ICHEP '98

Volume II

Proceedings of the 29th International Conference on

HIGH ENERGY PHYSICS

ICHEP '98

Volume II

Proceedings of the 29th International Conference on

HIGH ENERGY PHYSICS

Vancouver, Canada   23 – 29 July 1998

 TRIUMF

Editors

Alan Astbury
David Axen
Jacob Robinson

TRIUMF, Vancouver

 **World Scientific**
Singapore • New Jersey • London • Hong Kong

Published by

World Scientific Publishing Co. Pte. Ltd.

P O Box 128, Farrer Road, Singapore 912805

USA office: Suite 1B, 1060 Main Street, River Edge, NJ 07661

UK office: 57 Shelton Street, Covent Garden, London WC2H 9HE

British Library Cataloguing-in-Publication Data
A catalogue record for this book is available from the British Library.

ISBN 981-02-3772-3 (Set)
ISBN 981-02-3955-6 (Vol. 1)
ISBN 981-02-3956-4 (Vol. 2)

Printed in the United States of America.

TABLE OF CONTENTS

PLENARY SESSIONS

PARALLEL SESSIONS

Pa-02 Neutrino and Non Accelerator Experiments
Neutrino Oscillations; Solar Neutrinos; Double Beta-decay
(Convener: *William Louis*)

Pa-04 DIS, Low x; Structure Functions, Spin Structure Functions
(Conveners: *Masahiro Kuze, Michael Vetterli, Arie Bodek*)

Pa-05 Low Q^2, Soft Phenomena, Two Photon Physics
(Conveners: *Pierre Marage, Sampa Bhadra*)

VOLUME II

Pa-06 CP Violation and Rare Decays of K, mu, and tau
(Conveners: *Conrad Kleinknecht, Jack Ritchie*)

Pa-08 Heavy Hadrons: Lifetimes: Mixing, Rare Decays
(Conveners: *Michael Danilov, Manfred Paulini*)

Pa-09 Light Hadron Spectroscopy (Glueballs, Exotics, States with c Quarks)
(Conveners: *Saj Alam, Rolf Landau*)

Pa-10 Searches for New Particles at Accelerators
(Conveners: *Hendrik Weerts, Robert McPherson*)

Pa-17 Superstrings
(Convener: *Costas Kounnas*)

Pa-18 Lattice Gauge Theory
(Conveners: *Kenneth Bowler, Keh-Fei Liu*)

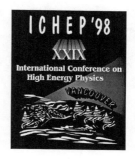

Parallel Session 6

CP Violation and Rare Decays of K, mu, and tau

Convenors: Konrad Kleinknecht (Mainz, NA48)

Jack Ritchie (Texas, AGS-E871)

RESULTS OF THE SINDRUM II EXPERIMENT

P. WINTZ

Phys. Inst. University of Zurich, 8057-Zurich, Switzerland
E-mail: Peter.Wintz@psi.ch

Tests of lepton-flavor conservation (LFC) are very sensitive probes of new physics beyond the Standard Model. The most stringent upper limit on a LF-nonconserving decay mode now comes from the search for $\mu^- \to e^-$ conversion on titanium with the SINDRUM II[1] spectrometer at PSI: $\Gamma(\mu^- Ti \to e^- Ti^{g.s.})/\Gamma(\mu^- Ti\ capture) < 6.1 \times 10^{-13}$ (90% CL). This new limit improves on the best previous result by a factor of 7. A search for the charge-changing $\mu^- \to e^+$ conversion yielded new upper limits which improve on our previous results by a factor of 2.5.

1 Introduction

The strict conservation of lepton flavor (LFC), as build into the Standard Model, is primarily based on experimental evidence. In most of the proposed extensions LFC-violating decays are expected to occur at some level and an observation of these processes would be an unambiguous indication of new physics beyond the Standard Model. Until now no signal for any LF-nonconserving decay mode has been observed and the experimental tests have resulted in upper limits only. Nevertheless these very low limits put strong constraints on the allowed parameters space of the various models [3,4,5].

2 $\mu \to e$ Conversion in Nuclei

Neutrinoless $\mu \to e$ conversion in muonic atoms, $\mu^-(A,Z) \to e^-(A,Z)$ with A mass number and Z atomic number, gives the best constraints on LFC-violation in a large variety of models [6–14]. For the conversion leaving the nucleus in its ground state the nucleons act coherently, which boosts the conversion probability relative to the rate of the dominant process of nuclear μ capture.

When negative muons stop in matter, they quickly get captured and form muonic atoms, which mostly reach their ground state before the bound μ decays $\mu^-(A,Z) \to e^- \bar{\nu}_e \nu_\mu (A,Z)$ or is captured by the nucleus $\mu^-(A,Z) \to \nu_\mu(A,Z-1)$. The capture probability increases with Z, to $f_{capt}^{Ti} = 85.3\%$ in the case of muonic titanium corresponding to a lifetime of 329 ns [15]. In the usual convention the branching ratio for $\mu \to e$ conversion is quoted relative to the rate for nuclear μ capture:

$$B_{\mu e} \equiv \frac{\Gamma_{\mu^- Ti \to e^- Ti^{g.s.}}}{\Gamma_{\mu^- Ti\ capture}}$$

The ratio $B_{\mu e}$ is predicted [16,17] to increase with Z. The same calculations predict the coherent fraction of the conversion to be larger than 80% for all nuclear systems. Earlier calculations [18,19,20] estimated a maximum of $B_{\mu e}$ in the region around $Z = 30$ so that most of

the experiments have been performed on medium-heavy nuclei. A general model-independent analysis [21] requires measurements on at least two nuclear targets with significantly different values for the normalized neutron excess $(N-Z)/(N+Z)$.

The nonconservation of total lepton number may lead to the charge-changing conversion $\mu^-(A,Z) \to e^+(A,Z-2)$. Since the nucleus changes from initial to final state there is no coherent enhancement in this case.

3 The SINDRUM II Experiment

At the $\mu E1$ beamline at PSI data were taken on titanium [22,24,25] and lead [23]. For the search for $\mu^- \to e^-$ conversion on Ti an experimental sensitivity below 10^{-12} was achieved, which is now the highest sensitivity on any LFC-violating decay. In the Ti data also a search for $\mu^- \to e^+$ conversion has been performed.

The SINDRUM II experiment aims at a final one event sensitivity around 10^{-14} for $B_{\mu e}$. The pion beamline $\pi E5$ with the new Pion-Muon-Converter (PMC) solenoid at its end will deliver the worldwide first high-intense μ^- beam with a negligible π^- contamination which is required to reach the proposed sensitivity (see Sect. 3.3).

Last year the PMC could be brought into operation but only with limited fieldstrength. Nevertheless a search for $\mu^- \to e^-$ conversion on a gold target was carried out with a temporarily changed beam concept. The $\pi E5$ channel was tuned to select lower momentum cloud muons for which the PMC fieldstrength was sufficient. Meanwhile the PMC can be operated at full field strength and the next $\mu \to e$ measurement starts in Oct. 1998.

3.1 $\mu \to e$ Signature

The signature of coherent $\mu \to e$ conversion is a single electron with the characteristic energy $E_{\mu e}$ given by the muon mass minus the muon binding energy E_b and the kinetic energy E_r of the recoiled atom:

$$E_{\mu e} = m_\mu c^2 - E_b(A, Z) - E_r(A, Z) \quad .$$

In case of titanium $E_{\mu e} = 104.3$ MeV with $E_b = 1.27$ MeV and $E_r = 0.12$ MeV. The conversion energy $E_{\mu e}$ coincides with the kinematical endpoint of electrons from bound-μ decay which is the dominant intrinsic background source.

3.2 Intrinsic Background

The energy spectrum of electrons from bound-muon decay has been calculated for a number of muonic atoms [26,27,28] and the results are shown in Figure 1. The spectra drop exponentially above $m_\mu c^2/2$ which is the endpoint energy of electrons from free-μ decay. Radiative corrections have not been applied and will lead to a reduction of the rate in the upper end of the spectra by 20 to 40% [26]. The probability to get an electron with energy

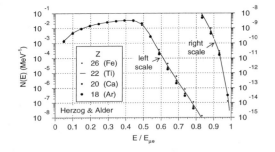

Figure 1: Normalized electron energy spectrum of bound-μ decay for muonic atoms Ar, Ca, and Fe [26]. The spectrum of Ti resulted from these distributions by interpolation.

within 2% of the endpoint is around 5×10^{-14}. Taking into account that the probability of bound-μ decay per captured μ is around 17%, this yields a background well below 10^{-14}. Simulation confirms that a Gaussian energy resolution of around 2 MeV (FWHM) is sufficient to remain free of the background from bound-μ decay down to the proposed sensitivity.

Electron background from radiative μ capture is less significant. Due to the mass difference of initial and final nucleus the kinematical endpoint of the energy spectrum is lowered. In case of Ti with Sc as final nucleus the shift is 4 MeV so that the endpoint is well below the $\mu \to e$ conversion energy $E_{\mu e}$. For the whole measured energy spectrum the contribution of electrons from μ capture compared to the bound-μ decay background is lower than a few percent. However, radiative μ capture produces the only intrinsic positron background in the search for $\mu^- \to e^+$ conversion (see Sec. 4.2).

3.3 Prompt and Cosmic Ray Background

Prompt background may be caused by beam e^- scattering off the target and by e^- from radiative π^- capture and subsequent pair production. The scattered electrons may originate from π^0 decays in the pion production target or from μ^- decays in flight at the end of the beamline. In the first case their momentum distribution reaches up to the selected beam momentum. Electrons from μ decay in flight may have significantly higher momenta and to avoid this background completely the beam momentum would have to be below 76 MeV/c.

Negatively charged pions at rest are immediately captured by a nucleus. With a probability of about 1% a photon is emitted with an energy above 100 MeV. These photons may convert in the target into an e^+e^- pair. If the positron remains unnoticed such events are a background source.

There are numerous processes in which cosmic rays produce e^- in the target with an energy around $E_{\mu e}$. In almost all cases additional particles may be observed, like high-energy μ or associated e^+. This background source can be reduced by passive shielding. A cosmic background run with beam switched off allows to study and identify the characteristic event topologies.

3.4 Spectrometer

The SINDRUM II spectrometer (Fig. 2) consists of a set of cylindrical detectors inside a superconducting solenoid. The μ^- beam enters the setup on the axis and is mod-

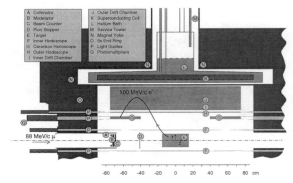

Figure 2: Vertical cross-section through the SINDRUM II spectrometer.

erated before reaching the Ti target at the center. The incoming μ^- go through a scintillation counter used to recognize prompt background events in the spectrometer and to monitor the beam intensity. The purpose of the second thinner moderator in front of the target is to absorb most of the remaining beam pions.

At the field strength of 1.27 T all electrons or positrons leaving the target with transverse momenta below 105 MeV/c are contained radially inside the magnet (see Fig. 3). The particles of interest follow helical trajectories, making at least one turn before reaching a Čerenkov hodoscope at either end of the spectrometer.

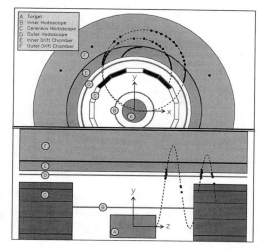

Figure 3: Example of a reconstructed electron trajectory in the $x - y$ and $y - z$ plane.

Two radial drift chambers are used for the spacial reconstruction of the trajectories. The average resolution of the inner drift chamber is 150 μm (σ_{xy}) in the $x-y$ plane and 2 mm (σ_z) in z direction. The main purpose of the larger outer chamber is to recognize cosmic ray induced events by coming from outside into the spectrometer or by correlated signals from additional particles. The spacial resolution of this chamber is restricted to the $x - y$ plane and of the order of a few millimeters.

The trigger for data readout is mainly based on the expected hit pattern in the outer hodoscope and inner drift chamber from a $\mu \rightarrow e$ simulation. In addition, signals in the inner hodoscope and a Čerenkov hodoscope are required.

The time signals of the scintillator and Čerenkov hodoscopes belonging to a trajectory are used to determine the particle charge and direction. The first signal in the outer plastic-scintillator hodoscope defines the spectrometer reference time which is used to start the drift time measurements and to identify prompt background with the help of the beam counter. The second inner plastic-scintillator hodoscope helps to recognize background in particular from $\gamma \rightarrow e^+e^-$ conversions in the target.

3.5 Track Reconstruction and Momentum Resolution

In the event reconstruction helical trajectories are searched and the particle momentum is fitted to the first two track elements in the inner drift chamber (see Fig. 3). Inhomogeneities of the magnetic field, energy loss, and multiple scattering along the particle trajectory outside the target are taken into account. The trajectory is followed backward to the point of closest approach to the spectrometer axis. This point is taken as the particle origin.

To check the momentum calibration and resolution the beamline was tuned to select π^+ which were stopped in a low-mass polystyrene target inside the spectrometer. Fig. 4 shows the measured energy distribution of the positrons from $\pi^+ \rightarrow e^+\nu_e$ decay. Compared to the $\mu \rightarrow e$ measurement the magnetic field of the spectrom-

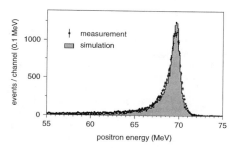

Figure 4: Energy spectrum of positrons from $\pi^+ \rightarrow e^+\nu_e$ decay in the low-mass polystyrene target. The width of the measured distribution is 1.5 MeV (FWHM).

eter was reversed and scaled down to the lower momentum. As may be seen the observed width is slightly larger than expected from simulation. Since the energy resolution for the $\mu \rightarrow e$ measurement is dominated by the spread in the energy loss inside the heavier Ti target the agreement is much better there. For this target a simulation of $\mu \rightarrow e$ conversion yields an energy resolution of 2.3 MeV which is sufficient to remain free from bound-μ decay background (see Sec. 3.2).

4 Results of the Experiment

4.1 Search for $\mu^- \rightarrow e^-$ Conversion on Titanium

The μ^- flux at the $\mu E1$ beamline rose as a function of the momentum and reached a value of 1.2×10^7 $\mu^- s^{-1}$ at the selected momentum of 88 MeV/c. This was the highest rate for the operation of the beam counter scintillator and the momentum was low enough regarding the background from electron scattering. During a measurement live time of 50.4 days with beam switched on

3.09×10^{13} μ^- stopped in the target with a π^- contamination around 10^{-7}.

In total 4.0×10^6 events fulfilled the trigger conditions and were recorded from which $\simeq 170000$ electrons originating in the target could be reconstructed. Events caused by radiative π^- capture and scattered beam e^- were removed by a 20 ns wide prompt veto on signals in the beam counter. The relative large width was caused by the flight-time spread of low-energy π^- between second moderator and target.

The criteria to recognize cosmic background was checked and optimised by studying the data measured with beam switched off. To identify cosmic ray background several tests were developed looking for additional signals in the various detectors. If at least three of these tests were positive an event was declared as cosmic ray background. A specific background source survived these tests. A cosmic photon could enter the spectrometer through the unshielded cryogenic service tower and make asymmetric e^+e^- pair production in the target. If the e^+ had not enough momentum to reach the inner hodoscope only signals from the electron were seen. The only way to remove this background was a 3% cut in the geometrical acceptance.

The final electron sample consisted of $\simeq 21000$ events, mainly from the intrinsic background of bound-μ decay. Radiative μ^- capture and unrecognized scattered beam e^- contributed at the level of 1% and 5%, respectively. Fig. 5 shows the resulting electron energy distributions and a $\mu \to e$ simulation assuming a branching ratio of $B_{\mu e} = 4 \times 10^{-12}$. The endpoint of the final

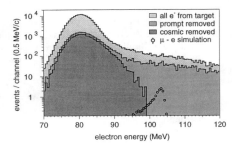

Figure 5: Energy distribution of the reconstructed electrons from the target and of the remaining events after suppression of prompt and cosmic ray background. A GEANT simulation of $\mu \to e$ conversion electrons shows the expected amount assuming $B_{\mu e} = 4 \times 10^{-12}$.

energy distribution is at 99 MeV. At higher energies the prompt and cosmic ray background is completely removed. Above 99 MeV one expects 83% of all $\mu \to e$ events but no $\mu \to e$ candidate event has been found.

In Fig. 6 various kinematical distributions of selected events from bound-μ decay in the final sample are compared with a GEANT simulation of this event type. The shift of the r_{ca} distribution to a mean value around 2.5 cm is caused by the steeply falling energy spectrum (see Fig. 1) and the vanishing acceptance towards lower values around 75 MeV. The z_{ca} distribution reflects the range of the stopped μ^- which has been shifted to the front of the target to reduce the π^- contamination. Electrons emitted at $\Theta \simeq 90^o$ have a low probability to reach one of the Čerenkov hodoscopes which explains the dip in the Θ distribution. The simulated distributions are normalised to the observed number of events.

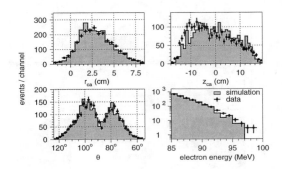

Figure 6: Kinematical distributions of electrons from bound-μ decay for data and simulation. r_{ca} and z_{ca} are the radial distance and corresponding z-coordinate of the closest point of the trajectory to the spectrometer axis. Θ is the angle of the trajectory relative to the z axis.

Since no $\mu \to e$ conversion candidate has been found the measurement results in a new upper limit on the branching ratio

$$ B_{\mu e} < \frac{n_{max}^{\mu e} \, (1 + n_{max}^{\mu e} \cdot \sigma_r^2/2)}{N_{stop}^{\mu} \cdot f_{capt}^{Ti} \cdot \epsilon^{\mu e}} \ . $$

$n_{max}^{\mu e} = 2.3$ is the upper limit on the expected number of events with 90% confidence, $N_{stop}^{\mu} = (3.09 \pm 0.14) \times 10^{13}$ is the number of stopped μ^-, and $f_{capt}^{Ti} = (0.853 \pm 0.003)$ is the μ^- capture probability for Ti [15]. $\epsilon^{\mu e}$ is the probability that a $\mu \to e$ conversion in the target leads to the observation of an e^- with an energy above 99 MeV and σ_r is the relative uncertainty of the denominator [29].

The probability $\epsilon^{\mu e}$ is the product of the geometrical $\mu \to e$ acceptance ϵ_ω and the efficiencies of the trigger ϵ_{trig}, track reconstruction ϵ_{track}, and selection cuts ϵ_{select} to suppress background. The acceptance $\epsilon_\omega = 0.418$ of the spectrometer was obtained by a $\mu \to e$ simulation whereas the efficiency factors and their energy dependencies were estimated from the events of bound-μ decay. Extrapolation to the $\mu \to e$ energy region yielded

$\epsilon_{trig} = 0.751$, $\epsilon_{track} = 0.864$, and $\epsilon_{select} = 0.538$ including the energy cut of 99 MeV to suppress the bound-μ decay background (see Fig. 5). Using $\epsilon^{\mu e} = (0.146 \pm 0.012)$ and $\sigma_r = 0.11$ the new upper limit on the $\mu \to e$ branching ratio is (with 90% CL)

$$B_{\mu e} < 6.1 \times 10^{-13}$$

which improves on the previous result [22] by a factor of 7.

4.2 Search for $\mu^- \to e^+$ Conversion on Titanium

Since there is no charge decision in the $\mu \to e$ trigger also positrons were recorded which allows to search for $\mu^- Ti \to e^+ Ca$ conversion. The positron energy depends on the mass difference of the nuclei and varies only slightly for the different Ti isotopes if Ca is in ground state. But unlike $\mu \to e$ conversion, one can expect an excitation of the final nucleus described by a giant dipole resonance with mean energy and width of 20 MeV in case of Ca. This leads to a broad energy spectrum of the positrons and has severe consequences for the background discrimination. Fortunately, the intrinsic background is much lower for e^+ than e^-.

Fig. 7 shows the energy distribution of the final data sample which consists mainly of positrons from radiative μ capture and some misidentified scattered beam electrons unrecognized by the beam counter[30]. The endpoint

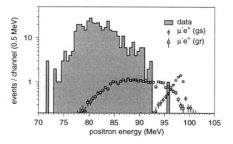

Figure 7: Energy distribution of the final positron data sample. A simulation of $\mu^- e^+$ conversion with Ca in ground state (gs) and with a giant resonance excitation (gr) shows the expected signals using $B_{\mu e}^{gs} = 4.3 \times 10^{-12}$ and $B_{\mu e}^{gr} = 8.9 \times 10^{-11}$.

of the positron spectrum is at 92.3 MeV with an isolated event at 95.7 MeV. Also shown are the distributions resulting from $\mu^- \to e^+$ simulations for ground state and giant resonance final states. The measured distribution with beam switched off shows also one event in the interesting energy region so that the event at 95.7 MeV can be explained by cosmic ray background. Therefor, above 92.3 MeV no candidate for $\mu^- \to e^+$ conversion is found

and the new upper limits are (with 90% CL)

$$B_{\mu^- e^+}^{gs} < 1.7 \cdot 10^{-12} \quad and \quad B_{\mu^- e^+}^{gr} < 3.6 \cdot 10^{-11}$$

which improve on the previous limits [22] by a factor of 2.5.

Acknowledgments

The experiment is supported in part by the BMFT, Germany, under contract number 06AC651 and by the Swiss National Science Foundation.

References

1. PSI proposal R-87-03 (1987).
   SINDRUM II collaboration: W. Bertl, T. Buchsteiner, G. Cahsor, V. Djordjadze, C. Dohmen, J. Egger, S. Eggli, R. Engfer, C. Findeisen, H. Haan, E. Hermes, J. Hofmann, W. Honecker, J. Kaulard, T. Kozlowski, B. Krause, G. Kurz, J. Kuth, M. Maas, F. Muheim, C. Niebuhr, G. Otter, S. Playfer, H. Pruys, F. Riepenhausen, F. Rosenbaum, M. Rutsche, A. van der Schaaf (spokesman), A. Schuengel, M. Starlinger, and P. Wintz.
   III. Phys. Inst. B RWTH Aachen (Germany) – Paul Scherrer Institute, Villigen (Switzerland) – Phys. Inst., University of Zurich (Switzerland).
2. Y. Ohama, *Phys. Rev.* D **57**, 6594 (1998).
3. J.D. Vergados, *Phys. Rep.* **133**, 1 (1986).
4. A. v.d. Schaaf, *Prog. Part. Nucl. Phys.* **31**, 1 (1993).
5. P. Depommier and C. Leroy, *Rep. Prog. Phys.* **58**, 61 (1995).
6. G. Altarelli et al., *Nucl. Phys.* B **125**, 285 (1977).
7. O. Shanker, *Nucl. Phys.* B **206**, 253 (1982).
8. R.N. Cahn and H. Harari, *Nucl. Phys.* B **176**, 135 (1980).
9. J. Bernabeu et al., *Nucl. Phys.* B **409**, 69 (1993).
10. L.N. Chang et al., *Phys. Rev.* D **50**, 4598 (1994).
11. T.S. Kosmas and J.D. Vergados, *Phys. Rep.* **264**, 251 (1996).
12. G. Barenboim and M. Raidal, *Nucl. Phys.* B **484**, 63 (1996).
13. J.E. Kim et al., *Phys. Rev.* D **56**, 100 (1997).
14. M. Raidal and A. Santamaria, *Phys. Lett.* B **421**, 250 (1998).
15. T. Suzuki et al., *Phys. Rev.* C **35**, 2212 (1987).
16. H.C. Chiang et al., *Nucl. Phys.* A **559**, 526 (1993).
17. T.S. Kosmas et al., *Phys. Rev.* C **56**, 526 (1997).
18. S. Weinberg and G. Feinberg, *Phys. Rev. Lett.* **3**, 111 (1959).
19. T.S. Kosmas and J.D. Vergados, *Phys. Lett.* B **217**, 19 (1989).
20. T.S. Kosmas and J.D. Vergados, *Nucl. Phys.* A **510**, 641 (1990).
21. O. Shanker, *Phys. Rev.* D **20**, 1608 (1979).
22. SINDRUM II Collab., C. Dohmen et al., *Phys. Lett.* B **317**, 631 (1993).
23. SINDRUM II Collab., W. Honecker et al., *Phys. Rev. Lett.* **76**, 200 (1996).
24. P. Wintz, *Ph.D. thesis*, RWTH Aachen (1995).
25. S. Eggli, *Ph.D. thesis*, University of Zurich (1995).
26. F. Herzog and K. Alder, *Helv. Phys. Acta* **53**, 53 (1980).
27. R. Watanabe et al., *Atomic Data and Nuclear Data Tables* **54**, 165 (1993).
28. O. Shanker and R. Roy, *Phys. Rev.* D **55**, 7307 (1997).
29. R. D. Cousins and V. L. Highland, *Nucl. Instr. Meth.* A **320**, 331 (1992).
30. SINDRUM II Collab., J. Kaulard et al., *Phys. Lett.* B **422**, 334 (1998).

τ decays to hadrons at LEP

Francisco Matorras

Universidad de Cantabria, Av. los Castros s/n, 39005 Santander SPAIN
DDELPHI collaboration
E-mail: Francisco.Matorras@cern.ch

A large variety of τ hadronic decays have been studied by the LEP experiments. Latest results are described here, including channels with up to seven charged particles, final states with several π^0 and channels with charged and neutral kaon identification. General agreement with previous measurements was found and significant improvement in the precision was achieved in many cases.

1 Introduction

The analysis of the whole LEP1 data sample during the last years is still producing lot of new results on τ physics and particularly in the subject of it hadronic decays. In this paper I will briefly summarise the new measurements submitted to this conference by two LEP collaborations (DELPHI and OPAL) [1,2,3,4,5,6] as well as the main analysis tools used. The studies described here cover most of the possible channels:

- one prong decays with several π^0 and no kaon/pion separation [1,4]

- three prong decays with several π^0 and no kaon/pion separation [1]

- topological Branching Ratios [1]

- five and seven prong decays with π^0 and no kaon/pion separation [5,6]

- One and three prong decays with pion/kaon separation [2,3]

- modes including neutral kaons [1]

It is important to remember that despite the moderate statistics of the LEP experiments (about 100000 $\tau^+\tau^-$ per experiment) most measurements are competitive because the systematic errors are very small, taking advantage of the very good separation of $\tau^+\tau^-$ from $q\bar{q}$ events leading to a clean and unbiased samples with high efficiency and to the powerful LEP detectors, whose response is very well known after seven years of running. To illustrate this point, in all summary tables the new results are shown together with the current world average [7].

2 One prong decays

In these analyses no charged kaon/pion separation was performed and results are quoted in terms of a generic

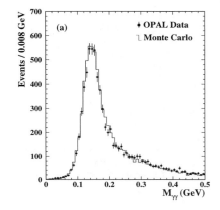

Figure 1: $\gamma\gamma$ invariant mass distribution for π^0 candidates in OPAL.

hadron h^- (a π^- or a K^-). The main issue for the measurement of these channels is, after leptonic rejection, the π^0 counting. For low energy, typically $E_{\pi^0} < 8 - 10$ GeV, the LEP detectors resolve the two γ from the π^0 decays and from its invariant mass (figure 1) they can be identified. For higher energies, the γ are too close in the calorimeters and produce a single shower. In the OPAL case all high energetic showers are considered as π^0, while for DELPHI a further identification is done. Taking advantage of the DELPHI electromagnetic calorimeter granularity it is possible to observe the substructure of the shower being able to identify those coming from two close photons. An example of a 25 GeV π^0 is shown in figure 2.

DELPHI analyses channels with none, one, two and three or more π^0, the preliminary results are shown in table 1. OPAL analyses none, one and two or more π^0, the results are final and are also shown in table 1. Note that OPAL results are already included in the world averages.

There is some confusion regarding the definition of

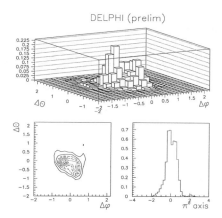

DELPHI (prelim)

Figure 2: DELPHI example of π^0 identification in single a shower for a 25 GeV simulated π^0. The figures show different projections of the shower, where the structure of two sub-showers is evident.

Table 1: One prong hadronic BR (in %) without kaon/pion separation. Note that for the mode $h^- \geq 2\pi^0\nu$ there are two entries accounting for different definitions of the channel, see text for explanation. The first column summarises DELPHI results, the second OPAL's and the third the current world averages[7].

Mode	DELPHI	OPAL	PDG98
$h^-\nu$	11.76 ± 0.14	11.98 ± 0.21	11.79 ± 0.12
$h^-\pi^0\nu$	25.88 ± 0.31	25.89 ± 0.34	25.84 ± 0.14
$h^-2\pi^0\nu$	9.37 ± 0.37		9.23 ± 0.14
$h^- \geq 2\pi^0\nu$	10.55 ± 0.24	9.61 ± 0.41	10.49 ± 0.16
$h^- \geq 2\pi^0\nu$	10.85 ± 0.24	9.91 ± 0.41	10.79 ± 0.16
$h^- \geq 3\pi^0\nu$	1.14 ± 0.25		1.40 ± 0.11

the channels with two or more π^0. While DELPHI considers all kind of modes including K^0 as a background for these channels, OPAL accepts as signal the case of $K^0 \to 2\pi^0$. For comparison, for the channel $h^- \geq 2\pi^0\nu$ measured by both collaborations the results are converted from one definition to another according to world averages for the channels including K^0 and two entries are shown in the table (the first excluding that contribution and the second including it).

Good agreement is found between measurements, and only a deviation of about two standard deviations is seen on the $h^- \geq 2\pi^0\nu$ mode. DELPHI measurement of the $h^-\nu$ is currently the world's most precise.

Table 2: DELPHI preliminary Branching Ratios (in %) for modes involving 3 charged hadrons without kaon/pion separation. For comparison current world averages[7] are shown. Note that for the mode $3h^-\nu$ there are two entries accounting for different definition of the channel, see text for explanation.

Mode	DELPHI	PDG98
$2h^-h^+\nu(K^0exc)$	9.27 ± 0.14	9.57 ± 0.10
$2h^-h^+\nu(K^0inc)$	9.65 ± 0.14	9.96 ± 0.10
$2h^-h^+\pi^0\nu$	4.72 ± 0.14	4.50 ± 0.09
$2h^-h^+ \geq 2\pi^0\nu$	0.49 ± 0.09	0.54 ± 0.04

Table 3: DELPHI preliminary topological Branching Ratios (in %). For comparison current world averages[7] are shown.

Mode	DELPHI	PDG98
one prong	85.12 ± 0.15	85.39 ± 0.33
three prongs	14.79 ± 0.15	14.63 ± 0.25

3 Three prong decays and topological Branching Ratios

For the study of these modes, in addition to the π^0 identification outlined above, the critical point is the reconstruction of three very close tracks and the rejection of secondaries. This is achieved with the help of the silicon micro strip vertex detectors. DELPHI preliminary results on three prong decays with none, one and two or more π^0 are listed on table 2.

As for the case of $h^- \geq 2\pi^0\nu$, for the $2h^-h^+\nu$ mode, the current measurements follow two different definitions, including the $K^0 \to \pi^-\pi^+$ or not. Both cases are shown in the table. Again, the agreement is perfect except in the $2h^-h^+\nu$ modes, were the new result is at almost 2 standard deviations from the world average. On the other hand, this result confirms higher values of the Branching Ratio (above 9%) as obtained by CLEO, OPAL or ALEPH and in clear disagreement with ARGUS result of $7.6 \pm 0.1 \pm 0.5$ (not including $K^0 \to \pi^-\pi^+$).

Similarly the topological Branching Ratios to one or three charged particles and any number of neutrals were obtained. Preliminary results are listed in table 3. In this case K^0 is treated as a neutral even when decaying to $\pi^+\pi^-$ inside the tracking detectors. Both DELPHI numbers have a strong statistical correlation but have no explicit constraint and the systematics are partially independent. These results improve by almost a factor two the current world average.

4 Five and seven prong decays

OPAL has studied the decays to five charged hadrons and none or one π^0. It relies on an extremely good re-

Table 4: OPAL final Branching Ratios (in %) for modes with five charged particles. For comparison current world averages[7] are shown (OPAL numbers included). Also shown the OPAL upper bound at 95% C.L. for seven prong decays. It is compared with the from world average from $3h^-2h^+2\pi^0\nu$ at 90% C.L.

Mode	OPAL	PDG98
$3h^-2h^+\nu$	0.091 ± 0.015	0.075 ± 0.007
$3h^-2h^+\pi^0\nu$	0.027 ± 0.019	0.022 ± 0.005
7 prongs	$< 1.8\ 10^{-5}$	$< 1.1\ 10^{-4}$

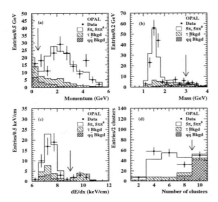

Figure 3: OPAL distributions for several kinematic and electron ID variables used in the five prong selection: minimum momentum (top left), invariant mass (top right), dE/dx (bottom left) and number of clusters in the calorimeter (bottom right)

jection of secondaries, mainly γ conversions, performed using electron identification variables. Also several kinematic variables can be used to separate signal from background (momentum, mass...) as shown in figure 3.

A similar analysis has been extended to search for any possible decay to seven charged hadrons and any number of accompanying π^0. One of the most powerful variables is shown in figure 4, consisting on combination for all tracks of probabilities of being electrons. For true decays to seven hadrons it has a value close to 0, while for the background, where one or more conversion have occurred it is displaced to higher values. No events are found, while 0.5 ± 1.0 were expected from background. Therefore an upper limit is set at 95% confidence level. Results are summarised in table 4. Since no other measurement of seven prongs is available it is compared with an upper limit from $3h^-2h^+\pi^0\nu$ at 90%.

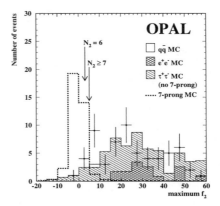

Figure 4: OPAL distribution for the main electron rejection variable used in the 7 prong search.

5 Decays including kaons

The charged kaon identification is one of the most challenging subjects in τ physics, having to reject a background 50 times larger from decays whose only difference is the existence of a pion instead of a kaon and at the same time to keep a reasonably high efficiency due to the small fraction of these decays.

OPAL kaon/pion separation relies on the ionization measurement in the tracking devices, dE/dx. Its response is shown in figure 5 for different types of particles. In particular the kaon and pion signals are separated from 1.5 to 4 standard deviations in the whole momentum spectra (3.5 to 45 GeV).

The DELPHI dE/dx separation is somewhat poorer, between 1.5 and 2 standard deviations (figure 6) but on the contrary has the advantage of the existence of a Ring Imaging Cherenkov Counter (RICH). It works in two modes, the veto mode, for tracks of less than 8.5 GeV where the kaons are recognised because they are below the Cherenkov threshold and therefore no photons are emitted, this allows a pion rejection at the level of 80. Above that threshold a Cherenkov ring can be reconstructed for both particles. The combined effect of RICH and dE/dx gives a separation of at least 2 standard deviations and up to 10 for particles close to threshold as shown in figure 6.

In both cases the identification is based on a so called *pull* variable built as the difference of the measured signal and the expected for a π (equivalently it can be defined for K hypothesis) divided by the experimental error. This variable has to be a Gaussian centred at 0 and

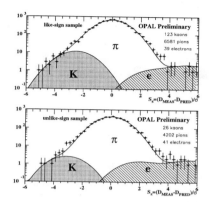

Figure 7: OPAL pull variables for the different types of events. The total MonteCarlo is represented by the white histogram, the Kaon component by the shaded histogram, the electron component by the hatched histogram and the data by the dots.

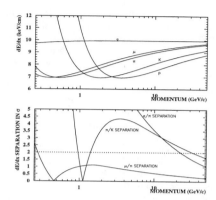

Figure 5: OPAL kaon/pion discrimination using dE/dx: $< dE/dx >$ signal as a function of momentum for different particles (top) and separation in number of standard deviations (bottom)

Table 5: OPAL preliminary Branching Ratios (in %) for modes involving charged kaons. For comparison current world averages[7] are shown.

Mode	OPAL	PDG98
$K^- K^+ \pi^- \geq 0\pi^0 \nu$	0.036 ± 0.053	0.23 ± 0.04
$K^- \pi^+ \pi^- \geq 0\pi^0 \nu$	0.358 ± 0.100	0.31 ± 0.06

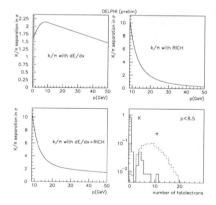

Figure 6: DELPHI kaon/pion separation for TPC dE/dx only (top left), RICH ring only (top right) ring and dE/dx combined (bottom left) in number of standard deviations and number of photoelectrons for the RICH in veto mode (bottom right).

with a unit width for π, while it will be displaced by the amount given in figures 5 and 6 for K. It is very important for the selection and especially for the systematic errors to reduce and control all possible non-gaussian tails of these distributions. The good behaviour of these distributions is shown in figures 7 and 8, where simulation and data are compared. The results are summarised in tables 5 and 6.

The K_S^0 are identified through its decays to $\pi^+\pi^-$ far from the interaction point, the K_L^0 as a neutral deposition on the hadronic calorimeters. The results are summarised in table 7.

Again, good agreement is found except in the $K^- K^+ \pi^- \geq 0\pi^0 \nu$ mode, where a deviation of almost 3σ is found. In most of the channels involving charged kaon identification the precision achieved is comparable with the best existing measurements, while the improvement is more modest for the channels with K^0.

6 Conclusions

Three years after the end of LEP I data taking period, at the Z^0 resonance, still a lot of new interesting and competitive results in τ physics are being produced and in particular on hadronic Branching Fractions. However, since most analysis are becoming final and no new data

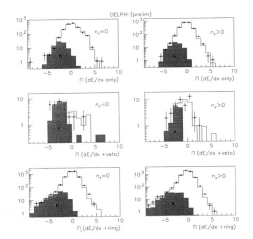

Figure 8: DELPHI pull variables for the different types of events: events with TPC dE/dx only (top), RICH veto and dE/dx (middle), RICH ring and dE/dx (bottom). Right/Left hand side figures correspond to the sample with/without γ. The total MonteCarlo sample is represented by the white histogram, the kaon component by the shaded histogram and the data by the dots.

Table 6: DELPHI preliminary Branching Ratios (in %) for one prong modes where pion/kaon separation was performed. For comparison current world averages[7] are shown.

Mode	DELPHI	PDG98
$K^-\nu$	0.703 ± 0.083	0.71 ± 0.05
$K^- \geq 1\pi^0\nu$	0.744 ± 0.102	0.60 ± 0.05
$K^- \geq 0\pi^0 \geq 0K^0\nu$	1.800 ± 0.121	1.66 ± 0.10
$\pi^-\nu$	11.06 ± 0.16	11.08 ± 0.13
$\pi^-\pi^0\nu$	25.22 ± 0.33	25.32 ± 0.15

Table 7: DELPHI preliminary Branching Ratios (in %) for one prong modes involving K^0. For comparison current world averages[7] are shown.

Mode	DELPHI	PDG98
$K^0 X^-\nu$	1.94 ± 0.22	1.66 ± 0.09
$K^*(892)\nu$	1.50 ± 0.33	1.28 ± 0.08
$h^- K^0\nu$	1.05 ± 0.31	0.99 ± 0.08
$h^- K^0\pi^0\nu$	0.51 ± 0.39	0.55 ± 0.05
$h^- K^0_S K^0_L\nu$	0.13 ± 0.12	0.06 ± 0.01

will be taken, very few improvements are expected in the future.

LEP has demonstrated, maybe unexpectedly, to be an optimal tool for the study of these decays despite the moderate statistics compared to other lower energy experiments, taking advantage of the clean environment and the precise and very well understood detectors.

In various channels, the quoted preliminary numbers will improve the current world averages significantly. No large inconsistencies between measurements are found, there is agreement at the level of two standard deviation between all the new results presented in this conference and the world averages, except, maybe, on the channel $K^- K^+ \pi^- \geq 0\pi^0\nu$ where OPAL's result differs by almost three sigma with the current world average and also other preliminary results from DELPHI and CLEO. However, it is better not to draw any conclusion before all these measurements become final.

An interesting exercise is to combine all the available information of the topological Branching Ratios, assuming the value for decays to five prongs or more from the world average of 0.097 ± 0.007. Under this assumption both one and three prong measure the same quantity and can be combined. DELPHI measurements quoted here have a high statistical correlation, and although the systematic errors are partially independent only one of them will be considered for this average. On the contrary, the current world averages are dominated by measurements done by different experiments and therefore are treated as independent. In such a way the result $B_1 = 0.999 - B_3 = 85.19 \pm 0.12$ is obtained. This can be compared with the world average fit taking into account all the exclusive decays $B_1 = 85.30 \pm 0.13$, with a perfect agreement showing the completeness of the τ decays or in other words that despite the increasing precision the "missing one prong problem" does not exist anymore.

References

1. "A measurement of the τ hadronic decays", DELPHI, this conf.

2. "A study of charged kaon production in one prong τ decays", DELPHI, this conf.

3. "Determination of τ branching ratios to three prong final states with charged kaons", OPAL, this conf.

4. "Measurement of the τ one prong hadronic branching ratios", OPAL, this conf.

5. "Measurement of the τ branching ratio to five charged particles", OPAL, this conf.

6. "An upper limit on the branching ratio for τ decays into seven charged particles", OPAL, this conf.

7. "Review of Particle Physics", Particle Data Group, *Eur. Phys. J. C3, 1-794 (1998)*

SPECTRAL FUNCTIONS AND THE STRONG COUPLING CONSTANT α_s IN TAU DECAYS FROM OPAL

A. STAHL

Physikalisches Institut, Nussallee 12, 53115 Bonn, Germany
E-mail: stahl@physik.uni-bonn.de

A measurement of the strong coupling constant α_s from τ decays by the OPAL collaboration is presented. The result using the contour improved third order calculation is $\alpha_s\left(m_\tau^2\right) = 0.348 \pm 0.009 \pm 0.019$ or $\alpha_s\left(m_{Z^0}^2\right) = 0.1219 \pm 0.0010 \pm 0.0017$.

1 Introduction

The τ lepton is the only lepton heavy enough to decay into hadrons. It offers a unique possibility to study the strong interaction and measure α_s at a mass scale $s = m_\tau^2$ where α_s is large enough, so that QCD corrections become sizable and therefore easier to measure, but not too large so that non-perturbative corrections are still small.

In analogy to the R-Ratio in e^+e^- annihilation R_τ is defined as

$$R_\tau := \frac{\Gamma\left(\tau \to \nu_\tau\, hadrons\right)}{\Gamma\left(\tau \to e\, \nu_e \nu_\tau\right)}. \tag{1}$$

Naively one expects that ratio to be three, as the quarks come in three colours. But the ratio is altered by QCD corrections. These QCD corrections depend on α_s and therefore the deviation of R_τ from 3 determines α_s.

The perturbative corrections have been calculated to third order in α_s [1] and the SVZ-approach [2] is used to organize the non-perturbative corrections which are finally measured from the data simultaneous with α_s. There are also some small electroweak corrections [3]. (For theoretical aspects see [4].) R_τ is given by

$$R_\tau = 3\left(|V_{ud}|^2 + |V_{us}|^2\right) S_{ew}$$
$$\left(1 + \delta_{\text{pert}} + \delta_{\text{non-pert}} + \delta_{ew}\right). \tag{2}$$

The perturbative correction δ_{pert} is by far the largest correction ($\approx 20\,\%$), whereas the others are on the level of 1 %.

R_τ can be determined purely from the leptonic branching ratios of the τ, by requiring the branching ratios to sum up to unity ($b_e + b_\mu + b_{\text{had}} = 1$). Then

$$R_\tau = \left(1 - b_e - b_\mu\right) / b_e \tag{3}$$

and from the world averages one gets $\alpha_s\left(m_\tau^2\right) = 0.35 \pm 0.03$ [5]. Although this is a valid result, it is not the end of the presentation. It is the advantage of τ decays and the purpose of this talk, to demonstrate the various ways in which the result can be cross checked.

2 Spectral Functions

The experimental input to these checks are the spectral functions. Figures 1 and 2 show the vector and axial vector spectral function respectively as measured by the OPAL collaboration [6]. This measurement has been restricted to the Cabibbo allowed fraction of hadronic τ decays. The data presented has already been corrected for experimental defects like the finite resolution, efficiency or background. The spectral functions are related to the Spin 1 two-point correlators through

$$v\left(s\right) = 2\pi\, \text{Im}\, \Pi_V^{(J=1)}, \qquad a\left(s\right) = 2\pi\, \text{Im}\, \Pi_A^{(J=1)} \tag{4}$$

and these are related to R_τ through

$$\left(R_\tau\right)_{V/A} = \frac{12\pi S_{ew}|V_{ud}|^2}{m_\tau^2}$$

$$\int_0^{m_\tau^2} ds \left(1 - \frac{s}{m_\tau^2}\right)^2 \left[\left(1 + \frac{2s}{m_\tau^2}\right)\Pi_{V/A}^{(J=1)} + \Pi_{V/A}^{(J=0)}\right] \tag{5}$$

The Spin 0 correlators are simple: the axial vector part only contains the pion pole and the vector part vanishes.

3 Spectral Moments

The value of α_s extracted from R_τ only depends on the integral of the hadronic spectrum, not on its shape. Non-perturbative corrections however alter the shape. Therefore it is possible to measure the non-perturbative corrections from the shape of the spectral functions through appropriate moments [7]

$$\left(R_\tau^{kl}\right)_{V/A} = \int_0^{m_\tau^2} ds \left(1 - \frac{s}{m_\tau^2}\right)^k \left(\frac{s}{m_\tau^2}\right)^l \left(\frac{dR_\tau}{ds}\right)_{V/A}. \tag{6}$$

In the SVZ approach [2] the non-perturbative corrections are organized as a series of terms with increasing mass dimensions. Each term is a product of the expectation value of a local operator, a so-called condensate, a Wilson coefficient, a factor $(1/m_\tau)^n$ to compensate for the mass dimension of the condensate and another coefficient depending on k and l, for example

$$g^{kl}\frac{1}{m_\tau^4}C_{V/A}(\mu)\left\langle\frac{\alpha_s}{\pi}GG\right\rangle. \qquad (7)$$

This example would be a correction of mass dimension 4. The local operator is the gluon condensate, describing the creation of a pair of gluons from the vacuum at a certain point in space-time. The Wilson coefficient $C_{V/A}(\mu)$ depends on the possible ways how such a process could be introduced into τ decays. The scale μ separates the non-perturbative effects included in the condensate from the perturbative effects in the Wilson coefficient. The Wilson coefficients (and also the g^{kl}) can be calculated, but not the condensates. Only condensates with an even dimension of mass can be defined and it is important to realize, that there is no dimension 2 condensate, i.e. the leading non-perturbative correction is absent.

As the coefficients g^{kl} are different for different dimensions and different moments kl, it is possible to disentangle the various corrections by taking a sufficient number of moments ($kl = 00, 10, 11, 12,$ and 13 were used). The result is

$$\begin{aligned}
\left\langle\tfrac{\alpha_s}{\pi}GG\right\rangle &= (0.001 \pm 0.008) \text{ GeV}^4\\
\delta_V^6 &= 0.026 \pm 0.004\\
\delta_A^6 &= -0.020 \pm 0.004\\
\delta_V^8 &= -0.008 \pm 0.002\\
\delta_A^8 &= 0.004 \pm 0.002
\end{aligned}$$

4 Perturbative Corrections

As already mentioned, the perturbative corrections have been calculated to third order in α_s [1], and even some of the higher terms are known. Nevertheless the uncertainty introduced by the missing higher order corrections is the dominating systematic uncertainty. It is estimated by varying the first unknown coefficient of the perturbative series (K_4), the renormalisation scheme, and the renormalisation scale. The best guess for K_4 is 25. It is varied by ± 50. The renormalisation scheme is varied by changing β_2 from 0 to twice its $\overline{MS}$ value, and the scale is modified between $0.4\,m_\tau^2$ and $2\,m_\tau^2$.

The strong coupling α_s can be calculated from a fixed order perturbative series, i.e. by dropping any term which is more than 3rd order in α_s. This is called Fixed Order Perturbation Theory (FOPT). Alternatively one might try to improve the result by partially including higher orders. If the full information from the running of α_s is kept, the calculation is called Contour Improved Perturbation Theory (CIPT) [9]. Another approach, the Renormalon Chains Perturbation Theory (RCPT) resums the large β_0 coefficients to all orders [10]. The variation of α_s due to the partial inclusion of higher orders gives another estimate of the theoretical uncertainties, independent from the one described above. The results for α_s are

$$\begin{aligned}
0.324 \pm 0.015 &\quad \text{FOPT}\\
0.348 \pm 0.021 &\quad \text{CIPT}\\
0.306 \pm 0.012 &\quad \text{RCPT}
\end{aligned}$$

The three numbers are in agreement with each other within the errors and this confirms the estimate of the theoretical uncertainties.

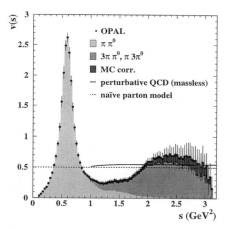

Figure 1: The vector spectral function measured from τ decays.

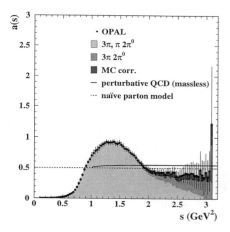

Figure 2: The axial vector spectral function from τ decays.

5 Global Quark Hadron Duality

The QCD calculations are carried out using free quarks, whereas in reality the quarks are bound into hadrons. The statement that the calculation is meaningful nevertheless is non-trivial and is called the global quark hadron duality. It is assumed to hold if the process considered averages over a sufficient number of bound states. This assumption can be tested in τ decays by reducing the number of hadrons included, i.e. by measuring α_s from the vector or axial vector spectral functions separately. The results (CIPT) are

$$0.341 \pm 0.017 \quad \text{V only}$$
$$0.357 \pm 0.019 \quad \text{A only}$$

where the errors include only the contributions uncorrelated between the two numbers. They are in good agreement with each other and with the value measured from the sum of V and A: 0.348.

6 The Running of α_s

The running of α_s can be tested in τ decays by not extending the integral of eqn. 5 all the way to m_τ^2 but stopping at an intermediate value s_0. The measurement then gives $\alpha_s(s_0)$. The resulting running of α_s is shown in fig. 3. The dominating theoretical errors are omitted from the plot, because they would be fully correlated between α_s at different values of s_0. The data points follow the

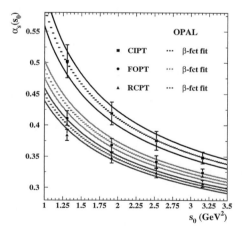

Figure 3: The running of α_s. Data points: measured values of $\alpha_s(s_0)$ with uncorrelated errors only. Graphs: predicted running of α_s, with α_s fitted to the three higher most data points. The spectral moments at $s_0 = m_\tau^2$ are used throughout the plot to fix the non-perturbative corrections.

predicted running nicely, especially for the CIPT values. That demonstrates, that the mainly perturbative calculation is applicable to values of s_0 even much lower than m_τ^2 and that the non-perturbative contributions are well under control.

7 QCD Sum Rules

Interesting tests of QCD can also be performed from the difference of the vector and axial vector spectral functions. QCD sum rules predict the value of the following integrals

$$\frac{1}{4\pi^2} \int_0^{s_0} ds \, f(s) \left[v(s) - a(s) \right] \tag{8}$$

in the limit of large s_0. There are several sum rules [11] with different weighting functions $f(s)$:

name	$f(s)$	QCD prediction
1st Weinberg	1	f_π^2
2nd Weinberg	s	0
DMO	$\frac{1}{s}$	$f_\pi^2 \frac{\langle r_\pi^2 \rangle}{3} - F_A$
Δm_π	$s \ln \frac{s}{\lambda^2}$	$-\frac{4\pi f_\pi^2}{3\alpha} \left(m_{\pi\pm}^2 - m_{\pi^0}^2 \right)$

Of particular interest is the Das-Mathur-Okubo (DMO) sum rule, because the $1/s$ weighting suppresses the high s_0 range, which is statistically less precisely measured in τ decays. The value of the integrals are displayed as a function of s_0 in fig. 4. The DMO integral has almost reached the perturbative regime at m_τ^2. This supports the perturbative treatment of R_τ for the α_s measurement, as one expects the sum $v(s) + a(s)$ to approach the perturbative regime much faster.

8 Conclusion

A new measurement of the strong coupling constant α_s by the OPAL collaboration has been summarized [6]. A value of $\alpha_s\left(m_\tau^2\right) = 0.348 \pm 0.009 \pm 0.019$ has been derived using the contour improved perturbative calculation, where the first error given is experimental and the second theoretical. Running this value up to $s = m_{Z^0}^2$, it shrinks by about a factor of 3, and the error comes down by an order of magnitude, making it a very competitive measurement: $\alpha_s\left(m_{Z^0}^2\right) = 0.1219 \pm 0.0010 \pm 0.0017$. Checks on the measurement have been presented, all of them supporting the reliability of the result.

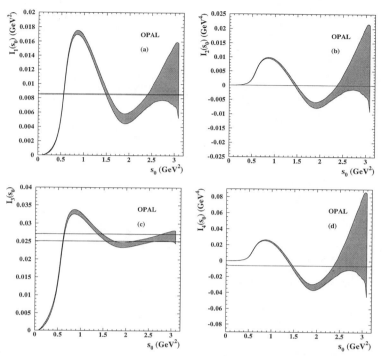

Figure 4: QCD sum rules as a function of the upper integration limit s_0. The shaded band indicates the value extracted from τ decays. The horizontal band represents the asymptotic QCD prediction. (a): 1st Weinberg, (b): 2nd Weinberg, (c): DMO and (d): the electromagnetic mass difference of the pions.

Acknowledgments

I would like to thank Sven Menke for many useful discussions.

References

1. K.G. Chetyrkin, A.L. Kataev and F.V. Tkachev, *Phys. Lett.* B **85**, 1979 (277),
   M. Dine and J. Sapirstein, *Phys. Rev. Lett.* **43**, 1979 (668),
   W. Celmaster and R.J. Gonsalves, *Phys. Rev. Lett.* **44**, 1980 (560),
   S.G. Gorishnii, A.L. Kataev and S.A. Larin, *Phys. Lett.* B **259**, 1991 (144).

2. M. A. Shifman, A. I. Vainshtein, and V. I. Zakharov, *Nucl. Phys.* B **147**, 1979 (385, 448 and 519).

3. W.J. Marciano and A. Sirlin, *Phys. Rev. Lett.* **61**, 1988 (1815).

4. S. Narison and A. Pich, *Phys. Lett.* B **211**, 1988 (183),
   E. Braaten, S. Narison and A. Pich, *Nucl. Phys.* B **373**, 1992 (581).

5. A. Pich, *Nucl. Phys.* B (Proc. Suppl.) **55 C**, 1997 (3).

6. OPAL-collab., K. Ackerstaff et al., CERN-EP-98-102, submitted to *Euro. Phys. Jour.* C.

7. F. Le Diberder and A. Pich, *Phys. Lett.* B **289**, 1992 (165).

8. T. van Ritbergen , J.A.M. Vermaseren, and S.A. Larin, *Phys. Lett.* B **400**, 1997 (379).

9. F. Le Diberder and A. Pich, *Phys. Lett.* B **286**, 1992 (147).

10. M. Neubert, *Nucl. Phys.* B **463**, 1996 (511).

11. T. Das, V.S. Mathur, and S. Okubo, *Phys. Rev. Lett.* **18**, 1967 (761), *Phys. Rev. Lett.* **19**, 1967 (895)
    S. Weinberg, *Phys. Rev. Lett.* **18**, 1967 (506),
    T. Das et al., *Phys. Rev. Lett.* **18**, 1967 (759)

Measurement of the Tau Dipole Moments at LEP

P. García–Abia

Institut für Physik der Universität Basel, CH–4056–Basel, Switzerland
E-mail: Pablo.Garcia.Abia@cern.ch

From the analysis of $e^+e^- \to \tau^+\tau^-(\gamma)$ events collected by the four experiments at LEP between 1990 and 1995, a measurement of the electromagnetic and weak dipole moments of the tau lepton (τ) has been performed. No indication of CP violation or new physics beyond the Standard Model has been observed.

1 Introduction

In the Standard Model[1] (SM) leptons are assumed to be pointlike particles. This has been proved experimentally for electrons and muons[2]. Before LEP, precise data were missing about the heaviest lepton, the tau (τ), and only limits on its anomalous magnetic moments were given at $q^2 \neq 0$[3] and, with very low statistics in the first stages of LEP, at $q^2 = 0$[4].

At LEP, high statistics data samples of $e^+e^- \to Z \to \tau^+\tau^-(\gamma)$ processes allow to measure the electromagnetic and weak dipole moments of the τ.

The electromagnetic anomalous magnetic and electric dipole moments are measured at $q^2 = 0$ from the radiation of real photons by the τ in $Z \to \tau\tau\gamma$ events. Weak dipole moments are obtained from spin measurements using the τ decay products in $Z \to \tau^+\tau^-$ events.

Large anomalous magnetic dipole moments would be a signal for substructure of the τ. A measurement of non-zero electric (electromagnetic or weak) dipole moments is an unambiguous signal of CP violation[5].

In this paper, I review the measurements of the electromagnetic and weak dipole moments of the τ lepton performed by the LEP experiments. I will introduce the fundamental concepts of the topic and will outline the basic ideas of the respective analyses. The results hereby quoted include the full statistics collected in the first phase of running of the LEP collider.

2 Electromagnetic Dipole Moments

In general a photon may couple to a τ lepton through its electric charge, magnetic dipole moment or electric dipole moment. This coupling may be parametrised using a matrix element in which the usual γ^μ is replaced by a more general Lorentz-invariant form[6],

$$\Gamma^\mu = F_1(q^2)\gamma^\mu + F_2(q^2)\frac{i}{2m_\ell}\sigma^{\mu\nu}q_\nu - F_3(q^2)\sigma^{\mu\nu}\gamma^5 q_\nu.$$

The q^2-dependent form-factors, $F_i(q^2)$, have familiar interpretations for $q^2 = 0$: $F_1(0) \equiv Q_\tau$ is the electric charge; $F_2(0) \equiv a_\tau$ is the anomalous magnetic moment ($a_\tau \equiv (g_\tau - 2)/2$); and $F_3(0) \equiv d_\tau/Q_\tau$, where d_τ is the electric dipole moment. In the SM a_τ is non-zero due to loop diagrams and is predicted to be $a_\tau^{SM} = 0.001\,177\,3(3)$[7]. A non-zero value of d_τ is forbidden by both P invariance and T invariance. Assuming CPT invariance, observation of a non-zero value of d_τ would imply CP violation.

2.1 Data Analysis

Anomalous electromagnetic moments would influence the production of high energy isolated photons in $e^+e^- \to \tau^+\tau^-\gamma$ events, compared to the SM initial and final state radiation processes[8] (figure 1). In this calculation, the cross section contains terms linear and quadratic in a_τ, which come from the interference with the SM amplitude.

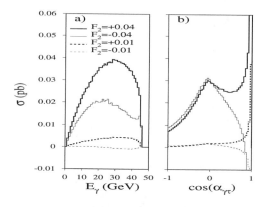

Figure 1: Effect of the additional contribution to the Standard Model cross section of the process $e^+e^- \to \tau^+\tau^-\gamma$, due to non zero values of F_2, in the a) photon energy and b) $\tau\gamma$ opening angle distributions.

The OPAL[9] and L3[10] experiments selected $e^+e^- \to \tau^+\tau^-\gamma$ events containing at least one high energy isolated photon. The samples used in the analysis collected during the 1990-1995 LEP runs at the Z peak, corresponding to an integrated luminosity of about 100 pb^{-1} per experiment, consisted of 1429 events for OPAL and 1590 for L3. A typical $\tau\tau\gamma$ event is shown in figure 2.

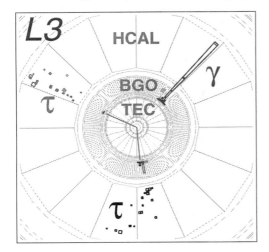

Figure 2: Cross view of a typical $e^+e^- \to \tau^+\tau^-\gamma$ event as seen in the L3 Detector. The Vertex chamber (TEC) and the calorimeters (BGO, HCAL) are also drawn.

The anomalous moments were determined from fits to the distributions of the photon energy (OPAL, L3) and isolation angle (L3), allowing for the SM backgrounds. Figure 3 shows the energy and angular photon spectra from L3 data, together with the SM expectation and the expected signal for a $a_\tau = 0.1$ contribution.

2.2 Results

The LEP results on electric and anomalous magnetic dipole moments are summarised in figure 4.

Detailed studies on the possible systematic uncertainties were performed by the two collaborations [9,10]. The most important sources of systematic error are those affecting the normalisation of the $\tau\tau\gamma$ sample: uncertainty in the $e^+e^- \to \tau^+\tau^-$ cross section, event selection cut and photon reconstruction efficiencies. Studies were performed on the effect of multiple photon radiation and of the theoretical modelling of the $e^+e^- \to \tau^+\tau^-\gamma$ process in the fitted distributions. They were shown to have a negligible effect on the result of the fit. The overall systematic uncertainty is at the level of the statistical error.

A limit is established by combining the individual likelihood functions, taking appropriate care of the correlated systematic errors. The LEP limits at the 95% CL on electric and anomalous magnetic dipole moments are:

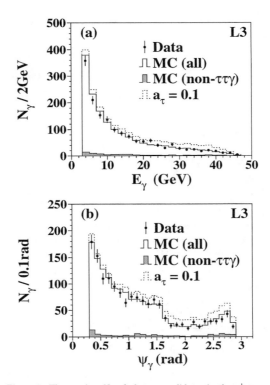

Figure 3: The number N_γ of photon candidates in the $e^+e^- \to \tau^+\tau^-\gamma$ sample (L3) as a function of (a) photon energy E_γ and (b) the isolation angle Ψ_γ with respect to the τ. The points with error bars denote the data and the solid histograms denote the Monte Carlo predictions, assuming the SM values of a_τ and d_τ. For illustration, the dashed histograms show how the distributions would appear for $a_\tau = 0.1$. Both the increase in the total cross section and the relatively greater importance of photons with large E_γ and Ψ_γ are evident.

$$-0.05 < a_\tau < 0.05$$
$$(-3 < d_\tau < 3) \times 10^{-16} \quad e \cdot cm.$$

The SM predictions are consistent with the measured values. Therefore, these give no indication of either CP violation or new physics beyond the SM.

3 Weak Dipole Moments

To be sensitive a to CP asymmetry in $Z \to f\bar{f}$ production, access to the spin of the final state fermions is required. At LEP, this can be done by exploiting the kinematics of the decay products of the short lived τ leptons.

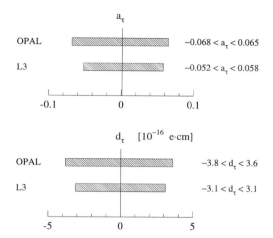

Figure 4: LEP results on the electromagnetic dipole moments a_τ and d_τ.

Different analysis methods were used by the different experiments. L3[11] based its measurement in the analysis of the spin of single τ decays. ALEPH[12], DELPHI[13] and OPAL[14] used the spin correlations of both τ decays.

All the experiments selected $e^+e^- \to \tau^+\tau^-$ events applying standard cuts. The analysis of L3 was restricted to π and ρ final states. The other collaborations used most of the available τ decay channels. A list of the data samples and the τ decay channels analysed by each experiment is shown in table 1.

Table 1: Data samples and τ decay channels analysed by the LEP experiments.

Experiment	Data	#decays	Decay channels
ALEPH	1990–95	83 000	e, μ, π, ρ, a_1
DELPHI	1992–95	22 000	e, μ, π, ρ, a_1 (3pr)
OPAL	1991–95	70 000	e, μ, π, ρ, a_1 (3pr)
L3	1991–95	9 000	π, ρ

3.1 L3 Method

This analysis exploited the polarisation of single τ decays. Events containing only $\tau \to \pi\nu$ and/or $\tau \to \rho\nu$ decays were used. The τ flight direction was reconstructed in these final states. Special azimuthal asymmetries were defined which were sensitive to the transverse and normal components of the polarisation of each single τ [15],

allowing for the measurement of both d_τ^w and a_τ^w. The results on a_τ^w are:

$$\text{Re}\,(a_\tau^w) = (0.0 \pm 1.6 \pm 2.3\,) \times 10^{-3}$$
$$\text{Im}\,(a_\tau^w) = (-1.0 \pm 3.6 \pm 4.3\,) \times 10^{-3}.$$

This is the first measurement of a_τ^w. Results on d_τ^w are summarised, together with the other LEP results, in section 3.3. Details of the analysis can be found elsewhere [11,16].

3.2 ALEPH, DELPHI and OPAL Method

These experiments tested the CP invariance of τ pair production using CP-odd observables constructed from the measured momenta and energies of the τ decay products. The optimal observables are given by:

$$\mathcal{O}^{\text{Re}} = \frac{\mathcal{M}_{\text{CP}}^{\text{Re}}}{\mathcal{M}_{\text{SM}}} \quad, \quad \mathcal{O}^{\text{Im}} = \frac{\mathcal{M}_{\text{CP}}^{\text{Im}}}{\mathcal{M}_{\text{SM}}} ,$$

where $\mathcal{M}_{\text{SM}}$ is the squared SM matrix element of the $e^+e^- \to \tau^+\tau^-$ process and $\mathcal{M}_{\text{CP}}^{\text{Re}}$ and $\mathcal{M}_{\text{CP}}^{\text{Im}}$ are the CP violating contributions to the squared amplitude [17]. These matrix elements are computed from an estimation of the τ spin and flight directions.

The mean value of the optimal observables can be expressed in terms of d_τ^w as follows:

$$\langle \mathcal{O}^{\text{Re}} \rangle_{AB} = \frac{m_Z}{e} \cdot c_{AB} \cdot \text{Re}\,(d_\tau^w)$$
$$\langle \mathcal{O}^{\text{Im}} \rangle_{AB} = \frac{m_Z}{e} \cdot f_{AB} \cdot \text{Im}\,(d_\tau^w) ,$$

where c_{AB} and f_{AB} are the sensitivities for a given final state $\tau^+\tau^- \to AB + n\nu$. They are calculated performing detailed Monte Carlo studies, including a complete simulation of the detectors.

With this definition of optimal observables, any significant observed deviation from zero points unambiguously to CP violation. As an example, the values of $\text{Re}\,(d_\tau^w)$ measured by the OPAL collaboration for the different final states are shown in figure 5. Details of the analyses from the different experiments can be found elsewhere [12,13,14].

3.3 Results

The LEP results on the weak electric dipole moment are summarised in figure 6. The OPAL and L3 results are final, while those of ALEPH and DELPHI are preliminary. The L3 results on the weak anomalous magnetic moment are given in section 3.1.

Detailed studies on the possible systematic uncertainties were performed by all the collaborations. The

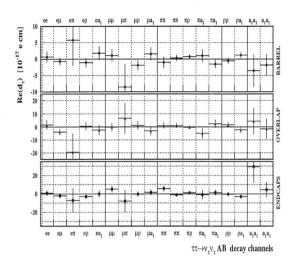

Figure 5: Real part of the weak dipole moment for the different decay topologies and detector regions (OPAL). Decay topologies which appear twice refer to different data periods (1991–92 and 1993–95, respectively).

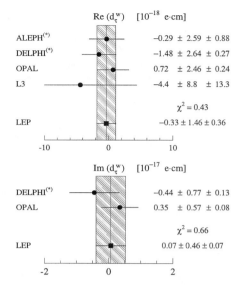

Figure 6: LEP results on the weak electric dipole moment d_τ^w. The OPAL and L3 results are final, the ALEPH and DELPHI ones are preliminary[(*)].

most important sources of systematic error are the alignment, homogeneity and CP invariance of the detectors. The latter has been checked to be at the level of 10^{-19} e·cm by artificially destroying the spin correlations of the final state in an event by event basis. The overall systematic uncertainty is well below the statistical error.

A limit is established by combining the individual values assuming gaussian errors and considering the systematic errors completely uncorrelated. The LEP limits at the 95% CL on the real and imaginary parts of the weak electric dipole moment are:

$$(-3.3 < \text{Re} \ (d_\tau^w) < 2.7 \ ) \times 10^{-18} \quad e \cdot cm$$
$$(-0.9 < \text{Im} \ (d_\tau^w) < 1.0 \ ) \times 10^{-17} \quad e \cdot cm$$
$$|d_\tau^w| \quad < 0.7 \ \times 10^{-17} \quad e \cdot cm.$$

The SM predictions are consistent with the measured values both of a_τ^w and d_τ^w. As a consequence, no indication of either CP violation or new physics beyond the SM has been found.

Acknowledgements

I would like to thank my colleagues from the LEP collaborations, in particular W. Lohmann, T. Paul, L. Taylor and N. Wermes.

References

1. S.L. Glashow, *Nucl. Phys.* **22**, 579 (1961);
   S. Weinberg, *Phys. Rev. Lett.* **19**, 1264 (1967);
   A. Salam, *Elementary Particle Theory*, Ed. N. Svartholm, Stockholm, "Almquist and Wiksell" (1968), 367.
2. E. R. Cohen and B. N. Taylor, *Rev. Mod. Phys.* **59**, 1121 (1987).
3. D. J. Silverman and G. L. Shaw, *Phys. Rev. D* **27**, 1196 (1983);
   R. Escribano and E. Massó, *Phys. Lett.* B **395**, 369 (1997).
4. J. A. Grifols and A. Méndez, *Phys. Lett.* B **255**, 611 (1991), Erratum *ibid* B **259**, 512 (1991).
5. W. Bernreuther and O. Nachtmann, *Phys. Rev. Lett.* **63**, 2787 (1989);
   N. Wermes, Invited talk at TAU96, Colorado (1996); also Bonn University preprint BONN-HE-96-10.
6. S. Weinberg, *The Quantum Theory of Fields I*, Cambridge University Press, Cambridge, (1995) 452.
7. M. A. Samuel, G. Li and R. Mendel, *Phys. Rev. Lett.* **67**, 668 (1991), Erratum *ibid* **69**, 995 (1992);
   F. Hamzeh and N.F. Nasrallah, *Phys. Lett.* B **373**, 211 (1996).

8. J. Biebel and T. Riemann, *Z. Phys.* C **76**, 53 (1997);
   S.S. Gau, T. Paul, J. Swain, and L. Taylor, *Nucl. Phys.* B **523**, 439-449 (1998);
   T. Paul and Z. Wąs, HEP-PH/9801301.

9. OPAL Collaboration, K. Ackerstaff *et al.*, *Phys. Lett.* B **431**, 188-198 (1998).

10. L3 Collaboration, M. Acciarri *et al.*, CERN-EP/98-045, *Phys. Lett.* B accepted (June 1998).

11. L3 Collaboration, M. Acciarri *et al.*, *Phys. Lett.* B **426**, 207-216 (1998).

12. ALEPH Collaboration, D. Buskulic *et al.*, *Phys. Lett.* B **346**, 371-378 (1995); ALEPH Collaboration, Paper PA08-030 contributed to ICHEP96, Warsaw (1996).

13. DELPHI Collaboration, M.-C. Chen *et al.*, Paper #321 submitted to HEP97, Jerusalem (1997).

14. OPAL Collaboration, K. Ackerstaff *et al.*, *Z. Phys.* C **74**, 403-412 (1997).

15. J. Bernabéu, G. A. González-Sprinberg and J. Vidal, *Phys. Lett.* B **326**, 168 (1994);
   U. Stiegler, *Z. Phys.* C **57**, 511 (1993).

16. E. Sánchez, Invited talk at TAU96, Colorado (1996).

17. D. Atwood and A. Soni, *Phys. Rev.* D **45**, 2405 (1992);
   M. Diehl and O. Nachtmann, *Z. Phys.* C **62**, 397 (1994);
   P. Overmann, Dortmund University preprint DO-TH 93-24 (1993).

FIRST SEARCH FOR CP VIOLATION IN TAU LEPTON DECAY

Karl M. Ecklund

Representing the CLEO Collaboration

Laboratory of Nuclear Studies,Cornell University, Ithaca, New York 14853
E-mail: kme@mail.lns.cornell.edu

We have performed the first search for CP violation in tau lepton decay. CP violation in lepton decay does not occur in the minimal standard model but can occur in extensions such as the multi-Higgs doublet model. It appears as a characteristic difference between the τ^- and τ^+ decay angular distributions for the semi-leptonic decay modes such as $\tau^- \to K^0\pi^-\nu$. We define an observable asymmetry to exploit this and find no evidence for any CP violation.

1 Introduction and Motivation

Despite observation of CP violation in the kaon system over 30 years ago, [1] its origin is still not understood. In the Standard Model (SM), CP violation arises from a complex phase in the CKM Matrix [2] which describes the coupling of the quarks to the W boson in flavor changing charged currents. However, the observed baryon-antibaryon asymmetry of the universe is much larger than expected from the SM. [3] There is no CP violation in the lepton sector of the SM, but in some extensions to the SM a new charged scalar boson may give rise to CP violation in lepton decays. A search for CP violation in the lepton sector may lead to a better understanding of the matter-antimatter asymmetry of the universe.

In this paper I report on a search for CP violation in tau lepton decays undertaken by the CLEO collaboration. The paper is organized as follows. In Section 2 I continue the discussion of CP violation in tau decays, followed by a description of the experimental technique in Section 3. Results and interpretation of the search are discussed in Section 4.

2 CP Violation in τ decays

One possible mechanism for CP violation in tau decays is the interference of two amplitudes with both a CP-odd phase difference and a CP-even phase difference. [a] In the SM, semi-leptonic tau decays proceed via W exchange. An additional amplitude for tau decay occurs through a charged Higgs scalar in multi-Higgs doublet extensions to the Standard Model. [4] Any extension to the SM that has a new charged scalar particle would provide a similar mechanism. Figure 1 shows the two amplitudes that would contribute to tau decay in this case. If these amplitudes have a relative CP-odd phase θ_{cp} (which arises from the complex coupling to the charged scalar) and a CP-even phase difference δ (which arises from final state

---
[a] Similar to the mechanism expected to lead to CP violation in the decay $B^0 \to J/\psi K_S^0$, where two amplitudes, $B^0 \to J/\psi K_S^0$ and $B^0 \to \bar{B}^0 \to J/\psi K_S^0$, interfere causing CP violation.

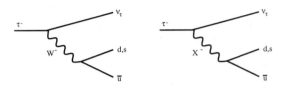

Figure 1: Tau decay mediated by W vector boson and hypothetical charged scalar (X) exchange.

interactions), a CP-violating asymmetry results from the interference term. For τ^- decays we have

$$A^- = |A_1 + A_2|^2 = A_1^2 + A_2^2 + 2A_1A_2\cos(\delta - \theta_{cp}), \quad (1)$$

and for τ^+ decays

$$A^+ = |A_1 + A_2|^2 = A_1^2 + A_2^2 + 2A_1A_2\cos(\delta + \theta_{cp}). \quad (2)$$

The net result is an asymmetry

$$A^- - A^+ = 4A_1A_2\sin\delta\sin\theta_{cp} \quad (3)$$

As long as the relative phases are nontrivial ($\sin\delta, \sin\theta_{cp} \neq 0$) there is a CP-violating asymmetry. The CP-odd phase comes from the non-Standard Model scalar coupling constant. The CP-even phase δ from the strong interaction is likely to be nontrivial since the W exchange is p-wave and the scalar X exchange must be s-wave. For tau decays to final states with two hadrons, final state interactions may lead to a strong phase difference δ. In the particular case $\tau^- \to K_S^0\pi^-\nu_\tau$, the standard model p-wave is dominated by the $K^*(892)$ resonance, yielding a strong phase $\delta \approx \pi/2$ compared to the s-wave exchange. Selecting a final state with an s-quark rather than a d-quark also increases any expected signal from scalar exchange with Higgs-like mass dependent couplings.

In order to observe the CP-violating asymmetry the matrix elements for the two amplitudes, which depend on kinematic variables, must be evaluated and an observable

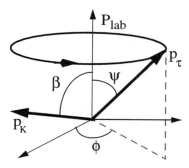

Figure 2: Definition of kinematic variables in hadronic rest frame. The direction of the laboratory frame as viewed from the hadronic rest frame is $\vec{p}_{\text{lab}}$ and β is the angle between the direction of the K^0 ($\vec{p}_K$) and $\vec{p}_{\text{lab}}$. ψ is the angle between the tau flight direction ($\vec{p}_\tau$) and $\vec{p}_{lab}$.

calculated while integrating over unobserved variables. Kühn and Mirkes [5] have evaluated the matrix elements in the rest frame of the hadronic system ($\vec{p}_\pi + \vec{p}_K = 0$) and find the interference term to be proportional to $\cos\beta\cos\psi$ where the angles are defined in Figure 2. In the hadronic rest frame, the tau flight direction is fully determined apart from an unknown azimuthal angle ϕ due to the unmeasured neutrino in the final state. In performing the integration over unobserved kinematic variables, they have also integrated over the unknown angle ϕ. The angles β and ψ are fully determined given the final state hadron momenta.

Using the full result from Kühn and Mirkes, the CP asymmetry may be written as a function of the angles ψ and β.

$$A_{\text{theory}}(\cos\beta\cos\psi) = Kg\sin\delta\sin\theta_{cp}\cos\beta\cos\psi, \quad (4)$$

where K depends on form factors and g is the ratio of scalar to vector coupling strength (i.e. g is in units of $G_F/2\sqrt{2}$). Experimentally CP violation would be observed as an asymmetry in the number of τ^- and τ^+ decays as a function of the angles β and ψ. Note that the asymmetry vanishes when integrating over β and ψ. We define the experimentally observable asymmetry in terms of the number of observed τ^- and τ^+ decays in an interval of $\cos\beta\cos\psi$:

$$A_{\text{obs}}^{\text{sample}}(\cos\beta\cos\psi) = \frac{N^+(\cos\beta\cos\psi) - N^-(\cos\beta\cos\psi)}{N^+(\cos\beta\cos\psi) + N^-(\cos\beta\cos\psi)} \quad (5)$$

3 Experimental Method

We wish to collect a clean sample of $\tau^- \rightarrow K_S^0\pi^-\nu_\tau$ decays in order to measure the asymmetry. However, the presence of background in the sample will dilute the observed asymmetry. The observed sample will contain backgrounds each with fraction $f_{\text{mode}}^{\text{sample}}$ of the total sample and with theoretical asymmetry α_{mode} relative to $K^0\pi^-\nu_\tau$. Finite resolution of the detector may dilute the observed asymmetry by a factor D_{det}, and differences in detection efficiency for τ^- and τ^+ events may bias the asymmetry measurement by the amount $A_{\text{det}}^{\text{sample}}$. Assuming any such bias to be small we may write the observed asymmetry as follows:

$$A_{\text{obs}}^{\text{sample}} = \sum_{\text{mode}} f_{\text{mode}}^{\text{sample}}\alpha_{\text{mode}}D_{\text{det}}A_{\text{theory}}^{\text{sig}} + A_{\text{det}}^{\text{sample}}. \quad (6)$$

3.1 CLEO Experiment

The data used in this analysis have been collected from e^+e^- collisions at a center of mass energy ($\sqrt{s}$) of 10.6 GeV with the CLEO II detector at the Cornell Electron Storage Ring (CESR). The total integrated luminosity of the data sample is 4.8 fb^{-1}, corresponding to the production of 4.4×10^6 $\tau^+\tau^-$ events. This analysis makes use of the CLEO II drift chambers for charged particle tracking and the CsI crystal calorimeter for photon and π^0 identification. The CLEO II detector has been described in detail elsewhere. [6]

3.2 Event Selection and Reconstruction

We reconstruct $\tau^\pm \rightarrow K_S^0h^\pm\nu$ events where the K_S^0 decays to two charged pions and $h^\pm$ may be any charged track due to the absence of particle identification for the range of relevant momenta. We first select tau pair events using a 1-prong tag in the opposite hemisphere from the $K_S^0h^\pm\nu$ candidate. We require a total of four tracks with zero net charge. The tag track must be separated by at least 90° from any other track. To reduce backgrounds from two photon events and beam gas interactions, we require momentum transverse to the beam axis $p_T > 0.025E_{\text{beam}}(E_{\text{beam}} = \sqrt{s}/2)$ and $|\cos\theta| < 0.90$, where θ is the polar angle with respect to the beam direction. The tag track must satisfy $p_T > 0.05E_{\text{beam}}$ and $|\cos\theta| < 0.80$. We further suppress backgrounds by requiring the net missing momentum to be greater than $0.03E_{\text{beam}}$ and have $|\cos\theta_{\text{miss}}| < 0.95$.

We make use of the calorimeter to reduce contamination from other tau decays on the signal side. We allow up to one neutral pion in the tag hemisphere, but veto any events with additional showers of $E > 350$ MeV or

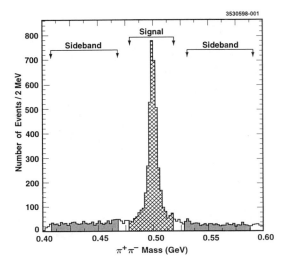

3530598-001

Figure 3: The $\pi^+\pi^-$ invariant mass distribution for the final data sample.

Table 1: Observed asymmetries in signal (Sig) and sideband (Side) regions and their difference (Sub=Signal-Side).

	$A_{\mathrm{obs}}(\cos\beta\cos\psi < 0)$	$A_{\mathrm{obs}}(\cos\beta\cos\psi > 0)$
Sig	0.058 ± 0.023	$0.024 \quad \pm \quad 0.021$
Side	0.049 ± 0.030	$0.034 \quad \pm \quad 0.033$
Sub	0.009 ± 0.038	$-0.010 \quad \pm \quad 0.039$

small variations in the selection criteria. The samples exhibit an overall rate asymmetry not expected from CP-violating interference effects. [b] In addition to CP violation, a non-zero asymmetry can arise from either a statistical fluctuation or a difference in detection efficiency for positive and negatively charged particles ($A_{\mathrm{det}}^{\mathrm{sample}}$). A Monte Carlo simulation is used to estimate the expected CP violation in terms of the extended standard model scalar coupling parameters. The sideband sample is used to empirically estimate the asymmetry due to detector effects.

photon candidates above 300 (100) MeV in the tag (signal) hemisphere. These shower vetos reduce backgrounds from tau decays containing an undetected K_L^0 or π^0.

A K_S^0 vertex is reconstructed by pairing oppositely charged tracks and defining the vertex at the intersection of the two tracks in the x-y plane (transverse to the beam direction). In order to be consistent with the decay of a long-lived K^0, we require the vertex point to have a separation of at least 5 mm from the interaction point (IP). In addition the z separation of the two tracks must be less than 12 mm at their intersection in the x-y plane, and the reconstructed K_S^0 momentum vector must have a distance of closest approach to the IP less than 2 mm. The reconstructed invariant mass assuming the tracks to be pions must be within 20 MeV/c^2 of the K^0 mass. We define a sideband region 30–90 MeV/c^2 above the K^0 mass for use as a control sample. Figure 3 shows the K_S^0 candidate invariant mass distribution.

4 Results and Interpretation

4.1 Asymmetry Measurement

Using the selected events we measure the asymmetry for both signal and sideband in two intervals of $\cos\beta\cos\psi$, $A_{\mathrm{obs}}^{\mathrm{sig,side}}(\cos\beta\cos\psi < 0)$ and $A_{\mathrm{obs}}^{\mathrm{sig,side}}(\cos\beta\cos\psi > 0)$, given in Table 1. Both signal and sideband exhibit similar non-zero asymmetries but with low statistical significance. The measured asymmetries are insensitive to

4.2 Monte Carlo Simulation of Expected CP Violation

Under the assumption of a multi-Higgs doublet model, we estimate the expected CP-violating asymmetry, $A_{\mathrm{theory}}^{K_S^0\pi^-\nu_\tau}$, for a pure $\tau^- \to K_S^0\pi^-\nu_\tau$ sample using the KORALB Monte Carlo [7] to generate τ-pairs. It has been modified to include a scalar Higgs coupling in addition to the Standard Model W boson coupling for the signal $K_S^0\pi^-\nu_\tau$ mode. We model the s-wave as a nonresonant two body decay. For the p-wave we use a relativistic Breit-Wigner with energy dependent width for the $K^*(892)$ resonance. We normalize the Breit-Wigner to have equal amplitude to the non-resonant s-wave at $q^2 = 0$, where q is the four momentum transfer. For kinematically allowed q^2, the p-wave dominates, and the average strong phase difference is $\langle \delta_{strong} \rangle = \pi/2$. The GEANT code [8] is used to simulate detector response and assumes equal detection efficiencies for positive and negatively charged particles. We calculate $A_{\mathrm{theory}}^{K_S^0\pi^-\nu_\tau}(\cos\beta\cos\psi < 0) = -0.033g\sin\theta_{cp}$ and $A_{\mathrm{theory}}^{K_S^0\pi^-\nu_\tau}(\cos\beta\cos\psi > 0) = +0.033g\sin\theta_{cp}$ for a pure $\tau^- \to K_S^0\pi^-\nu_\tau$ signal. The dilution from detector reso-

[b]In the most general case we do not expect an overall rate asymmetry. The angular dependence of an s, p, d etc.,-wave amplitude is given by the Legendre polynomials $P_{m=0,1,2}(x)$ so that the angular dependence of a CP-violating interference term is given by $P_m(x)P_n(x)$ and since

$$\int_{-1}^{+1} dx P_m(x)P_n(x) = \frac{2}{2n+1}\delta_{mn}$$

we can only have a non-zero rate asymmetry if $m = n$ in which case the strong phases of the two amplitudes will be equal ($\sin\delta_{strong} = 0$) so that the overall rate asymmetry will be zero.

Table 2: Signal and sideband mode composition. $f_{\text{mode}}^{\text{sig,side}}$ is the fraction of the total signal or sideband sample for a particular mode. α_{mode} is the approximate magnitude of asymmetry expected relative to the $\tau^- \to K_S^0 \pi^- \nu_\tau$ mode. The last column gives the dilution factor expected from backgrounds when the measured asymmetry in the sideband control sample is subtracted from the measured asymmetry in the signal sample, $D_{\text{bkg}} = \sum_{\text{mode}} (f_{\text{mode}}^{\text{sig}} - f_{\text{mode}}^{\text{side}})\alpha_{\text{mode}} = 0.48$.

Tau Mode	α_{mode}	$f_{\text{mode}}^{\text{sig}}$	$f_{\text{mode}}^{\text{side}}$	$(f_{\text{mode}}^{\text{sig}} - f_{\text{mode}}^{\text{side}})\alpha_{\text{mode}}$
$K_S^0 \pi^- \nu_\tau$	1	0.525 ± 0.057	0.043 ± 0.005	0.4820 ± 0.0570
$K_S^0 K^- \nu_\tau$	1/20	0.124 ± 0.036	0.009 ± 0.003	0.0060 ± 0.0020
$a_1^- \nu_\tau$	1/80	0.106 ± 0.003	0.620 ± 0.013	-0.0064 ± 0.0002
$K_S^0 \pi^- \pi^0 \nu_\tau$	1/4	0.066 ± 0.016	0.006 ± 0.002	0.0150 ± 0.0040
$K_S^0 K_L^0 \pi^- \nu_\tau$	1/80	0.055 ± 0.018	0.003 ± 0.001	$0.0007 \pm 0\ .0002$
$K_S^0 K^- \pi^0 \nu_\tau$	1/20	0.030 ± 0.008	0.003 ± 0.001	0.0014 ± 0.0004
$\pi^+ \pi^- \pi^- \pi^0 \nu_\tau$	1/20	0.028 ± 0.002	0.167 ± 0.007	-0.0070 ± 0.0004
$K^- \pi^+ \pi^- \nu_\tau$	1/4	0.008 ± 0.003	0.043 ± 0.007	-0.0090 ± 0.0020
others	0	0.012 ± 0.002	0.071 ± 0.017	0
$q\bar{q}$	0	0.044 ± 0.003	0.037 ± 0.003	0
Total	-	1.00 ± 0.07	1.00 ± 0.03	0.48 ± 0.06

lution effects is negligible due to the high precision of the tracking: $D_{\text{det}} = 1.0$.

4.3 Simulation of Expected Backgrounds

From Equation (6) we see that to compare $A_{\text{theory}}^{K_S^0 \pi^- \nu_\tau}$ to $A_{\text{obs}}^{\text{sample}}$ we must take into account the diluting effect of backgrounds ($f_{\text{mode}}^{\text{sample}}, \alpha_{\text{mode}}$) since the signal region is not pure $K_S^0 \pi^- \nu_\tau$. We use the KORALB Monte Carlo to estimate the contributions to the signal and sideband samples from other tau decay modes, and the Lund Monte Carlo[9] to estimate the contamination from $e^+ e^- \to q\bar{q}$ events. Table 2 gives the estimated signal and sideband compositions by mode. The backgrounds arise from our inability to distinguish kaons and pions in the desired momentum range, lack of K_L^0 identification, particles that fall outside the fiducial region of the detector, and charged track mismeasurement. We note that the signal and sidebands are composed of different modes, and it is unlikely that both samples would exhibit a similar CP asymmetry since the strong phases, and possibly the coupling strengths, are different for each mode.

4.4 Sideband and Detection Asymmetries

It is also necessary to estimate the asymmetry expected from charge dependent detection efficiencies represented in Equation (6) by $A_{\text{det}}^{\text{sample}}$. We use the $m(\pi\pi)$ sideband region, where we expect little CP violation, to estimate the bias in the asymmetry measurement due to detector effects. Both signal and sideband samples satisfy the same kinematic selection criteria so we expect

$$A_{\text{det}}^{\text{sig}} = A_{\text{det}}^{\text{side}}. \qquad (7)$$

We expect a number of charge dependent detection efficiencies to affect the measured asymmetries. Studies of pions from an independent $K_S^0 \to \pi^+ \pi^-$ sample indicate that at low momentum the reconstruction efficiency for π^+ is slightly greater than π^- and also the reconstruction of a K_S^0 in close proximity to a π^+ is slightly more efficient than for a π^-. The hadronic interaction of charged pions and kaons with the CsI crystals can produce fake electromagnetic clusters which can then be used to veto the event. The cross-sections for these interactions are different for positive and negative charged hadrons and cause charge dependent detection inefficiencies. All of these effects are more pronounced at lower momentum (< 1 GeV) and thus for $\cos \beta \cos \psi < 0.0$ since the pion from $\tau^- \to K_S^0 \pi^- \nu_\tau$ tends to be of lower momentum in this region.

The sidebands may be used as a control sample to estimate these combined effects in our signal region in a simple empirical way providing we assume that any CP-violating effects are suppressed in the sideband modes. Table 2 gives the expected CP-violating asymmetry α_{mode} relative to the $\tau^- \to K_S^0 \pi^- \nu_\tau$ signal mode for both signal and sideband samples. Two effects cause the expected CP asymmetry in the background modes to be less than in the signal mode. First from the mass dependence of the Higgs coupling and second due to the dilution of the p-wave nature of the standard model final state. For example, the $\tau^- \to \pi^- \pi^+ \pi^- \nu_\tau$ mode is dominated in the standard model decay by an s-wave $\tau^- \to a_1^- \nu_\tau \to \rho^0 \pi^- \nu_\tau$ intermediate state which dilutes the $s - p$ wave interference by a factor of ≈ 4 in addition to a mass suppression of m_u/m_s relative to the $K_S^0 \pi^-$ mode. From Table 2 we see that the sideband should have negligible asymmetry with respect to the signal under the assumption of a mass dependent coupling and can be used as a control sample to subtract the charge dependent detector asymmetries common to both signal and sideband. Using Equations (6) and (7):

$$
\begin{aligned}
A_{\text{obs}}^{\text{sub}} &= A_{\text{obs}}^{\text{sig}} - A_{\text{obs}}^{\text{side}} \\
&= \sum_{\text{mode}} (f_{\text{mode}}^{\text{sig}} - f_{\text{mode}}^{\text{side}}) \alpha_{\text{mode}} A_{\text{theory}}^{K_S^0 \pi^- \nu_\tau} \\
&= D_{\text{bkg}} A_{\text{theory}}^{K_S^0 \pi^- \nu_\tau}
\end{aligned} \tag{8}
$$

Taking D_{bkg} from Table 2 we see that if a true CP violation exists the subtracted quantity should still exhibit a significant asymmetry but diluted by a factor 0.48.

4.5 Corrected CP Asymmetry Results

From Table 1 the measured subtracted asymmetry is $A_{\text{obs}}^{\text{sub}}(\cos\beta\cos\psi < 0) = 0.009 \pm 0.038$, $A_{\text{obs}}^{\text{sub}}(\cos\beta\cos\psi > 0) = -0.010 \pm 0.039$ which is consistent with no CP violation. This can be compared with a revised Monte Carlo estimate that takes into account the background dilution factor, $D_{\text{bkg}} A_{\text{theory}}^{K_S^0 \pi^- \nu_\tau}(\cos\beta\cos\psi < 0) = -0.016g\sin\theta_{cp}$, $D_{\text{bkg}} A_{\text{theory}}^{K_S^0 \pi^- \nu_\tau}(\cos\beta\cos\psi > 0) = 0.016g\sin\theta_{cp}$, to give the constraint $-1.7 < g\sin\theta_{cp} < 0.6$ at the 90 % confidence limit.

4.6 Cross-check of Sideband CP Asymmetry

To cross check our assumption of suppressed CP violation in the sidebands we measure the asymmetry in an independent high-purity high-statistics data sample of the dominant sideband mode, $\tau^- \to a_1^- \nu_\tau$, using the selection criteria in Balest et al.[10] We find $A_{\text{obs}}^{a_1}(\cos\beta\cos\psi < 0) = -0.0013 \pm 0.0047$, $A_{\text{obs}}^{a_1}(\cos\beta\cos\psi > 0) = -0.0023 \pm 0.0047$ giving no evidence for CP violation. The higher track momentum and cluster veto thresholds combined with the absence of a K_S^0 requirement from this sample removes the contribution to the asymmetry from charge dependent detection inefficiencies but a true CP-violating effect should remain. We note that by measuring the CP-violating asymmetry in the dominant sideband mode as zero our results are approximately valid for a non-mass dependent coupling. However, we cannot fully relax this assumption due to the difficulty of empirically isolating a sample of each background mode in which to measure the asymmetry.

5   Conclusion

We have performed the first search for CP violation in tau lepton decay. We find no evidence for CP violation and constrain the coupling strength g (in units of $G_F/2\sqrt{2}$) and phase θ_{cp} of a new CP-violating mass-dependent scalar interaction, $-1.7 < g\sin\theta_{cp} < 0.6$ at the 90% confidence limit, assuming a non-resonant amplitude for the scalar decay. At the forthcoming B-factory experiments we anticipate substantial improvements in sensitivity both from the increased statistical precision and detector improvements. The addition of K_L^0 detection, K^-/π^- separation and improved precision tracking will significantly decrease the backgrounds which dilute the asymmetries.

Acknowledgments

CLEO gratefully acknowledges the effort of the CESR staff in providing us with excellent luminosity and running conditions. This work was supported by the National Science Foundation, the U.S. Department of Energy, Research Corporation, the Natural Sciences and Engineering Research Council of Canada, the A.P. Sloan Foundation, the Swiss National Science Foundation, and the Alexander von Humboldt Stiftung.

References

1. J.H.Christenson, J.W. Cronin, V.L. Fitch, and R. Turlay, *Phys. Rev. Lett.* **13**, 138 (1964).
2. N. Cabbibo, *Phys. Rev. Lett.* **10**, 531 (1963); M. Kobayashi and T. Maskawa, *Prog. Theor. Phys.* **49**, 652 (1973).
3. E. W. Kolb and M. S. Turner, *The Early Universe*, (Addison Wesley, New York, 1994).
4. S. Weinberg, *Phys. Rev. Lett.* **37**, 657 (1976).
5. J.H. Kühn and E. Mirkes, *Phys. Lett.* B **398**, 407 (1997); Y.S. Tsai, *Nucl. Phys.* B (Proc. Suppl.) **55C**, 293 (1997).
6. Y. Kubota et al., *Nucl. Instrum. Methods* A **320**, 66 (1992).
7. KORALB (v. 2.2)/TAUOLA (v 2.4): S. Jadach and Z. Was, *Comput. Phys. Commun.* **36** , 191 (1985) and *ibid.* **64**, 267 (1991); S. Jadach, J.H. Kühn and Z. Was, *Comput. Phys. Commun.* **64**, 275 (1991), *ibid.* **70**, 69 (1992), and *ibid.* **76**, 361 (1993).
8. **GEANT 3.15**: R. Brun et al., CERN DD/EE/84-1
9. **JETSET 7.3**: T. Sjostrand and M. Bengtsson, *Comput. Phys. Commun.* **43**, 367 (1987).
10. R. Balest et al., *Phys. Rev. Lett.* **75**, 3809 (1995).

On A New Class of Models for Soft CP Violation

David Bowser-Chao[1], Darwin Chang[2,3], and Wai-Yee Keung[1]

[1] *Physics Department, University of Illinois at Chicago, IL 60607-7059, USA*
[2] *NCST and Physics Department, National Tsing-Hua University, Hsinchu 30043, Taiwan, R.O.C.*
[3] *Institute of Physics, Academia Sinica, Taipei, R.O.C.*

We elaborate on a new class of models proposed recently by us and compare with another class proposed by Georgi and Glashow(GG). The models can be roughly classified as the righted-handed (or left-handed) models in our (or GG's) case. Both classes of models use softly broken CP symmetry to suppress tree level KM phase as well as the strong CP phase. The measured value of the CP-violating parameter ϵ are accounted for by employing a new heavy sector of scalars and vectorial fermions. The models can be milliweak or superweak in nature depending on the scale of the heavy sector. We review the phenomenology of the right-handed models and compare with the left-handed models.

(Contribution to International Conference on High Energy Physics, 1998, Triumf, presented by DC)

Introduction

Recently, two new classes of models [1,2] of soft CP violation have been proposed as alternatives to the standard KM model [3]. The models aim at reproducing the attractive characters of KM model and at the same time solve the long-standing strong CP problem.

These models share the features of imposing soft (or spontaneous) CP violation in order to avoid tree level KM phase as well as the tree level strong CP θ phase. To account for the measured value of the CP-violating parameter ϵ of the neutral kaon system, the models employ a new heavy sector of scalars and vectorial fermions. The first class of models, that we shall roughly classified as the right-handed (RH) models [1], uses a heavy sector that couples only to right-handed down type quarks, which are $SU_L(2)$ singlets. Instead, the alternative left-handed (LH) models [2] uses new particles that couple only to the ordinary left-handed quarks, which are $SU_L(2)$ doublet. In this report, we review the phenomenology of the RH models and make comments on the LH models as we go for comparison. The more extensive presentation is under preparation [4]. Note that the ideas in these directions were presented by Barr [5] and his collaborators some time ago.

The heavy sector of a typical model for our purposes requires two additional Higgs singlets, $h_\alpha (\alpha = 1, 2)$ of charge q_h and a vectorial pair of heavy fermions, $Q_{L,R}$, of electromagnetic charge $-\frac{1}{3} + q_h$. One can also choose to have two pairs of fermions and only one heavy Higgs singlet. In addition, One can choose to assign the scalars or the fermions to be carrying the color such that together they will couple to the right-handed down type quarks, d_{Ri}. In case of neutral, colorless fermions (the neutrinos), it is not even necessary to have vectorial pairs. One can also use the freedom in choosing charge q_h to avoids fractionally charged hadrons. Most of the phe-

nomenology mention below are more or less independent of the choice of q_h and color. Relevant new terms in the Lagrangian are:

$$\begin{aligned}
\mathcal{L}_{h_i} = & \left[(g\lambda_{i\alpha}\bar{Q}_L d_{iR} h_\alpha + M_Q \bar{Q}_L Q_R) + \text{h.c.} \right] \\
& - (m^2)_{\alpha\beta} h_\alpha{}^\dagger h_\beta - \kappa_{\alpha\beta}(\phi^\dagger\phi - |\langle\phi\rangle|^2) h_\alpha^\dagger h_\beta \\
& - \kappa'_{\alpha\beta\gamma\delta} h_\alpha^\dagger h_\beta h_\gamma^\dagger h_\delta ,
\end{aligned} \tag{1}$$

where ϕ is the Standard Model Higgs doublet. The soft breaking of CP symmetry implies a special basis where λ, κ, κ' and the SM Yukawa couplings are real. If fermions carry color, we also require (see below) that dim-3 couplings, M_Q, are real to avoid tree level contribution to θ. This leaves only a single CP violating parameter: $\text{Im}(m^2)_{12}$. We can diagonalize $(m^2)_{\alpha\beta}$ by a unitary matrix $U_{\alpha i}$ which in general is complex: $h_\alpha = U_{\alpha i} H_i$, with H_i the mass eigenstates. The quark-Higgs interaction in the mass eigenstate basis becomes

$$\mathcal{L}_{QqH} = g \sum_{q=d,s,b} \xi_{qj} (\bar{Q}_L q_R) H_j^- + \text{h.c.} , \tag{2}$$

where $H_i = U_{\alpha i}^* h_\alpha$ is the mass eigenstates, $\xi_{qj} \equiv \lambda_{q\alpha} U_{\alpha j}$. The Yukawa couplings ξ_{qj} are defined relative to the gauge coupling g of $SU_L(2)$. The rephasing-invariant measure of CP violation are

$$\mathcal{A}_{ij}^{qq'} = \lambda_{q\alpha}\lambda_{q'\beta} U_{\beta i} U_{\alpha j}^* = \xi_{qi}^* \xi_{q'j} \tag{3}$$

with $(q, q' = d, s, b)$ and $i, j = 1, 2$. For flavor-conserving amplitudes like EDM, we define the counterpart $\mathcal{B}_{ij} = \kappa_{\alpha\beta} U_{\beta i} U_{\alpha j}^*$. $\mathcal{A}_{ij}^{qq}$ as well as $\mathcal{B}_{ij}$ are Hermitian in indices $i.j$. Thus CP is broken only in the off-diagonal terms. As a result, they contribute to CP violation only when both the light and heavy charged Higgs are involved in a diagram at the lowest order. It is also possible to avoid the CP violating part of the coupling κ if one chooses to have only one heavy Higgs singlet and breaks CP symmetry in the dimension-3 heavy fermion mixing terms instead as long as these heavy fermions are colorless. The contribution related to parameter κ' in the last term of (1) occurs only at the higher loop level and generally can be ignored.

Before continuing, we like to emphasize again that if the new fermions carry color it is necessary to impose CP symmetry on their bare masses also in order to avoid tree level strong CP problem. When there are more than one pairs of vectorial fermions, one can potentially make the mass matrix Hermitian, however that would require some additional symmetry to be implemented.

Constraint from ϵ

With CP broken only softly, the CKM matrix is real at tree level. Leading contribution to the CP violating parameter ϵ is due to the box diagram with only heavy particles in loop.

$$
\begin{array}{ccc}
d_R & Q_L & s'_R \\
\end{array}
$$

We evaluate all possible diagrams and match them with the low energy effective Hamiltonian which has the form

$$\mathcal{H}^{\Delta S=2} = \frac{G_F^2 m_W^2}{16\pi^2} \sum_{I=R,L} C^I_{\Delta S=2}(\mu) O^I_{\Delta S=2}(\mu) , \quad (4)$$

with $\quad O^{R,L}_{\Delta S=2} = \bar{s}\gamma_\mu(1 \pm \gamma_5)d \, \bar{s}\gamma^\mu(1 \pm \gamma_5)d$. (5)

The $W^\pm$ diagrams yield a purely real Wilson coefficient $C^L_{\Delta S=2}(\mu)$; CP violation is due solely to the operator $O^R_{\Delta S=2}$ rather than $O^L_{\Delta S=2}$, in contrast to the KM model, because the complex coefficient $C^R_{\Delta S=2}(\mu)$ is generated by the charged Higgs. At the scale $\mu = M_Q$, we have

$$C^R_{\Delta S=2}(M_Q) = 2\xi_{d1}\xi^*_{s1}\xi_{d2}\xi^*_{s2}\frac{m_W^2}{M_Q^2}\frac{f(x_2)-f(x_1)}{x_2-x_1}$$

$$+ \sum_{i=1,2}(\xi_{di}\xi^*_{si})^2\frac{m_W^2}{M_Q^2}\frac{df}{dx}(x_i) , \quad (6)$$

with $f(h) = (1-h)^{-2}(1+2h+h^2+h^2\ln h)$.

The real part of the diagram can contribute to part of the $K_L - K_s$ mass difference while the imaginary gives rise to ϵ. This is analyzed in detail in Ref. [1]. For illustration here, we can take the (decoupling) limit $m_2 \gg m_1$ and assume $m_1 = M_Q$ for simplicity. Demanding that the imaginary part of the box diagram gives enough contribution to ϵ while the corresponding real part gives just

a fraction $\mathcal{F}$ of the mass difference Δm_K, we obtain[1] the constraints,

$$\text{Im}\left(\mathcal{A}_{sd}/(0.058)^2\right)^2 R_Q^2 = 1 , \quad (7)$$

$$\text{Re}\left(\mathcal{A}_{sd}/(0.058)^2\right)^2 R_Q^2 = 156\mathcal{F} ; \quad (8)$$

where $R_Q = 300 \text{ GeV}/M_Q$. The reasonable constraint $|\mathcal{F}| < 1$ can be easily satisfied. It is important to emphasize that, in RH models, the heavy particle box diagrams induce a right-handed four fermion operator, contrary to the LH models and the KM model in which the leading CP violating operators are left-handed.

Constraints from (ϵ'/ϵ)

The leading contribution to the direct CP violating parameter ϵ' is due to the gluonic penguin diagrams with only the heavy particles in the loop.

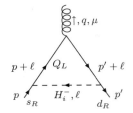

The CP violating piece of the effective Hamiltonian is parametrized as

$$\mathcal{H}^{\Delta S=1} = (G_F/\sqrt{2})\tilde{C}(\bar{s}T^a\gamma_\mu(1+\gamma_5)d) \times \sum_q(\bar{q}T^a\gamma^\mu q) .$$

At the electroweak scale, the Wilson coefficient is

$$\tilde{C} = -\alpha_s \sum_i \frac{\xi_{di}\xi^*_{si}}{6\pi}\frac{m_W^2}{M_Q^2}F\left(\frac{m_{H_i}}{M_Q^2}\right) ,$$

$$F(h) = \left[\frac{(2h-3)h^2\ln h}{(1-h)^4} + \frac{7-29h+16h^2}{6(1-h)^3}\right] .$$

The electromagnetic penguin and long distance effects can contribute to the imaginary part of the $\Delta I = 3/2$ amplitude and give a small contribution. This is analyzed in detail in Ref. [1]. In our decoupling limit, the result is

$$\epsilon'/\epsilon = -1.9 \times 10^{-5}\text{Im}(\mathcal{A}_{sd}/(0.058)^2)R_Q^2 \quad (9)$$

$$= \pm 1.9 \times 10^{-5}\left(\sqrt{(156\mathcal{F})^2+1} - 156\mathcal{F}\right)^{\frac{1}{2}}R_Q/\sqrt{2} ,$$

using the constraints in Eq.(8). For $R_Q = 1$ and $\mathcal{F} \approx 0$ (or -0.3), $\epsilon'/\epsilon = 1.4 \times 10^{-5}$ (or 1.3×10^{-4}), which is roughly the same order of magnitude as the KM model. One can of course makes the model more superweak[6] by setting the scale higher (and R_Q smaller) and $\mathcal{A}_{sd}$ larger. For $\mathcal{A}_{sd}$ of order one, M_Q is roughly 100 TeV.

Constraints from Strong CP θ_{QCD} and the induced KM phase

In both RH and LH models, $\theta_{\rm QCD}$ is only induced starting at the two-loop level, via generation of complex down-flavor quark masses as long as the coupling κ exists. A typical diagram is shown in Fig. 1 in Ref. [1]; this effect does not require more than one flavor of down-quark. However, it does require both charged Higgs to be involved.

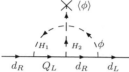

Roughly, $\theta_{\rm QCD} \sim g^2 I \, {\rm Im}(\mathcal{A}_{12}^{dd} \, \mathcal{B}_{12})/(16\pi^2)^2$. The factor I, of order one, is given by the integral

$$I = \int_0^1 \frac{dz}{(1-z)} \int_0^{1-z} dx \times$$

$$\left[\frac{zm_{H_2}^2 + xM_Q^2 + ym_{H_1}^2}{zm_{H_2}^2 + xM_Q^2 + ym_{H_1}^2 - z(1-z)M_{\phi^0}^2} \cdot \right.$$

$$\left. \log \frac{z(1-z)M_{H^0}^2}{zm_{H_2}^2 + xM_Q^2 + ym_{H_1}^2} - (m_{H_1} \leftrightarrow m_{H_2}) \right] ,$$

with the Feynman parameters $y = 1-x-z$. The integral I vanishes at the degenerate case $m_{H_2} = m_{H_1}$, but its size approaches [4] 1 as $m_{H_2} \to \infty$. This non-decoupling phenomena is not surprising because, in the large m_{H_2} limit, the CP is a broken symmetry. Numerically, the present constraint, $\theta_{\rm QCD} < 10^{-9}$, can easily be accommodated. In addition, there are also three loop diagrams due to the gluonic contribution. A typical graph is shown below.

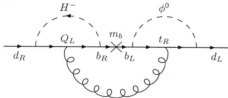

This contribution is independent of the coupling κ. However, unless κ happens to be very small, it may not be competitive with the two loop contributions because of the KM angle suppression and the additional loop factor [4]. Both the 2-loop and the 3-loop contributions also exist generically in LH models. However, they are typically numerically smaller in that case. The two loop contributions disappear (in both RH and LH models) of course when only one complex scalar boson is used as mentioned earlier.

Since CP is broken at higher energy, a non-vanishing KM phase η (defined in the Wolfenstein parametrization) [7] can in general be loop induced. It can originate either from the loop-induced complex mass matrix or from that of the complex kinetic energy terms. The contribution from complex mass matrix is similar to the analysis of θ and is therefore small. The contribution from complex kinetic terms is induced at the one loop level in LH models as analyszed in Ref. [2]. It is in general also suppressed by some small KM mixing angles and small mass ratios and therefore numerically tiny enough to be ignored phenomenologically. In contrast, in RH models, since the KM phase is related only to the rotation of the left-handed quarks, such contribution will not arise until at the two loop level as given in the following figure. Therefore induced η is even smaller [4].

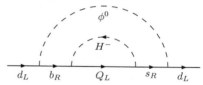

Constraints from electric dipole moments

A down-flavor quark EDM, however, is generated at the two-loop (or three-loop) level, in parallel with the generation of complex down quark masses discussed above. The typical contribution is given by diagrams similar to those for θ, except with an external photon attached to internal charged lines. An estimate of the two loop contribution gives EDM which is consistent with the current experimental bound. The contribution from chromo-electric dipole moment of gluon (the Weinberg operator) won't arise until three loop level (even with κ coupling) and therefore expected to be small. The electron couples only indirectly with the CP violating sector, so its EDM vanishes at two loops and the three loop contributions are insignificantly small.

B^0–$\bar{B}^0$ Mixing, $b \to s\gamma$ and Other Constraints

Another (much weaker) constraint to be considered is that from the $B_{s,d}^0$ mass splitting [1]. Using the usual estimate of strong form factors involved and the experimental value for ΔM_{B^0}, we have

$$\delta(\Delta M_{B^0})/\Delta M_{B^0} = 1.1 \times 10^{-3} \, R_Q^2 \, {\rm Re}\left(\mathcal{A}_{bd}/0.058^2\right)^2 ,$$

so even taking $\mathcal{A}_{bd} = (0.15)^2$, the fractional contribution is only about 5% for $M_Q = 300$ GeV.

In RH models, the operator induced by the exotic sector has helicity opposite to the SM contribution, the two do not interfere in the rate. In the decoupling limit with $\mathcal{F}_{b \to s\gamma} \equiv \delta B(b \to s\gamma)/B(b \to s\gamma)_{\rm SM}$, we have

$$\mathcal{F}_{b \to s\gamma} = 6.4 \times 10^{-6} \left| R_Q^2 \cdot \frac{0.0389}{V_{tb} V_{ts}^*} \cdot \frac{\mathcal{A}_{bs}}{(0.058)^2} \right|^2 .$$

Furthermore, the relevant parameter $\mathcal{A}_{bd}$ is not subjected to constraints from ϵ or ϵ'. If it is of the same size as $\mathcal{A}_{sd}$, the deviation from the SM would be negligible and the future B factory would observe only a collpased unitarity KM angle [8]. However, if $\mathcal{A}_{bd} \gg \mathcal{A}_{sd}$, the triangle can looks substantially different from that predicted by standard KM model. It is worthwhile to point out that, in LH models, the exotic contribution gives rise to operator that will interefere with that of SM and therefore the model is more severely constrained [4].

Decays of New Particles

In the generic RH model, h and Q can be assigned a new conserved quantum number which guarantees a stable lightest exotic particle, either H_1 or Q. However, when $q_h = -1$, an interaction, $h_\alpha L_i L_j$, is allowed, which can lead to H^- (on-shell or off-shell) decays into $l^- \nu$. Even so, lepton number is still conserved, just as in the SM, since Q and H will naturally carry lepton number ($L = \pm 2$). Another way for H to decay is to introduce a second Higgs doublet and let H couple to two different Higgs doublets. In that case H can decay into a neutral Higgs, plus a charged Higgs which in turn decays into ordinary quarks and leptons.

Spontaneous broken CP symmetry

To show how the above softly broken CP symmetry can in fact originate from a spontaneously broken one, one can first add a CP-odd scalar, a, which develops a non-zero vacuum expectation value (VEV) and breaks CP. However, this scalar will in general couple to $\bar{Q}_L Q_R$ and give rise a complex tree level M_Q and, therefore, a tree level $\theta_{\rm QCD}$. To avoid this, one can add another scalar singlet, s, which is CP-even and impose a discrete symmetry which changes the signs of both a and s and nothing else. As a result, a term such as $ia\bar{Q}\gamma_5 Q$ is forbidden. The only additional term relevant for CP violation is $i \left[s \, a \left(h_1{}^\dagger h_2 - h_2{}^\dagger h_1 \right) \right]$ which generates a complex $(m^2)_{12}$ after both s and a develop VEVs. The extra neutral Higgs bosons will mix with the SM Higgs, but since s and a do not couple to fermions directly, they have tiny scalar-pseudoscalar coupling to fermions only at loop level. As a result, the CP phenomenology considered below applies equally well to both the softly and spontaneously broken versions of our model.

Interplay between Strong and Weak CP Phase

The RH models provide a good example to look at the subtlety, raised in Ref. [9], involving the interplay of the CP phases between the strong and weak interactions. For our purposes, we can focus on the reduced effective theory which contains CP-conserving Standard Model-type interactions and vanishing $\theta_{\rm QCD}$, plus the new, induced superweak interaction with the strength $C_{\Delta S=2}^R$ defined in Eq. (4). We shall consider, for the sake of argument, the scenario in which the up quark is massive while m_d is zero. Without the new $C_{\Delta S=2}^R$ interaction the parameter $\theta_{\rm QCD}$ is then unphysical, with CP a good symmetry. (By the usual argument, with a massless quark present — in this case, the d quark — the right-handed component of that quark can be rotated to absorb $\theta_{\rm QCD}$ via the axial anomaly, while otherwise leaving the lagrangian invariant.) With the addition of the induced interaction $\mathcal{H}^{\Delta S=2}$, however, $\theta_{\rm QCD}$ becomes physical, as can be seen by considering the following cases:

(a) If $C_{\Delta S=2}^R$ is complex and $\theta_{\rm QCD} = 0$, CP is violated. In this case, the correct (non-zero) value for ϵ can be calculated without complication; all hadronic matrix elements (modulo absorptive contributions) can correctly be assumed to be real. It is illuminating to consider the calculation of ϵ in another basis, which is obtained by a phase rotation of d_R such that $C_{\Delta S=2}^R$ becomes real and $\theta_{\rm QCD}$ non-zero. Since the two theories are the same, one must arrive at the same result for ϵ. One can thus draw a rather surprising conclusion: $\theta_{\rm QCD}$ *can also, in certain situations, contribute to* ϵ.

In fact, from the way we obtain the $\theta_{\rm QCD}$ contribution to ϵ in this example, one realizes that there is an important subtlety here. *The actual contribution from* $\theta_{\rm QCD}$ *to* ϵ *is correlated to the explicit mechanism of CP violation*, which in our current example is the superweak $C_{\Delta S=2}^R$. A related result is that when $\theta_{\rm QCD}$ is not zero, how each hadronic matrix element develops a phase *also depends on the particular electroweak mechanism of CP violation in the theory*. In the present case, the CP violating coupling also happens to be the chiral symmetry breaking phase.

Another lesson one learns is that the usual argument which concludes that the contribution of $\theta_{\rm QCD}$ to CP-violating quantities such as the neutron electric dipole moment (edm) must be proportional to $m_u m_d$, is not strictly correct, a counterexample is offered by the simplified model presented above. The role of m_d is replaced by the coupling $C_{\Delta S=2}^R$. Of course, $C_{\Delta S=2}^R$ breaks the chiral symmetry associated with d quark, so that the d quark will certainly pick up mass at some (probably higher-loop) level, but the point is that the $C_{\Delta S=2}^R$ coupling plays a much more direct role in the contribution of θ_{QCD} than even the induced m_d!

Now we come to an apparent paradox whose resolution gives even further insight into the interplay between strong and weak CP phases.

We parenthetically noted above that if redefinition of the quark phases generates an imaginary part for the quark mass matrix *not proportional* to the identity matrix, the low-energy meson states must be suitably reinterpreted to ensure stability of the vacuum around which we carry out perturbation theory. This redefinition explicitly reintroduces the phase(s) rotated from the couplings into certain hadronic matrix elements, to ensure rephasing invariance. If both m_d, m_u vanish, then arbitrary rotation of the corresponding right-handed quarks seems to have no effect on vacuum stability, since the mass matrix is left real and diagonal (only m_s non-zero). Then, apparently, all phases may be arbitrarily rotated away, and with them, any possibility of CP violation. Specifically, consider the following variant of the two cases already considered:

(b) Let $C_{\Delta S=2}^R$ be complex, but take θ_{QCD} to be zero. If *both* m_u, m_d are strictly zero, is CP conserved or violated? At first glance, one might claim the phase in $C_{\Delta S=2}^R$ to be unphysical, since a combined phase rotation of the form $u_R \rightarrow e^{-i\delta} u_R$ and $d_R \rightarrow e^{i\delta} d_R$ can make $C_{\Delta S=2}^R$ real and maintain $\theta_{QCD} = 0$. It is very tempting to claim that CP violation is proportional to m_u for a small up quark mass and further that there is no CP violation when $m_u \rightarrow 0$ because the phase of $C_{\Delta S=2}^R$ then becomes absorbed.

This conclusion is *incorrect*, however, because we have ignored the vacuum degeneracy in the case of massless u and d quarks. Different choices of vacua would give different CP violation. It is true that there exists one very special vacuum where CP is conserved. However, a general vacuum posseses chiral condensate with a phase uncorrelated to that of $C_{\Delta S=2}^R$, and thus CP violation usually occurs, *even if* $C_{\Delta S=2}^R$ is real (since what is important is the relative phase between the vacuum and $C_{\Delta S=2}^R$). This idea can be demonstrated directly in the chiral effective lagrangian approach. The chiral field Σ (3×3 unitary matrix) can be perturbed around a vacuum configuration diag($e^{-i\phi}$, $e^{i\phi}$, 1). If $C_{\Delta S=2}^R$ is turned off, the strong interaction is independent of ϕ because of the chiral symmetry. However, with $C_{\Delta S=2}^R$, the phase ϕ has physical meaning and has implications with respect to CP violation. Now we include effects of the real quark masses $m_u \neq 0$ and $m_d \neq 0$. Their net effect is simply to pick out a particular vacuum, with Σ =diag(1, 1, 1). In this case of vacuum alignment, a complex $C_{\Delta S=2}^R$ is necessary, but also sufficient, for CP violation, since again it is the relative phase between $C_{\Delta S=2}^R$ and the vacuum

that is important. In some sense, the (possibly infinitesimal) up and down quark masses *enforce* CP violation, in the particular case that $C_{\Delta S=2}^R$ is real, whereas in the massless case, CP violation is still generally expected via the vacuum phase.

Conclusion

We have reviewed a new (RH) class of models whose CP violation is solely mediated by exotic Higgs bosons and fermions that couple to the right handed quarks, and compared with another (LH) class of models whose exotic particles couple to the left-handed quarks. The model naturally prevents the strong CP problem. Both classes of models are surprisingly similar to the KM model in the sense that the CP-breaking mechanism is seemingly milliweak(if the exotic particle scale is chosen to be as low as possible), while its phenomenology (as studied here) is quite superweak-like. The phenomenological distinction between the two will likely be made clear in experiments planned for the B factory. A careful and detailed analysis of such issues is clearly necessary and is in progress [4].

D. B.-C. and W.-Y. K. are supported by a grant from the DOE of USA, and D. C. by the NSC of R.O.C. We thank H. Georgi, S. Glashow, P. Frampton, R. Mohapatra, and L. Wolfenstein for very useful discussions.

References

1. D. Bowser-Chao, D. Chang, and W.-Y. Keung, Phys. Rev. Lett. **81**, 2028 (1998).
2. H. Georgi and S. Glashow, hep-ph/9807399. See also, P. Frampton and M. Harada, hep-ph/9809402.
3. M. Kobayashi and T. Maskawa, Prog. Theor. Phys. **49**, 652 (1973).
4. D. Bowser-Chao, D. Chang, and W.-Y. Keung, in preparation.
5. S. Barr and A. Zee, Phys. Rev. Lett. **55**, 2253 (1985); J. Nieves, Nucl. Phys. **B189**, 189 (1981); Phys. Lett **164B**, 85 (1985); S. Barr, Phys. Rev. **D34**, 1567 (1986); S. Barr and E. M. Freire Phys. Rev. **D41**, 2129 (1990).
6. L. Wolfenstein, Phys. Rev. Lett. **13**, 562 (1964);
7. L. Wolfenstein, Phys. Rev. Lett. **51**, 1945 (1983); L.–L. Chau and W.–Y. Keung, Phys. Rev. Lett. **53**, 1802 (1984).
8. S. Mele, hep-ph/9810333; see also R. Barbieri, L. Hall, A. Stocchi, and N. Weiner, Phys. Lett. **B425** 119 (1998) or hep-ph/9712252.
9. D. Bowser-Chao, D. Chang, and W.-Y. Keung, Chin. J. Phys. **35**, 842 (1997), hep-ph/9803275.

MEASUREMENT OF DIRECT CP-VIOLATION WITH THE NA48 EXPERIMENT AT THE CERN SPS

H. BLÜMER

CERN, EP-Division, CH-1211 Geneva 23, Switzerland;
Johannes Gutenberg-Universität Mainz, Institut für Physik, D-55099 Mainz, Germany;
E-mail: hans.bluemer@cern.ch

The NA48 experiment at the CERN SPS uses simultaneous, nearly collinear beams of long-lived and short-lived neutral kaons to measure the direct CP-violation parameter ε'/ε using the double ratio method to an overall accuracy of $2 \cdot 10^{-4}$, three times better than previous results. The detector has been installed and commissioned in 1995 and 1996. First physics data were recorded during 42 days in fall 1997 yielding more events than the previous experiment NA31. The talk presents the apparatus performance, data quality, current status of the physics analysis and ongoing activities. The experiment has performed another data run from May to September 1998, which has given a substantial increase in statistics.

Introduction

CP-violation remains an interesting and very active search field even 34 years after its discovery [1]. The Superweak Model explains CP violation by a new force acting on the extremely sensitive interferometric system of neutral kaons [2], and predicts $\varepsilon'/\varepsilon \equiv 0$. The Standard Model of electroweak interactions accounts for CP-violation in the Cabbibo-Kobayashi-Maskawa matrix of three generations (CKM matrix). Here, CP-violation can have contributions from state mixing in the kaon wave functions or directly from the weak decay amplitudes. The latter effect leads to interference between the amplitudes involving strangeness changes $\Delta S=0$ and $\Delta S=2$. The SM calculation is difficult due to hadronic uncertainties and imperfect knowledge of the CKM matrix elements. Different calculations [3] give $2 \cdot 10^{-4} < \mathrm{Re}(\frac{\varepsilon'}{\varepsilon}) < 15 \cdot 10^{-4}$.

Our present experimental knowledge is based on measurements of the NA31 and E731 collaborations at CERN and FNAL, giving $\mathrm{Re}\frac{\varepsilon'}{\varepsilon} = (23 \pm 6.5) \cdot 10^{-4}$ and $\mathrm{Re}\frac{\varepsilon'}{\varepsilon} = (7.5 \pm 5.9) \cdot 10^{-4}$, respectively [4, 5]. Given the fundamental nature of the problem it is important that these investigations are continued.

The NA48 collaboration [6] aims at a precision measurement of direct CP-violation to an overall accuracy of. $\delta(\mathrm{Re}\frac{\varepsilon'}{\varepsilon}) = 2 \cdot 10^{-4}$ using the double ratio,

$$R = 1 - 6\,\mathrm{Re}\frac{\varepsilon'}{\varepsilon} = \left|\frac{\eta_{00}}{\eta_{+-}}\right|^2 = \frac{\Gamma(K_L \to \pi^0\pi^0)}{\Gamma(K_S \to \pi^0\pi^0)} / \frac{\Gamma(K_L \to \pi^+\pi^-)}{\Gamma(K_S \to \pi^+\pi^-)}$$

Beam layout

To achieve this level of accuracy, the double ratio R is measured using all four decay modes simultaneously. Nearly collinear beams are used, where K_S and K_L are produced concurrently and are distinguished by tagging of the protons producing the K_S component. Since both beams coincide in space at the detector position all detection efficiencies cancel in principle in the double ratio if the result is evaluated independently for different kaon momenta. Accidental activities and the corresponding losses and gains of good events as well as time-dependant detector efficiencies cancel also since all four decay modes are observed concurrently. A schematic layout of the beam is shown in Figure 1, a detailed description has been given elsewhere [7].

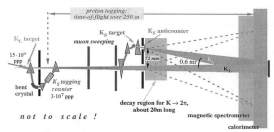

Figure 1: Schematic layout of the combined neutral kaon beams of the NA48 experiment, not to scale.

Detector and trigger

Charged particle tracks are detected in a spectrometer consisting of four drift chambers and a magnet providing a transverse momentum kick of 250 MeV/c. While the decay volume is evacuated to 10^{-4} mbar the drift chambers are enclosed by Helium to reduce multiple scattering. The invariant mass resolution is 2.5 MeV/c² for both K_L and K_S decays. A scintillator hodoscope triggers on charged particles and provides the reference time for these decays. Two-track events are reconstructed by an asynchronous, pipelined system of dedicated processors and are required to have an invariant mass consistent with the kaon mass.

Neutral particles are measured in a homogenous Liquid-Krypton (LKr) calorimeter providing excellent resolutions for energy, position and time of electromagnetic showers. The calorimeter is an ionisation chamber equipped with a tower-like readout structure of 13,500 cells of transverse dimensions 2×2 cm^2. The neutral particle trigger reconstructs the lifetime of kaons from the energies and positions of electromagnetic clusters imposing the kaon mass constraint. A range of $c\tau<5.5$ $c\tau_s$ is accepted. The majority of $K_L \to 3\pi^0$ events have missing photons and are therefore rejected due to their much larger apparent lifetimes.

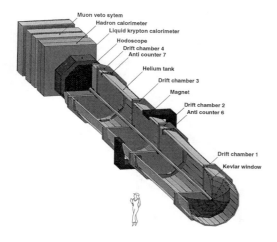

Figure 2: three-dimensional view of the NA48 detector generated by the GEANT simulation program

Anticounter rings, a hadronic calorimeter and a muon veto system complement the detector. Figure 2 shows a three-dimensional view of the detector generated by the GEANT simulation program.

The entire readout electronics is based on fast analog-to-digital converters operating at 25 MHz sampling rate. This clock is derived from a 960 MHz signal, which drives the proton tagging detector located 250 m upstream.

Experience and data quality for the 1997 run

The complete detector was operated to collect first physics data during 42 days in 1997. During this period, 25TB of data for 1.1 billion triggers were recorded and processed in real-time. A sample of 650,000 $K_L \to 2\pi^0$ events has been

obtained after off-line cuts for $0<c\tau<4c\tau_s$ and $70<E_K<170$GeV.

The LKr calorimeter

The LKr calorimeter performed well at 1.5kV instead of 3kV to avoid high voltage breakdown. A 4cm-wide vertical strip was disconnected from the high voltage. A small space charge correction needed to be applied for the high-intensity region near the beam pipe. The performance of the calorimeter was checked by comparing the electromagnetic energy E to the momentum p measured for well-identified electron tracks from K_{e3} decays. The linearity is better than $2\cdot10^{-3}$ in E/p and the resolution is better than 1% for energies greater than 20GeV. The invariant mass of neutral pions can be reconstructed from photon pairs with 1.1 MeV resolution.

Tagging

The vertical beam separation is only 72mm at the beginning of the decay volume. For charged mode decays, the vertex position allows an independent K_S-K_L assignment. In the neutral mode, this is not possible; NA48 relies on tagging the protons that produce the K_S beam.

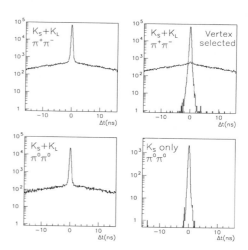

Figure 3: Event time with respect to the time of protons registered in the tagging detector 250m upstream. For charged mode decays a vertex selection has been applied, $K_S := y_v > 4$cm, $K_L := y_v < 4$cm. For the neutral mode, the K_L beam has been switched off to produce a pure K_S time distribution.

997

If the event time in the detector (hodoscope or LKr calorimeter) minus the time of the nearest proton in the tagging counter [8] is less than 2ns, the event is identified as a K_S decay; otherwise it is taken as a K_L decay. Since the tagger is blind to the final state, it can't make ε'/ε artificially non-zero.

The tagging method worked very well. Inefficiencies were found to be a few 10^{-4}. The misidentification of K_L events due to the high rate in the tagger is about 11% and can be measured to very high accuracy. Thus, we expect the tagging method to contribute less than 10^{-4} to the uncertainty on ε'/ε.

Charged decays

Background decays are first suppressed using efficient and K_S-K_L blind cuts and then statistically subtracted. For $K_L \to \pi^+\pi^-$ decays, the most important backgrounds are K_{e3} (background/signal=188) and $K_{\mu3}$ (background/signal=135). The number of events as a function of transverse momentum-squared, ($p_t'^2$, corrected for the finite target size) and invariant mass of the two-track system is used to perform the background subtraction. In the $p_t'^2$-$m_{\pi\pi}$ plane two rectangular regions are defined to contain signal and background events, respectively.

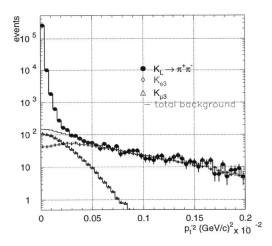

Figure 4: Charged mode decays: signal and background distribution in transverse momentum-squared, corrected for the finite target size.

Samples of well identified K_{e3} and $K_{\mu3}$ events are then used to determine the background level inside the signal region

fitting the relative amount of K_{e3} and $K_{\mu3}$. This is done in bins of momentum.

Figure 4 shows the global result projected on the $p_t'^2$ axis for 40% of the 1997 data integrated over a mass interval of ± 7.5 MeV/c^2 around the kaon mass. In this example a charged background of $(2.6 \pm 0.4_{\text{stat}}) \cdot 10^{-3}$ is subtracted in the signal region $p_t'^2 < 200$ (MeV/c)^2.

Neutral decays

In the neutral mode the kaon decay vertex is reconstructed from the positions and energies of photon showers in the LKr calorimeter, imposing the kaon mass constraint. Using this vertex, pairs of photons are combined to find the two invariant 2-photon masses $m_{\gamma\gamma}$ that are closest to the neutral pion mass. The dominant neutral background is from $K_L \to 3\pi^0$ decays and is subtracted using the data distribution in the $m_{\gamma1\gamma2} - m_{\gamma3\gamma4}$ plane.

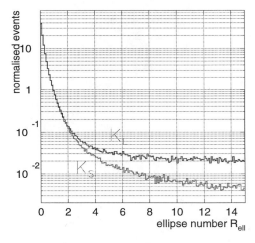

Figure 5: Signal and background subtraction in the R_{ell} variable for neutral mode decays. The K_S curve still includes the mistagged K_L component.

The quantity R_{ell} is proportional to the chisquare value for finding an event at any point in the $m_{\gamma1\gamma2} - m_{\gamma3\gamma4}$ plane using the known resolutions in $m_{\gamma\gamma}$. $K_L \to 3\pi^0$ decays have a flat distribution in R_{ell}. In the example shown the background is $3 \cdot 10^{-3}$ in a signal region $R_{ell} < 1.5$ without lifetime weighting. Since most of the $K_L \to 3\pi^0$ events reconstruct at large lifetime values due to their missing energy, lifetime weighting reduces the neutral background further by a factor of three.

Weighting

We intend to weight all K_L decays with a momentum-dependent factor $w_L = \exp\left(-z m_K (1 - \frac{z}{\tau_L}) / p c \tau_S\right)$, making use of the known values of kaon mass and lifetimes. In this way, the vertex distributions are made identical and the acceptances $A_{L,S}(z,p)$ for K_L and K_S decays as a function of decay vertex z and momentum cancel out in z. The energy spectra of the recorded decays become equal to ±10%. The effective loss of statistical power is estimated to be only 15% for a fiducial decay volume of $2c\tau_S$ (about 10m at 100 GeV/c). This is more than outweighed by keeping the acceptance correction as small as a few per-mille on the double ratio.

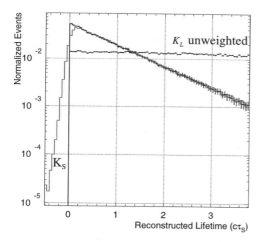

Figure 6: Reconstructed lifetime for charged mode decays. The K_S-line shows the resolution-smeared cut-off due to the anticounter, which defines the start of the fiducial volume.

Rare decays

In addition to the precision measurement of direct CP-violation the NA48 experiment has acquired large samples of rare kaon decays. The channel $K_L \to ee\gamma$ can be used to measure the form factor α_K for pseudoskalar and vector meson intermediate states [9],

$$A(K_L \to \gamma\gamma^*) = A(K_L \to \gamma\gamma^*)_{P-P} \pm \alpha_K \cdot A(K_L \to \gamma\gamma^*)_{V-V}.$$

The form factor can be fitted using the invariant electron-positron mass distribution. A preliminary analysis is shown in Figure 7, where the result $\alpha_K = -0.29 \pm 0.04$ includes

radiative corrections and the statistical error is based on 12,000 events analysed.

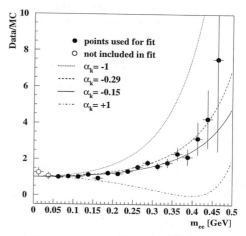

Figure 7: Ratio of data and MC simulated distributions of invariant electron-positron mass in $K_L \to ee\gamma$ decays.

The decay $K_L \to ee\gamma\gamma$ is interesting as a background to the process $K_L \to \pi^0 e^+ e^-$, which may proceed largely through direct CP-violation, and as a test for radiative corrections [10]. The data sample of 1997 contains 106 events in the signal region, considerably more than previously available. The preliminary branching ratio is found to be

$$\Gamma(K_L \to ee\gamma\gamma)/\Gamma(K_L \to all) = (4.6 \pm 0.7_{stat} \pm 1.4_{sys}) \cdot 10^{-7},$$

where cut-off parameters $E_\gamma > 5\,MeV$, $m_{e\gamma} > 1\,MeV$ were used.

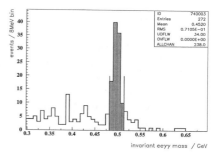

Figure 8: 106 $K_L \to ee\gamma\gamma$ events in the signal region.

Run98

A number of upgrades were implemented to increase the event yield:

- the liquid-Krypton calorimeter was repaired successfully; all channels hold 3kV and essentially no space charge effect is visible even at increased beam intensity;
- faster on-line processors were used for filtering $K \to \pi^+\pi^-$ events;
- data read-out and event building was given to a 26-units cluster of commercial PCs;
- the link to the CERN computing centre was upgraded to sustain 18 MB/s aggregate throughput;
- a light, carbon fibre beam pipe was installed to reduce photon conversions into the detector;
- η^0 decays produced in a special target were used to check the energy and length scales;
- the beam intensity was increased by 40% with respect to 1997.

The improved trigger system allowed to record events with four charged tracks. In 20 days analysed, about 250 candidate events of the type $K_L \to \pi^+\pi^-e^+e^-$ were found. We expect to have nearly 1000 events on tape for a precise measurement of the CP-violating asymmetry in the decay planes of electrons and pions [11].

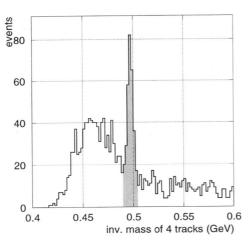

Figure 10: invariant mass of 4-track events using relaxed conditions to select $K_L \to \pi^+\pi^-e^+e^-$.

All these improvements worked well and additional 2.25 million $K_L \to 2\pi^0$ events were recorded in 116 days of physics data taking.

Summary and Outlook

The combined data sample of 1997 and 1998 of the NA48 experiment is approaching 3 million $K_L \to 2\pi^0$ events. The data of 1997 are currently being analysed and will provide a preliminary value of ε'/ε which has a similar weight as the entire NA31 result. At the same time, the 1998 data sample is being reprocessed using final calibrations. Another long physics run is scheduled for 1999. We are very confident to reach the initial goal of $\delta(\frac{\varepsilon'}{\varepsilon}) \le 2 \cdot 10^{-4}$.

References

[1]   J. H. Christenson, J. W. Cronin, V. L. Fitch, R. Turlay, Phys. Rev. Lett. 13(4) (1964) 214.

[2]   L. Wolfenstein, Phys. Rev. Lett. 13 (1964) 562.

[3]   See e.g., C. Caso et al., The European Physics Journal C3 (1998) 1, and references therein; A. Buras, Proc. Workshop on K Phys., Orsay, France, 30 May-4 June, 1997, p. 459; T. Hambye et al., DO-TH 97/28.

[4]   G. D. Barr et al., Phys. Lett. B 317 (1993) 233.

[5]   L. K. Gibbons et al., Phys. Rev. Lett. 70 (1993) 1199.

[6]   NA48 is a collaboration of Cagliari, Cambridge, CERN, Dubna, Edinburgh, Ferrara, Firenze, Mainz, Orsay, Perugia, Pisa, Saclay, Siegen, Torino, Warszawa, Wien; G. D. Barr et al., Proposal of the NA48 collaboration, CERN/SPSC/90-22, SPSC/P253, 20 July 1990.

[7]   C. Biino et al., CERN-SL-98-033-EA; C. Biino et al., CERN-SL-98-041-EA.

[8]   T. Beier et al., Nucl. Instr. Phys. Res. A 360 (1995) 390.

[9]   G. D. Barr et al., Phys. Lett. B 240 (1990) 283; K. E. Ohl et al., Phys. Rev. Lett. 65 (1990) 1407; E. Cheu et al., Proposal for KTeV run 1999, FNAL Dec. 9, 1997.

[10] T. Nakaya et al., Phys. Rev. Lett. 73 (1994) 2169; W. M. Morse et al., Phys. Rev. D45 (1992) 36.

[11] J. Adams et al., Phys. Rev. Lett. 80 (1998) 4123.

STATUS OF THE DIRECT CP VIOLATION MEASUREMENT FROM KTEV AND OTHER RESULTS

E. CHEU

Dept. of Physics, University of Arizona
Tucson, AZ 85721
e-mail: elliott@physics.arizona.edu

The primary goal of the new KTeV-E832 experiment at Fermilab is the measurement of ϵ'/ϵ from the two-pion decays of the neutral kaons (K_L and K_S). A non-zero effect is expected in the Standard Model but has so far not been established. Data were collected between November 1996 and July 1997; the statistical samples and background rejections exceed those of previous experiments. Effort has concentrated upon analyzing approximately 20% of the total ϵ'/ϵ data sample. New results on $K_L \to \pi^0\gamma\gamma$ and $K_L \to \pi^0 e^+ e^-\gamma$ are also presented.

1 Introduction

Although CP violation through mixing was discovered in the K meson system over 30 years ago, the existence of *direct* CP violation remains an open question. Within the context of the Standard Model, one expects direct CP violation, characterized by ϵ', to be very small [1], on the order of 10^{-4} to 10^{-3}. The superweak model of Wolfenstein predicts that the fundamental interaction is a $\Delta S = 2$ coupling and therefore direct CP violation should be exactly zero. No conclusive experimental evidence exists for direct CP violation since the most recent measurements of $Re(\epsilon'/\epsilon)$ are in nominal disagreement [2,3]. The Fermilab E731 experiment measured, $Re(\epsilon'/\epsilon) = (7.4 \pm 5.2 \pm 2.9) \times 10^{-4}$ while the CERN NA31 experiment measured, $Re(\epsilon'/\epsilon) = (23.0 \pm 3.5 \pm 6.0) \times 10^{-4}$. The KTeV experiment was built to help resolve this discrepancy.

The decay $K_L \to \pi^0\gamma\gamma$ recently has been the subject of much theoretical interest. The measured branching ratio [4,5] for this decay is almost three times higher than the predictions from $O(p^4)$ chiral perturbation calculations. More recent $O(p^6)$ calculations [6,7] predict a branching ratio consistent with the measured value. The new calculations also predict distinct shapes for the $m_{\gamma\gamma}$ and $y = |E_3 - E_4|/M_K$ distributions, where E_3 and E_4 are the energies of photons in the kaon rest frame that are not associated with the π^0. The shapes of these distributions have yet to be confirmed by experiment due to the low statistics available to the two previous measurements, which recorded approximately 50 events. Therefore, a high statistics measurement of $K_L \to \pi^0\gamma\gamma$ will help determine the validity of of the chiral perturbation predictions.

Study of the $K_L \to \pi^0\gamma\gamma$ mode is also relevant to understanding the $K_L \to \pi^0 e^+ e^-$ decay. The decay, $K_L \to \pi^0 e^+ e^-$, contains a direct CP violating amplitude and measuring this mode would provide another means for probing direct CP violation. The $K_L \to \pi^0 e^+ e^-$ mode also contains a CP conserving amplitude in which the two electrons couple to two photons that is related to the decay, $K_L \to \pi^0\gamma\gamma$. The branching ratio for the CP conserving amplitude is quite sensitive to the parameter, a_v, which can be measured using $K_L \to \pi^0\gamma\gamma$ decays.

The decay $K_L \to \pi^0 e^+ e^-\gamma$ probes much of the same physics as the decay $K_L \to \pi^0\gamma\gamma$. It proceeds through a CP conserving two-photon process and therefore the $e^+ e^-\gamma$ system is similar to the $\gamma\gamma$ system in the $K_L \to \pi^0\gamma\gamma$. Somewhat unexpectedly, the branching ratio for $K_L \to \pi^0 e^+ e^-\gamma$ is higher than that for $K_L \to \pi^0 e^+ e^-$. In the case of $K_L \to \pi^0 e^+ e^-$ a single photon coupling to the $e^+ e^-$ system is CP violating while it is CP allowed in the $K_L \to \pi^0 e^+ e^-\gamma$ case. Current calculations [8] predict the branching ratio to be 2.3×10^{-8}.

2 The KTeV Experiment

The KTeV detector is shown in Figure 1. The KTeV spectrometer consists of four drift chambers, two located upstream and two downstream of an analyzing magnet. Each of the drift chambers contains two planes offset by half of a drift cell for measuring a track's horizontal position and two planes for measuring the vertical position. The KTeV calorimeter [9] consists of 3100 pure CsI crystals that are 27 X_0 or 50 cm long. The crystals in the center of the calorimeter have a transverse cross section of $(2.5 \text{ cm})^2$ while the crystals towards the outside of the calorimeter have a cross section of $(5.0 \text{ cm})^2$. To reduce backgrounds from events with missing photons, the KTeV detector contains a system of 11 photon vetoes with a coverage out to 100 mrad. The KTeV detector has a muon system capable of detecting muons above 7 GeV/c. In the $Re(\epsilon'/\epsilon)$ measurement the muon system is primarily used to reject $K_L \to \pi\mu\nu$ events.

The $Re(\epsilon'/\epsilon)$ measurement requires collecting both CP violating and CP conserving $K \to \pi^+\pi^-$ and $K \to \pi^0\pi^0$ decays. In the KTeV experiment all four modes

Table 1: Comparison of the E731 and KTeV experiments.

	E731	KTeV
BEAM		
μ flux/proton	4×10^{-5}	2×10^{-7}
max protons/spill	2×10^{12}	5×10^{12}
CsI radiation	450	50
(rad/E12/week)		
REGENERATOR		
Inelastic bkg	$\sim 1.8\%$	$\sim 0.3\%$
CALORIMETER		
γ resolution (20 GeV)	3.5%	0.65%
Nonlinearity	10%	0.4%
π/e rejection	~ 50	~ 400
SPECTROMETER		
PT kick (MeV/c)	200	400
p resolution	0.5%	0.25%
PHOTON VETOES		
(photo electrons/MeV)	0.02	0.2
RESOLUTION		
$\pi^0\pi^0$ (MeV/c^2)	5.5	1.5
$\pi^+\pi^-$ (MeV/c^2)	3.5	1.6

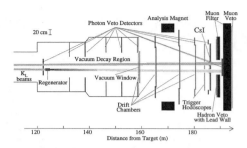

Figure 1: Plan view of the KTeV detector.

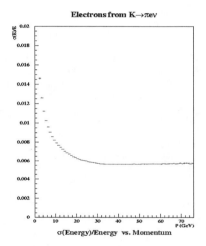

Figure 2: KTeV CsI calorimeter energy resolution for electrons as a function of the electron momentum.

are collected simultaneously, which helps to reduce the systematics due to accidental activity and due to time dependent conditions during the data taking. Approximately 125 meters downstream of the target, two nearly parallel K_L beams enter the KTeV detector. One of the beams is allowed to strike a regenerator of approximately 1.8 interaction lengths. The regenerator is composed of scintillator to help veto events that inelastically scatter in the regenerator. Although the regeneration amplitude is only a few percent, decays downstream of the regenerator are dominated by $K_S \to \pi\pi$ because $K_L \to \pi\pi$ decays are suppressed by CP violation and because the decay region corresponds to only a small fraction of the K_L lifetime. The regenerator moves from one beam to the other once per minute, thereby reducing systematics related to asymmetries in the beams or the detector. A comparison of relevant parameters of the E731 and KTeV experiments is shown in Table 1.

3 The Measurement of $Re(\epsilon'/\epsilon)$

The KTeV experiment collected approximately four million CP violating neutral decays between November 1996 and July 1997. As noted in Table 1, the KTeV detector performed very well. The calorimeter achieved a resolution for electrons well below 1% as shown in Figure 2. For reconstructed $\pi^0\pi^0$ decays, the kaon mass resolution is 1.5 MeV/c^2. In the energy range from 3 GeV to 60 GeV, the calorimeter is linear to better than 0.5%. During the run the center of the array received approximately

2.5 kRad and experienced a loss of light of approximately 0.5% per week. However, the change in linearity of affected crystals was very small.

The single hit resolution of the charged spectrometer was approximately 100 μm. Some chamber degradation was encountered in the region of the chambers through which the beam passed. After nominal offline cuts, the mass resolution for $\pi^+\pi^-$ events is approximately 1.6 MeV/c^2, and the background level from all sources, including scattered kaons, is approximately 0.1%.

To correct for the difference between the K_L and K_S acceptance, the KTeV experiment relies upon a detailed Monte Carlo. The main inputs to the Monte Carlo are the kaon momentum spectrum, the kaon transverse momentum spectrum and the detector apertures. Details

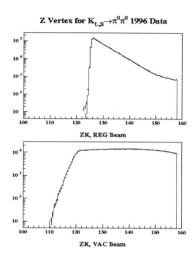

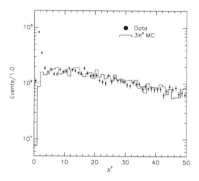

Figure 4: The photon shape parameter. The dots are the data and the histogram is the $3\pi^0$ Monte Carlo. The excess at low χ^2 is due to $K_L \to \pi^0\gamma\gamma$ events.

Figure 3: The $K_L \to \pi^0\pi^0$ decay vertex postions for a) the regenerator beam and b) the vacuum beam. The crosses represent the data and the histogram the $2\pi^0$ Monte Carlo.

of the Monte Carlo are checked using high statistic kaon decays such as $K_L \to \pi^0\pi^0\pi^0$ and $K_L \to \pi e\nu$. Figure 3 shows the decay vertex position for the regenerator and vacuum beams in neutral mode events. Note that the Monte Carlo simulation is overlaid on the data distribution and is nearly indistinguishable from the data. Systematic effects due to acceptance can be determined through the ratio of decay vertex distributions between data to Monte Carlo

Currently, the experiment plans to present a preliminary result based upon 20% of the total data set. The errors from this analysis will be smaller than those of either the E731 or the NA31 result. During 1999 the KTeV experiment will continue to record data for the $Re(\epsilon'/\epsilon)$ measurement. This run should increase the $Re(\epsilon'/\epsilon)$ data set by a factor of two. After combining the upcoming data with the current data, the KTeV experiment should achieve a combined statistical and systematic error on the order of 1×10^{-4}.

4   The Decay $K_L \to \pi^0\gamma\gamma$

To reconstruct $K_L \to \pi^0\gamma\gamma$ events, we isolate a sample of events that contain only four clusters in the CsI calorimeter. Assuming the kaon mass, we reconstruct the decay vertex position. This position is then used in reconstructing a π^0 from the six possible $\gamma\gamma$ pairings.

We choose the $\gamma\gamma$ pair whose mass reconstructs closest to the nominal π^0 mass. We veto events with detected energy in the photon veto system and only keep those events that reconstruct in the vacuum beam. $K_L \to \pi^0\pi^0$ events that were recorded in the same trigger as the $K_L \to \pi^0\gamma\gamma$ events are used to normalize the events. Misreconstructed $K_L \to \pi^0\pi^0$ events are a possible background and are removed by reconstructing the other two $\pi^0\pi^0$ pairings and discarding events that have two good π^0s.

The background in this measurement is dominated by $K_L \to \pi^0\pi^0\pi^0$ decays. Many of these events are removed by restricting the decay vertex region since $3\pi^0$ decays with missing photons reconstruct downstream of the true vertex position. The remaining $3\pi^0$ background results from events in which one or more photons fuse together in the CsI calorimeter. These events can be reduced by making requirements on the photon shape. We developed a shape χ^2 variable that is sensitive to whether a cluster is consistent with a single cluster. This shape χ^2 variable is shown in Figure 4. As can be seen, the Monte Carlo simulation of the background matches the data well except for χ^2 values below about four. The lowest values of the χ^2 distribution correspond to $\pi^0\gamma\gamma$ events since these events do not have fused photons. By making a tight requirement on this distribution, we are able to isolate a relatively clean sample of $K_L \to \pi^0\gamma\gamma$ events.

Figure 5 shows the final $m_{\gamma\gamma}$ distribution for all events after making the photon shape χ^2 requirement. The Monte Carlo is the $O(p^6)$ chiral perturbation theory prediction and matches the data very well. In our

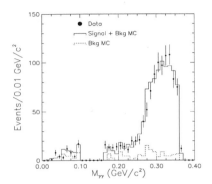

Figure 5: The $\gamma\gamma$ mass distribution for $K_L \to \pi^0\gamma\gamma$ candidates. The dots are the data, the histogram the sum of the background and signal Monte Carlo and the dashed histogram is the normalized background Monte Carlo.

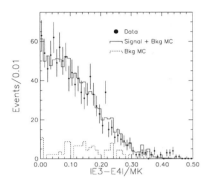

Figure 6: The y Dalitz parameter for $K_L \to \pi^0\gamma\gamma$ candidate events. The dots are the data, the histogram the sum of the background and signal Monte Carlo and the dashed histogram is the normalized background Monte Carlo.

analysis we find approximately 800 $K_L \to \pi^0\gamma\gamma$ events, an order of magnitude improvement over the previous world sample. Using this sample of events, we calculate a branching ratio[a] of $(1.78\pm0.06\pm0.10)\times10^{-6}$ to be compared with the previous measurement of $(1.70 \pm 0.3) \times 10^{-6}$. One should note that this sample of events demonstrates for the first time the existence of a low mass tail in the $m_{\gamma\gamma}$ distribution. The NA31 experiment had set a limit of $\Gamma(m_{\gamma\gamma} < 0.240 \text{ GeV/c}^2)/\Gamma(m_{\gamma\gamma} \text{ all}) < 0.09$. The y Dalitz distribution is shown in Figure 6. We performed a fit to the branching ratio, the $m_{\gamma\gamma}$ and y distributions and find $a_v = -0.84\pm0.06$. This value of a_v corresponds to a CP conserving branching ratio for $K_L \to \pi^0 e^+ e^-$ of approximately $3\text{-}4\times10^{-12}$ which is about a few orders of magnitude higher than predictions based upon the $O(p^4)$ calculations. During the 1999 run, we plan to continue to collect $K_L \to \pi^0\gamma\gamma$ events and hope to more than double the current sample.

5   The Decay $K_L \to \pi^0 e^+ e^- \gamma$

We reconstruct $K_L \to \pi^0 e^+ e^- \gamma$ events by requiring two oppositely charged tracks in the charged spectrometer and three neutral clusters in the CsI calorimeter. Using the charged vertex, we choose the $\gamma\gamma$ combination that reconstructs closest to the nominal π^0 mass. $K_L \to \pi^0 e^+ e^- \gamma$ events are selected by requiring the $e^+ e^- \gamma\gamma$ mass to lie within approximately 6.5 MeV/c^2

---

[a]The number presented in this contribution is somewhat higher than the result reported at the ICHEP98 conference. An error in the acceptance calculation lead to the lower value initially reported.

of the nominal kaon mass. Because the neutral vertex resolution is often better than the charged vertex resolution, we determine the kaon decay vertex from the π^0 vertex reconstruction.

The main backgrounds to this decay mode result from $K_L \to \pi^0\pi^0$ and $K_L \to \pi^0\pi^0\pi^0$ decays where one of the π^0s decays via $\pi^0 \to e^+ e^- \gamma$. In the case of the $K_L \to \pi^0\pi^0$ decays the mass reconstructs at the kaon mass and the $e^+ e^- \gamma$ mass reconstructs near the nominal π^0 mass. These events are removed by requiring the $e^+ e^- \gamma$ mass to be below 110 MeV/c^2 or above 160 MeV/c^2. In a small fraction of events, the wrong π^0 was chosen and the $e^+ e^- \gamma$ mass does not reconstruct at the π^0 mass. To remove these events, we examine the two other π^0 pairings and remove events where the $e^+ e^- \gamma$ mass for that pairing reconstructs near the nominal π^0 mass.

Since the $K_L \to \pi^0\pi^0\pi^0$ branching ratio is quite large, $3\pi^0$ decays potentially are a very large background. In order for these events to contribute to the final sample, two photons must either miss the detector or fuse with other photons in the calorimeter. A number of cuts were developed to help reduce this background. First, the transverse momentum of the kaon (p_t) was required to be less than 0.018 GeV/c. This requirement helps to remove events with missing energy. We then imposed a requirement on the photon shape variable described above to reduce events with one or more fused clusters. To further remove events in which one or more photons miss the detector, we imposed a requirement on the missing momentum of the kaon. After all of the cuts described

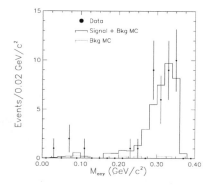

Figure 7: The $e^+e^-\gamma$ mass distribution for $K_L \to \pi^0 e^+ e^- \gamma$ candidates. The dots are the data, the histogram the signal plus background Monte Carlo and the filled histogram is the background Monte Carlo.

above, the background from $K_L \to \pi^0\pi^0\pi^0$ events is negligible.

After all of the cuts have been imposed we find 40 candidate events with an estimated background of 1.1 events. The remaining background comes from misreconstructed $K_L \to \pi^0\pi^0$ events. Background from $K_L \to \pi^0\pi^0\pi^0$ events is negligible. Figure 7 shows the $e^+e^-\gamma$ mass distribution for the candidate events. As expected, the $e^+e^-\gamma$ mass distribution closely resembles the $\gamma\gamma$ mass distribution in Figure 5. However, existence of a low-mass tail in the $K_L \to \pi^0 e^+ e^- \gamma$ events is somewhat inconclusive. To calculate the acceptance, we employed a $O(p^6)$ chiral perturbation calculation and found the acceptance to be 0.64%. For normalization, we used $K_L \to \pi^0\pi^0, \pi^0 \to e^+e^-\gamma$ decays. Although these events are CP violating and the Dalitz branching ratio is only about 1%, we reconstructed over 44,000 events. From our sample of 40 $K_L \to \pi^0 e^+ e^- \gamma$ events, we calculate the branching ratio to be, $\mathrm{BR}(K_L \to \pi^0 e^+ e^- \gamma)$ $= (2.22 \pm 0.35 \pm 0.08) \times 10^{-8}$. This should be compared to the chiral perturbation prediction of 2.3×10^{-8} which is well above the expected branching ratio for $K_L \to \pi^0 e^+ e^-$. This is the first evidence for the decay, $K_L \to \pi^0 e^+ e^- \gamma$.

6 Conclusions

The KTeV experiment had a very successful run between November 1996 and July 1997. The experiment plans to present a preliminary result based upon 20% of this data set. This result should be a large improvement over the

results of the E731 or the NA31 experiments. The KTeV experiment will collect data during 1999 with the goal of more than doubling the current data set. Combining all of the data from the KTeV experiment should allow us to achieve total error on $Re(\epsilon'/\epsilon)$ of 1×10^{-4}

The KTeV experiment also has new results on the decays $K_L \to \pi^0\gamma\gamma$ and $K_L \to \pi^0 e^+ e^- \gamma$. The sample of $K_L \to \pi^0\gamma\gamma$ events is an order of magnitude increase over the previous world sample. These events allow us to demonstrate the existence of a low-mass tail in the $\gamma\gamma$ mass distribution. Also, our fit to a_v indicates that the CP conserving component of $K_L \to \pi^0 e^+ e^-$ may be comparable to the CP violating component. The KTeV experiment made the first measurement of the decay $K_L \to \pi^0 e^+ e^- \gamma$. The measured branching ratio for this mode is consistent with chiral perturbation predictions and somewhat surprisingly above that for $K_L \to \pi^0 e^+ e^-$. Both decays, $K_L \to \pi^0\gamma\gamma$ and $K_L \to \pi^0 e^+ e^- \gamma$ provide new information relevant to the search for $K_L \to \pi^0 e^+ e^-$. Data collected in 1999 will allow us to continue study of these interesting modes.

Acknowledgments

The KTeV experiment is a collaboration of the following twelve institutions: Arizona, Chicago, Colorado, Elmhurst, Fermilab, Osaka, Rice, Rutgers, UCLA, UCSD, Virginia and Wisconsin. We would like to thank the DOE, the NSF and the US-Japan Foundation for their support. We would also like to acknowledge the excellent support of the Fermilab technical staff.

References

1. A.J. Buras and R. Fleischer, "Quark Mixing, CP violation and Rare Decays After the Top Quark Discovery," Heavy Flavors II (World Scientific, 1997).
2. L.K. Gibbons *et. al.*, Phys. Rev. Lett. **70** 1203 (1993).
3. G.D. Barr *et. al.*, Phys. Lett. **B317**, 233 (1993).
4. G.D. Barr *et. al.*, Phys. Lett **B284**, 440 (1992).
5. V. Papadimitriou *et. al.*, Phys. Rev. **D44** 573 (1991).
6. J.F. Donoghue and F. Gabbiani, Phys. Rev. **D51**, 2187 (1995).
7. G. D'Ambrosio and J. Portoles, Nucl. Phys **B492** 417 (1997).
8. J.F. Donoghue and F. Gabbiani, Phys. Rev. **D56**, 1605 (1997).
9. A.J. Roodman, "The KTeV Pure CsI Calorimeter," Proceedings of the VII International Conference on Calorimetry (World Scientific, 1998).

RESULTS ON CP, T, CPT SYMMETRIES WITH TAGGED K^0 AND $\bar{K}^0$ BY CPLEAR

P. KOKKAS

CERN/EP, CH-1211 Geneva 23, Switzerland

for the CPLEAR collaboration:
A. Angelopoulos[1], A. Apostolakis[1], E. Aslanides[11], G. Backenstoss[2], P. Bargassa[13], O. Behnke[17], A. Benelli[2], V. Bertin[11], F. Blanc[7,13], P. Bloch[4], P. Carlson[15], M. Carroll[9], E. Cawley[9], M.B. Chertok[3], M. Danielsson[15], M. Dejardin[14], J. Derre[14], A. Ealet[11], C.A. Eleftheriadis[16], L. Faravel[7], W. Fetscher[17], M. Fidecaro[4], A. Filipčič[10], D. Francis[3], J. Fry[9], E. Gabathuler[9], R. Gamet[9], H.- J. Gerber[17], A. Go[4], A. Haselden[9], P.J. Hayman[9], F. Henry-Couannier[11], R.W. Hollander[6], K. Jon-And[15], P.-R. Kettle[13], P. Kokkas[4], R. Kreuger[6], R. Le Gac[11], F. Leimgruber[2], I. Mandić[10], N. Manthos[8], G. Marel[14], M. Mikuž[10], J. Miller[3], F. Montanet[11], A. Muller[14], T. Nakada[13], B. Pagels[17], I. Papadopoulos[16], P. Pavlopoulos[2], G. Polivka[2], R. Rickenbach[2], B.L. Roberts[3], T. Ruf[4], M. Schäfer[17], L.A. Schaller[7], T. Schietinger[2], A. Schopper[4], L. Tauscher[2], C. Thibault[12], F. Touchard[11], C. Touramanis[9], C.W.E. Van Eijk[6], S. Vlachos[2], P. Weber[17], O. Wigger[13], M. Wolter[17], D. Zavrtanik[10] and D. Zimmerman[3].

[1]University of Athens, Greece; [2]University of Basle, Switzerland; [3]Boston University, USA; [4]CERN, Geneva, Switzerland; [5]LIP and University of Coimbra, Portugal; [6]Delft University of Technology, Netherlands; [7]University of Fribourg, Switzerland; [8]University of Ioannina, Greece; [9]University of Liverpool, UK; [10]J. Stefan Inst. and Phys. Dep., University of Ljubljana, Slovenia; [11]CPPM, IN2P3-CNRS et Université d'Aix-Marseille II, France; [12]CSNSM, IN2P3-CNRS, Orsay, France; [13]Paul Scherrer Institut (PSI), Switzerland; [14]CEA, DSM/DAPNIA, CE-Saclay, France; [15]Royal Institute of Technology, Stockholm, Sweden; [16]University of Thessaloniki, Greece; [17]ETH-IPP Zürich, Switzerland

We report the results of the CPLEAR experiment on CP-, T- and CPT-symmetries in the neutral kaon system. CPLEAR has experimentally determined, for the first time, the violation of T invariance by a direct method using semileptonic decays. The CPT symmetry is tested through the parameters $\mathrm{Re}(\delta)$ with a precision of a few 10^{-4} and $\mathrm{Im}(\delta)$, from the Bell-Steinberger relation, with a precision of 10^{-5}. This allows the mass equality between the K^0 and $\bar{K}^0$ to be tested down to the level of 10^{-19} GeV.

1 Introduction

At present, the neutral-kaon system remains a most precise laboratory where to measure the set of parameters which describe the exactness of discrete symmetries CP, T and CPT. CPLEAR has successfully developed a novel experimental approach[1] for the measurement of these parameters. The method is based on the measurement of time-dependent decay-rate asymmetries for the main decay modes, in the range 0 to 20 τ_S, between particles (K^0) and antiparticles ($\bar{K}^0$). The strangeness of the neutral kaons is tagged at the time of production and of decay (for the semileptonic decays).

2 The CPLEAR experiment

The CPLEAR experiment uses initially pure K^0 and $\bar{K}^0$ states produced in $p\bar{p}$ annihilations via the reactions

$$p\bar{p} \rightarrow \begin{array}{c} K^-\pi^+K^0 \\ K^+\pi^-\bar{K}^0 \end{array} \qquad (1)$$

each with a branching ratio of $\approx 2 \times 10^{-3}$. The initial strangeness of the neutral kaon is tagged by the charge of the accompanying kaon.

The CPLEAR detector is described elsewhere[2] and only a brief outline is presented here. It had a cylindrical geometry and was mounted inside a solenoid of length 3.6 m and internal radius 1 m, which produced a magnetic field of 0.44 T parallel to the $\bar{p}$ beam. The experiment used an intense 200 MeV/c antiproton beam ($\approx 10^6$ $\bar{p}$/s) from the Low Energy Antiproton Ring (LEAR) at CERN, which stopped in the target at the centre of the detector. The target, consisting of a 7-cm radius sphere filled with gaseous hydrogen at 16-bar pressure, was replaced in mid 1994 by a 1.1-cm radius, 27-bar, cylindrical target surrounded by a 1.5-cm radius, cylindrical, proportional chamber (PC0).

The tracking of the annihilation products was performed by two layers of proportional chambers, six layers of drift chambers and two layers of streamer tubes. A scintillator-Cherenkov-scintillator (S1-C-S2) sandwich (PID) provided input to a fast trigger system to identify the charged kaon, based on energy loss, time of flight and Cherenkov light measurements. An 18-layer, lead/gas-sampling electromagnetic calorimeter completed the detector.

Because of the small branching ratio of the desired annihilation channels, Eq. (1), and the high beam intensity, a multi-level trigger system[2], based on custom-made hardwired processors, was used to provide fast and efficient background rejection.

3 Measurements

The CPLEAR method allows detecting interference effects as a function of the decay time, in the various decay channels of the K^0 and $\bar{K}^0$ mesons. The time-dependent asymmetries do have the form:

$$A_f(\tau) \equiv \frac{R(\bar{K}^0 \to f)(\tau) - R(K^0 \to f)(\tau)}{R(\bar{K}^0 \to f)(\tau) + R(K^0 \to f)(\tau)} \quad (2)$$

where the final state f can be $\pi^+\pi^-$, $\pi^0\pi^0$, $\pi^+\pi^-\pi^0$, $\pi^0\pi^0\pi^0$ or $\pi^\pm e^\mp \nu$. The method of symmetrical production of particles and antiparticles in the proton-antiproton annihilation at rest, Eq. (1), and the identical detection of their decay products, has the advantage of minimizing the systematic errors. This method should allow the detection of CP breakdown in channels other than the two-pion channel, thus giving a global overview of the whole CP problem. In Table 1 we present all important CPLEAR measurements while in the following subsections we focus on the results on the $\pi^+\pi^-$ and on the semileptonic decay modes which are to be published shortly.

Table 1: The CPLEAR measurements.

Parameter	Measurement
$\|\eta_{+-}\|$	$(2.254 \pm 0.024_{\text{stat}} \pm 0.024_{\text{syst}})10^{-3}$
ϕ_{+-}	$(43.6 \pm 0.5_{\text{stat}} \pm 0.2_{\text{syst}})^o$
$\|\eta_{00}\|$	$(2.5 \pm 0.3_{\text{stat}} \pm 0.2_{\text{syst}})10^{-3}$
ϕ_{00}	$(42.0 \pm 5.6_{\text{stat}} \pm 1.9_{\text{syst}} \pm 0.4_{\Delta m})^o$
$\text{Re}(\eta_{+-0})$	$(-2 \pm 7 \text{ stat} \, ^{+4}_{-1} \text{ syst})10^{-3}$
$\text{Im}(\eta_{+-0})$	$(-2 \pm 9 \text{ stat} \, ^{+2}_{-1} \text{ syst})10^{-3}$
$\text{Re}(\eta_{000})$	$0.18 \pm 0.14_{\text{stat}} \pm 0.06_{\text{syst}}$
$\text{Im}(\eta_{000})$	$0.15 \pm 0.20_{\text{stat}} \pm 0.03_{\text{syst}}$
Δm	$(529.5 \pm 2.0_{\text{stat}} \pm 0.3_{\text{syst}})10^7 \hbar s^{-1}$
$\text{Re}(x)$	$(-1.8 \pm 4.1_{\text{stat}} \pm 4.5_{\text{syst}})10^{-3}$
A_T	$(6.6 \pm 1.3_{\text{stat}} \pm 1.0_{\text{syst}})10^{-3}$
$\text{Re}(\delta)$	$(3.0 \pm 2.6_{\text{stat}} \pm 0.6_{\text{syst}})10^{-4}$

3.1 The K^0 ($\bar{K}^0$) $\to \pi^+\pi^-$ decay mode

CP violation in the $\pi^+\pi^-$ decay mode (CP=+1) of neutral kaons manifests itself in the occurence of $K_L \to \pi^+\pi^-$ decays and is commonly described by the parameter η_{+-}.

The asymmetry

$$A_{+-}(\tau) \equiv \frac{R(\bar{K}^0 \to \pi^+\pi^-)(\tau) - \alpha R(K^0 \to \pi^+\pi^-)(\tau)}{R(\bar{K}^0 \to \pi^+\pi^-)(\tau) + \alpha R(K^0 \to \pi^+\pi^-)(\tau)}$$
$$= -2 \frac{|\eta_{+-}| e^{\frac{1}{2}(\Gamma_S - \Gamma_L)\tau} \cos(\Delta m \tau - \phi_{+-})}{1 + |\eta_{+-}|^2 e^{(\Gamma_S - \Gamma_L)\tau}} \quad (3)$$

is well suited to extract both the magnitude $|\eta_{+-}|$ and the phase ϕ_{+-} of η_{+-}, where $\Gamma_{S,L}$ are the $K_{S,L}$ decay widths, Δm the $K_L - K_S$ mass difference and α is a normalization factor which includes the ratio of the $\bar{K}^0$ over the K^0 tagging efficiency. The details of the analysis procedure can be found in [3]. Figure (1) shows the asymmetry $A_{+-}(\tau)$ made out of 7×10^7 events selected in the decay-time region 1-20 τ_S.

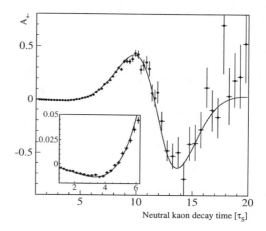

Figure 1: The time-dependent asymmetry $A_{+-}(\tau)$. The solid line is the result of our fit. The inset displays the data at short decay times with a refined binning.

A fit of Eq. (3) to the data gives

$$|\eta_{+-}| = (2.254 \pm 0.024_{\text{stat}} \pm 0.024_{\text{syst}}) \times 10^{-3}$$
$$\phi_{+-} = 43.63^\circ \pm 0.54^\circ_{\text{stat}} \pm 0.23^\circ_{\text{syst}} (\pm 0.42^\circ_{\Delta m}) . \quad (4)$$

The above value of ϕ_{+-} is the most precise result from a single experiment.

3.2 The semileptonic decay mode

In the semileptonic decays of the neutral kaons there are four independent decay rates, depending on the strangeness of the kaon (K^0 or $\bar{K}^0$) at the production time, $t = 0$, and on the charge of the decay lepton (e^+ or e^-):

$$R_+(\tau) \equiv R\left[K^0_{t=0} \to e^+\pi^-\nu_{t=\tau}\right]$$
$$\overline{R}_-(\tau) \equiv R\left[\bar{K}^0_{t=0} \to e^-\pi^+\overline{\nu}_{t=\tau}\right]$$
$$R_-(\tau) \equiv R\left[K^0_{t=0} \to e^-\pi^+\overline{\nu}_{t=\tau}\right] \quad (5)$$
$$\overline{R}_+(\tau) \equiv R\left[\bar{K}^0_{t=0} \to e^+\pi^-\nu_{t=\tau}\right] .$$

According to the $\Delta S = \Delta Q$ rule, the first two processes (R_+ and $\overline{R}_-$) are directly allowed, while the other two (R_- and $\overline{R}_+$) occur only after a $\Delta S = 2$ transition.

The above four rates can be parametrized as a function of the mixing parameters ϵ (T-violation parameter) and δ (CPT-violation parameter), defined as

$$\epsilon = \frac{\Lambda_{\overline{K}^0,K^0} - \Lambda_{K^0,\overline{K}^0}}{2(\lambda_L - \lambda_S)} \qquad \delta = \frac{\Lambda_{\overline{K}^0,\overline{K}^0} - \Lambda_{K^0,K^0}}{2(\lambda_L - \lambda_S)} \qquad (6)$$

where Λ_{ij} are the elements of the effective Hamiltonian Λ and $\lambda_{L,S} = m_{L,S} - \frac{i}{2}\Gamma_{L,S}$ are the Λ's eigenvalues, with $m_{L,S}$ and $\Gamma_{L,S}$ the masses and decay widths for the K_L and K_S states. The decay amplitudes for the four rates can be written as [4,5]

$$\begin{aligned}
\langle e^+\pi^-\nu|\Lambda|K^0\rangle &= a + b \\
\langle e^-\pi^+\overline{\nu}|\Lambda|\overline{K}^0\rangle &= a^* - b^* \\
\langle e^-\pi^+\overline{\nu}|\Lambda|K^0\rangle &= c + d \\
\langle e^+\pi^-\nu|\Lambda|\overline{K}^0\rangle &= c^* - d^*
\end{aligned} \qquad (7)$$

where the parameters c and d describe a possible violation of the $\Delta S = \Delta Q$ rule, b and d represent a possible CPT violation. The imaginary parts of a, b, c and d are all T-violating.

We also introduce the following $\Delta S = \Delta Q$ violation rule parameters

$$x = \frac{c^* - d^*}{a + b} , \qquad \overline{x} = \frac{c^* + d^*}{a - b} \qquad (8)$$

$$x_+ = \frac{x + \overline{x}}{2} , \qquad x_- = \frac{x - \overline{x}}{2} \qquad (9)$$

and the CPT-violation parameter

$$y = -\frac{b}{a} . \qquad (10)$$

The values of the above parameters under different hypothesis for the symmetry properties in the semileptonic decay amplitudes are shown in Table 2.

Table 2: The values of the parameters x, $\overline{x}$, x_+, x_- and y under different hypothesis for the symmetry properties in the semileptonic decay amplitudes.

Symmetry	Parameter
$\Delta S = \Delta Q$ exact	$x = \overline{x} = x_+ = x_- = 0$
CPT exact	$x = \overline{x} = x_+, \ x_- = 0,$ $y = 0$
T exact	$x, \ \overline{x}, \ y = $ real
CP exact	$x = \overline{x} \ x_- = 0$ $y = $ imaginary

With the four semileptonic decay rates, Eq. 5, it is possible to construct several asymmetries which are sensitive to different parameters of the kaon mixing matrix (Δm, ϵ and δ).

The asymmetry

$$\begin{aligned}
A_{\Delta m}(\tau) &\equiv \frac{\overline{R}_- + R_+ - \overline{R}_+ - R_-}{\overline{R}_- + R_+ + \overline{R}_+ + R_-} \\
&= \frac{e^{-1/2(\Gamma_S + \Gamma_L)\tau}\cos(\Delta m\tau)}{(1 + \mathrm{Re}(x))e^{-\Gamma_S\tau} + (1 - \mathrm{Re}(x_+))e^{-\Gamma_L\tau}}
\end{aligned} \qquad (11)$$

is sensitive to Δm and $\mathrm{Re}(x)$. For this measurement we take into account a possible violation of the $\Delta S = \Delta Q$ rule, but assume CPT to be exact in the semileptonic decay amplitudes (see Table 2). Figure 2 shows the asymmetry $A_{\Delta m}$. From a fit with Eq. (11) we obtain [6]

$$\begin{aligned}
\Delta m &= (0.5295 \pm 0.0020_{\text{stat}} \pm 0.0003_{\text{syst}}) \times 10^{10}\hbar s^{-1} \\
\mathrm{Re}(x) &= (-1.8 \pm 4.1_{\text{stat}} \pm 4.5_{\text{syst}}) \times 10^{-3}
\end{aligned} \qquad (12)$$

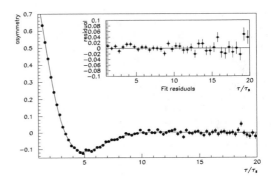

Figure 2: The asymmetry $A_{\Delta m}$ vs the decay time. The solid line represents the result of our fit.

Our Δm measurement gives the best single experiment result and has the same accuracy as the present world average. The $\mathrm{Re}(x)$ measurement improves the present limit on a possible $\Delta S = \Delta Q$ rule violation by a factor of three.

Among the highlights of the CPLEAR experiment are certainly the direct test of T and CPT-violation. The asymmetry

$$A_T(\tau) = \frac{R(\overline{K}^0 \to K^0) - R(K^0 \to \overline{K}^0)}{R(\overline{K}^0 \to K^0) + R(K^0 \to \overline{K}^0)} \qquad (13)$$

compares the rate of a neutral kaon produced as a $\overline{K}^0$ and decaying as a K^0 with the rate of the time reversed process, a kaon produced as a K^0 and decaying as a $\overline{K}^0$. Any asymmetry in this case is a clear sign of T violation [7]. In the CPLEAR experiment we measure the decay-time asymmetry

$$A_T^{exp}(\tau) = \frac{\overline{R}_+ - R_-}{\overline{R}_+ + R_-} \qquad (14)$$

which in the limit of CPT conservation in the semileptonic decay and of the validity of the $\Delta S = \Delta Q$ rule, is identical with the time-reversal asymmetry, Eq. 13. Figure 3 shows the asymmetry A_T^{exp} for a decay-time interval $1\tau_S \leq \tau \leq 20\tau_S$. The data points scatter around a positive and constant offset from zero, with the average being [8]

$$< A_T^{exp} >= (6.6 \pm 1.3_{stat} \pm 1.0_{syst}) \times 10^{-3} . \quad (15)$$

The above measurement of $< A_T^{exp} >$ is the first mea-

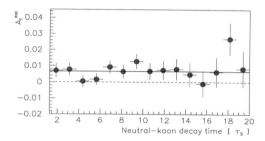

Figure 3: The asymmetry A_T^{exp} vs the decay time (in units of τ_S). The solid line represents the average $< A_T^{exp} >$ while the dotted line represents the zero.

surement of a departure from time-reversal invariance.

Finally the asymmetry

$$A_\delta(\tau) \equiv \frac{\overline{R}_+ - R_-}{\overline{R}_+ + R_-} + \frac{\overline{R}_- - R_+}{\overline{R}_- + R_+} =$$
$$= 2 \frac{\text{Im}(x)e^{-\frac{1}{2}(\Gamma_S + \Gamma_L)\tau}\sin(\Delta m\tau)}{E_+(\tau) - e^{-\frac{1}{2}(\Gamma_S + \Gamma_L)\tau}\cos(\Delta m\tau)}$$
$$+ \frac{[2\text{Im}(x) + 4\text{Im}(\delta)]e^{-\frac{1}{2}(\Gamma_S + \Gamma_L)\tau}\sin(\Delta m\tau)}{E_+(\tau) + e^{-\frac{1}{2}(\Gamma_S + \Gamma_L)\tau}\cos(\Delta m\tau)} \quad (16)$$
$$+ \frac{-4\text{Re}(\delta)E_-(\tau)}{E_+(\tau) + e^{-\frac{1}{2}(\Gamma_S + \Gamma_L)\tau}\cos(\Delta m\tau)} + 4\text{Im}(\delta)$$

with

$$E_\pm(\tau) = \frac{(e^{-\Gamma_S\tau} \pm e^{-\Gamma_L\tau})}{2} \quad (17)$$

is sensitive [9] for large decay-times ($\tau >> \tau_S$) to $\text{Re}(\delta)$ while for short decay-times to $\text{Im}(\delta)$ and $\text{Im}(x)$. We have assumed no CPT violation in the semileptonic decay amplitudes. Figure 4 shows the asymmetry A_δ. From a fit with Eq. (16) we obtain [9]

$$\text{Re}(\delta) = (\quad 3.0 \pm 2.6_{stat} \pm 0.6_{syst}) \times 10^{-4} ,$$
$$\text{Im}(\delta) = (-1.2 \pm 1.2_{stat} \pm 0.3_{syst}) \times 10^{-2} , \quad (18)$$
$$\text{Im}(x_+) = (\quad 0.8 \pm 0.9_{stat} \pm 0.3_{syst}) \times 10^{-2} .$$

The correlation of $\text{Re}(\delta)$ with $\text{Im}(\delta)$ and $\text{Im}(x_+)$ is weak (the correlation coefficients being -0.26 and 0.16, respectively) while the parameters $\text{Im}(\delta)$ and $\text{Im}(x_+)$ are

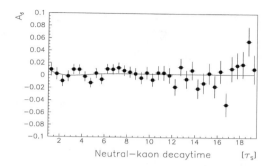

Figure 4: The asymmetry A_δ vs the decay time. The solid line represents the result of our fit.

strongly correlated with each other (the correlation coefficient is -0.98). Our result on $\text{Re}(\delta)$ improves by two orders of magnitude the limits obtained by an analysis [10] of previously available measurements of semileptonic decays.

4   An indirect test of CPT

The precise values of the various parameters of discrete symmetries which we extract from the CPLEAR experiment, can be used in the determination of $\text{Re}(\epsilon)$ and $\text{Im}(\delta)$ through the Bell-Steinberger relation [11],

$$-i\text{Im}(\delta) + \text{Re}(\epsilon) = \frac{1}{2(i\Delta m + \frac{1}{2}\gamma)}$$
$$\times (\sum (|A_S|^2 \eta_{\pi\pi}) + \sum (|A_L|^2 \eta_{3\pi}^*)$$
$$2[-i(\text{Im}(x_+) + \text{Im}(\delta) + \text{Re}(\epsilon) + \text{Re}(y)]|f|^2) \quad (19)$$

with

$$\gamma = \Gamma_S + \Gamma_L$$
$$|A_S|^2 = BR(K_S \rightarrow \pi\pi)\Gamma_S$$
$$|A_L|^2 = BR(K_L \rightarrow \pi\pi\pi)\Gamma_L \quad (20)$$
$$|f|^2 = BR(K_L \rightarrow \pi\ell\nu)\Gamma_L .$$

Using the values from CPLEAR together with PDG values, we obtain

$$\text{Im}(\delta) = (-1.9 \pm 2.0) \times 10^{-5}$$
$$\text{Re}(\epsilon) = (164.7 \pm 1.6) \times 10^{-5} \quad (21)$$

where this value has been obtained assuming $\eta_{+-0} = \eta_{000}$. If we drop this assumption the error on $\text{Im}(\delta)$ increases by a factor 4. The above result on $\text{Im}(\delta)$ improves by three orders of magnitude the previous limit [10] on this parameter.

Using our results on $\text{Re}(\delta)$ and $\text{Im}(\delta)$ we are able to set a limit to the K^0 and $\bar{K}^0$ mass and width difference, finding

$$|m_{K^0} - m_{\bar{K}^0}| = (-2.2 \pm 2.3) \times 10^{-18} \text{ GeV}$$
$$|\Gamma_{22} - \Gamma_{11}| = (4.1 \pm 4.9) \times 10^{-18} \text{ GeV} . \quad (22)$$

Assuming CPT invariance in the decay, ($\Gamma_{11} = \Gamma_{22}$), we decrease the mass-difference uncertainty by one order of magnitude, leading to

$$|m_{K^0} - m_{\bar{K}^0}| = (-2.6 \pm 2.8) \times 10^{-19} \text{ GeV} . \quad (23)$$

5 Conclusions

The CPLEAR experiment measured CP, T and CPT parameters in the neutral kaon system with an unprecedented accuracy. T violation was observed for the first time at the level of 4σ. The experiment yields the most precise values for ϕ_{+-}, Δm, x, η_{+-0} and η_{000}. The measured limit on the CPT violation parameters $\text{Re}(\delta)$ is two orders of magnitude lower than any previous measurement, while the limit on $\text{Im}(\delta)$ is by one order of magnitude lower than the limit for $\text{Re}(\delta)$. The mass difference between K^0 and $\bar{K}^0$ was probed to be zero with an accuracy of 10^{-19} GeV.

References

1. E. Gabathuler and P. Pavlopoulos, in *Strong and weak CP violation at LEAR*, Proc. Workshop on physics at LEAR with low energy cooled antiprotons, eds. U. Gastaldi and R. Klapisch (Plenum, New York, 1982) p. 747.
2. R. Adler *et al.*, The CPLEAR collaboration, *Nucl. Instrum. Methods* **376**, 76 (1996).
3. R. Adler *et al.*, The CPLEAR collaboration, *Phys. Lett.* B **363**, 243 (1995).
4. C.D. Buchanan *et al.*, *Phys. Rev.* D **45**, 4088 (1992).
5. L. Maiani, *CP and CPT violation in Neutral Kaon Decays*, in The Second DAΦNE Physics Handbook, eds. L. Maiani et al. (INFN, Frascati, 1995) p. 3. The parametrizations presented here are equivalent to ours but formulated with a slightly different notation.
6. A. Angelopoulos *et al.*, The CPLEAR collaboration, *Measurement of the* $K_L - K_S$ *mass difference using semileptonic decays of tagged neutral kaons*, submitted to *Phys. Lett.* B.
7. P.K. Kabir, *Phys. Rev.* D **2**, 540 (1970); A. Aharony, *Nuovo Cimento* **3**, 791 (1970).
8. A. Angelopoulos *et al.*, The CPLEAR collaboration, in *First direct observation of Time-Reversal non invariance in the neutral-kaon system*, submitted to *Phys. Lett.* B.
9. A. Angelopoulos *et al.*, The CPLEAR collaboration, in *A determination of the CPT violation parameter* $\text{Re}(\delta)$ *from the semileptonic decay of strangeness-tagged neutral kaons*, submitted to *Phys. Lett.* B.
10. V. Demidov et al., *Phys. Atom. Nucl.* **58**, 968 (1995).
11. J.S. Bell and J. Steinberger, Weak interactions of kaons, in Proc. of the Oxford International Conference on elementary Particles, eds. R.G. Moorhouse et al. (Rutherford Laboratory, Chilton, England, 1965), p. 195.

Investigation of CP violation in $B^0 \rightarrow J/\psi K_S^0$ decays at LEP with the OPAL detector

E.Barberio on behalf of the OPAL Collaboration.

CERN-EP, Geneva 1211 CH
E-mail: E.Barberio@cern.ch

An investigation of CP violation is performed using a total of 24 candidates for $B^0 \rightarrow J/\psi K_S^0$ decay, with a purity of about 60%. These events were selected from 4.4 million hadronic Z^0 decays recorded by the OPAL detector at LEP. An analysis procedure, involving techniques to reconstruct the proper decay times and tag the produced b-flavours, B^0 or $\bar{B}^0$, is developed to allow a first direct study of the time dependent CP asymmetry that, in the Standard Model, is $\sin 2\beta$. The result is

$$\sin 2\beta = 3.2^{+1.8}_{-2.0} \pm 0.5 \,,$$

where the first error is statistical and the second systematic.

1 Introduction

CP violation was observed in 1964 in K^0 decays[1], and, so far, this phenomenon has been seen only in the K system, however it is also expected to occur in the B system. CP violation can be accommodated in the Standard Model, provided that the CKM matrix elements are allowed to be complex. It is therefore important to investigate CP violation in the B system, where the relation between CKM matrix elements and CP violation can be tested.

Here is presented a study of CP violation in $B^0 \rightarrow J/\psi K_S^0$ decays. This decay mode is considered a 'golden' channel for CP violation studies[2,3,4], since the final state is a CP eigenstate which is experimentally favourable for reconstruction because the J/ψ and K_S^0 are narrow states and $J/\psi \rightarrow \ell^+\ell^-$ decays give a distinctive signature. In addition, CP violation in this channel is dominated by diagrams having a single relative phase, allowing a clean extraction of the phase of a CKM matrix element. In the Standard Model, the expected time-dependent rate asymmetry, $A(t)$, is given by[4]

$$A(t) \equiv \frac{B^0(t) - \bar{B}^0(t)}{B^0(t) + \bar{B}^0(t)} = -\sin 2\beta \sin \Delta m_{\mathrm{d}} t \,, \quad (1)$$

where the parameter Δm_{d} is the mass difference between the two B^0 mass eigenstates and $B^0(t)$ ($\bar{B}^0(t)$) represents the rate of produced B^0's ($\bar{B}^0$'s) decaying to $J/\psi K_S^0$ at a given proper decay time, t. Constraints on the CKM matrix, including measurements of CP violation in the K system, imply that the Standard Model expectation for $\sin 2\beta$ lies in the range 0.3–0.9[2].

At LEP, due to the small branching ratios of $B^0 \rightarrow J/\psi K_S^0$ and $J/\psi \rightarrow \ell^+\ell^-$, only a handful of these decays may be seen. Nonetheless, it is interesting to try to extract $\sin 2\beta$ because of the fundamental importance of this quantity, and the fact that no direct measurement has been performed so far. It is important to maximise the reconstruction efficiency and to determine the b-flavour at production with a minimum error rate.

In contrast, the proper time resolution is not critical, since the frequency of B^0 oscillation is easily resolved. In this analysis, $B^0 \rightarrow J/\psi K_S^0$ decays are reconstructed and their decay proper times are measured. The production flavour (B^0 or $\bar{B}^0$) of each candidate is determined using a combination of jet and vertex charge techniques. The CP-violating amplitude, $\sin 2\beta$, is extracted using an unbinned maximum likelihood fit to the proper-time distribution of the selected data.

2 $B^0 \rightarrow J/\psi K_S^0$ reconstruction

The selection of $B^0 \rightarrow J/\psi K_S^0$ decays is described in[5]. The efficiency for reconstructing the decay $B^0 \rightarrow J/\psi K_S^0 \rightarrow \ell^+\ell^-\pi^+\pi^-$ is estimated to be $19.5 \pm 1.6\%$ where the error is due to Monte Carlo statistics. The distribution of reconstructed mass is shown in Figure 1. In the mass region 5.0–5.6 GeV/c^2, 24 candidates are selected, with an overall purity of $60 \pm 8\%$, where the error is statistical. The background is estimated from a fit to the data. This fit also assigns the event-by-event background probabilities, f_{bac}, used in the fit for $\sin 2\beta$.

The B^0 decay length, l_{B^0}, is determined from the distance between the average beam spot position and the J/ψ vertex. The B^0 momentum, p_{B^0}, is taken from the constrained fit of the $J/\psi K_S^0$ system, and the proper decay time is then calculated as

$$t_{\mathrm{rec}} = l_{B^0} \cdot \frac{M_{B^0}}{p_{B^0}} \,, \quad (2)$$

where M_{B^0} is the B^0 mass. The uncertainty on t_{rec}, is typically $\sigma_t = 0.1\,\mathrm{ps}$.

3 Tagging the produced b-flavour

Information from the rest of the event is used to determine the production flavour of each candidate. The

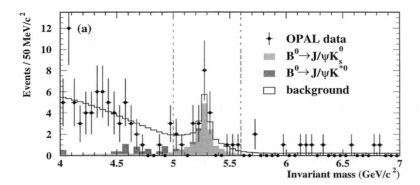

Figure 1: Mass distribution of $J/\psi K_S^0$ candidates. Data are shown by the points with error bars, and the fit is shown by the open histogram. The estimated contribution from the $B^0 \to J/\psi K_S^0$ signal is shown by the cross hatched histogram, the one from $B^0 \to J/\psi K^{*0}$ is shown by the hatched histogram.

weighted track charge sum, or 'jet charge', gives information on the charge and hence b-flavour, of the primary quark initiating the jet within which the $J/\psi K_S^0$ candidate is isolated. Additionally, since the Z^0 decays into quark-antiquark pairs, measuring the b-flavour of the other b quark produced in the event also provides information on the production flavour.

In this analysis, three different pieces of information are used to determine the B^0 production flavour: (a) the jet charge of the highest energy jet other than that containing the B^0 candidate (Q_{opp}), (b) the jet charge of the jet containing the B^0 candidate, excluding the tracks from the J/ψ and K_S^0 decays (Q_{same}), and (c) secondary vertices reconstructed in jets in the hemisphere opposite to the B^0 candidate. Approximately 40% of Monte Carlo $B^0 \to J/\psi K_S^0$ events has such an accepted secondary vertex in the opposite hemisphere.

For events with such a selected secondary vertex (9 of the 24 B^0 candidates), a neural network is constructed to tag the produced B^0 or $\bar{B}^0$, combining Q_{opp}, Q_{same} and the vertex charge. The output variable Q_B represents the effective produced b-flavour for each candidate ($Q_B = +1$ or -1 for pure B^0 or $\bar{B}^0$, respectively) and the tagging dilution. The average value of $|Q_B|$ is 0.38 for such events, according to Monte Carlo.

For events without such a selected secondary vertex, only the jet charge information is available. In this case the two jet charges are combined linearly to form $Q_{2jet} = Q_{same} - 1.43 \cdot Q_{opp}$, where the factor of 1.43 is optimised to minimise the mistag. For these events, the average value of $|Q_B|$ is 0.31.

4 Fit and results

In order to quantify the CP asymmetry in the data, an unbinned maximum likelihood fit is constructed, using for each $B^0 \to J/\psi K_S^0$ candidate the following inputs: t_{rec}, σ_t, Q_B and f_{bac} and $\Delta m_d = 0.467 \pm 0.022^{+0.017}_{-0.015}$ ps^{-1} [6]. The background is dominated by $b\bar{b}$ events and is assumed to have no CP asymmetry. Possible bias due to this assumption is treated as a systematic error. Fitting the data for the single parameter $\sin 2\beta$, gives:

$$\sin 2\beta = 3.2^{+1.8}_{-2.0} \; .$$

Figure 2 shows the $-\Delta \log \mathcal{L}$ value as a function of $\sin 2\beta$ from the fit. The estimator A, of the $B^0 \to J/\psi K_S^0$

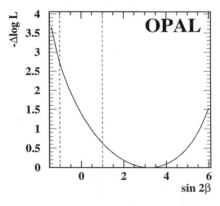

Figure 2: $-\Delta \log \mathcal{L}$ value as a function of $\sin 2\beta$, the physical region indicated by the dotted lines.

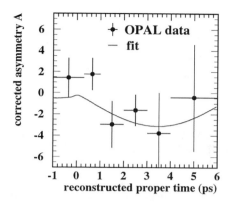

Figure 3: Distribution of the corrected asymmetry, A, versus t_{rec} for the $J/\psi K_S^0$ data with the fit result superimposed.

asymmetry (corrected for the average dilution in each time bin), is shown in Figure 3, with the fit result superimposed on the data points. The large observed values of A, typically exceeding the physical range of the asymmetry, are due to the tagging dilution factors and, to a lesser extent, the background fraction.

The main sources of systematic error and their effect on the measurement of $\sin 2\beta$ are listed in Table 1. The fit result is sensitive to the level and possible CP asymmetry of the background, the accuracy of the decay time reconstruction and the production flavour tagging dilution. Many of the sources of error have a statistical component, and many of them scale with the fitted value of $\sin 2\beta$. The systematic error would thus decrease in an analysis with higher statistics. A number of consistency checks is also performed, and the results are in [5].

Source	$\delta(\sin 2\beta)$
Background level (data statistics)	$+0.06$ -0.07
Background shape	± 0.32
Signal shape	± 0.13
Background asymmetry	± 0.03
Proper time reconstruction	± 0.01
Jet and vertex charge modelling	± 0.30
Jet charge offsets	$+0.14$ -0.08
Vertex charge performance	± 0.01
Δm_d value	± 0.10
B^0 lifetime	± 0.01
Background lifetime (± 0.4 ps)	$+0.01$ -0.02
Total	± 0.50

Table 1: Sources of systematic error

5 Conclusion

The time dependent CP asymmetry in the decays $B^0 \to J/\psi K_S^0$ is measured using data collected with the OPAL detector at LEP between 1990 and 1995. From 24 $B^0 \to J/\psi K_S^0$ candidates with a purity of about 60%, the CP violation amplitude, $\sin 2\beta$ in the Standard Model, is found to be

$$\sin 2\beta = 3.2^{+1.8}_{-2.0} \pm 0.5 \, ,$$

where the first error is statistical and the second systematic.

This is the first direct study of the CP asymmetry in the $B^0 \to J/\psi K_S^0$ system. The result is highly statistically limited, although it does yield information on the CP asymmetry, which is expected to be large.

References

1. J.H. Christenson, J.W. Cronin, V.L. Fitch and R. Turlay, *Phys. Rev.* D **13**, 1964 (138).

2. For recent reviews see :
   M. Neubert, Int. J. Mod. Phys. **A 11** (1996) 4173;
   A. Ali and D. London, Nucl. Phys. Proc. Suppl. **54 A** (1997) 297;
   A.J. Buras and R. Fleischer, Heavy Flavours II, World Scientific (1997), eds. A.J. Buras and M. Linder;
   Y. Nir, Proceedings of the 18th International Symposium on Lepton Photon Interactions Hamburg, Germany, July 28 - August 1 1997;
   I. Dunietz, Fermilab-CONF-97/278-T.

3. I.I. Bigi and A.I. Sanda, Nucl. Phys. **B 193** (1981) 85;
   A.B. Carter and A.I. Sanda, *Phys. Rev.* D **23**, 1981 (1567).

4. I.Dunietz and J.Rosener, *Phys. Rev.* D **34**, 1986 (1404);
   Ya.I. Azimov, N.G. Uraltsev and V.A.Khoze, *Sov.Journ.Nucl.Phys.* **45**, 1987 (878).

5. OPAL Collaboration, K. Ackerstaff *et al.*, CERN-EP/98-001, accepted by *Eur. J. Phys.* **C**

6. OPAL Collaboration, K. Ackerstaff *et al.*, *Z. Phys.* C **76**, 1997 (417).

MEASUREMENT OF $B^0/\overline{B}^0 \to J/\psi K_S^0$ DECAY ASYMMETRY USING SAME-SIDE B-FLAVOR TAGGING

J. TSENG

Massachusetts Institute of Technology
for the CDF Collaboration

We present a measurement of the time-dependent asymmetry in the rate for $\overline{B}^0$ versus B^0 decays to $J/\psi K_S^0$. The data sample consists of 198 ± 17 $B^0/\overline{B}^0$ decays collected by the CDF detector in $p\overline{p}$ collisions at the Fermilab Tevatron, where the initial b flavor has been determined by a same-side flavor tagging technique. This asymmetry is interpreted in the Standard Model as a measurement of the CP-violation parameter $\sin(2\beta)$. Our analysis results in $\sin(2\beta) = 1.8 \pm 1.1(stat) \pm 0.3(syst)$.

1 Introduction

Since its discovery over thirty years ago, CP violation has remained an elusive effect experimentally observed in only a few signatures among kaon decays. The fact that it lies at or near the heart of fundamental questions in physics—including not only questions of what lies beyond the Standard Model, but also such expansive questions as the origin of the matter-antimatter asymmetry of the universe—only intensifies the precise exploitation of known signatures as well as the search for its manifestations wherever they may be found.

A popular mechanism for explaining CP violation lies in the relationship between the weak and mass eigenstates of the different generations of quarks. This relationship is parameterized in the Standard Model with the unitary Cabibbo-Kobayashi-Maskawa (CKM) quark mixing matrix, [1]

$$\begin{pmatrix} d' \\ s' \\ b' \end{pmatrix} = \begin{pmatrix} V_{ud} & V_{us} & V_{ub} \\ V_{cd} & V_{cs} & V_{cb} \\ V_{td} & V_{ts} & V_{tb} \end{pmatrix} \begin{pmatrix} d \\ s \\ b \end{pmatrix}, \quad (1)$$

where d, s, and b are the mass eigenstates and d', s', and b' are the weak interaction eigenstates. With three generations, this matrix possesses a physical complex phase capable of accommodating CP violation. It is interesting that the original 1973 proposal of a third quark generation to explain CP violation predates by a year the unexpected discovery of the charmonium states which served to complete the *second* generation.

This article concerns the extension of the search for CP violation to the third generation, a topic which has engendered considerable experimental interest. One reason for such interest is that these measurements directly address the unitarity of the CKM matrix. For instance, one of the unitary constraints can be written as follows:

$$V_{ud}V_{ub}^* + V_{cd}V_{cb}^* + V_{td}V_{tb}^* = 0. \quad (2)$$

This equation can be represented as a triangle in the complex plane; the measurement of enough angles and sides of this triangle, thus overconstraining its construction, constitutes a fundamental test of the Standard Model understanding of CP violation.

The present analysis concerns the measurement of the angle

$$\beta \equiv \arg\left(-\frac{V_{cd}V_{cb}^*}{V_{td}V_{tb}^*}\right) \quad (3)$$

which is accessible through the relative decay rates of B^0 and $\overline{B}^0$ to the common CP eigenstate $J/\psi K_S^0$. A potentially large time-dependent asymmetry, cleanly related to this angle, arises from the interference between the direct decay path (*e.g.*, $B^0 \to J/\psi K_S^0$) and the mixed decay path (*e.g.*, $B^0 \to \overline{B}^0 \to J/\psi K_S^0$),

$$\mathcal{A}_{CP} \equiv \frac{\overline{B}^0(t) - B^0(t)}{\overline{B}^0(t) + B^0(t)} = \sin(2\beta)\sin(\Delta m_d t), \quad (4)$$

where $B^0(t)$ and $\overline{B}^0(t)$ are the numbers of decays to $J/\psi K_S^0$ at proper time t given that the produced meson (at $t = 0$) was a B^0 or $\overline{B}^0$, respectively [2]. The effect of mixing enters through Δm_d, the mass difference between the two B^0 mass eigenstates. Indirect evidence, interpreted in the light of Standard Model constraints, imply $0.30 \leq \sin(2\beta) \leq 0.88$ at 95% C.L. [3]. Thus the direct measurement of $\sin(2\beta)$ also provides a test of the Standard Model. The OPAL collaboration has reported $\sin(2\beta) = 3.2^{+1.8}_{-2.0} \pm 0.5$, also using $B^0/\overline{B}^0 \to J/\psi K_S^0$ decays [4].

This article summarizes the CDF result, which may be found in [5]. Many analysis details may also be found in [6] and [7].

2 Data Sample

The sample of $B^0/\overline{B}^0 \to J/\psi K_S^0$ decays used in this analysis was collected by CDF at the Fermilab Tevatron $p\overline{p}$ collider operating at $\sqrt{s} = 1.8$ TeV. The candidates

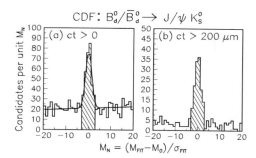

Figure 1: The normalized mass distribution of the $B^0/\overline{B}^0 \rightarrow J/\psi K_S^0$ candidates with $ct > 0$ and $ct > 200$ μm. The curve is the gaussian signal plus linear background from the full likelihood fit.

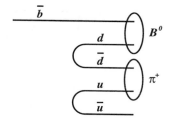

Figure 2: A simple picture of a $\overline{b}$ quark hadronizing into a B^0 meson. The "leading" pion is a π^-. The opposite correlation holds for a $\overline{B}^0$ meson.

are selected from data taken in the 1992-96 run, during which CDF accumulated an integrated luminosity of 110 pb^{-1}. The J/ψ is reconstructed via its decay to two oppositely-charged muons. Both muon tracks are required to have been reconstructed in the silicon microstrip detector, thereby obtaining a precise measurement of its decay vertex. The other pairs of oppositely-charged tracks are then searched for those consistent with a $K_S^0 \rightarrow \pi^+\pi^-$ hypothesis, where the K_S^0 is significantly displaced from the J/ψ vertex, which is presumed to be the B decay vertex. Each K_S^0 candidate is then combined with the J/ψ candidate in a four-particle fit which also includes constraints pointing the K_S^0 back to the J/ψ vertex, and the $J/\psi K_S$ system as a whole to the primary interaction vertex. The mass calculated by the fit, M_{FIT}, has a typical resolution of $\sigma_{FIT} \sim 9$ MeV/c^2. The proper decay length, ct, has a typical resolution of ~ 50 μm.

Figure 1 (left) shows the distribution of positive-lifetime candidate events in normalized mass $M_N \equiv (M_{FIT} - M_0)/\sigma_{FIT}$, where M_0 is the central value of the B^0 mass peak (5.277 GeV/c^2). The ratio of signal to background in the signal region is approximately 1.6, where the background has been modeled as a linear distribution. The corresponding negative-lifetime distribution shows no discernible signal peak. Also shown in the plot is the result of the maximum likelihood fit, described later, which takes into account both mass and lifetime information for both positive and negative lifetimes. The fit yields 198 ± 17 mesons, with a gaussian RMS of 1.39 ± 0.11, which is similar to corresponding widths in other $B \rightarrow J/\psi K$ normalized mass distributions [6].

It is interesting to note that since the CP asymmetry varies in time as $\sin(\Delta m_d t)$, it reaches its maximum close to a proper decay length of 1000 μm. Hence it is

instructive to examine the normalized mass distribution for long-lived events. Events with $ct > 200$ μm are shown in the right plot of Figure 1. The decay length requirement substantially reduces the background, and thus the measurement of the amplitude of the time-dependent CP asymmetry is largely insensitive to it.

3 Same-side Flavor Tagging

Once the sample of B's is obtained, the next step is to ascertain ("tag") whether they were B^0's or $\overline{B}^0$'s (their "flavor") when they were produced. The method used here is a "same-side" technique, *i.e.*, it aims to examine particles produced in association with the reconstructed B, in contrast with "opposite-side" techniques which attempt to identify the flavor of the other b hadron in order to infer the initial flavor of the first one. In particular, the present technique relies upon the correlation between the flavor of the reconstructed B and the charge of a nearby particle. This idea was first proposed in order to take advantage of the fact that the b quark may first hadronize to a B^{**} state, whose decay products would be the B^0 as well as the "tagging" pion [8]:

$$B^{**-} \rightarrow \overline{B}^0 \pi^- \tag{5}$$
$$B^{**+} \rightarrow B^0 \pi^+. \tag{6}$$

A $\overline{B}^0$ would thus be associated with a π^-, and a B^0 with a π^+. The same correlation is expected to exist between the B and the "leading" pion from fragmentation, as shown in Figure 2, and the present analysis utilizes both sources of correlation.

The criteria for the "nearby" track are as follows: the track must lie in an $\eta - \phi$ cone of half-angle 0.7 around the B direction, where $\eta \equiv -\ln[\tan(\theta/2)]$ is the pseudorapidity, θ is the polar angle relative to the proton beam direction, and ϕ is the azimuthal angle around the beam line. The tag must also have a minimum momentum transverse to the beam, p_T, of 400 MeV/c, and it should

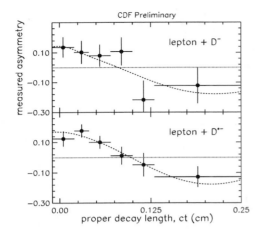

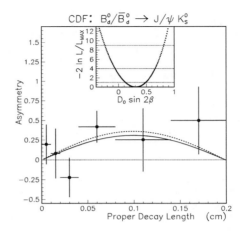

Figure 3: The measured **tagging asymmetries as a function** of ct for $B \to \ell D^{(*)-}X$ data (points). The dashed curve is the fit to the data, taking into account contributions to the sample from charged B's.

Figure 4: The sideband-subtracted tagging asymmetry as a function of the reconstructed $J/\psi K_S^0$ proper decay length (points). The dashed curve is the result of a simple binned χ^2 fit with the function $\mathcal{A}_0 \sin(\Delta m_d t)$. The solid curve is the likelihood fit result, for which the inset shows a scan through the log-likelihood function as $\mathcal{D}_0 \sin(2\beta)$ is varied about the best fit value.

be consistent with having come from the primary interaction vertex. If there is more than one "nearby" track, the one with the smallest p_T^{rel} is selected as the one tagging track, where p_T^{rel} is defined as the component of its momentum transverse to the momentum of the combined B+particle system.

The tagging algorithm, based as it is upon physical processes that happen before the B decay, should be applicable to other decay modes, and indeed, it has been applied successfully to the observation of $B^0 - \overline{B}^0$ time-dependent mixing and measurement of Δm_d using $B \to \ell D^{(*)}X$ decays [6,9], as shown in Figure 3. The algorithm has also been applied to a lower-statistics sample of $B^0 \to J/\psi K^{*0}$ decays, which are kinematically similar to the $J/\psi K_S^0$ events used in this analysis, yielding behavior consistent with mixing. The amplitude of the mixing oscillation in the $\ell D^{(*)}$ and $J/\psi K^{*0}$ data provides a direct measurement of the "dilution," $\mathcal{D}_0$, which is related to the "mistag" probability, $\mathcal{P}_{mistag}$:

$$\mathcal{D}_0 \equiv \frac{N_{RS} - N_{WS}}{N_{RS} + N_{WS}} = 1 - 2\mathcal{P}_{mistag} \qquad (7)$$

where N_{RS} is the number of events with the correct sign correlation and N_{WS} the others. In the case of SST, $\mathcal{D}_0 \sim 20\%$ [6]. As noted below, knowledge of the dilution is necessary for the extraction of $\sin(2\beta)$.

4 Tagging Asymmetry

The above same-side tagging technique tags approximately 65% of the $B^0/\overline{B}^0 \to J/\psi K_S^0$ events, which is a typical efficiency for this tagging algorithm. Figure 4 shows the sideband-subtracted asymmetry in bins of ct, where the asymmetry is calculated by counting the sideband-subtracted number of positive tags, N^+, and negative tags, N^-, in each bin:

$$\mathcal{A}(ct) = \frac{N^-(ct) - N^+(ct)}{N^-(ct) + N^+(ct)}. \qquad (8)$$

The signal region has been defined as $|M_N| < 3$, and the sideband region as $3 < |M_N| < 20$, for the purposes of counting. The events in the signal region generally prefer negative tags (i.e., a positive asymmetry), whereas events in the sideband regions favor positive tags (negative asymmetry). As noted before, however, the signal purity is high at large ct, and the sideband subtraction is a correspondingly small effect.

Two fits are shown in Figure 4. The dashed curve gives the results of a simple χ^2 fit of the function $\mathcal{A}_0 \sin(\Delta m_d t)$ to the binned asymmetries, where Δm_d has been fixed to its 1996 world-averaged [10] value of 0.474 ps^{-1}. In the absence of other charge asymmetries in the tagging, the amplitude, $\mathcal{A}_0 = 0.36 \pm 0.19$, measures $\mathcal{D}_0 \sin(2\beta)$. The amplitude of the time-dependent asymmetry is dominated by the asymmetries at large ct.

The solid curve is the result of an unbinned maximum likelihood fit which incorporates both signal and background distributions in M_N and ct. Sideband and negative-lifetime events are included to help constrain the background distributions. The likelihood function also incorporates resolution effects and corrections for systematic biases, such as the small inherent charge asymmetry favoring positive tracks resulting from the wire plane orientation in the main drift chamber. However, it is clear from the fact that the solid and dashed curves lie close to one another that these corrections do nothing dramatic; the result is dominated by the sample size. Also shown in the Figure 4 inset is the relative log-likelihood as a function $\mathcal{D}_0 \sin(2\beta)$. It is very close to parabolic, indicating gaussian errors.

Before ascribing the above asymmetry entirely to CP violation, one must eliminate other sources of charge asymmetry which could mimic the signature. The charge asymmetry of the main drift chamber has already been mentioned; it has been measured in an independent sample of inclusive $B \rightarrow J/\psi X$ decays and corrected for in the maximum likelihood fit. Backgrounds from other B decays, such as $B^0 \rightarrow J/\psi K^{*0}$, $K^{*0} \rightarrow K_S^0 \pi^0$ where the π^0 has not been reconstructed, have been considered and found to be negligible. The high signal purity at large ct also limits contributions to the asymmetry from backgrounds which are present in the sidebands. In addition, the analysis has been applied to the larger samples of $B^0 \rightarrow J/\psi K^{*0}$ and $B^+ \rightarrow J/\psi K^+$ decays in order to look for systematic effects, but none have been found.

We determine the systematic uncertainty on $\mathcal{D}_0 \sin(2\beta)$ by shifting the central value of each fixed input parameter to the likelihood fit by $\pm 1\sigma$ and refitting to find the shift in $\mathcal{D}_0 \sin(2\beta)$. Varying the B^0 lifetime shifts the central value by ± 0.001. The parameterization of the intrinsic charge asymmetry of the detector is also varied, yielding a $^{+0.016}_{-0.019}$ uncertainty. The largest shift is due to Δm_d, which gives a $^{+0.029}_{-0.025}$ shift. These shifts are added in quadrature, giving $\mathcal{D}_0 \sin(2\beta) = 0.31 \pm 0.18(stat) \pm 0.03(syst)$.

5 Extracting $\sin(2\beta)$

As mentioned above, the dilution $\mathcal{D}_0$, which reduces the amplitude of the CP asymmetry, can be measured in other data samples, including that of $B^0 \rightarrow J/\psi K^{*0}$ decays and the much larger $B \rightarrow \ell D^{(*)} X$ sample. These different dilution measurements can be extrapolated to the kinematic range appropriate for the $J/\psi K_S^0$ data and then combined. The extrapolation is performed with a Monte Carlo simulation based upon a version of the PYTHIA event generator tuned to CDF data [6,11]; the necessary adjustments to the different dilutions are on

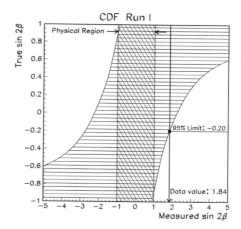

Figure 5: 95% confidence belts for the $\sin(2\beta)$ result. The vertical line represents the actual measurement of $\sin(2\beta)$, and its intersection with the confidence belts gives the confidence interval.

the 10% level at most, and it is found that the appropriate dilution for the $J/\psi K_S^0$ data is $\mathcal{D}_0 = 0.166 \pm 0.018 \pm 0.013$. The first uncertainty is due to the uncertainties in the contributing dilution measurements. The second uncertainty is due to the Monte Carlo extrapolation and is calculated by varying parameters of the Monte Carlo model.

Using this value of $\mathcal{D}_0$, it is found that

$$\sin(2\beta) = 1.8 \pm 1.1(stat) \pm 0.3(syst), \qquad (9)$$

where the dilution uncertainty has been added to the systematic uncertainty. The central value is unphysical since the amplitude of the measured raw asymmetry exceeds the measured dilution.

The result may be phrased in terms of confidence intervals. The present analysis follows the frequentist construction of [12], which gives proper confidence intervals even for measurements in the unphysical region, as is the case here. The confidence interval is shown in Figure 5. It is found that the measurement corresponds to an exclusion of values of $\sin(2\beta)$ below -0.20 at 95% C.L. It can also be seen from the Figure that if the true value of $\sin(2\beta)$ was 1, the median expectation of an exclusion for this analysis technique in the 1992-96 CDF data would be for $\sin(2\beta) < -0.89$ at 95% C.L. This expectation of a limit is a measure of experimental sensitivity. The fact that the present limit is higher reflects the unphysical value that was actually measured.

It is interesting to note that since this result has been obtained using only a single tagging algorithm, then as long as $\mathcal{D}_0 \neq 0$, the exclusion of $\sin(2\beta)$ from this result is independent of the actual value of $\mathcal{D}_0$. Given that $\mathcal{D}_0 > 0$, the same prescription as above for calculating limits yields an exclusion of negative values of $\sin(2\beta)$ at 90% C.L.

6 Conclusion

This article summarizes the CDF result on $\sin(2\beta)$, finding the value

$$\sin(2\beta) = 1.8 \pm 1.1(stat) \pm 0.3(syst) \qquad (10)$$

using only a single-track same-side flavor tagging algorithm on a sample of 198 ± 17 $B^0/\overline{B}^0 \to J/\psi K_S^0$ decays. This measurement corresponds to an exclusion of values of $\sin(2\beta) < -0.20$ at 95% C.L. This result establishes the feasibility of measuring CP asymmetries in B decays at a hadron collider. Upgrades to the Tevatron collider and to the CDF detector, as well as the application of other tagging methods, should reduce the uncertainty on $\sin(2\beta)$ to about 0.08, or by more than an order of magnitude.

Acknowledgements

The author would like to thank his fellow collaborators for the pleasure of representing them and their work, with special thanks to Ken Kelley, Petar Maksimovic, Gerry Bauer, and Paris Sphicas.

References

1. N. Cabibbo, *Phys. Rev. Lett.* **10**, 531 (1963); M. Kobayashi and T. Maskawa, *Prog. Theor. Phys.* **49**, 652 (1973).
2. A.B. Carter and A.I. Sanda, *Phys. Rev. Lett.* **45**, 952 (1980); *Phys. Rev.* D **23**, 1567 (1981); I.I. Bigi and A.I. Sanda, *Nucl. Phys.* B **193**, 85 (1981).
3. A. Ali, DESY Report 97-256, to be published in *Proc. of the First APCTP Workshop*, Seoul, South Korea.
4. OPAL Collaboration, K. Ackerstaff *et al.*, CERN-EP/98-001, to be published in *E. Phys. J. C.*
5. CDF Collaboration, F. Abe *et al.*, FERMILAB-PUB-98/189-E, submitted to *Phys. Rev. Lett.*.
6. CDF Collaboration, F. Abe *et al.*, FERMILAB-PUB-98/188-E, submitted to *Phys. Rev.* D.
7. K. Kelley, Ph.D. dissertation, Massachusetts Institute of Technology, 1998. CDF dissertations may be found on-line at `http://www-cdf.fnal.gov/grads/thesis_complete.html`.
8. M. Gronau, A. Nippe, J.L. Rosner, *Phys. Rev.* D **47**, 1988 (1993); M. Gronau and J.L. Rosner, *Phys. Rev.* D **49**, 254 (1994).
9. CDF Collaboration, F. Abe *et al.*, *Phys. Rev. Lett.* **80**, 2057 (1998); P. Maksimovic, Ph.D. dissertation, Massachusetts Institute of Technology, 1997.
10. Particle Data Group, R.M. Barnett *et al.*, *Phys. Rev.* D **54**, 1 (1996).
11. D. Vučinić, Ph.D. dissertation, Massachusetts Institute of Technology, 1998.
12. G.J. Feldman and R.D. Cousins, *Phys. Rev.* D **57**, 3873 (1998).

REVIEW OF FUTURE CP-VIOLATION EXPERIMENTS IN B-MESON DECAYS

T. NAKADA

CERN, CH-1211 Geneva 23, Switzerland
and
PSI, CH-5232 Villigen-PSI, Switzerland
E-mail: tatsuya.nakada@cern.ch

Experimental prospects for studying CP violation in B-meson decays are reviewed. Comparisons are made for various options: experiments at e^+e^- storage rings operating at the $\Upsilon(4S)$ energy, HERA and the Tevatron will produce results in the near future. They will have a good chance to discover CP violation in B-meson decays. On a longer time scale, experiments at the LHC will aim at much more accurate measurements to make a precision test of the Standard Model in CP violation. There exists also a proposal to build a dedicated experiment to study heavy flavour physics at the Tevatron.

1 Introduction

The Standard Model has so far been very successful in describing the world of elementary particles in a consistent way. The validity of the theory has been tested up to the quantum correction level and all the precision measurements performed at both high and low energies show no sign of deviation from the Standard Model predictions, except a few intriguing experimental results in the neutrino sector.

Along with mass generation and the Higgs particle, CP violation is one of the few remaining untested areas of the Standard Model. Since the discovery in 1964 [1], the neutral kaon system remains the only place where CP violation is observed. Although the effect can be accommodated within the Standard Model, it is not excluded that a mechanism beyond the Standard Model is responsible.

CP violation is one of the three necessary conditions to generate matter-antimatter asymmetry in the universe [2]. The Standard Model, however, does not seem to be capable of generating a sufficient amount of CP violation to explain the observed dominance of matter in our universe [3]. This calls for new sources of CP violation beyond the Standard Model.

In the Standard Model, CP violation is described in the framework of the 3×3 unitary mass-mixing matrix (CKM-matrix) [4]. Such a matrix can have a nontrivial phase which satisfies a condition [5], $\mathrm{Im}(V_{ij}V_{lk}V_{ik}^*V_{lj}^*) \neq 0$, necessary to generate CP violation where V_{ij} is an element of the CKM matrix.

The CKM matrix is uniquely defined by four parameters. One of the most commonly used parameterizations was introduced by Wolfenstein [6], with parameters λ, A, ρ and η. CP violation requires $\eta \neq 0$. The parameter λ is the sine of the Cabibbo angle [7] and is measured to be ~ 0.22 from kaon and hyperon decays. Given λ, A can be obtained from $|V_{cb}|$ measured in B-meson decays produced by the $b \rightarrow c + W$ process. For ρ and η, a combination of $|V_{ub}|$ and $|V_{td}|$ is needed. Currently, both elements are obtained using B-mesons: $|V_{ub}|$ is determined from B-meson decays produced by the $b \rightarrow u + W$ process and $|V_{td}|$ is obtained from the frequency of B^0-$\overline{B^0}$ oscillations. With current uncertainties on $|V_{ub}|$ and $|V_{td}|$, $\eta = 0$ is not excluded [a]. Determinations of $|V_{ub}|$ and $|V_{td}|$ will continue to improve with more experimental data and better theoretical understanding in the future.

It has to be noted that the element $|V_{td}|$ could also be measured from the rare $K^+ \rightarrow \pi^+ \nu \overline{\nu}$ decays [8]. The CP violating decay $K_L \rightarrow \pi^0 \nu \overline{\nu}$ [8] is dominated by the contribution from the interplay between the decays and oscillations, and the branching fraction is proportional to η. Its theoretical uncertainty is considered to be small.

As a first step, "first generation" of CP violation experiments in the B-meson systems will become operational around the turn of the century. If the Standard Model is largely responsible for the observed CP violation in the neutral kaon system, they will discover CP violation in $B_d \rightarrow J/\psi K_S$ decays [9]. A consistency test can then be made by comparing the observed CP asymmetry with the Standard Model prediction. If they disagree, this would mean a sign of new physics and opens up a wide rage of questions which have to be answered by "second generation" of experiments which can measure CP violation in many more decay channels with much higher precision.

Even if they agree, this could well be due to a numerical accident generated by new physics. New physics naturally introduces additional flavour-changing neutral currents. They can affect the oscillations of K^0-$\overline{K^0}$, B^0-$\overline{B^0}$ and B_s^0-$\overline{B_s^0}$, B-meson decay amplitudes dominated by the penguin and box diagrams and CP violation in K- and B-meson (and possibly D-meson) systems. Therefore, η and ρ extracted from the processes involving B^0-$\overline{B^0}$ and K^0-$\overline{K^0}$ oscillations no longer correspond to the

---

[a] A non-zero value of η ($\eta > 0$) is only obtained if the CP violation parameter in the neutral kaon system, ϵ, is taken into account.

real values. Thus, experiments are needed which measure enough different CP asymmetries with high precision so that a real consistency test can be done.

In addition, collection of data which help further understanding of B-meson decays has to be continued to improve the precision on the CKM matrix elements.

For the B-meson system, "second generation" CP experiments with the following capabilities are then needed:

- Being able to measure B_s^0-$\overline{B}_s^0$ oscillations even if the oscillation frequency is far beyond what is expected from the Standard Model.

- Being able to measure CP asymmetries in many different decay modes, including those of B_s-mesons, where the Standard Model predictions have little theoretical uncertainty: some examples are $B_d \rightarrow J/\psi K_S$, $B_d \rightarrow D^* \pi$, $B_d \rightarrow \rho\pi$, $B_d \rightarrow \phi K_S$, $B_s \rightarrow J/\psi\phi$, $B_s \rightarrow D_s K$, $B_s \rightarrow K^+ K^-$.

- Being able to study very rare decay modes generated by the penguin and box diagrams, such as $B_s \rightarrow \mu\mu$, at least down to the level expected by the Standard Model.

In the following, experimental prospects to pursue such studies are reviewed.

2 Future Experiments

2.1 Experiments before 2005

CESR is the only e^+e^- collider currently working at $\Upsilon(4S)$ studying decays of B-mesons. The machine luminosity has been constantly improved and has now reached a level of $4 \times 10^{32} \mathrm{cm}^{-2}\mathrm{s}^{-1}$ with a final goal of $1.7 \times 10^{33} \mathrm{cm}^{-2}\mathrm{s}^{-1}$ by the turn of the century[10]. There even exists an idea to increase its luminosity beyond $10^{34} \mathrm{cm}^{-2}\mathrm{s}^{-1}$ by constructing an additional ring. This may allow them to observe $B_d \rightarrow \pi^0\pi^0$ decays which provides a crucial information to understand the effect of gluonic penguin diagrams in the study of CP violation for $B_d \rightarrow \pi^+\pi^-$ decays.

The current detector at CESR, CLEO, is undergoing a major upgrade[10] (Figure 1) replacing the vertex detector and drift chamber. The upgrade also includes installation of a Ring Imaging Cherenkov Counter which will allow clean separation of $B_d \rightarrow K^\pm\pi^\mp$ and $B_d \rightarrow \pi^+\pi^-$ decays.

CLEO will stay as a front-runner of B physics for some time. By accumulating statistics, they look for CP violation in decay amplitudes. However, they cannot observe CP violation generated by the interplay between the decays and oscillations such as B^0 and $\overline{B}^0 \rightarrow J/\psi K_S$ as explained later. For this purpose, asymmetric e^+e^-

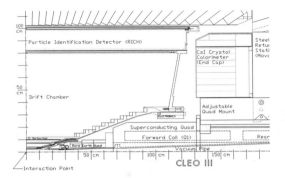

Figure 1: Upgrade plan for the CLEO detector.

colliders are being constructed, KEKB at KEK[11] and PEP-II at SLAC[12].

Table 1 summarises some parameters of the two e^+e^- B-meson factories. KEKB is being constructed in the existing TRISTAN tunnel and PEP-II in the PEP tunnel. Two beams with moderately different energies collide in both machines producing the $\Upsilon(4S)$ boosted. This is essential for the CP violation study. At the $\Upsilon(4S)$, B^0-$\overline{B}^0$ are produced in a state with an orbital angular momentum of one. Due to quantum coherency of this state, one can show that the difference in the decay time between the two B_d-mesons must be measured in order to obtain a visible signal of CP violation. With a stationary $\Upsilon(4S)$, only the sum of the decay times can be measured since the production points of B_d-mesons given by the e^+e^- collision points are not well defined compared with the average flight path of B_d-mesons. With a boosted $\Upsilon(4S)$, measuring the decay time difference becomes possible as illustrated in Figure 2.

In the PEP-II design, the two beams collide head-on. Then the two beams are separated by dipole magnets in order to avoid parasitic collisions. Separation has to be done quickly in order to increase the number of bunches, which leads to high luminosity. For quick separation of the beams after the collision, the dipoles have to be placed very close to the interaction point,

Table 1: List of some machine parameters for PEP-II and KEKB.

Machine	PEP-II	KEKB
$E_{\mathrm{H}}/E_{\mathrm{L}}$ [GeV]	9/3.1	8/3.5
$\mathcal{L}$ [cm^{-2}s^{-1}]	3×10^{33}	10^{34}
Bunch spacing [m]	1.26	0.6
Crossing angle	$0°$	± 11 mrad

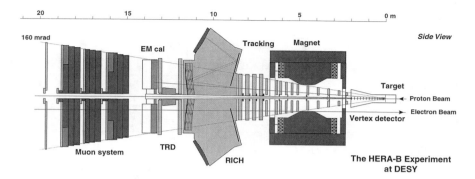

Figure 3: A schematic view of the HERA-B detector.

which makes the detector design difficult. In the KEK design, the two beams collide with a small angle. In this scheme, the beams are automatically separated after the collision. Therefore, no dipole is needed close to the interaction point and the bunch spacing can be reduced, i.e. the number of bunches can be increased. However, crossing bunches with a finite angle may introduce an instability of the beams which limits the luminosity.

Both machines have only one interaction region with a detector in order to obtain the highest luminosities.

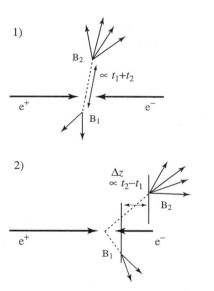

Figure 2: Illustration of $\Upsilon(4S) \to B^0\overline{B}^0$ followed by decays of the two B mesons for 1) a symmetric collider and 2) an asymmetric collider.

BELLE [11] is placed at KEKB and BaBar [12] at PEP-II. They are equipped with Si strip vertex detectors and high performance particle-identification devices. They expect to start taking data in Spring 1999.

The HERA-B experiment [13] (Figure 3) uses the halo of the HERA proton beam and internal targets placed inside the beam pipe. It runs parasitically to the other HERA experiments. The internal targets are made of eight thin metal wires ($\sim 50~\mu$m diameter). Since it is a fixed target experiment, B-mesons are boosted with an average decay length of ~ 6 mm. The collaboration plans to start taking data with the full detector sometime in 1999.

The CDF experiment has already demonstrated its potential by reconstructing the largest sample of $B_d \to J/\psi K_S$ decays [14] (Figure 4) used to measure a CP asymmetry [b]. Both CDF [16] and D0 [17] are upgrading their detector which will enhance their capability for studying CP violation in B-meson decays in the next run starting in 2000. In particular, D0 will introduce a magnet.

Some of the experimental conditions for those three different approaches are listed in Table 2. For all the experiments, the micro-vertex detector is a crucial part of the spectrometer. It is essential for reducing the background and measuring the B-meson decay time. The second point is particularly important for BaBar and BELLE to observe CP violation.

In addition to lepton identification, the capability to separate kaons from pions introduced by BaBar, BELLE, CLEO and HERA-B is another important point. This can be seen for the reconstruction of the $B_d \to \pi^+\pi^-$ decay. Dangerous backgrounds in this channel are $B_d \to K^\pm\pi^\mp$ (also $B_s \to K^\pm\pi^\mp$ and K^+K^- for hadron machines). Since they are real two-body B-meson decays,

---

[b]OPAL [15] also gives a measurement of CP asymmetry in $B_d \to J/\psi K_S$. For both cases, measured asymmetries are compatible with zero.

Table 2: Experimental conditions for e⁺e⁻ machines, HERA and the Tevatron.

Machines	e^+e^-			HERA	Tevatron
	CESR	KEKB	PEP-II		
Experiments	CLEO-III	BELLE	BaBar	HERA-B	CDF, D0
Reactions	$e^+ + e^-$ at $\Upsilon(4S)$			p + Metal wires at $\sqrt{s}$ =40 GeV	$p + \bar{p}$ at $\sqrt{s}$ =1.8 TeV
$\sigma_{b\bar{b}}$	∼1 nb			~ 760 nb	$\sim 60\ \mu$b
$\sigma_{b\bar{b}}/\sigma_{hadronic}$	$\sim 2 \times 10^{-1}$			$\sim 10^{-6}$	$\sim 10^{-3}$
b-hadrons	B_u, B_d			$B_u, B_d, B_s, B_c,$ all b-baryons	$B_u, B_d, B_s, B_c,$ all b-baryons
In one event	only $B\overline{B}$			many other particles	many other particles
Geometry of detector	central symmetric	central, slightly asymmetric		very forward fixed target	central symmetric
Particle ID	$p/K/\pi/\mu/e$			$p/K/\pi/\mu/e$	hadron/μ/e

the micro-vertex detectors will not help to remove them, and the mass resolutions of spectrometers are not sufficient to separate those channels using invariant masses

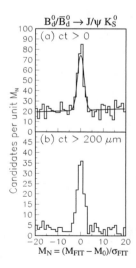

Figure 4: Invariant mass distribution for $J/\psi\,K_S$ measured by the CDF experiment with different cuts on the flight distance.

with different mass hypotheses. Furthermore, the efficiency for the flavour tag can be increased by using the kaon tag.

BaBar, BELLE and CLEO are equipped with electromagnetic calorimeters made of CsI crystals. Their excellent energy resolution and clean environment at the $\Upsilon(4S)$ decays allow the experiments to enhance the decay channels to be investigated by including final states with multi π^0's.

Among the future experiments, CP violation studies in the B_s-meson system are unique for those at hadron machines. Good decay time resolutions and the large number of produced B_s mesons allow them to have better sensitivities in detecting B_s^0-$\overline{B}_s^0$ oscillations than those achieved by the LEP experiments. However, their expected sensitivities still do not cover the whole range of the oscillation frequencies allowed by the Standard Model.

2.2 Experiments after 2005

The $b\bar{b}$ cross section for proton-proton collisiona at LHC ($\sqrt{s} = 14\ TeV$) is expected to be $\sim 500\ \mu b$ [18]. Even with a modest luminosity of 10^{32} to 10^{33} cm^{-2}s^{-1}, more than 10^{11} B-mesons will be produced in 10^7 seconds. The cross section ratio, $\sigma_{b\bar{b}}/\sigma_{inelastic}$ is predicted to be $\sim 5 \times 10^{-3}$, which is similar to $\sigma_{c\bar{c}}/\sigma_{inelastic}$ in the current fixed target charm experiments. Therefore, LHC is a promising machine for studying CP violation in B-meson decays with

Experiment	general purpose detectors ATLAS CMS	dedicated experiment LHCb
Reactions	p + p at $\sqrt{s}$ =14 TeV	
$\sigma_{b\bar{b}}$	$\sim$500 μb	
$\sigma_{b\bar{b}}/\sigma_{\mathrm{hadronic}}$	$\sim$5 $\times$ 10^{-3}	
$\mathcal{L}$ for B physics	10^{33} cm^{-2}s^{-1}	2×10^{32} cm^{-2}s^{-1}
Acceptance	central region $\|\eta\| < 2.5$ $\|\eta\| < 2.4$	forward region $1.6 < \eta < 5.3$
Early level trigger with a reduction factor > 100	high-p_t μ high-p_t μ or e	medium-p_t μ, e or hadrons followed by secondary vertex trigger
Particle ID	hadron/μ/e	p/K/π/μ/e

high statistics.

Three experiments are under construction for proton-proton collisions at LHC. ATLAS [19] and CMS [20] are two general-purpose experiments designed to look for Higgs and supersymmetric particles: LHCb [21] is a dedicated experiment for the CP violation study. Experimental conditions for the three experiments are summarised in Table 3. All the detectors are naturally equipped with high performance vertex detectors.

Since ATLAS and CMS are designed to look for particles produced in a very hard collision, detectors cover the central region. For the initial phase of the LHC operation where the machine luminosity is $\sim 10^{33}$ cm^{-2}s^{-1}, they intend to do physics with B-mesons. The b-quark events are triggered by the high transverse momentum (p_t) lepton trigger by reducing the threshold value.

LHCb chose the forward geometry mainly due to the following reasons:

- The b-quark production is peaked in the forward region where both b and $\bar{b}$ fly in the same direction. Therefore, a single arm spectrometer with a modest angular coverage of up to $\sim$ 300 mrad can detect the decay products of both b-hadrons for 10 to 20% of b$\bar{b}$ events. This keeps the cost of the detector low.

- B-hadrons produced in the forward direction are on average faster than those in the central region. Their average momentum is about 80 GeV/c, corresponding to a mean decay length of $\sim$7 mm. Therefore, a good decay time resolution can be obtained for reconstructed B-mesons.

- In the forward region, momenta are mainly carried by the longitudinal components. Therefore, the threshold value for the p_t trigger can be set low for electrons, muons and hadrons; around 1.5 GeV/c. This makes the p_t trigger more efficient than in the central region.

- The detector can be built in an open geometry which allows easy installation and maintenance.

As already demonstrated by CDF, general purpose collider experiments can reconstruct well B-meson final states containing lepton pairs, such as J/ψ K$_S$, J/$\psi\phi$ and $\ell^+\ell^-$. As an example, Figure 5 shows the simulated invariant mass distribution for reconstructed B$_d \to$ J/ψ K$_S$ decays with the CMS detector.

An especially good mass resolution and particle identification become essential for the reconstruction of hadronic final states. As an example, Figure 6 shows the reconstructed B$_s \to$ D$_s$K decay with and without particle identification for the LHCb experiment. The major background B$_s \to$ D$_s\pi$ decay can be successfully eliminated by the RICH and excellent mass resolution, although its branching fraction is about 15 times larger than that of B$_s \to$ D$_s$K. This is estimated by taking into account the Cabibbo suppression and form factor difference.

The importance of K/π separation can be also demonstrated in the B$_d \to \pi^+\pi^-$ decays. As discussed previously, the major background comes from other two body decay modes of the B-meson. Figure 7 shows the simulated $\pi^+\pi^-$ invariant mass distributions of reconstructed B$_d$ mesons for ATLAS. The contribution from the real B$_d \to \pi^+\pi^-$ decays shown as the crossed his-

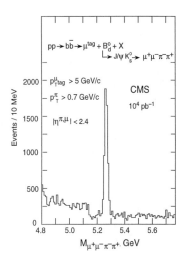

Figure 5: Simulated invariant mass distribution of $B_d \rightarrow J/\psi K_S$ decays for the CMS detector.

togram is completely buried in the background. The background peaking at the B_d-meson mass is due to other two-body B-meson decays and the combinatorial background is flat under the mass peak.

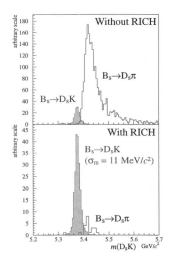

Figure 6: Simulated $D_s(KK\pi)K$ invariant mass distributions for reconstructed $B_s \rightarrow D_s K$ decays in LHCb. The shaded histogram is the true $B_s \rightarrow D_s K$ decays. The background channel, $B_s \rightarrow D_s \pi$, is assumed to have 15 times larger branching fraction than that of the signal.

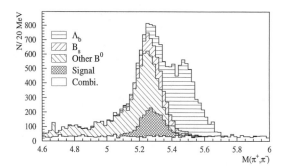

Figure 7: Simulated $\pi^+\pi^-$ invariant mass distributions of reconstructed $B_d \rightarrow \pi^+\pi^-$ decays for ATLAS with various background contributions.

In some two-body decay modes like $B_d \rightarrow K^\pm \pi^\mp$, a CP violation effect from the decay amplitudes is expected. The size of CP asymmetry could be as large as ~ 0.01. Expected statistical accuracies of CP asymmetry measurements in this decay mode by the LHC experiments are of the same order. Therefore, this effect is not negligible. Furthermore, a discovery of CP violation in $B_d \rightarrow K^\pm \pi^\pm$ decays by itself is very important. Such studies are clearly only possible with the LHCb detector which can separate kaons from pions in all the necessary momentum range.

An excellent decay time resolution is essential for B_s-meson studies. ATLAS and CMS will improve the sensitivities to observe B_s^0-$\bar{B}_s^0$ oscillations compared to CDF, D0 and HERA-B, for example. However, LHCb having expected decay time resolution of 43 fs and 86k reconstructed and initial flavour tagged $B_s \rightarrow D_s \pi$ decays is needed to observe B_s^0-$\bar{B}_s^0$ oscillations even if the oscillation frequency is well above the Standard Model expectation.

BTeV [22] (Figure 8) is a dedicated heavy flavour experiment being proposed at the Tevatron. The experiment is based on a forward spectrometer, similar to

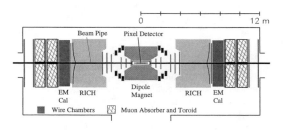

Figure 8: A schematic view of the BTeV detector.

LHCb, but with a double arm configuration. In its detector design, a dipole magnet is at the interaction region and a vertex detector with Si pixels is placed inside the magnetic field. This vertex detector is used in the very first level of the trigger in order to select events with detached vertices. The spectrometer is also equipped with tracking chambers, RICH's, electromagnetic calorimeters and muon systems. BTeV's physics goal and capability is very similar to that of LHCb. In addition, they intend to focus on charm physics as well.

3 Conclusions

CP violation will stay as one of the most important subjects of elementary particle physics for sometime. Together with continuous strong theoretical interests, the experimental programme already laid down beyond 2005 will keep the field very active and competitive.

In addition to the well established experimental programme in the kaon system, CP violation experiments in B-meson systems are rapidly developing. The clean environment of e^+e^- machines and the very high statistics available at hadron machines provide complementary physics programmes in the B-meson systems.

We are all hoping to see within the next ten years a sign of new physics which will explain why we have not annihilated soon after the big bang.

Acknowledgements

The author thanks the ATLAS, BaBar, BELLE, BTeV, CDF, CLEO, CMS, D0, HERA-B and LHCb collaborations for providing essential information for this presentation. O. Schneider and R. Forty are especially acknowledged for his careful reading of this manuscript.

References

1. J. H. Christenson et al., *Phys. Rev. Lett.* **13**, 138 (1964).
2. A. D. Sakharov, *JETP Lett.* **6** 21 (1967).
3. See for example M. B. Gavela et al., *Modern Phys. Lett.* **9A** 795 (1994).
4. M. Kobayasi and K. Maskawa, *Prog. Theor. Phys.* **49** 652 (1973).
5. C. Jarlskog,*Phys. Rev. Lett.* **55**, 1039 (1985) and *Z. Phys.* C **C29**, 491 (1985).
6. L. Wolfenstein, *Phys. Rev. Lett.* **51**, 1945 (1983).
7. N. Cabibbo, *Phys. Rev. Lett.* **10**, 531 (1963).
8. See for example A. Buras and R. Fleischer, TUM-HEP-275/97, 1997.
9. For a predicted value of CP asymmetries in $B_d \rightarrow J/\psi\,K_S$ decays, see for example F. Parodi, these proceedings.
10. Talk given by A. Wolf at Beauty 97, Proceedings *Nucl. Instrum. Methods Phys. Res.* A **408**, 58 (1998), Ed. P. E. Schlein.
11. Talk given by A. Bonder at Beauty 97, Proceedings *Nucl. Instrum. Methods Phys. Res.* A **408**, 64 (1998), Ed. P. E. Schlein.
12. Talk given by D. Hitlin at Beauty 97.
13. Talk given by T. Lhose at Beauty 97, Proceedings *Nucl. Instrum. Methods Phys. Res.* A **408**, 154 (1998), Ed. P. E. Schlein.
14. J. Tseng, these proceedings and F. Abe et al., FERMILAB-PUB-98-189-E, 1998.
15. E. Barberio, these proceedings and K. Ackerstaff et al., CERN-EP-98-001, 1998.
16. Talk given by N. Lockyer at Beauty 97, Proceedings *Nucl. Instrum. Methods Phys. Res.* A **408**, 94 (1998), Ed. P. E. Schlein.
17. Talk given by M. Wayne at Beauty 97, Proceedings *Nucl. Instrum. Methods Phys. Res.* A **408**, 103 (1998), Ed. P. E. Schlein.
18. Talk given by M. Mangano, Workshop "Theory of LHC Processes", Geneva, February 1997.
19. Talk given by P. Eerola, LHCC Open session, March 1998.
20. Talk given by D. Denegri, LHCC Open session, March 1998.
21. S. Amato et al., LHCb Technical Proposal, CERN/LHCC 98-4, February 1998.
22. A. Kulyavtsev et al., Paper 92, submitted to this conference, BTeV-pub-98/3, 1998.

BNL E871 RARE KAON DECAY SEARCHES: $K_L \to \mu e, K_L \to ee, K_L \to \mu\mu$

Mark Bachman

Department of Physics and Astronomy, University of California at Irvine, Irvine, CA 92697-4575
E-mail: mbachman@uci.edu

E871 at the Brookhaven Alternating Gradient Synchrotron (BNL AGS) searched for the rare decays of the neutral kaon, $K_L \to \mu e, K_L \to ee, K_L \to \mu\mu$. Data were taken during two run periods in 1995 and 1996. Over 6,000 $K_L \to \mu\mu$ events were seen in the data. The E871 collaboration has concluded its analysis of these decays and results are being prepared for publication. Preliminary results are: $B(K_L \to \mu e) < 4.8 \times 10^{-12}(90\%\text{CL})$, $B(K_L \to ee) = (8.8^{+5.7}_{-4.1}) \times 10^{-12}$, $B(K_L \to \mu\mu) < (7.23 \pm 0.22) \times 10^{-9}$ Experimental technique and analysis is discussed.

1 Introduction

1.1 Physics motivation

Experiment 871 at the Brookhaven Alternating Gradient Synchrotron (BNL AGS) searched for the rare decays of the neutral kaon, $K_L \to \mu e, K_L \to ee, K_L \to \mu\mu$. These decays probe the limits of our understanding of the Standard Model.

In the limit of massless neutrinos, the $K_L \to \mu e$ decay is forbidden under conservation of lepton flavor, and an observed signal would indicate physics beyond the Standard Model. Current estimates of neutrino masses and mixing in the Standard Model leads to predicted rates several orders of magnitude below current experimental sensitivities[1].

Many popular extensions to the Standard Model predict non-conservation of lepton flavor at levels significantly higher than those from massive neutrinos in the Standard Model. These include horizontal gauge interactions[2], left-right symmetry[3], technicolor[4], compositeness[5], and supersymmetry[6]. The E871 result is the most sensitive search for lepton flavor violation in the kaon system to date, and is consistent with the Standard Model.

The $K_L \to ee$ decay is strongly suppressed by the GIM mechanism[7] and the helicity structure of the $V - A$ interaction. New physics not subject to these effects could be revealed by an unexpected rate in this decay mode. Recent Standard Model predictions in the framework of chiral perturbation theory[8] indicate a decay rate approximately 9×10^{-12}. This is significantly greater than the absorptive contribution arising from the intermediate state, $K_L \to \gamma\gamma \to ee$, estimated[9] to be 3×10^{-12}. The E871 measurement represents the first observation of the $K_L \to ee$ decay and is consistent with this prediction, ruling out significant non-Standard Model contributions to this decay.

The $K_L \to \mu\mu$ decay rate is analogous to that of $K_L \to ee$. It is known that the prominent contribution comes from the absorptive part of the decay amplitude, deduced from $K_L \to \gamma\gamma$ measurement to be $(7.07 \pm 0.18) \times 10^{-9}$, the so-called unitarity bound. Contributions to the dispersive part are not yet understood, but could include short range effects involving flavor changing neutral currents, and the CKM matrix element, V_{td}. The uncertainty in the measurement of this rate (approximately 7%) limits the ability to discern the short range effects, and a better measurement should help isolate the short range contributions to the decay. Of course, a measurement below the unitarity bound would indicate new physics. The E871 result is above the unitarity bound and consistent with the the world average.

1.2 Basic experimental considerations

All three decay modes are kinematically similar and their experimental considerations are likewise similar. The basic requirements are: high kaon flux and the ability to handle associated high rates, high acceptance, excellent background rejection, and excellent control of systematics (for $K_L \to \mu\mu$).

With the exception of $K_L \to \pi\pi$, all backgrounds are three-body or four-body decays. The kinematic signature is based on the two body decay, which is easily identified by a reconstructed momentum which points in the same direction of the parent kaon, and by an invariant mass which is 497.672 MeV/c². We refer to the component of the momentum sum which points away from the parent kaon direction as the transverse momentum, p_T. For a two-body decay, $p_T = 0$. To reject the two-pion decay, particle identification counters are required. Good vertex reconstruction is also necessary to reject against track scatters and accidental pile-up events.

2 Measurement

2.1 Beam line

For E871, a 24 GeV proton beam incident on a 1.4 interaction length platinum target produced the neutral beam, collimated at 3.75° to maximize the K_L^0 yield while minimizing the n/K_L^0 ratio. Thin foil photon converters and subsequent sweeping magnets downstream removed charged particles and reduced the γ flux. The beam collimation was approximately 4×16 mrad^2; the proton intensity was typically 1.5×10^{13} per 1.2–1.6 s pulse, which repeated every 3.2–3.6 s. The resulting kaon intensity was 2×10^8 K_L^0 per pulse, within a momentum range of 2 GeV to 16 GeV. Of these, 7.5% decayed in an evacuated decay volume between 9.5 m to 20.75 m from the target. The n/K_L^0 ratio was measured to be 8 ± 3.

2.2 Spectrometer and trigger hodoscope

The detector system, shown in figure 1, consisted of a two arm spectrometer downstream of the decay volume with two dipole magnets and twelve straw and drift chamber tracking stations. Following this were a series of particle identification counters (PID) used to provide redundant identification of electrons and muons. All detectors were finely segmented to accommodate high rates, typically 700 kHz in the front straw tracking detectors. A large beam stop was installed in the first spectrometer magnet to absorb neutral beam within the detector system itself. This reduced the rates in the downstream counters and allowed them to be placed such that their active elements spanned the center of the beam line.

To reduce data rates from background events, the spectrometer magnetic fields were set at opposite polarities, providing net transverse momentum kicks of 418 MeV/c and 216 MeV/c. These values were such to bend the tracks from two-body decays at the Jacobian peak to be nearly parallel to the beam line when exiting the spectrometer. The baseline trigger (L0), based on hits in the segmented scintillator hodoscope (TSC), used hits which corresponded to tracks nearly parallel to the beam line. The TSC also defined the event time.

Eight straw stations, four per side, were composed of 5 mm diameter kapton straws, nearly 1 m in length. Downstream, four conventional wire drift chambers, 1 cm cell size, completed the tracking system. The average ef-

ficiencies and resolutions were 96% and 98% per wire, 160 μm and 120 μm, for the straws and drift chambers, respectively. The total mass of the spectrometer system, which included helium bags between the tracking stations, amounted to 1.5×10^{-2} radiation lengths.

2.3 Particle identification counters

Particle identification was performed by four different detectors. For electron identification, a gas Čerenkov detector and a lead glass calorimeter were employed. For muon identification, a muon hodoscope and a muon rangefinder were used.

The Čerenkov threshold detector (CER) was filled with hydrogen gas at atmospheric pressure and was segmented into 4×4 arrays of mirror-phototube pairs on each side of the beam line. The average photoelectron yield was 5.5 for electrons. The threshold momentum for electrons was 0.031 GeV/c; the efficiency for electrons was 0.977 ± 0.001. The threshold momentum for pions was 8.396 GeV/c; the probability for misidentifying pions below threshold momentum was 0.0019 ± 0.0002.

The lead glass calorimeter (PBG) consisted of 216 lead glass blocks arranged in two layers, with 3.5 radiation length converter blocks in front of 10.5 radiation length absorber blocks. The measured energy resolution for electrons was $\sigma/E = 0.015 + 0.062/\sqrt{E(\text{GeV})}$. The efficiency for electrons was 0.987 ± 0.004; the probability for misidentifying pions was 0.0093 ± 0.0004. Used together, the CER and PBG provided strong confidence in electron identification.

The muon hodoscope (MHO) was downstream of the electron counters and consisted of six planes of vertical and horizontal scintillation counters sandwiched between iron and marble and aluminum absorbing slabs. This was overlapped and followed by the muon rangefinder (MHO) which consisted of proportional wire chambers separated throughout the 20.5 m stack at 5% muon ranges. Muon identification was accomplished by forming tracks in the hodoscope and rangefinder, then comparing the deduced muon momentum and time-of-flight to the spectrometer results. The efficiency of the muon identification was approximately 0.96 per muon, with a pion acceptance of about 0.05.

The semi-leptonic decays $K_L \rightarrow \pi\mu\nu$ ($K_{\mu 3}$) and $K_L \rightarrow \pi e\nu$ (K_{e3}) were used to calibrate the particle identification counters.

2.4 Data acquisition

The baseline L0 trigger required a coincidence of signals in all six TSC planes. A parallelism condition was imposed, requiring that the tracks be within 31 mrad of

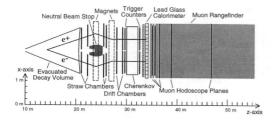

Figure 1: Plan view of the E871 apparatus. The origin is at the production target.

the beam line, in order to reject against three-body decays.

A second trigger (L1) used information from the CER and MHO counters to identify the probable decay mode for each event. The muon signal was taken from the hodoscope position corresponding to 1 GeV energy loss. Spatial correlation between the TSC and MHO or CER were required. Events not consistent with one of the three decay modes were rejected. In addition, a special trigger (*minbias*) required that only L0 be satisfied; this trigger was prescaled by 1000.

Data were collected, buffered, then streamed directly into eight dual-ported memory boards in a VME-based, Silicon Graphics V35 computer system. A software trigger (L3) ran on the processors and analyzed data in real time, performing tracking and kinematic analysis. The L3 software was able to make a decision within approximately 1.1 ms and rejected over 95% of the events.

Events were collected over the period of two years, 1995 and 1996. In all, over 1.7 Tbytes of data were written to tape.

3 Analysis

3.1 Data reduction

Offline analysis of the data was performed in three passes. The first two passes (pass-1 and pass-2) were performed primarily to reduce the data set to a manageable size. Pattern recognition was performed to associate hits in the spectrometer with tracks. Accurate hit assignment was facilitated by a maximum likelihood algorithm which employed parameterized calibrations of noise and inefficiencies. Events were selected if the invariant mass exceeded 470 MeV/c^2 and if either $p_T < 40$ MeV or $\theta_c < 4.5$ mrad, where θ_c is the angle between the kaon direction (determined from the target and vertex position) and the direction of the momentum sum of the two tracks. A second pass was used to perform detailed particle identification analysis and separate the data into three streams corresponding to the three decay modes.

Pass-3 used full magnetic field maps and fitting algorithms to perform kinematic analysis of the events. Two fitters were employed in an effort to cross-check the analysis. One fitter (QT) performed independent calculations of particle momenta using the front and back sections of the spectrometer. A χ^2 was constructed based on the front-back agreement from this fitter and served as useful cut for removing tracks containing pion decays. The second fitter (FT) did a fit to the full spectrometer at once, minimizing a global χ^2 at the event vertex, correctly accounting for multiple scattering and detector resolutions. The FT mass resolution was better (1.13 MeV/c^2 vs. 1.26 MeV/c^2 for QT, for $K_L \to \pi\pi$). An event was written to file if its invariant masses was greater than 485 MeV/c^2 and the square of its transverse momentum (p_T^2) was less than 900 (MeV/c)^2

In an attempt to perform a "blind" analysis, events near the signal region were excluded from being written to file; thus, experimenters had no access to the signal events when selecting cuts. This was done to minimize the possibility of biasing the results by a priori expectations. For $K_L \to \mu e$ and $K_L \to ee$, the excluded zone was 490 MeV/$c < M_{+-} < 505$ MeV/c and $p_T^2 < 100$ (MeV/c)^2. After analysis cuts were finalized, excluded events were extracted from the data tapes. For $K_L \to \mu\mu$, no signal events were excluded. However, a software prescale was applied to the $K_L \to \pi\pi$ normalization sample, and the value of this prescale kept secret until the analysis was completed.

Experimenters studied the results of pass-3 to determine potential backgrounds and the effects of their cuts. For all decay modes, cuts were made to ensure that tracks and vertices were of good quality, vertices fell within allowed regions, track and PID times were reasonable, and momenta were consistent throughout the spectrometer. All analyses required that the two event fitters (FT and QT) showed good agreement.

3.2 $K_L \to \mu e$ backgrounds and cuts

For the $K_L \to \mu e$ analysis, the main background was from the semi-leptonic decay, $K_L \to \pi e\nu$, where the pion decayed or was misidentified as a muon. This process is kinematically constrained such that the invariant mass must reconstruct to 489.3 MeV/c^2, which is 8.4 MeV below the kaon mass. However, it was found that a small fraction of these events resulted in particles (mostly electrons) which elastically scattered from the front vacuum wall and first straw chamber (0.12% and 0.23% radiation lengths, respectively). Large angle front scatters which happen to be in the decay plane and scatter away from the beam have the feature that their vertex quality remains good, and their apparent invariant mass is increased beyond the kinematic limit. To estimate the effect of this process, large-scale Monte Carlo jobs were run, simulating the experiment and subsequent analysis at 2,500 times the actual sensitivity. The simulated distributions of kinematic variables matched observed results extremely well when the front scattering was taken into account, and poorly when it was not.

Several cuts were used to reduce the effects of this background. A special variable, $p_T^\parallel$ was used, defined as the component of transverse momentum (p_T) which lies in the decay plane. This variable identified backgrounds well, since scatters in the decay plane were the dominant background. The E871 analysis imposed the condition that $p_T^{\parallel}$ be small.[a] Unfortunately, a tight cut on this quantity also significantly reduced signal acceptance. In addition, a momentum asymmetry, defined by the relative difference in momentum between the muon and the electron, $(p_\mu - p_e)/(p_\mu + p_e)$, was required to be less than 0.5. The electron momentum was required to be above 1.0 GeV.

A second source of potential background came from accidental pile-up of $K_L \to \pi e\nu$ and $K_L \to \pi\mu\nu$ events. These were removed by timing cuts, vertex quality cuts, and by explicitly looking for extra track fragments in the spectrometer. While accidental pile-up was reduced below 0.01 for the final E871 signal box, it represents a significant hurdle to performing this measurement at higher sensitivity.

3.3 $K_L \to ee$ backgrounds and cuts

For the $K_L \to ee$ analysis, background came primarily from the physics processes, $K_L \to e^+e^-\gamma$ and $K_L \to e^+e^-e^+e^-$. To reduce the $K_L \to e^+e^-e^+e^-$ background, special algorithms were developed to search for and reject events with short partial tracks in the front of the spectrometer. No cut could be made against the $K_L \to e^+e^-\gamma$ background; tightening of the signal region was the only option for reducing it.

Backgrounds were calculated in the region defined by $476 < M_{ee} < 490$ MeV/c^2, and $p_T^2 < 400$ (MeV/c)^2. The Monte Carlo prediction, using absolute normalization, was 24 ± 11 for $K_L \to e^+e^-e^+e^-$ and 38 ± 8 for $K_L \to e^+e^-\gamma$. The uncertainties in these results arise from the uncertainty in the $K_L \to e^+e^-\gamma$ form factor, and the uncertainty in the $K_L \to e^+e^-e^+e^-$ partial track cut efficiency. There were 43 observed events in the same region.

A cut was made on momentum asymmetry $(p_+ - p_-)/(p_+ + p_-) < 0.55$. In addition, to reduce background from accidental pile-up of two $K_L \to \pi e\nu$ events, extra track cuts, track timing, and vertex quality were tightened similarly to $K_L \to \mu e$.

3.4 $K_L \to \mu\mu$ backgrounds and cuts

For the $K_L \to \mu\mu$ analysis, background came primarily from the $K_L \to \pi\mu\nu$ events which reconstructed near the kaon mass. Since this analysis was not a "discovery mode" analysis, cuts were loosened to improve signal acceptance, allowing more background in the signal region. A detailed study of efficiencies and systematics was performed to reduce the error in the final result. It was found that several of the cut efficiencies were correlated.

---

[a] A correlated cut between $p_T^{\parallel}$ and p_e was made: $p_T^{\parallel} < (3p_e - 2.5) \times 10^{-3}$, where p_e and $p_T^{\parallel}$ are in GeV/c.

For this reason, a large weighting Monte Carlo was performed to estimate the final acceptance function.

Since large systematics resulted from poorly known detector response at the edges of the track momentum distribution, a cut was made on muon momentum to be within 1.05 GeV/c and 6.5 GeV/c. The loss in signal was more than offset by the reduction in systematic error.

4 Results

4.1 General remarks

After cuts were finalized, and the signal region defined, the blind analysis was complete and experimenters looked in the signal region for events. Although the analyses required agreement between FT and QT fitters, the FT results (which had better mass resolution) were used for calculating the final numbers. All results were normalized to the measured $K_L \to \pi\pi$ yield, using the PDG's average for $B(K_L^0 \to \pi\pi) = (2.067 \pm 0.035) \times 10^{-3}$. Relative acceptances between the decay modes and the normalization sample were carefully calculated and included calculation of all efficiencies. Effects specific to pions, such as decay and hadronic interactions were estimated and included in the acceptance calculations. Corrections were made for $K_L - K_S$ interference and inner bremsstrahlung.

4.2 $K_L \to \mu e$ results

Large scale optimization of the cut parameters was performed in order to produce the highest possible sensitivity with a background estimate of 0.1. The resulting cuts yielded a contour-box curve in the mass and p_T^2 plane. For $M_{\mu e} < M_K$, the signal ellipse is defined by $[p_T^2/(20(\text{MeV/c})^2)]^2 + [\Delta M/2.4(\text{MeV/c}^2)]^2 < 1$, where $\Delta M = M_{\mu e} - M_K$. For $M_{\mu e} > M_K$, signal is defined by the box $\Delta M < 4$ MeV/c^2 and $p_T^2 < 20$(MeV/c)^2. This can bee seen in figure 2. In addition, a $p_T^{\parallel}$ cut was imposed, as previously mentioned.

No events were observed in this region. The resulting upper limit at the 90% confidence level is

$$B(K_L \to \mu^{\pm} e^{\mp}) < 4.8 \times 10^{-12}.$$

This represents the most sensitive search for $K_L \to \mu^{\pm} e^{\mp}$ to date.

4.3 $K_L \to ee$ results

A discovery zone was defined to be an ellipse in the mass and p_T^2 plane, corresponding to about 2.5σ in each, as seen in figure 3. This region was chosen to have an expected background of 0.2 events. Four events were found in the discovery ellipse. These were carefully examined

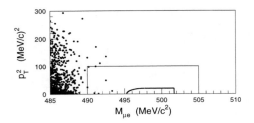

Figure 2: p_T^2 versus $M_{\mu e}$ for $K_L \to \mu e$ data. The large box shows the exclusion region; the small curved zone bounds the signal region.

for anomalies; none were found and these were declared a discovery of $K_L \to e^+ e^-$. The probability of 0.2 background events fluctuating to four events in this zone is 6×10^{-5}.

A maximum likelihood fit over the background region, modeling backgrounds and signal, was performed to extract a final branching ratio. This was found to be

$$B(K_L \to e^+ e^-) = 8.8^{+5.7}_{-4.1} \times 10^{-12}.$$

This result is consistent with recent predictions based on the Standard Model. This is the first measurement of the decay, $K_L \to e^+ e^-$, and is the smallest branching ratio yet measured.

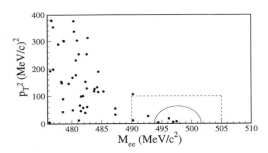

Figure 3: p_T^2 versus M_{ee} for $K_L \to ee$ candidates. The dashed box shows the exclusion region; the solid curve bounds the signal region.

4.4 $K_L \to \mu\mu$ results

A fit to the $K_L \to \mu\mu$ data yielded 6224 ± 80 events on a background of 55 ± 5 events, as seen in figure 4. Various background models were considered, including flat background, backgrounds from high χ^2 events, and

backgrounds from events with poor particle identification. All gave essentially the same result, and their deviations from each other were included in calculations of systematic error.

A preliminary ratio of branching fractions is determined to be

$$\frac{B(K_L^0 \to \mu\mu)}{B(K_L^0 \to \pi\pi)} = (3.50 \pm 0.11) \times 10^{-6}.$$

Normalizing to $K_L \to \pi\pi$ yields

$$B(K_L^0 \to \mu^+ \mu^-) = (7.23 \pm 0.22) \times 10^{-9}.$$

This is a 3% relative error, dominated by preliminary conservative systematic error estimates. Final results should yield less than 2% error. This value is consistent with the current world average for $B(K_L^0 \to \mu^+ \mu^-)$.

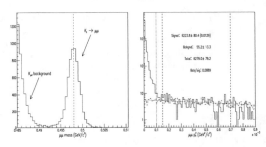

Figure 4: Invariant mass and p_T^2 for $K_L \to \mu\mu$. There are over 6200 events under the signal curve.

References

1. P. Langacker, S. U. Sankar, and K. Schilcher, *Phys. Rev.* D **38**, 2821 (1988).
2. R. N. Cahn and H. Harari, *Nucl. Phys.* B **176**, 135 (1980).
3. Z. Gagyi-Palffy, A. Pilaftsis, and K. Schilcher, *Nucl. Phys.* B **513**, 517 (1998).
4. S. Dimopoulous and J. Ellis, *Nucl. Phys.* B **182**, 505 (1981).
5. J. Pati and H. Stremnitzer, *Phys. Lett.* B **172**, 441 (1986).
6. B. Mukhopadhyaya and A. Raychaudhuri, *Phys. Rev.* D **42**, 3215 (1990).
7. S. L. Glashow, J. Iliopoulos, and L. Maiani, *Phys. Rev.* D **2**, 1285 (1970).
8. G. Valencia, *Nucl. Phys.* B **517**, 339 (1998); D. Gomez Dumm and A. Pich, *Phys. Rev. Lett.* **80**, 4633 (1998).
9. L. M. Sehgal, *Phys. Rev.* **183**, 1511 (1969).

Study of $K^+ \to \pi^+ e^+ e^-$ and $K^+ \to \pi^+ \mu^+ \mu^-$ decays in E865 at the AGS

H. Ma[1] R. Appel[6] G.S. Atoyan[2] B. Bassalleck[5] D. Bergman[8] D.N. Brown[6] N. Cheung[6] S. Dhawan[8]
H. Do[8] J. Egger[3] S. Eilerts[5] C. Felder[6] H. Fischer[5] M. Gach[6] W. Herold[3]
V.V. Issakov[2] H. Kaspar[3] D.E. Kraus[6] D. M. Lazarus[1] L. Leipuner[1] P. Lichard[6] J. Lozano[8]
J. Lowe[5] W. Majid[8] W. Menzel[4] S. Pislak[7] A.A. Poblaguev[2] V.E. Postoev[2] A.L. Proskurjakov[2] P. Rehak[1]
P. Robmann[7] A. Sher[6] J.A. Thompson[6] P. Trüol[7] H. Weyer[4] D. Wolfe[5] M.E. Zeller[8]

[1] *Brookhaven National Laboratory,* [2] *Institute for Nuclear Research, Moscow*
[3] *Paul Scherrer Institute,* [4] *University of Basel,* [5] *University of New Mexico,*
[6] *University of Pittsburgh,* [7] *University of Zürich,* [8] *Yale University*

10,000 $K^+ \to \pi^+ e^+ e^-$ events were studied in E865 at the AGS, yielding a preliminary result of the form factor parameter $\lambda = 0.2 \pm 0.02$. The measurement shows that the chiral perturbation theory of $\mathcal{O}(p^4)$ is insufficient to describe the data. 400 $K^+ \to \pi^+ \mu^+ \mu^-$ events have been observed. The $M_{\mu\mu}$ distribution confirms the form factor measured in the $K^+ \to \pi^+ e^+ e^-$ decay.

1 Introduction

The decay $K^+ \to \pi^+ e^+ e^-$ (πee) is a rare decay which involves a flavor changing neutral current (FCNC). Its rate was first calculated assuming the decay to occur through a short distance $s \to d\gamma$ transition. Since then, however, it has been realized that the decay mechanism is dominated by long-distance effects, and many calculations have been made to describe it, including parameter-free vector dominance model [1]. Chiral perturbation theory (CHPT), having been quite successful at modeling many decay modes of the light mesons is currently one of the most favored approaches. It has been employed to predict the decay rate and invariant electron-positron mass (M_{ee}) distribution for πee. Published results of measurements of these quantities [2] were consistent with next to leading order predictions of the theory [3], albeit statistically only to one and a half standard deviations. We report here the results of a new measurement, with significantly improved statistics and greater acceptance over the M_{ee} range than previously published, which we find to be in disagreement with these predictions.

The decay $K^+ \to \pi^+ \mu^+ \mu^-$, which was first observed by experiment E787 at the AGS [4], explores the same physics. Assuming the dominance of one-photon exchange process, the difference between the two decays lies only in the available phase space. A comparison of the two decays allows close examination of the decay mechanism.

2 The experiment

The experiment was performed at the Brookhaven National Laboratory's AGS. The experimental apparatus, a schematic drawing of which is shown in Figure 1 , was constructed to search for the decay $K^+ \to \pi^+ \mu^+ e^-$ [5,6]. It resided in a 6 GeV/c unseparated beam containing about 10^8 K^+ and 2×10^9 π^+ and protons per 1.6 second AGS pulse. Downstream of a 5 m evacuated decay volume a spectrometer formed by proportional chambers P1-P4 surrounding the dipole magnet 120D36 determined the momenta and trajectories of the decay products. The dipole magnet 48D48 separated the trajectories by charge with negative particles going mainly to the left and positive to the right.

Particle identification was accomplished with atmospheric-pressure Čerenkov counters filled with hydrogen gas on the left (C1L and C2L), and methane on the right (C1R and C2R); a shower calorimeter of the Shashlik design, consisting of of 600 modules, each 11.4 cm by 11.4 cm by 15 radiation lengths in depth, arrayed 30 horizontally and 20 vertically; and a muon range stack consisting of 24 planes of proportional tubes situated between iron plates. With these components electrons were identified as having light in the appropriate Čerenkov counters and energy in the calorimeter consistent with the measured momentum of the trajectory. Muons were identified as having no light in the Čerenkov counters, energy in the calorimeter consistent with having come from a minimum ionizing particle, and penetration in the range stack either consistent with that expected by the measured momentum of the trajectory, or completely through the stack if the measured momentum was greater than 1.5 GeV/c. Pions were identified in the same manner as muons except their range was required to be shorter than that expected from the measured momentum.

This apparatus was similar to that of E777 [2] but with a larger acceptance, improved particle identification capabilities and spectrometer resolution.

The primary trigger for the experiment is based on three charged particle hits using hodoscope D (an array of 20 "slats" between P1 and P2, with 10 on either side of the beam line), hodoscope A (an array of 30 "slats" be-

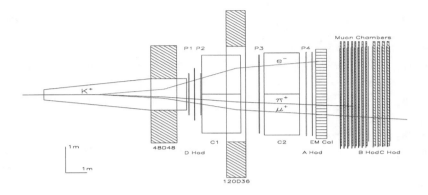

Figure 1: The plan view of the E865 detector

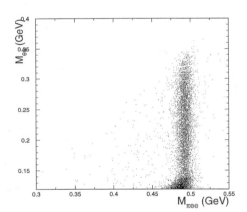

Figure 2: e^+e^- invariant mass versus $\pi^+e^+e^-$ invariant mass for $K^+ \to \pi^+e^+e^-$ candidates.

tween P4 and the shower counter, with 15 on either side of the beam line), and the shower counter. The dominant trigger rate at this stage comes from accidentals, and $K^+ \to \pi^+\pi^+\pi^-$ decays.

For $K^+ \to \pi^+e^+e^-$ decays, Čerenkov signals were required on each side of the detector. This trigger was dominated by events originating from the decay chain $K^+ \to \pi^+\pi^0$; $\pi^0 \to e^+e^-\gamma$ ($K_{\pi 2Dal}$), where the typical e^+e^- invariant mass was low. Taking advantage of the fact that the low mass e^+e^- pair were not separated

vertically in the non-bending plane, a "high mass" trigger was configured by requiring a vertical separation of calorimeter hits associated with the Čerenkov counter response.

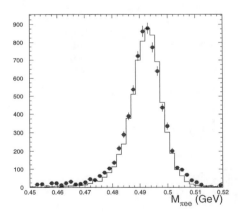

Figure 3: $\pi^+e^+e^-$ invariant mass for $K^+ \to \pi^+e^+e^-$ candidates with $M_{ee} > 150$ MeV.

For $K^+ \to \pi^+\mu^+\mu^-$ decays, muons were identified by the coincidence of the muon trigger counters, B and C. One coincidence on each side of the detector was required. This trigger was dominated by accidentals at the operating intensity.

The $K^+ \to \pi^+e^+e^-$ data were taken in 1995 and

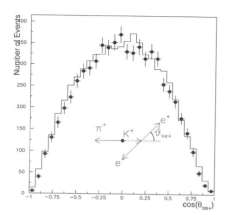

Figure 4: $cos\theta$ distribution of $K^+ \to \pi^+e^+e^-$ events in comparison with Monte Carlo simulation with pure vector interaction.

1996, running typically at 7×10^{12} protons/pulse for about 25 weeks. The $K^+ \to \pi^+\mu^+\mu^-$ data were collected in 1997, at a reduced intensity because of the trigger rate, for about 6 weeks.

Events were selected which contained unambiguous identification of a positive pion, and the lepton pair, whose trajectories were consistent with having come from a common vertex located within the 5 m long evacuated decay volume. The reconstructed kaon was required to have come from the target upstream of the beam magnets.

3 The $K^+ \to \pi^+e^+e^-$ decay

Figure 2 displays the M_{ee} versus the invariant mass of the three final state particles, $M_{\pi ee}$, for the $K^+ \to \pi^+e^+e^-$ candidates. The cluster of events below 130 MeV are the background from $K_{\pi 2Dal}$, where the $M_{\pi ee}$ is just below M_K because of a missing soft photon. Figure 3 shows $M_{\pi ee}$ for events with $M_{ee} > 0.15$ GeV in comparison with Monte Carlo. The background is insignificant. About 10,000 events are selected after all the cuts.

In general for a vector interaction the $q^2 = M_{ll}^2$ distribution for $\pi^+l^+l^-$ can be described by:

$$\frac{d\Gamma(q^2, cos\theta)}{dq^2 dcos\theta} = \frac{G^2\alpha^2 M_K^3}{16\pi}\lambda^{3/2}(1, q^2/M_K^2, m_\pi^2/M_K^2) \times$$
$$(1 - 4m_l^2/q^2)^{1/2}\left[1 - (1 - 2m_l^2/q^2)cos^2\theta\right] \times$$
$$|f(q^2)|^2 \tag{1}$$

where $\lambda(a, b, c) = a^2 + b^2 + c^2 - 2(ab + bc + ac)$, G is

a constant, m_l is the lepton mass and $f(q^2)$ is the form factor, a dimensionless function of q^2.

In order to confirm that the data should be fit with a vector spectrum we display in Figure 4 the $cos\theta$ distribution, where θ is the angle between the positron and pion momentum vectors in the e^+e^- center of mass frame. The data is compared with a simulation assuming a vector interaction.

The parameters of $f(q^2)$ can be determined by fitting the observed M_{ee} spectrum to that resulting from a simulation of such data after normalizing to the $K_{\pi 2Dal}$ sample for the same number of incident kaons. Because $K_{\pi 2Dal}$ has the same charged particles in the final state, much of the detector efficiency cancels in the normalization.

In analogy to the formulation of the semi-leptonic decay $K^+ \to \pi^0 e^+\nu_e$ (K_{e3}) the function $f(q^2)$ can be expressed as

$$f(q^2) = f_+(1 + \lambda\frac{q^2}{M_\pi^2}) \tag{2}$$

For K_{e3}, $\lambda = 0.0289$ has been measured.

From a fit to a partial sample of the πee data, the preliminary results are Br($K^+ \to \pi^+e^+e^-$)= $(2.69\pm0.2) \times 10^{-7}$ and $\lambda = 0.20 \pm 0.02$; Figure 5 displays the M_{ee} distribution compared to the Monte Carlo simulation using the fitted form factor.

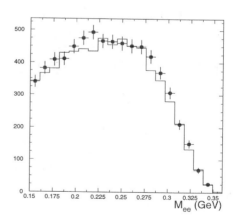

Figure 5: M_{ee} distribution for accepted $K^+ \to \pi^+e^+e^-$ events in comparison with Monte Carlo simulation with $\lambda = 0.2$.

From the appropriately normalized M_{ee} distribution one can also determine the parameters for a CHPT formulation [3]. In this case, the form factor is explicitly expressed as

$$f(q^2) = -(\phi_K(q^2) + \phi_\pi(q^2) + w_+) \qquad (3)$$

with ϕ_K and ϕ_π being analytic functions of q^2 and w_+ being a parameter to be determined from the data. Although this formulation can fit the M_{ee} spectrum well, it can not fit the branching ratio and the M_{ee} spectrum simultaneously. Figure 6 shows the experimental data in the plane of BR$_{\pi ee}$ and w_+. The previous data points [2,7] are also shown. The solid curve is the prediction from reference [3], while the dotted line shows a more recent revision of such calculation [8]. From this, we conclude that CHPT of $\mathcal{O}(p^4)$ is insufficient to describe the $K^+ \to \pi^+ e^+ e^-$ data.

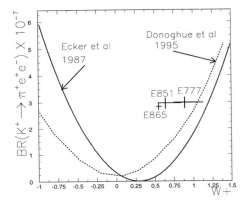

Figure 6: Experimental data on Branching Ratio versus w_+ plane. The solid curve came from reference 2 and the dashed curve came from reference 8. The previous measurements came from experiment E777 [2] and E851 [7].

4   The $K^+ \to \pi^+ \mu^+ \mu^-$ decay

The main background for the $K^+ \to \pi^+ \mu^+ \mu^-$ decay comes from $K^+ \to \pi^+ \pi^+ \pi^-$ decay with subsequent $\pi^\pm \to \mu^\pm \nu$ decays. This background satisfies the same particle identification requirement. Although in general the $\pi\mu\mu$ mass reconstructed from such background tends to be below the kaon mass, the pion decays in flight inside the spectrometer magnet can cause significantly false measurements extending $M_{\pi\mu\mu}$ into the signal region. To reject such background, a likelihood function is constructed based on back-tracking the reconstructed kaon to the target, track χ^2 and the vertex quality. Figure 7 displays the likelihood versus the $\pi\mu\mu$ invariant mass. The resolution of the detector is sufficient to separate the signal from the background.

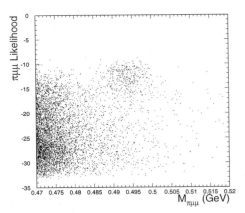

Figure 7: The likelihood versus the $\pi\mu\mu$ invariant mass for $K^+ \to \pi^+ \mu^+ \mu^-$ candidates. The $K^+ \to \pi^+ \mu^+ \mu^-$ events center around the kaon mass with high likelihood value. The background comes from $K^+ \to \pi^+ \pi^+ \pi^-$ decays, followed by $\pi^\pm \to \mu^\pm \nu$ decays.

Figure 8 displays $\pi\mu\mu$ invariant mass after requiring a good likelihood function. The background is modeled by an exponential function determined below the kaon mass. The signal to background ratio is about 10:1. Approximately 400 events have been observed.

Similar to the study with $K^+ \to \pi^+ e^+ e^-$, we can also check the angular distribution of the pion in the $\mu\mu$ center of mass frame, as shown in Figure 9. The data is consistent with the vector dominance.

Unlike the previous experimental measurement of this decay [4], this experiment covered the full $M_{\mu\mu}$ invariant mass range with good acceptance. The spectrum was measured and valuable form factor information can be extracted. Figure 10 shows the $\mu\mu$ invariant mass distribution, in comparison with the Monte Carlo simulation with $\lambda = 0.2$ and $\lambda = 0$. Although the statistics are low, the data prefer the solution with $\lambda = 0.2$.

For determining the branching ratio, the $K^+ \to \pi^+ \mu^+ \mu^-$ data can be normalized to the observed $K^+ \to \pi^+ \pi^+ \pi^-$ events, which is kinematically very similar to the signal. $K^+ \to \pi^+ \pi^+ \pi^-$ events were selected from a highly prescaled minimum biased trigger without any online particle identification requirement. This analysis is still in progress.

5   Summary

A high statistics study of $K^+ \to \pi^+ e^+ e^-$ events reveals the form factor's strong dependence on q^2, characterized by a large λ. This behavior, together with the branch-

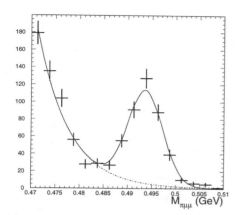

Figure 8: $\pi\mu\mu$ invariant mass after a cut on the likelihood function.

ing ratio measurment, can not be modeled by CHPT of $\mathcal{O}(p^4)$. Further theoretical study is needed, possibly investigating the high order contributions. The decay is consistent with vector dominance.

The $K^+ \rightarrow \pi^+\mu^+\mu^-$ data sample, twice as many events as the previously published data, provides the first measurement of the $\mu\mu$ invariant mass. The comparison with the $K^+ \rightarrow \pi^+e^+e^-$ data will be of great interest.

References

1. P. Lichard, *Phys. Rev.* D **55**, 5385 (1997)
2. C. Alliegro *et al*, Phys. Rev. Lett. **68**, 278 (1992).
3. G. Ecker, A. Pich, E. de Rafael, *Nucl. Phys.* B **291**, 692 (1987)
4. Phys. Rev. Lett. **79** 4756 (1997)
5. *Experiment E865 at BNL, A search for the decay $K^+ \rightarrow \pi^+\mu^+e^-$* , Ph .D. thesis, University of Zürich, 1997
6. *A search for the decay $K^+ \rightarrow \pi^+\mu^+e^-$* , Ph .D. thesis, Yale University, 1997
7. A. L. Deshpande, *A study of the decay of positively charged kaon into a positively charged pion, a positron and an electron, and a measurement of the decay of a neutral pion into a positron and an electron*, Ph.D. thesis, Yale University, 1995
8. J. Donoghue, F. Gabbiani, *Phys. Rev.* D **51**, 2187 (1995)

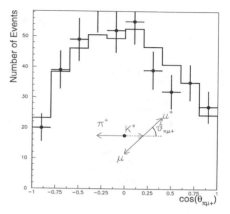

Figure 9: $cos\theta$ distribution of $K^+ \rightarrow \pi^+\mu^+\mu^-$ events in comparison with Monte Carlo simulation with pure vector interaction.

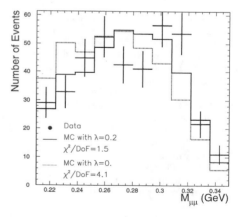

Figure 10: $\mu\mu$ invariant mass distribution of the accepted events after background subtraction, in comparison with MC simulation of the $K^+ \rightarrow \pi^+\mu^+\mu^-$ decay with $\lambda = 0.2$ and $\lambda = 0.$.

KTeV at Fermilab: New K_L^0 and π^0 Rare Decay Results from Phase II of Experiment 799

A.R. Barker

for the KTeV Collaboration

Department of Physics, University of Colorado, Boulder, CO, 80309-0390 USA
E-mail: tonyb@cuhep.Colorado.edu

The second phase of Fermilab Experiment 799 collected data during 1997, in order to search for a variety of rare K_L decays, with particular emphasis of those relevant to CP violation. The analysis of these data, representing over 200 billion K_L decays, is currently underway. We report here preliminary findings from a number of rare decay analyses. In each case, the sensitivity of E799 is at least an order of magnitude greater than that of any published result. Specifically, we describe our observation of the muonic Dalitz decay $K_L^0 \to \mu^+\mu^-\gamma$, and report a preliminary decay branching ratio for the Double Dalitz decay $K_L \to e^+e^-e^+e^-$. Both of these modes can provide insight into the $K_L\gamma\gamma$ form factor, which is needed to better understand the long-distance contribution to $K_L \to \mu^+\mu^-$. We also report on the Double Dalitz decay of the π^0, where we have over ten thousand events in which the decaying pion is produced in the decay $K_L \to \pi^0\pi^0\pi^0$. This decay can be used to measure the full $\pi^0\gamma\gamma$ form factor. In addition, we present a new measurement of the branching ratio for the rare decay $\pi^0 \to e^+e^-$, also obtained using $K_L \to \pi^0\pi^0\pi^0$ events. This decay tests unitarity in QED and provides another window into the structure of the $\pi^0\gamma\gamma$ coupling. We also discuss the rare decay $K_L \to \pi^+\pi^-e^+e^-$. We report here on preliminary results of an analysis of the full E799 data set, including a new branching ratio measurement as well as our observation of a new CP- and T-violating asymmetry in the decay angular distributions of this final state. Finally, we report two new upper limits from searches for the rare decay $K^0 \to \pi^0\nu\bar\nu$, which represents the future direction of direct CP-violation studies in rare K_L decays.

1 Introduction

1.1 The KTeV Project

The KTeV project at Fermilab is an effort to learn more about the origins of CP violation in the neutral kaon system. KTeV encompasses two experiments: E832 and E799 Phase II. E832 is the successor to FNAL E731, and is dedicated to making a detailed study of $K_{L,S} \to \pi\pi$ decays in order to extract the value of the direct CP violation parameter ϵ'/ϵ as well as other CP-violation and mixing parameters in the neutral kaon system. Despite having a sensitivity of better than 10^{-3} to Re(ϵ'/ϵ), the best experiments to date, E731 and NA31 at CERN, have not be able to conclusively resove the question of whether ϵ' is different from zero. Although the NA31 final result was three sigma from zero, E731 was only one sigma from zero, and the average two sigma. Moreover, the two results differed from each other at the 1.7σ significance level. Dissatisfaction with this situation led to a new round of ϵ' experiments. KTeV experiment E832 and the new CERN experiment, NA48, both aim to measure Re(ϵ'/ϵ) to a precision of better than 2×10^{-4}. Recent theoretical calculation in the Standard model come up with estimate for Re(ϵ'/ϵ) in the range of 2 to 10×10^{-4}, so this level of precision may well be adequate to see a definitely non-zero value, if the origin of CP violation lies in the Standard Model. For more details about that status of E832, the reader is referred to the contribution to this Conference by Prof.E.Cheu on that subject.

In recent years, theoretical predictions of the magnitude of Re(ϵ'/ϵ) have become substantially smaller, due to the large mass of the top quark. When the top quark

is heavy, there is substantial destructive interference between some of the amplitudes contributing to ϵ' in the Standard Model. Some of these direct CP-violating amplitudes correspond to so-called electromagnetic Penguin graphs, which describe an $s \to d$ transition mediated by a loop consisting of a virtual W boson and a virtual u, c, or t quark. Either of these virtual particles can couple to a photon or a Z boson, which materializes into a $\bar dd$ or $\bar uu$ pair, leading to a $K^0 \to \pi\pi$ decay. However, if the photon or Z couples instead to a lepton pair, the diagram can describe the rare decay $K_L \to \pi^0 ll$, where the lepton l can be an electron, a muon, or any flavor of neutrino. Because the CP-conserving and indirect CP-violating (through $K - \overline K$ mixing) contributions to these processes range from small (in the case of charged leptons), to nearly non-existent (for neutrinos), direct CP-violating amplitudes like those described by these Penguin graphs should play an important role in these rare processes.

1.2 The Rare K_L Decay Experiment E799

The first rare-K decay experiment at Fermilab, E799 Phase I, searched for evidence of these decays. No signal was found and upper limits of less than 5×10^{-9} (90% C.L.) were published[1] for the $\pi^0e^+e^-$ and $\pi^0\mu^+\mu^-$ modes. New searches for these modes will be carried out by E799 Phase II. For both modes, these is a problematic background from radiative Dalitz decays, $K_L \to ll\gamma\gamma$, where the photons accidentally combine to form a π^0 mass.

A search was also made for the experimentally very difficult mode $K^0 \to \pi^0\nu\bar\nu$, which does not suffer from

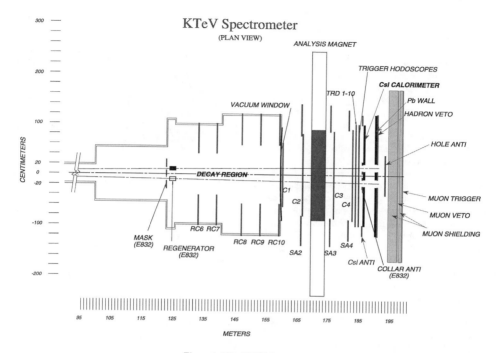

Figure 1: The KTeV Apparatus.

such a background. The experimental sensitivity to this mode is much poorer, and an upper limit of 5.8×10^{-5} was reported [2] from E799 Phase I. Nevertheless, the $\pi^0 \nu \bar{\nu}$ is especially interesting because in the Standard Model it proceeds almost entirely through a direct CP-violating amplitude involving no unknown hadronic matric elements. Consequently, the imaginary part of V_{td} in the CKM matrix can be extracted from a measurement of the branching ratio for $K_L \to \pi^0 \nu \bar{\nu}$ with very little theoretical uncertainty.

1.3 The KTeV Apparatus

Phase II of Experiment E799 is part of the KTeV project, and uses an almost entirely-new apparatus, which is shown schematically in Figure 1. The vertical scale on the figure is exaggerated to show the details of the apparatus, which, as the horizontal scale shows, is roughly 100 meters long, including the large evacuated decay volume. There are two parallel neutral beams, because in the ϵ' experiment E832, a regenerator is placed in one of the beams in order to produce a K_S component in that beam. The $\pi\pi$ decays in the regenerator beam are then compared to those in the vacuum beam in order

to extract $\text{Re}(\epsilon'/\epsilon)$ and other parameters of the neutral kaon system. In E799, the regenerator is removed, as are upstream absorbers in the regenerator beam. Both beams then provide a maximum flux of K_L for rare decay searches.

Charged particles are removed from the beams by sweeping magnets, the last of which is some 93 meters downstream of the target, just off the left edge of Figure 1. The vacuum decay region begins at this point, and extends downstream to a large window just before the first of four drift chambers, some 66 meters away. Decay particles produced at angles too large to lie within the acceptance of the main part of the apparatus are caught by a series of lead/scintillator veto detectors, spaced every few meters along the decay region as well as between the vacuum window and the main CsI calorimeter. These vetoes allow us to reject events containing particles that would otherwise escape the detector, in order to reduce non-exclusive backgrounds.

Charged particles are detected, and their momenta measured, by a series of four drift chambers, two upstream and two downstream of an analyzing magnet that imparted a 200 MeV/c transverse momentum kick. A series of eight transition radiation detectors downstream of the last drift chamber provided excellent π-e separation, achieving pion rejection of approximately 150:1 at 90% efficiency.

The heart of the apparatus is the 3100-crystal pure CsI calorimeter. The calorimeter consisted of 2.5 × 2.5 cm^2 crystals in the area near the beam, and 5.0 × 5.0 cm^2 crystals farther out. All the crystals were some 27 radiation lengths long, and an energy resolution of better than 1% was achieved for electrons above 10 GeV. The calorimeter was read out in a series of 19-ns time slices for each crystal, which allowed rejection of background caused by out-of-time accidental energy. The small transverse size of the crystals allowed an excellent position resolution of approximately 1 mm to be achieved by analysing the energy sharing between crystals. The calorimeter was used both to measure photon momenta and to identify electrons by comparing the deposited energy to the measured momenta in the drift chambers.

Dowstream of the CsI calorimeter, there was a three-meter thick steel muon filter, followed by a scintillator plane and another one-meter thick steel slab, followed finally by a scintillator hodoscope. This part of the detector allowed us to identify muons and separate them from pions that didn't shower in the CsI calorimeter.

A variety of triggers were defined which used signals from the CsI, the various scintillator planes, and the chambers to identify interesting events. Events were read out and passed to the online computers at a rate of about 10 kHz. Filter software reconstructed these events and selected roughly 10% for further offline analysis. E799 took data in two periods: one in the winter of 1997, and another in the summer. The total number of kaon decays in the decay volume during both runs was approximately 250 billion.

2 Rare π^0 Decays

Although the main purpose of the experiment is to study rare kaon decays, the common decay $K_L \to \pi^0\pi^0\pi^0$ provides a copious source of clean π^0 mesons, which allows us to search for a variety of rare π^0 decays. The most common type of event collected by E799's 2-electron/n-cluster trigger is $K_L \to \pi^0\pi^0\pi_D^0$, where π_D^0 indicates a π^0 undergoing the Dalitz decay $\pi^0 \to e^+e^-\gamma$. This decay has been well-measured, but the Double Dalitz decay $\pi^0 \to e^+e^-e^+e^-$ has been little-studied since it was used by Samios in 1962 to determine the parity of the π^0. We can observe this decay by searching for events containing four photons and four electrons. These events can come from either $K_L \to \pi^0\pi_D^0\pi_D^0$ or $K_L^0 \to \pi^0\pi^0\pi_{DD}^0$, where the π_{DD}^0 indicates a Double Dalitz decay. This is clearly seen in Figure 2, which shows the 4e mass for these events. The peak is due to π_{DD}^0 decays, while the

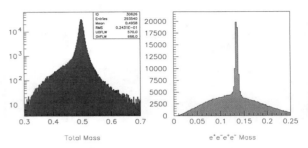

$\pi^0 \to e^+e^-e^+e^-$ Analysis using $K_L \to \pi^0\pi^0\pi^0$

Figure 2: The 4e mass in 4e4γ events.

continuum comes from double-Single Dalitz events.

We separate these two by forming a mass χ^2 for both hypotheses, considering all possible grouping of the final state particles. Finally, we eliminate background from photon conversions in the vacuume window or material of the first chamber by requiring that the smallest e^+e^- invariant mass exceed 2MeV. With this done, we estimate that there are a total of 52,446 $K_L \to \pi^0\pi_D^0\pi_D^0$ events, compared to 10,506 π_{DD}^0 events. We use the former sample for normalization, and, after correcting for the experimental acceptance, we find

$$\frac{\Gamma(\pi^0 \to e^+e^-e^+e^-)}{\Gamma(\pi^0 \to \gamma\gamma)} = (3.31 \pm 0.04 \pm 0.22) \times 10^{-5}.$$

We have also measured the rare decay $\pi^0 \to e^+e^-$, by looking for events with four photons and one e^+e^- pair. Backgrounds to the signal include $\pi^0\pi^0\pi_D^0$ events with one missing soft photon and $\pi^0\pi^0\pi_{DD}^0$ events with one soft e^+e^- pair having been swept out of the detector by the analysis magnet. We reduce the latter background by rejecting any event with in-time hits in the upstream drift chambers. In the E799 winter data, we see both of these backgrounds are seen at the predicted level, but in addition a clear peak containing 163 events is seen in the e^+e^- mass region between 132 and 138 MeV. In this region, we estimate only 11 background events, corresponding to over 150 signal events. On this basis, we report a preliminary branching ratio of

$$B(\pi^0 \to e^+e^-) = (6.16 \pm 0.52 \pm 0.26) \times 10^{-8}.$$

This measurement, with 10% precision, represents a substantial improvement over the previous measurements by BNL E851 and E799 Phase I, and is the first measurement that differs significantly from the lower bound of about 4×10^{-8} set by unitarity. The addition of the summer data should increase the statistics in this analysis by about 70%.

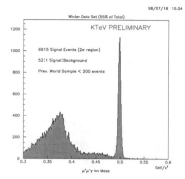

Winter Data Set (55% of Total)

KTeV PRELIMINARY

6915 Signal Events (2σ region)

52:1 Signal:Background

Prev. World Sample < 200 events

$\mu^+\mu^-\gamma$ Inv Mass

Figure 3: The total mass in $\mu^+\mu^-\gamma$ events.

3   Rare K_L^0 Decays

The two rare π^0 decays discussed in the previous section probe the coupling between the π^0 and two (possibly virtual) photons. The analogous K_L coupling can be studied by measuring the single Dalitz decays $K_L \to e^+e^-\gamma$, $K_L \to \mu^+\mu^-\gamma$, and the Double Dalitz modes $K_L \to e^+e^-e^+e^-$ and $K_L \to e^+e^-\mu^+\mu^-$. E799 has collected data for all these decays. Figure 3 shows the mass peak we see for $K_L \to \mu^+\mu^-\gamma$ in the winter data.

There are about 7000 events in the winter data, and about 5000 more are expected in the summer data. The signal to background ratio is over 50:1, so these events will allow us to make an accurate measurement of the $K_L\gamma^*\gamma$ form factor in the high-x region.

We also have a signal for the Double Dalitz decay $K_L \to e^+e^-e^+e^-$. The best previous measurement of this mode comes from E799 Phase I, in which some 27 events were identified. In the mass spectrum of events containing two electrons, two positrons, and no photons, we see backgrounds at masses below the kaon mass. However, as in other exclusive analyses, these backgrounds can be nearly eliminated by requiring that the total momentum of the observed final state particles be parallel to the line of flight of the decaying kaon. So we require that $P_t^2 < 300\,\mathrm{MeV}^2$, where P_t^2 is the square of the momentum component perpendicular to the line of flight, in addition to the a 2 MeV cut on the minimum e^+e^- invariant mass to remove most of the background from photon conversions. After these cuts, we see a very clean signal of some 252 events, over an estimated background of 10 events. Based on this signal, we extract a preliminary branching ratio

$$B(K_K \to e^+e^-e^-e^-) = (4.14 \pm 0.27 \pm 0.31) \times 10^{-8}.$$

This agrees well with the previous measurement [3] from E7899 Phase I, but is substantially more precise.

3.1   The Decay $K_L \to \pi^+\pi^-e^+e^-$

Based on a sample of about one-day's worth of E799 data, KTeV has already published the first measurement [4] of the branching ratio for the rare decay $K_L \to \pi^+\pi^-e^+e^-$. In the small data sample used for the paper, we saw about 35 examples of this decay. Now, with the full data set, we see a signal of nearly 3000 events. This decay is particularly interesting because we have observed a CP- and T-violating angular asymmetry in the final state.

The main background to this decay comes from the decay $K_L \to \pi^+\pi^-\pi_D^0$, where the photon from the π^0 Dalitz decay is not detected. If the photon is soft, mass of the remaining particles can be close to the kaon mass, and the event may be confused with a signal event. To remove this background, we assume that the particles seen come from a $\pi^+\pi^-\pi_D^0$ event. Then, based on the on the charged particle, we can calculate the square of the longitudinal momentum of the hypothesized π^0 in the K_L center of mass, called $P_\parallel^2$. If the hypothesis that the event is a $K_L \to \pi^+\pi^-\pi_D^0$ event is correct, then this quantity cannot be negative, except for some small smearing due to resolution effects. However, for events which are actually $K_L \to \pi^+\pi^-e^+e^-$, the quantity $P_\parallel^2$ is usually negative. So by requiring that it be negative, we can eliminate almost all of the dominant background while preserving most of the signal.

In view of the high statistics available for this mode with the full E799 dataset, the main concern in extracting the branching ratio for $K_L \to \pi^+\pi^-e^+e^-$ is the systematic error resulting from background subtraction. In order to minimize this error, the branching ratio analysis uses a very tight cut on $P_\parallel^2$, which reduces the number of candidate signal events to 708, of which 63 are estimated to be background. Based on the estimated total of 645 signal events with these tight cuts, we find after correcting for experimental acceptance a branching ratio

$$B(K_L \to \pi^+\pi^-e^+e^-) = (3.32 \pm 0.14 \pm 0.28) \times 10^{-7},$$

in good agreement with the theoretical predictions. [5]

To understand the interesting effects seen in the final state angular distributions, consider the decay $K_L \to \pi^+\pi^-\gamma$. Two different amplitudes contribute to this decay: a CP-violating amplitude describing the decay $K_L \to \pi^+\pi^-$ with inner bramsstrahlung from one of the pions, in which the photon is usually soft; and a (mostly) CP-conserving M1 direct emission amplitude in which the photon tends to be hard. In the region of moderate photon energy, both amplitudes are present at similar levels. Because one is CP-violating and the other is CP-conserving, there is a CP-violating interference term in the partial width. However, the CP violation manifests itself only in the polarization of the photon, which is normally not measured.

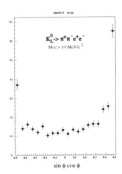

Figure 4: The angular asymmetry in $K_L \to \pi^+\pi^-e^+e^-$

However, when the photon is virtual and undergoes internal conversion to an e^+e^- pair, the angular distributions of the leptons are related to the virtual photon polarization, and the asymmetry should reveal itself. In particular, if ϕ is the angle between the normals to the hadronic and leptonic planes in the $\pi^+\pi^-e^+e^-$ decay, as measured in the K_L center of mass frame, then the partial width for the decay is predicted to have a distribution given by

$$\frac{d\Gamma(K_L \to \pi^+\pi^-e^+e^-)}{d\phi} = \Gamma_1 + \Gamma_2 \cos^2\phi + \Gamma_3 \cos\phi\sin\phi.$$

The last term violates both CP and T since $\sin\phi\cos\phi$ is odd under either CP or T separately, though it is even under CPT. An integrated asymmetry can be defined by

$$\mathcal{A} = \frac{N(\sin\phi\cos\phi > 0) - N(\sin\phi\cos\phi < 0)}{N(\sin\phi\cos\phi > 0) + N(\sin\phi\cos\phi < 0)}.$$

If Γ_3 were zero, then we would find $\mathcal{A} = 0$. The theoretical models [5] predict that the asymmetry resulting from the usual CP violation in $K_L \to \pi^+\pi^-$ should be large, about 14%.

Figure 4 shows the distribution of $\sin\phi\cos\phi$ that we observe. There is clearly a CP- and T-violating large asymmetry. In fact, the raw asymmetry seen in the data is

$$\mathcal{A}_{raw} = 23.2 \pm 2.3\%.$$

We have studied the acceptance of the apparatus and find that it has no intrinsic asymmetry, so this measurement shows that we have observed a new CP- and T-violating phenomenon with more than 10σ significance. However, the asymmetry in the decay itself may cause a secondary asymmetry in the experimental acceptance, so that the raw asymmetry is different from the "true" asymmetry that would be measured in the absence of acceptance effects. We are currently studying how best to extract a corrected asymmetry from these data.

3.2 The Decay $K_L \to \pi^0\bar{\nu}\nu$

The rare K_L decay that is the most theoretically attractive is at the same time the most experimentally difficult. The only amplitude contributing to $K_L \to \pi^0\bar{\nu}\nu$ is one that violates CP directly, and the rate for this process can be cleanly related to the parameters of the CKM matrix. On the other hand, the experimental signature is but a single π^0. Given the vast number of common K_L decays producing π^0's there are a huge number of events in which all the other particles are lost, and only a single π^0 is seen. In order to minimize these backgrounds, a highly efficient and hermetic system of photon vetoes is needed. Even then, one must require that the single π^0 have a large transverse momentum in order to beat down the remaining backgrounds from $K_L \to \pi^0\pi^0\pi^0$ as well as $\Lambda \to n\pi^0$ and $\Xi^0 \to \Lambda\pi^0$.

However, with the large beams normally used by E799, the transverse location of the kaon decay producing the single π^0 is only poorly determined, and the transverse momentum of the π^0 cannot be calculated well enough. This problem can be solved in two ways, both of which have been used by KTeV. First, the beams can be tightened to a so-called "pencil beam", which permits a reasonably good calculation of the transverse momentum of the π^0 from the two observed photons, at the cost of dramatically reducing the kaon flux. This was done for a special run of one day. We saw one event consistent with $K_L \to \pi^0\bar{\nu}\nu$ in those data, which we think probably comes from a neutron interacting in the vacuum window. Assuming that this was a signal event, we found an upper limit

$$B(K_L \to \pi^0\bar{\nu}\nu) < 1.6 \times 10^{-6} \qquad (90\% \text{ C.L., } \gamma\gamma \text{ mode})$$

The other approach to the problem of measuring transverse momentum is to use the Dalitz decay of the π^0. Then the vertex is accurately determined from the e^+e^- tracks. Unfortunately, this strategy costs a factor of 80 in sensitivity because of the small Dalitz branching ratio. However, this method is compatible with large beams, and we have applied it to the full E799 dataset, obtaining a somewhat better upper limit:

$$B(K_L \to \pi^0\bar{\nu}\nu) < 5.9 \times 10^{-7} \qquad (90\% \text{ C.L., Dalitz mode})$$

References

1. D.A.Harris *et al*, *Phys. Rev. Lett.* **71**, 3914 (1993) and *Phys. Rev. Lett.* **71**, 3918 (1193).
2. M.Weaver *et al.*, *Phys. Rev. Lett.* **72**, 3758 (1994).
3. P.Gu *et al.*, *Phys. Rev. Lett.* **72**, 3000 (1994).
4. J.Adams *et al.*, *Phys. Rev. Lett.* **80**, 4123 (1998).
5. L.M.Sehgal and M.Wanninger, *Phys. Rev.* D **46**, 1035 (1992).

SEARCH FOR TIME-REVERSAL VIOLATION IN $K^+ \to \pi^0 \mu^+ \nu$ DECAY

M. Abe[a], M. Aoki[b], I. Arai[c], Y. Asano[a], T. Baker[d], M. Blecher[e], M.D. Chapman[b], D. Dementyev[f],
P. Depommier[g], M. Grigorjev[f], P. Gumplinger[h], M. Hasinoff[i], R. Henderson[h], K. Horie[c], W.S. Hou[j],
H.C. Huang[j], Y. Igarashi[c], T. Ikeda[c], A. Ivashkin[f], J. Imazato[b], J-M. Lee[k], L.S. Lee[l], G.Y. Lim[l], J.H. Khang[k],
W. Keil[h], M. Khabibullin[f], A. Khotjantsev[f], Y. Kudenko[f], Y. Kuno[b], J.A. Macdonald[h], D.R. Marlow[m],
C.R. Mindas[m], O. Mineev[f], C. Rangacharyulu[d], S.K. Sahu[j], S. Sekikawa[c], H.M. Shimizu[b], S. Shimizu[n],
K. Shibata[c], Y.-M. Shin[d], Y.H. Shin[k], K.S. Sim[l], A. Suzuki[c], T. Tashiro[c], A. Watanabe[c], D. Wright[h], T. Yokoi[o]
(KEK-PS E246 Collaboration)

a) Institute of Applied Physics, University of Tsukuba, Ibaraki 305, Japan,
b) IPNS, High Energy Accelerator Research Organization (KEK), Ibaraki 305, Japan,
c) Institute of Physics, University of Tsukuba, Ibaraki 305, Japan,
d) Department of Physics, University of Saskatchewan, Saskatchewan, Canada S7N 0W0,
e) Department of Physics, Virginia Polytechnic Institute and State University, VA 24061-0435, U.S.A.,
f) Institute for Nuclear Research, Moscow 117312, Russia,
g) Laboratoire de Physique nucleaire, University de Montreal, Montreal, Canada H3C 3J7,
h) TRIUMF, Vancouver, British Columbia, Canada V6T 2A3,
i) Department of Physics and Astronomy, University of British Columbia, Vancouver, Canada V6T 1Z1,
j) Department of Physics, National Taiwan University, Taipei, Taiwan,
k) Department of Physics, Yonsei University, Seoul 120-749, Korea,
l) Department of Physics, Korea University, Seoul 136-701, Korea,
m) Department of Physics, Princeton University, NJ 08544, U.S.A.,
n) Department of Applied Physics, Tokyo Institute for Technology, Tokyo 152, Japan,
o) Department of Physics, University of Tokyo, Tokyo 113, Japan.

presented by Y. Kuno
E-mail: yoshitaka.kuno@kek.jp

A new experiment E246 at the 12-GeV proton synchrotron at KEK (KEK-PS), which is searching for the T-violating transverse muon polarization in $K^+ \to \pi^0 \mu^+ \nu$ decay, is described. The search is sensitive to new mechanisms of T or CP violation beyond the Standard Model. The preliminary result of the partial data samples analyzed so far is presented.

1 INTRODUCTION

The triple-vector correlation is odd under the time-reversal operation and measurement of such a correlation would provide a promising testing ground to scrutinize the time-reversal invariance. One such example concerning K decay is the transverse muon spin polarization $P_\mu^\perp$ in $K^+ \to \pi^0 \mu^+ \nu$ ($K_{\mu3}^+$) decay,[1] where $P_\mu^\perp$ is defined as the component of muon spin polarization normal to the decay plane, determined by the μ^+ and π^0 momentum vectors. It is given by

$$P_\mu^\perp \equiv \frac{\vec{s}_{\mu+} \cdot (\vec{p}_{\pi^0} \times \vec{p}_{\mu+})}{|\vec{p}_{\pi^0} \times \vec{p}_{\mu+}|}, \tag{1}$$

where $\vec{s}_{\mu+}$ is the muon spin vector and $\vec{p}_{\mu+}$ and $\vec{p}_{\pi^0}$ are the momentum vectors of the muon and neutral pion, respectively. Since the T-reversal operation changes the sign of $P_\mu^\perp$, a non-zero value of $P_\mu^\perp$ would signal T-violation. If CPT invariance is obeyed, T-violation implies CP-violation.

A search for $P_\mu^\perp$ in $K_{\mu3}^+$ decay is sensitive to new mechanisms of CP violation beyond the Standard Model. The physics motivation of new CP violation arises from the observed baryon asymmetry in the universe, which cannot be explained by the CP violation in the Standard Model alone.[2] Therefore there must be new additional sources of CP violation. Furthermore, recent theoretical progress of electroweak baryogenesis suggests that new CP violation sources might exist at the electroweak scale which can be accessible experimentally.

2 $P_\mu^\perp$ in $K_{\mu3}^+$ Decay

As a clean search for T-violating phenomena, the measurement of $P_\mu^\perp$ in $K_{\mu3}^+$ decay has several striking advantages. First, the final-state electromagnetic interaction (FSI), which would otherwise mimic a fake T-odd effect, is negligible, of the order of 10^{-6}, in $K_{\mu3}^+$ decay.[3] This is due to the fact that only one charged particle exists in the final state. However, this small FSI is not always the case; for instance, in triple correlations in nuclear β decays and $P_\mu^\perp$ in $K_L^0 \to \pi^- \mu^+ \nu$ ($K_{\mu3}^0$) decay, the FSI is predicted[4] to be as large as 10^{-3}. This advantage in $K_{\mu3}^+$ decay allows a wider window to search for T-violation mechanisms, free of FSI-induced background.

Second, $P_\mu^\perp$ in $K_{\mu3}^+$ decay has no contribution from the CKM phase in the minimal Standard Model at the tree level, and higher-order contributions are extremely small ($\sim 10^{-6}$). This implies that the observation of a non-zero $P_\mu^\perp$ value would be a definite signature of new physics beyond the minimal Standard Model.

In $K_{\mu3}^+$ decay, the hadronic matrix element can be described as

$$< \pi|J|K > = f_+^K(q^2)(\tilde{p}_K + \tilde{p}_\pi) + f_-^K(q^2)(\tilde{p}_K - \tilde{p}_\pi) \quad (2)$$

where $\tilde{p}_K$ and $\tilde{p}_\pi$ are the four momenta of the kaon and pion, respectively. $f_+^K(q^2)$ and $f_-^K(q^2)$ are the form factors of the hadronic matrix elements as a function of momentum transfer squared (q^2). Time-reversal invariance requires that the phases of f_+^K and f_-^K are relatively the same; in another words, if the parameter $\xi(q^2) \equiv f_-^K(q^2)/f_+^K(q^2)$ is defined, ξ should be a real number. Conversely, a non-zero value of Imξ would indicate T-violation. Based on eq.(2), $P_\mu^\perp$ can be calculated as a function of the energies of the muon (E_μ) and neutral pion (E_{π^0}).[5] Then, $P_\mu^\perp$ is given by[6]

$$P_\mu^\perp \cong \text{Im}\xi \left(\frac{m_\mu}{m_K} \right) \frac{|\vec{p}_\mu|}{[E_\mu + |\vec{p}_\mu|\vec{n}_\mu \cdot \vec{n}_\nu - m_\mu^2/m_K]} \quad (3)$$

in the K^+ rest frame. $P_\mu^\perp$ is proportional to Imξ with the kinematic factor which depends on the phase space of $K_{\mu3}$ decay sampled. It is $P_\mu^\perp \sim 0.3\times$ Imξ for E246.

The present experimental upper limit on $P_\mu^\perp$ in $K_{\mu3}^+$ decay, which was obtained from the previous experiment using in-flight K^+ decays at Brookhaven National Laboratory (BNL),[7] is $P_\mu^\perp = (-4.2 \pm 6.7) \times 10^{-3}$ in the K^+ rest frame. This yielded Im$\xi = -0.016 \pm 0.025$, or $|\text{Im}\xi| \leq 0.049$ at 90 % confidence level.[a]

3 Theoretical Predictions of $P_\mu^\perp$ in $K_{\mu3}^+$ Decay

In general[8] neither an effective vector (V) nor axial-vector (A) interaction introduce $P_\mu^\perp$, but only effective scalar (S) or pseudoscalar (P) interactions give a non-zero $P_\mu^\perp$. Some examples of theoretical models are given below.

3.1 Three Higgs Doublet Models

$P_\mu^\perp$ is sensitive to the extension of the Higgs sectors, such as multi Higgs-doublet models, which need at least

[a]Since the $P_\mu^\perp$ measurements in $K_{\mu3}^0$ decay might have FSI effects, only the $K_{\mu3}^+$ data is quoted instead of the combined result.

three Higgs-doublets to generate CP violation.[9] The three Higgs-doublet model (3HDM) has two charged Higgs particles H_i^+ (i=1,2). $P_\mu^\perp$ in $K_{\mu3}^+$ decay arises from the interference between the W^+-exchange and H^+-exchange diagrams. The former contributes mostly to $f_+(q^2)$ and the latter only to $f_-(q^2)$. Imξ in $K_{\mu3}^+$ decay in the 3HDM is given by[6]

$$\text{Im}\xi = \text{Im}(\alpha_1\beta_1^*) \cdot \left(\frac{v_2}{v_3} \right)^2 \cdot \left(\frac{m_K}{m_{H_1^+}} \right)^2, \quad (4)$$

where Im$(\alpha_1\beta_1^*)$ is a measure of the magnitude of CP-violation in 3HDM. m_K and $m_{H_1^+}$ are the masses of the kaon and charged Higgs, respectively. v_i ($i = 1, 2, 3$) are the three Higgs vacuum-expectation-values (V.E.V.). Although there have been no estimations of the magnitudes of the V.E.V.s, one possible scenario[10] is that the three V.E.V.s are proportional to the fermion masses which they couple to: $i.e.$ $v_1 : v_2 : v_3 \sim m_b : m_t : m_\tau$. This scenario gives a large $P_\mu^\perp$ in semileptonic K decays because of a large value of v_2/v_3 ($\sim m_t/m_\tau$) of about 100. Then, $P_\mu^\perp$ in $K_{\mu3}^+$ decay is the most sensitive among the other constraints (from the neutron electric dipole moment, $b \to s\gamma$ decay, and $B \to \tau\nu_\tau X$). The details are described elsewhere.[11]

3.2 Supersymmetric models with squark family mixing

The supersymmetric (SUSY) contribution to $P_\mu^\perp$ in $K_{\mu3}^+$ decay could be significant when the squark family mixing at the quark-squark-gluino couplings is large.[12] In general, there exist four squark-flavor mixing matrices denoting the couplings of quarks and their corresponding squarks ($\tilde{u}_L$, $\tilde{u}_R$, $\tilde{d}_L$, $\tilde{d}_R$). They contribute to a non-zero $P_\mu^\perp$ in $K_{\mu3}^+$ decay if their relative phase is not zero. Considering these experimental constraints, they found that $P_\mu^\perp$ can be as high as 7×10^{-3} when the maximally-allowed values are taken. It should be noted that the minimum SUSY model[13] without large squark family mixing predicts $P_\mu^\perp$ of 10^{-6}.

3.3 Leptoquark models or Supersymmetric model with R-parity breaking

Leptoquark models are an attractive candidate for new physics beyond the Standard Model.[14] One of the manifestations of leptoquark models is a supersymmetric extension with R-parity breaking. In these models, $P_\mu^\perp$ in $K_{\mu3}$ decay can be introduced by slepton ($\tilde{l}_i$) or down-type squark ($\tilde{d}_i$) exchange diagrams. For the squark exchange case, Imξ can be given by

$$\text{Im}\xi = \sum_k \frac{\text{Im}[\lambda'_{21k}(\lambda'_{22k})^*]}{4\sqrt{2}G_F \sin\theta_c(m_{\tilde{d}_k})^2} \cdot \frac{m_K^2}{m_\mu m_s} \quad (5)$$

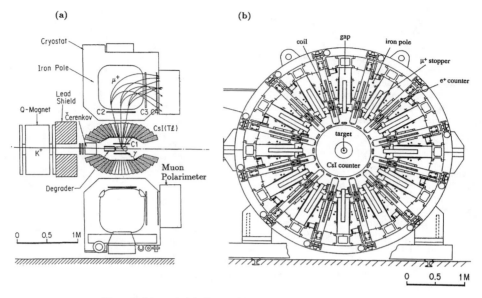

Figure 1: Schematic (a) side and (b) end views of the E246 detector

where λ'_{ijk} are the coupling constants in the R-parity breaking superpotential (with i, j, k being family indices), G_F is the Fermi coupling constant, θ_c is the Cabbibo angle, and m_s is a strange quark mass. In this case, $\mathrm{Im}[\lambda'_{21k}(\lambda'_{22k})^*]$ is a measure of CP violation in this model.[b] Since the upper limits on $\mathrm{Im}[\lambda'_{21k}(\lambda'_{22k})^*]$ are determined by the previous $P_\mu^\perp$ measurement, $P_\mu^\perp$ could appear just below the limit.[15]

4   KEK Experiment E246 Detector

KEK-PS E246 employs K decays at rest, in contrast to the previous experiments which used in-flight K decays. There are several significant advantages of the use of the K decays at rest. For instance, it allows clean and precise determination of K-decay kinematics. This, together with the carefully-designed detector, will reduce systematic errors.

4.1   E246 Detector

Schematic side and end views of the E246 detector are shown in Fig.1. Incident K^+s of 650 MeV/c, which are produced by a 12-GeV proton beam from the KEK-PS, are slowed down in the degrader, and stopped in a target

located at the center of the detector. The K^+-stopping target consists of an array of 256 plastic-scintillating fibers with 5×5 mm^2 square cross-section. After exiting the target, the muons from K^+ decays enter the spectrometer (a superconducting toroidal magnet) where their momenta are analyzed. The magnet has 12 identical magnet-gaps with accurate 30° rotational symmetry. To track the muons, four sets of wire chambers are installed, one of which is located at the entrance and two at the exit of each of the 12 magnet gaps. The fourth one is a cylindrical drift chamber surrounding the target. In addition to these, there is an array of 32 thin ring-shaped plastic scintillator (ring counter) which surrounds the target to provide an additional coordinate of the track along the beam axis.

Photons from the π^0 decays are detected with a highly-segmented photon detector, consisting of an array of 768 thallium-doped CsI crystals with silicon PIN photodiode readout[16] in which each crystal points to the K^+ stopping target. The assembly covers about 75 % of 4π and has a beam hole for allowing the K^+s into the target, and 12 holes for μ^+s going out into the magnetic spectrometer. In addition to the conventional ADC/TDC readout, transient digitizers based on switched capacitor arrays are instrumented for each crystal.[17]

The μ^+s exiting the spectrometer are stopped in a muon polarimeter installed at the exit of each magnet

---

[b]For the slepton exchange case, it can be replaced by $\mathrm{Im}[\lambda'_{i12}(\lambda_{2i2})^*]$.

gap. The muon polarimeter consists of a muon stopper made of pure aluminum, and e^+ counters located at clockwise (cw) and counter-clockwise (ccw) sides of the stopper. The e^+ counters detect e^+s in $\mu^+ \to e^+ \nu \bar{\nu}$ decay ($\tau_\mu = 2.2$ μsec). The muon polarization is deduced from the asymmetric angular distribution of e^+s, which are emitted preferentially along the μ^+ spin direction. The direction of the fringing magnetic field is parallel to $P_\mu^\perp$ so as to hold $P_\mu^\perp$, and at the same time it precesses the in-plane polarization components to average out.

4.2 $P_\mu^\perp$ Measurement and Systematic Cancellation

In E246, the $K_{\mu3}^+$ events which have the π^0 moving along the detector axis, either forward or backward from the target, will be accepted as "gold-plated events" For those events, the decay plane can be set radially from the detector axis. $P_\mu^\perp$ is directed azimuthally in a *screw-sense* around the detector axis, as in Fig.2. $P_\mu^\perp$ would then manifest itself as a difference in the e^+ counts between the cw and ccw counters from the muon stopper. By summing the cw- and ccw- counts of all the 12 sectors, $P_\mu^\perp$ is given by

$$\frac{\sum_{i=1}^{12} N_i(cw)}{\sum_{i=1}^{12} N_i(ccw)} \cong 1 \pm 2\alpha P_\mu^\perp, \qquad (6)$$

where $N_i(cw)$ and $N_i(ccw)$ are total e^+ counts at the cw and ccw counters at the ith sector, respectively. α is the effective analyzing power (defined as the ratio of the observed asymmetry to the muon polarization).

There are two important techniques for suppressing bias asymmetries in E246. First, summing over all magnet sectors would cancel any non-screw type biases. For instance, a fake asymmetry from the e^+-counter inefficiency, which would have opposite signs in its two adjacent sectors since the same e^+ detector acts as the cw counter in one sector and the ccw counter in the neighboring sector, would be cancelled by this summing. A shift of the K^+ stopping distribution at the target would introduce an asymmetric muon stopping distribution within the muon stopper, resulting in spurious asymmetries with the same sign in one hemisphere but with the opposite sign in the other hemisphere. Again, this can be cancelled after summing over all sectors.

The second is the comparison of the $K_{\mu3}^+$ samples with forward π^0 and backward π^0 directions. Since $P_\mu^\perp$ is of the same magnitude but opposite in sign between these two samples as shown in Fig.2 and any biased asymmetries are likely to be independent of π^0 directions, their comparison would enable us to reduce the systematic errors significantly. A double ratio can be formed by these two samples as follows:

$$\frac{[\sum_{i=1}^{12} N_i(cw)/\sum_{i=1}^{12} N_i(ccw)]_{fwd}}{[\sum_{i=1}^{12} N_i(cw)/\sum_{i=1}^{12} N_i(ccw)]_{bwd}} \cong 1 + 4\alpha P_\mu^\perp. \qquad (7)$$

It increases the asymmetry by a factor of two. In addition, by using the $K_{\mu3}^+$ events with the π^0s moving transverse to the detector axis, we will be able to examine the zero-asymmetry level.

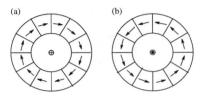

Figure 2: Schematic asymmetry directions in the 12 magnet sectors for (a) forward π^0 events and (b) backward π^0 events.

4.3 $K_{\mu3}^+$ Event Selection

E246 has analyzed the data taken from May, 1996 to December, 1997. The $K_{\mu3}^+$ event selection consists of identification of K^+ decays at rest in the K^+ stopping target, momentum selection (100 MeV/c < P_μ < 190 MeV/c), muon identification by time-of-flight, muon trajectory matching with hits in the target and the ring counter, π^0 mass cut, and a reconstructed kinematics consistent with $K_{\mu3}^+$ decay. A 8-μsec offline time window of e^+s from μ^+ decay was set 15 nsec after the K^+ decay timing to eliminate K_{e3} decay. After further selection of the forward and backward π^0s, a total number of gold-plated $K_{\mu3}^+$ samples of 1.3 M events is obtained. Also $K_{\mu3}$ samples with only one photon from π^0 decay detected (E_γ >70 MeV) are also accumulated to increase the total statistics, since the high energy photon in asymmetric π^0 decay carries most of the information of the original π^0 direction. This one-photon sample contains about 0.6 M events.

Major physics backgrounds are $K^+ \to \pi^+ \pi^0$ with π^0 decay in flight, $K^+ \to \pi^0 e^+ \nu$, and $K^+ \to \pi^+ \pi^0 \pi^0$ which are estimated to be 4 %, < 0.1 %, and < 0.5%, respectively, in these event samples. These background decays only dilute the asymmetry, but do not introduce any spurious asymmetries.

The analyzing power α was experimentally obtained from the $K_{\mu3}^+$ samples with the π^0 moving perpendicular to the detector axis. For those events, the e^+ asymmetry associated with the inplane muon polarization (P_μ^N) can be measured by the polarimeter. It is then compared with the expected P_μ^N which was calculated with the known value of Reξ. Then $\alpha = 0.198$ was determined. Since the stopping distributions of muons at the polarimeter were determined to be almost identical for the samples for P_μ^N and $P_\mu^\perp$, the same α value was used for the $P_\mu^\perp$ measurement.

4.4 Studies of Systematic Errors

Extensive studies of the possible systematic errors was done before examining $P_\mu^\perp$, based on real data as much as possible. Among many studies, two examples are given. One is the measurement of the e^+ asymmetry without any π^0 direction tagging. As mentioned earlier, a vanishing e^+ asymmetry is expected after summing over all sectors. It was confirmed at a level of $< 10^{-3}$, which is further reduced by the cancellation by forward and backward π^0s.[c] Another study is to examine distributions of the angle between the $K_{\mu 3}^+$ decay plane and the mid-plane of the magnet gap, where any deviation of the distribution from the expected would introduce bias asymmetry from the inplane muon polarization. A possible contribution to $P_\mu^\perp$ of $< 10^{-3}$ was determined. Further cancellation by the comparison of forward and backward π^0 events are estimated for each bias source separately. As a result of all the examinations, a combined systematic error of less than 1.0×10^{-3} is obtained.

4.5 Preliminary results

The very *preliminary* E246 result[18] of $P_\mu^\perp$ in $K^+ \rightarrow \pi^0 \mu^+ \nu$ decay is $P_\mu^\perp = (-2.5 \pm 5.7(stat) \pm 1.0(sys)) \times 10^{-3}$, which yields Im$\xi = (-0.77 \pm 1.9(stat) \pm 0.3(sys)) \times 10^{-2}$. At this moment, the statistical error dominates the systematic error. $P_\mu^\perp$ in $K^+ \rightarrow \mu^+ \nu \gamma$ decay, in which the decay plane is determined by the muon and photon momentum vectors, is also being investigated at E246.

5 Summary

E246 completed its first phase data taking in July, 1998. The total data are more than twice the data analyzed so far. A naive extrapolation of the final statistical sensitivity of Imξ to the whole data is about 1.1×10^{-2}, which would be about 2.5 times better than the previous BNL K^+ experiment. A request for a beam-time extension of an additional 66 days has been approved at KEK-PS. Together with the KEK-PS intensity upgrade, an ultimate sensitivity improvement of about a factor five is possible.

References

1. J.J. Sakurai, *Phys. Rev.* **109**, 980 (1958).

---

[c]This cancellation factor is about a few tens, estimated from real data.

2. L. Mclerran, M. Shaposhnikov, N. Turok, and M. Voloshin, Phys. Lett. **B 25 6**, 451 (1991); N. Turok and M. Voloshin, Phys. Lett. **B 256**, 451 (1991); N. Turok and J. Zadrozny, Nucl. Phys. **B 358**, 471 (1991); M. Dine, P. Huet, R. Singleton, and L. Susskind, Phys. Lett. **B 257**, 351 (1991).

3. A.R. Zhitnitskii, *Yad. Fiz.* **31**, 1014 (1980) [*Sov. J. Nucl. Phys*, **31**, 529 (1980)].

4. G.S. Adkins, *Phys. Rev.* D **28**, 2885 (1983) and references therein.

5. N. Cabbibo and A. Maksymowicz, *Phys. Lett.* **9**, 352 (1964); *Phys. Lett.* **11**, 360 (1964); *Phys. Lett.* **14**, 2 (1966); S.W. MacDowell, *Nuovo Cimento* **9**, 258 (1958).

6. G. Bélanger and C.Q. Geng, *Phys. Rev.* D **44**, 2789 (1991).

7. M.K. Campbell *et al.*, *Phys. Rev. Lett.* **47**, 1032 (1981); S.R. Blatt *et al.*, *Phys. Rev.* D **27**, 1056 (1983).

8. M. Leurer, *Phys. Rev. Lett.* **62**, 1967 (1989); P. Castoldi, J.-M. Frère and G.L. Kane, *Phys. Rev.* D **39**, 2633 (1989).

9. S. Weinberg, *Phys. Rev. Lett.* **37**, 657 (1976) and as a recent review, H.-Y. Cheng, *Int. J. Mod. Phys.* A **7**, 1059 (1992) and references therein.

10. R. Garisto and G. Kane, *Phys. Rev.* D **44**, 2038 (1991).

11. Y. Kuno, *Nucl. Phys. B (Proc. Suppl.)* **37A**, 87 (1994); Y. Kuno, *Chinese Journal of Physics* **32**, 1015 (1994).

12. G.-H. Wu and J.N. Ng, *Phys. Lett.* B **392**, 93 (1997).

13. E. Christova and M. Fabbrichesi, *Phys. Lett.* B **315**, 113 (1993).

14. G. Altarelli, J. Ellis, G.F. Giudice, S. Lola and M.L. Mangano, *Nucl. Phys.* B **506**, 3 (1997).

15. M. Fabbrichesi and F. Vissani, *Phys. Rev.* D **55**, 5334 (1997).

16. D.V. Dementyev *et al.*, *Nucl. Instrum. Methods* A **379**, 499 (1996).

17. R.L. Wixted *et al.*, *Nucl. Instrum. Methods* A **386**, 483 (1997).

18. T. Yokoi, Ph.D. thesis, University of Tokyo, 1998, unpublished; C.R. Mindas, Ph.D. thesis, Princeton University, 1998, unpublished; A. Ivashkin, Ph.D. thesis, Institute for Nuclear Research, 1998, unpublished.

STUDY OF THE RARE K_S^0, K_L^0, K^+, K^- DECAYS AT ϕ RESONANCE WITH THE CMD-2 DETECTOR

CMD-2 collaboration[1].
Presented by E.P.SOLODOV

The integrated luminosity $\approx 15pb^{-1}$ has been collected with the CMD-2 detector at the VEPP-2M collider for ϕ meson study (20×10^6 of ϕ's). The ϕ meson is a good source of the neutral and charged kaons and kaon decays in flight are detected with the complete reconstruction in the CMD-2 detector. A latest analysis of the $K_S^0 K_L^0$ coupled decays, decays of charged kaons and first observation of the semileptonic decay of K_S^0 based on 25% of available data are presented in this paper.

1 Introduction

As it was realized at the very early steps of the ϕ meson studies at the colliding beam machines, $K_S^0 K_L^0$ pairs can be used for studying CP and CPT violation. These suggestions including studies of quantum mechanical correlations were discussed [2,3] for experiments at VEPP-2M [4], an electron-positron collider at the Budker Institute of Nuclear Physics in Novosibirsk, Russia and carefully reviewed in paper [5]. Some preliminary results in $K_S^0 K_L^0$ coupled decays study with the CMD-2 detector based on relatively small data sample were published [6].

It should be mentioned that the decay of ϕ into two neutral kaons is pure $K_S^0 K_L^0$ state and the detection of one kaon at long distance predetermines the presence of K_S^0 decay within few millimetres from the beam providing an unique opportunity to study rare decays of pure K_S^0.

The combination of intensive decay of ϕ into two slow charged kaons and good 4π detector gives a possibility to study charged kaon decays and interactions using the recoil particle as a tag. All major decay modes are well detecting in the CMD-2 detector and their relative rates could be measured with relatively low systematic errors. These branching ratios were measured by optical spark chamber experiments [7] in 70's and were not consistent to each other.

The CMD-2 detector has been described in more detail elsewhere [3,8]. A 3.4 cm diameter vacuum beam pipe is made of Be with a 0.077 cm wall thickness and may be considered as a target for studies of the kaon nuclear interaction. As it was shown [9] about 50% of K_L^0 interacting in the barrel CsI calorimeter can be used as a tag for rare K_S^0 decays search.

The integrated luminosity of 4 pb^{-1} corresponding to about 5.5×10^6 produced ϕ's (about 25% of available data) has been analysed around ϕ. In this paper we present recent results in study of coupled $K_S^0 K_L^0$ decays and interactions including better measurement of regeneration cross section, measurements of charged kaon decay rates and the observation of the $K_S^0 \rightarrow \pi e \nu$ decay.

2 Study of $K_S^0 K_L^0$ coupled decays

Candidates were selected from a sample in which two vertices, each with two opposite charge tracks, were found within 15 cm from the beam axis and all tracks were reconstructed. An example of such event is shown in Figure 1.

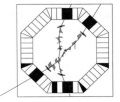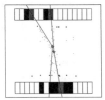

Figure 1: Display of $\phi \rightarrow K_S^0 K_L^0$ event with coupled decay.

The cuts on invariant mass $470 MeV/c^2 < M_{inv} < 525 MeV/c^2$ and missing momentum $80 MeV/c < P_{mis} < 140 MeV/c$ with an additional requirement to have another reconstructed vertex in the P_{mis} direction select $K_S^0 \rightarrow \pi^+ \pi^-$ events in one of the vertices. In this case K_L^0 is expected to be in the other one.

Figure 2a shows the decay length distribution for selected K_S^0's with the expected 0.55 ± 0.02 cm average decay length. The decay radius distribution for the K_L^0's with two charged tracks in final state is shown in Figure 2b. At the region from 0 to 7 cm where reconstruction efficiency is uniform, a bump with 266 ± 32 events is seen starting at the 1.7 cm radius corresponding to K_L^0 interactions with nuclei in the Be tube and DC material (Cu-Ti wires and Ar gas). The remain events are representing K_L^0 decaying in flight.

To select candidates to $K_L^0 \rightarrow \pi^+ \pi^-$ events additional cut requiring the invariant mass of two tracks from a K_L^0 vertex to be in the range of 470-525 MeV/c^2 was applied. The obtained distribution is presented in Figure 2d together with the fit function where all parameters except the number of events are fixed at the values obtained from the distribution in Figure 2b. The number of events under the peak drops down to 72 ± 12 events.

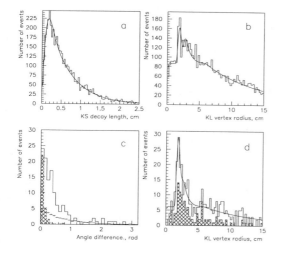

Figure 2: a. Decay length for K_S^0; b. Decay radius for K_L^0; c. Plane angular distributions for "tube" events (histogram), K_L^0 semileptonic decays (points with errors and K_S^0 two pions decays (shaded); d. Decay radius for K_L^0s after M_{inv} cut and after K_S^0 selecting cut (shaded).

One can apply stronger requirements for these events to satisfy $K_L^0 \to \pi^+\pi^-$ kinematics within detector resolution, i.e. $80MeV/c < P_{mis} < 140MeV/c$ and K_S^0 vertex being in the P_{mis} direction. This selection is illustrated in Figure 2d by shaded histogram. The peak at the Be tube survives with 38 ± 11 events and 65 K_L^0 decays in flight and interactions in DC material remain. About 8 CP violation decays of $K_L^0 \to \pi^+\pi^-$ are expected but cannot be identified because of K_L^0 semileptonic decays background and nuclear interactions.

The 72 ± 12 events found in the peak at Be pipe after the invariant mass cut and efficiency corrections are interpreted as regeneration of K_L^0 into K_S^0 with its decay into $\pi^+\pi^-$. Another 194 ± 32 events in the peak region represents nuclear interactions with two visible charged particles.

Using simulated efficiencies for estimation of full number of K_L^0 passed through beryllium pipe, the following cross sections for regeneration and visible inelastic scattering have been obtained:
$\sigma_{reg}^{Be} = 50 \pm 9 \pm 5$ mb.
$\sigma_{inel}^{vis} = 106 \pm 20$ mb.

The sources of the inelastic scattering events are the reactions with Σ and Λ production. To estimate the total cross section, the relative weight of these reactions was taken 0.21 from the CERN GEANT code (NUCRIN). With the ratio $\sigma_{inel}/\sigma_{tot} = 0.52$ taken from paper [10], one can estimate $\sigma_{tot}^{Be} = 854 \pm 161 \pm 240$ mb.

The histogram in Figure 2c shows the plane angular distribution for the regenerated at the beam pipe K_S^0 relative to the initial K_L^0 direction. Angular distribution for the semileptonic decays of K_L^0 is shown by dots with errors. The obtained angular distribution is wider than in the case of coherent regeneration which can be illustrated by the shown shaded distribution for original K_S^0 decays at the same distance from the beam. There is no evidence for the coherent contribution to the regenerated events.

In Figure 3a the measured regeneration cross section is plotted together with the theoretical calculations [10] for Be and Cu. The comparison of the calculated regeneration cross sections for these two different materials shows, that at momenta below 200 MeV/c one cannot scale them by a simple $A^{2/3}$ dependence. The experimental angular

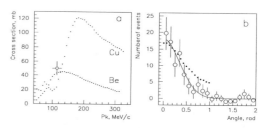

Figure 3: a. Experimental regeneration cross section and theoretical calculations for Be and Cu; b. Plane angular distribution of the regenerated K_S^0 with a fit function (solid line) and theoretical prediction (dots);

distribution of the regenerated K_S^0 after background subtraction is presented in Figure 3b together with gaussian fit function and theoretical prediction [10] and seems to be more narrow.

The obtained regeneration cross section for Be for low momentum kaons gives a possibility to estimate regeneration background for KLOE experiments at DAΦNE and helps to subtract it.

3   Charged kaon decays study

With the 120 MeV/c momentum the charged kaons from ϕ decays have about 50 cm decay length. Within 10 cm radius about 6% of charged kaon decays are completely reconstructed. Most of the major charged kaon decays have one charged particle in the final state with some number of neutrals (neutrino and π^0). Only $K^\pm \to \pi^\pm\pi^+\pi^-$ with $5.59\pm0.05\%$ decay probability has three charged particles in the final state. This decay is also detected but is not included in the present analysis.

Candidates were selected by presence of two charged tracks in the DC - one track coming from the beam axis with 120 MeV/c momentum and high dE/dx and second track with low dE/dx signal, acollinearity angle greater than 0.15 radian and impact parameter greater than 0.2 cm. The decay point was reconstructed by prolongation of kaon track to intersection with second track which was assumed to be from decay of another kaon. Under these conditions about 50000 tagged kaon decays were selected with decay radius less than 9 cm. The mo-

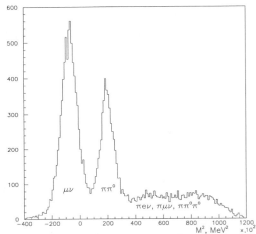

Figure 4: Missing squared mass distribution for selected events.

mentum and the direction of the decayed kaon are completely determined by tagging kaon and missing mass can be calculated assuming decay particle to be pion. Figure 4 presents distribution of calculated squared missing masses in the detected two track systems. Peaks are corresponding to two body decays $K^{\pm} \rightarrow \mu^{\pm}\nu$ and $K^{\pm} \rightarrow \pi^{\pm}\pi^{0}$. Using number of events under the peaks and with the efficiencies obtained by simulation the following ratio has been obtained:
$\Gamma(K^{\pm} \rightarrow \mu^{\pm}\nu)/\Gamma(K^{\pm} \rightarrow \pi^{\pm}\pi^{0}) = 0.340 \pm 0.008 \pm 0.017$.
This result should be compared with one listed in PDG tables[7]:
$\Gamma(K^{\pm} \rightarrow \mu^{\pm}\nu)/\Gamma(K^{\pm} \rightarrow \pi^{\pm}\pi^{0}) = 0.3316 \pm 0.0032$.

The broad distribution in Figure 4 represents the three body decays of charged kaons to $\pi^{0}\mu\nu$, $\pi^{0}e\nu$ and $\pi\pi^{0}\pi^{0}$. The reconstruction of π^{0}'s from detected photons can help to separate channels mentioned above. In

the present analysis the different calorimeter response for electrons and mesons was used. The charged particles

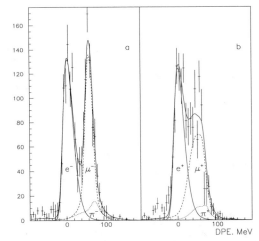

Figure 5: The DPE parameter distributions for negative (a) and positive (b) decay particles.

from three body decays have low momenta and stop in the barrel CsI calorimeter. It was found that average energy deposition for mesons is close to their kinetic energy and a parameter
DPE $= P \cdot c - \Delta E - E_{cluster}$
has weak dependence on momentum and gives relatively sharp peak at 60 MeV. In this parameter P is the particle momentum measured in DC, ΔE - ionization energy loss in material in front of CsI calorimeter, $E_{cluster}$ - energy deposition in CsI-cluster matched with the particle track. The shape of the DPE was studied with pions and muons from two body decays and the functions fitting DPE for positive and negative mesons were found. The average value of the DPE parameter for electrons is close to zero and electrons could be separated from mesons in the DPE distribution.

Figures 5a,b present the DPE distribution for negative and positive particles for events with squared missing mass $M_{miss} > 3.5 \cdot 10^{4}$ selecting three body decays. Lines present gaussian fit functions used for separation of electrons from mesons. Electrons from semileptonic decays are relatively good separated and 1503 ± 52 events have been found. Muon and pion distributions overlaps and more sophisticated analysis including information from detected photons is under preparation now to separate all decay channels.

With the found number of electrons the following rel-

ative decay rate has been found:
$\Gamma(K^{\pm} \to e^{\pm}\pi^0\nu)/(\Gamma(K^{\pm} \to \mu^{\pm}\nu) + \Gamma(K^{\pm} \to \pi^{\pm}\pi^0)) = 0.0578 \pm 0.0020 \pm 0.0020$.

This result should be compared to the values 0.0601 ± 0.0015 (average) or 0.0570 ± 0.0008 (fit with S=1.4) listed in the tables[7].

4 Observation of the $K^0_S \to \pi e\nu$ decay

Candidates were selected by requirement of cluster in the CsI barrel calorimeter and two prong vertex in the DC detected at the other side of the beam axis. The vertex was required to be inside vacuum pipe but at least 0.2 cm from the beam. Under these conditions about 46500 of K^0_S decays into $\pi^+\pi^-$ mode were selected. These events were used for normalisation. To separate three

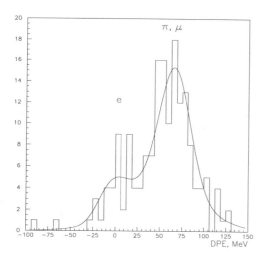

Figure 6: The DPE parameter distribution for selected K^0_S three body decays.

body decays from major decay modes the additional cuts on opening angle and sum of particle energies calculated from measured momenta was applied. The distribution over DPE parameter described in the previous section for the selected events is shown in Figure 6. The data were fitted using DPE shapes for e, μ and π found from $e^+e^- \to e^+e^-, \mu^+\mu^-$ events with beam energy 195 MeV and from $K^0_S \to \pi^+\pi^-$ events. As the result of the fit the number of the electrons $N_e = 20.0 \pm 6.8$ has been found. The ratio of N_e to the number of pions and muons $N_\mu + N_\pi$ was found to be 1:7.5 and was in good agreement with the expectation taking into account major K^0_S

three body decays[7]. The corresponding branching ratio was found to be
$B(K^0_S \to \pi e\nu)=(7.8 \pm 2.7) \times 10^{-4}$.

This result is consistent with one presented in PDG tables[7] obtained by calculation from K^0_L semileptonic rate and K^0_S lifetime assuming $\Delta S = \Delta Q$.

Acknowledgements

This work is supported in part by the grant INTAS 96-0624.

References

1. For the full list of CMD-2 collaboration see article "Investigation of the rare ϕ radiative decays with the CMD-2 detector at VEPP-2M collider" in this Proceedings.
2. V.N.Bayer, Russian JETP **17**, 446 (1973).
3. G.A.Aksenov *et al.*, Preprint BudkerINP 85-118, Novosibirsk, 1985.
4. V.V. Anashin *et al.*, Preprint BudkerINP 84-114, Novosibirsk, 1984.
5. J.L. Rosner, I. Dunietz, J. Hauser *Phys. Rev.* D **35**, 2166 (1987).
6. R.R. Akhmetshin *et al.*, *Phys. Lett.* B **398**, 423 (1997).
7. *European Physical Journal* **C3**, *1 (1998)*.
8. E.V. Anashkin *et al.*, ICFA Instrumentation Bulletin 5 (1988), p.18.
9. R .R. Akhmetshin *et al.*, Preprint BudkerINP 95-62, Novosibirsk, 1995.
10. R. Baldini, A. Michetti, Preprint LNF-96/008 (1996).

A NEW SEARCH FOR DIRECT CP VIOLATION IN HYPERON DECAYS

K. S. NELSON

Dept. of Physics, University of Virginia, 205 McCormick Rd., Charlottesville, Virginia, USA

Representing the HyperCP Collaboration

The HyperCP experiment [1] (Fermilab E871) is performing a search for CP violation in Ξ and Λ hyperon decays by comparing the slopes of the p and $\bar{p}$ angular distributions in the Λ and $\bar{\Lambda}$ helicity frames. A difference between these slopes would signal direct CP violation. Theoretical predictions of the slope asymmetry range from 0 up to about 10^{-3}. The expected experimental sensitivity using data from a run in 1997 is $\sim 2 \times 10^{-4}$. A future run planned for 1999 will push the sensitivity below 10^{-4}.

1 Introduction

Since the discovery of CP violation in 1964, observations of the effect have been made only in a few decay modes of the K_L meson. The source of CP violation in these decays is still unknown and the disentanglement of the $\Delta S = 1$ (direct) and $\Delta S = 2$ (mixing) contributions remains an active subject of investigation [2].

In the standard model CP violating effects arise from a single complex phase in the CKM matrix. If our theoretical prejudices are correct then CP violation should also be seen in other weak decays, *e.g.* in the decays of B mesons. The non-leptonic decays of hyperons offer an alternate and unique venue to look for CP violation outside of the K system. Any such effects would necessarily signal *direct* CP violation. The connection between hyperon CP violating observables and standard model parameters is not clean at the moment due to uncertainties in evaluating hadronic matrix elements and strong final state phase shifts. On the other hand, recent developments in high rate detectors and data acquisition systems have made it feasible in a reasonably short run to reach sensitivities where some model calculations predict CP violating effects to occur.

2 Theory of non-leptonic hyperon decays

The non-leptonic decay of a spin-$\frac{1}{2}$ hyperon into a specific daughter spin-$\frac{1}{2}$ baryon and pion proceeds by an admixture of S- and P-waves. The transition rate [3] is completely described by three mode-specific parameters [a] governing the angular distribution of the baryon;

$$\alpha = \frac{2\mathrm{Re}(S^*P)}{|S|^2 + |P|^2},$$
$$\beta = \frac{2\mathrm{Im}(S^*P)}{|S|^2 + |P|^2}, \qquad (1)$$
$$\gamma = \frac{|S|^2 - |P|^2}{|S|^2 + |P|^2},$$

---

[a] Only two of these are independent since $\alpha^2 + \beta^2 + \gamma^2 = 1$.

and the decay rate into that mode, Γ .

The parameter α expresses the parity violation in the decay of a polarized hyperon in the case when the spin of the daughter baryon is not observed. The angular distribution is given by:

$$\frac{1}{N_d}\frac{dN_d}{d\Omega} = \frac{1}{2\pi}\left(1 + \alpha_p \vec{P}_p \cdot \hat{p}_p\right) \qquad (2)$$

where $\vec{P}_p$ is the parent polarization and $\hat{p}_p$ is the unit momentum vector of the daughter baryon. Each of the three parameters in turn are associated with a component of the polarization of the daughter baryon along one of three orthogonal axes in its rest frame:

$$\vec{P}_d =$$

$$\frac{(\alpha_p + \vec{P}_p \cdot \hat{p}_d)\hat{p}_d + \beta_p(\vec{P}_p \times \hat{p}_d) + \gamma_p(\hat{p}_d \times (\vec{P}_p \times \hat{p}_d))}{(1 + \alpha_p \vec{P}_p \cdot \hat{p}_d)}. \qquad (3)$$

Note that in the decay of an *unpolarized* hyperon the daughter is produced in a helicity state with polarization given by the parent's α parameter. The parameter β is associated with the component of the daughter's polarization which is perpendicular to the decay plane defined by the directions of the parent's polarization and the daughter's momentum. Assuming CP conservation in the decay, these parameters of the antihyperon decay (indicated by over bar) are related to those of the hyperon decay by:

$$\bar{\alpha} = -\alpha$$
$$\bar{\beta} = -\beta . \qquad (4)$$

3 Signatures for Hyperon CP Violation

Donoghue et al. [4] have considered several asymmetries sensitive to CP odd effects that can be formed from the above mentioned observables. Three, which are of inter-

Table 1: Some predictions for A_Ξ and A_Λ.

Model	A_Ξ $[10^{-4}]$	A_λ $[10^{-4}]$	Ref.
CKM	$-(0.1 - 1)$	$-(0.1 - 0.5)$	5
Weinberg	≈ -3.2	≈ -0.25	4
Multi-Higgs	≈ 0	≈ 0	5
LR (mixing)	< 1	< 7	6

est here, are:

$$A = \frac{\alpha + \overline{\alpha}}{\alpha - \overline{\alpha}}$$
$$B = \frac{\beta + \overline{\beta}}{\beta - \overline{\beta}} \qquad (5)$$
$$\Delta = \frac{\Gamma - \overline{\Gamma}}{\Gamma + \overline{\Gamma}} .$$

Their model independent analysis reveals a hierarchy of these asymmetries, $B \sim 10A \sim 100\Delta$. The advantage of pursuing a measurement of a potentially large value of B is somewhat offset by the experimental difficulties in measuring both the parent and daughter polarizations to the required accuracy. Although A is suppressed relative to B by small final state interaction phase shifts, it has the advantage of requiring precise knowledge of only the parent polarization. It is mainly for this reason that the only asymmetry reported in the literature up to now is A_Λ.

Calculations of the CP asymmetries in hyperon decays are not considered to be reliable at the present time due to the uncertainty in evaluating the hadronic matrix elements and incomplete knowledge of some (*e.g.* $\Lambda\pi$) final state phase shifts. The magnitude of CP violating effects are also model dependent, as indicated by Table 1 which lists recent predictions for the CP asymmetries A_Ξ and A_Λ.

4 Previous Experimental Searches

In order to measure the asymmetry A, one needs samples of hyperons and antihyperons possessing precisely known polarizations. Previous experimental approaches with $\Lambda/\overline{\Lambda}$ hyperons made use of exclusive production,[8,9] where symmetry arguments could be used to infer equal Λ and $\overline{\Lambda}$ polarizations, or assumed an equal dependence of polarization on kinematic quantities[7]. The sensitivities of these approaches were limited by statistics. The best limit of $\delta A_\Lambda \approx 2 \times 10^{-2}$, coming from PS185, is still about two orders of magnitude above the theoretical predictions in Table 1.

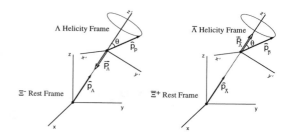

Figure 1: Λ helicity frame.

5 The HyperCP Experiment

The HyperCP experiment obtains Λ and $\overline{\Lambda}$ hyperons of precisely known polarization from the decay of Ξ^- and $\overline{\Xi}^+$ hyperons which are copiously produced at zero degrees in hadronic collisions. Assuming parity conservation of the strong interaction, such Ξ hyperons are unpolarized, and from Eq. 3 the daughter Λ and $\overline{\Lambda}$ hyperons are found in helicity states with polarizations of $P_\Lambda = \alpha_\Xi$ and $P_{\overline{\Lambda}} = \alpha_{\overline{\Xi}}$. In Ξ^- decay the angular distribution of the p in the Λ rest frame (see Fig. 1), whose z axis is defined by the direction opposite to the Ξ^- direction, is given by:

$$\frac{1}{N_p}\frac{dN_p}{d\Omega} = \frac{1}{2\pi}(1 + \alpha_\Lambda \vec{P}_\Lambda \cdot \hat{p}_p)$$
$$= \frac{1}{2\pi}(1 + \alpha_\Lambda \alpha_\Xi \hat{p}_\Lambda \cdot \hat{p}_p)$$
$$= \frac{1}{2\pi}(1 + \alpha_\Lambda \alpha_\Xi \cos\theta). \qquad (6)$$

A similar expression is obtained for the $\overline{p}$ distribution in $\overline{\Xi}^+$ decay. If these decays violate CP symmetry then $\alpha_\Lambda \alpha_\Xi \neq \alpha_{\overline{\Lambda}} \alpha_{\overline{\Xi}}$, *i.e.* the the slopes of the two distributions are different. The asymmetry that HyperCP measures is thus a composite slope asymmetry;

$$A_{\Xi\Lambda} \equiv \frac{\alpha_\Xi \alpha_\Lambda - \alpha_{\overline{\Xi}} \alpha_{\overline{\Lambda}}}{\alpha_\Xi \alpha_\Lambda + \alpha_{\overline{\Xi}} \alpha_{\overline{\Lambda}}} \approx A_\Xi + A_\Lambda . \qquad (7)$$

It is not possible to distinguish, with this measurement alone, whether it is the Ξ or Λ decay or both which is responsible for an asymmetry. Also note that cancellation is unlikely since most models predict that A_Ξ and A_Λ have the same sign.

A plan view of the HyperCP spectrometer is shown in Fig. 2. Protons of $800\,\text{GeV}$ energy are directed onto a target which is immediately followed by a curved collimator having $4.88\,\mu r$ acceptance embedded in a $6\,\text{m}$ long

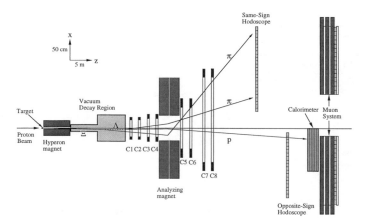

Figure 2: Plan view of the HyperCP apparatus.

dipole magnet. Charged particles of $\sim 157\,\text{GeV/c}$ follow the central orbit of the collimator which bends upward (out of the page) by 19.5 mrad. At this momentum the Ξ to charged particle rate is nearly maximal. Following the vacuum decay volume is a conventional magnetic spectrometer employing high-rate, narrow pitch wire chambers. The spectrometer terminates downstream in a muon system to allow rare and forbidden hyperon decays to be studied. The spectrometer analyzing magnet provides a sufficient momentum kick that the proton from the Λ decay is always deflected to one side of the spectrometer and the two pions from the Ξ and Λ decays are deflected to the other side. Furthermore, both are well separated at the last chamber from the intense charged beam emanating from the collimator. To go from Ξ^- to $\overline{\Xi}^+$ running the polarities of both the collimator magnet and the spectrometer analyzing magnet are reversed. Then, up to the small difference between particle and anti particle cross sections in normal matter, the apparatus and running conditions are CP invariant. In typical running a primary beam intensity on target of 7.5×10^9 protons per second yields a secondary beam of $\sim 20\,\text{MHz}$ issuing from the exit of the collimator. A simple yet selective trigger is formed by requiring the coincidence of at least one charged particle in scintillator hodoscopes on each side of the spectrometer downstream of the analyzing magnet. A calorimeter on the "proton" side is used to suppress triggers due to muons and secondary interactions of the collimator beam with material in the spectrometer. The raw data are recorded with a high rate data acquisition system [10] which is capable of transporting 20 MB/s of data to tape.

6   Status of the Analysis

From the data collected in the 4 month run in 1997 we expect to reconstruct 1.6 billion $\Xi \to \Lambda\pi \to p\pi\pi$ decays and 280 million $K^\pm \to 3\pi$ (all charged) decays, as well as a large number of $\Omega \to \Lambda K$ and $K_S \to \pi^+\pi^-$ decays. This quantity of events translates to a statistical uncertainty on the combined asymmetry of $\delta A_{\Xi\Lambda} \approx 2 \times 10^{-4}$.

The first pass analysis involving particle trajectory reconstruction and some minimal topological fitting to various hypotheses is expected to be completed by the middle of 1999. Preliminary analysis of a few percent of the data set has been completed at this time. Fig. 3 shows the $p\pi^-\pi^-$ invariant mass from negative running and the $\bar{p}\pi^+\pi^+$ mass from positive running. Without corrections, the resolution is about $1.6\,\text{GeV/c}^2$. The good agreement between the Ξ^- and $\overline{\Xi}^+$ mass peaks and the small underlying background demonstrates the degree of CP symmetry of the apparatus and the running conditions.

To measure the CP asymmetry $A_{\Xi\Lambda}$ to $\sim 10^{-4}$ will require an understanding and control of biases that would affect the Ξ^- and $\overline{\Xi}^+$ differently and nonuniformly over the $\cos\theta$ distribution. One strength of the analysis technique rests on the fact that the polar axis of the Λ helicity frame changes from event to event so that acceptance differences localized to a particular part of the apparatus do not map onto a particular region of the $\cos\theta$ distribution. To illustrate this effect we plot in Fig. 4 angular distributions of the proton with respect to the (fixed) laboratory y axis from two Ξ^- data sets taken at different production angles ($\pm 2.5\,\text{mrad}$) such that the produced Ξ^- polarizations are parallel and antiparallel to this axis. The

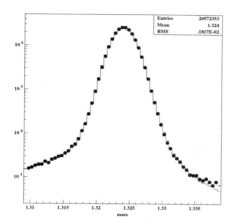

Figure 3: The $\Lambda\pi$ invariant mass for positive and negative Ξ trigger events.

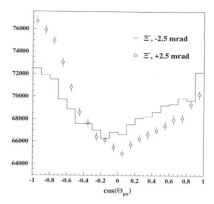

Figure 4: The proton $\cos\theta$ distributions along the y-axis in the Λ rest frame for Ξ^- events from ± 2.5 mrad production angles.

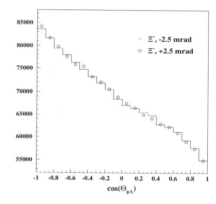

Figure 5: The proton $\cos\theta$ distributions in the Λ helicity frame for Ξ^- events from ± 2.5 mrad production angles.

two data sets clearly have different and non-uniform acceptances. Nevertheless, in the lambda helicity frame the two $\cos\theta$ distributions are essentially identical, as seen in Fig. 5. Studies have shown that without corrections most biases taken individually should produce false asymmetries in $A_{\Xi\Lambda}$ of less that 10^{-4}.

7 Future Prospects

The HyperCP collaboration is preparing for a second run in 1999 where modest upgrades to the apparatus and improvements in the accelerator duty factor should enable a four-fold increase in statistics resulting in a sensitivity in $A_{\Xi\Lambda}$ of better than 1×10^{-4}. This is at a level where some models predict CP violating effects will begin to occur.

Acknowledgements

We would like to thank the Fermilab staff and management for helping us build the apparatus and beam in less than one year and for providing good running conditions. This work is supported in part by the U.S. Department of Energy.

References

1. The HyperCP collaboration is: R.A. Burnstein, A. Chakravorty, A. Chan, Y.C. Chen, W.S. Choong, K. Clark, M. Crisler, E.C. Dukes, C. Durandet, J. Felix, G. Gidal, H.R. Gustafson, C. Ho, T. Holmstrom, M. Huang, C. James, M. Jenkins, D.M. Kaplan, L.M. Lederman, N. Leros, M.J. Longo, F. Lopez, G. Lopez, W. Luebke, K.B. Luk, K.S. Nelson, J.P. Perroud, D. Rajaram, H.A. Rubin, J. Sheng, M. Sosa, P.K. Teng, B. Turko, J. Volk, C.G. White, S.L. White, C. Yu, Z. Yu, and P. Zyla.

2. see the talks by H. Bluemer and E. Cheu, these proceedings.

3. see for example *Weak Interactions of Quarks and Leptons*, E.D. Cummins and P.H. Bucksbaum, Cambridge University Press (1983).

4. J.F. Donoghue and S. Pakvasa, Phys. Rev. Lett. **55** (1985) 162; J.F. Donoghue, X.-G. He, and S. Pakvasa, Phys. Rev. **D34** (1986) 833.

5. X.-G. He and S. Pakvasa, in Proceedings of the 8th Meeting Division of Particles and Fields of the American Physical Society, edited by S. Sei-

del (World Scientific, Singapore, 1995), pp. 984 – 990.

6. D. Chang, X.-G. He, and S. Pakvasa, Phys. Rev. Lett. **20** (1995) 3927.

7. P. Chauvat *et al.*, Phys. Lett. **B163** (1985) 273.

8. M.H. Tixier *et al.*, Phys. Lett. **B212** (1988) 523.

9. P.D. Barnes *et al.*, Phys. Rec **C54** (1996) 1877.

10. D.M. Kaplan *et al.*, Proceedings of the Sixth Annual LeCroy Conference on Electronics for Particle Physics, Chestnut Ridge, NY, May 28–29, 1997, ed. by G.J. Blanar and R.L. Sumner, LeCroy Research Systems.

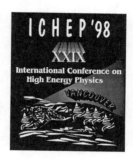

Parallel Session 7

Production and Decay of Heavy Quarks and Onia

Convenors: Mark Strovink (LBNL, DO)

Mark Wise (Caltech, Theory)

UPDATE ON $b \to s\gamma$ AND $b \to s l^+ l^-$ FROM CLEO

T.SKWARNICKI

Syracuse University, Department of Physics, Syracuse, NY 13244, USA
E-mail: tomasz@physics.syr.edu

(Representing The CLEO Collaboration)

Improved measurement of inclusive $b \to s\gamma$ rate, $(3.15 \pm 0.35 \pm 0.41) \times 10^{-4}$, and updated searches for $b \to s l^+ l^-$ in inclusive and exclusive modes are presented.

1 Measurement of $b \to s\gamma$

1.1 Brief History of $b \to s\gamma$ Measurements

The first evidence for $b \to s\gamma$ was obtained by CLEO in 1993 via observation of $B \to K^* \gamma$ in a sample of 1.6×10^6 $b\bar{b}$ pairs.[1] The branching ratio for this exclusive process was measured with precision of $\pm 1.7 \times 10^{-5}$. Using enlarged sample of 2.6×10^6 $b\bar{b}$ pairs, CLEO improved the precision of this measurement to $\pm 1.0 \times 10^{-5}$, with about equal contribution from the statistical and systematic uncertainties.[2]

Theoretical predictions for exclusive branching ratios suffer from large uncertainties in estimates of hadronization probabilities. Therefore, measurements of the inclusive $b \to s\gamma$ rate are more interesting, but also more difficult experimentally. The inclusive $b \to s\gamma$ rate was first measured by CLEO in 1994-95[3] in a sample of 2.2×10^6 $b\bar{b}$ pairs with precision of $\pm 6.7 \times 10^{-5}$, limited by the statistical fluctuations in the background level ($\pm 5.7 \times 10^{-5}$). Recently also the ALEPH experiment has measured the inclusive rate[4], however, with a larger error of $\pm 10.8 \times 10^{-5}$ due to smaller number of $b\bar{b}$ pairs (0.8×10^6).

In this article we present update of the CLEO measurement, based on increased statistics (3.3×10^6 $b\bar{b}$ pairs) and improved analysis techniques.[5] The statistical precision of the measurement is substantially improved, $\pm 3.5 \times 10^{-5}$, and for the first time becomes smaller than the systematic error, $\pm 4.1 \times 10^{-5}$.

The experimental errors should be compared to theoretical uncertainty in the Standard Model predictions[6] of $\pm 3.3 \times 10^{-5}$.

1.2 Signal Selection Tools

Photons produced in $b \to s\gamma$ decays are quasi-monochromatic ($E_\gamma \sim m_b/2 \sim 2.5$ GeV). Binding of b quark to the spectator quark in the B meson and QCD corrections smear the photon energy. Further smearing in the laboratory frame is produced by small, but non-zero momentum of B mesons from $\Upsilon(4S) \to B\bar{B}$ decays ($P_B \sim 0.34$ GeV). In this analysis we require 2.1 GeV$< E_\gamma < 2.7$ GeV.

Since the dominant $b \to c$ decays produce softer photons, B meson backgrounds are small. Furthermore, these backgrounds are dominated by π^0 and η decays, thus they can be controlled from the data themselves by explicit π^0 and η reconstruction.

Much larger background comes from continuum production in which photon energies are not limited by the B meson mass. There are two types of continuum background; π^0 and η decays and initial state radiation. Both types of continuum background are subtracted using below-$\Upsilon(4S)$ data. Fluctuations in on- and off-resonance continuum background yield dominate the statistical error on the $b \to s\gamma$ signal, thus suppression of the continuum background is the main experimental challenge.

We veto reconstructed π^0 and η mesons to reduce the continuum and B backgrounds. Further suppression of the continuum backgrounds is achieved by use of topological differences in continuum and signal event shapes (Figure 1). The B mesons are produced right above the kinematic threshold, thus they decay almost at rest, and produce spherical events. Continuum quark-pairs are produced well above their kinematic thresholds, thus the quarks hadronize into back-to-back jets. Initial state radiation makes continuum events more spherical, however, their back-to-back nature can be restored by boosting the event to the rest frame of the system recoiling against the energetic photon. Various shape variables are used to discriminate between the signal and two types of continuum background. A sample of shape variables is displayed in Figure 2. Each variable contains some separation of the signal and backgrounds, but the separation is not complete. There are also strong correlations between the shape variables. For the most efficient background suppression, we use neural network technique to combine 8 different shape variables into just one, $\mathbf{r}_{\text{SHAPE}}$. By definition, $\mathbf{r}_{\text{SHAPE}}$ peaks at $+1$ for the signal events and at -1 for the continuum background. Use of event shape information alone results in large signal efficiency ($\sim 50\%$) but also in relatively large background level.

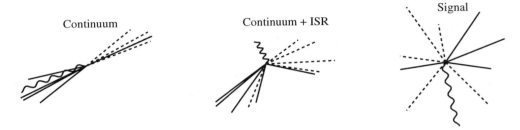

Figure 1: Continuum background and $b \to s\gamma$ signal event shapes.

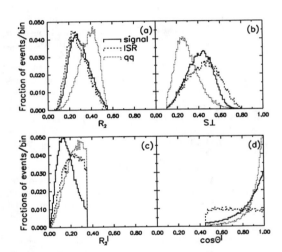

Figure 2: Distributions of various shape variables for simulations of $b \to s\gamma$ signal (solid), $q\bar{q}$ continuum (dotted) and the continuum with Initial State Radiation (dashed). (a) R_2 – the normalized second Fox-Wolfram moment. (b) $S_\perp$ – sum of the magnitudes of the momenta transverse to the photon direction. (c) R'_2 – the normalized second Fox-Wolfram moment evaluated in the rest frame of the system recoiling against the photon. (d) $\cos\theta'$ – the cosine of the angle between the photon and the thrust axis of the rest of the event in the boosted frame.

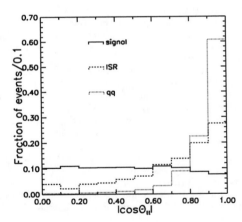

Figure 3: Distribution of the angle between the thrust axes for the reconstructed B meson candidate and the rest of the event for simulations of the $b \to s\gamma$ signal (solid), $q\bar{q}$ continuum (dotted) and the continuum with Initial State Radiation (dashed).

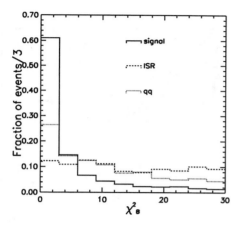

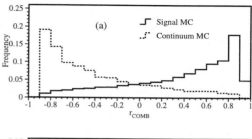

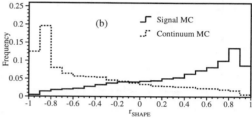

Figure 4: Distribution of χ_B^2, implementing the beam energy and B meson mass constraints, for simulations of the $b \to s\gamma$ signal (solid), $q\bar{q}$ continuum (dotted) and the continuum with Initial State Radiation (dashed).

Figure 5: Distribution of the neural net variables, (a) $\mathbf{r}_{\mathrm{COMB}}$ and (b) $\mathbf{r}_{\mathrm{SHAPE}}$, for Monte Carlo samples of $b \to s\gamma$ signal (solid histograms) and continuum background (dashed histograms) for: (a) events with reconstructed B meson candidate; (b) events with no reconstructed B meson candidate.

More stringent continuum background suppression can be achieved using B reconstruction technique, which requires detection of the other B decays products (i.e. of the X_s system) in $B \to X_s\gamma$ decays. Then, beam energy constraint, $E_{X_s\gamma} = E_{beam}$, and B mass constraint, $m_{X_s\gamma}^2 = E_{beam}^2 - P_{X_s\gamma}^2 = m_B$, can be used. Cosine of the angle between the thrust direction for the $X_s\gamma$ system and the thrust direction for the rest of event, $\cos\theta_{tt}$, also proves to be a sensitive variable, since the signal has a flat distribution, whereas the continuum strongly peaks at ± 1 (see Figure 3). Even limited kaon identification provides further rejection of continuum events. When applied to low multiplicity modes, like $B \to K^*\gamma$, the B reconstruction results in good signal efficiency and excellent background suppression. The B reconstruction can also be employed to measure inclusive rate by allowing a large variety of possible final states. For X_s system, we select a kaon ($K^\pm$ or $K_s \to \pi^+\pi^-$) and $1 - 4$ pions including up to one π^0. Signal efficiency is lower because of neglected decay modes and the large number of particles to be reconstructed in the considered modes. Due to large combinatorics, background suppression is not as good as in low multiplicity exclusive modes, but much more stringent than using event shape information alone.

For the best effectiveness in background suppression, B reconstruction variables, $\chi_B^2 \equiv (m_{X_s\gamma} - m_B)^2/\sigma^2 + (E_{X_s\gamma} - E_{beam})^2/\sigma^2$ (Figure 4) and $\cos\theta_{tt}$ (Figure 3), and event shape variable, $\mathbf{r}_{\mathrm{SHAPE}}$, are combined using neural network algorithm into one variable, $\mathbf{r}_{\mathrm{COMB}}$. Since only

about 20% of signal events can be reconstructed, we use $\mathbf{r}_{\mathrm{SHAPE}}$ alone for the rest of events. Distributions of $\mathbf{r}$ variables are shown for the signal and continuum Monte Carlo events in Figure 5.

1.3 Signal Extraction Technique

For the most efficient use of data, we do not cut on $\mathbf{r}$ variables, but unfold signal amplitude using the expected signal and background shapes.

For convenience we use event weighting rather than fitting. The event weight is a function of $\mathbf{r}$, and is defined as $w(\mathbf{r}) = s(\mathbf{r})/\Delta^2 s(\mathbf{r})$, where $s(\mathbf{r})$ is the expected signal yield assuming $\mathcal{B}(b \to s\gamma) = 3 \times 10^{-4}$, and $\Delta s(\mathbf{r})$ is the expected error on the signal yield. The latter is calculated from $\Delta^2 s(\mathbf{r}) = s(\mathbf{r}) + (1 + \alpha)b(\mathbf{r})$. Here $b(\mathbf{r})$ is the expected continuum background yield, and α is the luminosity scaling factor (~ 2) between the on- and off-resonance samples to account for subtraction of the off-resonance data. Weighted event yields, $N_w = \sum_{data} w(\mathbf{r})$, and weighted efficiency, $\epsilon_w = \sum_{MC} w(\mathbf{r})/N_{MC}^{generated}$, are calculated. Finally we calculate:

$$\mathcal{B}(b \to s\gamma) = (N_w^{on} - \alpha N_w^{off} - N_w^{B\bar{B}})/(2N_{b\bar{b}}\epsilon_w)$$

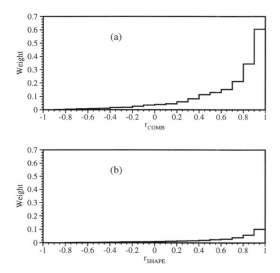

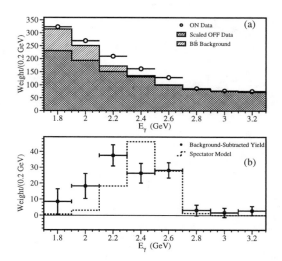

Figure 6: Event weights as function of the neural net variables for: (a) events with reconstructed B meson candidate; (b) events with no reconstructed B meson candidate.

Figure 7: Photon energy spectra. (a) On-resonance (points), scaled off-resonance (double-hatched), and background from $B\bar{B}$ (single hatched). (b) Background subtracted data (points) and Monte Carlo prediction for the shape of the $b \to s\gamma$ signal (dashed curve) for the model of Ali and Greub [7] with $<m_b> = 4.88$ GeV and $P_F = 250$ MeV/c. Error bars indicate statistical uncertainties only.

The weighting procedure automatically takes care of proper combination of results from the reconstructed and unreconstructed events. As illustrated in Figure 6, reconstructed events receive much larger weights because of better background suppression.

To estimate the B meson background, $N_w^{B\bar{B}}$, we measure π^0 and η high momentum yields in the data. We use Monte Carlo only to predict a fraction of the π^0 and η yields contaminating the photon spectrum (we use a transfer matrix indexed by E_γ and $P_{\pi^0,\eta}$). We also rely on Monte Carlo to predict small background contributions from sources other than π^0 and η.

1.4 Results

Inclusive photon spectrum from B decays is shown in Figure 7. In the $2.1 - 2.7$ GeV interval we obtain: $N_w^{on} = 500.5 \pm 7.4$, $\alpha N_w^{off} = 382.3 \pm 7.0$, $N_w^{B\bar{B}} = 26.1 \pm 1.5$, thus $N_w^{signal} = 92.2 \pm 10.3$ (statistical significance of 9 standard deviations!). Using model of Ali and Greub [7] to estimate number of photons outside the $2.1 < E_\gamma < 2.7$ GeV range and using detector simulation to estimate selection losses we calculate $\epsilon_w = (4.4 \pm 0.4 \pm 0.3) \times 10^{-2}$. The first error is systematic due to detector simulation and modeling of X_s hadronization. The second error is theoretical uncertainty in the predicted $b \to s\gamma$ photon

spectrum. From these numbers we obtain:

$$\mathcal{B}(b \to s\gamma) = (3.15 \pm 0.35 \pm 0.32 \pm 0.26) \times 10^{-4}$$

where the first error is statistical and the other two errors are explained above. Model dependence can be reduced by calculating the rate observed within the $2.1 - 2.7$ GeV photon energy interval:

$$\mathcal{B}(b \to s\gamma)_{2.1<E_\gamma<2.7\text{GeV}} = (2.97 \pm 0.33 \pm 0.30 \pm 0.21) \times 10^{-4}$$

This results is compared to the theoretical prediction and to the previous measurements in Table 1. The agreement with the Standard Model prediction is very good.

There are number of improvements in our analysis compared to our previously published result. We have used 50% more data. We have made more efficient use of information in the data by using expected signal and backgrounds shapes in χ_B^2 and $\cos\theta_{tt}$ variables, and by integrating the B reconstruction and event shape approaches. We have also improved estimates of $B\bar{B}$ backgrounds, which gave us confidence to extend the signal region in E_γ from $2.2 - 2.7$ GeV to $2.1 - 2.7$ GeV in order to decrease our dependence on the theoretical predictions for the photon spectrum. The last change made

Table 1: Branching ratio measurements for $b \to s\,\gamma$ decays.

Year	Experiment	$N_{b\bar{b}}$	Error		Value
			Stat.	Sys.	
		$\times 10^6$		$\times 10^{-6}$	
Exclusive $B \to K^*\gamma$					
1993	CLEO [1]	1.5	± 15	± 9	44
1996	CLEO [2]	2.6	± 8	± 6	42
Inclusive $b \to s\,\gamma$					
1994/95	CLEO [3]	2.2	± 57	± 35	232
1997/98	ALEPH [4]	0.8	± 80	± 72	311
1998	CLEO [5]	3.3	± 35	± 41	315
Standard Model [6]			± 33		328

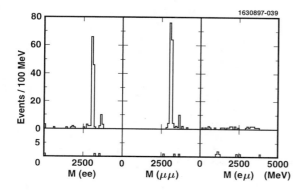

Figure 8: $M(l^+l^-)$ distributions for the on- (top) and off-resonance (bottom) data in the search for inclusive $b \to s\,l^+l^-$ decays.

the biggest difference in the central value of the measured branching ratio. As argued by Neubert in the other contribution to this conference, this may be understood on theoretical grounds. [8]

2 Search for $b \to s\,l^+l^-$

2.1 Inclusive search for $b \to s\,l^+l^-$

CLEO has search for $b \to s\,l^+l^-$ decays using the inclusive B reconstruction technique. [9] The resulting di-electron and di-muon mass spectra are shown in Figure 8. Prominent ψ and ψ' peaks are seen which stem from $b \to c$ decays. The remaining of events are consistent with the background estimates (unlike in $b \to s\,\gamma$ process the background here is predominantly from $B\bar{B}$ events). The upper limits are summarized in Table 2.

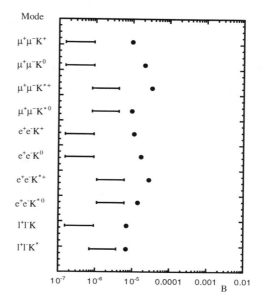

Figure 9: Comparison of the obtained 90% C.L. upper limits on branching ratios for various exclusive $b \to s\,l^+l^-$ decay modes (circles) and the theoretical predictions (solid lines).

2.2 Search in exclusive modes

We are also presenting an updated search for for exclusive $B \to K\,l^+l^-$ and $B \to K^*l^+l^-$ decays based on 3.3×10^6 $b\bar{b}$ pairs. [10] Exclusive B reconstruction and event shape cuts are used. All together we have used 12 different final states: ($K^\pm$ or $K_s \to \pi^+\pi^-$)+ ($\pi^\pm$ or π^0)+ (e^+e^- or $\mu^+\mu^-$). Combining various channels together we obtain the limits listed in Table 2. Limits in individual channels[a] are graphically displayed in Figure 9, where they are also compared to the range of theoretical predictions.

3 Summary

We have presented improved measurement of the inclusive $b \to s\,\gamma$ rate. The measured branching ratio is in excellent agreement with the theoretical predictions imposing limitations on the possible extensions of the Standard Model.

Searches for $b \to s\,l^+l^-$ decays yield only upper limits so far. The inclusive limit is a factor ~ 6 away

---

[a]The numerical results can be found in the paper submitted to this conference. [10]

Table 2: Searches for $b \to s\, l^+l^-$ decays.

Experiment	$\mathcal{B}$ in units of 10^{-5}		
	e^+e^-	$\mu^+\mu^-$	$^\dagger\ l^+l^-$
Exclusive $B \to K\, l^+l^-$			
CLEO [10]	< 0.9	< 0.8	< 0.7
CDF [11]		< 1.0	
Standard Model [12]	$0.03 - 0.07$		
Exclusive $B \to K^*l^+l^-$			
CLEO [10]	< 1.3	< 1.1	$^\ddagger\ < 0.7$
CDF [11]		< 2.5	
Standard Model [12]	$0.1 - 0.4$		
Inclusive $b \to s\, l^+l^-$			
CLEO [10]	< 5.7	< 5.6	< 4.2
D0 [13]		< 32.0	
Standard Model [14]	0.8 ± 0.2	0.6 ± 0.1	0.7 ± 0.2

† By definition $\mathcal{B}(l^+l^-) = (\mathcal{B}(e^+e^-) + \mathcal{B}(\mu^+\mu^-))/2$.
‡ With $m_{l^+l^-} > 0.5$ GeV cut to minimize virtual
$b \to s\,\gamma$ contribution.

from the Standard Model predictions. The limit on $B \to K\, l^+l^-$ is a factor ~ 10 away, and the limit on $B \to K^*l^+l^-$ is within a factor of ~ 3 from the upper range of theoretical calculations.

The updates on inclusive $\mathcal{B}(b \to s\,\gamma)$, and searches for $B \to K\, l^+l^-$ and $B \to K^*l^+l^-$ are preliminary. A factor of two more data is already being processed in the CLEO analysis pipeline. Thus, further updates are expected in near future.

Acknowledgements

I would like to thank Gerald Eigen, Jesse Ernst, Steve Glenn, Ed Thorndike and Sheldon Stone for help in preparation of this article.

References

1. R. Ammar et al. (CLEO Collaboration), Phys. Rev. Lett. **71**, 674 (1993).
2. R. Ammar et al. (CLEO Collaboration), Report No. CLEO CONF 96-05, contributed paper to The 28th International Conference on High Energy Physics, Warsaw, Poland (1996).
3. M.S. Alam et al. (CLEO Collaboration), Phys. Rev. Lett. **74**, 2885 (1995).
4. R. Barate et al. (ALEPH Collaboration), Phys. Lett. B **429**, 169 (1998).
5. S. Glenn et al. (CLEO Collaboration), Report No. CLEO CONF 98-17, contributed paper to The 29th International Conference on High Energy Physics, Vancouver, Canada (1998), ICHEP98 1011.
6. K. Chetyrkin et al., Phys. Lett. B **400**, 206 (1997); Phys. Lett. B **425**, 414 (1998); see also Reference 8.
7. A. Ali and C. Grueb, Phys. Lett. B **259**, 182 (1991).
8. M. Neubert, these proceedings, Report No. CERN-TH-98-301, hep-ph/9809377.
9. Glenn S et al. (CLEO Collaboration), Phys. Rev. Lett. **80**, 2289 (1998).
10. R. Godang et al. (CLEO Collaboration), Report No. CLEO CONF 98-22, contributed paper to The 29th International Conference on High Energy Physics, Vancouver, Canada (1998), ICHEP98 1012.
11. F. Abe et al. (CDF Collaboration), Phys. Rev. Lett. **76**, 4675 (1996).
12. A. Ali and T. Mannel, Phys. Lett. B **264**, 447 (1991); A. Ali, Nucl. Instrum. Methods A **384**, 8 (1996); C. Greub, A. Ioannisian and D. Wyler, Phys. Lett. B **346**, 149 (1994); P. Ball and V.M. Braun, Phys. Rev. D **58**, 4016 (1998); D. Melikov, N. Nikitin and S. Simula, Phys. Lett. B **410**, 290 (1997); see also Reference 10.
13. B. Abbott et al. (D0 Collaboration), Phys. Lett. B **423**, 419 (1998).
14. A. Ali, G. Hiller, L.T. Handoko and T. Morozumi, Phys. Rev. D **55**, 4105 (1997).

THEORETICAL STATUS OF $B \to X_s\gamma$ DECAYS

M. NEUBERT

Theory Division, CERN, CH-1211 Geneva 23, Switzerland
E-mail: Matthias.Neubert@cern.ch

We review the theoretical understanding of the branching ratio and photon-energy spectrum in $B \to X_s\gamma$ decays at next-to-leading order in QCD, including consistently the effects of Fermi motion. For the Standard Model, we obtain $\mathrm{B}(B \to X_s\gamma) = (3.29 \pm 0.33) \times 10^{-4}$ for the total branching ratio, and $\mathrm{B}(B \to X_s\gamma) = (2.85^{+0.34}_{-0.40}) \times 10^{-4}$ if a cut $E^{\mathrm{lab}}_\gamma > 2.1\,\mathrm{GeV}$ is applied on the photon energy, as done in the recent CLEO analysis. A precise measurement of the photon spectrum would help reducing the theoretical uncertainty and yield important information on the momentum distribution of b quarks inside B mesons.

1 Introduction

About three years ago, the CLEO Collaboration reported the first measurement of the inclusive branching ratio for the radiative decays $B \to X_s\gamma$, yielding[1] $\mathrm{B}(B \to X_s\gamma) = (2.32 \pm 0.57 \pm 0.35) \times 10^{-4}$. At this Conference, this value has been updated to[2] $\mathrm{B}(B \to X_s\gamma) = (3.15 \pm 0.35 \pm 0.32 \pm 0.26) \times 10^{-4}$, where the first error is statistical, the second systematic, and the third accounts for model dependence. The ALEPH Collaboration has reported a measurement of the corresponding branching ratio for b hadrons produced at the Z resonance, yielding[3] $\mathrm{B}(H_b \to X_s\gamma) = (3.11 \pm 0.80 \pm 0.72) \times 10^{-4}$. Theoretically, the two numbers are expected to differ by at most a few percent. Taking the weighted average gives

$$\mathrm{B}(B \to X_s\gamma) = (3.14 \pm 0.48) \times 10^{-4}\,. \tag{1}$$

Being rare processes mediated by loop diagrams, radiative decays of B mesons are potentially sensitive probes of New Physics beyond the Standard Model, provided a reliable calculation of their branching ratio can be performed. The theoretical framework for such a calculation is set by the heavy-quark expansion, which predicts that to leading order in $1/m_b$ inclusive decay rates agree with the parton-model rates for the underlying decays of the b quark[4–7]. The leading nonperturbative corrections have been studied in detail and are well understood. The prediction for the $B \to X_s\gamma$ branching ratio suffers, however, from large perturbative uncertainties if only leading-order expressions for the Wilson coefficients in the effective weak Hamiltonian are employed[8–10]. Therefore, it was an important achievement when the full next-to-leading order calculation of the total $B \to X_s\gamma$ branching ratio in the Standard Model was completed, combining consistently results for the matching conditions[11–13], matrix elements[14,15], and anomalous dimensions[16]. The leading QED and electroweak radiative corrections were included, too[17,18]. As a result, the theoretical uncertainty was reduced to a level of about 10%, which is slightly less than the current experimental error. During the last year, the next-to-leading

order analysis was extended to the cases of two-Higgs-doublet models[19,20] and supersymmetry[21], so that accurate theoretical predictions are now at hand also for the most popular extensions of the Standard Model.

The fact that only the high-energy part of the photon spectrum in $B \to X_s\gamma$ decays is accessible experimentally introduces a significant additional theoretical uncertainty[18]. For instance, in the new CLEO analysis reported at this Conference[2] a lower cut on the photon energy of 2.1 GeV in the laboratory is imposed, which eliminates about three quarters of phase space in this variable. After reviewing the theoretical status of calculations of the total $B \to X_s\gamma$ branching ratio, we thus discuss to what extent the effects of a photon-energy cut-off can be controlled theoretically, pointing out the importance of measurements of the photon spectrum for reducing the theoretical uncertainty in the extraction of the total branching ratio.

2 $B \to X_s\gamma$ branching ratio

The starting point in the analysis of $B \to X_s\gamma$ decays is the effective weak Hamiltonian

$$H_{\mathrm{eff}} = -\frac{4G_F}{\sqrt{2}} V^*_{ts} V_{tb} \sum_i C_i(\mu_b) O_i(\mu_b)\,. \tag{2}$$

The operators relevant to our discussion are

$$\begin{aligned}
O_2 &= \bar{s}_L \gamma_\mu c_L\, \bar{c}_L \gamma^\mu b_L\,,\\
O_7 &= \frac{e\, m_b}{16\pi^2}\, \bar{s}_L \sigma_{\mu\nu} F^{\mu\nu} b_R\,,\\
O_8 &= \frac{g_s m_b}{16\pi^2}\, \bar{s}_L \sigma_{\mu\nu} G^{\mu\nu}_a t_a b_R\,.
\end{aligned} \tag{3}$$

To an excellent approximation, the contributions of other operators can be neglected. The renormalization scale μ_b in (2) is chosen of order m_b, so that all large logarithms reside in the Wilson coefficients $C_i(\mu_b)$. For inclusive decays, the relevant hadronic matrix elements of the local operators O_i can be calculated using the heavy-quark expansion[5–7]. The complete theoretical prediction for the

$B \to X_s \gamma$ decay rate at next-to-leading order in α_s has been presented for the first time by Chetyrkin et al. [16]. It depends on a parameter δ defined by the condition that the photon energy be above a threshold given by $E_\gamma > (1 - \delta) E_\gamma^{\max}$. The prediction for the $B \to X_s \gamma$ branching ratio is usually obtained by normalizing the result for the corresponding decay rate to that for the semileptonic rate, thereby eliminating a strong dependence on the b-quark mass. We define

$$
R_{\text{th}}(\delta) = \frac{\Gamma(B \to X_s \gamma)|_{E_\gamma > (1-\delta) E_\gamma^{\max}}}{\Gamma(B \to X_c e \bar{\nu})}
$$
$$
= \frac{6\alpha}{\pi f(z)} \left| \frac{V_{ts}^* V_{tb}}{V_{cb}} \right|^2 K_{\text{NLO}}(\delta), \qquad (4)
$$

where $f(z) \approx 0.542 - 2.23(\sqrt{z} - 0.29)$ is a phase-space factor depending on the quark-mass ratio $z = (m_c/m_b)^2$. The fine-structure constant α is renormalized at $q^2 = 0$, as is appropriate for real-photon emission [17]. The quantity $K_{\text{NLO}}(\delta)$ contains the next-to-leading order corrections. In terms of the theoretically calculable ratio $R_{\text{th}}(\delta)$, the $B \to X_s \gamma$ branching ratio is given by $B(B \to X_s \gamma) = 0.105 N_{\text{SL}} R_{\text{th}}(\delta)$, where $N_{\text{SL}} = B(B \to X_c e \bar{\nu})/10.5\%$ is a normalization factor to be determined from experiment. To good approximation $N_{\text{SL}} = 1$. The current experimental situation of measurements of the semileptonic branching ratio of B mesons and their theoretical interpretation are reviewed in Refs. 22,23.

In the calculation of the quantity $K_{\text{NLO}}(\delta)$ we consistently work to first order in the small parameters α_s, $1/m_Q^2$ and α/α_s, the latter ratio being related to the leading-logarithmic QED corrections. The structure of the result is

$$
K_{\text{NLO}}(\delta) = \sum_{\substack{i,j=2,7,8 \\ i \le j}} k_{ij}(\delta, \mu_b) \, \text{Re}\left[C_i(\mu_b) \, C_j^*(\mu_b) \right], \qquad (5)
$$

where the Wilson coefficients $C_i(\mu_b)$ are expanded as

$$
C_i^{(0)}(\mu_b) + \frac{\alpha_s(\mu_b)}{4\pi} C_i^{(1)}(\mu_b) + \frac{\alpha}{\alpha_s(\mu_b)} C_i^{(\text{em})}(\mu_b) + \dots. \qquad (6)
$$

The coefficients $C_i^{(k)}(\mu_b)$ are complicated functions of the ratio $\eta = \alpha_s(m_W)/\alpha_s(\mu_b)$, which also depend on the values $C_i(m_W)$ of the Wilson coefficients at the weak scale. In the Standard Model, these inital conditions are functions of the mass ratio $x_t = (m_t/m_W)^2$. Whereas the leading-order coefficients $C_i^{(0)}(\mu_b)$ are known since a long time [8-10], the next-to-leading terms in (6), which must be kept for the coefficient $C_7(\mu_b)$, have been calculated only recently. The expression for $C_7^{(1)}(\mu_b)$ can be found in eq. (21) of Ref. 16, and the result for $C_7^{(\text{em})}(\mu_b)$ is given in eq. (11) of Ref. 18.

Explicit expressions for the functions $k_{ij}(\delta, \mu_b)$ in (5) can be found, e.g., in Ref. 18, where we have corrected some mistakes in the formulae for real-gluon emission used by previous authors. (The corrected expressions are also given in the Erratum to Ref. 16.) Bound-state corrections enter the formulae for the coefficients k_{ij} at order $1/m_Q^2$ and are proportional to the hadronic parameter [24] $\lambda_2 = \frac{1}{4}(m_{B^*}^2 - m_B^2) \approx 0.12 \,\text{GeV}^2$. Most of them characterize the spin-dependent interactions of the b quark inside the B meson [5-7]. However, a peculiar feature of inclusive radiative decays is the appearance of a correction proportional to $1/m_c^2$ in the coefficient k_{27}, which represents a long-distance contribution arising from $(c\bar{c})$ intermediate states [25-29].

2.1 Definition of the total branching ratio

The theoretical prediction for the $B \to X_s \gamma$ branching ratio diverges in the limit $\delta \to 1$ because of a logarithmic soft-photon divergence of the $b \to sg\gamma$ subprocess, which would be canceled by an infrared divergence of the $O(\alpha)$ corrections to the process $b \to sg$. We have argued that a reasonable definition of the "total" branching ratio is to use an extrapolation to $\delta = 1$ starting from the region $\delta \sim 0.5$–0.8, where the theoretical result exhibits a weak, almost linear dependence on the cutoff [18]. The extrapolated value so defined agrees, to a good approximation, with the result obtained by taking $\delta = 0.9$, and hence we define the total branching ratio using this particular value of the cutoff.

The theoretical result is sensitive to the values of various input parameters. For the (one-loop) quark pole masses we take $m_c/m_b = 0.29 \pm 0.02$, $m_b = (4.80 \pm 0.15)\,\text{GeV}$, and $m_t = (175 \pm 6)\,\text{GeV}$. The corresponding uncertainties in the branching ratio are, respectively, $^{+5.9}_{-5.0}\%$, $\mp 1.0\%$, and $\pm 1.6\%$. We use the two-loop expression for the running coupling $\alpha_s(\mu)$ with the initial value $\alpha_s(m_Z) = 0.118 \pm 0.003$, which induces an uncertainty of $\pm 2.7\%$. For the ratio of the CKM parameters in (4) we take the value $|V_{ts}^* V_{tb}|/|V_{cb}| = 0.976 \pm 0.010$ obtained from a global analysis of the unitarity triangle [30]. This gives an uncertainty of $\pm 2.1\%$. Finally, we include an uncertainty of $\pm 2.0\%$ to account for next-to-leading electroweak radiative corrections [17]. The theoretical uncertainty arising from the variation of the renormalization scale will be addressed below. We find an uncertainty of $\pm 6.3\%$ from the variation of the scale μ_b, and of $^{+2.2}_{-1.5}\%$ from the variation of the scale $\bar{\mu}_b$ entering the expression for the semileptonic decay rate in the denominator in (4). Adding the different errors in quadrature gives a total uncertainty of $^{+10}_{-9}\%$ (adding them linearly would lead to the more conservative estimate of $^{+24}_{-22}\%$). For the total $B \to X_s \gamma$ branching ratio in the Standard Model we obtain

$$\mathrm{B}(B \to X_s\gamma) = (3.29 \pm 0.33) \times 10^{-4} N_{\mathrm{SL}}, \qquad (7)$$

in good agreement with the experimental value in (1).

2.2 Sensitivity to New Physics

Possible New Physics contributions would enter the theoretical prediction for the $B \to X_s\gamma$ branching ratio through non-standard values of the Wilson coefficients of the dipole operators O_7 and O_8 at the weak scale m_W. To explore the sensitivity to such effects, we normalize these coefficients to their values in the Standard Model and introduce the ratios $\xi_i = C_i(m_W)/C_i^{\mathrm{SM}}(m_W)$ with $i = 7, 8$. In the presence of New Physics, the parameters ξ_7 and ξ_8 may take (even complex) values different from 1. Similarly, New Physics may induce dipole operators with opposite chirality to that of the Standard Model, i.e. operators with right-handed light-quark fields. If we denote by C_7^R and C_8^R the Wilson coefficients of these new operators, expression (5) can be modified to include their contributions by simply replacing $C_i C_j^* \to C_i C_j^* + C_i^R C_j^{R*}$ everywhere. We thus define two additional parameters $\xi_i^R = C_i^R(m_W)/C_i^{\mathrm{SM}}(m_W)$ with $i = 7, 8$, which vanish in the Standard Model. Since the dipole operators only contribute to rare flavour-changing neutral current processes, there are at present rather weak constraints on the values of these parameters [31]. On the other hand, we assume that the coefficient C_2 of the current–current operator O_2 takes its standard value, and that there is no similar operator containing right-handed quark fields. With these definitions, the $B \to X_s\gamma$ branching ratio can be decomposed as [18]

$$
\begin{aligned}
\frac{1}{N_{\mathrm{SL}}} \mathrm{B}(B \to X_s\gamma)\big|_{E_\gamma > (1-\delta)E_\gamma^{\max}} \\
= B_{22}(\delta) + B_{77}(\delta)\left(|\xi_7|^2 + |\xi_7^R|^2\right) + B_{88}(\delta)\left(|\xi_8|^2 + |\xi_8^R|^2\right) \\
+ B_{27}(\delta)\,\mathrm{Re}(\xi_7) + B_{28}(\delta)\,\mathrm{Re}(\xi_8) \\
+ B_{78}(\delta)\left[\mathrm{Re}(\xi_7\xi_8^*) + \mathrm{Re}(\xi_7^R\xi_8^{R*})\right].
\end{aligned}
\qquad (8)
$$

In Table 1, the values of the components $B_{ij}(\delta)$ obtained with $\delta = 0.9$ are shown for different choices of the renormalization scale. Assuming $N_{\mathrm{SL}} = 1$, the Standard Model result for the total branching ratio is given by $B_{\mathrm{SM}} = \sum_{ij} B_{ij}$. The most important contributions are B_{27} and B_{22} followed by B_{77}. The smallness of the remaining terms shows that there is little sensitivity to the coefficient $C_8(m_W)$ of the chromo-magnetic dipole operator. Once the parameters ξ_i and ξ_i^R are calculated in a given New Physics scenario, the result for the $B \to X_s\gamma$ branching ratio can be derived using the numbers shown in the table. For the remainder of this talk, however, we assume the validity of the Standard Model.

Table 1: Values of the coefficients B_{ij} (for $\delta = 0.9$) in units of 10^{-4}, for different choices of μ_b. The coefficient $B_{88} = 0.015$ in all three cases

μ_b	B_{27}	B_{22}	B_{77}	B_{28}	B_{78}	$\sum_{ij} B_{ij}$
$m_b/2$	1.265	1.321	0.335	0.179	0.074	3.188
m_b	1.395	1.258	0.382	0.161	0.083	3.293
$2m_b$	1.517	1.023	0.428	0.132	0.092	3.206

2.3 Perturbative uncertainties

The components $B_{ij}(\delta)$ in (8) are formally independent of the renormalization scale μ_b. Their residual scale dependence results only from the truncation of perturbation theory at next-to-leading order and is conventionally taken as an estimate of higher-order corrections. Typically, the different components vary by amounts of order 10–20% as μ_b varies between $m_b/2$ and $2m_b$. The good stability is a result of the cancelation of the scale dependence between Wilson coefficients and matrix elements achieved by a next-to-leading order calculation.

Previous authors [13,16,19,20] who estimated the μ_b dependence of the total $B \to X_s\gamma$ branching ratio in the Standard Model found a more striking improvement over the leading-order result, namely a variation of only $^{+0.1}_{-3.2}\%$ as compared with $^{+27}_{-20}\%$ at leading order. However, the apparent excellent stability observed at next-to-leading order is largely due to an accidental cancelation between different contributions to the branching ratio. A look at Table 1 shows that the residual scale dependence of the individual contributions B_{ij} is much larger than that of their sum, which determines the total branching ratio in the Standard Model. Note, in particular, the almost perfect cancelation of the scale dependence between B_{27} and B_{22}, which is accidental since the magnitude of B_{27} depends on the top-quark mass, whereas B_{22} is independent of m_t. In such a situation, the apparent weak scale dependence of the sum of all contributions is not a good measure of higher-order corrections. Indeed, higher-order corrections must stabilize the different components B_{ij} individually, not only their sum. The variation of the individual components as a function of μ_b thus provides a more conservative estimate of the truncation error than does the variation of the total branching ratio. For each component, we estimate the truncation error by taking one half of the maximum variation obtained by varying μ_b between $m_b/2$ and $2m_b$. The truncation error of the sum is then obtained by adding the individual errors in quadrature. In this way, we find a total truncation error of $\pm6.3\%$, which is more than a factor of 2 larger than the estimates obtained by previous authors. An even larger truncation error could be justified given that the choice of the range of variation of μ_b is ad hoc, and that the scale dependence of the various components is not symmetric around the point $\mu_b = m_b$.

3 Partially integrated branching ratio and photon-energy spectrum

Whereas the explicit power corrections to the functions k_{ij} are small, an important nonperturbative effect not included so far is the motion of the b quark inside the B meson caused by its soft interactions with the light constituents. It leads to a modification of the photon-energy spectrum, which must be taken into account if a realistic cutoff is imposed [18,32]. This so-called "Fermi motion" can be included in the heavy-quark expansion by resumming an infinite set of leading-twist contributions into a shape function $F(k_+)$, which governs the light-cone momentum distribution of the heavy quark inside the B meson [33-35]. The physical decay distributions are obtained from a convolution of parton-model spectra with this function. In the process, phase-space boundaries defined by parton kinematics are transformed into the proper physical boundaries defined by hadron kinematics.

The shape function is a universal characteristic of the B meson governing the inclusive decay spectra in processes with massless partons in the final state, such as $B \to X_s\gamma$ and $B \to X_u \, \ell\,\nu$. However, this function does not describe in an accurate way the distributions in decays into massive partons such as [34,35] $B \to X_c \ell \nu$. Unfortunately, therefore, the shape function cannot be determined using the lepton spectrum in semileptonic decays, for which high-precision data exist. On the other hand, there is some useful theoretical information on the moments $A_n = \langle k_+^n \rangle$ of the function $F(k_+)$, which are related to the forward matrix elements of local operators [33]. In particular, $A_1 = 0$ vanishes by the equations of motion (this condition defines the heavy-quark mass), and $A_2 = \frac{1}{3}\mu_\pi^2$ is related to the kinetic energy of the heavy quark inside the B meson. For our purposes it is sufficient to adopt the simple form $F(k_+) = N\,(1-x)^a e^{(1+a)x}$ with $x = k_+/\bar{\Lambda} \le 1$, where $\bar{\Lambda} = m_B - m_b$. This ansatz is such that $A_1 = 0$ by construction. The parameter a can be related to the second moment, yielding $\mu_\pi^2 = 3\bar{\Lambda}^2/(1+a)$. Thus, the b-quark mass (or $\bar{\Lambda}$) and the quantity μ_π^2 (or a) are the two parameters of our function. We take $m_b = 4.8\,\mathrm{GeV}$ and $\mu_\pi^2 = 0.3\,\mathrm{GeV}^2$ as reference values, in which case $a \approx 1.29$.

Let us denote by $B_{ij}^{\mathrm{p}}(\delta_{\mathrm{p}})$ the various components in (8) calculated in the parton model, where the cutoff δ_{p} is defined by the condition that $E_\gamma \ge \frac{1}{2}(1-\delta_{\mathrm{p}})m_b$. Then the corresponding physical quantities $B_{ij}(\delta)$ with δ de-

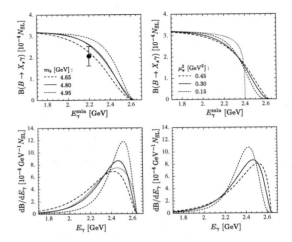

Figure 1: Partially integrated $B \to X_s\gamma$ branching ratio (upper plots) and photon-energy spectrum (lower plots) for various choices of shape-function parameters

fined such that $E_\gamma \ge \frac{1}{2}(1-\delta)m_B$ are given by [18,32]

$$B_{ij}(\delta) = \int_{m_B(1-\delta)-m_b}^{m_B-m_b} \mathrm{d}k_+\, F(k_+)\, B_{ij}^{\mathrm{p}}\left(1 - \frac{m_B(1-\delta)}{m_b + k_+}\right).$$
(9)

This relation is such that $B_{ij}(1) = B_{ij}^{\mathrm{p}}(1)$, implying that the total branching ratio is not affected by Fermi motion. The effect is, however, important for realistic values of the cutoff δ.

As an illustration of the sensitivity of our results to the parameters of the shape function, the upper plots in Figure 1 show the predictions for the partially integrated $B \to X_s\gamma$ branching ratio as a function of the energy cutoff $E_\gamma^{\mathrm{min}} = \frac{1}{2}(1-\delta)m_B$. In the left-hand plot we vary m_b keeping the ratio $\mu_\pi^2/\bar{\Lambda}^2$ fixed. The gray line shows the result obtained using the same parameters as for the solid line but a different functional form given by $F(k_+) = N\,(1-x)^a e^{-b(1-x)^2}$. For comparison, we show the data point $\mathrm{B}(B \to X_s\gamma) = (2.04 \pm 0.47) \times 10^{-4}$ obtained in the original CLEO analysis [1] with a cutoff at $2.2\,\mathrm{GeV}$. In the right-hand plot, we keep $m_b = 4.8\,\mathrm{GeV}$ fixed and compare the parton-model result (gray dotted curve) with the results corrected for Fermi motion using different values for the parameter μ_π^2. This plot illustrates how Fermi motion fills the gap between the parton-model endpoint at $m_b/2$ and the physical endpoint at $m_B/2$. (The true endpoint is actually located at $[m_B^2 - (m_K + m_\pi)^2]/2m_B \approx 2.60\,\mathrm{GeV}$, i.e. slightly below $m_B/2 \approx 2.64\,\mathrm{GeV}$.) Comparing the two plots, it is evident that the uncertainty due to the value of the

Table 2: Partially integrated $B \to X_s \gamma$ branching ratio for different values of the cutoff on the photon energy

$E_{\gamma,\min}^{\text{lab}}$	$B(B \to X_s\gamma)$ $[10^{-4}N_{\text{SL}}]$
2.2 GeV	$2.56 \pm 0.26\,^{+0.31}_{-0.36}$
2.1 GeV	$2.85 \pm 0.29\,^{+0.18}_{-0.27}$
2.0 GeV	$3.01 \pm 0.30\,^{+0.09}_{-0.18}$

b-quark mass is the dominant one. Variations of the parameter μ_π^2 have a much smaller effect on the partially integrated branching ratio, and also the sensitivity to the functional form adopted for the shape function turns out to be small. This behaviour is a consequence of global quark–hadron duality, which ensures that even partially integrated quantities are rather insensitive to bound-state effects. The strong remaining dependence on the b-quark mass is simply due to the transformation by Fermi motion of phase-space boundaries from parton to hadron kinematics.

Taking the three curves in the left-hand upper plot for a representative range of parameters and applying a small correction for the boost from the B rest frame to the laboratory as appropriate for the CLEO analysis, we show in Table 2 the predictions for the partially integrated branching ratio for three different values of the cutoff on the photon energy. The first error refers to the dependence on the various input parameters discussed previously, and the second one accounts for the uncertainty associated with the description of Fermi motion. In general, this second error can be reduced in two ways. The first possibility is to lower the cutoff on the photon energy. A first step in this direction has already been taken in the new CLEO analysis reported at this Conference [2], in which the cutoff has been lowered from 2.2 to 2.1 GeV. If a value as low as 2 GeV could be achieved, the theoretical predictions would become rather insensitive to the parameters of the shape function. To what extent this will be possible in future experiments will depend on their capability to reject the background of photons from other decays. The second possibility is that future high-precision measurements of the photon spectrum will make it possible to adjust the parameters of the shape function from a fit to the data. For the purpose of illustration, the photon spectra corresponding to the various parameter sets are shown in the lower plots in Figure 1. Such a determination of the shape-function parameters from $B \to X_s \gamma$ decays would not only help to reduce the theoretical uncertainty in the determination of the total branching ratio, but would also enable us to predict the lepton spectrum in $B \to X_u \ell \nu$ in a model-independent way [33]. This may help to reduce the theoretical uncertainty in the value of $|V_{ub}|$. A detailed analysis of the

photon spectrum will therefore be an important aspect in future analyses of inclusive radiative B decays.

4 Conclusions

The inclusive radiative decays $B \to X_s \gamma$ play a key role in testing the Standard Model and probing the structure of possible New Physics. A reliable theoretical calculation of their branching ratio can be performed using the operator product expansion for inclusive decays of heavy hadrons combined with the twist expansion for the description of decay distributions near phase-space boundaries. To leading order in $1/m_b$ the decay rate agrees with the parton-model rate for the underlying quark decay $b \to X_s \gamma$. With the completion of the next-to-leading order calculation of the Wilson coefficients and matrix elements of the operators in the effective weak Hamiltonian, the perturbative uncertainties in the calculation of this process have been reduced to a level of about 10%. Bound-state corrections to the total decay rate are suppressed by powers of $1/m_b$ and can be controlled in a systematic way.

A more important effect is the Fermi motion of the heavy quark inside the meson, which is responsible for the characteristic shape of the photon-energy spectrum in $B \to X_s \gamma$ decays. It leads to the main theoretical uncertainty in the calculation of the branching ratio if a restriction to the high-energy part of the photon spectrum is imposed. Fermi motion is naturally incorporated in the heavy-quark expansion by resumming an infinite set of leading-twist operators into a non-perturbative shape function. The main theoretical uncertainty in this description lies in the value of the b-quark mass. Other features associated with the detailed functional form of the shape function play a minor role. The value of m_b and other shape-function parameters could, in principle, be extracted from a precise measurement of the photon-energy spectrum, but also gross features of this spectrum such as the average photon energy would provide valuable information [18]. For completeness, we note that besides the photon-energy spectrum also the invariant hadronic mass distribution in radiative B decays can be studied. Investigating the pattern of individual hadron resonances contributing to the spectrum, one can motivate a simple description of the hadronic mass spectrum with only a single parameter, the $B \to K^* \gamma$ branching ratio, to be determined from experiment [18].

Possible New Physics contributions would enter the theoretical predictions for the $B \to X_s \gamma$ branching ratio and photon spectrum through the values of parameters ξ_i and ξ_i^R, which are defined in terms of values of Wilson coefficients at the scale m_W. This formalism allows one to account for non-standard contributions to the magnetic and chromo-magnetic dipole operators, as well as operators with right-handed light-quark fields. Quite generally, New Physics would not affect the shape of the photon spectrum[18] but could change the total branching ratio by a considerable amount. This implies that the analysis of the photon-energy and hadronic mass spectra, which is crucial for the experimental determination of the total branching ratio, can be performed without assuming the correctness of the Standard Model. On the other hand, the total branching ratio will provide a powerful constraint on the structure of New Physics beyond the Standard Model as experimental data become more precise.

Acknowledgments

The work reported here has been done in a most pleasant collaboration with Alex Kagan, which is gratefully acknowledged.

1. M.S. Alam et al. (CLEO Collaboration), Phys. Rev. Lett. **74**, 2885 (1995).
2. T. Skwarnicki (CLEO Collaboration), Talk no. 702 presented at this Conference.
3. R. Barate et al. (ALEPH Collaboration), Phys. Lett. B **429**, 169 (1998).
4. J. Chay, H. Georgi and B. Grinstein, Phys. Lett. B **247**, 399 (1990).
5. I.I. Bigi, N.G. Uraltsev and A.I. Vainshtein, Phys. Lett. B **293**, 430 (1992) [E: **297**, 477 (1993)]; I.I. Bigi, M.A. Shifman, N.G. Uraltsev and A.I. Vainshtein, Phys. Rev. Lett. **71**, 496 (1993); B. Blok, L. Koyrakh, M.A. Shifman and A.I. Vainshtein, Phys. Rev. D **49**, 3356 (1994) [E: **50**, 3572 (1994)].
6. A.V. Manohar and M.B. Wise, Phys. Rev. D **49**, 1310 (1994).
7. A.F. Falk, M. Luke and M.J. Savage, Phys. Rev. D **49**, 3367 (1994).
8. B. Grinstein, R. Springer and M.B. Wise, Nucl. Phys. B **339**, 269 (1990).
9. M. Ciuchini et al., Phys. Lett. B **316**, 127 (1993).
10. A.J. Buras, M. Misiak, M. Münz and S. Pokorski, Nucl. Phys. B **424**, 374 (1994).
11. K. Adel and Y.P. Yao, Phys. Rev. D **49**, 4945 (1994).
12. C. Greub and T. Hurth, Phys. Rev. D **56**, 2934 (1997).
13. A.J. Buras, A. Kwiatkowski and N. Pott, Phys. Lett. B **414**, 157 (1997); Nucl. Phys. B **517**, 353 (1998).
14. A. Ali and C. Greub, Phys. Lett. B **361**, 146 (1995); see also the extended version in Preprint DESY 95-117 [hep-ph/9506374].
15. C. Greub, T. Hurth and D. Wyler, Phys. Lett. B **380**, 385 (1996); Phys. Rev. D **54**, 3350 (1996).
16. K. Chetyrkin, M. Misiak and M. Münz, Phys. Lett. B **400**, 206 (1997) [E: **425**, 414 (1998)].
17. A. Czarnecki and W.J. Marciano, Phys. Rev. Lett. **81**, 277 (1998).
18. A.L. Kagan and M. Neubert, Preprint CERN-TH/98-99 [hep-ph/9805303], to appear in Eur. Phys. J. C.
19. M. Ciuchini, G. Degrassi, P. Gambino and G.F. Giudice, Nucl. Phys. B **527**, 21 (1998).
20. F.M. Borzumati and C. Greub, Preprint ZU-TH-31-97 [hep-ph/9802391].
21. M. Ciuchini, G. Degrassi, P. Gambino and G.F. Giudice, Preprint CERN-TH/98-177 [hep-ph/9806308].
22. P. Drell, Preprint CLNS-97-1521 [hep-ex/9711020], to appear in the Proceedings of the 18th International Symposium on Lepton–Photon Interactions, Hamburg, Germany, July 1997.
23. M. Neubert, Preprint CERN-TH/98-2 [hep-ph/9801269], to appear in the Proceedings of the International Europhysics Conference on High Energy Physics, Jerusalem, Israel, August 1997.
24. A.F. Falk and M. Neubert, Phys. Rev. D **47**, 2965 (1993).
25. M.B. Voloshin, Phys. Lett. B **397**, 275 (1997).
26. A. Khodjamirian, R. Rückl, G. Stoll and D. Wyler, Phys. Lett. B **402**, 167 (1997).
27. Z. Ligeti, L. Randall and M.B. Wise, Phys. Lett. B **402**, 178 (1997).
28. A.K. Grant, A.G. Morgan, S. Nussinov and R.D. Peccei, Phys. Rev. D **56**, 3151 (1997).
29. G. Buchalla, G. Isidori and S.J. Rey, Nucl. Phys. B **511**, 594 (1998).
30. A.J. Buras, Proceedings of the 28th International Conference on High-Energy Physics, Warsaw, Poland, July 1996, edited by Z. Ajduk and A.K. Wroblewski (World Scientific, Singapore, 1997), pp. 243.
31. For a review, see: J. Hewett, Talk no. 1617 at this Conference.
32. A.L. Kagan and M. Neubert, Preprint CERN-TH/98-1 [hep-ph/9803368], to appear in Phys. Rev. D.
33. M. Neubert, Phys. Rev. D **49**, 3392 and 4623 (1994).
34. I.I. Bigi, M.A. Shifman, N.G. Uraltsev and A.I. Vainshtein, Int. J. Mod. Phys. A **9**, 2467 (1994); R.D. Dikeman, M. Shifman and N.G. Uraltsev, Int. J. Mod. Phys. A **11**, 571 (1996).
35. T. Mannel and M. Neubert, Phys. Rev. D **50**, 2037 (1994).

b - QUARK FRAGMENTATION MEASUREMENTS

U. GASPARINI

Dipartimento di Fisica Univ. di Padova, Via Marzolo 8, 35100 Padova, Italy
E-mail: ugo.gasparini@pd.infn.it

Recent measurements on b-quark fragmentation properties from DELPHI and OPAL experiments at LEP collider and from CDF experiment at Tevatron collider are reported.

1 Introduction

The fractions of weakly decaying b-hadrons produced in the fragmentation (non-perturbative) process of a b-quark generated in high energy collisions at e^+e^- and pp Colliders are not predicted by theory. They are defined as the ratio between the yield of different species of B-mesons or baryons and the total number of hadrons containing a b-quark :

$$f_{B_d^0,B^+,B_s^0,\Lambda_b} = \frac{N(B_d^0,B^+,B_s^0,\Lambda_b)}{N(b-hadrons)} \qquad (1)$$

Their knowledge and the comparison/tuning of phenomenological models describing the quark fragmentation process with experimental data is very important for precision studies of other basical physical quantities (e.g. the CKM matrix elements V_{cb}, V_{ub}, the $B - \bar{B}$ mixing parameters Δm_d, Δm_s), to reduce their systematic uncertainty. Moreover, branching fractions like $BR(B_s, \Lambda_b \to l\nu X)$, $BR(B_s, \Lambda_b \to (D_s, \Lambda_c)l\nu X)$, which give complementary information w.r.t. lifetime measurements on the weak decay mechanism, are experimentally accessible via the products $f_{B_s,\Lambda_b} \cdot BR(B_s, \Lambda_b \to l\nu X)$, $f_{B_s,\Lambda_b} \cdot BR(B_s, \Lambda_b \to (D_s, \Lambda_c)l\nu X)$.

Until now, $f_{B_s^0,\Lambda_b}$ rates were determined from measurements of the exclusive semi-leptonic decays [1,2] : $f_{B_s^0} \cdot BR(B_s \to D_sl\nu X) = (0.86^{+0.20}_{-0.22})\%$, $f_{\Lambda_b} \cdot BR(\Lambda_b \to \Lambda_cl\nu X) = (0.89 \pm 0.25)\%$. Using inclusive semi-leptonic branching fractions and lifetime measurements the following values can be derived [3]: $f_{B_s} = (12.0^{+4.5}_{-3.4})\%$, $f_{\Lambda_b} = (10.1^{+3.9}_{-3.1})\%$. Assumptions on the quantities $\Gamma_{D_s}^{sl} = \frac{BR(B_s \to D_sl\nu X)}{BR(B_s \to l\nu X)}$, $\Gamma_{\Lambda_b}^{sl} = \frac{BR(\Lambda_b \to \Lambda_cl\nu X)}{BR(\Lambda_b \to l\nu X)}$ (inferred by the corresponding measured quantities in the $B_{u,d}$ sector) are also needed in the computation (see ref. [4] for a more detailed discussion). The results reported above are then limited by the small statistics and the D_s, Λ_c branching fraction uncertainties, and suffer from large systematics due to the assumptions used. Finally, the $B_{d,u}$ production rates, assumed to be equal, are usually derived from the normalization relation : $f_{B_d^0} + f_{B^+} + f_{B_s^0} + f_{\Lambda_b} = 1$.

The B_s production rate can be more precisely derived in an alternative way from mixing measurements of the average mixing parameter $\overline{\chi}$ measured at LEP and the χ_d mixing parameter of the B_d^0 system measured at Y(4S) resonance and at LEP from time-dependent measurements. Using the relation :

$$\overline{\chi}_{LEP} = f_{B_d^0}\chi_d + f_{B_s^0}\chi_s \qquad (2)$$

and the normalization relation quoted above (keeping $\chi_s = 0.5$ and assuming, again, $f_{B_d^0} = f_{B^+}$) the value [3] :

$$f_{B_s} = \frac{2\overline{\chi} - \chi_d(1 - f_{\Lambda_b})}{1 - \chi_d} = (10.2 \pm 1.7)\% \qquad (3)$$

is derived.

New methods have been recently developed by DELPHI[5] and OPAL [6] experiments at LEP collider to determine the fragmentation rates to $B_{d,u}$, B_s and b-baryons using inclusive selections, which will be described in this talk. New results from CDF experiment [7] at Tevatron, based on exclusive reconstruction of semi-leptonic decays, will be also reported. Finally, a study of the charge multiplicity in $b\bar{b}$ events as a function of the CM energy up to the highest energies available at LEP ($\sqrt{s} = 183 GeV$) [8] will be discussed, comparing the result with the QCD prediction.

2 Fragmentation measurements with inclusive methods

Inclusive methods recently developed in the analysis of LEP data use the presence of 'companion' particles like kaons and protons or Λ 's in the same jet of the reconstructed decay vertex of a b-hadron. As explained below, these accompaining particles may be signatures for B_s and b-baryon production in the b-fragmentation process. In addition, a 'vertex charge' variable properly defined for the decay vertex candidate can discriminate between neutral and charged decaying b-hadrons.

2.1 B_s production rate

To measure the production rate of B_s meson, the 'companion method' exploit the 'bottomness'-strangeness correlation inside jets between the secondary vertex and an

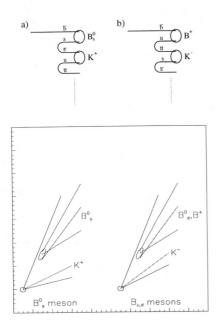

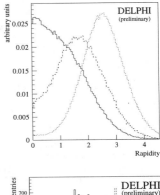

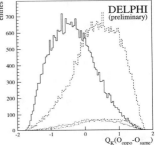

Figure 1: Correlation between the bottomness of b-hadron and strangeness of accompaining kaon produced at primary vertex.

Figure 2: Up: rapidity distribution of identified kaons in simulated events for B decay products (dotted line), 'rank 1' (see text) primary kaons (dashed line) and higher rank particles (full line). Down: distribution of the discriminating variable defined in the text for events with $B_{u,d}$ (dashed line) and B_s mesons (full line).

identified kaon at primary vertex. This correlation (in absence of oscillations of neutral B-mesons due to mixing) is different for B_s^0 and $B_{u,d}$ mesons, as depicted in Fig.1.

A b-tagging algorithm was used by DELPHI [5] to select events with relatively high efficiency (50%) and b-purity (95%); in addition, a secondary vertex separated by more than 3σ from the primary interation and an identified kaon by the RICH detector at primary vertex were required.

The rapidity distribution of the identified kaon in simulated events, which is sensitive to the order of appearence of the kaon in the fragmentation chain ('rank'), shows a clear separation between particles from primary and secondary vertex, as can be seen in Fig.2. A 'bottomness variable' O in each hemisphere of the event was defined; this variable is a neural-net output based on variables like jet charge and (eventually) lepton properties, trained on simulated events to give 1 (-1) for $\bar{b}(b)$-quark initiated jets. To minimize the dilution effect on the correlation due to $B - \bar{B}$ mixing, the variable $Q_K \cdot (O_{opp} - O_{same})$ was considered, were Q_K is the charge of the identified kaon at primary vertex and $O_{opp(same)}$ is the 'bottomness variable' in the opposite

(same) hemisphere w.r.t. the kaon. A binned-likelihood fit to the bi-simensional distribution of kaon rapidity vs $Q_K \cdot (O_{opp} - O_{same})$ was performed, giving the result on the production rate [5] :

$$f'_{B_s} = (12.0 \pm 1.4(stat) \pm 2.5(syst))\%. \qquad (4)$$

It must be stressed that f'_{B_s} is the production rate summed over B_s, B_s^* and B_s^{**} states, which must be converted into f_{B_s} (the production rate of *weakly decaying* B_s meson). Assuming a production rate $P_{b \to B_s^{**}}/P_{b \to B_s} = 0.27 \pm 0.08$ (compatible with the corresponding rate measured in the $B_{u,d}$ sector), the value :

$$f_{B_s} = (8.8 \pm 1.0(stat) \pm 1.8(syst) \pm 1.0(B^{**}rate))\%. \quad (5)$$

is obtained.

2.2 Neutral and charged B-mesons rates

A method to separate neutral from charged B mesons was developed by DELPHI [5], based on a 'secondary vertex charge' defined as :

$$Q_B = \Sigma(Q_i P_{Bi}) \qquad (6)$$

where Q_i was the charge of the track and P_{Bi} was the probability that it originated from a b-hadron decay rather than from fragmentation. Input variables to define this probability were the probabilities that the track fits to the primary and secondary vertex, the energy and the rapidity of the track with respect to the jet axis. Assuming binomial statistics, an error on this quantity was defined: $\sigma_{Q_B} = \sqrt{\Sigma P_B (1 - P_B)}$; this quantity tends to be small if all tracks are well classified, having values of P_B close to 0 or 1, and it is strongly correlated to the hemisphere multiplicity N_{hemi}. Cutting on small values of the variable $s_{Q_B} = \sigma_{Q_B} / < \sigma_{Q_B}(N_{hemi}) >$, hemispheres with good separation of B-decay and fragmentation tracks were selected, thus allowing a reliable determination of the B-dacay hadron charge. Parameters in the simulation, mainly lifetimes of B-hadrons, had to be adjusted. The real data distribution together with the fit result and the simulation prediction for neutral, positively and negatively charged b-hadrons is shown in Fig.3. The result was :

$$BR(b \to X_B^0) = (57.8 \pm 0.5(stat) \pm 1.0(syst))\%$$
$$BR(b \to X_B^\pm) = (42.2 \pm 0.5(stat) \pm 1.0(syst))\%.$$

Assuming a 7% fraction of charged b-baryons as predicted by JETSET fragmentation model and using the B_s^0 meson production rate quoted above, lead to the result for the production rate of $B_{u,d}$ mesons [5]:

$$f_{B_u^+} = f_{B_d^0} = (41.7 \pm 1.3)\% \qquad (7)$$

2.3 b-baryons production rate

The 'companion method' was used by OPAL experiment to separate b-baryons from B mesons [6]. Events were selected by requiring a b-tagging with high purity (94%) and the presence of a reconstructed fast Λ ('tagging Λ' with $p > 4\ GeV/c$). Differently from the analysis for the B_s rate measurement, in this case the 'companion particle' (an anti-proton or an additional $\bar\Lambda$ in the same hemisphere) may originate either from the primary or the secondary vertex. Due to local baryon number conservation, the two cases discriminate between the production of a b-baryon decaying in the mode $b - baryon \to \Lambda X$ and a B-meson underlaying the decay $B \to \Lambda \bar{p}(\bar\Lambda)X$. The momentum spectrum of the companion anti-baryon distinguishes between the b-baryon production (in this case the spectrum is softer) from the B-meson production case. A

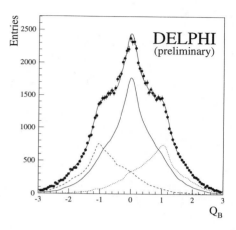

Figure 3: Vertex charge distribution for the data (points with error bars) with the fit result superimposed. The shapes for neutral (solid line), negatively (dashed) and positively (dotted) charged b-hadrons obtained in the simulation and used in the fit are also shown.

binned maximum likelihood fit to the bi-dimensional distribution of the momentum of the tagging Λ vs the momentum of the companion anti-baryon was performed. The result was [6] :

$$f(b \to \Lambda_b) \cdot BR(\Lambda_b \to \Lambda X) = (3.93 \pm 0.46(stat) \pm 0.37(syst))\%$$

$$f(b \to B) \cdot BR(B \to \Lambda \bar{p} X) = (1.94 \pm 0.28(stat) \pm 0.24(syst))\%.$$

Assuming $BR(\Lambda_b \to \Lambda X) = BR(\Lambda_b \to \Lambda_c X) \cdot BR(\Lambda_c \to \Lambda X)$, using $BR(\Lambda_c \to \Lambda X) = (35 \pm 11)\%$ and neglecting the $BR(\Lambda_b \to pDX)$ mode, the value:

$$f(b \to \Lambda_b) = (11.2 \pm 3.9)\%$$

can be derived.

3 Production rates using exclusive semi-leptonic decays

Measurement of $f_{u,d,s,b-baryons}$ were made by CDF using the exclusive semi-electronic decays [7] :

$$B^+ \to \bar{D}^0 e^+ \nu X$$
$$B^0 \to D^{(*)-} e^+ \nu X$$
$$B_s^0 \to D_s^- e^+ \nu X$$
$$\Lambda_b \to \Lambda_c^- e^+ \nu X$$

with the weakly decaying charm states exclusively reconstructed in $K^+\pi^-$, $K^+\pi^-\pi^-$, $K^+K^-\pi^-$ and $\bar{p}K^+\pi^-$ mode respectively, and the D^{*-} reconstructed in the

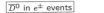

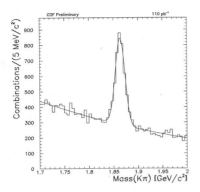

Figure 4: $K\pi$ invariant mass distribution for D^0 vertex candidates accompining a high p_T electron.

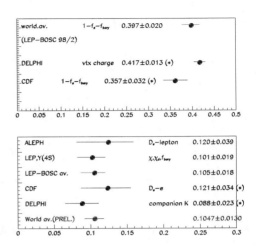

$\bar{D}^0\pi^-$ mode (charged conjugate states are implied). As an example, the $(K\pi)$ invariant mass of D^0 vertex candidates accompaining a high p_T electron is shown in Fig.4. The $D_s\ell$ and $\Lambda_c\ell$ samples are pure samples of B_s and Λ_b states, while D^0 (D^-/D^{*-}) mainly tag B^+ (B_d^0) mesons. However, in the B^+, B_d^0 case the situation is complicated due to the presence of $D^{**0} \to D^-\pi^+$, $D^{**-} \to D^0\pi^-$ and $D^{*-} \to D^0\pi^-$ decays which dilute the above correspondence, and correction to this simple assumption must be done using the measured relevant branching fractions. A simultaneous fit to the event yields in the 5 channels considered with the ratios f_d/f_u, $f_s/(f_d + f_u)$ and $f_{bary}/(f_d + f_u)$ as free parameters gave the result:

$$f_d/f_u = (87.0 \pm 24.5)\%$$
$$f_s/(f_d + f_u) \cdot BR(D_s \to \phi\pi) = (0.643 \pm 0.116)\%$$
$$f_{bary}/(f_d + f_u) \cdot BR(\Lambda_c \to pK\pi) = (1.3 \pm 0.3)\%$$

By fitting the above relative fractions, important systematic uncertainties (e.g. the electron trigger and reconstruction efficiency) cancel out, and only small topology-dependent correction factors had to be included in simulation studies. Using the relevant D_s, Λ_c branching fractions and the usual normalization relation $\Sigma f_i = 1$, the following production rates were derived [7] :

$$f_d = (31.6 \pm 5.6)\%$$
$$f_u = (36.3 \pm 5.0)\%$$
$$f_s = (12.1 \pm 3.4)\%$$
$$f_{bary} = (16.6 \pm 5.5)\%$$

A summary of existing measurements of b fragmentation rates is given in Fig.5. The recent measurements reported in this talk are marked as 'new', and a preliminary world average including the new measurements is also reported.

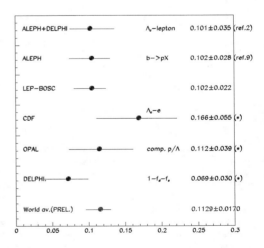

Figure 5: Summary of available results on B, B_s, and b-baryon production rates in b-quark fragmentation. The results discussed in this talk are marked by a $*$. The preliminary world averages quoted include these new results.

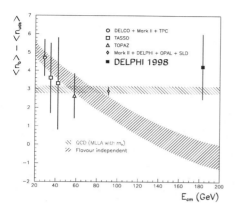

Figure 6: Difference between the average charge multiplicity in b-quark and light-quark initiated jets, measured as a function of the CM energy of the $q\bar{q}$ pair.

4 b-jets properties

The average charge multiplicity in $b\bar{b}$ events was measured by DELPHI up to the highest LEP energies ($\sqrt{s} = 183\ GeV$)[8] . The difference of the multiplicity between b and light quark initiated jets, $\delta_{bq} = <n_b> - <n_q>$, is predicted to be energy independent by perturbative QCD (in the limit $E_{CM} << m_b << \Lambda_{QCD}$). Previous measurements at lower energies were marginally discriminating between QCD and models with quark mass-independent hadronization, which predict a decreasing behaviour of δ_{bq} as a function of E_{CM}. The charge multiplicity measured in e^+e^- collisions at $\sqrt{s} = 183\ GeV$ was[8] :

$<n_b> = 29.54 \pm 0.79(stat) \pm 0.63(syst)$,

while in hadronic events was:

$<n_{hadr}> = 26.58 \pm 0.24(stat) \pm 0.54(syst)$.

Assuming the Standard Model values for R_c, R_b, R_{uds} and the PYTHIA prediction for the multiplicity in $c\bar{c}$ events, $<n_c> = 27.6 \pm 1.5$, the value:

$\delta_{bq} = 4.23 \pm 1.09(stat) \pm 1.20(syst) \pm 0.67(<n_c>)$

was obtained. This value is in fair agreement with the QCD prediction extrapolated from lower energies measurements and excludes flavour independent hadronizatin models, as illustrated by Fig.6.

5 Summary and conclusions

The fragmentation rates of b-quarks have been extensively studied at e^+e^- and pp Colliders; they are important inputs to reduce the systematic uncertainties in the measurement of fundamental electro-weak physical constants in the b-quark sector. New inclusive methods developped by DELPHI and OPAL experiments at LEP to determine the B_s^0 and b-baryons production rates, avoiding the limitation from the uncertainty on exclusive D_s, Λ_c branching fractions, have been presented. New results from CDF experiment using exclusive reconstruction of final states in semi-electronic decays were also reported.

The preliminary world average values reported here:

$$f_{B_s} = (10.47 \pm 1.30)\%, f_{b-bary} = (11.29 \pm 1.70)\% \quad (8)$$

significantly improve the previous available results (cfr. PDG'96: $f_{B_s} = (11.2^{+1.8}_{-1.9})\%, f_{b-bary} = (13.2 \pm 4.1)\%$).

A first direct measurement of $f_{B_{u,d}}$ from DELPHI experimenmt was reported, and a quantitative check of the QCD prediction on the fragmentation process in light and b quark jets, as a function of the CM energy, was presented.

All results are preliminary; further improvements by extending the new methods to other LEP experiments are expected in the near future.

References

1. D.Buskulic et al., ALEPH Collaboration, *Phys. Lett.* B **361**, 221 (1995).
2. D.Buskulic et al., ALEPH Collaboration, *Phys. Lett.* B **357**, 685 (1995); P.Abreu et al., DELPHI Collaboration, *Z. Phys.* C **68**, 375 (1995).
3. LEP Working group on B oscillations, LEPBOSC 98/2; Web page $http://www.cern.ch/LEPBOSC/PDG-1998/$.
4. D.Buskulic et al., ALEPH Collaboration, *Phys. Lett.* B **359**, 236 (1995).
5. DELPHI Collaboration, 'A direct measurement of the branching fractions of the b-quark into strange, neutral and charged B-mesons', contributed paper n.232 to this Conference.
6. K.Ackerstaff et al.,OPAL Collaboration, *Z. Phys.* C **74**, 423 (1996).
7. W.Taylor, CDF Collaboration, 'Bottom quark fragmentation fractions in $p\bar{p}$ collisions at $\sqrt{s} = 1.8$ TeV',APS/AAPT Meeting, Columbus,Ohio, April 1998 and contributed paper n.979 to this Conference.
8. DELPHI Collaboration, 'Fragmentation properties of b quarks compared to light quarks in $q\bar{q}$ events at $\sqrt{s} = 183\ GeV$', contributed paper n.141 to this Conference.
9. D.Buskulic et al., ALEPH Collaboration, European Physical Journal C **5**, 205 (1997).

Leptoproduction of J/ψ

Thomas Mehen

California Institute of Technology, Pasadena, CA 91125
E-mail: mehen@theory.caltech.edu

I review the status of the extraction of NRQCD color-octet J/ψ production matrix elements. Large theoretical uncertainties in current extractions from hadropoduction and photoproduction are emphasized. Leptoproduction of J/ψ is calulated within the NRQCD factorization formalism. Color-octet contributions dominate the cross section, allowing for a reliable extraction of $\langle \mathcal{O}_8^{J/\psi}(^1S_0)\rangle$ and $\langle \mathcal{O}_8^{J/\psi}(^3P_0)\rangle$. A comparison with preliminary data from the H1 collaboration shows that leading order color-octet mechanisms agree with the measured total cross section for $Q^2 > 4m_c^2$, while the color-singlet model underpredicts the cross section by a factor of 5.

The NRQCD factorization formalism of Bodwin, Braaten, and Lepage[1] has emerged as a new paradigm for computing the production and decay rates of heavy quarkonia. This formalism provides a rigorous theoretical framework which systematically incorporates relativistic corrections and ensures the infrared safety of perturbative calculations.[2] In the NRQCD factorization formalism, cross sections for the production of a quarkonium state H are written as

$$\sigma(H) = \sum_n \frac{c_n(\alpha_s, m_Q)}{m_Q^{d_n-4}} \langle 0|\mathcal{O}_n^H|0\rangle, \qquad (1)$$

where m_Q is the mass of the heavy quark Q. The short-distance coefficients, c_n, are associated with the production, at distances of order $1/m_Q$ or less, of a $Q\bar{Q}$ pair with quantum numbers indexed by n (angular momentum, $^{2S+1}L_J$, and color, 1 or 8). They are computable in perturbation theory. In Eq. (1), $\langle 0|\mathcal{O}_n^H|0\rangle$ are vacuum matrix elements of NRQCD operators:

$$\langle 0|\mathcal{O}_n^H|0\rangle \equiv \sum_X \sum_\lambda \langle 0|\mathcal{K}_n^\dagger|H(\lambda) + X\rangle\langle H(\lambda) + X|\mathcal{K}_n|0\rangle, \qquad (2)$$

where $\mathcal{K}_n$ is a bilinear in heavy quark fields which creates a $Q\bar{Q}$ pair in an angular-momentum and color configuration indexed by n. The bilinear combination $\mathcal{K}_n^\dagger\mathcal{K}_n$ has energy dimension d_n. The production matrix elements describe the evolution of the $Q\bar{Q}$ pair into a final state containing the quarkonium H plus additional hadrons (X) which are soft in the quarkonium rest frame. Throughout the remainder of this talk, a shorthand notation will be used in which the vacuum matrix elements are written as $\langle \mathcal{O}_{(1,8)}^H(^{2S+1}L_J)\rangle$.

The NRQCD matrix elements obey simple scaling laws[3] with respect to v, the relative velocity of the Q and $\bar{Q}$ in the quarkonium bound state. Therefore, Eq. (1) is a double expansion in v and α_s. Since the NRQCD matrix elements are sensitive only to large distance scales, they are independent of the short-distance process in which the Q and $\bar{Q}$ are produced. Thus, the NRQCD matrix

elements are universal parameters which can be extracted from one experiment and used to predict production cross sections in other processes.

Prior to the innovations presented in Ref. [1], most J/ψ production calculations took into account only the hadronization of $c\bar{c}$ pairs initially produced in a color-singlet $^3S_1$ state, as parameterized by the NRQCD matrix element $\langle \mathcal{O}_1^{J/\psi}(^3S_1)\rangle$. An important aspect of the NRQCD formalism is that, in addition to the color-singlet contribution, it allows for the possibility that a $c\bar{c}$ pair produced in a color-octet state can evolve non-perturbatively into a J/ψ. The most important color-octet matrix elements are $\langle \mathcal{O}_8^{J/\psi}(^3S_1)\rangle$, $\langle \mathcal{O}_8^{J/\psi}(^1S_0)\rangle$, and $\langle \mathcal{O}_8^{J/\psi}(^3P_J)\rangle$, which are suppressed by v^4 relative to the leading color-singlet matrix element. They describe the non-perturbative evolution of a color-octet $c\bar{c}$ pair in either a $^3S_1$, $^1S_0$, or ^3P_J angular momentum state into a J/ψ. Using heavy quark spin symmetry relations, it is possible to express all three P-wave matrix elements in terms of one: $\langle \mathcal{O}_8(^3P_J)\rangle = (2J+1)\langle \mathcal{O}_8(^3P_0)\rangle + O(v^2)$. Thus, at this order in the v expansion, there are three independent color-octet matrix elements.

The NRQCD factorization formalism has enjoyed considerable phenomenological success. Most notably, the large $p_\perp$ production of J/ψ at hadron colliders, which is underpredicted by over an order of magnitude in the color-singlet model (CSM), can be easily accounted for by color-octet production mechanisms.[4] The CSM cannot account for the observed branching ratios for $B \rightarrow J/\psi + X$, $B \rightarrow \psi' + X$. Inclusion of color-octet mechanisms removes this discrepancy.[5,6] Color-octet mechanisms also improve the understanding of J/ψ production in Z^0 decay.[7]

The NRQCD factorization formalism has yet to be conclusively proven as the correct theory of quarkonium production. One important test of the formalism is the verification of the universality of the NRQCD matrix elements. Another important prediction, which has yet to be verified, is the polarization of J/ψ at large $p_\perp$ at hadron colliders. Large $p_\perp$ production of J/ψ is dominated by fragmentation of a (nearly) on-shell gluon into a $^3S_1^{(8)}$ $c\bar{c}$ pair, which inherits the gluon's transverse polarization. Because of heavy quark spin symmetry, soft gluons emitted as the $c\bar{c}$ pair hadronizes into the quarkonium state do not dilute this polarization.[8,9] Therefore, at large $p_\perp$, J/ψ are expected to be almost 100% transversely polarized. This gets significantly diluted at lower $p_\perp$ due to nonfragmentation production.[10,11]

Polarization is a particularly interesting test of quarkonium production because it can distinguish between NRQCD and the other remaining model of quarkonium production, the color evaporation model (CEM).[12] In this model, the cross section for producing a J/ψ is proportional to the total production rate for $c\bar{c}$ pairs with invariant mass less than the $D\bar{D}$ threshold:

$$\sigma_{J/\psi} = \rho_{J/\psi} \int_{2m_c}^{2m_D} dm_{c\bar{c}} \frac{d\sigma_{c\bar{c}}}{dm_{c\bar{c}}}, \qquad (3)$$

where $\rho_{J/\psi}$ is a universal factor which represents the fraction of $c\bar{c}$ pairs which hadronize into a J/ψ. Because this model includes color-octet production mechanisms, it is consistent with hadron collider data as well as data from Z^0 decay.[12] A prediction of this model is that the emission of soft gluons in the hadronization washes out any polarization of the $c\bar{c}$ produced in the short distance process. This prediction is at odds with heavy quark spin symmetry arguments. The CEM would be a reasonable model if the relative velocity, v, in the quarkonium state were too large to serve as a useful expansion parameter. From the existing estimates of the color-octet matrix elements, this does not appear to be the case. The leading color-singlet matrix element is well determined: $\langle \mathcal{O}_1^{J/\psi}(^3S_1) \rangle = 1.1 \pm 0.1, /\text{GeV}^3$. Extractions of $\langle \mathcal{O}_8^{J/\psi}(^3S_1) \rangle$, $\langle \mathcal{O}_8^{J/\psi}(^1S_0) \rangle$, and $\langle \mathcal{O}_8^{J/\psi}(^3P_0) \rangle/m_c^2$ suffer from considerable uncertainty but they are known to be of order $10^{-2}\,\text{GeV}^3$. These values are consistent with the v^4 suppression expected on the basis of NRQCD scaling laws.

To quantitatively predict polarization as a function of $p_\perp$, color-octet matrix elements need to be accurately determined. At the present time, extractions from production at hadron colliders suffer from considerable theoretical uncertainty. Beneke and Krämer[11] extract $\langle \mathcal{O}_8^{J/\psi}(^3S_1) \rangle = 1.2^{+1.2}_{-0.7} \times 10^{-2}\,\text{GeV}^3$, where statistical errors, and errors due to variation of the renormalization scale and the parton distribution functions, have been added in quadrature. The error is dominated by scale uncertainty. Other extractions[13,14] of $\langle \mathcal{O}_8^{J/\psi}(^3S_1) \rangle$ are consistent within these errors. $\langle \mathcal{O}_8^{J/\psi}(^1S_0) \rangle$, $\langle \mathcal{O}_8^{J/\psi}(^3P_0) \rangle$ are even more poorly determined since hadroproduction is only sensitive to a particular linear combination. The extracted value of this linear combination is also much more sensitive to the choice of parton distribution function and the effects of intital and final state radiation.[15,16] Extractions from lower energy hadroproduction[17] suffer similar from problems. These calculations are extremely sensitive to the charm quark mass because of the small energy scale of the process. Corrections due to higher orders in perturbation theory and higher twist effects are expected to be large.

Recently, progress has been made in analysis of B decays and Z^0 decay. A next-to-leading order analysis of B decay[6] establishes a bound on the following linear combination of color-octet matrix elements:

$$\langle \mathcal{O}_8^{J/\psi}(^1S_0) \rangle + 3.1 \frac{\langle \mathcal{O}_8^{J/\psi}(^3P_0) \rangle}{m_c^2} < 2.3 \times 10^{-2}\,\text{GeV}^3. \quad (4)$$

The individual matrix elements, $\langle \mathcal{O}_8^{J/\psi}(^1S_0) \rangle$ and $\langle \mathcal{O}_8^{J/\psi}(^3P_0) \rangle$ remain undetermined. Production of J/ψ in Z^0 decays is dominated by gluon fragmentation and is therefore sensitive to $\langle \mathcal{O}_8^{J/\psi}(^3S_1) \rangle$. A recent analysis,[18] which sums logarithms of $2M_{J/\psi}/M_Z$ and $2E_{J/\psi}/M_Z$ to all orders, extracts the following effective matrix element:

$$\langle \hat{\mathcal{O}}_8^{J/\psi}(^3S_1) \rangle = \sum_H \langle \hat{\mathcal{O}}_8^H(^3S_1) \rangle \text{Br}(H \to J/\psi + X)$$
$$= 1.9 \pm 0.5 \pm 1.0 \times 10^{-2}\,\text{GeV}^3. \qquad (5)$$

The first error is statistical, the second theoretical. The effective $\langle \hat{\mathcal{O}}_8^{J/\psi}(^3S_1) \rangle$ includes feeddown from higher mass charmonium resonances (H) because it is not possible to seperate prompt J/ψ experimentally.

Next-to-leading order calculations of photoproduction would seem to indicate that NRQCD predicts a large excess of J/ψ at large z which is not observed in the data.[19] (Here $z = E_{J/\psi}/E_\gamma$, where energies are measured in the proton rest frame.) This led to the conclusion that the color-octet matrix elements extracted from the Tevatron were an order of magnitude too large to be consistent with photoproduction data from the HERA experiments. However, this enhancement is actually an artifact of the next-to-leading order perturbative calculation. It is worth examining this process in more detail because it is indicative of some of the pitfalls one may encounter when trying to compare NRQCD with data.

The leading order diagrams are shown in Fig. 1, and results in the following perturbative cross section:

$$\frac{d\sigma}{dz} = \sigma_0 \delta(1-z), \qquad (6)$$

$$\sigma_0 = \frac{4\pi^3 \alpha_s \alpha}{s_{\gamma p} m_c^3} G(4m_c^2/s_{\gamma p}) \times \qquad (7)$$

$$\left(\langle \mathcal{O}_8^{J/\psi}(^1S_0) \rangle + 7 \frac{\langle \mathcal{O}_8^{J/\psi}(^3P_0) \rangle}{m_c^2} \right)$$

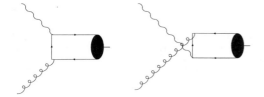

Figure 1: Leading order diagrams for color-octet photoproduction of J/ψ

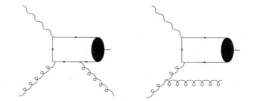

Figure 2: Next-to-leading order diagrams for color-octet photoproduction of J/ψ

$G(x_g)$ is the gluon distribution function of the proton. $s_{\gamma p}$ is the center of mass energy squared of the photon-proton system. The delta function in Eq.(6) is an artifact of the perturbative expansion which will be smeared by nonperturbative emission of soft gluons as the $c\bar{c}$ pair hadronizes into the final state hadron. How this nonperturbative smearing can be accounted for within the NRQCD formalism will be discussed below. For now, we concern ourselves with perturbative corrections.

The next-to-leading order diagrams are shown in Fig. 2. In the limit in which the final state gluon is soft, the cross section is given by: [20]

$$
\begin{aligned}
\frac{d\sigma}{dz} &= \int dx_g G(x_g) \frac{\sum |M_{LO}|^2}{16\pi x_g s_{\gamma p}} \\
&\times C_A g_s^2 \left(\frac{2P \cdot g}{P \cdot k \, g \cdot k} - \frac{4m_c^2}{(P \cdot k)^2} \right) \\
&= \int_\rho^z dx G(\rho/x) \frac{\sum |M_{LO}|^2}{4m_c^2 s_{\gamma p}} C_A \alpha_s \frac{z-x}{(1-x)^2(1-z)} \\
&= -\sigma_0 \frac{C_A \alpha_s}{\pi} \frac{\ln(1-z)}{1-z} + ...
\end{aligned}
\tag{8}
$$

M_{LO} is the leading order matrix element from Fig. 1. P, g, and k are the four momentum of the J/ψ, the initial state gluon, and the final state gluon, respectiverly. $\rho = 4m_c^2/s_{\gamma p}$ and $x = \rho/x_g$. In the last line of Eq.(8), we take $x = 1$ inside the argument of the structure function in order to obtain an analytic expression. This approximation is valid up to subleading logs since the leading logs come from the region $x \to z \to 1$. In higher orders, we expect to find terms of the form $\alpha_s^n \ln^{2n-1}(1-z)/(1-z)$ which need to be resummed before theory can be compared with experimental data. The divergence of the cross section as $z \to 1$ is an artifact of the next-to-leading order calculation. Once the resummation is performed, the cross section will be well-behaved as $z \to 1$.

There are also important nonperturbative corrections[21] near $z = 1$. For quarkonium production near the boundaries of phase space, it is sometimes the case that contributions from NRQCD operators which are higher order in v are enhanced by kinematic factors. This results in the breakdown of the NRQCD expansion. The crux of the problem is that in the perturbative QCD part of the matching calculation one uses twice the heavy quark mass instead of the quarkonium mass to compute the phase space for the production of the quarkonium meson. The difference between $2m_Q$ and M_H is a v^2 correction, which is ignored in leading-order calculations. However, at the boundaries of phase space this difference becomes important, and it is necessary to sum an infinite number of NRQCD matrix elements. This resummation leads to a universal distribution function called a shape function which replaces the delta function in Eq. (6). Because of the universality of the shape functions, it may be possible in the future to test NRQCD by comparing shape functions extracted from different quarkonium production processes. This would require more precise data than is currently available.

As $z \to 1$, the photoproduction cross section becomes sensitive to the effects of soft gluon radiation. Within pertubation theory this sensitivity is signalled by the presence of large logarithms which must be resummed. The v expansion in Eq. (1) also breaks down and an infinite set of operators become relevant. For this reason, it is impossible to extract NRQCD matrix elements from this distribution. When leptoproduction is discussed below, the analysis will focus on observables for which shape function corrections are not kinematically enhanced, because it is only from these distributions where one can hope to obtain a reliable extraction of NRQCD matrix elements.

In Ref.[22], leptoproduction of J/ψ is examined and found to be a useful process from which to extract NRQCD matrix elements. This process is similar to photoproduction, however, now the photon is off-shell. At leading order, the cross section is again given by the diagrams in Fig. 1:

$$
\frac{d\sigma}{dQ^2} = \int dy \int dx_g \, G(x_g) \, \delta(x_g y s - (2m_c)^2 - Q^2)
$$

$$\times \frac{2\alpha_s(\mu)\alpha^2 e_c^2 \pi^2}{Q^2(Q^2+(2m_c)^2)m_c} \left\{ \frac{1+(1-y)^2}{y} \left[\langle \mathcal{O}_8^\psi(^1S_0) \rangle \right. \right.$$

$$\left. + \frac{3Q^2+7(2m_c)^2}{Q^2+(2m_c)^2} \frac{\langle \mathcal{O}_8^\psi(^3P_0) \rangle}{m_c^2} \right]$$

$$\left. - \frac{8(2m_c)^2 Q^2}{x_g s(Q^2+(2m_c)^2)} \frac{\langle \mathcal{O}_8^\psi(^3P_0) \rangle}{m_c^2} \right\}, \qquad (9)$$

where s is the electron-proton center-of-mass energy squared. The momentum fraction of the virtual photon relative to the incoming lepton is $y \equiv P_p \cdot q / P_p \cdot k$, where P_p is the proton four-momentum, q is the photon four-momentum, and k is the incoming lepton four-momentum, and $Q^2 \equiv -q^2$.

There are several advantages to extracting NRQCD matrix elements from this process. First of all, note that the relative importance of $\langle \mathcal{O}_8^\psi(^1S_0) \rangle$ and $\langle \mathcal{O}_8^\psi(^3P_0) \rangle$ changes as a function of Q^2. Thus, it is possible to fit the differential cross section as a function of Q^2 and extract both of these matrix elements seperately. Second, the scale for the coupling is set by $\mu^2 \approx Q^2 + (2m_c)^2$, so as Q^2 increases, perturbative corrections should be increasingly suppressed relative to photoproduction. In Ref.[22], next-to-leading order graphs with real gluon emission were computed and found to give a small contribution to the total cross section for $Q^2 > 4\,\text{GeV}^2$. The uncertainty due to varying the renormalization scale is less than 10% for $Q^2 > 4\,\text{GeV}^2$. This is a great deal smaller than the corresponding uncertainty in hadroproduction. This indicates that higher order perturbative corrections to the leading order color-octet production mechanism will be small. Higher twist corrections to the parton model are also expected to be small as they are suppressed by powers of Q^2.

The leading color-singlet mechanism requires the emission of an additional hard gluon. (See the first diagram in Fig. 2.) This $\alpha_s(Q)$ suppression compensates for the v^4 suppression of the color-octet matrix elements, and it turns out that for $Q^2 > 4\,\text{GeV}^2$, the color-octet contribution is roughly 4 times bigger than the leading color-singlet process. However, color-singlet mechanisms will dominate in the inelastic ($z < 1$) region, where there is no $O(\alpha_s)$ color-octet contribution. In the elastic region, one also expects production of J/ψ via diffractive processes. At large Q^2, diffractive leptoproduction can be studied using perturbative QCD. In Ref.[23], a perturbative analysis of diffractive leptoproduction predicts the cross section to fall as $1/(Q^2+(2m_c)^2)^3$, as compared to $1/(Q^2+(2m_c)^2)^2$ for the color-octet mechanism. Therefore, at sufficiently large Q^2, the diffractive contribution should be negligible correction to our calculation.

One must also consider the possibility of kinematic enhancement of higher order v^2 corrections. An analysis of the Q^2 distribution in Ref.[22] shows that the higher

order in v^2 corrections associated with the shape function are suppressed in the large Q^2 limit. Therefore, this distribution is calculable in NRQCD. However, the z distribution still suffers from large endpoint corrections, and a perturbative calculation of this distribution cannot be compared with experiment near $z = 1$, even at large Q^2.

The largest errors in the calculation of the Q^2 distribution are associated with uncertainty in the charm quark mass. Varying m_c between 1.3 GeV and 1.7 GeV, results in an error of $^{+60\%}_{-25\%}$ at $Q^2 = 10\,\text{GeV}^2$. This error decreases slightly as Q^2 is increased.

Once color-octet matrix elements are extracted, the polarization of the J/ψ can be predicted without introducing any new parameters. Calculations of the polarization of J/ψ produced via the leading order color-octet mechanism as a function of Q^2 are given in Ref.[22]. A precise measurement of the polarization of leptoproduced J/ψ could provide an excellent test of the NRQCD factorization formalism.

Preliminary results from HERA,[24] indicate that $d\sigma/dQ^2$ is well described by NRQCD for $Q^2 > (2m_c)^2$. This is in contrast with CSM calculations which underpredict the cross section by a factor of ~ 5. So this appears to be strong evidence in favor of color-octet mechanisms as understood within the NRQCD factorization formalism. However, the shape of the rapidity distribution at large rapidity is not well reproduced by NRQCD. This could be a consequence of the effects of soft gluon emission which are not accounted for in the leading order calculations of Ref.[22]. Analysis of soft gluon effects and extraction of NRQCD matrix elements will be performed in a future publication.[25]

Acknowledgements

This work was done in collaboration with Sean Fleming. This work was supported by the National Science Foundation under grant number PHY-94004057, by the Departmnent of Energy under grant number DE-FG03-92-ER40701, and by a John A. McCone Fellowship.

References

1. G.T. Bodwin, E. Braaten, and G.P. Lepage, Phys. Rev. D **51** 1125 (1995).
2. H.W. Huang and K.T. Chao, Phys. Rev. D **55** 244, (1997); E. Braaten and Y.Q. Chen, Report No. OHSTY-HEP-T-97-001, hep-ph/9701242 (unpublished).
3. G.P. Lepage, L. Magnea, C. Nakhleh, U. Magnea, and K. Hornbostle, Phys. Rev. D **46**, 4052 (1992).
4. E. Braaten and S. Fleming, Phys. Rev. Lett. **74** 3327 (1995).

5. P. Ko, J. Lee, and H.S. Song, Phys. Rev. **D53**, 1409 (1996).

6. M. Beneke, F. Maltoni, and I. Z. Rothstein, hep-ph/9808360.

7. P. Cho, Phys. Lett. **B368** 171 (1996).

8. P. Cho and M. Wise, Phys. Lett. **B346** 129 (1995).

9. M. Beneke and I. Z. Rothstein, Phys. Lett. **B372** 157 (1996), **B389** 769 (1996).

10. A. Leibovich, Phys. Rev. **D56** 4412 (1997).

11. M. Beneke and M. Krämer, Phys. Rev. **D55**, 5269 (1997).

12. J.F. Amundson, O.J.P. Éboli, E. M. Gregores, and F. Halzen, Phys. Lett. B **372**, 127 (1996), *ibid* **390** 323 (1997); O.J.P. Éboli, E. M. Gregores, and F. Halzen, Phys. Lett. B **395**, 113 (1997); G. A. Schuler and R. Vogt, Phys. Lett. B **387**, 181 (1996).

13. P. Cho and A. Leibovich, Phys. Rev. D **53**, 150 (1996); **53**, 6203 (1996).

14. M. Cacciari, M. Greco, M.L. Mangano, and A. Petrelli, Phys. Lett. **356B**, 553 (1995).

15. B. A. Kniehl and G. Kramer, hep-ph/9803256.

16. B. Cano-Coloma and M.A. Sanchis-Lozano, Nucl. Phys. **B508**, 753 (1997); Phys. Lett. **B406**, 232 (1997).

17. M. Beneke and I. Z. Rothstein, Phys. Rev. **D54** 2005 (1996), **D54** 7082 (1996).

18. C.G. Boyd, A. Leibovich, and I. Rothstein, hep-ph/9810364.

19. M. Cacciari and M. Krämer, Phys. Rev. Lett. **76**, 4128 (1996); P. Ko, J. Lee, and H.S. Song, Phys. Rev. **D54**, 4312 (1996).

20. F. Maltoni, M.L. Mangano, and A. Petrelli, Nucl. Phys. **B519**, 361 (1998).

21. M. Beneke, I. Z. Rothstein, and M. B. Wise, Phys. Lett. **B408**, 373 (1997).

22. S. Fleming and T. Mehen, Phys. Rev. **D57**, 1846 (1998).

23. S. J. Brodsky, L. Frankfurt, J. F. Gunion, A.H. Mueller, and M. Strikman, Phys. Rev. D **50**, 3134 (1994);
M.G. Ryskin, R.G. Roberts, A.D. Martin, and E.M. Levin, Report No. RAL-TR-95-065, hep-ph/9511228.

24. H1 Collaboration, preprint submitted to ICHEP 98.

25. S. Fleming and T. Mehen, in preparation.

New Results on Heavy $Q\bar{Q}$ Pair Production Close to Threshold

A. H. Hoang

Department of Physics, University of California, San Diego,
La Jolla, CA 92093-0319, USA
E-mail: hoang@einstein.ucsd.edu

We review the theoretical ideas and tools required to arrive at a next-to-next-to-leading order (NNLO) description of the heavy quark-antiquark production cross section in e^+e^- annihilation for the case that the center of mass kinetic energy of the quarks is larger than $\Lambda_{\rm QCD}$. In this case the NNLO cross section can be calculated with purely perturbative methods. We present details of the calculation and discuss one application, the determination of the bottom quark mass from Υ sum rules.

1 Introduction

Within the last year significant progress has been achieved in the perturbative description of heavy quark-antiquark ($Q\bar{Q}$) pairs in the kinematic regime close to threshold using the concepts of effective field theories. In this talk we report on some of these developments from the point of view of practical applications. There have also been many interesting new results concerning the systematics of the effective field theory description of $Q\bar{Q}$ pairs at threshold which would deserve to be presented in detail but can only be mentioned peripherally due to lack of space.

To be definite, we consider the total cross section $\sigma^{thr} \equiv \sigma(e^+e^- \to Q\bar{Q} + \text{anything})$ for the c.m. energies $\sqrt{s} \approx 2M_Q$, M_Q being the heavy quark mass. In this kinematic regime the $Q\bar{Q}$ dynamics is characterized by the fact that the heavy quarks have small c.m. velocities. Considering the strong dependence of the velocity ($v = \sqrt{1 - 4M_Q^2/s}$) in the threshold regime on M_Q, it is obvious that, at least in principle, accurate calculations of the cross section may lead to precise determinations of the quark mass. The situation, however, is not simple due to the fact that the $Q\bar{Q}$ system in the nonrelativistic regime is governed by (at least) three scales. This makes this system particularly complicated. The three scales are M_Q, the quark mass, $M_Q v$, the relative momentum, and $M_Q v^2$, the kinetic energy in the c.m. frame. This has three important consequences which are symptomatic for all nonrelativistic $Q\bar{Q}$ systems:

1. Because $v \ll 1$, the three scales are widely separated ($M_Q \gg M_Q v \gg M_Q v^2$). The theoretical tools required to describe the $Q\bar{Q}$ pair strongly depend on whether the scales each are smaller or larger then $\Lambda_{\rm QCD}$. The most transparent situation arises if $M_Q v^2 > \Lambda_{\rm QCD}$ because in this case the theoretical tools resemble most closely those of QED bound state calculations. In fact, for the application mentioned in this talk it is ensured that this condition is satisfied.

2. Because ratios of the three scales arise, the de-

scription of the nonrelativistic $Q\bar{Q}$ pair involves a double expansion in α_s and v. This means that the standard multi-loop expansion in α_s breaks down. The most prominent indication of this fact is the so called *Coulomb singularity* which corresponds to a singular $(\alpha_s/v)^n$-behavior in the n-loop correction to the amplitude $\gamma \to Q\bar{Q}$ for $v \to 0$. The latter singularity is caused by the 00-component of the gluon propagator and, in Coulomb gauge, is directly related to the exchange of longitudinally polarized gluons. This singularity has to be treated by resumming diagrams involving longitudinal gluon exchange to all orders in α_s. The Coulomb singularity exists in the nonrelativistic limit, but there are also power-like and logarithmic divergences in v which are suppressed by powers of the velocity. At this point we would like to specify more clearly what "resummation" means in this context. Strictly speaking it would mean a resummation of the perturbation series in α_s to all orders, where the respective coefficients are expanded up to a certain power in v. This means that the resummation would be carried out in the (formal) limit $\alpha_s \ll v \ll 1$. The resulting series would then (uniquely) define analytic functions which could be continued to the region of interest $|v| \sim \alpha_s$. Typical structures like $Q\bar{Q}$ bound states can only be observed after this continuation. Of course, this would be a highly cumbersome and inefficient method. It is therefore mandatory to reformulate the problem in terms of wave equations. The solutions of these wave equations are equivalent to the results of the resummation method. As a matter of convenience we will also call the wave equation method "resummation" for the rest of this talk.

3. From point 1 we can conclude that the c.m. velocity of the heavy quarks should satisfy the condition $v > (\Lambda_{\rm QCD}/M_Q)^{1/2}$ to ensure that also the scale $M_Q v^2$ is perturbative. In addition, relativistic corrections are not suppressed by factors of π like multi-loop corrections. This means that relativistic corrections can be quite sizeable. The corrections of $\mathcal{O}(v^2)$, called NNLO from now on, can be estimated to be of order $20 - 30\%$ for $b\bar{b}$ and

5% for $t\bar{t}$. Thus, the calculation of higher order relativistic corrections is mandatory in order to achieve sufficient theoretical accuracy and to test the reliability of the perturbative description itself. Obviously the perturbative treatment works better if M_Q is large.

In order to arrive at a reliable theoretical description of σ^{thr} we have to go through two steps: first, we have to address the question how to organize the calculation systematically keeping in mind points 1-3, and, second, we have to actually carry out the calculation itself. In Section 2 we will briefly address the systematics and in Section 3 we will present the calculation of the photon mediated cross section at NNLO in the nonrelativistic expansion. Section 4 is devoted to the determination of the bottom quark mass.

2 Systematics and NRQCD

A very economical approach to systematically deal with the problems described previously is to take advantage of the separation of the scales M_Q, $M_Q v$ and $M_Q v^2$ using the concepts of effective field theories. In the following we outline the conceptual steps to arrive at a NNLO description of a nonrelativistic $Q\bar{Q}$ pair for $\Lambda_{QCD} < M_Q v^2$ without going very far into formal considerations. It is the basic idea of the effective field theory approach to integrate out momenta above the scales relevant for the nonrelativistic dynamics of the $Q\bar{Q}$ pair. Doing this, one always has to keep in mind the relation of each of the scales to Λ_{QCD}. (If M_Q were of order or even smaller than Λ_{QCD}, a nonrelativistic expansion would be meaningless in the first place.)

Suppose that M_Q is larger than Λ_{QCD}. In this case we can integrate out momenta of order M_Q because they are not responsible for the nonrelativistic dynamics of the $Q\bar{Q}$ pair. Starting from QCD we then arrive at an effective field theory in which the heavy quarks and the gluons interacting with them only carry momenta below M_Q. This forces us to introduce different fields for heavy quark and antiquark. The resulting theory is called non-relativstic QCD (NRQCD) [1] and its Lagrangian reads

$$
\begin{aligned}
\mathcal{L}_{NRQCD} = {} & -\frac{1}{2}\,\mathrm{Tr}\,G^{\mu\nu}G_{\mu\nu} + \sum_{\text{light quarks}} \bar{q}\,i\!\not{\!D}\,q \\
& + \psi^\dagger\left[iD_t + a_1\frac{\boldsymbol{D}^2}{2\,M_Q} + a_2\frac{\boldsymbol{D}^4}{8\,M_Q^3} + \frac{a_3\,g}{2\,M_Q}\,\boldsymbol{\sigma}\boldsymbol{B}\right. \\
& \left. + \frac{a_4\,g}{8\,M_Q^2}(\boldsymbol{D}\boldsymbol{E} - \boldsymbol{E}\boldsymbol{D}) + \frac{i\,a_5\,g}{8\,M_Q^2}\boldsymbol{\sigma}(\boldsymbol{D}\times\boldsymbol{E} - \boldsymbol{E}\times\boldsymbol{D})\right]\psi + \dots .
\end{aligned}
$$

(1)

The gluonic and light quark degrees of freedom are described by the conventional relativistic Lagrangian, whereas the heavy quark and antiquark are described by the Pauli spinors ψ and χ, respectively. For convenience all color indices are suppressed. Only those terms relevant for the NNLO cross section are displayed, where

we have omitted the antiquark bilinears. The latter can be obtained through charge conjugation symmetry. The effects coming from momenta or order M_Q are encoded in the short-distance coefficients $a_1,\dots,a_5$. They can be determined as a perturbative series in α_s at the scale $\mu_{hard} = M_Q$ through the matching procedure. If Λ_{QCD} were of order of or even larger than $M_Q v$ (which is essentially the situation for $c\bar{c}$) this would be all we could do perturbatively. For Λ_{QCD} smaller than $M_Q v$, however, one can go further and also integrate out gluonic (and light quark) momenta of order $M_Q v$. The resulting theory has been called "potential NRQCD" (PNRQCD) in Ref. [2] and is characterized by the fact that its Lagrangian contains spatially non-local four-fermion interactions which are nothing else than instantaneous (static) $Q\bar{Q}$ potentials. The "short-distance" coefficients of the corresponding operators describe the gluonic (and light quark) effects from momenta of order $M_Q v$ and can be calculated perturbatively at the scale $\mu_{soft} = M_Q v$. To NNLO (i.e. including potentials suppressed by at most α_s^2, α_s/M_Q or $1/M_Q^2$ relative to the Coulomb potential) the relevant $Q\bar{Q}$ potentials read ($a_s \equiv \alpha_s(\mu_{soft})$, $C_A = 3$, $C_F = 4/3$, $T = 1/2$, $\tilde{\mu} \equiv e^\gamma\,\mu_{soft}$, $r \equiv |\vec{r}|$)

$$
\begin{aligned}
V_c(\vec{r}) = {} & -\frac{C_F\,a_s}{r}\left\{1 + \left(\frac{a_s}{4\pi}\right)\left[2\beta_0\ln(\tilde{\mu}\,r) + a_1\right] + \left(\frac{a_s}{4\pi}\right)^2\left[\right.\right. \\
& \left.\left. \beta_0^2\left(4\ln^2(\tilde{\mu}\,r) + \frac{\pi^2}{3}\right) + 2\left(2\beta_0 a_1 + \beta_1\right)\ln(\tilde{\mu}\,r) + a_2\right]\right\},
\end{aligned}
$$

(2)

$$
\begin{aligned}
V_{BF}(\vec{r}) = {} & \frac{C_F\,a_s\,\pi}{M_Q^2}\left[1 + \frac{8}{3}\vec{S}_t\vec{S}_{\bar{t}}\right]\delta^{(3)}(\vec{r}) + \frac{C_F\,a_s}{2\,M_Q^2\,r}\left[\vec{\nabla}^2\right. \\
& \left. + \frac{1}{r^2}\vec{r}(\vec{r}\vec{\nabla})\vec{\nabla}\right] - \frac{3\,C_F\,a_s}{M_Q^2\,r^3}\left[\frac{1}{3}\vec{S}_t\vec{S}_{\bar{t}} - \frac{1}{r^2}\left(\vec{S}_t\vec{r}\right)\left(\vec{S}_{\bar{t}}\vec{r}\right)\right] \\
& + \frac{3\,C_F\,a_s}{2\,M_Q^2\,r^3}\vec{L}\,(\vec{S}_t + \vec{S}_{\bar{t}}),
\end{aligned}
$$

(3)

$$
V_{NA}(\vec{r}) = -\frac{C_A\,C_F\,a_s^2}{2\,M_Q\,r^2},
$$

(4)

where $\vec{S}_t$ and $\vec{S}_{\bar{t}}$ are the quark and antiquark spin operators, $\vec{L}$ is the angular momentum operator and $\beta_{0,1}$ are the one- and two-loop beta-functions. The constants $a_{1,2}$ have been calculated in Refs. [3,4]. V_c is the Coulomb (static) potential and V_{BF} the Breit-Fermi potential known from positronium. V_{NA} is a purely non-Abelian potential generated through non-analytic terms in the one-loop vertex corrections to the Coulomb potential involving the triple gluon vertex. The remaining dynamical gluon (light quark) fields can only carry momenta of order $M_Q v^2$ and describe radiation and retardation effects. If $\Lambda_{QCD} < M_Q v^2$ one can show that those retardation effects are of NNNLO in σ^{thr} using arguments known from QED and taking into account how the gluon self coupling scales with v for gluonic momenta of order $M_Q v^2$. [This only works because the $Q\bar{Q}$ pair is produced in a color singlet state!] This means that retardation effects (and the scale $M_Q v^2$) can be ignored at NNLO and that the $Q\bar{Q}$ dynamics can be described by a two-body positronium-like Schrödinger equation of the form ($E \equiv \sqrt{s} - 2M_Q$)

$$\left(-\frac{\vec{\nabla}^2}{M_Q} - \frac{\vec{\nabla}^4}{4\,M_Q^3} + \left[V_c(\vec{r}) + V_{\mathrm{BF}}(\vec{r}) + V_{\mathrm{NA}}(\vec{r}) \right] - E \right)$$
$$\times\, G(\vec{r}, \vec{r}', E) = \delta^{(3)}(\vec{r} - \vec{r}'), \qquad (5)$$

containing the heavy quark kinetic energy up to NNLO and the instantaneous potentials (2)–(4). In Eq. (5) M_Q is defined as the pole mass. If Λ_{QCD} were of order or even larger than $M_Q v^2$, on the other hand, the coupling of the radiation gluon with the heavy quark would become of order one and retardation effects would be NNLO. In this case a perturbative calculation of σ^{thr} would be impossible at NNLO because retardation effects would be non-perturbative. For this reason we have to make sure that the condition $\Lambda_{\mathrm{QCD}} < M_Q v^2$ is always satisfied, if the NNLO expression for σ^{thr} shall be trusted.

3 Calculation of the Total Cross Section

For simplicity we only consider the photon mediated total cross section. The inclusion of the Z exchange is trivial for the vector current contributions. The contributions from the axial-vector current can be easily implemented at NNLO because the axial-vector current produces the $Q\bar{Q}$ pair in a P-wave state which leads to a suppression $\propto v^2$ relative to the vector current contribution. We start from the fully covariant expression for the total cross section ($R_{Q\bar{Q}} \equiv \sigma^{thr}/\sigma(e^+e^- \to \mu^+\mu^-)$)

$$R_{Q\bar{Q}}(q^2) = \frac{4\pi Q_b^2}{q^2} \mathrm{Im}\left[-i \int dx\, e^{iq.x} \langle 0|T j_\mu^b(x)\, j^{b\,\mu}(0)|0\rangle \right]$$
$$\equiv \frac{4\pi Q_b^2}{q^2} \mathrm{Im}\left[-i \langle 0|T \tilde{j}_\mu^b(q)\, \tilde{j}^{b\,\mu}(-q)|0\rangle \right], \qquad (6)$$

and expand the electromagnetic current which produces/annihilates the $Q\bar{Q}$ pair with c.m. energy $\sqrt{q^2}$ in terms of $^3S_1$ NRQCD currents up to dimension eight ($i = 1, 2, 3$)

$$\tilde{j}_i(q) = b_1 \left(\tilde{\psi}^\dagger \sigma_i \tilde{\chi} \right) - \frac{b_2}{6M_Q^2} \left(\tilde{\psi}^\dagger \sigma_i (-\tfrac{i}{2}\overset{\leftrightarrow}{\boldsymbol{D}})^2 \tilde{\chi} \right) + \dots,$$
$$\tilde{j}_i(-q) = b_1 \left(\tilde{\chi}^\dagger \sigma_i \tilde{\psi} \right) - \frac{b_2}{6M_Q^2} \left(\tilde{\chi}^\dagger \sigma_i (-\tfrac{i}{2}\overset{\leftrightarrow}{\boldsymbol{D}})^2 \tilde{\psi} \right) + \dots, (7)$$

where the constants b_1 and b_2 are short-distance coefficients normalized to one at the Born level. Only the spatial components of the currents contribute at the NNLO level. Inserting expansion (7) back into Eq. (6) leads to the nonrelativistic expansion of the NNLO cross section

$$R_{\mathrm{NNLO}}^{thr}(E) = \frac{\pi Q_Q^2}{M_Q^2} C_1(\mu_{\mathrm{hard}}, \mu_{\mathrm{fac}}) \mathrm{Im}\left[\mathcal{A}_1(E, \mu_{\mathrm{soft}}, \mu_{\mathrm{fac}}) \right]$$
$$-\frac{4\pi Q_Q^2}{3M_Q^4} C_2(\mu_{\mathrm{hard}}, \mu_{\mathrm{fac}}) \mathrm{Im}\left[\mathcal{A}_2(E, \mu_{\mathrm{soft}}, \mu_{\mathrm{fac}}) \right] + \dots, (8)$$

where

$$\mathcal{A}_1 = i\,\langle 0| (\tilde{\psi}^\dagger \vec{\sigma} \tilde{\chi})\, (\tilde{\chi}^\dagger \vec{\sigma} \tilde{\psi})|0\rangle, \qquad (9)$$
$$\mathcal{A}_2 = \tfrac{1}{2} i\,\langle 0| (\tilde{\psi}^\dagger \vec{\sigma} \tilde{\chi})\, (\tilde{\chi}^\dagger \vec{\sigma} (-\tfrac{i}{2}\overset{\leftrightarrow}{\boldsymbol{D}})^2 \tilde{\psi}) + \mathrm{h.c.}|0\rangle. \quad (10)$$

The cross section is expanded in terms of a sum of absorptive parts of nonrelativistic current correlators, each of them multiplied by a short-distance coefficient. In fact, the right-hand side (RHS) of Eq. (8) just represents an application of the factorization formalism proposed in [5]. The second term on the RHS of Eq. (8) is suppressed by v^2, i.e. of NNLO. This can be seen explicitly by using the equations of motion from the NRQCD Lagrangian, which relates the correlator $\mathcal{A}_2$ directly to $\mathcal{A}_1$,

$$\mathcal{A}_2 = M_Q\, E\, \mathcal{A}_1. \qquad (11)$$

Relation (11) has also been used to obtain the coefficient $-4/3$ in front of the second term on the RHS of Eq. (8). The nonrelativistic current correlators $\mathcal{A}_{1,2}$ contain the resummation of the singular terms mentioned previously. They depend on the renormalization scale μ_{soft} through the potentials (2)–(4). The constants C_1 and C_2 (which are also normalized to one at the Born level), on the other hand, describe short-distance effects and, therefore, depend on the hard scale μ_{hard}. They only represent a simple power series in α_s (where the coefficients contain numbers and logarithms of M_Q, μ_{fac} and μ_{hard}) and do not contain any resummations in α_s. At NNLO they have to be calculated up to order α_s^2 because we count α_s/v of order one for a perturbative nonrelativistic $Q\bar{Q}$ system.

The nonrelativistic correlators $\mathcal{A}_{1,2}$ are calculated by determining the Green function of the Schrödinger equation (5) where V_{BF} is evaluated for the $^3S_1$ configuration. The NNLO relation between the correlator $\mathcal{A}_1$ and Green function reads

$$\mathcal{A}_1 = 6\,N_c \left[\lim_{|\vec{r}|,|\vec{r}'| \to 0} G(\vec{r}, \vec{r}', E) \right]. \qquad (12)$$

Eq. (12) can be quickly derived from the facts that $G(\vec{r}, \vec{r}', E)$ describes the propagation of a bottom-antibottom pair which is produced and annihilated at relative distances $|\vec{r}|$ and $|\vec{r}'|$, respectively, and that the $Q\bar{Q}$ pair is produced and annihilated through the electromagnetic current at zero distances. Therefore $\mathcal{A}_1$ must be proportional to $\lim_{|\vec{r}|,|\vec{r}'| \to 0} G(\vec{r}, \vec{r}', E)$. The correct proportionality constant can then be determined by considering production of a free (i.e. $\alpha_s = 0$) $Q\bar{Q}$ pair in the nonrelativistic limit. (In this case the Born cross section in full QCD can be easily compared to the imaginary part of the Green function of the free nonrelativistic Schrödinger equation.) The correlator $\mathcal{A}_2$ is determined from $\mathcal{A}_1$ via relation (11). We would like to emphasize that the zero-distance Green function on the RHS of Eqs. (12) contains UV divergences from the higher dimensional NNLO effects which have to be regularized. In the actual calculations carried out in Ref. [6] we have imposed the explicit short-distance cutoff μ_{fac}, called factorization scale. This is the reason why the correlators also depend on μ_{fac}. One way to solve Eq. (5) is to start from the well known Green function $G_c^{(0)}$ of the non-relativistic Coulomb problem and to incorporate all the higher order terms via first and second order Rayleigh-Schrödinger time-independent perturbation theory.

To determine the short-distance constant C_1 up to $\mathcal{O}(\alpha_s^2)$ we can expand expression (8) in the (formal) limit $\alpha_s \ll v \ll 1$ (for $\mu_{\text{soft}} = \mu_{\text{hard}}$) up to $\mathcal{O}(\alpha_s^2)$ and demand equality (i.e. match) to the total cross section obtained at the two-loop level in full QCD keeping terms up to NNLO in an expansion in v. In this limit fixed multi-loop perturbation theory (i.e. an expansion in α_s) as well as the nonrelativistic approximation (i.e. a subsequent expansion in v) are feasible. Because the full QCD cross section is independent of μ_{fac}, C_1 also depends on μ_{fac}. We have called this kind of matching calculation at the level of the final result "direct maching". The consistency of the effective field theory approach ensures that C_1 only contains contributions from momenta of order M_Q and does not have any terms singular in v. Details of this calculation and references regarding the important calculations of the two-loop cross section in full QCD at NNLO in the velocity expansion can be found in Ref. [6].

4 Bottom Quark Mass from Υ Mesons

Due to causality, derivatives of the electromagnetic current-current correlator with respect to q^2 at $q^2 = 0$ are directly related to the total photon mediated cross section of bottom quark-antiquark production in e^+e^- annihilation,

$$P_n \equiv \frac{4\pi^2 Q_b^2}{n! \, q^2} \left(\frac{d}{dq^2} \right)^n \Pi_\mu^{\ \mu}(q) \Big|_{q^2=0} = \int \frac{ds}{s^{n+1}} R_{b\bar{b}}(s). \tag{13}$$

Assuming global duality, P_n can be either calculated from experimental data for the total cross section in e^+e^- annihilation[a] or theoretically using quantum chromodynamics (QCD). It is the basic idea of this sum rule to set the moments calculated from experimental data, P_n^{ex}, equal to those determined theoretically from QCD, P_n^{th}, and to use this relation to determine the bottom quark mass (and the strong coupling) by fitting theoretical and experimental moments for various values of n. At this point it is important to set the range of allowed values of n for which the moments P_n^{th} can be trusted

---

[a] At the level of precision in this work the Z mediated cross section can be safely neglected.

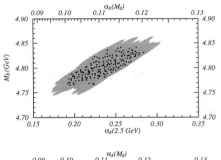

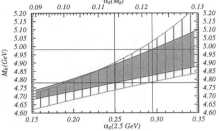

Figure 1: (a) Result for the allowed region in the M_b^{pole}-α_s plane for the unconstrained fit using the moments at NNLO. The gray shaded region represents the allowed region. The dots represent points of minimal χ^2 for a large number of models. Experimental errors are included at the 95% CL level and visible as the narrow gray ellipses. (b) Result for the allowed M_b^{pole} values for a given value of α_s. The gray shaded region corresponds to the allowed ranges for the NNLO analysis and the striped region for the NLO analysis. Experimental errors are included at the 95% CL level. It is illustrated how the allowed range for M_b^{pole} at NNLO is obtained if $0.114 \leq \alpha_s(M_z) \leq 0.122$ is taken as an input.

using the NNLO cross section calculated in the preceding section. As mentioned, we have to make sure that $M_b\langle v \rangle^2 > \Lambda_{\text{QCD}}$ where $\langle v \rangle$ is the average c.m. bottom quark velocity. One can show that $\langle v \rangle \sim 1/\sqrt{n}$ for large n which means that n should be sufficiently smaller than $15 - 20$. It is interesting that the same conclusion can be drawn from the Poggio-Quinn-Weinberg argument that the effective energy smearing range contained in the moments should be larger than Λ_{QCD} [13]. On the other hand, n has to be large enough that the use of the cross section at threshold is justified because the regime close to threshold dominates P_n only in this case. In our analysis [6] we have taken the range $4 \leq n \leq 10$. Larger values of n increase the possibility of large systematic errors. To determine the allowed range for the bottom quark mass we have fitted P_n^{th} to P_n^{ex} for various sets of n's. It turns out that the theoretical errors are much larger than the

Table 1: Recent determinations of bottom quark masses using QCD sum rules for the Υ mesons. $m_b(m_b)$ refers to the $\overline{\text{MS}}$ mass. NLO refers to analyses including corrections of order α_s to the nonrelativistic limit and NNLO to analyses including corrections or order α_s^2, $\alpha_s v$ and v^2. No order is indicated for Ref. 8 because the bound state poles have not been taken into account for P_n^{th} in that analysis. The numbers in the brackets refer to the value of $\alpha_s(M_z)$ obtained in those analyses where a simultaneous fit for mass and QCD coupling has been carried out. All the other analyses have taken α_s as an input. In our analysis (Ref. 6) the errors are not written as Gaussian errors but as allowed ranges.

	order	n	M_b^{pole}	$m_b(m_b)$
Ref. [6]	NLO	$4 - 10$	$4.64 - 4.92$ [.086 − .132]	
	NNLO	$4 - 10$	$4.74 - 4.87$ [.096 − .124]	$4.17 - 4.35$
	NNLO	$4 - 10$	$4.78 - 4.98$	$4.16 - 4.33$
Ref. [7]	NLO	$8 - 20$	$4.827 \pm .007$	
Ref. [8]		$8 - 20$	$4.604 \pm .014$ [.109 ± .001]	$4.13 \pm .06$ [.120$^{+.010}_{-.008}$]
Ref. [9]	NLO	$10 - 20$	$4.75 \pm .04$ [.118$^{+.007}_{-.008}$]	
Ref. [10]	NNLO	$10 - 20$	$4.78 \pm .04$	
Ref. [11]	NNLO	$14 - 18$		$4.20 \pm .10$
Ref. [12]	NNLO	$8 - 12$	$4.80 \pm .06$	

experimental ones. The dominant theoretical errors come from the dependence of P_n^{th} on the scale μ_{soft}. We have combined experimental and theoretical errors by using the "scanning" method [14] in which a large number of statistical fits is carried out for various "reasonable" choices for the scales, each called "a model". It is believed that this method represents a conservative way to combine experimental and large theoretical errors. Our result for a simultaneous fit for the bottom pole mass and α_s is displayed in Fig. 1a. In Fig. 1b the result for the pole mass is displayed if α_s is taken as an input. It is conspicuous that the extracted values for the pole mass are quite different for both methods. This variation could be explained from the fact that the pole mass is not defined beyond an accuracy of Λ_{QCD} due to its strong infrared sensitivity. [15,16]. It therefore seems to be more advantageous to extract a short-distance mass like the $\overline{\text{MS}}$ mass. Taking into account the strong correlations between pole mass and α_s and using the two-loop conversion formula between pole and $\overline{\text{MS}}$ mass we find very good agreement in the mass determination for both methods (see Tab. 1). Using the moments at NLO we also found errors which were much more conservative than the ones given in the NLO analysis of Ref. [7]. In Tab. 1 we give a compilation of all recent sum rule determinations of bottom quark masses based on experimental data from the Υ mesons.

5 Conclusions

During the last year there has been significant progress in the understanding of perturbative heavy quark-antiquark systems in the kinematic regime close to threshold. Using the concepts of effective field theories our knowledge has increased at the conceptual level and a number of previously unknown NNLO corrections have been determined. In this talk we have reviewed the ideas involved to perturbatively calculate and then apply the NNLO corrections of the total $Q\bar{Q}$ production cross section at threshold.

Acknowledgments

I am grateful to P. Labelle, T. Teubner and S. M. Zebarjad for their collaboration on topics related to this talk. This work is supported in part by the U.S. Department of Energy under contract No. DOE-FG03-97ER40546.

References

1. W. E. Caswell and G. E. Lepage, *Phys. Lett.* B **167**, 437 (1986).
2. A. Pineda and J. Soto, *Nucl. Phys. Proc. Suppl.* **64**, 428 (1998).
3. W. Fischler, *Nucl. Phys.* B **129**, 157 (1977); A. Billoire, *Phys. Lett.* B **92**, 343 (1980).
4. M. Peter, *Phys. Rev. Lett.* **78**, 602 (1997); *Nucl. Phys.* B **501**, 471 (1997).
5. G. T. Bodwin, E. Braaten, and G. P. Lepage, *Phys. Rev.* D **51**, 1125 (1995).
6. A. H. Hoang, UCSD/PTH 98-02 [hep-ph/9803454].
7. M. B. Voloshin, *Int. J. Mod. Phys.* **A 10**, 2865 (1995).
8. M. Jamin and A. Pich, *Nucl. Phys.* B **507**, 334 (1997).
9. J. H. Kühn, A. A. Penin, and A. A. Pivovarov, TTP/98-01 [hep-ph/9801356].
10. A. A. Penin and A. A. Pivovarov, TTP/98-13 [hep-ph/9803363].
11. K. Melnikov and A. Yelkovsky, TTP/98-17 [hep-ph/9805270].
12. A. A. Penin and A. A. Pivovarov, INR-98-0986 [hep-ph/9807421].
13. E. C. Poggio, H. R. Quinn, and S. Weinberg, *Phys. Rev.* D **13**, 1958 (1976).
14. A. J. Buras, proceedings of the workshop *Symposium of Heavy Flavours*, Santa Barbara, July 7–11, 1997 [hep-ph/9711217].
15. A. H. Hoang, M. C. Smith, T. Stelzer and S. Willenbrock, UCSD/PTH 98-13 [hep-ph/9804227].
16. M. Beneke, CERN-TH/98-120 [hep-ph/9804241].

PROPERTIES OF b-QUARK PRODUCTION AT THE TEVATRON

F. STICHELBAUT

State University of New York at Stony Brook, Stony Brook, NY 11794, USA
E-mail: stichel@fnal.gov

We report on results obtained by the DØ and CDF collaborations on the properties of b-quark production in $p\bar{p}$ collisions at $\sqrt{s} = 1.8$ TeV. The DØ experiment has measured the inclusive production cross section for muons originating from b-quark decays in the forward region, corresponding to the rapidity range $2.4 < | y^\mu | < 3.2$. Combined with previous measurements in the central region, these results permit a comparison of the rapidity dependence of the b-quark production cross section with the predictions of next-to-leading order QCD. The CDF experiment has achieved the first direct measurement of $b\bar{b}$ rapidity correlations at a hadron collider, with the ratio of the cross sections obtained in two rapidity bins. This measurement is a direct probe of the high-x gluon content of the proton.

1 Introduction

The production of b quarks in $p\bar{p}$ collisions provides a benchmark process for the study of perturbative QCD. Predictions of the QCD theory at the next-to-leading order (NLO) have been available for a decade [1]. They describe the production of heavy flavors by three major mechanisms: flavor creation, with s-channel gluon or t-channel quark exchange, t-channel gluon exchanges with subsequent gluon splitting, and, flavor excitation. The latter two mechanisms dominate the $\mathcal{O}(\alpha_s^3)$ contribution to the cross section and are of about the same size as the leading order (LO) contribution.

The NLO QCD predictions show a large dependence on the choice of the renormalization and factorization scale, μ. This scale is usually choosen as $\mu = \mu_o \equiv \sqrt{m_b^2 + (p_T^b)^2}$, where p_T^b is the transverse momentum of the b-quark and m_b its mass. The theoretical uncertainties are obtained by varying μ between $\mu_0/2$ and $2\mu_0$. This large scale dependence is a symptom of large next-to-NLO (NNLO) corrections. The theoretical predictions are also affected by the choice of m_b, with a central value set to 4.75 GeV/c^2, and varied between 4.5 and 5.0 GeV/c^2 for the uncertainties.

In the last few years, the inclusive b-quark production cross section has been measured in the central rapidity region $| y | < 1$ by CDF [2] and DØ [3] using various data samples. These measurements were found to agree in shape with the NLO QCD predictions over the studied transverse momentum range, $6 < p_T^b < 40$ GeV/c, but to be a factor 2 to 3 larger than the central value obtained with $m_b = 4.75$ GeV/c^2 and $\mu = \mu_0$.

The DØ experiment has now measured the differential muon cross section due to b-quark decays in the forward region, $2.4 < | y^\mu | < 3.2$, as a function of the muon transverse momentum, p_T^μ. This analysis is presented in section 2.

The study of $b\bar{b}$ correlations is a fundamental test of QCD at the next-to-leading order. It allows the comparison of the shapes of kinematic distributions rather than their more uncertain normalizations. Previous studies of the $b\bar{b}$ correlations performed by CDF [4] and DØ [5] focused on the difference in azimuthal angle between the two muons coming from $b\bar{b}$ decays. These results show consistently higher values than the prediction of the NLO QCD theory. The shape of the opening angle between the two muons is found to be consistent with the theory within the uncertainties. However, another CDF analysis [6] of the correlations between the b and $\bar{b}$ quarks in μ–jet events shows discrepancies between the predicted and observed shape of the azimuthal opening angle between the two quarks, indicating that a simple K factor is insufficient to explain the difference between data and theory.

Recently, CDF has extended its correlation studies to the forward region, measuring the $b\bar{b}$ production cross section when one quark is produced in the forward region $(1.8 < | \eta^b | < 2.6)$, and the other in the central pseudorapidity range $(| \eta^b | < 1.5)$. This measured cross section was found to be higher than the QCD prediction by a factor $2.4^{+0.6}_{-0.5}$, with the uncertainty reflecting the experimental error only. In order to reduce the systematic uncertainties affecting both the data and the theory, they now measure the cross section in two rapidity bins and take the ratio. This analysis is described in section 3.

2 Forward μ and b-Quark Production

The DØ analysis of the b-quark production in the forward region is based on the small angle muon spectrometer [7] consisting of magnetized iron toroids and drift tube stations on each side of the interaction region with pseudorapidity coverage of $2.2 < | \eta^\mu | < 3.3$ for a single muon. The data were collected in special runs during the 1994-95 Tevatron run. Using only events corresponding to a single interaction, the effective integrated luminosity for these runs is 82 ± 7 nb^{-1}.

Muons are selected in the rapidity range $2.4 < |y^\mu| < 3.2$, with momentum $p^\mu < 150$ GeV/c and transverse momentum $p_T^\mu > 2$ GeV/c. Additional cuts are applied to ensure a good momentum measurement and an energy deposition in the calorimeter consistent with that from a minimum ionizing particle, leading to a final sample of 6709 events. The contamination due to cosmic rays and hadronic punch-through are estimated to be less than 1%. The contribution from W and Z decays is determined by both DØ data and ISAJET[8] Monte Carlo (MC) simulation and is also negligible. The muon detection efficiency is obtained from MC events with full detector simulation. It is cross-checked with muon unbiased data.

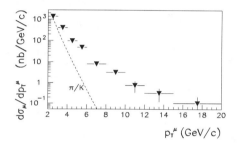

Figure 1: The inclusive muon cross section, per unit of rapidity, in the forward region as a function of p_T^μ. The dashed line represents the expected contributions from π/K decays.

The measured inclusive muon cross section is shown in Fig. 1 as a function of p_T^μ. It is computed per unit of rapidity and only the statistical errors are shown. The systematic uncertainties vary between 14 and 18%, the major error sources being the limited MC statistics and the momentum unfolding.

The two major contributions to the inclusive muon cross section consist of in-flight decays of pions and kaons, and, b- and c-quark decays. The expected π/K contribution is obtained using ISAJET. The p_T spectrum of charged particles predicted by ISAJET was found to be in agreement with the CDF measurement in the central region. As shown in Fig. 1, this contribution dominates the inclusive muon cross section for low p_T^μ but falls more rapidly than the data, the excess being attributed to the heavy flavors decays. After subtraction of the π/K contribution, the muon cross section is multiplied by the b-quark fraction, f_b, defined as the ratio of the muon yield due to b-quark decays to the total muon yield from b- and c-quark decays. This ratio is determined using NLO QCD predictions for b- and c-quark production cross sections. DØ uses events containing a muon associated to a jet to compare the b-quark fraction obtained from the data against the MC predictions. The presence of the

jet allows the determination of the origin of the muon by looking at its transverse momentum relative to that of the jet. From a sample of 31,000 events collected during the entire 1994-95 Tevatron run ($\int \mathcal{L}dt = 90$ nb^{-1}), they find an excellent agreement between data and MC simulation for the b-quark fraction $vs.$ p_T^μ.

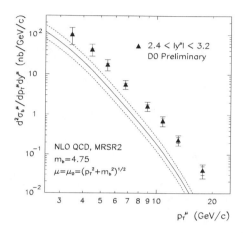

Figure 2: p_T spectrum of forward muons from b decays compared to NLO QCD prediction.

The differential muon cross section from b-quark production and decay is presented in Fig. 2. The systematic uncertainties on this measurement include those of the inclusive muon cross section together with uncertainties due to the f_b estimation $(10-3)$% and the π/K subtraction $(14-1)$%. In Fig. 2, the data are compared to the NLO QCD prediction obtained using the Monte Carlo simulation HVQJET[9]. This MC relies on MNR calculations[1] for the production of the $b/\bar{b}$ partons and ISAJET for the fragmentation to B mesons and the decay of these mesons to muons. The "partonic events" are generated with $m_b = 4.75$ GeV/c, $\mu = \mu_0$, and the MRSR2[10] parton distribution functions. The theoretical uncertainty is determined by varying the parameters m_b and μ from 4.5 and 5.0 GeV/c^2 and from $\mu_0/2$ to $2\mu_0$, respectively. The disagreement between the data and the model prediction is patent and increases with p_T^μ, from around 2.5 at low p_T^μ to more than 5 when $p_T^\mu > 8$ GeV/c. A slightly better agreement in shape is obtained when using the MRSA'[11] parton distribution functions but the data remain larger than the theory by a factor close to 4.

These measurements can be combined with previous DØ results[3] obtained in the central region ($|y^\mu| < 0.8$) to study the rapidity dependence of the b-produced muon

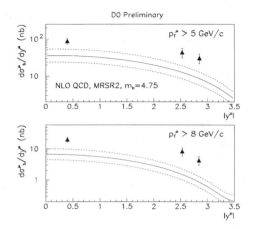

Figure 3: b-produced muon cross section $vs.$ rapidity compared to the NLO QCD prediction.

cross section. The differential muon cross section from b decay as a function of rapidity is shown in Fig. 3 for two p_T^μ ranges, $p_T^\mu > 5$ GeV/c and $p_T^\mu > 8$ GeV/c. The NLO QCD prediction obtained with HVQJET does not exhibit the same rapidity dependence as the data: for $p_T^\mu > 5$ GeV/c, the ratio data/theory is equal to 2.5 ± 0.5 in the central region and increases to 3.6 ± 0.6 in the forward region. The uncertainty on these ratios only reflects the experimental errors. For $p_T^\mu > 8$ GeV/c, these numbers become 3.1 ± 0.6 and 5.1 ± 1.1 in the central and forward regions, respectively.

3 $b\bar{b}$ Rapidity Correlations

Using their forward muon data set collected during the 1994–95 Tevatron run, CDF recently measured the production cross section for $b\bar{b}$ pairs with one quark in the forward region ($1.8 < | \eta^{b_1} | < 2.6$), and the other in central pseudorapidity range ($| \eta^{b_2} | < 1.5$) [12]. The resulting cross section

$$\sigma(p\bar{p} \to b_1 b_2 X) = 6.49 \pm 0.63(\text{stat})^{+1.43}_{-1.23}(\text{syst}) \text{ nb,}$$

valid for $p_T^{b_1, b_2} > 25$ GeV/c, is a factor $2.4^{+0.6}_{-0.5}$ larger than the NLO QCD prediction (2.7 nb) obtained with the standard parameter values and the MRSA$'$ [11] parton distribution functions.

In order to reduce the systematic uncertainties on both the data and the theory, CDF now measures the cross section in two rapidity bins and takes the ratio. In this way, several of the large uncertainties affecting

the absolute measurement of the forward-central cross section cancel in the ratio. Moreover, the dependence of the theory prediction on the choice of the μ scale is reduced.

The event selection requires a central b-quark jet with $E_T > 26$ GeV and $| \eta^{\text{jet}} | < 1.5$, accompanied by a second b-quark decaying into a muon and a jet. This jet must have an $E_T > 15$ GeV and the events are classified as central-central or forward-central depending upon the muon pseudorapidity range, $| \eta^\mu | < 0.6$ or $2.0 < | \eta^\mu | < 2.6$, respectively. For the μ-jet combination, the b-quark decay is identified on the basis of the muon momentum transverse to the μ-jet direction, p_T^{rel}. The central b-quark jet is identified through a secondary vertex tag. A minimal azimuthal opening angle of 60° between the two tags is required to remove events from the region of poor acceptance. Based on 80 pb^{-1} of data collected in 1994-1995, the final data sample contains 382 forward-central and 7544 central-central events.

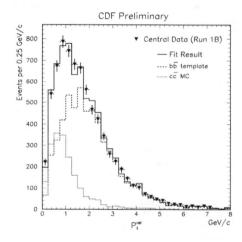

Figure 4: p_T^{rel} distribution for the muons in central-central events together with the fit result.

To extract the signal fraction in each sample, the p_T^{rel} of the muon and the pseudo-$c\tau$ of the b-jet are fitted simultaneously using a binned maximum likelihood method. The template histograms for the signal and the dominant background sources ($c\bar{c}$ production and $b\bar{b}$ production where one b tag is a fake) are obtained either from MC simulation or directly from the data. The p_T^{rel} distribution obtained for central-central events is shown in Fig. 4, together with the fit result. The results of the fits are summarized in Table 1. The signal fraction is about 75% in the forward-central events and 60% in the central-central events.

Table 1: Composition of the data samples.

Source (b-tag/μ-tag)	forward-central	central-central
Real b / Real b	0.739 ± 0.073	0.582 ± 0.021
c / $\bar{c}$	0.123 ± 0.089	0.169 ± 0.021
Real b / Fake	$0.034^{+0.087}_{-0.034}$	0.085 ± 0.026
Fake / Real b	0.104 ± 0.030	0.165 ± 0.009

The ratio of the cross sections is obtained with the relation:

$$R_{\text{data}} = \frac{\sigma(p\bar{p} \to b_1 b_2 X; \; 2.0 < |\,y_{b_1}\,| < 2.6)}{\sigma(p\bar{p} \to b_1 b_2 X; \; |\,y_{b_1}\,| < 0.6)} \quad (1)$$

where $p_T(b_1, b_2) > 25$ GeV/c, $|\,y_{b_2}\,| < 1.5$, and, $\delta\phi(b\bar{b}) > 60°$. It is calculated using the number of signal events in both samples and the ratio of the total efficiency for central-central and forward-central events. Due to the much smaller kinematic acceptance of the forward toroids relative to the central detector, this ratio is equal to 5.4. The result is

$$R_{\text{data}} = 0.361 \pm 0.041(\text{stat})^{+0.011}_{-0.023}(\text{syst}),$$

in good agreement with the NLO QCD prediction obtained with MRSA':

$$R_{\text{theory}} = 0.338^{+0.014}_{-0.097}.$$

The dominant systematic errors on R_{data} come from uncertainty in the energy scale (3%), acceptance (3%), fragmentation (3%), and background assumption (5%). The uncertainty on R_{theory} results from changing the μ scale between $2\mu_0$ and $\mu_0/2$. The evolution of the cross section ratio with the rapidity range of the muon coming from the b-quark semileptonic decay is shown in Fig. 5. The measurement is in agreement with both the LO and NLO QCD predictions.

With the cross section definition given in (1), the forward-central events arise from collisions between one parton with an average $x \approx 0.025$ and a second parton at $x \approx 0.25$. The cross section ratio should thus be sensitive to the gluon distribution in the proton at high x values, i.e. in a region where this gluon distribution is not very well known [13].

Figure 6 shows a comparison of the R_{data} measurement with R_{theory} obtained using the parton distribution sets MRSR1(2) [10] and CTEQ4HJ [14]. The data point and theory curves are normalized to MRSA' and are presented as a function of the rapidity of the b-quark decaying to μ-jet. The data point is in good agreement

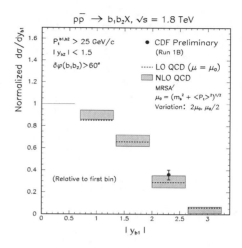

Figure 5: Evolution with rapidity of the cross section ratio.

with the MRS sets and slightly disfavors the CTEQ4HJ distribution.

4 Summary

DØ has measured the b-produced muon cross section as a function of the muon rapidity. Compared with the NLO QCD calculations, the data are a factor 2.5 ± 0.5 larger than the central prediction in the region $|\,y^\mu\,| < 0.8$ for $p_T^\mu > 5$ GeV/c. This discrepancy increases to a factor 3.6 ± 0.6 in the forward region, $2.4 < |\,y^\mu\,| < 3.2$. Recent theoretical developments attempt to account for this increase of the data/theory discrepancy with rapidity. New calculations based on a variable flavor number scheme [15] predict an increase of the forward b-quark production cross section by a factor of 1.2 - 1.5 with respect to the standard NLO QCD calculations corresponding to a fixed flavor number scheme. A stiffer b-quark fragmentation function would result in a 30 to 50% increase of the B-meson production cross section for transverse momentum larger than 10 GeV/c when compared to that obtained with the usual Peterson fragmentation function [16].

CDF has measured the ratio of the forward-central to the central-central cross sections. This ratio $R_{\text{data}} = 0.361 \pm 0.041(\text{stat})^{+0.011}_{-0.023}(\text{syst})$ is in good agreement with the NLO QCD prediction. Although this measurement should show some sensitivity to the gluon distribution in the proton at large x values, more data are needed before being able to distinguish between various parton distributions.

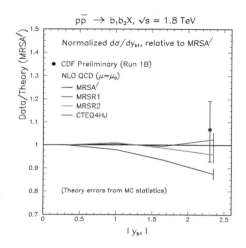

Figure 6: Comparison of the cross section ratio with various parton distribution functions.

Acknowledgements

I would like to thank the organizers for inviting me to this very pleasant conference.

References

1. M. Mangano, P.Nason and G. Ridolfi, *Nucl. Phys.* B **373**, 295 (1992).
2. F. Abe *et al.*, *Phys. Rev. Lett.* **71**, 500, 2396, 2537 (1993); *Phys. Rev. Lett.* **75**, 1451 (1995); *Phys. Rev. Lett.* **79**, 572 (1997).
3. S. Abachi *et al.*, *Phys. Rev. Lett.* **74**, 3548 (1995).
4. F. Abe *et al.*, *Phys. Rev.* D **55**, 2546 (1997).
5. D. Fein, FERMILAB-Conf-97/008-E (1997).
6. F. Abe *et al.*, *Phys. Rev.* D **53**, 1051 (1996).
7. DØ Collaboration, S. Abachi *et al.*, *Nucl. Instrum. Methods* **A338**, 185 (1994).
8. F. Paige and S.D. Protopopescu, BNL-Report BNL-38034 (1986).
9. M. Baarmand, private communication.
10. A.D. Martin, R.G. Roberts and W.J. Stirling, *Phys. Lett.* B **387**, 419 (1996).
11. A.D. Martin, R.G. Roberts and W.J. Stirling, *Phys. Rev.* D **47**, 867 (1993).
12. D. Zieminska, in: Proceedings of the XXXIIIrd Rencontres de Moriond '98, QCD and Hadronic Interactions.
13. J. Huston *et al.*, the CTEQ Collaboration, hep-ph/9801444.
14. H. L. Lai *et al.*, the CTEQ Collaboration, *Phys. Rev.* D **55**, 1280 (1997).
15. F.I. Olness, R.J. Scalise, and Wu-Ki Tung, FERMILAB-Pub-97/428-T, hep-ph/9712494.
16. M.L. Mangano, hep-ph/9711337.

FIRST OBSERVATION OF OPEN b PRODUCTION AT HERA

Georgios TSIPOLITIS

Institute for Particle Physics, ETH Zurich, CH-8093 Zurich, Switzerland
E-mail: yorgos@mail.desy.de

The first observation of open b production in ep collisions is reported. The measurement was carried out using data taken by the H1 experiment at HERA at a centre of mass energy of $\sqrt{s} = 300$ GeV. Events containing muons and jets were used to select a data sample enriched in b events. The cross section is found to be a factor ~ 5 higher than calculations based on LO-QCD models.

1 Introduction

The study of heavy quark production in electron–proton scattering provides an important testing ground for QCD. At the electron–proton collider HERA with an ep centre of mass energy $\sqrt{s} = 300$ GeV, heavy quarks are produced predominantly by the photon gluon fusion process $\gamma g \to q\bar{q}$ $(q = c, b)$, where a real or virtual photon emitted by the electron and a gluon from the proton generate a heavy quark pair.

The kinematics of the ep interactions are described by three independent variables, the centre of mass energy $\sqrt{s}$, the photon four momentum squared $q^2 = -Q^2$ and either one of the scaling variables $y = (q \cdot P/l \cdot P)$, the inelasticity of the ep-interaction, or the Bjorken-x $x = Q^2/(2P \cdot l)$, where P and l denote the four-momenta of the proton and the electron, respectively. The $\gamma - p$ centre of mass energy squared is given by $W_{\gamma p}^2 = W^2 \approx ys - Q^2$. When y is calculated with the Jacquet-Blondel method[1], it is denoted as y_{JB}.

The dominant contribution to heavy quark production is due to the exchange of an almost real photon (*photoproduction*), where the negative squared four-momentum transfer carried by the photon is $Q^2 \approx 0$. The heavy quarks hadronize and are then either detected as *"open charm (beauty)"*, i.e with charm(beauty) hadrons, or alternatively as *"hidden charm(beauty)"* states such as $J/\psi(\Upsilon)$. The mean photon-proton center-of-mass (cms) energy $W_{\gamma p} \sim 200$ GeV is roughly an order of magnitude larger than at previous fixed-target experiments.

Heavy quark production at HERA is dominated by charm quarks. The total cross section of open b production at HERA, $\sigma(ep \to b\bar{b}X)$, is expected to be about two orders of magnitude smaller than $\sigma(ep \to c\bar{c}X)$. The predictions for open b production at HERA for the NLO QCD calculations [2,3,4] are between 4.7 and 10 nb depending on the assumptions for the mass of the b quark and the parton distribution function (pdf) used. The expectation of the AROMA[5] Monte Carlo event generator, which is based on LO-QCD matrix elements with parton showers, is 3.8 nb for a mass of the b quark, $m_b = 4.75$ GeV/c^2 and the MRS(G) pdf. The first obser-

vation of open b production at HERA is reported here. The semileptonic decays of b quarks into muons were used as a tool to find a sample of events enriched in b decays. Two independent methods were used in the analysis. The first uses events with one muon candidate and two jets whereas the second uses events with two muon candidates.

The measurement is made in a visible range determined by the leptons from the heavy quark decay. The NLO calculations stop at the parton level and do not produce the decay leptons. Therefore, the result is compared to the LO-QCD models where this information is available. An estimate for the NLO-QCD contributions can be extracted from the prediction for the total cross section, $\sigma(ep \to b\bar{b}X)$, which are given above.

2 Open Beauty Photoproduction

In the following, two methods used to identify b events in the H1 based on data collected with the H1 detector[6] during the 1995–96 running periods, corresponding to an integrated luminosity of $\mathcal{L} = 8.3$ pb^{-1}.

2.1 Single Muon Analysis

Events are selected with the requirement of at least two jets with $E_T > 6$ GeV in the pseudorapidity range $|\eta| < 2.5$. The jets were identified using a cone algorithm with radius $r = \sqrt{\Delta\eta^2 + \Delta\phi^2} < 1$. In addition it was required that no electron candidate was found with $\theta_e < 177.8°$, which limits the data sample to photoproduction events with $Q^2 < 1$ GeV^2. To identify b events the semileptonic b decays into muons are used where at least one muon is identified in the central region of the detector $(35° < \theta^\mu < 130°)$ where a transverse momentum $p_{T,lab}^\mu > 2$ GeV/c is required. The muon has to be found within one of the jets. Beauty and charm events are separated on a statistical basis using the transverse momentum of the muon $p_{T,rel}^\mu$ relative to the thrust axis of the jet defined as the direction in which T is maximum, with

$$T = \max \frac{\sum |p_i^L|}{\sum |p_i|}$$

i = all the particles in the jet except the muon,

where p_i is the momentum and p_i^L its longitudinal component with respect to the thrust axis. Because the b quark is heavier than the c quark it is expected that the $p_{T,rel}^\mu$ distribution tends to higher values for the former. The shapes of the $p_{T,rel}^\mu$ distributions for beauty and charm are taken from the AROMA Monte Carlo event generator.

The main problem is to identify the background contribution in the event sample. The background consists of events with two jets and with one of the hadrons identified as a fake muon in the detector. To solve this problem it is essential to know the probability that a hadron h fakes a muon. The probability functions $\mathcal{P}_h^\mu(p, \theta)$, where $h = \pi, K, p$, are parametrized as a function of the momentum and the polar angle of the hadron.

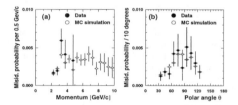

Figure 1: Momentum (a) and polar angle (b) distribution of pion misidentification probabilities $P_\pi^\mu(p)$ and $P_\pi^\mu(\theta)$ describing the probability that a pion is identified as muon in the instrumented iron return yoke. The filled points show the misidentification probability for pions obtained from data; the open points give the corresponding quantities for a Monte Carlo simulation.

These probability functions can be determined directly from the data by studying K_S^0, ϕ and Λ decays as a clean source of pions, kaons and protons. However, the available data do not allow the determination of these functions to sufficient precision. Therefore, after it was checked that the detector simulation describes the data very well (Figure 1), large Monte Carlo samples with single pion, kaon and proton events were generated, passed through the full detector simulation and were used to determine the probability functions $\mathcal{P}_h^\mu(p, \theta)$.

To verify that the method of finding the fake muons works, the sample of pions from the K_S^0 decays in the data is used. In this sample, the number of pions misidentified as muons is known. Then for each of the pions of the sample the probability that it fakes a muon is calculated and is used as weight to plot the θ, p distributions.

As can be seen in Figure 2, the agreement of the distributions expected after applying the $\mathcal{P}_\pi^\mu(p, \theta)$ on the data with the fake muons actually found is very good, both in shape and magnitude. Similar tests have been performed for the case of the kaons using events of elastic ϕ production. The misidentification probability, $\mathcal{P}_h^\mu(p, \theta)$, varies with the polar angle but it does not exceed $6 \cdot 10^{-3}$ in the case of pions and $2 \cdot 10^{-2}$ in the case of kaons. For protons the misidentification probability was found to be lower than $2 \cdot 10^{-3}$.

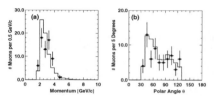

Figure 2: Momentum (a) and polar angle (b) distributions for fake muons due to a pion sample. The pions are identified via the decay $K_S^0 \to \pi^+\pi^-$. The filled points show the muon yield as measured in data. The solid histogram gives the estimate of the muon yield as obtained by assigning $P_\pi^\mu(p, \theta)$ as weight to every pion and summing the weights over the entire pion sample. The measured yield is 64 fake muons; the estimate amounts to 69.8.

The knowledge of the $\mathcal{P}_h^\mu(p, \theta)$ functions allows the calculation of the background directly from the data both in shape and absolute magnitude. For this purpose, events with two jets with $E_T > 6$ GeV were selected. The expected background in the $p_{T,rel}^\mu$ distribution was calculated from all the hadrons, passing the $p_{T,lab}^\mu$ and polar angle requirements. The assignment of a hadron to a pion, kaon or proton was done according to the fractions of these particles as given by JETSET[7]. This hypothesis is supported by the measurements of the SLD[8] and DELPHI[9] collaborations in e^+e^- annihilation.

After the contribution of the fake muons in the $p_{T,rel}^\mu$ spectrum is estimated both in shape and absolute magnitude, the contributions of beauty and charm in addition to the fixed background are fitted to the measured $p_{T,rel}^\mu$ distribution. The result is shown in Figure 3.

The relative composition of the data sample is determined from the fit to the $p_{T,rel}^\mu$ distribution and amounts to $f_b = 51.4 \pm 4.4\%$ (beauty), $f_c = 23.5 \pm 4.3\%$ (charm), and $f_{fake} = 23.5\%$ (background, fixed). Distributions of other variables, like the muon transverse momentum and polar angle as well as the jet transverse energy and polar angle distributions, are very well described with this data composition (Figure 4). The polar angle distributions of the muon and the jet containing the muon

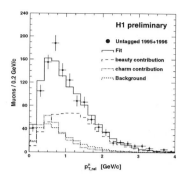

Figure 3: The measured $p^{\mu}_{T,rel}$ distributions and the result of the fit (solid line) are shown. The contributions of beauty (long dashed line) charm (dotted line) and the background (dashed line) are also shown.

Figures 4b,d, indicate that the data are produced mainly in the forward region of the detector.

The open beauty cross section is determined from the number of muons N^{μ}_b attributed by the fit to b decays which are in the visible range $Q^2 < 1\,\text{GeV}^2$, $0.1 < y_{JB} < 0.8$, $p^{\mu}_{T,lab} > 2.0\,\text{GeV}/c$, $35° < \theta^{\mu} < 130°$:

$$\sigma_{vis}(ep \to b\bar{b}X) = [0.93 \pm 0.08(stat.)\,^{+0.21}_{-0.12}(syst.)]\,\text{nb},$$

(H1 preliminary)

where the first uncertainty is statistical and the second systematic. The estimated visible beauty cross section of AROMA amounts to 0.19 nb, which is roughly a factor of five below the measured value. This measurement has been cross-checked using the HERWIG [10] Monte Carlo event generator. The difference from AROMA is of the order of 5% and is included in the systematic error. Other sources of systematic uncertainties are the branching fractions $\mathcal{B}(b \to X\mu\nu)$, the background estimation, the energy scale and the trigger efficiency. The visible charm cross section has been measured with the same method and shows good agreement with the expectation. Using AROMA for the extrapolation to the full kinematic range, the total charm cross section agrees to well within the errors with the total charm cross section measured using the reconstruction of D^* mesons [11].

2.2 A Dimuon Analysis

An alternative method to search for b production is based on like-sign dimuon events. As it can be seen in Figure 5 the only physics channel, in the lower orders, that can produce like-sign dimuons at HERA is b production. The number of like-sign dimuon events, $N_{\pm\pm}$, can be written

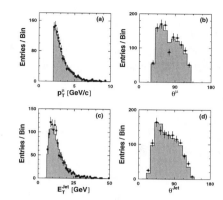

Figure 4: Comparison of measured kinematic distributions (full dots) to the sum of the different components (hatched histogram). (a) The transverse momentum of the muons, (b) the polar angle distribution of the muons, (c) the transverse energy of the jet containing the muon and (d) the polar angle distribution of the jets containing the muon.

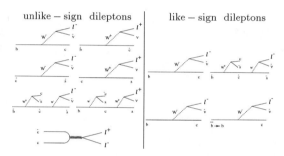

Figure 5: Sources of dilepton events

as:

$$N_{\pm\pm} = N_{bb} + N_{bf} + N_{cf} + N_{ff}$$

where N_{bb} is the "real" number of dimuons both coming from b decays, N_{bf} (N_{cf}) the number of events where one b(c) decays via a semi-muonic channel and one like-sign hadron misidentified as a muon, and N_{ff} is the number of events where both muon candidates are due to hadron misidentification. Note that $N_{bf} + N_{cf} + N_{ff}$ is nothing but the data sample with the single muon candidates combined with a hadron faking a muon. Therefore, one can use the single muon event sample in combination with the $\mathcal{P}^{\mu}_h(p, \theta)$ to estimate the background. Requiring two muons, each one having $p^{\mu}_{T,lab} > 2.0\,\text{GeV}/c$ and $35° < \theta^{\mu} < 130°$, gives 9 events in which like sign dimuons are found ; the expected background is 3.0 ± 0.15 events. This number of events is translated into a cross section in the visible range $(0.1 < y_{JB} < 0.9,$ $p^{\mu}_{T,lab\ 1(2)} > 2.0\,\mathrm{GeV}/c,\ 35^\circ < \theta^\mu < 130^\circ)$ and found to be

$$\sigma_{vis}(ep \to b\bar{b}X) = 55 \pm 30(stat.) \pm 7(syst.)\,\mathrm{pb},$$

(H1 preliminary)

The expectation of AROMA is 17 pb. The main source of the systematical uncertainty is due to the errors on the measurements of the branching fractions $\mathcal{B}(b \to c\mu\nu)$ and $\mathcal{B}(b \to c \to s\mu\nu)$.

3 Conclusion

The open b production cross section has been observed for the first time at HERA with the H1 detector using semi-muonic decays of the b quarks. The cross section found to be a factor of ~ 5 higher than an expectation based on LO-QCD. Using events with two like-sign muons, the cross section is larger than the LO-QCD expectation, though the statistical uncertainty of the this latter measurement is large.

Acknowledgments

It is my pleasure to thank the organizers for a very stimulating conference.

References

1. F. Jacquet and A. Blondel, *Proceedings of the Study of an ep facility for Europe*, Ed. U. Amaldi, DESY 79/48, 391.

2. R.K. Ellis, and P. Nason, *Nucl. Phys.* B **312**, 551 (1989).

3. J. Smith and W. L. van Neerven, *Nucl. Phys.* B **374**, 36 (1992).

4. S. Frixione *et al*, *Phys. Lett.* B **348**, 633 (1994).

5. G. Ingelman, J. Rathsman and G. A. Schuler, *Comp. Phys. Com.* **101**, 135 (1997).

6. H1 Collab.,*Nucl. Instrum. Methods* A **386**, 310 (1997).

7. T. Sjöstrand and M. Bengtsson, *Comp. Phys. Com.* **43**, 367 (1987).

8. SLD Collab., "Production of $\pi^\pm$, $K^\pm$, K^0, K^{*0}, ϕ, p and Λ^0 in Hadronic Z decays", contributed paper # 287 to the International Europhysics Conference on HEP, Aug. 19-26, 1997, Jerusalem, Israel.

9. DELPHI Collab., "$\pi^\pm$, $K^\pm$, p and $\bar{p}$ production in $Z^0 \to q\bar{q}$, $Z^0 \to b\bar{b}$ and $Z^0 \to u\bar{u}, d\bar{d}, s\bar{s}$", contributed paper # 541 to the International Europhysics Conference on HEP, Aug. 19-26, 1997, Jerusalem, Israel.

10. G. Marchesini *et al*, *Comp. Phys. Com.* **67**, 465 (1992).

11. H1 Collaboration, *Nucl. Phys.* B **472**, 32 (1996).

Top Quark Mass from CDF

W.-M. Yao
For the CDF Collaboration

Lawrence Berkeley National Laboratory, One Cyclotron Road, Berkeley, CA 94720, USA
E-mail: wmyao@lbl.gov

We report on the most recent measurements of the top quark mass in three decay channels, performed by the CDF Collaboration at the Tevatron collider. We combine these results to obtain a mass value of 176.0 ± 6.5 GeV/c^2.

1 Introduction

A precise measurement of the top quark mass is an important ingredient in testing the consistency of the standard model with experimental data. In addition, precise W and top mass measurements can provide information on the mass of the Higgs boson, which is a remnant of the mechanism that gives rise to spontaneous electroweak symmetry breaking.

Within the framework of the Standard Model the top quark decays almost exclusively into a real W boson and a b quark. The observed event topology is then determined by the decay modes of the two W bosons, which can be classified into three decay channels. Decays of W boson to τ leptons are not explicitly included in the study except when they subsequently decay to an electron or a muon.

- Dilepton Channel: About 5% of the time both W bosons decay to $e\nu$ or $\mu\nu$.

- Lepton + Jets Channel: In 30% of the cases, one W boson decays to $e\nu$ or $\mu\nu$, and the other to a $q\bar{q}'$ pair.

- All Hadronic Channels: Finally 44% of the final states involve the hadronic decay of both W bosons.

Using 109 pb^{-1} of data accumulated by the CDF experiment at the Fermilab Tevatron from 1992 through 1995, we report an improved measurement of the top quark mass by combining the results from the three decay channels [1, 2, 3].

2 Top Mass Measurement from Lepton + jets

The advantage of measuring a top quark mass in the Lepton + Jets channel [1] is its relatively larger branching ratio and the ability to full reconstruct the top mass on an event-to-event basis. We select events containing a single isolated electron (muon) with E_T $(P_T) > 20$ GeV (GeV/c) in the central region and missing transverse energy $\not{E}_T > 20$ GeV. At least four jets are required in each event, three of which must have an ob-

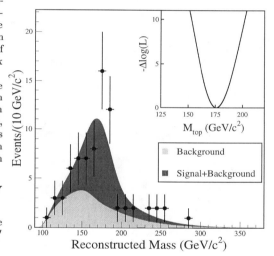

Figure 1: Reconstructed-mass distribution of the four mass subsamples combined. The data (points) are compared with the result of the combined fit (dark shading) and with the background component of the fit (light shading). The inset shows the variation of the combined negative log-likelihood with M_{top}.

served $E_T > 15$ GeV and $|\eta| < 2$ in a cone radius $\Delta R = \sqrt{\Delta\eta^2 + \Delta\phi^2} = 0.4$. In order to increase the acceptance, we relax the requirements on the fourth jet to be $E_T > 8$ GeV and $|\eta| < 2.4$, provided one of four leading jets is tagged by the Silicon Vertex tagging (SVX) or Soft Lepton tagging (SLT) algorithms. SVX tags are only allowed on jets with observed $E_T > 15$ GeV, while SLT tags are allowed on jets with $E_T > 8$ GeV. If no such tag is present, the fourth jet must satisfy the same E_T and η requirements as the first three. The above selection defines our mass sample, which contains 83 events.

Table 1: Subsamples used in the lepton + jets top quark mass analysis, expected background fractions, and the measured top quark mass.

Data Sample	# Events	$x_b(\%)$	Top Mass (GeV/c^2)
SVX Double	5	5^{+4}_{-2}	170.1 ± 9.3
SVX Single	15	13^{+5}_{-4}	178.0 ± 7.9
SLT	14	40^{+9}_{-9}	142.1^{+33}_{-14}
No Tag	42	56^{+14}_{-17}	180.8 ± 9.0
Combined	76	-	175.9 ± 4.8

Measurement of the top quark mass begins by fitting each event in the sample to the hypotheses of $t\bar{t}$ production followed by decay in the lepton + jets channel ($t\bar{t} \rightarrow W^+bW^-\bar{b} \rightarrow (l^+\nu b)(q\bar{q}'\bar{b})$). There are twelve distinct ways of assigning the four leading jets to the four partons b, $\bar{b}$, q and $\bar{q}'$. In addition, there is a quadratic ambiguity in the determination of the longitudinal component of the neutrino momentum. This yields up to twenty-four different configurations for reconstructing an event according to the $t\bar{t}$ hypothesis. We require that SVX and SLT-tagged jets to be assigned to b-partons and choose the configuration with lowest χ^2. Events with lowest $\chi^2 > 10$ are rejected.

A maximum-likelihood method is used to extract a top mass measurement from a sample of events which have been reconstructed according to the $t\bar{t}$ hypothesis. An essential ingredient of the likelihood functions are the probability density $f_s(M_{rec}; m_t)$ to reconstruct a mass M_{rec} from a $t\bar{t}$ events if the true top mass is m_t, the probability density $f_b(M_{rec})$ for reconstructing a mass M_{rec} from a background events, and the background fraction x_b constrained directly from the data. The likelihood is then maximized with respect to m_t, x_b and the parameters that define the shapes of f_s and f_b.

The precision of the top quark mass measurement is expected to increase with the number of observed events, the signal-over-background ratio, and the narrowness of the reconstructed-mass distribution. Monte Carlo studies show that an optimum way to partition the sample consists of subdividing the events into the four statistically independent subsamples shown in Table 1. The table also shows the numbers of events, the expected background fraction x_b, and the fitted top mass m_t.

The reconstructed-mass distribution of the sum of the four subsamples is plotted in Figure 1. The inset shows the shape of the corresponding sum of negative log-likelihoods as a function of top mass. From this we measure $m_t = 175.9 \pm 4.8$ GeV/c^2, where the uncertainty corresponds to a half-unit change in the negative log-likelihood with respect to its minimum.

We list the systematic uncertainties in Table 2. The systematic error due to hard gluon radiation uncertainty has been revised since publication. The uncertainty is now estimated using events generated with the PYTHIA Monte Carlo[4] to isolate the effects on the top mass due to initial and final state radiation jets. In summary, we have measured the top quark mass to be $175.9 \pm 4.8 \pm 5.3$. This is the most precise determination of the top mass in a single channel.

3 Top Mass Measurement from Dilepton

We report an improved measurement of the top quark mass using dilepton events[2] originating predominantly from $t\bar{t} \rightarrow W^+bW^-\bar{b} \rightarrow (l^+\nu b)(l^-\bar{\nu}\bar{b})$, where $l = e$ or μ. This measurement supersedes our previously reported result in the dilepton channel[5]. The previous result was obtained by comparing data with Monte Carlo simulation of $t\bar{t}$ events for two kinematic variables, the b-jet energies and the invariant masses of the lepton and b-jet systems.

We apply the same event selection criteria as those employed in the previous mass analyses of the dilepton channel[5]. We obtain a sample of eight candidate events. The expected background of 1.3 ± 0.3 events consists of events in which a track or a jet is misidentified as a lepton (0.29 events), Drell-Yan production (0.35 events), WW production (0.24 events), $Z \rightarrow \tau\tau$ decays (0.26 events) and $Z \rightarrow \mu\mu$ decays in which μ tracks are mismeasured (0.20 events).

Each candidate event is reconstructed according to the $t\bar{t}$ decay hypothesis in the dilepton channel:

$$t \rightarrow W^+b \rightarrow \ell_1^+\nu_1 b$$
$$\bar{t} \rightarrow W^-\bar{b} \rightarrow \ell_2^-\bar{\nu}_2\bar{b}$$

The two highest E_T jets in the event are assumed to be the b-jets from top decays. We assume the b-jet mass to be 5 GeV/c^2. Because the system is underconstrained due to the two unmeasured neutrinos, we use a weighting technique to determine a function, $f(m_t)$, from which we extract a top mass value[6]. We proceed as follows. We assume a top quark mass (m_t) and the two neutrino η values (η_1, η_2) and solve for the neutrino momenta, up to a four-fold ambiguity (two $P_z(\nu)$ choices for each ν) for each of the two jet charged-lepton pairings. We then assign a weight to each solution by comparing $\not{E}_T^p$, the sum of the neutrino transverse momenta for that solution, to $\not{E}_T^m$, the measured missing transverse energy after proper correction:

$$f(m_t, \eta_1, \eta_2) = exp\left(-\frac{(\not{E}_T^p - \not{E}_T^m)^2}{2\sigma^2}\right)$$

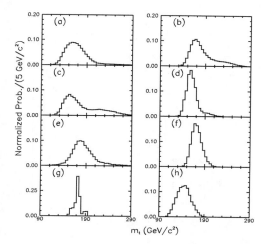

Figure 2: Weight distribution normalized to unity as a function of m_t for the eight dilepton top candidate events (a-h).

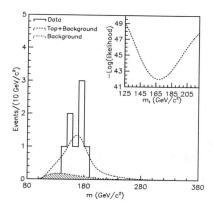

Figure 3: Reconstructed top mass for the eight dilepton events (solid). Background distribution (shaded, 1.3 events) and top Monte Carlo (6.7 events) added to background (dashed). The likelihood distribution as a function of the top mass, is shown in the inset.

where σ is the $\not{E}_T$ resolution for that event.

For each choice of m_t, η_1, and η_2 we take into account the detector resolution for jets and leptons by sampling (*i.e.* fluctuating) the measured quantities many times according to their resolutions.

For each assumed top mass value we use several (100) pairs of (η_1, η_2) values, chosen from distributions obtained from the HERWIG Monte Carlo predictions[7]. They are consistent with independent Gaussian distributions with $\sigma = 1.0$ in units of η. The weight is summed for all samplings as well as over all η_1, η_2 values and all the eight possible combinations; thus for each event at each top mass, m_t, we evaluate an overall weight:

$$f(m_t) = \sum_{\eta_1, \eta_2, E_{T_i}, \ell_1, \ell_2} f(m_t, \eta_1, \eta_2)$$

where E_{T_i} refers to all the jets in the event. We then compute the weight as a function of the top mass in 2.5 GeV/c^2 steps in the range 90-290 GeV/c^2.

The $f(m_t)$ distribution for each of the eight candidate events, normalized to unity, is shown in Figure 2. For each event, i, we use this distribution to determine a top mass estimate, m_i, by averaging the values of m_t corresponding to values of $f(m_t)$ greater than $f(m_t)_{max}/2$ on either side of the maximum. The m_i distribution

for the eight events is shown in Figure 3, together with the Monte Carlo expectation for background alone, and top plus background normalized to the data. The inset shows the negative log-likelihood as a function of the top mass, from which we determine a top quark mass value of 167.4 ± 10.3 (stat) GeV/c^2. Including the systematic uncertainties presented in Table 2, the top quark mass is measured to be $167.4 \pm 10.3 \pm 4.8$ for the events in the dilepton channel.

4 Top Mass Measurement from all Hadronic

In this analysis [3] we select $t\bar{t}$ events in which both W bosons decay into quark-antiquark pairs, leading to an all hadronic final state. The study of this channel, with a branching ratio of about 4/9, complements the leptonic modes and the mass measurement, takes advantage of a fully reconstructed final state, but suffers a very large QCD multijet background. To reduce this background, events are required to have at least one identified SVX b-jet and to pass strict kinematic criteria that favor $t\bar{t}$ production and decay.

To determine the top quark mass, full kinematic reconstruction is applied to the sample of events with 6 or more jets, $E_T > 15$ GeV, $|\eta| < 2$, and at least one b-tagged jet. Events are reconstructed to the $t\bar{t} \rightarrow W^+ b W^- \bar{b}$ hypothesis, where both W bosons decay into a quark pair, with each quark associated to one of the

Table 2: Systematic uncertainties on the top mass measurements (GeV/c^2)

Source	Dilep	Lep+jets	All Hadr.
Jet E_T	3.8	4.4	5.0
Gluon Radiation	3.1	2.6	1.8
Background	0.3	1.3	1.7
PDF, MC	1.1	0.5	1.0
Total	4.8	5.3	5.7

six highest E_T jets. All the combinations are tried except the SVX-tagged jet is assigned to b-partons and the combination with lowest $\chi^2 < 10$ is chosen. The data sample consist of 136 events, of which 108 ± 9 events are expected to come from background. The reconstructed 3-jet mass distribution is shown in Figure 4. The inset shows the shape of the log-likelihood as a function of top mass. From this we measure a top quark mass of $186.0 \pm 10.0 \pm 5.7$ GeV/c^2. The revised systematic uncertainties are shown in Table 2. The overall systematic error for the all-hadronic channel is reduced from 12.0 to 5.7 GeV/c^2.

5 A Combined Top Mass Measurement from CDF

The results for the three channels are combined with standard methods [8] to yield an overall CDF mass measurement. The three statistical errors are taken as uncorrelated, while the systematic errors are assumed to be either entirely correlated or uncorrelated between any two channels. The primary systematic error, that due to jet energy uncertainty, is taken as entirely correlated among all channels, as is the systematic error due to the Monte Carlo model used (mostly due to initial and final state radiation). The combined result is

$$m_t = 176.0 \pm 6.5 \text{ GeV/}c^2$$

including both statistical and systematic errors. In Table 3 we show the combined value with separate statistical and systematic errors. They are obtained by defining the combined statistical error as the sum in quadrature of the weighted individual statistical errors, and the systematic error as the difference in quadrature of the total and statistical errors. The relative contributions from the three channels are 67% for lepton plus jets, 18% for dileptons and 15% for all-hadronic.

In Run II with 2 fb^{-1} of integrated luminosity, we expect more than 1000 single tagged and about 600 double tagged $t\bar{t}$ events. It will allow us to measure the top quark mass down to approximately 2 GeV/c^2 precision.

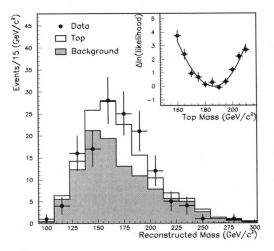

Figure 4: Reconstructed mass distribution for events with at least one b-tag. Also shown are the background distribution (shaded) and $t\bar{t}$ Monte Carlo events added to background (hollow). The inset shows the log-likelihood and the fit used to determine the top mass.

Table 3: Summary of top mass measurements with the CDF detector

Channel	Top Mass (GeV/c^2)
Lep + Jets	$175.9 \pm 4.8 \pm 5.3$
Dilepton	$167.4 \pm 10.3 \pm 4.8$
All Hadronic	$186.0 \pm 10.0 \pm 5.7$
Combined	$176.0 \pm 4.0 \pm 5.1$

Acknowledgements

We thank the Fermilab staff and the technical staffs of the participating institutions for their contributions. This work was supported by the U.S. Department of Energy and National Science Foundation, the Italian Istituto Nazionale di Fisica Nucleare, the Ministry of Science, Culture, and Education of Japan, the Natural Sciences and Engineering Research Council of Canada, the National Science Council of the Republic of China, and the A. P. Sloan Foundation.

References

1. F. Abe *et al.* *Phys. Rev. Lett.* **80**, 2767 (1998).
2. F. Abe *et al.* "An Improved Measurement of the Top Quark Mass with Dilepton Events in the Collider Detector at Fermilab", To be submitted to *Phys. Rev. Lett.*.
3. *F. Abe et al.* Phys. Rev. Lett. **79**, *1992 (1997)*.
4. *T. Sjöstrand,* Comput. Phys. Commun. **82**, *74 (1994). Pythia version 5.7 was used in this analysis.*
5. *F. Abe et al.* Phys. Rev. Lett. **80**, *2779 (1998)*.
6. *K. Kondo,* J. Phys. Soc. Jpn. **57**, *4126 (1988) and ibid.* **60**, *836 (1991). R.H. Dalitz and G.R. Goldstein,* Phys. Rev. *D* **45**, *1531 (1992). B. Abbott et al.,* Phys. Rev. Lett. **80**, *2063 (1998)*.
7. *G. Marchesini and B. Webber,* Nucl. Phys. *B* **310**, *461 (1988). HERWIG version 5.6 was used in this analysis.*
8. *Particle Data Group, R.M. Barnett et al.,* Phys. Rev. *D* **54**, *1 (1996), p. 161.*

DIRECT MEASUREMENT OF THE TOP QUARK MASS AT DØ

E. BARBERIS

for the DØ collaboration

Lawrence Berkeley National Laboratory, 1 Cyclotron Rd., Berkeley, CA 94720, USA
E-mail: EBarberis@lbl.gov

A summary of results on the direct measurement of the top quark mass (m_t) at the Tevatron by the DØ collaboration is presented. This includes measurements in the dilepton and lepton+jets channels, which yield $m_t = 173.3 \pm 5.6$ *(stat)* ± 5.5 *(sys)* and $m_t = 168.4 \pm 12.3$ *(stat)* ± 3.6 *(sys)* GeV/c^2 respectively, a well as a combined DØ result: $m_t = 172.1 \pm 5.2$ *(stat)* ± 4.9 *(sys)* GeV/c^2, and the world average of direct top mass measurements from the Tevatron (CDF and DØ): $m_t = 173.8 \pm 5.0$ *(tot)* GeV/c^2.

1 Top production and overview

The top quark is produced at the Tevatron mainly through $q\bar{q}$ annihilation processes. Assuming standard model couplings, all top quarks decay via $W + b$ and the $t\bar{t}$ decay channels are set by the W decay branching ratios, as summarized in the chart below (Figure 1).

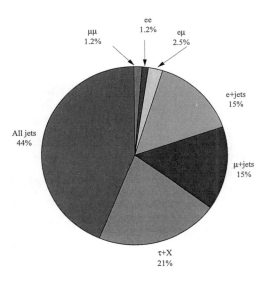

Figure 1: Branching ratios for $t\bar{t}$ decay channels.

Since the cross section for background processes, *i.e.* for the production of two or more jets, is $\sim 10^6$ larger than the $t\bar{t}$ production cross section, systematic uncertainties counteract statistical precision and the most weight in the top mass measurement comes from lepton+jets and dilepton channels, where a "clean" isolated lepton (e or μ) is required.

2 Top mass measurement in the lepton+jets channel (DØ)

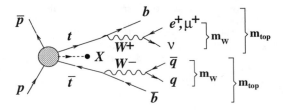

Figure 2: $t\bar{t} \rightarrow l$+jets final state

The lepton+4 jets final state (Figure 2) is a well-defined problem; all final state variables (18) are measured except for $p_z(\nu)$. With the additional three constraints: $m(q_i\bar{q}_j) = m_W$, $m(l\nu) = m_W$, and $m(\bar{b}_iq_i\bar{q}_j) = m(b_il\nu)$, the problem is twice over-constrained and a 2C fit can be performed to obtain a value for the top quark mass m_{fit}.

There are, however, a few complications arising from jet combinatorics and QCD radiation. There are 12(6) possible assignments of the 4 jets to $b\bar{b}q\bar{q}$ if 0(1) jet is b-tagged. There are two solutions for $p_z(\nu)$. The lowest χ^2 permutation is chosen (with $\chi^2 < 10$). When the fitted 4 jets correspond to the 4 quarks this choice has a probability of $\sim 40\%$ to be correct. Extra jets are radiated and only $\sim 50\%$ of the time the 4 fitted ones (4 highest E_T jets with $|\eta| < 2$) correspond to the 4 quarks to be fit. Due to both of these effects, the m_{fit} distribution is distorted from the true top mass (m_t) distribution, although still strongly correlated with it.

2.1 The Likelihood fit procedure

The top mass is derived from m_{fit} and extracted from a data sample which is a mixture of signal and background, according to the following procedure. Event selection is applied, based on the requirement of the final state particles signature (e/μ, $\not{E}_T(\nu)$, jets). Events with and without b-tagging[a] are separately examined. After cuts, 77 events are obtained, of which approximately 51 are estimated to be background events.

For each candidate event two quantities are computed: m_{fit}, from the two constraint fit, and a top probability D. For each event we now have a measurement in the 2D space defined by (m_{fit}, D), which is a function of m_t. The shape of these measurements is defined for data, background models, and signal Monte Carlo for several input values of the top mass m_t. Signal and background are fitted to the data using a 2D Poisson likelihood fit and a likelihood curve as a function of m_t, $L(m_t)$, is thus defined. A quadratic fit to $-\ln L(m_t)$ yields m_t and the statistical error.

2.1.1 The top probability D

D is the output of a discriminant which uses multivariate techniques and whose inputs are four kinematic variables $x_1, ..x_4$:

- x_1 = missing E_T
- x_2 = aplanarity
- x_3 = measures the event centrality
- x_4 = measures the extent to which jets are clustered together (sensitive to gluon radiation)

D is sensitive to the discrimination between signal and background and is weakly correlated to the top mass. Figure 3 shows the output of two discriminants, D_{LB} and D_{NN}, both based on $x_1, ..x_4$ but constructed using two different techniques (respectively a low mass bias likelihood and neural networks), for signal and background.

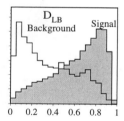
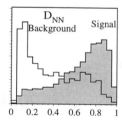

Figure 3: The "low bias" likelihood and the neural net discriminant outputs for signal and background samples.

Negligible correlation exists between D and m_{fit} as shown in Figure 4 for (a) signal, (b) background, and (c) data events.

---
[a] Here b-tagging refers to soft lepton b-tagging, which requires the presence of a muon consistent with $b \to \mu + X$.

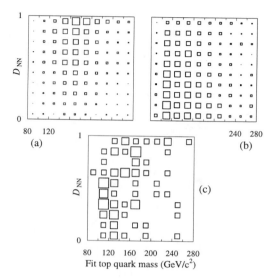

Figure 4: Discriminant as a function of fitted top mass for (a) signal, (b) background, and (c) data samples. The area of each box is proportional to the number of events in that bin.

2.1.2 Fit to the true top mass likelihood ln $L(m_t)$

Once a quadratic fit is performed to the true top mass likelihood (Figure 5), the value obtained by $D\emptyset$ in the lepton+jets channel[1,2] is:

$$m_t = 173.3 \pm 5.6 \ (stat) \pm 5.5 \ (sys) \ \text{GeV/c}^2 \quad (1)$$

The different components of the systematic error are summarized in Table 1.

As a final cross check, the top mass is determined with a different technique which uses 3C fits at a fixed set of

Table 1: Summary of systematic errors in the ℓ+jets channel.

Source of uncertainty	Error in GeV/c^2
Jet energy scale	4.0
$t\bar{t}$ (QCD radiation)	1.9
background (VECBOS)	2.5
Noise/Multiple Interactions	1.3
MC statistics	0.85
LB/NN diff.	0.8
Likelihood fit	1.0

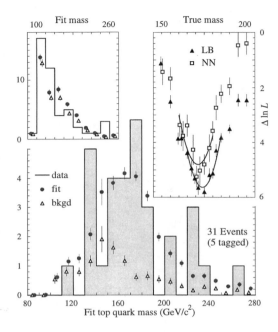

Figure 5: The distribution of m_{fit} for both a signal-enriched and a background-enriched sample and L vs. m_t for the two different definitions of D.

m_{fit} values for each event. At each m_{fit}, the jet permutation which gives the smallest $\chi^2/2$ is chosen, $\chi^2/2$ is summed over all events and plotted vs m_{fit}. After background subtraction, a parabola is fitted to obtain the "raw" m_t with its error. Monte Carlo experiments are used to relate this to the "true" m_t. This technique yields $m_t = 176.0 \pm 7.9(stat) \pm 4.8(sys)$ GeV/c^2, in good agreement with the main analysis.

3 Top mass measurement in the dilepton channel (DØ)

In the dilepton channel, the final state consists of two charged leptons, two neutrinos and two jets (18 variables). The four-vectors of the jets and leptons are measured together with the two components of the $\not{E}_T$ and the same three constraints used in the ℓ+jets case apply to the dilepton channel as well. The system has unconstrained kinematics and in this case we do need to assume a value for m_t.

DØ has introduced a strategy[3] which leads to a substantial advance in reducing the statistical and systematic uncertainties in this measurement. The basic idea is to assume a value for the top mass and then estimate the

likelihood of observing a determined final state in the detector. A weight is assigned to the event, based on the agreement between the final state observables and our hypothesis on the top mass. Two methods are used to determine the probability of an observed final state as a function of m_t:

- Matrix Element weighting (MWT): takes into account the parton distribution functions and the transverse momenta of the leptons from the W decays.

- Neutrino Weighting (νWT): is based on the agreement between the calculated neutrino p_T and the observed $\not{E}_T$.

To obtain the final weight, we sum over solutions (there is a four-fold ambiguity in pairing the leptons and b-jets that originate from the same top quark), jets assignment (including the ambiguity related to ISR and FSR), and versions of the same event smeared according to resolution. The weight function vs. m_t for each of the six dilepton candidates is shown in Figure 6 for both the MWT and the νWT methods.

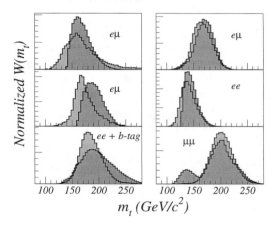

Figure 6: Weight distributions as a function of m_t for each of the dilepton candidates for the MWT (dark shaded area) and the νWT (light shaded area) methods.

In order to find the value of top mass most consistent with the entire data sample, $t\bar{t}$ Monte Carlo and data weight functions are compared at different values of m_t. A maximum likelihood fit is performed with respect to the number of signal events, n_s, and background events, n_b, for each m_t. The weight functions enter the fit as four-dimensional vectors of weight fractions in fixed bins of m_t. This improves the precision of the result, as it conveys more information than the simple use of the

Table 2: Summary of systematic errors in the dilepton channel.

Source of uncertainty	Error in GeV/c²
Jet energy scale	2.4
Multiple interactions	1.3
Generator (signal,bkg)	1.1,1.8
MC statistics	0.3
Likelihood fit	1.1

peak value. A quadratic is then fitted to the nine points nearest to the minimum of the likelihood curve to obtain the value of the top mass (shown in Figure 7).

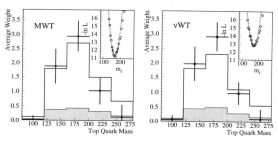

Figure 7: The average weight as a function of m_t for data (dots), signal+background (histogram), and background (shaded histogram). The inset show the likelihood vs. m_t.

The combined result in the dilepton channel from the two weighting methods is:

$$m_t = 168.4 \pm 12.3 \ (stat) \pm 3.6 \ (sys) \ \text{GeV/c}^2 \qquad (2)$$

The value of the statistical errors is supported by Monte Carlo experiments, and the summary of systematic uncertainties is shown in Table 2.

The DØ result from the lepton+jets and dilepton channels combined is:

$$m_t = 172.1 \pm 5.2 \ (stat) \pm 4.9 \ (sys) \ \text{GeV/c}^2 \qquad (3)$$

4 World average of direct top mass measurements (CDF and DØ)

We have so far shown that is possible to obtain a very precise measurement of the top quark mass m_t from individual channels, but in order to determine m_t with best accuracy it is important to use all possible decay modes, combine the results, use several methods to cross check techniques and systematics and, ultimately, combine experiments. This leads to the combined Tevatron effort

(CDF+DØ) in determining the value of the top quark mass.

The method is sketched in here. We want to measure a quantity Q=m_t. In the simplest case we have two experiments a, b and therefore two estimates for this quantity $Q_{a,b}$. The statistical error for each experiment is given by $T^{a,b}$. The systematic errors $y_i^{a,b}$ for each experiment are classified into categories i=1,....,N uncorrelated among themselves. The total systematic error is: $Y^{a,b} = \sqrt{\sum_{i=1}^N (y_i^{a,b})^2}$, and the total error per experiment is: $S^{a,b} = \sqrt{(Y^{a,b})^2 + (T^{a,b})^2}$.

The correlation coefficient ρ between two experiments can be expressed in terms of the mutual correlation coefficients ρ_i for the systematic error category i, the systematic errors in each category and the total errors for each experiment as:

$$\rho = \frac{\sum_{i=1}^N \rho_i y_i^a y_i^b}{S^a S^b} \qquad (4)$$

In the basis $\{a, b\}$, in terms of ρ, the covariance matrix is:

$$\mathcal{S} = \begin{pmatrix} (S^a)^2 & \rho S^a S^b \\ \rho S^a S^b & (S^b)^2 \end{pmatrix} \qquad (5)$$

From the inverse of the covariance matrix a χ^2 is defined for the hypothesis that both experiments measure the same best fit quantity $\langle Q \rangle$. A χ^2 minimization yields $\langle Q \rangle$ and the curvature of χ^2 vs $\langle Q \rangle$ yields the total error $S_{\langle Q \rangle}$.

The formalism can be extended to the case of combining more than two experimental results (in this case the five results obtained in separate channels by the two detectors, as listed in Table 3). The correlation coefficients between each two of the five measurements are shown in Figure 8.

Although some degree of correlation is observed between measurements within one detector, the correlation between the two detectors appears to be very small.

The final Tevatron result is:

$$\begin{aligned} m_t = 173.8 &\pm 3.2 \ (stat) \pm 3.9 \ (sys) \ \text{GeV/c}^2 \\ &= 173.8 \pm 5.0 \ (tot) \ \text{GeV/c}^2 \end{aligned} \qquad (6)$$

with χ^2 probability of 79%. Table 3 summarizes the results in the separate channels for both detectors and the combined result within each detector.

The relative weight of the different channels in the average in shown in Figure 9.

5 Summary

The DØ collaboration has measured the mass of the top quark in both the lepton+jets and the dilepton channels,

Correlation in top mass

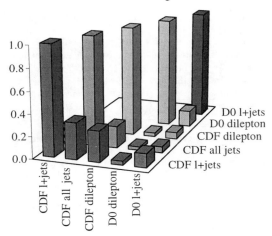

Figure 8: Correlation coefficients in the measurement of the top mass for different channels and two detectors (CDF and DØ)

Relative weight in top mass average

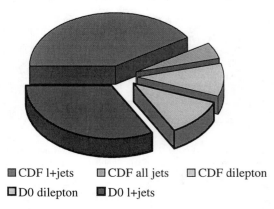

■ CDF l+jets   ■ CDF all jets   □ CDF dilepton
▥ D0 dilepton   ■ D0 l+jets

Figure 9: Relative weight of different measurement channels in the top mass average.

4. CDF Collaboration, F. Abe *et al.*, *Phys. Rev. Lett.* **80**, 2767 (1998).
5. Wei-Ming Yao, these proceedings.

resulting in a combined value of $m_t = 172.1 \pm 5.2\ (stat) \pm 4.9\ (sys)$ GeV/c^2. The average of the Tevatron measurements for the top mass is $m_t = 173.8 \pm 5.0\ (tot)$ GeV/c^2. With a fractional error of less than 3%, the top mass is to date the best known quark mass.

References

1. DØ Collaboration, B. Abbott *et al.*, *Phys. Rev. Lett.* **79**, 1197 (1997).
2. DØ Collaboration, B. Abbott *et al.*, *Phys. Rev. D* **58**, 052001 (1998).
3. DØ Collaboration, B. Abbott *et al.*, *Phys. Rev. Lett.* **80**, 2063 (1997).

Table 3: Summary of CDF and DØ top mass values (in GeV/c^2).

ℓ+jets (CDF) [4]	$m_t = 175.9 \pm 4.8(stat) \pm 4.9(sys)$
dilepton (CDF) [5]	$m_t = 167.4 \pm 10.3(stat) \pm 4.8(sys)$
all jets (CDF) [5]	$m_t = 186.0 \pm 10.0(stat) \pm 8.2(sys)$
combined (CDF)	$m_t = 175.3 \pm 4.1(stat) \pm 5.0(sys)$
ℓ+jets (DØ)	$m_t = 173.3 \pm 5.6(stat) \pm 5.5(sys)$
dilepton (DØ)	$m_t = 168.4 \pm 12.3(stat) \pm 3.6(sys)$
combined (DØ)	$m_t = 172.1 \pm 5.2(stat) \pm 4.9(sys)$

GLUON RADIATION IN TOP MASS RECONSTRUCTION: EFFECT OF HADRONIC W DECAYS

L.H. ORR

Department of Physics and Astronomy, University of Rochester, Rochester NY 14627-0171, USA
E-mail: orr@pas.rochester.edu

T. ANDRE

Department of Physics and Astronomy, University of Rochester, Rochester NY 14627-0171, USA and
Department of Physics, University of Chicago, Chicago IL 60637, USA
E-mail: troy@hep.uchicago.edu

T. STELZER

Department of Physics, University of Illinois, Urbana, IL 61801, USA
E-mail: stelzer@pobox.hep.uiuc.edu

Top quark events in hadron collisions often contain additional hadronic jets from gluon bremsstrahlung off the quarks and gluons in the hard subprocesses. Such extra jets must be taken into account in attempts to reconstruct the momentum of the top quark from those of its decay products. We have performed a complete calculation of gluon radiation in top production and decay at the Fermilab Tevatron including hadronically decaying W bosons. In this talk we discuss the effect of gluon radiation on the reconstructed mass of the top quark, using various top mass reconstruction scenarios. Implications for the LHC are briefly discussed.

1 Introduction

Measuring the top quark mass at hadron colliders requires reconstructing its momentum from its decay products. Radiated gluons in top events can complicate the reconstruction process, because for example the jets from gluons can be indistinguishable from the jets in top decays. It is important to account correctly for these gluons because future top mass measurements will be dominated by systematic effects due to gluon radiation.

Given a top event with an extra jet from a radiated gluon, what should we do with the extra jet? In particular, should the extra jet be combined with the W and b quark to reconstruct m_t? The answer depends on where the gluon originated. If it was radiated from an initial state quark, then it is a correction to the production process that is not part of the top decay, and it should be ignored. If the gluon was radiated from one of the b quarks from the t or $\bar{t}$ decay, then it is itself part of the decay and should be included in the reconstruction. Suppose the gluon was radiated by the top quark itself – is it associated with top production or decay? In fact it can be either, depending on when the top quark went on shell.

The point is that in a given event we cannot usually distinguish between the possibilities (even apart from the fact that they interfere), so we must consider top production and decay together in our treatment of gluon radiation. This has been done for top production and decay at the Tevatron[1] and LHC[2], without radiation from hadronic decays of the W bosons. But in the detection modes in which at least one of the top quarks can be fully reconstructed from its decay products — the lepton + jets and all-jets modes — one or both of the W bosons decays to quarks, which can themselves radiate. Radiation from hadronically decaying W's in top events was treated in the soft gluon approximation in[3]. The soft approximation serves as a useful guide to the distribution of gluons and the relative importance of the various contributions, but it does not incorporate exact kinematics and cannot be used to study mass reconstruction.

In this talk we present the results for an exact calculation of gluon radiation in top production and decay at the Tevatron with hadronic W decays fully taken into account.

2 Gluon Distributions

We have calculated the cross section for $p\bar{p} \to b\bar{b}q\bar{q}'l\nu j$ and $p\bar{p} \to b\bar{b}q\bar{q}'q\bar{q}'j$ where j is an extra radiated jet. This tree-level calculation is exact at $\mathcal{O}(\alpha_s^3)$ and contains all spin correlations, top width effects, and interferences. The center of mass energy is 1.8 TeV, and top and bottom masses are 175 and 5 GeV, respectively. Helicity amplitudes are computed with the assistance of the MADGRAPH package[4]. The results shown below are for the $q\bar{q}$ initial state that dominates in top production at the Tevatron; we have done the calculation for the gg

and qg initial states[5] but do not show those results here.

We apply the following kinematic cuts to all final-state jets (which in this parton-level calculation are quarks and gluons) and to the charged lepton:

$$E_{Tj}, E_{Tl} \geq 15 \text{ GeV} ,$$
$$|\eta_j|, |\eta_l| \leq 2.5 ,$$
$$\Delta R_{jj}, \Delta R_{jl} \geq 0.4 . \qquad (1)$$

These are meant to mimic experimental cuts, so that the partons are likely to appear in the detectors with enough angular separation to be distinguishable as separate particles. They also protect the theoretical cross section from the soft and collinear singularities that appear at tree level.

In the distributions we present below we will decompose the cross section into contributions from radiation associated with various parts of the process. These contributions are:

i. *Production-stage radiation,* which comes from the initial quarks or internal gluon line, or from the top (or antitop) quark *before* it goes on shell.

ii. *Decay-stage radiation* that is part of the t or $\bar{t}$ decay; this is further subdivided into

    a. Decay-tb radiation from either of the b's or from either of the t's *after* they go on shell.

    b. Decay-W radiation from the decay products of hadronically decaying W bosons.

We make these distinctions in the parton-level calculation based on kinematics to see how the various contributions behave; this cannot of course be done for a given event in the experiments. In principle the production-stage and decay-stage contributions can interfere with each other (with the exception of the decay-W radiation, which cannot interfere with the other processes because the W is a color singlet). And although we do include all interferences in our calculation, in practice the production-decay interference is very small for gluon energy thresholds large compared to the top width $\Gamma_t = 1.5$ GeV, as in the present case.

Figure 1 shows the distribution in pseudorapidity of the extra jet at the Tevatron for the lepton + jets case, *i.e.* for a single hadronically decaying W. The production-stage radiation, shown as a dashed histogram, has the broadest distribution, populating most of the accessible rapidity range. The two decay-stage contributions are more centrally peaked. The decay-tb contribution (dotted histogram) is slightly larger, as it accounts for radiation from both the t and $\bar{t}$, but the decay-W contribution (dot-dashed histogram) from a single W is similar in size and shape to the decay-tb, as was found in the soft approximation[3]. The central region of the detector is populated by all three contributions,

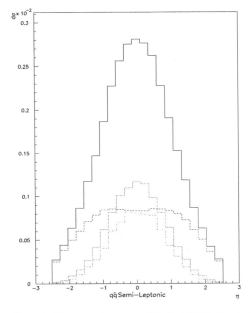

Figure 1: The extra jet pseudorapidity (η_j) distribution (solid histogram) and its decomposition in terms of production (dashed histogram), decay-tb (dotted histogram) and decay-W (dot-dashed histogram) emission contributions, for the lepton + jets mode.

which means that distinguishing them will be challenging at best.

Results for the all-hadronic mode, where both W bosons decay to quarks, are similar, the main difference being that the decay-W contribution approximately doubles in size.

The transverse energy distributions of the radiated jet are shown in Figure 2 for the lepton + jets (top) and all-hadronic modes. The spectra look quite similar for the various contributions, and extend to large values of transverse energy.

3 Top Mass Reconstruction

Given the difficulty of distinguishing extra jets radiated in the production stage from those from decay, we can ask what effect the extra jet has on mass reconstruction. First, however, we have to ask which jet *is* the extra one. It seems reasonable to assume that the gluon jet is the one with the lowest E_T, given the infrared singularity that characterizes emitted radiation. While it is true that of the final state jets, the gluon has the softest E_T

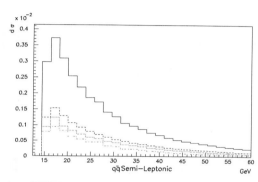

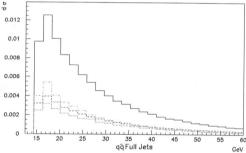

Figure 2: The extra jet transverse energy distribution (solid histogram) and its decomposition in terms of production (dashed histogram), decay-tb (dotted histogram) and decay-W (dot-dashed histogram) emission contributions, for the lepton + jets mode (top plot) and all-hadronic (bottom plot) mode.

spectrum, the gluon has the lowest E_T only just over half the time (just under half for the all-jets mode). Still, there is no method that is obviously better for identifying the extra jet.

We reconstruct the top mass from the final state partons without assuming we know which jet is which, but we omit the jet with the lowest E_T. For the lepton + jets mode we do the following.

(1.) Drop the lowest E_T jet. This leaves four jets.
(2.) Find the jet pair with invariant mass closest to m_W.
(3.) Solve for the neutrino four-momentum using the charged lepton momentum and W mass constraint.
(4.) Combine each of the W's with each of the remaining jets to give m_t and $m_{\bar{t}}$. Choose the combination that minimizes the t-$\bar{t}$ mass difference.

The m_t distributions obtained from this procedure are shown in Figure 3. The top plot shows the distribution for the top with the hadronically decaying W, and the bottom corresponds to leptonic W decay. In both cases we see a peak at the correct central value, where the procedure resulted in the correct mass. We also see smooth, reasonably flat high and low tails from wrong combinations, in addition to bumps in the low tails corresponding to the omission of jets that were part of the decays. Note that these bumps appear in the contributions from decay-stage radiation, when *all* of the jets should be included in the mass reconstruction. Finally, the leptonically decaying W gives a sharper top mass distribution because with no radiation from the W decay, there are fewer wrong combinations.

The results shown in Figure 3 are meant to be illustrative and should not be taken as a direct representation of distributions measured in experiment. In particular, this calculation is at the parton level; we have not included backgrounds; and these distributions only include events with a radiated gluon. The effects of hadronization, energy resolution and detector effects, and background will certainly make things worse. However there are certainly ways to improve on this simplistic analysis as well. For example, although we minimized mass differences we did not cut on them explicitly. Figure 4 shows the magnitude of the t-$\bar{t}$ mass difference on an event-by-event basis for the case shown above. It suggests that an absolute cut on the mass difference could reduce the tails in the mass distributions. Interestingly, b-tagging, *i.e.* assuming we can identify b jets, does not improve the mass distributions much. This is because we do not include backgrounds.

The results for the all-jets mode are similar, except that with both W's decaying to quarks there are two more jets in each event, leading to more possible wrong combinations and a corresponding increase in the tails.

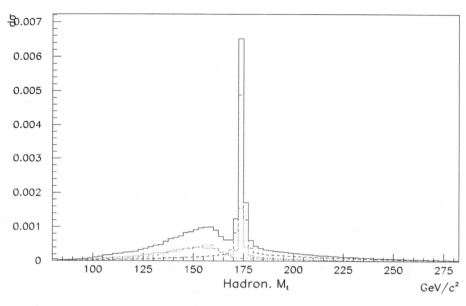

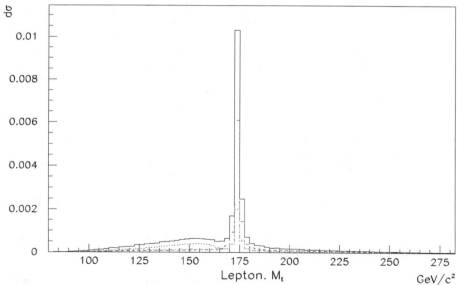

Figure 3: The reconstructed top mass distributions (solid histograms) in the lepton + jets mode for the hadronically decaying W (top plot) and the leptonically decaying W, and their decomposition in terms of production, decay-tb and decay-W emission contributions.

4 Conclusions

In top quark physics, as statistics improve, systematic effects associated with gluon radiation will dominate measurements of the top mass. We have added hadronic W decays to analyses of gluon radiation in top production and decay and presented some initial results here. We find that the contribution from radiation from a single hadronically decaying W is nearly as large as and comparable in shape to the remaining decay-stage radiation from both the t and $\bar{t}$. The presence of radiation from both the top production and decay stages complicates the reconstruction of the top momentum from its decay products and hence complicates the measurement of the top mass. Further analysis is in progress for the Tevatron and LHC [5].

Acknowledgements

Work supported in part by the U.S. Department of Energy, under grant DE-FG02-91ER40685 and by the U.S. National Science Foundation, under grant PHY-9600155.

References

1. L.H. Orr, T. Stelzer, and W.J. Stirling, *Phys. Rev. D* **52**, 124 (1995); *Phys. Lett.* **B354** (1995) 442.
2. L.H. Orr, T. Stelzer, and W.J. Stirling, *Phys. Rev. D* **56**, 446 (1997).
3. B. Masuda, L.H. Orr, and W. J. Stirling, *Phys. Rev. D* **54** (1996) 4453.
4. T. Stelzer and W.F. Long, *Comp. Phys. Commun.* **81** (1994) 357.
5. T. Andre, L.H. Orr, and T. Stelzer, in preparation.

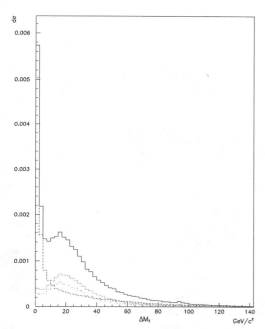

Figure 4: The distribution in magnitude of the t-$\bar{t}$ mass difference for the events in the previous figure.

DØ ALL-HADRONIC TOP DECAY AND TOP CROSS SECTION SUMMARY

M. A. C. Cummings

M.S. 352, Expt. D0, Fermilab, Batavia, IL 60510, USA
E-mail: MACC@fnal.gov

A description of the analysis of $t\bar{t}$ production in $p\bar{p}$ collisions at $\sqrt{s} = 1.8$ TeV from 110 pb^{-1} of data collected in the all hadronic decay channel with the DØ detector at Fermilab. A neural network analysis yielded a cross section of 7.1 ± 2.8 (stat.) ± 1.5 (syst.), at a top quark mass of 172.1 GeV/c^2 in this channel. Also presented is a summary of all DØ $t\bar{t}$ cross sections. The combined DØ result for the $t\bar{t}$ production cross section is 5.9 ± 1.2 (stat.) ± 1.1 (syst.) pb for m_t=172.1 GeV/c^2.

1 Introduction

The Standard Model predicts that, at Tevatron energies, top quarks are produced primarily in $t\bar{t}$ pairs, and that each top quark decays into a b quark and a W boson. The expected branching fraction for both W bosons in a $t\bar{t}$ event decaying into quark-antiquark pairs is 44%. The signal for these pure hadronic, or "all-jets", $t\bar{t}$ events is six reconstructed jets. The main background is from QCD multijet events that arise from a 2→2 parton process producing two energetic ("hard") leading jets and less energetic ("soft"), radiated gluon jets.

2 Apparatus

The DØ detector is described elsewhere[1]. We use calorimetry and tracking information for jet and muon reconstruction, respectively, and employ both for electron identification. The muons in this analysis are used for b-tagging of jets, and are restricted to the pseudorapidity range $|\eta| \leq 1.0$. The calorimeter covers the range $|\eta| \leq 4.2$, with an energy resolution of $80\%/\sqrt{E(\text{GeV})}$ for reconstructed jets[1,2]. The fine segmentation of the calorimeter (0.1×0.1 in $\Delta\eta \times \Delta\phi$), allows details of jet shape to be used in this analysis.

3 Event Selection

The multijet data sample was selected using a hardware trigger and an online filter requiring five reconstructed jets with pseudorapidity $|\eta| < 2.5$ and transverse energy $E_T > 10$ GeV, a total transverse energy (H_T) of > 115 or 120 GeV (depending on run conditions) and jet quality requirements. The data sample after the initial cuts is 600,000 events. With about 200 expected top events in this channel, the background overwhelms the signal by a factor of ≈ 3000 (see Fig. 1), and discrimination from many variables is required. For this reason, neural networks are an integral part of our analysis.

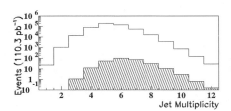

Figure 1: Jet Multiplicity. Data from Run Ib (histogram) and $t\bar{t}$ HERWIG events for m_t=175 GeV/c^2 (shaded histogram). Signal is overwhelmed by background by three to four orders of magnitude.

4 Analysis

The analysis proceeded in four stages: initial cuts, muon-tag requirement, background modeling, and neural network calculation. Events with an isolated muon or electron were excluded to keep the sample independent of the other top samples. Events were required to have at least six $\mathcal{R}$=0.3 cone jets and less than nine $\mathcal{R}$=0.5 cone jets. We generally used $\mathcal{R}$=0.3 cone jets because of the greater reconstruction efficiency in that cone size, but used $\mathcal{R}$=0.5 cone jets to calculate mass-related variables. We required that at least one jet have an associated muon that satisfied muon quality criteria and was kinematically consistent with originating from a b quark decay within the jet. As about 20% of $t\bar{t}$ all-jets events have such a "b-tagged" jet, compared to approximately 3% of the QCD multijet background, the background/signal is reduced by almost a factor of ten. Of the 280,000 events surviving the offline cuts, 3853 have at least one b-tagged jet. These tagged events comprise the data sample used in the analysis.

The very large background to signal ratio in the untagged data represents an almost pure background sample. With a correction for the very small $t\bar{t}$ component expected, and with a method of assigning a muon tag to the untagged event, the background model can be deter-

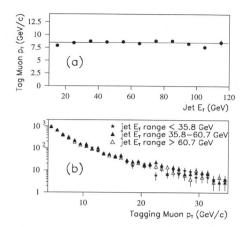

Figure 2: (a) Mean muon p_T (dots) versus tagged jet E_T and (b) muon p_T distributions for three jet E_T ranges (chosen to be equally populated) for data events. The line in (a) is the average of the points. No correlation is observed between the muon p_T and the jet E_T.

mined directly from the data. Separate sets of untagged data with an added muon tags were used for background neural network response training and background modeling. HERWIG[5] $t\bar{t}$ events were used for the $t\bar{t}$ signal response training.

Correctly assigning a muon tag to a jet in an untagged event was critical to our background calculation. We derived a "tag rate" from the entire multijet data set, which was defined as the probability for any individual jet to have a tagging muon. We chose a function that factorized into two pieces: ϵ, the detector efficiency dependent on run number and η of the jet (to account for chamber aging), and $f(E_T)$, the probability that a jet of tranverse energy E_T has a tagging muon.

We established that the p_T of the tagging muon and the E_T of the tagged jet (uncorrected for the muon and neutrino energy effects) are uncorrelated. This is illustrated in Fig. 2. Therefore, the muon p_T spectrum factors out of the tag rate function, and can be generated independently. Applying the tag rate function to every jet in the untagged data sample, and generating a muon p_T to those jets determined as tagged, produced the background model sample.

The analysis used two neural networks. The inputs to the first neural network (NN1) were ten correlated variables involving kinematic and topological properties of the events. The output of this neural network was used as an input variable to a second and final neural network (NN2) whose three other inputs were variables describing other event features. The four input variables in NN2 were less correlated than the ten kinematic input variables in NN1. The top quark production cross section is calculated from the output of NN2. Both networks were trained to force output near 1 for $t\bar{t}$ events and near 0 for QCD multijet events using the back-propagation learning algorithm in JETNET[4] and one layer of hidden nodes. The network training was done with event samples that are independent of the data sample used for the measurement, and with the tag rate function described above, we had a good model for the tagged multijet background. The $t\bar{t}$ cross section is derived from the excess of events in the NN2 output data distribution over the NN2 modeled background output.

We need to exploit many sources of discrimination between signal and background. Compared with the QCD multijet background, $t\bar{t}$ events are typically more energetic, have their energy more uniformly distributed among the jets, are more isotropic, and have their jets distributed at smaller η. At least two variables describing each of these qualities (total energy, jet energy distribution, event shape, and rapidity distribution) were defined for each event:

1. H_T: Total scalar E_T.

2. $\sqrt{\hat{s}}$: Total invariant mass.

3. E_{T_1}/H_T: E_T fraction carried by the leading jet.

4. H_T^{3j}: Total scalar E_T of non-leading jets.

5. N_{jets}^A: The threshold-weighted number of jets.

6. $E_{T_{5,6}}$: Average E_T of the fifth and sixth jets.

7. $\mathcal{A}$: The aplanarity, calculated from the normalized momentum tensor.

8. $\mathcal{S}$: The sphericity, calculated from the normalized momentum tensor.

9. $\mathcal{C}$: The centrality, $\mathcal{C} = H_T/H_E$, where H_E is the sum of all the jet total energies.

10. $<\eta^2>$: The E_T-weighted mean square of jet η.

These ten variables are the inputs into the first neural network (NN1), whose output is used as an input variable into the second neural network (NN2). The three other inputs into NN2 are:

11. p_T^μ: The transverse momentum of the tagging muon.

The p_T^μ distribution is harder for tagged jets in $t\bar{t}$ events than for tagged jets in QCD multijet events.

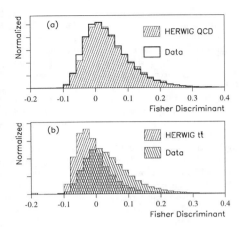

Figure 3: Distributions of $\mathcal{F}$ for, (a) data (predominantly background) and HERWIG QCD, and (b) data and HERWIG $t\bar{t}$ events.

12. $\mathcal{M}$: The mass-likelihood variable. This variable is defined as $\mathcal{M} = (M_{W_1} - M_W)^2/\sigma_W^2 + (M_{W_2} - M_W)^2/\sigma_W^2 + (m_{t_1} - m_{t_2})^2/\sigma_t^2$, with the parameters M_W, σ_W, and σ_t set to 80, 16 and 62 GeV/c^2, respectively. M_{W_i} and M_{t_i} refer to the jet combinations that best define the W boson and top quark masses in an event.

The mass likelihood variable $\mathcal{M}$ is a χ^2-like object minimized when there are two invariant masses consistent with the W mass, and two candidate top quark masses that are identical.

13. $\mathcal{F}$: The jet-width Fisher discriminant. This is defined as $\mathcal{F}_{\mathrm{jet}} = (\sigma_{\mathrm{jet}} - \sigma_{\mathrm{quark}}(E_T))^2/\sigma_{\mathrm{quark}}^2(E_T) - (\sigma_{\mathrm{jet}} - \sigma_{\mathrm{gluon}}(E_T))^2/\sigma_{\mathrm{gluon}}^2(E_T)$, where $\sigma_{\mathrm{quark}}^2(E_T)$ and $\sigma_{\mathrm{gluon}}^2(E_T)$ are mean square jet widths calculated from Monte Carlo, for quarks and gluons respectively, as a functions of jet E_T.

It has been demonstrated that quark jets are, on average, narrower than gluon jets [6,7]. The Fisher discriminant, based on the η-ϕ RMS jet widths is calculated for the four narrowest jets in the event, and indicates whether the jets were most probably "quark-like" ($t\bar{t}$) or "gluon-like" (QCD multijet). Using quark-gluon jet width differences as a physics analysis tool is a unique feature of our analysis[3]. Fig. 3 shows the signal/background separation in the Fisher discriminant.

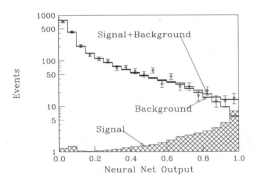

Figure 4: The distribution in NN2 output (on a log scale) for data (diamonds + error bars) and the fits for expected signal and background are shown. The signal was modeled with HERWIG for m_t=180 GeV/c^2. The errors shown are statistical.

5 Results

The NN2 output distributions for data, modeled background and HERWIG top signal are plotted in Fig. 4. The cross section is obtained from a simultaneous fit of the data to the background and HERWIG shapes, with the background normalization (A_{bkg}) and the top pair cross section ($\sigma_{t\bar{t}}$) as free parameters. The result of this fit is also shown in Fig. 4. Because of the preponderance of background at the low end of NN2, we use the region NN2 > 0.1 for our quoted results.

The values of the cross section and background normalization are obtained from similar plots with HERWIG $t\bar{t}$ events generated at several different top masses. The results are shown in the graph of Fig. 5, along with the value for the DØ leptonic channels. Interpolating to top quark mass as measured by DØ [8] ($m_t = 172.1 \pm 7.1$ GeV), we obtain $\sigma_{t\bar{t}} = 7.1 \pm 2.8$ (stat.) ± 1.5 (sys.) pb. Contributions to the systematic error on the cross section are described in detail elsewhere[3].

As a check we did a straight counting analysis (assuming a background normalization of 1). For an NN2 output threshold of 0.85 (chosen to minimize the errors on the cross section) we had 44 events with an expected background of 25.3 $\pm$ 3.1 for an excess of 15.9 $\pm$ 2.6 events. This corresponds to a cross section of 7.3 $\pm$ 3.3 $\pm$ 1.6 pb at the DØ mass, consistent with our result above.

Previous DØ measurements of the $t\bar{t}$ production cross section in the dilepton and semi-leptonic channels [10] gives an average cross section of 5.6 $\pm$ 1.4 (stat) $\pm$ 1.2 (syst) pb at m_t=172.1 GeV/c^2, in very good agreement with that from the all-jets channel. Contributions are (1) $\sigma_{t\bar{t}} = 6.4 \pm 3.4$ from the dilepton channel, (2) $\sigma_{t\bar{t}}$

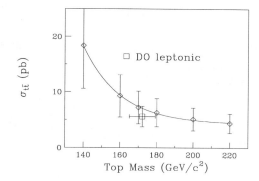

Figure 5: The $t\bar{t}$ cross section extracted through fitting the shapes of the distributions in neural network output to data, shown as a function of top quark mass. Error bars are statistical only. For reference, the DØ $t\bar{t}$ cross section and top quark mass from leptonic channels [10] is shown in the figure (open square).

$= 4.1 \pm 2.1$ from the topological lepton plus jets channel and $\sigma_{t\bar{t}} = 8.3 \pm 3.6$ from the tagged semileptonic channel. We can combine the all-jets cross section with these results, assuming that statistical errors are uncorrelated, and the systematic errors each have correlation coefficients as appropriate. The combined DØ result for the $t\bar{t}$ production cross section is 5.9 ± 1.2 (stat) ± 1.1 (syst) pb for $m_t = 172.1$ GeV/c^2.

References

1. DØ Collaboration, S. Abachi , Nucl. Instrum. Methods Phys. Res. A **338**, 185 (1994).
2. DØ Collaboration, S. Abachi , Phys. Rev. D **52**, 4877 (1995).
3. DØ Collaboration, N. Amos , submitted to Phys. Rev. D, FERMILAB-Pub-98/130-E, (1998).
4. C. Peterson and T. Rögnvaldsson, *JETNET 3.0 - A Versatile Artifical Neural Network Package*, CERN-TH.7135/94 (1994).
5. G. Marchesini , Comput. Phys. Commun. **67**, 465 (1992).
6. The AMY Collaboration, Y.K. Kim , Phys. Rev. Lett. **63**, 1772 (1989).
7. The OPAL Collaboration, G. Alexander , Phys. Lett. B **265**, 462 (1991).
   The OPAL Collaboration, G. Alexander , Z.Phys. **C68**, 179 (1995).
8. DØ Collaboration, B. Abbott , submitted to Phys. Rev. D, FERMILAB-Pub-98/031-E, (1998).
9. CDF Collaboration, F. Abe , Phys. Rev. Lett. **79**, 1992 (1997).
10. DØ Collaboration, S. Abachi , Phys. Rev. Lett. **79**, 1203 (1997).
    DØ Collaboration, B. Abbott , to be submitted to Phys. Rev. D, (1998).

Top Quark Production and Decay Measurements from CDF

K. Tollefson

University of Rochester, Department of Physics and Astronomy, Rochester, NY 14627, USA
E-mail: kirsten@fnal.gov

The CDF collaboration is completing a number of studies on the top quark, based on samples collected during Run I of the Tevatron Collider. The production and decay properties of the top quark are being examined in most of the $t\bar{t}$ decay channels, and many of these results have recently been published. The study of the top quark has moved beyond measurements of its mass and production cross section, to detailed studies of W polarization in top decays, single top production, branching fractions, the $W - t - b$ coupling, and searches for rare decays.

1 Introduction

The existence of the top quark was firmly established in early 1995 when the two collider experiments at the Tevatron, CDF and DØ, observed a significant number of events in excess of the background [1]. The characteristics of these events were consistent with $t\bar{t}$ production and decay. Initial measurements of the top quark mass indicated a mass [2] of approximately 175 GeV/c^2, nearly 40 times more massive than the b quark. It is essential to fully explore this remarkable quark for both its standard model properties and for hints of physics beyond the standard model. Since its discovery, the 1992-1996 Tevatron collider run has concluded and analysis of the current dataset is nearly complete. This paper reviews the status of our understanding of top quark production and decay.

During Run I the Tevatron operated at a center-of-mass energy of 1.8 TeV with a peak luminosity near 2×10^{31} $cm^{-2}sec^{-1}$. The CDF results described below are based on 110 pb^{-1} of integrated luminosity. A detailed description of the CDF detector can be found elsewhere [3].

2 Identification of Top Quark Samples

At the Tevatron, the dominant form of top quark production is $t\bar{t}$ pair production via $q\bar{q}$ annihilation. A NLO calculation indicates an expected production cross section [4] for $t\bar{t}$ in the range of 4.7-5.5 pb for $M_{top} = 175$ GeV/c^2. This production process is rather rare, having a cross section which is nine orders of magnitude less than the total inelastic cross section. It is also possible to produce events which contain only one top quark (single top production) although the expected cross section [5], as well as the acceptance for such decays, is smaller than that for $t\bar{t}$ events. For a top quark mass of 175 GeV/c^2, the top quark width is expected to be about 1.8 GeV/c^2, so the top quark decays before hadronizing. In the standard model, a t or $\bar{t}$ quark decays almost 100% of the time to a real W boson and a b quark. Figure 1 shows the Feynman diagram for top quark production and stan-

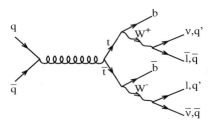

Figure 1: The tree-level Feynman diagram for top quark production by $q\bar{q}$ annihilation and standard model top quark decay.

dard model top quark decay. The subsequent decays of the W bosons lead to many possible final states. CDF classifies top events into three categories according to the decay channels of the two W bosons; the lepton plus jets channel, the dilepton channel, and the all-hadronic channel.

2.1 The Lepton Plus Jets Channel

In this channel, one W decays leptonically ($W \rightarrow \ell\nu$) while the other decays hadronically ($W \rightarrow q\bar{q}'$). Events are characterized by one high P_T lepton, missing energy from the neutrino, and normally four jets, each created by the hadronization of a final state quark (q, $\bar{q}'$, b, $\bar{b}$). The lepton (e or μ) plus jets channel represents about 30% of the $t\bar{t}$ decays but suffers from a large background of W plus multijet production. CDF requires the presence of three or more jets, an isolated electron or muon and missing transverse energy [6]. Even after these selection criteria the signal-to-background ratio is 1 to 6. The amount of background is further reduced by identifying the bottom quark(s) from the t or $\bar{t}$ decay using one of two methods. The first method attempts to identify the semileptonic decay of the b quark, $b \rightarrow \ell X$ or $b \rightarrow c \rightarrow \ell X$, with ℓ being a muon or electron. This method is referred to as soft lepton tagging or SLT. The second b-tagging method uses the precision tracking of the silicon vertex (SVX) detector and the long-lifetime

Table 1: Event summaries for various $t\bar{t}$ decay channels. The data samples are based on 110 pb^{-1} of integrated luminosity. The combined cross section result does not include the τ dilepton measurement.

Channel	Mode	Data	Background	$\sigma(t\bar{t})$ (pb)
Lepton plus jets	SVX b-tag	34	9.2 ± 1.5	$6.2^{+2.1}_{-1.7}$
	SLT b-tag	40	22.6 ± 2.8	$9.2^{+4.3}_{-3.6}$
Dilepton	$ee, \mu\mu$, or $e\mu$	9	2.4 ± 0.5	$8.2^{+4.4}_{-3.4}$
	$e\tau$ or $\mu\tau$	4	2.0 ± 0.4	$15.6^{+16.3}_{-10.3}$
All Hadronic	1 SVX b-tag	187	142 ± 12	$9.6^{+4.4}_{-3.6}$
	2 SVX b-tag	157	120 ± 18	$11.5^{+7.7}_{-7.0}$
Combined				$7.6^{+1.8}_{-1.5}$

of B hadrons to locate decay vertices that are displaced from the primary interaction point. This method is called silicon vertex tagging or SVX.

2.2 The Dilepton Channel

In this channel, both W's decay leptonically. Events are characterized by two high P_T leptons of opposite charge, substantial missing energy from two neutrinos, and two jets. CDF searches mainly for dilepton events with electrons or muons (ee, $\mu\mu$ or $e\mu$) [7]. This channel has a better signal-to-background ratio than the lepton plus jets channel but it suffers from low statistics since it only represents about 5% of all $t\bar{t}$ decays.

CDF has also looked for dilepton events containing τ leptons [8]. The sample is restricted to events with one W boson decaying to an electron or muon, and the other decaying to a τ lepton. In principle this should double the number of dilepton events since 5% of all $t\bar{t}$ decays are $e\tau$ or $\mu\tau$. However the τ selection is less efficient than the e or μ selection, resulting in a total τ dilepton acceptance which is about 5 times smaller than that for the ee, $\mu\mu$ or $e\mu$ events.

2.3 The All Hadronic Channel

In this channel, both W's decay hadronically and the final state appears as 6 jets [9]. The all-hadronic channel accounts for 44% of all $t\bar{t}$ decays and has the advantage of no missing energy since there are no neutrinos in the final state. However, this channel is dominated by background from QCD multijet production. To reduce the background CDF requires 5 or more jets with either one SVX b-tagged jet and kinematic requirements, or two SVX b-tagged jets and looser kinematic requirements.

3 The $t\bar{t}$ Production Cross Section

The number of observed events and expected backgrounds in each of the channels described previously is summarized in Table 1. Given these numbers, the acceptance in each channel, and the luminosity, it is a simple calculation to determine the $t\bar{t}$ production cross section. Table 1 also lists the cross section measurements in each channel. Currently, the accuracies of the cross section measurements are limited statistically. These measurements can be compared to the NLO theoretical predictions from QCD of ~ 5 pb [4]. A measured cross section that is significantly higher than the QCD prediction could indicate new production mechanisms beyond the standard model. In addition, a comparison of the cross section extracted from different decay channels provides a test of the decays open to the top quark. The measurements from the lepton plus jets, dilepton, and all-hadronic channels, were combined, properly accounting for correlated errors, resulting in a CDF combined cross section [6] of [a]:

$$\sigma(t\bar{t}) = 7.6^{+1.8}_{-1.5} \ pb \qquad (1)$$

4 Properties of the Top Quark

Once the $t\bar{t}$ events have been found, it is important to use these events to explore the properties of the top quark. There are many different measurements that can be made in the top quark sector of the standard model [10]. In addition to the measurement of the $t\bar{t}$ production cross section described above, and the measurement of the top quark mass described elsewhere in these proceedings, there are a variety of other measurements being made and several of these are presented below.

---

[a]The τ dilepton cross section measurement is not included in the combined result.

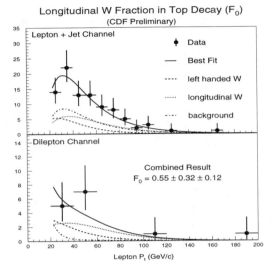

Figure 2: Fit of the CDF lepton P_T spectra in the dilepton and lepton plus jets samples to a mixture of $t\bar{t}$ events with longitudinal and left-handed W decays and background.

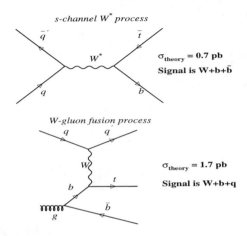

Figure 3: Feynman diagrams for single top production.

4.1 Measurement of the Polarization of W's in Top Quark Decay

The standard model predicts that the polarization of W bosons in top quark decay be either left-handed or longitudinal. For a top quark mass of 175 GeV/c, the expected fraction of longitudinal W's in top quark decay is:

$$F_0 = \frac{M_{top}^2}{2M_W^2 + M_{top}^2} = 70\ \%. \tag{2}$$

CDF measures this fraction by using the shape of the lepton P_T spectra in the lepton plus jets and dilepton data samples. This technique takes advantage of the fact that a charged lepton from a left-handed W tends to move opposite the W direction, while that from a longitudinal W tends to be perpendicular to the W direction. In the lab frame, this results in a P_T spectrum for leptons from longitudinal W's which is significantly harder than those from left-handed W's.

In the dilepton sample, the e^+e^- and $\mu^+\mu^-$ events are removed, due to an inability to reliably model the background P_T spectrum for these events. The lepton plus jets sample is broken up into three distinct samples depending on their b-tagging status, in a manner similar to that used in the CDF mass analysis[12].

The lepton P_T distribution observed in the data is shown in Figure 2, along with a fit to a mixture of $t\bar{t}$ events with longitudinal W decays, $t\bar{t}$ events with left-

handed W decays, and background. The result from a simultaneous fit to the dilepton and lepton plus jets channels is:

$$F_0 = 0.55 \pm 0.32(stat) \pm 0.12(syst). \tag{3}$$

This result has been corrected for the kinematic acceptance difference between top quarks which decay to longitudinal W's versus those that decay to left-handed W's. The largest systematic effect comes from the uncertainty in the shape of the background in the lepton plus jets sample.

4.2 Single Top Production

Single top quarks can be produced by two processes: the s-channel W^* process or the W-gluon fusion process. Figure 3 shows the Feynman diagrams for these processes and their theoretical cross sections[5]. A single top event is characterized by a W and two jets. In the W^* process the two jets are from b decays whereas for W-gluon one jet is from a b quark and the other from a light quark. Currently CDF only has preliminary results from the W-gluon fusion process. Events are required to have an electron or muon from the leptonically decaying W and exactly two jets, one of which must be identified as a b candidate by either an SVX or SLT tag.

The W-gluon events are distinguished from background by looking at a combination of kinematic properties. The pseudo-rapidity[b] distribution, η, of the light

---

[b]Pseudo-rapidity, η, is defined as the $-log(tan(\theta/2))$ where θ is

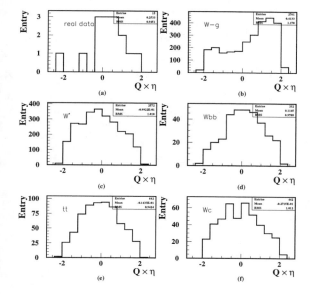

Figure 4: The lepton charge, Q, times pseudo-rapidity, η, distributions for a) $W+2$ jet data, b) W-gluon fusion single top, c) W^* single top, d) $Wb\bar{b}$ Monte Carlo, e) $t\bar{t}$ Monte Carlo, and f) Wc Monte Carlo events.

quark (non-tagged) jets tends to be positive for top quarks and negative for anti-top quarks. By taking the product of the lepton charge, Q, and η a distribution which is more asymmetric for W−gluon events than background events is formed. Figure 4 shows the Q times η distributions for W-gluon events and several backgrounds. A binned maximum likelihood is used to fit this distribution to extract an upper limit on single top production cross section in the W−gluon fusion channel of:

$$\sigma(W - \text{gluon}) \quad < \quad 15.4 \ pb \quad (95\% \ \text{C.L.}) \quad (4)$$

4.3 BF($t \to Wb$)/BF($t \to Wq$)

In the standard model, the top quark is expected to decay to a W boson and a b quark with a branching fraction near 1. Since each $t\bar{t}$ event should have two b quarks, a prediction for the number of events with 0, 1, or 2 b-tags can be made if the b-tagging efficiency is known. By comparing the prediction to the observed tagging multiplicities in both the dilepton and the lepton plus jets

the polar angle measured relative to the positive z axis (taken as the direction of the outgoing proton beam) assuming a z-vertex position of zero.

decay channels, CDF has extracted the preliminary result:

$$R_b = \frac{BF(t \to Wb)}{BF(t \to Wq)} = 0.99 \pm 0.29 \quad (5)$$

4.4 Rare Decays

Although the decay $t \to Wb$ dominates, the standard model predicts the top quark can decay into other final states such as $t \to \gamma q$, $t \to Zq$, $t \to WWc$ and $t \to WZb$, which for $M_{top} \sim 175$ GeV/c^2 is near threshold. However, the branching fraction to these final states is expected to be extremely small, BF $\sim 10^{-10}$. These decays occur through higher order diagrams involving loops and triboson vertices. CDF has made a search for the decay mode $t \to \gamma q$ by looking for two classes of events. The first type has a final state of a W boson ($W \to \ell\nu$), a photon, and jet activity. The second type has a final state with a high energy photon and four or more jets with one jet tagged as a b quark. In addition, a search has been made for the leptonic decay of a Z boson in place of the photon. Using these data samples, CDF establishes a limit on the branching fractions [11]

$$BF(t \to \gamma q) \quad < \quad 3.3\% \quad (95\% \ \text{C.L.}) \quad (6)$$

$$BF(t \to \gamma Z) \quad < \quad 33\% \quad (95\% \ \text{C.L.}). \quad (7)$$

The primary difference between the limits of these two channels is the small branching fraction from demanding $Z \to \ell\ell$. Although these limits are far from the expected standard model prediction, they are the first direct limits of these branching ratios. Even with the large statistics available in Run II, the limit of these rare decay modes will only approach $\approx 10^{-3}$, still far from the standard model expectations.

4.5 Br($t \to l\nu b$)

If top decays via the process $t \to Xb$ instead of $t \to Wb$, where the decay modes of X are very different from those of a W boson, we would expect to obtain a value for the branching ratio $Br(t \to l\nu b)$ which is different that the standard model expectation of 1/9. The method used by CDF is to take the ratios of the $t\bar{t}$ production cross section measurements times standard model branching fraction using the lepton plus jets, dilepton, and all-hadronic samples. Each ratio can be written as a function of the branching ratio $Br(t \to l\nu b)$, and a likelihood technique is used to extract a measurement of this branching fraction from each ratio. The three results obtained are shown in Table 2. The weighted average of the two best results (the all-hadronic to lepton plus jets and the dilepton to all-hadronic ratios) give the CDF published[6] result of:

$$Br(t \to l\nu b) = 0.094 \pm 0.024 \quad (8)$$

Table 2: The measured value for $Br(t \rightarrow l\nu b)$ using the three ratios of measured cross section times standard model branching fractions.

Ratio	$Br(t \rightarrow l\nu b)$
Dilepton/Lepton plus jets	0.127 ± 0.044
All Hadronic/Lepton plus jets	0.083 ± 0.031
Dilepton/All Hadronic	0.104 ± 0.022

The correlated uncertainty due to the common channel is accounted for, as well as the effect of the contamination of decays across different channels.

5 Summary

The first explorations of the top quark have been made at the Tevatron. With the discovery of the top quark complete, the attention of physicists has turned towards measuring the properties of the top quark. The measurements include the production cross section, the measurement of the W polarization in top decays and branching ratios, limits on single top production and rare decays and the measurement of the top quark mass. In all respects the top quark appears to be what the standard model predicts. Perhaps the only surprise is the accuracy to which the experiments have been able to measure the top quark mass [12]. Currently, most measurements are statistically limited and for that reason the collaboration is looking forward to Run II with great anticipation.

Acknowledgments

The results presented here are the work of the CDF collaboration at Fermilab. I wish to thank all of my colleagues who contributed to the success of this research and the Fermilab Accelerator, Computing and Research Divisions for their support. This work was supported in part by the Department of Energy and the National Science Foundation.

References

1. F. Abe *et al.*, *Phys. Rev. Lett.* **74**, 2626 (1995); S. Abachi *et al.*, *Phys. Rev. Lett.* **74**, 2632 (1995).
2. F. Abe *et al.*, *Phys. Rev.* D **50**, 2966 (1994).
3. F. Abe *et al.*, *Nucl. Instrum. Methods* A **271**, 387 (1988).
4. E. Laenen, J. Smith, W.L. van Neerven, *Phys. Lett.* B **321**, 254 (1994); Catani, Mangano, Nason and Trentadue, CERN Preprint, CERN-TH/96-21 hep-ph/9602208; E. Berger, H. Contopanagos, Argonne National Lab Preprint, ANL-HEP-CP-96-51.
5. M.C. Smith, S. Willenbrock, *Phys. Rev.* D **54**, 6696 (1996); T. Stelzer, Z. Sullivan, S. Willenbrock, *Phys. Rev.* D **56**, 5919 (1997); T. Tait, C.P. Yuan, hep-ph/9710372.
6. F. Abe *et al. Phys. Rev. Lett.* **80**, 2773 (1998).
7. F. Abe *et al. Phys. Rev. Lett.* **80**, 2779 (1998).
8. F. Abe *et al. Phys. Rev. Lett.* **79**, 3585 (1997).
9. F. Abe *et al. Phys. Rev. Lett.* **79**, 1992 (1997).
10. *Future ElectroWeak Physics at the Fermilab Tevatron: Report of the tev_2000 Study Group*, Editors D. Amidei and R. Brock Fermilab–Pub–96/082.
11. F. Abe *et al. Phys. Rev. Lett.* **80**, 2525 (1998).
12. F. Abe *et al. Phys. Rev. Lett.* **80**, 2767 (1998).

CHARGE ASYMMETRY OF HEAVY QUARKS AT HADRON COLLIDERS*

J. H. KÜHN[a)] and G. RODRIGO[b)]

a) *Institut für Theoretische Teilchenphysik, Universität Karlsruhe, Germany*
b) *INFN-Sezione di Firenze, Italy*

A sizeable difference in the differential production cross section of top and antitop quarks, respectively, is predicted for hadronically produced heavy quarks. It is of order α_s and arises from the interference between charge odd and even amplitudes respectively. For the TEVATRON it amounts up to 15% for the differential distribution in suitable chosen kinematical regions. The resulting integrated forward-backward asymmetry of 4–5% could be measured in the next round of experiments. At the LHC the asymmetry can be studied by selecting appropriately chosen kinematical regions. Furthermore, a slight preference at LHC for centrally produced antitop is predicted, with top quarks more abundant at large positive and negative rapidities.

Heavy flavor production at hadron colliders is one of the most active fields of current theoretical and experimental studies. Large event rates, combined with improved experimental techniques, allow for detailed investigations of the properties of heavy quarks and their production mechanism at the same time. While charm production with a quark mass around 1.5 GeV is barely accessible to perturbative QCD calculations, bottom and *a fortiori* top production should be well described by this approach.

Theoretical and experimental results [1,2] for the cross section of hadronic top production are well consistent with this expectation. Obviously, in view of the large QCD coupling, the inclusion of higher order QCD corrections in these calculations is mandatory for a successful comparison. Recent studies have, to a large extent, concentrated on the predictions of the total cross section and a few selected one particle inclusive distributions. In this paper a different issue of heavy flavor production is investigated, namely the charge asymmetry, which is sensitive toward a specific subclass of virtual and real radiative corrections.

Evaluated in Born approximation the lowest order processes relevant for heavy flavor production

$$q + \bar{q} \to Q + \bar{Q} \,, \tag{1}$$

$$g + g \to Q + \bar{Q} \,, \tag{2}$$

do not discriminate between the final quark and antiquark, thus predicting identical differential distributions also for the hadronic production process. However, radiative corrections involving either virtual or real gluon emission lead to a sizeable difference between the differential quark and antiquark production process and hence to a charge asymmetry [3,4] which could be well accessible experimentally.

This asymmetry has its origin in two different reactions: radiative corrections to quark-antiquark fusion

(Fig. 1) and heavy flavor production involving interference terms of different amplitudes contributing to gluon-quark scattering

$$g + q \to Q + \bar{Q} + q \,, \tag{3}$$

a reaction intrinsically of order α_s^3. The contribution from quark gluon scattering to the asymmetry has been shown to be small [4] and will be ignored in this review. Gluon fusion remains of course charge symmetric. In both reactions (1) and (3) the asymmetry can be traced to the interference between amplitudes which are relatively odd under the exchange of Q and $\bar{Q}$. In fact, as shown below in detail, the asymmetry can be understood in analogy to the corresponding one in QED reactions and is proportional to the color factor d_{abc}^2. In contrast, the non-Abelian contributions, in particular those involving the triple gluon coupling, lead to symmetric pieces in the differential cross section.

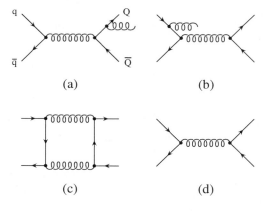

Figure 1: Origin of the QCD charge asymmetry in hadroproduction of heavy quarks: interference of final-state (a) with initial-state (b) gluon bremsstrahlung plus interference of the box (c) with the Born diagram (d). Only representative diagrams are shown.

*Presented by J. H. Kühn.

Let us briefly discuss a few important aspect of this calculation. The box amplitude for $q\bar{q} \to Q\bar{Q}$ is ultraviolet finite and the asymmetric contribution to the cross section of order α_s^3 is therefore not affected by renormalization, an obvious consequence of the symmetry of the lowest order reaction. The same line of reasoning explains the absence of initial state collinear singularities in the limit $m_q \to 0$ which would have to be absorbed into the (symmetric) lowest order cross section. Infrared singularities require a more careful treatment. They are absent in the asymmetric piece of the process in eq. (3). However, real and virtual radiation (Fig. 1), if considered separately, exhibit infrared divergences, which compensate in the sum, corresponding to the inclusive production cross section.

The charge asymmetry in the partonic reactions (1) and (3) implies for example a forward-backward asymmetry of heavy flavor production in proton-antiproton collisions. In particular, it leads to a sizeable forward-backward asymmetry for top production which is dominated by reaction (1), and can, furthermore, be scrutinized by studying $t\bar{t}$ production at fixed longitudinal momenta and at various partonic energies $\hat{s}$. However, the charge asymmetry can also be observed in proton-proton collisions at high energies. In this case one has to reconstruct the $t\bar{t}$ restframe and select kinematic regions, which are dominated by $q\bar{q}$ annihilation or flavor excitation $gq \to t\bar{t}X$. Alternatively, one may also study the difference in the one-particle inclusive rapidity distribution of top versus antitop, which again integrates to zero.

The analysis of these effects allows to improve our understanding of the QCD production mechanism. At the same time it is important for the analysis of single top production through Wb fusion. This reaction is charge asymmetric as a consequence of weak interactions. Although the final states in single top production and hadronic $t\bar{t}$ production are different and should in principle be distinguishable, it is nevertheless mandatory to control the charge asymmetry from both sources.

As shown in [3], the dominant contribution to the charge asymmetry originates from $q\bar{q}$ annihilation, namely from the asymmetric piece in the interference between the Born amplitude for $q\bar{q} \to Q\bar{Q}$ (Fig. 1d) and the one loop corrections to this reaction (Fig. 1c), which must be combined with the interference term between initial state and final state radiation (Fig. 1a,1b).

However, only QED like terms are relevant for the charge asymmetric piece [4]. The QCD asymmetry is thus obtained from the QED results by the replacement

$$\alpha_{QED}Q_qQ_Q \to \frac{d_{abc}^2}{16N_CT_FC_F}\alpha_s = \frac{5}{12}\alpha_s . \qquad (4)$$

Let us note in passing that diagrams involving the triple gluon coupling lead to charge symmetric terms.

The differential charge asymmetry in the inclusive cross section

$$q + \bar{q} \to Q + X , \qquad (5)$$

at the partonic level can then be defined through

$$\hat{A}(\cos\hat{\theta}) = \frac{N_t(\cos\hat{\theta}) - N_{\bar{t}}(\cos\hat{\theta})}{N_t(\cos\hat{\theta}) + N_{\bar{t}}(\cos\hat{\theta})} , \qquad (6)$$

where $\hat{\theta}$ denotes the top quark production angle in the $q\bar{q}$ restframe and $N(\cos\hat{\theta}) = d\sigma/d\Omega(\cos\hat{\theta})$. Since $N_{\bar{t}}(\cos\hat{\theta}) = N_t(-\cos\hat{\theta})$ as a consequence of charge conjugation symmetry, $\hat{A}(\cos\hat{\theta})$ can also be interpreted as a forward-backward asymmetry of top quarks (Fig. 2).

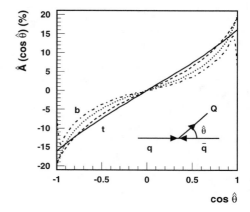

Figure 2: Differential charge asymmetry in top quark pair production for fixed partonic center of mass energies $\sqrt{\hat{s}} = 400$ GeV (solid), 600 GeV (dashed) and 1 TeV (dotted). We also plot the differential asymmetry for b-quarks with $\sqrt{\hat{s}} = 400$ GeV (dashed-dotted).

The integrated charge asymmetry

$$\bar{A} = \frac{N_t(\cos\hat{\theta} \geq 0) - N_{\bar{t}}(\cos\hat{\theta} \geq 0)}{N_t(\cos\hat{\theta} \geq 0) + N_{\bar{t}}(\cos\hat{\theta} \geq 0)} , \qquad (7)$$

is shown in Fig. 3 as a function of $\sqrt{\hat{s}}$. With a typical value around $6 - 8.5\%$ it should be well accessible in the next run of the TEVATRON.

In addition to the pure QCD amplitudes also a mixed QCD-electroweak interference term will lead to an asymmetric contribution to the $q\bar{q}$ process [3,4]. This leads to an increase of the asymmetry as given by pure QCD by a factor 1.09.

The asymmetry can in principle be studied experimentally in the partonic restframe, as a function of $\hat{s}$, by

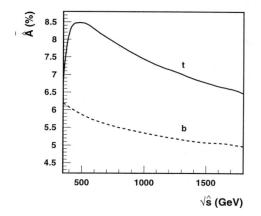

Figure 3: Integrated charge asymmetry as a function of the partonic center of mass energy for top and bottom quark pair production.

scale, $\mu = m_t/2$ and $\mu = 2m_t$, have been considered and the factor 1.09 is included. An increase in the center of mass energy to 2 TeV leads to a slight decrease of our prediction to $4.6 - 5.5\%$.

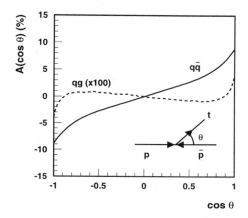

Figure 4: Differential charge asymmetry in the proton-antiproton restframe, $\sqrt{s} = 2$ TeV, using the CTEQ-1 structure function, $\mu = m_t$. The contributions from $q\bar{q}$ and qg (plus $\bar{q}g$) initiated processes are shown separately.

measuring the invariant mass of the $t\bar{t}$ system plus an eventually radiated gluon. It is, however, also instructive to study the asymmetry in the laboratory frame by folding the angular distribution with the structure functions[5,6]. For proton-antiproton collisions it is convenient to consider the forward-backward asymmetry as function of the production angle in the center of mass system. The differential asymmetry for $\sqrt{s} = 2$ TeV is shown in Fig. 4 which displays separately the contribution from $q\bar{q}$ and qg (plus $\bar{q}g$) initiated reactions. The denominator includes both $q\bar{q}$ and gg initiated processes in lowest order. The numerator is evidently dominated by quark-antiquark annihilation.

At this point we have to emphasize that both numerator and denominator are evaluated in leading order (LO). The next-to-leading (NLO) corrections to the $t\bar{t}$ production cross section are known to be large[7], around 30% or even more. In the absence of NLO corrections for the numerator we nevertheless stay with the LO approximation in both numerator and denominator, expecting the dominant corrections from collinear emission to cancel. However, from a more conservative point of view an uncertainty of around 30% has to be assigned to the prediction for the asymmetry.

For the total charge asymmetry at $\sqrt{s} = 1.8$ TeV we predict

$$\bar{A} = \frac{N_t(\cos\theta \geq 0) - N_{\bar{t}}(\cos\theta \geq 0)}{N_t(\cos\theta \geq 0) + N_{\bar{t}}(\cos\theta \geq 0)} = 4.8 - 5.8\% \,, \quad (8)$$

where different choices of the structure function and different choices of the factorization and renormalization

Top-antitop production in proton-proton collisions at the LHC is, as a consequence of charge conjugation symmetry, forward-backward symmetric if the laboratory frame is chosen as the reference system. However, by selecting the invariant mass of the $t\bar{t}(+g)$ system and its longitudinal momentum appropriately, one can easily constrain the parton momenta such that a preferred direction is generated for quark-antiquark reactions.

For some of the more extreme kinematic regions, namely large x and/or large $\hat{s}$, a sizeable difference between top and antitop production can be observed at the LHC[4]. From this it may seem that the reconstruction of both t, $\bar{t}$ and even the gluon is required for the study of the charge asymmetry in pp collisions. However, also the difference between the single particle inclusive distribution of t and $\bar{t}$ respectively may provide evidence for the charge asymmetry. Production of $t\bar{t}(g)$ with negative x is dominated by initial $\bar{q}$ with small x_1 and q with large x_2. The charge asymmetry implies that $Q(\bar{Q})$ is preferentially emitted into the direction of $q(\bar{q})$. The same line of reasoning is applicable for positive x, with $Q(\bar{Q})$ again preferentially emitted in the direction of $q(\bar{q})$, and the role of x_1 and x_2 reversed. In total this leads to a slight preference for centrally produced antiquarks and quarks slightly dominant in the forward and backward direction, i.e., at large positive and negative rapidities.

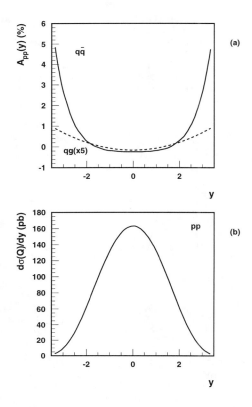

2. P. Tipton, "Experimental top quark physics," *Proceedings of the ICHEP 96, Warsaw, Poland* (World Scientific, Singapore, 1996), pg.123.
3. J. H. Kühn and G. Rodrigo *Phys. Rev. Lett.* **81** (1998) 49.
4. J. H. Kühn and G. Rodrigo, `hep-ph/9807420`.
5. A. D. Martin, R. G. Roberts, and W. J. Stirling *Phys. Lett.* **B387** (1996) 419, `hep-ph/9606345`.
6. H. L. Lai *et. al.* *Phys. Rev.* **D55** (1997) 1280, `hep-ph/9606399`.
7. R. Bonciani, S. Catani, M. L. Mangano and P. Nason, "NLL resummation of the heavy quark hadroproduction cross-section", `hep-ph/9801375`.

Figure 5: Rapidity distribution of charge asymmetry (a) and total cross section at Born order (b) of top quark production in proton-proton collisions, $\sqrt{s} = 14$ TeV and $\mu = m_t$. Contributions from $q\bar{q}$ fusion and flavor excitation, $qg(\bar{q}g)$, are shown separately. Laboratory frame (CTEQ-1).

The differential charge asymmetry

$$A_{pp}(y) = \frac{\dfrac{dN(Q)}{dy} - \dfrac{dN(\bar{Q})}{dy}}{\dfrac{dN(Q)}{dy} + \dfrac{dN(\bar{Q})}{dy}} \, , \qquad (9)$$

is shown in Fig. 5a for top quark production at the LHC ($\sqrt{s} = 14$ TeV). As expected, a sizeable charge asymmetry is predicted in the region of large rapidity. It remains to be seen, if the low event rates in these extreme regions will permit the observation of this effect. The quark-gluon process is again negligible.

1. S. Catani, "QCD at high-energies," `hep-ph/9712442`, and references therein.

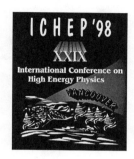

Parallel Session 8

Heavy Hadrons: Lifetimes: Mixing, Rare Decays

Convenors: Michael Danilov (ITEP)

Manfred Paulini (LBNL, CDF)

B^+, B_d^0 AND b-BARYON LIFETIMES

S. WILLOCQ

Stanford Linear Accelerator Center, P.O. Box 4349, Stanford, CA 94309, USA
E-mail: willocq@slac.stanford.edu

We review recent B^+, B_d^0 and b-baryon lifetime measurements performed by the LEP, SLD and CDF collaborations. Lifetime ratios of $\tau(B^+)/\tau(B_d^0) = 1.070 \pm 0.027$ and $\tau(b \text{ baryon})/\tau(B_d^0) = 0.77 \pm 0.04$ are obtained using all existing measurements. The ratio between charged and neutral B meson lifetimes is in good agreement with theory but the ratio between b-baryon and B meson lifetimes remains somewhat lower than expected.

1 Introduction

The study of exclusive b-hadron lifetimes provides an important test of our understanding of b-hadron decay dynamics. Lifetimes are especially useful to probe the strong interaction effects arising from the fact that b quarks are not free particles but are confined inside hadrons. In the naive spectator model, the b quarks are treated as if they were free and one therefore expects $\tau(B^+) = \tau(B_d^0) = \tau(B_s^0) = \tau(\Lambda_b)$. However, this picture does not hold in the case of charm hadrons for which the lifetimes follow the pattern $\tau(D^+) \simeq 2.3 \ \tau(D_s) \simeq 2.5 \ \tau(D^0) \simeq 5 \ \tau(\Lambda_c^+)$. These factors are predicted to scale with the inverse of the heavy quark mass squared and the b-hadron lifetimes are thus expected to differ by only 10-20%. Using the Heavy Quark Expansion, Bigi et al.[1] predict $\tau(B^+)/\tau(B_d^0) = 1 + 0.05 \ (f_B/200 \text{ MeV})^2$, where f_B is the B-meson decay constant ($f_B = 200 \pm 40$ MeV), and $\tau(\Lambda_b)/\tau(B_d^0) \simeq 0.9$. However, Neubert and Sachrajda[2] argue that a more theoretically conservative approach yields $0.8 < \tau(B^+)/\tau(B_d^0) < 1.2$ and $0.85 < \tau(\Lambda_b)/\tau(B_d^0) < 1.0$.

Precise knowledge of exclusive b-hadron lifetimes is required for accurate measurements of $|V_{cb}|$ and B mixing, and is also an important input parameter for $Z^0 \to b\bar{b}$ electroweak measurements.

2 B^+ and B_d^0 Lifetimes

The LEP, SLD and CDF collaborations have taken advantage of their precision vertex detectors and of the significant boost for b hadrons produced in high energy e^+e^- and $\bar{p}p$ collisions to measure exclusive b-hadron lifetimes. Three main analysis techniques have been used to measure B^+ and B_d^0 lifetimes. The first method relies on fully reconstructed B decays (e.g. $B \to J/\psi \, K$). This is the ideal method for a lifetime measurement since there is little or no modelling uncertainty in the B energy and the sample composition. However, exclusive branching ratios for B decays are typically small (10^{-4} to 10^{-3}) which severely limits the statistics available at current

facilities. The second and most utilized method selects semileptonic decays of the type $B \to D^{(*)} l \nu X$, where the $D^{(*)}$ meson is fully reconstructed. Sample composition can be controlled from the data using the charge correlation between the lepton and the $D^{(*)}$ meson. A complication arises from decays of the type $B \to D^{**} l \nu$ which spoil the B^+ and B_d^0 purity of the respective $\overline{D^0} l^+$ and $D^{*-} l^+$ samples, and whose rates are not well known.

The CDF collaboration has finalized a study[3] based on the full Run-I data sample and corresponding to an integrated luminosity of 110 pb^{-1}. A "B^+" sample consisting of $\overline{D^0} l^+$ pairs is selected with fully reconstructed $\overline{D^0} \to K^+ \pi^-$ decays. Similarly, a "B^0" sample consisting of $D^{*-} l^+$ pairs is selected by reconstructing the decays $D^{*-} \to \overline{D^0} \pi^-$ where $\overline{D^0} \to K^+ \pi^- (\pi^0)$ or $\overline{D^0} \to K^+ \pi^- \pi^+ \pi^-$. The B decay vertex is then formed by intersecting the lepton and $D^{(*)}$ trajectories.

A fit using decay length and momentum information for the $\overline{D^0} l^+$ and $D^{*-} l^+$ samples yields $\tau(B^+) = 1.637 \pm 0.058(\text{stat})^{+0.045}_{-0.043}(\text{syst})$ ps, $\tau(B_d^0) = 1.474 \pm 0.039(\text{stat})^{+0.052}_{-0.051}(\text{syst})$ ps, and $\tau(B^+)/\tau(B_d^0) = 1.110 \pm 0.056(\text{stat})^{+0.033}_{-0.030}(\text{syst})$. Contamination from $B \to D^{**} l \nu$ decays is estimated to be 10-15% and constitutes the dominant systematic uncertainty in the lifetime ratio.

A third method for lifetime measurements relies on inclusive topological vertexing, pioneered by the DELPHI and SLD collaborations. Here, the charged particle topology of the decays is reconstructed and the separation between charged and neutral b hadrons is achieved simply using the sum of the charges of all tracks associated with a secondary vertex. This method has the advantage of large statistics but requires good control in the detailed simulation of b hadron production and decay.

The SLD collaboration has updated its topological vertexing analysis[4] with data taken during the first part of the 1997-98 run. A set of 49,664 B decay candidates is selected with an efficiency of 50% and a purity of 98%. Separation between B^+ and B_d^0 decays is performed on the basis of the total charge Q_{tot} of tracks associated with the secondary vertex (see Fig. 1). The charged (neutral) sample consists of 30,028 (19,636) decays with $|Q_{tot}| =$

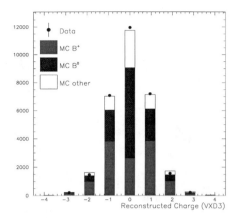

Figure 1: Distribution of the vertex charge for the SLD 1997-98 data (points) and Monte Carlo simulation (histograms) indicating the contributions from charged and neutral B mesons. The category "MC other" contains mostly neutral b hadrons: B_s^0 and b baryons.

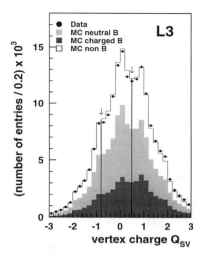

Figure 2: Distribution of the vertex charge for L3 data (points) and Monte Carlo simulation (histograms) for charged and neutral B mesons.

$1, 2, 3$ ($Q_{tot} = 0$). The charge separation is enhanced somewhat by taking into account the dependence upon the reconstructed vertex mass and the b-quark charge at production (using techniques developed for the study of time-dependent B^0–$\overline{B^0}$ mixing). An effective $B^+ : B_d^0$ ($B_d^0 : B^+$) separation of 2.6 : 1 is then obtained in the charged (neutral) sample.

The lifetimes are extracted with a simultaneous fit to the decay length distributions of the charged and neutral samples. Combining with previous data, corresponding to a total sample of 400,000 hadronic Z^0 decays, the lifetimes are $\tau(B^+) = 1.686 \pm 0.025(\text{stat}) \pm 0.042(\text{syst})$ ps, $\tau(B_d^0) = 1.589 \pm 0.026(\text{stat}) \pm 0.055(\text{syst})$ ps, and $\tau(B^+)/\tau(B_d^0) = 1.061^{+0.031}_{-0.029}(\text{stat}) \pm 0.027(\text{syst})$. These are currently the most precise determinations of the B^+ and B_d^0 lifetimes. The dominant contribution to the lifetime measurement error arises from the uncertainty in the b-fragmentation function. Specifically, the range of scaled b-hadron energy was taken to be $\langle x_E \rangle_b = 0.700 \pm 0.011$, which translates into an uncertainty of ± 0.035 ps in both B^+ and B_d^0 lifetimes. This uncertainty cancels out in the lifetime ratio since all b-hadrons are assumed to have the same fragmentation function. It should be noted that recent measurements of $\langle x_E \rangle_b$, including an analysis by SLD using the same topological technique,[5] find a somewhat larger value for $\langle x_E \rangle_b \simeq 0.72$ (see also the L3 measurement below). Such a value would shift the above lifetimes down by about 0.064 ps.

The L3 collaboration has also developed an inclusive topological vertexing technique, first applied to measure the average b-hadron lifetime.[6] The vertexing algorithm uses the 3-D impact parameters and rapidity of tracks to reconstruct 3 vertices per event corresponding to the one primary and two secondary vertices expected in $Z^0 \to b\bar{b}$ decays. Here, the lifetime is extracted from either the secondary vertex decay length or the impact parameters of tracks attached to the secondary vertex. The latter has the advantage of having a reduced dependence on the b fragmentation uncertainty. Since the two different variables have different sensitivities to this uncertainty, they can be combined to yield very precise determinations of both the average b-hadron lifetime $\tau_b = 1.556 \pm 0.010(\text{stat}) \pm 0.017(\text{syst})$ ps and the average scaled b-hadron energy $\langle x_E \rangle_b = 0.709 \pm 0.004(\text{stat+syst})$.

L3 extended this technique to the study of B^+ and B_d^0 lifetimes.[7] From a sample of 2×10^6 hadronic Z^0 decays, the analysis selects 890,506 secondary vertices. The separation between charged and neutral decays is then obtained by forming the vertex charge Q_{SV} defined as the product of the weighted sum of track charges and the sign of the Jet Charge, where the weight represents the probability to belong to the secondary vertex. Fig. 2 shows the vertex charge distribution and the cuts used to define the charged ($Q_{SV} > 0.5$) and neutral ($-0.8 < Q_{SV} < 0.5$) samples. For $Q_{SV} > -0.8$, the sample is 69% pure in b hadrons. The $B^+ : B_d^0$ ($B_d^0 : B^+$) separation is estimated to be 1.25 : 1 (1.10 : 1) in the charged (neutral) sample. To reduce the b-fragmentation uncertainty, the lifetimes are extracted using weighted average track impact parameters and a b tag is used in the opposite hemisphere

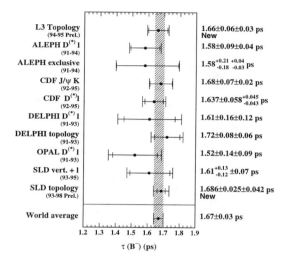

L3 Topology (94-95 Prel.)	1.66±0.06±0.03 ps **New**
ALEPH D$^{(*)}$ l (91-94)	1.58±0.09±0.04 ps
ALEPH exclusive (91-94)	1.58$^{+0.21}_{-0.18}$ $^{+0.04}_{-0.03}$ ps
CDF J/ψ K (92-95)	1.68±0.07±0.02 ps
CDF D$^{(*)}$ l (92-95)	1.637±0.058$^{+0.045}_{-0.043}$ ps
DELPHI D$^{(*)}$ l (91-93)	1.61±0.16±0.12 ps
DELPHI topology (91-93)	1.72±0.08±0.06 ps
OPAL D$^{(*)}$ l (91-93)	1.52±0.14±0.09 ps
SLD vert. + l (93-95)	1.61$^{+0.13}_{-0.12}$ ±0.07 ps
SLD topology (93-98 Prel.)	1.686±0.025±0.042 ps **New**
World average	1.67±0.03 ps

$\tau\,(\mathrm{B}^-)$ (ps)

Figure 3: Measurements of the B^+ lifetime.

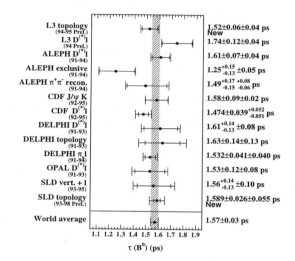

L3 topology (94-95 Prel.)	1.52±0.06±0.04 ps **New**
L3 D$^{(*)}$ l (94 Prel.)	1.74±0.12±0.04 ps
ALEPH D$^{(*)}$ l (91-94)	1.61±0.07±0.04 ps
ALEPH exclusive (91-94)	1.25$^{+0.15}_{-0.13}$ ±0.05 ps
ALEPH π$^+$π$^-$ recon. (91-94)	1.49$^{+0.17}_{-0.15}$ $^{+0.08}_{-0.06}$ ps
CDF J/ψ K (92-95)	1.58±0.09±0.02 ps
CDF D$^{(*)}$ l (92-95)	1.474±0.039$^{+0.052}_{-0.051}$ ps
DELPHI D$^{(*)}$ l (91-93)	1.61$^{+0.14}_{-0.13}$ ±0.08 ps
DELPHI topology (91-93)	1.63±0.14±0.13 ps
DELPHI π l (91-93)	1.532±0.041±0.040 ps
OPAL D$^{(*)}$ l (91-93)	1.53±0.12±0.08 ps
SLD vert. + l (93-95)	1.56$^{+0.14}_{-0.13}$ ±0.10 ps
SLD topology (93-98 Prel.)	1.589±0.026±0.055 ps **New**
World average	1.57±0.03 ps

$\tau\,(\mathrm{B}^0)$ (ps)

Figure 4: Measurements of the B^0_d lifetime.

to suppress the background. As a result, the lifetimes are found to be $\tau(B^+) = 1.662 \pm 0.056(\mathrm{stat}) \pm 0.025(\mathrm{syst})$ ps, $\tau(B^0_d) = 1.524 \pm 0.055(\mathrm{stat}) \pm 0.037(\mathrm{syst})$ ps, and $\tau(B^+)/\tau(B^0_d) = 1.09 \pm 0.07(\mathrm{stat}) \pm 0.03(\mathrm{syst})$.

The measurements presented above have been combined with all previous measurements (see Figs. 3-5) to yield the following world averages:

$$\tau(B^+) = 1.67 \pm 0.03 \text{ ps}, \tag{1}$$

$$\tau(B^0_d) = 1.57 \pm 0.03 \text{ ps}, \tag{2}$$

$$\tau(B^+)/\tau(B^0_d) = 1.070 \pm 0.027. \tag{3}$$

It is interesting to note that the recent progress in inclusive topological techniques has allowed a reduction of about 25% in overall uncertainty since the last summer conferences. Furthermore, the measurements are becoming precise enough to begin to measure a difference between B^+ and B^0_d lifetimes.

3 b-baryon Lifetime

As mentioned earlier, the lifetime of b baryons is expected to be about 10% shorter than that of B^0_d mesons. However, measurements over the past few years have indicated that the effect may be as large as 20-25% which remains somewhat difficult to accommodate. Measurements of b-baryon lifetimes are challenging since b baryons represent only about 10% of all b hadrons produced in $Z^0 \to b\bar{b}$ decays and the properties of b baryons

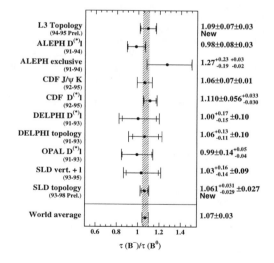

L3 Topology (94-95 Prel.)	1.09±0.07±0.03 **New**
ALEPH D$^{(*)}$ l (91-94)	0.98±0.08±0.03
ALEPH exclusive (91-94)	1.27$^{+0.23}_{-0.19}$ $^{+0.03}_{-0.02}$
CDF J/ψ K (92-95)	1.06±0.07±0.01
CDF D$^{(*)}$ l (92-95)	1.110±0.056$^{+0.033}_{-0.030}$
DELPHI D$^{(*)}$ l (91-93)	1.00$^{+0.17}_{-0.15}$ ±0.10
DELPHI topology (91-93)	1.06$^{+0.13}_{-0.11}$ ±0.10
OPAL D$^{(*)}$ l (91-93)	0.99±0.14$^{+0.05}_{-0.04}$
SLD vert. + l (93-95)	1.03$^{+0.16}_{-0.14}$ ±0.09
SLD topology (93-98 Prel.)	1.061$^{+0.031}_{-0.029}$ ±0.027 **New**
World average	1.07±0.03

$\tau\,(\mathrm{B}^-)/\tau\,(\mathrm{B}^0)$

Figure 5: Measurements of the ratio between B^+ and B^0_d lifetimes.

are not well known. Therefore, most measurements have concentrated on semileptonic decays and have relied on charge correlations between Λ_c^+-lepton or Λ-lepton pairs to enhance the signal fraction and control the sample composition.

The OPAL collaboration has finalized a study [8] of partially reconstructed $\Lambda_b \to \Lambda_c^+ l^- \overline{\nu} X$ decays with $\Lambda_c^+ \to p K^- \pi^+$ or $\Lambda_c^+ \to \Lambda l^+ \nu X$ decays in a total sample of 4.4×10^6 hadronic Z^0 events. The $\Lambda_c^+ l^-$ signal is estimated to be 129 ± 25 events and the Λ_b lifetime extracted from the reconstructed decay length distribution is $\tau(\Lambda_b) = 1.29^{+0.24}_{-0.22}(\text{stat}) \pm 0.06(\text{syst})$ ps.

The DELPHI collaboration released a preliminary study of the same modes using a sample of 3.6×10^6 hadronic Z^0 events. [9] Charge-correlations allow the signal fraction to determined from the data to be $f_{signal} = (56 \pm 6)\%$. A lifetime fit to the reconstructed proper time distribution yields $\tau(\Lambda_b) = 1.17^{+0.20}_{-0.18}(\text{stat})^{+0.04}_{-0.05}(\text{syst})$ ps. DELPHI also studied more inclusive final states consisting of Λ-lepton and proton-lepton pairs. These have the advantage of increasing the statistical sensitivity of the measurement but the sample composition is more difficult to control which leads to higher systematic uncertainties. The proton-lepton analysis is unique and proceeds by applying an inclusive reconstruction of b-hadron semileptonic decays which relies on both vertexing and kinematical information. Then, vertices containing an opposite-sign proton-lepton pair are selected, where the proton is required to be the fastest hadron in the vertex and to be positively identified by the RICH particle identification system. A rejection factor of ~ 10 is achieved for both pion/proton and kaon/proton separation over most of the momentum range of interest (3 to 20 GeV/c). This analysis is only applied to the 1994-95 data sample, corresponding to 2×10^6 hadronic Z^0 decays, since the RICH was not fully operational before 1994. The b-baryon lifetime is then extracted from the reconstructed proper time distribution of the proton-lepton sample (Fig. 6): $\tau(b \text{ baryon}) = 1.19 \pm 0.14(\text{stat}) \pm 0.07(\text{syst})$ ps with $f_{signal} = (47 \pm 5)\%$ as estimated from the data. A study of Λ-lepton pairs yields $\tau(b \text{ baryon}) = 1.16 \pm 0.20(\text{stat}) \pm 0.09(\text{syst})$ ps with $f_{signal} = (35 \pm 8)\%$ as estimated from the data.

Measurements of the b-baryon lifetime are summarized in Fig. 7. Averaging Λ_c^+-lepton with more inclusive Λ-lepton and proton-lepton measurements yields the following world average:

$$\tau(b \text{ baryon}) = 1.21 \pm 0.05 \text{ ps.} \qquad (4)$$

4 Summary

B^+, B_d^0 and b-baryon lifetimes have been measured by the LEP, SLD and CDF collaborations. Recent progress

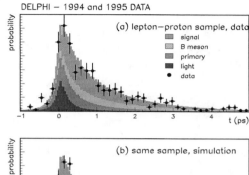
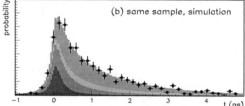

DELPHI – 1994 and 1995 DATA

(a) lepton–proton sample, data
- signal
- B meson
- primary
- light
- data

(b) same sample, simulation

Figure 6: Reconstructed proper time distribution for the DELPHI proton-lepton analysis for 1994-95 data (points) and the various sample components (histograms).

in the precision of B^+ and B_d^0 lifetimes has stemmed from the application of inclusive topological techniques and the addition of new data collected by SLD. As seen in Fig. 8, lifetime differences are small and the observed hierarchy $\tau(\Lambda_b) < \tau(B_s^0) < \tau(B_d^0) < \tau(B^+)$ is consistent with predictions based on the Heavy Quark Expansion. [a] The measurements are becoming precise enough to begin to see a difference between B^+ and B_d^0 lifetimes, the significance being at the $2.6\,\sigma$ level. The b-baryon lifetime remains significantly low which continues to spur theoretical activity.

Further improvements are expected in the near future from SLD with the inclusion of the full 1997-98 data sample, corresponding to an increase of $\sim 40\%$ in statistics. In the longer term, the next step in precision will come from experiments at the B Factories and the Tevatron.

Acknowledgments

I wish to thank Claire Shepherd-Themistocleous, Hans-Gunther Moser and Juan Alcaraz from the LEP B Lifetime Working Group for updating the world averages. I have also benefited from interesting discussions with Claire Bourdarios and Franz Muheim. John Jaros and Su Dong are thanked for their proofreading. This work

[a] A review of B_s^0 lifetime measurements was presented by A. Ribon at this conference.

was supported in part by Department of Energy contract DE–AC03–76SF00515.

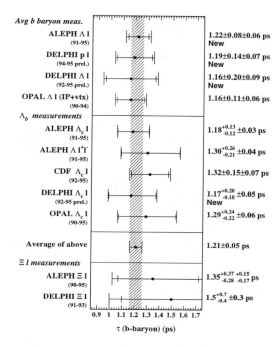

Figure 7: Measurements of the b-baryon lifetime.

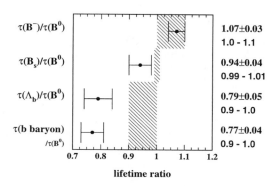

Figure 8: World averages for various b-hadron lifetime ratios. The hatched bands indicate the approximate range of predictions. [1]

References

1. I. I. Bigi *et al.*, in *B Decays*, ed. S. Stone (World Scientific, New York, 1994), p. 132.

2. M. Neubert and C.T. Sachrajda, *Nucl. Phys.* B **483**, 339 (1997).

3. CDF Collaboration, *Improved Measurement of the B^- and $\overline{B^0}$ Meson Lifetimes using Semileptonic Decays*, FERMILAB-Pub-98/167-E, Paper ICHEP98-PA08/731 contributed to the XXIXth International Conference on High Energy Physics, 23-29 July 1998, Vancouver, Canada.

4. SLD Collaboration, *Measurement of the B^+ and B^0 Lifetimes using Topological Vertexing at SLD*, SLAC-PUB-7868, Paper ICHEP98-PA08/180 contributed to the XXIXth International Conference on High Energy Physics, 23-29 July 1998, Vancouver, Canada.

5. SLD Collaboration, *A Preliminary Improved Measurement of the B Hadron Energy Distribution in Z^0 Decays*, SLAC-PUB-7826, Paper ICHEP98-PA03/261 contributed to the XXIXth International Conference on High Energy Physics, 23-29 July 1998, Vancouver, Canada.

6. L3 Collaboration, *Phys. Lett.* B **416**, 220 (1998), Paper ICHEP98-PA08/554 contributed to the XXIXth International Conference on High Energy Physics, 23-29 July 1998, Vancouver, Canada.

7. L3 Collaboration, *Upper Limit on the Lifetime Difference of Short- and Long-Lived B_s^0 Mesons*, L3 Note 2281, Paper ICHEP98-PA08/557 contributed to the XXIXth International Conference on High Energy Physics, 23-29 July 1998, Vancouver, Canada.

8. OPAL Collaboration, *Phys. Lett.* B **426**, 161 (1998), Paper ICHEP98-PA08/317 contributed to the XXIXth International Conference on High Energy Physics, 23-29 July 1998, Vancouver, Canada.

9. DELPHI Collaboration, *Measurement of the Lifetime of b-baryons*, DELPHI 98-72 CONF 140, Paper ICHEP98-PA08/156 contributed to the XXIXth International Conference on High Energy Physics, 23-29 July 1998, Vancouver, Canada.

MEASUREMENTS OF THE B_s LIFETIME AND SEARCH FOR A LIFETIME DIFFERENCE $\frac{\Delta\Gamma}{\Gamma}$

A. RIBON

Padova University and I.N.F.N. Padova

The various B_s^0 lifetime measurements, both at LEP and at the Tevatron, are reviewed. Recent upper limits on the difference of the lifetimes for the two B_s mass-eigenstates are also presented.

1 Introduction

The measurements of B lifetimes are important for the determination of elements of the Cabibbo-Kobayashi-Maskawa matrix. Moreover, measurements of the lifetimes of B-hadrons probe decay mechanisms beyond the simple spectator quark decay model. Lifetime differences can arise from unequal amplitudes for the annihilation and W-exchange diagrams, as well as from final state Pauli interference effects. In the case of charm mesons, such differences have been observed to be quite large $(\tau(D^+)/\tau(D^0) \sim 2.5)$. Among bottom hadrons, the lifetime differences are expected to be smaller due to the heavier bottom quark mass. Phenomenological models predict a difference between the B^+ and B^0 meson lifetimes of the order of 5%, and very similar B^0 and B_s^0 lifetimes [1]. Whereas in the $B_d^0 - \overline{B}_d^0$ system the relative difference between the width of the two mass eigenstates is expected to be less than 1%, for the $B_s^0 - \overline{B}_s^0$ system $\Delta\Gamma/\Gamma$ has been predicted to be sizeable: $10 - 30\,\%$ [2]. In the Standard Model, the ratio $\Delta\Gamma_s/\Delta m_s$ is given by a kinematical factor and by QCD corrections which have been calculated recently [3]:

$$\left(\frac{\Delta\Gamma}{\Delta m_s}\right)_{B_s^0} = (5.6 \pm 2.6) \times 10^{-3} \qquad (1)$$

Therefore, from a direct measurement of $\Delta\Gamma_s$ an indirect Δm_s measurement can be obtained. Finally, from the ratio $\Delta m_d/\Delta m_s$ it is possible to determine the ratio $|V_{td}/V_{ts}|$ of the Cabibbo-Kobayashi-Maskawa matrix elements.

From the experimental point of view, the direct measurements of Δm_s and $\Delta\Gamma_s$ are complementary: the higher the Δm_s the more difficult the oscillation $B_s^0 - \overline{B}_s^0$ measurement, whereas the simpler the $\Delta\Gamma_s$ measurement, because a larger $\Delta\Gamma_s$ results from Eq.1.

The paper is organized as follows. We first discuss the various B_s lifetime measurements, in the various channels: semileptonic, more inclusive, exclusive. Then we discuss the recent $\Delta\Gamma/\Gamma$ searches. Finally we conclude by summarizing all of the results.

2 B_s^0 Lifetime Measurements

2.1 Channels

The possible channels that can be used for a B_s lifetime measurement are the following:

- Exclusive channels: $J/\psi\phi$, $D_s^{(*)+}D_s^{(*)-}$, etc.: this is in principle the best approach, but at the moment is statistically limited.

- Semileptonic channel: $D_s^- \ell^+$ [a] where the D_s^- can be fully or partially reconstructed. With the current available statistics this approach gives the best measurement of B_s lifetime both at LEP and at CDF.

- More inclusive channels: $D_s^- + \text{hadron(s)}^+$, inclusive D_s^-, $\phi - \ell^-$ correlations. This approach has been used only at LEP, because a clean event topology and good particle identification are necessary. However, these analyses are less powerful than the semileptonic ones but can be combined to improve the sensitivity. Recently [12] a completely inclusive analysis has been done by L3 for the $\Delta\Gamma/\Gamma$ search.

2.2 Semileptonic channel

CDF measurement

Let us consider in some detail the CDF measurement of the B_s lifetime using the semileptonic decay $B_s^0 \to D_s^- \ell^+ \nu$, where the D_s^- candidates are reconstructed in four decay modes:

1. $D_s^- \to \phi\pi^-$, $\phi \to K^+K^-$
2. $D_s^- \to K^{*0}K^-$, $K^{*0} \to K^+\pi^-$
3. $D_s^- \to K_s^0 K^-$, $K_s^0 \to \pi^+\pi^-$
4. $D_s^- \to \phi\mu^-\nu$, $\phi \to K^+K^-$

For the first three decay modes the reconstruction is based on a single lepton trigger data set with a principal lepton p_t threshold of about $8\,GeV/c$, while the semileptonic D_s^- decay mode is based on a dimuon data sample

---

[a] Charge conjugate modes are always implied.

obtained with a trigger requirement of $m(\mu\mu) < 2.8\,GeV$ and a muon p_t threshold of about $2\,GeV/c$. The total number of signal events, out of a data sample of about $110\,pb^{-1}$ of $p\bar{p}$ collisions at $\sqrt{s} = 1.8\,TeV$, is ~ 600: this is the world largest sample of semileptonic B_s^0 decays.

In the figure below all four mass spectra are shown. Notice that for the semileptonic D_s^- decay mode the invariant mass distribution is for the pair K^+K^-: the peak is the ϕ signal. In all four decay modes, the shaded histograms are the wrong-sign combinations, which means: same sign $D_s^{\pm}\ell^{\pm}$ events for the three hadronic D_s^- decay modes, and same sign $K^{\pm}K^{\pm}$ or $\mu^{\pm}\mu^{\pm}$ events for the semileptonic mode. As expected, there is no evidence for a signal in the wrong-sign mass spectrum. Notice that, in the $D_s^- \rightarrow \phi\pi^-$ mode, there is also evidence for the Cabibbo-suppressed decay $D^- \rightarrow \phi\pi^-$.

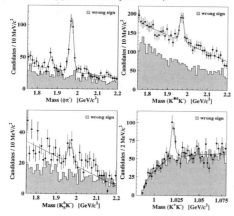

In the $D_s^- \rightarrow K^{*0}K^-$ and $D_s^- \rightarrow K_s^0 K^-$ decays there is significant background from $B_d^0 \rightarrow D^-\ell^+X$ decays with $D^- \rightarrow K^{*0}\pi^-$ or $D^- \rightarrow K_s^0\pi^-$, where the pion is misassigned as a kaon. This kind of background is more relevant at CDF than at LEP experiments, because the particle identification at CDF provides a pion/kaon separation (through dE/dx) of only about one standard deviation in this momentum range. The fraction of the signal coming from real D_s^- decays has been measured in two different ways. First, because of the large difference between the D_s^- and D^- lifetimes, the D_s^- and D^- fractions can be determined from a two-component fit of the proper decay length distribution, where the lifetimes of D_s^- and D^- are fixed to the world average. The second method is a simultaneous fit to the D_s^- and D^- mass distributions, where the D^- mass distribution has been created from the data by switching the mass assignment on the third track of the D_s^- candidate from a kaon to

a pion. Each of the two mass distributions is fit with a Gaussian for the corresponding signal, a linear background, plus the shape of the events coming from the reflection as determined with Monte Carlo.

The fraction of signal coming from real D_s^- decays is obtained from the weighted average of the two independent measurements, which gives about 70 % for both decay modes.

For all of the four decays modes there are the following possible sources of real physics backgrounds:

1. $B_d^0 \rightarrow D_s^{(*)-}D^{(*)+}X$ and $B^- \rightarrow D_s^{(*)-}D^{(*)0}X$ with the $D^{(*)}$ decaying semi-leptonically;

2. $B_s \rightarrow D_s^{(*)-}D_s^{(*)+}$ with one $D_s^{(*)}$ decaying semi-leptonically;

3. $B^0/B^- \rightarrow D_s^- K\ell^+\nu X$

The last decay has not yet been established, therefore the fraction is assumed negligible, but a small possible contamination is taken into account as systematic error. By using the branching ratios from the PDG, and the relative efficiencies from Monte Carlo, the fraction of the first two type of backgrounds have been estimated to be only a few percent of the signal.

The proper decay length is determined as follows:

$$t = F_{corr} \cdot \frac{M(B_s) \cdot L_{xy}^B}{p_t(D_s\ell)} \qquad (2)$$

where L_{xy}^B is the decay length of the B_s^0 in the transverse plane, defined as the distance between the beam spot and the B_s^0 vertex. The latter is determined from the intersection of the lepton with the D_s direction. The correction factor F_{corr} is determined from Monte Carlo to take into account that the B_s^0 is not fully reconstructed, hence the momentum of the system $D_s^0\ell$ must be corrected to obtain the B_s^0 momentum.

A simultaneous unbinned likelihood fit of the four samples is used to extract the B_s^0 lifetime. The probability distribution of the signal consists of an exponential function, convoluted with the distribution of the correction factor and the resolution function. The combinatorial background is parametrized with three components: a Gaussian for the prompt component, and two exponentials convoluted with a Gaussian for the left ($c\tau < 0$) and right ($c\tau > 0$) tails. For each sample, the proper time distribution of the events in the signal mass region is fitted simultaneously with the one of the sideband mass regions and the wrong-sign combinations, to model the combinatorial background. For the D^- reflection contamination in the $K^{*0}K^-$ and K_sK^- samples, the lifetime is fixed to world average B_d^0 lifetime. For the physics background $B \rightarrow D_s^{(*)}D_s^{(*)}X$ and $B_s \rightarrow D_s^{*-}D_s^{*+}$ the lifetimes are fixed to the effective ones as measured in the Monte Carlo. The final result, shown in the figure below is:

$$\tau_s = 1.36 \pm 0.09(stat) \, ^{+0.06}_{-0.05}(syst) \, ps \qquad (3)$$

This is the most precise B_s^0 lifetime measurement from a single experiment.

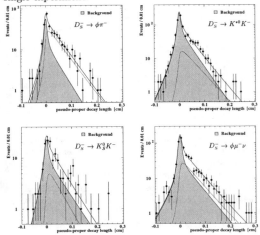

LEP measurements

- ALEPH [5]: four million hadronic Z^0 decays collected from 1991 to 1995 have been used in this analysis. Leptons (electrons or muons) are combined with opposite sign D_s^- candidates reconstructed in seven different decay modes: $\phi\pi^-; K^{*0}K^-; K_s^0 K^-; \phi\pi^+\pi^-\pi^-; K^{*0}K^{*-}; \phi e^-\nu; \phi\mu^-\nu$ where: $\phi \to K^-K^+; K^{*0} \to K^+\pi^-; K_s^0 \to \pi^+\pi^-$. 277 candidates in the signal mass window have been selected: the B_s^0 estimated purity is larger than 60%. The measured B_s^0 lifetime is:

$$\tau_s = 1.54 \, ^{+0.14}_{-0.13}(stat) \pm 0.04(syst) \, ps \qquad (4)$$

- DELPHI [9]: four million hadronic Z^0 decays collected from 1991 to 1995 have been used in this analysis. 278 $D_s^- \ell^+$ signal events reconstructed in eight decay modes: $\phi\pi^-; K^{*0}K^-; K_s^0 K^-; \phi\pi^-\pi^+\pi^-; \phi\pi^-\pi^0; K^{*0}K^{*-}; \phi e^-\nu; \phi\mu^-\nu$. The measured B_s^0 lifetime is:

$$\tau_s = 1.47 \, ^{+0.13}_{-0.12}(stat.) \pm 0.03(syst.) \, ps \qquad (5)$$

- OPAL [11]: 4.4 million hadronic Z^0 decays collected from 1990 to 1995 have been used in this analysis. Leptons (electrons or muons)

are combined with opposite sign D_s^- candidates reconstructed in five different decay modes: $\phi\pi^-; K^{*0}K^-; K_s^0 K^-; \phi e^-\nu; \phi\mu^-\nu$, where: $\phi \to K^-K^+; K^{*0} \to K^+\pi^-; K_s^0 \to \pi^+\pi^-$. In total 199 ± 26 $D_s^-\ell^+$ candidates in the mass peak are observed: of these, 172 are attributed to B_s^0. The B_s^0 lifetime results:

$$\tau_s = 1.50 \, ^{+0.16}_{-0.15}(stat) \pm 0.04(syst) ps \qquad (6)$$

2.3 More inclusive channels

D_s^- **hadron(s)**$^+$

- ALEPH [6]: four million hadronic Z^0 decays collected from 1991 to 1995 have been used in this analysis. One or more tracks (excluding leptons to keep this analysis statistically independent of the $D_s^-\ell^+$ channel) are combined with D_s^- candidates reconstructed in the following channels: $\phi\pi^-; K^{*0}K^-; K_s^0 K^-; \phi e^-\nu; \phi\mu^-\nu$, where: $\phi \to K^-K^+; K^{*0} \to K^+\pi^-; K_s^0 \to \pi^+\pi^-$. Also the semileptonic channel $D_s^-\ell^+$, where the D_s^- is reconstructed in the mode $\phi\rho^-$, $\phi \to K^+K^-$, $\rho^- \to \pi^-\pi^0$, $\pi^0 \to \gamma\gamma$, has been included in this analysis. The total number of data candidates in the signal mass region is 1620, with an estimated B_s^0 purity of 22%. The measured B_s^0 lifetime results:

$$\tau_s = 1.47 \pm 0.14(stat) \pm 0.08(syst) \, ps \qquad (7)$$

- DELPHI [8]: one track (not lepton) h^+ is combined with D_s^- candidates reconstructed in the two channels: $\phi\pi^-; K^{*0}K^-$, where: $\phi \to K^-K^+; K^{*0} \to K^+\pi^-$. The measured B_s^0 lifetime results:

$$\tau_s = 1.52 \, ^{+0.23}_{-0.22}(stat) \pm 0.12(syst) \, ps \qquad (8)$$

inclusive D_s^-

- OPAL [10]: 3.7 million hadronic Z^0 decays collected from 1991 and 1995 have been used in this analysis. Only the D_s^- mesons are reconstructed, without any additional accompanying track, in the two channels: $\phi\pi^-; K^{*0}K^-$, where: $\phi \to K^-K^+; K^{*0} \to K^+\pi^-$. The total number of candidates in the mass peak is 911 ± 83, of which about 57% are expected to be from B_s^0 decay. The B_s^0 lifetime results:

$$\tau_s = 1.72 \, ^{+0.20}_{-0.19}(stat) \, ^{+0.18}_{-0.17}(syst) \, ps \qquad (9)$$

- DELPHI [8]: only the D_s^- mesons are reconstructed in the two channels: $\phi\pi^-; K^{*0}K^-$, where: $\phi \to K^-K^+; K^{*0} \to K^+\pi^-$. The total number of data candidates in the mass peak is 342 ± 41, with an estimated B_s^0 purity of 55%. The B_s^0 lifetime results:

$$\tau_s = 1.60 \pm 0.26(stat) \, {}^{+0.13}_{-0.15}(syst) \, ps \qquad (10)$$

$\phi\ell$ correlations

The DELPHI collaboration[8] also measures τ_s in a more inclusive sample requiring a lepton together with a ϕ meson in the same jet. The estimated B_s^0 purity is about 50%. The measured B_s^0 lifetime results:

$$\tau_s = 1.76 \pm 0.20(stat) \, {}^{+0.15}_{-0.10}(syst) \, ps \qquad (11)$$

2.4 Exclusive channels

$J/\psi\phi$

The only fully reconstructed B_s^0 channel used so far is the $J/\psi\phi$ final state by CDF[13]. Out of $110\,pb^{-1}$ of $p\bar{p}$ collisions, 58 ± 12 signal events have been reconstructed from the dimuon trigger data set. The lifetime fit, performed simultaneously to the mass fit, gives:

$$\tau_s = 1.34 \, {}^{+0.23}_{-0.19}(stat) \, \pm 0.05(syst) \, ps \qquad (12)$$

$D_s^{(*)+}D_s^{(*)-}$

Recently the ALEPH Collaboration[7] has used the $\phi\phi X$ final state to partially reconstruct the decay $B_s^0 \to D_s^{(*)+}D_s^{(*)-}$. From about four million hadronic Z^0 decays (1992-1995) 32 ± 17 pure signal events have been collected, and the lifetime fit gives:

$$\tau_s = 1.42 \pm 0.23(stat) \pm 0.16(syst) \, ps \qquad (13)$$

By assuming that only the CP-even component contributes to the channel $D_s^{(*)+}D_s^{(*)-}$ (indeed it is a mixture of CP-even and CP-odd states, but it is expected to be CP-even dominated), and therefore that the above τ_s measurement gives the lifetime of the light eigenstate (the CP violation is negligible); and by assuming that the world τ_s average is the mean of the lifetimes of the light and heavy mass eigenstates, then it results:

$$\frac{\Delta\Gamma}{\Gamma} = (24 \pm 35)\% \qquad (14)$$

3 $\Delta\Gamma/\Gamma$ search

There are essentially three ways to measure $\Delta\Gamma/\Gamma$:

1. The lifetime distribution of the semileptonic sample $B_s^0 \to D_s^- \ell^+ \nu$ can be fitted with a sum of two exponentials, one for the light and one for the heavy states:

$$const \cdot \left(e^{-\Gamma_L t} + e^{-\Gamma_H t}\right) \qquad (15)$$

where the normalization takes into account the fact that the width, not the branching ratio, for the semileptonic channel is the same for the two mass eigenstates. By rewriting Γ_L and Γ_H in terms of $\Gamma = (\Gamma_L + \Gamma_H)/2$ and $\Delta\Gamma = \Gamma_L - \Gamma_H$ then $\Delta\Gamma/\Gamma$ can be fitted.

Using this method, and by fixing $\tau_s = \tau_d$ (because from the theory we expect the two lifetimes to be the same within 1%), both DELPHI and CDF set an upper limit at $95\%\,C.L.$:

$$\frac{\Delta\Gamma}{\Gamma} < \begin{array}{ll} 0.46 & (DELPHI) \\ 0.83 & (CDF) \end{array} \qquad (16)$$

2. The exclusive channels, like $J/\psi\phi$ or $D_s^{(*)+}D_s^{(*)-}$, are a mixture of the CP-even and CP-odd states. These two components can be separated with an angular analysis[4] and then from the lifetime fit of the CP-even (CP-odd) component the lifetime of the light (heavy) eigenstate can be determined. In practice this method is difficult and suffers from low statistics, expecially for the CP-odd component because these exclusive decays are CP-even dominated. A more realistic approach is described below.

3. This approach combines the two previous ones as follows. From the exclusive decays, by assuming that the CP-even component is completely dominant, the lifetime of the light eigenstate is measured. The measurement of the B_s^0 lifetime in the semileptonic channel gives instead the weighted average of the lifetimes of the light and heavy states, where the weights are given by the corresponding branching ratios. By combining these two measurements, $\Delta\Gamma/\Gamma$ can be extracted. A simplified version of this method has been used by ALEPH, as described in the previous section.

A completely new, totally inclusive technique has been used recently by the L3 collaboration[12] to search for $\Delta\Gamma/\Gamma$. The main idea is that value of $\Delta\Gamma \neq 0$ causes significant deviations in the *inclusive* b−hadron decay time distribution as compared to a single average b lifetime. The deviations occur at long time scales. The analysis was performed on the 1994-1995 data sample which consists of ~ 2 million hadronic Z^0 decays. After requiring two secondary vertices, the analysis combines a measurements of the b decay time, based on the impact parameter of tracks reconstructed at the secondary vertices, and of the b hadron charge, based on the weighted average of the charge of tracks reconstructed at the secondary vertices. By splitting the sample in *neutral* and *charged* according to the b hadron charge, and in *b-enriched* and *b-reduced* according to the decay length in the opposite hemisphere, and then by performing a grand-fit in which all possible sources with non-zero lifetime are taken properly into account, the following upper limit was set:

$$\Delta\Gamma/\Gamma < 0.67 \qquad (95\% C.L.) \qquad (17)$$

4 Summary

The table below summarizes all of the $\Delta\Gamma/\Gamma$ searches:

Exp.	Meas.	Limit (95% CL)	Ch.
DELPHI	-	0.46	$D_s^- \ell^+$
L3	$0.00^{+0.30}_{-0.00}$	0.67	incl.
ALEPH	0.24 ± 0.35	-	$D_s^{(*)} D_s^{(*)}$
CDF	$0.49^{+0.25}_{-0.49}$	0.83	$D_s^- \ell^+$

The table below summarizes all of the τ_s measurements [b].

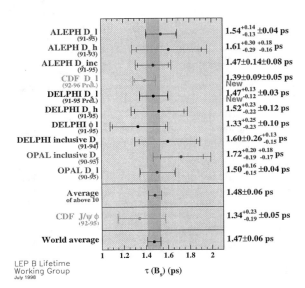

LEP B Lifetime
Working Group
July 1998

$\tau\,(B_s)$ (ps)

M.B. Voloshin, N.G. Uraltsev, V.A. Khoze and M.A. Shifman, *Sov. J. Nucl. Phys.* **46**, 112 (1987); A. Datta, E.A. Paschos and U. Turke, *Phys. Lett.* B **196**, 382 (1987); R. Aleksan, A. Le Yaouanc, L. Oliver, O. Pene and J.C. Raynal, *Phys. Lett.* B **316**, 567 (1993); I.I. Bigi, CERN-TH.7282/94 (1994).

3. I. Dunietz, *Phys. Rev.* D **52**, 1995 (3048); M. Beneke, G. Buchalla and I. Dunietz, *Phys. Rev.* D **54**, 1996 (4419).

4. A.S. Dighe, I. Dunietz, H.J. Lipkin and J.L. Rosner, *Phys. Lett.* B **369**, 144 (1996);

5. D. Buskulic *et al.*, ALEPH Coll., *Phys. Lett.* B **377**, 205 (1996).

6. D. Buskulic *et al.*, ALEPH Coll., CERN-PPE/97-157 (1997).

7. ALEPH Coll., contributed paper to the XXIX International Conference on High Energy Physics, July 23-29, 1998, Vancouver, Canada.

8. P. Abreu *et al.*, DELPHI Coll., *Z. Phys.* C **71**, 11 (1996).

9. DELPHI Coll., contributed paper to the XXIX International Conference on High Energy Physics, July 23-29, 1998, Vancouver, Canada.

10. K. Ackerstaff *et al.*, OPAL Coll., CERN-PPE/97-095 (1997).

11. K. Ackerstaff *et al.*, OPAL Coll., CERN-PPE/97-159 (1997).

12. L3 Coll., contributed paper to the XXIX International Conference on High Energy Physics, July 23-29, 1998, Vancouver, Canada.

13. F. Abe *et al.*, CDF Coll., *Phys. Rev.* D **57**, 5382 (1998).

14. F. Abe *et al.*, CDF Coll., FERMILAB-Pub-98/172-E, submitted to Phys.Rev.D.

References

1. M.B. Voloshin and M.A. Shifman, *Sov. Phys. JETP* **64**, 698 (1986); I.I. Bigi, B. Blok, M. Shifman, N. Uraltsev, A. Vainshtein, CERN-Th.7132/94 (1994); I.I. Bigi, UND-HEP-95-BIG02 (1995); M. Neubert, *Int. J. Mod. Phys.* A **11**, 4173 (1996).
2. A.J. Buras, W. Slominsky and H. Steger, *Nucl. Phys.* B **245**, 369 (1984);

[b]Notice that the value reported here for the CDF measurement in the semileptonic channel is slightly older than the updated value quoted in this paper.

STATUS OF $D^0 - \overline{D}^0$ MIXING AND DOUBLY CABIBBO-SUPPRESSED D^0 DECAY RATE MEASUREMENTS

M. D. SOKOLOFF

Physics Department, ML11; University of Cincinnati; Cincinnati, OH 45221-0011; USA
E-mail: sokoloff@physics.uc.edu

$D^0 - \overline{D}^0$ mixing occurs when the weak eigenstates of the $D^0 - \overline{D}^0$ system have either masses or widths which differ. Such mixing is predicted to occur at very low levels in the Standard Model, but may occur at rates approaching the current experimental limits in models of physics beyond the Standard Model. The ALEPH experiment has results of a new search for mixing and doubly Cabibbo-suppressed decays using the experimental signature $D^{*+} \to \pi^+ D^0$; $D^0 \to K^+\pi^-$. Their results are compared to earlier results from the E791 and CLEO experiments. E791 has measured the decay rates for $D^0 \to^- K\pi^+$ and for $D^0 \to K^-K^+$ and directly derived a limit on the weak eigenstate width difference. Finally, CLEO 2.5 has acquired new data which promises to produce more sensitive measurements of mixing physics.

1 Introduction

$D^0 - \overline{D}^0$ mixing occurs when the weak eigenstates of the $D^0 - \overline{D}^0$ system have either masses or widths which differ. Such mixing is predicted to occur at very low levels in the Standard Model, but may occur at rates approaching the current experimental limits in models of physics beyond the Standard Model. The ALEPH experiment has results of a new search for mixing and doubly Cabibbo-suppressed decays using the experimental signature $D^{*+} \to \pi^+ D^0$; $D^0 \to K^+\pi^-$. Their results are compared to earlier results from the E791 and CLEO experiments. E791 has measured the decay rates for $D^0 \to K\pi$ and for $D^0 \to KK$ and directly derived a limit on the weak eigenstate width difference. Finally, CLEO 2.5 has acquired new data which promises to produce more sensitive measurements of mixing physics.

1.1 Some Formalism

The time evolution of a D^0 or $\overline{D}^0$ is determined by Schrodinger's Equation:

$$i\frac{\partial}{\partial t}\begin{pmatrix} |D^0\rangle \\ |\overline{D}^0\rangle \end{pmatrix} = \begin{bmatrix} M - i\Gamma/2 & M_{12} - i\Gamma_{12}/2 \\ M_{12}^* - i\Gamma_{12}^*/2 & M - i\Gamma/2 \end{bmatrix} \begin{pmatrix} |D^0\rangle \\ |\overline{D}^0\rangle \end{pmatrix}.$$

With the Hamiltonian written as

$$\mathbf{H} = a\,\mathbf{I} + b_x\,\sigma_{\mathbf{x}} + b_y\,\sigma_{\mathbf{y}} + b_z\,\sigma_{\mathbf{z}},$$

$$CP \Longleftrightarrow \sigma_{\mathbf{x}} = \begin{pmatrix} 0 & 1 \\ 1 & 0 \end{pmatrix}.$$

Diagonalizing $\mathbf{H}$, Schrodinger's Equation has solutions of the form $e^{-i\omega t} = e^{-i(m - i\gamma/2)t}$ with eigenvalues

$$m_{H,L} = M \pm Re\left[(M_{12} - i\Gamma_{12})(M_{12}^* - i\Gamma_{12}^*)\right]$$

$$\gamma_{H,L} = \Gamma \mp Im\left[(M_{12} - i\Gamma_{12})(M_{12}^* - i\Gamma_{12}^*)\right]$$

for the "light" and "heavy" eigenstates

$$|D_L\rangle = p|D^0\rangle + q|\overline{D}^0\rangle$$
$$|D_H\rangle = p|D^0\rangle - q|\overline{D}^0\rangle$$

with

$$\frac{q}{p} = \frac{1 + \epsilon_B}{1 - \epsilon_B} = \left(\frac{M_{12}^* - i\Gamma_{12}^*}{M_{12} - i\Gamma_{12}}\right)^{\frac{1}{2}} = e^{2i\phi_M}.$$

We define the following variables which relate to experimental measurements.

$$\Delta m = m_H - m_L\,; \qquad x = \Delta m/\Gamma$$
$$\Delta\Gamma = \gamma_h - \gamma_L\,; \qquad y = \Delta\Gamma/2\Gamma$$

1.2 $D^0 - \overline{D}^0$ Mixing and DCSD Rates

Searching for $D^0 - \overline{D}^0$ mixing and/or D^0 doubly Cabibbo-suppressed decays requires knowing the strong eigenstate at birth (production) and again at death (decay). Two approaches have been used experimentally.

- The MARK III experiment[1] searched for mixing using coherent $D^0 - \overline{D}^0$ production at the $\Psi(3770)$. If one D decays as a particle, then the other D can decay to the same final state *only* if it mixes into the particle state.

- Most experiments (including ALEPH, E791, and CLEO) have used the strong decay $D^{*+} \to \pi^+ D^0$ to identify the strong eigenstate at birth.

The strong eigenstate at death can be identified either

- unambiguously – using a semileptonic decay such as $D^0 \to \overline{D}^0 \to K^+\ell^-\nu$, or

- ambiguously – using a hadronic decay such as $D^0 \rightarrow \overline{D}^0 \rightarrow K^+\pi^-$. These Cabibbo-favored decays can be confused with the unmixed doubly Cabibbo-suppressed decays $D^0 \rightarrow K^+\pi^-$, which complicates the search for mixing when using hadronic final states.

The decay to "wrong-sign" final states takes the form[2]

$$\Gamma[D^0(t) \rightarrow f] = e^{-\Gamma t}|\langle f|H|\overline{D}^0\rangle_{\mathrm{CF}}|^2 \left|\frac{q}{p}\right|^2$$
$$\times \left[|\lambda|^2 + \tfrac{1}{4}(x^2+y^2)(t/\tau)^2 + (y\Re(\lambda) + x\Im(\lambda))(t/\tau)\right]$$

where
$$\lambda \equiv \frac{p}{q}\frac{\langle f|H|D^0\rangle_{\mathrm{DCS}}}{\langle f|H|\overline{D}^0\rangle_{\mathrm{CF}}}$$

and we define the following shorthand for use later:

$$r_{\mathrm{dcsd}} = |\lambda|^2$$
$$r_{\mathrm{mix}} = \tfrac{1}{2}(x^2+y^2)$$
$$r_{\mathrm{int}} = (y\Re(\lambda) + x\Im(\lambda))$$

2   Results from the ALEPH Collaboration

The ALEPH Collaboration submitted new results to this conference[3]. The data were extracted from four million Z decays collected in their detector from 1991 - 1995. They used the decay chain $D^{*+} \rightarrow \pi^+ D^0$ to identify the strong eigenstate at birth. They used the decays $D^0 \rightarrow K^-\pi^+$ as normalization, and searched for the decay chains $D^0 \rightarrow \overline{D}^0 \rightarrow K^+\pi^-$ and $D^0 \rightarrow K^+\pi^-$. They identified their signal by selecting D-meson candidates with $|M(K\pi) - M(D^0)| < 30$ MeV/c^2 and looking at the $D^{*+} - D^0$ mass difference distribution $\Delta M = M(K\pi\pi^+) - M(K\pi)$ which has very good resolution. In selecting $K\pi$ candidates, ALEPH required first that only one of the mass hypothesis assignments produced a candidate within the D^0 window considered. Subsequently, they required the dE/dx measured in at least 50 of the 338 layers of the TPC to be consistent with the assigned particle hypotheses. A more complete description of the data selection (and analysis) is available in their publication, from which this discussion has been extracted.

The right- and wrong-sign data from ALEPH are shown in Figs. (1a) and (1b). Within a ΔM window from 143.5 MeV/c^2 to 147.5 MeV/c^2 they find the numbers of right-sign and wrong-sign events to be $N_{\mathrm{RS}} = 1038.8 \pm 32.5 \pm 4.4$ and $N_{\mathrm{WS}} = 21.3 \pm 6.1 \pm 3.5$ after the combinatorial background is subtracted. Of the 21.3 wrong-sign events, 2.2 ± 1.0 are real physics background events whose proper time distribution is assumed

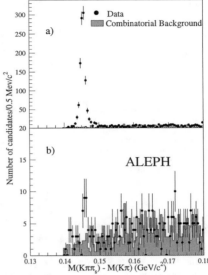

Figure 1: Mass-difference distribution (a) for candidates of the decay channel $D^{*+} \rightarrow D^0\pi^+$, $D^0 \rightarrow K^-\pi^+$ and (b) candidates of the decay channel $D^{*+} \rightarrow D^0\pi^+$, $D^0 \rightarrow K^+\pi^-$. The dots with error bars are data while the hatched histogram represents the distribution of the combinatorial background.

to be exponential with the D^0 lifetime. This leaves a signal of $19.1 \pm 6.1 \pm 3.5$ $K^+\pi^-$ events which result from a combination of $D^0 - \overline{D}^0$ mixing and doubly Cabibbo-suppressed D^0 decays.

The proper time distributions of the mixing and DCSD signals differ, which allows the separation of the two. Assuming small mixing and neglecting CP-violating terms, the time evolution detailed for D^0 decay above becomes

$$N(D^0 \rightarrow K^+\pi^-)(t) \propto$$
$$\left[r_{\mathrm{dcsd}} + \sqrt{2r_{\mathrm{mix}}r_{\mathrm{dcsd}}}\cos\phi(t/\tau) + \tfrac{1}{2}r_{\mathrm{mix}}(t/\tau)^2\right]e^{-t/\tau}$$

The second term describes the interference between the mixing and DCSD amplitudes, and $\cos\phi$ is the phase angle parameterizing this interference.

The proper decay time of a D^0 candidate is calculated from the distance between the primary vertex and the D^0 decay vertex projected along the direction of the flight of the D^0, using the reconstructed momentum and mass of the candidate. The average resolution of the proper decay time is ≈ 0.1 ps and is dominated by the uncertainty of the D^0 vertex ($\approx 40\mu m$). The distribution of proper time in the signal region for the wrong-sign sample is shown in Fig. 2a and for the right-sign sample

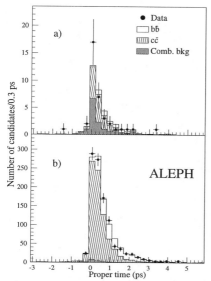

Figure 2: Proper time distribution for (a) the $D^0 \to K^+\pi^-$ candidates, and (b) for the $D^0 \to K^-\pi^+$ candidates. The dots with error bars are the data. The histograms are the contributions of $c\bar{c}$, $b\bar{b}$, and background events resulting from the unconstrained fit when no interference between the mixing and DCSD amplitudes is assumed.

is shown in Fig. 2b. The data are indicated as dots with error bars. The histograms indicate the contributions from $c\bar{c}$ events, $b\bar{b}$ events, and background events, determined using a binned maximum likelihood fit. The direct $c\bar{c} \to D^0 X$ fraction (rather than the $b\bar{b} \to c\bar{c} \to D^0 X$) is found to be $(77.0 \pm 2.7 \pm 0.5)\%$ in fitting the right-sign sample. The background in the wrong-sign sample is a sum of combinatorial background, whose proper time dependence is measured in the data from the side band regions in the ΔM plot, and a physics background contribution whose proper time distribution is assumed to be exponential with a decay constant given by the D^0 lifetime.

Fitting the proper time distribution of the wrong-sign sample with the interference term set to zero ($\cos\phi = 0$), gives $N_{\text{dcsd}} = 20.8^{+8.4}_{-7.4} \pm 4.0$ and $N_{\text{mix}} = -2.0 \pm 4.4 \pm 1.1$. The fitted value of N_{mix} is outside the physical region, implying that no mixing is observed. The 95% CL upper limit on the number of mixed events is found to be $N_{\text{mix}} < 9.6$, from which the mixing limit is found to be $r_{\text{mix}} < 0.92\%$ at the 95% CL. Setting $N_{\text{mix}} = 0$, the doubly Cabibbo-supressed decay rate for $D^0 \to K^+\pi^-$ is found to be $r_{\text{dcsd}} = (1.77^{+0.60}_{-0.56} \pm 0.31)\%$.

The effect of interference was studied by fitting the

data with fully constructive ($\cos\phi = +1$) and fully destructive ($\cos\phi = -1$) interference. The respective upper limits were found to be $r_{\text{mix}} < 0.96\%$ at the 95% CL and $r_{\text{mix}} < 3.6\%$ at the 95% CL.

These results can be compared with previously published results from other experiments. CLEO measured the rate of right-sign and wrong-sign $D^0 \to K\pi$ events with no decay-time information. Assuming no mixing, they reported[4] $r_{\text{dcsd}} = (0.77 \pm 0.25 \pm 0.25)\%$. E791 searched for right- and wrong-sign D^0 decays to $K\pi$ and $K\pi\pi$[5]. Assuming no mixing, they reported $r_{\text{dcsd}} = (0.68^{+0.34}_{-0.33} \pm 0.07)\%$ for the $K\pi$ final state. Allowing full interference between mixing and DCSD amplitudes, and allowing CP violation in the interference term but not allowing CP violation in the mixing or DCSD terms themselves, they reported $r_{\text{mix}} < 0.85\%$ at the 90% CL. From a mixing analysis using semileptonic D^0 decays[6] rather than hadronic D^0 decays (which eliminates the DCSD amplitudes and the possibility of their interfering with mixing amplitudes), E791 reported $r_{\text{mix}} < 0.50\%$ at the 90% CL.

3 E791's Direct Measurement of $\Delta\Gamma$

Fermilab experiment E791 has studied the reduced proper lifetime distributions of a $D^0 \to K^-K^+$ sample and of a $D^0 \to K^-\pi^+$ sample selected from the same data set using very similar criteria. Assuming that the weak eigenstates are CP even and odd, and that the $D^0 \to K^-\pi^+$ decay is an equal CP admixture, the width difference can be extracted directly:

$$\Gamma(D^0 \to K^-K^+) - \Gamma(D^0 \to K^-\pi^+)$$
$$\approx \Gamma_1 - \tfrac{1}{2}(\Gamma_1 + \Gamma_2)$$
$$= \tfrac{1}{2}(\Gamma_1 - \Gamma_2)$$
$$= \tfrac{1}{2}\Delta\Gamma$$

To determine $\Delta\Gamma$ experimentally, one may either measure $\Gamma(D^0 \to K^-K^+)$ and $\Gamma(D^0 \to K^-\pi^+)$ separately and then take the difference, or one may directly compare the number of decays observed at any decay time. This latter procedure tends to reduce the effect of common systematic uncertainties in reconstruction acceptance × efficiency. In any bin of reduced proper lifetime,

$$n(D^0 \to K^-K^+) = A_{KK}\, e^{-\Gamma_1 t}$$
$$n(D^0 \to K^-\pi^+) = A_{K\pi}\, e^{-\frac{1}{2}(\Gamma_1 + \Gamma_2)t}$$
$$\Rightarrow \ln\left[\frac{n(D^0 \to K^-K^+)}{n(D^0 \to K^-\pi^+)}\right] = \ln\left(\frac{A_{KK}}{A_{K\pi}}\right) - \frac{\Delta\Gamma}{2}t$$

Thus, the slope of the logarithm of the ratio of the number of $D^0 \to K^-K^+$ events compared to the number of $D^0 \to K^-\pi^+$ events as a function of reduced proper lifetime is directly proportional to $\Delta\Gamma$.

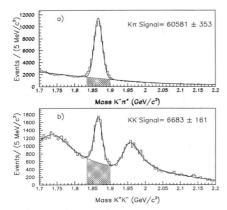

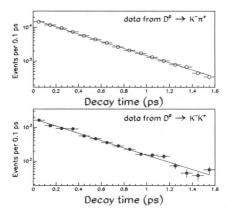

Figure 3: (a) The $D^0 \to K^-\pi^+$ sample used in E791's $\Delta\Gamma$ measurement; (b) $D^0 \to K^-K^+$ sample used in E791's $\Delta\Gamma$ measurement;

Figure 4: (a) The reduced proper decay time distribution for E791's $D^0 \to K^-\pi^+$ data, weighted for acceptance. (b) The reduced proper decay time distribution for E791's $D^0 to K^-K^+$ data, corrected for Čerenkov efficiency and weighted for acceptance.

The value of (limit on) y determined from this measurement can be compared to the limit on y extracted from the most stringent limit on r_{mix}, that reported by E791 from its search for mixing using semileptonic D^0 decays:

$$r_{\mathrm{mix}} = \frac{1}{2}(x^2 + y^2) < 0.50\%$$

$$\Rightarrow y < 0.1$$

The $D^0 \to K^-\pi^+$ and $D^0 \to K^-K^+$ samples used for this analysis are shown in Figs. 3a and 3b. Similar plots are created for each bin of reduced proper decay time (using the length beyond the minimum required to survive the analysis selection criteria). Because Čerenkov identification was used to improve the statistical significance of the $D^0 \to K^-K^+$ signal, each $D^0 \to K^-K^+$ event was weighted to correct for efficiency. In addition, the net signal in each bin of reduced proper decay time was corrected for acceptance. Assuming CP conservation, the K^-K^+ final state is an even eigenstate, and the decay $D^0 \to K^-K^+$ is exactly exponential. The decay $D^0 \to K^-\pi^+$ is a superposition of two exponentials with very similar decay constants. The difference between the components is small enough (based on previously measured limits on r_{mix}) that fitting the $D^0 \to K^-\pi^+$ reduced proper decay time distribution as a single exponential will produce a very good estimate of the average of the two inverse lifetimes. Accordingly, the distributions of the net signals as functions of reduced proper decay time, shown in Figs. 4a and 4b, were fit as exponentials. The inverse lifetimes were found to be Γ [for $D^0 \to K\pi$] $= (2.456 \pm 0.018)$ ps^{-1} and Γ [for $D^0 \to KK$] $= (2.472 \pm 0.066)$ ps^{-1}

where the errors are statistical only. The corresponding lifetimes are $\tau = (0.407 \pm 0.003)$ ps [for $D^0 \to K\pi$] and $\tau = (0.404 \pm 0.011)$ ps [for $D^0 \to KK$]. The final systematic errors are expected to be roughly equal to the statistical errors for the $D^0 \to K\pi$ measurement. For comparison, the Particle Data Group[7] reports $\tau(D^0) = 0.415 \pm 0.004$ ps.

The decay width difference was determined directly by comparing the numbers of $D^0 \to K^-K^+$ and $D^0 \to K^-\pi^+$ events, after the corrections described above, in bins of reduced proper decay time. Fitting the ratio, the error on the slope gives $\Delta\Gamma = (-0.022 \pm 0.073)$ ps^{-1} (statistical error only). From this, one finds, $y = \Delta\Gamma/(2\Gamma) = -0.005 \pm 0.015 \pm$ syst. The final systematic error is expected to be approximately the same size as the statistical error. It thus appears that the constraint on $\Delta\Gamma$ from this direct measurement will be significantly more stringent than those from mixing measurements to date.

4 First data from CLEO 2.5

CLEO 2.5 is currently (September, 1998) taking data. They had 6 fb^{-1} of data on tape at the time of the conference (late July, 1998), of which 4 fb^{-1} had been skimmed and examined. They expect $8 - 9$ fb^{-1} by the end of the run. Some charm data is shown in Figs. 6 and 7.

The horizontal and vertical beam resolutions are $\sigma_x \approx 350\mu$m and $\sigma_y \approx 10\mu$m. The plots show very good signals for D^0 from $D^{*+} \to \pi^+, D^0$ which can be used for mixing/DCSD and lifetime difference studies. To improve mass resolutions, the D^0 momentum vectors

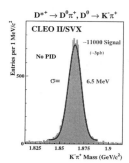
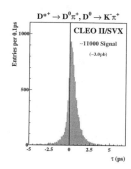
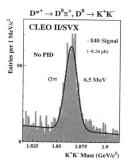
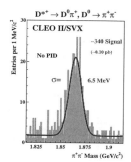
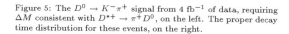
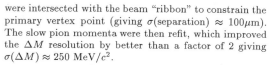

Figure 5: The $D^0 \to K^-\pi^+$ signal from 4 fb^{-1} of data, requiring ΔM consistent with $D^{*+} \to \pi^+ D^0$, on the left. The proper decay time distribution for these events, on the right.

Figure 6: The $D^0 \to K^- K^+$ signal from 4 fb^{-1} of data, requiring ΔM consistent with $D^{*+} \to \pi^+ D^0$, on the left. The $D^0 \to \pi^-\pi^+$ signal from 4 fb^{-1} of data, requiring ΔM consistent with $D^{*+} \to \pi^+ D^0$, on the right.

were intersected with the beam "ribbon" to constrain the primary vertex point (giving σ(separation) $\approx 100\mu m$). The slow pion momenta were then refit, which improved the ΔM resolution by better than a factor of 2 giving $\sigma(\Delta M) \approx 250$ MeV/c^2.

In studying $D^0 \to K^-\pi^+$, the D^{*+} momentum was required to be greater than 1.5 GeV/c. The plot on the left-hand side of Fig. 5 shows a clean signal of approximately 11,000 events after requiring $|\Delta M| < 500$ keV/c^2. Approximately 90% of the signal comes from $c\bar{c}$ events and 10% from $b\bar{b}$ events. The decay time distribution of these D^0's is shown in the plot on the right-hand side of Fig. 6. This extrapolates to $\approx 25,000$ events for the full run which can be compared to ≈ 1000 from ALEPH and ≈ 5000 from E791.

Fig. 6 shows clean $D^0 \to K^- K^+$ and $D^0 \to \pi^-\pi^+$ signals, again originating from $D^{*+} \to \pi^+ D^0$. For these samples, the D^{*+} momentum was required to be greater than 1.5 GeV/c, and essentially 100% of the signal comes from $c\bar{c}$ events. The 840 event $D^0 \to K^- K^+$ signal extrapolates to ≈ 1800 clean events for the full run. This can be compared to N_{eff} [$\equiv (\mathcal{S}/\sigma)^2$] ≈ 1700 from E791 (which does not require the D^0 to come from a D^{*+} for its lifetime difference analysis). If CLEO chooses to combine its $D^0 \to K^- K^+$ and its $D^0 \to \pi^-\pi^+$ data for a lifetime difference study (both are CP-even eigenstates), the sample will be 40% larger. And it should be possible to increase the sample again by using D^0's which are not D^{*+} decay products.

It is also worth noting that the CLEO 2.5 signals shown here used no particle identification. Altogether, CLEO 2.5 should have significantly improved measurements of $D^0 - \overline{D}^0$ mixing and DCSD rates and very competitive direct measurements of $\Delta\Gamma$.

Acknowledgements

Reporting analyses performed by other individuals and other experiments is always challenging. I would like to express my special appreciation to Mario Aleppo (from ALEPH), K. C. Peng (from E791), Simon Kwan (from E791), and David Asner (from CLEO) for their help in preparing this report. Any mistakes in understanding and conveying the results belong solely to the author. This work was supported in part by the U.S. National Science Foundation.

References

1. G. E. Gladding, "$D^0 - \overline{D}^0$ Mixing, The Experimental Situation," in the Proceedings of the International Symposium on the Production and Decay of Heavy Flavors, Stanford (1988).
2. G. Blaylock, A. Seiden, and Y. Nir, Phys. Lett. **B355**, 555 (1995).
3. ALEPH Collaboration (R. Barate *et al.*), CERN-EP-98-118 and submitted to Phys. Lett. **B** (1998).
4. CLEO Collaboration (D. Cinabro *et al.*), Phys. Rev. Lett. **72**, 1406 (1994).
5. E791 Collaboration (E. M. Aitala *et al.*), Phys. Rev. **D57**, 13 (1998).
6. E791 Collaboration (E. M. Aitala *et al.*), Phys. Rev. Lett. **77**, 2384 (1996).
7. Review of Particle Properties, C. Caso *et al.*, European Physical Journal **C3** (1998).

NEARBY RESONANCES AND $D^0 - \bar{D}^0$ MIXING

E. GOLOWICH

Department of Physics and Astronomy, University of Massachusetts, Amherst, MA 01003, USA
E-mail: gene@het.phast.umass.edu

The talk begins with an overview of the current theoretical understanding of $D^0 - \bar{D}^0$ mixing. Following this, a description is given of very recent work involving the dynamical effect of nearby resonances on the mixing amplitude. Finally, some concluding remarks are made regarding possible future work in this area.

1 Introduction

1.1 Motivation

The primary purpose of this talk is to describe a recently published work by Alexey Petrov and myself which shows how the presence of resonances with mass nearby the D^0 meson can influence the mixing process $D^0 \leftrightarrow \bar{D}^0$.[1] We envisage the resonance mechanism as a D^0 meson undergoing a weak transition to a resonance R which propagates and then itself undergoes a weak mixing to $\bar{D}^0$. There are several points which make this possibility of particular interest:

1. The resonance mechanism, if operative, can be available only to D mesons. The light kaon lies below the resonance region and the heavy $B_{d,s}$ mesons lie above it.

2. The dynamical mechanism of resonant enhancement constitutes an explicit violation of the quark-hadron duality assumption and could influence power counting rules built into the heavy quark effective theory (HQET) estimate of Δm_D[2,3], which assumes a large energy gap between m_c and the scale Λ_{QCD} at which hadron dynamics is active.

3. If a resonance is viewed as a single-particle intermediate state, then its contribution will be favored over that of multibody intermediate states by the $1/N_c$ counting rules.

1.2 Meson Mixing in General

It is important to understand the rather different roles played by theory in analyzing $D^0 - \bar{D}^0$ mixing as compared to that undergone by K^0, B_d^0 and B_s^0 mesons. For the latter three, the Standard Model mixing is large and theoretically relatively well understood. For these cases, one studies mixing to better constrain the parameter space of the CKM matrix. To be sure, theoretical uncertainties like the precise values of B-parameters (*e.g.* B_K, B_{B_d}) are nuisances that negatively impact on phe-

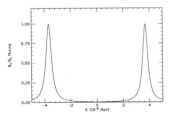

Figure 1: $B_s - \bar{B}_s$ Mixing profile.

nomenological determination of the $\{V_{ij}\}$, but one hopes with time to gradually overcome such limitations.

In order to better comprehend the qualitative difference between $D^0 - \bar{D}^0$ mixing and the others, consider Figs. 1,2. In Fig. 1, I have plotted the Lorentzian profiles of the two mass eigenstates for the B_s system.[a] The mass difference Δm_{B_s} is simply the distance between the two peaks and the difference in decay rates $\Delta \Gamma_{B_s}$ is surmised from the relative widths of the two peaks. Although the latter may be somewhat hard to discern, I have allowed for $\Delta \Gamma_{B_s}/\Gamma_{B_s} \simeq 0.2$ as predicted by theory.[4] The mixing expected in the B_s system is the largest of all the mesons and accounts for the most striking aspect of Fig. 1, namely how far apart the peaks are relative to the width of either peak. This considerable mixing effect is the very reason why $B_s - \bar{B}_s$ mixing is difficult to observe – the oscillation frequency is very large.

Contrast this with the situation for $D^0 - \bar{D}^0$ mixing as depicted in Fig. 2. As with the case of $B_s - \bar{B}_s$ mixing, I have had to use a theoretical estimate since no experimental signal has been observed. It would appear that some kind of error has been made in preparing this diagram, as there seems to be just one curve. In actuality two curves have indeed been drawn. However, they are so close together that on a scale in which the width Γ_D gracefully occupies the space it is not possible to separate them with the eye. This, more than any other means

---

[a]Since $B_s - \bar{B}_s$ mixing not yet been experimentally detected, the value $x_s \equiv \Delta m_{B_s}/\Gamma_{B_s} \simeq 17$ has been assumed for the sake of argument.

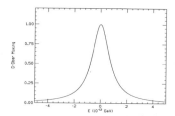

Figure 2: $D^0 - \bar{D}^0$ Mixing profile.

that comes to mind, vividly demonstrates just how small $D^0 - \bar{D}^0$ mixing (both Δm_D and $\Delta\Gamma_D$) is expected to be in the Standard Model. Of course, this has not inhibited experimental searches for $D^0 - \bar{D}^0$ mixing, nor should it. After many years of effort by a number of groups, the most recent lower bound for the mixing[5] is several orders of magnitude larger, $|\Delta m_D|^{\rm expt} < 1.3 \cdot 10^{-13}$ GeV, than the signal expected from Standard Model effects.

There is a second point we must not forget – that calculations in the charm region tend to be *inherently* difficult. To see why, it pays again to constrast the situation with much lower and higher energies. For kaon physics, chiral symmetry is a powerful constraint and one can generally employ chiral perturbation theory (ChPT) to get reliable results. B-meson physics relies upon HQET as its guiding principle. However, the mass region of charm physics is clearly too high for ChPT and evidently too low (at least for the electroweak sector) for HQET.

It seems to me that given (i) the wide gap between existing experimental limits and Standard Model estimates together with (ii) the difficulty of quantifying $D^0 - \bar{D}^0$ mixing effects in the Standard Model lead to the conclusion — *the real purpose of doing $D^0 - \bar{D}^0$ searches is to find new physics and the real job of the theorist is to establish as best as possible an upper bound for Δm_D and $\Delta\Gamma_D$ in the Standard Model.*

2 $D^0 - \bar{D}^0$ Mixing in the Standard Model

For $B_d - \bar{B}_d$ and $B_s - \bar{B}_s$ mixing the familiar box-diagram contributions, which explicitly involve the quark degrees of freedom, provide good quantitative descriptions of the effect. For $K^0 - \bar{K}^0$ mixing the relative size of the (quark-level) box diagrams and (hadronic) long-distance processes is not so clear. However, it has become reasonably evident that the box contribution provides at least a rough first approximation and perhaps rather better.

For $D^0 - \bar{D}^0$, the situation is clouded, although my bet is that (quark-level) box diagrams do *not* provide a reliable guide to the size of the effect. At any rate, it is convenient to categorize various possibilities by studying

the $D^0 - \bar{D}^0$ mass matrix. From standard perturbation theory, the $ij^{\rm th}$ element can be represented as

$$\left[M - i\frac{\Gamma}{2}\right]_{ij} = \frac{1}{2m_D}\langle D_i^0|\mathcal{H}_W^{\Delta C=2}|D_j^0\rangle$$
$$+ \frac{1}{2m_D}\sum_I \frac{\langle D_i^0|\mathcal{H}_W^{\Delta C=1}|I\rangle\langle I|\mathcal{H}_W^{\Delta C=1\dagger}|D_j^0\rangle}{m_D^2 - m_I^2 + i\epsilon} \quad . \quad (1)$$

The first term in the above mass matrix expansion corresponds to the contribution of local $\Delta C = 2$ 'box' and 'dipenguin' operators. These are small in the Standard Model[6,7,8]. Next come the bilocal contributions which are induced by the insertion of two $\Delta C = 1$ operators. This class of terms might be enhanced by various non-perturbative effects, and therefore is of considerable interest. As follows from Eq. (1), one introduces a sum over all possible n-particle intermediate states allowed by the corresponding quantum numbers. For these continuum contributions, the summation in the second term of Eq. (1) takes the form of an integral over the energy variable. There will be a unitarity cut in the complex energy plane lying along the real axis and beginning at the two-pion threshold. The contribution from charged pseudoscalar two-body intermediate states was originally considered in Refs. [7,9] and estimated to be potentially large. However, it remains very difficult to reliably determine the total effect associated with $n \geq 2$ intermediate states due to the many decay modes present, each having unknown final state interaction (FSI) phases. Below, we discuss these contributions in somewhat more detail.

2.1 Short-distance Box Contribution

Correct expressions for the box-components of Δm_D and $\Delta\Gamma_D$, unadorned by QCD radiative corrections, are

$$\Delta m_D^{\rm box} = \frac{2X_D}{3\pi^2}\frac{(m_s^2 - m_d^2)^2}{m_c^2}\left[1 - \frac{5}{4}\frac{B_D'}{B_D}\frac{m_D^2}{(m_c + m_u)^2}\right]$$

$$\simeq -1.9 \cdot 10^{-17} \text{ GeV} ,$$

$$\Delta\Gamma_D^{\rm box} = \frac{4X_D}{3\pi} \cdot \frac{(m_s^2 - m_d^2)^2}{m_c^2} \cdot \frac{m_s^2 + m_d^2}{m_c^2}$$

$$\times \left[1 - \frac{5}{2}\frac{B_D'}{B_D}\frac{m_D^2}{(m_c + m_u)^2}\right] ,$$

$$\simeq 0.39 \Delta m_D^{\rm box} ,$$

where $X_D \equiv \xi_d^2 B_D G_F^2 m_D F_D^2$ and B_D, B_D' are D-meson 'B-parameters' arising from the D^0-to-$\bar{D}^0$ matrix elements of chiral four-quark operators. Until recently, B_D and B_D' represented a major source of numerical uncertainty, even as to the sign of the effect. However, a lattice determination[10] of both B-parameters has alleviated this source of numerical ambiguity, leaving the value of m_c as perhaps the least well-determined quantity. The above values correspond to the choice $m_c = 1.3$ GeV.

2.2 Dipenguin Contribution

It has long been understood that the effect of two penguin operators (dipenguin) can also occur as a four-quark local operator.[11] This effect has been studied for both the $K^0 - \bar{K}^0$ and $B_d - \bar{B}_d$ mixings, but until recently not for $D^0 - \bar{D}^0$ mixing. This gap in the literature has been filled by a very recent calculation of Petrov who finds[8]

$$|\Delta m_D^{\text{di-pen}}| \leq 10^{-17} \text{ GeV} .\qquad(2)$$

Thus, the dipenguin does not lead to a surprisingly large contribution.

2.3 HQET Analysis

The effect of QCD radiative corrections on the box contribution to Δm_D has been studied in the context of the HQET. Georgi gave the original formulation of HQET to $D^0 - \bar{D}^0$ mixing,[2] and analysis of the QCD effects appeared soon thereafter.[3] HQET represents the most natural means for computing QCD radiative corrections for heavy quark processes, as otherwise large logarithmic factors are avoided. There are several steps in this approach: (i) obtain an effective hamiltonian in terms of local operators (OPE), (ii) run the energy scale (RG equations) from $\mu \simeq M_W$ down to the value of interest ($\mu \simeq m_c$ in our case) and (iii) obtain accurate values for the matrix elements of the local operators. For application to Δm_D, operators of dimension four, six and eight were considered.[3] The main uncertainites here involve convergence of the OPE (i.e. is the c-quark really 'heavy'?) and evaluation of the multi-quark matrix elements. The leading-order HQET contribution to $\Delta\Gamma_D$ has been calculated by Petrov and myself,[12]

$$\Delta m_D^{\text{HQET}} = \Delta m_D^{(4)} + \Delta m_D^{(6)} + \Delta m_D^{(8)} + \dots$$

$$\simeq (0.9 \to 3.5) \cdot 10^{-17} \text{ GeV} .$$

$$\Delta\Gamma_D^{\text{HQET}} = \Delta\Gamma_D^{(4)} + \dots \simeq 1.83\,\Delta\Gamma_D^{\text{box}} .$$

2.4 Long-distance Multiparticle Contributions

For this class of contributions, only hadronic degrees of freedom appear. We briefly comment on the original work done on this topic and then consider a toy estimate based in part on a more recent effort to understand a large number of D decays.

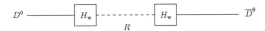

Figure 3: Contribution of Resonance R to the D^0-to-$\bar{D}^0$ matrix element.

Charged Pseudoscalar Component

Realizing that the large observed SU(3) breaking large in D decays could imply important long-distance contributions to $D^0 - \bar{D}^0$ mixing, the UMass group[7] and Wolfenstein[9] considered the processes

$$D^0 \to \begin{bmatrix} K^-\pi^+ \\ \pi^+\pi^- \\ K^+K^- \\ K^+\pi^- \end{bmatrix} \to \bar{D}^0$$

as a potentially important mixing mechanism. Individual terms were indeed found to be large.

Extended Analysis

One can contemplate extending the above by using a more comprehensive study of D decays[13] as input,

$$D^0 \to \begin{bmatrix} PP \\ PV \\ VV \\ SP \end{bmatrix} \to \bar{D}^0$$

where P, S, V stand for pseudoscalar, scalar and vector mesons. For example, employing the estimate[13] $\Delta\Gamma_D/\Gamma_D \simeq (1.5 + i0.0014) \cdot 10^{-3}$ and making the ansatz $|\Delta m_D^{\text{disp}}| \sim 0.2\,\Delta\Gamma_D$ which is typical of individual terms in a dispersive treatment of mixing, one arrives at the 'guesstimate' $|\Delta m_D^{\text{disp}}| \sim \mathcal{O}(10^{-16} \text{ GeV})$. However, this little exercise should not be accepted as a real calculation.

3 The Mechanism of Nearby Resonances

Among the terms in Eq. (1) will be those arising from single-particle intermediate states. Consider specifically the contributions from resonances,

$$\Delta m_D\Big|_{\text{tot}}^{\text{res}} = \frac{1}{2m_D} \sum_R Re\, \frac{\langle D_L|\mathcal{H}_W|R\rangle\langle R|\mathcal{H}_W^\dagger|D_L\rangle}{m_D^2 - m_R^2 + i\Gamma_R m_D}$$

$$- (D_L \to D_S) .\qquad(3)$$

The resonance mechanism is depicted in Fig. 3. The pseudoscalar 0^{-+} (scalar 0^{++}) intermediate states have $CP = -1$ ($CP = +1$) and contribute respectively to D_L (D_S). If the mass of the resonance is not too far from the D-meson mass, an interesting effect occurs. To highlight the dependence on the resonance mass, temporarily consider just the energy denominator in Eq. (3).

Table 1: Magnitudes of Pseudoscalar Resonance Contributions.

Table: Magnitudes of Pseudoscalar Resonance Contributions.

Resonance	$|\Delta m_D| \times 10^{-16}$ (GeV)	$|\Delta\Gamma_D| \times 10^{-16}$ (GeV)
$K(1460)$	$\sim 1.24\ (f_{K(1460)}/0.025)^2$	$\sim 0.88\ (f_{K(1460)}/0.025)^2$
$\eta(1760)$	$(0.77 \pm 0.27)\ (f_{\eta(1760)}/0.01)^2$	$(0.43 \pm 0.53)\ (f_{\eta(1760)}/0.01)^2$
$\pi(1800)$	$(0.13 \pm 0.06)\ (f_{\pi(1800)}/0.01)^2$	$(0.41 \pm 0.11)\ (f_{\pi(1800)}/0.01)^2$
$K(1830)$	$\sim 0.29\ (f_{K(1830)}/0.01)^2$	$\sim 1.86\ (f_{K(1830)}/0.01)^2$

The contribution to the energy denominator of a light bound state (*e.g.* pions or kaons) of mass m is $P(m^2) \equiv 1/(m_D^2 - m^2) = m_D^{-2} + \mathcal{O}(m^2/m_D^2)$, which amounts to a suppression factor of order m_D^{-2}. By contrast, the energy denominator (*cf* Eq. (3)) for a resonance of mass m_R and width Γ_R will yield

$$\Delta m_D \Big|_{R}^{res} \propto \frac{m_D^2 - m_R^2}{(m_D^2 - m_R^2)^2 + \Gamma_R^2 m_D^2} \ ,$$

$$\Delta \Gamma_D \Big|_{R}^{res} \propto -\frac{\Gamma_R m_D}{(m_D^2 - m_R^2)^2 + \Gamma_R^2 m_D^2} \ . \quad (4)$$

The following array points out the expected behavior as the mass of the single-particle intermediate state is varied:

	$\Delta m_D^{(R)}$	$\Delta \Gamma_D^{(R)}$
Bound State	$\mathcal{O}(m_D^{-2})$	0
Light Resonance	$\mathcal{O}(m_D^{-2})$	$\mathcal{O}(m_D^{-2})$
Nearby Resonance	$\mathcal{O}((\Gamma_R m_D)^{-1})$	$\mathcal{O}((\Gamma_R m_D)^{-1})$

It turns out that for both Δm_D and $\Delta \Gamma_D$, the m_D^{-2} dependence which would appear for a very light resonance has been replaced by $\Gamma_R m_D$. Thus, for a resonance sufficiently near the D meson the possibility exists for an enhancement factor of order $m_D/\Gamma_R \simeq 5 \to 15$ for both Δm_D and $\Delta \Gamma_D$ relative to an unenhanced pole contribution. Actually it even makes sense to broaden the term 'nearby resonance' to include $m_R > 1$ GeV since the m_D^{-2} suppression mentioned above will be largely overcome.

To continue with this line of thought would require more detailed calculation. However, we do note:

1. In calculating the matrix element $\langle R|\mathcal{H}_W^\dagger|D^0\rangle$ in vacuum saturation, the contribution of resonance R to mixing will be proportional to its squared decay constant f_R^2.[b]

---

[b]It has been argued that contributions from $Q\bar{Q}$ resonances with $J^P = 0^+$ will be suppressed as they occur in P-waves and thus have vanishing wave function at the origin. This point is being studied and results will be announced elsewhere.

2. The decay constants $\{f_R\}$ will be smaller than the more common ground state decay constants like f_π. We interpret $J^P = 0^-$ $Q\bar{Q}$ resonances with masses nearest the D as second radial excitations. We include also *first* radial excitations in our study due to their larger decay constants but omit radial excitations above the second.

3. One should try to incorporate full SU(3) resonance multiplets into the description in order to take full account of possible cancellations.

4. There is no reason to limit the discussion to resonant intermediate states to just $\bar{Q}Q$ composites. Hybrid states of the $\bar{Q}QG$ type [14] should also be considered.

All these points are considered more fully elsewhere.[1] We summarize the situation in Table 1, which demonstrates that individual resonance contributions are quite sizable and that $\Delta \Gamma_D$ need not be small compared to Δm_D.

4 Conclusion

Some points I have made in this talk are:

1. Among contributions to $D^0 - \bar{D}^0$ mixing arising from Standard Model effects, the resonance mechanism discussed here is likely to be important. However, it is not currently possible to provide a definitive measure of this mechanism without additional knowledge about meson resonances in the $1 \to 2$ Gev mass range.

2. The area of experimental meson spectroscopy could supply mass and decay data for the $J^P = 0^\pm$ resonances. Completion of SU(3) multiplets would be of great value, as would classification of resonances into standard $\bar{Q}Q$ or hybrid $\bar{Q}QG$ type.

3. The lattice-gauge community could supply numerical estimates of both decay constants of excited mesons and also, as a test of vacuum saturation, matrix elements like $\langle a_0|H_W|D^0\rangle$.

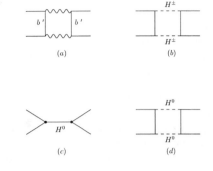

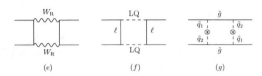

Figure 4: Effects of new physics on $\bar{D}^0 D^0$ Mixing: (a) extra $Q = -1/3$ quark b', (b) charged higgs scalars $H^\pm$, (c-d) tree, box contributions of flavor-changing neutral higgs scalars H^0, (e) left-right symmetric W-boson W_R, (f) leptoquark LQ, (g) gluino $\tilde{g}$ and squarks $\tilde{q}_{1,2}$.

4. It cannot be emphasized too strongly that $D^0 - \bar{D}^0$ mixing affords an excellent area for the exploration of new physics signals. This is because, according to my best judgement of this difficult subject, Standard Model contributions remain well below the current experimental bound. Experiments performed at either e^+e^- or hadron colliders would enhance our knowledge of particle physics.

5. A subject deserving serious thought is the effect of resonant mixing on possible CP-violation signals in the D system. [15]

Acknowledgements

The work described in this talk was supported in part by grants from the National Science Foundation and the Department of Energy.

References

1. E. Golowich and A.A. Petrov, *Phys. Lett.* B **427**, 172 (1998).
2. H. Georgi, Phys. Lett., **B297** (1992) 353.
3. T. Ohl *et al*, Nucl. Phys. **B403** (1993) 605.
4. R. Aleksan. A. LeYaouanc, L. Oliver, O. Pène and J-C Reynard, Phys. Lett., **B316** (1993) 567; M. Beneke, G. Buchalla and I. Dunietz, Phys. Rev., **D54** (1996) 4419.
5. E.M. Aitala *et al* (E791 collaboration), Phys. Rev. **D57** (1998) 13.
6. A. Datta and M.Khambakhar, Zeit. Phys., **C27** (1985) 515.
7. J. Donoghue, E. Golowich, B. Holstein and J. Trampetic, Phys. Rev., **D33** (1986) 179.
8. A.A. Petrov, Phys. Rev. **D56** (1997) 1685.
9. L. Wolfenstein, Phys. Lett., **B164** (1985) 170.
10. R. Gupta, T. Bhattacharya and S. Sharpe, *Phys. Rev.* D **55**, 4036 (1997).
11. J. Donoghue, E. Golowich and G. Valencia, Phys. Rev. **D33** (1986) 1387.
12. E. Golowich and A.A. Petrov, unpublished.
13. F. Buccella, M. Lusignoli and A. Pugliese, Phys. Lett., **B379** (1996) 249.
14. F.E. Close and P. Page Nucl. Phys. **B443** (1995) 233; Phys. Rev. **D52** (1995) 1706.
15. E. Golowich and A.A. Petrov, to appear.

OSCILLATIONS OF THE B_d^0 MESONS

VALERY P. ANDREEV[†]

Department of Physics and Astronomy,
Louisiana State University,
Baton Rouge, LA 70803, USA
e-mail address: Valeri.Andreev@cern.ch

The status of the time-dependent $B_d^0 - \bar{B}_d^0$ mixing measurements is reviewed. The new precise world average value of the B_d^0 meson oscillation frequency corresponds to a mass difference Δm_d between the two B_d^0 mass eigenstates of

$$\Delta m_d^{\text{world}} = 0.471 \pm 0.016 \text{ ps}^{-1}.$$

1   Introduction

As in the case of K-mesons, oscillations between particle-antiparticle states are expected in neutral B-mesons. In the Standard Model, the mechanism causing mixing is a second-order weak interaction through box diagrams (Figure 1).

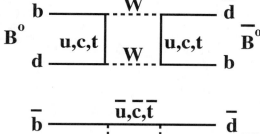

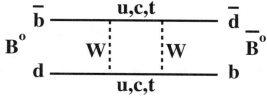

Figure 1: The Standard Model mechanism of mixing in the B_d^0 meson system.

The flavour eigenstates B_d^0 ($\bar{b}d$) and $\bar{B}_d^0$ ($b\bar{d}$) are linear combinations of the mass eigenstates B_1 and B_2. Neglecting the decay width difference between B_1 and B_2 and the effects of CP violation, both expected to be small, the probability P to find a B_d^0 decaying at proper time t, provided it was produced as $\bar{B}_d^0$ at $t = 0$, is given by

$$P(\bar{B}_d^0 \to B_d^0) = \frac{1}{\tau} e^{-\frac{t}{\tau}} \left(\frac{1 - \cos \Delta m_d\, t}{2} \right)$$

[†]on leave from Petersburg Nuclear Physics Institute

where τ is the lifetime of the B_d^0 meson. A measurement of the oscillation frequency thus gives a direct measurement of the mass difference Δm_d between the two mass eigenstates.

The phenomenon of B^0-$\bar{B}^0$ mixing is well established by experiment [1,2,3]. The time dependence of mixing for B_d^0 mesons has been measured at LEP, SLC and Tevatron using different techniques [4,5,6,7,8,9].

2   Method

In order to tag a B candidate as mixed or unmixed, it is necessary to determine its flavour state both at production (initial state) and at decay (final state). Several techniques are applied for flavour tagging. One can use the charge of the lepton from semileptonic B decay; a negatively charged lepton is produced in b flavoured meson decay. The charge of $D^{*\pm}$ mesons from B_d^0 decays is used as well in semi-exclusive analyses. For fully inclusive analyses, final state tagging techniques include jet charge (the momentum- or rapidity-weighted sum of charges of the particles) and charge dipole methods. SLD benefits from the high SLC beam polarization. The large forward backward asymmetry for $Z \to b\bar{b}$ decays is used as an efficient ($\epsilon \sim 100$ %) tag of the initial state flavour based on the polar angle of the B candidate. The initial state can also be tagged by the charge of a track from the primary vertex; it is correlated with the production state of the B if that track is a decay product of a B^{**} state, or it is the first particle in the fragmentation chain (leading particle) [5]. The combinations of those tags from the same side to the decay of a B meson candidate or from the opposite side are used to classify events as mixed or unmixed.

To study the time dependence of mixing one needs to estimate the production and the decay vertex positions of the B hadron candidate. The proper time t is related to the decay length in the laboratory system l by $t = l m_{\mathrm{B}}/p_{\mathrm{B}}$ where m_{B} and p_{B} are the mass and momentum of the b-hadron. The resolution can be parametrized as $\sigma_t = \sqrt{a_1^2 + a_2^2 t^2}$, where a_1 depends on the error on decay length l (typically 100-300 μm) and a_2 on the error on the Lorentz boost factor (typically 10-20%). Impact parameters of leptons are also used for the mixing time dependence measurements [6,7]. In this method the time of decay is measured indirectly and statistically from the impact parameters of the reconstructed lepton trajectories with respect to the beam axis.

The maximum likelihood fit is performed on the selected sample. The likelihood function is constructed for the like- and unlike-sign events and normally has very complicated structure.

3 Results

A total of 24 different analyses have been performed to measure Δm_d, of which 2 are new and 2 have been updated since last summer 1997. The different measurements are illustrated on the example of the contributed papers to the ICHEP98 conference [Abstracts 371,372,552,727]. In Figure 2 the measurement [7] by the L3 collaboration of the fraction of like-sign dilepton events is plotted as a function of proper time (Figure 1a) or the lepton impact parameter (Figure 1b) and compared to the fit result. The expected variation with proper time is clearly observed. The flavour of the b quark at production time is determined from the charge of the lepton in the opposite hemisphere (lepton - lepton combination).

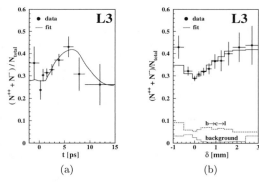

(a)   (b)

Figure 2: The L3 lepton-lepton analysis. Ratio of same-sign dilepton events to the total number of dilepton events versus the measured proper time (a) or versus the lepton impact parameter (b) compared to the fit result.

The measurement [8] by the OPAL collaboration using lepton -jet charge combination is shown in Figure 3. Jet charge technique gives much bigger statistics than lepton tag but tagging purity is smaller. The purity of the jet

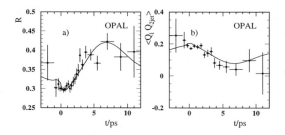

Figure 3: The OPAL lepton - jet charge analysis. a) the right-sign fraction R versus proper time t for $|Q_{2jet}| > 2$ and b) $< Q_l Q_{2jet} >$ product versus t for $|Q_{2jet}| < 2$.

charge tagging is the major systematic uncertainty for this method.

A flavour tagging technique based on the correlations of B meson flavour with the charge of the leading particle from a primary vertex (same side tag) [12] is pioneered in the CDF measurement [5]. The B flavour at decay is determined by reconstructing the semileptonic decays $B^0 \to l^+ D^{(*)-} X$ and $B^+ \to l^+ \bar{D}^0 X$. In Figure 4 the asymmetry between right-sign initial-final state flavour combinations and wrong-sign combinations $A(t) = (N_{RS}(t) - N_{WS}(t))/(N_{RS}(t) + N_{WS}(t))$ is plotted for decay signatures: $l^+ \bar{D}^0$ (dominated by B^+), $l^+ D^-$ and $l^+ D^{(*)-}$ (dominated by B^0). The dashed line is the result of the fit. The time dependence (oscillation) in the neutral B signatures is present, giving the Δm_d measurement.

In Fig. 5 the individual measurements of Δm_d and their combined value are shown. The procedure for combining of the available Δm_d measurements has been developed by the LEP B oscillations working group [10]. The measurements of Δm_d are scaled to a common set of physics input parameters and then combined taking into account the common systematic uncertainties among the analyses. Following the procedure discussed in [10], a combined value of

$$\Delta m_d^{\text{LEP+CDF+SLD}} = 0.477 \pm 0.010 \,(\text{stat}) \pm 0.013 \,(\text{syst}) \text{ps}^{-1}$$

is found. The dominant uncertainty on Δm_d is now the systematics. This combined value of Δm_d from time dependent mixing measurements, and the values of $\tau_{B_d^0}$ and $\chi_d^{\Upsilon(4S)}$ [11], may be used to determine $\chi_d^{\text{LEP+CDF+SLD+}\Upsilon(4S)} = 0.175 \pm 0.009$, or equivalently $\Delta m_d^{\text{world}} = 0.471 \pm 0.016 \text{ ps}^{-1}$.

The Δm_d average can be used to improve the knowledge of the fraction of B_s^0 mesons in a sample of weakly decaying b-hadrons. Assuming $f_{B_d} = f_{B^+}$ and $f_{B_d} + f_{B^+} + f_{B_s} + f_{b-\text{baryon}} = 1$, the values of χ_s, $f_{b-\text{baryon}}$ and the b-lifetimes from [11] and $\bar{\chi}$ from [13], the world

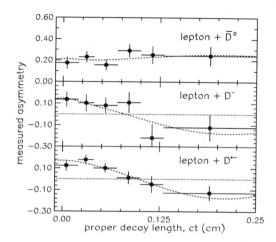

Figure 4: The CDF leading pion - lepton-D^*/D analysis. The top plot, dominated by B^+, shows no time dependence. The other two plots are mostly due to B_d^0 contribution, giving the Δm_d measurement.

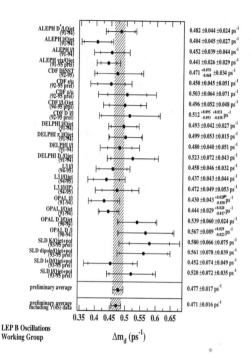

LEP B Oscillations Working Group

Figure 5: Summary of the Δm_d measurements.

average of Δm_d leads to $f_{B_s} = (10.6 \pm 1.6)\%$. Averaging this value of f_{B_s} from mixing with that from branching ratio measurements $f_{B_s} = (11.6^{+3.4}_{-2.9})\%$ [10], the fraction of B_s^0 mesons in a sample of weakly decaying b-hadrons is found to be $f_{B_s} = (10.8 \pm 1.4)\%$. This further implies that $f_{B_d} = f_{B^+} = (39.5^{+1.3}_{1.4})\%$.

4 Constraint on CKM mixing matrix and new physics

The result of Δm_d measurement may be used to determine the Cabibbo-Kobayashi-Maskawa (CKM) matrix element V_{td}. Assuming the dominant contribution to the mass difference arises from the matrix element of a box diagram of Figure 1 with W bosons and top quark as sides, B_d^0 meson oscillation frequency can be expressed [14] as

$$\Delta m_d = |V_{td} V_{tb}^*|^2 \, \frac{G_F^2}{6\pi^2} \, m_{B^0} \, m_W^2 \, F\left(\frac{m_t^2}{m_W^2}\right) f_{B^0}^2 \, B_{B^0} \, \eta_B$$

where F is the known function of m_t^2/m_W^2 ratio. The product $f_{B^0}^2 B_{B^0} \eta_B$ is coming from QCD involvement in the weak decay of the B_d^0 meson and consists of a B^0 decay constant, a non-perturbative correction factor (bag factor) and a perturbative QCD correction, respectively. The V_{tb} is assumed to be equal to 1. The oscillation

frequency Δm_d is now measured to a precision of 3%, but the uncertainty on QCD factors limits the extracted value of V_{td} to an accuracy of 20% [11]. Nevertheless, this may be used to constrain the nature of a possible CP violation in B decays. In the Wolfenstein parametrization [15] of the CKM flavour mixing matrix, the V_{td} can be expressed as $V_{td} = A \lambda^3 (1 - \rho - i\eta)$, where λ is equal to the sine of the Cabibbo angle and the complex parameter $i\eta$ accounts for the CP violating phase. Unitarity condition on CKM matrix, involving the V_{td} matrix element, is geometrically represented in the complex plane $(\eta - \rho)$ by the "unitarity triangle". In Figure 6a this triangle [‡] is shown [16] with constraint bands $(\pm\sigma)$ from the measurements of Δm_d, the CP violation in the K^0 system ϵ_K and, the $|V_{ub}|/|V_{cb}|$ ratio. The power of the co nstraint from the measurement of the mixing in the B meson system can be increased by measuring the ratio of the oscillation frequencies of B_d^0 and B_s^0 mesons. In the ratio many common factors cancel, and lattice QCD pre-

---

[‡] $\bar{\rho} = \rho(1 - \frac{\lambda^2}{2}), \bar{\eta} = \eta(1 - \frac{\lambda^2}{2})$

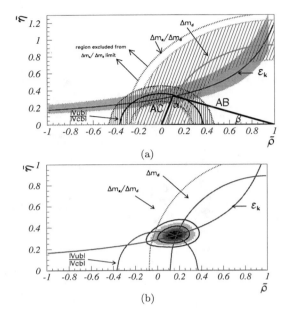

(a)

(b)

Figure 6: The unitarity triangle in the $(\bar\eta - \bar\rho)$ plane. a) The constraints on Δm_d, ϵ_K and $|V_{ub}|/|V_{cb}|$ are shown as bands corresponding to $\pm\sigma$. b) The allowed region for vertex A. The contours at 68% and 95% CL are shown.

dicts then a QCD correction factor to the precision of the order of 10%. There is no measurement up to now of the mass difference Δm_s in the system of B_s^0 mesons, a lower limit $\Delta m_s > 12.4$ps^{-1} at 95% CL is set [17]. The limit on Δm_s allows to restrict further the possible domain for the vertex position, A in Figure 6, of the unitarity triangle. This is shown in Figure 6a as the region excluded at 95% CL by the $\Delta m_s/\Delta m_d$ ratio limit. The comparison of the calculated CKM matrix elements with the values based on direct measurements is used to place restrictions on new physics [16]. Those three constraints were used to fit the position of the vertex A. The result of the fit is presented in Figure 6b, where the vertex position and the 68% and 95% CL contours are shown. It is clear from the figure that the measurements of such different physical quantities as the mixing of the B_d^0 mesons, the CP violation in K^0 mesons and the ratio of charmless decays to decays with charm, $|V_{ub}|/|V_{cb}|$, are very compatible with each other (the curves cross at the same point). It means there is no big change of corresponding CKM matrix elements due to new physics.

5 Impact on CP-violation measurements

The importance of a precise measurement of the B_d^0 oscillation frequency Δm_d can be appreciated because of the impact it has for the searches of CP-violation in the neutral B meson decays. CP-violation is expected in the B decays [18]. In the system of neutral B mesons, CP-violation can arise from interference between the two paths to a given final state f, that is without or with mixing $B \to f$ or $B \to \bar{B} \to f$. Large effects may be seen in the difference in decay rates of B_d^0 and $\bar{B}_d^0$ to the CP eigenstate $J/\psi K_s^0$ [18], which corresponds to a decay asymmetry

$$A_{CP}(t) \equiv \frac{\bar{B}_d^0(t) - B_d^0(t)}{\bar{B}_d^0(t) + B_d^0(t)} = \sin(2\beta)\sin(\Delta m_d t).$$

This relationship shows that CP-violation effect, $\sin(2\beta)$, has to be measured on top of the time dependence from the mixing, $\sin(\Delta m_d t)$. The results of the first attempts to measure time dependent asymmetries of $B_d^0/\bar{B}_d^0 \to J/\psi K_s^0$ have been presented at the ICHEP98 conference by the CDF [19] and the OPAL [20] collaborations. For the time being, the world pool of such events consists of 198 events by CDF and 24 events with a purity of about 60% by OPAL. In Figure 7a the CDF measurement is shown for the mass distribution of $J/\psi K_s^0$ candidates, where one can see clearly identified $B_d^0/\bar{B}_d^0 \to J/\psi K_s^0$ candidates. The background is highly suppressed for large decay lengths (Figure 7b). The same side tag (leading "pion") method has been applied for initial state tagging. The asymmetry as a function of proper decay length is shown in Figure 7c together with the result of the fit to the function $D_0 \sin(2\beta)\sin(\Delta m_d t)$, where D_0 is dilution factor. For $\sin(2\beta)$ CDF finds $1.8 \pm 1.1 \pm 0.3$. The central value is unphysical since the amplitude of the measured asymmetry is larger than D_0 (the statistics are not big enough yet), but for this example we would like to underline that the largest systematic uncertainty is due to Δm_d. There will be some improvement on the statistical error on Δm_d in the future, but the major improvement on Δm_d precision will come from an improved knowledge of the Δm_d fit input parameters (b-hadron fractions, b-lifetimes, decay branching ratios, fragmentation, etc.)

6 Conclusion

The new precise world average value of the B_d^0 meson oscillation frequency corresponds to a mass difference Δm_d between the two B_d^0 mass eigenstates of

$$\Delta m_d^{\text{world}} = 0.471 \pm 0.016 \text{ ps}^{-1}.$$

This result provides an important constraint on the CKM mixing matrix and will serve as the basic input parameter for future CP-violation studies. This measurement is

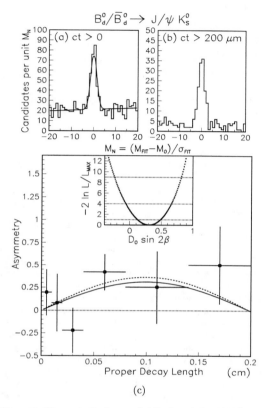

Figure 7: The normalized mass distribution of the $J/\psi K_s^0$ candidates of CDF experiment with $ct > 0$ (a) and $ct > 200\mu m$ (b). (c) shows the sideband-subtracted tagging asymmetry as function of the reconstructed $J/\psi K_s^0$ proper decay length versus fit to $A_0 \sin(\Delta m_d t)$.

used to improve the knowledge of the fraction of weakly decaying B_s^0 mesons in a sample of weakly decaying b-hadrons, giving a result of $f_{B_s} = (10.8 \pm 1.4)\%$.

Acknowledgements

I would like to thank my colleagues of the L3 Collaboration and my colleagues of the LEP B Oscillations Working Group for the cooperation.

References

1. UA1 Collaboration, Phys. Lett. **B 186** (1987) 247.
2. ARGUS Collaboration, Phys. Lett. **B 192** (1987) 245; Phys. Lett. **B 324** (1994) 249; CLEO Collaboration, Phys. Rev. Lett. **71** (1993) 1680.
3. ALEPH Collaboration, Phys. Lett. **B 284** (1992) 177;
DELPHI Collaboration, Phys. Lett. **B 332** (1994) 488;
L3 Collaboration, Phys. Lett. **B 335** (1994) 542;
OPAL Collaboration, Z. Phys. **C 60** (1993) 199.
4. ALEPH Collaboration, Z. Phys. **C 75** (1997) 397.
5. CDF Collaboration, Phys. Rev. Lett. 80 (1998) 2057.
6. DELPHI Collaboration, Z. Phys. **C 76** (1997) 579.
7. L3 Collaboration, CERN-EP/98-028, Eur. Phy. J. **C** (in press).
8. OPAL Collaboration, Z. Phys. **C 72** (1996) 377; Z. Phys. **C 76** (1997) 401; Z. Phys. **C 76** (1997) 417.
9. SLD Collaboration SLAC-PUB-7228,7229,7230, contributed papers PA08-026A,026B,026C, to ICHEP'96 Warsaw.
10. The LEP B Oscillations Working Group (note LEPBOSC 98/3). Combinations of all available measurements, including preliminary results, are updated regularly and may be found at: http://www.cern.ch/LEPBOSC/ .
11. Particle Data Group, Eur. Phys. J. **C 3** (1998) 1.
12. M. Gronau, A. Nippe and J. Rosner, Phys. Rev. **D 47** (1993) 1988.
13. The LEP Collaborations: ALEPH, DELPHI, L3 and OPAL, the LEP Electroweak Working Group and the SLD Heavy Flavour Group, *A Combination of Preliminary Electroweak Measurements and Constraints on the Standard Model*, CERN-PPE/97-154.
14. A.J. Buras and R. Fleischer, *Quark Mixing, CP Violation and Rare Decays After the Top Quark Discovery*, preprint hep-ph/9704376 (1997), to be published in *Heavy Flavours II*, World Scientific (1997), Eds. A.J. Buras and M. Lindner.
15. L. Wolfenstein, Phys. Rev. Lett. **51** (1983) 1945.
16. P. Paganini, F. Parodi, P. Roudeau, and A. Stocchi, CERN-OPEN-97-028 (1997).
F. Parodi, P. Roudeau, and A. Stocchi, *Constraints on the Parameters of the V_{CKM} matrix at the end of 1997*, preprint hep-ph/9802289 (1998).
17. F. Parodi, these proceedings.
18. A.B. Carter and A.I. Sanda, Phys. Rev. Lett. **45** (1980) 952; Phys. Rev. **D 23** (1981) 1567;
I.I. Bigi and A.I. Sanda, Nucl. Phys. **B 193** (1981) 85; Nucl. Phys. **B 281** (1987) 41.
19. FERMILAB-PUB-98/189-E and J. Tseng, these proceedings.
20. CERN-EP/98-001 and E. Barberio, these proceedings.

B_s^0 MIXING, LIMITS ON Δm_s

F. Parodi

Dipartimento di Fisica, Università di Genova and INFN, Via Dodecaneso 33, IT-16146 Genova, Italy
E-mail: Fabrizio.Parodi@ge.infn.it

The experimental status of the B_s^0-$\overline{B_s^0}$ oscillations is presented. Several B_s^0 mixing analyses from different experiments (LEP, SLD, CDF) are reviewed and the world average is given. The world average of these measurements implies $\Delta m_s > 12.4 \ ps^{-1}$ at 95 % C.L..
Implications for the CKM matrix elements are discussed.

1 Introduction

In these proceedings I will review the status of the search for B_s^0 oscillations. After a short introduction on the general features of these analyses I will concentrate on the new results sent to this Conference [1,2,3,4,5,6,7]. The world average of these measurements will be also discussed [8]. The last part is devoted to the impact of the present limit on Δm_s on the determination of the CKM matrix elements [9].

2 B^0–$\overline{B^0}$ mixing in the Standard Model

In the Standard Model, B_q^0–$\overline{B_q^0}$ (q = d,s) mixing is a direct consequence of second order weak interactions.
Neglecting CP-violation effects, the mass eigenstates of the B^0–$\overline{B^0}$ system, $B_{S,L} = (B^0 \pm \overline{B^0})/\sqrt{2}$ [a] are also CP eigenstates with masses $m_{S,L} = m \pm \Delta m/2$ and decay width $\Gamma_{S,L} = \Gamma \pm \Delta\Gamma/2$.
Starting with a B_q^0 meson produced at time t=0, the probabilities to observe a B_q^0 or a $\overline{B_q^0}$ decaying at the proper time t are, neglecting CP violating effects:

$$\mathcal{P}_{B_q^0 \to B_q^0, \bar{B}_q^0} = \frac{\Gamma_L \Gamma_S}{2\Gamma} e^{-\Gamma t}(\cosh((\Delta\Gamma/2)t) \pm \cos \Delta m_q t)$$

Since no measurement of $\Delta\Gamma$ exists and theory predicts $\Delta\Gamma \ll \Delta m$, the oscillation analyses neglect $\Delta\Gamma$ simplifying the mixing probability density to the function $\Gamma e^{-\Gamma t}(1 \pm \cos \Delta m_q t)/2$.
The mass differences depend on the CKM parameters via the following relations:

$$\Delta m_q = \frac{G_F^2}{6\pi^2} \mid V_{tb} \mid^2 \mid V_{tq} \mid^2 m_{B_q} f_{B_q}^2 B_{B_q} \eta_B m_t^2 F(\tfrac{m_t^2}{m_W^2})$$
$$\Delta m_d \propto A^2 \lambda^6[(1-\rho)^2 + \eta^2]$$
$$\Delta m_s \propto A^2 \lambda^4$$

where A, λ, ρ and η are the parameters appearing in the Wolfenstein parametrisation of the CKM matrix.
ρ and η are the most uncertain parameters. The interest

in measuring Δm_q consists in constraining these two parameters.
Δm_d has a direct dependence on ρ and η but its impact is limited by the uncertainty on the hadronic factor $f_{B_d} \sqrt{B_{B_d}}$.
Δm_s does not depend on ρ and η but it contains the corresponding hadronic factor for B_s^0 ($f_{B_s} \sqrt{B_{B_s}}$). Since most of the theoretical uncertainties cancel in the ratio of the factors for B_s^0 and B_d^0, the $\Delta m_s/\Delta m_d$ ratio could be very efficient in constraining the (ρ, η) plane provided that Δm_s is known (or limited to some interval).
To give some orders of magnitude on the oscillation frequencies the values $A \simeq 0.8$ and $\lambda \simeq 0.22$ can be assumed: the ratio $\Delta m_s/\Delta m_d$ turns out to be roughly 20. Since $\Delta m_d \simeq 0.5 \ ps^{-1}$ and $\tau(B_s^0) \simeq 1.6 \ ps$ the B_s^0 meson is expected to oscillate about three times in one lifetime. This rapid oscillation represents a challenge for the experiments, nevertheless I will show that the world average is already exploring a substantial part of the Δm_s allowed range in the Standard Model.

3 Methods for B_s^0 oscillation analyses

Due to the expected large value of Δm_s the integrated mixing parameter $\chi_s = 1/2(\Delta m^2/(\Delta m^2 + \Gamma^2))$ is very close to 0.5 and it does not give any information on the oscillation frequency: time-dependent analyses are needed.
A time-dependent analysis requires the measurement of the decay proper time and the tag of the B candidate as mixed or unmixed. For this purpose it is necessary to determine the particle-antiparticle state both at the production and at the decay time.
The statistical significance ($\mathcal{S}$) of an oscillation signal can be approximated by [12]:

$$\mathcal{S} = \sqrt{N/2} f_{B_s^0}(2\epsilon - 1)e^{-(\Delta m_s \sigma_t)^2/2}$$

where

---
[a] S and L indicate the "short" and "long" lifetime eigenstates.

N   = number of candidates
$f_{B_s^0}$ = B_s^0 purity of the selected sample
ϵ = right-tag probability
σ_t = proper time resolution

The significance is a decreasing function of Δm_s controlled by the parameter σ_t.

This parameter turns out to be crucial in exploring high oscillation frequencies. The proper time resolution $\sigma_t = \sqrt{(\sigma_L m/\langle p\rangle)^2 + (\sigma_p/p)^2 t^2}$ receives contribution from the decay length resolution σ_L and from the relative momentum resolution σ_p/p; the second term grows with the proper time.

Another crucial parameter is the "right-tag" probability, that is the probability of correctly assigning mixed and unmixed events.

The global factor $(2\epsilon - 1)$ in $\mathcal{S}$ includes the contribution from the decay tag (ϵ_{dec}) and from the production tag (ϵ_{prod}) ($(2\epsilon - 1) = (2\epsilon_{dec} - 1)(2\epsilon_{prod} - 1)$).

The decay tag depends heavily on the decay chain used to reconstruct the B_s^0 (high p_t leptons, completely reconstructed D_s^+,...) and it is usually quite pure (in the semileptonic channels the main contamination coming from $b \to c \to \bar{\ell}$ is reduced to about 10 % using a p_t cut).

The production tag is less dependent on the procedure used to select B candidates and it uses several quantities that can be divided into two groups: same side variables that tag the $\bar{b}$ quark contained in the reconstructed B meson (jet–charge, spectator kaon), opposite side variables that tag the initial state of the other b quark produced in the event (jet–charge, high p_t lepton, kaons at the secondary vertex,...). These variables are usually combined to give the so called "combined tag" of the initial state.

4 Techniques used to set limits on Δm_s

No oscillation signal has been seen so far in the B_s^0 system and all B_s^0 analyses set lower limits.

The original method used for this purpose, the likelihood with respect to the minimum, was found to be inaccurate and quite complicated when combining different experiments.

A new method [12] has been developed, in which the oscillation amplitude $\mathcal{A}$ is measured, at fixed value of Δm_s, using a maximum likelihood fit based on the modified probability function $\Gamma e^{-\Gamma t}(1 \pm \mathcal{A}\cos\Delta m_s t)$. $\mathcal{A}$ could be thought as a normalised Fourier amplitude and it is equal to 1 when $\Delta m_s = \Delta m_s^{true}$ and to 0 when Δm_s is far enough from Δm_s^{true} (typically $|\Delta m_s - \Delta m_s^{true}| > 2\Gamma$). The measurements of $\mathcal{A}$ are Gaussian with r.m.s. $\sigma_{\mathcal{A}} = 1/\mathcal{S}$. As a result a value of Δm_s can be excluded at 95% C.L. if $\mathcal{A}(\Delta m_s) + 1.645\,\sigma_{\mathcal{A}} < 1$.

This method allows to combine different experiments and

systematics. It gives also an unbiased estimator of the goodness of the analysis: the sensitivity.

The sensitivity is defined as the largest Δm_s value that would be excluded if all the amplitudes were put at 0.

The Amplitude method is the method used by the LEP B oscillations working group to combine results on Δm_s.

5 Δm_s at LEP

The data sample used by each LEP experiment to study B_s^0 oscillations consists in about 4 million hadronic Z^0 decays collected between 1991 and 1995.

Different channels have been studied: they are summarized in Tab. 1 starting from the most inclusive to the most exclusive channel. Each analysis will be labelled using the convention: "decay tag/production tag".

The experiments labelled in bold have provided new

Method of analysis	Experiment		
Inclusive lepton/Comb	**A**	**D**	O
Exclusive D_s and/or $D_s h$/Comb	A	**D**	
$D_s \ell$/Comb	A	**D**	
Exclusive B_s^0/Comb		**D**	

Table 1: *LEP: summary of the channels used in Δm_s studies. The following labels are used: A=ALEPH,D=DELPHI,O=OPAL*

results for this Conference.

5.1 Inclusive lepton channel

Samples of inclusive semileptonic b decays are selected by requiring the presence, in one hemisphere, of an identified lepton (electron or muon) with a high p_t (in order to enrich the sample in direct $b \to \ell$ decays). The charm vertex is inclusively reconstructed.

This channel is characterised by high statistics ($N \sim$ some tens of thousands) but it suffers from a quite low B_s^0 purity ($\sim 10\%$). The proper time resolution and the initial state tagging are also challenging in such an inclusive topology.

ALEPH has shown that it is possible to get a very good proper time resolution by carefully selecting the initial sample ($\sigma_t(t < 1\ ps) < 0.27\ ps$).

The combined tag of the initial state allows to have a quite good effective tagging purity (70%).

Both ALEPH and DELPHI tried also different techniques to improve the effective B_s^0 purity.

DELPHI, for instance, combines, using the the likelihood ratio method, different variables to enhance the $b \to \ell$ contribution ($p_t(\ell)$, $p(\ell)$, b-tagging, B-energy) and the B_s^0 fraction inside the $b \to \ell$ (spectator kaon, vertex charge, number of secondary kaons). The effective increase in the B_s^0 purity is 1.32.

The Δm_s results obtained with the inclusive lepton samples are summarized in Tab. 2.

The ALEPH inclusive lepton analysis has the best sensi-

Experiment	95% C.L. limit	Sensitivity	Sens. '97
ALEPH	9.6	9.5	10.5
DELPHI	4.6	6.4	3.5
OPAL(ℓ/Q_{jet})	2.9	not updated	4.8

Table 2: LEP: summary of the results obtained with the samples of inclusive leptons (unit in ps^{-1}).

tivity among all other results on B_s^0 oscillations (Fig. 1).

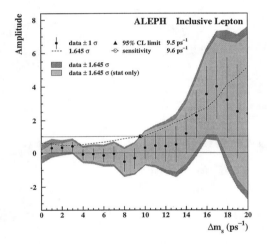

Figure 1: *ALEPH: amplitude spectrum for the inclusive lepton analysis.*

5.2 $D_s\ell$ channel

$D_s\ell$ candidates are reconstructed by pairing completely reconstructed D_s mesons with high p_t leptons of opposite charge.

The excess of events under the D_s peak is dominated by the signal (i.e. $B_s^0 \to D_s^-\ell^+\nu X$), taking into account the combinatorial background, the B_s^0 purity is about 60%. Since all the charged decay products of the B_s^0 are completely isolated, a quite good accuracy is achieved on the proper time resolution and the initial state tagging profits from the use of all the information in the event, giving effective tagging purities as high as 78% (DELPHI). The only weak point is the statistics, limited essentially by

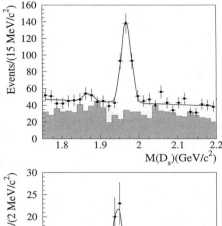

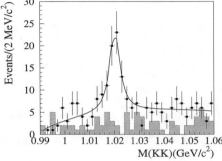

Figure 2: *DELPHI: invariant mass distributions for $D_s\ell$ candidates in hadronic (upper plot) and semileptonic (lower plot) decay modes. Shaded histograms represent the wrong sign combination.*

the accessible number of D_s decay channels.

A large effort has been put by ALEPH and DELPHI to reconstruct as many D_s decay channels as possible. Fig. 2 shows the sample selected in DELPHI based on six hadronic and two semileptonic D_s decay modes ($D_s^+ \to \phi\pi^+, \phi\pi^+\pi^0, \phi\pi^+\pi^-\pi^+, \overline{K^{0*}}K^+, \overline{K^{0*}}K^{*+}, K_S^0K^+$; $D_s^+ \to \phi e^+\nu_e, \phi\mu^+\nu_\mu$).

The results obtained from the $D_s\ell$ samples are summarized in Tab. 3

Experiment	95% C.L. limit	Sensitivity	Sens. '97
ALEPH	7.6	not updated	6.7
DELPHI	7.4	8.4	8.0

Table 3: LEP: summary of the results obtained with the $D_s\ell$ samples (unit in ps^{-1}).

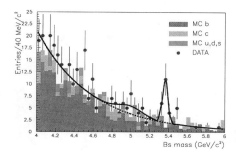

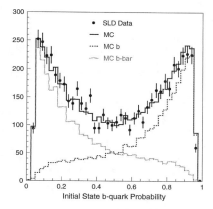

Figure 3: *DELPHI: invariant mass of* B_s^0 *candidates reconstructed in the decay modes* $\mathrm{D}^0\mathrm{K}\pi$ *and* $\mathrm{D}_s\pi$

5.3 Exclusive B_s^0 signals

The use of exclusively reconstructed B_s^0 mesons has been recently proposed by the DELPHI Collaboration.

The idea behind this analysis is that, to explore high Δm_s values, a very good proper time resolution is essential.

The exclusive B_s^0 reconstruction offers this opportunity since, for this type of events, the error on the momentum determination is negligible giving a constant proper time resolution of about 0.08 ps (1/20 of lifetime !).

Fig. 3 shows the candidates selected by DELPHI in the two decay modes $\mathrm{B}_s^0 \to \mathrm{D}_s^- \pi^+$ and $\mathrm{B}_s^0 \to \overline{\mathrm{D}}^0 \mathrm{K}^- \pi^+$. The signal yield has been evaluated to be 17 ± 8 over a background of 35%.

This channel alone has a quite low sensitivity because of the small statistics available. Nevertheless the error on the amplitude varies slowly with Δm_s and its contribution to high values of Δm_s is remarkable.

Considering, as an example, the various analyses done in DELPHI, the exclusive B_s^0 appears to be competitive (has smaller $\sigma_{\mathcal{A}}$) with the inclusive lepton sample at Δm_s=10 ps^{-1} and with the $\mathrm{D}_s\ell$ at Δm_s=20 ps^{-1}. The combined DELPHI sensitivity moves from 9.7 ps^{-1} to 10.4 ps^{-1} by adding the exclusive B_s^0 analysis.

The impact could be even more important if other LEP Collaborations will perform similar analyses.

Concluding the paragraph on LEP analyses, I would like to point out that, even if the LEP contribution dominates the world average (Sec. 8), improvements are still expected at LEP from a wider sharing of the existing analyses (see Tab. 1).

6 Δm_s at SLC

The SLD has presented two analyses at this Conference. The data sample consists of about 250,000 hadronic Z^0

Figure 4: *SLD: distribution of initial state b-quark probability for DATA and MC.*

decays collected with the upgraded vertex detector between January 1996 and March 1998.

Both analyses benefit of the initial state tagging based on opposite hemisphere variables (similar to those used at LEP) and on beam polarisation. SLC performs the best initial state tagging with an effective purity of 85%. Fig. 4 shows the initial state tagging as a function of the discriminant variable for DATA and MC.

The first analysis is called "lepton+D/Comb" and is quite similar to the inclusive lepton performed by the LEP Collaborations with the advantage of a very precise proper time resolution due to the Vertex Detector placed very close to the interaction point ($\sigma_t(t < 1\ ps) = 0.13\ ps$).

The second analysis, called "Charge Dipole/Comb", aims to perform a more inclusive reconstruction. Both the B and D vertices are inclusively reconstructed giving access to the full sample of hadronic decays. The final state tagging is then performed using the "Charge Dipole" $\delta Q \equiv D_{\mathrm{BD}} \times SIGN(Q_{\mathrm{D}} - Q_{\mathrm{B}})\ (\delta Q > 0(< 0) \to b(\bar{b}))$.

The average purity of this tag is 69%.

The lepton+D and the Charge Dipole analyses were combined taking into account correlated systematic errors.

Two regions are excluded at 95% C.L.: $\Delta m_s < 1.7\ ps^{-1}$ and $3.3 < \Delta m_s < 5.0\ ps^{-1}$. The sensitivity of the combined analysis is 4.2 ps^{-1}.

The slope of $\sigma_{\mathcal{A}}$, thanks to the very good proper time resolution, is quite small: the overall sensitivity is expected to improve significantly by adding the rest of the 1998 statistics (some 150,000 Z^0) and other analysis techniques.

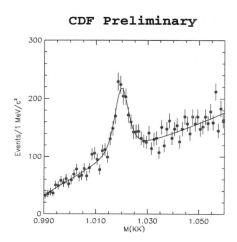

Figure 5: *CDF: K^+K^- invariant mass spectrum in the selected dilepton triggers*

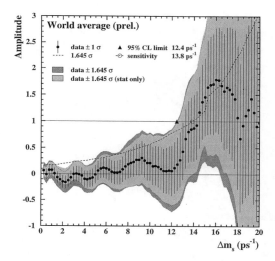

Figure 6: *a) LEP B Oscillations Working Group: combined B_s^0 oscillation amplitude as a function of Δm_s*

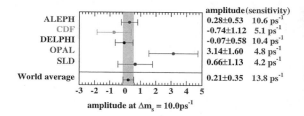

Figure 7: *b) Amplitudes at 10 ps^{-1} averaged per experiment*

CDF has performed a search for B_s^0-$\overline{B_s^0}$ oscillations using a sample of B_s^0 semileptonic decays collected using dilepton triggers during the 1992-95 run of the Tevatron. The B_s^0 meson is reconstructed in the $\phi\ell$ final state and the initial state is tagged using the second lepton in the trigger.

The selected sample contains 1068 ϕ candidates over the combinatorial background (Fig. 5) with an estimated B_s^0 purity of 61%. The production tag given by the lepton has a purity of about 76%.

This analysis excludes at 95% C.L. all the Δm_s values below 6 ps^{-1} and has a sensitivity of 5.1 ps^{-1}.

CDF expects to improve its result by adding other tagging methods in the $\phi\ell$ analysis and by performing a new analysis ($D_s\ell$).

8 Combined result on Δm_s

The LEP B Oscillations Working Group has combined the results of the various analyses taking into account correlated systematics. The overall limit is then deduced from the combined amplitude spectrum (Fig. 6)

All the values below 12.4 ps^{-1} are excluded at 95 % C.L.. The combined sensitivity is 13.8 ps^{-1}.

Using LEP data alone, the combined limit is 11.5 ps^{-1} and the sensitivity is 12.9 ps^{-1}.

The amplitude in the region between 14.5 ps^{-1} and 16.5 ps^{-1} is more than 1.645 standard deviations away

from 0. This region is above the sensitivity and the points are highly correlated. The most significant deviation at 15 ps^{-1} (2σ) is consistent with zero at 2.3 % C.L..

No observation can be claimed.

In Fig. 7 the values of the amplitudes at 10 ps^{-1}, averaged for each experiment, are given. Sensitivities are also reported.

9 Implications on the CKM matrix elements

In order to appreciate the importance of the B_s^0 oscillations study it is useful to analyse the impact of this measurement on the determination of the CKM unitarity triangle in the framework of the Standard Model.

The present analysis is an update of the results presented

Parameter	Value
A	0.835 ± 0.029
$\frac{|V_{ub}|}{|V_{cb}|}$	0.093 ± 0.014
Δm_d	$(0.471 \pm 0.016) \ ps^{-1}$
Δm_s	$> 12.4 \ ps^{-1}$
$\overline{m_t}(m_t)$	$(167 \pm 6) \ GeV/c^2$
B_K	0.86 ± 0.07
$f_{B_d}\sqrt{B_{B_d}}$	$(204 \pm 35) \ MeV$
$\xi = \frac{f_{B_s}\sqrt{B_{B_s}}}{f_{B_d}\sqrt{B_{B_d}}}$	1.11 ± 0.06

Table 4: *Input parameters.*

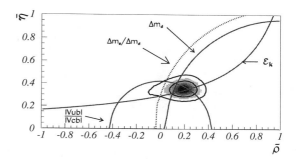

Figure 8: *$\bar{\rho}$–$\bar{\eta}$ probability distribution. The 68 % and 95 % C.L. contours are shown.*

in [10,11] using recent measurements submitted to this Conference (the improved limit on Δm_s and the new V_{ub} and V_{cb} results) and the progress obtained by lattice QCD evaluations.

The convention adopted for the CKM parameters is the Wolfenstein parametrisation in which the four independent parameters are λ, A, $\bar{\rho}$ and $\bar{\eta}$ [b]. Since λ is precisely measured ($\lambda = 0.2205 \pm 0.0018$) and A is the second best known parameter ($A\lambda^2 = V_{cb}$) the analysis will concentrate on the determination of the allowed region in the $(\bar{\rho}, \bar{\eta})$ plane.

Four constraints are used: the measurements of $|\epsilon_K|$, $\left|\frac{V_{ub}}{V_{cb}}\right|$, Δm_d and the Δm_s amplitude spectrum. Further details on the method can be found in [10].

The input values for each parameter entering into the definition of the constraints are given in Tab. 4. The region of the $(\bar{\rho}, \bar{\eta})$ plane selected by those constraints has been obtained assuming Gaussian errors. A similar study based on flat distributions for the parameters B_K, $f_{B_d}\sqrt{B_{B_d}}$, ξ and $\left|\frac{V_{ub}}{V_{cb}}\right|$ was also done [11] and gave rather similar results.

The contours corresponding to 68% and 95% confidence levels are shown in Fig. 8.

The measured values for the two parameters are:

$$\bar{\rho} = 0.189 \pm 0.074, \quad \bar{\eta} = 0.354 \pm 0.045$$

The experimental constraints are compatible and do not show evidence for New Physics inside the measurement errors.

The constraint on Δm_s cuts the left part of the plane, allowing to improve considerably the precision on $\bar{\rho}$; this is clearly shown by the results obtained without the Δm_s constraint:

$$\bar{\rho} \in [-0.35, +0.34] \quad \text{at 95 \% C.L.}$$
$$\bar{\rho} \in [0.00, +0.30] \quad \text{at 68 \% C.L.}$$
$$\bar{\eta} = 0.350 \pm 0.054$$

[b]$\bar{\rho}$ and $\bar{\eta}$ are related to the original ρ and η parameters introduced by Wolfenstein: $\bar{\rho}(\bar{\eta}) = \rho(\eta)(1 - \frac{\lambda^2}{2})$ [13]

It is of interest to derive, using the $(\bar{\rho}, \bar{\eta})$ probability distribution, the central values and the uncertainties on the quantities $\sin 2\alpha$ and $\sin 2\beta$ which will be measured directly at future facilities like HERA-B and B factories:

$$\sin 2\alpha = -0.15 \pm 0.30 \quad \sin 2\beta = 0.73 \pm 0.08$$

The value of $\sin 2\beta$ is rather precisely determined, with an accuracy already at the level expected after the first years of running at B factories.

Quite good precision has been also achieved in the determination of the γ angle:

$$\gamma = (62 \pm 10)^o$$

Removing the constraint from the measured limit on Δm_s the probability density for Δm_s can be deduced (Fig. 9). The predicted range at 68 % C.L. is $[9.5, 17] \ ps^{-1}$ and the preferred value is $13 \ ps^{-1}$.

The present analyses are exploring the region where the signal is expected. The combined analysis excludes already a large fraction of this distribution ($\Delta m_s < 12.4 \ ps^{-1}$) and has a three sigma significance for $\Delta m_s = 9.7 \ ps^{-1}$ indicating that if Δm_s was below $10 \ ps^{-1}$ it should have been measured.

10 Conclusions

Several new analyses on Δm_s have been presented at this Conference by ALEPH, DELPHI, SLD and CDF.

The world average combined analysis has a sensitivity of $13.8 \ ps^{-1}$ and excludes, at 95 % C.L., all values below $12.4 \ ps^{-1}$.

This limit falls in the high probability region of the Δm_s probability distribution predicted by the Standard Model and therefore it is quite efficient in constraining the CKM parameters.

The present result could still be improved by adding new statistics (SLD) and new analyses (LEP, SLD, CDF).

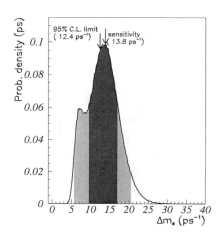

Figure 9: Δm_s *probability distribution. The dark-shaded and clear-shaded intervals correspond to 68 % and 95 % C.L. regions respectively.*

Acknowledgements

I would like to thank the representatives of the various collaborations who provided the results for this review. I am thankful to Achille Stocchi for his help in preparing the talk and to Carlo Caso and Patrick Roudeau for the useful suggestions received during the writing of this paper. The work of my colleagues of the LEP B Oscillations Working Group is also gratefully acknowledged.

References[c]

1. ALEPH Collaboration, "Search for B_s^0 oscillations using inclusive lepton events", ICHEP98-PA08-948.
2. ALEPH Collaboration, "Study of B_s^0 oscillations and lifetime using fully reconstructed D_s decays", ICHEP98-PA08-954.
3. DELPHI Collaboration, "Study of B_s^0-$\overline{B_s^0}$ oscillations and B_s^0 lifetime measurement using $D_s\ell$ candidates", ICHEP98-PA08-235.
4. DELPHI Collaboration, "Study of B_s^0-$\overline{B_s^0}$ oscillations using inclusive leptons emitted at large P_t", ICHEP98-PA08-236.
5. DELPHI Collaboration, "Study of B_s^0-$\overline{B_s^0}$ oscillations using exclusive B_s^0 signals and D_s-hadrons final states", ICHEP98-PA08-237.

6. SLD Collaboration, "Time Dependent B_s^0-$\overline{B_s^0}$ Mixing Using Inclusive and Semileptonic B Decays at SLD", ICHEP-PA08-184.
7. CDF Collaboration, "Search for B_s^0-$\overline{B_s^0}$ Flavour Oscillations in $p\bar{p}$ Collisions with CDF at $\sqrt{s} = 1.8$ TeV", ICHEP-PA08-732.
8. LEP B Oscillations Working Group, "Combined Results on B^0 Oscillations: Results for Summer 1998 Conferences", LEPBOSC 98/3.
9. F. Parodi, P. Roudeau and A. Stocchi, "Present determination of the CKM unitarity triangle", ICHEP-PA06-586.
10. P. Paganini, F. Parodi, P. Roudeau and A. Stocchi, " Measurement of the ρ and η parameters of the V_{CKM} matrix and perspectives", LAL 97-79, hep-ph/9711261, accepted by Physica Scripta.
11. F. Parodi, P. Roudeau and A. Stocchi, "Constraints on the parameters of the V_{CKM} matrix at the end of 1997", hep-ph/9802289.
12. H.-G. Moser and A. Roussarie, *Nucl. Instrum. Methods* A **384**, 491 (1997).
13. A.J. Buras, M.E. Launtenbacher and G. Ostermaier, *Phys. Rev.* D **50**, 3433 (1994).

[c]Abbreviation:
ICHEP98-... = contributed paper to the 29th International Conference on High Energy Physics, 22-29 July 1998, Vancouver, Canada.

SEMILEPTONIC B DECAYS AT THE Z^0

PAULINE GAGNON

Centre for Research in Particle Physics, Carleton University, Ottawa, Canada K1S 5B6
E-mail: pauline.gagnon@cern.ch

Over the past few years, precise measurements of b hadron lifetimes have uncovered lifetime differences larger than what is theoretically expected between b baryons and B mesons. Meanwhile, a long-standing disagreement between experimental results has emerged for the b hadron semileptonic branching fraction when measured at the $\Upsilon(4S)$ or near the Z^0 resonance. This has resulted in an increased interest in the study of semileptonic decays as an alternative approach to the understanding of the mechanics of b hadron decays. This paper first reviews two recent measurements made near the Z^0 resonance for the semileptonic branching fractions of the average b baryons. The results support the evidence of a much shorter lifetime for b baryons than for B mesons. In a second part, new techniques used to determine $BR(b \to \ell X)$ are described. Preliminary results indicate that progress is being made towards the understanding of the experimental discrepancy between the $\Upsilon(4S)$ and Z^0 measurements.

1 Introduction

Measurements of the semileptonic decays of b hadrons are important in testing our understanding of the dynamics of heavy quark physics. While theoretical calculations [1,2] agree with the experimental measurements within errors, the situation has been more problematic on the experimental front. The measurements obtained at different centre-of-mass energy have disagreed for several years. The semileptonic branching fraction for B mesons has been measured at the $\Upsilon(4S)$ resonance to be $BR_{sl}^B = (10.45 \pm 0.21)\%$. [3] Using all results obtained above the B meson production threshold, one obtains $BR_{sl}^b = (11.02 \pm 0.21)\%$ [4] taking into account the correlated errors with all other electroweak measurements. The "b" superscript indicates that the high-energy data correspond to a mixture of $B^\pm$, B^0, B_s and b baryons as opposed to $B^\pm$ and B^0 only as seen at the $\Upsilon(4S)$ resonance, which is represented by the "B" superscript. [a]

The experimental result obtained near the Z^0 resonance needs to be corrected to account for the presence of b baryons and B_s in the sample. Assuming the semileptonic width Γ_{sl} to be the same for all b hadrons (as suggested by the result of [5,6] described in section 2.3), and given that the semileptonic branching ratio is defined as $\Gamma_{sl}/\Gamma_{total} = \tau_{sl}\Gamma_{sl}$, one obtains the needed relation:

$$BR(B \to \ell X) = \frac{\tau_B}{\tau_b} \cdot BR(b \to \ell X). \quad (1)$$

To remove the difference between the $\Upsilon(4S)$ and Z^0 results would require that τ_B/τ_b be 0.948 ± 0.026, whereas current lifetime measurements indicate that the b baryon lifetime is less than the B meson lifetime, and hence that τ_B/τ_b is expected to be greater than one. [2] The current world average is 1.027 ± 0.018 [3] where the error on τ_B dominates.

By measuring the b-baryon semileptonic branching fraction, one can check if τ_{Λ_b} really is much smaller than τ_B, independent of any bias that the lifetime measurement could have introduced. This measurement can also sort out if the presence of b-baryons is causing the difference between the $BR_{sl}^{\Upsilon(4S)}$ and $BR_{sl}^{Z^0}$ measurements.

New methods that measure $BR(b \to \ell X)$ with higher precision and less model-dependence can help resolve the discrepancy between experimental measurements. In particular, using lifetime b-tagging techniques and devising methods aimed at extracting $BR(b \to \ell X)$ only, eliminates the correlation with the R_b measurements (as opposed to previous methods [7]) and provides more reliable results.

2 Semileptonic branching fractions for Λ_b

Measuring the inclusive semileptonic branching fraction $BR(\Lambda_b \to \ell X)$ is a very difficult task since it is nearly impossible to get a pure, inclusive sample of Λ_b [b] at the denominator. Instead, two groups at LEP have measured closely related quantities, namely $R_{\Lambda\ell} = BR(\Lambda_b \to \Lambda\ell X)/BR(\Lambda_b \to pX)$ at OPAL [5] and $R_{p\ell} = BR(\Lambda_b \to p\ell\bar{\nu}X)/BR(\Lambda_b \to \Lambda X)$ at ALEPH. [6] In both cases, ℓ refers to electrons and muons combined. Since the decay $(\Lambda_b \to \Lambda_c X)$ is expected to dominate both for inclusive and semileptonic decays, then $R_{\Lambda\ell} \sim R_{p\ell} \sim BR(\Lambda_b \to \ell X)$. Both analyses use the baryon-lepton charge correlation to select the events needed at the numerator.

---

[a] This notation is used throughout the text.

[b] Λ_b denotes all weakly decaying b baryons found near the Z^0 resonance.

2.1 The $R_{\Lambda\ell}$ measurement

For the denominator of $R_{\Lambda\ell}$,[5] ($\Lambda_b \to \Lambda X$) events are found by seeking Λ-baryon pairs in b-tagged events. The "direct" Λ is assumed to come from the ($\Lambda_b \to \Lambda_c \to \Lambda$) decay chain, whereas the "companion" baryon (either $\bar{p}$ or $\bar{\Lambda}$) is produced to locally conserved baryon number[c]. This sample contains contributions from different sources: the signal events ($\Lambda_b \to \Lambda\bar{p} X$), (B $\to \Lambda\bar{p} X$) and other backgrounds, mostly Λ-baryon pairs produced during the fragmentation process in b$\bar{b}$ events. The fraction of selected events coming from one of these three sources is obtained by fitting to the Λ-baryon pair momentum spectra. No assumptions need to be made about branching fractions or the relative abundance of a particular hadron. The results of the fit to the shapes of the momentum spectra are shown separately for the direct Λ and the companion baryon in figure 1.

To evaluate the numerator, ($\Lambda_b \to \Lambda\ell X$) events are selected by making use of the $\Lambda\ell$ correlations. Semileptonic Λ_b decay produce ($\Lambda\ell^-$) pairs whereas leptons coming from semileptonic B decays can be paired with a Λ or $\bar{\Lambda}$ from fragmentation. Therefore, the number of ($\Lambda_b \to \Lambda\ell X$) is estimated by subtracting wrong-sign ($\Lambda\ell^+$) pairs from the right-sign ($\Lambda\ell^-$) sample. Small corrections are made using the Monte Carlo predictions to account for imbalances in the wrong-sign background due to contributions from B mesons. The final result is:
$$R_{\Lambda\ell} = \frac{\mathrm{BR}(\Lambda_b \to \Lambda\ell X)}{\mathrm{BR}(\Lambda_b \to \Lambda X)} = (7.0 \pm 1.2(stat) \pm 0.7(syst))\%$$

2.2 The $R_{p\ell}$ measurement

In this analysis,[6] the number of Λ_b decaying into protons is estimated in a few steps. First, events are classified according to their primary quark flavour (b$\bar{b}$, c$\bar{c}$ and uds) by assigning a weight to each hemisphere for each different flavour. This is achieved using a lifetime b-tagging technique. Next, the information on the deposited energy per unit length, dE/dx, is used to set a probability for each track of being a proton, kaon or pion. By performing a fit to the data using these various probability weights, the momentum spectrum of protons, kaons and pions is obtained for b$\bar{b}$, c$\bar{c}$ and uds events. Furthermore, additional information from the impact parameter and the cosine of the angle between the track and the thrust axis allows tracks coming from b hadron decays to be distinguished from those produced in the fragmentation process within the b$\bar{b}$ sample. The momentum spectrum of protons coming from direct b hadron decays in b$\bar{b}$ events is shown in figure 2.

Finally, integrating the total spectrum from figure 2 gives N_p, the total number of protons produced in b hadron decays. The shape of the momentum spectrum

[c]Charge-conjugation is implied throughout the text

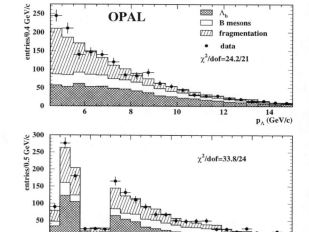

Figure 1: Results of the fit where the contributions from the three types of Monte Carlo events are compared with data for the direct Λ and companion baryon momentum spectra. The gap in the companion momentum spectrum corresponds to the region where the dE/dx information is too ambiguous to be used to identify protons.

as predicted by the Monte Carlo is used to extrapolate the spectrum below the minimum momentum cut of 4 GeV/c, after normalisation to the data. N_p is given by

$$N_p = (f_{\Lambda_b}\mathrm{BR}(\Lambda_b \to pX) + (1 - f_{\Lambda_b})\mathrm{BR}(B \to pX))N_b \tag{2}$$

where f_{Λ_b} is the fraction of b quarks forming Λ_b's at hadronisation. N_b is the number of b hadron decays given by $2R_b \cdot N_{\mathrm{mh}}$, with N_{mh} the number of all multihadronic events initially selected. The value for $\mathrm{BR}(B \to pX)$ is taken from CLEO. Some assumptions are made to extract f_{Λ_b}: f_{B_s}, the fraction of b quarks forming B_s[3]; $\mathrm{BR}(B_s \to pX) = (8.0 \pm 4.0)\%$ and $\mathrm{BR}(\Lambda_b \to pX) = (58 \pm 6)\%$. This gives

$$f_{\Lambda_b} = (10.2 \pm 0.7 \pm 2.2 \pm 1.6 \ (\mathrm{syst. \ BR}))\% \tag{3}$$

in close agreement with $(10.2^{+3.9}_{-3.1})\%$, the current best estimate from the Particle Data Group.[3]

The numerator is evaluated by selecting proton-lepton pairs having the proper charge correlation. A high momentum lepton is required to enhance the b-purity of selected pairs and reduce contributions from backgrounds such as protons coming from the fragmentation process. The number of protons coming from Λ_b semileptonic decays is obtained from integrating the momentum spec-

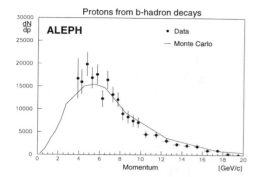

Figure 2: Momentum spectrum for protons from b hadron decays for data and Monte Carlo. The Monte Carlo spectrum is normalised to the data.

Table 1: Combined result for $\mathrm{BR}(\Lambda_b \to \ell X)$

OPAL	$\frac{\mathrm{BR}(\Lambda_b \to \Lambda \ell X)}{\mathrm{BR}(\Lambda_b \to \Lambda X)} = (7.0 \pm 1.2 \pm 0.7)\%$
ALEPH	$\frac{\mathrm{BR}(\Lambda_b \to p\ell X)}{\mathrm{BR}(\Lambda_b \to pX)} = (8.0 \pm 1.2 \pm 1.4)\%$
combined	$\mathrm{BR}(\Lambda_b \to \ell X) = (7.4 \pm 1.1)\%$

trum of the selected protons:

$$f_{\Lambda_b}\mathrm{BR}(\Lambda_b \to p\ell X) = (4.72 \pm 0.66 \pm 0.44) \; 10^{-3} \quad (4)$$

By rearranging eq. 2, one gets an expression for $f_{\Lambda_b}\mathrm{BR}(\Lambda_b \to pX)$ in terms of f_{Λ_b} and other branching fractions. This is solved by inserting the value of f_{Λ_b} derived in eq. 3. Finally, taking the ratio with eq. 4 gives $R_{p\ell} = \frac{\mathrm{BR}(\Lambda_b \to p\ell X)}{\mathrm{BR}(\Lambda_b \to pX)} = (8.0 \pm 1.2 \; (stat) \pm 1.4 \; (syst))\%$. The systematic errors take into account the dependency of f_{Λ_b} on $\mathrm{BR}(\Lambda_b \to pX)$.

2.3 Summary for semileptonic Λ_b decays

Combining the two results from table 1 gives $\mathrm{BR}(\Lambda_b \to \ell X) = (7.4 \pm 1.1)\%$, which is clearly less than for B mesons with $\mathrm{BR}_{sl}^B = (10.45 \pm 0.21)\%$ ($\Upsilon(4S)$). This supports the evidence of a much shorter Λ_b lifetime.

By definition, $\mathrm{BR}_{sl} = \tau \Gamma_{sl}$. One can write such an expression for Λ_b, B mesons and the average b hadron. The ratio of the semileptonic width of two different b hadron species will be the same as the ratio of their lifetimes if all b hadron species have the same semileptonic width, which is indeed the case:

$$\frac{\mathrm{BR}_{sl}^{\Lambda_b}}{\mathrm{BR}_{sl}^B} = 0.71 \pm 0.11 \text{ and } \frac{\tau_{\Lambda_b}}{\tau_B} = 0.75 \pm 0.05$$

$$\frac{\mathrm{BR}_{sl}^{\Lambda_b}}{\mathrm{BR}_{sl}^b} = 0.68 \pm 0.10 \text{ and } \frac{\tau_{\Lambda_b}}{\tau_b} = 0.77 \pm 0.05.$$

3 Measuring BR(b → ℓX) at LEP

The main difficulty in this measurement is to distinguish leptons coming from "direct" decays, (b → ℓ), from those coming from "cascade" decays, (b → c → ℓ). One important difference between these two decays is the amount of transverse momentum (measured with respect to the jet axis) carried by the lepton. Leptons from b hadron decays receive on average more p_T than those coming from cascade decays, due to the larger b quark mass.

Unfortunately, the precise modelling of the lepton momentum is not known and one must rely on different approximations to estimate the exact spectrum shape. The LEP experiments have agreed to use the ACCMM model for central value and the IGSW and IGSW** models [8] to simulate a $\pm 1\sigma$ variation. This problem particularly affects the LEP measurements since one must apply a minimum momentum cut on the leptons to reject fake leptons. To extract a value for BR(b → ℓX), one must extrapolate the lepton momentum spectrum below the minimum momentum selection cut where the exact momentum shape is not known, introducing large model dependences. This uncertainty is the largest contribution to the systematic error.

Prior to the introduction of lifetime b-tagging techniques, high-p_T leptons were used to enrich the b$\bar{\text{b}}$ content of the sample. This extra momentum cut further increases the model-dependence. Several attempts have been made to design new analysis techniques that would rely less on the lepton modelling. The most recent ones are reviewed here. [9,10,11] All are still preliminary.

These analyses share common features: each event is divided into two hemispheres using the thrust axis. An enriched b$\bar{\text{b}}$ sample is selected using a lifetime b-tagging technique. The b purity, typically greater than 90%, is extracted from the data using events with one or both hemispheres b-tagged. The c$\bar{\text{c}}$ and uds efficiencies are taken from the Monte Carlo, and the experimental values are used for R_b and R_c. Each hemisphere opposite a b-tagged hemisphere is searched for prompt leptons. The selected lepton samples contain mostly direct decays (b → ℓ), and cascade decays (b → c → ℓ). There are also small contributions from primary charm and uds events, (b → J/Ψ → $\ell^+\ell^-$), (b → τ → ℓ), fake leptons, and (b → $\bar{\text{c}}$ → ℓ). This last category denotes events where the b quark emits a W^- which decays into a $\bar{\text{c}}s$ pair. The $\bar{\text{c}}$ then decays semileptonically, giving a ℓ^- instead of a ℓ^+ from cascade decays. Several different approaches are used to disentangle the specific contributions from these various sources. These are detailed below for each individual analysis.

3.1 Updated result from DELPHI

Electrons and muons are selected with above 3 GeV/c from all data collected in 1994 − 1995. A fit to the shape of the combined momentum $p_c = \sqrt{p_T^2 + p^2/100}$ is performed, comparing data to the Monte Carlo. Different momentum spectrum shapes are used from six types of events: $(b \to \ell)$, $(b \to c \to \ell)$, $(b \to \bar{c} \to \ell)$, fakes, $(c \to \ell)$, and $(b \to J/\Psi \to \ell^+\ell^-)$.[9] Additional constraints are obtained from b-tagged events containing a lepton in each hemisphere, using the charge correlation between the two leptons. The fit has four free fit parameters: the fractions $f(b \to \ell)$ and $f(b \to c \to \ell)$, as well as ϵ_b and $\bar{\chi}_b$, where ϵ_b is the Petersen fragmentation parameter and $\bar{\chi}_b$ is the effective mixing parameter for b hadrons. All other contributions are kept fixed. A maximum likelihood fit to the single lepton and dilepton samples yields the fractions of events from $(b \to \ell)$ and $(b \to c \to \ell)$. The results of the fit can be seen in figure 3 where the contributions from the different Monte Carlo samples are superimposed and compared to the data.

The main source of dileptons with opposite charge in figure 3(b) is from two direct b decays: $(b \to \ell^-)$ in one hemisphere and $(\bar{b} \to \ell^+)$ in the other. In the presence of $B^0 - \bar{B}^0$ mixing, a fraction of these dileptons from direct decays have the same charge, as seen in figure 3(c). This fraction is given by $2\bar{\chi}_b(1 - \bar{\chi}_b)$. Same charge dileptons mostly come from events with one direct b decay and one cascade decay but misidentified leptons and mixing introduce other combinations as well.

The results of the fit are:

BR(b → ℓX) $(10.65 \pm 0.11 \pm 0.23^{+0.43}_{-0.27})\%$
BR(b → c → ℓX) $(7.91 \pm 0.23 \pm 0.40^{+0.52}_{-0.65})\%$
$\bar{\chi}_b$ $(0.128 \pm 0.013 \pm 0.003^{+0.007}_{-0.006})\%$
$< x_E >$ 0.712 ± 0.003 (stat),

where $< x_E >$ is the mean fractional energy carried by b hadrons, determined from ϵ_b. By leaving ϵ_b as a free fit parameter, the error from the fragmentation model is greatly reduced. The last error represents all modelling errors, and is mostly due to the uncertainty in the shape of the lepton momentum spectra.

3.2 New result from OPAL

Jet shape and muon[d] momentum variables in b-tagged events are used as input variables to a neural network trained to select prompt muons coming from direct b decays.[10] By combining the information from several variables, more discrimination power is obtained than when using momentum information alone. Five kinematic variables are used: 1) the muon candidate momentum, 2) the muon candidate transverse momentum calculated w.r.t.

---

[d]Only muons are analysed here. Electrons will be included for the final publication.

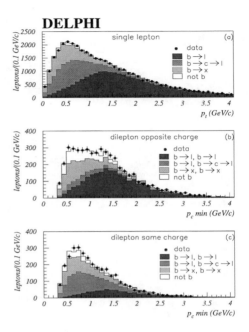

Figure 3: Comparison of the momentum spectra between data and Monte Carlo (after renormalisation to the data). (a) Transverse momentum distribution for single electrons and muons. Combined momentum distribution for dilepton in jets from opposite hemispheres, with (b) opposite charge and (c) same charge. In (b) and (c), p_c min refers to the minimum combined momentum for the two leptons.

the jet, 3) the energy of the jet containing the muon candidate, 4) the energy of the sub-jet containing the muon, and 5) the scalar sum of transverse momenta of charged tracks for the muon jet.

Muons are selected above 3 GeV/c to reduce contributions from fake muons. This analysis uses data with centre-of-mass energies within 3 GeV of the Z^0 peak collected during the 1992 − 1995 running period when the silicon microvertex detector was fully operational. Monte Carlo events having similar distributions for the neural network output variable are grouped together and three sub-samples are formed: $(b \to \mu)$ and $(b \to J/\Psi \to \mu^+\mu^-)$, $(b \to c \to \mu)$ and $(b \to \bar{c} \to \mu)$, and all other backgrounds. The data is compared to these three Monte Carlo shapes to extract the sample composition. A χ^2 fit to the neural network output variable is performed with two free parameters: $f([b, J/\Psi] \to \mu)$, the fraction of events from the first Monte Carlo sub-

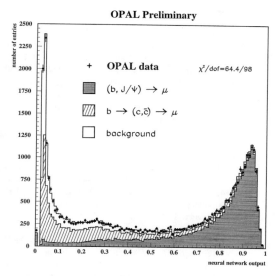

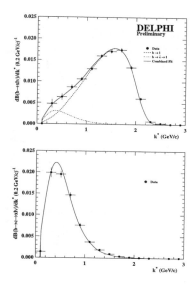

Figure 4: Results of the fit after applying small systematic corrections. The χ^2 is calculated including statistical errors only as an indicator of the goodness-of-fit.

Figure 5: The lepton k^* spectrum for the b enriched sample. Contributions to the $\ell-$b right-sign sample are from $(b \to \ell)$ and $(b \to \bar{c} \to \ell)$ events in the top plot and from $(b \to c \to \ell)$ events in the bottom plot (wrong-sign sample).

sample, and $f(b \to [c, \bar{c}] \to \mu)$, the fraction of events from the second sub-sample. The result is shown in figure 4.

After removing a small contribution from $(b \to J/\Psi \to \ell^+\ell^-)$ and making corrections to account for sources affecting the shape of the neural network output variable (b species content, charm content, fake muon rate, contribution from $(b \to \tau \to \mu)$, etc.), the result is:

BR$(b \to \mu X) = (10.86 \pm 0.08 \pm 0.22^{+0.46}_{-0.31} \text{ (model)})\%$

The modelling error, which is shown separately, includes contributions from the lepton spectrum modelling uncertainty as well as from the fragmentation model.

3.3 New approach from DELPHI

A new method [11] is being developed by DELPHI but is not yet included in the LEP average since sources of systematic uncertainties are still being investigated. This method takes full advantage of the separation power offered by combining several neural networks. After b-tagging the event, all tracks are classified as originating from a b hadron decay or from the fragmentation process using a neural network algorithm. Tracks associated with the b decay vertex are used to evaluate the b hadron vertex charge, $Q_B = \sum Q_i p_i$. This quantity then becomes one major input to a second neural network trained to separate hemispheres containing b and $\bar{b}$ jets. Cutting on either end of this neural network yields samples enriched

(depleted) in b and $\bar{b}$ jets.

Further separation between charged and neutral B mesons is achieved with yet one more neural network which uses the vertex charge Q_B and the output of the previous network as main inputs. A cut is applied on this last neural network output at either end of the distribution to select a B$^+$ (B$^-$) enriched (depleted) sample. For these three neural networks, additional inputs are the jet charge, the b-tagging probability, and the charge of kaons identified as B decay products, among other variables.

Leptons are selected if they have a high probability of coming from a b hadron decay, as determined by the first neural network mentioned above. These leptons are boosted using a Lorentz transformation to obtain k^*, their momentum in the b rest frame, using the momentum associated with the b decay vertex. The k^* resolution is not affected by the missing neutrino. The lepton charge and the b jet (B meson) charge are used to form the right and wrong-sign samples. A fit to the lepton momentum k^* allows to separate the contributions from $(b \to \ell)$ and $(b \to \bar{c} \to \ell)$, as is shown in the top part of figure 5. The contribution from $(b \to c \to \ell)$ appears in the wrong-sign sample and is shown in the bottom part of figure 5.

The results of the various fits to the $b/\bar{b}$ jet samples and B$^+$/B$^-$ samples are given in table 2. Results are

Table 2: Results for BR(b → ℓX) and BR(B$^+$ → ℓX)

BR(b → ℓX)	$(10.32 \pm 0.12 \pm 0.24^{-0.21}_{+0.31}\%)$
BR(b → c → ℓX)	$(0.94 \pm 0.16 \pm 0.27^{-0.50}_{+0.33}\%)$
BR(b → c̄ → ℓX)	$(7.43 \pm 0.16 \pm 0.28^{-0.24}_{+0.18}\%)$
BR(B$^+$ → ℓX)	$(9.89 \pm 0.28 \pm 0.24^{-0.21}_{+0.31}\%)$
BR(B$^+$ → c → ℓX)	$(1.33 \pm 0.35 \pm 0.27^{-0.50}_{+0.33}\%)$
BR(B$^+$ → c̄ → ℓX)	$(5.98 \pm 0.31 \pm 0.28^{-0.24}_{+0.18}\%)$

Table 3: Results for BR(b → ℓX) currently used in the electroweak global fit by the LEP Working Group for the average presented at this conference. The first error is from statistics, the second from systematics and the third one from modelling sources.

ALEPH [12]	$(11.01 \pm 0.23 \pm 0.28 \pm 0.11)\%$
L3 [13]	$(10.85 \pm 0.12 \pm 0.47)\%$
OPAL [10]	$(10.86 \pm 0.08 \pm 0.22^{+0.46}_{-0.31}\%$
DELPHI [9]	$(10.65 \pm 0.11 \pm 0.23^{+0.43}_{-0.27})\%$
LEP EW global fit	$(10.89 \pm 0.24)\%$

from data collected in 1994 − 1995.

3.4 Summary for BR(b → ℓX)

Prior to the introduction of the new techniques presented in section 3, most measurements of BR(b → ℓX) were made by fitting simultaneously for R_b and BR(b → ℓX) after selecting a sample of prompt leptons to enhance the bb̄ content. [7] These techniques had the distinct disadvantage of introducing strong correlations (−0.95) between these two quantities. Consequently, the LEP Electroweak Working Group has decided to no longer include these older measurements in the global electroweak fit. The fit is now performed using only dedicated analyses to evaluate R_b and BR(b → ℓX) separately, each one using lifetime b-tagging techniques.

The new LEP average is BR(b → ℓX) = (10.89 ± 0.24)%. The contribution to the systematic error due to the modelling of the lepton spectrum is ±0.13. Note that in the global electroweak fit, there is an additional constraint on the BR(b → ℓX) error imposed by the asymmetry results. When this constraint is removed, the error from modelling increases from ±0.13 to ±0.18, for a total error of ±0.26 with essentially no effect on the central value. Further studies of the sensitivity of BR(b → ℓX) to the averaging procedure is underway to improve the reliability of the LEP combined branching ratio results. All results used in the fit are listed in table 3. Further developments in these new techniques should soon close the gap between the $BR_{sl}^{Z^0}$ and $BR_{sl}^{\Upsilon(4S)}$ measurements.

Acknowledgements

Work supported by a grant from the Natural Sciences and Engineer Research Council of Canada. I would like to thank Ursula Becker from Aleph, and Marta Calvi and Jong Yi from Delphi for providing the material reviewed in this paper. Thanks also to Chris Jones for his essential contribution to the OPAL analysis. Special thanks to Hans Mes and Peter Igo-Kemenes who made it possible for me to participate in this conference.

References

1. E. Bagan, P. Ball, V.M. Braun, and P. Gosdzinsky, *Nucl. Phys.* B **432**, 3 (1994); *Phys. Lett.* B **342**, 362 (1995); E. Bagan, P. Ball, B. Fiol, and P. Gosdzinsky, *Phys. Lett.* B **351**, 546 (1995);

2. M. Neubert, C.T. Sachrajda, *Nucl. Phys.* B **483**, 339 (1997)

3. Review of Particle Physics, C. Caso *et al.* (Particle Data Group), *Eur. Phys. J.* C **3**, 1 (1998)

4. The LEP Collab. (ALEPH, DELPHI, L3 and OPAL), the LEP Electroweak Working Group, the SLD Heavy Flavour Group, CERN-PPE/97-154 (1997); March 1998 update for the XXXIIInd Rencontres de Moriond, Les Arcs, France.

5. R. Akers *et al.* (OPAL Collab) *Z. Phys.* C **74**, 423 (1997) ICHEP'98 contributed paper #375

6. R.Barate *et al.* (ALEPH Collab.) *Eur. Phys. J.* C **5**, 205 (1998) ICHEP'98 contributed paper #971

7. D. Buskulic *et al.* (ALEPH Collab.) PLB **384**, 414 (1996); P. Abreu *et al.* (DELPHI Collab.) *Z. Phys.* C **66**, 323 (1995); M. Acciari *et al.* (L3 Collab.) *Phys. Lett.* B **335**, 542 (1994); R. Akers *et al.* (OPAL Collab.) *Z. Phys.* C **60**, 199 (1993)

8. The LEP Collab. (ALEPH, DELPHI, L3 and OPAL) *Nucl. Instrum. Methods* **378**, 101 (1996); CERN-PPE/96-183 (1997).

9. (DELPHI Collab) DELPHI 98-122 CONF 183, 22 June 1998, ICHEP'98 contributed paper #129

10. (OPAL Collab) OPAL Physics Note 334, 10 March 1998. ICHEP'98 contributed paper #370

11. (DELPHI Collab) DELPHI 98-135 CONF 196, 22 June 1998, ICHEP'98 contributed paper #233

12. (ALEPH Collab) Paper contributed to the International Europhysics Conference on High Energy Physics, Brussels, Belgium 27 July-2 August 1995 EPS95 Ref. eps0404

13. M. Acciarri *et al.* (L3 Collab) *Z. Phys.* C **71**, 379 (1996)

Selected Results on $b \to c\ell\nu$ from CLEO

Ronald A. Poling

School of Physics and Astronomy, University of Minnesota, Minneapolis, MN 55116, USA
E-mail: poling@umn.edu

Recent results on semileptonic B-meson decays from the CLEO experiment are reported. Data on exclusive decay processes continue to improve, with a new study of $B \to D\ell\nu$ and the first results from updated measurements of $B \to D^*\ell\nu$. Inclusive measurements have entered a new phase, with the focus now being tests of the theory and determination of theoretical parameters used in the extraction of the CKM elements.

1 Introduction and Theoretical Background

Semileptonic processes continue to be one of the most powerful probes of heavy-flavor physics. In recent years studies of semileptonic B decays have focused increasingly on the determination of the CKM parameters V_{cb} and V_{ub}. Initially restricted to inclusive analyses, these measurements have evolved to concentrate on exclusive decays. The reasons for this have been twofold: growing data samples and improved experimental techniques have rendered the exclusive modes accessible, and the development and application of Heavy Quark Effective Theory (HQET) provide what many (not all) theorists believe is a superior route to the CKM couplings through exclusive measurements.

While the CKM measurements have become the principal objective, studies of semileptonic B decays provide excellent opportunities to learn about hadronic physics as well. In fact it is inescapable. To probe the dominant semileptonic decay $b \to c\ell\nu$ at the quark level, for example, we must disentangle that process from the obscuring (but interesting) effects of the strong interaction at the meson level. From the beginning of heavy-flavor studies this disentangling has been facilitated by theoretical *models*, beginning with spectator calculations,[1] and continuing with potential models.[2] In recent years we have begun to move beyond these phenomenological models to the application of QCD *theory*, with the development of HQET and the computational tools of the Operator Product Expansion (OPE).[3] The result is the ability to express a number of the observables of semileptonic B decays in a rigorous and systematic expansion in powers of α_s and Λ_{QCD}/m_Q, where m_Q is the heavy-meson mass.[4,5] The power of the procedure arises from the controlled corrections (none at order $1/m_Q$) and quantitatively estimable errors.

The parameters of the expansion are as follows:

$$\lambda_1 = -\langle B|\bar{b}_v(i\vec{D})^2 b_v|B\rangle, \text{ and} \qquad (1)$$

$$\lambda_2 = \frac{1}{3}\langle B|\bar{b}_v\sigma_{\mu\nu}G^{\mu\nu}b_v|B\rangle, \qquad (2)$$

where b_v is the heavy quark field and $G^{\mu\nu}$ is the chromomagnetic field operator. Intuitively, λ_1 and λ_2 can be thought of as the negative of the average momentum-squared of the b quark and the energy of the hyperfine interaction of the spin of the b with the light degrees of freedom, respectively. Introduction of a third parameter, $\bar{\Lambda}$, the energy of the light degrees of freedom, makes it possible to eliminate the quark mass m_b in favor of the meson masses m_B and m_B^*:

$$m_B = m_b + \bar{\Lambda} - \frac{\lambda_1}{2m_b} - d_M\frac{\lambda_2}{2m_b}, \qquad (3)$$

where $d_P = 3$ and $d_V = -1$ are substituted for the pseudoscalar and vector meson states. The mass difference between the B and B^* mesons thus leads to $\lambda_2 \simeq 0.2$ GeV^2.

Determination of the theoretical parameters λ_1 and $\bar{\Lambda}$ requires detailed measurements of heavy-meson decays. In the absence of such data, their values have been guessed. This is tough, as is reflected by the range of values that have been used in the past: $\lambda_1 \simeq -0.1$ GeV^2 (spectator model), $\lambda_1 \simeq -0.6$ GeV^2 (QCD sum rules) and $\lambda_1 \geq -0.36$ GeV^2 (quantum mechanics), to give a few examples.

There is a clear challenge here for the experimenters. HQET/OPE allows the determination of $|V_{cb}|$ from inclusive semileptonic B decay, given the values of λ_1 and $\bar{\Lambda}$. It allows the determination of $|V_{cb}|$ from exclusive modes like $B \to D^*\ell\nu$, if we have λ_1. Both determinations depend on the validity of the theoretical approach. A program of testing the theory and measuring $\bar{\Lambda}$ and λ_1 is a *high* priority.

2 CLEO Measurements of Exclusive Semileptonic B Decays

Since it was brought into operation in 1989, the CLEO II detector has been used to make a series of increasingly precise measurements of semileptonic B decays. Highlights have included a study of $B \to D^*\ell\nu$,[6] which produced a measurement of $|V_{cb}| = (39.5\pm3.6)\times10^{-3}$, where

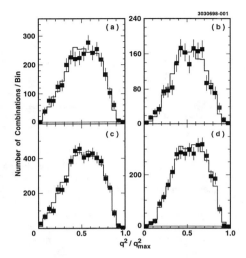

Figure 1: Background-subtracted $\tilde{q}^2/q^2_{max}$ distributions for data (points with error bars) and the fit results for four different subsets of the CLEO II data sample: electrons (a and c) and muons (b and d), each for two generations of tracking code. Note that $\tilde{q}^2$ is the estimated q^2 determined from the soft-pion and lepton four-momenta.

the combined statistical and systematic uncertainty is ±9%. This effort continues, and an overall precision of ±5% is expected to be achieved with data collected by the end of CLEO II in early 1999. This determination is based on fitting the measured distribution in $d\Gamma/dq^2$ within the framework of HQET. To refine and test the validity of this approach, detailed measurements of $B \to D^*\ell\nu$, in particular of the ratios of the form factors, have been made.[7] These have generally supported HQET, although there is much room for improvement in the precision.

With the continuing growth of the CLEO II data sample, and the development of improved analysis procedures, efforts on $B \to D^*\ell\nu$ have been renewed. A first installment of this work has been presented to this conference, in the form of a new measurement of the q^2 distribution in $\bar{B}^0 = D^{*+}\ell^-\bar{\nu}$ based on the technique of partial reconstruction.[8] This measurement is complementary to and has higher efficiency than a full-reconstruction analysis. A total $B\bar{B}$ sample of 3.3 million events yields $\sim 25,000$ candidates. The resulting distributions are shown in Fig. 1, and the result of a fit constraining the form-factor ratios to CLEO's measurements gives

$$\hat{\rho}^2 = 0.54 \pm 0.05 \pm 0.10 \pm 0.02.$$

While this result is consistent with the world average, it is just barely so. With the completion of the updated analyses it is to be hoped that the picture will become

more clear.

Although attention was initially entirely on $B \to D^*\ell\nu$, more recently there has also been work on $B \to D\ell\nu$.[9] The CKM measurement, $|V_{cb}| = (38.5 \pm 5.3) \times 10^{-3}$, is less precise than for the $B \to D^*\ell\nu$ analysis, but this mode represents an important piece of the puzzle and the consistency is encouraging.

CLEO has presented a new preliminary measurement of the exclusive semileptonic B decay to $D\ell\nu$.[10] This is an independent analysis of the same data sample as was used for the published measurement (3.3 million $B\bar{B}$ events). In contrast with that analysis (full reconstruction with "neutrino detection"), this analysis has higher efficiency, different systematics, and better overall precision.

The approach is to select $D-\ell$ pairs from the decay of the same B meson, a sample that includes the processes $\bar{B} \to D\ell\bar{\nu}$, $\bar{B} \to D^*\ell\bar{\nu}$, $\bar{B} \to D^{**}\ell\bar{\nu}$, and $\bar{B} \to D^{(*)}\pi\ell\bar{\nu}$. Both charged and neutral $\bar{B}$'s are measured. The components are separated with the kinematic variable:

$$cos\theta_{B-D\ell} = \frac{E_B E_{D\ell} - m_B^2 - m_{D\ell}^2}{2|\vec{p_B}||\vec{p_{D\ell}}|} \quad (4)$$

This quantity has physical values (-1 to 1) for correctly measured $D\ell\bar{\nu}$ signal, and is pushed to smaller values when more than the neutrino is missing. The resulting distributions include backgrounds from several sources: random $K\pi(\pi)$ combinations, $D - \ell$ combinations from different B's, non-signal $D - \ell$ descending from the same B, fake leptons, and continuum. The total background correction is approximately 50% for $D^0\ell$ combinations and approximately 75% for $D^+\ell$. The resulting samples consist of $\sim 3.2K$ $D^0\ell$ and $\sim 2.3K$ $D^+\ell$, and are shown in Fig. 2. These distributions are fitted to determine the contribution from each mode. The parameterizations in the fits are from the ISGW2 model[2] for $\bar{B} \to D\ell\bar{\nu}$ and $\bar{B} \to D^{**}\ell\bar{\nu}$, from calculations using CLEO's measured form factors for $\bar{B} \to D^*\ell\bar{\nu}$, and from the model of Goity and Roberts[11] for the nonresonant decays.

To extract branching fraction, form-factor and CKM information, the differential decay width is measured by fitting in eight bins of w, the standard kinematic variable of HQET,

$$w = \frac{(m_B^2 + m_D^2 - q^2)}{2m_B m_D}. \quad (5)$$

The differential decay width is given by

$$\frac{d\Gamma}{dw} = \frac{G_F^2 |V_{cb}|^2}{48\pi}(m_B + m_D)^2 m_D^3(w^2 - 1)^{3/2} F_D(w)^2, \quad (6)$$

where $F_D(w)$ is the form factor. There is a practical complication in this measurement. The actual measured quantity is not w, but rather is an approximation $\tilde{w}$, which is smeared by B motion and resolution effects. The

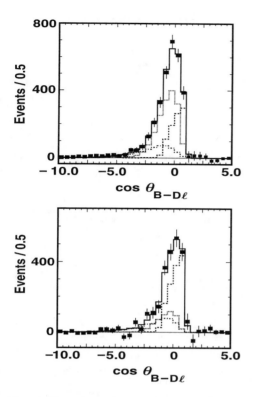

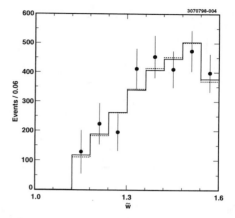

Figure 3: The sum of $B^- \to D^0\ell\bar\nu$ and $\bar B^0 \to D^+\ell\bar\nu$ yields as a function of $\widetilde{w}$, for the data (solid squares), and as computed using three form factors: linear expansion in w (dashed histogram), ISGW2 (dotted histogram) and following Boyd *et al.* (solid histogram).

$$\mathcal{B}(\bar B^0 \to D^0\ell\bar\nu) = (2.25 \pm 0.18 \pm 0.18 \pm 0.13)\% \qquad (8)$$

The uncertainties are statistical, systematic, and that associated with the input B lifetimes and D branching fractions. The values are consistent with previous measurements, and the precision is somewhat improved. A fit assuming a linear form factor gives a slope parameter $\rho_D^2 = 0.81 \pm 0.14 \pm 0.09$. The value of $|V_{cb}|$ is also determined from the $\widetilde{w}$ distribution, with additional theoretical uncertainty due to the value of the form factor at $w = 1$. Using $F_D(1) = 1 \pm 0.1$, we find $|V_{cb}| = 0.048 \pm 0.006 \pm 0.003 \pm 0.001 \pm 0.005$, where the first three errors are the same as for the branching fraction, and the fourth is due to the theoretical uncertainty in $F_D(1)$. This result is somewhat larger than the previous measurement from $B \to D\ell\nu$. It is consistent with studies of $B \to D^*\ell\nu$, although with somewhat inferior precision.

3 CLEO Measurements of Inclusive Semileptonic B Decays

For most of the history of B-decay studies the only measurements of semileptonic decays were inclusive measurements. By virtue of enormous statistical power, they are still of great interest. CLEO II data have been used to determine the B semileptonic branching fraction of $(10.5 \pm 0.5)\%$, and the shapes of the primary and secondary lepton spectra, which are shown in Fig.4.[14] More precise measurements of the momentum spectrum and other details of $B \to X_c\ell\nu$ will continue to be a rich

Figure 2: The $\cos\theta_{B-D\ell}$ distribution for $B^- \to D^0 X\ell\bar\nu$ (top) and $\bar B^0 \to D^+ X\ell\bar\nu$ (bottom). The data (solid squares) are overlaid with simulated $B \to D\ell\bar\nu$ decays (short-dashed histogram), $B \to D^*\ell\bar\nu$ decays (dotted histogram), $B \to D^{**}\ell\bar\nu + D^{(*)}\pi\ell\bar\nu$ decays (long dash-dotted histogram), and the total (solid histogram). Normalizations are provided by the fit.

distribution in $\widetilde{w}$ is shown in Fig. 3. The smearing of w into $\widetilde{w}$ is accounted for in the fitting procedure with a Monte Carlo-determined response matrix. The form factor is parameterized both in the usual way (linear, linear plus quadratic), and by alternative forms constrained by dispersion relations.[12,13] (For details on these see the contributed paper.[10]) The results of the fits are shown in Fig. 3.

There are several sources of systematic error that must be accounted for, including the B mass/momentum scale, the modeling of $D^{**}\ell\nu$ and $D^{(*)}\pi\ell\bar\nu$, and the $D^*\ell\nu$ form factors. The branching fractions are extracted by integrating $\frac{d\Gamma}{dw}$ over w and using B-lifetime data to be independent of the charged/neutral production fractions at the $\Upsilon(4S)$. The results are as follows:

$$\mathcal{B}(B^- \to D^0\ell\bar\nu) = (2.33 \pm 0.19 \pm 0.19 \pm 0.14)\% \qquad (7)$$

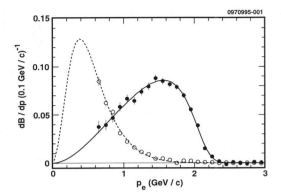

Figure 4: Primary $(B \to X_c e\nu)$ and secondary $(B \to D \to Xe\nu)$ electron spectra in $B\bar{B}$ events, from CLEO's lepton-tagged analysis.

source of information about the dynamics of B decay.

CLEO has presented a preliminary analysis of the moments of both the lepton-energy spectrum and the hadronic recoil mass-squared spectrum in $B \to X_c \ell\nu$.[15] The combination of HQET and the OPE provides quantitative predictions of these moments in terms of the theoretical parameters $\bar{\Lambda}$ and λ_1. By measuring the moments of both distributions, we have an opportunity to determine the values of the QCD parameters while making a consistency check of the theoretical framework. The ultimate goal is improved precision in determining $|V_{cb}|$.

CLEO has measured the first two moments of the recoil mass-squared distribution in semileptonic B decays without explicitly reconstructing the final state hadrons X_c. Like several of our recent measurements of semileptonic decays, this study is based on "detecting" the neutrino with careful determinations of missing momentum and energy. Events are chosen with leptons in the momentum range $1.5 - 2.4$ GeV/c. The missing mass is required to be consistent with a single neutrino, and events are restricted to those where this determination is reliable by demanding no extra leptons and a net charge of zero. Finally, events are required to be topologically consistent with $B\bar{B}$ to suppress continuum background. This investigation has been made with the full CLEO II data sample of 3.4 million $B\bar{B}$ events.

The objective is to determine the missing mass squared:

$$M_X^2 = (E_B - E_\ell - E_\nu)^2 - (\vec{p}_B - \vec{p}_\ell - \vec{p}_\nu)^2 \quad (9)$$

$$M_X^2 = M_B^2 + M_{\ell\nu}^2 - 2E_B E_{\ell\nu} + 2|\vec{p}_B||\vec{p}_{\ell\nu}|cos\theta_{\ell\nu,B} \quad (10)$$

Since we do not know the direction of the B momentum we must approximate by dropping the final term in

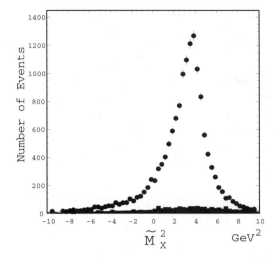

Figure 5: Measured $\widetilde{M_X^2}$ distributions, for on-resonance data (points) and scaled off-resonance data (shaded histogram).

Eq. 10:

$$\widetilde{M_X^2} = M_B^2 + M_{\ell\nu}^2 - 2E_B E_{\ell\nu} \quad (11)$$

Fig. 5 shows the distributions of $\widetilde{M_X^2}$ for $\Upsilon(4S)$ decays and continuum events. Because of the smearing caused by the neglect of the last term in Eq. 10 and the resolution of the missing momentum and energy determinations, the expected shape of this distribution does not reveal the detailed structure: sharp resonant peaks on a nonresonant background. It nevertheless should yield the overall shape of the spectrum, and allow a precise determination of the moments.

Two methods are used to determine the moments from the background-subtracted M_X^2 distribution. The distribution is fitted directly with Monte Carlo prediction of the contributions of expected $B \to X_c\ell\nu$ processes: $D\ell\nu$, $D^*\ell\nu$ and a mixture of charmed mesons of masses above that of D^* and nonresonant modes. The mixture of these that best fits the distribution is then used to determine the moments. The second procedure is to measure the moments of the raw distribution, and then to use Monte Carlo simulations to correct for experimental bias. In both cases the moments are computed about the square of the spin-averaged D/D^* mass, $\bar{M}_D^2 = (1.975~GeV)^2$.

Results from the two procedures are quite consistent and have comparable errors. For what follows we use the averages computed with equal weighting: $\langle M_X^2 - \bar{M}_D^2 \rangle = (0.286 \pm 0.023 \pm 0.080)$ GeV^2 and $\langle (M_X^2 - \bar{M}_D^2)^2 \rangle = (0.911 \pm 0.066 \pm 0.309)$ GeV^2.

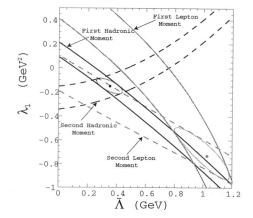

Figure 6: Bands in $\bar{\Lambda} - \lambda_1$ space defined by the measured first and second moments of the hadronic mass-squared and lepton energy. The intersections of the centerlines of the mass-squared and lepton-energy bands are shown as dots. The bands and ellipses represent one-sigma limits.

The HQET parameters are extracted from the moments by fitting to parameterizations due to Falk *et al.* and Gremm *et al.* [4] The resulting constraints on $\bar{\Lambda}$ and λ_1 are shown in Fig. 6. The best solution is given by $\bar{\Lambda} = (0.33 \pm 0.02 \pm 0.08)$ GeV and $\lambda_1 = -(0.13 \pm 0.01 \pm 0.06)$ GeV^2.

Voloshin [5] and Gremm *et al.* [16] have proposed the determination of $\bar{\Lambda}$ and λ_1 from the moments of the lepton-energy spectrum. Along with the analysis of the hadronic mass-squared moments, CLEO has presented preliminary measurements [15] of the lepton-energy moments, using the previously measured electron spectrum (Fig. 4). [14] This analysis is based on the first two moments, the mean lepton energy $\langle E_\ell \rangle$ and the spread about the mean $\langle (E_\ell - \langle E_\ell \rangle)^2 \rangle$.

By necessity the lepton-energy spectrum is measured in the lab frame, with a minimum momentum requirement of 0.6 GeV/c. Corrections must be applied for electromagnetic radiative effects, experimental resolution (including radiation in the material of the detector), the boost from the B rest frame into the lab frame, and charmless decays $b \to u\ell\nu$. In addition, theoretical models must be used to correct for the unmeasured part of the spectrum.

The procedure for computing the true moments of the underlying B-decay spectrum from the data is first to calculate raw moments from the measured momentum spectrum, and then to apply Monte Carlo-determined corrections for each of the effects listed above (radiation, boost, etc.). Each correction is evaluated by calculating the moments for a simulated sample, then turning off the specific effect under study, and taking the changes in the moment values to be the corrections. In the course of this study it was verified that the procedure yields the true moments (which were known, since this was Monte Carlo) to excellent precision.

The results are quite sensitive to the momentum dependence of the detection efficiency and to the details of several physics backgrounds. We consider leptons from charmed mesons D^0, $D^\pm$ and D_s produced in "upper-vertex" processes ($b \to c\bar{c}s$), leptons from charmonium mesons from B decay, leptons from τ's from $B \to X\tau\nu$, semileptonic decays of Λ_c's from B decays, π^0 and η Dalitz decays, γ conversions, fake leptons, and secondary $b \to c \to s\ell\nu$ leptons. There are also small uncertainties in the procedures for correcting the raw moments as described above. Evaluation of systematic errors is not yet complete, and several have been estimated quite conservatively for this preliminary result. Overall, the largest single systematic uncertainty is that due to upper-vertex charm.

The preliminary results for the measured moments are as follows:

$$\langle E_\ell \rangle = (1.36 \pm 0.01 \pm 0.02) \text{ GeV, and} \qquad (12)$$

$$\langle (E_\ell - \langle E_\ell \rangle)^2 \rangle = (0.19 \pm 0.004 \pm 0.005) \text{ GeV}^2, \quad (13)$$

where all systematic errors have been combined in quadrature.

The constraints on λ_1 and $\bar{\Lambda}$ that are extracted from these moment measurements, following Voloshin, are sumperimposed on the results from the hadronic mass-squared moments in Fig. 6. Note that there is no common intersection among the one-sigma bands for the four inputs, and the disagreement appears to be outside the errors. Possible explanations for the disagreement include experimental error, the need to go beyond order $1/m_Q^2$ in the expansions, or perhaps a more serious problem with the theory.

Given values for $\bar{\Lambda}$ and λ_1, we can determine $|V_{cb}|$ from measurements of the B lifetime and semileptonic decays. If we were to accept the values and precision of the parameters as measured from the hadronic moments, we would find a value of $|V_{cb}|$ consistent with current determinations, and with a theoretical uncertainty of $\sim 2\%$. Substituting the values favored by the lepon moments, on the other hand, would result in a 4.5% (10%) increase in the $|V_{cb}|$ value from the exclusive (inclusive) data. Using the parameter values from the hadronic moments and

Eq. 3 to determine the b-quark mass, we would obtain $m_b = 4.97 \pm 0.10$ GeV where this is the pole mass in $\overline{MS}$ at one loop. Using the results from the lepton moments, the value for m_b would be lower by 0.7 GeV. Clearly we will benefit from further improvements in the precisions of both moments, full analyses of systematic errors, and a definitive test of the consistency of the theory with both the lepton spectrum and the recoil mass-squared distribution.

4 Conclusion

CLEO has continued to make significant progress in the study of exclusive $B \to X_c \ell \nu$. While new results are consistent with previous measurements, the level of consistency is not as stunning as has been the case in the past, and the completion of a new comprehensive analysis of $B \to D^* \ell \nu$ is eagerly anticipated. New directions for studying inclusive semileptonic decays are currently being developed. Preliminary tests of HQET/OPE calculations, and the determination of the parameters of the theory, are not yet conclusive, but seem to suggest that a conflict is brewing.

These studies will continue to be pursued aggressively. CLEO has more than doubled its $B\bar{B}$ data sample compared to the analyses reported here, and running continues with CESR luminosity now above $\sim 6 \times 10^{32}/(cm^2 s)$. Beginning in 1999 the next phase of the project will commence, as CESR Phase III and CLEO III will begin the push toward B-factory luminosity with a significantly upgraded detector.

Acknowledgements

I would like to thank the organizers and convenors of ICHEP '98 for the opportunity to participate in a most enjoyable and informative conference. The CLEO research described in this report has been made possible by the outstanding efforts of the CESR accelerator staff, and continuing support from the U.S. Department of Energy and the National Science Foundation.

References

1. A. Ali and E. Pietarinen, *Nucl. Phys.* B **154**, 519 (1979); G. Altarelli *et al.*, *Nucl. Phys.* B **208**, 365 (1982).
2. N. Isgur *et al.*, *Phys. Rev.* D **39**, 799 (1989); D. Scora and N. Isgur, *Phys. Rev.* D **52**, 2783 (1995).
3. J. Chay, H. Georgi and B. Grinstein, *Phys. Lett.* B **247**, 399 (1990); I. I. Bigi, N. G. Uraltsev and A. I. Vainshtein, *Phys. Lett.* B **293**, 430 (1992); I. I. Bigi, M. A. Shifman, N. G. Uraltsev, and A. I. Vainshtein, *Phys. Rev. Lett.* **71**, 496 (1993); A. V. Manohar and M. B. Wise, *Phys. Rev.* D **49**, 1310 (1994); A. Falk, M. Luke and M. Savage, *Phys. Rev.* D **49**, 3367 (1994); A. Falk, hep-ph/9610363 (15 October 1996).
4. A. Falk, M. Luke and M. Savage, *Phys. Rev.* D **53**, 2491 (1996); A. Falk, M. Luke and M. Savage, *Phys. Rev.* D **53**, 6316 (1996); M. Gremm and A. Kapustin, hep-ph/9603448 (10 March 1997).
5. M. Voloshin *et al.*, *Phys. Rev.* D **51**, 4934 (1995).
6. B. Barish *et al.* (CLEO), *Phys. Rev.* D **51**, 1014 (1995).
7. J. E. Duboscq *et al.* (CLEO), *Phys. Rev. Lett.* **76**, 3898 (1996).
8. M. S. Alam *et al.* (CLEO), contributed paper ICHEP 98-853, also available as Cornell LNS preprint CLEO CONF 98-13.
9. M. Athanas *et al.* (CLEO), *Phys. Rev. Lett.* **79**, 2208 (1997).
10. M. Artuso *et al.* (CLEO), contributed paper ICHEP 98-856, also available as Cornell LNS preprint CLEO CONF 98-12.
11. J. L. Goity and W. Roberts, *Phys. Rev.* D **51**, 3459 (1995).
12. C. G. Boyd, B. Grinstein and R. F. Lebed, *Phys. Rev.* D **56**, 6895 (1997).
13. I. Caprini, L. Lellouch and M. Neubert, CERN-TH/97-91, hep-ph/9712417.
14. B. Barish *et al.* (CLEO), *Phys. Rev. Lett.* **76**, 1570 (1996).
15. J. Bartelt *et al.* (CLEO), cntributed paper ICHEP 98-1013, also available as Cornell LNS preprint CLEO CONF 98-12.
16. M. Gremm *et al.*, *Phys. Rev. Lett.* **77**, 20 (1996); M. Gremm, I. Stewart, *Phys. Rev.* D **55**, 1226 (1996); M. Gremm, I. Kapustin, *Phys. Rev.* D **55**, 6924 (1997).

EXPERIMENTAL REVIEW ON $|V_{ub}|$

Philippe Rosnet

Laboratoire de Physique Corpusculaire de Clermont-Ferrand
Université Blaise Pascal / IN2P3 - CNRS
E-mail: rosnet@clermont.in2p3.fr

This paper presents an experimental review of the Cabibbo-Kobayashi-Maskawa (CKM) matrix element V_{ub}. From the data samples recorded by the $\Upsilon(4S)$ and the Z resonance experiments, both inclusive and exclusive analysis of the charmless semileptonic decays of b hadrons $b \to ul\nu_l$ give measurements of $|V_{ub}|$. Combining all the results, the value obtained is: $|V_{ub}| = (3.56 \pm 0.55) \times 10^{-3}$.

1 Introduction

The quarks are mixed under the weak interaction and the phenomena is described by the CKM matrix [1]:

$$V_{CKM} = \begin{pmatrix} V_{ud} & V_{us} & V_{ub} \\ V_{cd} & V_{cs} & V_{cb} \\ V_{td} & V_{ts} & V_{tb} \end{pmatrix} .$$

Using the parametrisation of Wolfenstein [2], one gets:

$$V_{CKM} \approx \begin{pmatrix} 1 - \frac{\lambda^2}{2} & \lambda & A\lambda^3(\rho - i\eta) \\ -\lambda & 1 - \frac{\lambda^2}{2} & A\lambda^2 \\ A\lambda^3(1 - \rho - i\eta) & -A\lambda^2 & 1 \end{pmatrix} ,$$

valid up to $\mathcal{O}(\lambda^4)$.

The unitarity of the matrix leads to $V_{ud}V_{ub}^* + V_{cd}V_{cb}^* + V_{td}V_{tb}^* = 0$, represented by the unitarity triangle on figure 1. One of these sides is characterised by the element V_{ub}. In the framework of the Standard Model, the understanding of the CP violation requires a detailed study of the whole triangle, namely the sides and the angles. In this optic, a precise measurement of V_{ub} is required.

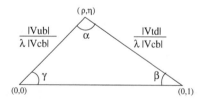

Figure 1: Unitarity triangle in the complex plane (ρ, η).

2 Semileptonic b → u decays

From a theoretical and experimental point of view, only semileptonic $b \to u$ transitions are studied to extract $|V_{ub}|$ (see Fig. 2). In practice, two approaches can be used to derive the matrix element from the measurement of the branching ratio:

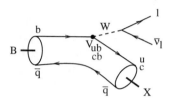

Figure 2: Semileptonic B decay, where X represent the hadronic system produced.

- In the case of exclusive final states ($X_u = \pi$ or ρ or η ...), the experimental reconstruction is relatively easy, but the bad knowledge of the form factors implies a large model dependence.

- In the inclusive study, where all the final states are taken into account ($X_u = \pi$, $\eta^{(\prime)}$, ρ, ω, ..., $\pi\pi$, $\pi\pi\pi$, ...), the experimental reconstruction is more difficult, but the theoretical prediction is quite accurate [3] :

$$|V_{ub}| = 0.00465 \sqrt{\frac{Br(b \to X_u l\nu_l)}{0.002}}$$
$$\sqrt{\frac{1.55ps}{\tau_b}}(1 \pm 0.025_{\text{pert}} \pm 0.03_{m_b}) ,$$

which corresponds to 4 % of precision.

Unfortunately, the $b \to c$ transitions are dominant by approximatively two orders of magnitude, with $X_c = D$, D^*, D^{**}, $D^*\pi$, ... Therefore a good discrimination between the two types of decays ($b \to u$ and $b \to c$) is needed.

Inclusively, the $b \to X_u l\nu_l$ transitions are characterised by a low hadronic mass spectrum (M_X) and a high lepton energy end-point in the centre of mass of the b hadron (E_l^*), see Fig. 3. With a perfect reconstruction and 100 % of efficiency, we can select around 90 % of $b \to u$ decays with the M_X spectrum and only 10 % from

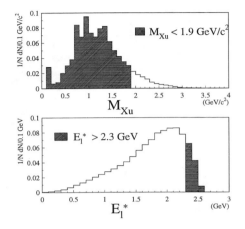

Figure 3: Hadronic mass (upper plot) and lepton energy (lower plot) for inclusive $b \to X_u l \nu_l$ decays. For each plot, the hatched part shows the region forbidden to the $b \to X_c l \nu_l$ decays.

the E_l^* distribution, illustrating the advantage to use the properties of the hadronic system.

At the $\Upsilon(4S)$, the two B mesons are produced almost at rest and their decay products are completely mixed. In this case, it is difficult to distinguish the particles coming from the two B's. On the other hand, at LEP, the situation is different because the two b quarks are boosted and then the decay products of the b hadrons are well separated. In these conditions, it is possible to reconstruct inclusively the hadronic system coming from the semileptonic b decay.

3 $\Upsilon(4S)$ experiments

3.1 Inclusive studies

The first evidence of $b \to u l \nu_l$ transitions has been done, next to the $e^+e^- \to \Upsilon(4S) \to B\bar{B}$ accelerators, by observing an excess of events in the lepton energy spectrum above the $b \to c$ end-point. ARGUS (at DORIS) and CLEO (at CESR) experiments [5] give the following values of the ratio $|V_{ub}|/|V_{cb}|$:

$$\text{ARGUS} \to \frac{|V_{ub}|}{|V_{cb}|} = 0.130 \pm 0.017_{stat+syst} \pm 0.047_{model}$$
$$\text{CLEO} \to \frac{|V_{ub}|}{|V_{cb}|} = 0.080 \pm 0.008_{stat+syst} \pm 0.021_{model}.$$

With the value [4] $|V_{cb}| = 0.0392 \pm 0.0027$, one gets the values of $|V_{ub}|$ given in table 5 (called respectively ARGUS and CLEO end-point).

3.2 Exclusive studies

An other way to observe $b \to u l \nu_l$ transition consists to reconstruct a B meson in a specific channel. First, upper limits on the vector modes $\rho l \nu_l$ and $\omega l \nu_l$ have been given by CLEO [6], and afterwards a measurement of the exclusif branching ratio have been performed from the excess of events observed in the reconstructed B mass [6]:

$$\begin{cases} Br(B^0 \to \pi^+ l^- \bar{\nu}_l) = (1.8 \pm 0.4_{stat} \pm 0.3_{syst} \\ \qquad\qquad\qquad \pm 0.2_{model}) \times 10^{-4} \\ Br(B^0 \to \rho^+ l^- \bar{\nu}_l) = (2.5 \pm 0.4_{stat} \pm^{0.5}_{0.7 syst} \\ \qquad\qquad\qquad \pm 0.5_{model}) \times 10^{-4} \end{cases},$$

where the first error is statistical, the second is from the estimation of the residual background and the third is from the modelling of the signal.

By combining these results, the value of $|V_{ub}|$ is determined, see table 5 (called CLEO exclusif).

A new exclusive study has been done by the CLEO experiment [7], with $N(e^+e^- \to \Upsilon(4S) \to B\bar{B}) = 3.3 \times 10^6$ (compare to 2.8×10^6 in the previous analysis). The analysis concerns the vector mode $B \to \rho(\to \pi\pi)l\nu_l$ and $B \to \omega(\to \pi\pi\pi)l\nu_l$, with a lepton momentum greater than 1.7 GeV/c, while the pseudo-scalar mode $B \to \pi l\nu_l$ is used only as a check and to estimate its contribution as background in the ρ channel. Assuming the isospin symmetry $\Gamma_{\rho^\pm} = 2\Gamma_{\rho^0} = 2\Gamma_\omega$, the fraction of $B^0 \to \rho^+ l^- \bar{\nu}_l$ is fitted in the plane $M_{\pi\pi(\pi)}$ versus ΔE (see Fig. 4), where $\Delta E = E_{\pi\pi(\pi)} + E_l + |\vec{p}_{miss}| - E_{beam}$. In the fit, the lepton energy is divided in three regions: $1.7 < E_l < 2.0$ GeV poor in $b \to u$ and used to normalized the background, $2.0 < E_l < 2.3$ GeV mixing $b \to u$ and $b \to c$ decays, and $2.3 < E_l < 2.7$ GeV rich in $b \to u$ transitions. The result

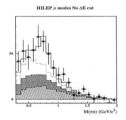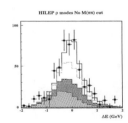

Figure 4: $M_{\pi\pi}$ (left plot) and ΔE (right plot) for the ρ channel in the lepton energy range $2.3 < E_l < 2.7$ GeV. For each plot, the squares represent the data; and the histograms the Monte Carlo with the signal (in white), the cross-feed components of the signal (dotted line), the background from $b \to u l \nu_l$ non-signal modes (double-hatched region), and the background from $b \to c l \nu_l$ (single-hatched region).

Table 1: Result of the branching ratio for the $B^0 \to \rho^+ l^- \bar{\nu}_l$ obtained by CLEO with different models, where the first error is statistical while the second comes from the model itself.

Model	$Br(B^0 \to \rho^+ l^- \bar{\nu}_l) \times 10^4$
Quark model (average)	$2.7 \pm 0.4 \pm 0.4$
Lattice (UKQCD)	$2.8 \pm 0.4 \pm 0.1$
LCSR (Ball/Braun)	$3.2 \pm 0.5 \pm 0.4$
Ligeti/Wise + E791	2.5 ± 0.4

of the fit gives:

$$Br(B^0 \to \rho^+ l^- \bar{\nu}_l) = (2.8 \pm 0.4_{stat} \pm 0.4_{syst} \\ \pm 0.6_{model}) \times 10^{-4} \, ,$$

where the modelling error comes from the dispersion of the result with respect to the different exclusive models used [7], see table 1.

The excess of events observed after the fit (see Fig. 5) and the efficiency for each channel, in the region $2.0 < E_l < 2.7$ GeV, are:

$$\begin{cases} \rho^\pm \longrightarrow (116 \pm 18) \text{ events with } \epsilon = 3.9\,\% \\ \rho^0 \longrightarrow (114 \pm 17) \text{ events with } \epsilon = 9.1\,\% \\ \omega \longrightarrow (53 \pm 8) \text{ events with } \epsilon = 3.7\,\% \end{cases}$$

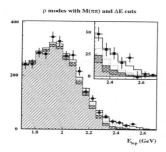

ρ modes with M($\pi\pi$) and ΔE cuts

E_{lep} (GeV)

Figure 5: Lepton energy for the ρ mode after the cuts: $|M_{\pi\pi} - M_\rho| \le 150$ MeV/c^2 and $|\Delta E| \le 500$ MeV. The differents contribution are the same than for the previous figure.

From the measured branching ratio, the value of $|V_{ub}|$ (largely statistically independent from the previous analysis) given in table 5 (called CLEO vector) is extracted.

4 Z experiments

For all the $\Upsilon(4S)$ results, the modelling of the $b \to u l \nu_l$ transitions is the dominant error on $|V_{ub}|$. It comes from the fact that in the exclusive searches the prediction of the associated form factors suffers from large uncertainties, and in the end-point studies only a small part of the

lepton energy phase space is considered. The idea is then to develop at the Z pole an inclusive analysis including the information provided by the hadronic system, which is in theory very powerful (see Fig. 3).

4.1 ALEPH measurement

The ALEPH study [8] is based on 3.6×10^6 $Z \to q\bar{q}$ events. The principle of the analysis is the following:

- A lepton (electron or muon) is identified with a momentum greater than 3.0 GeV/c. A b-tag is applied on the opposite hemisphere, the neutrino is reconstruct with the help of the missing momentum of the event ($\vec{p}_{\nu_l} = \vec{p}_{miss}$ and $E_{\nu_l} = |\vec{p}_{\nu_l}|$), and the hadronic system X is selected by using two neural networks (one for charged tracks and one for photons) to discriminate the b tracks with respect to the fragmentation tracks. All the decay products of the b hadrons (l, ν_l and X) allow to calculate its four-momentum.

- Afterwards, discriminating variables between $b \to u l \nu_l$ and $b \to c + l$ decays (direct $b \to c l \nu_l$, cascade $b \to c \to l$, $c \to l$, $b \to \bar{c} \to l$, $J/\psi \to l$, miss-identifications, ...) are constructed in the centre of mass of the b hadron: invariant mass, sphericity, lepton energy, ... 20 such variables are combined inside a neural network called NN_{bu} to improve the separation between $b \to u$ and $b \to c$ transitions, see Fig. 6. The advantage of this method is to have similar efficiency for neutral and charged modes as explained by figure 6.

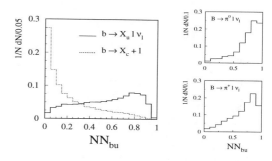

Figure 6: Neural network output NN_{bu} (left plot) to discriminate the $b \to u l \nu_l$ transitions from the $b \to c + l$ background. NN_{bu} for the $B \to \pi l \nu_l$ modes (right plots): for the π^0 (upper plot) and the $\pi^\pm$ (lower plot) channels.

- Finally the shape of the neural network NN_{bu} is fitted to extract $Br(b \to X_u l \nu_l)$.

Figure 7 shows the comparison between data and simulation for NN_{bu}. An excess of 303 ± 88 events appears in the signal region ($NN_{bu} > 0.6$) which is attributed to $b \to X_u l \nu_l$ transitions. A fit to this part of the neural

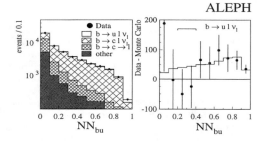

Figure 7: Comparison of the neural network output NN_{bu} (left plot) for data (points) and the simulation (histograms) with the signal, the direct $b \to c l \nu_l$ background, the cascade $b \to c \to l$ and the other sources of leptons ($c \to l$, $b \to \bar{c} \to l$, $J/\psi \to l$, miss-identifications, ...). Difference data - Monte Carlo without $b \to u$ transitions (right plot).

network (corresponding to an efficiency of 50 % for the signal) gives the inclusive branching ratio:

$$Br(b \to X_u l \nu_l) = (1.73 \pm 0.55_{stat} \pm 0.51_{syst} \\ \pm 0.21_{model}) \times 10^{-3} ,$$

where the $b \to c + l$ systematic errors come from four main sources and are summarised in table 2.

In this analysis, the signal is described with an hybrid model [8] (see Fig. 3) by taking resonant states (25 %) at low hadronic recoil $E_u < 1.6\,\text{GeV}$ with ISGW2 model [8] and multi-body states (75 %) at large recoil $E_u \geq 1.6\,\text{GeV}$ with the DSU model[8] (Dikeman-Shifman-Uraltsev). The corresponding systematic errors are estimated by varying the different parameters of the model [8]. The small error obtained (see table 2) is due to the relatively small sensitivity of the neural network output to the modelling, as explained in figure 8.

Several checks were made to confirm the evidence of inclusive $b \to X_u l \nu_l$ transitions in this analysis: by comparing the NN_{bu} output of Data and Monte Carlo for selected $b \to X_c l \nu_l$ decays; by checking the stability of the result with respect to the fit procedure; by studying the effect of the variables used; and by searching for an evidence of signal for $NN_{bu} > 0.9$ with vertexing method, in the $M_{X l \nu_l}$ distribution, and by a scan of the events (see Ref. [8] for more details).

Table 2: Systematic errors on the branching ratio $Br(b \to X_u l \nu_l)$ for the ALEPH analysis.

Source of error	$\Delta Br(b \to X_u l \nu_l)$
b hadron production	$\pm 0.16 \times 10^{-3}$
b hadron decay	$\pm 0.31 \times 10^{-3}$
c hadron decay	$\pm 0.37 \times 10^{-3}$
lepton identification	$\pm 0.08 \times 10^{-3}$
Hybrid model parameter	$\pm 0.08 \times 10^{-3}$
Exclusive model	$\pm 0.05 \times 10^{-3}$
inclusive model	$\pm 0.18 \times 10^{-3}$
Λ_b modelling	$\pm 0.04 \times 10^{-3}$

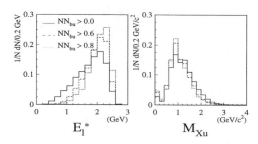

Figure 8: Lepton energy (left plot) and invariant mass (right plot) generated for the $b \to X_u l \nu_l$ transitions for three different cuts on NN_{bu}.

Finally, with the help of theoretical predictions [3], one obtains the result of $|V_{ub}|$ presented in the table 5 (called ALEPH inclusif).

4.2 DELPHI measurement

The DELPHI experiment has presented in the conference a new inclusive analysis [9] based on 2.8×10^6 hadronic Z events. All the decay products of the b hadron: the hadronic system X, the lepton l and the neutrino ν_l are reconstructed; and the background coming from the $b \to c + l$ decays is suppressed with cuts like $E_{Xl} > 12\,\text{GeV}$ and $M_{Xl} > 2.0\,\text{GeV/c}^2$. Then, two samples are defined by using the vertexing conditions of the lepton with respect to the b decay vertex and with the presence of kaon candidates in the lepton hemisphere. Figure 9 shows the reconstructed invariant mass M_X for the $b \to u$ depleted and enriched samples. By cutting at $M_X = 1.6\,\text{GeV/c}^2$ in the two samples, four sub-samples are defined. Then a fit of the reconstructed lepton energy in the b hadron rest frame E_l^* is done simultaneously in the four sub-samples to determine the ratio $|V_{ub}|/|V_{cb}|$. The result obtained

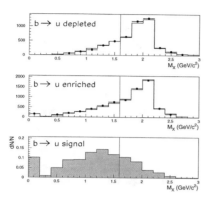

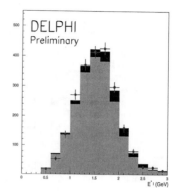

Figure 9: Comparison of the hadronic invariant mass M_X between the Data (point) and the simulation (histogram) for the $b \to u$ depleted sample (upper plot) and the $b \to u$ enriched sample (middle plot). The lower plot shows the distribution for $b \to u$ simulated events.

Figure 10: Lepton energy E_l^* for the $b \to u$ enriched sample with $M_X < 1.6\,\mathrm{GeV/c^2}$. The points represent the Data while the histograms the simulation contributions with the signal (in dark) and the $b \to c + l$ background (in light).

Table 3: Systematic errors on the ratio $|V_{ub}|/|V_{cb}|$ for the DELPHI analysis.

Source of error	$\Delta|V_{ub}|/|V_{cb}|$
b hadron production	± 0.0052
b hadron decay	± 0.0083
c hadron decay	± 0.0065
detector effects	± 0.0101
Signal efficiency	± 0.0012
m_b	± 0.0057
$<p_b^2>$	± 0.0010
b kinematic model	± 0.0010
Hadronisation model	± 0.0060

is:

$$\frac{|V_{ub}|}{|V_{cb}|} = 0.104 \pm 0.012_{stat} \pm 0.015_{syst} \pm 0.009_{model} \; ,$$

where the error are summarised in table 3. After the fit, the excess of events in the lepton energy spectrum (see Fig. 10) is 205 ± 56 for the $b \to u$ enriched sample with $M_X < 1.6\,\mathrm{GeV/c^2}$ (with 11.2 % of signal efficiency).

With the value of $|V_{cb}|$ given in Ref.[4], one obtains the value of $|V_{ub}|$ given in table 5 (called DELPHI inclusif).

4.3 L3 measurement

An other new inclusive study[10] was presented by the L3 experiment with $1.8 \times 10^6\ Z \to q\bar{q}$ events. The procedure of the analysis is similar to the ALEPH one, except that only the two most energetic tracks of the lepton hemisphere are selected to reconstruct the hadronic system. Then discriminating variables based on the lepton and these two tracks are built (see Fig. 11). A total of 8 such

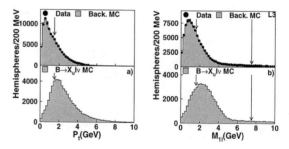

Figure 11: Lepton transverse momentum with respect to the jet axis (left plots), and invariant mass between the lepton and the most energetic track (right plots). For each variable, the upper plot show a comparison between the Data (point) and the background simulation (histogram), while the lower plot the expected signal. The arrows indicate the cuts applied on the variables.

variables are used to reduce the background by a factor 160. Finally, the number of remaining events is used to determined the inclusive branching ratio. The excess of 81 events observed over 576 Data (with 1.5 % of signal efficiency) leads to the result:

$$Br(b \to X_u l \nu_l) = (3.3 \pm 1.0_{stat} \pm 1.66_{syst} \\ \pm 0.55_{model}) \times 10^{-3} \; ,$$

Table 4: Systematic errors on the branching ratio $Br(b \to X_u l \nu_l)$ for the L3 analysis.

Source of error	$\Delta Br(b \to X_u l \nu_l)$
b hadron production	$\pm 0.68 \times 10^{-3}$
b hadron decay	$\pm 1.42 \times 10^{-3}$
detector effects	$\pm 0.52 \times 10^{-3}$
MC statistics	$\pm 0.06 \times 10^{-3}$
Exclusive π rate	$\pm 0.19 \times 10^{-3}$
lepton spectrum	$\pm 0.04 \times 10^{-3}$
π spectrum	$\pm 0.27 \times 10^{-3}$
Λ_b rate	$\pm 0.43 \times 10^{-3}$

where the errors are given in table 4. In this analysis, the signal is described by the ACCMM model [10] (with $p_f = 298\,\text{MeV/c}$ and $m_u = 150\,\text{MeV/c}^2$), except for the $B \to \pi l \nu_l$ channel where an exclusive approach is used (Burdman-Kambor model [10]). As shown in figure 12, a large fraction of the phase space is considered by the analysis explaining the small modelling error.

The cuts on each variable have been changed to check

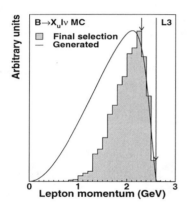

Figure 12: Comparison of the lepton momentum between the generated (line) and selected (histogram) $b \to X_u l \nu_l$ events. To compare the arrows define the range considered by the end-point analysis of CLEO.

the stability of the result. Furthermore, the 8 variables have been combined inside a neural network and the analysis redone, see Fig. 13. The result obtained is in good agreement with the previous one.

Finally, the value of the matrix element $|V_{ub}|$ is determined with the help of the theoretical prediction [3], see table 5 (called L3 inclusif).

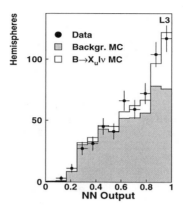

Figure 13: Comparison of the neural network output between the Data (points) and the simulation (histograms) with the signal and background contributions.

Table 5: Summary of all the values of the CKM matrix element $|V_{ub}|$.

Experimental result	$|V_{ub}|$ with statistical, systematic and modelling errors ($\times 10^3$)
ARGUS end-point	$5.1 \pm 0.6 \pm 0.2 \pm 1.8$
CLEO end-point	$3.1 \pm 0.2 \pm 0.3 \pm 0.8$
CLEO exclusif	$3.3 \pm 0.2 \pm^{0.3}_{0.4} \pm 0.7$
CLEO vector	$3.2 \pm 0.3 \pm^{0.2}_{0.3} \pm 0.6$
ALEPH inclusif	$4.16 \pm 0.70 \pm 0.64 \pm 0.31$
DELPHI inclusif	$4.08 \pm 0.47 \pm 0.69 \pm 0.35$
L3 inclusif	$6.0 \pm 0.9 \pm 1.6 \pm 0.6$

5 Comparison and discussion

In summary, seven values of the CKM matrix element $|V_{ub}|$ are now available and presented in table 5.

An average value, obtained with a blue technique (consisting to minimise the total error), gives the following result:

$$|V_{ub}| = (3.55 \pm 0.21_{stat} \pm 0.28_{syst} \pm 0.43_{model}) \times 10^{-3} \,,$$

with the dominant error coming from the modelling of the $b \to u l \nu_l$ decays.

A comparison of the different values of $|V_{ub}|$ is given on figure 14. All the values are in good agreement; and we see clearly that the results coming from the $\Upsilon(4S)$ experiments are dominated by the modelling error, while the results of the Z experiments are dominated by the statistical and systematic errors.

Considering all the other constraints [4] [11] on the unitarity triangle:

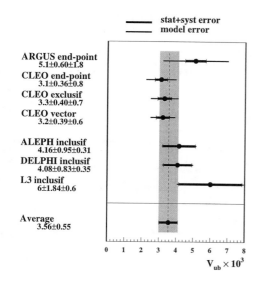

Figure 14: Comparison of the $|V_{ub}|$ measurements.

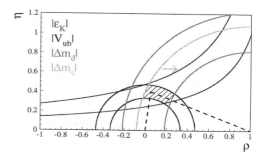

Figure 15: Constraints on the unitarity triangle, with the allowed region in hatched.

- $|\epsilon_K| = (2.280 \pm 0.013) \times 10^{-3}$ and $B_K = (0.75 \pm 0.15)$,

- $\Delta m_d = (0.471 \pm 0.016)\,ps^{-1} \Rightarrow |V_{td}| = (8.8 \pm 1.6)) \times 10^{-3}$,

- $\Delta m_s > 12.4\,ps^{-1} \Rightarrow \frac{|V_{ts}|}{|V_{td}|} > 4.2$ at 95 % C.L.;

one obtains the result presented on figure 15.

To conclude, these new results on the charmless semileptonic b hadron decays lead to an average value of the CKM matrix element (with 68 % of C.L. and with no gaussian error):

$$|V_{ub}| = (3.56 \pm 0.55) \times 10^{-3},$$

that is 15 % of accuracy.

References

1. N. Cabibbo, Phys. Rev. Lett. 10 (1963) 531;
   M. Kobayashi and T. Maskawa, Prog. Theor. Phys. 42 (1973) 652;
2. L. Wolfenstein, Phys. Rev. Lett. 51 (1983) 233.
3. N. Uraltsev, Int. Jour. Mod. Phys. A11 (1996) 515;
   I. Bigi, R. Dikeman and N. Uraltsev, hep-ph/9706520.
4. A. Pich, Nucl. Phys. B (Proc. Suppl. WIN97) 66 (1998) 456;
   K. Berkelman, Nucl. Phys. B (Proc. Suppl. WIN97) 66 (1998) 447.
5. ARGUS Collaboration, Phys. Lett. B 234 (1990) 409;
   ARGUS Collaboration, Phys. Lett. B 255 (1991) 297;
   CLEO Collaboration, Phys. Rev. Lett. 64 (1990) 64;
   CLEO Collaboration, Phys. Rev. Lett. 71 (1993) 4111.
6. CLEO Collaboration, Phys. Rev. Lett. 70 (1993) 2681;
   CLEO Collaboration, Phys. Rev. Lett. 77 (1996) 5000.
7. CLEO Collaboration, ICHEP98 # 855.
8. ALEPH Collaboration, CERN-EP/98-067 (May 5, 1998), ICHEP98 # 933.
9. DELPHI Collaboration, ICHEP98 # 241.
10. L3 Collaboration, CERN-EP/98-097 (June 15, 1998), ICHEP98 # 561.
11. V. Andreev, *Oscillations of the B_d^0 mesons*, these proceedings;
    F. Parodi, B_s *mixing and limit on* Δm_s, these proceedings.

NEW MEASUREMENTS OF $|V_{ub}|$ and $|V_{cb}|$ WITH DELPHI AT LEP

M. BATTAGLIA

Helsinki Institute of Physics, P.O. Box 9 FIN 00014, University of Helsinki, Finland
E-mail: Marco.Battaglia@cern.ch

The $|V_{ub}|$ and $|V_{cb}|$ elements in the CKM mixing matrix have been measured by inclusive reconstruction of semileptonic B decays with the DELPHI data recorded at LEP, at $\sqrt{s}$ energies around the Z^0 peak, from 1992 to 1995. The measurement of $|V_{ub}|$ is based on a novel technique that uses the reconstructed mass of the secondary hadronic system produced in semileptonic B decays. Since $b \rightarrow u\ell\nu$ transitions are characterized by hadronic masses smaller than those of D mesons produced in $b \rightarrow c\ell\nu$ transitions, events with a reconstructed value of M_X below the D mass have been selected. $|V_{ub}|/|V_{cb}|$ has been extracted by a fit to the shape of the spectrum of the lepton energy in the B rest frame and found to be 0.104 ± 0.012 (stat) ± 0.015 (syst) ± 0.009 (model). A new precise measurement of $|V_{cb}|$ has been obtained from the extrapolation to zero recoil of the differential rate of $\bar{B}^0 \rightarrow D^{*+}\ell\nu$ decays. These have been reconstructed by inclusively tagging the soft pion from $D^{*+} \rightarrow D^0\pi^+$. The result has been found to be $|V_{cb}| = (41.19 \pm 1.52 \text{ (stat)} \pm 1.77 \text{ (syst exp)} \pm 1.36 \text{ (syst th)}) \times 10^{-3}$. From these results, $|V_{ub}| = 4.28 \pm 0.86 \times 10^{-3}$ has been obtained.

1 Introduction

The $|V_{ub}|$ and $|V_{cb}|$ elements in the Cabibbo-Kobayashi-Maskawa mixing matrix determine the rates of the weak decays of the b quark through tree level $b \rightarrow u$ and $b \rightarrow c$ transitions and non-spectator annihilation diagrams. As these are not predicted by the Standard Model, it is important to perform precise experimental determinations of their values. This has important implications in the study of the weak decays of the b quark and in the perspective of the interpretation of CP violation in the B system.

Recent theoretical progress has provided ways to reduce the model dependencies in the extraction of $|V_{ub}|$ and $|V_{cb}|$ from the analysis of semileptonic B decays. The preliminary results of new measurements of $|V_{ub}|$ and $|V_{cb}|$ inspired by these developments are reported [1]. Both analyses are based on the data collected by DELPHI at LEP at centre-of-mass energies around the Z^0 peak. These consist of 2.78 M selected hadronic Z^0 decays in the combined 1993-95 statistics used for the V_{ub} analysis and 3.04 M including also the 1992 data used for the V_{cb} analysis. These data sets have been enriched in $Z^0 \rightarrow b\bar{b}$ events by applying a b-tag algorithm based on the measured track impact parameters.

In the reconstruction of semileptonic B decays, candidate leptons have been tagged in the momentum interval 2 GeV/c $< p <$ 30 GeV/c. Muons have been identified by the hits associated in the muon chambers while electron candidates have been selected using a Neural Network based on the response of the HPC e.m. calorimeter and on the measured specific ionization in the TPC. Only events with one lepton were considered in the analyses. To reduce the background from misidentified leptons and $c \rightarrow \ell$ decays, the lepton was required to satisfy the cut $p_t^{in} > 0.5$ GeV/c for the V_{ub} analysis and $p_t^{out} > 1.0$ GeV/c for the V_{cb} analysis. For the re-

construction of the secondary hadronic system, charged kaons have been identified by the combination of the response of the DELPHI Ring Imaging CHerenkov (RICH) detectors and the dE/dx in the TPC. K_s^0 have been reconstructed in their $\pi^+\pi^-$ decay mode. $\pi^0 \rightarrow \gamma\gamma$ decays have been tagged either by observing photon pairs compatible with the π^0 mass or by the e.m. cluster shape reconstructed in the HPC calorimeter.

2 Determination of $|V_{ub}|/|V_{cb}|$

The measurement of the branching ratio for the decay $b \rightarrow u\ell\nu$ provides the most precise way to determine the $|V_{ub}|$ element in the CKM mixing matrix. However the extraction of $|V_{ub}|$ from the yield of leptons above the $b \rightarrow c\ell\nu$ endpoint is subject to large systematic uncertainties. Also the derivation of $|V_{ub}|$ from exclusive semileptonic decays, recently observed by CLEO [2], contrary to the case for $|V_{cb}|$ in $B \rightarrow D^*\ell\nu$, has a significant model dependence.

The extraction of $|V_{ub}|$ from the shape of the invariant mass of the hadronic system recoiling against the lepton in $b \rightarrow u\ell\nu$ transitions was proposed several years ago [3] and it has been recently the subject of new theoretical calculations [4,5]. The proposed method starts from the observation that this hadronic system has an invariant mass lower than the charm mass for the majority of $b \rightarrow u\ell\nu$ decays (see Figure 1). The model dependence in predicting the fraction of $b \rightarrow u\ell\nu$ decays with invariant mass below a given cut value is expected to be smaller than those for the estimate of the lepton spectrum in the end-point region and for the exclusive final state branching ratios. For this analysis, the model dependence has been studied using $b \rightarrow u\ell\nu$ decays simulated with a dedicated decay generator, implementing different prescriptions for the initial state kinematics and the resonance decomposition of the hadronic final state [6].

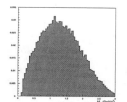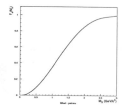

Figure 1: Spectrum of hadronic mass M_X for $b \to u\ell\nu$ decays at the parton level (left) and the fraction $F_u(M_X)$ of $b \to u$ transitions giving an hadronic system below a given mass value (right).

The measurement of the ratio $|V_{ub}|/|V_{cb}|$ reduces the experimental systematic uncertainties since selection efficiencies cancel out in first approximation. Finally the systematics due to the description of charm decays have been reduced by defining samples enriched and depleted in $b \to u$ transitions using the secondary vertex topology and identified kaons

The analysis is based on the reconstruction of the secondary hadronic system emitted in the semileptonic B decay. This reconstruction has been performed inclusively. For each particle in the same hemisphere as the lepton, the probability to be a B decay product has been evaluated on the basis of kinematical and topological variables. An iterative vertex fit algorithm defined charged candidate B decay products consistent with originating at a common secondary vertex. Identified K^0 and π^0 candidates with large B probability values were then tested for compatibility with this secondary hadronic system. The average total (charged) reconstructed secondary multiplicity is 3.64 (2.74) and the mass distribution shows a peak corresponding to the D mass and an asymmetric tail at lower values due to partially reconstructed decays. In order to identify these decays, the total mass $M_{X\ell\nu}$ of the candidate B decay has been estimated from the invariant mass of the system formed by the secondary hadronic system, the lepton and the neutrino reconstructed by the missing energy. In fully reconstructed $B \to X\ell\nu$ decays, the distribution of this mass peaks at 5.0 GeV/c^2 and has a resolution of 0.9 GeV/c^2 while, in partially reconstructed decays, it peaks at 4.5 GeV/c^2 with a tail at lower masses. Therefore decays with $M_{X\ell\nu} < 3.0$ GeV/c^2 have been removed and, for decays with $M_{X\ell\nu} < 4.5$ GeV/c^2 and $M_X < 1.6$ GeV/c^2, the measured hadronic mass M_X has been rescaled by the ratio $M_B/M_{X\ell\nu}$. In order to further remove partially reconstructed D decays, hemispheres with charged particle tracks not associated to the secondary vertex but incompatible with originating from the primary vertex have been rejected. Also decays with two identified leptons in the same hemisphere have

been removed, since double semileptonic decays $b \to c\ell\nu$, $c \to s\ell\nu$ result in a low mass hadronic system. In addition, an inclusive search for $D^* \to D\pi$, similar to that presented in the next section, has been performed. The mass M_X of secondary hadronic systems compatible with containing a reconstructed D^* has been set to 2.01 GeV/c^2. The resulting hadronic mass distribution is shown in Figure 2.

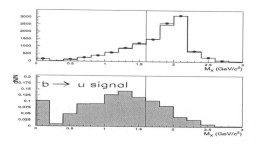

Figure 2: Distribution of scaled hadronic mass M_X for $b \to X\ell\nu$ candidate decays. The upper plot shows simulated decays (full histogram) compared with real data (points with error bars). The lower plot shows the simulation prediction for signal $B \to X_u\ell\nu$. The lines indicate the mass cut used in the analysis.

In order to further reduce the background due to $b \to c$ transitions, the decay topology and the presence of identified kaons in the same hemisphere as the lepton have been used to define samples enriched or depleted in $b \to u$ transitions, independently of the reconstructed hadronic mass. The impact parameter of each tagged lepton associated with a reconstructed secondary vertex has been computed w.r.t. this vertex and a sign has been attributed using the lifetime convention. In $b \to c\ell\nu$ decays the secondary vertex corresponds to the D decay vertex. Therefore $b \to c$ semileptonic transitions preferentially give leptons with negative impact parameters, since the lepton comes from the B decay vertex located behind the D vertex. On the contrary, in the $b \to u\ell\nu$ transitions the secondary vertex coincides with the lepton production point and its sign is either positive or negative with equal probabilities (see Figure 3). Decays with a reconstructed secondary vertex and lepton impact parameter $d_{sec}^\ell < -0.015$ cm have been assigned to the "$b \to u$ depleted" class. The detection of a strange particle in the semileptonic B decay has been also used to separate $b \to c\ell\nu$ decays from $b \to u$ transitions where the production of strange particle is suppressed. Decays with an identified $K^\pm$ or K_s^0 with $p > 2.5$ GeV/c in the same hemisphere as the lepton have also been assigned to the "$b \to u$ depleted" class.

The lepton energy E_ℓ^* has been computed in the B

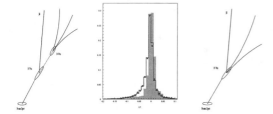

Figure 3: Distribution of lepton impact parameter computed w.r.t. the secondary vertex and expected impact parameter configuration for $b \to c$ (left) and $b \to u$ (right) decays . The real data (dots with error bars) and the simulated background (histogram) show an enhancement at negative values due to the charm lifetime. The expected distribution for the signal (filled histogram) is centered around zero.

rest frame defined by the reconstructed B energy and direction. The resolution on the E_ℓ^* reconstruction has been studied on simulated events and found to be 14 % for 81 % of the selected decays.

Candidate semileptonic B decays have been further selected by requiring a the lepton momentum to exceed 3.5 GeV/c and the invariant mass of the secondary hadronic system and the lepton to exceed 2.0 GeV/c^2. Decays in which the lepton charge had a sign equal to that of the hadronic system have also been discarded.

These criteria selected 12134 decays in real data and 11898 in the simulated backgrounds, with an estimated efficiency for $b \to u\ell\nu$ with $\ell = e$ or μ, of 0.165 ± 0.004. The background composition has been studied on simulation and found to consist of $b \to c\ell\nu$ decays (91%), cascade $b \to c \to s\ell\nu$ decays (7%) and $c \to \ell$ decays and misidentified hadrons (2%) for $E_\ell^* > 1.0$ GeV.

The selected decays have been divided into four independent classes according to the reconstructed secondary hadronic mass M_X and the $b \to u$ enrichment criteria. A M_X upper cut at 1.6 GeV/c^2 has been chosen to ensure a reduced model dependence in the extraction of $|V_{ub}|$[5,6] while being sufficiently below the D mass to suppress the bulk of $b \to c\ell\nu$ decays.

The numbers of events selected in the real data and of expected background events in the four classes are summarised in Table 1. For decays selected in the "low M_X, $b \to u$ enriched" class that is expected to contain 68% of the $b \to u$ signal, an excess of 205 ± 56 events has been found in the real data compared to the expected background. No significant excess has been observed in the other classes.

As a cross-check, the analysis has been repeated using both anti-b tagged events and decays with same-sign lepton and hadronic vertex combinations. Both these samples are expected to be depleted in signal but are

sensitive to possible discrepancies between real data and simulation in the description of backgrounds. The numbers of selected decays in the "low M_X, $b \to u$ enriched class" show no excess in the real data compared with the expected backgrounds.

Table 1: Number of events selected in the real data and from expected backgrounds. These backgrounds have been rescaled by the normalisation factor obtained from the fit and its error has been included.

Data/Back.	$b \to u$ enriched	$b \to u$ depleted
Low M_X	2292±48/2087±30	1081±33/1143±19
High M_X	5017±71/4901±58	3744±61/3648±45

The numbers of events in each decay class and their E_ℓ^* distributions have been used to extract the value of $|V_{ub}|/|V_{cb}|$ by a simultaneous binned maximum likelihood fit. In the fit, the overall simulation to real data normal-

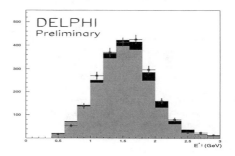

Figure 4: E_ℓ^* distribution for the decays in the $b \to u$ enriched class with $M_X < 1.6$ GeV/c^2. Real data are indicated by the points with error bars, the background by the light shaded histogram and the $b \to u\ell\nu$ signal by the dark shaded histogram normalised to the fitted fraction of events.

isation and the value of $|V_{ub}|/|V_{cb}|$ have been left free. By leaving the normalization free, the systematic uncertainties from the lepton identification and other sources have been significantly reduced by absorbing their effects in this overall normalisation factor. The result of the fit is $\frac{|V_{ub}|}{|V_{cb}|} = 0.104 \pm 0.012$ (stat), with the normalisation factor 0.985 ± 0.010. The stability of the result has been checked by changing the cut on M_X. This cut has been varied from 1.25 GeV/c^2 to 1.75 GeV/c^2, corresponding to a variation of the signal-to-background ratio from 0.11 to 0.06. The result of the fit has been found to be stable. Several cuts used in the selection of the events have been varied or dropped and the fit has been repeated. The results have been found to agree within the errors.

Several sources of systematic uncertainties have been considered and their contributions evaluated as summarised in Table 2. In conclusion the value of the ra-

Table 2: Estimate of systematic errors on $|V_{ub}|/|V_{cb}|$.

Source	Syst. Error
Charm Decays	$\pm$ 0.0090
B Production and Decays	$\pm$ 0.0076
Detector Systematics	$\pm$ 0.0100
$b \to u\ell\nu$ Model	$\pm$ 0.0085
Total	$\pm$ 0.018

tio $|V_{ub}|/|V_{cb}|$ has been measured to be: $\frac{|V_{ub}|}{|V_{cb}|} = 0.104 \pm 0.012$ (stat) $\pm$ 0.015 (syst) $\pm$ 0.009 (model).

A search for fully reconstructed few prong $b \to u\ell\nu$ decays has been performed using the events selected in the $b \to u$ enriched class with reconstructed hadronic system consisting of either a single charged particle or two particles with total zero charge and at most one neutral. An excess of events has been observed for hadronic masses below 1.3 GeV/c^2 in agreement with that expected from $b \to u\ell\nu$ decays. An example of fully reconstructed $B^0 \to \pi^+\mu^-\nu$ decay is shown in Figure 5.

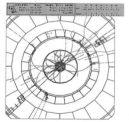

Figure 5: A candidate $B^0 \to \pi^+\mu^-\nu$ decay.

3 Determination of $|V_{cb}|$

The determination of $|V_{cb}|$ is based on the measurement of the differential decay rate $d\Gamma/dw$ for the process $\bar{B}^0 \to D^{*+}\ell\nu$ where w is the product of the $\bar{B}^0$ and D^{*+} velocities. This rate is proportional to $|V_{cb}|^2 \times \mathcal{F}(w)$ with $\mathcal{F}(w)$ being the $B \to D^{*+}$ transition form factor. In the Heavy Quark limit, at $w \to 1$ the D^{*+} is emitted at rest and the form factor normalisation is $\mathcal{F} = 1$. Corrections to the Heavy Quark limit have been computed to second order. $|V_{cb}|$ can be extracted from the extrapolation of the decay rate at zero recoil and there have been several determinations of $|V_{cb}|$ using this technique. Due to the phase space suppression at zero recoil, the data have to be extrapolated from larger w values and the parametrisation of the form factor used introduces a model uncer-

tainty. Recently a new parametrisation, including constraints on the form of the QCD functions and the relationship between the curvature coefficient and the heavy meson radius, that reduces this uncertainty, has been suggested [7]. The new DELPHI analysis, which also profits from improved resolution on w and from larger statistics obtained by an inclusive reconstruction technique of $D^{*+}\ell\nu$ decays, adopts these theoretical improvements in the extraction of $|V_{cb}|$.

Charm hadron candidates have been inclusively reconstructed from particles selected in the same jet as the lepton, using an iterative algorithm similar to that described for the previous analysis. The inclusive charm vertex has been constrained to the D^0 mass. The resulting D^0 trajectory has been extrapolated to intercept the lepton candidate to define the B decay vertex. To tag the D^{*+} decay, an additional pion π^* with charge opposite to that of the lepton has been added to the candidate D^0 vertex and the mass difference $\Delta m = M_{D^0\pi} - M_{D^0}$ computed. The resulting Δm distribution shows an excess of events at $\Delta m < 0.165$ GeV/c^2 corresponding to genuine D^{*+} decays.

The w variable has been computed as $w = (M_{\bar{B}^0}^2 + M_{D^{*+}}^2 - q^2)/(2M_{\bar{B}^0}M_{D^{*+}})$ with $q^2 = (p_{\bar{B}^0} - p_{D^{*+}})^2$. The squared recoil mass μ has been used to improve the estimate on the B direction by imposing the constraint $\mu^2 = M_\nu^2 = 0$. The resulting resolution on the w determination has been found to be described by a central gaussian with $\sigma_w = 0.125$ and small additional tails.

In the analysis of the differential rate for the exclusive $D^{*+}\ell\nu$ decay, it is important to evaluate the background contributions. The combinatorial background was determined from the real data using same-charge $\pi^*\ell$ combinations and subtracted from the selected events, leaving 7075 $\pm$ 164 candidate D^{*+} decays. The remaining backgrounds are due to processes containing a D^{*+} meson other than $\bar{B}^0 \to D^{*+}\ell\nu$. Among these backgrounds, the contribution from decays into orbitally excited D^{**} states represent a dominant source of systematical uncertainty. Therefore this fraction has been directly measured on the selected inclusive D^{*+} sample using the measured squared recoil mass μ^2. Due to the additional particles produced in the decay of $D^{**} \to D^{*+}$, the μ^2 value is shifted to positive values in semileptonic decays including a D^{**} compared to those with a directly produced D^{*+} meson. The fraction of D^{**} meson was extracted by a fit to the measured μ^2 distribution leaving the D^{**} fraction free and fixing the contribution to the other background sources to their simulation estimates. The shape of the μ^2 distribution for $D^{**}\ell\nu$ was taken from HQET predictions [8] and that for $D^{*+}\ell\nu$ from the ISGW model [9]. The result of the fit has been found to agree with that obtained from the direct D^{**} reconstruction and the combined result for the fraction of D^{**}

decays to the sum of D^{**} and D^* corresponds to $R^{**} = 0.19 \pm 0.08$.

The value of $|V_{cb}|\mathcal{A}_1(1)$ has been extracted by a χ^2 fit to the differential distribution of the number of events dN/dw. The product $|V_{cb}|\mathcal{A}_1(1)$, where $\mathcal{A}_1(w)$ is a new function introduced in the parametrisation of the $\mathcal{F}_1(w)$ form factor [7] and the parameter $\rho_{\mathcal{A}_1}$ have been left free in the fit. The distribution for the combinatorial background has been obtained directly from the data using same sign $\pi\ell$ combinations, that for D^{**} meson from theory [8] and the others from the simulation. The contribution from signal $D^{*+}\ell\nu$ events has been recomputed at each step of the minimisation by reweighting each accepted event by the ratio of the present value of the fitting function to that at the generation. The result is: $|V_{cb}|\mathcal{A}_1(1) = (37.48 \pm 1.38) \times 10^{-3}$ and $\rho_{\mathcal{A}_1} = 1.38 \pm 0.13$. In addition the branching ratio $\mathrm{BR}(\bar{B}^0 \to D^{*+}\ell\nu)$ was determined to be $(5.11 \pm 0.12)\%$ by integrating the the differential decay width.

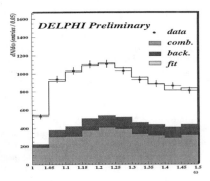

Figure 6: Fit to the dN/dw distribution. The dots with error bars represent the real data, the open hisogram the fitted genuine D^* constribution, the light shaded histogram the combinatorial background and the dark shaded histogram the contributions from other backgrounds sources including the D^{**} decays.

The fit has been repeated assuming different expansions of $\mathcal{A}(w)$. In particular using a linear extrapolation the value $|V_{cb}|\mathcal{A}_1(1) = (37.28 \pm 1.24) \times 10^{-3}$ has been obtained. Several sources of systematics errors have been investigated (see Table 3).

Table 3: Estimate of systematic errors on $|V_{cb}|$.

Source	Syst. Error ($\times 10^{-3}$)
D Decays	± 0.61
B Production and Decays	± 1.14
Detector Systematics	± 0.96
Total	± 1.38

Using $\mathcal{A}(1) = \mathcal{F}(1) = 0.91 \pm 0.03$ the value $|V_{cb}| = (41.19\pm 1.52 \text{ (stat)} \pm 1.77 \text{ (syst exp)} \pm 1.36 \text{ (syst th)}) \times 10^{-3}$ is obtained.

4 Conclusions

New determinations of $|V_{ub}|/|V_{cb}|$ and $|V_{cb}|$ have been obtained in the study of semileptonic B decays using inclusive techniques by the DELPHI experiment at LEP. Both analyses profited from recent theoretical progress to reduce the model systematics in the extraction of the two CKM elements governing the decays of b quarks. The preliminary results are: $|V_{ub}|/|V_{cb}| = 0.104 \pm 0.012 \text{ (stat)} \pm 0.015 \text{ (syst)} \pm 0.009 \text{ (model)}$ and $|V_{cb}| = (41.19\pm1.52 \text{ (stat)} \pm1.77 \text{ (syst exp)} \pm1.36 \text{ (syst th)}) \times 10^{-3}$. From these results the value of $|V_{ub}| = (4.28 \pm 0.86) \times 10^{-3}$ has been derived.

Acknowledgements

I would like to thank my DELPHI colleagues that have contributed to the measurements reported: P.M. Kluit, M. Margoni, E. Piotto and F. Simonetto. I am also grateful to I. Bigi and M. Neubert for discussions of the analysis methods and of the estimate of the model dependent systematics.

References

1. M. Battaglia, P.M. Kluit and E. Piotto (DELPHI Collaboration), DELPHI 98-97 CONF 165, ICHEP-98 Paper 241
   M. Margoni and F. Simonetto (DELPHI Collaboration), DELPHI 98-140 CONF 201, ICHEP-98 Paper 238.

2. C.P. Jessop et al. (CLEO Collaboration) CLEO CONF 98-18, ICHEP-98 Paper 855 and J. Alexander, in these Proceedings.

3. V. Barger, C.S. Kim and R.J.N. Phillips, *Phys. Lett.* B **251**, 629 (1990).

4. A.F. Falk, Z. Ligeti and M.B. Wise, *Phys. Lett.* B **406**, 225 (1997).

5. I. Bigi, R.D. Dikeman and N. Uraltsev, *Eur. Phys. J.* C **4**, 453 (1998).

6. M. Battaglia, DELPHI 98-42 PHYS 772.

7. C.G. Boyd, B. Grinstein and R.F. Lebed, *Phys. Rev.* D **56**, 6895 (1997); I. Caprini, L. Lellouch and M. Neubert, CERN-TH/97-91 and hep-ph/9712417 submitted to *Nucl. Phys.* B.

8. V. Morenas et al., *Phys. Rev.* D **56**, 5668 (1997).

9. N. Isgur and M. Wise, *Phys. Lett.* B **232**, 113 (1989).

A CONSTITUENT QUARK-MESON MODEL FOR HEAVY MESON DECAYS

ALDO DEANDREA

Centre de Physique Théorique, CNRS Luminy, Case 907, F-13288 Marseille Cedex 9, France
E-mail: deandrea@cpt.univ-mrs.fr

I describe a model for heavy meson decays based on an effective quark-meson lagrangian. I consider the heavy mesons S with spin and parity $J^P = (1^+, 0^+)$, H with $J^P = (1^-, 0^-)$ and T^μ with $J^P = (2^+, 1^+)$, i.e. S and P wave heavy-light mesons. The model is constrained by the known symmetries of QCD in the $m_Q \to \infty$ for the heavy quarks and chiral symmetry in the light quark sector. Using a very limited number of free parameters it is possible to compute several phenomenological quantities, e.g. the leptonic B and B^{**} decay constants; the three universal Isgur-Wise form factors: ξ, $\tau_{3/2}$, $\tau_{1/2}$, describing the semi-leptonic decays $B \to D^{(*)}\ell\nu$, $B \to D^{**}\ell\nu$; the strong and radiative D^* decays; the weak semi-leptonic decays of B and D into light mesons: π, ρ, A_1. An overall agreement with data, when available, is achieved.

1 The Model

The model described in the present paper is based on an effective constituent quark-meson lagrangian containing both light and heavy degrees of freedom. It is constrained by the known symmetries of QCD, i.e. chiral symmetry and heavy quark symmetry for the heavy quarks in the limit $m_Q \to \infty$ [1].

The model can be thought of as an intermediate approach between a pure QCD calculation and an effective theory for heavy and light mesons retaining only the symmetries of the problem [2].

It conjugates the symmetry approach of effective lagrangians with well motivated dynamical assumptions on chiral symmetry breaking and confinement. The effective quark-meson interaction can be for example deduced from partial bosonization of an extended Nambu Jona-Lasinio (NJL) model [3]. In the following this dynamical information will be implemented in an effective lagrangian and the few remaining free parameters will be fixed by data, thus allowing a number of predictions based on symmetry and on the implemented dynamics.

This will allow to calculate the parameters of the effective heavy meson theory without solving the non-perturbative QCD problem. In this simplified approach one can hope to describe the essential part of the QCD behavior, at least in a limited energy range, and extract useful information from it. The model is suitable for the description of higher spin heavy mesons as they can be included in the formalism in a very easy way [4] (see also [5]). On the contrary the inclusion of higher order corrections, albeit possible, requires the determination of new free parameters, which proliferate as new orders are added to the expansion. In this sense the model allows a simple and intuitive approach to heavy-meson processes if it is kept at lowest order, while it loses part of its predictive power if corrections have to be included.

1.1 Heavy meson field

In order to implement the heavy quark symmetries in the spectrum of physical states the wave function of a heavy meson has to be independent of the heavy quark flavor and spin. It can be characterized by the total angular momentum s_ℓ of the light degrees of freedom. To each value of s_ℓ corresponds a degenerate doublet of states with angular momentum $J = s_\ell \pm 1/2$. The mesons P and P^* form the spin-symmetry doublet corresponding to $s_\ell = 1/2$ (for charm for instance, they correspond to D and D^*).

The negative parity spin doublet (P, P^*) can be represented by a 4×4 Dirac matrix H, with one spinor index for the heavy quark and the other for the light degrees of freedom.

An explicit matrix representation is:

$$H = \frac{(1 + \not{v})}{2} [P^*_\mu \gamma^\mu - P\gamma_5] \tag{1}$$

$$\bar{H} = \gamma_0 H^\dagger \gamma_0 \ . \tag{2}$$

Here v is the heavy meson velocity, $v^\mu P^*_{a\mu} = 0$ and $M_H = M_P = M_{P^*}$. Moreover $\not{v}H = -H\not{v} = H$, $\bar{H}\not{v} = -\not{v}\bar{H} = \bar{H}$ and $P^{*\mu}$ and P are annihilation operators normalized as follows:

$$\langle 0|P|Q\bar{q}(0^-)\rangle = \sqrt{M_H} \tag{3}$$

$$\langle 0|P^{*\mu}|Q\bar{q}(1^-)\rangle = \epsilon^\mu \sqrt{M_H} \ . \tag{4}$$

The formalism for higher spin states was introduced by Falk and Luke [6]. I shall consider only the S and P-waves of the system $Q\bar{q}$. The heavy quark effective theory predicts two distinct multiplets, one containing a 0^+ and a 1^+ degenerate state, and the other one a 1^+ and a 2^+ state. In matrix notation, analogous to the one used for the negative parity states, they are described by

$$S = \frac{1 + \not{v}}{2}[P^{*\prime}_{1\mu}\gamma^\mu\gamma_5 - P_0] \tag{5}$$

and

$$T^\mu = \frac{1 + \not{v}}{2}\left[P_2^{*\mu\nu}\gamma_\nu - \sqrt{\frac{3}{2}}P_{1\nu}^*\gamma_5\left(g^{\mu\nu} - \frac{1}{3}\gamma^\nu(\gamma^\mu - v^\mu)\right)\right] .$$

(6)

These two multiplets have $s_\ell = 1/2$ and $s_\ell = 3/2$ respectively, where s_ℓ is conserved together with the spin s_Q in the infinite quark mass limit because $\vec{J} = \vec{s}_\ell + \vec{s}_Q$.

1.2 Meson-Quark Interaction

The light degrees of freedom, i.e. the light quark fields χ and the pseudo-scalar $SU(3)$ octet of mesons π are introduced using the chiral lagrangian:

$$\mathcal{L}_{\ell\ell} = \bar{\chi}(iD^\mu\gamma_\mu + g_A\mathcal{A}^\mu\gamma_\mu\gamma_5)\chi - m\bar{\chi}\chi$$
$$+ \frac{f_\pi^2}{8}\partial_\mu\Sigma^\dagger\partial^\mu\Sigma.$$

(7)

Here $D_\mu = \partial_\mu - i\mathcal{V}_\mu$, $\xi = \exp(i\pi/f_\pi)$, $\Sigma = \xi^2$, $f_\pi = 130$ MeV and

$$\mathcal{V}^\mu = \frac{1}{2}(\xi^\dagger\partial^\mu\xi + \xi\partial^\mu\xi^\dagger)$$
$$\mathcal{A}^\mu = \frac{i}{2}(\xi^\dagger\partial^\mu\xi - \xi\partial^\mu\xi^\dagger) .$$

(8)

The term with g_A is the coupling of pions to light quarks; it will not be used in the sequel. It is a free parameter, but in NJL model $g_A = 1$.

One can introduce a quark-meson effective lagrangian involving heavy and light quarks and heavy mesons. At lowest order one has:

$$\mathcal{L}_{h\ell} = \bar{Q}_v iv \cdot \partial Q_v - \left(\bar{\chi}(\bar{H} + \bar{S} + i\bar{T}_\mu\frac{D^\mu}{\Lambda_\chi})Q_v + h.c.\right)$$
$$+ \frac{1}{2G_3}\text{Tr}[(\bar{H} + \bar{S})(H - S)] + \frac{1}{2G_4}\text{Tr}[\bar{T}_\mu T^\mu]$$

(9)

where the meson fields H, S, T have been defined is section 1.1, Q_v is the effective heavy quark field, G_3, G_4 are coupling constants and Λ_χ (= 1 GeV) has been introduced for dimensional reasons. Lagrangian (9) has heavy spin and flavor symmetry. This lagrangian comprises three terms containing respectively H, S and T. Note that the fields H and S have the same coupling constant. In doing this one assumes that this effective quark-meson lagrangian can be justified as a remnant of a four quark interaction of the NJL type by partial bosonization.

1.3 Cut-Off Prescription

The cut-off prescription is the way in which part of the dynamical information regarding QCD is introduced in the model, this is why it is crucial and is part of the definition of the model. As the heavy mesons are described

consistently with HQET, the heavy quark propagator in the loop contains the residual momentum k which arises from the interaction with the light degrees of freedom. It is natural to assume an ultraviolet cut-off on the loop momentum of the order of $\Lambda \simeq 1$ GeV, even if the heavy quark mass is larger than the cut-off.

In the infrared the model is not confining and its range of validity can not be extended below energies of the order of Λ_{QCD}. In practice one introduces an infrared cut-off μ, to take this into account.

The cut-off prescription is implemented via a proper time regularization (a different choice is followed in [7]). After continuation to the Euclidean space it reads, for the light quark propagator:

$$\int d^4 k_E \frac{1}{k_E^2 + m^2} \to \int d^4 k_E \int_{1/\mu^2}^{1/\Lambda^2} ds\, e^{-s(k_E^2 + m^2)}$$

(10)

where μ and Λ are infrared and ultraviolet cut-offs.

Reasonable values are $\Lambda \simeq 1$ GeV, $m \simeq \mu \simeq 10^2$ MeV. The cut-off prescription is similar to the one in [3], with $\Lambda = 1.25$ GeV; the numerical results are not strongly dependent on the value of Λ. The constituent mass m in the NJL models represents the order parameter discriminating between the phases of broken and unbroken chiral symmetry and can be fixed by solving a gap equation, which gives m as a function of the scale mass μ for given values of the other parameters. In the second paper of Ebert et al. [3] the values $m = 300$ MeV and $\mu = 300$ MeV are used and we shall assume the same values. As shown there, for smaller values of μ, m is constant (=300 MeV) while for much larger values of μ, it decreases and in particular it vanishes for $\mu = 550$ MeV.

2 Analytical and Numerical Results

2.1 Decay constants

The leptonic decay constants $\hat{F}$ and $\hat{F}^+$ are defined as follows:

$$\langle 0|\bar{q}\gamma^\mu\gamma_5 Q|H(0^-, v)\rangle = i\sqrt{M_H}v^\mu\hat{F}$$

(11)

$$\langle 0|\bar{q}\gamma^\mu Q|S(0^+, v)\rangle = i\sqrt{M_S}v^\mu\hat{F}^+ .$$

(12)

and they can be computed by a loop calculation, where the heavy meson interacts with the heavy and light quarks (via the interaction introduced in (9)) and then those interact with the current. The result is

$$\hat{F} = \frac{\sqrt{Z_H}}{G_3}$$

(13)

$$\hat{F}^+ = \frac{\sqrt{Z_S}}{G_3}$$

(14)

where Z_H and Z_S are the field renormalization constants. Detailed results can be found in Deandrea et al. [4]. Values

Table 1: Renormalization constants and couplings. Δ_H in GeV; G_3, G_4 in GeV^{-2}, Z_j in GeV^{-1}.

Δ_H	$1/G_3$	Z_H	Z_S	Z_T	$1/G_4$
0.3	0.16	4.17	1.84	2.95	0.15
0.4	0.22	2.36	1.14	1.07	0.26
0.5	0.345	1.14	0.63	0.27	0.66

Table 2: $\hat{F}$ and $\hat{F}^+$ for various values of Δ_H. Δ_H in GeV, leptonic constants in GeV$^{3/2}$.

Δ_H	$\hat{F}$	$\hat{F}^+$
0.3	0.33	0.22
0.4	0.34	0.24
0.5	0.37	0.27

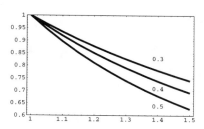

Figure 1: Isgur-Wise form factor at different Δ values.

for the renormalization constants and couplings can be read in Table 1 for three values of the parameter Δ_H.

The numerical results for the decay constants can be found in Table 2. Neglecting logarithmic corrections, $\hat{F}$ and $\hat{F}^+$ are related, in the infinite heavy quark mass limit, to the leptonic decay constant f_B and f^+. For example, for $\Delta_H = 400$ MeV, one obtains from Table 2:

$$f_B \simeq 150 \text{ MeV} \tag{15}$$
$$f^+ \simeq 100 \text{ MeV} . \tag{16}$$

2.2 Semi-leptonic Decays and Form Factors

As an example of the quantities that can be analytically calculated in the model, one can examine the Isgur-Wise function ξ:

$$\langle D(v') \mid \bar{c}\gamma_\mu(1-\gamma_5)b|B(v)\rangle = \sqrt{M_B M_D}$$
$$\times C_{cb}\xi(\omega)(v_\mu + v'_\mu) \tag{17}$$

$$\langle D^*(v',\epsilon)|\bar{c}\gamma_\mu(1-\gamma_5)b|B(v)\rangle = \sqrt{M_B M_{D^*}} C_{cb}\,\xi(\omega)$$
$$\times[i\epsilon_{\mu\nu\alpha\beta}\epsilon^{*\nu}v'^\alpha v^\beta - (1+\omega)\epsilon^*_\mu + (\epsilon^* \cdot v)v'_\mu] \tag{18}$$

where $\omega = v \cdot v'$ and C_{cb} is a coefficient containing logarithmic corrections depending on α_s; within our approximation it can be put equal to 1: $C_{cb} = 1$. We also note that, in the leading order we are considering here $\xi(1) = 1$.

One finds [3]:

$$\xi(\omega) = Z_H \left[\frac{2}{1+\omega}I_3(\Delta_H) + \left(m + \frac{2\Delta_H}{1+\omega}\right)I_5(\Delta_H, \Delta_H, \omega) \right] . \tag{19}$$

The ξ function is plotted in Fig. 1.

The integrals I_3, I_5 can be found in the Appendix.

One can compute in a similar way the form factors describing the semi-leptonic decays of a meson belonging to the fundamental negative parity multiplet H into the positive parity mesons in the S and T multiplets. Examples of these decays are

$$B \to D^{**}\ell\nu \tag{20}$$

where D^{**} can be either a S state (i.e. a 0^+ or 1^+ charmed meson having $s_\ell = 1/2$) or a T state (i.e. a 2^+ or 1^+ charmed meson having $s_\ell = 3/2$).

The decays in (20) are described by two form factors $\tau_{1/2}, \tau_{3/2}$ [8] which can be computed by a loop calculation similar to the one used to obtain $\xi(\omega)$. The result is

$$\tau_{1/2}(\omega) = \frac{\sqrt{Z_H Z_S}}{2(1-\omega)}\Big[I_3(\Delta_S) - I_3(\Delta_H)$$
$$+ (\Delta_H - \Delta_S + m(1-\omega)) I_5(\Delta_H, \Delta_S, \omega)\Big] \tag{21}$$

and

$$\tau_{3/2}(\omega) = -\frac{\sqrt{Z_H Z_T}}{\sqrt{3}}$$
$$\times \Big[m\Big(\frac{I_3(\Delta_H) - I_3(\Delta_T) - (\Delta_H - \Delta_T) I_5(\Delta_H, \Delta_T, \omega)}{2(1-\omega)}$$
$$- \frac{I_3(\Delta_H) + I_3(\Delta_T) + (\Delta_H + \Delta_T) I_5(\Delta_H, \Delta_T, \omega)}{2(1+\omega)}\Big)$$
$$- \frac{1}{2(-1-\omega+\omega^2+\omega^3)}\Big(- 3 S(\Delta_H, \Delta_T, \omega)$$
$$- (1-2\omega) S(\Delta_T, \Delta_H, \omega) + (1-\omega^2) T(\Delta_H, \Delta_T, \omega)$$
$$- 2(1-2\omega) U(\Delta_H, \Delta_T, \omega)\Big)\Big] \tag{22}$$

where the integrals S, T, U are defined in the Appendix.

The numerical results are reported in Table 3. For a comparison with other calculations of these form factors see [9].

Table 3: Form factors and slopes. Δ_H in GeV.

Δ_H	$\xi(1)$	ρ_{IW}^2	$\tau_{1/2}(1)$	$\rho_{1/2}^2$	$\tau_{3/2}(1)$	$\rho_{3/2}^2$
0.3	1	0.72	0.08	0.8	0.48	1.4
0.4	1	0.87	0.09	1.1	0.56	2.3
0.5	1	1.14	0.09	2.7	0.67	3.0

Table 4: Theoretical and experimental D^* branching ratios (%). Theoretical values are computed with $\Delta_H = 0.4$ GeV.

Decay mode	Br (%)	Exp
$D^{*0} \to D^0 \pi^0$	65.5	61.9 ± 2.9
$D^{*+} \to D^0 \pi^+$	71.6	68.3 ± 1.4
$D^{*+} \to D^+ \pi^0$	28.0	30.6 ± 2.5

An important test of our approach is represented by the Bjorken sum rule, which states

$$\rho_{IW}^2 = \frac{1}{4} + \sum_k \left[|\tau_{1/2}^{(k)}(1)|^2 + 2|\tau_{3/2}^{(k)}(1)|^2 \right] . \tag{23}$$

Numerically we find that the first excited resonances, i.e. the S and T states ($k = 0$) practically saturate the sum rule for all the three values of Δ_H.

2.3 Strong decays

The model can be used to calculate strong coupling constants, such as those concerning the decays:

$$H \to H\pi \tag{24}$$
$$S \to H\pi \tag{25}$$

The constant $g_{D^* D\pi}$ is related to the strong coupling constant of the effective meson field theory g appearing in the heavy meson effective lagrangian [2]

$$\mathcal{L} = ig\mathrm{Tr}(\overline{H}H\gamma^\mu\gamma_5\mathcal{A}_\mu) + \left[ih\,\mathrm{Tr}(\overline{H}S\gamma^\mu\gamma_5\mathcal{A}_\mu) + \mathrm{h.c.}\right] \tag{26}$$

by the relation

$$g_{D^* D\pi} = \frac{2m_D}{f_\pi} g \tag{27}$$

valid in the $m_Q \to \infty$ limit.

Numerically one gets

$$g = 0.456 \pm 0.040 \tag{28}$$

where the central value corresponds to $\Delta_H = 0.4$ GeV and the lower (resp. higher) value corresponds to $\Delta_H = 0.3$ GeV (resp. $\Delta_H = 0.5$ GeV). In an analogous way one obtains

$$h = -0.85 \pm 0.02 \tag{29}$$

The details of the calculation can be found in Deandrea et al. [4]. Once the coupling constants are calculated, it is possible to make predictions for branching ratios in strong heavy mesons decays. The results are given in Table 4 and are in good agreement with experimental data.

3 Conclusions

Starting from an effective lagrangian at the level of mesons and constituent quarks, one can calculate meson transition amplitudes by evaluating loops of heavy and light quarks. In this way it is possible to compute the Isgur-Wise function, the form factors $\tau_{1/2}$ and $\tau_{3/2}$, the leptonic decay constant $\hat{F}$ and $\hat{F}^+$, and many other quantities, such as radiative and strong couplings and decays which are only briefly mentioned in this short note (see for details [4] and [10]). The agreement with data, when available, is very good in most cases. The model is able to describe a number of essential features of heavy meson physics in a simple and compact way.

Acknowledgments

I acknowledge the support of a "Marie Curie" TMR research fellowship of the European Commission under contract ERBFMBICT960965. The subject described in this paper is based on work in collaboration with R. Gatto, G. Nardulli, A. Polosa and in the early stage of the work with N. Di Bartolomeo. I wish to thank them all. Centre de Physique Théorique is Unité Propre de Recherche 7061.

Appendix

The integrals used in the paper are listed in this appendix. A more exhaustive list of integrals with proper time regularization useful for calculations in the model described here can be found in the appendix of [4].

$$I_1 = \frac{iN_c}{16\pi^4} \int^{reg} \frac{d^4k}{(k^2 - m^2)} = \frac{N_c m^2}{16\pi^2} \Gamma(-1, \frac{m^2}{\Lambda^2}, \frac{m^2}{\mu^2})$$

$$I_3(\Delta) = -\frac{iN_c}{16\pi^4} \int^{reg} \frac{d^4k}{(k^2 - m^2)(v \cdot k + \Delta + i\epsilon)}$$

$$= \frac{N_c}{16\pi^{3/2}} \int_{1/\Lambda^2}^{1/\mu^2} \frac{ds}{s^{3/2}} e^{-s(m^2 - \Delta^2)} \left(1 + \mathrm{erf}(\Delta\sqrt{s}) \right)$$

where m is the constituent light quark mass of the order of 300 MeV and

$$\Gamma(\alpha, x_0, x_1) = \int_{x_0}^{x_1} dt \, e^{-t} \, t^{\alpha-1}$$

is the generalized incomplete gamma function and erf is the error function. In order to keep I_5 and I_6 in a short form one can introduce the function:

$$\sigma(x, \Delta_1, \Delta_2, \omega) = \frac{\Delta_1 (1-x) + \Delta_2 x}{\sqrt{1 + 2(\omega-1)x + 2(1-\omega)x^2}}$$

Finally one has:

$$I_5 (\Delta_1, \Delta_2, \omega) = \int_0^1 dx \frac{1}{1 + 2x^2(1-\omega) + 2x(\omega-1)} \times$$
$$\left[\frac{6}{16\pi^{3/2}} \int_{1/\Lambda^2}^{1/\mu^2} ds \, \sigma \, e^{-s(m^2-\sigma^2)} \, s^{-1/2} (1 + \mathrm{erf}(\sigma\sqrt{s})) + \right.$$
$$\left. \frac{6}{16\pi^2} \int_{1/\Lambda^2}^{1/\mu^2} ds \, e^{-s(m^2-2\sigma^2)} \, s^{-1} \right]$$

$$I_6 (\Delta_1, \Delta_2, \omega) = I_1 \int_0^1 dx \frac{\sigma}{1 + 2x^2(1-\omega) + 2x(\omega-1)}$$
$$- \frac{N_c}{16\pi^{3/2}} \int_0^1 dx \frac{1}{1 + 2x^2(1-\omega) + 2x(\omega-1)} \times$$
$$\int_{1/\Lambda^2}^{1/\mu^2} \frac{ds}{s^{3/2}} e^{-s(m^2-\sigma^2)} \left\{ \sigma[1 + \mathrm{erf}(\sigma\sqrt{s})] \times \right.$$
$$\left. [1 + 2s(m^2 - \sigma^2)] + 2\sqrt{\frac{s}{\pi}} e^{-s\sigma^2} \left[\frac{3}{2s} + (m^2 - \sigma^2) \right] \right\}$$

$$S(\Delta_1, \Delta_2, \omega) = \Delta_1 I_3(\Delta_2) + \omega (I_1 + \Delta_2 I_3(\Delta_2))$$
$$+ \Delta_1{}^2 I_5(\Delta_1, \Delta_2, \omega)$$
$$T(\Delta_1, \Delta_2, \omega) = m^2 I_5(\Delta_1, \Delta_2, \omega) + I_6(\Delta_1, \Delta_2, \omega)$$
$$U(\Delta_1, \Delta_2, \omega) = I_1 + \Delta_2 I_3(\Delta_2) + \Delta_1 I_3(\Delta_1)$$
$$+ \Delta_2 \Delta_1 I_5(\Delta_1, \Delta_2, \omega) \tag{30}$$

References

1. For further references see the review papers: H. Georgi, contribution to the *Proceedings of TASI 91*, R.K. Ellis ed., World Scientific, Singapore, 1991; B. Grinstein, contribution to *High Energy Phenomenology*, R. Huerta and M.A. Peres eds., World Scientific, Singapore, 1991; N. Isgur and M. Wise, contribution to *Heavy Flavours*, A. Buras and M. Lindner eds., World Scientific, Singapore, 1992; M. Neubert, *Phys. Rep.* **245** 259 (1994).
2. R. Casalbuoni et al., *Phys. Rep.* **281** 145 (1997).
3. D. Ebert, T. Feldmann, R. Friedrich and H. Reinhardt, *Nucl. Phys.* B **434**, 619 (1995); D. Ebert, T. Feldmann and H. Reinhardt, *Phys. Lett.* B **388**, 154 (1996).
4. A. Deandrea, N. Di Bartolomeo, R. Gatto, G. Nardulli, A.D. Polosa, *Phys. Rev.* D **58**, 034004 (1998).
5. T. Feldmann, hep-ph/9606451 and also PhD Thesis (in German) available at http://wptu38.physik.uni-wuppertal.de/~feldmann/pub/thesis.ps
6. A. Falk and M. Luke *Phys. Lett.* B **292**, 119 (1992).
7. W. H. Bardeen and C. T. Hill, *Phys. Rev.* D **49**, 409 (1994). B. Holdom and M. Sutherland, *Phys. Rev.* D **47**, 5067 (1993), B. Holdom and M. Sutherland, *Phys. Lett.* B **313**, 447 (1993), B. Holdom and M. Sutherland, *Phys. Rev.* D **48**, 5196 (1993), B. Holdom, M. Sutherland and J. Mureika, *Phys. Rev.* D **49**, 2359 (1994), M. Sutherland, B. Holdom, S. Jaimungal, and R. Lewis, *Phys. Rev.* D **51**, 5053 (1995).
8. N. Isgur and M. B. Wise, *Phys. Rev.* D **43**, 819 (1991).
9. V. Morenas et al., *Phys. Rev.* D **56**, 5668 (1997).
10. A. Deandrea, R. Gatto, G. Nardulli, A.D. Polosa, BARI-TH/98-314

TOWARDS THE EXTRACTION OF THE CKM ANGLE γ

ROBERT FLEISCHER

Theory Division, CERN, CH-1211 Geneva 23, Switzerland
E-mail: Robert.Fleischer@cern.ch

The determination of the angle γ of the unitarity triangle of the CKM matrix is regarded as a challenge for future B-physics experiments. In this context, the decays $B^\pm \to \pi^\pm K$ and $B_d \to \pi^\mp K^\pm$, which were observed by the CLEO collaboration last year, received a lot of interest in the literature. After a general parametrization of their decay amplitudes, strategies to constrain and determine the CKM angle γ with the help of the corresponding observables are reviewed. The theoretical accuracy of these methods is limited by certain rescattering and electroweak penguin effects. It is emphasized that the rescattering processes can be included in the bounds on γ by using additional experimental information on $B^\pm \to K^\pm K$ decays, and steps towards the control of electroweak penguins are pointed out. Moreover, strategies to probe the CKM angle γ with the help of $B_s \to K\overline{K}$ decays are briefly discussed.

1 Introduction

Among the central targets of the B-factories, which will start operating in the near future, is the direct measurement of the three angles α, β and γ of the usual, non-squashed, unitarity triangle of the Cabibbo–Kobayashi–Maskawa matrix (CKM matrix). From an experimental point of view, the determination of the angle γ is particularly challenging, although there are several strategies on the market, allowing – at least in principle – a theoretically clean extraction of γ (for a review, see for instance Ref. 1).

In order to obtain direct information on this angle in an experimentally feasible way, the decays $B^+ \to \pi^+ K^0$, $B_d^0 \to \pi^- K^+$ and their charge conjugates appear very promising.[2-4] Last year, the CLEO collaboration reported the observation of several exclusive B-meson decays into two light pseudoscalar mesons, including also these modes.[5] So far, only results for the combined branching ratios

$$\text{BR}(B^\pm \to \pi^\pm K) \equiv$$
$$\frac{1}{2}\left[\text{BR}(B^+ \to \pi^+ K^0) + \text{BR}(B^- \to \pi^- \overline{K^0})\right] \quad (1)$$

$$\text{BR}(B_d \to \pi^\mp K^\pm) \equiv$$
$$\frac{1}{2}\left[\text{BR}(B_d^0 \to \pi^- K^+) + \text{BR}(\overline{B_d^0} \to \pi^+ K^-)\right] \quad (2)$$

have been published, with values at the 10^{-5} level and large experimental uncertainties.

A particularly interesting situation arises if the ratio

$$R \equiv \frac{\text{BR}(B_d \to \pi^\mp K^\pm)}{\text{BR}(B^\pm \to \pi^\pm K)} \quad (3)$$

is found to be smaller than 1. In this case, the following allowed range for γ is implied:[6]

$$0° \le \gamma \le \gamma_0 \quad \lor \quad 180° - \gamma_0 \le \gamma \le 180°, \quad (4)$$

where γ_0 is given by

$$\gamma_0 = \arccos(\sqrt{1 - R}). \quad (5)$$

Unfortunately, the present data do not yet provide a definite answer to the question of whether $R < 1$. The results reported by the CLEO collaboration last year give $R = 0.65 \pm 0.40$,[5] whereas an updated analysis, which was presented at this conference, yields $R = 1.0 \pm 0.4$.[7] Since (4) is complementary to the presently allowed range of $41° \lesssim \gamma \lesssim 134°$ arising from the usual fits of the unitarity triangle,[8] this bound would be of particular phenomenological interest (for a detailed study, see Ref. 9). It relies on the following three assumptions:

   i) $SU(2)$ isospin symmetry can be used to derive relations between the $B^+ \to \pi^+ K^0$ and $B_d^0 \to \pi^- K^+$ QCD penguin amplitudes.

   ii) There is no non-trivial CP-violating weak phase present in the $B^+ \to \pi^+ K^0$ decay amplitude.

   iii) Electroweak (EW) penguins play a negligible role in the decays $B^+ \to \pi^+ K^0$ and $B_d^0 \to \pi^- K^+$.

Whereas (i) is on solid theoretical ground, provided the "tree" and "penguin" amplitudes of the $B \to \pi K$ decays are defined properly,[10] (ii) may be affected by rescattering processes of the kind $B^+ \to \{\pi^0 K^+\} \to \pi^+ K^0$.[11-15] As for (iii), EW penguins may also play a more important role than is indicated by simple model calculations.[3,13] Consequently, in the presence of large rescattering and EW penguin effects, strategies more sophisticated[16,17] than the "naïve" bounds sketched above are needed to probe the CKM angle γ with $B \to \pi K$ decays. Before turning to these methods, let us first have a look at the corresponding decay amplitudes.

2 The General Description of $B^\pm \to \pi^\pm K$ and $B_d \to \pi^\mp K^\pm$ within the Standard Model

Within the framework of the Standard Model, the most important contributions to the decays $B^+ \to \pi^+ K^0$ and $B_d^0 \to \pi^- K^+$ arise from QCD penguin topologies. The

$B \to \pi K$ decay amplitudes can be expressed as follows:

$$A(B^+ \to \pi^+ K^0) = \lambda_u^{(s)}(P_u + P_{\text{ew}}^u + \mathcal{A})$$
$$+ \lambda_c^{(s)}(P_c + P_{\text{ew}}^c) + \lambda_t^{(s)}(P_t + P_{\text{ew}}^t) \qquad (6)$$

$$A(B_d^0 \to \pi^- K^+) = -\left[\lambda_u^{(s)}(\tilde{P}_u + \tilde{P}_{\text{ew}}^u + \tilde{\mathcal{T}})\right.$$
$$\left. + \lambda_c^{(s)}(\tilde{P}_c + \tilde{P}_{\text{ew}}^c) + \lambda_t^{(s)}(\tilde{P}_t + \tilde{P}_{\text{ew}}^t)\right], \qquad (7)$$

where P_q, $\tilde{P}_q$ and P_{ew}^q, $\tilde{P}_{\text{ew}}^q$ denote contributions from QCD and electroweak penguin topologies with internal q quarks ($q \in \{u,c,t\}$), respectively, $\mathcal{A}$ is related to annihilation topologies, $\tilde{\mathcal{T}}$ is due to colour-allowed $\bar{b} \to \bar{u}u\bar{s}$ tree-diagram-like topologies, and $\lambda_q^{(s)} \equiv V_{qs}V_{qb}^*$ are the usual CKM factors. Because of the tiny ratio $|\lambda_u^{(s)}/\lambda_t^{(s)}| \approx 0.02$, the QCD penguins play the dominant role in Eqs. (6) and (7), despite their loop suppression.

Making use of the unitarity of the CKM matrix and applying the Wolfenstein parametrization[18] yields

$$A(B^+ \to \pi^+ K^0) = -\left(1 - \frac{\lambda^2}{2}\right)\lambda^2 A\left[1 + \rho\,e^{i\theta}e^{i\gamma}\right]\mathcal{P}_{tc},$$
$$(8)$$

where

$$\mathcal{P}_{tc} \equiv |\mathcal{P}_{tc}|\,e^{i\delta tc} = (P_t - P_c) + (P_{\text{ew}}^t - P_{\text{ew}}^c) \qquad (9)$$

and

$$\rho\,e^{i\theta} = \frac{\lambda^2 R_b}{1 - \lambda^2/2}\left[1 - \left(\frac{\mathcal{P}_{uc} + \mathcal{A}}{\mathcal{P}_{tc}}\right)\right]. \qquad (10)$$

In these expressions, δ_{tc} and θ denote CP-conserving strong phases, $\mathcal{P}_{uc}$ is defined in analogy to Eq. (9), $\lambda \equiv |V_{us}| = 0.22$, $A \equiv |V_{cb}|/\lambda^2 = 0.81 \pm 0.06$, and $R_b \equiv |V_{ub}/(\lambda V_{cb})| = 0.36 \pm 0.08$. The quantity $\rho\,e^{i\theta}$ is a measure of the strength of certain rescattering effects, as will be discussed in more detail in Section 4.

If we apply the $SU(2)$ isospin symmetry of strong interactions, implying

$$\tilde{P}_c = P_c \quad \text{and} \quad \tilde{P}_t = P_t, \qquad (11)$$

the QCD penguin topologies with internal top and charm quarks contributing to $B^+ \to \pi^+ K^0$ and $B_d^0 \to \pi^- K^+$ can be related to each other, yielding the following amplitude relations (for a detailed discussion, see Ref. 10):

$$A(B^+ \to \pi^+ K^0) \equiv P \qquad (12)$$
$$A(B_d^0 \to \pi^- K^+) = -[P + T + P_{\text{ew}}], \qquad (13)$$

which play a central role to probe the CKM angle γ. Here the "penguin" amplitude P is *defined* by the $B^+ \to \pi^+ K^0$ decay amplitude, the quantity

$$P_{\text{ew}} \equiv -|P_{\text{ew}}|e^{i\delta_{\text{ew}}} = -\left(1 - \frac{\lambda^2}{2}\right)\lambda^2 A$$
$$\times \left[\left(\tilde{P}_{\text{ew}}^t - \tilde{P}_{\text{ew}}^c\right) - \left(P_{\text{ew}}^t - P_{\text{ew}}^c\right)\right] \qquad (14)$$

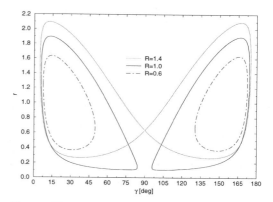

Figure 1: The contours in the $\gamma - r$ plane for $|A_0| = 0.2$ and for various values of R in the case of $\rho = \epsilon = 0$.

is essentially due to electroweak penguins, and

$$T \equiv |T|e^{i\delta_T}e^{i\gamma} = \lambda^4 A R_b\left[\tilde{\mathcal{T}} - \mathcal{A} + \left(\tilde{P}_u - P_u\right)\right.$$
$$\left. + \left(\tilde{P}_{\text{ew}}^u - \tilde{P}_{\text{ew}}^t\right) - \left(P_{\text{ew}}^u - P_{\text{ew}}^t\right)\right]e^{i\gamma} \qquad (15)$$

is usually referred to as a "tree" amplitude. However, owing to a subtlety in the implementation of the isospin symmetry, the amplitude T does not only receive contributions from colour-allowed tree-diagram-like topologies, but also from penguin and annihilation topologies.[10,16] It is an easy exercise to convince oneself that the amplitudes P, T and P_{ew} are well-defined physical quantities.[16]

In the parametrization of the $B^{\pm} \to \pi^{\pm}K$ and $B_d \to \pi^{\mp}K^{\pm}$ observables, it turns out to be very useful to introduce the quantities

$$r \equiv \frac{|T|}{\sqrt{\langle|P|^2\rangle}}, \quad \epsilon \equiv \frac{|P_{\text{ew}}|}{\sqrt{\langle|P|^2\rangle}}, \qquad (16)$$

with $\langle|P|^2\rangle \equiv (|P|^2 + |\overline{P}|^2)/2$, as well as the CP-conserving strong phase differences

$$\delta \equiv \delta_T - \delta_{tc}, \quad \Delta \equiv \delta_{\text{ew}} - \delta_{tc}. \qquad (17)$$

In addition to the ratio R of combined $B \to \pi K$ branching ratios defined by Eq. (3), also the "pseudo-asymmetry"

$$A_0 \equiv \frac{\text{BR}(B_d^0 \to \pi^- K^+) - \text{BR}(\overline{B_d^0} \to \pi^+ K^-)}{\text{BR}(B^+ \to \pi^+ K^0) + \text{BR}(B^- \to \pi^- \overline{K^0})} \qquad (18)$$

plays an important role to probe the CKM angle γ. Explicit expressions for R and A_0 in terms of the parameters specified above are given in Ref. 16.

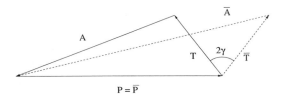

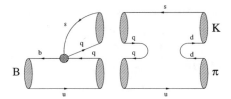

Figure 2: Triangle construction with $A \equiv A(B_d^0 \to \pi^- K^+)$ and $\overline{A} \equiv A(\overline{B_d^0} \to \pi^+ K^-)$ to determine the CKM angle γ in the case of $\rho = \epsilon = 0$, where $P \equiv A(B^+ \to \pi^+ K^0) = A(B^- \to \pi^- \overline{K^0}) \equiv \overline{P}$.

Figure 3: Illustration of rescattering processes of the kind $B^+ \to \left\{ F_q^{(s)} \right\} \to \pi^+ K^0$. The shaded circle represents insertions of the usual current–current operators $Q_{1,2}^q$ ($q \in \{c, u\}$).

3 Strategies to Constrain and Determine the CKM Angle γ with the Help of $B^\pm \to \pi^\pm K$ and $B_d \to \pi^\mp K^\pm$ Decays

The observables R and A_0 provide valuable information about the CKM angle γ. If in addition to R also the asymmetry A_0 can be measured, it is possible to eliminate the strong phase δ in the expression for R, and contours in the $\gamma - r$ plane can be fixed;[16] these are shown in Fig. 1 for $|A_0| = 0.2$ and for various values of R. These contours correspond to a mathematical implementation of a simple triangle construction,[2] which is illustrated in Fig. 2. In both Figs. 1 and 2, rescattering and EW penguin effects have been neglected for simplicity. A detailed study of their impact can be found in Refs. 16 and 17.

In order to determine the CKM angle γ, the quantity r, i.e. the magnitude of the "tree" amplitude T, has to be fixed. At this step, a certain model dependence enters. In recent studies based on "factorization", the authors of Refs. 3 and 4 came to the conclusion that a future theoretical uncertainty of r as small as $\mathcal{O}(10\%)$ may be achievable. In this case, the determination of γ at future B-factories would be limited by statistics rather than by the uncertainty introduced through r, and $\Delta\gamma$ at the level of $10°$ could in principle be achieved. However, since the properly defined amplitude T (see Eq. (15)) does not only receive contributions from colour-allowed "tree" topologies, but also from penguin and annihilation processes,[10,16] it may be shifted sizeably from its "factorized" value so that $\Delta r = \mathcal{O}(10\%)$ may be too optimistic.

Interestingly, it is possible to derive bounds on γ that do *not* depend on r at all.[6] To this end, we eliminate again the strong phase δ in the ratio R of combined $B \to \pi K$ branching ratios. If we now treat r as a "free" variable, while keeping (ρ, θ) and (ϵ, Δ) fixed, we find that R takes the following minimal value:[16]

$$R_{\min} = \kappa \sin^2 \gamma + \frac{1}{\kappa} \left(\frac{A_0}{2 \sin \gamma} \right)^2. \qquad (19)$$

In this expression, which is valid *exactly*, rescattering and

EW penguin effects are described by

$$\kappa = \frac{1}{w^2} \left[1 + 2 \left(\epsilon w \right) \cos \Delta + \left(\epsilon w \right)^2 \right], \qquad (20)$$

with

$$w = \sqrt{1 + 2\rho \cos\theta \cos\gamma + \rho^2}. \qquad (21)$$

An allowed range for γ is related to $R_{\min}$, since values of γ implying $R_{\exp} < R_{\min}$ are excluded ($R_{\exp}$ denotes the experimentally determined value of R). This range can also be read off from the contour in the $\gamma - r$ plane corresponding to the measured values of R and A_0, as can be seen in Fig. 1.

The theoretical accuracy of these contours and of the associated bounds on γ is limited by rescattering and EW penguin effects, which will be discussed in the following two sections. In the "original" bounds on γ derived in Ref. 6, no information provided by A_0 has been used, i.e. both r and δ were kept as "free" variables, and the special case $\rho = \epsilon = 0$ has been assumed, implying $\sin^2 \gamma < R_{\exp}$. Note that a measurement of $A_0 \neq 0$ allows us to exclude a certain range of γ around $0°$ and $180°$.

4 The Role of Rescattering Processes

In the formalism discussed above, rescattering processes are closely related to the quantity ρ (see Eq. (10)), which is highly CKM-suppressed by $\lambda^2 R_b \approx 0.02$ and receives contributions from penguin topologies with internal top, charm and up quarks, as well as from annihilation topologies. Naïvely, one would expect that annihilation processes play a very minor role, and that penguins with internal top quarks are the most important ones. However, also penguins with internal charm and up quarks lead, in general, to important contributions.[19] Simple model calculations, performed at the perturbative quark level, do not indicate a significant compensation of the large CKM suppression of ρ through these topologies. However, these crude estimates do not take into account certain rescattering processes,[11-15] which may play an important role and can be divided into two classes:[10]

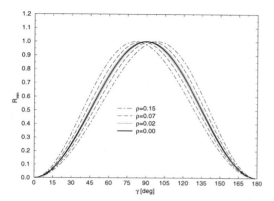

Figure 4: The effect of final-state interactions on $R_{\min}$ for $A_0 = 0$. The curves for a given value of ρ correspond to $\theta \in \{0°, 180°\}$ and represent the maximal shift from $\rho = 0$.

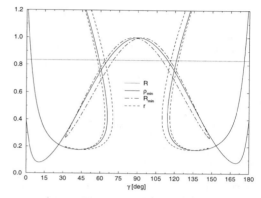

Figure 5: Illustration of the strategy to control rescattering effects in $R_{\min}$ and the contours in the $\gamma - r$ plane through the decay $B^{\pm} \to K^{\pm} K$ within a simple model (for details, see Ref. 17).

i) $B^+ \to \{\overline{D^0} D_s^+, \overline{D^0} D_s^{*+}, \ldots\} \to \pi^+ K^0$

ii) $B^+ \to \{\pi^0 K^+, \pi^0 K^{*+}, \ldots\} \to \pi^+ K^0$,

where the dots include also intermediate multibody states. These processes are illustrated in Fig. 3. Here the shaded circle represents insertions of the usual current–current operators

$$Q_1^q = (\bar{q}_\alpha s_\beta)_{V-A} (\bar{b}_\beta q_\alpha)_{V-A}$$
$$Q_2^q = (\bar{q}_\alpha s_\alpha)_{V-A} (\bar{b}_\beta q_\beta)_{V-A} , \qquad (22)$$

where α and β are colour indices, and $q \in \{c, u\}$. The rescattering processes (i) and (ii) correspond to $q = c$ and u, respectively.

If we look at Fig. 3, we observe that the final-state-interaction (FSI) effects of type (i) can be considered as long-distance contributions to penguin topologies with internal charm quarks, i.e. to the P_c amplitude. They may affect BR($B^{\pm} \to \pi^{\pm} K$) significantly. On the other hand, the rescattering processes characterized by (ii) result in long-distance contributions to penguin topologies with internal up quarks and to annihilation topologies, i.e. to the amplitudes P_u and $\mathcal{A}$. They play a minor role for BR($B^{\pm} \to \pi^{\pm} K$), but may affect assumption (ii) listed in Section 1, thereby leading to a sizeable CP asymmetry, A_+, as large as $\mathcal{O}(10\%)$ in this mode.[12-15] The point is as follows: while we would have $\rho \approx 0$ if rescattering processes of type (i) played the dominant role in $B^+ \to \pi^+ K^0$, or $\rho = \mathcal{O}(\lambda^2 R_b)$ if both processes had similar importance, ρ would be as large as $\mathcal{O}(10\%)$ if the FSI effects characterized by (ii) would dominate $B^+ \to \pi^+ K^0$ so that $|\mathcal{P}_{uc}|/|\mathcal{P}_{tc}| = \mathcal{O}(5)$. This order of magnitude is found in a recent attempt to evaluate rescattering processes of the kind $B^+ \to \{\pi^0 K^+\} \to \pi^+ K^0$ with the

help of Regge phenomenology.[14] A similar feature is also present in other approaches to deal with these FSI effects.[12,13] Therefore, we have arguments that rescattering processes may play an important role.

A detailed study of their impact on the constraints on γ arising from the $B^{\pm} \to \pi^{\pm} K$ and $B_d \to \pi^{\mp} K^{\pm}$ observables was performed in Ref. 16. While these effects, which are included in the formalism discussed above through the parameter κ (see Eq. (20)), are minimal for $\theta \in \{90°, 270°\}$ and only of second order, they are maximal for $\theta \in \{0°, 180°\}$. In Fig. 4, these maximal effects are shown for various values of ρ in the case of $A_0 = 0$. Looking at this figure, we observe that we have negligibly small effects for $\rho = 0.02$, which was assumed in Ref. 6 in the form of point (ii) listed in Section 1. For values of ρ as large as 0.15, we have an uncertainty for γ_0 (see Eqs. (4) and (5)) of at most $\pm 10°$.

The FSI effects can be controlled through experimental data. A first step towards this goal is provided by the CP asymmetry A_+. It implies an allowed range for ρ, which is given by $\rho_{\min} \leq \rho \leq \rho^{\max}$, with

$$\rho_{\min}^{\max} = \frac{\sqrt{A_+^2 + (1 - A_+^2) \sin^2 \gamma} \pm \sqrt{(1 - A_+^2) \sin^2 \gamma}}{|A_+|}. \qquad (23)$$

In order to go beyond these constraints, $B^{\pm} \to K^{\pm} K$ decays – the $SU(3)$ counterparts of $B^{\pm} \to \pi^{\pm} K$ – play a key role, allowing us to include the rescattering processes in the contours in the $\gamma - r$ plane and the associated constraints on γ completely, as was pointed out in Refs. 16 and 17 (for alternative strategies, see Refs. 10 and 14). As a by-product, this strategy moreover gives an allowed region for ρ, and excludes values of γ within ranges around

$0°$ and $180°$. It is interesting to note that $SU(3)$ breaking enters in this approach only at the "next-to-leading order" level, as it represents a correction to the correction to the bounds on γ arising from rescattering processes. Moreover, this strategy also works if the CP asymmetry A_+ arising in $B^+ \to \pi^+ K^0$ should turn out to be very small. In this case, there may also be large rescattering effects, which would then not be signalled by sizeable CP violation in this channel.

Following Ref. 17, this approach to control the FSI effects is illustrated in Fig. 5 by showing the contours in the $\gamma-r$ plane and the dependence of $R_{\min}$ on the CKM angle γ. Here the simple model advocated by the authors of Refs. 12 and 13 was used to obtain values for the $B \to \pi K$, KK observables by choosing a specific set of input parameters (for details, see Ref. 17). The value of $R = 0.83$ arising in this case is represented in Fig. 5 by the dotted line. It is an easy exercise to read off the corresponding allowed range for γ from this figure.

Since the "short-distance" expectation for the combined branching ratio $BR(B^\pm \to K^\pm K)$ is $\mathcal{O}(10^{-6})$,[20] experimental studies of $B^\pm \to K^\pm K$ appear to be difficult. These modes have not yet been observed, and only upper limits for $BR(B^\pm \to K^\pm K)$ are available.[5,7] However, rescattering effects may enhance this quantity significantly, and could thereby make $B^\pm \to K^\pm K$ measurable at future B-factories.[16,17] Another important indicator of large FSI effects is provided by $B_d \to K^+K^-$ decays,[21] for which stronger experimental bounds already exist.[5,7]

Although $B^\pm \to K^\pm K$ decays allow us to determine the shift of the contours in the $\gamma-r$ plane arising from rescattering processes, they do not allow us to take into account these effects also in the determination of γ, requiring some knowledge on r, in contrast to the bounds on γ. As we have already noted, this quantity is not just the ratio of a "tree" to a "penguin" amplitude, which is the usual terminology, but has a rather complex structure and may in principle be considerably affected by FSI effects. However, if future measurements of $BR(B^\pm \to K^\pm K)$ and $BR(B_d \to K^+K^-)$ should not show a significant enhancement with respect to the "short-distance" expectations of $\mathcal{O}(10^{-6})$ and $\mathcal{O}(10^{-8})$, respectively, and if A_+ should not be in excess of $\mathcal{O}(1\%)$, a future theoretical accuracy of r as small as $\mathcal{O}(10\%)$ may be achievable.

5   The Role of Electroweak Penguins

The modification of $R_{\min}$ through EW penguin topologies is described by $\kappa = 1 + 2\epsilon \cos\Delta + \epsilon^2$. These effects are minimal and only of second order in ϵ for $\Delta \in \{90°, 270°\}$, and maximal for $\Delta \in \{0°, 180°\}$. In the case of $\Delta = 0°$, which is favoured by "factorization",

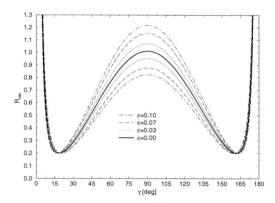

Figure 6: The effect of electroweak penguins on $R_{\min}$ for $|A_0| = 0.2$. The curves for a given value of ϵ correspond to $\Delta \in \{0°, 180°\}$ and represent the maximal shift from $\epsilon = 0$.

the bounds on γ get stronger, excluding a larger region around $\gamma = 90°$, while they are weakened for $\Delta = 180°$. In Fig. 6, the maximal EW penguin effects are shown for $|A_0| = 0.2$ and for various values of ϵ. The EW penguins are "colour-suppressed" in the case of $B^+ \to \pi^+K^0$ and $B^0_d \to \pi^-K^+$; estimates based on simple calculations performed at the perturbative quark level, where the relevant hadronic matrix elements are treated within the "factorization" approach, typically give $\epsilon = \mathcal{O}(1\%)$.[20] These crude estimates may, however, underestimate the role of these topologies.[3,13]

An improved theoretical description of the EW penguins is possible, using the general expressions for the corresponding four-quark operators and performing appropriate Fierz transformations. Following these lines,[16] we arrive at the expression

$$\frac{\epsilon}{r} e^{i(\Delta - \delta)} \approx \frac{3}{2\lambda^2 R_b} \left[\frac{C_1'(\mu)C_{10}(\mu) - C_2'(\mu)C_9(\mu)}{C_2'^2(\mu) - C_1'^2(\mu)} \right] a \, e^{i\omega} \tag{24}$$

with $C_1'(\mu) \equiv C_1(\mu) + 3C_9(\mu)/2$ and $C_2'(\mu) \equiv C_2(\mu) + 3C_{10}(\mu)/2$, where $C_{1,2}(\mu)$ are the Wilson coefficients of the current–current operators specified in Eq. (22), and $C_{9,10}(\mu)$ those of the EW penguin operators

$$\begin{aligned} Q_9 &= \tfrac{3}{2}(\bar{b}_\alpha s_\alpha)_{\mathrm{V-A}} \sum_{q=u,d,c,s,b} c_q \, (\bar{q}_\beta q_\beta)_{\mathrm{V-A}} \\ Q_{10} &= \tfrac{3}{2}(\bar{b}_\alpha s_\beta)_{\mathrm{V-A}} \sum_{q=u,d,c,s,b} c_q \, (\bar{q}_\beta q_\alpha)_{\mathrm{V-A}} \, . \end{aligned} \tag{25}$$

The combination of Wilson coefficients in Eq. (24) is essentially renormalization-scale-independent and changes only by $\mathcal{O}(1\%)$ when evolving from $\mu = M_W$ down to $\mu = m_b$. Employing $R_b = 0.36$ and typical values for the

Wilson coefficients yields[16]

$$\frac{\epsilon}{r} e^{i(\Delta - \delta)} \approx 0.75 \times a\, e^{i\omega}. \qquad (26)$$

The quantity $a\, e^{i\omega}$ is given by

$$a\, e^{i\omega} \equiv \frac{a_2^{\text{eff}}}{a_1^{\text{eff}}}, \qquad (27)$$

where a_1^{eff} and a_2^{eff} correspond to a generalization of the usual phenomenological colour factors a_1 and a_2 describing the "strength" of colour-suppressed and colour-allowed decay processes, respectively.[16] Comparing experimental data on $B^- \to D^{(*)0}\pi^-$ and $\overline{B_d^0} \to D^{(*)+}\pi^-$, as well as on $B^- \to D^{(*)0}\rho^-$ and $\overline{B_d^0} \to D^{(*)+}\rho^-$ decays gives $a_2/a_1 = \mathcal{O}(0.25)$, where a_1 and a_2 are – in contrast to a_1^{eff} and a_2^{eff} – real quantities, and their relative sign is found to be positive. For $a = 0.25$, we obtain a value of ϵ/r that is larger than the "factorized" result

$$\left. \frac{\epsilon}{r} e^{i(\Delta - \delta)} \right|_{\text{fact}} = 0.06 \qquad (28)$$

by a factor of 3. A detailed study of the effects of the EW penguins described by Eq. (24) on the strategies to probe the CKM angle γ discussed in Section 3 was performed in Ref. 16. There it was also pointed out that a first step towards the experimental control of the "colour-suppressed" EW penguin contributions to the $B \to \pi K$ amplitude relations (12) and (13) is provided by the decay $B^+ \to \pi^+\pi^0$. More refined strategies will certainly be developed in the future, when better experimental data become available.

6 Probing γ with $B_s \to K\overline{K}$ Decays

In this section, we focus on the modes $B_s \to K^0\overline{K^0}$ and $B_s \to K^+K^-$, which are the B_s counterparts of the $B_{u,d} \to \pi K$ decays discussed above, where the up and down "spectator" quarks are replaced by a strange quark. Because of the expected large B_s^0–$\overline{B_s^0}$ mixing parameter $x_s \equiv \Delta M_s / \Gamma_s = \mathcal{O}(20)$, experimental studies of CP violation in B_s decays are regarded as being very difficult. In particular, an excellent vertex resolution system is required to keep track of the rapid oscillatory $\Delta M_s t$ terms arising in tagged B_s decays. These terms cancel, however, in the untagged B_s decay rates defined by

$$\Gamma[f(t)] \equiv \Gamma(B_s^0(t) \to f) + \Gamma(\overline{B_s^0}(t) \to f), \qquad (29)$$

where one does not distinguish between initially, i.e. at time $t = 0$, present B_s^0 and $\overline{B_s^0}$ mesons. In this case, the expected sizeable width difference[22] $\Delta\Gamma_s \equiv \Gamma_H^{(s)} - \Gamma_L^{(s)}$ between the mass eigenstates B_s^H ("heavy") and B_s^L ("light") of the B_s system may provide an alternative route to explore CP violation.[23] Several strategies

were proposed to extract CKM phases from experimental studies of such untagged B_s decays.[23-25]

In Ref. 24, it was pointed out that the modes $B_s \to K^0\overline{K^0}$ and $B_s \to K^+K^-$ probe the CKM angle γ. Their decay amplitudes take a form completely analogous to Eqs. (12) and (13), and the corresponding untagged decay rates can be expressed as follows:[26]

$$\Gamma[\dot{K}^0\overline{K^0}(t)] = R_L\, e^{-\Gamma_L^{(s)}t} + R_H\, e^{-\Gamma_H^{(s)}t} \qquad (30)$$

$$\Gamma[K^+K^-(t)] = \Gamma[K^0\overline{K^0}(0)]\left[a\, e^{-\Gamma_L^{(s)}t} + b\, e^{-\Gamma_H^{(s)}t}\right]. \qquad (31)$$

Since we have $a + b = R_s$, where R_s corresponds to the ratio R of the combined $B \to \pi K$ branching ratios (see Eq. (3)), bounds on γ similar to those discussed in Sections 1 and 3 can also be obtained from the untagged $B_s \to K\overline{K}$ observables. Moreover, a comparison of R and R_s provides valuable insights into $SU(3)$ breaking.

A closer look shows, however, that it is possible to derive more elaborate bounds from the untagged $B_s \to K\overline{K}$ rates:[26]

$$\frac{|1 - \sqrt{a}|}{\sqrt{b}} \leq |\cot\gamma| \leq \frac{1 + \sqrt{a}}{\sqrt{b}}, \qquad (32)$$

corresponding to the allowed range

$$\gamma_1 \leq \gamma \leq \gamma_2 \quad \vee \quad 180° - \gamma_2 \leq \gamma \leq 180° - \gamma_1 \qquad (33)$$

with

$$\gamma_1 \equiv \operatorname{arccot}\left(\frac{1 + \sqrt{a}}{\sqrt{b}}\right), \quad \gamma_2 \equiv \operatorname{arccot}\left(\frac{|1 - \sqrt{a}|}{\sqrt{b}}\right). \qquad (34)$$

Besides a sizeable value of $\Delta\Gamma_s$ and non-vanishing observables a and b, the bound (32) does not require any constraint on these observables such as $R_s = a + b < 1$, which is needed for Eqs. (4) and (5) to become effective.

As in the $B \to \pi K$ case, the theoretical accuracy of these constraints, which make use only of the general amplitude structure arising within the Standard Model and of the $SU(2)$ isospin symmetry of strong interactions, is also limited by certain rescattering processes and contributions arising from EW penguins. In Eq. (32), these effects are neglected for simplicity. The completely general formalism, taking also into account these effects, is derived in Ref. 26, where also strategies to control them through experimental data are discussed.

In order to go beyond these constraints and to determine γ from the untagged $B_s \to K\overline{K}$ observables, the magnitude of an amplitude T_s, which corresponds to T (see Eq. (15)), has to be fixed, leading to hadronic uncertainties similar to those in the $B \to \pi K$ case. Such an input can be avoided by considering the contours in the γ–$r_{(s)}$ and γ–$\cos\delta_{(s)}$ planes, and applying the $SU(3)$ flavour symmetry to relate r_s to r and $\cos\delta_s$ to $\cos\delta$,

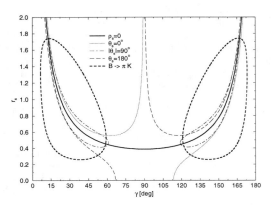

Figure 7: The contours in the $\gamma - r_{(s)}$ plane for $a = 0.60$, $b = 0.15$ and $\rho_s = 0.15$, $\epsilon_s = 0$. The $B \to \pi K$ contours correspond to $R = 0.75$, $A_0 = 0.2$ and $\rho = \epsilon = 0$.

respectively.[26] The contours in the $\gamma - r_{(s)}$ plane are illustrated in Fig. 7. Using the formalism presented in Refs. 16, 17 and 26, rescattering and EW penguin effects can be included in these contours. As a "by-product", also values for the hadronic quantities $r_{(s)}$ and $\cos \delta_{(s)}$ are obtained, which are of special interest to test the factorization hypothesis.

Provided a tagged, time-dependent measurement of $B_s \to K^0 \overline{K^0}$ and $B_s \to K^+ K^-$ can be performed,[27] it would be possible to extract γ in such a way that rescattering effects are taken into account "automatically".[26] To this end, the $B_s \to K\overline{K}$ observables are sufficient, and the theoretical accuracy of γ would only be limited by EW penguins. Let me finally note that the $B_s \to K\overline{K}$ decays represent also an interesting probe for certain scenarios of physics beyond the Standard Model.[26]

7 Conclusions

On the long and winding road towards the extraction of the CKM angle γ, the decays $B^{\pm} \to \pi^{\pm} K$ and $B_d \to \pi^{\mp} K^{\pm}$ are expected to play an important role. An accurate measurement of these modes, as well as of $B \to KK$ and $B \to \pi\pi$ decays to control rescattering and EW penguin effects, is therefore an important goal of the future B-factories. At present, data for these decays are already starting to become available, and the coming years will certainly be very exciting. The modes $B_s \to K^0 \overline{K^0}$ and $B_s \to K^+ K^-$ also offer interesting strategies to probe the CKM angle γ. Here the width difference $\Delta\Gamma_s$ may provide an interesting tool to accomplish this task. In order to investigate B_s decays, experiments at hadron machines appear to be most promising.

References

1. R. Fleischer, *Int. J. Mod. Phys.* **A12**, 2459 (1997).
2. R. Fleischer, *Phys. Lett.* **B365**, 399 (1996).
3. M. Gronau and J.L. Rosner, *Phys. Rev.* **D57**, 6843 (1998).
4. F. Würthwein and P. Gaidarev, preprint CALT-68-2153 (1997) [hep-ph/9712531].
5. CLEO Collaboration (R. Godang *et al.*), *Phys. Rev. Lett.* **80**, 3456 (1998).
6. R. Fleischer and T. Mannel, *Phys. Rev.* **D57**, 2752 (1998).
7. CLEO Collaboration (M. Artuso *et al.*), preprint CLEO CONF 98-20, ICHEP98 858; J. Alexander, these proceedings.
8. A. Buras, preprint TUM-HEP-299/97 (1997) [hep-ph/9711217].
9. Y. Grossman *et al.*, *Nucl. Phys.* **B511**, 69 (1998).
10. A.J. Buras, R. Fleischer and T. Mannel, preprint CERN-TH/97-307 (1997) [hep-ph/9711262], to appear in *Nucl. Phys.* **B**.
11. L. Wolfenstein, *Phys. Rev.* **D52**, 537 (1995).
12. J.-M. Gérard and J. Weyers, preprint UCL-IPT-97-18 (1997) [hep-ph/9711469].
13. M. Neubert, *Phys. Lett.* **B424**, 152 (1998).
14. A.F. Falk *et al.*, *Phys. Rev.* **D57**, 4290 (1998).
15. D. Atwood and A. Soni, *Phys. Rev.* **D58**, 036005 (1998).
16. R. Fleischer, preprint CERN-TH/98-60 (1998) [hep-ph/9802433], to appear in *Eur. Phys. J.* **C**.
17. R. Fleischer, preprint CERN-TH/98-128 (1998) [hep-ph/9804319], to appear in *Phys. Lett.* **B**.
18. L. Wolfenstein, *Phys. Rev. Lett.* **51**, 1945 (1983).
19. A.J. Buras and R. Fleischer, *Phys. Lett.* **B341**, 379 (1995); R. Fleischer, *Phys. Lett.* **B341**, 205 (1994); M. Ciuchini *et al.*, *Nucl. Phys.* **B501**, 271 (1997).
20. For a recent study, see A. Ali, G. Kramer and C.-D. Lü, preprint DESY 98-041 (1998) [hep-ph/9804363]; A. Ali, these proceedings.
21. M. Gronau and J.L. Rosner, preprint EFI-98-23 (1998) [hep-ph/9806348].
22. For a recent calculation of $\Delta\Gamma_s$, see M. Beneke *et al.*, preprint CERN-TH/98-261 (1998) [hep-ph/9808385].
23. I. Dunietz, *Phys. Rev.* **D52**, 3048 (1995).
24. R. Fleischer and I. Dunietz, *Phys. Rev.* **D55**, 259 (1997).
25. R. Fleischer and I. Dunietz, *Phys. Lett.* **B387**, 361 (1996).
26. R. Fleischer, preprint CERN-TH/97-281 (1997) [hep-ph/9710331], to appear in *Phys. Rev.* **D**.
27. C.S. Kim, D. London and T. Yoshikawa, *Phys. Rev.* **D57**, 4010 (1998).

OBSERVATION OF THE B_c MESON AT CDF

VAIA PAPADIMITRIOU

for the CDF Collaboration

Texas Tech University, Department of Physics, Box 41051, Lubbock, TX 79409, USA

E-mail: vaia@fnal.gov

We report on the observation of bottom-charm mesons via the decay mode $B_c^{\pm} \to J/\psi l^{\pm} \nu$ in 1.8 TeV $p\bar{p}$ collisions using the CDF detector at the Fermilab Tevatron. A fit of background and signal contributions to the $J/\psi l$ mass distribution yielded $20.4^{+6.2}_{-5.5}$ events from B_c mesons. We measured the B_c^+ mass to be $6.40 \pm 0.39(stat.) \pm 0.13(syst.)$ GeV/c^2 and the B_c^+ lifetime to be $0.46^{+0.18}_{-0.16}(stat.) \pm 0.03(syst.)$ ps. The measured production cross section times branching ratio for $B_c^+ \to J/\psi l^+ \nu$ relative to that for $B^+ \to J/\psi K^+$ is $0.132^{+0.041}_{-0.037}(stat.) \pm 0.031(syst.)^{+0.032}_{-0.020}(lifetime)$.

1 Introduction

The B_c^+ meson is the lowest-mass bound state of quarkonium states containing a charm quark and a bottom anti-quark [1]. Since this pseudoscalar ground state has non-zero flavor, it has no strong or electromagnetic decay channels. It is the last such meson predicted by the Standard Model and the only meson with two different heavy quarks. It decays weakly yielding a large branching fraction to final states containing a J/ψ [2,3,4,5]. Non-relativistic potential models predict a B_c mass in the range 6.2–6.3 GeV/c^2 [6,7]. In these models, the c and $\bar{b}$ are tightly bound in a very compact system and have a rich spectroscopy of excited states.

The production of B_c mesons has been calculated in perturbative QCD. At transverse momenta p_T large compared to the B_c mass the dominant process is that in which a $\bar{b}$ is produced by gluon fusion in the hard collision and fragmentation provides the c. At lower p_T a full α_s^4 calculation [8] shows that the dominant process is one in which both the $\bar{b}$ and c quarks are produced in the hard scattering. These and other calculations [8,9,10,11,12] provide inclusive production cross sections along with distributions in p_T and other kinematic variables.

We expect three major contributions to the B_c decay width: $\bar{b} \to \bar{c}W^+$ with the c as a spectator, leading to final states like $J/\psi \pi$ or $J/\psi \ell \nu$; $c \to sW^+$, with the $\bar{b}$ as spectator, leading to final states like $B_s \pi$ or $B_s \ell \nu$; and $c\bar{b} \to W^+$ annihilation, leading to final states like $D K$, $\tau \nu_\tau$ or multiple pions. Since these processes lead to different final states, their amplitudes do not interfere. When phase space and other effects are included, the predicted lifetime is in the range 0.4–1.4 ps [2,13,14,15,16,17]. Because of the wide range of predictions, a B_c lifetime measurement is a test of the different assumptions made in the various calculations. Several authors have also calculated the B_c partial widths to semileptonic final states [2,3,4,5,18].

Limits on the B_c production have been placed by various experimental searches at LEP [19,20,21]. A prior CDF search placed a limit on B_c production in the $B_c^+ \to J/\psi \pi^+$ decay mode [22].

We report here on the first observation of B_c mesons produced in 1.8 TeV $p\bar{p}$ collisions at the Fermilab Tevatron collider using a 110 pb^{-1} data sample collected with the CDF detector. A more detailed description of this work can be found in Ref. [23]. We searched for the decay channels $B_c^+ \to J/\psi \mu^+ \nu$ and $B_c^+ \to J/\psi e^+ \nu$ followed by $J/\psi \to \mu^+ \mu^-$. A Monte Carlo calculation of B_c production and decay to $J/\psi \ell \nu$ showed that, for an assumed B_c mass of 6.27 GeV/c^2, 93% of the $J/\psi \ell$ final state particles would have $J/\psi \ell$ masses with $4.0 < M(J/\psi \ell) < 6.0$ GeV/c^2. We refer to this as the signal region, but we accepted candidates with $M(J/\psi \ell)$ between 3.35 and 11 GeV/c^2.

We have described the CDF detector in detail elsewhere [24,25]. The tracking system for CDF gives a transverse momentum resolution $\delta p_T/p_T = [(0.0009 \times p_T)^2 + (0.0066)^2]^{1/2}$, where p_T is in units of GeV/c. The average track impact parameter resolution relative to the beam axis is $(13 + (40/p_T))$ μm in the plane transverse to the beam [26]. An online di-muon trigger and subsequent offline selection yielded a sample of about 196,000 $J/\psi \to \mu^+ \mu^-$ mesons.

2 Event Selection

We searched for the B_c through $B_c^+ \to J/\psi \ell^+ \nu$ decays. These decays have a very simple topology: a decay point for $J/\psi \to \mu^+ \mu^-$ displaced from the primary interaction point and a third track emerging from the same decay point. This J/ψ + track sample included $B_c^+ \to J/\psi e^+ \nu$, $B_c^+ \to J/\psi \mu^+ \nu$, $B^+ \to J/\psi K^+$, and background from various sources. We subjected the three tracks to a fit that constrained the two muons to the J/ψ mass and that constrained all three tracks to originate from a common point. A measure of the time between production and decay of a B_c candidate is the quantity

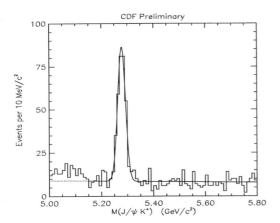

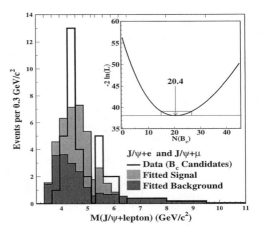

Figure 1: The distribution of masses of $J/\psi K^\pm$ candidates. The solid curve represents a least squares fit to the data between 5.15 and 5.8 GeV/c² consisting of a Gaussian signal above a linear background. The area of the Gaussian contribution is 290 ± 19 events.

Figure 2: Histogram of the $J/\psi\,\ell$ mass that compares the signal and background contributions determined in the likelihood fit to the combined data for $J/\psi\,e$ and $J/\psi\,\mu$. Note that the mass bins, indicated by tick marks at the top, vary in width. The total B_c contribution is $20.4^{+6.2}_{-5.5}$ events. The inset shows the behavior of the log-likelihood function $-2\ln(L)$ vs. the number of B_c mesons.

ct^*, defined as

$$ct^* = \frac{M(J/\psi\,\ell) \cdot L_{xy}(J/\psi\,\ell)}{|p_T(J/\psi\,\ell)|} \qquad (1)$$

where L_{xy} is the distance between the beam centroid and the decay point of the B_c candidate in the plane perpendicular to the beam direction and projected along the direction of the $J/\psi\,\ell$ combination in that plane, $M(J/\psi\,\ell)$ is the mass of the tri-lepton system, and $p_T(J/\psi\,\ell)$ is its momentum transverse to the beam. Our average uncertainty in the measurement of ct^* is 25 μm. We required $ct^* > 60$ μm.

$B^+ \to J/\psi\,K^+$ candidates were identified by a peak in the $\mu^+\mu^-K^+$ mass distribution centered at $M(B^+) = 5.279$ GeV/c² with an r.m.s. width of 14 MeV/c². (See Fig. 1.) The peak contained 290 ± 19 events after correction for background. Events within 50 MeV/c² of $M(B^+)$ were eliminated as $B_c^+ \to J/\psi\,\ell^+\,\nu$ candidates.

Electrons were identified by the association of a charged-particle track with $p_T > 2$ GeV/c and an electromagnetic shower in the calorimeter. Additional information for identifying electrons was obtained from specific ionization in the tracking chambers and from the shower profile in proportional chambers embedded in the electromagnetic calorimeter. Muons from J/ψ decay were identified by matching a charged-particle track with $p_T > 2$ GeV/c to a track segment in muon drift chambers outside the central calorimeter (5 to 9 interaction lengths

thick depending on angle). The third muon was required to have a transverse momentum exceeding 3 GeV/c and to pass through an additional three interaction lengths of steel to produce a track segment in a second set of drift chambers. We found 23 $B_c^+ \to J/\psi\,e^+\,\nu$ candidates of which 19 were in the signal region, and we found 14 $B_c^+ \to J/\psi\,\mu^+\,\nu$ candidates of which 12 were in the signal region.

3   Background Determination

The major contributions to backgrounds in the sample of B_c candidates come from misidentification of hadron tracks as leptons (i.e. false leptons) and from random combinations of real leptons with J/ψ mesons. There are three significant sources of false leptons: hadrons that reach the muon detectors without being absorbed; hadrons that decay in flight into a muon in advance of entering the muon detectors; and hadrons that are falsely identified as electrons. In one type of random combination, electrons from photons that convert to e^+e^- pairs in the material around the beam line or from Dalitz decay of π^0 contribute to a "conversion background" when the other member of the pair remains undetected. The other type of random combination involves a B that has decayed into a J/ψ and an associated $\overline{B}$ that has decayed semileptonically (or through semileptonic decays of its daughter hadrons) into a muon or an electron. The displaced J/ψ and the lepton can accidentally appear to originate from a common point. A number of other backgrounds [23] were found to be negligible. From a combination of data and Monte Carlo calculations, we determined the J/ψ + track mass distribution for each of the sources of background. As a check of our background calculations, we verified that we are able to predict the number of events and mass distribution in an independent, background-rich sample of same-charge, low-mass lepton pairs. (See Fig. 27 in Ref. [23].)

The topology for candidate events was verified by applying all selection criteria except the requirement that the third track intersect the J/ψ vertex. The impact parameter distribution between the third track and the J/ψ vertex has a prominent peak at zero, demonstrating that, for most candidate events, the three tracks arise from a common vertex. (See Fig. 28 in Ref. [23].)

In Table 1 we summarize the results of the background calculation and of a simultaneous fit for the muon and electron channels to the mass spectrum over the region between 3.35 and 11 GeV/c^2. Figure 2 shows the mass spectra for the combined $J/\psi\,e$ and $J/\psi\,\mu$ candidate samples, the combined backgrounds and the fitted contribution from $B_c^+ \rightarrow J/\psi\,\ell^+\,\nu$ decay. The fitted number of B_c events is $20.4^{+6.2}_{-5.5}$.

4 Test of Significance and Determination of the B_c Lifetime and Mass

To test the significance of this result, we generated a number of Monte Carlo trials with the statistical properties of the backgrounds, but with no contribution from B_c mesons. These were subjected to the same fitting procedure to determine contributions consistent with the signal distribution arising from background fluctuations. The probability of obtaining a yield of 20.4 events or more is 0.63×10^{-6}, equivalent to a 4.8 standard-deviation effect.

To check the stability of the B_c signal, we varied the value assumed for the B_c mass. We generated signal templates, i.e. Monte Carlo samples of $B_c^+ \rightarrow J/\psi\,\ell^+\,\nu$, with various values of $M(B_c)$ from 5.52 to 7.52 GeV/c^2. The signal template for each value of $M(B_c)$ and the background mass distributions were used to fit the mass spectrum for the data. This study established that the magnitude of the B_c signal is stable over the range of theoretical predictions for $M(B_c)$, and the dependence of the log-likelihood function on mass yielded $M(B_c) = 6.40 \pm 0.39\,\text{(stat.)} \pm 0.13\,\text{(syst.)}$ GeV/c^2.

We obtained the mean proper decay length $c\tau$ and hence the lifetime τ of the B_c meson from the distribution of ct^*. We used only events with $4.0 < M(J/\psi\,\ell) < 6.0$

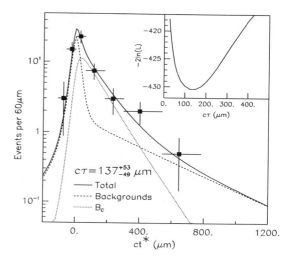

Figure 3: The distribution in ct^* for the combined $J/\psi\,\mu$ and $J/\psi\,e$ data along with the fitted curve and contributions to it from signal and background. The inset shows the log-likelihood function vs. $c\tau$ for the B_c.

GeV/c^2, and we changed the threshold requirement on ct^* from $ct^* > 60\,\mu m$ to $ct^* > -100\,\mu m$ for this lifetime measurement. This yielded a sample of 71 events, 42 $J/\psi\,e$ and 29 $J/\psi\,\mu$. We determined a functional form for the shapes in ct^* for each of the backgrounds. To these, we added a resolution-smeared exponential decay distribution for a B_c contribution, parameterized by its mean decay length $c\tau$. Because of the missing neutrino, the proper decay length ct for each event differs from ct^* of Eq. 1. We convoluted the exponential in ct with the distribution of ct^*/ct derived from Monte Carlo studies. Finally, we incorporated the data from each of the candidate events in an unbinned likelihood fit to determine the best-fit value of $c\tau$. The data and the signal and background distributions are shown in Fig. 3, and the result is:

$$c\tau = 137^{+53}_{-49}\,\text{(stat.)} \pm 9\,\text{(syst)}\ \mu m \qquad (2)$$

$$\tau = 0.46^{+0.18}_{-0.16}\,\text{(stat.)} \pm 0.03\,\text{(syst.)}\ \text{ps} \qquad (3)$$

From the 20.4 B_c events and the 290 $B^+ \rightarrow J/\psi\,K^+$ events, we calculated the B_c production cross section times the $B_c^+ \rightarrow J/\psi\,\ell^+\,\nu$ branching fraction $\sigma \cdot BR(B_c^+ \rightarrow J/\psi\,\ell^+\,\nu)$ relative to that for the topologically similar decay $B^+ \rightarrow J/\psi\,K^+$. Systematic uncertainties arising from the luminosity, from the J/ψ trigger efficiency, and from the track-finding efficiencies cancel in the ratio. Our Monte Carlo calculations yielded the

Table 1: B_c Signal and Background Summary

	3.35 < $M(J/\psi\,\ell)$ < 11.0 GeV/c^2	
	$J/\psi\,e$ Events	$J/\psi\,\mu$ Events
False Electrons	4.2 ± 0.4	
Undetected Conversions	2.1 ± 1.7	
False Muons		11.4 ± 2.4
$B\bar{B}$ bkg.	2.3 ± 0.9	1.44 ± 0.25
Total Background (predicted)	8.6 ± 2.0	12.8 ± 2.4
(from fit)	9.2 ± 2.0	10.6 ± 2.3
Predicted $N(B_c \to J/\psi\,e\,\nu)/N(B_c \to J/\psi\,\ell\,\nu)$	0.58 ± 0.04	
e and μ Signal (derived from fit)	$12.0^{+3.8}_{-3.2}$	$8.4^{+2.7}_{-2.4}$
Total Signal (fitted parameter)	$20.4^{+6.2}_{-5.5}$	
Signal + Background	21.2 ± 4.3	19.0 ± 3.5
Candidates	23	14
Probability for background alone to fluctuate to the apparent signal of 20.4 events	0.63×10^{-6}	

values for the efficiencies that do not cancel. The detection efficiency for $B_c^+ \to J/\psi\,\ell\,\nu$ depends on $c\tau$ because of the requirement that $ct^* > 60$ μm, and we quote a separate systematic uncertainty because of the lifetime uncertainty. We assumed that the branching fraction is the same for $B_c^+ \to J/\psi\,e^+\,\nu$ and $B_c^+ \to J/\psi\,\mu^+\,\nu$. We multiply the 20.4 events by a factor 0.85 ± 0.15 to correct for other B_c decay channels such as $B_c \to \psi'\,\ell\,\nu$ [23]. We find

$$\mathcal{R}(J/\psi\,\ell\,\nu) \equiv \frac{\sigma(B_c) \cdot BR(B_c \to J/\psi\,\ell\,\nu)}{\sigma(B) \cdot BR(B \to J/\psi\,K)} =$$

$$0.132^{+0.041}_{-0.037}\,\text{(stat.)} \pm 0.031\,\text{(syst.)}$$

$$^{+0.032}_{-0.020}\,\text{(lifetime)}, \qquad (4)$$

for B_c^+ and B^+ with transverse momenta $p_T > 6.0$ GeV/c and rapidities $|y| < 1.0$. This result is consistent with limits from previous searches [19,20,21]. Figure 4 compares phenomenological predictions with our measurements of $c\tau$ and $\mathcal{R}(J/\psi\,\ell\,\nu)$. Within experimental and theoretical uncertainties, they are consistent.

5 Conclusions

In conclusion, we report the observation of B_c mesons through their semileptonic decay modes, $B_c \to J/\psi\,\ell X$ where ℓ is either an electron or a muon. We measured the B_c mass and the product of its production cross section times semileptonic branching fraction, which confirm phenomenological expectations. We measured a B_c lifetime consistent with calculations in which the decay width is dominated by the decay of the charm quark.

The increase of statistics expected in Run II of the Tevatron Collider (increase in yield by at least a factor of 30), combined with refinement in technique and investigation of additional decay channels for the B_c meson should allow us to measure its mass, lifetime and production cross section with much better accuracy than we did with the currently available data. It should also allow us to measure ratios of branching ratios of the B_c for various decay channels.

Acknowledgements

We thank the Fermilab staff and the technical staffs of the participating institutions for their vital contributions. This work was supported by the U.S. Department of Energy and National Science Foundation; the Italian Istituto Nazionale di Fisica Nucleare; the Ministry of Education, Science and Culture of Japan; the Natural Sciences and Engineering Research Council of Canada; the National Science Council of the Republic of China; the A. P. Sloan Foundation; and the Swiss National Science Foundation.

References

1. References to a specific state imply the charge-conjugate state as well.
2. M. Lusignoli and M. Masetti, *Z. Phys.* C **51**, 549 (1991).
3. N. Isgur, D. Scora, B. Grinstein, M. B. Wise, *Phys. Rev.* D **39**, 799 (1989).

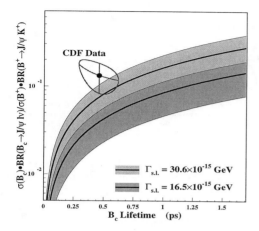

Figure 4: The point with 1-standard-deviation contour shows our measured value of the $\sigma \cdot BR$ ratio plotted at the value we measure for the B_c lifetime. The shaded region represents theoretical predictions and their uncertainty corridors for two different values of the semileptonic width $\Gamma_{s.l.} = \Gamma(B_c \to J/\psi l\nu)$ based on Refs. 2 and 4. The other numbers assumed in the theoretical predictions are $V_{cb} = 0.041 \pm 0.005$ [27], $\sigma(B_c^+)/\sigma(\bar{b}) = 1.3 \times 10^{-3}$ [9], $\frac{\sigma(B^+)}{\sigma(\bar{b})} = 0.378 \pm 0.022$ [27], $BR(B^+ \to J/\psi K^+) = (1.01 \pm 0.14) \times 10^{-3}$ [27].

4. D. Scora and N. Isgur, *Phys. Rev.* D **52**, 2783 (1995).

5. C. H. Chang and Y. Q. Chen, *Phys. Rev.* D **49**, 3399 (1994).

6. W. Kwong and J. Rosner, *Phys. Rev.* D **44**, 212 (1991).

7. E. Eichten and C. Quigg, *Phys. Rev.* D **49**, 5845 (1994).

8. C. H. Chang and Y. Q. Chen and R. J. Oakes, *Phys. Rev.* D **54**, 4344 (1996).

9. M. Lusignoli and M. Masetti and S. Petrarca, *Phys. Lett.* B **266**, 142 (1991).

10. E. Braaten, K. Cheung and T. C. Yuan, *Phys. Rev.* D **48**, R5049 (1993).

11. C. H. Chang and Y. Q. Chen, *Phys. Rev.* D **48**, 4086 (1993).

12. M. Masetti and F. Sartogo, *Phys. Lett.* B **357**, 659 (1995).

13. I.I. Bigi, *Phys. Lett.* B **371**, 105 (1996).

14. M. Beneke and G. Buchalla, *Phys. Rev.* D **53**, 4991 (1996).

15. S. S. Gershtein *et al.*, *Int. J. Mod. Phys.* A6, 2309 (1991).

16. P. Colangelo *et al.*, *Z. Phys.* C **57**, 43 (1993).

17. C. Quigg, Fermilab-CONF-93/267(SSCL-SR-1225) (1994).

18. Myoung-Taek Choi and Jae Kwan Kim, *Phys. Rev.* D **53**, 6670 (1996).

19. P. Abreu *et al.*, The DELPHI Collaboration, *Phys. Lett.* B **398**, 207 (1997).

20. K. Ackerstaff *et al.*, The OPAL Collaboration, *Phys. Lett.* B **420**, 157 (1998).

21. R. Barate *et al.*, The ALEPH Collaboration, *Phys. Lett.* B **402**, 213 (1997).

22. F. Abe *et al.*, The CDF Collaboration, *Phys. Rev. Lett.* **77**, 5176 (1996).

23. F. Abe *et al.*, The CDF Collaboration, Fermilab-Pub-98/121-E, submitted to *Phys. Rev. D*, April 1998.

24. F. Abe *et al.*, The CDF Collaboration, *Nucl. Instrum. Methods* A **271**, 387 (1988).

25. F. Abe *et al.*, The CDF Collaboration, *Phys. Rev.* D **50**, 2966 (1994).

26. D. Amidei *et al.*, The CDF Collaboration, *Nucl. Instrum. Methods* A **350**, 73 (1994).

27. "Review of Particle Physics", R. M. Barnett *et al.*, *Phys. Rev.* D **54**, 1 (1996).

RARE HADRONIC B DECAYS

J. ROY

University of Colorado, Department of Physics, Campus Box 390, Boulder,CO,80309.

(for the CLEO Collaboration)

New results in the investigation of rare charmless hadronic decays of B mesons by the CLEO Collaboration are reported. For some decay modes recent data collected with an upgraded detector are included. Earlier observations of decays of the type $B \to K\pi$ and $B \to \eta'K$ are confirmed and the first observation of $B^+ \to K^+\pi^0$ is presented. Also, evidence for the decay $B^+ \to \eta_c K^+$ is shown. Theoretical implications based on some of these results are briefly discussed.

1 Introduction

Rare hadronic B decays are usually defined as decays with fully hadronic charmless final states that involve a transition from a b quark to a u,s, or d quark. The $b \to u$ transition occurs at tree level (Fig. 1) and probes the V_{ub} element of the CKM matrix[2], which accounts for the mixing between the quark flavor eigenstates and the quark weak interaction eigenstates. The $b \to s$ and $b \to d$ transitions cannot occur at tree level in the Standard Model and are effective flavor-changing neutral current transitions. They can occur at the one-loop level because of the different masses of the virtual quarks in the loop, and are therefore sensitive to several CKM matrix elements. The diagram on the right of Figure 1 shows an example of the Feynman diagram for such a transition. It is commonly referred to as a "gluonic penguin", where the virtual quark emits a gluon.

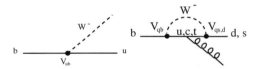

Figure 1: Feynman diagrams for $b \to u$ transition (left) and $b \to s, d$ transitions (right).

One of the key points in the interest in rare hadronic B decays is that for some final states both tree-level and penguin transitions can contribute to the overall decay amplitude. This can lead to CP-violating interference effects that are potentially large, as opposed to semileptonic decays where the CP asymmetries are expected to be out of reach of experimental measurements for quite some time. Another interesting aspect is that the presence of one loop transitions allows for the possibility of new physics to contribute to these decays. The observation of unexpectedly large decay rates could be an indication of physics beyond the Standard Model.

In this talk, I will present new results from the CLEO collaboration in the investigation of rare hadronic B decays. Several of these results are updates to previously reported results [3,4]

2 Detector and data sample

The CLEO detector is located on the CESR storage ring on the Cornell University campus. The storage ring is tuned for the reaction $e^+e^- \to \Upsilon(4S) \to B\overline{B}$. For two-thirds of the time the center-of-mass energy is set to the peak of the $\Upsilon(4S)$ resonance, and the other third of the time it is set approximately $60\ MeV$ below the resonance in order to study background from continuum quark production.

The CLEO-II [5] detector underwent some changes in 1995 and is now referred to as CLEO 2.5. The innermost wire chamber (closest to the beam pipe) was removed and replaced with a silicon vertex detector around a smaller beam pipe. This allows for better tracking and charm identification, although the latter aspect has not yet been exploited in the analyses described here. CLEO also changed the gas used in the central drift chamber from argon-ethane (50:50) to helium-propane (60:40). This improved both the momentum and specific ionization (dE/dx) resolution for charged particles by about 15%. In addition, all CLEO-II data was reprocessed with a more uniform procedure, better tracking algorithm and better calibration. Most of the analyses presented here made use of an on-resonance integrated luminosity of 3.11 fb^{-1} collected with the CLEO-II detector and of an additional 2.45 fb^{-1} collected with the CLEO 2.5 detector. The total corresponds to 5.8×10^6 $B\overline{B}$ pairs. For all decays discussed here charge conjugation is implied.

3 Analysis techniques

All analyses start with the application of preliminary cuts to select charged tracks and photons. Photon pairs are used to reconstruct π^0's and η's. Neutral kaons are identified via $K_S^0 \rightarrow \pi^+\pi^-$. For B decays to charmless hadrons, $q\bar{q}$ pair production from the continuum is the dominant background, for which the best rejection tool is the event shape. Since at CESR the B mesons are produced almost at rest in the lab, their decay products are distributed more or less isotropically, whereas $q\bar{q}$ background has a more jet-like distribution of decay products.

Two quantities based on the event shape are used in the analyses. The first is the cosine of the angle θ_T between the thrust axis of the candidate B and the thrust axis of the remainder of the tracks and photons in the event. For some analyses, the sphericity axis is used instead of the thrust axis. The left side of Figure 2 shows the distribution of $\cos\theta_T$ for simulated signal, which is more or less flat, and for continuum background, strongly peaked at ± 1. The other variable is a Fisher discriminant formed of 11 variables, nine of which are the sum of the energy of tracks and photons in concentric 10 degree cones around the candidate's thrust axis. The other two variables are the angle between the thrust axis of the candidate and the beam axis, and the angle between the direction of the B candidate and the beam axis. For some analyses the latter quantity is replaced with R_2, the ratio of the second to the zeroth Fox-Wolfram moments. The right side of figure 2 shows the distribution of signal (left curve) and continuum background (right curve) after a cut has been applied on $\cos\theta_T$ to remove a large fraction of continuum background.

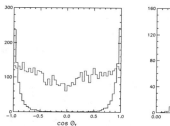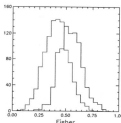

Figure 2: Distribution of $\cos\theta_T$ (left) and Fisher discriminant (right) for simulated signal and continuum background. See text for details. Vertical scales are arbitrary.

Candidate B mesons are identified with kinematics. The quantity $\Delta E = E_{\mathrm{sig}} - E_{\mathrm{beam}}$, where E_{sig} is the energy of the candidate B and E_{beam} is the energy of the beam, should be centered on zero for real signal. For charged tracks a hypothesis has to be made as to the particle type. In two-body decays where a single track is ambiguous, a pion assumption for a real kaon results in a shift of the mean of the ΔE distribution of approximately

43 MeV. The other quantity used is the beam-constrained mass, defined as $M_B = \sqrt{E_{\mathrm{beam}}^2 - p_B^2}$. The energy of the B candidate is assumed to be equal to the beam energy, which results in an order of magnitude improvement in the resolution.

Other quantities used in the analyses are the dE/dx for high momentum tracks, which provides approximately 2σ separation between kaons and pions at 2.5 GeV/c, the invariant mass of resonances in the final state, and the helicity angle of vector mesons in cases where we have a pseudoscalar meson that decays to a pseudoscalar and a vector meson. All these quantities help in the identification of signal events and also in the rejection of background, including other B decays. The latter has been estimated to be negligible in all cases.

The results presented below are obtained with two different types of analyses : the simple cut and count method, where cuts are applied on the quantities described above, and unbinned maximum likelihood analyses, which also use the same quantities as input but in a much broader range, allowing for increased efficiency. The probability distribution functions (PDFs) used in the likelihood analyses to describe each input variable are obtained from data for background and from a Monte Carlo simulation for signal. The fit returns the number of signal and background events corresponding to the maximum of the likelihood function.

4 Results

4.1 $B \rightarrow K\pi/\pi\pi$

Decays of the type $B \rightarrow K\pi/\pi\pi$ are the simplest two-meson charmless hadronic B decays. $B \rightarrow K\pi$ decays are expected to be dominated by penguin amplitudes, while $B \rightarrow \pi\pi$ decays are expected to be dominated by tree-level amplitudes. It should be noted that some decay modes, such as $B^+ \rightarrow K^0\pi^+$ have no tree-level contributions, and are therefore important in the investigation of penguin amplitudes. Table 1 shows the results from maximum likelihood analyses. For the final states $K^0\pi^0$, $\pi^0\pi^0$, and $K^0\overline{K}^0$ the results are for CLEO-II only and were previously published [3]; all others are for the CLEO-II and CLEO 2.5 data samples, and are preliminary only.

From this table, we see that significant signals are observed for the decay modes $B^0 \rightarrow K^+\pi^-$, $B^+ \rightarrow K^0\pi^+$, and $B^+ \rightarrow K^+\pi^0$. The branching fractions are all very similar to each other :

$$\mathcal{B}(B^0 \rightarrow K^+\pi^-) = (1.4 \pm 0.3 \pm 0.1) \times 10^{-5} \quad (1)$$
$$\mathcal{B}(B^+ \rightarrow K^+\pi^0) = (1.5 \pm 0.4 \pm 0.3) \times 10^{-5} \quad (2)$$
$$\mathcal{B}(B^+ \rightarrow K^0\pi^+) = (1.4 \pm 0.5 \pm 0.2) \times 10^{-5}. \quad (3)$$

For the first two the addition of CLEO 2.5 data increased the significance of earlier observations. Figure 3 shows

Table 1: Results for $B \to K\pi/\pi\pi$ decays. Shown for each decay mode are the detection efficiency, the event yield, the significance, and the branching fraction or 90% confidence level upper limit. For modes with a quoted branching fraction the significance includes systematic effects.

Mode	Eff. (%)	Yield	Signif.	$\mathcal{B}(10^{-5})$
$K^+\pi^-$	53	$43.1^{+9.0}_{-8.2}$	6.7σ	$1.4 \pm 0.3 \pm 0.1$
$K^+\pi^0$	42	$38.1^{+9.7}_{-8.7}$	4.7σ	$1.5 \pm 0.4 \pm 0.3$
$K^0\pi^+$	15	$12.3^{+4.7}_{-3.9}$	4.2σ	$1.4 \pm 0.5 \pm 0.2$
$K^0\pi^0$	8	$4.1^{+3.1}_{-2.4}$	2.2σ	< 4.1
$\pi^+\pi^-$	53	$11.5^{+6.3}_{-5.2}$	2.9σ	< 0.84
$\pi^+\pi^0$	42	$14.9^{+8.1}_{-6.9}$	2.3σ	< 1.6
$\pi^0\pi^0$	29	$2.7^{+2.7}_{-1.7}$	2.4σ	< 0.9
K^+K^-	53	$0.0^{+1.6}_{-0.0}$	0.0σ	< 0.24
$K^+\overline{K^0}$	15	$1.8^{+2.6}_{-1.4}$	1.5σ	< 0.93
$K^0\overline{K^0}$	5	0		< 1.7

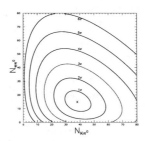

Figure 4: Maximum likelihood fit result for $B^+ \to K^+\pi^0$ and $B^+ \to \pi^+\pi^0$.

the projections onto ΔE and M_B for these two decay modes, with clear signals where expected. To make the projections cuts are applied to variables not shown, and the overall likelihood fit curve is scaled to take into account the effect of these cuts and then overlayed on the histogram. For $B^+ \to K^+\pi^0$, this result is the first ever observation. Figure 4 shows the likelihood fit result for this mode, as well as for the mode $B^+ \to \pi^+\pi^0$, which is searched for simultaneously since the only difference is the identity of the charged track. Contours in 1σ increments indicate the significance of the result. Figure 5 shows the projection of the data for ΔE and M_B, clearly indicating the presence of a signal.

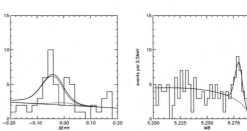

Figure 5: Projections of the data for $B^+ \to K^+\pi^0$ and $B^+ \to \pi^+\pi^0$. For the ΔE projection the solid line shows the scaled overall fit, the dotted line shows the $B^+ \to K^+\pi^0$ contribution and the dashed line the $B^+ \to \pi^+\pi^0$ contribution. For the M_B projection the dotted line shows the background contribution to the likelihood fit.

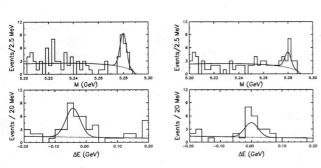

Figure 3: Projections of the data for $B^0 \to K^+\pi^-$ (left) and $B^+ \to K^0\pi^+$ (right). The solid line shows the scaled maximum likelihood fit and the dotted line shows the background contribution only.

Other observations can also be made from these results. There is still no significant signal for decay modes of the type $B \to \pi\pi$, in particular $B^+ \to \pi^+\pi^-$, for which we get :

$$\mathcal{B}(B^0 \to \pi^+\pi^-) < 8.4 \times 10^{-6}(90\%C.L.). \qquad (4)$$

This mode is favored for future B factories to study the angle α from the CKM unitarity triangle. There is also no signal for $B \to KK$ modes, although these are expected to have very small branching fractions within the Standard Model.

Finally, we can use the measurements for $B^0 \to K^+\pi^-$ and $B^+ \to K^0\pi^+$ to update the Fleischer-Mannel bound [6]. This bound places a constraint on the angle γ from the CKM unitarity triangle based on the quantity

$$R \equiv \frac{\Gamma(B^0 \to K^+\pi^-)}{\Gamma(B^+ \to K^0\pi^+)}. \qquad (5)$$

There are various conditions for the validity of this constraint, some of which were addressed at this conference [7]. Regardless of these considerations, this bound is only useful if $R < 1$. Based on earlier CLEO results, a value of $R = 0.6 \pm 0.4$ was reported [3]. From our present numbers, we get the preliminary result

$$R = 1.0 \pm 0.4. \qquad (6)$$

Although this is certainly not the last experimental word on this topic, it appears that the most recent data significantly reduce the usefulness of this bound.

4.2 $B \to \eta' X_s$

The gluonic penguin decay $b \to sg^*$ is expected to make a significant contribution to the total B decay rate. The presence of high momentum η' mesons in B decays is a potential signature for this decay. The analysis used CLEO-II data only and looked for the inclusive decay $B \to \eta' X_s$ with a partial B reconstruction technique in which the final state is identified as $\eta' K^{\pm} X$, where X represents up to four pions, one of which can be neutral. The η' is identified through its $\eta' \to \eta\pi^+\pi^-$ decay, and the continuum background is taken into account by subtracting the scaled off-resonance contribution. Figure 6 shows the $\eta\pi^+\pi^-$ invariant mass for the momentum bin $2.0 < p_{\eta'} < 2.7$ GeV/c. The top and bottom plots show the on and off-resonance data respectively, clearly indicating a signal. The on-resonance excess is measured to be 39.0 ± 11.6 events.

The spectrum of the invariant mass of X_s, shown in Figure 7, is indicative of the nature of the observed signal. The first observation is that there are several events at the kaon mass, but none at the K^* mass. Background from color-suppressed decays $B \to D\eta'$ and $B \to D^*\eta'$ is possible, but both would peak in M_{X_s}, which is not supported by the data. The best fit to this spectrum, shown on the figure as a dashed line, is obtained with a

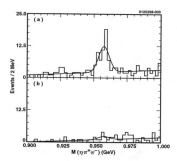

Figure 6: $\eta\pi^+\pi^-$ invariant mass spectrum for $2.0 < p_{\eta'} < 2.7$ GeV/c for a) on-resonance data and b) unscaled off-resonance data.

model in which the η' QCD anomaly plays an important role through $b \to sg^*, g^* \to g\eta'$ [8].

For the excess indicated above, we obtain the following result :

$$\mathcal{B}(B \to \eta' X_s) = (6.2 \pm 1.6 \pm 1.3^{+0.0}_{-1.5}) \times 10^{-4}. \qquad (7)$$

This branching fraction is unexpectedly large, and it is difficult for theory, even after the inclusion of the η' QCD anomaly, to account for it.

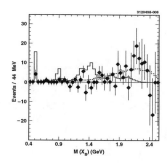

Figure 7: Background subtracted M_{X_s} invariant mass spectrum for $2.0 < p_{\eta'} < 2.7$ GeV/c (points with error bars). The solid line shows the result of a simulation of two-body $b \to sq\bar{q}$ and the dashed line shows a simulation with three-body $b \to sg^*, g^* \to g\eta'$.

4.3 $B \to \eta' K$

CLEO also searched for the exclusive decay $B \to \eta' K$ using both the $\eta' \to \eta\pi^+\pi^-$ and $\eta' \to \rho\gamma$ decays. Significant signals from CLEO-II data were previously reported for both the charged and neutral B decays [4]. Figure 8 shows the result of a maximum likelihood analysis simultaneously searching for the decays $B^+ \to \eta' K^+$ and $B^+ \to \eta'\pi^+$ with the CLEO-II and CLEO 2.5 data samples. The axes are in units of branching fraction and they show a clear signal for the mode $B^+ \to \eta' K^+$ with more than 12σ statistical significance. From this we obtain the following branching fractions :

$$\mathcal{B}(B^+ \to \eta' K^+) = (7.4^{+0.8}_{-1.3} \pm 1.0) \times 10^{-5} \qquad (8)$$

$$\mathcal{B}(B^+ \to \eta' \pi^+) < 1.2 \times 10^{-5}(90\%CL) \qquad (9)$$

Figure 9 shows the projections of the data for ΔE and M_B individually for the two η' decay modes. There is a clear signal in each case.

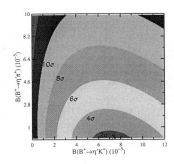

Figure 8: Preliminary maximum likelihood fit result for $B^+ \to \eta' K^+$ and $B^+ \to \eta' \pi^+$ for CLEO-II and CLEO 2.5 data. Contours are in 2σ increments.

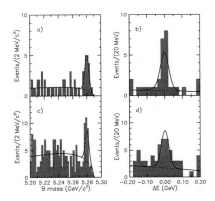

Figure 9: Projections of the data for $B^+ \to \eta' K^+$ for CLEO-II and CLEO 2.5 data for a) and b) $\eta' \to \eta\pi^+\pi^-$ and c) and d) $\eta' \to \rho\gamma$.

Figure 10 shows the result of the maximum likelihood fit for CLEO-II and CLEO 2.5 data for the decay mode $B^0 \to \eta' K^0$, also indicating a significant signal. The preliminary branching fraction is

$$\mathcal{B}(B^0 \to \eta' K^0) = (5.9^{+1.8}_{-1.6} \pm 0.9) \times 10^{-5}. \qquad (10)$$

This decay mode is particularly interesting because it could be used by B factory experiments to measure the

angle β from the CKM unitarity triangle. This would be complementary to investigations performed with the decay $B^0 \to \psi K_S^0$ [9]. Figure 11 shows the projections for ΔE and M_B, again indicating the presence of a signal where expected. Table 2 gives detailed numbers for the maximum likelihood fit results for $B \to \eta' K$.

Based on these results and the ones from section 4.1, we can verify the validity of a recently proposed sum rule [10]. Neglecting electroweak penguin and so-called "hairpin" amplitudes, the following sum rule can be obtained :

$$\Gamma(B^+ \to \eta' K^+) + \Gamma(B^+ \to \eta K^+) = \qquad (11)$$
$$\Gamma(B^+ \to K^+\pi^0) + \Gamma(B^+ \to K^0\pi^+).$$

If we neglect $B^+ \to \eta K^+$, which currently[4] has a branching fraction upper limit of 1.4×10^{-5}, the left-hand side of the equation is equal to $(7.4^{+1.3}_{-1.6}) \times 10^{-5}$ and the right hand side is only $(2.9 \pm 0.7) \times 10^{-5}$, perhaps indicating that the assumptions made in deriving the sum rule need to be revised.

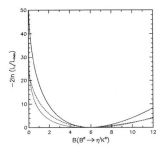

Figure 10: Preliminary maximum likelihood fit result for $B^0 \to \eta' K^0$ for CLEO-II and CLEO 2.5 data. The dotted line corresponds to $\eta' \to \eta\pi^+\pi^-$, the dashed line corresponds to $\eta' \to \rho\gamma$, and the solid line shows the result for the combined fit.

Table 2: Results for $B \to \eta' K$ decays. Shown for each of the sub-decay modes of the η' are the event yield, the efficiency, and the efficiency corrected for branching fractions of final state particles. Also shown is the statistical significance of the combined result for each decay mode.

Final state	Fit events	$\epsilon(\%)$	$\epsilon\mathcal{B}_s(\%)$	Stat.sig.(σ)
$\eta' K^+$				12.7
$\eta'_{\eta\pi\pi}K^+$	$18.4^{+5.0}_{-4.3}$	29	5.0	
$\eta'_{\rho\gamma}K^+$	$50.2^{+9.9}_{-9.0}$	36	10.8	
$\eta' K^0$				7.3
$\eta'_{\eta\pi\pi}K^0$	$5.4^{+2.8}_{-2.1}$	28	1.6	
$\eta'_{\rho\gamma}K^0$	$12.7^{+5.0}_{-4.2}$	33	3.4	

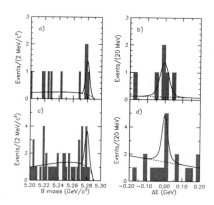

Figure 11: Projections of the data for $B^+ \to \eta' K^0$ for CLEO-II and CLEO 2.5 data for a) and b) $\eta' \to \eta\pi^+\pi^-$ and c) and d) $\eta' \to \rho\gamma$.

4.4 $B \to \eta_c K$

One of the possible explanations for the large $B \to \eta' K$ branching fractions is the intrinsic charm content of the η'. This lead to a search for the decay $B \to \eta_c K$, which should be very similar to hadronic B decays including a J/ψ. Theoretical expectations for the branching fractions for this decay are in the 10^{-3} range.

The analysis used CLEO-II data only and looked for the η_c with the decay $\eta_c \to \phi\phi$. The latter has a rather small branching fraction (0.71%), but provides an extremely clean experimental signature. Figure 12 shows the events remaining in the ΔE—M_B plane after cuts on all other variables. There are two events in the region where signal is expected, and essentially no background. The final event yield of $2.0^{+1.8}_{-1.1}$ was obtained with a maximum likelihood analysis. The probability that the observed signal is a fluctuation of the background is 4.3×10^{-5}. If this yield is interpreted as a signal, we obtain the following preliminary result :

$$\mathcal{B}(B^+ \to \eta_c K^+) = (1.5^{+1.4}_{-0.9} \pm 0.2 \pm 0.6) \times 10^{-3}. \quad (12)$$

The last quoted error is that due to the η_c branching fraction, and it clearly dominates the overall systematic uncertainty.

5 Summary

In summary, with the addition of data collected with a few upgrades to the detector, the CLEO collaboration has confirmed their earlier observations of the decays $B^0 \to K^+\pi^-$, $B^+ \to K^0\pi^+$, $B^+ \to \eta' K^+$, and $B^0 \to \eta' K^0$, and made the first observation of the decay $B^+ \to K^+\pi^0$. These results allow for interesting tests of

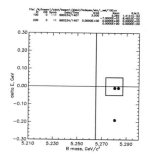

Figure 12: ΔE versus M_B distribution of events after cuts on all other variables for $B^+ \to \eta_c K^+$. The solid box shows the region where signal is expected.(CLEO-II only, preliminary)

recent theoretical developments, notably the Fleischer-Mannel bound and a sum rule proposed by Lipkin. A large rate was observed for the inclusive decay $B \to \eta' X_s$, which could be a signature for the gluonic penguin amplitude. This unexpected result has thus far eluded a complete theoretical explanation. CLEO also finds evidence for the decay $B^+ \to \eta_c K^+$ at a rate comparable to theoretical expectations.

Analyses are currently underway for several other charmless hadronic decay modes, and since even more data will be available soon from the CLEO 2.5 detector, we can expect this field to continue to be very active in the months to come.

Acknowledgments

The CLEO Collaboration gratefully acknowledges the effort of the CESR staff in providing us with excellent luminosity and running conditions. This work was supported by the National Science Foundation, the U.S. Department of Energy, Research Corporation, the Natural Sciences and Engineering Research Council of Canada, the A.P. Sloan Foundation, the Swiss National Science Foundation, and the Alexander von Humboldt Stiftung.

References

1. This talk covered abstracts 857, 858, and 860. See also M. Artuso *et al.* (CLEO Collaboration), CLEO CONF 98-20, ICHEP98-858, July 1998; T.E. Browder *et al.* (CLEO Collaboration), CLEO CONF 98-02, ICHEP98-857, July 1998; B.H. Behrens *et al.* (CLEO Collaboration), CLEO CONF 98-09, ICHEP98-860, July 1998; S. Chan *et al.* (CLEO Collaboration), CLEO CONF 98-24, ICHEP98-1073, July 1998.

2. M. Kobayashi and T. Maskawa, Prog. Theor. Phys. **49**, 652 (1973).
3. R. Godang *et al.* (CLEO Collaboration), *Phys. Rev. Lett.* **80**, 3456 (1998).
4. B.H. Behrens *et al.* (CLEO Collaboration), *Phys. Rev. Lett.* **80**, 3710 (1998).
5. Y. Kubota *et al.* (CLEO Collaboration), *Nucl. Instrum. Methods* A **320**, 66 (1992).
6. R. Fleischer and T. Mannel, *Phys. Rev.* D **57**, 2752 (1998).
7. R. Fleischer, Talk #813 at this conference.
8. D. Atwood and A. Soni, *Phys. Lett.* B **405**, 150 (1997).
9. D. London and A. Soni, *Phys. Lett.* B **407**, 61 (1997).
10. H.J. Lipkin, hep-ph/9708253, May 1998.

CHARMLESS AND DOUBLE-CHARM B DECAYS AT SLD

M. DAOUDI

Stanford Linear Accelerator Center, Stanford, CA 94309, USA
E-mail: daoudi@slac.stanford.edu

We present new results from the study of B decays at SLD: a measurement of the inclusive double-charm branching fraction in B-hadron decays and a search for the rare exclusive charmless decays $B^+ \to \rho^0\pi^+(K^+)$ and $B^+ \to K^{*0}\pi^+(K^+)$. Using a novel technique which consists of counting charged kaons produced at the secondary (B) vertex and the tertiary (D) vertex, we measure $\mathcal{B}(B \to D\bar{D}X) = 0.188 \pm 0.025(stat) \pm 0.059(syst)$. Another technique, based on the comparison of the charge of the D vertex to the flavor of the B hadron at production, yields a consistent result. In the search for rare exclusive B^+ decays, no candidates were found and competitive branching ratio upper limits are derived.

1 Introduction

The motivation for the analyses presented here lies in trying to shed some light into a long-standing B decay puzzle. For some time, the B semileptonic branching ratio (B_{SL}) and the charm yield in B decays (n_C) appeared to be too low compared to the theoretical expectation. While this is no longer the case at LEP, this discrepancy seems to persist at the $\Upsilon(4S)$, as illustrated in Fig 1 [1,2].

One way of reconciling a small B_{SL} is by having a large B hadronic width Γ_{Had}, which in turn can be due to a large $b \to c\bar{c}s$ partial width. By the same token, a large $B \to D\bar{D}X$ branching ratio can result in a large charm yield. This is shown in the following equations:

$$B_{SL} = \frac{\Gamma_{SL}}{\Gamma_{SL} + \Gamma_{Had}}$$

$$\Gamma_{Had} = \Gamma(b \to c\bar{u}d) + \Gamma(b \to c\bar{c}s)$$
$$+ \Gamma(b \to sg) + \Gamma(b \to u)$$

$$n_C = 1 + f_{D\bar{D}} - f_{NOC}.$$

Here, $f_{D\bar{D}}$ and f_{NOC} represent the B double-charm and no-open-charm branching fractions, respectively.

Another possible scenario is to have a large contribution from the $b \to sg$ and $b \to u$ transitions. For example, A. Kagan and J. Rathsman suggest that the $b \to sg$ decay could possibly be enhanced to a branching ratio as large as 10% [3]. An inclusive search for $b \to sg$ is currently underway at SLD [4]. Preliminary results from that analysis were covered in A. Litke's talk [5] at this conference.

In the present paper, results are given on the measurement of the inclusive branching fraction $\mathcal{B}(B \to D\bar{D}X)$ and a search for the rare exclusive decays $B^+ \to \rho^0\pi^+(K^+)$ and $B^+ \to K^{*0}\pi^+(K^+)$. These analyses benefit from the very good tracking and vertexing efficiency and resolution of SLD provided by a precision 3-D CCD pixel vertex detector. In addition, the high polarization of the SLC electron beam permits an excellent $b/\bar{b}$ separation, while the Cherenkov Ring Imaging Detector

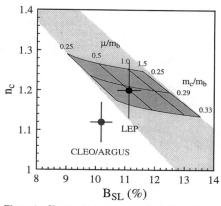

Figure 1: Charm yield vs. semileptonic branching ratio, from Ref. 1. Theoretical bounds are defined by the quark mass ratio m_c/m_b and the renormalization scale μ.

(CRID) provides good particle identification. The total SLD data sample of $\sim 550k$ hadronic Z^0's, collected between 1993 and 1998, was used in the rare B^+ decay search, whereas $\sim 250k$ hadronic Z^0's were analyzed in the B double-charm fraction measurement.

2 Measurement of $\mathcal{B}(B \to D\bar{D}X)$

The $B \to D\bar{D}X$ decay channel is governed mainly by the $b \to c\bar{c}s$ quark transition illustrated by the spectator diagram of Fig. 2, where in addition to the D meson from the $b \to c$ transition, a second so-called "wrong-sign" D is produced in the $W^- \to \bar{c}s$ decay. Previous measurements [6,7,8] of the branching ratio for this decay yielded a value of $\sim 20\%$ with an uncertainty $\simeq \pm 4\%$.

At SLD, the measurement is performed using an inclusive approach based on selecting B decays with two well separated vertices. This is achieved, in part, by requiring that the χ^2 probability that all tracks in the B

Table 1: Number of charged kaons in B- and D-vertex and their ratio.

	$N_{K^-,B}$	$N_{K^+,B}$	$N_{K^-,D}$	$N_{K^+,D}$	$N_{K,B}/N_{K,D}$
Data	758	559	1168	746	0.688 ± 0.025
M.C. $D\bar{D}$	3115	2339	2380	1880	1.280 ± 0.026
M.C. "Not-$D\bar{D}$"	5386	3974	11489	6549	0.519 ± 0.007

decay chain originate from a single vertex be $< 5\%$. We refer to the vertex closest to the interaction point as the "B-vertex" and the one furthest as the "D-vertex", as in the standard $b \to c$ decays. Note that for $B \to D\bar{D}X$ decays, the so-called "B-vertex" is rather likely to be one of the two D mesons. The two unique techniques we make use of to extract the double-charm fraction $f_{D\bar{D}}$ rely essentially on this fact. They are described below.

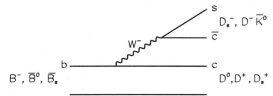

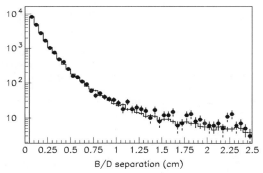

Figure 2: Tree-level diagram for $b \to c\bar{c}s$.

Figure 3: Reconstructed distance between B and D vertices. The dots represent the data and the histogram the Monte Carlo.

Fig. 3 shows the reconstructed distance between the B and D vertices. Good agreement between the data and the Monte Carlo is seen. The Monte Carlo simulation of the production and decay of B mesons at the Z^0 resonance is based on the JETSET 7.4 generator [9] and the CLEO B decay model [10].

2.1 The Kaon Counting Method

In this analysis we count the number of charged kaons identified by the CRID that are attached to the B-vertex and the D-vertex. We then compare the ratio of these two numbers in the data to that from two Monte Carlo samples: one containing exclusively double-charm B decays, and one made up of all other B decay channels. Since kaons are copiously produced in the $c \to s$ transition, one expects to find substantially more kaons in the D-vertex for single-charm B decays. This is not the case for double-charm B decays where one of the D mesons may be reconstructed as the B-vertex. The ratio of the number of $K^\pm$'s in the B-vertex to that in the D-vertex is shown in the last column of Table 1. It can be seen clearly that this ratio in the data has a double charm component, whose fraction we deduce to be

$$f_{D\bar{D}} = 0.188 \pm 0.025(stat).$$

This is after having applied a correction for the difference in efficiency for selecting single-charm (ϵ_D) and double-charm ($\epsilon_{D\bar{D}}$) decays of the B. We estimate the ratio of these efficiencies to be $\epsilon_{D\bar{D}}/\epsilon_D = 1.34 \pm 0.03$.

An interesting cross check is provided by the K^+/K^- asymmetry at the D-vertex (see Table 1). For a single-charm B decay, the sign of the kaon there is $-$ ($+$) if it originated from the $b \to c \to s$ ($\bar{b} \to \bar{c} \to \bar{s}$) decay chain. For double-charm decays, the D-vertex may correspond to the wrong-sign D meson which produces a kaon with opposite sign. Thus the kaon charge asymmetry in this case is different. The $B \to D\bar{D}X$ fraction we extract from this asymmetry is $f_{D\bar{D}} = 0.29 \pm 0.11(stat)$.

Note that, in this discussion, the $K^\pm$ sign is given by the kaon charge multiplied by the sign of the b-quark charge, as determined by tagging the b or $\bar{b}$ flavor of the B meson at production. The initial state b flavor tagging at SLD relies mainly on the large polarized forward-backward asymmetry for $Z^0 \to b\bar{b}$ decays. It is complemented by other tags using information from the opposite hemisphere to the selected B vertex. These tags include the momentum-weighted track charge sum (jet charge), the dipole charge for a topologically reconstructed vertex and the charge of an identified kaon or

Table 2: Systematic errors in the measurement of $f_{D\bar{D}}$ for the kaon counting method.

	σ_{syst}
Detector Systematics	
Tracking Efficiency	±0.003
Tracking Resolution	±0.007
$K^{\pm}$ Id. Efficiency	< 0.001
$\pi^{\pm}$ Fake Rate	±0.007
Track Misassignment	±0.042
Physics Systematics	
$udsc$ Background	±0.004
B Multiplicity	±0.015
B_s Fraction	±0.011
Λ_b Fraction	±0.007
$\mathcal{B}(B_{u,d} \to D^0 X)$	±0.002
$\mathcal{B}(B_{u,d} \to D^+ X)$	±0.002
$\mathcal{B}(B_{u,d} \to D_s^+ X)$	±0.004
$B \to D\bar{D}$ Model	±0.015
Charm Multiplicity	±0.032
TOTAL	±0.059

a high transverse-momentum lepton. Further details can be found in Ref. 11.

The analyzing power of the kaon counting technique is somewhat reduced experimentally due to the misassignment of tracks between the B and D vertices. In order not to rely entirely on the Monte Carlo simulation, the track misassignment rate is calibrated in the data using semileptonic $b \to l$ and $b \to c \to l$ decays. This is done by comparing the lepton sign in these decays to the initial-state b-quark flavor (sign). Wrong combinations of the two are the result of an initial-state flavor mistag (whose rate is well determined) and/or a misassignment of the lepton to the B- or D-vertex. We measure the track assignment efficiency to be $\epsilon_{T.A.} = 0.70 \pm 0.03$. The same analysis performed on the Monte Carlo gives a consistent result. The statistical error in this efficiency translates into a systematic uncertainty of ±0.042 in $f_{D\bar{D}}$.

Another important source of systematic error in this measurement is the uncertainty in the charm decay multiplicity, which is not very well determined experimentally [12]. Other systematic errors are listed in Table 2. Combining all the components in quadrature we get a total systematic error of ±0.059. Thus, our preliminary measurement of the B double-charm branching fraction using the kaon counting method is

$$f_{D\bar{D}} = 0.188 \pm 0.025(stat) \pm 0.059(syst).$$

2.2 D Charge Method

In this analysis, we compare the sign of the reconstructed D-vertex to the initial-state flavor of the B meson. For example, when a B^- decays to a single charmed meson, we expect the D-vertex to be positively charged (see Fig. 2). Whereas, for a double-charm decay of the same B^-, the reconstructed D-vertex can be either positively or negatively charged, since it may correspond to either a right- or wrong-sign D meson. This is, of course, without taking into account decays of the B^- into a neutral D meson, for simplicity. The distribution of the D-vertex charge multiplied by the b-quark charge sign of the B meson at production is shown in Fig. 4. There is good agreement between the data, represented by the dots, and the Monte Carlo, by the solid histogram. The Monte Carlo is a sum of a pure single-charm (shaded histogram) component and double-charm component. Note that the latter is asymmetric.

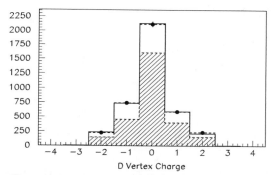

Figure 4: D-vertex charge (multiplied by sign of b-quark charge). The various components are explained in the text.

Similarly to the kaon counting method, this technique is also sensitive to the misassignment of tracks between the B and D vertices. Furthermore, the large D^0 production rate degrades its analyzing power.

We extract the B double-charm branching fraction by fitting the data distribution in Fig. 4 to the two Monte Carlo distributions corresponding to the D-charge for single- and double-charm decays. The result of the χ^2 fit is

$$f_{D\bar{D}} = 0.195 \pm 0.045(stat),$$

which is consistent with the above value obtained with the kaon counting method. The systematic error for this measurement is currently being evaluated.

3 Search for Charmless B^+ Decays

We have searched for the exclusive three-prong decays $B^+ \to \rho^0 \pi^+ (K^+)$, $\rho^0 \to \pi^+ \pi^-$; and $B^+ \to K^{*0} \pi^+ (K^+)$,

$K^{*0} \to K^+\pi^-$. These are mediated by the tree-level $b \to u$ and the one-loop penguin $b \to d(s)$ transitions. Theoretical predictions for the corresponding branching ratios are in the $10^{-6} - 10^{-5}$ range [13,14,15]. The best experimental limits come from CLEO [16] and are about a factor $5 - 10$ larger. These are based on the observation of a small number of events (e.g. 8 observed decays for an expected background of 3 in the $\rho\pi$ channel). For the decays $B^+ \to \rho^0\pi^+$, $K^{*0}\pi^+$, the DELPHI [17] experiment observes a total of 3 decays, with a background estimated at 0.15. The resulting branching ratio is $\mathcal{B}(B \to \rho(K^*)\pi) = (1.7^{+1.2}_{-0.8} \pm 0.2) \times 10^{-4}$.

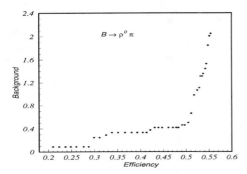

Figure 5: Background vs. efficiency evolution as the cut on the discriminator function output is varied.

While CLEO and DELPHI have used relatively large samples of B^+'s, their efficiency is small ($\approx 6.0\%$). At SLD, with our high resolution vertexing capability, we achieve an efficiency as high as 50% within our acceptance. With a sample of $\sim 100k$ B^+'s, we can set comparable limits to CLEO and DELPHI.

Table 3: Efficiency and background estimate for each of the four final states in the $B^+ \to 3$-prong search (not including the branching ratio for $\rho^0 \to \pi^+\pi^-$ and $K^{*0} \to K^+\pi^-$).

	Efficiency	Estimated Background
$\rho^0\pi^+$	0.50	0.61
$\rho^0 K^+$	0.45	0.61
$K^{*0}\pi^+$	0.52	0.37
$K^{*0}K^+$	0.46	0.37

The analysis consists of three steps. First, we form a three-prong vertex with charge = ± 1. We apply track quality, background suppression, and decay kinematics requirements. The latter include B mass and energy cuts, and ρ^0 and K^{*0} mass and center-of-mass helicity angle cuts. No kaon identification is required. In the second stage of the analysis, we construct a discriminator

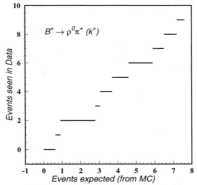

Figure 6: Number of observed decays in the data vs. expected background from the Monte Carlo, as the cut of the discriminator function is relaxed.

function using eight kinematic and vertexing variables, each with a parametrization for signal and background. Finally, we determine the optimal cut on the output variable of this discriminator function. This is done, relying on Poisson statistics, by minimizing the expected branching ratio upper limit as a function of efficiency, for each of the decay channels, under the assumption that the true branching fraction is 0. The variation of the expected background vs. efficiency for the $\rho\pi$ channel is shown in Fig. 5. The optimal discriminator function output cut for this channel can be inferred from the figure, corresponding to an efficiency of $\approx 50\%$. It is important to note that this study is performed using only a Monte Carlo sample for the background, independently of the data.

Table 4: Systematic uncertainty in the efficiency for each of the $B^+ \to 3$-prong final states.

	$\rho^0\pi^+$	$\rho^0 K^+$	$K^{*0}\pi^+$	$K^{*0}K^+$
Mass Resolution	-0.3%	-0.9%	-0.4%	+0.1%
Tracking Effic.	-0.6	-0.5	-0.4	-0.2
Vertex Resolution	-2.5	-1.7	-1.0	-0.9
b Fragmentation	±0.4	±0.0	±0.3	±0.4
Total Uncertainty	±3.8%	±3.1%	±2.1%	±1.4%

No events are observed in any of the four decays we searched for in the data. The amount of background we estimate for each channel is listed in Table 3, together with the individual efficiencies. The number of observed decays in the data is compared to the expected background as the cut on the discriminator function is relaxed. We find that the two track one another very well, providing a nice cross check of the analysis. This is illustrated in Fig. 6.

Table 4 contains the systematic uncertainty in the efficiency for each of the decay channels. The first three components in the table correspond to the shift in absolute efficiency observed in the Monte Carlo when it is smeared according to transverse momentum resolution, tracking efficiency, and impact parameter resolution, respectively. The last component corresponds to the uncertainty associated with the measurement of the average B-hadron energy at the Z^0. Taking into account these systematic errors and the uncertainty in the estimate of the number of B^+'s contained in the data ($\pm 1.1\%$), we set 90% C.L. upper limits. These are given in Table 5. They are consistent with the CLEO limits. For the combined $\rho^0\pi^+$ and $K^{*0}\pi^+$ final states, the agreement between our result and DELPHI's branching ratio measurement, given above, is only at the $\leq 10\%$ confidence level.

Table 5: Preliminary 90% CL upper limits (B.R.$\times 10^5$) for the four exclusive final states in the $B^+ \to$ 3-prong search. Corresponding CLEO and DELPHI limits are also given.

	CLEO	DELPHI	SLD Preliminary
$B^+ \to \rho^0\pi^+$	5.8	16	8.2
$B^+ \to \rho^0 K^+$	1.4	12	9.2
$B^+ \to K^{*0}\pi^+$	3.9	39	11.9
$B^+ \to K^{*0}K^+$			13.5

4 Conclusions

We have presented a measurement of the $B \to D\bar{D}X$ branching ratio using two novel techniques. These rely on the high efficiency inclusive topological vertex reconstruction of the SLD detector, as well as its excellent initial-state b / $\bar{b}$ flavor tagging, and its very good particle identification capability. The first method, called the kaon counting technique, gives the following preliminary result:

$$\mathcal{B}(B \to D\bar{D}X) = 0.188 \pm 0.025(stat) \pm 0.059(syst).$$

This is comparable in precision and in good agreement with recent measurements by other experiments. The systematic error, which is conservatively estimated at this stage of the analysis, is dominated by the track misassignment rate between secondary and tertiary vertices, and by the charm decay multiplicity. It is expected to be improved significantly.

The second method, dubbed the D-charge technique, gives a consistent result. This analysis is not as mature yet, but it provides a good cross check for the kaon counting technique. The two measurements will be combined, taking into account common systematics, and the overall error in the double-charm branching ratio will be significantly decreased. An additional improvement in the statistical error is expected with the inclusion of a data sample not yet analyzed.

We have also shown the results of a search for rare exclusive charmless $B^+ \to$ 3-prong decays. Here also, the high resolution tracking and vertexing of SLD greatly enhances the experimental sensitivity to these modes. An overall efficiency of 50% is achieved. No candidates were observed in the data for the $\rho^0\pi^+(K^+)$ and $K^{*0}\pi^+(K^+)$ final states, and competitive upper limits are derived.

Other interesting analyses in B decays are underway at SLD, in particular a measurement of the B semileptonic branching fraction and a search for the $b \to sg$ transition. Together with the measurements presented here, a significant contribution toward a better understanding of the semileptonic branching ratio vs. charm yield puzzle will be made.

Acknowledgements

I would like to thank my colleagues Glen Crawford, Per Reinertsen, and Bruce Schumm for their contribution.

References

1. M. Neubert, CERN-TH/98-2, hep-ph/9801269.
2. M. Feindt, IEKP-KA/98-4, hep-ph/9802380.
3. A.L. Kagan and J. Rathsman, hep-ph/9701300.
4. SLD Collaboration (K. Abe et al.), SLAC-PUB-7896, July 1998.
5. A. Litke, elsewhere in these proceedings.
6. CLEO Collaboration (T.E. Coan et al.), Phys. Rev. Lett. **80**, 1150 (1998).
7. ALEPH Collaboration (R. Barate et al.), Eur. Phys. J. C **4**, 387 (1998).
8. DELPHI Collaboration (P. Abreu et al.), Phys. Lett. B **426**, 193 (1998).
9. T. Sjöstrand, Comp. Phys. Comm. **82**, 74 (1994).
10. CLEO B decay model provided by P. Kim and the CLEO Collaboration.
11. SLD Collaboration (K. Abe et al.), SLAC-PUB-7885, July 1998.
12. Mark-III Collaboration (D. Coffman et al.), Phys. Lett. B **263**, 135 (1991).
13. N.G. Deshpande and J. Trampetic, Phys. Rev. D **41**, 895 (1990).
14. L.-L. Chau et al., Phys. Rev. D **43**, 2176 (1991).
15. A. Deandrea et al., Phys. Lett. B **320**, 170 (1994).
16. CLEO Collaboration (A. Anastassov et al.), CLEO CONF 97-24, EPS97-334.
17. DELPHI Collaboration (W. Adam et al.), Z. Phys. C **72**, 207 (1996).

TOWARDS A THEORY OF CHARMLESS NON-LEPTONIC TWO-BODY B DECAYS

A. ALI

Deutsches Elektronen-Synchrotron DESY, Hamburg, Germany
E-mail: ali@x4u2.desy.de

We address a number of theoretical and phenomenological issues in two-body charmless non-leptonic B decays in the context of a factorization model. A classification of the exclusive decays involving tree and penguin amplitudes is reviewed. The role of QCD anomaly in exclusive decays involving an η or η' is elucidated and comparison is made with the existing data. We argue that the factorization approach accounts for most of the observed two-body B decays.

1 Introduction

The standard theoretical framework to study B decays is based on an effective Hamiltonian, obtained by integrating out the top quark and $W^{\pm}$ fields, which allows to separate the short- and long-distance-contributions in these decays using the operator product expansion. QCD perturbation theory is used in deriving the renormalization group improved short-distance contributions and in evaluating the matrix elements at the parton level. The long-distance part in the two-body hadronic decays $B \to M_1 M_2$ involves the transition matrix elements $\langle M_1 M_2 | O_i | B \rangle$, where O_i is a four-quark or magnetic moment operator. Calculating these matrix elements from first principle is a true challenge and a quantitative theory of exclusive B decays is not yet at hand. Hence, some assumption about handling the hadronic matrix elements is at present unavoidable. We assume factorization, in which soft final state interactions (FSI) are ignored, and hence the hadronic matrix elements in the decays $B \to M_1 M_2$ factorize into a product of theoretically more tractable quantities[1].

The rationale of factorization lies in the phenomenon of colour-transparency[2] in which a pair of fast moving (energetic) quarks in a colour-singlet state effectively decouples from long-wavelength gluons. In two-body B-decays, the decay products have each an energy E_i of $O(m_B/2)$, which is large enough (compared to Λ_{QCD}) for the above argument to hold. Phenomenologically, the factorization framework does remarkably well in accounting for the observed non-leptonic two-body B-decays involving the current-current operators $O_{1,2}^c$, inducing $b \to c$ transitions[3]. The decays $B \to h_1 h_2$, where h_1 and h_2 are light hadrons, are more complex as they involve, apart from the current-current, also QCD- and electroweak penguin-induced amplitudes. However, in simpler circumstances where a single (Tree or Penguin) amplitude dominates, it should be possible to make predictions for the decays $B \to h_1 h_2$ in the factorization framework which are on the same theoretical footing as the two-body B-decays governed by the operators $O_{1,2}^c$.

We review some selected $B \to h_1 h_2$ decays. The underlying theoretical framework and the results presented here are based on the work done in collaboration with Greub[4], Chay, Greub and Ko[5], and Kramer and Lü[6,7]. A comparison with the available data from the CLEO collaboration, some of which has been updated at this conference[8], is also made.

2 Effective Hamiltonian Approach to B Decays

The effective Hamiltonian for the $\Delta B = 1$ transitions can be written as:

$$H_{eff} = \frac{G_F}{\sqrt{2}} [V_{ub} V_{uq}^* (C_1 O_1^u + C_2 O_2^u)$$

$$+ V_{cb} V_{cq}^* (C_1 O_1^c + C_2 O_2^c) - V_{tb} V_{tq}^* \sum_{i=3}^{12} C_i O_i], (1)$$

where G_F is the Fermi coupling constant, V_{ij} are the CKM matrix elements, C_i are the Wilson coefficients and $q = d, s$. The counting of these operators is as follows: $O_{1,2}^u$ and $O_{1,2}^c$ are the current-current (Tree) operators which induce the $b \to u$ and $b \to c$ transitions, respectively; $O_3, ..., O_6$ are the QCD-penguin operators, $O_7, ..., O_{10}$ are the Electroweak-penguin operators, and $O_{11}(O_{12})$ is the electromagnetic (chromo-magnetic) dipole operator. Their precise definition can be seen elsewhere[6]. The Wilson coefficients depend (in general) on the renormalization scheme and the scale μ at which they are evaluated. However, the physical matrix elements $\langle h_1 h_2 | H_{eff} | B \rangle$ are obviously independent of both the scheme and the scale. Hence, the dependencies in the Wilson coefficients must be compensated by a commensurate calculation of the hadronic matrix elements in a non-perturbative framework, such as lattice QCD. Presently, this is not a viable strategy as the calculation of the matrix elements $\langle h_1 h_2 | O_i | B \rangle$ is beyond the scope of the current lattice technology. However, perturbation theory comes to (partial) rescue, with the help of which one-loop matrix elements can be rewritten in terms of the tree-level matrix elements of the operators and the effective coefficients C_i^{eff}, which are scheme- and (largely) scale-independent:

$$\langle sq'\bar{q}'|\mathcal{H}_{eff}|b\rangle = \sum_{i,j} C_i^{eff}(\mu)\langle sq'\bar{q}'|O_j|b\rangle^{\text{tree}}. \quad (2)$$

The effective coefficients multiplying the matrix elements $< sq'\bar{q}'|O_j^{(q)}|b >^{\text{tree}}$ may be expressed as [4,6]:

$$C_1^{eff} = C_1 + \frac{\alpha_s}{4\pi}\left(r_V^T + \gamma_V^T \log\frac{m_b}{\mu}\right)_{1j} C_j + \cdots,$$

$$C_2^{eff} = C_2 + \frac{\alpha_s}{4\pi}\left(r_V^T + \gamma_V^T \log\frac{m_b}{\mu}\right)_{2j} C_j + \cdots,$$

$$C_3^{eff} = C_3 - \frac{1}{2N}\frac{\alpha_s}{4\pi}(C_t + C_p + C_g)$$
$$+ \frac{\alpha_s}{4\pi}\left(r_V^T + \gamma_V^T \log\frac{m_b}{\mu}\right)_{3j} C_j + \cdots,$$

$$C_4^{eff} = C_4 + \frac{1}{2}\frac{\alpha_s}{4\pi}(C_t + C_p + C_g)$$
$$+ \frac{\alpha_s}{4\pi}\left(r_V^T + \gamma_V^T \log\frac{m_b}{\mu}\right)_{4j} C_j + \cdots,$$

$$C_5^{eff} = C_5 - \frac{1}{2N}\frac{\alpha_s}{4\pi}(C_t + C_p + C_g)$$
$$+ \frac{\alpha_s}{4\pi}\left(r_V^T + \gamma_V^T \log\frac{m_b}{\mu}\right)_{5j} C_j + \cdots,$$

$$C_6^{eff} = C_6 + \frac{1}{2}\frac{\alpha_s}{4\pi}(C_t + C_p + C_g)$$
$$+ \frac{\alpha_s}{4\pi}\left(r_V^T + \gamma_V^T \log\frac{m_b}{\mu}\right)_{6j} C_j + \cdots,$$

$$C_7^{eff} = C_7 + \frac{\alpha_{\text{ew}}}{8\pi}C_e + \cdots,$$

$$C_8^{eff} = C_8$$

$$C_9^{eff} = C_9 + \frac{\alpha_{\text{ew}}}{8\pi}C_e + \cdots,$$

$$C_{10}^{eff} = C_{10}. \quad (3)$$

Here, r_V^T and γ_V^T are the transpose of the matrices derived by Buras et al. [9] and Ciuchini et al. [10]. The functions C_t, C_p and C_e generate (perturbative) strong interaction phases, essential for CP violation in B decays [11]. They are functions of k^2, the off-shell virtuality in the process $g(k^2) \to q\bar{q}$, CKM parameters, quark masses and the scale μ [4,6].

A number of remarks on C_i^{eff} is in order. First of all, the scale- and scheme-dependence in C_i mentioned above are now regulated. However, there are still scheme-independent but process-specific terms omitted in Eq. (3) indicated by the ellipses [4]. The specific constant matrix r_V^T used in our work [4,6,7] in defining C_i^{eff} has been obtained in the Landau gauge using an off-shell scheme in the calculation of the virtual corrections [9,10]. This raises

the spectre of C_i^{eff} becoming gauge dependent [12]. A remedy of these related problems is a perturbative formulation, in which the real and virtual corrections to the matrix elements are calculated in the NLL approximation in a physical (on-shell) scheme. The gauge-dependence in C_i^{eff} will then cancel in much the same way as in inclusive decays [13]. However, for exclusive decays, this procedure will bring in a certain cut-off dependence of C_i^{eff} due to the bremsstrahlung contribution, for which only a limited part of the phase space can be included in C_i^{eff}. This sensitivity has to be treated as a theoretical systematic error. Next, as already stated, the coefficients C_i^{eff} are functions of k^2. In the factorization approach, there is no model-independent way to keep track of this dependence. So, one has to model the k^2-dependence or use data to fix it. At present one varies this parameter in some reasonable range [14], $\frac{m_b^2}{4} \lesssim k^2 \lesssim \frac{m_b^2}{2}$, and includes this uncertainty in the estimates of the branching ratios. The k^2-related uncertainty in the CP-asymmetries in some cases is prohibitively large [15,7]. Clearly, more theoretical work and data are needed on these aspects.

3 Factorization Ansatz for $B \to h_1 h_2$

The factorization Ansatz for the decays $B \to h_1 h_2$ is illustrated below on the example of the u-quark contribution in the operator O_5 in the decay $B^- \to K^-\omega$ [4], where

$$O_5^{(u)} = (\bar{s}\gamma_\mu(1-\gamma_5)b)(\bar{u}\gamma^\mu(1+\gamma_5)u). \quad (4)$$

There are two diagrams which contribute to this decay (see, Fig. 3 in ref. 4). Calling the contributions by D_1 and D_2, the factorization approximation for D_1 is readily obtained:

$$D_1 = \langle\omega|\bar{u}\gamma^\mu(1+\gamma_5)u|0\rangle\langle K^-|\bar{s}\gamma_\mu(1-\gamma_5)b|B^-\rangle$$
$$= \langle\omega|\bar{u}u_-|0\rangle\langle K^-|\bar{s}b_-|B^-\rangle, \quad (5)$$

where $\bar{q}q'_- = \bar{q}\gamma_\mu(1-\gamma_5)q'$. To get D_2, one has to write the operator $O_5^{(u)}$ in its Fierzed form:

$$O_5^{(u)} = -2(\bar{u}_\beta(1-\gamma_5)b_\alpha)(\bar{s}_\alpha(1+\gamma_5)u_\beta)$$
$$= -2\left[\frac{1}{N_c}(\bar{u}(1-\gamma_5)b)(\bar{s}(1+\gamma_5)u)\right.$$
$$\left.+\frac{1}{2}(\bar{u}(1-\gamma_5)\lambda b)(\bar{s}(1+\gamma_5)\lambda u)\right], \quad (6)$$

where λ represents the $SU(3)$ colour matrices and N_c is the number of colours. Now, in the factorization approximation the second term in the square bracket does not contribute and one retains only the color-singlet contribution. This example illustrates the general structure of the matrix elements in the factorization approach. Thus, generically, one has

$$\langle |h_1 h_2| \mathcal{H}_{eff}|B \rangle \simeq \left[C_{2i-1}^{eff} \langle I \otimes I \rangle + \frac{1}{N_c} C_{2i}^{eff} \langle I \otimes I \rangle \right. $$
$$\left. + C_{2i-1}^{eff} \langle 8 \otimes 8 \rangle \right] + \left[C_{2i-1}^{eff} \leftrightarrow C_{2i}^{eff} \right] . \tag{7}$$

The factorization approximation amounts to discarding the $\langle 8 \otimes 8 \rangle$ contribution and compensating this by the parameters a_{2i} and a_{2i-1} ($i = 1, ..., 5$):

$$a_{2i-1} = C_{2i-1}^{eff} + \frac{1}{N_c} C_{2i}^{eff} \quad , \quad a_{2i} = C_{2i}^{eff} + \frac{1}{N_c} C_{2i-1}^{eff} . \tag{8}$$

These phenomenological parameters have to be determined by experiment. A particularly simple parametrization is obtained by replacing $1/N_c$ in Eq. (7) by a phenomenological parameter ξ:

$$1/N_c \to \xi .$$

With this parametrization, a variety of decays such as $B \to (J/\psi, \psi')(K, K^*)$ and $B \to (D, D^*)(\pi, \rho)$ etc. yield a universal value of ξ. Further, the parameter a_1 in these decays comes out close to its perturbative value, obtained by setting $N_c = 3$, and the experimental phase of a_2/a_1 is found to agree with the one based on factorization[3]. It is, therefore, tempting to extend this simplest parametrization to all ten parameters a_i in the decays $B \to h_1 h_2$. For a different point of view on the parametrization of the penguin amplitudes, see the talk by Cheng[17] and the papers by Ciuchini et al.[18].

The numerical values of a_i are given in Table 1 for three representative values of N_c (equivalently ξ) for the quark level transitions $b \to s$ $[\bar{b} \to \bar{s}]$. They are evaluated for $k^2 = m_b^2/2$, $\mu = m_b/2$ with $m_b = 4.88$ GeV, $\alpha_s(M_Z) = 0.118$, $m_t(m_t) = 168$ GeV and wherever necessary, the CKM-Wolfenstein parameters are set to $A = 0.81$, $\rho = 0.12$ and $\eta = 0.34$, corresponding to the present best fits[16].

A number of observations on the entries in Table 1 is in order:

- Only the coefficients a_1, a_4, a_6 and a_9 are stable against N_c-variation, i.e., they are of $O(1)$ as $N_c \to \infty$, with their relative magnitudes reflecting the SM dynamics (quark masses and mixing angles). The rest a_2, a_3, a_5, a_7, a_8 and a_{10} being of $O(1/N_c)$ are unstable against the variation of N_c.

- The coefficients a_1, a_2, a_4, a_6 and a_9 can be determined by measuring the ratios of some selected branching ratios. This has been studied extensively

in the paper with Kramer and Lü[6], where detailed formulae and their (reasonably accurate) approximate forms are given. These ratios will be helpful in testing the predictions of the factorization approach in forthcoming experiments.

- The QCD-penguin coefficient a_3 and the electroweak-penguin coefficients a_7, a_8 and a_{10} are numerically very small. Hence, it will be difficult to measure them.

3.1 Classification of Factorized Amplitudes

In the context of Tree-decays, a classification of the two-body decay amplitudes was introduced by Stech and co-workers[1]. These classes, concentrating now on the $B \to h_1 h_2$ decays, are the following:

- Class-I, involving decays in which only a charged meson can be generated directly from a colour-singlet current, as in $B^0 \to \pi^+ \pi^-$. For this class, $\mathcal{M}(B \to h_1 h_2) \propto a_1$.

- Class-II, involving decays in which the meson generated from the weak current is a neutral meson, like in $B^0 \to \pi^0 \pi^0$. For this class, $\mathcal{M}(B \to h_1 h_2) \propto a_2$.

- Class-III, involving the interference of class-I and class-II decay amplitudes, $\mathcal{M}(B \to h_1 h_2) \propto a_1 + r a_2$, where r is a process-dependent (but calculable in terms of form factors etc.) constant. Some examples are $B^\pm \to \pi^\pm \pi^0$, $B^\pm \to \pi^\pm \rho^0$ and $B^\pm \to \pi^\pm \omega$.

This classification has been extended to the decays involving penguin operators[6]. In the $B \to h_1 h_2$ decays, one now has two additional classes:

- Class-IV, involving decays whose amplitudes contain one (or more) of the dominant penguin coefficients a_4, a_6 and a_9, with constructive interference among them. Their amplitudes have the generic form:

$$\mathcal{M}(B^0 \to h_1^\pm h_2^\mp) \simeq \alpha_1 a_1 + \sum_{i=4,6,9} \alpha_i a_i + ..., \tag{9}$$
$$\mathcal{M}(B^0 \to h_1^0 h_2^0) \simeq \alpha_2 a_2 + \sum_{i=4,6,9} \alpha_i a_i + ...,$$
$$\mathcal{M}(B^\pm \to h_1^\pm h_2^0) \simeq \alpha_1 (a_1 + r a_2) + \sum_{i=4,6,9} \alpha_i a_i + ...,$$

with the second (penguin-induced) term dominant in each of the three amplitudes. The ellipses indicate possible contributions from the coefficients a_3, a_5, a_7, a_8 and a_{10} which can be neglected for

Table 1: Effective coefficients a_i for the $b \to s$ [$\bar{b} \to \bar{s}$] transitions; the entries for $a_3,...,a_{10}$ are to be multiplied with 10^{-4}.

	$N_c = 2$	$N_c = 3$	$N_c = \infty$
a_1	0.99 [0.99]	1.05 [1.05]	1.16 [1.16]
a_2	0.25 [0.25]	0.053 [0.053]	-0.33 [-0.33]
a_3	$-37- 14i$ $[-36-14i]$	48 [48]	$218+ 29i$ $[215+ 29i]$
a_4	$-402-72i$ $[-395-72i]$	$-439-77i[-431-77i]$	$-511-87i[-503-87i]$
a_5	$-150-14i[-149-14i]$	$-45[-45]$	$165+ 29i$ $[162+ 29i]$
a_6	$- 547-72i[-541-72i]$	$-575-77i[-568-77i]$	$-630-87i[-622-87i]$
a_7	$1.3-1.3i$ $[1.4-1.3i]$	$0.5-1.3i[0.5-1.3i]$	$-1.2-1.3i$ $[-1.1- 1.3i]$
a_8	$4.4-0.7i$ $[4.4-0.7i]$	$4.6-0.4i[4.6-0.4i]$	5.0[5.0]
a_9	$-91- 1.3i$ $[-91-1.3i]$	$-94-1.3i[-94-1.3i]$	$-101-1.3i[-101- 1.3i]$
a_{10}	$-31-0.7i$ $[-31-0.7i]$	$-14-0.4i$ $[-14-0.4i]$	20 [20]

this class of decays. The coefficients α_j are process-dependent and contain the CKM matrix elements, form factors etc.

Examples of Class-IV decays are quite abundant. They include decays such as $B^{\pm} \to K^{\pm}\pi^0$, $B^{\pm} \to K^{\pm}\eta^{(\prime)}$, which involve $a_1 + ra_2$ as the tree amplitude, and $B^0 \to K^0\pi^0$, $B^0 \to K^0\eta^{(\prime)}$ (and charged conjugates), which involve a_2 from the tree amplitude. Finally, the pure-penguin decays, such as $B^{\pm} \to \pi^{\pm}K^0$, $B^{\pm} \to K^{\pm}\bar{K}^0$ etc. naturally belong here. Several of these decays have been measured by the CLEO collaboration[8].

- Class-V, involving decays with strong N_c-dependent penguin coefficients a_3, a_5, a_7 and a_{10}, interfering significantly with one of the dominant penguin coefficients. Decays in which the dominant penguin coefficients interfere destructively are also included here.

Examples of this class are: $B^{\pm} \to \pi^{\pm}\phi$, $B^0 \to \pi^0\phi$, $B^0 \to \eta^{(\prime)}\phi$. In all these cases, the amplitudes are proportional to the linear combination $[a_3 + a_5 - 1/2(a_7 + a_9)]$. Examples of decays whose amplitudes are proportional to the dominant penguin coefficients interfering destructively are: $B^{\pm} \to K^{\pm}\phi$, $B^0 \to K^0\phi$. The above five classes exhaust all cases.

One expects that only class-I and class-IV decays (and possibly some class-III decays) can be predicted with some reasonable theoretical accuracy (typically a factor 2). In almost all of the decays studied in these classes[4,6], the variation of the branching ratios with N_c is not very marked. Hence, the effective coefficients extracted from data should come out rather close to their perturbative QCD values. Decays in other classes involve, in most cases, large and delicate cancellations, and hence their decay rates[6] (and CP asymmetries) are difficult to be predicted reliably[7]. An example is the decay $B^{\pm} \to \omega K^{\pm}$, which has been measured by the CLEO collaboration[8] but whose decay rate varies by more than

an order of magnitude in the range $0 < \xi < 0.5$ in the present approach[4,6].

4 Charm Content of the η and η' and the Role of QCD Anomaly in the Decays $B \to K\eta^{(\prime)}, K^*\eta^{(\prime)}$

Before comparing this framework with data, we discuss the decays $B \to K\eta'(\eta)$ and $B \to K^*\eta'(\eta)$, which have received lot of interest lately[19]. To be specific, we concentrate on the decay $B^{\pm} \to K^{\pm}\eta'(\eta)$.

In the factorization approach, the matrix element for the decay $B^{\pm} \to K^{\pm}\eta'$ can be expressed as

$$\mathcal{M} = -\frac{G_F}{\sqrt{2}} V_{cb} V_{cs}^* a_2 \langle \eta'(q)|\bar{c}\gamma_\mu\gamma_5 c|0\rangle \langle K(p')|\bar{s}\gamma^\mu b|B(p)\rangle . \tag{10}$$

Defining

$$\langle \eta'(q)|\bar{c}\gamma_\mu\gamma_5 c|0\rangle \equiv -if_{\eta'}^{(c)} q_\mu , \tag{11}$$

the quantities $f_{\eta'}^{(c)}$ and $f_{\eta}^{(c)}$ measure the charm content of the η' and η, respectively. Using this notation, the coupling constant f_{η_c}, defined analogously,

$$\langle \eta_c(q)|\bar{c}\gamma_\mu\gamma_5 c|0\rangle \equiv -if_{\eta_c} q_\mu , \tag{12}$$

can be used to normalize them. The constants $f_{\eta'}^{(c)}$ and $f_{\eta}^{(c)}$ have been determined in a variety of ways. Here the following two methods are reviewed[4,5].

- $f_{\eta'}^{(c)}$ and $f_{\eta}^{(c)}$ via the η-η'-η_c mixing[4]

This is a purely phenomenological approach. One admits a small $|c\bar{c}\rangle$ admixture in the $SU(3)$-singlet state vector $|\eta_0\rangle$, characterized by θ_{cc}. In the small-$\tan\theta_{cc}$ limit, and using one-mixing-angle (θ) formalism for the (η-η') complex, one can write down the following relations[4]:

$$f_{\eta'}^{(c)} \simeq \cos\theta \tan\theta_{cc} f_{\eta_c} ,$$
$$f_{\eta}^{(c)} \simeq \sin\theta \tan\theta_{cc} f_{\eta_c} . \tag{13}$$

Using the observed decay width [20]

$$\Gamma(\eta_c \to \gamma\gamma) = \frac{4(4\pi\alpha)^2 f_{\eta_c}^2}{81\pi m_{\eta_c}} = 7.5^{+1.6}_{-1.4} \text{ KeV} , \qquad (14)$$

one obtains $f_{\eta_c} = 411$ MeV from the central value. The mixing angle θ_{cc} can be determined from the ratio of the following radiative J/ψ-decays [20]:

$$\frac{\mathcal{B}(J/\psi \to \eta_c\gamma)}{\mathcal{B}(J/\psi \to \eta'\gamma)} = \frac{(1.3 \pm 0.4) \times 10^{-2}}{(4.31 \pm 0.30) \times 10^{-3}}$$

$$\simeq \left(\frac{k_{\eta_c}}{k_{\eta'}}\right)^3 \frac{1}{\cos^2\theta \tan^2\theta_{cc}} . \qquad (15)$$

This leads to a value $|\theta_{cc}| \simeq 0.014$, yielding [4]:

$$|f_{\eta'}^{(c)}| = |\cos\theta \tan\theta_{cc} f_{\eta_c}| \simeq 5.8 \text{ MeV} ,$$

$$|f_{\eta}^{(c)}| = |\sin\theta \tan\theta_{cc} f_{\eta_c}| \simeq 2.3 \text{ MeV} . \qquad (16)$$

Note that the signs of $f_{\eta'}^{(c)}$ and $f_{\eta}^{(c)}$ are not determined in this method. In the two-angle mixing formalism for the $(\eta$-$\eta')$ complex [21], the angle θ in the expressions for $f_{\eta'}^{(c)}$ and f_{η_c} gets replaced by θ_0, the angle in the singlet sector. Since $|\theta_0| < |\theta|$ (typical values are: $\theta \simeq -22°$ and $\theta_0 = -(4-9)°$), the value of f_{η_c} is reduced, yielding $|f_{\eta}^{(c)}| \simeq 1$ MeV.

- $f_{\eta'}^{(c)}$ and $f_{\eta}^{(c)}$ via QCD Anomaly [5]

In this case, the matrix elements are modeled by annihilating the charm-anticharm quark pair into two gluons, effecting the decay $b \to sgg$, followed by the transition $gg \to \eta^{(')}$ (see Fig. 1 in ref. 5). The first part of this two-step process, i.e., $b \to sgg$, has been worked out by Simma and Wyler [22], and their result can be transformed in the language of the effective theory:

$$H_{\text{eff}}^{gg} = -\frac{\alpha_s}{2\pi} a_2 \frac{G_F}{\sqrt{2}} V_{cb} V_{cs}^* \Delta i_5 \left(\frac{q^2}{m_c^2}\right) O_{sgg} , \qquad (17)$$

where the non-local operator O_{sgg} is given by

$$O_{sgg} = \frac{1}{k_1 \cdot k_2} G_a^{\alpha\beta} (D_\beta \tilde{G}_{\alpha\mu})_a \bar{s}\gamma^\mu(1-\gamma_5)b , \qquad (18)$$

with $\tilde{G}_{\mu\nu} = \frac{1}{2}\epsilon_{\mu\nu\alpha\beta}G^{\alpha\beta}$ and $q^2 = (k_1+k_2)^2 = 2k_1 \cdot k_2$, where k_1 and k_2 are the momenta of the two gluons. The function $\Delta i_5(z)$ is given by

$$\Delta i_5(z) = -1 + \frac{1}{z}\left[\pi - 2\tan^{-1}(\frac{4}{z}-1)^{1/2}\right]^2 , \text{ for } 0 < z < 4 . \qquad (19)$$

One can expand this function in q^2/m_c^2, which makes it clear that the leading term in H_{eff}^{gg} induces a power $(1/m_c^2)$ correction. In fact, in this form H_{eff}^{gg} becomes

the chromo-magnetic analogue of the corresponding operator in the decay $B \to X_s\gamma$ discussed by Voloshin [23]. Then, on using the equation of motion and the U_1 axial-anomaly:

$$\langle\eta^{(')}|\frac{\alpha_s}{4\pi}G_a^{\alpha\beta}\tilde{G}_{\alpha\beta,a}|0\rangle = m_{\eta^{(')}}^2 f_{\eta^{(')}}^u , \qquad (20)$$

one gets [5]:

$$f_{\eta'}^{(c)} \simeq -3.1[-2.3] \text{ MeV} , \qquad (21)$$

$$f_{\eta}^{(c)} \simeq -1.2[-0.9] \text{ MeV} , \qquad (22)$$

corresponding to the value $m_c = 1.3$ GeV [$m_c = 1.5$ GeV]. The two calculations give (within a factor 2) consistent results, with the anomaly method determining both the magnitudes and signs. The charm contents of the η' and η are, however, found to be small in both approaches.

5 Comparison with Data and Outlook

In Table 2, the branching ratios, averaged over the charge-conjugated modes, for the decays $B \to PP$ involving two pseudoscalar mesons are shown. The entries in this table [6] have been calculated using the BSW-model [1] and [Lattice QCD/QCD sum rule] form factors. Experimental numbers from CLEO [8] are shown in the last column.

We conclude this contribution with a number of remarks.

- All five decays measured by the CLEO collaboration shown in Table 2 are penguin-dominated class-IV decays. The estimates based on the factorization model are in reasonable agreement with data, except perhaps for the decay $B^+ \to K^+\eta'$ for which experiment lies (approximately) a factor 2 higher. All upper limits are in accord with the estimates given here. Thus, QCD-penguins in B decays are measurably large but not anomalous.

- It is fair to conclude that the QCD-improved factorization framework discussed here provides a first step towards understanding exclusive two-body B decays. However, there are many open theoretical questions and more work is needed. In particular, most class-V decays is a hrad nut to crack.

- Finally, non-leptonic B decays provide new avenues to determine the CKM parameters. At present no quantitative conclusions can be drawn as the experimental errors are large, but potentially some of these decays will provide complementary information on the CKM parameters [24] to the one from the unitarity constraints.

Table 2: $B \to PP$ Branching Ratios (in units of 10^{-6}).

Channel	Class	$N_c = 2$	$N_c = 3$	$N_c = \infty$	Exp.
$B^0 \to \pi^+\pi^-$	I	9.0 [11]	10.0 [12]	12 [15]	< 15
$B^0 \to \pi^0\pi^0$	II	0.35 [0.42]	0.12 [0.14]	0.63 [0.75]	< 9.3
$B^0 \to \eta'\eta'$	II	0.05 [0.07]	0.02 [0.02]	0.09 [0.10]	< 47
$B^0 \to \eta\eta'$	II	0.19 [0.22]	0.08 [0.10]	0.29 [0.34]	< 27
$B^0 \to \eta\eta$	II	0.17 [0.20]	0.10 [0.11]	0.24 [0.29]	< 18
$B^+ \to \pi^+\pi^0$	III	6.8 [8.1]	5.4 [6.4]	3.0 [3.6]	< 20
$B^+ \to \pi^+\eta'$	III	2.7 [3.2]	2.1 [2.5]	1.1 [1.4]	< 12
$B^+ \to \pi^+\eta$	III	3.9 [4.7]	3.1 [3.7]	1.9 [2.2]	< 15
$B^0 \to \pi^0\eta'$	V	0.06 [0.07]	0.07 [0.09]	0.11 [0.13]	< 11
$B^0 \to \pi^0\eta$	V	0.20 [0.24]	0.23 [0.27]	0.30 [0.36]	< 8
$B^+ \to K^+\pi^0$	IV	9.4 [11]	10 [12]	12 [15]	15 ±4 ± 3
$B^0 \to K^+\pi^-$	IV	14 [16]	15 [18]	18 [21]	14 ± 3 ± 2
$B^0 \to K^0\pi^0$	IV	5.0 [5.9]	5.7 [6.8]	7.4 [8.9]	< 41
$B^+ \to K^+\eta'$	IV	21 [25]	25 [29]	35 [41]	$74^{+8}_{-13} \pm 9$
$B^0 \to K^0\eta'$	IV	20 [24]	25 [29]	35 [41]	$59^{+18}_{-16} \pm 9$
$B^+ \to K^+\eta$	IV	2.0 [2.3]	2.4 [2.7]	3.4 [3.9]	< 14
$B^0 \to K^0\eta$	IV	1.7 [1.9]	2.0 [2.2]	2.6 [3.0]	< 33
$B^+ \to \pi^+K^0$	IV	14 [17]	16 [20]	22 [26]	14 ± 5 ± 2
$B^+ \to K^+\bar{K}^0$	IV	0.82 [0.95]	0.96 [1.1]	1.3 [1.5]	< 21
$B^0 \to K^0\bar{K}^0$	IV	0.79 [0.92]	0.92 [1.1]	1.2 [1.4]	< 17

References

1. M. Bauer and B. Stech, *Phys. Lett.* B **152**, 380 (1985); M. Bauer, B. Stech and M. Wirbel, *Z. Phys.* C **34**, 103 (1987).

2. J.D. Bjorken, in *New Developments in High Energy Physics*, ed. E.G. Floratos and A. Verganelakis, Nucl. Phys. (Proc. Suppl.) **11**, 321 (1989).

3. M. Neubert and B. Stech, preprint CERN-TH/97-99 [hep-ph 9705292], in *Heavy Flavors*, Second Edition, ed. A.J. Buras and M. Lindner (World Scientific, Singapore).

4. A. Ali and C. Greub, *Phys. Rev.* D **57**, 2996 (1998).

5. A. Ali, J. Chay, C. Greub and P. Ko, *Phys. Lett.* B **424**, 161 (1998).

6. A. Ali, G. Kramer and C.D. Lü, report DESY 98-041, hep-ph/9804363, Phys. Rev. D (in press).

7. A. Ali, G. Kramer and C.D. Lü, report DESY 98-056, hep-ph/9805403, Phys. Rev. D (in press).

8. J. Alexander (these proceedings); R. Godang et al. (CLEO Collaboration), *Phys. Rev. Lett.* **80**, 3456 (1998); B.H. Behrens et al. (CLEO Collaboration), *Phys. Rev. Lett.* **80**, 3710 (1998); T. Bergfeld et al. (CLEO Collaboration), *Phys. Rev. Lett.* **81**, 272 (1998).

9. A.J. Buras et al., *Nucl. Phys.* B **370**, 69 (1992).

10. M. Ciuchini et al., *Z. Phys.* C **68**, 239 (1995).

11. M. Bander, D. Silverman and A. Soni, *Phys. Rev. Lett.* **43**, 242 (1979); J. M. Gerard and W. S. Hou, *Phys. Rev.* D **43**, 2909 (1991); H. Simma, G. Eilam and D. Wyler, *Nucl. Phys.* B **352**, 367 (1991).

12. A.J. Buras and L. Silvestrini, report TUM-HEP-315/98, hep-ph/9806278.

13. M. Beneke, F. Maltoni and I.Z. Rothstein, report CERN-TH-98-240, hep-ph/9808360.

14. N.G. Deshpande and J. Trampetic, *Phys. Rev.* D **41**, 2926 (1990); H. Simma and D. Wyler, *Phys. Lett.* B **B272**, 395 (1991).

15. G. Kramer, W. F. Palmer and H. Simma, *Nucl. Phys.* B **428**, 77 (1994); *Z. Phys.* C **66**, 429 (1995).

16. A. Ali, report DESY 97-256, hep-ph/9801270.

17. H.-Y. Cheng (these proceedings).

18. M. Ciuchini, E. Franco, G. Martinelli and L. Silvestrini, *Nucl. Phys.* B **501**, 271 (1997); M. Ciuchini et al., *Nucl. Phys.* B **512**, 3 (1998).

19. For review, see K. Lingel, T. Skwarnicki and J.G. Smith, preprint COLO-HEP-395, SLAC-PUB-7796, HEPSY 98-1, hep-ex/9804015.

20. R.M. Barnett et al., *Phys. Rev.* D **54**, 1 (1996).

21. H. Leutwyler, preprint hep-ph/9709408; P. Herrera-Sikoldy *et al.*, *Phys. Lett.* B **419**, 326 (1998); T. Feldmann and P. Kroll, *Eur. Phys. J.* C **5**, 327 (1998).

22. H. Simma and D. Wyler, *Nucl. Phys.* B **344**, 283 (1990).

23. M.B. Voloshin, *Phys. Lett.* B **B397**, 275 (1997).

24. R. Fleischer (these proceedings).

NONFACTORIZABLE EFFECTS IN CHARMLESS B DECAYS AND B MESON LIFETIMES

HAI-YANG CHENG

Institute of Physics, Academia Sinica, Taipei, Taiwan 115, Republic of China
E-mail: phcheng@ccvax.sinica.edu.tw

Status of nonfactorizable effects in hadronic charmless B decays is reviewed. Implications of new CLEO measurements on $B^0 \to \pi^+\pi^-$ and $B \to \eta' K$ are discussed. Nonfactorizable effects due to color octet 4-quark operators are calculated using renormalization group improved QCD sum rules. The resultant B-meson lifetime ratio $\tau(B^-)/\tau(B_d)$ agrees with experiment.

1 Generalized Factorization

The nonleptonic two-body decays of mesons are conventionally evaluated under the factorization hypothesis. In the factorization approach, the decay amplitude is expressed in terms of factorizable hadronic matrix elements multiplied by some combinations of Wilson coefficient functions. To be more specific, the factorization hypothesis assumes that the 3-body hadronic matrix element $\langle h_1 h_2|O|M\rangle$ for the decay $M \to h_1 h_2$ is approximated as th product of two matrix elements $\langle h_1|J_{1\mu}|0\rangle$ and $\langle h_2|J_2^\mu|M\rangle$. However, it is known that this approach of naive factorization fails to describe the decays proceeding through the (class-I) color-suppressed internal W-emission diagrams, though it is at work for decay modes dominated by (class-II) external W-emission diagrams. This implies that it is necessary to take into account nonfactorizable contributions to the decay amplitude in order to render the color suppression of internal W-emission ineffective.

Because there is only one single form factor involved in the class-I or class-II decay amplitude of $B \to PP$, PV decays, the effects of nonfactorization can be lumped into the effective parameters a_1 and a_2: [1]

$$a_{1,2}^{\mathrm{eff}} = c_{1,2}^{\mathrm{eff}} + c_{2,1}^{\mathrm{eff}}\left(\frac{1}{N_c} + \chi_{1,2}\right), \qquad (1)$$

where χ_i are nonfactorizable terms and receive main contributions from color-octet current operators. Since $|c_1/c_2| \gg 1$, it is evident from Eq. (1) that even a small amount of nonfactorizable contributions will have a significant effect on the color-suppressed class-II amplitude. If $\chi_{1,2}$ are universal (i.e. process independent) in charm or bottom decays, then we have a generalized factorization scheme in which the decay amplitude is expressed in terms of factorizable contributions multiplied by the universal effective parameters $a_{1,2}^{\mathrm{eff}}$. For $B \to VV$ decays, this new factorization implies that nonfactorizable terms contribute in equal weight to all partial wave amplitudes so that $a_{1,2}^{\mathrm{eff}}$ can be defined. It should be stressed that, contrary to the naive one, the improved factorization does incorporate nonfactorizable effects in a process

independent form. Phenomenological analyses of two-body decay data of D and B mesons indicate that while the generalized factorization hypothesis in general works reasonably well, the effective parameters $a_{1,2}^{\mathrm{eff}}$ do show some variation from channel to channel, especially for the weak decays of charmed mesons [1,2,3]. An eminent feature emerged from the data analysis is that a_2^{eff} is negative in charm decay, whereas it becomes positive in the two-body decays of the B meson [1,4,5]:

$$a_2^{\mathrm{eff}}(B \to D\pi) \sim 0.20 - 0.28,$$
$$\chi_2(B \to D\pi) \sim 0.12 - 0.19. \qquad (2)$$

Phenomenologically, it is often to treat the number of colors N_c as a free parameter to model the nonfactorizable contribution to hadronic matrix elements and its value can be extracted from the data of two-body nonleptonic decays. Theoretically, this amounts to defining an effective number of colors N_c^{eff}, called $1/\xi$ in [6], by $1/N_c^{\mathrm{eff}} \equiv (1/N_c) + \chi$. It is clear from (2) that

$$N_c^{\mathrm{eff}}(B \to D\pi) = 1.8 - 2.2 \approx 2. \qquad (3)$$

2 Nonfactorizable Effects in Hadronic Charmless B Decays

What are the nonfactorizable effects in hadronic charmless B decays [7]? We note that the effective Wilson coefficients appear in the factorizable decay amplitudes in the combinations $a_{2i} = c_{2i}^{\mathrm{eff}} + \frac{1}{N_c}c_{2i-1}^{\mathrm{eff}}$ and $a_{2i-1} = c_{2i-1}^{\mathrm{eff}} + \frac{1}{N_c}c_{2i}^{\mathrm{eff}}$ $(i = 1, \cdots, 5)$. As discussed in Sec. 1. nonfactorizable effects in the decay amplitudes of $B \to PP$, VP can be absorbed into the parameters a_i^{eff}. This amounts to replacing N_c in a_i^{eff} by $(N_c^{\mathrm{eff}})_i$. Explicitly,

$$a_{2i}^{\mathrm{eff}} = c_{2i}^{\mathrm{eff}} + \frac{1}{(N_c^{\mathrm{eff}})_{2i}}c_{2i-1}^{\mathrm{eff}},$$
$$a_{2i-1}^{\mathrm{eff}} = c_{2i-1}^{\mathrm{eff}} + \frac{1}{(N_c^{\mathrm{eff}})_{2i-1}}c_{2i}^{\mathrm{eff}}. \qquad (4)$$

It is customary to assume in the literature that $(N_c^{\mathrm{eff}})_1 \approx (N_c^{\mathrm{eff}})_2 \cdots \approx (N_c^{\mathrm{eff}})_{10}$; that is, the nonfactorizable term is usually assumed to behave in the same way in tree and penguin decay amplitudes. A closer investigation shows that this is not the case. We have argued in[8] that nonfactorizable effects in the matrix elements of $(V-A)(V+A)$ operators are different from that of $(V-A)(V-A)$ operators. One reason is that the Fierz transformation of the $(V-A)(V+A)$ operators $O_{5,6,7,8}$ is quite different from that of $(V-A)(V-A)$ operators $O_{1,2,3,4}$ and $O_{9,10}$. Hence, we will advocate that

$$N_c^{\text{eff}}(LL) \equiv \left(N_c^{\text{eff}}\right)_{1,2,3,4,9,10},$$
$$N_c^{\text{eff}}(LR) \equiv \left(N_c^{\text{eff}}\right)_{5,6,7,8}, \tag{5}$$

and that $N_c^{\text{eff}}(LR) \neq N_c^{\text{eff}}(LL)$. In principle, N_c^{eff} can vary from channel to channel, as in the case of charm decay. However, in the energetic two-body B decays, N_c^{eff} is expected to be process insensitive as supported by data[5].

2.1 Nonfactorizable effects in spectator amplitudes

To study $N_c^{\text{eff}}(LL)$ in spectator amplitudes, we focus on the class-III decay modes sensitive to the interference between external and internal W-emission amplitudes. Good examples are the class-III modes: $B^{\pm} \to \omega\pi^{\pm}$, $\pi^0\pi^{\pm}$, $\eta\pi^{\pm}$, $\pi^0\rho^{\pm}$, $\cdots$, etc. Considering $B^{\pm} \to \omega\pi^{\pm}$, we find that the branching ratio is sensitive to $1/N_c^{\text{eff}}$ and has the lowest value of order 2×10^{-6} at $N_c^{\text{eff}} = \infty$ and then increases with $1/N_c^{\text{eff}}$. The 1997 CLEO measurement yields[9]

$$\mathcal{B}(B^{\pm} \to \omega\pi^{\pm}) = \left(1.1^{+0.6}_{-0.5} \pm 0.2\right) \times 10^{-5}. \tag{6}$$

Consequently, $1/N_c^{\text{eff}} > 0.35$ is preferred by the data[8]. Because this decay is dominated by tree amplitudes, this in turn implies that $N_c^{\text{eff}}(LL) < 2.9$. If the value of $N_c^{\text{eff}}(LL)$ is fixed to be 2, the branching ratio for positive ρ, which is preferred by the current analysis[10], will be of order $(0.9 - 1.0) \times 10^{-5}$, which is very close to the central value of the measured one. Unfortunately, the significance of $B^{\pm} \to \omega\pi^{\pm}$ is reduced in the recent CLEO analysis and only an upper limit is quoted[11]: $\mathcal{B}(B^{\pm} \to \pi^{\pm}\omega) < 2.3 \times 10^{-5}$. Nevertheless, the central value of $\mathcal{B}(B^{\pm} \to \pi^{\pm}\omega)$ remains about the same as (6). The fact that $N_c^{\text{eff}}(LL) \sim 2$ is preferred in charmless two-body decays of the B meson is consistent with the nonfactorizable term extracted from $B \to (D, D^*)(\pi, \rho)$ decays: $N_c^{\text{eff}}(B \to D\pi) \approx 2$. Since the energy release in the energetic two-body decays $B \to \omega\pi$, $B \to D\pi$ is of the same order of magnitude, it is thus expected that $N_c^{\text{eff}}(LL)|_{B \to \omega\pi} \approx 2$.

In analogue to the decays $B \to D^{(*)}(\pi, \rho)$, the interference effect of spectator amplitudes in class-III charmless B decay can be tested by measuring the ratios:

$$R_1 \equiv 2 \frac{\mathcal{B}(B^- \to \pi^-\pi^0)}{\mathcal{B}(\overline{B}^0 \to \pi^-\pi^+)},$$

$$R_2 \equiv 2 \frac{\mathcal{B}(B^- \to \rho^-\pi^0)}{\mathcal{B}(\overline{B}^0 \to \rho^-\pi^+)},$$
$$R_3 \equiv 2 \frac{\mathcal{B}(B^- \to \pi^-\rho^0)}{\mathcal{B}(\overline{B}^0 \to \pi^-\rho^+)}. \tag{7}$$

The ratios R_i are greater (less) than unity when the interference is constructive (destructive). Hence, a measurement of R_i (in particular R_3)[8], which has the advantage of being independent of the Wolfenstein parameters ρ and η, will constitute a very useful test on the effective number of colors $N_c^{\text{eff}}(LL)$.

During this conference, CLEO has reported the updated limits on $B^0 \to \pi^+\pi^-$ and $B^- \to \pi^-\pi^0$:[12]

$$\mathcal{B}(B^0 \to \pi^+\pi^-) < 0.84 \times 10^{-5},$$
$$\mathcal{B}(B^- \to \pi^-\pi^0) < 1.6 \times 10^{-5}. \tag{8}$$

In particular, the limit on $B^0 \to \pi^+\pi^-$ is improved by a factor of 2. It appears that this decay provides a stringent constraint on the form factor $F_0^{B\pi}$. Irrespective of the values of N_c^{eff}, the predicted branching ratio for $B^0 \to \pi^+\pi^-$ will easily exceed the current limit if $F_0^{B\pi}(0) \gtrsim 0.30$. Note that the decay rate of $B^0 \to \pi^+\pi^-$ increases slightly with $N_c^{\text{eff}}(LL)$ as it is dominated by the tree coefficient a_1. For $F_0^{B\pi}(0) = 0.30$, we find $N_c^{\text{eff}}(LL) \lesssim 0.20$.

2.2 Nonfactorizable effects in penguin amplitudes

The penguin amplitude of the class-VI mode $B \to \phi K$ is proportional to the QCD penguin coefficients $(a_3 + a_4 + a_5)$ and hence sensitive to the variation of $N_c^{\text{eff}}(LR)$ since a_4 is N_c^{eff}-stable, but a_3 and a_5 are N_c^{eff}-sensitive. Neglecting W-annihilation and space-like penguin diagrams, we find[8] that $N_c^{\text{eff}}(LR) = 2$ is evidently excluded from the present CLEO upper limit[11]

$$\mathcal{B}(B^{\pm} \to \phi K^{\pm}) < 0.5 \times 10^{-5}, \tag{9}$$

and that $1/N_c^{\text{eff}}(LR) < 0.23$ or $N_c^{\text{eff}}(LR) > 4.3$. A similar observation was also made in[13]. The branching ratio of $B \to \phi K^*$, the average of ϕK^{*-} and ϕK^{*0} modes, is also measured recently by CLEO with the result[11]

$$\mathcal{B}(B \to \phi K^*) = \left(1.1^{+0.6}_{-0.5} \pm 0.2\right) \times 10^{-5}. \tag{10}$$

We find that the allowed region for $N_c^{\text{eff}}(LR)$ is $4 \gtrsim N_c^{\text{eff}}(LR) \gtrsim 1.4$. This is in contradiction to the constraint $N_c^{\text{eff}}(LR) > 4.3$ derived from $B^{\pm} \to \phi K^{\pm}$. In fact, the factorization approach predicts that $\Gamma(B \to \phi K^*) \approx \Gamma(B \to \phi K)$ when the W-annihilation type of contributions is neglected. The current CLEO measurements (9) and (10) are obviously not consistent with the prediction based on factorization. One possibility is that generalized factorization is not applicable to $B \to VV$.

Therefore, the discrepancy between $\mathcal{B}(B \to \phi K)$ and $\mathcal{B}(B \to \phi K^*)$ will measure the degree of deviation from the generalized factorization that has been applied to $B \to \phi K^*$. It is also possible that the absence of $B \to \phi K$ events is a downward statistical fluctuation. At any rate, in order to clarify this issue and to pin down the effective number of colors $N_c^{\text{eff}}(LR)$, we urgently need measurements of $B \to \phi K$ and $B \to \phi K^*$, especially the neutral modes, with sufficient accuracy.

2.3 $B \to \eta' K$ decays

The published CLEO results [14] on the decay $B \to \eta' K$

$$\mathcal{B}(B^\pm \to \eta' K^\pm) = \left(6.5^{+1.5}_{-1.4} \pm 0.9\right) \times 10^{-5},$$
$$\mathcal{B}(B^0 \to \eta' K^0) = \left(4.7^{+2.7}_{-2.0} \pm 0.9\right) \times 10^{-5}, \quad (11)$$

are several times larger than previous theoretical predictions [15,16,17] in the range of $(1-2) \times 10^{-5}$. It was pointed out last year by several authors [18,19,20] that the decay rate of $B \to \eta' K$ will get enhanced because of the small running strange quark mass at the scale m_b and sizable $SU(3)$ breaking in the decay constants f_8 and f_0. Ironically, it was also realized last year that [19,18] the above-mentioned enhancement is partially washed out by the anomaly effect in the matrix element of pseudoscalar densities, an effect overlooked before. Specifically,

$$\langle \eta' | \bar{s}\gamma_5 s | 0 \rangle = -i \frac{m_{\eta'}^2}{2m_s} \left(f_{\eta'}^s - f_{\eta'}^u \right), \quad (12)$$

where the QCD anomaly effect is manifested by the decay constant $f_{\eta'}^u$. Since $f_{\eta'}^u \sim \frac{1}{2} f_{\eta'}^s$, it is obvious that the decay rate of $B \to \eta' K$ induced by the $(S-P)(S+P)$ penguin interaction is suppressed by the anomaly term in $\langle \eta' | \bar{s}\gamma_5 s | 0 \rangle$. As a consequence, the net enhancement is not large. If we treat $N_c^{\text{eff}}(LL)$ to be the same as $N_c^{\text{eff}}(LR)$, as assumed in previous studies, we would obtain typically $\mathcal{B}(B^\pm \to \eta' K^\pm) = (2-3) \times 10^{-5}$ (see the dot-dashed curve in Fig. 1).

What is the role played by the intrinsic charm content of the η' to $B \to \eta' K$? It has been advocated that the new internal W-emission contribution coming from the Cabibbo-allowed process $b \to c\bar{c}s$ followed by a conversion of the $c\bar{c}$ pair into the η' via two gluon exchanges is potentially important since its mixing angle $V_{cb}V_{cs}^*$ is as large as that of the penguin amplitude and yet its Wilson coefficient a_2 is larger than that of penguin operators. The decay constant $f_{\eta'}^c$, defined by $\langle 0 | \bar{c}\gamma_\mu\gamma_5 c | \eta' \rangle = i f_{\eta'}^c q_\mu$, has been calculated theoretically [21,22] and extracted phenomenologically from the data of $J/\psi \to \eta_c \gamma$, $J/\psi \to \eta' \gamma$ and of the $\eta\gamma$ and $\eta'\gamma$ transition form factors [18,23]; it lies in the range -2.3 MeV $\leq f_{\eta'}^c \leq -18.4$ MeV. The sign of $f_{\eta'}^c$ is crucial for the η' charm content contribution. For a negative $f_{\eta'}^c$, its contribution to $B \to \eta' K$ is constructive

for $a_2 > 0$. Since a_2 depends strongly on $N_c^{\text{eff}}(LL)$, we see that the $c\bar{c} \to \eta'$ mechanism contributes constructively at $1/N_c^{\text{eff}}(LL) > 0.28$ where $a_2 > 0$, whereas it contributes destructively at $1/N_c^{\text{eff}}(LL) < 0.28$ where a_2 becomes negative. In order to explain the abnormally large branching ratio of $B \to \eta' K$, an enhancement from the $c\bar{c} \to \eta'$ mechanism is certainly welcome in order to improve the discrepancy between theory and experiment. This provides another strong support for $N_c^{\text{eff}}(LL) \approx 2$. If $N_c^{\text{eff}}(LL) = N_c^{\text{eff}}(LR)$, then $\mathcal{B}(B \to \eta' K)$ will be *suppressed* at $1/N_c^{\text{eff}} \leq 0.28$ and enhanced at $1/N_c^{\text{eff}} > 0.28$ (see the dashed curve in Fig. 1 for $f_{\eta'}^c = -15$ MeV). If the preference for N_c^{eff} is $1/N_c^{\text{eff}} \lesssim 0.2$ (see e.g. [24]), then it is quite clear that te contribution from the η' charm content will make the theoretical prediction even worse at small $1/N_c^{\text{eff}}$! On the contrary, if $N_c^{\text{eff}}(LL) \approx 2$, the $c\bar{c}$ admixture in the η' will always lead to constructive interference irrespective of the value of $N_c^{\text{eff}}(LR)$ (see the solid curve in Fig. 1).

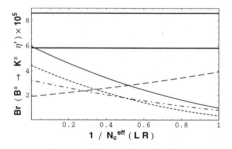

Figure 1: The branching ratio of $B^\pm \to \eta' K^\pm$ as a function of $1/N_c^{\text{eff}}(LR)$ with $N_c^{\text{eff}}(LL)$ being fixed at the value of 2 and $\eta = 0.34$, $\rho = 0.16$. The charm content of the η' with $f_{\eta'}^c = -15$ MeV contributes to the solid curve but not to the dotted curve. The anomaly contribution to $\langle \eta' | \bar{s}\gamma_5 s | 0 \rangle$ is included. For comparison, predictions for $N_c^{\text{eff}}(LL) = N_c^{\text{eff}}(LR)$ as depicted by the dot-dashed curve with $f_{\eta'}^c = 0$ and dashed curve with $f_{\eta'}^c = -15$ MeV is also shown. The solid thick lines are the preliminary updated CLEO measurements (13) with one sigma errors.

At this conference we learned that a recent CLEO reanalysis of $B \to \eta' K$ using a data sample 80% larger than in previous studies yields the preliminary results [25,12]:

$$\mathcal{B}(B^\pm \to \eta' K^\pm) = \left(7.4^{+0.8}_{-1.3} \pm 1.0\right) \times 10^{-5},$$
$$\mathcal{B}(B_d \to \eta' K^0) = \left(5.9^{+1.8}_{-1.6} \pm 0.9\right) \times 10^{-5}, \quad (13)$$

suggesting that the original measurements (11) were not an upward statistical fluctuation. This favors a slightly larger $f_{\eta'}^c$ in magnitude. For $N_c^{\text{eff}}(LL) = 2$ and $f_{\eta'}^c = -15$ MeV, which is consistent with all the known theoretical and phenomenological constraints, we show in Fig. 1 that

$\mathcal{B}(B^\pm \to \eta' K^\pm)$ at $1/N_c^{\text{eff}}(LR) \leq 0.2$ is enhanced considerably from $(2.5-3) \times 10^{-5}$ to $(4.6-5.9) \times 10^{-5}$. In addition to the η' charm content contribution, $N_c^{\text{eff}}(LL) \approx 2$ leads to constructive interference in the spectator amplitudes of $B \to \eta' K$ and an enhancement in the term proportional to $2(a_3 - a_5)X_u^{(BK,\eta')} + (a_3 + a_4 - a_5)X_s^{(BK,\eta')}$.

In the B_s system, we [26] find that $B_s \to \eta\eta'$, the analogue of $B^0 \to \eta' K^0$, has also a large branching ratio of order 2×10^{-5}.

2.4 Summary

Tree-dominated rare B decays, $B^\pm \to \omega\pi^\pm$ and $B^0 \to \pi^+\pi^-$, favor a small $N_c^{\text{eff}}(LL)$, namely $N_c^{\text{eff}}(LL) \approx 2$. The constraints on $N_c^{\text{eff}}(LR)$ derived from the penguin-dominated decays $B^\pm \to \phi K^\pm$ and $B \to \phi K^*$, which tend to be larger than $N_c^{\text{eff}}(LL)$, are not consistent with each other. Our analysis of $B \to \eta' K$ clearly indicates that $N_c^{\text{eff}}(LL) \approx 2$ is favored and $N_c^{\text{eff}}(LR)$ is preferred to be larger. The preliminary updated CLEO measurements of $B \to \eta' K$ seem to imply that the contribution from the η' charm content is important and serious.

3 Final-state interactions and $B \to \omega K$

The CLEO observation [11] of a large branching ratio for $B^\pm \to \omega K^\pm$

$$\mathcal{B}(B^\pm \to \omega K^\pm) = \left(1.5^{+0.7}_{-0.6} \pm 0.2\right) \times 10^{-5}, \quad (14)$$

is difficult to explain at first sight. Its factorizable amplitude is of the form

$$A(B^- \to \omega K^-) \propto (a_4 + Ra_6)X^{(B\omega,K)} \quad (15)$$
$$+ \left(2a_3 + 2a_5 + \frac{1}{2}a_9\right)X^{(BK,\omega)} + \cdots,$$

with $R = -2m_K^2/(m_b m_s)$, where ellipses represent for contributions from W-annihilation and space-like penguin diagrams. It is instructive to compare this decay mode closely with $B^- \to \rho K^-$

$$A(B^- \to \rho^0 K^-) \propto (a_4 + Ra_6)X^{(B\rho,K)} + \cdots. \quad (16)$$

Due to the destructive interference between a_4 and a_6 penguin terms, the branching ratio of $B^\pm \to \rho^0 K^\pm$ is estimated to be of order 5×10^{-7}. The question is then why is the observed rate of the ωK mode much larger than the ρK mode ? By comparing (15) with (16), it is natural to contemplate that the penguin contribution proportional to $(2a_3 + 2a_5 + \frac{1}{2}a_9)$ accounts for the large enhancement of $B^\pm \to \omega K^\pm$. However, this is not the case: The coefficients a_3 and a_5, whose magnitudes are smaller than a_4 and a_6, are not large enough to accommodate the data unless $N_c^{\text{eff}}(LR) < 1.1$ or $N_c^{\text{eff}}(LR) > 20$ (see Fig. 9 of [8]).

So far we have neglected three effects in the consideration of $B^\pm \to \omega K^\pm$: W-annihilation, space-like penguin diagrams and final-state interactions (FSI). It turns out that FSI may play a dominant role for $B^\pm \to \omega K^\pm$. The weak decays $B^- \to K^{*-}\pi^0$ via the penguin process $b \to su\bar{u}$ and $B^- \to K^{*0}\pi^-$ via $b \to sd\bar{d}$ followed by the quark rescattering reactions $\{K^{*-}\pi^0, K^{*0}\pi^-\} \to \omega K^-$ contribute constructively to $B^- \to \omega K^-$ (see Fig. 2), but destructively to $B^- \to \rho K^-$. Since the branching ratios for $B^- \to K^{*-}\pi^0$ and $K^{*0}\pi^-$ are large, of order $(0.5-0.8) \times 10^{-5}$, it is conceivable that a large branching ratio for $B^\pm \to \omega K^\pm$ can be achieved from FSI via inelastic scattering. Moreover, if FSI dominate, it is expected that $\mathcal{B}(B^\pm \to \omega K^\pm) \approx (1+\sqrt{2})^2\mathcal{B}(B^0 \to \omega K^0)$.

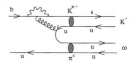

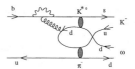

Figure 2: Contributions to $B^- \to K^-\omega$ from final-state interactions via the weak decays $B^- \to K^{*-}\pi^0$ and $B^- \to K^{*0}\pi^-$ followed by quark rescattering.

4 Nonspectator Effects and B Meson Lifetimes

In the heavy quark limit, all bottom hadrons have the same lifetimes, a well-known result in the parton picture. With the advent of heavy quark effective theory and the OPE approach for the analysis of inclusive weak decays, it is realized that the first nonperturbative correction to bottom hadron lifetimes starts at order $1/m_b^2$ and it is model independent. However, the $1/m_b^2$ corrections are small and essentially canceled out in the lifetime ratios. The nonspectator effects such as W-exchange and Pauli interference due to four-quark interactions are of order $1/m_Q^3$, but their contributions can be potentially significant due to a phase-space enhancement by a factor of $16\pi^2$. As a result, the lifetime differences of heavy hadrons come mainly from the above-mentioned nonspectator effects.

The four-quark operators relevant to inclusive nonleptonic B decays are

$$O_{V-A} = \bar{b}_L \gamma_\mu q_L\, \bar{q}_L \gamma^\mu b_L,$$

$$O_{S-P} = \bar{b}_R q_L \, \bar{q}_L b_R \, ,$$
$$T_{V-A} = \bar{b}_L \gamma_\mu t^a q_L \, \bar{q}_L \gamma^\mu t^a b_L \, ,$$
$$T_{S-P} = \bar{b}_R t^a q_L \, \bar{q}_L t^a b_R \, . \qquad (17)$$

From which one can follow [27] to define four hadronic parameters $B_1, B_2, \varepsilon_1, \varepsilon_2$ relevant to our purposes:

$$\frac{1}{2m_B} \langle \overline{B} | O_{V-A(S-P)} | \overline{B} \rangle \equiv \frac{f_B^2 m_B}{8} B_{1(2)} \, ,$$
$$\frac{1}{2m_B} \langle \overline{B} | T_{V-A(S-P)} | \overline{B} \rangle \equiv \frac{f_B^2 m_B}{8} \varepsilon_{1(2)} \, . \qquad (18)$$

Under the factorization approximation, $B_i = 1$ and $\varepsilon_i = 0$. To the order of $1/m_b^3$, the B-hadron lifetime ratios are given by

$$\frac{\tau(B^-)}{\tau(B_d^0)} = 1 + \left(0.043 B_1 + 0.0006 B_2 - 0.61 \varepsilon_1 + 0.17 \varepsilon_2 \right) \, ,$$
$$\frac{\tau(B_s^0)}{\tau(B_d^0)} = 1 + (-1.7 \times 10^{-5} B_1 + 1.9 \times 10^{-5} B_2$$
$$-0.0044 \varepsilon_1 + 0.0050 \, \varepsilon_2) \, . \qquad (19)$$

It is clear that even a small deviation from the factorization approximation $\varepsilon_i = 0$ can have a sizable impact on the lifetime ratios.

We have derived in heavy quark effective theory the renormalization-group improved QCD sum rules [28] for the hadronic parameters B_1, B_2, ε_1, and ε_2. The results are [28]

$$B_1(m_b) = 0.96 \pm 0.04, \ \ B_2(m_b) = 0.95 \pm 0.02,$$
$$\varepsilon_1(m_b) = -0.14 \pm 0.01, \ \ \varepsilon_2(m_b) = -0.08 \pm 0.01, (20)$$

to the zeroth order in $1/m_b$. The resultant B-meson lifetime ratios are $\tau(B^-)/\tau(B_d) = 1.11 \pm 0.02$ and $\tau(B_s)/\tau(B_d) \approx 1$, to be compared with the world averages [29]: $\tau(B^-)/\tau(B_d) = 1.07 \pm 0.03$ and $\tau(B_s)/\tau(B_d) = 0.94 \pm 0.04$. Therefore, our prediction for $\tau(B^-)/\tau(B_d)$ agrees with experiment.

References

1. H.Y. Cheng, *Phys. Lett.* B **395**, 345 (1994).
2. A.N. Kamal, A.B. Santra, T. Uppal, and R.C. Verma, *Phys. Rev.* D **53**, 2506 (1996).
3. H.Y. Cheng, *Z. Phys.* C **69**, 647 (1996).
4. H.Y. Cheng and B. Tseng, *Phys. Rev.* D **51**, 6295 (1995).
5. M. Neubert and B. Stech, CERN-TH/97-99 [hep-ph/9705292].
6. M. Bauer, B. Stech, and M. Wirbel, *Z. Phys.* C **34**, 103 (1987).
7. For a review of CLEO measurements on charmless B decays, see K. Lingel, T. Skwarnicki, and J.G. Smith, hep-ex/9804015.
8. H.Y. Cheng and B. Tseng, hep-ph/9803457, to appear in Phys. Rev. D.
9. CLEO Collaboration, M.S. Alam *et al.*, CLEO CONF 97-23 (1997).
10. F. Parodi, P. Roudeau, and A. Stocchi, hep-ph/9802289.
11. CLEO Collaboration, T. Bergfeld *et al.*, *Phys. Rev. Lett.* **81**, 272 (1998).
12. CLEO Collaboration, J. Roy, invited talk presented at the XXIX International Conference on High Energy Physics, Vancouver, July 23-28, 1998.
13. N.G. Deshpande, B. Dutta, and S. Oh, OITS-644 [hep-ph/9712445].
14. CLEO Collaboration, B.H. Behrens *et al.*, *Phys. Rev. Lett.* **80**, 3710 (1998).
15. L.L. Chau, H.Y. Cheng, W.K. Sze, H. Yao, and B. Tseng, *Phys. Rev.* D **43**, 2176 (1991); D **58**, (E)019902 (1998).
16. G. Kramer, W.F. Palmer, and H. Simma, *Z. Phys.* C **66**, 429 (1995); *Nucl. Phys.* B **428**, 77 (1994).
17. D.S. Du and L. Guo, *Z. Phys.* C **75**, 9 (1997); D.S. Du, M.Z. Yang, and D.Z. Zhang, *Phys. Rev.* D **53**, 249 (1996).
18. A. Ali and C. Greub, *Phys. Rev.* D **57**, 2996 (1998).
19. A.L. Kagan and A.A. Petrov, UCHEP-27 [hep-ph/9707354].
20. N.G. Deshpande, B. Dutta, and S. Oh, *Phys. Rev.* D **57**, 5723 (1998); A. Datta, X.G. He, and S. Pakvasa, *Phys. Lett.* B **419**, 369 (1998).
21. A. Ali, J. Chay, C. Greub, and P. Ko, *Phys. Lett.* B **424**, 161 (1998).
22. F. Araki, M. Musakhanov, and H. Toki, hep-ph/9803356 and hep-ph/9808290.
23. A.A. Petrov, *Phys. Rev.* D **58**, 054003 (1998); T. Feldmann, P. Kroll, and B. Stech, hep-ph/9802409; T. Feldmann and P. Kroll, hep-ph/9711231; J. Cao, F.G. Cao, T. Huang, and B.Q. Ma, hep-ph/9807508.
24. A. Ali, G. Kramer, and C.D. Lü, hep-ph/9804363.
25. CLEO Collaboration, B.H. Behrens *et al.*, CLEO CONF 98-09 (1998).
26. Y.H. Chen, H.Y. Cheng, and B. Tseng, IP-ASTP-10-98.
27. M. Neubert and C.T. Sachrajda, *Nucl. Phys.* B **483**, 339 (1997).
28. H.Y. Cheng and K.C. Yang, hep-ph/9805222.
29. S. Willocq, invited talk presented at the XXIX International Conference on High Energy Physics, Vancouver, July 23-28, 1998.

B/B̄ FLAVOUR TAGGING AND DOUBLY CHARMED B DECAYS IN ALEPH

R. BARATE

LAPP, Chemin de Bellevue, BP110, F74941, Annecy-le-Vieux, CEDEX, France
E-mail: barate@lapp.in2p3.fr

This contribution concerns three contributed papers that share the common feature of analysing fully- (or almost fully-) reconstructed B decays coming from a sample of four million hadronic Z decays collected with the ALEPH detector at LEP. In the first paper [1], 404 charged and neutral B mesons decaying in standard modes are fully reconstructed and used to look for resonant structure in the $B\pi^{\pm}$ system. In the framework of Heavy Quark Symmetry (HQS), the mass of the B_2^* state and the relative production rate of the B^{**} system are measured. In the same sample of B mesons, significant $B\pi^{\pm}$ charge-flavour correlations are observed. In the second paper [2], a search for doubly-charmed B decays with both charmed mesons reconstructed is performed. A clear signal is observed in the channels $b \rightarrow D_s\bar{D}(X)$ and $b \rightarrow D\bar{D}(X)$ providing the first direct evidence for doubly-charmed b decays involving no D_s production. Evidence for associated K_S^0 and $K^{\pm}$ production in the decays $B \rightarrow D\bar{D}(X)$ is also presented and some candidates for completely reconstructed B meson decays $B \rightarrow D_s\bar{D}(n\pi)$, $B \rightarrow D\bar{D}K_S^0$ and $B \rightarrow D\bar{D}K^{\pm}$ are observed. Furthermore, candidates for the two-body Cabibbo suppressed decays $B^0 \rightarrow D^{*-}D^{*+}$ and $B^- \rightarrow D^{(*)0}D^{(*)-}$ are also observed. One $B_s^0 \rightarrow D_s^+D_s^-$ event is reconstructed, which can be only the short-lived CP even eigenstate. In the third paper [3], the B_s decay to $D_s^{(*)+}D_s^{(*)-}(X)$ is observed, tagging the final state with two ϕ in the same hemisphere. It corresponds mostly to the short-lived CP even eigenstate. A preliminary value of the B_s short lifetime is obtained.

1 Resonant Structure and Flavour Tagging in the $B\pi^{\pm}$ System

1.1 Introduction

Fully reconstructed B meson decays are used to extract a precise mass of the B_2^* state and to obtain the B/B̄ signature at the decay point. Using the π from B^{**} decay or the nearest π from fragmentation, direct tagging of the initial B just before it oscillates and decays is possible. It could be used in future CP violation experiments.

1.2 B meson and associated pion selection

Charged and neutral B mesons are fully reconstructed in various exclusive modes. Eighty percent are in the mode[a] $B \rightarrow \bar{D}^*(X)$, where X is a charged π, ρ or a_1, and 20% are of the form $B \rightarrow J/\psi(\psi')X$, where X is a charged K or a neutral K^*. In addition, charged B candidates are reconstructed in the channels $B^- \rightarrow D^{*0}\pi^-$ and $D^{*0}a_1^-$, with a missing soft γ or π^0 from the $D^{*0} \rightarrow D^0\gamma$ or $D^0\pi^0$ decays. In total, 238 charged and 166 neutral B candidates are reconstructed with purities of $(84 \pm 3)\%$ and $(86 \pm 3)\%$ respectively.

The neighbouring pion is selected using the P_L^{max} algorithm which chooses the track with the highest projected momentum along the B direction (and a $B\pi$ mass below 7.3 GeV/c).

1.3 Resonant Structure in the $B\pi^{\pm}$ System

Using the pion selected with the P_L^{max} algorithm, the right sign and wrong sign $B\pi$ mass distributions are made

---

[a]Throughout this paper, charge conjugate decay modes are always implied.

and B^{**} signals are extracted. The gain of one order of magnitude in mass resolution compared to previous inclusive experiments, due to the quality of exclusive B decays, allows a more precise measurement of the masses if one uses the Heavy Quark Symmetry parameters [1]. HQS predicts 4 resonances giving 5 correlated Breit Wigner (3 narrow and 2 wide) in the $B\pi$ mass distributions. Here only the overall mass scale and the total number of signal events are left free. An unbinned likelihood fit (Fig. 1) gives :

$$M(B_2^*) = (5739^{+8}_{-11}(\text{stat})^{+6}_{-4}(\text{syst})) \text{ MeV/c}^2$$

$$f_{B^{**}} \equiv \frac{\mathcal{B}(b \rightarrow B^{**} \rightarrow B^{(*)}\pi)}{\mathcal{B}(b \rightarrow B_{u,d})} = (31 \pm 9(\text{stat})^{+6}_{-5}(\text{syst}))\%.$$

The result for the mass of the B_2^* state is somewhat low compared to the predicted value of 5771 MeV/c^2 .

1.4 B/B̄ Flavour Tagging

The sign of the neighbouring pion (from B^{**} or fragmentation) tags the B/B̄ nature at production : a π^+ right-sign tags a B^0, etc. The P_L^{max} tagging algorithm efficiencies for neutral and charged B's are :

$$\epsilon_{tag}^N = (89 \pm 3(\text{stat}) \pm 2(\text{syst}))\% ,$$

$$\epsilon_{tag}^C = (89 \pm 2(\text{stat}) \pm 1(\text{syst}))\%.$$

The asymmetry $\mathcal{A}^N$ between the number of right-sign and wrong-sign tags, $\mathcal{A}^N = (N_{rs} - N_{ws})/(N_{rs} + N_{ws})$, is shown as a function of the B decay proper time, t, in Fig. 2a. The sinusoidal mixing of B^0 into $\bar{B}^0$ gives rise to the excess of wrong-sign tags (negative value of $\mathcal{A}^N$) at high proper times.

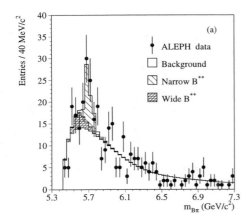

Figure 1: The $B\pi$ mass spectrum from data (points with error bars) and fit (histogram). The fit includes the expected background plus contributions from the narrow and wide B^{**} states.

The mistag rate for neutral B's, ω_{tag}^N, is measured from the oscillation amplitude, with Δm_d fixed to the world average of $0.474 \pm 0.031\,\mathrm{ps}^{-1}$. An unbinned likelihood fit gives:

$$\omega_{tag}^N = (34.4 \pm 5.5(\mathrm{stat}) \pm 1.0(\mathrm{syst}))\,\% \, ;$$

similarly, an unbinned likelihood fit for ω_{tag}^C, the mistag rate for charged B's gives :

$$\omega_{tag}^C = (26.0 \pm 3.6(\mathrm{stat}) \pm 0.7(\mathrm{syst}))\,\%$$

showing that this method gives good tagging performance, whilst being very efficient. The corresponding fits to the data are displayed in Fig. 2.

2 Doubly charmed B decays

2.1 Introduction

The final states with 2 charm mesons allow a precise study of the b quark decay into $c\bar{c}s$ and give a direct access to the average number of charm quarks per b decay n_c. The study of 3-body B meson decay in $\bar{D}DK$ gives many results on the diagrams, possible resonances and dynamics. Added to the Cabbibo suppressed $D\bar{D}$, standard $D_s^+\bar{D}$, and new $D_s^+\bar{D}n\pi^\pm$ decays, one obtains the total double charm contribution to B meson decay.

2.2 $D\bar{D}$ selection

The charmed mesons are searched for in the decay modes $D^0 \to K^-\pi^+$, $D^0 \to K^-\pi^+\pi^-\pi^+$,

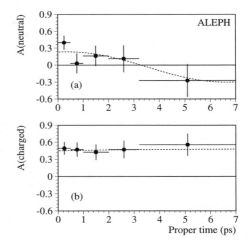

Figure 2: The right-sign/wrong-sign asymmetries in the data as a function of the proper decay time. The dashed curves display the charge asymmetries determined from the unbinned likelihood fits.

$D^+ \to K^-\pi^+\pi^+$, $D^{*+} \to D^0\pi^+$, $D_s^+ \to \phi\pi^+ (\phi \to K^-K^+)$ and $D_s^+ \to \bar{K}^{*0}K^+$ ($\bar{K}^{*0} \to K^-\pi^+$). For D^0 mesons from D^{*+} decay, the decay mode $D^0 \to K^-\pi^+\pi^0$ is also used.

The pairs of D candidates must be in the same hemisphere and the two D candidates are required to form a vertex with a probability of at least 0.1%. To maintain a good acceptance for the $B \to D\bar{D}X$ signal whilst rejecting the backgrounds and minimizing the model dependence of the selection efficiencies, a cut $d_{BD}/\sigma_{BD} > -2$ (>0) is applied on the D^0, D_s^+ (D^+) decay length significance (defined in Fig. 3). The decay length significance of the $D\bar{D}$ vertex is also required to satisfy the condition $d_B/\sigma_B > -2$.

To obtain the number of real $D\bar{D}$ events, standard tables are made (for instance $m(K\pi)_1/m(K\pi)_2$) and the combinatorial background is subtracted linearly.

2.3 Inclusive b quark decays in $D_s\bar{D}(X)$ or $D\bar{D}(X)$

After acceptance corrections, the different branching fractions obtained are given in Table 1.

The inclusive branching fraction of b quarks to $D_sD(X)$ is measured to be

$$\mathcal{B}(b \to D_sD^0, D_sD^\pm(X)) = \left(13.1^{+2.6}_{-2.2}(\mathrm{st})^{+1.8}_{-1.6}(\mathrm{sy})^{+4.4}_{-2.7}(\mathcal{B}_D)\right)\%,$$

in good agreement with previous measurements of the inclusive branching fraction of the B mesons to D_s.

For the first time, doubly-charmed B decays involving no D_s production are observed. The corresponding

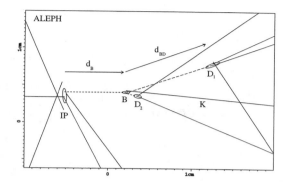

Figure 3: Display of a decay $B^0 \to D^- D^0 K^+$ reconstructed in the ALEPH detector (real data).

Table 1: Summary of the different branching fractions measured in this analysis. The modes involving a D^{*+} (lowest part of the table) are also included in the upper part results as a subsample of the modes involving a D^0 or a D^+.

Channel	$\mathcal{B}(\%)$
$b \to D^0 D_s^- (X)$	$9.1^{+2.0}_{-1.8}{}^{+1.3}_{-1.2}{}^{+3.1}_{-1.9}$
$b \to D^+ D_s^- (X)$	$4.0^{+1.7}_{-1.4} \pm 0.7{}^{+1.4}_{-0.9}$
Sum $b \to D^0 D_s^-, D^+ D_s^- (X)$	$13.1^{+2.6}_{-2.2}{}^{+1.8}_{-1.6}{}^{+4.4}_{-2.7}$
$b \to D^0 \bar{D}^0 (X)$	$5.1^{+1.6}_{-1.4}{}^{+1.2}_{-1.1} \pm 0.3$
$b \to D^0 D^-, D^+ \bar{D}^0 (X)$	$2.7^{+1.5}_{-1.3}{}^{+1.0}_{-0.9} \pm 0.2$
$b \to D^+ D^- (X)$	$< 0.9\%$ at 90%C.L.
Sum $b \to D^0 \bar{D}^0, D^0 D^-, D^+ \bar{D}^0 (X)$	$7.8^{+2.0}_{-1.8}{}^{+1.7}_{-1.5}{}^{+0.5}_{-0.4}$
$b \to D^{*+} D_s^- (X)$	$3.3^{+1.0}_{-0.9} \pm 0.6{}^{+1.1}_{-0.7}$
$b \to D^{*+} \bar{D}^0, D^0 D^{*-} (X)$	$3.0^{+0.9}_{-0.8}{}^{+0.7}_{-0.5} \pm 0.2$
$b \to D^{*+} D^-, D^+ D^{*-} (X)$	$2.5^{+1.0}_{-0.9}{}^{+0.6}_{-0.5} \pm 0.2$
$b \to D^{*+} D^{*-} (X)$	$1.2^{+0.4}_{-0.3} \pm 0.2 \pm 0.1$

inclusive branching fractions are

$$\mathcal{B}(b \to D^0 \bar{D}^0, D^0 D^\pm (X)) = \left(7.8^{+2.0}_{-1.8}(\text{st})^{+1.7}_{-1.5}(\text{sy})^{+0.5}_{-0.4}(\mathcal{B}_D)\right)\%$$

Hence a significant fraction of the doubly-charmed B decays leads to no D_s production. For the average mixture of b hadrons produced at LEP, the sum over all the decay modes above yields:

$$\mathcal{B}(b \to D_s D^0, D_s D^\pm, D^0 \bar{D}^0, D^0 D^\pm (X)) =$$
$$\left(20.9^{+3.2}_{-2.8}(\text{stat})^{+2.5}_{-2.2}(\text{syst})^{+4.5}_{-2.8}(\mathcal{B}_D)\right)\%.$$

Adding the small hidden charm and charmed baryon contributions, this measurement is in good agreement with the recent ALEPH measurement of the average number of charm quarks per b decay :

$n_c = 1.230 \pm 0.036(\text{stat}) \pm 0.038(\text{syst}) \pm 0.053(\mathcal{B}_D)$.

Some corresponding $D^{(*)}\bar{D}^{(*)}$ mass spectra are given in Fig. 4.

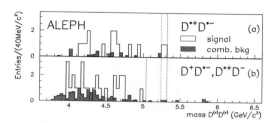

Figure 4: Unshaded histogram: the $D\bar{D}$ mass spectra of the selected $b \to D\bar{D}(X)$ candidates (a) $D^{*+}D^{*-}$ (b) $D^\pm D^{*\mp}$. The channels are mutually exclusive. Shaded histogram: the $D\bar{D}$ mass distribution of the events in the sidebands of the D_1 or D_2 mass spectra, normalised to the expected number of combinatorial background events.

2.4   Cabbibo suppressed $B \to \bar{D}D$ decays

As can be seen in Fig. 4a, two candidates for the Cabibbo suppressed decay $B_d^0 \to D^{*+}D^{*-}$ are observed. Asking for no charged track at the $D\bar{D}$ vertex, the background is reduced to 0.10 ± 0.03 event. The corresponding branching fraction is measured to be

$$\mathcal{B}(\bar{B}_d^0 \to D^{*+}D^{*-}) = \left(0.23^{+0.19}_{-0.12} \pm 0.04 \pm 0.02(\mathcal{B}_D)\right)\%.$$

One candidate for the Cabibbo suppressed decay $B^- \to D^{*-}D^0$, with both D vertices well separated from the reconstructed B decay point, is also observed [4] and limits on branching fractions are obtained [2].

2.5   Semi inclusive $B \to \bar{D}DK(X)$ decays

Reconstructing a K_S^0 and $K^\pm$ compatible with the $D\bar{D}$ vertex, and making the $\bar{D}DK$ invariant mass (without using the soft pion from a D^*), 3 peaks separated by about 150 MeV/c^2 appear, corresponding to the three 3-body B meson decay modes (Fig. 5). At a lower mass, the events correspond to more than 3-body decay modes (mainly $D^{(*)}\bar{D}^{(*)}K\pi$). After background subtraction, the total number of events is 32.2 ± 7.9, and in the 3-body region 21.2 ± 5.5. Hence, one sees that the three-body decays $B \to \bar{D}^{(*)}D^{(*)}K$ are a large part (about 70%) of the inclusive doubly-charmed $B \to \bar{D}^{(*)}D^{(*)}K(X)$ decays.

2.6   Exclusive $B \to \bar{D}DK$ decays

Asking now for no other charged track at the $\bar{D}DK$ vertex (and using the soft pion from a D^* when available), a clear B signal appear (Fig. 6) with 9 $\bar{D}^{(*)}D^{(*)}K_S^0$ events and 9 $\bar{D}^{(*)}D^{(*)}K^\pm$ events above 5.04 GeV/c^2. One of these events [4] was displayed in Fig. 3.

No evidence for resonant decays $B \to \bar{D}^{(*)}D_{s1}^+(2535)$ is found.

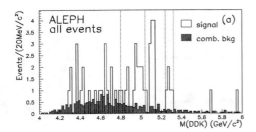

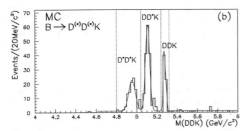

Figure 5: The $D^0\bar{D}^0K$, D^0D^-K or D^+D^-K mass of $D\bar{D}$ events with a reconstructed K_S^0 or a $K^\pm$ for (a) ALEPH data (b) simulated three-body decays $B \to D^{(*)}\bar{D}^{(*)}K$. The π^+ from $D^{*+} \to D^0\pi^+$, even if reconstructed, are not used in the mass.

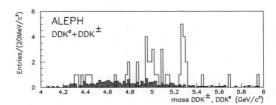

Figure 6: Invariant mass $m(D\bar{D}K)$ for events with one identified K and no other additional track from the $D\bar{D}K$ vertex. D can be either a D^0, a D^+ or a D^{*+}.

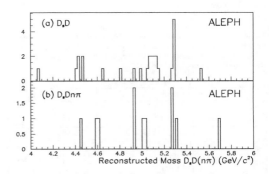

Figure 7: Invariant mass $m(D_s^+\bar{D}(n\pi^\pm))$ reconstructed for ALEPH data (a) $D_s^+\bar{D}$ (b) $D_s^+\bar{D}n\pi^\pm$, $n \geq 1$. The peak close to 5.1 GeV/c^2 is due to events with one missing neutral from decays $D^* \to D\pi^0, \gamma$ or or $D_s^{*+} \to D_s^+\gamma$.

B and $\bar{B}$ decays give different final states and, for instance in a $\bar{B}$ decay, the D coming from the b quark decay (called D_b) can be distinguished from the $\bar{D}$ coming from the W decay (D_W): in most of the events, the invariant mass $m(D_bK)$ tend to be higher than $m(D_WK)$ (and hence the momentum $p(D_b)$ in the B rest frame is higher than $p(D_W)$).

The diagrams contributing to these decays can be divided into 3 classes (Table 2): External W (E), Internal W (I) or color suppressed (cf $B \to J/\psi K$) and interference (EI); as can be seen in Table 2, some I type events are also observed.

Using isospin symmetry, the different branching fractions are given in Table 2 and the sum is :

$$\mathcal{B}(B \to \bar{D}^{(*)}D^{(*)}K) = \left(7.1^{+2.5}_{-1.5}(\text{st})^{+0.9}_{-0.8}(\text{sy}) \pm 0.5(\mathcal{B}_D)\right)\%.$$

2.7 $B \to \bar{D}DKX$ decays

Compared to the semi-inclusive result of Sec. 2.5 or to the inclusive b results of Sec. 2.3 , scaled by a factor $1/2f_{B_d^0} = 1.3$ to account for $b \to \bar{B}^0, B^-$, one sees that $\mathcal{B}(B \to \bar{D}^{(*)}D^{(*)}KX)$ should be about 3%.

2.8 $B \to D_s^+\bar{D}(n\pi^\pm)$ decays

Using the events with a $D_s^+\bar{D}$ mass above 5.04 GeV/c^2 in Fig. 7a, the branching fraction of B^0 and B^+ mesons into doubly-charmed two-body decay modes is also measured and gives

$$\mathcal{B}(B \to D_s^{(*)+}\bar{D}^{(*)}) = \left(5.6^{+2.1}_{-1.5}(\text{st})^{+0.9}_{-0.8}(\text{sy})^{+1.9}_{-1.1}(\mathcal{B}_D)\right)\%,$$

in good agreement with previous measurements of the same quantity .

For the first time, some candidates for completely reconstructed decays $B^0, B^+ \to \bar{D}^{(*)}D_s^+ n\pi^\pm$ ($n \geq 1$) are also observed (Fig. 7b). A measurement of the branching fraction for many-body decays $B^0, B^+ \to \bar{D}^{(*)}D_s^+ X$ is performed, leading to

$$\mathcal{B}(B \to D_s^{(*)\pm}D^{(*)}X) = \left(9.4^{+4.0}_{-3.1}(\text{st})^{+2.2}_{-1.8}(\text{sy})^{+2.6}_{-1.6}(\mathcal{B}_D)\right)\%.$$

Table 2: Summary of the various branching fractions B → DD̄K measured in this analysis.

Diagram	Channel (B^0, B^+)	Number of candidates	$\mathcal{B}(B \to D^{(*)}\bar{D}^{(*)}K)$ $(B^0/B^+$ average$)$
E	$D^-D^0K^+$, $\bar{D}^0D^+K^0$	3	$1.7^{+1.2}_{-0.8} \pm 0.2 \pm 0.1\%$
E	$(D^{*-}D^0 + D^-D^{*0})K^+$, $(\bar{D}^{*0}D^+ + \bar{D}^0D^{*+})K^0$	5	$1.8^{+1.0}_{-0.8} \pm 0.3 \pm 0.1\%$
E	$D^{*-}D^{*0}K^+$, $\bar{D}^{*0}D^{*+}K^0$	1	$< 1.3\%$
I	$D^0D^0K^+$, $D^+D^-K^+$	1	$< 2.0\%$
I	$(\bar{D}^0D^{*0} + \bar{D}^{*0}D^0)K^0$, $(D^{*+}D^- + D^+D^{*-})K^+$	1	$< 1.6\%$
I	$\bar{D}^{*0}D^{*0}K^0$, $D^{*+}D^{*-}K^+$	1	$< 1.5\%$
EI	$D^+D^-K^0$, $\bar{D}^0D^0K^+$	1	$< 1.9\%$
EI	$(D^{*+}D^- + D^+D^{*-})K^0$, $(\bar{D}^{*0}D^0 + \bar{D}^0D^{*0})K^+$	4	$1.6^{+1.0}_{-0.7} \pm 0.2 \pm 0.1\%$
EI	$D^{*+}D^{*-}K^0$, $\bar{D}^{*0}D^{*0}K^+$	1	$< 3.0\%$
Sum E	$D^{(*)-}D^{(*)0}K^+$, $\bar{D}^{(*)0}D^{(*)+}K^0$	9	$3.5^{+1.7}_{-1.1}{}^{+0.5}_{-0.4} \pm 0.2\%$
Sum I	$\bar{D}^{(*)0}D^{(*)0}K^0$, $D^{(*)+}D^{(*)-}K^+$	3	$0.8^{+1.0}_{-0.4}{}^{+0.2}_{-0.1} \pm 0.1\%$
Sum EI	$D^{(*)+}D^{(*)-}K^0$, $\bar{D}^{(*)0}D^{(*)0}K^+$	6	$2.8^{+1.6}_{-1.0}{}^{+0.4}_{-0.3} \pm 0.2\%$
E+I+EI	Sum DD̄K	5	$2.3^{+1.5}_{-0.9}{}^{+0.3}_{-0.3} \pm 0.2\%$
E+I+EI	Sum DD̄*K + D*D̄K	10	$3.8^{+1.6}_{-1.1}{}^{+0.5}_{-0.4} \pm 0.2\%$
E+I+EI	Sum D*D̄*K	3	$1.0^{+1.3}_{-0.6}{}^{+0.2}_{-0.2} \pm 0.1\%$
E+I+EI	Sum $D^{(*)}\bar{D}^{(*)}K$	18	$7.1^{+2.5}_{-1.5}{}^{+0.9}_{-0.8} \pm 0.5\%$

2.9 Conclusion on B meson decays

Summing the results of Sec. 2.4, 2.6, 2.7 and 2.8, one sees that B → DD̄(X) and B → D_s^+D̄(X) are a big part (about 25%) of B mesons decays.

2.10 $B_s^0 \to D_s^+D_s^-$ decay

One event [4] is reconstructed with a $D_s^+D_s^-$ mass at the B_s^0 mass, on a negligible background, with $D_s^+ \to \bar{K}^{*0}K^+$ and $D_s^- \to \phi\pi^-$. This can be only the pure CP even B_s^0 short state as explained in Sec. 3.1 below.

3 Width difference in the $B_s - \bar{B}_s$ system

3.1 Introduction

Most of the channels common to B_s and $\bar{B}_s$ are CP even (cf $D_s^+D_s^-$ ...). Hence the B_s short is the CP even state. Here a direct measurement of the B_s short lifetime is made using the mostly CP even decay modes $B_s^0 \to D_s^{(*)+}D_s^{(*)-}(X)$ with $D_s^+ \to \phi X$, that is a $\phi\phi X$ final state [5].

3.2 $\phi\phi$ selection

The same method as for double charm (Sec. 2.2) is applied (table $m(K^+K^-)_1/m(K^+K^-)_2$). The number of $\phi\phi$ events measured in this way is $N_{\phi\phi} = 50 \pm 15$, taking into account the non linear background shape near the K^+K^- threshold. In these events, 20% of charm contamination is expected from Monte Carlo;

the contribution of $\phi\phi$ coming from B_d and B_u decays is evaluated using $\mathcal{B}(B \to D_s^{(*)\pm}D^{(*)}(X))$ measured in Sec. 2.8. After subtracting these events, the number of $B_s^0 \to D_s^{(*)+}D_s^{(*)-}(X)$ found is $N_{sig} = 32 \pm 17$ (over a background of 78%). This number corresponds to a $B_s^0 \to D_s^{(*)+}D_s^{(*)-}(X)$ branching ratio of approximately 10%.

3.3 B_s short lifetime

The $\phi\phi$ vertex (which has a resolution of about $200\mu m$) is a good approximation of the B_s vertex due to the short D_s lifetime. An unbinned maximum likelihood fit using background parametrisation from the sidebands gives a preliminary lifetime $\tau_s = 1.42 \pm 0.23 \pm 0.16$ ps for this eigenstate. Using the world average B_s lifetime, $\bar{\tau} = 1.61 \pm 0.10$ ps, it would correspond [5] to $\Delta\Gamma/\Gamma = (24 \pm 35)\%$ where the statistical and systematic errors are combined.

References

1. Paper # 946, Preprint CERN EP/98-017, submitted to *Phys. Lett.* B.
2. Paper # 941, Preprint CERN EP/98-037, submitted to *The European Physical Journal* C.
3. Paper # 1054, Preprint ALEPH 98-064.
4. Color displays of the events shown during the talk can be found in http://alephwww.cern.ch/.
5. See also A. Ribon, Talk # 802, these proceedings.

Production of Orbitally Excited B Mesons in Z Decays

Steven Goldfarb

(On Behalf of the L3 Collaboration)

University of Michigan, CERN-EP, 1211 Genève 23, Suisse
E-mail: Steven.Goldfarb@cern.ch

We measure the mass, decay width and production rate of orbitally excited B mesons in 1.25 million hadronic Z decays registered by the L3 detector in 1994 and 1995. B meson candidates are inclusively reconstructed and combined with charged pions produced at the event primary vertex. An excess of events above the expected background is observed in the $B\pi$ mass spectrum near 5.7 GeV. These events are interpreted as resulting from the decay $B^{**} \to B^{(*)}\pi$, where B^{**} denotes a mixture of $l = 1$ B meson spin states. The masses and decay widths of the B_2^* ($j_q = 3/2$) and B_1^* ($j_q = 1/2$) resonances and the relative production rate for the combination of all spin states are extracted from a fit to the mass spectrum.

1 Introduction

Detailed understanding of the resonant structure of orbitally excited B mesons provides important information regarding the underlying theory. A symmetry (Heavy Quark Symmetry) arises from the fact that the mass of the b quark is large relative to Λ_{QCD}. In this approximation, the spin of the heavy quark ($\vec{s}_Q$) is conserved independently of the total angular momentum ($\vec{j}_q = \vec{s}_q + \vec{l}$) of the light quark. Excitation energy levels are thus degenerate doublets in total spin and can be expressed in terms of the spin-parity of the meson J^P and the total spin of the light quark j_q.

In this model, the $l = 0$ mesons, for which $j_q = 1/2$, have two possible spin states: a pseudo-scalar P, corresponding to $J^P = 0^-$ and a vector V, corresponding to $J^P = 1^-$. If the spin of the heavy quark is truly conserved independently, the relative production rate of these states is $V/(V + P) = 0.75$.[a] Recent measurements of this rate for the B system [7–10] agree well with this ratio.

For the orbitally excited or $l = 1$ mesons one expects two sets of degenerate doublets: one corresponding to $j_q = 1/2$ and the other to $j_q = 3/2$. For the $j_q = 1/2$ states, there is one possible spin combination corresponding to $J^P = 0^+$ and three corresponding to $J^P = 1^+$. For the $j_q = 3/2$ states, there are three spin combinations corresponding to $J^P = 1^+$ and five corresponding to $J^P = 2^+$. Rules for the decay of these states to the $1S$ states are determined by spin-parity conservation [2, 5]. For the dominant two-body decays, the $j_q = 1/2$ states can decay via an $L = 0$ transition (S-wave) and their decay widths are expected to be broad in comparison to those of the $j_q = 3/2$ states which must decay via an $L = 2$ transition (D-wave). Table 1 presents the nomenclature of the various spin states for $l = 1$ B mesons

---

[a]Corrections due to the decay of higher excited states are predicted to be small.

Table 1: Spin states of the $l = 1$ mesons with their predicted production rates and decay modes.

j_q	J^P	Production	$l = 1$ B decay	Transition
1/2	0^+	1 state	$B_0^* \to B\pi$	S-wave
1/2	1^+	3 states	$B_1^* \to B^*\pi$	S-wave
3/2	1^+	3 states	$B_1 \to B^*\pi$	D-wave
3/2	2^+	5 states	$B_2^* \to B^*\pi, B\pi$	D-wave

containing either a u or d quark, with the predicted production rates and two-body decay modes.

Predictions for the masses and decay widths of the four degenerate spin states are based on an effective QCD theory, called Heavy Quark Effective Theory (HQET), in which corrections are expressed as perturbations in powers of Λ_{QCD}/m_Q [5]. Such corrections, which can be relatively large for c hadrons are much smaller for the b hadrons, since $m_b \ll \Lambda_{QCD}$. Excited B meson spectroscopy is thus an excellent proving ground for the theory.

Mass measurements for $l = 1$ D mesons exist only for the D-wave decays of the D_1 and D_2^* resonances. A lack of corresponding data for the D_1^* and D_0^* states may indicate that the S-wave decays are too broad to be resolved in the mass spectra. As a result, precise HQET predictions for the B system which depend on input from the D system exist only for the B_1 and B_2^*. For example, Eichten, *et al.* [3] predict $M_{B_1} \approx 5759\,\mathrm{MeV}, \Gamma_{B_1} \approx 21\,\mathrm{MeV}, M_{B_2^*} \approx 5771\,\mathrm{MeV}, \Gamma_{B_2^*} \approx 25\,\mathrm{MeV}$. Several independent studies [1, 3, 4] predict the masses of the broad states to be lower than those of the narrow states, while others [5, 6] support the opposite "spin-orbit inversion" hypothesis. The decay widths of these S-wave states are generally predicted to be in the range $100 - 300\,\mathrm{MeV}$.

Recent analyses at LEP combining a charged pion produced at the primary event vertex with an inclusively reconstructed B meson [10–12] have measured an average mass of $M_{B^{**}} = 5700 - 5730\,\mathrm{MeV}$, where B^{**} indicates a mixture of all $l = 1$ spin states. An analysis [13] combining a primary charged pion with a fully reconstructed B meson, measures $M_{B_2^*} = (5739 \, ^{+8}_{-11}(\text{stat}) \, ^{+6}_{-4}(\text{syst}))$ MeV by performing a fit to the mass spectrum which fixes the mass differences, widths and relative rates of all of the spin states according to the predictions of Eichten, et al.. [3]

The analysis presented here is based on the combination of primary charged pions with inclusively reconstructed B mesons. Several new analysis techniques make it possible to improve on the resolution of the $B\pi$ mass spectrum and to unfold this resolution from the signal components. As a result, measurements are obtained for masses and widths of D-wave B_2^* decays and of S-wave B_1^* decays.

2 Event Selection

2.1 Selection of $Z \to b\bar{b}$ decays

The data were registered by the L3 detector [14] in 1994 and 1995 and correspond to an integrated luminosity of 90 pb^{-1} with LEP operating at the Z mass. Hadronic Z decays are selected by making use of their characteristic energy distributions and high multiplicity [15]. In addition, all events are required to have an event thrust direction satisfying $|\cos\theta| < 0.74$, where θ is the polar angle; to contain an event primary vertex reconstructed in three dimensions; to contain at least two calorimetric jets, each with energy greater than 10 GeV; and to pass stringent detector quality criteria for the vertexing, tracking and calorimetry. A total of 1,248,350 events pass this selection. A cut on a $Z \to b\bar{b}$ event discriminant based on track DCA significances [16] yields a b-enriched sample of 176,980 events.

A sample of 6 million hadronic Z decays have been generated with Jetset 7.4 [17] to study the content of the selected data. These events are passed through the L3 simulation program [18], including full realistic detector resolution and efficiency effects. Performing the same event reconstruction and hadronic selection on these events as on the data yields 3.5 million Monte Carlo $Z \to q\bar{q}$ decays. From this sample, the $Z \to b\bar{b}$ event purity is determined to be $\pi_{b\bar{b}} = 0.828$.

2.2 Selection of $B^{**} \to B^{(*)}\pi$ decays

Secondary decay vertices and primary event vertices are reconstructed in three dimensions by an iterative procedure such that a track can be a constituent of no more than one of the vertices. A calorimetric jet is selected for analysis as a B candidate if it is one of the two most energetic jets in the event, if a secondary decay vertex has been reconstructed from tracks associated with that

jet, and if the decay length of that vertex with respect to the event primary vertex is greater than 3σ, where σ is the estimated error of the measurement.

The decay of a B** to a B$^{(*)}$ meson and a pion is carried out via a strong interaction and thus occurs at the primary event vertex. In addition, the predicted masses for the $l = 1$ states correspond to relatively small Q values, so that the decay pion (π^{**}) direction is forward with respect to the B meson direction. We take advantage of these decay kinematics by requiring that, for each B meson candidate, there correspond at least one track which is a constituent of the event primary vertex and which is located within 90 degrees of the jet axis. A total of 60,205 track-jet pairs satisfy these criteria.

To further decrease background, typically due to charged fragmentation particles, only the track with the largest component of momentum in the direction of the jet is selected. This method has been found [13,19] to be an efficient means to improve the purity of the signal. In addition, background due to charged pions from $D^* \to D\pi$ decays is reduced by requiring the track to have a transverse momentum with respect to the jet axis larger than 100 MeV. These selection criteria are satisfied by 48,022 $B\pi$ pairs with a b hadron purity of $\pi_B = 0.942$.

2.2.1 B meson direction reconstruction

The direction of the B candidate is estimated by taking a weighted average in the θ (polar) and ϕ (azimuthal) coordinates of directions defined by the vertices and by particles with a high rapidity relative to the jet axis. A numerical error-propagation method [20] makes it possible to obtain accurate estimates for the uncertainty of the angular coordinates measured from vertex pairs. These errors, as well as the error for the decay length measurement used in the secondary vertex selection, are calculated for each pair of vertices from the associated error matrices and determine the weight for the vertex-defined coordinate measurements.

Particles coming from the decay of b hadrons produced in Z decays have a characteristically high rapidity relative to the original direction of the hadron when compared to that of particles coming from fragmentation. A cut on the particle rapidity distribution is thus a powerful tool for selecting the B meson decay constituents [8]. A second estimate for the direction of the B is obtained by summing the momenta of all charged and neutral particles (excluding the π^{**} candidate) with rapidity $y > 1.6$ relative to the original jet axis. Estimates for the uncertainty of the coordinates obtained by this method are determined from simulated B meson decays as an average value for all events and determine the weight for the rapidity-defined coordinate measurements. The final B direction coordinates are taken as the error-weighted averages of these two sets of coordinates.

The resolution for each coordinate is parametrized by a two-Gaussian fit to the difference between the reconstructed and generated values. For θ, the two widths are $\sigma_1 = 18\,\mathrm{mrad}$ and $\sigma_2 = 34\,\mathrm{mrad}$ with 68% of the B mesons in the first Gaussian. For ϕ, the two widths are $\sigma_1 = 12\,\mathrm{mrad}$ and $\sigma_2 = 34\,\mathrm{mrad}$ with 62% of the B mesons in the first Gaussian.

2.2.2 B meson energy reconstruction

The energy of the B meson candidate is estimated by taking advantage of the known center of mass energy at LEP to constrain the measured value. The energy of the B meson from this method [11] can be expressed as

$$E_{\mathrm{B}} = \frac{M_{\mathrm{Z}}^2 - M_{\mathrm{B}}^2 + M_{\mathrm{recoil}}^2}{2M_{\mathrm{Z}}} \quad , \tag{1}$$

where M_{Z} is the mass of the Z boson and M_{recoil} is the mass of all particles in the event other than the B. To determine M_{recoil}, the energy and momenta of all particles in the event with rapidity $y < 1.6$, including the π^{**} candidate (regardless of its rapidity), are summed and $M_{\mathrm{recoil}}^2 = E_{y<1.6}^2 - p_{y<1.6}^2$. Fitting the difference between reconstructed and generated values for the B meson energy with an asymmetric Gaussian yields a maximum width of 2.8 GeV.

3 Analysis of the Bπ Mass Spectrum

The combined Bπ mass is defined as

$$M_{\mathrm{B}\pi} = \sqrt{M_{\mathrm{B}}^2 + m_{\pi}^2 + 2E_{\mathrm{B}}E_{\pi} - 2p_{\mathrm{B}}p_{\pi}\cos\alpha} \quad , \tag{2}$$

where M_{B} and m_{π} are set to 5279 MeV and 139.6 MeV, respectively, and α is the measured angle between the B meson and π^{**} candidates.

3.1 Background function

The background distribution is estimated from the Monte Carlo data sample, excluding $\mathrm{B}^{**} \to \mathrm{B}^{(*)}\pi$ decays, and fit with a six-parameter threshold function given by

$$p_1 \times (x - p_2)^{p_3} \times e^{(p_4 \times (x-p_2) + p_5 \times (x-p_2)^2 + p_6 \times (x-p_2)^3)} \quad . \tag{3}$$

Parameters p_2 through p_6 are fixed to the shape of the simulated background, but the overall normalization factor p_1 is allowed to float freely in order to obtain a correct estimate of the contribution of the background to the statistical error of the signal.

3.2 Signal function

To examine the underlying structure of the signal, it is necessary to unfold effects due to detector resolution. The π^{**} candidates are expected to have typical momenta of a few GeV. In this range, the single track momentum resolution is no more than a few percent with an angular resolution better than 2 mrad. The dominant sources of uncertainty for the mass measurement are thus the B meson angular and energy resolutions. Monte Carlo studies confirm that these two components are dominant and roughly equal in magnitude. This analysis thus concentrates on unfolding the effects of these components by parametrizing and removing their contribution to the mass resolution.

3.2.1 Signal resolution and efficiency

Dependence of the Bπ mass resolution and selection efficiency on Q value is studied by generating signal events at several different values of B** mass and Breit-Wigner width. The simulated events are passed through the same event reconstruction and selection as the data. The resulting Bπ mass distributions are each fit with a Breit-Wigner function convoluted with a Gaussian resolution (Voigt function) and the detector resolution is extracted by fixing the Breit-Wigner width to the generated value.

The Gaussian width is found to increase linearly from 20 MeV to 60 MeV in the B** mass range $5.6 - 5.8$ GeV. This increase with Q value is mainly due to the angular component of the uncertainty, which increases as a function of the opening angle α. The resolution is parametrized as a linear function of the B** mass from a fit to the extracted widths. Similarly, the selection efficiency is found to increase slightly with Q value and the dependence is parametrized with a linear function.

Agreement between data and Monte Carlo for the B meson energy and angular resolution is confirmed by analyzing B$^* \to$ Bγ decays selected from the same sample of B mesons. The photon selection for this test is the same as that described in reference [7]. A B* meson decays electromagnetically and hence has a negligible decay width compared to the detector resolution. As in the case of the Bπ mass resolution, the B meson energy and angular resolution are the dominant components of the reconstructed Bγ mass resolution. Fits to the $M_{\mathrm{B}\gamma} - M_{\mathrm{B}}$ spectra are performed with the combination of a Gaussian signal and the background function described above. For the Monte Carlo, the Gaussian mean value is found to be $M_{\mathrm{B}\gamma} - M_{\mathrm{B}} = (46.5 \pm 0.6(\mathrm{stat}))$ MeV with a width of $\sigma = (11.1 \pm 0.7(\mathrm{stat}))$ MeV. The input generator mass difference is 46.0 GeV. For the data, the Gaussian mean value is found to be $M_{\mathrm{B}\gamma} - M_{\mathrm{B}} = (45.1 \pm 0.6(\mathrm{stat}))$ MeV with a width of $\sigma = (10.7 \pm 0.6(\mathrm{stat}))$ MeV. Good agreement between the widths of the data and Monte Carlo signals provides confidence that the B energy and angular resolution are well understood and simulated.

3.2.2 Combined signal

According to spin-parity rules one expects to resolve mass resonances of five possible B^{**} decay modes: $B_2^* \to B\pi$, $B_2^* \to B^*\pi$, $B_1 \to B^*\pi$, $B_1^* \to B^*\pi$ and $B_0^* \to B\pi$. No attempt is made to tag subsequent $B^* \to B\gamma$ decays, as the efficiency for selecting the soft photon is relatively low. As a result, the effective $B\pi$ mass for a decay to a B^* meson is shifted down by the 46 MeV $B^* - B$ mass difference.

The five resonances are fit with five Voigt functions, with the relative production fractions determined by spin counting rules. The Gaussian convolutions to the widths are determined by the resolution function. Additional physical constraints are applied to the mass differences and relative widths in order to obtain the most information possible from the data sample.

Predictions for the mass differences $M_{B_2^*} - M_{B_1}$ and $M_{B_1^*} - M_{B_0^*}$ depend on several factors, including the b and c quark masses and, in some cases, input from experimental data of the D meson system. In general, the values are predicted to be roughly equal and in the range $5 - 20$ MeV. [1,3–6]. We constrain both of the mass differences to 12 MeV. The Breit-Wigner widths are predicted to be functions of the decay Q values and of the probability for $B^{**} \to B^{(*)}\rho$ decays. Both of these contributions depend on mass and are accounted for by using an iterative procedure to determine the constraints at the measured masses. In the final fit, we constrain $\Gamma_{B_0^*} = \Gamma_{B_2^*}$ and $\Gamma_{B_1} = 0.84\Gamma_{B_2^*}$. No constraint is applied to the difference between the mass of the narrow states and that of the broad states, nor to their relative decay widths.

3.3 Fit results

Monte Carlo events for each of the expected B^{**} decays are generated then passed through the simulation and reconstruction programs and the $B\pi$ event selection. The resulting mass spectra are combined with background then fit with the signal and background functions under the constraints described above. Mass values and decay widths for the B_2^* and B_1^* resonances and the overall normalization are extracted from the fit and found to agree well with the generated values. All differences lie within the statistical error and have no systematic trend.

The data $B\pi$ mass spectrum is fit with the combined signal and background functions, allowing the normalization parameters to float freely. The resulting fit, shown in Figure 1, has a χ^2 of 57 for 60 degrees of freedom. A total of 2673 events occupy the signal re-

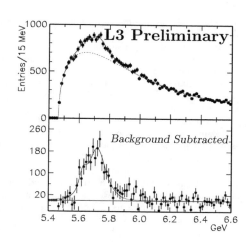

Figure 1: Fit to the data $B\pi$ mass distribution with the five-peak signal function and the background function described in the text.

Table 2: Sources of systematic uncertainty and their estimated contributions to the errors of the measured values.

Sources	$M_{B_2^*}$	$\Gamma_{B_2^*}$	$M_{B_1^*}$	$\Gamma_{B_1^*}$	f^{**}
b purity	—	—	—	—	± 0.02
$B_{u,d}$ fraction	—	—	—	—	± 0.03
background	± 2	± 9	± 3	± 9	± 0.05
M constraints	± 3	± 7	± 3	± 7	< 0.01
Γ constraints	< 1	± 2	< 1	± 2	< 0.01
resolution	± 2	± 9	± 1	± 9	± 0.01
efficiency	< 1	± 1	< 1	± 1	± 0.01
Total	± 4	± 15	± 4	± 15	± 0.06

gion corresponding to a relative $B_{u,d}^{**}$ production rate of $\sigma(B_{u,d}^{**})/\sigma(B_{u,d}) = 0.39\pm0.05(\text{stat})$. The mass and width of the B_2^* are found to be $M_{B_2^*} = (5770 \pm 6(\text{stat}))$ MeV and $\Gamma_{B_2^*} = (23 \pm 26(\text{stat}))$ MeV and the mass and width of the B_1^* are found to be $M_{B_1^*} = (5675 \pm 12(\text{stat}))$ MeV and $\Gamma_{B_1^*} = (76 \pm 28(\text{stat}))$ MeV.

3.4 Systematic uncertainty

Sources of systematic uncertainty and their estimated contributions to the errors of the measured values are summarized in Table 2. The b hadron purity of the sample is varied from 91% to 96%. The fraction of b quarks hadronizing to $B_{u,d}$ mesons is taken to be 79% and is varied between 74% and 83% in accordance with the recommendations of the LEP B Oscillation Working Group [21]. These variations effect only the overall B^{**} production fraction.

Systematic effects due to background modelling are studied by varying the shape parameters of the background function and by performing the fit with other background functions to study the effect on the measured values. Contributions to the error due to modelling of the signal are estimated for the mass and width constraints: the $M_{B_2^*} - M_{B_1}$ and $M_{B_1^*} - M_{B_0^*}$ mass differences are varied in the range $6 - 18\,\text{MeV}$ and the $\Gamma_{B_1}/\Gamma_{B_2^*}$ and $\Gamma_{B_0^*}/\Gamma_{B_1^*}$ ratios are varied between 0.8 and 1. Effects due to uncertainty in the resolution and efficiency functions are estimated by varying the slopes and offsets of the linear parametrizations.

Three-body decays of the type $B_2^* \to B\pi\pi$ have been generated and passed through the simulation and reconstruction programs and the $B\pi$ event selection. $B\pi$ pairs, for which only one of the pions is tagged, are studied as a possible source of resonant background. The resulting reflection is found to contribute insignificantly to the background in regions of small Q value. Similarly, generated $B_s^{**} \to BK$ decays, where the K is mistaken for a π are found to contribute only slightly to the low Q value region and their effects are included in the background modelling uncertainty contribution.

4 Conclusion

We measure for the first time the masses and decay widths of the B_2^* ($j_q = 3/2$) and B_1^* ($j_q = 1/2$) mesons. From a constrained fit to the $B\pi$ mass spectrum, we find

$$M_{B_2^*} = (5770 \pm 6(\text{stat}) \pm 4(\text{syst}))\,\text{MeV}$$
$$\Gamma_{B_2^*} = (23 \pm 26(\text{stat}) \pm 15(\text{syst}))\,\text{MeV}$$
$$M_{B_1^*} = (5675 \pm 12(\text{stat}) \pm 4(\text{syst}))\,\text{MeV}$$
$$\Gamma_{B_1^*} = (76 \pm 28(\text{stat}) \pm 15(\text{syst}))\,\text{MeV}$$

The relative $B_{u,d}^{**}$ production rate, including all $l = 1$ spin states, is measured to be

$$\frac{\text{Br}(b \to B_{u,d}^{**} \to B^{(*)}\pi)}{\text{Br}(b \to B_{u,d})} = 0.39 \pm 0.05(\text{stat}) \pm 0.06(\text{syst})$$

where isospin symmetry is employed to account for decays to neutral pions. These results provide strong support for Heavy Quark Effective Theory.

Acknowledgements

I wish to thank Alan Astbury for inviting me to present these results at ICHEP98. I also thank Chris Quigg for several informative discussions on B^{**} production and Franz Muheim for editing the presentation and contributed papers.

References

1. M. Gronau, A. Nippe and J.L. Rosner, Phys. Rev. **D 47** (1993) 1988;
   M. Gronau and J.L. Rosner, Phys. Rev. **D 49** (1994) 254.
2. J.L. Rosner, Comments Nucl. Part. Phys. **16** (1986) 109.
3. E.J. Eichten, C.T. Hill and C. Quigg, Phys. Rev. Lett. **71** (1993) 4116;
   E.J. Eichten, C.T. Hill and C. Quigg, Fermilab-Conf-94/118-T (1994).
4. A.F. Falk and T. Mehen, Phys. Rev. **D 53** (1996) 231.
5. N. Isgur, Phys. Rev. **D 57** (1998) 4041.
6. D. Ebert, V.O. Galkin and R.N. Faustov, Phys. Rev. **D 57** (1998) 5663.
7. L3 Collaboration, Phys. Lett. **B 345** (1995) 589;
   L3 Collaboration, Contribution to EPS Conference, Brussels EPS95 (1995).
8. DELPHI Collaboration, Z. Phys. **C 68** (1995) 353.
9. OPAL Collaboration, Z. Phys. **C 74** (1997) 413.
10. ALEPH Collaboration, Z. Phys. **C 69** (1996) 393.
11. OPAL Collaboration, Z. Phys. **C 66** (1995) 19.
12. DELPHI Collaboration, Phys. Lett. **B 345** (1995) 598.
13. ALEPH Collaboration, Phys. Lett. **B 425** (1998) 215.
14. L3 Collaboration, Nucl. Instr. and Meth. **A 289** (1990) 35;
    L3 Collaboration, Physics Reports **236** (1993) 1.
15. L3 Collaboration, Phys. Lett. **B 307** (1993) 237.
16. J. Branson et al., L3 Note 2108, June 24, (1997). This is an L3 note available on request from: L3 Secretariat, CERN-PPE, 1211 Genève 23, Suisse.
17. T. Sjöstrand and M. Bengtsson, Comput. Phys. Commun. **43** (1987) 367;
    T. Sjöstrand, Comp. Phys. Comm. **82** (1994) 74;
    T. Sjöstrand, Lund and CERN Preprints LU TP 95-20/CERN-TH.7112/93 (revised, 1995).
18. The L3 detector simulation is based on GEANT Version 3.15; see R. Brun et al., GEANT 3, CERN DD/EE/84-1 (revised 1987) and the GHEISHA program (H. Fesefeld, RWTH Aachen Report PITHA 85/02 (1985)) for the simulation of hadronic interactions.
19. CDF Collaboration, Phys. Rev. Lett. **80** (1998) 2057.
20. J. Swain and L. Taylor, *Numerical Construction of Likelihood Distributions and the Propagation of Errors*, Nucl. Instr. and Methods **411** (1998) 153.
21. The LEP B Oscillation Working Group, V. Andreev et al., LEPBOSC/97-01 (1997).

MEASUREMENT OF b-BARYON POLARIZATION IN Z⁰ DECAYS

RICHARD J. VAN KOOTEN

Physics Department, Indiana University, Bloomington, IN 47405, USA
e-mail: rickv@paoli.physics.indiana.edu

THE OPAL COLLABORATION

European Center for Particle Physics Research (CERN), EP Division, CH-1211 Geneva 23, Switzerland,
e-mail: opal-physics-coordinators@opal-lb.cern.ch

In the Standard Model, b quarks produced in e^+e^- annihilation at the Z^0 peak have a large average longitudinal polarization of -0.94. Some fraction of this polarization is expected to be transferred to b-flavored baryons during hadronization. The average longitudinal polarization of weakly decaying b baryons, $\langle P_L^{\Lambda_b} \rangle$, is measured in approximately 4.3 million hadronic Z^0 decays collected with the OPAL detector between 1990 and 1995 at LEP. Those b baryons that decay semileptonically and produce a Λ baryon are identified through the correlation of the baryon number of the Λ and the electric charge of the lepton. In this semileptonic decay, the ratio of the neutrino energy to the lepton energy is a sensitive polarization observable. The neutrino energy is estimated using missing energy measurements. From a fit to the distribution of this ratio, the value

$$\langle P_L^{\Lambda_b} \rangle = -0.56^{+0.20}_{-0.13} \pm 0.09$$

is obtained, where the first error is statistical and the second systematic.

1 Introduction

The results presented here are described in more detail elsewhere[1].

According to the Standard Model, the process $e^+e^- \to Z^0 \to b\bar{b}$ gives rise to b quarks that are longitudinally polarized[2,3] with a large average value of $\langle P_L^b \rangle = -0.94$ for a weak mixing angle of $\sin^2 \theta_W = 0.23$. If subsequent hadronization to a Λ_b^0 baryon is considered for example, the light u and d quarks form a spin-0 system and the spin of the Λ_b^0 should be carried entirely by the b quark as shown in Fig. 1. In the heavy-quark limit, an important prediction of Heavy Quark Effective Theory (HQET) is that the degrees of freedom of the b quark are decoupled from the spin-0 light diquark so that the Λ_b^0 should retain almost 100% of this polarization[3] with only a slight reduction of about 3% due to hard gluon emission during hadronization[4]. This can be contrasted with spin-0 pseudoscalar B mesons where the polarization information of the b quark is lost or vector B* meson where the polarization information is not expected to be observable[5]. In b baryon production, it is also possible for the b quark to combine with a spin-1 uu, ud, or dd diquark system to form the higher mass baryonic states Σ_b and Σ_b^* that are expected to decay strongly to $\Lambda_b^0 \pi$ leading to substantial reduction in the polarization averaged over the weakly decaying b baryons[5,6]. Measuring the Λ_b polarization[a] therefore provides a test of HQET

and information about heavy baryon hadronization and nonperturbative corrections to spin transfer in fragmentation.

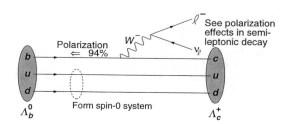

Figure 1: Polarization of a b quark from Z^0 decay in a Λ_b^0 baryon.

The sign of measured polarization also gives information on the chirality of the b quark coupling to the weak charged current[7], particularly interesting because of recent analyses testing for the presence of sizeable right-handed components in the $b \to c$ charged current coupling[8].

In the weak semileptonic decay $\Lambda_b \to X_c \ell^- \bar{\nu}_\ell X$, both the charged lepton and neutrino energy spectrum are sensitive to $\langle P_L^{\Lambda_b} \rangle$ [2,3,9]. Assuming that b decays proceed via the usual left-handed current (i.e., $(V - A)$ coupling), in the rest frame of the Λ_b, charged leptons ℓ^- tend to be emitted antiparallel to the spin of the Λ_b and the $\bar{\nu}_\ell$'s

---

[a]The symbol Λ_b will be used to refer to all b baryons that decay weakly, including for example the Λ_b^0 and the b quark in combination with a spin-1 us or ds diquark forming a Ξ_b. The symbol Λ_b^0 is used to denote the particular ground state b baryon with quark content (bud).

parallel to the Λ_b spin. In the laboratory frame, polarization then implies a harder lepton energy spectrum and softer neutrino energy spectrum compared to the unpolarized case. However, there are substantial uncertainties in the exact shape of these spectra due to uncertainties in fragmentation, the ratio of the quark masses m_c/m_b, and QCD corrections[10,11] such that the extraction of the average polarization from the spectra alone is problematic. These systematic effects partially cancel in the ratio of the average of the lepton energy to the average of the neutrino energy or in the ratios of higher moments of the energy spectra[12].

A measurement is described of the average longitudinal polarization of weakly decaying b baryons, $\langle P_L^{\Lambda_b} \rangle$, using about 4.3 million multihadronic Z^0 decays collected by the OPAL detector [13] from 1990 to 1995. To ensure a data sample with reasonably large statistics, events containing decay chains of the form[b] $\Lambda_b \to \Lambda_c^+ \ell^- \bar{\nu}_\ell X$ followed by $\Lambda_c^+ \to \Lambda X$ are selected without reconstructing the intermediate charm state. The correlation of a Λ with a negatively-charged lepton ($\ell = $ e or μ) or a $\bar{\Lambda}$ with a positively-charged lepton can indicate the presence of a semileptonic b-baryon decay, as used previously by OPAL[14,15]. The other charge combinations ($\bar{\Lambda}$-ℓ^- and Λ-ℓ^+) are used to characterize the background. The neutrino energy E_ν was estimated using missing energy measurements in the hemisphere containing the Λ-ℓ^- pair. To extract the polarization of the b baryon, a fit was then made to the distribution of the ratio E_ν/E_ℓ formed event by event. Fitting to the shape of this distribution should lead to an estimate with lower statistical variance than the use of the average sample energies $\langle E_\nu \rangle/\langle E_\ell \rangle$ as employed in the previously published measurement of $\langle P_L^{\Lambda_b} \rangle$ [16].

To model the signal, samples of simulated data were generated of Z^0 hadronic decays that included $\Lambda_b^0 \to \Lambda_c^+ \ell^- \bar{\nu}_\ell X$ followed by $\Lambda_c^+ \to \Lambda \pi^+ X$ and $\Lambda \to p\pi^-$ using a modified JETSET decay routine incorporating polarized Λ_b^0 decay with QCD corrections[11]. Samples of 40 000 events each were generated at twelve different values of $\langle P_L^{\Lambda_b} \rangle$.

2 Data and event selection

The selection of b-baryon decays using Λ-lepton pairs is similar to the one previously used to measure the average b-baryon lifetime and production rate[15] but uses improved electron identification, a different jet definition, different lepton kinematic cuts, and the inclusion of an additional π^+ to exploit the charge correlation of Λ-π^+ in the decay chain $\Lambda_b \to \Lambda_c^+ \to \Lambda \pi^+ X$. Details of the event selection can be found elsewhere[1]. Firstly,

---

[b]Charge conjugate processes are implied throughout.

electrons and muons having large momentum and large transverse momentum with respect to their associated jet (including the lepton candidate) were identified. Λ baryons were then reconstructed and combined with π^+'s that were consistent with coming from a common vertex formed by the Λ and lepton candidate. In the signal process the π^+ most commonly comes from the Λ_c^+ decay with a decay length from the point of Λ_b decay which is small compared with the typical Λ-lepton vertex resolution. Requirements were placed on the Λ-lepton combination followed by requirements to improve the missing energy determination and its applicability in estimating the neutrino energy in the semileptonic decay of the Λ_b.

The invariant mass distribution of the $p\pi^-$ combination is shown in Fig. 2 for $\Lambda \ell^- \pi^+$ (right-sign) and $\Lambda \ell^+ \pi^+$ (wrong-sign) combinations after all other selection requirements have been applied. A total of 912 right-sign and 316 wrong-sign $\Lambda \ell \pi^+$ combinations are selected with an overall b baryon purity of 69% in the right-sign sample. The 596 excess right-sign combinations can be attributed mainly to b-baryon decays. The number of observed wrong-sign combinations are used after small corrections as an estimate of the level of the background as well as to estimate the shape of the background in the distribution of the ratio E_ν/E_ℓ.

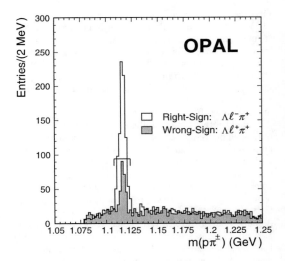

Figure 2:   Invariant mass distribution of $p\pi^-$ combinations in the right-sign $\Lambda \ell^- \pi^+$ sample (open histogram) superimposed by wrong-sign $\Lambda \ell^+ \pi^+$ (shaded histogram) combinations.

3 Neutrino and lepton energy measurement

In the semileptonic decay of the Λ_b, the neutrino energy was estimated by the measured missing energy in the signal hemisphere, which can be defined by

$$E_\nu = E_{\text{miss}}^{\text{hemi}} = E_{\text{beam}} + E_{\text{corr}} - E_{\text{vis}}^{\text{hemi}},$$

where E_{beam} is the beam energy and $E_{\text{vis}}^{\text{hemi}}$ the visible energy in the signal hemisphere. If we assume a two-body decay of the Z^0, with each hemisphere considered as one body, then E_{corr} is a correction term calculated assuming that the two hemisphere momenta are equal:

$$E_{\text{corr}} = (M_{\text{sig}}^2 - M_{\text{recoil}}^2)/4E_{\text{beam}}.$$

M_{sig} and M_{recoil} are the measured invariant masses of the signal hemisphere and opposite recoil hemisphere, respectively. The term E_{corr} uses the beam energy constraint to improve the resolution for the missing energy.

The agreement of the quantity $E_{\text{miss}}^{\text{hemi}}$ between data and Monte Carlo events containing identified leptons following a preselection with relaxed p and p_t cuts to enhance statistics is shown in Fig. 3(a). To further test and calibrate the measurement of this important quantity, event samples from both data and Monte Carlo simulation were prepared that were enhanced in b quarks using a lifetime tag[17] in one hemisphere Such control samples were used to determine an additional small linear correction to the reconstructed neutrino energy. Independent samples of events with hemispheres failing the b-tag requirement also show good agreement between data and Monte Carlo.

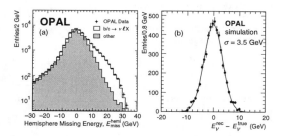

Figure 3: (a) Reconstructed missing energy in data (points with error bars) and Monte Carlo simulated events in $q\bar{q}$ events after a preselection with relaxed p and p_t requirements; the open histogram shows the missing energy for events containing b and c hadrons decaying semileptonically, and the shaded histogram is for all other processes. (b) Neutrino energy resolution from reconstructed E_ν^{rec} and true E_ν^{true} in Monte Carlo signal events with $\Lambda_b^0 \to \Lambda_c^+ \ell^- \bar{\nu}_\ell X$.

In the signal Monte Carlo samples, the average neutrino energy resolution was observed to be 3.5 GeV as

shown in Fig. 3(d). For the electron and muon candidates the momentum was used as an estimate of the lepton energy. The distributions of lepton spectra in the control samples described above were also compared between data and Monte Carlo showing good agreement. The average values are $\langle E_\ell \rangle = 10.18 \pm 0.03$ GeV for the data and $\langle E_\ell \rangle = 10.11 \pm 0.03$ GeV for the Monte Carlo simulation for the b-tagged sample containing identified leptons. For the unpolarized signal process, $\langle E_\ell \rangle = 10.03$ GeV.

The distributions of the measured electron and muon energies and of the reconstructed neutrino energies for the selected right-sign $\Lambda \ell^- \pi^+$ and wrong-sign $\Lambda \ell^+ \pi^+$ samples in the data are shown in Fig. 4.

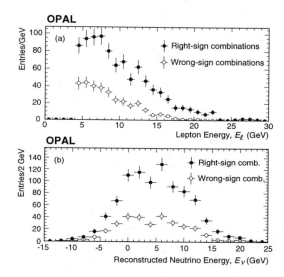

Figure 4: (a) Lepton (electron and muon) and (b) reconstructed neutrino energy spectra for the hemispheres containing selected right-sign $\Lambda \ell^- \pi^+$ combinations with the spectra for wrong-sign combinations overlaid.

4 Background estimate

The wrong-sign sample is used as an estimate of the level and shape of the background in the right-sign sample. The component of the background due to random combinations of tracks and fakes due to misidentified leptons or Λ baryons is expected to contribute equally to right- and wrong-sign samples, as verified by Monte Carlo simulations.

A larger source of background is the combination of a Λ baryon from the fragmentation process with a genuine lepton from the semileptonic decay of b or c hadrons. In the framework of string fragmentation, the background of a fragmentation $\bar{\Lambda}\ell^-$ combination will preferentially enter the wrong-sign sample. Table 4 gives the background composition of the right- and wrong-sign sample as predicted by Monte Carlo simulation and shows the resulting imbalances as detailed elsewhere[15].

Source	RS	WS
Signal: Λ and ℓ from b baryon	629 ± 30	
Fragmentation Λ plus:		
ℓ from B meson	94 ± 9	77 ± 8
ℓ from c baryon	9 ± 3	3 ± 2
ℓ from b baryon	5 ± 4	66 ± 6
ℓ from c meson	32 ± 6	43 ± 7
Exclusive Backgrounds:	32 ± 7	27 ± 6
Combinatorials and Fakes:	111 ± 8	100 ± 8

Table 1: Monte Carlo predictions of the composition among the 912 right-sign (RS) combinations and the 316 wrong-sign combinations. Errors are due to Monte Carlo statistics, branching ratio uncertainties, and modelling systematic errors.

5 Fitting procedure and result

The b-baryon polarization was extracted by comparing the reconstructed distribution of E_ν/E_ℓ in the data to spectra estimated from fully simulated Monte Carlo events corresponding to various values of $\langle P_L^{\Lambda_b}\rangle$ and satisfying all the selection criteria. A binned maximum likelihood fit was used to extract $\langle P_L^{\Lambda_b}\rangle$ by determining which Monte Carlo spectrum gives the best description of the data. The right-sign and wrong-sign spectra were fit simultaneously.

The event-by-event distribution of E_ν/E_ℓ was formed, as shown in Fig. 5(a), and the right-sign and wrong-sign distributions were fitted. The resulting curve of $-\log\mathcal{L}$ versus $\langle P_L^{\Lambda_b}\rangle$, offset so the maximum value of $\log\mathcal{L}$ is zero, is shown in Fig. 5(b) indicating a measurement of $\langle P_L^{\Lambda_b}\rangle = -0.56^{+0.20}_{-0.13}$, and a value of zero polarization ruled out at the 95% confidence level (CL), considering only the statistical errors. The χ^2 per degree of freedom is 0.91.

The sources of systematic error that have been considered are summarized in Table 2 and are described in more detail elsewhere[1]. The systematic uncertainties from each source were added in quadrature to obtain an estimated total systematic error of ±0.09 on $\langle P_L^{\Lambda_b}\rangle$.

Further checks were performed to search for other systematic effects. Wide variations were made on the selection cuts to change the b quark content and fragmentation Λ baryons in the sample and in each case $\langle P_L^{\Lambda_b}\rangle$ redetermined. Observed variations were consistent with the statistical errors on the uncorrelated fractions of the different samples. A fit was also made to the reconstructed

Source of Uncertainty	$\Delta\langle P_L^{\Lambda_b}\rangle$
E_ν resolution	±0.02
E_ν reconstruction	±0.05
E_ℓ scale and shape	±0.03
Selection criteria	±0.02
Background fraction and shape	±0.04
b fragmentation	±0.03
Λ_c polarization	±0.02
$b \to \tau$	±0.01
Fitting method	±0.03
Theoretical uncertainty (form factor modelling, QCD corrections, m_c/m_b)	±0.03
Total	±0.09

Table 2: Summary of systematic uncertainties in the measurement of $\langle P_L^{\Lambda_b}\rangle$.

E_ν/E_ℓ distribution in the inclusive lepton control sample from the data and a value of $\langle P_L\rangle = -0.043\pm0.019$ (stat.) was measured. Although B mesons are expected to be unpolarized, a fraction of the b hadrons are predicted to be b baryons in this sample, and although this total observed polarization is not necessarily expected to be zero, it should be small, as observed. A further cross check was made with a data sample containing candidates for the exclusive decay $B^0 \to D^{*-}\ell^+\nu_\ell$ selected from events where a D^{*-} and a lepton of opposite charge were found in the same jet[18]. A value consistent with zero was found as expected for a sample with a b hadron content almost exclusively from B mesons.

6 Discussion and summary

The average polarization of b baryons in Z^0 decays at OPAL has been measured to be:

$$\langle P_L^{\Lambda_b}\rangle = -0.56^{+0.20}_{-0.13} \pm 0.09,$$

where the first error is statistical and the second systematic. This level of polarization is larger than, but consistent with, the published value of $\langle P_L^{\Lambda_b}\rangle = -0.23^{+0.24}_{-0.20}$ (stat.) $^{+0.08}_{-0.07}$ (syst.)[16] and preliminary $\langle P_L^{\Lambda_b}\rangle = -0.30^{+0.19}_{-0.16}$ (stat.) $\pm$ 0.06 (syst.)[19] measured by the ALEPH Collaboration and a preliminary result from DELPHI of $\langle P_L^{\Lambda_b}\rangle = -0.08^{+0.35}_{-0.29}$ (stat.) $^{+0.18}_{-0.16}$ (syst.)[20].

Including systematic errors, this OPAL measurement implies bounds on the longitudinal polarization of b baryons of $-0.13 \geq \langle P_L^{\Lambda_b}\rangle \geq -0.87$ at 95% CL, therefore disfavoring full observed average polarization of -0.94. This is the first measurement to exclude zero polarization of b baryons at larger than 95% CL providing direct evidence that the b quark is longitudinally polarized in the decay $Z^0 \to b\bar{b}$.

A simple model[5] can be used to predict the total observed b baryon polarization after depolarization of those b baryons proceeding through intermediate states involving strong decays: $b \rightarrow \Sigma_b^{(*)} \rightarrow \Lambda_b$. A central value of $\langle P_L^{\Lambda_b} \rangle = -0.68$ is predicted and varying input parameters as suggested by measurements[21], a range of predictions between -0.54 and -0.88 is found. In this model, the measured value of $\langle P_L^{\Lambda_b} \rangle$ is consistent with no depolarization during fragmentation.

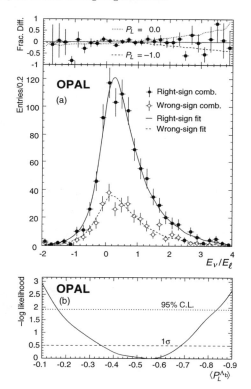

Figure 5: (a) Bottom: distribution of E_ν/E_ℓ for each right-sign $\Lambda \ell^- \pi^+$ combination in the data (solid circles) overlaid by the distribution for each of the wrong-sign combinations (open circles). The solid and dashed lines show the result of the likelihood fit to the right-sign and wrong-sign distribution, respectively. Top: fractional difference $(n_i^{RS,data} - n_i^{RS,fit})/n_i^{RS,fit}$ compared to the analogous distributions for the prediction for $\langle P_L^{\Lambda_b} \rangle = 0.0$ (dotted line) and -1.0 (dashed line). (b) Negative log likelihood as a function of $\langle P_L^{\Lambda_b} \rangle$ for the fit.

References

1. OPAL Collab., G. Abbiendi *et al.*, CERN-EP/98-119, 20 July 1998, to be published in *Phys. Lett. B*. Reported as talk #823 of abstract #320.

2. F.E. Close, J.G. Körner, R.J.N. Phillips, and D.J. Summers, *J. Phys. G* **18** (1992) 1716.

3. T. Mannel and G. Schuler, *Phys. Lett.* B **279** (1992) 194.

4. J.G. Körner, A. Pilaftsis, and M. Tung, *Z. Phys.* C **63** (1994) 575; M. Tung, *Phys. Rev.* D **52** (1995) 1353; S. Groote, J.G. Körner, and M. Tung, *Z. Phys.* C **74** (1997) 615.

5. A.F. Falk and M.E. Peskin, *Phys. Rev.* D **49** (1994) 3320.

6. J. Körner, *Nucl. Phys. B, Proc. Suppl.* **50** (1996) 130.

7. J.F. Amundson, J.L. Rosner, M. Worah, and M.B. Wise, *Phys. Rev.* D **47** (1993) 1260; M. Gronau, in *B Decays*, 2nd Edition, ed. by S. Stone, World Scientific (1994).

8. T. Rizzo, *Right-Handed Currents in B Decay Revisited*, SLAC-PUB-7738, March 1998, to be published in *Phys. Rev. D*.

9. B. Mele and G. Altarelli, *Phys. Lett.* B **299** (1993) 345; B. Mele, *Mod. Phys. Lett.* A **9** (1994) 1239.

10. I.I. Bigi, M. Shifman, N.G. Uraltsev, and A. Vainshtein, *Phys. Rev. Lett.* **71** (1993) 496; A. Czarnecki, M. Jeżabek, J.G. Körner, and J.H. Kühn, *Phys. Rev. Lett.* **73** (1994) 384.

11. A. Czarnecki and M. Jeżabek, *Nucl. Phys.* B **427** (1994) 3.

12. G. Bonvicini and L. Randall, *Phys. Rev. Lett.* **73** (1994) 392; C. Diaconu, M. Talby, J.G. Körner, and D. Pirjol, *Phys. Rev.* D **53** (1996) 6186.

13. OPAL Collab., K. Ahmet *et al.*, *Nucl. Instrum. Methods* A **305** (1991) 275; P.P. Allport *et al.*, *Nucl. Instrum. Methods* A **346** (1994) 476.

14. OPAL Collab., P.D. Acton *et al.*, *Phys. Lett.* B **281** (1992) 394.

15. OPAL Collab., R. Akers *et al.*, *Z. Phys.* C **69** (1996) 195.

16. ALEPH Collab., D. Buskulic *et al.*, *Phys. Lett.* B **365** (1996) 437.

17. OPAL Collab., R. Akers *et al.*, *Z. Phys.* C **65** (1995) 17.

18. OPAL Collab., K. Ackerstaff *et al.*, *Phys. Lett.* B **395** (1997) 128.

19. ALEPH Collab., conference proceedings 28th Inter. Conf. on HEP, Warsaw, Poland, ICHEP96 PA01-072.

20. DELPHI Collab., conference proceedings 28th Inter. Conf. on HEP, Warsaw, Poland, ICHEP96 PA01-049 and DELPHI 96-95 CONF 24.

21. OPAL Collab., P.D. Acton *et al.*, *Phys. Lett.* B **291** (1992) 503.

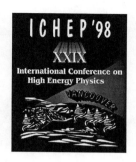

Parallel Session 9

Light Hadron Spectroscopy (Glueballs, Exotics, States with c Quarks)

Convenors: Saj Alam (SUNY, CLEO)

Rolf Landua (CERN)

STUDY OF D** AND D*' PRODUCTION IN B AND C JETS, WITH THE DELPHI DETECTOR

C. BOURDARIOS

Université de Paris Sud, Laboratoire de l'Accélérateur Linéaire, Bât. 200, B.P. 34, FR-91898 ORSAY CEDEX
E-mail: claire.bourdarios@cern.ch

Using D^{*+} mesons exclusively reconstructed in the DELPHI detector at LEP, orbital and radial excitations of non strange charmed mesons are studied. The multiplicities of the two narrow D_1^0 and D_2^{*0} orbital excitations are measured in $Z^0 \to c\bar{c}$ and $Z^0 \to b\bar{b}$ decays. Preliminary results are obtained on the production of broad D** states, using B meson semi-leptonic decays. A narrow signal of 66 ± 14 events is observed in the $(D^{*+}\pi^+\pi^-)$ final state, interpreted as the first evidence of the predicted $D^{*'}$ radial excitation.

1 Introduction

For mesons containing heavy and light quarks ($Q\bar{q}$), and in the limit where the heavy quark mass is much larger than the typical QCD scale, the spin $\vec{s_Q}$ of the heavy quark decouples from other degrees of freedom. Thus, for strong decays, the total (spin+orbital) angular momentum $\vec{j_q} = \vec{s_q} + \vec{L}$ of the light component is conserved. This heavy quark symmetry, together with quark potential models used for lower mass mesons, allows the masses and decay widths of heavy mesons to be predicted [1].

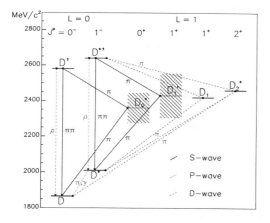

Figure 1: Spectroscopy of non-strange D mesons. The shaded areas show predicted widths for these states. For clarity the expected D_1 and D_2^* decays involving a ρ meson or $\pi\pi$ pairs are not shown.

The present knowledge of charmed meson spectroscopy is summarized in Figure 1. The well established D and D^* mesons [2] correspond to the two degenerate lev-els of the (L=0, $j_q = 1/2$) state. The two (L=1, $j_q = 3/2$) states have been clearly observed [2], because they have narrow decay widths of about 20 MeV/c^2. The measured masses of the $D_1^0(2420)$ and $D_2^{*0}(2460)$ agree within 20 MeV/c^2 with the prediction of the models. Section 3 presents a measurement of their production rate in $c\bar{c}$ and $b\bar{b}$ jets.

The (L=1, $j_q = 1/2$) states decay through a S wave and are expected to have large decay widths. Up to now, they have not been observed directly, but their total production rate is measured using B meson semi-leptonic decays (section 4).

In addition to these orbital excitations, radial excitations of heavy mesons are forseen. The D' and $D^{*'}$ are expected to have masses of 2.58 GeV/c^2 and 2.64 GeV/c^2 respectively, with a 10-25 MeV/c^2 uncertainty on the mass predictions [3]. They are expected to decay, in S wave, into $D^{(*)}\pi\pi$. Section 5 presents the first evidence for the $D^{*'}$ meson, observed in the decay mode $(D^*\pi\pi)$.

2 D** and D*' reconstruction

DELPHI [4] is a multipurpose LEP detector, with special emphasis on precise vertex and charged tracks momentum reconstruction, and particle identification. The micro-vertex detector provides 3 Rϕ and 2 Z hits per track, with intrisic resolutions of 7.6 and 9 μm. For muons of 45 GeV/c momentum, a resolution of $\sigma(p)/p$ of $\pm 3\%$ is obtained, and the precision of the track extrapolation to the beam collision point is 26 ± 2 μm. Kaon and pion identification is performed using a Ring Imaging CHerenkov detector, and the ionisation loss in the TPC, which is the main tracking device. A total of 3.4 million hadronic events is obtained from the 1992-1995 data, at center-of-mass energies close to the Z^0 mass.

2.1 D* reconstruction

All the decay channels considered here involve the $D^{*+} \to D^0\pi_*^+$ decay, followed by $D^0 \to (K^-\pi^+)$ or

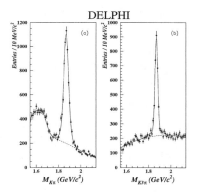

DELPHI

Figure 2: $K\pi$ and $K3\pi$ invariant mass distributions in (a) the $D^{*+} \to (K^-\pi^+)\pi_*^+$ and (b) the $D^{*+} \to (K^-\pi^+\pi^-\pi^+)\pi_*^+$ decay channels.

$D^0 \to (K^-\pi^+\pi^-\pi^+)$. [a]

To reconstruct the D^0 decay final state, all $(K^-\pi^+)$ and $(K^-\pi^+\pi^-\pi^+)$ combinations are tried to fit a secondary vertex in space. Kinematical and track selection cuts are described in detail in [5]. Kaon candidates are considered if they have a momentum larger than 1 GeV/c and, in the $K3\pi$ channel, a loose kaon identification is required. The D^0 momentum and invariant mass are computed from the momenta of the decay products. Then, all charged particles with momentum between 0.4 GeV/c and 4.5 GeV/c and charge opposite to that of the kaon candidate are used as pion candidates for the $D^{*+} \to D^0\pi_*^+$ decay. In the $K\pi$ ($K3\pi$) channel, events are selected if the mass difference $(M_{K\pi\pi_*} - M_{K\pi})$ (resp. $(M_{K3\pi\pi_*} - M_{K3\pi})$) is within ± 2 (± 1) MeV/c^2 of the nominal value $(M_{D^*} - M_{D^0})$. The D* candidates must have an energy fraction $X_E(D^*) = E(D^*)/E_{beam}$ greater than 0.25. Figure 2 shows the distribution of the M($K\pi$) and M($K3\pi$) invariant masses for the selected events. The fitted D^0 masses and widths are 1868 $\pm$ 1 (1869 $\pm$ 1) and 19 $\pm$ 1 (12 $\pm$ 2) MeV/c^2. The reconstructed D^0 mass is required to lie within $\pm$ 40 ($\pm$ 30) MeV/c^2 of the nominal D^0 mass: 4661 $\pm$ 88 (2164 $\pm$ 65) D* candidates are selected in the $K\pi$ ($K3\pi$) channels. The selection efficiency is estimated, using the simulation, to be 21% (8%).

2.2 D_1^0, D_2^{*0} and $D^{*'}$ reconstuction

Similar selection criteria and vertex reconstruction are used to reconstruct narrow orbitally and radially excited states.

---

[a]Throughout this paper, charge-conjugate states are always implied, and the π from D* decay is denoted as π_*.

In the case of D_1^0 and D_2^{*0} decaying into $D^{*+}\pi^-$, a pion with a charge opposite that of the D^{*+} is added, and the $D^0\pi_*^+\pi^-$ vertex is fitted. All combinations are tried, provided the pion candidate has a momentum larger than 1.0 (1.5) GeV/c in the $K\pi$ ($K3\pi$) channel. The reconstruction efficiency is 14% (6%) in the $K\pi$ ($K3\pi$) channels.

In the case of D$^{*'}$ decaying into $D^{*+}\pi^+\pi^-$, all pairs of oppositely charged pions are used to fit a $D^0\pi^+\pi^-$ vertex. The pion candidates are required to have a momentum larger than 0.6(1.0) GeV/c, and those compatible with a kaon according to particle identification are rejected. For a signal of mass 2640 MeV/c^2, the reconstruction efficiency is 4% (2%) in the $K\pi$ ($K3\pi$) channels.

In both cases, the precision on the invariant mass reconstruction is improved by correcting for a 4 MeV/c^2 shift observed in the D^0 mass, by using:

$$M(D^*\pi) = M_{(D^0\pi_*\pi)} - M_{(D^0\pi_*)} + m_{D^*}$$
$$M(D^*\pi\pi) = M_{(D^0\pi_*\pi\pi)} - M_{(D^0\pi_*)} + m_{D^*} \quad (1)$$

where m_{D^*} is the nominal D^{*+} mass. The simulation predicts a resolution of about 6 MeV/c^2 on the mass reconstruction, for both radial and orbital excitations.

2.3 Selection of $b\bar{b}$ and $c\bar{c}$ samples

Due to the relatively long lifetimes of charmed and bottomed particles, heavy flavour events are characterized by the presence of secondary vertices. The probability $\mathcal{P}$ that all tracks detected in the event come from the primary vertex is small: for $b\bar{b}$ events, a purity of 90% is archieved, with an efficiency of 60%, by requiring $\mathcal{P} \leq 10^{-2}$.

Charmed mesons from $Z^0 \to b\bar{b}$ events are distinguished from those in $c\bar{c}$ events by considering both their energy and lifetime informations. Bottom quarks fragment into a B hadron, which subsequently decays into a D^{*+} meson, whereas in $c\bar{c}$ events charmed mesons are directly produced in the fragmentation process. This difference in the hadronization leads to a smaller energy fraction of $X_E(D^*)$ for $b\bar{b}$ events. Also, due to the b quark lifetime, the apparent flight of the D^0 meson is greater than the true decay length. Its measured proper time distribution is larger than the mean B meson lifetime, 1.6 ps, compared to a true D^0 lifetime of 0.4 ps.

By combining these variables, $b\bar{b}$ and $c\bar{c}$ samples are selected, with high purities: 92 % for $b\bar{b}$, 89 % for $c\bar{c}$.

In the $b\bar{b}$ sample, the combinatorial background is higher, but is reduced by 50% using the kaon identification, and also by asking that the impact parameter of the additional pion is positive, ie that the intersection of the pion and D^* directions is on the same side of the primary vertex as the D^* vertex. As a consequence, the ratio of efficiencies $\epsilon(D^*\pi)/\epsilon(D^*)$ is 52% for $b\bar{b}$, compared to 62% for $c\bar{c}$.

3 Study of narrow orbital excitations

Figure 3 shows the $M(D^*\pi)$ invariant mass distribution obtained for the sum of the $b\bar{b}$ and $c\bar{c}$ samples [6]. A clear excess of $(D^{*+}\pi^-)$ pairs is observed between 2.4 and 2.5 GeV/c^2, corresponding to the two overlapping contributions of the D_1^0 and D_2^{*0}. They are fitted by two Breit-Wigner functions, whose widths are fixed to the measured world average [2], convoluted with the experimental resolution. A total signal of (361 ± 58) $D_1^0 + D_2^{*0}$ events is fitted, out of which (65 ± 10) % is assigned to D_1^0. The masses are left free in the fit, and the result is $M_{D_1^0} = 2425 \pm 3(stat)$ MeV/c^2 and $M_{D_2^{*0}} = 2461 \pm 6(stat)$ MeV/c^2, ie consistent with the world averages [2]. The helicity distributions are consistent with the production of a $J^p = 1^+$ and $J^p = 2^+$ states.

Figure 4 shows the same mass distribution, but for the $b\bar{b}$ and $c\bar{c}$ samples separately. The same fit is performed, but both D_1^0 and D_2^{*0} masses and widths are fixed to the world average. The result of the fit is (97 ± 26) D_1^0 and (69 ± 27) D_2^{*0} in the $b\bar{b}$ sample, (141 ± 26) D_1^0 and (104 ± 26) D_2^{*0} in the $c\bar{c}$ sample. In order to measure the D_1^0 and D_2^{*0} production rates, these results are unfolded from the reconstruction efficiencies, signal purities, D_1^0 and D_2^{*0} decay widths into $D^{*+}\pi^-$. The errors quoted below are statistical only. Systematic errors are still under study, but smaller than the statistical errors.

The $c\bar{c}$ sample provides direct information on the charm fragmentation. Results are:

$$f(c \rightarrow D_1^0) = 1.9 \pm 0.4 \ (stat) \ \%$$
$$f(c \rightarrow D_2^{*0}) = 4.3 \pm 1.3 \ (stat) \ \% \tag{2}$$

Both results are in agreement with previous LEP and CLEO measurements. For the D_1^0, the result is also in agreement with theoretical calculations [7], which predicted 1.7%. For the D_2^{*0}, the result is high compared to the expectation (2.4 %), but has large errors. A more precise measurement would need to use the $D_2^{*0} \rightarrow D^+\pi^-$ channel, which is forbidden for the D_1^0.

For the $b\bar{b}$ sample, results are:

$$f(b \rightarrow D_1^0) = 2.2 \pm 0.6 \ (stat) \ \%$$
$$f(b \rightarrow D_2^{*0}) = 4.8 \pm 2.0 \ (stat) \ \% \tag{3}$$

This shows that the charm fragmentation properties are similar, although in a different environment.

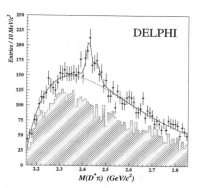

Figure 3: Invariant mass distributions $(D^{*+}\pi^-)$ (dots) and $(D^{*+}\pi^+)$ (hatched histogram). The mass computation and the fit are explained in the text.

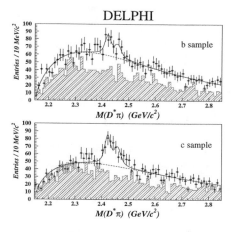

Figure 4: Invariant mass distributions $(D^{*+}\pi^-)$ (dots) and $(D^{*+}\pi^+)$ (hatched histogram) for the separate $b\bar{b}$ abd $c\bar{c}$ samples. The mass computation and the fit are explained in the text.

4 Study of broad orbital excitations in B meson semileptonic decays

In B meson semileptonic decays, only 60% to 70% of the final states are described by $D\ell\bar{\nu}_\ell$ and $D^*\ell\bar{\nu}_\ell$. The remaining contribution is attributed to D^{**}. The total production rate, including broad orbital excitations, can be measured using the impact parameter of the pion, denoted π_{**}, emitted in the decay chain $B \to D^{**}X \to (D^*\pi_{**})X'$.

Events are selected if a lepton with momentum larger than 3 GeV/c^2 is identified, and if its transverse momentum relative to the D^{*+} is larger than 0.5 GeV/c^2. The kaon candidates in the D^0 decay must have the same charge as the lepton. 459 $\pm$25 (288 $\pm$ 19) events are selected in the $K\pi(K3\pi)$ channel. All remaining tracks, of charge opposite to that of the D^{*+}, are π_{**} candidates. Kinematical and selection cuts are described in [8]. The background due to fake D^{*+} associated to a true lepton ℓ^- is subtracted by using events in the tail of the D^* invariant mass distribution. The contribution of true D^{*+} associated to a fake lepton is subtracted using $D^*\ell$ pairs with the wrong sign combination. The remaining 111 $\pm$ 16 events are due to true b semileptonic decays into $D^{*+}\ell^-X$ final state, associated with a π_{**} candidate either from D^{**} decay, or from jet fragmentation. The shape of the two contributions are shown in figure 5. They are used to fit the π_{**} impact parameter distribution shown in figure 6. From the result of the fit, the following branching ratio is obtained:

$$BR(B^- \to (D^{*+}\pi^-)\ell^-\bar{\nu}X) \\ = 1.15 \pm 0.17(stat) \pm 0.14(syst) \% \quad (4)$$

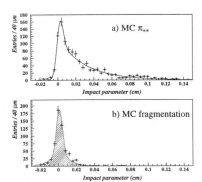

Figure 5: Impact parameter relative to the primary interaction vertex in simulated B semileptonic decays for a) π_{**} from D^{**} decay and b) charged particles from jet fragmentation.

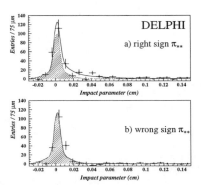

Figure 6: Impact parameter relative to the primary interaction vertex in real data for a) right charge π_{**}^- and b) wrong charge π_{**}^+ candidates. The hatched area is the contribution from jet fragmentation.

This result significantly improves a previous DELPHI measurement, and is in agreement with other LEP measurements.

The $(D^{*+}\pi_{**})$ invariant mass is also reconstructed and used to fit, in the way described in the previous section, the D_1^0 and D_2^{*0} narrow resonances. A signal of 26.7 $\pm$ 8.2 D_1^0 is fitted, and the corresponding production rate is:

$$BR(B^- \to D_1^0\ell^-\bar{\nu}X) \\ = 0.72 \pm 0.22(stat) \pm 0.13(syst) \% \quad (5)$$

For the D_2^{*0} state, 14.8 $\pm$ 7.7 events are fitted, ie a signal significance smaller than 2 σ. More data would be necessary to estimate the corresponding branching ratio.

5 Evidence for a narrow radial excitation

Figure 7 shows the invariant mass distribution obtained when two pions of opposite charges are added to the D^* candidate, and using the sum of the two $b\bar{b}$ and $c\bar{c}$ samples. An excess of 66 $\pm$ 14 (stat) events is observed in the $(D^{*+}\pi^+\pi^-)$ combination. The signal is fitted by a Gaussian distribution of free parameters: the χ^2 per degree of freedom is 60/59, and would be 91/62 if the Gaussian was removed. About $(57 \pm 10)\%$ of the signal is selected in the $c\bar{c}$ sample.

The fitted mass is 2637 $\pm$ 2 (stat) $\pm$ 6 (syst) MeV/c^2. It is thus consistent with the predictions for the $D^{*'}$ radial excitation[3]: 2640 MeV/c^2. Other L=2 states are predicted, but with masses higher by at least 50 MeV/c^2.

The width of the fitted Gaussian is $7 \pm 2 \text{ MeV}/c^2$, ie compatible with the detector resolution. Therefore, only an upper limit is derived: the full width of the signal is smaller than $15 \text{ MeV}/c^2$ at 95 % C.L. There is no natural explanation of such a small value, neither for the $D^{*'}$ nor for higher orbital excitations [9].

Various checks were performed. Varying the background shape and the kinematical cuts has no effect within statistics. No peculiar double counting was noticed, and the signal is stable when the π^* is added to the $D^0 \pi^+ \pi^-$ tracks in the vertex fit. As explained above, mass shifts are studied using D_1^0 and D_2^{*0} narrow states, and a conservative systematic error of $6 \text{ MeV}/c^2$ is attached to the mass measurement.

The production rate of this signal can be compared with that of the D_1^0 and D_2^{*0} narrow states:

$$\frac{<N_{D^{*'}}> \times Br(D^{*'} \to D^* \pi^+ \pi^-)}{\sum_{J=1,2} <N_{D_J^{(*)}}> \times Br(D_J^{(*)} \to D^* \pi)} \quad (6)$$
$$= 0.49 \pm 0.18(stat) \pm 0.10(syst)$$

Most of the systematic uncertainties cancel in this ratio. The quoted systematics is due to the Monte-Carlo statistics, and to the uncertainties on widths and on the kaon rejection. This result is compatible, within its large errors, with the value obtained using the thermodynamical models already mentioned for orbital states [5,7].

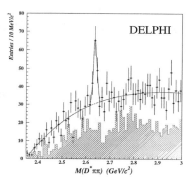

Figure 7: Invariant mass distributions $(D^{*+}\pi^+\pi^-)$ (dots) and $(D^{*+}\pi^-\pi^-)$ (hatched histogram). The mass computation and the fit are explained in the text.

The measured multiplicities in $b\bar{b}$ events are consistent with the ones in $c\bar{c}$ events, both for the D_1^0 and D_2^{*0} .

The total D^{**} production rate, involving a D^{*+} in the final state, is measured in B meson semileptonic decays.

A narrow signal is observed in the $(D^{*+}\pi^+\pi^-)$ final state, at the mass M = 2637 $\pm$ 2 (stat) $\pm$ 6 (syst) MeV/c^2 , interpreted as the first evidence of the predicted $D^{*'}$ meson.

Acknowledgements

I am very grateful to D.Bloch and P.Roudeau for their help while preparing this talk, and to the DELPHI collaboration for choosing me to give it.

References

1. N.Isgur and M.B. Wise, Phys. Rev. Lett. **66**(1991)1130 and ref. therein.
2. Particle Data Group, "Review of Particle Properties", Euro. Phys. J. **C3**(1998)1.
3. S. Godfrey and N. Isgur, Phys. Rev. **D32**(1985)189.
   D.Ebert, R.N. Faustov and V.O. Galkin, "Mass spectrum of orbitally and radially excited heavy-light mesons in the relativistic quark model", preprint HUB-EP-97/90, hep-ph/9712318(1997).
4. DELPHI collab., P. Aarnio et al., Nucl. Instr. Meth. **A303**(1991)233.
   DELPHI collab., P. Aarnio et al., Nucl. Instr. Meth. **A378**(1996)57.
5. DELPHI collab., P. Abreu et al., Phys. Lett. **B426**(1998)231-242
   and contribution to ICHEP'98 number 483.
6. Contribution to ICHEP'98 number 240, DELPHI 98-128 CONF 189.
7. F. Becattini, J. Phys. **G23**(1997)1933.
8. Contribution to ICHEP'98 number 239, DELPHI 98-119 CONF 180.
9. D.Melikhov and O.Péne, HEP-PH/9809308.

6 Conclusion

Using about 7000 exclusively reconstructed D^* mesons, the D_1^0 and D_2^{*0} multiplicities are measured in $c\bar{c}$ events, and found to be consistent with theoretical calculations.

Recent Preliminary Results on D Meson Decays from the BES

Gang Rong

Institute of High Energy Physics, Chinese Academy of Sciences, P.O. Box 918-1, Beijing 100039, China

Representing the BES Collaboration

We report on a first measurement of the branching fractions for the decays $D^{+(0)} \to \overline{K^{*0}}/K^{*0}X, \phi X$, where X can be any unidentified particles, using a data sample of 22.3 pb^{-1} collected at a center-of-mass energy of 4.03 GeV at the BEPC e^+e^- storage ring. We have also searched for the decay $D^+ \to \phi e^+ \nu$, and place a limit on its branching fraction based on non-observation.

1 Introduction

Measurement of the inclusive decay branching fractions of D mesons provides general information about D decay properties. It may be used to determine characteristics of decay modes which have not been observed yet through measurements of exclusive decay modes. For instance, the measurement of the inclusive decay branching fractions could provide evidence for new decay modes. At present the values of the inclusive $\overline{K^{*0}}/K^{*0}$ and ϕ decay branching fractions of the D^a mesons are not available[1]. These inclusive measurements would be beneficial to more precise measurements of B meson decay properties in the experiments such as LEP, CDF and B factories, where the values of the inclusive $\overline{K^{*0}}/K^{*0}$ and ϕ decay branching fractions may be needed as inputs to simulate the D decays and to subtract charm background in the measurements of inclusive B mesons decay properties. The inclusive branching fraction for $D \to \phi X$ is important for studies of gluonic penguin ($b \to sg$) decay of B mesons[1], as well as for time-dependent $B_s^0 \overline{B_s^0}$ oscillation measurements that use the $\phi \ell$ pair to tag the B_s^0 meson.

This paper reports the first measurement of the inclusive Cabibbo-favored $\overline{K^{*0}}$ decay branching fractions, inclusive ϕ decay branching fractions of the charged and the neutral D mesons based on an analysis of the 22.3 pb^{-1} data collected with BES[2] in e^+e^- annihilation at $\sqrt{s} = 4.03$ GeV at BEPC[3].

2 Method

At $\sqrt{s} = 4.03$ GeV charm mesons are produced as
$$e^+e^- \to D_s^+ D_s^-, D^+ D^-, D^0 \overline{D^0},$$
$$D^+ D^{*-}, D^0 \overline{D^{*0}},$$
$$D^{*+} D^{*-}, D^{*0} \overline{D^{*0}}.$$
Since the D^{*+} can decay to $\pi^+ D^0$ and $\pi^0(\gamma)D^+$, it is impossible to know whether a D^+ or D^0 appears in the side recoiling against a tagged $\overline{D}$ meson. But if we combine single and double tags for the two mesons to mea-

---

[a]Throughout this paper, charged conjugation is implied.

sure their inclusive K^{*0} or any other particle P decay branching fractions together, called combinative double tag method (CDTM)[4,5], we can obtain the values of the decay branching fractions as shown bellow separately,

$$BF(\overline{D^0} \to K^{*0}X) = f((1-\beta)\frac{N_{D_{tag}^0}^{K^{*0}X}}{N_{D_{tag}^0}} - \eta \frac{N_{D_{tag}^+}^{K^{*0}X}}{N_{D_{tag}^+}}), \quad (1)$$

and

$$BF(D^- \to K^{*0}X) = f((1-\eta)\frac{N_{D_{tag}^+}^{K^{*0}X}}{N_{D_{tag}^+}} - \beta \frac{N_{D_{tag}^0}^{K^{*0}X}}{N_{D_{tag}^0}}), \quad (2)$$

where $f = 1/(\epsilon(1-\beta-\eta))$, the values $\eta = 0.2116$ and $\beta = 0.5235$ were calculated based on doubly tagged events (see below), $N_{D_{tag}^0}$ is the number of the singly tagged D^0 and $N_{D_{tag}^+}$ the number of the singly tagged D^+, and $N_{D_{tag}^0}^{K^{*0}}$ and $N_{D_{tag}^+}^{K^{*0}}$ are the numbers of the inclusive K^{*0} events appeared in the system recoiling against the D^0 and D^+ tags, respectively.

If we examine the total charge[5] in the recoil against the tagged D mesons and consider a balance of the charge to the charm flavor quantum number, we can get additional information about the decay modes.

In this paper we use the CDTM to determine the inclusive K^{*0} branching fractions[4], and use the CDTM and the recoil charge method together[5] to measure the inclusive ϕ branching fractions of the D^0 and the D^+ mesons.

3 Single Tags

In this analysis, the singly tagged neutral and charged D mesons were reconstructed in hadronic decay modes: $K^+\pi^-$, $K^+\pi^-\pi^-\pi^+$, and $K^+\pi^-\pi^-$. For a candidate 2(or 3 or 4)-charged-track combination, each track must be well reconstructed and consistent with coming from the interaction region. In addition, to optimize momentum resolution and charged particle identification, a geometry cut ($|\cos\theta| \leq 0.8$, where the θ is polar angle of the track) is applied. For the charged particle mass assignment, one (or two) level(s) of particle identification is (are) applied. Firstly a combined particle confidence level calculated using the dE/dx and TOF measurements is required to be greater than 1% for $\pi(K)$ hypothesis; then a normalized likelihood ratio $LR(K) > 50\%$ is required for the K hypothesis. If a track passes the particle identification cuts, the $\pi(K)$ mass is assigned to the track.

After the particle mass assignments, the invariant masses of different $Kn\pi$ (n=1 or 2 or 3) combinations for the three hadronic decay modes $\overline{D^0} \rightarrow K^+\pi^-$, $K^+\pi^-\pi^-\pi^+$ and $D^- \rightarrow K^+\pi^-\pi^-$ are calculated using momenta components corrected for a small shift in the magnetic field and for ionization loss in the material of the Main Drift Chamber[2]. In order to reduce combinatorial background in the singly tagged D meson sample, a roughly 3σ cut on the momenta is used. In this analysis, only the D mesons from the $e^+e^- \rightarrow \overline{D}D^*, D^*\overline{D}^*$ reactions are selected by using the D momentum cut[5]. Figures 1(a), 1(b) and 1(c) show the invariant mass distributions for events in which the combinations passed the momentum cuts for the three hadronic decay modes. The spectra were fit with a Gaussian function and a third order polynomial background. After we corrected the number of events for double counting[5], the fits give $4207 \pm 196 \pm 92$, $2716 \pm 158 \pm 88$ neutral and $2234 \pm 120 \pm 49$ charged $\overline{D}$ mesons for the $K^+\pi^-$, $K^+\pi^-\pi^-\pi^+$ and $K^+\pi^-\pi^-$ modes, respectively, where the first errors are statistical and the second are systematic due to variations in the mass fits.

4 Double Tags

4.1 Inclusive $D \rightarrow \overline{K^{*0}}/K^{*0}X$

The inclusive $\overline{K^{*0}}$ meson is reconstructed in its $\overline{K^{*0}} \rightarrow K^-\pi^+$ mode in the system recoiling against both the $\overline{D}$ tagged region and $\overline{D}$ sideband. The region which is within $\pm 3\sigma_{M_D}$ of the D mass for each of the three modes is defined as signal region, while regions of $\sim \pm 1.5\sigma_{m_D}$ both above and below the tag region for each of the three modes are defined as sideband control regions to be used to estimate the numbers of background events in the tag region. Figure 2(c) shows the distribution of the subtracted-sideband-background invariant masses of the $K^-\pi^+$ pairs selected in the recoils. The mass distribution reflects the net events from the recoils of the tagged $\overline{D}$. Breaking down the events in the figure 2(c) into two groups for the tagged neutral and the charged $\overline{D}$ events separately, we obtained figures 2(a) and 2(b). After correcting for double counting, we obtained 224 ± 27 inclusive Cabibbo-favored $\overline{K^{*0}}$ events from the total $\overline{D}$ tags, 158 ± 22 inclusive Cabibbo-favored $\overline{K^{*0}}$ events from

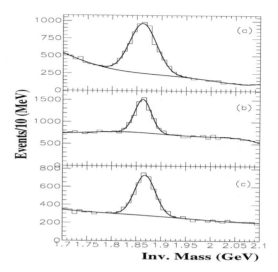

Figure 1: Distribution of invariant masses of the selected (a) $K^+\pi^-$, (b) $K^+\pi^-\pi^-\pi^+$, and (c) $K^+\pi^-\pi^-$ combinations.

the neutral $\overline{D}$ tags and 66 ± 16 inclusive Cabibbo-favored $\overline{K^{*0}}$ events from the charged $\overline{D}$ tags. Figure 3 shows the distribution of the subtracted-sideband-background invariant masses of the $K^+\pi^-$ pairs selected in the recoils. It reflects the net events from the recoils of the tagged $\overline{D}$. After correcting for double counting, we obtain 37.1 ± 15.7 inclusive Cabibbo-suppressed K^{*0} candidate events in total.

4.2 Inclusive $D \rightarrow \phi X$

Figure 4(c) shows the invariant mass distribution of the K^+K^- pairs selected for the events whose $Kn\pi$ invariant masses are within $\pm 2.5\sigma_{M_D}$ of the D masses. Breaking down the events in the figure 4c into the neutral and charged D events separately, we obtained figure 4(a) and 4(b). The K^+K^- mass region 1.00-1.04 GeV is chosen as the signal region, and the intervals 0.98 - 1.00 GeV and 1.04 - 1.15 GeV are considered as background regions for the ϕ. The $Kn\pi$ mass regions from 1.7 to 2.1 GeV , except for regions within $\pm 3\sigma_{M_D}$ of the fitted D masses, are defined as sideband background control regions for the D mesons. As shown in Figure 4(c), 15 events are found as $D\phi$ candidates, and 14 events are selected as D-'ϕ side band' events. Using the D side band events, a total of 0.5 ± 0.5 background events has been estimated as the background among the D candidates. Subtracting the background contributions to both the

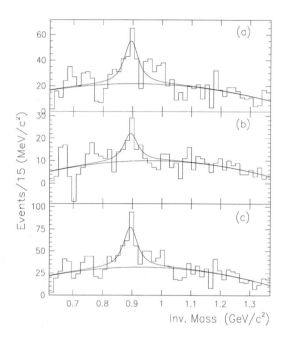

Figure 2: Distribution of the invariant masses of $K^-\pi^+$ combinations for Cabibbo-favored decay modes of D mesons in the system recoiling against the tagged (a) neutral $\overline{D}$, (b) charged $\overline{D}$, and (c) combined the neutral and charged $\overline{D}$ mesons together.

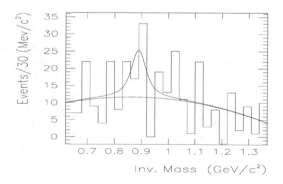

Figure 3: Distributions of the invariant masses of $K^+\pi^-$ combinations for Cabibbo-suppressed decay modes of the D mesons from the system recoiling against the tagged neutral and charged $\overline{D}$ mesons together.

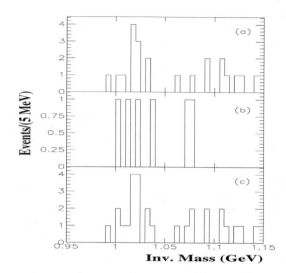

Figure 4: Distributions of the invariant masses of K^+K^- combinations in the system recoiling against the tagged (a) neutral D, (b) charged D, and (c) combined the neutral and charged D mesons together.

D and the ϕ, we obtain 10.2 ± 4.0 inclusive $D^{0(+)} \to \phi X$ events, 6.6 ± 3.5 $D^{0(+)} \to \phi X$ events in the recoil of D^0 tags, and 3.6 ± 2.0 $D^{0(+)} \to \phi X$ events in the recoil of D^+ tags. These correspond 3.7 ± 4.7 $D^0 \to \phi X$ events, and 6.5 ± 5.5 $D^+ \to \phi X$ events in the recoils.

By using the CDTM and recoil charge method together, we obtained 6.7 ± 4.5 $D^0 \to \phi X$, 3.5 ± 3.6 $D^+ \to \phi X$ events, and 6803 ± 303 D^0 and 2251 ± 77 D^+ in the recoil side[5]. These numbers are used to determine inclusive ϕ branching fractions.

4.3 Search for the decay $D^+ \to \phi e^+ \nu$

Among the 15 ϕ candidates observed in the recoil side of the events, 4 are accompanied by at least one charged track which has $|cos\theta| < 0.85$. Each of these tracks is checked against the electron identification procedure by using information from the dE/dx measurement. None of the accompanying tracks is identified as an electron.

4.4 Ratios of D^0 and D^+ in the Recoil of Tags

In order to determine the inclusive $\overline{K^{*0}}/K^{*0}$ decay branching fractions using formula (1) and (2), the ratio $R_{D^0_{tag}}^{\overline{D^0}}$ ($R_{D^+_{tag}}^{\overline{D^0}}$) of the number of the $\overline{D^0}$ produced separately in the recoils to the numbers of the D^0 (D^+) in the singly tagged sample must be determined. To

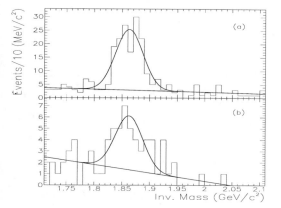

Figure 5: Invariant mass distributions of the combined $K^+\pi^-$ and $K^+\pi^-\pi^-\pi^+$ combinations together observed in the recoils against the D^0 (a) and the D^+ (b) tags.

obtain them, a search for the doubly tagged events $\overline{D^0} \to K^+\pi^-, K^+\pi^-\pi^-\pi^+$ in the recoil side of the tagged D^0 and D^+ was performed. Figure 5 shows the invariant mass distribution of the combined $K^+\pi^-$ and $K^+\pi^-\pi^-\pi^+$ combinations together observed in the recoils against the (a) D^0 and the (b) D^+ tags. The mass spectra were fitted with a Gaussian plus a first-order polynomial to parameterize the background. The standard deviation of the Gaussian distribution was fixed at 23.8 MeV/c^2. The fit gives $N^{\overline{D^0}}_{D^0_{tag}} = 140 \pm 12$, and $N^{\overline{D^0}}_{D^+_{tag}} = 30 \pm 6$. A Monte Carlo study shows that the efficiencies to observe the doubly tagged events are $\epsilon_{\overline{D^0} \to K^+\pi^-} = 0.4234$ and $\epsilon_{\overline{D^0} \to K^+\pi^-\pi^-\pi^+} = 0.1258$. These lead to $R^{\overline{D^0}}_{D^+_{tag}} = 0.5235 \pm 0.1047$ and $R^{\overline{D^0}}_{D^0_{tag}} = 0.7884 \pm 0.0732$, giving $\beta = R^{\overline{D^0}}_{D^+_{tag}}$ and $\eta = 1 - R^{\overline{D^0}}_{D^0_{tag}}$.

5 Results

5.1 Inclusive $\overline{K^{*0}}/K^{*0}$ decay branching fractions of the charged and neutral D mesons

Inserting the numbers of the inclusive $\overline{K^{*0}}$ events observed in the recoil, the numbers of the charged and the neutral $\overline{D}$ tags, the efficiency $\epsilon = 0.156 \pm 0.0078$ estimated by Monte Carlo simulation, and the ratios β and η into equations (1) and (2), we determine the inclusive $\overline{K^{*0}}$ decay branching fractions of the charged and the neutral D mesons to be

$$BF(D^+ \to \overline{K^{*0}}X) = (27.7 \pm 14.1 \pm 4.4)\%,$$

and

$$BF(D^0 \to \overline{K^{*0}}X) = (11.1 \pm 5.1 \pm 3.2)\%.$$

To obtain the average inclusive Cabibbo-favored $\overline{K^{*0}}$ and Cabibbo-suppressed K^{*0} decay branching fractions of the neutral and the charged D mesons combined together, the 210.5 ± 24.9 and 37.1 ± 15.7 events from the $9157 \pm 279 \pm 136$ D decays have to be corrected with the efficiency ϵ. We obtain the average inclusive Cabibbo-favored $\overline{K^{*0}}$ decay branching fraction of

$$BF(D \to \overline{K^{*0}}X) = (15.7 \pm 1.9 \pm 1.4)\%,$$

and we place a 90% C.L. upper limit including systematic uncertainty for the Cabibbo-suppressed K^{*0} decay branching fraction at

$$BF(D \to K^{*0}X) < 4.1\%.$$

Treating our K^{*0} candidates as a signal we obtained the Cabibbo-suppressed K^{*0} decay branching fraction of

$$BF(D \to K^{*0}X) = (2.6 \pm 1.1 \pm 0.2)\%.$$

In the values of the decay branching fractions, the first errors are statistical and the second are systematic. The systematic errors arise from uncertainties ($\pm0.1\%$, $\pm1.2\%$, $\pm0.06\%$ and $\pm0.06\%$) in the numbers of singly tagged D mesons due to uncertainty in background, uncertainties ($\pm1.3\%$, $\pm2.9\%$, $\pm0.8\%$ and $\pm0.1\%$) in the efficiency and uncertainties ($\pm2.5\%$, $\pm3.0\%$, $\pm0.0\%$ and $\pm0.0\%$) in η and β, uncertainties ($\pm1.6\%$, $\pm0.9\%$, $\pm1.2\%$ and $\pm0.2\%$) in variation in the sidebands (for the neutral, the charged, the total of the neutral and the charged D mesons for Cabibbo-favored, and for Cabibbo-suppressed decays, respectively.) The combined effect of these sources is obtained by adding the uncertainties in quadrature, which yields total systematic errors of $\pm3.2\%$, $\pm4.4\%$, $\pm1.4\%$, and $\pm0.2\%$ for each of the cases.

5.2 Inclusive ϕ decay branching fractions of the charged and neutral D mesons

The measurement of $BF(D^{0(+)} \to \phi X)$ is obtained from the number of neutral or charged D events ($N^{D^{0(+)}}_{REC}$) appeared in the recoils of the tagged D sample, number of inclusive ϕ events ($N^{D^{0(+)} \to \phi X}_{REC}$), and the inclusive efficiency ($\epsilon_{D \to \phi X}$) as follows

$$BF(D^{0(+)} \to \phi X) = \frac{N^{D^{0(+)} \to \phi X}_{REC}}{N^{D^{0(+)}}_{REC} \times \epsilon_{D \to \phi X}}. \quad (3)$$

Inserting the number of the inclusive ϕ events, the number of the neutral or the charged D events appeared in the recoils determined by using the CDTM and the recoil charge method together as shown in subsection 4.2, the inclusive efficiency $\epsilon_{D\rightarrow\phi X} = 0.084 \pm 0.004$, including the branching fraction for $\phi \rightarrow K^+K^-$, as estimated by Monte Carlo simulations, into the equation (3), we obtained the inclusive ϕ decay branching fractions for the neutral, charged, and a mixture of $(75\pm5)\%$ neutral and $(25\pm3)\%$ charged D mesons as in the data sample to be

$$BF(D^0 \rightarrow \phi X) = (1.2 \pm 0.8 \pm 0.1)\%,$$

$$BF(D^+ \rightarrow \phi X) = (1.9 \pm 1.9 \pm 0.2)\%,$$

and

$$BF(D \rightarrow \phi X) = (1.3 \pm 0.5 \pm 0.1)\%.$$

Where the first errors are statistical and second systematic. The systematic errors arise from uncertainties ($\pm0.05\%$, $\pm0.08\%$ and $\pm0.05\%$) in the numbers of singly tagged D mesons due to the choice of a background function and fit interval for the single tag sample, uncertainties ($\pm0.1\%$, $\pm0.2\%$ and $\pm0.1\%$) in the inclusive ϕ efficiency. The combined effect of these sources is obtained by adding in the uncertainties in quadrature, which yields a total systematic errors.

5.3 Exclusive semi-electronic decay branching fraction

A 90% C.L. Possion upper limit, $N^{90\%C.L.}_{D^+\rightarrow\phi e^+\nu} = 2.3$, is obtained for $D^+ \rightarrow \phi e^+\nu$ based on 0 candidate event. Using the total number, $N^{D^+}_{REC}$, of the D^+ in the recoil side of the tagged D mesons, and the detection efficiency $\epsilon^{MC}_{D^+\rightarrow\phi e^+\nu}$, we have

$$BF(D^+ \rightarrow \phi e^+\nu) = \frac{N^{90\%C.L.}_{D^+\rightarrow\phi e^+\nu}}{\epsilon^{MC}_{D^+\rightarrow\phi e^+\nu} \times N^{D^+}_{REC}}.$$

The detection efficiency $\epsilon^{MC}_{D^+\rightarrow\phi e^+\nu}$ is estimated to be $\epsilon^{MC}_{D^+\rightarrow\phi e^+\nu} = 0.0652$ by Monte Carlo simulation. A 90% C. L. upper limit that includes a 1σ systematic error is set at

$$BF(D^+ \rightarrow \phi e^+\nu) < 1.6\%.$$

6 Summary

By using the double tag method, we have measured the inclusive Cabibbo-favored $\overline{K^{*0}}$ decay branching fractions of the charged and the neutral D mesons of

$$BF(D^+ \rightarrow \overline{K^{*0}}X) = (27.7 \pm 14.1 \pm 4.4)\%,$$

and

$$BF(D^0 \rightarrow \overline{K^{*0}}X) = (11.1 \pm 5.1 \pm 3.2)\%.$$

At the same time we have measured an average inclusive Cabibbo-favored $\overline{K^{*0}}$ decay branching fraction of the D mesons to be

$$BF(D \rightarrow \overline{K^{*0}}X) = (15.7 \pm 1.9 \pm 1.4)\%,$$

and placed a 90% C.L. upper limit for the Cabibbo-suppressed K^{*0} decay branching fraction at

$$BF(D \rightarrow K^{*0}X) < 4.1\%$$

for the mixture of the neutral and the charged D mesons as in the data sample. If we treat our K^{*0} candidates as a signal, we obtain a Cabibbo-suppressed K^{*0} decay branching fraction of $BF(D \rightarrow K^{*0}X) = (2.6 \pm 1.1 \pm 0.2)\%$.

Combining the CDTM and the recoil charge method together, we measured the inclusive ϕ branching fractions of

$$BF(D^0 \rightarrow \phi X) = (1.2 \pm 0.8 \pm 0.1)\%,$$
$$BF(D^+ \rightarrow \phi X) = (1.9 \pm 1.9 \pm 0.2)\%,$$
$$BF(D \rightarrow \phi X) = (1.3 \pm 0.5 \pm 0.1)\%,$$

and set a new C.L.=90% upper limit at

$$BF(D^+ \rightarrow \phi e^+\nu) < 1.6\%$$

based on non-observation of $D^+ \rightarrow \phi e^+\nu$ event.

Comparing with the sums of the existing measurements on the exclusive D^0 and D^+ decays containing a ϕ in the final states, our measured values of the inclusive ϕ decay branching fractions of the charged and the neutral D mesons indicate that there is little room for additional ϕ containing decay modes of the D^0 and the D^+ mesons.

References

1. Particle Data Group, C. Caso et al., Eur. Phys. J. **C3** (1998) 1.
2. J.Z. Bai et al, Nucl. Instr. and Methods A 344(1994)319.
3. M. H. Ye and Z.P. Zheng, Proceeding of the XIV International Symposium on Lepton and Photon Interaction, Stanford, California, 1989 (World Scientific, Singapore, 1990)
4. G. Rong (BES Collaboration), "Measurements of Inclusive Cabibbo-favored $\overline{K^{*0}}$ and Cabibbo-suppressed K^{*0} Decay Branching Fractions of Charged and Neutral D Mesons",TALK 903, Contributed Paper to the ICHEP'98,Pa-678(1).
5. G. Rong (BES Collaboration), "Measurement of Inclusive ϕ Decay Branching Fractions of the D^0 and D^+ Mesons and a Search for $D^+ \rightarrow \phi e^+\nu$", TALK 903, Contributed Paper to the ICHEP'98,Pa-678(2).

OPAL Results on the Production of Excited Charm Mesons in Hadronic Z^0 Decays

Peter W. Krieger

Dept. of Physics, Carleton University, 1125 Colonel By Drive, Ottawa, Canada, K1S 5B6
E-mail: krieger@physics.carleton.ca

Using data collected with the OPAL detector at LEP, final results are presented on the production of orbitally-excited charm and charm-strange mesons in hadronic Z^0 decays. Also presented are the preliminary results of a search for a radial excitation of the $D^{*\pm}$. All analyses are based on the full 1990-1995 LEP I data sample recorded using the OPAL detector, approximately 4.3 million multihadronic events.

1 Introduction

For each $c\bar{q}$ system, where $\bar{q}$ is a light-antiquark, there should be four orbitally-excited states corresponding to the addition of a single unit of orbital angular momentum to spin 0 or spin 1. These states have spin-parity $J^P = 0^+$, 1^+, 1^+, and 2^+. The decay widths of these charm and charm-strange states are expected to be saturated by two-body decays to $D^{(*)}\pi$ and $D^{(*)}K$, respectively. Spin-parity conservation then dictates the allowed decay modes for each state, as well as the allowed partial waves.

The two 1^+ states mix to produce the physical states. In the heavy-quark limit, this results in one state degenerate in mass with the 2^+ state, decaying as a pure D-wave, and another degenerate with the 0^+ state, decaying as a pure S-wave. The states decaying as D-waves are expected and observed to be quite narrow ($\Gamma \sim 20$ MeV/c^2). The spectroscopy of these states is now well established [1]. The remaining states are expected to be quite broad ($\Gamma \sim 150$ MeV/c^2); none has yet been observed. Studies of P-wave charm meson production at LEP energies provides information on the fragmentation process in the charm sector. This allows, for instance, more complete studies of the vector to pseudoscalar production ratio, V/(V+P). Such a study has been performed by OPAL [2]. In the b-quark sector, studies at LEP currently provide the only existing measurements of the inclusive production of orbitally-excited charm mesons in b-hadron decays. OPAL results on D_1^0, D_2^{*0} and D_{s1}^+ production in hadronic Z^0 decays are final and have been published [3]. These results are summarized sections 2 and 3. Preliminary results were presented at this conference by the ALEPH [4] and DELPHI [5] collaborations.

At somewhat higher masses, one expects meson states corresponding to radial excitations of the $c\bar{q}$ system [6]. The lowest-mass states should correspond to excitations of the D and D^*. Evidence for such a state was recently published by the DELPHI collaboration [7]. The results of a similar search by OPAL are presented in Section 5.

2 Production of $D_J^{(*)0}$ Mesons

The OPAL reconstruction of orbitally-excited charm mesons was performed using the decay sequence

$$D_J^{(*)0} \rightarrow D^*(2010)^+ \pi^-$$
$$\quad\quad \hookrightarrow D^0 \pi^+$$
$$\quad\quad\quad\quad \hookrightarrow K^- \pi^+ \quad\quad (1)$$

which provides the cleanest channel since D^{*+} mesons can be reconstructed on a relatively low background. D^{*+} candidates with scaled energy $x_E > 0.2$ were combined with pion candidates having momentum $p > 2.0$ GeV/c. Charm and bottom-enriched samples were obtained using vertexing and energy information. The $c\bar{c}$-enriched sample was obtained by requiring $x_E(D^{*+}\pi^-) > 0.5$ and a small apparent proper time for the D^0 decay. The $b\bar{b}$- enriched sample was obtained by reconstructing the b-decay vertex topology in events containing D^{*+} candidates, including a fit to the two-track D^0 vertex, and making requirements on the decay length significance of the b-vertex with respect to the primary vertex and of the D^0 vertex with respect to the b-vertex.

The mass-difference distributions obtained from the $c\bar{c}$ and $b\bar{b}$-enriched samples are shown in Figure 1. The flavour-separated production rates for the D_1^0 and D_2^{*0} were obtained from a simultaneous fit to the two distributions which accounts for the charm contribution to the $b\bar{b}$-enriched sample and *vice versa*. In order to measure the individual production rates for the D_1^0 and D_2^{*0}, the observed $D_J^{(*)0}$ signal must be separated into contributions from the two states. This was achieved by parameterizing the signal as the sum of two Breit-Wigners convoluted with the experimental mass resolution, with the masses and widths of the D_1^0 and D_2^{*0} components fixed to their nominal values [1]. Information from the angular distribution of the $D_J^{(*)0}$ decay products was also used: the narrow $J^P = 1^+$ and 2^+ states have distinct distributions of $\cos\alpha$, where α is defined as the angle between the π^- from the $D_J^{(*)0}$ decay and the π^+ from the D^{*+} decay, in the rest frame of the D^{*+}. In the heavy-quark

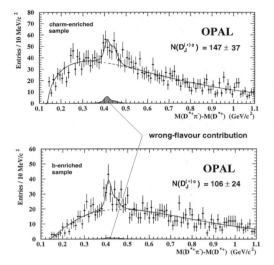

Figure 1: Mass-difference distribution for D*+π− combinations in the cc̄-enriched and bb̄-enriched samples. In each case, the result of the simultaneous fit is overlaid as a solid line, and the fitted wrong-flavour contribution is shown as a shaded distribution.

limit these are of the form $1 + 3\cos^2\alpha$ for the D_1^0 and $\sin^2\alpha$ for the D_2^{*0}, regardless of any spin alignment of the initial state [8]. Experimental results are consistent with these predictions [9]. The inclusion of the helicity-angle information in the fit to the two distributions improves the separation of the D_1^0 and D_2^{*0} signal components. The fit yields

$$\overline{n}_{Z^0 \to c\bar{c} \to D_J^{(*)0}}(x_E > 0.5) \cdot \mathrm{Br}(D_J^{(*)0} \to D^{*+}\pi^-)$$
$$= \left(5.4\ ^{+1.4+0.6}_{-1.3-0.8}\right) \times 10^{-3} \qquad (2)$$

$$\overline{n}_{Z^0 \to b\bar{b} \to D_J^{(*)0}}(x_E > 0.2) \cdot \mathrm{Br}(D_J^{(*)0} \to D^{*+}\pi^-)$$
$$= \left(16.1\ ^{+3.7+2.0}_{-3.6-1.8}\right) \times 10^{-3}. \qquad (3)$$

The fractions of these signals attributed to the D_1^0 state, in cc̄ and bb̄ events, respectively, are

$$f_1^{c\bar{c}} = 0.56 \pm 0.15\ ^{+0.03}_{-0.04} \qquad (4)$$

and

$$f_1^{b\bar{b}} = 0.77\ ^{+0.16}_{-0.14} \pm 0.04. \qquad (5)$$

In each case, the quoted errors are statistical and systematic, respectively. For a description of the contributions to the systematic errors, see reference [3].

The flavour-separated rates (2) and (3) are for a limited region of x_E. Monte Carlo was used to extrapolate to all energies. Production rates were then calcu-lated by making assumptions about the $D_J^{(*)0}$ branch-ing fractions: the value of $\mathrm{Br}(D_1^0 \to D^{*+}\pi^-)$ is con-strained to be 0.65 by phase-space and isospin symme-try; isospin considerations along with measurements [1] of $\mathrm{Br}(D_2^{*0} \to D^+\pi^-)/\mathrm{Br}(D_2^{*0} \to D^{*+}\pi^-) = (2.3 \pm 0.6)$ yield $\mathrm{Br}(D_2^{*0} \to D^{*+}\pi^-) = (0.21 \pm 0.04)$. Accounting for these branching fractions yields the results

$$f(c \to D_1^0) = 0.021 \pm 0.007 \pm 0.003 \qquad (6)$$
$$f(c \to D_2^{*0}) = 0.052 \pm 0.022 \pm 0.013 \qquad (7)$$
$$f(b \to D_1^0) = 0.050 \pm 0.014 \pm 0.006 \qquad (8)$$
$$f(b \to D_2^{*0}) = 0.047 \pm 0.024 \pm 0.013. \qquad (9)$$

3 Production of $D_{s1}^+(2536)$ Mesons

In the charm-strange sector, despite the theoretical ex-pectation that the widths of the two narrow $L = 1$ states be about the same, the width of the $D_{s1}^+(2536)$ is observed to be much smaller than D_{s2}^{*+} width of 16 MeV/c². This narrow width ($\Gamma < 2.3$ MeV/c² at 90% CL) [1] is attributed to the phase-space factor, given the state's proximity to the D^*K threshold, since this is assumed to be the only available decay channel. We first discuss a partial recon-struction of the decay chain

$$D_{s1}^+(2536) \to D^{*0}K^+$$
$$\hookrightarrow D^0(\pi^0, \gamma)$$
$$\hookrightarrow K^-\pi^+. \qquad (10)$$

The low energy release in the D^{*0} decay means that non-observation of the neutral particle does not greatly affect the decay kinematics. A narrow peak arising from the decay sequence (10) should be observable in the mass-difference distribution, $M(D^0K^+) - M(D^0)$.

This analysis requires efficient background suppres-sion; this is obtained primarily with energy and momen-tum requirements. D^0K^+ combinations were required to have a scaled energy larger than 0.57. Due to the missing neutral particle from the D^{*0} decay, this corresponds to an effective scaled-energy cut at 0.6 for the $D_{s1}^+(2536)$. For this reason, the OPAL analysis is sensitive primarily to the production of this state in charm fragmentation.

Figure 2a shows the mass-difference distribution ob-tained from analysis of the OPAL data. A clear peak is seen near threshold. Overlaid as a solid line is the result of a fit to the signal and background. The signal width used in the fit has been fixed to the expected detector res-olution. The fitted mass difference, 527.3 ± 2.2 MeV/c², is consistent with the PDG value [1]. The fitted signal contribution is 28.7 ± 8.3 events, which corresponds to

$$\bar{n}_{Z^0 \to D_{s1}^+} \cdot \mathrm{Br}(D_{s1}^+ \to D^{*0}K^+)_{x_E > 0.6} =$$
$$(1.9 \pm 0.5 \pm 0.2) \times 10^{-3}. \qquad (11)$$

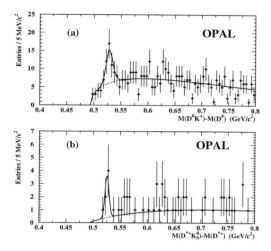

Figure 2: Mass-difference distributions for D_{s1}^+ candidates reconstructed in the final state a) $D^0 K^+$ and b) $D^{*+} K_s^0$. In each case the result of the fit overlaid as a solid line.

The $D_{s1}^+(2536)$ has also been reconstructed in its other available final state using the decay sequence

$$D_{s1}^+(2536) \to D^{*+} K_s^0$$
$$\hookrightarrow D^0 \pi^+ \qquad (12)$$
$$\hookrightarrow K^- \pi^+$$

which provides a better signal-to-background ratio, due the the availability of clean samples of D^{*+} and K_s^0 mesons, but reduced statistics due to the presence of additional branching fractions and reconstruction efficiencies. Figure 2b shows the mass-difference distribution obtained from the OPAL data. A small signal is visible near threshold. The fit yields $5.9^{+2.8}_{-2.3}$ events which corresponds to

$$\bar{n}_{Z^0 \to D_{s1}^+} \cdot Br(D_{s1}^+ \to D^{*+} K^0)_{x_E > 0.6} =$$
$$(1.0^{+0.5}_{-0.4} \pm 0.1) \times 10^{-3}. \qquad (13)$$

Assuming that $D^{*0} K^+$ and $D^{*+} K^0$ saturate the available final states, we can sum (11) and (13) to obtain

$$\bar{n}_{Z^0 \to D_{s1}^+ (x_E > 0.6)} = (2.9^{+0.7}_{-0.6} \pm 0.2) \times 10^{-3}. \qquad (14)$$

Using Monte Carlo to correct for small contributions from $b\bar{b}$ events and for extrapolation to lower energies yields

$$f(c \to D_{s1}^+(2536)) = 0.016 \pm 0.004 \pm 0.003. \qquad (15)$$

4 Summary of P-wave Charm Meson Production at LEP

A summary of the LEP results on the production of orbitally-excited charm and charm-strange mesons is given in Table 1. Preliminary results from the ALEPH and DELPHI collaborations are taken from their contributed papers, which are summarized in these proceedings. Besides the hadronization fractions and inclusive b branching ratios, Table 1 also shows the fraction of the $D_J^{(*)0} \to D^{*+}\pi^-$ signal which is attributed to the $J^P = 1^+$ state. As the production mechanisms in $b\bar{b}$ and $c\bar{c}$ events are different, there is no reason to expect these fractions to be the same in the two classes of events. Also shown are the predictions of the model of Pei [10], which describes LEP data using a simple formula inspired by string fragmentation, and of the model of Becattini [11] which predicts hadron abundances in e^+e^- collisions based on thermodynamic assumptions.

Overall the agreement between the three sets of results is good. However, the OPAL result for $f(b \to D_1^0)$ is somewhat higher than the corresponding results from ALEPH and DELPHI. This has some impact on the discussion of the search for the $D^{*\prime\pm}$, discussed in the next section. Additionally, ALEPH obtains only a limit on $f(b \to D_2^{*0})$ while both OPAL and DELPHI see evidence for a $f(b \to D_2^{*0})$ value which equals or exceeds that of $f(b \to D_1^0)$. This situation is similar to that which exists for $B \to D_J^{(*)0} \ell \nu_\ell$ decays in which some results indicate a suppression of the $J^P = 2^+$ state [12].

5 Search for a Radial Excitation of the $D^{*\pm}$

The DELPHI collaboration has recently published [7] evidence for a radial excitation of the $D^{*\pm}$. In the $D^{*+}\pi^+\pi^-$ final state they observe a narrow resonant structure; 66 ± 14 events at a mass of $2637 \pm 2 \pm 6$ MeV/c^2. This consistent with theoretical predictions [6] for a radial excitation (2^3S_1) of the D^{*+}. The most recent prediction is $M(D^{*\prime\pm}) = 2629$ MeV/c^2. The uncertainty of the mass predictions is assumed to be about 20 MeV/c^2, based on comparison of measured and predicted masses in the P-wave sector. The natural width of the observed state is $\Gamma < 15$ MeV/c^2 at the 90% CL. As a systematic check of their selection, DELPHI perform a similar reconstruction of $D_J^{(*)0} \to D^{*+}\pi^-$ decays. Their final result is quoted as a relative production rate:

$$R' \equiv \frac{\bar{n}_{Z^0 \to D^{*\prime\pm}(2637)} \cdot Br(D^{*\prime+} \to D^{*+}\pi^+\pi^-)}{\sum_{J=1,2} \bar{n}_{Z^0 \to D_J^{(*)0}} \cdot Br(D_J^{(*)0} \to D^{*+}\pi^-)}$$
$$= 0.49 \pm 0.18 \pm 0.10. \qquad (16)$$

OPAL has also searched for this state. D^{*+} candidates were reconstructed in the $D^{*+} \to D^0\pi^+$, $D^0 \to K^-\pi^+$

Fraction (%)	OPAL[3]	ALEPH[4]	DELPHI[5]	Pei[10]	Becattini[11]
$b \to D_1^0$	$5.0 \pm 1.4 \pm 0.6$	2.3 ± 0.7	2.0 ± 0.6	–	–
$b \to D_2^{*0}$	$4.7 \pm 2.4 \pm 1.3$	< 2.0 (95% CL)	4.8 ± 2.0	–	–
$c \to D_1^0$	$2.1 \pm 0.7 \pm 0.3$	1.6 ± 0.5	1.9 ± 0.4	3.5	1.7
$c \to D_2^{*0}$	$5.2 \pm 2.2 \pm 1.3$	4.7 ± 1.0	4.7 ± 1.3	5.0	2.4
$f_1^{b\bar{b}}$	0.77 ± 0.16	> 0.8 (95% CL)	0.58 ± 0.12	–	–
$f_1^{c\bar{c}}$	0.56 ± 0.15	0.47 ± 0.09	0.58 ± 0.08	–	–
$c \to D_{s1}^+$	$1.6 \pm 0.4 \pm 0.3$	$0.77 \pm 0.20 \pm 0.08$	–	1.1	0.54
$c \to D_{s2}^{*+}$	–	$1.3 \pm 0.5 \pm 0.2$	–	1.7	–
$b \to D_{s1}^+$	–	$1.1 \pm 0.3 \pm 0.2$	–	–	–
$b \to D_{s2}^{*+}$	–	$2.2 \pm 0.8 \pm 0.5$	–	–	–

Table 1: A summary of LEP results on the production of orbitally-excited charm and charm-strange mesons. Also shown are the model predictions of Pei and of Becattini. The results from ALEPH and DELPHI are preliminary. Uncertainties on the DELPHI results are statistical only. f_1 values for DELPHI are calculated from the numbers of events provided in Table 2 of the cited contributed paper.

decay sequence. Charm and bottom-enriched samples of hadronic Z^0 decays were obtained using energy and vertexing requirements. The expected signal size was estimated[a] by calculating the OPAL product efficiency times branching fraction ($\epsilon \cdot$ Br) times the number of analyzed events (N_{MH}), relative to DELPHI:

$$\frac{\epsilon \cdot N_{MH} (\text{OPAL})}{\sum_{i=1,2} \epsilon_i \cdot Br_i \cdot N_{MH} (\text{DELPHI})} \cong 1.1 \pm 0.1. \quad (17)$$

Making a conservative assumption of an expected number of signal events equal to the number observed by DELPHI, the expected signal plus background distribution from the OPAL search is shown in Figure 3a. The signal contribution is clearly visible above a modest background. The signal region indicated by the arrows is 2.59-2.67 GeV/c^2 which includes the $\pm 2\sigma$ range of both the theoretical predictions and the DELPHI result.

Applying this analysis to the OPAL data yields the invariant-mass distribution shown in Figure 3b. The background distribution was obtained by fitting the invariant-mass distribution with the signal region excluded. The number of signal events, calculated by subtracting the integrated interpolated background function in the signal region, is -32 ± 24. The statistical error includes two contributions, one from the Gaussian variance of the number of entries in the mass window (± 21), the other from the uncertainty on the background integral (± 11), computed by propagating the error from the fit parameters.

As a systematic check of the OPAL analysis, a similar selection was used to reconstruct $D_J^{(*)0} \to D^{*+}\pi^-$ decays, both in OPAL data and in a Monte Carlo data sample

[a] Using the published R' value from DELPHI and the published OPAL $D_J^{(*)0}$ results would yield a higher estimate for the expected signal, since the OPAL results for the production of $D_J^{(*)0}$ mesons lead to a larger value for the denominator in (16).

in which the $D_J^{(*)0}$ production fractions were tuned to the published OPAL results. No significant bias was observed. Systematic errors on the reconstruction efficiency arise due to modelling uncertainties, Monte Carlo statistics, tracking resolution fractions and uncertainty on the width of the $D^{*\pm}$. The possibility of decays via an intermediate L=1 state was also considered. Based on the absence of a signal, limits are set on the production of this state in hadronic Z^0 decays, and separately for its production in $Z^0 \to c\bar{c}$ and $Z^0 \to b\bar{b}$ events. The 95% CL upper limit on the number of signal events is 32.8 which corresponds to

$$f(Z^0 \to D^{*\prime\pm}(2629)) \times Br(D^{*\prime+} \to D^{*\prime+}\pi^+\pi^-) \quad (18)$$
$$< 2.1 \times 10^{-3} \quad (95\% \text{ C.L.}).$$

The corresponding limits for production of this state in $Z^0 \to c\bar{c}$ and $Z^0 \to b\bar{b}$ events are given below:

$$f(c \to D^{*\prime\pm}(2629)) \times Br(D^{*\prime+} \to D^{*+}\pi^+\pi^-) \quad (19)$$
$$< 1.2 \times 10^{-2} \quad (95\% \text{ C.L.})$$

$$f(b \to D^{*\prime\pm}(2629)) \times Br(D^{*\prime+} \to D^{*+}\pi^+\pi^-) \quad (20)$$
$$< 1.0 \times 10^{-2} \quad (95\% \text{ C.L.}).$$

The reconstruction of $D_J^{(*)0} \to D^{*+}\pi^-$ decays, with a similar selection, done as a systematic check, also provides the denominator for the ratio R', in which numerous systematic errors cancel. The resulting OPAL limit is

$$R'(\text{OPAL}) < 0.21 \quad (95\% \text{ C.L.}). \quad (21)$$

The CLEO collaboration has also searched for the production of this state in e^+e^- collisions at centre-of-mass energies in the region of the $\Upsilon(4S)$ [13]. They apply a hard cut on the scaled momentum of their candidates and thus are sensitive only to production in charm

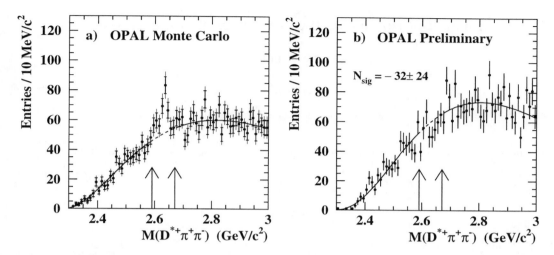

Figure 3: Invariant-mass distributions for selected $D^{*+}\pi^+\pi^-$ combinations. a) shows the expected distribution from OPAL Monte Carlo. The signal normalization was obtained by assuming a number of observed events equal to the number observed by DELPHI, as described in the text. b) shows the distribution obtained from analysis of the OPAL data. In each case the arrows indicate the signal region. Overlaid as a solid line is the result of the fit to the distribution, excluding the signal region. The interpolation of the fitted background function is shown in the signal region as a dotted line. In a) the thick error bars represent the statistical error obtained from the full Monte Carlo sample. The lighter (larger) error bars have been scaled to correspond to statistics of the OPAL data.

fragmentation. No evidence for a signal was observed. CLEO has a large data sample with $\sim 10^3$ reconstructed $D_J^{(*)0} \to D^{*+}\pi^-$ decays. The absence of a signal in the $D^{*+}\pi^+\pi^-$ final state yields

$$R'(\text{CLEO}) < 0.16 \quad (90\% \text{ C.L.}). \qquad (22)$$

6 Summary of $D^{*/\pm}$ Search Results

In summary, the published DELPHI evidence for $D^{*/\pm}$ production in hadronic Z^0 decays cannot be confirmed with the OPAL data. No evidence is observed in either $c\bar{c}$ or $b\bar{b}$-enriched samples of hadronic Z^0 decays. A similar search, performed by the CLEO collaboration [13], finds no evidence for the production of this state in charm fragmentation at $\sqrt{s} \approx 10.6$ GeV. Both the OPAL and CLEO results are preliminary.

References

1. Particle Data Group, R.M.Barnett et al., Phys. Rev. D54 (1996) 1.
2. OPAL Collaboration, K. Ackerstaff et al., CERN-EP/98-006.
3. OPAL Collaboration, K. Ackerstaff et al., Z. Phys C76 (1997) 425.
4. ALEPH Collaboration, ICHEP'98 contributed papers 943 and 944; S.Wasserbaech, these proceedings.
5. DELPHI Collaboration, ICHEP'98 contributed paper 240; C. Bourdarios, these proceedings.
6. S. Godfrey and N. Isgur, Phys. Rev. D32 (1985) 189; D. Ebert, V.O. Galkin and R.N. Faustov, Phys. Rev. D57 (1998) 5663
7. DELPHI Collaboration, P. Abreu et al., Phys. Lett. B426 (1998) 231.
8. J. L. Rosner, Comments Nucl. Part. Phys. 16 (1986) 109.
9. CLEO Collaboration, P. Avery et al., Phys.Lett. B331 (1994) 236.
10. Yi-Jin Pei, Z. Phys. C72 (1996) 39.
11. F. Becattini, J. Phys. G23 (1997) 1933.
12. ALEPH Collaboration, D. Buskulic et al., Z. Phys. C73 (1997) 601; OPAL Collaboration, R. Akers et al., Z. Phys. C67 (1995) 57; CLEO Collaboration, A. Anastassov et al., Phys. Rev. Lett. 80 (1998) 4127; ARGUS Collaboration, H. Albrecht et al., Z. Phys. C57 (1993) 533.
13. I. Shipsey, these proceedings.

LEPTONIC DECAYS OF THE D_s MESON
AND PRODUCTION OF ORBITALLY EXCITED D AND D_s MESONS

STEVEN WASSERBAECH

Dept. of Physics, University of Washington, PO Box 351560, Seattle, WA 98195-1560, USA
E-mail: wasser@u.washington.edu

Representing the ALEPH Collaboration

The purely leptonic decays $D_s \to \mu\nu$ and $D_s \to \tau\nu$ are studied in a sample of four million hadronic Z decays collected in 1991–1995 with the ALEPH detector at LEP. A linear discriminant analysis is used to measure the branching fraction $B(D_s \to \mu\nu) = (0.64 \pm 0.08[\text{stat}] \pm 0.21[\text{syst}] \pm 0.15[\phi\pi])\%$ (preliminary), where the last uncertainty is from the branching fraction $B(D_s \to \phi\pi)$. Within the Standard Model, this result corresponds to $f_{D_s} = (284 \pm 18[\text{stat}] \pm 49[\text{syst}] \pm 34[\phi\pi])$ MeV. Preliminary results are also reported from an investigation of orbitally excited ($L = 1$) charm mesons in $Z \to c\bar{c}$ and $b\bar{b}$ events. Significant production of such mesons is observed, and the production rates are measured.

1 Outline

In this talk I will discuss two preliminary analyses by the ALEPH Collaboration. The first, described in Section 2, is a study of purely leptonic decays of the D_s meson.[1] The second, presented in Section 3, is an investigation of orbitally excited ($L = 1$) charmed mesons.[2]

2 Leptonic decays of the D_s Meson

2.1 Introduction

The purely leptonic decay of the D_s^+ ($c\bar{s}$) meson (Fig. 1) is the second-generation analogue of π^+ ($u\bar{d}$) decay.[a] The decay rate for $D_s \to \ell\nu$ is proportional to $f_{D_s}^2$, where the decay constant f_{D_s} characterizes the $c\bar{s}$ wavefunction at zero separation.

The decay constants of the pseudoscalar mesons D, D_s, B, etc. can be predicted by means of lattice QCD or other methods. The values of f_B and f_{B_s} are needed in order to extract CKM matrix elements from measurements of B-$\bar{B}$ mixing, but f_B and f_{B_s} are difficult to measure; at the present time it is necessary to rely on the theoretical predictions. A measurement of f_{D_s} can serve as a useful check on the lattice calculations.

The D_s leptonic branching fractions are given by

$$B(D_s \to \ell\nu) = \tau_{D_s} \frac{G_F^2}{8\pi} f_{D_s}^2 |V_{cs}|^2 M_{D_s} M_\ell^2 \left(1 - \frac{M_\ell^2}{M_{D_s}^2}\right)^2,$$

[a] Charge conjugate states are implied throughout this article.

Figure 1: The decay $D_s \to \ell\nu$.

Table 1: Recent measurements of f_{D_s}.

Experiment	Decay	f_{D_s} (MeV)
E653[4]	$D_s \to \mu\nu$	$194 \pm 35 \pm 20 \pm 14$
CLEO[5]	$D_s \to \mu\nu$	$280 \pm 17 \pm 25 \pm 34$
L3[6]	$D_s \to \tau\nu$	$309 \pm 58 \pm 33 \pm 38$
DELPHI[7]	$D_s \to \tau\nu$	330 ± 95 (prelim)
Average		264 ± 36

where τ_{D_s} is the D_s lifetime and the M's are the masses of the D_s and the final state charged lepton. The factor M_ℓ^2 reflects the helicity suppression of the nonradiative leptonic decay. (For the purposes of the present analysis, the effect of radiative decays is expected to be rather small.[3]) The tree-level formula predicts the branching fractions for D_s decays to $e\nu$, $\mu\nu$, and $\tau\nu$ to be in the ratio 2.4×10^{-5} to 1 to 9.75.

Some recent measurements of leptonic D_s decays are listed in Table 1. The average, calculated after adjusting the measurements to[8] $B(D_s \to \phi\pi) = (3.6 \pm 0.9)\%$, is $f_{D_s} = 264 \pm 36$ MeV. This central value corresponds to $B(D_s \to \mu\nu) = 0.56\%$ and $B(D_s \to \tau\nu) = 5.4\%$.

2.2 Analysis procedure

The decays $D_s \to \mu\nu$ and $D_s \to \tau\nu$ are studied in a sample of four million hadronic Z decays collected in 1991–1995 with the ALEPH detector[9] at LEP. The event selection is optimized to select $e^+e^- \to c\bar{c}$ events containing the decay $D_s \to \mu\nu$. The analysis is also sensitive to $D_s \to \tau\nu$ decays, which are treated as signal. The preselection of hadronic Z decays is performed according to the standard ALEPH selection based on charged tracks. The cut on the minimum number of reconstructed charged tracks coming from the area of the interaction point is tightened to reduce background from

$e^+e^- \rightarrow \tau^+\tau^-$ events. The event thrust axis is required to satisfy $|\cos\theta_{\text{thrust}}| < 0.8$ to select events within the acceptance of the vertex detector. A loose muon identification algorithm based on muon chamber hits and the digital pattern information from the hadron calorimeter is used to select muon candidates.

A kinematic fit is then performed in order to improve the resolution on the missing momentum, assumed to arise from the undetected neutrino from $D_s \rightarrow \mu\nu$. In this fit, the missing mass is constrained to be zero and the D_s displacement direction (from the primary event vertex to some point on the muon track) is constrained to be parallel to the D_s momentum direction. The energies of the reconstructed charged and neutral particles are varied in the fit, while their directions are held constant. The event primary vertex and the D_s decay vertex are also allowed to vary within their uncertainties. The energy resolutions for charged, neutral electromagnetic, and neutral hadronic objects are parametrized from simulated events. The fitted energy of the D_s candidate (the muon and the neutrino) is required to be greater than 15 GeV.

To reduce background from $b\bar{b}$ events, selections are made based on a set of rapidity and impact parameter dependent hemisphere variables. An existing ALEPH lifetime tag[10] was modified to include a dependence on the rapidity of the charged tracks in the event. Events from $c\bar{c}$ are expected to have higher rapidity tracks relative to the jet axes than $b\bar{b}$ events because of the lower mass of c hadrons compared to b hadrons. Additionally, if one excludes the identified muon, signal events will have only fragmentation tracks in the muon hemisphere, and can be distinguished from other $c\bar{c}$ events. The tracks in each hemisphere are divided according to whether their rapidity is greater or less than a certain value. The probabilities that the sets of high and low rapidity tracks in each hemisphere originate from the Z production point are then calculated. Cuts are made on the probabilites for the low rapidity tracks in each hemisphere, and on the high rapidity tracks' probability in the muon hemisphere.

Finally, to distinguish signal from the remaining primarily $c\bar{c}$ and $b\bar{b}$ background events a hard cut is made on the energy of the fitted D_s candidate (>25 GeV). The leptonic branching fractions are extracted by means of a fitting procedure which is designed to distinguish between the two principal background components separately. Two linear discriminant variables are created. One is constructed specifically to separate $c\bar{c} \rightarrow D_s \rightarrow \mu\nu$ events from $b\bar{b}$ background events (mainly containing semileptonic b and c decays), and another to separate them from $c\bar{c}$ background events (mainly containing semileptonic c decays). The linear discriminant variables U_b and U_c are linear combinations of variables

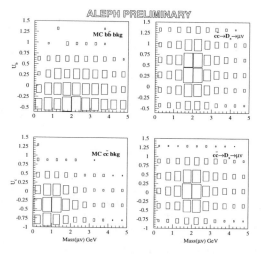

Figure 2: Unnormalized U_b vs $M(\mu\nu)$ (top row) and U_c vs $M(\mu\nu)$ (bottom row) distributions for backgrounds (left) and $c\bar{c} \rightarrow D_s \rightarrow \mu\nu$ signal (right). The same cuts used in the fitting procedure are applied here. The top row shows how the U_b variable discriminates between $b\bar{b}$ background events and the signal. The bottom row shows that $c\bar{c}$ background events are distinguished from the signal both in the U_c and $M(\mu\nu)$ distributions.

which optimally distinguish between two distributions. Each linear discriminant variable is a linear combination of fifteen variables. Plots of the distributions for the $c\bar{c} \rightarrow D_s \rightarrow \mu\nu$ events and the respective background events are shown in Fig. 2.

The most discriminating ingredients of U_b are

- b tag neural network outputs from both hemispheres;

- $|\cos\theta_{\text{thrust}}|$;

- the missing energy in the muon hemisphere;

- the fitted muon momentum.

The most discriminating ingredients of U_c are

- c tag neural network output from the muon hemisphere;

- the fitted muon momentum;

- the fitted neutrino momentum;

- the output of a b tagging algorithm, based on lifetime and mass information.

The procedure to fit for the $D_s \rightarrow \mu\nu$ (plus $D_s \rightarrow \tau\nu$) component in the data consists of fitting the $\mu\nu$ invariant mass distribution in 36 (6 × 6) bins in the (U_b, U_c) plane simultaneously. Fitting functions are generated by separately fitting the mass distributions for each event type in each slice. One to three Gaussians are used depending on the number of events in the slice. The fits for each event type are normalized to one another according to the Monte Carlo predictions.

In the maximum likelihood fit to the data, the backgrounds are divided into two categories, $b\bar{b}$ events and all other events (including $udsc$ events and the few remaining $e^+e^- \rightarrow \tau^+\tau^-$ events).

A single fitting function describing the signal is obtained by combining the decays $D_s \rightarrow \mu\nu$ and $D_s \rightarrow \tau\nu$ from $c\bar{c}$ and $b\bar{b}$ events. These four components are combined in their expected proportions; in particular, it is assumed that $B(D_s \rightarrow \tau\nu)/B(D_s \rightarrow \mu\nu) = 9.75$. These processes are all treated as signal because the yields are all proportional to $f_{D_s}^2$. Small contributions are included in the signal from the Cabibbo suppressed decays $D^+ \rightarrow \mu\nu$ and $D^+ \rightarrow \tau\nu$, which are nearly indistinguishable from D_s leptonic decays.

The normalizations of the combined signal and the two background categories are free to vary in the fit, while the shapes are held fixed.

Signal and background distributions, projected onto the $M(\mu\nu)$ axis, are shown in Fig. 3. The largest signal

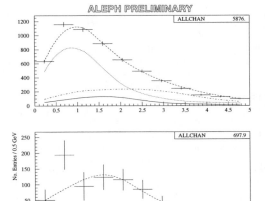

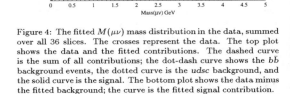

Figure 4: The fitted $M(\mu\nu)$ mass distribution in the data, summed over all 36 slices. The crosses represent the data. The top plot shows the data and the fitted contributions. The dashed curve is the sum of all contributions; the dot-dash curve shows the $b\bar{b}$ background events, the dotted curve is the $udsc$ background, and the solid curve is the signal. The bottom plot shows the data minus the fitted background; the curve is the fitted signal contribution.

components are $c\bar{c} \rightarrow D_s \rightarrow \mu\nu$ (25% of the total signal) and $c\bar{c} \rightarrow D_s \rightarrow \tau\nu$ (52%). The contribution from D^+ leptonic decays is 6% of the total.

2.3 Results

The results of the three-dimensional maximum likelihood fit to the 1991–1995 ALEPH data are plotted in Fig. 4, projected onto the $M(\mu\nu)$ axis. This fit is used to measure the branching fraction $B(D_s \rightarrow \mu\nu)$ and to extract the decay constant f_{D_s}. The fitted number of signal events is 747 ± 93, yielding $B(D_s \rightarrow \mu\nu) = (0.670 \pm 0.083[\text{stat}])\%$. The goodness of fit is assessed by generating and fitting fake data samples; the distribution of likelihood values from the fake samples shows that the fit to the data has a confidence level of 0.60.

Additional projections of the fitted three-dimensional space are given in Fig. 5. Here the (U_b, U_c) plane is divided into four regions by the lines $U_b = 0.45$ and $U_c = 0.1$, and the $M(\mu\nu)$ distribution in each of the four regions is plotted separately. The upper-left plot is dominated by background events, and the fitted histogram agrees with the data points in all but the lowest mass bin. The lower-right plot is enriched with signal events, and the data is again well represented by the fitted distribution including signal and background.

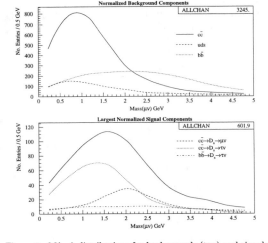

Figure 3: $M(\mu\nu)$ distributions for backgrounds (top) and signals (bottom), after integrating over U_b and U_c. The top plot shows the uds, $c\bar{c}$, and $b\bar{b}$ background components. In the bottom plot, the solid curve represents the total of all D_s and D^+ signal components. The other curves indicate the $M(\mu\nu)$ distributions from some of the largest signal contributions: the dashed curve is $c\bar{c} \rightarrow D_s \rightarrow \mu\nu$, dotted is $c\bar{c} \rightarrow D_s \rightarrow \tau\nu$, and dot-dashed is $b\bar{b} \rightarrow D_s \rightarrow \tau\nu$.

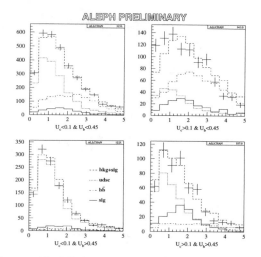

Figure 5: The $M(\mu\nu)$ mass distribution in four regions of the (U_b, U_c) plane. The crosses are the data. The solid histogram shows the fitted signal contribution, while the dotted and dot-dashed histograms represent the $udsc$ and $b\bar{b}$ background contributions, respectively.

As a cross check, the data are refitted without the constraint on the ratio $B(D_s \to \tau\nu)/B(D_s \to \mu\nu)$. The results are $B(D_s \to \mu\nu) = (0.596 \pm 0.246)\%$ and $B(D_s \to \tau\nu)/B(D_s \to \mu\nu) = 11.8 \pm 7.0$. The latter value is to be compared to the expected ratio of 9.75.

2.4 Systematic uncertainties

The contributions to the systematic uncertainty on the branching fraction $B(D_s \to \mu\nu)$ are listed in Table 2.

The three-dimensional fitting procedure applied to fake data samples is found to produce a bias of $+4.8\%$ on the fitted number of signal events. A correction for this offset is applied to the result obtained from the data. The total uncertainty on the measured leptonic branching ratios related to the statistics of the various fully-simulated Monte Carlo samples (signals and backgrounds) is $\pm 8.7\%$.

The production fractions of charmed mesons in $b\bar{b}$ and $c\bar{c}$ events are renormalized according to the most recent ALEPH measurements. Applying the corrections to the Monte Carlo background distributions produces a shift in the fitted leptonic branching fraction. The production fractions are also varied according to their experimental uncertainties. The sum in quadrature of the variations in the resulting branching fraction measurement plus half the shift is taken as the uncertainty on the branching fraction measurement.

The effects of the charm meson production fractions and other physics parameters on the numbers of produced D_s's and D^+'s are also taken into account as shown in Table 2. A large contribution to the systematic uncertainty on $B(D_s \to \mu\nu)$ is due to the uncertainties on $f(c \to D_s)$ and $f(b \to D_s)$, which are dominated by the 25% relative uncertainty[8] on $B(D_s \to \phi\pi)$. In effect we are measuring the ratio $B(D_s \to \mu\nu)/B(D_s \to \phi\pi)$, and the uncertainty on $B(D_s \to \phi\pi)$ enters when the result is expressed in terms of $B(D_s \to \mu\nu)$ or f_{D_s}.

The uncertainty in the signal efficiency due to the muon identification efficiency and the uncertainty on the fragmentation parameter for the signal events is estimated. The muon efficiency is reduced by 20% in the Monte Carlo for muons with $P < 3\,\text{GeV}$ and the data is refitted. The uncertainty due to fragmentation is estimated by lowering the energy cut on the D_s candidates by 2%. The resulting change in the fraction of signal events predicted by Monte Carlo is taken as the uncertainty.

Other significant systematic errors remain to be estimated. The fragmentation parametrization must be varied. The composition and properties of the principal background components must be understood. A relative uncertainty on $B(D_s \to \mu\nu)$ of 30% is assigned as a conservative estimate of the effects related to the shapes of the background distributions.

A check on the procedure will also be performed by running the analysis to search for $D_s \to e\nu$, which is expected to be extremely small due to helicity suppres-

Table 2: Summary of systematic uncertainties on $B(D_s \to \mu\nu)$. Entries labeled "$[\phi\pi]$" are contributions due to the uncertainty on $B(D_s \to \phi\pi)$.

Source	Rel. uncertainty (%)
Fit bias $\pm$ MC statistics	$+4.8 \pm 8.7$
Hadron production (bkg)	± 6.4
Physics parameters (signal):	
$f(c \to D_s)$	$\pm 13.3 \pm 19.5[\phi\pi]$
$f(c \to D^+)$	± 0.5
$f(b \to D_s)$	$\pm 1.9 \pm 4.1[\phi\pi]$
$f(b \to D^+)$	± 0.1
$B(D^+ \to \mu\nu)/B(D_s \to \mu\nu)$	± 0.3
$B(D^+ \to \tau\nu)/B(D_s \to \mu\nu)$	± 0.2
Signal efficiency:	
μ identification	± 1.0
Fragmentation	± 4.0
Background shape:	
($\sim 5\times$ hadron prod. uncert.)	± 30.0
Total bias and uncertainty	$+5 \pm 34 \pm 24[\phi\pi]$

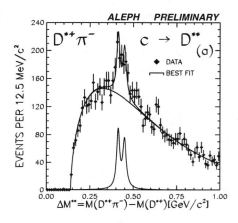
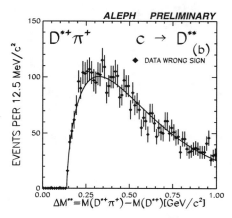

Figure 6: Mass difference distributions $\Delta M^{**} = M(D^*\pi) - M(D^*)$ in the charm-enriched sample: (a) opposite-sign combinations $(D^{*+}\pi^-)$ and (b) same-sign combinations $(D^{*+}\pi^+)$. The solid curves show the results of a fit for the background function along with Breit-Wigner functions for the D_1 and D_2^* in (a).

sion. The signal efficiency can be checked by analyzing $D_s \to \phi\pi$ decays. The reason for the bias in the fitting routine must be understood, and the effects of detector resolution must be examined.

In summary, significant signals for $D_s \to \mu\nu$ and $D_s \to \tau\nu$ are observed in $Z \to c\bar{c}$ events collected with the ALEPH detector at LEP. A preliminary measurement of

$$B(D_s \to \mu\nu) = (0.64 \pm 0.08_{[stat]} \pm 0.21_{[syst]} \pm 0.15_{[\phi\pi]})\%$$

is obtained. This central value corresponds to $B(D_s \to \tau\nu) = 6.2\%$, under the assumption $B(D_s \to \tau\nu)/B(D_s \to \mu\nu) = 9.75$ which is used in the fit to the data. The decay constant is calculated to be

$$f_{D_s} = (284 \pm 18_{[stat]} \pm 49_{[syst]} \pm 34_{[\phi\pi]}) \text{ MeV}.$$

3 Orbitally excited charm mesons

It is desirable to measure the relative probabilities for producing various spin states in charm hadronization and in $b \to c$ decays.

In the $c\bar{q}$ system (where $q = u, d, s$), we expect

- two states with $L = 0$: D and D^*;
- four states with $L = 1$: D_0^*, D_1, D_1, D_2^*, decaying via the strong interaction.

Due to spin-parity conservation, one D_1 and the D_2^* can only decay via d wave and hence are expected to be narrow. The D_0^* and the other D_1 can decay via s wave and hence should be wide. All of the narrow states with

$q = u$, d, and s have been observed[8], as indicated in Table 3.

ALEPH has investigated the reactions

$$D_1^0 \to D^{*+}\pi^-,$$
$$D_2^{*0} \to D^{*+}\pi^-, D^+\pi^-,$$
$$D_2^{*+} \to D^0\pi^+,$$
$$D_{s1}^+ \to D^{*+}K^0, D^{*0}K^+,$$
$$D_{s2}^{*+} \to D^0K^+.$$

The results presented herein were obtained from a sample of 4.1 million hadronic Z decays collected between 1991 and 1995. Only a brief summary of the analysis strategy is given here; more details are available in the contributed papers[2].

The analysis of the $D\pi$ and $D^*\pi$ channels proceeds as follows. (The procedure in the DK and D^*K studies is similar.) First, D^0, D^+, and D^{*+} candidates are reconstructed. Then a search for $D\pi$ or $D^*\pi$ resonances is performed in the mass difference distribution. Subsamples enriched in $Z \to c\bar{c}$ and $Z \to b\bar{b}$ events are selected by means of cuts on (1) the D meson energy, (2) the presence of tracks not from the Z decay vertex and not from the D^0 or D^+ candidate, and (3) the decay length significance of the $D^{(*)}\pi$ vertex. The $c\bar{c}$ and $b\bar{b}$ purities are measured by applying a b tagging algorithm to the hemisphere opposite the charm meson candidate.

Examples of the mass difference plots are shown in Figs. 6 and 7. Evidence for the narrow D_1 and D_2^* states is observed in the c-enriched sample, whereas only the D_1 gives a significant signal in the b-enriched sample.

The following preliminary results are obtained:

$$\frac{\Gamma(Z \to c/\bar{c} \to D_1^0/\bar{D}_1^0 X)}{\Gamma(Z \to q\bar{q})} B(D_1^0 \to D^{*+}\pi^-) = (0.37 \pm 0.075_{\text{stat}} \pm 0.06_{\text{syst}})\%$$

$$\frac{\Gamma(Z \to c/\bar{c} \to D_2^{*0}/\bar{D}_2^{*0} X)}{\Gamma(Z \to q\bar{q})} B(D_2^{*0} \to D^{*+}\pi^-) = (0.41 \pm 0.075_{\text{stat}} \pm 0.07_{\text{syst}})\%$$

$$\frac{\Gamma(Z \to c/\bar{c} \to D_2^{*0}/\bar{D}_2^{*0} X)}{\Gamma(Z \to q\bar{q})} B(D_2^{*0} \to D^{+}\pi^-) = (0.64 \pm 0.15_{\text{stat}} \pm 0.11_{\text{syst}})\%$$

$$\frac{\Gamma(Z \to c/\bar{c} \to D_2^{*\pm} X)}{\Gamma(Z \to q\bar{q})} B(D_2^{*+} \to D^{0}\pi^+) = (0.71 \pm 0.16_{\text{stat}} \pm 0.13_{\text{syst}})\%$$

$$\frac{\Gamma(Z \to b/\bar{b} \to D_1^0/\bar{D}_1^0 X)}{\Gamma(Z \to q\bar{q})} B(D_1^0 \to D^{*+}\pi^-) = (0.66 \pm 0.16_{\text{stat}} \pm 0.12_{\text{syst}})\%$$

$$\frac{\Gamma(Z \to b/\bar{b} \to D_2^{*0}/\bar{D}_2^{*0} X)}{\Gamma(Z \to q\bar{q})} B(D_2^{*0} \to D^{*+}\pi^-) < 0.13\% \quad (95\% \text{ CL})$$

$$\frac{\Gamma(Z \to b/\bar{b} \to D_2^{*\pm} X)}{\Gamma(Z \to q\bar{q})} B(D_2^{*+} \to D^{0}\pi^+) < 0.55\% \quad (95\% \text{ CL})$$

$$\text{n}(c \to D_{s1}^{\pm})\, B(D_{s1}^{+} \to D^{*+}K^0) = (0.27 \pm 0.09_{\text{stat}} \pm 0.02_{\text{syst}})\%$$

$$\text{n}(c \to D_{s1}^{\pm})\, B(D_{s1}^{+} \to D^{*0}K^+) = (0.51 \pm 0.18_{\text{stat}} \pm 0.06_{\text{syst}})\%$$

$$\text{n}(c \to D_{s2}^{*\pm})\, B(D_{s2}^{*+} \to D^{0}K^+) = (0.58 \pm 0.24_{\text{stat}} \pm 0.06_{\text{syst}})\%$$

$$\text{n}(b \to D_{s1}^{\pm})\, B(D_{s1}^{+} \to D^{*+}K^0) = (0.48 \pm 0.21_{\text{stat}} \pm 0.11_{\text{syst}})\%$$

$$\text{n}(b \to D_{s1}^{\pm})\, B(D_{s1}^{+} \to D^{*0}K^+) = (0.57 \pm 0.27_{\text{stat}} \pm 0.08_{\text{syst}})\%$$

$$\text{n}(b \to D_{s2}^{*\pm})\, B(D_{s2}^{*+} \to D^{0}K^+) = (1.0 \pm 0.4_{\text{stat}} \pm 0.2_{\text{syst}})\%.$$

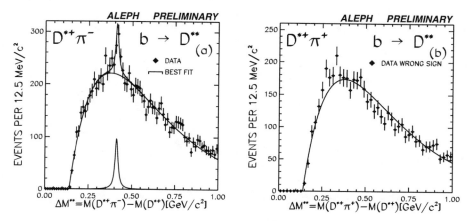

Figure 7: Mass difference distributions $\Delta M^{**} = M(D^*\pi) - M(D^*)$ in the b-enriched sample: (a) opposite-sign combinations ($D^{*+}\pi^-$) and (b) same-sign combinations ($D^{*+}\pi^+$). The solid curves show the results of a fit with a D_1 signal in the opposite-sign distribution.

Table 3: The narrow $L = 1$ charm mesons.

State	Mass (MeV/c^2)	Width (MeV/c^2)	Decays
D_1^0 $(c\bar{u})$	2422.2 ± 1.8	$18.9^{+4.6}_{-3.5}$	$D^*\pi$
D_2^{*0} $(c\bar{u})$	2458.9 ± 2.0	23 ± 5	$D^*\pi$, $D\pi$
D_1^+ $(c\bar{d})$	2427 ± 5	28 ± 8	$D^*\pi$
D_2^{*+} $(c\bar{d})$	2459 ± 4	25^{+8}_{-7}	$D^*\pi$, $D\pi$
D_{s1}^+ $(c\bar{s})$	2535.35 ± 0.34	<2.3	D^*K
D_{s2}^{*+} $(c\bar{s})$	2573.5 ± 1.7	15^{+5}_{-4}	D^*K, DK

In order to draw some conclusions about the production probabilities for the $L = 1$ states, we assume that the D_1 decays to $D^*\pi$ and the D_2^* decays to $D\pi$ and $D^*\pi$ only. We then obtain

$$n(c \to D_1) = (3.2 \pm 0.9)\%$$

$$n(c \to D_2^*) = (9.4 \pm 1.9)\%$$

$$n(b \to D_1) = (4.6 \pm 1.4)\%$$

$$n(b \to D_2^*) < 3.9\% \quad (95\% \text{ CL}).$$

Further assuming that the D_{s1} decays to D^*K only and $B(D_{s2}^{*+} \to D^0K^+) = 0.45$, we find

$$n(c \to D_{s1}) = (0.77 \pm 0.20_{\text{stat}} \pm 0.08_{\text{syst}})\%$$

$$n(c \to D_{s2}^*) = (1.3 \pm 0.5_{\text{stat}} \pm 0.2_{\text{syst}})\%$$

$$n(b \to D_{s1}) = (1.1 \pm 0.3_{\text{stat}} \pm 0.2_{\text{syst}})\%$$

$$n(b \to D_{s2}^*) = (2.2 \pm 0.8_{\text{stat}} \pm 0.5_{\text{syst}})\%.$$

These results show that production of the narrow $L = 1$ charm states is substantial in $Z \to c\bar{c}$ and $b\bar{b}$ events.

References

1. ALEPH Collaboration, contributed paper 937 to this conference.
2. ALEPH Collaboration, contributed papers 943 and 944 to this conference.
3. D. Atwood, G. Eilam, and A. Soni, Mod. Phys. Lett. A **11**, 1061 (1996).
4. FNAL E653 Collaboration, Phys. Lett. B **382**, 299 (1996).
5. CLEO Collaboration, Phys. Rev. D **58**, 32002 (1998).
6. L3 Collaboration, Phys. Lett. B **396**, 327 (1997).
7. DELPHI Collaboration, contributed paper 455 to the International Europhysics Conference on High Energy Physics, Jerusalem, Israel, August 1997.
8. R.M. Barnett et al. (Particle Data Group), Phys. Rev. D **54**, 1 (1996).
9. ALEPH Collaboration, Nucl. Instrum. Methods A **294**, 121 (1990); B. Mours et al., Nucl. Instrum. Methods A **379**, 101 (1996).
10. ALEPH Collaboration, Phys. Lett. B **313**, 535 (1993).

FIRST CHARM HADROPRODUCTION RESULTS FROM SELEX

THE SELEX COLLABORATION[1], presented by J. S. RUSS

Physics Department, Carnegie Mellon University, Pittsburgh, PA 15213, USA
E-mail: russ@cmphys.phys.cmu.edu

The SELEX experiment (E781) at Fermilab is a 3-stage magnetic spectrometer for the high statistics study of charm hadroproduction out to large x_F using 600 GeV Σ^-, p and π beams. The main features of the spectrometer are:

- high precision silicon vertex system
- broad-coverage particle identification with TRD and RICH
- 3-stage lead glass photon detector

Preliminary results on differences in hadroproduction characteristics of charm mesons and Λ_c^+ for $x_F \geq 0.3$ are reported. For baryon beams there is a striking asymmetry in the production of baryons compared to antibaryons. Leading particle effects for all incident hadrons are discussed.

1 Introduction

Understanding charm hadroproduction at fixed-target energies has been a difficult theoretical problem because of the complexities of renormalization scale, of parton scale, and of hadronization corrections. The recent review by Frixione, Mangano, Nason, and Ridolfi summarizes the theoretical situation, using data through 1996.[2] More recent data from Fermilab E791 (500 GeV π^- beam) greatly improves the statistical precision on charm meson production by pions, but E791 has not yet reported absolute cross sections or compared yields between charm species. In this first report of the SELEX hadroproduction results, we compare our pion results at 580 GeV with those from E791 as well as comparing SE-LEX pion data with our proton data at 550 GeV and Σ^- data at 620 GeV mean momenta. All SELEX data were taken in the same spectrometer with the same trigger. We limit this report to data having $x_F \geq 0.3$, where the spectrometer acceptance is essentially constant with x_F for all final states.

2 The Experiment

SELEX used the Fermilab Hyperon beam in negative polarity to make a mixed beam of Σ and π in roughly equal numbers. In positive polarity, protons comprised 92% of the particles, with π^+ making up the balance. The beam was run at 0 mrad production. The experiment aimed especially at understanding charm production in the forward hemisphere and was built to have good mass and vertex resolution for charm momenta from 100-500 GeV/c. The spectrometer is shown in Figure 1.

Interactions occurred in a target stack of 5 foils: 2 Cu and 3 C. Total target thickness was 5% of Λ_{int} for protons. Each foil was spaced by 1.5 cm from its neighbors. Decays occurring inside the volume of a target were

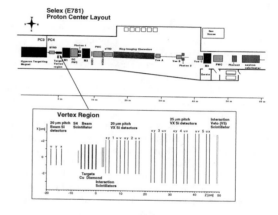

Figure 1: E781 Layout

rejected in this analysis. Interactions were selected by a scintillator trigger. The charm trigger was very loose, requiring only ≥ 4 charged tracks in a forward $10°$ cone and ≥ 2 hits in a hodoscope after the second analyzing magnet. We triggered on about 1/3 of all inelastic interactions.

A major innovation in E781 was the use of online selection criteria to identify reconstructable events. This experiment uses a RICH counter to identify p, K, or π after the second analyzing magnet. A computational filter used only these RICH-identifiable tracks to make a full vertex reconstruction in the vertex silicon and downstream PWCs. It selected events that had evidence for a secondary vertex. This reduced the data size (and offline computation time) by a factor of nearly 8 at a cost of about a factor of 2 in charm written to tape, as normalized from a study of unfiltered K_s^0 and Λ^0 decays. Most of the charm loss came from selection cuts that are independent of charm species or kinematic variables. No bias is expected from the filter. Filter operation depends on stable track reconstruction and detector alignment. These features were monitored online and were extremely stable throughout the run.

3 Charm Selection

All data reported here result from a preliminary pass through the data, using a production code optimized for speed but not efficiency. Final yields will be higher than these preliminary results. However, our simulations indicate that the inefficiency does not affect the kinematic features of the results for $x_F \geq 0.3$. For all final states, the charm selection required that the primary vertex lie within the target region and that the secondary vertex occur before the start of the VX silicon. At our high energy, this latter cut removed a number of $D^\pm$ events which can be recovered later.

In this analysis secondary vertices were reconstructed when the vertex χ^2 for the ensemble of tracks was inconsistent with a single primary vertex. All combinations of tracks were investigated, and every secondary vertex candidate was tested against a reconstruction table that listed acceptable particle identification tags for a charm candidate, track selection criteria necessary (RICH identification for a proton, for example), and any other selections, e.g., minimum significance cut for primary/secondary vertex separation. Selected events were written to output files and the essential reconstruction features for each identified secondary vertex were saved in a PAW-like output structure for quick pass-II analysis. All data shown here come from analysis using this reduced output.

3.1 System performance for charm

Vertex resolution is a critical factor in charm experiments. The primary and secondary longitudinal vertex resolution for all data in a typical run of the experiment are shown in Figure 2. The lower plot shows the primary vertex distribution overlaid on rectangles that represent the physical placement of the 5 targets. The average relativistic transformation factor from lab time to proper time for charm states in these data is 100. This spatial resolution corresponds to about a 20 fs proper time resolution for lifetime studies.

Another important factor in charm studies at large x_F is having good charm mass resolution at all momenta. Figure 3 shows that the measured width of the $D^0 \rightarrow K^- + \pi^+$ is about 10 MeV for all x_F.

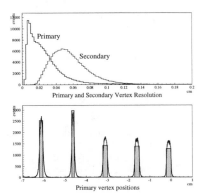

Figure 2: Typical Primary and Secondary Vertex Error Distributions

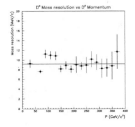

Figure 3: D^0 Mass Resolution versus D^0 Momentum

Finally, we depend on the RICH to give correct identification of K and p decay prongs. Figure 4 shows the π/K separation in interaction data for 100 GeV/c tracks, a typical momentum for prongs from our charm states. The RICH gives π/K separation up to 165 GeV/c (2σ confidence level).[3]

4 Overall Charm Features at Large x_F

Previous high-statistics charm production results from pions [4] and protons [5] have emphasized central production, although both NA32 and E791 have presented results for $x_F \geq 0.5$. SELEX and E769 are the only high energy experiments reporting results from three different beam particles with identical systematics. The important features of the SELEX data can be seen at a glance in Figure 5 for the charged states $D^\pm$, Λ_c^+, and $\overline{\Lambda_c^-}$ produced respectively by Σ^-, π^-, and proton beams. The pion data show comparable particle and antiparticle

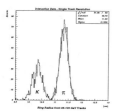

Figure 4: RICH K and π Response at 100 GeV/c

yields both for charm mesons and for charm baryons, as reported by NA32 at lower energy. [4]

It remains a surprising feature of hadroproduction that one finds significant antibaryon production from pions even at $x_F \geq 0.5$. The source of the antiquark pair which combines with the charmed antiquark has been the subject of considerable theoretical speculation. The pion provides a $\bar{u}$ valence quark which can contribute in some models. No present model gives an adequate description. There is good agreement for the $D^{\pm}$ production asymmetry integrated over $x_F \geq 0.3$ between these preliminary results and the E791 results [4]. E791 has not published Λ_c^+ asymmetry results. Their observations are consistent with those shown here.[6]

The relative efficiencies for each beam particle are almost the same in this x_F region, so that one can quote the ratio of the cross sections even though we have not yet determined absolute yields. The normalization between different incident hadrons depends on the number of incident beam particles for each data sample and on the total inelastic cross section for each beam particle. We use 34 mb for the proton inelastic cross section, 27 mb for Σ^-, and 22 mb for π^- to compare yields for different beam particles. For these data the relative yields of selected charmed states, normalized to pion production, are given in Table 1. No errors are included in this preliminary analysis. Note that this table does not directly provide information about the relative yields for the different charmed states.

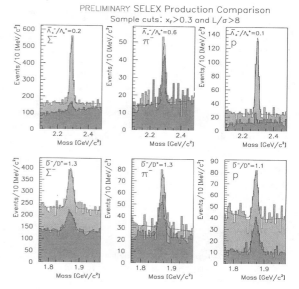

Figure 5: Charm and Anticharm mass distributions for Σ^-, π^-, and p beams in modes $\Lambda_c^+ \to pK^-\pi^+$ or c.c. and $D^+ \to K^-\pi^+\pi^+$ or c.c.

is the observation that baryon beams are very effective charm baryon producers, at least at large x_F. Also, for the states listed here, the Σ^- beam has yields comparable to pions, except for the non-leading case of the D^+. We have not yet compiled the yields for the c-s-q baryons, where we expect the Σ^- beam relative yields will large.

The previous table gave the relative efficacy of each beam particle for producing a given charm state at large x_F. It does *not* compare relative yields of the different charm states for the same beam. As can be seen from Figure 5, there are strong asymmetries. These are tabulated in Table 2. Again, errors are omitted at this stage of analysis.

Table 1: Relative Charmed Particle Yields for $x_F \geq 0.3$ versus beam type

Relative Charmed Particle Yields	p	π^-	Σ^-
$\overline{\Lambda_c^-}$	0.25	1.0	1.1
Λ_c^+	0.8	1.0	1.2
D^-	0.4	1.0	0.8
D^+	0.2	1.0	0.4

Perhaps the most surprising result from this table

Table 2: Charmed Particle Antiparticle Ratios for $x_F \geq 0.3$ versus beam type

Yield Ratio	p	π^-	Σ^-
$\overline{\Lambda_c^-}/\Lambda_c^+$	0.1	0.6	0.2
D^-/D^+	1.1	1.2	1.3

Table 2 shows for both baryon beams there are striking differences in production asymmetries for charm baryons compared to the pion beam. For charm mesons, that is not the case. Baryon beams, which have no valence antiquarks, show strong suppression of antibaryon production, compared to pions. This feature was not observed by NA27 in 400 GeV pp collisions. They reported comparable baryon/antibaryon production but had only a few events, all in the central region. No other proton data exist for charm baryons. The WA89 results for charm baryon production by Σ^- are consistent with our findings. [7]

The D^- and Λ_c^+ are leading hadrons in the sense that all 3 beam hadrons *may* contribute at least one valence quark to the final state. The large difference in the Λ_c^+ asymmetry between the meson beam (largely symmetric) and the baryon beams (very asymmetric) is a new issue for charm hadroproduction analysis, which has assumed that there is a universal baryon/meson fraction for all incident hadrons. [2]

5 Summary

The SELEX experiment complements previous charm hadroproduction experiments by exploring different regions of production phase space and by using different beams. The early results already show some noteworthy new features of charm production. Further studies of different states and details of single- and double-differential charm production distributions are underway and will be reported at meetings in the fall.

Further analysis will extend the x_F coverage down to about 0.1, to enhance overlap with other experiments and to increase statistics. Also, other charm baryon states are being analyzed and results will be reported later.

1.. Carnegie Mellon University, Fermilab, University of Iowa, University of Rochester, University of Hawaii, University of Michigan-Flint, Petersburg Nuclear Physics Institute (Russia), Institute for Theoretical and Experimental Physics (Moscow), Institute for High Energy Physics (Protvino), Moscow State University, University of Sao Paulo, Centro Brasileiro de Pesquisas Fisicas, Universidade Federal de Paraiba, Insitute of High Energy Physics (Beijing), University of Bristol, Tel Aviv University, Max Planck Institut fuer Kernphysik (Heidelberg), University of Trieste and INFN, University of Rome and INFN, Universidad Autonoma de San Luis Potosi, Bogazici University

2. S. Frixione, M. Mangano, P. Nason, and G. Ridolfi, "Heavy Quark Production" in "Heavy Flavours II", A. J. Buras and M. Lindner, eds., World Scientific Publishing Co, Singapore (1997); see also preprint hep-ph/9702287

3. Fermilab-Pub-98/299-E, submitted to Nucl Instr & Methods.

4. NA32: 230 GeV; S. Barlag, et al., Z. Phys. **C39**(1988)451; *ibid* **C49**(1991)555; Phys. Lett. **B257** (1991)519
   E769: 250 GeV; G.A. Alves, et al., Phys. Rev. Lett. **69** (1992)3147; *ibid* **77** (1996)2388; 2392
   WA92: 330 GeV; M. I. Adamovich, Phys. Lett. **B348** (1995)256; *ibid* **B385**(1996)487; Nucl. Phys. **B495**(1997)3
   E791: 500 GeV; E. M. Aitala, et al., Phys. Lett. **B411**(1997)230; *ibid* **B403**(1997)185; *ibid* **B371**(1996)157

5. E769: 250 GeV; G.A.Alves, et al., Phys. Rev. Lett. **77** (1996)2388;2392
   NA27: 400 GeV; M. Aguilar-Benitez, et al., Z. Phys. **C40**(1988)321

6. Simon Kwan, private communication

7. M. I. Adamovich, et al., submitted to Eur. Phys. J. C; preprint hep-ex/9803021

FURTHER EVIDENCE ON GLUONIC STATES FROM THE WA102 EXPERIMENT

A.V.Singovsky

*Laboratoire de Annecy-le-Vieux de Physique des Particules, Annecy-le-Vieux, France
and IHEP, Protvino, Russia
E-mail: Alexander.Singovski@cern.ch*

representing the WA102 Collaboration [1]

A study of central meson production is a convenient way of investigating gluonic matter. In addition to the expected enrichment of glueball production via the Double Pomeron process, it also shows some kinematic peculiarity which can be used to separate all glueball candidates from undisputed $q\bar{q}$ mesons. The difference in transverse momentum (dP_T) of the exchanged particles can be considered as a possible glueballs-$q\bar{q}$ filter.

1 Introduction

For more than a decade experimental attempts to detect glueballs, in particular the lightest scalar glueball, have not yet brought a final success. Glueball discovery remains an experimental challenge although the existence of such objects is certain from the QCD point of view.

The situation can be well illustrated by the scalar glueball case. The scalar glueball mass predictions made by different models [2] are rather consistent and point to about 1500 MeV with typically 100-150 MeV errors and about 200 MeV spread over different predictions. This mass region is very well investigated by several generations of experiments and one can take as a realistic approximation that all scalar states between 1 and 2 GeV are already observed. The number of established scalars is greater than what is needed to fill the 0^{++} quark model nonet. This fact is considered as a clear indication of the existence of the scalar gluonic state in 1-2 GeV mass region. But none of the observed scalar mesons can be unambiguously identified as a glueball although some of them have exotic properties. This can be understood if the scalar glueball is strongly mixed with the quark neighbours and instead of pure gg state we have a certain fraction of glue in several $q\bar{q}$ scalar states. Hence the subject of "glueball spectroscopy" is not a search for new particles with fully exotic properties but the detailed investigation of all meson states in the given mass region in preferably all decay modes in order to quantify glue-induced discrepancies with the standard quark model predictions.

Mixed $q\bar{q}$-gg states can be produced via their quark or glue component in practically any hadroproduction reaction. However it is preferable for glueball spectroscopy experiments to use some of the glue-rich production mechanisms, like $p\bar{p}$-annihilation or Double Pomeron exchange (DPE) to ensure the presence of the subject of

investigation in the collected data sample.

2 The WA102 experiment

The WA102 experiment was designed to study exclusive final states formed in the reaction

$$pp \rightarrow p_f X^0 p_s,$$

where the subscripts f and s refer to the fastest and slowest particles in the laboratory frame respectively and X^0 represents the central system. Such reactions are expected to be mediated by double exchange processes where both Pomeron and Reggeon exchange can occur. Assuming the Pomeron is a colour singlet gluonic system all meson states, pure $q\bar{q}$ or gg and mixed $q\bar{q}$-gg, should be produced in such a reaction. Theoretical predictions [3] of the evolution of the different exchange mechanisms with centre of mass energy, $\sqrt{s}$, suggest that

$$\sigma(RR) \sim s^{-1},$$
$$\sigma(RP) \sim s^{-0.5},$$
$$\sigma(PP) \sim \text{constant},$$

where RR, RP and PP refer to Reggeon-Reggeon, Reggeon-Pomeron and Pomeron-Pomeron exchange respectively. Hence the glueball-contained states produced via DPE should have a production cross section energy dependence different from the pure quark states produced via Double Reggeon exchange. This effect, specific to central production, gives an additional very useful tool for the identification of gluonic states. But the most promising peculiarity of the central production is the recently observed kinematic filter.

3 A kinematic filter

The basis of the possible glueball-$q\bar{q}$ filter suggested recently by the WA102 collaboration is the experimental observation that the central production of meson states

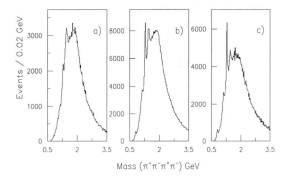

Figure 1: The $\pi^+\pi^-\pi^+\pi^-$ mass spectrum for a) $dP_T < 0.2$ GeV, b) $0.2 < dP_T < 0.5$ GeV and c) $dP_T > 0.5$ GeV

is sensitive to the kinematic variables representing production particles - Reggeons and/or Pomerons. The variable currently used for analysis is the difference of the exchange particles transverse momentum, defined as

$$dP_T = \sqrt{(P_{y1} - P_{y2})^2 + (P_{z1} - P_{z2})^2}$$

where Py_i, Pz_i are the y and z components of the momentum of the ith exchanged particle in the pp centre of mass system. Although other relevant variables, as for example, the angle between exchange particles can be used as well.

It was found that the mass spectra of the centrally produced systems like 2π, $2K$ or 4π depend dramatically on dP_T. This effect is illustrated in fig 1. In the $\pi^+\pi^-\pi^+\pi^-$ mass spectrum the $f_1(1285)$ signal disappears as dP_T goes to zero while the two structures at 1.5 and 2.0 GeV regions, attributed to $f_0(1500)$ and $f_2(1930)$, remain.

The full setup Monte-Carlo simulation, including trigger, detector acceptance, reconstruction procedure and applied off-line cuts shows practically no or very small dependence of the acceptance as a function of dP_T for the given channel and hence the observed resonance production dP_T dependence cannot be explained as acceptance effects. It is also not an artifact of the WA102 experiment since the effect was confirmed by the NA12/2 experiment.

4   Summary of the dP_T filtering

In order to understand the nature of the kinematic filtering, production of all resonances detected by WA102 experiment was investigated as a function of dP_T. Due

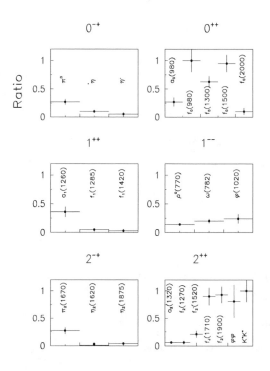

Figure 2: The ratio of the amount of resonance with $dP_T < 0.2$ GeV to the amount with $dP_T > 0.5$ GeV.

to the limited statistics dP_T dependence was presented as a relative production cross-section in three intervals: less than 0.2 GeV, 0.2 to 0.5 GeV and more than 0.5 GeV. Results are summarized in the Table 1.

It can be noticed that all undisputed $q\bar{q}$ states, i.e. $\rho^0(770)$, $f_2(1270)$, $f_1(1285)$, $f_2'(1525)$ etc., are suppressed at low dP_T, whereas all glueball candidates $f_0(1300)$, $f_0(1500)$, $f_J(1710)$ and $f_2(1930)$ survive. Figure 2 shows the ratio of the number of events for $dP_T < 0.2$ GeV to the number of events for $dP_T > 0.5$ GeV for each resonance considered. All states can be arranged into three groups by the value of this ratio: all undisputed $q\bar{q}$ states which can be produced via DPE have the ratio value about 0.1. States with I=1 or G parity negative, which can not be produced by DPE, have slightly higher value around 0.25. All states with suspected gluonic com-

Table 1: Resonance production as a function of dP_T expressed as a percentage of its total contribution

J^{PC}	Resonance	$dP_T < 0.2$ GeV	$0.2 < dP_T < 0.5$ GeV	$dP_T > 0.5$ GeV
0^{-+}	π^0	12 ± 2	44 ± 2	44 ± 2
	η	6 ± 2	34 ± 2	60 ± 3
	η'	3 ± 2	32 ± 2	64 ± 3
0^{++}	$a_0(980)$	14 ± 4	35 ± 4	51 ± 7
	$f_0(980)$	22 ± 2	56 ± 3	22 ± 3
	$f_0(1300)$	20 ± 2	48 ± 2	32 ± 4
	$f_0(1500)$	23 ± 2	53 ± 3	24 ± 4
	$f_0(2000)$	5 ± 3	43 ± 5	52 ± 5
1^{++}	$a_1(1260)$	13 ± 3	51 ± 4	36 ± 3
	$f_1(1285)$	3 ± 1	35 ± 2	61 ± 4
	$f_1(1420)$	2 ± 2	38 ± 2	60 ± 4
1^{--}	$\rho(770)$	8 ± 2	38 ± 2	54 ± 3
	$\omega(782)$	10 ± 2	40 ± 2	49 ± 3
	$\phi(1020)$	10 ± 3	48 ± 3	42 ± 4
2^{-+}	$\pi_2(1670)$	11 ± 2	48 ± 4	40 ± 4
	$\eta_2(1620)$	2 ± 1	42 ± 6	54 ± 5
	$\eta_2(1875)$	1 ± 1	36 ± 7	63 ± 7
2^{++}	$a_2(1320)$	4 ± 4	35 ± 3	61 ± 5
	$f_2(1270)$	4 ± 2	25 ± 2	71 ± 3
	$f_2'(1520)$	11 ± 3	37 ± 3	52 ± 4
	$f_J(1710)$	26 ± 3	45 ± 2	29 ± 4
	$f_2(1900)$	26 ± 2	46 ± 3	28 ± 4
	$\phi\phi$	22 ± 5	51 ± 4	27 ± 4
	$K^*(892)\bar{K}^*(892)$	23 ± 3	54 ± 3	23 ± 2

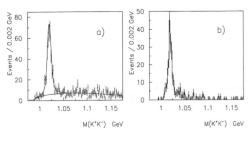

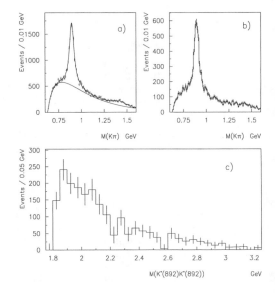

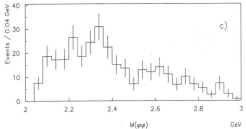

Figure 3: a) the K^+K^- mass spectrum (four combinations per event) for the data sample with at least three central particles identified as kaons, b) the K^+K^- effective mass of one K^+K^- combination if the other lie in ϕ mass band, c) the $\phi\phi$ effective mass spectrum

Figure 4: a) the $K^\pm\pi^\mp$ mass spectrum, b) the $K^\pm\pi^\mp$ mass spectrum after selecting the $K^\mp\pi^\pm$ to lie in $K^*(892)$ mass band, c) the $K^*(982)\bar{K}^*(982)$ effective mass spectrum

ponent have a significantly higher value of this ratio.

The separation of states by this ratio is valid for all resonances considered. It is interesting to note that enigmatic $f_0(980)$ has a "glueball-like" ratio value as well as the $\phi\phi$ and $K^*(982)\bar{K}^*(982)$ systems.

4.1 $\phi\phi$ and $K^*(982)\bar{K}^*(982)$ systems

It was recently suggested that the observed dP_T effect could be due to the fact that the production mechanism is through the fusion of two vector particles [4,5] and, hence it is interesting to study the dP_T dependence of systems decaying to two vectors. Two such systems, $\phi\phi$ and $K^*(982)\bar{K}^*(982)$ have been studied. The $\phi\phi$ system has the added interest that the observation of glueball candidates has been claimed in this channel [7].

The $\phi\phi$ system decaying to $K^+K^-K^+K^-$ produced in central pp interaction was studied [6] with a factor ten increase over previously published data samples. Kaons where identified by the threshold Cherenkov detector. The K^+K^- combinatorial mass and the $\phi\phi$ effective mass spectrum are shown in fig. 3. It was found that $\phi\phi$ production dominates in the $K^+K^-K^+K^-$ central pro-

duction at 450 GeV and the ratio of the ϕK^+K^- to $\phi\phi$ cross sections was estimated to be $\sigma(\phi K^+K^-)/\sigma(\phi\phi) \approx 0.8$. The $K^+K^-K^+K^-$ system angular distributions where analyzed to determine the spin-parity of the intermediate $\phi\phi$ states. The analysis shows the best chi-squared for $J^{PC} = 2^{++}$ (with $J^{PC} = 0^{++}$ and $J^{PC} = 0^{+-}$ ruled out).

A study of the reaction $pp \to p_f p_s (K^+K^-\pi^+\pi^-)$ shows evidence for the $K^*(982)\bar{K}^*(982)$ system production [8]. The $K\pi$ combinatorial mass and $K^*(982)\bar{K}^*(982)$ mass spectra are shown in fig. 4. The angular analysis suggests $J^P = 2^+$ for the $K^*(982)\bar{K}^*(982)$ intermediate states.

Although the effect of kinematic filtering is well established experimentally, the underlying physics behind it is not fully understood. Close and Kirk [9] have suggested that low dP_T can favour gluon coupling of the exchange particles (Pomerons or Reggeons) into the final state, enriched by gg in this case. It was also pointed out that the results may be explained if the particles exchanged in the formation of the central system carry non-zero spin [4,5]. Hence the results may have implication for the spin structure of the Pomeron. The $\phi\phi$ and $K^*(982)\bar{K}^*(982)$

data fit well to this scheme.

5 Conclusions

A study of central pp interactions performed by the WA102 experiment show that central production of meson states is sensitive to the kinematic variables representing the exchanged particles, in particular dP_T. This possible glueball-$q\bar{q}$ filter together with the production cross section energy dependence could bring to "glueball spectroscopy" new information capable of resolving the long-standing glueball puzzle.

The ratio of the production cross section at low and high dP_T intervals separate all undisputed $q\bar{q}$ states from glueball candidates. The underlying physics behind the dP_T effect is still not fully understood, although some models are already proposed.

References

1. The WA102 Collaboration: D. Barberis, W. Beusch, F.G. Binon, A.M. Blick, F.E. Close, K.M. Danielsen, A.V. Dolgopolov, S.V. Donskov, B.C. Earl, D. Evans, B.R. French, T. Hino, S. Inaba, A.V. Inyakin, T. Ishida, A. Jacholkowski, T. Jacobsen, G.V. Khaustov, T. Kinashi, J.B. Kinson, A. Kirk, W. Klempt, V. Kolosov, A.A. Kondashov, A.A. Lednev, V. Lenti, S. Maljukov, P. Martinengo, I. Minashvili, K. Myklebost, T. Nakagawa, K.L. Norman, J.M. Olsen, J.P. Peigneux, S.A. Polovnikov, V.A. Polyakov, V. Romanovsky, H. Rotscheidt, V. Rumyantsev, N. Russakovich, V.D. Samoylenko, A. Semenov, M. Sené, R. Sené, P.M. Shagin, H. Shimizu, A.V. Singovsky, A. Sobol, A. Solovjev, M. Stassinaki, J.P. Stroot, V.P. Sugonyaev, K. Takamatsu, G. Tchlatchidze, T. Tsuru, M. Venables, O. Villalobos Baillie, M.F. Votruba, Y. Yasu.

2. G. Bali et al. (UKQCD), *Phys. Lett.* B **309**, 378 (1993);
   D. Weingarten, hep-lat/9608070;
   J. Sexton et al. *Phys. Rev. Lett.* **75**, 4563 (1995).

3. S.N. Ganguli and D.P. Roy *Phys. Rep.* **67**, 203 (1980).

4. F.E. Close *Phys. Lett.* B **419**, 387 (1998)

5. P. Castoldi, R. Escribano and J.-M. Frere *Phys. Lett.* B **425**, 359 (1998).

6. D. Barbetis et al., *Phys. Lett.* B **432**, 436 (1998).

7. A.Etkin et al. *Phys. Lett.* B **201**, 568 (1988).

8. D. Barbetis et al., hep-ex/9807021, to be published in Phys. Lett.

9. F.E. Close and A. Kirk *Phys. Lett.* B **397**, 333 (1997).

EXOTIC MESON PRODUCED IN ANTIPROTON-NEUTRON ANNIHILATION: $\hat{\rho} \to \eta\pi$

W. DÜNNWEBER (for the Crystal Barrel Collaboration)

Universität München, D-85748 Garching, Germany
E-mail: duennweb@physik.uni-muenchen.de

We report on an exotic resonance, decaying into $\eta\pi$, observed in the $\bar{p}n$ annihilation channel $\eta\pi^-\pi^0$. Antiprotons from LEAR were stopped in a liquid deuterium target. Using the Crystal Barrel Detector, 53×10^3 events of the type $\bar{p}d \to \eta\pi^-\pi^0 p$ with proton spectator momentum < 100 MeV/c were fully reconstructed. The $\eta\pi$ P-wave, carrying quantum numbers ($J^{PC} = 1^{-+}$) not accessible for a quark-antiquark system, is produced with a large relative rate (11%) in this highly selective annihilation channel. Together with the other two dominant resonances, $\rho(770)$ and $a_2(1320)$, it generates a simple pattern in the Dalitz plot of the intensity distribution. The observed interferences pin down the resonance characteristics of the $\eta\pi$ P-wave. A partial wave analysis yields the Breit-Wigner parameters $m = (1400 \pm 20 \pm 20)MeV/c^2$ and $\Gamma = (310 \pm 50^{+50}_{-30})MeV/c^2$.

1 Introduction

Mesons with exotic quantum numbers $J^{PC} = 0^{--}, 0^{+-}, 1^{-+}, 2^{+-}, ...$ have necessarily non-$q\bar{q}$ structure due to the generalized Pauli principle. Glueballs and quark-gluon hybrids, but also diquonia ($qq\bar{q}\bar{q}$) or meson molecules can attain these quantum numbers. Unlike the 0^{++} glueball, mesons with exotic quantum numbers cannot mix with ordinary ($q\bar{q}$) mesons. Thus the identification of a meson with exotic J^{PC} gives unambiguous evidence for other or more constituents than one quark and one antiquark. The present contribution reports on Crystal Barrel's recent observation[1] of an exotic meson with $J^{PC} = 1^{-+}$ decaying into $\eta\pi$.

The $\eta\pi$ system is attractive for the exotics search since its P-wave must carry the non-$q\bar{q}$ quantum numbers $J^{PC} = 1^{-+}$ (G= -). In the experiment it is produced in the annihilation

$$\bar{p}n \to \pi^-\pi^0\eta$$

The crucial question is whether a resonance exists in the $\eta\pi$ P-wave (Fig.1). Utilizing a different production mechanism, previous work[2-4] on pion-induced reactions on nucleons reported resonant behaviour of the $\eta\pi$ P-wave around 1400 MeV/c^2. While some of these and other experiments seemed to be mutually inconsistent, the recent detailed results from BNL[4] are in agreement with the more qualitative results from VES[3]. Clearly, the first exotic meson with an exotic address deserves various inquiries by alternative experiments before it reaches full respectability.

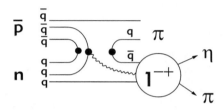

Figure 1: Quarkline diagram of $\bar{p}n$ annihilation into π and the exotic 1^{-+} resonance ("$\hat{\rho}$"), which may be a hybrid (as shown here) or a ($qq\bar{q}\bar{q}$) state or a meson molecule.

2 Experimental Results and Partial Wave Analysis

Antiprotons with a momentum of 200 MeV/c from the LEAR facility at CERN were stopped in a liquid deuterium target. The Crystal Barrel Detector is equipped for charged particle tracking in a magnetic field of 1.5 T and proton spectroscopy with a highly granular CsI calorimeter, both with close-to-4π geometry[1]. (After closure of LEAR it is presently being rebuilt at ELSA in Bonn.)

From a total of 8.2 million 1-prong events, corresponding to about 10^8 $\bar{p}d$ events, a sample of 52576 events of the type $\bar{p}d \to \pi^-\pi^0(\gamma\gamma)\eta(\gamma\gamma)p$ with a proton spectator momentum <100 MeV/c was fully reconstructed and kinematically selected. Background from misidentified events of other origin contributes about 0.5%. The momentum cut was chosen to guaranty the spectator role of the proton, i.e. the negligibility of final state interactions with the produced mesons. This was confirmed by the observed independence of our results on variations of this cut and by the symmetry of the intensity distribution with respect to $\pi^0\pi^-$ exchange.

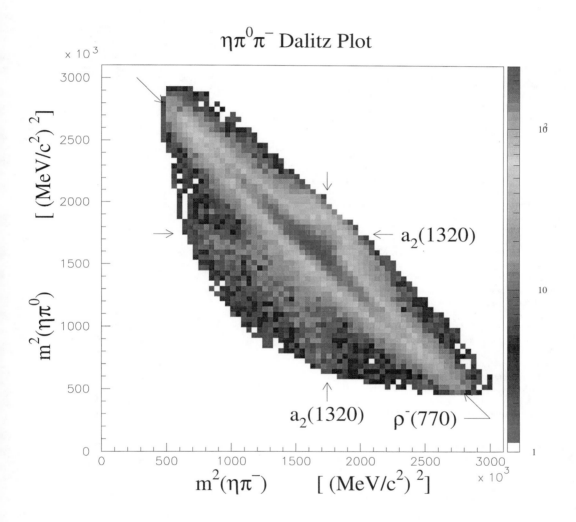

Figure 2: Dalitz plot of the 52576 $\eta\pi^0\pi^-$ events from $\bar{p}d \to \eta\pi^0\pi^- p_{spectator}$. The experimental intensity distribution was binned and acceptance corrected.

The experimental intensity distribution is displayed as a Dalitz plot in Fig.2. A simple pattern is observed which is dominated by a diagonal $\pi^-(770)$ band and two broad orthogonal bands in the region of the $a_2^{-/0}$ (1320). The latter show large modulations typical of interference effects.

The partial wave analysis of this intensity distribution assumes intermediate states of $\pi^-\pi^0$ resonances with a recoiling η or $\eta\pi$ resonances with recoiling π. Selection rules, based on J, P, G, and I conservation, strongly restrict the model space of intermediate states of the present annihilation channel. All allowed known or candidate resonances with nominal mass inside or close to the phase space boundary were tried in the fits, using relativistic Breit-Wigner resonance parametrizations[1]. A simple model space (A) containing only the $\rho^-(770)\eta$, $a_2(1320)\pi$ and $(\eta\pi)_P\pi$ intermediate states for each of the two incoherent initial $\bar{p}n$ states, $^3S_1$ and $^1P_1$, is sufficient for a good fit ($\chi^2/N_{dof} = 506/391 = 1.29$). The resonant $\eta\pi$ P-wave contributes a significant fraction, summing up to 11% plus the interferences. Its intensity distribution, including the interferences with the other two dominant resonances, shows a characteristic pattern covering the whole phase space (Fig.3). In addition to the accumulation of intensity in the central region of the DP, its most significant contributions arise from interference with the ρ^- and a_2 resonances. Its fitted mass of about $1400 MeV/c^2$ and width of about $300 MeV/c^2$ are not significantly affected by various extensions of the model space. With a model space that does not include the $\eta\pi$ P-wave but all other allowed known resonances or candidates (a_0, excited ρ^- and a_2) no satisfactory fits are obtained and the χ^2 distribution gives evidence for missing interferences (Fig.4). It is noted that conservation laws do not allow a_0 production for the dominant $^3S_1$ initial state. In contrast, the fit model A, which includes the $\eta\pi$ P-wave, yields structureless χ^2 distributions with merely statistical fluctuations[1].

3 Discussion of the Resonant Behaviour of the $\eta\pi$ P-wave

The interference of the $\eta\pi$ P-wave with both the ρ^- and the a_2 resonances pins down the resonance characteristics. The relative phase of the latter two resonances is well determined by their crossing in the DP. It is close to 0^0 for the $^3S_1$ initial state. Both probe the $\eta\pi$ phase motion in different regions. Constructive and destructive interference on opposite sides of the ρ^- band center is visible in Fig.3. Moving along a parallel just below the ρ^- band, one observes the rise and the

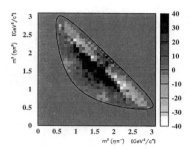

Figure 3: Intensity distribution of the $\eta\pi$ P-wave, as obtained by substracting from the experimental intensity the ρ^- and a_2 contributions of fit model A.

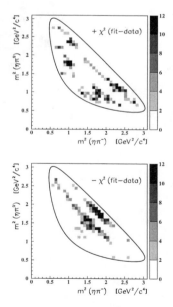

Figure 4: Deviations between the data and a fit that does not include the $\eta\pi$ P-wave but all other allowed resonances. Upper panel: fit exceeds data, lower panel: reverse.

fall of the constructive term, which reflects the almost complete phase rotation of the $\eta\pi$ resonance. Moving along an $m^2(\eta\pi) = (1.7 - 1.8)GeV^2/c^4$ line, one finds in addition a minimum and a maximum close to the lower and upper phase space boundaries, respectively. These interferences arise from the overlap with the a_2. The interference with the ρ^- resonance constitutes the essential difference from studies of pion-induced reactions[2-4] which are confined to the interference with a dominant a_2.

As an alternative model of the $\eta\pi$ P-wave, an effective range amplitude is found to yield convergent or divergent fits in the range of scattering parameters that characterize resonant or nonresonant behaviour, respectively. The resonant solution is practically identical to the Breit-Wigner fit amplitude. Its phase motion shows the typical resonance behaviour in an Argand diagram (Fig.5).

The opening of another channel under a broad resonance may create an apparently narrower structure, a so-called cusp. Using an amplitude parametrization that treats this coupled-channels effect in an approximative fashion, we have investigated a possible threshold effect of the $f_1\pi$ system with threshold mass 1420 MeV/c^2. With the rather large $\Gamma_{\eta\pi} = 1GeV/c^2$ a passable fit is obtained (with χ^2 increasing by 42 with respect to fit model A), yielding a resonance mass close to 1400 MeV/c^2. However, the corresponding $f_1\pi$ strength could not be detected, so far. Since additional 1^{-+} resonant structure is found[4] in the mass range 1600 - 1700 MeV/c^2 in $\rho\pi$ and is also indicated[3] in $\eta'\pi$, we investigated the possibility that a broad resonance in that range produces a cusp at 1400 MeV/c^2 in the present channel. However, the corresponding Flatté fit clearly fails to account for the data: χ^2 increases by 120 with respect to fit model A for m = 1700, $\Gamma = 1000$ MeV/c^2. However, all observations are consistent with the existence of two 1^{-+} resonances, one at 1400 MeV/c^2, the other at 1600 - 1700 MeV/c^2.

To summarize, we have extracted the following parameters for the $\hat\rho$ resonance with $J^{PC} = 1^{-+}$ from a simple Breit Wigner fit (model A) that accounts well for the data:
m = $(1400 \pm 20_{stat} \pm 20_{syst})MeV/c^2$.
$\Gamma = (310 \pm 50_{stat} + 50/ - 30_{syst})$ MeV/c^2.
Additional resonance structure above $\approx 1.7 GeV/c^2$ is not excluded, because of the present phase space limitation. It may be that a 1^{-+} doublet of dominant hybrid and molecular type structure will be required to reconcile all data. Finally, an interesting entrance channel selectivity is noted: in the $\bar{p}p \to \eta\pi^0\pi^0$ annihilation channel, the

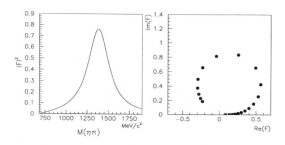

Figure 5: Lhs: $(\eta\pi)_P$ effective range amplitude[6] fitted to the data. Rhs:Corresponding Argand plot, showing the imaginary versus the real part of the effective range amplitude . The range from m = 690 to 1800 MeV/c^2 is divided into equal Δm steps. An almost complete anticlockwise phase rotation is observed.

$J^{PC} = 1^{-+}$ resonance shows up with a smaller relative intensity than in the present channel, but with consistent resonance parameters[7]. In the atomic S-wave, which is dominant in the annihilation at rest, only the $^1S_0$ $\bar{p}p$ state and the $^3S_1$ $\bar{p}n$ state can decay into the present $\eta\pi\pi$ channels. The observed entrance channel selectivity is probably due to different angular momentum configurations arising from the different initial states.

References

1. A.Abele *et al.*(Crystal Barrel Collaboration), Phys.Lett. **B423** (1998) 175.
2. D.Alde et al., Phys.Lett.**B205** (1988) 397.
3. G.M.Beladidze et al.,Phys.Lett.**B313** (1993) 276.
4. D.R.Thompson *et al.*,Phys.Rev.Lett.**79** (1997) 1630; D.P. Weygand, these proceedings
5. S.U.Chung et al., Ann.d.Physik 4 (1995) 404.
6. K.Hüttmann, Diploma thesis, Univ. München (1997)
7. C.Amsler *et al.*,(Crystal Barrel Collaboration), Phys.Lett. **B333** (1994), and to be published.

Evidence for a $J^{PC} = 1^{-+}$ Exotic Mesons from BNL E852

D. P. Weygand

for the E852 collaboration

Jefferson Lab, Newport News, Virginia 23606, USA
and
Rensselear Polytechnic Institute, Troy, New York 12180, USA

The exclusive reactions $\pi^- p \to \eta \pi^- p$ and $\pi^- p \to \pi^+ \pi^- \pi^- p$ at 18 GeV/c have been studied at the MultiParticle Spectrometer at Brookhaven National Laboratory. Partial wave analyses have been performed on the 47,200 $\eta \pi^- p$ events and 250,000 $\pi^+ \pi^- \pi^- p$ events. In $\eta \pi^-$ a $J^{PC} = 1^{-+}$ exotic state of mass $(1370 \pm 16 \, {}^{+50}_{-30})$ MeV/c^2, and $\Gamma = (385 \pm 40 \, {}^{+65}_{-105})$ MeV/c^2 is seen to interfere with the $a_2(1320)$ with resonance-like phase motion, while in $\pi^+ \pi^- \pi^-$ a $J^{PC} = 1^{-+}$ exotic wave is observed with resonance-like behavior at a mass of $1593 \pm 8^{+29}_{-47}$ MeV/c^2 and $\Gamma = 168 \pm 20^{+150}_{-12}$ MeV/c^2 showing interference with the $a_1(1260)$, the $a_2(1320)$, and the $\pi_2(1670)$.

1 Introduction

Progress has been made in recent years in the theoretical description of hadrons lying outside the scope of the constituent quark model. QCD predicts the existence of multi-quark $q\bar{q}q\bar{q}$ and hybrid $q\bar{q}g$ mesons as well as purely gluonic states, referred to collectively as exotic mesons. Irrefutable experimental evidence for an exotic meson would be the observation of a state with $J^{PC} = 0^{--}, 0^{+-}, 1^{-+}, 2^{+-}$, etc., as a $q\bar{q}$ pair cannot have these quantum numbers.

Theoretical predictions for the mass of the lightest $J^{PC} = 1^{-+}$ hybrid meson are based on various models. The flux tube model [1] predicts 1^{-+} states between 1.8 and 2.0 GeV/c^2. Similar results are obtained with lattice QCD in the quenched approximation [2]. Earlier bag model estimates suggest somewhat lower masses in the 1.3–1.8 GeV/c^2 range [3]. QCD sum-rule predictions vary widely between 1.5 GeV/c^2 and 2.5 GeV/c^2 [4]. The diquark cluster model [5] predicts the 1^{-+} state to be at 1.4 GeV/c^2. Finally, the constituent gluon model [6] predicts that light exotic masses should lie between 1.8 and 2.2 GeV/c^2. Most of these models predict that the decay modes of the 1^{-+} isovector hybrid meson will be dominated by $b_1(1235)\pi$ and $f_1(1285)\pi$, with small (but non-negligible) $\rho\pi$ and $\eta\pi$ decay probability [7,8]. The decay of this state to $\eta\pi$ may be prohibited based purely on symmetry arguments [9].

In this note we present experimental evidence for two isovector 1^{-+} exotic mesons produced in the reactions $\pi^- p \to \eta \pi^- p$ (reaction 1), and $\pi^- p \to \pi^+ \pi^- \pi^- p$ (reaction 2). Experiment E852 was performed at the Multi-Particle Spectrometer facility at Brookhaven National Laboratory (BNL). The experimental apparatus is described elsewhere [10,11]. The experiment was performed with a π^- beam of momentum 18.3 GeV/c incident on a 30 cm liquid hydrogen target. The trigger was based on the requirement of either one (reaction 1) or three (reaction 2) forward-going charged tracks and one charged recoil track. In addition, the η sample was enhanced in reaction 1 by use of a processor which only triggered events with more than a neutral pion mass in the lead glass calorimeter. Thirty million triggers for reaction 1 and seventeen million trigger for reaction 2 were recorded by the experiment. Event selection reduced this number to a final sample of 47,200 events of reaction 1 and 250,000 events of reaction 2 used in the final partial-wave analyses. Strict cuts were applied to insure that all data were exclusively proton-recoil events.

The partial-wave analyses for both reactions was performed using a program developed at BNL [12]. For reaction 2 each event is considered in the framework of an isobar model: an initial decay of a parent particle into a $\pi\pi$ isobar and an unpaired pion followed by the subsequent decay of the isobar. Each partial wave α is characterized by the quantum numbers $J^{PC} L M^\epsilon$ in the case of reaction 1, and $J^{PC}[isobar]LM^\epsilon$ — here J^{PC} are spin, parity and C-parity of the partial wave; M is the absolute value of the spin projection on the quantization axis; ϵ is the reflectivity (and corresponds to the naturality of the exchanged particle); L is the orbital angular momentum between the η and the pion in reaction 1, and between the isobar and the unpaired pion in the case of reaction 2.

The spin-density matrix is parameterized in terms of the complex production amplitudes $V_\alpha^{k\epsilon}$ for wave α with reflectivity ϵ. These amplitudes are determined from an extended maximum likelihood fit. The index k corresponds to the different possibilities at the baryon vertex and determines the rank of the spin-density matrix. In general this rank does not exceed two for the proton-recoil reaction (from proton spin-flip and spin-non-flip diagram contributions).

The experimental acceptance is taken into account by means of Monte Carlo normalization integrals as described in [12]. Relativistic Breit-Wigner functions with standard Blatt-Weisskopf factors were used in the description of the $\rho(770)$, $f_2(1270)$ and $\rho_3(1690)$ isobars. The $\pi^+\pi^-$ S-wave parameterization was based on a K-matrix formalism [13]. Different K-matrix parameterizations [14,15] were tested, as was a simple Breit-Wigner description of f_0 isobars. The 1^{-+} signal was found to be insensitive to the particular choice of 0^{++} $\pi\pi$ isoscalar parameterization. The results presented here were obtained in a fit with the parameterization based on the modified "M" solution [14].

$$\pi^- p \to \eta\pi^- p$$

From 47 million triggers 47,200 events were reconstructed which were consistent with reaction 1.

The partial-wave analysis for reaction 1 was performed in 40 MeV/c^2 mass bins between 0.98 and 1.82 MeV/c^2 [16]. For each mass bin all events with $0.1 < -t < 0.95$ (GeV/c)^2 were included in the fit. The rank of the density matrix was fixed at 1.

Four of the more critical moments $H(LM)$ are shown in Fig. 1. These moments, as functions of the production amplitudes, is given below[a].

$$H(30) = \frac{3}{7\sqrt{5}} 2Re\{\sqrt{3}P_0 D_0^* - P_- D_-^* - P_+ D_+^*\}$$
$$H(32) = \frac{1}{7}\sqrt{\frac{3}{2}} 2Re\{P_- D_-^* - P_+ D_+^*\}$$
$$H(40) = \frac{2}{7}|D_0|^2 - \frac{4}{21}\left(|D_-|^2 + |D_+|^2\right)$$
$$H(42) = \frac{\sqrt{10}}{21}\left(|D_-|^2 - |D_+|^2\right)$$

The signs of $H(40)$ and $H(42)$ indicate the dominance of D_+. The non-zero $H(30)$ and $H(32)$ moments indicate that there is some P_+ wave in the data. The match between the data points and the fit results are a measure of the quality of the fit.

The acceptance-corrected numbers of events predicted by the PWA fit for the D_+ and P_+ waves and their phase difference $\Delta\Phi(D_+ - P_+)$ are shown as a function of $M(\eta\pi^-)$ in figure 2. There are eight ambiguous solutions in the fit [17,18]., each of which leads to the same angular distribution. We show the range of fitted values for these ambiguous solutions in the vertical rectangular bar at each mass bin, and the maximum extent of their errors is shown as the error bar. The $a_2(1320)$ is clearly observed in the D_+ partial wave (Fig. 2a). A broad peak

is seen in the P_+ wave at about 1.4 GeV/c^2 (Fig. 2b). $\Delta\Phi(D_+ - P_+)$ increases through the $a_2(1320)$ region, and then decreases above about 1.5 GeV/c^2 (Fig. 2c). The intensities for the waves of negative reflectivity (not shown) are generally small and are all consistent with zero above about 1.3 GeV/c^2.

Fits were also carried out on Monte Carlo events generated with a pure D_+ wave to determine whether P_+-wave structure could be artificially induced by acceptance effects, resolution, or statistical fluctuations. We do find that some P_+ intensity can be induced by resolution and/or acceptance effects. Such "leakage" leads to a P_+ wave that mimics the generated D_+ intensity (and in our case would therefore have the shape of the $a_2(1320)$) with a $\Delta\Phi(D_+ - P_+)$ that is independent of mass. Neither property is seen in our result.

In an attempt to understand the nature of the P_+ wave observed in our experiment, we have carried out a mass-dependent fit to the results of the mass-independent amplitude analysis. The fit has been carried out in the $\eta\pi$ mass range from 1.1 to 1.6 GeV/c^2. The input quantities to the fit included, in each mass bin, the P_+-wave intensity; the D_+-wave intensity; and the $D_+ - P_+$ phase difference. Each of these quantities was taken with its error and correlation coefficients obtained from the amplitude analysis. In this fit, we have assumed that the D_+-wave and the P_+-wave decay amplitudes are resonant and have used relativistic Breit-Wigner forms for these amplitudes. We introduce a constant relative production phase between the P_+-wave and D_+-wave amplitudes. The parameters of the fit included the D_+-wave mass, width and intensity; the P_+-wave mass, width and intensity; and the $D_+ - P_+$ production phase difference. One can view this fit as a test of the hypothesis that the correlation between the fitted P-wave intensity and its phase (as a function of mass) can be fit with a resonant Breit-Wigner amplitude.

Results of the fit are shown as the smooth curves in Fig. 2a, b, and c. The mass and width of the $J^{PC} = 2^{++}$ state (Fig. 2a) are (1317 $\pm$1 $\pm$2) MeV/c^2 and (127 $\pm$2 $\pm$2) MeV/c^2 respectively. (The first error given is statistical and the second is systematic.) The mass and width of the $J^{PC} = 1^{-+}$ state as shown in Fig. 2b are (1370 $\pm$16 $^{+50}_{-30}$) MeV/c^2 and (385 $\pm$40 $^{+65}_{-105}$) MeV/c^2 respectively [b]. Shown in Fig. 2d are the Breit-Wigner phase dependences for the $a_2(1320)$ (line 1) and the P_+ waves (line 2); the fitted $D_+ - P_+$ production phase difference (line 3); and the fitted $D_+ - P_+$ phase difference (line 4). (Line 4, which is identical to the fitted curve shown in Fig. 2c, is obtained as line 1 $-$ line 2 + line 3.)

---

[a]The partial waves in the two pseudo-scalar system are labeled by the orbital angular momentum of the decay. The subscript $+$ indicates the natural-parity exchange wave with $|J_z| = 1$ in the Gottfried-Jackson frame, while the subscript $-$ describes the unnatural-parity exchange wave with $|J_z| = 1$. The subscript 0 is the wave with $|J_z| = 0$, which is an unnatural-parity exchange wave.

[b]The systematic errors are based on the range of values allowed by taking into account the ambiguous solutions previously discussed.

$\pi^- p \to \pi^+ \pi^- \pi^- p$

In reaction 2 the $a_1(1260)$, $a_2(1320)$ and $\pi_2(1670)$ resonances dominate the three-pion spectrum. The two-body mass spectrum shows the $\rho(770)$ and $f_2(1270)$ isobars.

The partial-wave analysis for reaction 2 was performed in 40 MeV/c^2 mass bins with $0.05 < -t < 1.0$ (GeV/c)^2 [19]. The rank of the density matrix was fixed at 1. Goodness-of-fit was determined by comparison of the experimental moments $H(LMN)$ with those predicted by the PWA fit. These moments are the integrals of the $D_{MN}^L(\alpha, \beta, \gamma)$ functions of three Euler angles taken over the experimental or predicted angular distributions. It was determined that a minimal set of 21 partial waves was required in order to achieve a reasonable agreement between the experimental and predicted moments. This set takes into account all relevant decay modes of the known resonances. It includes three 0^-+ waves, four 1^{++} waves, three 1^-+ waves, two 2^{++} waves, seven 2^-+ waves, one 3^{++} wave, and a non-interfering isotropic background wave. The 1^-+ waves were found to be essential for the description of the moments.

The acceptance-corrected numbers of events for the major non-exotic spin-parity states predicted by the PWA fit are shown in Fig. 3. The $J^{PC} = 1^{++}$ wave is dominant and accounts for almost half of the total number of events. It consists mostly of the $a_1(1260)$ meson and has a large contribution from the Deck-type background [20]. The $a_2(1320)$ is prominent in the $J^{PC} = 2^{++}$ waves, and the $\pi_2(1670)$ is seen in the $J^{PC} = 2^-+$ waves. The $J^{PC} = 0^-+$ spectrum is quite complex.

The intensities of the exotic waves are shown in Fig. 4. All three $1^-+[\rho(770)]P$ waves with $M^\epsilon = 0^-, 1^-, 1^+$ show similar structure, a broad enhancement at 1.1–1.4 GeV/c^2 and a peak at 1.6–1.7 GeV/c^2. This structure is similar to that reported for the $\rho\pi$ channel by the VES collaboration [21]. It is not present in the $1^-+[f_2(1270)]D1^+$ wave (not shown) which is consistent with zero.

The phase difference between the $1^-+[\rho(770)]P1^+$ wave and all other significant natural parity exchange waves indicates a resonance-like increase in the phase of the 1^-+ in the region 1.5–1.7 GeV/c^2 Some of these phase differences are shown in Figs. 5.

Extensive studies have been made to test the stability of the results with respect to the assumptions made in the analysis. It was found that no significant change in the 1^-+ waves takes place by inclusion of rank 2 in the spin density matrix, by different choice of the $\pi\pi$ S-wave parameterization, by exclusion of the events from the regions with a relatively large uncertainty in the instrumental acceptance, or by making PWA fits in restricted regions of t.

The finite acceptance of the apparatus can cause 'leakage' from one wave to another in the analysis. The impact of leakage into the 1^-+ waves was studied by generating Monte Carlo events with a spin-density matrix without any 1^-+ waves, then processing the sample through the analysis chain. Considerable leakage from the non-exotic waves to the 1^-+ waves is evident below 1.4 GeV/c^2 as seen in Fig. 4. An additional study has identified the $1^{++}[\rho(770)]S0^+$ wave as a primary source of this leakage at small values of the three-pion effective mass. However, above 1.4 GeV/c^2 the $J^{PC} = 1^-+$ waves are not subject to leakage.

To determine the resonance parameters, a series of two-state χ^2 fits of the $1^-+[\rho(770)]P1^+$ and $2^-+[f_2(1270)]S0^+$ waves as a function of mass were performed. Both waves are parameterized with relativistic Breit-Wigner forms including Blatt-Weisskopf barrier factors. In addition to Breit-Wigner phases, a production phase difference which varies linearly with mass is assumed. The fit yields $\chi^2 = 25.8$ for 22 degrees of freedom, with the production phase difference between the two waves being almost constant throughout the region of the fit. The fitted mass and width of the 1^-+ state are M=$1593 \pm 8^{+29}_{-47}$ MeV/c^2 and Γ=$168 \pm 20^{+150}_{-12}$ MeV/c^2 (See Fig. 6). The error values correspond to statistical and systematic uncertainties, respectively. The systematic errors were estimated by fitting the PWA results obtained for different sets of partial waves and different rank of the PWA fit.

Conclusions

The E852 collaboration has now provided credible evidence for the existence of two new $J^{PC} = 1^-+$ exotic, isovector mesons. A partial wave analysis of the reaction $\pi^- p \to \eta\pi^- p$ observes the $a_2(1320)$ in natural parity exchange, and in addition we observe a $J^{PC} = 1^-+$ exotic state, also in natural parity exchange. A mass-dependent fit has determined the mass of this resonance to be (1370 $\pm 16 \, ^{+50}_{-30}$) MeV/c^2 and Γ =(385 $\pm 40 \, ^{+65}_{-105}$) MeV/c^2. In addition, a partial wave analysis was performed on the reaction $\pi^- p \to \pi^+ \pi^- \pi^- p$. Here we see all the expected three pion resonances, the $a_1(1260)$, the $a_2(1320)$, and the $\pi_2(1670)$. In addition, we observe a $J^{PC} = 1^-+$ exotic state which shows phase motion against all the observed states. A mass-dependent fit, performed with 2 states, the $\pi_2(1670)$ and this state, yields a mass of $1593 \pm 8^{+29}_{-47}$ MeV/c^2 and a width of $168 \pm 20^{+150}_{-12}$ MeV/c^2.

References

1. N. Isgur and J. Paton, Phys. Rev. D **31**, 2910 (1985).
2. C. Bernard *et al.*, Phys. Rev. D **56**, 7039 (1997); P. Lacock *et al.*, Phys. Lett. B **401**, 308 (1997).

3. T. Barnes and F. E. Close, Phys. Lett. B **116**, 365 (1982).

4. I. I. Balitsky, D. I. Dyakonov, and A. V. Yung, Z.Phys. C **33**, 265 (1986); J. I. Latorre, P. Pascual, and S. Narison, Z.Phys. C **34**, 347 (1987); J. Govaerts et al., Nucl. Phys. B **284**, 674 (1987).

5. Y.Uehara et al., Nucl. Phys. A **606**, 357 (1996).

6. S. Ishida, H. Sawazaki, M. Oda, and K. Yamada, Phys. Rev. D **47**, 179 (1992).

7. N. Isgur, R. Kokoski, and J. Paton, Phys. Rev. Lett. **54**, 869 (1985).

8. Frank E. Close and Philip R. Page, Nucl. Phys. B **443**,233-254 (1995).

9. H. J. Lipkin, Phys. Lett. B **219**, 99 (1989).

10. N. M. Cason et al., in *Proceedings of the VI International Conference on Hadron Spectroscopy, Manchester, England, 1995,* edited by M. C. Birse, G. D. Lafferty and J. A. McGovern (World Scientific, Singapore, 1996), p.55.

11. B. B. Brabson et al., Nucl. Inst. and Meth. A **332**, 419 (1993); T. Adams et al., Nucl. Inst. and Meth. A **368**, 617 (1996); Z. Bar-Yam et al., Nucl. Inst. and Meth. A **386**, 235 (1997); R. R. Crittenden et al., Nucl. Inst. and Meth. A **387**, 377 (1997).

12. J. P. Cummings and D. P. Weygand, BNL-64637

13. S. U. Chung et al., Annalen der Physik **4**, 404 (1995).

14. K. L. Au, D. Morgan and M. R. Pennington, Phys. Rev. D **35**, 1633 (1987).

15. C. Amsler et al., Phys. Lett. B **342**, 443 (1995).

16. D.R. Thompson et al., Phys. Rev. Lett. **79**, 1630 (1997).

17. S. U. Chung, Phys. Rev. D **56**, 7299 (1997).

18. S. A. Sadovsky. "On the Ambiguities in the Partial-Wave Analysis of $\pi^- p \to \eta \pi^0 n$ Reaction", Inst. for High Energy Physics IHEP-91-75, unpublished (1991).

19. G. S. Adams et al., submitted to Phys. Rev. Lett (1998).

20. G. Ascoli et al., Phys. Rev. D **8**, 3894 (1973).

21. Yu. P. Gouz et al., in *Proceedings of the XXVI International Conference on High Energy Physics, Dallas, 1992,* edited by J. R. Sanford (American Institute of Physics, 1993), Vol.1, p.572.

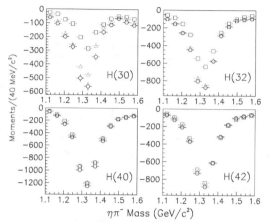

Figure 1: Four of the $H(LM)$ uncorrected moments. The circles are the data points, the triangles are calculated from the results of the fit, and the squares represent a data set with only the D_+ wave included.

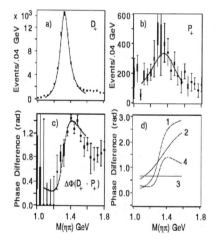

Figure 2: Results of the $\eta \pi^-$ partial wave amplitude analysis. Shown are a.) the fitted intensity distributions for the D_+ and b.) the P_+ partial waves, and c.) $\Delta \Phi(D_+ - P_+)$, their phase difference. The range of values for the eight ambiguous solutions is shown by the central bar and the extent of the maximum error is shown by the error bars. Also shown as curves in a.), b.), and c.) are the results of the mass dependent analysis described in the text. The lines in d.) correspond to (1) the fitted D_+ Breit-Wigner phase, (2) the fitted P_+ Breit-Wigner phase, (3) the fitted $D_+ - P_+$ relative production phase, and (4) the overall $D_+ - P_+$ phase difference as shown in c.) but with a different scale.

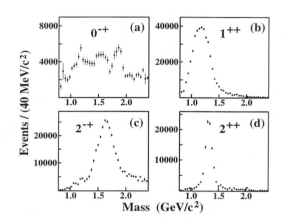

Figure 3: Combined intensities of all (a) 0^{-+} waves, (b) 1^{++} waves, (c) 2^{-+} waves, (d) 2^{++} waves.

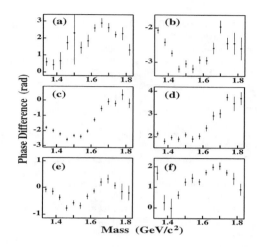

Figure 5: Phase difference between the $1^{-+}[\rho(770)]P1^+$ wave and (a) the $0^{-+}[f_0(980)]S0^+$ wave, (b) the $2^{++}[\rho(770)]D1^+$ wave, (c) the $1^{++}[\rho(770)]S0^+$ wave, (d) the $1^{++}[\rho(770)]S1^+$ wave, (e) the $2^{-+}[\rho(770)]P0^+$ wave, (f) the $2^{-+}[f_2(1270)]D0^+$ wave.

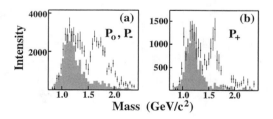

Figure 4: Wave intensities of the $1^{-+}[\rho(770)]P$ exotic waves: (a) the $M^\epsilon = 0^-$ and 1^- waves combined, (b) the $M^\epsilon = 1^+$ wave. The PWA fit to the data is shown as the points with error bars and the shaded histograms show estimated contributions from all non-exotic waves due to leakage.

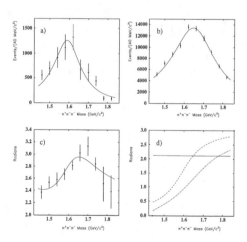

Figure 6: A coupled mass-dependent Breit-Wigner fit of the $1^{-+}[\rho(770)]P1^+$ and $2^{-+}[f_2(1270)]S0^+$ waves. (a) $1^{-+}[\rho(770)]P1^+$ wave intensity. (b) $2^{-+}[f_2(1270)]S0^+$ wave intensity. (c) Phase difference between the $1^{-+}[\rho(770)]P1^+$ and $2^{-+}[f_2(1270)]S0^+$ waves. (d) Phase motion of the $1^{-+}[\rho(770)]P1^+$ wave (1), $2^{-+}[f_2(1270)]S0^+$ wave (2), and the production phase between them (3).

THE 1.4 GEV $J^{PC} = 1^{-+}$ STATE AS AN INTERFERENCE OF A NON–RESONANT BACKGROUND AND A RESONANCE AT 1.6 GEV

A. DONNACHIE

Department of Physics and Astronomy, University of Manchester, Manchester M13 9PL, UK
E-mail: ad@a3.ph.man.ac.uk

P. R. PAGE

T-5, MS-B283, Los Alamos National Laboratory, P.O. Box 1663, Los Alamos, NM 87545, USA
E-mail: prp@t5.lanl.gov

We investigate theoretical interpretations of the 1.4 GeV J^{PC} exotic resonance reported by the E852 collaboration. A K–matrix analysis shows that the 1.4 GeV enhancement in the E852 $\eta\pi$ data can be understood as an interference of a non–resonant Deck–type background and a resonance at 1.6 GeV.

Evidence for a $J^{PC} = 1^{-+}$ isovector resonance $\hat{\rho}(1405)$ at 1.4 GeV in the reaction $\pi^- p \to \eta\pi^- p$ has been published recently by the E852 collaboration at BNL [1]. The mass and width quoted are $1370 \pm 16^{+50}_{-30}$ MeV and $385 \pm 40^{+65}_{-105}$ respectively. These conclusions are strengthened by the claim of the Crystal Barrel collaboration that there is evidence for the same resonance in $p\bar{p}$ annihilation with a mass of $1400 \pm 20 \pm 20$ MeV and a width of $310 \pm 50^{+50}_{-30}$ MeV [2], consistent with E852. However, the Crystal Barrel state is not seen as a peak in the $\eta\pi$ mass distribution, but is deduced from interference in the Dalitz plot. Since the J^{PC} of this state is "exotic", i.e. it implies that it is *not* a conventional meson, considerable excitement has been generated, particularly because the properties of the state appear to be in conflict with theoretical expectations.

In addition there are two independent indications of a more massive isovector $J^{PC} = 1^{-+}$ exotic resonance $\hat{\rho}(1600)$ in $\pi^- N \to \pi^+\pi^-\pi^- N$. The E852 collaboration recently reported evidence for a resonance at $1593 \pm 8^{+29}_{-47}$ MeV with a width of $168 \pm 20^{+150}_{-12}$ MeV [3]. These parameters are consistent with the preliminary claim by the VES collaboration of a resonance at 1.62 ± 0.02 GeV with a width of 0.24 ± 0.05 GeV [4]. In both cases a partial wave analysis was performed, and the decay mode $\rho^0\pi^-$ was observed. There is also evidence for $\hat{\rho}(1600)$ in $\eta'\pi$ peaking at 1.6 GeV [5]. It has been argued that the $\rho\pi$, $\eta'\pi$ and $\eta\pi$ couplings of this state qualitatively support the hypothesis that it is a hybrid meson, although other interpretations cannot be entirely eliminated [6].

Recent flux–tube and other model estimates [7] and lattice gauge theory calculations [8] for the lightest 1^{-+} hybrid support a mass substantially higher than 1.4 GeV and often above 1.6 GeV [6]. Further, on quite general grounds, it can be shown that an $\eta\pi$ decay of 1^{-+} hybrids is unlikely [9]. There is thus an apparent conflict between

experimental observation and theoretical expectation as far as the 1.4 GeV peak is concerned.

The purpose of the present paper is to propose a resolution of this apparent conflict. We suggest a mechanism whereby an appropriate $\eta\pi$ decay of a hybrid meson can be generated and argue that there is only one $J^{PC} = 1^{-+}$ isovector exotic, the lower–mass signal in the E852 experiment being an artefact of the production dynamics. We demonstrate explicitly that is possible to understand the 1.4 GeV peak observed in $\eta\pi$ as a consequence of a 1.6 GeV resonance interfering with a non–resonant Deck–type background with an appropriate relative phase. We do *not* propose that there should necessarily be a peak at 1.4 GeV; but that if experiment unambiguously confirms a peak at 1.4 GeV, it can be understood as a 1.6 GeV resonance interfering with a non–resonant background.

1 Interference with a non–resonant background

The current experimental data on the 1.6 GeV state is consistent with mass predictions and decay calculations for a hybrid meson [6,10]. This then leaves open the interpretation of the structure at 1.4 GeV.

There are two basic problems to be solved. Firstly it is necessary to find a mechanism which can generate a suitable $\eta\pi$ width for the hybrid. Then having established that, it is necessary to provide a mechanism to produce a peak in the cross section which is some way below the real resonance position.

The $\eta\pi$ peak in the E852 data spans the $\rho\pi$ and $b_1\pi$ thresholds, so we propose a Deck–type model [11] as a source of a non–resonant $\eta\pi$ background. We then show that, within the K–matrix formalism, interference between this background and a resonance at 1.6 GeV can account for the E852 $\eta\pi$ data.

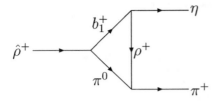

Figure 1: Decay of $\hat{\rho}$ to $\eta\pi$ via final state interactions.

1.1 $\eta\pi$ width of a 1.6 GeV state

Although the $\eta\pi$ width of a hybrid is suppressed by symmetrization selection rules [9] which operate on the quark level and have been estimated in QCD sum rules to be tiny (~ 0.3 MeV) [12], long distance contributions to this width are possible.

An essential ingredient is the presence of an allowed dominant decay which can couple strongly to the channel of interest. In the flux–tube model $b_1\pi$ is such a dominant decay [10], and it is strongly coupled to $\eta\pi$ by ρ exchange (see Figure 1). Diagrams like that in Fig. 1 are expected to make the $\eta\pi$ width more appreciable.

1.2 Non–resonant $\eta\pi$ Deck background

The 1.4 GeV peak in the $\eta\pi$ channel occurs in the vicinity of the $\rho\pi$ and $b_1\pi$ thresholds, and it is therefore natural to consider these as being responsible in some way for the $\eta\pi$ peak. The Deck mechanism [11] is known to produce broad low–mass enhancements for a particle pair in three–particle final states, for example in $\pi p \to (\rho\pi)p$. In this latter case, the incident pion dissociates into $\rho\pi$, either of which can then scatter off the proton [13]. At sufficiently high energy and presumed dominance of the exchange of vacuum quantum numbers (pomeron exchange) for this scattering one obtains the "natural parity change" sequence $\pi \to 0^-, 1^+, 2^- \ldots$ (the Gribov–Morrison rule [14]). However if the scattering involves the exchange of other quantum numbers then additional spin–parity combinations can be obtained, including $J^P = 1^-$. This can be seen explicitly in ref. [11] for the reaction $\pi p \to (\rho\pi)p$ in which the full πp scattering amplitude was used, so that the effect of exchanges other than the pomeron are automatically included. The J^P sequence from the "natural parity change" dominates due to the dominant contribution from pomeron exchange, but other spin-parity states are present at a non–negligible level. The Reggeised Deck effect can simulate resonances, both in terms of the mass distribution and the phase [11,15]. It can produce circles in the Argand plot, the origin of which is the Regge phase factor $\exp[-i\frac{1}{2}\pi\alpha(t_R)]$.

It is also important to note that rescattering of the lighter particle from the dissociation of the incident beam particle is not a prerequisite, and indeed both can contribute [13]. We suggest that in our particular case the relevant processes are (from left to right in Figure 2)

1. $\pi \to b_1\omega$, $\omega p \to \pi p$ giving a $b_1\pi$ final state.

2. $\pi \to \pi\rho$, $\rho p \to \eta p$ giving a $\eta\pi$ final state.

3. $\pi \to \rho\pi$, $\pi p \to \pi p$ and $\rho p \to \rho p$ giving a $\pi\rho$ final state.

For each of these processes the rescattering will be predominantly via ρ (natural parity) exchange to give the required parity in the final state. Obviously process (ii) produces a final $\eta\pi$ state directly, but for (i) and (iii) the $b_1\pi$ and $\rho\pi$ final states are required to rescatter into $\eta\pi$.

The characteristic mass–dependence is a peak just above the threshold. Thus there are three peaks from our proposed mechanism: a sharp peak just above the $\eta\pi$ threshold; a broader one at about 1.2 GeV from the $\rho\pi$ channel; and a very broad one at about 1.4 GeV from the $b_1\pi$ channel. The first of these is effectively removed by experimental cuts, but the net effect of the two latter is to produce a broad peak in the $\eta\pi$ channel. Thus invoking this mechanism does provide an explanation of the larger width of the $\eta\pi$ peak at 1.4 GeV in the E852 data compared to that of the $\rho\pi$ peak at 1.6 GeV. Because of the resonance–like nature of Deck amplitudes it is also possible in principle to simulate the phase variation observed. However as there are Deck amplitudes and the 1.6 GeV resonance, presumably produced directly, it is necessary to allow for interference between them. We use the K–matrix formalism to calculate this, and also to demonstrate that the Deck mechanism is essential to produce the 1.4 GeV peak.

1.3 K–matrix with P–vector formalism

It is straightforward to demonstrate that within the K–matrix formalism it is impossible to understand the $\eta\pi$ peak at 1.4 GeV as due to a 1.6 GeV state if only resonant decays to $\eta\pi$, $\rho\pi$ and $b_1\pi$ are allowed despite the strong threshold effects in the two latter channels [a]. We find that for a $b_1\pi$ width of ≈ 200 MeV and $\eta\pi$ and $\rho\pi$ widths in the region $1 - 200$ MeV there is no shift of the peak. However, when a non–resonant $\eta\pi$ P–wave is introduced, the interference between this and the 1.6 GeV state can appear as a 1.4 GeV peak in $\eta\pi$.

We have seen that the non–resonant $\eta\pi$ wave can have significant presence at the $b_1\pi$ or $f_1\pi$ threshold

---

[a]The use of $b_1\pi$ is not critical here: any channel with a threshold near 1.4 GeV will suffice.

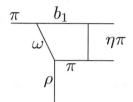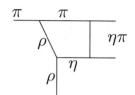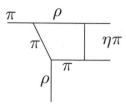

Figure 2: Deck background production in $\eta\pi$.

(called the "P+S" threshold), e.g. 1.368 GeV for $b_1\pi$, because of the substantial "width" generated by the Deck mechanism. Since the hybrid is believed to couple strongly to "P+S" states due to selection rules [16], the interference effectively shifts the peak in $\eta\pi$ down from 1.6 GeV to 1.4 GeV. It is not necessary for the 1.6 GeV resonance to have a strong $\eta\pi$ decay. It is significant that the E852 experiment finds $\hat\rho$ at $1370 \pm 16^{+50}_{-30}$ MeV, near the $b_1\pi$ threshold, but not at 1.6 GeV. It is possible for a state to peak near the threshold of the channel to which it has a strong coupling, assuming that the (weak) channel in which it is observed has a significant non–resonant origin.

We follow the K–matrix formalism in the P–vector approach as outlined in [17,18]. We assume there to be a $\hat\rho$ with $m_{\hat\rho} = 1.6$ GeV as motivated by the structure observed in $\rho\pi$ [3]. The problem is simplified to the case where there is decay to two observed channels i.e $\eta\pi$ and $\rho\pi$, and one unobserved $P + S$ channel. These channels are denoted 1, 2 and 3 respectively. The production amplitudes and the amplitude after final–state interactions are grouped together in the 3-dimensional P– and F–vectors respectively. In order to preserve unitarity [17] we assume a real and symmetric 3×3 K–matrix. The amplitudes after final–state interactions and production are related by [17]

$$F = (I - iK)^{-1}P \qquad (1)$$

We define the widths as

$$\Gamma_i = \gamma_i^2 \, \Gamma_{\hat\rho} \frac{B^2(q_i)}{B^2(q_i^{\hat\rho})}\rho(q_i) \qquad i = 1, 2 \qquad (2)$$

$$\Gamma_3 = \gamma_3^2 \, \Gamma_{\hat\rho} \, \rho(q_3) \qquad (3)$$

where q_i is the breakup momentum in channel i from a state of effective mass w, and $q_i^{\hat\rho}$ is the breakup momentum in channel i from a state of effective mass $m_{\hat\rho}$. The kinematics is taken care of by use of the phase space factor

$$\rho(q) = \frac{2q}{w} \qquad (4)$$

and the P–wave angular momentum barrier factor

$$B^2(q) = \frac{(q/q_R)^2}{1 + (q/q_R)^2} \qquad (5)$$

where the range of the interaction is $q_R = 1$ fm $= 0.1973$ GeV.

We assume the experimental width in $\rho\pi$ of $\Gamma_{\hat\rho} = 168$ MeV [3] to be the total width of the state[b]. We adopt the flux–tube model of Isgur and Paton and use the $\rho\pi$ and $b_1\pi$ widths which it predicts for a hybrid of mass 1.6 GeV. Since the model predicts that the branching ratio of a hybrid to $b_1\pi$ is $59 - 74$ % and to $f_1\pi$ is $12 - 16$ % [10], we obtain the $P + S$–wave width to be $120 - 150$ MeV. Analysis of the data shows that the $\rho\pi$ branching ratio of $\hat\rho(1600)$ is 20 ± 2 % [6], corresponding to a $\rho\pi$ width of $30 - 37$ MeV. This is consistent with flux–tube model predictions of $9 - 22$ % [10]. For the simulation we use a $b_1\pi$ width of 120 MeV, a $\rho\pi$ width of 34 MeV, and an $\eta\pi$ width of 14 MeV, well within the limits set by the doorway calculation. We neglect other predicted modes of decay since we restrict our analysis to three channels.

The K–matrix elements are

$$K_{ij} = \frac{m_{\hat\rho}\sqrt{\Gamma_i\Gamma_j}}{m_{\hat\rho}^2 - w^2} + c_{ij} \qquad (6)$$

where c_{ij} includes the possibility of an unknown background.

In the simulation we assume that the Deck terms can be treated as conventional resonances. This is not necessary, but is done to reduce the number of free parameters. We assume that the $\eta\pi$ Deck amplitude is produced predominantly via the $b_1\pi$ and $\rho\pi$ channels, and so is modelled as a resonance at a mass $m_{b1} = 1.32$ GeV and a width $\Gamma_{b1} = 300$ MeV. This width fits the E852 data at low $\eta\pi$ invariant masses (see Figure 3a). The

[b]It is found that our results in Fig. 3 are very similar even for a width of 250 MeV.

$\rho\pi$ background is assumed to peak at a mass $m_{b2} = 1.23$ GeV with a width $\Gamma_{b2} = 400$ MeV, which when plotted as an invariant mass distribution effectively peaks at ~ 1.15 GeV, in agreement with detailed Deck calculations in the 1^{++} wave [11].

We incorporate the $\eta\pi$ and $\rho\pi$ Deck background by putting $c_{ij} = 0$ except for

$$c_{11} = \frac{m_{b1}\Gamma_{b1}}{m_{b1}^2 - w^2} \qquad c_{22} = \frac{m_{b2}\Gamma_{b2}}{m_{b2}^2 - w^2} \qquad (7)$$

The widths are defined analogously to Eq. 2 as

$$\Gamma_{bi} = \gamma_{bi}^2 \, \Gamma_{\hat{\rho}} \, \frac{B^2(q_i)}{B^2(q_i^b)} \rho(q_i) \qquad i = 1, 2 \qquad (8)$$

where q_i^b is the breakup momentum from a state of effective mass m_{bi} (for $i = 1, 2$).

The production amplitudes are given by

$$P_i = \frac{m_{\hat{\rho}} V_{\hat{\rho}} \sqrt{\Gamma_i \Gamma_{\hat{\rho}}}}{m_{\hat{\rho}}^2 - w^2} + c_i \qquad (9)$$

where the (dimensionless) complex number $V_{\hat{\rho}}$ measures the strength of the production of $\hat{\rho}$. We take $c_3 = 0$ and

$$c_1 = \frac{m_{b1} V_{b1} \sqrt{\Gamma_{b1} \Gamma_{\hat{\rho}}}}{m_{b1}^2 - w^2} \qquad c_2 = \frac{m_{b2} V_{b2} \sqrt{\Gamma_{b2} \Gamma_{\hat{\rho}}}}{m_{b2}^2 - w^2} \qquad (10)$$

where the complex numbers V_{bi} gives the production strengths of the Deck background in channel i.

The results of this fit are shown in Fig. 3 and clearly provide a good description of the $\eta\pi$ data [1,18].

We briefly discuss the results. Fig. 3a indicates a steep rise for low invariant $\eta\pi$ masses, and a slow fall for large $\eta\pi$ masses. This naturally occurs because of the presence of the resonance at 1.6 GeV in the high mass region, which shows as a shoulder in our fit. Figure 3b reproduces the experimental slope and phase change in $\eta\pi$ [18]. One might find this unsurprising, since the background changes phase like a resonance. However, we have confirmed, by assuming a background that has constant phase as a function of $\eta\pi$ invariant mass, that the experimental phase shift is still reproduced. The experimental phase shift is hence induced by the resonance at 1.6 GeV.

Without the inclusion of a dominant $P + S$–wave channel the $\eta\pi$ event shape clearly shows two peaks, one at 1.3 GeV and one at 1.6 GeV, which is not consistent with the data [1]. The phase motion is also more pronounced in the region between the two peaks than that suggested by the data [18]. The rôle of the dominant $P + S$–channel is thus that at invariant masses between the two peaks, the formalism allows coupling of the strong $P + S$ channel to $\eta\pi$, so that the $\eta\pi$ appears stronger than it

would otherwise, interpolating between the peaks at 1.3 and 1.6 GeV, consistent with the data [1]. A dominant $P + S$ decay of the $\hat{\rho}$ is hence suggested by the data.

2 Discussion

We have argued that on the basis of our current understanding of meson masses it is implausible to interpret the 1.4 GeV peak seen in the $J^{PC} = 1^{-+}$ $\eta\pi$ channel by the BNL E852 experiment as evidence for an exotic resonance at that mass. We acknowledge that this is not a proof of non–existence and note the Crystal Barrel claim for the presence of a similar state at $1400 \pm 20 \pm 20$ MeV in the reaction $p\bar{p} \to \eta\pi^+\pi^-$. However this is not seen as a peak and is inferred from the interference pattern on the Dalitz plot. It has not been observed in other channels in $p\bar{p}$ annihilation at this mass, which is required for confirmation. So at present we believe that the balance of probability is that the structure does not reflect a real resonance.

Given this view, it is then necessary to explain the data and in particular the clear peak and phase variation seen by the E852 experiment. Additionally the observation of the peak only in the $\eta\pi$ channel, which is severely suppressed by symmetrization selection rules, requires justification. We have dealt with these two questions in reverse order. We first suggest final–state interactions can generate a sizable $\eta\pi$ decay. We then suggest that the E852 $\eta\pi$ peak is due to the interference of a Deck–type background with a hybrid resonance of higher mass, for which the $\hat{\rho}$ at 1.6 GeV is an obvious candidate. This mechanism also provides the natural parity exchange for the former which is observed experimentally. The parametrization of the Deck background is found not to be critical.

A key feature in our scenario is the presence of the large "$P + S$" amplitude which drives the mechanism. This should be observable both as a decay of the 1.6 GeV state and as a lower–mass enhancement due to the Deck mechanism. Depending on the relative strength of these two terms the resulting mass distribution could be considerably distorted from a conventional Breit–Wigner shape as the Deck peak is broad and the interference could be appreciably greater than in the $\rho\pi$ channel.

References

1. D.R. Thompson *et al.* (E852 Collab.), *Phys. Rev. Lett.* **79** (1997) 1630.
2. A. Abele *et al.* (Crystal Barrel Collab.), *Phys. Lett.* **B423** (1998) 175; W. Dünnweber (Crystal Barrel Collab.), *Proc. of HADRON'97* (Upton, N.Y., August 1997), p. 309, eds. S.-U. Chung, H.J. Willutzki.

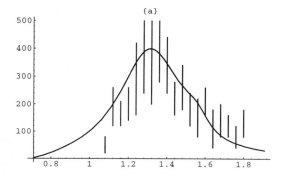

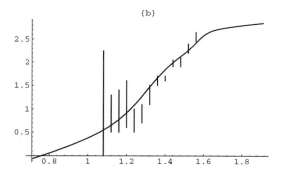

Figure 3: Results of the K–matrix analysis. (a) The events ($|F_1|^2$) in $\eta\pi$ as compared to experiment [1]; (b) The phase (of F_1) in $\eta\pi$ compared to experiment [18]. The invariant mass w is plotted on the horisontal axis in GeV. When the phase is plotted it is in radians, with the overall phase *ad hoc*. The parameters of the simulation are $m_{\hat{\rho}} = 1.6$ GeV, $\Gamma_{\hat{\rho}} = 168$ MeV [3], $\gamma_1 = 0.31$, $\gamma_2 = 0.52$, $\gamma_3 = 1.49$, $m_{b1} = 1.32$ GeV, $m_{b2} = 1.23$ GeV, $\gamma_{b1} = 1.53$, $\gamma_{b2} = 2.02$, $V_{b1}/V_{\hat{\rho}} = 2.05e^{2.77i}$, $V_{b2}/V_{b1} = 0.35e^{1.6i}$. $V_{\hat{\rho}}$ sets the overall magnitude and phase, which is not shown. None of the ratios of production strengths should be regarded as physically significant, since the K–matrix formalism allows for the introduction of additional parameters in the modelling of the backgrounds, which would change the values of these ratios. The plots shown here are only weakly dependent on the $\rho\pi$ parameters γ_{b2} and V_{b2}. The parameters have been chosen to fit both the $\eta\pi$ data [1] and the preliminary $\rho\pi$ data [3]. Experiment has not been able to eliminate the possibility that the low mass peak in $\rho\pi$ is due to leakage from the a_1. The background amplitude in $\rho\pi$ is being used as a means of parametrising all forms of background into the $\rho\pi$ channel, including leakage or Deck.

3. N.M. Cason (E852 Collab.), *Proc. of Intersections between Nuclear and Particle Physics*, p. 471 (Big Sky, Montana, May 1997), ed. T.W. Donnelly, American Institute of Physics; D.P. Weygand and A.I. Ostrovidov (E852 Collab.), *Proc. of HADRON '97* (Upton, N.Y., August 1997), p. 313, 263, eds. S.-U. Chung, H.J. Willutzki; A.I. Ostrovidov *et al.* (E852 Collab.), *in preparation*.

4. Yu.P. Gouz. *et al.* (VES Collab.), *Proc. of the 26th ICHEP* (Dallas, 1992), p. 572, ed. J.R. Sanford.

5. D. Ryabchikov (E852 Collab.), "Study of the reaction $\pi^- p \to \eta\pi^+\pi^-\pi^- p$ in E852 experiment", *Proc. of HADRON'97* (Upton, N.Y., August 1997), p. 527, eds. S.-U. Chung, H.J. Willutzki; Yu.P. Gouz. *et al.* (VES Collab.), *Proc. of the 26th ICHEP* (Dallas, 1992), p. 572, ed. J.R. Sanford.

6. P.R. Page, *Phys. Lett.* **B415** (1997) 205.

7. T. Barnes, F.E. Close and E.S. Swanson, *Phys. Rev.* **D52** (1995) 5242.

8. C. Michael *et al.*, *Nucl. Phys.* **B347** (1990) 854; *Phys. Lett.* **B142** (1984) 291; *Phys. Lett.* **B129** (1983) 351; *Liverpool Univ. report* LTH-286 (1992), *Proc. of the Workshop on QCD : 20 Years Later* (Aachen, 9–13 June 1992)

9. P.R. Page, *Phys. Lett.* **B401** (1997) 313; F. Iddir *et al.*, *Phys. Lett.* **B207** (1988) 325; H.J. Lipkin, *Phys. Lett.* **219** (1989) 99

10. P.R. Page, *Proc. of HADRON'97* (Upton, N.Y., August 1997), p. 121, eds. S.-U. Chung, H.J. Willutzki, hep-ph/9711241; P.R. Page, E.S. Swanson, A.P. Szczepaniak, *in preparation*.

11. G. Ascoli *et al. Phys. Rev.* **D8** (1973) 3894; *ibid.* **D9** (1974) 1963.

12. J.I. Latorre *et al.*, *Z. Phys.* **C34** (1987) 347; S. Narison, "QCD spectral sum rules", Lecture Notes in Physics, Vol. 26 (1989), p. 369; J. Govaerts, F. de Viron, *Phys. Rev. Lett.* **53** (1984) 2207.

13. L. Stodolsky, *Phys. Rev. Lett.* **18** (1967) 973.

14. V. Gribov, *Sov. J. Nucl. Phys.* **5** (1967) 138; M.R.O. Morrison, *Phys. Rev.* **165** (1968) 1699.

15. M.G. Bowler *et al.*, *Nucl. Phys.* **B97** (1975) 227; M.G. Bowler, *J. Phys.* **G5** (1979) 203.

16. P.R. Page, *Phys. Lett.* **B402** (1997) 183.

17. I.J.R. Aitchison, *Nucl. Phys.* **A189** (1972) 417; S.-U. Chung *et al.*, *Ann. der Phys.* 4 (1995) 404.

18. S.-U. Chung, Private communication; *Proc. of Jefferson Lab / NCSU Workshop on Hybrids and Photoproduction Physics* (1997).

SEARCHES FOR GLUEBALL CANDIDATES IN $\gamma\gamma$ COLLISIONS AT LEP AND CESR

ALVISE FAVARA

California Institute of Technology, HEP 256-48, Pasadena, CA 91125, USA
also at
CERN, EP, 1211 Geneva 23, Switzerland
E-mail: Alvise.Favara@cern.ch

Recent results on resonance formation in $\gamma\gamma$ collisions are presented. Searches for the two best glueball candidates, the $f_0(1500)$ and the $f_J(2220)$, in the $\pi^+\pi^-$ and in the $K_S^0 K_S^0$ channels are reviewed. No evidence for the $f_0(1500)$ or the $f_J(2220)$ signal is reported and new compelling upper limits, at 95% C.L., on their $\Gamma_{\gamma\gamma}$ are set. These limits support the interpretation of the $f_0(1500)$ and the $f_J(2220)$ in terms of gluon bound states.

1 Introduction

Quantum Chromodynamics (QCD) is, by now, a well established theory for the description of strong interactions between quarks and gluons. Many experimental confirmations of this theory have been obtained in the high energy regime as well as in hadron spectroscopy.

The distinctive feature of QCD as a non-Abelian gauge theory is the self-interaction of gluons. Indirect experimental evidence of this fact has been obtained from the study of multi-jet events in experiments at high energy colliders. Yet the main consequence of self-interactions among gluons is the existence of a new form of matter [1], called glueball, which is a bound state of two or more gluons. Unfortunately, glueballs have not been identified so far and theoretical predictions in the non-perturbative regime are not precise enough to guide the experimental search.

The discovery of glueballs is heavily related to a complete understanding of the $q\bar{q}$ mass spectrum. Glueball production is expected to be enhanced in gluon rich environments, such as in $p\bar{p}$ and pp collisions and in J/ψ decays, and suppressed in two-photon collisions as gluons have no electric charge.

1.1 Two-photon Collisions

Resonance formation in two-photon collisions follows the scheme shown in Figure 1 and its cross section is given by:

$$\sigma(\gamma^*\gamma^* \to R) = 8\pi(2J+1)\frac{\Gamma_{\gamma\gamma}\Gamma_R}{(m_R^2 - W_{\gamma\gamma}^2)^2 + m_R^2\Gamma_R^2} \quad (1)$$

where $W_{\gamma\gamma}$ is the two-photon invariant mass, J the spin of the resonance, m_R its mass, $\Gamma_{\gamma\gamma}$ its two-photon width and Γ_R its total width. The two-photon width, being proportional to the fourth power of the constituent quark charges, expresses the flavour content of a resonance. In

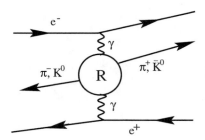

Figure 1: Resonance formation in $\gamma\gamma$ collisions.

two-photon collisions, resonances with spin 0 or 2 [a] and with positive C-parity can be produced.

The role of two-photon collisions in glueball searches is to anti-tag known resonances. For this purpose a useful glueball figure of merit, called stickiness, is defined [2]. It measures the ratio of the colour charge to the electric charge for a given resonance, R. Its expression is given by:

$$S_R = N_l \left(\frac{m_R}{k_{J/\psi \to \gamma R}}\right)^{2l+1} \frac{\Gamma(J/\psi \to \gamma R)}{\Gamma(R \to \gamma\gamma)} \sim \frac{|<R|gg>|^2}{|<R|\gamma\gamma>|^2} \quad (2)$$

where $\Gamma(R \to \gamma\gamma)$ is the two-photon width, $\Gamma(J/\psi \to \gamma R)$ the J/ψ partial width into γR, m_R and l the mass and spin of the resonance, k the photon energy in $J/\psi \to \gamma R$ decays, and N_l is a normalization factor such that $S_{f_2(1270)} = 1$.

1.2 Scalar and Tensor Meson Nonets

In this paper we report on searches in the $\pi^+\pi^-$ and in the $K_S^0 K_S^0$ channels where only scalar and tensor mesons can be observed. According to theoretical predictions [3] based on lattice calculations, the scalar glueball should

---

[a] Photons are quasi-real if e^+e^- are undetected, hence spin 1 is highly suppressed.

have a mass of 1.6 ± 0.15 GeV, and the tensor glueball a mass of 2.2 ± 0.2 GeV.

A full understanding of the scalar nonet has not yet been achieved. Many broad and interfering resonances in the $1.0 \div 1.5$ GeV region make the identification of the scalar nonet members very difficult. In addition, it is likely that among those resonances the scalar glueball is present. Unfortunately, the glueball state can mix with $q\bar{q}$ states loosing its specific features, like flavour blind couplings. It is still under discussion [4] whether the $s\bar{s}$ state and the scalar glueball have been identified or not. The $s\bar{s}$ state is supposed to be in the region $1.5 \div 1.8$ GeV, while the $f_0(1500)$ is thought to be the scalar glueball [5].

The situation is much cleaner in the tensor nonet where all the members have found their corresponding physical resonance. The anomaly is related to the observation of the $f_J(2220)$, also known as $\xi(2230)$, by Mark III [6] and BES [7], which is probably a spin 2 resonance. The statistical significance of this resonance is now large and its narrow width ($\Gamma \simeq 20$ MeV) as well as its flavour blind couplings suggest an interpretation as the tensor glueball.

2 Searches in the $\pi^+\pi^-$ Final State

2.1 Search for the $f_0(1500)$ at LEP

This analysis [8] uses 160.9 pb^{-1} of integrated luminosity collected from 1990 to 1995 at $\sqrt{s} \simeq 91$ GeV with the ALEPH detector. Events in the $\pi^+\pi^-$ channel should satisfy the following criteria: only 2 tracks originating from the interaction point and with opposite charges; small total transverse momentum; small visible energy and mass to reject Z decays; no energy in the forward detectors to select quasi-real photon collisions. In addition muons, electrons and kaons are efficiently rejected by means of the muon chambers and the energy loss measurement in the tracking chamber.

The $\pi^+\pi^-$ invariant mass spectrum for the selected data is shown in Figure 2. This spectrum is used to search for the process $\gamma\gamma \to f_0(1500) \to \pi^+\pi^-$. The clear peak in Figure 2 is due to the $f_2(1270)$. The mass resolution ($\sigma_M \simeq 10$ MeV) and the detection efficiency for the $f_2(1270)$ and the $f_0(1500)$ ($\varepsilon_{f_0(1500)} \simeq 14\%$) are evaluated by means of a Monte Carlo simulation. The background level in the region of the $f_0(1500)$ signal is estimated by extrapolating the fit to the mass spectrum into the region $1.38 \div 1.62$ GeV. The fitted Breit-Wigner parameters for the $f_2(1270)$ are 1211 ± 4 MeV for the mass and 211 ± 18 MeV for the width, revealing a large interference with the $\pi^+\pi^-$ continuum (the PDG [4] values are 1275 ± 1 MeV and 185 ± 4 MeV respectively) which has not been taken into account in the simulation.

No evidence for the $f_0(1500)$ signal is observed. From

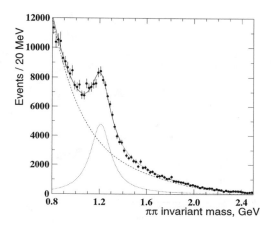

Figure 2: $\pi^+\pi^-$ mass spectrum for the selected data in ALEPH. The fit to the data includes a polynomial for the $\pi^+\pi^-$ continuum and a Breit-Wigner for the $f_2(1270)$.

a likelihood fit in the mass window $1.38 \div 1.62$ GeV, a limit at 95% C.L. is set on $\Gamma_{\gamma\gamma} \times BR(f_0(1500) \to \pi^+\pi^-) < 33.6$ eV. Including also systematic uncertainties, the limit is set to $\Gamma_{\gamma\gamma} < 170$ eV. This value is used to set a lower limit on the stickiness of the $f_0(1500)$: $S_{f_0(1500)} > 13$ at 95% C.L.

2.2 Search for the $f_J(2220)$ at CESR

This analysis [9] uses 4.77 fb^{-1} of integrated luminosity collected at $\sqrt{s} \simeq 10.6$ GeV with the CLEO detector. The selection of events from the process $\gamma\gamma \to \pi^+\pi^-$ is very similar to that described in Section 2.1. In this analysis the small visible energy requirement is applied to reject Υ decays.

In order to perform a Monte Carlo simulation of the process $\gamma\gamma \to f_J(2220) \to \pi^+\pi^-$, the $f_J(2220)$ resonance is assumed to have spin 2. For helicity 0, events are in the very forward region and the detection efficiency is only 13.1%, while for helicity 2, events are in the central region and the efficiency is 26.9%. The final efficiency of 24.9% is obtained using a ratio of 1:6 between the two helicity states. The $\pi^+\pi^-$ invariant mass spectrum for the selected data is shown in Figure 3. In the same Figure, the hatched histogram shows the simulated signal distribution, where resolution effects ($\sigma_M \simeq 12$ MeV) are included. The background level in the region of the $f_J(2220)$ signal is estimated by extrapolating the fit to the mass spectrum into the window $2200 \div 2268$ MeV.

No excess over the expected background is observed in the $f_J(2220)$ mass region. In this mass window, a likelihood fit is performed to obtain an upper limit at 95% C.L. of 2.5 eV on $\Gamma_{\gamma\gamma} \times BR(f_J(2220) \to \pi^+\pi^-)$. In

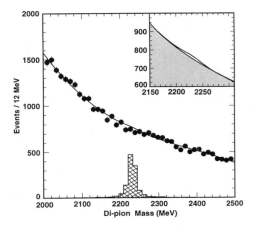

Figure 3: $\pi^+\pi^-$ mass spectrum for the selected data in CLEO. In the insert the two fits: without signal and with the signal yield corresponding to the upper limit.

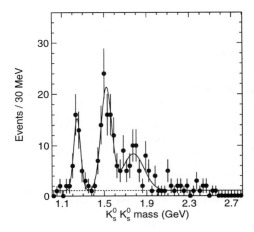

Figure 4: $K_S^0 K_S^0$ mass spectrum for the selected data in L3. Three Gaussians, and a constant for the background, are used to fit the peaks of the $f_2(1270) - a_2(1320)$, the $f_2'(1525)$ and the $f_J(1710)$.

this fit, the final signal distribution is the convolution of different signal distributions obtained spanning the mass and the width of the $f_J(2220)$ within their errors. The upper limit on $\Gamma_{\gamma\gamma} \times BR(f_J(2220) \rightarrow \pi^+\pi^-)$ is used to set a lower limit on the stickiness of the $f_J(2220)$: $S_{f_J(2220)} > 73$ at 95% C.L.

3 Searches in the $K_S^0 K_S^0$ Final State

3.1 Search for the $f_J(2220)$ at LEP

The L3 collaboration presented an analysis [10] of the reaction $\gamma\gamma \rightarrow K_S^0 K_S^0$, with $K_S^0 \rightarrow \pi^+\pi^-$, based on 143 pb^{-1} of integrated luminosity at $\sqrt{s} \simeq 91$ GeV and 52 pb^{-1} at $\sqrt{s} \simeq 183$ GeV. The selection is similar to that described in Section 2.1. Events with four charged tracks, small visible energy and small total transverse momentum are selected. In addition the presence of two secondary vertices is required. The final criterion to select a clean sample of K_S^0 pairs is $\sum_i (M_{(\pi^+\pi^-)_i} - M_{K_S^0})^2 < (40 \text{ MeV})^2$, where $\sigma_{M_{(\pi^+\pi^-)}} = 9.6$ MeV.

Figure 4 shows the 253 events selected in data from the reaction $\gamma\gamma \rightarrow K_S^0 K_S^0$, where the background contamination is expected to be negligible. The spectrum is dominated by the $f_2'(1525)$, but a peak at approximately 1250 MeV and an enhancement at approximately 1750 MeV are also clearly visible. No apparent excess is detected in the $f_J(2220)$ mass region.

The small peak at 1250 MeV is the result of the $f_2(1270) - a_2(1320)$ destructive interference, in agreement with the theoretical prediction for the $K_S^0 K_S^0$ final state. The enhancement at 1750 MeV could be evidence of the $s\bar{s}$ state of the scalar meson nonet. A firm evi-

dence of this state, which would be strongly coupled to $K\bar{K}$, would support the interpretation of the $f_0(1500)$ as the scalar glueball. From a fit of the spectrum in Figure 4 the mass and the width of the $f_2'(1525)$ and of a second resonance are measured and their values are respectively: 1520 ± 7 MeV and 119 ± 15 MeV, 1770 ± 20 MeV and 200 ± 50 MeV. In the existing spectrum of known resonances, the $f_J(1710)$ is the one which best matches the observation of L3 in the 1750 MeV region. The measured mass and width, and the limited statistics do not allow a firm conclusion, but from now on we will refer to this second peak as the $f_J(1710)$.

According to the Monte Carlo simulation the detection efficiency for the $f_2'(1525)$ is 3.9% for helicity 0 and 7.6% for helicity 2. The theoretical prediction that helicity-2 states should be dominant is verified studying the K_S^0 polar angle distribution in the two-photon centre-of-mass system ($\cos\theta^*$). Data events in the selected mass window and Monte Carlo events generated with helicity 2 or helicity 0 hypothesis are compared and the result is shown in Figure 5. The χ^2 values for the two hypotheses are, respectively, 7 and 81 for nine degrees of freedom. This result confirms the prediction that helicity 2 states should be dominant. Using the helicity 2 hypothesis a value of 85 ± 12 eV is obtained for the $\Gamma_{\gamma\gamma} \times BR(f_2'(1525) \rightarrow K\bar{K})$.

Even though the statistics for the $f_J(1710)$ are very limited an attempt is made to establish its spin. The $f_J(1710)$ could be a radially excited state of the $f_2'(1525)$ or the $s\bar{s}$ state of the scalar meson nonet. Figure 6 shows the K_S^0 polar angle distributions in the two-photon

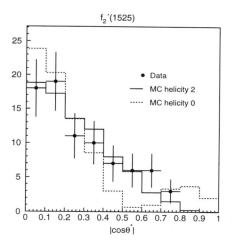

Figure 5: K_S^0 polar angle distributions in the two-photon centre-of-mass system, for L3 data and Monte Carlo in the $f_2'(1525)$ mass region.

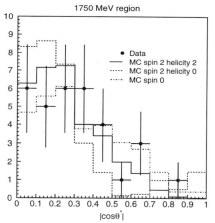

Figure 6: K_S^0 polar angle distributions in the two-photon centre-of-mass system, for L3 data and Monte Carlo in the $f_J(1710)$ mass region.

centre-of-mass system for data events in the selected mass window, and for Monte Carlo events generated with spin 2 helicity 2, spin 2 helicity 0 and spin 0. Having χ^2 values of 15, 55 and 6 for nine degrees of freedom, both the spin 2 helicity 2 and the spin 0 interpretations are allowed.

The process $\gamma\gamma \to f_J(2220) \to K_S^0 K_S^0$ has been simulated assuming a spin 2 for the $f_J(2220)$ resonance. The detection efficiency for helicity-2 states, which are as-

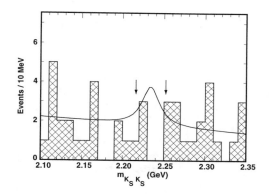

Figure 7: $K_S^0 K_S^0$ mass spectrum for the selected data in the $f_J(2220)$ mass region in CLEO. The fit to the data includes the signal yield corresponding to the upper limit.

sumed to be dominant, is 16.5%. The mass resolution for K_S^0 pairs is approximately 60 MeV. The upper limit on the production of the $f_J(2220)$ is obtained counting events in the window 2234 ± 120 MeV. The background level is estimated by an extrapolation of the values of the 2σ wide sidebands. The upper limit at 95% C.L. on $\Gamma_{\gamma\gamma} \times BR(f_J(2220) \to K_S^0 K_S^0)$ is set at 4.1 eV. This upper limit is converted into a lower limit on the stickiness of the $f_J(2220)$: $S_{f_J(2220)} > 28.6$ at 95% C.L.

3.2 Search for the $f_J(2220)$ at CESR

There is a similar analysis [11] performed by CLEO using 3.0 fb^{-1} of integrated luminosity collected at $\sqrt{s} \simeq 10.6$ GeV. Events from the reaction $\gamma\gamma \to K_S^0 K_S^0$, with $K_S^0 \to \pi^+\pi^-$, are selected as previously described for the L3 analysis. The main difference is in the $\pi^+\pi^-$ mass requirement, which is $\sum_i (M_{(\pi^+\pi^-)_i} - M_{K_S^0})^2 < (10 \text{ MeV})^2$, where $\sigma_{M_{(\pi^+\pi^-)}} = 3.3$ MeV.

For signal simulation the $f_J(2220)$ is assumed to have spin 2. The detection efficiency for helicity-0 states is 7% while for helicity-2 states is 15%. The final efficiency is obtained according to the ratio 1:6 between the two helicity states. The estimated mass resolution for K_S^0 pairs is 9 MeV.

No enhancement over the expected background is observed in the signal mass region shown in Figure 7. The mass window to be considered in the calculation of the upper limit on the production of the $f_J(2220)$, is optimized according to ε^2/b. The efficiency (ε) depends on the $f_J(2220)$ parameters, and the background (b) is evaluated from a fit in the region $2.05 \div 2.35$ GeV excluding 40 MeV around the expected signal mass. The upper limit at 95% C.L. on $\Gamma_{\gamma\gamma} \times BR(f_J(2220) \to K_S^0 K_S^0)$ is 1.3 eV. Varying the values of the $f_J(2220)$ parameters by 1σ

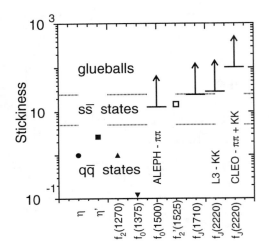

Figure 8: Stickiness values for several resonances. The recent 95% C.L. lower limits from ALEPH, CLEO and L3 are also shown.

this limit can range between 1.2 eV and 1.8 eV. This upper limit is converted into a lower limit on the stickiness of the $f_J(2220)$: $S_{f_J(2220)} > 82$ at 95% C.L.

4 Summary and Conclusions

New results on resonance formation in two-photon collisions have been reviewed. Searches for the scalar glueball candidate $f_0(1500)$ and for the tensor glueball candidate $f_J(2220)$ do not find any evidence for a signal both in the $\pi^+\pi^-$ and in the $K_S^0 K_S^0$ final states. In the L3 analysis there may be evidence for an $f_J(1710)$ signal, but this interpretation needs confirmation.

To more clearly identify the glueball nature of a resonance we can combine experimental upper limits on its $\Gamma_{\gamma\gamma}$ with its production rate in J/ψ radiative decays. This combination is expressed in the stickiness, defined in Eq. (2), which is to some extent equivalent to the ratio of the colour charge to the electric charge. Small values of stickiness are expected for mixed $u\bar{u} - d\bar{d}$ mesons. Larger values are expected for pure $s\bar{s}$ states, while even larger values are expected for glueballs which can couple to $\gamma\gamma$ only by creation of a virtual $q\bar{q}$ pair.

Figure 8 shows stickiness values for several resonances. For some resonances only upper limits on $\Gamma_{\gamma\gamma}$ exist, which result in lower limits on the stickiness. In Figure 8 indicative domains for the stickiness of glueballs, $s\bar{s}$ and $q\bar{q}$ states are shown. Resonances like the η and η' are in the $q\bar{q}$ domain, while the $f_2'(1525)$ is thought to be the $s\bar{s}$ state of the tensor meson nonet. So far only upper limits have been published [4] on the $\Gamma_{\gamma\gamma}$ of the $f_J(1710)$,

leaving open both the interpretation as the scalar glueball and as the $s\bar{s}$ state of the scalar meson nonet. If the L3 signal can be interpreted as the $f_J(1710)$, this would lead to a stickiness value very close to the present 95% C.L. lower limit, and in agreement with the $s\bar{s}$ state hypothesis. Concerning the two best glueball candidates, the $f_0(1500)$ and the $f_J(2220)$, the new results are summerized below:

- new limit on $S_{f_0(1500)} > 13$ at 95% C.L. by ALEPH
- new limit on $S_{f_J(2220)} > 28.6$ at 95% C.L. by L3
- new limit on $S_{f_J(2220)} > 102$ at 95% C.L. obtained by CLEO combining their analyses of the $\pi^+\pi^-$ and $K_S^0 K_S^0$ channels [9].

No firm conclusions can be drawn yet, but there are strong indications that the $f_J(2220)$ is the tensor glueball. Our improved knowledge of the $f_0(1500)$ and the $f_J(1710)$, suggests that the $f_0(1500)$ might be the scalar glueball and the $f_J(1710)$ the scalar $s\bar{s}$ state.

In the coming years LEP will take advantage of an integrated luminosity forty times higher (4 experiments with 500 pb^{-1} each) than that presented in this paper. CLEO as well, will take new data and we can foresee that the final word will be said about glueballs.

References

1. H. Fritzsch *et al.*, *Nuovo Cimento* A **30**, 393 (1975).
2. M. Chanowitz in *VI International Workshop on Photon-Photon Collisions*, ed. R. Lander (World Scientific, Singapore, 1984).
3. G. Bali *et al.*, *Phys. Lett.* B **309**, 378 (1993).
   H. Chen *et al.*, *Nucl. Phys.* B **34**, 357 (1994).
   J. Sexton *et al.*, *Phys. Rev. Lett.* **75**, 4563 (1995).
   A. Szczepaniak *et al.*, *Phys. Rev. Lett.* **76**, 2011 (1996).
4. Particle Data Group, *Eur. Phys. Journal* C **3**, 1 (1998).
5. C. Amsler and F.E. Close, *Phys. Rev.* D **53**, 295 (1996).
6. Mark III Collab., R.M. Baltrusaitis *et al.*, *Phys. Rev. Lett.* **56**, 107 (1986).
7. BES Collab., J.Z. Bai *et al.*, *Phys. Rev. Lett.* **76**, 3502 (1996).
8. ALEPH Collab., R. Barate *et al.*, Contrib. paper 907, ICHEP 98, Vancouver, 1998.
9. CLEO Collab., R. Godang *et al.*, Contrib. paper 872, ICHEP 98, Vancouver, 1998.
10. L3 Collab., M. Acciarri *et al.*, Contrib. paper 522, ICHEP 98, Vancouver, 1998.
11. CLEO Collab., R. Godang *et al.*, *Phys. Rev. Lett.* **79**, 3829 (1997).

THE HADRONIC STRUCTURE IN $\tau \to 3\pi\nu$ DECAYS

A. ANDREAZZA

CERN, EP Division, CH-1211 Geneva 23, Switzerland
E-mail: Attilio.Andreazza@cern.ch

DELPHI COLLABORATION

The hadronic structure of the decay of the τ lepton to three pions, $\tau \to 3\pi\nu_\tau$, has been studied using DELPHI LEP-1 data. The invariant mass distributions of the 3π system have been fitted to the models of Kühn and Santamaria, Isgur Morningstar and Reader, and Feindt. The 3π and $\pi^+\pi^-$ mass spectra are compared to the models. Qualitative agreement is observed at lower masses but for higher masses anomalous behaviour is observed, consistent with the existence of a hitherto unseen decay mode of the τ through a radial excitation of the a_1 meson.

1 Introduction

The structure of the decay of the τ to three charged pions, $\tau \to 3\pi\nu_\tau$, has been the subject of much theoretical and experimental effort. It allows studies of the weak hadronic current and light-meson spectroscopy. The present paper describes the results of a recent study performed by the DELPHI Collaboration.[1]

In the decay $\tau \to 3\pi\nu_\tau$, the 3π system is expected to have the quantum numbers $J^{PC} = 1^{++}$, with some small contribution of 0^{-+}, and to be produced predominantly via the a_1 resonance. However, problems still exist, both in the determination of the mass and width of the a_1, and in the mechanism by which the τ decays to three pions. Measurements of the a_1 meson parameters in hadronic interactions obtain lower values for the mass and width than those from τ decays. Values obtained from τ data alone also vary, depending on the model and decay mechanism assumed.

The models of Kühn and Santamaria[2] (KS), Isgur Morningstar and Reader[3] (IMR), and Feindt[4] have been reasonably successful in describing the decay $\tau \to 3\pi\nu_\tau$. All assume that the decay proceeds predominantly through the a_1 resonance.

The KS model allows the τ to decay to a_1 which then decays simply to a mixture of $\rho\pi$ and $\rho(1450)\pi$ which exist in the lowest dimensional Born state (approximately the S-wave state). The resonances are parameterised as Breit-Wigner functions with energy dependent widths. Constant form-factors are employed in describing the $a_1 \to \rho\pi$ and $\rho \to \pi\pi$ vertices.

The IMR model allows the τ to decay to the axial-vector a_1 and the pseudo-scalar $\pi(1300)$. The a_1 decays through both D-wave and S-wave $\rho\pi$ states, as well as through the K^*K channel. The $\pi(1300)$ decays to $\rho\pi$ and through a broad $\pi\pi$ S-wave state. In addition, the model makes use of a three parameter polynomial background term which was found necessary in order to "take into account many possible small effects ... e.g., the low-mass

tails of radial excitations of the a_1 and ρ".[3] IMR has energy dependent form-factors. These affect the mass dependent width of the a_1 in different ways: in KS the width increases with s; in IMR it increases to a maximum at $s \approx 1.4$ after which it decreases. It is principally this effect which causes the large differences in the masses and widths of the a_1 measured in τ decays.[5]

The Feindt model is similar to the KS and IMR models but has a more general form. It can be made to resemble either KS or IMR. This is useful in evaluating the effects of different combinations of decay channels.

2 $\tau \to 3\pi\nu_\tau$ selection

This analysis uses data taken between 1992 and 1995 by the DELPHI experiment, corresponding to about 170,000 τ pairs. The selection of $\tau \to 3\pi\nu_\tau$ candidates starts by identifying decays of the Z to two τ leptons produced back to back, one of which decays to three charged particles, while the other decays to one or three charged particles.

To reject $\tau \to \pi\pi\pi(n\pi^0)\nu_\tau$ decays, three charged track decays having two or more neutral energy deposits greater than 500 MeV or a single deposit greater than 2 GeV are rejected. The event is also removed if any track has muon chamber hits or deposits in the hadron calorimeter consistent with a minimum ionising particle. Such events may originate from four-fermion processes.

The track and vertex reconstruction algorithm is the same as that used in the DELPHI determination of the τ lifetime,[6] with tight requirements on both track and vertex reconstruction quality. In total, 7180 events remain after all cuts.

The efficiency inside $|\cos\theta| < 0.73$ was found to be 36% with a background of 20%. At high s, the sample is essentially pure.

The invariant mass squared s of the 3π system and its reconstruction error σ_s are calculated at the vertex assuming the pion mass for all the particles. The error

	KS	IMR with poly.	IMR no poly.
m_{a_1} (MeV)	$1255 \pm 7 \pm 6$	$1207 \pm 5 \pm 8$	1217 ± 7
Γ_{a_1} (MeV)	$587 \pm 27 \pm 21$	$478 \pm 3 \pm 15$	475 ± 3

	MF-KS	MF-IMR
m_{a_1} (MeV)	$1264 \pm 8 \pm 4$	$1196 \pm 4 \pm 5$
Γ_{a_1} (MeV)	$547 \pm 25 \pm 23$	$425 \pm 14 \pm 8$
κ	$.75 \pm .06 \pm .02$	$.50 \pm .06 \pm .02$
ϕ (rad)	$6.1 \pm .2 \pm .0$	$3.2 \pm .2 \pm .0$

Table 1: The fitted a_1 parameters for various models as described in the text. The first error is statistical; the second is due to systematic effects and is presented for the KS and IMR models, as well as for their MF equivalents in which the effect of an a_1' has been included.

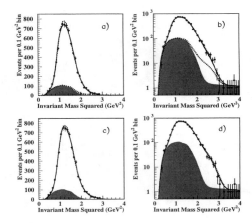

Figure 1: Distribution of 3π invariant mass squared with various models superimposed. The points are the data; the shaded area is the background contribution. In a) the dashed curve is the KS fit; the solid curve the IMR fit; the dotted curve is the contribution of the polynomial background. Plot b) is as for a) but on a logarithmic scale. In c) the dashed curve is the MF-KS fit; the solid curve is the MF-IMR fit. Plot d) is as for c) but on a logarithmic scale.

on $\sqrt{s}$ is typically between 10 and 35 MeV.

3 KS and IMR models

This section describes the fitting of the distribution of the invariant mass squared of the three pions to the models of KS and IMR and the extraction of the a_1 parameters. The Dalitz plot projections are compared with the predictions of both models. The distribution of the invariant mass squared s of the three pions is fitted to the models by maximising a binned likelihood function. The predicted distribution is the sum of signal and background contributions. The shape and proportion of the background are taken from simulation. The signal is obtained from the theoretical model distribution corrected for resolution and efficiency effects.

Results for the following models are presented:

- the KS model, with parameters and masses as given in the first row of Table 1 of their paper.[2] The mass and width of the a_1 meson are left free.

- the IMR model, where the prescription for their 'preferred' fit is followed but the KK^* contributions are not included as they do not contain 3π in the final state. The effect of the KK^* channel, where the final state kaons are mis-identified as pions, is considered below when estimating the backgrounds. The strong on-shell form factors are taken from the model predictions. The mass of the a_1 meson and the three terms characterising the polynomial background are left as free parameters. The a_1 width is calculated from the a_1 mass in accordance with the model.

The fits are shown in Fig. 1a) and b); the a_1 parameters are given in Table 1.

The data are in reasonably good agreement with the KS model. The χ^2 over the 27 bins from $s = 0.5$ GeV^2 to $s = 3.2$ GeV^2 is 32. Close to the endpoint, however, the data lie somewhat higher than the model predicts. The IMR model is also in reasonable agreement with the data. The χ^2 over the 27 bins from $s = 0.5$ GeV^2 to $s = 3.2$ GeV^2 is 36. In both models the obtained a_1 mass and width agree with previous measurements.[7,8]

The IMR polynomial term contributes 2% of the total 3π decay rate. However, it contributes one third above $s = 2.5$ GeV^2. This could be an indication of the existence of a higher mass resonance. Results for the fit without the polynomial background term are given in Table 1. A poorer fit is obtained, especially in the end region where a χ^2 of 40 is found for the 12 bins between $s = 2.0$ GeV^2 and 3.2 GeV^2.

The modelling of the background from other τ decays was investigated and gave systematic errors 5 and 8 MeV for the mass, and 18 and 3 MeV for the width for the KS and IMR models respectively.

The detector resolution was checked by reconstructing the D^0 mass peak from a data sample of D^* mesons which decay via $D^* \to D^0 \pi \to K\pi\pi$. Errors from these sources of 3 and 2 MeV for the mass, and 11 and 1 MeV for the width were found for KS and IMR respectively. An error of 15 MeV is included on the a_1 width in the IMR model, due to variations in the form factors.

Fig. 2 shows the Dalitz distributions, and Fig. 3 the $\sqrt{s_1}$ and $\sqrt{s_2}$ mass spectra, for various s ranges. The

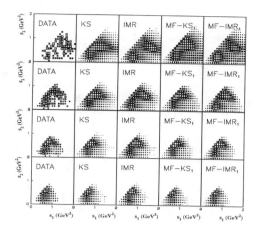

Figure 2: Dalitz plots in sequential s ranges, for the data and the models described in the text. From top to bottom the s ranges (in GeV^2) are: $2.3 < s < 3.2$; $2.0 < s < 2.3$; $1.7 < s < 2.0$; $1.4 < s < 1.7$. The different columns contain: 1) data; 2) KS model; 3) IMR model; 4) MF-KS model; 5) MF-IMR model.

quantities s_1 and s_2 are the invariant mass squared values for the two unlike-sign two-particle combinations, with $s_1 > s_2$.

In the lower s ranges displayed, the s_1 and s_2 structures apparent in the data appear to be well reproduced by both models.

For the high s bin the ρ bands of the KS model are not reproduced in the data. The double peaked structure of the IMR model appears to provide a much better qualitative agreement with the data. This double peaked structure is due to the energy dependent form-factors which, for IMR, create the $\rho\pi$ predominantly in a D-wave state at high values of s, whereas KS assumes an approximate S-wave state at all energies. However, the IMR model does not predict the normalisation in the high s region; an excess of events is observed for $s > 2.3$ GeV^2 and the Dalitz projection in Fig. 3a) shows an excess of events at $\sqrt{s_1} \approx 1.25$ GeV.

Taking into account phase space effects, the enhancement observed at $\sqrt{s_1}$ of about 1.25 GeV for s above 2.3 GeV^2 could be explained by a decay channel of the τ to a resonance of mass similar to or greater than the τ mass which then decays to 3π through the intermediate state of a pion plus a particle of mass 1.25 GeV or greater.

An excellent candidate for this resonance is the a_1', which is a radial excitation of the a_1 meson. It is predicted by Isgur and Kokoski[9] to have a mass of 1820 MeV and to decay to the $\rho(770)\pi$ D-wave state in preference to the S-wave state. Such a resonance has probably already

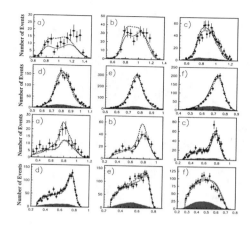

Figure 3: Distributions for $\sqrt{s_1}$ and $\sqrt{s_2}$ for various s bins. The top six plots are for $\sqrt{s_1}$; the bottom six are for $\sqrt{s_2}$. The six s regions (in units of GeV^2) are: a) $2.3 < s < 3.2$; b) $2.0 < s < 2.3$; c) $1.7 < s < 2.0$; d) $1.4 < s < 1.7$; e) $1.1 < s < 1.4$; f) $0.8 < s < 1.1$. The points represent the data; the dotted line is the KS fit; the solid line is the IMR fit; the shaded histogram is the background. Units on the horizontal axis are in GeV.

been seen by the VES collaboration,[10] which reported a clear signal in the D-wave state (and an enhancement of equal size in the S-wave), for a $J^{PC} = 1^{++}$ particle with a mass of about 1700 MeV and a width of about 300 MeV. If this exists, then there is every reason to expect it to be present in τ decays, and in fact Iizuka et al.[11] and Shuryak and Kapusta[12] have already postulated such a particle.

4 Feindt model and a_1' resonance

The extra versatility of the Feindt model is now used to introduce an a_1' resonance. The Feindt model is made to look like the previous two models and the fits repeated to check the consistency of the two approaches. With this achieved, the Feindt model is then modified so that, in addition to the decays through the a_1, the τ is allowed to decay through an a_1'. The complex Breit-Wigner function (BW) which described the amplitude of the a_1 resonance, is modified by an admixture of a_1': $BW_a \to BW_a + \kappa e^{i\phi} BW_{a_1'}$. It was not possible to fit for the mass and width of the BW which describes the a_1', as the data had little sensitivity to these values. The mass and width were fixed to be 1700 MeV and 300 MeV respectively, following the VES results. The a_1' was allowed to decay with equal probability into the S-wave $\rho(1450)\pi$ state and the D-wave $\rho(770)\pi$ state.

For each model a further fit (called respectively MF-KS and MF-IMR for the KS and IMR models) was made

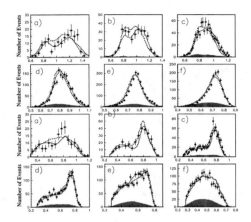

Figure 4: As Fig. 3, for the MF-KS and MF-IMR fits. The points are data; the dotted line is the MF-KS fit; the solid line is the MF-IMR fit; the shaded histogram is the background.

where the mass and width of the a_1 were left as free parameters as well as κ and ϕ.

The results of the fits are given in Table 1. The fitted widths of the a_1 are both lower than those obtained before adding an a_1', This can be attributed to the addition of an a_1'. This could explain the discrepancy between the parameters of the a_1 measured in τ decay and in hadronic production. The fits to the s distribution are shown in Figs. 1c) and d). The χ^2 values over the 27 bins from $s = 0.5$ GeV^2 to $s = 3.2$ GeV^2 are 28 for MF-KS and 25 for MF-IMR, which are better than those for previous fits. In the high s region, the χ^2 values for the 12 bins between $s = 2.0$ GeV^2 and $s = 3.2$ GeV^2 improve to 13 for MF-KS and 11 for MF-IMR.

The Dalitz plots are shown in the last two columns of Fig. 2 and the $\pi^+\pi^-$ mass spectra in Fig. 4. MF-IMR shows the same characteristics in the high s plots as the data. The agreement of MF-KS is less striking, but better than KS. In the lower s bins, the level of agreement is similar to that of the fit without an a_1'.

5 Conclusions

Below $s = 2.3$ GeV^2, the 3π and 2π mass distributions are found to be in reasonable agreement with the KS and IMR models. showing that the decay proceeds predominantly through a_1 and ρ resonances.

Above $s = 2.3$ GeV^2, an enhancement is observed. This effect is not described by the models and leads us to hypothesise the existence of a hitherto unobserved decay channel in τ decays consistent with the decay chain $\tau \to a_1' \nu_\tau$.

The KS model, although having the merit of simplicity and providing a reasonable description of the shape of the s distribution, has three problems. Firstly, the large fitted width of the a_1 leads to a branching ratio prediction at variance with the experimentally measured value. Secondly, the fit to the data at high s values is not particularly good. Thirdly, the distribution over the Dalitz plot at high s differs from the data.

The IMR model gives a fair description of the data, although in the high s region this is in large part due to the polynomial background term. This is clearly a weak point of the model, as it does not provide a physical explanation for the behaviour.

A better description of the data is obtained using the Feindt model to extend the models of KS and IMR by including the effect of the a_1'. Reasonable assumptions for the mass and width of the a_1' and its decay mechanism are taken from other experimental and theoretical work. With this extra resonance included, a much improved fit to the data is obtained. The best description, especially in the high s region, is provided by MF-IMR which extends the formalism of IMR by the inclusion of an a_1'. Unfortunately, not enough data are available to unambiguously identify the resonances which may exist in this region. Further theoretical and experimental work is required.

References

1. P. Abreu et al., DELPHI coll., *Phys. Lett.* **B426** (1998) 411.

2. J. H. Kühn and A. Santamaria, *Z. Phys.* **C48** (1990) 445.

3. N. Isgur, C. Morningstar and C. Reader, *Phys. Rev.* **D39** (1989) 1357.

4. M. Feindt, *Z. Phys.* **C48** (1990) 681.

5. P.R. Poffenberger, *Z. Phys.* **C71** (1996) 579.

6. P. Abreu et al., DELPHI coll., *Phys. Lett.* **B365** (1996) 448.

7. R. Akers et al., OPAL coll., *Z. Phys.* **C67** (1995) 45;
   K. Ackerstaff et al., OPAL coll., *Z. Phys.* **C75** (1997) 593.

8. H. Albrecht et al., ARGUS coll., *Z. Phys.* **C58** (1993) 61.

9. R. Kokoski and N. Isgur, *Phys. Rev.* **D35** (1987) 907.

10. D. V. Amelin et al., VES Collaboration, *Phys. Lett.* **B356** (1995) 595.

11. J. Iizuka, H. Koibuchi and F. Masuda, *Phys. Rev.* **D39** (1989) 3357.

12. J. I. Kapusta and E. V. Shuryak, *Phys. Rev.* **D49** (1994) 4694.

INVESTIGATION OF THE RARE ϕ RADIATIVE DECAYS WITH THE CMD-2 DETECTOR AT VEPP-2M COLLIDER

R.R.AKHMETSHIN, G.A.AKSENOV, E.V.ANASHKIN, M.ARPAGAUS, V.A.ASTAKHOV, V.M.AULCHENKO, V.S.BANZAROV, L.M.BARKOV, N.S.BASHTOVOY, S.E.BARU, A.E.BONDAR, D.V.BONDAREV, D.V.CHERNYAK, A.G.CHERTOVSKIKH, A.S.DVORECKY, S.I.EIDELMAN, G.V.FEDOTOVICH, N.I.GABYSHEV, A.A.GREBENIUK, D.N.GRIGORIEV, V.F.KAZANIN, B.I.KHAZIN, A.V.KLIMENKOV, I.A.KOOP, L.M.KURDADZE, A.S.KUZMIN, I.B.LOGASHENKO, P.A.LUKIN, A.P.LYSENKO, Yu.I.MERZLYAKOV, V.S.OKHAPKIN, E.A.PEREVEDENTSEV, T.A.PURLATZ, N.I.ROOT, A.A.RUBAN, N.M.RYSKULOV, Yu.M.SHATUNOV, M.A.SHUBIN, B.A.SHWARTZ, V.A.SIDOROV, A.N.SKRINSKY, V.P.SMAKHTIN, I.G.SNOPKOV, E.P.SOLODOV, P.Yu.STEPANOV. A.I.SUKHANOV, Yu.V.YUDIN, S.G.ZVEREV

Budker Institute of Nuclear Physics, Novosibirsk, 630090, Russia

J.P.MILLER, B.L.ROBERTS

Boston University, Boston, MA 02215, USA

J.A.THOMPSON,

University of Pittsburgh, Pittsburgh, PA 15260, USA

S.K.DHAWAN, V.W.HUGHES

Yale University, New Haven, CT 06511, USA

In experiments with the CMD-2 detector at VEPP-2M collider the integrated luminosity about 15 pb^{-1} has been collected around ϕ. The search of rare decays of ϕ was performed in the $\pi^+\pi^-$, $\pi^+\pi^-\gamma$, $\pi^0\pi^0\gamma$, $\pi^0\eta\gamma$, $\eta'\gamma$, $\mu^+\mu^-$, $\mu^+\mu^-\gamma$ final states. About 25% of available data has been analysed and some preliminary results are presented here. The new upper limits for CP,P-violation decay $\eta \rightarrow \pi^+\pi^-$ and C-violation decay $\phi \rightarrow \rho\gamma$ have been found.

1 Introduction

The CMD-2 detector has been described in more detail elsewhere [1,2]. The detector is in operation since 1992 and he total integrated luminosity about 25 pb^{-1} have been collected in the whole energy range of the VEPP-2M collider [3]. The BGO endcap calorimeter has been installed in 1996 increasing the solid angle for photons from 70% to 95% of 4π.

The integrated luminosity $\approx 15pb^{-1}$ has been collected around ϕ mass and the results based on part of available data were published in papers[4,5,6,7]. In this paper we present a recent progress in study of rare ϕ decays based on about 25% of available data.

2 Study of $\phi \rightarrow \pi^+\pi^-$ and $\phi \rightarrow \mu^+\mu^-$ decays

The ϕ signal can be seen as an interference pattern in the $e^+e^- \rightarrow \pi^+\pi^-$ or in the $e^+e^- \rightarrow \mu^+\mu^-$ cross sections. The event candidates were selected by requirement of two collinear charged tracks in the DC with the sum of the energy depositions in two CsI clusters associated with

them to be less than 450 MeV to remove Bhabha events. The latter were used for luminosity determination.

To separate pions and muons the inner muon range system which covered the polar angle between 0.85 and 2.25 radians was used. Muons with 500 MeV/c momenta leave about 90 MeV energy deposition in the CsI calorimeter and have close to 100% probability to reach inner muon system while pions have about 35% of probability to interact and stop in the CsI. The correlation between energy deposition in the CsI calorimeter and probability to reach muon system was used to determine muon system efficiency which was measured to be 96.2 ± 0.3%.

The requirement of both particles to hit the inner muon system selects muons with about 10% of pion admixture. The remaining events mostly are pions with small admixture of muons due to muon system inefficiency.

The probability for pions to stop in the CsI vs. polar angle was studied for positive and negative particles and the obtained numbers were used to subtract pions from muon sample and pions from muon sample.

The visible cross sections vs. energy for pions and

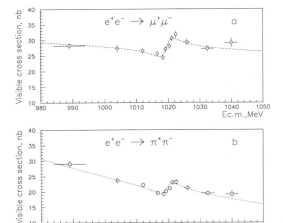

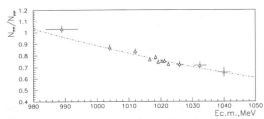

Figure 1: a. Visible cross section for $e^+e^- \rightarrow \mu^+\mu^-$ Line represents the fit with ϕ signal. b. Visible cross section for $e^+e^- \rightarrow \pi^+\pi^-$. Line represents the fit with ϕ signal.

Figure 2: Ratio of number of pions to number of muons vs. energy.

muons are presented in Figure 1. The signal from ϕ is clearly seen as well as a different energy dependence of the pion and muon pair production due to pion formfactor.

To get $\phi \rightarrow \pi^+\pi^-$ and $\phi \rightarrow \mu^+\mu^-$ branching ratios the obtained cross sections were fitted by the functions proportional to $1/E^2$ for muons and $|F_\pi(E)|^2/E^2$ for pions multiplied by additional term describing vacuum polarisation by ϕ resonance [8]. The following results have been obtained:

$(B(\phi \rightarrow \mu^+\mu^-) \cdot B(\phi \rightarrow e^+e^-))^{1/2} = (2.77 \pm 0.35) \times 10^{-4}$,

$B(\phi \rightarrow \mu^+\mu^-) = (2.57 \pm 0.70) \times 10^{-4}$,

$B(\phi \rightarrow \pi^+\pi^-) = (1.60 \pm 0.49) \times 10^{-4}$.

For the last two branchings above the value of $B(\phi \rightarrow e^+e^-) = (2.99 \pm 0.08) \times 10^{-4}$ was taken from the PDG tables[10].

To extract the amplitude of the direct ϕ decay into two pions the ratio of number of pions to muons was calculated. This ratio cancels the influence of ϕ to photon propagator and interference signal of direct amplitude can be seen. This ratio is shown in Figure 2. No signal is seen and fit gives the upper limit
$B(\phi \rightarrow \pi^+\pi^-)_{direct} < 5 \times 10^{-6}$ at 90% C.L.

3 Search for $\phi \rightarrow f_0(980)\gamma$

3.1 $\phi \rightarrow \pi^+\pi^-\gamma$ channel

The event candidates were selected by a requirement of two charged tracks in the DC and one or two photons with energy greater than 20 MeV in the CsI calorimeter. All particles were required to have the polar angle in between 0.85 and 2.25 radians for the bremsstrahlung processes suppression which gave the main contribution to the observed events.

The main background for the studied process was $\phi \rightarrow \pi^+\pi^-\pi^0$ decay when one of the photons from π^0 decay escaped detection. To reduce this background a constrained fit with requirement of energy and momentum conservation was used. As a result the 3π background was removed for the gamma energy range 20 MeV$< E_\gamma <$160 MeV, where maximum signal from f_0 was expected. The muon range system was used to separate $\mu^+\mu^-\gamma$ contamination as described in previous section.

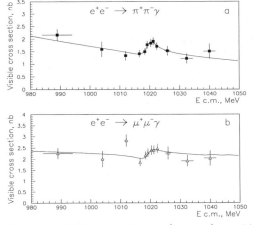

Figure 3: a. Visible cross section for $e^+e^- \rightarrow \pi^+\pi^-\gamma$ with $20 < E_\gamma < 120$ MeV. Line is a fit with $B(\phi \rightarrow \pi^+\pi^-\gamma) = 0.38 \times 10^{-4}$ and relative phase 0.7 radian. b. Visible cross section for $e^+e^- \rightarrow \mu^+\mu^-\gamma$.

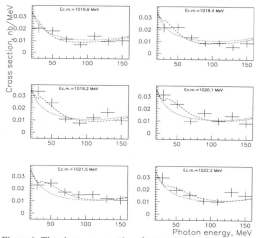

Figure 4: The photon spectra for $\pi^+\pi^-\gamma$ events at "ϕ" region normalised to the integrated luminosity. Lines are theoretical prediction for four quark model with 2.4×10^{-4} branching ratio(dashed) and pure bremsstrahlung spectra(dotted).

Under these conditions 7309 events of $\pi^+\pi^-\gamma$ and 7590 of $\mu^+\mu^-\gamma$ events were selected. The detection efficiency was found by simulation to be 17%. The experimental cross sections vs. centre of mass energies are presented in Figures 3a,b.

The signal from $\phi \to \pi^+\pi^-\gamma$ decay was searched as an interference patterns in the cross section vs. energy behaviour of the $e^+e^- \to \pi^+\pi^-\gamma$ process (Fig. 3) and in the photon spectra of the selected events (Fig. 4) at ϕ peak. Using a simple "model independent" fit by sum of power function representing $e^+e^- \to \pi^+\pi^-\gamma$ process and Breit-Wigner from ϕ the following results have been obtained:
$Br(\phi \to \pi^+\pi^-\gamma) = (0.38 \pm 0.16) \times 10^{-5}$,
$Br(\phi \to \mu^+\mu^-\gamma) = (1.3 \pm 0.6) \times 10^{-5}$.

The result above can be interpreted as $\phi \to f_0\gamma$ decay with the branching ratio $Br(\phi \to f_0\gamma) = (0.57 \pm 0.24) \times 10^{-5}$. That value is close to one predicted for two quark structure of f_0 in paper [11]. The four quark model of f_0 structure suggested by [11] is also in good agreement with observed data as seen in Figure 4. The fit gives:
$Br(\phi \to f_0\gamma) = (2.0 \pm 0.9) \times 10^{-4}$
with the relative phase to bremsstrahlung background 0.7 ± 0.3 radian. This relative phase corresponds to strong destructive interference and reduce visible signal from ϕ decay in case of four quark model.

The $\pi^+\pi^-$ events with photons in the 100-300 MeV range after 3π background subtraction were used to

search for C violation process $\phi \to \rho\gamma$ and a new upper limit was obtained:
$B(\phi \to \rho\gamma) < 3 \times 10^{-4}$ at 90% CL.
The $\pi^+\pi^-$ events with photons in the narrow range around 362 MeV were used to search for CP,P violation decay of η into two charged pions and the new limit is:
$B(\eta \to \pi^+\pi^-) < 3 \times 10^{-4}$ at 90% CL.

3.2 $\phi \to \pi^0\pi^0\gamma$ channel

The decay $\phi \to f_0\gamma$ with $f_0 \to \pi^0\pi^0$ has no bremsstrahlung background and can help to determine the f_0 structure. The observation of this decay was reported by SND group in 1997 [12].

Events with 5 photons detected both in the CsI and BGO calorimeter with no charged particles were selected.

The constrained fit with requirement of momentum and energy conservation removes the main background from $K_S \to \pi^0\pi^0$ decays when K_L makes fifth cluster in the calorimeter. To select $\pi^0\pi^0\gamma$ events the constrained fit finding two best combinations of photon pairs with π^0 masses was performed.

The main background to events of interest comes from the processes $e^+e^- \to \omega\pi^0$ with $\omega \to \pi^0\gamma$ decay and from $\phi \to \eta\gamma$, $\eta \to \pi^0\pi^0\pi^0$ events with two lost photons.

The invariant mass of $\pi^0\gamma$ for selected events is shown in Figure 5a. The number of $\omega\pi^0$ events was found to be 172±18 with the $\pi^0\gamma$ mass 785±3 MeV close to table value [10]. Background subtraction from $\eta \to \pi^0\pi^0\pi^0$ events was performed according to simulation.

The invariant mass of $\pi^0\pi^0$ system is shown in Figure 5b. The distribution demonstrates the increasing of event number with higher invariant mass (lower photon energy) and after background subtraction 47±13 $\pi^0\pi^0\gamma$ events have been found. Taking into account 14% detection efficiency the branching ratio
$B(\phi \to \pi^0\pi^0\gamma) = (0.98 \pm 0.26 \pm 0.20) \times 10^{-4}$
has been obtained. Systematic error comes from uncertainty in background subtraction.

The obtained value is in contradiction with visible branching ratio obtained for $\pi^+\pi^-\gamma$ mode

If $\phi \to \pi^+\pi^-\gamma$ decay goes through f_0 intermediate state the branching ratio with charged pions should be two times larger then ones for neutrals. The experimental ratio of visible branchings was found to be 0.4±0.2. So only assuming the strong destructive interference in charged mode these two channels could be described in one model. Using the four quark model suggested in paper[11] a combined fit of photon spectra from $\pi^0\pi^0\gamma$ events and $\pi^+\pi^-\gamma$ was performed. This fit gives the following f_0 parameters, coupling constants and branching ratio:
$M_{f_0} = 965 \pm 8$ MeV, $g^2_{KK}/4\pi = (2.34 \pm 1.17)$ GeV^2,

$g^2_{\pi\pi}/4\pi = (0.58 \pm 0.25)\ GeV^2,$

$B(\phi \to f_0\gamma) = (3.4 \pm 0.8) \times 10^{-4}.$

The influence of other resonances (ρ, σ) was estimated[13,14] to be about 10-15%.

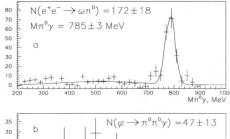

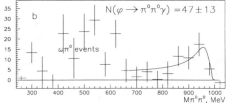

Figure 5: a. The $\pi^0\gamma$ invariant mass for 5 photons sample; b. The $\pi^0\pi^0$ invariant mass for 5 photons sample. Line is the theoretical prediction for f_0 production in case of four quark model for 3.4×10^{-4} branching ratio.

The obtained results and interpretations are in good agreement with the previously published by SND group[12,15].

4 Search for $\phi \to \eta\pi^0\gamma$

This channel also can be searched in the 5 photon mode with η decays into two photons. The observation of this decay mode was reported in 1997 by SND group[12]. After kaon background subtraction as described above the additional constrained fit finding one best combination of photon pairs with π^0 mass was performed. The invariant masses of coupled combinations of the rest 3 photons are shown in Figure 6. On the bump of background events from $\omega\pi^0$ and from three pion η decays the clear peak at η mass with 40 ± 13 events is seen. With 14% detection efficiency obtained by simulation the branching ratio

$B(\phi \to \eta\pi^0\gamma) = (1.09 \pm 0.28 \pm 0.20) \times 10^{-4}$

has been obtained. The systematic error corresponds to uncertainty in the background subtraction.

Statistic used in this analysis is not sufficient to study $\eta\pi^0$ invariant mass and make a conclusion about a_0 intermediate state in this decay.

The result in $\eta\pi^0\gamma$ mode is in good agreement with published by SND group[12,15].

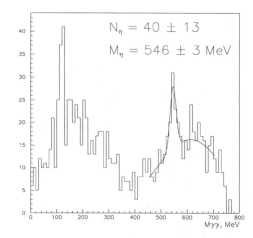

Figure 6: The $\gamma\gamma$ invariant mass spectrum for 3 combinations of "free" photons when one pair gives π^0 mass and fixed by constrained fit.

5 Search for $\phi \to \eta'\gamma$

The first observation of this decay was reported by CMD-2 collaboration last year[7,9]. The reported branching ratio was based on data sample, corresponding to 5.5 millions of ϕ decays. Using the same data sample, but adding information from BGO endcap calorimeter, the number of events identified as $\phi \to \eta'\gamma$ decays was increased.

The decay $\phi \to \eta'\gamma$ was searched in the mode, when η' decays into $\pi^+\pi^-\eta$ and $\eta \to \gamma\gamma$. This final state as well as one from much more intensive ϕ decay into $\eta\gamma$ has 2 charged particles and 3 photons. The events with all these particles detected were used for the constrained fit and as a result 1518 ± 39 events of $\eta\gamma$ were selected.

The decay into $\eta\gamma$ is the basic background for the $\eta'\gamma$ process and after anti–$\eta\gamma$ cut the scatter plot of the invariant mass of two hardest photons M_{12} versus the weakest photon energy ω_3 was studied. For $\eta'\gamma$ events M_{12} should be close to η mass 547.5 MeV, while ω_3 is a monochromatic 60 MeV photon. The Figure 7 presents the result together with simulation of $\phi \to \eta'\gamma$. The $9.2^{+3.5}_{-2.9}$ $\eta'\gamma$ events were found as a result of a fit taking into account detector resolutions and background distributions. Using number of events and the efficiencies obtained from the simulation, $\varepsilon_{\eta\gamma} = 16.6\%$ and $\varepsilon_{\eta'\gamma} = 12.2\%$, the following result has been obtained:

$$Br(\phi \to \eta'\gamma) = 1.35^{+0.55}_{-0.45} \cdot 10^{-4}.$$

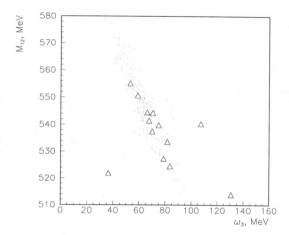

Figure 7: Invariant mass M_{12}(MeV) vs. ω_3(MeV) after constrained fit. Dots are simulation, triangles - experiment.

6 Summary

The CMD-2 detector is taking data at the VEPP-2M collider in Novosibirsk. Data collection was performed at the ϕ meson region with the total luminosity integral about 15 pb^{-1} About 25% of available data have been analysed. The new results in rare ϕ decays study
$B(\phi \to \mu^+\mu^-) = (2.57 \pm 0.70) \times 10^{-4}$,
$B(\phi \to \pi^+\pi^-) = (1.60 \pm 0.49) \times 10^{-4}$,
$B(\phi \to \pi^+\pi^-\gamma) = (0.38 \pm 0.16) \times 10^{-4}$,
$B(\phi \to \pi^0\pi^0\gamma) = (0.98 \pm 0.33) \times 10^{-4}$,
$B(\phi \to \eta\pi^0\gamma) = (1.09 \pm 0.35) \times 10^{-4}$,
$B(\phi \to \eta'\gamma) = (1.35^{+0.55}_{-0.45}) \times 10^{-4}$,
$B(\phi \to \mu^+\mu^-\gamma) = (1.27 \pm 0.68) \times 10^{-5}$
have been obtained. The results above for $\pi\pi\gamma$ can be interpreted as the $\phi \to f_0\gamma$ process with the f_0 decay into two charged or neutral pions. The data are better described in the frame of the four quark model and combined fit for neutral and charged pions gives:
$B(\phi \to f_0\gamma) = (3.4 \pm 0.8) \times 10^{-4}$, $M_{f_0} = 965 \pm 8$ MeV.

The $\pi^+\pi^-$ events with photons in the 100-300 MeV range were used to search for C violation process $\phi \to \rho\gamma$ and a new upper limit was obtained:
$B(\phi \to \rho\gamma) < 3 \times 10^{-4}$ at 90% CL.
The $\pi^+\pi^-$ events with photons in the narrow range around 362 MeV were used to search for CP,P violation decay of η into two charged pions and the new limit is:
$B(\eta \to \pi^+\pi^-) < 3 \times 10^{-4}$ at 90% CL.

References

1. G.A.Aksenov *et al.*, Preprint BudkerINP 85-118, Novosibirsk, 1985.
2. E.V. Anashkin *et al.*, ICFA Instrumentation Bulletin 5, 1988, p.18.
3. V.V. Anashin *et al.*, Preprint Budker INP 84-114, Novosibirsk, 1984.
4. R.R. Akhmetshin *et al.*, *Phys.Lett.* B **364**, 199 (1995).
5. R.R. Akhmetshin *et al.*, *Phys.Lett.* B **398**, 423 (1997).
6. R.R. Akhmetshin *et al.*, *Phys.Lett.* B **415**, 452 (1997).
7. R.R. Akhmetshin *et al.*, *Phys.Lett.* B **415**, 445 (1997).
8. E.Kuraev and V.Fadin, *Sov. Journal of Nucl. Phys.* **41**, 466 (1985).
9. R.R. Akhmetshin *et al.*, Proceedings of the 7th International Conference on Hadron Spectroscopy, Upton, NY, 1997, p.778.
10. Review of Particles Physics, The European Physical Journal, V3, Number 1-4, 1998.
11. N.N.Achasov, V.Gubin and E.P.Solodov, *Phys. Rev.* D **55**, 2672 (1997).
12. M.N.Achasov *et al.*, Proceedings of the 7th International Conference on Hadron Spectroscopy, Upton, NY, 1997, p.26.
13. N.N.Achasov and V.Gubin, *Phys. Rev.* D **57**, 1987 (1998).
14. N.N.Achasov and V.Gubin, Proceedings of the 7th International Conference on Hadron Spectroscopy, Upton, NY, 1997, p.574.
15. M.N.Achasov *et al.*, Preprint Budker INP 98-65, Novosibirsk, 1998.

RADIATIVE WIDTH OF THE a_2 MESON.

The SELEX Collaboration [1]
presented by
V. KUBAROVSKY

Institute for High Energy Physics, Protvino, Russia 142284
E-mail: vpk@fnal.gov

We present data on coherent production of the $(\pi^+\pi^-\pi^-)$ system by 600 GeV/c pion beam in the reaction $\pi^- + A \rightarrow A + (\pi^+\pi^-\pi^-)$ for the C, Cu and Pb targets. The Primakoff formalism was used for extracting the radiative width of the a_2 meson. We obtain a preliminary value $\Gamma(a_2^- \rightarrow \pi^-\gamma) = 225 \pm 25(\text{stat}) \pm 45(\text{syst})$ keV.

1 Introduction

Radiative decay widths of mesons and baryons are powerful tools for understanding the structure of elementary particles and for constructing a dynamical theory of hadronic systems. Straightforward predictions for radiative widths make possible the direct comparison of experimental data and theory.

The small value of the branching ratios of radiative decays makes it difficult to measure them directly, if there is a large background from the strong decay channels with $\pi^0 \rightarrow \gamma\gamma$ in the final state with one lost photon. For this reason most experimental data for these decays have been obtained using production reactions in the Coulomb field of the nuclei. Studying the inverse reaction (the so called Primakoff [2] formalism) $\gamma + \pi^- \rightarrow M^-$ provides a relatively clean method for the determination of the radiative widths.

The Primakoff reaction works better when the energy of the beam particles is increased. However, very good spatial resolution is needed to extract the signal at small t where the electromagnetic processes dominate over the strong interaction.

SELEX (E781) had a beam energy of 600 GeV and had a high resolution vertex detector which made it possible to explore the features of the Primakoff reaction.

The differential cross section for the Primakoff reaction

$$\pi^- + Z \rightarrow Z + a_2^-, \quad a_2^- \rightarrow \pi^-\pi^-\pi^+ \tag{1}$$

can be written as a function of the mass of the final state (M) and the square of the four-momentum transfer to the nucleus (t) as follows

$$\frac{d\sigma}{dt\,dM} = 8\pi\alpha Z^2(2J_{a_2}+1)\Gamma_\gamma \left(\frac{M}{M^2-m_\pi^2}\right)^3 g_\gamma(M)g_{a_2}(M)$$

$$\frac{2}{\pi}\frac{m_{a_2}^2\Gamma_{a_2}}{(M^2-m_{a_2}^2)^2+m_{a_2}^2 g_{a_2}^2(M)\Gamma_{a_2}^2}\frac{t-t_{min}}{t^2}|F(t)|^2 \tag{2}$$

Here α is the fine-structure constant, Z is the charge of the nucleus, $F(t)$ is the electromagnetic form factor of

the nucleus, J_{a_2} is the spin of the a_2 meson and t_{min} is the minimum four-momentum transfer:

$$t_{min} \approx (M^2 - m_\pi^2)^2/4P_{beam}^2, \tag{3}$$

where P_{beam} is the beam momentum. In our case $P_{beam} = 600$ GeV and for the a_2 meson with the mass 1.32 GeV we have

$$t_{min} \approx 2\cdot 10^{-6} GeV^2 \tag{4}$$

The t-distribution for the Primakoff reaction is characterized by a sharp peak with the maximum near $t = 2\cdot t_{min}$. The influence of the nuclear form factor is small in the region of $t < 0.001$ GeV^2 where we expect the Primakoff signal. It is less than 5% for the Pb target and 3% for the Cu target.

The mass-dependent total width and the partial widths $\Gamma(a_2 \rightarrow all)$, $\Gamma(a_2 \rightarrow \pi\gamma)$ and $\Gamma(a_2 \rightarrow \rho\pi)$ can be rewritten in terms of resonance widths as follows: $\Gamma(a_2 \rightarrow \pi\gamma) = \Gamma_\gamma g_\gamma(M)$ and $\Gamma(a_2 \rightarrow \rho\pi) = \Gamma_\rho g_\rho(M)$, where $g_\gamma(m_{a_2}) = g_\rho(m_{a_2}) = 1$. We used $g(a_2 \rightarrow all) = g(a_2 \rightarrow \rho\pi)$.

For the energy dependence of the widths we used

$$g_\gamma(M) = \left(\frac{k}{k_0}\right)^3 \frac{2k_0^2}{k^2+k_0^2} \tag{5}$$

$$g_{a_2}(M) = \left(\frac{q}{q_0}\right)^5 \frac{2q_0^2}{q^2+q_0^2} \tag{6}$$

where k and q are the momenta in the $\pi\gamma$ and $\rho\pi$ frames, respectively.

Numerically integrating over M and t from t_{min} up to 0.001 GeV^2 we get the cross section for the reaction (1) in this t-interval

$$\sigma_{\text{Primakoff}} = \int_{t_{min}}^{0.001} \frac{d\sigma}{dt\,dM}dt\,dM = \Gamma(a_2 \rightarrow \pi\gamma)\cdot C, \tag{7}$$

where $C = 27.2,\ 626,\ 4870\ (mb/GeV)$ for the C, Cu and Pb targets respectively.

1296

Table 1: Statistics of the experiment.

Target	Number of Events
Carbon	2 760 523
Copper	1 997 972
Lead	549 092

2 Cuts and Statistics of the experiment

The following cuts were applied in order to extract the reaction

$$\pi^- + A \rightarrow A + (\pi^- \pi^- \pi^+) \tag{8}$$

from the exclusive trigger stream of SELEX data:

1. The beam transition radiation detector shows that the beam particle is a pion.

2. For the quasielastic peak, the difference between the beam particle momentum and the sum of momenta of three secondaries must be less than 17.5 GeV/c.

3. The z-position of the interaction must be in the vicinity of a C, Cu or Pb target.

4. The energy deposition in the Photon-1 detector must be less than 2 GeV.

5. The average distance between the point of interaction and all tracks (beam and secondaries) must be less than 20, 25 and 75 μm for the C, Cu and Pb targets.

The event statistics are presented in Table 1.

3 Primakoff Production of the a2 Meson

The p_T^2 distribution of the (3π)-system for the Cu target is shown in Fig. 1a as an example. The distribution was fitted by the sum of two exponentials. The slope for the coherent production was found to be close to the previously published data [3]. The p_T^2 distributions for all targets have an enhancement in the region of small p_T^2. The two exponentials fit yields a slope for the second term greater than 1000 GeV^{-2} which corresponds to the Monte Carlo estimate for the Primakoff reaction.

Two p_T^2 regions were defined (the shaded regions in Fig. 1a). In the first one ($p_T^2 < 0.001$ GeV^2) the Primakoff mechanism dominates. We used the second one $0.0015 < p_T^2 < 0.0035$ GeV^2 for the estimation of the background from the coherent production of the (3π)-system.

The $\pi\pi\pi$ mass spectra for these two regions are shown in Fig. 1b for the Cu target. The result of the

subtraction of these histograms is shown in Fig. 1c. A clear a_2 meson signal is seen. The distributions were fitted using Eq. (2) integrated over t. The mass and width of the resonance were determined using the combined data from all targets: $M_{a_2} = 1304 \pm 5$ MeV, $\Gamma_{a_2} = 121 \pm 20$ MeV. The mass and the width of the peaks are in good agreement with the mass and width of the a_2 meson. The number of a_2 events was found to be 1587 ± 480, 5170 ± 590 and 2945 ± 400 for the C, Cu an Pb targets, respectively.

To make sure that the a_2 meson is produced via the Primakoff mechanism we divided the data into five p_T^2-bins each 0.00025 GeV^2 and repeated the subtraction procedure. We got 5 histograms for each target similar to Fig. 1c. After the fit of the mass spectra and the determination the number of a_2 mesons in each bin the p_T^2 distribution was extracted and shown in Fig. 1d. The slope was determined for every target. It is consistent with the Primakoff distribution smeared by the resolution of the experimental setup.

4 Monte Carlo

A simple Monte Carlo program was written to simulate the p_T^2-distribution for the Primakoff production of the a_2 meson. The p_T resolution had these different values for the different targets: $\sigma(p_{Tx}) = \sigma(p_{Ty}) = 16$, 17 and 20 MeV for the C, Cu and Pb targets, respectively.

The subtraction procedure efficiency was evaluated using the Monte Carlo simulation:

$$\varepsilon(subtr.) = \frac{\int\limits_0^{0.001} \frac{d\sigma'}{dt} dt - k \int\limits_{0.0015}^{0.0035} \frac{d\sigma'}{dt} dt}{\int\limits_0^{0.001} \frac{d\sigma}{dt} dt} \tag{9}$$

where σ is the Primakoff cross section, σ' is the Primakoff cross section smeared by the experimental resolution and k is the normalization factor. These efficiencies were found to be $\varepsilon = 0.697$, 0.630 and 0.488 for the C, Cu and Pb targets.

5 Absolute Normalization

The Primakoff approach gives the possibility to determine the radiative width only in the case when the absolute cross section of the reaction (1) is measured. This is the crucial point for the experiment. At the present stage of analysis we use the cross section for the reaction (8) from E272 [3]. We determined the product of $(L \cdot \varepsilon)$ and didn't need to evaluate the reconstruction efficiency ε alone. Here L is the luminosity. The experiment E272 was done in a beam with energy 200 GeV. We assumed

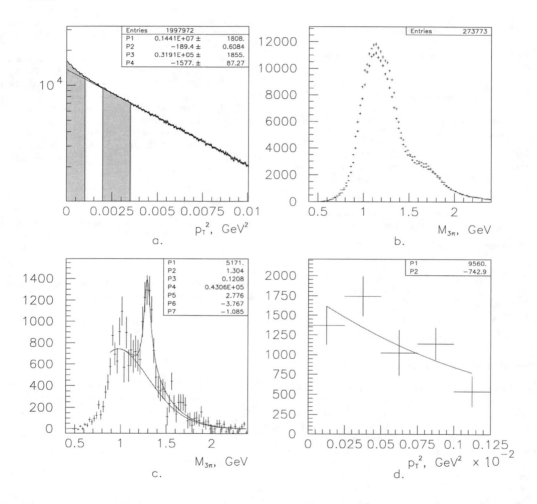

Figure 1: Primakoff production of the a_2 meson on Copper target in the reaction: $\pi^- + Cu \rightarrow Cu + (\pi^-\pi^+\pi^-)$. a. p_T^2 distribution of (3π)-system. The Primakoff signal is clearly seen in the region of small p_T^2 . b. $M_{3\pi}$ distribution. The first curve(light points) is for events with $p_T^2 < 0.001$ GeV^2, the second one is for events in the band $0.002 < p_T^2 < 0.0035$ GeV^2. c. Result of subtraction of two curves (see b). d. p_T^2 distribution of the a_2 meson.

that the coherent cross section of the reaction (8) is independent of the beam energy. The simplest estimation based on Regge theory confirms this assumption.

We determined the absolute cross section in the p_T^2 region close to zero to avoid the acceptance corrections:

$$(L \cdot \varepsilon) = \frac{\int_{t_{min}}^{t_{max}} \frac{dN}{dt} dt}{\int_{t_{min}}^{t_{max}} \frac{d\sigma}{dt} dt} \qquad (10)$$

where $\frac{dN}{dt}$ is the differential t-distribution from SELEX data, $\frac{d\sigma}{dt}$ is the differential cross section from E272 data, $t_{min} = 0, \ 0.001, \ 0.0015 \ \text{GeV}^2$ and $t_{max} = 0.020, \ 0.020, \ 0.006 \ \text{GeV}^2$ for three targets. $(L \cdot \varepsilon) = 881558, \ 149302, \ 17831$ events/mb for the C, Cu and Pb targets, respectively.

6 Radiative Width of the a_2 Meson

The radiative width of the a_2 meson was determined using the expression

$$\Gamma(a_2^- \to \pi^- \gamma) = \frac{N a_2}{C \cdot (L\varepsilon) \cdot CG \cdot BR \cdot \varepsilon(subtr.)} \qquad (11)$$

where $CG(a_2 \to \rho\pi)$=0.5 is the Clebsch-Gordon coefficient, $BR(a_2 \to \rho\pi)$=0.701 is the branching ratio, C is determined by numerical integration (see Eq. (7)), $(L\varepsilon)$ is the product of the luminosity and the reconstruction efficiency (see Eq. (10)). The radiative width was determined for all three targets: C, Cu and Pb. The results are shown in Fig. 2. As can be seen from this figure the results are consistent with each other, confirming the Coulomb production of the a_2 meson. The average value over all targets is

$$\Gamma(a_2^- \to \pi^- \gamma) = 225 \pm 20(\text{stat}) \pm 45(\text{syst}) \ \text{keV} \qquad (12)$$

The major sources of uncertainty in this result are the normalization procedure[3] and the errors in the determination of the number of a_2 events. The PDG value for the $\Gamma(a_2^- \to \pi^- \gamma) = 295 \pm 60$ keV is shown in Fig. 2 as well.

7 Conclusion

Based on a weighted average over the C, Cu and Pb targets of the values for $\Gamma(a_2^- \to \pi^- \gamma)$ we present the preliminary value of $225 \pm 20(\text{stat}) \pm 45(\text{syst})$ keV for the radiative width of the a_2 meson. Our result has the best world statistical error[4,5] for the radiative width of the a_2 meson. The systematic error can be reduced in the future by accurate measurement of the $(\pi^+ \pi^- \pi^-)$ coherent cross section using SELEX data.

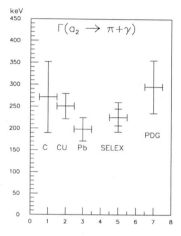

Figure 2: The radiative width $\Gamma(a_2^- \to \pi^- \gamma)$ for the C, Cu and Pb targets. The average of SELEX result (with statistical and systematic errors) and the PDG value are shown as well.

References

1. SELEX Collaboration: Carnegie-Mellon University, Fermilab, University of Iowa, University of Rochester, University of Hawaii, University of Michigan-Flint, Ball State, Petersburg Nuclear Physics Institute, ITEP (Moscow), IHEP (Protvino), Moscow State University, University of São Paulo, Centro Brasileiro de Pesquisas Fisicas (Rio de Janeiro), Universidade Federal da Paraiba, IHEP (Beijing), University of Bristol, Tel Aviv University, Max-Planck-Institut für Kernphysik (Heidelberg), University of Trieste, University of Rome "La Sapienza", INFN, Universidad Autonoma de San Luis Potosí, Bogazici University.

2. H.Primakoff, *Phys. Rev.* **81**, 899 (1951).

3. M.Zielinski *et al.*, *Z. Phys.* C **16**, 197 (1983).

4. S.Cihangir *et al.*, *Phys. Lett.* B **117**, 119 (1982).

5. *Phys. Rev. D* **54**, (1996)

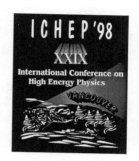

Parallel Session 10

Searches for New Particles at Accelerators

Convenors: Hendrik Weerts (MSU D0)

Robert McPherson (IPP, OPAL)

Standard Model Higgs at LEP

P. A. McNamara

Department of Physics, University of Wisconsin, 1150 University Ave, Madison, WI, 53706, USA
E-mail: Peter.McNamara@cern.ch

The Standard Model of elementary particle physics predicts and requires the existence of a neutral scalar Higgs boson. Recent fits to electroweak data suggest that it may be visible at current LEP energies. Searches for Higgs production at LEP at center of mass energies up to 183 GeV have been performed, and a combined limit on the mass of the Higgs boson is set at 89.8 GeV/c^2 at 95% CL. Preliminary searches for the Higgs boson at center of mass energy 189 GeV are also examined.

1 Introduction

The Standard Model[1] is very successful in describing the electroweak interactions of elementary particles. In the model, the Higgs mechanism[2] introduces a doublet of complex scalar fields, which allow the $W^{\pm}$ and Z^0 bosons to acquire mass via spontaneous symmetry breaking. An unavoidable consequence of the Higgs mechanism is the addition of a neutral colorless scalar particle, the Higgs boson, to the spectrum of elementary particles.

Although the mass of the Higgs boson is not predicted by the theory, the dependence of several electroweak observables on the Higgs mass allows a fit to electroweak data to give a prediction of the Higgs mass.[3] Figure 1 shows the prediction of this fit, which places an upper limit on the mass of the Higgs boson, requiring $m_H < 262$ GeV/c^2 at 95% CL, with preferred values close to the current LEP exclusion limits.

LEP is the Large Electron Positron collider located in Geneva, Switzerland. There are four experiments at LEP: ALEPH; DELPHI; L3; and OPAL. The LEP collider has been operating at center-of-mass energies above the W^+W^- threshold since 1996. A summary of the luminosities collected at each energy is given in Table 1. The latest results, from $\sqrt{s} = 183$ and 189 GeV, are discussed below.

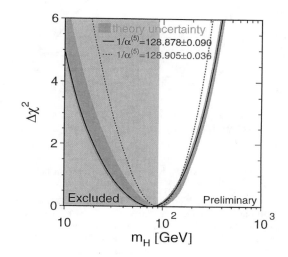

Figure 1: Higgs mass prediction from electoweak fit results. The width of the band represents an estimate of the theoretical error due to missing higher order corrections. The vertical band shows the 95% CL exclusion limit on m_H from direct searches.

2 Search Strategy

In e^+e^- collisions at LEP2 energies, the Higgs is predominantly produced in the "Higgs-Strahlung" process, $e^+e^- \to Z^\star \to HZ$. As Figure 2 shows, the HZ production cross section drops rapidly with Higgs mass, falling below 0.05 pb at threshold.[4] Thus, it is primarily with increases in the center of mass energy that Higgs mass sensitivity is extended at LEP.

Figure 3 shows that for Higgs masses below about 100 GeV/c^2, the Higgs branching ratio to $b\bar{b}$ is about 85%.[4] The predominance of b quarks in the final state means that b-tagging (the identification of b-jets, and rejection of udscg-jets) is a very important component of the Higgs boson searches. Each of the experiments has devoted significant effort to developing high purity b-tags

Table 1: Luminosities collected at LEP.

Year	$\sqrt{s}$ (GeV)	Lumi./Exp. (pb^{-1})
1989–1995	91.2	180
Fall 1995	130–136	6
1996	161–172	20
1997	181–184	57
	136	6
summer 1998	189	35

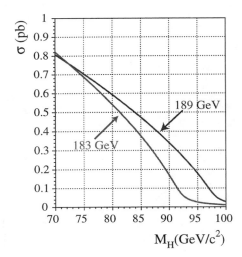

Figure 2: HZ production cross section for e^+e^- collisions with $\sqrt{s} = 183$ and 189 GeV.

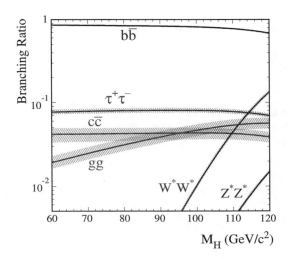

Figure 3: Branching Ratios for Standard Model Higgs boson. Below 100 GeV/c^2, the Higgs branching ratio to b quarks is approximately 85%.

in order to achieve the best possible seperation between signal and background.

The identification of b-jets relies on the fact that b hadrons have a much larger mass than other hadrons, and a long lifetime. The b hadron typically travels three millimeters before decaying, giving rise to a distinctive "Displaced Vertex" topology, which can be searched for using silicon vertex detectors. Jets from decays of b hadrons will also typically have a larger mass than jets from light quark decays, and may contain leptons with high transverse momentum relative to the jet direction.

Multivariate techniques, such as neural networks, likelihoods and linear discriminant analyses, are used by each of the experiments to improve the performance of the Higgs searches. The use of these techniques has been found by the experiments to be very helpful in improving the performance of the analyses without introducing significant additional systematic concerns. The techniques have been used both to improve the event selection in individual channels and to improving the tools most useful to the event selection. The ALEPH b-tag neural network, for example, incorporates six b-tag variables, considering lifetime information, jet shapes, and the presence of high transverse momentum leptons. This network incorporates the information from the several variables in an optimal way in order to achieve a b-jet efficiency 8% better than the most powerful single algoritm in the network when light quark rejection is at 90%.[5]

3 Search Channels

Higgs searches are done using four topologies, making use of the distinct characteristics of different HZ final states. The predominance of b decays in the final states means that most of the search channels will concentrate on b$\bar{b}$ final states. The different topologies are distinguished primarily by the decay products of the Z boson, implicitly assuming the Higgs decay to b$\bar{b}$ in most cases, with decays to τ leptons searched for using a seperate selection. The individual experiments classify the search channels differently, and use different techniques to distinguish Higgs signal from known standard model background processes, so it is not possible to describe them in detail here. The general strategies are determined by the Higgs topologies, however, and are the same for all experiments. Table 2 gives a summary of the topologies, which are described in more detail below.

Table 2: Summary of Higgs Topologies.

Channel	Branching Ratio
H$q\bar{q}$	64%
H$\nu\bar{\nu}$	20%
H$\ell^+\ell^-$	6.7%
H$\tau^+\tau^-$,$\tau^+\tau^- q\bar{q}$	8.9%

The HZ → Hqq̄ channel comprises about 64% of the total Higgs signal. The signature for signal in this channel is four hadronic jets, two of which are b-tagged. B-tagging eliminates much of the background, especially from WW → qq̄qq̄ events, but additional constraints are placed on events to further reduce background in this channel. Kinematic information, such as the fact that the non b-tagged jets are expected to have an invariant mass close to 91 GeV/c^2, and information about jet-jet angles are used to reduce background coming from qq̄gg events. The background contribution from ZZ → bbqq̄ is impossible to reduce, but it can be dealt with using background subtraction.

The HZ → Hνν̄ channel makes up approximately 20% of the total signal. The signature for this channel is a large missing mass and two b-tagged hadronic jets. Requiring the Higgs jets to be acoplanar and located in the central region of the detector reduces much of the background from radiative returns to the Z^0 resonance and other events with particles missing down the beamline. The b-tagging again plays an important role in reducing background, as less than 20% of the remaining background is expected to contain b decays.

The HZ → Hℓ⁺ℓ⁻ channel makes up approximately 6.7% of the total Higgs signal, half from electrons, half from muons. The signature is two energetic leptons and two b-tagged hadronic jets. By requiring the leptons to be isolated and energetic, most of the background from non-ZZ sources can be eliminated. B-tagging can then be applied to distinguish ZZ events from the HZ.

The ττ channels make up the remaining 8.9% of the signal cross section. There are two ττ channels, HZ → Hττ, and HZ → ττqq̄. The primary challenge in the ττ channels is in identifying τ candidates in a hadronic environment. The signatures for these channels are two τ candidates and two hadronic jets: b-tagged in the Hττ channel; not b-tagged in the ττqq̄ channel. The τ candidates are required to be isolated and energetic in order to reduce the probability of fake τ candidates, and to reduce the remaining background, the event is then required to be kinematically consistent with an HZ decay.

4 Results

The searches for the various HZ final states have been performed by the LEP experiments, and the data analyzed for center of mass energies up to 183 GeV, and a first look has been taken at data from summer 1998 at center of mass energy 189 GeV. No evidence for Higgs production is seen by any of the four experiments, whose results are summarized below.

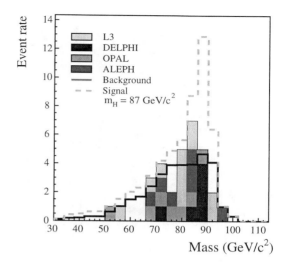

Figure 4: Higgs candidate mass distribution from 183 GeV data. In addition to the masses of observed candidates, the expected mass distribution for background events, and the expected mass distribution given a Higgs mass of 87 GeV/c^2 is shown. Only the 11 most significant L3 candidates are shown.

4.1 Results from 183 GeV

In 1997, LEP collected an integrated luminosity of approximately 57 pb^{-1} per experiment with a center of mass energy near 183 GeV.

The observations from this sample were consistent with standard model background predictions. For a total of 47.8 events expected by the four experiments, 40 were observed. Figure 4 shows the mass distributions for these candidates, which are summarized in Table 3.

Table 3: Events Observed at 183 GeV.

Experiment	Num. Observed	Num. Expected
ALEPH[6]	7	7.2
DELPHI[7]	6	7.5
L3[8]	18	24.2
OPAL[9]	9	8.9

The good agreement between data and standard model background suggests that the Higgs boson has not been discovered in this data sample. Accordingly, each collaboration has placed a lower bound on the mass of the Higgs boson based upon their data sample. Where DELPHI, L3 and OPAL have chosen to perform background

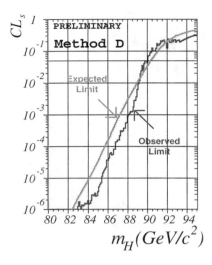

Figure 5: LEP combined limit from 183 GeV data.

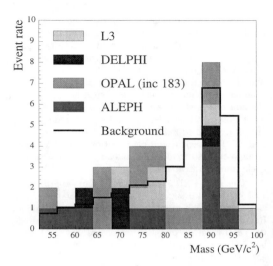

Figure 6: Higgs mass distribution from 189 GeV data.

subtraction, ALEPH performs no background subtraction on this sample. The limits set by the four experiments are consistent with the expectations based on their individual sensitivities, and are summarized in Table 4.

Table 4: Individual Limits at 183 GeV.

Experiment	Limit Observed (GeV/c^2)	Limit Expected (GeV/c^2)
ALEPH[6]	87.9	85.5
DELPHI[7]	85.7	86.5
L3[8]	87.6	86.8
OPAL[9]	88.3	86.2

The LEP Higgs working group has performed a preliminary combination of the results from the individual experiments from $\sqrt{s} = 183$ GeV. The combination was repeated using four different statistical techniques to verify the robustness of the result. The limits produced by the four techniques are in very good agreement, with a spread of only 150 MeV/c^2, implying that the limits produced are quite robust. The lowest of these mass limits is taken as the combined limit on the mass of the Higgs boson, giving a lower limit on the Higgs boson mass, $m_H > 89.8$ GeV/c^2 at 95% CL.[10] Figure 5 shows the limit curve obtained using method D.

4.2 Preliminary Results from 189 GeV

In 1998, LEP has been running with center of mass 189 GeV. As of mid-july, luminosities of approximately 35 pb^{-1} per experiment were collected. Using this sample, a first look for the Higgs boson at this new energy has been performed by each of the four experiments.

Figure 6 shows the mass distribution of the candidates observed at 189 GeV. With 39.4 events expected by the four experiments together, 40 were observed, in good agreement with expectation. Rather than an excess, which might signal the presence of a Higgs, there appears to be a deficit of events in the region around 90 GeV/c^2. The number of candidates observed by each experiment is summarized in Table 5.

Table 5: Candidates Observed at 189 GeV.

Experiment	Num. Observed	Num. Expected
ALEPH[11]	10	8.0
DELPHI[12]	4	4.8
L3[13]	22	18.0
OPAL[14]	4	8.6

The LEP collaborations have computed very preliminary limits based on this sample. As Table 6 shows, the limits are again broadly in line with what would be expected, based on the sensitivity of the individual exper-

1306

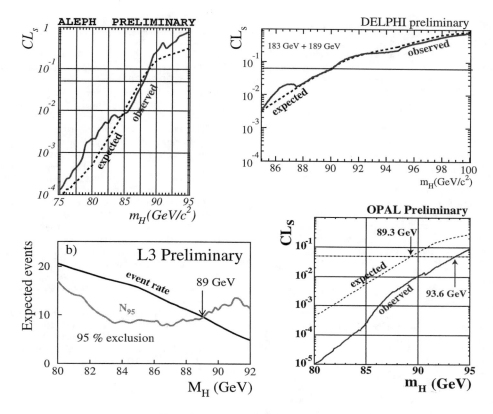

Figure 7: Higgs Limits from 189 GeV data.

iments. The OPAL limit, however, is significantly higher than expected, most likely as a result of the downward fluctuation in the number of candidates observed. In fact, OPAL estimates the probability of getting a better limit to be only 3.1%.[14] Figure 7 shows the limit curves for the individual experiments.

Table 6: Individual Limits at 189 GeV.

Experiment	Limit Observed (GeV/c^2)	Limit Expected (GeV/c^2)
ALEPH[11]	88.0	87.5
DELPHI[12]	89.6	89.5
L3[13]	89.0	88.6
OPAL[14]	93.6	89.3

5 Prospects

Extrapolating from current analysis performance, it is possible to estimate the combined LEP sensitivity to Higgs boson production using a model of the current LEP Higgs searches.[15] This year, if LEP collects the expected 150 pb^{-1} per experiment at 189 GeV, a Higgs mass limit of 97 GeV/c^2 is expected in the absence of a signal. Similarly, one can estimate that LEP will be sensitive to a Higgs discovery this year if the mass is below 94 GeV/c^2. By the end of LEP, the collaborations hope to collect total luminosities of 200 pb^{-1} per experiment at $\sqrt{s} = 200$ GeV. It is estimated that with a sample this size, LEP would be sensitive to a 5σ discovery of a Higgs signal with mass up to 107 GeV/c^2. Similarly, in the absence of a signal, it is estimated that with this sample, LEP should be able to exclude Standard Model Higgs bosons with masses up to 109 GeV/c^2.

Conclusion

Although the standard model predicts the existence of a Higgs boson with a mass near the region of current LEP sensitivity, data collected at center of mass energies up to 189 GeV show no evidence of HZ production. A preliminary lower limit of 89.8 GeV/c^2 on the Higgs boson mass is set using data from the four LEP experiments from energies up to 183 GeV together. In the next few years, LEP hopes to extend this limit by almost 20 GeV/c^2, if the Higgs is not discovered in the meantime.

Acknowledgements

I would like to thank S. Armstrong, R. Clare, F. Di Lodovico, M. Felcini, E. Gross, P. Igo-Kemenes, P. Janot, S. de Jong, W. Murray, F. Teubert, D. Treille, S. L. Wu and the entire LEP Higgs Working Group for their assistance. This work is supported by US Department of Energy, grant DE-FG0295-ER40896.

References

1. S. Glashow, *Partial-symmetries of weak interactions*, Nucl. Phys. **22** (1961) 579;
   S. Weinberg, *A Model of Leptons*, Phys. Rev. Lett. **19** (1967) 1264;
   A. Salam, *Weak and electromagnetic interactions*, in *Elementary Particle Theory*, ed. N. Svartholm (Almqvist and Wiksells, Stockholm, 1968).

2. P. W. Higgs, *Broken Symmetries, Massless Particles and Gauge Fields*, Phys. Lett. **12** (1964) 132;
   F. Englert and R. Brout, *Broken Symmetry and the Mass of Gauge Vector Mesons*, Phys. Lett. **13** (1964) 321;
   G. S. Guralnik, C. R. Hagen, T. W. Kibble, *Global Conservation Laws and Massless Particles*, Phys. Rev. Lett. **13** (1964) 585;
   J. F. Gunion, H. E. Haber, G. Kane, S. Dawson, *The Higgs Hunter's Guide* (Addison-Wesley, 1990) and references therein.

3. M. Grünewald, *these proceedings*.

4. E. Gross, B. A. Kniehl, and G. Wolf, *Production and Decay of the Standard Model Higgs Boson at LEP-200* Z. Phys. **C63** (1994) 417;
   erratum: *ibid*, **C66** (1995) 321.

5. ALEPH Collaboration, R. Barate, *et al*, *Search for the neutral Higgs bosons of the MSSM in e^+e^- collisions at $\sqrt{s}$ from 130 to 172 GeV*, Phys. Lett. **B412** (1997) 173.

6. ALEPH Collaboration, *Preliminary Limits from Searches for Neutral Higgs Bosons in e^+e^- Collisions at Centre-of-mass Energies of 181–184 GeV*, ALEPH 980029 CONF 98-017, 1998.

7. DELPHI Collaboraton, *Search for neutral Higgs bosons in e^+e^- collisions at $\sqrt{s} = 183$ GeV*, DELPHI 98-95 CONF 163, 1998.

8. L3 Collaboration, *Search for the Standard Model Higgs Boson in e^+e^- Collisions at $\sqrt{s} = 183$ GeV*, CERN-EP/98-52, 1998.

9. OPAL Collaboration, *Search for Higgs Bosons in e^+e^- Collisions at $\sqrt{s} \approx 183$ GeV*, OPAL PN-366, 1998.

10. The LEP Higgs Working Group, *Bounds on SM and neutral MSSM Higgs boson masses using combined data of the four LEP experiments collected at $\sqrt{s}$ up to 183 GeV*, ALEPH 98-069 PHYSIC 98-028, DELPHI 98-144 PHYS 790, L3 Note 2310, OPAL TN-558, 1998.

11. ALEPH Collaboration, *Search for Neutral Higgs bosons in e^+e^- collisions at $\sqrt{s} = 189$ GeV*, ALEPH 98-075 CONF 98-043, 1998.

12. DELPHI Collaboration, *Updates of DELPHI results on searches with 189 GeV data*, DELPHI 98-137, CONF 198, 1998.

13. L3 Collaboration, *Preliminary results on New Particle Searches at $\sqrt{s} = 189$ GeV*, L3 Note 2309, 1998.

14. OPAL Collaboration, *Search for Higgs Bosons at $\sqrt{s} = 189$ GeV*, OPAL PN-361, 1998.

15. E. Gross, A. Read, D Lellouch, *Prospects for the Higgs Boson Search in e^+e^- Collisions at LEP 200*, CERN-EP/98-094, 1998.

SEARCHES FOR HIGGS BOSONS BEYOND THE STANDARD MODEL AT LEP

K. DESCH

EP Division, CERN, CH–1211 Geneva 23, Switzerland
E-mail: Klaus.Desch@cern.ch

Searches for various types of Higgs bosons in models beyond the Standard Model (SM) are reported. They are based on data taken by the four LEP experiments ALEPH, DELPHI, L3 and OPAL at centre-of-mass energies up to 189 GeV. The searches comprise neutral and charged Higgs bosons in models with two Higgs doublets, in particular the Minimal Supersymmetric Standard Model (MSSM). Non standard decay modes of Higgs bosons like decays into photons and invisible decays are also covered. None of the searches revealed a significant excess over the expectation from SM background processes. Therefore new limits on Higgs boson masses and other model parameters are reported.

1 Introduction

In the Standard Model (SM) the Higgs mechanism allows the particles to acquire masses while through spontaneous symmetry breaking the underlying gauge theory remains renormalisable. As a consequence a complex scalar field doublet has to be introduced giving rise to one physical scalar particle, the Higgs boson. Recent search results on the SM Higgs boson are reported in [1].

Many models beyond the Standard Model also predict one or more Higgs bosons. Models with two Higgs field doublets where one doublet couples to up-type quarks and the other to down-type quarks and charged leptons (Type II Higgs Doublet Models) predict five physical Higgs bosons, three being neutral (h, H, A) and two charged (H^+, H^-). In particular supersymmetric models require two Higgs field doublets. The Minimal Supersymmetric Standard Model (MSSM) is considered an attractive extension of the SM and Higgs boson searches within this model are a major topic of this contribution.

In the SM, the Higgs boson decays predominantly into the heaviest particles kinematically possible, i.e. b-quarks and τ leptons for the mass range currently accessible. Beyond the SM, this feature might be modified and different Higgs boson decays might be dominant. Especially decays into photon pairs and decays into massive invisible particle (e.g. a pair of neutralinos in the MSSM) are also investigated by the four LEP experiments.

The results summarised here are reported in the following contributed papers on searches for neutral Higgs bosons in the MSSM and in Type II Higgs Doublet Models [3], on charged Higgs boson searches [4], on searches for invisibly decaying Higgs bosons [5] and on Higgs bosons decaying into photons [6]. A summary on Higgs physics at LEP2 can be found elsewhere [2].

Since 1995 the energy of LEP has been increased in steps up to 189 GeV in 1998. This increase opens the possibility to search for Higgs bosons in a previously unaccessible mass range. In this contribution, partly preliminary results from the four LEP experiments ALEPH, DELPHI, L3 and OPAL are reported from searches at centre-of-mass energies between 130 GeV and 189 GeV. The analyses are based on integrated luminosities per experiment of approximately 5 pb^{-1} at 130-136 GeV, 10 pb^{-1} at 161 GeV, 10 pb^{-1} at 170-172 GeV and 55 pb^{-1} at 183 GeV. In many cases very preliminary updates of the analyses are reported using approximately 35 pb^{-1} of data per experiment at 189 GeV from the current LEP run [7].

2 MSSM Higgs Bosons

In the neutral Higgs sector of the MSSM three Higgs bosons are predicted, the lighter h and the heavier H being CP-even and A being CP-odd. The h boson can be produced in the same process as the Standard Model Higgs boson $(e^+e^- \rightarrow hZ)$. Furthermore in the MSSM there exists the pair production process of h and A: $e^+e^- \rightarrow hA$. These two processes show a complementarity in their cross sections:

$$\sigma_{hZ} = \sin^2(\beta - \alpha)\,\sigma_{HZ}^{SM} \tag{1}$$
$$\sigma_{hA} = \cos^2(\beta - \alpha)\,\bar{\lambda}\,\sigma_{\nu\bar{\nu}}^{SM}. \tag{2}$$

Here, $\tan\beta$ denotes the ratio of the vacuum expectation values of the two Higgs doublets and α is the mixing angle of the CP-even fields, $\bar{\lambda}$ is a kinematic factor and σ_{HZ}^{SM} ($\sigma_{\nu\bar{\nu}}^{SM}$) are the SM cross sections for HZ ($\nu\bar{\nu}$) production.

Thus, in regions where one process is suppressed, the other is enhanced. While for large regions of the MSSM parameter space both h and A predominantly decay into $b\bar{b}$, in some regions the decays into $\tau^+\tau^-$, $c\bar{c}$ or even gg become important. Furthermore, if kinematically possible also the decay into pairs of supersymmetric particles, e.g. into neutralinos $\chi_0^1\chi_0^1$ is possible, giving rise to invisible Higgs decays. Also, for $2m_{A^0} < m_{h^0}$, the decay $h^0 \rightarrow A^0 A^0$ is possible.

2.1 Search channels

The search channels for the $e^+e^- \to hZ$ process are the same as for the SM process. The corresponding analyses reported in [1] are (in some cases after reoptimisations including dedicated searches for the $h^0 \to A^0 A^0$ case) adopted for the MSSM search. The pair production process is searched for in the $e^+e^- \to h^0 A^0 \to b\bar{b}b\bar{b}$ and $e^+e^- \to h^0 A^0 \to b\bar{b}\tau^+\tau^-$ final states. In some analyses, the $h^0 \to A^0 A^0$ case is also treated in the $e^+e^- \to A^0 A^0 A^0 \to b\bar{b}b\bar{b}b\bar{b}$ final state.

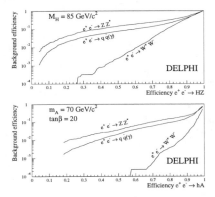

Figure 1: b-tagging: signal efficiency for the $h^0 Z^0 \to b\bar{b}q\bar{q}$ and $h^0 A^0 \to b\bar{b}b\bar{b}$ processes versus the background efficiency for major backgrounds (DELPHI).

In most cases the analyses adopted use highly efficient multivariate techniques like likelihood analyses or artificial neural network (ANN) analyses. The tagging of b-flavoured jets exploiting the relatively long lifetime of b hadrons plays a major role in all analyses. b-tagging proceeds predominantly through the reconstruction of displaced vertices using the excellent spatial resolution of the silicon micro-vertex detectors of all LEP experiments. As an example the background suppression potential of DELPHI is shown in Fig. 1. Furthermore, the tagging of τ leptons in a multi-hadronic environment has made significant process recently.

The searches for the $b\bar{b}b\bar{b}$ ($b\bar{b}\tau^+\tau^-$) final states have typical signal efficiencies of 50–60% (25–45%) at very low levels of the remaining SM background (30–60 fb for $b\bar{b}b\bar{b}$ and 2–20 fb for $b\bar{b}\tau^+\tau^-$). The remaining background consists mainly of four–fermion final states like $W^+W^-$and Z^0Z^0pair production, the latter partially being an irreducible background.

2.2 Limits on the MSSM parameter space

None of the four LEP experiments observed an excess over the SM background expectation. This negative search result was used to constrain the allowed parameter space of the MSSM. All four experiments interpret the results within a constrained MSSM (*benchmark scan*). The assumptions in this benchmark scan are: unification of the scalar–fermion masses m_0 and at the grand unification (GUT) scale. m_0 and the SU(2) gaugino mass term M_2 are fixed at 1 TeV. The supersymmetric Higgs mass parameter μ is set to -100 GeV and the scalar-top mixing parameter A_t is either set to 0 (no scalar-top mixing) or $\sqrt{6}$ TeV (maximal scalar-top mixing). Furthermore the top quark mass is fixed at its measured value. These benchmark scans are motivated and defined in the LEP2 *yellow report* [2]. The obtained limits from the four LEP experiments on the masses of the h^0 and A^0 bosons are listed in table 1.

Table 1: Mass limits for h^0 and A^0 at 95% CL within the MSSM benchmark scan for $\tan\beta > 1$ for the four LEP experiments. The limits obtained including data at 189 GeV are very preliminary.

	183 GeV		189 GeV	
	m_{h^0}	m_{A^0}	m_{h^0}	m_{A^0}
	(GeV)	(GeV)	(GeV)	(GeV)
ALEPH	72.2	76.1	79.7	79.7
DELPHI	74.4	75.3	77.5	78.3
L3	70.7	71.0	74.0	74.0
OPAL	70.5	72.0	73.5	74.5

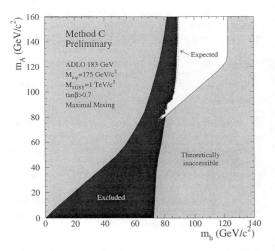

Figure 2: MSSM benchmark scan: excluded region in (m_{A^0}, m_{h^0})-plane at 95% CL (combination of the four LEP experiments).

2.3 LEP-wide combination

The LEP Higgs working group has performed a first preliminary combination of the MSSM Higgs searches for this conference. The combination has been done in the benchmark scan scenario for data taken at centre-of-mass energies up to and including 183 GeV. The statistical methods [8] C and D have been used to obtain limits. Upper limits of 77 GeV on m_{h^0} and 78 GeV on m_{A^0} have been derived at the 95% CL (see Fig. 2). A range $0.8 < \tan\beta < 2.1$ can be excluded at 95% CL for the case of no scalar-top mixing and $m_{top} = 175$ GeV (Fig. 3). This result disfavours the low-tan β-solution of the infrared fixed-point scenario. However, for higher values of m_{top} the exclusion is less stringent.

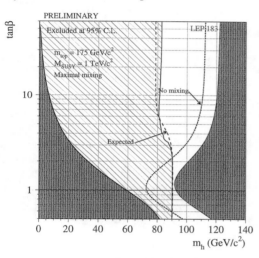

Figure 3: MSSM benchmark scan: excluded region in $(\tan\beta, m_{h^0})$-plane at 95% CL (combination of the four LEP experiments).

2.4 General MSSM scans

Three of the LEP collaborations (ALEPH, DELPHI, OPAL) have also performed an interpretation of the search results in a less constrained parameter space of the MSSM. In these general MSSM parameter scans, all relevant model parameters (m_0, M_2, μ, A_t) are allowed to vary within a wide range. Also the case of $\tan\beta < 1$ is investigated as well as a variation of the top quark mass within its current measurement error is considered. The three experiments consistently observe unexcluded parameter sets with values for m_{h^0} and m_{A^0} below the benchmark scan exclusions. Some of those parameter sets can be excluded using external constraints from $b \rightarrow s\gamma$ and $\Delta\rho$ measurements, and combination with direct search

results for supersymmetric particles. Also some of the unexcluded parameter sets might exhibit non-physical behaviour (charge and colour breaking (CCB) minima of the MSSM lagrangian). However, even after rejection of these parameter sets, unexcluded points remain and the absolute limits on m_{h^0} and m_{A^0} are lower than in the benchmark case (Fig. 4). ALEPH concludes that still for a large fraction ($> 99.9\%$) of the explored parameter sets the limits obtained from the benchmark scan remain valid. The additional unexcluded points have typically simultaneously large values of $|A_t|$ and $|\mu|$. They are currently subject to more detailed investigation.

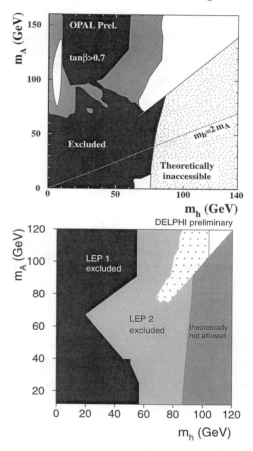

Figure 4: General MSSM parameter scan: excluded region at 95% CL in the (m_{A^0}, m_{h^0}) plane from OPAL (upper figure) and DELPHI (lower figure). For OPAL, the grey area is only excluded when a criterion to reject CCB minima is applied. For DELPHI, in the dotted area some parameter sets are excluded, some are not.

3 Type-2 Two Higgs Doublet Model Interpretation

The searches for the processes $e^+e^- \to h^0 Z^0$ and $e^+e^- \to h^0 A^0$ have been used by DELPHI and OPAL to set mass limits in Type-2 Two Higgs Doublet Models using the complementarity of the cross section in eqs. 1 and 2. In Fig. 5 the DELPHI exclusion is shown in the (m_{A^0}, m_{h^0})-plane. Under the assumption that of dominating b-decays of the Higgs bosons, which is valid for $\tan \beta \geq 1$, the light grey area can be excluded at 95% CL. The dark grey area can be excluded even if non-b-decays are dominant (for $\tan \beta < 1$).

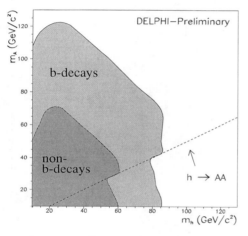

Figure 5: Exclusion at 95% CL in the (m_{A^0}, m_{h^0})-plane for Type-2 Two Higgs Doublet models.

4 Charged Higgs Bosons

In the MSSM at tree level, charged Higgs bosons are predicted to be heavier than the W bosons. Radiative corrections do not change this prediction significantly. Therefore a discovery of a charged Higgs boson lighter than the W boson would put strong constraints on the MSSM parameter space.

4.1 Search channels

At LEP2, charged Higgs bosons are pair–produced through the process $e^+e^- \to H^+ H^-$. It is assumes that charged Higgs bosons decay either hadronically, $H^+ \to q\bar{q}'$, or leptonically, $H^+ \to \tau^+ \nu_\tau$ with unknown branching ratios but assuming $\mathrm{BR}(H^+ \to q\bar{q}') + \mathrm{BR}(H^+ \to \tau^+ \nu_\tau) = 1$. Therefore, three final

states have to be considered: the leptonic channel, $e^+e^- \to H^+ H^- \to \tau^+ \nu_\tau \tau^- \bar{\nu}_\tau$, the semi-leptonic channel, $e^+e^- \to H^+ H^- \to \tau \nu_\tau q\bar{q}'$, and the hadronic channel, $e^+e^- \to H^+ H^- \to q\bar{q}' q\bar{q}'$. The pair production of W bosons is a large and for $m_{H^\pm} \sim m_{W^\pm}$ irreducible background for all charged Higgs boson final states.

4.2 Mass limits

None of the LEP experiments have reported an excess over the expected SM background. This result is used to set limits on the charged Higgs boson mass. The partially preliminary lower mass limits from the four LEP experiments using data up to 183 GeV are listed in table 2. The ALEPH result for data up to 183 GeV energy is shown in Fig. 6. An updated result from DELPHI using data up 189 GeV energy yields a limit of 59.2 GeV.

Table 2: Mass limits on charged Higgs bosons at 95% CL independent of $\mathrm{BR}(H^+ \to \tau^+ \nu_\tau)$ for data up to 183 GeV centre-of mass energy from the four LEP experiments.

	$m_{H^\pm} >$
ALEPH	59 GeV
DELPHI	57 GeV
L3	56.5 GeV
OPAL	56 GeV

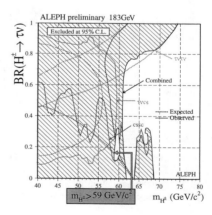

Figure 6: Charged Higgs search: ALEPH exclusion for data up 183 GeV centre-of-mass energy.

5 Invisible Higgs Decays

The process $e^+e^- \to h^0 Z^0$ where the Higgs boson decays invisibly is searched for in the channels $Z^0 \to \ell^+ \ell^-$

(leptonic channel) and $Z^0 \to q\bar{q}$ (hadronic channel). For BR($h^0 \to$ invisible) = 1 and an assumed SM production cross section, the obtained mass limits at 95% CL are listed in Tab. 3. Fig. 7 shows the number of events excluded at 95% CL together with the number of expected events for BR($h^0 \to$ invisible) = 1 for the L3 analysis.

Table 3: Mass limits on invisibly decaying Higgs bosons at 95% CL for BR($h^0 \to$ invisible) = 1 from the four LEP experiments.

	$m_{h^0} >$	$\sqrt{s}$	
ALEPH	80 GeV	$\leq$ 183 GeV	$Z^0 \to \ell^+\ell^-, Z^0 \to bb$
DELPHI	84.9 GeV	$\leq$ 189 GeV	
L3	83.6 GeV	$\leq$ 183 GeV	
OPAL	81 GeV	$\leq$ 183 GeV	only $Z^0 \to q\bar{q}$

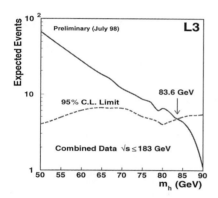

Figure 7: Invisibly decaying Higgs search: exclusion for BR($H^0 \to$ invisible)=1 from L3.

6 Higgs And Photons

There are models in which the lightest CP even Higgs boson does not couple to fermions [9] (*fermiophobic Higgs*). For the mass range accessible at LEP2 this leads to large BR($h^0 \to \gamma\gamma$). The OPAL analysis for this models searches for the process $e^+e^- \to h^0 Z^0 \to \gamma\gamma Z^0$ in all decay modes for the Z^0. No excess has been observed. The obtained Higgs boson mass limit in the fermiophobic Higgs model is 92.6 GeV for data up to 189 GeV centre-of-mass energy (Fig. 8).

A DELPHI analysis performs a search for Higgs boson final states involving one, two or three photons in the final state. The negative search results are interpreted as limits on effective $HZ\gamma$ and $H\gamma\gamma$ couplings [10]. Under the assumption that all unconstrained effective couplings in

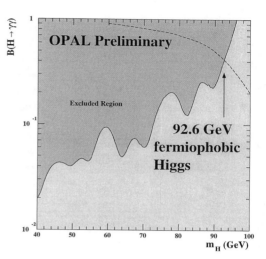

Figure 8: Search Higgs Bosons decaying into photons: exclusion in the fermiophobic Higgs model (OPAL). The full line is the 95% CL upper limit on BR($H \to \gamma\gamma$) assuming a SM production rate. The dashed line is the predicted BR($H \to \gamma\gamma$) in the fermiophobic Higgs model.

the model are equal, the exclusion shown in Fig. 9 is obtained.

7 Summary

The four LEP collaborations have reported updated results on searches for Higgs bosons beyond the SM. Those searches cover a wide range of possible extensions of the SM. A detailed interpretation of the searches in the framework of the MSSM has been worked out. Since many of the searches look for certain topologies rather than specific models, the searches are also prepared for the unexpected. By the time of writing these proceedings the delivered luminosity from LEP at 189 GeV energy has already exceeded 100 pb^{-1}. Together with the good prospects for 1999 and 2000, Higgs searches will remain exciting.

Acknowledgments

I would like to thank many of my colleagues from the four LEP collaborations who provided me with information, plots, and help. Especially the support of the LEP Higgs working group is thankfully acknowledged.

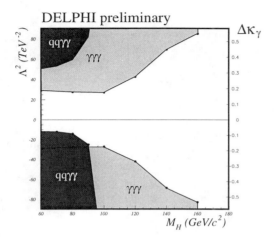

DELPHI preliminary

Figure 9: Limits on anomalous HZγ and Hγγ couplings as a function of the Higgs boson mass (DELPHI). The dark grey area is excluded by the qq̄γγ search, the light grey area is excluded by the γγγ search.

References

1. P. McNamara, *The search for the Standard Model Higgs Boson at LEP*, these proceedings.

2. E. Accomando *et al.*in *Physics at LEP 2*, CERN 96-01, 1996, p.351.

3. ALEPH Collaboration, *Limits from Searches for the Neutral Higgs bosons in e^+e^- collisions at centre-of-mass energies of 181-184 GeV*, Contributed paper, Abstract #896;
   DELPHI Collaboration, *Search for neutral Higgs bosons in e+e- collisions at $\sqrt{s}$ = 183 GeV*, Contributed paper, Abstract #200;
   DELPHI Collaboration, *MSSM Parameter Scan at $\sqrt{s}$ = 183 GeV*, Contributed paper, Abstract #209;
   DELPHI Collaboration, *Interpretation of the search for Neutral Higgs Bosons in two-Doublet Models*, Contributed paper, Abstract #350;
   L3 Collaboration, *Search for Neutral Higgs Bosons of the Minimal Supersymmetric Standard Model in e^+e^- Interactions at $\sqrt{s}$ = 130 -183 GeV*, Contributed paper, Abstract #487;
   OPAL Collaboration, *Search for Higgs Bosons at $\sqrt{s}$ = 183 GeV*, Contributed paper, Abstract #355.
   The LEP Higgs Working Group, *Higgs Boson Searches: Combined Results from the Four LEP Collaborations Using Data Collected at 183 GeV Energy*, Contributed paper, Abstract #582.

4. ALEPH Collaboration, *Search for charged Higgs bosons in e^+e^- collisions at centre-of-mass energies of 181-184 GeV*, Contributed paper, Abstract #900;
   ALEPH Collaboration, *Limits from searches of Higgs bosons in the MSSM with a parameter scan*, Contributed paper, Abstract #897;
   DELPHI Collaboration, *Search for Charged Higgs Bosons at LEP 2*, Contributed paper, Abstract #214;
   L3 Collaboration, *Search for Charged Higgs Bosons in e^+e^- Collisions at Centre-of-Mass Energies between 130 and 183 GeV*, Contributed paper, Abstract #489;
   OPAL Collaboration, *Search for Higgs Bosons at $\sqrt{s}$ = 183 GeV*, Contributed paper, Abstract #356.

5. ALEPH Collaboration, *Search for invisible decays of the Higgs boson in e^+e^- collisions at centre-of-mass energies of 181-184 GeV*, Contributed paper, Abstract #902;
   DELPHI Collaboration, *A search for invisible Higgs bosons produced in e^+e^- interactions at LEP2 energies*, Contributed paper, Abstract #218;
   L3 Collaboration, *Missing Mass Spectra in Hadronic Events from e^+e^- Collisions at $\sqrt{s}$ = 183 GeV and Limits on Invisible Higgs Decays*, Contributed paper, Abstract #491;
   OPAL Collaboration, *Search for Higgs Bosons at $\sqrt{s}$ = 183 GeV*, Contributed paper, Abstract #357;

6. DELPHI Collaboration, *Search for the Higgs boson in events with isolated photons at LEP II*, Contributed paper, Abstract #216;
   OPAL Collaboration, *Search for Higgs Bosons Decaying in Photons at $\sqrt{s}$ = 183 GeV*, Contributed paper, Abstract #353.

7. ALEPH Collaboration, *Search for Neutral Higgs Bosons in e^+e^- collisions at $\sqrt{s}$ = 189 GeV*, Contributed paper, Abstract #905;
   DELPHI Collaboration, *Updates of DELPHI results with 189 GeV data*, Contributed paper, Abstract #219;
   L3 Collaboration, *Preliminary Results on New Particles Searches from the L3 Experiment at $\sqrt{s}$ = 189 GeV* , Contributed paper, Abstract #485;
   OPAL Collaboration, *Search for Higgs bosons in e+e- collisions at $\sqrt{s}$ = 189 GeV*, Contributed paper, Abstract #1066.

8. *Lower bound on the Standard Model Higgs boson from combining the results of the four LEP experiments*, CERN-EP/98-046.

9. A. G. Akeroyd, *Phys. Lett.* B **368**, 89 (1996);
   A. Stange, W. Marciano, and S. Willenbrock, *Phys. Rev.* D **49**, 1354 (1994).

10. K.Hagiwara, R.Szalapski, D. Zeppenfeld, *Phys. Lett.* B **318**, 155 (1993).

MSSM AND HIGGS SEARCH AT THE TEVATRON

Juan A. Valls

Rutgers, The State University of New Jersey, P.O. Box 849, Piscataway, NJ 08904, USA
E-mail: valls@fnal.gov

In this paper a summary of present CDF and DØ results on supersymmetry and Higgs searches at Tevatron is presented. Analysis include results from a variety of signatures: missing E_T and jets (or leptons and jets) for squark/gluino searches, trileptons and missing E_T for gaugino searches and, for the first time, charmed-tagged jets and missing E_T as a new signature for stop quark pair production. We show also results for various searches of standard and non-standard model Higgs bosons in hadron collisions. The seeked signature is Higgs associated production with a vector boson. Different channels involve jets and leptons from vector boson decays and b-tagged jets or photons from Higgs decays.

1 Introduction

The Fermilab Tevatron $p\bar{p}$ collider has provided collisions at $\sqrt{s} = 1.8$ TeV during two recent running periods. In Run IA (1992-93) a total accumulated 20 pb^{-1} of data was collected by CDF and 15 pb^{-1} by DØ, and in Run IB (1995-96) 109 pb^{-1} and 90 pb^{-1} were collected by CDF and DØ respectively. The results presented here correspond to partial and complete analysis of this large sample of data.

Hints for supersymmetric (SUSY) particles like squarks, gluinos, and gauginos have been searched in classical channels at hadron colliders, including missing E_T ($E\!\!\!/_T$) + jets or $E\!\!\!/_T$ + jets + leptons. Other SUSY searches with signatures involving photons in the final state are reported in a different review[1] within these proceedings. For the first time, CDF reports a search for direct stop pair production using c-tagged jets + $E\!\!\!/_T$, improving the sensitivity reached in previous analysis.

Although with limited sensitivity, Higgs boson searches including new channels and signatures have been performed and combined. Search techniques have been developed and proved to be promising for the next Tevatron run, when an approximate twenty-fold increase in the total integrated luminosity is expected.

Results are interpreted in general unification scenarios along the line of supergravity (SUGRA), which mediates the interaction needed for supersymmetry breaking. This results in a great simplification on the number of parameters at the unified energy scale, which are defined to be M_0 for the common scalar (squark and slepton) masses, and $M_{1/2}$ for the common gaugino masses. Three other parameters define the Higgs sector of the model: $\tan\beta$, the ratio of the vacuum expectation values of the two Higgs doublets, A_0, the universal trilinear coupling constant, and the sign of μ, the mixing parameter in the Higgsino mass matrix.

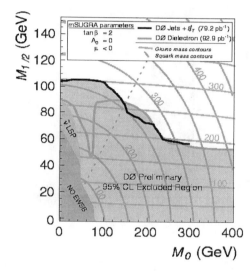

Figure 1: DØ 95% CL excluded region in the $M_0 - M_{1/2}$ plane.

2 SUSY Searches

2.1 Squark and Gluino Searches

Both CDF and DØ have performed searches for squarks and gluinos in events with jets + $E\!\!\!/_T$. The signature arises from the ultimate decays of these sparticles into jets and lightest neutralinos. The analysis reported here corresponds to an update of the previous DØ[2] results with 79.2 pb^{-1} of Run IB data. CDF results[3] for 19 pb^{-1} of Run IA data will also be briefly summarized.

The DØ search requires one jet with $E_T > 115$ GeV and two more jets with $E_T > 25$ GeV. A $E\!\!\!/_T > 75$ GeV as well as a total scalar sum of jet E_T's (excluding the leading jet) > 100 GeV are required. Further cuts require the $E\!\!\!/_T$ to be uncorrelated in ϕ with any jet. After these cuts, remaining standard model (SM) backgrounds are $t\bar{t}$, W/Z and QCD production. The DØ analysis is interpreted in the context of a minimal SUGRA model with fixed $\tan\beta$, A_0, and sign of μ. The results are shown in Figure 1 as a function of M_0 and $M_{1/2}$. All models with $M_{\tilde{q}} < 250$ GeV/c^2 are excluded. For models with $M_{\tilde{q}} = M_{\tilde{g}}$, a common mass below 260 GeV/c^2 is excluded.

The limits derived from the Run IA CDF analysis are shown in Figure 2 as the hashed region area in the $M_{\tilde{q}} - M_{\tilde{g}}$ plane. In this case a modified SUGRA-inspired framework is used to interpret the results. The model

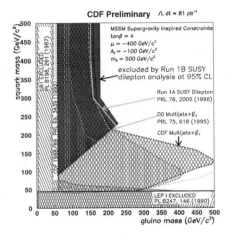

Figure 2: CDF 95% CL excluded region in the $M_{\tilde{q}} - M_{\tilde{g}}$ plane from the lepton + $\not{E}_T$ and the LS dilepton analysis.

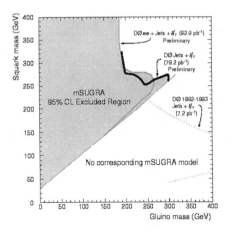

Figure 3: DØ 95% CL excluded region in the $M_{\tilde{q}} - M_{\tilde{g}}$ plane.

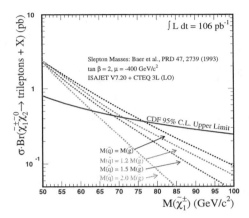

Figure 4: CDF 95% CL limits on $\sigma_{\tilde{\chi}_1^\pm \tilde{\chi}_2^0} BR(\tilde{\chi}_1^\pm \tilde{\chi}_2^0 \to 3l+X)$ versus $\tilde{\chi}_1^\pm$ mass for representative points in the MSSM parameter space.

is specified by using as input parameters $M_{\tilde{q}}$, $M_{\tilde{g}}$, M_A, $\tan\beta$, and the magnitud and sign of μ. At the 95% CL CDF excludes common squark and gluino masses $M_{\tilde{q}} = M_{\tilde{g}} < 216$ GeV/c^2, and $M_{\tilde{g}} < 173$ GeV/c^2, independent of squark masses. DØ has also produced an experimental limit in the $M_{\tilde{q}} - M_{\tilde{g}}$ plane using the 79.2 pb^{-1} of Run IB data. Their results are shown in Figure 3. The results of these analysis do not change substantially as parameters are varied within the theoretical framework.

The limits from the jets + $\not{E}_T$ analysis can be extended by searching for signatures with isolated lepton pairs. These arise from cascade decays of gluinos to quark pairs via charginos/neutralinos which decay to leptons. These channels produce relatively clean experimental signals, in particular if the lepton pairs are required to be like-sign (LS). Latest CDF LS dilepton (e, μ) results [4] and DØ ee results [5] from these channels in the $M_{\tilde{q}} - M_{\tilde{g}}$ plane are shown in Figures 2 and 3 respectively.

2.2 Associated Gaugino Pair Production

Both CDF [6] and DØ [7] have searched for associated chargino-neutralino pair production $\tilde{\chi}_1^\pm \tilde{\chi}_2^0$ using the complete Run I data sample. Assuming SUGRA constrains the chargino can decay to $\tilde{\chi}_1^\pm \to \tilde{\chi}_1^0 l\nu$ and the neutralino to $\tilde{\chi}_2^0 \to l^+ l^-$, giving rise to very distinct signatures with three leptons + $\not{E}_T$ and small SM backgrounds. Only the CDF analysis will be reported here. The search include four channels: $e^+ e^- e^\pm$, $e^+ e^- \mu^\pm$, $\mu^+ \mu^- e^\pm$, and $\mu^+ \mu^- \mu^\pm$. No candidates are found while 0.3 events are expected

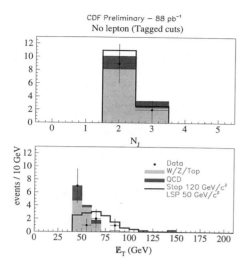

CDF Preliminary – 88 pb^{-1}
No lepton (Tagged cuts)

N_J

events / 10 GeV

- Data
  W/Z/Top
  QCD
  Stop 120 GeV/c^2
  LSP 50 GeV/c^2

$\not{E}_T$ (GeV)

Figure 5: Jet multiplicity (top) and $\not{E}_T$ (bottom) distributions for the pretagged $t\bar{t}$ sample for data, SM backgrounds and signal.

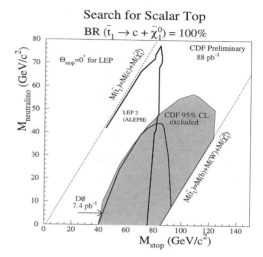

Search for Scalar Top
BR $(\tilde{t}_1 \to c + \tilde{\chi}_1^0) = 100\%$

$M_{neutralino}$ (GeV/c^2)

CDF Preliminary
88 pb^{-1}

$\Theta_{stop}=0°$ for LEP

$M(\tilde{t}_1)=M(c)+M(\tilde{\chi}_1^0)$

LEP 2
(ALEPH)

CDF 95% CL
excluded

$M(\tilde{t}_1)=M(b)+M(W)+M(\tilde{\chi}_1^0)$

D0
7.4 pb^{-1}

M_{stop} (GeV/c^2)

Figure 6: CDF 95% CL excluded region in the $M_{\tilde{t}} - M_{\tilde{\chi}_1^0}$ plane.

from background processes like Drell-Yan (plus a fake lepton), $b\bar{b}$, $c\bar{c}$ and diboson events.

Results are shown in Figure 4 as 95% CL upper limits on the sum of the branching ratio times cross section for the four channels: $\sigma_{\tilde{\chi}_1^\pm \tilde{\chi}_2^0} BR(\tilde{\chi}_1^\pm \tilde{\chi}_2^0 \to 3l + X)$. This represents a lower limit of $M_{\tilde{\chi}_1^\pm} > 81.5$ GeV/c^2 and $M_{\tilde{\chi}_2^0} > 82.2$ GeV/c^2 for $\tan\beta = 2$, $\mu = -600$ GeV/c^2 and $M_{\tilde{q}} = M_{\tilde{g}}$.

2.3 Direct Stop Quark Pair Production

As a consequence of the large top mass, the MSSM predicts a large splitting between the two top quark superpersymetry partners, with the lightest one significantly lighter than the top. This stop could be then observable at Tevatron.

CDF has recently finished a preliminar analysis for direct $\tilde{t}\bar{\tilde{t}}$ production using the decays $\tilde{t} \to c\tilde{\chi}_1^0$ with 88.6 pb^{-1} of data from Run IB. The signature is two acolinear jets from charmed quarks, significant $\not{E}_T$, and the absence of leptons in the final state. The analysis requires 2 or 3 jets with $E_T > 15$ GeV and a $\not{E}_T > 40$ GeV. The $\not{E}_T$ is required to be uncorrelated in ϕ with any jet. After these cuts the main background sources arise from SM W/Z + jets, QCD and $t\bar{t}$ events. Figure 5 shows the jet multiplicity and $\not{E}_T$ distributions for data and the different background contributions after the selection cuts. To further reduce backgrounds, a c-tagged jet probability algorithm is, for the first time, utilized

in CDF with acceptable efficiencies. After this cut 11 events are left to be compared to a background estimate of 13.5 ± 4.

The 95% CL exclusion region in the $M_{\tilde{t}} - M_{\tilde{\chi}_1^0}$ parameter space is shown in Figure 6. A lower limit of $M_{\tilde{t}} > 120$ GeV/c^2 for $M_{\tilde{\chi}_1^0} = 38$ GeV/c^2 is obtained. The analysis significantly extends the limit as compared to previous analysis from LEP and D0[8].

Further sensitivity to stop quark pair production is achieved by studying the decays $\tilde{t} \to \tilde{\chi}_1^\pm b$, $l\nu b\tilde{\chi}_1^0$ and $\tilde{\chi}_1^\pm \to \tilde{\chi}_1^0 l\nu$ with signatures including b jets, leptons and $\not{E}_T$. This analysis is in progress within the CDF collaboration. Finally stop quarks could also be detectable through top decays $t \to \tilde{t}\tilde{\chi}_1^0$ and subsequent $\tilde{t} \to b\tilde{\chi}_1^\pm$, $bl\nu\tilde{\chi}_1^0$. Results from this channel have already been presented by the CDF collaboration[9].

3 Higgs Searches

3.1 Standard Model Higgs

The standard model provides the simplest mechanism for spontaneous symmetry breaking through the introduction of a scalar field doublet. This leaves a single observable scalar particle, the Higgs boson, with unknown mass but fixed couplings to other particles.

At Tevatron, one of the Higgs production mechanism more likely to be observed is associated production $V + H$ with $V = W, Z$. Both CDF and D0 have searched for this channel using different signatures. Both experiments

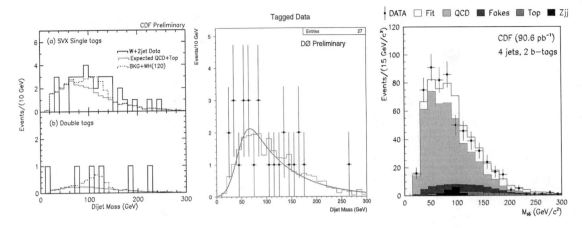

Figure 7: $b\bar{b}$ invariant mass distributions for the CDF and D∅ SM Higgs selections: $WH \to l\nu b\bar{b}$ (left, middle plots) and $VH \to jjb\bar{b}$ ($V = W, Z$, right plot) compared to the expected backgrounds.

look for an isolated high p_T lepton + $\not{E}_T$ to identify the W decay, and jets with b-tags to identify the $b\bar{b}$ Higgs decay [10]. CDF also searches in high jet multiplicity events from hadronic decays of W and Z bosons with the requirement of at least two b-tagged jets [11]. No deviation from the expected SM background contributions is observed in the reconstructed $b\bar{b}$ invariant mass distributions. A likelihood fit is then made to the shape of the observed distributions using a combination of signal and different SM background sources. Figure 7 shows the $b\bar{b}$ invariant mass distribution for the CDF and D∅ $l +$ jets WH selection, as well as for the CDF all-hadronic VH selection. In all cases, the data is compared to the expected SM background sources.

D∅ has also searched for invisible $\nu\bar{\nu}$ decays of Z bosons in association with a Higgs boson. Muon-tagged jets are utilized to identify b-quark decays of the Higgs, and a $\not{E}_T > 35$ GeV cut is further required to reject most of the SM backgrounds.

Figure 8 shows the 95% CL upper limit results on $\sigma(WH)BR(H \to b\bar{b})$ and $\sigma(ZH)BR(H \to b\bar{b})$ for the individual CDF and D∅ results. Figure 9 shows the 95% CL upper limit results on $\sigma(VH)\beta(H \to b\bar{b})$ with $V = W, Z$ for the CDF all-hadronic analysis. The same figure shows the combined limit obtained from the all-hadronic and lepton + jets results. The sensitivity of these searches is limited by statistics to a cross section approximately two orders of magnitude larger than the predicted cross section for standard model Higgs production. For the next Tevatron run we hope for an approximately twenty-fold increase in the total integrated luminosity and a substantial increase in the total acceptances

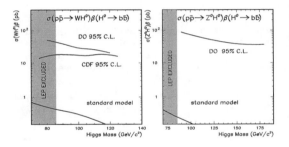

Figure 8: CDF and D∅ 95% CL upper limits on $\sigma(WH)BR(H \to b\bar{b})$ (left) and $\sigma(ZH)\beta(H \to b\bar{b})$ (right).

by improving the single and double b-tagging efficiencies, and the use of a more efficient, dedicated Higgs trigger.

3.2 MSSM Higgs

A more complex symmetry breaking mechanism occurs in the MSSM, when five observable scalar states are predicted: two charged Higgs particles ($H^{\pm}$), two CP-even scalars (h^0, H^0) and one CP-odd pseudoscalar (A^0). In order to describe the MSSM Higgs sector one has to introduce two additional parameters to describe the properties of the scalar particles and their interactions with gauge bosons and fermions: $\tan\beta$ and the mixing angle α in the neutral CP-even sector.

CDF has presented results on direct and indirect

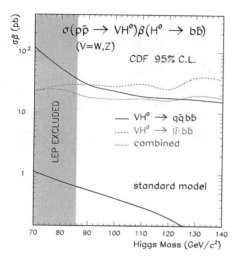

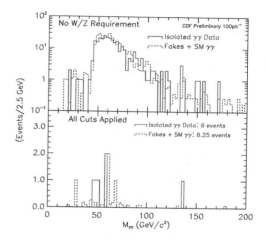

Figure 9: CDF 95% CL upper limits on $\sigma(VH)BR(H \to b\bar{b})$ with $V = W, Z$ from the individual and combined $VH \to l\bar{\nu}b\bar{b}$, $q\bar{q}b\bar{b}$ channels.

Figure 10: CDF diphoton invariant mass distribution with photon selection cuts only before (top) and after (bottom) the W/Z selection cuts.

Higgs production via top quark decays $t \to H^+ b$ with $H^+ \to \tau\nu$ showing sensitivity to the high and low $\tan\beta$ regions in the $M_{H^\pm}$-$\tan\beta$ plane [12].

More recently, CDF has started an analysis seeking for neutral MSSM Higgs particles produced via the associated production $b\bar{b}\Phi$ with $\Phi = h^0, H^0, A^0$ and $\Phi \to b\bar{b}$. The signature is high multiplicity jets with at least two of them tagged as b-quark jets. In the MSSM, the Yukawa couplings between the Higgs scalars and the b quarks are enhanced for large $\tan\beta$ values with respect to the standard model. With basic parameter choices for the SUSY scale and the stop mixing, CDF derived 95% CL lower mass limits for the neutral Higgs sector of the MSSM as a function of $\tan\beta$. These are preliminary results still not presented by the time of this conference.

3.3 Higgs Decaying to Two Photons

Finally, several extended Higgs models allow a light neutral scalar Higgs, with standard model strength couplings to vector bosons but suppresed couplings to fermions. Such a "bosophilic" Higgs decays dominantly to $\gamma\gamma$ for masses below 90 GeV/c^2 and is most easily detected in the associated production mode with a vector boson.

Both CDF and DØ have analyzed the complete Run I data sample seeking for diphoton signatures + leptons, $\not{E}_T$, or jets. The CDF diphoton sample consists

of events with two isolated central ($|\eta| < 1.0$) photon candidates with $E_T > 25$ GeV. DØ uses a similar sample with slightly different thresholds. These high p_T photon samples suffer from significant backgrounds from jets misidentified as photons, calculated from independent fake control samples with modified isolation requirements. CDF requires further the presence of a high p_T lepton (e, μ), or $\not{E}_T$, or two jets, covering all possible decay channels of the vector bosons. 6 evenys are left with a predicted background of 6.2 ± 2.1. DØ only requires the presence of two additional jets. They find 4 candidates with an expected background of 6.0 ± 2.1 events. Figures 10 and 11 show the diphoton invariant mass distributions before and after the vector boson selection cuts, compared to SM background expectations for CDF and DØ respectively.

CDF results for the 95% CL upper limits on $BR(H \to \gamma\gamma)$ assuming SM production cross section for $W/Z + H$ are shown in Figure 12. The DØ results are presented in Figure 13 as 95% CL upper limits on the production cross section times branching ratio compared to the full "bosophilic" cross section.

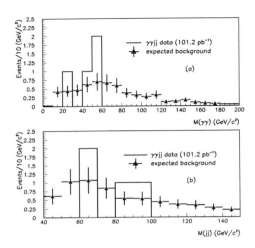

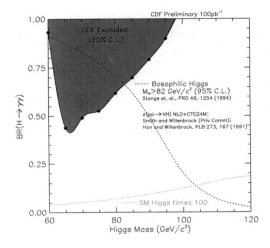

Figure 11: DØ (a) diphoton invariant mass and (b) jet mass distributions compared with expected backgrounds.

Conclusions

A summary of the present undergoing analysis results from the CDF and DØ collaborations at the Tevatron has been presented. A broad range of signatures corresponding to very different channels have been investigated for hints of new physics in the framework of supersymmetry and several Higgs models including the SM, the MSSM Higgs sector, and "bosophilic" Higgs scenarios. No evidence for any of the channels covered has been found with the present statistics and 95% CL limits on different model parameters have been stablished. The analysis of the Run I data is still not completed and several new results will be reported in the near future.

The Tevatron is scheduled to operate again in Run II by the fall of 1999 with substantial luminosity improvements and an energy reach increased to $\sqrt{s} = 2$ TeV. A twenty-fold increase in statistics is expected by the first years of operation. The Tevatron will be able to significantly probe a large fraction of the expected SUSY mass range. With sufficient integrated luminosity provided by further Tevatron runs, a light intermediate-mass Higgs in the range $80 < M_H < 130$ GeV/c^2 or event above this threshold, will also be accesible.

Figure 12: CDF 95% CL upper limit on $BR(H \rightarrow \gamma\gamma)$ as a function of the Higgs mass compared with "bosophilic" and SM branching ratio predictions.

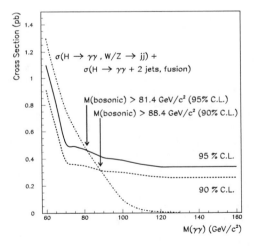

Figure 13: DØ 90% and 95% CL upper limits on the total production cross section times branching ratio for associated production of vector bosons and Higgs decaying to $\gamma\gamma$.

Acknowledgements

I am very grateful to all members of the CDF and DØ exotics physics group at Fermilab for providing me the different material shown in this paper.

References

1. M. Paterno, *Gauge Mediated SUSY Searches at the Tevatron*, these proceedings.
2. B. Abbott *et al.*, Fermilab-Conf-97/357-E.
3. F. Abe *et al.*, Phys. Rev. **D56**, 1357 (1997).
4. F. Abe *et al.*, Phys. Rev. Lett. **76**, 2006 (1996). J.Done, Fermilab-Conf-96-371-E, *Proc. of 1996 DPF, Minneapolis, 1996*.
5. S. Abachi *et al.*, Fermilab-Conf-96-427-E, *Proc. of XXVIII ICHEP, Warsaw, Poland, 1996*.
6. F. Abe *et al.*, Phys. Rev. Lett. **76**, 4307 (1996).
7. B. Abbott *et al.*, Fermilab-Pub-97/201-E.
8. S. Abachi *et al.*, Phys. Rev. Lett. **76**, 2222 (1996).
9. P.J. Wilson, Fermilab-Conf-97-241-E, *Proc. of Les Rencontres de Physique de la Vallee d'Aosta, La Thuile, 1997*.
10. F. Abe *et al.*, Phys. Rev. Lett. **79**, 3819 (1997). S. Abachi *et al.*, *Proc. of XXVIII ICHEP, Warsaw, Poland, 1996*.
11. F. Abe *et al.*, Fermilab-Pub-98-252, Aug. 1998. Submitted to Phys. Rev. Lett.
12. F. Abe *et al.*, Phys. Rev. Lett. **79**, 357 (1997). Fermilab-Conf-97-213-E, *Proc. of the XII Hadron Coolider Physics, SUNY, Stony Brook, 1997*.

SUSY WITH NEUTRALINO LSP AT LEP

P. REBECCHI

CERN, CH-1211 Geneva 23, Switzerland
E-mail: Pierpaolo.Rebecchi@cern.ch

A review of the main results obtained by the four LEP experiments on the search for new particles predicted by supersymmetric theories is reported, based mainly on the recent data collected at centre of mass energies of 189 GeV. The analyses presented in this note are carried out in the framework of the Minimal Supersymmetric Standard Model (MSSM) where the lightest supersymmetric particle (LSP) is assumed to be the lightest neutralino. No evidence of such signal is found, although candidate events consistent with the expectations from Standard Model processes are observed. Limits at 95 % C.L. on the production cross-sections and on the particle masses obtained by the four LEP experiments and bounds on the parameters of MSSM are reported.

1 Introduction

An overview of the main results from the searches performed at LEP [1] is presented, for new particles predicted by the Minimal Supersymmetric Standard Model (MSSM). The results obtained by the four LEP experiments [2] are based mainly on the analyses of the $\sim 40 \text{pb}^{-1}$ of data taken per experiment during the beginning of the '98 high energy run of LEP. Some results combining all LEP energies are also reported.

The searches presented in this note are carried out in the framework of the MSSM where the lightest supersymmetric particle (LSP) is assumed to be the lightest neutralino. In this model, ordinary particles and s-particles are distinguished by their R-parity, a multiplicative quantum number, which is assumed to be conserved to ensure leptonic and baryonic quantum number conservation. The parameters of MSSM relevant to the present searches are the masses M_1 and M_2 of the gaugino sector, the universal mass m_0 of the scalar lepton sector, the Higgs mass parameter μ, and the ratio $\tan\beta$ of vacuum expectation values of the two Higgs doublets. The GUT relation $M_1 = \frac{5}{3}\tan^2\theta_W M_2 \approx 0.5 M_2$ at the electroweak scale in not always assumed to be true.

Searches for different s-particles have discovery power in different regions in the MSSM space: chargino and neutralino searches play an important role for either large M_2 and $|\mu| << M_2$ (the chargino is mainly higgsino) or for small M_2, $|\mu| >> M_2$ and large m_0 (the chargino is mainly gaugino and the sneutrino is heavy); sleptons and neutralino searches play an important role for small M_2, $|\mu| >> M_2$ and small m_0 (the chargino is mainly gaugino and the sneutrino is light).

Three important phenomenological aspects of the considered model must be underlined: the lightest supersymmetric particle (LSP) is stable and escapes detection; each s-particle must eventually decay into a state which contains an odd number of LSPs; and, in collider experiments, s-particles can only be produced in even numbers (usually two at a time). The first direct consequence of these three aspects is that the main important variable in s-particles searches is ΔM, defined as the difference $M_{s-particle}$ - $M_{\tilde{\chi}^0}$, where $M_{s-particle}$ is the mass of the s-particle searched for. In fact, close to the kinematic limit, the visible energy of the event is roughly twice ΔM: for low values of ΔM (5-10 GeV/c^2) signal events are very similar to two-photon interactions, for high values of ΔM (50-80 GeV/c^2) signal events are very similar to four fermions events.

1.1 Importance of ΔM in s-particle search

The direct dependence of the efficiency in detecting s-particles on ΔM can be seen in Figure 1. The figure shows iso-curves of chargino detection efficiency in the $(M_{\tilde{\chi}^\pm}, M_{\tilde{\chi}^0})$ plane.

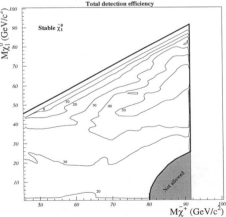

DELPHI $\tilde{\chi}^+\tilde{\chi}^-$ efficiencies (183 GeV)

Figure 1: Chargino pair production total detection efficiencies (%) at 183 GeV in the $(M_{\tilde{\chi}^\pm}, M_{\tilde{\chi}^0})$ plane.

The highest efficiency is obtained when $M_{\tilde{\chi}^\pm} \sim 2 M_{\tilde{\chi}^0}$, so for moderate values of ΔM (~ 40 GeV/c^2); due to the selection criteria against background from two-photon interactions the efficiency decreases when the chargino is degenerate in mass with the neutralino, for low values of ΔM (< 10 GeV/c^2); on the other side, when the neutralino becomes light the signal is more similar to WW background and so the efficiency is reduced by the selection criteria used to suppress 4-fermion backgrounds: this happens for high values of ΔM (> 70 GeV/c^2). Therefore the efficiency is very sensitive to the mass of the searched s-particle and to ΔM. Efficiencies are estimated from the Monte Carlo simulation of a huge range of $M_{s-particle}$ and ΔM values and then parametrised and extended to all MSSM space.

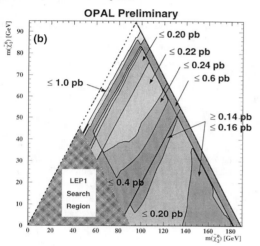

Figure 2: The contours of the 95% C.L. upper limit for $e^+e^- \to \tilde{\chi}_2^0 \tilde{\chi}_1^0$ production cross-sections at $\sqrt{s} = 188.7$ GeV are shown assuming Br($\tilde{\chi}_2^0 \to \tilde{\chi}_1^0 Z^{0(*)}$)=100%. The region for which $M_{\tilde{\chi}_2^0} + M_{\tilde{\chi}_1^0} < M_Z$ is not considered in this analysis.

Another phenomenological variable which depends on ΔM is the average length of flight of the searched s-particle. Different searches are optimised for different ranges of ΔM:

- 3 GeV/$c^2 < \Delta M$: the s-particle decays promptly and the standard search is applied. The main signature is the missing energy and different selection criteria are optimised for different ΔM regimes.

- 500 MeV/$c^2 < \Delta M < 3$ GeV/c^2: the secondary vertex of the s-particle decay is not yet detectable. Only events with ISR photons are selected in order to reduce the high cross-section of two-photon interactions background.

- 150 MeV/$c^2 < \Delta M < 500$ MeV/c^2: the decay length is short and the s-particle still decays inside the

detectors. Main signatures are the high impact parameter and the presence of kinks.

- $\Delta M < 150$ MeV/c^2: the s-particle decays outside the detector and stable heavy charged particles are searched.

2 Experimental results

Several candidates have been observed in data but in agreement with Standard Model processes. No evidence of signal from s-particles has been found. Therefore the experimental results are: limits at 95 % C.L. on the production cross-sections and on the s-particle masses, new constraints on the parameters of MSSM.

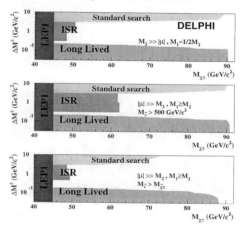

Figure 3: Regions in the plane $(M_{\tilde{\chi}_1^\pm}, \Delta M)$ excluded at 95% C.L., for $\sqrt{s}$=189 GeV, using: the standard search for high ΔM charginos; the search for soft particles accompanied by ISR; the search for long lived charginos. The three scenarios are the ones which allow low ΔM: the lightest chargino is higgsino; the lightest chargino is gaugino and $M_{\tilde{\nu}} > 500$ GeV/c^2; the lightest chargino is a gaugino and $M_{\tilde{\nu}} > M_{\tilde{\chi}_1^\pm}$.

2.1 Neutralinos and charginos

Direct production of $\tilde{\chi}_2^0 \tilde{\chi}_1^0$ and of $\tilde{\chi}^+ \tilde{\chi}^-$ has been searched at LEP. Since the characteristics of neutralino and chargino events mainly depends on ΔM, different selection criteria have been applied. Model-independent upper limits are obtained at 95% C.L. for the production cross-section. Exclusion limits are determined from the observed number of events, the signal detection efficiencies and their uncertainties, and the number of background events expected and their uncertainties. To obtain a limit on a given $(M_{\tilde{\chi}_2^0}, M_{\tilde{\chi}_1^0})$ or $(M_{\tilde{\chi}_1^\pm}, M_{\tilde{\chi}_1^0})$ point, the independent analyses performed for different ΔM regimes are combined.

Figure 2 shows the OPAL limit for $\tilde{\chi}_2^0\tilde{\chi}_1^0$ production assuming that $\tilde{\chi}_2^0$ decays dominantly to $\tilde{\chi}_1^0 Z^{0(*)}$. The best cross-section limit is ~ 0.14 pb. The worst case happens for low values of ΔM (<5 GeV/c^2) where the best limit is of the order of 1 pb.

Similar model-independent upper limits are obtained on $\tilde{\chi}_1^+\tilde{\chi}_1^-$ production cross-section. the best limit is ~ 0.15 pb, obtained for $M_{\tilde{\chi}_1^\pm}\sim 2M_{\tilde{\chi}_1^0}$ and $M_{\tilde{\chi}_1^\pm}>70$ GeV/c^2, assuming that Br($\tilde{\chi}_1^\pm \to \tilde{\chi}_1^0 W$)$=100\%$. The worst case occurs for low values of ΔM, where a limit of ~ 3 pb is reached for $\Delta M\sim 3$ GeV/c^2.

ALEPH PRELIMINARY

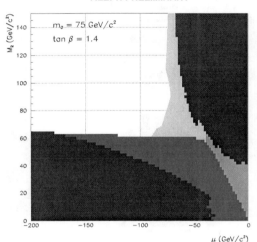

Figure 4: Regions in the (μ, M_2) plane covered by the combination of (from light to dark gray) neutralino search at LEP2, neutralino search at LEP1, selectron search and chargino search at LEP2, for $\tan\beta=\sqrt{2}$, $m_0=75$ GeV/c^2 and $\sqrt{s}=189$ GeV.

The model-independent limits can be translated into mass limits following the MSSM assumptions introduced in section 1. These limits depend tightly on ΔM and on $M_{\tilde{\nu}}$ (so on m_0). For $m_0 > 300$ GeV/c^2 and $\Delta M>10$ GeV/c^2 the minimal non excluded chargino mass ranges from ~ 92 GeV/c^2 for a chargino dominantly gaugino to the kinematic limit for a chargino dominantly higgsino. For an arbitrary value of m_0 the lower mass limit decreases to 57.2 GeV/c^2, due to a lower cross-section. For lower values of ΔM (<10 GeV/c^2) the chargino is purely higgsino and the mass limit is independent of m_0, but in this case the efficiency is lower due to the selection criteria used to reduce background from two-photon interactions. The lower mass limit obtained by DELPHI for $\Delta M=5$ GeV/c^2 is 89.9 GeV/c^2. More sophisticated analyses are needed for lower values

of ΔM as explained in Section 1.1. Figure 3 shows the chargino mass limit from DELPHI as a function of ΔM obtained by the different analyses, in different regions of the (μ, M_2) plane, for different relations between M_1 and M_2, and for different values of $M_{\tilde{\nu}}$. In conclusion the lower chargino mass limit, valid when the chargino is dominantly gaugino, the $\tilde{\nu}$ is heavy and $\Delta M>0.5$ GeV/c^2 is 62.2 GeV/c^2.

2.2 Lower mass limit on the LSP

In order to obtain an absolute lower mass limit on the LSP ($\tilde{\chi}_1^0$) it is necessary to combine chargino, neutralino and slepton searches. The commonly used method consists in converting the negative results from searches for charginos and neutralinos into exclusion regions in (M_2, μ) plane for different $\tan\beta$ values and determine the minimal allowed LSP mass as a function of $\tan\beta$. The exclusion regions obtained from chargino and neutralino searches depend however on the sneutrino and selectron mass values, respectively. The searches for sleptons are therefore used to set a minimal allowed sneutrino mass value. The scan over the sneutrino mass is performed starting from the value allowed by the slepton searches. The GUT relation with a common m_0 is used to relate selectron and sneutrino masses and to obtain a minimal allowed sneutrino mass for given $\tan\beta$ and M_2 values. The LEP1 search for $\tilde{\chi}_2^0\tilde{\chi}_1^0$ production with $\tilde{\chi}_2^0 \to \tilde{\chi}_1^0\gamma$ was also included in order to cover regions of the (M_2, μ) plane not excluded by the LEP2 gaugino searches (for $M_2 \sim |\mu|$, $\mu < 0$ and $\tan\beta<1.5$).

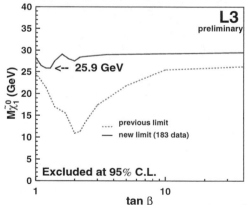

Figure 5: Lower limit on $M_{\tilde{\chi}_1^0}$ as a function of $\tan\beta$ and for any value of m_0, at $\sqrt{s}=183$ GeV.

An example of the exclusion regions resulting from searches by ALEPH for charginos, neutralinos and sleptons are shown in Figure 4.

The lower limit on the LSP mass obtained by L3 is

Table 1: Lower mass limits on sleptons masses at $\sqrt{s}$=183 GeV. The results obtained at $\sqrt{s}$=189 GeV are shown in italic.

Experiment	$M_{\tilde{e}_R}$	$M_{\tilde{\mu}_R}$	$M_{\tilde{\tau}_R}$
ALEPH	*84* GeV/c^2 for ΔM>15 GeV/c^2, $\tan\beta$=2	*78* GeV/c^2 for ΔM>15 GeV/c^2	*65* GeV/c^2 for ΔM>15 GeV/c^2
DELPHI	81.4 GeV/c^2 for $M_{\tilde{\chi}_1^0}$<35 GeV/c^2, $\tan\beta$=1.5	70.3 GeV/c^2 for $M_{\tilde{\chi}_1^0}$<35 GeV/c^2	*70* GeV/c^2 for ΔM>15 GeV/c^2
L3	80 GeV/c^2 for ΔM>15 GeV/c^2, $\tan\beta$=1.41	69 GeV/c^2 for ΔM>15 GeV/c^2	
OPAL	78 GeV/c^2 for ΔM>2 GeV/c^2, $\tan\beta$=1.5	*73* GeV/c^2 for ΔM>3 GeV/c^2	*69.4* GeV/c^2 for ΔM>10 GeV/c^2

shown in Figure 5, as a function of $\tan\beta$, for any value of m_0, and at $\sqrt{s}$=183 GeV. The absolute minimum is 25.9 GeV/c^2.

Similar values on the lower mass limit on the LSP are obtained by the other LEP experiments and they are summarised in Table 2.

Table 2: Lower mass limits on the LSP ($\tilde{\chi}_1^0$), for any value of m_0 and for high values of m_0 (> 200 GeV/c^2), for $\tan\beta{\geq}1$.

Exp.	$\sqrt{s}$	Any m_0	High m_0
ALEPH	189 GeV	26.0 GeV/c^2	32.0 GeV/c^2
DELPHI	189 GeV	23.4 GeV/c^2	30.6 GeV/c^2
L3	183 GeV	25.9 GeV/c^2	30.1 GeV/c^2
OPAL	189 GeV	25.4 GeV/c^2	32.8 GeV/c^2

2.3 Sleptons

Cross section limits and exclusion limits at 95% C.L. are determined in the same way as for neutralinos and charginos, but in the $(M_{\tilde{\ell}}, M_{\tilde{\chi}_1^0})$ plane. Figure 6 shows the right-handed selectron exclusion mass limit obtained by ALEPH for $\sqrt{s}$=189 GeV/c^2. The excluded limit is the best for $\tilde{\chi}_1^0$ masses smaller than 40 GeV/c^2 and becomes worse when the selectron is mass-degenerate with the LSP.

The numerical mass bounds on the slepton masses set by the four LEP experiments are listed in Table 1, for different values of ΔM and $\tan\beta$.

The results from the four LEP experiments at $\sqrt{s}$=183 GeV were combined by the LEP SUSY working group. For a fixed $\sqrt{s}$, mass limits are extended by 5-6 GeV/c^2 for $\tilde{e}_R$ and $\tilde{\tau}_R$ and by 2-3 GeV/c^2 for $\tilde{\mu}_R$. These results are superseded by the mass limits obtained by a single LEP experiment at $\sqrt{s}$=189 GeV.

2.4 Squarks

No evidence for $\tilde{t}_1\bar{\tilde{t}}_1$ and $\tilde{b}_1\bar{\tilde{b}}_1$ production has been observed in data. In order to calculate mass limits, the number of signal events passing the event selections is determined as a function of $M_{\tilde{q}_1}$, $M_{\tilde{\chi}_1^0}$ and the mixing angle $\theta_{\tilde{q}}$. The results obtained with all centre-of-mass

energies data ($\sqrt{s}$=130-189 GeV) were used to calculated these limits.

Figure 7 shows the OPAL 95% C.L. excluded regions in the $(M_{\tilde{b}_1}, M_{\tilde{\chi}_1^0})$ plane, for different values of the mixing angle between the left and right components. There is a good agreement between the expected and the obtained exclusion limit.

The numerical $\tilde{t}_1$ and $\tilde{b}_1$ mass bounds for the four LEP experiments are listed in Table 3

The results from the four LEP experiments at $\sqrt{s}$=183 GeV were combined by the LEP SUSY working group. For a fixed $\sqrt{s}$, mass limits are extended by 5-6 GeV/c^2.

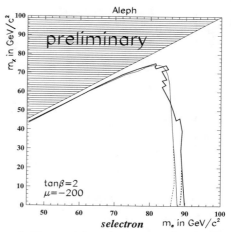

Figure 6: The $\tilde{e}_R$ mass limit for $\tan\beta$=2, μ=-200 GeV/c^2 and $\sqrt{s}$=189 Gev. The actual (thick line) and expected (thin line) mass limits are shown for selectrons in the MSSM assuming Br($\tilde{e}_R \to e\tilde{\chi}_1^0$)=100% (full lines). The dashed curves show the effect of cascade decays assuming no efficiency for these decays.

2.5 Anomalous production of photon pairs

Pairs of acoplanar photons events with missing energy are expected from pair production of $\tilde{\chi}_2^0$ with each $\tilde{\chi}_2^0$

Table 3: Lower mass limits on stop and sbottom masses at $\sqrt{s}$=183 GeV, for $\theta_{\tilde{q}}$=0. The results obtained at $\sqrt{s}$=189 GeV are shown in italic.

Experiment	$M_{\tilde{t}}$ $\tilde{t} \to c\tilde{\chi}$	$M_{\tilde{t}}$ $\tilde{t} \to bl\tilde{\nu}$	$M_{\tilde{b}}$ $\tilde{b} \to b\tilde{\chi}$
ALEPH	*84* GeV/c^2 for ΔM>10 GeV/c^2	*88* GeV/c^2 for ΔM>10 GeV/c^2	*84* GeV/c^2 for ΔM>10 GeV/c^2
DELPHI	78.4 GeV/c^2 for ΔM>10 GeV/c^2	82.0 GeV/c^2 for ΔM>15 GeV/c^2	78.6 GeV/c^2 for ΔM>10 GeV/c^2
L3	81 GeV/c^2 for ΔM>10 GeV/c^2		80 GeV/c^2 for ΔM>20 GeV/c^2
OPAL	*88.3* GeV/c^2 for ΔM>10 GeV/c^2	*87.3* GeV/c^2 for ΔM>10 GeV/c^2	*85.5* GeV/c^2 for ΔM>10 GeV/c^2

decaying into $\tilde{\chi}_1^0\gamma$. The Standard Model backgrounds are $e^+e^- \to \nu\bar{\nu}\gamma\gamma$ and $e^+e^- \to \gamma\gamma(\gamma)$.

The LEP SUSY working group combined the spectra of the recoil mass of acoplanar photons events coming from the four LEP experiments. The results are shown in Figure 8. A good agreement with SM expectations is obtained. Since no signal has been observed, exclusion limits at 95% C.L. on the production cross-section are set in the ($M_{\tilde{\chi}_2^0}$,$M_{\tilde{\chi}_1^0}$) plane. Limits range from ~0.02 pb for $M_{\tilde{\chi}_2^0}$>80 GeV/c^2 and ΔM>5 GeV/c^2, to 0.13 pb for $M_{\tilde{\chi}_2^0}$~47 GeV/c^2 and $M_{\tilde{\chi}_1^0}$~15 GeV/c^2.

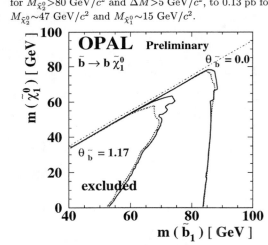

Figure 7: The 95% C.L. excluded regions in the ($M_{\tilde{b}_1}$,$M_{\tilde{\chi}_1^0}$) plane, at $\sqrt{s}$=189 GeV, for a mixing angle of the $\tilde{b}_1$ assumed to be 0.0 and 1.17 rad, assuming that the $\tilde{b}_1$ decays into $b\tilde{\chi}_1^0$. The solid lines show the actual limits and the dotted lines show the limits calculated only with the expected number of background events. $\theta_{\tilde{b}}$ is defined as $\tilde{b}_1 = \tilde{b}_L \cdot \cos\theta_{\tilde{b}} + \tilde{b}_R \cdot \sin\theta_{\tilde{b}}$. if $\theta_{\tilde{b}}$=1.17 rad, $\tilde{b}_1$ decouples from the Z^0.

Acknowledgements

I would like to thank the four LEP experiments ALEPH, DELPHI, L3 and OPAL for the results and plots they sent me in order to show them in ICHEP98. I'm extremely great-full to Sergio Navas and Luc Pape for their help in preparing my presentation and these proceedings.

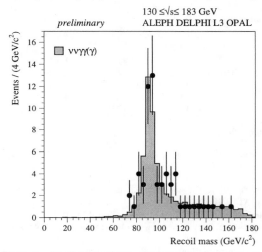

Figure 8: Recoil mass of two acoplanar photons events obtained combining the spectra of the four LEP experiments, for $\sqrt{s}$=130 GeV to $\sqrt{s}$=183 GeV.

References

1. "LEP Storage Ring", Technical Notebook **CERN** (1989).
2. ALEPH Coll.,Nucl. Instr. and Meth. **A360** (1995) 481;
   DELPHI Coll.,Nucl. Instr. and Meth. **A378** (1996) 57;
   L3 Coll.,Nucl. Instr. and Meth. **A289** (1990) 35;
   OPAL Coll.,Nucl. Instr. and Meth. **A305** (1991) 275.

SEARCHES FOR SUPERSYMMETRY AT HERA

Zhiqing ZHANG

Laboratoire de l'Accélérateur Linéaire,
IN2P3-CNRS et Université de Paris-Sud, BP 34, F-91898 Orsay Cedex
E-mail: zhangzq@lal.in2p3.fr

Supersymmetric signals were searched for in e^+p collisions collected by H1 and ZEUS in 1994 → 1997 corresponding to an integrated luminosity of about $40\,\text{pb}^{-1}$ per experiment at a center-of-mass energy of 300 GeV. Within the framework of minimal supersymmetric extensions to the Standard Model which conserve R-parity, ZEUS found no evidence for the production of selectron and squark decaying directly into the lightest neutralino. H1 performed a search for direct single production of squarks of R-parity violating supersymmetry. The sensitivity to the corresponding Yukawa couplings was shown to be only weakly dependent on the free parameters of the minimal supersymmetric Standard Model and the reach in the mass-coupling plane extended to domains unexplored in other direct or indirect searches.

1 Introduction

Supersymmetry (SUSY) is one of the most likely ingredients for a theory beyond the Standard Model (SM). In particular the Minimal Supersymmetric extension of the Standard Model (MSSM) describes as well as the SM all experimental data, and in addition it offers among its appealing consequences solutions for the cancellation of quadratic divergences occurring in the scalar Higgs sector of the SM and models beyond the SM.

SUSY relates fermions to bosons and predicts for each SM particle a partner with spin differing by half a unit. For example selectrons $\tilde{e}_L$, $\tilde{e}_R$ are scalar partners of electrons e_L, e_R, and similarly squarks $(\tilde{u}_L, \tilde{d}_L)$, $\tilde{u}_R$, $\tilde{d}_R$ are the partners of up and down quarks. Two Higgs doublets with vacuum expectation values v_2, v_1 are necessary to generate masses for up-type quarks (v_2) and for down-type quarks and charged leptons (v_1). The partners of the gauge bosons $W^\pm, Z^0, \gamma$ and the two Higgs doublets are called gauginos and higgsinos. They can mix and form two charged mass eigenstates $\chi^\pm_{1,2}$ (charginos) and four neutral mass eigenstates $\chi^0_{1,2,3,4}$ (neutralinos).

Since supersymmetric particles are not observed at the masses of their SM partners, SUSY must be broken. In the MSSM, this breaking is achieved by adding extra mass parameters M_2 and M_1 for the $SU(2)$ and $U(1)$ gauginos. Thus the masses of charginos and neutralinos depend on $M_1, M_2, \tan\beta \equiv v_2/v_1$ and the higgsino mass parameter μ.

The MSSM is constructed to conserve R-parity (R_p): for a particle of spin S, the multiplicative quantum number $R_p \equiv (-1)^{3B+L+2S}$ distinguishes particles ($R_p = +1$) from SUSY particles ($R_p = -1$). Here B and L are baryon and lepton number, respectively. This implies that supersymmetric particles can only be produced in pairs and that the lightest supersymmetric particle (LSP), which is generally assumed to be χ^0_1, is stable. At HERA the dominant MSSM process is the produc-

tion of a selectron and a squark via a t-channel exchange of a neutralino $ep \to \tilde{e}\tilde{q}X$. The $\tilde{e}$ and $\tilde{q}$ can then decay into any lighter gaugino and their SM partners. The decay involving χ^0_1 gives an experimentally clean signature of missing transverse energy plus an electron[a] and a hadronic system. Based on the e^+p collisions taken in 1994 → 1997 corresponding to an integrated luminosity of $46.6\,\text{pb}^{-1}$, ZEUS has performed such a search[1] (Sec. 2).

The most general SUSY theory which preserves gauge invariance of the SM allows, however, for R_p violating ($\slashed{R}_p$) Yukawa couplings $\lambda, \lambda', \lambda''$ between one scalar squark or slepton and two SM fermions:

$$W_{\slashed{R}_p} = \lambda_{ijk}L_iL_j\overline{E}_k + \lambda'_{ijk}L_iQ_j\overline{D}_k + \lambda''_{ijk}\overline{U}_i\overline{D}_j\overline{D}_k. \quad (1)$$

where $i,j,k = 1,2,3$ are generation indices, $L_i(Q_i)$ are the lepton (quark) $SU(2)_L$ doublet superfields and $\overline{E}_i(\overline{D}_j, \overline{U}_j)$ are the electron (down and up quark) $SU(2)_L$ singlet superfields. Of particular interest for HERA are the $\slashed{R}_p$ terms $\lambda'L_iQ_j\overline{D}_k$ as HERA provides both leptonic and baryonic quantum numbers in the initial state. The resonant squarks at HERA are thus singly produced (in contrast to the MSSM) in the s-channel with masses up to the kinematic limit of $\sqrt{s} \simeq 300$ GeV. From the theoretical understanding of unification, there is no clear preference between R_p-conservation and $\slashed{R}_p$, it is thus mandatory to experimentally search for both possibilities. Based on an integrated luminosity of $37\,\text{pb}^{-1}$, H1 has searched for direct production of squarks via $\slashed{R}_p$ Yukawa coupling λ' by taking into account various possible $\slashed{R}_p$ decays and gauge decays of the squarks[2]. Such couplings could lead to leptoquark-like final states (Sec. 3) or to explicit manifestation of lepton flavour violation (Sec. 4).

The SM deep inelastic scattering (DIS) processes, neutral current (NC) and charged current (CC) interactions, become backgrounds for the searches considered

---

[a] Unless specified, an electron in the following can be either an electron or a positron.

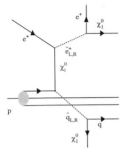

Figure 1: Selectron-squark production via neutralino exchange and the subsequent decays into the lightest supersymmetric particle χ_1^0.

here. The most commonly used DIS kinematic variables are Q^2, x and y, with $Q^2 = -q^2$, q being the four-momentum of the exchanged gauge bosons (γ, Z, W), x is the momentum fraction of the proton carried by the struck quark in the quark parton model, and y is related to Q^2 and x by $y = Q^2/(xs)$ with $0 < x, y < 1$. Experimentally, the kinematics for a NC DIS event is over-constrained as both H1 and ZEUS detectors measure not only the scattered electron but also the hadronic final state, while the kinematics for a CC DIS event can only be reconstructed from the hadronic information alone.

2 Search for $\tilde{e}$ and $\tilde{q}$ within MSSM

Within the MSSM, the production of a $\tilde{e}$ and a $\tilde{q}$ is the lowest order process in which supersymmetric particles could be produced at HERA (Fig. 1). The cross section depends on the MSSM parameters $M_1, M_2, \tan\beta, \mu$, and on the masses of the produced particles. The branching ratios for the decays $\tilde{e} \to e\chi_1^0$ and $\tilde{q} \to q\chi_1^0$ depend on the same MSSM parameters. To reduce the number of free parameters, the following assumptions are made: (i) $m_{\tilde{e}_L} = m_{\tilde{e}_R}, m_{\tilde{q}_L} = m_{\tilde{q}_R} = m_{\tilde{q}}(\tilde{q} \neq \tilde{t})$[b] (ii) $M_1 = 5/3 \tan^2\theta_W M_2$, and (iii) $m_{\tilde{q}} < m_{\tilde{g}}$ such that the decay $\tilde{q} \to q\tilde{g}$ is kinematically forbidden.

The main selection cuts are defined according to the event signature mentioned in Sec.1: (1) an isolated e with $p_t^e > 4\,\text{GeV}$ ($> 10\,\text{GeV}$ if $\theta_e < 0.35$), (2) a hadronic system with $p_t^h > 4\,\text{GeV}$, (3) a missing transverse momentum with $\not{P}_t > 10\,\text{GeV}$, and (4) a set of final cuts $E - p_z < 50\,\text{GeV}$, $\not{P}_t > 14\,\text{GeV}$, $(E - p_z)/\not{P}_t < 1$ determined from an optimization procedure. The acceptance is close to zero for small mass differences $\Delta m =$

---
[b]The mass eigenstates $\tilde{e}_{1,2}(\tilde{q}_{1,2})$ are generally different from the interaction eigenstates $\tilde{e}_{L,R}(\tilde{q}_{L,R})$. The degree of mixing being proportional to the lepton (quark) masses is however small and is neglected here, since the contribution from the top quark is negligible.

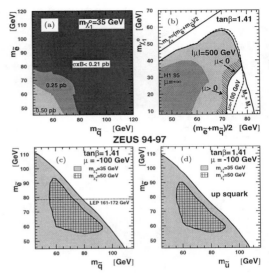

Figure 2: Upper limits at 95% CL on $\sigma \times B$ (a) and excluded regions at 95% CL for degenerate $\tilde{e}$ and $\tilde{q}$ (b,c) and for the up squark alone (d).

$\min(m_{\tilde{e}} - m_{\chi_1^0}, m_{\tilde{q}} - m_{\chi_1^0})$ and reaches a plateau for $\Delta m > 10\,\text{GeV}$. The level of the plateau increases from 25% at $m_{\tilde{e}} = m_{\tilde{q}} = 40\,\text{GeV}$ to about 50% at $m_{\tilde{e}}$ or $m_{\tilde{q}} = 120\,\text{GeV}$.

One event survived the selection criteria and was identified as containing a high-Q^2 positron with associated $\not{P}_t$ in the calorimeter due to two muons in the final state. The expected SM background is $1.99^{+0.57}_{-0.84}$ events from five considered SM reactions: W production processes (dominant contribution), NC DIS events (second dominant contribution), CC DIS events, lepton pair production (l^+l^-) and photoproduction (negligible).

The resulting upper limits on the $\tilde{e}, \tilde{q}$ production cross section times the branching ratios $(\sigma \times B)$ for the decay to the lightest neutralino χ_1^0 at 95% CL are shown in Fig. 2(a). When compared with the theoretical value from model calculations, exclusion areas in the parameter space of the MSSM are derived (Fig. 2(b)). For large $|\mu|$ the excluded region reaches $(m_{\tilde{e}} + m_{\tilde{q}})/2 = 77\,\text{GeV}$ for a 40 GeV neutralino. This limit worsens at lower neutralino masses, because new decay channels to charginos and next to the lightest neutralino open and compete with the direct decay to χ_1^0. In the limit $M_2 \gg M_1$, the charginos and the next to lightest neutralino masses increase leaving only the direct decay channel to χ_1^0 open. The excluded region is limited by the small cross section of the process at large $(m_{\tilde{e}} + m_{\tilde{q}})/2$, while for large neutralino masses it is limited by the efficiency that falls to

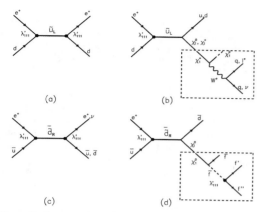

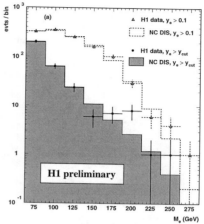

Figure 3: Lowest order s-channel diagrams for first generation squark production at HERA followed by (a), (c) $\not{R}_p$ decays and (b), (d) gauge decays. In (b) and (d), the emerging neutralino or chargino might subsequently undergo $\not{R}_p$ decays of which examples are shown in the dashed boxes for (b) the χ_1^+ and (d) the χ_1^0.

Figure 4: Mass spetra for eq topology. The subscript e in y_e and $M_e (= \sqrt{sx})$ indicates that the event kinematics is reconstructed with the measured energy and angle of the scattered electron.

zero as $\Delta m \to 0$. A large variation in $\tan\beta$ results only in slight changes. In the range $45 < (m_{\tilde{e}} + m_{\tilde{q}})/2 < 85\,\text{GeV}$ the up quark contribution to the cross section ranges between 70% to 90% because it dominates the parton densities at high x. The limits on the $\tilde{u}$ mass alone (Fig. 2(d)), assuming all other squarks to be much heavier, are only $\sim 2\,\text{GeV}$ below the limit obtained for degenerate squark masses (Fig. 2(c)). The mass limits substantially improve those limits previously published by H1[3] due to the seven-fold increase in the integrated luminosity. These limits are at the same level as those obtained from LEP[4c] and are complementary to those obtained from Tevatron[6] as they investigate different regions of the MSSM parameter space[d].

3 Search for $\tilde{q}$ within $\not{R}_p$-SUSY

In contrast to the MSSM, in $\not{R}_p$-SUSY, the squarks can be singly produced and they can decay not only via their gauge couplings to a quark/antiquark and a neutralino/chargino (Fig. 3(b,d)) but also via their Yukawa coupling into SM fermions (leptoquark-like, Fig. 3(a,c)). Moreover, the LSP (again assumed to be χ_1^0), which is no longer stable, decays via λ'_{1jk} into a quark, an antiquark and a lepton.

With e^+p collisions, HERA is best sensitive to couplings λ'_{1j1} among the nine possible couplings λ'_{1jk}, where mainly $\tilde{u}_L^j$ squarks are produced via processes involving

[c]The Limits from LEP have been improved recently[5].
[d]For example, in contrast to Tevatron, the gluinos are assumed here to be heavy.

a valence d quark. On the contrary, future HERA e^-p data will allow to better probe couplings λ'_{11k} and $\tilde{d}_R^k$ squarks.

Depending on whether the produced squarks undergo a $\not{R}_p$ decay or a gauge decay, there are many different final state event topologies, e.g. (a) a lepton plus a jet, (b) a neutrino plus a jet, (c) a right sign lepton plus multijets, and (d) a wrong sign lepton plus multijets.

Topology (a) is indistinguishable from a NC DIS event on an event-by-event basis. Statistically, however, one expects for the signal a resonant peak in the mass distribution and a flat distribution in y ($1/y^2$ distribution for NC DIS events). For this reason, a mass dependent y cut is chosen and it allows to reduce significantly the NC DIS background (Fig. 4). The other selection cuts are basically the same as in the high Q^2 paper[7]. The total efficiency varies between 35% at 75 GeV, 50% at 200 GeV and 70% at 250 GeV. In total 312 events are observed, which is in good agreement with the SM expectation of 306 ± 23 events. Data are found to be well described by MC, although an excess of events still remains in the mass range $200 \pm 12.5\,\text{GeV}$ where 8 events are observed while 3.01 ± 0.54 are expected. This clustering is however less significant than that observed with 1994 $\to$ 1996 data alone[7].

Topology (b) has only a low sensitivity with the e^+ beam since the produced squarks $\tilde{d}_R^{k*}$ couples to a sea quark $\bar{u}$ from the proton (Fig. 3(c)), the density of which is small at high x. Thus it will not be considered in the following.

The main SM background for topology (c) is also

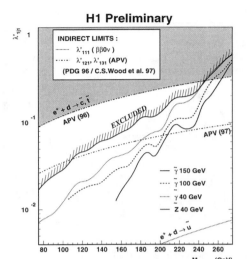

Figure 5: Exclusion upper limits at 95% CL for the couplings λ'_{1j1} as a function of squark mass, for various masses and mixtures of the χ_1^0; also represented are the most stringent indirect limits on λ'_{1j1}.

from NC DIS where QCD radiation leads to multijets. Two main cuts are $y > 0.4$ and $\theta_e < 110°$. The latter cut exploits the fact that for high y NC DIS events, one hard jet is usually scattered in the backward region of the calorimeter (the proton beam direction is defined as the forward direction), on the contrary jets coming from a squark will be boosted in the forward direction independent of y. The efficiency varies from about 30% for a 100 GeV squark decaying into a 40 GeV χ_1^0 to about 50% when these masses are set respectively at 200 GeV and 80 GeV. In total 289 candidates are observed, which is in good agreement with the mean SM background of 285.7 ± 28.0 expected from NC DIS and photoproduction (the latter has a contribution of less than 3%).

Topology (d) has such a striking final state that it is essentially background free. Indeed, when the negative charge of the lepton track with good quality is required in addition to the cuts applied for topology (c), only one event survives, which is compatible with 0.49 ± 0.2 events coming from NC DIS.

Under the assumption that only one of the Yukawa couplings λ'_{1j1} dominates[e], mass dependent upper limits on these couplings are derived by combining these three topologies (Fig. 5). The sensitivity on λ'_{1j1} for squark masses below about 200 GeV is better by roughly a fac-

[e]This is not unreasonable as in the SM the top quark Yukawa coupling is almost a factor 40 larger than the bottom Yukawa coupling.

tor 2 for a $\tilde{\gamma}$-like χ_1^0 than for a χ_1^0 dominated by its zino component due to the higher part of the total branching actually being analyzed in the three considered topologies. The sensitivity achieved increases with the $\tilde{\gamma}$ mass due to the higher efficiency in topology (c). For electromagnetic coupling strengths $\lambda'_{1j1} = \sqrt{4\pi\alpha_{em}} \simeq 0.3$, squark masses up to 262 GeV are excluded at 95% CL. This reach in the mass-coupling plane extends beyond that covered by other collider experiments, [2] e.g. (i) in their leptoquark-like searches, D0 and CDF rule out $\tilde{u}_L^j$ squark masses below 200 GeV for a branching ratio $B(\tilde{u}_L^j \to e^+q) \geq 50\%$ (the excluded mass domain is lowered to about 110 GeV when $B(\tilde{u}_L^j \to e^+q) \leq 10\%$, which can be natually small in R_p-SUSY), (ii) the light stop mass limit is estimated to be $130 - 150$ GeV in models in which R_p is violated by a λ'_{13k} coupling from the specific SUSY searches performed by D0 and CDF. The results are also compared to the best indirect limits. The most stringent constraint comes from the non-observation of neutrinoless double beta decay but only concerns coupling λ'_{111}. For couplings λ'_{121} and λ'_{131}, the limits derived by H1 are comparable or more stringent than the best indirect limits coming from Atomic Parity Violation.

4 Search for lepton flavor violation within R_p-SUSY

Under the assumption that one product of couplings $\lambda'_{1j1} \times \lambda'_{ljk}(l = 2, 3)$ is non-vanishing and donimates over all remaining couplings, namely squarks $\tilde{u}_L^j$ are produced via coupling λ'_{1j1} and decay via λ'_{ljk}, striking explicit lepton flavour violation processes $e^+d \to \tilde{u}_L^j \to \mu^+(\tau^+)d^k$ may occur.

H1 has observed 5 events of the type $e^+p \to \mu^\pm X$. [8] None of these events survive as soon as the kinematic constraints of a $2 \to 2$ body process is imposed.[f]

H1 has also searched for squarks coupling to a third generation lepton leading to τq final states. No candidate is observed while 0.8 ± 0.3 events are expected from SM processes. The efficiency ranges from $\sim 10\%$ for a 100 GeV squark to $\sim 25\%$ for squark masses above 200 GeV. Under a further assumption that the gauge decays of squarks are forbidden (so that the only squark decay modes are $\tilde{u}_L^j \to e^+d$ and $\tilde{u}_L^j \to \tau^+d^k$), exclusion limits at 95% CL on λ'_{3jk} as a function of squark masses have been derived by fixing the production coupling λ'_{1j1} to a given value (Fig. 6). A similar analysis to that presented here has been published by the ZEUS Collaboration with an integrated luminosity of about 3 pb^{-1} using

[f]On the other hand, part of the observed muon events are kinematically compatible [9] as proceeding through the process $e^+d \to \tilde{t}_1 \to \tilde{b}_1 W^+$ with subsequent decays $\tilde{b}_1 \to d\bar{\nu}_e$, $W^+ \to \mu^+\nu_\mu$ as suggested by T. Kon et al. [10].

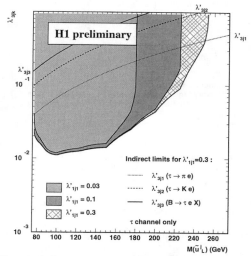

Figure 6: Exclusion upper limits at 95% CL for the coupling λ'_{3jk} as a function of squark mass, for several fixed values of λ'_{1j1} (greyed domains). The regions above the full, dashed and dash-dotted curves correspond to the best relevant indirect limits.

1994 e^+p data[11]. When the two couplings are both equal to 0.03, the analysis presented here extends the squark mass range by about 65 GeV. The only relevant indirect limits [12] come from the process $\tau \to \pi e, \tau \to Ke$ and $B \to \tau eX$.[g] H1 direct limits improve these constraints by typically one order of magnitude.

5 Summary and Outlook

SUSY processes have been searched for at HERA using full e^+p data taken in 1994 $\to$ 1997. Within the MSSM, ZEUS has found no evidence for the production of selectron and squarks decaying directly into the lightest neutralino. Excluded regions reach $(m_{\tilde{e}} + m_{\tilde{q}})/2 = 77$ GeV at 95% CL for $m_{\chi^0_1} = 40$ GeV and large values of the MSSM parameter $|\mu|$. The process is dominated by the $\tilde{u}$ contribution and the exclusion limit is 75 GeV when only the $\tilde{u}$ squark is considered. These mass values are still far away from the HERA kinematic limit and are currently limited by the available integrated luminosity.

H1 has searched for within R-parity violating SUSY direct production of single resonant squarks decaying via R-parity violating as well as gauge couplings. Notwith-

---

[g]Note that better indirect limits on couplings λ'_{3jk} alone exist. However these only concern the $\tilde{d}_R$ and can thus be evaded assuming e.g. $\tilde{u}^j_L$ to be much lighter than other squarks, which could be achieved especially in the case of a light stop ($j = 3$).

standing an excess at masses around 200 GeV in channel with a positron and one jet as the final states, no significant evidence for the production of squarks was found and mass dependent limits on the R-parity violating couplings were derived. Squarks with masses up to 262 GeV are excluded at 95% CL for a strength of the Yukawa coupling of $\alpha_{\rm em}$. The limits extend beyond the domain covered by other collider experiments, and for some R-parity violating couplings, are better than the most stringent indirect constraints.

Lepton flavour violation processes have been sought. No candidates satisfying a $2 \to 2$ body process kinematics are found. The resulting H1 direct limits on λ'_{3jk} improve the only relevant indirect limits by typically one order of magnitude.

HERA will provide e^-p collisions in 1998 and probably also in 1999 with an expected integrated luminosity of $50\,{\rm pb}^{-1}$. The proton beam energy will increase by more than 10% from 820 GeV to 920 GeV. In two years from now, there will be a major luminosity upgrade, which will result in a factor of more than 5 increase in the peak luminosity providing a total integrated luminosity of $1\,{\rm fb}^{-1}$ in the period 2000 $\to$ 2005 corresponding to a yearly luminosity of $150\,{\rm pb}^{-1}$ per experiment. Therefore new potential for SUSY searches or discovery is expected in the next years from HERA.

Acknowledgements

It is pleasure to thank all members of the H1 and ZEUS collaborations, whose efforts allow me to present these results. I am grateful to M. Corradi, M. Kuze, E. Perez and Y. Sirois for their help in preparing the talk.

References

1. ZEUS Collab., paper 759 to ICHEP'98; DESY 98-069, accepted by *Phys. Lett.* **B**.
2. H1 Collab., paper 580 to ICHEP'98.
3. H1 Collab., *Phys. Lett.* B**380**, 461 (1996).
4. See e.g. ALEPH Collab., *Phys. Lett.* B**407**, 377 (1997); OPAL Collab., *Z. Phys.* C**75**, 409 (1997).
5. P. Rebecchi, talk given at ICHEP'98.
6. D0 Collab., *Phys. Rev. Lett.* **75**, 618 (1995); CDF Collab., *Phys. Rev.* D **56**, 1357 (1997); *Phys. Rev. Lett.* **69**, 3439 (1992).
7. H1 Collab., *Z. Phys.* C**74**, 191 (1997).
8. H1 Collab., DESY-98-063, Submitted to *Eur. Phys. J.* C; C. Diaconu, talk given at ICHEP'98.
9. E. Perez, talk given at SUSY'98.
10. T. Kon *et al.*, *Mod. Phys. Lett.* A**12**, 3143 (1997).
11. ZEUS Collab., *Z. Phys.* C**73**, 613 (1997).
12. S. Davidson *et al.*, *Z. Phys.* C**61**, 613 (1994).

Closing the Light Gluino Window Using $\pi^+\pi^-$ Pairs Produced in a Neutral Beam

Amitabh Lath*

Rutgers University, Piscataway NJ 08855

We have performed an extensive search for spontaneous appearance for $\pi^+\pi^-$ pairs in a neutral beam, for invariant masses of the charged pions as low as the kinematic limit of $2M_\pi$. The main source of events with two charged particles are decays of neutral kaons, and detailed understanding of the resultant invariant mass shape constrains most hypothetical processes that produce a charged pion pair and an unobservable particle. In particular, we place a limit on the decay $R^0 \to \pi^+\pi^-\tilde{\gamma}$ for any $M_{R^0} - M_{\tilde{\gamma}} \geq 2M_\pi$. In the most interesting R^0 mass range, ≤ 3 GeV/c^2, we can exclude lifetimes from 5×10^{-10} seconds to as high as 10^{-4} seconds.

1 Introduction

Several theories of supersymmetry, including those which solve the SUSY-CP problem by eliminating dimension 3 operators [1], predict light gauginos and heavy squarks. In such theories, the super-partners of the massless bosons remain light. In fact, light masses for gluinos($\tilde{g}$) and photinos($\tilde{\gamma}$) arise naturally in many SUSY models, but have been ignored without adequate evidence for their exclusion. In most such models, the gluino and photino masses are expected to be ≤ 1.0 GeV/c^2. The gluino should form bound states with normal quarks and gluons (g), the lightest of which is called the R^0, a spin $\frac{1}{2}$ $g\tilde{g}$ bound state.

Estimates of the mass and lifetime of the R^0 vary from 1 to 3 GeV/c^2 and 10^{-10} to 10^{-5} s respectively [2]. The ratio of masses $M_{R^0}/M_{\tilde{\gamma}} \equiv$ r is expected to be $1.3 \leq r \leq 1.8$, based on particle physics [2] and cosmological [3] arguments.

1.1 Previous Searches

A previous direct search for the R^0 by this collaboration is described in [4]. That result, based on 5% of the data collected by KTeV in 1996, excluded the R^0 with the constraint $M_{R^0} - M\tilde{\gamma} \geq 0.648$ GeV/c^2. However, that previous KTeV result, along with several related indirect searches, [5][6][7][8], failed to exclude the R^0 at the required level of sensitivity in the region of primary interest, $r \leq 1.4$ [9].

1.2 R^0 Production and Decay

R^0 production in pN collisions has been discussed in [4]. It is important to note that the production may proceed through conventional QCD [11]. A light R^0 can therefore be produced in large numbers in pN collisions. The R^0 is expected to decay mainly into $\rho\tilde{\gamma}$, but may decay into $\eta\tilde{\gamma}$ or $\pi^0\tilde{\gamma}$ if C-parity were to be violated. A neutral beam fixed target experiment with a large decay volume downstream of the target is therefore the ideal place to conduct a search for the R^0. In this analysis, we concentrate on the dominant decay, $R^0 \to \rho\tilde{\gamma}$, $\rho \to \pi^+\pi^-$, in which the $\tilde{\gamma}$ escapes undetected.

2 The KTeV Experiment

The KTeV experiment is described in [10], and as used in the R^0 search in [4]. The neutral beam, extracted from a BeO target with 800 GeV/c protons incident, was allowed to propagate for approximately 100 m; then decay in a volume approximately 50 m long. The products of the decays were detected by a spectrometer consisting of four drift chambers with single-hit resolution of $\leq 100\mu$m and a 3100 element pure CSI calorimeter with electromagnetic energy resolution of $\leq 1\%$. In addition, a series of muon hodoscopes downstream of the calorimeter were used to identify tracks as muons.

3 Data Analysis

The data used in this analysis were collected during the 1996 run of KTeV (FNAL E832). The trigger and analysis cuts used are similar to those described in [4]. To detect a possible R^0 signal we examined decays with two charged particles; specifically the shape of the invariant mass distribution, with the assumption that the particles were pions ($M_{\pi^+\pi^-}$). Backgrounds consisted of $K_L \to \pi^\pm l^\mp \nu$ ($l = e, \mu$) decays with leptons misidentified as pions (semi-leptonic decays); $K_L \to \pi^+\pi^-$ and $K_L \to \pi^+\pi^-\gamma$ decays; as well as $K_L \to \pi^+\pi^-\pi^0$ decays with undetected π^0's.

Semi-leptonic decays were identified and rejected in a manner similar to that described in [4]. Tracks made by electrons were identified as such since the energy they deposited in the calorimeter matched the momentum measured by the spectrometer. Muon tracks were identified by extrapolating the track from the spectrometer to one or more hits in the muon hodoscopes. The $K_L \to \pi^+\pi^-$ decays were rejected by requiring the transverse momentum squared of the $\pi^+\pi^-$ with respect to the beam direction (P_t^2) to be greater than 0.001 (GeV/c)^2. The $K_L \to$

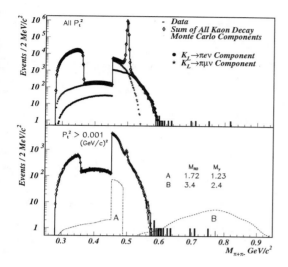

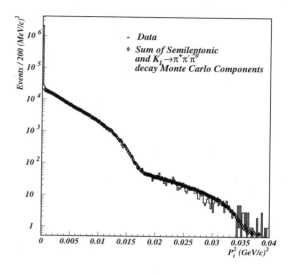

Figure 1: $M_{\pi^+\pi^-}$ distribution with all but P_t^2 and $K_L \to \pi^+\pi^-\pi^0$ specific cuts (top), and all cuts (bottom). The solid line is data, and diamonds are the sum of all kaon decay simulations. The data with $M_{\pi^+\pi^-} \leq 0.45$ GeV/c^2 is from a data stream with a trigger prescale. The contribution of $K_L \to \pi^\pm e^\mp \nu$ (dots) and $K_L \to \pi^\pm \mu^\mp \nu$ (stars) decays is also shown. In the bottom plot, the distributions due to two sample R^0's of lifetime $\tau = \tau(K_L)$, and 1/10 of the pQCD predicted flux are shown (dash, dot-dash), labelled A and B. The masses are in GeV/c^2.

Figure 2: P_t^2 distribution for the data, with all cuts but P_t^2 and $K_L \to \pi^+\pi^-\pi^0$ specific cuts. The data (solid) has a spike at zero due to $K_L \to \pi^+\pi^-$ decays. The diamonds show the sum of the semileptonic and $K_L \to \pi^+\pi^-\pi^0$ decays.

$\pi^+\pi^-\gamma$ and $K_L \to \pi^+\pi^-\pi^0$ decays were rejected by identifying photons in the calorimeter from these decays. Additional cuts in the region $M_{\pi^+\pi^-} \leq 0.36$ GeV/c^2 on energy deposited in the calorimeter further rejected the $K_L \to \pi^+\pi^-\pi^0$ decays ($K_L \to \pi^+\pi^-\pi^0$ specific cuts).

Figure 1 (top) shows the $M_{\pi^+\pi^-}$ distribution for all the data, without the $K_L \to \pi^+\pi^-\pi^0$ specific and P_t^2 cuts. The data (solid) and the appropriately weighted sum of all K_L decay components (diamonds) are shown. The contributions of the semi-leptonic kaon decays (dots, and stars) are also shown separately. There are $\sim 2.1 \times 10^6$ CP-violating $K_L \to \pi^+\pi^-$ decays in the peak at M_K. An edge is evident at $M_{\pi^+\pi^-} = 0.45$ GeV/c^2. The data with $M_{\pi^+\pi^-} \leq 0.45$ GeV/c^2 is from a data stream collected with a trigger prescale, since the main charged-track data stream concentrated on $K_L \to \pi^+\pi^-$ decays, and filtered out events with $M_{\pi^+\pi^-} \leq 0.45$ GeV/c^2. The CP-conserving $K_L \to \pi^+\pi^-\pi^0$ decays are evident at $M_{\pi^+\pi^-} \leq 0.36$ GeV/c^2, and the kinematic limit is evident at $2M_\pi = 0.28$ GeV/c^2. The data and kaon decay

simulation are in agreement for over six orders of magnitude.

Figure 1 (bottom) shows the $M_{\pi^+\pi^-}$ distribution for the data (solid) with all the cuts and the sum of the K_L decay simulations (diamonds) are shown. The $K_L \to \pi^+\pi^-$ peak is significantly reduced due to the P_t^2 cut, and the $K_L \to \pi^+\pi^-\pi^0$ specific cuts reduce the number of events at $M_{\pi^+\pi^-} \leq 0.36$ GeV/c^2. Two sample R^0 distributions are shown (dash, dot-dash), to illustrate their effect on the $M_{\pi^+\pi^-}$ distribution. The masses of the sample R^0 shown are 1.72 and 3.4 GeV/c^2, and the respective photino masses are 1.23 and 2.4 GeV/c^2, for the distributions labelled A and B. The number of the sample R^0 in the distributions is set at 1/10th of that expected due to the flux predicted by perterbative QCD (pQCD) [11][12], at a lifetime equal to that of the K_L.

In the data, there are 12 events with $M_{\pi^+\pi^-} \geq 0.6$ GeV/c^2 that are not simulated by K_L decays. The number of events is consistent with that expected from interactions of the beam with residual gas in the vacuum decay volume. The sample R^0 distributions shown in figure 1 illustrate the effect of $R^0 \to \pi^+\pi^-\tilde{\gamma}$ decay on the $M_{\pi^+\pi^-}$ shape. Since the shape due to R^0 decay is significantly different from those due to kaon decays, we can

Table 1: Results for various R^0, $\bar{\gamma}$ combinations from the MINUIT fit. The first two columns are the R^0, $\bar{\gamma}$ masses; the third column is the 90% C.L. upper limit for the number of R^0 decays found in the data; the fourth column is the 90% C.L. upper limit for the number of R^0 produced at the target (for $\tau = \tau(K_L)$). The fifth and sixth columns are the 90% confidence level upper limit for the R^0/K^0 flux at the target, and the prediction for this flux ratio using pQCD.

$\mathbf{M_{R^0}}$ GeV/c^2	$\mathbf{M_{\bar{\gamma}}}$ GeV/c^2	Observed (90% C.L. Limit)	Produced (90% C.L. Limit)	$\mathbf{R^0/K^0}$ Flux	Expected Flux (pQCD)
3.40	2.40	13.7	50×10^3	1.3×10^{-7}	5.2×10^{-5}
1.72	1.23	929.8	7.2×10^6	1.9×10^{-5}	2.1×10^{-3}
1.61	1.15	455.8	13×10^6	3.5×10^{-5}	2.7×10^{-3}
1.16	0.83	78.5	4.6×10^6	1.2×10^{-5}	6.6×10^{-3}
1.07	0.69	236.7	11.9×10^6	3.2×10^{-5}	8.74×10^{-3}
0.93	0.60	76.0	4.1×10^6	1.1×10^{-5}	11.9×10^{-3}
0.55	0.25	23.1	2.5×10^6	6.7×10^{-6}	27.4×10^{-3}

conduct a search for the R^0 by inspecting the $M_{\pi^+\pi^-}$ shape and its difference from that expected from summing the $M_{\pi^+\pi^-}$ shape from the various kaon decays. The data shows no deviation in the $M_{\pi^+\pi^-}$ shape that could indicate a contribution from an R^0 decay. Quantitative limits on R^0 were obtained from a maximum likelihood fit explained below.

Figure 2 shows the P_t^2 distribution for all the data. The CP violating $K_L \to \pi^+\pi^-$ events are evident in the first bin, which covers the range 0 to 200 (MeV/c)^2. The solid distribution is the data, while the diamonds are the sum of the semileptonic and $K_L \to \pi^+\pi^-\pi^0$ decay Monte Carlo simulations. The data at higher P_t^2 is well matched by a combination of semileptonic and $K_L \to \pi^+\pi^-\pi^0$ decays. The magnitudes for the kaon decay components were fixed by the fit to $M_{\pi^+\pi^-}$ shown in figure 1.

3.1 Fitting for R^0

We used the MINUIT [13] fitting program to perform a maximum-likelihood fit to the $M_{\pi^+\pi^-}$ shape from data, using shapes from $K_L \to \pi e\nu$, $K_L \to \pi\mu\nu$, $K_L \to \pi^+\pi^-$, $K_L \to \pi^+\pi^-\pi^0$, and $R^0 \to \pi^+\pi^-\bar{\gamma}$ from Monte Carlo simulations. The amplitudes for all the simulated $M_{\pi^+\pi^-}$ shapes were allowed to vary independently, as was the amplitude of the trigger prescale for $M_{\pi^+\pi^-} \leq 0.45$ GeV/c^2. The agreement between data and Monte Carlo has an overall χ^2/degree of freedom of $\sim 194/148$ for the region 0.28 GeV/c^2 $\leq M_{\pi^+\pi^-} \leq 0.58$ GeV/c^2.

Approximately 70 different R^0, $\bar{\gamma}$ combinations were used in the fits. All fits yielded R^0 components consistent with zero. An upper limit for a given R^0 was determined by evaluating the maximum-likelihood curve at the 90% confidence interval, after shifting the peak of the curve to zero if it lay in a negative region. Table 1 lists the 90% confidence level upper limit on the number of certain R^0 found in the data, as well as the acceptance corrected number (assuming the R^0 lifetime to be equivalent to the

kaon lifetime) which yields the upper limit on the number of R^0 produced at the target. The last two columns list the 90% confidence level upper limit for the R^0/K^0 flux ratio, and pQCD predicted flux ratio. Using the 2.1×10^6 $K_L \to \pi^+\pi^-$ events observed, we determined that 37.7×10^{10} K_L of all energies exited the absorbers.

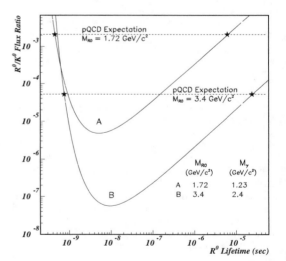

Figure 3: Upper limits with 90% confidence level on the R^0/K^0 flux ratio as a function of R^0 lifetime, for two M_{R^0}, $M_{\bar{\gamma}}$ combinations. The dotted lines show the pQCD expectation for the flux ratio, and the stars mark the corresponding lifetime limits.

3.2 R^0 Flux Upper Limits

Figure 3 shows the variation of the 90% C.L. upper limit on the R^0/K^0 flux ratio with the R^0 lifetime for two R^0, $\tilde{\gamma}$ combinations. The dashed lines indicate the flux ratio expected due to pQCD, and the intersection of the flux prediction and limit, indicated by the stars, correspond to the lifetime limits for that particular R^0. Particles with shorter lifetimes decay too close to the target to be visible in the detector, while those with much longer lifetimes exit the detector without decaying.

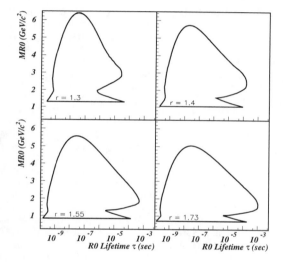

Figure 4: R^0 mass-lifetime regions excluded at 90% confidence level, assuming pQCD flux, for values of r = 1.3 (top left), 1.4 (top right), 1.55 (bottom left) and 1.73 (bottom right). The lower edges are due to the kinematic limit of $M_{R^0} - M_{\tilde{\gamma}} = 2M_\pi$.

We can now use the R^0/K^0 flux expectation from perturbative QCD to exclude a range of lifetimes for a given R^0, $\tilde{\gamma}$ combination. Figure 4 shows R^0 lifetimes excluded at 90% confidence level for a given mass. Contours are shown for r = 1.3, 1.4, 1.55, and 1.73.

The region of particular interest, 1.4 GeV/$c^2 \leq M_{R^0} \leq 2.2$ GeV/c^2, is now accessible. The lower line is due to the kinematic limit of $M_{R^0} - M_{\tilde{\gamma}} = 2M_\pi$, below which the decay cannot proceed. We note that in the theoretically interesting regions of M_{R^0} and r, our exclusion covers lifetimes as low 3×10^{-10} seconds, and as high as 10^{-3} seconds.

4 Conclusion

The analysis presented in this paper excludes most R^0 masses, over six decades in lifetime. In particular the region of primary interest, $M_{R^0}/M_{\tilde{\gamma}} \leq 1.4$ and $M_{R^0} \leq 2.2$ GeV/c^2, which was not addressed by previous direct and indirect searches, is now excluded. We thus definitively close the light gluino window. Our null result constrains most SUSY models in which gauginos remain massless at tree-level and obtain masses from loop diagrams. More generally, our understanding of the $M_{\pi^+\pi^-}$ shape will constrain many future models that require a light hadronic component in their decays.

Acknowledgements

We thank Glennys Farrar for suggesting this search and for discussions concerning this work, and along with Rocky Kolb, for pointing out the cosmological significance of this search.

References

[*] Representing the KTeV Collaboration; email: lath@physics.rutgers.edu

1. R. Mohapatra and S. Nandi, Phys. Rev. Lett. **79**, 181 (1997); Z. Chacko et al., Phys. Rev. D**56**, 5466 (1997); S. Raby, Phys. Rev. D**56**, 2852 (1997).

2. G.R. Farrar, Phys. Rev. Lett. **76**, 4111 (1996); G.R. Farrar, Phys. Rev. D **51**, 3904 (1995);

3. D.J.H. Chung, G.R. Farrar, and E.W. Kolb, Phys Rev. D **56**, 6096 (1997); G.R. Farrar and E.W. Kolb, Phys. Rev. D **53**, 2990 (1996).

4. J. Adams et al., Phys Rev Lett **79**, 4083 (1997).

5. I. Albuquerque et al., Phys. Rev. Lett. **78**, 3252 (1997).

6. F. Csikor and Z. Fodor, Phys. Rev. Lett. **78**, 4335 (1997).

7. R. Barate et al., Report No. CERN-PPE-97-002.

8. R.H. Bernstein et al., Phys. Rev. D **37**, 3103 (1988).

9. G. R. Farrar, Nucl. Phys. Proc. Suppl. **62**, 485 (1998) (available as eprint hep-ph/9710277).

10. K. Arisaka et al., Report No. FERMILAB-580-1992; L.K. Gibbons et al., Phys. Rev. D **55** 6625 (1997).

11. S. Dawson, E. Eichten, and C. Quigg, Phys. Rev. D **31**, 1581 (1985).

12. C. Quigg (private communication).

13. F. James, M. Roos, Comput. Phys. Commun. **10**, 343 (1975).

DIRECT SEARCH FOR LIGHT GLUINOS

M. M. Velasco

CERN, CH-1211 Geneve 23, Switzerland
E-mail: mayda.velasco@cern.ch
On behalf of the NA48 Collaboration

We present the results for a direct search of light gluinos through the appearance of $\eta \to 3\pi^0$ with high transverse momentum in the vacuum tank of the NA48 experiment at CERN. We find one event within a lifetime range of $10^{-9} - 10^{-3}$ s and another one between $10^{-10} - 10^{-9}$ s. Both events are consistent with the expected background from neutrons in the beam, produced by 450 GeV protons impinging on the Be targets, which interact with the residual air in the tank. From these data we give limits on the production of the hypothetical $g\tilde{g}$ bound state, the R^0 hadron, and its $R^0 \to \eta\tilde{\gamma}$ decay in the R^0 mass range between 1 and 5 GeV.

1 Motivation

Recent theoretical work [1] has proposed a class of supersymmetric models in which the gluino ($\tilde{g}$) and the photino ($\tilde{\gamma}$) are expected to have small masses and the photino is stable and an ideal candidate for dark matter. In such models there is a hypothetical spin-1/2 gluon-gluino ($g\tilde{g}$) bound state, the R^0 hadron. This strongly interacting particle is expected to have a mass of a few GeV and a lifetime between 10^{-10} and 10^{-6} seconds. For these reasons the NA48 experiment [2] (see Fig. 1), designed to measure the CP violation parameter $\Re(\epsilon'/\epsilon)$ using high intensity K_L and K_S beams, is a suitable experiment to look for R^0's produced by a 450 GeV proton beam impinging on a Be target.

We have searched for $R^0 \to \eta\tilde{\gamma}$ through the appearance of $\eta \to 3\pi^0$ with high transverse momentum in the decay volume of this experiment, under the assumption that the $\tilde{\gamma}$ is not detectable. The data were collected in about 3 weeks of data taking during 1997.

As shown in Fig. 1, the NA48 experiment has two nearly collinear K_S and K_L beams which operate concurrently [3]. The beams are produced from 1.1×10^{12} and 3.4×10^7 protons impinging on the 40 cm long K_L and K_S Be targets, respectively, every 14.4 seconds in a burst that is 2.4 seconds long. Decays occurring in the K_S and K_L beamline are distinguished by a tagging scintillator hodoscope which is positioned in the proton beam producing the K_S beam by measuring the time of flight between the tagging scintillator hodoscope and the main detector. This dual beamline design offers a wide lifetime range ($10^{-10} - 10^{-3}$ s) in the R^0 search.

2 Search description

A dedicated trigger derived from the ϵ'/ϵ neutral trigger, based on the liquid krypton electromagnetic calorimeter (LKr)[4] information, was implemented in order to select $3\pi^0$ events with high transverse momentum (high-P_T).

The complete 'neutral' trigger system is described in [5]. The high-P_T trigger decision was based on the calculated total electromagnetic energy E_{LKr}, the first moment of the energy m_1, the energy center-of-gravity COG, and the counted number of clusters in each projection. The resulting high-P_T trigger had a high background rejection power with losses less than 25% of the geometrically accepted signal. This was achieved by requiring $E_{LKr} \geq 40$ GeV, $m_1 \geq 1500$ GeV cm, COG ≥ 20 cm from the beam axis, and four or more distinguishable clusters in at least one of the two projections. The COG requirement limits the transverse momentum to more than 0.15 GeV for ηs of momentum greater then 75 GeV., while the first moment requirement rejects backgrounds from $K_L \to 3\pi^0$, where some of the photons escape detection. The largest measured loss (15%) for fully contained events is due to the overlapping of clusters in both projections. For the final trigger to be issued, the neutral trigger signal required to be in anti-coincidence with the muon veto and the ring-shape array of photon detectors appearing as "anti-counters". The high-P_T trigger was downscaled by a factor of two. The resulting trigger rate was below 100 triggers/burst out of a total of 13,000 triggers/burst handled by the data acquisition system during the data taking using 1.5×10^{12} protons/burst on the K_L target.

3 Event selection

The six photons in an event are used to reconstruct $3\pi^0$'s that have to come from a common vertex. The photons must have energies above 2 GeV, a time difference between them which is smaller than 1.5 ns, to be within the defined LKr fiducial volume, and to have no associated track in the drift chambers. In addition, it is required that there is no activity in the hadronic calorimeter and that the energy of the η's is greater than 95 GeV. The reconstructed π^0 masses, $m_i(d)$, are used as a constraint in a fit to minimize the χ^2 as a func-

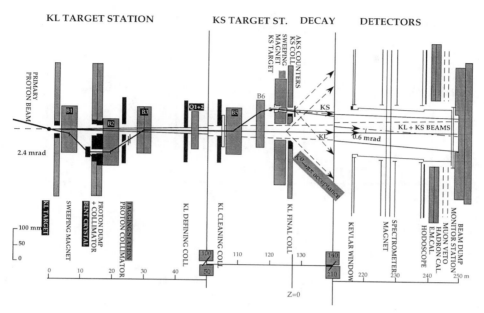

Figure 1: *Schematic drawing of the NA48 beamline.*

tion of the longitudinal vertex position, d, without an assumption on the mass of the parent particle, that is, $\chi^2(d) = \sum_{i=1}^{3}(m_{\pi^0} - m_i(d))^2/\sigma_i^2$. The typical error on the π^0 mass σ_i is around 1.2 MeV. The $\chi^2(d)$ was required to be smaller than 8, which corresponds to a confidence level (CL) larger than 98.2%.

4 Results

The masses of the selected high-P_T events are shown in Fig. 2 as a function of the best fit longitudinal vertex position. We find around 150 $K_L \rightarrow 3\pi^0$ events and around 35 $\eta \rightarrow 3\pi^0$ events in the d-region between - 300 and 9600 m. The integrated beam corresponds to 1.2×10^{17} and 2.1×10^{12} protons impinging on the K_L and K_S target, respectively. The COG distribution for the $K_L \rightarrow 3\pi^0$ events is consistent with simulations made for elastic and quasi-elastic interactions in the AKS and beam cleaning collimators shown in Fig. 1.

Since the η has a very short lifetime, its decay vertex practically coincides with the position at which it was produced. Therefore, the expected R^0 signature is an η with high-P_T in the vacuum region right after the last collimators and the AKS counter. The fiducial region be-

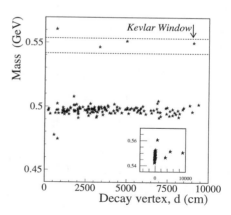

Figure 2: *Reconstructed high transverse momentum* K_L *and* η *particle decays into* $3\pi^0$ *in the vacuum region. The two horizontal lines define the mass window for* η *candidates. The inserted plot includes the* η *events produced by beam interactions in elements of the beamline before the allowed fiducial decay volume.*

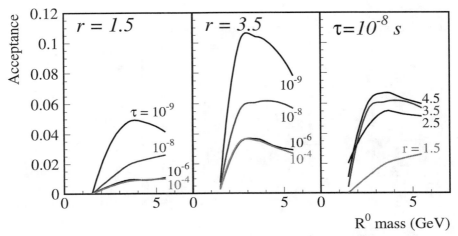

Figure 3: *Calculated geometrical acceptance for several values of the mass ratio $r=m_{R^0}/m_{\tilde{\gamma}}$ and R^0 lifetime as a function of the R^0 mass.*

gins around 6 m downstream of the K_S target, and ends at around 96 m downstream at the Kevlar window, see Fig. 1. The vertex resolution for $\eta \to 3\pi^0$ events is about 70 cm. Therefore, in order to reduce the background from η's produced in the collimators and the AKS counter, only events with a vertex which is at least 200 cm away from the AKS counter position were accepted. As shown in Fig. 2, there are three events that survive all the above cuts and they are in the mass window of ± 6 MeV, which corresponds to $\pm 3\sigma$ of the mass resolution for $3\pi^0$ events. The vertex of the most downstream event is consistent with the position of the Kevlar window, and therefore is excluded from the analysis. The two remaining events are identified as one particle coming from the K_L Be target and the other one from the K_S Be target, by comparing the time of the event as defined by the LKr system and the K_S proton tagging system [6]. A time difference smaller than 1.5 ns is required for an association with the K_S target. We assign one event produced in the K_L beamline and the other one in the K_S beamline. The tagging system has been found to have an efficiency greater than 99.9%, but the rate of protons in the tagger is of the order of 30 MHz, which gives a probability greater than 10% of having an event from the K_L beamline identified as coming from the K_S due to accidental coincidence.

5 Backgrounds

The main background is due to diffractive neutron interactions in the remaining air in the 6-9×10^{-5} mbar

vacuum region. There are also about 10^9 photons/burst coming from the K_L beamline, but they do not contribute to the background because their mean energy is only 30 GeV. The expected mean energy for the neutrons in the K_L and K_S beamlines is around 190 GeV and 100 GeV, and the rates are expected to be 2×10^8 and 1.5×10^4 per burst, respectively.

The background estimates are based on a special run taken with a charged 75 GeV pion beam where $\pi^- N \to \eta X$ events were recorded, and where $\sigma(\pi^- N \to \eta X)$ was measured from eight hours of data taking in which 10^7 pions/burst hit a 6 cm thick CH_2-target at the nominal SPS cycle time. In these data only $\eta \to 3\pi^0$ events with η energies above 95 GeV had trigger requirements similar to those of the high-P_T $3\pi^0$ trigger. For this reason, a minimum energy requirement of 95 GeV in the analysis was applied for the events shown in Fig. 2. However, we do not see additional high-P_T η events if this requirement is relaxed.

The estimates for high-P_T η production in interactions of neutrons in the vacuum tank are found from the ratios of cross sections between $\sigma(\pi^- N \to \eta X)/\sigma(nN \to \eta X)$. We find that our sample should contain about 0.4 and $1.0 \times (10^{-5})$ η's in the K_L and the K_S beamline, respectively. The background estimates for the K_S beamline can be cross checked using the η events produced in the collimators and the 2 mm Iridium crystal (AKS) [7]. The AKS is located around 6 m after the K_S target, and is used to detect K_S decaying before this point. As shown in the inserted plot in Fig. 2 there are about 30 events produced in the region of the AKS and collimators. Ac-

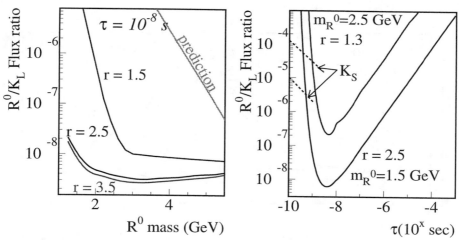

Figure 4: *Upper limits at 95% confidence level on the flux ratio between R^0 and K_L production in p^+-Be interactions assuming a 100% branching ratio for this decay mode. A small improvement (not shown in the figure) in the exclusion at small lifetime is obtained from K_S events.*

cording to the tagging system there are 13 of them are in the K_S beamline a which implies an expected background of 0.4×10^{-5} η events produced by all particles in that beamline. This is consistent with the πN estimates.

As discussed above, there is a 10% probability that an event produced in the K_L beamline is assigned to the K_S instead, and a 40% probability of having an η event in our data sample that was produced by neutrons in the K_L beamline. As a consequence, the probability of having two K_L events and that one of them is tagged as from the K_S beamline is 1.2%. Therefore, we conclude that all events, both in the K_S and the K_L beamline, are consistent with background expectation.

6 Limits and comparison

As shown in Fig. 3, the detector acceptance for $\eta \to 3\pi^0 (BR \simeq 32\%)$ events produced from $R^0 \to \eta\tilde{\gamma}$ decaying within the fiducial volume, shows a strong dependence on the mass ratio between the R^0 and the $\tilde{\gamma}$, $r = m_{R^0}/m_{\tilde{\gamma}}$. At long lifetimes and/or for $r - 1 \gg m_\eta/m_{\tilde{\gamma}}$ the acceptance becomes almost independent of r. We have assumed that the R^0 energy spectrum is the same as that of Λ production measured by this experiment. The sensitivity to the energy spectrum is weak, and it is seen only at short lifetimes, where the sensitivity drops very quickly as a function of R^0 mass.

The angular acceptance in the K_L and the K_S beamline is 0.15 and 0.375 mrad, respectively. The maxi-

mum difference in the angular acceptance between neutral kaons and $\pi^\pm, K^\pm, p, \overline{p}$ was found using the results from Ref. [8] and used as an upper estimate of the expected difference in 'collimator' acceptance between R^0 and neutral kaons. We concluded that the R^0 'collimator' acceptance will be smaller than that of kaons by about 4% and 23% in the K_L and the K_S beamline, respectively.

The expected interaction rate for $R^0 N$ is expected to be between 10% to 100% of the pN cross section [9]. This means that ratio of the absorption probabilities in the Be target of R^0 and kaons can be between 0.75 and 1.

It is more conservative to not apply a background subtraction, but to evaluate the limits based on one signal event in each beam. In addition, it is assumed that the the branching ratio of $R^0 \to \eta\tilde{\gamma}$ is equal to 100%. After taking all the above in consideration the resulting limits on the flux ratio between $R^0 \to \eta\tilde{\gamma}$ and K_L production at a 95% CL are shown in Fig.4 (a)-(b). The fall-off in the sensitivity for small R^0 mass on the left side of plot (a) is due to the loss in phase space for the η to be produced, while the drop in the right side in (b) is due to the decrease in detected events as the lifetime increases. The numbers of expected K_L and K_S at the exit of the last collimator are 2×10^7 and 2×10^2, respectively. This means that there are also enough protons in the K_S target in order to improve the limits at low lifetimes. The K_S data gives an upper bound of 10^{-5} and 10^{-4} for $r = 2.5$ and $r = 1.3$, respectively.

These results can be summarized for r=2.2 by plot-

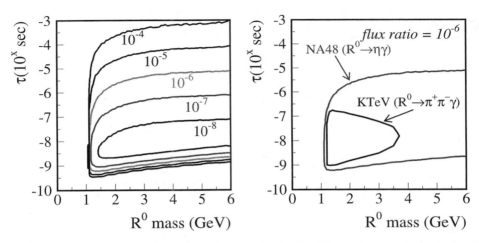

Figure 5: *Upper limits at 95% CL on the flux ratio between R^0 and K_L production in p-Be interactions for r=2.2. The second plot shows the 10^{-6} contours limits given by the analysis presented here and by the KTeV collaboration. In both cases it was assumed a 100% branching ratio in the analysed decay more, $R^0 \to \eta\tilde{\gamma}$.*

ting the contours of the upper limits at 95% CL. These are shown in Fig. 5 where they are compared with the best limit at 90% CL from the direct search for $R^0 \to \pi^+\pi^-\tilde{\gamma}$ by the KTeV collaboration [10]. The two searches are complementary and not necessarily comparable because of the different decay modes. Both analyses show their results by setting the indicated branching ratio equal to 100%. The two-body R^0 decays are suppressed due to approximate C invariance in SUSY QCD [1], while the three-body decays are not. The π^0, η and R^0 have $C = +1$, while $C = -1$ for photinos. Nevertheless, our limits on the R^0/K_L flux ratio are stringent on the R^0 production even if the branching ratio of $R^0 \to \eta\tilde{\gamma}$ is of the order of 10^{-2}.

Though R^0 production cross sections are quite model dependent, and the theoretical uncertainties in the estimates of the R^0 branching ratio into $R^0 \to \eta\tilde{\gamma}$ and the productions cross sections of R^0's are rather large, Nevertheless, the available perturbative QCD calculations [11] imply that the R^0/K_L flux ratio goes as $0.14e^{-2.7m_{R^0}}$. This implies that R^0s with low mass are excluded by these results even if the branching ratio for $R^0 \to \eta\tilde{\gamma}$ is at the level of 1%, as shown in Fig. 5 for an R^0 lifetime of 10^{-8}.

In conclusion, limits are given on the upper values for the R^0/K_L flux ratio in an R^0 mass and lifetime region between 1-5 GeV and $10^{-10} - 10^{-3}$ seconds, respectively. As shown in Fig. 4, depending on the value for the branching ratio of $R^0 \to \eta\tilde{\gamma}$, the 95% CL on the upper value on the R^0/K_L flux ratio could be as low as 6×10^{-9} for a R^0 with a mass of 1.5 GeV, a lifetime of 6×10^{-9} seconds and r=2.5.

References

1. G. Farrar, Phys. Rev. **D51** , 3904 (1995); G. Farrar, Phys. Rev. Lett. **76**, 4111 (1996).
2. G. Barr, *et al.*, CERN/SPSC/90-22 (1990).
3. C. Binno, N. Double, L. Gatignon, P. Grafstrom, and H. Wahl, CERN-SL-98-033 (EA).
4. G. Barr, *et al.*, Nucl. Instr. and Meth. A **370**, (1996) 413.
5. B. Gorini, *et al.*, IEEE Trans. Nucl. Sci. **45**, 1771-1775 (1998).
6. T. Beier, *et al.*, Nucl. Instr. and Meth. A **360**, (1995) 390-394.
7. C. Biino, Workshop on Channelling, Aarhus, Denmark, July 10-14, 1995.
8. H.W. Atherton, *et al.*, CERN Yellow Report 80-07 (1980).
9. S. Nussinov, Phys. Rev. **D57**, 7006 (1998).
10. J. Adams, Phys. Rev. Lett. **79**, 4083 (1997).
11. S. Dawson, E. Eichten, and C. Quigg, Phys. Rev. D **37**, 1581(1985); C. Quigg, S.V. Somalwar, (pri. comm.).

CONTACT INTERACTIONS AT LEP AND LIMITS FROM SM PROCESSES

MARCO PIERI

INFN Sezione di Firenze, L.go E. Fermi, 2, 50125 Firenze, Italy
E-mail: Marco.Pieri@cern.ch

From 1995 LEP collected data above the Z peak at centre of mass energies between 130 and 189 GeV. Using the measurements of cross sections and asymmetries of the two fermion production process different scenarios beyond the Standard Model have been investigated. We will review the results of the four LEP experiments on the search for contact interactions, R-parity violating sneutrinos and squarks, leptoquarks and Z' bosons.

1 Introduction

The results on the measurements of two fermion production at LEP and their interpretations in terms of new physics are reviewed. At energies above the Z peak the sensitivity of this process to additional contributions is enhanced and limits on their presence can be derived if no signal is found.

2 Two Fermion Production at LEP

Two fermion production, $e^+e^- \rightarrow f\bar{f}$, occurs in the Standard Model through the two tree level diagrams shown in Figure 1. The cross sections and forward-backward asymmetries of this process would be modified by the presence of new physics. All LEP experiments have measured these quantities when the final state fermions are $f = e, \mu, \tau, q^{1,2,3,4}$ using the data summarized in Table 1. All results obtained at $\sqrt{s} = 189$ GeV are preliminary.

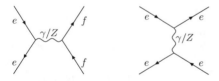

Figure 1: Feynman diagrams for the process of two fermion production.

$\sqrt{s}$(GeV)	$\mathcal{L}$/exp. (pb^{-1})	Year
130	~ 6	1995 + 1997
136	~ 6	1995 + 1997
161	~ 10	1996
172	~ 10	1996
183	~ 55	1997
189	~ 30–40	1998

Table 1: Data collected at LEP above the Z peak.

As an example in Figure 2 the cross section measurements from ALEPH are shown. ALEPH, DELPHI and

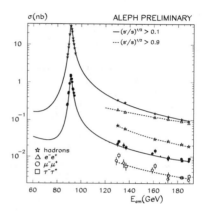

Figure 2: ALEPH measurements of cross sections of two fermion production.

OPAL also measured cross sections and asymmetries for heavy quark production. The interesting measurements for the search for new physics are those corresponding to non-radiative events which occur at high effective centre of mass energy. All measured quantities are found in good agreement with the Standard Model predictions and no signal of new phenomena is observed.

3 Contact Interactions

Contact interactions provide a general description of new physics which occurs at a higher energy scale. They are characterized by the coupling strength g and by the energy scale Λ which is equivalent to the mass of the exchanged particle. Even if the energy scale Λ is much larger that the centre of mass energy, contact interactions could affect observables and produce appreciable deviations from the Standard Model predictions.

The effective Lagrangian for contact interactions in

the $e^+e^- \to f\bar{f}$ process can be written as [5]:

$$\mathcal{L} = \frac{1}{1+\delta_{ef}} \sum_{i,j=\text{L,R}} \eta_{ij} \frac{g^2}{\Lambda^2} (\bar{e}_i \gamma^\mu e_i)(\bar{f}_j \gamma^\mu f_j)$$

where δ_{ef} is 1 for the e^+e^- final state and 0 otherwise and η_{ij} $(i,j = \text{L,R})$ define the helicity structure of the interaction. For simplicity we usually assume $|\eta_{ij}| = 0, 1$. In addition, when quoting limits on Λ, we conventionally choose $g^2/4\pi = 1$.

In presence of contact interactions the differential cross section becomes:

$$\frac{d\sigma}{d\cos\theta} = \frac{d\sigma^{\text{SM}}}{d\cos\theta} + c_2(s,\cos\theta)\frac{1}{\Lambda^2} + c_4(s,\cos\theta)\frac{1}{\Lambda^4}$$

where c_2 is the interference between SM and new physics and c_4 is the pure contact interaction term.

To determine the models of contact interactions the values of η_{ij} have to be defined. The models investigated at LEP are shown in Table 2. We should note that the models VV, AA, LL+RR, and LR+RL, which are parity conserving are the most interesting at LEP as they are not subject to limits from atomic parity violation experiments. For the others $\Lambda \lesssim 15$ TeV is already excluded [6]. Figure 3 shows the effect of the presence of contact inter-

Model	η_{LL}	η_{RR}	η_{LR}	η_{RL}
LL	± 1	0	0	0
RR	0	± 1	0	0
LR	0	0	± 1	0
RL	0	0	0	± 1
VV	± 1	± 1	± 1	± 1
AA	± 1	± 1	∓ 1	∓ 1
LL+RR	± 1	± 1	0	0
LR+RL	0	0	± 1	± 1
LL$-$RR	± 1	∓ 1	0	0

Table 2: Helicity structure of the models of contact interactions considered.

actions in the LL model for the $e^+e^- \to q\bar{q}$ cross section and $e^+e^- \to \mu^+\mu^-$ asymmetry. It is apparent that the effect becomes larger at higher centre of mass energies. From fits to the measurements limits on the presence of contact interactions are derived in the different models. They are expressed as 95% C.L. lower limits on the scale Λ when contact interactions couple to the different final state fermions. The limits Λ_+ (Λ_-) correspond to the upper (lower) sign combination in Table 2. The limits from the four experiments for contact interactions [1,2,3,4] affecting the channels: $f\bar{f}$, $q\bar{q}$ and l^+l^- are reported in Table 3. DELPHI and L3 have included the results from $\sqrt{s} = 189$ GeV data.

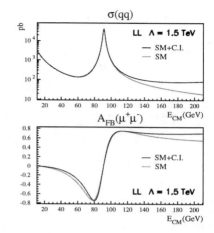

Figure 3: Effect of LL contact interactions with $\Lambda = 1.5$ TeV on the $q\bar{q}$ cross section and $\mu^+\mu^-$ asymmetry.

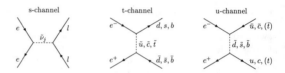

Figure 4: Feynman diagrams of R-parity violating supersymmetric particle exchange

4 R-Parity Breaking Scalar Fermions and Leptoquarks

In Supersymmetric theories with R-parity breaking the superpotential contains L and B violating trilinear Yukawa couplings [7]:

$$W_{R\!\!\!/} = \lambda_{ijk} L_L^i L_L^j \bar{E}_R^k + \lambda'_{ijk} L_L^i Q_L^j \bar{D}_R^k$$

where L and Q are left-handed doublets of leptons and quarks, E and D are right-handed singlets of charged leptons and down-type quarks and ijk are the generation indices. Given that $\lambda_{ijk} \neq 0$ only for $i < j$, at least two different generations are coupled in the purely leptonic vertices and the number of free couplings in the leptonic vertices is reduced from 27 to 9. Depending on these couplings supersymmetric particles can be exchanged in the s, t or u channels as shown in Figure 4.

4.1 Scalar Neutrinos

The scenario with scalar neutrinos is particularly interesting because they could be produced in the s-channel and induce large deviations in the cross sections and

Table 3: Limits on Λ_+ (Λ_-) in TeV at 95 % CL for $f\bar{f}$, l^+l^- and $q\bar{q}$ final states.

$f\bar{f}$	ALEPH		DELPHI		L3		OPAL	
	Λ_+	Λ_-	Λ_+	Λ_-	Λ_+	Λ_-	Λ_+	Λ_-
LL	6.4	7.5	—	—	6.7	4.3	5.8	5.2
RR	5.4	6.0	—	—	6.6	4.3	5.7	5.0
LR	4.9	4.8	—	—	4.2	5.8	5.7	5.4
RL	5.6	5.0	—	—	4.7	6.1	5.5	6.1
VV	9.7	10.4	—	—	9.9	8.8	9.5	9.7
AA	8.5	10.1	—	—	10.6	5.4	8.4	8.1
LL+RR	8.2	9.5	—	—	9.2	5.8	7.4	7.4
LR+RL	7.5	6.8	—	—	6.1	8.3	7.7	7.9

l^+l^-	ALEPH		DELPHI		L3		OPAL	
	Λ_+	Λ_-	Λ_+	Λ_-	Λ_+	Λ_-	Λ_+	Λ_-
LL	5.1	5.7	4.7	5.6	6.8	4.2	5.2	5.3
RR	5.0	5.4	4.4	5.3	6.5	4.1	5.0	5.1
LR	4.9	4.8	4.5	4.7	4.2	5.9	5.6	5.2
RL	4.9	4.8	5.1	4.7	3.2	5.9	5.6	5.2
VV	9.1	9.7	9.0	9.6	9.8	8.9	9.6	9.3
AA	7.4	8.1	6.3	7.8	10.8	5.4	7.7	8.3
LL+RR	7.1	7.9	—	—	9.4	5.9	7.2	7.4
LR+RL	7.1	7.0	—	—	8.4	5.9	7.8	7.3

$q\bar{q}$	ALEPH		DELPHI		L3		OPAL	
	Λ_+	Λ_-	Λ_+	Λ_-	Λ_+	Λ_-	Λ_+	Λ_-
LL	5.6	6.4	—	—	4.1	2.7	4.4	2.8
RR	4.1	4.5	—	—	2.9	3.7	3.0	3.9
LR	3.0	3.2	—	—	3.2	2.4	3.3	3.6
RL	2.3	4.0	—	—	2.4	4.6	2.5	4.9
VV	6.7	7.4	—	—	4.0	5.4	4.1	5.7
AA	7.4	8.2	—	—	6.0	3.7	6.3	3.8
LL+RR	6.9	7.7	—	—	4.2	3.7	4.4	3.1
LR+RL	2.9	4.5	—	—	3.0	5.2	3.1	5.5

λ^2	e^+e^-	$\mu^+\mu^-$	$\tau^+\tau^-$
λ_{121}^2	$\tilde{\nu}_\mu$ (t,s)	$\tilde{\nu}_e$ (t,s)	
λ_{131}^2	$\tilde{\nu}_\tau$ (t,s)		$\tilde{\nu}_e$ (t)
$\lambda_{121}\lambda_{233}$			$\tilde{\nu}_\mu$ (s)
$\lambda_{131}\lambda_{232}$		$\tilde{\nu}_\tau$ (s)	

Table 4: Sneutrinos contributing to the different leptonic final states (and in which exchange channel, t or s,) when one or two λ are non-vanishing.

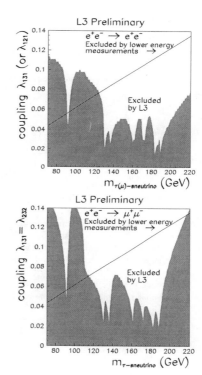

Figure 5: L3 preliminary limits on the sneutrino couplings with the assumptions described in the text.

asymmetries. To derive limits on the couplings, for simplicity we restrict ourselves to two cases:

- we assume that only one coupling is non vanishing;

- we assume that two couplings are different from zero and equal.

In Table 4 we show which scalar neutrinos can be exchanged and in which channels in the different cases.

The L3 limits are shown in Figure 5, in case of only one coupling different from zero Λ_{121} and Λ_{131} values above 0.05 are excluded at 95% C.L. for a sneutrino mass $130 \lesssim m_{\tilde{\nu}} \lesssim 200$ GeV.

4.2 Scalar Quarks and Leptoquarks

Scalar quarks and leptoquarks can be exchanged in the t or u-channel leading to hadronic final states. A large variety of leptoquarks is predicted in different theories and they can be classified according to reference [8]. Figure 6 shows the Feynman diagrams of these processes. The t-channel is associated with the transition $e^- \to q$ and leptoquarks with $F = 0$ are involved, the u-channel is associated with the transition $e^- \to \bar{q}$ and leptoquarks with $F = 2$ are involved, being F the leptoquark fermion number $F = 3B + L$.

The cross section for leptoquark exchange is given

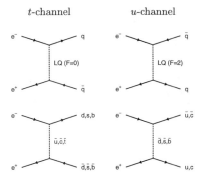

t-channel   u-channel

Figure 6: Feynman diagrams for squark and leptoquark exchange.

by [7]:

$$\frac{d\sigma}{d\cos\theta} = \frac{N_c}{128\pi s}\sum_{i,k=L,R}\rho_{ik}|f_{ik}|^2$$

where ρ_{ik} are the spin density matrix elements and f_{ik} are the helicity amplitudes.

If one coupling $|\lambda'_{1jk}|$ is much larger than the others (which are neglected) the limits for R-parity violating scalar quark exchange are almost identical to those for the leptoquarks. Down-type right handed scalar quarks correspond to leptoquarks S_0 while up-type left handed scalar quarks correspond to leptoquarks $\tilde{S}_{1/2}$. Figure 7 shows the limits on scalar leptoquark couplings g_L or g_R obtained by OPAL when assuming that only one of the two is non-zero.

If the leptoquark mass is such that $m_{LQ} \gg \sqrt{s}$ this process is equivalent to contact interactions with:

$$\frac{g^2_{L,R}}{m^2_{LQ}} \sim \frac{1}{\Lambda^2_{ij}}$$

and the limits obtained above apply.

5   Indirect Search for Z′

The presence of a Z′ would modify the neutral current Lagrangian by an additional term:

$$\mathcal{L} = eA_\mu J^\mu_\gamma + gZ_\mu J^\mu_Z + g'Z'_\mu J^\mu_{Z'}$$

The Z′ parameters are the mass $M_{Z'}$, the Z–Z′ mixing angle θ_M and its couplings to the fermions. The Z′ couplings are defined in GUT theories, such as E_6 models [9] and left-right symmetric models [10], or can be left free in a model independent search.

In the E_6 models the Z′ current is function of the parameter θ_6:

$$J^\mu_{Z'} = J^\mu_\chi \cos\theta_6 + J^\mu_\psi \sin\theta_6$$

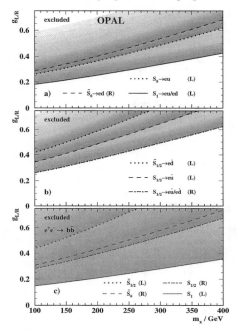

Limits on the coupling for Scalar Leptoquarks

Figure 7: OPAL upper limits on scalar leptoquark couplings.

and specific models are obtained for given value of the parameter:

- χ model; $\theta_6 = 0$,
- ψ model; $\theta_6 = \pi/2$,
- η model; $\theta_6 = -\arctan\sqrt{5/3}$.

In the left-right symmetric models the current is:

$$J^{LR} = \alpha_{LR}J^\mu_{3R} - \frac{1}{2\alpha_{LR}}J^\mu_{B-L}$$

and the LR model is defined for a value of $\alpha_{LR} = 1.53$.

L3 also consider the Sequential Standard Model (SSM), where the Z′ has the same couplings as the Z and it is only heavier.

New preliminary results using 189 GeV data have been reported by DELPHI and L3 [2,3,11]. Mixing angles larger than ~ 0.03 radians are excluded for any $M_{Z'}$ in the different models. Figure 8 shows the exclusions from L3 in the plane $M_{Z'}$—θ_M, while preliminary limits on $M_{Z'}$, valid for any value of the mixing angle, are reported in Table 5. The indirect lower limit of 820 GeV on $M_{Z'}$

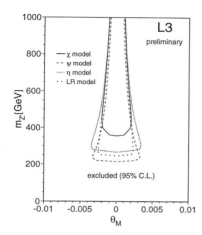

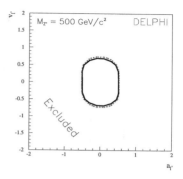

Figure 9: DELPHI model independent limits on the Z' couplings to the leptons. The solid line indicates the exclusion region using all data, the dotted line when only data up to 183 GeV are used.

Figure 8: L3 limits on the Z' mass as funcion of the mixing angle in different models.

	$m_{Z'}$ (GeV)	
Z' model	DELPHI	L3
χ	315	390
ψ	330	230
η	240	285
LR	270	250
SSM	–	820

Table 5: Preliminary 95% C.L. limits on the Z' mass in the different models.

in the SSM improves that coming from direct searches at the Tevatron.

Limits are also obtained on the Z' couplings to leptons in a model independent analysis which assumes lepton universality. In Figure 9 the DELPHI results for $M_{Z'} = 500$ GeV are shown. The two curves in the figure show the difference if $\sqrt{s} = 189$ GeV data are included or not in the fit.

6 Conclusions

New preliminary results from all LEP experiments on the measurements of $e^+e^- \to f\bar{f}$ cross sections and asymmetries based on ≈ 30–40 pb^{-1}/exp. at 189 GeV have been reported. Al results are found to be in agreement with the Standard Model predictions. No signal of new physics has been observed and improved limits, using $\sqrt{s} = 189$ GeV data have been derived by DELPHI and L3. Lower limits on contact interaction scale in the range 2 – 10 TeV have been reported as well as improved limits on R-parity violating supersymmetric particle and leptoquark couplings. The best up to date indirect limit on the Se-

quential Standard Model Z' mass of 820 GeV has been obtained by L3.

Further increase of LEP energy in the coming years will allow to extend these searches to even higher energies with the possibility of observing effects due to new phenomena.

References

1. ALEPH Collab., ICHEP'98 **906**, (1998),
   ALEPH Collab., ICHEP'98 **908**, (1998).
2. DELPHI Collab., ICHEP'98 **439**, (1998),
   DELPHI Collab., ICHEP'98 **441**, (1998),
   DELPHI Collab., ICHEP'98 **643**, (1998).
3. L3 Collab., *Phys. Lett.* B **370**, 195 (1996),
   L3 Collab., *Phys. Lett.* B **484**, 361 (1997),
   L3 Collab., ICHEP'98 **484**, (1998),
   L3 Collab., ICHEP'98 **510**, (1998),
   L3 Collab., ICHEP'98 **513**, (1998).
4. OPAL Collab., *Euro. Phys. J.* C **2**, 441 (1998),
   OPAL Collab., ICHEP'98 **1067**, (1998).
5. E. Eichten et al., *Phys. Rev. Lett.* **50**, 811 (1983).
6. C. S. Wood et al., Science **275**, 1759 (1997),
   V.Barger et al., *Phys. Lett.* B **404**, 147 (1997),
   N. Di Bartolomeo, M. Fabbrichesi, *Phys. Lett.* B **406**, 237 (1997).
7. J. Kalinowski et al., *Z. Phys.* C **74**, 595 (1997),
   J. Kalinowski et al., *Phys. Lett.* B **406**, 314 (1997).
8. A. Djouadi et al. *Z. Phys.* C **46**, 679 (1990).
9. For a review see e.g. J. L. Hewett, T. G. Rizzo, *Phys. Rev.* **183**, 193 (1989).
10. L. S. Durkin, P. Langacker, PLB **166**, 436 (1986),
    For a review see e.g. R. N. Mohapatra, "Unifications and Supersymmetries", Springer, New York (1989).
11. L3 Collab., ICHEP'98 **511**, (1998).

CONTACT INTERACTIONS AT HERA

L. STANCO

I.N.F.N. Padova, Via Marzolo 8, I-35131 Padova, Italy
E-mail: stanco@pd.infn.it

Deep–inelastic scattering, $e^+p \to e^+X$, has been studied at HERA at 300 GeV center–of–mass energy by the H1 and ZEUS experiments. The data were collected during 1994–1997 and correspond to 37.0 and 46.2 pb^{-1} of luminosity for H1 and ZEUS, respectively. The measurements were compared to the Standard Model expectation and analyzed to search for *lepton–quark* contact interactions. No significant signal is observed and limits on the compositeness scales are computed for different coupling models. Where already available, limits are compared to those from other experiments.

1 Introduction

The HERA ep collider has extended the kinematic range available for the study of deep–inelastic scattering (DIS) by two orders of magnitude in Q^2, the negative square of the four–momentum transfer between the lepton and proton. Positrons with $E_e = 27.5$ GeV collide with 820 GeV protons at a center of mass energy of $\sqrt{s} = 300$ GeV. In terms of Bjorken–x and y, the fraction of lepton's energy lost in the rest frame of the proton, Q^2 is $\simeq x \cdot y \cdot s$, allowing the study of DIS up to values of Q^2 of several 10^4 GeV^2.

This new kinematic domain opens a window to search for indications of physics processes beyond the Standard Model (SM). The measured DIS cross section is compared to the prediction by the Standard Model which describes the ep interaction via the neutral and charged electroweak currents. A wide class of hypothesized new interactions at characteristic mass scales in the TeV range would modify the DIS cross sections in a way which can be parameterized by effective four–fermion contact interactions (CI). Compositeness or leptoquark structures can then be indirectly inferred at the *low* energy scale of HERA.

The neutral–current e^+p DIS data have been collected with the two detectors, H1 and ZEUS, during the 1994–1997 runs and correspond to an integrated luminosity of 37.0 and 46.2 pb^{-1} for H1 and ZEUS, respectively. It represents a tenfold statistics increase compared to the previous analysis done by H1 with 1994 data only[1].

This paper is organized as follows: after a description of the theoretical aspects, the relevant techniques of the selection of the data and fitting of differential cross section are described. Determination of the limits on the compositeness energy scale are given in the subsequent section, followed by a brief summary and some perspective considerations.

The data selection is also extensively reported in the two contributed papers to this conference by H1[2] and ZEUS[3], on the measurement of the inclusive cross section for neutral and charged current interaction at high Q^2.

H1 and ZEUS results on Contact Interaction search[4] have been submitted to this conference.

2 Theoretical Aspects

In full analogy to the Fermi's description of the weak force at low energy through an effective four–fermion interaction, non Standard Model physics processes at mass scales much beyond the HERA center–of–mass energy can be approximated in their low energy limit by $lq \to lq$ CI. Examples include the exchange of new, heavy objects with mass $M > \sqrt{s}$ such as leptoquarks or new vector bosons[5], and the exchange of common constituents between the lepton and quark in models of compositeness[6,7].

The simplest $lq \to lq$ contact interaction that conserve $SU(3)_C \times U(1)$ symmetry can be represented for each quark flavor as additional terms in the Standard Model Lagrangian $\mathcal{L}_{SM}$. For chirally invariant neutral current vector interactions one finds[5]:

$$\mathcal{L} = \mathcal{L}_{SM} + \frac{g^2}{\Lambda^2} \sum_{q=u,d} \{ \eta^q_{LL}(\bar{e}_L\gamma^\mu e_L)(\bar{q}_L\gamma^\mu q_L)$$
$$+ \eta^q_{LR}(\bar{e}_L\gamma_\mu e_L)(\bar{q}_R\gamma^\mu q_R) + \eta^q_{RL}(\bar{e}_R\gamma_\mu e_R)(\bar{q}_L\gamma^\mu q_L)$$
$$+ \eta^q_{RR}(\bar{e}_R\gamma_\mu e_R)(\bar{q}_R\gamma^\mu q_R) \}$$

where the indices L and R denote the left–handed and right–handed fermion helicities, g is the coupling, Λ is the effective mass scale and the η determine the relative size of the individual terms and the sign of the positive or negative interference with the Standard Model current. Strong limits beyond the HERA sensitivity have already been placed on the scalar and tensor interactions[5] and therefore neglected in the present analysis. Conventionally, g is chosen to be equal to 4π while $\eta = 0, \pm 1$ in each of the 8 terms.

To pare down the corresponding long list of possibilities, the following considerations are taken in:

- The SM invariance $SU(3)_C \times SU(2)_L \times U(1)_Y$ implies $\eta^u_{iL} = \eta^d_{iL}$, $i = L, R$. ZEUS has considered scenarios which violate this relation only for *up* quarks which dominate the high–x cross section at HERA.

- Very strong limits on Λ come from atomic parity violation measurements from Cs[8]. These limits are avoided if purely chiral CI are excluded, i.e. if $\eta^q_{LL} + \eta^q_{LR} - \eta^q_{RL} - \eta^q_{RR} = 0$. Within this constraint the ZEUS analysis explored the 24 possible scenarios for Contact Interaction.

- VV, AA and VA were also considered by H1 who preferred to analyze also the pure chiral terms to provide easy access to further combinations of them. See table 1 for the full list.

Table 1: The 24+8 scenarios of contact interaction considered by the two experiments, H1 and ZEUS. Each row of this table represents two scenarios corresponding to $a = +1$ and $a = -1$.

	η^u_{LL}	η^u_{LR}	η^u_{RL}	η^u_{RR}	η^d_{LL}	η^d_{LR}	η^d_{RL}	η^d_{RR}
VV	$+a$	$+a$	$+a$	$+a$	$+a$	$+a$	$+a$	$+a$
AA	$+a$	$-a$	$-a$	$+a$	$+a$	$-a$	$-a$	$+a$
VA	$+a$	$-a$	$+a$	$-a$	$+a$	$-a$	$+a$	$-a$
X1	$+a$	$-a$	0	0	$+a$	$+a$	0	0
X2	$+a$	0	$+a$	0	$+a$	0	$+a$	0
X3	$+a$	0	0	$+a$	$+a$	0	0	$+a$
X4	0	$+a$	$+a$	0	0	$+a$	$+a$	0
X5	0	$+a$	0	$+a$	0	$+a$	0	$+a$
X6	0	0	$+a$	$-a$	0	0	$+a$	$-a$
U5	0	$+a$	0	$+a$	0	0	0	0
U1	$+a$	$-a$	0	0	0	0	0	0
U4	0	$+a$	$+a$	0	0	0	0	0
	η_{LL}	η_{LR}	η_{RL}	η_{RR}				
LL	$+a$	0	0	0				
LR	0	$+a$	0	0				
RL	0	0	$+a$	0				
RR	0	0	0	$+a$				

The effect of the addition of a contact interaction on the NC DIS cross section varies depending on the scenario but can be divided into two portions: a term proportional to $1/\Lambda^4$ that enhances the cross section at high Q^2 and a second term proportional to $1/\Lambda^2$ caused by interference with the SM amplitude that can either enhance or suppress the cross section at intermediate Q^2.

3 Experimental Aspects

Data from both experiments were selected and analyzed as for the Neutral Current DIS cross section measurements[2,3].

In H1, the cross section is determined from a purely inclusive measurement of the final state positron with energy E'_e and polar angle θ_e. The kinematic quantities are calculated via $Q^2 = 4E_eE'_e\cos(\theta_e/2)$ and $y = 1 - E'_e/E_e\sin^2(\theta_e/2)$. The absolute energy uncertainty varies between 1% in the backward region ($\theta_e > 90^0$ wrt proton direction) and 3% in the forward region, the scattering angle is known to within 3 mrad. The data are corrected for detector effects and QED radiation and represent the cross section within the kinematic phase space $180 < Q^2 < 30,000$ GeV^2, $y < 0.9$ and $E'_e > 11$ GeV.

In ZEUS, the final state is determined by the polar angles of the positron (θ_e) and the *current* hadronic system (γ_H), with a corresponding systematic error on the computed Q^2 of 2.1 - 3.0 %, in the $400 < Q^2 < 51,200$ GeV^2, $y < 0.95$, $E'_e > 10$ GeV phase space. The quoted ZEUS systematic error on luminosity is less than 1.5%.

For both experiments the global systematic error is around 3% against an error on the Standard Model prediction around 7% due to the choice of the parton distribution functions.

Good agreement was found in both analysis between the measured differential cross sections $d\sigma/dQ^2$, $d\sigma/dx$, $d\sigma/dy$ and the respective Standard Model prediction, with the exception of a slight excess at very high $Q^2 > 15,000$ GeV^2. H1 performed a χ^2 fit to the $d\sigma/dQ^2$ ($180 < Q^2 < 30,000$ GeV) distribution as compared to the expectation of the Standard Model. Using the CTEQ4D parton parameterizations[9] in the DIS scheme, the fit yields $\chi^2 = 11.9/14$ $d.o.f$ with a normalization of $\int d\sigma^{SM}/\int d\sigma^{Data} = 0.988^a$. The CI analysis from H1 is based on a comparison of the measured cross section to that predicted by SM+CI Lagrangians. The dominant correlated systematic errors, i.e. the experimental uncertainty of the lepton energy scale and the scattering angle, are taken into account in the fit procedure, which analyzes the differential cross section in terms of the coupling coefficients $\pm 4\pi/\Lambda^2$ and leaves the sign of interference free.

The CI analysis from ZEUS is based on a comparison of measured event distributions of the kinematic variables Q^2, x and y to the corresponding results of a MC simulation provided by standard SM generators and a weight $d\sigma^{SM+CI}/d\sigma^{SM}$ to consider the CI scenarios. This reweighting procedure accounts correctly for correlations between the possible presence of a CI signal and the pattern of acceptance losses and migrations.

ZEUS applied two independent statistical methods to extract a log–likelihood function $L(a/\Lambda)^b$ from the comparison of measured and simulated event distribu-

---

^{a}An overall uncertainty of 1.8% is quoted by H1 due to the luminosity determination. It should be noted that the luminosity error is lower than the one quoted in [2] due to an improved understanding of the systematic uncertainties.

$^b a = \pm 1$ represents the two scenarios for positive and negative interference of CI with the Standard Model.

tions. In the first method, the likelihood is determined from the binned Q^2 distribution of events, using Poissonian statistics. Systematic uncertainties, including the effects of different parton distribution functions, are included in L under the assumption that they are fully correlated between bins and that the probability density for their magnitude has a Gaussian shape. In the second method the likelihood is computed from the unbinned bidimensional distribution in the $x - y$ plane. By applying the two methods to MC samples, they were checked to provide unbiased and consistent estimates of the CI parameters.

The most probable value Λ_0 for each CI scenario is extracted from the likelihood maximum. MC experiments (i.e. statistically independent MC data samples corresponding to the data luminosity) were used to calculate 95% C.L. CI limits by determining the value of Λ at which 95% of such MC experiments produce most likely values of $1/\Lambda^2$ larger than found in the data. In such cases where $1/\Lambda_0^2 \neq 0$ the MC experiments are also used to estimate the probabilities that a statistical fluctuation in an experiment with SM cross section would produce a value of Λ_0 as small or smaller than that obtained from the data.

H1 followed a similar procedure to give limits on the compositeness scale parameters, which were derived by fitting Λ^+ and Λ^- with the appropriate interference signs and taking the corresponding changes in χ^2 with respect to the fit to the Standard Model. Choosing different parton distributions for the cross section calculation, e.g MRST or GRV [10], does not change the shape of the Q^2 spectrum significantly, but rather the normalization by up to 1.2%. The limits on $\Lambda^\pm$ change by $\sim 0.2 - 0.8$ TeV

in either direction compared to the CTEQ4D parton densities. The most conservative values of $\Lambda^\pm$ obtained from a variation of the parton distributions are quoted in the results.

4   Results

The χ^2 distributions obtained from H1 are shown in fig. 1. An example of the negative log–likelihood functions derived with the two different methods by ZEUS, is reported in fig. 2.

Fit results derived from the Q^2 cross section (H1) and the ratio of events from data and events expected from SM Monte Carlo (ZEUS) are given in fig. 3 and 4, respectively.

All Λ_0 results, their 1σ confidence intervals, the 95% C.L. limits and the probabilities for SM fluctuations are summarized in table 2 and fig. 5. Corresponding limits which have recently been reported by other experiments are also shown.

The values of Λ_0 range from about 2 TeV to infinity. The lower 95% C.L. limits of Λ range from 1.5 TeV to almost 5 TeV. For all the cases the Standard Model prediction, i.e. $1/\Lambda = 0$, is compatible with the result found within two standard deviations. The probabilities of observing a Standard Model fluctuation are typically in the range 20 - 40% and reach a value as low as 4.5% only for the X2($a = -1$) CI scenario in the ZEUS analysis.

Note that the LL and RR models and the LR and RL models are almost indistinguishable in deep–inelastic e^+p scattering. The pure chiral couplings analyzed by H1 prefer negative values of the interference. This is a consequence of the trend of the data (fig. 3 and 4) to be slightly low around $Q^2 \simeq 4,000 - 12,000$ GeV^2 followed by an upward fluctuation at higher Q^2, which favors a

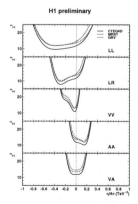

Figure 1: Distributions of χ^2 versus $\eta/4\pi$ ($\eta = \pm 4\pi/\Lambda^2$) from H1 fits to various CI models using different parton distributions. The pure chiral models, like LL and LR are almost indistinguishable in DIS.

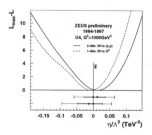

Figure 2: A comparison of the negative log–likelihood function derived with the two different methods used by ZEUS for the U4 scenario. The horizontal bars at the bottom indicate the most likely values of η/Λ^2 (dots), the 1σ confidence intervals (thick lines) and the 95% C.L. limits calculated from the log–likelihood curves (thin lines). The upper and lower bars are for the unbinned and the binned methods, respectively.

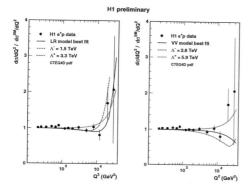

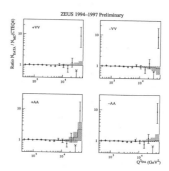

Figure 3: NC cross section $d\sigma/dQ^2$ normalized to the SM expectation using CTEQ4D. parton distributions. H1 data (dots) are compared with fits to the LR and VV scenarios and with 95% C.L. contributions of $\Lambda^{\pm}$. The overall normalization uncertainty of 1.8% is not shown.

Figure 4: Examples of CI fit results from ZEUS. The number of events expected from the SM was evaluated using the CTEQ4 parton distribution. The best fit is plotted as a dark line, the lighter lines are the $\pm 1\sigma$ limits. For $+VV$ the best fit is the SM. For $-VV$ and $-AA$, the lower 1σ limit is also the SM.

negative interference term. This trend is present in the data of both H1 and ZEUS.

5 Conclusions and Outlook

The neutral current reaction $e^+p \to e^+X$ has been measured at high momentum transfers Q^2 between a few hundred and a few ten thousand GeV^2 with the two detectors H1 and ZEUS at the HERA collider. The differential cross section $d\sigma/dQ^2$ is well described over six orders of magnitude by the Standard Model expectation. The data were analyzed to search for new phenomena mediated through $(\bar{e}e)(\bar{q}q)$ contact interactions with 37.0 and 46.2 pb^{-1} of luminosity by H1 and ZEUS, respectively.

No significant signal for contact interactions was found. The most likely values of the compositeness mass scale Λ and 95% C.L. limits on $1/\Lambda$ have been determined for 32 different CI scenarios.

In all cases the 95% C.L. limits include the Standard Model (i.e. S.M. differs at most two standard deviations from each CI scenario). The lower limits on Λ are typically between 2 and 5 TeV. The results show a sensitivity to contact interactions which is similar to that reported by other experiments. ZEUS is also setting limits on scenarios which were not studied by the others.

In the short term, HERA is going to provide $\approx$ 50 pb^{-1} of e^-p data with the 1998/99 run. This will allow to set stronger limits on CI compositeness scale in the 5 TeV range for many more of different scenarios.

In the long term, after the year 2000 the increase of the instantaneous luminosity will provide $\approx$ 1 fb^{-1} in 5 years for e^+p and e^-p interactions with significant longitudinal polarization. The latter will allow to address

the study of the helicity structure of the couplings and, possibly, to get hints for physics beyond the Standard Model, finally.

Acknowledgements

I like to acknowledge support from the two collaborations, H1 and ZEUS, whose people were of great help in preparing the presentation to ICHEP98 and the present paper.

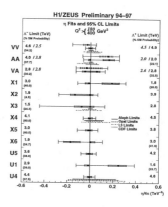

Figure 5: A summary of the contact interaction fit results. Each scenario of CI is presented in one row (slanted: H1). The dots (upper: H1) denote the most likely values of $\pm 1/\Lambda^2$ and have 1σ errors represented by the double error bars (dashed: H1). The thin error bars represent the 95% C.L. limits. The CI limits reported by other experiments are indicated as single lines where available.

Table 2: Results from this paper compared to those from other experiments, where available. All limits corresponds to 95% C.L.

Class	a	Fit ZEUS Λ_0 (TeV)	Limit ZEUS Λ (TeV)	Limit H1 Λ (TeV)	Limit CDF[11] Λ (TeV)	Limit OPAL[12] Λ (TeV)	Limit ALEPH[13] Λ (TeV)	Limit L3[14] Λ (TeV)
VV	$+1$	∞	4.9	4.5	3.5	4.1	6.7	3.2
VV	-1	>10	4.6	2.5	5.2	5.7	7.4	3.9
AA	$+1$	2.5	2.0	2.0	3.8	6.3	7.4	4.3
AA	-1	>10	4.0	3.8	4.8	3.8	8.2	2.9
VA	$+1$	4.2	2.8	2.6	–	–	–	–
VA	-1	4.6	2.8	2.8	–	–	–	–
$X1$	$+1$	2.3	1.8	–	–	–	–	–
$X1$	-1	8.8	3.0	–	–	–	–	–
$X2$	$+1$	∞	3.9	–	–	–	–	–
$X2$	-1	2.7	1.9	–	–	–	–	–
$X3$	$+1$	∞	2.8	–	–	4.4	6.9	3.2
$X3$	-1	3.7	1.5	–	–	3.8	7.7	2.8
$X4$	$+1$	∞	4.5	–	–	3.1	2.9	2.4
$X4$	-1	>10	4.1	–	–	5.5	4.5	3.7
$X5$	$+1$	∞	3.8	–	–	–	–	–
$X5$	-1	8.2	3.0	–	–	–	–	–
$X6$	$+1$	>10	3.0	–	–	–	–	–
$X6$	-1	2.4	1.9	–	–	–	–	–
$U5$	$+1$	∞	4.2	–	–	–	–	–
$U5$	-1	8.7	3.6	–	–	–	–	–
$U1$	$+1$	2.0	1.6	–	–	–	–	–
$U1$	-1	8.2	2.9	–	–	–	–	–
$U4$	$+1$	∞	4.6	–	–	2.0	2.1	1.8
$U4$	-1	>10	4.4	–	–	2.3	2.6	2.2

Class	a	Limit H1 Λ (TeV)	Limit OPAL[12] Λ (TeV)	Limit ALEPH[13] Λ (TeV)	Limit L3[14] Λ (TeV)	Limit A.P.V.[8] up Λ (TeV)	Limit A.P.V.[8] down Λ (TeV)
LL	$+1$	2.4	4.4	5.6	3.0	7.4	7.9
LL	-1	1.2	2.8	6.4	2.1	11.7	12.3
LR	$+1$	3.0	3.3	3.0	2.4	7.4	7.9
LR	-1	1.5	3.6	3.2	2.6	11.7	12.3
RL	$+1$	3.0	2.5	2.3	2.0	11.7	12.3
RL	-1	1.6	4.9	4.0	2.3	7.4	7.9
RR	$+1$	2.4	3.0	4.1	2.3	11.7	12.3
RR	-1	1.2	3.9	4.5	2.7	7.4	7.9

References

1. H1 Coll., S.Aid et al., *Phys. Lett.* B353 (1995) 578.
2. H1 Coll., *Measurement of Inclusive Cross Sections for Neutral and Charged Interactions at High Q^2*, paper #533, submitted to this conference.
3. ZEUS Coll., *Measurement of High Q^2 Neutral–Current DIS Cross Section at HERA*, paper #752, submitted to this conference.
4. H1 Coll., *Search for Contact Interactions in Neutral Current Scattering at HERA*, paper #584, submitted to this conference.
   ZEUS Coll., *Investigation of eeqq Contact Interactions in Deep–Inelastic e^+p Scattering at HERA*, paper #753, submitted to this conference.
5. P.Haberl et al., Proc. Workshop 'Physics at HERA', eds. W.Buchmüller and G.Ingelman, DESY–Hamburg (1991), vol. 2, pag. 1133.
6. R.Rückl, *Phys. Lett.* B129 (1983) 363.
7. E.Eichten et al., *Phys. Rev. Lett.* 50 (1983) 811.
8. A.Deandrea, *Phys. Lett.* B409 (1997) 277.
   L.Giusti, A.Strumia, *Phys. Lett.* B410 (1997) 229.
   C.S.Wood et al., *Phys. Lett.* B409 (1997) 1759
9. H.L.Lai et al., *Phys. Rev.* D55(1997) 1280.
10. PDFLIB, CERN program library W5051.
    A.D.Martin, R.G.Roberts, W.J.Stirling and R.S.Thorne, hep–ph/9803445.
11. CDF Coll., *Phys. Rev. Lett.* 79 (1987) 2198.
12. OPAL Coll., CERN–EP/98–108, subm. to EPJ, ICHEP98 # 264 (1998).
13. ALEPH Coll., ALEPH98–060, ICHEP98 # 906 (1998).
14. L3 Coll., CERN–EP/98–31 (1998).

LEPTOQUARK SEARCHES

E. GALLO

INFN Firenze, L.go E. Fermi 2, 50125 Firenze, Italy
E-mail: gallo@desy.de

This paper reports on preliminary results on direct searches of leptoquarks at LEP2 and at HERA. No clear signal is observed over the Standard Model expectation. For a Yukawa coupling of the order of the electromagnetic strength, limits are set on various types of first generation leptoquarks up to masses of $\simeq 280$ GeV.

1 Introduction

Various theories beyond the Standard Model predict the existence of leptoquarks (LQ), particles which couple to both leptons and quarks. Leptoquarks are coloured, they have both baryon (B) and lepton (L) numbers, fractional charge, and can have spin 0 or spin 1. This talk reports on direct searches for first generation LQs from LEP2 and HERA[1]. Assuming flavour and chirality conservation, there are 14 types of first generation LQs; the branching ratios to $eq, \nu_e q$ are given in the model of Buchmüller et al.[2]. In the analyses here described, the leptoquarks are produced from the fusion of the electron [a] with a quark ($F = 3B + L = 2$) or an antiquark ($F = 0$). The production cross section is proportional to the Yukawa coupling to the lepton-quark pair, λ^2, and to the quark momentum density in the initial beam. The decay width, in leading order, is also proportional to λ^2 and is generally very small (< 1 MeV).

2 Single Leptoquark Production at LEP2

Leptoquarks can be produced singly at LEP2, allowing searches up to the kinematical limit. The largest contribution to the cross section comes from the process in which one of the electrons/positrons from the beams emits a quasi-real photon, and a quark from the resolved component of the photon fuses with the other beam to form a leptoquark. The LQ decays then to an electron and a jet or to a neutrino and a jet, giving a very distinct signature, with either a lepton+jet or a monojet. The initial electron which emits the photon is assumed to escape detection down the beam pipe.

DELPHI has looked for such signatures in the data collected at $\sqrt{s} = 183$ GeV, corresponding to an integrated luminosity of 47.7 pb^{-1}. They find one candidate event in each channel $e + jet$ and $\nu + jet$, compatible with the expected background (coming mainly from $q\bar{q}$ and WW). They present then upper limits on the mass of the various types of leptoquarks M_{LQ}, as a function

of the Yukawa coupling λ. For $\lambda = \sqrt{4\pi\alpha_{em}} \simeq 0.3$, that is of the same order of the electromagnetic strength, DELPHI excludes, at 95% CL, scalar LQs in the range $134-161$ GeV and vector LQs in the range $149-171$ GeV.

Similar searches at lower $\sqrt{s}$ are presented by OPAL[1].

3 Leptoquark Searches at HERA

HERA is the ideal place to look for leptoquarks. They are produced in the s-channel from the fusion of the electron/positron beam with a valence or sea quark from the proton. Their distinct signature is a peak in the $x_{Bjorken}$ distribution at a value $x \simeq M_{LQ}^2/s$, where s is the ep center of mass energy squared ($\sqrt{s} = 300$ GeV in 94-97). The events are in principle indistinguishable from normal neutral current (NC) or charged current DIS events, in which there is the t-channel exchange of a γ, Z or W. However they have a different inelasticity y distribution ($d\sigma/dy \simeq$ flat for scalar LQs, $d\sigma/dy \simeq 1/y^2$ at fixed x for DIS) or, alternatively, they have a different angular distribution. Denoting with θ^* the positron decay angle in the LQ rest frame relative to the proton beam direction, where $\cos\theta^* = 2y - 1$, the prediction for scalar leptoquarks is a flat distribution in $\cos\theta^*$ and for vector LQs a distribution $(1 - \cos\theta^*)^2$, whereas DIS events peak at low values of $\cos\theta^*$.

In the 94-96 NC e^+p data, both HERA experiments observed an excess over the Standard Model prediction at high Q^2 and high x[3]. Here an analysis is presented based on the 94-97 e^+p NC data, based on approximately twice the integrated luminosity (37 pb^{-1} for H1, 47 pb^{-1} for ZEUS).

3.1 Results from H1

H1 looks for LQ resonances decaying in $e^+ + jet$, selecting events with a scattered positron in the final state and with high momentum transfer between the initial and final state positrons ($Q^2 > 2500$ GeV^2). They obtain 1297 events, in good agreement with the Standard Model expectation (based on a DIS neutral current MC, with

---

[a]In the case of positron, the $F = 0$ leptoquarks are formed from the quark, the $F = 2$ from the antiquark.

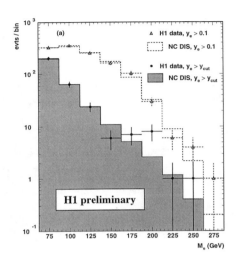

Figure 1: The H1 mass distribution reconstructed with the electron method.

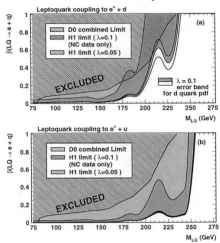

Figure 2: The H1 LQ limits in the plane mass-branching ratio in $e + q$.

the MRSH structure function parametrization), where 1276 ± 98 events are expected. In order to reconstruct the mass, they use the electron method, where the kinematics is reconstructed from the energy and angle of the scattered positron. They performed a new energy calibration of their calorimeter which gives a better control on the systematic errors. In the regions close to the calorimeter cracks they use the double angle method, where the angle of the scattered positron and the angle of the hadronic system coming from the struck quark are used to measure the mass.

Figure 1 shows the reconstructed mass M_e for all selected events with $y_e > 0.1$ (triangles), compared to the DIS MC expectation (dashed line), showing a very good agreement between data and MC. In order to look for LQ resonances and optimize the signal significance with respect to the DIS NC background, they apply a $y_e > y_{cut}$ selection, where y_{cut} varies with M_e and decreases with increasing mass. The mass distribution after this cut is shown in fig. 1, where the data (black dots) are compared to the DIS MC (shaded area). An excess is visible in the mass range $200\pm25/2$ GeV, where, for $y_e > 0.4$, 8 events are observed in the data and 3.0 ± 0.5 are expected; 5 of these events come from the 94-96 sample, 3 come from the 97 sample, which corresponds to more than half of the total integrated luminosity. The 97 data alone, therefore, do not give a significant excess in this particular mass interval. H1 then assumes a statistical

fluctuation and puts upper limits on LQ production. The limits are presented, at 95% CL, in the plane $\lambda - M_{LQ}$ assuming the branching ratios in e^+q from ref.[2], and are most sensitive for $F = 0$ leptoquarks, as the positron couples to valence quarks which have the higher parton density at high x. For $\lambda \simeq 0.3$, scalar LQs with $F = 2$ are excluded up to masses of $200 - 250$ GeV, depending on the LQ type, and scalar LQs with $F = 0$ are excluded up to masses in the range $260 - 275$ GeV.

Alternatively, they fix the Yukawa coupling λ and present limits on the mass versus the branching ratio $\beta(LQ \to eq)$, as shown in fig. 2. For small β, i.e. $\beta(LQ \to e^+u) \simeq 0.1$, and small $\lambda \simeq 0.1$, LQ production is excluded at 95% CL up to masses around 210 GeV. The limit is very competitive with respect to the published Tevatron results[4], which give a limit only up to masses around 110 GeV for these small values of the coupling and branching ratio. On the other hand the plot shows that, for small λ and β, there is still a margin for discovery at HERA: it is therefore important to look at other decay channels, such as those involving for instance lepton flavour violation, as discussed also in the H1 papers[1].

3.2 Results from ZEUS

ZEUS searches for a LQ signal in the neutral current sample at very high transverse energy ($E_T > 60$ GeV).

ZEUS 1994-97 Preliminary

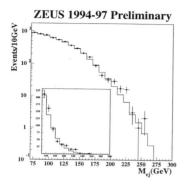

ZEUS 1994-97 Preliminary

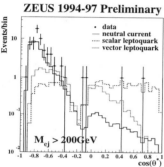

Figure 3: The ZEUS mass and $\cos(\theta^*)$ distributions. The inset in the upper plot shows the high mass part in linear scale. The dip is the $\cos\theta^*$ distribution corresponds to a fiducial cut between calorimeter boundaries.

In total 7255 events are selected, in good agreement with the Standard Model expectation of 7121 events (a DIS neutral current MC with the CTEQ4 parametrization is used here). The invariant mass M_{ej} of the scattered positron and the highest E_T jet in the event is calculated: in this way the sensitivity to initial state QED and QCD radiation is removed. The M_{ej} distribution is shown for the selected events in the upper plot in fig. 3, where the data (dots) are compared to the DIS MC expectation (solid histogram). A modest excess is visible at high mass, where, for $M_{ej} > 200\ GeV$, 68 events are observed, while 43^{+14}_{-12} are expected.

In order to investigate the nature of this excess, the $\cos\theta^*$ distribution is plotted for the events which have a mass $> 200\ GeV$, and compared to the expectations of a DIS MC and of a LQ MC. The distribution is shown in the lower plot in fig. 3: the events peak at small values of $\cos\theta^*$, as expected from normal NC events, while a scalar or a vector LQ signals have a flatter distribution.

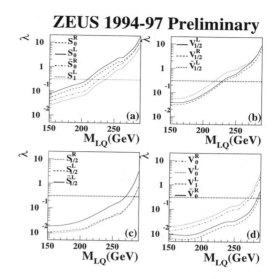

Figure 4: The ZEUS LQ limits in the plane $\lambda - M_{LQ}$.

For $\cos\theta^* > -0.5$, 7 events are observed while 4.3 are expected; 5 events are from the 94-96 sample, and only 2 come from the 97 sample, which corresponds to about half of the integrated luminosity. ZEUS then assumes no significant evidence and puts upper limits on the LQ cross section. In order to optimize the limits, a varying $\cos\theta^*$ cut was applied for the different masses.

Assuming the branching ratios $\beta(LQ \to eq)$ as given by ref.[2], the limits are presented for the 14 types of lepto-quarks (scalar and vector) in the plane $\lambda - M_{LQ}$ (fig. 4). The region above the lines is excluded at 95% CL. For a Yukawa coupling of the order of the electromagnetic coupling, $F = 2$ leptoquarks are excluded up to masses in the range $200 - 240$ GeV (upper two plots) and $F = 0$ leptoquarks up to masses of $255 - 280$ GeV (lower two plots).

4 Summary

We have reported on preliminary results on direct searches for first generation leptoquarks at LEP2 and HERA. No significant signal is found above the Standard Model predictions. Limits are set by HERA up to masses of 280 GeV (250 GeV) for $F = 0$ ($F = 2$) lepto-quarks, for a Yukawa coupling $\lambda \simeq \sqrt{4\pi\alpha_{em}}$. For $F = 0$ the limits are three times better than the ones previously published[5]. For small branching ratios to e^+q and for high masses, HERA is very competitive compared to Tevatron. The 1998-1999 HERA running with electron beam will allow to investigate, with increased proton energy, leptoquarks with $F = 2$, where the electron couples to valence quarks in the proton.

Acknowledgements

I would like to thank Robert McPherson (OPAL), Clara Matteuzzi and Antonio Onofre (DELPHI), and my H1 and ZEUS colleagues, in particular Wuyi Liu and Yves Sirois, for the help in preparing this talk and these proceedings.

References

1. DELPHI Collaboration, Paper #211 submitted to this conference; OPAL Collaboration, Paper #342 submitted to this conference; H1 Collaboration, Papers #579 and #580 submitted to this conference; ZEUS Collaboration, Paper #754 submitted to this conference.
2. W. Buchmüller, R. Rückl and D. Wyler, *Phys. Lett.* B **191**, 442 (1987).
3. H1 Collaboration, C. Adloff et al., *Z. Phys.* C **74**, 191 (1997); ZEUS Collaboration, J. Breitweg et al., *Z. Phys.* C **74**, 207 (1997).
4. D0 Collaboration, B. Abbott et al., *Phys. Lett.* B **79**, 1997 (4321); D0 Collaboration, B. Abbott et al., *Phys. Lett.* B **80**, 1998 (2051).
5. H1 Collaboration, T. Ahmed et al., *Phys. Lett.* B **369**, 173 (1996).

Searches for Exotic Particles at the Tevatron[1]

Carla Grosso-Pilcher

For the CDF and DØ Collaborations

The Enrico Fermi Institute, The University of Chicago, 5640 S. Ellis Avenue, Chicago, IL 60637, USA
E-mail: carla@hep.uchicago.edu

This paper reports on recent searches for exotic particles by the CDF and DØ Collaborations. The results are derived from the data collected during the 1992-95 Run I at the Fermilab Tevatron. Limits are presented on new heavy gauge bosons, leptoquarks, monopoles and technicolor particles.

1 Introduction

The standard model has proved remarkably successful at describing the experimental observations at presently available energies. However, unsatisfactory explanations of some basic questions, notably the particle masses, have led to the development of extensions of the standard model. Many of these models predict the existence of new particles, some of which could be observed at the Tevatron. Both the CDF and DØ collaborations have searched for a number of these particles in the data collected during the 1992-95 Run I at a center of mass energy of 1.8 TeV, corresponding to an integrated luminosity of the order of 100 pb^{-1}. No evidence has been observed, therefore limits have been set on the production cross section times the branching ratio to a given channel as a function of the candidate particle mass.

New limits in a number of channels have been presented at this conference. This paper summarizes the present mass limits for new gauge bosons, leptoquarks, technicolor states, and monopoles, as well as for compositeness scales.

2 Heavy Gauge Boson Searches

Heavy gauge bosons, in addition to the standard model W and Z, are predicted by many extensions of the standard model and Grand Unified Theories[2]. These particles may have decay modes analogous to the ones of the standard model of W and Z. The models specify the coupling strengths, but make no predictions on the masses of the new bosons. In the absence of an observed signal, experiments set limits on the production cross section times branching ratio as a function of the new boson mass, from which mass limits are obtained in the context of a given model. Present searches via leptonic or quark decays have produced no observation of such particles. A recent search for $W' \to \mu\nu$ by the CDF collaboration that set a mass limit of 663 GeV/c^2 based on the Run I data set has been presented at this conference. The $\mu\nu$ transverse mass spectrum shown in Fig. 1 is consistent

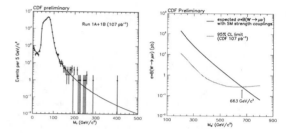

Figure 1: Left: Transverse mass distribution of the $\mu\nu$ system (dots) and the standard model expectation (line). Right: CDF upper limit on $\sigma \cdot B(W' \to \mu\nu)$ and the theoretical expectation for SM couplings.

with the expectation from W decay plus background, indicated by the line. The experimental limit on the cross section, with the theoretical expectation assuming standard model coupling strengths, is also shown in Fig. 1.

Table 1: New Gauge Bosons Mass Limits

	95% C.L. Limit (GeV/c^2)	$\int \mathcal{L}dt$ (pb^{-1})
$Z' \to e^+e^-$	> 655	CDF (110)
$Z' \to \mu^+\mu^-$	> 590	CDF (107)
$Z' \to l^+l^- (e^+e^-, \mu^+\mu^-)$	> 690	CDF (107)
$Z_\phi, Z_\eta, Z_\chi, Z_I \to l^+l^-$	>580,610,585,555	CDF (110)
$Z_{LR}, Z_{ALRM} \to l^+l^-$	> 620,590	CDF (110)
$W' \to e\nu$	>720	DØ (1a,1b)
$W'_R \to eN_R$ (heavy N_R)	> 650	DØ (1a,1b)
$W'_R \to eN_R$ (light N_R)	> 549	DØ (1a,1b)
$W' \to \mu\nu$	> 663	CDF (107)
$W' \to WZ$	> 560	CDF (90)
$W' \to jj$	300<$M_{W'}$< 420	CDF (90)

Current mass limits for $Z' \to ee/\mu\mu$ in the contest of specific models[3] and for $W' \to (e/\mu)\nu/WZ/jj$[4] are

Table 2: Compositness Scale Limits (DØ)

Model	Λ^+ (GeV)	Λ^- (GeV)
LL	3300	4200
RR	3300	4000
LR	3400	3600
RL	3300	3700
VV	4900	6100
AA	4700	5500

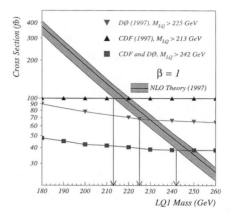

Figure 2: 95% C.L. upper cross section limits from CDF (triangles), DØ (inverted triangles), and combined (squares) leptoquark analyses. The band shows the NLO theoretical cross section; arrows correspond to the respective 95% C.L. lower mass limits.

summarized in Table 1.

3 Quark-Lepton Compositness

If quarks and leptons have a substructure, the constituent's contact interaction can contribute an additional amplitude to the standard cross section for the Drell-Yan process $q\bar{q} \rightarrow l^+ l^-$. A deviation of the dilepton invariant mass distribution, M_{ll}, from the one predicted by Drell-Yan could indicated a compositness of the scattering partons. Previous searches performed with ee and $\mu\mu$ final states have set limits on the value of the compositness scale Λ in the range of 2.5 to 4.2 TeV[5], depending on the model. DØ has presented a new analysis of the dielectrons mass spectrum collected during Run I (120 pb^{-1}) to obtain the limits for the $e - q$ contact interaction scale shown in Table 2, where the +/- signs correspond to constructive or destructive interference, respectively.

4 Leptoquarks Searches

The apparent symmetry between leptons and quarks has suggested, in many extensions of the standard model, the existence of particles that carry both lepton and quark quantum numbers and couple directly to both leptons and quarks[6]. Theoretical models give no predictions on the values of the masses, but limits on these leptoquark states can be obtained from direct searches at the colliders, or indirectly from four-fermion interactions that could be induced by these particles. The absence of flavor-changing neutral currents and lepton family-violations, in particular, suggest leptoquarks that couple only within a single generation[7]. Leptoquarks may be directly pair produced at the Tevatron via the strong interactions, by $q\bar{q}$ annihilation or gluon fusion. The production cross section is therefore independent of the unknown $LQ - lq$ coupling.

Searches for all three generations of leptoquarks have been performed by the CDF and DØ collaborations. They have now combined their results in the search for a first generation scalar leptoquark[8,9], with the resulting mass limit of 242 GeV/c^2 for the branching ratio

to the ej channel $\beta = 1$. The combined cross section limit, with the theoretical calculation[10], is shown in Fig. 2. The DØ Collaboration has also extended the search for first generation leptoquarks to the vector case, using all final states corresponding to combinations of the decay mode into ej or νj. In this case the LQ cross section depends on the model and the mass limit is presented for typical choices of the couplings. The limits in M_{LQ} vs β space for different couplings are shown in Fig. 3. A summary of leptoquark mass limits, as a function of the unknown branching ratio β to the $(e/\mu/\tau)jet$ channel is shown in Table 3[11]. Although generational crossing is disfavoured by experimental constraints, it is not theoretically excluded. Bound states of leptons and quarks of different generations are allowed in some models, like the Pati-Salam model[12]. If such states exist, they could mediate decays of the type $B^o \rightarrow e\mu$, which are strictly forbidden in the standard model. This decay mode has been used by CDF to search for these leptoquarks in mass ranges well above those allowed by direct production at the Tevatron energies. The invariant mass distribution for oppositely-charged $e\mu$ pairs, originating from a vertex displaced from the primary interaction, has been searched for events consistent with the B_s (0 events observed) and B_d (1 event observed) masses. From the 95% limit on the branching ratios in 102 pb^{-1}, mass limits on Pati-Salam leptoquarks of 19.3 TeV/c^2 and 20.4 TeV/c^2 have been obtained for the B_s and B_d cases, respectively[13].

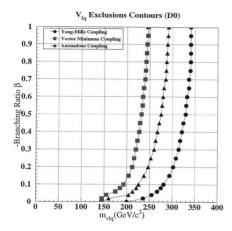

Figure 3: First generation vector LQ mass limits as function of the branching ratio β into the ej channel for Yang-Mills, anomalous, and minimal vector coupling.

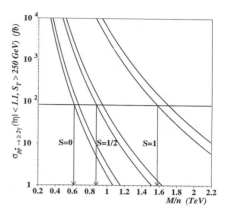

Figure 4: Theoretical cross section times acceptance for production of monopoles of spin 0, 1/2 and 1 as a function of M/n, where M is the monopole mass and n is an integer. The bands represent the uncertainty on the cross section. The horizontal line represents the DØ experimental cross section limit.

Table 3: Leptoquarks Mass Limits

channel	β	$M_{LQ}(GeV/c^2)$	$(\int \mathcal{L}dt(pb^{-1}))$
first generation scalar			
$eejj$	1	213	CDF (110)
$eejj$	1	225	DØ (123)
$e(e/\nu)jj$	0.5	204	DØ (115)
$\nu\nu jj$	1	79	DØ (7.4)
$eejj$	1	242	CDF/DØ
first generation vector (Yang-Mills couplings)			
$eejj$	1	340	DØ (123)
$e(e/\nu)jj$	0.5	329	DØ (115)
$\nu\nu jj$	0	200	DØ (7.4)
second generation scalar			
$\mu\mu jj$	1	202	CDF (110)
$\mu\mu jj$	0.5	160	CDF (110)
$\mu\mu jj$	1	185	DØ
$\mu\mu jj$	0.5	140	DØ
third generation scalar			
$\tau\tau jj$	1	99	CDF (110)
$\nu\nu b\bar{b}$	0	94	DØ (1a,1b)
third generation vector (Yang-Mills couplings)			
$\tau\tau jj$	1	225	CDF (110)
$\nu\nu b\bar{b}$	0	216	DØ (1a,11b)

5 Monopole Search

DØ has searched for evidence of magnetic monopoles[14] in 70 pb^{-1} of diphoton data. Magnetic monopoles would allow rescattering of two photons via a box diagram, producing a distinct signature of two central, high energy photons. No events are observed with $S_T = \sum_i E_T^{\gamma i} > 250$ GeV, from which a 95% limit on the cross section, $\sigma(p\bar{p} \rightarrow \geq 2\gamma)|_{S_T>250GeV,|\eta^\gamma|<1.1} < 83fb$, is derived. Figure 4 shows the theoretical cross section for three values of the monopole spin times the experimental acceptances, compared to the observed $\sigma_{lim}^{95\%}$. The 95% C.L. mass limits are 610 GeV/c^2 (S=0), 870 GeV/c^2 (S=1/2) and 1580 GeV/c^2 (S=1).

6 Technicolor Searches

Technicolor models offer an alternative explanation for the mechanism of electroweak symmetry breaking. While in many extensions of the standard model the symmetry is spontaneously broken by fundamental scalar fields, the Higgs bosons, that give masses to the standard model W's and Z's, in technicolor models the symmetry is dynamically broken[15]. These models predict the existence of a number of new, heavy fermions interacting via the strong technicolor gauge interaction to form new boson bound states. Technicolor models have evolved to extended technicolor and topcolor-assisted technicolor to account for the fermion masses and the large value of the top quark mass. Recent models[16] predict techniparticles light enough to be produced at Tevatron en-

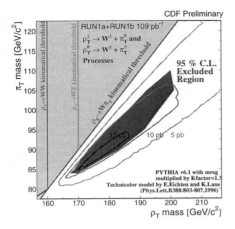

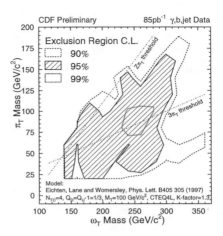

Figure 5: The 95% C.L. excluded region in the M_{ρ_T}, M_{π_T} plane. The lines represent production cross section contours.

Figure 6: The 95% excluded region in the M_{ω_T}, M_{π_T} plane.

ergies. CDF has searched for technirho, ρ_T, and techniomega, ω_T, via their decay to technipions, π_T, and standard model bosons. The decays $\rho_T^\pm \to W^\pm \pi_T^o$ and $\rho_T^o \to W^\pm \pi_T^\mp$ would have the largest cross section for $M_W + M_{\pi_T} \leq M_{\rho_T} \leq 2M_{\pi_T}$. The π_T is expected to mostly decay to $b\bar{b}$ or $b\bar{c}$ states, therefore the final state will be characterized by the presence of two heavy quarks, identified by a vertex displaced from the primary interaction, while the W is identified by its leptonic decay into $(e/\mu)\nu$. The search requires a high transverse momentum, isolated lepton, large missing transverse energy, $\not{E}_T$, and two b-tagged jets. The presence of technicolor would be characterized by peaks in the $b\bar{b}$ ($\to \pi_T$) and the Wjj ($\to \rho_T$) mass distributions. A W mass constraint on the lepton and $\not{E}_T$ has been imposed for the M_{Wjj} calculation. The data are consistent with the expected background. Cross section limits for each M_{ρ_T}, M_{π_T} mass combination are obtained by counting events in a $\pm 3\sigma$ window around the given masses. Figure 5 shows the 95% C.L. excluded region.

Similarly, CDF has searched for ω_T via the $\gamma\pi_T$ decay channel, requiring an isolated photon and two jets, with at least one jet tagged as a b-jet. The distributions of the b-jet invariant mass and of $M_{\gamma,b,jet} - M_{b,jet}$ show no evidence of resonance production. The 95% C.L. excluded region in the M_{π_T}, M_{ω_T} plane is shown in Fig. 6.

7 Search for Resonances in $b\bar{b}$ States

New resonant states with widths narrower than the experimental resolution would manifest themselves in dijets mass spectra as peaks on top of a smooth distribution from the dominant QCD production. Restricting the

search to production of b quark pairs reduces this background while increasing the sensitivity to production of new particles that couple preferentially to the third generation. Such exotic states would include a color-octect technirho ρ_{T8}[17], topgluon, topcolor Z'[18], and new gauge bosons. CDF has presented preliminary results from a study of the $M_{b\bar{b}}$ distribution from 87 pb^{-1} of data[19], shown in Fig. 7. The spectrum has been fitted to a smooth background plus a resonance shape, for a range of values of the resonance mass and width. The 95% C.L. upper limit on the cross section times branching ratio into $b\bar{b}$ pairs is shown in Fig. 8 as a function of the resonance mass for narrow states, and as an excluded region in the mass and width parameters for topgluons. The present data are not sensitive enough to exclude the models considered in this analysis, but the cross section limit can be applied to any state narrower than the experimental resolution. This procedure appears promising for similar searches in the coming run of the Tevatron.

8 Conclusions

Searches for exotic particles with the data available from the Tevatron Run I have not found any signals. New limits on the mass of a number of new particles have been presented at this conference:

- $M_{Z'}, M_{W'} > \approx 700$ GeV/c^2
- Compositeness scale $\Lambda > (3.3 - 6.1)$ TeV
- Leptoquarks:
  - $M_{LQ1}^{scalar} > 242$ GeV/c^2 ($\beta = 1$)
  - $M_{LQ1}^{vector} > 340$ GeV/c^2 (Yang-Mills couplings)
  - Pati-Salam: $M_{LQ} > 20$ TeV/c^2

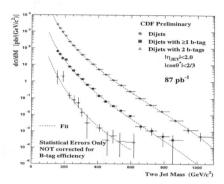

Figure 7: Mass spectra for inclusive dijet (dots), events with one b-jet (squares), and events with 2 b-jets (triangles).

- Dirac Monopole: $M/n > 610$ - 1580 GeV/c^2

- Walking Technicolor: limits on $\rho_T - \pi_T$, $\omega_T - \pi_T$

- Limits on topgluon, ρ_{T8} from $b\bar{b}$ mass spectrum

Run II, with an expected increase of luminosity of a factor of 20 and a center of mass energy of 2 TeV, will allow for either a discovery or significant increase of the present limits.

9   Acknowledgments

I would like to thank members of the CDF and DØ collaborations for providing the material shown in this paper, particularly the authors of the submitted contributions.

10   References

1. This paper includes results from contributions 591, 594, 655, 657, 674 and 706.
2. G. G. Ross, *Grand Unified Theories* (Cambridge U.P., Cambridge,1987) and references therein. R. N. Mohapatra, *Unification and Supersymmetry* (Springer,New York,1992) and references therein.
3. F. Abe *et al.*, *Phys.Rev.Lett.* **79**, 2192 (1997).
4. S. Abachi *et al.*, *Phys.Rev.Lett.* **76**, 3271 (1996); F. Abe *et al.*, *Phys.Rev.Lett.* **74**, 2990 (1995).
5. F. Abe *et al.*, *Phys.Rev.Lett.* **79**, 2198 (1997).
6. J.C. Pati and A. Salam, *Phys.Rev.D* **19**, 275 (1974); H. Georgi and S. Glashow, *Phys.Rev.Lett.* **32**, 438 (1974); L.F. Abott and E. Farhi, *Nucl.Phys.B* **189**, 547 (1981); E. Eichten *et al.*, *Phys.Rev.Lett.* **50**, 811 (1983); E. Witten,

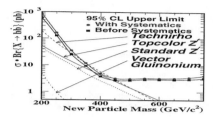

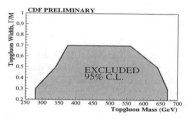

Figure 8: Top: 95% C.L. limit on $\sigma \times BR(X \to b\bar{b})$ as function of M_X, compared to the expected cross sections from typical models. Bottom: Togluon excluded region as a function of mass and width.

Nucl.Phys.B **258**, 75 (1985); M. Dine *et al.*, *Nucl.Phys.B* **259**, 519 (1985); J. Breit *et al.*, *Phys.Lett.B* **158**, 33 (1985); S. Pakvasa, Int.J. Mod. Phys. **A2**, 1317 (1987); J.L. Hewett and S. Pakvasa, *Phys.Rev.D* **37**, 3165 (1988);

7. W. Buchmüller and D. Wyler, *Phys.Lett.B* **177**, 337 (1986).
8. F. Abe *et al.*, *Phys.Rev.Lett.* **79**, 4327 (1997).
9. B. Abbott *et al.*, *Phys.Rev.Lett.* **79**, 4321 (1997).
10. M. Krämer, T. Plehn, M. Spira, and P.M. Zerwas, *Phys.Rev.Lett.* **79**, 341 (1997).
11. F. Abe *et al.*, submitted to *Phys.Rev.Lett.*; F. Abe *et al.*, *Phys.Rev.Lett.* **78**, 2906 (1997); B. Abbott *et al.*, *Phys.Rev.Lett.* **81**, 38 (1998).
12. J. Pati and A. Salam, *Phys.Rev.D* **10**, 275 (1974).
13. F. Abe *et al.*, submitted to *Phys.Rev.Lett.*, FERMILAB-PUB-98/263-E.
14. P.A.M. Dirac, *Proc.R.Soc. London*, **A**, 60(1931).
15. S. Weinberg, *Phys.Rev.D* **13**, 974 (1976); **19**, 1277 (1979); L . Susskind, *Phys.Rev.D* **20**, 2619 (1979).
16. E.Eichten, K.Lane, and J.Womersley, *Phys.Lett.B* **405**, 305 (1997); E. Eichten and K. Lane, *Phys.Lett.B* **388**, 803 (1996).
17. E. Eichten and K. Lane, *Phys.Lett.B* **388**, 129 (1994)
18. C.T. Hill, *Phys.Lett.B* **345**, 438 (1995) and C.T. Hill and S.J. Parke, *Phys.Rev.D* **49**, 4454 (1994)
19. F. Abe *et al.*, submitted to *Phys.Rev.Lett.*, FERMILAB-PUB-98/290-E.

SEARCHES FOR HEAVY AND EXCITED FERMIONS AT $\sqrt{s}$ =183–189 GEV AT LEP2

R. J. TEUSCHER

CERN, EP Division, CH-1211 Geneva 23, Switzerland
E-mail: Richard.Teuscher@cern.ch

The LEP collaborations have searched for signals of new physics in e^+e^- collisions at centre-of-mass energies of 183-189 GeV. Searches for unstable neutral and charged heavy leptons, N and $L^\pm$, and for excited states of neutral and charged leptons, ν^*, e^*, μ^* and τ^*, are reported. First results from 1998, using the highest energy data ever taken in e^+e^- collisions, are given.

1 Beyond the Standard Model

Physics beyond the Standard Model (SM) is searched for along two broad paths, via (i) unification or (ii) compositeness. In the first approach, which includes SUSY and Grand Unified models such as E_6 and SO(10), the particles of the SM are taken to be fundamental and pointlike. Experimental consequence of unification may include the existence of new fermions, such as heavy charged leptons $L^\pm$, or heavy neutral leptons N. In contrast to unification, the second path to new physics suggests that the particles of the SM are composite at some scale $\Lambda \geq 1$ TeV. Experimental consequences include the existence of excited fermions, which couple to the known fermions via vector bosons. This report includes searches for new physics following both paths described above.

2 Previous Searches

Lower limits on the masses of heavy leptons have been obtained in e^+e^- collisions at centre-of-mass energies of $\sqrt{s} \sim M_Z$ [2,5] and recent searches at $\sqrt{s} = 172$ GeV [6,7] and $\sqrt{s} = 130\text{-}183$ GeV [8] have improved these limits. Excited leptons have been sought at $\sqrt{s} \sim M_Z$ [9], $\sqrt{s} = 172$ GeV [7], and at the HERA ep collider [12]. This report concentrates on the newest data from the LEP collaborations, recorded in 1997 and 1998, at $\sqrt{s} = 183\text{-}189$ GeV. All results are preliminary and at the 95% confidence level.

3 Searches for Heavy Leptons

Heavy neutral leptons are particularly interesting in light of recent evidence for SM neutrinos having mass [3,4]. In this case, the see-saw mechanism predicts that the mass of heavy neutral leptons are given by $m_{N_\ell} = m_\ell^2/m_{\nu_\ell}$, where $\ell =$ e, μ, or τ. For example, if $m_{\nu_e} \sim 3$ eV, then the above relation gives $m_{N_e} \approx 90$ GeV, which is within the reach of LEP 2. Such heavy neutral leptons, which can be either Dirac or Majorana particles, can be pair-produced at LEP, via the process $e^+e^- \to N\bar{N}$, followed by a flavour-mixing decay into a light lepton, $N \to eW$, $N \to \mu W$ or $N \to \tau W$. The mixing angle-squared, ζ^2, of

Channel		Mass Limit (GeV)			
		O	O	L	D
		183+189	183	183	183
$N \to eW$	Dirac	89.5	87.3	86.3	76.5
	Major.	77.7	75.7	72.5	-
$N \to \mu W$	Dirac	89.9	86.4	87.7	79.5
	Major.	78.3	74.0	74.9	-
$N \to \tau W$	Dirac	80.1	77.0	75.4	60.5
	Major.	62.6	58.1	63.6	-
$L^\pm \to \nu_\ell W^\pm$		91.3	84.0	85.0	78.3
$L^\pm$ long-lived		92.6	89.5	93.5*	-

Table 1: Summary of mass limits for heavy leptons from DELPHI (D), L3 (L), and OPAL (O). The number 183 indicates that the data are taken at $\sqrt{s} = 183$ GeV, while the numbers 183+183 indicate that the data taken at $\sqrt{s} = 183\text{-}189$ GeV are combined. *The L3 limit for $L^\pm$ includes data taken at $\sqrt{s} = 189$ GeV.

the heavy lepton with the standard leptons, is limited to the range $\mathcal{O}(10^{-12})< \zeta^2 < 0.005$, where the lower bound is given by the decay radius $< \mathcal{O}(1$ cm) to which the searches are sensitive, and the upper bound is constrained by precision electroweak data. [13]

The typical event topology resulting from the decay of heavy neutral leptons, assuming 100% CC-decays, is two isolated leptons and jets. In the case of Majorana heavy neutral leptons, the final-state leptons can be of equal charge. From a search using 57 pb^{-1} of data at $\sqrt{s} = 183$ GeV and 37 pb^{-1} of data at $\sqrt{s} = 189$ GeV, the OPAL collaboration[1] observes 8, 6, and 66 candidates in the $\ell =$ e, μ, or τ channels respectively. This is in agreement with the expected SM background of 8.9, 5.8, and 80.4 events, which arise mainly from $\ell\ell qq$ and $qqqq$ production. As shown in Table 1, the resulting 95% C.L. mass limits for heavy neutral leptons range from 60 to 89 GeV. These limits already represent an improvement of about 10 GeV over the 1998 PDG limits[2], and an increase of about 40 GeV over the searches from LEP 1.

If charged heavy leptons exist, they can be pair-produced at LEP via the process $e^+e^- \to L^+L^-$, with charged-current decays of the $L^\pm$. Typical final-state topologies depend on the decay of the W-boson, and include four jets and missing transverse momentum (p_T^{miss}), or two jets, one lepton, and p_T^{miss}. The SM backgrounds are WW pair production, with WW → qqqq and WW → ℓνqq production. No excess is observed by the LEP experiments in the data from $\sqrt{s}$ = 183-189 GeV, and the resulting mass limits, ranging from 80-90 GeV, are shown in Table 1.

Searches for stable or long-lived charged heavy leptons, L^+L^-, involve topologies with two back-to-back charged tracks. Since such particles will be massive and not strongly-interacting, no electromagnetic or hadronic showers should be associated to the tracks. In addition, the measured energy-loss, dE/dx, of the tracks should lie outside the range of that expected from SM particles. No candidates are observed in the LEP data from $\sqrt{s}$ = 183-189 GeV, and the resulting limits of $O(90)$ GeV are given in Table 1.

The last topology for heavy leptons is a search for stable or long-lived neutral leptons, $e^+e^- \to L^+L^-$ with $L^- \to L^0W^-$ (where L^0 is a stable or long-lived neutral heavy lepton which decays outside the detector). This production is possible if L^0 is the lighter member of an SU(2) doublet which does not mix appreciably with the three known lepton generations. The typical signature is a pair of acoplanar particles and missing transverse momentum. The signal leads to a very low visible energy if the mass difference, $\Delta M \equiv M_{L^\pm} - M_{L^0}$, is small. From the data at $\sqrt{s}$ = 183 GeV, the L3 and OPAL collaborations exclude long-lived neutral heavy leptons for a range of masses close to the kinematic limit, and down to $\Delta M = 3 - 5$ GeV. The exclusion region in the $(M_{L^\pm}, M_{L^0})$ plane is shown in Figure 1, obtained by the L3 Collaboration[1].

4 Searches for Excited Fermions

Excited fermions F^* are new massive states of fermions expected in compositeness models. They are assumed to have the same electroweak SU(2) and U(1) gauge couplings to the vector bosons, g and g', as the SM fermions, but are expected to be grouped into both left- and right-handed weak isodoublets with vector couplings. The existence of the right-handed doublets is required to protect the ordinary light leptons from radiatively acquiring a large anomalous magnetic moment via the F^*FV interaction[14], where V is a γ, Z, or $W^\pm$ vector boson.

We use the effective Lagrangian[14]

$$\mathcal{L} = \frac{1}{2\Lambda}\overline{F^*}\sigma^{\mu\nu}\left[gf\frac{\tau}{2}\mathbf{W}_{\mu\nu} + g'f'\frac{Y}{2}B_{\mu\nu} + g_sf_s\frac{\lambda}{2}\mathbf{G}_{\mu\nu}\right]F_L$$

which describes the generalized de-excitation of the F^* states. The matrix $\sigma^{\mu\nu}$ is the covariant bilinear tensor, τ are the Pauli matrices, λ are the Gell-Mann matrices,

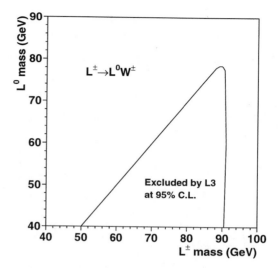

Figure 1: Excluded region for long-lived heavy neutral leptons in the $(M_{L^\pm}, M_{L^0})$ plane.

g_s is the strong coupling constant, $\mathbf{W}_{\mu\nu}$, $B_{\mu\nu}$, and $\mathbf{G}_{\mu\nu}$ represent the fully gauge-invariant field tensors, and Y is the weak hypercharge. The parameter Λ has units of energy and can be regarded as the "compositeness scale," while f, f', and f_s are the coupling constants associated with the different gauge groups. For excited leptons, the final-state topologies, branching ratios, and efficiencies depend on the relative values of f and f'. For simplicity, we will interpret our results using two example complementary coupling assignments, $f = f'$ and $f = -f'$. In the first case, for $f = f'$, photonic decays are expected to dominate for charged excited leptons, while for excited neutrinos, charged-current decays are expected to dominate. The reverse is true for the second case, $f = -f'$, in which photonic decays of excited neutrinos are possible. A third case, $f = 0$, is also studied, in this scenario there is no $t-$channel production of excited neutrinos.

Excited leptons could be produced in pairs in e^+e^- collisions via the process $e^+e^- \to F^*\bar{F}^*$. In the searches for pair-production of excited leptons with charged-current decays, the topologies are similar to those for pair-production of heavy leptons, and the same analyses are used. In the searches for the pair-production of charged excited leptons with photonic decays, the final states involve two leptons and two photons, while for excited neutrinos the final states include two photons and missing transverse momentum. In all of the searches for the pair-production of excited leptons, no excesses are observed by the LEP collaborations in the data from

Type	Coupling	V	Mass Limit (GeV)			
			O 183+189	O 183	L 189	D 183
e^*	$f = f'$	γ	94.1	91.4	93.7	90.7
μ^*	$f = f'$	γ	94.1	91.4	93.7	90.7
τ^*	$f = f'$	γ	94.0	91.3	93.6	89.7
e^*	$f = -f'$	$W^\pm$	91.3	86.0	-	81.3
μ^*	$f = -f'$	$W^\pm$	91.3	86.0	-	81.3
τ^*	$f = -f'$	$W^\pm$	91.3	86.0	-	81.3
ν_e^*	$f = f'$	$W^\pm$	92.8	91.1	-	87.3
ν_μ^*	$f = f'$	$W^\pm$	93.6	90.9	-	88.0
ν_τ^*	$f = f'$	$W^\pm$	90.5	87.0	-	81.0
ν_e^*	$f = -f'$	γ	93.2	91.0	85.3*	90.0
ν_μ^*	$f = -f'$	γ	93.2	91.0	85.3*	90.0
ν_τ^*	$f = -f'$	γ	93.2	91.0	85.3*	90.0

Table 2: Summary of mass limits for excited leptons from DELPHI (D), L3 (L), and OPAL (O). The number 183 indicates that the data are taken at $\sqrt{s} = 183$ GeV, while the numbers 183+183 indicate that the data taken at $\sqrt{s} = 183$-189 GeV are combined. The column "V" indicated whether the dominant decay channel is via a charged-current ($W^\pm$) or photonic (γ) decay. *The L3 limit for ν^* at $\sqrt{s} = 189$ GeV is lower due to a candidate event.

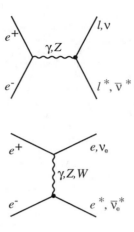

Figure 2: Feynman diagrams for the single-production of excited leptons via the s-channel and t-channel.

$\sqrt{s} = 183$-189 GeV. The resulting lower limits on the masses for excited leptons are summarized in Table 2. From this table, one concludes that charged and neutral excited leptons with charged-current or photonic decays to ordinary leptons are excluded by LEP at the 95% CL below approximately the mass of the Z^0.

Excited leptons could be produced singly, via the coupling to ordinary fermions, as shown in Figure 2. In this case, the discovery reach in e^+e^- collisions extends up to the full centre-of-mass energy. The possible topologies included charged-current decays ($\ell^*\ell \to \nu\ell W$ and $\nu^*\nu \to \nu\ell W$), neutral-current decays ($\ell^*\ell \to \ell\ell Z$ and $\nu^*\nu \to \nu\nu Z$), and photonic decays ($\ell^*\ell \to \ell\ell\gamma$ and $\nu^*\nu \to \nu\nu\gamma$). No excesses are observed in the data taken at $\sqrt{s} = 183$-189 GeV. The resulting limits on f/Λ from the OPAL collaboration[1] are shown in Figure 3.

Another possibility for $f = -f'$ is the t-channel exchange of an excited electron. In this case, the final topology consists of two photons. Since the e^* may be a virtual particle, limits on the coupling versus mass beyond the centre-of-mass energy may be probed. As shown in Figure 4, the DELPHI collaboration uses this search at $\sqrt{s} = 183$ GeV to set limits on $\lambda/m_{e^*} = f/(\sqrt{2}\Lambda)$, where λ is the Yukawa coupling of the excited electron to the SM electron, and m_{e^*} is the mass of the excited electron.

The last case considered for excited neutrinos is the case $\nu^*\nu \to \nu\nu\gamma$. The final topology is a single photon and missing transverse momentum. While no direct mass reconstruction is possible, the kinematics of the photon can be used to restrict the range of excited neutrino masses consistent with the event. The numbers

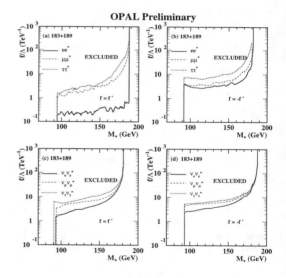

Figure 3: 95% confidence level upper limits on the ratio of the coupling to the compositeness scale, f/Λ, as a function of the excited lepton mass. (a) shows the limits on e^*, μ^* and τ^* with $f = f'$, (b) shows the limits on e^*, μ^* and τ^* with $f = -f'$, (c) shows the limits on ν_e^*, ν_μ^* and ν_τ^* with $f = f'$, and (d) shows the limits on ν_e^*, ν_μ^* and ν_τ^* with $f = -f'$. The regions above and to the left of the curves are excluded by the single and pair production searches, respectively.

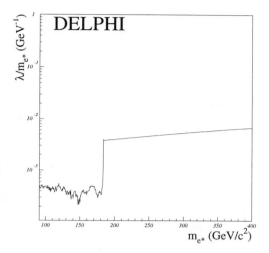

Figure 4: 95% confidence level upper limits on the ratio of the coupling to the excited electron mass, $\lambda/m_{e^*} = f/(\sqrt{2}\Lambda)$ as a function of the excited electron mass. The regions above the curves are excluded by the single production searches for $e^+e^- \to \gamma\gamma$.

and energy distributions of the data taken at $\sqrt{s} = 183$ GeV are consistent with the expectation from SM $\nu\bar{\nu}\gamma$ production. Limits on λ/m_{ν^*} versus the mass of the excited neutrino, from the L3 collaboration[1], are shown in Figure 5. These include limits for both $f' = 0$, and $f = 0$, in which there is no t-channel production of excited neutrinos.

Finally, the single production of excited quarks is searched for. The two decays considered are photonic decays, $q^* \to q\gamma$, and gluonic decays, $q^* \to qg$. Final states include one photon and two jets, or three jets. From the data taken at $\sqrt{s} = 161\text{-}183$ GeV, the DELPHI collaboration observes 196 and 140 candidates in these two channels respectively, in agreement with 140 and 152 events expected from SM processes. The resulting limits on λ/m_{ν^*} versus the mass of the excited quark are shown in Figure 6.

Conclusion

No evidence for neutral and charged heavy leptons, or excited leptons has been found in the mass ranges accessible to LEP 2. Using only about six weeks of data taken in 1998 at $\sqrt{s} = 189$ GeV, the LEP collaborations have already increased the mass limits on these particles by up to 10 GeV, compared to previous searches. The future LEP program will reach up to 500 pb^{-1} at energies up to $\sqrt{s} = 200$ GeV, which still allows for a window of discovery.

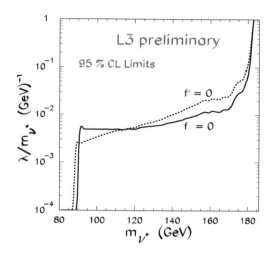

Figure 5: 95% confidence level upper limits on the ratio of the coupling to the excited neutrino mass, $\lambda/m_{\nu^*} = f/(\sqrt{2}\Lambda)$ as a function of the excited neutrino mass. The regions above the curves are excluded by the single production searches for $e^+e^- \to \nu\nu\gamma$.

Acknowledgements

I would like to thank the CERN SL Division for their efficient operation of the LEP accelerator, and the LEP Collaborations for providing me with their results.

References

1. DELPHI Collaboration, *"Search for composite and exotic fermions at LEP 2"*, DELPHI 98-69 CONF 137, ICHEP 98 Contributed Paper 215, June 1998; L3 Collaboration, *"Search for Heavy Neutral and Charged Leptons in e^+e^- Annihilation at $\sqrt{s} = 183$ GeV"*, L3 Note 2293, ICHEP 98 Contributed Paper 497, June 1998; L3 Collaboration, *"Search for Excited Leptons Decaying Radiatively at $\sqrt{s} = 183$ GeV"*, L3 Note 2217, ICHEP 98 Contributed Paper 498, June 1998; L3 Collaboration, *"Preliminary results on New Particle Searches at $\sqrt{s} = 189$ GeV"*, L3 Note 2309, ICHEP 98 Contributed Paper 485, July 1998; OPAL Collaboration, *"Search for Unstable Heavy and Excited Leptons in e^+e^- Collisions at $\sqrt{s} = 181\text{-}184$ GeV"*, OPAL Physics Note PN 343, ICHEP 98 Contributed Paper 283, June 1998;

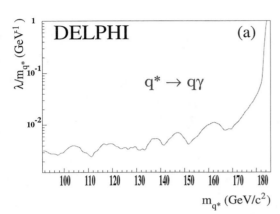

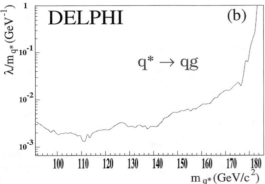

Figure 6: 95% confidence level upper limits on the ratio of the coupling to the excited quark mass, $\lambda/m_{q^*} = f/(\sqrt{2}\Lambda)$ as a function of the excited quark mass. Figure (a) gives the limits from searches for photonic decays of excited quarks, while figure (b) shows limits for hadronic decays. The regions above the curves are excluded.

OPAL Collaboration, *"New Particle Searches in e^+e^- Collisions at $\sqrt{s} = 189$ GeV"*, OPAL Physics Note PN 362, ICHEP 98 Contributed Paper 1065, July 1998.

2. "Review of Particle Physics" D. Haidt *et al.*, Eur. Phys. J. C1 (1998).

3. Super-Kamiokande Collaboration, Y. Fukuda *et al.*, Phys.Rev.Lett. 81 (1998) 1562-1567.

4. MACRO Collaboration, M. Ambrosio *et al.*, hep-ex/9809006, 9 Sep 1998.

5. ALEPH Collaboration, D. Decamp *et al.*, Phys. Lett. B236 (1990) 511;
OPAL Collaboration, M.Z. Akrawy *et al.*, Phys. Lett. B240 (1990) 250;
OPAL Collaboration, M.Z. Akrawy *et al.*, Phys. Lett. B247 (1990) 448;
L3 Collaboration, B. Adeva *et al.*, Phys. Lett. B251 (1990) 321;
DELPHI Collaboration, P. Abreu *et al.*, Phys. Lett. B274 (1992) 230.

6. L3 Collaboration, M. Acciarri *et al.*, Phys. Lett. B412 (1997) 189.

7. OPAL Collaboration, K. Ackerstaff *et al.*, Eur. Phys. J. C1 (1998) 45.

8. OPAL Collaboration, K. Ackerstaff *et al.*, "Search for Stable and Long-Lived Massive Charged Particles in e^+e^- Collisions at $\sqrt{s} = 130$-183 GeV", CERN-EP/98-039, submitted to Phys. Lett. B.

9. OPAL Collaboration, M.Z. Akrawy *et al.*, Phys. Lett. B257 (1990) 531;
ALEPH Collaboration, R Barate *et al.*, CERN-EP/98-022, submitted to Eur. Phys. J. C;
ALEPH Collaboration, D. Decamp *et al.*, Phys. Lett. B250 (1990) 172;
DELPHI Collaboration, P. Abreu *et al.*, Z. Phys. C53 (1992) 41;
L3 Collaboration, M. Acciarri *et al.*, Phys. Lett. B353 (1995) 136.

10. OPAL Collaboration, K. Ackerstaff *et al.*, Phys. Lett. B391 (1997) 197.

11. DELPHI Collaboration, P. Abreu *et al.*, Phys. Lett. B393 (1997) 245;
L3 Collaboration, M. Acciarri *et al.*, Phys. Lett. B401 (1997) 139.

12. H1 Collaboration, I. Abt *et al.*, Nucl. Phys. B396 (1993) 3;
ZEUS Collaboration, M. Derrick *et al.*, Z. Phys. C65 (1994) 627;
H1 Collaboration, S. Aid *et al.*, Nucl. Phys. B483 (1997) 44.

13. E. Nardi, E. Roulet, and D. Tommasini, Phys. Lett. B344 (1995) 225.

14. F. Boudjema, A. Djouadi and J.L. Kneur, Z. Phys. C57 (1993) 425.

Events with high energy isolated leptons and missing transverse momentum and excited fermion searches at HERA

Cristinel DIACONU

Centre de Physique des Particules de Marseille, 13288 Marseille cedex 09, France
E-mail: diaconu@cppm.in2p3.fr

The observation of events containing isolated leptons and missing transverse momentum in e^+p collisions at HERA is reported. H1 observed one event containing an electron e^- and five events containing muons $\mu^\pm$. ZEUS observed three events containing positrons e^+. All events have significant missing transverse momentum. The main Standard Model explanation is the production of $W^\pm$ bosons.

Excited fermion searches with H1 and ZEUS detectors are also presented. Several decay channels giving different experimental signatures are analysed. No experimental evidence for excited fermions is found. Limits on production and decay parameters versus the excited fermion mass are extracted for mass domains up to 250 GeV.

1 Introduction

At HERA 27.5 GeV positrons collide with 820 GeV protons resulting in a center of mass energy of 300 GeV. This configuration open an unexplored domain for new physics searches. This note presents two analysis performed by H1 and ZEUS collaborations on data samples corresponding to an integrated luminosity of 40 pb^{-1}/experiment.

In the first part the observation of events with high energy isolated leptons and missing transverse momentum is reported. The search for excited fermion production and decay is described in the second part.

2 Observation of events containing high energy isolated leptons and missing transverse momentum

In the frame of the Standard Model, events containing a high energy lepton together with significant missing energy are issued at HERA from the W production and decay in e-p collisions. The same final state topology can also provide a signature for the processes beyond the Standard Model like production of excited fermions or supersymmetric particles.

Both HERA collaborations have searched for events with high energy isolated leptons and missing transverse momentum[1,2]. The charged leptons are directly identified in compact detectors like H1 and ZEUS. The presence of non-detected particles can only be deduced from a transverse momentum imbalance.

2.1 H1 analysis

The H1 analysis[1] is based on the inclusive search for events with the calorimetric transverse momentum (P_T^{calo}) above 25 GeV. This condition ensures a good

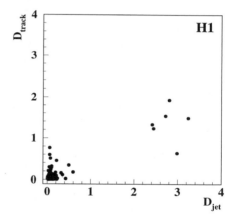

Figure 1: High energy track isolation with respect to the tracks (D_{track}) versus the isolation with respect the hadronic jets (D_{jet}) in the events with $P_T^{calo} > 25$ GeV.

understanding of experimental efficiencies and drastically reduces the contributions of neutral current deep inelastic scattering (NC-DIS) and photoproduction processes. Among the selected sample, 124 events contain also a high energy track with a transverse momentum $P_T > 10$ GeV. The isolation of those tracks with respect to the calorimetric deposits and to other tracks is quantified using the Cartesian distance in the $\eta - \phi$ plane, where η stand for pseudorapidity and ϕ is the azimuth in a right reference frame having its positive z direction along the incoming proton and the origin in the event interaction vertex. The isolation of high P_T tracks with respect to the closest track (D_{track}) and to the closest hadronic jet

(D_{jet}) is presented in the figure 1. A majority of high energy tracks present small isolation values corresponding to tracks inside hadronic jets. Six events have isolated high energy tracks ($D_{jet} > 1.0$ and $D_{track} > 0.5$). A loose lepton identification using calorimetric deposit shape for $e^{\pm}$ and the calorimeter and the muon detector information for $\mu^{\pm}$ reveals the leptonic nature of the isolated tracks: one event contain an e^- and five events contain muons: $2\mu^+$ $2\mu^-$ and one very energetic μ associated to a stiff track whose sign cannot be determined.

In all events a hadronic shower has been detected in the calorimeters. The leptonic signature has been investigated in details and found to be consistent with the assigned hypothesis. In particular, the path of calorimetric deposits for muon candidates sampled over at least seven interaction lengths has been found compatible with a minimum ionizing particle.

All events present significant imbalance in total transverse momentum, reconstructed using calorimetric information for the electron event and using in addition the muon track information for the muon events. The missing transverse momentum points to the presence of an undetected particle. Additional evidence supporting this hypothesis is provided by the acoplanarity ($\Delta\phi$) observed in most events between the hight P_T lepton and the hadronic system.

The processes within the Standard Model which may yield events with an isolated high P_T lepton and missing transverse momentum have been investigated by applying the selection procedure explained above. Events with more than one isolated lepton of the same generation have not been taken into account as they do not correspond to the topology of the observed events.

The main Standard Model contribution in both electron and muon channels is provided by the W boson production in e-p collisions. The subsequent W leptonic decay $W \to \ell\nu$ provide both an isolated high energy charged lepton and missing transverse momentum associated to the undetected neutrino. This process is studied using EPVEC generator [3]. The selection efficiencies are 33% in the electron channel and 10% in the muon channel. The expected number of events corresponding to an integrated luminosity of 37 pb^{-1} are 1.7 ± 0.5 in the electron channel and 0.5 ± 0.1 in the muon channel.

The lower acceptance in the muon channel is mainly due to the cut on the calorimetric transverse momentum at 25 GeV which selects events with a high P_T hadronic jet associated to the recoiling quark. This topology is disfavoured in the W production process, where the P_T distribution of the recoil hadronic jet peak at zero with and exponential tail towards higher values described by a perturbative computation in EPVEC.

Other standard Model contributions can yield events with isolated leptons and missing transverse momentum.

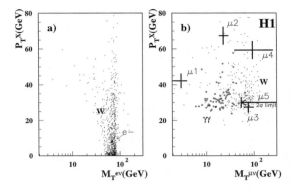

Figure 2: Comparison of the observed event (crosses) to the Standard Model contributions – W production(points) and photon-photon process (small circles) – in the plane defined by the hadronic transverse momentum P_T^X and the lepton-neutrino transverse mass $M_{\ell\nu}$. The Monte Carlo sample correspond to a luminosity a factor 500 higher than the data.

For instance, the NC-DIS process can produce events containing an isolated positron e^+ and missing transverse momentum due to energy measurement fluctuations. The muon events topology may also be produced in photon-photon muon pair production, with one of the muons escaping down the beam pipe. All the other contributions have been estimated to bee small or negligible.

The analysis of the Standard Model processes concluded to an expected number of events of 2.4 ± 0.5 events in the electron channel (1e$^-$ event observed) and 0.8 ± 0.2 $\mu^{\pm}$ events in the muon channel (5 events observed).

The expectation being dominated by the W production process, in the figure 2 a comparison of the observed events and the W simulation is done in terms of the transverse momentum of the hadronic system P_T^X versus the transverse mass of the lepton-neutrino system $M_T^{\ell\nu}$. The electron event and two of the muon events are consistent with the Jacobian peak around the W mass. Three of the muon events only marginally agree with the W interpretation. None of the muon events is consistent with the muon pair production in photon-photon collisions also represented in the figure 2.

2.2 ZEUS analysis

The search for events with isolated leptons and missing transverse momentum has been done by the ZEUS collaboration in the frame of the W production analysis [2]. The selection in the electron channels is based on the requirement of a missing transverse momentum above 20 GeV for events with an identified $e^{\pm}$ having

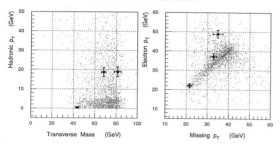

ZEUS 1994−97 PRELIMINARY

Figure 3: Comparison of the 3 e^+ observed events (crosses) to the W production in the planes defined by the hadronic transverse momentum P_T^X and the lepton-neutrino transverse mass $M_{\ell\nu}$ (left) and electron transverse momentum versus missing transverse momentum (right). The Monte Carlo sample correspond to a luminosity a factor 1000 higher than the data.

Table 1: H1 and ZEUS results in the isolated lepton and missing P_T event searches.

Analysis	$e^{\pm}$ channel		$\mu^{\pm}$ channel	
	DATA	MC	DATA	MC
Strategy **H1** 36.5 pb^{-1}	$P_T^{calo} > 25 P_T^e > 10$ $1\ e^-$	2.4 ± 0.5	$P_T^{calo} > 25 P_T^\mu > 10$ $5\ \mu^{\pm}$	0.8 ± 0.2
Strategy **ZEUS** 46.6 pb^{-1}	$P_T^{calo} > 20 P_T^e > 10$ $3\ e^+$	3.0 ± 0.7	$P_T^{calo} > 30 P_T^\mu > 5$ $0\ \mu^{\pm}$	0.8 ± 0.3

a transverse momentum higher than 10 GeV. The NC-DIS events are rejected by removing from the sample the events with backward electrons ($\theta_e < 2$ rad.) or with electrons aligned to the hadronic system in the transverse plane ($\Delta\phi_{e-hadrons} > 0.3$ rad.). The efficiency to select W events was estimated to be 38% using EPVEC[3]. With those cuts three events have been selected in a data sample corresponding to an integrated luminosity of 46.6 pb^{-1}. The Standard Model expectation has been calculated to be 3 events, the W production accounting for 2 events. All observed data events contain a positron (e^+) and the kinematics is consistent with the W production and decay, as can be seen in the figure 3. Using those events, ZEUS collaboration computed the W production cross section $\sigma(e^+p \to e^+WX) = 1.0^{+1.0}_{-0.7} \pm 0.3(syst.)$

The selection of W events in the muon channel is based on the requirement of both calorimetric and missing transverse momentum above 15 GeV. The muon is then identified as a track reconstructed in the tracking system with a momentum higher than 5 GeV and corresponding to a minimum ionizing pattern in the calorimeters. The selection efficiency for W events is 11%. No events are selected in the ZEUS data for an expectation of 1.2 events (0.6 from W production). A limit on the W production cross section has been calculated in this analysis: $\sigma(e^+p \to e^+WX) < 4.4$ pb at 95% C.L. .

In order to compare to the H1 analysis, ZEUS searched for events having an identified muon and high calorimetric transverse momentum. For $P_T^{calo} > 20$ GeV 1 event is observed in the data for 2.3 events expected and no event is observed with $P_T^{calo} > 30$ GeV while 0.8 events are expected.

In conclusion, ZEUS collaboration observe three W-like events in the $e^{\pm}$ channel and no event with high

energy isolated muon and missing transverse momentum.

2.3 Comparison H1/ZEUS

Table 1 summarize the results obtained by H1 and ZEUS in searching for events containing high energy isolated leptons and missing transverse momentum. The observed and expected numbers of events are in agreement in the electron channel. In the muon channel, H1 analysis reveals an excess of events compared to the expectation. This excess is not present in the ZEUS data. Three of the H1 events are kinematically atypical for any Standard Model interpretation and can only be explained with the present data sample by a statistical fluctuation. Beyond the Standard Model possible explanations of such topology includes production and decay of excited quarks or supersymmetric particles [4].

3 Excited fermion searches at HERA

The excited fermions are one of the possible manifestations for the substructure of the nowadays fundamental particles. The production and decay of excited fermions F^* are described in terms of a phenomenological Lagrangian [5]:

$$L_{FF^*} = \frac{e}{\Lambda} \sum_V^{\gamma,Z,W} \bar{F}^* \sigma^{\mu\nu} (c_V - d_V \gamma^5) F \partial_\mu V_\nu + h.c.$$

where c_V and d_V are coupling constants at the $F \leftrightarrow F^*$ vertex labelled for each vector boson V, and Λ is the compositeness scale. The precise measurement of muon $g-2$ and the absence of electron and muon electric dipole moment implies that $c_V = d_V$ for compositeness scales less than 10–100 TeV. This fact lead to an interaction Lagrangian where the coupling constants of excited

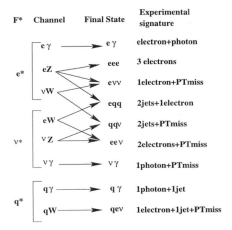

Figure 4: Analyzed decay modes and their experimental signatures for the excited fermion searches.

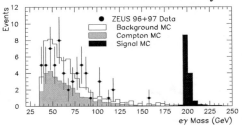

Figure 5: Distribution of the $e\gamma$ invariant mass for selected $e^* \rightarrow e\gamma$ candidates for the data (solid points) and for the total background from NC-DIS and QED Compton (open histogram). The contribution from the QED Compton is shown separately by the shaded histogram. An example of e^* signal with the mass of 200 GeV is also shown as a dark histogram in arbitrary normalization.

fermions to Standard Model gauge bosons are described in terms of three constants corresponding to the three gauge groups: f for $SU(2)_L$, f' for $SU(1)_Y$ and f_s for $SU(3)_C$. For specific assumptions relating f, f' and f_s the branching ratios can be predicted and the cross sections are described by a single parameter (e.g. f/Λ).

At HERA, excited fermions can be produced via t-channel exchange of a gauge boson: γ or Z for e^* and q^* production and $W^\pm$ for ν^* production. The excited states decay to known fermions and gauge bosons. The search for excited fermions is therefore a search for a fermion-boson resonance.

H1 and ZEUS collaboration have published excited fermion analysis [6] based on integrated luminosities of 2.75 pb^{-1} and 9.4 pb^{-1}, respectively. The results presented here summarize the new multi-channel H1 analysis [7] of a data sample corresponding to 37 pb^{-1} and ZEUS update [8] of the excited electron channel $e^* \rightarrow e\gamma$ with similar luminosity.

The decay channels analyzed and their experimental signatures are illustrated in the figure 4. The same final state may correspond to different excited fermion flavours and decay.

As an example, we will shortly describe the excited electron search in the channel $e^* \rightarrow e\gamma$. The branching ratio is relatively high (predicted to be 30% for e^* masses above 150 GeV) and the experimental signature particularly clear: two high energy electromagnetic clusters. The Standard Model contributions are NC-DIS and Compton $e\gamma$ scattering (QED Compton).

ZEUS $e^* \rightarrow e\gamma$ analysis require two electromagnetic clusters of transverse momentum above 10 GeV (30 GeV if the polar angle of the cluster is below 0.3 rad) consistent with a electron-photon final state and low activity in the tracking system (except the track associated to the electron). The last condition mainly selects the elastic and quasi-elastic production of excited electrons. The acceptance for the signal ranges from 60% for e^* masses of 100 GeV to 75% at higher masses.

H1 $e^* \rightarrow e\gamma$ analysis is splitted in the exclusive and inclusive strategies. The inclusive analysis demand the sum of the energies of the two clusters to be above 20 GeV and the residual energy deposited in the calorimeters to be below 5 GeV. The inclusive strategy, designed to improve the efficiency for masses above 50 GeV, ask for the transverse momentum of the two e.m. clusters to be above 10 and 20 GeV, respectively, relaxing in the same time the empty detector condition. This strategy selects also inelastic production of e^*. The efficiency ranges in the H1 analysis from 65% for e^* masses of 50 GeV and increases to about 80-85% at higher masses.

The invariant mass of the two electromagnetic clusters obtained in ZEUS analysis $e^* \rightarrow e\gamma$ is showed in the figure 5. No significant excess of events characteristic of a peak for e^* production is observed.

For all channels analyzed (see figure 4), good agreement between data and Standard Model expectation is observed. Limits at 95% confidence level (C.L.) are obtained for the parameters related to the excited fermions production and decay. This is done first in a model independent manner by extracting limits on the product between the production cross-section σ^* and the branching ratio Br^* corresponding to a particular final state. In the case of the excited electrons (figure 6), the best limits are obtained in the channel $e^* \rightarrow e\gamma$ thanks to

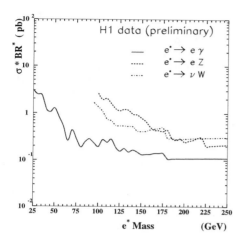

Figure 6: Limits on $\sigma^* Br^*$ for excited electron analysis in three decay modes. Upper regions are excluded at 95% C.L..

the high selection efficiency. Mass domains from 25 to 250 GeV are explored. For the highest mass region a limit $\sigma^* Br^* < 0.11$ pb at 95% C.L. is obtained in the H1 analysis. A similar result is obtained in the ZEUS update of that channel, improving by a factor of 4 the previous result.

For excited neutrino, H1 analysis set upper limits on $\sigma^* Br^*$ ranging from 0.2 pb for ν^* masses of 50 GeV to 0.3 pb at masses around 250 GeV. In the same mass range, the upper limits on $\sigma^* Br^*$ for the excited quarks are 0.3 to 0.2 pb.

Using the model described above and making assumptions for the coupling constants f, f' and f_s, limits on the ratio of couplings and the compositeness scale can be obtained. In the case of the excited electron, assuming equal strengths for the electromagnetic and weak couplings $f = f'$, upper limits for the ratio f/Λ from 0.6×10^{-3} to 1.0×10^{-2} are obtained for a e^* mass range between 60 GeV and 250 GeV. Assuming $f/\Lambda = 1/M_{e*}$, excited electron states with masses below 222 GeV are excluded at 95% C.L..

In the case of the excited neutrino, the assumption $f = -f'$ allows the radiative decay $\nu^* \to \nu\gamma$. The exclusion region of f/Λ, obtained in H1 analysis, is above values varying between 0.3×10^{-2} to 9×10^{-2} for a ν^* mass interval of 60 to 185 GeV.

The model dependent excited quark analysis suppose no strong coupling $f_s = 0$. This assumption is motivated by the results obtained in $p\bar{p}$ collisions at TEVATRON[9]. The excited quarks are produced in that case in a quark-

gluon fusion mechanism which requires strong q^* coupling ($f_s \neq 0$). Excited quarks with masses in the range from 80 to 300 GeV are excluded for $f = f' = f_s$ values greater than 0.2 for the ansatz $\Lambda = M(q^*)$. Assuming, as explained above, only electroweak mechanisms for q^* production and equal strength for the electromagnetic and weak couplings $f = f'$, the exclusion region, obtained in the H1 analysis for masses between 60 to 230 GeV, correspond to values of f/Λ above 0.8×10^{-3} to 1.9×10^{-2}.

Excited fermions could be produced in e^+e^- collisions at LEP[10] or in $p\bar{p}$ interactions at TEVATRON[9]. LEP limits are more stringent up to masses corresponding to the center of mass energy, but HERA set limits beyond LEP kinematical reach for e^* and ν^*. In the case of excited quarks, HERA complements the TEVATRON results for the case of electroweak production mechanism.

Acknowledgements

The help in preparing this talk and the provision of information and plots by several H1 and ZEUS members are greatly appreciated.

References

1. C.Adloff et al., H1 Collaboration, preprint DESY 98-063, submitted to *Eur. Phys. Jour.* C.
2. ZEUS Collaboration, ICHEP98, contributed paper 756.
   T.Matsushita, these proceedings.
3. U.Baur, J.A.M. Vermaseren, D.Zeppenfeld, *Nucl. Phys.* B **375**, 3 (1992).
4. Z.Zhang, these proceedings.
   T.Kon et al., *Mod. Phys. Lett.* A **12**, 3143 (1997).
   E.Perez, talk at SUSY98, Oxford, 11-17, July 1998.
5. K.Hagiwara, S. Komamiya, D. Zeppenfeld, *Z. Phys.* C **29**, 115 (1985).
6. H1 Collaboration, S.Aid et al., *Nucl. Phys.* B **483**, 44 (1997),
   ZEUS Collaboration, J.Breitweg et al., *Z. Phys.* C **76**, 631 (1997). .
7. H1 Collaboration, ICHEP98, contributed paper 581.
8. ZEUS Collaboration, ICHEP98, contributed paper 758.
9. CDF Collaboration, F.Abe et al., *Phys. Rev.* D **55**, 5263 (1997).
   D0 Collaboration, I Bertram, FERMILAB-CONF-96/389-E.
10. R.Teuscher, these proceedings.

Search for Neutral Heavy Leptons in the NuTeV Experiment at Fermilab

R. B. DRUCKER[6]*, T. ADAMS[4], A. ALTON[4], S. AVVAKUMOV[7], L. de BARBARO[5], P. de BARBARO[7], R. H. BERNSTEIN[3], A. BODEK[7], T. BOLTON[4], J. BRAU[6], D. BUCHHOLZ[5], H. BUDD[7], L. BUGEL[3], J. CONRAD[2], R. FREY[6], J. FORMAGGIO[2], J. GOLDMAN[4], M. GONCHAROV[4], D. A. HARRIS[7], R. A. JOHNSON[1], S. KOUTSOLIOTAS[2], J. H. KIM[2], M. J. LAMM[3], W. MARSH[3], D. MASON[6], C. McNULTY[2], K. S. McFARLAND[3,7], D. NAPLES[4], P. NIENABER[3], A. ROMOSAN[2], W. K. SAKUMOTO[7], H. SCHELLMAN[5], M. H. SHAEVITZ[2], P. SPENTZOURIS[2], E. G. STERN[2], B. TAMMINGA[2], M. VAKILI[1], A. VAITAITIS[2], V. WU[1], U. K. YANG[7], J. YU[3] and G. P. ZELLER[5]

Presented by R. B. DRUCKER

[1] *University of Cincinnati, Cincinnati, OH 45221*
[2] *Columbia University, New York, NY 10027*
[3] *Fermi National Accelerator Laboratory, Batavia, IL 60510*
[4] *Kansas State University, Manhattan, KS 66506*
[5] *Northwestern University, Evanston, IL 60208*
[6] *University of Oregon, Eugene, OR 97403*
[7] *University of Rochester, Rochester, NY 14627*

Preliminary results are presented from a search for neutral heavy leptons in the NuTeV experiment at Fermilab. The upgraded NuTeV neutrino detector for the 1996-1997 run included an instrumented decay region for the NHL search which, combined with the NuTeV calorimeter, allows detection in several decay modes ($\mu\mu\nu$, $\mu e\nu$, $\mu\pi$, $e\pi$, and $ee\nu$). We see no evidence for neutral heavy leptons in our current search in the mass range from 0.3 GeV to 2.0 GeV decaying into final states containing a muon.

1 Introduction

Many extensions to the Standard Model incorporating non-zero neutrino mass predict the existence of neutral heavy leptons (NHL). See Refs. [1] and [2] for discussions and references concerning massive neutrinos. The model considered in this paper is that of Ref. [1] in which the NHL is an iso-singlet particle that mixes with the Standard Model neutrino. Figure 1 shows the Feynman diagrams for the production and decay of such an NHL.

The upgraded NuTeV detector includes a Decay Channel designed specifically to search for NHL's and provides a significant increase in sensitivity over previous searches.

2 The Experiment

The NuTeV calorimeter is described elsewhere[3]; only the features essential to this analysis are described here. The calorimeter consists of 84 layers of 10 cm steel plates and scintillating oil counters. A multi-wire gas drift chamber is positioned at every 20 cm of iron for particle tracking and shower location.

The decay channel is an instrumented decay space upstream of the calorimeter. The channel is 30 m long and filled with helium using 4.6 m diameter plastic bags. The helium was used to reduce the number of neutrino interactions in the channel. Drift chambers are positioned at three stations in the decay channel to track the NHL decay products. Figure 2 shows a schematic diagram of the decay channel. A 7 m × 7 m scintillating plastic "veto wall" was constructed upstream of the decay channel in order to veto on any charged particles entering the experiment.

3 Event Selection

Figure 2 also shows an example of the event signature for which we are searching. The characteristics of an NHL event are a neutral particle entering the channel and decaying in the helium region to two charged (and possibly an additional neutral) particles. The charged particles must project to the calorimeter and at least one must be identified as a muon.

To select events for this analysis we triggered on energy deposits of at least 2.0 GeV in the calorimeter and required no veto wall signal. We then require that there be two well-reconstructed tracks in the decay channel that form a vertex in the helium well away from the edges of the channel and the tracking chambers. The event vertex was required to be at least 3σ away from the fiducial volume edges, where σ is the resolution of the vertex position measurement. By requiring two tracks and separation from the tracking chambers we greatly reduce the number of background events from neutrinos interacting in the decay channel materials. For all the cuts a vertex constrained fit is used in which the two tracks are required to come from a single point in space. The

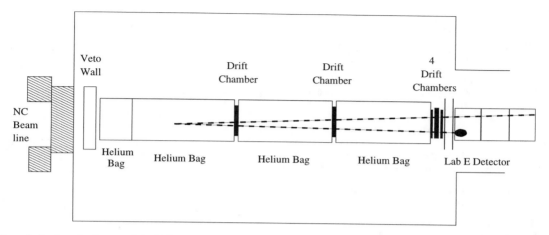

Figure 2: A schematic diagram of the NuTeV decay channel. The beam enters from the left, and at the far right is the NuTeV neutrino target. An example of an NHL decay to $\mu\pi$ is also shown. The event appears as two tracks in the decay channel, a long muon track in the calorimeter and a hadronic shower.

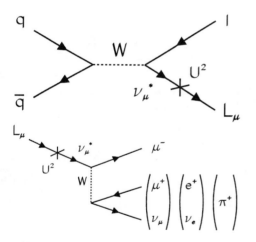

Figure 1: Feynman diagrams showing the production (from meson decay) and decay of neutral heavy leptons (L_μ). Decay via the Z^0 boson is also allowed, but not shown.

vertex resolution depends on the opening angle of the tracks, but it is typically 25 cm along on the beam axis and 2.5 cm transverse.

The two decay tracks are required to project to the calorimeter and to match (in position) with particles identified in the calorimeter. At least one of the two particles must be identified as a muon, because for this analysis we only consider decay modes with at least one

muon. In order to insure good particle identification and energy measurement, we require all muons in the event to have energy greater than 2.0 GeV and all electrons or hadrons to have energy greater than 10.0 GeV. These energy cuts also reduce backgrounds from cosmic rays and neutrino interactions.

To further reduce acceptance for background events, additional kinematic cuts are applied. NHL decays are expected to have a small opening angle; therefore, the decay particles are required to have slopes p_x/p_z and p_y/p_z less than 0.1 (p_z is the momentum component along the direction of the incoming beam, p_x and p_y are the transverse components). We are only considering NHL's produced by kaon and charmed meson decays in this analysis; therefore, NHL's with mass above 2.0 GeV are not considered. We require the transverse mass[a] of the event to be less than 5.0 GeV in order to restrict ourselves to this lower mass region. Finally, in order to reduce neutrino-induced events even further we form the quantities x_{eff} and W_{eff} by assuming that: i) the event is a neutrino charged current interaction ($\nu N \to \mu N' X$), ii) that the highest energy muon comes from the neutrino-W vertex, and iii) the missing transverse momentum in the event is carried by the final state nucleon. We require $x_{\text{eff}} < 0.1$ and $W_{\text{eff}} > 2.0$ GeV.

---

[a]The transverse mass is $p_T + \sqrt{p_T^2 + m_V^2}$, where p_T is the component of the total momentum of the two charged tracks perpendicular to the beam direction (i.e. the "missing transverse momentum"), and m_V is the invariant mass of the two charged tracks.

4 NHL Monte Carlo

Figure 3 shows a schematic of the NuTeV beamline. The experiment took 2.5×10^{18} 800 GeV protons from the Fermilab Tevatron on a BeO target. Secondaries produced from the target are focused in the decay pipe with a central momentum of 250 GeV. The decay pipe is 0.5 km long, and the center of the decay pipe is 1.5 km from the center of the decay channel. Non-interacting protons, wrong-sign and neutral secondaries are dumped into beam dumps just beyond the BeO target. NHL's would be produced in decays of kaons and pions in the decay pipe, as well as from charmed hadron decays in the primary proton beam dumps. Pion decays do not contribute significantly to this analysis, as they cannot produce NHL's in the mass range of our search.

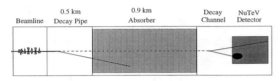

Figure 3: A schematic diagram of the NuTeV beamline. The 800 GeV proton beam from the Fermilab Tevatron enters from the left. NHL's are produced from the decays of kaons and pions in the decay pipe and from the decays of charm hadrons in the beam dumps.

The production of kaons is simulated using the Decay Turtle[4] program. The simulation of kaon decays to NHL's includes the effects of mass both in decay phase space and in helicity suppression. The production of charmed hadrons in the beam dump are simulated using a Monte Carlo based on the production cross sections reported in Ref. [5]. For this analysis we only generate muon flavored NHL's. Figure 4 shows examples of the momentum distribution of NHL's produced by the NuTeV beamline. For a 1.45 GeV mass NHL, the average momentum is ∼140 GeV. For a 0.35 GeV mass NHL the average momentum is ∼100 GeV.

The simulation of NHL decays uses the model of Ref. [6]. The polarization of the NHL is also included in the decay matrix element[7]. The decay products of the NHL are run through a full Geant detector simulation to produce simulated raw data which is then run through our analysis software.

5 Results

We observe no events which pass our event selection cuts. The number of expected background events are approximately 0.5. The largest background is 0.4 events expected from neutrino interactions in the decay channel

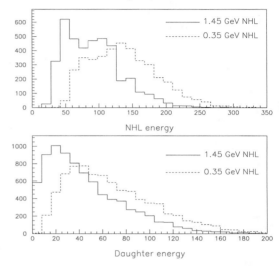

Figure 4: The upper plot shows the energy distributions for Monte Carlo NHL's with mass 1.45 GeV and 0.35 GeV. The lower plot shows the energy of the decay products of the NHL.

helium. This estimate was made using the Lund Monte Carlo[8] to simulate neutrino–nucleon interactions. In order to present a conservative limit, we assume an expected background of zero events (this is only a small change in the resulting limits).

In order to demonstrate the acceptance and reconstruction efficiency of the experiment, we loosened several cuts in order to examine the neutrino interactions in the decay channel material. We removed the cuts on the event vertex position (allowing events at the positions of the chambers), and allow events with more than 2 tracks. No calorimeter cuts (matching to particles, or energy cuts) were applied, and no x_{eff} or W_{eff} cuts were applied. Figure 5 shows the distribution of the event vertex along the beam axis. The peaks correspond to the positions of the tracking chambers. The plot also shows the neutrino interactions in the helium gas between the chambers. The number of events seen is consistent with expectations. This study demonstrates that the channel and our tracking reconstruction are working well.

Figure 6 shows our limits on the NHL–neutrino coupling, $U_{2\mu}^2$, as a function of the mass of the NHL. The results of previous experiments[9,10,11,12,13] are shown for comparison. Our result is a significant increase in sensitivity in the range from 0.3 GeV to 2.0 GeV. These limits

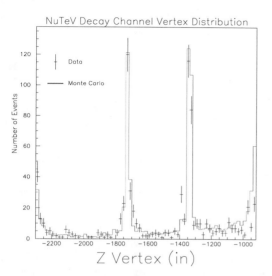

Figure 5: The Z vertex distribution for neutrino interaction events in the NuTeV decay channel. The points are data and the lines are Monte Carlo. The peaks correspond to the positions of the drift chambers.

are for muon flavored NHL's and only include their decay modes containing a muon. The limits do not yet include the effects of systematic uncertainties.

6 Conclusions

We have shown new preliminary limits from a search for muon flavored neutral heavy leptons from the NuTeV experiment at Fermilab. In the future we plan to expand our search to include masses greater than 2.0 GeV as well as masses less than 0.3 GeV (perhaps to a final range of ~ 0.020 GeV to ~ 10.0 GeV). We will also expand our search to include electron flavored NHL's and all NHL decay modes ($\mu\mu\nu$, $\mu e \nu$, $\mu\pi$, $e\pi$, and $ee\nu$).

Acknowledgements

This research was supported by the U.S. Department of Energy and the National Science Foundation. We would also like to thank the staff of Fermliab for their substantial contributions to the construction and support of this experiment during the 1996–97 fixed target run.

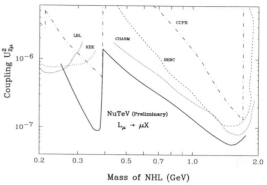

Figure 6: Preliminary limits from NuTeV on the coupling, $U^2_{2\mu}$, of neutral heavy leptons (NHL) to the Standard Model left-handed muon neutrino as a function of NHL mass. Only the μX decay modes of the NHL are included in this first search. The limits are 90% confidence and are based on zero observed events with zero expected background events. The limits do not yet include effects from systematic uncertainties

References

1. Michael Gronau, C.N. Leung and Jonathan L. Rosner, *Phys. Rev.* D **29**, 2539 (1984).

2. Particle Data Group, *Eur. Phys. J.* C **3**, 1 (1998).

3. W. Sakamoto, *et al.*, *Nucl. Instrum. Methods* **A294**, 179 (1990).

4. David C. Carey, Karl L. Brown, F.C. Iselin, SLAC-0246 (1982).

5. Stefano Frixione, *et al.*, *Nucl. Phys.* B **431**, 453 (1994).

6. Loretta M. Johnson, Douglas W. McKay and Tim Bolton, *Phys. Rev.* D **56**, 2970 (1997).

7. Joseph A. Formaggio, *et al. Phys. Rev.* D **57**, 7037 (1998)

8. G. Ingelman, *et al.*, DESY HERA Workshop 1366 (1991).

9. S.R. Mishra *et al.*, *Phys. Rev. Lett.* **59**, 1397 (1987).

10. A.M. Cooper-Sarkar *et al.*, *Phys. Lett.* B **160**, 207 (1985).

11. J. Dorenbusch *et al.*, *Phys. Lett.* B **166**, 473 (1986).

12. T. Yamazaki *et al.*, Procedings of Neutrino 84 (1985).

13. C.Y. Pang *et al.*, *Phys. Rev.* D **8**, 1989 (1973).

SUSY SEARCH WITH PHOTONIC EVENTS AT LEP WITH $\tilde{G}$ AS LSP AND $\tilde{\chi}_1^0$ AS NLSP

JEAN FAY

Institut de Physique Nucléaire de Lyon, IN2P3/CNRS and Université Claude Bernard

We present results from SUSY scenarios with the gravitino as LSP and the neutralino as NLSP. The searches, performed by the 4 LEP experiments in 1997 at $\sqrt{s}$=183 GeV, include slepton pair production, chargino pair production and photon(s) plus missing energy final states. Preliminary data obtained in 1998 at $\sqrt{s}$=189 GeV are also presented. No evidence from new physics has been found and limits on the parameters of the models have been derived.

1 Introduction

Usually, the supersymmetry breaking mechanism is supposed to occur at very high energy and the gravitino is too heavy to be produced at LEP.

On the other hand, in Gauge Mediated Susy Breaking (GMSB) models[1], the supersymmetry breaking takes place at much lower energy scale $\sqrt{F}$ and the gravitino $\tilde{G}$ which mass reads:

$$m_{\tilde{G}} = \frac{F}{\sqrt{3}\, M_P} \simeq 2.5 \text{ eV} \left(\frac{\sqrt{F}}{100\text{TeV}}\right)^2$$

becomes the Lightest Susy Particle (LSP).

The Next to Lightest Susy Particle (NLSP) can be either a right handed scalar lepton[2] $\tilde{\ell}_R^\pm$ or the neutralino $\tilde{\chi}_1^0$. In this latter case, which is the subject of this paper, the search for $\tilde{\ell}$, charginos $\tilde{\chi}^\pm$ and $\tilde{\chi}^0$ must be optimized to take the $\tilde{\chi}_1^0$ decay into account. In most cases, the branching ratio $\tilde{\chi}_1^0 \to \gamma\tilde{G}$ is near 100 %.

The neutralino decay length:

$$D.L. \simeq (17\mu m) \frac{1}{\kappa} \sqrt{\frac{E_{\tilde{\chi}_1^0}^2}{m_{\tilde{\chi}_1^0}^2}} \left(\frac{m_{\tilde{G}}}{1 \text{ eV}}\right)^2 \left(\frac{m_{\tilde{\chi}_1^0}}{100 \text{ GeV}}\right)^{-5}$$

can be non negligible. Nevertheless, unless otherwise stated, we will consider that $\tilde{\chi}_1^0$ is short lived and decays near the interaction region.

ALEPH has performed a search for slepton, DELPHI and OPAL for chargino, while the 4 LEP experiments were looking for neutralino.

The main results come from data taken at $\sqrt{s}$=183 GeV. Preliminary updates at 189 GeV have already been obtained.

2 Search for slepton

The production of $\tilde{\mu}$ and $\tilde{\tau}$ proceeds via γ or Z exchange while $\tilde{e}$ can also be produced via $\tilde{\chi}^0$ exchange in the t channel. The slepton dominantly decays in its partner lepton and $\tilde{\chi}_1^0$:

A dedicated search has been developed with acoplanar leptons plus photons.[3] In the case of e, μ, there must be at least 2 photons with energies greater than 3 GeV and 2 identified leptons. Additional cuts are applied in the selectron search to remove radiative Bhabha events. For acoplanar τ, the two photons must have at least 1 GeV and a τ tagging algorithm is applied.

When the mass difference $m_{\tilde{\ell}}$-$m_{\tilde{\chi}_1^0}$ decreases below 1 GeV, the efficiency of the previous selections becomes too low and the selection is modified in order to search for two acoplanar photons + X. Both photons must have an energy $E_\gamma > 3$ GeV. Other cuts are applied on invariant mass from tracks and photons and on additionnal energy.

The results obtained from data taken at $\sqrt{s}$=161 (11.1 pb^{-1}), 172 (10.6 pb^{-1}) and 183 GeV (57.0 pb^{-1}) are displayed in Table 1.

Table 1: Efficiencies, measured and expected number of events.

	ϵ (%)	data	background
$\tilde{e}$	30 - 50	0	0.6
$\tilde{\mu}$	45 - 70	0	0.8
$\tilde{\tau}$	20 - 35	1	1.1
$\gamma + X$	15 - 50	3	2.7 ($\nu\bar{\nu}\gamma(\gamma)$)

As there is no observation of any excess of events, these results lead to the exclusion regions (figure 1). The limit on $m_{\tilde{\ell}}$ is derived assuming the 3 sleptons have the same mass and combining $\tilde{e}$ and $\tilde{\mu}$ results.

The following mass limits at 95% confidence level are derived:

$m_{\tilde{e}_R}$: 77 GeV, $m_{\tilde{\mu}_R}$: 77 GeV, $m_{\tilde{\tau}_R}$: 52 GeV

$m_{\tilde{\ell}_R}$: 82 GeV (for degenerated slepton masses)

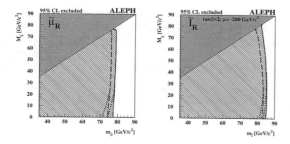

Figure 1: Mass limits for right-handed sleptons in GMSB models.

3   Search for chargino

When the neutralino is the NLSP, the chargino decay is the following:

$$e^+e^- \longrightarrow \tilde{\chi}_1^+ \tilde{\chi}_1^-$$

with $\tilde{\chi}_1^0 W^{(*)-}$, where $\tilde{\chi}_1^0 \to q\bar{q}'$ or $\ell^-\bar{\nu}_\ell$, $\gamma\tilde{G}$; and $\tilde{\chi}_1^0 W^{(*)+}$, where $\tilde{\chi}_1^0 \to q\bar{q}'$ or $\ell^+\nu_\ell$, $\gamma\tilde{G}$.

The final states include isolated photons and missing energy plus jets or jets and isolated leptons or two isolated leptons.

DELPHI[4] makes no distinction on the event topology and searches for $\gamma\gamma + X$: it requires at least 1 photon with energy $E_\gamma > 5$ GeV and angle $15° < \theta_\gamma < 165°$ and at least 1 track. Additional cuts are put on $\not{p}_t$, E_{vis} and acoplanarity, dependent on $\Delta M = m_{\tilde{\chi}_1^+} - m_{\tilde{\chi}_1^0}$.

OPAL[5] divides the events into two categories: a high multiplicity ($N_{tracks} > 4$) topology, in this case, the event must contain at least 1 photon with $E_\gamma > 15$ GeV, and a low multiplicity ($2 < N_{tracks} \leq 4$) topology where there must be at least 1 photon with $E_\gamma > 5\%\, E_{beam}$. Cuts on p_t, E_{vis}, acoplanarity, γ isolation are added to remove the background.

The results obtained at $\sqrt{s} = 183$ GeV and preliminary ones at $\sqrt{s} = 189$ GeV are presented in Table 2.

Table 2: Efficiencies, number of events observed in data and expected from background.

	ϵ (%)	data	backgr.
DELPHI	183 GeV 50.6 pb^{-1}		
$\Delta M > 10$ GeV	35 - 45	3	4.9
$5 < \Delta M < 10$ GeV	40 - 55	0	0.9
$\Delta M < 5$ GeV	40 - 50	1	0.5
total		4	6.3
	189 GeV 34 pb^{-1}		
$\Delta M > 10$ GeV		2	3.0
$5 < \Delta M < 10$ GeV		0	0.5
$\Delta M < 5$ GeV		1	0.5
OPAL	183 GeV 56.8 pb^{-1}		
high mult.	15 - 70	2	2.1
low mult.	40 - 65	4	4.9
total	30 - 50	6	7.0
	189 GeV 32 pb^{-1}		
high mult.		2	2.2
low mult.		2	2.6

DELPHI computes the chargino cross section limits for different chargino masses (figure 2) and extracts the lower limit at 95 % confidence level for the chargino mass: 90.5 GeV in the non-degenerate case for a heavy sneutrino and 90.6 GeV in the degenerate one. The preliminary limits obtained at $\sqrt{s} = 189$ GeV are respectively 93.6 and 93.7 GeV.

DELPHI $\tilde{\chi}_1^+\tilde{\chi}_1^-$ limits at 183 GeV

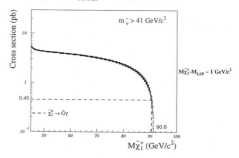

Figure 2: Expected cross section versus the chargino mass, in the degenerate case ($\Delta M \sim 1$ GeV), for different SUSY parameter values.

Production cross section limit contours are set by OPAL (figure 3).

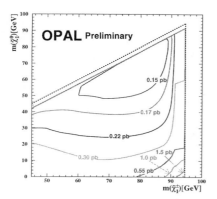

Figure 3: $\tilde{\chi}^+\tilde{\chi}^-$ production cross section upper limits (95 % confidence level) obtained at $\sqrt{s}$=189 GeV.

OPAL also presents exclusion region for non zero $\tilde{\chi}_1^0$ lifetimes (figure 4). In this case, one or both $\tilde{\chi}_1^0 \to \gamma\tilde{G}$ decay can occur outside the detector and escapes detection.

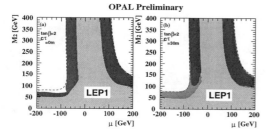

Figure 4: Excluded region for different $\tilde{\chi}_1^0$ lifetime $c\tau$. Dark shaded region is excluded from this analysis. Medium shaded region is excluded from standard $\tilde{\chi}^+\tilde{\chi}^-$ searches.

4 One and two photon final states

GMSB models predict final states with 1γ + missing energy $e^+e^- \to \tilde{\chi}_1^0\tilde{G} \to \gamma\tilde{G}\tilde{G}$ or 2γ + missing energy $e^+e^- \to \tilde{\chi}_1^0\tilde{\chi}_1^0 \to \gamma\gamma\tilde{G}\tilde{G}$.

These final states are also predicted by a no-scale supergravity model (LNZ)[6] which separates local ($\sqrt{F}$) and global SUSY breaking (m_0, $m_{3/2}$) and predicts the $\tilde{G}$ to be the LSP. The $\tilde{\chi}_1^0$ is the NLSP and is almost purely bino. This model has only 2 free parameters, the $\tilde{G}$ and $\tilde{\chi}_1^0$ masses.

The 4 LEP experiments are looking for photon(s) plus missing energy [7] and the LEP SUSY working group (LSWG) performs a combination of the individual results.

4.1 One photon final states

The table 3 shows the results from data taken at 183 GeV. The other backgrounds ($\gamma\gamma(\gamma)$, $\mu\mu\gamma$, $\ell\ell\nu\bar{\nu}\gamma$) and the cosmic rays contamination are negligible.

Preliminary results obtained at $\sqrt{s}$=189 GeV are shown in figure 5.

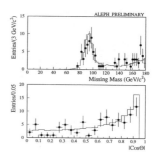

Figure 5: Missing mass spectrum and photon angular distribution. The full line shows the simulation obtained from KORALZ Monte Carlo.

Figure 6 displays the combination made by the LEP SUSY working group of all the 1 γ results.

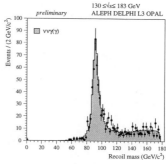

Figure 6: Recoil mass spectrum obtained by the 4 LEP collaborations in 1 γ events, compared to KORALZ predictions (from LSWG).

No evidence is observed for new physics and limits are derived. For this, additional cuts are applied: for each $\tilde{\chi}_1^0$ mass, the photon energy E_γ must be consistent with the kinematics derived from the expected signal.

Table 3: One photon search.

	ALEPH	DELPHI			L3		OPAL
Luminosity (pb^{-1})	58.5	50.2	44.3	52.0	55.3		54.5
θ_γ range (°)	18 - 162	45 - 135	11 - 30 150 - 169	3.8 - 8 172 - 176.2	43 - 137 14 - 36 144 -166	45 - 135	15 - 165
Energy cuts (GeV)	$p_t >$ .0375$\sqrt{s}$	$E_\gamma >$ .06$\sqrt{s}$	$E_\gamma >$ .2$\sqrt{s}$	$E_\gamma >$ .6$\sqrt{s}$	$E_\gamma > 5$	$10 > E_\gamma > 1.3$ $p_t > 1.3$	$E_\gamma >$ .05E_{beam}
Data	195	54	58	28	195	144	191
			140				
MC $\nu\bar{\nu}\gamma(\gamma)$	187	59.5	48.1	24.7	203.2	28.2	201.3
			132.3				
MC $e^+e^-\gamma$	0	0	3.1	2.2	1.2	107.4	0
			5.3				

DELPHI, L3 (figure 7) and OPAL have derived upper limits on $e^+e^- \to \tilde{\chi}_1^0\tilde{G} \to \gamma\tilde{G}\tilde{G}$ cross section.

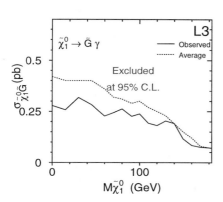

Figure 7: Upper limits at 95 % C.L. for the cross section of the process $e^+e^- \to \tilde{\chi}_1^0\tilde{G} \to \gamma\tilde{G}\tilde{G}$.

In the LNZ model, L3 has also computed the exclusion region in the plane of the 2 parameters of the model (figure 8).

If the gravitino is very light, the cross section for the process $e^+e^- \to \tilde{G}\tilde{G}\gamma$ can be sizeable. This scenario, with all other SUSY particles too heavy to be produced, has been recently studied. [8] The total cross section is given by:

$$\sigma = \frac{\alpha \, s^3}{320\pi^2|F|^4} \, I(E_{\gamma,\mathrm{min}}, |\cos\theta_\gamma|_{\mathrm{max}})$$

where integral I is dependent on the cuts on photon en-

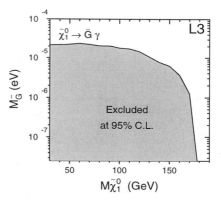

Figure 8: Upper limits at 95 % C.L. for the cross section of the process $e^+e^- \to \tilde{\chi}_1^0\tilde{G} \to \gamma\tilde{G}\tilde{G}$.

ergy and angle. One can derive a lower limit on the gravitino mass:

$$m_{\tilde{G}} > 3.8 \; 10^{-6}\mathrm{eV} \left[\frac{\sqrt{s}(\mathrm{GeV})}{200}\right]^{\frac{3}{2}} \left[\frac{I}{\sigma_{\mathrm{exp}}(\mathrm{pb})}\right]^{\frac{1}{4}}$$

Mass limits have been set by three experiments:

	$m_{\tilde{G}}$ (eV)
ALEPH	8.3 10^{-6}
DELPHI	6.6 10^{-6}
L3	8.0 10^{-6}

4.2 Two photon final states

The results from data collected at $\sqrt{s}$=183 GeV are displayed in table 4. The other backgrounds are negligible.

Table 4: Search for acoplanar two photon final states.

	ALEPH	DELPHI	L3	OPAL
θ_γ range (°)	18 - 162	γ_1 : 10 - 170 γ_2 : 25 - 155	43 - 137 14 - 36 144 -166	15 - 165
Energy cuts (GeV)	$E_\gamma > 1$	$E_\gamma > .05E_{beam}$	$E_\gamma > 1$	$E_\gamma > .05E_{beam}$ $E_\gamma > 1.5$ if $|\cos\theta| < 0.8$
Acopl. (°)	3.	3.	2.4	2.5
Data	9	10	14	9
MC ($\nu\bar{\nu}\gamma\gamma$)	10.8	10.2	13.3	8.2

Figure 9, from the LEP SUSY working group, displays the events selected by the 4 experiments.

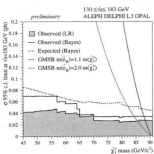

Figure 9: Recoil mass spectrum obtained by the 4 LEP collaborations in acoplanar 2 γ events (from LSWG).

No anomaly is found and limits are derived, imposing additional cuts to improve signal to background ratio. An upper limit on $\gamma\gamma\tilde{G}\tilde{G}$ cross section has been computed by the LEP SUSY working group (figure 10).

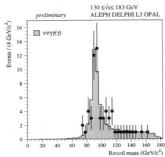

Figure 10: Upper limit cross section for the process $e^+e^- \rightarrow \tilde{\chi}_1^0\tilde{\chi}_1^0 \rightarrow \gamma\gamma\tilde{G}\tilde{G}$ (from LSWG).

The interpretation of the CDF event in $q\bar{q} \rightarrow \tilde{e}\tilde{e} \rightarrow e^+e^-\tilde{\chi}_1^0\tilde{\chi}_1^0 \rightarrow e^+e^-\gamma\gamma\tilde{G}\tilde{G}$ leads to specific region in the $m_{\tilde{e}}$, $m_{\tilde{\chi}_1^0}$ plane. Exclusion region (figure 11) in this plane can be derived from 2 γ plus missing energy results.

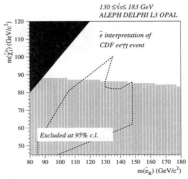

Figure 11: Exclusion mass plot from 2 γ and missing energy data (from LSWG).

5 Conclusion

No evidence for production of events compatible with Susy models predicting a $\tilde{G}$ as LSP and $\tilde{\chi}_1^0$ as NLSP has been reported by any of the four LEP experiments. CDF event interpretation in $\tilde{e}$ $\tilde{e}$ is almost ruled out.

References

1. P. Fayet, *Phys. Lett.* B117 (1982) 423,
C. Kolda *Nucl. Phys.* B Proc. Suppl (1998) 62.
2. G. Wolf, these Proceedings.
3. ALEPH Coll., CERN-EP/98-077, ICHEP98#950.
4. DELPHI Coll., DELPHI 98-66, ICHEP98#201.
5. OPAL Coll., OPAL PN-332, ICHEP98#332.
6. J.L. Lopez et al., *Phys. Rev.* D55 (1997) 5813.
7. ALEPH Coll., *Phys. Lett.* B429 (1998) 201, ICHEP98#953,
DELPHI Coll., DELPHI 98-76, ICHEP98#206,
DELPHI Coll., DELPHI 98-59, ICHEP98#207,
L3 Coll., L3 Note 2289, ICHEP98#494,
OPAL Coll., OPAL PN-351, ICHEP98#275.
8. A. Brignole et al., CERN-TH/97-339.

Searches in Light Gravitino Scenarios with Sfermion NLSP at LEP

G. Wolf

CERN, CH-1211 Geneva 23, Switzerland
E-mail: Gustavo.Wolf@cern.ch

A review of the main results obtained by the four LEP experiments on the search for new particles predicted in theories of Supersymmetry with light gravitinos is reported. It is also assumed that sfermions are the next-to-lightest supersymmetric particles. The report is based on data collected at LEP 2 centre-of-mass energies, $\sqrt{s} = 133$ to 189 GeV. No evidence for such signal is observed, although candidate events consistent with the expectations from Standard Model processes are observed. Limits at 95% C.L. on the production cross-section and on the particle masses obtained by the four LEP experiments are reported.

1 Introduction

An overview of the main results from the searches for new particles performed at LEP [1] in the framework of Light Gravitino Scenarios (LGS) is presented. The results obtained by the four LEP collaborations [2] are mainly based on the combination of all LEP 2 energies, ranging from $\sqrt{s} = 133$ to 183 GeV, and are in some cases updated to include the first ~ 40 pb^{-1} of data taken per experiment during the beginning of the '98 run of LEP.

In models including supersymmetry (SUSY), it is often assumed that the messengers of supersymmetry breaking couple to the observable sector with interactions of gravitational strength and that the SUSY breaking scale in the hidden sector is of the order of 10^{11} GeV. An alternative possibility is that supersymmetry is broken at some lower scale (below 10^7 GeV), and that the ordinary gauge interaction acts as the messenger of supersymmetry breaking [3,4].

In this case, the gravitino is naturally the lightest supersymmetric particle (LSP) and the lightest Standard Model superpartner is the next to lightest supersymmetry particle (NLSP). Thus, the NLSP is unstable and decays to its Standard Model (SM) partner and a gravitino.

Since the gravitino couplings are suppressed compared to electroweak and strong interactions, decays to the gravitino are in general only relevant for the NLSP and therefore the production of pairs of supersymmetric particles at high energy colliders would generally take place through Standard Model couplings. One exception to this rule is the process $e^+e^- \to Z^*/\gamma^* \to \tilde{G}\tilde{\chi}_1^0$, which will be discussed later.

Although most of the attention has been focused on the case where the neutralino is the NLSP, it is also possible that the NLSP is any other sparticle, and in particular a charged slepton [5,6,7,8]. Moreover, when left-right sfermion mixing occurs [9], the corresponding $\tilde{\tau}$ state, $\tilde{\tau}_1$, becomes the NLSP.

Throughout this work, it is assumed that the $\tilde{\tau}_1$ is the NLSP, and that the lightest neutralino, $\tilde{\chi}_1^0$, is the

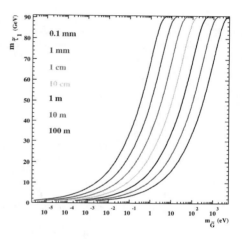

Figure 1: $\tilde{\tau}_1$'s mean decay length as a function of stau and gravitino masses for $\sqrt{s} = 183$ GeV/c^2.

next-to-NLSP (NNLSP). Its mean decay length is:

$$L = 1.76(E^2/m_{\tilde{\tau}_1}^2 - 1)^{\frac{1}{2}} \left(\frac{m_{\tilde{\tau}_1}}{100 \, \text{GeV}/c^2} \right)^{-5} \left(\frac{m_{\tilde{G}}}{1 \, \text{eV}/c^2} \right)^2 \mu\text{m}, \tag{1}$$

and depends strongly on the $\tilde{\tau}_1$ mass, $m_{\tilde{\tau}}$, on the mass of the gravitino, $m_{\tilde{G}}$, and of the energy of the $\tilde{\tau}_1$, E. Figure 1 shows the $\tilde{\tau}_1$'s mean decay length as a function of stau and gravitino masses for $\sqrt{s} = 183$ GeV/c^2.

The decay length of the stau will determine the final topology of the type of sparticle production being searched for. The following three sections deal respectively with searches for supersymmetric particles decaying promptly after production, in the middle of the detector, and outside the detector.

2 Searches for Promptly Decaying Sparticles

In this section, all the produced particles will be treated as decaying at the vertex, with the exception of the gravitino, which is stable. Four kinds of searches are possible in this case. Two of them, are already covered by searches in the MSSM framework:

2.1 Single Neutralino Production

For the case of ultralight gravitino scenarios ($m_{\tilde{G}} < \mathcal{O}(10^{-2} \text{ eV}/c^2)$), the production cross section of $e^+e^- \rightarrow \tilde{\chi}_1^0 \tilde{G}$ is big enough for this channel to be searched for. The signature is the same as the MSSM joint production of the lightest and *next-to*-lightest neutralinos, in the low m_0 scenario with massless lightest neutralino and the *next-to*-lightest neutralino playing the role of the LGS lightest neutralino.

2.2 Sleptons

The signature of stau pair production consists of two taus and two gravitinos. This signature and its associated kinematics are the same as in the case of MSSM stau pair production with massless neutralino. The same holds for the case of direct decays of selectrons (smuons) into and electron (muon) and a gravitino. For centre-of-mass energies up to 183 GeV, the four combined LEP experiments 95% C.L. limits on sleptons are: $m_{\tilde{\tau}} > 75 \text{ GeV}/c^2$, $m_{\tilde{\mu}} > 71 \text{ GeV}/c^2$ and $m_{\tilde{e}} > 85 \text{ GeV}/c^2$. The case in which selectron (smuon) decays through cascade decays ($\tilde{e} \rightarrow \tau \nu_\tau \nu_e \tilde{G}$ or $\tilde{e} \rightarrow \tau e \tau \tilde{G}$) is still not treated by any of the collaborations.

2.3 Charginos

In this case, the charginos would decay into a stau and a tau neutrino, with further decay of the stau into a tau and a gravitino. This case is equal to MSSM chargino production with massless neutralino and low m_0.

2.4 Neutralinos

The signature for the lightest neutralino pair production is four τs and missing energy. Only the DELPHI collaboration reports on this search. The main SM backgrounds come from the production of final states containing taus; and from $\gamma\gamma$ processes for the case of low mass difference between those of the stau and the neutralino. For centre-of-mass energies between 161 and 183 GeV, DELPHI expects 0.78±0.16 events and observes two. Figure 2 shows the 95 % C.L. upper limit on $\tilde{\tau}_1$ pair production as a function of stau and neutralino masses.

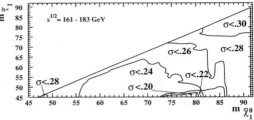

Figure 2: 95 % C.L. upper limit on $\tilde{\tau}_1$ pair production as a function of stau and neutralino masses for $\sqrt{s} = 161 - 183$ GeV/c^2.

3 Searches for Long Lived Particles

As can be seen from fig. 1, if the mass of the gravitino is between 10 eV and a KeV and the mass of the stau is between 70 and 90 GeV/c^2, its mean decay length will be between 1 cm and 10 m. It is in this context that the following searches are framed.

3.1 Sleptons

ALEPH and DELPHI present results on two kinds of slepton searches.

- If the sleptons decay before the tracking devices (O(1 mm) < Mean Decay Length < O(40 cm)), the search is oriented to look for displaced vertices or tracks with big impact parameter (IP).

- If the sleptons decay inside the tracking devices (O(10 cm) < Mean Decay Length < O(2 m)), searches are oriented too look for tracks with kinks (for an illustration, see fig. 3).

In the case of the IP searches, one of the mean backgrounds is cosmic rays. Fig. 4 illustrates how they are efficiently eliminated by requiring the ratio between the IPs of the two most energetic tracks in the event to be negative and between certain values (-1.5 and -0.5 for DELPHI). When the mean life-time is bigger, tracks with kinks are searched for. Specific cuts are designed to deal with the main sources of background: tracks badly reconstructed, hadronic interactions with the material in the detector, and tracks of charged particles that emit a hard photon. Slepton track candidates are also required to have hits in the vertex detectors. Quality cuts on the kinks are applied. ALEPH expect 0.32 events from background estimates at $\sqrt{s} = 183$ GeV, and observes one event. DELPHI expects $1.47^{+0.63}_{-0.32}$ at $\sqrt{s} = 161 - 183$ GeV and observes none in the data.

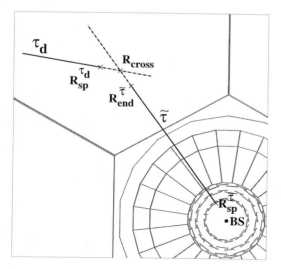

Figure 3: Sketch illustrating the reconstruction of a secondary vertex, shown in the plane perpendicular to the beam. The stau track (labelled with $\tilde{\tau}$) and the track of the decay product of the tau (labelled with τ_d) are extrapolated (dashed line). The extrapolated tracks define a crossing point at radius R_{cross}. $R_{sp}^{\tilde{\tau}}$ and $R_{end}^{\tilde{\tau}}$ are the radii of the first and last measured points of the $\tilde{\tau}$ track. $R_{sp}^{\tau_d}$ is the first measured point of track selected as the τ decay product. All the radii are measured with respect to the beam spot (BS).

3.2 Charginos

Chargino pair production has the same final signature than the slepton pair production: each chargino would decay into a stau and a neutrino, and each stau would in turn decay producing a track with big impact parameter, or with a kink.

3.3 Neutralinos

Neutralinos decay into a stau and a tau. The signature for this kind of events would be between two and four normal charged tracks, and one or two tracks with big impact parameter or a kink. None of the LEP collaborations reports on this search.

4 Search for Stable Sparticles

When the mean decay length of the sleptons exceeds 10 meters, only a very small fraction of them will decay inside the tracking volume of the detectors. For all practical purposes, sleptons are considered to be stable. All four LEP experiments report results for the slepton pair production channel. The chargino pair production channel looks identical to it, with one exception that will be

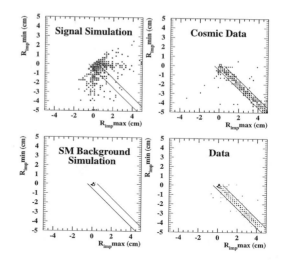

Figure 4: Impact parameters of the two most energetic tracks for Signal, Cosmic Rays recorded while LEP was not running, SM background simulation, and Data taken during LEP operation.

Table 1: Number of expected and observed events, and 95 % C.L. lower mass limits coming from stable slepton searches combining results from $\sqrt{s} = 130 - 183$ GeV runs.

	ALEPH	DELPHI	L3	OPAL
Expected	0.6	0.7	0.3	2
Observed	0	0	0	1
$\tilde{\tau}_R >$	81	81	80	82

pointed up in the next paragraph. The neutralino pair production channel is not yet investigated by any of the LEP Collaborations.

Stable sleptons are produced back to back in the transverse plane (charginos will in general be acoplanar due to the presence on neutrinos in their decays), do not interact strongly and are massive. Thus, such events should not contain neither hadronic nor electromagnetic showers. The main search tool is dE/dx analysis combined with qualitative and kinematical cuts based on the properties previously described. The results of these searches for $\sqrt{s} = 130 - 183$ GeV, and the combination for the four LEP Collaborations are shown in table 1 and fig. 5. The characterisation as $\tilde{\mu}$ could be changed for $\tilde{\tau}$ given that both particles share the same characteristics when both are stable. The combined 95 % C.L. lower mass limit is 86.5 GeV/c^2.

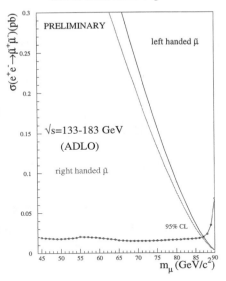

ADLO limits stable sleptons

Figure 5: 95 % C.L. lower mass limits on stable sleptons combining results from the four LEP experiments at $\sqrt{s} = 130 - 183$ GeV.

5   Combination of different searches

Once the whole range of possible gravitino masses is searched for for a given channel, results can be expressed as a function of the mass of the gravitino. As an example, fig. 6 shows ALEPH's 95% C.L. lower limits on the mass of the stau right.

Results can also be expressed within a certain model as exclusion plots in the mass space. DELPHI gives an interpretation in the frame of a very general model by Cheung et al.[6]. Figure 7 shows the combination of three searches for the case of sparticles decaying at the vertex for $\sqrt{s} = 161 - 189$ GeV. Above the diagonal line, the lightest neutralino is the NLSP, and neutralino pair production gives two acoplanar photons and missing energy as signature[10]. Below the diagonal line, the stau is the NLSP. Two searches are used to excluded different parts of the mass space: the search for neutralino pair production, and the MSSM search for staus with massless neutralino, as explained in subsection 2.2. Results are also shown for the case of gaugino-like neutralinos, and $n = 2$, where n is the number of messenger generations. An increase in n would in general tip the region in mass space allowed by theory towards the zone below the diagonal. As an example, for $n = 4$ there is no mass combination with $m_{\tilde{\tau}} > 80$ GeV/c^2 and $m_{\tilde{\tau}} > m_{nuno}$. The regions excluded by neutralino searches will also diminish when neutralinos are higgsino-like.

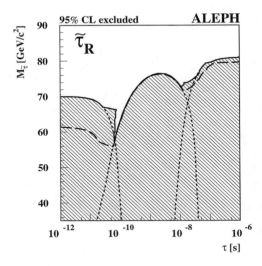

Figure 6: ALEPH's 95 % C.L. lower limits on the mass of the stau right as a function of stau's mean life-time.

Acknowledgements

I would like to express my gratefulness to the members of the four LEP Collaborations who gave me their plots and results to help me in preparing my talk and this proceedings.

References

1. *LEP Storage Ring*, Technical Notebook **CERN** (1989).

2. ALEPH Coll., Nucl. Instr. and Meth. **A360** (1995) 481;
   DELPHI Coll., Nucl. Instr. and Meth. **A378** (1996) 57;
   L3 Coll., Nucl. Instr. and Meth. **A289** (1990) 35;
   OPAL Coll., Nucl. Instr. and Meth. **A305** (1991) 275.

3. M. Dine, W Fischler and M. Srednicki, *Nucl. Phys.* B18981575 ; S. Dimopoulos and S. Raby, *Nucl. Phys.* B19281353 ; M. Dine and W. Fischler, *Phys. Lett.* B11082227 ; M. Dine and M. Srednicki, *Nucl. Phys.* B20282238 ; L. Alvarez-Gaumé, M. Claudson and M. Wise, *Nucl. Phys.* B2078296 ; C. Nappi and B. Ovrut, *Phys. Lett.* B11382175 .

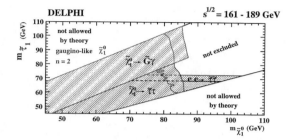

Figure 7: Areas excluded at 95% C.L. for $n = 2$ and gaugino-like neutralinos for the case of staus decaying at the vertex in the $m_{\tilde{\chi}_1^0}$ vs. $m_{\tilde{\tau}_1}$ plane. The positive-slope dashed area is excluded by the search for lightest neutralino pair production. The negative-slope dashed area is excluded by the search for $\tilde{\chi}_1^0 \rightarrow \gamma\tilde{G}$, and the point-hatched area by the direct search for stau pair production in the MSSM framework. The wiggled curve that indicates the limit from the search for neutralino pair production, is due to a rapid variation in the 95% C.L. limit on the production cross section. In all cases, the first limit line corresponds to the limit up to $\sqrt{s} = 183$ GeV, and the second to the preliminary update including $\sim 34pb^{-1}$ of data at $\sqrt{s} = 189$ GeV.

4. M. Dine and W. Fischler, *Nucl. Phys.* B20482346 ; S. Dimopoulos and S. Raby, *Nucl. Phys.* B21983479.

5. J. A. Bagger, K. Matchev, D. M. Pierce and R. Zhang, *Phys. Rev.* D55973188.

6. D. A. Dicus, B. Dutta, S. Nandi, *Phys. Rev.* D56975748 ; D. A. Dicus, B. Dutta, S. Nandi, *Phys. Rev. Lett.78973055.*

7. F. Borzumati, *hep-ph/9702307.*

8. G.F. Giudice, R. Rattazzi, *hep-ph/9801271,* Submitted to Phys. Rev. D.

9. A. Bartl et. al., *Z. Phys. C73(1997) 469.*

10. J. Fay, review # 1015, these proceedings.

SUSY Searches at the Tevatron Using Photons

M. PATERNO

Department of Physics
University of Rochester, Rochester, NY 14627-0251, USA
E-mail: paterno@fnal.gov
for the CDF and DØ Collaborations

We report on recent searches by the CDF and DØ collaborations for physics beyond the Standard Model, using events with one or more photons with high transverse momentum. These searches were conducted using the complete 1992–1996 data sample, corresponding to an integrated luminosity in excess of 100 pb^{-1} per experiment. In all cases, the searches have null results, with the possible exception of a single $ee\gamma\gamma \not{E}_T$ candidate seen by the CDF collaboration. These null results are interpreted to exclude parts of the parameter spaces of several models of supersymmetry.

1 Introduction

Supersymmetry (SUSY) is a symmetry relating fermions and bosons.[1] For every particle in the Standard Model, it introduces a partner with the same quantum numbers, but with spin differing by one half. Suggestions [2,3,4] have been made that SUSY may result in signatures containing one or more high transverse momentum photons. Some of this interest is a result of the observation of one event containing two high transverse energy photons, two electron candidates, and a large imbalance in transverse momentum ($\not{E}_T$) by the CDF collaboration.[5] We present a summary of recent analyses by the CDF and DØ collaborations searching for such signatures. In these analyses, we assume that R-parity is conserved. (R-parity [6] is a multiplicative quantum number, $+1$ for Standard Model particles and -1 for their superpartners.) Therefore, superparticles are produced in pairs, and the lightest supersymmetric particle (LSP) is stable.

In the minimal supersymmetric extention of the Standard Model (MSSM), the gaugino-Higgsino sector (excluding gluinos) is described by four parameters: M_1 (the U(1) gaugino mass parameter), M_2 (the SU(2) gaugino mass parameter), μ (the Higgsino mass parameter), and $\tan(\beta)$ (the ratio of the vacuum expectation values of the two Higgs doublets). The MSSM can be further constrained by demanding unification of the gaugino masses M_2 and M_1 at some large GUT scale, which provides the relation $M_1 = \frac{5}{3} M_2 \tan^2 \theta_W$. It can be sill further constrained by the requirement that the magnitude of μ (but not its sign) be determined by electroweak symmetry breaking. Each of these possibilities is considered by at least one of the following analyses.

Photons appear in a variety of SUSY models. For example, in models in which SUSY is broken by a gauge interaction (GMSB models), the LSP is the gravitino ($\widetilde{G}$). The mass of the gravitino is of order $4.2 \times 10^{-5} (\Lambda/500)^2$ eV, where Λ is the SUSY breaking scale in GeV. For a typical SUSY breaking scale of 100 TeV, this yields a gravitino mass of a few electron volts. In many GMSB models, the next lightest superpartner is the lightest neutralino, $\widetilde{\chi}_1^0$. In such models, the signature of SUSY production is the production of two $\widetilde{\chi}_1^0$s (either directly, or through the decay of other SUSY particles), with the subsequent decay $\widetilde{\chi}_1^0 \rightarrow \gamma\widetilde{G}$; the observable effect is the production of two high transverse momentum photons and missing transverse momentum.

Photons can also appear in models without GMSB (in which the $\widetilde{\chi}_1^0$ is the LSP), for example, in the radiative decay $\widetilde{\chi}_2^0 \rightarrow \gamma\widetilde{\chi}_1^0$. Such a model has been proposed [4] as a possible explanation for the CDF $ee\gamma\gamma\not{E}_T$ candidate. Within the MSSM, the radiative decay $\widetilde{\chi}_2^0 \rightarrow \gamma\widetilde{\chi}_1^0$ dominates in the region of parameter space defined by $50 \leq M_1 \sim M_2 \leq 100$ GeV/c^2, $1 \leq \tan(\beta) \leq 3$, and $-65 \leq \mu \leq -35$ GeV/c^2.[7]

The SUSY searches presented here fall into two types: diphoton searches, covered in Section 2, and photon + $\not{E}_T$ searches, covered in Section 3.

2 SUSY Searches Using Two Photons

2.1 CDF Results

The CDF collaboration has published [5] a search for events with two high transverse energy photons. Using a data sample corresponding to an integrated luminosity of 85 pb^{-1}, they searched for events with two isolated photons with $E_T > 12$ GeV in the central region of production ($|\eta^\gamma| < 1.0$). Each photon candidate was required to have no associated track with $p_T > 1$ GeV/c, and no more than one associated track with $p_T < 1$GeV/c, a shower shape consistent with that of a single photon, and no other photon candidate within the same 15° segment of the calorimeter. To maintain the projective geometry of the calorimeter, the primary collision vertex was required to be within 60 cm of the center of the detector. To suppress cosmic rays, events were required to have no central hadronic calorimeter tower with $E > 1$ GeV out of time with the collision. For events in which both photon candidates had $E_T > 22$ GeV, the fiducial and isolation cuts were loosened. The efficiency for identifying an isolated photon was measured with a control sample of $Z \to ee$ decays to be $68 \pm 3\%$ for the 12 GeV selection criteria, and $84 \pm 4\%$ for the 22 GeV criteria. The above selections yield a diphoton sample of 2239 events; the fraction of this total measured to contain two prompt photons is $15 \pm 4\%$.

In order to minimize the effect on $\not{\!\!E}_T$ of fluctuations in jet measurements, events with a jet with uncorrected $E_T > 10$ GeV within $10°$ in azimuth of the $\not{\!\!E}_T$ were rejected. Standard Model processes that contribute to this background contain no intrinsic $\not{\!\!E}_T$, and the expected distribution of $\not{\!\!E}_T$ should therefore correspond to the resolution of the detector. Figure 1 shows the $\not{\!\!E}_T$ distribution of both the $E_T > 12$ GeV and $E_T > 25$ GeV samples, compared with that predicted from the detector resolution. Neither distribution shows a significant excess of events beyond the prediction. Note that the events in the high-threshold sample are a subset of those in the low-threshold sample. The one event containing large $\not{\!\!E}_T$ (55 ± 7 GeV) appears in both plots. For a photon threshold of 12 GeV and $\not{\!\!E}_T > 35$ GeV, 0.5 ± 0.1 events were expected from SM sources.

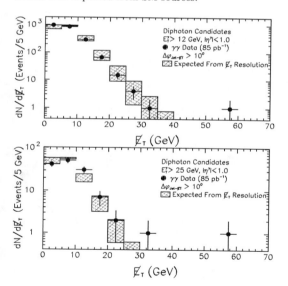

Figure 1: $\not{\!\!E}_T$ spectra for events with two central photons, from the CDF experiment. The top (bottom) plot contains events with both photons with E_T above 12 (25) GeV.

The event at high $\not{\!\!E}_T$ has, in addition to the two high E_T photons and large $\not{\!\!E}_T$, a central electron and an electromagnetic cluster in the plug calorimeter that passes the electron selection criteria used for Z-boson identification.[8] The total p_T of the 4-cluster system is 48 ± 2 GeV/c, opposite to the $\not{\!\!E}_T$, and in good agreement with the measured magnitude, implying the imbalance in p_T is intrinsic to the 4-cluster system. Although the cluster passes all standard electron selection criteria, there is no track in the silicon vertex system pointing directly at it, as would be expected if the cluster were due to an electron. Interpretation of the cluster as coming from an isolated photon, the hadronic decay of a τ, or a jet, while possible, each have expectations less than a few percent. The origin of the cluster remains unclear.

CDF used these data to set limits on two SUSY models, one in which the photons appear from the radiative decay $\widetilde{\chi}_2^0 \to \gamma \widetilde{\chi}_1^0$, and one model with light gravitinos, in which the photons result from the decay $\widetilde{\chi}_1^0 \to \gamma \widetilde{G}$. Comparison with SUSY predictions was done using simulated event samples created with SPYTHIA.[9]

To investigate the $\widetilde{\chi}_2^0 \to \gamma \widetilde{\chi}_1^0$ model of Ref. 3, with $m_{\widetilde{\chi}_1^0} = 36.6$ GeV/c^2 and $m_{\widetilde{\chi}_2^0} = 64.6$ GeV/c^2, CDF simulated both direct production and cascade decays, yielding a prediction of 2.4 events passing the selection criteria of $E_T^\gamma > 12$ GeV and $\not{\!\!E}_T > 35$ GeV. In the data, one event passes these requirements; consequently, this model could not be ruled out at the 95% confidence level (C.L.). Without background subtraction, the resulting 95% C.L. upper limit on the cross section is 1.1 pb.

In the model with a light gravitino, diphoton production is dominated by $\widetilde{\chi}_1^\pm \widetilde{\chi}_2^0$ and $\widetilde{\chi}_1^\pm \widetilde{\chi}_1^\mp$ states. In the MSSM, further constrained to have the magnitude of μ determined by electroweak symmetry breaking, and by requiring gauge unification for the SU(2) and U(1) symmetries, the parameter space important to the gaugino sector (excluding gluinos) is spanned by M_2, $\tan(\beta)$, and the sign of μ. Variation of these parameters over the range $1 < \tan(\beta) < 25$, $M_2 < 200$ GeV/c^2, for both signs of μ, yields the range of limits on cross sections shown in Figure 2.

2.2 DØ Results

The DØ collaboration has published[10] a search for events with two photons of large transverse energy and large $\not{\!\!E}_T$. This study was conducted in the context of a model with a light gravitino, with the assumption of gaugino mass unification at the GUT scale.

The $\widetilde{\chi}_1^0$ is assumed to be short-lived, and to decay to $\gamma \widetilde{G}$ in the volume of the detector, with branching fraction of 100%. Since R-parity conservation is assumed, the noninteracting $\widetilde{G}$ is stable. Pair production of gauginos thus yields events with high E_T photons and large $\not{\!\!E}_T$, with or without jets.

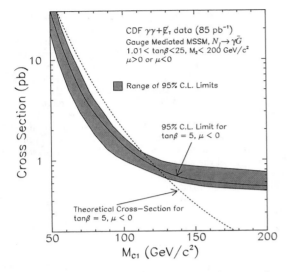

Figure 2: From CDF, the 95% C.L. upper limits on the cross section for the production of $\widetilde{\chi}_1^\pm \widetilde{\chi}_1^\mp$ and $\widetilde{\chi}_1^\pm \widetilde{\chi}_2^0$ in a model with a light gravitino. The shaded region shows the range of limits on cross sections as the parameters are varied within the limits $1 < \tan(\beta) < 25$ and $M_2 < 200$ GeV/c^2, for both signs of μ. The solid (dashed) line shows the limit (predicted cross section) for the lowest value of $m_{\widetilde{\chi}_1^\pm}$ that is excluded: $m_{\widetilde{\chi}_1^\pm} < 120$ GeV/c^2 for $\tan(\beta) = 5$ and $\mu < 0$.

The data for this analysis were collected during the 1992–1996 run of the Tevatron collider and represent an integrated luminosity of 106.3 ± 5.6 pb^{-1}. Photon identification consisted of a two-step process: the selection of isolated electromagnetic (EM) clusters, and the rejection of electrons. EM clusters were defined as calorimeter energy clusters with at least 95% of the energy of the cluster deposited in the EM section of the calorimeter. Also, photon EM clusters were required to have the longitudinal and transverse shower shapes consistent with that of a single photon or electron, and were required to be isolated in the calorimeter. Electrons were then suppressed by rejecting those clusters having either a reconstructed track or a large number of tracking-chamber hits in the road between the calorimeter cluster and the event vertex.

To be selected as a $\gamma\gamma\not{E}_T$ candidate, an event was required to have two identified photons, one with $E_T > 20$ GeV and the second with $E_T > 12$ GeV, each in the pseudorapidity interval $|\eta| < 1.2$ or $1.5 < |\eta| < 2.0$, and $\not{E}_T > 25$ GeV. Two events satisfied all requirements.

The principal backgrounds are from Standard Model processes with misidentified photons or mismeasured $\not{E}_T$, or both. The background due to mismeasured $\not{E}_T$

was determined from the data using events with EM clusters satisfying looser requirements than used for photon identification, and for which at least one cluster failed the shower profile requirement; these events are expected to have a $\not{E}_T$ resolution similar to that of the diphoton sample. By normalizing the number of events in this sample with $\not{E}_T < 20$ GeV to that observed in the diphoton sample, DØ obtained a background estimate of 2.1 ± 0.9 due to $\not{E}_T$ mismeasurement for $\not{E}_T > 25$ GeV.

Other background sources include events with real $\not{E}_T$ and with an electron misidentified as a photon. Using the probability for an electron to be misidentified as a photon ($0.45 \pm 0.08\%$, as determined from $Z \to ee$ data), DØ estimated a total of 0.2 ± 0.1 events from such sources, for a total background expectation of 2.3 ± 0.9 events. The $\not{E}_T$ distribution of the data (before imposition of the $\not{E}_T$ cut) and the expectation from background are compared with that of two simulated SUSY samples in Figure 3.

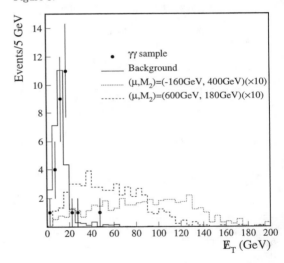

Figure 3: $\not{E}_T$ distribution of the $\gamma\gamma$ and background samples from the DØ experiment. The number of events with $\not{E}_T < 20$ GeV in the background sample is normalized to that in the $\gamma\gamma$ sample. Also shown are the expected distributions (multiplied by 10) from two representative points in the SUSY parameter space, with $\tan(\beta) = 2$.

DØ modeled chargino and neutralino production and decay using SPYTHIA. They explored the parameter space in the (μ, M_2) plane, while keeping $\tan(\beta)$ fixed, generating $\widetilde{\chi}_i^0 \widetilde{\chi}_j^0$, $\widetilde{\chi}_i^0 \widetilde{\chi}_j^\pm$ and $\widetilde{\chi}_i^\pm \widetilde{\chi}_j^\mp$ events for many points in the (μ, M_2) mass plane. The total signal efficiency varied from $\sim 0.01\%$ to $\sim 26\%$, depending on the masses of $\widetilde{\chi}_1^\pm$ and $\widetilde{\chi}_1^0$. With two events observed and 2.3 ± 0.9 events expected from background, the result was used to calculate a 95% C.L. upper limit on the cross section for chargino and neutralino production; the upper limit varied from several hundred pb for light charginos or neutralinos to 0.18 pb for heavy charginos or neutralinos. Figure 4 shows this limit expressed in the (μ, M_2) parameter plane, for values of $\tan(\beta)$ from 1.05 to 100. DØ obtains a lower limit for the mass of $\widetilde{\chi}_1^{\pm}$ of 150 GeV/c^2(and of 77 GeV/c^2 for $\widetilde{\chi}_1^0$), at the 95% C.L.

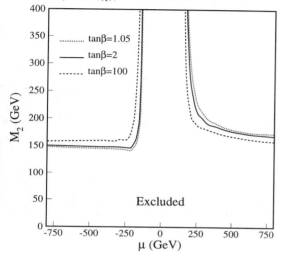

Figure 4: From DØ, the 95% C.L. bounds in the (μ, M_2) plane for $\tan(\beta) = 1.05$ (dotted line), $\tan(\beta) = 2$ (solid line), and $\tan(\beta) = 100$ (dashed line).

3 SUSY Searches Using One Photon

3.1 CDF Results

The CDF collaboration has preliminary results from a search for production of light scalar top quarks $\tilde{t}_1$ in a model including radiative decay of the $\widetilde{\chi}_2^0$. In such a model, the chargino would decay via $\widetilde{\chi}_1^{\pm} \to b\tilde{t}_1 \to b(c\widetilde{\chi}_1^0)$, and so $\widetilde{\chi}_1^{\pm}\widetilde{\chi}_2^0$ production would have the signature $\gamma bc\not{\!\!E}_T$. The analysis was not presented at this conference; a description may be found in Ref. 11.

3.2 DØ Results

DØ has submitted for publication [12] a search for events with one high transverse energy photon, two or more jets, and large missing E_T, motivated principally by recent suggestions that SUSY may result in signatures involving one or more photons in association with multiple jets and $\not{\!\!E}_T$.[3,4] DØ interprets the result in terms of squark and

gluino production in the context of SUSY models with a dominant $\widetilde{\chi}_2^0 \to \gamma \widetilde{\chi}_1^0$ decay.

Within the MSSM, the radiative decay $\widetilde{\chi}_2^0 \to \gamma \widetilde{\chi}_1^0$ dominates[7] in the region of parameter space defined by $50 \leq M_1 \sim M_2 \leq 100$ GeV/c^2, $1 \leq \tan(\beta) \leq 3$, and $-65 \leq \mu \leq -35$ GeV/c^2.

The data used in this analysis were collected with the during the 1992–1996 run of the Tevatron, and represent an integrated luminosity of 99.4 ± 5.4 pb^{-1}. To be selected as a $\gamma \not{\!\!E}_T + 2$ jets candidate, an event was required to have at least one identified photon with $E_T > 20$ GeV and $|\eta| < 1.1$, or $1.5 < |\eta| < 2.0$, and two or more jets reconstructed with cones of radius $\mathcal{R} = 0.5$, having $E_T > 20$ GeV and $|\eta| < 2.0$. The $\not{\!\!E}_T$ distribution for events satisfying these requirements is compared with the prediction from Standard Model backgrounds and two simulated SUSY samples in Figure 5. The turn-on at low $\not{\!\!E}_T$ is caused by a trigger requirement of $\not{\!\!E}_T > 14$ GeV; to avoid this region, $\not{\!\!E}_T > 25$ GeV was required in the analysis. A total of 318 events satisfied all requirements.

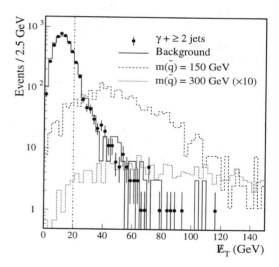

Figure 5: $\not{\!\!E}_T$ distribution for $\gamma + 2$ jet data from DØ (solid circles) compared with predicted background (solid histogram). The number of events in the background is normalized to the $\gamma + 2$ jets sample for $\not{\!\!E}_T < 20$ GeV, the region to the left of the vertical dot-dashed line. Also shown (as dashed and dotted histograms) are the distributions expected from SUSY production for $m_{\tilde{q}} = m_{\tilde{g}} = 150$ GeV/$c^2$and 300 GeV/c^2.

The principal backgrounds to the signal are QCD direct photon and multijet events, with mismeasured $\not{\!\!E}_T$ and either a real photon or a jet misidentified as a photon. Smaller backgrounds include W + jets events, in which the electron from $W \to e\nu$ is misidentified as a photon, and W + jets events, with $W \to \ell\nu$ and one of the jets misidentified as a photon. The expected number of events from these backgrounds was estimated from data. Above $\not{E}_T = 25$ GeV, the total expectation is 320 ± 30 events; 318 events are observed in the data.

To optimize the selection criteria for the detection of a signal from SUSY, DØ simulated squark and gluino pair production, and production in association with charginos and neutralinos, using SPYTHIA, with $M_1 = M_2 = 60$ GeV/c^2, $\tan(\beta) = 2$, and $\mu = -40$ GeV/c^2, yielding $m_{\widetilde{\chi}_1^0} = 34$ GeV/c^2, $m_{\widetilde{\chi}_2^0} = 60$ GeV/c^2, and a branching fraction B of 100% for $\widetilde{\chi}_2^0 \to \gamma\widetilde{\chi}_1^0$. Defining H_T as the scalar sum of the E_T of all jets with $E_T > 20$ GeV and $|\eta| < 2.0$, the sensitivity to SUSY was increased by selecting $\not{E}_T$ and H_T cutoffs to maximize the ratio of signal efficiency to uncertainty in the expected background. To ensure sensitivity for both high and low squark and gluino masses, the optimization was performed for $m_{\tilde{q}} = m_{\tilde{g}} = 150$ GeV/c^2 and 300 GeV/c^2; the optimum $(\not{E}_T, H_T)$ values were (35, 100) and (45, 220) at these two points, respectively. The lower cutoffs were used for squark and gluino masses below 200 GeV/c^2, and the higher values above those masses. The number of events passing these cutoffs were 60 and 5, with 75 ± 15 and 8 ± 6 events expected from Standard Model processes.

Having observed no excess, DØ used this result to set a 95% C.L. upper limit on $\sigma(p\bar{p} \to \tilde{q}/\tilde{g} \to \widetilde{\chi}_2^0 + X) \times B(\widetilde{\chi}_2^0 \to \gamma\widetilde{\chi}_1^0)$. Figure 6 shows the limit for the case $m_{\tilde{q}} = m_{\tilde{g}}$, together with the cross section to leading order, calculated with SPYTHIA. This result excludes squarks and gluinos of equal mass below 310 GeV/c^2 at the 95% C.L. For the case of heavy squarks, the resulting limit on the gluino mass is 240 GeV/c^2; in the case of heavy gluinos, the resulting limit on the squark mass is 240 GeV/c^2. These limits constrain the models proposed in Ref. 4, but do not exclude all of them.

4   Conclusion

The CDF and DØ experiments have performed searches for supersymmetry in channels involving one or more photons. Although the CDF collaboration has one remarkable event that is difficult to explain, no evidence for supersymmetry has yet been found. In the absence of such evidence, both collaborations have established new cross section and mass limits in the context of various SUSY models.

Both collaborations are now preparing for the next run of the Tevatron collider, with a greatly increased luminosity. It is clear that SUSY searches involving photons will remain a topic of great interest into the next data collection period.

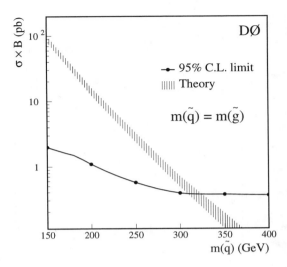

Figure 6: From the DØ experiment, the upper limit at the 95% C.L. on $\sigma \times B$ as a function of $m_{\tilde{q}/\tilde{g}}$, assuming equal squark and gluino masses. The hatched band represents the range of expected cross sections for different sets of MSSM parameters, consistent with the constraints $B(\widetilde{\chi}_2^0 \to \gamma\widetilde{\chi}_1^0) = 100\%$, and $m_{\widetilde{\chi}_2^0} - m_{\widetilde{\chi}_1^0} > 20$ GeV/c^2.

Acknowledgements

The author thanks members of the CDF and DØ collaborations for help in the preparation of this paper.

References

1. Yu.A. Golfand and E.P. Likhtman, *JETP Letters* **13**, 323 (1971); D.V. Volkov and V.P. Akulov, *Phys. Lett.* B **46**, 109 (1973); J. Wess and B. Zumino, *Nucl. Phys.* B **70**, 39 (1974).

2. H.E. Haber, G.L. Kane and M. Quiros, *Phys. Lett.* B **160**, 297 (1985).

3. S. Dimopoulos, S. Thomas and J.D. Wells, *Phys. Rev.* D **54**, 3283 (1996); S. Dimopoulos, M. Dine, S. Raby and S. Thomas, *Phys. Rev. Lett.* **76**, 3494 (1996); K.S. Babu, C. Kolda and F. Wilczek, *Phys. Rev. Lett.* **77**, 3070 (1996); J.L. Lopez, D.V. Nanopoulos and A. Zichichi, *Phys. Rev. Lett.* **77**, 5168 (1996); S. Ambrosanio *et al.*, *Phys. Rev.* D **54**, 5395 (1996); H Baer, M. Brhlik, C.-H. Chen and X. Tata, *Phys. Rev.* D **55**, 4463 (1997).

4. S. Ambrosanio *et al.*, *Phys. Rev. Lett.* **76**, 3498 (1996); S. Ambrosanio *et al.*, *Phys. Rev.* D **55**, 1372 (1997).

5. F. Abe *et al.* (CDF Collaboration), *Phys. Rev.*

Lett. **81**, 1791 (1998); F. Abe *et al.* (CDF Collaboration), hep-ex/9806034, submitted to *Phys. Rev. D.*

6. G.R. Farrar and P. Fayet, *Phys. Lett.* B **76**, 575 (1978).

7. S. Ambrosanio and B. Mele, *Phys. Rev.* D **55**, 1399 (1997), and erratum, *Phys. Rev.* D **56**, 3157 (1997).

8. F. Abe *et al.* (CDF Collaboration), *Phys. Rev.* D **52**, 2624 (1995).

9. T. Sjöstrand, *Comp. Phys. Commun.* **82**, 74 (1994); S. Mrenna, *Comp. Phys. Commun.* **101**, 232 (1997).

10. B. Abbott *et al.* (DØ Collaboration), *Phys. Rev. Lett.* **80**, 442 (1998).

11. D. Toback, "CDF Searches for New Phenomena", FERMILAB-CONF-98/183-E.

12. B. Abbott *et al.* (DØ Collaboration), hep-ex/9808010v3, submitted to *Phys. Rev. Lett.*

SEARCH FOR SUSY WITH $\not{R}_p$ AT LEP

G. GANIS

Max Planck Institut für Physik, Föhringer Ring 6, 80805 München, GERMANY
E-mail: Gerardo.Ganis@cern.ch

Searches for supersymmetry at LEP allowing for $\not{R}_p$ are reviewed. The results are compared with the R_p conserving scenario.

1 Introduction

The searches for supersymmetry performed at LEP are guided by a minimal supersymmetric extension of the Standard Model (MSSM) [1], based on a minimal number of additional degrees of freedom and the addition of soft supersymmetry breaking terms, i.e. those breaking the mass degeneracy inside the supermultiplets preserving the relevant advantages of the theory. Furthermore, in order to insure at tree level lepton (L) and baryon (B) number conservation - which, contrarily to the Standard Model, in supersymmetry is not guaranteed by gauge invariance - the conservation of the so called R_p parity is required [2], R_p being defined as $(-1)^{3B+L+2S}$ with S the particle spin.

The so far negative results of these searches have pushed the experimental groups to consider less minimal models. Among these, those obtained allowing some degree of R_p violation are particularly interesting because they predict significant modifications of the phenomenology of the reactions searched for.

This report summarises the most recent results produced by the LEP collaborations on the subject, focusing on the main ideas. Details can be found in [3]. ALEPH results are based on ~ 200 pb^{-1} collected at $\sqrt{s} = 91 - 189$ GeV; L3 results on ~ 76 pb^{-1} collected at $\sqrt{s} = 161 - 183$ GeV; DELPHI and OPAL results on ~ 56 pb^{-1} collected at $\sqrt{s} = 183$ GeV.

2 Basic Model and assumptions

The basic model used as guideline contains the minimal set of additional particles: two charginos, $\chi^{\pm}, \chi_2^{\pm}$, partners of $W^{\pm}$ and $H^{\pm}$; four neutralinos, χ, χ', χ'', χ''', partners of γ, Z, h, H, A; seven complex scalars per family, squarks ($\tilde{u}_L, \tilde{d}_L, \tilde{u}_R, \tilde{d}_R$), sleptons ($\tilde{l}_L^-, \tilde{l}_R^-$), sneutrinos ($\tilde{\nu}$), partners of quarks, charged leptons and neutrinos. The gluinos (the eight partners of gluons) and the gravitino (partner of graviton) are assumed to play no rôle at the energies involved. The masses of the Higgs bosons and of the supersymmetric partners are determined in terms of μ, the Higgs doublet mixing term, $\tan \beta$, the Higgs *vev* ratio, M_1 and $M_2 = 3/5 \cot^2 \theta_W M_1$,

the soft supersymmetry breaking terms associated with $U(1)_Y$ and $SU(2)_L$, and M_0, a common GUT scale mass for the scalar partners of the matter fermions.

The $\not{R}_p$ lagrangian terms can be cast in the form [4]

$$\frac{\lambda_{ijk}}{2} L_i L_j E_k^c + \lambda'_{ijk} L_i Q_j D_k^c + \frac{\lambda''_{ijk}}{2} U_i^c D_j^c D_k^c + \mu_i L_i H_u \quad (1)$$

where i, j, k are generation indices, L, Q and E, U, D are, respectively, the lepton and quark $SU(2)$-doublet and $SU(2)$-singlet superfields, and H_u is the Higgs doublet giving mass to the up-like fermions. Symmetry properties of the λ_{ijk}'s and λ''_{ijk}'s reduce the total number of new independent complex parameters to 48. The main phenomenological consequence of the μ_i terms are contributions to neutrino masses and oscillations [5], and are therefore not considered in LEP analyses.

The complexity of the new scenario is reduced by assuming that only one coupling at a time is non-zero. This is a sufficient (although not necessary) condition to cope with the constraints derived from low energy experiments on the λ couplings [a] or on their products [6] [b].

In e^+e^- environments the most straightforward consequence of the $\not{R}_p$ terms is a modification of the final states deriving from pair production of supersymmetric particles. Non vanishing $\not{R}_p$ couplings allow the Lightest Supersymmetric Particle (LSP) to decay into Standard Model (SM) particles. Consequently, usual cosmological arguments demanding a neutral and weakly interacting LSP [7] do not apply any longer. The nature of the LSP determines the sensitivity in λ of the searches. In LEP analyses it is required that the LSP decays within ~ 1 cm from the interaction point. This implies $|\lambda| > 10^{-4}$ if the lightest neutralino is the LSP, or $|\lambda| > 10^{-7}$ if one of the sfermions is the LSP[c]. Lower $|\lambda|$ values give either a long-lived LSP, behaving as in the R_p conserving

---

[a] Except when the family structure is specified, in the following the symbol "λ" will denote any of the Yukawa couplings.
[b] Typical upper limits on $|\lambda|$'s are of the order of a few percent; some products are much more constrained: for example, present lower limits on proton lifetime require $\lambda'_{11k}\lambda''_{11k} < 10^{-22}$.
[c] Assuming the GUT relation for gaugino masses an LSP chargino exists only for $M_{\chi^{\pm}} < M_Z/2$, a mass region ruled out by the Z-lineshape measurement; however, it is noted that the developed searches cover also chargino LSP topologies.

Figure 1: Typical sensitivity regions in the $(|\lambda|, M_{\tilde{f}})$ plane, for current experiments and different experimental techniques. Here "e^+e^- 2f" and "Single Prod" stand, respectively, for two-fermion cross sections and single χ production in e^+e^- interactions. The gaugino curves come from calculating the χ lifetime assuming that M_χ is small enough for production at LEP.

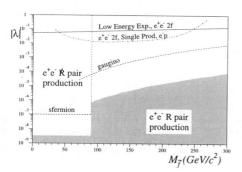

$|\lambda|$

Low Energy Exp., e^+e^- 2f

e^+e^- 2f, Single Prod, e^-p

gaugino

e^+e^- $\not{R}$ pair production

sfermion

e^+e^- R pair production

$M_{\tilde{f}}(GeV/c^2)$

Table 1: New direct decay modes of SUSY particles mediated by $\not{R}_p$ couplings. Here i, j, k are generations indices. For lines with more than one particle in the first column, the relevant decays are indicated by $\rightarrow$ in the second; for example, $\tilde{d}_{jL}$ decays via LQD into $\bar{\nu}_i d_k$.

LLE (λ_{ijk})	
χ	$\rightarrow \bar{\nu}_i l_j^+ l_k^-, \bar{\nu}_j l_i^+ l_k^- \nu_i l_j^- l_k^+, \nu_j l_i^- l_k^+$
χ^+	$\rightarrow \nu_i \nu_j l_k^+, l_i^+ l_j^+ l_k^-, l_i^+ \nu_j \nu_k, \nu_i l_j^+ \nu_k$
$\tilde{l}_{iL}^-, \tilde{l}_{kR}^-, \tilde{\nu}_i$	$\rightarrow \nu_j l_k^-, \rightarrow \nu_i l_j^-, \nu_j l_i^-, \rightarrow l_j^+ l_k^-$
LQD (λ'_{ijk})	
χ	$\rightarrow l_i^- u_j \bar{d}_k, l_i^+ \bar{u}_j d_k, \nu_i d_j \bar{d}_k, \bar{\nu}_i \bar{d}_j d_k$
χ^+	$\rightarrow \nu_i u_j \bar{d}_k, l_i^+ \bar{d}_j d_k, l_i^+ \bar{u}_j u_k, \bar{\nu}_i \bar{d}_j u_k$
$\tilde{d}_{kR}, \tilde{d}_{jL}, \tilde{u}_{jL}$	$\rightarrow \bar{\nu}_i d_j, l_i^- u_j, \rightarrow \bar{\nu}_i d_k, \rightarrow l_i^+ d_k$
$\tilde{l}_{iL}^-, \tilde{\nu}_i$	$\rightarrow \bar{u}_j d_k, \rightarrow d_j \bar{d}_k$
UDD (λ''_{ijk})	
χ	$\rightarrow u_i d_j d_k, \bar{u}_i \bar{d}_j \bar{d}_k$
χ^+	$\rightarrow u_i u_j d_k, u_i d_j u_k, \bar{d}_i \bar{d}_j \bar{d}_k$
$\tilde{u}_{iR}, \tilde{d}_{kR}, \tilde{d}_{jR}$	$\rightarrow \bar{d}_j \bar{d}_k, \rightarrow \bar{u}_i \bar{d}_j, \rightarrow \bar{u}_i \bar{d}_k$

case, or a medium-lived LSP, decaying inside the detector volumes and giving topologies not yet addressed [d]. In the case of a neutralino LSP, for most of the couplings the decay length constraint can be satisfied only if $M_\chi > 10$ GeV/c^2.

The typical experimental situation is summarised in Figure 1 as a function of the relevant parameters: the coupling λ and the mass $M_{\tilde{f}}$ of the scalar coupling to the $\not{R}_p$ vertex. The sensitivity of low energy experiments to large λ values decreases almost linearly as $M_{\tilde{f}}$ increases. As it will be discussed briefly in Sect. 4, other signatures can be studied at e^+e^- colliders which can be sensitive to lower values of $|\lambda|$, for $M_{\tilde{f}}$ below the effective centre of mass energy of the experiment. The same is valid for leptoquark searches at ep colliders [9]. Depending on the MSSM parameters, pair production in e^+e^- can be sensitive to much lower $|\lambda|$ values. Also indicated is the region possibly covered by R_p conserving searches.

In the next section the searches for topologies arising from pair production are discussed in some detail. Single production, example of an alternative signature studied at LEP, is discussed in Sect. 4.

3 Pair Production

As in the case of R_p conservation, supersymmetric particles are mostly expected to be pair-produced via s-channel exchange of γ, Z and, for gaugino and slepton pairs, t-channel exchange of the relevant slepton or gaug-

[d]Searches for heavy particles decaying inside the detector volume have been performed by ALEPH and DELPHI [8] but not yet interpreted in the $\not{R}_p$ scenario.

ino. $\not{R}_p$ couplings involving electrons allow additional t-channel contributions which, given the existing limits on the couplings, are expected to be small and neglected. Therefore, total production cross sections follow the usual pattern: typically large for gauginos, allowing in some cases to push the sensitivity very close to the kinematic limit; smaller for sfermions, where the sensitivity is limited by the integrated collected luminosity.

3.1 Decay phenomenology and final state topologies

The decay modes of the supersymmetric particles mediated by the $\not{R}_p$ couplings (the so called *direct* decay modes) are given in table 1. Depending on the size of the couplings and on the nature of the LSP, *indirect* decays, i.e. those proceeding in a first stage via the lightest neutralino, can still play a rôle. Direct and indirect decays are treated separately since they can lead to topologies selected with different strategies and performances.

Table 2 gives the possible topologies expected from pair production. It is noted that, despite of the LSP instability, in many cases final state neutrinos can carry away a significant fraction of energy.

3.2 Selections

The complexity of the situation depicted in Table 2 renders necessary to develop many selections, one for every corner of the space of the relevant topological and dynamical quantities.

New signals must be disentangled from the expectations from the Standard Model. The most problematic

Table 2: Topologies arising from pair production; here l stands for any charged lepton, q for any quark and $\not{E}$ indicates the presence of undetected particles.

	LLE	LQD	UDD
$\chi^+\chi^-$	$6l$, $4l+\not{E}$, $2l+\not{E}$, $6l+\not{E}$, $5l+2q+\not{E}$, $4l+4q+\not{E}$	$2nq+2l$, $2nq+\not{E}$, $2nq+l+\not{E}$ (n=2,3), $8q+2l$, $8q+ml+\not{E}$ (m≤1)	$6q$, $8q$, $10q$, $8q+l+\not{E}$, $6q+2l+\not{E}$
$\chi\chi$, $\chi\chi'$ $\chi'\chi'$	$2ml+\not{E}$ (m=2,3,4), $4q+4l+\not{E}$, $2q+2ml+\not{E}$ (m=2,3)	$2nq+2ml$ $\binom{n=2,3,4}{m=1,2,3}$, $2nq+ml+\not{E}$ $\binom{n=2,3,4}{m<6}$	$2nq$ (n=3,4,5), $2mq+\not{E}$ (m=3,4), $2nq+2ml$ $\binom{n=3,4}{m=1,2}$, $6q+2l+\not{E}$
$\tilde{l}^+\tilde{l}^-$	$2l+\not{E}$, $6l+\not{E}$	$4q$, $4q+4l$, $4q+ml+\not{E}$ (m=2,3)	$6q+2l$
$\tilde{\nu}\tilde{\nu}^*$	$4l$, $4l+\not{E}$	$4q$, $4q+2l+\not{E}$, $4q+ml+\not{E}$ (m=0,1)	$6q+\not{E}$
$\tilde{q}\tilde{q}^*$	$2q+4l+\not{E}$	$2q+2l$, $6q+2l$, $2nq+ml+\not{E}$ $\binom{n=1,3}{m=0,1}$	$4q$, $6q$, $8q$

standard events to reject are the 4-fermion events proceeding via heavy gauge bosons (W,Z), which, in some cases, result in irreducible contamination to be dealt with statistical methods.

Final states mediated by LLE couplings are characterised by a large number of charged leptons and neutrinos. The type of events expected is quite unusual for the SM; consequently they can be tagged efficiently by mainly looking at the number of identified leptons. Hadronic *jets* from quark hadronization can occur in case of indirect decays. In this case the signal events are tagged by requiring the identified leptons to be isolated from the hadronic system and to carry a large fraction of visible energy.

Direct LQD decays of sleptons or gauginos can lead, respectively, to four jet final states or four jet and two charged leptons final states; in both these cases the equal reconstructed mass constraint can be exploited to enhance the discriminating power. Other decays lead to final states which always contain quarks and, depending on the process and the coupling structure, charged leptons or neutrinos. Selections therefore require hadronic activity, lepton identification and isolation, missing energy. Selection efficiencies are generally lower than in the LLE case.

Direct UDD decays of sleptons or gauginos produce pure hadronic multi-jet events, with no missing energy, suffering huge backgrounds from two-quark and four-quark SM processes. Mass reconstruction and sphericity-like variables are generally exploited for final discrimination. Indirect decay modes can lead either to fully hadronic final states or to mixed topologies with charged leptons and/or missing energy, which can be selected with relatively higher efficiency. Figure 2 shows an example of variable used in these selections.

Table 3 gives an overlook of the typical selections developed by the experimental groups and of their performances.

Figure 2: Example of variables used in the selections: reconstructed χ and $\chi^\pm$ masses in DELPHI 10-jet analysis. Data are shown in black dots. The light grey histogram (green in colour) corresponds to the expected background. The dark grey (red in colour) histogram shows the signal $(\chi^+\chi^- \rightarrow 4j\chi\chi \rightarrow 10j$, $M_\chi^\pm = 68$ GeV/c^2 and $M_\chi = 38$ GeV/c^2) normalised to 2 pb.

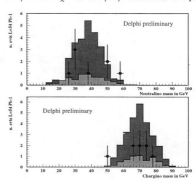

Table 3: Typical selections developed by the experiments (denoted with their initial) in searching for pair production of supersymmetric particles. Here L indicates some degree of charged lepton identifications, J hadronic activity, and $\not{E}$ missing energy.

	Exp	Efficiency
LLE		
Acoplanar Leptons	ADLO	~40%
"4L", "6L" (w/o $\not{E}$)	ADO	~50%,~70%
"4L", "6L" + $\not{E}$	ADLO	~30%,~50%
"$L+J$" + $\not{E}$	ADLO	~35%
LQD		
"4J" (w/o $\not{E}$)	ADO	~50%
"4J" + $\not{E}$	AD	~30%
"$L+J$" (w/o $\not{E}$)	ADO	~30%
"$L+J$" + $\not{E}$	ADO	~20%
UDD		
"6J,8J,10J"	AD	~15%
Multi-J+$\not{E}$	A	~25%
"$L+J$" (w/o $\not{E}$)	A	~50%

Figure 3: Event selected by the 4 lepton OPAL LLE search. The most likely Standard Model interpretation is $ZZ \to e^+e^-\tau^+\tau^-$.

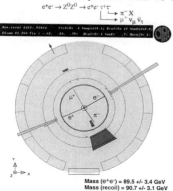

Mass (e^+e^-) = 89.5 +/- 3.4 GeV
Mass (recoil) = 90.7 +/- 3.1 GeV

Figure 4: Limits in the (μ, M_2) plane from ALEPH LLE analysis.

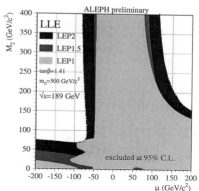

3.3 Results

As expected by the irreducibility of some of the background processes, the developed analyses select candidates. An example is shown in Figure 3. However, the number and dynamics of the selected events is in good agreement with the SM expectations.

The negative results of the searches are used to constrain the parameter space. Given the large amount of $\not{R}_p$ couplings, a conservative approach is adopted to reduce the parameter dependence of the limits by scanning the space of the relevant parameters for the most conservative configuration; for λ and λ' couplings this corresponds, in general, to final states with τ's.

Gauginos

For charginos and neutralinos the derived upper limits on the production cross section are interpreted in the usual (μ, M_2) plane. The ALEPH result for LLE couplings is shown in Figure 4, where the first results from the 189 GeV data have also been included. For large values of M_0 the chargino cross section is large enough for the excluded region to come very close to the kinematic limit, while neutralino cross sections are small and can only help when the collected luminosities are large [e]. Lowering M_0, chargino (neutralino) cross sections move down (up) because of the negative (positive) interference between t-channel sneutrino (selectron) and s-channel exchanges. The neutralino processes and in particular the visibility of $\chi\chi$ act naturally as a backup for the loss of sensitivity of $\chi^+\chi^-$ [f].

Charged sleptons

In the case of charged sleptons decaying directly into an acoplanar lepton pair (LLE) or four jets (LQD), the upper limit on the cross section is translated into a lower limit on the mass. For example, in the upper-left part of Figure 5 it can be seen that a $\tilde{\tau}_R^-$ decaying directly in $\tau^-\nu_e$ is excluded by DELPHI if its mass is below 62 GeV/c^2, while from Figure 6 the ALEPH search for four jet excludes $\tilde{\mu}_L$ with masses below 61 GeV/c^2. Indirect decays, dominating mainly when χ is the LSP, give rise to exotic signatures, like six charged leptons and missing energy (LLE), four quarks and four charged leptons (LQD), or six quarks and two leptons (UDD); these are generally selected with better efficiency; the mass lower limits, expressed usually in the plane $(M_{\tilde{l}}, M_\chi)$, are therefore better than in the case of direct decays. The lower part of Figure 5 gives the DELPHI exclusion region for $\tilde{\mu}$ and $\tilde{\tau}$ decaying indirectly via λ_{133} or λ_{122}. It can be noticed that no unexcluded "corridor" is present near $\Delta M = M_{\tilde{l}} - M_\chi \sim 0$ as in the equivalent plot for the R_p conserving case. This is a general feature of $\not{R}_p$ searches due to the fact that χ is visible. However, it should be noted that the efficiency is generally lower for small ΔM. Therefore for cross section limited channels the corridor could still appear.

[e]This is the case, for instance, for LEP 1 results that, as can be seen in Figure 4 begin to be superseded only by the most recent data collected at 189 GeV.

[f]The interplay works particularly well for LLE, since the topologies arising from $\chi\chi$ are exotic and can be selected with high effective efficiency. For LQD, the ALEPH analysis with 161 and 172 GeV data has shown that the most conservative scenario corresponds to dominance of $\chi\chi \to qqqq\nu\nu$, for which the efficiency was not enough to cover, for low M_0, the mixed region $\mu \sim -M_2$; assuming similar performances the large luminosity collected at 183 is expected to smooth out the M_0 dependence. The full M_0 analysis under UDD dominance has not yet been performed, but arguments similar to LQD are expected to hold.

Figure 5: Limits from the charged slepton DELPHI LLE analysis.

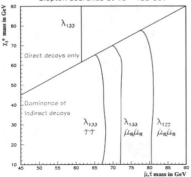

Table 4: Summary of lower limits on sfermion masses from pair production, in some of the most conservative scenarios. Experiments are denoted by their initial. DELPHI limits for $\tilde{t}_1$ have been recalculated in the most conservative scenario. Latest limits assuming R_p conservation (labelled as R_p) are also shown for comparison; these are taken from the LEP SUSY working group[9] for $\tilde{t}_1$ and $\tilde{\tau}_R$, and from the Z lineshape measurement for $\tilde{\nu}$. Values in GeV/c².

	Coupling	Decay	A	D	O	R_p
$\tilde{\nu}$	LLE	indirect	62	62		~43
	LLE	direct	66	62		
	LQD	direct	60	60		
$\tilde{\tau}_R$	LLE	direct	61	61	66	72
	LLE	indirect	70	67		
	LQD	direct	62	71		
$\tilde{t}_1$	LQD	direct		~71	72	83.5
	UDD	direct	63	~72		

Sneutrinos

Direct decays mediated by LLE give the four charged lepton signature selected with high efficiency; the worst case are $\tilde{\nu}_\mu$'s decaying via λ_{233} into $\tau\tau$, for which ALEPH gets a lower mass limit of 66 GeV/c². LQD direct decays give the four jet signature whose non evidence translates in a lower mass limit of 61 GeV/c², as shown in Figure 6. Indirect decays lead in this case to less exotic signatures and therefore to worse absolute mass limits. For UDD, improvements on the Z-lineshape LEP 1 limit are obtained either for $\tilde{\nu}_e$ in the gaugino region, where constructive t-channel interference enhances the cross section, or considering the three sneutrinos mass degenerate.

Squarks

Direct squark decays via UDD lead to four jet final states. As an example, again from Figure 6, ALEPH excludes $\tilde{u}_R$'s of masses below 69 GeV/c². LQD direct decays give final states with two charged leptons and two quarks, which are selected quite efficiently. The OPAL result for $\tilde{t}_1 \to \tau^- d_i$, the most conservative case, is shown in Figure 7; for $\tilde{t}_1$ decoupling from the Z the lower mass limit is 72 GeV/c². Topologies arising from indirect decays are selected with lower efficiencies, and therefore give worse absolute lower mass limits.

It should be noted that hadron colliders are in better position for squark searches, especially for LQD couplings, where the lepton tag can be used to enhance the selection efficiency. First results in this direction have been presented at this conference by D0 and CDF[11].

3.4 Summary of Pair-Production results

The picture that emerges from the study of pair production with $\not{R}_p$ shows that, in general, the situation is at

Figure 6: Cross section upper limits from ALEPH LQD 4-jet analysis.

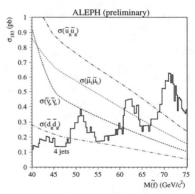

least as good as in the R_p conserving case. For gauginos, the (μ, M_2) analysis for gauginos leads to similar conclusions. The typical sfermion limits are summarised in Table 4; taking into account that the combination of the LEP experiments would bring more or less the same gain as in the R_p conserving case, i.e. 5-10 GeV, the sensitivity reached allowing for $\not{R}_p$ is very similar to the one obtained forbidding it.

4 Other Signatures

Dominant $\not{R}_p$ couplings whose structure involves the first lepton family can manifest themselves in e^+e^- collisions in alternative ways.

For example, the $e^+e^- \to e^+e^-$ cross section is expected to receive a contribution from s-channel exchange of $\tilde{\nu}_j$ if $\lambda_{1j1} \neq 0$; and t-channel contributions to the

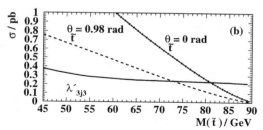

Figure 7: Cross section upper limits from OPAL LQD search for $\tilde{t}_1 \to \tau^- d_i$.

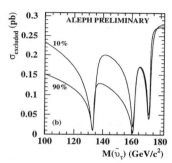

Figure 8: 95% C.L. cross section upper limits for single resonant sneutrino production via λ_{131} coupling in ALEPH. The rescaling factor σ_{183}/σ' was calculated explicitly and naturally leads to dips at each centre-of-mass energy except 183 GeV.

hadronic cross section are expected from non-zero λ'_{1jk} couplings. The study of these deviations is part of the electroweak physics at LEP 2 analyses, and thoroughly discussed in [12].

A new analysis has been presented by ALEPH at this conference looking for single neutralino production which would occur in the case of non-zero λ_{1j1} couplings. The full reaction would be $e^+e^- \to \tilde{\nu}_j \to \nu_j \chi \to \nu_j e^+ e^- \nu_j$ or $\nu_j e^+ l_j^- \nu_e$, leading to acoplanar e^+e^-, $e^\pm\mu^\mp$ or $e^\pm\tau^\mp$ final states. ALEPH has translated the negative results of the acoplanar lepton search into cross section upper limits for these processes. The selection efficiency depends upon the flavour of the final state leptons, the sneutrino mass and the ratio of the neutralino and sneutrino masses. In Figure 8 the excluded cross section for the coupling λ_{131} is shown; the best and worst case neutralino masses (expressed as a fraction of the sneutrino mass) are plotted. Once a point in the MSSM parameter space is chosen, the cross section limit can be interpreted as an upper limit on $|\lambda_{131}|$; for $\tilde{\nu}$ masses in the range of the LEP energies these limits in general significantly improve on low energy bounds.

5 Conclusions

The LEP collaborations have submitted at this conference the results of several searches for supersymmetry allowing for some degree of R_p violation. Overall, these searches cover, with good efficiency and purity, most of the signatures expected in these scenarios. The resulting sensitivity is, in general, as good as in the R_p conserving case, making the negative results of searches for supersymmetry at LEP not invalidated by $\not{R}_p$.

Acknowledgements

I would like to thank the contact persons of the LEP collaborations for their prompt help, in particular I.Fleck and S.Braibant (OPAL), and S.Katsanevas, M.Besancon and C.Berat (DELPHI). I am also indebted with my ALEPH colleagues for providing material and helpful comments during the preparation of the talk, in particular to J.Coles, P.Coyle, D.Fouchez and M.Williams. Finally I would like to thank V.Büscher for careful proof-reading of the manuscript.

References

1. An example of such minimal model can be found in H.E.Haber, G.L.Kane, *Phys. Rep.* **117**, 75 (1985). For phenomenological implications in e^+e^- colliders see the supersymmetry section in *Physics at LEP 2*, CERN 96-01, Vol. 1.
2. G.Farrar, P.Fayet, *Phys. Lett.* B **76**, 575 (1978).
3. ICHEP 98 abstracts **949** and **951** (ALEPH), **210** (DELPHI), **496** (L3), **278** and **279** (OPAL).
4. S.Weinberg, *Phys. Rev.* D **26**, 287 (1982); N.Sakai, T.Yanagida, *Nucl. Phys.* B **197**, 83 (1982); S.Dimopoulos, S.Raby, F.Wilczek, *Phys. Lett.* B **212**, 133 (1982).
5. R.Hempfling, UCDPHY-96-36, 1997, hep-ph/9702412.
6. For a recent review see G. Bhattacharyya, talk given at *Physics Beyond the Standard Model: Beyond the Desert*, Tegernsee, Germany, 8-14 June 1997, hep-ph/9709395.
7. J.Ellis et al.,*Nucl. Phys.* B **238**, 453 (1984).
8. ALEPH Coll., CERN EP/98-077; DELPHI Coll., ICHEP 98 abstract **205**.
9. ICHEP 98 abstracts **579** and **580** (H1), **759** (ZEUS).
10. see http://www.cern.ch/LEPSUSY/.
11. ICHEP 98 abstracts **653** (CDF), **588** (D0).
12. ICHEP 98 abstracts **906** (ALEPH), **441** (DELPHI), **512** and **513** (L3), **264** (OPAL).

Searches for R-Parity Violating SUSY at the Tevatron

Lee Sawyer

Physics Program, Louisiana Tech University, Ruston, LA 71272, USA
E-mail: sawyer@phys.latech.edu

On Behalf of the CDF and DØ Collaborations

We present searches for R-parity violating supersymmetry (SUSY) signals at the Fermi National Accelerator Laboratory's Tevatron $p\bar{p}$ collider. These searches were conducted by the CDF and DØ experiments, and cover a variety of models. Results are shown for both models inspired by the apparent excess of high Q^2 events at HERA, as well as more general R-parity violating supergravity models. No evidence for these signals is found, allowing a variety of limits to be set.

1 Introduction

Supersymmetry (SUSY)[1] describes a class of extensions to the Standard Model which predict a new supersymmetric partner particle for each current particle in the Standard Model. Most SUSY models are constructed such that there is a* conserved multiplicative quantum called "R-parity", R= $(-1)^{3B+L+2S}$; and $B \equiv$ Baryon number, $L \equiv$ Lepton number, and $S \equiv$ spin of the particle. The particles of the Standard Model thus have R= +1, while SUSY particles would have R= −1. Strict conservation of R-parity leads to many of the familiar properties of SUSY models; namely, a lightest supersymmetric particle (LSP) that is absolutely stable, and (due to cosmological considerations) is also uncharged and colorless and therefore would escape detection, leading to an apparent momentum imbalance in the final state. Conversely, nonconservation of R-parity would lead in general to violation of either baryon number or lepton number or both. Limits on B- and L-violating process such as proton decay generally lead to the assumption that R-parity must be conserved in any SUSY extension to the Standard Model.

However, small R-parity violations can be incorporated into SUSY models, while adhering to previous limits[2]. R-parity violation can occur through Yukawa coupling terms in the superpotential

$$\lambda_{ijk} L_i L_j \bar{E}_k + \lambda'_{ijk} L_i Q_j \bar{D}_k + \lambda''_{ijk} \bar{U}_i \bar{D}_j \bar{D}_k \quad (1)$$

where L and Q refer to the doublet lepton and quark superfields; E, U, and D are the singlet superfields; and ijk run over the number of generations. There are in general 45 terms corresponding to the terms λ_{ijk} and λ'_{ijk}. While R-parity conserving SUSY models set these quantities to zero, R-parity violation is still possible if only some of the coupling are nonzero. Thus it is possible to have, for example, B conservation but L violation. To reduce the parameter space of R-parity SUSY searches, the following assumptions are often made : 1) Only one R-parity violating coupling is nonzero, 2) This R-parity

violating coupling is strong enough to lead to decay of the LSP well within the volume of the detector, and 3) The R-parity violating term is small enough that all other features of R-parity conserving SUSY (such as pair production of SUSY particles) still hold.

2 Searches for R-parity Violating SUSY at the Tevatron

We will describe searches for R-parity violating SUSY performed at the Fermi National Accelerator Laboratory's Tevatron $p\bar{p}$ collider, based on data collected during the 1992-1995 run by the CDF and DØ experiments. Details of these two detectors are given elsewhere and are well-known in the high energy physics community; nevertheless it is worth pointing out that these two experiments are based on complimentary detectors, with different capabilities which drive different search strategies. Thus, although both experiments perform R-parity violating SUSY searches in the dielectron + jets channel, details of the analyses as well as presentation of search limits are quite different.

3 R-Parity Violation in Gluino Pair Production

The CDF collaboration searches for the process[3]

$$p\bar{p} \to \tilde{g}\tilde{g} \to (c\tilde{c}_L)(c\tilde{c}_L) \Rightarrow c(e^{\pm}d)c(e^{\pm}d). \quad (2)$$

(Here and in subsequent descriptions of decays the R-parity violating decay will be indicated by the symbol "$\Rightarrow$".) This process is of particular interest, as it involves the same R-parity violating coupling invoked to explain the apparent excess of positron plus jet events at high Q^2, recorded by the experiments at the HERA ep collider[4]. Such events can be explained as the production and decay of a charm squark: $\tilde{c}_L \Rightarrow e + d$, assuming a nonzero value of the R-parity violating coupling λ'_{121}.

The CDF search requires events with 2 like-sign electrons, having $E_T > 15$ GeV and within the pseudorapidity range $|\eta| < 1.1$, as well as two or more jets with $E_T > 15$ GeV in $|\eta| < 2.4$. These electrons are required to pass standard identification cuts for lateral shower shape, be well-matched with charged tracks from the same primary vertex, and be isolated from one another such that $\Delta R = \sqrt{\Delta\phi^2 + \Delta\eta^2} > 0.4$. The jets are defined using a cone algorithm with cone size $\Delta R = 0.7$, and are required to be isolated from each another and from either of the electrons by $\Delta R > 0.7$. Since the LSP is expected to decay, no significant missing E_T is expected; events are therefore vetoed if $\not{E}_T/\sqrt{\Sigma E_T} > 5$ GeV$^{1/2}$.

No events were found in 107 pb^{-1} of data (by comparison, there were 166 opposite-sign dielectron + jet events passing the same cuts). This is consistent with the expected Standard Model backgrounds from $t\bar{t}$ and $b\bar{b}/c\bar{c}$ production. Signal acceptances were modeled using the ISAJET [5] Monte Carlo and the CDF detector simulation. The HERA-inspired model of Choudhury and Raychaudhuri [6] was used to set mass relations among the squarks. In particular, events were assumed to be due to the decay of a charm squark of mass $M(\tilde{c}_L) = 200$ GeV (consistent with the invaiant mass distribution of the HERA events), and $M(\tilde{g}) > M(\tilde{c}_L)$ while the masses of the other squarks were allowed to vary. These acceptances were then used to calculate the 95% C.L. limit on the process $p\bar{p} \to \tilde{g}\tilde{g} \cdot Br(\tilde{g}\tilde{g} \to e^\pm e^\pm + \geq 2 \text{ jets})$. CDF excludes $\sigma \cdot Br \geq 0.18$ pb, independent of the mass of the $\tilde{g}$.

Figure 1 shows the resulting limit in the plane of $(M(\tilde{g}), M(\tilde{q}))$, where $M(\tilde{q})$ is the mass of the other degenerate up-type and right-handed down-type squarks. Contours are drawn corresponding to two different values of the branching fraction for the decay $\tilde{c}_L \Rightarrow ed$. The loss of sensitivity near $M(\tilde{q} \approx 260$ GeV is due to the bottom squark $\tilde{b}_L$ becoming lighter than the $\tilde{c}_L$, causing the decay $\tilde{g} \to b\bar{b}_L$ to dominate and the decay $\tilde{g} \to \bar{c}\tilde{c}_L$ to be suppressed, since there are no R-parity violating decays for the $\tilde{b}_L$ in this model.

4 R-Parity Violation in Squark Pair Production

Using the same data selection described above, CDF also places limits on the process

$$p\bar{p} \to \tilde{q}\tilde{\bar{q}} \to (q\chi_1^0)(\bar{q}\chi_1^0) \Rightarrow q(dce^\pm)\bar{q}(dce^\pm) \qquad (3)$$

where the R-parity violating decay $\tilde{\chi}_1^0 \Rightarrow dce^\pm$ is mediated by an off-shell charm squark. In order to take into account the possibility of large mass splittings in the top squark sector, two different scenarios are considered: 1) $\tilde{q}\tilde{\bar{q}}$ production of 5 mass degenerate squark flavors, and 2) Production of light $\tilde{t}_1\tilde{t}_1$ pairs. Events are modeled assuming that $M(\tilde{\chi}_1^\pm) > M(\tilde{q}) > M(\tilde{\chi}_1^0)$ (which suppresses the decay $\tilde{q} \to \tilde{\chi}_1^\pm$) and $M(\tilde{\chi}_1^\pm) > M(\tilde{t}_1) - M(b)$ (which

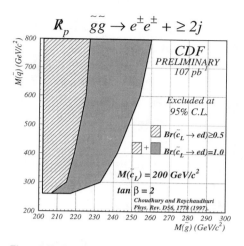

Figure 1: Exclusion region in the $(M(\tilde{g}), M(\tilde{q}))$ plane for the search for R-parity violating decays in gluino pair production, for two different values of the branching fraction for the R-parity violating decay $\tilde{c}_L \Rightarrow ed$.

ensures that $Br(\tilde{t}_1 \to c\tilde{\chi}_1^0) = 100\%$). Equal values were assumed for the branching fractions $Br(\tilde{\chi}_1^0 \Rightarrow q\bar{q}'e)$ and $Br(\tilde{\chi}_1^0 \Rightarrow q\bar{q}'\nu)$.

The resulting 95% C.L. exclusion regions are shown in Fig. 2 (for the degenerate squark case) and Fig. 3 (for the case of $\tilde{t}\tilde{t}$ production). In each case, $\sigma \cdot Br \geq 0.81$ pb to 0.20 pb are ruled out, as a function of the masses. In the degenerate squark case, quarks of mass less than 200 to 260 GeV are ruled out, depending on the mass of the gluino and neutralino. In the top squark case, $\tilde{t}_1$ of mass less than 120 GeV is ruled out for light neutralinos, while $\tilde{t}_1$ of mass less than 135 GeV is ruled out for heavy neutralinos.

5 Searches for R-Parity Violation in SUGRA-Motivated Models

The DØ experiment performs a general search for R-parity violating neutralino decays, within the context of supergravity (SUGRA) motivated models [7]. Only the R-parity violating coupling λ'_{122} is assumed to be significant, leading to the decay $\tilde{\chi}_1^0 \Rightarrow e^\pm cs$. Otherwise the SUGRA framework [8] is assumed, with production cross sections, branching fractions, and mass relationships fully specified by the parameters m_0 (the common scalar mass at the unification energy), $m_{1/2}$ (the common fermion mass at unification), A_0 (the trilinear coupling constant), $\tan\beta$ = ratio of Higgs vacuum expectation values, and the sign of μ (the Higgsino mass parameter).

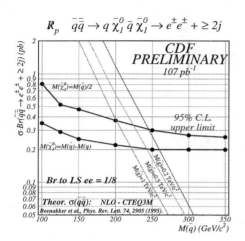

$$\mathcal{R}_p \quad \tilde{q}\bar{\tilde{q}} \to q\, \tilde{\chi}_1^0\, \bar{q}\, \tilde{\chi}_1^0 \to e^{\pm}e^{\pm} + \geq 2j$$

CDF
PRELIMINARY
107 pb^{-1}

$M(\tilde{\chi}_1^0)=M(\tilde{q})/2$

95% C.L.
upper limit

$M(\tilde{\chi}_1^0)=M(\tilde{q})-M(q)$

$M(\tilde{g})=0.2\ TeV/c^2$
$M(\tilde{g})=1\ TeV/c^2$
$M(\tilde{g})=0.5\ TeV/c^2$

Br to LS ee = 1/8

Theor. $\sigma(\tilde{q}\bar{\tilde{q}})$: NLO - CTEQ3M
Beenakker et al., Phys. Rev. Lett. 74, 2905 (1995).

$M(\tilde{q})\ (GeV/c^2)$

Figure 2: Upper limits at 95% C.L. on cross section times branching fraction for the search for R-parity violating decays in squark pair production. All squarks are considered to be mass degenerate except for $\tilde{t}_1$.

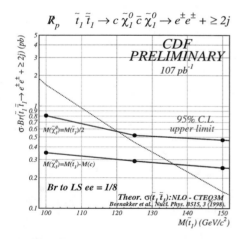

$$\mathcal{R}_p \quad \tilde{t}_1\bar{\tilde{t}}_1 \to c\, \tilde{\chi}_1^0\, \bar{c}\, \tilde{\chi}_1^0 \to e^{\pm}e^{\pm} + \geq 2j$$

CDF
PRELIMINARY
107 pb^{-1}

95% C.L.
upper limit

$M(\tilde{\chi}_1^0)=M(\tilde{t}_1)/2$

$M(\tilde{\chi}_1^0)=M(\tilde{t}_1)-M(c)$

Br to LS ee = 1/8

Theor. $\sigma(\tilde{t}_1\bar{\tilde{t}}_1)$:NLO - CTEQ3M
Beenakker et al., Nucl. Phys. B515, 3 (1998).

$M(\tilde{t}_1)\ (GeV/c^2)$

Figure 3: Upper limits at 95% C.L. on cross section times branching fraction for the search for R-parity violating decays in $\tilde{t}_1\bar{\tilde{t}}_1$ pair production.

Table 1: Summary of backgrounds to the DØ SUGRA-motiveated search.

Background Process	Expected Contribution to 96 pb^{-1}
Drell-Yan$\to ee$	$0.42 \pm 0.15 \pm 0.16$
$Z \to \tau\tau \to ee$	$0.08 \pm 0.02 \pm 0.02$
$t \to ll \to ee$	$0.08 \pm 0.01 \pm 0.03$
Instrumental	$1.27 \pm 0.13 \pm 0.24$
Total	$1.8 \pm 0.2 \pm 0.3$

The DØ search proceeds by requiring events with two or more electrons, where the leading electron has $E_T > 15$ GeV and the next-to-leading electron has $E_T > 10$ GeV; these electrons are required to be within the pseudorapidity range $|\eta| < 1.1$ or $1.5 < |\eta| < 2.5$. In addition, the events were required to have four or more jets, having $E_T > 15$ GeV within $|\eta| < 2.5$, and the events were also required to have a "hardness" $H_T \equiv \Sigma E_T^{jet} + \Sigma E_T^{e} > 150$ GeV. Electrons must pass standard identification cuts on shower profile, be isolated, and have a match to a central track. Jets are reconstructed using a cone algorithm, with $\Delta R = 0.5$. Backgrounds from the process $Z \to ee$ are reduced by requiring that the invariant mass of the electron pair fall within $M_{ee} < 76$ GeV or $M_{ee} > 106$ GeV.

In a data sample of 96 pb^{-1}, two events were found which satisfied the cuts described above. Background estimates were made using a GEANT[9]-based detector simulation for the processes $t\bar{t}, WW$, Drell-Yan, and $Z \to \tau\bar{\tau} \to ee$. Instrumental backgrounds from jets misidentified as electrons are estimated from data. The expected level of background from each source is summarized in Table 1. The total estimated background was found to $1.8 \pm 0.2 \pm 0.3$ events, where the first error is statistical and the second error is due to systematic uncertainites.

Interpreting the two events as a null result, DØ sets limits in the $(m_0, m_{1/2})$ plane, assuming $\tan\beta = 2, A_0 = 0$ and $\mu < 0$. Signal acceptances were calculated from events generated at 125 points with different values af m_0 and $m_{1/2}$. The top quark mass was fixed at 170 GeV, and all SUSY processes were generated at each point in the parameter space. The branching fractions for R-parity violating neutralino decay was added to ISAJET in a separate subroutine, using the formulae in Dreiner, et al. [10]. Figure 4 shows the resulting 95% C.L. exclusion region. The prominent dip near $m_0 = 70 - 80$ GeV is in a region where $M(\tilde{\chi}_2^0) > M(\tilde{\nu})$, leading to the decay $\tilde{\chi}_2^0 \to \tilde{\chi}_1^0 \nu\bar{\nu}$, which suppresses the rate of dielectron events. Overlaid on this limit curve are the corresponding contours for $M(\tilde{g})$ and $M(\tilde{q})$. In general, this search rules out gluinos of mass below 252 GeV, and squarks

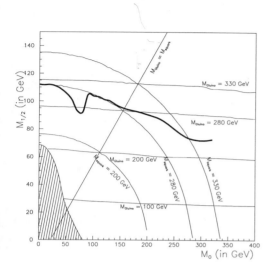

Figure 4: Exclusion contour (dark line) for the DØ SUGRA-motivated search. Results are shown in the $(m_0, m_{1/2})$ plane, assuming $\tan\beta = 2$, $A_0 = 0$, $\mu < 0$, and the R-parity violating neutralino decay $\tilde\chi_0^1 \Rightarrow e^\pm cs$.

Table 2: Summary of limits from the R-parity violating SUSY searches described in this paper.

Search Channel	Limits
CDF HERA-Inspired	$\sigma * Br \geq 0.18$ pb
1) $\tilde g \tilde g \to (c\tilde c_L)(c\tilde c_L)$ $\Rightarrow c(e^\pm d)c(e^\pm d)$	$M(\tilde c_L) = 200$ Gev/c^2 excluded for range of $M(\tilde q)$ and $M(\tilde g)$
2) $\tilde q \tilde{\bar q} \to (q\chi_1^{\tilde 0})(\bar q \chi_1^{\tilde 0})$ $\Rightarrow q(dce^\pm)\bar q(dce^\pm)$	$M(\tilde t_1) > 135$ GeV/c^2 $M(\tilde q > 260$ GeV/c^2 (heavy $\tilde\chi_1^0$)
DØ SUGRA	Exclusion in $M_{1/2}/M_0$ plane
$\tilde\chi_1^0 \Rightarrow e^\pm cs$ (all initial states)	$M(\tilde q) = M(\tilde g) > 283$ Gev/c^2 $M(\tilde q) > 252$ GeV/c^2 and $M(\tilde g) > 232$ GeV/c^2

of mass less than 232 GeV, within the model described above. The limit on equal mass gluinos and squarks is 283 GeV.

6 Conclusions

Both the CDF and DØ experiments at the Tevatron have searched for evidence of R-parity violating SUSY events in $p\bar p$ collisions. The data have been interpreted in a a variety of frameworks, both those inspired by the excess of high Q^2 events at HERA as well as more general SUGRA-motivated models. No evidence for R-parity violating SUSY is seen in roughly 100 pb^{-1} of data taken by each experiment. Limits from the various searches are summarised in Table 2.

Both experiments are currently extending the searches described in this talk to channels with muons in the final state. The combination of theoretical interest and increased luminosity in the upcoming Run II of the Tevatron promises to make R-parity violating SUSY a continuing topic of interest in the search for new phenomena at the Tevatron.

Acknowledgements

The author would like to express his thanks to the members of the CDF and DØ collaborations for their assistance in preparing this presentation.

References

1. For reviews of SUSY, see X. Tata, in *The Standard Model and Beyond*, edited by H. Kim (World Scientific, Singapore, 1991); H. Niles, Phys. Rep. **110**, 1 (1984); H. Haber and G. Kane, Phys. Rep. **117**, 75 (1995).
2. For a complete review of R-parity violating searches, including limits on the λ_{ijk} and $\lambda_{ijk}{\prime}$ couplings, see H. Dreiner, "An Introduction to Explicit R-parity Violation", hep-ph/9707435 (v2), 24 Jul, 1997 (to be published in *Perspectives in Supersymmetry*, G. Kane, ed., World Scientific.)
3. M. Chertok, J. Done, and T. Kamon (for the CDF Collaboration), "Search for R Parity Violating Supersymmetry using Like-sign Dielectrons at CDF", submitted to this conference.
4. C. Adloff, *et al.*(H1 Collaboration), Z. Phys. C**74**, 191 (1997); J. Breitwig, *et al.*(ZEUS Collaboration), Z. Phys. C**74**, 207 (1997).
5. H. Baer, F. Paige, S. Protopopescu, and X. Tata, "Simulating Supersyummetry with ISAJET 7.0/ISASUSY 1.0", in *Physics at Current Accelerators and Supercolliders*, edited by J. Hewitt, A. White, and D. Zeppenfield, Argonne National Laboratory (1993).

6. D. Choudhury and S. Raychaudhuri, Phys. Lett. **B 401**, 54 (1997); D. Choudhury and S. Raychaudhuri, Phys. Rev. **D 56**, 1 778 (1997).

7. B. Abbott, *et al.*(DØ Collaboration), "Search for R-parity Violating Supersymmetry in the Dielectron Channel", submitted to this conference.

8. H. Baer and X. Tata, Phys. Rev. **D 47**, 2739 (1993); M. Dress and M. Nojiri, Nucl. Phys. **B 369**, 54 91992); L. Ibanez, C. Lopez, and C. Munoz, Nucl. Phys. **B 256**, 218 (1985).

9. R. Brun and F. Carminati, CERN Program Library Long Writeup W5013, 1993 (unpublished).

10. H. Dreiner, M .Guchait, and D. Roy, Phys. Rev. **D 49**, 3270 (1994).

ASPECTS OF HIGGS PHYSICS AND PHYSICS BEYOND THE STANDARD MODEL AT LHC AND e^+e^- LINEAR COLLIDERS

W. KILIAN

Institut für Theoretische Physik, Universität Heidelberg, D–69120 Heidelberg, Germany
E-mail: W.Kilian@thphys.uni-heidelberg.de

P. M. ZERWAS

Deutsches Elektronen-Synchrotron DESY, D–22603 Hamburg, Germany
E-mail: zerwas@desy.de

Recent developments in prospects of searching for Higgs particles and testing their properties at the LHC and at TeV e^+e^- linear colliders are summarized. The discovery limits of supersymmetric particles at the LHC are presented and the accuracy is explored with which the fundamental SUSY parameters in the context of supergravity models can be determined at high-luminosity linear colliders. Finally, new discovery limits for gauge bosons in left-right symmetric models at the LHC are presented.

1 The physical basis

1. Recent results from high-precision measurements of electroweak observables at LEP, SLC and elsewhere strongly support the hypothesis that the electroweak symmetries are broken through the Higgs-mechanism and that a light fundamental Higgs boson is realized in Nature[1]. Moreover, the perturbative expansion of the theory up to the GUT scale, backed by the observed value of the electroweak mixing angle, favors a Higgs mass in the intermediate range $M_H \lesssim 180$ GeV. This is the most difficult range to explore at the LHC, yet it is evident now that the lower part of this range can be covered in the $H \to \gamma\gamma$ decay channel[2] while the upper part is accessible in the four-lepton decay $H \to ZZ^* \to 4\ell^\pm$, both with high significance between $S = 8$ and 10.

Once Higgs bosons are discovered, their properties must be explored to establish the Higgs mechanism *sui generis* as the basic mechanism for the breaking of the electroweak symmetries. This program can be carried out in three consecutive steps. (i) The external quantum numbers J^{PC} must be determined[3]. (ii) The generating of vector-boson and fermion masses through the Higgs mechanism can be scrutinized by measuring the HVV and Hff couplings, *nota bene* the Htt Yukawa coupling (of the heaviest matter particle in the Standard Model) to the Higgs particle. The bremsstrahlung of Higgs bosons in the process $e^+e^- \to t\bar{t}H$ offers a method for measuring this fundamental coupling directly[4,5]. (iii) The Higgs potential which provides the operational basis for the Higgs mechanism, must finally be reconstructed. The strength and the form of the potential define the Higgs mass and the trilinear and quadrilinear self-couplings of the Higgs particle. The prediction for the trilinear coupling can be tested in Higgs-pair production at the LHC in the process $pp \to HH$, and at e^+e^- colliders in Higgs-strahlung $e^+e^- \to ZHH$, for instance, where the Higgs pair is emitted in the decay of a virtual H boson[6].

2. Extending the Standard Model to a supersymmetric theory provides a natural way to keep light fundamental Higgs bosons stable in the background of large GUT scales. The search for supersymmetric particles is therefore a very important endeavor at existing and future accelerators. The ultimate discovery limits, in the next two decades, of squarks and gluinos will be set by the LHC[7]. TeV e^+e^- linear colliders will play the same role in the non-colored sector for charginos/neutralinos and sleptons[8].

Moreover, the high-luminosity version of e^+e^- colliders allows to carry out very accurate measurements of the sparticle properties[8]. These measurements can be used to explore the structure of the basic supersymmetric theory, in particular the mechanism responsible for the breaking of supersymmetry. Since SUSY breaking mechanisms are (in general) generated at scales of the order of the Planck scale, these machines open windows to physics scenarios in which gravity is an integral part of the system. Thus, they prepare the basis for the unification of the four fundamental forces.

3. Recent experimental evidence that neutrinos are massive particles has renewed interest in grand unified theories such as $SO(10)$, which incorporate right-handed neutrino fields in a natural way. Even though the typical scales of gauge bosons associated with right-handed symmetries and heavy neutrino masses may likely be close to the GUT scale, it is nevertheless an interesting problem to search for new LR degrees of freedom[9], in particular at the LHC which defines the energy frontier of laboratory accelerators in the near future.

Table 1: Cross section at the LHC, branching ratio and acceptance for $H \to \gamma\gamma$ with $M_H = 100$ GeV, left column; background cross sections for the same invariant $\gamma\gamma$ mass, right column.

Signal σ		Backgrounds $d\sigma/dm_{\gamma\gamma}$	
$\sigma(pp \to H + X)$	56.3 pb	quark annihil.	92 fb/GeV
BR($H \to \gamma\gamma$)	1.53×10^{-3}	gluon fusion	167
acceptance	51.9 %	isol. bremsstr.	120
$\sigma \times$ BR	86.2 fb	total	379 fb/GeV

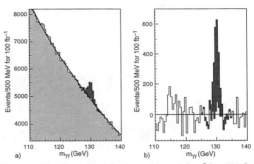

Figure 1: The Higgs signal in the $\gamma\gamma$ channel for $\int \mathcal{L} = 100$ fb^{-1} before (a) and after (b) background subtraction [2].

2 Higgs physics

2.1 The Higgs two-photon channel at the LHC

The SM Higgs boson in the lower part of the intermediate mass range $M_H \lesssim 150$ GeV, can be searched for at the LHC in the $H \to \gamma\gamma$ decay mode. The Higgs bosons are primarily produced in gluon-gluon fusion. However, the signal-to-background ratio improves from typical values $S/B \sim 1/15$ for small-$p_\perp$ Higgs production to $S/B \sim 1/4$ if a cut is applied to the minimal transverse momentum, balanced by a recoiling gluon or quark jet [10]:

$$pp \longrightarrow H(\to \gamma\gamma) + \text{jet} \qquad (1)$$

Important subprocesses are the parton reactions $g + q/g \to H + q/g$ in which the Higgs boson is formed in the fusion of the initial-state gluon and a virtual gluon emitted from the quark/gluon spectator line.

Besides the reducible γ backgrounds from decaying hadrons, such as $\pi^0 \to \gamma\gamma$, irreducible backgrounds are generated in QCD Compton processes like $q + g \to q + \gamma\gamma$; their size can readily be predicted theoretically [10]. It is more difficult to control background processes in which photons are fragments of quarks or gluons. These events must be eliminated by vetoing the accompanying parent jets. Under realistic experimental conditions the impurity probabilities are of the order $P[\gamma/q_{\text{veto}}] \sim 2 \times 10^{-4}$ and $P[\gamma/g_{\text{veto}}] \sim 0.3 \times 10^{-4}$ for quark and gluon parents, respectively.

Selecting large-transverse momentum events allows to search for Higgs bosons in the $\gamma\gamma$ channel [9] also in high-luminosity LHC runs with $\mathcal{L} = 100$ fb^{-1}/year in which many events pile up. Exploiting recoiling high-$p_\perp$ tracks, z_{vertex} can be nevertheless reconstructed very accurately so that a high $\gamma\gamma$ mass resolution can be achieved also in this environment: $\sigma_{\gamma\gamma} \approx 690$ MeV in the CMS ECAL, for instance. With a transverse-momentum cut of 2 GeV on the tracks, the average transverse momentum of the jet formed by the recoiling hadrons is of order $\langle p_\perp^{\text{jet}} \rangle \approx 40$ GeV.

The typical size of signal and background cross sections is shown in Table 1, referring to the illustrations in

the Figs.1a/b. The significance $S = N_S/\sqrt{N_B}$ varies for $\int \mathcal{L} = 100$ fb^{-1} from about 8 to 12 to 8 if the Higgs mass is increased from 100 to 130 to 150 GeV, i.e.

$$S > 8 \quad \text{for} \quad M_H = [100, 150 \text{ GeV}] \qquad (2)$$

Even though theoretical refinements could change the estimates somewhat, it can nevertheless be concluded that a sufficient buffer does exist for the discovery of the SM Higgs boson in the $\gamma\gamma$ resonance channel at the LHC.

2.2 The parity of Higgs bosons

The external quantum numbers of the Higgs boson can be studied in production and decay processes. For heavy enough Higgs bosons, spin correlations in Higgs decays to top-quark pairs can be exploited to measure the parity [3]. Denoting the couplings

$$\langle H|t\bar{t}\rangle = (M_t/v)[a + i\gamma_5 \tilde{a}] \qquad (3)$$

the coefficient $a \neq 0$ describes the scalar component, the coefficient $\tilde{a} \neq 0$ the pseudoscalar component of the state. If both are non-zero, CP is violated in the interactions.

The couplings determine the spin correlations between the final-state top and anti-top quark. Near threshold, the correlations are given by $\langle s_t s_{\bar{t}} \rangle = +1/4$ and $-3/4$ for scalar and pseudoscalar Higgs decay matrix elements, corresponding to spin triplets and singlets. The continuum prediction for the spin correlation in gg collisions at threshold is again $-3/4$. The correlation may be analyzed at the LHC through lepton-angular distributions in semileptonic t and $\bar{t}$ decays [3]. The size of the angular correlations is given by $\langle \cos \theta_{\ell^+\ell^-} \rangle = 0.396, 0.383$ and 0.402 in the continuum, scalar and pseudoscalar decays, respectively, for $M_H = 400$ GeV. The statistical error with which $\langle \cos \theta_{\ell^+\ell^-} \rangle$ can be measured is estimated to be $\Delta = 0.001$ for $\int \mathcal{L} = 100$ fb^{-1}. Thus, the small differences between the resonance values and the continuum are significant.

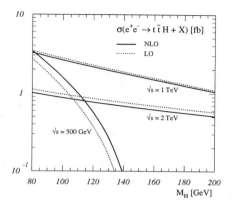

Figure 2: The cross section of Higgs bremsstrahlung in $e^+e^- \rightarrow t\bar{t}H$ for three different collider energies [4].

2.3 The Htt Yukawa coupling

The masses of the fundamental particles are generated in the Higgs mechanism by interactions with the non-zero Higgs field in the ground state. Within the Standard Model, the Yukawa couplings of the leptons and quarks with the Higgs particle are therefore uniquely determined by the particle masses:

$$g_{Hff} = M_f/v \quad : \quad v = (\sqrt{2}\,G_F)^{-1/2} \quad (4)$$

The measurement of these couplings provides a fundamental experimental test of the Higgs mechanism. Several methods have been proposed in the literature to test Yukawa couplings of the Higgs boson [8]. Since the mass of the top quark is maximal in the fermion multiplets of the Standard Model, the Htt coupling ranks among the most interesting predictions of the Higgs mechanism. This coupling can be tested indirectly by measuring the $H\gamma\gamma$ and Hgg couplings, which are mediated by virtual top-quark loops. If the Higgs boson is very heavy, the decay mode $H \rightarrow t\bar{t}$ can be exploited. On the other hand, if the Higgs boson is light, $M_H \sim 100 \ldots 200$ GeV, the radiation of Higgs bosons off top quarks lends itself as a basic mechanism for measuring the Htt coupling. The prospects of operating e^+e^- linear colliders at high luminosities[a] render the radiation process

$$e^+e^- \rightarrow t\bar{t}H \quad (5)$$

a suitable instrument to carry out this measurement.

The Higgs boson can be radiated off the final-state top quarks as well as off the intermediate Z boson; due to

[a]In TESLA designs, integrated luminosities of $\int \mathcal{L} = 0.3$ ab^{-1} and 0.5 ab^{-1} per year are planned at c.m. energies $\sqrt{s} = 0.5$ TeV and 0.8 TeV, respectively.

the heavy top mass the first mechanism is by far leading. QCD rescattering corrections[4,5] increase the cross section near the threshold while for high energies the positive contributions from gluon radiation are overwhelmed by negative Htt vertex corrections. The two limits can be paraphrased by K factors:

$$K_{\text{thr}} \approx 1 + \frac{\alpha_s}{\pi}\frac{64}{9}\frac{\pi M_t}{\sqrt{(s^{1/2} - M_H)^2 - 4M_t^2}} \quad (6)$$

$$K_\infty \approx 1 - 3\frac{\alpha_s}{\pi} \quad (7)$$

The complete cross section including Higgs emission from the Z line, has been evaluated in Ref. 4. Characteristic values of the cross section are presented in Fig.2 for three e^+e^- energy values $\sqrt{s} = 500$ GeV, 1 TeV and 2 TeV, as a function of the Higgs mass.

While the phase-space suppression is gradually lifted with rising energy, the lower part of the intermediate Higgs mass range is experimentally accessible already at high-luminosity e^+e^- colliders for $\sqrt{s} = 500$ GeV; for $\sqrt{s} = 1$ TeV the entire intermediate Higgs mass range can be covered. For a typical size $\sigma \sim 1$ fb of the cross section, about 10^3 events are generated when an integrated luminosity $\int \mathcal{L} \sim 1$ ab^{-1} is reached within two to three years of running. This should provide a sufficiently large sample for detailed experimental studies of this process. Since Higgs radiation from the top quarks is dominant, the sensitivity to the Htt Yukawa coupling is nearly quadratic, $\sigma(e^+e^- \rightarrow t\bar{t}H) \propto g_{Htt}^2/4\pi$, thus being very high.

2.4 Higgs self-couplings

The electroweak symmetries are spontaneously broken in the Standard Model and related theories through a potential in the scalar sector for which the minimum is realized at a non-vanishing value of the fields,

$$V = \frac{\lambda}{2}\left[|\varphi|^2 - \frac{v^2}{2}\right]^2 \quad (8)$$

Expanding the field φ about the ground-state value $\langle\varphi\rangle_0 = v/\sqrt{2}$, the physical Higgs mass $M_H = \sqrt{2\lambda}\,v$, and trilinear and quadrilinear self-couplings of the physical Higgs fields are generated:

$$V = \frac{1}{2}M_H^2 H^2 + \frac{1}{2}(M_H^2/v)H^3 + \frac{1}{8}(M_H^2/v^2)H^4 \quad (9)$$

The fundamental Higgs potential can thus be reconstructed by measuring the trilinear and quadrilinear self-couplings of the Higgs boson.

The trilinear self-coupling of the Higgs boson determines the production of pairs of Higgs particles at e^+e^-

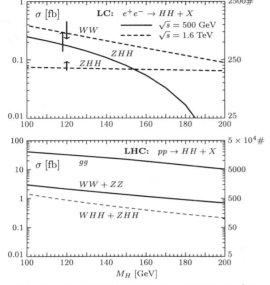

Figure 3: The cross sections for Higgs pair production at e^+e^- colliders and at LHC in various channels. The event rates on the first and second right scale correspond to integrated luminosities $\int \mathcal{L} = 2.5$ ab^{-1} and 500 fb^{-1}, respectively.

colliders and at the LHC[6]. Several subprocesses are relevant in this context:

$e^+e^- \rightarrow HH + X$:

    double Higgs-strahlung $e^+e^- \rightarrow Z + HH$
    WW fusion $e^+e^- \rightarrow \bar{\nu}_e \nu_e + HH$

$pp \rightarrow HH + X$:

    double Higgs-strahlung $q\bar{q} \rightarrow W/Z + HH$
    WW/ZZ fusion $q\bar{q} \rightarrow q\bar{q} + HH$
    gluon fusion $gg \rightarrow HH$

In all subprocesses, the two Higgs bosons can either be emitted from the W/Z lines of the first two subprocesses and the t line of the third subprocess, or from the splitting of a virtual Higgs boson H_{virt} generated, for instance, in Higgs-strahlung $e^+e^- \rightarrow Z + H_{\mathrm{virt}}$ followed by $H_{\mathrm{virt}} \rightarrow HH$.

The size of the cross sections for the different channels is illustrated in Figs.3a/b. Since all the processes are of higher order in the electroweak coupling, the cross sections are small, posing severe background problems at the LHC, and requiring very high luminosities at e^+e^- linear colliders. The sensitivity to the trilinear coupling λ_{HHH} is indicated by arrows which point to the variation of the cross sections if the coupling is modified, $ad\ hoc$,

from $1/2$ to $2 \times \lambda_{HHH}$. The shifts are distinctly larger than the statistical fluctuations.

3 Supersymmetry

There are strong indications that the Standard Model is embedded in a supersymmetric theory. Besides the doubling of the basic Higgs doublet, SUSY doubles the spectrum of the SM particles in the minimal supersymmetric extension of the Standard Model (MSSM). The SM leptons and quarks are associated with scalar sleptons and squarks; the gluons with spin-$\frac{1}{2}$ gluinos; the partners of the electroweak gauge bosons and Higgs bosons mix to form two charginos and four neutralinos. The LHC is the machine proper to search for colored squarks and gluinos while lepton colliders are suitable machines to discover the non-strongly interacting particles, charginos/neutralinos and sleptons.

Apart from variants, essentially two scenarios have been developed to induce the breaking of supersymmetry: mSUGRA and gauge-mediated supersymmetry breaking. Most phenomenological analyses have been performed so far in the minimal supergravity model mSUGRA. In this approach, the breaking is triggered by gluino condensation in a shadow-world, transferred by gravitational interactions to the eigen-world. Soft SUSY breaking terms are generated this way near Planck scale distances. Evolving the universal gaugino and scalar mass parameters from the GUT scale down to the low electroweak scale, the mass parameters split eventually, generating the Higgs mechanism when the mass parameter (squared) in the Higgs sector becomes negative. This scenario is described by five parameters, with masses and couplings being universal at the GUT scale: the scalar mass M_0; the gaugino mass $M_{1/2}$; the trilinear coupling A_0; $\tan\beta$, the ratio of the VEVs in the Higgs sector; and sgn(μ), the sign of the higgsino parameter. The more than one hundred phenomenological SUSY parameters at the electroweak scale can all be expressed in terms of those five fundamental parameters, leading to many relations for testing the scheme.

3.1 Ultimate discovery limits at the LHC

Sparticles can be searched for at the LHC in a variety of channels. While the classical signature of squark and gluino production is the observation of "*multi-jets* + $E_\perp^{miss}$", the search for isolated leptons, together eventually with combinations of the other signatures, have proven extremely successful, too[7]:

$$pp \rightarrow \text{sparticles} \rightarrow \text{lepton(s)} + \text{(jets)} + E_\perp^{\mathrm{miss}} \qquad (10)$$

This set of signatures can be applied to the search of squarks/gluinos, charginos/neutralinos, and sleptons.

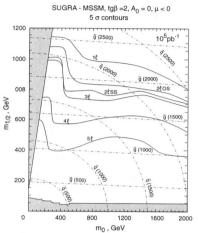

SUGRA - MSSM, tgβ =2, A₀ = 0, μ < 0
5 σ contours

Figure 4: Contours for the discovery of squarks and gluinos at the LHC in leptonic channels [7].

a) <u>Squarks and gluinos</u>

Squark and gluino masses can be expressed in terms of the fundamental parameters by two simple approximate relations (for moderate values of $\tan\beta$) in the mSUGRA scenario:

$$M_{\tilde{q}} \approx \sqrt{M_0^2 + 6M_{1/2}^2} \qquad (11)$$

$$M_{\tilde{g}} \approx 2.5\, M_{1/2} \qquad (12)$$

Depending on the relative magnitude of $M_{\tilde{q}}$ vs. $M_{\tilde{g}}$, the SUSY particles first decay via mutual cascades, before leptonic channels become relevant, e.g.

$$pp \;\to\; \tilde{g}\,\tilde{g}/\tilde{g}\tilde{q}/\tilde{q}\tilde{q}$$

$$\quad \hookrightarrow \bar{q} + \tilde{q}$$
$$\qquad\quad \hookrightarrow q + \ell^{\pm}\nu_\ell + \tilde{\chi}_1^0$$
$$\quad \hookrightarrow \bar{q} + \tilde{q}$$
$$\qquad\quad \hookrightarrow q + \ell^{\pm}\nu_\ell + \tilde{\chi}_1^0$$

As shown in Fig.4, ultimate discovery limits [7] of

$$M_{\tilde{q}} \lesssim 2 \text{ to } 2.5 \text{ TeV} \qquad (13)$$

$$M_{\tilde{g}} \lesssim 2 \text{ to } 2.5 \text{ TeV} \qquad (14)$$

deep in the TeV range, can be reached for leptonic signatures. Similar bounds are found if the parameters $\tan\beta$, A_0 and μ are altered.

b) <u>Charginos/neutralinos and sleptons</u>

The mSUGRA mass relations (for the lightest members of the two species) can be cast in the form:

$$M_{\tilde{\chi}_1^0} \approx 0.5\, M_{1/2} \qquad\qquad M_{\tilde{\chi}_1^\pm} \approx M_{1/2}$$
$$M_{\tilde{\chi}_2^0} \approx M_{1/2}$$

and

$$M_{\tilde{\ell}_R}^2 = M_0^2 + 0.15\, M_{1/2}^2 - s_w^2 M_Z^2 \cos 2\beta$$
$$M_{\tilde{\ell}_R}^2 = M_0^2 + 0.52\, M_{1/2}^2 - \tfrac{1}{2}(1 - 2s_w^2) M_Z^2 \cos 2\beta$$
$$M_{\tilde{\nu}_L}^2 = M_0^2 + 0.52\, M_{1/2}^2 + \tfrac{1}{2} M_Z^2 \cos 2\beta$$

Apart from cascade decays, these particles can be generated through the Drell-Yan mechanism:

$$pp \to \tilde{\chi}_i\tilde{\chi}_j \quad \text{and} \quad \tilde{\ell}\tilde{\ell}$$

in quark-antiquark collisions leading eventually to clean non-jetty events. From the abundant leptonic decay states, bounds [7] of

$$\text{charginos/neutralinos}: M_{1/2} \lesssim 180 \text{ GeV}$$
$$\text{sleptons} \qquad\qquad : M_{\tilde{\ell}} \; \lesssim 350 \text{ GeV}$$

can be set in the high-luminosity runs at LHC.

The mass limits accessible in the Drell-Yan mechanism can be doubled if suitable cascade decays are realized in the SUSY models.

3.2 Discovery limits and ultimate precision at LC

Discovery limits for spin-$\tfrac{1}{2}$ $\tilde{\chi}$ states and spin-0 $\tilde{\ell}, \tilde{\nu}_\ell$ states in e^+e^- linear colliders coincide nearly with the kinematical limits, i.e.

$$M_{\tilde{\chi}_1^\pm} \lesssim \sqrt{s}/2 \qquad \text{etc.}$$
$$M_{\tilde{\ell}, \tilde{\nu}_\ell} \lesssim \sqrt{s}/2 \qquad \text{etc.}$$

Thus, in a 2 TeV collider, chargino/neutralino and slepton masses up to ~ 1 TeV [and in some unpaired $\tilde{\chi}_i\tilde{\chi}_j$ channels even beyond] can be observed.

However, the second target of e^+e^- linear colliders is the measurement to high precision of the phenomenological SUSY mass parameters etc. This is possible in a high-luminosity machine, such as TESLA, which allows to perform many threshold scans consecutively. Anticipating [8] about 50 fb^{-1} for the high-precision mass measurement at spin-$\tfrac{1}{2}$ thresholds which rise steeply $\sim \beta$, and about 100 fb^{-1} for spin-0 thresholds which rise more slowly $\sim \beta^3$, more than a dozen independent channels can be analyzed if a total luminosity $\int\mathcal{L} \sim 1.5$ ab^{-1} can be provided by the machine.

In such an experimental program, the masses can be determined at the per-mille level [8]:

$$M_{\tilde{\chi}_1^\pm} = 138 \text{ GeV} \pm 100 \text{ MeV}$$
$$M_{\tilde{\mu}_R} = 132 \text{ GeV} \pm 300 \text{ MeV} \qquad \text{etc.}$$

Based on these measurements, the fundamental mSUGRA parameters can be extracted with very high accuracy [11]; for example,

$$\tan\beta = 3 \pm 0.01 \quad M_0 = 100 \text{ GeV} \pm 120 \text{ MeV}$$
$$A_0 = 0 \pm 5 \text{ GeV} \quad M_{1/2} = 200 \text{ GeV} \pm 130 \text{ MeV}$$

These high-precision measurements at the per-mille level will allow us to test stringently scenarios of supersymmetry breaking. Since these mechanisms are generated at the GUT/Planck scale, eventually involving gravity, high-luminosity e^+e^- linear colliders appear indispensable facilities for opening windows to physics scales where gravity plays an integral role, thus providing a bridge to the unification of all four forces.

4 Left-right symmetric gauge theories

Left-right symmetric gauge theories connect the right-chiral leptons, charged and neutral, and the right-chiral quarks by the absorption or emission of $W_R^\pm$ gauge bosons. Moreover, neutral gauge bosons Z' exist in these scenarios, heavier than the $W_R^\pm$ by the ratio $(2\cos^2\theta_w/\cos 2\theta_w)^{1/2} \approx 1.7$. The left and right degrees of freedom mix in general to form mass eigenstates W, W' and Z, Z'; however, these mixing effects will in general be neglected.

To generate a spectrum of very light and very heavy neutrinos, a see-saw mechanism may be operative which is driven by large Majorana masses associated with the right-handed neutrinos. The light as well as the heavy mass eigenstates are Majorana neutrinos ν and N, marked by a family index each.

The gauge bosons $W_R^\pm$ and Z' can be produced in the Drell-Yan mechanism:

$$pp \to W_R \text{ and } Z' \tag{15}$$

The W_R bosons are produced in collisions of right-handed quarks and left-handed antiquarks. Similarly Z', yet also in quark/antiquark beams with reverse handedness. The particles decay into right/left-handed quarks/antiquarks following the same rule, yet also to charged leptons and heavy neutrinos, if kinematically possible:

$$W_R^- \to q\bar{q}' \quad \text{and} \quad N_\ell\ell^-$$
$$Z' \to q\bar{q} \quad \text{and} \quad \ell^+\ell^- \oplus N_\ell N_\ell$$

The quark decays dominate strongly. The Majorana neutrinos N decay, if mixing effects are neglected, into charged leptons of either sign with equal probability,

$$N_\ell \to \ell^\pm + jj \tag{16}$$

where a right-handed quark current is coupled to the right-handed lepton current by virtual W_R exchange.

The signatures of the events[9] are spectacular isolated leptons $\ell^\pm$ plus two jets, clustering at the $N_\ell = (\ell jj)$ mass, and a second charged lepton, clustering together with N_ℓ at the $W_R = (\ell\ell jj)$ mass. Similarly the Z'

Table 2: Discovery limits of gauge bosons and heavy Majorana neutrinos in LR symmetric theories at the LHC[9].

$pp \to$	$M(W_R/Z')$	$M(N)$
$W_R \to \ell N_\ell$	6.4 TeV	3.3 TeV
$Z' \to N_\ell N_\ell$	4.5 TeV	1.7 TeV

chains. As expected, the W_R channel provides the highest discovery limit[9], shown in Table 2.

Thus, left-right symmetric phenomena can be probed at the LHC in the multi-TeV mass range, in the gauge-boson as well as the neutrino sector.

Acknowledgements

We are grateful to R. McPherson and H. Weerts for the invitation to this talk. Thanks should go to D. Barney, W. Bernreuther and L. Rurua for providing us with unpublished material. We also thank our collaborators G. Blair, A. Djouadi and M. Muhlleitner for allowing us to include unpublished results. Help in preparing the manuscript is acknowledged to S. Günther and T. Plehn.

References

1. D. Karlen, *these proceedings*.
2. D. Barney [CMS], *Contribution to ICHEP98*; see also *CMS, The Electromagnetic Calorimeter, Technical Design Report*, CMS TDR4.
3. W. Bernreuther, M. Flesch and P. Haberl, Aachen Report PITHA 97/34 (RV).
4. S. Dittmaier, M. Krämer, Y. Liao, M. Spira and P.M. Zerwas, DESY 98-111 [hep-ph/9808433].
5. S. Dawson and L. Reina, BNL–HET–98/27 [hep-ph/9808443].
6. A. Djouadi, W. Kilian, M. Muhlleitner and P.M. Zerwas, *Contribution to ICHEP98*.
7. L. Rurua [CMS], *Contribution to ICHEP98*; see also *Discovery Potential for Supersymmetry at CMS*, CMS Note 1998/006.
8. E. Accomando *et al.*, Phys. Rep. **299**, 1 (1998).
9. J. Collot and A. Ferrari [ATLAS], *Contribution to ICHEP98*.
10. M.N. Dubinin, V.A. Ilyin and V.I. Savrin, *Proceedings, Workshop Samara 1997* [hep-ph/9712335]; S. Abdullin *et al.*, INPMSU 98–13/514.
11. G. Blair, DESY/ECFA LC Workshop, *private communication*.

Parallel Session 11

Particle Astrophysics (Dark Matter Searches, Extensive Air Shower, Space and Underground Experiments)

Convenors: Lawrence Sulak (Boston)

Thomas Gaisser (Bartol)

THE DETECTION OF GRAVITATIONAL WAVES
LIGO

Barry C. Barish

California Institute of Technology, Pasadena, CA 91125 USA

The Laser Interferometer Gravitational-Wave Observatory (LIGO) is being developed to directly observe gravitational waves originating from astrophysical sources. A Caltech-MIT collaboration is building two widely separated detectors which will be used in coincidence to search for sources from compact binary systems, spinning neutron stars, supernovae and other astrophysical or cosmological phenomena that emit gravitational waves. The construction of LIGO is well underway and preparations are being made for commissioning phase.

1 Introduction

Einstein first predicted gravitational waves in 1916 as a consequence of the general theory of relativity. In this theory, concentrations of mass (or energy) warp space-time and changes in the shape of such objects cause a distortion that propagates through the Universe at the speed of light. Although gravitational waves have yet to be observed directly, strong indirect evidence resulted from the beautiful experiment of Hulse and Taylor[1]. They studied the neutron star binary system PSR1913+16 and observed by using pulsar timing the gradual speed up of the ~ 8 hour orbital period of this system. This speed up of about 10 seconds was tracked accurately over about 14 years and the result is in very good quantitative agreement with the predictions of general relativity.

Of course, the motivation for direct detection of gravitational waves, based on the empirical desire to 'see these waves.' However, such studies also have enormous potential both to study the nature of gravity in a new regime and to probe the Universe in a fundamentally new way. It is tempting to draw an analogy with the neutrino, where it was 'indirectly observed' by Pauli and Fermi in the 1930's as the explanation for the apparent non-conservation of energy and angular momentum in nuclear beta decay. Decades of rich physics have followed, which were first focussed on the goal of direct detection. Since the direct detection by Reines and Cowan, neutrino physics has been a rich subject, both for studies of the properties of neutrinos themselves (e.g. the question of neutrino mass remains an important topic) and as a probe of the constituent nature of nucleons.

LIGO[2] is designed to directly detect gravitational ways using the technique of laser interferometry. The arms of the interferometer are arranged in an L-shaped pattern that will measure changes in distance between suspended test masses at the ends of each arm. The basic principle is illustrated in figure 1. A gravitational wave produces a distortion of the local metric such that one axis of the interferometer is stretched while the orthogonal direction shrinks. This effect oscillates between the two arms with the frequency of the gravitational wave. Thus,

$$\Delta L = \Delta L_1 - \Delta L_2 = hL$$

where h is the gravitational strain or amplitude of the gravitational wave. Since the effect is linearly proportional to L, the interferometer should have arm length as long as is practical and for LIGO that is 4 km, yielding an expected strain sensitivity of $h \sim 10^{-21}$ for the initial interferometers now being installed.

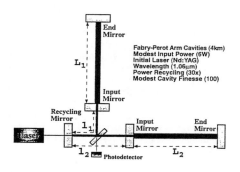

Figure 1: Initial LIGO interferometer configuration

Construction of LIGO is well underway at the two observatory sites: Hanford, Washington and Livingston, Louisiana and the commissioning of the detectors will begin in 2000. The first data run is expected to begin in 2002 at a sensitivity of h ~ 10^{-21}. Incremental technical improvements that will lead to a better sensitivity of h ~ 10^{-22} are expected to follow shortly, and the facility will allow further improved second generation interferometers when they are developed with sensitivity of h ~ 10^{-23}. It is also important to note that all the detectors in the world of comparable sensitivity will be used in a worldwide network to make the most sensitive and reliable detection.

2 Sources of Gravitational Waves

There are a large number of processes in the Universe that could emit detectable gravitational waves. Interferometers like LIGO, will search for gravitational waves in the frequency range f ~ 10Hz- 10KHz. It is worth noting that there are proposals to put interferometers in space which would be complementary to the terrestrial experiments, as they are sensitive to much lower frequencies (f < 0.1 Hz), where there are known sources like neutron binaries or rotating black holes. For LIGO, characteristic signals from astrophysical sources will be sought by recording time-frequency data and examples of such signals include the following:

Chirp Signals: The inspiral of compact objects such as a pair of neutron stars or black holes will give radiation that increases in amplitude and frequency as they move toward the final coalescence of the system. This characteristic chirp signal can be characterized in detail, depending on the masses, separation, ellipticity of the orbits, etc. A variety of search techniques, including comparisons with an array of templates will be used for this type of search. The expected rate of such events is expected to be a few per year within about 200 Mpc from neutron star pairs. The rate is more uncertain for black hole pairs, but due to the heavier masses they make a large signal which will allow a deeper search into the Universe for a given LIGO sensitivity.

Burst Signals: The gravitational collapse of stars (e.g. supernovae) will lead to emission of gravitational radiation. Type I supernovae involve white dwarf stars and are not expected to yield substantial emission. However, Type II collapses can lead to strong radiation, if the core

collapse is sufficiently non-axisymmetric. Estimates of the strengths indicate detection might be possible out to the Virgo Cluster, which would yield rates of one or more per year. The detection will require identifying burst like signals in coincidence from multiple interferometers.

Periodic Signals: Radiation from rotating non-axisymmetric neutron stars will produce periodic signals in the detectors. The gravitational wave frequency is twice the rotation frequency, which is typically within the LIGO sensitivity band for known neutron stars. Neutron stars spin down partially due to emission of gravitation waves. Searches for signals from identified neutron stars will involve tracking the system for many cycles, taking into account the Doppler shift for the motion of the Earth around the Sun, and effects of spin-down of the pulsar. Both targeted searches for known pulsars and general sky searches are anticipated.

Stochastic Signals: Signals from gravitational waves emitted in the first instants of the early universe (t ~10^{-43} sec) can be detected through correlation of the background signals from two or more detectors. Some models of the early Universe can result in detectable signals. Observations of this early Universe gravitational radiation would provide an exciting new cosmological probe.

3 The LIGO Facilities

The LIGO facilities at Hanford, WA and Livingston, LA each have a 4 km L shaped vacuum enclosure, which is 1 meter in diameter. Vacuum is required to reduce scattering off residual molecules that bounce off the walls and are modulated by the small shaking of the vacuum walls modulating this background. In addition, we have installed baffles on the walls to reduce scattering. The large diameter of the tube is both to minimized scattering and to provide the ability to house multiple interferometers within the same facility. Each facility has a 4 km interferometer with test masses housed in vacuum chambers at the vertex and the ends of the L shaped arms. At Hanford, there will also be a 2 km interferometer implemented in the same vacuum system, allowing a triple coincidence requirement. The overall vacuum system is capable of achieving pressures of 10^{-9} torr. We presently have both arms at each site installed vacuum tight. We are beginning to bake the tube to reach the desired high vacuum. We expect to have the entire vacuum system complete, all control systems operational and at high vacuum before the end of 1999.

The initial detector for LIGO is a Michelson interferometer with a couple of special features:

1. The arms are Fabry-Perot cavities to increase the sensitivity by containing multiple bounces and effectively lengthening the interferometer arms. The number of bounces is set to not exceed half the gravitational wave wavelength (e.g. ~ 30 bounces).

2. The interferometers are arranged such that the light from the two arms destructively interferes in the direction of the photodetector, thus producing a dark port. However, the light constructively interferes in the direction of the laser and this light is 're-used' by placing a recycling mirror between the laser and beam splitter. This mirror forms an additional resonant cavity by reflecting this light back into the interferometer, effectively increasing the laser power and thereby the sensitivity of the detector.

Much work has been done over the past decade to demonstrate this configuration and the detailed techniques and required sensitivities in smaller scale laboratory prototypes. This includes experiments on a 40 m prototype interferometer at Caltech, which is a scale model of LIGO, which has provided an excellent test bed to study sensitivity, optics, controls and even some early work on data analysis, noise characterization, etc. We also have built a special interferometer (PNI) at MIT, which has successfully demonstrated our required phase sensitivity, the limitation to the sensitivity at high frequencies.

LIGO is limited in practice by three noise sources:

1. At low frequencies (~10 Hz to 50 Hz), the limitation in sensitivity is set by the level of seismic noise in the system. We employ a seismic isolation system to control this noise that consists of a four-layer, passive vibration isolation stack having stainless steel plates separated by constrained-layer damped springs. This system is contained within large vacuum chambers. The stack supports an optical platform from which the

test mass is suspended. The combination of the seismic isolation stack and the test mass suspension give an isolation from ground motion at the relevant frequencies of about 10 orders of magnitude. The possibility of a more elaborate isolation system and/or the addition of active isolation exists for the future, as well as improved suspension systems. Improvements in this area are planned early in the future improvement program of LIGO.

2. In the middle range of frequencies (~ 50 Hz to 200 Hz) the principle effect limiting the sensitivity is thermal noise. This noise comes partially from the suspension system, where there are violin wire resonances from the steel suspension fibers. However the principal noise source is from the vibrational modes of the test masses. This noise is reduced by the choice of test masses, presently fused silica, to have a very high Q thereby dissipating most of its noise out of our frequency band. The test masses will also improve in the future by using higher Q fused silica, better wires and bonding techniques for the wires, and perhaps even new materials for the test masses, like Sapphire.

Figure 2 shows the expected sensitivity curves vs. frequency for LIGO, along with possible signals discussed above.

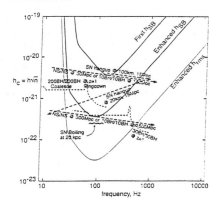

Figure 2: Sensitivity of enhanced LIGO showing various gravitational source signals

3. At the higher frequencies (200 Hz to 5 kHz) the main limitation comes from shot noise and the sensitivity is limited by the power of the laser (or the effective photostatistics). The

initial laser (Nd:YAG) is designed and produced for LIGO, using a master oscillator/power amplifier configuration, which yeild a 10 watt high quality output beam. We also have developed a system to pre-stabilize this laser in power and frequency. Again, we expect to incorporate higher power lasers in the future as they become available.

4   Conclusions and Prospects

The LIGO interferometer parameters have been chosen such that our initial sensitivity will be consistent with estimates needed for possible detection of known sources. Although the rate for these sources have large uncertainty, improvements in sensitivity linearly improve the distance searched for detectable sources, which increases the rate by the cube of this improvement in sensitivity. So, anticipated future improvements will greatly enhance the physics reach of LIGO and for that reason a vigorous program for implementing improved sensitivities is integral to the design and plans for LIGO.

We are now entering into the final year of the construction of the LIGO facilities and initial detectors.   We have formed the scientific collaboration that will organize the scientific research on LIGO. This collaboration already consists of more than 200 collaborators from 22 institutions. By early in the next millenium we will turn on and begin the commissioning of these detectors. We anticipate that we will reach a sensitivity of $h \sim 10^{-21}$ by the year 2002. At that point, we plan to enter into the first physics data run ($\sim$ 2 years) to search for sources. This will be the first search for gravitational waves with sensitivity where we might expect signals from known sources.   Following this run in 2004, we will begin incremental improvements to the detector interleaved with further data runs. We expect to reach a sensitivity of $h \sim 10^{-22}$ within the next 10 years, making direct detection of gravitational waves within that time frame reasonably likely.

5   References

1. R.A. Hulse and J.H. Taylor. *Astrophys. J.*, **324**, 355, (1975) and J.H.. Taylor. *Rev Mod. Phys.*, **66**, 1994

2. A. Abramovici et al, *Science*, **256**, 325, (1992).

STATUS OF THE GRAVITATIONAL WAVE DETECTOR VIRGO

presented by
R. FLAMINIO

Laboratoire d'Annecy-le-Vieux de Physique des Particules,
Chemin de Bellevue BP110, Annecy-le-Vieux, F-74941, France
E-Mail: flaminio@lapp.in2p3.fr

Y.Acker[f], C.Arnault[f], P.Astone[k], D.Babusci[c], R.Barillet[f], F.Barone[e], G.Barrand[f], M.Barsuglia[f], F.Bellachia[a], J.L.Beney[f], M.Bermond[a], M.Bernardini[j], J.P.Berthet[f], R.Bilhaut[f], C.Boccara[h], D.Boget[a], A.Bozzi[j], S.Braccini[j], C.Bradaschia[j], A.Brillet[f], V.Brisson[f], F.Bronzini[k], J.Cachenaut[f], G.Cagnoli[i], E.Calloni[e], B.Caron[a], T.Carron[a], C.Casciano[j], D.Castellazzi[a], F.Cavalier[f], G.Cella[j], R.Chiche[f], F.Chollet[a], A.Ciampa[j], F.Cleva[f], J.P.Coulon[f], E.Cuoco[j], G.Curci[j], S.Cuzon[f], E.D'Ambrosio[j], J.B.Daban[h], G.Daguin[a], V.Dattilo[j], P.Y.David[a], M.Davier[f], G.De Carolis[j], R.De Salvo[j], M.Dehamme[f], L.Derome[a], M.Dialinas[f], L.Di Fiore[e], A.Di Virgilio[j], C.Drezen[a], A.Dubois[h], D.Dufournaud[a], C.Eder[f], D.Enard[j], A.Errico[j], H.Fang[c], G.Feng[j], I.Ferrante[j], F.Fidecaro[j], R.Flaminio[a], S.Frasca[k], F.Frasconi[j], A.Gaddi[j], L.Gammaitoni[i], P.Ganau[d], F.Garufi[e], M.Gaspard[f], A.Gennai[j], G.Gennaro[j], L.Giacobone[a], A.Giazotto[j], G.Giordano[c], C.Girard[a], P.Gleyzes[h], A.Grado[e], X.Grave[a], H.Heitmann[f], P.Hello[f], R.Hermel[a], P.Heusse[f], A.Hrisoho[f], M.Iannarelli[c], J.M.Innocent[f], E.Jules[f], J.Kovalik[f], P.La Penna[j], J.C.Lacotte[a], B.Lagrange[d], J.C.Le Marec[a], M.Leliboux[h], B.Lieunard[a], V.Loriette[h], G.Losurdo[j], J.C.Lucenay[f], J.M.Mackowski[d], M.Maggiore[j], E.Majorana[k], C.N.Man[f], S.Mancini[j], P.T.Manh[f], F.Marchesoni[i], J.A.Marck[f], P.Marin[f], F.Marion[a], J.C.Marrucho[f], L.Massonnet[a], G.Matone[c], L.Matone[f], M.Mazzoni[b], C.Mehmel[a], M.Mencik[f], C.Michel[d], L.Milano[e], R.Morand[a], N.Morgago[d], B.Mours[a], P.Mugnier[a], R.Nahoum[f], F.Palla[j], C.Palomba[k], H.B.Pan[j], F.Paoletti[j], A.Pasqualetti[j] , R.Passaquieti[j], D.Passuello[j], D.Pelat[f], M.Perciballi[k], L.Pinard[d], R.Poggiani[j], P.Popolizio[j], M.Punturo[i], P.Puppo[k], F.Raffaelli[j], P.Rapagnani[k], S.Rapisarda[j], A.Reboux[f], V.Reita[f], A.Remillieux[d], F.Ricci[k], J.P.Roger[h], P.Roudier[f], V.Sannibale[a], S.Solimeno[e], R.Sottile[a], R.Stanga[b], R.Taddei[j], M.Taubman[f], M.Taurigna[f], E.Turri[c], D.Verkindt[j], A.Vicere[j], J.Y.Vinet[f], M.Yvert[a], Z.Zhou[j]

a - LAPP Annecy	b - INFN Firenze	c - INFN Frascati
d - IPN Lyon	e - INFN Napoli	f - LAL Orsay
g - LAS Orsay	h - ESPCI Paris	i - INFN Perugia
j - INFN Pisa	k - INFN Roma	

The gravitational interaction is described by Einstein's theory of General Relativity. According to the theory, gravitational waves are emitted by accelerated masses and should be observable as small changes of the space-time metric. The detection of these signals will constitute a basic step forward in the study of the gravitational force under strong field conditions. VIRGO is one of the two collaborations in the world that are now building new kilometer-scale detectors. The status of the project as well as the main experimental difficulties and data analysis issues are presented.

1 Introduction

VIRGO [1,2] is a collaborative effort between the CNRS in France and the INFN in Italy. It involves scientists from Firenze, Frascati, Napoli, Perugia, Pisa and Rome in Italy and from Annecy, Lyon, Orsay and Paris in France. The goal of the project is the construction of a Michelson interferometer of 3 km arm length, aimed at detecting gravitational waves in the frequency range from 10 Hz to a few kHz. The detector, currently under construction, is located in Cascina, Italy, 10 km from Pisa.

The radiation of gravitational waves from accelerating masses is predicted by general relativity. According to the theory, gravitational waves are perturbations of the space-time metric that propagate at the speed of light and that are transverse to the propagation direction. As for electro magnetic waves, there are two independent polarization states, but the wave force field has a quadrupolar symmetry and the transformation properties under rotation are those of a spin 2 wave (see figure 1). The wave force field causes an antisymmetric change of the distance between free masses placed along two orthogonal directions transverse to the propagation direction of the wave (see figure 1). The relative difference in the displacements along two orthogonal directions $\delta L/L$ is equal the gravitational wave amplitude h.

The gravitational wave force field is such that as it propagates across a Michelson interferometer having free suspended masses as mirrors, the interferometer arm

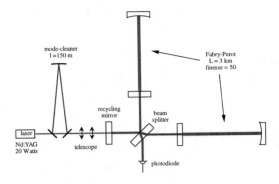

Figure 2: *VIRGO: interferometer optical scheme.*

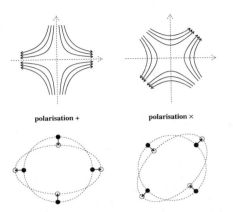

Figure 1: *Gravitational wave force field (above) and the effect on a set of masses placed on a circle (below) for the two polarization states.*

lengths are changed anti-symmetrically and a variation of the light power transmitted at the interferometer output port will be observed[5]. Interferometric detectors like VIRGO are based on this principle.

The amplitude h of the gravitational wave is roughly proportional to the non-spherical component of the source kinetic energy ϵ_{ns}^{kin} and decreases linearly with the source distance d:

$$h \simeq \frac{G}{c^4} \frac{\epsilon_{ns}^{kin}}{d} \qquad (1)$$

The small coupling constant makes the generation of detectable gravitational waves on earth impossible at present. On the other hand many astrophysical process are believed to be sources of gravitational waves[3]. A process involving the conversion of 0.01 solar masses into gravitational waves at the distance of the VIRGO cluster (10 Mpc) would produce a gravitational wave with an amplitude of about h=10^{-21} on earth. Such a gravitational wave would cause a displacement of about 10^{-18} m between two free masses placed 1 km apart. The detection of such a tiny displacement is now possible.

Processes involving compact objects in fast motion like coalescing binaries, supernovae explosions and rotating neutron stars are some of the astrophysical phenomena that should produce gravitational waves. The period slow down of the binary pulsar PSR 1913+16[4] confirms the theoretical prediction and represents the first indirect observation of a process involving the emission of gravitational waves. A few such systems have been detected in the galaxy. It is expected that in a radius of 200 Mpc a few binary pulsars reach the coalescing phase each year. These events should produce chirp-like gravitational waves with an amplitude large enough to be detected. The detection of such a signal represents a unique method to study the physics inside a neutron star and to test the theory of gravitation in strong field conditions.

2 Detector overview

The VIRGO interferometer optical scheme is shown in figure 2. The change in length of the interferometer arms induced by a gravitational wave is proportional to the length of the arms. As a consequence, the sensitivity of such a detector increases linearly with such lengths. Practical problems and cost limit the length of such a detector to a few km. The signal is increased by placing a Fabry-Perot in each arm of the interferometer to bounce the light back and forth many times along the arms. Thus the optical path can be increased up to more than 100 km.

In order to detect the tiny displacement induced by the expected gravitational waves the mirror should be completely isolated from all external perturbations. The effect of seismic noise will be reduced by suspending all the mirrors to a cascade of seven pendulums, each 1m in length, called Super-Attenuator[7] (see figure 3). Vertical compliance is added to each stage by means of prestressed metal cantilever blades whose spring constant is softened with magnetic anti-springs. Thus each stage has resonant frequencies below 1 Hz in all six degrees of freedom. This system will provide an attenuation of the seismic vibration by more than ten orders of magnitude at 10 Hz.

Acoustic noise as well as air index of refraction fluctuations may also limit the sensitivity of such a detector. To avoid these effects the whole interferometer should be placed under vacuum. The VIRGO vacuum system will consist of two 3 km long and 1.2 m diameter tubes and several 10 m high towers containing the required suspensions (see figure 4). The pressure level along the light path should be kept below 10^{-9} mbar.

The ultimate limitation to the sensitivity of a laser interferometer gravitational wave detector is due to photon shot-noise. In order to reduce its effect of this noise the light power impinging on the beam splitter should be as large as possible. Current technology allows the construction of lasers delivering a power of the order of tens of Watts with the required frequency and power stability. By adding an additional mirror in between the laser and the beam splitter it is possible to "recycle" the unused light reflected towards the laser thus increasing the light power impinging on the beam splitter. This technique, called power, recycling [6] is equivalent to using a more powerful laser.

The VIRGO collaboration will use these techniques in order to reach a power of more than 10 kW inside the Fabry-Perot cavities. This should allow to reduce the shot noise level to an equivalent strain spectrum noise of $3 \ 10^{-23}/\sqrt{\text{Hz}}$ at 100 Hz. The required light source stability will be obtained by using a stabilized injection locked Nd:YAG laser delivering 20 Watts.

In order to increase the amount of recycled light, the losses due to scattering and/or absorption inside the mirrors and on the reflecting coating layers should be kept as small as possible. All the optics should respect very stringent specifications in order to keep losses down to a few ppm. This is especially true for the Fabry-Perot mirrors. Furthermore these mirrors have to be relatively large in order to deal with the beam dimensions (more than 10 cm diameter at the end mirror) and quite massive in order to reduce the effect of pendulum thermal noise. The VIRGO interferometer will use as mirrors cylinders of fused silica 35 cm in diameter and 10 or 20 cm thick. The mirror surfaces will be polished in their central area to a precision of $\lambda/100$. The coating of all large optics is developed inside the VIRGO experiment by the Lyon group.

All the optical cavities will be kept at resonance by a complex system of servos using the signals read by several photodiodes and acting on the mirror position. These servos will also keep the interferometer output on the null interference condition in order to maximize the detector sensitivity. Other independent control loops will act on the suspension in order to reduce the low frequency oscillation amplitude and on the laser in order to keep the

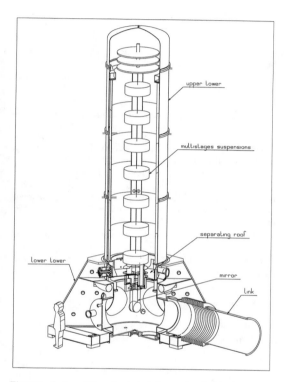

Figure 3: *A view of the seismic isolation system inside the vacuum tank.*

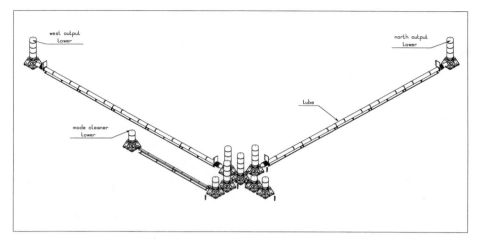

Figure 4: *The VIRGO vacuum system; each 'tower' contains an optical element suspended to a seismic isolator.*

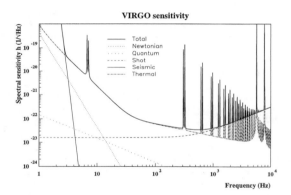

Figure 5: *The expected VIRGO spectral sensitivity*

frequency and the power at the required level of stability.

3   Detector sensitivity

The VIRGO expected sensitivity expressed in terms of an equivalent strain spectrum is shown in figure 5.

The use of the Super-Attenuator should keep the seismic noise effects at a negligible level above a few hertz. From a few Hz up to 100 Hz the sensitivity will be limited by pendulum thermal noise. The effect of the pendulum thermal noise decreases as the frequency increases and around 100 Hz the effect of mirror internal thermal noise is expected to be the main limitation to

the detector sensitivity. Around 200 Hz the sensitivity reaches the lowest level of about $3\ 10^{-23}/\sqrt{\text{Hz}}$.

Above a few hundred Hz the light storage time in the arms becomes non negligible and the interferometer response decreases. As a consequence the shot noise level gets larger and it becomes the main limitation to the sensitivity.

Various resonances peaks are visible above 300 Hz. These peaks in the noise spectrum are due to the excitation by thermal noise of the violin modes in the mirror suspension wires.

4   Status and Schedule

The VIRGO experiment has been jointly approved by INFN and CNRS in 1994.

The construction of the infrastructure is organized in two steps. During the first step the central area is built. Central area includes the main building (the place where the laser, the detection system, most of the optics and the assembly labs are installed) and the control building (where the offices and the control room are placed). Then the 3 km long arms and the terminal buildings will be built and the tube installed. The central area is now near completion and the tube production will be started by the end of the year.

The detector start-up follows the construction evolution. As soon as the central building will become available (by the end of the year) the vacuum chambers will be installed and a Michelson interferometer with 6m long arms will be suspended under vacuum. Even if shorter this interferometer will use the final suspensions, control system, vacuum and data acquisition thus providing a good test bed to check most of the technical and scientific choices made by VIRGO.

As soon as the arms will become available the detector will gradually move to the final 3 km long configuration. The full detector will become operational in 2002.

5 Signal analysis

Different kind of gravitational wave signals are expected to be emitted by astrophysical sources.

Pulsars and more generally rotating neutron stars are expected to produce continuous and periodical gravitational waves. Most of these are foreseen to radiate at frequencies around 1 Hz but some of them are known to rotate at higher frequency and thus are in the VIRGO bandwidth. Due to the earth rotation around the sun, such a signal will be Doppler shifted. Such a Doppler shifted signal can, if strong enough, be detected with a single detector.

Impulsive signals with rather well known waveforms are expected to be produced by coalescing binaries. The detection of such a waveform in the detector output signal will also be the evidence of a gravitational wave. Techniques as matched filtering can be used in this case.

Impulsive signals are also expected from supernovae (galactic or extra-galactic). The waveforms are mostly unknown in this case. A search for such a signal will be limited by non Gaussian events due to environment and technical noises. To reduce this rate a network of sensitive environmental noise detectors are used but the evidence for the detection can only come from the synchronous detection of the same signal in different detectors.

The rather large amounts of data produced by such a detector (about 5 Mbyte/s) requires on-line triggering to be used in order to reduce the data stored on tape and to help off-line data analysis [12]. The goal will be to strongly reduce the amount of information coming from the additional environmental sensors that is the major contribution to the total amount of data stored on tape.

A network of gravitational wave detectors based on laser interferometer is currently being built in the world. Apart from VIRGO, two other km scale detectors are being built in the U.S. by the LIGO project [8]. Shorter prototypes are being built in Germany by the GEO collaboration [9] and in Japan by the TAMA project [10] in order to test new technologies for second generation interferometers. A common data format [11] is being developed and adopted by the various projects in order to help coincidence analysis as soon as the detectors will become operational.

References

1. VIRGO Coll., Final Conceptual Design (1992)
2. VIRGO Coll., Final Design, Version 0 (1995)
3. K.S.Thorne, Gravitational Radiation in: 300 Years of Gravitation, Cambridge University Press (ed.), Cambridge (1987) 330-452
4. J.H.Taylor and J.M.Weisberg, The Astrophysical Journal **253** (1982) 908-920
5. M.E.Gersenshtein and V.I.Pustovoit, Sov. Phys. - JETP **16**, (1963), 433
   R.Weiss, Q.Prog. Rep. Lab. Electron. MIT **105** (1972) 54
6. R.W.P.Drever, Gravitational Radiation, les Houches 1982, eds. N.Deruelle and T.Piran (North-Holland Amsterdam) (1983) 321
7. A.Giazotto, Physics Reports **182** (1989) 365-425
8. A.Abramovici et al., Science 256 (1992) 325-333
9. J.Hough et al., GEO 600: Current Status and Some Aspects of the Design, Proceedings of the TAMA International Workshop on Gravitational Detection, see ref 11
10. T.Tsubono et al., TAMA project, Proceedings of the TAMA International Workshop on Gravitational Detection, see ref 11
11. LIGO Data Group and VIRGO Data Acquisition Group, LIGO-T970130-B-E and VIRGO-SPE-LAP-5400-102 (1997)
    B.Mours, A Common Data Format for Gravitational Wave Interferometers, Proceedings of the TAMA International Workshop on Gravitational Detection, eds. K.Tsubono M.-K.Fujimoto, K.Kuroda, Universal Academy Press, Tokyo, Japan (1997) 31
12. R.Flaminio et al., Astroparticle Physics 2 (1994) 235-248

Search for Rare Particles with the MACRO Detector

Fabrizio Cei for the MACRO Collaboration

University of Michigan, Department of Physics, 2071 Randall Laboratory, 500 East University, Ann Arbor, MI 48109-1120, USA
E-mail: cei@pooh.physics.lsa.umich.edu

We present the results of the search for rare particles (magnetic monopoles, nuclearites, WIMPs and LIPs) with the MACRO detector. For magnetic monopoles (the main goal of the experiment) our limit is ~ 0.4 times the Parker bound for $10^{-4} \leq \beta \leq 10^{-1}$.

1 Introduction

MACRO [1,2] [a] at the Gran Sasso Laboratory is a large area underground detector devoted to the search for rare events in the cosmic radiation. It is optimized to search for GUT magnetic monopoles (MM), but can also perform many observations relevant to astrophysics, nuclear, particle and cosmic ray physics. The main MACRO physics items are the study of atmospheric neutrinos and oscillations, the search for MMs, the study of the high energy underground muons, the primary cosmic ray composition, the measurement of the muon residual energy spectrum, the search for low energy stellar collapse neutrinos and the high energy neutrino astronomy.

Here we present the results of the search for rare particles: magnetic monopoles and nuclearites, weakly interacting massive particles and lightly ionizing particles. All these searches gave (until now) null results, setting significant limits on the fluxes of these rare particles.

2 The MACRO detector

The MACRO detector consists of six supermodules (total sizes $77 \times 12 \times 9$ m^3), each one divided in a lower and an upper part. The lower part is made by ten horizontal planes of limited streamer tubes, interleaved with seven rock absorber layers, and two liquid scintillation counter layers on the top and bottom. The lateral walls are closed by four "vertical detectors", formed by a liquid scintillator layer sandwiched between two sets of streamer tubes (three planes each). The upper part is made by two "vertical detectors" on the East and West faces and by a roof with one layer of scintillators sandwiched between two planes of tubes; this part is left open on the North and South faces to house the electronics. A nuclear track detector [3] is located horizontally in the middle of the lower part and vertically on the East and North walls.

The scintillation counters are equipped with specific triggers for rare particles, muons and stellar gravitational collapse neutrinos and by 200 MHz WFDs. The streamer

---

[a] For the MACRO author list see the D. Michael paper in these proceedings

tubes are read by 8-channel cards which discriminate the signals and send the analog information (time development and total charge) to an ADC/TDC system; the discriminated signals form two different chains of TTL pulses, which are the inputs for the streamer tube Fast and Slow Particle Triggers.

3 Magnetic Monopoles

Massive ($M_M \sim 10^{17}$ GeV/c^2) magnetic monopoles arise spontaneously in Grand Unified Theories (GUTs) [4] of electroweak and strong interactions. Magnetic monopoles of such a large mass cannot be produced with accelerators and must be searched in the cosmic radiation. The MACRO experiment was designed to be sensitive to monopoles at a flux level well below the Parker Bound [5] $\Phi_M \lesssim 10^{-15}$ cm^{-2} s^{-1} sr^{-1} in the monopole velocity range $4 \times 10^{-5} < \beta < 1$. The use of three subdetectors (liquid scintillators, streamer tubes and nuclear track detector) ensure redundancy of information, multiple cross-checks and independent signatures for possible monopole candidates. The results reported here are obtained using the various subdetectors in a stand-alone and in a combined way. All the limits refer to monopoles with unit Dirac magnetic charge ($g = 137/2$ e), catalysis cross section $\sigma < 10$ mb (we do not consider the monopole induced nucleon decay) and isotropic flux (we consider monopoles with enough kinetic energy to traverse the Earth); the last condition sets a β-dependent mass threshold of $\sim 10^{17}$ GeV for $\beta \sim 5 \times 10^{-5}$ and lower (down to 10^{10} GeV) for faster monopoles.

3.1 Searches with the scintillator subdetector

The energy loss, arrival time and velocity of a particle passing in the scintillator system are measured by using the total charge and shape of the photomultiplier pulses.

In the low velocity region ($1.8 \times 10^{-4} < \beta < 3 \times 10^{-3}$) we studied the photomultiplier waveforms, looking for the wide, flat and small amplitude signals (or long trains of single photoelectrons) expected for a monopole. No candidates were found in two independent analyses [6,7]; the flux upper limits (90 % C.L.) are 5.6 and 4.1×10^{-15} cm^{-2} s^{-1} sr^{-1} (curves "A" and "B" in fig. 1).

In the medium velocity range ($1.2 \times 10^{-3} < \beta < 10^{-1}$) the monopoles are searched using the data collected between October 1989 and March 1998 by the stellar gravitational collapse trigger PHRASE [8]. The events selected in this β range are rejected since their pulse width is smaller than the expected counter crossing time or since the light produced is lower than that expected for a monopole [9]. The flux upper limit at 90 % C.L. is 4.3×10^{-16} cm^{-2} s^{-1} sr^{-1} (curve "D" in fig. 1). The technique is fully discussed in [10].

Finally, in the high velocity range ($\beta > 0.1$) the data collected by the muon trigger ERP [8] are used. All events are rejected since the measured energy deposit in two counter layers is much lower than the energy loss [9] expected for a fast monopole. The 90 % C.L. flux upper limit is 4.4×10^{-15} cm^{-2} s^{-1} sr^{-1} (curve "C" in fig. 1) [10].

3.2 Search with the streamer tubes subdetector

The MACRO streamer tubes [2,11] are filled with a mixture of He (73 %) and n-pentane (27 %); the Helium was chosen to allow the detection of slow ($\beta \lesssim 10^{-3}$) monopoles by the Drell-Penning effect [12]. The hits on the streamer tubes system and the charge collected on each tube provide measurements of track, velocity and energy loss of an ionizing particle [11]. The data were collected from February 1992 to October 1997. The monopole analysis is based on the search for clean single tracks of well reconstructed velocity; it was checked that the trigger selection and analysis procedure are velocity independent. No candidated survived; the flux upper limit (90 % C.L.) is $\Phi < 4.5 \times 10^{-16}$ cm^{-2} s^{-1} sr^{-1} for $1.1 \times 10^{-4} < \beta < 5 \times 10^{-3}$ (curve "Streamer" in fig. 1) [7,13].

3.3 Search with the nuclear track subdetector

The MACRO nuclear track subdetector is made by three layers of LEXAN and three layers of CR39; its total surface is 1263 m^2 and its acceptance for fast monopoles is ≈ 7100 m^2 sr. A calibration of the CR39 with slow and fast ions showed that its response depends on the restricted energy loss only [3]. The track-etch subdetector is used in a stand-alone or in a "triggered" mode by the streamer tubes and the scintillator systems. A total surface of 181 m^2 was etched, with an average exposure time of 7.23 years; the flux upper limits (90 % C.L.) are ~ 0.88 and $\sim 1.3 \times 10^{-15}$ cm^{-2} s^{-1} sr^{-1} at $\beta \sim 1$ and $\beta \sim 10^{-4}$ respectively (curves "CR39" in fig. 1) [7,13].

3.4 Combined search

The fast monopoles are expected to release a huge amount of energy by ionization/excitation without pro-

ducing showers for $\beta < 0.99$. A monopole search based on the energy deposition only can be seriously affected by a large background due to showering cosmic rays, but such a background is efficiently rejected by a combined analysis. This analysis uses at the same time the data collected by the scintillators and by the streamer tubes, requiring large photomultiplier pulses, an isolated single track and a high streamer charge per unit path length. Only few (~ 5/year) events survive and are looked for in the appropriate track-etch sheets. In 667 days of live time no candidates were found; the 90 % C.L. flux upper limit is 1.5×10^{-15} cm^{-2} s^{-1} sr^{-1} for $5 \times 10^{-3} < \beta < 0.99$ (curve "E" in fig. 1) [14].

3.5 Conclusions about magnetic monopoles

We show in fig. 1 the flux upper limits obtained using the various MACRO subdetectors. Since each subdetector can rule out, within its acceptance and sensitivity, a potential candidate from the others, we obtain a global MACRO limit (curve "GLOBAL" in fig. 1), as an "OR" combination of the separate results. The prescriptions used for this combination are described in [7]. In fig. 2 we compare the MACRO combined result with

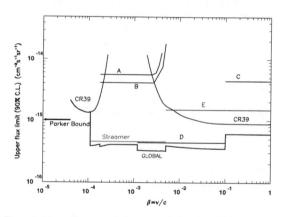

Figure 1: Magnetic monopole flux upper limits obtained using the various MACRO subdetectors.

that obtained by other experiments [15,16,17,18,19,20,21] and with the Parker Bound. Our limit is at the level of 0.3 times the Parker Bound for $\beta > 10^{-4}$ and is the best existing for $10^{-4} < \beta < 5 \times 10^{-2}$.

4 Nuclearites

The results obtained using the liquid scintillator and the nuclear track subdetectors can be, at least in part, extrapolated to the search for nuclearites [22], hypothesized

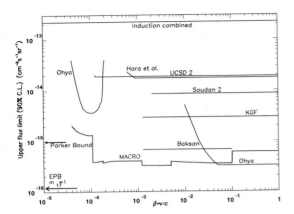

Figure 2: Magnetic monopole flux upper limits obtained by MACRO and by other experiments.

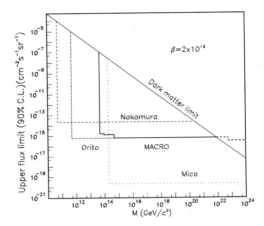

Figure 3: Nuclearite flux upper limits obtained by MACRO and by other experiments; the dark matter bound is also shown.

nuggets of strange quark matter (the streamer tubes are not sensitive to nuclearites because of the low density of the filling gas [27]). By studying the mechanism of nuclear energy loss [23] it was shown that the scintillators are sensitive to nuclearites [26] down to $\beta \lesssim 10^{-4}$ and the CR39 [27] down to $\beta \approx 10^{-5}$. The MACRO flux upper limits are $\sim 4 \times 10^{-16}$ cm^{-2} s^{-1} sr^{-1} for $M_N > 0.1$ g and about 2 times higher for $M_N < 0.1$ g (the lighter nuclearites cannot traverse the Earth and then only the down-going flux must be considered). Assuming a nuclearite velocity at the ground level $\beta = 2 \times 10^{-3}$ we compared our limit with the results of other experiments[19,24,25] and with the dark matter bound; our results look competitive in a large range of nuclearite masses (10^{14} GeV/$c^2 \lesssim M_N \lesssim 10^{21}$ GeV/c^2) (see fig. 3 [b].) For further details on this search see [27].

5   WIMPs

The *Weakly Interacting Massive Particles* (WIMPs) are important candidates for the Cold non-baryonic part of the Dark Matter in the Universe [28]. Between the various Cold Dark Matter candidates subject to weak interactions one of the most promising is the supersymmetric neutralino $\tilde{\chi}$ [29].

In supersymmetric theories where the R parity is conserved there exists a lightest stable supersymmetric

---

[b]The limit is prolonged above the Dark matter bound to show the transition to an isotropic flux for $M > 0.1$ g $\left(\approx 4.5 \times 10^{22} \text{ GeV}/c^2\right)$

particle (LSP), which is the natural candidate for the Dark Matter since its expected density is close to the critical one: $\Omega_{LSP} \sim 1$. In many theories the LSP is the neutralino $\tilde{\chi}$, the simplest linear combination of the gaugino and higgsino eigenstates. The $\tilde{\chi}$ mass depends on the supersymmetric parameters, as, for instance, the gaugino and higgsino mass parameters M_1, M_2 and μ and the ratio of the Higgs doublet vacuum expectation values $\tan \beta$. These parameters are constrained by accelerator searches and a lower limit on $m_{\tilde{\chi}}$ was set by LEP 2 data: $m_{\tilde{\chi}} > 20 \div 30$ GeV [30]. The search for WIMPs in underground detectors can probe complementary regions of the parameter space.

5.1   WIMP searches in MACRO

The WIMPs are indirectly searched in MACRO by using upward going muons. A WIMP intercepting a celestial body can lose its energy and be trapped in the core of this body and annihilate with an other WIMP; the decay of the annihilation products produces high-energy ν's which can be detected in an underground detector as upward going muons. Then, a statistically significant excess of upward going muons from the direction of a celestial body can be a hint for a WIMP-WIMP annihilation in the core of that body. Some possible traps were proposed and some calculations of the WIMP annihilation rate in the Earth and Sun and of the corresponding upward going muon fluxes were performed [29,31,32]. This indirect search achieves a better signal to noise ratio for high WIMP masses, since the higher the WIMP mass, the more the upgoing muon follows the parent neutrino direction. The technique was already used by other experiments [33,34,35].

The upward going muons in MACRO are selected by the time-of-flight technique, requiring a track in the streamer tubes and a time of flight between the scintillation counter layers consistent with a $\beta \sim 1$ particle from below. Details on the selection criteria can be found in [36].

5.2 Search for WIMP annihilation in the Earth

Upward going muons from WIMP annihilation in the Earth core are searched in angular cones ($3 \div 30°$ wide) around the vertical. We used 517 events, collected in 3.1 years of live time, requiring a minimum crossing of 200 g cm^{-2} of rock absorber. After background subtraction the number of selected events is $487 \pm 22_{stat}$, to be compared with $653 \pm 111_{theor}$ expected from a Monte Carlo calculation. Most of the deficit lies around the vertical, where the WIMP annihilation signal is expected, but also where the efficiency and acceptance of the apparatus are best known. To set a conservative limit on the WIMP flux we assumed that the number of measured and expected events are equal and normalized the expected distribution to the factor 0.85, corresponding to the ratio between measured and expected events for $\theta > 30°$. With this prescription our limits on the WIMP flux from the Earth range from 0.4 to 2.3×10^{-14} cm^{-2} s^{-1} for angular windows from 3 to 30°.

5.3 Search for WIMP annihilation in the Sun

In the search for WIMPs from the Sun, since the background for moving sources is lower than for steady sources, we used an enlarged sample (762 events) which includes also muons partially contained in the apparatus (produced by neutrino interactions in the absorber in the MACRO lower part) and muons crossing < 200 g cm^{-2} of rock absorber. The simulation was obtained by the data themselves to include properly the effects of the partially contained events and the arrival times were extracted randomly during the whole measurement time to take into account possible drifts of detection efficiency. The angular distribution of the upward going muon events do not show any significant excess around the Sun direction. Then, we set upper limits on the WIMP flux from the Sun ranging from 1.7 to 6.0×10^{-14} cm^{-2} s^{-1} in the angular window $3 \div 30°$.

5.4 Limits on the WIMP fluxes and $\tilde{\chi}$ mass from the angular distributions

As already stated, the angle between the upward going muon and the neutrino from the WIMP annihilation depends mainly on the WIMP mass. We performed a full

Monte Carlo calculation of the expected angle between the upward going muon and the Earth or Sun directions for neutralino masses from 60 GeV to 1000 GeV. This calculation uses the shape of the upward going muon signals from $\tilde{\chi} - \tilde{\chi}$ annihilation in the Earth and the Sun computed in [31], propagates the muons through the rock to the apparatus (with the cross sections given in [37]) and includes the angular smearing produced by the experimental resolution. As expected, the angle becomes narrower when $m_{\tilde{\chi}}$ increases, but also in the less favourable case more than 90 % of the signal is contained in an angular cone $\theta < 15°$ around the selected direction. For each $m_{\tilde{\chi}}$ we determine the cone which collects the 90 % of the expected signal and define a corresponding 90 % C.L. limit on the upward going muon flux from $\tilde{\chi} - \tilde{\chi}$ annihilation. These limits for the Sun are shown in fig. 4 and compared with the fluxes computed in [31], varying some supersymmetric parameters ($M_1, \mu, \tan\beta$ etc.).

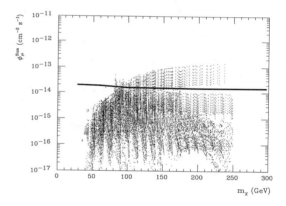

Figure 4: Upward going muon flux from the Sun vs $m_{\tilde{\chi}}$ computed in[31]. The solid line is the 90 % C.L. MACRO flux limit.

Fig. 4 shows that, thanks to the improved statistics and exposure, our data start to constrain the supersymmetric theoretical models [38].

6 LIPs

Fractionally charged particles have been actively searched for since many years, but without success; the detection of such particles would be a proof of their existence and/or of a lack of the confinement hypothesis under some circumstances. The family of the possible fractionally charged particles includes the quarks [39] (with charge $|e|/3$ and $2/3 |e|$) and a variety of other particles predicted by Grand Unified Theories, with charges ranging from $1/5$ $|e|$ up to $2/3$ $|e|$[40,41,42]. Particles with fractional charge deposit less energy than particle with unit charge since the energy loss is proportional to the square of the charge; such particles are called *Lightly Ionizing Particles* (LIPs).

A specialized trigger was developed in MACRO for the LIP search; this trigger uses the low energy events collected by the PHRASE system and performs a four-fold coincidence between the signals coming from three counters (each one in a different scintillator layer) and from the streamer tubes. The LIP trigger is sensitive down to $|Q/e| = 1/5$; the corresponding energy loss in the MACRO counters is $\Delta E \approx 1.6$ MeV. The LIP trigger gives the stop to the WFD system; the WFD data provide high quality measurements of the energy and timing of the events. For each LIP event, the maximum energy loss in the three scintillator layers is computed and compared with that expected for a fractionally charged particle. The efficiency as a function of the charge is obtained by a Monte Carlo simulation which takes into account the trigger behaviour close to the threshold and the cosmic muon background. No candidate survived out of 1.2 million triggers; the 90 % C. L. upper limit for an isotropic flux is $\Phi \leq 9.2 \times 10^{-15}$ cm^{-2} s^{-1} sr^{-1}. This limit is shown in fig. 5 (solid line) and compared with that set by other collaborations[43,44].

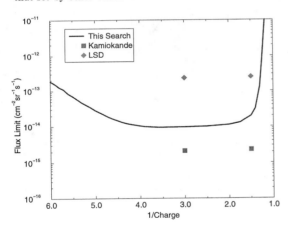

Figure 5: 90 % C.L. flux upper limits for LIP set by MACRO (solid line) and by other experiments as a function of the LIP charge.

We stress the fact that our experiment is the first one to be sensitive down to $|Q/e| = 1/5$ and that we can perform a high quality search thanks to the combined signature provided by the streamer tubes and the scintillators: only few events over 1.2 millions needed a hand scanning (instead of one event over few thousands in Kamiokande). This search is discussed in detail in [45].

References

1. MACRO Coll. NIM **A264** (1988) 18
2. MACRO Coll. NIM **A324** (1993) 337
3. MACRO Coll. LNGS **94/115** (1994)
4. J. Preskill Phys. Rev. Lett. **43** (1979) 1365
5. E. M. Parker et al. Phys. Rev. **D26** (1982) 1926
6. MACRO Coll. Phys. Rev. Lett. **72** (1994) 608
7. MACRO Coll. Phys. Lett. **B406** (1997) 249
8. MACRO Coll. Astropart. Phys. **1** (1992) 11
9. S. Ahlen & G. Tarlé Phys. Rev. **D27** (1983) 688
10. MACRO Coll. Astropart. Phys. **6** (1997) 113
11. MACRO Coll. Astropart. Phys. **4** (1995) 33
12. S. Drell et al. Phys. Rev. Lett. **50** (1983) 644
13. MACRO Coll. MACRO/PUB **98/3** (1998)
14. MACRO Coll. INFN/AE-**97/19** (1997)
15. S. Bermon et al. Phys. Rev. Lett. **64** (1990) 839
16. K. N. Buckland et al. Phys. Rev. **D41** (1990) 2726
17. J. L. Thron et al. Phys. Rev. **D46** (1992) 4846
18. Baksan Coll. 21st ICRC, Adelaide (1990) 83
19. S. Orito et al. Phys. Rev. Lett. **66** (1991) 1951
20. Kolar Coll. 21st ICRC, Adelaide (1990) 95
21. T. Hara et al. 21st ICRC, Adelaide (1990) 79
22. E. Witten Phys. Rev. **D30** (1984) 272
23. A. De Rújula & S. Glashow Nature **312** (1984) 734
24. S. Nakamura et al. Phys. Lett. **B263** (1991) 529
25. P. B. Price Phys. Rev. **D38** (1988) 3813
26. MACRO Coll. Phys. Rev. Lett. **69** (1992) 1860
27. MACRO Coll. INFN/AE-**97/20** (1997)
28. J. R. Primack DARK'96, Heidelberg (1996)
29. G. Jungman et al. Phys. Rep. **267** (1996) 1955
30. OPAL Coll. accepted by Z. Phys. **C**; ALEPH Coll. Z. Phys. **C72** (1996) 549
31. A. Bottino et al. Astropart. Phys. **3** (1995) 65
32. L. Bergström et al. Phys. Rev. **D55** (1997) 1765
33. Baksan Coll. Nucl. Phys. **B48** (Proc. Suppl.) (1996) 83
34. Kamiokande Coll. Phys. Rev. **D48** (1993) 5505
35. IMB Coll. Phys. Lett. **B188** (1987) 388
36. D. Michael this proceedings; MACRO Coll. submitted to Phys. Lett.
37. P. Lipari et al. Phys. Lett. **74** (1995) 4384
38. MACRO Coll. in preparation
39. M. Gell-Mann Phys. Lett. **8** (1964) 214
40. P. H. Frampton & T. Kephart Phys. Rev. Lett. **49** (1982) 1310
41. S. M. Bar et al. Phys. Rev. Lett. **50** (1983) 317
42. F. xiao Dong et al. Phys. Lett. **129B** (1983) 405
43. LSD Coll. Astropart. Phys. **2** (1994) 29
44. Kamiokande Coll. Phys. Rev. **D43** (1991) 2843
45. C. W. Walter Ph. D. thesis, California Institute of Technology (1997); MACRO Coll. in preparation

A SEARCH FOR DARK MATTER USING CRYOGENIC DETECTORS (CDMS)

D. A. BAUER (for the CDMS collaboration)

Department of Physics, University of California, Santa Barbara, CA 93106, USA
E-mail: dan_bauer@slac.stanford.edu

The Cryogenic Dark Matter Search experiment uses germanium and silicon detectors cooled to cryogenic temperatures in a direct search for weakly-interacting massive particles in our Galaxy. The novel detectors allow a high degree of background rejection by discriminating between electron and nuclear recoils through the simultaneous measurement of the energy deposited in phonons (heat) and ionization. Exposures of a few kilogram-days from initial runs of our experiment yield (preliminary) upper limits on the WIMP-nucleon cross section that are comparable to much longer runs of other experiments. Current and future data promise significant improvement, primarily due to improved detectors and reduced backgrounds from surface radioactivity.

1 Searching for WIMPS

Observations of stars and galaxies over a large range of distance scales indicate that most of the matter in the universe is "dark," seen only through its gravitational effects,[1,2] and that much of this dark matter is non-baryonic and "cold."[3] The best model for producing observed structure in the universe over three orders of magnitude in distance scale has 20% hot dark matter, 5% baryons, and 75% cold dark matter, with critical mass density $(\Omega_m = 1)$ [4]. Massive, stable (or long-lived) particles with weak-scale couplings, such as may exist under supersymmetry or other extensions to the standard model, could provide about the implied amount of non-baryonic cold dark matter.[5,6]

If weakly-interacting massive particles (WIMPs) exist, they would now make up a major component of the dark matter in our own galactic halo.[7] For a standard halo comprised of WIMPs with a Maxwellian velocity distribution characterized by $v_{rms} = 270$ km/s and a mass density of 0.4 GeV/cm^3, the expected WIMP flux at Earth is enormous, $\sim 10^7/m_\chi$ cm^{-2}, where m_χ is the WIMP mass in GeV. However, expected WIMP-nucleon scattering cross sections less than 10^{-41} cm^2 imply rates for WIMP-nuclear scattering ranging from 0.001 to 1 event per kilogram of detector per day, and the expected recoil energy is as low as 1 keV.[6,8] Direct detection of WIMPs will require large detector mass, low energy thresholds, efficient nuclear recoil detection, and long counting times.

2 CDMS at Stanford

Thus far, WIMP direct-detection experiments have been limited by irreducible backgrounds, primarily photons and electrons from radioactive contamination or activation. New techniques are needed to actively discriminate against such backgrounds without losing efficiency for the nuclear-recoil signature of WIMPs. One method for achieving this is to cool semi-conducting crystals to cryogenic temperatures (< 100 mK), where it becomes possible to detect the heat (phonons) liberated when particles interact in the crystal. Since nuclear recoils are less ionizing than electromagnetic backgrounds, one can reject such backgrounds by measuring both charge and phonon signals.[9,10] Following a decade-long development effort, the Cryogenic Dark Matter Search (CDMS) collaboration[a] began running such detectors in a low-background environment at Stanford in 1996. Our early data runs yield preliminary upper bounds on the WIMP-nucleon cross section that are comparable to much longer exposures of other experiments, illustrating the power of this technique.

2.1 Description of the experiment

The primary tools of this experiment are novel particle detectors which operate at 20 mK. Key to the experiment is the simultaneous measurement of the energy ΔE deposited in a scattering event both in phonon-mediated (heat) signals and in ionization (charge). The ionization measurement is made by applying a small (a few V/cm) bias voltage across the two sides of the cm-thick semiconductor targets. Electron-hole pairs are collected efficiently throughout the bulk of the detectors, resulting in FWHM energy resolutions as good as 640 eV. The division of the surface contacts into an inner region and an outer region provides some information on the radial position of the scattering event. Unfortunately, trapping

---

[a]The CDMS collaboration consists of: D. S. Akerib, T. A. Perera, R. W. Schnee from Case Western Reserve University; M. B. Crisler, R. Dixon, S. Eichblatt from Fermi National Accelerator Laboratory; E. E. Haller, R. R. Ross, A. Smith from Lawrence Berkeley National Laboratory; P. D. Barnes Jr from Lawrence Livermore National Laboratory; K. D. Irwin from NIST, Boulder; T. Shutt from Princeton University; B. A. Young from Santa Clara University; F. Lipschultz, B. Neuhauser from San Francisco State University; B. Cabrera, R.M. Clarke, P. Colling, A. K. Davies, S. W. Nam from Stanford University; A. Da Silva, R. J. Gaitskell, S. R. Golwala, J. Jochum, B. Sadoulet, A. L. Spadafora from the Center for Particle Astrophysics, University of California, Berkeley; and D. A. Bauer, D. O. Caldwell, A. H. Sonnenschein, S. Yellin from the University of California, Santa Barbara.

sites near detector surfaces result in a 10–30-μm-thick "dead layer" where charge collection is incomplete.[11] This makes the detectors vulnerable to surface contamination backgrounds, particularly from low energy electrons and photons. Recently, we have developed new ionization contacts which produce a much smaller dead layer, greatly reducing the effect of backgrounds at low energy. This technology will be tested on the next Ge detectors to be used at Stanford. If successful, it will be the basis for all future ionization contacts on both Ge and Si.

The CDMS detectors employ two distinct methods for performing the phonon-mediated measurement of the recoil energy ΔE. One technology uses two neutron-transmutation-doped (NTD) germanium thermistors eutectically bonded to a 1.2-cm-thick 6-cm-diameter 165-g cylindrical crystal of high-purity germanium. When the device is in contact with a 20 mK bath, monitoring the thermistor resistance gives the temperature rise $\Delta T = C^{-1}\Delta E$, where $C \sim 1$ keV/μK is the detector heat capacity. The resulting energy measurement has a FWHM resolution of 650 eV at 10 keV.[12] The use of two NTDs permits the rejection of events that originate in an NTD and would otherwise mimic the small ionization of a nuclear recoil. Implemented on Ge, we refer to this readout technology as BLIP.

The other technology uses quasiparticle-trap-assisted electrothermal-feedback transition-edge sensors (QETs) to detect non-equilibrium phonons before they have time to thermalize.[13] Tungsten meanders on a surface of a cooled 1-cm-thick cylindrical detector are held in the middle of their superconducting transitions by electrothermal feedback; a voltage bias V causes Joule heating (V^2/R), which increases at lower temperatures or resistances R, and decreases at higher temperatures. Deposited energy drives the tungsten towards normal conduction, producing a current signal. The time integral of this signal is proportional to the deposited energy, which is measured to 650 eV (FWHM) in our 100-gram silicon targets; the technology is now being transferred to germanium targets. Since the phonon collection time is fast (a few microseconds), relative-timing information from the four sensors on a device allows a two-dimensional determination of the event position to a few millimeters. Timing information also provides the ability to reject events on the top and bottom surfaces of the detectors, where the charge dead layer may otherwise compromise the detector discrimination capability. In our current implementation on Si, this technology is labelled FLIP.

Figure 1 shows the simultaneous phonon and ionization measurements for a 165 g Ge BLIP and a 100 g Si FLIP detector during gamma (^{60}Co) and neutron (^{252}Cf) source calibration runs. The plots clearly show the gamma/nuclear recoil discrimination capability of the detectors. The neutron calibration data show that the ratio of the ionization yield for nuclear recoils versus electron recoils is slightly higher in Si ($\sim 1/2$) than in the Ge ($\sim 1/3$). The relative ionization yield is not linear at low energies and, in general, must be measured as a function of recoil energy. The neutron recoil spectrum in Si is 2.5 times higher in energy than that of Ge from a similar incident neutron spectrum. This kinematic effect makes Si a good choice for measuring the neutron background and would allow us to measure the mass of incident WIMPs by comparing their spectra in Si and Ge. The discrimination for BLIP shown in Figure 1 corresponds to an effective gamma background contamination of less than 1% for recoil energies greater than 20 keV, with nuclear recoil acceptances of at least 99%. FLIP has much better background rejection (5×10^{-5} for recoil energies greater than 30 keV) due to the use of timing information, although the nuclear recoil acceptance is lower ($\sim 50\%$) at the present stage of analysis.

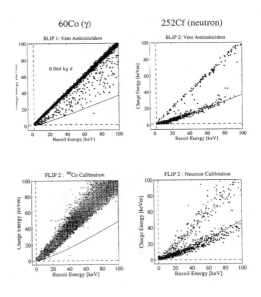

Figure 1: Scatter plots of the ionization measurement versus the recoil energy measurement for the 165 g Ge BLIP and 100 g FLIP detectors, obtained during calibration runs at the Stanford Underground Facility (SUF). The ionization measurements are normalized to electron equivalent energy. The curves represent fits to the region of nuclear recoil events.

The remainder of the experimental apparatus consists of specialized low-activity detector-housing modules mounted in a shielded cryostat made from a set of nested copper cans. The cans are cooled by conduction through a set of concentric horizontal tubes extending in a dog-leg from a dilution refrigerator. An external, 15-cm-thick lead shield reduces the flux of background photons by a factor of ∼1000, while 25 cm of polyethylene shielding reduces the flux of neutrons by a factor of ∼100.[14] Samples of all materials internal to the shield are carefully screened in a low-background HPGe counting facility for radioactive contaminants. Further shielding close to the detectors is achieved with 1 cm of ancient, ultra-low-activity lead, which has a low concentration of ^{210}Pb, a beta-emitter. Due to the complexity of the detectors and cryostat, the first phase of the experiment is being performed at a shallow site at Stanford University at a depth of 17 meters water-equivalent (mwe). This over-burden is enough to eliminate the hadronic cosmic-ray flux. However, the overburden reduces the cosmic-ray muon flux by a factor of only 5, requiring further rejection of backgrounds with a hermetic plastic-scintillator muon veto. The veto efficiency for direct muon hits in the detectors has been measured to be 99.995%. The ratio of nuclear recoils in FLIP which are not veto associated to those which are veto coincident implies a lower limit on the veto efficiency for cosmic-induced neutrons of 99%.

2.2 Results

Several data runs have been taken in the low-background facility over the past two years, indicating that the experiment can successfully operate over months-long timescales with energy resolutions comparable to those of calibration data. The rates of photon and neutron backgrounds have been consistent with or less than the expected levels, confirming that our goals at the shallow site are attainable and that our screening procedures are effective in limiting these sources of background.

Figure 2 shows ionization energy versus recoil energy for events during part of the latest background run of two 165 g Ge BLIPs and one 100 g Si FLIP at Stanford. The main diagonal band in each case is from gamma events. The off-axis events just below the gamma band for the BLIP detector arise from the interaction of electrons, most of which were ejected from surrounding material by muons, in the surface dead layer of the Ge detector.

The population of events lying along the nuclear recoil line below 20 keV recoil is believed to be due to ∼ 10^6 atoms of tritium on the surface of the Ge BLIP detectors. The smaller number of events (∼ 10 per live day in the range 20-100 keV) at higher energies close to the line could be due to electrons emitted from surface contamination such as ^{40}K from accidental sweat contamination, or ^{137}Cs, ^{14}C, and ^{210}Pb from radon plating.

Ultimately, it is our intention to use the gamma cali-

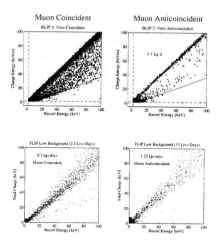

Figure 2: Ionization energy versus recoil energy in a 165 g Ge and a 100 g Si FLIP during low background running at Stanford after all cuts, including timing cuts for FLIP.

bration data in order to estimate the number of misidentified gammas that lie near to the nuclear recoil band in the muon anticoincident background data, and so subtract their contribution. However, at present the events that appear in the nuclear recoil band in Ge BLIP background data sets are dominated by surface contamination. Given the overlap of the electron event distribution with the measured nuclear recoil region these events effectively constrain the nuclear recoil (dark matter) limit set from the BLIP data. Recent data from the 100 g Si FLIP have shown that the rise time of pulses produced by surface electrons is significantly faster than those from gammas and neutrons. Use of this timing information produces the much-cleaner background spectrum for the Si FLIP shown in Figure 2. These distributions also show clear evidence of neutrons produced by cosmics in the co-incident data, but very few events in the nuclear recoil band in the veto anti-coincident set. Two events remain in the 30-60 keV energy range, where we expect 0.75 events from "punch-through" neutrons which originate outside of the muon veto.

2.3 WIMP Limits

The fundamental WIMP-nucleon cross section for coherent spin-independent elastic scattering allows direct comparisons between previous experiments and the preliminary and anticipated results of the CDMS experiment. Figure 3 shows the results of the first analysis of the latest sample of CDMS data using the 165 g Ge and 100 g Si detectors, together with recent limits from Rome NaI [15], UK NaI [16], Milan TeO_2 [17], Modane Al_2O_3 [18], and Ge diode [19] experiments. The CDMS results are preliminary and based on data that represent less than 7% of the total data anticipated from the experiment at Stanford. Above the Ge noise threshold of 2 keV total recoil energy, but below 18 keV, the Ge spectrum is dominated by backgrounds we expect to remove in upcoming data runs. Because the Si FLIP has additional rejection power against these low-energy surface backgrounds due to timing, it achieves better limits for low-mass WIMPs than does the Ge, in spite of the five-times larger cross section for WIMP-nucleon scattering in Ge. The current CDMS results are competitive with other existing limits because of the detectors' extraordinary capability to reject a large number of background events by simultaneously measuring ionization and thermal energy.

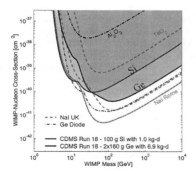

Figure 3: Existing limits set by previous experiments. The Ge diode limit is the envelope of the UCSB/LBL/UCB, COSME-2, Canfranc, and Heidelberg-Moscow experiments; the other published limits are referred to in the text. Also shown are preliminary limits from CDMS with a 100 g Si detector with 1.0 kg-d exposure and a threshold of 5 keV, and a 165 g Ge detector with 6.9 kg-d exposure and a threshold of 3 keV.

The data from the first CDMS data at Stanford have been encouraging, yielding both competitive dark matter limits and a deeper understanding of backgrounds. For future runs, improved cleanliness procedures, increased detector self-shielding, and the implementation of improved charge collection technologies and phonon timing

information all should help minimize the surface-electron background and restore the full effectiveness of our discrimination technique for both Ge and Si. Once this has been accomplished, we expect to be limited by the cosmogenic neutron backgrounds at the Stanford site with an exposure of about 100-kg-d. To obtain this exposure we will instrument two silicon and four germanium devices with QET readout and six germanium devices with NTD readout, for a total of 200 g of silicon and 2 kg of germanium. Comparison of distributions from the Ge and Si will provide information on the backgrounds, especially neutrons. Multiple scattering of neutrons in the detector arrays will also provide a handle for background subtraction.

3 CDMS at Soudan

Ultimately, cosmogenic neutron backgrounds will limit the sensitivity of the experiment at the Stanford site. In order to take full advantage of these advanced detectors, we intend to continue the experiment at the Soudan Mine in northern Minnesota. The new CDMS II experiment would exploit the experience from CDMS I and utilize similar detector designs. The improved sensitivity will be obtained by fully-filling the cryogenic detector volume with CDMS detectors, increasing the active detector mass by an order of magnitude, and by operating CDMS II in the low background environment of the Soudan mine, which will decrease the cosmic ray induced background rates. The 2090 mwe overburden at Soudan attenuates cosmic-ray muons by some 5 orders of magnitude, which will greatly reduce cosmogenic activity in the apparatus and hence the neutron background.

Figure 4 shows our expected ultimate sensitivities for the Stanford and Soudan phases of CDMS. The Stanford curve is based on using six, 250 g Ge detectors with an exposure of 100 kg-days (CDMS I), ambient backgrounds of approximately 3 events/(keV-kg-day), and intrinsic background rejection of 99%. Following background subtraction, the resulting sensitivity is 0.01 events/(keV-kg-day) at the Stanford site, good enough to allow CDMS I to discover relic neutralino WIMPs over a significant part of the allowed parameter space. However, CDMS II at Soudan, with a total exposure of 10,000 kg-days, will improve on those limits by at least an order of magnitude. For spin-independent couplings, CDMS II will have better sensitivity than any other experiment for all but the lowest mass WIMPs, which seem to be ruled out by experiments at LEP. Of course, the hope is to obtain an unambiguous detection of WIMP dark matter sometime within the next few years!

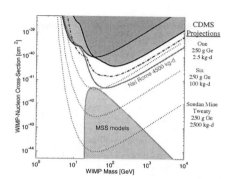

CDMS
Projections

One
250 g Ge
2.5 kg-d

Six
250 g Ge
100 kg-d

Soudan Mine
Twenty
250 g Ge
2500 kg-d

Figure 4: Projections for CDMS reach in WIMP-nucleon cross section versus WIMP mass. The upper dotted line is where we would be right now at Stanford if we had a single 250 g Ge detector with the FLIP readout technology. The middle dotted line is the goal for CDMS-I at the Stanford site, with six 250 g Ge detectors. The bottom dotted line is the projection for Ge detectors at Soudan. The best current limits are shown for comparison along with the envelope of possible MSSM predictions (the peak corresponds to 3 events/(kg-day) for Ge).

Acknowledgements

This work is supported by the Center for Particle Astrophysics, an NSF Science and Technology Center operated by the University of California, Berkeley, under Cooperative Agreement No. AST-91-20005, and by the Department of Energy under contracts DE-AC03-76SF00098, DE-FG03-90ER40569, and DE-FG03-91ER40618.

We gratefully acknowledge the skillful and dedicated efforts of the technical staffs at LBNL, Stanford University, UC Berkeley, and UC Santa Barbara.

References

1. V. Trimble, *Ann. Rev. Astron. Astrophys.* **25**, 425 (1987).
2. E.W. Kolb and M.S. Turner, *The Early Universe* (Addison-Wesley, Reading, 1988).
3. K.A. Olive, astro-ph/9707212.
4. E. Gawiser and J. Silk, *Science* **280** , 1405 (1998).
5. B.W. Lee and S. Weinberg, *Phys. Rev. Lett.* **39**, 165 (1977).
6. G. Jungman, M. Kamionkowski and K. Griest, *Phys. Rep.* **267**, 195 (1996).
7. E.I Gates and M.S. Turner, *Phys. Rev. Lett.* **72**, 2520 (1994).
8. P.F. Smith and J.D. Lewin, *Phys. Rep.* **187**, 203 (1990).
9. Reviewed in N. Booth, B. Cabrera and E. Fiorini, *Ann. Rev. Nucl. Part. Sci.* **46**, 471 (1996).
10. T. Shutt *et al.*, *Nucl. Phys.* **51B**, 318 (1996).
11. T. Shutt *et al.*, in *Proceedings of the VIIth International Workshop on Low Temperature Detectors*, ed. S. Cooper (Munich, Max Planck Institute of Physics, 1997) (LTD7), 224.
12. T. Shutt *et al.*, *Phys. Rev. Lett.* **69**, 3425 (1992); R.J. Gaitskell *et al.*, LTD7 *ibid.* 221.
13. R.M. Clarke *et al.*, LTD7 *ibid.* 229; A.K. Davies *et al.*, LTD7 *ibid.* 227.
14. A. Da Silva *et al.*, *Nucl. Instrum. Meth.* **A354**, 553 (1995).
15. R. Bernabei *et al.*, *Phys. Lett.* **B389**, 757-766 (1996).
16. P. F. Smith et al., *Phys. Lett.* **B379**, 299-308 (1996).
17. A. Alessandrello et al., *Nucl. Inst. Meth.* **A370**, 241 (1996).
18. A. de Bellefon, et al., *Nucl. Inst. Meth.* **A370**, 230 (1996)
19. M. Beck et al., *Phys. Lett.* **B336**, 141 (1994), E. Garcia *et al.*, *Phys. Rev.* **D51**, 1458 (1995), D. O. Caldwell et al., *Phys. Rev. Lett.* **61**, 510 (1988), and A. Morales (private communication).

SEARCH FOR $P \to K^+ \bar{\nu}$ WITH SUPER-KAMIOKANDE

Y. HAYATO

for the Super-Kamiokande collaboration

Institute of Particle and Nuclear Studies, High Energy Accelerator Research Organization,
1-1 Oho, Tsukuba, Ibaraki, 305-0801, Japan
E-Mail: hayato@neutrino.kek.jp

We present results of a search for the proton decay mode $p \to \bar{\nu} K^+$ using data from a 33 kton·year exposure of the Super–Kamiokande detector. Two decay modes of the kaon, $K^+ \to \mu^+ \nu_\mu$ and $K^+ \to \pi^+ \pi^0$, are studied. The data are consistent with the background expected from atmospheric neutrinos. Therefore we set the preliminary lower limit of the partial lifetime for the proton $(\tau/B(p \to \bar{\nu}_\mu K^+))$ to be 6.8×10^{32} years at 90% confidence level.

One of the most impressive predictions of Grand Unified Theories (GUTs) is baryon number violation. The minimal SU(5) GUT predicts the dominant decay mode to be $p \to e^+ \pi^0$ with a lifetime prediction of no more than $\sim 10^{31}$ years [1]. The current experimental lower limit is about 100 times longer than this [2][3]. Furthermore, the weak mixing angle predicted by this model does not agree with the experimental value and the three running coupling constants of the strong and electroweak forces do not meet exactly at a single point [4][5].

The minimal supersymmetric (SUSY) SU(5) GUT makes a prediction for the weak mixing angle which is much closer to experimental results and predicts the lifetime of the proton decay into $e^+ \pi^0$ to be more than four orders of magnitude longer than in non-SUSY minimal SU(5) GUT [7]. The minimal SUSY SU(5) model predicts the proton decay mode $p \to \bar{\nu} K^+$ to be dominant with the partial lifetime prediction varying from $\mathcal{O}(10^{29})$ to $\mathcal{O}(10^{35})$ yr [7], within the observable range of Super–Kamiokande. In this letter, we report the result of the search for proton decay through the channel $p \to \bar{\nu} K^+$ in 535 live-days of data in Super–Kamiokande, corresponding to a 33 kton·year exposure. The two dominant decay modes of the kaon were studied, $K^+ \to \mu^+ \nu_\mu$ and $K^+ \to \pi^+ \pi^0$. Two separate search methods were used to search for $K^+ \to \mu^+ \nu_\mu$.

The Super–Kamiokande detector is a ring imaging water Cherenkov detector located 1000 m (2700 m water equivalent) below the peak of Mt. Ikenoyama near Kamioka, Japan. 50 kilotons of ultra pure water is held within a stainless steel tank of height 39.3 m and diameter 41.4 m. The tank is optically separated into two regions, the inner and outer detectors, by a Photo Multiplier Tube (PMT) support structure. The inner detector has a height of 36.2 m and a diameter of 33.8 m. The outer detector completely surrounds the inner detector and is used to identify incoming and outgoing particles. On the wall of the inner detector, there are 11,146 50 cm inward facing PMTs which cover 40% of the surface.

The outer detector is lined with 1885 20 cm outward facing PMTs equipped with 60×60 cm² wavelength-shifter plates to increase the collection efficiency of Cherenkov photons.

In this analysis, we use the fully contained (FC) event sample which is identical to that used for the atmospheric ν analysis [9]. These events have vertices which were reconstructed inside the 22.5 kiloton fiducial volume of the detector, defined to be 2 m from the PMT support structure. All particles must deposit all of their energy within the inner detector. After FC event selection, events are reconstructed determining the vertex position, number of Cherenkov rings, particle type (e-like or μ-like), momentum, and number of decay electrons. More details of the detector, event selection, and event reconstruction are described in Ref. 3 and Ref. 9.

The absolute scale of the momentum reconstruction was checked with several calibration sources such as the electrons from a LINAC [11], decay electrons from cosmic-ray muons which stopped in the detector volume (stopping muons), dE/dx energy deposited by stopping muons, and the reconstructed mass of π^0s generated by atmospheric neutrino interactions. The error in the absolute energy scale is estimated to be less than $\pm 2.5\%$. The time variation of the energy scale was checked with muon decay electrons and was found to vary by only $\pm 0.5\%$ over the exposure period.

The momentum of the K^+ from $p \to \bar{\nu} K^+$ is 340 MeV/c and is below the threshold momentum for its production of Cherenkov light in water. Candidate events for this decay mode are therefore identified through the decay products of the K^+. The dominant K^+ decay modes are $K^+ \to \mu^+ \nu_\mu$ and $K^+ \to \pi^+ \pi^0$, with branching ratios of 63.5% and 21.2%, respectively. The hadronic cross section of the low momentum K^+ is known to be small, so more than 90% of them stop before decaying. In this paper, we therefore search for K^+ decays at rest.

We now describe the first method to search for $p \to \bar{\nu} K^+$; $K^+ \to \mu^+ \nu_\mu$. The momentum of the μ^+ from the decay of the stopped K^+ is 236 MeV/c. If a proton in the $p_{3/2}$ state of ^{16}O decays, the remaining ^{15}N nucleus is left in an excited state. This state quickly decays, emitting a prompt 6.3 MeV γ-ray. The signal from the decay particles of the K^+ should be delayed relative to that of the γ-ray due to its lifetime ($\tau = 12$ ns). The probability of a 6.3 MeV γ-ray emission from oxygen is estimated to be 41% [14]. By requiring this prompt γ-ray, almost all of the background events are eliminated.

To determine the signal region and to estimate the detection efficiency, a total of 2000 $p \rightarrow \bar{\nu} K^+$; $K^+ \rightarrow \mu^+ \nu_\mu$ Monte Carlo (MC) events were generated. The same reduction and reconstruction as for the data were applied to these events. For these proton decay MC events, 96% were identified as one ring. The resolution of the vertex fitting was 52 cm and the probability of particle misidentification was 3.8%. Based on this simulation, a signal region of the reconstructed momentum of the μ^+ was defined to be between 215 MeV/c and 260 MeV/c. Candidate events for this analysis were therefore selected by the following criteria: A1) one ring A2) μ-like A3) with one decay electron, A4) 215 MeV/c < momentum(μ) < 260 MeV/c, A5) misfit event rejection and A6) search for the prompt γ-ray. If the vertex position of an event was misreconstructed, the distribution of the timing residuals, (photon arrival time) - (time of flight), is wide. These events were rejected with A5. The prompt γ-ray signal is defined as the number of hit PMTs between 12 ns and 120 ns before the μ signal. Criterion A6 required the number of hit PMTs to be more than 7. Figures 1-(a,b,c) show the number of hit PMTs within the timing region of the prompt γ-ray for (a) proton decay MC events, (b) simulated atmospheric neutrino events, (c) real data. As a result, the detection efficiency is estimated to be 21%. Including the branching ratio of $K^+ \rightarrow \mu^+ \nu_\mu$ (63.5%), ratio of the bound proton in Oxygen to all the protons in H$_2$O (80%) and the emission probability of the 6.3 MeV γ-ray (41%), the total detection efficiency is estimated to be 4.4%. The background contamination is estimated to be 0.38 events/33 kt·yr.

When the same criteria were applied to the real data, no events were observed (Figure 1-(c)).

The partial lifetime (τ/B yr) was calculated:

$$\tau_\beta = (\Lambda \times \epsilon B_m)/N_{cand}, \tag{1}$$

where Λ is the exposure, ϵB_m is the detection efficiency(4.4%) and N_{cand} is upperlimit of the number of candidate events. The lower limit of the partial lifetime obtained for this mode is 2.1×10^{32} yr at 90% C.L. (from Eqn. 1).

The detection efficiency of the prompt γ-ray tagging method is rather small. Therefore, as the second method, we search for an excess of events by monoenergetic 236MeV/c μ^+ produced by the stopped K^+ decay

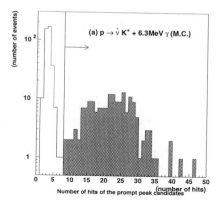

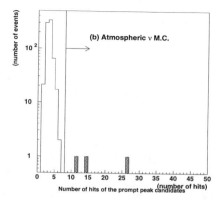

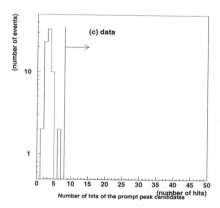

Figure 1: Number of hit PMTs within 12ns for the prompt peak candidate for $p \rightarrow \bar{\nu} K^+, K^+ \rightarrow \mu^+ \nu_\mu$ Monte Carlo (a) with 6.3MeV γ-ray, (b) 225kt·yr equivalent atmospheric neutrino Monte Carlo events and (c) 33kt·yr data of Super-Kamiokande.

for events with no observed prompt γ-ray. For this analysis, the selection criteria are: A1) to A4) described above and A7) no prompt γ-ray signal. The detection efficiency of this mode was estimated to be 63%.

To estimate the excess of proton-decay signal, the number of events in three momentum regions, 200 MeV/c to 215 MeV/c, 215 MeV/c to 260 MeV/c, and 265 MeV/c to 300 MeV/c, were summed separately for $p \to \bar{\nu}K^+$; $K^+ \to \mu^+\nu_\mu$ MC, atmospheric ν MC, and data. The χ^2 method was applied to fit parameters. The χ^2 function is defined as follows:

$$\chi^2(a,b) = \sum_{i=1}^{3} \frac{[N_i^{data} - (a \cdot N_i^{atm\nu} + b \cdot N_i^{pdcy})]^2}{N_i^{data}}, \quad (2)$$

where a and b are the fit parameters, $N_i^{data}, N_i^{pdcy}, N_i^{atm\nu}$ are the numbers of events of real data, proton decay MC and atmospheric neutrino MC in each momentum region i, respectively.

The uncertainty in the absolute scale of the reconstructed momentum was estimated to be less than 2.5%. When we shifted the reconstructed momentum by 2.5% and fit the parameters, the data were still consistent with no excess of $p \to \bar{\nu}K^+$ events. However the 90% C.L. upper limit on the number of candidate was slightly increased by 1.2 events in 33kt· yr. Therefore, we took the fitted parameters with the reconstructed momentum shifted by 2.5%. The minimum χ^2, $\chi^2_{min} = 1.9 \times 10^{-3}$, was in the unphysical region ($b \cdot N_2^{pdcy} = -12.5$). The χ^2_{min} in the physical region ($b \cdot N_2^{pdcy} = 0$) was 1.0. The data are therefore consistent with no excess of $p \to \bar{\nu}K^+$ events. The 90% C.L. upper limit on the number of proton decay events(N_{cand}) was obtained by requiring $\chi^2 - \chi^2_{physical} = 3.6$ based on the prescription described in Ref. [8]. The 90% upper limit on the number of candidates was estimated to be 14.6. The momentum distribution of the events which satisfy criteria A1) to A3) is shown in Figure 2. Also shown is the expected signal for the proton partial lifetime of 3.0×10^{32} years at 90% C.L. The lower limit of the partial lifetime of $p \to \bar{\nu}K^+$; $K^+ \to \nu_\mu\mu^+$ using the above method is 3.0×10^{32} years at 90% C.L.

Finally, we describe the search for $p \to \bar{\nu}K^+$; $K^+ \to \pi^+\pi^0$. The π^0 and π^+ from the K^+ decay at rest have equal and opposite momenta of approximately 205 MeV/c. The π^0 decay γs completely reconstruct this momentum. The π^+, barely over Cherenkov threshold with $\beta \approx .86$, emits very little Cherenkov radiation. However it does decay into a muon ($\pi^+ \to \mu^+\nu_\mu$) which decays into a positron ($\mu^+ \to e^+\nu_e\bar{\nu}_\mu$). Detection of this positron is possible. Furthermore, a small amount of Cherenkov radiation from the π^+ can be detected in the direction opposite that of the π^0. To quantify this we define "backwards charge" (Q_b) as the sum of the photoelectrons detected by the PMTs which lie within a

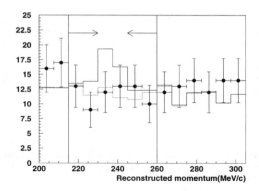

Figure 2: Reconstructed momentum distribution. Solid(dotted) line shows the estimated 90% C.L. number of proton decay+atmospheric ν(atmospheric ν) events, the black points with error-bars are the data with statistical errors.

40° cone whose axis is the opposite direction of the reconstructed direction of the π^0. This charge is corrected for light attenuation in the water, angular dependence of photon acceptance, and photocathode coverage.

To determine the signal region and estimate detection efficiency, a total of 1000 $p \to \bar{\nu}K^+$; $K^+ \to \pi^+\pi^0$ Monte Carlo events were generated, 771 of which had vertices which reconstructed inside the fiducial volume of the detector. For these 205 MeV/c π^0 events, the resolution of vertex fitting is 29 cm and 66% of the events are identified as 2-ring events.

The selection criteria for this type of event were determined, B1) 2 e-like rings, B2) with 1 decay electron, B3) 85 MeV/c^2 < mass$_{\gamma\gamma}$ < 185 MeV/c^2, B4) 175 MeV/c^2 < momentum$_{\gamma\gamma}$ < 250 MeV/c^2 and B5) 40 p.e. < Q_b < 100 p.e. Criteria B1),B3), and B4) search for π^0 with the monochromatic momentum expected. Criterion B2) is a search for the decay of the π^+ into muon into positron. Criterion B5) searches for Cherenkov light from the π^+. The π^0 mass resolution is determined to be 135±21 MeV/c^2. By passing the proton decay Monte Carlo events through these selection criteria, the detection efficiency was determined to be 31%. The dominant contribution to the inefficiency was the inefficiency of detection of two γ-rays from the decay of the π^0. Including the kaon branching ratio of 21.2% into $\pi^+\pi^0$, the total detection efficiency for this mode is estimated to be 6.5%.

Charged current interactions such as $\nu_\mu N \to \mu N'\pi^0$ from atmospheric neutrinos can imitate a kaon decay mode of this type. To obtain the number of background events, the flux was normalized using the result of the atmospheric neutrino analysis. The factors are 0.74 for

ν_μ charged current events and 1.17 for any other neutrino induced events [9]. The selection criteria were applied to the sample of atmospheric neutrino Monte Carlo and the number of background events expected is estimated to be 0.7 events/33kt·yr. Figure 3-a and b show $|\vec{p}_{\pi^0}|$ vs. Backwards Charge for proton decay MC and atmospheric neutrino MC, respectively.

The selection criteria for this decay mode were applied to the data. Figure 3-c shows the results of the final two cuts. No events passed. The single event which lies close to the cuts was examined manually with an event display. We found no additional evidence that it could be a signal event that fell outside of the cuts. The backwards charge appears to be a fragment of one of the rings and not from a small collapsed ring. Based on these numbers, the lower limit of the partial lifetime of $K^+ \to \pi^+\pi^0$ is estimated to be 3.1×10^{32} yr at the 90% C.L. (Eqn. 1).

The main sources of systematic errors are: 1) uncertainty in the momentum reconstruction, 2) uncertainty in the detection efficiency of the decay electron from the stopped muon, 3) uncertainty in the atmospheric neutrino fluxes and interaction cross sections used in the Monte Carlo simulation.

For the $K^+ \to \pi^+\pi^0$ mode, the detection efficiency changed by less than 1% if the momentum criterion was changed by 2.5%. For the $K^+ \to \mu^+\nu_\mu$ mode, this systematic error was already considered in the fitting. The systematic error of the detection efficiency of decay electrons from muons is estimated to be 1.5%[9] by comparing the fraction of cosmic-ray muon events with decay electrons between MC simulation and real data. These uncertainties contributed to the uncertainty on the limit via the detection efficiency; however the uncertainties were small enough that the lower limits of the partial lifetimes for each mode were unchanged.

The ambiguity of the estimation of number of background events comes from the uncertainty of the neutrino flux and the interaction cross sections. This ambiguity was estimated by comparing the number of events of the normalized atmospheric neutrino background with the real data at each reduction step. For the $K^+ \to \pi^+\pi^0$ mode, the data and atmospheric ν Monte Carlo agreed within statistical errors at each step. In searching for the excess of 236 MeV/c μ^+s, the shape of the momentum distribution and the absolute normalization was left as a free parameter in the fitting. The uncertainties of neutrino flux and the interaction cross-sections therefore did not affect the lower limit of the partial lifetime.

The combined 90% C.L. upper limit for the number of proton decay candidates (x_{limit}) was calculated by integrating the likelihood function to the 90% probability

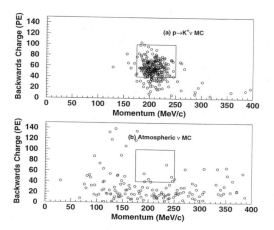

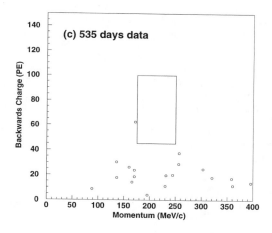

Figure 3: Backwards charge versus π^0 momentum for $p \to K^+\bar{\nu}_\mu$; $K^+ \to \pi^+\pi^0$ Monte Carlo, 225kt·yr equivalent atmospheric ν Monte Carlo, and 33kt·yr data of Super-Kamiokande.

level:

$$\frac{\int_0^{x_{limit}}[\prod_{i=1}^n P(N_{obs}^i, N_{bgd}^i(x))]dx}{\int_0^{\infty}[\prod_{i=1}^n P(N_{obs}^i, N_{bgd}^i(x))]dx} = 0.90, \qquad (3)$$

$$N_{bgd}^i(x) = X_{BG}^i + \frac{\epsilon B_m^i}{\sum_{j=1}^n [\epsilon B_m^j]}x, \qquad (4)$$

where n is the number of independent observations, $P(N, x)$ is the probability function of Poisson statistics, $N_{obs}(i)$ is the number of observed candidates, $X_{BG}(i)$ is the number of estimated background events, and $\epsilon B_m(i)$ is the detection efficiency multiplied by the meson branching ratio. The superscript i stands for the i-th observed mode. The lower limit of partial lifetime was calculated with

$$\tau/B = \frac{1}{x_{limit}}(\sum_{j=1}^n \epsilon B_m(j))\Lambda, \qquad (5)$$

where Λ is the exposure.

Finally, the combined lower limit of the partial lifetime for $p \to \bar{\nu}K^+$ was obtained $\tau/B(p \to K^+\bar{\nu}) \geq 6.8 \times 10^{32}$ yr (90% C.L.). This limit is more than six times longer than the previous best limit (1.0×10^{32} yr.(90% C.L.)) by the Kamiokande Collaboration [15].

In this letter, we have reported the result of a search for proton decay into $\bar{\nu}K^+$ in a 33 kt·yr exposure of the Super–Kamiokande detector. The data is consistent with the background expected from atmospheric neutrinos and no evidence for proton decay was observed. We set the preliminary lower limit of partial lifetime for $p \to \bar{\nu}K^+$ to be 6.8×10^{32} yr at 90% C.L. This limit will help to constrain SUSY GUT models.

We gratefully acknowledge the cooperation of the Kamioka Mining and Smelting Company. The Super-Kamiokande experiment was built from and has been operated with, funding by the Japanese Ministry of Education, Science, Sports and Culture, and the United States Department of Energy.

1. H.Georgi and S.L.Glashow, Phys. Rev. Lett. 32(1974), 438.
   For a review, see P.Langacker, Phys. Rep. 72(1981), 185.
2. R.Becker-Szendy et al., Phys. Rev. D42(1990), 2974.
3. M.Shiozawa et al., Search for Proton Decay via $p \to e^+\pi^0$ in a Large Water Cherenkov Detector, Phys. Rev. Lett.(submitted)
4. P.Langacker et al., Phys. Rev. D44(1991), 817.
5. Particle Data Group, Phys. Rev. D54(1996), 85.
6. J.Wess and B.Zumino, Phys. Lett. 49B(1974), 52.
   J.Iliopoulos and B.Zumino, Nucl. Phys. B76(1974), 310.

S.Ferrara and O.Piguet, Nucl. Phys. B93(1975), 261.
   A.A.Slavnov, Nucl. Phys. B97(1975), 155.
7. N.Sakai and T.Yanagida, Nucl. Phys. B197(1982), 533.
   S.Weinberg, Phys. Rev. D26(1982), 287.
   J.Ellis et al., Nucl. Phys. B202(1982), 43.
   P.Nath et al., Phys. Rev. D32(1985), 2348.
   P.Nath et al., Phys. Rev. D38(1988), 1479.
   J.Hisano et al., Nucl. Phys. B402(1993), 46.
   K.S.Babu, J.C.pati, F.Wilczek, Phys. lett. B 423 (1998) 337.
8. Particle Data Group, Review of Particle Physics, Section: Errors and confidence intervals – Bounded physical region, June 1996: R.M. Barnett et al., Phys. Rev. D54(1996), 375 .
9. Y.Fukuda et al., , Phys. Lett. B433(1998), 9.
10. Y. Fukuda et al., Evidence for oscillation of atmospheric neutrinos, Phys. Rev. Lett. 81(1998), 1562
11. M. Nakahata et al., Calibration of Super-Kamiokande using an electron LINAC, NIM (submitted)
12. M.Honda et al., Phys. Rev. D52(1995), 4985.
13. W.A.Mann et al., Phys. Rev. D34 (1986), 2545.
14. H.Ejiri, Phys. Rev. C48 (1993), 1442.
15. K.S.Hirata et al., Phys.Lett. B220(1989), 308.

STATUS AND RECENT RESULTS FROM THE HEGRA AIR SHOWER EXPERIMENT

G. HEINZELMANN for the HEGRA Collaboration

II. Inst. Exp. Physik, Universität Hamburg, Luruper Chausseee 149, 22761 Hamburg, Germany
E-mail: goetz.heinzelmann@desy.de

The HEGRA experiment investigates the relativistic (non thermal) universe by measuring air showers initiated in the atmosphere by photons and charged particles from the cosmos, using the imaging atmospheric Cherenkov light technique (≥ 500 GeV) and the showerfront sampling technique (≥ 20 TeV). A unique feature of HEGRA is the simultaneous observation of an air shower by several telescopes (stereoscopic observation). The status and recent results are presented, especially the measurement of photons ($E_\gamma \geq 500$ GeV) from the BL Lac object Markarian 501, which showed strong outbursts in 1997, being then the strongest TeV γ-source in the sky and which yielded high statistics and good quality data unprecedent in the field of the TeV γ-astronomy. The measured energy spectrum of Mkn 501 extends from 500 GeV to well beyond 10 TeV. Also the measurement of the energy spectrum and mass composition of charged cosmic rays in the PeV energy region (the knee) is briefly addressed.

1 Introduction and Experimental Method

The HEGRA (High Energy Gamma Ray Astronomy) air shower experiment explores processes of the violent, non thermal universe. The measurement of the cosmic primaries is indirect, via the showers generated in the atmosphere, since a direct measurement of primary photons ($E \geq 500$ GeV) and primary nuclei ($E \geq 0.5$ PeV) from satellites is not possible, due to their small detection areas in view of the low rates involved. HEGRA exploits two detection methods: the imaging atmospheric Cherenkov technique with 6 atmospheric Cherenkov telescopes ($E \geq 500$ GeV) and the showerfront sampling technique ($E \geq 20$ TeV) with three types of arrays (scintillator, wide angle Cherenkov counter AIROBICC and Geiger-towers). In both detection techniques HEGRA has pioneered new methods. The experiment is located at the Observatorio del Roque de los Muchachos on the Canary Island of La Palma (2200 m a.s.l., 28.8⁰N, 17.9⁰W) covering an area of about 200m · 200m (Figure 1).

1.1 Imaging Atmospheric Cherenkov Technique (IACT)

The showers in the atmosphere, initiated by the primary photons or nuclei, generate atmospheric Cherenkov light which is recorded during night using a mirror which focuses the Cherenkov light onto a multipixel (photomultiplier) camera in the focal plane of the mirror (Figure 2). Due to the multipixel camera the image of the airshower can be analyzed and the direction and energy be determined as well as γ-induced showers be separated from hadron initiated ones (which constitute in general a severe background). This imaging analysis technique was pioneered by the Whipple experiment, and initiated a breakthrough in the TeV γ-astronomy with the observation of the Crab nebula in 1989.[1] The Cherenkov light pool at detection level has a radius of ≈ 100m and thus the detection area is of the order of $3 \cdot 10^4 \text{m}^2$.

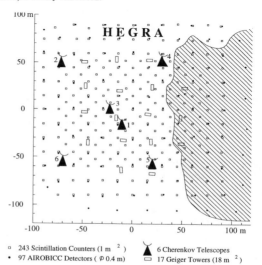

□ 243 Scintillation Counters (1 m²)	▲ 6 Cherenkov Telescopes
• 97 AIROBICC Detectors (φ 0.4 m)	▭ 17 Geiger Towers (18 m²)

Figure 1: *The layout of the HEGRA experiment at La Palma (Canary island). The hatched region was destroyed by a bushfire October 1997. The AIROBICC counters in this area have been reinstalled. In 1998 6 Cherenkov telescopes, 182 scintillation counters and 97 AIROBICC counters are operational.*

A substantial further improvement has been achieved by the HEGRA experiment using a system[2] of up to 4 (in future 5) telescopes to measure simultaneously the images of the same air shower from different viewing angles; the so called stereoscopic observation. With this unique technique HEGRA achieves[3] (on an event by event basis) an angular resolution of $\leq 0.1^0$, an energy resolution of better than 20 % and an excellent separation of gamma- from hadron-initiated showers, resulting in a flux sensitivity of $S/\sqrt{B} = 5\sigma$ in 1 h of observation for a flux corresponding to 1/4 the Crab. The low energy threshold of 500 GeV in view of the mirror size is related

Figure 2: *One of the HEGRA atmospheric Cherenkov telescopes used for the stereoscopic observations (segmented mirror area $8.5m^2$, 271 pixel camera, 4.3^0 field of view)*

to the background suppression due to the simultaneous measurement with several telescopes.

In total HEGRA operates 6 imaging Cherenkov telescopes: The prototype telescope 1 (since 1991) [4] [5] is used in stand alone mode and has been upgraded 1997 from $5m^2$ to $10m^2$ mirror area using a new type of Aluminum mirrors. The prototype telescope 2 (since 1992) is being modified to the same type as telescopes 3-6 which are used as a system for the stereoscopic observations. These system telescopes [3] have a mirror area of 8.5 m^2 each, and a 271 pixel (PMT) camera of 0.25^0pixel size, resulting in a field of view of 4.3^0. One of these telescopes is placed in the center and 3 (in future, 4) at the corners of a 100m · 100m square area (Fig. 1)corresponding to the size of the Cherenkov light pool. The system telescopes 3-6 were completed in fall 1996.

1.2 Showerfront Sampling Technique

At energies above ≈ 20 TeV the particles of an air shower reach the ground. Since the particles are concentrated in a so called showerfront the arrival time and particle density can be sampled with an array [6] of scintillation counters and the direction and showersize be deduced. This is the classical method of airshower observation. At HEGRA 243 scintillation counters have been placed on an 200m · 200m area in a 15 m grid and a denser grid in the center.

Also the atmospheric Cherenkov photons are concentrated in a corresponding (thinner) showerfront and the Cherenkov light front is sampled by an array of open photomultipliers (20 cm diameter, attached to a Winston cone), placed in a 30m grid (AIROBICC),[7] also with a densely instrumented part.

The energy thresholds of the arrays for vertical incident primary photons is 20 (11) TeV, requiring 14 (6) counters in coincidence for scintillators (AIROBICC).

The angular resolution $\sigma_\theta(68\%)$is $0.8^0(0.3^0)$ for events triggered with the corresponding arrays. For part of the time, data were collected also with 17 Geiger-towers [8] (measuring the electromagnetic energy at detection level and the direction of muons) and with the driftchamber modules of the CRT [9] (Cosmic Ray Tracking) detector. Exploiting the differences of electromagnetic and hadronic air showers, the measurements with the arrays allow also a separation of gamma- and hadron-showers and a coarse determination of the the type of primary nuclei.

After a bush fire October 1997, which destroyed about 30 % of the arrays, there are now, after a restoration, 97 AIROBICC counters and 182 scintillation counters in operation.

2 Results and Discussion

2.1 Imaging Atmospheric Cherenkov Technique

After the breakthrough achieved in 1989 with the observation of the Crab nebula by the imaging atmospheric Cherenkov technique (Whipple collaboration [1]), the Crab nebula has been observed by several experiments and is now the standard candle of TeV γ-astronomy. The Crab nebula, a Supernova remnant (a plerion) is supposed to be an electron accelerator generating photons by synchrotron radiation and inverse Compton effect. In contrast, the search for the galactic accelerators of nuclei, supposed to be associated with the shock waves of shell type supernova remnants has not yet been successful and this remains one of the main aims of the TeV γ-astronomy. Promising candidates have been searched for without clear success up to now. In view of the sparse galactic TeV sources detected, it was therefore a great surprise that the extragalactic sources Markarian 421 (z=0.031) and Markarian 501 (z=0.033) have been observed 1992 and 1994 first by the Whipple instrument [10] [11] and then by the prototype telescopes 1 and 2 of

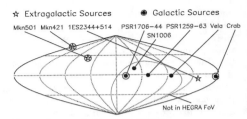

Figure 3: *The position of the known TeV γ-sources in galactic coordinates. The region of the sky not observable from the HEGRA site (La Palma) is indicated. Only sources observed with more than 5σ are shown.*

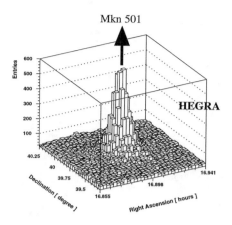

Figure 5: *Distribution of all events reconstructed in the $1^0 x 1^0$ solid angle region centered on the Mkn 501 direction. A prominent γ-ray excess can be seen above a flat background (stereoscopic observation, hardware threshold, no software cuts)*

the HEGRA experiment [12][13] as the next prominent TeV sources in the northern sky.

The sources known in the TeV sky as of today are shown in Figure 3 in galactic coordinates. From the northern sky only the Crab nebula (galactic source) and Mkn 421 and Mkn 501 have been observed by several experiments with high significance ($\geq 5\sigma$). (The source 1ES 2344+514 has only been observed [18] by Whipple during the winter 95/96). The BL Lac objects, such as Mkn 421 and Mkn 501 form a subclass of the active galactic nuclei (AGN). They are elliptical galaxies with two opposite side jets, one of which is pointing to the observer, supposedly powered from a central, super massive black hole surrounded by an accretion disk. The radiation is therefore not expected to be isotropic but rather beamed and boosted in the relativistic jet pointing to the observer. Mkn421 and Mkn501 at a redshift of z $\approx$ 0.03 are amongst the nearest objects of this class ($\approx$ 500 Mio. light years).

It was a fortunate coincidence that, short after the completion of the system of Cherenkov telescopes for the stereoscopic observations end of 1996, the BL Lac Mkn 501 showed strong outbursts during the whole observation period, from March to October 1997 when it was the strongest γ-source in the TeV sky and was therefore being monitored by several experiments. [14] These outbursts measured with the stereoscopic method allowed HEGRA the collection of data of unprecedented precision and statistics in the field of TeV γ-astronomy. First results have been presented in [15]. No strong flares have been observed in 1998.

Markarian 501 signal 1997

In 110 hours of of observation between March and October 1997 an excess of 38.000 events as shown in Fig. 5 from the direction of Mkn 501 have been measured by HEGRA with the system of telescopes. The figure displays the number of events in equatorial coordinates. The mean location of the excess photons coincides with the Mkn 501 position within 0.01^0[19]. Already without software cuts for a gamma-hadron separation a sample of photons is obtained that is almost free of background.

Markarian 501 lightcurve and variability

The unexpected variation of the flux amplitudes 1997 (flux at 2 TeV) are shown in Fig. 4 from March 97 until October 97. The gaps are due to the moon shine periods during which no stereoscopic observations were performed. The mean flux of Mkn 501 exceeds the mean flux of the Crab nebula, the next strongest source at TeV energies, by a factor of three. The daily averages range from a fraction of the Crab to about 10 Crab with the largest flare at 26th/27th June 1997. With the prototype telescope 1 also measurements during partial moon shine have been performed, leading to the most complete record of the flaring states of Mkn 501.[17]

A detailed study of the time scales of the flux variability from night to night and also on sub-day time scales have been performed with the HEGRA data of 1997 and variation time scales between 5h and 15h have been obtained. Due to causality and light travel arguments the size and the Dopplerfactors involved can be estimated and with certain assumptions the Doppler factors range between 5 and 50.[16] (The Doppler factor gives the ratio of observed to emitted energy). Although the variability time scales favor an electron acceleration process a hadronic acceleration (with a subsequent electron population due to π^0 production and decay) cannot be ruled out on the basis of the variation time scales, since the variability can be due to geometric effects.

In contrast to the strong variation of the flux amplitudes by more than a factor 10, the shape of the corresponding energy spectra are surprisingly similar. No correlation of the spectral shape (1-5 TeV) with the flux amplitudes is observed. This has been studied [16] in great detail by HEGRA making use of the ability to determine the energy spectra on a night by night basis due to the high sensitivity of the system of telescopes.

Markarian 501 energy spectrum

One of the most important measurements concerns the shape of the energy spectrum and its extent towards high energies. In Figure 6 the mean energy spectrum between 1 and 10 TeV is shown, averaged for nights with a high

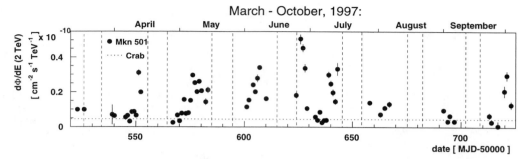

Figure 4: *The flares of Mkn 501 during 1997. The mean daily flux amplitudes for Mkn 501 from March to October 1997 (flux at 2 TeV) as a function of days. For comparison the Crab flux amplitude is also given. (No measurements during moon shine periods, which are indicated).*

flux amplitude ($\geq 3 \cdot 10^{-11} \mathrm{cm}^{-2} \mathrm{s}^{-1} \mathrm{TeV}^{-1}$ at 2 TeV). The energy spectrum is of unprecedented precision and shows a gradual steepening which is more pronounced considering the complete energy spectrum measured, which extends from 500 GeV well beyond 10 TeV [16] and where the systematic uncertainties are still under investigation.

The shape and extent of the energy spectrum is of great interest especially in the context of the following two questions: (1) The underlying acceleration process in the relativistic jets. A steepening is expected e.g. in models considering synchrotron radiation of electrons and the inverse Compton effect which upscatters the photons into the TeV energy range. On the other hand if the spectrum extends to very high energies a proton acceleration as the basic process seems to be likely with a subsequent electromagnetic part due to π^0 production and decay. (2) The spectrum is also of great interest in connection with the density of the diffuse infrared intergalactic background radiation, since the TeV photons interact due to pair production $\gamma_{TeV} + \gamma_{IR} \to e^+ e^-$ on the infrared background fields and are absorbed. This absorption may lead to a steepening or break in the energy spectrum and may be used as an indirect measurement of the density of these background fields (which are difficult to measure from satellites due to the foreground radiation of our galaxy). This has already been discussed in connection with the Mkn 421 spectrum. The density of these background fields are related to the epoch of star formation and are therefore of great importance for cosmological questions e.g. structure formation and models of cold and cold+hot dark matter.[20]

X-ray/TeV correlation for Markarian 501

Correlated flares in the keV and TeV energy bands are expected if both components are due to synchrotron and inverse Compton radiation of the same population of ultra relativistic electrons in a relativistic jet.

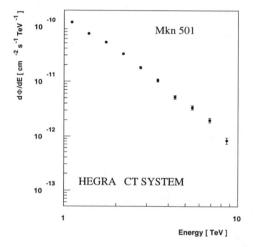

Figure 6: *Mean differential energy spectrum of Mkn 501 between 1 and 10 TeV averaged for nights with a high flux ($\geq 3 \cdot 10^{-11} \mathrm{cm}^2 \mathrm{TeV}^{-1}$ at 2 TeV). The energy spectrum from 500 GeV -1 TeV and above 10 TeV is still under investigation.*

The Rossi X-ray Timing Explorer satellite (RXTE), which has been monitoring Mkn 501 since January 1996 with the all sky monitor, showed a dramatic increase for the count rate in June/July 1997. Using the data base of RXTE, a weak correlation between the 2-12 keV count rate of RXTE and the TeV flux amplitudes in the HEGRA data has been found, [16] supporting the view of Synchrotron emission and inverse Compton scattering.

2.2 Scintillator and AIROBICC Arrays

The showerfront sampling method with the arrays permits the monitoring of a large fraction of the sky (1 sr) simultaneously ($E \geq 20$ TeV). However in spite of improved methods and sensitivities of this technique achieved with the HEGRA scintillator and AIROBICC arrays, no clear signal of a γ-source with a significance above 5σ has been observed for the region of the sky accessible from La Palma, resulting in typical flux limits of $F_\gamma(E \geq 15\text{TeV}) < 10^{-13}\text{cm}^{-2}\text{s}^{-1}$ for a large number of possible source candidates. Only the Crab nebula, the standard candle of TeV γ-astronomy has been detected with ≈ 3.5 sigma in the data of 1992/93 and Her X-1 with a similar significance in 1994/95. Flux limits obtained [21] for TeV counterparts of Gamma Ray Bursts are in agreement with recent measurements in the optical waveband indicating that GRBs are at cosmological distances.

The search for intergalactic cascades related to the cosmic ray event of highest energy ($3 \cdot 10^{20}$ eV)[22] yielded from 2.5 years of scintillator data an interesting excess. Interpreting this excess as being due to an intergalactic cascade, a distance of less than 28 Mpc [23] is obtained for a source assumed to be constantly emitting.

Besides the gamma astronomy with the arrays the measurement of the energy spectrum and composition of charged cosmic rays (nuclei) in the energy range from 0.4 PeV to 10 PeV is a further central theme of the physics with the arrays. This energy range is only accessible to the airshower technique. It is of specific interest, since the highest energies attainable in the models of galactic accelerators and the escape of galactic cosmic rays from our galaxy is expected to occur within this energy range. Also the measured energy spectrum is known to change the slope in the PeV region (the knee), although the measurements in details are still conflicting.

New methods have been developed at HEGRA for the reconstruction of the primary energy (without prior knowledge of the type of primary) and a coarse determination of primary mass [24] [25]. These methods are based on the fact, that the penetration depth, i.e. the position of the shower maximum in the atmosphere, can be remarkably well determined with a precision of about 1 rad. length from the lateral intensity distribution of the Cherenkov light as measured with the AIROBICC counters. This leads to an energy resolution of 30 % . Since the detectors were optimized for the γ-astronomy as a primary goal for the HEGRA experiment (and not for the determination of the primary energy over several orders of magnitude) a good understanding of the systematic effects is necessary and presently under way. Preliminary results [26] show, that the energy spectrum steepens in the PeV region i.e. the exponent of dN/dE $\propto E^{-\alpha}$ changes by about $\Delta\alpha = 0.3 - 0.5$ consistent with other experiments and the mass composition does not change abruptly in this energy range.

3 Summary

The flares in 1997 of the BL Lac object Mkn 501 (one of the only three known extragalactic TeV γ-sources) measured with the HEGRA Cherenkov telescopes yielded data of unprecedented precision in the field of TeV γ-astronomy. The full analysis of the data will supply us with important clues for the understanding of the relativistic universe.

Acknowledgements

We thank the Instituto de Astrofisica de Canarias (IAC) for supplying excellent working conditions at La Palma. HEGRA is supported by the BMBF (Germany) and CYCIT (Spain).

References

1. T.C. Weekes *et al.*, *Astrophys. J.* **342**, 379 (1989)
2. F. Aharonian *et al.*, *Astropart. Phys.* **6**, 343 (1997)
3. A. Daum *et al.*, *Astropart. Phys.* **8**, 1 (1997)
4. R. Mirzoyan *et al.*, *NIM* **351**, 513 (1994)
5. G. Rauterberg *et al.*, *Proc. 24th ICRC* **3**, 412, (1995)
6. H. Krawczynski *et al.*, *NIM* **383**, 431 (1996)
7. A. Karle *et al.*, *Astropart. Phys.* **3**, 321 (1995)
8. W. Rhode *et al.*, *NIM* **378**, 399 (1996)
9. K. Bernlöhr *et al.*, *NIM* **369**, 293 (1996)
10. M. Punch *et al.*, *Nature* **477**, 358 (1992)
11. J. Quinn *et al.*, *Astrophys. J.* **L83**, 456 (1996)
12. D. Petry *et al.*, *A&A* **L13**, 311 (1996)
13. S.M. Bradbury *et al.*, *A&A* **L5**, 320 (1997)
14. R.J. Protheroe *et al.*, 25th ICRC (1997) and astro-ph/9710118
15. F. Aharonian *et al.*, *A&A* **L5**, 327 (1997)
16. F. Aharonian *et al.*, submitted to *A & A* and astro-ph/9808296
17. B.C. Raubenheimer *et al.*, submitted to *Astroparticle Phys*
18. M. Catanese *et al.*, APJ in press, astro-ph/9712325
19. G. Pühlhofer *et al.*, *Astropart. Phys.* **8**, 101 (1997)
20. D. Macminn and J.R. Primack, *Space. Sci.Rev.* **75**, 413 (1996)
21. L. Padilla *et al.*, submitted to *A & A* 1998
22. D.J. Bird *et al.*, *Astrophys. J.* **441**, 1 (1995)
23. D. Horns *et al.*, to appear in Proc. 19th ICVHECR 1998
24. A. Lindner, *Astropart. Phys.* **8**, 235 (1998)
25. J. Cortina *et al.*, *J. Phys.* G **23**, 1733 (1997)
26. A. Lindner *et al.*, to appear in Proc. 19th ICVHECR 1998

TeV Gamma-Ray Astronomy at the Whipple Observatory

R.W. Lessard

for the Whipple and VERITAS collaborations

Department of Physics, Purdue University, West Lafayette, IN 47097, USA
E-mail: lessard@physics.purdue.edu

The Whipple collaboration operates a 10 m diameter optical reflector for gamma-ray astronomy at the Fred Lawrence Whipple Observatory situated on Mt. Hopkins in southern Arizona. The telescope will be described and a review of recent results will be given. Dramatic advances in Very High Energy (VHE) gamma-ray astronomy utilizing the imaging atmospheric Cherenkov technique (IACT) have greatly accelerated the development of new ground-based facilities. A next generation atmospheric Cherenkov observatory is also described which uses the Whipple Observatory gamma-ray telescope as a prototype. An array of imaging telescopes, named the Very Energetic Radiation Imaging Telescope Array System (VERITAS), will be deployed such that they will permit the maximum versatility and will give the highest sensitivity in the 50 GeV - 50 TeV band (with maximum sensitivity from 100 GeV to 10 TeV). In this band critical measurements of nature's accelerators will be made.

1 Introduction

The Whipple collaboration operates a 10 m diameter optical reflector, shown in Figure 1, for gamma-ray astronomy at the Fred Lawrence Whipple observatory situated on Mt. Hopkins in southern Arizona (elevation 2.3 km). A camera, consisting of 331 photomultiplier tubes (PMTs) mounted in the focal plane of the reflector, records images of atmospheric Cherenkov radiation from air showers produced by gamma rays and cosmic rays. By making use of the distinctive differences in the angular distribution of light and orientation of the shower images, a gamma-ray signal can be extracted from the large background of hadronic showers. Using this technique, a background rejection power of several hundred can be obtained. Monte Carlo simulations have shown that this analysis results in an effective energy threshold of approximately 300 GeV and an effective area of $\approx 3.5 \times 10^8 \text{cm}^2$ [1].

Figure 1: The Whipple 10 m gamma ray telescope.

2 The Crab Nebula

The Whipple collaboration broke new ground in 1989 with the first detection at high significance of TeV emission from the Crab Nebula [2] (see Figure 2). Nearly a decade of observations of the Crab Nebula has established it as a steady source, making it the standard candle for TeV gamma-ray astronomy.

The energy spectrum of TeV gamma-rays from the Crab Nebula is shown in Figure 3 and is found to be in agreement with a Synchrotron-Self-Compton (SSC) mechanism of emission [3]. Despite clear observation of a steady signal from the Nebula, pulsed emission from the Crab pulsar has not been observed.

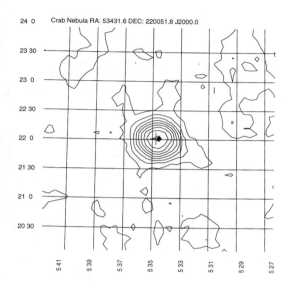

Crab Nebula RA: 53431.6 DEC: 220051.8 J2000.0

Figure 2: TeV detection of the Crab Nebula by the Whipple Observatory's 10 m gamma-ray telescope. Contours represent the statistical significance of excess events between observations centered on the Crab Nebula and a corresponding control region. Contours are in steps of one standard deviation. The cross depicts the pointing uncertainty of the telescope and the filled circle shows the position of the Crab Nebula.

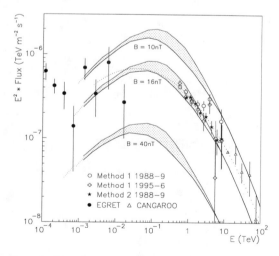

Figure 3: Energy spectrum of TeV gamma-rays from the Crab Nebula [3]. The SSC model fit for several values of the magnetic field intensity are given by the solid curves and shaded regions.

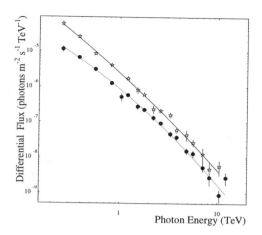

Figure 4: Energy spectra of TeV gamma-rays from Markarian 421, open stars, and Markarian 501, closed circles [9].

3 Active Galactic Nuclei

The detection of TeV gamma-rays from the Active Galactic Nucleus (AGN) Markarian 421 in 1992 [4] opened the field of ground based gamma-ray astronomy to extragalactic sources. A comprehensive search by the Whipple collaboration for very high energy (VHE) gamma rays from other AGNs with properties similar to Markarian 421 led to the subsequent detection of two additional BL Lacertae type objects: Markarian 501 in the spring of 1995 [5] and 1ES2344 in the fall of 1995 [6]. These AGNs are not significant sources of photons for EGRET on the Compton Gamma Ray Observatory (CGRO) so they are the first gamma ray sources discovered by a ground based instrument.

The energy spectrum has been measured for Markarian 421 and 501 and are found to differ significantly. The spectrum of Markarian 421 is consistent with a single power law [7]. The spectrum of Markarian 501, on the other hand, is not well represented by a single power law and requires an additional curvature term for the fit[8].

Intensive monitoring of Markarian 421 and 501 has revealed extreme variability on time scales ranging from 15 minutes to months. The two most extreme episodes of gamma ray variability from an AGN were detected from Markarian 421 in 1996 [10]. The first episode, observed on May 7, 1996, produced the largest flux of gamma rays ever recorded from a VHE source, 10 times the steady Crab rate and 20 times its average for the previous month. The second episode occurred 8 days later and showed both the rise and fall of the flare with a doubling time < 15 minutes. The light curves of these

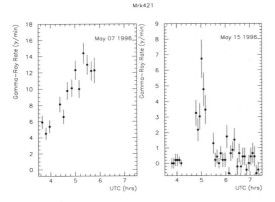

Figure 5: Observations of extreme variability in the AGN Markarian 421 [10].

episodes are shown in Figure 5.

These observations imply large Doppler factors in the jet and very small emission regions, not much larger than the Schwarzchild radius of the putative black hole at the center of the AGN.

For Markarian 421, the TeV gamma ray emission is best characterized as a series of flares with no baseline emission above the sensitivity limit of the telescope. In contrast, Markarian 501, even during 1995 when its flux was 8% that of the Crab Nebula, does appear to have a baseline emission but one which varies dramatically from year to year [11]; its average flux was 130% that of the Crab Nebula in 1997, making it the brightest source in the known TeV sky only two years after being discovered as the weakest source (see Figure 6).

The extreme variability of the VHE emitting BL Lac objects has generated intense interest including several multi-wavelength campaigns. Recent multi-wavelength observations of Markarian 501, initiated in response to a very high emission state in TeV gamma rays, revealed correlations between the TeV gamma ray flux and the X-ray flux. The observations, shown in Figure 7 also indicated a correlation with 50 - 500 keV photons detected at a very high flux level by OSSE on CGRO. EGRET did not detect a significant excess [12].

The multi-wavelength spectral energy distribution for Markarian 421, shown in Figure 8, reveals that the maximum power is output in the TeV band. The fitted curve in Figure 8 is from a single component synchrotron-self-Compton model of emission, although other models involving external sources of soft photons for inverse-Compton (IC) processes or proton initiated cascades within the jet, fit the data equally well [13].

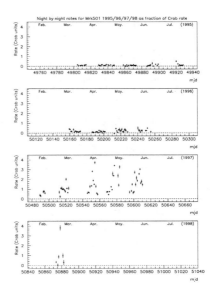

Figure 6: Long term variability of Markarian 501 [11].

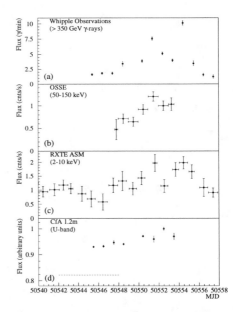

Figure 7: Correlated emission from Markarian 501 [12]

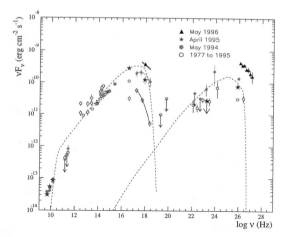

Figure 8: Multi-wavelength spectral energy distribution for Markarian 421. The fitted curve is a single SSC model [13].

Table 1: Characteristics of VERITAS.

Energy range	50 GeV - 50 TeV
	100 GeV - 10 TeV (max sens.)
Effective area	$> 10^5 m^2$
Flux sensitivity	1% of the Crab at 100 GeV
Angular resolution	$0°.05$ $(3')$
Source localization	$0°.005$ $(18'')$
Energy resolution	10%
Field of view	$5°$

4   VERITAS

The dramatic results that have come from VHE gamma-ray observations with existing telescopes justifies increasing efforts in this area of high energy astrophysics. The Very Energetic Radiation Imaging Telescope Array System (VERITAS) is the logical extension of the imaging atmospheric Cherenkov technique. The power of arrays of telescopes, demonstrated by the HEGRA group using an array of six small telescopes and the success of the Whipple 10 m telescope provides a solid design upon which the next generation ground based gamma-ray observatory can be forged.

4.1   Proposed Design

The proposed design for VERITAS will have the basic telescopes modeled on the Whipple 10 m telescope with a focal length of 12 m and wide field cameras of 499 pixels each. The array will consist of seven such telescopes, all capable of independent or coincident operation. The telescope layout is shown in Figure 9 and the design characteristics are given in Table 1. A somewhat similar array has been proposed by the Heidelberg MPI group.

The preferred location of VERITAS is a flat area within the Whipple Observatory complex. There are two sites near the Base camp (elevation 1.3km) where there is ample space for development as well as easy access to roads, power, etc. Southern Arizona has been shown to be an excellent site for these kinds of astronomical investigations with an impressive record of clear nights.

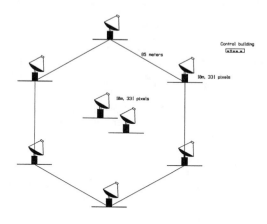

Figure 9: The array configuration for VERITAS.

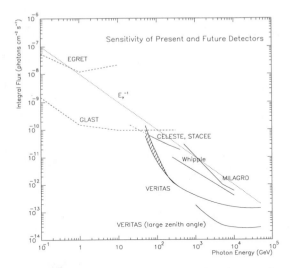

Figure 10: Predicted sensitivity of VERITAS. Also shown are the known sensitivities of EGRET and Whipple and the predicted sensitivities of the MILAGRO, STACEE, CELESTE and GLAST experiments, all of which are at various stages of development. The exposure for VERITAS, STACEE and CELESTE is 50 hours. The exposure for EGRET, GLAST and MILAGRO is one year of sky survey operation.

4.2 Flux Sensitivity

The parameters of the array are chosen to give optimum flux sensitivity in the 100 GeV - 10 TeV range which has proven to be rich in scientific returns. The predicted flux sensitivity is shown in Figure 10; it is seen to be a factor of 10 better than any other detector in this range. In these two decades of energy the major background comes from hadron-initiated air showers for which successful identification methods have been developed. At the lower end single muons become a major background but these can be removed by the coincident requirement in several telescopes. Also at lower energies, the cosmic electron background constitutes an irreducible isotropic background. Over these two decades of energy, the angular and energy resolutions will be pushed to their limits.

4.3 Operating Modes

The two outstanding limitations of the present generation atmospheric Cherenkov telescopes are: (1) poor sensitivity at low energies which is limited in part by the background of muon events and in part by angular resolution; and (2) the ability to look at more than one small part of the sky. This is especially frustrating for variable sources which require long term monitoring. VERI-

TAS will address both of these limitations since the array will increase the angular resolution, discriminate against muons and it will be capable of a variety of operating modes.

1. In the most sensitive mode, all telescopes will be operated in parallel and will thus achieve the maximum sensitivity. Based on a simple extrapolation from previous designs, an energy threshold of 50 GeV will be achieved. This will be most important for the discovery of new AGNs. This mode will achieve an energy resolution of 8% and an angular resolution of 0°.05.

2. In the survey mode, all nine telescopes will be operated independently to search for new sources and/or to monitor known sources. In this mode, each telescope will have a greater sensitivity than the existing state-of-the-art 10 m telescope. It will be possible to complete a sky survey (at the sensitivity of the Crab Nebula) for the entire sky in one year. A deeper survey of the Galactic Plane (at a level of 10% that of the Crab Nebula) could be completed in two years.

3. In the intermediate mode, the telescopes will be operated in clusters of three and four with an energy threshold of 75 GeV.

4. Operation at large zenith angles gives greatly increased sensitivity at higher energies since the collection area increases by almost the same factor as the energy threshold and the background is greatly reduced.

Acknowledgements

Gamma-ray astronomy at Purdue University is supported by a grant from the U.S. Department of Energy.

References

1. G. Mohanty et al., Astropart. Phys. 9, 15 (1998).
2. T.C. Weekes et al., ApJ 342, 379 (1989).
3. A.M. Hillas et al., ApJ 503, 744 (1998).
4. M. Punch et al., Nature 358, 477 (1992).
5. J. Quinn et al., ApJ 456, L83 (1996).
6. M. Catanese et al., ApJ 501, 616 (1998).
7. J.A. Zweerink et al., ApJ 490, L141 (1997).
8. F.W. Samuleson et al., ApJ 501, L17 (1998)
9. F. Krennrich et al., submitted to ApJ.
10. J.A. Gaidos et al., Nature 383, 319 (1996)
11. J. Quinn et al., to be submitted.
12. M. Catanese et al., ApJ 487, L143 (1997).
13. J.H. Buckley et al., ApJ 472, L9 (1996)

GLAST

T. H. Burnett

(on behalf of the GLAST Collaboration)

University of Washington, Physics 351560, Seattle WA 98195, USA
E-mail: tburnett@u.washington.edu

The Gamma-ray Large Area Space Telescope, GLAST, is a satellite-based experiment under development to measure the cosmic gamma-ray flux in the energy range 20 MeV to 300 GeV (with acceptance up to 1 TeV) with unprecedented precision and sensitivity. With a planned launch in 2005, GLAST will open a new and important window on a wide variety of high energy phenomena, including black holes and active galactic nuclei; gamma-ray bursts; supernova remnants; and searches for new phenomena such as supersymmetric dark matter annihilations and primordial black hole evaporation. In addition to the physics, we discuss the instrument design, the novel collaboration of particle physicists and high energy astrophysicists, and the mission status.

1 Introduction

As the highest-energy photons, gamma rays have an inherent interest to astrophysicists and particle physicists studying high energy nonthermal processes. Gamma-ray telescopes complement those at other wavelengths, especially radio, optical, and X-ray, providing the broad, multiwavelength coverage that has become such a powerful aspect of modern astrophysics. EGRET, the high energy telescope on the Compton Gamma Ray Observatory, has led the way in such an effort, contributing to braoadband studies of blazars, gamma-ray bursts, pulsars, solar flares, and diffuse radiation. New development is underway for the next significant advance in high energy gamma-ray astrophysics. GLAST, which will have 30 times the sensitivity of EGRET at 100 MeV, and more at higher energies, including the largely unexplored 30-300 GeV band. The following sections describe the science goals, instrument technologies, and international collaboration for GLAST.

Some key scientific parameters for GLAST are shown in Table 1[1,8]:

2 Scientific Goals

2.1 Blazars

Blazars are thought to be active galactic nuclei consisting of accretion-fed supermassive black holes with jets of relativistic particles directed nearly toward our line of sight. The formation, collimation, and particle acceleration in these powerful jets remain important open questions. Many blazars are seen as bright, highly-variable gamma-ray sources, with the high-energy gamma rays often dominating the luminosity[5]. For this reason, the gamma rays provide a valuable probe of the physics under these extreme conditions, especially when studied as part of multiwavelength campaigns (e.g. Shrader and Wehrle[12]). With its wide field of view and high sensitivity, GLAST will enable blazar studies with far better resolution and time coverage than was possible with EGRET.

Blazars are often very distant objects, and the extension of the gamma-ray spectrum into the multi-GeV range opens the possibility of using blazars as cosmological probes. In the energy range beyond that observed by EGRET but accessible to GLAST, the blazar spectra should be cut off by absorption effects of the extragalactic background light produced by galaxies during the era, of star formation. A statistical sample of high-energy blazar spectra at different redshifts may provide unique information on the early universe (Macminn and Primack[7]).

2.2 Gamma-Ray Bursts

The recent breakthrough associating gamma-ray bursts with distant galaxies (e.g. Djorgovski[3]) has changed the focus of gamma-ray burst research from the question of where they are to the questions of what they are and how they work. The power source and emission mechanisms for gamma-ray bursts remain mysteries. The high-energy gamma radiation seen from some bursts by EGRET[2] indicates that GLAST will provide important information about these questions. With its large field of view, GLAST can expect to detect over 100 bursts per year at GeV energies, allowing studies of the high-energy component of the burst spectra.

2.3 Search for Dark Matter

One of the leading candidates for the dark matter now thought to dominate the universe is a stable, weakly-interacting massive particle (WIMP). One candidate in supersymmetric extensions of the standard model in particle physics is the neutralino, which might annihilate into gamma rays in the 10-300 GeV range covered by GLAST (see Jungman, Kamionkowski and Griest[6], for a

Table 1: Some GLAST parameters.

Energy Range	20 MeV - 300 GeV
Energy Resolution	$10\%(> 100\text{MeV})$
Effective Area	$8000\text{cm}^2(> 1\text{GeV})$
Field of view	> 2.5 steradians
Source location determination	30 arc seconds to 5 arc minutes
Source sensitivity (1 yr)	$3 \times 10^{-9}(\text{E} > 100\text{MeV})/\text{cm}^2$ s
Mission life	> 2 years

general discussion of dark matter candidates). The good energy resolution possible with the GLAST calorimeter will make a search for such WIMP annihilation lines possible.

2.4 Pulsars

A number of young and middle-aged pulsars have their energy output dominated by their gamma- ray emission. Because the gamma rays a,re directly related to the particles accelerated in the pulsar magnetospheres, they give specific information about the physics in these high magnetic and electric fields. Models based on the EGRET-detected pulsars make specific predictions that will be testable with the larger number of pulsars that GLAST's greater sensitivity will provide[14].

2.5 Supernova Remnants and the Origin of Cosmic Rays

Although a near-consensus can be found among scientists that the high-energy charged particle cosmic rays originate in supernova remnants (SNR), the proof of that hypothesis has remained elusive. Some EGRET gamma-ray sources appear to be associated with SNR, but the spatial and spectral resolution make the identifications uncertain[4]. If SNR do accelerate cosmic rays, they should produce gamma rays at a level that can be studied with GLAST, which will be able to resolve some SNR spatially.

2.6 Diffuse Gamma Radiation

Within the Galaxy, GLAST will explore the diffuse radiation on scales from molecular clouds to galactic arms, measuring the product of the cosmic ray and gas densities. The extragalactic diffuse radiation may be resolved; GLAST should detect all the blazars suspected of producing this radiation. Any residual diffuse extragalactic

gamma rays would have to come from some new and unexpected source.

2.7 Unidentified Sources and New Discoveries

Over half the sources seen by EGRET in the high-energy gamma-ray sky remain unidentified with known astrophysical objects. Some may be radio-quiet pulsars, some unrecognized blazars, and some are likely to be completely new types of object (for a recent discussion, see ref [9]). In general, the EGRET error boxes are too large for spatial correlation, and the photon density is too small for detailed timing studies. Both these limitations will be greatly alleviated with GLAST. In particular, the combination of GLAST with the next generation of X-ray telescopes should resolve a large part of this long-standing mystery. The new capabilities of GLAST will surely produce unanticipated discoveries, just as each previous generation of gamma-ray telescope has done.

3 The GLAST Baseline Design

Any high energy gamma-ray telescope operates in the range where pair production is the dominant interaction process; therefore, GLAST shares some design heritage with SAS-2, COS-B, and EGRET: it has a plastic anticoincidence system, a tracker with thin plates of converter material, and an energy measurement system. See Figure 1 for the Baseline design. What GLAST benefits from most is the rapid advance in semiconductor and computing technology since the previous gamma-ray missions. The silicon revolution affects GLAST in several ways:

3.1 Multi-Layer Si Strip Tracker

The tracker consists of solid-state devices instead of a gas/wire chamber. The baseline design for GLAST uses Si strip detectors with 195μ m pitch offering significantly better track resolution with no expendable gas or high

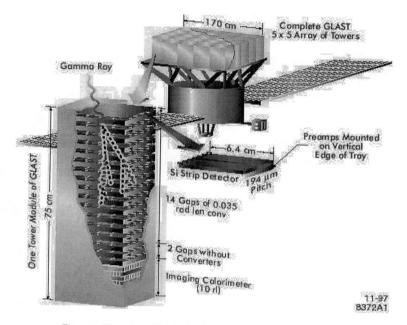

Figure 1: View of the GLAST baseline design, showing the spacecraft, an exploded tower and a Si detector "tray".

voltage discharge required. Low-power application specific integrated circuits (ASICs) allow readout of approximately 10^6 channels of tracker with only 260 W.

The 77m^2 of Si strip detectors planned for GLAST will be the largest silicon strip detector system ever made. Since manufacturers (Hamamatsu, Micron and others) have decided to move to 6-inch wafers, we can expect improved cost/performance.

3.2 On-board Computing

On-board computing, which was extremely limited in the Compton Observatory era, is now possible on a large scale. The 32-bit, radiation-tolerant processors now available allow software to replace most of the hardware triggering of previous missions and also enable considerable on-board analysis of the tracker data to enhance the throughput of useful gamma-ray data. In particular, it makes a very loose trigger possible, accepting all charged cosmic rays for later software rejection. The modular design, currently planned to have 25 autonomous towers, allows multiprocessing, using all 25 processors in parallel.

3.3 Monte Carlo Simulation

An important component of the design process has been Monte Carlo simulations that have been used to predict the efficiency and resolution (which were verified by the 1997 beam test[11]), and to validate the triggering and on-board processing schemes.

4 Plan and Schedule

Two prototype Si strip detectors have been made. The first prototype was a 6-cm-detector. The second prototype (Fig. 2), which had redundancy and bypass strips, showed superb characteristics, with fewer than 0.03% strips bad. A third prototype, made from 6-inch-wafers, is being requested.

A 'mini-tower' consisting of a stack of the 6 cm Si strip detectors, a CsI(T1) calorimeter, and a plastic scintillator anticoincidence, was tested at a tagged gamma-ray beam at SLAC in Fall, 1997[11]. A full prototype tower will be built by summer, 1999, for the Fall, 1999, beam test. We are expecting full production of the GLAST hardware to begin in 2001. We are currently waiting for approvals from DOE and NASA. Applications for support are pending in France, Italy, and Japan.

GLAST is planned as a facility-class mission involving a unique international collaboration with roughly equal participation from the particle physics and astrophysics communities. Currently, scientists from the United States, Japan, France, Germany, and Italy are involved in the development effort. GLAST is currently listed as a candidate for a new start at NASA, with a possible launch in 2005. Further information about GLAST can be found at http://www-glast.stanford.edu/.

References

1. Bloom, E.D., Sp. Sci. Rev. **75**, 109 (1996).
2. Dingus, B.L., Ap. 6 Sp. Sci. **231**, 187 (1995)
3. Djorgovski, S.G. et al., IA U Circ. , 6655 (1997)
4. Esposito, J.A. et al., ApJ **461**, 820 (1996)).
5. Hartman, R.C., Collmar, W., von Montigny, C., K Dermer, C.D., in Proc. Fourth Compton Symposium, ed. C.D. Dermer, M.S. Strickman, J.D. Kurfess, pp. 307 – 327, AIP CP410, Woodbury, NY (1997).
6. Jungman, G., Kamionkowski, M., k Griest, K., Phys. Reports **267**, 195 (1996).
7. Macminn, D. k Primack, J.R., Sp. Sci. Rev. **75**, 413 (1996)
8. Michelson, P.F., SPIE **2806**, 31 (1996).
9. Mukherjee, R., Grenier, I.A., k Thompson, D.J., in Proc. Fourth Compton Symposium, ed. C.D.
10. Dermer, M.S. Strickman, J.D. Kurfess, pp. 394 – 406, AIP CP410, Woodbury, NY (1997).
11. Ritz, S.M., et al., NIM, submitted (1998).
12. Shrader, C.R. k Wehrle, A.E., in Proc. Fourth Compton Symposium, ed. C.D. Dermer, M.S. Strickman,
13. J.D. Kurfess, pp. 328 – 343, AIP CP410, Woodbury, NY (1997).
14. Thompson, D.J., Harding, A.K., Hermsen, W., Sz Ulmer, M.P., Proc. Fourth Compton Symposium, ed.
15. C.D. Dermer, M.S. Strickman, J.D. Kurfess, pp. 39 – 56, AIP CP410, Woodbury, NY (1997).

INITIAL RESULTS FROM THE AMANDA HIGH ENERGY NEUTRINO DETECTOR

S.W. BARWICK, J. BOOTH, J. KIM, P. MOCK, R. PORRATA, D. ROSS, W. WU, G. YODH, S. YOUNG

Dept. of Physics, University of California-Irvine, Irvine, CA, USA

T. MILLER

Bartol Research Institute, University of Delaware, Newark, DE, USA

A. BIRON, S. HUNDERTMARK, H. LEICH, M. LEUTHOLD, P. NIESSEN, S. RICHTER, T. SCHMIDT, U. SCHWENDICKE, C. SPIERING, P. STEFFEN, O. STREICHER, T. THON, C. WIEBUSCH, R. WISCHNEWSKI

DESY-Institute for High Energy Physics, Zeuthen, Germany

S. CARIUS, P. LINDAHL

Dept. of Physics, Kalmar University, Sweden

E. ANDRÉS, L. BERGSTRÖM, A. BOUCHTA, E. DALBERG, J. EDSJO, P. EKSTRÖM, A. GOOBAR, P. HULTH, J. RODRIGUEZ, C. WALCK

Dept. of Physics, Stockholm University, Stockholm, Sweden

R. BAY, Y. HE, D. LOWDER, P. MIOCINOVIC, P. B. PRICE, W. RHODE, M. SOLARZ, K. WOSCHNAGG

Dept. of Physics, University of California-Berkeley, Berkeley, CA, USA

D. COWEN, M. NEWCOMER

Dept. of Physics, University of Pennsylvania, Philadelphia, PA, USA

M. CARLSON, T. DEYOUNG, F. HALZEN, R. HARDTKE, G. HILL, J. JACOBSEN, V. KANDHADAI, A. KARLE, I. LIUBARSKY, R. MORSE, P. ROMENESKO, S. TILAV

Dept. of Physics, University of Wisconsin, Madison, WI, USA

W. CHINOWSKY, D. NYGREN, G. PRZYBYLSKI, G. SMOOT, R. STOKSTAD

Lawrence Berkeley Laboratory, Berkeley, CA, USA

O. BOTNER, A. HALLGREN, P. LOAIZA, P. MARCINIEWSKI, C. PÉREZ DE LOS HEROS, H. RUBINSTEIN

University of Uppsala, Uppsala, Sweden

The AMANDA collaboration has deployed a 10-string array (AMANDA-B10) of 302 photomultiplier tubes at the South Pole at a depth of 1.5 to 1.9 km. Simulations indicate that AMANDA-B10 has an effective area of $\sim 10,000$ m^2 for TeV-scale neutrino-induced muons, with an angular resolution of 2.5° and good rejection ($> 10^5$) of down-going atmospheric muons. The primary purpose of the array is to search for astrophysical sources of high energy neutrinos, but its science reach extends to particle physics. Data from the first phase of the construction, AMANDA-B4, has been used to measure the optical quality of the ice, to determine geometric spacing, and to reconstruct tracks from atmospheric muons. The reliability of the detector simulations has been addressed by comparing the results to AMANDA-B4 data. Construction of the AMANDA-II upgrade began January 1998 with the deployment of three strings to a depth of 2350m. Simulations predict that AMANDA-II will have an effective area of $\sim 5 \times 10^4$ m^2 (depending on energy) and angular resolution of $\sim 1°$.

1 Introduction[a]

The Antarctic Muon and Neutrino Detector Array AMANDA is a multi-purpose instrument. While opti-

mized to search for astrophysical sources of high energy neutrinos, its science missions address questions in particle physics, astronomy and astrophysics, cosmology and cosmic ray physics[1]. Its deployment creates new opportunities for glaciology[2]. The rich scientific potential of the AMANDA detector is due in part to the flexible ar-

---

[a]Talk presented at the *International Conference on High Energy Physics (ICHEP 98)*, Vancouver, Canada, July 1998.

chitecture, which creates the opportunity to optimize the sensitivity over broad range of neutrino energies. In addition, of all known high energy particles, only neutrinos convey information from the edge of the universe or from the central engine of cosmic accelerators. Gamma rays come from a variety of objects, both galactic (Crab Nebula) and extragalactic (blazars) with energies to 20 TeV (or higher). However, absorption of gamma rays by extragalactic infrared radiation may limit the seeing to distances less than 100 Megaparsecs. Cosmic rays are accelerated to energies as high as 10^{20} eV, but their range is limited by absorption by the cosmic microwave photons. At the highest energies, cosmic rays and neutrinos may originate from gamma ray bursts (GRBs), active galaxies, or topological defects such as cosmic strings.

A high energy neutrino can be detected only if it converts to a charged lepton. Muon detection is the usual mode of operation, but neutrino detectors have the capability to observe electrons (and taus) via the Cherenkov photons generated by electromagnetic and hadronic cascades. The muon is detected by distributing photon sensors (PMTs) over the largest posssible volume of transparent medium and recording the arrival times and intensities of the Cherenkov wavefront. To have sufficient sensitivity to attack the fundamental problems, it is generally agreed that the volume of the detector should be at least 1 km^3. Great care must be taken to design the system to reject the background of ordinary downgoing atmospheric muons, whose flux is more than a factor 10^5 larger at AMANDA depths than the irreducible background of atmospheric neutrinos. The low levels of background light and cold ambient temperatures result in relately small rates of random background signals from AMANDA OMs. Consequently, AMANDA can detect a burst of low energy neutrinos ($/sim10$ MeV) by measuring an increase in the dark noise rates, averaged over all OMs with good stability.

AMANDA-B10 has been optimized to detect neutrinos with energies in excess of 1 TeV, but the sensitivity of detector at lower energies is sufficient to observe ~100 atmospheric neutrinos above 100 GeV per year[6] , thereby providing a convenient source for calibration.

The status of the AMANDA project can be summarized as follows:

- Construction of the first generation AMANDA detector[3] was completed in the austral summer 96–97. It consists of 302 optical modules, located on 10 separate strings, that are deployed to depths between 1500–2000 m; see Fig. 1. An optical module (OM) consists of an 8 inch photomultiplier tube (R5912-02) encapsulated in a glass pressure sphere and mounting hardware. Analog signals are sent to the surface via electrical cables

in AMANDA-B10. The conservative design has resulted in an *in-situ* failure rate of only 3 %. System level robustness is a consequence of architectural redundancy. Each sensor and calibration tool are connected to the surface by an individual cable, eliminating single-point failure concerns.

- Data taken with the first 4 strings (a total of 80 OM's), deployed in January of 1996 to assess the optical properties of the deep ice, have been analysed. This partial detector will be referred to as AMANDA-B4. Nearly vertical up-going muons are found at a rate that is statistically consistent with the expected flux of atmospheric neutrinos, athough complete background assessment is still in progress. Simulations and data agree at various levels of refinement - from a crude check of hardware trigger rate to a more sophisticated examination of the muon angular distributions as the reconstruction criteria are made more restrictive. Absolute event rates agree to within a factor of 3 at a rejection level of 10^5.

- The commissioning phase of the full detector is now completed (July '98) and analysis of data from 1997 is in progress. Final calibration of array geometry, cable-dependent time delays, and PMT performance was completed after the return of the first year of full operation. First-look analysis indicates that events can be extracted with trajectories in the upward direction. A more extensive evaluation of background and detector performance for a variety of signal classes is currently in progress.

2 Installation of AMANDA-II strings

AMANDA-II is an approved and funded expansion of the AMANDA-B array. The proposed array consists of 11 additional strings of OMs arranged concentrically around AMANDA-B10. Simulations predict that AMANDA-II will have an effective detection area of $\sim 5 \times 10^4$m^2 (depending on energy; significantly less for atmospheric neutrinos and somewhat larger for PeV-scale neutrinos) and angular resolution of $\sim 1°$ (again, depending slightly on energy). Construction of the AMANDA-II upgrade began in January 1998 with the deployment of three strings to a depth of 2350m. Each string contained 42 OMs that were positioned along the lowest kilometer of cable. Thus, these strings serve as a full-scale prototype for a planned expansion to a kilometer-scale array of sensors called IceCube.

The deployment of AMANDA-II strings in 1998 addressed both science and R&D goals. First, the optical properties of the ice at depths above and below

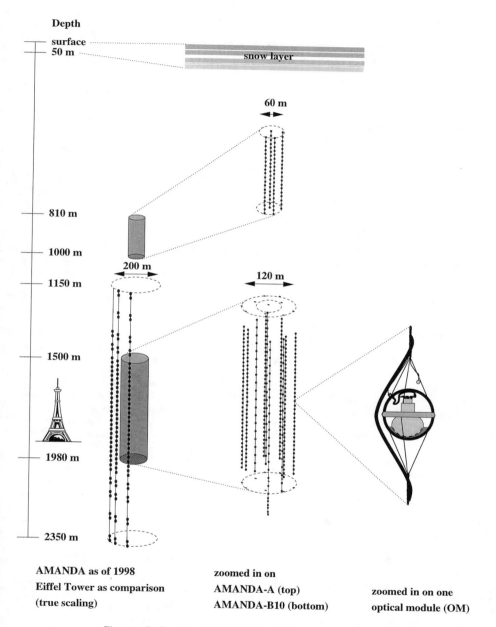

Depth

surface
50 m

snow layer

60 m

810 m

1000 m

200 m

1150 m

120 m

1500 m

1980 m

2350 m

AMANDA as of 1998
Eiffel Tower as comparison
(true scaling)

zoomed in on
AMANDA-A (top)
AMANDA-B10 (bottom)

zoomed in on one
optical module (OM)

Figure 1: Configuration of Antarctic Muon And Neutrino Detector Array (AMANDA) in 1998.

AMANDA-B10 were measured. These results will be used to optimize the depth and spacing of the remainng eight strings of AMANDA-II sensors. Second, the longer lever arms of the new strings provides crucial data to verify simulation results on event topologies not readily obtained by AMANDA-B10.

Research and development of more capable technologies is a strong component of the AMANDA-II program. For example, a pair of TV cameras were lowered into the last hole. The resulting images visually confirm the exceptional clarity of the ice deduced from calibration measurements. The fidelity of signal transmission was dramatically improved by transmitting analog signals from the PMT to the surface over optical fiber and electrical cable simultaneously. Optical transmission of signals, using an LED, eliminates the distortion of the PMT waveform while preserving many aspects of the conservative design features introduced by analog signal transmission over electrical cable . The high fidelity of signal reproduction at the surface improves the double-pulse resolution by an order of magnitude. Reconstruction should benefit from better identification of multiphoton signals, and from reduced cross-talk. Time-delay calibration procedures are simplified so fewer manpower resources are required.

The robustness of the optical fiber cables and connectors was improved. One combination of fiber and connector technologies produced a 90% survival rate. Based on the success of the optical technologies, the remaining AMANDA-II OMs will transmit analog optical and electrical signals to the surface.

In addition, new technologies were introduced to reduce problems associated with more complex string design. Operational factors, such as the status and limitations of several new deployment strategies, were assessed. New mechanical designs and integration procedures were introduced and tested that maintained an acceptable duration for string assembly despite an increased number of OMs per string (42), larger separations between OMs(as large as 60m in some cases), an extended length of active sensors (more than 1 km), and more complex certification procedures due to the increased complexity of the optical signal transmission. The next iteration of string design should realize similar reduction in the time to assemble and certify the AMANDA string. Fuel usage and manpower requirments were another concern. While the increased depth of the holes taxed the capabilities of the ice-drilling equipment, the holes were drilled straight to within a meter of tranverse displacement. The success of the drilling operations and the encouraging performance of the optical fiber technologies suggest that strings containing 60-80 sensors, as currently envisioned for IceCube, can be deployed within the time contraints imposed by the deployment process.

Anticipating the requirements of the full AMANDA-II array, the data acquisition system was upgraded to VME-based readout of the front-end electronics. Macintosh computers were replaced by powerful workstations running LINUX and the acquisition sofware was completely re-written. Data is acquired at 90 kB/s and automatically archived. A new satellite link transfers $\sim$ 20% of the data from the South Pole daily - a significant improvement in bandwidth over previous years.

3   Commissioning of AMANDA-B10

After one year of operation, a complete calibration of the B10 array was performed. Interstring laser pulses are used to determine the geometry of the detector. In conjunction with telemetry from the drill, the OMs have been positioned with an absolute precision of better than 1 meter. Mapping the detector has been by far the most challenging aspect of the calibration of this novel instrument. A precise knowledge of the location of the optical sensors is crucial for track reconstruction. Therefore, two completely independent methods were developed for the final determination of the geometry. One method makes use of drill data. A variety of sensors are installed in the drill to determine its speed and direction during drilling. Every second a data string is transmitted to the control system and recorded. This data is combined with survey information at the surface to determine the transverse string position. The absolute depth of the strings is deduced from a pressure sensor. The final positioning of the strings is obtained by a laser calibration system. Laser pulses (532 nm) are transmitted with optical fibers to every optical module on strings 1–4, and to every second module on strings 5–10. After the timing calibration is completed, the laser calibration provides time of flight measurements to determine the distances between strings and a check on possible vertical offsets. More than a hundred laser runs were globally fit both to determine the geometry and to verify the timing calibration. Figure 2 shows a horizontal cross-section of the position of the strings at a depth midway between the top and bottom of the array. The two methods of determining the geometry agree to within 1 m, so the position error of the optical sensors matches the time resolution of the sensors.

Time offsets measure the relative time for the signal to propagate from the optical module to the electronics. Since the cables are 1.5–2.0 km in length, the propagation time is $\sim$ 10 microseconds. The challenge is to measure this value to an accuracy of 5 ns. Commercial and special purpose optical time domain reflectometers measured the propagation time of 532nm optical pulses. Offset times show excellent stability over a two year period, agreeing to better than 10ns (rms), although exhibiting a small

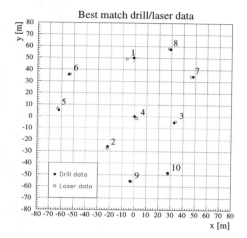

Figure 2: Top view of array geometry. Open circles = geometry determined by YAG laser. Filled circles = geometry determined by survey, drill, and pressure sensor data.

systematic drift of 10ns. These results are consistent with expectation based on the temperature stability of the bulk ice.

4 Reconstruction of muon tracks

The reconstruction of muon tracks that traverse optical media in which the scattering of Cherenkov photons cannot be ignored represents one of the major challenges facing the AMANDA collaboration. Reconstruction at the necessary levels of accuracy have been achieved by maximizing a likelihood function which matches the observed and expected time of arrival values[7]. Studies of the optical properties of deep polar ice reveal that the absorption length is 100 m or more, depending on depth[2]. With such a large absorption length, scattering becomes a critical issue. The scattering length is 22–30 m, depending on depth (preliminary). A typical event triggers 20 OM's. Of these more than 5 photons are, on average, "not scattered." They are referred to as direct photons, i.e. photons which arrive within time residuals of $[-15; 25]$ ns relative to the calculated time it takes for unscattered Cherenkov photons to reach the OM from the reconstructed muon track. The choice of residual allows for delays of slightly scattered photons.

The reconstruction method was tested with AMANDA-B4 data (see Fig. 3), where the measured arrival directions of background cosmic ray muon tracks, reconstructed with 5 or more unscattered photons, are compared to simulated events using a similar analysis

chain. The bottom panel of Fig. 3 imposes a cut which requires that the track, reconstructed from timing information, traces the spatial positions of the OM's in the trigger. The power of this cut, especially for events recorded with only 4 strings, is instructive.

The results confirm the expectation based on simulation of the AMANDA-B4 detector that less than one in 10^5 triggered events is misreconstructed as originating below the detector[4]. We conclude from Fig. 3 and numerous additional checks that the agreement between real and simulated data is adequate, bolstering our confidence in the reliability of studies which predict the performance of the AMANDA-B10 array.

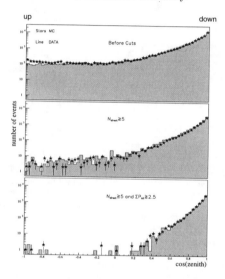

Figure 3: Reconstructed zenith angle distribution of muons: AMANDA-B4 data and simulation. The relative normalization has not been adjusted at any level. Less than one in 10^5 triggered events is reconstructed in the upward direction.

Visual inspection reveals that the misreconstructed tracks are mostly showers, radiated by muons or initiated by electron neutrinos, which are misreconstructed as up-going tracks of muon neutrino origin. They can be readily identified on the basis of the characteristic nearly isotropic distribution of the OM amplitudes, and by the fact that the direct hits occur over a short distance near the origin of the shower, rather than spread over a longer muon track.

We have verified the angular resolution of AMANDA-B4 by reconstructing muon tracks registered in coincidence with a surface air shower array SPASE[5]. Figure 4 demonstrates that the zenith angle distribution of the coincident SPASE-AMANDA cosmic ray beam reconstructed by the surface array is quantitatively reproduced by reconstruction of the muons in AMANDA. Note that that small 3 degree difference in mean zenith angle was predicted by simulation. The errors in the zenith angle analysis are biased to directions parallel to the string axis. This effect is reduced in the full array.

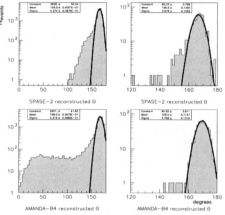

Figure 4: Zenith angle distributions of cosmic rays triggering AMANDA and the surface air shower array SPASE. AMANDA reconstruction of muon direction is compared to the air shower direction using SPASE. AMANDA events are selected by requiring signals on 2 or more strings (left), and 5 or more direct photons (right).

5 Conclusions

Data from the first phase of construction, AMANDA-B4, has been used to measure the optical quality of ice, geometric spacing *in situ*, and angular distributions of atmospheric muons. The reliability of the detector simulations has been confirmed by comparing to angular distributions as the acceptance criteria become more selective. Independent confirmation of the angular resolution was obtained from events that simultaneously trigger the SPASE air shower array and AMANDA.

During the six month commissioning phase following the first year of AMANDA-B10 operation, system calibration of the array geometry, propagation constants, and gain drifts was completed. Concurrently, software was developed to reduce the 0.5 TB of data by a factor of 10 by filtering events that were readily identified as atmopheric muons. The data was processed through the filter at NERSC/LBL and distributed to the collaboration in July 1998.

We are excited by the prospect of scanning the sky with a new tool specifically designed to detect high energy neutrinos from cosmological sources. As high energy neutrino, gamma ray, and gravity wave detectors are commissioned to search for powerful transient phenomema, we may soon see the birth of multi-messenger astronomy.

Acknowledgements

The AMANDA collaboration is indebted to the Polar Ice Coring Office and to Bruce Koci for the successful drilling operations, and to the National Science Foundation (USA), DESY (Germany), the Swedish National Research Council, the K.A. Wallenberg Foundation and the Swedish Polar Research Secretariat. S.W.B is supported in part by the University of California–Irvine and in part by NSF grants OPP-9512196 and PHY-9722641. F.H. is supported in part by the U.S. Department of Energy and in part by the University of Wisconsin Research Committee.

References

1. For a review, see T.K. Gaisser, F. Halzen and T. Stanev, *Phys. Rep.* **258**(3), 173 (1995); R. Gandhi, C. Quigg, M.H. Reno and I. Sarcevic, *Astropart. Phys.*, **5**, 81 (1996).

2. The AMANDA collaboration, *Science* **267**, 1147 (1995).

3. S.W. Barwick *et al.*, *The status of the AMANDA high-energy neutrino detector*, in Proceedings of the 25th International Cosmic Ray Conference, Durban, South Africa (1997).

4. S. Tilav *et al.*, *First look at AMANDA-B data*, in Proceedings of the 25th International Cosmic Ray Conference, Durban, South Africa (1997).

5. S.W. Barwick *et al*, Proceedings of the 22nd International Cosmic Ray Conference, Dublin (Dublin Institute for Advanced Studies, 1991), Vol. 4, p. 658.

6. R. Bay *et al.*, *The AMANDA Neutrino Telescope*, in Proceedings of the 18th International Conference on Neutrino Physics and Astrophysics (Neutrino 98), Takayama, Japan (1998).

7. C. Wiebusch *et al.*, *Muon reconstruction with AMANDA-B*, in Proceedings of the 25th International Cosmic Ray Conference, Durban, South Africa (1997).

8. T. Miller *et al.*, *Analysis of SPASE-AMANDA coincidence events*, in Proceedings of the 25th International Cosmic Ray Conference, Durban, South Africa (1997).

9. R. Bay *et al*, The AMANDA collaboration, Physics Reports **306**, to be published; A. Bouchta, University of Stockholm, PhD thesis (1998)

ANTARES

J. J. AUBERT

on behalf of the Antares Collaboration

*Centre de Physique des Particules de Marseille, Université de la Méditerranée, IN2P3-CNRS,
Case 907, 163 Avenue de Luminy, 13288 Marseille Cedex 09, FRANCE
E-mail: aubert@cppm.in2p3.fr*

A demonstrator for a large scale deep sea neutrino detector is under construction and long term measurements of a site close to
Toulon (Mediterranean sea) are in progress. We report on the encouraging results received to date.

1 Introduction

To ultimately achieve a km scale detector for very high
energy cosmic neutrinos, necessitates a phased development programme. For a final detector of ten thousand
optical sensors installed on hundred strings (figure 1,
right), the first phase would be a demonstrator consisting
of two or three strings (figure 1, left). A second step may
consist in a thousand optical modules arranged in fifteen
strings. This second step should allow the understanding of long term operations and to measure the neutrino
flux. The first phase will validate technological choices
and allow for a correct cost estimate.

Long term site measurements are mandatory to understand properly deep sea conditions. This phase is approved and financed.

The collaboration benefits appreciably from previous
work from Amanda [1], Baïkal [2], Dumand [3], Nestor [4] and
D. Nygren et al. [5].

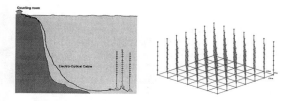

Figure 1: A possible demonstrator (left) ; a possible detector configuration (right).

2 Detector principle

The principle has been suggested by Markov in the
1960's. The earth is used as a muon filter. A three dimensional matrix of optical sensors, installed in a transparent medium, collects the Čerenkov light emitted by
the muon induced in a ν_μ interaction (figure 2). For
contained events, the hadronic shower contributes to the
amount of light observed. Since the muon range increases

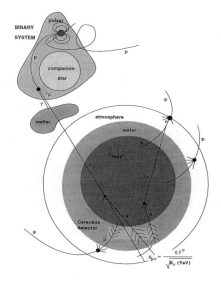

Figure 2: Neutrino detection principle.

with the muon energy, the effective volume of a detector
increases with the neutrino energy and partially compensates for the corresponding expected decrease of the flux.
The muon track parameters are determined using the
light arrival time on photomultiplier tubes. There is no
analytical solution, one needs a minimisation technique
to get the parameters. The angular accuracy will depend, for a given geometry, on the medium properties
(absorption, diffusion), the time arrival accuracy fixed
by the light diffusion and for long diffusion length by
the time accuracy of our system. The optimal ulti-
mate accuracy has to be smaller than the θ_{ν_μ} angle. It
should be noted that in practice, for $E_\mu \geq 0.1\ E_\nu$ or
$E_\mu \geq 100$ GeV, the θ_{ν_μ} angle is smaller than the number
usually quoted, namely $\frac{1.5^\circ}{\sqrt{E_{TeV}}}$ by a factor 3. For 100 TeV
neutrino (for $E_\mu > 100$ GeV), $\theta_{\nu_\mu} = 0.05^\circ$. The muon en-

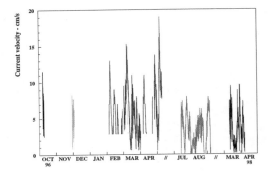

Figure 3: Current intensity versus time in the Toulon site.

correlation (>40 m). We have also seen that bioluminescence activity strongly depends on the current and on the seasons (figure 4, right). No conclusion can be drawn in comparing the Corsica and Toulon sites. The variability is too great and the data have to be taken over a longer period.

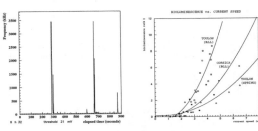

Figure 4: An example of bioluminescence signal (left); bioluminescence (in arbitrary units) as a function of the current for three separate time intervals (right).

ergy is determined by measuring dE/dx with the amount of light received on the photomultipliers. $\frac{dE}{dx} = a + bE$ above 500 GeV/1 TeV. This allows an accuracy of $0.3/0.4$ on $\log E$.

3 Site measurements - Mediterranean sea

The program to measure the current behaviour, the bioluminescence noise, the long term biofouling effect and the optical properties of deep sea water is well-advanced. Most of the tests have been done in Toulon, some of them in Corsica (close to Calvi), the limitation in the latter case is the overall amount of technical support available.

3.1 Current

The figure 3 shows the deep sea current at 2300 m in Toulon area over 18 months. The mean value is around 4 cm/s with one peak at 19 cm/s. This shows that the variability is very large on a timescale of a few days and for year to year. Our measurements are in agreement with results from C. Millot [6].

3.2 Bioluminescence

This is a well-known effect whereby short pulses (msec to few secondes) of high intensity are seen from just above the K^{40} background to saturation of the 8" phototube (EMI) used (figure 4, left). Since one monitors the rate of each optical module, the bioluminescence is only a potential deadtime, not a background for the experiment. In fact, the electronics implemented allows the mean deadtime per photomultiplier to be less than 3 % with a maximum of 5 %. It has been measured that the bioluminescence noise is eliminated (~ 100 Hz) when a two-photoelectron threshold is required. We have observed short distance correlations but no long distance

3.3 Biofouling

When an optical module stays in ocean water, the sedimentation occurs on the upper part (Baïkal measurements[2],) and, on a longer time scale, biofouling can induce a kind of glue which will deteriorate the optical properties. Biofouling [7] depends mainly on the density of bacteria. This density is lower in deep water than in shallow water. We have illuminated an optical module with LED diodes and monitored the absorption with pin diode installed at 0°, 20, 40, 50, 60, 70, 80 and 90° (0° is the top of the optical module, 90° is horizontal). Measurements taken over three months show a rather strong sediment deposit at 0° decreasing to 40°. The cleaning of the sediment with a rapid current variation is also visible. Measurements taken over eight months (figure 5) show an overall 3% attenuation at 90°. The attenuation observed originates from the source and detector, thus giving an overall loss of 1.5% in eight months.
These results are compatible with the in-situ measurement of the bacterial activity.

3.4 Optical properties of deep sea water

We have instrumented a long stiff support of 33 m on which a movable light source and a fixed optical module detector have been mounted. With a continuous, uniform LED source, one should measure a position dependent intensity $e^{-D/\Lambda} \frac{1}{D^2}$ where D is the source to detector distance and Λ is the attenuation length of water. Figure 6 shows the results for water measurement and calibration in air on a log scale. $\Lambda_{attenuation} = 41 \pm 1_{stat} \pm 1_{syst} m$ at a wavelength of 465 nm. The scattering length is measured

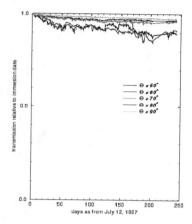

Figure 5: Biofouling effect versus time.

by the arrival time of a fast pulsed source on a 1" photomultiplier. This measurement has been made in July 1998. Current analysis of the data indicate a scattering length for Rayleigh-type scattering in excess of 100 m. It should be noted that a 150 m scattering length and a 55 m absorption length combine to give a 40 m attenuation length. However encouraging these results may be, the wavelength, depth and time dependances still need to be studied.

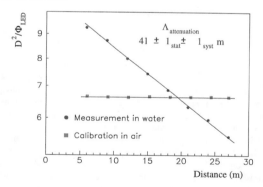

Figure 6: Attenuation measurement in situ.

3.5 Some conclusions about site measurements

Apart from the tests described, we have taken sediment samples in situ, performed a geological site study, and we will make a visual submarine survey this winter. Our present understanding is that the Toulon site has no un-

favourable characteristics which would prevent the possibility to operate a second phase there.

4   The Demonstrator

We are building a demonstrator consisting of two strings which will be interconnected electrically by submarine. The first string is mainly intended to understand the technology and deployment issues and consists of 32 optical modules (OMs). The optical modules are 8" hemispheric photomultipliers from Hamamatsu. They are arranged (figure 7) in pairs and installed at 15 m intervals starting at 100 m from the sea-bed. This string is connected to a 40 km long 4 fibre optical cable. There are also 11 instrumented containers equipped with tiltmeters and compasses. Four acoustic beacons will be used for the position measurement of the OMs. In practice, only 8 PMTs will be mounted and the signal will be transferred directly to the fibre. The first string is ready, first series of preliminary tests are complete and the first deployment in shallow water (400 m) has been performed during the time of the conference. Further work is now continuing. The electro-optical cable was laid a few months ago. The second string will be orientated towards an economically viable solution.

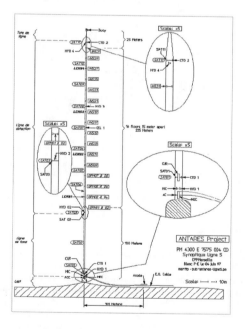

Figure 7: First string: conceptual design.

4.1 Detector alignment

Since such a detector will be in fact a telescope, one needs an absolute measurement. This has been achieved with a long base acoustic device (2 x 2 km²) and with more than thousand measurements on boat with Differential GPS. An absolute accuracy of 1 m has been achieved. One should also monitor the exact position ot the optical module with time. For this purpose, a short base acoustic device (300 m) has been built and the beacon installed on the string will give a relative accuracy of 20 cm. Tiltmeters and compasses provide redundancy and will allow the calibration of our system.

4.2 Submarine connections

IFREMER is organising a connection test in winter 98 with a manned submarine, "The Nautile" which will be used to connect cables stored on a drum between two supports (anchors). This operation, a routine one for IFREMER, is well described in figure 8. The performance of the connector has been made in a high pressure test facility at IFREMER Brest.

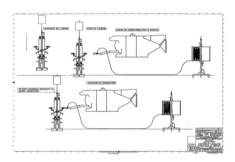

Figure 8: Schematic of a submarine connection.

4.3 String 2

The fabrication of the first string and the deployment operation have provided us with invaluable experience. String 2 will implement a simpler design to reduce costs and facilitate easier operation in the sea. It will be a modular structure of two or three optical modules with only one electromechanical cable to provide support and to carry the signals. The laboratory study is well-advanced and the industrial fabrication should start soon for deployments by the end of 1999.

5 The Next step

5.1 A detector to measure the TeV-PeV neutrino flux

We aim at a next step consisting of ≈ 1000 optical modules on ≈ 15 strings giving us around 0.1 km² detector for a TeV, PeV muons. As an example, we report here the on possibilities of a device of 15 strings each of 64 OM grouped by 4 every 15 m, the distance between the strings being 100 m with the first OM 100 m above the sea bottom. Figure 9 displays the arrival time of the Čerenkov light for different muon energies. In sea water, the late pulse are due to the electromagnetic shower, not to the scattering in the medium.

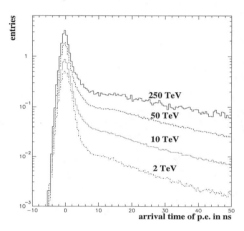

Figure 9: Time spread for different muons energies.

Figure 10 shows the effective surface area of such a detector requesting at least 2 level coincidence in more than 2 strings. The effective size is in the range of 0.1 km² allowing a first measurement of very high energy cosmic neutrinos if any. The pixel size of the reconstructed events is such that 50% of them are within a radius of 0.17°. After reconstruction and energy determination, we have checked that a E^{-2} spectrum and a E^{-3} spectrum are not smeared out. We are testing different geometries to optimize the effective surface, the angular resolution and the rate of misreconstructed cosmic rays (not problematic at a depth of 2400 m).

5.2 Atmospheric neutrino oscillations searches with ANTARES

Following the results of Superkamiokande neutrino oscillations [8], predicting a mass difference of $\Delta m^2 \sim 10^{-2}/10^{-3}$ eV², we are studying the detection possibilities with the proposed strings. Quasivertical atmospheric

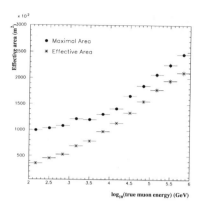

Figure 10: Maximal area that could be achieved (≥ 6 hits on ≥ 3 strings). Effective area with the present trigger (≥ 4 coincidences on ≥ 2 strings.)

neutrinos cross 13000 km of the Earth, the muon energy can be reconstructed by range in the 5-50 GeV domain with a good accuracy. This will probe the oscillation in a clean way in the expected range (see figure 11).

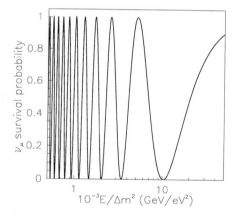

Figure 11: Survival probability for atmospheric muon neutrino crossing all the earth.

Acknowledgements

Antares Collaboration: Aslanides-E[1], Amram-P[2], Arpesella-C[1], Aubert-JJ[1], Azoulay-R[3], Basa-S[1], Benhammou-Y[4], Bernard-F[1], Berthier-R[3], Bertin-V[1], Billault-M[1], Biller-S[5], Blanc-F[6], Blanc-PE[1], Bland-RW[3], Blondeau-F[3], Bottu-N[3], Boulesteix-J[2], Brooks-B[5], Brunner-J[1], Calzas-A[1], Carloganu-C[1], Carmona Flores-E[7], Carr-J[1], Carton-PH[3], Cartwright-S[10], Cases-R[7], Cassol-F[1], Charles-F[4], Charles-J[4], Cooper-S[5], De Botton-[3], Desages-FE[3], Destelle-JJ[1], Dispau-G[3], Engelen-J[8], Feinstein-F[1], Fopma-J[5], Fuda-JL[9], Goret-P[3], Gosset-L[3], Gournay-JF[3], Hernandez-JJ[7], Hubaut-F[1], Hubbard-R[3], Huss-D[4], Jaquet-M[1], Jelley-N[5], Kajfasz-E[1], Kudryavtsev-V[10], Lachartre-D[3], Lafoux-H[3], Lamare-P[3], Languillat-JC[3], Laugier-JP[3], Le Provost-H[3], Le Van Suu-A[1], Loiseau-D[3], Loucatos-S[3], Marcelin-M[2], Martin-L[1], Mazeau-B[3], Mazure-A[11], McMillan-J[10], Meessen-C[1], Millot-C[6], Mols-P[3], Montanet-F[1], Moorhead-M[5], Moscoso-L[3], Navas-S[1], Nooren-G[8], Olivetto-C[1], Palanque-Delabrouille-N[3], Pallares-A[4], Payre-P[1], Perrin-P[3], Poinsignon-J[3], Potheau-R[1], Qian-Z[1], Raymond-M[1], Roberts-J[10], Sacquin-Y[3], Schuller-JP[3], Schuster-N[10], Stolarczyk-T[3], Tabary-A[3], Talby-M[1], Tao-C[1] Thompson-L[10], Triay-R[12], Valdy-P[9], Velasco-J[7], Vigeolas-E[1], Vignaud-D[3], Vilanova-D[3], Wark-D[5], Zuniga-J[7].

References

1. *Upgrade of Amanda B towards Amanda II. Proposal PRC 97/05 Desy-Zeuthen.*

2. *The Baïkal Neutrino Telescope NT-200, Baïkal 92-03, Nov 92.*

3. *Proposal to construct a Deep-Ocean Laboratory for the study of High Energy Neutrino Astrophysics and Particle Physics*, HDC-2-88, July 88, Aug. 94.

4. *Nestor International Workshop, Oct. 19-21, 1993, Pylos, Greece*, ed. L.K. Resvanis.

5. **H.J. Crawford** et al., *Stage R & D for a km-scale Neutrino Observatory*, unpublished proposal, 97.
   D. Nygren, *PMT Waveforms Captived by LBNL/JPL Digital Optical Modules in the Ice*, unpublished, March 97.

6. **C. Millot**, *Circulation in the Western Mediterranean Sea, J. of Marine Systems, 1998*, in press.

7. **E. L'Hostis**, *Détection et caractérisation de biofilms par méthodes élecrochimiques*, PhD Univ. Paris VI, Oct. 96.

8. *M. Takita's talk, plenary talk, this conference.*

[1] Centre de Physique des Particules de Marseille
[2] Observatoire de Marseille
[3] DAPNIA-DSM, CEA de Saclay
[4] Groupe de Recherche des Hautes Energies, Mulhouse
[5] Oxford University, Dept. of Physics
[6] Centre d'Océanologie de Marseille
[7] CSIC, Universitad de Valencia
[8] NIKHEF, Amsterdam
[9] IFREMER, La Seyne Sur Mer and Toulon
[10] Sheffield University, Dept. of Physics and Astronomy
[11] Laboratoire d'Astronomie Spatiale de Marseille
[12] Centre de Physique Théorique de Marseille

ALPHA MAGNETIC SPECTROMETER AMS
REPORT ON THE FIRST FLIGHT IN SPACE JUNE 2–12

U. BECKER

AMS COLLABORATION
Massachusetts Institute of Technology,
Laboratory of Nuclear Science,
77 Mass. Ave, Cambridge, MA 02139, USA
E-mail: becker@mitlns.mit.edu

Launched by the shuttle Discovery, AMS took data during 140 revolutions around the earth. The detector, the counting rates, specific events of p, *e*, He, C, Fe, and p̄ are presented as well as preliminary results from the few % of the total data which were downlinked during the flight. All data were saved on disc and are being processed now.

1 AMS Experiment

1.1 For the space station Alpha

The AMS experiment[1] is scheduled to measure charged particles on the space station. One possible detector configuration is shown in Figure 1. Six planes of Silicon microstrip detectors measure the momentum of the particles, being triggered by four layers of scintillation hodoscopes, which determine the velocity of the particle and magnitude of charge as well. At the top of the detector, 24 layers of TRD select electrons from other particles, RICH counter underneath identifies nuclei and isotopes, and a calorimeter on the bottom measures the electromagnetic energy. (γ's can be detected by conversion, too.)

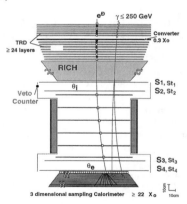

Figure 1: AMS detector for the space station Alpha.

The detector is constructed by the AMS collaboration[2] under DOE responsibility, while NASA provides for the flight[3]. It follows many balloon, and non-magnet space-based experiments[4] as well as ground and underground-based experiments[5] which provided most of the data about the origin of cosmic rays to date. AMS is a large aperture magnetic spectrometer for accurate measurements of primary cosmic ray spectra of >0.1 GV in space. Among the physics interests are:

- Search for dark matter, which may constitute as much as 90% of the universe. In particle physics, SUSY particles can reveal themselves in annihilation by: $\chi\bar{\chi} \rightarrow \bar{p} + ...$, $\chi\bar{\chi} \rightarrow e^+ ...$ as characteristic bumps in the spectra.

- Search for antimatter: in the form of H̄e to 10^{-9} and C̄ to 10^{-8} level. Although the non-observation of $e^+e^- \rightarrow \gamma\gamma(511\text{ keV})$ makes large amounts of antimatter unlikely in our cluster of galaxies, there are 10^8 other clusters we have no data from. Detection is by that of negatively charged nuclei which requires a magnet. In particle physics a "matter only" universe is hard to understand, since the Big Bang created equal amounts of matter and antimatter. A "matter only" universe requires the three Sacharov conditions to be fulfilled *simultaneously*:

  - Baryon number violation. So far no p decay is observed to the level of $2 \cdot 10^{32}$ years[5].

  - CP (or T) violation. - The presently observed CP violation is many magnitudes too small.

  - Non-equilibrium, so that an imbalance is not remixed to 50:50. This demands the Higgs to couple only weakly, but then its mass $m_H < 60$ GeV, in conflict with recent measurements[6] of $m_H > 89$ GeV.

- Primary cosmic rays. They will provide information on galactic propagation. 10^9 isotopes of D, He, Li, Be, B, C... will be recorded.

1.2 AMS for the precursor flight June 2-12, 1998

For NASA the flight was primarily scheduled for MIR supply deliveries and astronaut return. For AMS it was an engineering run to provide space experience for the dectector, which was constructed in an HEP experiment-like manner. Figure 2 shows the detector, and the responsibilities of various countries, and its location in the shuttle Discovery, where it is supported by the Unique Support Structure of NASA.

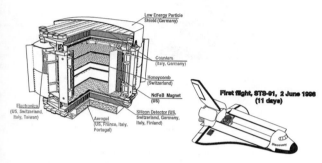

Figure 2: AMS detector for the shuttle flight, showing main components with country of origin. Location in Discovery.

1.3 Detector components

The magnet was constructed in China from 64 premagnetized columns of Fe Nd B material with > 48 Gauss Oersted magnetization[1,2]. This is > 10 times higher than conventional ferrite materials as Prof. S.C.C.Ting realized. Most of the Nd is found in China. The ring shape assembly in Figure 3a produces a homogeneous field of 1.3 kGauss with a low fringe field and vanishing dipole moment, as required by NASA to avoid gyroscopic motion in the earth field. Before and after a 17g centrifuge test it was exposed to a vibration test, Figure 3b, showing no change in amplitudes at different frequencies.

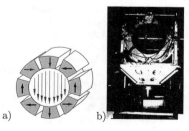

Figure 3: a) Schematic of Magnet. b) Magnet being subjected to a vibration test.

The counters consist of x, y layers with 14 scintillation pieces each, located above and below the magnet. Viewed by 3 photomultipliers at each end for redundancy, they register the TOF (time of flight) with 115ps design accuracy. Mounted in carbon fibre casings with lightight air pressure exits, the counters overlap to avoid inefficiencies, see Figure 4.

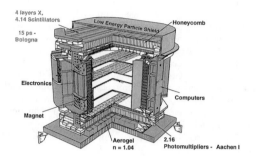

Figure 4: 3-dim representation of the components for the shuttle flight.

Along the inner magnet wall, 16 veto counters register "illegal" tracks. The cylindrical scintillators have inlaid lightfibres, which guide the light to photomultipliers on each end. With ~28 photoelectrons they have very high efficiency. The six planes of Silicon strip detectors are given in the photograph of Figure 5 during assembly in a precision jig at ETH in Zuerich using very precise quartz spacers. The tracker has:

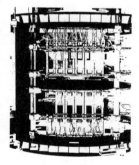

Figure 5: Silicon tracker assembly on six C-fibre honeycomb plates.

- six planes equipped with 'ladders' of double sided Si strip detectors, as seen in Figure 4,

- an accuracy of 10μm in the bending plane and 30μm in the other plane, providing $\Delta p/p = 7\%$ at 10 GeV limited by multiple scattering,

- 180,000 channels with signal/noise>17 for a MIP with a <400W power allowance,

- space approved and tested precise honeycomb panels of <3.2% X_0, yet stable to $< 10\mu m$,

- a precise special C-Fibre structure for mounting.

The preinstalled and tested structure was inserted into the magnet, secured by special Ti-feet and then surveyed. Alignment is checked using double infrared laser rays penetrating all six planes, generating signals directly in the Silicon wafers.

The electronics were specially built for low power and space conditions and mounted in crates thermally connected to the magnet. Time of flight and pulse height are registered for each counter, Cerenkov counter and anti-counter. The trigger was formed from a coincidence of the four scintillation layers.

The Cerenkov counter underneath discriminates electrons from protons with <4.2 GeV. It consists of 164 cells, arranged in two layers of $(11cm)^3$ Aerogel blocks viewed by one selected photomultiplier. At the bottom and top a 15mm carbon fibre shield protects the detector from copious low energy particles and micrometeorites.

2   Flight June 2-12

2.1   Preparations and launch

The assembled detector was tranferred to KSC, Florida, inserted into the NASA unique support structure, and tested with cosmic rays in all stations and positions of the mounting procedure into the shuttle. Figure 6a depicts the time resolution achieved. The photon yield in the Cerenkov counter for equivalent proton momentum is shown in Figure 6b.

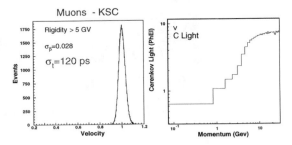

Figure 6: a) Time distribution for muons > 5 GeV. b) Light yield in the Cerenkov counter.

Well guarded by friendly personnel and occasional alligators, tests and flight simulations continued till the last day before the launch, Figure 7.

Figure 7: Launch of shuttle Discovery from KSC, June 2, 5:06 local time.

Two hours after the start, the thermosensors showed expected normal conditions for the detector, using the NASA slow communication link (S-band). However the fast Ku band transmitter on the shuttle could not be made operational. Therefore the data were all stored on disc as planned under the supervision of astronaut Dr Chang-Diaz. In addition the Air Force permitted the downlinking of data for a few minutes each time the shuttle passed over one of their stations, with the subsequent relaying of data to the control center in Houston. The data presented are from this sample corresponding to a few % of the total data acquired. Detailed calibrations in feedback could not be carried out.

2.2   Monitor data and conditions

Immediately after startup of AMS, the trigger rate showed the expected pattern, for the flight at 51.6° inclination at 173 nmi above ground, Figure 8.

With a 90 min. period and minima when crossing the equator, where the earth field deflects most particles, and maxima when approaching the magnetic poles, modulated by the beat effect from the different geographic poles, the rate shows exactly the ~9 min transition through the South Atlantic Anomalty at the expected positions. A laser calibration run showed no difference of the Silicon location under gravity on the pad at KSC when compared to the weightless flight environment as shown in Figures 9a and b. The data taking consisted of a short period before and a 75 h period after MIR docking with controlled attitude of AMS looking into space, in total 105h.

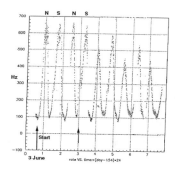

Figure 8: Coincidence rate of the four triggering planes vs time starting 6h after launch.

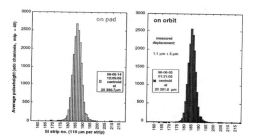

Figure 9: a) Intensity profile of laser on Si strips at KSC. b) In space 1.1μm different.

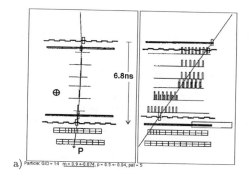

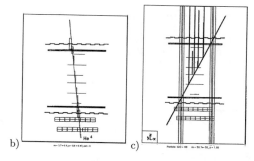

Figure 10: a) Normal proton, bending view and non-bending view. b) He in bending view. c) Fe candidate.

3 Data and Preliminary Results

3.1 Recognition of different events

Figure 10a gives a 'normal' proton event. From the transit time of 6.8ns, β is derived, and from the curvature, $p = +0.5$GeV and a positive charge of unity. Using the relation $p^2 + m^2 = E^2 = (p/\beta)^2$, a value of $m = 0.9$ GeV is determined. The size of the boxes signifies the dE/dx of a slow proton. The side view determines the track uniquely through the TOF hodoscopes. The Silicon strips have been linked in the non-bending plane to reduce the number of channels. For electrons, the dE/dx corresponds to that of a minimal ionizing particle, the Cerenkov counter gives a large signal, the momentum a negative bend and the mass = 0, within errors. Figure 10b shows a 4 times minimal track ionization ($\sim Z^2$) and the mass determination agrees also with the mass of a He nucleus. There are clean candidates for Li, Be, and C, as well. Amusing maybe is Figure 10c showing ionization which is well off-scale. From mass determination it could be Fe, in which case the signal is $\sim$600 times that of a proton track!

3.2 Preliminary Evaluation

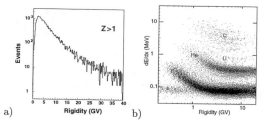

Figure 11: a) Spectrum of Z>1 particles vs. rigidity. b) Energy loss in the tracker as function of rigidity.

Taking the distribution of all events which have an ionization of $Z > 1$ against the rigidity (p/Z) results in the spectrum of Figure 11a. Uncalibrated as it is, it displays roughly the expected E^{-2} behaviour. Figure 11b is a scatterplot of all energy losses in the Silicon tracker against the rigidity. Bands corresponding to light particles, protons, Helium, Li,..., C are visible.

3.3 Antiproton Candidates

The lower part of Figure 12 gives 3-dim projections of a particle with proton mass, but negative momentum (charge). The upper part depicts further, different antiproton candidates. With about one dozen candidates the total sample cannot be estimated, since these come from "flyover downlink data", which natually deteriorate with the visibility of the shuttle as it disappears over the horizon, hence the fraction is hard to determine.

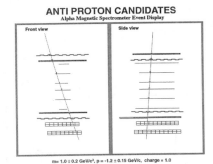

ANTI PROTON CANDIDATES
Alpha Magnetic Spectrometer Event Display

m= 1.0 ± 0.2 GeV/c², p = -1.2 ± 0.15 GeV/c, charge = 1.0

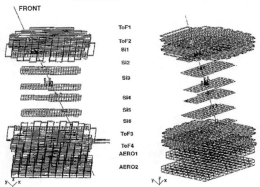

Figure 12: Antiproton candidates.

Reconstructing the mass distribution for particles with $Z = 1$, 2, and 6 results in the spectrum of Figure 13. The sample was taken during MIR docking and contains π's possibly generated in solar panels. Despite that and the uncalibrated analysis, the p, D, He, and C contributions are clearly visible and appear in the expected ratios[7] for the time in the solar cycle.

4 Summary

The first flight of the large aperture magnetic spectrometer was successful, AMS functioned properly. Despite

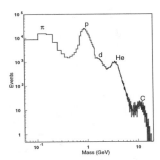

Figure 13: Mass spectrum of particles with Z=1 or 2 or 6.

the failure of the fast satellite data link all the data was recorded and retrieved from discs. A few % of the data were downlinked. They confirm correct operation. The post mission test verified this as well. The optimising calibration which was not possible during the flight will be carried out in a heavy ion beam soon. Evaluation of the full data sample has started. Most importantly, valuable space experience has been gained.

It is a pleasure to thank NASA, the astronauts, and AMS colleagues for most enjoyable, effective help and the DOE for support.

References

1. "Alpha Magnetic Spectrometer" (AMS) for Extraterrestrial Study of Antimatter, Matter, and missing Matter on the International Space Station Alpha, Proposal Dec 1994, S.C.C. Ting et. al.
2. The AMS Experiment in "Experimental Results and Future Opportunities". S.C.C Ting. Proceed. of 17th International Symposium on Lepton and Photon Interactions Aug 1995, p.724
3. NASA flight STS 91 for this precursor flight, STS 120 for ISSA scheduled 5/2002
4. M. Spiro "Experiments in Particle Astrophysics", and references therein, these proceedings 1998, also S. Ahlen et. al. NIM A350 (1994) 2316
5. L. Sulak "Review of p-decay", M. Takita "Recent results from Super-Kamiokande" and references therein, these proceedings 1998
6. D. Karlen "Experimental Status of the Standard Model", this conference. ibid.
7. For example: M. S. Longair "High energy Astrophysics" chapters 9,10 Cambridge University Press 2.ed 1997

BESS MEASUREMENT OF COSMIC-RAY ANTIPROTON SPECTRUM AND SEARCH FOR ANTIMATTER

S. ORITO, T. MAENO, H. MATSUNAGA, K. ABE, K. ANRAKU, Y.ASAOKA, M. FUJIKAWA, M. IMORI, N. MATSUI, M. MOTOKI, T. SAEKI, T. SANUKI, Y. SHIKAZE, T. SONODA, I. UEDA, K. YOSHIMURA

University of Tokyo, Tokyo 113-0033, JAPAN
E-mail: oriton@icepp.s.u-tokyo.ac.jp

Y. MAKIDA, J. SUZUKI, K. TANAKA, A. YAMAMOTO, T. YOSHIDA

High Energy Accelerator Research Organization (KEK), Tsukuba, Ibaraki 305-0801, JAPAN

H. MATSUMOTO, T. MITSUI, M. NOZAKI, M. SASAKI

Kobe University, Kobe, Hyogo 657-8501, JAPAN

J. MITCHELL, A. MOISEEV, J. ORMES, R. STREITMATTER

National Aeronautics and Space Administration, Goddard Space Flight Center, Greenbelt, MD 20771, USA

J. NISHIMURA, Y. YAJIMA, T. YAMAGAMI

The Institute of Space and Astronautical Science (ISAS), Sagamihara, Kanagawa 229-8510, JAPAN

E. S. SEO

University of Maryland, College Park, MD 20742, USA

M. ISHINO

Kyoto University, Kyoto 606-8502, Japan

The absolute fluxes of the cosmic-ray antiproton are measured at solar minimum in the energy range 0.2 to 3.2 GeV, based on 460 antiprotons unambiguously detected by BESS spectrometer during its '95 and '97 balloon flights from Lynn Lake, Canada. The large number of events, the low enough rigidity cut-off during the entire flight, and the wide energy range covered, all together allow us to study the antiproton spectrum, for the first time, in a quantitative way. We have detected a clear peak in the spectrum around 2 GeV and measured its flux to 10 % accuracy. The position and the absolute flux of the peak agree with the prediction of the Standard Leaky Box model for the "secondary" antiprotons, i.e., produced by the high energy cosmic rays interacting with the interstellar gas, when we utilize the interstellar proton flux deduced from the BESS measurement of the proton spectrum. At low energies below 1 GeV, we observe an excess antiproton flux over the simple Standard Leaky Box prediction. This might indicate that the propagation mechanism needs to be modified, or might suggest a contribution of low-energy antiproton component from novel sources such as evaporating primordial black holes or the annihilating neutralino dark matter. Data from '98 and future flights are expected to clarify the issue. On the antimatter search, a new upper limit of 1.7×10^{-6} is obtained for the antihelium/helium ratio. We aim to reach in year 2001 a sensitivity level of 1×10^{-7}.

1 BESS Spectrometer and Flights

The construction of BESS detector was initiated in 1988. The basic concept [1] was a compact high-resolution spectrometer which possesses a large enough acceptance to perform sensitive searches for antinuclei and for the low-energy antiprotons from novel primary sources. Various new detector technologies developed for collider experiments are incorporated to the spectrometer.

Figure 1 shows the cross-sectional view of BESS spec-

trometer in its '97 configuration. A uniform field of 1 Tesla is produced by a thin (4 g/cm^2) superconducting coil, through which particles can pass without too much interactions. The magnetic-field region is filled with the central tracking volume. This geometry results in a large acceptance of 0.3 m^{2}Sr, which is a factor 30 larger than those of previous cosmic-ray spectrometers. The tracking in the central region is performed by fitting up to 28 hit-points in the drift chambers, resulting in a magnetic-

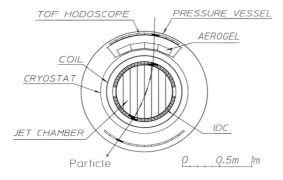

Figure 1: BESS spectrometer.

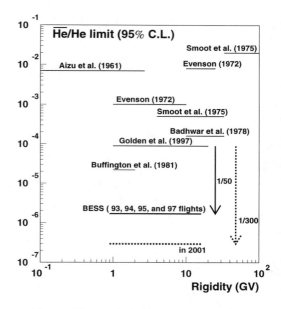

He/He limit (95% C.L.)

Figure 2: The upper limits of antihelium/helium ratio.

rigidity (R) resolution of 0.5 % at 1 GV/c. The continuous and redundant 3-dimensional tracking enables us to recognize multi-track events and tracks having interactions or scatterings. The upper- and lower-scintillator hodoscopes measure the time-of-flight (TOF) of particles with 70 ps resolution and also provides two measurements of dE/dx. In addition, dE/dx of the particle in the drift chamber gas is obtained as a truncated mean of the integrated charges of the hit-pulses. The first-level trigger is provided by a coincidence between the top and the bottom scintillators, with the threshold safely set at 1/3 of pulse height from minimum ionizing particles. The second-level trigger, which utilizes the hit-patterns of the hodoscopes and of the inner drift chambers (IDC), first rejects the null- and multi-track events and makes a rough rigidity-determination to select negatively-charged particles. In addition, one of every 60 first-level triggers is recorded, irrespective of the second-level trigger condition, to build a sample of unbiased triggers from which the efficiencies are determined.

We had successful balloon flights of 16 to 22 hours each in '93, '94, '95, and '97, all from Lynn Lake, Canada. The place is close enough to the North magnetic pole so that the rigidity cut-off during the flight remained below 0.5 GV/c (corresponding to 0.125 GeV proton kinetic energy). Most of scientific data were collected at altitudes of about 36 km (residential atmosphere of 5 g/cm^2).

2 Antihelium Search[2]

Simple off-line selections are applied simultaneously for negative- and positive-rigidity events, requiring a single track which is fully contained in the fiducial tracking region with acceptable track-qualities such as the χ^2 of the track fitting. The velocity ($\beta \equiv v/c$) determined by the time-of-flight and the three measured dE/dx are required to be helium- (antihelium) like as functions of

the magnetic-rigidity (R). These selections eliminates all backgrounds and produce a pure sample of helium (antihelium) with 80 % of efficiency. We observe no events in the negative rigidity region below 16 GV/c. The events with higher rigidities are consistent to be the spill-over of the positive rigidity heliums. The corresponding upper limits on the antihelium/helium ratio is obtained after correcting for the expected losses of antihelium and helium in the air and in the instrument. The resultant upper limit (1.7×10^{-6} at 95 % C.L.) is compared with previous limits in Figure 2. We aim to reach a sensitivity of 1×10^{-7} in the year 2001 by accumulating enough flight time.

3 Antiproton Spectrum

The strict limit on the antihelium abundance is the most direct evidence for Galaxy and nearby part of Universe being composed solely of particles. If so, antiprotons in cosmic rays must have been pair-created together with protons by some elementary-particle processes. One such process which should certainly exist at some level is the interaction of high-energy cosmic rays with interstellar gas. The energy spectrum of antiprotons from this "secondary" process is expected to show a characteristic peak around 2 GeV and sharp decreases of the flux below and above the peak, a generic feature which reflects the kinematics of the antiproton production. The absolute peak flux and details of the spectrum are expected to provide important and unique informations on the propagation process of the cosmic rays.

One can also conceive novel elementary-particle processes such as the annihilation of neutralino dark matter or the evaporation of primordial black holes [3]. The antiprotons from these "primary" sources are expected to show very soft energy spectra [4], peaking toward the lower energies, and would exhibit large solar modulations [5], thus distinguishable in principle from the secondary antiproton component. The antiprotons from the primordial black holes are especially interesting as an unique window to search for such exotic object and to investigate (or at least to set a limit on) the density fluctuations in the Early Universe at extremely small mass scale of $10^{-19} M_\odot$ which would correspond to the horizon mass at temperature of about 10^{11} GeV.

However the detection of the secondary peak and the search for low-energy primary antiproton component has been a difficult experimental endeavor due to the huge backgrounds and to the extremely small flux especially at low energies. The first "mass-identified" detection of cosmic-ray antiprotons was reported by BESS [6] in the low-energy region (0.3–0.5 GeV) using its '93 flight data, which was followed by IMAX [7] and CAPRICE [8] detections at higher energies. The BESS '95 extended its energy range and measured the spectrum [9] at the solar minimum, based on 43 antiprotons unambiguously detected in the energy range 0.2–1.4 GeV. The antiproton spectrum appears to be flat below 1 GeV and does not exhibit the steep decline generally expected to the secondary antiprotons, although the statistical significance is very marginal.

In '97 flight, we succeeded to collect a factor 2.7 times more statistics at low energies than '95 flight, by the combination of longer flight-time, less dead-time and higher trigger- and selection-efficiencies. The energy range of the antiproton identification was extended to 3.2 GeV by improving the TOF resolution and by installing a threshold Aerogel ($n = 1.032$) Čerenkov counter [10], which reject $e^\pm$ background by a factor 5000 while having 97 % efficiency for protons and antiprotons. The off-line analysis selects events with a single track in the fiducial region with acceptable track qualities. The three measured dE/dx are required to be proton (antiproton) like as function of R. The Čerenkov output is then required to be below a threshold. The overall off-line selection efficiency was about 85 % slightly depending on the energy.

Figure 3 shows the β^{-1} versus R plot for the surviving events. We see a clean band of 425 antiprotons at the position exactly opposite to the protons. The antiproton sample is mass-identified and background-free, as the cleanness of the band demonstrates and various

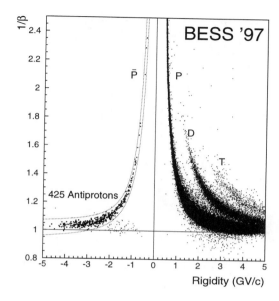

Figure 3: The identification of 425 antiprotons (BESS '97).

background-studies and consistency-checks indicate. To check against the re-entrant albedo background, we confirmed that the trajectories of all low-energy antiprotons can be traced numerically through the Earth's geomagnetic field back to the outside of geomagnetic sphere.

We obtain the antiproton energy spectrum at the top of the atmosphere (TOA) in the following way: The TOA energy of each event is calculated by tracing back the particle through the detector material and the air. The geometrical of the spectrometer acceptance can be calculated reliably due to simple geometry and uniform magnetic field. The efficiency of the second-level trigger as well as the off-line selection efficiency are determined by using the unbiased trigger sample. The survival probability of the antiprotons through the air and instrument is evaluated with estimated relative accuracy of about 10 % by GEANT/GHEISHA simulation, which incorporate the detailed material distribution and realistic detector performance as well as the correct antiproton-nuclei cross sections. We subtract the expected number of atmospheric antiprotons, produced by the collisions of cosmic rays in the air. The subtraction amount to 9 ± 2 %, 15 ± 3 % and 19 ± 5 %, respectively, at 0.25, 0.7 and 2 GeV, where the errors correspond to the maximum difference among the three independent recent calculations.

The resultant '97 spectrum is very consistent with the '95 spectrum in the overlapping energy range (0.2

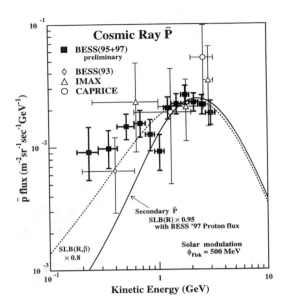

Figure 4: Energy spectrum of cosmic-ray antiprotons. Lines represent Standard Leaky Box calculations for secondary antiprotons, normalized by the peak flux.

to 1.4 GeV). The solar activities at the time of the two flights were both close to the minimum as shown by world neutron monitors and by the low-energy proton spectra measured by BESS. The combined ('95+'97) spectrum is shown in Figure 4 together with previous measurements which have very much larger errors.

We detect a clear peak around 2 GeV, i.e. exactly at the position expected for "secondary" antiprotons, and measure the peak flux to ±10 % accuracy. It is remarkable that a simple Standard Leaky Box (SLB) calculation [11] (solid line) appears to predict approximately correct peak flux when one utilize as input the interstellar proton flux [12] deduced from BESS '97 measurement of the proton spectrum. The SLB calculation utilizes the escape length $\lambda(R)$ determined as function of rigidity from the data on Boron/Carbon ratio, and incorporates important details such as the subsequent (tertiary) interactions of the antiprotons with the interstellar gas. Therefore the predicted shape of the spectrum is rather constrained within this SLB scheme.

At energies below 1 GeV, our data show a significant excess over the SLB prediction, while data points at higher energies are well reproduced by the SLB calculation. If one takes the SLB calculation seriously, the low-energy excess might indicate [4,13] a contribution of

primary antiprotons from novel sources such as the evaporating primordial black holes. However one can also elaborate on the propagation. For example, if one suppose the escape length $\lambda(R, \beta)$ which also depends on the particle velocity, the low-energy part of the spectrum could change. An extreme case [11] of such trial is shown (after scaling by a factor 0.8) as the dotted line, which could explain the data though marginally.

BESS '98 data, which we took at the very last stage of the solar-minimum phase, will double the statistics and greatly help us to clarify the situation. Furthermore, as previously pointed out [5], rapid increase of solar activity toward year 2001 will drastically suppress any primary antiproton component while only modestly affecting the secondary antiproton spectrum, thusby allowing us to draw a firm conclusion on whether we are seeing the primary antiprotons from novel sources and at the same time to determine the propagation mechanism of cosmic rays.

Acknowledgements

Sincere thanks are given to NASA and NSBF for balloon launch. The analysis was performed by using the computing facilities at ICEPP, Univ. of Tokyo. This experiment was supported by Grant-in-Aid from Monbusho in Japan and by NASA in the USA.

References

1. S. Orito, in *Proceedings of the ASTROMAG Workshop*, eds. J. Nishimura, K. Nakamura, and A. Yamamoto, KEK Report 87-19, 1987.
2. T. Saeki *et al.*, *Phys. Lett.* B **422**, 319 (1998).
3. S. W. Hawking, *Commun. Math. Phys.* **43**, 199 (1975).
4. K. Maki, T. Mitsui, and S. Orito, *Phys. Rev. Lett.* **76**, 3474 (1996). .
5. T. Mitsui, K. Maki, and S. Orito, *Phys. Lett.* B **389**, 169 (1996).
6. K. Yoshimura *et al.*, *Phys. Rev. Lett.* **75**, 3792 (1995); A. Moiseev *et al.*, *Astrophys. J.* **474**, 479 (1997).
7. J. W. Mitchell *et al.*, *Phys. Rev. Lett.* **76**, 3057 (1996).
8. M. Boezio *et al.*, *Astrophys. J.* **487**, 415 (1997).
9. H. Matsunaga *et al.*, *Phys. Rev. Lett.*, (in print).
10. Y. Asaoka *et al.*, *Nucl. Instrum. Methods*, (in print).
11. T. Mitsui, S. Orito, and J. Nishimura, (in preparation).
12. T. Sanuki *et al.*, (to be published).
13. S. Orito, K. Maki, and T. Mitsui, (in preparation).

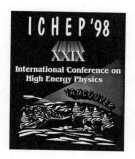

Parallel Session 13

Heavy Ion Collisions at High Energies

Convenors: Lars Leistam (CERN, NA57)

PROBING THE FINAL AND INITIAL STATE IN ULTRARELATIVISTIC HEAVY ION COLLISIONS – RESULTS FROM THE WA98 EXPERIMENT

T. PEITZMANN

University of Münster, 48149 Münster, Germany
E-mail: peitzmann@ikp.uni-muenster.de

WA98 COLLABORATION

Recent results of the WA98 experiment studying 158·A GeV ^{208}Pb+^{208}Pb collisions at the CERN SPS are presented. Transverse momentum spectra of neutral pions emitted near midrapidity are studied. They exhibit an invariance of the spectral shape and a simple scaling of the yield with the number of participating nucleons for centralities with greater than about 30 participating nucleons. It is shown that neutral pion spectra are not compatible with very low freeze-out temperatures. Directed flow of protons near target rapidity is observed in semi-central collisions, the magnitude being smaller than observed at AGS energies and than RQMD model predictions. Preliminary results on photon production indicate an excess of direct photons over the hadronic decay background.

1 Introduction

The WA98 experiment[1] searches for signatures of a phase transition to a quark-gluon plasma consists of large acceptance hadron and photon spectrometers. The experiment is designed to study a variety of observables in heavy-ion reactions via inclusive distributions and also event-by-event. Like its predecessor, the WA80 experiment, it is unique in its capabilities to measure photons and neutral mesons in these reactions. It consists of hadronic calorimetry, multiplicity detectors for charged particles and for photons, the Plastic Ball detector identifying particles in the target rapidity region, the 10,000 module lead glass photon spectrometer LEDA to measure photons and neutral mesons and two magnetic spectrometer arms including time of flight detectors to measure negative and positive particles.

In this paper we will present results from reactions of 158·A GeV ^{208}Pb+^{208}Pb. We will concentrate on high precision neutral pion spectra, on studies of directed flow and the production of direct photons. The neutral meson spectra are mainly influenced by thermal and chemical freeze-out in the final state. Directed flow is sensitive to the equation of state of hadronic matter and might carry information on a possible softening of it due to a phase transition. Direct photons, however, are produced throughout the history of the collisions and are not influenced by hadronic rescattering. Important contributions come from the hot initial state and from a possible mixed phased, therefore allowing to extract information on the initial state parameters, like the initial temperature.

2 Neutral Mesons

The measured neutral pion spectrum from central Pb+Pb reactions (10% of min.bias cross section) as a function of $m_T - m_0$ is shown in Fig. 1. The data are compared to predictions of the string model Monte Carlo generators FRITIOF 7.02[2] and VENUS 4.12[3]. As already observed in S+Au reactions[4], both generators fail to describe the data at large m_T. The FRITIOF prediction is more than an order of magnitude lower at high m_T while VENUS significantly overpredicts the data. Alternatively, it has recently been shown that perturbative QCD calculations, including initial state multiple scattering and intrinsic p_T[5], are able to describe the preliminary WA98 data at intermediate and high p_T. This prediction is included in Fig. 1 as a solid line. The surprisingly good agreement has been interpreted as an indication for unexpectedly small effects of parton energy loss[5].

To study the centrality dependence of the spectral shape in a manner which is independent of model or fit function we have used the truncated mean transverse momentum $\langle p_T(p_T^{min}) \rangle$, where

$$\langle p_T(p_T^{min}) \rangle = \left(\int_{p_T^{min}}^{\infty} p_T \frac{dN}{dp_T} dp_T \middle/ \int_{p_T^{min}}^{\infty} \frac{dN}{dp_T} dp_T \right) - p_T^{min}.$$

(1)

The lower cutoff $p_T^{min} = 0.4$ GeV/c is introduced to avoid systematic errors from extrapolation to low p_T.

Figure 2 shows $\langle p_T(p_T^{min}) \rangle$ as a function of the average number of participants N_{part} for 158·A GeV ^{208}Pb+Pb collisions. For comparison, $\langle p_T(p_T^{min}) \rangle$ values for 200·A GeV S+Au[4] and from a parametrization of pp data are also included. Together these data show the general trend of a rapid increase of $\langle p_T(p_T^{min}) \rangle$ compared to pp results for small system sizes. For N_{part} greater than about 30 the mean transverse momentum appears to attain a limiting value of ≈ 280 MeV/c. VENUS 4.12 calculations show a qualitatively similar behaviour, although the values of $\langle p_T(p_T^{min}) \rangle$ are somewhat lower than the experimental data. The simple implementation

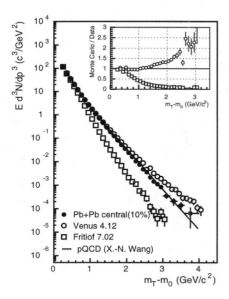

Figure 1: Transverse mass spectra of neutral pions in central collisions of 158·A GeV Pb+Pb. Invariant yields per event are compared to calculations using the FRITIOF 7.02 and VENUS 4.12 Monte Carlo programs. Predictions of a pQCD calculation [5] are included as a solid line. The inset shows the ratios of the results of the Monte Carlo codes to the experimental data.

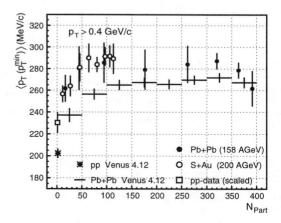

Figure 2: Truncated mean transverse momentum $\langle p_T(p_T^{min})\rangle$ of π^0 mesons as defined by Eq. 1 plotted as a function of the average number of participants N_{part}. The solid circles correspond to the 8 E_T based centrality selections for Pb+Pb. The open square shows $\langle p_T(p_T^{min})\rangle$ extracted from a parametrization of pp data scaled to the same cms-energy, the open circles the results for S+Au collisions at 200 AGeV [4]. For comparison, results from VENUS 4.12 are included as histograms for Pb+Pb collisions and as a star for pp. A cut parameter $p_T^{min} = 0.4\,\text{GeV}/c$ was used.

of rescattering which is used in this model seems to be strong enough to lead to a saturation for semi-peripheral collisions as in the experimental data. One should, however, keep in mind that VENUS 4.12 does not correctly describe pion production at high p_T (see figure 1).

More detailed information about the centrality dependence of the pion spectral shape and yield is shown in Fig. 3 where the neutral pion yield per event has been parameterized as $Ed^3N/dp^3 \propto N_{part}^{\alpha(p_T)} \cdot \sigma_0(p_T)$. The results for $N_{part} \geq 30$ are well described by this scaling with an exponent $\alpha(p_T) \approx 1.3$, independent of p_T. Consistent with the previous discussion, the results indicate a constant spectral shape over the entire interval of measurement from $0.5 \geq p_T \geq 3$ GeV/c. The observed $N_{part}^{4/3}$ scaling for symmetric systems implies a scaling with the number of nucleon collisions, as confirmed by a similar analysis. However, this scaling does not extrapolate from the pp results. On the contrary, when comparing semi-peripheral Pb+Pb collisions with pp collisions the exponent α increases over the entire p_T interval, confirming the very different spectral shapes.

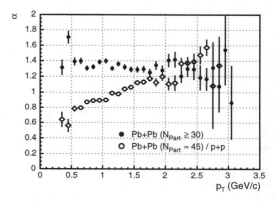

Figure 3: The exponent $\alpha(p_T)$ of the dependence of the π^0 yield on the average number of participants N_{part} plotted as a function of the transverse momentum for 158·A GeV Pb+Pb. The solid circles are calculated based on the centrality selections with $N_{part} \geq 30$. The open circles are calculated based on the ratio of the semi-peripheral data ($N_{part} \approx 45$) to a parameterization of pp data.

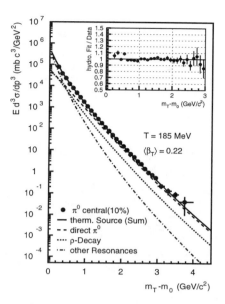

Figure 4: Transverse mass spectra of neutral pions in central collisions of 158·A GeV Pb+Pb. The invariant cross section is compared to a fit using a hydrodynamical model [6] including transverse flow and resonance decays, with the direct production and the contribution of ρ decays and all other resonances shown separately. The ratio of the fit to the data is shown in the inset.

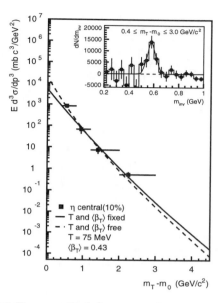

Figure 5: The cross section for eta meson production as a function of $m_T - m_0$ together with hydrodynamical fits with parameters fixed at the same values as obtained for the pions (solid line) and fits with all parameters varying freely (dashed line). The horizontal error bars correspond to the bin width. Here, the inset shows the invariant mass spectrum in the eta region integrated for $0.4 \, \text{GeV}/c^2 \leq m_T - m_0 \leq 3.0 \, \text{GeV}/c^2$.

The measured neutral pion cross section from central Pb+Pb reactions has also been compared with a hydrodynamical model [6] including transverse flow and resonance decays (Fig. 4). This model can provide an excellent description of the neutral pion spectra with a temperature $T = 185 \text{MeV}$ and an average flow velocity of $\langle \beta_T \rangle = 0.225$. These values are very similar to the parameters obtained with comparable fits to neutral pion spectra in central reactions of ^{32}S+Au [4]. Contrary to other analyses with charged pions [7,8] the neutral pion spectra can constrain the parameters significantly, the 2σ lower limit on the temperature being $T^{low} = 171 \text{MeV}$ and the corresponding upper limit on the flow velocity $\langle \beta_T^{upp} \rangle = 0.265$.

Fig. 5 shows the cross section of η mesons as a function of $m_T - m_0$ for central collisions of Pb+Pb. The poorer statistics of the transverse mass spectrum does not allow to fix hydrodynamical parameters. A free fit (dashed line) yields $T = 75 \text{MeV}$ and $\langle \beta_T \rangle = 0.43$. This is a much lower temperature than for pions (there a constrained fit with $\langle \beta_T \rangle = 0.43$ would require $T = 108 \text{MeV}$), but a description using the same parameters as extracted from the neutral pions (solid line) is also acceptable. Using this description of the spectra with the

same temperature and flow velocity one can obtain the ratio of the integrated yields $\eta/\pi^0 = 0.081 \pm 0.013$ which yields an optimum chemical temperature within the same model of $T \approx 160 \text{MeV}$. The ratio is however relatively insensitive to the temperature – it is compatible with temperatures in the range of $T = 120 - 200 \text{MeV}$.

3 Directed Flow

We determine the reaction plane as the azimuthal direction, Φ, of $\vec{P_T}$, the total transverse momentum vector of fragments (p, d, and t) detected in the target rapidity region in the Plastic Ball detector. In order to study how well the flow direction can be defined in our experiment, we divide each event randomly into two equal sized subevents and determine a fragment flow direction for each subevent, Φ_a and Φ_b. The relative angle distributions are fitted with the form

$$\frac{1}{2\pi N} \frac{dN}{d(\Phi_a - \Phi_b)} = 1 + A_1 \cdot \cos(\Phi_a - \Phi_b), \quad (2)$$

where the asymmetry parameter A_1 quantifies the strength of the correlation between the subevent flow di-

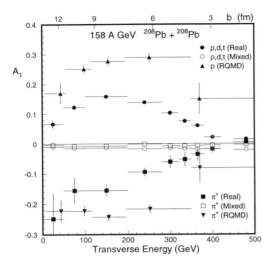

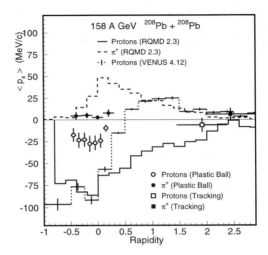

Figure 6: The centrality dependence of the strength of the correlation between the total transverse momentum directions of pairs of subevents. The circles indicate the results for subevents consisting of target fragments (p,d,t). The squares indicate the results for one subevent of target fragments and the other subevent of π^+ in the target rapidity region. Triangles are results from RQMD model calculations. The open symbols are for subevents constructed from mixed events. The horizontal bars indicate the E_T bin intervals (or impact parameter intervals for RQMD).

Figure 7: The average transverse momentum projected onto the reaction-plane for semi-central 158·A GeV ^{208}Pb + ^{208}Pb collisions (Note $y_{cm} = 2.9$). The errors include uncertainties of the fit procedure and the event plane resolution correction. RQMD model calculations (b=8-10 fm) and VENUS model calculations (b=5-12 fm) are also shown.

rections.

The dependence of the A_1 parameter for target fragments on centrality, as determined by the measured transverse energy (E_T), is shown in Fig. 6. The strength of the correlation between the flow directions of the two subevents increases with centrality and reaches a maximum value for semi-central collisions with $b \approx 8$ fm. Also shown in Fig. 6 is the strength of the correlation between the total transverse momentum directions of one subevent constructed of all p, d, and t, and another subevent constructed of π^+, all identified in the Plastic Ball. A clear anti-correlation, or anti-flow [12], is observed between the fragment and π^+ flow directions. This behaviour has been observed at incident energies from 1·A GeV to SPS energies and has been explained as resulting from preferential absorption of the pions emitted in the target spectator direction [10,11,12,13].

A conventional directed flow analysis has been performed [14], in which the average transverse momentum with respect to the reaction plane $\langle p_x \rangle$ is evaluated as a function of rapidity. This is done for semi-central colli-

sions ($100 < E_T < 200$ GeV) where the largest azimuthal asymmetry is observed (see Fig. 6). The distribution $d^3N/dp'_x dp'_y dy$ is constructed for protons and π^+ in the Plastic Ball and in the tracking arm, where the new axis p'_x corresponds to the reaction plane determined event-by-event using all remaining fragments measured in the Plastic Ball (then reflected, $p'_x \rightarrow -p'_x$, to correspond to the projectile fragment direction, according to convention). At each rapidity the average transverse momentum in the reaction plane, $\langle p'_x \rangle$, is calculated from fits to the experimental distributions.

After correction for the event-plane resolution, the $\langle p_x \rangle$ for protons and π^+ are plotted as a function of rapidity in Fig. 7. As expected from Fig. 6, the π^+ show an anti-flow relative to the proton flow. Summing over the Plastic Ball acceptance $\langle p_x \rangle$ values of $5.3 \pm 1.7, -22.7 \pm 7.1, -51.1 \pm 16.0$, and -67.6 ± 18.5 MeV/c are obtained for π^+, p, d, and t, respectively. The observed scaling with fragment mass for p, d, and t indicates emission sources with a common collective motion.

In Fig. 7 the measured results are compared to RQMD 2.3 [15] and VENUS 4.12 predictions for similar impact parameter range. The RQMD calculation, in cascade mode, overpredicts the observed proton flow by about a factor of three. The VENUS predictions show a similar disagreement in the target rapidity region. The results suggest a significant softness in the nuclear response.

4 Direct Photons

The extraction of direct photons is a very difficult task, because they have to be obtained from the inclusive photon yield, which has to be fully corrected for efficiency and contamination by charged hadrons and neutrons, by a subtraction of the dominating hadronic decay photons. For this purpose the most important hadrons decaying into photons, the π^0 and η, have been measured within the same data sample as used for the direct photon extraction. The analysis procedure has to be checked carefully for systematic errors. Preliminary results indicate that for the first time in ultrarelativistic heavy ion reactions a net excess of photons over the hadronic decay background is observed. After a variety of further checks, which are currently being performed, including a complete independent reanalysis, the direct photon cross section will allow to test different scenarios for the early evolution in these reactions and will thereby yield unique information on the properties of the initial state.

5 Summary

We have analyzed the centrality dependence of high precision transverse momentum spectra of neutral pions from 158·A GeV Pb+Pb collisions. The neutral pion spectra show an increasing deviation from pp results with increasing centrality, indicating the importance of multiple scattering effects. However, for centralities with more than about 30 participating nucleons, the shape of the transverse momentum spectrum becomes invariant over the interval $0.5 \le p_T \le 3$ GeV/c. In this interval the pion yield scales like $N_{part}^{1.3}$, or like the number of nucleon collisions, for this range of centralities. Since the amount of rescattering increases with centrality, the invariance of the spectral shape with respect to the number of rescatterings, most naturally suggests a dominantly thermal emission process. It will be important to determine whether cascade models which reproduce the observed invariant spectral shape will support the interpretation as an "effective" thermalization due to significant rescattering.

We have used the transverse mass spectra to estimate hadronic freeze-out parameters in central collisions of 158·A GeV Pb+Pb. It turns out that neutral pion spectra measured over a large range in m_T can provide important constraints. Even with the additional uncertainty regarding hard scattering contributions the

high m_T part of the spectrum yields an upper bound for the transverse flow velocity of $\langle \beta_T^{upp} \rangle = 0.265$ (98% CL) within the model used. We have shown that the relative production of η and π^0 is compatible with the assumption of a simultaneous thermal and chemical freeze-out at $T \approx 185$MeV. There appears to be a large discrepancy between the neutral pion spectra discussed here and the results by NA49 from HBT of negative particles [7]. It has to be investigated, whether these two analyses do really probe the same information.

The directed flow of protons and π^+ has been studied in 158·A GeV ^{208}Pb + ^{208}Pb collisions. It is largest for impact parameters ≈ 8 fm, which is considerably more peripheral than observed at lower incident energies. The π^+ directed flow is in the direction opposite to the protons, similar to observations at 11·A GeV energy [10]. The magnitude of the proton directed flow, as measured by the maximum average transverse momentum projected onto the reaction plane, is much less than cascade mode RQMD model predictions, which underpredict the proton flow at AGS energies. It is also much less than VENUS model predictions. The results suggest a soft nuclear response at SPS energies.

Preliminary results on direct photon production indicate an excess over hadronic decay photons. Once the ongoing analysis is finalized the direct photon cross section allows to extract important information on the initial state in these reactions.

References

1. WA98 Collaboration, CERN/SPSLC 91-17 and CERN/SPSLC 95-35.
2. B. Andersson, G. Gustafson, and H. Pi, Z. Phys. C **57**, 485 (1993).
3. K. Werner, Phys. Rep. **232**, 87 (1993).
4. R. Albrecht et al., nucl-ex/9805007, to appear in Eur. Phys. J. C.
5. X.-N. Wang, 1998, preprint hep-ph/9804384.
6. U.A. Wiedemann and U. Heinz, Phys. Rev. C **56**, 3265 (1997).
7. H. Appelshäuser et al., Eur. Phys. J. C **2**, 661–670 (1998).
8. I. G. Bearden et al., Phys. Rev. Lett. **78**, 2080 (1997).
9. F. Becattini, Z. Phys. **C69** (1996) 485–492.
10. J. Barrette, et al., Phys. Rev. C **56**, 3254 (1997).
11. T.C. Awes, et al., Phys. Lett. B **381**, 29 (1996).
12. A. Jahns, et al., Phys. Rev. Lett. **72**, 3463 (1994).
13. A. Kugler, et al., Acta Phys. Pol. **25**, 691 (1994).
14. P. Danielewicz and G. Odyniec, Phys. Lett. **157B**, 146 (1985).
15. H. Sorge, Phys. Rev. C **52**, 3291 (1995).

INTERMEDIATE MASS DIMUONS IN NA38/NA50

C.Soave[11] - NA50 collaboration

M.C. Abreu[6,a], B. Alessandro[11], C. Alexa[3], R. Arnaldi[11], J. Astruc[8], M. Atayan[13], C. Baglin[1], A. Baldit[2], M. Bedjidian[12], F. Bellaiche[12], S. Beolè[11], V. Boldea[3], P. Bordalo[6,b], A. Bussière[1], V. Capony[1], L. Casagrande[6], J. Castor[2], T. Chambon[2], B. Chaurand[9], I. Chevrot[2], B. Cheynis[12], E. Chiavassa[11], C. Cicalò[4], M.P. Comets[8], S. Constantinescu[3], J. Cruz[6], A. De Falco[4], N. De Marco[11], G. Dellacasa[11,c], A. Devaux[2], S. Dita[3], O. Drapier[12], B. Espagnon[2], J. Fargeix[2], S.N. Filippov[7], F. Fleuret[9], P. Force[2], M. Gallio[11], Y.K. Gavrilov[7], C. Gerschel[8], P. Giubellino[11], M.B. Golubeva[7], M. Gonin[9], A.A. Grigorian[13], J.Y. Grossiord[12], F.F. Guber[7], A. Guichard[12], H. Gulkanyan[13], R. Hakobyan[13], R. Haroutunian[12], M. Idzik[11,d], D. Jouan[8], T.L. Karavitcheva[7], L. Kluberg[9], A.B. Kurepin[7], Y. Le Bornec[8], C. Lourenço[5], M. Mac Cormick[8], P. Macciotta[4], A. Marzari-Chiesa[11], M. Masera[11], A. Masoni[4], S. Mehrabyan[13], S. Mourgues[2], A. Musso[11], F. Ohlsson-Malek[12,e], P. Petiau[9], A. Piccotti[11], J.R. Pizzi[12], W.L. Prado da Silva[11,f], G. Puddu[4], C. Quintans[6], C. Racca[10], L. Ramello[11,c], S. Ramos[6,b], P. Rato-Mendes[11], L. Riccati[11], A. Romana[9], S. Sartori[11], P. Saturnini[2], E. Scomparin[5,g], S. Serci[4], R. Shahoyan[6,h], S. Silva[6], C. Soave[11], P. Sonderegger[5,b], X. Tarrago[8], P. Temnikov[4], N.S. Topilskaya[7], G.L. Usai[4], C. Vale[6], E. Vercellin[11], N. Willis[8].

[1] LAPP, CNRS-IN2P3, Annecy-le-Vieux, France.
[2] LPC, Univ. Blaise Pascal and CNRS-IN2P3, Aubière, France.
[3] IFA, Bucharest, Romania.
[4] Università di Cagliari/INFN, Cagliari, Italy.
[5] CERN, Geneva, Switzerland.
[6] LIP, Lisbon, Portugal.
[7] INR, Moscow, Russia.
[8] IPN, Univ. de Paris-Sud and CNRS-IN2P3, Orsay, France.
[9] LPNHE, Ecole Polytechnique and CNRS-IN2P3, Palaiseau, France.
[10] IReS, Univ. Louis Pasteur and CNRS-IN2P3, Strasbourg, France.
[11] Università di Torino/INFN, Torino, Italy.
[12] IPN, Univ. Claude Bernard and CNRS-IN2P3, Villeurbanne, France.
[13] YerPhI, Yerevan, Armenia.
a) also at FCUL, Universidade de Lisboa, Lisbon, Portugal
b) also at IST, Universidade Técnica de Lisboa, Lisbon, Portugal
c) Dipartimento di Scienze e Tecnologie Avanzate, II Facoltà di Scienze, Alessandria, Italy
d) now at Faculty of Physics and Nuclear Techniques, University of Mining and Metallurgy, Cracow, Poland
e) now at ISN, Univ. Joseph Fourier and CNRS-IN2P3, Grenoble, France
f) now at UERJ, Rio de Janeiro, Brazil
g) on leave of absence from Università di Torino/INFN, Torino, Italy
h) on leave of absence of YerPhI, Yerevan, Armenia

The NA38/NA50 experiments have measured the dimuon production in ultrarelativistic proton–nucleus and nucleus–nucleus collisions at the CERN SPS for different systems and energies. We present here the analysis of the intermediate mass region ($1.5 < M < 2.5 \ GeV/c^2$) for the 450 GeV/c p - $A(A = Al, Cu, Ag, W)$, 200 $A \ GeV/c$ S - U and 158 $A \ GeV/c$ Pb - Pb data sets. We show that, after background subtraction, the intermediate mass region can be simply described as a superposition of dimuons from the semi–leptonic decays of charmed meson pairs and DY. No other contribution, as thermal dimuons, has been taken into account. We find that the open charm yield in nucleus–nucleus collisions exceeds a linear extrapolation from proton–nucleus, while in proton–nucleus collisions it is consistent with previous FNAL and CERN experimental results. The intermediate mass region enhancement increases with the system size and c.m. energy up to a factor of $\simeq 3$ in central Pb - Pb collisions. For events in the same mass range, a comparison of the Pb - Pb PYTHIA generated and experimental p_T differential distributions is also shown.

1 Introduction

The NA50 experiment at the CERN SPS studies the muon pair production in nucleus–nucleus collisions to probe the behaviour of nuclear matter under extreme conditions of energy density. Dimuons generated in the interaction volume interact weakly with the hot surrounding medium; they can thus escape undisturbed, carrying out informations about the state of the matter where they were generated [1]. The dimuon mass spectrum can be roughly subdivided in three main regions :

the Low Mass Region (LMR) below the ϕ resonance, the Intermediate Mass Region (IMR) between the ϕ and the J/ψ and the High Mass Region (HMR) above the J/ψ. While the latter can be seen as a simple superposition of high mass DY, J/ψ and ψ', the former originate from different sources (hadron Dalitz decays, charmed meson semi–leptonic decays and DY) with individual shapes and relative contributions not immediatly deductable from the data.

In this paper we study the IMR for the $450\ GeV/c\ p$ - $A(A = Al, Cu, Ag, W)$, $200\ A\ GeV/c\ S$ - U and $158\ A\ GeV/c\ Pb$ - Pb data sets. We assume the intermediate mass region to be a superposition of dimuons from DY, charmed hadron decays and combinatorial background. Thermal dimuons, which are expected to appear in nucleus–nucleus collisions, have not been taken into account. Monte–Carlo techniques have been used to generate the DY and open charm mass shapes. The Monte–Carlo generated shapes have then been used to fit the data in order to determine their relative contributions. The opposite–sign background shape has been deduced from the like–sign experimental distributions. The background relative contribution has been fixed with a Monte–Carlo simulation. A similar procedure has been used by the HELIOS-3 collaboration [2]. In the previous NA38/NA50 IMR analysis [3,4] the combinatorial background contribution was estimated without using the Monte–Carlo techniques detailed in paragraph 4.

2   The experimental set–up

The NA38/NA50 experimental apparatus is based on the NA10 muon spectrometer [5]. It consists of a toroidal deflecting magnet, 8 $MWPC$'s and 4 trigger hodoscopes. It is separated from the target region by a 4 m long hadron absorber made of Carbon followed by 0.8 m of Iron. A segmented active target allows the primary vertex identification thanks to a system of quartz blades located after each subtarget. A beam hodoscope (BH) tags incident ions and rejects the beam pile-up. An electromagnetic calorimeter, EMC, measures the neutral transverse energy. In NA38 this was the only way of estimating the collision centrality. In NA50 two more detectors have been included, a multiplicity detector, MD [6], and a zero degree calorimeter, ZDC [7].

In the following the most important differences between the NA38 and NA50 set–ups are listed. For a more detailed description see ref.8,9.

2.1 S - U $200\ A\ GeV/c$ (NA38)

In the S - U set–up, the segmented active target was made of 12 U subtargets with a total thickness of 0.2 λ_i. The centrality of the collision was measured by an electromagnetic calorimeter located in front of the hadron absorber and covering the pseudorapidity interval $1.7\ <\eta<\ 4.1$. The beam hodoscope and the target detectors were made of plastic scintillator. Muon pairs were detected in the pseudorapidity interval $2.8\ <\eta<\ 4.0$. The field in the toroidal magnet was $1.2\ Tm$. The beam intensity was about $10^8\ S\ ions/burst$, with a 5 s spill.

2.2 p - A $450\ GeV/c$ and Pb - Pb $158\ A\ GeV/c$ (NA50)

In the Pb - Pb set–up, the segmented active target is made of 7 Pb subtarget with a total thickness of $0.3\ \lambda_i$ [10]. The pseudorapidity coverage of the EMC has been shifted to $1.1 < \eta < 2.3$. The beam hodoscope and the active target detectors have been re–built using quartz in order to stand the high radiation level. The field in the toroidal magnet is $2.1\ Tm$ to improve the mass resolution and reduce the rate of soft background muons. A zero degree calorimeter is located along the beam axis in the hadron absorber internal core. It measures the energy of the projectile spectators of the collision and thus estimates the collision centrality. The beam intensity is $5 \cdot 10^7\ Pb\ ions/burst$, with a 5 s spill.

A similar set–up has been adopted for the p - A data taking. The beam intensity is about $3 \cdot 10^9\ protons/burst$, with a $2.5\ s$ spill. The active target is replaced by a single target, with no vertex identification system. Four nuclear targets have been used, namely Al, Cu, Ag and W, with thicknesses ranging from $32\ g/cm^2$ (Al) to $80\ g/cm^2$ (W).

3   Data reduction and centrality selection

For both proton–nucleus and nucleus–nucleus samples, the selected events must have two and only two fully reconstructed tracks in the muon spectrometer; furthermore, the two tracks must point to the target. These quality cuts are very effective since they reject more than 50% of the triggered events. For nucleus-nucleus collisions, the BH rejects the beam pile-up. In S - U data reduction, the information given by the target detectors has been used to select events with an interaction in one of the subtargets. This method has a low efficiency for peripheral collisions. For this reason, in Pb - Pb data reduction a less restrictive selection has been applied: since events originating from the target region must be strongly correlated in the E_T vs. E_{ZDC} plane, only the events lying in a 2σ region around the average correlation are retained. In order to reject dimuons from kinematical regions where the acceptance is very low, only events with $-0.5 < cos\theta_{CS} < 0.5$ have been accepted, where θ_{CS} is the polar angle of the muons relative to the beam axis in the rest frame of the dimuon. For the same reason, the rapidity cuts $0 < y_{cm} < 1$ for S - U and Pb - Pb

and $-0.52 < y_{cm} < 0.48$ for p - A collisions have been applied.

The S - U and Pb - Pb data sets have been subdivided in 5 and 9 centrality classes respectively corresponding to different E_T bins. The average number of participants and impact parameter for each centrality class have been estimated by means of a geometrical model [11], taking into account realistic nuclear densities. The results have been further checked using a simulation based on the VENUS 4.02 [12] event generator.

4 Combinatorial background

One of the most important contributions to the IMR opposite–sign dimuon spectrum is the background from pion and kaon decays. In p - A collisions at $450\, GeV/c$, its contribution to the IMR amounts to 85% and to 90% in the most central Pb - Pb collisions at $158\, A\, GeV/c$. The background subtraction from the opposite–sign dimuon mass spectrum is usually made using the like–sign dimuon mass distributions. This is due to the fact that π and K decays contribute both to the opposite and to the like–sign dimuon spectra. While the opposite–sign distribution is the superposition of many different sources, like–sign muon pairs are purely due to combinatorial background.

For p - A and S - U data the following formula has been used to estimate the opposite–sign background from the like–sign dimuon distributions,

$$N^{+-} = 2\sqrt{N^{++}N^{--}} \qquad (1)$$

In Pb - Pb, where a better degree of accuracy is needed because of the higher background contribution, the opposite–sign distributions have been obtained with mixed event techniques [13]. Eq.1 has then been used for normalization. In fact, eq.1 is strictly valid if there are no acceptance biases introduced by the apparatus and if the $K's$ and $\pi's$ produced in the collision are not charge correlated. While the first condition is satisfied in NA38/NA50 thanks to a suited off-line cut, the second one can be unsatisfied in p - A and peripheral A - B collisions where the produced particle multiplicity is not high enough. For this reason, eq.1 has to be multiplied by a factor $R \geq 1$. A detailed Monte–Carlo simulation has been carried out to compute the background R factor in each particular case. For each of the analysed systems $\simeq 10^5$ VENUS events have been generated and the produced $K's$ and $\pi's$ stored. Each K and π has been tracked through the apparatus and the corresponding decay probabilities into muons, P_{μ_i} with $i = 1, N_\pi + N_K$, have been computed. Within each event, $K's$ and $\pi's$ have been exhaustively combined in opposite and like–sign pairs and subsequently

forced to decay. Each dimuon has been given a relative weight equal to the product of its parents decay probabilities $P_{\mu\mu_{i,j}} = P_{\mu_i}P_{\mu_j}$. The dimuon kinematical variables have been computed and the same kinematical cuts as imposed on the experimental data have been applied. With the opposite–sign and like–sign surviving dimuons, the Monte–Carlo background R factor has been evaluated as

$$R^{MC} = \frac{\sum_{n=1}^{N}\sum_{i,j} P_{\mu\mu_{i,j}}^{+-}}{2\sqrt{(\sum_{n=1}^{N}\sum_{i,j} P_{\mu\mu_{i,j}}^{++})(\sum_{n=1}^{N}\sum_{i,j} P_{\mu\mu_{i,j}}^{--})}} \qquad (2)$$

where N is the number of generated VENUS events.

For A - B systems, where R is expected to depend on the collision centrality, several calculations have been worked out for different values of the impact parameter b. The R vs. b correlation for S - U and Pb - Pb have been derived. For each centrality class, the average value $\langle R^{MC}\rangle$ has been computed as a weighted average of $R^{MC}(b)$ over the b distribution of the events in that class. As expected, $\langle R^{MC}\rangle$ turns out to be compatible with 1 for central Pb - Pb collisions and larger than 1 for peripheral ones ($\langle R^{MC}\rangle$=1.035 for $\langle b\rangle = 10.7$ corresponding to the most peripheral bin).

The NA50 p - A data have been collected with a high beam intensity and no pile-up rejection. Particles produced in consecutive events can thus pile-up and be seen as originating from a single event. This smears the $K's$ and $\pi's$ charge correlation and reduces the R factor value. Introducing this effect in our simulation, we obtain $\langle R^{MC}\rangle = 1.05$ for p-W collisions at the experimental beam intensity while $\langle R^{MC}\rangle = 1.21$ when pile-up effects are not taken into account.

5 Data analysis

5.1 Introduction

We describe the IMR as a simple superposition of dimuons from charmed meson (and baryon) pair decays and DY, after background subtraction. In order to exclude the contribution of the low mass resonances, the fit starting point has been fixed at $1.5\, GeV/c^2$. The DY and open charm mass shapes are taken from PYTHIA event generator. The J/ψ and ψ' mass shapes have also been generated, using the method explained in ref.11. The fit of the mass spectra extends up to $8\, GeV/c^2$ and allows the determination of the DY and open charm contributions as free parameters. The pure DY high mass region ($4.5 < M < 8.0\, GeV/c^2$) constrains the DY normalisation. It has been shown [14] that the high mass DY, after isospin correction, scales with A in p - A and with $A \times B$ in A–B collisions. The open charm cross-section

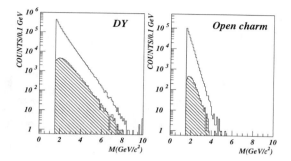

Figure 1: *Generated (plain histogram) and reconstructed (filled histogram) DY and open charm dimuon invariant mass spectra for the Pb - Pb set-up.*

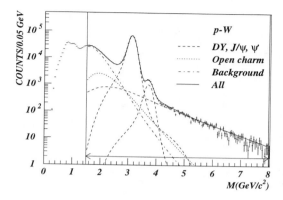

Figure 2: *Fit to the p–W invariant mass spectrum, with the background contribution fixed using R^{MC}. The double arrow indicates the fitted mass range.*

also scales with A in p - A interactions [15] and is expected to scale as $A \times B$ in A–B collisions.

5.2 DY and open charm mass shapes

The DY and open charm mass shapes have been obtained using PYTHIA [16] with the MRS A set of PDFs [17], with m_c=1.5 GeV/c^2 and $\langle k_T^2 \rangle$=0.8 GeV^2. For each process, a large sample of PYTHIA events has been generated. The generation extends over a kinematical window wider than the experimental one in order to account for multiple scattering smearing effects. The generated dimuons have been tracked and reconstructed through the apparatus following the same procedure used for experimental data. In fig.1 the DY and open charm generated and reconstructed mass spectra are shown for the Pb - Pb set–up and for $M > 1.5$ GeV/c^2. The acceptance for a single process is defined as the ratio of reconstructed to generated events in the experimentally covered phase space. For the Pb - Pb set–up and for $M > 1.5$ GeV/c^2, the resulting DY and open charm acceptances are $A_{DY} = 0.028$ and $A_{D\overline{D}} = 0.011$. For p - A a similar result has been obtained as the Pb - Pb and p - A set–ups are practically identical. The S - U set–up has acceptances $\simeq 4$ times higher due to the significantly lower magnetic field.

5.3 Fit of the mass spectra

The four analysed 450 GeV/c p - $A(A = Al, Cu, Ag, W)$ sets of data have been taken with the same experimental set–up. Moreover, as previously said, it has been shown that the DY [18,14] and open charm cross sections [15] scale with A in p - A collisions. The four data sets are therefore fitted simultaneously with the constraint that the ratio $\sigma_{D\overline{D}}/\sigma_{DY}$, left as a free parameter in the fit, is the same in the four samples of events. The J/ψ, ψ' and DY

normalizations and the J/ψ mass and width are also left free. The combinatorial background R factors have been either imposed using the results of the simulation, or left as additional free parameters. In both cases a satisfactory description of the mass spectra has been achieved even if, as expected, a better χ^2 is obtained when leaving R free. The normalization of the open charm contribution varies by $\sim 20\%$ between the two fitting procedures; this provides an evaluation of the systematic error due to the combinatorial background subtraction. In fig.2, the resulting fit to the p–W opposite sign mass spectrum is shown with all the contributing processes. The Pb - Pb and S - U spectra have been fitted with the same procedure as adopted for the p - A data samples, without any constraint on the $\sigma_{D\overline{D}}/\sigma_{DY}$ ratio. The values of the combinatorial background R factor have been computed accordingly to the procedure outlined in section 4. The resulting fits satisfactorily describe the mass spectra for the 5 S - U and the 9 Pb - Pb centrality bins, but the open charm contribution increases, with respect to DY, with the size of the system and the centrality of the collision. We find a factor $\simeq 1.5$ enhancement from peripheral to central S - U and $\simeq 2$ from peripheral to central Pb - Pb collisions. Fig.3, shows the Pb–Pb fit for a peripheral and central bins.

6 Results

We can now compare our measured open charm yield with direct measurement of charmed hadron production. In a recent paper [19], results of various CERN and FNAL open charm hadro–production experiments have been re-

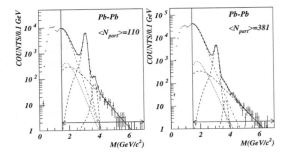

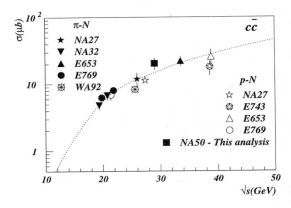

Figure 3: *Fit to peripheral and central Pb - Pb spectra. The DY, J/ψ and ψ' contributions are shown as dashed lines, the open charm as a dotted line, the background as a dashed–dotted line.*

Figure 4: *Cross sections for forward ($x_F > 0$) $c\bar{c}$ production compared with the value extrapolated from this analysis.*

viewed. The $\sqrt{s}$ dependence of D meson cross sections has been found to be well described by PYTHIA within a factor ~ 2, which can be attributed to residual systematic errors. The D meson differential distributions, as well as the lepton distributions from their semi–leptonic decays measured in several experiments, are in good agreement with PYTHIA calculations. To match the experimental points, the absolute normalization provided by PYTHIA must be scaled up by an appropriate K–factor.

Since PYTHIA reproduces the dilepton spectra from open charm and since the open charm and DY shapes have been here obtained with PYTHIA, we can extract the open charm cross-section from the open charm dimuon yield as measured from our fits. The resulting open charm cross section per nucleon for forward $c\bar{c}$ production as obtained from the p - A data sets is shown in fig.4, where it is compared with results from direct open charm measurements. The NA50 open charm cross section per nucleon extracted from the present analysis is in good agreement with the results from other experiments. The dotted line in fig.4 represents the PYTHIA prediction, scaled up by an ad-hoc K–factor. The S - U and Pb - Pb open charm cross sections per nucleon can be similarly computed and compared, at the relevant energies, with the points of fig.4. Either the dotted line in fig.4, or the same line re-scaled to match the NA50 p - A results, can be taken as a reference. In fig.5 the open charm enhancement with respect to the second reference, i.e. to the $c\bar{c}$ cross section normalized to the NA50 p - A point, is shown versus the number of participant nucleons N_{part} and for all the studied systems. The enhancement clearly increases with the collision centrality and the system size. In central Pb - Pb collisions the dimuon yield from open charm decays is 3 times higher than expected from a linear extrapolation of the observed

p - A yield.

6.1 The IMR dimuons p_T spectra

To support the hypothesis of an open charm enhancement, we further check the p_T differential distributions where the DY and open charm contributions can be easily distinguished. The Monte–Carlo p_T spectra have been obtained in the same way as for the mass differential distributions. The DY, open charm and J/ψ contributions have been fixed using the results of the fits to the mass spectra, including the enhanced open charm. Fig.6 shows the p_T dimuon distributions, in the mass range $1.5 < M_{\mu\mu} < 2.5 \ GeV/c^2$, measured in peripheral and central Pb - Pb collisions. With no additional hypothesis, the superposition of the PYTHIA generated DY, open charm and J/ψ contributions, normalized according to the results of the fit to the mass spectra, are in good agreement with the data.

7 Conclusions

The intermediate mass muon pair continua measured by the NA38/NA50 experiment in p - A, S - U and Pb - Pb collisions are well described as a superposition of dimuons from charmed meson (and baryon) semi-leptonic decays and DY, after combinatorial background subtraction. While in p - A collisions the open charm cross-section correctly scales with A, in S - U and Pb - Pb the dimuon production from open charm is enhanced with respect to the QCD expectations. The enhancement factor increases linearly with the number of participants and a factor 3 enhancement is found in central Pb - Pb colli-

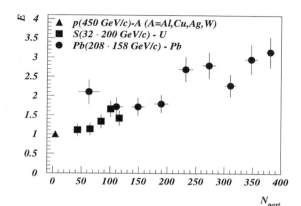

Figure 5: *The enhancement factor E of the open charm dimuon component with respect to p - A collisions versus the number of participant nucleons N_{part}.*

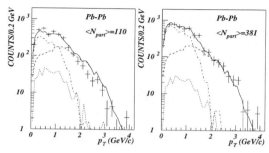

Figure 6: *The p_T distribution of IMR dimuons for peripheral and central Pb - Pb collisions. The DY (dashed line), J/ψ (dotted line) and $D\overline{D}$ (dashed-dotted line) individual contributions are shown separately, together with their sum (solid line).*

sions. The analysis of the intermediate mass region p_T spectra provides further support to the hypothesis that the excess dimuons originate from open charm decays.

References

1. For an overview on the subject see e.g., Ruuskanen P.V., *Nucl.Phys.***A544**(1992)169c
2. Angelis A.L.S. *et al.* (HELIOS/3 Collaboration), preprint CERN-EP/98-92
3. Lourenço C. *et al.* (NA38 Collaboration), *Nucl.Phys.***A566**(1994)77c
4. Scomparin E. *et al.* (NA50 Collaboration), *Nucl.Phys.***A610**(1996)331c
5. Anderson L. *et al.*, *Nucl.Inst.Meth.* **223**(1984)26
6. S.Beolè, Pd.D. Thesis, Torino, Italy, 1998.
7. Arnaldi R. *et al.*, *Nucl.Inst.Meth.* **A411**(1998)1
8. Baglin C. *et al.* (NA38 Collaboration), *Phys.Lett.***B220**(1989)471
9. Abreu M.C. *et al.* (NA50 Collaboration), *Phys.Lett.***B410**(1997)327
10. Bellaiche F. *et al.*, *Nucl.Inst.Meth.* **A398**(1997) 180
11. Fleuret F., Thesis, Ecole Polytechinique, Palaiseau (1997)
12. Werner K., *Phys.Rep.* **232**(1993)87
13. Constantinescu S. *et al.*, preprint IPNO-DRE-96-01
14. Abreu M.C. *et al.* (NA50 Collaboration), *Phys.Lett.***B410**(1997)337
15. Leitch M.J. *et al.* (E789 Collaboration), *Phys.Rev.Lett.* **72**(1994)2542
16. Sjostrand T., *Computer Physics Commun.* **82**(1994)74
17. Martin A.D. *et al.*, *Phys.Rev.* **D51**(1995)4756
18. Alde D.M. *et al.* (E772 Collaboration), *Phys.Rev.Lett.***64**(1990)2479
19. Braun-Munzinger P. *et al.*, *Eur.Phys.J.***C1** (1998)123

RESULTS ON CHARMONIUM STATES IN Pb-Pb INTERACTIONS

Presented by SÉRGIO RAMOS, *LIP-Lisbon*

NA50 COLLABORATION

C. BAGLIN, A. BUSSIÈRE, V. CAPONY
LAPP, CNRS-IN2P3, Annecy-le-Vieux, France

A. BALDIT, J. CASTOR, T. CHAMBON, I. CHEVROT, A. DEVAUX, B. ESPAGNON,
J. FARGEIX, P. FORCE, S. MOURGUES, P. SATURNINI
LPC, Univ. Blaise Pascal and CNRS-IN2P3, Aubière, France

C. ALEXA, V. BOLDEA, S. CONSTANTINESCU, S. DITA
IFA, Bucharest, Romania

C. CICALÒ, A. DE FALCO, P. MACCIOTTA, A. MASONI, G. PUDDU, S. SERCI,
P. TEMNIKOV, G.L. USAI
Università di Cagliari/INFN, Cagliari, Italy

C. LOURENÇO, E. SCOMPARIN[g], P. SONDEREGGER[b]
CERN, Geneva, Switzerland

M.C. ABREU[a], P. BORDALO[b], L. CASAGRANDE, J. CRUZ,
C. QUINTANS, S. RAMOS[b], R. SHAHOYAN[h], S. SILVA, C. VALE
LIP, Lisbon, Portugal

S.N. FILIPPOV, Y.K. GAVRILOV, M.B. GOLUBEVA, F.F. GUBER,
T.L. KARAVITCHEVA, A.B. KUREPIN, N.S. TOPILSKAYA
INR, Moscow, Russia

J. ASTRUC, M.P. COMETS, C. GERSCHEL, D. JOUAN, Y. LE BORNEC,
M. MAC CORMICK, X. TARRAGO, N. WILLIS
IPN, Univ. de Paris-Sud and CNRS-IN2P3, Orsay, France

B. CHAURAND, F. FLEURET, M. GONIN, L. KLUBERG, P. PETIAU, A. ROMANA
LPNHE, Ecole Polytechnique and CNRS-IN2P3, Palaiseau, France

C. RACCA
IReS, Univ. Louis Pasteur and CNRS-IN2P3, Strasbourg, France

B. ALESSANDRO, R. ARNALDI, S. BEOLÈ, E. CHIAVASSA, N. DE MARCO,
G. DELLACASA[c], M. GALLIO, P. GIUBELLINO, M. IDZIK[d], A. MARZARI-CHIESA,
M. MASERA, A. MUSSO, A. PICCOTTI, W.L. PRADO DA SILVA[f], L. RAMELLO[c],
P. RATO-MENDES, L. RICCATI, S. SARTORI, C. SOAVE, E. VERCELLIN
Università di Torino/INFN, Torino, Italy

M. BEDJIDIAN, F. BELLAICHE, B. CHEYNIS, O. DRAPIER, J.Y. GROSSIORD,
A. GUICHARD, R. HAROUTUNIAN, F. OHLSSON-MALEK[e], J.R. PIZZI
IPN, Univ. Claude Bernard and CNRS-IN2P3, Villeurbanne, France

M. ATAYAN, A.A. GRIGORIAN, H. GULKANYAN, R. HAKOBYAN, S. MEHRABYAN
YerPhI, Yerevan, Armenia

a) also at FCUL, Universidade de Lisboa, Lisbon, Portugal
b) also at IST, Universidade Técnica de Lisboa, Lisbon, Portugal
c) Dipartimento di Scienze e Tecnologie Avanzate, II Facoltà di Scienze, Alessandria, Italy
d) now at Faculty of Physics and Nuclear Techniques, University of Mining and Metallurgy, Cracow, Poland
e) now at ISN, Univ. Joseph Fourier and CNRS-IN2P3, Grenoble, France
f) now at UERJ, Rio de Janeiro, Brazil
g) on leave of absence from Università di Torino/INFN, Torino, Italy
h) on leave of absence from YerPhI, Yerevan, Armenia

We present cross-sections for J/ψ, ψ' and Drell-Yan production in lead-lead interactions at 158 GeV/nucleon. The Pb-Pb data, when compared with previous results obtained with lighter target or projectiles, show a similar behaviour for Drell-Yan, but exhibit an anomalous J/ψ suppression, which increases with centrality.

1 Introduction

NA50 is a CERN fixed target experiment designed for the study of Pb-Pb collisions in the North Area High Intensity Facility of the SPS accelerator. It is the upgrade of the previous NA38 experiment which took data from 1986 to 1992, using proton, oxygen and sulphur beams.

NA50 is devoted to the detection of the Quark-Gluon Plasma (QGP) formation. It studies charmonium production, considered as an unumbiguous signature [1].

2 Experimental setup

The NA50 detector [2] consists mainly of a muon spectrometer, an active target formed by several lead subtargets, an electromagnetic calorimeter, a forward hadronic calorimeter and a charged multiplicity detector.

The muon spectrometer is based on eight multi-wire proportional chambers and a toroidal air-gap magnet. It covers the pseudo-rapidity interval $2.8 < \eta_{lab} < 4.0$. Four highly segmented scintillator hodoscopes perform the dimuon trigger, whereas two other are used for trigger efficiency measurements. The hadron absorber, located between the target region and the muon spectrometer, consists of a 60 cm long BeO preabsorber, followed by 400 cm of carbon and 80 cm of iron. An iron wall at the end of the spectrometer, before the last trigger hodoscope, ensures that the triggering particles are indeed muons.

The active target assembly is formed by a set of seven individual lead subtargets, 1 or 2 mm thick, with total interaction length L_{int}, as shown in Table 1. Each subtarget is followed by pairs of quartz blades, located on the left and the right of the beam axis, used to identify the vertex interaction and fight spectator fragment reinteractions. Two anti-halo quartz counters, located immediately upstream from the subtargets, sign preinteractions.

A beam hodoscope (BH), subdivided in sixteen quartz counters, located 22 m upstream from the target, is used to count the incoming lead ions and to reject beam pile-up. It is followed downstream by an interaction detector (BHI) which tags interactions in the BH itself.

The electromagnetic calorimeter, located 32 cm downstream from the target centre, sorrounds the BeO preabsorber, and thus covers a pseudo-rapidity interval outside the muon spectrometer acceptance: $1.1 < \eta_{lab} < 2.3$. It is subdivided in four rings and six sextants of roughly equal pseudo-rapidity ranges and

Table 1: Characteristics of the data samples

Period	L_{int}	Intensity	Nb. triggers	Nb. J/ψ
1996	30%	$5.4 \cdot 10^7$	$168 \cdot 10^6$	$275 \cdot 10^3$
1995	17.5%	$3.2 \cdot 10^7$	$58 \cdot 10^6$	$48 \cdot 10^3$

is made of scintillating optical fibres embedded in lead converter in a 1:2 volume ratio ($< L_{rad} > = 0.93$ cm). Its resolution is 5% for central collisions. The neutral transverse energy, E_T, is obtained by subtracting the charged particle contribution, which is estimated (by Monte-Carlo) to amount to 40%.

The zero degree calorimeter is a very forward detector ($\eta_{lab} \geq 6.2$) measuring the energy of the beam spectator fragments, E_{ZDC} (by means of the Čerenkov light produced in the fibres). It is located 165 cm dowstream from the target centre, near the beam axis, protected against secondaries produced in the interaction by a long conical Cu collimator. It is made of SiO_2 optical fibres embedded in a tantalum converter in a 1:17 volume ratio. Its resolution is 7% for 32.7 TeV incident ^{208}Pb nuclei.

The multiplicity detector is a two plane highly segmented silicon microstrip device, measuring charged particle multiplicities within the pseudo-rapidity range $1.5 < \eta_{lab} < 3.5$. It has not been used in the present analysis.

3 Data analysis

Data have been collected with two different Pb target configurations, in 1995 and 1996, with a lead beam of 158 GeV/nucleon incident momentum. Details of the two data samples are given in Table 1.

The event selection criteria are as follows: the BH detects only one incident ion; no preinteraction is detected by the BHI or the halo counters; only two tracks are reconstructed in the fiducial region of the spectrometer; each image track, conceptually obtained from the real track by reversing the magnetic field, is also accepted by the apparatus so that muon acceptance is charge independent; interaction in one of the subtargets and no reinteraction, as assigned by the active target algorithm (in 1995); and, because of the target algorithm low efficiency for peripheral collisions use, for this type of events, of a contour cut based on a 2 σ correlation between E_T and E_{ZDC} reinforced with a strong cut on the distance between each muon track and the beam axis (in 1996).

The purpose of this analysis is to study the dimuon yield in our high mass region, which originates from the J/ψ and ψ' decays as well as from the Drell-Yan mechanism.

The kinematical domain used is defined by: $2.92 \leq y_{lab} \leq 3.92$ (i.e., $0 \leq y_{cms} \leq 1$) and $|\cos\theta_{CS}| < 0.5$, leading, in the mass region of interest, to acceptances of the order of 15%. The J/ψ mass resolution is 96 MeV (3.1%) and increases to 104 MeV (3.3%) for the most peripheral events recovered by the $E_T - E_{ZDC}$ contour cut.

In this analysis the background and the open charm ($D\bar{D}$) contributions are also taken into account. The background is mainly due to uncorrelated π and K decays into muons. It is computed from the like-sign mass distributions dN/dM^{++} and dN/dM^{--} using the relation

$$\frac{dN^{Bg}}{dM} = 2R \cdot \left[\sqrt{\left(\frac{dN^{++}}{dM} \cdot \frac{dN^{--}}{dM}\right)_{field+}} + \sqrt{\left(\frac{dN^{++}}{dM} \cdot \frac{dN^{--}}{dM}\right)_{field-}} \right],$$

where R is a factor depending on the type of the collision. For ion induced reactions, $R = 1$.

The opposite sign muon pair invariant mass distribution is fitted according to the following procedure in order to determine the amounts of its different components. The shapes of the muon pairs originating from the J/ψ and ψ' decays and from the Drell-Yan mechanism are obtained from a simulation of the NA50 detector using the same reconstruction and selection criteria as used for the real data. Drell-Yan contribution is calculated at the leading order and uses the MRS43 structure functions [3], which take into account the $\bar{u}/\bar{d}$ asymmetry as measured by NA51 experiment [4]. The $D\bar{D}$ shape is taken from Pythia and its amplitude is previously fixed by fitting the data in the range $1.9 < M_{\mu\mu} < 2.9$ GeV/c², allowing for a charm-like excess [5]. Finally, the amplitudes of J/ψ, ψ' and Drell-Yan contributions are obtained from a fit to the experimental data (above $M_{\mu\mu} > 3.05$ GeV/c²), according to:

$$\frac{dN^{+-}}{dM} = A_{J/\psi} \cdot \frac{dN^{J/\psi}}{dM} + A_{\psi'} \cdot \frac{dN^{\psi'}}{dM} + A_{DY} \cdot \frac{dN^{DY}}{dM} + \frac{dN^{D\bar{D}}}{dM} + \frac{dN^{Bg}}{dM},$$

with five free parameters: $A_{J/\psi}$, $A_{\psi'}$, A_{DY}, $M_{J/\psi}$ and $\sigma_{J/\psi}$; the ψ' mass and width are functions of the corresponding J/ψ values.

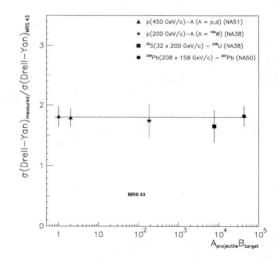

Figure 1: The Drell-Yan cross-section as a function of A·B, in terms of the experimental K factor

4 Results

Let us first concentrate on Drell-Yan production. Figure 1 shows for different systems, as a function of the product of projectile and target atomic mass numbers, A·B, the measured Drell-Yan cross-section normalized to the theoretical value, computed at the leading order using the MRS43 structure functions. This ratio is the so-called K-factor and is a measure of the higher order corrections needed to account for the data.

The different proton and neutron content of the interacting nuclei are taken into account, by referring all data to p-p collisions, using the relation:

$$\sigma_{DY}^{AB}(corr) = \sigma_{DY}^{AB}(meas) \times \frac{AB \cdot \sigma_{DY}^{pp}(th)}{\sigma_{DY}^{AB}(th)} .$$

All the results are compatible and lead to the average value K = 1.78 ± 0.09, in agreement with expectations.

Using the usual power law behaviour to parametrize the nuclear dependence of hard processes

$$\sigma_{DY}^{AB} = (AB)^{\alpha} \cdot \sigma_{DY}^{pp} ,$$

a fit to the data, corrected as explained above, leads to $\alpha = 1.002 \pm 0.011$, in perfect agreement with the expected value $\alpha = 1$.

That is, from p-p up to Pb-Pb the Drell-Yan cross-section behaves normally and is proportional to the number of elementary nucleon-nucleon collisions (i.e., the

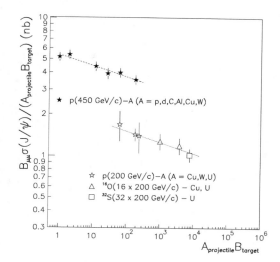

Figure 2: J/ψ cross-section per nucleon-nucleon collision, as a function of A·B, for 450 GeV/c and 200 GeV/c incident projectiles, for several systems ranging from p-p to S-U

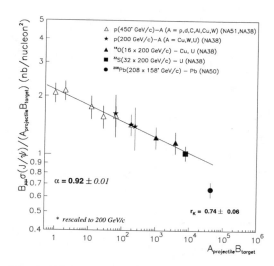

Figure 3: J/ψ cross-section per nucleon-nucleon collision, as a function of A·B, for several systems ranging from p-p to Pb-Pb, having rescaled all incident momenta to 200 GeV/c

product A·B). Thus, it is useful to use it as a reference to the study of other systems.

In order to compare our new measurements on Pb-Pb with results obtained with proton and other ion induced reactions, and besides the isospin correction discussed above, the measured cross-sections are all rescaled to the same incident beam momentum according to:

$$\sigma_{DY}^{corr}(E_f) = \sigma_{DY}^{meas}(E_i) \times \frac{\sigma_{DY}^{th}(E_f)}{\sigma_{DY}^{th}(E_i)} \quad .$$

We turn now to the study of J/ψ. Figure 2 shows the J/ψ cross-section per nucleon-nucleon collision as a function of A·B, for the two data sets, p-A collisions at 450 GeV/c, and proton, oxygen and sulphur induced reactions at 200 GeV/c. The power law behaviour fit gives compatible values, namely $\alpha_{450} = 0.92 \pm 0.02$ and $\alpha_{200} = 0.91 \pm 0.04$. The common trend observed from p-p to S-U allows us to merge all the data on the same curve after energy rescaling (Figure 3). An overall fit leads to $\alpha = 0.92 \pm 0.01$. According to some authors, this behaviour may be interpreted as due to an absorption in nuclear matter of the pre-resonant $c\bar{c}g$ state [7].

Figure 3 also shows that the Pb-Pb cross-section, rescaled to 200 GeV, lies well below the value expected from the pattern of lighter interacting nuclei. This anomalous suppression can be quantified by the ratio between the measured and expected value, $R_K = 0.74 \pm 0.06$, which includes a 7% systematic error.

The J/ψ cross-section per nucleon-nucleon collision can also be estimated from the ratio of the J/ψ to Drell-Yan cross-sections. This ratio is almost free of systematic errors, which are common to both samples (only 1.5% left). In Figure 4 the two samples collected in 1995 and 1996 are shown separately, as a function of E_T . They are significantly different in statistics but are fully compatible with each other. A threshold is clearly seen near 40 GeV in the 1996 sample which, thanks to statistics, can be studied with a much narrower binning.

A common description is needed in order to put together the NA50 data on Pb-Pb with previous NA38 and NA51 data obtained with lighter systems. We define L as the path length of the pre-resonant $c\bar{c}g$ state through nuclear matter. In the framework of a nuclear geometrical model, L can be related to a given impact parameter b. As b and E_T are also related, E_T bins just correspond to different L values. In a given E_T bin, L is computed by averaging the path length over the production point within the nuclei, and the impact parameter.

The J/ψ over Drell-Yan cross-sections' ratio, through its decay into muons, denoted as $B_{\mu\mu}\sigma_{J/\psi}/\sigma_{DY}$, is shown in Figure 5, as a function of L. All 450 GeV/c and 200 GeV/c data have been rescaled to 158 GeV/c using the Schuler parametrization [6]. A good overall fit from p-p to S-U interactions is obtained with an absorption model parametrized as $exp(-\rho L \sigma_{abs})$, where $\rho = 0.17$ nucleons/fm³ is the standard nuclear density. The

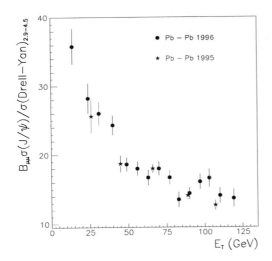

Figure 4: The ratio of J/ψ to Drell-Yan cross-sections as a function of E_T, for two data sets, in Pb-Pb collisions

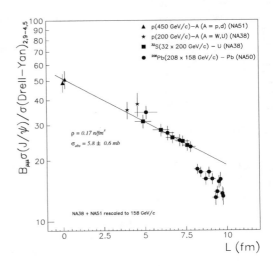

Figure 5: The ratio of J/ψ to Drell-Yan cross-sections as a function of L, for several systems ranging from p-p to Pb-Pb

fit leads to an absorption cross-section $\sigma_{abs} = 5.8 \pm 0.6$ mb, compatible with theoretical predictions [7]. Whereas the more peripheral Pb-Pb points lie on the absorption curve, the more central ones show a clear departure at $L \simeq 8$ fm, suggesting the onset of another J/ψ suppression mechanism. In fact, several authors claim that this behaviour can only be explained in the framework of QGP formation [8].

Finally, we turn to the study of ψ' production. Because of its larger radius as compared to J/ψ, ψ' should have a lower threshold for breakup. Figure 6 shows as a function of A·B, for several systems ranging from p-p to Pb-Pb, the ψ' over J/ψ cross-sections' ratio, $B_{\mu\mu}\sigma_{\psi'}/B_{\mu\mu}\sigma_{J/\psi}$, where $B_{\mu\mu}$ are the branching ratios of J/ψ and ψ' into muons. While ψ' and J/ψ show the same behaviour in proton-nucleus collisions, their ratio has much lower values in ion induced interactions. The ψ' over Drell-Yan ratio, that is, $B_{\mu\mu}\sigma_{\psi'}/\sigma_{DY}$, is plotted in Figure 7 as a function of L, for different proton and ion induced reactions. In proton-nucleus collisions ψ' follows the same nuclear absorption curve as J/ψ, supporting the assumption of a $c\bar{c}g$ state interacting in the target nucleus. But, for S-U and Pb-Pb interactions, ψ' shows an additional suppression, which may be attributed to another mechanism. Several authors have indeed tried to explain it by means of a further ψ' absorption by co-movers (hadrons produced in the interaction and accompaining the resonance) [9]. However, this kind of mechanism can hardly account for sudden pattern changes, as

observed for J/ψ.

5 Conclusions

We have studied the Drell-Yan, J/ψ and ψ' production cross-sections for various systems, ranging from p-p to Pb-Pb interactions. The Drell-Yan production is proportional to the number of collisions, and is thus used as a reference to the study of J/ψ and ψ'. J/ψ follows a nuclear absorption pattern for p-A, S-U, and Pb-Pb peripheral collisions. It shows a clear departure from this absorption trend when centrality increases, exhibiting a sharp decrease at $L \simeq 8$ fm. ψ' behaves, as J/ψ, according to the same absorption curve for proton-nucleus data, but shows a stronger suppression for sulphur and lead induced interactions, which begins at lower L values. This suggests that the suppression mechanism for J/ψ and ψ' are different.

References

1. T. Matsui and H. Satz, *Phys. Lett.* B **178**, 416 (1986).
2. NA50 Collaboration, M.C. Abreu et al., *Phys. Lett.* B **410**, 327 (1997), *Phys. Lett.* B **410**, 337 (1997) and references therein.
3. A.D. Martin et al., *Phys. Lett.* B **306**, 145 (1993).
4. NA51 Collaboration, A. Baldit et al., *Phys. Lett.* B **332**, 244 (1994).

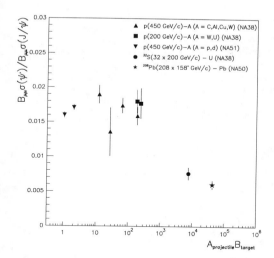

Figure 6: The ratio of ψ' to J/ψ cross-sections as a function of A·B, for several systems ranging from p-p to Pb-Pb

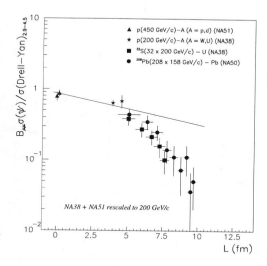

Figure 7: The ratio of ψ' to Drell-Yan cross-sections as a function of L, for several systems ranging from p-p to Pb-Pb

5. See C. Soave's contribution to these Proceedings.

6. G.A. Schuler, CERN-TH 7170/94 (HEP-PH/9403387) (1994).

7. See for instance, D. Kharzeev and H. Satz, *Phys. Lett.* B **366**, 316 (1996).

8. D.Kharzeev, *Nucl. Phys.* A **610**, 418c (1996); C.Y. Wong, *Nucl. Phys.* A **610**, 434c (1996); J.P. Blaizot, J.Y. Ollitraut, *Phys. Rev. Lett.* **77**, 1703 (1996); N. Armesto et al., *Phys. Rev. Lett.* **77**, 3736 (1996); D.Kharzeev et al., *Z. Phys.* C **74**, 307 (1997); M. Nardi and H. Satz, hep-ph/9805247.

9. S. Gavin and R. Vogt, *Phys. Rev. Lett.* **78**, 1006 (1997); A. Capella et al, *Phys. Lett.* B **393**, 431 (1997); W. Cassing and C.M. Ko, *Phys. Lett.* B **396**, 39 (1997).

HIGH ENERGY Pb+Pb COLLISIONS AS SEEN BY THE NA44 EXPERIMENT

D. Hardtke[h] for the NA44 Collaboration

I.G.Bearden[1], H. Bøggild[1], J. Boissevain[2], J. Dodd[3], B. Erazmus[4], S. Esumi[5,c], C.W. Fabjan[6], D. Ferenc[7], D.E. Fields[2,d], A. Franz[6,e], J. Gaardhøje[1], M. Hamelin[10], A.G. Hansen[1], O. Hansen[1], D. Hardtke[8,h], H. van Hecke[2], E.B. Holzer[6], T.J. Humanic[8], P. Hummel[6], B.V. Jacak[11], R. Jayanti[8], K.Kaimi[5], M. Kaneta[5], T. Kohama[5], M. Kopytine[11], M. Leltchouk[3], A. Ljubicic, Jr.[7,e], B. Lörstad[9], N. Maeda[5,f], R.Malina[6], A. Medvedev[3], M. Murray[10],H.Ohnishi[5], G. Paic[8], S.U. Pandey[8,g], F. Piuz[6], J. Pluta[4], V. Polychronakos[12], M. Potekhin[3], G. Poulard[6], D.Reichhold[8], A. Sakaguchi[5,b], J. Simon-Gillo[2], J. Schmidt-Sørensen[9], W. Sondheim[2], M.Spegel[6], T. Sugitate[5], J.P. Sullivan[2], Y. Sumi[5], W.J. Willis[3], K.L. Wolf[10], N. Xu[2,h], and D.S.Zachary[8]

[1] Niels Bohr Institute, DK-2100, Copenhagen, Denmark.
[2] Los Alamos National Laboratory, Los Alamos, NM 87545, USA.
[3] Columbia University, New York, NY 10027, USA.
[4] Nuclear Physics Laboratory of Nantes, 44072 Nantes, France.
[5] Hiroshima University, Higashi-Hiroshima 739, Japan.
[6] CERN, CH-1211 Geneva 23, Switzerland.
[7] Rudjer Boskovic Institute, Zagreb, Croatia.
[8] Ohio State University, Columbus, OH 43210, USA.
[9] University of Lund, S-22362 Lund, Sweden.
[10] Texas A&M University, College Station, TX 77843, USA.
[11] State University of New York, Stony Brook, NY 11794, USA.
[12] Brookhaven National Laboratory, Upton, NY 11973, USA.
[b] now at Osaka University, Toyonaka, Osaka 560-0043, Japan.
[c] now at Heidelberg University, D-69120 Heidelberg, Germany.
[d] now at University of New Mexico, Albuquerque, NM 87131, USA.
[e] now at Brookhaven National Laboratory, Upton, NY 11973, USA.
[f] now at Florida State University, Tallahassee, FL 32306, USA.
[g] now at Wayne State University, Detroit, MI 48202, USA.
[h] now at Lawrence Berkeley National Laboratory, Berkeley, CA 94720, USA.

NA44 is a focusing spectrometer at the CERN SPS that measures single-particle inclusive spectra and two-particle correlations near mid-rapidity. Recent data taken using a 158 GeV/nucleon Pb beam incident on a Pb target will be presented. These data will be compared to data taken using other projectile-target combinations. The data show that the emission volume in Pb+Pb collisions is larger than in lighter projectile-target systems, that a large degree of collective radial flow is created in these collisions, and that different particle species probe different regions of the freeze-out volume.

1 Introduction

The goal of relativistic heavy-ion collision research is to look for phenomena that cannot be explained in terms of a superposition of simple nucleon-nucleon collisions. It is hoped that at sufficient bombarding energy, the system will evolve through a state where the quarks and gluons are no longer bound in hadrons but instead form a Quark-Gluon plasma. Inevitably, this plasma hadronizes, and the produced hadrons interact via elementary collision processes. Finally, the hadrons freeze-out and can be detected.

The final-state hadrons produced in nucleus-nucleus collisions do not directly probe the early stages of the system, but the final state one- and two- particle hadron spectra can be sensitive to the dynamical evolution of the system integrated over the collision history. Through a measurement of the particle spectra, it is possible to infer whether the system reaches a stage of kinetic equilibrium and estimate the freeze-out temperature and radial expansion velocity. Measurements of the two-particle correlation functions can be used to infer the freeze-out volume and duration of particle emission. Composite particles, such as deuterons and anti-deuterons, can be used to constrain both the final state temperature and expansion velocity by looking at the inverse slopes of their momentum spectra. In addition, the absolute yield of composite particles is sensitive to the freeze-out volume because they are produced primarily through the coalescence of nucleons during the later stages of the collision.

NA44 is a focusing spectrometer that measures charged hadrons produced in p-A and A-A collisions at the CERN SPS. The spectrometer has a very small acceptance centered around mid-rapidity, typically measuring only one or two of the thousands of produced hadrons in a central Pb+Pb collision. The experiment, however, is able to trigger on particles of different species using two threshold Cerenkov counters, and has excellent particle identifical using a combination of three time-of-flight detectors, a Threshold Imaging Cerenkov Detector,[1] and a Uranium calorimeter. More detail about the experiment can be found elsewhere.[2,3]

2   Single-Particle Inclusive Spectra

NA44 has measured the mid-rapidity single particle inclusive spectra for $\pi^\pm$, $K^\pm$, p, $\bar{p}$, d, $\bar{d}$, and t. The particle spectra tend to be nearly exponential, and are parameterized by,

$$\frac{1}{m_T}\frac{dN}{dm_T} = Ae^{-m_T/T}, \tag{1}$$

where $m_T = \sqrt{m^2 + p_T^2}$ is the transverse mass and T is the inverse slope parameter. NA44 showed that the π, K and p inverse slope parameters increase nearly linearly with particle mass in A+A collisions, and that the magnitude of this effect increases with projectile-target system size.[4] This has been interpreted as evidence for collective radial expansion, and on the basis of the single-particle spectra a kinetic freeze-out temperature of 140±7(138±5) MeV and a transverse expansion velocity of 0.41±0.11(0.28±0.10) c has been estimated for central Pb+Pb (S+S) collisions.[4] Fig. 1 shows that the nearly linear increase with particle mass continues up to the deuteron for both the S+S and Pb+Pb systems. It was shown recently[7] that the deuteron slope is sensitive to both the collective radial flow and the density profile of the source. A box type density distribution leads to a linear increase in the inverse slope parameter with particle mass, while a Gaussian density profile leads to a less than linear increase of the inverse slope parameter with increasing particle mass. The difference in predicted slope parameters between the box and Gaussian density profiles increases with increasing particle mass, so we must wait until the deuteron analysis is finalized before drawing conclusions about the nucleon density profile.

Deuterons and anti-deuterons are not produced in the primary nucleon-nucleon collisions, but are instead thought to be produced primarily through the coalescence of nucleons in the final-state. The probability for the production of a deuteron (anti-deuteron) is related to the final state nucleon (anti-nucleon) density. Hence a useful quantity is the coalescence parameter B_A, which relates the probability for producing a composite particle

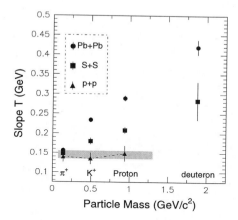

Figure 1: The inverse slope parameters near mid-rapidity ($y = 0$) for π^+, K^+, p, and d for p+p, S+S, and Pb+Pb collisions. The p+p results are taken from Alper et al.[5] and Guettler et al.[6]. The S+S and Pb+Pb π^+, K^+, and p inverse slopes are published NA44 data[4], while the deuteron inverse slopes are preliminary.

of nucleon number A to the proton yield:

$$B_A = \frac{\frac{dN_A}{dy}}{\left(\frac{dN_p}{dy}\right)^A}. \tag{2}$$

Fig. 2 shows a compilation of the B_2 for various energies and various projectile-target combinations along with the preliminary NA44 data for deuterons and anti-deuterons. B_2 is constant for deuterons in proton-induced collisions at all energies. In nucleus-nucleus collisions, B_2 for deuterons decreases with increasing system size and projectile energy. Note that the B_2 parameter for anti-deuterons measured by NA44 is significantly smaller than that measured by NA52.[18] The NA44 measurement is for semi-central collisions and the mean transverse momentum of the anti-deuteron sample is 1 GeV/c. In contrast, the NA52 measurement is for minimum bias collisions and anti-deuterons with $p_T = 0$.

Under certain assumptions, the coalescence parameter B_2 can be related to the volume of the source. In one such model[8] that assumes a spherically symmetric Gaussian source density $\rho(r) = e^{-r^2/2R_G^2}$, the Gaussian source radius can be related to B_2:

$$R_G^3 = \frac{3}{4}(\pi)^{3/2}(\hbar c)^3 \frac{m_d}{m_p^2}\frac{1}{B_2}. \tag{3}$$

The B_2 parameter is inversely proportional to the volume in this formulation. The NA44 preliminary mea-

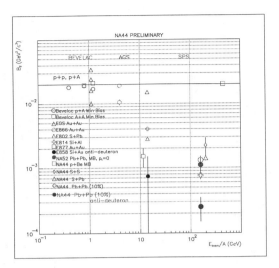

Figure 2: Compilation of B_2 measured in p-A and A-A collisions. [10,11,12,13,14,15,16,17,18] The filled points are for deuterons and the open points are anti-deuterons. The NA44 data are preliminary. The compilation comes from Johnson[9].

surements of the B_2 coalescence parameter for deuterons and anti-deuterons in central Pb+Pb collisions at 158 GeV/nucleon suggest that the anti-deuterons probe a larger freeze-out volume than the deuterons. This could be the result of anti-nucleon annihilations. This would be consistent with the large number of protons observed at mid-rapidity by NA44.[19]

3 Two-Particle Correlations

NA44 has measured two-particle (HBT[20]) correlations for $\pi^+\pi^+$, $\pi^-\pi^-$, and K^+K^+. The two-particle correlation function is defined as

$$C_2(\vec{k}_1, \vec{k}_2) = \frac{R(\vec{k}_1, \vec{k}_2)}{B(\vec{k}_1, \vec{k}_2)}, \tag{4}$$

where $R(\vec{k}_1, \vec{k}_2)$ is the number of particle pairs with single particle momenta $\vec{k}_1$ and $\vec{k}_2$ observed in the real data, and $B(\vec{k}_1, \vec{k}_2)$ is number of pairs that would be expected in the absence of quantum statistical correlations. The background distribution is found by using mixed events. For pairs of identical bosons, the correlation function shows an enhancement at small momentum difference. The width of this enhancement is inversely proportional to size of the particle emitting source.

The correlation function is typically parameterized as a function of the pair momentum difference $\vec{Q}$ and the

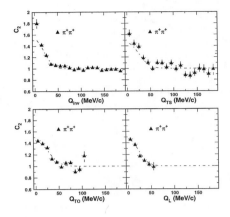

Figure 3: The low-p_T ($\langle p_T \rangle \approx 170$ MeV) $\pi^+\pi^+$ correlation function as measured in Pb+Pb collisions by NA44. Shown is the one-dimensional parameterization ($Q_{inv} = \sqrt{|\vec{Q}|^2 - Q_0^2}$) and the projections of the three-dimensional parameterization. The projections of the Gaussian fits are also shown.

average pair momentum $\vec{K}$:

$$C_2(\vec{Q}, \vec{K}) = 1 + \lambda e^{-R_{TO}^2(\vec{K})Q_{TO}^2 - R_{TS}^2(\vec{K})Q_{TS}^2 - R_L^2(\vec{K})Q_L^2}. \tag{5}$$

Here, the longitudinal direction (Q_L) is defined as the direction along the beam axis, the outward direction (Q_{TO}) is defined as the direction transverse to the beam direction and parallel to the total transverse momentum of the pair, and the sideward direction (Q_{TS}) is defined as the direction perpendicular to the beam axis and perpendicular to the total transverse momentum of the pair. R_{TO}, R_{TS}, and R_L are the Gaussian radius parameters, and λ is the strength of the correlation as the momentum difference Q goes to 0. This parameterization is used since for a static Gaussian source the difference between the outward and sideward radius parameters is related to the duration of particle emission, and this is expected to be large if there is a first order phase transition from the Quark-Gluon Plasma.[21]

A typical correlation function is shown in Fig. 3. The correlation functions are typically well described by a Gaussian. A summary of the measured Pb+Pb radius and lambda parameters are shown in Fig. 4 as a function of the average transverse mass of the pair. A detailed description of the analysis can be found elsewhere.[2,3]

The Gaussian radius of a Pb nucleon is 3.2 fm, so the low-m_T pion radii indicate that source undergoes significant expansion before particle freeze-out. The large-m_T kaon radii, however, are smaller than the size of the

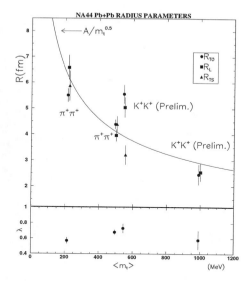

Figure 4: Pb+Pb HBT radius and intercept parameters as a function of transverse mass for pions[3] and kaons. The kaon results are preliminary. Also drawn is the expectation of $A/\sqrt{m_T}$ scaling.

anti-deuterons. Within a coalescence model, these measurements suggest a larger source size for anti-deuterons compared to deuterons. NA44 has also measured the two-particle correlation functions of pions and kaons at several values of the pair transverse mass. The HBT radii are larger than the projectile size, indicating expansion before freeze-out. The three-dimensional radius parameters do not follow the simple $1/\sqrt{m_T}$ that was observed by NA44 in S+Pb collisions. The intercept parameter grows with increasing transverse mass, which suggests a reduced resonance contribution to the particle sample with increasing m_T.

Acknowledgements

The NA44 Collaboration wishes to thank the staff of the CERN PS-SPS accelerator complex for their excellent work. We thank the technical staff at CERN and the collaborating institutes for their valuable contributions. We are also grateful for the support given by the Science Research Council of Denmark; the Japanese Society for the Promotion of Science, and the Ministry of Education, Science and Culture, Japan; the Science Research Council of Sweden; the US Department of Energy; and the National Science Foundation.

projectile. This illustrates the importance of transverse expansion on the measured radii. Higher transverse momentum particles see a smaller region of the source.[3]

For S+Pb collisions at 200 GeV/nucleon, NA44 observed that the three-dimensional radius parameters for π^+ and K^+ follow a common $A/\sqrt{m_T}$ scaling.[22] This is no longer the case in Pb+Pb collisions. The R_{TO} and R_L radius parameters for the low-m_T kaons are larger than the pion radius parameters at a similar m_T.

Fig. 4 also shows that the λ parameter increases as a function of transverse mass. This is what would be expected if the reduction of λ from unity is caused entirely by the contributions of particles from resonance decay to the pair sample. As the m_T is increased, fewer particles come from resonance decay, so the λ parameter should approach unity.

4 Conclusions

NA44 has measured single-particle inclusive spectra and two-particle correlations in Pb+Pb collisions at 158 GeV/nucleon. The mass systematics of the single-particle inverse slope parameters show that a large collective radial flow is created in these collisions, and that the magnitude of this flow is larger Pb+Pb collisions than in S+S collisions at a comparable energy. NA44 has also measured the absolute yield of deuterons and

References

1. C.W. Fabjan *et al.*, *Nucl. Instrum. Methods* A**367**m 240 (1995); A. Bream *et al.*, CERN-PPE/97-120.
2. H. Bøggild *et al.* (NA44 collaboration), *Phys. Lett.* B**302** (1993) 510.
3. I.G. Bearden *et al.* (NA44 collaboration), *Phys. Rev.* C**58** (1998) 1656.
4. I.G. Bearden *et al.* (NA44 collaboration), *Phys. Rev. Lett.* **78** (1997) 2080.
5. B. Alper *et al.*, *Nucl. Phys.* B**100** (1975) 237.
6. K. Guettler *et al.*, *Nucl. Phys* B**116** (1976) 77.
7. A. Polleri, J.P. Bondorf, and I.N. Mishustin, *Phys. Lett.* B**419** (1998) 19.
8. W.J. Llope *et al.*, *Phys. Rev.* C**52** (1995) 2004; S. Pratt, Private Communication.
9. S. Johnson, Ph.D. thesis, SUNY-Stony Brook (1997).
10. S. Nagamiya *et al.*, *Phys. Rev.* C**24** (1981) 971.
11. M.C. Lemaire *et al.*, *Phys. Lett.* **37** (1976) 667.
12. R.L. Auble *et al.*, *Phys. Rev.* C**28** (1983) 1552.
13. S. Wang *et al.*, *Phys. Rev. Lett.* **74** (1995 2648.
14. J. Barrette *et al.*, *Phys. Rev.* C**50** (1994) 1077; *ibid*, *Phys. Rev.* C**50** (1994) 3047; *ibid*, *Phys. Rev. Lett.* **70** (1993) 1763.
15. T. Abbot *et al.*, *Phys. Rev.* C**50** (1994) 1024.
16. J. Simon-Gillo *et al.*, *Nucl. Phys* A**590** (1995) 483c.

17. J.W. Cronin *et al.*, *Phys. Rev.* **D11** (1975) 3105.
18. G. Appelquist *et al.*, *Phys. Rev. Lett.* **76** (1996) 3907; *ibid, Phys. Lett.* **B376** (1996) 245.
19. I.G. Bearden *et al.* (NA44 collaboration), *Phys. Lett.* **B388** (1996) 431.
20. R. Hanbury Brown and R.Q. Twiss, *Nature* **178** (1956) 1046.
21. S. Pratt, *Phys. Rev.* **D33** (1986) 1314; G. Bertsch and G.E. Brown, *Phys. Rev.* **C40** (1989) 1830.
22. H. Beker *et al.* (NA44 collaboration), *Phys. Rev. Lett.* **74** (1995) 209.

OVERVIEW OF HADRONIC OBSERVABLES MEASURED BY NA49 AT THE CERN/SPS IN CENTRAL $^{208}Pb + Pb$ COLLISIONS AT 158 GeV/NUCLEON

P. FOKA

for the NA49 collaboration

IKF, August Euler Str. 6, D60486 Frankfurt, Germany
E-mail: yiota.foka@cern.ch

Results on single and two-particle distributions measured by NA49 experiment in central $^{208}Pb + Pb$ collisions at 158 GeV/nucleon are presented. Their relevance concerning the observation and characterisation of an extended equilibrated deconfined state of partonic matter, the quark gluon plasma (QGP) is discussed. Available data support that sufficient energy density for a phase transition of normal hadronic matter to a deconfined state is reached at the early stage of such collisions. The evolution of the created system is probed via hadronic observables, and information on transverse expansion, chemical and thermal freeze-out is obtained. Comparisons of particle abundances for different collision systems and energies suggest that $^{208}Pb + Pb$ collisions cannot be described as a superposition of independent elementary collisions.

1 Heavy Ion Physics and NA49 Experiment

Heavy ion physics studies strongly interacting matter at extreme conditions of energy density in the region where Lattice QCD predicts [1] a phase transition from hadronic to deconfined chiraly restored quark gluon plasma [2]. However, the exact nature of non-perturbative QCD in hot and dense systems created in heavy ion collisions is not known. Experimental input is thus crucial for theoretical models attempting to describe such strongly interacting matter. The aim being, through the parallel advances of theory and experiment, to establish the equation of state and to study the phase diagram of QCD matter and of quarks and gluons in a deconfined medium. Thus heavy ion physics explores and tests QCD on its natural scale, Λ_{QCD}, and addresses the understanding of confinement and chiral-symmetry breaking.

In the case that QGP has been formed in the early stage of heavy ion collisions, partonic matter will eventually hadronise as the system expands, dilutes and cools. The yield of different particle species is then determined at the moment of chemical freezeout. The system keeps expanding till strong interactions cease. At this thermal freezeout stage the spectral shape of different particles is determined. This is the stage that the two-particle correlations are established and can give information about the expansion dynamics of the final hadronic phase, which is expected to differ considerably with the presence or absence of a prior plasma phase.

NA49 aims to provide experimental information on the evolution of the system which could then contribute to infer the initial conditions and establish an equation of state. The experiment was designed as a large acceptance hadron spectrometer [3] aiming to characterise the hadronic final state as completely as possible. About 1200 charged particles are detected in a single central $^{208}Pb + Pb$ collision by its four large vol-

ume Time Projection Chambers (TPCs) and the four Time-of-Flight (TOF) walls covering 80% of all produced charged particles. Two of the TPCs in the magnetic field provide momentum determination with an accuracy of $\Delta p/p^2 \sim 0.3 \ 10^{-4}$ and identification of secondary vertices. The other two are 4 m long to measure the energy loss to better than 4.5%, permitting particle determination in the relativistic rise. The TOF walls, achieving 60 ps resolution, are used to complement particle identification at mid-rapidity.

Particle identification via dE/dx and TOF measurement provides the experiment with identified π^+, π^-, K^+, K^-, p, $\bar{p}$ and deuterons. Reconstruction of decay topologies provides Λ, $\overline{\Lambda}$, K_S^0, K^+, K^- and recently Ξ cascade measurements. ϕ have been measured via invariant mass reconstruction of identified $K^+ K^-$ pairs. The multitude of measured observables over a large acceptance provide a comprehensive view of the hadronic final state.

Transverse energy production at mid-rapidity and energy flow studied in the NA49 calorimeters, have confirmed that energy densities of $\sim 3 \ GeV/fm^3$, have been reached in central $^{208}Pb + Pb$ collisions [4,5,6].

The analysis of a sample of the 5% most central interactions selected by triggering on the energy deposited in the forward calorimeter is presented here. This continues and complements systematic studies of ^{32}S induced reactions by NA35, the predecessor experiment of NA49.

2 Hadronic Observables

2.1 Stopping and Energy Deposition

Information on the degree of stopping and energy deposition at the initial stage of the collision can be obtained from the distribution of net-baryons, defined as the difference of the baryon and antibaryon distribution,

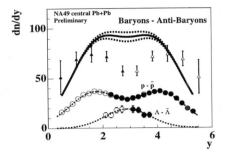

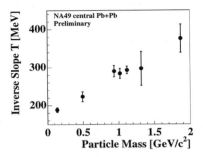

Figure 1: Rapidity distributions of net-protons, net-Λ and net-baryons in central $^{208}Pb + Pb$. The net-baryon distribution in central $^{32}S + S$ (triangles) is scaled up by the ratio of participant nucleons in each system.

Figure 2: Systematics of inverse slope parameters of identified π, K, p, ϕ, Λ, Ξ and d in central $^{208}Pb + Pb$ collisions as a function of particle mass.

in the final state. The baryon-antibaryon rapidity distribution for $^{208}Pb + Pb$ is shown in Fig. 1 (solid line) together with the ones for net protons, $(p - \bar{p})$ and net-Λ, $(\Lambda - \bar{\Lambda})$. It is obtained from the sum $2.1 \times (p - \bar{p})$ plus $1.6 \times (\Lambda - \bar{\Lambda})$, where the scaling factors account for neutrons and higher mass hyperons respectively.

The net-baryon distribution is compared to the properly scaled NA35 central $^{32}S + S$ data. The $^{208}Pb + Pb$ distribution is characterised by a plateau which extends over two units of rapidity, indicating large stopping power and a rather sharp drop at high and low rapidity. The $^{32}S + S$ data show a higher yield near target and beam rapidity and a dip close to central rapidity. A mean rapidity shift for the $^{208}Pb + Pb$ system of $< \Delta y > = 1.78$ units of rapidity as compared to $< \Delta y > = 1.63$ for the $^{32}S + S$ indicates larger stopping power for the heavier system, which results in a larger amount of deposited energy. Already in the $^{32}S + S$ system a 25% higher energy loss of the incoming nucleons was observed as compared to pp collisions.

The redistribution of the available energy, into particle production, thermal and ordered motion can be revealed by the data as discussed in the following.

2.2 Dynamic Evolution

NA49 has provided systematic differential measurements of transverse mass distributions in rapidity bins for negative hadrons and participant protons [3]. Single particle spectra for a variety of identified particles have been measured around mid-rapidity. Fig. 2 shows as a summary the inverse slop parameters obtained by fitting the transverse mass spectra to the expression, $\frac{1}{m_T}\frac{dn}{dm_T} = C \exp\left(-\frac{m_T}{T}\right)$ as function of the mass of the observed hadron. The spectra become harder with increasing par-

ticle mass which is indicated by the almost linear increase of their inverse slope parameter, as shown in Fig. 2, up to a value of ~ 380 MeV for deuterons. Clearly, inverse slope parameters of this magnitude cannot be interpreted as "real" temperatures of a hadronic system, particularly when recalling the Hagedorn limiting temperature of ~ 165 MeV for a hadronic source [7].

This observed dependence and the fact that particles of similar mass show very similar inverse slope parameters can be accommodated by introducing into the simple thermal model a transverse velocity field, common to all particles, assuming that the system develops radially ordered flow [8]. Then the further observation that the inverse slope parameter increases with the size of the colliding system for heavy particles, as has been shown for p, Λ, $\bar{\Lambda}$, K_S^0, can be attributed to the development of higher flow in the heavier systems.

This expansion pattern is a manifestation of the Doppler effect. For hadrons emerging from the decoupling fireball created in nuclear collisions the energy spectra are shifted according to the local expansion velocities relative to the observer. The transverse expansion is directed towards the detectors and the measured particle momentum spectra are hence "blue-shifted".

It has been shown [9,10,11,12] that with the choice of an appropriate velocity profile, the inverse slope systematics can be described with two parameters, a common thermal freeze-out temperature T_f, and a flow velocity parameter $\beta_\perp$, that characterises the strength of the flow velocity field common to all particles. However with large uncertainty particularly in the determination of $\beta_\perp$ [9].

The extraction of transverse velocity using transverse mass spectra alone is in fact ambiguous, since a similar effect results from thermal motion, the temperature of the decaying source being in principle unknown,

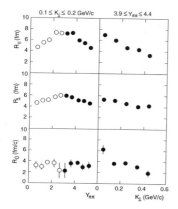

Figure 3: Radius HBT parameters in bins of rapidity and transverse momentum of the pair.

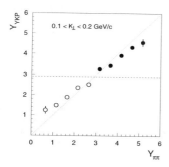

Figure 4: Rapidity of the pion-emitting source element Y_{YKP} as a function of the average rapidity of the emitted pion pairs $Y_{\pi\pi}$. Filled circles are data, open circles their reflection about mid-rapidity.

and therefore flow velocity and real temperature cannot be uniquely disentangled. However, quantum statistical Bose-Einstein correlations of identical pions can provide independent information.

The theoretical framework for this measurement has been considerably refined recently [13] and has been demonstrated [10] that it is possible to extract dynamical parameters and disentangle the spatial and temporal properties of an expanding source when the correlation function is parametrised properly by the Yano-Koonin-Podgoretsky formalism [14]. According to this formalism the source is described as a succession of local source elements along the longitudinal axis, each one of them described by its spatio-temporal extent and average longitudinal velocity, the advantage being that the extracted

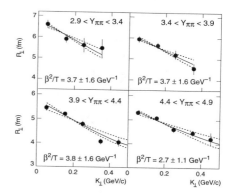

Figure 5: Dependence of the transverse correlation length $R_\perp$ on the average transverse momentum $K_\perp$ of the pion pair in four successive rapidity intervals. The fit to the data determines the ratio of the flow velocity parameter $\beta_\perp$ and the thermal freeze-out temperature T_f. Dashed lines show the variation of the fit function one standard deviation from the best fit values.

radius HBT-parameters do not depend on the longitudinal velocity of the system.

The parametrisation is then fitted to the measured 3-dimensional correlation function in bins of rapidity and transverse momentum [15]. High statistics, highly differential measurements fix the HBT-parameters with negligible statistical errors, mapping the system in different kinematical areas as shown in Fig. 3.

There is a clear dependence of the longitudinal and transverse HBT-radius parameters $R_\parallel$ and $R_\perp$ on the rapidity and transverse momentum of the pair, while R_0 appears to be constant. The source at mid-rapidity appears locally isotropic with both the longitudinal and transverse HBT-radius parameters $R_\parallel$ and $R_\perp$ being 6–7 fm and reducing towards higher rapidities. The duration of particle emission as measured by the parameter R_0 is $\Delta\tau \sim 3.5$ fm/c. The time profile of pion emission is assumed to be gaussian. The overall duration of expansion is estimated to be $\tau_0 \sim 8$ fm/c near mid-rapidity, decreasing slightly to $\tau_0 \sim 6$ fm/c at $y = 5$. The emission of pions fades away at ~ 18 fm/c. It is estimated that the source expands tranvsersely by a factor of about 2.5 between initial maximum energy density and final hadronic decoupling stage.

The dynamical picture of the source in the longitudinal direction is illustrated in Fig. 4 by the almost linear dependence of the rapidity of the source element, Y_{YKP}, on the average rapidity of the emitted pion pairs, $Y_{\pi\pi}$. A spherical static fireball at mid-rapidity would be revealed by the lack of such correlation, whereas an infinite boost-invariant source would exhibit a strict linear correlation[16]. The data show a boost-invariant lon-

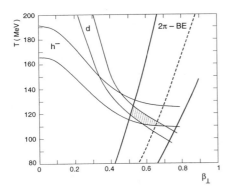

Figure 6: Allowed regions of thermal freeze-out temperature T_f and transverse flow velocity parameter $\beta_\perp$ derived from fit constraints of single-particle m_T spectra and two-particle correlation analysis.

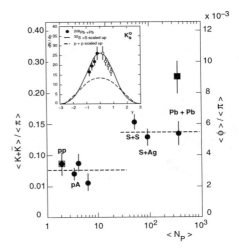

Figure 7: Strangeness enhancement in S and Pb induced collisions as measured by K and ϕ yields relative to pion yield as a function of the size of colliding systems.

gitudinal expansion with rather small deviations at high rapidities as expected for a beam of finite energy.

The dependence of the two-particle correlation function on the average pair momentum reflects the correlation between position and momentum of particles in an expanding source. Transverse expansion is then demonstrated in the dependence of the transverse radius HBT-parameter on the transverse momentum of the pair in different rapidity bins as shown in Fig. 5. Information on the strength of the transverse velocity field can be extracted by fitting the source function to the correlation data. A value of $\beta_\perp^2/T_f = (3.7 \pm 1.6)$ GeV^{-1} is determined where $\beta_\perp$ is the transverse flow velocity and T_f the thermal freeze-out temperature. But again the expansion velocity and thermal freeze-out temperature cannot be determined independently.

However by fitting the transverse mass spectra and two particle correlations within the same model of a 3-dimensional expanding source we can now tightly constrain $\beta_\perp$ and T_f at the same time, as shown in Fig. 6. The fit results in sets of solutions $(\beta_\perp, T_f)$ but with different constraint for the two pion correlation and the transverse mass spectra of negative hadrons and deuterons. The allowed regions in the plane of freeze-out temperature T versus transverse surface velocity $\beta_\perp$ are shown by the bands in Fig. 6 which correspond to one σ variation from the best fit values. These complementary observations define a rather narrow area of $(\beta_\perp, T_f)$ parameters from which we extract the "true" freeze-out temperature T_f=120 MeV and the "true" transverse expansion velocity $\beta_\perp$=0.55.

Theoretical models without collective expansion before hadrons are formed fail to reproduce such a transverse growth.

2.3 Particle Abundances

The character of the system at an early stage of the evolution may further be understood through the study of the relative particle abundances which are defined at the chemical freeze-out point. NA49 results on particle composition are used to study hadron chemistry in a model independent way and further constrain models.

Fig. 7 shows the dependence of particle ratios on the number of participant nucleons which essentially measures the size of the system. The left scale measures the ratio of kaon mean multiplicity to the total pion mean multiplicity, shown by dots, resulting from 4π integrated yields. This is a measure of the $(s + \bar{s})/(q + \bar{q})$ content. Since K_S^0 carry $\sim 80\%$ of the produced strangeness one can infer a strangeness enhancement by almost a factor of two in nucleus-nucleus collisions relative to elementary systems, which remains almost the same for the $^{32}S + S$ and $^{208}Pb + Pb$ systems. The recent preliminary analysis of ϕ production with 4π-integrated yields from combined TOF and dE/dx measurements has provided the ratio of ϕ to the total pion yield shown in Fig. 7 by rectangles with the scale at the right side. It shows an increase by a factor of ~ 2.6 relative to pp.

This enhancement is also observed in the $\bar{\Lambda}/\bar{p}$ ratio measured at mid-rapidity and verified by recent Ξ production studies. It is shown that double-strange hadrons are further enhanced over single-strange hadrons with respect to elementary systems. The measurement of multi-strange hadron production increases the confidence that

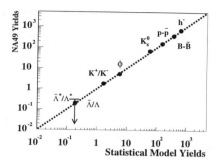

Figure 8: Comparison between particle yields observed by NA49 and the fit of Becattini's statistical model.

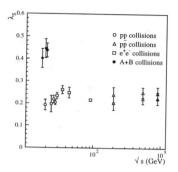

Figure 9: Strangeness suppression factor calculated in statistical model by Becattini for elementary and nucleus-nucleus collisions.

the observed strangeness enhancement originates from the partonic stage and is not due to hadronic interactions in the later stage.

In the insert of Fig. 7 is shown the rapidity distributions of K_S^0 for the $^{208}Pb + Pb$ system by points, and for $^{32}S + S$ and pp by lines, the last two being scaled up by appropriate factors accounting for the number of participant nucleons in each system. The $^{208}Pb + Pb$ distribution is in agreement with the $^{32}S + S$ scaled up distribution, and both are enhanced relative to the pp scaled up distribution. It is thus seen that the increase of strange particle production in S and Pb induced reactions relative to pp occurs mainly around mid-rapidity.

The rapidity distribution of negative hadrons in $^{32}S + S$, scaled up by the same factor as the one used for the K_S^0 distribution in Fig. 7, reproduces the h^- distribution in $^{208}Pb + Pb$ system. Since h^- represent the majority of produced particles, and K_S^0 the majority of strange particles one has an indication that neither negative hadron nor strange particle production accounts for the higher deposited energy in $^{208}Pb + Pb$ relative to $^{32}S + S$ system.

The NA49 observed particle yields are further compared with a fit by a statistical hadron gas model[17] which studies in particular the question whether full chemical equilibration is achieved in these collisions. The model yields very good agreement with the data, as shown in Fig. 8, only if one allows for partial strangeness saturation measured by the strangeness saturation factor γ_s, with a value of 0.6. This is significantly lower than unity as expected for a hadron gas in full chemical equilibrium. The model gives a hadronisation temperature $\sim$ 190 MeV, different from the kinetic freeze-out temperature $T_f \sim 120$ MeV. The same model also describes the particle yields in elementary p+p, $e^+ + e^-$ and $p + \bar{p}$ col-

lisions, showing a universal hadronisation temperature of 180 MeV. This indicates that the hadronisation process in elementary and nucleus-nucleus collisions is similar.

Within the same model the strangeness suppression factor λ_s, defined as the ratio of newly produced strange quarks to non-strange quarks is calculated in a consistent way for different experiments[17]. It is shown in Fig. 9 that for elementary collisions λ_s remains constant with a value of $\sim$ 0.25 while for central nucleus-nucleus collisions a value of $\lambda_s \sim 0.45$ is obtained with no significant change from $^{32}S + S$ to $^{208}Pb + Pb$.

This strangeness enhancement, unique to heavy-ion collisions, reaches a saturation already in central $^{32}S + S$ collisions, and this occurs at a level that it is not compatible with an ideal hadron gas calculation in full chemical equilibrium. This effect is compatible[18] with a scenario in which the strangeness content of the system is determined in a globally equilibrated partonic phase and remains unchanged during hadronisation and freeze-out.

3 Summary and Conclusions

The NA49 experiment has measured the hadronic final state in central $^{208}Pb + Pb$ collisions. An energy density of $\sim$ 3 GeV/fm^3 is inferred for the early stage of the collision, well within the regime predicted for a phase transition. The stopping power and energy deposition as extracted by net-baryon distributions in the final state is higher for the heavier system relative to $^{32}S + S$ and pp collisions.

The systematics of inverse slope parameters of the transverse mass spectra of identified particles and the dependence of the transverse HBT radius parameter on the pair transverse momentum indicate collective transverse flow. The simultaneous analysis of single and two-

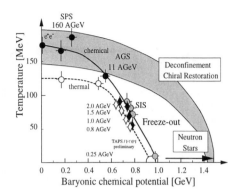

Figure 10: Phase diagram of temperature versus baryochemical potential with experimental points superimposed.

particle spectra within the framework of a hydrodynamicaly motivated 3-dimensional model of the expanding source fix the transverse expansion velocity $\beta_\perp$=0.55 and the thermal freeze-out temperature T_f=120 MeV. The overall expansion time is $\sim$ 8 fm/c, and the transverse source size at freeze-out is about 2.5 times its primordial size. The duration of pion emission is 3-4 fm/c and fades away at about 18 fm/c. A longitudinal velocity of $\sim 0.9c$ is estimated at the highest rapidity region of $Y_{\pi\pi}$.

The observed particle yields are well reproduced by a statistical model which assumes a hadron gas in full chemical equilibrium freezing out at the same temperature for all species, T=190 MeV after allowing for partial strangeness saturation. This hadronisation temperature is very close to the 180 MeV extracted within the same model for the elementary processes p+p, $e^+ + e^-$ and $p+\bar{p}$, indicating a universal hadronisation mechanism. A qualitative distinction of nucleus-nucleus collisions is an enhanced strangeness production quantified by increase of the strangeness suppression factor to $\lambda_s \sim 0.45$ as compared to $\lambda_s \sim 0.25$ for the elementary interactions.

In a phase diagram of temperature versus baryochemical potential we find that the observed values of the hadronic chemical freeze-out of central $^{208}Pb + Pb$ collisions lies in the region where models predict the transition from hadronic to chiral restored, deconfined partonic matter. Such a phase diagram is shown in Fig. 10 where data points from SIS, AGS and SPS experiments are combined. It illustrates the difference between chemical freeze-out of particle abundances and thermal freeze-out of momentum spectra. At lower SIS energies where the reaction zone expands rather slowly there is indication of common thermal and chemical freeze-out, while in the large and fast expanding systems at SPS chemical freeze-out occurs earlier and at higher temperature than

thermal freeze-out.

Results from the SPS heavy ion program reveal that very dense matter and high energy densities are produced leading to phenomena beyond extrapolation of elementary systems. The SPS chemical freeze-out point is located close to the region where the phase transition is expected to occur. It is therefore possible that a rapidly cooling system has indeed crossed the phase transition boundary evolving away from its hotter initial state. To be able to prove conclusively that the system has gone through a phase transition systematic studies on the dependence on the energy and mass of the system are required.

References

1. E. Laermann, Nucl. Phys. **A610** (1996) 1,
   E. Laermann, Proceedings of the Hirschegg Workshop on QCD Phase transitions, 1997
2. J. C. Collins and M. J. Perry, Phys. Rev. Lett. **34** (1975) 1353,
   E. V. Shuryak, Phys. Rep. **C61** (1980) 71 and **C115** (1984) 151
3. P. Jones et al. (NA49 Collaboration), Nucl. Phys. **A610** (1996) 188
4. T. Alber et al, NA49 collaboration, Phys. Rev. Lett. 75 (1995) 3814.
5. T. Wienold et al, NA49 collaboration, Nucl. Phys. A610 (1996) 76C.
6. R. Stock Nucl. Phys. **A630** (1998) 535
7. R. Hagedorn, Ref.TH.3684-CERN (1983).
8. J. Sollfrank et al., Z. Phys. **C52** (1991) 593, E. Schnedermann et al., Phys. Rev. **C48** (1993) 2462
9. B. Kämpfer, preprint FZR-149, hep-ph/9612336 preprint
10. U. Heinz, Nucl. Phys. **A610** (1996) 264.
11. U. Wiedemann et al., nucl-th/9611031 preprint.
12. J. Alam, J. Cleymans, K. Redlich, H. Satz, nucl-th/970742 preprint.
13. S. Pratt, Phys. Rev. **D33** (1986) 1314, A. N. Maklin and Y. M. Sinyukov, Z. Phys. **C39** (1988) 69; G. Bertsch, Nucl. Phys. **A498** (1989) 173 S. Chapman, J. R. Nix and U. Heinz, Phys. Rev. **C52** (1995) 2694, S. Pratt, Phys. Rev. **D33** (1986) 1314.
14. F. B. Yano and S. E. Koonin, Phys. Lett. B78 (1978) 556; M. I. Podgoretskii, Sov. J. Nucl. Phys. 37 (1983) 272.
15. H. Appelshauser et al. (NA49 Collaboration) Eur. Phys. J. C 2, 661-670 (1998).
16. GIBS Collaboration, Phys. Lett. B397 (1997) 30.
17. F. Becattini, M. Gaździcki, J. Sollfrank, hep-ph/9710529
18. M. Gaździcki, J. Phys. **G23** (1997) 1881

RELATIVISTIC MULTIPARTICLE PROCESSES IN THE CENTRAL RAPIDITY REGION AT ASYMPTOTICALLY HIGH ENERGIES IN NUCLEAR COLLISIONS

A. I. MALAKHOV

Joint Institute for Nuclear Research, 141980, Dubna, Moscow reg., Russia
E-mail: malakhov@lhe.jinr.ru

The principles of symmetry and self-similarity have been used to obtain an explicit analytical expression for inclusive cross sections of production of particles, nuclear fragments and antinuclei in relativistic nuclear collisions in the central rapidity region (y=0). The result is in agreement with available experimental data. It is shown that the effective number of nucleons participating in nuclear collisions decreases with increasing energy and the cross section tends to a constant value equal both for particles and antiparticles. The analysis of the obtained results makes it possible to conclude that the hopes for obtaining dense and hot matter in heavy ultrarelativistic nuclear collisions will not be realized.

In the present paper, using the approaches based on the application of the laws of symmetry and self-similarity and the correlation depletion principle in relativistic nuclear physics we answer the following questions:

1. What is the effective number of nucleons, that participate in nucleus-nucleus collisions, depending upon the measured parameters of inclusive particles?

2. What is the asymptotic behavior of the effective number of nucleons depending upon the collision energy?

The regularities in the behavior of the cross sections in the central rapidity region we have observed are tested using the available in literature experimental material and predictions for the collider energy region of nuclear collisions have been made. [1]

At the Joint Institute for Nuclear Research, starting with the first papers on cumulative effect [2] the notion of the minimal number of nucleons participating in the nucleus-nucleus collision

$$I + II \rightarrow 1 + 2 + \dots \qquad (1)$$

was introduced. There these quantities were denoted as N_I and N_{II}. N^{min} corresponds to the minimal number of nucleons which should take part in the transfer of the momentum to an observed fast particle according to the laws of conservation.

The cumulative effect is defined as the production of particles in the region

$$(u_1 u_{II}) > (u_I u_{II}) \gg 1,$$

when

$$N \geq N^{min} = \frac{m_1}{m_0} \cdot \frac{(u_1 u_{II})}{(u_I u_{II})} > 1, \qquad (2)$$

where

$$u_i = p_i / m_i = \left\{ \frac{E_i}{m_i}; \frac{\vec{p}_i}{m_i} \right\},$$

E_i is the energy, $\vec{p}_i$ and m_i are the three-dimensional momentum and the mass of an i-th particle, respectively, m_1 the inclusive particle mass, m_0 the nucleon mass.

Generalization of these ideas given in ref. [3] consists in the introduction of a self-similarity parameter

$$\Pi = min[\frac{1}{2} \sqrt{(u_I N_I + u_{II} N_{II})^2}]. \qquad (3)$$

The quantities N_I and N_{II} become measurable if we take into account the law of conservation of four-momentum in the form:

$$(N_I m_0 u_I + N_{II} m_0 u_{II} - m_1 u_1)^2 = (N_I m_0 + N_{II} m_0 + \Delta)^2 \qquad (4)$$

neglecting the relative motion of all the remaining not detected particles.

Δ is the mass of the particles providing conservation of the barion number, strangeness and other quantum numbers. For antinuclei and K^- mesons (the case of antimatter formation [3]) $\Delta = m_1$, for nuclear fragments $\Delta = -m_1$. For particles produced without accompanying antiparticles (π mesons, jets and others) $\Delta = 0$.

Eq. (4) can be written in the form

$$N_I N_{II} - \Phi_I N_I - \Phi_{II} N_{II} = \Phi_\delta, \qquad (5)$$

where we introduce relativistic invariant dimensionless quantities:

$$\Phi_I = \left[\frac{m_1}{m_0} (u_I u_1) + \frac{\Delta}{m_0} \right] / [(u_I u_{II}) - 1],$$

$$\Phi_{II} = \left[\frac{m_1}{m_0} (u_{II} u_1) + \frac{\Delta}{m_0} \right] / [(u_I u_{II}) - 1],$$

$$\Phi_\delta = (\Delta^2 - m_1^2) / [2 m_0^2 (u_I u_{II}) - 1].$$

We write Eq. (5) in the form

$$[(N_I / \Phi_{II}) - 1] \cdot [(N_{II} / \Phi_I) - 1] = 1 + [\Phi_\delta / (\Phi_I \Phi_{II})], \quad (6)$$

and find the self-similarity parameter Π as a solution of the equations

$$d(u_I N_I + u_{II} N_{II})^2/dF_I = 0,$$

$$d(u_I N_I + u_{II} N_{II})^2/dF_{II} = 0.$$

Here we introduce the variables

$$F_I = (N_I/\Phi_{II}) - 1,$$

$$F_{II} = (N_{II}/\Phi_I) - 1, \qquad (7)$$

$$F_I \cdot F_{II} = 1 + [\Phi_\delta/(\Phi_I \Phi_{II})].$$

These equations are symmetric with respect to the replacement of F_I by F_{II} and Φ_I by Φ_{II} and for $\Phi_I = \Phi_{II}$ they coincide. From here for the minimal value of Π we have the following solution valid for $(u_1 u_I) = (u_1 u_{II})$, i.e. in the central rapidity region:

$$F_I = F_{II} = F; \qquad F^2 = 1 + (\Phi_\delta/\Phi^2).$$

Making use of Eq. (7) we get $(N_I/\Phi) - 1 = (N_{II}/\Phi) - 1$ and consequently $N_I = N_{II} = N$. Thus, we obtain

$$N = N_I = N_{II} = (1 + F)\Phi = \left[1 + [1 + (\Phi_\delta/\Phi^2)]^{1/2}\right]\Phi,$$

$$\Pi = \frac{1}{2}\sqrt{2N^2 + 2N^2(u_I u_{II})} = \frac{N}{\sqrt{2}}\sqrt{1 + (u_I u_{II})}$$

$$= N \cosh Y, \qquad (8)$$

$$\Pi = N \cosh Y. \qquad (9)$$

Taking into account that

$$(u_I u_{II}) = \cosh(2Y),$$

$$(u_I u_1) = \frac{m_T}{m_1}\cosh(-Y - y) = \frac{m_T}{m_1}\cosh(Y + y),$$

$$(u_{II} u_1) = \frac{m_T}{m_1}\cosh(Y - y),$$

at $y = 0$ we get

$$(u_I u_1) = (u_{II} u_1) = \frac{m_T}{m_1}\cosh Y.$$

Here

$$m_T = \sqrt{m_1^2 + p_T^2}.$$

Then

$$\Phi = \frac{1}{m_0} \cdot [m_T \cosh Y + \Delta] \cdot (1/2\sinh^2 Y),$$

$$\Phi_\delta = (\Delta^2 - m_1^2)/(4m_0^2 \sinh^2 Y). \qquad (10)$$

From Eq. (8), by using Eq. (10) we get

$$N = [1 + \sqrt{(\Phi_\delta/\Phi^2) + 1}] \cdot [\frac{m_T}{m_0}\cosh Y + \frac{\Delta}{m_0}]$$

$$\times [1/(2\sinh^2 Y)]. \qquad (11)$$

Now we consider the asymptotic behavior of the self-similarity parameter with increasing interaction energy. Employing Eq. (10) we obtain at

$$s/(2m_I m_{II}) \approx (u_I u_{II}) = \cosh 2Y \to \infty,$$

$$\Phi_\delta/\Phi^2 = \frac{\Delta^2 - m_1^2}{m_T^2} \cdot \frac{\sinh^2 Y}{[\cosh Y + (\Delta/m_T)]^2} \to \frac{\Delta^2 - m_1^2}{m_T^2},$$

$$\Phi \cosh Y = \left(\frac{m_T}{m_0}\cosh Y + \frac{\Delta}{m_0}\right) \cdot \frac{\cosh Y}{2\sinh^2 Y} \to \frac{m_T}{2m_0}.$$

Hence it follows that at $\cosh Y \to \infty$ (in the collider energy region) the self-similarity parameter Π assumes the finite value

$$\Pi_\infty = \frac{m_T}{2m_0}\left[1 + \sqrt{1 + (\Delta^2 - m_1^2)/m_T^2}\right]. \qquad (12)$$

As is seen from Eq. (11), the effective number of the nucleons involved in the reaction $N \to 0$ at $\cosh Y \to \infty$. In this connection, we may say with certainty that the hopes for obtaining dense and hot matter (in any case, for detecting it by fast inclusive particles) in ultrarelativistic nuclear collisions are not feasible. Our conclusion is based on earlier carried out tests of the dependence of the invariant differential cross section upon the self-similarity parameter Π:

$$\frac{d^2\sigma}{m_T dm_T dy} = 2\pi C_1 A_I^{1/3 + N_I/3} A_{II}^{1/3 + N_{II}/3}$$

$$\times \exp[-\Pi/C_2], \qquad (13)$$

where $C_1 = 1.9 \cdot 10^4 \; mbGeV^{-2}C^3 st^{-1}$, $C_2 = 0.125 \pm 0.002$ in a wide range of variables $y, m_T, \cosh Y$ and for different inclusive particles [4].

An additional possibility of testing Eq. (13) appeared as a result of the CERN SPS experiments on Pb + Pb collisions at an energy of 160 A·GeV which corresponds to $\cosh Y = 8.971$. It is especially interesting to estimate the effect of the transition to heavy nuclei on the number of the nucleons in the initial state participating in production of inclusive particles.

Predictions of the cross sections for antimatter production in relativistic nuclear collisions and, in particular, quantitative predictions of the cross sections for the reaction

$$Pb + Pb \to \bar{p} + \dots \qquad \text{and} \qquad Pb + Pb \to \bar{d} + \dots$$

at the energy of 160 A·GeV are given in ref. [3] over the whole rapidity region. However a computor method for solution of the equations

$$\frac{d\Pi}{dN_I} = 0 \quad \text{and} \quad \frac{d\Pi}{dN_{II}} = 0$$

suggested there makes it possible to obtain the results in the form of numerical tables and diagrams. The analytical representation for Π enables us to draw the following new conclusions:

1. There exists the limiting value of Π described by Eq. (12).

2. For $\Phi_\delta = 0$ the expression for Π is factorized and proportionality of m_1 to the inclusive particle mass makes it possible to test in details the self-similarity laws. From Eq. (11) and Eq. (9) it follows that the cross section Eq. (13) exponentially quickly decreases with increasing m_1. In particular, this implies that the probability of observing even light antinuclei and fragments in the region $y = 0$ is insignificantly small.

3. A yield of strange particles in the central region (for $y = 0$) increases with increasing collision energy $\cosh 2y$ Figure 1.

4. The effective number of nucleons involved in the reaction decreases with increasing $\cosh Y$ (Eq. (11)) Figure 2.

5. A strong factorizable dependence of Π on $m_T = \sqrt{m_1^2 + p_T^2}$ we have discovered explains the observed m_T scaling.

At $p_T = m_T/m_0 \geq 5$ (see Eq. (9) and Eq. (11)) the cross section becomes very small and the transition form the exponential dependence to the power one (hard scattering) should be observed.

We give a number of estimates of the validity of the above-mentioned equations on the basis of the available in literature experimental data. We consider the ratio of the yield of antideuterons to that of deuterons at an energy of 160 GeV per nucleon in the collisions of Pb nuclei which have been realized at the SPS.

$$(u_I u_{II}) = \cosh 2Y = E_I/m_I \approx 160 \ GeV/m_0 \approx 160,$$

$$2Y = 5.7683, \qquad Y = 2.884,$$

$$\cosh Y = 8.971, \qquad \sinh Y = 8.915.$$

For d and anti-d (and, generally, nucleus-antinucleus) $\Delta^2 - m_1^2 = 0$. Let us consider the case $m_T = m_1$ ($p_T = 0$)

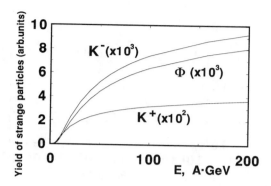

Figure 1: Yield of strange particles in the central region (for y=0) as a function of the collision energy.

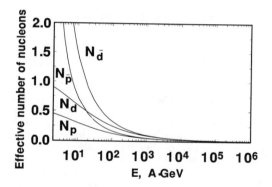

Figure 2: The effective number of nucleons involved in the nuclear interaction as a function of the collision energy.

corresponding to the upper estimate of the cross section Eq. (13). Then for the deuteron we have

$$\Pi_d = [\frac{m_1}{m_0} \cosh Y - \frac{m_1}{m_0}] \cdot \frac{\cosh Y}{\sinh^2 Y} = 1.799.$$

and for the antideuteron

$$\Pi_{\bar{d}} = [\frac{m_1}{m_0} \cosh Y + \frac{m_1}{m_0}] \cdot \frac{\cosh Y}{\sinh^2 Y} = 2.251.$$

By inserting these numbers in Eq. (13) we find the ratio of the production cross sections for antideuterons and deuterons

$$\bar{d}/d = \exp[-(\Pi_{\bar{d}} - \Pi_d)/C_2] = 0.027.$$

In the asymptotical region $\Pi_{\bar{d}} = \Pi_d$ (according to Eq. (12)) and the ratio of the production cross section

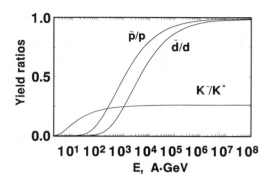

Figure 3: Predictions of the production cross section ratios for antiparticles to particles versus collision energy.

for deuterons to that for antideuterons is close to unity. Similar estimates for K^-/K^+ and $\bar{p}/p$ give:

$$K^-/K^+ \approx 0.25, \qquad \bar{p}/p \approx 0.16$$

Preliminary data of the NA52 and NA44 experiments on measurements of the yields of antiparticles produced as a result of the interactions of Pb nuclei of an energy of 160 GeV per nucleon with a Pb target at the CERN SPS [5,6] and of Au+Au interactions of an energy of 11 GeV per nucleon in E866 experiment at the BNL [7] have recently been obtained. In these papers, in particular, the rapidity distributions for the ratios of the $\bar{p}/p$, K^-/K^+ and $\bar{d}/d$ yields have been presented. The ratios for the particle and antiparticle yields in the central rapidity region $y = 0$ measured in these experiments are in agreement with our abovementioned upper estimates or are close to the latter in magnitude. For the sake of comparison the results of our estimates and the experimental values are tabulated:

Table 1: Comparison of results of calculations and experimental data.

Ratios of the yields	$\bar{p}/p$	$\bar{d}/d$	K^-/K^+
Calculation (160 A · GeV)	0.16	0.027	0.25
NA52 (160 A · GeV)	≈ 0.1	≈ 0.01	≈ 0.2
NA44 (160 A · GeV)	≈ 0.08	–	≈ 0.4
Calculation (11 A · GeV)	0.00039	–	0.11
E866 (11 A · GeV)	≈ 0.0003	–	≈ 0.2

Our predictions of the ratios of the production cross section for antiparticles to particles are presented in Figure 3.

Acknowledgements

I express my sincere gratitude to Prof. A.M.Baldin for stimulating of this work and I.S.Baldina and I.I.Migulina for help in preparatin of manuscript.

The work is supported by the Department of the Physics of Relativistic Multiparticle Systems of the Lebedev Physical Institute and a grant of the Russian Fund of Fundamental Researches № 96-02-18728.

References

1. A.M.Baldin and A.I.Malakhov, *JINR Rapid Communications* 1[87]-98, 5 (1998).
2. A.M.Baldin, *Phys.Part.Nucl.* 8(3), 429 (1977).
3. A.A.Baldin, *JINR Rapid Communications* 4[78]-96, 61 (1996).
4. A.M.Baldin and A.A.Baldin, *Phys.Part.Nucl.* (to be published), *Preprint JINR* P2-97-309, (1997).
5. G.Ambrosini *et al.*, *Nucl.Phys.* A 610, 306c (1996).
6. I.G.Bearden *et al.*, *Nucl.Phys.* A 610, 175c (1996).
7. L.Ahle *et al.*, *Nucl.Phys.* A 610, 139c (1996).

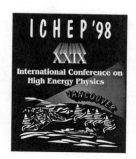

Parallel Session 14

Experimental Techniques

Convenors: Horst Oberlack (MPI Munich, ATLAS)

Robert Brown (Rutherford CMS)

THE COMPASS EXPERIMENT

The COMPASS collaboration, presented by L. Schmitt

Physik-Department, E18, Technische Universität München, D-85748 Garching, Germany
E-mail: Lars.Schmitt@ph.tum.de

COMPASS is a new experiment under construction at the CERN SPS aimed at the study of structure and spectroscopy of hadrons with multiple types of high intensity beams.
A polarised muon beam is used for deep inelastic scattering off polarised proton and deuteron targets to study the spin structure of nucleons. In particular open charm production is used to measure the gluon contribution to the nucleon spin. Other items are the study of Λ^0 polarisation and longitudinal and transverse quark polarisation. The physics program connected to the hadron beams includes Primakoff scattering of pions and kaons to study their polarisabilities, central production of hybrids and glueballs in a proton beam on a liquid hydrogen target and production and leptonic and semi-leptonic decays of charmed hadrons.
The COMPASS experiment uses a double forward spectrometer for best momentum resolution. Both spectrometer parts are equipped with RICH detectors, electromagnetic and hadronic calorimeters and muon filters for particle identification. Highlights of the COMPASS apparatus are presented.

1 The Physics of COMPASS

The COMPASS experiment covers a wide range of topics in hadron physics: From polarisabilities of hadrons studied in Primakoff scattering to centrally produced glueballs and exotic mesons, from charm production and semi-leptonic and leptonic charm decays in hadron hadron collisions to open charm production in deep inelastic muon scattering important questions of perturbative and non-perturbative QCD are addressed by our experiment [1].

1.1 Physics with the Muon Beam

The experimental program using a muon beam comprises the study of nucleon structure and this continues the investigations undertaken at SLAC, DESY and CERN. The CERN experiment EMC [2] observed, that the contribution of valence quarks to the nucleon spin is smaller than expected. This result has been confirmed by other experiments at SLAC, DESY and CERN. It is therefore necessary to measure all individual parton spin-distribution functions to fully understand the spin-dependent structure functions of protons and neutrons.

A main focus of COMPASS is to measure the contributions of gluons to the nucleon spin. This can be achieved by means of the photon-gluon-fusion process (see fig.1). If the spin of the nucleon translates into a polarisation of the interacting gluon it may be detected using a polarised muon as a probe exploiting the spin dependence of the underlying interaction. The change of target polarisation allows the measurement of an asymmetry that is related to the gluon spin.

A rather clean signature for photon-gluon-fusion is the production of a $c\bar{c}$-pair. The experiment is therefore

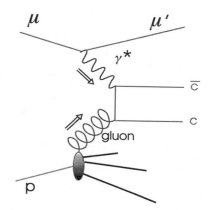

Figure 1: Photon-gluon-fusion to measure the gluon polarisation $\Delta G/G$ in the nucleon

designed to efficiently reconstruct D^0 mesons, the most abundantly produced charm hadrons.

As additional channel a pair of charged light hadrons at high p_T can be a signature for a photon-gluon-fusion process enhancing the sample after further kinematic cuts. COMPASS expects to measure $\Delta G/G$ to approximately $\pm10\%$ with charm mesons and to around $\pm5\%$ with light hadron pairs.

In addition strange sea-quarks may contribute to the nucleon spin. This is studied by reconstructing the polarisation of Λ and $\overline{\Lambda}$ produced in deep inelastic scattering of polarised muons.

For the first time the transverse spin distribution functions will be measured. For systematic studies the longitudinal spin distributions will be measured as well.

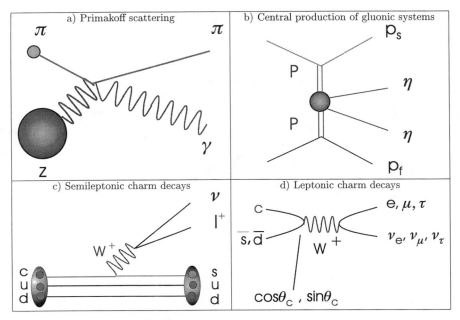

Figure 2: Processes relevant to the physics programs with hadron beams

1.2 Physics with Hadron Beams

The very forward region in the scattering of hadrons off heavy targets is dominated by the Compton process in the Coulomb field of the target nucleus (fig. 2a). This process, Primakoff scattering, gives access to the electric and magnetic polarisabilities of the incoming hadrons predicted by chiral perturbation theory. A meson for example is supposed to have an induced dipole moment associated to its quark structure determined by the strong interaction between the quarks. Existing measurements for the pion can be improved substantially and the kaon polarisabilities can be measured for the first time.

Other predictions of chiral perturbation theory like the amplitude of $\gamma \to 3\pi$ can be tested with Primakoff scattering as well.

In addition the Primakoff process allows the production of vector mesons and possibly exotic non-$q\bar{q}$ states.

Central collisions of a proton beam on a proton target provide a gluon-rich environment in the double pomeron exchange mechanism (fig. 2b). This is assumed to efficiently produce glueballs and exotic non-$q\bar{q}$ mesons predicted by QCD. A candidate for the scalar glueball ground state 0^{++} is the f_0 at 1500 MeV/c observed by the Crystal Barrel experiment at LEAR[3].

COMPASS will detect many decay modes of this state and in addition access the mass region above 2GeV where the first excited state, the tensor glueball 2^{++} is predicted around 2.3GeV by lattice QCD.

Furthermore it is assumed that exotic non $q\bar{q}$-states, especially hybrid mesons consisting of a $q\bar{q}$-pair and a gluon can be produced in central collisions.

The results obtained with the central production process can be cross-checked with the Coulomb excitation process in the same experiment using a high Z target. Glueball candidates should be absent in the final states produced by this process.

Finally charm hadron decays allow the study of the influence of strong interaction and spectator effects on the weak decay of hadrons at a scale where QCD predictions are already possible (unlike in the light quark sector) but are still sizable (unlike in the beauty sector).

Semileptonic decays of charm hadrons (fig. 2c) probe predictions of Heavy Quark Effective Theory. The decay constants of D and D_s can be measured in purely leptonic decays (fig. 2d) of these mesons, a statistical error of about $\pm 20\%$ and $\pm 10\%$ respectively can be reached by COMPASS.

A further goal of the charm program is to look for doubly charmed baryons never observed so far.

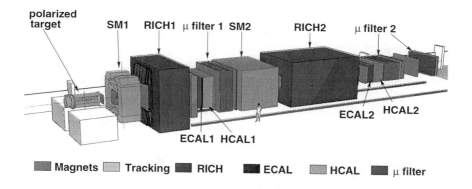

polarized target | SM1 | RICH1 | μ filter 1 | SM2 | RICH2 | μ filter 2

ECAL2 HCAL2

ECAL1 HCAL1

■ Magnets ■ Tracking ■ RICH ■ ECAL ■ HCAL ■ μ filter

Figure 3: The experimental setup of COMPASS

2 The Experimental Setup

The COMPASS apparatus is a two stage magnetic spectrometer to obtain a good momentum resolution from $1 - 2$ GeV up to 150GeV. Both stages will be equipped with RICH detectors for identification of charged particles, electromagnetic and hadronic calorimeters and muon filters. Tracking detectors will be multi-staged using silicon microstrips in the beam area and gaseous strip detectors with larger acceptance immediately around. Behind the first magnet straw tubes provide the large area tracking in lower intensity regions. Further downstream standard MWPC can be employed. Scintillating fibre detectors are used as fast beam detectors for timing and triggering .

The experiment will use muon beams of 100 and 200 GeV at a rate of 10^8 particles per second for the deep inelastic scattering program studying the nucleon spin structure. For Primakoff scattering 280GeV pion and Kaon beams are used at 5×10^6 particles per second, for central production 5×10^7 protons per second and for the charm program 5×10^7 pions per second.

The apparatus of COMPASS will have 250,000 channels which are read out by a fast DAQ with minimum deadtime. At a trigger rate of $10 - 100$kHz and an event size around 30kB a data rate of 600MB $-$ 6GB per SPS spill is reached leading to 300TB of data per year on tape.

3 Targets used in COMPASS

COMPASS will employ different targets according to the various physics programs.

3.1 The Polarised Target

For the study of the spin structure of the nucleons a polarised beam and a polarised target are needed. The muon beam has a natural polarisation of about 80% due to the decay kinematics of the initial pions.

The target consists of a dilution refrigerator with a microwave cavity included in a superconducting solenoid magnet. As material ammonia is used as proton-rich target and ^{6}LiD as deuteron-rich target to extract the neutron structure functions.

This material is polarised by dynamic nuclear polarisation in the magnetic field by means of microwaves. After a polarisation of about 80% for NH_3 or 50% for ^{6}LiD is achieved at 0.5K, it is frozen by cooling to 50mK to be maintained for several hundred hours.

3.2 The Charm Target

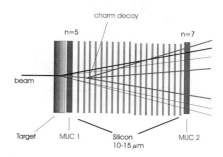

Figure 4: Target for charm production with fine pitch silicon microstrip detectors

The charm target consists of a thin copper plate closely followed by trigger counters and an array of silicon microstrip detectors. These detectors have an extremely fine pitch of around $12\mu m$ and are narrowly spaced at a distance of only 2mm along the beam axis. This setup allows to fully reconstruct a charm decay by measuring also a track piece of the charm hadron itself. For leptonic and semi-leptonic decays, where the neutrino carries away momentum, only the additional information from the charm track allows the kinematical reconstruction of the decay.

3.3 The Target for Central Production

For the central production of mesons and gluon-rich states a liquid hydrogen target is used along with a proton beam at about 300GeV. This target is surrounded by two concentric cylindrical barrels composed of scintillator slabs to detect the slow recoil proton emitted in the studied process. Further trigger counters detect the fast forward proton deflected at a small angle w.r.t. the beam.

4 Detectors of COMPASS

4.1 Tracking Detectors

The tracking detectors of COMPASS will be subjected to very high particle rates posing stringent requirements on readout speed, rate resistance and radiation hardness. For this reason in the upstream parts of the experiment hybrid tracking stations are being developed consisting of very radiation hard inner detectors surrounded by larger detectors with bigger pitch and lower occupancy.

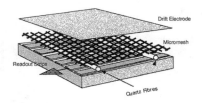

Figure 5: Schematic view of the Micromesh Gaseous detector

The stations between the target and the first magnet consist of double-sided silicon microstrip detectors operated at liquid nitrogen temperature to reduce the effect of radiation damage [4]. These detectors of about $6.3 \times 6.3 cm^2$ are mounted in front of micromesh gaseous detectors ("Micromegas") of $40 \times 40 cm^2$ arranged in a 2×2 mosaic to cover $80 \times 80 cm^2$. This new type of detector implements a fine copper mesh with a 25 μm grid $50 - 100\mu m$ above an anode plane with strip structure to achieve an additional gas amplification in a high electric field. This detector promises high rate

capacity at good resolution and is developed at Saclay [5].

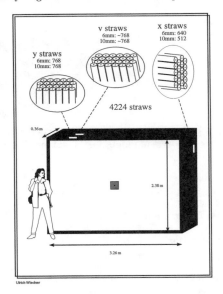

Figure 6: A COMPASS straw chamber

The tracking after the first magnet is mainly performed by straw tubes. They are packed in chambers of $3.2 \times 2.4 m^2$ size and contain straws with 6mm diameter in the inner region and 10mm in the outer region which are arranged in three projections. First testbeam measurements gave promising results concerning resolution and performance.

The straw chambers are augmented by GEM chambers [6] (Gas Electron Multiplier) which cover the inner high rate part of the acceptance. These new detectors employ one or two copper coated foils with $70\mu m$ large holes spaced at $140\mu m$ as amplification stage about 1mm above an anode plane with strip structure. As the Micromegas these detectors provide better rate resistance than ordinary wire chambers.

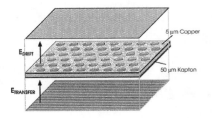

Figure 7: Principle of a GEM detector

4.2 The RICH

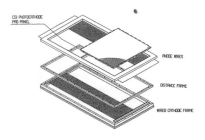

Figure 8: The RICH photon detector with CsI cathode

Identification of charged particles over a large momentum range is best achieved using a ring-imaging Cherenkov detector. COMPASS employs a RICH detector with a 3m long C_4F_{10}-radiator for good $\pi-K$ separation in the momentum range from $3-65\text{GeV}/c$. Wire chambers with CsI-coated cathodes are used as photon detectors. The cathode is segmented in pads of 8mm size giving a two-dimensional information. The 69000 channels are read out by GASSIPLEX preamplifier/multiplexer chips.

4.3 The Calorimeters

Very good photon detection is required in particular by the hadron program. Therefore two electromagnetic calorimeters are constructed. As building blocks of the first calorimeter mainly lead glass modules of the GAMS calorimeter previously used in the WA102 experiment are employed. The cells are arranged in a $4 \times 2.9\text{m}^2$ array with a central hole of $61 \times 107\text{cm}^2$.

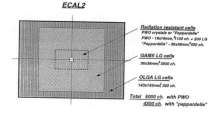

Figure 9: Setup of the second electromagnetic calorimeter

The second electromagnetic calorimeter of $3.7 \times 1.85\text{m}^2$ uses existing lead glass blocks in the outer region whereas the inner region will be constructed of radiation hard material. For this part either PWO crystals or a new type of lead/scintillator sandwich are considered.

The two hadronic calorimeters are mostly used for triggering purposes to indicate production of a minimum amount of high energy hadrons in the studied reactions. The first calorimeter of $4 \times 3\text{m}^2$ size is composed of interleaved lead and scintillator plates, the second one of $4 \times 2.2\text{m}^2$ is made of steel and scintillator plates with WLS fibre readout.

For all calorimeters a new common readout electronics is developed to match the requirements of the high trigger rate and the fast DAQ.

5 Conclusions

COMPASS is a challenging experiment addressing a rich program of physics. It is currently under construction at the EHN2 beamline of the SPS at CERN, with a planned commissioning in the year 2000. Detector prototypes are presently being tested and the development of readout electronics and detectors is in its decisive stage.

The experiment is approved to run until the year 2004, but the richness of the physics program suggests a further prolongation well into the LHC era.

Acknowledgements

We acknowledge the support by the Bundesministerium für Bildung und Forschung, Germany, under contract 06 TM 896 I.

References

1. The COMPASS Collaboration, *COMPASS, A Proposal for a COmmon Muon and Proton Apparatus for Structure and Spectroscopy*, CERN/SPSLC 96-14, SPSLC/P297, 1996
2. The European Muon Collaboration, J. Ashman et al., *Phys. Lett.* B206, 364 (1988); *Nucl. Phys.* B328, 1 (1989)
3. The Crystal Barrel Collaboration, C. Amsler et al., *Phys. Lett.* B342, 433 (1995)
4. V. Palmieri et al., *Evidence for charge collection efficiency recovery in heavily irradiated silicon detectors operated at cryogenic temperatures*, BUHE-98-06, subm. to *Nucl. Instrum. Methods*
5. Y. Giomataris et al., Nucl. Instrum. Methods A376, 29 (1996); Y. Giomataris, *Development of fast gaseous detector "Micromegas"*, DAPNIA-98-01, presentation at WCC'98, Vienna, subm. to Nucl. Instrum. Methods
6. F. Sauli et al., *Nucl. Instrum. Methods* A386, 531 (1997); F. Sauli, *Gas detectors: Recent developments and future perspectives*, CERN-EP/98-51, invited opening presentation at WCC'98, Vienna, subm. to *Nucl. Instrum. Methods*

STATUS OF THE BABAR DETECTOR

F. C. PORTER

Physics Department, Caltech, Pasadena, CA 91125, USA
E-mail: fcp@hep.caltech.edu

FOR THE BABAR COLLABORATION

The BaBar experiment, under construction at the PEP-II e^+e^- colliding beam storage ring, is motivated by the desire to study rare B decays, and especially CP violation in the B system. The requirements for, design, and status of the construction of the various detector systems are summarized. Much of the construction is complete, and BaBar has entered the installation and check-out phase. Problems with potential schedule impact have been agressively addressed, and it is planned to complete construction and roll onto the beam line in time for the start of physics collisions in April 1999, as originally scheduled.

1 Introduction

The B meson is especially interesting because of its long lifetime, which means that standard model processes are sufficiently suppressed that mecahnisms beyond the standard model may be observable. An area of particular importance is the study of CP violation via the large observed mixing in the B^0 system. To study CP violation via mixing in $B^0\bar{B}^0$ produced at the $\Upsilon(4S)$ in e^+e^- collisions, a center-of-mass which is moving in the laboratory is required, in order that the time evolution of the $B^0 - \bar{B}^0$ decays can be measured. With this approach, an error on, for example, $\sin 2\beta$ of order 0.12 in the $J/\psi(\ell^+\ell^-)K_S(\pi^+\pi^-)$ mode, and 0.076 averaged over several modes, is expected for a 30 fb^{-1} dataset. The PEP-II accelerator [1] at SLAC is a new asymmetric storage ring collider for this purpose, and the BaBar detector [2] is under construction to take data on collisions at PEP-II.

The BaBar collaboration consists of approximately 600 physicists and engineers from 72 institutions in 9 countries (Canada, China, France, Germany, Italy, Norway, Russia, United Kingdom, and USA). Despite the wide geographical distribution, this has functioned remarkably well, with different countries contributing major portions of the detector.

2 PEP-II

The PEP-II accelerator is designed to produce e^+e^- collisions at center of mass energies in the Υ region, with a moving center of mass, $\beta\gamma = 0.56$ at the $\Upsilon(4S)$. The design luminosity is $\mathcal{L} = 3\times10^{33}$ cm^{-2}s^{-1}. The storage ring consists of two rings, a "high energy ring" (HER), carrying electrons at approximately 9 GeV, located below a "low energy ring" (LER), carrying positrons at about 3.1 GeV. Collisions are head-on, with trajectories separated magnetically as the beams emerge from the interaction point (IP). The bunch collision frequency is 238 MHz.

Table 1: PEP-II Parameters.

Parameter	HER	LER	Units
Energy	9	3.1	GeV
I	0.75	2.14	A
# bunches	1658	1658	
σ_x^*	155	155	μm
σ_y^*	4.7	4.7	μm
σ_z^*	1.0	1.0	cm
$\sigma_{E_{CM}}$	5.2		MeV
$\mathcal{L}_{\text{peak}}$	3×10^{33}		cm^{-2}s^{-1}

Table 1 summarizes some of the interesting accelerator parameters.

PEP-II construction is essentially complete, and the accelerator is being commissioned. The HER commissioning began in May 1997. It has performed well, and has met already several of the key design goals, such as bunch current, lifetime, and number of bunches. The LER commissioning began in July 1998, and first collisions were observed, in beam-beam disruption, and in tune crosstalk, also in July 1998. Further commissioning will take place in Fall 1998, prior to disassembly of the interaction region in January 1999 for detector roll-on.

3 Detector Requirements

- Vertex resolution: Good vertex resolution is crucial for the CP-violation studies, because of the requirement to measure the decay time difference between the two B^0 decays. The scale of the requirement is set by the B average flight path of $\sim 260\mu$m at the PEP-II asymmetry. BaBar is designed to have an impact parameter resolution of

$$\sigma_z = \sigma_{xy} = [50/p_t(\text{GeV}) \oplus 15] \ \mu\text{m},$$

where the $\oplus$ denotes quadratic combination.

- Tracking resolution: Good tracking resolution is important for reconstruction of the final decay states, in order to suppress backgrounds. BaBar is designed to have a track momentum resolution of

$$\sigma_{p_t}/p_t = [0.21 + 0.14 p_t (\mathrm{GeV})] \, \%,$$

for $p_t > 200$ MeV, in the drift chamber.

- Photon detection and measurement: A substantial fraction of the decays of the particles of interest contain π^0's or other neutral particles. Thus, photon measurement is important in the reconstruction of many exclusive states. The BaBar calorimeter is designed to have a photon energy resolution of

$$\sigma_E/E = [1/E(\mathrm{GeV})^{\frac{1}{4}} \oplus 1.2] \, \%,$$

and an angular resolution of

$$\sigma_\theta = [3/\sqrt{E(\mathrm{GeV})} \oplus 2] \, \mathrm{mr}.$$

- Particle identification: A key part of the anticipated CP violation measurements involves the tagging of the B^0 flavor (i.e., B or $\bar{B}$), using both leptons and kaons. Pion-kaon separation is also important in distinguishing exclusive modes such as $B \to \pi\pi$ from $B \to K\pi$, for which kinematic separation is marginal. Particle identification up to about 4 GeV momentum is required, and BaBar is designed to achieve this, using dE/dx, Cerenkov, and calorimetric measurements.

- Hermeticity: Large solid angle coverage is of course important for efficiently reconstructing multi-particle final states. It is also important in studies which require measuring the "missing" momentum, e.g., reconstructing the neutrino in $B \to \ell\nu$ decays. The instrumented flux return (Ifr) serves the dual roles of muon identifier and hadron calorimeter, in particular for K_L^0 and other neutral hadrons.

- The detector and computing must be able to handle a trigger rate to tape of order 100 Hz, of which approximately 10 Hz is single-photon hadronic tirggers. The event reconstruction should be nearly real time, to provide rapid data quality feedback, and timely physics analysis. The simulation tools must contain sufficient detail and quality to model efficiencies and backgrounds in large datasets.

4 Status of Detector Systems

The overall BaBar detector is of a by-now standard large solid angle solenoidal design. The various elements are

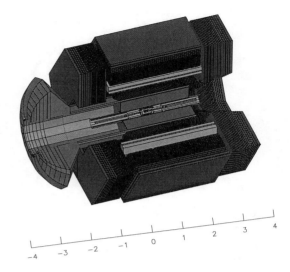

Figure 1: Cut-away view of the BaBar detector, as seen by the simulation program. The scale is in meters. At smallest radii are the PEP-II beam components and silicon vertex tracker, followed by the drift chamber and the DIRC quartz bars. Outside the DIRC is the CsI calorimeter, followed by cylindrical RPCs, the solenoid coil, and the instrumented flux return. The large object at the left is the water stand-off box for the DIRC.

illustrated in Figure 1. Note that the detector is somewhat asymmetric, to match the asymmetry in the beam energies.

4.1 Silicon Vertex Tracker

The silicon vertex tracker (Svt) consists of five layers of double-sided silicon strip tracking, fabricated on $300 \, \mu$m silicon, with strip readout pitch ranging from 50 to 210 microns. The radii of the layers range from 3.3 cm to 14.4 cm, and cover polar angles between 17.2 and 150 degrees. The provision of five layers permits this to be used as an independent tracker, besides providing for precision vertex measurement. This helps, for example, in the detection of the slow pion from $D^* \to D$ decays. The digitization uses a time-over-threshold custom integrated circuit, the "AToM" chip (for "A Time-over-threshold Machine"),[3] with 128 channels, sparsification, and serial readout.

The support structure for the Svt, a low-mass carbon fiber structure which mounts on the permanent magnet bends near the IP, is complete. The silicon strip wafers are all in hand, and have been tested to be of good quality in terms of pinholes and leakage. Wirebonding (570k

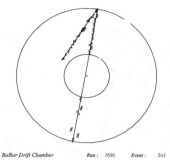

BaBar Drift Chamber   Run : 1690   Event : 503

Figure 2: View of a cosmic ray from commissioning of the BaBar drift chamber. Approximately 50% of the chamber is instrumented; the gaps in the lower track and the termination in the track away from the IP are due to uninstrumented regions. The stereo hits are plotted at wire center, hence deviations from the fitted tracks indicate z position. The tracks are drawn as reconstructed by the reconstruction software – the track through the IP is actually reconstructed as two tracks.

bonds) is in production. Single module tests have been performed in a CERN test beam, and in conjunction with the PEP-II commissioning. There have been some difficulties with the front end readout electronics (AToM IC and high density interconnect hybrid) which have taken up available float in the schedule, although the Svt installation is still scheduled for January 1999.

4.2 Drift Chamber

The drift chamber (Dch) [4] is a 40-layer hexagonal-cell device with cell diameters of approximately 1.5 cm, 276 cm in length, and occupying radii between 23.6 and 81 cm. The layers are arranged in axial-U-V $(40 - 70 \text{ mrad})$ superlayers of 4 layers each. A low-Z gas consisting of 80% helium and 20% isobutane is used. The readout of the 7104 sense wires provides both time and amplitude, giving dE/dx information for particle identification. The $B^0 \to \pi^+\pi^-$ mass resolution is expected to be 20-23 MeV over 80% of 4π. A capacitive system is used to monitor the relative positions of the Svt and the final bend magnets, and of the Dch and the support tube containing the Svt and bends.

Tests of a full-length prototype give resolution in agreement with design, with better than $100\,\mu$m at the minimum, $130\,\mu$m averaged over a cell, and dE/dx resolution extrapolating to 6.8% for the full 40-layer Dch. The chamber stringing is complete, and it is currently being commissioned with cosmic rays. Figure 2 shows a cosmic ray track observed in the drift chamber, prior to installation in the solenoid scheduled for late August. The gas system is under construction at the interaction region.

4.3 Cerenkov Particle Identification System

The particle identification system is a novel Cerenkov ring imaging device called a "DIRC", for "Direct Imaging Ring Cerenkov". Quartz bars serve as the Cerenkov radiator, and also as light guides which carry the ring image out the ends of the bars via total internal reflection. At the end of the quartz bar, the light exits into a large water volume, providing a lever arm of about 1.2 meter before the ring is detected by a dense array of 10,752 photomultipliers. Each of the 144 quartz bars is 4.9 m long, constructed by gluing 4 shorter bars together. The radial thickness of a bar is 1.7 cm. The DIRC system presents about 14% of a radiation length (at 90°) to particles before they reach the calorimeter, following slightly over 5% from the components at smaller radii, averaged over azimuth.

The water volume ("Stand-Off Box", or SOB) and photomultiplier array are complete. A substantial prototype test [5] was made in a CERN test beam, and the pattern recognition software is well-advanced. The quartz bars are in production, although the vendor has been having difficulties leading to delay. The installation schedule is being kept flexible in order to accomodate the delays without incurring a delay in the final detector readiness. The most critical aspect is that a third of the bars must be installed before the detector rolls onto the beam-line.

4.4 Electromagnetic Calorimeter

As good photon energy resolution is important, the electromagnetic calorimeter (Emc) consists of an array of CsI(Tl) crystals, 5760 in the barrel arranged in 48 z slices of 120 crystals, and 820 in the forward calorimeter arranged in 8 rings of 80-120 crystals. Each crystal is observed by two photodiodes, with two detector-mounted amplifier ICs, for redundancy. Crystals are supported in carbon fiber compartments, and mounted from the rear, in the barrel, to a cylindrical aluminum strongback in groups of 3×7 crystals. The crystals range from approximately 16 to 17.5 X_0, with the longer lengths concentrated in the forward region in the lab frame. A relatively high energy radioactive calibration source is generated via neutron activation of a flourine-based liquid (3M Fluorinert FC-77), eventually yielding a 6.13 MeV line in $^{16}O^*$ decays. When a calibration is desired, the activated fluid is pumped through panels on the inner surface of the calorimeter. The design is for a statistical uncertainty in a crystal of 0.25%, in a 15 minute calibration.

A 25 crystal prototype calorimeter array was tested in a beam at PSI.[6] The barrel calorimeter is complete, and was installed in the solenoid on July 10, 1998. All channels have been determined to be functional via pulser and ^{228}Th (2.6 MeV γ) tests. The alignment of the barrel calorimeter has been completed. The source calibration panels are installed in the barrel. All but 25 of the forward calorimeter crystals are in hand, with the final delivery due mid-August.

4.5 Solenoid Magnet

A 1.5 Tesla superconducting solenoid provides the magnetic field for momentum analysis in the tracking volume. There is a bucking coil located where the DIRC bars exit the flux return, in order to reduce the field at the DIRC photomultipliers to an acceptable level.

The magnet is complete and installed in the flux return at the interaction hall. The net magnetic axial force has been determined to be acceptable and in approximate agreement with modeling. Magnetic mapping has been performed at both 1 and 1.5 T. The DIRC bucking coil is found to perform its design function, and the field strength at the final focus quadrupoles is acceptable.

4.6 Instrumented Flux Return

The instrumented flux return [7] consists of 18-19 layers of resistive plate chambers (RPCs) interspersed with the iron of the flux return for the magnet. The thickness of the iron plates increases from 2 cm to 10 cm with distance from the IP. The total iron thickness is 65 cm in the barrel, and 60 cm in the endcaps. The RPCs consist of a 2 mm gap between 2 mm thick Bakelite plates, on the outer surface of which high resistivity graphite layers are painted. The graphite is followed by insulating PVC film, then aluminum strips to pick up induced signals for readout. The minimum muon momentum required to be detected in the Ifr is 500 MeV. In addition, there is a 2-layer (each with two views) cylindrical RPC between the Emc and the solenoid. There are approximately 50,000 strips in the Ifr.

The iron for the flux return is complete and installed. The RPCs are complete, and nearly all installed in the detector.

4.7 Electronics and Data Acquisition

In addition to the Svt chips, there are 6 custom integrated circuits in BaBar: *(i)* Dch amplifier/shaper/discriminator; *(ii)* Dch ELEFANT (for "ELEctronics For Amplitude 'N Timing") time/amplitude digitizer with 1 ns TDC binning and 15 MHz FADC sampling (6-bit resolution with an 8-bit range); *(iii)* DIRC amplifier/shaper/discriminator; *(iv)* DIRC time-to-digital and fifo IC; *(v)* The Emc preamplifier/shaper, mounted one/photodiode on the crystals; *(vi)* The CARE (for "Custom Auto-Range Encoding") amplifier and range-selecting chip for the

Emc. All of these chips are complete or in production. There are approximately 40 different circuit boards, nearly all complete or in production.

The data acquisition for the experiment proceeds from the front end digitization through VME single-board computer modules, called read-out modules (ROMs). These modules have a detector-specific "personality card" (in fact, there are only two varieties of this card for BaBar). Upon receipt of a level one trigger, the data is transferred to a commercial PowerPC-based single board computer (also part of the ROM) running the VxWorks operating system. From the approximately 150 ROMs, data is transferred over a switched 100 BaseT network to the online farm, consisting of Sun Ultra 60 computers. The events are assembled in this process, and the level three trigger (there is no "level two" trigger) is run on the farm, reducing a design level 1 trigger rate of 2 kHz to a level 3 tape rate of 100 Hz. The raw event size is anticipated to be slightly over 30 kbyte/event, including backgrounds. Detector control and monitoring is based on the EPICS toolkit.

The level 1 trigger makes decisions based on tracks in the drift chamber and energy clusters in the calorimeter. It must generate its own event time. The level 3 trigger has all of the event information available as required. The trigger is essentially fully efficient for $B\bar{B}$ events, with high efficiency for $\tau^+\tau^-$ events which can be traded off for background rejection if required by running conditions.

The basic data acquisition chain has been demonstrated in the drift chamber cosmic ray commissioning, though there is still much to do before the system will be ready for colliding beams, such as multicrate readout and partitioning of the detector permitting parallel independent readout chains. The installation of the first contingent of online farm CPUs will take place in August.

4.8 Offline Computing

The offline computing [8] consists of several components, for event simulation, event reconstruction, physics analysis, data storage, *etc.* In the planning for the experiment, an object-oriented methodology, with implementation in C++, was adopted. Code management and distribution for the widespread development team is implemented in a software release tools package, based on CVS (Concurrent Versions System) and AFS (Andrew File System). Several Unix platforms (DEC/OSF, HP/UX, IBM/AIX, and Sun/Solaris) are currently supported, subject to ongoing evaluation.

The detector simulation currently is based on the Fortran GEANT3 package; however, we are in the midst of developing a GEANT4 (C++) implementation, and plan for this to serve both our fast- and detailed-

Table 2: Near-term BaBar Schedule.

Task	Date
Barrel calorimeter install in coil	Jul 1998
Drift chamber installation in coil	Aug 1998
Install DIRC SOB	Late Sep 1998
Install forward calorimeter	Oct 1998
Cosmic ray commiss. with magnet	Nov-Dec 1998
Break IR beamline for roll-on	Jan 1999
Install Svt	Jan 1999
Roll-on to beam line	Jan-Mar 1999
First colliding beam data	Apr 1999

simulation needs, starting at the end of 1998. The reconstruction code has essentially complete functionality, and is in the process of undergoing a second major round of testing under production conditions. For both the "conditions" (*e.g.*, detector configuration, calibration data) database and the event store, we are developing a solution based on the commercial Objectivity object oriented database management system. For the event store, this is being layered on the HPSS (High Performance Storage System, IBM) package to manage the underlying disk/tape files. The size of the overall event store is of order 100 Tbyte/year.

5 Schedule

Table 2 summarizes highlights of the remaining schedule for BaBar installation, commissioning, and roll-on, with colliding beam data to start in April 1999. The cosmic ray commissioning run at the end of 1998 will be with a complete detector, excepting the Svt, which will be installed in the coil in January, and the DIRC, which will not yet have its complete array of quartz bars.

6 Conclusion

The construction of BaBar has not been without its technical problems and delays. However, the collaboration has been very agressive in dealing with issues in order to keep to a schedule in which colliding beam data begins in April of 1999. Much of the detector is complete, and we are in the installation and commissioning phase. Many activities remain to be finished, but we are so far on track for an April 1999 start. We are therefore actively planning for the "first year physics".

Acknowledgements

Credit goes to the collaborators and collaborating institutions on BaBar. This work was supported in part by the US Department of Energy under grant DE-FG03-92-ER40701 (Caltech) and contract DE-AC03-76-SF00515 (SLAC).

References

1. "An Asymmetric B Factory Based on PEP: Conceptual Design Report", edited by M. Zisman, SLAC-372 (1991); updated as SLAC-418 (1993).
2. D. Boutigny *et al.*, Technical Design Report for the BaBar Detector, SLAC-R-95-457, March 1995; BaBar web pages at: http://www.slac.stanford.edu/BFROOT/doc/www/bfHome.html.
3. I. Kipnis *et al.*, *IEEE Trans. Nucl. Sci.* **44** (1997) 289.
4. G. Sciolla *et al.*, SLAC-PUB-7779, May 1998, Submitted to *Nucl. Instrum. Methods*.
5. H. Staengle *et al.*, *Nucl. Instrum. Methods* A **397** (1997) 261.
6. R. J. Barlow *et al.*, SLAC-PUB-7887, July 1998, Submitted to *Nucl. Instrum. Methods* A.
7. F. Anulli *et al.*, INFN-AE-97-30, July 1997, Submitted to *Nucl. Phys. Proc. Suppl.*
8. N. Geddes, *Comp. Phys. Comm.* **110** (1998) 38.

The HERA-B Experiment: Overview and Concepts

R. Mankel

Humboldt University Berlin, Institut für Physik, EEL, Invalidenstr 110, D-10115 Berlin, Germany
E-mail: mankel@ifh.de

For the HERA-B Collaboration

The key idea of the HERA-B experiment is to convert the HERA proton ring into a B factory by suspending a wire target into the outer reaches of the beam. The detector will have to cope with on average four inelastic proton-nucleus interactions, which are recorded with a complicated spectrometer setup including silicon micro-strip detectors, micro-strip gaseous chambers and honeycomb drift chambers. The high interaction rate and the dense event structure present also challenges to triggering and event reconstruction which are not unlike those at the LHC. This article gives an overview of the experiment, with some detailed emphasis on the target, the trigger and the track reconstruction concept.[1].

1 Introduction

Precision measurements in exclusive decays of b hadrons represent an excellent means to probe fundamental aspects of the standard model and, possibly, find indications for new physics. Small branching ratios of the key decay modes require rich sources of B mesons. While e^+e^- collisions at the $\Upsilon(4S)$ resonance produce b quarks with a cross section of $\mathcal{O}(1nb)$, the beauty cross section in hadronic collisions is eg. expected to be about four orders of magnitude higher at LHC energies. It is therefore conceivable that dedicated experiments at high energy hadron machines will finally explore the most elusive phenomena in b physics, provided that the huge experimental challenges of these interactions can be mastered.

HERA-B can be viewed as a prototype of a dedicated hadronic b physics experiment. It is based on the idea of turning the HERA proton ring at DESY into a b factory by inserting a wire target into the halo of the proton beam. The energy of the HERA proton beam has recently been increased from 820 to 920 GeV which is expected to improve the b production cross section by about 40%. The HERA-B collaboration consists of $\approx$ 280 physicists from 14 countries. The project was approved in 1995, and the detector is currently in the state of installation and commissioning. The detailed status of the detector hardware is reported elsewhere in these proceedings[2].

One of the most attractive physics goals is the search for CP violation in the B system. The design of the HERA-B experiment is particularly motivated by studying the asymmetry in the *golden decay* $B^0 \rightarrow J/\psi K_S^0$, which allows to measure the angle β in the well-known *unitarity triangle*. In the following, the emphasis will be on the technology mandatory for studying this channel.

Figure 1: Photograph of HERA-B target wires.

2 The Internal Target

In order to provide high target efficiency and at the same time separate multiple interaction vertices as good as possible, the HERA-B target consists of thin wires with a typical thickness of 50 by 500 μm, which is adjusted to the vertex resolution of the detector. The wires are mounted in forks which are moved with stepping motors (see fig. 1). The delicate interplay between target and beam has been explored intensively during the last years. This has lead to the development of an automatic multi-wire steering which is designed to stabilize the interaction rate and to react appropriately to changes in the beam conditions.

The operation of the target is illustrated in a run with four wires in fig. 2. At the beginning, the wires are moved close to the beam one by one. Evidently, an interaction rate above 30 MHz is reached without problems. In the following, the automatic target steering keeps the rate stable in spite of the slowly decreasing proton beam current, by accordingly moving the wires closer to the beam. It is an essential observation that the target operation does not cause problematic background for the other HERA experiments; this can be seen from the bottom diagram of fig. 2 where the trigger rate of the ZEUS forward neutron calorimeter (FNC) and the HERMES proton veto (pV) is monitored: the dotted lines indicate the background level defining "good running conditions" for these experiments.

The target efficiency, ie. the fraction of protons leaving the beam which interacts in the HERA-B target was found to be better than 60%, which satisfies the requirements of the HERA-B proposal and indicates an efficient use of the proton fill.

3 The Detector

In accordance to the kinematics of beauty production in fixed target pN interactions, the HERA-B detector [3,4] is designed as a forward spectrometer with excellent vertex resolution and particle identification capabilities. The design interaction rate of 40 MHz requires the use of technologies which are also common with LHC experiments. A top view of the HERA-B spectrometer is shown in fig. 3. The main components are:

- a silicon strip microvertex detector with 50 μm readout pitch

- the spectrometer magnet and the main tracking system which consists of an inner part (<20 cm from the beam) with micro-strip gaseous chambers (equipped with GEMs), and an outer part based on honeycomb drift chamber technology

- dedicated pad and gas pixel chambers for the high-p_T pretrigger

- a ring imaging Čerenkov counter using C_4F_{10} as radiator and multi-cell photo multiplier tubes

- a transition radiation detector based on straw chamber technology covering the inner region

- a lead/tungsten-scintillator electromagnetic calorimeter with shashlik readout

- a muon system with four superlayers of tube, pad and gas pixel chambers, embedded in steel and concrete absorber shields

The detailed setup and implementation status of these detector compenents are discussed in another contribution to these proceedings [2].

4 The Trigger System

A highly selective trigger system is required to reduce the event rate of 10 MHz, determined by the bunch frequency

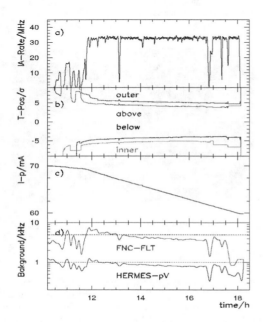

Figure 2: Interaction rate, target wire positions (distance from the beam in units of the beam width), machine current and backgrounds for ZEUS and HERMES in a HERA-B target run.

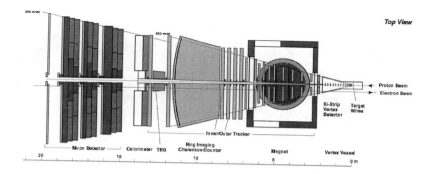

Figure 3: Top view of the HERA-B detector layout. The proton beam enters from the right.

of the HERA proton ring, to about 20 Hz which can be accomodated for temporary storage, without losing significant part of the signal. An overview of the trigger scheme developed for this task is shown in tab. 1.

4.1 The First Level Trigger

In order to achieve a high background rejection already at an early stage, it is necessary to exploit the invariant mass of the lepton pair from the J/ψ decay[a] already on the first trigger level. Four superlayers of inner and outer tracker between magnet and calorimeter (plus part of the muon system) are used to search for track candidates. The search is initiated by a pretrigger seed which is provided by either an energy deposition in the calorimeter of high transverse energy, a twofold coincidence of pads in the two hindmost superlayers of the muon system (fig. 4). For the decay $B^0 \to \pi^+\pi^-$, a special pretrigger for high-p_T tracks is generated by a threefold coincidence of pads in the magnet. The seed provides an initial set of track parameters, and a *region of interest* (ROI) on the next superlayer upstream. If a coincidence of hits in the three projections ($0°$ and $\pm 5°$) is found, the track parameters are updated and used to calculate the ROI in the next superlayer. The hardware implementation of this process is based on *track finding units* (TFU), each of which searches the hit coincidences in a ROI in the part of the tracker for which it is responsible. The updated track parameters are then passed as messages to the appropriate TFUs in the next superlayer (see fig. 5). For those candidates emerging after processing all superlayers, a *track parameter unit* (TPU) estimates the momentum from the apparent deflection by the magnetic field. The *trigger decision unit* (TDU) evaluates invariant masses of track pairs and finally generates the FLT decision. Accepted events are moved to the *second level buffer*. It is worth

---

[a]or of the pion pair in case of $B^0 \to \pi^+\pi^-$

noting that, while the detector data are transferred to the FLT processors in tune with the bunch clock, the data flow between the TFUs is asynchronous, and the time needed for a track candidate to pass the chain depends on the amount of congestion. The maximum latency allowed for the FLT is $12\mu s$, which is determined by the bunch crossing rate and the pipeline depth of 128.

A special software tool capable of a clock-level simulation of the FLT is under development. It is also used for cross-checking the electronics of the trigger boards.

4.2 The Second Level Trigger

The second level trigger is designed to achieve its background suppression by evaluating more tracking information on the lepton (or pion) candidates delivered by the first level trigger. In a first step, hits in other layers of the field-free part of the main tracker are located within the first level trigger's ROI, and the measured drift distances are taken into account. Track candidates surviving this check are extrapolated to the entrance of the magnet (avoiding the costly location of hits within the field) and propagated towards the target with the hits of the vertex detector. Finally, a detached vertex (or a non-vanishing impact parameter) is required.

The hardware implementation of the second level trigger is based on a farm of diskless Pentium-II PCs running Linux as operating system. Presently, 100 nodes are already installed. The data are received from the second level buffer through a fast switch based on *SHARC* digital signal processors, which has recently been demonstrated to work well. The same nodes are also the platform for event building and third level triggering.

Table 1: Overview of HERA-B trigger levels

Trigger level	Data/Method	Input rate	Reduction factor	Time scale
Pretrigger L1	ECAL, μ system, p_T pads e/μ "tracking" in 4 SL, p_T cut, mass cut	10 MHz	1/200	12 μs
L2	+ drift times, magnet traversal, vertexing	50 kHz	1/100	5 ms
L3	full track & vertex fit, + VD tracks, part.id.	500 Hz	1/10	200 ms
L4	+ full reconstruction, physics selection	50 Hz	1/2.5	4 s
Output rate to tape:		20 Hz		

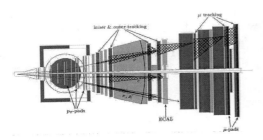

Figure 4: Track search scheme in the first level trigger. The search proceeds from right to left.

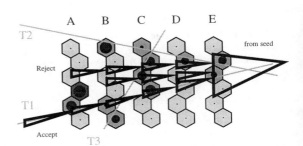

Figure 6: Schematic illustration of the Concurrent Track Evolution technique. The hexagons represent honeycomb tracker cells, the circles drift-isochrones of hits. The wedge-shaped areas symbolize search windows of concurrently propagated track candidates.

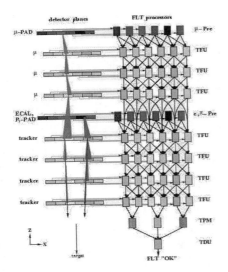

Figure 5: Layout of first level trigger processors, and connection to corresponding detector parts.

5 Track Reconstruction

In view of the high track density, which leads to occupancies up to 20% in the outer tracker chambers, reliable track pattern recognition becomes a crucial issue. The track finding concept developed for HERA-B proceeds in the following way: in the first pass, straight-line track segments are found in the "pattern tracker", which consists of 4 superlayers in the field-free area between magnet and RICH. In the second step, track candidates are continued through the chambers within the highly inhomogeneous magnetic field. Finally, the tracks are matched to track segments in the vertex detector and propagated to the devices behind the RICH.

5.1 Pattern tracker

Pattern recognition in the main tracker uses the *Concurrent Track Evolution* technique[5], which is a *track following method* based on the Kalman filter[6,7,8,9]. This method evaluates the *available paths* for a track candidate *concurrently* to find the optimal solution, but also

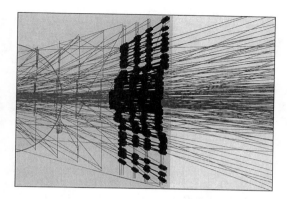

Figure 7: Display of a simulated event with seven superimposed interactions. The Monte Carlo tracks are shown as grey lines. The tracks found by the pattern recognition algorithm are shown as thick dark lines, with the hits used indicated by crosses.

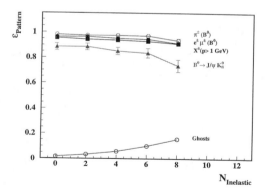

Figure 8: Mean efficiency as a function of the number of superimposed inelastic interactions, for particles related to the golden B decay, and ordinary charged tracks ($X^{\pm}$). The relative rate of ghosts is also shown.

keeps the combinatorics at a reasonable level (illustrated in fig. 6). This is achieved by evaluating a quality estimator function for all candidates in each step and applying relative and absolute cuts. In the pattern tracker, the algorithm works on the chambers of the $0°$ projection first, the stereo layers are considered in a subsequent step. The display of a simulated event with seven interactions in the same bunch crossing is displayed in fig. 7. The efficiency of this algorithm as a function of the number of inelastic interactions superimposed to the one producing the heavy flavour is shown in fig. 8. For the nominal four inelastic interactions corresponding to 40 MHz of interaction rate, the resulting mean pattern recognition efficiency for particles from the golden B decay is better than 96%, the average number of ghosts is 4.4 per event at this stage.

5.2 Magnet tracker

A three-dimensional extension of the Concurrent Track Evolution technique is used to continue the track candidates into the magnet tracker [10]. The track parameter transport is achieved with a fifth-order Runge-Kutta method with adaptive stepsize control. The magnet propagation efficiency obtained with the standard HERA-B event simulation is found to be $\geq 97\%$ for tracks from the golden B decay, resulting in a combined main tracker pattern recognition efficiency of 83% for the $J/\psi K_S^0 \rightarrow \mu^+\mu^-\pi^+\pi^-$ final state. The geometrical acceptance of the combined magnet and pattern tracker amounts to 34%. The ghost rate is reduced to 0.9 per event.

5.3 Track fitting

The momentum resolution of the spectrometer has been studied after applying a Kalman filter-based track fit, which included corrections for multiple scattering and energy loss [11]. The resulting relative momentum resolution without pattern recognition effects (i.e. locating the hits using Monte Carlo information) is shown in fig. 9. Multiple scattering effects dominate the resolution below 20 GeV of momentum. In fig. 9a, the improvement due to the coordinate resolution of the vertex detector (squares) compared to magnet and pattern tracker alone (circles) is clearly visible above 30 GeV.

The influence of pattern recognition effects has been investigated for muons from the golden B decay, using only main tracker hits. The "ideal" momentum resolution discussed above for these muons is $\delta p/p = 8.7 \cdot 10^{-3}$. The raw output of the pattern recognition step determines the momentum already with a precision of $8.7 \cdot 10^{-3}$. A subsequent iterative refit improves this resolution further to $8.1 \cdot 10^{-3}$. Even without exploiting

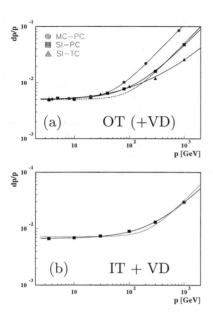

Figure 9: Relative momentum resolution for different ranges of superlayers in the detector, for a particle traversing vertex detector and outer tracker (a), and for one passing vertex detector and inner tracker (b). "MC–PC" stands for the outer magnet and pattern tracker chambers, the range "SI–PC" contains in addition the silicon vertex detector, and in "SI–TC" the trigger chambers behind the RICH have been added.

further improvement options, as outlier removal or inclusion of vertex detector hits, one concludes that pattern recognition effects have a visible but limited effect on the HERA-B momentum resolution.

6 Summary

HERA-B is the prototype of a *hadronic B factory*. The main goal of the experiment is to test fundamental aspects of the standard model by investigating CP violation in the B system. To achieve this goal, the experiment is built on technology which is also important for the LHC era. Routinely operation of the internal target, providing the necessary interaction rates without disturbing other HERA experiments has been achieved. A sophisticated triggering scheme has been devised. Track reconstruction techniques suitable to cope with the high track density have been developed.

Acknowledgements

I would like to express my gratitude to K. Ehret, E. Gerndt and H. Kolanoski, whose abstracts are also covered here, for their help in preparing the presentation. This work was supported by the Bundesministerium für Bildung, Wissenschaft, Forschung und Technologie under contract number 05 7 BU 35I (5).

References

1. This article covers the abstracts #168, #449, #746 and #891 of session 14.
2. K. Lau, (these proceedings).
3. T. Lohse et al., *An Experiment to Study CP Violation in the B System Using an Internal Target at the HERA Proton Ring* (Proposal), DESY-PRC 94/02 (1994).
4. E. Hartouni et al., *An Experiment to Study CP Violation in the B System Using an Internal Target at the HERA Proton Ring* (Design Report), DESY-PRC 95/01.
5. R. Mankel, *Nucl. Instrum. Methods* A **395**, 169 (1997).
6. R.E. Kalman, Trans. ASME, J. Basic Engineering (1960);
   R. Battin, Am. Rocket Soc. **32**, 1681 (1962);
   Am. Rocket Soc. 32 1681 (1962);
   R.E. Kalman and R.S. Bucy, Trans. ASME, J. Basic Engineering (1962);
7. R. Frühwirth, *Nucl. Instrum. Methods* A **262**, 444 (1987).
8. P. Billoir and S. Qian, *Nucl. Instrum. Methods* A **294**, 219 (1990).
9. R. Mankel, *Application of the Kalman Filter Technique in the HERA-B Track Reconstruction*, HERA-B Note 95-239 (1995).
10. R. Mankel and A. Spiridonov, *The Concurrent Track Evolution Algorithm: Extension for Track Finding in the Inhomogeneous Magnetic Field of the HERA-B Spectrometer*, DESY 98-142 (1998).
11. R. Mankel, *ranger – a Pattern Recognition Algorithm for the HERA-B Main Tracking System, Part IV: The Object-Oriented Track Fit*, HERA-B Note 98-079 (1998).

THE RUN II UPGRADE TO THE COLLIDER DETECTOR AT FERMILAB

B. L. WINER

Department of Physics, The Ohio State University, Columbus, OH USA
E-mail: winer@mps.ohio-state.edu

The CDF collaboration is preparing for Run II at the Tevatron by making significant detector upgrades. These upgrades will enhance CDF's ability to study the top quark, search for CP violation, and search for extensions beyond the standard model. The center piece of a new integrated tracking system is an open-cell drift chamber (COT) that is ∼3 meters long and covers between 40 and 137 cm in the radial direction. Residing near the beam pipe will be a 5 layer silicon-strip micro-vertex detector (SVX II) to provide precision tracking and secondary vertex identification near the interaction point. Between the SVX II and COT are additional layers of silicon to enhance stand-alone tracking in the silicon tracking systems. New calorimeters using scintillating tile-fiber technology have been built and are being installed. The calorimeters cover approximately $1.1 < |\eta| < 3.6$. Located at a depth of approximately 6 radiation lengths is a shower maximum detector to measure the precise position of electron and photon showers and provide rejection against π^{o}'s. The muon detection system will be extended to provide additional acceptance both in the central and forward directions. Finally, new front-end electronics and a new data aquisition and trigger system are being constructed to handle the 132 ns bunch spacing of an upgraded Tevatron. The front-end electronics will be fully pipelined with on board buffering for 42 beam crossings. The trigger is separated into three levels (stages). The first level is synchronous with a 5.5 μs decision time and up to a 40 KHz accept rate. The second level is asynchronous with a 20 μs decision time and a 300 Hz accept rate. The third and final stage of the trigger is a farm of processors that reconstruct high level event quantities prior to the final decision to write the event to mass storage. The status of these upgrades will be reviewed and the future schedule summarized.

1 Introduction

The Collider Detector at Fermilab (CDF) was first commissioned in 1985. It has taken data during Tevatron runs over more than a 10 year period. During that time CDF has undergone several upgrades. However, the current upgrades planned for Run II at the Tevatron are the most substantial changes ever made to the detector [1]. For that reason the detector is being given a new name, CDF II. Run II at the Tevatron will see the commissioning of the Main Injector Facility which will allow the instantaneous luminosity in the Tevatron to reach $2 \times 10^{32} cm^{-1} s^{-1}$, a factor of 10 higher than Run I. The integrated luminosity is expected to reach 2 fb^{-1}, a factor of 20 above the current datasets. At this high luminosity it is important to keep the number of multiple interactions as low as possible. This will be achieved by reducing the spacing between proton and antiproton bunches from 3.5 μsec to 396 ns and eventually to 132 ns.

Run II is expected to yield a number of important physics measurements and opportunities. Continued study of the top quark and its properties will be the sole territory of the Tevatron experiments. The uncertainty on the top quark mass is anticipated to be as low as 1-2 GeV/c^2 [1,2]. The measurement of the top quark mass in conjunction with a precise measurement of the W boson mass deepens our understanding the electroweak symmetry breaking in the standard model. The precision on the W mass is expected to reach $\approx$ 40 MeV/c^2 [1,2]. Besides these electroweak measurements, CDF will continue exploring the B sector of the standard model. The corner stone of this exploration will be the investigation of CP violation in the B system. In addition, the

high b-quark production rate at the Tevatron provides an opportunity to search for rare decays. Finally, the Tevatron will provide the highest center-of-mass energy collisions in the world, and therefore it will be important to search for extensions beyond the standard model.

The physics program for CDF during Run II is broad. To take advantage and extend the physics opportunities, CDF will be upgraded substantially. The detector configuration for Run II is shown in Figure 1. Some of these upgrades are necessitated by the changes in the accelerator complex. For example, the shortening of the bunch spacing and the increase in the luminosity requires the complete replacement and redesign of the trigger and data aquisition systems and the front-end electronics. However, some detector subsystems will remain for Run II. The central electromagnetic and hadronic calorimeters will remain intact along with the wall hadronic calorimeter. The existing muon system will remain, although its coverage will be extended. The entire central tracking system will be replaced. This includes the large drift chamber and the silicon microstrip vertex detector. Finally, the electromagnetic and hadronic calorimeter in the endplug regions will be replaced. In the sections below, we review each of the major upgrades. The time scale for these upgrades is short. Most of the parts will be installed by mid to late 1999. The current schedule has colliding beams starting in early 2000.

2 Tracking System

Precision charged tracking with excellent momentum resolution has always been a high priority for the CDF ex-

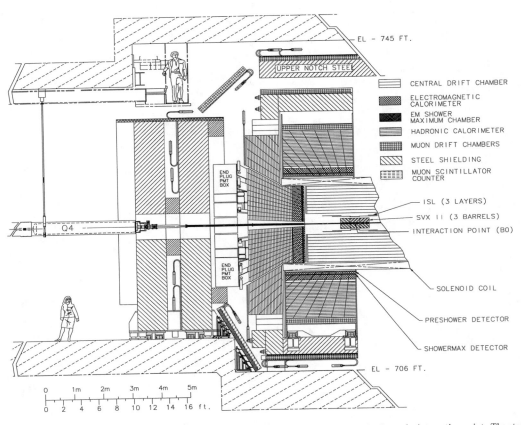

CENTRAL DRIFT CHAMBER

ELECTROMAGNETIC CALORIMETER

EM SHOWER MAXIMUM CHAMBER

HADRONIC CALORIMETER

MUON DRIFT CHAMBERS

STEEL SHIELDING

MUON SCINTILLATOR COUNTER

EL - 745 FT.

UPPER NOTCH STEEL

END PLUG PMT BOX

END PLUG PMT BOX

Q4

ISL (3 LAYERS)

SVX II (3 BARRELS)

INTERACTION POINT (B0)

SOLENOID COIL

PRESHOWER DETECTOR

SHOWERMAX DETECTOR

EL - 706 FT.

0   1m   2m   3m   4m   5m

0   2   4   6   8   10   12   14   16 ft.

Figure 1: A cross section of 1/2 of the CDF II Detector. The detector is left-right symmetric about the interaction point. The steel from the Run I forward muon toroid has been moved closer to the interaction region.

1520

periment. An entirely new tracking system is being developed to meet the challenges of the Run II environment. The tracking systems can bee seen in Figure 1. The entire system is placed in magnetic field generated by a superconducting solenoid at $r \approx 1.5$ m. Closest to the beam pipe is a 5 layer silicon microstrip vertex detector (SVXII). Between $r = 20$ cm and $r = 30$ cm are additional layers of silicon microstrip detectors called the Intermediate Silicon Layers (ISL). Finally, the main tracking detector is the Central Outer Tracker (COT), a drift chamber approximately 3 meters long and extending from $r = 40$ cm to $r = 137$ cm. The COT supplies complete coverage of tracks for $|\eta| < 1.0$ For the region $1.0 < |\eta| < 2.0$, where the COT has only partial coverage, the Intermediate Silicon Layers provide charged particle sampling along with SVXII. For tracks passing completely through all 3 tracking systems, the momentum resolution is expected to be $\Delta P_T / P_T^2 < 0.1\%$ for $P_T > 4$ GeV/c.

2.1 Silicon Vertex Detector (SVXII)

The 5 layer silicon-microstrip vertex detector builds upon the experience gained with the SVX and SVX' systems during Run I. Besides one additional layer, a number of important improvements have been made. First, the electronics for the front end of SVXII will be more radiation hard than previous versions. It is expected that the readout chips (SVX3 Chip) will withstand 2-4 Mrad of dose. This corresponds to about 3-5 fb^{-1} of integrated luminosity for the inner most layer. Second, the length of SVXII is nearly twice as long as the SVX and SVX' detectors. SVXII will cover $\approx 95\%$ of the luminous region. Third, the silicon layers are double sided, providing both r-ϕ and r-z information for the tracks. The r-z information is obtained from 2 layers with strips at $1.2°$ stereo angle and 3 layers with $90°$ stereo angle. Three dimensional standalone tracking will be possible with SVXII.

An important feature of the SVX3 readout chip is its ability to acquire data while digitizing and reading out the results of a previous beam crossing. This dramatically reduces the deadtime caused by this system. In addition, the SVXII system has two high speed readout paths. One supplies data to the normal data aquisition path for mass storage. The other path supplies the hit information to the Level 2 trigger. Having this information available at the trigger level allows for a trigger requirement on track impact parameter (see Section 5). Therefore, it provides the opportunity for CDF to trigger on hadronic b-quark decays and extend its reach in areas such as the study of CP violation in the B system.

2.2 Intermediate Silicon Layers (ISL)

The Intermediate Silicon Layers provide an important link between the SVXII track hits and the tracks reconstructed by the Central Outer Tracker (COT). The system consists of a complete layer of silicon at $r \approx 20$ cm and a partial layer at $r \approx 28$ cm and $1 < |\eta| < 2$. Many components of the ISL are similar to the components of the SVXII system. The readout uses SVX3 chips, the DAQ beyond the port cards is the same as SVXII, however, the ladders, hybrids are somewhat simpler than SVXII. The parameters of the ISL were optimized to (i) best bridge the gap between the SVXII and COT and (ii) extend precision tracking into the plug calorimeter to assist with lepton identification and b-quark tagging at low angles ($1 < |\eta| < 2$). The SVXII and the ISL provide tracking in the region $1 < |\eta| < 2$ with $\Delta P_T / P_T^2 = 0.4\%$ ($P_T > 5$ GeV/c).

2.3 Central Outer Tracker (COT)

The Central Outer Tracker is based on the Central Tracking Chamber (CTC) from Run I. However, a number of changes have been made to correct short comings of the CTC, to respond to changing accelerator conditions, and to simplify the construction. Like the CTC, the COT consists of 8 "superlayers". Four of these layers have axial sense wires to provide $r - \phi$ measurements. Four of the superlayers have wires tilted at a $\sim 3°$ stereo angle to provide z positions of charged particles. Each superlayer is divided into many identical cells. The total number of cells is 2520, four times more cells than the CTC. A cell has an average maximum drift distance of 0.88 cm with a maximum drift time which is less than the beam crossing time. Therefore, there is no "pile-up" of events in the chamber. A cell has 12 sense wires and 17 potential wire which are 40 micron diameter gold plated tungsten. The total number of sense wires in the COT is 30,240. The cathodes for the cells are sheets of gold plated mylar. The cells are tilted at $35°$ for the Lorentz angle. In the construction phase, wire planes and cathode sheet planes are prestrung and attached to printed-circuit boards that snap into place on the COT endplates. At the time of these proceedings the endplates of the COT have been delivered to Fermilab. The endplates have been aligned and insertion of the wire and cathode planes is about to begin.

3 Plug Upgrade

In Run I, the calorimeters in the forward-backward region ($4.2 < |\eta| < 1.1$) were separated into two pieces on each side. These calorimeters were gas proportion chambers between lead and iron absorber. The resolution and readout speed of these detectors are no longer sufficient. These calorimeters are being replaced with a single fast, hermetic, scintillating tile-fiber calorimeter. The calorimeter resides in the position of the old plug calorimeter and the coverage is extended from 10° down to 3° from the beam line. The electromagnetic calorimeter is 21 radiation length deep and uses 4 mm lead plates. A scintillating fiber position detector is located at shower maximum in the EM calorimeter. The hadronic calorimeter reuses the existing steel absorber from the previous plug calorimeter. Additional layers of stainless steel absorber are used to extend the depth of the calorimeter to 6.6 interaction lengths and fill the absorber gap between 10° and 3°. Since the plug provides the return flux for the magnetic field, stainless steel was used so that detailed remapping of the magnetic field could be avoided.

The construction of the new plug calorimeter is complete and final installation and calibration are underway. Part of the new calorimeter has been studied in a test beam at Fermilab. The resolution of the electromagnetic calorimeter was determined to be $\sigma(E)/E = 16\%/\sqrt{E}$. For the hadronic calorimeter the energy resolution was found to be $\sigma(E)/E = 75\%/\sqrt{E} \oplus 5.3\%$. These resolutions are substantially better than the previous plug calorimeter. The improved resolution along with the improved tracking into the plug region should dramatically improve lepton identification and b-quark tagging in this region.

4   Muon System Upgrade

The muon detection system for CDF has been improved several times over the last few years and improvements are continuing for Run II. The existing system consists of the CMU ($|\eta| < 0.6$), CMP ($|\eta| < 0.6$ behind more shielding), and CMX ($0.6 < |\eta| < 1.0$). These existing systems are being improved in a number of ways. First, the CMP and the CMX had several azimuthal regions where coverage was not present due to the obstruction of mechanical and structural components of the detector or collision hall. These gaps in coverage are being filled for Run II. The coverage of the CMP system is being increased by 17% while the coverage for the CMX is being increased by 45%. In addition, shielding for the CMX system is being added to reduce accidental triggers from the scatter of low energy particles near the beam line. This shielding should help control the muon trigger rates from this system. Finally, an additional muon detector subsystem is being added. This system is called the Intermediate Muon system (IMU). It covers the region of forward-backward going muons ($1.0 < |\eta| < 2.0$) and covers 270° in azimuthal angle. The IMU is located on the Run I toroid steel which has been pushed closer to the interaction region. The system includes scintillation

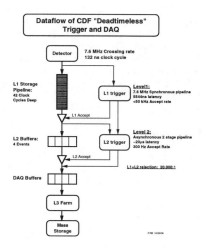

Figure 2: Block diagram for CDF II data aquisition and trigger systems.

counters over the region $1.0 < |\eta| < 1.5$ that are used to include this system in trigger decisions. Including all the enhancements to the muon system, the total coverage for muons increases by a factor of 2 when compared to Run I.

5   Data Aquisition and Trigger Systems

Although changes in the accelerator operation necessitate the complete redesign of the data aquisition and trigger systems, desire to improve the capabilities of the trigger system in order to extend the physics reach of CDF is also a large motivation. A schematic diagram of the trigger and data aquisition system are shown in Figure 2. The design of the system assumes a 7.6 MHz crossing rate of the beams, therefore the "fundamental clock" for operation of the electronics is 132ns. One of the most important features of the system is the fact that the readout is pipelined at all stages. This will keep the deadtime of the system to a minimum. The system is designed to be 90% live at 90% of the maximum bandwidth. The pipelines of the front-end electronics will be 42 clock cycles deep. This provides a 5.5 μsec latency time for a synchronous Level 1 trigger (see below). Upon a Level 1 accept, events are read from the Level 1 pipeline into one of four Level 2 buffers. The Level 2 trigger makes an asynchronous trigger decision. Upon a Level 2 accept, the data is collected in DAQ buffers that reside in 12 VME crates. Event building is performed via a network switch between these buffers and the Level 3 PC Farm. At Level 3 full data reconstruction is performed to make

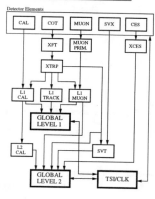

RUN II TRIGGER SYSTEM

Figure 3: A block diagram for the Level 1 and Level 2 trigger systems for CDF II.

a final decision whether or not to write the event to mass storage for offline analysis.

The structure of the Level 1 and Level 2 trigger system is shown in Figure 3. The trigger system follows the structure of the Run I trigger, in that it is highly flexible and programmable and therefore capable of responding to changes in the accelerator or the physics goals. The Level 1 trigger combines primitives from tracking (COT), muons, and energy depositions in the calorimeters to form a global Level 1 decision. The processing at Level 1 is entirely synchronous. A new decision is supplied every 132 ns. The accept rate into Level 2 is < 40 KHz. One of the important additions to the Level 1 system is the COT track information. Previously this information was only available at Level 2. Tracks with measured P_T and ϕ at this early stage are important for keeping the accept rates as low as possible while still maintaining excellent efficiency for important physics signatures. Level 2 consists of Alpha processor nodes that examine the events in more detail. Level 2 is asynchronous and has a latency of approximately 20 μs. The most important addition to the Level 2 trigger is the presence of hit data from the SVXII detector. A Silicon Vertex Trigger (SVT) will use Level 1 COT tracks as seeds for linking hits in the silicon system. The resulting track will have an improved momentum resolution and an impact parameter measurement. The performance for this device is expected to be $\Delta P_T / P_T^2 = 0.3\%$ with a 35 micron resolution on the impact parameter. This feature will allow trigging on displaced tracks and is a tremendous advantage for the study of the b quark.

The construction of the trigger and data aquisition system is well underway. Prototypes of nearly all the components have been built. Some components have moved into the production phase and others are expected to enter production soon. The DAQ and trigger systems will be installed in the later half of 1999. Initial commissioning of the subsystems will take place with cosmic rays in late 1999 and early 2000.

6 Summary

The upgrades to CDF are extensive and ambitious. Many of the detector components are being completely redesigned and replaced. All aspects of the upgrades are well underway with many detector components moving into the production and construction stages. The detector should be ready for collider running by early 2000. Many components will be ready by the fall of 1999. The upgrades significantly enhance the physics potential for Run II at the Tevatron and all aspects of the physics program will benefit from the upgrades.

Acknowledgements

I would like to thank all of my collaborators whom are working on the upgrade for Run II. Their dedication, inspiration, and perspiration will truly make Run II a success. The upgrade to CDF is supported by the U.S. Department of Energy; the National Science Foundation, the Italian Istituto Nazionale di Fisica Nucleare; the Ministry of Education, Science and Culture of Japan; the Natural Sciences and Engineering research Council of Canada; the National Science Council of the Republic of China; the Swiss National Science Foundation; and the A.P. Sloan Foundation.

References

1. The CDF II Collaboration, *The CDF II Detector: Technical Design Report*, Fermilab-Pub-96-390-E, 1996.
2. Editors D Amidei, R. Brock *Future ElectroWeak Physics at the Fermilab Tevatron: Report of the TeV2000 Study Group*, Fermilab-Pub-96/082, 1996.

THE DØ DETECTOR UPGRADE AND JET ENERGY SCALE

MARIA TERESA ROCO
For the DØ Collaboration

MS 357 Fermilab, P.O. Box 500, Batavia, IL. 60510-0500, U.S.A.
E-mail: roco@fnal.gov

A major detector upgrade project is presently underway for the DØ experiment in preparation for Run 2 of the Fermilab Tevatron collider scheduled to begin in the year 2000. The upgrade is driven by the DØ Run 2 physics goals and by the higher event rates and backgrounds expected in the new high luminosity environment. A general overview of the DØ upgrade is given in the first part of this report. The second part describes the calibration of jets observed with the DØ detector at the transverse energy and pseudorapidity range $E_T > 8$ GeV and $|\eta| < 3$.

1 The DØ Detector Upgrade

The Fermilab Tevatron collider luminosity upgrade program involves the replacement of the existing Main Ring accelerator with the Main Injector and the construction of a new antiproton storage ring within a common tunnel. The ultimate performance goal of the collider upgrade for Run 2 is to achieve luminosities up to $2 \times 10^{32} cm^{-2} s^{-1}$, more than an order of magnitude increase over the previous run. In addition, the center-of-mass energy will increase to 2 TeV and the bunch spacing will be reduced to 396 ns and eventually to 132 ns, substantially shorter than 3.5 μs in Run 1.

The higher luminosities and much shorter bunch crossing time require some major modifications to the DØ detector to optimize its physics capabilities in the new Main Injector era. A major element of the detector upgrade[1] involves the replacement of the inner tracking systems and the installation of a 2 Tesla superconducting solenoid. The new tracking system includes a scintillating fiber tracker, a silicon vertex detector and preshower detectors in the central and forward regions. Improvements to the muon detector are required to handle the higher event rates and backgrounds expected in Run 2. These include the installation of fast trigger elements and the addition of shielding material to reduce the detector occupancy. Electronic upgrades are necessary to handle the smaller bunch spacing and to provide pipelining for the front-end signals coming from the calorimeter, muon and tracking systems. A sophisticated trigger control system is planned to integrate the new detector trigger elements and front-end electronics, and to reduce the raw event rates to a manageable level.

1.1 Tracking System

The momenta of charged particles are determined from their curvature in the uniform 2 Tesla axial magnetic field provided by a superconducting solenoid. The 2.8 m long solenoid has a two layer coil with a mean radius of 60 cm. It encloses the scintillating fiber tracker and the silicon microstrip vertex detector. The fiber tracker provides charged particle track reconstruction, momentum measurement within $|\eta| < 2$ and a fast first level track triggering with good resolution. It consists of eight concentric barrels of scintillating fiber doublet layers, one layer with fibers parallel to the beam line and another layer with $\pm 2^o$ stereo. It has an active length of 2.6 m and a radial coverage 20 $cm < r < 50$ cm. There are approximately 77,000 fibers which are connected to the photodetectors via 11 m long clear waveguides. The optical signals from the fibers are detected by visible photon light counters (VLPC). VLPCs are arsenic-doped silicon diodes which operate at temperatures between 8-10 ^oK. A large sample of the VLPCs were tested in a cosmic ray test stand[2]. The sample exhibited $\sim$ 80% quantum efficiency, relatively high gain and less than 0.1% noise occupancy when operated at full efficiency. The doublet hit efficiency for cosmic ray tracks is better than 99.9% and a position resolution $\sim 100 \mu m$ is obtained.

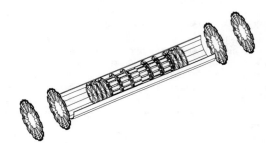

Figure 1: The silicon microstrip tracker consists of disks interspersed with barrels.

The main factors which constrain the design of the silicon microstrip tracker include[3]: the extended luminous region in z, three dimensional track reconstruction capabilities with transverse impact parameter resolutions better than 30 μm, radiation hardness to cope with the

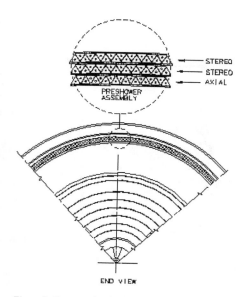

Figure 2: Cross sectional end view of the tracking system showing the central preshower assembly.

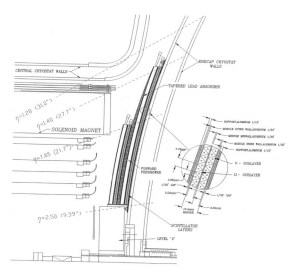

Figure 3: Side view of the tracking system showing the forward preshower assembly.

expected Run 2 delivered luminosity of 2-5 fb^{-1}, and electronics and readout which can operate reliably with bunch crossing intervals down to 132 ns. The long luminous region sets the length scale and motivates a hybrid design which consists of disk and barrel modules shown in Fig. 1. Larger disks at $z = \pm 94$ cm, ± 126 cm provide improved momentum resolution up to $\eta = \pm 3$. The barrel and disk modules are supported by a double-walled carbon-fiber/epoxy half cylinder which aids in maintaining the precise relative alignment and supports the detector cabling and cooling services. Each barrel contains four concentric layers of silicon detectors ranging from 2.6 cm to 10 cm. The basic detector unit is the ladder, which consists of two 300 μm thick wafers, 6 cm × 2.1 cm, positioned end to end and electrically connected by microwirebonds. The wafers are held in position by two longitudinal rails consisting of Rohacell foam and carbon fiber sandwich. Four inner barrels around $z = 0$ have layers of double-sided silicon ladders with either a 2° or 90° stereo angle. The outer barrels have single-sided ladders in layers 1,3 while layers 2,4 are double-sided with 2° stereo. The ladders are supported by beryllium bulkheads with very tight tolerances since they establish the precision of the alignment. By means of an integrated coolant channel, the bulkheads also provide cooling for the electrical components mounted at one end of the ladder. To avoid excessive radiation damage, the detectors are operated at temperatures between 5-10 ^oC. The

ladders, as well as the fiber tracker VLPC system, are read out by radiation hard CMOS custom integrated circuits, SVX-IIe chips [4], optimized for 396 ns and 132 ns bunch crossing intervals. This new 128 channel device incorporates preamplifiers, 32 cell analog pipelines, 8 bit ADCs, and sparsification. The SVX-II chip is highly flexible, with programmable test pulse patterns, ADC ramp, pedestal, bandwidth and polarity.

The cylindrical central preshower detector, shown in Fig. 2, is designed to enhance the identification of electrons and photons, provide precise position measurements in the region $|\eta| < 1.2$, and correct the calorimeter electromagnetic energy for the effects of the solenoid. It is installed in the small gap between the solenoid coil and the central calorimeter cryostat at a radius of 72 cm. It consists of three layers of 7 mm base triangular scintillating strips with wavelength-shifting fiber readout. The fast energy and position measurements enable the use of the preshower information at the trigger level. Simulation studies show that the central preshower information allows the electron trigger rate to be reduced by a factor of 3-5 without a loss in efficiency. There are also forward preshower detectors, shown in Fig. 3, mounted on the face of each end calorimeter cryostat. Like the central preshower detector, they are intended to improve the electron identification and triggering capabilities by making precise position measurements of particle trajectories using dE/dx and showering information.

1.2 Muon Detector

The upgraded muon system is designed to provide excellent muon identification and muon triggering capabilities for $|\eta| < 2$. For $|\eta| < 1$, the existing proportional drift tubes (PDT) are retained and a faster gas mixture is used to reduce the drift time from 750 ns to 450 ns. The front-end electronics are replaced to allow a deadtimeless operation. A new layer of scintillation counters are installed between the calorimeter and the inner PDT layer. They are used, together with the fiber tracker information, to reject out-of-time backgrounds and to enable a reduced trigger threshold for low p_T muons down to 2 GeV/c. For the forward muon system covering $1 < |\eta| < 2$, the PDTs are replaced with planes of plastic mini-drift tubes (MDTs). Three layers of scintillator pixel counters, shown in Fig. 4, provide time information and matching of muon tracks with the fiber tracker. The minimum pixel size, $\Delta\eta = 0.1$ and $\Delta\phi = 4.5^o$, is dictated by the requirement of efficient muon identification and triggering down to $p_T = 3$ GeV/c. The muon detector upgrade also includes the addition of shielding material extending from the rear of the calorimeter to the accelerator tunnel to substantially reduce the background arising from the scattered beam fragments exiting the calorimeter and the beam pipe.

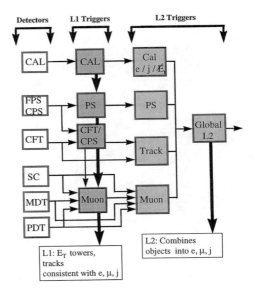

Figure 5: $L1$ and $L2$ block diagram.

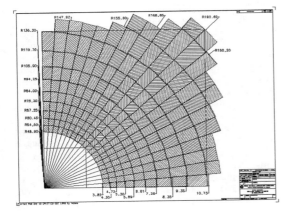

Figure 4: One quadrant of a forward muon scintillation pixel plane.

1.3 Trigger and Data Acquisition Systems

A new trigger system is designed to deal with the higher event rates in Run 2. The system includes three hardware trigger levels, $L0, L1, L2$, and a software trigger $L3$. $L0$, which serves primarily as a luminosity monitor, provides an inelastic collision trigger. $L1$ reduces

the inelastic trigger rate by a factor of 10^3 while maintaining a high trigger acceptance for leptons and jets. Fig. 5 shows the $L1$ and $L2$ block diagram. The new $L1$ trigger includes the calorimeter, central fiber tracker (CFT), central (CPS) and forward preshower (FPS) detectors, muon scintillation counters and chambers. For deadtimeless operation, all $L1$ triggers are pipelined and buffered. $L1$ provides a trigger decision in 4.2 μs and has an accept rate of 5-10 KHz. The number of triggers at $L1$ is 128 and fractional prescales may be available. The $L2$ system consists of a set of detector-specific preprocessors which prepare the information to be sent to the global processors. The result of the preprocessing algorithm is a list of candidate objects from each detector such as muons, electrons, jets, tracks and secondary vertices. The $L2$ system introduces a minimal amount of deadtime ($< 1\%$). It makes a trigger decision within 400 μs with an accept rate of 1 KHz, providing a rejection factor of 10. The $L3$ trigger system uses a farm of standard high-performance commercial processors to run event filtering algorithms. $L3$ makes a trigger decision in 100 ms and the accept rate is 20-50 Hz.

The data acquisition system is basically unchanged from Run 1. The architecture has been designed with parallel and redundant paths which provide a high degree of reliability in case of a component failure. VME buffer drivers (VBD) drive the raw data via high-speed data cables which feed a new data distributor system where further buffering is performed. The data is passed to a farm of event building and filtering $L3$ processing nodes. Its output goes to another VBD and data cable system which feeds the data to a host interface node. This node transfers the data to data-logging and monitoring cluster nodes via a standard network connection. Event reconstruction is performed on the Fermilab processor farm system capable of matching the 20-50 Hz data acquisition rate. Processed data are then stored on a robotic tape system and made available for analysis. In Run 2 the expected nominal event size is roughly 250 KBytes.

1.4 Calorimeter Upgrade

To maintain the excellent performance of the calorimeter for running conditions expected in Run 2, an upgrade of the calorimeter system is required. The uranium liquid-argon calorimeter remains unchanged but modifications to the front-end electronics are necessary. The shorter bunch spacing and high luminosity conditions require a re-optimization of the noise arising from the electronics, pile-up and uranium radioactivity. This can be achieved by shortening the effective integration time and reducing the intrinsic noise of the preamplifier. Pile-up effects in the calorimeter are minimized by reducing the peak sampling time from 2.2 μs to 400 ns. The increased sensitivity to noise and signal reflections in the signal cables due to the shorter shaping times require new preamplifier hybrids as well as impedance matched cables which have better noise performance. Switched capacitor arrays perform the necessary pipelining of the calorimeter signals to provide time to generate a trigger decision.

2 The Jet Energy Scale in the DØ Calorimeters

The major source of systematic uncertainty in both the DØ inclusive jet cross section and top quark mass measurements is the jet energy scale. The calorimeters[5] are the primary tool for jet measurements at DØ. An accurate understanding of the calorimeter energy calibration is essential since most physics measurements at the Tevatron involve events with jets. After a brief summary of the characteristics and performance of the DØ calorimeters, the different corrections involved in the jet energy calibration are enumerated below[6].

2.1 DØ Calorimeters

The DØ experiment has a hermetic, finely-segmented, thick, radiation-hard calorimeter based on the detection of ionization in liquid argon. The central (CC) and end (EC) sections, shown in Fig. 6, contain approximately 7 and 8 interaction lengths of material, ensuring the containment of nearly all particles except neutrinos and high

p_T muons. The hermetic calorimeter consists of plates of depleted uranium in the electromagnetic (EM) and fine hadronic (FH) sections, and either copper or stainless steel in the coarse hadronic (CH) sections. The segmentation is $\Delta \eta \times \Delta \phi = 0.1 \times 0.1$ (0.05×0.05 at the EM shower maximum).

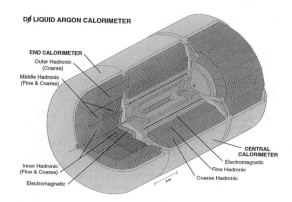

Figure 6: View of the DØ calorimeters.

From previous test beam measurements[7], the energy resolution for single electrons and pions, with energy E in GeV, are given by $\sigma(E)/E \sim 15\%/\sqrt{E}$ and $50\%/\sqrt{E}$, respectively. The DØ calorimeters are nearly compensating with e/π ratio less than 1.05 above 30 GeV.

2.2 Jet Energy Calibration

Jet energies measured in the calorimeter are distorted by particle energy losses in the material in front of the calorimeter, detector noise, phenomena which affect the response such as detector boundaries, non-uniformities and e/π, and jet reconstruction and resolution effects. The goal of the jet energy calibration is to obtain the particle-level or true jet energy from the observed jet energy measured in the calorimeter. Energy corrections are derived using the reconstructed Run 1 data sets at center-of-mass energies 1800 GeV and 630 GeV. A fixed cone algorithm is used to reconstruct jets from cell energy depositions in the calorimeter. The particle level jet energy, E_{jet}^{ptcl}, is obtained from

$$E_{jet}^{ptcl} = \frac{E_{jet}^{meas} - E_O}{R_{jet} \cdot S}. \tag{1}$$

E_O is an offset correction. It subtracts the excess energy not associated with the high p_T interaction which includes the detector noise, the effects of previous crossings (pile-up), the underlying event contributed by spectator partons to the high p_T interaction and the contribution from additional $p\bar{p}$ interactions. R_{jet} is a measure of the calorimeter energy response to jets due to e/π and energy losses in the calorimeter cracks. S compensates for the net energy flow through the jet cone boundary during the shower development in the calorimeter. The jet energy calibration is performed for several cone sizes, $\mathcal{R}$, from 0.3 to 1. Representative plots shown below are typically for central 0.7 cone jets.

2.3 Offset Correction

The total offset correction is measured as a transverse energy density in $\eta - \phi$ space and can be written as $D_O = D_{ue} + D_\theta$. The first term is the contribution from the physics underlying event, or the energy associated with the spectator partons. It is measured as the average transverse energy density in minimum bias events, which are dominated by hard core interactions. The second term accounts for the detector noise, pile-up effects, and energy from additional $p\bar{p}$ interactions. Pile-up increases with luminosity since it depends on the number of interactions in the previous crossing. D_θ is determined from a zero bias data sample, recorded on arbitrary beam crossings, which may or may not contain a hard interaction. The top plot in Fig. 7 shows the η dependence of D_{ue} for two center-of-mass energies 1800 and 630 GeV. D_{ue} is independent of the luminosity and the number of $p\bar{p}$ interactions in the event. The η and luminosity dependence of D_θ are shown in the bottom plot of Fig. 7.

2.4 Response and Showering Corrections

The overall calorimeter response to particles is less than unity. This is due to its non-linear response to low energy particles, inactive material, detector boundaries, and module-to-module non-uniformities. The $\gamma - jet$ data sample is a useful calibration tool in determining the calorimeter response to jets. These events consist of a photon balanced in p_T by one or more jets. For a $\gamma - jet$ event in the DØ detector, any non-zero E_T gives a measure of the overall transverse energy imbalance due to the differences in calorimeter response to photons and jets. In this case the jet response can be derived from

$$R_{jet} = 1 + \frac{\vec{\not{E}}_T \cdot \hat{n}_{T_\gamma}}{E_{T_\gamma}} \qquad (2)$$

where E_{T_γ} and $\hat{n}_{T_\gamma}$ are the transverse energy and direction of the photon, respectively.

R_{jet} includes η-dependent corrections arising from detector inhomogeneities and boundaries. Energy dependent corrections are also included to account for the jet

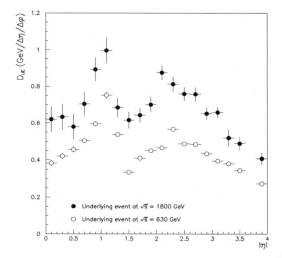

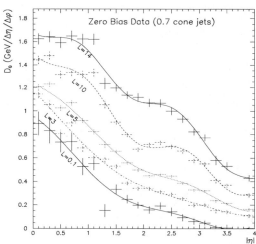

Figure 7: (Top) The underlying event density D_{ue} plotted as a function of η for the center-of-mass energies 1800 GeV and 630 GeV. (Bottom) The eta dependence of D_θ for different luminosities in units of $10^{30} cm^{-2} s - 1$.

E_T threshold of 8 GeV, jet reconstruction inefficiencies and finite jet energy resolutions. The use of the CC data is limited to energies below 120 GeV due to the rapidly falling photon cross section. Data from the EC are used to extend the energy reach up to 300 GeV. Monte Carlo information is also included at high energy to constrain the jet energy response extrapolated from the data. The variation of R_{jet} with the jet energy, shown in Fig. 8 for $\mathcal{R} = 0.7$ cone jets, is independent of the cone size and is the same for the 1800 GeV and 630 GeV data sets. The top plot is the same as the bottom plot except the jet energy axis is logarithmic to show the behavior of R_{jet} at low jet energies. The solid curve shown is a fit to the data with the associated error band. The fit has a functional form $R_{jet} = a + b \cdot \ln(E) + c \cdot \ln(E)^2$.

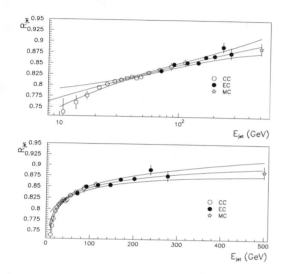

Figure 8: The dependence of R_{jet} on the jet energy, from the data using the central (CC) and end (EC) calorimeters, with 0.7 cone jets. The solid curve is the fit with the associated error band.

The last jet energy correction applied corrects for the effects of showering. It compensates for the energy associated with particles emitted inside the cone but deposited outside the cone as the shower develops. S is less than one which implies that the effect of showering is a net flow of energy from inside to outside the cone. The energy outside the cone may be associated with jet fragmentation outside the cone or gluon radiation. It may also be attributed to detector noise, pile-up and underlying event. The out-of-cone jet energy from gluon radiation and fragmentation effects, determined using the HERWIG [8] Monte Carlo at particle-level, is subtracted from the out-of-cone jet energy measured in the data. Showering losses are derived using jet energy density profiles using both the data and particle-level Monte Carlo. These losses have been shown [6] to have a strong dependence on the jet cone size $\mathcal{R}$, the jet energy, and η. They are independent of the center-of-mass energy for jets of the same energy and pseudorapidity.

2.5 Total Jet Energy Correction Factor

Figure 9 shows the total jet energy correction and the associated uncertainties as a function of the uncorrected jet energy for $|\eta| = 0$. The overall correction factor in the central calorimeter is 1.16 ± 0.015 and 1.12 ± 0.023 at 70 and 400 GeV, respectively. In general, the correction factors are about 15% above 30 GeV and almost flat all the way to E_T of 200 GeV. At lower jet energies, smaller cone sizes, and larger pseudorapidities, the corrections and errors increase. The total error is dominated by the contribution from showering corrections at large values of η.

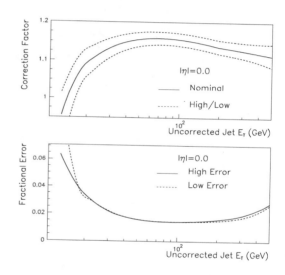

Figure 9: The total jet energy correction and associated errors versus the uncorrected jet energy.

3 Summary

The DØ experiment has performed a detailed evaluation of the jet energy scale based primarily on data taken during the 1992-1996 $p\bar{p}$ collider run at center-of-mass energies $\sqrt{s} = 1800$ GeV and 630 GeV. The energy corrections compensate for spectator interactions, detector noise, calorimeter response, and showering losses for the transverse energy and pseudorapidity range $E_T > 8$ GeV and $|\eta| < 3$. The correction procedure has been verified, using a HERWIG Monte Carlo sample with the full DØ detector simulation, which shows that the jet energy is corrected to the particle-level within the quoted errors.

This report also presented a general overview of the DØ upgrade. All aspects of the upgrade are well underway with many detector components moving into the production and construction stages. The upgraded DØ detector is scheduled to be complete by the spring of the year 2000.

References

1. DØ Collab., FERMILAB-PUB-96-357 (1996).
2. D. Adams *et al.*, IEEE Trans. Nucl. Sci. **43**, 1146 (1996).
3. DØ Collab., The DØ Silicon Tracker Technical Design Report, Fermilab (1995).
4. R. Yarema *et al.*, FERMILAB-TM-1892 (1994).
5. DØ Collab., *Nucl. Instrum. Methods* A **338**, 185 (1994).
6. DØ Collab., FERMILAB-PUB-97-330 (1998), submitted to *Nucl. Instrum. Methods* A.
7. J. Kotcher for the DØ Collab., FERMILAB-CONF-95-007 (1995).
8. G. Marchesini *et al.*, hep-ph/9607393 (1996).

THE KLOE CALORIMETRIC SYSTEM FOR NEUTRAL PARTICLE DETECTION AND TRIGGERING

F. Cervelli for the KLOE collaboration [a]

Dipartimento di Fisica dell'Università e Sezione INFN, Pisa

The KLOE experiment at the DAΦNE ϕ-factory in Frascati is devoted to study the CP violation in the neutral Kaon decays. The e.m. calorimeter of the experiment is arranged as a barrel closed at both ends with two end-caps. The barrel is organized in 24 modules 4.3 m long; each end-cap is made of 32 modules of different length. The modules are structured as a fine sampling lead-scintillating fibers calorimeter. At the test beam this calorimeter has shown linearity from 20 MeV to 300 MeV, good energy resolution and excellent time resolution. The calorimetry system is completed by two compact detectors (QCAL) placed close to the interaction point and surrounding the focalization quadrupoles. Their purpose is to increase the hermeticity and to improve the background rejection. The installation of the e.m. calorimeter ended in December 1997. Afterward, cosmic rays were used for intensive calibration studies of the whole calorimeter, collecting a large amount of data by means of the DAQ system of the experiment. In this article the first results of these calibration studies are reported.

1 Introduction

The main goal of the KLOE experiment at the DAΦNE ϕ–factory is to study the CP violation in the neutral kaon decays with an accuracy of $\mathcal{O}(10^{-4})$ in $\Re(\epsilon'/\epsilon)$ [1,4]. The identification of the neutral decays ($K_{L,S} \rightarrow \pi^0\pi^0 \rightarrow 4\gamma$'s) in the photon energy range 20 MeV $< E_\gamma <$ 280 MeV, the reconstruction of the decay path K_S, $K_L \rightarrow \pi^0\pi^0$ to a precision of ~ 1 cm and the rejection of the $K_L \rightarrow \pi^0\pi^0\pi^0$ background at the 10^3 level, make the electromagnetic calorimeter one of the most demanding elements of the detector. The reconstruction of the neutral vertex $K^0 \rightarrow \pi^0$'s $\rightarrow 4\gamma$'s in KLOE will be performed by measuring the impact point and the arrival time of each photon on the calorimeter surface. Given the low speed of the K^0 's ($\beta \sim 0.2$),

[a] M. Adinolfi, F. Ambrosino, A. Aloisio, A. Andryakov, A. Angeletti, A. Antonelli, C. Bacci, R. Baldini-Ferroli, G. Barbiellini, G. Bencivenni, S. Bertolucci, C. Bini, C. Bloise, V. Bocci, F. Bossi, P. Branchini, L. Bucci, G. Cabibbo, A. Calcaterra, R. Caloi, G. Carboni, P. Campana, G. Capon, M. Carboni, C. Carusotti, G. Cataldi, F. Ceradini, F. Cervelli, F. Cevenini, G. Chiefari, P. Ciambrone, S. Conticelli, E. De Lucia, R. De Sangro, P. De Simone, G. De Zorzi, S. Dell'Agnello, A. Denig, A. Di Domenico, S. Di Falco, A. Doria, F. Donno, E. Drago, V. Elia, L. Entesano, O. Erriquez, A. Farilla, G. Felici, A. Ferrari, M. L. Ferrer, G. Finocchiaro, D. Fiore, C. Forti, G. Foti, A. Franceschi, P. Franzini, A. Galli, M. L. Gao, C. Gatti, P. Gauzzi, S. Giovannella, V. Golovatyuk, E. Gorini, F. Grancagnolo, E. Graziani, P. Guarnaccia, U. v. Hagel, H. G. Han, S. W. Han, M. Incagli, L. Ingrosso, W. Kim, W. Kluge, V. Kulikov, F. Lacava, G. Lanfranchi, J. Lee-Franzini, T. Lomtadze, T. Lomtadze, C. Luisi, A. Martini, W. Mei, L. Merola, R. Messi, S. Miscetti, A. Moalem, S. Moccia, F. Murtas, M. Napolitano, A. Nedosekin, P. Pagès, M. Palutan, L. Paoluzi, E. Pasqualucci, L. Passalacqua, M. Passaseo, A. Passeri, V. Patera, E. Petrolo, G. Petrucci, D. Picca, M. Piccolo, A. Pintus, G. Pirozzi, L. Pontecorvo, M. Primavera, F. Ruggieri, P. Santangelo, E. Santovetti, G. Saracino, R. D. Schamberger, C. Schwick, B. Sciascia, A. Sciubba, F. Scuri, I. Sfiligoi, S. Sinibaldi, T. Spadaro, S. Spagnolo, E. Spiriti, C. Stanescu, L. Tortora, E. Valente, P. Valente, G. Venanzoni, S. Veneziano, D. Vettoretti, S. Weseler, Y. G. Xie, C. D. Zhang, J. Q. Zhang, P. P. Zhao

a time resolution of ~ 50 ps$/\sqrt{E(\text{GeV})}$ and a measurement of the photon conversion point with an accuracy of ~ 1 cm are required. The rejection factor of $\mathcal{O}(10^3)$ for the $K_L \rightarrow \pi^0\pi^0\pi^0$ background requires a calorimeter with a good energy resolution (~ 5 %$/\sqrt{E(\text{GeV})}$) and hermetic coverage of the fiducial volume. Finally, the calorimeter should provide a fast and unbiased first level trigger and some particle identification to aid rejection of $K_{\mu 3}$ decays.

2 The KLOE electromagnetic calorimeter

The KLOE calorimeter consists of a central part, the *barrel*, and two *endcaps* [2,5]. The barrel is organised in 24 modules of trapezoidal cross section, 4.3 m long, $\sim$ 60 cm wide and 23 cm thick. They form a cylindrical shell of 4 m inner diameter and cover a polar region $49° < \theta < 131°$. The fibers run parallel to the beams. Each endcap is made of 32 modules of different lengths, 23 cm thick, and running vertically along the chords of the circle inscribed in the barrel. In each module the two ends are bent outwards and become parallel to the barrel. The resulting C-shape provides a hermetic coverage of the polar region $9° < \theta < 46°$.

Each calorimeter module is built by glueing 1 mm diameter fibers between grooved lead planes obtained by plastic deformation of 0.5 mm thick lead foils. When layers are superimposed the fibers are located at the corners of quasi-equilateral triangles with 1.35 mm fiber pitch. The resulting structure has a fiber:lead:glue volume ratio of 48:42:10, an average density of $\bar{\rho} \sim 5$ g/cm^3 and a mean radiation length $\overline{X_0} \sim 1.6$ cm. The final composite has a considerable stiffness and can be easily machined. The small lead foil thickness ($< 0.1\ \overline{X_0}$) results in a quasi-homogeneous structure and in a high efficiency for low energy photons ($\sim 99\%$ at 20 MeV).

The readout granularity is $\sim (4.4 \times 4.4)$ cm^2 for a total number of 4880 read-out channels. The granularity is such to define five radial planes in each module. The light is collected at the two module ends by means of light guides coupled with photomultipliers (PM's). Each light guide consists of a tapered mixing part and a Winston cone concentrator, to maximize the light collection efficiency. Since the calorimeter is inside a 0.6 T magnetic field and in the PM area the residual magnetic field is up to 2 kGauss with an angle with respect to PM axis up to 25°, fine mesh PM's, Hamamatsu R5946, specially designed for KLOE, are used [5].

A first full size barrel module was tested [6-7] at Paul Scherrer Institute (Zurich) with e^-, μ^- and π^- beams of momentum 100 MeV/c $< p <$ 450 MeV/c. Calibration was done using the energy release of minimum ionizing particles impinging perpendicularly at module center (*MIP*).

The energy response was linear and independent of the beam incidence angle. Signals produced by e.m. showers were equivalent to 31 MIP/GeV, corresponding to a sampling fraction of $\sim$13%. The light yield was $\sim$ 1330 N_{pe}/GeV, the energy resolution resulted to be $\sigma(E)/E = 4.7\%/\sqrt{E(\text{GeV})}$ and it was fully dominated by sampling fluctuations [6,7].

The arrival time of a particle impinging on the calorimeter surface was obtained by a proper energy weighting of the times measured by the cells which belonged to the shower. The time resolution scaled with energy as $\sigma_t = 55$ ps/$\sqrt{E(\text{GeV})}$, showing to be fully dominated by the light yield [6,7].

The coordinate along the fiber direction, the z-coordinate, was measured using the time difference ΔT at the two module ends and the effective light propagation speed (v_f) in the fibers, through the relation $z = \frac{1}{2} v_f \times \Delta T$. v_f was valued to be about 17.0 cm/ns, in good agreement with the refractive index of the fiber core ($n = 1.6$) and the bounce angle ($\theta \sim 21°$) of the light travelling in the fiber. The z resolution was $\sigma_z = 1.24$ cm/$\sqrt{E(\text{GeV})}$.

All the produced modules were tested in Frascati at a Cosmic Ray Stand [10]: performances became better and better along production and a final time resolution, $\sigma_t = 50$ ps/$\sqrt{E(\text{GeV})}$ can be quoted.

3 The calorimeter trigger system

Due to the high peak luminosity of the DAΦNE accelerator (10^{33} cm^{-2}s^{-1}) event and background rates at KLOE are higher than in any other e^+e^- collider experiment. The trigger must run continuously, since the high bunch crossing rate of 370 MHz excludes a trigger decision to be formed between bunch crossings. Given the average event size of 5 Kbytes, maximal data rates of

25 Mbytes/s are expected from physics events. Various background sources are entering the detector with a rate of several hundred MHz. Main sources are machine background (beam gas interactions, Touschek effect), cosmic ray muons, and Bhabha scattering. The trigger is required to have an efficiency larger than 99% for all events relevant to investigate CP violation, while rejecting the background so to keep the final data rate at the level of 10 kHz. The adopted trigger is a two level system, which uses information both from the e.m. calorimeter and the central drift chamber.

To achieve a high trigger purity, the following differences between physics and background events are exploited:

- Calorimeter hits of background events are concentrated in the endcaps, since the particles have a low transverse momentum with respect to the beam.

- Background events have drift chamber hits predominantly in the region close to the beam pipe, whereas in physics events they are distributed all over the chamber volume.

- Background events have relatively low multiplicities in the calorimeter and in the drift chamber.

Specifically, for triggering purposes, the calorimeter is divided into 200 overlapping sections comprising 20 to 30 adjacent photomultipliers on each readout side of the calorimeter. Analogue signals of these sectors are summed for each side separately, and are then applied to two programmable trigger thresholds. The lower threshold is needed to trigger on small energy deposits by low energy particles, the higher to form a veto for Bhabha events. A trigger decision is taken within a time window on the basis of the number and location of sectors with signals above the two aforementioned thresholds.

The complete hardware of the ECAL trigger has been installed at the beginning of 1998. It was tested with cosmic rays and by pulsing the ECAL P.M.'s preamplifiers at high frequencies. Pulse rates up to 200 kHz were triggered for several hours without errors. This corresponds to a frequency about 10 times larger than expected at DAΦNE with the maximum luminosity. At present the system is used to trigger on cosmic rays events at a rate of about 3-5 kHz (depending on trigger's conditions). With these data the tigger efficiency as a function of the deposited energy can be studied for every sector. A first analysis shows that it behaves according to project's expectations; on average, the uncertainty on the threshold's value is of the order of or less than the uncertainties due to the calorimeter's resolution.

4 Calibration studies

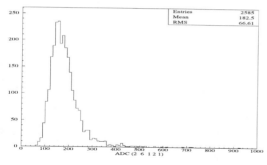

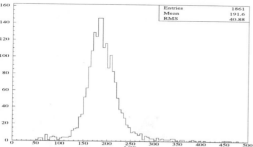

Figure 1: a) Distribution of m.i.p.s energy deposition in one calorimeter cell in ADC counts b) Distribution of MIP peaks for the barrel calorimeter

After the installation of the calorimeter modules and of the final front end electronics, the use of the KLOE DAQ [8] and trigger [9] systems allows routinely acquisition of cosmic rays events for fast measurements of the main calibration parameters. The trigger requires at least one calorimeter sector with an energy deposition of $\sim 1/2$ MIP (typical trigger rate ~ 3 Khz). The DAQ system is able to record these events rate with no effort once the appropriate zero suppression is performed (typical event size of 400 Bytes for a total data throughput of 1.2 Mbytes/sec).

Fig. 1.a shows a typical charge distribution of an ADC channel for tracks passing within 20 cm from the middle of the module. A sample of clean minimum ionizing particles (m.i.p.s) is selected requiring the track to be almost orthogonal to the module's longitudinal face; the impact angle is determined using the calorimeter information only by calculating, for each radial plane, the three coordinates (x,y,z) of the charge center of gravity and then by fitting the reconstructed points with a straight line. Only tracks with incidence angle $\theta < 23°$ with respect to the normal incidence are selected. Fitting these distribution with a gaussian we determine the energy deposition in a cell (MIP) for a m.i.p.

In Fig. 1.b the distribution of the MIP values for most of the barrel PM's is reported after the HV setting determined at the Cosmic Ray Stand [10], has been downloaded. One MIP roughly corresponds to 180 ADC counts with an equalization of ~ 15 %. The equalization in response reached at the CRS was at a level of ~ 2 %. This discrepancy is due to the fact that, before mounting the final set of PM's, all ligth guides were cleaned and a new layer of optical grease was added to obtain a good optical contact. At present a process is in progress to readjust the HV setting and tidy up the equalization.

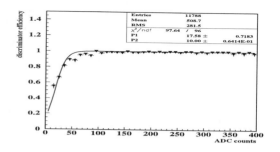

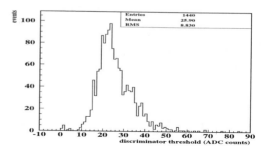

Figure 2: a) TDC firing efficiency (ϵ) as function of pulse height in ADC counts b) Distribution of discriminator thresholds in ADC counts

The discriminators threshold for each channel was evaluated by looking at the ratio (ϵ) between the number of events firing the TDC channel and the ones firing the corresponding ADC channel as a function of the ADC content. An example of such a distribution for a single PM channel is shown in fig. 2.a: the threshold can be derived from the ADC value at $\epsilon = 0.5$. The distribution of the threshold settings for all the TDC channels is reported in fig. 2.b: the average threshold corresponds to $\sim 1/6$ of a MIP.

The difference $\Delta T = T_a - T_b$ between the arrival

PRECISION LUMINOSITY MEASUREMENT WITH THE OPAL SILICON-TUNGSTEN CALORIMETERS

SILVIA ARCELLI

for the OPAL collaboration

University of Maryland, College Park, Maryland 2074, USA

Present address: CERN-PPE, CH 1211 Geneva 23, Switzerland.
E-mail: Silvia.Arcelli@cern.ch

A preliminary, high precision measurement of the luminosity of the LEP colliding beams for the LEP I data recorded between 1993 and 1995 with the OPAL experiment is presented. The measurement is based on the OPAL Silicon-Tungsten Luminosity Monitor, which detects electrons from small-angle Bhabha scattering at angles between approximately 25 and 58 mrad. The overall experimental uncertainty on the measured luminosity is 3.3×10^{-4}. Such uncertainty contributes negligibly to the uncertainty on the OPAL cross section measurements near the Z^0 resonance.

1 Introduction

The reference process used at LEP to monitor the luminosity of the colliding beams is the small-angle $e^+e^- \to e^+e^-$ Bhabha scattering, which offers the following advantages:

- due to the steepness of the Bhabha angular distribution, by instrumenting a region of the detector at suffincently low angle, the counting rate from small angle Bhabha events can be made larger than the rate of Z^0 production. In this way the contribution of the luminosity measurement to the statistical error of the measured cross sections is minimised.

- being dominated by the pure QED t-channel photon exchange, the small angle Bhabha cross section is largely independent of the electroweak parameters and can be calculated theoretically with high precision.

- it is essentially free from background from other physics processes.

As many of the electroweak measurements performed in electron-positron annihilations at the Z^0 resonance [1] rely on cross sections measurements, a very precise determination of the luminosity is required at LEP. In particular, for an efficient use of the $5 \cdot 10^6$ Z^0 samples which were collected by each experiment, the luminosity must be measured with an accuracy of better than 0.1%.

In 1993 OPAL installed a new luminometer which employs calorimeters with tungsten absorber and silicon sampling. In this document the determination of the OPAL luminosity for the 1993, 1994 and 1995 LEP runs at the Z^0 is described.

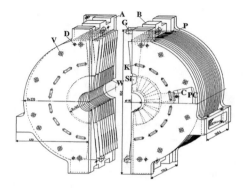

Figure 1: The isometric view of one of the SiW calorimeters.

2 The detector

The SiW Luminometer consists of two higly homogeneus, cylindrical calorimeters encircling the beam pipe at approximately ±2.5 m from the interaction point and covering the polar angles between approximately 25 and 58 mrad. One such calorimeter is shown in figure 1. Each calorimeter is made of 18 tungsten plates, interleaved with 19 layers of highly segmented silicon diode detectors. The first 14 tungsten plates are 1 X_0 thick, while the last 4 are 2 X_0 thick, for a total of 22 radiation lengths (X_0).

Each active layer consists of 32 large, thick-film, ceramic hybrid carrying a 64-pad silicon wafer diode and four AMPLEX chips for the detector readout. The pads have a radial pitch of 2.5 mm, and are arranged in an $r-\phi$ geometry. From the charge induced on the pads the position and the energy of the 45 GeV electrons showers are reconstructed with an inefficiency lower than 10^{-5}. The calorimetric approach, together with the fine radial and longitudinal granularity of the detector, allows the reconstruction of precise and continuous radial coordinates and the separation of multiple clusters above 5 mrads.

3 Analysis strategy and systematic errors

The luminosity is measured by counting Bhabha events in two asymmetrical fiducial acceptances defined by the following "Narrow-Wide" angular cuts:

- SWITR   7.7 cm $< R_R <$ 12.7 cm
                      6.7 cm $< R_L <$ 13.7 cm

- SWITL   7.7 cm $< R_L <$ 12.7 cm
                      6.7 cm $< R_R <$ 13.7 cm,

where R_R (R_L) is the radial coordinate calculated with respect to the beam axis in the right (left) calorimeter. The reference luminosity is then defined as:

$$\mathcal{L} = \frac{1}{2} \frac{(N_{\text{SWitR}} + N_{\text{SWitL}})}{\sigma_{\text{SWitRL}}}, \quad (1)$$

where $N_{\text{SWitR,SWitL}}$ are the number of events accepted by the SWITR, SWITL selections. The BHLUMI Monte-Carlo [2], convoluted with a detailed parametrization of the detector, was used to calculate σ_{SWitRL}, the average of the Bhabha cross sections in the two acceptances.

One crucial aspect in the analysis is that the forward-peaked $1/\theta^3$ Bhabha spectrum requires the inner edge of the acceptance to be defined with very high precision. More specifically, the uncertainty on the inner radius of the angular acceptance propagates into a systematic uncertainty on the luminosity according to the relation:

$$\frac{\Delta L}{L} \approx \frac{\Delta r_{\text{inner}}}{25 \,\mu\text{m}} \times 10^{-3}. \quad (2)$$

Therefore, both the radial geometry of the detector and any possible bias affecting the radial coordinate must be known with an accuracy better than 25 μm to reach a 0.1% precision in the luminosity.

Another potentially large source of uncertainty in the measurement is the sensitivity of the Bhabha acceptance to the movements of the interaction point and the other beam parameters. The azimuthal symmetry of the detector, coupled with the careful definition of the event selection, minimizes such effects ensuring that they influence the reference acceptance only in second-order.

An uncertainty finally arises from the off-momentum beam particles background, whose accidental coincidences can mimic a Bhabha event. Additional conditions on the acoplanarity, acollinearity, and the energies

of the electron and the positron are applied to strongly suppress such background. The sensitivity to the modelling of the detector response at low energy, which could be introduced by the energy requirements, is significantly reduced by the acollinearity cut, which causes the acceptance for radiative events to be determined by geometry rather than by the energy requirements.

In the following each of the sources of systematic uncertainty mentioned above will be discussed in more detail.

3.1 Detector Metrology

A very careful metrology of the detector, both during assembly and during operation conditions, was performed to establish the absolute radial dimensions of the detector. Taking into account mechanical deformations, temperature effects, and the precision of the optical apparatus used for the survey, the inner radius of the luminometer acceptance is determined with an accuracy of 4.4 μm. This uncertainty in the acceptance radius contributes a systematic error of 1.4×10^{-4} to the OPAL luminosity measurement during 1993-1995. A much smaller uncertainty of 5×10^{-5} arises from the axial metrology of the detector.

3.2 Radial Coordinate

In the reconstruction of the radial shower coordinate, shown in figure 2 for the left and the right calorimeters, the pad having the maximum energy and the two pads adjacent to it in 9 consecutive layers (from 2 to 11 X_0) are considered. The information from the different planes is combined and projected onto a reference plane in the detector at 7 X_0. Such a procedure allows for the reconstruction of a continous coordinate with an average resolution of 160 μm. The crucial point in the measurement is however to minimize any bias in the shower radius at the inner edge of the acceptance.

A bias is actually expected due to the $r - \phi$ geometry of the pads, which will cause the true position of electrons depositing equal equal charge in two nearby pads to lie at a smaller radius than the physical pad boundary. This bias depends sensitively on the lateral shower profile, and was directly measured for 45 GeV electrons in a test beam, with an SiW calorimeter module fully-equipped in depth, and a four-plane, double-sided Si micro-strip telescope with a resolution of better than 3μm for individual tracks. A beam of 100 GeV muons and the sensitivity of the SiW electronics to individual mips was used to align the telescope to the detector.

From the test beam measurements the bias of the pad maximum is parameterized in terms of a variable related to shower size, the pad-maximum transition width,

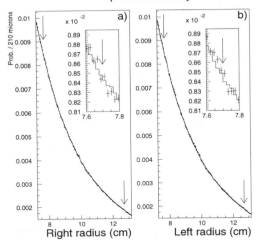

Figure 2: The distribution of the radial coordinates R_R and R_L in the right (a) and left (b) calorimeter, for the data (points) and the MonteCarlo (histogram). The narrow angular cuts defining the acceptance used in the measurement are shown as arrows. The wide angular cuts have already been applied.

and this variable is then measured in the data to directly determine the bias of the radial coordinate used in the luminosity measurement. The combined statistical and systematic uncertainty on such bias is $3.8\,\mu m$, which translates into an uncertainty on the luminosity of 1.2×10^{-4}.

3.3 Beam Parameters

Although the Bhabha selection is chosen to minimize the effects of the LEP beam parameters on the luminosity measurement, the eccentricity of the beam with respect to the detector, which is typically of order 1 mm, still produce a second-order variation of several 10^{-4} in the measured luminosity. The eccentricities, which are a function of the offset of the interaction point and the tilt of the beam direction with respect to the detector axis, can be measured on a fill-by-fill basis from the azimuthal modulation of the intensity of Bhabha events and from the Bhabha acollinearity distribution, shown in figure 3, with a precision of few microns. A correction is then applied to the measured luminosity.

The finite beam sizes and divergences at the OPAL interaction region produce variations in the acceptance used for the luminosity of approximately 1×10^{-4}. By using measurements of the longitudinal and transverse size

of the luminous region from the OPAL central detector and micro-vertex detector, as well from the acollinearity distribution of Bhabha events, such parameters are also measured and incorporated into the simulation used to extract the final acceptance.

The overall uncertainty deriving from the uncomplete knowledge of the LEP beam parameters results in a systematic error on the luminosity of 0.57×10^{-4}.

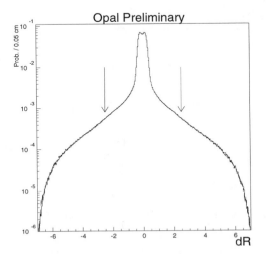

Figure 3: The acollinearity distribution of Bhabha events, after all cuts except that on the acollinearity, for data (points) and MonteCarlo (histogram). The double-peak structure in the core of the distribution is generated by the eccentricity of the beam with respect to the detector.

3.4 Beam Background

Accidental coincidences of beam background clusters in the right and left calorimeters can produce events which will potentially pass the Bhabha selection. Because the trigger signals from every beam crossing are digitized and recorded, it is possible to form a trigger which requires a coincidence when one side is delayed by one full revolution of LEP with respect to the other. The background fractions measured with this method are between 0.1 and 0.6×10^{-4}, with a systematic uncertainty smaller than 0.1×10^{-4}.

The accidental overlap of a background cluster with a Bhabha event can also change the values of reconstructed quantities, modifying the acceptance. This effect has been evaluated both by incorporating the measured background into the detector simulation and by convoluting the measured background with data events. The Bhabha acceptance varies by less than 10^{-4}. A systematic error of 0.75×10^{-4} is assigned, which covers the difference between the two methods.

3.5 Detector Simulation

The OPAL SiWdetector simulation is based on a detailed parametrization of the detector response in energy, radius and azimuth, which was derived from test beam measurements and the data itself.

The energy response in terms of sensitivity to material upstream the detector and effects of non linearity has been carefully studied using the kinematical properties of radiative Bhabha events and the very precise radial information. The uncertainties related to these tests result in a systematic error on the luminosity of 1.8×10^{-4}. A systematic of 1×10^{-4} is also assigned to account for uncertainties in the implementation of the clustering algorythm in the MonteCarlo. Finally, the uncertainties from the simulation of the radial and azimuthal coordinates are found to produce a negligible effect on the luminosity.

4 Results and Conclusions

The luminosity of the colliding beams at LEP for the OPAL 1993-95 data has been measured with a (preliminary) overall experimental systematic error of:

$$d\mathcal{L}/\mathcal{L} \; = \; 3.3 \times 10^{-4}.$$

This precision represents a two-order of magnitude improvement compared to luminosity measurements done below the Z^o peak [3], and more than a factor of 10 improvement compared to previously published OPAL luminosity measurement [1]. In the overall uncertainty, the current theoretical error of 11×10^{-4} associated to the Bhabha cross section is dominant [a] To quantify the impact of the luminosity uncertainty on the electroweak observables which are derived from lineshape measurements, the ratio of the Z^0 invisible width to the leptonic width can be considered:

$$R_{\mathrm{inv}} \; = \; \left(\frac{12\pi}{\sigma_{\ell^+\ell^-}^{\mathrm{pole}} m_Z^2} \right)^{\frac{1}{2}} - R_{\mathrm{had}} - 3$$

where $\sigma_{\ell^+\ell^-}^{\mathrm{pole}}, m_Z$ and R_{had} are respectivly the leptonic pole cross section, the mass of the Z^0 and the ratio of the hadronic to the leptonic pole cross sections. This quantity is predicted with very high accuracy in the Standard Model and is sensitive to the production of any new invisible particles coupling to the Z^0 at the tree level. The individual contributions to the uncertainty on R_{inv} are listed in table 1. The experimental uncertainty on the luminosity clearly does not represent a limiting factor in the precision of the measurement.

Table 1: The individual contribution to the relative error on R_{inv} from the experimental and theoretical uncertainty on the luminosity, in units of 10^{-4}. For comparison, the contribution from the experimental error on hadrons and leptons channels are also listed.

Quantity	Error (stat) ($\times 10^{-4}$)	Error (Sys) ($\times 10^{-4}$)	Error on R_{inv} ($\times 10^{-4}$)
hadrons	6	7	32
leptons	17	13	21
Luminosity	3	3	11
Bhabha Theory	-	11	27
total			48

References

1. OPAL Collaboration, Akers et al., Z. Phys. **C61** (1994) 19;
   ALEPH Collaboration, Buskulic et al., Z. Phys. **C62** (1994) 539;
   DELPHI Collaboration, Abreu et al., Nucl. Phys. **B418** (1994) 403;
   L3 Collaboration, Acciari et al., Z. Phys. **C62** (1994) 551.
2. S. Jadach et al., Comp. Phys. Com. **102** (1997) 229.
   S. Jadach et al., Comp. Phys. Com. **70** (1992) 305.
3. HRS Collaboration, D. Bender et al., Phys. Rev. **D31** (1985) 1;
   MAC Collaboration, E. Fernandez et al., Phys. Rev. **D31** (1985) 1537;
   CELLO Collaboration, H.-J. Behrend et al., Phys. Lett. **B183** (1987) 407;
   JADE Collaboration, W Bartel et al., Phys Lett. **B160** (1985) 337;
   B. Adeva et al., Phys. Rev. **D34**, (1986) 681;
   TASSO Collaboration, M. Althoff, et al., Phys Lett. **B138** (1984) 441.

---

[a] A substantial improvement in the theoretical error was announced at this conference by Prof. B.Ward, these proceedings.

THE ATLAS ELECTROMAGNETIC CALORIMETER AND THE ATLAS LEVEL-1 CALORIMETER TRIGGER

G.SAUVAGE[a]

L.A.P.P., Chemin de Bellevue BP110, 74941 Annecy le Vieux Cedex, France
E-mail:gilles.sauvage@lapp.in2p3.fr

P.BRIGHT-THOMAS[b]

School of Physics and Astronomy, University of Birmingham, Edgbaston, Birmingham B15 2TT, England
E-mail: pbt@hep.ph.bham.ac.uk

A short description of the ATLAS liquid argon electromagnetic calorimeter is given. The construction of modules 0 for the central and end cap calorimeters as well as for the central preshower is reviewed. These modules are prototypes of the serial construction. The architecture and the technologies of the level-1 calorimeter trigger are described. The trigger algorithms are presented and the expected performance is summarised.

Introduction[ab]

The Large Hadron Collider (LHC) at CERN will provide proton-proton collisions at a centre of mass energy of 14 TeV with a design luminosity of 10^{34} cm^{-2}s^{-1}. This luminosity and the 25ns bunch crossing time impose severe constraints on the calorimeters of the ATLAS experiment[1]. Below are described the ATLAS electromagnetic calorimeters[2] and the ATLAS level-1 calorimeter trigger[3].

1 The ATLAS electromagnetic calorimeter

1.1 Choice of the liquid argon calorimetry

The liquid argon sampling calorimetry has several advantages in the LHC environment : its intrinsic radiation hardness, the full hermeticity apart from cryostat walls in η, the possibility of a high segmentation, the uniformity, the large dynamic range and the long term stability of the response facilitating the calibration. The use of the accordion geometry for the lead absorbers and the electrodes allow a fast readout scheme as required by the 25ns bunch crossing time. This new geometry has been fully studied by the RD3 collaboration[4]. It has been shown that the following energy, position and angular resolutions can be obtained for the ATLAS electromagnetic calorimeter : $\sigma_E/E \leq 10\%/\sqrt{(E(\text{GeV}))} \oplus 0.7\% \oplus 0.27/E(\text{GeV})$
$\sigma_r \leq 8$ mm$\sqrt{(E(\text{GeV}))}$ $\sigma_\theta \leq 40$ mrad$/\sqrt{(E(\text{GeV}))}$
These resolutions allow the detection of H $\rightarrow \gamma\gamma$ in the low mass $\sim 90 \leq M_H \leq 130$ GeV range.

The ATLAS electromagnetic calorimeter (figure 1) is composed of a central calorimeter and two end-caps

---
[a]Talk 1408 (abstracts 348 and 749) presented by G.Sauvage
[b]on behalf of the Level-1 Calorimeter Trigger Collaboration

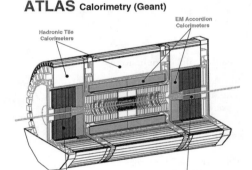

ATLAS Calorimetry (Geant)

Figure 1: The ATLAS calorimeters.

calorimeters housed in 3 different cryostats. The end-cap cryostats house also the hadronic end-cap calorimeters and the forward calorimeters.

1.2 The barrel calorimeter and the barrel presampler

Two half barrels cover the $-1.4 \leq \eta \leq 0$. and $0. \leq \eta \leq 1.4$ ranges. Each half barrel (figure 2), about 3.2m long and with an internal and external radii of 1.5m and 2.1m, is composed of 16 modules covering a $\delta\phi = 22.5°$ range. Each module contains 64 accordion absorbers and 64 electrodes. An half barrel is supported by 7 stainless steel rings (each ring is segmented in 16 ring pieces).

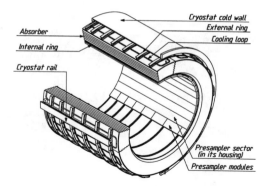

Figure 2: The electromagnetic barrel calorimeter (general layout).

On the internal face of each barrel module are fixed two presamplers sectors. The aim of the presampler is to measure the energy lost in the dead material in front of the calorimeter (cryostat walls, inner solenoid, inner detector). Each sector is composed of sub-sectors covering each a $\delta\eta = .2$ range.

1.3 The end-cap calorimeters

Two end-cap calorimeters cover the $1.375 \leq |\eta| \leq 3.2$ ranges. One end-cap (figure 3) is composed of two concentric wheels ($1.4 \leq |\eta| \leq 2.5$ and $2.5 \leq |\eta| \leq 3.2$ respectively). The internal and external radii for the outer (inner) wheel are .6m and 2.1m (.3m and .6m) and the thickness is about .5m. Each end-cap is composed of 8 modules covering a $\delta\phi = 45°$ range. Each module has 96 (32) absorbers and electrodes for the outer (inner) wheel. The supporting structure is composed of 3 front rings and 3 rear rings. A presampler, with the same aim as in the barrel, covering the $1.5 \leq \eta \leq 1.8$ range is placed in front of the outer wheel.

1.4 Readout electronics

The granularity in ϕ is obtained by ganging electrodes with summing boards installed directly on the electrodes. The granularity in η is obtained by suitably drawing pads or strips on the electrodes. The strips and pads drawing allow 3 readouts in depth. The number of readout channels is about 3400 for a barrel module and 3900 for an end-cap module, with a total of 173000 channels for the electromagnetic calorimeters. Motherboards connected to the summation boards, contain precise resistors for the distribution of the calibration signals and are connected to the cables going to the preamplifiers and shapers situated outside the cryostats ("0T electronics").

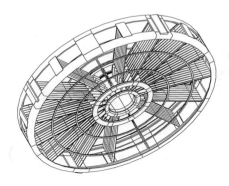

Figure 3: Schematic view of an electromagnetic end-cap calorimeter.

The front-end crates are situated in cracks in the hadronic tile calorimeters. The front-end boards, placed in these crates, provide the amplification and the shaping of the analog pulses, the summation of the calorimeter cells by trigger tower within each depth layer, the storage of the signals in an analog memory to allow the level-1 trigger latency, the digitisation of the selected pulses and the transmission on optical fibres of the multiplexed digital results. The front-end crates contain also the calibration boards with precision pulsers, the tower builder boards where the last stage of analog summation to form trigger tower signals for the level-1 trigger is performed and some other complementary boards. Prototypes of the different boards have been produced and will be used for the beam tests.

1.5 Modules 0 production

Before going to the production of all the modules, it has been decided to build modules 0 identical to the series modules. The aim is to validate all the changes and improvements made since the RD3 prototypes.

The absorbers consist of a lead sheet sandwiched between 2 stainless-steel plates to get a high mechanical strength, the gluing being done by a layer of "prepreg" adhesive on each face. Due to the different geometry of the absorbers (length and shape) new folding presses and gluing presses have been built. Once folded, the barrel absorbers are glued on 2 G10 bars which will define precisely its azimuthal location. The outer G10 bars are screwed on the outer stainless steel ring-pieces. The inner G10 bar is screwed on internal G10 ring pieces on which is also fixed the presampler. The fabrication of the end-cap absorbers has also needed new folding presses. The fabrication of the absorbers for the modules 0 is now finished. The geometry of the absorber has carefully been checked. In particular the thickness of the lead sheets is well inside the tolerances. The electrodes consist of 3 layers of copper insulated by 2 kapton polyimide sheets. The design of the about 2m long electrodes has also led to develop new folding presses. The first flat electrodes produced in the industry have been received and folded.

The assembly of modules 0 is under way using assembly jigs which will be used for the assembly of the series modules. The production of the sectors 0 of the presampler has been done. These modules 0 will be tested in an electron beam. For the series production, a fraction of the 32 barrel modules and of the 16 end-cap modules will be calibrated with an electron beam. The other modules will undergo electrical and HV tests at the liquid argon temperature before doing the assembly of the barrel and end-caps calorimeters.

2 The ATLAS level-1 calorimeter trigger

The ATLAS experiment at the LHC will have a multi-level trigger, in order to reduce the 40 MHz bunch-crossing rate (an inelastic pp interaction rate of $\sim$1 GHz at design luminosity) to a rate of $\sim$100 Hz for events written to disk. The first stage of this rate reduction, from 40 MHz to $\sim$75 kHz, is performed by the Level-1 trigger[3], which reaches its decision on each event within the 2.5 microsecond latency during which full detector data from each beam-crossing are retained in the detector front-end pipelines. Almost half of this latency is due to the propagation times of trigger inputs and trigger decision to and from the detector respectively. The tightly constrained time budget is met by the use of fast pipelined trigger processors at Level-1.

The Level-1 trigger is divided into separate calorimeter and muon systems. The calorimeter trigger inputs are taken from $\sim$7200 calorimeter trigger towers, which are coarse granularity sums of cells from the five ATLAS calorimeters. The muon trigger takes its inputs from dedicated fast muon trigger chambers. Both systems send multiplicities of trigger features, such as isolated electrons, photons and muons, to the Central Trigger Processor (CTP). The CTP forms its trigger decision from the OR of up to 96 combinations of trigger feature mul-

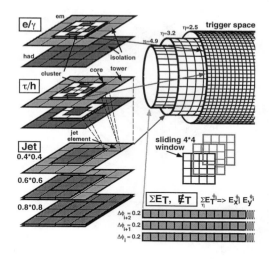

Figure 4: Level-1 calorimeter algorithms.

tiplicities at a range of E_T thresholds. When an event is accepted by the Level-1 trigger, all data for that event is secured by reading from the front-end pipelines into de-randomiser buffers.

The two trigger systems also generate pointers to "Regions of Interest" (RoIs) surrounding each trigger feature. Full-precision data from these regions, which amounts to a few per cent of the total data for each event, are used by the Level-2 trigger to improve the identification of trigger features and thereby achieve a further trigger rate reduction by a factor of $\sim$100. Finally, the ATLAS Event Filter performs full event reconstruction, using all detector data, which makes a further rate reduction to $\sim$100 Hz possible. The total data rate for events passed by the Event Filter will be $\sim$100 Mbyte/s.

The features produced by the Level-1 calorimeter trigger are:

- isolated high-E_T electromagnetic clusters (corresponding to electrons or photons);

- isolated high-E_T hadronic clusters (corresponding mostly to hadrons from τ decay);

- high-E_T jets of particles;

- global sums of transverse energy (both the scalar sum of total transverse energy and the vector sum of missing transverse energy.)

The dominant background to all features comes from QCD events.

The analogue trigger tower inputs are shaped pulses of several hundred nanoseconds duration, which are reduced to a single digital value synchronised with the bunch-crossing of origin for that pulse (termed bunch-crossing identification, or BCID). The projective trigger towers come separately from the electromagnetic and hadronic calorimeters. The cluster algorithms (e/γ and τ/hadron) are performed in the region $|\eta| \leq 2.5$, for which the trigger towers are of granularity $\delta\eta * \delta\phi = 0.1 * 0.1$. The jet and energy sum algorithms operate within $|\eta| \leq 3.2$ and 4.9 respectively, and take their inputs as jet elements formed (from the trigger towers) on a granularity of $\delta\eta * \delta\phi = 0.2 * 0.2$ (electromagnetic and hadronic energies are summed in each 0.2*0.2 element).

The algorithms which identify the four types of trigger features are illustrated in figure 4. The cluster and jet algorithms operate within multiple 4*4 windows, sliding by one tower/element in each direction (the windows are implemented in parallel in the trigger processor hardware). The algorithm for isolated electromagnetic (e/γ) clusters is based on identifying the maximum 2*1 cluster within the central 2*2 of the electromagnetic layer, whilst requiring that the energies deposited within both the surrounding electromagnetic isolation frame and the whole hadronic 4*4 region behind the cluster are below separate isolation thresholds. The τ/hadron cluster algorithm is similar, but the central cluster also includes energy from the central 2*2 of the hadronic layer, again with separate isolation regions in each layer.

The jet algorithm also operates within a 4*4 window and implements three jet cluster sizes in parallel: 0.4*0.4, 0.6*0.6 and 0.8*0.8. The larger jet clusters provide better high-E_T inclusive performance, while the smaller clusters provide better noise rejection and spatial resolution and thus are better suited to low-E_T multi-jet separation. The global E_T sums are formed by summing jet elements in η, resolving these sums into x and y components for combination in the missing energy sum. The combination of global energy sums with other trigger features allows lower trigger thresholds, maintaining high efficiency for many physics processes, whilst achieving good rejection of the dominant QCD background. Each class of trigger feature is discriminated at between 4 and 8 E_T thresholds, and the multiplicities of the local trigger features (clusters and jets) at each threshold are sent to the CTP.

The performance of the trigger algorithms, and the pre-processing of the trigger inputs, has been evaluated using detailed detector simulations, optimising the parameters of each algorithm to lower trigger rates whilst maintaining efficiency. The 2*1 cluster size of the e/γ algorithm provides greater jet rejection than a 3- or 4-tower cluster, whilst independent electromagnetic and hadronic isolation thresholds allow an O(10) rate reduction compared to a cluster threshold alone. The larger jet cluster size allows for a lower high-E_T inclusive jet rate than for the smaller clusters, whilst the multiplicity of low-E_T jet RoIs passed on to Level-2 for triggered events is lowest for the smallest cluster size. The noise rejection achieved by the trigger tower pre-processing significantly improves the missing E_T resolution. At the nominal LHC luminosity of $10^{34} \mathrm{cm}^{-2}\mathrm{s}^{-1}$ the e/γ inclusive trigger with a threshold of $\sim$30 GeV will be the largest single component of the $\sim$40 kHz total Level-1 trigger rate. This reflects both the difficulty of rejecting jets from the dominant QCD background using only reduced granularity information, and the importance of isolated electromagnetic clusters in the triggering of many important physics processes.

The key property of the calorimeter trigger is the "fan-out" of each tower or jet element to participate in 16 neighbouring 4*4 trigger windows. The density of the processors is maximised by separating the linear pre-processing step from the parallel processing of the trigger algorithms themselves. The calorimeter trigger is divided into the Pre-processor, which synchronises and digitises the calorimeter trigger inputs, plus a Cluster Processor (e/γ and τ/hadron algorithms) and Jet/Energy-sum Processor acting in parallel.

A ϕ-quadrant topology is adopted throughout, with a single modularity in the Pre-processor and processors, which simplifies both design and maintenance. Each quadrant has two Pre-processor crates (electromagnetic and hadronic towers) and a crate for each of the Cluster and Jet/Energy-sum Processors. The processor modules in a crate map onto a one-dimensional array of sub-spaces within the quadrant. The sharing of towers which appear simultaneously in neighbouring sub-spaces is performed by nearest-neighbour communication over a fast quasi-serial backplane. Towers which appear in sub-spaces in neighbouring quadrants are duplicated at the Pre-processor and sent in parallel to the processor crates for each quadrant.

In order to minimise the numbers of modules and crates, and thus maximise the use of short-path intra-crate fan-out, high density processors are achieved by the use of custom ASICs and MCMs (multi-chip modules). Fast serial communication is used between the Pre-processor and the processors, and over the backplanes between processor modules, including serial inputs to the processing ASICs and FPGAs. The serial links between Pre-processor and processors are error-encoded to minimise fake triggers, and this encoding is maintained across the cluster processor backplane. The key technology elements of the calorimeter trigger — ASIC cluster-finding, digital bunch-crossing identification, Gbit/s serial link communication and fast backplanes — have been successfully demonstrated in testbeams during 1992–97, which included a "complete slice" test of Calorimeter + Pre-processor + Cluster Processor + CTP.

3 Conclusion

The construction of modules 0 of the different parts of the electromagnetic calorimeter are under way. These modules will be tested in a beam soon. The series construction of all modules will follow.

The level-1 calorimeter trigger is now entering the detailed design phase of its components, with production scheduled to begin in 2001.

Acknowledgements

Up to date information has been kindly given to the authors by many of their colleagues in the ATLAS LAr and the ATLAS level-1 trigger communities. We thank them all here.

References

1. ATLAS Collaboration, *ATLAS Technical Proposal*, CERN/LHCC/94-43, 15 december 1994.
2. ATLAS Collaboration, *Calorimeter Performance*, CERN/LHCC/96-40, 15 december 1996 and *Liquid Argon Calorimeter Technical Design Report*, CERN/LHCC/96-41, 15 december 1996.
3. ATLAS Collaboration, *First-Level Trigger Technical Design Report*, CERN/LHCC/98-14, 30 june 1998.
4. B.Aubert et al. (RD3 collaboration), *Nucl. Instr. Meth.* A**309**, 438-449 (1991), A**321**, 467-478 (1992), A**325**, 116-128 (1993), A**330**, 405-415 (1993),
   D.M.Gingrich et al. (RD3 collaboration), *Nucl. Instr. Meth.* A**355**, 290-306 (1995), A**355**, 290-306 (1995), A**385**, 45-47 (1997), A**389**, 398-408 (1997),
   Z.Ajaltouni et al. (ATLAS collaboration - Calorimetry and Data Acquisistion), *Nucl. Instr. Meth.* A**387**, 333-351 (1997)

Test of CsI(Tl) crystals for the Dark Matter Search

H.J.Kim, S.K.Kim, E.Won, H.J.Ahn

Department of Physics, Seoul National University, Seoul 151-742, Korea
E-mail: hjkim@hep1.snu.ac.kr, skkim@hep1.snu.ac.kr, eiwon@hep1.snu.ac.kr, hjahn@hep1.snu.ac.kr

Y.D.Kim

Department of Physics, Sejong University, Seoul, Korea
E-mail: ydkim@kunja.sejong.ac.kr

Searches for weakly interacting particles(WIMP) can be based on the detection of nuclear recoil energy in CsI(Tl) crystals. We demonstrate that low energy gamma rays down to few keV is detected with CsI(Tl) crystal detector. A clear peak at 6 keV is observed using X-ray source. In addition, we also show that alpha particles and gamma rays can be clearly separated using the different time characteristics of the crystal.

1 Introduction

Several evidence from a variety of sources indicate the universe contains a large amount of dark matter. [1] The most strong evidence for the existence of dark matter comes from galactic dynamics. There is simply not enough luminous matter observed in spiral galaxies to account for the observed rotational curves. [2] Among several dark matter candidates, the most prominent candidate is the weakly-interacting massive particles(WIMP). The leading candidate WIMP is perhaps the neutralino, the lightest super-symmetric particle such as photinos, Higgsinos and Z-inos. These particles typically have masses between 10 GeV and a few TeV and couple to ordinary matter only with weak-scale interactions. These particles would interact via elastic scattering on nuclei and could be detected by measuring the recoil energy of nucleus. [3] Recently, a great deal of attention has been drawn on crystal detectors for their possible use for detection of some possible candidates for the dark matter in galactic halos. Especially, NaI(Tl) crystal has been used for the dark matter search and one of the most stringent limit for direct detection of weakly interacting massive particle(WIMP) has been established using the NaI(Tl) crystal detector. [4] With NaI(Tl) crystal they achieved as low threshold as 6keV and relatively good separation of recoil energy spectrum from the ionizing particle spectrum using the difference of decay time of scintillation light.

It has been noted by several authors that CsI(Tl) crystal might have better performance for recoil/ionizing particle separation [5]. Although the light yield is relatively lower than NaI(Tl) crystal, the better separation can be more advantageous for WIMP search. Also CsI(Tl) crystal has higher density, much less hygroscopicity than NaI(I), as shown in Table I. The spin-independent cross section of WIMP is larger for CsI(Tl) than NaI(Tl) be-

Table 1: Comparison of CsI(Tl) and NaI(Tl) characteristics.

Quantity	CsI(Tl)	NaI(Tl)
Density(g/cm^3)	4.53	3.67
Decay constant(ns)	~1000	~230
Peak emission(nm)	550	415
Light yield(relative)	85	100
Hygroscopicity	slight	strong

cause CsI(Tl) has a compound with 2 heavy nuclei while spin-dependent cross section will be comparable. Moreover hundreds of tons of CsI(Tl) crystals are already being used for several detectors in high energy experiment. [6] Thus fabricating large amount of crystals will be quite feasible. We have studied the possibility of CsI(Tl) crystal for this application.

2 Experimental setup

We prepared a 3cm×3cm×3cm CsI(Tl) crystal with all surfaces polished. Photomultiplier tubes of 2 inch diameter(Hamamtsu H1161) are attached on two opposite end surfaces. The cathode plane of PMT covers almost all the area of the crystal surface attached. The other sides are wrapped with 1.5 μm aluminized foil. It is necessary to use thin foil so that low energy X-ray signal is not blocked. Signals from PMTs are then amplified using a home-made AMP(x8) with low noise and high slew rate. Trigger signals are amplified with ORTEC AMP(x200). Discriminators are set at the level of single photo-electron signal. By using LED, we can clearly demonstrate that single photo-electron signal is well above the electronic noise. In order to suppress the background from dark current, for each signal we delay the signal by 100 ns and

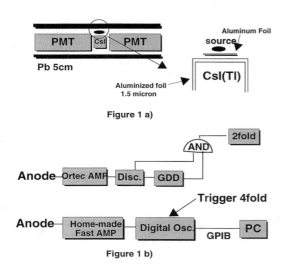

Figure 1 a)

Figure 1 b)

Figure 1: a) Schematics of the experimental setup and b) trigger logic.

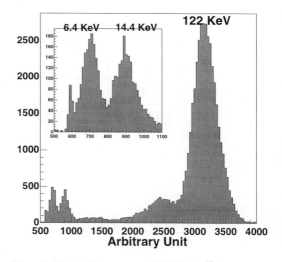

Figure 2: Pulse height spectrum of CsI(Tl) for ^{57}Co source. The left top plot is zoomed pulse height spectrum for the low energy X-ray.

then form a self coincidence, which force that at least two photo-electron occurs within 200 ns. Then coincidence of those two signals are made for the final trigger decision. In this way the trigger caused by the accidental noises are suppressed by a great amount. With this condition our effective threshold is four photo-electrons, which corresponds roughly to 40 photons produced in the crystal. Using the widely accepted light yield of CsI(Tl), ~50000 /MeV, our threshold can be interpreted as 2keV. The crystal and PMTs are located inside blocks of 5cm thick lead in order to stop the environmental background. A digital oscilloscope is used for the data taking with GPIB interface to a PC with LINUX system. Entire data taking and analysis has been done using the ROOT package recently developed at CERN[7]. Schematics of the experimental setup is shown in Figure 1 a) and the trigger logic in Figure 1 b). The digital oscilloscope we used for our experiment samples the signal at 1 Gs/sec with 8 bit pulse height information and two channels are read out simultaneously. Full pulse shape informations are saved for the further analysis.

3   Result

We have performed measurement of x-rays, gamma-rays, and alpha particles using various radioactive sources with the setup described in the previous section. The distribution of total charge collected by x-rays and gamma rays

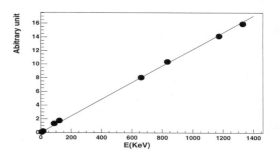

Figure 3: Linearity of CsI(Tl). The solid line shows the linear fit to data.

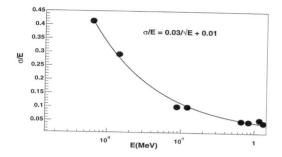

Figure 4: The energy resolution of CsI(Tl) with one-side PMT. The solid curve shows the best fit to data.

from the ^{57}Co source is given in Figure 2. The highest peak is from the gamma ray of 122 keV. Shown in left side of broad distribution of pulses are Compton edge. The energy resolution at 122 keV is about 7%. Also, the X-ray peak at 6.4 and 14.4 keV are clearly seen with energy resolution of 30 and 20%, respectively. This resolution is not much worse than that of NaI(Tl) crystal. Many calibration sources such as ^{57}Co, ^{109}Cd, ^{137}Cs, ^{54}Mn and ^{60}Co are used for the determination of linearity and resolution. The pulse shape is quite linear at high energy as shown in Figure 3 but there is some deviation at low energy. It turns out that the variation in the response function near the L and K-shell of Cs and I causes non-linearity at X-ray region within 20%.[8] Also the NaI(Tl) has similar tendency. The pulse shape is linear within 10 % upto low energy X-ray region if these effects are considered.

The Figure 4 shows resolution of CsI(Tl) crystal with one side PMT. The best fit of the resolution is

$$\frac{\sigma}{E(MeV)} = \frac{0.03}{\sqrt{E(MeV)}} \oplus 0.01, \qquad (1)$$

and it becomes

$$\frac{\sigma}{E(MeV)} = \frac{0.02}{\sqrt{E(MeV)}} \oplus 0.01, \qquad (2)$$

when we add both side of PMT signal.

A high ionization loss produces a higher density of free electrons and holes which favors their recombination into loosely bound systems and results in fast timing component. On the other hand, the singly free electrons and holes are captured successively, resulting in the excitation of certain meta-stable states and produce slow timing component. By using this characteristics, we may be able to separate X-ray background from the high ionization loss produced by WIMP. To demonstrate this difference, we measured signals produced by alpha particles using ^{241}Am source. Energy of the alpha particle

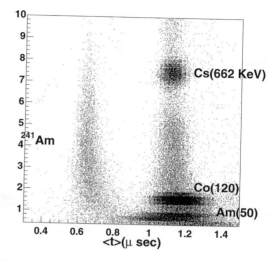

Figure 5: Decay time distribution of CsI(Tl) crystal with ^{241}Am and γ sources.

is 5.5 MeV and the energy of entering the detector was controlled by the thickness of thin aluminum foil in front of the crystal. Although alpha particle at this energy stops in the crystal, the visible energy seen by the PMT is about 75% of the energy. This is due to the quenching factor for alpha particles and agrees with what was observed by other experiments[9]. We show the two dimensional histogram of mean time vs. integrated charge in Figure 5. The mean time is the pulse hight weighted time average, defined as

$$<t> = \frac{\sum t_i \times q_i}{\sum q_i} \qquad (3)$$

, where q_i is the amplitude of the pulse at the channel time t_i up to $4\mu s$. It is approximately the same as decay time of the crystal. Two clear bands in the Figure 5 indicate that we can make good separation between alpha particle and gamma rays. The low energy of gamma ray from the ^{241}Am source is 59.5 keV. Decay time for alpha particles is ~700 ns while for gamma rays ~1100 ns. Two peaks are well separated by more than 3 sigma in this energy region.

4 Conclusion

We demonstrated that CsI(Tl) crystal can be used to measure low energy gamma rays down to few keV. In addition, a good separation of alpha particle from gamma rays has been achieved. As we expect recoiled ions in the crystal would behave similar to alpha particles, this mean time difference would be very useful to differenciate WIMP signals from backgrounds. The background study and neutron response on CsI(Tl) are underway and pilot experiment with a large amount crystals will be lunched in a near future.

References

1. G.G. Raffelt, hep-ph/9712538, Dec 1997.
2. K.G. Begeman, A.H. Broeils, and R.H. Sanders, *Mon. Not. R. Astr. Soc. 249*, 523 (1991).
3. M.W. Goodman, E.Witten, *Phys. Rev.* D **31**, 3059 (1985).
4. P.F. Smith *et al.*, *Phys. Lett.* B **379**, 299 (1996), R. Bernabei *et al.*, *Phys. Lett.* B **389**, 757 (1996).
5. J.B. Birks, *Theory and practice of scintillation counter*, (Pergamnon press, Oxford, 1964).
6. E. Blucher *et al.* (CLEO), *Nucl. Instrum. Methods* **A235**, 319 (1985), M.T.Cheng *et al.* (BELLE), *Technical Design Report*, KEK Report 95-1, 1995.
7. R. Brun and F. Rademakers, *ROOT - An Object Oriented Data Analysis Framework*, *Proceedings AIHENP'96 Workshop*, Lausanne, Sep. 1996, *Nucl. Instrum. Methods* **A389**, (1997).
8. D. Aitken, B.L. Leron, G. Yenicay, H.R. Zulliger, *Trans. Nucl. Sci. NS-14, No. 2*, 468 (1967).
9. Y.K. Akimov, *Phys. Part. Nucl. 25*, 92, 1994, W.R. Leo, *Techniques for Nuclear and Particle Physics Experiments*, (Springer-Verlag, Berlin, 1993).

THE DRIFT CHAMBER AND THE DATA ACQUISITION SYSTEM OF THE KLOE EXPERIMENT

F. CERADINI for the KLOE collaboration [a]

Physics Department of the University Roma Tre and INFN, Rome, Italy
E-mail: filippo.ceradini@roma1.infn.it

The tracking detector of the KLOE experiment is a large volume drift chamber designed to contain most of the decays of K_L produced at the DAΦNE φ-factory. The design, construction technique and the first tests of the chamber are presented. The Data Acquisition System of the experiment is also described.

1 Introduction

The main goal of the KLOE experiment [1] at DAΦNE, the Frascati φ-factory, is the precise measurement of the CP-symmetry violation parameters in the decay of neutral kaons produced at the φ resonance, $\phi \rightarrow K_L K_S$. The experiment is made of a large volume drift chamber (DC) [2], where most of the kaons decays occur, and a hermetic calorimeter [3], both immersed in a 0.6 T uniform magnetic field. At the design luminosity of 5 10^{32} cm^{-2} s^{-1} the foreseen trigger rate [4] is of ∼ 10 kHz. A Data Acquisition System (DAQ) [5] with a bandwith of 50 MByte/s and the required reliability for a ∼ 10^{-4} precision measurement has been built.

[a] M. Adinolfi, F. Ambrosino, A. Aloisio, A. Andryakov, A. Angeletti, A. Antonelli, C. Bacci, R. Baldini-Ferroli, G. Barbiellini, G. Bencivenni, S. Bertolucci, C. Bini, C. Bloise, V. Bocci, F. Bossi, P. Branchini, L. Bucci, G. Cabibbo, A. Calcaterra, R. Caloi, G. Carboni, P. Campana, G. Capon, M. Carboni, C. Carusotti, G. Cataldi, F. Ceradini, F. Cervelli, F. Cevenini, G. Chiefari, P. Ciambrone, S. Conticelli, E. De Lucia, R. De Sangro, P. De Simone, G. De Zorzi, S. Dell'Agnello, A. Denig, A. Di Domenico, S. Di Falco, A. Doria, F. Donno, E. Drago, V. Elia, L. Entesano, O. Erriquez, A. Farilla, G. Felici, A. Ferrari, M. L. Ferrer, G. Finocchiaro, D. Fiore, C. Forti, G. Foti, A. Franceschi, P. Franzini, A. Galli, M. L. Gao, G. Gatti, P. Gauzzi, S. Giovannella, V. Golovatyuk, E. Gorini, F. Grancagnolo, E. Graziani, P. Guarnaccia, U. v. Hagel, H. G. Han, S. W. Han, M. Incagli, L. Ingrosso, W. Kim, W. Kluge, V. Kulikov, F. Lacava, G. Lanfranchi, J. Lee-Franzini, T. Lomtadze, C. Luisi, A. Martini, W. Mei, L. Merola, R. Messi, S. Miscetti, A. Moalem, S. Moccia, F. Murtas, M. Napolitano, A. Nedosekin, P. Pagès, M. Palutan, L. Paoluzi, E. Pasqualucci, L. Passalacqua, M. Passaseo, A. Passeri, V. Patera, E. Petrolo, G. Petrucci, D. Picca, M. Piccolo, A. Pintus, G. Pirozzi, L. Pontecorvo, M. Primavera, F. Ruggieri, P. Santangelo, E. Santovetti, G. Saracino, R. D. Schamberger, C. Schwick, B. Sciascia, A. Sciubba, F. Scuri, I. Sfiligoi, S. Sinibaldi, T. Spadaro, S. Spagnolo, E. Spiriti, C. Stanescu, L. Tortora, E. Valente, P. Valente, G. Venanzoni, S. Veneziano, D. Vettoretti, S. Weseler, Y. G. Xie, C. D. Zhang, J. Q. Zhang, P. P. Zhao

2 The construction of the Drift Chamber

The long decay path of the K_L ($\lambda_L \sim 3.5$ m) demands for a very large sensitive volume: ∼ 4 m diameter and ∼ 3.4 m length. Since the angular distribution of the charged decay products is isotropic, this volume is uniformely filled with 58 cell layers in a *all-stereo* configuration. The 12,582 drift cells are approximately square: ∼ 2 × 2 cm^2 in the inner 12 layers, and ∼ 3 × 3 cm^2, in the outer layers [6]. The ratio of field to sense wires is 3:1, for a total of 52,140 wires, including the inner and outer guard layers. The stereo angle increases approximately with the square root of the layer radius from 60 to 150 mrad.

To minimize the K_L regeneration in the inner wall, the conversion of low energy photons in front of the calorimeter and the multiple scattering contribution to the momentum resolution of charged particles, very thin carbon fiber walls for the DC mechanics, as well as aluminum field wires and a ultra-ligth gas mixture ($90\%He - 10\%C_4H_{10}$) have been chosen. The choice of aluminum for the field and guard wires (tempered Al-5056, silver-plated, 80 μm diameter) allows to sensibly reduce the load on the End-Plates (EP). To obtain a good gain and a small sagitta over the ∼ 3.6 m length, a 25 μm diameter gold-plated tungsten wire was chosen for the sense wires. The overall thickness of the DC, including front-end electronics, is 0.1 X_o.

The optimization of the cell design and the study of the DC performance have been done with several beam tests of different chamber prototypes [7]. The results of these tests showed that a spatial resolution of 130 μm over most of the drift cell can be obtained.

The whole DC mechanics is made of carbon fiber. Two spherical EP's of ∼ 9.7 m curvature radius, and 8 mm thickness, with a 30 μm Cu foil glued on the external surface for ground shield, are kept apart by 12 carbon fiber rods. An external ring is coupled to each

EP through 48 screws, to allow recovering the EP deformation under the wire tension load during the wires stringing. The gas sealing of the chamber is ensured by the inner cylinder (25 cm radius, 0.7 mm thickness) and 12 external panels.

The correct DC stringing required a systematic measurement of the linear density and Young modulus of each wire spool. In addition, in order to maintain the cell shape as uniform as possible, i.e. equal sagging of Al and W wires, we had to take into account the creep of aluminum wires under load and define the proper value of the mechanical tension. For this, we have done a long-term creep test with several Al wires suspended with different loads. With a load of 80 g, a total elongation of about 1.5 mm has been measured for the Al wires after 1.5 years, corresponding to a tension loss of approximately 10 g.

The wires are held in place with the crimping method: each wire is inserted through the hole of a metallic pin and is fixed by flattening the pin. Special care was taken in the choice of the pin, that should match the material of the wire to ensure a good mechanical coupling and avoid wire damages along the crimping. We have used copper pins, with an outer diameter of 1 mm and a hole of 150 μm, to hold both W and Al wires. The coupling is ensured by crimping the pin by means of a pneumatic clamping jaw.

3 Chamber stringing

The DC has been strung in a *class 1000* clean room, temperature and humidity controlled (21±1 ^{o}C and 50±2 % respectively), by means of a computer controlled semi-automatic system[8], composed of three main parts:

- a special rig was supporting and rotating the chamber EP's with an accuracy of 100 μm at 2 m radius. The EP's, connected by the 12 rods, were fixed by means of flanges to a central rotating iron tube; the coupling between the flanges and the central tube was done by means of linear bearings that allowed the EP's to freely move axially to account for the EP's deformation under the wire load;

- a 3-axes robot brought the wire from one EP to the other with an accuracy of 500 μm;

- two automatic stations: one equipped with a wire supply system and a pneumatic crimping jaw, the other equipped with a wire tensioning torquemeter and a crimping jaw. The position accuracy on the 3-axes was 100 μm.

The operations of feeding the wire to the robot and of inserting the wire into the feed-through were done manually.

After stringing one complete layer, systematic mechanical and electrical tests were done on the wires. To check the mechanical tension we used an electrostatic method[9]: the mechanical oscillation of the wire was excited applying an alternating voltage (1 kV) to the wire with respect to nearby wires. The resonance frequency was detected measuring the variation of the wire capacitance. The main advantage of this method is that all wires can be measured at any time. We also used this method to monitor the EP deformation and its recovery, to re-measure the tension of a large sample of wires at the end of the chamber stringing and to monitor the effect of the gas over-pressure onto the EP's. In addition, for all sense and guard wires we measured the leakage current at a voltage of 3 kV. The measurement was done in air with a relative humidity of 50%. About 1% of the wires failed these tests and were replaced.

The recovery of the EP deformations, which, for the wire tension load of $\sim$ 3.5 tons, amounted to about 3 mm per EP, was done acting on the external ring. The 48 screws used to connect the ring to the EP are monitored by means of strain gauges. By tensioning the ring a radial force is applied to the EP causing a longitudinal displacement of the EP opposite to the wire tension. Tests with static loads, up to $\sim$ 5 tons, showed that this method allows to keep the average residual deformation down to 0.5 mm. The recovery operation was done by tensioning the external ring every 300 kg of wire load. The deformation and its recovery were determined by measuring the variation of the wire mechanical tension on a sample of wires on different layers.

All wires were strung at a nominal sagitta of 250 μm. The average nominal tension was 80 g for aluminum and 40 g for tungsten wires. Additional corrections to these values were applied during stringing to account for:

- the not vanishing EP deformation residual. This (negative) correction is a function of the layer radius and is of the order of few grams for tungsten and slightly larger for aluminum;

- the creep effect of aluminum wires, resulting in a positive correction of about 10 g.

After the last EP deformation recovery, we measured the tension of about 1500 W and 500 Al wires, on different layers. The wire tensioning procedure allowed to achieve a good cell shape uniformity with sagitta$_{Al}$ $\sim$ sagitta$_W$ at the level of 10 μm.

At the end of the stringing the number of broken wires was 0.07% of the total: all these wires have been succesfully replaced, before closing the chamber, with a very delicate manual procedure.

Figure 1: The KLOE Drift Chamber.

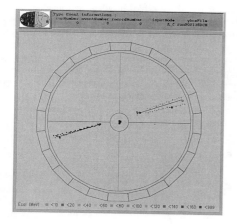

Figure 2: A typical cosmic ray event.

4 Chamber tests

The Drift Chamber is shown in Fig. 1 at the end of the stringing operation. After completion of the construction, the DC has been sealed and tested for gas leaks. Gas tightness is crucial for the experiment since the DC is surrounded by $\sim$ 5000 photomultipliers for the readout of the calorimeter whose glass windows are not fully tight to He atoms. Then the DC has been tested at the nominal high voltage setting for about one month and partially equipped with TDC's to check the quality of reconstructed tracks when triggered with a cosmic ray hodoscope. A typical event recorded during the test is shown in Fig. 2. The solid lines represent the track fit through the hit points of the two stereo views. The circles are centered on the hit wire and their radii are proportional to the distance of the track from the wire.

The DC was then transported in the KLOE Assembly Hall, installed inside the calorimeter and surveyed. At present, the chamber is fully instrumented and a complete test with cosmic rays triggered by the calorimeter is in progress for the final debugging of the whole system before moving the experiment in the DAΦNE Hall.

5 The Data Acquisition architecture

The DAQ system is designed to collect data from about 25,000 front-end electronics (FEE) channels (5,000 calorimeter ADC's and TDC's, 13,000 drift camber

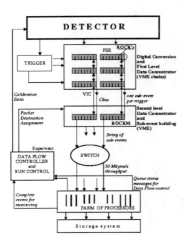

Figure 3: The DAQ system architecture.

first level of concentration, data from the FEE boards are organized in sub-events flagged by the trigger counter.

The FEE crates are organized in chains that end in standard VME crates housing a chain controller, the ROCK-manager (ROCKM) [10], and a CPU with a FDDI interface. The ROCKM reads out the data from a chain of ROCK's via a custom-bus (C-bus) in a token-ring fashion and organizes data in strings of sub-events ready for transfer. The CPU's continuously read data from the ROCKM's through the VME-bus and transfer sub-events via FDDI. At this second level of data concentration the measured bandwidth reaches ~ 9 MByte/s per chain. A Vertical Interconnect commercial bus (VIC) is used to connect all the crates of a chain allowing the VME CPU's to program, test and debug the FEE boards.

A commercial high-speed FDDI switch with bridge functionality is used to provide parallel paths between the VME processors and the on-line farm processors in a scalable way. At present the system is made of 10 chains: 4 for the calorimeter, 4 for the drift chamber and 2 for the trigger. Both the number of ports dedicated to the second level VME processors and to the on-line farm processors can be increased according to the needs of the experiment. A maximum bandwidth of 130 MByte/s can be attained using all available ports of the switch.

The address of the CPU of the on-line farm that receives a string of sub-events is assigned by a VME control processor, the Data Flow Controller (DFC), that is connected to the second level crates via another VIC. By mainteining address tables in all second level crates, the DFC guarantees that all sub-events originating from the same event are sent to the same on-line farm processor.

All FEE crates and second level crates are located close the detector in the DAΦNE Hall. The whole experiment is connected to the counting room via ten FDDI optical links and few other links for the slow controls. The DAQ chains of the calorimeter and of the trigger system are fully operational since few months taking data with cosmic rays at a rate of ~ 3 kHz.

6 Conclusions

The KLOE Drift Chamber is the largest wire tracking detector with very thin walls ever built. Its design is innovative in many aspects: the choice of materials, the drift cell configuration and the gas mixture. With the help of a dedicated semi-automatic machine the stringing of the 52,000 wires was completed in 11 months. The chamber has been installed in the experiment and has taken data with cosmic rays triggered by the calorimeter.

A powerful and efficient Data Acquisition System has been built whose performance and reliability have been tested with a continuous operation for few months. Both

TDC's and the trigger data) at a rate of ~ 10 kHz, half due to $\phi-$decays and half to down-scaled Bhabha scattering, cosmic rays and machine backgrounds. With an estimated occupancy of 5%, a bandwidth of 50 MByte/s is required. Since particles transit times in the detector and signal transfer are much longer than the DAΦNE bunch crossing period of 2.7 ns, the DAQ should run asynchronous. To limit systematic errors to the required level the DAQ should introduce a very small dead time, ensuring that it does not depend on the event configuration, and should minimize errors in data read-out and transfer.

The scheme of the read-out and data acquisition is shown in Fig. 3. Data collected by the FEE are first buffered through a high-speed two-level data concentrator, then transferred to a on-line processor farm via FDDI and finally written to tape. The FEE is organized in VME boards whose input stage is locked for a fixed time of 2.2 μs after reception of a first-level trigger signal: at a rate of 10 kHz, the dead-time is 2%. Each FEE crate houses a hardwired Read Out Controller (ROCK) [10]. The crate uses a custom-designed bus (AUX-bus) optimized for the relatively low occupancy of the FEE's channels with hardwired implementation of the sparse data scan. The AUX-bus can sustain a bandwidth of about 50 MByte/s. The FEE boards and the ROCK's have local buffers such that there is no need for the read-out to run synchronous with the trigger signal. At this hardware and software tools meet the severe specifications of the experiment program.

Acknowledgements

I would like to thank all my friends of the KLOE Collaboration and in particular G.Bencivenni and E.Graziani for their help and advice in preparing this presentation.

References

1. The KLOE Collaboration, *The KLOE Detector, Technical Proposal*, LNF-93/002 (1993).
2. The KLOE Collaboration, *The KLOE Drift Chamber, Addendum to the Technical Proposal*, LNF-94/028 (1994).
3. F.Cervelli, *The KLOE electromagnetic calorimeter*, ICHEP98 contribution No.634.
4. The KLOE Collaboration, *The KLOE Trigger System, Addendum to the Technical Proposal*, LNF-96/043 (1996).
5. The KLOE Collaboration, *The KLOE Data Acquisition System, Addendum to the Technical Proposal*, LNF-95/014 (1995).
6. G.Finocchiaro *et al.*, *The KLOE Drift Chamber Geometry*, KLOE note 157 (1996).
7. G.Finocchiaro, Nucl. Instrum. Methods A **360**, 48 (1995); F.Lacava, Nucl. Phys. B **54**, 327 (1997); A.Andryakov *et al.*, Nucl. Instrum. Methods A **404**, 248 (1998).
8. G.Bencivenni *et al.*, *An Automated Facility for Stringing the KLOE Drift Chamber*, KLOE note 149 (1995).
9. A.Andryakov *et al.*, *Electrostatic digital method of wire tension measurement for the KLOE Drift Chamber*, submitted to Nucl. Instrum. Methods.
10. R.Aloisio *et al.*, *A Fast Readout System for the KLOE Experiment*, KLOE note 88 (1994).

PROGRESS TOWARDS CLEO III: THE SILICON TRACKER AND THE LIF-TEA RING IMAGING CHERENKOV DETECTOR

M. ARTUSO

Syracuse University, Syracuse, NY 13244-1130, USA
E-mail: artuso@suhep.phy.syr.edu

We describe the two major components of CLEO III: the Silicon Vertex Detector and the Ring Imaging Cherenkov Detector (RICH). The Silicon Vertex Detector is a four layer barrel-style device which spans the radial distance from 2.5 cm to 10.1 cm and covers 93% of the solid angle. It is being constructed using double-sided silicon sensors read out by front end electronics devices especially designed for this application. The RICH system consists of LiF radiators and multiwire proportional chambers containing a mixture of CH4 and TEA gases. The radiators are both flat and "sawtooth." Results from a test beam run of final CLEO III RICH modules will be presented, as well as test beam data on sensors to be employed in the Silicon Vertex Tracker.

1 Introduction

The Cornell e^+e^- collider (CESR) is currently being upgraded to a luminosity in excess of $1.7 \times 10^{33} \text{cm}^{-2}\text{s}^{-1}$. In parallel the CLEO III detector is undergoing some major improvements.

One key element of this upgrade is the construction of a four layer double-sided silicon tracker. This detector spans the radial distance from 2.5 cm to 10.1 cm and cover 93% of the solid angle surrounding the interaction region. The outermost layer is 55 cm long and will present a large capacitive load to the front-end electronics. The innermost layer must be capable of sustaining large singles rates typical of a detector situated near an interaction region.

A novel feature of CLEO III is a state of the art particle identification system that will provide excellent hadron identification at all the momenta relevant to the study of the decays of B mesons produced at the $\Upsilon(4S)$ resonance. The technique chosen is a proximity focused Ring Imaging Cherenkov detector (RICH) [1] in a barrel geometry occupying 20 cm of radial space between the tracking system and the CsI electromagnetic calorimeter.

The physics reach of CLEO III is quite exciting: the increased sensitivity of the upgraded detector, coupled with the higher data sample available, will provide a great sensitivity to a wide variety of rare decays, CP violating asymmetries in rare decays and precision measurements of several Cabibbo-Kobayashi-Maskawa matrix elements.

2 Vertex Detector Design

The barrel-shaped CLEO III Silicon Tracker (Si3) is composed of 447 identical sensors combined into 61 ladders. The sensors are double sided with $r\phi$ readout on the n side and z strips on the p side. The strip pitch is 50 μm on the n side and 100 μm on the p side. Readout hybrids are attached at both ends of the ladders, each reading out

half of the ladder sensors. More details on the detector design can be found elsewhere. [2] Sensors and front end electronics are connected by flex circuits that have traces with a 100 μm pitch on both sides, being manufactured by General Electric, Schenectady, New York.

All the layers are composed of identical sensors. In order to simplify the sensor design, the detector biasing resistors and the coupling capacitors have been removed from the sensor into a dedicated R/C chip, mounted on the hybrid. Another key feature in the sensor design is the so called "atoll" geometry of the p-stop barriers, using isolated p-stop rings surrounding individual n-strips. Furthermore a reverse bias can be applied to the p-stop barriers through a separate electrode. Thus the parasitic capacitance associated with these insulation barriers can be significantly reduced with a corresponding reduction of the sensor noise in the frequency range of interest. [3]

The middle chip in the readout chain is the FEMME preamplifier/shaper VLSI device. It has an excellent noise performance. At the shaping time of $2\mu s$, well matched to the CLEO III trigger decision time, its equivalent noise charge is measured as:

$$ENC = 149e + 5.5e/pF \times C_{in}, \tag{1}$$

giving satisfactory noise performance also with the high input capacitances in the outer layer sensors. More details on the design and performance of this device can be found elsewhere. [4] The last chip in the readout chain is the SVX_CLEO digitizer and sparsifier. [3] Both these chips are manufactured utilizing radiation -hard CMOS technology from Honeywell.

3 The Si3 test beam run

The silicon sensors have been tested in several test beam runs that took place at CERN in the last few months. The sensors, flex circuits, and R/C chips used were the same ones planned for the final system. However

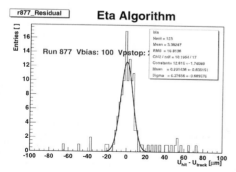

Figure 1: Residual distribution with η algorithm at $V_{bias}=100$ V and $V_{pstop} = 20V$.

the readout electronics used was not the combination of FEMME + SVX_CLEO, but the low noise VA2 chip produced by IDE AS, Norway [5] with digitization implemented in the remote data acquisition system.

The data were collected inserting the Si3 sensor in the test beam set-up used by the RD42 collaboration [6] to test their diamond sensors. A 100 GeV π beam was used and the silicon sensor was inserted in a silicon telescope composed of 8 microstrip reference planes defining the track impact parameters with a precision of about 2 μm. Two data sets were collected. The first one contains 300,000 events with tracks at normal incidence and different bias points. The second one consists of about 200,000 events at $\theta = 0$ and $\theta = 10°$. The probability distribution of the variable η defined as $\eta = Q_L/(Q_L + Q_R)$ (referring to the relative location of two adjacent strips where at least one of them recorded a hit) has been mapped and has been used in a non-linear charge weighting to reconstruct the track hit location. [7] Fig. 1 shows the hit resolution achieved on the n-side with a detector bias of 100 V and a p-stop reverse bias of 20 V. A value of σ of 6 μm is quite impressive for a track at normal incidence. The expectation for a 50 μm strip pitch is about 9 μm, as the charge spreading due to diffusion does not provide appreciable signal in the neighboring strip unless the track impact point is at the periphery of the strip. The higher resolution achieved in this case is attributed to an increase in charge spread due to the reverse biased p-stops. In fact the ability to modulate the reverse bias of the p-stops has allowed the tuning of this charge sharing for optimum resolution and may be of interest for other applications.

4 Description of the CLEO III RICH System

The CLEO III RICH system is based on the 'proximity focusing' approach, in which the Cherenkov cone, produced by relativistic particles crossing a LiF radiator, is let to expand in a volume filled with gas transparent to ultraviolet light before intersecting a photosensitive detector where the coordinates of the Cherenkov photons are reconstructed. The photodetector is a multiwire proportional chamber filled with a mixture of Triethylamine (TEA) gas and Methane. The TEA molecule has good quantum efficiency (up to 35%) in a narrow wavelength interval between 135 and 165 nm and an absorption length of only 0.5 mm.

The position of the photoelectron emitted by the TEA molecule upon absorption of the Cherenkov photon is detected by sensing the induced charge on an array of 7.6mm × 8 mm cathode pads. The probability distribution for the charge in the avalanche initiated by a single photoelectron is exponential at low gain. This feature implies that a low noise front end electronics is crucial to achieve good efficiency. A dedicated VLSI chip, called VA_RICH, based on a very successful chip developed for solid state application, has been designed and produced for our application at IDE AS, Norway. We have acquired and characterized all the hybrids necessary to instrument the whole RICH detector, a total of 1800 hybrids containing two VA_RICH chips each, corresponding 230,400 readout channels. We have fully characterized all of them and for moderate values of the input capacitance C_{in}, the equivalent noise charge ENC is found to be about:

$$ENC = 130e^- + (9e^-/pF) \times C_{in}. \qquad (2)$$

The traces that connect the cathode pads with the input of the preamplifier in the VA_rich are rather long and the expected value of ENC in absence of other contribution is of the order of 200 e^-.

The charge signal is transformed into a differential current output transmitted serially by each hybrid to a remote data acquisition board, where the currents are transformed into voltages by transimpedance amplifiers and then digitized by a 12 bit flash-ADC capable of digitizing the voltage difference at its input. The data boards perform several additional complex functions, like providing the power supply and bias currents necessary for the VA_RICH to be at its optima working point. In addition, the digital component of these boards provides sparsification, buffering and memory for pedestal and threshold values.

If a track crosses the LiF radiator at normal incidence, no light is emitted in the wavelength range detected by TEA, due to total internal reflection. In order to overcome this problem a novel radiator geometry has been proposed. [8] It involves cutting the outer surface of the radiator like the teeth of a saw, and therefore is referred to as "sawtooth radiator". A detailed simulation of several possible tooth geometries has been performed and a tooth angle of 42° was found out to be close to optimal and technically feasible. [8]

There are several technical challenges in producing these radiators, including the ability of cutting the teeth with high precision without cleaving the material and polishing this complex surface to yield good transmission properties for the ultraviolet light. One of the goals of the test beam run described below was to measure the performance of sawtooth radiators and we were able to produce two full size pieces working with OPTOVAC in North Brookfield, Mass. The light transmission properties of these two pieces were measured relative to a plane polished sample of LiF and found to be very good. [9]

5 Test beam results

Two completed CLEO III RICH modules were taken to Fermilab and exposed to high energy muons emerging from a beam dump. Their momentum was ≥ 100 GeV/c The modules were mounted on a leak tight aluminum box with the same mounting scheme planned for the modules in the final RICH barrel. One plane radiator and the two sawtooth radiators were mounted inside the box at a distance from the photodetectors equal to the one expected in the final system. Two sets of multiwire chambers were defining the μ track parameters and the trigger was provided by an array of scintillator counters. The data acquisition system was a prototype for the final CLEO system.

The beam conditions were much worse than expected: the background and particle fluxes were about two order of magnitude higher than we expect in CLEO and in addition included a significant neutron component that is going to be absent in CLEO. Data were taken corresponding to different track incidence angles and with tracks illuminating the three different radiators. For the plane radiator, we were able to configure the detector so that the photon pattern would appear only in one chamber. For the sawtooth radiator a minimum of three chambers would have been necessary to have full acceptance.

The study of this extensive data sample has been quite laborious and the full set of results is beyond the scope of this paper. The results from two typical runs will be summarized in order to illustrate the expected performance from our system: the first case will involve tracks incident at 30° to the plane radiator and the second tracks incident at 0° on a sawtooth radiator.

The fundamental quantities that we study to ascertain the expected performance of the system are the number of photons detected in each event and the an-

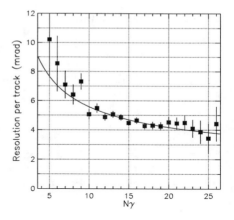

Figure 2: The Cherenkov angular resolution per track as a function of the number of detected photons (background subtracted) for a plane radiator with tracks at 30° incidence.

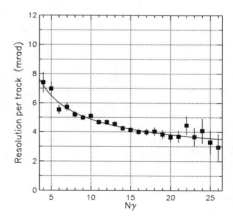

Figure 3: The Cherenkov angular resolution per track as a function of the number of detected photons (background subtracted) for a sawtooth radiator with tracks at normal incidence.

Table 1: Summary on the performance of the RICH modules in two typical test beam runs. The symbol γ refers to single photon distributions and the symbol t refers to quantities averaged over the photons associated with a track.

Parameter	Plane Radiator (30°)	Sawtooth Radiator (0°)
σ_γ	13.5 mr	11.8 mr
$< N_\gamma >$	15.5	13.5
σ_t	4.5 mr	4.8 mr
$\sigma_t(MC)$	3.9 mr	3.8 mr
$\sigma_t(CLEO)$	4.0 mr	2.9 - 3.8 mr

gular resolution per photon. The measured distributions are compared with the predictions of a detailed Monte Carlo simulation, including information on the CH_4-TEA quantum efficiency as a function of wavelength, ray tracing, crystal transmission, etc.. It includes also a rudimentary model of the background, attributed to out of time tracks. The agreement between Monte Carlo and data is quite good. The average number of detected photoelectrons is about 14 after background subtraction and the angular resolution per photon is about 13.5 ± 0.2 mr. The average number of photons, after background subtraction, is 13.5, with a geometrical acceptance of only 55% of the final system. In this case it can be seen that the expected resolution is slightly better than the one achieved. In order to estimate the particle identification power in our system, we need to combine the information provided by all the Cherenkov photons in an event. We can use the resolution on the mean Cherenkov angle per track as an estimator for the resolving power in the final system.

Table 1 shows a summary of the predicted and achieved values of these variables in the two data sets discussed in this paper and in the corresponding Monte Carlo simulation, as well as the expectations for the final system. Fig. 2 shows the measured resolution per track σ_t as a function of the background subtracted number of photons detected for the flat radiator and Fig. 2 shows the corresponding curve for the sawtooth radiator. The curves are fit to the parameterization $(a/\sqrt{N_{ph}})^2 + b^2$.

The data shown, although preliminary, show a good understanding of the system and give confidence that the predicted level of efficiency versus resolution will be achieved in CLEO III. An active analysis program is under way to study the additional data available.

6 Conclusions

Both the major systems for the CLEO III detectors are well under way. Test runs data support the expectations of excellent performance of both the silicon tracker and the Ring Imaging Cherenkov detector. This will lead to a quite exciting physics program expected to start in 1999.

7 Acknowledgements

The author would like to thank her colleagues in the CLEO III Si3 and RICH groups for their excellent work reported in this paper. Especially noteworthy were P. Hopman, H. Kagan, I. Shipsey and M. Zoeller for their help in collecting the information relative to the Si3 tracker, S. Anderson, S. Kopp, E. Lipeles R. Mountain, S. Schuh, A. Smith, T. Skwarnicki, G. Viehhauser that were instrumental to a successful test beam run. Special thanks are due to C. Bebek and S. Stone for their help throughout the length of the RICH project.

References

1. R. Arnold et al. Nucl. Instrum. Methods **A314**, 465 (1992).
2. I. Shipsey et al. Nucl. Instrum. Methods **A386**, 37 (1997).
3. G. Brandenburg et al. in Proceedings of the 3rd International Symposium on the Development and Application of Semiconductor Tracking Detectors, Melbourne, Dec. 1997.
4. H. Kagan et al. Nucl. Instrum. Methods **A383**, 189 (1996).
5. E. Nygard et al. Nucl. Instrum. Methods **A301**, 506 (1991).
6. P. Weilhammer et al. Nucl. Instrum. Methods **A409**, 264 (1998).
7. E. Belau et al. Nucl. Instrum. Methods **124**, 253 (1993).
8. A. Efimov and S. Stone Nucl. Instrum.. Methods **A371**, 79 (1996).
9. M. Artuso et al. ICFA Instr. Bullettin **15**, 3 (1997)

STATUS OF HERA-B SUB-DETECTORS & OPERATIONAL EXPERIENCE

KWONG LAU

Physics Department, University of Houston, Houston, TX 77204-5506, USA
E-mail: Lau@uh.edu

For the HERA-B Collaboration[*]

The HERA-B detector, a multi-particle spectrometer specially designed to study beauty decays in search of evidence for *CP* violation, is at the final stage of its construction. The ring imaging Čerenkov detector, the electromagnetic calorimeter, and parts of the silicon vertex detector, the inner tracker, and the muon system are installed for the 1998 run which is now underway. The initial physics goal, in conjunction with the commissioning of the various sub-detectors, is to measure the production cross section of beauty hadrons. The status of the HERA-B sub-detectors is reviewed in this talk.

1 Introduction

HERA-B is a fixed-target experiment at DESY dedicated to measuring the details of *CP* violation in the beauty system. The beauty hadrons are produced by collisions of 920 GeV halo protons in the HERA ring with an internal multi-wire target. The decay products of the beauty hadrons, whose decay distances are boosted by a large Lorentz factor of about

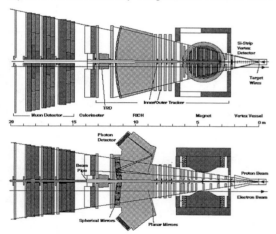

Figure 1. The top (upper drawing) and side (lower drawing) views of the HERA-B detector. Particles produced by interactions of the 920 GeV protons (entering from right) with an internal target (not shown) traverse in succession a vertex detector, a series of tracking chambers, a ring imaging Čerenkov detector, an electromagnetic calorimeter, and a muon detector.

20 in the lab frame, are analyzed by a fine-grain state-of-the-art magnetic spectrometer constructed around the beampipe. Schematic drawings of the top and side views of the HERA-B detector are shown in Figure 1. The momenta of charged particles are analyzed by a dipole magnet and a tracking system consisting of a multi-layer radiation-hard silicon microstrip vertex detector, a microstrip gas chamber inner tracker, and a honeycomb wire outer tracker, which covers the angular acceptance from 10 to 300 mrad in the lab frame. *CP* violation measurements in HERA-B rely on faithful flavor tagging via efficient identification of hadrons and leptons. Charged particles are identified by a gas ring imaging Čerenkov (RICH) detector, a shower sampling electromagnetic calorimeter, and a muon detector. A transition radiation detector supplements the electron identification in the small angle region. Three layers of straw tubes with pad readouts located inside the magnet are planned for the triggering of high p_t hadrons. HERA-B has been successfully operated in parallel with other experiments at the HERA storage ring with no measurable adverse effects [1]. The year-round operation of HERA-B, coupled with a high design instantaneous interaction rate of 40 MHz, renders HERA-B a unique hadron beauty and charm factory, with capabilities (such as cm long decay lengths for beauty hadrons) unmatched by beauty factories based on asymmetric electron-positron storage rings.

A general overview of the HERA-B experiment, including its sophisticated multi-level triggering strategy, is given by Rainer Mankel in a separate talk in this conference

---

[*] HERA-B is a collaboration of some 300 Ph.D.s from the following 34 institutions in 14 countries: NIKHEF, U. of Texas at Austin, U. of Barcelona, Institute for High Energy Physics (Beijing), Tsinghua U., U. of Cincinnati, Humboldt U., U. of Bologna, U. of Coimbra, Wayne State U., U. of Dortmund, Northwestern U., DESY, U. of Hamburg, Max-Planck-Institute at Heidelberg, U. of Heildelberg, U. of Houston, Institute for Nuclear Research (Ukraine), Niels Bohr Institute, Institute J. Stefan & U. of Ljubljana, U. of California at Los Angeles, Lund U., U. of Mannheim, Institute of Theoretical and Experimental Physics (Moscow), DUBNA, Moscow Physical Engineering Institute, Max-Planck-Institute at München, U. of Oslo, U. of Siegen, St. Petersburg Nuclear Institute, Brookhaven National Laboratory, U. of Utrecht, DESY Zeuthen, and U. of Zürich.

Table 1. Summary of HERA-B sub-detectors and their status

Sub-detector	Detector technology and basic unit	Status
Vertex detector	25 μm pitch double-sided silicon with 25 μm pitch readout by Helix chip	1/4 installed in '98, complete in '99
Inner tracker	300 μm pitch microstrip gas chamber (MSGC) with GEM foil	8 modules installed in '98, complete in '99
Outer tracker	5/10 mm honeycomb drift chambers	Production in '98, ready in '99
RICH	Gas radiator with PMTs	Ready in '98
ECAL	W/Pb scintillator sandwich with wavelength shifting fiber readout	Ready in '98
Muon	Drift and pixel chambers in hadron absorber	3/4 in '98, complete in '99
High p_t chambers	Resistive straw tubes with pad readout	1 test chamber in '98, complete in '99
TRD	Fiber radiator with straw readout	80% in '98, ready in '99

[2]. In this talk, the status of the various subsystems of HERA-B and their operational experience is reviewed. Special emphases are given to the tracking and particle identification systems. The 1998 run is presently underway. The configuration and status of the hardware for the 1998 run is described. The status of the various sub-detectors is summarized in Table 1.

2 The HERA-B Tracking Systems

2.1 The Silicon Vertex Detector

The HERA-B silicon vertex detector (SVD) [3] consists of 11 super-layers of silicon strip detectors. The super-layers are organized as modules, each having 2 double-sided 300 μm thick silicon wafers fabricated with 25 μm pitch strips. The strips are tilted by 2.5° on both sides for resolving tracking ambiguities, and are readout at 50 μm intervals. A super-layer consists of 4 modules, one at each quadrant centered at the beam pipe. Each SVD module is encapsulated in an RF shield, placing the SVD in a secondary vacuum. Figure 2 is the side view of a module, showing the cabling

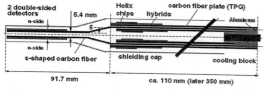

Figure 2. A schematic side view of a silicon vertex detector module. See text for details.

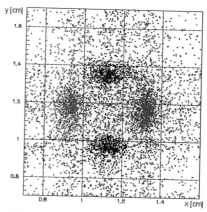

Figure 3. Distribution of interaction points reconstructed from three silicon planes in the 1997 run. The images of the four target wires are clearly discernible. Resolutions obtained from the residuals in each plane are 30 and 300 μm, transverse and along the beam respectively.

and support structures for readout and cooling.

In order to maximize the resolution and to minimize the material to the beam, the silicon detector planes are oriented perpendicular to the beam. The modules are mounted on roman pots, allowing the silicon planes to be moved out the beam during injection to avoid unintended radiation damage. During data taking, the SVDs are located as close as 1 cm from the beam, tracking charged particles at a peak fluence of 3×10^{14}/cm^2. The detector planes are cooled to about 10° C to minimize the effects of radiation damage. The readout electronics, based on HELIX chips located at about 10 cm from the silicon detector, are subject to much lower radiation dose. In-situ tests conducted in the 1997 run with three detector planes clearly demonstrated the ability of the SVD to locate the primary interaction points in the target (see Figure 3) to the requisite precision, namely 30 and 300 μm, transverse and along the beam respectively, with more than 98% efficiency per plane [3]. The expected lifetime of the SVD is at least one HERA-B year. Silicon planes and electronics aged by radiation will be exchanged during the annual shutdown. Twelve modules of the SVD have been installed in the lower and outer quadrants of the SVD for the 1998 run. The system is expected to be completed in 1999.

2.2 Microstrip Gas Detector

The charged particle tracking of HERA-B in the intermediate angular region is done by microstrip gas chambers (MSGCs). The construction and operation principle of a MSGC is illustrated in Figure 4. Ionization electrons from

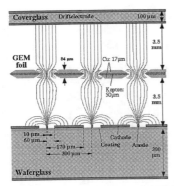

Figure 4. A schematic illustrating the construction of a MSGC with a GEM foil. See text for details.

charged particles are drifted toward the 10 μm wide anodes deposited on a wafer-glass, where the avalanches take place. The granularity (300 μm anode pitch) and radiation hardness of MSGCs are suited to this application.

Early aging test results, however, showed that conventional MSGCs cannot withstand the high counting rate at HERA-B, and that the anodes are easily damaged by heavily ionizing charged particles. The final design of the MSGC employs a gas electron multiplier (GEM) foil, which cured the problems by distributing the gain in two stages with a gain of about 150 in the anode stage [4]. The effect of the magnetic field on the chamber (the chambers are placed inside the dipole magnet) is compensated for by tilting the chambers by the Lorentz angle. The projected resolutions in the bending and non-bending directions are 80 and 800 μm, respectively. The layout of the inner tracker is similar to that of the SVD with 4 modules, one per quadrant, assembled around the beampipe. The lifetime of the MSGCs is expected to be at least one HERA-B year. The complete system consists of 10 superlayers. Eight modules have been installed for the 1998 run, and we expect to complete the system in 1999.

2.3 The Outer Tracker

The outer tracker of HERA-B covers the angular region up to about 300 mrad with conventional wire drift chambers. The chamber cathodes, made of gold-plated Pokalon, are folded and glued to adjacent layers to form a rigid lightweight honeycomb structure. Drift cells of 5 and 10 mm diameters are deployed in 11 super-layers for near complete coverage of charged particles. A typical $B_d \rightarrow J/\psi\, K_S \rightarrow e^+e^-\pi^+\pi^-$ Monte Carlo event superimposed in the outer tracker is shown in Figure 5. Extensive aging tests have been conducted with different cathode materials, gas mixtures, and operating voltages. The test results showed that these chambers could survive at least one and a half years of high-intensity running at HERA-B. The design of the outer tracker has been finalized, and mass production is in progress. Several prototypes have been deployed in the 1998 run for further aging studies. The plan is to have a complete outer tracking system in the 1999 run.

3 The Particle Identification Systems

3.4 RICH

The RICH consists of a 2.5 m long gas radiator, two sets of

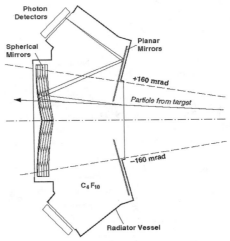

Figure 6. Schematic side-view of the RICH showing the placement of the various components. Cerenkov photons emitted by a charged particle and their paths to the photon detector are indicated.

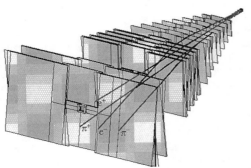

Figure 5. Layout of the outer tracker in HERA-B. A Monte Carlo $B_d \rightarrow J/\psi$ $K_S \rightarrow e^+e^-\pi^+\pi^-$ event is shown for illustration.

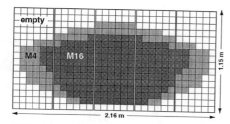

Figure 7. Distribution of base-boards (groups of 4 PMTs) for the 5 super-modules on one of the two photon detectors. The cells in blue are in the inner region and the cells in orange are in the outer region of the photon detector. Two empty outer super-modules are not shown.

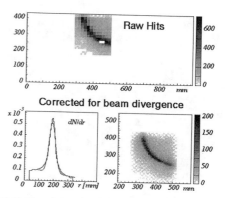

Figure 8. Cerenkov photons observed in one supermodule of the RICH in a beam test with 5 GeV electrons in a radiator filled with Argon.

spherical and plane mirrors, and two photon detectors. A schematic side-view of the RICH is shown in Figure 6. The radiator vessel, located at about 11 m downstream of the target, was made with stainless steel plates with 1 mm aluminum particle entrance and exit windows. Cerenkov light exits the vessel through 2 mm thick UV grade Plexiglas windows. The vessel will be filled with 100 m^3 of perfluorobutane (C_4F_{10}), a gas chosen for its low dispersion (5% in our spectral acceptance from 300 to 480 nm) and high refractive index ($n = 1 + 1.35 \times 10^{-3}$) under atmospheric pressure. The Cerenkov opening angle for $\beta = 1$ particles is 52 mrad.

The radiator gas will be circulated in a closed system with liquefaction stages for cleaning and buffering. Requirements on gas purity are modest because the photon detectors operate mostly in the visible part of the light spectrum.

The photon detectors are equipped with two kinds of multi-anode photomultipliers (PMTs), 1500 16-anode PMTs (Hamamatsu model R5900-00-M16) in the inner region and 750 4-anode PMTs (Hamamatsu model R5900-03-M4) in the outer region. The PMTs are arranged according to occupancy and resolution requirements as indicated in Figure 7, where the inner and outer regions are defined. The PMTs have a continuous 18×18 mm^2 bi-alkali photo cathode with metal channel dynodes (12 stages for the M16 and 10 for the M4). Both types of PMT have very low noise, and allow efficient single photon detection at a nominal high voltage of 650–850 V [5]. To reduce the total number of PMTs needed, a two-lens telescopic light collection system was designed to focus the Cerenkov light onto the PMT cathode surface, reducing the effective linear cell size by about a factor of two [6]. Each light-guide consists of a 35.3 mm^2 square plano-convex field lens ($f = 95$ mm) and a circular 32 mm Ø biconvex collector lens ($f = 30$ mm). The telescope has been optimized for good acceptance for photon incidence angles up to ±150 mrad. Both lenses are aspherical and are injection

molded to optical quality from UV transparent acrylic. The efficiency of the telescopes has been measured to be 65%, with 15% reflection, 15% absorption and about 5% geometric losses.

Custom-designed printed circuit boards (base-boards) have been built to provide mechanical, electrical, and electronic interfaces to the photomultipliers [7]. Each base-board, 70×70 mm^2 in area, has four independent high voltage divider chains, one for each PMT which is mounted on a custom-built socket. The base-board contains also four 16-channel readout-cards.

The upper and lower photon detectors are each constructed as seven 1.1×0.4 m^2 super-modules, each containing a grid made from soft iron sheet that serves as a magnetic shield and a mounting structure for the PMTs. The complete system contains 23,808 9×9 mm^2 and 3008 18×18 mm^2 cells, covering a total area of 2.9 m^2.

A full size super-module prototype has been tested at a 5 m long argon-filled radiator setup in a 3 GeV electron beam at DESY. The results are shown in Figure 8. The 21 photons detected for each $\beta = 1$ particle in this test translates to an expected yield of 34 ±2 photons per β=1 particle in the HERA-B RICH. From this number we calculated a kaon identification efficiency of 85% between 10 and 40 GeV/c kaon momenta.

The gas system for the RICH is near completion. Commissioning of the RICH with air in the radiator tank is currently taking place. Cerenkov rings in coincidence with the proton beam have been observed with the expected number of photons. The RICH radiator vessel is currently being flushed with dry nitrogen. The plan is to fill the tank with C_4F_{10} by the end of November.

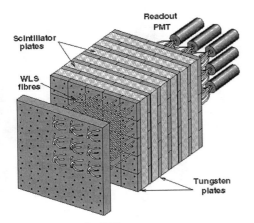

Figure 10. A schematic of an ECAL tower.

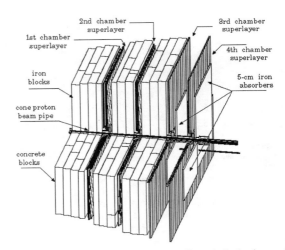

Figure 9. A cut-out view of the muon system. Shown in the drawing are four super-layers of muon chambers interspersed in steel and concrete absorbers.

3.5 The Electromagnetic Calorimeter

The electromagnetic calorimeter (ECAL) is designed to provide energy and position measurements for photons and electrons with an energy resolution of about $10\%/\sqrt{E}$. The ECAL is constructed with converter-scintillator sandwich towers with a total of 23 radiation lengths. The light in the scintillators is read out by PMTs via 1.2 mm diameter wavelength shifting optical fibers running longitudinally. The converters in the inner and outer regions are W and Pb, respectively. A schematic of an ECAL tower is shown in Figure 10. A neutral pion signal was seen in the 1997 run data, which has a width that is consistent with the expected energy resolution. The electron mis-identification probability is expected to be at the 1% level. The ECAL is completely installed for the 98 run. Commissioning of the full ECAL detector and systematic calibration work is currently underway.

Electron identification in the small angle region is augmented by a transition radiation detector (TRD). Thirty-two layers of 6 mm straw tubes operating in a Xenon-based gas mixture are interspersed in fiber radiators, which detect the X-rays from electrons with high efficiency.

3.6 Muon System

The muon system is designed to recognize muons at the trigger level, and to provide a J/ψ trigger based on the muon signals. The muon detector consists of four super-layers of wire chambers, whose design and granularity depend on the distance from the beam, embedded in about 18 interaction lengths of hadron absorber in the form of steel blocks in the inner region and steel-concrete blocks in the outer region. A cut-out view of the muon system is shown in Figure 9. In the inner region where the particle density is high, the tracking is done by pixel chambers whose wires are oriented along the beam direction. The cell size of about 1 cm is matched to the particle density in this region. In the outer region, the active detectors are drift tubes with pad and/or strip readouts. One super-layer of muon chambers was tested in a beam. The measured chamber occupancy with the beam on is in accordance with expectation.

4 The 1998 run

HERA-B is currently taking data with several major subdetectors in place. The hardware configuration of the detector for the 1998 run is shown in Figure 11. The RICH and the ECAL are completely installed. Parts of the SVD, the inner tracker, and the muon detector are in place. One quarter of the SVD and eight quadrants of the inner tracker have been installed. One muon super-layer has been installed at the beginning of the run, and two more are expected to be installed by the end of November in 1998. We will start filling the RICH radiator vessel with C_4F_{10} in early November of 98. Commissioning and calibration work with the RICH and ECAL is in progress. One high p_t prototype module and some TRD chambers will be installed during the 1998 run.

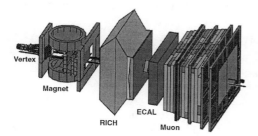

Figure 11. Hardware configuration of the HERA-B detector for the 1998 run.

Acknowledgement

The author thanks members of the HERA-B collaboration for their assistance in the preparation of this talk. He is particularly indebted to R. Harr, K.T. Knöpfle, and J. Pyrlik for providing useful information and plots and for systems for which they are experts. Support from the US DOE for the author's participation in HERA-B is acknowledged.

References

1 T. Lohse, Nucl. Instr. and Meth. A 408 (1998) 154.
2 R. Mankel, these proceedings.
3 For a recent review, see C.Bauer *et al.*, talk presented at the 6th International Workshop on vertex Detectors (VERTEX 97), Rio de Janeiro, September 1997.
4 B. Schmidt, talk presented at the 36th workshop of the INFN ELOISA-TRON Project on New Detectors, Erice, November 1997.
5. P. Krizan *et al.*, Nucl. Instr. and Meth. A 394 (1997) 27.
6 J. Rosen *et al.*, Nucl. Instr. and Meth. A 408 (1998) 191
7 J. Pyrlik, talk presented at the 3rd International Conference on Hyperons, Charm and Beauty Hadrons, Genoa, June 1998

RECENT RESULTS ON THE DEVELOPMENT OF RADIATION-HARD DIAMOND DETECTORS

J. S. CONWAY

Department of Physics and Astronomy, Rutgers Univ., 136 Frelinghuysen Rd., Piscataway, NJ, 08854-0819
e-mail: conway@physics.rutgers.edu

for the RD42 Collaboration:

W. Adam[1], C. Bauer[2], E. Berdermann[3], P. Bergonzo[4], F. Bogani[5], E. Borchi[6], A. Brambilla[4], M. Bruzzi[6],
C. Colledani[7], J. Conway[8], W. Dabrowski[9], J. DaGraca[8], P. Delpierre[10], A. Deneuville[11], W. Dulinski[7], B. van Eijk[12],
A. Fallou[4], F. Fizzotti[13], F. Foulon[4], M. Friedl[1], K. K. Gan[14], E. Gheeraert[11], E. Grigoriev[9], G. Hallewell[10],
R. Hall-Wilton[15], S. Han[14], F. Hartjes[12], J. Hrubec[1], D. Husson[7], D. Jamieson[16], H. Kagan[14], D. Kania[14], J. Kaplon[9],
C. Karl[17], R. Kass[14], K.T. Knoepfle[2], M. Krammer[1], A. Logiudice[13], R. Lu[13], P.F. Manfredi[18], C. Manfredotti[13],
R.D. Marshall[4], D. Meier[9], M. Mishina[19], A. Oh[17], L.S. Pan[14], V.G. Palmieri[20], M. Pernicka[1], A. Peitz[8], S.Pirollo[6],
R. Plano[8], P. Polesello[13], S. Prawer[16], K. Pretzl[20], M. Procario[21], V. Re[18], J.L. Riester[7], S. Roe[9], D. Roff[15],
A. Rudge[9], J. Russ[21], S. Schnetzer[8], S. Sciortino[6], S.V. Somalwar[8], V. Speziali[18], H. Stelzer[3], R. Stone[8], B. Suter[21],
R.J. Tapper[15], R. Tesarek[8], G.B. Thomson[8], M. Trawick[14], W. Trischuk[22], E. Vittone[13], A.M. Walsh[8], R. Wedenig[9],
P. Weilhammer[9], C. White[23], H. Ziock[24], M. Zoeller[14]

[1] *Institut für Hochenergiepphysik der Österreich, Akademie d. Wissenschaftern, Vienna, Austria*
[2] *MPI für Kernphysik, Heidelberg, Germany*
[3] *GSI, Darmstadt, Germany*
[4] *LETI, Saclay, France*
[5] *LENS, Florence, Italy*
[6] *Univ. of Florence, Florence, Italy*
[7] *LEPSI, IN2P3/CNRS-ULP, Strasbourg, France*
[8] *Rutgers University, Piscataway, NJ, USA*
[9] *CERN, Geneva, Switzerland*
[10] *CPPM, Marseille, France*
[11] *LEPES, Grenoble, France*
[12] *NIKHEF, Amsterdam, Netherlands*
[13] *Univ. of Torino, Torino, Italy*
[14] *The Ohio State University, Columbus, Ohio*
[15] *Bristol Univ., Bristol, UK*
[16] *Univ. of Melbourne, Melbourne, Australia*
[17] *II.Inst. für Exp. Physik, Hamburg, Germany*
[18] *Univ. di Pavia Dipartimento di Elettronica, Pavia, Italy*
[19] *Fermilab, Batavia, Illinois, USA*
[20] *Lab. für Hochenergiephysik, Bern, Switzerland*
[21] *Carnegie-Mellon University, Pittsburgh, Pennsylvania*
[22] *Univ. of Toronto, Toronto, ON, Canada*
[23] *Illinois Institute of Technology, Chicago, IL, USA*
[24] *LANL, Los Alamos, NM, USA*

Abstract

Charged particle detectors made from chemical vapor deposition (CVD) diamond have radiation hardness greatly exceeding that of silicon-based detectors. The CERN-based RD42 collaboration has developed and tested CVD diamond microstrip and pixel detectors with an eye to their application in the intense radiation environment near the interaction region of hadron colliders. This paper presents recent results from tests of these detectors.

1 Diamond as a detector material

Chemical vapor deposition (CVD) diamond is a poly-crystalline form of diamond grown in thin films on sub-strates. The material possesses the remarkable proper-ties of single-crystal diamond, including its mechanical strength and hardness, very high electrical resistivity, and high thermal conductivity. As such, it represents a very attractive material for application as a charged particle detection medium. Indeed, diamond has been used for many years for particle detection, and in the past several years the RD42 Collaboration[1] at CERN has developed microstrip and pixel tracking detectors using diamond as the detection medium.

The main challenge in the application of CVD dia-mond to charged particle detection lies in the facts that in diamond a minimizing ionizing particle liberates an average of 3600 electron-hole pairs per 100 μm material traversed, and the liberated charges do not travel the full extent of the crystal lattice before being trapped at impu-rity or lattice dislocation sites. This limits the magnitude of the electrical signal induced on the microstrip or pixel electrodes, which must be distinguished from noise for efficient operation of such a device.

In the past several years, manufacturers have steadily improved the purity and crystal quality of CVD diamond such that the charge collection distance, the average dis-tance by which electrons and holes separate in the dia-mond under the applied electric field, has improved by several orders of magnitude. Furthermore, by removing material from the substrate face of the diamond film the collection distance can be improved further. Nearly all diamond samples presently studied by RD42 have had substrate material removed.

The signal size in diamond is also observed to in-crease after exposure to ionizing radiation; it is believed that the liberated charge diffuses in the lattice until trapped at impurity sites or lattice dislocations, passivat-ing these traps. Unless noted, the results of tests of detec-tors discussed here are from diamond in this "pumped" state, and at applied electric fields of 1 V/μm.

2 Diamond microstrip detectors

Diamond microstrip detectors can be fabricated by straightforward techniques of metallization, typically us-ing chromium and gold, or titanium and gold. RD42 has fabricated and tested both small-scale (1 cm × 1 cm) and larger-scale (2 cm × 4 cm) detetors, metallized with 50-μm-pitch microstrips. These detectors are made from di-amond samples which show promise based on their char-acteristics in bench tests using simple circular metallized electrodes, and a ^{90}Sr beta source.

The beam tests, carried out in a 100-GeV/c pion

beam at CERN, utilize a set of eight silicon detectors equipped with VA2 readout chips.[2] These detectors, four situated upstream and four downstream of the diamond detectors under study, provide four space measurements each in the x and y directions transverse to the beam. The intrinsic position resolution of the silicon detectors is such that after removing tracks with large-angle scat-ters, the resulting position resolution of tracks projected onto the diamond plane is about 1-2 microns. This reso-lution is limited by the accuracy of the diamond detector alignment.

Equipping the diamond detectors with VA2 readout chips, which have very low noise (typically under 100 e^- ENC at 2 μsec integration time), allows detailed study of the characteristics of the diamond devices. Figure 1 shows the pulse height distribution (obtained from the sum of the pulse heights on the three strips nearest the extrapolated track) for one of the most promising sam-ples tested so far, which has 432 μm thickness, and is 1 cm × 1 cm in area. The distribution shows a most probable charge near 5000 e^-, and an average charge collected of about 8000 e^-. The region near zero shows that the distribution starts near 1000 e^-, but a fit of the overall distribution to a Landau shape convoluted with a Gaussian does not describe the data well.

Figure 2 shows the distribution of diamond hit po-sition minus extrapolated track position for one of the first 2 cm × 4 cm detectors. A Gaussian fit to this dis-tribution reveals a sigma of about 10 μm. (In this case the diamond hit is not used in the track fit.) This shows that the detector construction techniques developed for the smaller scale diamond microstrip detectors can be successfully applied to larger-area detectors.

In a real application in a hadron collider the read-out times required are much shorter; at the LHC the bunch crossing times are 25 nsec. RD42 has performed tests of diamond detectors equipped with the SCT128A readout chip, based on a radiation-hard DMILL bipo-lar technology.[3] This chip, which typically performs with about 700 e^- noise ENC, samples the signal at the req-uisite 40 MHz. Figure 3 shows the distribution of pulse heights on a 432-μm-thick diamond microstrip detector equipped with an SCT128A. The ratio of peak signal size to average single-channel noise for this detector is about 7.

3 Diamond pixel detectors

The innermost tracking detectors at the LHC are pixel devices, due to the stringent demands on channel occu-pancy and radiation hardness that the very high luminos-ity imposes. Present designs call for silicon-based detec-tors with either square or elongated rectangular metal-lized pixel patterns, with readout chips with arrays of

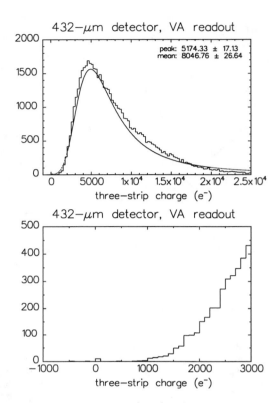

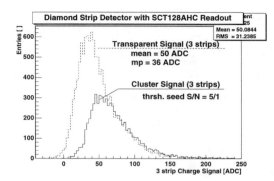

Figure 3: Pulse height distribution of CVD diamond microstrip detector with SCTA128 readout chip.

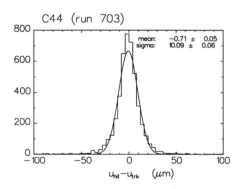

Figure 1: Pulse height distribution (above) and detail of region near zero pulse height (below) for a 432-μm-thick diamond microstrip detector.

Figure 2: Position resolution of large-area (2 cm × 4 cm) CVD diamond microstrip detector with VA2 readout chip.

preamplifier inputs arranged with matching geometry. These devices will survive the first phase of LHC operations, but at no smaller radius than about 6-7 cm.

Diamond-based pixel detectors, an order of magnitude more radiation hard (see next section), offer an attractive alternative to silicon devices. RD42 demonstrated in late 1996 a diamond detector with a pixel array read out by a VA2 microstrip readout chip via a fanout fabricated with metallized glass. This showed that diamond pixel detectors could be made to work, and attention in the past year has turned to fabricating actual diamond pixel detectors using development versions of the pixel readout chips from ATLAS and CMS.

The readout chip inputs can be bonded to the pixel array using a technique similar to that employed with silicon. Small "bumps" of indium are vapor-deposited on the metallized surface, using a mask and then the readout chip is aligned and pressed into place and heated to 170°C. The most successful test so far was made using a diamond sample metallized with the 50 μm × 536 μm pattern of the ATLAS/3 readout chip, shown in figure 4. The indium bump bonding process, performed in industry at Boeing, resulted in a 100% bump yield, and in a beam test, 97% of the channels in the properly functioning section of the readout chip were operational. This can be seen in figure 5 which shows the frequency of hits in each pixel in the detector. Ignoring the two "hot" columns of pixels (columns 5 and 7 in the figure), and focusing on just the first four columns, we find only 6 out of 256 channels with no hits.

The pulse height distribution of this detector, measured in the "unpumped" state, shown in figure 6, has a sharp edge at 3500 electrons, reflecting the threshold for reading out a hit pixel. The distribution of charge in the absence of any applied threshold, determined from the same sample configured as a microstrip detector with

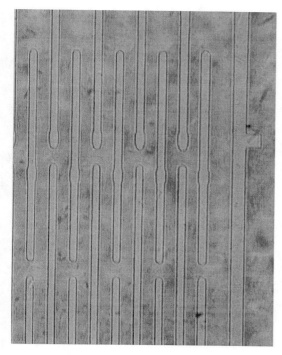

Figure 4: ATLAS/3 Pixel pattern metallized on diamond.

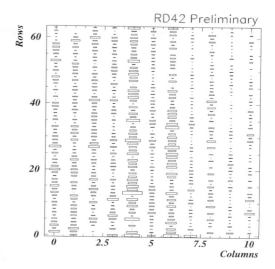

Figure 5: Frequency of pixel hits in CVD diamond pixel detector with ATLAS/3 readout.

VA2 readout, actually extends down to about 1000 e^-, after "pumping" with ionizing radiation. Thus though the efficiency with a threshold of 3500 e^- is about 25%, after pumping and with a reduced threshold of 2000 e^- the efficiency of this device would be about 85%.

The position resolution of the diamond pixel detector appears in figure 7, and a fit to the central portion of the distribution gives a sigma of roughly 15 μm, as expected for digital position resolution.

4 Radiation hardness of diamond

The radiation hardness of CVD diamond has been tested with exposures to electromagnetic radiation (^{60}Co), 300 MeV/c pions, 24 GeV/c protons, and 1 MeV neutrons, up to fluences of well over 10^{15}. The result for exposures to protons appears in figure 8, which shows that after exposures of 5×10^{15} p/cm^2, the average pulse height decreases by about 40%. A similar decrease is seen after exposure to 1.3×10^{15} neutrons/cm^2; an exposure to 1.13×10^{15} pions/cm^2 results in a decrease of about 30% in average pulse height. We note that this exposure level represents the charged hadron dose anticipated[4] after about 5 years

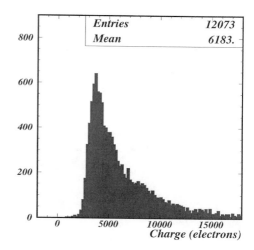

Figure 6: Distribution of collected charge in CVD diamond pixel detector with ATLAS/3 readout.

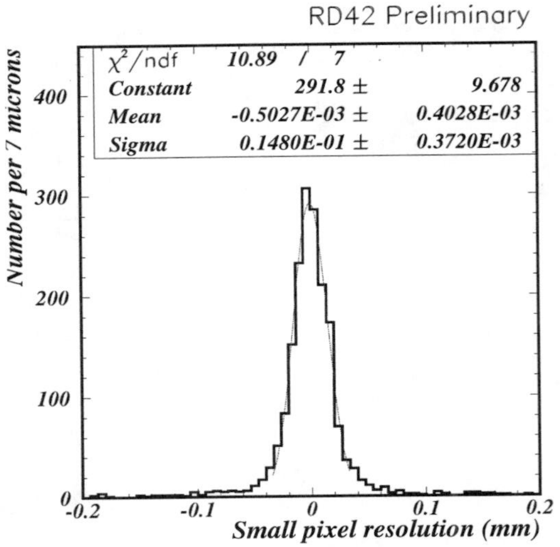

RD42 Preliminary

χ^2/ndf	10.89 / 7	
Constant	291.8 ±	9.678
Mean	-0.5027E-03 ±	0.4028E-03
Sigma	0.1480E-01 ±	0.3720E-03

Number per 7 microns

400

300

200

100

0

-0.2   -0.1   0   0.1   0.2

Small pixel resolution (mm)

Figure 7: Position resolution transverse to the long direction of the pixels for a CVD diamond pixel detector with ATLAS/3 readout.

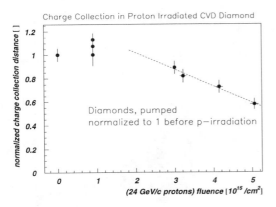

Figure 8: The reduction of charge collection distance as a function of proton fluence.

of operation at 8 cm radius of the interaction region of the LHC.

5 Sumamry

In the past year the RD42 Collaboration has advanced the development of CVD-diamond detectors in several ways: The first large-area (2 cm × 4 cm) detector was successfully constructed and operated. Average signal sizes of over 8000 e^- were achieved in detectors of 432 μm thickness, and when equipped with a fast (25-nsec) radiation-hard readout chip, had a 7:1 peak signal to single-strip noise ratio. The first beam test of an indium bump-bonded diamond-based pixel detector with an actual pixel readout chip performed with about 97% of the channels operational, and gave the expected position resolution. Most importantly, the radiation hardness of high quality diamond detectors was demonstrated to meet that required for application near the interaction region of the LHC and other high-luminosity hadron colliders.

References

1. "Development of Diamond Tracking Detectors for High Luminosity Experiments at the LHC," CERN/LHCC 98-20, Status Report RD42, June 1998.

2. O. Toker, *et al.*, Nucl. Instrum. Meth. A340, 1994, 527-579.

3. F. Anghinolfi, *et al.*, IEEE Trans. Nucl. Sci.44:298-302, 1997.

4. CMS Technical Proposal, CERN/LHCC 94-38.

THE ATLAS TRANSITION RADIATION TRACKER AND RECENT DEVELOPMENTS IN P+N SILICON STRIP DETECTORS FOR LHC

G. F. TARTARELLI

CERN, EP Division, CH-1211 Geneva 23, Switzerland
E-mail: tarta@mail.cern.ch

The ATLAS Collaboration is building, as a part of its tracking system, a 370,000 straw drift tube detector, the Transition Radiation Tracker, that provides flexible level-2 triggering, robust pattern recognition and good electron/pion separation. We discuss the final design choices that will ensure reliable operation over the experiment lifetime in the harsh LHC environment, with particular emphasis on high counting rate and detector ageing issues. We also discuss recent developments in single-sided p+n silicon strip detectors optimised for capacitively-coupled read-out with single strip threshold binary electronics. Results on high voltage operation capability and on the use of a novel biasing technique, making use of implanted bias resistors, after irradiation up to a fluence of 3×10^{14} cm^{-2} 24 GeV protons, are presented.

1 Introduction

Detectors designed for use at the Large Hadron Collider (LHC) will have to meet unprecedented standards due to the harsh environment in which they will be operated. The requirement that the detectors be fully functional up to the design LHC luminosity (10^{34} cm^{-2} s^{-1}) and for the entire lifetime of the experiment (10 years) poses serious challenges to physicists and engineers.

Some of the problems that detectors will face at LHC, like radiation damage for silicon detectors and ageing for gaseous detectors, have been already encountered in past experiments and have generated long series of studies. However, the scale with which some of these effects will appear at the LHC is such that new efforts and new design solutions have been necessary.

This paper describes some of these efforts by examining two particular detectors. In the first part of this paper, the Transition Radiation Tracker, being built as a part of the tracking system of the ATLAS detector, is presented. The second part describes how issues connected to the operation of microstrip silicon detectors in a heavily damaging radiation environment have been addressed by the development of single-sided p+n silicon detectors designed at the MPI Semiconductor Lab (Munich, Germany).[1]

2 The Transition Radiation Tracker

The Transition Radiation Tracker (TRT) is a drift tube system that, together with silicon strip and pixel detectors at lower radii, constitutes the ATLAS Inner Detector.[2] The TRT is meant to provide robust tracking information in order to ease the pattern recognition in the LHC environment, to increase the momentum resolution by providing track measurement points up to the solenoid radius, and to be used at the trigger level. In addition, by integrating the functionality of a Transition

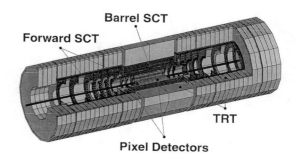

Figure 1: Schematic 3D view of the ATLAS Inner Detector.

Radiation Detector, the TRT also provides stand-alone electron/pion separation.

In Sec. 2.1 a brief description of the ATLAS Inner Detector is given. The geometry and the design of the TRT are briefly summarised in Sec. 2.2 and 2.3. Finally, in Sec. 2.4, 2.5 and 2.6, more details on issues connected to the high-rate operation and to the ageing of the detector are presented.

2.1 The ATLAS Inner Detector

The ATLAS detector[3] is a multi-purpose detector being built to study the physics of pp collisions provided by the LHC. It has been designed to work up to the highest LHC luminosity (10^{34} cm^{-2} s^{-1}) providing as many signatures as possible using leptons, photons, missing transverse energy, jets and b-jets. The detector has both forward-backward and azimuthal symmetry and consists of a tracking system contained inside a solenoid magnet, a calorimetry system using both liquid argon and tile scintillator techniques and a muon spectrometer embedded inside three air-core toroid magnets.

The ATLAS Inner Detector is designed to reconstruct charged particles and measure accurately their momenta and impact parameters. It also provides π/e separation meant to extend the one provided by the calorimeter. The Inner Detector also plays an important role in the trigger system, providing a Level-2 track trigger which is crucial for B-physics analyses during the first years of running at low luminosity (10^{33} cm^{-2} s^{-1}) and for high-p_T lepton triggers at high luminosity.

The Inner Detector (see Figure 1) radially extends between a 2.5 cm radius beryllium beam-pipe and a 1.15 m radius superconducting solenoid magnet generating a 2 Tesla magnetic field parallel to the beam axis. The total length of the detector is 6.8 m which translates in a pseudorapidity coverage of $|\eta| < 2.5$.[4] The solenoid is shorter than the detector length (it is about 5.3 m long) causing the field to be non-uniform at the two far ends. In the *barrel* region, covering $|\eta| < 0.7$, the detectors are arranged as cylindrical layers around the beam-line. In the two *end-cap* regions ($0.7 < |\eta| < 2.5$), the detectors are arranged as disks or wheels with a radial geometry. The Inner Detector is divided into three sub-systems using three different technologies: the Pixel Detector, the Semi-Conductor Tracker (SCT) and the TRT.

The Pixel Detector[5] makes uses of about 140 million silicon pixels (each pixel cell is 50 μm long in the $R\phi$ direction and 300 μm in z) to provide three precise spatial measurements per charged particle. It consists of three layers in the barrel region and five disks in each end-cap region. The barrel layers are located at radii of 4.3 cm, 10.1 cm and 13.2 cm. The pixel cells will be read out by DC-coupled rad-hard electronics. The read-out will be binary, but low-resolution (4–5 bits) digital charge information will be available via a Time-Over-Threshold measurement.

The pixel system is surrounded by 4 barrel layers and 9 end-cap wheels (on each side of the barrel region) of silicon microstrip detectors, that constitute the SCT. Each SCT layer is made of basic detecting units called *modules*. Each module is made of pairs of 80 μm pitch and 12 cm long single-sided strip detectors glued together back to back. One detector is given a relative rotation (*stereo angle*) of, alternatively, 40 mrad and -40 mrad, with respect to the other one which has axial (radial) strips in the barrel (end-cap) region. The strips are read-out by AC-coupled single-threshold binary electronics. The SCT detector has about 6.2 million read-out channels and provides four precision measurements per track over $|\eta| < 2.5$.

Both the pixel and the SCT detectors are contained inside the TRT which is described in more detail in the next sections.

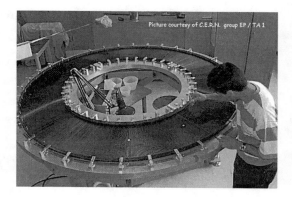

Figure 2: Insertion of a straw inside the end-cap TRT module-0.

2.2 Design of the Transition Radiation Tracker

The TRT consists of 370,000 proportional tubes or *straws*. Each straw is a 4 mm diameter Kapton® cylindrical tube. The inner surface of the straw, which acts as a cathode and is kept at high voltage, is covered by a 0.2 μm thick layer of aluminium (to increase its conductivity) in turn protected by a 6 μm thick layer of carbon-doped polyimide. In the centre of the straw is a 30 μm thick gold-plated tungsten anode wire from which the output signal is read-out. The straws are filled with a ternary gas mixture based on xenon (80%) with the addition of CF_4 (20%) and CO_2 (10%). The total electron drift-time in the gas is about 40 ns.

The use of xenon is motivated by the double functionality of the TRT as both a tracking and a transition radiation detector, in which, a novel feature, the layers of straws are interleaved with the radiator material. Charged particles with a relativistic γ factor in excess of about 1000 will produce Transition Radiation (TR) photons in the radiator, which can then be detected with high efficiency in the xenon gas through photoelectric absorption. The read-out is therefore performed using double-threshold binary electronics. A low-threshold setting (about 200 eV) is used to detect the signal (on average 2 keV) left by charged MIPs ionising the gas mixture; the high-threshold setting (about 6.5 keV) is used to detect TR photons, which have an average energy deposition of about 8–10 keV. The first threshold setting allows drift-time measurements with a resolution as good as 120 μm (see Sec. 2.5) and the second one is expected to provide electron/hadron separation with a hadron rejection in the range 15–2000 (depending on the particle p_T and η) at low luminosity, for a 90% electron efficiency. Using 20 GeV pion and electron beams at the CERN SPS, a 78-cm long TRT prototype (corresponding

Table 1: Expected straw counting rates at design LHC luminosity.

Position (cm)	Low threshold		High threshold		Streamers	
	total	neutrons	total	neutrons	total	neutrons
Barrel (R=108)	6 MHz	0.8 MHz	0.4 MHz	0.2 MHz	11 kHz	8.5 kHz
Barrel (R=64)	18 MHz	1.1 MHz	1.1 MHz	0.2 MHz	21 kHz	12 kHz
Endcap (z=180)	7 MHz	0.7 MHz	0.5 MHz	0.1 MHz	4 kHz	<1 kHz
Endcap (z=330)	19 MHz	1.9 MHz	1.1 MHz	0.3 MHz	10 kHz	<1 kHz

to $|\eta| \sim 1.2$ in ATLAS) has achieved a rejection factor of 100 against pions. [6]

In the barrel region, the straws are 150-cm long and are parallel to the beam. There are 73 layers of straws occupying the radial region between 56 cm and 107 cm. The straws are arranged in ϕ as *modules* of rhomboidal section and laid out radially into three rings of 32 modules each. In each module, the straws are embedded in polypropylene/polyethylene fibre material acting as radiator. The wires are split in the centre and read out at both ends, thus reducing the occupancy but doubling the number of electronic channels. The innermost 6 layers are active only over the last 36 cm along z, in order to reduce the occupancy but nevertheless achieve some improvement in the number of measurement points in the barrel/end-cap transition region. In total, there are 52,544 straws and 105,088 read-out channels in the barrel part.

In the end-caps, the straws are radial and are arranged in 18 units (per side) called *wheels*, for a total of 224 layers of straws on each side. The straws extend between radii of 64 cm and 103 cm, except in the last 6 wheels (64 layers) where the straws extend between radii of 48 cm and 103 cm. The radiator consists of 15–20 μm thick foils made of polypropylene and regularly spaced by 200–300 μm, distributed between the straw layers. Contiguous layers of straws are given a rotation in the ϕ direction in order to obtain a uniform coverage for tracks: on average, a charged particle generated in the interaction region will cross 36 straws over the full pseudorapidity range (this figure is $\sim$30% lower in the barrel/end-cap transition region). The total number of end-cap straws (and read-out channels) is 319,488.

2.3 Present Status of the Project

The TRT has recently started construction. The straws are currently in production for the complete detector: the 500 kg of 25-μm thick Kapton® film needed for the complete detector have been already purchased and the coating of 80% of the film has been succesfully applied. At the same time, the procurement for the other components has started. The assembly (see Figure 2) of the final pre-production prototypes (the *module-0's*) is going

on right now (September '98). The mass production will start in the second half of '99.

2.4 Operating Environment

At design luminosity, the straw counting rates can be very high, as shown in Table 1. The most critical regions are the barrel straws at small radii and the end-cap straws at large η where the counting rates can approach 20 MHz. About 50–60% of the total rates are due to interactions in the Inner Detector itself and in the beam-pipe. About 10% of the counting rate is due to neutron interactions generating photons in the MeV range. The counting rate from streamers is much smaller.

If the particle rates are translated into radiation levels, all TRT materials should be validated to survive an ionising dose of 6 (4.7) Mrad and charged-hadron fluxes of 2.0×10^{14} (1.0×10^{14}) cm^{-2} 1 MeV n$_{eq}$ at the position of the straws (FE electronics).

2.5 High-rate Operation

The design choices and the operating environment briefly depicted in the previous section put stringent requirements on the TRT front-end electronics. The need to operate with a 200 eV low-level threshold requires careful consideration of both extrinsic and intrinsic noise in the design of the FE chip. At the same time, the threshold must be uniform and stable even at the highest counting rates. Due to the fact that xenon-based gas mixtures are characterised by a very long ion tail (lasting up to 60 μs), the implementation of an AC-coupled ion tail cancellation and baseline restoration circuitry has been necessary.

A rad-hard version of the TRT analog FE chip, the ASDBLR, has been produced and tested in the laboratory and in a high-energy pion beam at the CERN SPS. The relation between the input signal amplitude (in units of energy) and the discriminator threshold (in units of current) is plotted in Figure 3. This figure shows that a threshold current of 47 μA corresponds to a 200 ± 25 eV energy threshold for all 16 channels measured. The high-threshold uniformity is even better. Moreover, at the low-threshold value, the noise counting rate has been

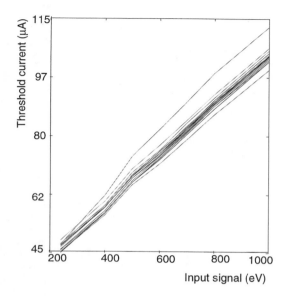

Figure 3: Low-level threshold current as a function of the input signal for 16 different channels.

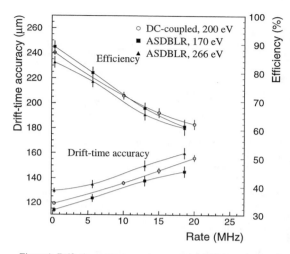

Figure 4: Drift-time measurement accuracy and efficiency (see text) as a function of the straw counting rate for two different low-level thresholds. Results for a DC-coupled, discrete, reference circuit are also shown.

measured to be less than 10 kHz. The drift-time accuracy and efficiency as a function of the counting rate are shown in Figure 4. The drift-time resolution has been measured to be 113 (145) μm with an efficiency of 89 (62)% at 0 (18.6) MHz rate. This efficiency is defined as the fraction of times the measured drift-time is within 2.5 standard deviations of the true drift-time. For comparison, the basic straw hit efficiency is greater than 97%. A new, improved rad-hard version of the chip will be submitted to DMILL in September '98.

2.6 Ageing

Gaseous proportional chambers operated in a high-radiation environment can have their performance seriously compromised by *ageing* effects. By ageing is meant a degradation with time of the performance of the chamber that can appear as one or more of the following effects: loss of gain, worsening of the energy resolution, discharges, sparking and so on. The cause of this behaviour has to be ascribed to the complex chemical phenomena that take place in the avalanche (an environment similar to the one studied by plasma chemistry) leading to the creation of a variety of molecular species (oxides, polymers and others) that can accumulate as deposits or layers on the electrode surfaces.

The ageing processes that can appear in the chamber depend on the design choices: type of gas used, anode and cathode material and operating conditions. However, even if the initial design choices are correct, ageing processes can be initiated by the presence of contaminants already present in the gas mixture or appearing at a later stage as a consequence of natural or radiation-induced outgassing of all the detector materials in contact with the gas.

The ageing processes are proportional to the quantity of ions produced in the avalanche, so a suitable unit for these studies is the total accumulated charge per unit length of the detector. The TRT is expected to accumulate about 6 C/cm after 10 years of operation at the LHC. A series of tests done with small-scale prototypes (prototypes with short straws) have proven that the straw cathode shows no deterioration up to 18 C/cm (the electrical and mechanical properties of the cathode are basically unchanged and the protective layer is still intact). For what concerns the anode wire, etching of the gold layer has been observed with some wires. With the best candidates, no drop of the signal has been observed within a few percent up to 8 C/cm. A total of 14 wires from different vendors is under study now in order to select the best one for mass production.

A large-scale setup, containing about 100 straws of 1 m length that can be irradiated along 60 cm in order to reproduce conditions much closer to those that will be encountered at the LHC, has been assembled and operated for ageing measurements with high statistics. The impact of streamers is under test with a 100 MeV α-beam in Karlsruhe. These particles have an energy deposition

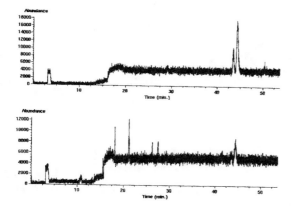

Figure 5: Total ion chromatograms from samples of Araldit® AW 106, obtained at 60 °C before (top) and after (bottom) irradiation.

in the straws in the range 108–190 keV. About 0.1 C/cm, i.e. 10% of the total dose expected from streamers at the LHC, have already been accumulated at this time.

All materials and glues to be used in the construction of the TRT must undergo a strict validation of their properties before and after irradiation. In addition to their mechanical and electrical properties, their outgassing properties and compatibility with the gas mixture need to be understood. For these studies, all the samples under consideration have been irradiated up to a dose of 50 Mrad (which is about 10 times higher than what expected) using a source of Cobalt 60.

The outgassing study is performed in two steps. In a first step, the intrinsic tendency of the materials to outgas is studied by keeping each sample in a helium atmosphere and then analysing the resulting gas with the help of a gas chromatograph connected to a mass spectrometer. The measurement is performed at ambient temperature and is then repeated at 60 °C in order to speed up the outgassing (although the proportionality factor relating time to temperature is unknown). This test is used to pre-select a number of candidates more likely to satisfy the detector requirements and to reduce the number of candidates that need to be tested in the second step. The second step consists of a true ageing test in which the standard TRT gas mixture is flushed through samples of materials and then sent to an instrumented straw which is exposed to a high irradiation dose using an X-ray source. The electrical signal from this straw is then compared to the one of a reference straw operated at a much lower dose in order to distinguish changes due to ageing processes from changes in the operating conditions (temperature, pressure and so on).

The first step of this procedure has been already completed and has already allowed to rule out some candidate materials and glues. Figure 5 shows, as an example, the chromatograms before and after irradiation for Araldit® AW 106, a glue commonly used in high-energy physics applications. Clear outgassing peaks of various organic compounds are visible and, as a consequence, this glue will not be used in TRT construction. The second step of the material validation program is still in progress.

3 Recent Developments in p⁺n Silicon Strip Detectors for the LHC

Silicon detectors at the LHC will be exposed to extremely high radiation levels. After 10 years of operation a detector at 30 cm from the beam-line will have been exposed to fluences in the range 1–2×10^{14} cm^{-2} 1 MeV n$_{eq}$ and to an ionising dose of about 10 Mrad.

The main consequences are an increase of the oxide charge at the Si-SiO$_2$ interface and a damage of the bulk material. The first effect will cause locally enhanced electric fields and will increase the possibility of discharges and breakdown. The second effect will cause an increase of the leakage current (I_L). Moreover, the silicon doping will change from n to p-type (the so-called *type inversion*) and the p-type doping will slowly increase with time even after the end of the exposure (*reverse annealing*): both these effects cause a steady increase of the depletion voltage (V_d).

Both the I_L and V_d variations are a function of the temperature and the dependence is such that they are reduced by keeping the silicon at low temperature. But, while the leakage current is a strong function of the operating temperature, the depletion voltage depends on the full thermal history of the detector (both during and after irradiation). For this reason one would like to always keep and operate the silicon at the lowest possible temperature (but not much lower than about −10 °C for the simplicity of the detector module design).

It is important to accurately estimate the depletion voltage required after 10 years of running in order to ensure that it is not too large. Too high depletion voltages would increase the risk of detector breakdown, would require to keep the silicon at too low temperatures, and might cause *thermal runaway* of the detector. Thermal runaway is a closed-loop process in which, at high bias voltages, the current in the detector is so high that the power dissipation increases and so does the temperature, causing an increase of the leakage current. This, in turn, gives rise to a further increase of the power dissipation in the detector and so on. The result is that the detector cooling system is not able anymore to keep the detector cool and the silicon temperature increases without control. As a consequence, it is important to have detec-

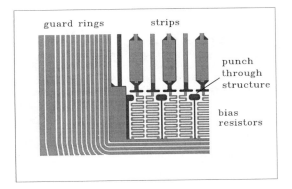

Figure 6: View of the p$^+$n silicon detector corner region. The strip bias resistors and the edge protection structure are visible.

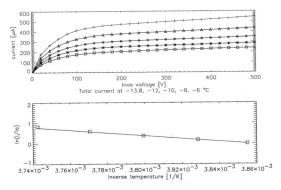

Figure 7: Detector current versus bias voltage measured at various temperatures (-6, -8, -10, -12 and -13.8 °C from top to bottom) after irradiation (top). Logarithm of the current normalised at -13.8 °C versus the inverse temperature (bottom). Data (empty squares) is superimposed to the theoretical prediction (line).

tors with low leakage currents, that do not require too high bias voltages for depletion after irradiation (about 350–400 V as a maximum) and that in their operating range are exempt from discharges and breakdowns.

3.1 Detector Layout and Design Choices

The p$^+$n detectors developed at MPI have been designed in order to cope with the conditions described above. They are p$^+$n 280 μm-thick detectors with 80 μm strip pitch and 80 μm read-out pitch. The width of the strip implant is 22 μm and is covered by a 16 μm metal line. Great attention has been put in the design of the detector, and especially in the design of its guard ring structure (16 rings with increasing gap widths towards the edge), in order to keep the electrical fields to a minimum everywhere. In this way, the detector is less prone to breakdown and can be operated at higher voltages. A schematic view of the detector is given in Figure 6.

A novel biasing technique, using *implanted resistors*, has been used as an alternative to the more common polysilicon biasing. In this configuration, biasing is done through the bulk by means of a highly doped ion implanted region connecting the bias ring on one side to the strip implant on the other side. The implanted resistors are put in parallel to punch-through structures, which become operational only in the case of radiation bursts, limiting the voltage drop across the resistors (see Figure 6). The advantage of this configuration is that both the noise problems of punch-through biasing after high irradiation doses and the complications and the cost of polysilicon biasing are avoided. Polysilicon biasing requires the additional production steps of polysilicon deposition, two photolithographic steps and two implantations while detectors with implanted resistors are produced by using only five photolithographic steps.

3.2 Detector Properties and Results of the Tests

A very high strip yield, ranging from 99.3% to 100% has been obtained for the eight detectors produced. The strip capacitance to the four closest neighbours has been measured to be 0.8–0.9 pF/cm. Adding 0.3 pF/cm for the capacitance towards the back-plane, this leads to a total strip capacitance of 1.1–1.2 pF/cm.

The detectors have shown low leakage currents before and after irradiation. Before irradiation, leakage currents have been measured to be less than 1 μA for the 64×63.6 mm^2 device and all the detectors could be biassed up to 1000 V while keeping the current still below 20 μA. The current-to-voltage characteristic, after irradiation at the CERN PS with 24 GeV protons up to a fluence of 2.1×10^{14} cm^{-2}, is shown in Figure 7. The detector current has been measured at various temperatures after a warm-up period at 23 °C of 64 hours. Also shown in the same figure is the logarithm of the leakage current at V_{bias}=400 V, normalised to the value at -13.8 °C, as a function of the inverse of the temperature. The dependence is linear as expected from processes of generation/recombination in the bulk of the detector. This shows that other processes, like avalanche generation, are essentially absent.

The implanted resistors performed very well after irradiation: they introduced no additional noise and only a moderate increase of the resistance (20% at most) was observed (see Figure 8).

Irradiated detectors have been micro-bonded together in order to obtain 12-cm long strips and have been connected to non-irradiated electronics operating with a 25 ns effective shaping time. The detectors have shown full depletion at about 400 V, stable noise figures as a

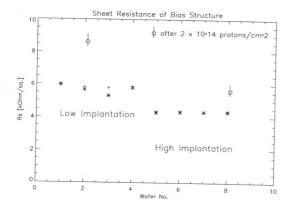

Figure 8: Resistance of two implanted resistors, measured on each of the eight wafers produced, before (stars and crosses) and after (circles) a dose of 2×10^{14} p/cm^2. The first four wafers were produced with a higher doping concentration.

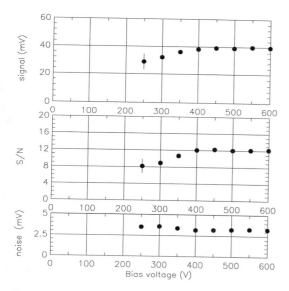

Figure 9: Signal, signal-to-noise ratio (S/N) and noise measured on a 12-cm long detector irradiated to 3×10^{14} p/cm^2 and connected to electronics with an effective shaping time of 25 ns.

function of the bias voltage and a signal-to-noise ratio of about 12, as shown in Figure 9.

4 Conclusions

The ATLAS Transition Radiation Tracker has entered the construction phase and is scheduled to enter mass production in mid-1999. The analog front-end electronics meets all requirements for low noise, threshold uniformity and operation up to counting rates of about 20 MHz. A careful material validation and ageing study program is ongoing.

Single-sided p$^+$n microstrip detectors developed at MPI meet all requirements for successful operation at the LHC. Laboratory and beam tests have shown that they have excellent and reproducible I-V characteristics and provide good S/N both before and after irradiation. The novel technique of implanted resistors for strip biasing has also shown no noise problems and small resistor increase after irradiation. These detectors represent a simple, high-yield and cost-effective solution for use at the LHC.

Acknowledgements

This paper presents results obtained by the ATLAS TRT Collaboration and by the ATLAS Silicon group in MPI. The author wishes to thank D. Froidevaux (CERN) and H.-G. Moser (MPI, Munich) for careful reading of the manuscript and for providing some of the material presented.

References

1. H.-G. Moser *et al.*, "Single-sided p$^+$n Silicon Strip Detectors for Large-scale Experiments in the Harsh Radiation Environment of Future Accelerators," contributed paper to this Conference.
2. The ATLAS Collaboration, "Inner Detector TDR Vol. 1," CERN/LHCC/97-16 (1997).
   The ATLAS Collaboration, "Inner Detector TDR Vol. 2," CERN/LHCC/97-17 (1997).
3. The ATLAS Collaboration, "Technical Proposal," CERN/LHCC/94-43 (1994).
4. The pseudorapidity is defined by $\eta \equiv -\ln(\tan\theta/2)$ where θ is the polar angle measured from the positive z axis. ATLAS uses a coordinate system with the x axis in the horizontal plane towards the centre of the LHC ring, y in the vertical direction, and z to form a right-handed coordinate system.
5. The ATLAS Collaboration, "Pixel Detector TDR," CERN/LHCC/98-13 (1998).
6. T. Åkesson *et al.*, *Nucl. Instrum. Methods* A412 (1998), 200–215.

The RICH system and vertex detector of LHCb

G. Wilkinson

European Laboratory for Particle Physics (CERN), CH-1211, Geneva 23, Switzerland

The LHCb experiment will make precise studies of CP violation in the B meson sector. Good proper time resolution and π/K separation are vital to this programme. The design and present status of the LHCb vertex detector and RICH system are described.

1 Introduction

The study of CP violation through rare decay asymmetries in the B sector is an established priority of particle physics over the coming two decades. LHCb is an experiment approved for operation at the LHC which will pursue this programme with very high statistical precision. Due to the large b production cross-section in pp collisions at $\sqrt{s} = 14$ TeV, a flux of 10^6 $b\bar{b}$ pairs will be created per year at the chosen LHCb running luminosity of 2×10^{32} cm^{-2}s^{-1}. To exploit these enormous statistics requires high performance instrumentation. The vertex detector and RICH system are central to the success of LHCb. Discussion of the prospects of LHCb and other B physics experiments can be found in [1].

2 LHCb overview

The LHCb experiment is a forward spectrometer operating in collider mode. The layout is shown schematically in figure 1. The vertex detector and both RICHes are labelled. More information on the LHCb experiment may be found in [2].

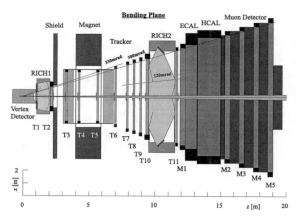

Figure 1: A schematic of the LHCb experiment seen from above. The interaction point is within the vertex detector region.

3 Vertex Detector

3.1 Detector requirements

The ability to be sensitive to decay vertices of B mesons displaced from the primary interaction vertex, and to reconstruct the proper lifetime of each meson is a fundamental requirement of the LHCb experiment. The vertex detector must contribute both in the triggering and in the offline reconstruction:

- Trigger – it is intended to use the vertex detector at the second stage of triggering to select B events by identifying interactions with tracks from secondary vertices. This decision will be made at an input rate of 1 MHz with a maximum latency of 256 μs, and a suppression factor of 25 necessary. The vertex detector will also be used at higher triggering levels.

- Offline reconstruction – analyses will need to reconstruct the proper time of B meson decays, and will require impact parameter determination and vertexing in order to reject combinatorial background. B_s studies in particular demand proper time resolutions << 1 ps in order to be sensitive to the rapid oscillation frequency.

3.2 Detector design

The vertex detector will consist of a sequence of stations placed transverse to the beam direction. During coasts the stations will be lowered to within ~ 1 cm of the beam within Roman pots to permit a first measured point very close to the interaction region. This arrangement is one of the main advantages of the forward spectrometer geometry. A total of 17 stations are envisaged, displaced along 1 m. The number and distribution of the stations is chosen to ensure that the majority of tracks seen downstream in the spectrometer will traverse at least 3 vertex detector planes, and to account for the ~ 5 cm longitudinal spread in the interaction region. A schematic of the vertex detector is shown in figure 2.

Silicon strip detectors are a suitable technology for the LHCb environment. Each station will consist of two separate detectors of thickness 150 μm to read orthogonal coodinates which will be r and ϕ respectively, with a $\pm 5°$

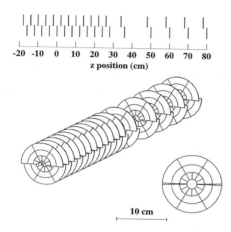

Figure 2: A schematic of the LHCb vertex detector, showing distribution of stations along the beampipe direction.

stereo angle in the ϕ strips. This coordinate scheme has been chosen as being the natural geometry for fast (r, z) impact parameter calculation in a forward spectrometer, an important consideration in the trigger. The stations will appear as discs, being made of azimuthal slices of detector spanning $61°$ with an inner radius of 1 cm and an outer radius of 6 cm. The $1°$ overlap between slices will allow for internal alignment of planes. The detectors will be divided into various regions of differing strip pitch $(40 - 100\,\mu\text{m})$ chosen to optimise occupancy per channel. With these pitches, hit precisions of $6 - 10\,\mu\text{m}$ are expected for double channel clusters. A diagram of both r and ϕ detectors is shown in figure 3.

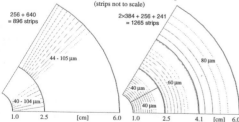

Figure 3: The LHCb r and ϕ strip vertex detector elements.

The detectors will be read out using a double-metal technology, as pioneered by DELPHI[3], in which the signal from a given strip is routed out over the top of the other structures via a $1\,\mu\text{m}$ thick aluminium strip isolated from the rest of the detector by a thin SiO_2 or polyamide dielectric layer. Thus the readout electronics can be placed outside the acceptance, thereby minimising the multiple scattering.

Calculations of the expected particle flux around the LHCb interation region, and knowledge of the response of strip detectors to radiation, indicate that it may be necessary to replace the detectors every few years. The open geometry of the experiment makes this a practical operation.

3.3 Detector performance

Using the described design a trigger algorithm has been devised and evaluated using a full GEANT simulation, including secondary interactions. The algorithm proceeds by finding tracks in the $r - z$ projection with a significant impact parameter to a reconstructed primary vertex. Three dimensional information is then invoked to search for secondary vertices made from these tracks. The result of this search is used to define a probability that the event does not contain a b-hadron. From placing cuts on this probability the performance curves in figure 4 are obtained, for a variety of final states used in the CP violation studies. It can be seen that with a background rentention of 4%, signal efficiencies of $\sim 50\%$ are achieved.

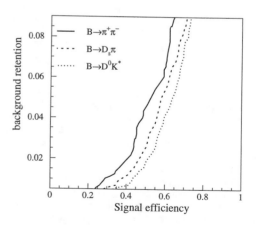

Figure 4: Performance of the second stage LHCb vertex trigger.

Full simulation studies have been performed in a variety of decay modes. In particular LHCb's potential for observing and measuring B_s mixing have been evaluated using the channel $B_s^0 \to D_s^- \pi^+$. A proper time resolution of 43 ± 2 fs is observed. With this performance and one year of data LHCb expects to be able to observe mixing if $x_s < 75$ and set a limit of $x_s > 91$ if no signal is observed.

3.4 Present R & D

Prototype r and ϕ 'n on n' detectors with a size and strip pitches close to that of the final design have been studied in a beam environment. An array of such detectors was used to observe tracks from collisions between the beam and metal targets. The position of these targets was then reconstucted much as secondary vertices will be at the LHC. Resolutions were obtained compatible with expectation.

Present and future studies involve fast readout electronics and irradiation tests. A question specific to LHCb concerns the effect of the $1/r^2$ dependence of the radiation across the detector, which is a particular consequence of the forward geometry.

4 RICH system

4.1 Detector requirements

Pion/kaon discrimination is a mandatory attribute of a B physics experiment. Many important final states have background of the same topology, but with one or more tracks being a kaon (pion) instead of a pion (kaon). At best the background will dilute the observed CP asymmetry of the signal; at worse it will carry unknown CP asymmtry itself, and thus contaminate the measurement, preventing a meaningful interpretation. An example where the latter is expected to occur is the determination of the unitarity angle α from the channel $B_d^0 \to \pi^+\pi^-$, in which the other two body decays $B_d^0 \to K^+\pi^-$, $B_s^0 \to K^+\pi^-$ and $B_s^0 \to K^+K^-$ will outnumber the signal by a factor of ~ 2. Another example is the measurement of the angle γ from the channel $B_s^0 \to D_s K$, where background comes from the ~ 10 times more common $B_s^0 \to D_s\pi$ decays. These backgrounds can only be removed if the experiment has hadron identification capabilities.

Hadron identification can also be used to improve the *flavour tagging* of the experiment. CP asymmetry measurements require that the b flavour at birth of the decaying meson be known. This can be established on a statistical basis by determining the charge of a kaon from the accompanying meson in the event. This *kaon tag* complements the more conventional *lepton tag*, which

is limited by the low semi-leptonic branching ratio. This is particularly important for LHCb, where many events without leptons are triggered by the presence of a high transverse momentum cluster in the hadron calorimeter.

The momentum regime over which hadron identification is necessary is $1 < p < 150$ GeV/c. The higher momentum limit is determined by background suppression requirements in decays such as $B_d^0 \to \pi^+\pi^-$; the low momentum limit comes from the rather soft spectrum of kaon tag candidates. Pion/kaon discrimination over such a broad interval can only be achieved with a RICH system.

4.2 Detector design

In order to cover the necessary momentum range LHCb will be equipped with two RICH counters: an upstream counter before the magnet, intended for lower momentum, wide angle tracks, and a downstream counter, in front of the calorimeters, optimised for the high momentum tracks. A schematic of the upstream RICH is shown in figure 5. Tracks traverse two radiators: ~ 5 cm of aerogel and ~ 95 cm of C_4F_{10}. The Cherenkov rings from both are focussed with spherical mirrors tilted by ± 250 mrad onto photodetector planes to the left and the right of the beampipe outside the spectrometer acceptance.

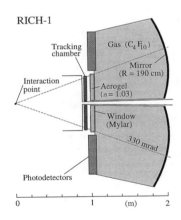

Figure 5: Schematic of the upstream LHCb RICH, seen from above.

Tracks entering the downstream RICH will traverse ~ 180 cm of CF_4. Again the rings are transported to photodetector planes outside the acceptance, by a tilted focussing mirror at an angle of ± 370 mrad, and a intemediate flat mirror.

Table 1: Specifications of the LHCb RICH system.

	Aerogel	C_4F_{10}	CF_4
Refractive index	1.03	1.0014	1.0005
$P_{thresh}(\pi)$ [GeV/c]	0.6	2.6	4.4
$P_{thresh}(K)$ [GeV/c]	2.0	9.3	15.6
θ_C [mrad]	242	53	32
$N_{p.e.}$	15	55	30
$\sigma_{\theta_C}^{p.e.}$ [mrad]	1.45	1.10	0.35

In table 1 is listed for each radiator the refractive index, pion and kaon momentum thresholds, Cherenkov angle per saturated track, expected number of photoelectrons after all loses per saturated track and single photoelectron Cherenkov angle resolution.

4.3 Photodetector evaluation and development

The most important issue in the RICH design is the choice of photodetector. A technology is required which will meet many criteria:

- UV and visible single photon efficiency.

- Spatial resolution of ~ 1 mm.

- Efficient coverage of $\sim 3\,m^2$.

- LHC compatible readout.

- Channel cost ≤ 10 CHF ($350k$ channels in total).

There are two candidate solutions: multianode photomultipliers (MAPMTs) and hybrid photodetectors (HPDs).

MAPMTs operate in the same manner as conventional photomultipliers, but have many detection cells within a single vacuum envelope. A commercial device with 64 channels will soon be available which will meet LHCb's needs, but with significant dead area around the outside of the device. This problem could be solved by an array of optical lenses mounted in front of the tubes. MAPMTs are currently being evaluated, both in the laboratory and in the beamline.

In an HPD a liberated photoelectron is accelerated over a large potential difference, typically 15 kV, onto a pixelated silicon detector at the base of the tube. Electrostatic focussing can be used to map from a large entrance window to a smaller area detector. HPDs are commercially available and have been studied by the LHCb collaboration in a prototype RICH counter. Figure 6 shows Cherenkov rings integrated over many events from a 10 GeV pion beam passing through aerogel (wide ring) and C_4F_{10} (central arc) radiators. The detectors are proximity focussed HPDs with hexagonal 2 mm pixels.

The photoelectron yield and observed resolution are in agreement with expectations.

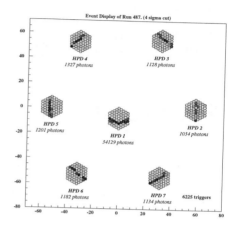

Figure 6: Aerogel and C_4F_{10} rings seen by commercial HPDs in RICH prototype. The scale is in mm.

The HPDs currently available, however, are inadequate for LHCb's needs with too small a fraction of live area and unsuitable readout possibilities. Therefore two parallel development projects, the so-called 'pad' and 'pixel' HPDs, are underway at CERN to explore how these problems may be solved.

In the pad HPD, electrodes are used to demagnify by a factor of 2.5 onto 1 mm silicon pads, giving an active-area fraction of 81%. These pads are readout by routing lines which carry the signal to discrete readout chips within the tube. This project is well advanced with recent successes including the fabrication of high efficiency UV and visible light sensitive photocathodes.

The electrostatics in the pixel HPD both demagnify by a factor of 4 and 'cross-focus' the image – an attribute that provides good resolution and robustness against external electric and magnetic fields. The detector, consisting of true pixels of $500\,\mu m$ size, is bump-bonded to a readout chip. Development of the pixel HPD has proceeded in collaboration with industry and is an extension of the successful 'ISPA-tube'[4] which demonstrated the feasibility of the approach.

4.4 Detector performance

Full GEANT simulations have been performed of the LHCb RICH system, assuming photodetectors with effective pixel size, quantum efficiency and live area consistent with those of the HPD development goals. Typical hit pixel multiplicites per event are found to be 1400 in the upstream RICH and 600 in the downstream RICH.

A pattern recognition algorithm has been developed which considers all RICH and track data in the event and forms an event likelihood. Maximising the likelihood leads to high efficiency, high purity particle assignments. Figure 7 shows the number of sigma separation between the pion and kaon hypotheses obtained for true pions, as a function of momentum.

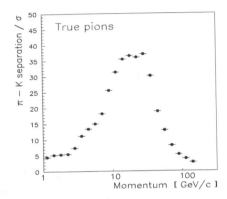

Figure 7: Number of sigma separation between pion and kaon hypotheses after full pattern recognition for true pions, as a function of momentum on a logarithmic scale. Each bin shows the mean separation obtained over many events.

With hadron identification of such high performance the physics goals set out in section 4.1 can be met. Kaons can be identified for use in the flavour tag with negligible contamination from other particles. Background can be rejected with high efficiency in the final states, as is illustrated by figure 8 which shows the reconstructed $B_d^0 \to \pi^+\pi^-$ mass peak with no hadron identification, and with RICH information, demanding that the particles be compatible with a pion or lighter particle. With the pion criteria and a mass cut of $\pm 30 \, \text{MeV}/c^2$ the signal to background becomes 18.0, having started at a value of 0.4. There is $< 10\%$ loss in signal events from the RICH cut. That background which remains comes mainly from events with tracks above $150 \, \text{GeV}/c$.

5 Conclusions

The LHCb experiment has an exciting programme of precision measurements of CP violation in the B meson sector to which the vertex and RICH detectors are central. Detector designs have been conceived which meet the

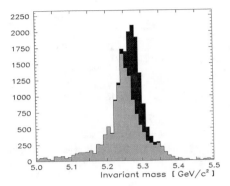

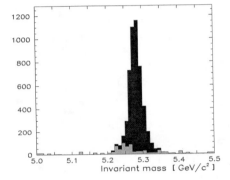

Figure 8: $B_d^0 \to \pi^+\pi^-$ mass peak before (top) and after (bottom) the RICH requirement. The dark shading is the signal and the light shading the two body background.

physics requirements, and the development of the technologies needed to constuct these detectors are at an advanced stage.

Acknowledgements

I am grateful to my colleagues in the LHCb vertex detector and RICH groups for their assistance in preparing this talk.

References

1. T. Nakada, these proceedings.
2. LHCb Technical proposal, CERN/LHCC 98-4.
3. V. Chabaud *et al.*, *Nucl. Instrum. Methods* A **368**, 314 (1996).
4. T. Gys *et al.*, *Nucl. Instrum. Methods* A **335**, 386 (1995).

Parallel Session 15

Field Theory - Perturbative and Non Perturbative

Convenors: Andrzej Buras (TU, Munich)

Harry C.S. Lam (McGill University)

GLOBAL QUANTIZATION OF YANG-MILLS THEORY

H. HÜFFEL and G. KELNHOFER

Institut für Theoretische Physik, Universität Wien, Boltzmanngasse 5, A-1090 Vienna, Austria
E-mail: helmuth.hueffel@univie.ac.at

Recently a modified Faddeev-Popov path integral density for the Yang-Mills field was derived using a generalized stochastic quantization method. The modification of the path integral consists in the presence of specific finite contributions of the pure gauge degrees of freedom. Due to the Gribov problem the gauge fixing can be defined only locally and the whole space of the gauge potentials has to be partitioned into cells, each corresponding to a different gauge fixing choice. We prove that in any of the overlap regions the expectation values of gauge invariant observables are manifestly equal. As a consequence a global path integral density for Yang-Mills fields can be defined by summing over all cells, which is manifestly independent of the specific choice of cells and gauge fixing conditions, respectively.

1 Stochastic quantization of Yang–Mills theory

In a recent paper [1] we performed the stochastic quantization [2,3,4] of Yang–Mills theory in configuration space and derived a generalized Faddeev–Popov path integral density. This result was obtained from the equilibrium limit of the associated Fokker–Planck equation after a specific modification of both its original drift and diffusion terms [5,6,7]; we gave a geometrical interpretation of our procedure.

We would like to first sketch the main result of [1]. In order to properly discuss *global* aspects of the quantization of Yang–Mills fields the theory of fiber bundles is indispensable. Let $P(M, G)$ be a principal fiber bundle with structure group G over the compact Euclidean space time M. Let $\mathcal{A}$ denote the space of all irreducible connections on P and let $\mathcal{G}$ denote the gauge group, which is given by all vertical automorphisms on P reduced by the centre of G. Then $\mathcal{G}$ acts freely on $\mathcal{A}$ and defines a principal $\mathcal{G}$-fibration $\mathcal{A} \xrightarrow{\pi} \mathcal{A}/\mathcal{G} =: \mathcal{M}$ over the space $\mathcal{M}$ of all inequivalent gauge potentials with projection π.

According to the Parisi–Wu approach for the stochastic quantization of the Yang–Mills theory in terms of a Langevin equation we have

$$dA = -\frac{\delta S}{\delta A} ds + dW. \tag{1}$$

Here S denotes the Yang–Mills action without gauge symmetry breaking terms and without accompanying ghost field terms, s denotes the extra time coordinate ("stochastic time" coordinate) with respect to which the stochastic process is evolving, dW is the increment of a Wiener process.

We separate the Yang-Mills fields into gauge independent and gauge dependent degrees of freedom. Due to the Gribov ambiguity [8], however, the principal $\mathcal{G}$-bundle $\mathcal{A} \to \mathcal{M}$ is not globally trivializable. In order to define a local section, we choose a fixed background connection $A_0 \in \mathcal{A}$ and consider a sufficient small neighbourhood

$U(A_0)$ of $\pi(A_0)$ in $\mathcal{M}$. Then the subspace

$$\Gamma = \{B \in \pi^{-1}(U(A_0))/D_{A_0}^\star(B - A_0) = 0\} \tag{2}$$

defines a local section of $\mathcal{A} \to \mathcal{M}$; here $D_{A_0}^\star$ is the adjoint operator of the covariant derivative D_{A_0} with respect to A_0. Our analysis is performed on the isomorphic trivial principal $\mathcal{G}$-bundle, $\Gamma \times \mathcal{G} \to \Gamma$, where the isomorphism is given by the map

$$\chi : \Gamma \times \mathcal{G} \to \pi^{-1}(U(A_0)), \qquad \chi(B, g) := B^g \tag{3}$$

with $B \in \Gamma$, $g \in \mathcal{G}$ and B^g denoting the nonabelian gauge transformation of B by g. The inverse map χ^{-1} is given by the expression

$$\chi^{-1} : \pi^{-1}(U(A_0)) \to \Gamma \times \mathcal{G}, \ \chi^{-1}(A) := (A^{\omega(A)^{-1}}, \omega(A)), \tag{4}$$

where $\omega : \pi^{-1}(U(A_0)) \to \mathcal{G}$ is uniquely defined by the requirement that $A^{\omega(A)^{-1}} \in \Gamma$, i.e.

$$D_{A_0}^\star(A^{\omega(A)^{-1}} - A_0) = 0. \tag{5}$$

When the Parisi–Wu Langevin equation (1) is transformed into the adapted coordinates $\Psi = (B, g)$ we have to use Ito's stochastic calculus obtaining

$$d\Psi = \left(-G^{-1}\frac{\delta S}{\delta \Psi} + \frac{1}{\sqrt{\det G}}\frac{\delta(G^{-1}\sqrt{\det G})}{\delta \Psi}\right) ds + E dW \tag{6}$$

The vielbein E, the metric G, its inverse G^{-1} and its determinant $\det G$ were discussed in detail in [1]. We just recall that the determinant of G is given by

$$\det G = \det(R_g^\star R_g)(\det \mathcal{F}_B)^2(\det \Delta_{A_0})^{-1} \tag{7}$$

where $\sqrt{\det(R_g^\star R_g)}$ implies an invariant volume density on $\mathcal{G}$, with R_g denoting the right logarithmic derivative of the identity on $\mathcal{G}$; $\mathcal{F}_B = D_{A_0}^\star D_B$ is the Faddeev–Popov operator and $\Delta_{A_0}^{-1}$ is the inverse of the covariant Laplacian $\Delta_{A_0} = D_{A_0}^\star D_{A_0}$.

It was shown in [1,5,6,7] that the above procedure can be generalized by adding specific terms to both the drift and the Wiener increment term of the Langevin equation (1), yet leaving all expectation values of gauge invariant observables unchanged. As a consequence there is induced a new metric $\widetilde{G}$ for the stochastic process, with respect to which the gauge fixing surface becomes orthogonal to the gauge orbits.

In specific we found the generalized Langevin equation

$$d\Psi = \left[-\widetilde{G}^{-1}\frac{\delta S_{\text{tot}}[\Psi]}{\delta\Psi} + \frac{1}{\sqrt{\det G}}\frac{\delta(\widetilde{G}^{-1}\sqrt{\det G})}{\delta\Psi} \right] ds$$
$$+ \widetilde{E} dW \tag{8}$$

where

$$S_{\text{tot}}[\Psi] = S[B] + S_{\mathcal{G}}[g]. \tag{9}$$

Here $S_{\mathcal{G}}[g]$ is an arbitrary damping function with the property that

$$\int_{\mathcal{G}} \mathcal{D}g \sqrt{\det(R_g^\star R_g)}\, e^{-S_{\mathcal{G}}[g]} < \infty. \tag{10}$$

We remark that the new (inverse) metric obeys $\widetilde{G}^{-1} = \widetilde{E}\widetilde{E}^\star$, where the vielbein $\widetilde{E}$ is defined in terms of its projections $\widetilde{E}^\Gamma$ and $\widetilde{E}^{\mathcal{G}}$ along the gauge fixing surface and the gauge orbit by $\widetilde{E}^\Gamma = E^\Gamma$ and $\widetilde{E}^{\mathcal{G}} = R_g^{-1}\Delta_B^{-1}D_B^\star \operatorname{ad}(g)$, respectively.

2 The modified Faddeev–Popov formula

Associated to the Langevin equation (8) there is a Fokker–Popov equation

$$\frac{\partial\rho[\Psi, s]}{\partial s} = L[\Psi]\rho[\Psi, s], \tag{11}$$

where the Fokker-Planck operator $L[\Psi]$ is appearing in just factorized form

$$L[\Psi] = \frac{\delta}{\delta\Psi}\widetilde{G}^{-1}\left[\frac{\delta S_{\text{tot}}[\Psi]}{\delta\Psi} - \frac{1}{\sqrt{\det G}}\frac{\delta(\sqrt{\det G})}{\delta\Psi} + \frac{\delta}{\delta\Psi} \right].$$
(12)

As a consequence of the fluctuation–disssipation theorem the equilibrium Fokker–Planck distribution $\rho^{\text{eq}}[\Psi]$ is obtained as [1]

$$\rho^{\text{eq}}[\Psi] = \frac{\sqrt{\det G}e^{-S_{\text{tot}}[\Psi]}}{\int_{\Gamma\times\mathcal{G}} \mathcal{D}B\mathcal{D}g\sqrt{\det G}e^{-S_{\text{tot}}}}$$

$$= \frac{\det \mathcal{F}_B e^{-S[B]}\sqrt{\det R_g^\star R_g}e^{-S_{\mathcal{G}}[g]}}{\int_\Gamma \mathcal{D}B \det \mathcal{F}_B e^{-S[B]}\int_{\mathcal{G}}\mathcal{D}g\sqrt{\det R_g^\star R_g}e^{-S_{\mathcal{G}}[g]}}.$$
(13)

It should be stressed that within the framework of stochastic quantization [2,3,4] the equilibrium Fokker–Planck distribution by definition is the conventional path integral density of field theory. Our result (13) is equivalent to the Faddeev–Popov prescription [9] for Yang–Mills theory but contains a modification due to the presence of the contributions of the gauge degrees of freedom. Although these *finite* contributions always cancel out when evaluated on gauge invariant observables their presence implies far reaching consequences.

3 Independence of the gauge fixing surface

Let us define a different local section, choosing a different background connection $A_0' \in \mathcal{A}$. We define Γ' as in (2) by

$$\Gamma' = \{B' \in \pi^{-1}(U(A_0'))\,|\,D_{A_0'}^\star(B' - A_0') = 0\} \tag{14}$$

and study the separation of the Yang–Mills fields into gauge independent and gauge dependent degrees of freedom in the overlap region $U(A_0) \cap U(A_0')$. Introducing ω' in analogy to (5) by

$$D_{A_0'}^\star(A^{\omega'(A)^{-1}} - A_0') = 0 \tag{15}$$

we have now

$$A = B^g = B'^{g'} \tag{16}$$

where

$$B = A^{\omega(A)^{-1}}, \qquad g = \omega(A) \tag{17}$$

and

$$B' = A^{\omega'(A)^{-1}} = B^{\omega'(B)^{-1}}, \qquad g' = \omega'(A) = \omega'(B)g. \tag{18}$$

Performing the change of variables

$$(B, g) \to A \to (B', g') \to (B', g'') \tag{19}$$

where g'' is the left translation

$$g'' = \omega(B')g' \tag{20}$$

the path integral density simply transforms as

$$\mathcal{D}B \det \mathcal{F}_B \frac{1}{\sqrt{\det \Delta_{A_0}}}e^{-S[B]} \times$$

$$\mathcal{D}g\sqrt{\det R_g^\star R_g}e^{-S_{\mathcal{G}}[g]}$$

$$= \mathcal{D}B' \det \mathcal{F}'_{B'}\frac{1}{\sqrt{\det \Delta_{A_0'}}}e^{-S[B']} \times$$

$$\mathcal{D}g'\sqrt{\det R_{g'}^\star R_{g'}}e^{-S_{\mathcal{G}}[\omega(B')g']}$$

$$= \mathcal{D}B' \det \mathcal{F}'_{B'} \frac{1}{\sqrt{\det \Delta_{A'_0}}} e^{-S[B']} \times$$

$$\mathcal{D}g'' \sqrt{\det R^{\star}_{g''} R_{g''}} e^{-S_{\mathcal{G}}[g'']}. \tag{21}$$

We remark that $\det \mathcal{F}'_{B'}$ is defined by

$$\det \mathcal{F}'_{B'} = D^{\star}_{A'_0} D_{B'}. \tag{22}$$

Consequently the expectation values of gauge invariant observables f are manifestly equal when performing the path integrations over the subspaces

$$\bar{\Gamma} = \{B \in \pi^{-1}(U(A_0) \cap U(A'_0)) | D^{\star}_{A_0}(B - A_0) = 0\} \tag{23}$$

and

$$\bar{\Gamma}' = \{B' \in \pi^{-1}(U(A_0) \cap U(A'_0)) | D^{\star}_{A'_0}(B' - A'_0) = 0\}, \tag{24}$$

respectively. Specifically we have

$$\int_{\bar{\Gamma}} \mathcal{D}B \det \mathcal{F}_B \frac{1}{\sqrt{\det \Delta_{A_0}}} f(B) e^{-S[B]} =$$

$$\int_{\bar{\Gamma}'} \mathcal{D}B' \det \mathcal{F}'_{B'} \frac{1}{\sqrt{\det \Delta_{A'_0}}} f(B') e^{-S[B']}. \tag{25}$$

4 Global quantization

Let $\{U_i\}$ be an open covering of the space $\mathcal{M}$ of all the inequivalent gauge potentials such that background gauge fields $A_0^{(i)} \in \mathcal{A}$ with $\pi(A_0^{(i)}) \in U_i \subset \mathcal{M}$ can be chosen with respect to which the definition of local sections is possible as discussed above. Furthermore let e_α be a partition of unity subordinate to $\{U_i\}$. For each α we choose a specific $A_0^{(i_\alpha)}$ from the above background fields such that $\pi(A_0^{(i_\alpha)}) \in supp\, e_\alpha$ and define local sections by

$$\Gamma_{(\alpha)} = \{B^{(\alpha)} \in \pi^{-1}(supp\, e_\alpha) | D^{\star}_{A_0^{(i_\alpha)}}(B^{(\alpha)} - A_0^{(i_\alpha)}) = 0\} \tag{26}$$

Finally we propose the definition of the global expectation value of a gauge invariant observable f by summing over all the partitions e_α such that

$$\langle f \rangle = \frac{\sum_\alpha \int_{\Gamma_{(\alpha)}} \mu_\alpha\, f(B^{(\alpha)})\, e^{-S[B^{(\alpha)}]} e_\alpha(B^{(\alpha)})}{\sum_\alpha \int_{\Gamma_{(\alpha)}} \mu_\alpha\, e^{-S[B^{(\alpha)}]} e_\alpha(B^{(\alpha)})} \tag{27}$$

with

$$\mu_\alpha = \mathcal{D}B^{(\alpha)} \det \mathcal{F}_{B^{(\alpha)}}^{(\alpha)} \frac{1}{\sqrt{\det \Delta_{A_0^{(i_\alpha)}}}} \tag{28}$$

and where

$$\det \mathcal{F}_{B^{(\alpha)}}^{(\alpha)} = D^{\star}_{A_0^{(i_\alpha)}} D_{B^{(\alpha)}}. \tag{29}$$

Due to the result of the previous section it is trivial to explicitly see the independence of the global expectation value $\langle f \rangle$ on the specific choice of the open covering $\{U_i\}$, on the specific choice of background gauge fields $\{A_0^{(i)}\}$, on the choice of the partition of unity e_α subordinate to $\{U_i\}$, and on the choice of the specific background gauge fields $A_0^{(i_\alpha)}$ associated to e_α.

Acknowledgments

G. Kelnhofer acknowledges support by "Fonds zur Förderung der wissenschaftlichen Forschung in Österreich", project P10509-NAW.

References

1. H. Hüffel and G. Kelnhofer, "Generalized Stochastic Quantization of Yang–Mills Theory", Ann. of Phys. 270 (1998) 231
2. G. Parisi and Wu Yongshi, Sci.Sin. 24 (1981) 483
3. P. Damgaard and H. Hüffel, Phys. Rep. 152 (1987) 227
4. M. Namiki, "Stochastic Quantization", Springer, Heidelberg, 1992
5. H. Hüffel, Nucl. Phys. B (Proc. Suppl.) 57 (1997) 209
6. H. Hüffel and G. Kelnhofer, Phys. Lett. B 408 (1997) 241
7. H. Hüffel and G. Kelnhofer, "Nonperturbative Stochastic Quantization of the Helix Model", Ann. of Phys. 266 (1998) 417
8. V. Gribov, Nucl. Phys. B139 (1978) 1
9. L. Faddeev and V. Popov, Phys. Lett. 25B (1967) 29

Vilkovisky-DeWitt Effective Potential and the Higgs-Mass Bound

Guey-Lin Lin and Tzuu-Kang Chyi

Institute of Physics, National Chiao-Tung University, Hsinchu, Taiwan, R.O.C.
E-mail: glin@cc.nctu.edu.tw

We compute the Vilkovisky-DeWitt effective potential of a simplified version of the Standard Electroweak Model, where all charged boson fields as well as the bottom-quark field are neglected. The effective potential obtained in this formalism is gauge-independent. We derive from the effective potential the mass bound of the Higgs boson. The result is compared to its counterpart obtained from the ordinary effective potential.

1 Introduction

The gauge dependence of the effective potential was first pointed out by Jackiw in early seventies[1]. This finding raised concerns on the physical significance of the effective potential. In a later work by Dolan and Jackiw [2], the effective potential of scalar QED was calculated in a set of R_ξ gauges. It was concluded that only the limiting unitary gauge gives sensible result on spontaneous symmetry-breaking. This difficulty was partially resolved by the work of Nielsen[3]. In his paper, Nielsen derived a simple identity characterizing the mean-field and the gauge-fixing-parameter dependences of the effective potential, namely,

$$(\xi \frac{\partial}{\partial \xi} + C(\phi, \xi) \frac{\partial}{\partial \phi}) V(\phi, \xi) = 0, \quad (1)$$

where ξ is the parameter appearing in the gauge-fixing term $L_{gf} = -\frac{1}{2\xi}(\partial_\mu A^\mu)^2$. The above identity implies that the local extrema of V for different ξ are located along the same characteristic curve on (ϕ, ξ) plane, which satisfies $d\xi = \frac{d\phi}{C(\phi,\xi)/\xi}$. Hence covariant gauges with different ξ are equally good for computing V. On the other hand, a choice of the multi-parameter gauge[2] $L_{gf} = -\frac{1}{2\xi}(\partial_\mu A^\mu + \sigma\phi_1 + \rho\phi_2)^2$ would break the homogeneity of Eq. (1)[3]. Hence effective potential calulated in this gauge has no physical significance.

Recently it was pointed out[11] that the Higgs mass bound as derived from the effective potential is gauge-dependent. The gauge dependence enters in the calculation of one-loop effective potential, a quantity that is crucial for the determination of the Higgs mass bound. Boyanovsky, Loinaz and Willey has proposed a resolution[4] to the gauge dependence of the Higgs mass bound. Their approach is based upon the *Physical Effective Potential* constructed as the expectation value of the Hamiltonian in physical states[5]. The effective potential of the abelian Higgs model is computed explicitly as an illustration. However, this formalism requires the identification of first-class constraints in the theory and a projection to the physical states. Such a procedure necessarily breaks the manifest Lorentz invariance of the theory. Consequently we expect it is highly non-trivial to apply this formalism to the Standard Model(SM).

In our work[6], we introduce the formalism of Vilkovisky and DeWitt [7,8] for constructing an gauge-independent effective potential, and therefore obtaining a gauge-independent lower bound for the Higgs mass. We present the idea with a toy model[13] which corresponds to neglect all charged boson fields in the SM. The generalization to the full SM is straightforward. In fact, the applicability of Vilkovisky-DeWitt formulation to non-abelian gauge theories has been extensively demonstrated in literatures[12].

The outline of this presentation is as follows. In Section 2, we briefly review the formalism of Vilkovisky and DeWitt using scalar QED as an example. We shall illustrate that the effective action of Vilkovisky and DeWitt is equivalent to the ordinary effective action constructed in the Landau-DeWitt gauge[14]. In Section 3, we calculate the effective potential of the simplified standard model, and the relevant renormalization constants of the theory using the Landau-DeWitt gauge. The effective potential is then extended to large vacuum expectation value of the scalar field by means of renormalization group analyses. In Section 4, the mass bound of the Higgs boson is derived and compared to that given by oridinary effective action. Section 5 is the conclusion.

2 Vilkovisky-DeWitt Effective Action of Scalar QED

The formulation of Vilkovisky-DeWitt efective action is motivated by the parametrization dependence of the ordinary effective action, which can be written generically as

$$\exp \frac{i}{\hbar} \Gamma[\Phi] = \exp \frac{i}{\hbar}(W[j] + \Phi^i \frac{\delta\Gamma}{\delta\Phi^i})$$
$$= \int [D\phi] \exp \frac{i}{\hbar}(S[\phi] + (\Phi^i - \phi^i) \cdot \frac{\delta\Gamma}{\delta\Phi^i}). (2)$$

The parametrization dependence of the ordinary effective action arises because the difference $\eta^i \equiv (\Phi^i - \phi^i)$ is not a vector in the field configuration space, hence the product $\eta^i \cdot \frac{\delta\Gamma}{\delta\Phi^i}$ is not a scalar under reparametrization. The remedy to this problem is to replace $-\eta^i$ with a two-point function $\sigma^i(\Phi, \phi)$ [7,8,9] which, at the point Φ, is tangent to the geodesic connecting Φ and ϕ. The precise form of $\sigma^i(\Phi, \phi)$ depends on the connection Γ^i_{jk} of the configuration space. It is easy to show that[10]

$$\sigma^i(\Phi, \phi) = -\eta^i - \frac{1}{2}\Gamma^i_{jk}\eta^j\eta^k + O(\eta^3). \quad (3)$$

For scalar QED described by the Lagrangian:

$$L = -\frac{1}{4}F_{\mu\nu}F^{\mu\nu} + (D_\mu\phi)^\dagger(D^\mu\phi) \\ - \lambda(\phi^\dagger\phi - \mu^2)^2, \quad (4)$$

with $D_\mu = \partial_\mu + ieA_\mu$ and $\phi = \frac{\phi_1 + i\phi_2}{\sqrt{2}}$, we have

$$\Gamma^i_{jk} = \left\{ {i \atop jk} \right\} + T^i_{jk}, \quad (5)$$

where $\left\{ {i \atop jk} \right\}$ is the Christoffel symbol of the field configuration space which has the following metric:

$$G_{\phi_a(x)\phi_b(y)} = \delta_{ab}\delta^4(x - y), \\ G_{A_\mu(x)A_\nu(y)} = -g^{\mu\nu}\delta^4(x - y), \\ G_{A_\mu(x)\phi_a(y)} = 0. \quad (6)$$

We note that the metric of the field configuration space is determined by the quadratic part of the Lagrangian according to the prescription of Vilkovisky[7]. In the above flat-metric, we have $\left\{ {i \atop jk} \right\} = 0$. However, the Christoffel symbols would be non-vanishing in the parametrization with polar variables ρ and χ such that $\phi_1 = \rho\cos\chi$ and $\phi_2 = \rho\sin\chi$. T^i_{jk} is a quantity pertinent to generators g^i_α of the gauge transformation. Explicitly, we have[7,10]

$$T^i_{jk} = -B^\alpha_j D_k g^i_\alpha + \frac{1}{2}g^\rho_\alpha D_\rho K^i_\beta B^\alpha_j B^\beta_k + j \leftrightarrow k, \quad (7)$$

where $B^\alpha_k = N^{\alpha\beta}g_{k\beta}$ with $N^{\alpha\beta}$ being the inverse of $N_{\alpha\beta} \equiv g^k_\alpha g^l_\beta G_{kl}$. In scalar QED, the generators g^i_α are given by

$$g^{\phi_a(x)}_y = -\epsilon^{ab}\phi_b(x)\delta^4(x - y), \\ g^{A_\mu(x)}_y = -\partial_\mu\delta^4(x - y). \quad (8)$$

with $\epsilon^{12} = 1$. The one-loop effective action of scalar QED can be calculated from Eq. (2) with the quantum fluctuations η^i replaced by $\sigma^i(\Phi, \phi)$. The result is written as[10]:

$$\Gamma[\Phi] = S[\Phi] - \frac{i\hbar}{2}\ln\det G + \frac{i\hbar}{2}\ln\det \tilde{D}^{-1}_{ij}, \quad (9)$$

where $S[\Phi]$ is the tree-level effective action; $\ln\det G$ arises from the function space measure $[D\phi] \equiv$

$\prod_x d\phi(x)\sqrt{\det G}$, and $\tilde{D}^{-1}_{ij}$ is the modified inverse-propagator:

$$\tilde{D}^{-1}_{ij} = \frac{\delta^2 S}{\delta\Phi^i\delta\Phi^j} - \Gamma^k_{ij}[\Phi]\frac{\delta S}{\delta\Phi^k}. \quad (10)$$

To study the symmetry-breaking behaviour of the theory, we focus on the effective potential which can be obtained from $\Gamma[\Phi]$ by setting the classical fields Φ's to constants.

The Vilkovisky-DeWitt effective potential of scalar QED has been calculated in various gauges and different parametrizations of scalar fields [14,10,15]. The results all agree with one another. In this work, we calculate the effective potential and other relevant quantities in Landau-DeWitt gauge[16], which is characterized by the gauge-fixing term: $L_{gf} = -\frac{1}{2\xi}(\partial_\mu B^\mu - ie\eta^\dagger\Phi + ie\Phi^\dagger\eta)^2$, with $\xi \to 0$. In L_{gf}, $B^\mu \equiv A^\mu - A^\mu_{cl}$, and $\eta \equiv \phi - \Phi$ are quantum fluctuations while A^μ_{cl} and Φ are classical fields. The advantage of performing calculations in the Landau-DeWitt gauge is that T^i_{jk} vanishes[14] in this case. In other words, Vilkovisky-DeWitt formalism coincides with the conventional one in the Landau-DeWitt gauge.

For computing the effective potential, we choose $A^\mu_{cl} = 0$ and $\Phi = \frac{\rho_{cl}}{\sqrt{2}}$, i.e. the imaginary part of Φ is set to zero. In this set of background fields, L_{gf} can be written as

$$L_{gf} = -\frac{1}{2\xi}\left(\partial_\mu B^\mu \partial_\nu B^\nu - 2e\rho_{cl}\chi\partial_\mu B^\mu + e^2\rho^2_{cl}\chi^2\right), \quad (11)$$

where χ is the quantum field defined by $\eta = \frac{\rho + i\chi}{\sqrt{2}}$. We note that $B_\mu - \chi$ mixing in L_{gf} is ξ dependent, and therefore would not cancell out the corresponding mixing term in the classical Lagrangian of Eq. (4). This induces the mixed-propagator such as $< 0|T(A_\mu(x)\chi(y))|0 >$ or $< 0|T(\chi(x)A_\mu(y))|0 >$. The Faddeev-Popov ghost Lagrangian is given by

$$L_{FP} = \omega^*(-\partial^2 - e^2\rho^2_{cl})\omega. \quad (12)$$

With each part of the Lagrangian determined, we are ready to compute the effective potential. Since we choose a flat-metric, the one-loop effective potential is completely determined by the modified inverse propagators $\tilde{D}^{-1}_{ij}$[17]. From Eqs. (4), (11) and (12), we arrive at

$$\tilde{D}^{-1}_{B_\mu B_\nu} = (-k^2 + e^2\rho^2_0)g^{\mu\nu} + (1 - \frac{1}{\xi})k^\mu k^\nu, \\ \tilde{D}^{-1}_{B_\mu\chi} = ik^\mu e\rho_0(1 - \frac{1}{\xi}), \\ \tilde{D}^{-1}_{\chi\chi} = (k^2 - m^2_G - \frac{1}{\xi}e^2\rho^2_0), \\ \tilde{D}^{-1}_{\rho\rho} = (k^2 - m^2_H), \\ \tilde{D}^{-1}_{\omega^*\omega} = (k^2 - e^2\rho^2_0)^{-2}, \quad (13)$$

1585

where we have set $\rho_{cl} = \rho_0$, which is a space-time independent constant, and defined $m_G^2 = \lambda(\rho_0^2 - 2\mu^2)$, $m_H^2 = \lambda(3\rho_0^2 - 2\mu^2)$. Using the definition $\Gamma[\rho_0] = (2\pi)^4\delta^4(0)V_{eff}(\rho_0)$ along with Eqs. (9) and (13), and taking the limit $\xi \to 0$, we obtain $V_{eff}(\rho_0) = V_{tree}(\rho_0) + V_{1-loop}(\rho_0)$ with

$$V_{1-loop}(\rho_0) = \frac{-i\hbar}{2}\int \frac{d^n k}{(2\pi)^n} \ln[(k^2 - e^2\rho_0^2)^{n-3}$$
$$\times (k^2 - m_H^2)(k^2 - m_+^2)(k^2 - m_-^2)], \quad (14)$$

where m_+^2 and m_-^2 are solutions of the quadratic equation $(k^2)^2 - (2e^2\rho_0^2 + m_G^2)k^2 + e^4\rho_0^4 = 0$. In the above equation, the gauge-boson's degree of freedom has been continued to $n - 3$ in order to preserve the relevant Ward identities. Our expression of $V_{1-loop}(\rho_0)$ agree with previous results obtained in the unitary gauge[15]. One could also calculate the effective potential in the *ghost-free* covariant gauges with $L_{gf} = -\frac{1}{2\xi}(\partial_\mu B^\mu)^2$. The cancellation of gauge-parameter dependence in the effective potential has been demonstrated in the case of massless scalar-QED with $\mu^2 = 0$[4,10]. It can be easily extended to the massive case.

It is instructive to rewrite Eq. (14) as

$$V_{1-loop}(\rho_0) = \frac{\hbar}{2}\int \frac{d^{n-1}\vec{k}}{(2\pi)^{n-1}}\left((n-3)\omega_B(\vec{k}) + \omega_H(\vec{k})\right.$$
$$\left. + \omega_+(\vec{k}) + \omega_-(\vec{k})\right), \quad (15)$$

where $\omega_B(\vec{k}) = \sqrt{\vec{k}^2 + e^2\rho_0^2}$, $\omega_H(\vec{k}) = \sqrt{\vec{k}^2 + m_H^2}$ and $\omega_\pm(\vec{k}) = \sqrt{\vec{k}^2 + m_\pm^2}$. One can see that V_{1-loop} is a sum of the zero-point energies of four excitations with masses $m_B \equiv e\rho_0$, m_H, m_+ and m_-. Since there are precisely four physical degrees of freedom in scalar QED, we see that Vilkovisky-DeWitt effective potential does exhibit a correct number of physical degrees of freedom.

3 Vilkovisky-DeWitt Effective Potential of the Simplified Standard Model

In this section, we compute the effective potential of the simplified standard model where charged boson fields and all fermion fields except the top quark field are discarded. The gauge interactions of this model are prescribed by the following covariant derivatives[13]:

$$D_\mu t_L = (\partial_\mu + ig_L Z_\mu - \frac{2}{3}ieA_\mu)t_L,$$
$$D_\mu t_R = (\partial_\mu + ig_R Z_\mu - \frac{2}{3}ieA_\mu)t_R,$$
$$D_\mu\phi = (\partial_\mu + i(g_L - g_R)Z_\mu)\phi, \quad (16)$$

where $g_L = (-g_1/2 + g_2/3)$, $g_R = g_2/3$ with $g_1 = g/\cos\theta_W$ and $g_2 = 2e\tan\theta_W$. Clearly this toy model

exhibits a $U(1)_A \times U(1)_Z$ symmetry where each $U(1)$ symmetry is associated with a neutral gauge boson. The $U(1)_Z$-charges of t_L, t_R and ϕ are related in such a way that the following Yukawa interactions are invariant under $U(1)_A \times U(1)_Z$:

$$L_Y = -y\bar{t}_L\phi t_R - y\bar{t}_R\phi^* t_L. \quad (17)$$

Since Vilkosvisky-DeWitt effective action coincides with ordinary effective action calculated in the Landau-DeWitt gauge, we hence calculate the effective potential in this gauge which has

$$L_{gf} = -\frac{1}{2\alpha}(\partial_\mu Z^\mu + \frac{ig_1}{2}\eta^\dagger\Phi - \frac{ig_1}{2}\Phi^\dagger\eta)^2$$
$$- \frac{1}{2\beta}(\partial_\mu A^\mu)^2, \quad (18)$$

with α, $\beta \to 0$. We note that A^μ and Z^μ are quantum fluctuations associated with the photon and the Z boson. Their classical backgrounds can be set to zero for computing the effective potential. Following the method of the previous section, we obtain

$$V_{VD}(\rho_0) = \frac{\hbar}{2}\int \frac{d^{n-1}\vec{k}}{(2\pi)^{n-1}}\left((n-3)\omega_Z(\vec{k}) + \omega_H(\vec{k})\right.$$
$$\left. + \omega_+(\vec{k}) + \omega_-(\vec{k}) - 4\omega_F(\vec{k})\right), \quad (19)$$

with $\omega_i(\vec{k}) = \sqrt{\vec{k}^2 + m_i^2}$ where $m_Z^2 = \frac{g_1^2}{4}\rho_0^2$, $m_\pm^2 = m_Z^2 + \frac{1}{2}(m_G^2 \pm m_G\sqrt{m_G^2 + 4m_Z^2})$ and $m_F^2 \equiv m_t^2 = \frac{y^2\rho_0^2}{2}$. Performing the integration in Eq. (19) and subtracting the infinities with $\overline{MS}$ prescription, we obtain

$$V_{VD}(\rho_0) = \frac{1}{64\pi^2}(m_H^4 \ln\frac{m_H^2}{\kappa^2} + m_Z^4 \ln\frac{m_Z^2}{\kappa^2}$$
$$+ m_+^4 \ln\frac{m_+^2}{\kappa^2} + m_-^4 \ln\frac{m_-^2}{\kappa^2} - 4m_t^4 \ln\frac{m_t^2}{\kappa^2})$$
$$- \frac{1}{128\pi^2}(3m_H^4 + 5m_Z^4 + 3m_G^4$$
$$+ 12m_G^2 m_Z^2 - 12m_t^4). \quad (20)$$

Although $V_{VD}(\rho_0)$ is obtained in the Landau-DeWitt gauge, we should stress that any other gauge with non-vanishing T^i_{jk} should lead to the same result. For later comparisons, let us write down the ordinary effective potential in the *ghost-free* Landau gauge[2] (equivalent to removing the scalar part of Eq. (18)):

$$V_L(\rho_0) = \frac{\hbar}{2}\int \frac{d^{n-1}\vec{k}}{(2\pi)^{n-1}}\left((n-1)\omega_Z(\vec{k}) + \omega_H(\vec{k})\right.$$
$$\left. + \omega_G(\vec{k}) - 4\omega_F(\vec{k})\right). \quad (21)$$

Performing the integrations in V_L and subtracting the infinities give

$$V_L(\rho_0) = \frac{1}{64\pi^2}(m_H^4 \ln\frac{m_H^2}{\kappa^2} + 3m_Z^4 \ln\frac{m_Z^2}{\kappa^2}$$

$$+ m_G^4 \ln \frac{m_G^2}{\kappa^2} - 4m_t^4 \ln \frac{m_t^2}{\kappa^2})$$
$$- \frac{1}{128\pi^2}(3m_H^4 + 5m_Z^4 + 3m_G^4 - 12m_t^4). \quad (22)$$

We remark that V_L differs from V_{VD} except at the point of extremum where $\rho_0^2 = 2\hat{\mu}^2$. At this point, one has $m_G^2 = 0$ and $m_\pm^2 = m_Z^2$ which leads to $V_{VD}(\rho_0 = 2\hat{\mu}^2) = V_L(\rho_0^2 = 2\hat{\mu}^2)$. That $V_{VD} = V_L$ at the point of extremum is a consequence of Nielsen identities[3].

To derive the Higgs mass bound from $V_{VD}(\rho_0)$ or $V_L(\rho_0)$, one encounters a breakdown of the perturbation theory at, for instance, $\frac{\lambda}{16\pi^2} \ln \frac{\lambda\rho_0^2}{\kappa^2} > 1$ for a large ρ_0. To extend the validity of the effective potential for a large ρ_0, the effective potential has to be improved by the renomalization group(RG) analysis. Let us denote the effective potential generically as V_{eff}. The renormalization-scale independence of V_{eff} implies the following equation[18,4]:

$$\left(-\mu(\gamma_\mu + 1)\frac{\partial}{\partial\mu} + \beta_{\hat{g}}\frac{\partial}{\partial\hat{g}}\right.$$
$$\left. - (\gamma_\rho + 1)t\frac{\partial}{\partial t} + 4\right) V_{eff}(t\rho_0^i, \mu, \hat{g}, \kappa) = 0. \quad (23)$$

where $\hat{g}$ denotes collectively the coupling constants λ, g_1, g_2 and y; ρ_0^i is an arbitrarily chosen initial value for ρ_0. Solving this differential equation gives

$$V_{eff}(t\rho_0^i, \mu_i, \hat{g}_i, \kappa) = \exp\left(\int_0^{\ln t} \frac{4}{1 + \gamma_\rho(x)} dx\right)$$
$$\times V_{eff}(\rho_0^i, \mu(t, \mu_i), \hat{g}(t, \hat{g}_i), \kappa), \quad (24)$$

where

$$t\frac{d\hat{g}}{dt} = \frac{\beta_{\hat{g}}(\hat{g}(t))}{1 + \gamma_\rho(\hat{g}(t))} \text{ with } \hat{g}(0) = \hat{g}_i, \quad (25)$$

and

$$\mu(t, \mu_i) = \mu_i \exp\left(-\int_0^{\ln t} \frac{1 + \gamma_\mu(x)}{1 + \gamma_\rho(x)} dx\right) \quad (26)$$

To fully determine V_{eff} at large ρ_0, we need to calculate β functions of λ, g_1, g_2 and y, and the anomalous dimensions γ_μ and γ_ρ. It has been demonstrated that the n-loop effective potential is improved by $(n+1)$-loop β and γ functions[19,20]. Since the effectve potential is calculated to the one-loop order, a consistent RG analysis requires the knowledge of β and γ functions up to two loops. As the computation of two-loop β and γ functions are quite involved, we will simply improve the tree-level effective potential with one-loop β and γ functions.

We have the following one-loop β and γ functions in the Landau-DeWitt gauge(we will set $\hbar = 1$ from this point on):

$$\beta_\lambda = \frac{1}{16\pi^2}\left(\frac{3}{8}g_1^4 - 3\lambda g_1^2 - 2y^4 + 4\lambda y^2 + 20\lambda^2\right),$$

$$\beta_{g_1} = \frac{g_1}{4\pi^2}\left(\frac{g_1^2}{16} - \frac{g_1 g_2}{18} + \frac{g_2^2}{27}\right),$$
$$\beta_{g_2} = \frac{g_2}{4\pi^2}\left(\frac{g_1^2}{16} - \frac{g_1 g_2}{18} + \frac{g_2^2}{27}\right),$$
$$\beta_y = \frac{y}{8\pi^2}\left(y^2 - \frac{3g_1^2}{8} + \frac{g_1 g_2}{12}\right),$$
$$\gamma_\mu = \frac{1}{2\pi^2}\left(\frac{3\lambda}{4} + \frac{3g_1^4}{128} - \frac{3g_1^2}{32} - \frac{y^4}{8\lambda} + \frac{y^2}{8}\right),$$
$$\gamma_\rho = \frac{1}{64\pi^2}\left(-5g_1^2 + 4y^2\right). \quad (27)$$

We stress that all the above functions except γ_ρ, the anomalous dimension of the scalar field, are in fact gauge-independent in the $\overline{MS}$ subtraction scheme. For γ_ρ in the Landau gauge, we have

$$\gamma_\rho = \frac{1}{64\pi^2}\left(-3g_1^2 + 4y^2\right). \quad (28)$$

4 The Higgs Mass Bound

The lower bound of the Higgs mass can be derived from the vacuum instability condition for the effective potential. To derive the mass bound, one begins with Eq. (24) which implies

$$V_{tree}(t\rho_0^i, \mu_i, \lambda_i) = \frac{1}{4}\chi(t)\lambda(t, \lambda_i)\left((\rho_0^i)^2 - 2\mu^2(t, \mu_i)\right)^2,$$
$$\quad (29)$$

with $\chi(t) = \exp\left(\int_0^{\ln t} \frac{4}{1+\gamma_{\rho_0}(x)} dx\right)$. Since $\mu(t, \mu_i)$ decreases as t increases, we have $V_{tree}(t\rho_0^i, \mu_i, \lambda_i) \approx \frac{1}{4}\chi(t)\lambda(t, \lambda_i)(\rho_0^i)^4$ for a sufficiently large t. Similarly, the one-loop effective potential $V_{1-loop}(t\rho_0^i, \mu_i, \hat{g}_i, \kappa)$ is also proportional to $V_{1-loop}(\rho_0^i, \mu(t, \mu_i), \hat{g}(t, \hat{g}_i), \kappa)$ with the same proportional constant $\chi(t)$ as in V_{tree}. Since we shall neglect all running effects in V_{1-loop}, we have $\hat{g}(t, \hat{g}_i) = \hat{g}_i$ and $\mu(t, \mu_i) = \frac{1}{t}\mu_i$ in V_{1-loop}. For a sufficiently large t, we can again approximate V_{1-loop} by its quartic terms. In the Landau-DeWitt gauge with the choice $\rho_0^i = \kappa$, we have

$$V_{VD} \approx \frac{(\rho_0^i)^4}{64\pi^2}\left[9\lambda_i^2\ln(3\lambda_i) + \frac{g_{1i}^4}{16}\ln(\frac{g_{1i}^2}{4}) - y_i^4\ln(\frac{y_i^2}{2})\right.$$
$$+ A_+^2(g_{1i}, \lambda_i)\ln A_+(g_{1i}, \lambda_i)$$
$$+ A_-^2(g_{1i}, \lambda_i)\ln A_-(g_{1i}, \lambda_i)$$
$$\left. - \frac{3}{2}(10\lambda_i^2 + \lambda_i g_{1i}^2 + \frac{5}{48}g_{1i}^4 - y_i^4)\right], \quad (30)$$

where $A_\pm(g_1, \lambda) = g_1^2/4 + \lambda/2 \cdot (1 \pm \sqrt{1 + g_1^2/\lambda})$. Similarly, the effective potential in the Landau gauge reads:

$$V_L \approx \frac{(\rho_0^i)^4}{64\pi^2}\left[9\lambda_i^2\ln(3\lambda_i) + \frac{3g_{1i}^4}{16}\ln(\frac{g_{1i}^2}{4}) - y_i^4\ln(\frac{y_i^2}{2})\right.$$
$$\left. + \lambda_i^2\ln(\lambda_i) - \frac{3}{2}(10\lambda_i^2 + \lambda_i g_{1i}^2 + \frac{5}{48}g_{1i}^4 - y_i^4)\right], \quad (31)$$

Combining the tree level and the one-loop effective potential, we arrive at

$$V_{eff}(t\rho_0^i, \mu_i, \hat{g}_i, \kappa) \approx \frac{1}{4}\chi(t)\left(\lambda(t, \lambda_i) + \Delta\lambda(\hat{g}_i)\right)(\rho_0^i)^4,$$

(32)

where $\Delta\lambda$ denotes one-loop corrections given by Eqs. (30) or (31). Let $t_{VI} = \rho_{VI}/\rho_0^i$. The condition for vacuum instability of the effective potential is then[21]

$$\lambda(t_{VI}, \lambda_i) + \Delta\lambda(\hat{g}_i) = 0.$$

(33)

We note that couplings $\hat{g}_i$ in $\Delta\lambda$ is evaluated at $\kappa = \rho_0^i$, which can be taken as the electroweak scale. Hence $g_{1i} \equiv g/\cos\theta_W = 0.67$, $g_{2i} \equiv 2e\tan\theta_W = 0.31$, and $y_i = 1$. The running coupling $\lambda(t_{VI}, \lambda_i)$ also depends on g_1, g_2 and y through β_λ, and γ_ρ shown in Eq. (27).

The strategy for solving Eq. (33) is to make an initial guess on λ_i, which enters into $\lambda(t)$ and $\Delta\lambda$, and repeatedly adjusting λ_i till $\lambda(t)$ completely cancells $\Delta\lambda$. For $t_{VI} = 10^2$ (or $\rho_0 \approx 10^4$ GeV) which is the new-physics scale reachable by LHC, we find $\lambda_i = 4.83 \times 10^{-2}$ for Landau-DeWitt gauge, and $\lambda_i = 4.8 \times 10^{-2}$ for Landau gauge. For a higher instability scale such as the scale of grand unification, we have $t_{VI} = 10^{13}$ or $\rho_0 \approx 10^{15}$ GeV. In this case, we find $\lambda_i = 3.13 \times 10^{-1}$ for both Landau-DeWitt and Landau gauges. The numerical agreement between λ_i's of two gauges can be attributed to an identical β function for the running of $\lambda(t)$, and a small difference in $\Delta\lambda$ between two gauges. We recall from Eq. (25) that the evolution of λ in two gauges will be different if effects of next-to-leading logarithm are taken into account. In that case, the difference in γ_ρ between two gauges give rise to different evolutions for λ. One may expect to see non-negligible differences in λ_i between two gauges for a large t_{VI}.

The critical value $\lambda_i = 4.83 \times 10^{-2}$ corresponds to a lower bound for the $\overline{MS}$ mass of the Higgs boson. Since $m_H = 2\sqrt{\lambda}\mu$, we have $(m_H)_{\overline{MS}} \geq 77$ GeV. For $\lambda_i = 3.13 \times 10^{-1}$, we have $(m_H)_{\overline{MS}} \geq 196$ GeV. To obtain the lower bound for the physical mass of the Higgs boson, finite radiative corrections must be added to the above bounds[11]. We will not pursue any further on these finite corrections since we are simply dealing with a toy model. However we like to point out that this finite correction is gauge-independent as ensured by Nielsen identities[3].

5 Conclusion

We have computed the one-loop effective potential of an abelian $U(1) \times U(1)$ model in the Landau-DeWitt gauge, which reproduces the result given by the gauge-independent Vilkovisky-DeWitt formalism. One-loop β and γ functions are also computed to facilitate the RG improvement of the effective potential. A gauge-independent lower bound for the Higgs self-coupling or equivalently the $\overline{MS}$ mass of the Higgs boson is derived. We compare this bound to that obtained by the ordinary effective potential computed in Landau gauge. The numerical values of both bounds are almost identical due to the leading-logarithmic approximation we have taken. A complete next-to-leading analysis as well as an extension of this work to the full standard model will be reported in future publications.

Acknowledgements

We thank W.-F. Kao for discussions. This work is supported in part by National Science Council of R.O.C. under grant numbers NSC 87-2112-M-009-038, and NSC 88-2112-M-009-002.

References

1. R. Jackiw, *Phys. Rev.* D **9**, 1686 (1974).
2. L. Dolan and R. Jackiw, *Phys. Rev.* D **9**, 2904 (1974).
3. N. K. Nielsen, *Nucl. Phys.* B **101**, 173 (1975).
4. D. Boyanovsky, W. Loinaz and R. S. Willey, *Phys. Rev.* D **57**, 100 (1998).
5. D. Boyanovsky, D. Brahm, R. Holman and D.-S. Lee, *Phys. Rev.* D **54**, 1763 (1996).
6. G.-L. Lin and T.-K. Chyi, NCTU-HEP-9804, hep-ph/9811213.
7. G. Vilkovisky, "The Gospel According to DeWitt" in *Quantum Theory of Gravity*, ed. S. M. Christensen (Adam Hilger, Bristol, 1983); *Nucl. Phys.* B **234**, 215 (1984).
8. See, for example, B. S. DeWitt in *Quantum Field Theory and Quantum Statistics: Essays in Honour of the 60th Birthday of E. S. Fradkin*, eds. I. A. Batalin, C. J. Isham and G. A. Vilkovisky (Hilger, Bristol, 1987), p. 191.
9. A mere replacement of $-\eta^i$ with $\sigma^i(\Phi, \phi)$, as suggested in Ref.[7], is not satisfactory for calculations beyond one-loop, because $\Gamma[\Phi]$ does not generate one-particle irreducible diagrams at the higher loops. A modified construction was given by DeWitt in Ref.[8]. Since we will be only concerned with one-loop corrections, we shall adhere to the current construction in our subsequent discussions.
10. See, for example, G. Kunstatter, in *Super Field Theories*, eds. H. C. Lee, V. Elias, G. Kunstatter, R. B. Mann, and K. S. Viswanathan(Plenum, New York, 1987), p. 503.
11. W. Loinaz and R. S. Willey, *Phys. Rev.* D **56**, 7416 (1997).

12. A. Rehban, *Nucl. Phys.* B **288**, 832 (1987); *Nucl. Phys.* B **298**, 726 (1988).

13. G.-L. Lin, H. Steger and Y.-P. Yao, *Phys. Rev.* D **44**, 2139 (1991).

14. E. S. Fradkin and A. A. Tseytlin, *Nucl. Phys.* B **234**, 509 (1984).

15. I. H. Russell and D. J. Toms, *Phys. Rev.* D **39**, 1735 (1989).

16. Although properties of this gauge are discussed in Ref.[14], authors of this paper used other gauge to calculate the effective potential of scalar QED.

17. In the current gauge, $\tilde{D}_{ij}^{-1} = D_{ij}^{-1}$ since Γ_{jk}^i vanishes. Furthermore, for ghost fields, it is the ghost propagator rather than its inverse that will appear in the effective action. This is due to the Grassmannian nature of ghost fields.

18. S. Coleman and E. Weinberg, *Phys. Rev.* D **7**, 1888 (1973).

19. B. Kastening, *Phys. Lett.* B **283**, 287 (1992).

20. M. Bando, T. Kugo, N. Maekawa and H. Nakano, *Phys. Lett.* B **301**, 83 (1993).

21. J. A. Casas, J. R. Espinosa and M. Quiros, *Phys. Lett.* B **342**, 171 (1995).

EXACT RESULTS IN SOFTLY BROKEN SUSY GAUGE THEORIES

D.I.KAZAKOV

Bogoliubov Laboratory of Theoretical Physics, Joint Institute for Nuclear Research,
141 980 Dubna, Moscow Region, RUSSIA
E-mail: kazakovd@thsun1.jinr.ru

It is shown that softly broken theory is equivalent to a rigid theory in external spurion superfield. This enables one to get the singular part of effective action in a broken theory from a rigid one by a simple redefinition of the couplings. This way one can reproduce all known results on the renormalization of soft couplings and masses in a softly broken theory. As an example the renormalization group functions for soft couplings and masses in the Minimal Supersymmetric Standard Model up to the three-loop level have been calculated. The method opens a possibility to construct a totally all loop finite N=1 SUSY gauge theory, including the soft SUSY breaking terms. Explicit relations between the soft terms, which lead to a completely finite theory in any loop order, are given.

1 Introduction

The gauge theory with softly broken supersymmetry has been widely studied. A powerful method which keeps supersymmetry manifest is the supergraph technique [1,2]. It is also applicable to softly broken SUSY models by using the "spurion" external superfields [3,4,5]. As has been shown by Yamada [6] with the help of the spurion method the calculation of the β functions of soft SUSY-breaking terms is a much simpler task than in the component approach.

In a recent paper [7] we have developed a modification of the spurion technique in gauge theories and have formulated the Feynman rules. We have shown that the ultraviolet divergent parts of the Green functions of a softly broken SUSY gauge theory are proportional to those of a rigid theory with the spurion fields factorized.

The general idea is that a softly broken supersymmetric gauge theory can be considered as a rigid SUSY theory imbedded into the external space-time independent superfield, so that all the parameters as the couplings and masses become external superfields.

The main Statement:

> Softly Broken $\approx$ Rigid SUSY Theory
> SUSY Theory   in External Field
>
> The Coupling g $\Rightarrow$ External Superfield Φ_0

Consequence: Singular part of effective action depends on external superfield, but not on its derivatives:

$$S_{eff}^{sing}(g) \Rightarrow S_{eff}^{sing}(\Phi_0, \cancel{D^2\Phi_0}, \cancel{\bar{D}^2\Phi_0}, D^2\cancel{\bar{D}^2\Phi_0})$$

This approach to a softly broken sypersymmetric theory allows us to use remarkable mathematical properties of $N = 1$ SUSY theories such as non-renormalization theorems, cancellation of quadratic divergences, etc. We show that the renormalization procedure in a softly broken SUSY gauge theory can be performed in exactly the same way as in a rigid theory with the renormalization constants being external superfields. They are related to the corresponding renormalization constants of a rigid theory by the coupling constants redefinition. This allows us to find explicit relations between the renormalizations of soft and rigid couplings.

Throughout the paper we assume the existence of some gauge and SUSY invariant regularization and a minimal subtraction procedure.

As an application of the above mentioned relations we consider a possibility of constructing totally finite supersymmetric theories including the soft breaking terms. This problem has already been discussed several times [8,9]. We show that by choosing the soft terms in a proper way one can reach complete all loop finiteness. Moreover, there is no new fine-tuning. The soft terms are fine-tuned in exactly the same way as the corresponding Yukawa couplings. [10]

2 Softly Broken $N = 1$ SUSY Gauge Theory

Consider pure $N = 1$ SUSY Yang-Mills theory with a simple gauge group. The Lagrangian of a rigid theory is given by

$$\mathcal{L}_{rigid} = \int d^2\theta \, \frac{1}{4g^2} \text{Tr} W^\alpha W_\alpha + \int d^2\bar{\theta} \, \frac{1}{4g^2} \text{Tr} \bar{W}^\alpha \bar{W}_\alpha \tag{1}$$

To perform a soft SUSY breaking, one can introduce a gaugino mass term

$$-\mathcal{L}_{soft-breaking} = \frac{m_A}{2}\lambda\lambda + \frac{m_A}{2}\bar{\lambda}\bar{\lambda}, \tag{2}$$

where λ is the gaugino field. To rewrite it in terms of superfields, let us introduce an external spurion superfield $\eta = \theta^2$, where θ is a Grassmannian parameter. The softly broken Lagrangian can now be written as

$$\mathcal{L}_{soft} = \int d^2\theta \, \frac{1}{4g^2}(1 - 2\mu\theta^2)\text{Tr} W^\alpha W_\alpha$$

$$+ \int d^2\bar{\theta} \, \frac{1}{4g^2}(1 - 2\bar{\mu}\bar{\theta}^2)\mathrm{Tr}\bar{W}^\alpha\bar{W}_\alpha. \qquad (3)$$

In terms of component fields the interaction with external spurion superfield leads to a gaugino mass equal to $m_A = \mu$, while the gauge field remains massless. This external chiral superfield can be considered as a vacuum expectation value of a dilaton superfield emerging from supergravity, however, this is not relevant to further consideration.

As has been shown in [7] the Feynman rules corresponding to the Lagrangian (3) needed for the calculation of the singular part of effective action are the same as in a rigid theory with the substitution:

$$g^2 \rightarrow \tilde{g}^2 = g^2\left(1 + \mu\theta^2 + \mu\bar{\theta}^2 + 2\mu^2\theta^2\bar{\theta}^2\right). \qquad (4)$$

Then the vector propagator in a softly broken theory is:

$$\langle V(x_1,\theta_1,\bar{\theta}_1)V(x_2,\theta_2,\bar{\theta}_2)\rangle_{soft} = \qquad (5)$$

$$\frac{\tilde{g}^2}{g^2}\langle V(x_1,\theta_1,\bar{\theta}_1)V(x_2,\theta_2,\bar{\theta}_2)\rangle_{rigid} + \text{irrelevant terms},$$

where by irrelevant terms we mean the ones decreasing faster than $1/p^2$ for large p^2. The same is true for the ghost fields

$$\langle G(z_1)\bar{G}(z_2)\rangle_{soft} = (\tilde{g}^2/g^2)\langle G(z_1)\bar{G}(z_2)\rangle_{rigid} + \text{i.t.},$$
$$(6)$$

where G stands for any ghost superfield.

Hence, to perform the analysis of the divergent part of the diagrams in a soft theory, one has to use the same propagators as in a rigid theory multiplied by the factor $\tilde{g}^2/g^2$. It is also obviously true for any vertex of the ghost-vector interactions of the softly broken theory. Each vertex of this type has to be multiplied by the inverse factor $g^2/\tilde{g}^2$. The situation is less obvious with the vector vertices, however it happens to be true in this case as well.

Thus, we see that any element of the Feynman rules for a softly broken theory coincides with the corresponding element of a rigid theory multiplied by a common factor which is a polynomial in the grassmann coordinates.

Consider now a rigid SUSY gauge theory with chiral matter. The Lagrangian written in terms of superfields looks like

$$\mathcal{L}_{rigid} = \int d^2\theta d^2\bar{\theta} \, \bar{\Phi}^i(e^V)_i^j\Phi_j + \int d^2\theta \, \mathcal{W} + \int d^2\bar{\theta} \, \bar{\mathcal{W}}, \qquad (7)$$

where the superpotential $\mathcal{W}$ in a general form is

$$\mathcal{W} = \frac{1}{6}\lambda^{ijk}\Phi_i\Phi_j\Phi_k + \frac{1}{2}M^{ij}\Phi_i\Phi_j. \qquad (8)$$

The SUSY breaking terms which satisfy the requirement of "softness" can be written as

$$-\mathcal{L}_{soft-breaking} = \left[\frac{1}{6}A^{ijk}\phi_i\phi_j\phi_k + \frac{1}{2}B^{ij}\phi_i\phi_j + h.c.\right]$$
$$+ (m^2)_j^i\phi_i^*\phi^j, \qquad (9)$$

Like in the case of a pure gauge theory the soft terms (9) can be written down in terms of superfields by using the external spurion field. The full Lagrangian for the softly broken theory can be written as

$$\mathcal{L}_{soft} = \int d^2\theta d^2\bar{\theta} \, \bar{\Phi}^i(\delta_i^k - (m^2)_i^k\eta\bar{\eta})(e^V)_k^j\Phi_j \quad (10)$$

$$+ \int d^2\theta \left[\frac{1}{6}(\lambda^{ijk} - A^{ijk}\eta)\Phi_i\Phi_j\Phi_k\right.$$

$$\left. + \frac{1}{2}(M^{ij} - B^{ij}\eta)\Phi_i\Phi_j\right] + h.c.$$

The Lagrangian (10) allows one to write down the Feynman rules for the matter field propagators and vertices in a soft theory.

$$\langle \Phi(z_1)_i\bar{\Phi}(z_2)^j\rangle_{soft} = \qquad (11)$$

$$(\delta_i^k + \frac{1}{2}(m^2)_i^k\eta\bar{\eta})\langle \Phi(z_1)_k\bar{\Phi}(z_2)^l\rangle_{rigid}(\delta_l^j + \frac{1}{2}(m^2)_l^j\eta\bar{\eta})$$
$$+\text{i.t.},$$

The vector-matter vertices, according to eq.(10), gain the factor $(\delta_i^j - (m^2)_i^j\eta\bar{\eta})$ so that if in a diagram one has an equal number of chiral propagators and vector-matter vertices the spurion factors cancel.

The chiral vertices of a soft theory, as it follows from eq.(10), are the same as in a rigid theory with the Yukawa couplings being replaced by

$$\lambda^{ijk} \rightarrow \lambda^{ijk} - A^{ijk}\eta, \quad \bar{\lambda}_{ijk} \rightarrow \bar{\lambda}_{ijk} - \bar{A}_{ijk}\bar{\eta}.$$

The structure of the UV counterterms in chiral vertices is similar to that of the vector vertices, but is simpler due to the absence of the covariant derivatives on external lines. Effectively the corrections to the propagator (11) may be associated with the chiral vertices, which allows one to reduce all soft term corrections to the modification of the couplings.

3 Renormalization of Soft versus Rigid Theory: the General Case

The external field construction described above allows one to write down the renormalization of soft terms starting from the known renormalization of a rigid theory without any new diagram calculation. The following statement is valid:[7]

Statement 1 *Let a rigid theory (1,8) be renormalized via introduction of the renormalization constants Z_i, defined within some minimal subtraction massless scheme. Then, a softly broken theory (3,10) is renormalized via introduction of the renormalization superfields $\tilde{Z}_i$ which are related to Z_i by the coupling constants redefinition*

$$\tilde{Z}_i(g^2, \lambda, \bar{\lambda}) = Z_i(\tilde{g}^2, \tilde{\lambda}, \bar{\tilde{\lambda}}), \qquad (12)$$

where the redefined couplings are

$$\tilde{g}^2 = g^2(1 + \mu\eta + \bar{\mu}\bar{\eta} + 2\mu\bar{\mu}\eta\bar{\eta}), \qquad (13)$$

$$\tilde{\lambda}^{ijk} = \lambda^{ijk} - A^{ijk}\eta \qquad (14)$$
$$+ \frac{1}{2}(\lambda^{njk}(m^2)^i_n + \lambda^{ink}(m^2)^j_n + \lambda^{ijn}(m^2)^k_n)\eta\bar{\eta},$$

$$\bar{\tilde{\lambda}}_{ijk} = \bar{\lambda}_{ijk} - \bar{A}_{ijk}\bar{\eta} \qquad (15)$$
$$+ \frac{1}{2}(\bar{\lambda}_{njk}(m^2)^n_i + \bar{\lambda}_{ink}(m^2)^n_j + \bar{\lambda}_{ijn}(m^2)^n_k)\eta\bar{\eta}$$

From eqs.(12) and (13-15) it is possible to write down an explicit differential operator which has to be applied to the β functions of a rigid theory in order to get those for the soft terms.

Consider first the gauge couplings α_i. One has

$$\alpha_i^{Bare} = Z_{\alpha i}\alpha_i \quad \Rightarrow \quad \tilde{\alpha}_i^{Bare} = \tilde{Z}_{\alpha i}\tilde{\alpha}_i, \qquad (16)$$

where $Z_{\alpha i}$ is the product of the wave function and vertex renormalization constants.

Though $\tilde{\alpha}_i$ and $\tilde{Z}_{\alpha i}$ are general superfields, one has to consider only their chiral or antichiral parts. The chiral part of eq.(16) is

$$\alpha_i^{Bare}(1 + m_{A_i}^{Bare}\eta) = \alpha_i(1 + m_{A_i}\eta)Z_{\alpha i}(\tilde{\alpha})|_{\bar{\eta}=0}.$$

Expanding over η one has

$$\alpha_i^{Bare} = \alpha_i Z_{\alpha i}(\alpha), \qquad (17)$$
$$m_{A_i}^{Bare}\alpha_i^{Bare} = m_{A_i}\alpha_i Z_{\alpha i}(\alpha) + \alpha_i D_1 Z_{\alpha i}, \qquad (18)$$

where the operator D_1 extracts the linear w.r.t. η part of $Z_{\alpha i}(\tilde{\alpha})$. Due to eqs.(13-15) the explicit form of D_1 is

$$D_1 = m_{A_i}\alpha_i\frac{\partial}{\partial\alpha_i}.$$

Combining eqs.(17) and (18) one gets

$$m_{A_i}^{Bare} = m_{A_i} + D_1 \ln Z_{\alpha i}. \qquad (19)$$

To find the corresponding β functions one has to differentiate eqs.(17) and (19) w.r.t. the scale factor having in mind that the operator D_1 is scale invariant. This gives

$$\beta_{\alpha i} = \alpha_i\gamma_{\alpha i}, \quad \beta_{m_{A i}} = D_1\gamma_{\alpha i}, \qquad (20)$$

where $\gamma_{\alpha i}$ is the logarithmic derivative of $\ln Z_{\alpha i}$ equal to the anomalous dimension of the vector superfield in some particular gauges.

One can make also the transition from a rigid to a broken theory at the level of the renormalization group equation. Namely take

$$\dot{\alpha} = \beta_\alpha \quad \Rightarrow \quad \dot{\tilde{\alpha}} = \beta_{\tilde{\alpha}}, \qquad (21)$$

and expand over θ^2. Then one immediately reproduces eq.(20) for m_A.

One can go even further and consider a solution of the RGE. Then one has in a rigid theory

$$\int^\alpha \frac{d\alpha'}{\beta(\alpha')} = \log\left(\frac{Q^2}{\Lambda^2}\right). \qquad (22)$$

Making a substitution $\alpha \to \tilde{\alpha}$ one has

$$\int^{\tilde{\alpha}} \frac{d\alpha'}{\beta(\alpha')} = \log\left(\frac{Q^2}{\tilde{\Lambda}^2}\right), \qquad (23)$$

where $\tilde{\Lambda} = \Lambda(1 + c\theta^2 + ...)$. Expanding over θ^2 one finds

$$\frac{m_A\alpha}{\beta(\alpha)} = const. \qquad (24)$$

This result is in complete correspondence with that of Ref.[11].

The same procedure can be applied for the other soft terms. This needs the modification of the operator D_1 to include the A parameter from the chiral vertex. As for the mass square terms, they need the second order differential operator. Relations between the rigid and soft terms renormalizations in general case are summarized below. Similar results on the soft terms renormalization have been obtained in ref.[12]

The Rigid Terms

$$\beta_{\alpha_i} = \alpha_i\gamma_{\alpha i} \qquad (25)$$

$$\beta_M^{ij} = \frac{1}{2}(M^{il}\gamma_l^j + M^{lj}\gamma_l^i) \qquad (26)$$

$$\beta_y^{ijk} = \frac{1}{2}(y^{ijl}\gamma_l^k + y^{ilk}\gamma_l^j + y^{ljk}\gamma_l^i) \qquad (27)$$

The Soft Terms

$$\beta_{m_{A i}} = D_1\gamma_{\alpha i} \qquad (28)$$

$$\beta_B^{ij} = \frac{1}{2}(B^{il}\gamma_l^j + B^{lj}\gamma_l^i) \qquad (29)$$
$$- (M^{il}D_1\gamma_l^j + M^{lj}D_1\gamma_l^i)$$

$$\beta_A^{ijk} = \frac{1}{2}(A^{ijl}\gamma_l^k + A^{ilk}\gamma_l^j + A^{ljk}\gamma_l^i) \qquad (30)$$
$$- (y^{ijl}D_1\gamma_l^k + y^{ilk}D_1\gamma_l^j + y^{ljk}D_1\gamma_l^i)$$

$$(\beta_{m^2})_j^i = D_2\gamma_j^i \qquad (31)$$

with

$$D_1 = m_{A_i}\alpha_i \frac{\partial}{\partial \alpha_i} - A^{ijk}\frac{\partial}{\partial y^{ijk}}, \qquad (32)$$

$$\bar{D}_1 = m_{A_i}\alpha_i \frac{\partial}{\partial \alpha_i} - A_{ijk}\frac{\partial}{\partial y_{ijk}} \qquad (33)$$

$$D_2 = \bar{D}_1 D_1 + m_{A_i}^2 \alpha_i \frac{\partial}{\partial \alpha_i} \qquad (34)$$

$$+ \frac{1}{2}(m^2)_n^a \left(y^{nbc}\frac{\partial}{\partial y^{abc}} + y^{bnc}\frac{\partial}{\partial y^{bac}} + y^{bcn}\frac{\partial}{\partial y^{bca}} \right.$$

$$\left. + y_{abc}\frac{\partial}{\partial y_{nbc}} + y_{bac}\frac{\partial}{\partial y_{bnc}} + y_{bca}\frac{\partial}{\partial y_{bcn}} \right)$$

4 Illustration

To make the above formulae more clear and to demonstrate how they work in practice, we consider the renormalization group functions in a general theory in one loop. We follow the notation of ref. [9] except that our β functions are half of those. Note that all the calculations in ref. [9] are performed in the framework of dimensional reduction and the $\overline{MS}$ scheme.

The gauge β functions and the anomalous dimensions of matter superfields in a massless scheme are the functions of dimensionless gauge and Yukawa couplings of a rigid theory.

In the one-loop order, the renormalization group functions of a rigid theory are (for simplicity, we consider the case of a single gauge coupling)[a]

$$\gamma_\alpha^{(1)} = \alpha Q, \quad Q = T(R) - 3C(G), \qquad (35)$$

$$\gamma_j^{i\,(1)} = \frac{1}{2}y^{ikl}y_{jkl} - 2\alpha C(R)_j^i, \qquad (36)$$

where $T(R)$, $C(G)$ and $C(R)$ are the Casimir operators. Using eqs.(28-31) we construct the renormalization group functions for the soft terms

$$\beta_{m_A}^{(1)} = \alpha m_A Q, \qquad (37)$$

$$\beta_B^{ij\,(1)} = \frac{1}{2}B^{il}(\frac{1}{2}y^{jkm}y_{lkm} - 2\alpha C(R)_l^j) \qquad (38)$$

$$+ M^{il}(\frac{1}{2}A^{jkm}y_{lkm} + 2\alpha m_A C(R)_l^j) + (i \leftrightarrow j),$$

$$\beta_A^{ijk\,(1)} = \frac{1}{2}A^{ijl}(\frac{1}{2}y^{kmn}y_{lmn} - 2\alpha C(R)_l^k) \qquad (39)$$

$$+ y^{ijl}(\frac{1}{2}A^{kmn}y_{lmn} + 2\alpha m_A C(R)_l^k)$$

$$+ (i \leftrightarrow j) + (i \leftrightarrow k),$$

$$[\beta_{m^2}]_j^{i\,(1)} = \frac{1}{2}A^{ikl}A_{jkl} - 4\alpha m_A^2 C(R)_j^i \qquad (40)$$

$$+ \frac{1}{4}y^{nkl}(m^2)_n^i y_{jkl} + \frac{1}{4}y^{ikl}(m^2)_j^n y_{nkl} + \frac{4}{4}y^{isl}(m^2)_s^k y_{jkl}.$$

[a]To simplify the formulas hereafter we use the following notation:
$\alpha_i = \frac{g_i^2}{16\pi^2}$, $y^{ijk} = \frac{\lambda^{ijk}}{4\pi}$, $A^{ijk} = \frac{A^{ijk}}{4\pi}$.

One can easily see that the resulting formulae coincide with those of ref. [9]. The same procedure works in higher orders of PT.

5 Soft Renormalizations in the MSSM

The general rules described in the previous section can be applied to any model, in particular to the MSSM. In the case when the field content and the Yukawa interactions are fixed, it is more useful to deal with numerical rather than with tensor couplings. Rewriting the superpotential (8) and the soft terms (9) in terms of group invariants, one has

$$\mathcal{W}_{SUSY} = \frac{1}{6}\sum_a y_a \lambda_a^{ijk}\Phi_i\Phi_j\Phi_k + \frac{1}{2}\sum_b M_b h_b^{ij}\Phi_i\Phi_j, \quad (41)$$

and

$$-\mathcal{L}_{soft} = \left[\frac{1}{6}\sum_a \mathcal{A}_a \lambda_a^{ijk}\phi_i\phi_j\phi_k + \frac{1}{2}\sum_b \mathcal{B}_b h_b^{ij}\phi_i\phi_j \right.$$

$$\left. + \frac{1}{2}m_{A_j}\lambda_j\lambda_j + h.c. \right] + (m^2)_i^j \phi^{*i}\phi_j, \qquad (42)$$

where we have introduced numerical couplings y_a, M_b, $\mathcal{A}_a$ and $\mathcal{B}_b$.

Usually, it is assumed that the soft terms obey the universality hypothesis, i.e. they repeat the structure of a superpotential, namely

$$\mathcal{A}_a = y_a A_a, \quad \mathcal{B}_b = M_b B_b, \quad (m^2)_j^i = m_i^2 \delta_j^i. \qquad (43)$$

The renormalization group β functions of a rigid theory are (for simplicity, we assume the diagonal renormalization of matter superfields)

$$\beta_{\alpha_j} = \beta_j \equiv \alpha_j \gamma_{\alpha_j}, \qquad (44)$$

$$\beta_{y_a} = \frac{1}{2}y_a \sum_i K_{ai}\gamma_i, \qquad (45)$$

$$\beta_{M_b} = \frac{1}{2}M_b \sum_i T_{bi}\gamma_i, \qquad (46)$$

where γ_i is the anomalous dimension of the superfield Φ_i, γ_{α_j} is the anomalous dimension of the gauge superfield (in some gauges) and numerical matrices K and T specify which particular fields contribute to a given term in eq.(41).

Applying the algorithm of the previous section the renormalizations of the soft terms are expressed through those of a rigid theory in the following way:

$$\beta_{m_{A_j}} = D_1 \gamma_{\alpha_j}, \qquad (47)$$

$$\beta_{A_a} = -D_1 \sum_i K_{ai}\gamma_i, \qquad (48)$$

1593

$$\beta_{B_b} = -D_1 \sum_i T_{bi}\gamma_i, \qquad (49)$$

$$\beta_{m_i^2} = D_2 \gamma_i, \qquad (50)$$

and the operators D_1 and D_2 now take the form

$$D_1 = m_{A_i}\alpha_i \frac{\partial}{\partial \alpha_i} - A_a Y_a \frac{\partial}{\partial Y_a}, \qquad (51)$$

$$D_2 = (m_{A_i}\alpha_i \frac{\partial}{\partial \alpha_i} - A_a Y_a \frac{\partial}{\partial Y_a})^2$$
$$+ m_{A_i}^2 \alpha_i \frac{\partial}{\partial \alpha_i} + m_i^2 K_{ai} Y_a \frac{\partial}{\partial Y_a}. \qquad (52)$$

where we have used the notation $Y_a \equiv y_a^2$.

To illustrate these rules, we consider as an example one loop renormalization of the MSSM couplings. Leaving for simplicity the third generation Yukawa couplings only, the superpotential is

$$\mathcal{W}_{MSSM} = (y_t Q^j U^c H_2^i + y_b Q^j D'^c H_1^i + y_\tau L^j E^c H_1^i$$
$$+ \mu H_1^i H_2^j)\epsilon_{ij}, \qquad (53)$$

where Q, U, D', L and E are quark doublet, up-quark, down-quark, lepton doublet and lepton singlet superfields, respectively, and H_1 and H_2 are Higgs doublet superfields. i and j are the $SU(2)$ indices.

The soft terms have a universal form

$$-\mathcal{L}_{soft-br} = \sum_i m_i^2 |\phi_i|^2 + (\frac{1}{2} \sum_a \lambda_a \lambda_a + (A_t y_t q^j u^c h_2^i$$
$$+ A_b y_b q^j d'^c h_1^i + A_\tau y_\tau l^j e^c h_1^i + B\mu h_1^i h_2^j + h.c.), \qquad (54)$$

where the small letters denote the scalar components of the corresponding superfields and λ_a are the gauginos. The $SU(2)$ indices are suppressed.

Renormalizations in a rigid theory in the one loop order are given by

$$\gamma_{\alpha_i}^{(1)} = b_i \alpha_i, \quad i = 1, 2, 3, \quad b_i = \frac{33}{5}, 1, -3,$$

$$\gamma_Q^{(1)} = Y_t + Y_b - \frac{8}{3}\alpha_3 - \frac{3}{2}\alpha_2 - \frac{1}{30}\alpha_1,$$

$$\gamma_U^{(1)} = 2Y_t - \frac{8}{3}\alpha_3 - \frac{8}{15}\alpha_1,$$

$$\gamma_D^{(1)} = 2Y_b - \frac{8}{3}\alpha_3 - \frac{2}{15}\alpha_1,$$

$$\gamma_L^{(1)} = Y_\tau - \frac{3}{2}\alpha_2 - \frac{3}{10}\alpha_1,$$

$$\gamma_E^{(1)} = 2Y_\tau - \frac{6}{5}\alpha_1,$$

$$\gamma_{H_1}^{(1)} = 3Y_b + Y_\tau - \frac{3}{2}\alpha_2 - \frac{3}{10}\alpha_1,$$

$$\gamma_{H_2}^{(1)} = 3Y_t - \frac{3}{2}\alpha_2 - \frac{3}{10}\alpha_1.$$

Consequently, the renormalization group β functions are

$$\beta_{\alpha_i}^{(1)} = b_i \alpha_i^2,$$

$$\beta_{Y_t}^{(1)} = Y_t(6Y_t + Y_b - \frac{16}{3}\alpha_3 - 3\alpha_2 - \frac{13}{15}\alpha_1),$$

$$\beta_{Y_b}^{(1)} = Y_b(Y_t + 6Y_b + Y_\tau - \frac{16}{3}\alpha_3 - 3\alpha_2 - \frac{7}{15}\alpha_1),$$

$$\beta_{Y_\tau}^{(1)} = Y_\tau(3Y_b + 4Y_\tau - 3\alpha_2 - \frac{9}{5}\alpha_1),$$

$$\beta_{\mu^2}^{(1)} = \mu^2(3Y_t + 3Y_b + Y_\tau - 3\alpha_2 - \frac{3}{5}\alpha_1).$$

This allows us immediately to write down the soft term renormalizations

$$\beta_{A_t}^{(1)} = 6Y_t A_t + Y_b A_b + \frac{16}{3}\alpha_3 m_{A_3} + 3\alpha_2 m_{A_2}$$
$$+ \frac{13}{15}\alpha_1 m_{A_1},$$

$$\beta_{A_b}^{(1)} = Y_t A_t + 6Y_b A_b + Y_\tau A_\tau + \frac{16}{3}\alpha_3 m_{A_3} + 3\alpha_2 m_{A_2}$$
$$+ \frac{7}{15}\alpha_1 m_{A_1},$$

$$\beta_{A_\tau}^{(1)} = 3Y_b A_b + 4Y_\tau A_\tau + 3\alpha_2 m_{A_2} + \frac{9}{5}\alpha_1 m_{A_1},$$

$$\beta_B^{(1)} = 3Y_t A_t + 3Y_b A_b + Y_\tau A_\tau + 3\alpha_2 m_{A_2} + \frac{3}{5}\alpha_1 m_{A_2},$$

$$\beta_{m_{A_i}}^{(1)} = \alpha_i b_i m_{A_i},$$

$$\beta_{m_Q^2}^{(1)} = Y_t(m_Q^2 + m_U^2 + m_{H_2}^2 + A_t^2) + Y_b(m_Q^2 + m_D^2$$
$$+ m_{H_1}^2 + A_b^2) - \frac{16}{3}\alpha_3 m_{A_3}^2 - 3\alpha_2 m_{A_2}^2 - \frac{1}{15}\alpha_1 m_{A_1}^2,$$

$$\beta_{m_U^2}^{(1)} = 2Y_t(m_Q^2 + m_U^2 + m_{H_2}^2 + A_t^2)$$
$$- \frac{16}{3}\alpha_3 m_{A_3}^2 - \frac{16}{15}\alpha_1 m_{A_1}^2,$$

$$\beta_{m_D^2}^{(1)} = 2Y_b(m_Q^2 + m_D^2 + m_{H_1}^2 + A_b^2)$$
$$- \frac{16}{3}\alpha_3 m_{A_3}^2 - \frac{4}{15}\alpha_1 m_{A_1}^2,$$

$$\beta_{m_L^2}^{(1)} = Y_\tau(m_L^2 + m_E^2 + m_{H_1}^2 + A_\tau^2) - 3\alpha_2 m_{A_2}^2$$
$$- \frac{3}{5}\alpha_1 m_{A_1}^2,$$

$$\beta_{m_E^2}^{(1)} = 2Y_\tau(m_L^2 + m_E^2 + m_{H_1}^2 + A_\tau^2) - \frac{12}{5}\alpha_1 m_{A_1}^2,$$

$$\beta_{m_{H_1}^2}^{(1)} = 3Y_b(m_Q^2 + m_D^2 + m_{H_1}^2 + A_b^2) + Y_\tau(m_L^2 + m_E^2$$
$$+ m_{H_1}^2 + A_\tau^2) - 3\alpha_2 m_{A_2}^2 - \frac{3}{5}\alpha_1 m_{A_1}^2,$$

$$\beta_{m_{H_2}^2}^{(1)} = 3Y_t(m_Q^2 + m_U^2 + m_{H_2}^2 + A_t^2) - 3\alpha_2 m_{A_2}^2$$
$$- \frac{3}{5}\alpha_1 m_{A_1}^2,$$

which perfectly coincide with those in the literature [13].

We calculate here also the two and three loop gaugino mass renormalization out of a corresponding gauge β functions. The RG β functions for the gauge couplings in the MSSM are

$$\beta_{\alpha_i} = b_i\alpha_i^2 + \alpha_i^2\left(\sum_j b_{ij}\alpha_j - \sum_f a_{if}Y_f\right) \qquad (55)$$

$$+\alpha_i^2\left[\sum_{jk} b_{ijk}\alpha_j\alpha_k - \sum_{jf} a_{ijf}\alpha_jY_f + \sum_{fg} a_{ifg}Y_fY_g\right],$$

where Y_f means Y_t, Y_b and Y_τ and the coefficients $b_i, b_{ij}, a_{if}, b_{ijk}, a_{ijf}$ and a_{ifg} are given in ref.[14].

For the gaugino masses we have

$$\beta_{m_{A_i}} = b_i\alpha_i m_{A_i} \qquad (56)$$

$$+\alpha_i\left(\sum_j b_{ij}\alpha_j(m_{A_i} + m_{A_j}) - \sum_f a_{if}Y_f(m_{A_i} - A_f)\right)$$

$$+\alpha_i\left[\sum_{jk} b_{ijk}\alpha_j\alpha_k(m_{A_i} + m_{A_j} + m_{A_k})\right.$$

$$-\sum_{jf} a_{ijf}\alpha_jY_f(m_{A_i} + m_{A_j} - A_f)$$

$$\left.+\sum_{fg} a_{ifg}Y_fY_g(m_{A_i} - A_f - A_g)\right] + ...$$

6 Finiteness of Soft Parameters in a Finite SUSY GUT

Consider now the application of the proposed formulae to construct totally finite softly broken theories.

For rigid N=1 SUSY theories there exists a general method of constructing totally all loop finite gauge theories proposed in refs.[15,16,17]. The key issue of the method is the one-loop finiteness. If the theory is one-loop finite and satisfies some criterion verified in one loop[18], it can be made finite in any loop order by fine-tuning of the Yukawa couplings order by order in PT. In case of a simple gauge group the Yukawa couplings have to be chosen in the form

$$Y_a(\alpha) = c_1^a\alpha + c_2^a\alpha^2 + ..., \qquad (57)$$

where the finite coefficients c_n^a are calculated algebraically in the n-th order of perturbation theory.

Suppose now that a rigid theory is made finite to all orders by the choice of the Yukawa couplings as in eq.(57). This means that all the anomalous dimensions and the β functions on the curve $Y_a = Y_a(\alpha)$ are identically equal to zero.

Consider the renormalization of the soft terms. According to eqs.(47-50) and (51,52) the renormalizations of the soft terms are not independent but are given by the differential operators acting on the same anomalous dimensions. One has either

$$\beta_{soft} \sim D_1\gamma(Y,\alpha) = (m_A\alpha\frac{\partial}{\partial\alpha} - A_aY_a\frac{\partial}{\partial Y_a})\gamma(Y_a,\alpha), \qquad (58)$$

or

$$\beta_{soft} \sim D_2\gamma(Y,\alpha) = [(m_A\alpha\frac{\partial}{\partial\alpha} - A_aY_a\frac{\partial}{\partial Y_a})^2$$

$$+m_A^2\alpha\frac{\partial}{\partial\alpha} + m_i^2K_{ai}Y_a\frac{\partial}{\partial Y_a}]\gamma(Y_a,\alpha), \quad (59)$$

where $\gamma(Y_a,\alpha)$ is some anomalous dimension.

From the requirement of finiteness

$$\gamma_i(Y_a(\alpha),\alpha) = 0, \qquad (60)$$

the Yukawa couplings $Y_a(\alpha)$ are found in the form (57).

To reach the finiteness of all the soft terms in all loop orders, one has to choose the soft parameters A_a and m_i^2 in a proper way. The following statement is valid:[10]

Statement 2 *The soft term β functions become equal to zero if the parameters A_a and m_i^2 are chosen in the following form:*

$$A_a(\alpha) = -m_A\alpha\frac{\partial}{\partial\alpha}\ln Y_a(\alpha), \qquad (61)$$

$$m_i^2 = -m_AK_{ia}^{-1}\frac{\partial}{\partial\alpha}\alpha A_a(\alpha) \qquad (62)$$

$$= m_A^2K_{ia}^{-1}\frac{\partial}{\partial\alpha}\alpha^2\frac{\partial}{\partial\alpha}\ln Y_a(\alpha),$$

where the matrix K_{ia}^{-1} is the inverse of the matrix K_{ai}.

This statement follows from the form of the operators D_1 and D_2. After substitution of solutions (61) and (62) into D_1 and D_2, D_1 becomes a total derivative over α and D_2 becomes a second total derivative.

Indeed, consider eq.(58). For A_a chosen as in eq.(61) the differential operator D_1 takes the form

$$D_1 = m_A\alpha\frac{\partial}{\partial\alpha} - A_aY_a\frac{\partial}{\partial Y_a}$$

$$= m_A(\frac{\partial\ln Y_a}{\partial\ln\alpha}\frac{\partial}{\partial\ln Y_a} + \frac{\partial}{\partial\ln\alpha}) = m_A\frac{d}{d\ln\alpha}.$$

Hence, since on the curve $Y_a = Y_a(\alpha)$ the anomalous dimension $\gamma(Y_a,\alpha)$ identically vanishes, so does its derivative

$$\frac{d}{d\ln\alpha}\gamma(Y_a(\alpha),\alpha) = 0.$$

The operator D_2 is the second derivative. Using eq.(62) one has

$$D_2 = (m_A \alpha \frac{\partial}{\partial \alpha} - A_a Y_a \frac{\partial}{\partial Y_a})^2 + m_A^2 \alpha \frac{\partial}{\partial \alpha} + m_i^2 K_{ai} Y_a \frac{\partial}{\partial Y_a}$$

$$= (m_A \alpha \frac{\partial}{\partial \alpha} - A_a Y_a \frac{\partial}{\partial Y_a})^2 - m_A \frac{\partial A_a}{\partial \ln \alpha} Y_a \frac{\partial}{\partial Y_a}$$

$$+ m_A (m_A \alpha \frac{\partial}{\partial \alpha} - A_a Y_a \frac{\partial}{\partial Y_a})$$

$$= m_A^2 \frac{d}{d \ln \alpha} + m_A^2 \frac{d^2}{d \ln^2 \alpha}.$$

The term with the derivative of A_a is essential to get the total second derivative over α, since in the bracket the derivative $\alpha \partial / \partial \alpha$ does not act on A_a by construction. Like in the previous case the total derivatives identically vanish on the curve $Y_a = Y_a(\alpha)$.

The solutions (61,62) can be checked perturbatively order by order. In the leading order one has

$$A_a = -m_A, \quad m_i^2 = \frac{1}{3} m_A^2, \tag{63}$$

since $\sum_a K_{ia}^{-1} = 1/3$. These relations coincide with the already known ones [8] and with those coming from supergravity [19] and supersting-inspired models [20]. There they usually follow from the requirement of finiteness of the cosmological constant and probably have the same origin. Note that since the one-loop finiteness of a rigid theory automatically leads to the two-loop one and hence the coefficients $c_2^a = 0$, the same statement is valid due to eqs.(61,62) for a softly broken theory. Namely, relations (63) are valid up to two-loop order in accordance with [9]. In higher orders, however, they have to be modified.

This way one can construct all loop finite N=1 SUSY GUT including the soft SUSY breaking terms with the fine-tuning given by exactly the same functions as in a rigid theory.

7 Discussion

Our approach is based on a consideration of the soft theory as a rigid one embedded into the external x−independent superfields, that are the charges and masses of the theory. The Feynman rules together with the operator constructions D_1 and D_2 are just the technical consequences of this approach.Apparently it has wider application. Any spontaneously broken theory may be treated this way.

Later there has been considerable activity exploring the renormalization group invariant relations between the soft term renormalizations [21,22,23]. They essentially use the explicit form of the differential operators D_1 and D_2 and their reduction to the total derivatives found in [10]. In particular the exact form for the correction due to the ϵ-scalar mass in component approach in the renormalization scheme corresponding to the NSVZ β function has been obtained [22]. The exact β function for the scalar masses in NSVZ scheme for softly broken SUSY QCD has been also found [23].

Refereneces

1. R. Delbourgo, Nuovo Cim. **25A** (1975) 646;
   A. Salam and J. Strathdee, Nucl. Phys. **B86** (1975) 142;
   K. Fujikawa and W. Lang, ibid. Nucl. Phys. **B88** (1975) 61.
2. M.T. Grisaru, M. Roček and W. Siegel, Nucl. Phys. **B59** (1979) 429.
3. L. Girardello and M.T. Grisaru, Nucl. Phys. **B194** (1982) 65.
4. J.A. Helayël-Neto, Phys. Lett. **135B** (1984) 78;
   F. Feruglio, J.A. Helayël-Neto and F. Legovini, Nucl. Phys. **B249** (1985) 533;
5. M. Scholl, Z. Phys. **C28** (1985) 545.
6. Y. Yamada, Phys. Rev. **D50** (1994) 3537.
7. L.A.Avdeev, D.I.Kazakov and I.N.Kondrashuk, Nucl.Phys. **B510** (1998) 289 (hep-ph/9709397).
8. D.R.T.Jones, L.Mezinchescu and Y.-P. Yao, Phys.Lett., **148B** (1984) 317.
9. I. Jack and D.R.T. Jones, Phys. Lett. **333B** (1994) 372.
10. D.I.Kazakov, Phys.Lett. **B421** (1998) 211 (hep-ph/9709465)
11. J. Hisano and M.A. Shifman, Phys.Rev. **D56** (1997) 5475 (hep-ph/9705417).
12. I. Jack and D.R.T. Jones, Phys.Lett. **B415** (1997) 383 (hep-ph/9709364).
13. see, for example,
    V. Barger, M.S. Berger and P. Ohmann, Phys. Rev. **D47** (1993) 1093;
    W. Boer, R. Ehret and D.I. Kazakov, Z. Phys. **C67** (1995) 667;
14. P.M. Ferreira, I. Jack and D.R.T. Jones, hep-ph/9605440, Phys. Lett. **387B** (1996) 80.
15. A.V.Ermushev, D.I.Kazakov and O.V.Tarasov, Nucl.Phys., **B281** (1987) 72.
16. D.R.T.Jones, Nucl.Phys. **B277** (1986) 153.
17. C.Lucchesi, O.Piguet and K.Sibold, Phys.Lett. **201B** (1988) 241.
18. D.I.Kazakov, Mod.Phys.Lett., **A9** (1987) 663.
19. H.-P.Nilles, Phys.Rep. **110** (1984) 1.
20. A.Brignole, L.E.Ibáñez and C.Muñoz, Nucl.Phys. **B422** (1994) 125.
21. I.Jack, D.R.T.Jones, A.Pickering, hep-ph/9712542.
22. I.Jack, D.R.T.Jones, A.Pickering, hep-ph/9803405.
23. T.Kobayashi, J.Kubo and G.Zoupanos, hep-ph/9802267.

HIGGS MASSES AND S-SPECTRA IN FINITE AND THE MINIMAL SUSY GAUGE-YUKAWA GUT

T. KOBAYASHI

Dept. of Physics, Univ. of Helsinki, FIN-00014 Helsinki, Finland

J. KUBO

Dept. of Physics, Kanazawa Univ. , Kanazawa 920-1192, Japan

M. MONDRAGON

Inst. de Física, UNAM, Apdo. Postal 20-364, México 01000 D.F., México

G. ZOUPANOS

Physics Dept., Nat. Technical University, GR-157 80 Zografou, Athens, Greece
and
Institut f. Physik, Humboldt-Universität, D10115 Berlin, Germany

The all-loop relations among couplings obtained in supersymmetric Gauge-Yukawa and Finite Unified Theories have been extended to include those appearing in the soft supersymmetry breaking (SSB) sector. A sum rule for the scalar of the SSB sector holding to all-orders permits the minimal supersymmetric and two Finite-Gauge-Yukawa $SU(5)$ models that have been examined to overcome the phenomenological problems related to the universality of the soft scalar masses while they exhibit a remarkable predictive power. The most characteristic features of both kind of models are: i) the old agreement of the top quark mass prediction remains unchanged, ii)the lightest Higgs boson mass is predicted to be ~ 120 GeV, iii) the s-spectrum starts above 400 GeV.

1 Introduction

The Standard Model (SM) is very successful in explaining the known particle interactions, but is full of parameters whose values are determined only experimentally. This fact can be interpreted as signaling the existence of a more fundamental theory, possibly at the Planck scale. The usual way of reducing the number of parameters in a theory is the addition of symmetries. Well known examples of this are supersymmetry (susy) and Grand Unified Theories (GUTs), and specifically $N = 1$ susy GUTs, which can explain some of the low energy free parameters of the SM.

In our recent studies[1,9] we have developed a complementary approach in the search for the fundamental theory. It consists in studying renormalization group invariant (RGI) relations among couplings holding at the GUT scale, and which can therefore be continued all the way to the Planck scale. This allows to perform a *reduction of couplings*, [10−12] thus rendering the resulting theory more predictive. The reduction of couplings is an essential ingredient in the search for finite unified theories (FUTs), since finiteness is based on the fact that it is possible to find RGI relations among couplings that keep finiteness in perturbation theory, even to all orders.[13] The RGI approach has already been proved successful in predicting the top quark mass (among others) in finite and in the minimal $SU(5)$ $N = 1$ susy GUTs. An important aspect of the RGI programme is that it is possible to guarantee the validity of the relations to all-orders in perturbation theory by studying the uniqueness of the relations at one-loop.[10]

Supersymmetry is an essential feature of the above described method, and thus it is important to understand its breaking in this framework. Given this, the search for RGI relations was naturally extended to the soft susy breaking (SSB) sector of these theories,[7,15−17] which involves parameters of dimensions one and two. In non-finite theories, the RGI method was successfully extendend to the dimensionful sector, if use of a mass independent renormalization scheme is assumed.[7] In finite theories, some SSB terms that preserve finiteness to one- and two-loops have already been known for some time,[15,16] and the all-loop finiteness can also be proved.[18] This solution requires that the soft scalar masses are the same at the GUT scale (universality), which leads to phenomenological problems, namely, it predicts a charged particle to be the lightest susy particle (LSP). More recently, though, it was proven that the universality condition can be relaxed to a sum rule for the soft scalar masses,[8] and that in terms of this sum rule also the SSB sector can be made finite to all-loops,[19] i.e. it is possible to find completely finite unified theories .

In here we show how to apply the RGI method to include the dimensionful sector in specific $SU(5)$ models, and to explore their phenomenological consequences, in particular as the mass spectra is concerned. Special attention is given to the prediction for the lightest Higgs mass m_h, the LSP, and the restrictions imposed by having a large $\tan \beta$.

2 Reduction of Couplings and Finiteness in $N = 1$ SUSY Gauge Theories

A RGI relation among couplings, $\Phi(g_1, \cdots, g_N) = 0$, has to satisfy the partial differential equation (PDE) $\mu \, d\Phi/d\mu = \sum_{i=1}^{N} \beta_i \, \partial \Phi / \partial g_i = 0$, where β_i is the β-function of g_i. There exist $(N - 1)$ independent Φ's, and finding the complete set of these solutions is equivalent to solve the so-called reduction equations (REs), $\beta_g \, (dg_i/dg) = \beta_i$, $i = 1, \cdots, N$, where g and β_g are the primary coupling and its β-function respectively. Using all the $(N - 1)$ Φ's to impose RGI relations, one can in principle express all the couplings in terms of a single coupling g. The complete reduction, which formally preserves perturbative renormalizability, can be achieved by demanding a power series solution, where the uniqueness of such a power series solution can be investigated at the one-loop level. The completely reduced theory contains only one independent coupling with the corresponding β-function.

It is clear by examining specific examples, that the various couplings in supersymmetric theories have easily the same asymptotic behaviour. Therefore searching for a power series solution to the REs is justified. This is not the case in non-supersymmetric theories.

Let us then consider a chiral, anomaly free, $N = 1$ globally supersymmetric gauge theory based on a group G with gauge coupling constant g. The superpotential of the theory is given by

$$W = \frac{1}{2} m^{ij} \, \Phi_i \, \Phi_j + \frac{1}{6} C^{ijk} \, \Phi_i \, \Phi_j \, \Phi_k , \qquad (1)$$

where m^{ij} and C^{ijk} are gauge invariant tensors and the matter field Φ_i transforms according to the irreducible representation R_i of the gauge group G.

The one-loop β-function of the gauge coupling g is given by

$$\beta_g^{(1)} = \frac{dg}{dt} = \frac{g^3}{16\pi^2} \left[\sum_i l(R_i) - 3 \, C_2(G) \right] , \qquad (2)$$

where $l(R_i)$ is the Dynkin index of R_i and $C_2(G)$ is the quadratic Casimir of the adjoint representation of the gauge group G. The β-functions of C^{ijk}, by virtue of the

non-renormalization theorem, are related to the anomalous dimension matrix γ_i^j of the matter fields Φ_i as:

$$\beta_C^{ijk} = \frac{d}{dt} C^{ijk} \qquad (3)$$

$$= C^{ijp} \sum_{n=1} \frac{1}{(16\pi^2)^n} \gamma_p^{k(n)} + (k \leftrightarrow i) + (k \leftrightarrow j) \quad (4)$$

At one-loop level γ_i^j is

$$\gamma_i^{j(1)} = \frac{1}{2} C_{ipq} C^{jpq} - 2 \, g^2 \, C_2(R_i) \delta_i^j , \qquad (5)$$

where $C_2(R_i)$ is the quadratic Casimir of the representation R_i, and $C^{ijk} = C_{ijk}^*$.

As one can see from Eqs. (2) and (5) all the one-loop β-functions of the theory vanish if $\beta_g^{(1)}$ and $\gamma_i^{j(1)}$ vanish, i.e.

$$\sum_i \ell(R_i) = 3 C_2(G) \quad \frac{1}{2} C_{ipq} C^{jpq} = 2\delta_i^j g^2 C_2(R_i) . \quad (6)$$

A very interesting result is that the conditions (6) are necessary and sufficient for finiteness at the two-loop level.[15,16]

The one- and two-loop finiteness conditions (6) restrict considerably the possible choices of the irreps. R_i for a given group G as well as the Yukawa couplings in the superpotential (1). Note in particular that the finiteness conditions cannot be applied to the supersymmetric standard model (SSM), since the presence of a $U(1)$ gauge group is incompatible with the condition (6), due to $C_2[U(1)] = 0$. This leads to the expectation that finiteness should be attained at the grand unified level only, the SSM being just the corresponding, low-energy, effective theory.

The finiteness conditions impose relations between gauge and Yukawa couplings. Therefore, we have to guarantee that such relations leading to a reduction of the couplings hold at any renormalization point. The necessary, but also sufficient, condition for this to happen is to require that such relations are solutions to the reduction equations (REs) to all orders. The all-loop order finiteness theorem of ref.[13] is based on: (a) the structure of the supercurrent in $N = 1$ SYM and on (b) the non-renormalization properties of $N = 1$ chiral anomalies [13] (there are also other proofs of finiteness [14]).

3 Sum rule of soft scalar masses and finiteness

The above described method of reducing the dimensionless couplings has been extended [7] in the soft supersymmetry breaking (SSB) dimensionful parameters of $N = 1$ supersymmetric theories. In addition it was found [20] that RGI SSB scalar masses in Gauge-Yukawa unified models satisfy a universal sum rule. Here we will briefly describe how to obtain the all-loop sum rule for the soft scalar masses.

The recent progress made using the spurion technique [21,22] leads to the following all-loop relations among SSB β-functions,[23,24]

$$\beta_M = 2\mathcal{O}\left(\frac{\beta_g}{g}\right) , \qquad (7)$$

$$\beta_h^{ijk} = \gamma^i{}_l h^{ljk} + \gamma^j{}_l h^{ilk} + \gamma^k{}_l h^{ijl}$$
$$\qquad -2\gamma^i_{1l}C^{ljk} - 2\gamma^j_{1l}C^{ilk} - 2\gamma^k_{1l}C^{ijl} , \qquad (8)$$

$$(\beta_{m^2})^i{}_j = \left[\Delta + X\frac{\partial}{\partial g}\right]\gamma^i{}_j , \qquad (9)$$

$$\mathcal{O} = \left(Mg^2\frac{\partial}{\partial g^2} - h^{lmn}\frac{\partial}{\partial C^{lmn}}\right) , \qquad (10)$$

$$\Delta = 2\mathcal{O}\mathcal{O}^* + 2|M|^2 g^2\frac{\partial}{\partial g^2}$$
$$\qquad + \tilde{C}_{lmn}\frac{\partial}{\partial C_{lmn}} + \tilde{C}^{lmn}\frac{\partial}{\partial C^{lmn}} , \qquad (11)$$

where $(\gamma_1)^i{}_j = \mathcal{O}\gamma^i{}_j$, $C_{lmn} = (C^{lmn})^*$, and

$$\tilde{C}^{ijk} = (m^2)^i{}_l C^{ljk} + (m^2)^j{}_l C^{ilk} + (m^2)^k{}_l C^{ijl}. \quad (12)$$

It was also found that the relation,[24]

$$h^{ijk} = -M(C^{ijk})' \equiv -M\frac{dC^{ijk}(g)}{d\ln g} , \qquad (13)$$

is all-loop RG-invariant. Using the Novikov-Shifman-Vainstein-Zakharov [25] (NSVZ) β function the all-loop RGI sum rule for the soft scalar masses was found,[19]

$$m_i^2 + m_j^2 + m_k^2 = |M|^2\left\{\frac{1}{1 - g^2 C(G)/(8\pi^2)}\frac{d\ln C^{ijk}}{d\ln g}\right.$$
$$\left. +\frac{1}{2}\frac{d^2\ln C^{ijk}}{d(\ln g)^2}\right\} + \sum_l \frac{m_l^2 T(R_l)}{C(G) - 8\pi^2/g^2}\frac{d\ln C^{ijk}}{d\ln g} . \quad (14)$$

In addition the exact β function for m^2 in the NSVZ scheme has been obtained [19] for the first time and is given by:

$$\beta_{m_i^2}^{\text{NSVZ}} = \left[|M|^2\left\{\frac{1}{1 - g^2 C(G)/(8\pi^2)}\frac{d}{d\ln g} + \frac{1}{2}\frac{d^2}{d(\ln g)^2}\right\}\right.$$
$$\left. +\sum_l \frac{m_l^2 T(R_l)}{C(G) - 8\pi^2/g^2}\frac{d}{d\ln g}\right]\gamma_i^{\text{NSVZ}} . \quad (15)$$

Surprisingly enough, the all-loop result (14) coincides with the superstring result for the finite case in a certain class of orbifold models [8] if $d\ln C^{ijk}/d\ln g = 1$.

4 Gauge-Yukawa-Unified Theories

In this section we will look at concrete $SU(5)$ models, where the reduction of couplings in the dimensionless and dimensionful sector has been achieved.

4.1 Finite Unified Models

A predictive Gauge-Yukawa unified $SU(5)$ model which is finite to all orders, in addition to the requirements mentioned already, should also have the following properties:

1. One-loop anomalous dimensions are diagonal, i.e., $\gamma_i^{(1)j} \propto \delta_i^j$.

2. Three fermion generations, $\overline{5}_i$ ($i = 1, 2, 3$), which obviously should not couple to the **24**. This can be achieved for instance by imposing $B - L$ conservation.

3. The two Higgs doublets of the MSSM should mostly be made out of a pair of Higgs quintet and anti-quintet, which couple to the third generation.

In the following we discuss two versions of the all-order finite model.
A: The well studied finite model [1] with the following matter content: $3\ \overline{5} + 3\ \mathbf{10} + 4\ (5 + \overline{5}) + \mathbf{24}$.
B: A slight variation of the model **A**.

The superpotential which describes the two models takes the form [1,8]

$$W = \sum_{i=1}^{3}\left[\frac{1}{2}g_i^u\ \mathbf{10}_i\mathbf{10}_i H_i + g_i^d\ \mathbf{10}_i\overline{5}_i\ \overline{H}_i\right]$$
$$\quad + g_{23}^u\ \mathbf{10}_2\mathbf{10}_3 H_4 + g_{23}^d\ \mathbf{10}_2\overline{5}_3\ \overline{H}_4 + g_{32}^d\ \mathbf{10}_3\overline{5}_2\ \overline{H}_4$$
$$\quad + \sum_{a=1}^{4}g_a^f\ H_a\ \mathbf{24}\ \overline{H}_a + \frac{g^\lambda}{3}\ (\mathbf{24})^3 , \qquad (16)$$

where H_a and $\overline{H}_a$ ($a = 1, \dots, 4$) stand for the Higgs quintets and anti-quintets.

The non-degenerate and isolated solutions to $\gamma_i^{(1)} = 0$ for the models $\{\mathbf{A}, \mathbf{B}\}$ are:

$$(g_1^u)^2 = \{\frac{8}{5}, \frac{8}{5}\}g^2 , \quad (g_1^d)^2 = \{\frac{6}{5}, \frac{6}{5}\}g^2 ,$$

$$(g_2^u)^2 = (g_3^u)^2 = \{\frac{8}{5}, \frac{4}{5}\}g^2 , \quad (g_2^d)^2 = (g_3^d)^2 = \{\frac{6}{5}, \frac{3}{5}\}g^2 ,$$

$$(g_{23}^u)^2 = \{0, \frac{4}{5}\}g^2 , \quad (g_{23}^d)^2 = (g_{32}^d)^2 = \{0, \frac{3}{5}\}g^2 ,$$

$$(g^\lambda)^2 = \frac{15}{7}g^2 , \quad (g_1^f)^2 = 0 , \qquad (17)$$

$$(g_2^f)^2 = (g_3^f)^2 = \{0, \frac{1}{2}\}g^2 , \quad (g_4^f)^2 = \{1, 0\}g^2 .$$

These models satisfy the requirements to be finite to all-orders.[13] After the reduction of couplings the symmetry of W is enhanced. [1,8]

The main difference of the models **A** and **B** is that three pairs of Higgs quintets and anti-quintets couple to the **24** for **B** so that it is not necessary to mix them with H_4 and $\overline{H}_4$ in order to achieve the triplet-doublet splitting after the symmetry breaking of $SU(5)$.

In the dimensionful sector, the sum rule gives us the following boundary conditions at the GUT scale: [8]

$$m_{H_u}^2 + 2m_{10}^2 = m_{H_d}^2 + m_{\overline{5}}^2 + m_{10}^2 = M^2 \quad \text{for } \mathbf{A}, (18)$$

$$m_{H_u}^2 + 2m_{10}^2 = M^2, m_{H_d}^2 - 2m_{10}^2 = -\frac{M^2}{3} ,$$

$$m_{\overline{5}}^2 + 3m_{10}^2 = \frac{4M^2}{3} \quad \text{for } \mathbf{B}, \quad (19)$$

where we use as free parameters $m_{\overline{5}} \equiv m_{\overline{5}_3}$ and $m_{10} \equiv m_{10_3}$ for the model $\mathbf{A}$, and m_{10} for $\mathbf{B}$, in addition to M.

4.2 The minimal supersymmetric $SU(5)$ model

Next let us consider the minimal supersymmetric $SU(5)$ model. The field content is minimal. Neglecting the CKM mixing, one starts with six Yukawa and two Higgs couplings. We then require GYU to occur among the Yukawa couplings of the third generation and the gauge coupling. We also require the theory to be completely asymptotically free. In the one-loop approximation, the GYU yields $g_{t,b}^2 = \sum_{m,n=1}^{\infty} \kappa_{t,b}^{(m,n)} h^m f^n g^2$ (h and f are related to the Higgs couplings). Where h is allowed to vary from 0 to 15/7, while f may vary from 0 to a maximum which depends on h and vanishes at $h = 15/7$. As a result, it was obtained [2]: $0.97\,g^2 \le g_t^2 \le 1.37\,g^2$, $0.57\,g^2 \le g_b^2 = g_\tau^2 \ll 0.97\,g^2$. It was found [2,9] that consistency with proton decay requires g_t^2, g_b^2 to be very close to the left hand side values in the inequalities.

In this model, the reduction of parameters implies that at the GUT scale the SSB terms are proportional to the gaugino mass. [7]

5 Predictions of Low Energy Parameters

Since the gauge symmetry is spontaneously broken below M_{GUT}, the finiteness conditions do not restrict the renormalization property at low energies, and all it remains are boundary conditions on the gauge and Yukawa couplings (17), the $h = -MC$ relation and the soft scalar-mass sum rules (18,19) at M_{GUT}. So we examine the evolution of these parameters according to their renormalization group equations, at two-loop for dimensionless parameters and at one-loop for dimensionful ones, with these boundary conditions. Below M_{GUT} their evolution is assumed to be governed by the MSSM. We further assume a unique supersymmetry breaking scale M_s which we take to be the average of the stop masses, so that below M_s the SM is the correct effective theory.

The predictions for the top quark mass M_t are ~ 183 and ~ 174 GeV in models $\mathbf{A}$ and $\mathbf{B}$ respectively, and ~ 181 GeV for the minimal $SU(5)$ model. Comparing

Table 1: A representative example of the predictions for the s-spectrum for the finite model $\mathbf{A}$ with $M = 1.0$ TeV, $m_{\overline{5}} = 0.8$ TeV and $m_{10} = 0.6$ TeV.

$m_\chi = m_{\chi_1}$ (TeV)	0.45	$m_{\tilde{b}_2}$ (TeV)	1.76
m_{χ_2} (TeV)	0.84	$m_{\tilde{\tau}} = m_{\tilde{\tau}_1}$ (TeV)	0.63
m_{χ_3} (TeV)	1.49	$m_{\tilde{\tau}_2}$ (TeV)	0.85
m_{χ_4} (TeV)	1.49	$m_{\tilde{\nu}_1}$ (TeV)	0.88
$m_{\chi_1^\pm}$ (TeV)	0.84	m_A (TeV)	0.64
$m_{\chi_2^\pm}$ (TeV)	1.49	$m_{H^\pm}$ (TeV)	0.65
$m_{\tilde{t}_1}$ (TeV)	1.57	m_H (TeV)	0.65
$m_{\tilde{t}_2}$ (TeV)	1.77	m_h (TeV)	0.122
$m_{\tilde{b}_1}$ (TeV)	1.54		

these predictions with the most recent experimental value $M_t = (175.6 \pm 5.5)$ GeV, and recalling that the theoretical values for M_t may suffer from a correction of less than $\sim 4\%$,[9] we see that they are consistent with the experimental data. In addition the value of $\tan\beta$ is obtained as $\tan\beta = 54$ and 48 for models $\mathbf{A}$ and $\mathbf{B}$ respectively, and $\tan\beta = 48$ for the minimal $SU(5)$ model.

In the SSB sector, besides the constraints imposed by reduction of couplings and finiteness, we also look for solutions which are compatible with radiative electroweak symmetry breaking, and that have the lightest supersymmetric particle (LSP) neutral. Furthermore, we also look at the constraints imposed by $BR(b \to s\gamma)$,[26] whose experimental value is $1 \times 10^{-4} < BR(b \to s\gamma) < 4 \times 10^{-4}$. The SM predicts $BR(b \to s\gamma) = 3.1 \times 10^{-4}$. The softly broken SUSY theory has contributions to this branching ratio due to the charged Higgs and the chargino other than the SM contribution. The contribution due to the charged Higgs field is always additive to the SM one. One property of our models is that the charged Higgs field is much lighter than the chargino field. Thus, the charged Higgs contribution is important in these models. This imposes a further restriction in our parameter space, namely $M \sim 1$ TeV if $\mu < 0$ for all three models. This restriction is less strong for the case that $\mu > 0$. For example, for the finite model B with $M = 1$ TeV and $m_{10} = 0.65$ TeV as shown in Table 1, we obtain $BR(b \to s\gamma) = 3.5 \times 10^{-4}$ for $\mu > 0$ and 4.1×10^{-4} for $\mu < 0$.

In Tables 1, 2, and 3 we present representative examples of the values obtained for the sparticle spectra in each of the models. The s-particle spectrum starts above 400 GeV, when we take into account the constraints explained above. Very relevant is also the prediction for the lightest Higgs mass. The physical mass for the lightest Higgs M_h, with one-loop corrections included,[27] is predicted to be ~ 120 GeV $\pm 10\%$ for all three models. This prediction is stable against changes in the free parameters.

Table 2: A representative example of the predictions of the s-spectrum for the finite model **B** with $M = 1$ TeV and $m_{10} = 0.65$ TeV.

$m_\chi = m_{\chi_1}$ (TeV)	0.45	$m_{\tilde{b}_2}$ (TeV)	1.70
m_{χ_2} (TeV)	0.84	$m_{\tilde{\tau}} = m_{\tilde{\tau}_1}$ (TeV)	0.47
m_{χ_3} (TeV)	1.30	$m_{\tilde{\tau}_2}$ (TeV)	0.67
m_{χ_4} (TeV)	1.31	$m_{\tilde{\nu}_l}$ (TeV)	0.88
$m_{\chi_1^\pm}$ (TeV)	0.84	m_A (TeV)	0.73
$m_{\chi_2^\pm}$ (TeV)	1.31	$m_{H^\pm}$ (TeV)	0.73
$m_{\tilde{t}_1}$ (TeV)	1.51	m_H (TeV)	0.73
$m_{\tilde{t}_2}$ (TeV)	1.73	m_h (TeV)	0.118
$m_{\tilde{b}_1}$ (TeV)	1.56		

Table 3: A representative example of the predictions of the s-spectrum for the minimal $SU(5)$ model with $M = 1.0$ TeV.

$m_\chi = m_{\chi_1}$ (TeV)	0.45	$m_{\tilde{b}_2}$ (TeV)	1.88
m_{χ_2} (TeV)	0.84	$m_{\tilde{\tau}} = m_{\tilde{\tau}_1}$ (TeV)	0.92
m_{χ_3} (TeV)	1.73	$m_{\tilde{\tau}_2}$ (TeV)	1.10
m_{χ_4} (TeV)	1.73	$m_{\tilde{\nu}_l}$ (TeV)	1.43
$m_{\chi_1^\pm}$ (TeV)	0.84	m_A (TeV)	0.70
$m_{\chi_2^\pm}$ (TeV)	1.73	$m_{H^\pm}$ (TeV)	0.70
$m_{\tilde{t}_1}$ (TeV)	1.69	m_H (TeV)	0.70
$m_{\tilde{t}_2}$ (TeV)	1.89	m_h (TeV)	0.120
$m_{\tilde{b}_1}$ (TeV)	1.70		

6 Conclusions

We have shown how to apply the RGI method to the supersymmetry breaking sector of Gauge-Yukawa and Finite Unified Theories. This way it is possible to find a sum rule for the scalar masses of the SSB sector, which holds at all orders, enhancing the predictivity of the theory. Particularly interesting is the fact that the finite unified theories, which could be made all-loop finite in the supersymmetric sector, can be made completely finite, i.e. including the SSB terms[8,19] Thus, the successful prediction for the top quark mass that the finite and Gauge-Yukawa unified theories gave, has now been complemented with a new important ingredient concerning the finiteness of the SSB sector of the theory. The sum rule for the scalar masses avoids the phenomenological problems related to the previously known "universal" solution. It is interesting to remark that this sum rule coincides with that of a certain class of string models in which the massive string modes are organized into $N = 4$ supermultiplets. Using the sum rule we can now determine the spectrum of realistic models in terms of just a few parameters. In addition to the successful prediction of the top quark mass the characteristic features of the spectrum are that 1) the lightest Higgs mass is predicted ~ 120 GeV and 2) the s-spectrum starts above 400 GeV. It is

also shown that there are no strong constraints placed on the parameter space of the three models by $BR(b \to s\gamma)$ for $\mu > 0$. For the case that $\mu < 0$, the value of the gaugino mass is constrained to be $M \geq 1$ TeV.

Therefore, the next important test of these Gauge-Yukawa and finite unified theories will be given with the measurement of the Higgs mass, for which the models show an appreciable stability in their prediction.

Acknowledgements

Supported partly by the mexican project CONA-CYT 3275-PE, by the EU projects FMBI-CT96-1212 and ERBFMRXCT960090, and by the Greek project PENED95/1170;1981.

1. D. Kapetanakis, M. Mondragón and G. Zoupanos, *Zeit. f. Phys.* **C60** 181 (1993); M. Mondragón and G. Zoupanos, *Nucl. Phys.* **B** (Proc. Suppl) **37C** 98 (1995).
2. J. Kubo, M. Mondragón and G. Zoupanos, *Nucl. Phys.* **B424** 291 (1994).
3. J. Kubo, M. Mondragón, N.D. Tracas and G. Zoupanos, *Phys. Lett.* **B342** 155 (1995).
4. J. Kubo, M. Mondragón, S. Shoda and G. Zoupanos, Nucl. Phys. **B469** 3 (1996).
5. J. Kubo, M. Mondragón, M. Olechowski and G. Zoupanos, *Gauge-Yukawa Unification and the Top-Bottom Hierarchy*, Proc. of the *Int. Europhysics Conf. on HEP*, Brussels 1995.
6. J. Kubo, M. Mondragón, M. Olechowski and G. Zoupanos, *Nucl. Phys.* **B479** 25 (1996).
7. J. Kubo, M. Mondragón and G. Zoupanos, *Phys. Lett.* **B389** 523 (1996).
8. T. Kobayashi, J. Kubo, M. Mondragón and G. Zoupanos, *Nucl. Phys. B*. **B511** 45 (1998).
9. For an extended discussion and a complete list of references see: J. Kubo, M. Mondragón and G. Zoupanos, *Acta Phys. Polon.* **B27** 3911 (1997).
10. W. Zimmermann, *Com. Math. Phys.* **97** 211 (1985); R. Oehme and W. Zimmermann *Com. Math. Phys.* **97** (1985) 569; R. Oehme, K. Sibold and W. Zimmermann, *Phys. Lett.* **B147** 117 (1984); **B153** 142 (1985); R. Oehme, *Prog. Theor. Phys. Suppl.* **86** 215 (1986).
11. J. Kubo, K. Sibold and W. Zimmermann, *Nucl. Phys.* **B259** (1985) 331; *Phys. Lett.* **B200** 185 (1989).
12. J. Kubo, *Phys. Lett.* **B262** 472 (1991).
13. C. Lucchesi, O. Piguet and K. Sibold, *Helv. Phys. Acta* **61** (1988) 321; *Phys. Lett.* **B201** 241 (1988).
14. A. Hanany, M.J Strassler, A.M. Uranga, *J.High Energy Phys.* **06** 011 (1998).

15. D.R.T. Jones, L. Mezincescu and Y.-P. Yao, *Phys. Lett.* **B148** 317 (1984).

16. I. Jack and D.R.T. Jones, *Phys.Lett.* **B333** 372 (1994).

17. I. Jack and D.R.T. Jones, *Phys. Lett.* **B349** 294 (1995).

18. D. I. Kazakov, *Finiteness of Soft Terms in Finite N=1 SUSY Gauge Theories,* hep-ph/9709465.

19. T. Kobayashi, J. Kubo and G. Zoupanos, *Further All-loop Results in Softly-broken Supersymmetric Gauge Theories,* hep-ph/9802667, to be published in *Phys. Lett.* **B**.

20. Y. Kawamura, T. Kobayashi and J. Kubo, *Phys. Lett.* **B705** 64 (1997).

21. R. Delbourgo, Nuovo Cim **25A** (1975) 646; A. Salam and J. Strathdee, Nucl. Phys. **B86** 142 (1975); K. Fujikawa and W. Lang, Nucl. Phys. **B88** (1975) 61; M.T. Grisaru, M. Rocek and W. Siegel, Nucl. Phys. **B59** 429 (1979).

22. L. Girardello and M.T. Grisaru, Nucl. Phys. **B194** 65 (1982); J.A. Helayel-Neto, Phys. Lett. **B135** 78 (1984); F. Feruglio, J.A. Helayel-Neto and F. Legovini, Nucl. Phys. **B249** (1985) 533; M. Scholl, Zeit. f. Phys. **C28** 545 (1985).

23. L.V. Avdeev, D.I. Kazakov and I.N. Kondrashuk, *Nucl. Phys.* B **510**, 289 (1998); D.I. Kazakov, *Phys. Lett.* B **421**, 211 (1998).

24. I. Jack, D.R.T. Jones and A. Pickering, *Phys. Lett.* B **426**, 73 (1998).

25. V. Novikov, M. Shifman, A. Vainstein and V. Zakharov, Nucl. Phys. **B229** 381 (1983); Phys. Lett. **B166** 329 (1986); M. Shifman, Int.J. Mod. Phys.**A11** 5761 (1996) and references therein.

26. S. Bertolini, F. Borzumati, A. Masiero and G. Ridolfi, Nucl. Phys. **B353** 591 (1991).

27. A.V. Gladyshev, D.I. Kazakov, W. de Boer, G. Burkart, R. Ehret,*Nucl. Phys.* **B498** 3 (1997); M. Carena, J.R. Espinosa, M. Quirós, C.E.M. Wagner, *Phys. Lett.* **B355** 209 (1995); H. E. Haber, R. Hempfling, A. H. Hoang, *Z.Phys.*C **75** 539 (1997); S. Heinemeyer, W. Hollik, G. Weiglein, preprint KA-TP-2-1998, Mar 1998. 9pp. e-Print Archive: hep-ph/9803277.

EFFECTIVE CHIRAL THEORY OF MESONS, COEFFICIENTS OF ChPT, AXIAL VECTOR SYMMETRY BREAKING

BING AN LI

Department of Physics and Astronomy, University of Kentucky, Lexington, KY 40506, USA
E-mail: li@ukcc.uky.edu

A phenomenological successful effective chiral theory of pseudoscalar, vector, and axial-vector mesons is introduced. Based on this theory all the 10 coefficients of ChPT are predicted. It has been found that a new symmetry breaking-axial-vector symmetry breaking is responsible for the mass difference between ρ and a_1 mesons. It is shown that in an EW theory without Higgs m_W and m_Z are dynamicallu generated by the combination of the fermion masses and the axial-vector symmetry breaking: $m_W = \frac{1}{\sqrt{2}} g m_t$, $m_Z = \rho m_W / cos\theta_W$ with $\rho \approx 1$, and $G_F = 1/(2\sqrt{2} m_t^2)$. Two gauge fixing terms of W and Z fields are dynamically generated too.

1 Effective chiral theory of mesons

The chiral perturbation theory is rigorous and phenomenologically successful in describing the physics of the pseudoscalar mesons at low energies($E < m_\rho$). Models attempt to deal with the two main frustrations that the ChPT is limited to pseudoscalar mesons at low energy and contains many coupling constants which must be measured.

I have proposed an effective chiral theory of pseudoscalar, vector, and axial-vector mesons[1]. It provides a unified study of pseudoscalar, vector, and axial-vector meson physics at low energies.

The ansatz made in this theory is that the meson fields are simulated by quark operators. For example,

$$\rho_\mu^i = -\frac{1}{g_\rho m_\rho^2} \bar{\psi} \tau_i \gamma_\mu \psi. \tag{1}$$

The ansatz can be tested. Applying PCAC, current algebra, and this expression to the decay $\rho \to \pi\pi$, under the soft pion approximation it is derived

$$\frac{1}{2} f_{\rho\pi\pi} g_\rho = 1 \tag{2}$$

which is just the result of the VMD.
As a matter of fact, using fermion operator to simulate a meson field has a long history:

1. More than six decades ago, Jordan et al observed

$$\frac{1}{\sqrt{\pi}} \partial_\mu \phi = \bar{\psi} \gamma_5 \gamma_\mu \psi \tag{3}$$

   in $1+1$ field theory.

2. A similar relation is proved in the bosonization of 1+1 field theory.

3. Use of quark operators to simulate meson fields has been already exploited in the Nambu-Jona-Lasinio(NJL) model, a model of four quark interactions.

4. Quark operators have been taken as interpolating fields in lattice gauge calculations.

The simulations of the meson fields by quark operators are realized by a Lagrangian which is constructed by chiral symmetry

$$\mathcal{L} = \bar{\psi}(x)(i\gamma \cdot \partial + \gamma \cdot v + \gamma \cdot a\gamma_5 - mu(x))\psi(x)$$
$$-\psi(\bar{x})M\psi(x) + \frac{1}{2} m_0^2 (\rho_i^\mu \rho_{\mu i} + K^{*\mu} K_\mu^* + \omega^\mu \omega_\mu + \phi^\mu \phi_\mu$$
$$+ a_i^\mu a_{\mu i} + K_1^\mu K_{1\mu} + f^\mu f_\mu + f_{1s}^\mu f_{1s\mu}) \tag{4}$$

$u = exp\{i\gamma_5(\tau_i \pi_i + \lambda_a K_a + \lambda_8 \eta_8 + \eta_0)\}$. To avoid double couting, there are no kinetic terms of mesons, which are generated dynamically. Using the least action principle, the relationships between meson fields and quark operators are found from the Lagrangian. Taking the case of two flavors as an example, from the least action principle we obtain

$$\frac{\Pi_i}{\sigma} = i(\bar{\psi} \tau_i \gamma_5 \psi + ix\bar{\psi} \tau_i \psi)/(\bar{\psi}\psi + ix\bar{\psi}\gamma_5\psi),$$
$$x = (i\bar{\psi}\gamma_5\psi - \frac{\Pi_i}{\sigma} \bar{\psi}\tau_i\psi)/(\bar{\psi}\psi + i\frac{\Pi_i}{\sigma} \bar{\psi}\tau_i\gamma_5\psi),$$
$$\rho_\mu^i = -\frac{1}{m_0^2} \bar{\psi}\tau_i\gamma_\mu\psi, \quad a_\mu^i = -\frac{1}{m_0^2} \bar{\psi}\tau_i\gamma_\mu\gamma_5\psi,$$
$$\omega_\mu = -\frac{1}{m_0^2} \bar{\psi}\gamma_\mu\psi, \quad f\mu = -\frac{1}{m_0^2} \bar{\psi}\gamma_\mu\gamma_5\psi, \tag{5}$$

where $\sigma + i\gamma_5\tau \cdot \Pi = ue^{-i\eta\gamma_5}$, $\sigma = \sqrt{1 - \Pi^2}$, and $x = tan\eta$. The pseudoscalar fields have very complicated quark structures. Substituting these expressions into the Lagrangian of two flavors, a Lagrangian of quarks is obtained. It is no longer a theory of four quarks. This model is different from NJL model.

Using path interal to integrate out the quark fields, the effective Lagrangian of mesons is derived.

$$\mathcal{L}_E = lndet\mathcal{D},$$
$$\mathcal{L}_{re} = \frac{1}{2} lndet(\mathcal{D}^\dagger \mathcal{D}),$$

$$\mathcal{L}_{im} = \frac{1}{2} ln det(\mathcal{D}/\mathcal{D}^\dagger).$$

The question is whether the masses, decay widths, and interactions of mesons can be described by the quark operators correctly. Taking masses as an example,

1. Pseudoscalar mesons

$$m_{\pi^\pm}^2 = \frac{4}{f_{\pi^\pm}^2}\{-\frac{1}{3}<\bar\psi\psi>(m_u+m_d) - \frac{F^2}{4}(m_u+m_d)^2\},$$

$$m_{\pi^0}^2 = \frac{4}{f_{\pi^0}^2}\{-\frac{1}{3}<\bar\psi\psi>(m_u+m_d) - \frac{F^2}{2}(m_u^2+m_d^2)\},$$

$$m_{K^+}^2 = \frac{4}{f_{K^+}^2}\{-\frac{1}{3}<\bar\psi\psi>(m_u+m_s) - \frac{F^2}{4}(m_u+m_s)^2\},$$

$$m_{K^0}^2 = \frac{4}{f_{K^0}^2}\{-\frac{1}{3}<\bar\psi\psi>(m_d+m_s) - \frac{F^2}{4}(m_d+m_s)^2\},$$

$$m_{\eta_8}^2 = \frac{4}{f_{\eta_8}^2}\{-\frac{1}{3}<\bar\psi\psi>\frac{1}{3}(m_u+m_d+4m_s)$$

$$-\frac{F^2}{6}(m_u^2+m_d^2+4m_s^2)\}, \qquad (6)$$

The formulas at the first order in quark masses are Gell-Mann, Oakes, and Renner chiral perturbation theory.

2. Vector mesons

$$m_\rho^2 = m_\omega^2 = 6m^2.$$

3. Axial-vector mesons

$$(1-\frac{1}{2\pi^2 g^2})m_a^2 = 6m^2 + m_\rho^2,$$

$$(1-\frac{1}{2\pi^2 g^2})m_f^2 = m_\rho^2 + m_\omega^2,$$

The masses of vector originate in dynamical chiral symmetry breaking. This is the reason why they are much heavier than pseudoscalars. The masses of the axial-vector mesons are resulted by a new symmetry breaking-axial-vector symmetry breaking.

The widths are

$$\Gamma_\rho = 142 MeV (Exp. 150 MeV),$$

$$\Gamma_a = 386 MeV (exp. \sim 400 MeV),$$

$$\Gamma_\omega = 7.7 MeV (exp. 7.49 MeV),$$

$$\Gamma_{f\to\rho\pi\pi} = 6.01 MeV (exp. 6.96(1\pm0.33)MeV)$$

This theory has following features:

1. The theory is chiral symmetric in the limit of $m_q \to 0$. The theory has dynamically chiral symmetry breaking(m),

2. VMD is a natural result

$$\frac{e}{f_v}\{-\frac{1}{2}F^{\mu\nu}(\partial_\mu\rho_\nu - \partial_\nu\rho_\mu) + A^\mu j^\mu\}.$$

3. Axial-vector currents are bosonized

$$-\frac{g_W}{4f_a}\frac{1}{f_a}\{-\frac{1}{2}F^{i\mu\nu}(\partial_\mu a_\nu^i - \partial_\nu a_\mu^i) + A^{i\mu} j_\mu^i\}$$

$$-\frac{g_W}{4}\Delta m^2 f_a A_\mu^i a^{i\mu} - \frac{g_W}{4}f_\pi A^{i\mu}\partial_\mu\pi^i,$$

Axial-vector symmetry breakin is taken part in.

4. The Wess-Zumino-Witten anomalous action is the leading term of the imaginary part of the effective Lagrangian,

5. Weinberg's first sum rule is satisfied analytically,

6. The constituent quark mass is introduced as m,

7. Theoretical results of the masses and strong, E&M, and weak decay widths of mesons agree well with data,

8. The form factors of pion, π_{l3}, K_{l3}, $\pi \to e\gamma\nu$, and $K \to e\gamma\nu$ are obtained and agree with data. For example,

$$<r^2>_\pi = 0.445 fm^2,$$

$$Exp. = 0.44 \pm 0.01 fm^2,$$

$$\rho - pole = 0.39 fm^2.$$

9. The theory has been applied to τ mesonic decays[3] successfully. For example,

$$B(\tau \to \eta 3\pi\nu) = 3.4 \times 10^{-4},$$

$$Exp. = (4.1 \pm 0.7 \pm 0.7) \times 10^{-4},$$

$$ChPT = 1.2 \times 10^{-6}.$$

10. $\pi\pi$ and πK scatterings are studied. Theory agrees with data,

11. The parameters of this theory are: m(quark condensate), g(universal coupling constant), and three current quark masses,

12. Large N_C expansion is natural in this theory. All loop diagrams of mesons are at higher orders in N_C expansion. So far all calculations are done at the three level,

13. A cut-off has been determined to be 1.6GeV. All the masses of mesons are below the cut-off. The theory is self consistent,

2 Coefficients of chiral perturbation theory

Chiral perturbation theory is the low energy limit of any successful effective meson theory

$$\mathcal{L} = \frac{f_\pi^2}{16} Tr D_\mu U D^\mu U^\dagger + \frac{f_\pi^2}{16} Tr\chi(U + U^\dagger)$$
$$+ L_1[Tr(D_\mu U D^\mu U^\dagger)]^2 + L_2(Tr D_\mu U D_\nu U^\dagger)^2$$
$$+ L_3 Tr(D_\mu U D^\mu U^\dagger)^2 + L_4 Tr(D_\mu U D^\mu U^\dagger) Tr\chi(U + U^\dagger)$$
$$+ L_5 Tr D_\mu U D^\mu U^\dagger(\chi U^\dagger + U\chi) + L_6[Tr\chi(U + U^\dagger)]^2$$
$$+ L_7[Tr\chi(U - U^\dagger)]^2 + L_8 Tr(\chi U^\dagger \chi U + \chi U^\dagger \chi U^\dagger)$$
$$- iL_9 Tr(F_{\mu\nu}^L D^\mu U D^\nu U^\dagger + F_{\mu\nu}^R D^\mu U^\dagger D^\nu U)$$
$$+ L_{10} Tr(F_{\mu\nu}^L U F^{\mu\nu R} U^\dagger). \tag{7}$$

Many models try to predict the coefficients of ChPT(see Table I)

The effective chiral theory of mesons is used to predict all the 10 coefficients.

1. The effective L of $\pi\pi$ scattering at low energy is derived. Two of the three coefficients are obtained

$$2(L_1 + L_2) + L_3 = \frac{1}{4}\frac{1}{(4\pi)^2}(1 - \frac{2c}{g})^4$$

$$L_2 = \frac{1}{4}\frac{c^4}{g^2} + \frac{1}{4}\frac{1}{(4\pi)^2}(1 - \frac{2c}{g})^2(1 - \frac{4c}{g} - \frac{4c^2}{g^2})$$
$$+ \frac{1}{8}(1 - \frac{2c}{g})\frac{c}{g}\{2gc + \frac{1}{\pi^2}(1 - \frac{2c}{g})\}$$

$$c = \frac{f_\pi^2}{2gm_\rho^2}. \tag{8}$$

2. A complete determination of $L_{1,2,3}$ is carried out from the effective L of πK scattering

$$-2L_1 + L_2 = 0,$$

$$L_3 = -\frac{3}{16}\frac{2c}{g}(1 - \frac{2c}{g})\{2gc + \frac{1}{\pi^2}(1 - \frac{2c}{g})\}$$
$$- \frac{1}{2}\frac{1}{(4\pi)^2}(1 - \frac{2c}{g})^2(1 - \frac{4c}{g} - \frac{8c^2}{g^2}) - \frac{3}{4}\frac{c^4}{g^2},$$

$$L_1 = \frac{1}{32}\frac{2c}{g}(1 - \frac{2c}{g})\{2gc + \frac{1}{\pi^2}(1 - \frac{2c}{g})\}$$
$$+ \frac{1}{8}\frac{1}{(4\pi)^2}(1 - \frac{2c}{g})^2(1 - \frac{4c}{g} - \frac{4c^2}{g^2}) + \frac{1}{4}\frac{c^4}{g^2}. \tag{9}$$

3. The coefficients L_{4-8} are determined from the quark mass expansions of m_π^2, m_K^2, m_η^2, f_π, f_K, and f_η

$$L_4 = 0, \quad L_6 = 0, \tag{10}$$

$$L_5 = \frac{f_{\pi 0}^2 f}{8m B_0}, \tag{11}$$

$$L_8 = -\frac{F^2}{16B_0^2}, \tag{12}$$

$$3L_7 + L_8 = -\frac{F^2}{16B_0^2}, \quad L_7 = 0, \tag{13}$$

where

$$B_0 = \frac{4}{f_{\pi 0}^2}(-\frac{1}{3}) < \bar{\psi}\psi > . \tag{14}$$

L_5 and L_8 are written as

$$L_5 = \frac{1}{32Q}(1 - \frac{2c}{g})\{(1 - \frac{2c}{g})^2(1 - \frac{1}{2\pi^2 g^2})$$
$$- (1 - \frac{2c}{g}) + \frac{4}{\pi^2}Q(1 - \frac{c}{g})\}, \tag{15}$$

$$L_8 = -\frac{1}{1536g^2 Q^2}(1 - \frac{2c}{g})^2, \tag{16}$$

where

$$Q = -\frac{1}{108g^4}\frac{1}{m^3} < \bar{\psi}\psi > . \tag{17}$$

Q is a function of the universal coupling constant g only. By fitting $\rho \rightarrow e^+ e^-$, g is determined to be 0.39. The numerical value of Q is 4.54.

4. L_9 and L_{10} are determined by $< r^2 >_\pi$ and the amplitudes of pion radiative decay, $\pi^- \rightarrow e^- \gamma\nu$

$$L_9 = \frac{f_\pi^2}{48} < r_\pi^2 >, \tag{18}$$

$$L_9 = \frac{1}{32\pi^2}\frac{R}{F^V}, \tag{19}$$

$$L_{10} = \frac{1}{32\pi^2}\frac{F^A}{F^V} - L_9, \tag{20}$$

$$L_9 = \frac{1}{4}cg + \frac{1}{16\pi^2}\{(1 - \frac{2c}{g})^2 - 4\pi^2 c^2\}, \tag{21}$$

$$L_{10} = -\frac{1}{4}cg + \frac{1}{4}c^2 - \frac{1}{32\pi^2}(1 - \frac{2c}{g})^2. \tag{22}$$

There are two parameters: g and f_π^2/m_ρ^2 in all the 10 coefficients, which have been determined already. The numerical values of the 10 coefficients predicted by present theory are shown in Table II.

3 Axial-vector symmetry breaking and W and Z masses

Pion, ρ and a_1 are made of u and d quarks. Pions are Goldstone bosons. Due to explicit chiral symmetry breaking by current quark masses pion mass is light. m_ρ is resulted in dynamical chiral symmetry breaking(quark condensate) and is much heavier than pions. It is well known that a_1 meson is the chiral partner of ρ meson. However, a_1 is much heavier(1.26GeV) than ρ meson (0.77GeV) is. Why?

Table 1: Coefficients obtained by various models

	Vectors	Quark			Nucleon loop	Linearσ model	ENJL
$(L_1+\frac{1}{2}L_3)\times 10^{-3}$	-2.1	-.8	2.1	1.1	-.8	-.5	
$L_1 \times 10^{-3}$							0.8
$L_2 \times 10^{-3}$	2.1	1.6	1.6	1.8	.8	1.5	1.6
$L_3 \times 10^{-3}$							-4.1
$L_4 \times 10^{-3}$							0.
$L_5 \times 10^{-3}$							1.5
$L_6 \times 10^{-3}$							0.
$L_7 \times 10^{-3}$							
$L_8 \times 10^{-3}$							0.8
$L_9 \times 10^{-3}$	7.3	6.3	6.7	6.1	3.3	.9	6.7
$L_{10} \times 10^{-3}$	-5.8	-3.2	-5.8	-5.2	-1.7	-2.0	-5.5

Table 2: Coefficients from this theory

$10^3 L_1$	$10^3 L_2$	$10^3 L_3$	$10^3 L_4$	$10^3 L_5$	$10^3 L_6$	$10^3 L_7$	$10^3 L_8$	$10^3 L_9$	$10^3 L_{10}$
.9 ± .3	1.7 ± .7	-4.4 ± 2.5	0. ± .5	2.2 ± .5	0. ± 0.3	-.4 ± .15	1.1 ± .3	7.4 ± .7	-6. ± .7

What kind of symmetry breaking is responsible for the mass diffrence between ρ and a_1 mesons?

There must be a new dynamical symmetry breaking which generates the mass difference between a_1 and ρ. It has been found out that the axial-vector coupling results a new dynamical symmetry breaking -axial-vector symmetry breaking which leads to

$$(1 - \frac{1}{2\pi^2 g^2})m_a^2 = 6m^2 + m_\rho^2.$$

Can this axial-vector symmetry breaking bring something to the EW theory?

A Lagrangian of EW interactions without Higgs is studied

$$\mathcal{L} = -\frac{1}{4}A_{\mu\nu}^i A^{i\mu\nu} - \frac{1}{4}B_{\mu\nu}B^{\mu\nu} + \bar{q}\{i\gamma \cdot \partial - M\}q$$
$$+\bar{q}_L\{\frac{g}{2}\tau_i\gamma \cdot A^i + g'\frac{Y}{2}\gamma \cdot B\}q_L + \bar{q}_R g'\frac{Y}{2}\gamma \cdot Bq_R$$
$$+\bar{l}\{i\gamma \cdot \partial - M_f\}l + \bar{l}_L\{\frac{g}{2}\tau_i\gamma \cdot A^i - \frac{g'}{2}\gamma \cdot B\}l_L$$
$$-\bar{l}_R g'\gamma \cdot Bl_R. \tag{23}$$

There are no $m_{W,Z}$ and there are fermion mass terms which break the charged part of the $SU(2)_L \times U(1)$ symmetry. This explicit symmetry breaking makes charged bosons, W, massive. However, there are still two U(1) symmetries. Z and photon are massless. In EW interactions there are axial-vector couplings, therefore, there are axial-vector symmetry breaking which contribute mass to both W and Z bosons. Using path integral to integrate out the fermion fields, the Lagrangian of boson fields is derived. After multiplicative renormalization following results are obtained

1.

$$m_W^2 = \frac{1}{2}g^2\{m_t^2 + m_b^2 + m_c^2 + m_s^2 + m_u^2 + m_d^2$$
$$+m_{\nu_e}^2 + m_e^2 + m_{\nu_\mu}^2 + m_\mu^2 + m_{\nu_\tau}^2 + m_\tau^2\}(24)$$

$$m_W = \frac{g}{\sqrt{2}}m_t. \tag{25}$$

Using the values $g = 0.642$ and $m_t = 180 \pm 12GeV$, it is found

$$m_W = 81.71(1 \pm 0.067)GeV, \tag{26}$$

which is in excellent agreement with data $80.33 \pm 0.15GeV$.

2.

$$G_F = \frac{1}{2\sqrt{2}m_t^2} = 0.96 \times 10^{-5}(1 \pm 0.13)m_N^{-2}.$$

3.

$$m_Z^2 = \rho m_W^2/cos^2\theta_W, \tag{27}$$

$$\rho = (1 - \frac{\alpha}{4\pi}f_4)^{-1},$$

$$f_4 = \frac{1}{3}N_G - \frac{2}{3}\sum_q f_3 + \frac{2}{3}\sum_l f_3.$$

$$f_3 = \frac{1}{2}\frac{1}{\sqrt{x}}ln\frac{1+\sqrt{x}}{1-\sqrt{x}}.$$

$$x = (\frac{m_2^2}{m_1^2})^2, \quad m_{1,2} = \frac{1}{2}(m_t \pm m_b).$$

$$\rho \sim 1,$$

$$m_Z^2 = m_W^2 / cos^2\theta_W.$$

4. The propagator of W field is derived

$$\frac{i}{q^2 - m_W^2}\{-g_{\mu\nu} + \frac{q_\mu q_\nu}{q^2}\} - \frac{i}{\xi_W q^2 - m_W^2}\frac{q_\mu q_\nu}{q^2}. \quad (28)$$

Changing the index W to Z, the propagator of Z boson field is obtained.

$$\xi_W = \frac{N_G}{(4\pi)^2}\frac{g^2}{3}, \quad \xi_Z = \frac{N_G}{(4\pi)^2}(1 - \frac{\alpha}{4\pi}f_4)^{-1}\frac{1}{3}(g^2 + g'),$$

$$N_G = 3N_C + 3.$$

1. $m_{W,Z}$ are correctly dynamically generated by axial-vector symmetry bre

2. The propagators of W and Z fields have no problems with renormalization.

4 Conclusions

1. The effective chiral theory is phenomenologically successful,

2. All the 10 coefficients of ChPT are predicted,

3. What can we learned about QCD from this theory:

   (a) Chiral symmetry,

   (b) Dynamical chiral symmetry breaking,

   (c) Large N_C expansion,

   (d) Simulations of meson fields by quark operators are working. It is worth to do theoretical study on whether the simulations are solutions of QCD at least in the limit of large N_C expansion,

4. The Lagrabgian is not closed. The quark condensate, m, should be related to gluons,

5. Axial-vector symmetry breaking provides an explanation to the mass difference between ρ and a_1,

6. $m_{W,Z}$ are dynamically generated from the fermion masses and the axial-vector symmetry breaking.

References

1. B.A.Li, *Phys. Rev.* D **52**, 5165 (1995);5251841995.
2. B.A.Li, Proc. of Intern. Europhys. Conf. on High Energy Phys., Brussels, Belgium, 27 Jul-2 Aug, 1995, edited by J.Lemonne *et al.*, p.225.
3. B.A.Li, —Journal*Phys. Rev.* D5514251997.
4. B.A.Li, hep-ph/9711397.

ANALYTIC PERTURBATIVE APPROACH TO QCD

K. A. MILTON

Department of Physics and Astronomy, The University of Oklahoma, Norman, OK 73019-0225 USA
E-mail: milton@mail.nhn.ou.edu

I. L. SOLOVTSOV, O. P. SOLOVTSOVA

Bogoliubov Laboratory of Theoretical Physics, Joint Institute for Nuclear Research, Dubna, 141980 Russia
E-mail: solovtso@thsun1.jinr.ru, olsol@thsun1.jinr.ru

A technique called analytic perturbation theory, which respects the required analytic properties, consistent with causality, is applied to the definition of the running coupling in the timelike region, to the description of inclusive τ-decay, to deep-inelastic scattering sum rules, and to the investigation of the renormalization scheme ambiguity. It is shown that in the region of a few GeV the results are rather different from those obtained in the ordinary perturbative description and are practically renormalization scheme independent.

1 Analytic Running Coupling Constant

The conventional renormalization-group resummation of perturbative series leads to unphysical singularities in the running coupling constant. For example, the usual QCD one-loop running coupling is [a]

$$\frac{\alpha^{\text{PT}}(Q^2)}{4\pi} = \frac{1}{\beta_0} \frac{1}{\ln(Q^2/\Lambda^2)}, \tag{1}$$

where $\beta_0 = 11 - 2n/3$, n being the number of quark flavors. This evidently has a singularity (Landau pole) at $Q^2 = \Lambda^2$, which is unphysical and inconsistent with the causality principle. Instead, we propose replacing perturbation theory (PT) by analytic perturbation theory (APT) [1,2] to enforce the correct analytic properties, for example, that the running coupling be regular except for a branch cut for $-Q^2 \geq 0$, by adopting the dispersion relation

$$\frac{\alpha^{\text{APT}}(Q^2)}{4\pi} = \frac{1}{\pi} \int_0^\infty d\sigma \frac{\rho(\sigma)}{\sigma + Q^2 - i\epsilon}, \tag{2}$$

where the imaginary part is given by the perturbative result, that is

$$\rho(\sigma) = \frac{1}{4\pi} \text{Im}\, \alpha^{\text{PT}}(-\sigma - i\epsilon). \tag{3}$$

This leads to a consistent spacelike coupling, which in one-loop is:

$$\alpha^{\text{APT}}(Q^2) = \frac{4\pi}{\beta_0} \left[\frac{1}{\ln(Q^2/\Lambda^2)} + \frac{\Lambda^2}{\Lambda^2 - Q^2} \right]. \tag{4}$$

The second, nonperturbative term, cancels the ghost pole.

---
[a] We use the notation $Q^2 = -q^2$, where $Q^2 > 0$ for spacelike momentum transfer.

The above defines the running coupling in the spacelike region. We can also define a timelike (or s-channel) coupling $\alpha_s(s)$,[2] which is related to the spacelike coupling through the following reciprocal relations:

$$\alpha_s(s) = -\frac{1}{2\pi i} \int_{s-i\epsilon}^{s+i\epsilon} \frac{dz}{z} \alpha(-z), \tag{5}$$

$$\alpha(Q^2) = Q^2 \int_0^\infty \frac{ds}{(s+Q^2)^2} \alpha_s(s), \tag{6}$$

where the contour in the first integral does not cross the cut on the positive real axis. In terms of the spectral function, then,

$$\frac{\alpha_s(s)}{4\pi} = \frac{1}{\pi} \int_s^\infty \frac{d\sigma}{\sigma} \rho(\sigma). \tag{7}$$

The spectral functions in 1-, 2-, and 3-loops are shown in Fig. 1, for $n = 3$. The areas under all these curves turn out to be the same, which implies a universal infrared fixed point α^*,

$$\frac{\alpha^*}{4\pi} = \frac{\alpha(0)}{4\pi} = \frac{\alpha_s(0)}{4\pi} = \frac{1}{\pi} \int_0^\infty \frac{d\sigma}{\sigma} \rho(\sigma) = \frac{1}{\beta_0}, \tag{8}$$

which is exact to all orders. [1,3] It is also possible to prove that symmetrical behavior of the timelike and spacelike couplings is inconsistent with the required analytic properties, that is, in any renormalization scheme

$$\alpha_s(s) \neq \alpha(Q^2), \quad s = Q^2. \tag{9}$$

This difference, which is important when the value of the running coupling is extracted from various experimental data, [4] is demonstrated in Fig. 2 for the $\overline{\text{MS}}$ scheme.

Because of the infrared fixed point, the running coupling implied by APT rises less rapidly for small Q^2 than

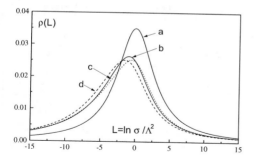

Figure 1: The spectral functions: (a) and (b) are the one- and two-loop results; (c) is the two-loop approximation obtained as the first iteration of the exact renormalization-group equation;[3] (d) is the three-loop result in the $\overline{\text{MS}}$ scheme.

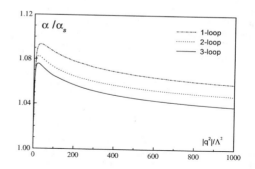

Figure 2: The ratio of spacelike and timelike values of the running coupling constant in the APT approach, at one, two, and three loops.

does the usual perturbative running coupling. This is illustrated[4] in Fig. 3.

Another interesting fact is that within APT Schwinger's conjecture[6] about the connection between the β and spectral functions is valid.[2] Indeed, the β function for the timelike coupling is

$$\beta_s = \frac{s}{4\pi}\frac{d\alpha_s}{ds} = -\frac{1}{\pi}\rho(s). \tag{10}$$

The above discussion assumes that the number of active flavors n is realistically small. However, if n is large enough, even the perturbative coupling can exhibit an infrared fixed point, at least in the two-loop level. This same fixed point defined by Eq. (8) occurs in the analytic approach and we have

$$\frac{\alpha^*}{4\pi} = \begin{cases} 1/\beta_0, & n \leq 8, \\ -\beta_0/\beta_1, & 9 \leq n \leq 16. \end{cases} \tag{11}$$

The transition between the nonperturbative and perturbative fixed point occurs for $n = 8.05$, as is shown in Fig. 4. This is qualitatively consistent with the phase transition seen, for example, in nonperturbative approaches and lattice simulations.[7]

2 Inclusive τ Decay within APT

The inclusive semileptonic τ decay ratio for massless quarks is given[8] in terms of the electroweak factor S_{EW}, the CKM matrix elements V_{ud}, V_{us}, and the QCD correction Δ_τ,

$$R_\tau = 3S_{\text{EW}}(|V_{ud}|^2 + |V_{us}|^2)(1 + \Delta_\tau). \tag{12}$$

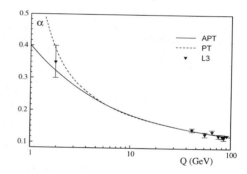

Figure 3: QCD evolution of the running coupling constants (defined in the spacelike region) compared to experimental data.[5] Here a matching procedure across the various quark thresholds has been applied in the timelike region.[4]

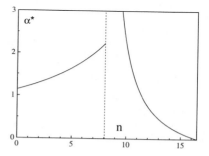

Figure 4: The analytic infrared fixed point α^* vs. n.

The correction Δ_τ may be written in terms of the functions $r(s)$ or $d(q^2)$, which are the QCD corrections to the imaginary part of the hadronic correlator Π: $\mathrm{Im}\,\Pi \sim 1+r$ and to the Adler D-function: $D = -q^2 d\Pi/dq^2 \sim 1+d$. The analytic properties of the Adler function allow us to write down the relations

$$d(q^2) = -q^2 \int_0^\infty \frac{ds}{(s-q^2)^2} r(s), \qquad (13)$$

$$r(s) = -\frac{1}{2\pi i} \int_{s-i\epsilon}^{s+i\epsilon} \frac{dz}{z} d(z). \qquad (14)$$

Because of the proper analytic properties (which are violated in the usual perturbative approach) the QCD contribution to the τ ratio is given by the two equivalent forms

$$\Delta_\tau = 2 \int_0^{M_\tau^2} \frac{ds}{M_\tau^2} \left(1 - \frac{s}{M_\tau^2}\right)^2 \left(1 + 2\frac{s}{M_\tau^2}\right) r(s) \quad (15)$$

$$= \frac{1}{2\pi i} \oint_{|z|=M_\tau^2} \frac{dz}{z} \left(1 - \frac{z}{M_\tau^2}\right)^3 \left(1 + \frac{z}{M_\tau^2}\right) d(z).$$

If one likes, one can think of r and d as effective running couplings in the timelike and spacelike regions, so as with the running couplings they may be expressed in terms of an effective spectral density,

$$d^{\mathrm{APT}}(q^2) = \frac{1}{\pi} \int_0^\infty \frac{d\sigma}{\sigma - q^2} \rho^{\mathrm{eff}}(\sigma), \qquad (16)$$

$$r^{\mathrm{APT}}(s) = \frac{1}{\pi} \int_s^\infty \frac{d\sigma}{\sigma} \rho^{\mathrm{eff}}(\sigma), \qquad (17)$$

possessing the same universal infrared limit as the running coupling.

In the ordinary perturbative approach, the function $d(q^2)$ may be expanded in terms of the running coupling and in the third order is

$$d^{\mathrm{PT}}(q^2) = a^{\mathrm{PT}}(-q^2) + d_1 \left[a^{\mathrm{PT}}(-q^2)\right]^2 + d_2 \left[a^{\mathrm{PT}}(-q^2)\right]^3, \qquad (18)$$

where we have introduced $a = \alpha_S/\pi$, and, numerically, for three active flavors, the coefficients are $d_1^{\overline{\mathrm{MS}}} = 1.6398$, $d_2^{\overline{\mathrm{MS}}} = 6.3710$. Such is not the case in APT; rather, d^{APT} is constructed from Eq. (16) with a spectral density obtained as the imaginary part of d^{PT} on the physical cut:

$$\rho^{\mathrm{eff}}(\sigma) = \rho_0(\sigma) + d_1 \rho_1(\sigma) + d_2 \rho_2(\sigma), \qquad (19)$$

where $\rho_n(\sigma) = \mathrm{Im}\,a^{n+1}(-\sigma-i\epsilon)$. We use the world average[9] value[b] $R_\tau = 3.633 \pm 0.031$. Our results are shown in Table 1, where for the sake of illustration we also show the PT results obtained by using the contour integral representation given in the second line of Eq. (15).

[b]This is consistent with the 1998 PDG value,[10] which we extract as $R_\tau = 3.650 \pm 0.016$.

Table 1: The APT and PT parameters in the $\overline{\mathrm{MS}}$ scheme extracted from τ-decay. The first two rows refer to NNLO calculations, the last two rows to NLO calculations. The errors in the last digits are shown in parentheses.

Method	$\Lambda_{\overline{\mathrm{MS}}}$ (MeV)	$\alpha(M_\tau^2)$	$d(M_\tau^2)$
APT	871(155)	0.3962(298)	0.1446(88)
PT	385(27)	0.3371(141)	0.1339(69)
APT	918(151)	0.3983(236)	0.1431(84)
PT	458(31)	0.3544(157)	0.1400(67)

Most remarkably, APT exhibits very little renormalization scheme (RS) dependence. The τ decay coefficients d_1 and d_2 are RS dependent, as are all but the first two beta function coefficients, defined by the renormalization group equation

$$\mu^2 \frac{da}{d\mu^2} = -\frac{b}{2} a^2 (1 + c_1 a + c_2 a^2). \qquad (20)$$

Because of the existence of RS invariants, one can investigate the sensitivity of the predicted value of the QCD correction to the choice of RS by varying d_1 and c_2 in a region where the degree of cancellation in the second RS invariant $\omega_2 = c_2 + d_2 - c_1 d_1 - d_1^2$ does not exceed a specified limit, taking, for example,

$$\frac{|c_2| + |d_2| + c_1|d_1| + d_1^2}{|\omega_2|} \le 2. \qquad (21)$$

That sensitivity is shown [11] in Fig. 5, based on $\Delta_\tau^{\overline{\mathrm{MS}}} = 0.1881$. Observe that the relative difference between the prediction of the lower corners of the domain defined by Eq. (21) is 0.8%, while in PT that difference is 5%. Note that the $\overline{\mathrm{MS}}$ scheme lies outside this domain, as does the so-called V scheme, the prediction of which differs by only 0.2% from the $\overline{\mathrm{MS}}$ value in APT, but by 66% in PT! To all intents and purposes, APT exhibits practically no RS dependence. (Because of the perturbative stability, the known three-loop level is quite adequate for these RS recalculations.)

3 Deep-Inelastic Scattering Sum Rules

At present, the polarized Bjorken and Gross–Llewellyn Smith deep inelastic scattering sum rules allow the possibility,[12] as with τ decay, of extracting the value of α_S from experimental data at low Q, here down to 1 GeV.

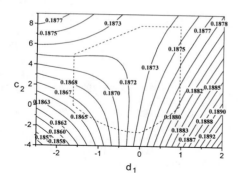

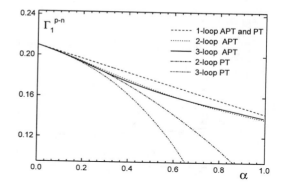

Figure 5: Contour plot of the APT correction Δ_τ at the three-loop order as a function of RS parameters d_1 and c_2. The dashed line indicates the boundary of the domain defined by Eq. (21).

Figure 6: Γ_1^{p-n} with 1-, 2-, and 3-loop QCD corrections vs. the coupling constant.

3.1 Bjorken Sum Rule

The Bjorken sum rule refers to the value of the integral of the difference between the polarized structure functions of the neutron and proton,

$$\Gamma_1^{p-n} = \int_0^\infty dx \left[g_1^p(x, Q^2) - g_1^n(x, Q^2) \right] \quad (22)$$

$$= \frac{1}{6} \left| \frac{g_A}{g_V} \right| \left[1 - \Delta_{\mathrm{Bj}}(Q^2) \right] , \quad (23)$$

where the prefactor in the second line is the parton-level description. In the conventional approach, with massless quarks, the QCD correction is given by a power series similar to Eq. (18), with coefficients for three active flavors being $d_1^{\overline{\mathrm{MS}}} = 3.5833$, $d_2^{\overline{\mathrm{MS}}} = 20.2153$. However, this description violates required analytic properties of the structure function moments, which, as has been argued,[13] follow from the existence of the Deser-Gilbert-Sudarshan integral representation.

Thus we adopt the analytic approach, which says instead

$$\Delta_{\mathrm{Bj}}^{\mathrm{APT}}(Q^2) = \delta^{(1)}(Q^2) + d_1 \delta^{(2)}(Q^2) + d_2 \delta^{(3)}(Q^2) , \quad (24)$$

where

$$\delta^{(k)}(Q^2) = \frac{1}{\pi^{k+1}} \int_0^\infty \frac{d\sigma}{\sigma + Q^2} \mathrm{Im} \left[\alpha_{\mathrm{PT}}^k(-\sigma - i\epsilon) \right] , \quad (25)$$

that is, Δ^{APT} is not a power series in α^{APT}.

Besides possessing the correct analyticity, the APT approach has two key properties in its favor. First, successive perturbative corrections are small, so that the two- and three-loop QCD results are nearly the same.

This is not the case in the conventional PT approach, as is shown[14] in Fig. 6. Second, the renormalization scheme dependence is again very small, so that various schemes which have the same degree of cancellation as the $\overline{\mathrm{MS}}$ scheme give the same predictions all the way down to $Q^2 \sim 1$ GeV^2. In contrast, it is impossible to make any reliable prediction for conventional PT, even if improved by the Padé approximant (PA) method,[15] for Q^2 below several GeV^2. This is illustrated[14] in Fig. 7. One can see that instead of RS unstable and rapidly changing PT functions, the APT predictions are slowly varying functions, which are practically RS independent.

3.2 Gross-Llewellyn Smith Sum Rule

A precisely similar analysis can be performed on the GLS sum rule, which refers to the integral

$$S_{\mathrm{GLS}} = \frac{1}{2} \int_0^1 dx \left[F_3^{\bar\nu p}(x, Q^2) + F_3^{\nu p}(x, Q^2) \right] \quad (26)$$

$$= 3 \left[1 - \Delta_{\mathrm{GLS}}(Q^2) \right] . \quad (27)$$

Again, the APT approach leads to perturbative stability, and to practically no renormalization scheme dependence down to very low Q^2.[20]

4 Conclusions

Our conclusions are four-fold.

- APT maintains correct analytic properties (causality), and allows for a consistent extrapolation between timelike (τ decay) and spacelike (DIS sum rules) data.

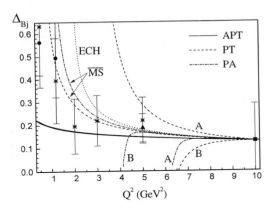

Figure 7: Renormalization scheme dependence of predictions for Δ_{Bj} vs. Q^2 for the APT and PT expansions. The solid curves, which are very close to each other, correspond to the APT result in the $\overline{MS}$, A, B, and ECH schemes. The PT evolution in the $\overline{MS}$, A, and B schemes are shown by dashed curves, the ECH scheme is indicated by a dotted curve, and the PA results in the $\overline{MS}$, A, and B schemes are denoted by dash-dotted curves. (The definition of the various schemes is given in our paper.[14]) The SMC data [16] is indicated by a square, the triangle is the E154 data,[17] circles are E143 data,[18] and the stars are recent E143 data.[19]

- Three loop corrections are much smaller than in PT; thus, there is perturbative stability.

- Renormalization scheme dependence is drastically reduced. The three-loop APT level is practically RS independent.

- The values of Λ are larger in the APT approach than in the PT approach. Yet these values are consistent between timelike and spacelike processes, and consistent with the data.

The work reported here is the beginning of a systematic attempt to improve upon the results of perturbation theory in QCD. In the future we will treat in detail the significance of power corrections which come from the operator product expansion, and examine the effect of finite mass corrections, which necessitate a more elaborate analytic structure.

Acknowledgements

This work was supported in part by grants from the US DOE, number DE-FG-03-98ER41066, from the US NSF, grant number PHY-9600421, and from the RFBR, grant 96-02-16126. Useful conversations with D. V. Shirkov and L. Gamberg are gratefully acknowledged. We dedicate this paper to the memory of our late colleague Mark Samuel.

References

1. D. V. Shirkov and I. L. Solovtsov, *JINR Rapid Comm.*, No. **2** [76]-96, p. 5 (1996); *Phys. Rev. Lett.* **79**, 1209 (1997).
2. K. A. Milton and I. L. Solovtsov, *Phys. Rev.* D **55**, 5295 (1997).
3. K. A. Milton, I. L. Solovtsov, and O. P. Solovtsova, *Phys. Lett.* B **415**, 104 (1997).
4. K. A. Milton and O. P. Solovtsova, *Phys. Rev.* D **57**, 5402 (1998).
5. L3 Collaboration, M. Acciarri *et al.*, *Phys. Lett.* B **404**, 390 (1997); **411**, 580 (1997).
6. J. Schwinger, *Proc. Natl. Acad. Sci. USA* **71**, 3024, 5047 (1974).
7. T. Appelquist, J. Terning, and L. C. R. Wijewardhana, *Phys. Rev. Lett.* **77**, 1214 (1996); V. A. Miransky and K. Yamawaki, *Phys. Rev.* D **55**, 5051 (1997); J. B. Kogut and D. R. Sinclair, *Nucl. Phys.* B **295**, 465 (1988); F. R. Brown, H. Chen, N. H. Christ, Z. Dong, R. D. Mawhinney, W. Shafer, and A. Vaccarino, *Phys. Rev.* D **46**, 5655 (1992); Y. Iwasaki, hep-lat/9707019; Y. Iwasaki, K. Kanaya, S. Kaya, S. Sakai, and T. Toshie, hep-lat/9804005.
8. E. Braaten, S. Narison, and A. Pich, *Nucl. Phys.* B **373**, 581 (1992).
9. Particle Data Group, R. M. Barnett *et al.*, *Phys. Rev.* D **54**, 1 (1996).
10. Particle Data Group, *Eur. Phys. J.* C **3**, 1 1998.
11. K. A. Milton, I. L. Solovtsov, and V. I. Yasnov, preprint OKHEP–98–01, hep-ph/9802262.
12. M. Albrow *et al.*, *Proc. of Snowmass Workshop 96: New Directions for High-Energy Physics*, hep-ph/9706470.
13. W. Wetzel, *Nucl. Phys.* B **139**, 170 (1978).
14. K. A. Milton, I. L. Solovtsov, and O. P. Solovtsova, to be published in Phys. Lett. B.
15. S. J. Brodsky, J. Ellis, E. Gardi, M. Karliner, and M. A. Samuel, *Phys. Rev.* D **56**, 6980 (1997), and references therein.
16. SMC Collaboration, A. Adams *et al.*, *Phys. Lett.* B **412**, 414 (1997).
17. E154 Collaboration, K. Abe *et al.*, *Phys. Lett.* B **405**, 180 (1997).
18. E143 Collaboration, K. Abe *et al.*, *Phys. Rev. Lett.* **78**, 815 (1997).
19. E143 Collaboration, K. Abe *et al.*, SLAC-PUB-7753, hep-ph/9802357.
20. K. A. Milton, I. L. Solovtsov, and O. P. Solovtsova, in preparation.

THE PERTURBATIVE QCD POTENTIAL AND THE $t\bar{t}$ THRESHOLD*

M. JEŻABEK[a], J. H. KÜHN[b], M. PETER[c], Y. SUMINO[b]**
and T. TEUBNER[d]

a) Institute of Nuclear Physics, Cracow, Poland, and
Department of Field Theory and Particle Physics, University of Silesia, Katowice, Poland
b) Institut für Theoretische Teilchenphysik, Universität Karlsruhe, Germany
c) Institut für Theoretische Physik, Universität Heidelberg, Germany
d) Deutsches Elektronen-Synchrotron (DESY), Hamburg, Germany

We include the full second-order corrections to the static QCD potential in the analysis of the $t\bar{t}$ threshold cross section. There is an unexpectedly large difference between the QCD potential improved by the renormalization-group equation in momentum space and the potential improved by the renormalization-group equation in coordinate space. This difference remains even at a fairly short distance $1/r \simeq 100$ GeV and its origin can be understood within perturbative QCD. We scrutinize the theoretical uncertainties of the QCD potential in relation to the $t\bar{t}$ threshold cross section. In particular there exists a theoretical uncertainty which limits our present theoretical accuracy of the $t\bar{t}$ threshold cross section at the peak to be $\delta\sigma_{\mathrm{peak}}/\sigma_{\mathrm{peak}} \gtrsim 6\%$ within perturbative QCD.

In this paper, based on Ref. [1], we report on our present theoretical understanding of the $t\bar{t}$ total cross section near the threshold. Up to now, all the $\mathcal{O}(\alpha_s)$ corrections (also leading logarithms) have been included in the calculations of various cross sections near threshold. In order to take into account the QCD binding effects properly in the cross sections, we have to systematically rearrange the perturbative expansion near threshold. Namely, we first resum all the leading Coulomb singularities $\sim (\alpha_s/\beta)^n$, take the result as the leading order contribution, and then calculate higher order corrections, which are essentially resummations of the terms $\sim \alpha_s^{n+1}/\beta^n$, α_s^{n+2}/β^n, ... It is also important to resum large logarithms arising from the large scale difference involved in the calculation [2]. This is achieved by (first) calculating the Green function of the non-relativistic Schrödinger equation with the QCD potential [3,2]. Conventionally both the coordinate-space approach developed in Refs. [2,4] and the momentum-space approach developed in Refs. [5,6] have been used in solving the equation by different groups independently. It has recently been found [7] that there are discrepancies in the results obtained from the two approaches reflecting the difference in the construction of the potentials in both spaces. It was argued that the differences are formally of $\mathcal{O}(\alpha_s^2)$ but their size turns out to be non-negligible.

Quite recently there has been considerable progress in the theoretical calculations of the second-order corrections to the cross section at threshold and the Coulombic bound-state problem. New contributions have been calculated analytically [8,9] and numerically [10] for QED bound-states. Very important steps have been accomplished in QCD as well. The full second-order correction

to the static QCD potential was computed in [11]. Also, the $\mathcal{O}(\alpha_s^2)$ total cross section is known now [12] in the region $\alpha_s \ll \beta \ll 1$ as a series expansion in β. All these results have to be included in the calculation of the full $\mathcal{O}(\alpha_s^2)$ corrections to the threshold cross section, which has just been completed (as far as the production process of top quarks are concerned) [13]. The full second-order corrections turned out to be anomalously large, which may suggest a poor convergence of the perturbative QCD in the $t\bar{t}$ threshold region.

In this paper, we incorporate the full $\mathcal{O}(\alpha_s^3)$ corrections (the second-order corrections to the leading contribution) to the static QCD potential into our analyses. In principle this is a step towards an improvement of the theoretical precision in our analysis of the $t\bar{t}$ threshold cross section. Then we scrutinize the problem of the difference between the momentum-space and the coordinate-space potentials. Contrary to our expectation, the inclusion of the above corrections does not reduce the difference of the cross sections significantly, and there still remains a non-negligible deviation. We find that there is a theoretical uncertainty within perturbative QCD which limits our present-day theoretical accuracy of the threshold cross section.

Let us now briefly explain the construction of our potentials in momentum space and in coordinate space, respectively. More detailed descriptions including formulas can be found in [1]. The large-momentum part of the momentum-space potential $V_{\mathrm{JKPT}}(q)$ is determined as follows. First the potential has been calculated up to $\mathcal{O}(\alpha_s^3)$ in a fixed-order calculation. The result is then improved using the three-loop renormalization group equation in momentum space. At low momentum, the potential is continued smoothly to a Richardson-like potential. On the other hand, the short-distance part of the

*Presented by J. H. Kühn.
**On leave of absence from Tohoku Univ., Sendai, Japan.

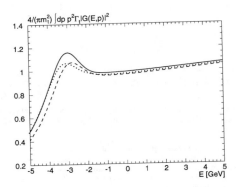

Figure 1: Comparison of the total cross sections (normalized to R) calculated from the different potentials: V_{JKPT} (solid), V_{SFHMN} (dashed), and V_{new} (dotted line). We set $\alpha_{\overline{\text{MS}}}(M_Z^2) = 0.118$, $m_t = 175$ GeV, and $\Gamma_t = 1.421$ GeV.

coordinate-space potential $V_{\text{SFHMN}}(r)$ is calculated by taking the Fourier transform of the fixed-order perturbative potential in momentum space, and then is improved using the three-loop renormalization group equation in coordinate space. At long distance, the potential is continued smoothly to a phenomenological ansatz. Thus, it is important to note that the two potentials are *not* the Fourier transforms of each other even in the large-momentum or short-distance region. They agree only up to the next-to-next-to-leading logarithmic terms of the series expansion in a fixed $\overline{\text{MS}}$ coupling. The difference begins with the non-logarithmic term in the three-loop fixed-order correction.

In Fig. 1 we show a comparison of the total cross sections (normalized to R) calculated from V_{JKPT} (solid) and from V_{SFHMN} (dashed line), without any weak or hard-gluon corrections:

$$R = \frac{4}{\pi m_t^2} \int_0^\infty dp\, p^2 |G(E, p)|^2 \, \Gamma_t. \tag{1}$$

For the physical parameters we used $\alpha_{\overline{\text{MS}}}(M_Z^2) = 0.118$, $m_t = 175$ GeV, and $\Gamma_t = 1.421$ GeV. We find that the two cross sections differ by 7.8% at the peaks and by 2.2% at $E = 5$ GeV.[a] Since the difference of the cross sections calculated from the next-to-leading order potentials is 8.6% at the peak and 2.4% at $E = 5$ GeV for the same value of $\alpha_s(M_Z^2)$, the cross sections have come closer only slightly after the inclusion of the second-order correction to the potential. The remaining difference is

[a]In this paper we are not concerned with those differences of the cross sections which can be absorbed into an additive constant to the potential $V(r)$, or equivalently, into a redefinition of the top quark mass. The theoretical uncertainty in the pole mass for our problem is discussed in [14].

much larger than what one would expect from an $\mathcal{O}(\alpha_s^3)$ correction relative to the leading order, which is not fully included in our analyses, even if we take into account the high sensitivity to the coupling [2], $\sigma_{\text{peak}} \propto \alpha_s^2$. The purpose of this paper is to understand the origin of this unexpectedly large difference.

In order to compare the asymptotic behavior of the potentials more clearly, we plot in Fig. 2(a) the coordinate-space effective couplings defined by

$$\bar{\alpha}_V(1/r) = (-C_F/r)^{-1}\, V(r) \tag{2}$$

for $V_{\text{JKPT}}(r)$ and $V_{\text{SFHMN}}(r)$ as solid and dashed lines, respectively. Contrary to our expectation, the difference of the couplings exceeds 3% even at very short distances, $1/r \simeq 100$ GeV.

Naturally the question arises: Why is there such a large discrepancy between the potential constructed in momentum space and that constructed in coordinate space? To answer this question, let us examine a relation connecting the effective coupling in coordinate space, defined by Eq. (2), and the effective coupling in momentum space, defined from the momentum-space potential as

$$\alpha_V(q) = \left(-4\pi C_F/q^2\right)^{-1} V(q). \tag{3}$$

The relation is derived from the renormalization group equation of $\alpha_V(q)$ and exact to all orders. In the asymptotic region where the couplings are small, it can be given in the form of an asymptotic series [14], which reads numerically

$$\bar{\alpha}_V(1/r) = \alpha_V + 1.225\, \alpha_V^3 + 5.596\, \alpha_V^4 + 32.202\, \alpha_V^5 + \cdots \tag{4}$$

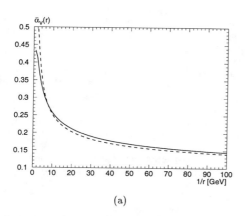

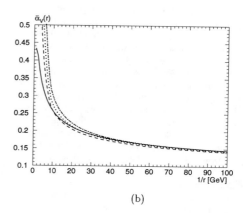

(a)   (b)

Figure 2: (a) Comparison of the coordinate-space effective charges defined from $V_{\rm JKPT}$ (solid) and from $V_{\rm SFHMN}$ (dashed line). (b) Comparison of the coordinate-space effective charges defined from the various terms of Eq. (4). See the text for the description of each curve.

for $n_f = 5$. On the right-hand-side, $\alpha_V = \alpha_V(q = e^{-\gamma_E}/r)$. All terms which are written explicitly are determined from the known coefficients of the β function, β_0^V, β_1^V, and β_2^V. At present, we can use the above relation consistently only at $O(\alpha_V^3)$ because we know the effective couplings only up to the next-to-next-to-leading order corrections in perturbative QCD, i.e. we know the relation between α_V and $\alpha_{\overline{\rm MS}}$ only up to $\mathcal{O}(\alpha_{\overline{\rm MS}}^3)$. Due to this limitation, essentially, the effective coupling $\bar{\alpha}_V$ defined from $V_{\rm SFHMN}$ is the right-hand-side of the above equation truncated at the $O(\alpha_V^3)$ term, while $\bar{\alpha}_V$ defined from $V_{\rm JKPT}$ is the right-hand-side including all terms. Numerically, the $O(\alpha_V^4)$ term and the $O(\alpha_V^5)$ term contribute as $+1.4\%$ and $+1.1\%$ corrections, respectively, for $\alpha_V = 0.1379$ (corresponding to $1/r = 100$ GeV). Therefore, these higher order terms indeed explain the difference of the effective couplings at small r. Fig. 2(b) shows several curves derived from the above relation:

1. The solid line is $\bar{\alpha}_V(1/r)$ defined from $V_{\rm JKPT}$.

2. The dashed curve is $\alpha_V + 1.225\,\alpha_V^3$, where $\alpha_V = \alpha_V(q = e^{-\gamma_E}/r)$ is calculated using the perturbative prediction in momentum space. This curve is essentially the same as $\bar{\alpha}_V(1/r)$ defined from $V_{\rm SFHMN}(r)$, since it is the next-to-next-to-leading order perturbative prediction for the coordinate-space coupling at short distances.

3. The dotted curve includes the next correction, $5.596\,\alpha_V^4$, which is in fact even larger than the $\mathcal{O}(\alpha_V^3)$ term below $1/r \sim 30$ GeV.

4. The dash-dotted curve includes the $\mathcal{O}(\alpha_V^5)$ term.

We observe that the agreement of both sides of Eq. (4) becomes better as we include more terms at small r, while it becomes worse at large r on account of the asymptoticness of the series. From the purely perturbative point of view, the discrepancy between our two potentials, $V_{\rm JKPT}$ and $V_{\rm SFHMN}$, in the asymptotic region thus seems real, an indication of large higher order corrections. When the third-order correction to the potential will be computed in terms of $\alpha_{\overline{\rm MS}}$ in the future, the $\mathcal{O}(\alpha_V^4)$ term will be treated consistently and the difference will reduce by 1.4% at $1/r \simeq 100$ GeV.

We may consider this difference of $\bar{\alpha}_V(1/r)$ at short-distances as an estimate of the higher order corrections on the basis of the following observations. First, the two potentials are equal up to the next-to-next-to-leading order, and there seems to be no reason *a priori* for considering one of the two to be more favorable theoretically. Secondly, if we apply the same method (the relation between $\bar{\alpha}_V$ and α_V) to estimate the size of the already known $\mathcal{O}(\alpha_s^3)$ correction, we obtain $\pi^2\beta_0^2/3 = 193.4$, which turns out to be a slight under-estimate of the true correction[11] $a_2 = 333.5$ $(n_f = 5)$.

Moreover, the above 3% uncertainty of $\bar{\alpha}_V(1/r)$ at $1/r \simeq 100$ GeV provides a certain criterion for the present theoretical uncertainty of the $t\bar{t}$ cross section. In fact, it would already limit the theoretical accuracy of $\bar{\alpha}_V(1/r)$ at longer distances to be not better than 3%. If we combine this with a naive estimate $\sigma_{\rm peak} \propto \bar{\alpha}_V^2$, we expect a theoretical uncertainty of the peak cross section to be $\delta\sigma_{\rm peak}/\sigma_{\rm peak} \gtrsim 6\%$. Therefore, the large discrepancy of the cross sections which we have seen turns out to be quite consistent with this estimated uncertainty.

One is tempted to include one more term of the series (4) to define a (new) coordinate-space potential despite our ignorance of the corresponding terms in the relation between α_V and $\alpha_{\overline{MS}}$, since this would apparently reduce the difference between the two effective couplings. In fact we did this exercise, but (to our surprise) it did not bring the cross section closer to the one calculated from the momentum-space potential V_{JKPT} in the peak region. This cross section calculated from the potential $V_{new}(r)$, which incorporates the $\mathcal{O}(\alpha_V^4)$ term of Eq. (4), is shown as a dotted curve in Fig. 1. This fact indicates that we are no longer able to improve the agreement of the cross sections by including even higher order terms, as we are confronting the problem of asymptoticness of the series.

Some indications can be obtained by looking into the nature of the perturbative expansion of each potential. Within our present knowledge of the static QCD potential, the perturbative series looks more convergent for the momentum-space potential than for the coordinate-space potential. To see this, one may compare the β functions of the effective couplings (the V-scheme couplings) in both spaces [11]. Numerically, the first three terms in the perturbative expansion read

- (momentum-space coupling)

$$\mu^2 \frac{d\alpha_V}{d\mu^2} = -0.6101\,\alpha_V^2 - 0.2449\,\alpha_V^3 - 1.198\,\alpha_V^4 + \cdots$$
(5)

- (coordinate-space coupling)

$$\mu^2 \frac{d\bar{\alpha}_V}{d\mu^2} = -0.6101\,\bar{\alpha}_V^2 - 0.2449\,\bar{\alpha}_V^3 - 1.945\,\bar{\alpha}_V^4 + \cdots$$
(6)

for $n_f = 5$. The first two coefficients are universal. The third coefficient depends on the scheme (the definition) of the coupling. As the third coefficients for the V-scheme couplings are quite large, the third term of the β function is comparable to the second term already for $\alpha_V = 0.20$ and for $\bar{\alpha}_V = 0.13$, respectively. The difference of the third coefficients between momentum space and coordinate space originates from the $\pi^2 \beta_0^2/3$ term. Although the magnitude of the third coefficients is of the same order, in practice it makes a certain difference whether an apparent convergence is lost at $\alpha_V = 0.20$ or $\bar{\alpha}_V = 0.13$ because there is a large scale difference between the two values. This indicates a worse convergence in coordinate space than in momentum space.

It is interesting to examine the level of uncertainties within the momentum-space approach or the coordinate-space approach by itself. Fig. 3 shows how the cross section changes when we vary the scale by a factor of

2 in each approach: from $\mu = q/\sqrt{2}$ to $\mu = \sqrt{2}q$ in Eq. (28) of [11] in the momentum-space approach (upper three curves), and from $\mu = \mu_2/\sqrt{2}$ to $\mu = \sqrt{2}\mu_2$ in Eq. (44) of [11] in the coordinate-space approach (lower three curves). For the momentum-space approach, the variation of the cross section is 2.2% at the peak and around 0.6% for c.m. energies above threshold. Meanwhile in the coordinate-space approach, the variation of the cross section amounts to 0.7% at the peak and 0.9% at larger c.m. energies. These results may be regarded as an internal consistency check for each approach.

So far we have examined the difference between the momentum-space potential and the coordinate-space potential in detail, and we have taken this difference as an estimate of the theoretical uncertainty of the QCD potential. This estimation may, however, be somewhat misleading. One might well argue that the difference is an artifact of our inadequate use of the perturbative expansion in describing the potentials related by Fourier transformation. For illustration, let us consider a hypothetical case where we know $\alpha_V(q)$ exactly. In this case, if we calculate $\bar{\alpha}_V(1/r)$ via (numerical) Fourier transformation of the momentum-space potential, in principle we can calculate $\bar{\alpha}_V(1/r)$ to any desired accuracy by investing more time. On the other hand, if we calculate $\bar{\alpha}_V(1/r)$ using the series on the right-hand side of Eq. (4) naively, there is a limitation in the achievable accuracy because the series is only asymptotic. Certainly such a limited accuracy does not reflect any theoretical uncertainty of $\bar{\alpha}_V(1/r)$. This nature should not be confused with our claim. We claim that there is a limitation within perturbative QCD in relating $\alpha_V(q)$ or $\bar{\alpha}_V(1/r)$ to $\alpha_{\overline{MS}}(M_Z^2)$, and estimate theoretical uncertainties in this relation using the difference of the potentials in the two spaces.

Let us comment on the relation between this work and the work [13] which has been completed very recently. Ref. [13] presents a fixed-order calculation (without log resummations), which includes the full second-order corrections to the $t\bar{t}$ threshold cross sections. While we include corresponding corrections only to the static QCD potential in our analyses, we also employ the renormalization-group improvement and thus resum large logarithms in the potential. In this sense the two works are complementary to each other. Both works give the common conclusion that the second-order corrections to the $t\bar{t}$ threshold cross section are large and may indicate a poor convergence of perturbative QCD, although they are based on qualitatively different arguments. The full set of the fixed-order $\mathcal{O}(\alpha_s^2)$ corrections to the cross section near threshold [13] are larger in size and even more scale dependent than the corrections to the potential alone. The theoretical uncertainty may therefore even be larger than indicated by the study in our paper.

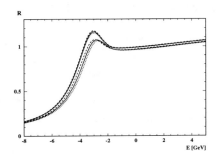

Figure 3: Comparison of the cross sections for different choices of the scale. The upper three curves are for the momentum-space approach: $\mu = q$ (solid), $\mu = \sqrt{2}q$ (dotted), and $\mu = q/\sqrt{2}$ (dashed line). The lower three curve are for the coordinate-space approach: $\mu = \mu_2 = \exp(-\gamma_E)/r$ (solid), $\mu = \sqrt{2}\mu_2$ (dotted), and $\mu = \mu_2/\sqrt{2}$ (dashed line).

Summary

- There is a difference between the potential constructed in momentum space and that constructed in coordinate space even at a fairly short-distance, $1/r \sim 100$ GeV. The difference can be understood within the framework of perturbative QCD. We already know that there is a large correction at $\mathcal{O}(\alpha_s^4)$ in the relation between the two potentials, although a consistent treatment is not possible until the full $\mathcal{O}(\alpha_s^4)$ corrections to the QCD potential are calculated.

- The above difference at short distances would limit the theoretical accuracy of the QCD potential at longer distances and thus provides a criterion for our present theoretical uncertainty of the $t\bar{t}$ cross section, $\delta\sigma_{\text{peak}}/\sigma_{\text{peak}} \gtrsim 6\%$.

- In addition, it seems that we are confronting the problem of the asymptoticness of the perturbative series in the calculation of the $t\bar{t}$ cross section, as the top quarks do not probe a region which is sufficiently deep in the potential. We may not be able to improve our theoretical precision even if the higher order corrections are calculated in perturbative QCD.

- We may, however, discuss which of the two approaches gives a more favorable result theoretically. Up to the second-order corrections, the perturbative series looks better convergent for the momentum-space potential than for the coordinate-space potential.

1. M. Jeżabek, J.H. Kühn, M. Peter, Y. Sumino and T. Teubner, *Phys. Rev.* **D58** (1998) 014006.
2. M.J. Strassler and M.E. Peskin, *Phys. Rev.* **D43** (1991) 1500.
3. V.S. Fadin and V.A. Khoze, *JETP Lett.* **46** (1987) 525; *Sov. J. Nucl. Phys.* **48** (1988) 309.
4. Y. Sumino, K. Fujii, K. Hagiwara, H. Murayama and C.-K. Ng, *Phys. Rev.* **D47** (1993) 56.
5. M. Jeżabek, J.H. Kühn and T. Teubner, *Z. Phys.* **C56** (1992) 653.
6. M. Jeżabek and T. Teubner, *Z. Phys.* **C59** (1993) 669.
7. M. Peter and Y. Sumino, University of Karlsruhe Report No. TTP97-27, hep-ph/9708223, to appear in *Phys. Rev.* **D**.
8. A.H. Hoang, *Phys. Rev.* **D56** (1997) 5851.
9. A.H. Hoang, P. Labelle and S.M. Zebarjad, *Phys. Rev. Lett.* **79** (1997) 3387.
10. G. Adkins, R.N. Fall and P.M. Mitrikov, *Phys. Rev. Lett.* **79** (1997) 3383.
11. M. Peter, *Phys. Rev. Lett.* **78** (1997) 602; *Nucl. Phys.* **B501** (1997) 471.
12. A. Czarnecki and K. Melnikov, University of Karlsruhe Report No. TTP97-54, hep-ph/9712222, and references therein.
13. A. Hoang and T. Teubner, University of California, San Diego Report No. UCSD/PTH 98-01, DESY 98-008, hep-ph/9801397.
14. M. Jeżabek, M. Peter and Y. Sumino, *Phys. Lett.* **B428** (1998) 352.
15. J.L. Richardson, *Phys. Lett.* **B82** (1979) 272.
16. M. Jeżabek, *Nucl. Phys.* (Proc. Suppl.) **37B** (1994) 197, and reference therein.

THE DEGREE OF INFRARED SAFETY OF QUARK-ANTIQUARK ANNIHILATION AT HIGH ENERGY TO ALL ORDERS

M. POLJŠAK

Jožef Stefan Institute, P.O.B. 3000, SI-1001 Ljubljana, Slovenia
E-mail: mat.poljsak@ijs.si

It is argued, using crossing symmetry and eikonal approximation, that the infrared divergent part of the Bloch-Nordsieck cross-section for quark-antiquark annihilation, calculated to all orders of perturbation theory, is suppressed at high c.m. energy $\sqrt{s}$ of the quark-antiquark pair by the same factor $O(m^4/s^2)$, where m is the quark mass, as at two-loops order.

1 Introduction

The cross-section for the unpolarized Drell-Yan (DY) process

$$A + B \to \ell\bar{\ell}(Q) + X,$$

may be written at large c.m. energy Q of the leptonic pair in a factorized form [1]

$$\sigma_{AB} = \sum_{ab} \int dx_a \, dx_b \, \Phi_{a/A}(x_a, \mu) \Phi_{b/B}(x_b, \mu)$$

$$\times \hat{\sigma}_{ab}\left(\frac{Q^2}{x_a x_b S}, \frac{Q}{\mu}, \alpha_S(\mu)\right)\left[1 + O\left(\left(\frac{\Lambda_{QCD}}{Q}\right)^2\right)\right].$$

Here S is the total c.m. energy of the incoming hadrons, functions $\Phi_{a/A}$, $\Phi_{b/B}$ describe the distributions of partons a, b inside incoming hadrons, $\hat{\sigma}_{ab}$ denotes a partonic cross-section, μ is the renormalization point, $\alpha_S(\mu) = \frac{g^2(\mu)}{4\pi}$ with $g(\mu)$ the QCD running coupling and Λ_{QCD} the QCD scale. The factorized form of the DY cross-section holds true, provided the partonic cross-sections are rendered free of infrared (IR) divergences by the Bloch-Nordsieck (BN) cancellation. This cancellation is known to be incomplete in QCD [2,3]: to fourth order of perturbative expansion in the strong coupling constant g, there is a subdominant IR divergence left over in the cross-section for the partonic process

$$q + \bar{q} \to \ell + \bar{\ell} + \text{soft gluons}, \qquad (1)$$

$$\hat{\sigma}_{q\bar{q}}^{(4)} = \sigma_B \alpha_S^2 C_F C_A \left(\frac{1}{\beta_L} - 1\right)\left(\frac{1}{2\beta_L} \ln \frac{1+\beta_L}{1-\beta_L} - 1\right)\frac{1}{d-4}$$
$$+ O(1), \quad d \to 4. \qquad (2)$$

Here, σ_B is the Born cross-section, C_F and C_A are the Casimir operators of the colour group SU(3) in fundamental and adjoint representation, respectively, d is the dimension of space-time continued analytically into the complex half-plane $\text{Re}\, d > 4$ in order to regularise IR divergences. The IR divergent part of the partonic cross-section $\hat{\sigma}_{q\bar{q}}$ depends also on the velocity of the antiquark, β_L, given in the frame of reference where the quark is at rest.

The IR divergent part of (2) behaves for small mass of the quark, m, like

$$\hat{\sigma}_{q\bar{q}}^{(4)} = O\left(m^4 \ln \frac{m^2}{s}\right), \quad m \to 0, \qquad (3)$$

where $s = x_a x_b S$ is the square of the c.m. energy of the incoming quark and antiquark, because the factor $(1/\beta_L - 1)$ in (2) is of the order $O(m^4/s^2)$ and the other factor depending on β_L of the order $O(\ln (m^2/s))$ while σ_B approaches a constant in the limit. So, to $O(g^4)$ the necessary condition for factorization of the DY cross-section is fulfilled at leading $(O(\ln m^2/s))$ and next-to-leading power $(O(m^2 \ln (m^2/s)/s), O(m^2/s))$ of the inverse of the large energy scale $\sqrt{s}$.

The question I ask in this paper is whether the same power of mass-suppression persits in the IR divergent part of $\hat{\sigma}_{q\bar{q}}^{(\infty)}$, i.e., to all orders of perturbative expansion in g.

The question has already been addressed by Frenkel, Gatheral and Taylor (FGT) [4]. They were able to establish suppression of $\hat{\sigma}_{q\bar{q}}^{(\infty)}$ by a factor of $O(\frac{m^2}{s})$. The question has also been dealt with by Bodwin [5] as well as Collins, Soper and Sterman [6]. Working to all orders of perturbation theory and assuming that the scale of hard scattering is much larger than all the masses, they showed not only that IR divergences in partonic cross-sections are suppressed by factors of $O(m^2/s)$, but also that there are no non-factorizing contributions to the leading-power DY cross-section coming from spectators. Factorization of the DY cross-section at leading power was proved to all orders also in a model for the incoming hadrons [7].

FGT [4] expressed their belief that suppression $O(\frac{m^4}{s^2})$ of IR divergences in the BN cross-section for quark-antiquark annihilation occurs to all orders. Basu, Ramalho and Sterman [8] suggested that not only do IR divergent contributions to the DY cross-section, but also IR finite contributions, factorise at next-to-leading power, $O(\frac{m^2}{s})$, as they do at leading power. Neither statement was substantiated, however, in these references.

Reasoning based on coherent state representation[9,10] of asymptotic states was also offered to suggest next-to-leading power factorization, but it seems to be inconclusive for two reasons: (i) it is doubtful whether the constructed initial states represent physical reality (cf., e.g., ref. [11]); (ii) the scattering amplitudes were expressed in terms of the quark velocity β_L as seen in the laboratory frame (which goes like $1 - O(\frac{m^4}{s^2})$ as $\frac{m^2}{s} \to 0$, in contradistinction to the quark velocity in the centre-of-mass frame, $\beta \equiv \beta_{CM} = 1 - O(\frac{m^2}{s})$, $\frac{m^2}{s} \to 0$) without paying due attention to the possibility that the scattering amplitude may not be an analytic function of β_L (This problem will be discussed in the following section).

An all-orders argument as to why the program of factorization can be extended to the level of next-to-leading power (but not to next-to-next-to-leading one) in many cases of inclusive cross-sections, including the DY one, was given by Qiu and Sterman [12]. Their argument implies that the answer to the question posed in the third paragraph of this section extends from $O(g^4)$ to all orders. The authors admit, however, that they "make no pretense of rigor", but believe that their "arguments are reasonably convincing, and compare favorably with the all-orders understanding of leading-power factorization in the 1980's". So, there seems to be some room for improvement upon their line of argument (based mainly on an analysis of the Landau equations), or for trying to explore alternative avenues to the same conclusions.

It is the purpose of this paper to point out the possibility of an alternative all-orders proof of the statement that IR divergences in the BN cross-section for quark-antiquark annihilation at high energy are suppressed by a factor of $O(\frac{m^4}{s^2})$.

The question of the degree of IR safety of quark-antiquark annihilation, in addition to being interesting for an exact proof of the DY factorization theorem, is of relevance also to recent advances in renormalon techniques. It has been shown [13,14] that single IR renormalon corrections contribute to the DY cross-section at $O((\Lambda_{QCD}/\sqrt{s})^k)$ with $k = 2$, i.e., at the same power where non-perturbative corrections (signalled in perturbation theory via uncancelled IR divergences [15]) would intervene if the quark-mass suppression factor of these IR divergences were not of $O(m^4/s^2)$ but only of $O(m^2/s)$.

2 Crossing symmetry and IR divergences

In search of the degree of IR safety of the Bloch-Nordsieck cross-section for quark-antiquark annihilation to all orders, we are interested in the dependence on the quark mass of the leading IR divergent part of the cross-section, which will be denoted by σ_{BN}^{IR}. This dependence is hidden in σ_B, which is known explicitly to be of $O(1)$ as

$m^2/s \to 0$, and in a function solely of

$$\beta_L = \frac{\sqrt{1 - 4m^2/s}}{1 - 2m^2/s} = 1 + O\left(\frac{m^4}{s^2}\right), \quad \frac{m^2}{s} \to 0. \quad (4)$$

The same degree of IR safety would therefore persist to higher than two-loop orders if σ_{BN}^{IR} were given by a sufficiently regular function of β_L. To get an idea of what should be understood by "sufficiently regular", observe that a term in σ_{BN}^{IR} proportional to $\sqrt{1 - \beta_L^2}$ is not acceptable, because it is of $O(m^2/s)$ as m^2/s tends to 0. Since this term is not an analytic function of β_L, one clue to the expected regularity is the analytic structure of σ_{BN}^{IR} as a function of β_L.

In order to learn something about the analytic properties of σ_{BN}^{IR} we may ask what should it look like as a function of the (anti)quark velocity in the c.m. frame of reference, $\beta = \sqrt{1 - 4m^2/s}$. This velocity is related to β_L by the formula $\beta_L = 2\beta/(1 + \beta^2)$. It is characteristic of this formula that its r.h.s. is invariant under the inversion

$$\beta \to 1/\beta. \quad (5)$$

The connection between analyticity and inversion (5) is that the latter is one of the conformal (more precisely, homographic) transformations of the complex β-plane. The question now arises: Should σ_{BN}^{IR} as a function of β be invariant under (5)?

As judged from the behaviour under (5) of the undesired term $\sqrt{1 - \beta_L^2}$ in σ_{BN}^{IR}, namely

$$\sqrt{1 - \beta_L^2} = \frac{1 - \beta^2}{1 + \beta^2} \xrightarrow{\beta \to \frac{1}{\beta}} \frac{\beta^2 - 1}{\beta^2 + 1} = -\sqrt{1 - \beta_L^2}, \quad (6)$$

as well as from the symmetry under (5) of the real part of the r.h.s. of eq. (2), the answer to the question should probably be in the affirmative. In the following, I am going to argue in favour of this suggestion on the grounds of crossing symmetry of scattering amplitudes.

First, I shall examine the BN cross-section for quark-antiquark annihilation to two-loop order. Using standard Feynman rules for QCD in the Coulomb gauge defined in the laboratory system where the quark is at rest, $p = (m, \mathbf{0})$, together with the eikonal approximation and averaging over spins of the quark and the antiquark, one can express the graph on fig. 1a algebraically in the form

$$\mathcal{A}_{eik} = 2ie^2 g^4 \beta_L^2 \, T_a T_b T_a T_b \, (p \cdot p' + 2m^2) \int \frac{d^3\mathbf{k}}{(2\pi)^3 |\mathbf{k}|}$$

$$\times \int \frac{d^4 k'}{(2\pi)^4 |\mathbf{k}'|^2} (1 - x^2) \frac{1}{k_0' - K' + i\frac{\epsilon}{p_0'}} \frac{1}{-k_0' + i\frac{\epsilon}{p_0}}$$

$$\times \frac{1}{-|\mathbf{k}| + K - i\frac{\epsilon}{p_0'}} \frac{1}{k_0' - K' - |\mathbf{k}| + K + i\frac{\epsilon}{p_0'}}, \quad (7)$$

where T_a are the generators of the gauge group SU(3) and the following definitions have been used

$$\beta_L = \frac{|\mathbf{p'}|}{|p_0'|}, \qquad K' = \frac{\mathbf{k'} \cdot \mathbf{p'}}{p_0'}, \qquad K = \frac{\mathbf{k} \cdot \mathbf{p'}}{p_0'} \equiv \beta_L x |\mathbf{k}|. \tag{8}$$

The diagram on fig. 1b gives

$$\mathcal{B}_{eik} = 2 e^2 g^4 \, T_a T_b T_a T_b \, (p \cdot p' + 2m^2) \int \frac{d^4 k}{(2\pi)^4} \frac{p_0 p_0'}{k^2 + i\epsilon}$$

$$\times \int \frac{d^4 k'}{(2\pi)^4 |\mathbf{k'}|^2} \left[-(\mathbf{p'})^2 + \frac{(\mathbf{p'} \cdot \mathbf{k})^2}{|\mathbf{k}|^2} \right] \frac{1}{p' \cdot k' + i\epsilon}$$

$$\times \frac{1}{-p \cdot k' + i\epsilon} \frac{1}{p' \cdot k + i\epsilon} \frac{1}{p' \cdot k' + p' \cdot k + i\epsilon}. \tag{9}$$

If on the r.h.s. of eq. (9) we perform integration over k_0, we get

$$\mathcal{B}_{eik} = 2 \, \mathrm{i} \, e^2 g^4 \beta_L^2 \, T_a T_b T_a T_b \, (p \cdot p' + 2m^2) \int \frac{d^3 \mathbf{k}}{(2\pi)^3 |\mathbf{k}|}$$

$$\times \int \frac{d^4 k'}{(2\pi)^4 |\mathbf{k'}|^2} (1 - x^2) \frac{1}{k_0' - K' + i\frac{\epsilon}{p_0'}} \frac{1}{-k_0' + i\frac{\epsilon}{p_0}}$$

$$\times \frac{1}{-|\mathbf{k}| - K + i\frac{\epsilon}{p_0'}} \frac{1}{k_0' - K' - |\mathbf{k}| - K + i\frac{\epsilon}{p_0'}}. \tag{10}$$

A comparison between eqs. (7) and (10) reveals that

$$\mathcal{A}_{eik} + \mathcal{A}_{eik}^* = \mathcal{B}_{eik} + \mathcal{B}_{eik}^*. \tag{11}$$

Therefore, $\sigma_{BN}^{IR} \propto \frac{1}{\beta_L} (\mathcal{A}_{eik} + \mathcal{A}_{eik}^* + \mathcal{B}_{eik} + \mathcal{B}_{eik}^*)$ is non-zero because either side of eq. (11) is non-zero (in fact, IR divergent).

Now, we shall verify that crossing of both the quark and the antiquark line (with $\ell\bar{\ell}$ lines) while leaving gluons intact has as a consequence that there are no IR divergences in the BN cross-section for the reaction

$$\ell + \bar{\ell} \to q + \bar{q} + \text{soft gluons}. \tag{12}$$

The (double) crossing transformation is effected by

$$p \to -p \quad \text{and} \quad p' \to -p'. \tag{13}$$

This transformation maps $\mathcal{A}_{eik}(p, p')$ into $\mathcal{A}_{eik}(-p, -p')$. From eq. (7) and definitions (8) we can see that the sole effect of the crossing transformation is to reverse the signs of imaginary parts of all denominators in $\mathcal{A}_{eik}(p, p')$. Because of the factor i (written as the second factor on the r.h.s. of eq. (7)), we can therefore write

$$\mathcal{A}_{eik}(-p, -p') = -\mathcal{A}_{eik}^*(p, p') \tag{14}$$

and consequently

$$\mathcal{A}_{eik}(-p, -p') + c.c. = -[\mathcal{A}_{eik}(p, p') + c.c.]. \tag{15}$$

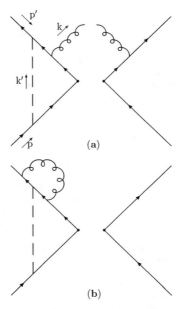

(a)

(b)

Figure 1: Real (a) and virtual (b) gluon contributions to the maximally non-abelian part of the cross-section for annihilation of a quark (having momentum p) and an antiquark (momentum p') into a virtual photon (denoted by a circle). The Coulomb propagator of a gluon is represented by a broken line, the transverse propagator by a curly, looping line. The right-hand parts of graphs (a) and (b) denote the complex conjugate of their pictures in a vertical mirror.

On the other hand, the crossing transformation (13) does not change the value of $\mathcal{B}_{eik}$ (eq. (9)). Therefore,

$$\mathcal{B}_{eik}(-p, -p') + c.c. = \mathcal{B}_{eik}(p, p') + c.c. \tag{16}$$

As a consequence of eqs. (15), (16) and (11), the IR divergent part of the BN cross-section for annihilation of a lepton pair into a quark, an antiquark and possibly soft gluons is vanishing: $\sigma_{BN}^{IR}(\ell + \bar{\ell} \to q + \bar{q} + \text{soft g}) \propto \frac{1}{\beta_L}[\mathcal{A}_{eik}(-p, -p') + \mathcal{B}_{eik}(-p, -p') + c.c.] = 0$.

3 Crossing symmetry implies symmetry under the inversion of quark velocity

From the expressions for $\mathcal{A}_{eik}$, eq. (7), $\mathcal{B}_{eik}$, eq. (10), it is evident that the ratio $\sigma_{BN}^{IR}/\sigma_B$ is a homogeneous function of the quark and antiquark momenta of degree zero if one ignores imaginary parts of denominators.

In deriving eqs. (7) and (10), I have employed the Coulomb gauge. On the other hand, if one wishes to establish the analytic properties of individual diagrams, one better uses a covariant gauge, because only then it might be simple to discern these properties. It is therefore important to notice that individual contributions to

1620

$\sigma_{BN}^{IR}/\sigma_B$ in a covariant gauge are also homogeneous of degree zero.

So far momenta of the quark and the antiquark were given in the laboratory system. For the discussion in this section it will be more appropriate to use the c.m. system. In this frame and in a covariant gauge, there are two further differences in how the expressions corresponding to $\mathcal{A}_{eik}$, eq. (7), and $\mathcal{B}_{eik}$, eq. (10), depend on p and p': (i) The denominator $-k_0' + \mathrm{i}\frac{\epsilon}{p_0}$ is replaced by the denominator $-k_0' - \frac{\mathbf{k}' \cdot \mathbf{p}}{p_0} + \mathrm{i}\frac{\epsilon}{p_0}$; (ii) The parameter β_L in eqs. (8) is replaced by β ($= \beta_{CM}$).

It is characteristic of the expressions so obtained that their dependence on p and p' is contained only in the parameter

$$\beta = \frac{|\mathbf{p}|}{|p_0|} = \frac{|\mathbf{p}'|}{|p_0'|} \qquad (17)$$

and in the imaginary parts of denominators, $\mathrm{i}\frac{\epsilon}{p_0}$ and $\mathrm{i}\frac{\epsilon}{p_0'}$. The values of the components of (on-shell) momenta p and p' are such that $\beta \in (0,1)$ and the imaginary parts of denominators are positive.

The question I would like to answer now is how to extend $\sigma_{BN}^{IR}/\sigma_B$ as a function of p and p' from the domain described above to the domain obtained by the crossing transformation $p' \to -p'$, $p \to p$. (The transformation $p' \to p'$, $p \to -p$ would be equally good for our purpose.)

There are two ways to do this:

(a) One can replace the parameter β in the expression for $\sigma_{BN}^{IR}/\sigma_B$, by the formula

$$\beta = \sqrt{1 - 4m^2/(p+p')^2}, \qquad (18)$$

which is equivalent to eq. (17) on the original domain of values of p and p', and then apply the crossing transformation referring to the momentum of the antiquark, i.e., $p' \to -p'$. As a result, the r.h.s. of eq. (18) is transformed into $\sqrt{1 - 4m^2/(p-p')^2}$. This expression satisfies, when momenta p and p' are on-shell, the following algebraic identity

$$\left[1 - 4m^2/(p-p')^2\right]^{\frac{1}{2}} = \left[1 - 4m^2/(p+p')^2\right]^{-\frac{1}{2}} \qquad (19)$$

whose r.h.s. equals $1/\beta$. Crossing transformation also changes the signs of the imaginary parts $\mathrm{i}\epsilon/p_0'$ of denominators.

(b) One can express the integrands as functions of $\frac{|\mathbf{p}|}{|p_0|}$, $\frac{|\mathbf{p}'|}{|p_0'|}$, $\mathrm{i}\frac{\epsilon}{p_0}$ and $\mathrm{i}\frac{\epsilon}{p_0'}$ in the same way as in the original domain, only now allowing the parameter p_0' to take negative values and lifting the constraints $\frac{|\mathbf{p}|}{|p_0|} < 1$, $\frac{|\mathbf{p}'|}{|p_0'|} < 1$.

When these two possible ways of analytic continuation of $\sigma_{BN}^{IR}/\sigma_B$ are considered together with the general theorem of complex analysis that analytic continuation of a function is unique, if it exists, we can draw the following conclusion: The ratio $\sigma_{BN}^{IR}/\sigma_B$, considered now

as a function of β taking values on the interval $(0, \infty)$, is even under the transformation (5), because the two proposed analytic extensions must coincide.

Two remarks are in order at this point:

1. As a check on the proposed symmetry to $O(g^4)$, we can verify from eq. (2) that the IR divergent part of the BN cross-section for the the process (1) (more precisely, the real part of $\sigma_{BN}^{IR}/\sigma_B(\beta)$) is indeed invariant under (5). Furthermore, we can avail ourselves of the calculation performed to the precisely required detail in ref.[16] to verify the symmetry also in the case of scattering of quarks off virtual photons,

$$q + \ell \to q + \ell + \text{soft gluons}. \qquad (20)$$

Although the published results for diagrams on fig. 2 in ref.[16] do not seem to be invariant under $\beta \to 1/\beta$ at first sight, when misprints are corrected[a] the symmetry is actually restored.

2. The argument presented above to $O(g^4)$ in favour of the statement that individual contributions (of pairs of Feynman diagrams or partial sums thereof) to the IR divergent part of the BN cross-section for processes (1), (12) and (20) with the Born cross-section factored out, are invariant under the inversion of the quark velocity in the c.m. frame depended upon the fact that they are homogeneous functions of quark momenta of order zero (for $d = 4$). It is important for an all-orders argument to recall that the condition of homogeneity is fulfilled also to higher orders of perturbative expansion in the strong coupling constant because of eikonal scaling (cf., e.g., ref.[4]). Consequently, the conclusion reached above also extends to all orders.

---

[a] As an example of typographycal error, notice that the value of the integral

$$\int_0^\pi \frac{(\sin \omega)^\epsilon d(\sin \omega)}{1 - \cos \omega} \qquad (21)$$

is quoted in ref.[16] (just before eq. (3.4) there) to be $\frac{2}{\epsilon} + 2\ln 2$, $\epsilon \to 0$, while in fact it equals minus this value. When this change in sign is taken into account, some terms in the expression for the sum of diagrams on figs. 2e and 2f also change sign so that the correct expression reads

$$d\sigma_c = -d\sigma_B \left[-\frac{g^4}{(2\pi)^2} \frac{(4\pi)^\epsilon}{[\Gamma(1 + \frac{1}{2}\epsilon)]^2} C_A C_F \frac{\Delta^{2\epsilon}}{\epsilon^2} \right] \left\{ -\left[\frac{2}{\epsilon} - 1 + 2\ln 2 \right] \right.$$
$$\times \left[\frac{1+\beta^2}{2\beta} \ln\left(\frac{1+\beta}{1-\beta}\right) - 1 \right] - \frac{1+\beta^2}{4\beta} \ln(1-\beta^2) \ln\left(\frac{1+\beta}{1-\beta}\right)$$
$$- \frac{1}{\beta} \ln\left(\frac{1+\beta}{1-\beta}\right) + \frac{1-\beta^2}{2\beta} \ln\left(\frac{1+\beta}{1-\beta}\right) + (1+\beta^2) \int_0^1 dx \frac{\ln(1-x^2)}{1 - \beta^2 x^2}$$
$$\left. + 2\ln 2 - 1 - \frac{(1+\beta^2)}{2} \int_0^1 dx \left[\frac{\ln(1-\beta x)}{1 + \beta x} + \frac{\ln(1+\beta x)}{1 - \beta x} \right] \right\} \qquad (22)$$

which can be verified, using the analytic properties of the dilogarithm function, to be invariant under $\beta \to \frac{1}{\beta}$.

4 How symmetry under the inversion of quark velocity constrains the degree of IR safety

Now, I would like to explore whether or not the symmetry under the inversion of quark velocity in any way restricts the quark mass dependence of the IR divergent part of the BN cross section for quark antiquark annihilation, σ_{BN}^{IR}. Because in this cross-section the quark mass always appears together with the square of the c.m. energy of the quark-antiquark pair, s, in the combination $z = \frac{4m^2}{s}$, from here onwards I shall trace the z-dependence of the cross-section instead of its m-dependence.

I shall assume that the IR divergent part of σ_{BN}^{IR} in general takes the following form as far as z-dependence is concerned:

$$\frac{\sigma_{BN}^{IR}}{\sigma_B}(z) = \text{Re} \sum_{n=0}^{\infty} z^n \left(c_{0n} + c_{1n}\ln z + c_{2n}\ln^2 z \right), \quad c_{00} = 0. \tag{23}$$

This particular form is suggested by the known z-dependence of low-order contributions of pairs of Feynman diagrams [2,3,4,17], as well as by the Froissart bound which restricts any cross-section to grow with increasing s not faster than $\ln^2 s$. The value $c_{00} = 0$ summarizes the result that there is no IR divergence present in the case of massless quarks [18]. I shall also assume that the coefficients c_{kn}, $k = 0, 1, 2$, are real, because the cross-section has to be real.

From the dependence of the (anti)quark velocity in the c.m. system on the quark mass, $\beta = \sqrt{1 - 4m^2/s} = \sqrt{1 - z}$, we can infer that the inversion of the (anti)quark velocity, eq. (5), is equivalent to the following transformation of the parameter z:

$$z \to -z/(1 - z). \tag{24}$$

The requirement that $\sigma_{BN}^{IR}/\sigma_B(z)$ be invariant under the transformation (24) can therefore be expressed by the equation:

$$\frac{\sigma_{BN}^{IR}}{\sigma_B}(z) = \frac{\sigma_{BN}^{IR}}{\sigma_B}\left(-\frac{z}{1-z}\right). \tag{25}$$

Since we are interested in the structure of $\sigma_{BN}^{IR}/\sigma_B(z)$ only for small values of z, we follow the z-dependence of the r.h.s. of eq. (25) only up to (and including) terms of $O(z^2)$ (modulo logarithmic factors of $\ln z$ and $\ln^2 z$), i.e.,

$$\text{Re}(\ln^2 z) \to \text{Re}(\ln^2 z) + (2z + z^2)\text{Re}(\ln z) - \pi^2 + z^2 + ...$$
$$\text{Re}(\ln z) \to \text{Re}(\ln z) + z + \tfrac{1}{2}z^2 + ...$$
$$z \to -z - z^2 + ... \tag{26}$$

Eq. (25) introduces, to the accuracy specified by the relations (26), the following constraints on the coeffi-

cients c_{kn} in eq. (23):

$$0 = -\pi^2 c_{20}, \; c_{10} = c_{10}, \; c_{20} = c_{20}, \; c_{21} = -c_{21}$$
$$c_{01} = -c_{01} + c_{10} + \pi^2 c_{21}, \; c_{11} = -c_{11} + 2c_{20}$$
$$c_{02} = c_{02} - c_{01} + \tfrac{1}{2}c_{10} + c_{20} + \pi^2 c_{21} - \pi^2 c_{22}$$
$$c_{12} = -c_{11} + c_{12} + c_{20} - 2c_{21}, \; c_{22} = c_{22} - c_{21} \tag{27}$$

From the known absence of the $\ln z$ term in $\sigma_{BN}^{IR}(z)$ to all orders [4], which in our notation amounts to the condition $c_{10} = 0$, and from eqs. (27) it follows that the coeficients c_{01}, c_{11}, c_{20}, c_{21} and c_{22} have to vanish, while the coefficients c_{12} and c_{02} are unconstrained.

This result makes it necessary that σ_{BN}^{IR} is suppressed at high energies by a factor of $O\left(m^4/s^2\right)$ to all orders, which is what I intended to demonstrate.

References

1. J.C. Collins, D.E. Soper and G. Sterman, in *Perturbative Quantum Chromodynamics*, ed. A.H. Mueller (World Scientific, Singapore 1989).
2. R. Doria, J. Frenkel and J.C. Taylor, *Nucl. Phys.* B **168**, 93 (1980) ; C. Di'Lieto *et al.*, *ibid.* **183**, 223 (1981).
3. A. Andraši *et al.*, *ibid.* **182**, 104 (1981).
4. J. Frenkel, J.G.M. Gatheral and J.C. Taylor, *Nucl. Phys.* B **233**, 307 (1984).
5. G. Bodwin, *Phys. Rev.* D **31**, 2616 (1985); *ibid.* **34**, 3932 (1986).
6. J.C. Collins, D.E. Soper and G. Sterman, *Nucl. Phys.* B **261**, 104 (1985); *ibid.* **308**, 833 (1988).
7. J. Frenkel, P.H. Sørensen and J.C. Taylor, *Z. Phys.* C **35**, 361 (1987);
   F.T. Brandt, J. Frenkel and J.C. Taylor, *Nucl. Phys.* B **312**, 589 (1989).
8. R. Basu, A.J. Ramalho and G. Sterman, *Nucl. Phys.* B **244**, 221 (1984).
9. Ch.A. Nelson, *Phys. Lett.* B **177**, 93 (1986).
10. S. Catani, M. Ciafaloni and G. Marchesini, *Phys. Lett.* B **168**, 284 (1986)
    Nucl. Phys. B **264**, 588 (1986).
11. (19 J.C. Taylor, *Phys. Rev.* D **54** , 2975 (1996).
12. J. Qiu and G. Sterman, *Nucl. Phys.* B **353**, 169 (1991).
13. M. Beneke and V.M. Braun, *Nucl. Phys.* B **454**, 253 (1995).
14. M. Beneke, *Renormalons*, hep-ph/9807443.
15. P.H. Sørensen and J.C. Taylor, *Nucl. Phys.* B **238**, 284 (1984).
16. J. Frenkel *et al. Nucl. Phys.* B **121**, 58 (1977).
17. M. Poljšak, *Phys. Lett.* B **388**, 813 (1996).
18. J. Frenkel, J.G.M. Gatheral and J.C. Taylor, *Nucl. Phys.* B **228**, 529 (1983);
    J.M.F. Labastida, *Nucl. Phys.* B **239**, 583 (1984).

PERTURBATIVE QCD- AND POWER-CORRECTED HADRON SPECTRA AND SPECTRAL MOMENTS IN THE DECAY $B \to X_s \ell^+ \ell^-$

A. ALI, G. HILLER

Deutsches Elektronen-Synchrotron DESY, Hamburg, Germany
E-mail: ali@x4u2.desy.de, ghiller@x4u2.desy.de

Leading order (in α_s) perturbative QCD and power ($1/m_b^2$) corrections to the hadronic invariant mass and hadron energy spectra in the decay $B \to X_s \ell^+ \ell^-$ are reviewed in the standard model using the heavy quark expansion technique (HQET). In particular, the first two hadronic moments $\langle S_H^n \rangle$ and $\langle E_H^n \rangle$, $n = 1, 2$, are presented working out their sensitivity on the HQET parameters λ_1 and $\bar{\Lambda}$. Data from the forthcoming B facilities can be used to measure the short-distance contribution in $B \to X_s \ell^+ \ell^-$ and determine the HQET parameters from the moments $\langle S_H^n \rangle$. This could be combined with complementary constraints from the decay $B \to X \ell \nu_\ell$ to determine these parameters precisely.

1 Introduction

The semileptonic inclusive decays $B \to X \ell^+ \ell^-$, where $\ell^\pm = e^\pm, \mu^\pm, \tau^\pm$ and X represents a system of light hadronic states, offer, together with the radiative electro-magnetic penguin decays $B \to X + \gamma$, presently the most popular testing grounds for the standard model (SM) in the flavour sector. In this contribution, we summarize the main steps in the derivation of the hadron spectra and hadron spectral moments in $B \to X_s \ell^+ \ell^-$ using perturbative QCD and the heavy quark expansion technique HQET [1,2,3], published recently by us [4,5]. This work, which incorporates the leading order (in α_s) perturbative QCD and power ($1/m_b^2$) corrections to the hadronic spectra, complements the derivation of the dilepton invariant mass spectrum and the forward-backward asymmetry of the charged lepton [6], calculated in the HQET framework some time ago by us in collaboration with T. Morozumi and L. Handoko [7]. (See, also Buchalla et al. [8].) Both the hadron and dilepton spectra are needed to distinguish the signal ($B \to X_s \ell^+ \ell^-$) from the background processes and in estimating the effects of the experimental selection criterion. We shall concentrate here on the short-distance contribution which can be extracted from data with the help of judicious cuts, such as those employed recently by the CLEO collaboration [9]. The residual effects from the resonant (long-distance) contributions have been studied in these distributions elsewhere [7,10], to which we refer for details and references to the earlier work.

We also underline the theoretical interest in measuring the first few hadronic spectral moments $\langle S_H^n \rangle$ and $\langle E_H^n \rangle$ ($n = 1, 2$). The former are sensitive to the HQET parameters $\bar{\Lambda}$ and λ_1; we work out this dependence numerically and argue that a combined analysis of the moments and spectra in $B \to X_s \ell^+ \ell^-$ and $B \to X \ell \nu_\ell$ will allow to determine the HQET parameters with a high precision. Since these parameters are endemic to a large class of phenomena in B decays, their precise knowledge is of great advantage in reducing the theoretical errors in the determination of the CKM matrix elements V_{td}, V_{ts}, V_{cb} and V_{ub}.

2 Kinematics and HQET Relations

We start with the definition of the kinematics of the decay at the parton level, $b(p_b) \to s(p_s)(+g(p_g)) + \ell^+(p_+) + \ell^-(p_-)$, where g denotes a gluon from the $O(\alpha_s)$ correction. The corresponding kinematics at the hadron level can be written as: $B(p_B) \to X_s(p_H) + \ell^+(p_+) + \ell^-(p_-)$. We define by q the momentum transfer to the lepton pair $q = p_+ + p_-$ and $s \equiv q^2$ is the invariant dilepton mass squared. We shall also need the variable u defined as $u \equiv -(p_b - p_+)^2 + (p_b - p_-)^2$. The hadronic invariant mass and the hadron energy in the final state is denoted by S_H and E_H, respectively; corresponding quantities at parton level are the invariant mass s_0 and the scaled parton energy $x_0 \equiv \frac{E_0}{m_b}$. From energy-momentum conservation, the following equalities hold in the b-quark, equivalently B-meson, rest frame ($v = (1, 0, 0, 0)$):

$$x_0 = 1 - v \cdot \hat{q}, \quad \hat{s}_0 = 1 - 2v \cdot \hat{q} + \hat{s},$$
$$E_H = m_B - v \cdot q, \quad S_H = m_B^2 - 2m_B v \cdot q + s , \quad (1)$$

where dimensionless variables with a hat are scaled by the b-quark mass, e.g., $\hat{s} = \frac{s}{m_b^2}$, $\hat{m}_s = \frac{m_s}{m_b}$ etc. Here, the 4-vector v denotes the velocity of both the b-quark and the B-meson, $p_b = m_b v$ and $p_B = m_B v$.

The relation between the B-meson and b-quark mass is given by the HQET mass relation $m_B = m_b + \bar{\Lambda} - 1/2m_b(\lambda_1 + 3\lambda_2) + \ldots$, where the ellipses denote terms higher order in $1/m_b$. The quantity λ_2 is known precisely from the $B^* - B$ mass difference, with $\lambda_2 \simeq 0.12$ GeV^2. The other two parameters are considerably uncertain at present [11,12] and are of interest here.

The hadronic variables E_H and S_H can be expressed in terms of the partonic variables x_0 and $\hat{s}_0$ by the following relations

$$E_H = \bar{\Lambda} - \frac{\lambda_1 + 3\lambda_2}{2m_B} + \left(m_B - \bar{\Lambda} + \frac{\lambda_1 + 3\lambda_2}{2m_B} \right) x_0 + \ldots,$$

$$S_H = m_s^2 + \bar{\Lambda}^2 + (m_B^2 - 2\bar{\Lambda}m_B + \bar{\Lambda}^2 + \lambda_1 + 3\lambda_2)(\hat{s}_0 - \hat{m}_s^2)$$
$$+ (2\bar{\Lambda}m_B - 2\bar{\Lambda}^2 - \lambda_1 - 3\lambda_2)x_0 + \ldots.$$

The dominant non-perturbative effect on the hadron spectra is essentially determined by the binding energy $\bar{\Lambda} = m_B - m_b + \ldots$, in terms of which one has the following transformation:

$$E_0 \to E_H = \bar{\Lambda} + E_0 + \ldots,$$
$$s_0 \to S_H = s_0 + 2\bar{\Lambda}E_0 + \bar{\Lambda}^2 + \ldots. \quad (2)$$

Thus, changing the variables of integration $(s_0, E_0) \to (s_H, E_0)$ and integrating over E_0 in the range $\sqrt{S_H} - \bar{\Lambda} < E_0 < 1/2m_B(S_H - 2\bar{\Lambda}m_B^2 + m_B^2)$, one gets an invariant hadron mass spectrum $d\Gamma/dS_H$ in the kinematic range $\bar{\Lambda}^2 < S_H < m_B^2$. In particular, already for the partonic decay $b \to s\ell^+\ell^-$ with $m_s = 0$, and hence $s_0 = 0$, one gets a non-trivial distribution in S_H for $\bar{\Lambda}^2 < S_H < \bar{\Lambda}m_B$. The kinematic boundary of the distribution $d\Gamma/dS_H$ is extended by the bremsstrahlung process $b \to s + g + \ell^+\ell^-$, where now $\bar{\Lambda}m_B < S_H < m_B^2$ (with $m_s = 0$). The $\mathcal{O}(\alpha_s)$ contribution leads to a double logarithmic (but integrable) singularity at $S_H = \bar{\Lambda}m_B$. Perturbation theory is valid for $\Delta^2 < S_H < m_B^2$, with $\Delta^2 > \bar{\Lambda}m_B$.

3 Matrix Element for $B \to X_s\ell^+\ell^-$ in the Effective Hamiltonian Approach

The effective Hamiltonian governing the decay $B \to X_s\ell^+\ell^-$ is given as [7]:

$$\mathcal{H}_{eff}(b \to s) = -\frac{4G_F}{\sqrt{2}} V_{ts}^* V_{tb} \quad (3)$$

$$\left[\sum_{i=1}^{6} C_i(\mu)O_i + C_7(\mu)\frac{e}{16\pi^2}\bar{s}_\alpha\sigma_{\mu\nu}(m_bR + m_sL)b_\alpha F^{\mu\nu} \right.$$

$$\left. + C_9(\mu)\frac{e^2}{16\pi^2}\bar{s}_\alpha\gamma^\mu Lb_\alpha\bar{\ell}\gamma_\mu\ell + C_{10}\frac{e^2}{16\pi^2}\bar{s}_\alpha\gamma^\mu Lb_\alpha\bar{\ell}\gamma_\mu\gamma_5\ell \right],$$

where G_F is the Fermi coupling constant, $L(R) = 1/2(1 \mp \gamma_5)$, and C_i are the Wilson coefficients. Note that the chromo-magnetic operator does not contribute to the decay $B \to X_s\ell^+\ell^-$ in the approximation which we use here.

The matrix element for the decay $B \to X_s\ell^+\ell^-$ can be factorized into a leptonic and a hadronic part as

$$\mathcal{M}(B \to X_s\ell^+\ell^-) = \frac{G_F\alpha}{\sqrt{2}\pi} V_{ts}^* V_{tb} \left(\Gamma^L{}_\mu L^{L\mu} + \Gamma^R{}_\mu L^{R\mu} \right), \quad (4)$$

with

$$L^{L/R}{}_\mu \equiv \bar{\ell}\gamma_\mu L(R)\ell, \quad (5)$$

$$\Gamma^{L/R}{}_\mu \equiv \bar{s} \left[R\gamma_\mu \left(C_9^{\text{eff}}(\hat{s}) \mp C_{10} + 2C_7^{\text{eff}}\frac{\hat{\slashed{A}}}{\hat{s}} \right) \right.$$
$$\left. + 2\hat{m}_s C_7^{\text{eff}}\gamma_\mu\frac{\hat{\slashed{A}}}{\hat{s}}L \right] b, \quad (6)$$

with $C_7^{\text{eff}} \equiv C_7 - C_5/3 - C_6$. The effective Wilson coefficient $C_9^{\text{eff}}(\hat{s})$ receives contributions from various pieces. The resonant $c\bar{c}$ states also contribute to $C_9^{\text{eff}}(\hat{s})$; hence the contribution given below is just the perturbative part:

$$C_9^{\text{eff}}(\hat{s}) = C_9\eta(\hat{s}) + Y(\hat{s}). \quad (7)$$

Here $\eta(\hat{s})$ and $Y(\hat{s})$ represent, respectively, the $\mathcal{O}(\alpha_s)$ correction [13] and the one loop matrix element of the Four-Fermi operators [14,15].

With the help of the above expressions, the differential decay width becomes on using $p_\pm = (E_\pm, \boldsymbol{p}_\pm)$,

$$d\Gamma = \frac{1}{2m_B}\frac{G_F^2\alpha^2}{2\pi^2}|V_{ts}^* V_{tb}|^2 \frac{d^3\boldsymbol{p}_+}{(2\pi)^3 2E_+}\frac{d^3\boldsymbol{p}_-}{(2\pi)^3 2E_-}$$
$$\times \left(W^L{}_{\mu\nu} L^{L\mu\nu} + W^R{}_{\mu\nu} L^{R\mu\nu} \right), \quad (8)$$

where $W_{\mu\nu}^{L,R}$ and $L_{\mu\nu}^{L,R}$ are the hadronic and leptonic tensors, respectively, and can be seen in the literature [7]. The hadronic tensor $W_{\mu\nu}^{L/R}$ is related to the discontinuity in the forward scattering amplitude, denoted by $T_{\mu\nu}^{L/R}$, through the relation $W_{\mu\nu}^{L/R} = 2\operatorname{Im}T_{\mu\nu}^{L/R}$. Transforming the integration variables to $\hat{s}$, $\hat{u}$ and $v \cdot q$, one can express the triple differential distribution in $B \to X_s\ell^+\ell^-$ as:

$$\frac{d\Gamma}{d\hat{u}\,d\hat{s}\,d(v \cdot q)} = \frac{1}{2m_B}\frac{G_F^2\alpha^2}{2\pi^2}\frac{m_b^4}{256\pi^4}|V_{ts}^* V_{tb}|^2$$
$$\times 2\operatorname{Im}\left(T^L{}_{\mu\nu} L^{L\mu\nu} + T^R{}_{\mu\nu} L^{R\mu\nu} \right). \quad (9)$$

Using Lorentz decomposition, the tensor $T_{\mu\nu}$ can be expanded in terms of three structure functions T_i,

$$T_{\mu\nu}^{L/R} = -T_1^{L/R} g_{\mu\nu} + T_2^{L/R} v_\mu v_\nu + T_3^{L/R} i\epsilon_{\mu\nu\alpha\beta} v^\alpha \hat{q}^\beta, \quad (10)$$

where the ones which do not contribute to the amplitude for massless leptons have been neglected.

4 Hadron Spectra in $B \to X_s\ell^+\ell^-$

We discuss first the perturbative $O(\alpha_s)$ corrections recalling that only the matrix element of the operator $O_9 \equiv e^2/(16\pi^2)\bar{s}_\alpha\gamma^\mu Lb_\alpha\bar{\ell}\gamma_\mu\ell$ is subject to such corrections. The corrected hadron energy spectrum in $B \to$

$X_s\ell^+\ell^-$ can be obtained by using the existing results in the literature on the decay $B \to X\ell\nu_\ell$ by decomposing the vector current in O_9 as $V = (V-A)/2 + (V+A)/2$. The $(V-A)$ and $(V+A)$ currents yield the same hadron energy spectrum [16] and there is no interference term present in this spectrum for massless leptons. So, the correction for the vector current case can be taken from the corresponding result for the charged $(V-A)$ case [13].

The $\mathcal{O}(\alpha_s)$ perturbative QCD correction for the hadronic invariant mass is discussed next. As already mentioned, the decay $b \to s + \ell^+ + \ell^-$ yields a delta function at $\hat{s}_0 = \hat{m}_s^2$ and hence only the bremsstrahlung diagrams $b \to s + g + \ell^+ + \ell^-$ contribute in the range $\hat{m}_s^2 < \hat{s}_0 \leq 1$. The resulting distribution $d\mathcal{B}(B \to X_s\ell^+\ell^-)/ds_0$ in the parton model in the $\mathcal{O}(\alpha_s)$ approximation and the Sudakov exponentiated form can be seen in our paper [5]. We remark that the Sudakov exponentiated double differential distribution for the decay $B \to X_u\ell\nu_\ell$ has been derived by Greub and Rey [17], which we have checked and used after changing the normalization for $B \to X_s\ell^+\ell^-$. The hadronic invariant mass spectrum $d\mathcal{B}(B \to X_s\ell^+\ell^-)/dS_H$, shown in Fig. 1 depends rather sensitively on m_b (or equivalently $\bar{\Lambda}$). An analogous analysis for the decay $B \to X_u\ell\nu_\ell$ has been performed earlier, with very similar qualitative results [18].

Next, we discuss the power corrections to the hadronic spectra. The structure functions $T_i^{L/R}$ in the hadronic tensor in Eq. (10) can be expanded in inverse powers of m_b with the help of the HQET techniques [1,2,3]. The leading term in this expansion, i.e., $\mathcal{O}(m_b^0)$, reproduces the parton model result [14,15]. In HQET, the next to leading power corrections are parameterized in terms of λ_1 and λ_2. After contracting the hadronic and leptonic tensors and with the help of the kinematic identities given in Eq. (1), we can make the dependence on x_0 and $\hat{s}_0$ explicit,

$$T^{L/R}{}_{\mu\nu} L^{L/R\mu\nu} = m_b{}^2 \left\{ 2(1 - 2x_0 + \hat{s}_0) T_1{}^{L/R} + \left[x_0^2 - \frac{1}{4}\hat{u}^2 - \hat{s}_0 \right] T_2{}^{L/R} \mp (1 - 2x_0 + \hat{s}_0)\hat{u} \, T_3{}^{L/R} \right\}.$$
(11)

By integrating Eq. (9) over $\hat{u}$, the double differential power corrected spectrum can be expressed as [5]:

$$\frac{d^2\mathcal{B}}{dx_0 \, d\hat{s}_0} = -\frac{8}{\pi} \mathcal{B}_0 \text{Im} \sqrt{x_0^2 - \hat{s}_0} \left\{ (1 - 2x_0 + \hat{s}_0) T_1(\hat{s}_0, x_0) + \frac{x_0^2 - \hat{s}_0}{3} T_2(\hat{s}_0, x_0) \right\} + \mathcal{O}(\lambda_i \alpha_s).$$
(12)

The structure function T_3 does not contribute to the double differential distribution and we do not consider it any further. The functions $T_1(\hat{s}_0, x_0)$ and $T_2(\hat{s}_0, x_0)$, together

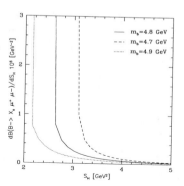

Figure 1: The differential branching ratio $d\mathcal{B}(B \to X_s\ell^+\ell^-)/dS_H$ in the hadronic invariant mass, S_H, shown for three values of m_b in the range where only bremsstrahlung diagrams contribute.

with other details of the calculations, have been given by us elsewhere [5].

The branching ratio for $B \to X_s\ell^+\ell^-$ is usually expressed in terms of the measured semileptonic branching ratio $\mathcal{B}_{sl}$ for the decay $B \to X_c\ell\nu_\ell$. This fixes the normalization constant $\mathcal{B}_0$ to be,

$$\mathcal{B}_0 \equiv \mathcal{B}_{sl} \frac{3 \alpha^2}{16\pi^2} \frac{|V_{ts}^* V_{tb}|^2}{|V_{cb}|^2} \frac{1}{f(\hat{m}_c)\kappa(\hat{m}_c)} ,$$
(13)

where $f(\hat{m}_c)$ is the phase space factor for $\Gamma(B \to X_c\ell\nu_\ell)$ and $\kappa(\hat{m}_c)$ accounts for both the $O(\alpha_s)$ QCD correction to the semileptonic decay width [19] and the leading order $(1/m_b)^2$ power correction [1]. The hadron energy spectrum can now be obtained by integrating over $\hat{s}_0$ with the kinematic boundaries: $max(\hat{m}_s^2, -1 + 2x_0 + 4\hat{m}_l^2) \leq \hat{s}_0 \leq x_0^2$, $\hat{m}_s \leq x_0 \leq \frac{1}{2}(1 + \hat{m}_s^2 - 4\hat{m}_l^2)$. The hadron energy spectrum $d\mathcal{B}(B \to X_s\ell^+\ell^-)/dE_0$ in the parton model (dotted line) and including leading power corrections (solid line) are shown in Fig. 2. For $m_b/2 < E_0 < m_b$ the two distributions coincide. Note that the $1/m_b^2$-expansion breaks down near the lower end-point of the hadron energy spectrum and at the $c\bar{c}$ threshold. Hence, only suitably averaged spectra are useful for comparison with experiments in these regions. Apart from these regions, the HQET and parton model spectra are remarkably close to each other.

5 Hadron Spectral Moments in $B \to X_s\ell^+\ell^-$

The lowest spectral moments in the decay $B \to X_s\ell^+\ell^-$ at the parton level are worked out by taking into account the two types of corrections discussed earlier, namely the leading power $1/m_b$ and the perturbative

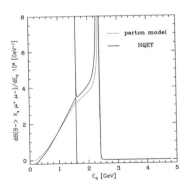

Figure 2: Hadron energy spectrum $d\mathcal{B}(B \to X_s \ell^+ \ell^-)/dE_0$ in the parton model (dotted line) and including leading power corrections (solid line). For $m_b/2 < E_0 < m_b$ the two distributions coincide.

$\mathcal{O}(\alpha_s)$ corrections. To that end, we define the moments for integers n and m:

$$\mathcal{M}_{l^+l^-}^{(n,m)} \equiv \frac{1}{\mathcal{B}_0} \int (\hat{s}_0 - \hat{m}_s^2)^n x_0^m \frac{d\mathcal{B}}{d\hat{s}_0 dx_0} d\hat{s}_0 dx_0 , \quad (14)$$

which obey $\langle x_0^m (\hat{s}_0 - \hat{m}_s^2)^n \rangle = \frac{B_0}{B} \mathcal{M}_{l^+l^-}^{(n,m)}$. These moments can be expanded as a double Taylor series in α_s and $1/m_b$:

$$\mathcal{M}_{l^+l^-}^{(n,m)} = D_0^{(n,m)} + \frac{\alpha_s}{\pi} C_9^2 A^{(n,m)}$$
$$+ \hat{\lambda}_1 D_1^{(n,m)} + \hat{\lambda}_2 D_2^{(n,m)} , \quad (15)$$

with a further decomposition of $D_i^{(n,m)}$, $i = 0, 1, 2$, into pieces from different Wilson coefficients:

$$D_i^{(n,m)} = \alpha_i^{(n,m)} C_7^{\text{eff}2} + \beta_i^{(n,m)} C_{10}^2 + \gamma_i^{(n,m)} C_7^{\text{eff}} + \delta_i^{(n,m)}. \quad (16)$$

The terms $\gamma_i^{(n,m)}$ and $\delta_i^{(n,m)}$ in Eq. (16) result from the terms proportional to $Re(C_9^{\text{eff}}) C_7^{\text{eff}}$ and $|C_9^{\text{eff}}|^2$ in Eq. (12), respectively. The explicit expressions for $\alpha_i^{(n,m)}, \beta_i^{(n,m)}, \gamma_i^{(n,m)}, \delta_i^{(n,m)}$ are given in our paper [5].

The leading perturbative contributions for the hadronic invariant mass and hadron energy moments can be obtained analytically,

$$A^{(0,0)} = \frac{25 - 4\pi^2}{9} , \quad A^{(1,0)} = \frac{91}{675} , \quad A^{(2,0)} = \frac{5}{486} ,$$
$$A^{(0,1)} = \frac{1381 - 210\pi^2}{1350} , A^{(0,2)} = \frac{2257 - 320\pi^2}{5400} . \quad (17)$$

The zeroth moment $n = m = 0$ is needed for the normalization; the result for $A^{(0,0)}$ was first derived by Cabibbo

and Maiani [19]. Likewise, the first mixed moment $A^{(1,1)}$ can be extracted from the results for the decay $B \to X \ell \nu_\ell$ [20] after changing the normalization, $A^{(1,1)} = 3/50$. For the lowest order parton model contribution $D_0^{(n,m)}$, we find, in agreement with [20], that the first two hadronic invariant mass moments $\langle \hat{s}_0 - \hat{m}_s^2 \rangle$, $\langle (\hat{s}_0 - \hat{m}_s^2)^2 \rangle$ and the first mixed moment $\langle x_0(\hat{s}_0 - \hat{m}_s^2) \rangle$ vanish: $D_0^{(n,0)} = 0$, for $n = 1, 2$ and $D_0^{(1,1)} = 0$.

Using the expressions for the HQET moments derived by us [5], we present the numerical results for the hadronic moments in $B \to X_s \ell^+ \ell^-$. The parameters used are : $m_s = 0.2$ GeV, $m_c = 1.4$ GeV, $m_b = 4.8$ GeV, $m_t = 175 \pm 5$ GeV, $\mu = m_b{}^{+m_b}_{-m_b/2}, \alpha_s(m_Z) = 0.117 \pm 0.005$, $\alpha^{-1} = 129$. We find for the short-distance hadronic moments, valid up to $\mathcal{O}(\alpha_s/m_B^2, 1/m_B^3)$:

$$\langle S_H \rangle = m_B^2 \Big(\frac{m_s^2}{m_B^2} + 0.093 \frac{\alpha_s}{\pi} - 0.069 \frac{\bar{\Lambda}}{m_B} \frac{\alpha_s}{\pi}$$
$$+ 0.735 \frac{\bar{\Lambda}}{m_B} + 0.243 \frac{\bar{\Lambda}^2}{m_B^2} + 0.273 \frac{\lambda_1}{m_B^2} - 0.513 \frac{\lambda_2}{m_B^2} \Big) ,$$

$$\langle S_H^2 \rangle = m_B^4 \Big(0.0071 \frac{\alpha_s}{\pi} + 0.138 \frac{\bar{\Lambda}}{m_B} \frac{\alpha_s}{\pi}$$
$$+ 0.587 \frac{\bar{\Lambda}^2}{m_B^2} - 0.196 \frac{\lambda_1}{m_B^2} \Big) , \quad (18)$$

$$\langle E_H \rangle = 0.367 m_B \Big(1 + 0.148 \frac{\alpha_s}{\pi} - 0.352 \frac{\bar{\Lambda}}{m_B} \frac{\alpha_s}{\pi} + 1.691 \frac{\bar{\Lambda}}{m_B}$$
$$+ 0.012 \frac{\bar{\Lambda}^2}{m_B^2} + 0.024 \frac{\lambda_1}{m_B^2} + 1.070 \frac{\lambda_2}{m_B^2} \Big) ,$$

$$\langle E_H^2 \rangle = 0.147 m_B^2 \Big(1 + 0.324 \frac{\alpha_s}{\pi} - 0.128 \frac{\bar{\Lambda}}{m_B} \frac{\alpha_s}{\pi} + 2.954 \frac{\bar{\Lambda}}{m_B}$$
$$+ 2.740 \frac{\bar{\Lambda}^2}{m_B^2} - 0.299 \frac{\lambda_1}{m_B^2} + 0.162 \frac{\lambda_2}{m_B^2} \Big) ,$$

where the numbers shown correspond to the central values of the parameters.

The dependence of the hadronic moments given in Eq. (18) on the HQET parameters λ_1 and $\bar{\Lambda}$ has been worked out numerically. In doing this, the theoretical errors on these moments following from the errors on the input parameters m_t, α_s and the scale μ have been estimated by varying these parameters in the indicated $\pm 1\sigma$ ranges, one at a time, and adding the individual errors in quadrature. The correlations on the HQET parameters λ_1 and $\bar{\Lambda}$ which follow from (assumed) fixed values of the hadronic invariant mass moments $\langle S_H \rangle$ and $\langle S_H^2 \rangle$ (calculated using $\bar{\Lambda} = 0.39$ GeV, $\lambda_1 = -0.2$ GeV^2 and $\lambda_2 = 0.12$ GeV^2) are shown in Fig. 3 (for the decay $B \to X_s \mu^+ \mu^-$). The $(\lambda_1 - \bar{\Lambda})$ correlation from the analysis of Gremm et al. [11] for the electron energy spectrum in $B \to X \ell \nu_\ell$ is shown as an ellipse in this figure. With the measurements of $\langle S_H \rangle$ and $\langle S_H^2 \rangle$ in the

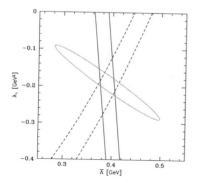

Figure 3: $\langle S_H \rangle$ (solid bands) and $\langle S_H^2 \rangle$ (dashed bands) correlation in (λ_1-$\bar{\Lambda}$) space for the decay $B \to X_s \ell^+ \ell^-$. The correlation from the analysis of the decay $B \to X \ell \nu_\ell$ by Gremm et al.[11] is shown as an ellipse.

decay $B \to X_s \ell^+ \ell^-$, one has to solve the experimental numbers on the l.h.s. of Eq. (18) for λ_1 and $\bar{\Lambda}$. It is, however, clear that the constraints from the decays $B \to X_s \ell^+ \ell^-$ and $B \to X \ell \nu_\ell$ are complementry. Using the CLEO cuts on hadronic and dileptonic masses[9], we estimate that $O(200)$ $B \to X_s \ell^+ \ell^- (\ell = e, \mu)$ events will be available per 10^7 $B\bar{B}$ hadrons[5]. So, there will be plenty of $B \to X_s \ell^+ \ell^-$ decays in the forthcoming B facilities to measure the correlation shown in Fig. 3.

Of course, the utility of the hadronic moments calculated above is only in conjunction with the experimental cuts which could effectively remove the resonant (long-distance) contributions. The optimal experimental cuts in $B \to X_s \ell^+ \ell^-$ remain to be defined, but for the cuts used by the CLEO collaboration we have studied the effects in the HQET-like Fermi motion (FM) model[21]. We find that the hadronic moments in the HQET and FM model are very similar and CLEO-type cuts remove the bulk of the $c\bar{c}$ resonant contributions[5].

In summary, we have calculated the dominant contributions to the hadron spectra and spectral moments in $B \to X_s \ell^+ \ell^-$ including contributions up to terms of $\mathcal{O}(\alpha_s/m_B^2, 1/m_B^3)$. We have presented the results on the spectral hadronic moments $\langle E_H^n \rangle$ and $\langle S_H^n \rangle$ for $n = 1, 2$ and have worked out their dependence on the HQET parameters $\bar{\Lambda}$ and λ_1. The correlations in $B \to X_s \ell^+ \ell^-$ are shown to be different than the ones in the semileptonic decay $B \to X \ell \nu_\ell$. This complementarity allows, in principle, a powerful method to determine them precisely from data on $B \to X \ell \nu_\ell$ and $B \to X_s \ell^+ \ell^-$ in forthcoming high luminosity B facilities.

References

1. J. Chay, H. Georgi and B. Grinstein, *Phys. Lett.* B **247**, 399 (1990); I.I. Bigi, N.G. Uraltsev and A.I. Vainshtein, *Phys. Lett.* B **293**, 430 (1992) [E. B**297**, 477 (1993)]; I.I. Bigi *et al.*, *Phys. Rev. Lett.* **71**, 496 (1993); B. Blok *et al.*, *Phys. Rev.* D **49**, 3356 (1994) [E. D **50**, 3572 (1994)].

2. A. Manohar and M. B. Wise, *Phys. Rev.* D **49**, 1310 (1994).

3. A. F. Falk, M. Luke and M. J. Savage, *Phys. Rev.* D **49**, 3367 (1994).

4. A. Ali and G. Hiller, *Phys. Rev.* D **58**, 071501 (1998).

5. A. Ali and G. Hiller, *Phys. Rev.* D **58**, 074001 (1998).

6. A. Ali, T. Mannel and T. Morozumi, *Phys. Lett.* B **273**, 505 (1991).

7. A. Ali, L. T. Handoko, G. Hiller and T. Morozumi, *Phys. Rev.* D **55**, 4105 (1997).

8. G. Buchalla, G. Isidori and S. -J. -Rey, *Nucl. Phys.* B **511**, 594 (1998).

9. S. Glenn et al. (CLEO Collaboration), *Phys. Rev. Lett.* **80**, 2289 (()1998.

10. A. Ali and G. Hiller, preprint DESY 98-031, hep-ph/9807418 (submitted to Phys. Rev. D).

11. M. Gremm, A. Kapustin, Z. Ligeti and M.B. Wise, *Phys. Rev. Lett.* **77**, 20 (1996).

12. M. Neubert, preprint CERN-TH/98-2, hep-ph/9801269; to be published in Proc. of the Int. Europhys. Conf. on High Energy Physics, Jerusalem, Israel, 19 - 26 August, 1997.

13. A. Czarnecki, M. Ježabek and J. H. Kühn, Acta. Phys. Pol. B20, 961 (1989); M. Ježabek and J. H. Kühn, *Nucl. Phys.* B **320**, 20 (1989).

14. A. J. Buras and M. Münz, *Phys. Rev.* D **D52**, 186 (1995).

15. M. Misiak, *Nucl. Phys.* B **393**, 23 (1993) [E. B**439**, 461 (1995)].

16. A. Ali, *Z. Phys.* C **1**, 25 (1979).

17. C. Greub and S.-J. Rey, *Phys. Rev.* D **56**, 4250 (1997).

18. A.F. Falk, Z. Ligeti and M.B. Wise, *Phys. Lett.* B **406**, 225 (1997).

19. N. Cabibbo and L. Maiani, *Phys. Lett.* B **79**, 109 (1978); Y. Nir, *Phys. Lett.* B **221**, 184 (1989).

20. A. F. Falk, M. Luke and M. J. Savage, *Phys. Rev.* D **53**, 2491 (1996).

21. A. Ali and E. Pietarinen, *Nucl. Phys.* B **B154**, 519 (1979); G. Altarelli et al., *Nucl. Phys.* B **208**, 365 (1982).

Dokshitzer-Gribov-Lipatov-Altarelli-Parisi Evolution and the Renormalization Group Improved Yennie-Frautschi-Suura Theory in QCD [a]

B.F.L. Ward

Department of Physics and Astronomy, University of Tennessee,
Knoxville, Tennessee 37996-1200, USA
SLAC, Stanford University, Stanford, California 94309
CERN, Theory Division, CH-1211 Geneva 23, Switzerland

S. Jadach

Institute of Nuclear Physics, ul. Kawiory 26a, PL 30-059 Cracow, Poland
CERN, Theory Division, CH-1211 Geneva 23, Switzerland

We show that the recently derived renormalization group improved Yennie-Frautschi-Suura (YFS) exponentiation of soft ($k_o \to 0$) gluons in QCD is fully compatible with the usual Dokshitzer-Gribov-Lipatov-Altarelli-Parisi (DGLAP) evolution of the structure functions of hadrons in the respective hadron-hadron hard interactions. We show how to implement the YFS exponentiation without double or over counting effects already implied by the DGLAP equation. In this way, we arrive at a theory which allows for the development of realistic, multiple gluon Monte Carlo event generators for hard hadron-hadron scattering processes in which the DGLAP evolved structure functions are correctly synthesized with the respective YFS exponentiated soft gluon effects in a rigorous way.

An important step in the implementation of the recently derived [1] renormalization group improved soft Yennie-Frautschi-Suura (YFS)[2] gluon exponentiation in QCD is the proper synthesization of these new infrared (IR) effects (where by infrared we do mean the gluon energy $k_o \to 0$ limit as opposed to the soft gluon regime analyzed by Dokshitzer *et al.*[3] in which $k_\perp \geq \frac{1}{R}$, $R \simeq$ hadron radius $\simeq 1/(200 MeV) \simeq \frac{1}{\Lambda_{QCD}}$) with the required Dokshitzer-Gribov-Lipatov-Altarelli-Parisi (DGLAP) [4] evolution of the respective hadron structure functions in computing any particular hard hadron-hadron scattering cross section, such as $p + \bar{p} \to t\bar{t} + X$, for example. Indeed, a naive use of the results in Ref. [1] would lead to double or over counting some of the same effects, rendering the precision of any such calculated cross section questionable at best. In what follows, we will show how one properly synthesizes the DGLAP structure function evolution with renormalization group improved soft gluon YFS QCD exponentiation derived in Ref. [1]. In this way, we develope the necessary remaining theoretical apparatus needed for the construction of realistic hadron-hadron hard scattering Monte Carlo event generators in which finite $P_\perp$ multiple soft gluon effects are taken into account at the level of the hard scattering subprocess amplitude while the proper structure function evolution is realized, on an event-by-event basis. Such a Monte Carlo event generator will appear elsewhere [5].

Specifically, let us recall the basic result from Ref. [1] for the YFS exponentiated differential cross section for the process $q + {}^{(}\bar{q}{}^{)\prime} \to q'' + {}^{(}\bar{q}{}^{)\prime\prime\prime} + n(G)$, where G is a soft gluon,

$$d\hat{\sigma}_{exp} = e^{\text{SUM}_{\text{IR}}(\text{QCD})} \sum_{n=0}^{\infty} \frac{1}{n!} \int \prod_{j=1}^{n} \frac{d^3 k_j}{k_j} \int \frac{d^4 y}{(2\pi)^4}$$
$$e^{+iy(p_1+p_2-q_1-q_2-\sum_j k_j)+D_{QCD}}$$
$$\times \bar{\beta}_n(k_1, \ldots, k_n) \frac{d^3 q_1 d^3 q_2}{q_1^0 q_2^0} \quad (1)$$

where

$$D_{QCD} = \int \frac{d^3 k}{k_0} \tilde{S}_{\text{QCD}} \left[e^{-iy \cdot k} - \theta(K_{max} - |\vec{k}|) \right] \quad . \quad (2)$$

Here, we have defined

$$\text{SUM}_{\text{IR}}(\text{QCD}) \equiv 2\alpha_s \bar{B}_{\text{QCD}} + 2\alpha_s \Re B_{\text{QCD}} \quad (3)$$

from Ref. [1] with the QCD YFS virtual and real genuinely infrared functions $\bar{B}_{\text{QCD}}$, B_{QCD}, and $\tilde{S}_{\text{QCD}}$ as defined therein[1]. The hard gluon residuals $\bar{\beta}_n(k_1, \ldots, k_n)$ are also as defined in Ref. [1], for example. At issue is how one synthesizes a calculation such as that in (1) with the fundamental structure functions for the incoming partons which are themselves solutions of the DGLAP equation and hence in general may have in themselves some of the effects in (1) already included?

Indeed, one might think that the usual formula

$$\sigma(p\bar{p} \to t\bar{t} + X) = \int \sum_{i,j} D_i(x_i) \bar{D}_j(x_j) d\hat{\sigma}_{exp}(x_i x_j s) dx_i dx_j \quad (4)$$

could be used to leading log accuracy when the existing structure functions $D_i(x_i)$, $\bar{D}_j(x_j)$ for partons i, j in $p, \bar{p}$, respectively, in the literature[6] are used. Here, x_i, x_j are the respective light-cone momentum fractions. But, this is not even true because the DGLAP equation is well known to contain all leading logs from the initial state radiation of gluons and all attendant collinear infrared effects. Thus, in using (4) naively one would double count some leading log and collinear infrared effects.

Specifically, in order to use the existing experience on the structure functions together with the YFS exponentiation improvement of perturbation theory given by (1), one has to identify those effects in the YFS formula which are already contained in the DGLAP equation for the structure functions and remove them. This we now do.

Since we have exponentiated the quantity $SUM_{IR}(QCD)$ in (1), we start by isolating from it and from all of its attendant parts in (1) that collinear infrared contribution which is generated by the DGLAP evolution of the structure functions. Viewing the DGLAP evolution as the complete LL series from the initial state radiative effects, we see that we must do the following to synthesize (1) with the existing structure functions[6] in (4):

- From the initial state part of $\tilde{S}_{QCD}$, we must remove that part which corresponds to the YFS limit of the DGLAP kernel, $\tilde{S}_{QCD}|_{DGLAP}$. This can be computed using the definitions in the original paper of Altarelli and Parisi[4] and we get the result that, if we wish to use (1) in (4), we should replace $\tilde{S}_{QCD}$ in (1) with $\tilde{S}_{QCD}^{nls}$ where

$$\tilde{S}_{QCD}^{nls} = \tilde{S}_{QCD} - \tilde{S}_{QCD}|_{DGLAP} \quad (5)$$

for

$$\tilde{S}_{QCD}|_{DGLAP}(k) = \frac{C_F \alpha_s}{4\pi^2} \left[\frac{\theta(z) \, 4(1-zv)^2}{(k_\perp^2 + z^2 v^2 m_q^2)} \right.$$
$$\times \left(1 - \frac{m_q^2 v^2 z^2}{k_\perp^2 + z^2 v^2 m_q^2} \right)$$
$$+ \frac{\theta(-z) \, 4(1+zv)^2}{(k_\perp^2 + z^2 v^2 m_q^2)}$$
$$\left. \times \left(1 - \frac{m_q^2 v^2 z^2}{k_\perp^2 + z^2 v^2 m_q^2} \right) \right] \quad (6)$$

with the identifications $zk = k_z$, $k^o = v\sqrt{s}/2, \vec{k} = (\vec{k}_\perp, k_z)$ and we have written the RHS of (6) for incoming q and $\bar{q}$ respectively for definiteness; for the $q\bar{q}'$ incoming case, for example, one would need to substitute $m_{q'}$ for m_q in the coefficient of $\theta(-z)$ on the RHS of (6). The superscript nls on $\tilde{S}_{QCD}$ on the LHS of (5) denotes that it is non-leading in the s-channel relative to DGLAP evo-

lution. Thus, $\tilde{S}_{QCD}^{nls}$ has the property that its integral over the gluon phase space has no collinear big logarithm terms proportional to $\frac{\alpha_s L}{\pi}$ or $\frac{\alpha_s L^2}{\pi}$, where $L = \ln s/m_q^2 - 1$ in the case of incoming $q\bar{q}$, for example. To obtain the result (6), one follows the derivation of the DGLAP kernel $P_{Gq}(\bar{z})$ in the paper of Altarelli and Parisi in Ref. [4] but one substitutes the YFS vertex $-ig\tau^a(2p+k)_\mu$ instead of the complete vertex $-ig\tau^a\gamma_\mu$ and one retains the quark mass corrections through order m_q^2 in the respective kinematics. Here, g is the QCD coupling constant and τ^a generate the quark (vector) representation of the QCD gauge group; p is the final 4-momentum of the respective radiating quark. We stress that $\tilde{S}_{QCD}^{nls}$ defined in this way is positive definite. The result (5), which corresponds to the replacement $\tilde{B}_{QCD}(K_{max}) \to \tilde{B}_{QCD}^{nls}(K_{max}) = \int \frac{d^3k}{k_o} \theta(K_{max} - |\vec{k}|) \tilde{S}_{QCD}^{nls}$, then avoids the double counting of real gluon radiation from the initial state that is already included in the DGLAP evolution of the structure functions.

- This brings us to the possible double counting of the virtual corrections represented by $\Re B_{QCD}$ in (1). We address this problem as follows. The DGLAP evolution equation generates the entire leading log series associated with initial state real and virtual gluon radiation. Thus, we should also remove from $\Re B_{QCD}$ all leading log effects associated with the initial state. Here, we stress that the definition of the big log L has to be the same as that used in the DGLAP evolution. In practice, this means that, in the formula for $SUM_{IR}(QCD)$ in eq.(9) of Ref. [1], for the initial state contribution corresponding to the case $A = s$ in the respective sum over channels in this formula, we should use the result

$$B_{tot}(s) = \frac{\pi^2}{3} - \frac{1}{2} \quad (7)$$

instead of the result given by eq.(10) in Ref. [1]. We denote this new result by $SUM_{IR}^{nls}(QCD)$. In addition, in the computation of the hard gluon residuals $\bar{\beta}_n$ in (1) one should remember again to remove from the real part of the integral over the first term in eq.(4) in Ref. [1] (this integral can be found in Refs. [7]), which is the initial state part of $\Re B_{QCD}$, all terms involving either $\frac{\alpha_s L}{\pi}$ and $\frac{\alpha_s L^2}{\pi}$, where $L \equiv \ln s/m_q^2 - 1$ in the $q\bar{q}$ incoming case, for example. The generalization to the $q\bar{q}'$ incoming case is immediate. Henceforth, we denote this new $\Re B_{QCD}$ with its initial state s-channel big logarithms removed as $\Re B_{QCD}^{nls}$. It is $\tilde{S}_{QCD}^{nls}$

and $\Re B_{QCD}^{nls}$ that should now be used in computing $\bar{\beta}_n$ in (1); we denote the resulting new hard gluon residual as $\bar{\beta}_n^{nls}$, where it is understood that the various finite order perturbative QCD differential cross sections in this definition of $\bar{\beta}_n^{nls}$ are the reduced, hard mass factorized cross sections.

When these steps are followed, we arrive at a new representation of the cross section in (1) in which $\tilde{S}_{QCD}^{nls}$, $SUM_{IR}^{nls}(QCD)$ and $\Re B_{QCD}^{nls}$ are substituted for $\tilde{S}_{QCD}$, $SUM_{IR}(QCD)$ and $\Re B_{QCD}$ respectively and we write this new version of the result in (1) as

$$d\hat{\sigma}'_{exp} = exp[SUM_{IR}^{nls}(QCD)] \sum_{n=0}^{\infty} \frac{1}{n!} \int \prod_{j=1}^{n} \frac{d^3 k_j}{k_j} \int \frac{d^4 y}{(2\pi)^4}$$
$$e^{+iy(p_1+p_2-q_1-q_2-\sum_j k_j)+D_{QCD}^{nls}}$$
$$\times \bar{\beta}_n^{nls}(k_1, \ldots, k_n) \frac{d^3 q_1 d^3 q_2}{q_1^0 q_2^0}. \tag{8}$$

We may now use this result (8) in (4) with the existing DGLAP evolved structure functions[6] without double counting any effects whatsoever. Moreover, the result (8) is finite and has a smooth limit as the initial state light quark masses $m_q \to 0$.

A related point is that, from the formulas for $\tilde{S}_{QCD}$ (eq.(5) in the first paper in Ref.[1]) and for $\tilde{S}_{QCD}|_{DGLAP}$ it is clear that the terms we are exponentiating in the factor $exp\{SUM_{IR}(QCD)\}$ in (1) are indeed the usual ones that all published calculations have treated to a finite order in α_s when they combine soft real radiation cross sections with virtual gluon radiation corrected ones to cancel the respective infrared singularities at $k_0 \to 0$ to the respective order in α_s; here, we treat this cancelation to all orders in α_s. Referring to Refs.[8,9,10,11,12], we see that the value of α_s that should be used in $SUM_{IR}(QCD)$ is $\alpha_s(\mu)$ where μ is the respective hard renormalization scale~factorization scale in the hard process- otherwise, the real gluon emission $k_0 \to 0$ singularity will not cancel the respective virtual $k_0 \to 0$ singularity: for illustration, we recall the result in Ref.[8] for the parton level Drell-Yan[13] process $q + \bar{q} \to \mu^+ \mu^- + G$ for a muon pair mass M with the IR gluon regulator mass $m_G \to 0$, eq.(5.3.17) in Ref.[8] (here, $\hat{\sigma}_0$ is the respective Born cross section),

$$\hat{\sigma}_{DY}(real) = \frac{2\alpha_s(M)}{3\pi} \hat{\sigma}_0 \{ \ln^2(m_G^2/M^2) + 3\ln(m_G^2/M^2) + \pi^2 \} \tag{9}$$

and its corresponding virtual correction, eq.(5.3.18) in Ref.[8],

$$\hat{\sigma}_{DY}(virtual) = \frac{2\alpha_s(M)}{3\pi} \hat{\sigma}_0 \{ -\ln^2(m_G^2/M^2) - 3\ln(m_G^2/M^2) - \frac{7}{2} - \frac{2\pi^2}{3} + \pi^2 \}, \tag{10}$$

the two of which sum to the total result, eq.(5.3.19) in Ref.[8],

$$\hat{\sigma}_{DY}(real) + \hat{\sigma}_{DY}(virtual) = \frac{2\alpha_s(M)}{3\pi} \hat{\sigma}_0 (\frac{4\pi^2}{3} - \frac{7}{2}), \tag{11}$$

which is independent of the infrared ($k_0 \to 0$) gluon regulator mass m_G – the $k_0 \to 0$ gluon infrared singularities regulated by m_G in (9) and (10) are cancelled in (11) *at the hard scattering coupling $\alpha_s(M)$*. We refer the reader to Refs.[9,10], to the second paper in Refs.[11], and to the second paper in Ref.[12] for further explicit examples of this choice of α_s in the cancelation of IR singularities at $k_0 \to 0$ in finite order perturbative QCD calculations. More recently, the reader may also see the analysis in Ref.[14] for further illustrations of this point. Finally, we recall the fundamental analysis of Altarelli and Parisi in Ref.[4] in which the regularized DGLAP kernel, $P_{qq}(z)$, generates the differential change in the total qq splitting function given by

$$d\Gamma_{qq}(z,t) = \frac{\alpha_s(Q)}{2\pi} P_{qq}(z) dt \tag{12}$$

and

$$P_{qq}(z) = C_F \left(\frac{1+z^2}{(1-z)_+} + \frac{3}{2}\delta(1-z) \right) \tag{13}$$

where the +function has the standard definition

$$\int_0^1 dz \frac{f(z)}{(1-z)_+} \equiv \int_0^1 dz \frac{f(z) - f(1)}{(1-z)} \tag{14}$$

for any appropriate test function $f(z)$, $C_F = 4/3$ is the respective quark color Casimir invariant and $t = \ln Q^2/m_q^2 - 1$ for example. Q is the respective hard scale. We focus here on the $\delta(1-z)$ part of $d\Gamma_{qq}(z,t)$ which realizes the cancellation of the $k_0 \to 0$ real and virtual gluon singularities in $d\Gamma_{qq}(z,t)$ and which enters into it with the squared coupling constant of $\alpha_s(Q)$, the coupling constant of the respective hard scale Q, in complete analogy with the value of α_s which we use in $SUM_{IR}(QCD)$. The correctness of the value of α_s multiplying the delta function in $d\Gamma_{qq}(z,t)$ is proven by agreement between the predictions of the DGLAP equation and the Wilson expansion for the deep inelastic structure functions (see Ref.[4,15,16], for example) and by the agreement of these predictions with the available data[17]. Thus, the value of α_s which we use in $SUM_{IR}(QCD)$ is well-founded in the published literature.

Up to this point in our discussion, we have focussed on how one uses the existing DGLAP evolved structure functions in conjunction with the renormalization group improved YFS theory to compute the rigorous hadron level cross section. It is also possible to go one step further, to use the explicit application of the YFS theory to the one-loop and single bremsstrahlung correction illustrated in Ref. [18] to exponentiate the IR singularities in the DGLAP equation itself. This calculation, which together with its application will be reported elsewhere [19], yields an entirely new approach to DGLAP evolution in which the splitting functions themselves are actually YFS exponentiated.

A final important issue that can be addressed is the relationship between our YFS exponentiated result (1) and the soft gluon resummation theory of Sterman [20] and of Catani and Trentadue [21]. We now turn to this relationship. We note that this latter resummation theory has recently been used in Refs. [22,23,24] at the respective leading log level to discuss the effects of soft gluons in the $p + \bar{p} \to t + \bar{t} + X$ cross section normalization.

Specifically, referring to the notation of Refs. [20,21,22,23,24] we can identify the resummed soft gluon cross section of the theory of Refs. [20,21] with the cross section in (1) via

$$\exp[\text{SUM}_{\text{IR}}(\text{QCD})] \sum_{n=0}^{\infty} \int \prod_{j=1}^{n} \frac{d^3 k_j}{k_j} \int \frac{d^4 y}{(2\pi)^4}$$

$$e^{iy \cdot (p_1 + p_2 - q_1 - q_2 - \sum k_j) + D_{\text{QCD}}}$$

$$* \bar{\beta}_n(k_1, \ldots, k_n) \frac{d^3 q_1}{q_1^0} \frac{d^3 q_2}{q_2^0} / d\Phi \xrightarrow[1-z \to 0]{} e^{E(1-z, \alpha_s)} d\hat{\sigma}_B / d\Phi, \tag{15}$$

where we note the limit $1 - z \to 0$ is the limit studied by Sterman, Catani and Trentadue in which the hard scattering invariant mass squared fraction z, defined by s'/s, approaches its maximum value 1, and where $E(1 - z, \alpha_s)$ is the resummation exponent estimated in Ref. [20] at the leading-log level and in Ref. [21] at the next-leading log level and used in Refs. [22,23,24] at the leading log level for the normalization of $\sigma(t\bar{t})$ at FNAL, $d\Phi$ is the respective Lorentz invariant phase space considered in Refs. [20,21] and $d\hat{\sigma}_B$ is the respective Born level cross section. For example, the well-known double integral over $\alpha_s(\bar{k}_\perp^2)$ in E is easily recovered from the lefthand side of (15) by approximating

$$D_{\text{QCD}}(y) \approx D_{\text{QCD}}(0)$$

$$= \int \frac{d^3 k}{k} \tilde{S}_{\text{QCD}}(k)(1 - \theta(K_{max} - k))$$

$$= \int \frac{dz}{1-z} \int \frac{dk_\perp^2}{k_\perp^2} \frac{\alpha_s(\bar{k}_\perp^2)}{\pi} C_F (1 - \theta((1-z)_{max}$$

$$- (1-z)))\{1 + \cdots\}, \tag{16}$$

so that for the leading log initial state part of $SUM_{IR}(\text{QCD}) + D_{\text{QCD}}(0)$ we recover exactly the double

integral over $\alpha_s(\bar{k}_\perp^2) = \alpha_s((1-z)k_\perp^2) \simeq \alpha_s((1-z)Q^2)$ proposed for E in Refs. [20,21] (the $\cdots$ in (16) stand for the remainder of $D_{\text{QCD}}(0)$ [1]). Here, in our infinite momentum frame (light-cone) kinematics, we identified $(1-z)_{max} = K_{max}/p_1^0$ and $Q^2 = s$. We note that, as the authors in Refs. [20,21] have pointed-out, in allowing α_s to run in (16) one resums a certain class of leading logs which in our exact series in $\alpha_s(Q^2)$ in (1) are generated by the infinite sum on n in (1). We note further that the remainder of E corresponds to the next leading log terms generated by the infinite sum on n in (1) over the $\bar{\beta}_n$ when one makes the approximation

$$\sum_{n=0}^{\infty} \frac{1}{n!} \int \prod_j^n \frac{d^3 k_j}{k_j} \delta(p_1 + p_2 - q_1 - q_2 - \sum k_j)$$

$$\bar{\beta}_n'(k_1, \cdots, k_n) \frac{d^3 q_1}{q_1^0} \frac{d^3 q_2}{q_2^0} / d\hat{\sigma}_B \equiv \mathcal{S}(\bar{\beta}') \approx \sum_{n=0}^{\infty} \frac{1}{n!} (E^{nll})^n \tag{17}$$

where we define E^{nll} to be the next leading log part of E and to be consistent we have defined $\bar{\beta}_n'$ to be that part of the $\bar{\beta}_n$ in our exact result (1) whose leading log content is not already resummed by integral over the running α_s in (16). We see that in our approach, the results of Refs. [20,21] for E^{nll} correspond to the approximation

$$(E^{nll})^n = \int \prod_j^n \frac{d^3 k_j}{k_j} \delta(p_1 + p_2 - q_1 - q_2 - \sum k_j)$$

$$\times \bar{\beta}_n'(k_1, \cdots, k_n) \frac{d^3 q_1}{q_1^0} \frac{d^3 q_2}{q_2^0} / d\hat{\sigma}_B. \tag{18}$$

We conclude that, at the parton level, the formulas in Refs. [20,21] are entirely contained in our formula (1) and that they represent an approximation to our result (1) in the limit that $1 - z \to 0$.

One of the important consequences of the identification of the Sterman-Catani-Trentadue exponent E in (15) is that we can address the issue of the presence or absence of renormalon [25] behavior associated with the leading log and next-leading log truncation of our exact result (1) as it is repeesented by E. Specifically, as the authors in Refs. [20,21,22,23,24] have noted, as $1 - z \to 0$ in (16), the argument of the running α_s would appear to approach the Landau pole, creating an ambiguity in the non-leading terms in the exponent, for example, of the famous renormalon type [25] *if one expands α_s in terms of this one-loop result*– we stress that the two and three loop formulas are known and they do not seem to support this pole in α_s [26]. On the other hand, we have argued in Ref. [1], by the uncertainty principle, that the truly long wavelength gluons, with wavelengths much larger than $1/\Lambda_{QCD}$, should decouple from the hard process under discussion here. Does this decoupling really happen or not? We look at the exact initial state formula for the sum

$$SUM_{IR}(QCD)|_{initial\ state} = \{2\alpha_s ReB_{QCD}$$
$$+2\alpha_s \tilde{B}_{QCD}(K_{max})\}|_{initial\ state}$$
$$= -\frac{C_F}{4\pi^2}\int \frac{d^3k}{k^0}\alpha_s(\bar{k}_\perp^2)\left(\frac{p_1}{p_1 k} - \frac{p_2}{p_2 k}\right)^2\theta(K_{max}-k)$$
$$+\Re\frac{iC_F}{4\pi^2}\int\frac{d^4k}{\pi}\left(\frac{1}{k^2 - m_G^2 + i\epsilon}\right)$$
$$\left(\frac{2p_1 + k}{k^2 + 2p_1 k + i\epsilon} + \frac{2p_2 - k}{k^2 - 2p_2 k + i\epsilon}\right)^2\alpha_s(\bar{k}_\perp^2)$$

$$(19)$$

where we have restored the running ansatz of Refs. [20,21] in their estimate of E. We stress that for consistency the same definition of α_s has to be used for both soft real and soft virtual gluons in order for the infrared singularities to cancel. Carrying out the virtual k^0 integral by standard contour methods in the upper complex k^0 plane, we get, for the residue of the gluon propagator $\equiv Res(gluon\ propagator)$, the contribution

$$2\alpha_s ReB_{QCD}|_{initial\ state,Res(gluon\ propagator)} =$$
$$\frac{C_F}{4\pi^2}\int\frac{d^3k}{k^0}\alpha_s(\bar{k}_\perp^2)\left(\frac{p_1}{p_1 k} - \frac{p_2}{p_2 k}\right)^2 \quad (20)$$

so that, when we combine this result with the real emission term in (19), we get finally the representation

$$SUM_{IR}(QCD)|_{initial\ state} = \frac{C_F}{4\pi^2}\int_{k^0\geq K_{max}}\frac{d^3k}{k^0}\alpha_s(\bar{k}_\perp^2)$$
$$\times\left(\frac{p_1}{p_1 k} - \frac{p_2}{p_2 k}\right)^2$$
$$-\frac{C_F}{2\pi}\Re\int\frac{d^3k}{\pi}\sum_{Res(fermion\ propagators)}\left(\frac{1}{k^2 - m_G^2 + i\epsilon}\right)$$
$$\times\left(\frac{2p_1 + k}{k^2 + 2p_1 k + i\epsilon} + \frac{2p_2 - k}{k^2 - 2p_2 k + i\epsilon}\right)^2\alpha_s(\bar{k}_\perp'^2),$$

$$(21)$$

where we have introduced the scale $\bar{k}_\perp'^2$ for α_s in the second term in this last equation to take into account that, as this term is not infrared divergent and is dominated by gluon momenta $\mathcal{O}(\sqrt{s}/2)$, $\bar{k}'_\perp$ should also be of this order for consistency. In the second term in (21), only the fermion propagator residues , $Res(a), a = fermion\ propagators$, enter into the respective sum over residues, as we indicate explicitly. We see in the first term in (21) that only gluons with energies exceeding the dummy parameter K_{max} (our result (1) is independent of K_{max} and $1 - z_{max}$) are actually involved in $SUM_{IR}(QCD)$ so that, if we use the resummation ansatz of Refs. [20,21] we do not encounter the regime

$(1 - z)s = \Lambda_{QCD}^2$ in our result for the LL part of E as we may take $(1 - z_{max}) \gg \frac{\Lambda_{QCD}^2}{s}$ without loss of content or predictive power. This is consistent with the earlier observation that (1), which is a series in $\alpha_s(Q^2)$, is consistent with the uncertainty principle and hence in fact is insensitive to the behavior of α_s at scales $\mathcal{O}(\Lambda_{QCD})$ when $Q^2 \gg \Lambda_{QCD}^2$.

In summary, we have shown how one rigorously synthesizes the powerful results of DGLAP evolution and renormalization group improved YFS exponentiation in QCD without double counting. Explicit Monte Carlo event generator data based on our prescription will appear elsewhere [5].

Acknowledgments

We would thank Profs. G. Veneziano and G. Altarelli for the support and kind hospitality of the CERN Theory Division, where a part of this work was performed. One of us (B.F.L. W.) thanks Prof. C. Prescott of SLAC for the kind hospitality of SLAC Group A while this work was completed.

References

1. D.B. DeLaney, S. Jadach, C. Shio, G. Siopsis, B. F. L. Ward, UTHEP-95-0102; Phys. Lett. **B342** (1995) 239; Phys. Rev. **D52** (1995) 108; see also F. A. Berends and W. T. Giele, Nucl. Phys. **B313**, 595(1989).

2. D. R. Yennie, S. C. Frautschi, and H. Suura, Ann. Phys. **13** (1961) 379.

3. Yu. L. Dokshitzer et al., Basics of perturbative QCD,(Editions Frontieres, Gif-Sur-Yvette, 1991).

4. G. Altarelli and G. Parisi, Nucl. Phys. **B126** (1977) 298; Yu. L. Dokshitzer, Sov. Phys. JETP **46** (1977) 641; L. N. Lipatov, Yad. Fiz. **20** (1974) 181; V. Gribov and L. Lipatov, Sov. J. Nucl. Phys. **15** (1972) 675, 938.

5. S. Jadach et al., in preparation.

6. A. D. Martin, R. G. Roberts and W. J. Stirling, DTP-96-44, 1996; A. D. Martin, W. J. Stirling and R. G. Roberts, Int. J. Mod. Phys. **A10** (1995) 2885; M. Gluck, E. Reya and M. Stratmann, Phys. Rev. D51 (1995) 3220; W. Tung, MSUHEP-60701, 1996; H. L. Lai et al., Phys. Rev. D51 (1995) 4763; andd references therein.

7. S. Jadach, W. Placzek and B.F.L. Ward, UTHEP-95-0801, 1995; Phys. Rev. D (1996) in press; B. F. L. Ward, ibid.**36** (1987) 939; and references therein.

8. R. D. Field, Applications of Perturbative QCD ,(Addison Wesley Publ. Co., Redwood City, 1989).

9. R.K. Ellis *et al.*, Nucl. Phys. **B173** (1980) 397.

10. R. K. Ellis and J. Sexton, Nucl. Phys. **B269** (1986) 445; and references therein.

11. P. Nason, S. Dawson, and R. K. Ellis, Nucl. Phys. **B303** (1988) 607; *ibid.* **B327** (1989) 49; *ibid.* **B335** (1990) 260.

12. W. Beenakker *et al.*, Phys. Rev. **D40** (1989) 54; W. Beenakker *et al.*, Nucl. Phys. **B351** (1991) 507.

13. S.D. Drell and T.M. Yan, Ann. Phys. **66**, 578 (1991).

14. S. Catani, in Proc. 1996 Cracow International Symposium on Radiative Corrections, ed. S. Jadach, Acta Phys. Pol., to be published.

15. K. G. Wilson, Phys. Rev. **D2**, 1473 (1970); *ibid.* **D2**, 1478 (1970); *ibid.* **D3**, 1818 (1971).

16. D.J. Gross and F. Wilczek, Phys. Rev. **D9**, 980 (1974); *ibid.***D8**, 3633 (1973); Phys. Rev. Lett. **30**, 1343 (1973); H.D. Politzer, Phys. Rev. Lett. **30**, 1346 (1973); Phys. Rep. **14**, 335 (1974); H. Georgi and H. D. Politzer, Phys. Rev. **D9**, 416 (1974).

17. See ,for example, J. Feltesse, in Proceedings of the XXVII International Conference on High Energy Physics, v.1, eds. I. G. Knowles and P. J. Bussey, (IOP Publ. Ltd., Bristol, 1995) p. 65; and, references therein.

18. S. Jadach, E. Richter-Was, B.F.L. Ward and Z. Was, Phys. Rev. **D44**, 2669(1991).

19. S. Jadach and B. F. L Ward, to appear.

20. G. Sterman, *Nucl. Phys.* **B281** (1987) 310.

21. S. Catani and L. Trentadue, *Nucl. Phys.* **B327** (1989) 323; *ibid.* **B353** (1991) 183.

22. E. Laenen, J. Smith, and W. van Neerven, Phys. Lett. **B321** (1994) 254; Nucl. Phys. **B369** (1992) 543.

23. E. Berger and H. Contopanagos, *Phys. Rev.* **D54** (1996) 3085.

24. S. Catani *et al.*, preprint CERN-TH/96-21, 1996.

25. G. 't Hooft, in The Whys of Subnuclear Physics, Erice, 1977, ed. A. Zichichi (Plenum, New York, 1979); Y. Firshman and A. R. White, Nucl. Phys. **B158** (1979) 221; J. C. Le Guillou and J. Zinn-Justin, eds., Large-order Behavior in Perturbation Theory, Current Physics – Sources and Comments, Vol. 7(North-Holland, Amsterdam, 1990); G. B. West, Phys. Rev. Lett. **67** (1991) 1388; L. S. Brown *et al.*, Phys. Rev. D**46** (1992) 4712; V. I. Zakharov, Nucl. Phys. **B385** (1992) 452; M. Beneke and V. I. Zakharov, Phys. Rev. Lett. **69** (1992) 2472; A. H. Mueller, QCD – 20 Years Later, Aachen, 1992, P. M. Zerwas and H. A. Kastrup, eds.,(World Scientific, Singapore,1993).

26. S. J. Brodsky, comment at the 1996 CTEQ QCD Symp., FNAL, 1996.

EFFECT OF ELECTROMAGNETISM IN NONLEPTONIC KAON DECAYS

E. GOLOWICH

Department of Physics and Astronomy, University of Massachusetts, Amherst, MA 01003, USA
E-mail: gene@het.phast.umass.edu

A precise understanding of the role of electromagnetism in the $K \to \pi\pi$ decays remains lacking to this day. At issue are a number of basic dynamical questions, such as the magnitude and origin of a $\Delta I = 5/2$ signal in these decays, the relative importance of $\Delta I = 3/2$ vs. electromagnetically-induced amplitudes in the $K^+ \to \pi^+\pi^0$ transition, etc. We report on a calculation which uses modern chiral techniques to analyze this problem.

1 Introduction

This talk represents a progress report on a project to calculate the effect of electromagnetism on $K \to \pi\pi$ decays. Although final results are not yet available, the overall architecture of the calculation is clear, and I can give some preliminary findings regarding the $K^+ \to \pi^+\pi^0$ transition. [1] Before describing the calculation itself, let me comment on why such a project is of interest.

1.1 Motivation: A Phenomenological Consideration

Consider the following simple fit to the $K \to \pi\pi$ decays. We take as inputs [2] the measured decay widths Γ_{+-}, Γ_{00} and Γ_{+0} which correspond to the transitions $K^0 \to \pi^+\pi^-$, $K^0 \to \pi^0\pi^0$ and $K^+ \to \pi^+\pi^0$. From these we obtain magnitudes of the associated decay amplitudes, $|a_{+-}|$, $|a_{00}|$ and $|a_{+0}|$.[a] To these we fit the isospin amplitudes $|a_{1/2}|$, $|a_{3/2}|$ and the two-pion phase shift $[\delta_{I=0}^{\pi\pi} - \delta_{I=2}^{\pi\pi}](m_K^2)$, where the $I = 0, 2$ labels refer to the two-pion final state isospin. This procedure is the simplest one possible in that we ignore the presence of electromagnetism and any $\Delta I = 5/2$ contribution. Upon performing the fit, we find

$$|a_{3/2}| = 0.0212 \text{ keV} \ ,$$
$$|a_{1/2}/a_{3/2}| = 22.2 \ ,$$
$$[\delta_{I=0}^{\pi\pi} - \delta_{I=2}^{\pi\pi}](m_K^2) = 56.7^o \pm 3.9^o \ . \quad (1)$$

The large numerical ratio in the second of the above relations is, of course, just the $\Delta I = 1/2$ rule as manifested in nonleptonic kaon decays.

So far, the analysis seems pedestrian and requires no further comment. However, let us compare the phase shift difference of Eq. (1) with those obtained more directly,

$$[\delta_{I=0}^{\pi\pi} - \delta_{I=2}^{\pi\pi}](m_K^2) = \begin{cases} 42^o \pm 4^o & (\pi\pi \text{ scatt.}) \\ 45^o \pm 6^o & (\text{ChPT}) \end{cases} . \quad (2)$$

We see that pion-pion phase shifts obtained either from a fit to pion-pion scattering [4] or from a determination

of the low energy pion-pion system from chiral perturbation theory [5] are several standard deviations beneath the determination of Eq. (1). The most promising avenue for addressing this contradiction is to incorporate electromagnetism into the fit of $K \to \pi\pi$ decays.

1.2 Motivation: A Theoretical Consideration

Previous lattice calculations of the $\Delta I = 3/2$ amplitude have tended to produce a value about twice the magnitude of the experimental value contained implicitly in Eq. (1).[b] In addition, lattice determinations of the B_K parameter have produced (via a procedure [7] based on chiral symmetry and SU(3) flavor symmetry) a magnitude for the $\Delta I = 3/2$ amplitude about twice that in Eq. (1).

Both these determinations deal with the true $\Delta I = 3/2$ amplitude. It is possible, due to contamination from electromagnetism, that the quantity $a_{3/2}$ of Eq. (1) is not the *true* $\Delta I = 3/2$ amplitude. In particular, since $|a_{1/2}|$ is so much larger than $|a_{3/2}|$, it is possible that electromagnetic corrections to $a_{1/2}$ may simulate an effect similar to the small $I = 3/2$ amplitude. These electromagnetic corrections are $\mathcal{O}(\alpha \cdot a_{1/2}) \sim a_{1/2}/137$, which can be comparable to a sizeable portion of the $I = 3/2$ amplitude $a_{3/2} \sim a_{1/2}/22$. The possibility then emerges that the relevant phenomenological amplitude (with electromagnetism removed) could differ significantly from that presently being used.

1.3 Final Introductory Comments

The literature on electromagnetic corrections to $K \to \pi\pi$ decays extends over many years [8,9,10,11]. Yet the issue has not yet received a definitive treatment. It is our aim to provide an analysis using the most up-to-date tools which hopefully will yield a reliable estimate of this effect. Although the remainder of this talk is devoted to analyzing $K^+ \to \pi^+\pi^0$ decay, we are planning a more comprehensive analysis of electromagnetic radiative corrections in kaon decays. [12] The full system, particularly

---

[a] We adopt the $K \to \pi\pi$ amplitudes Devlin and Dickey. [3]

[b] A recent lattice calculation is more promising in this regard. [6]

Figure 1: Various energy scales.

the decay $K^0 \to \pi^+\pi^-$, brings in several additional complications, such as the effect of Coulomb scattering on the final state, the violations of Watson's theorem from the mixing of final states and the induced $\Delta I = 5/2$ effect.

However, the decay $K^+ \to \pi^+\pi^0$ is particularly simple and can by itself clarify important issues. In the following, we focus on this one transition.

2 Chiral Analysis

A modern framework for structuring the calculation involves chiral lagrangians. The weak interactions involve left-handed currents only, and the nonleptonic hamiltonian has an octet and 27-plet component. The lowest-order lagrangian for the octet portion involves two derivatives,

$$\mathcal{L}_8 = g_8 \,\mathrm{Tr}\, \left(\lambda_6 D_\mu U D^\mu U^\dagger\right) \ ,$$
$$D_\mu U = \partial_\mu + iQ A_\mu \ , \tag{3}$$

where A_μ is the photon field, Q is the charge matrix and $|g_8| \simeq 7.8 \cdot 10^{-8} \, F_\pi^2$. Electromagnetic corrections involve both left-handed and right-handed effects and can lead to lagrangians which do not involve derivatives. The most important of these to the present work describes the leading electromagnetic correction to the weak interactions,

$$\mathcal{L}_{\mathrm{emw}} = g_{\mathrm{emw}} \,\mathrm{Tr}\, \left(\lambda_6 U Q U^\dagger\right) \ . \tag{4}$$

This is not to be confused with the lagrangian $\mathcal{L}_{\mathrm{ems}}$ which deals with (to lowest order) the electromagnetic shift,

$$\mathcal{L}_{\mathrm{ems}} = g_{\mathrm{ems}} \,\mathrm{Tr}\, \left(Q U Q U^\dagger\right) \ , \tag{5}$$

such that

$$\delta m_\pi^2 \equiv m_{\pi^+}^2 - m_{\pi^0}^2 = \frac{2}{F_\pi^2} \, g_{\mathrm{ems}} \ . \tag{6}$$

The calculation to follow amounts to a determination of g_{emw} and of its relation to $K^+ \to \pi^+\pi^0$ decay. For convenience, we work with the simpler $K^+ \to \pi^+$ transition in which the central object of interest is the amplitude $\mathcal{M}^{(K\pi)}$,

$$_{\mathrm{out}}\langle \pi^+(p') | K^+(p) \rangle_{\mathrm{in}} = i(2\pi)^4 \delta^{(4)}(p' - p) \ \mathcal{M}^{(K\pi)}(p^2) \ . \tag{7}$$

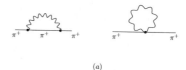

(a)

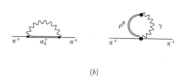

(b)

Figure 2: Electromagnetic mass shift of the pion.

In the chiral limit, this is related to g_{emw} via

$$\mathcal{M}^{(K\pi)}(0) = \frac{2}{F_\pi^2} \, g_{\mathrm{emw}} \ . \tag{8}$$

We calculate $\mathcal{M}^{(K\pi)}(0)$ in the next section and then turn to the full $K^+ \to \pi^+\pi^0$ amplitude, including effects that arise at the next chiral order ($\mathcal{O}(p^2)$) from the photon loop calculation.

3 Calculation of the leading chiral amplitude

To fully calculate the relevant amplitude, one needs to consider contributions from all energy scales. As depicted in Fig. 1, there are three distinct regions of the euclidean virtual photon momentum:

1. very low energies $Q^2 < \Lambda^2$ with $\Lambda \sim m_\rho$,

2. high energies with $Q^2 > m_c^2, \mu^2$, and

3. intermediate energies between these two regions.

The quantity μ is an energy scale that we shall take throughout as one which divides 'very low' from 'high' energy. It can be chosen such that $\mu > m_c$ or $\mu < m_c$ (as in Fig. 1). In the low energy regime, we use chiral techniques to obtain the leading effect. At high energies, the short distance analysis of QCD will be employed. The treatment of the intermediate energy region is driven by the need to match these two descriptions and can be modeled on physics already known for the case of electromagnetic mass shifts.

3.1 Long Distance Component

A long distance component which arises in a similar context occurs in the electromagnetic mass shift of the

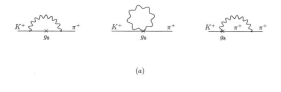

(a)

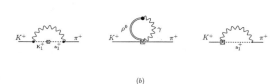

(b)

Figure 3: Weak transition $K^+ \to \pi^+$.

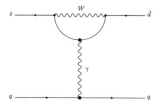

Figure 4: Electromagnetic penguin.

charged pion.[13] Diagrams like those in Fig. 2(a) lead to the mass shift

$$(\delta m_\pi^2)_{LD} = \frac{3\alpha}{4\pi} \int_0^{\Lambda^2} dQ^2 \quad , \tag{9}$$

where the integration is over the euclidean squared-momentum Q^2. The analogous long distance component of the kaon-to-pion transition can be calculated in the chiral limit using the chiral lagrangian of Eq. (3). From diagrams like those in Fig. 3(a), one finds after Wick rotation the matrix element

$$\mathcal{M}_{LD}^{(K\pi)} = -\frac{3\alpha g_8}{2\pi F_\pi^2} \int_0^{\Lambda^2} dQ^2 \quad , \tag{10}$$

where Λ represents the upper end of the low-energy region. The similarity of Eqs. (9),(10) can be inferred from the fact that in the latter calculation the weak vertex in the loop introduces a factor of the loop momentum Q^2 which compensates one of the two propagators, yielding an effect proportional to $(\delta m_\pi^2)_{LD}$.

At this stage, it appears as just a curiosity that the choice $\Lambda^2 = m_\rho^2$ provides an accurate description of the pion mass difference. However, we argue below that this is not an accident — that reliably known physics cuts off the integral above the rho mass and that similar considerations constrain $\mathcal{M}^{(K\pi)}(0)$. The similarity pointed out above between $(\delta m_\pi^2)_{LD}$ and $\mathcal{M}_{LD}^{(K\pi)}$ is the first indication of a more general, although approximate, relation

$$\mathcal{M}^{(K\pi)}(0) \simeq -2\frac{g_8}{F_\pi^2}\delta m_\pi^2 \tag{11}$$

or equivalently $(g_{\text{emw}}) \simeq -g_8 \, (\delta m_\pi^2)$

3.2 Short Distance Component

The short distance parts of the photon-exchange amplitudes between the four external quark fields of the weak interaction leads to an overall shift in the strength of the nonleptonic interaction. This is equivalent to a redefined 'nonleptonic' (NL) Fermi constant,[14]

$$G_{\text{NL}} = G_\mu \left[1 + \frac{2\alpha}{3\pi} \ln\left(\frac{M_W}{\mu}\right) \right] \quad , \tag{12}$$

where G_μ is the Fermi constant measured in muon decay. This shift does not lead to mixing of isospin amplitudes and is irrelevant for our purposes.

The only meaningful short distance physics which contributes to $\mathcal{M}_{SD}^{(K\pi)}$ involves the electromagnetic penguin of Fig. 4,

$$\mathcal{H}_{\text{EMP}} = -\frac{2}{9} \cdot \frac{G_F}{\sqrt{2}} \cdot \frac{\alpha}{\pi} \left[\xi_u \, I(m_c,\mu) + \xi_t \ln\frac{m_t^2}{m_c^2} \right] \hat{\mathcal{O}} \quad , \tag{13}$$

where

$$\hat{\mathcal{O}} = \bar{s}\gamma_\mu(1+\gamma_5)d \; \bar{q}Q\gamma^\mu q \quad , \tag{14}$$

and

$$I(m_c,\mu) \equiv \ln\frac{m_c^2 + \mu^2}{\mu^2} + \ldots \quad . \tag{15}$$

The ellipses in Eq. (15) represent finite non-logarithmic terms. The CKM dependence in Eq. (13) occurs in $\xi_k = V_{kd}^* V_{ks}$ $(k = u,t)$. Hereafter we drop the term proportional to ξ_t as it is about 1% in magnitude of the term proportional to ξ_u.

We estimate the kaon-to-pion matrix element of operator $\hat{\mathcal{O}}$ by passing to the chiral limit,

$$\langle \pi^+(0)|\hat{\mathcal{O}}|K^+(0)\rangle = -\frac{2}{3F_\pi^2}\langle 0| \bar{q}q |0\rangle^2 \quad , \tag{16}$$

so that

$$\mathcal{M}_{SD}^{(K\pi)} \simeq -\frac{\alpha}{4\pi} \cdot \frac{16 G_F \xi_u}{27\sqrt{2}} \frac{\langle 0| \bar{q}q |0\rangle^2}{F_\pi^2} \left[\ln\frac{m_c^2 + \mu^2}{\mu^2} + \ldots \right] . \tag{17}$$

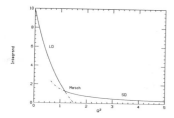

Figure 5: Depiction of Matching.

Anticipating the matching procedure between long and short distance components, we express $\mathcal{M}_{SD}^{(K\pi)}$ in the integral form

$$\mathcal{M}_{SD}^{(K\pi)} \simeq \int_{\mu^2}^{\infty} dQ^2 \, B_{SD} \left[\frac{m_c^2}{Q^2(Q^2+m_c^2)} + \cdots \right]$$

$$B_{SD} \equiv -\frac{\alpha}{4\pi} \cdot \frac{16 G_F \xi_u}{27\sqrt{2} F_\pi^2} \langle 0| \, \bar{q}q \, |0\rangle^2 \ . \tag{18}$$

Note that in the case of the electromagnetic mass shifts there is no local operator in the short distance expansion, in which case the high energy portion falls faster than $1/Q^2$

3.3 Intermediate Energies and Matching

A prototype for dealing with the intermediate energies is the pion electromagnetic mass difference. A rigorous approach would involve the sum rule of Das *et al*[15], in which δm_π^2 is expressed in terms of the difference of the experimental vector and axialvector spectral functions $(\rho_V - \rho_A)(s)$. This has been analysed successfully using experimental data and QCD constraints.[16] A simplified expression that captures the essential physics is obtained upon saturating ρ_V and ρ_A respectively with the vector ρ resonance and the axialvector resonance a_1. This yields

$$\delta m_\pi^2 = \frac{3\alpha}{4\pi} \int_0^\infty dQ^2 \, \frac{m_{a_1}^2}{Q^2+m_{a_1}^2} \cdot \frac{m_\rho^2}{Q^2+m_\rho^2} \ . \tag{19}$$

The long-distance amplitude given in Eq. (10) has been softened at values of Q^2 above the resonance region so that m_ρ and m_{a_1} act as the effective cutoff for the integral. This result is equally well reproduced by introducing resonance couplings to the effective lagrangian and imposing the Weinberg sum rules on the masses and couplings. This involves the diagrams of Fig. 2(b).

Something similar must happen to soften the weak interaction result. If we incorporate the effects of resonances in the intermediate energy region via diagrams like those in Fig. 3(b), we obtain an expression of the general form

$$\mathcal{M}_{LD}^{(K\pi)} + \mathcal{M}_{INT}^{(K\pi)} = -\frac{3\alpha g_8}{2\pi F_\pi^2} \int_0^{\mu^2} dQ^2 \tag{20}$$

$$\times \left[1 - \frac{B_V Q^2}{Q^2+m_V^2} - \frac{B_A Q^2}{Q^2+m_A^2} + \frac{C m_A^2 Q^2}{(Q^2+m_A^2)^2} \right] \ ,$$

where B_V, B_A and C contain unknown couplings from the weak interaction resonance lagrangians. It is, of course, possible to model these couplings. Instead, however, we constrain them by imposing requirements involving the transition to the high energy region. Most importantly, it is a consequence of the short distance analysis that the large Q^2 limit of the integrand does not contain any dependence which is constant in Q^2. This leads to the condition $B_V + B_A = 1$

We are now able to address the matching with the short distance result. If the charm mass were much higher, we would match the $1/Q^2$ tail of the integrand in Eq. (20) to the result of Eq. (18). However, this $1/Q^2$ tail is not valid above m_c^2, and m_c is too close to m_ρ, m_a for this to be done. Instead we simply require that the integrand of Eq. (20) equal the value of the short-distance amplitude at some scale $1 \geq \mu(\text{GeV}) \geq 2$ and treat the resulting variation as one of the uncertainties on the calculation. The matching appears schematically in Fig. 5, which depicts the integrands in the Q^2 integrals. We find the short distance integrand to be much smaller than that at long distances.

If m_ρ and m_a were equal this second constraint would uniquely determine the integrand in the intermediate energy region and would complete the matching. The difference between m_ρ and m_a leads to a slight further uncertainty. We have explored models for the resonance couplings[17] which weight the axialvector and vector resonances differently and find that this uncertainty is smaller than that associated with the matching scale.

From the results of the previous analysis, we obtain an expression for the electromagnetic penguin coupling g_{emw}, expressed in units of $g_8 F_\pi^2$,

$$\frac{g_{\text{emw}}}{g_8 F_\pi^2} = \frac{G_F F_\pi^4 \xi_u}{2\sqrt{2} g_8} \cdot \frac{\alpha}{4\pi} \mathcal{M}^{(K\pi)}(0) \ . \tag{21}$$

Below, we provide a range of determinations for $|g_{\text{emw}}|$ corresponding to different matching procedures. Because the relative sign of the long-distance and short-distance amplitudes is not known, we list (preliminary) results for either sign of g_8 (taking $m_c = 1.4$ GeV):

$$\left| \frac{g_{\text{emw}}}{g_8 F_\pi^2} \right| = \begin{cases} 0.108 \to 0.123 & (g_8 > 0) \\ 0.071 \to 0.087 & (g_8 < 0) \end{cases} \ . \tag{22}$$

4 The $K^+ \to \pi^+\pi^0$ transition

At last we turn to the $K^+ \to \pi^+\pi^0$ transition itself, first listing the distinct contributions,

$$a_{+0} = \left[\frac{\sqrt{3}}{2}(a_{3/2} + a_{3/2}^{\rm em}) + a_{1/2}^{\rm em\ (a)} + a_{1/2}^{\rm em\ (b)} \right] e^{i\delta_{I=2}^{\pi\pi}} \ , \tag{23}$$

where:

1. $a_{3/2}$ is the true $\Delta I = 3/2$ amplitude that we wish to learn about.

2. $a_{3/2}^{\rm em}$ contains all electromagnetic corrections to $a_{3/2}$ such as the K^+ self energy. We ignore this set of tiny $\mathcal{O}(\alpha \cdot a_{3/2})$ contributions.

3. $a_{1/2}^{\rm em\ (a)}$ is a term which we have not heretofore discussed. It too arises via 'electromagnetic leakage' from $a_{1/2}$ and is obtained from calculating the matrix element $\langle \pi^+\pi^0 | \mathcal{L}_8 | K^+ \rangle$,

$$a_{1/2}^{\rm em\ (a)} = -\frac{g_8}{F_\pi^3} \delta m_\pi^2 \ . \tag{24}$$

This contribution is easy to miss entirely as one does not expect $\mathcal{L}_8$ to couple to the final state since $\pi^+\pi^0$ carries isospin $I = 3/2$. Indeed, $a_{1/2}^{\rm em\ (a)}$ does vanish in the isospin limit, but only there.

4. $a_{1/2}^{\rm em\ (b)}$ is the contribution from $\mathcal{M}^{(K\pi)}$ which we have studied in detail in the previous section. It can be expressed as

$$a_{1/2}^{\rm em\ (b)} = \frac{g_8}{F_\pi^3} \delta m_\pi^2 + a_{1/2}^{\rm em\ (b')} \ , \tag{25}$$

where $a_{1/2}^{\rm em\ (b')}$ is a residual contribution much smaller in magnitude than the term proportional to δm_π^2.

5 Concluding Remarks

Our analysis has yielded important preliminary findings regarding the effect of electromagnetism on the $K \to \pi\pi$ decays. We have found two significant electromagnetic corrections to the $K^+ \to \pi^+\pi^0$ amplitude, called $a_{1/2}^{\rm em\ (a)}$ and $a_{1/2}^{\rm em\ (b)}$ in the above. As discussed in this talk, the latter ($a_{1/2}^{\rm em\ (b)}$) can be analyzed with chiral methods, and to a good numerical approximation we find it to be proportional to the pion electromagnetic mass difference (*cf* Eq. (25)). However, the leading term in $a_{1/2}^{\rm em\ (b)}$ is exactly canceled by $a_{1/2}^{\rm em\ (a)}$, arising from the effect of pion electromagnetic self-energies in the usual weak amplitude.

What is left are residual effects coming mainly from intermediate energy physics.

The end result is that the amplitude $a_{3/2}$ as extracted from the observed $K^+ \to \pi^+\pi^0$ decay rate is essentially the true $\Delta I = 3/2$ amplitude. In other words, 'what you see is what you get'. There will be more results available on the full set of $K \to \pi\pi$ decays in the near future.

Acknowledgements

The work described in this talk was supported in part by a grant from the National Science Foundation.

References

1. V. Cirigliano, J.F. Donoghue and E. Golowich, work in progress.
2. C. Caso *et al* (Particle Data Group), Euro. Phys. Jnl. **C3** (1998) 1.
3. T.J. Devlin and J.O. Dickey, Rev. Mod. Phys. **51** (1979) 237.
4. E. Chell and M.G. Olsson, Phys. Rev. **D48** (1993) 4076.
5. J. Gasser and U-G. Meissner, in Proc. Joint Intl. Lepton-Photon Symp. and Europhys. Conf. on High Energy Phys., Eds. S. Hegarty, K. Potter and E. Quercigh (World Scientific 1991).
6. JLQCD collaboration, KEK preprint 97-218, hep-lat/9711046 v2.
7. J.F. Donoghue, E. Golowich and B.R. Holstein, Phys. Lett. **B119** (1982) 412.
8. F. Abbud, B.W. Lee and C.N. Yang, Phys. Rev. Lett. **18** (1967) 980.
9. A.A. Belavin and I.M. Narodetskii, Sov. J. Nucl. Phys. **8** (1968) 568.
10. A. Neveu and J. Scherk, Phys. Lett. **B27** (1968) 384.
11. A.A. Bel'kov and V.V. Kostyuhkin, Sov. J. Nucl. Phys. **51** (1989) 326.
12. V. Cirigliano, J.F. Donoghue, E. Golowich and B.R. Holstein, work in progress.
13. J.F. Donoghue and A. Perez, Phys. Rev. **D55** (1997) 7075.
14. See Section 3 of Chapter IX in J.F. Donoghue, E. Golowich and B.R. Holstein, *Dynamics of the Standard Model*, (Cambridge University Press, Cambridge, England 1992).
15. T. Das, G.S. Guralnik, V.S. Mathur, F.E. Low and J.E. Young, Phys. Rev. Lett. **18** (1967) 759.
16. J.F. Donoghue and E. Golowich, Phys. Rev. **D49** (1994) 1513.
17. G. Ecker, J. Gasser, A. Pich and E. de Rafael, Nucl. Phys. **B321** (1989) 311.

THE THERMAL COUPLING CONSTANT IN THE VECTOR $\lambda\varphi^4_D$ MODEL

G.N.J.AÑAÑOS, A.P.C.MALBOUISSON AND N.F.SVAITER

Centro Brasileiro de Pesquisas Físicas-CBPF, Rua Dr.Xavier Sigaud 150, Rio de Janeiro, RJ 22290-180 Brazil

We re-examine the behavior at finite temperature of the $O(N)$-symmetric massive $\lambda\varphi^4$ model in a generic D-dimensional spacetime. In the cases $D = 3$ and $D = 4$, numerical analysis of the thermal behavior of the coupling constant is done for all temperatures. It results that the behavior of the thermal coupling constant is quite different in odd or even dimensional spacetimes. In $D = 3$, the thermal coupling constant decreases up to a positive minimum value diferent from zero and then grows up monotonically as the temperature increases. In the case $D = 4$, it is found that the thermal renormalized coupling constant tends in the high temperature limit to a positive constant asymptotic value.

In this note we investigate the behavior of the renormalized thermal coupling constant of the vector N-component $\lambda\varphi^4_D$ model. Withouth loss of generality, let us suppose that we are in the symmetric phase i.e $m_0^2 > 0$. To go beyond perturbation theory, we take the leading order in $\frac{1}{N}$, in which case we know that the the contributions come only from some classes of diagrams and that it is possible to perform summations over them. Proceeding in that way, we get for the thermal renormalized coupling constant an expression of the form,

$$\lambda(\beta) = \frac{\lambda_0}{1 - \lambda_0 L\left(m^2(\beta), \beta\right)}, \tag{1}$$

where

$$L\left(m^2(\beta), \beta\right) = -\frac{3}{2}\frac{1}{(2\pi)^{D/2}}\sum_{n=1}^{\infty}\left(\frac{m(\beta)}{\beta n}\right)^{\frac{D}{2}-2} \times$$
$$K_{\frac{D}{2}-2}\left(m(\beta)n\beta\right), \tag{2}$$

λ_0 is the zero-temperature renormalized coupling constant $m(\beta)$ is the thermally corrected mass and K_ν stands for the Bessel function of the third kind. As usual, β is the inverse temperature. For simplicity we have suppresed the subscript N everywhere.

Let us investigate firstly the thermal behavior of the coupling constant in a even dimensional spacetime. We use an integral representation of the Bessel function, which leads to the result,

$$L(m^2(\beta), \beta) = G(D)(m(\beta))^{D-4} \times$$
$$\int_1^\infty dt(t^2-1)^{\frac{D-5}{2}}\frac{1}{e^{m(\beta)\beta t}-1} \tag{3}$$

where

$$G(D) = -\frac{3}{2}\frac{1}{(2\sqrt{\pi})^{D-1}}\frac{1}{\Gamma(\frac{D-3}{2})}. \tag{4}$$

Note that since we are in an even-dimensional spacetime there are no poles in the Gamma function and $G(D)$

never vanishes. Defining

$$g(D, k) = G(D)(-1)^k C_{\frac{D-5}{2}}^k \tag{5}$$

it is not dificult to show that

$$L\left(m^2(\beta), \beta\right) = \beta^{-D}\sum_{k=0}^{\infty} g(D, k)\left(m(\beta)\beta\right)^{2k+2} \times$$
$$\int_{m(\beta)\beta}^\infty d\tau \frac{\tau^{D-5-2k}}{e^\tau - 1}. \tag{6}$$

We easily get the following expression

$$L\left(m^2(\beta), \beta\right) = \beta^{-D}\sum_{k=0}^{\infty} g(D, k)\left(m(\beta)\beta\right)^{2k+2} \times$$
$$I\left(m(\beta)\beta, D - 5 - 2k\right), \tag{7}$$

where $I(x, n)$ is the Debye integral. Then, substituting eq.(7) into eqs. (2) and (1) we get an expression for the high temperature thermal coupling constant in a even dimensional spacetime. In the case of a odd dimensional spacetime, the integral representation for the Bessel function in eq.(3) can not be used in particular in the cases $D = 3$ and $D = 1$. To obtain general formulas recovering these cases we use another integral representation of the Bessel function i.e.,

$$K_\nu(z) = \frac{1}{2}\left(\frac{z}{2}\right)^\nu \int_0^\infty dt\, e^{-t-\frac{z^2}{4t}}\, t^{-(\nu+1)} \tag{8}$$

which is valid for $|arg(z)| < \frac{\pi}{2}$ and $Re(z^2) > 0$. A straighforward calculation gives

$$L\left(m^2(\beta), \beta\right) = Q(D)m(\beta)^{D-4}\int_0^\infty dt\, e^{-t}\, t^{-\frac{D}{2}+1} \times$$
$$\left(\Theta_3(\pi, e^{-\frac{m^2\beta^2}{4t}}) - 1\right), \tag{9}$$

where the theta function $\Theta_3(z, q)$ is defined by [2]:

$$\Theta_3(z, q) = 1 + 2\sum_{n=1}^\infty q^{n^2}\cos(2nz). \tag{10}$$

and $Q(D) = -\frac{3}{2}\frac{1}{(2\sqrt{\pi})^D}$. Substituting $L\left(m^2(\beta),\beta\right)$ in eq.(1) we have a closed expression for the thermal renormalized coupling constant valid in the cases $D = 3$ and $D = 1$. We would like to stress that the behavior of the thermal renormalized coupling constant is quite different from the monotonically increasing in temperature behavior of the squared mass. Indeed the kind of thermal behavior of the coupling constant depends on the spacetime dimension. For $D = 3$ the coupling constant (as a function of the temperature) decreases until some minimum value and then start to increase (see fig.1). For $D = 4$ the thermal renormalized coupling constant tends to a constant value in the high temperature limit. See fig.(2). Our results recover those in Fendley's paper [3]. It should be emphasized that the form of the thermal corrections to the mass and coupling constant have been discussed using resummation methods. We have formulated the problem in a general framework, but our results are at leading order in the $\frac{1}{N}$ expansion. We have chosen this way of working in order to get answers as much as possible of a non-perturbative character. In what concerns the thermal mass behavior, the thermal renormalized squared mass is a monotonic increasing function of the temperature for any spacetime dimension. It is also possible to obtain a general formula for the critical temperature of the second order phase transition, valid for any dimension $D > 2$ provided the necessary renormalization procedure is done to circunvect singularities of the zeta-function. The values obtained for the critical temperatures in $D = 4$ and $D = 3$ agree with previous results (see comments below).

The behavior of the thermal coupling constant depends on the spacetime dimension. In $D = 3$ the renormalized coupling constant decreases until some positive minimum value and then starts to increase slightly as a function of the temperature. See fig.(1). This result seems to indicate that at a non-perturbative level, in the framework of the Vector N-component model at the large-N limit, the answer to the question raised in ref.[1] is negative: there is no first-order phase transition induced by the thermal coupling constant in $D = 3$. In $D = 4$ the thermal renormalized coupling constant in the high temperature limit tends to a constant value, $(\lambda_0 - (3\frac{\sqrt{6}}{8\pi})\lambda_0^{\frac{3}{2}}$, which coincides exactly with the result obtained by Fendley [3]).

It is interesting to note that the thermal behaviour of the coupling constant is very sensitive to the thermal behaviour of the mass. As an ilustration of this fact we exhibit in fig.(3) the general aspect of the coupling constant as function of the temperature for the same model we have treated here, but subjected to Wick ordering [4]. In this case all tadpoles are suppressed and the thermal behaviour of the coupling constant does not depend

at all on the mass thermal behaviour. We see that in this situation the coupling constant is a monotonic decreasing function of the temperature. The suppression of Wick ordering deeply changes this behaviour. We show in fig.(4) for $D = 3$ in the same scale the plots for $\lambda(T)$ with and without Wick ordering, respectively the lower and the upper curves. We see that in the region of temperatures where $\lambda_W(T)$ goes practically to zero, $\lambda(T)$ is pactically constant at a value slightly lower than the common zero-temperature coupling constant $\lambda(0)$. The growth of $\lambda(T)$ with the temperature presented in fig.(1) is in a much smaller scale for $\lambda(T)$ than in fig.(4). In fact this growth is "microscopic" in a scale where the Wick ordered coupling constant $\lambda_W(T)$ presents asymptotic thermal freedom.

A natural extension of this work should be to go beyond the $\frac{1}{N}$ leading order results using renormalization group ideas. Another possible direction is to introduce an abelian gauge field coupled to the N-component scalar field. In $D = 3$ a topological Chern-Simons term may be added, and also a φ^6 term. In this case we have shown using perturbative and semi classical techniques that the topological mass makes appear a richer phase structure introducing the possibility of first or second order phase transitions depending on the value of the topological mass [5]. The use of resummation methods to investigate the thermal behavior of the physical quantities could generalize these previous results to the N-component vector model. These will be subjects of future investigation.

Acknowledgements

We would like to thank M.B.Silva-Neto and C.de Calan by fruitful discussions. This paper was supported by Conselho Nacional de Desenvolvimento Cientifico e Tecnologico do Brazil (CNPq).

References

1. A.P.C.Malbouisson and N.F.Svaiter, Physica A **233**, 573 (1996).
2. Handbook of Mathemtical Functions, edited by M.Abramowitz and I.A.Stegun, Dover Inc.Pub. N.Y. (1965).
3. P.Fendley, Phys.Lett.B **196**, 175 (1987).
4. C.de Calan, A.P.C.Malbouisson and N.F.Svaiter, Mod Phys.Lett. A, **13**, 1757 (1998).
5. A.P.C.Malbouisson, F.S.Nogueira and N.F.Svaiter, Europhys.Lett. **41**, 547 (1998).

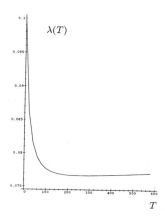

Figure 1: Coupling constant thermal behavior in dimension $D = 3$

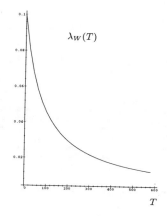

Figure 3: General aspect of coupling constant thermal behavior for the Wick ordered model.

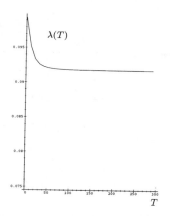

Figure 2: Coupling constant thermal behavior in dimension $D = 4$.

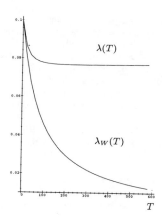

Figure 4: Compared thermal behaviors of the coupling constant for the Wick-ordered model and non Wick-ordered model.

FIXED POINT STRUCTURE OF PADÉ-SUMMATION APPROXIMATIONS TO THE QCD β-FUNCTION

V. Elias*, F. Chishtie

Department of Applied Mathematics, The University of Western Ontario, London, Ontario N6A 5B7, CANADA
* E-mail: zohar@apmaths.uwo.ca

T.G. Steele

Department of Physics and Engineering Physics, University of Saskatchewan, Saskatoon, Saskatchewan S7N 5C6
E-mail: steelet@sask.usask.ca

Padé-improvement of four-loop β-functions in massive ϕ^4 scalar field theory is shown to predict the known five-loop contribution with astonishing (0.2%) accuracy, supporting the applicability of Padé-summations for approximating all-orders $\overline{\text{MS}}$ QCD β-functions, as suggested by Ellis, Karliner, and Samuel. Surprisingly, the most general set of [2|2] approximants consistent with known two-, three-, and four-loop contributions to the QCD β-function with up to six flavours fail to exhibit any zeros that could be interpreted as positive infrared fixed points, *regardless* of the unknown five-loop term. When they occur, positive zeros of such [2|2] approximants are preceded by singularities, leading to a double-valued β-function that is decoupled entirely from the infrared region, similar to the β-function of SUSY gluodynamics.

Higher order terms of the QCD $\overline{\text{MS}}$ β-function

$$\mu^2 \frac{dx}{d\mu^2} \equiv \beta(x), \qquad (1a)$$

$$\beta(x) = -\sum_{i=0}^{\infty} \beta_i x^{i+2}, \qquad (1b)$$

$x \equiv \alpha_s(\mu)/\pi$ are, upon truncation, known to permit the occurrence of fixed points other than the ultraviolet fixed point at $x = 0$; e.g. the positive infrared fixed point (IRFP) which occurs for $9 \leq n_f \leq 16$ when the series for $\beta(x)$ in (1) is truncated after two terms [$\beta_0 = (11 - 2n_f/3)/4$; $\beta_1 = (102 - 38n_f/3)/16$; $x_{IRFP} = -\beta_0/\beta_1$]. However, the fixed points arising from such truncation are likely to be spurious, as the candidate-value for x_{IRFP} is sufficiently large for the highest-order term in the series $\beta(x_{IRFP})$ to be comparable in magnitude to lower terms [e.g.$|\beta_1 x^3| = |\beta_0 x^2|$]. In a recent paper, [1] Ellis, Karliner and Samuel predicted the coefficient β_3 via Padé approximant methods, and claimed that β_{0-2} and their prediction for β_3 yield a Padé summation of the β-function with a nonzero IRFP consistent with an earlier prediction by Mattingly and Stevenson. [2] This Mattingly-Stevenson scenario leads to the freezing-out of the coupling to a constant value in the infrared region, as shown schematically in Fig. 1.

Padé summation of the $\overline{\text{MS}}$ β-function identifies the infinte series

$$\beta(x) = -\beta_0 x^2 (1 + R_1 x + R_2 x^2 + R_3 x^2 + ...) \qquad (2)$$

($R_i \equiv \beta_i/\beta_0$) with a Padé approximant which incorporates the known UV asymptotics of the β-function,

$$\beta(x) \rightarrow \beta_{[N|M]}(x) \equiv -\beta_0 x^2 S_{[N|M]}(x), \qquad (3a)$$

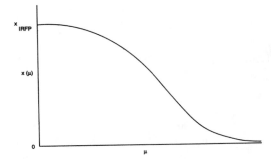

Figure 1: Mattingly-Stevenson scenario

whose Maclaurin expansion reproduces the known terms in the infinite series (2):

$$S_{[N|M]}(x) = \frac{1 + a_1 x + ... + a_N x^N}{1 + b_1 x + ... + b_M x^M}$$
$$= 1 + R_1 x + R_2 x^2 + R_3 x^3 + ... \qquad (3b)$$

An IRFP of the β-function would, in this approximation, necessarily be identified with a positive zero of $\beta_{[N|M]}$; i.e. a positive zero (x_{num}) of $1 + a_1 x + ... + a_N x^N$, the numerator of $S_{[N|M]}$, *provided* $S_{[N|M]}$ *remains positive for* $0 \leq x \leq x_{num}$. This latter requirement precludes the existence of a positive zero (x_{den}) of the denominator $1 + b_1 x + ... + b_M x^M$ that falls in the interval $0 \leq x \leq x_{num}$.

One cannot automatically dismiss the possibility of such a denominator zero occuring within the true QCD

β-function. The β-function of SU(N) SUSY gluodynamics is known *exactly* if no matter fields are present,[3] and exhibits precisely such a zero:

$$\beta(x) = -\frac{3Nx^2}{4}\left[\frac{1}{1 - Nx/2}\right]. \quad (4)$$

The β-function (4) has been discussed further by Kogan and Shifman.[4] If (4) is incorporated into (1a), the resulting Kogan-Shifman scenario (Fig. 2) for $x(\mu)$ is indicative of both a strong phase in the ultraviolet region (the upper branch of Fig. 2) as well as the existence of an infrared cut off (μ_c) on the domain of $x(\mu)$ that renders the infrared region $\mu < \mu_c$ inaccessible. [a]

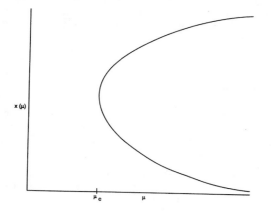

Figure 2: The Kogan-Shifman scenario $[x_{den} = x(\mu_c)]$

Without use of Padé summation methods, as described above, the known terms of the infinite series representation (2) for the $\overline{\text{MS}}$ QCD β-function offer little information as to whether the Mattingly-Stevenson (Fig. 1) or Kogan-Shifman (Fig. 2) scenario is appropriate for the evolution of the strong coupling. However, there is reason to believe that Padé summation representations of the β-function [eq. (3a)] may indeed be appropriate for quantum field theoretical calculations. Ellis, Gardi, Karliner, and Samuel [1,5] have argued that Padé-summations (3) converge to their perturbative series (2) as N and M increase for any such series dominated by a finite set of renormalon poles, consistent with the following asymptotic error formula for the difference between R_{N+M+1} in (2) and the value $R_{N+M+1}^{\text{Padé}}$ predicted via use of the $[N|M]$ approximant in (3b): [1,6]

$$\delta_{N+M+1} \equiv \frac{R_{N+M+1}^{\text{Padé}} - R_{N+M+1}}{R_{N+M+1}}$$

---

[a] We are assuming α_s to be real. The possibility of α_s being complex for $\mu < \mu_c$ is addressed in ref. 4.

$$= -\frac{M!A^M}{[N + M(1+a) + b]^M}. \quad (5)$$

In (5), $\{a, b, A\}$ are constants to be determined.

To demonstrate the utility of this asymptotic error formula, consider the known β-function for massive ϕ^4 scalar-field theory:[7]

$$\mathcal{L} = \frac{1}{2}(\partial_\mu\phi)(\partial^\mu\phi) + \frac{1}{2}m^2\phi^2 + g\left(\frac{16\pi^2}{4!}\right)(\phi^2)^2, \quad (6)$$

$$\beta(g) = 1.5g^2\left\{1 - \frac{17}{9}g + 10.8499g^2 - 90.5353g^3 + R_4g^4 + ...\right\}. \quad (7)$$

Using the first two terms of (7) to generate a $[0|1]$ approximant, as in (3), one would predict $R_2^{Padé} = (-17/9)^2$, in which case we see from (5) that

$$\delta_2 = \frac{(-17/9)^2 - 10.8499}{10.8499} = \frac{-A}{1 + (a+b)}. \quad (8)$$

Using the first three terms of (7) to generate a $[1|1]$ approximant, one would predict $R_4^{Padé} = (10.8499)^2/(-17/9)$, in which case

$$\delta_3 = \frac{[(10.8499)^2/(-17/9)] - (-90.5353)}{(-90.5353)}$$

$$= \frac{-A}{2 + (a+b)}. \quad (9)$$

Equations (8) and (9) are two equations for the unknown constants A and $(a+b)$, with solutions

$$A = [1/\delta_2 - 1/\delta_3]^{-1}, \ (a+b) = (\delta_2 - 2\delta_3)/(\delta_3 - \delta_2). \quad (10)$$

We can substitute (10) into (5) to determine R_4. The first three terms in the series (7) generate a $[2|1]$ approximant whose Maclaurin expansion (3b) predicts $R_4^{Padé} = (-90.5353)^2/(10.8499)$. Upon substitution of A, $(a+b)$, and $R_4^{Padé}$ into (5) we find that this asymptotic error formula *predicts* that

$$R_4 = R_4^{Padé}/(1 + \delta_4) \quad (11)$$
$$= R_4^{Padé}[3 + (a+b)]/[3 + (a+b) - A] = 947.8$$

The value of R_4 in (7) has been explicitly calculated[7] to be 949.5, in very close agreement with (11). The series of steps leading from (5) to (11), a methodological recipe first presented in ref. 6, has also been applied [1,8] to N-component scalar field theory, for which the Lagrangian (6) is modified such that $\phi \to \phi_a$, $\phi^2 \to \sum_{a=1}^{N}\phi^a\phi^a$, $N = \{2, 3, 4\}$. Agreement with calculated values of R_4 ($R_4 \equiv \beta_4/\beta_0$) remains within 3.5% for $N \leq 4$.[8]

This startling agreement suggests that Padé methodology may also be applicable to the QCD β-function, particularly in the $n_f = 0$ gluodynamic limit where

1. such methods are expected to be most accurate,[6]

2. comparison with the Kogan-Shifman scenario for SUSY gluodynmamics is most relevant.

For $n_f = 0$, the 4-loop $\overline{MS}$ QCD β-function, as defined by (1a), is given by [9]

$$\beta(x) = -\frac{11}{4}x^2[1 + 2.31818 + 8.11648x^2$$
$$+ 41.5383x^3 + \sum_{k=4}^{\infty} R_k x^k]. \quad (12)$$

The coefficients R_k are presently not known for $k \geq 4$. The first three terms in the series (12) are sufficient in themselves to determine the Padé approximants $S_{[1|2]}$ and $S_{[2|1]}$, as defined in (3). These approximants are

$$\beta_{[2|1]}(x) = -\frac{11}{4}x^2\left[\frac{1 - 2.7996x - 3.7475x^2}{1 - 5.1178x}\right], \quad (13)$$

$$\beta_{[1|2]}(x) = -\frac{11}{4}x^2\left[\frac{1 - 5.9672x}{1 - 8.2854x + 11.091x^2}\right]. \quad (14)$$

In both (13) and (14), the (first) positive denominator zero *precedes* the positive numerator zero: for (13), $x_{num} = 0.264 > x_{den} = 0.195$; for (14), $x_{num} = 0.168 > x_{den} = 0.151$. Consequently, x_{num} cannot be identified with the Mattingly-Stevenson IRFP in either case, as this zero is separated from the small x-region by a singularity past which the β-function switches sign. Indeed the ordering $0 < x_{den} < x_{num}$ is suggestive of a Kogan-Shifman scenario in which x_{num}, if taken seriously, is an *ultraviolet* fixed point (UVFP) characterizing the strong phase [*i.e.*, the upper branch of Fig 2].

We can apply the asymptotic error formula (5) to the series (12) in precisely the same way we applied it to (7). We then obtain an estimate $R_4 = 302.2$, analogous to (11). Using this value of R_4 in conjunction with the known terms of (12), it is possible to obtain a [2|2]-approximant β-function

$$\beta_{[2|2]}(x) = -\frac{11}{4}x^2\left[\frac{1 - 9.6296x + 4.3327x^2}{1 - 11.9477x + 23.913x^2}\right]. \quad (15)$$

The first positive numerator zero $x_{num} = 0.1092$ is again larger than the first positive denominator zero $x_{den} = 0.1063$, precluding the identification of x_{num} as the IRFP of the Mattingly-Stevenson scenario (Fig. 1). Instead, the β-function (15) is consistent with the Kogan-Shifman scenario of Fig. 2, with x_{num} again identified as a nonzero UVFP for the strong phase.

Curiously, the ordering $0 < x_{den} < x_{num}$ characterizes [2|2]-approximant β-functions *even if R_4 is allowed to be arbitrary*. The most general such β-function that

reproduces the first four terms of (12) [the first three being known] is

$$\beta_{[2|2]}(x) = -\frac{11}{4}x^2 \times \quad (16)$$
$$\left[\frac{1 + (13.403 - 0.076215R_4)x - (22.915 - 0.090166R_4)x^2}{1 + (11.084 - 0.076215R_4)x - (56.727 - 0.26685R_4)x^2}\right].$$

It is easy to verify the first positive numerator zero of (16) is always larger than the first positive denominator zero [Fig. 3], although these zeros become asymptotically close as $R_4 \to +\infty$. Thus, we see that the first positive zero of *any* [2|2] Padé approximant whose Maclaurin expansion reproduces the known terms of eq. (12) *cannot* be identified as an IRFP, nor is such Padé-summation indicative of a Fig. 1 scenario for the $\overline{MS}$ $n_f = 0$ β-function.

Remarkably, the same set of conclusions can be drawn for the physically interesting case of three light flavours. When $n_f = 3$, the 4-loop $\overline{MS}$ QCD β-function is given by [9]

$$\beta(x) = -\frac{9x^2}{4}\left[1 + (16/9)x + 4.471065x^2\right.$$
$$\left. + 20.99027x^3 + \sum_{k=4}^{\infty} R_k x^k\right], \quad (17)$$

with R_k not presently known for $k \geq 4$. The known terms in (17) determine [2|1] and [1|2] Padé-summation representations of the $n_f = 3$ β-function,

$$\beta_{[2|1]}(x) = -\frac{9x^2}{4}\left[\frac{1 - 2.91691x - 3.87504x^2}{1 - 4.69468x}\right], \quad (18)$$

$$\beta_{[1|2]}(x) = -\frac{9x^2}{4}\left[\frac{1 - 8.17337x}{1 - 9.95115x + 13.2199x^2}\right]. \quad (19)$$

The positive zero of $\beta_{[2|1]}(x)$ ($x = 0.2559$) occurs after the pole at $x = 0.2130$; the positive zero of $\beta_{[2|1]}(x)$ at $x = 0.1223$ similarly occurs after a pole at $x = 0.1194$. The most general [2|2] approximant consistent with (17) is

$$\beta_{[2|2]}(x) = -\frac{9x^2}{4} \times \quad (20)$$
$$\left[\frac{1 + (7.1945 - 0.10261R_4)x - (11.329 - 0.075643R_4)x^2}{1 + (5.4168 - 0.10261R_4)x - (25.430 - 0.25806R_4)x^2}\right].$$

The Maclaurin expansion of (20) reproduces the series in (17), including its (unknown) $R_4 x^4$ term. As was the case in (16), the first positive zero of the denominator of (20) is always seen to precede the first positive zero of the numerator, regardless of R_4. Thus the [1|2], [2|1] and most general possible [2|2]-approximant representations of the $n_f = 3$ $\overline{MS}$ β-function uphold the ordering

1644

$0 < x_{den} < x_{num}$, an ordering that precludes the identification of x_{num} with the IRFP of the Mattingly-Stevenson scenario. Moreover, [2|2]-approximant β-functions for arbitrary R_4 have been constructed [8] analogous to (16) and (20) for $n_f = \{4,5,6\}$, and for each of these, the $0 < x_{den} < x_{num}$ ordering persists regardless of R_4. A range for R_4 for which an ordering compatible with Fig. 1 $(0 < x_{num} < x_{den})$ is possible does not occur until $n_f = 7$.

As noted above, the ordering $0 < x_{den} < x_{num}$ suggests the occurrence of a double-valued QCD coupling constant, as is the case in SUSY gluodynamics (Fig. 2). Such a scenario is seen to decouple the infrared region $\mu < \mu_c$ from the domain of α_s, provided α_s is understood to be real. Such a scenario is also indicative of a strong phase at short distances [4] with possible implications for dynamical electroweak symmetry breaking, suggesting that QCD may even furnish its own "technicolour."

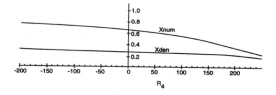

Figure 3: Dependence of the first positive numerator zero (x_{num}) and denominator zero (x_{den}) of (16) on R_4, the (presently unknown) five-loop coefficient of the $n_f = 0$ β-function.

Acknowledgements

VE and TGS are grateful for research support from NSERC, the Natural Sciences and Engineering Research Council of Canada.

References

1. J. Ellis, M. Karliner and M.A. Samuel, *Phys. Rev.* D **400**, 176 (1997).
2. A.C. Mattingly and P. M. Stevenson, *Phys. Rev.* D **49**, 437 (1994), and *Phys. Rev. Lett.* **69**, 1320 (1992).
3. V. Novikov, M. Shifman, A. Vainshtein, and V. Zakharov, *Nucl. Phys.* B **229**, 381 (1983), and *Phys. Lett.* B **166**, 329 (1986); D.R.T. Jones, *Phys. Lett.* B **123**, 45 (1983).
4. I. I. Kogan and M. Shifman, *Phys. Rev. Lett.* **75**, 2085 (1995).
5. M. Samuel, J. Ellis, and M. Karliner, *Phys. Rev. Lett.* **74**, 4380 (1995); J. Ellis, E. Gardi, M. Karliner, and M. A. Samuel, *Phys. Lett.* B **366**, 268 (1996) and *Phys. Rev.* D **54**, 6986 (1996).
6. J. Ellis, I. Jack, D.R.T. Jones, M. Karliner, and M. A. Samuel, *Phys. Rev.* D **57**, 2665 (1998).
7. H. Kleinert, J. Neu, V. Schulte-Frohlinde, K.G. Chetyrkin, and S. A. Larin, *Phys. Lett.* B **272**, 39 (1991).
8. V. Elias, T. G. Steele, F. Chishtie, R. Migneron, and K. Sprague, *Phys. Rev.* D , (to appear) hep-ph/9806324.
9. T. van Ritbergen, J.A.M. Vermaseren, and S.A. Larin, *Phys. Lett.* B **400**, 379 (1997).

Problem of Divergent Energy-Integrals in the Coulomb Gauge

G. Leibbrandt

Department of Mathematics and Statistics, University of Guelph, Guelph, Ontario, Canada N1G 2W1
and
Theoretical Physics Division, CERN, CH-1211 Genève 23, Switzerland
E-mail: gleibbra@msnet.mathstat.uoguelph.ca

Using the gauge-invariant technique of split dimensional regularization, we have computed the three-point vertex function in the noncovariant Coulomb gauge. The new regularization procedure is capable of handling both the conventional UV/IR divergences, as well as the divergences from the notorious energy-integrals. The relevant BRST identity holds exactly, despite the appearance of nonlocal integrals. Preliminary studies indicate that split dimensional regularization is also viable at the two-loop level.

1 Introduction and Review

For perturbative calculations in quantum field theory we require either a covariant gauge or a noncovariant gauge. If the gauge is *covariant*, such as the Lorentz gauge $\partial^\mu A_\mu^a(x) = 0, \mu = 0,1,2,3$, and a = 1,2,..., N^2 - 1, for SU(N), the corresponding Feynman integrals are generally computed by using Feynman's $i\epsilon$ - prescription, $\epsilon > 0$. If the gauge is *noncovariant*, such as the axial gauge[1] defined by $n^\mu A_\mu^a(x) = 0$, where n_μ is an arbitrary four-vector, $n^2 = 0$ or $n^2 \gtrless 0$, the axialtype integrals may be evaluated consistently by employing the n^* - prescription for the spurious poles of $(q \cdot n)^{-\beta}, \beta = 1, 2, \ldots$ [2,3].

However, if the gauge happens to be the noncovariant Coulomb gauge, evaluation of the associated integrals is wrought with fresh difficulties, especially at two and three loops [4]. In this talk we shall focus on the major problems afflicting the Coulomb gauge and suggest a general procedure for computing the integrals. The Coulomb gauge is defined by

$$\vec{\nabla} \cdot \vec{A}^a(x) = 0, \vec{\nabla} \equiv \vec{i}\partial_x + \vec{j}\partial_y + \vec{k}\partial_z, \quad (1)$$

where $\vec{A}^a$ is a gauge field, a = 1,2,...,8. Schwinger[5] had pointed out as early as 1962 that transition from the classical Coulomb-gauge Hamiltonian to the quantum Hamiltonian gives rise to additional terms, called V_1, V_2 terms[6]. The latter are *nonlocal* and lead to divergent, ambiguous integrals such as[7]

$$\int dq_0 \, q_0^2 \int \frac{d^3\vec{q}}{(q^2 + i\epsilon)(\vec{q} - \vec{p})^2}, \epsilon > 0, q^2 = q_0^2 - \vec{q}^2, \quad (2)$$

$$\int dp_0 \int dq_0 \int d^3\vec{p} \int d^3\vec{q} \frac{p_0}{p^2 + i\epsilon} \frac{q_0}{q^2 + i\epsilon} F(\vec{p}, \vec{q}), \quad (3)$$

which cause major technical problems in non-Abelian models. They are known[8] as energy-integrals, $\mathcal{E}$, because their divergences arise specifically from integrating over

dp_0 and dq_0. In 1987, Doust and Taylor concluded[9] that standard dimensional regularization did not regulate the divergences in the p_0-, q_0- integrations.

2 Previous Recipes. Split Dimensional Regularization

2.1 Earlier work

During the past 20 years, various recipes have been proposed for handling the notorious energy-integrals $\mathcal{E}$, and we thought it might be instructive to acquaint the reader with some of the more common recipes.

(i) Christ and Lee[6] write the Lagrangian density in the Coulomb gauge, $\mathcal{L}^{\text{Coul}}$, as a Weyl-ordered piece[6], plus two extra terms:

$$\mathcal{L}^{\text{Coul}} = \{\mathcal{L}^{\text{Coul}}\}_{\text{ordered}}^{\text{Weyl-}} + V_1 + V_2. \quad (4)$$

They suggest the following procedure for the ambiguous $\mathcal{E}$'s: Set the $\mathcal{E}$'s to zero, and then try to compute the *nonlocal* V_1, V_2 terms..

(ii) The basic idea of Cheng and Tsai[8] is to regulate the $\mathcal{E}$'s via a Θ - regulator, namely

$$\frac{1}{\vec{k}^2} \implies \frac{1}{\vec{k}^2 - \Theta^2 k_0^2}\Bigg|_{\Theta \to 0}, \quad (5)$$

and to demonstrate that *combinations* of Θ - regulated $\mathcal{E}$'s are equivalent to the V_1, V_2 terms of Christ and Lee.

(iii) Doust and Taylor[9] employ a *naive* $\mathcal{L}^{\text{Coul}}$, i.e. one that is not Weyl-ordered, and then attempt to combine Feynman graphs in such a way that a *combination* of $\mathcal{E}$'s yields a convergent expression. This technique works admirably for two-loop diagrams, but becomes inadequate at three loops.

(iv) In a recent publication by Baulieu and Zwanziger[10], the $\mathcal{E}$'s are likewise treated with a regulator.

2.2 Split dimensional regularization

Our approach [11] consists of using a naive $\mathcal{L}^{Coul}$ and trying to *regularize individual $\mathcal{E}$'s* with the help of *two* complex-dimensional regulating parameters σ, ω, i.e. we replace the measure $d^4q = dq_0\, d^3\vec{q}$ by

$$d^{2(\omega+\sigma)}q = d^{2\sigma}q_0\, d^{2w}\vec{q}, \tag{6}$$

taking the limits $\sigma \to (1/2)^+$, and $\omega \to (3/2)^+$ after all integrations have been executed. The new technique is called *split* dimensional regularization, and possesses two principal features:

(i) The UV/IR divergences manifest themselves as poles of $\Gamma(2-\sigma-\omega)$;

(ii) *all* one-loop Coulomb-gauge integrals are ambiguity-free, and some are *nonlocal*, as for example the integral

$$div \int \frac{d^4q}{q^2(q+p)^2(\vec{q}+\vec{p})^2} = -\frac{2}{p^2}\Gamma(2-\sigma-\omega)\Big|_{w\to(3/2)^+}^{\sigma\to(1/2)^+}, \tag{7}$$

where "div" denotes the *divergent part*.

3 The three-point function in split dimensional regularization

3.1 Calculation and Results

The purpose of this section is to apply split dimensional regularization to the quark self-energy $\Sigma(p)$ and the quark-quark-gluon vertex function $\Lambda_\mu(p',p)$ in the Coulomb gauge. The tools for this computation include the (naive) Lagrangian density $\mathcal{L}^{Coul}$,

$$\mathcal{L}^{Coul} = -\frac{1}{4}F_{\mu\nu}^{a\,2} - \frac{1}{2\alpha}(\vec{\nabla}\cdot\vec{A}^a)^2 + \mathcal{L}_{ghost} + L_{ext}, \alpha \to 0, \tag{8}$$

the gauge propagator $G_{\mu\nu}^{ab}(q)$,

$$G_{\mu\nu}^{ab}(q) = \frac{-i\delta^{ab}}{(2\pi)^4(q^2+i\epsilon)}\Big[g_{\mu\nu} + \frac{n^2 q_\mu q_\nu}{\vec{q}^2} - \frac{q\cdot n}{\vec{q}^2}(q_\mu n_\nu + q_\nu n_\mu)\Big], \epsilon > 0, \tag{9}$$

and the scalar ghost propagator $G_{\mu\nu}^{ghost}$,

$$G_{\mu\nu}^{ghost} = \frac{i\delta^{ab}}{(2\pi)^4\vec{q}^2}. \tag{10}$$

The three-gluon vertex has the same structure as in covariant gauges.

Evaluation of the three-point function is complicated by the appearance of new, divergent integrals with four and five propagators. The latter possess an intricate singularity structure that leads to *nonlocal* pole terms, as seen from the examples below [12] :

$$div \int \frac{d^4q}{(2\pi)^4 q^2(q+p)^2\vec{q}^2(\vec{q}+\vec{p})^2}$$
$$= +\frac{-4}{\vec{p}^2 p^2}\frac{\Gamma(2-\sigma-\omega)}{(4\pi)^2}\Big|_{\omega\to(3/2)^+}^{\sigma\to(1/2)^+}, \tag{11}$$

and

$$div \int \frac{d^4q}{(2\pi)^4 q^2(q+k)^2[(q+p)^2+m^2]\vec{q}^2(\vec{q}+\vec{k})^2}$$
$$= \frac{-2}{\vec{k}^2 k^2}\frac{\Gamma(2-\sigma-\omega)}{(4\pi)^2}$$
$$\times \Big[\frac{1}{p^2+m^2} + \frac{1}{(p+k)^2+m^2}\Big]\Big|_{\omega\to(3/2)^+}^{\sigma\to(1/2)^+}; \tag{12}$$

m is the quark mass, and the *nonlocalities* reside, respectively in $(\vec{p}^2 p^2)^{-1}$, and $(\vec{k}^2 k^2)^{-1}$. Despite the appearance of many such nonlocal integrals, the final expressions for $\Sigma(p)$ and $\Lambda_\mu(p',p)$ are not only *local* but surprisingly simple:

$$\Sigma(p) = \frac{1}{3\pi}\alpha_s(\not{p}-4m)\Gamma(2-\sigma-\omega), \tag{13}$$

$$\Lambda_\mu(p',p) = -\frac{1}{3\pi}\alpha_s\gamma_\mu\Gamma(2-\sigma-\omega), \tag{14}$$

with $\alpha_s \equiv \bar{g}^2/4\pi, g \equiv \bar{g}\mu^{2-\sigma-w}$, where $\sigma \to (1/2)^+, w \to (3/2)^+$, as usual; μ is the mass scale.

3.2 BRST identity

There is another surprise in the computation of Λ_μ which concerns the BRST identity relating Λ_μ and Σ. It was shown by Taylor [13,14], that the correct identity in the Coulomb gauge has the form

$$(p'-p)^\mu\Lambda_\mu(p',p) + E_1(p',p) + E_2(p',p) = -\Sigma(p') + \Sigma(p), \tag{15}$$

where E_1 and E_2 are 1 P I three-point functions, each consisting of one ghost propagator, one gluon propagator, and one quark propagator (see Figs 4a, b in ref. 12). E_1 has the nontrivial structure

$$E_1(p', p) = [factors] \int \frac{d^4 q}{(2\pi)^4 [(q-k)^2 + i\epsilon] \vec{q}^2}$$
$$\times q_i \gamma_j \frac{1}{(\not{p} + \not{q} - m + i\epsilon)}$$
$$\times [-\delta_{ij} + \frac{(q-k)_i (q-k)_j}{(\vec{q} - \vec{k})^2}]$$
$$\times (\not{p} - m), i, j = 1, 2, 3, k_\mu \equiv p_\mu' - p_\mu. \quad (16)$$

Utilizing the various integrals listed, for example, in refs. 11,12, we find that the *divergent* part of E_1 is actually zero:

$$div E_1(p', p) = 0; \quad (17)$$

similarly,

$$div E_2(p', p) = 0. \quad (18)$$

In summary, split dimensional regularization respects the "reduced" BRST identity

$$(p' - p)^\mu \Lambda_\mu(p', p) = -\Sigma(p') + \Sigma(p), \quad (19)$$

which is just the conventional identity for covariant gauges. We should stress that an identity resembling Eq. (15) has also been examined by Muzinich and Paige [15], and, more recently, by Newton [16]. (See in this connection also, ref. 17.)

4 What about two-loop graphs?

We have seen that split dimensional regularization is capable of regulating *all* one-loop energy-integrals in the Coulomb gauge. But how effective and reliable is the procedure at higher loops? To get an idea of the technical problems likely to arise in a two-loop calculation, we have taken a cursory look at the "sunset" diagram in Yang-Mills theory. This diagram contains a large quantity of ambiguous energy-integrals of the form

$$I_M = \int \frac{dq_0}{2\pi} \int \frac{dk_0}{2\pi} \frac{k_0}{k^2 + i\epsilon} \frac{q_0}{q^2 + i\epsilon},$$
$$q^2 = q_0^2 - \vec{q}^2, \epsilon > 0, \quad (20)$$

or, in Euclidean space,

$$I_E = i^4 \int \frac{dq_4}{2\pi} \int \frac{dk_4}{2\pi} \frac{k_4}{k_4^2 + \vec{k}^2} \frac{q_4}{q_4^2 + \vec{q}^2}. \quad (21)$$

Working in the framework of split dimensional regularization, we first define k_4, q_4 each over 2σ - dimensional complex space (labelling the "new" variables K_μ,

Q_μ respectively, $\mu = 1,2,3,...$), and then define $\vec{k}$, $\vec{q}$ each over 2ω -dimensional complex space (labelling the new vectors $\vec{K}$, $\vec{Q}$, respectively). Hence, I_E becomes

$$I_E = \int \frac{d^{2\sigma} Q \, Q_\mu}{(2\pi)^{2\sigma} (Q^2 + \vec{Q}^2)} \int \frac{d^{2\sigma} K \, K_\mu}{(2\pi)^{2\sigma} (K^2 + \vec{K}^2)},$$
$$Q^2 = Q_\mu Q^\mu, etc. \quad (22)$$

Each 2σ-dimensional integral is now well defined and can be evaluated by standard dimensional regularization. Thus,

$$\int \frac{d^{2\sigma} Q \, Q_\mu}{(2\pi)^{2\sigma} (Q^2 + \vec{Q}^2)} = 0, etc. \quad (23)$$

In the context of split dimensional regularization, the divergent part of the two-loop energy-integral (20) is, therefore, equal to zero:

$$div I_M = 0. \quad (24)$$

5 Conclusion

We have shown that the method of split dimensional regularization leads to ambiguity-free energy-integrals, and respects the BRST identity relating the quark self-energy $\Sigma(p)$ and the three-point function $\Lambda_\mu(p', p)$. Although a large portion of the Coulomb-gauge integrals have *nonlocal* pole parts, the final expressions for $\Sigma(p)$ and $\Lambda_\mu(p', p)$, as well as the gluon self-energy $\Pi_{\mu\nu}^{ab}(p)$, are strictly *local*. It remains to demonstrate that split dimensional regularization gives consistent results also in a general two-loop calculation and satisfies such basic concepts as BRST/Ward identities, Lorentz covariance, ghost number conservation, etc.

Acknowledgements

I am most grateful to John C. Taylor for his continued interest and sound advice concerning the Coulomb gauge. I have also benefitted from discussions with C. Becchi, M. Beneke, J.D. Bjorken, S. Brodsky, K. Haller, H. Hüffel, C. Newton, and T.T. Wu. It gives me great pleasure to thank Alvaro De Rújula, and the secretarial staff, for their hospitality during the summer of 1998 in the Theoretical Physics Division at CERN, where this paper was initiated. I should also like to thank my student, Jimmy Williams, for checking several of the more challenging integrals. This research was supported in part by the Natural Sciences and Engineering Research Council of Canada under Grant No. A8063.

References

1. G. Leibbrandt, *Noncovariant Gauges*, World Scientific Publishing Co., 1994.
2. S. Mandelstam, *Nucl. Phys.* B **213**, 149 (1983).
3. G. Leibbrandt, *Phys. Rev.* D **29**, 1699 (1984).
4. J.C. Taylor in *Physical and Non-standard Gauges*, eds. P. Gaigg, W. Kummer and M. Schweda, Lecture Notes in Physics, Vol. **361** (Springer, Berlin, 1990) p.137.
5. J. Schwinger, *Phys. Rev.* **127**, 324 (1962).
6. N.H. Christ and T.D. Lee, *Phys. Rev.* D **22**, 939 (1980).
7. P.J. Doust, *Ann. Phys.* (NY) **177**, 169 (1987).
8. H. Cheng and Er-Cheng Tsai, *Phys. Rev.* D **36**, 3196 (1987).
9. P.J. Doust and J.C. Taylor, *Phys. Lett.* B **197**, 232 (1987).
10. L. Baulieu and D. Zwanziger, private communication (1998).
11. G. Leibbrandt and J. Williams, *Nucl. Phys.* B **475**, 469 (1996).
12. G. Leibbrandt, *Nuc. Phys.* B (1998) in press.
13. The author is grateful to Professor J.C. Taylor for alerting him about the existence [14] of the ghost contributions $E_1(p', p)$ and $E_2(p', p)$ in Eq. (15), and for explaining to him their graphical representation in terms of 'ordinary' Feynman diagrams.
14. J.C. Taylor, *Nucl. Phys.* B **33**, 436 (1971).
15. I.J. Muzinich and F.E. Paige, *Phys. Rev.* D **4**, 378 (1971).
16. C. Newton, private communication (1997).
17. C. Newton and T.T. Wu, *Yang-Mills Ward Identities Without a Regulator: the Three-Point Function*, preprint Lund - MPh - 96/09 (1996).

NONLOCAL COLOR INTERACTIONS IN A GAUGE-INVARIANT FORMULATION OF QCD

Kurt Haller and Lusheng Chen

Department of Physics, University of Connecticut, Storrs, CT 06269
E-mail: khaller@uconnvm.uconn.edu

We construct a set of states that implement the non-Abelian Gauss's law for QCD. We also construct a set of gauge-invariant operator-valued quark and gluon fields by establishing an explicit unitary equivalence between the Gauss's law and the 'pure glue' part of the Gauss's law operator. This unitary equivalence enables us to use the 'pure glue' Gauss's law operator to represent the entire Gauss's law operator in a new representation. Since the quark field commutes with the 'pure glue' Gauss's law operator, it is a gauge-invariant field in this new representation. We use the unitary equivalence of the new and the conventional representations to construct gauge-invariant quark and gluon fields in both representations, and to transform the QCD Hamiltonian in the temporal gauge so that it is expressed entirely in terms of gauge-invariant quantities. In that form, all interactions between quark fields mediated by 'pure glue' components of the gluon field have been transformed away, and replaced by a nonlocal interaction between gauge-invariant color-charge densities. This feature — that, in gauge-invariant formulations, interactions mediated by pure gauge components of gauge fields have been replaced by nonlocal interactions — is shared by many gauge theories. In QED, the resulting nonlocal interaction is the Coulomb interaction, which is the Abelian analog of the QCD interaction we have identified and are describing in this work. The leading term, in a multipole expansion, of this nonlocal QCD interaction vanishes for quarks in color-singlet configurations, suggesting a dynamical origin for color confinement; higher order terms of this multipole expansion suggest a QCD mechanism for color transparency. We also show how, in an SU(2) model, this nonlocal interaction can be evaluated nonperturbatively.

1 Introduction

The program of the work we are discussing is designed to take advantage of a common thread that appears to underlie all gauge theories — non-Abelian as well as Abelian: When only gauge-invariant fields are used in constructing the Hamiltonian for a gauge theory, interactions between charged fields (electrical, color, etc.) and pure-gauge components of gauge fields cannot arise, and a nonlocal interaction between charge densities — the Coulomb interaction in QED — appears in their stead. Moreover, in gauge-invariant QED, the only interaction besides the Coulomb interaction is between the transverse (gauge-invariant) part of the gauge field and the current density; this interaction takes the form $-\int j_i(\mathbf{r}) A_{T\,i}(\mathbf{r})d\mathbf{r}$, where $A_{T\,i}$ is the transverse (gauge-invariant) part of the gauge field, and $\vec{j}(\mathbf{r}) = e\psi^\dagger(\mathbf{r})\vec{\alpha}\psi(\mathbf{r})$. Since the current density has a v/c dependence in the nonrelativistic limit, the Coulomb interaction is, by far, the most important electrodynamic force in the low-energy regime.

We will show that a very similar effect occurs in QCD. The main difference between the two cases is the fact that constructing gauge-invariant gauge fields is much more difficult in non-Abelian theories than in Abelian ones. Also, the nonlocal interaction in QCD is more complicated, but involves not only quark color-charge densities, but also gauge-invariant gluon fields. Significantly, in QCD, just as in QED, the only interaction other than the nonlocal one involves the gauge-invariant gauge (gluon) field and a quark color-current density that is the non-Abelian analog of the current density in QED, and that can be expected to also have a v/c

dependence in the nonrelativistic limit.

In this paper we will show how to implement the non-Abelian Gauss's law, how to construct gauge-invariant quark and gluon states, and how to express the QCD Hamiltonian in terms of these gauge-invariant fields. We will also discuss the physical implications of the formal results we obtain.

2 Implementing Gauss's law for QCD

The Gauss's law operator in QCD is given by

$$\hat{\mathcal{G}}^a(\mathbf{r}) = \overbrace{\partial_i\Pi_i^a(\mathbf{r}) + \underbrace{gf^{abc}A_i^b(\mathbf{r})\Pi_i^c(\mathbf{r})}_{J_0^a(\mathbf{r})}}^{D_i\Pi_i^a(\mathbf{r})=\text{'pure-glue' part}} + j_0^a(\mathbf{r}), \quad (1)$$

where $j_0^a(\mathbf{r}) = g\psi^\dagger(\mathbf{r})\frac{\lambda^a}{2}\psi(\mathbf{r})$ is the quark color-charge density, and $J_0^a(\mathbf{r})$ is the gluon color-charge density. To implement the non-Abelian Gauss's law, we must construct states that are annihilated by the Gauss's law operator, *i.e.* we must solve

$$\hat{\mathcal{G}}^a(\mathbf{r})|\hat{\Psi}\rangle = 0. \quad (2)$$

In QED, it is easy to implement Gauss's law, $\{\partial_i\Pi_i(\mathbf{r}) + j_0(\mathbf{r})\}|\hat{\xi}\rangle = 0$, because $\partial_i\Pi_i(\mathbf{r}) + j_0(\mathbf{r})$ and $\partial_i\Pi_i(\mathbf{r})$ are unitarily equivalent, and $\partial_i\Pi_i(\mathbf{r})|\xi\rangle = 0$ is simple to solve.[1] In QCD, this method for implementing Gauss's law is not available. $\hat{\mathcal{G}}^a(\mathbf{r})$ and $\partial_i\Pi_i^a(\mathbf{r})$ cannot be unitarily equivalent, because these two operator-valued quantities obey inequivalent commutator algebras; whereas $[\partial_i\Pi_i^a(\mathbf{r}), \partial_i\Pi_i^b(\mathbf{r}')] = 0$, $[\hat{\mathcal{G}}^a(\mathbf{r}), \hat{\mathcal{G}}^b(\mathbf{r}')] = igf^{abc}\hat{\mathcal{G}}^c(\mathbf{r})\delta(\mathbf{r} - \mathbf{r}')$.[2]

We will initially implement the 'pure glue' form of Gauss's law for QCD,

$$D_i \Pi_i^a(\mathbf{r})|\Psi\rangle = 0 , \qquad (3)$$

by constructing a state $|\Psi\rangle = \Psi|\phi\rangle$, for which $D_i\Pi_i^a(\mathbf{r})|\Psi\rangle = \{\partial_i\Pi_i^a(\mathbf{r}) + J_0^a(\mathbf{r})\}\Psi|\phi\rangle = 0$; $|\phi\rangle$ represents a state that is annihilated by $\partial_i\Pi_i^a(\mathbf{r})$ — the so-called 'Fermi' state.[3] We then seek to construct an operator Ψ, for which

$$[\partial_i\Pi_i^a(\mathbf{r}),\ \Psi] = -J_0^a(\mathbf{r})\,\Psi + B_Q^a(\mathbf{r}) . \qquad (4)$$

where $B_Q^a(\mathbf{r})$ is an operator that has $\partial_i\Pi_i^a(\mathbf{r})$ on its extreme right. Eq.(4) is essentially an operator differential equation, in which the commutator, $[\partial_i\Pi_i^a(\mathbf{r}),\ \Psi]$, is a derivative.

We have found a solution of this equation,[4] in which Ψ is expressed as

$$\Psi = \|\exp(\mathcal{A})\| \qquad (5)$$

with

$$\mathcal{A} = i\int d\mathbf{r}\ \overline{A_i^\gamma}(\mathbf{r})\ \Pi_i^\gamma(\mathbf{r}) , \qquad (6)$$

and with $\overline{A_i^\gamma}(\mathbf{r})$ represented as the series

$$\overline{\mathcal{A}_i^\gamma}(\mathbf{r}) = \sum_{n=1}^\infty g^n \mathcal{A}_{(n)i}^\gamma(\mathbf{r}) . \qquad (7)$$

The $\mathcal{A}_{(n)i}^\gamma(\mathbf{r})$ are nonlinear functionals of transverse and longitudinal parts of gauge fields, but are independent of the canonical momentum $\Pi_i^\gamma(\mathbf{r})$. The ordered product $\|\exp(\mathcal{A})\|$ is defined so that, in the nth order term, $\|(\mathcal{A})^n\|$, all functionals of the gauge field A_i^a are *to the left of* all functionals of the canonical momenta Π_j^b. We refer to $\overline{A_i^\gamma}(\mathbf{r})$ as the *resolvent gauge field*.

The requirement that $|\Psi\rangle$ implement Gauss's law can be translated into a condition on the resolvent gauge field. Before we formulate this condition, we first define the following quantities:

$$\mathcal{X}^\alpha(\mathbf{r}) = \frac{\partial_j}{\partial^2}A_j^\alpha(\mathbf{r}) \text{ and } \mathcal{R}_{(\eta)}^{\vec{\alpha}}(\mathbf{r}) = \prod_{m=1}^\eta \mathcal{X}^{\alpha[m]}(\mathbf{r}),$$

which are functionals of gauge fields; $\overline{\mathcal{Y}^\alpha}(\mathbf{r}) = \frac{\partial_j}{\partial^2}\overline{A_j^\alpha}(\mathbf{r})$ and $\mathcal{M}_{(\eta)}^{\vec{\alpha}}(\mathbf{r}) = \prod_{m=1}^\eta \overline{\mathcal{Y}^{\alpha[m]}}(\mathbf{r})$, which have structures similar to $\mathcal{X}^\alpha(\mathbf{r})$ and $\mathcal{R}_{(\eta)}^{\vec{\alpha}}(\mathbf{r})$ respectively, but which are functionals of the resolvent gauge fields. We also need to define:

$$f_{(\eta)}^{\vec{\alpha}\beta\gamma} = f^{\alpha[1]\beta b[1]}\ f^{b[1]\alpha[2]b[2]}\ f^{b[2]\alpha[3]b[3]} \times \cdots$$
$$\times f^{b[\eta-2]\alpha[\eta-1]b[\eta-1]}\ f^{b[\eta-1]\alpha[\eta]\gamma} ; \qquad (8)$$

these are chains of structure constants whose 'links' are summed over repeated indices.

The condition on the resolvent gauge field that is equivalent to implementing the 'pure glue' Gauss's law, is

$$ig\,f^{a\beta d}A_i^\beta(\mathbf{r})\int d\mathbf{r}'[\,\Pi_i^d(\mathbf{r}),\ \overline{A_j^\gamma}(\mathbf{r}')\,]\,V_j^\gamma(\mathbf{r}') +$$
$$i\int d\mathbf{r}'[\,\partial_i\Pi_i^a(\mathbf{r}),\ \overline{A_j^\gamma}(\mathbf{r}')\,]\,V_j^\gamma(\mathbf{r}') + g\,f^{a\mu d}A_i^\mu(\mathbf{r})\,V_i^d(\mathbf{r})$$
$$= \sum_{\eta=1}\frac{g^{\eta+1}B(\eta)}{\eta!}\,f^{a\beta c}f_{(\eta)}^{\vec{\alpha}c\gamma}\,A_i^\beta(\mathbf{r})\,\frac{\partial_i}{\partial^2}\left(\mathcal{M}_{(\eta)}^{\vec{\alpha}}(\mathbf{r})\,\partial_j V_j^\gamma(\mathbf{r})\right) -$$
$$\sum_{\eta=0}\sum_{t=1}(-1)^{t-1}g^{t+\eta}\frac{B(\eta)}{\eta!(t-1)!(t+1)}\ \times$$
$$f_{(t)}^{\vec{\mu}a\lambda}\,f_{(\eta)}^{\vec{\alpha}\lambda\gamma}\,\mathcal{R}_{(t)}^{\vec{\mu}}(\mathbf{r})\,\mathcal{M}_{(\eta)}^{\vec{\alpha}}(\mathbf{r})\,\partial_i V_i^\gamma(\mathbf{r})$$
$$-g\,f^{a\beta d}A_i^\beta(\mathbf{r})\sum_{\eta=0}\sum_{t=1}(-1)^t g^{t+\eta}\,\frac{B(\eta)}{\eta!(t+1)!}\ \times$$
$$f_{(t)}^{\vec{\mu}d\lambda}\,f_{(\eta)}^{\vec{\alpha}\gamma}\frac{\partial_i}{\partial^2}\left(\mathcal{R}_{(t)}^{\vec{\mu}}(\mathbf{r})\,\mathcal{M}_{(\eta)}^{\vec{\alpha}}(\mathbf{r})\,\partial_j V_j^\gamma(\mathbf{r})\right), \qquad (9)$$

where $B(\eta)$ is the ηth Bernoulli number, and $V_j^\gamma(\mathbf{r})$ represents any arbitrary vector field in the adjoint representation of SU(3).

We have solved Eq.(9);[5] The solution is given by

$$\int d\mathbf{r}\,\overline{A_j^\gamma}(\mathbf{r})V_j^\gamma(\mathbf{r}) = \sum_{\eta=1}^\infty \frac{ig^\eta}{\eta!}\int d\mathbf{r}\ \Big\{\ \psi_{(\eta)j}^\gamma(\mathbf{r}) +$$
$$f_{(\eta)}^{\vec{\alpha}\beta\gamma}\,\mathcal{M}_{(\eta)}^{\vec{\alpha}}(\mathbf{r})\,\overline{B_{(\eta)j}^\beta}(\mathbf{r})\ \Big\}\,V_j^\gamma(\mathbf{r}) , \qquad (10)$$

where

$$\psi_{(\eta)i}^\gamma(\mathbf{r}) = (-1)^{\eta-1}\,f_{(\eta)}^{\vec{\alpha}\beta\gamma}\,\mathcal{R}_{(\eta)}^{\vec{\alpha}}(\mathbf{r})\,\mathcal{Q}_{(\eta)i}^\beta(\mathbf{r}) \qquad (11)$$

$$\text{with } \mathcal{Q}_{(\eta)i}^\beta(\mathbf{r}) = [\,a_i^\beta(\mathbf{r}) + \frac{\eta}{(\eta+1)}\,x_i^\beta(\mathbf{r})\,] , \qquad (12)$$

and

$$\overline{B_{(\eta)i}^\beta}(\mathbf{r}) = a_i^\beta(\mathbf{r}) + \left(\delta_{ij} - \frac{\eta}{(\eta+1)}\frac{\partial_i\partial_j}{\partial^2}\right)\overline{A_j^\beta}(\mathbf{r}) ; \qquad (13)$$

a_i^α and x_i^α designate the transverse and longitudinal parts of the gauge field $A_i^\alpha(\mathbf{r})$ respectively. Eq.(10) calls for summations over the multiplicity index η, which labels the multiplicity of $\overline{\mathcal{Y}^\alpha}(\mathbf{r})$ factors in $\mathcal{M}_{(\eta)}^{\vec{\alpha}}(\mathbf{r})$, and the multiplicity of $\mathcal{X}^\alpha(\mathbf{r})$ factors in $\mathcal{R}_{(\eta)}^{\vec{\alpha}}(\mathbf{r})$. This multiplicity of factors makes Eq.(10) a nonlinear integral equation that specifies the resolvent gauge field $\overline{A_j^\gamma}(\mathbf{r})$ recursively. The possibility that this nonlinear equation has multiple solutions, and that these multiple solutions correspond to different physical regimes, is interesting, but as yet unexplored.

In our earlier work,[5] we proved that Eq.(10) is a solution of Eq.(9) — a proof that we have referred to as the 'fundamental theorem' — and that the resolvent gauge field, therefore, is the operator-valued field required for the implementation of the non-Abelian Gauss's law. We will see, in the remainder of this paper, that the resolvent gauge field plays a pivotal role in the construction of gauge-invariant fields and in the transformation of the QCD Hamiltonian to a functional of gauge-invariant fields.

3 Gauge-invariant quark and gluon fields

The observation that underlies the construction of gauge-invariant quark and gluon fields is that the Gauss's law operator, $\hat{\mathcal{G}}^a(\mathbf{r})$, and the 'pure glue' Gauss' law operator $\mathcal{G}^a(\mathbf{r}) = D_i \Pi_i^a(\mathbf{r})$ are unitarily equivalent. In earlier work,[5] we have shown that

$$\hat{\mathcal{G}}^a(\mathbf{r}) = \mathcal{U}_C \, \mathcal{G}^a(\mathbf{r}) \, \mathcal{U}_C^{-1} , \qquad (14)$$

where $\mathcal{U}_C = \exp(\mathcal{C}_0) \exp(\bar{\mathcal{C}})$ with $\mathcal{C}_0$ and $\bar{\mathcal{C}}$ given by

$$\mathcal{C}_0 = i \int d\mathbf{r} \, \mathcal{X}^\alpha(\mathbf{r}) j_0^\alpha(\mathbf{r}) , \quad \text{and} \qquad (15)$$

$$\bar{\mathcal{C}} = i \int d\mathbf{r} \, \overline{\mathcal{Y}^\alpha}(\mathbf{r}) j_0^\alpha(\mathbf{r}) . \qquad (16)$$

We are therefore free to interpret the 'pure glue' Gauss's law operator $\mathcal{G}^a(\mathbf{r})$ as the complete Gauss's law operator, $\hat{\mathcal{G}}^a(\mathbf{r})$, in a different, unitarily equivalent representation. We will refer to the conventional representation, in which $\hat{\mathcal{G}}^a(\mathbf{r})$ is the complete Gauss's law operator and $\mathcal{G}^a(\mathbf{r})$ the 'pure glue' Gauss's law operator, as the $\mathcal{C}$ representation; and the new representation, in which $\mathcal{G}^a(\mathbf{r})$ is the unitarily transformed *complete* Gauss's law operator, as the $\mathcal{N}$ representation. It is manifest that the spinor (quark) field $\psi(\mathbf{r})$ commutes with $\mathcal{G}^a(\mathbf{r})$. Since the Gauss's law operator is the generator of gauge transformations, $\psi(\mathbf{r})$ is a gauge-invariant spinor (quark) field in the $\mathcal{N}$ representation. To find the corresponding gauge-invariant quark field in the $\mathcal{C}$ representation, we apply the transformation that appears in Eq.(14), and obtain

$$\psi_{\mathsf{GI}}(\mathbf{r}) = \mathcal{U}_C \, \psi(\mathbf{r}) \, \mathcal{U}_C^{-1} = V_C(\mathbf{r}) \psi(\mathbf{r}) \qquad (17)$$

where

$$V_C(\mathbf{r}) = \exp\left(-ig\overline{\mathcal{Y}^\alpha}(\mathbf{r})\tfrac{\lambda^\alpha}{2}\right) \exp\left(-ig\mathcal{X}^\alpha(\mathbf{r})\tfrac{\lambda^\alpha}{2}\right) , \quad (18)$$

and the λ^h represent the Gell-Mann SU(3) matrices. $V_C(\mathbf{r})$ could be written as $\exp(-ig\mathcal{Z}^\alpha(\mathbf{r})\tfrac{\lambda^\alpha}{2})$, where the Baker-Hausdorff-Campbell theorem can be used to express $\mathcal{Z}^\alpha$ as a functional of $\mathcal{X}^\alpha$ and $\overline{\mathcal{Y}^\alpha}$. $V_C(\mathbf{r})$ takes the form of a unitary operator that carries out a gauge transformation on the spinor field $\psi(\mathbf{r})$; but, in this case, $\mathcal{Z}^\alpha$ is not a c-number function in the adjoint representation of SU(3), but is a complicated functional of the gauge field. Under a gauge transformation, precisely compensating transformations are made on $\psi(\mathbf{r})$

and $V_C(\mathbf{r})$, so that $\psi_{\mathsf{GI}}(\mathbf{r})$ is strictly gauge-invariant. When they appear In the $\mathcal{N}$ representation, the color-charge and color-current densities, $j_0^a(\mathbf{r}) = g\psi^\dagger(\mathbf{r})\tfrac{\lambda^a}{2}\psi(\mathbf{r})$ and $j_i^a(\mathbf{r}) = g\psi^\dagger(\mathbf{r})\alpha_i\tfrac{\lambda^a}{2}\psi(\mathbf{r})$ respectively, therefore are gauge-invariant, although, in the $\mathcal{C}$ representation, both of these quantities transform gauge-*covariantly*, as vectors in the adjoint representation of SU(3). In the $\mathcal{N}$ representation, the quark field $\psi(\mathbf{r})$ implicitly includes enough of the gluon field to achieve this gauge invariance.

We have used Eq.(17) to find the gauge-invariant gluon field

$$[A_{\mathsf{GI}\,i}^b(\mathbf{r}) \tfrac{\lambda^b}{2}] = V_C(\mathbf{r}) [A_i^b(\mathbf{r}) \tfrac{\lambda^b}{2}] V_C^{-1}(\mathbf{r}) + \tfrac{i}{g} V_C(\mathbf{r}) \, \partial_i V_C^{-1}(\mathbf{r}) , \qquad (19)$$

or, equivalently,[5]

$$A_{\mathsf{GI}\,i}^b(\mathbf{r}) = A_{T\,i}^b(\mathbf{r}) + [\delta_{ij} - \tfrac{\partial_i \partial_j}{\partial^2}]\overline{\mathcal{A}_j^b}(\mathbf{r}) . \qquad (20)$$

We observe that, as in QED, the gauge-invariant gauge field $A_{\mathsf{GI}\,i}^b(\mathbf{r})$ is transverse. But it is not merely the transverse part of the gauge field. In contrast to the gauge-invariant gauge field in QED, $A_{\mathsf{GI}\,i}^b(\mathbf{r})$ also involves the transverse part of the resolvent gauge field.

We have expanded $\psi_{\mathsf{GI}}(\mathbf{r})$ and $A_{\mathsf{GI}\,i}^b(\mathbf{r})$, and have verified that our gauge-invariant quark and gluon fields agree with the perturbative calculations of Lavelle, McMullan, et. al. to the highest order to which their perturbative calculations were available.[6,7]

4 A gauge-invariant QCD Hamiltonian

We have been able to express the QCD Hamiltonian

$$H_{QCD} = \int d\mathbf{r} \left\{ \tfrac{1}{2}\Pi_i^a(\mathbf{r})\Pi_i^a(\mathbf{r}) + \tfrac{1}{4}F_{ij}^a(\mathbf{r})F_{ij}^a(\mathbf{r}) + \right.$$
$$\left. \psi^\dagger(\mathbf{r}) \left[\beta m - i\alpha_i \left(\partial_i - igA_i^a(\mathbf{r})\tfrac{\lambda^\alpha}{2} \right) \right] \psi(\mathbf{r}) \right\} , \quad (21)$$

entirely in terms of gauge-invariant quark and gluon fields,[8] by systematically transforming it, term by term, from the $\mathcal{C}$ to the $\mathcal{N}$ representation. Under this transformation, $\psi(\mathbf{r}) \rightarrow V_C^{-1}(\mathbf{r})\psi(\mathbf{r})$, $\psi^\dagger(\mathbf{r}) \rightarrow \psi^\dagger(\mathbf{r})V_C(\mathbf{r})$, but the gauge field remains untransformed. Shifting $V_C(\mathbf{r})$ to the right and $V_C^{-1}(\mathbf{r})$ to the left until they encounter the $A_i^a(\mathbf{r})\tfrac{\lambda^\alpha}{2}$, turns the second line of Eq.(21) into $\int d\mathbf{r} \, \psi^\dagger(\mathbf{r}) \left(\beta m - i\alpha_i\partial_i\right) \psi(\mathbf{r}) + \tilde{H}_{j-A}$, with

$$\tilde{H}_{j-A} = - \int d\mathbf{r} \overbrace{g \, \psi^\dagger(\mathbf{r})\alpha_i\tfrac{\lambda^h}{2}\psi(\mathbf{r})}^{j_i^h(\mathbf{r})} A_{\mathsf{GI}\,i}^h(\mathbf{r}) . \qquad (22)$$

$F_{ij}^a(\mathbf{r})F_{ij}^a(\mathbf{r})$ remains untransformed in this process, but we found that this similarity transformation, when applied to $\Pi_i^a(\mathbf{r})\Pi_i^a(\mathbf{r})$, generates nonlocal interactions between color-charge densities which we will discuss below.[8]

The QCD Hamiltonian that results from the transformation to the $\mathcal{N}$ representation is

$$\tilde{H}_{QCD} = \int d\mathbf{r} \left\{ \tfrac{1}{2}\Pi_i^a(\mathbf{r})\Pi_i^a(\mathbf{r}) + \tfrac{1}{4}F_{ij}^a(\mathbf{r})F_{ij}^a(\mathbf{r}) + \right.$$

$$\left. \psi^\dagger(\mathbf{r})\left[\beta m - i\alpha_i\,\partial_i\right]\psi(\mathbf{r}) \right\} + \tilde{H}', \qquad (23)$$

where $\tilde{H}'$ describes interactions involving the gauge-invariant quark field. The parts of $\tilde{H}'$ relevant to the dynamics of quarks and gluons can be expressed as

$$\tilde{H}' = \tilde{H}_{j-A} + \tilde{H}_{LR}, \qquad (24)$$

where $\tilde{H}_{LR}$ is the nonlocal interaction

$$\tilde{H}_{LR} = H_{g-Q} + H_{Q-Q}. \qquad (25)$$

We will initially focus our attention on H_{Q-Q} in this report, since it illustrates the most important features of our work. A useful formulation of H_{Q-Q} can be given as [9]

$$H_{Q-Q} = \frac{1}{2}\int d\mathbf{r}d\mathbf{x}\, j_0^b(\mathbf{r})\mathcal{F}^{ba}(\mathbf{r},\mathbf{x})j_0^a(\mathbf{x}). \qquad (26)$$

where the Green's function $\mathcal{F}^{ba}(\mathbf{r},\mathbf{x})$ is represented as

$$\mathcal{F}^{ba}(\mathbf{r},\mathbf{x}) = \frac{\delta_{ab}}{4\pi|\mathbf{r}-\mathbf{x}|} +$$

$$g\,f^{\delta_{(1)}ba}\int\frac{d\mathbf{y}}{4\pi|\mathbf{r}-\mathbf{y}|}A_{\mathsf{Gl}\,i}^{\delta_{(1)}}(\mathbf{y})\,\partial_i\frac{1}{4\pi|\mathbf{y}-\mathbf{x}|} +$$

$$\cdots \qquad + \qquad \cdots \qquad +$$

$$(-1)^{(n-1)}g^n f^{\delta_{(1)}bs_{(1)}}f^{s_{(1)}\delta_{(2)}s_{(2)}}\cdots f^{s_{(n-1)}\delta_{(n)}a}\times$$

$$\int\frac{d\mathbf{y}_1}{4\pi|\mathbf{r}-\mathbf{y}_1|}A_{\mathsf{Gl}\,i}^{\delta_{(1)}}(\mathbf{y}_1)\,\partial_i\int\frac{d\mathbf{y}_2}{4\pi|\mathbf{y}_1-\mathbf{y}_2|}\times$$

$$A_{\mathsf{Gl}\,j}^{\delta_{(2)}}(\mathbf{y}_2)\,\partial_j\int\frac{d\mathbf{y}_3}{4\pi|\mathbf{y}_2-\mathbf{y}_3|}\cdots\int\frac{d\mathbf{y}_n}{4\pi|\mathbf{y}_{(n-1)}-\mathbf{y}_n|}\times$$

$$A_{\mathsf{Gl}\,\ell}^{\delta_{(n)}}(\mathbf{y}_n)\,\partial_\ell\frac{1}{4\pi|\mathbf{y}_n-\mathbf{x}|} + \cdots \qquad (27)$$

We make the following observations about $\mathcal{F}^{ba}(\mathbf{r},\mathbf{x})$: The initial term resembles the Coulomb interaction. The infinite series of further terms consists of chains through which the interaction is transmitted from one color-charge density to the other. Each chain contains a succession of 'links', which have the characteristic form

$$\text{link} = g f^{s_{(1)}\delta s_{(2)}}A_{\mathsf{Gl}\,j}^{\delta}(\mathbf{x})\,\partial_j\frac{1}{4\pi|\mathbf{x}-\mathbf{y}|}. \qquad (28)$$

The 'links' are coupled through summations over the $s_{(n)}$ indices and integrations over the spatial variables. As noted before, all the quantities in H_{Q-Q} are gauge-invariant — the color-charge density $j_0^b(\mathbf{r})$ as well as the

gauge field $A_{\mathsf{Gl}\,j}^{\delta}(\mathbf{x})$. H_{g-Q} — the other nonlocal interaction in $\tilde{H}_{LR}$ — couples quark to gluon color charge density. In H_{g-Q} the quark color-charge density is coupled, through the same Green's function $\mathcal{F}^{ba}(\mathbf{r},\mathbf{x})$, to a gauge-invariant expression describing 'glue'-color, which we have found to be [8,9]

$$\mathsf{K}_g^d(\mathbf{r}) = gf^{d\sigma e}\,\mathrm{Tr}\left[V_C^{-1}(\mathbf{r})\frac{\lambda^e}{2}V_C(\mathbf{r})\frac{\lambda^b}{2}\right]A_{\mathsf{Gl}\,i}^{\sigma}(\mathbf{r})\Pi_i^b(\mathbf{r}).$$

The chains that constitute $\mathcal{F}^{ba}(\mathbf{r},\mathbf{x})$ — coupled links, in which the nth order chain is a product of n links — suggest features closely associated with QCD: flux tubes, 'string'-like structures tying colored objects to each other, etc. But these rudimentary analogies only serve to direct our attention to potentially important features of this interaction — they do not, by themselves, demonstrate a physical effect.

With regard to $\tilde{H}_{j-A}$, we observe that the color-current density $j_i^h(\mathbf{r})$ has a configuration-space structure that is similar to that of the electric current density in QED. We can therefore reasonably expect that it, too, will manifest a v/c dependence for nonrelativistic quarks interacting with the purely transverse $A_{\mathsf{Gl}\,j}^{\delta}(\mathbf{x})$, and that $\tilde{H}_{LR}$ will be of predominant importance in the low-energy QCD regime.

5 Quark confinement and color transparency

The use of Eqs.(26) and (27) to explicitly evaluate an effective 'potential' between quark color-charge densities is precluded, at the present time, by the fact that we have not yet found an expression for the gauge-invariant gauge field $A_{\mathsf{Gl}\,j}^{\delta}(\mathbf{x})$. However, we can draw some important conclusions from the general form of Eqs.(26) and (27).

If we assume that the Green's function $\mathcal{F}^{ba}(\mathbf{r},\mathbf{x})$ varies smoothly as a function of $\mathbf{x}$, when evaluated in an appropriately chosen state, and that quarks, or configurations of quarks, are localized in wave packets whose size is small enough so that $\mathcal{F}^{ba}(\mathbf{r},\mathbf{x})$ varies only moderately over the space occupied by these wave packets, then we are entitled to make a 'color-multipole' expansion of the expression $\int \mathcal{F}^{ba}(\mathbf{r},\mathbf{x})j_0^a(\mathbf{x})d\mathbf{x}$ — which appears in Eq.(26) —about the point $\mathbf{x} = \mathbf{x}_0$, in the form

$$\int d\mathbf{x}\left\{\mathcal{F}^{ba}(\mathbf{r},\mathbf{x}_0) + X_i\partial_i\mathcal{F}^{ba}(\mathbf{r},\mathbf{x}_0) + \right.$$

$$\left. \frac{1}{2}X_iX_j\,\partial_i\partial_j\,\mathcal{F}^{ba}(\mathbf{r},\mathbf{x}_0) + \cdots\right\}j_0^a(\mathbf{x}) \quad (29)$$

where $X_i = (x-x_0)_i$ and $\partial_i = \partial/\partial x_i$. When we perform the integration in Eq. (29), the first term contributes $\mathcal{F}^{ba}(\mathbf{r},\mathbf{x}_0)\mathcal{Q}^a$, where $\mathcal{Q}^a = \int d\mathbf{x}\,j_0^a(\mathbf{x})$ (the integrated "color charge"). Since the color charge is the generator of rotations in SU(3) space, it will annihilate any multiquark state vector in a singlet color configuration. Multiquark packets in a singlet color configuration therefore are immune to the leading term of the nonlocal H_{Q-Q}. Color-singlet configurations of quarks are only subject to the color multipole terms, which act as color analogs to the Van der Waals interaction. The scenario that this model suggests is that the leading term in H_{Q-Q}, namely $Q^b \mathcal{F}^{ba}(\mathbf{r}_0, \mathbf{x}_0) Q^a$ for a quark color charge Q^a at $\mathbf{r}_0$ and another quark color charge Q^b at $\mathbf{x}_0$, may be responsible for the confinement of quarks and packets of quarks that are not in color-singlet configurations. Moreover, assuming that $\mathcal{F}^{ba}(\mathbf{r}, \mathbf{x}_0)$ varies only gradually within a volume occupied by quark packets, the effect of the higher order color multipole forces on a packet of quarks in a color-singlet configuration becomes more significant as the packet increases in size. As small quark packets move through gluonic matter, they will experience only insignificant effects from the multipole contributions to H_{Q-Q}, since, as can be seen from Eq. (29), the factors X_i, $X_i X_j$, $\cdots$, $X_{i(1)} \cdots X_{i(n)}$, keep the higher order multipole terms from making significant contributions to $\int d\mathbf{x}\, \mathcal{F}^{ba}(\mathbf{r}, \mathbf{x}) j_0^a(\mathbf{x})$ when they are integrated over small packets of quarks. As the size of the quark packets increases, the regions over which the multipoles are integrated also increases, and the effect of the multipole interactions on the color-singlet packets can become larger. This dependence on packet size of the final-state interactions experienced by color-singlet states — i.e. the increasing importance of final-state interactions as color-singlet packets grow in size — is in qualitative agreement with the characterizations of color transparency given by Miller and by Jain, Pire and Ralston.[10]

In making this argument, the following features of this gauge-invariant formulation must be borne in mind: The state vectors that can appropriately be chosen for such a representation must implement the non-Abelian Gauss's law applicable to QCD — i.e. the states would have to be annihilated by the Gauss's law operator, which is $\mathcal{G}$ in the $\mathcal{N}$ representation. That still allows us to apply quark creation operators to such states to form multiquark Fock states with orbitals whose configuration-space dependence is arbitrary as far as the gauge problem is concerned. These states are far from being Fock states in the gluon sector — they have to be complex enough to implement Gauss's law. But that fact does not inhibit us in using Fock states for multi-quark configurations, because, in the $\mathcal{N}$ representation, the quark fields contain enough gluon contributions to be gauge-invariant by themselves. The multi-quark states therefore have a status similar to multi-electron states in a gauge-invariant formulation of QED (e.g. in the Coulomb gauge), which are surrounded by a Coulomb field and therefore are gauge-invariant.

In order to test these ideas quantitatively, we have made use of Yang-Mills theory — the SU(2) version of this model — for which the structure constants $f^{\delta ba}$ are $\epsilon^{\delta ba}$.[9] We model $A^{\delta}_{\mathrm{Gl}\,i}(\mathbf{r})$, which, in this case, are transverse fields in the adjoint representation of SU(2), as the manifestly transverse "hedgehog" configuration

$$\langle A^{\delta}_{\mathrm{Gl}\,i}(\mathbf{r}) \rangle = \epsilon^{ij\delta} r_j \phi(r). \tag{30}$$

Although there is no reason to believe that the *ansatz* given in Eq. (30) follows from the dynamical equations that determine $A^{\delta}_{\mathrm{Gl}\,i}(\mathbf{r})$, it is a convenient choice for examining to what extent the structure of $\mathcal{F}^{ba}(\mathbf{r}, \mathbf{x})$ — as distinct from the precise form of $A^{\delta}_{\mathrm{Gl}\,i}(\mathbf{r})$ — enables us to nonperturbatively evaluate the infinite series given in Eq. (27). As we were able to show,[9] the perturbative evaluation of H_{Q-Q} for this SU(2) model could be bypassed, and the entire series in $\mathcal{F}^{ba}(\mathbf{r}, \mathbf{x})$ obtained from the solution of a sixth-order differential equation.

Acknowledgments

This research was supported by the Department of Energy under Grant No. DE-FG02-92ER40716.00.

References

1. P. A. M. Dirac, *Can. J. Phys.* 33, 650 (1955); K. Haller and E. Lim-Lombridas, *Found. of Phys.* 24, 217 (1994), and further references cited therein.
2. for example, R. Jackiw, *Rev. Mod. Phys.* 52, 661 (1980); K. Haller, *Phys. Rev.* D36, 1839 (1987).
3. E. Fermi, *Rev. Mod. Phys.* 4, 661 (1932).
4. M. Belloni, L. Chen and K. Haller, *Phys. Lett.* B373, 185 (1996).
5. L. Chen, M. Belloni and K. Haller, *Phys. Rev.* D55, 2347 (1997).
6. M. Lavelle and D. McMullan, *Phys. Lett.* B329 (1994) 68.
7. E. Bagan, M. Lavelle, B. Fiol, N. Roy and D. McMullan, *Constituent Quarks from QCD: Perturbation Theory and the Infrared*. hep-ph/9609330.
8. M. Belloni, L. Chen and K. Haller, *Phys. Lett.* B403, 316 (1997).
9. L. Chen and K. Haller, *Quark confinement and color transparency in a gauge-invariant formulation of QCD*, hep-th/9803250.
10. G. A. Miller, *Color Transparency — Color Coherent Effects in Nuclear Physics.* DOE/ER/41014-26-N97; nucl-th/9707040; P. Jain, B. Pire, and J. P. Ralston, *Physics Reports* 271, 67 (1996).

ON SCALAR FIELD THEORIES WITH POLYNOMIAL INTERACTIONS

J. BECKERS

Theoretical and Mathematical Physics, Institute of Physics (B5), University of Liège, B-4000 Liège 1 (Belgium)
E-mail: Jules.Beckers@ulg.ac.be

Toy models in Quantum Field Theory have already been visited and have enlightened realistic scalar field theories in particular. Here we come back on classical solutions of the *soliton* type and on their *stability* studied through *polynomial interactions*. We study differential realizations of quasi-exactly solvable operators having a hidden symmetry and show the fundamental interest of such nonlinear developments intimately connected with deformations of the angular momentum algebra.

1 Introduction

"Deformations of the angular momentum algebra" refer to technicalities introduced in mathematical physics dealing with main subjects like the study of the so-called "quantum groups or algebras" [1] and their irreducible representations, a subject which has already played an interesting role in Yang-Mills type gauge theories and conformal field theories in particular [2,3].

Let us *first* recall that this very important angular momentum theory is associated with the well known structure called the simple Lie algebra $sl(2,\Re)$ – one of the real forms of the Cartan A_1-algebra [4] – characterized by the commutation relations

$$[J_0, J_\pm] = \pm J_\pm \ , \tag{1}$$

$$[J_+, J_-] = 2J_0 \ , \tag{2}$$

dealing with the diagonal (J_0), raising (J_+) and lowering (J_-) operators.

Secondly, let us also point out its quantum or q-deformation denoted by $U_q(sl(2,\Re))$ with the corresponding commutation relations (1) but, instead of (2), with

$$[J_+, J_-] = [2J_0] \tag{3}$$

where the bracket $[x]$ refers to the usual definition [5]

$$[x] = \frac{q^x - q^{-x}}{q - q^{-1}} \ , \quad q = e^\gamma \ . \tag{4}$$

Here let us now be interested in a *third* type of algebra that we call a *polynomial* deformation of $sl(2,\Re)$: it is a "nonlinear Lie algebra", also characterized by the commutation relations (1) but, instead of (2) *or* of (3), by

$$[J_+, J_-] = P(J_0) \tag{5}$$

where $P(J_0)$ is a polynomial function of J_0 with a generally finite degree Δ. For example, if we take $\Delta = 3$ and

$$P(J_0) = 2J_0 + 8\beta J_0^3 \tag{6}$$

where β can be interpreted as a deformation parameter, we are concerned with the well known Higgs algebra [6] already recognized as a specific W-algebra [7] very well visited in connection with different physical applications [8].

From this kind of polynomial deformation of $sl(2,\Re)$, we want to direct our study towards *quasi-exactly-solvable* Hamiltonians [9] leading to discussions of *stability of solitons* in simple scalar field theories [10] and, more particularly, in connection with the so-called ϕ^6-model. Such nonlinear developments will thus appear as being intimately connected with polynomial deformations of the angular momentum algebra.

In section 2, we construct Hamiltonians of the type

$$H = \sum_{k=0}^{2} P_k(x) \ \frac{d^k}{dx^k} \tag{7}$$

where $P_k(x)$ are polynomials of arbitrary degrees in terms of the x-variable. Such operators belonging to the enveloping algebra of $sl(2,\Re)$ lead to polynomial self-interactions of the ϕ^6-model in particular and to stability of solitonic solutions which are considered in section 3.

2 On "quasi-exactly-solvable" Hamiltonians

A constructive way for realizing polynomial deformations of $sl(2,\Re)$ has recently been proposed [11] through differential operators given in the following forms

$$J_+ = x^N \ F(D), J_- = G(D) \ \frac{d^N}{dx^N}, J_0 = \frac{1}{N} \ (D + c) \tag{8}$$

where, in terms of a single variable x, $N = 1, 2, 3, ..., D$ is the dilation operator

$$D \equiv x \ \frac{d}{dx} \tag{9}$$

and c is a constant. Remember that through the Heisenberg algebra we evidently have

$$\left[\frac{d}{dx}, x \right] = 1 \tag{10}$$

as well as direct consequences like the following ones

$$[D, x^N] = N\, x^N \quad , \quad \left[\frac{d^N}{dx^N}, D\right] = N\, \frac{d^N}{dx^N} \quad . \quad (11)$$

If we consider the Higgs algebra, i.e. the polynomial deformation of $sl(2, \Re)$ characterized by the function $P(J_0)$ given by (??), we are able [11] to put forward the following expression for $N = 1$

$$J_+ = x(c_1 D^2 + c_2 D + c_3),$$
$$J_- = (c_4 D + c_5), \quad J_0 = D + c \quad (12)$$

and, for $N = 2$,

$$J_+ = x^2(d_1 D^2 + d_2 D + d_3),$$
$$J_- = d_4 \frac{d^2}{dx^2}, \quad J_0 = \frac{1}{2}\,(D + c). \quad (13)$$

They are *both* such that

$$[J_+, J_-] = f(J_0^3, J_0^2, J_0, \mathbf{1}) \quad (14)$$

in general, $c_1...c_5$ and $d_1...d_4$ being sets of real parameters. In fact, the above proposals (??) and (??) take care of the fact that we want to privilege Schrödingerlike Hamiltonians as operators at most of second orders with respect to space derivatives (according to quantization of classical (kinetic) energies at least). We thus expect Hamiltonians of the type

$$H = \alpha J_+ + \beta J_0 + \gamma J_0^2 + \delta J_- + \varepsilon \quad (15)$$

where α, β, γ, δ and ε are also real parameters. With the above realizations (??) and (??), we are led respectively to

$$H_1 = (A_1 x^3 + B_1 x^2 + C_1 x)\frac{d^2}{dx^2}$$
$$+ (D_1 x^2 + E_1 x + I)\frac{d}{dx} + F_1 x + G_1 \quad (16)$$

and

$$H_2 = (A_2 x^4 + B_2 x^2 + C_2)\frac{d^2}{dx^2}$$
$$+ (D_2 x^3 + E_2 x)\frac{d}{dx} + F_2 x^2 + G_2 \quad , \quad (17)$$

the subindices 1 and 2 referring thus to the $(N = 1, 2)$-contexts.

These Hamiltonians (??) and (??) are thus obtained on the required form (??) which can finally be put in the Schrödinger desired context, i.e.

$$H \equiv -\frac{d^2}{dy^2} + V(y) \quad (18)$$

with the help of standard methods (mainly changes of variables $x \to y$).

3 Toy models and polynomial interactions in scalar field theory

Let us here construct Hamiltonians like (??) and (??) but in the scalar field theory in $(1+1)$-dimensions supported by the Lagrangian density

$$\mathcal{L} = \mathcal{L}(\phi, \phi_t, \phi_x) = \frac{1}{2}\left(\frac{\partial \phi}{\partial t}\right)^2 - \frac{1}{2}\left(\frac{\partial \phi}{\partial x}\right)^2 - V(\phi) \quad (19)$$

in terms of the scalar field $\phi(x, t)$ and its partial derivatives with respect to time (ϕ_t) and with respect to x (ϕ_x). The principle of least action in the *stationary* context, i.e.

$$A(\phi) = \int_{-\infty}^{+\infty}\left\{\frac{1}{2}\left(\frac{\partial \phi}{\partial x}\right)^2 + V(\phi)\right\} dx \quad (20)$$

leads to the *static* field equation

$$\frac{\partial^2 \phi}{\partial x^2} - \frac{dV}{d\phi} = 0 \quad (21)$$

and to the first integral

$$K = 2V(\phi) - \phi_x^2 \quad . \quad (22)$$

By varying $\phi(x)$ near a solution (called $\overline{\phi}$) through small fluctuations η chosen such that

$$\phi(x) = \overline{\phi}(x) + \eta(x) \quad , \quad \eta(-\infty) = \eta(+\infty) = 0 \quad , \quad (23)$$

we immediately get the *stability* condition [10]

$$\left(-\frac{\partial^2}{\partial x^2} + \left.\frac{d^2 V}{d\phi^2}\right|_{\phi = \overline{\phi}}\right)\eta(x) = \omega^2 \eta(x) \quad (24)$$

as an interesting Schrödinger equation "associated with such a scalar field theory". It is precisely such a condition which could be strongly related to the eigenvalue- and eigenfunction-problems subtended by our Hamiltonians (??) and (??).

Indeed, let us suppose that $V(\phi)$ is an *even* function of ϕ and let us realize the following simple change of variables

$$z = \overline{\phi}^2(x) \quad . \quad (25)$$

The equation (??) then becomes

$$\left[P_{m+1}(z)\frac{d^2}{dz^2} + P_m(z)\frac{d}{dz} + P_{m-1}(z)\right]F(z) = \omega^2 F(z) \quad (26)$$

where here $P_m(z)$ refers to a polynomial of degree m in the variable z.

Its remarkable property is that the polynomials $P(z)$ do not depend on the form of the solution $\overline{\phi}$. They only

1656

depend on V and K. For instance, if V is quartic, it is well known that the equation (??) or equivalently here the equation (??) is a Lamé equation if $K > 0$ and a Pöschl-Teller equation if $K = 0$.

Let us now consider the polynomial self-interaction of the so-called interesting ϕ^6-model. To proceed further, our potential is thus parametrized by (??) and

$$V(\overline{\phi}) = \frac{1}{2}(z - A)(z - B)(z - C) + \frac{1}{2}K \qquad (27)$$

so that the polynomial $P_4(z)$, i.e. the coefficient multiplying the second order derivative, is simply

$$P_4(z) = 4z(z - A)(z - B)(z - C) \ . \qquad (28)$$

The corresponding $P_3(z)$ and $P_2(z)$ are then given by

$$P_3(z) = 2[4z^3 - 3z^2(A + B + C) \\ + 2z(AB + AC + BC) - ABC] \qquad (29)$$

and

$$P_2(z) = -15z^2 + 6z(A + B + C) \\ + AB + AC + BC \qquad (30)$$

so that our equation (??) is completely determined for the ϕ^6-model.

By entering these values, we finally get the stability condition on the form :

$$\{4z(z - A)(z - B)(z - C) \frac{d^2}{dz^2} \\ + 2[4z^3 - 3z^2(A + B + C) \\ + 2z(AB + AC + BC) - ABC] \frac{d}{dz} \\ - 15z^2 + 6z(A + B + C)\}F(z) \\ = (AB + AC + BC - \omega^2)F(z). \qquad (31)$$

This opens the discussion for getting families of solutions of the Heun equation [12] as already pointed out by Christ and Lee in their Section V [10]. Here, for example, the solution

$$F(z) = (A - z)^{1/2}(z - \frac{B + C}{2}), \ \omega^2 = 2(B - C)^2, (32)$$

exists only if $A = 2B + 2C$ and it generalizes the Christ and Lee results limited to the values $A = B = 1$ and $C = -\frac{1}{2}$.

We end the discussion here by remembering that all the above developments now are seen in a completely new context, i.e. the deformations of the angular momentum algebra subtended by differential nonlinear operators leading to Hamiltonians with polynomial interactions of special interest in scalar field theories.

Acknowledgments

Thanks are due to my colleagues Nathalie Debergh and Yves Brihaye for their enthusiastic collaboration in this research.

References

1. V. Chari and A.N. Pressley, "*A guide to quantum groups*" (Cambridge University Press, Cambridge, 1994);
   A. Klimyk and K. Schmudgen, "*Quantum groups and their representations*" (Springer Verlag, Berlin, 1997).
2. L. Alvarez-Gaume, C. Gomez and G. Sierra, *Phys.Lett.* **B220**, 142 (1989); *Nucl.Phys.* **B330**, 347 (1990).
   C. Gomez and G. Sierra, *Nucl.Phys.* **B352**, 791 (1991).
3. K. Schoutens, A. Sevrin and P. Van Nieuwenhuizen, *Nucl.Phys.* **B349**, 791 (1989); *Comm. Math. Phys.* **124**, 87 (1989); *Phys.Lett.* **B255**, 549 (1991).
4. J.F. Cornwell, "*Group Theory in Physics*", vol. 2 (Academic Press, 1984).
5. A.J. Macfarlane, *J.Phys.* **A22**, 4581 (1989).
   L.C. Biedenharn, *J.Phys.* **A22**, L873 (1989).
6. P.W. Higgs, *J.Phys.* **A12**, 309 (1979).
7. J. The Boer, F. Harmsze and T. Tjin, *Phys. Rep.* **272**, 139 (1996).
   F. Barbarin, E. Ragoucy and P. Sorba, *Nucl.Phys.* B**442**, 425 (1995).
8. M. Roček, *Phys.Lett.* **B255**, 554 (1991).
   D. Bonatsos and C. Daskaloyannis, *Phys.Lett.* **B307**, 100 (1993).
9. A.V. Turbiner, *Comm.Math.Phys.* **119**, (1988).
10. N.H. Christ and T.D. Lee, *Phys.Rev.* **D12**, 1606 (1975).
    M.A. Lohe, *Phys.Rev.* **D20**, 3120 (1979).
    R. Rajamaran, "*Solitons and Instantons*" (North-Holland Publ. Comp., Amsterdam, 1982).
11. J. Beckers, Y. Brihaye and N. Debergh, "On realizations of nonlinear Lie algebras by differential operators and some physical applications", *Liège preprint* (March 1998).
12. A. Erdelyi (Editor), "*Higher Transcendental Functions*" (Mc-Graw Hill, New York, 1955).

Parallel Session 16

Beyond the Standard Model - Theory

Convenors: Fridger Schrempp (DESY)

THE FERMION MASS PROBLEM AND THE ANTI-GRAND UNIFICATION MODEL

C. D. FROGGATT, M. GIBSON

Department of Physics and Astronomy, Glasgow University, Glasgow G12 8QQ, Scotland

H. B. NIELSEN

Niels Bohr Institute, Blegdamsvej 17, Copenhagen φ, Denmark

D. J. SMITH

Institut für Physik, Humboldt Universität Berlin, Invalidenstr. 110, 10115 Berlin, Germany

We describe the Anti-Grand Unification Model (AGUT) and the Multiple Point Principle (MPP) used to calculate the values of the Standard Model gauge coupling constants in the theory, from the requirement of the existence of degenerate vacua. The application of the MPP to the pure Standard Model predicts the existence of a second minimum of the Higgs potential close to the cut-off, which we take to be the Planck scale, giving our Standard Model predictions for the top quark and Higgs masses: $M_t = 173 \pm 5$ GeV and $M_H = 135 \pm 9$ GeV. We also discuss mass protection by chiral charges and present a fit to the charged fermion mass spectrum using the chiral quantum numbers of the maximal AGUT gauge group $SMG^3 \times U(1)_f$, where $SMG \equiv SU(3) \times SU(2) \times U(1)$. The neutrino mass and mixing problem is then briefly discussed for models with chiral flavour charges responsible for the charged fermion mass hierarchy.

1 Introduction

One of the outstanding problems in particle physics is to explain the observed pattern of quark-lepton masses and of flavour mixing. This is the problem of the hierarchy of Yukawa coupling constants in the Standard Model (SM), which range in value from of order 1 for the top quark to of order 10^{-5} for the electron. However there is no reason in the SM for the Higgs field to prefer to couple to one fermion rather than another; in fact one would expect them all to be of order unity. We suggest [1] that the natural resolution to this problem is the existence of some approximately conserved chiral charges beyond the SM. These charges, which we assume to be the gauge quantum numbers in the fundamental theory beyond the SM, provide selection rules forbidding the transitions between the various left-handed and right-handed quark-lepton states, except for the top quark. In order to generate mass terms for the other fermion states, we have to introduce new Higgs fields, which break the fundamental gauge symmetry group G down to the SM group. We also need suitable intermediate fermion states to mediate the forbidden transitions, which we take to be vector-like Dirac fermions with a mass of order the fundamental scale M_F of the theory. In this way effective SM Yukawa coupling constants are generated, which are suppressed by the appropriate product of Higgs field vacuum expectation values measured in units of M_F.

If we want to explain the observed spectrum of quarks and leptons, it is clear that we need charges which—possibly in a complicated way—separate the gen-

erations and, at least for $t - b$ and $c - s$, also quarks in the same generation. Just using the usual simple $SU(5)$ GUT charges does not help, because both (μ_R and e_R) and (μ_L and e_L) have the same $SU(5)$ quantum numbers. So we prefer to keep each SM irreducible representation in a separate irreducible representation of G and introduce extra gauge quantum numbers distinguishing the generations, by adding extra cross-product factors to the SM gauge group. In this talk we consider the maximal anomaly free gauge group of this type—the anti-grand unification (AGUT) group $SMG^3 \times U(1)_f$. In section 2 we discuss the structure of the AGUT model and the prediction of the values of the SM gauge coupling constants, using the so-called Multiple Point Principle (MPP). We apply this principle to the pure SM in section 3, assuming a desert up to the Planck scale, and obtain predictions for the top quark and SM Higgs particle masses. In section 4 we consider the Higgs fields responsible for breaking the AGUT gauge group and the structure of the resulting quark-lepton mass matrices, together with details of a fit to the observed spectrum. The problem of neutrino mass and mixing in models with approximately conserved chiral flavour charges are discussed in section 5. Finally we present our conclusions in section 6.

2 Anti-Grand Unification

In the AGUT model the SM gauge group is extended in much the same way as Grand Unified $SU(5)$ is often assumed; it is just that we assume another non-simple gauge group $G = SMG^3 \times U(1)_f$, where $SMG \equiv$

$SU(3) \times SU(2) \times U(1)$, becomes active near the Planck scale $M_{Planck} \simeq 10^{19}$ GeV. So we have a pure SM desert, without any supersymmetry, up to an order of magnitude or so below M_{Planck}. The existence of the $SMG^3 \times U(1)_f$ group means that, near the Planck scale, each of the three quark-lepton generations has got its own gauge group and associated gauge particles with the same structure as the SM gauge group. There is also an extra abelian $U(1)_f$ gauge boson, giving altogether $3 \times 8 = 24$ gluons, $3 \times 3 = 9$ W's and $3 \times 1 + 1 = 4$ abelian gauge bosons.

At first sight, this $SMG^3 \times U(1)_f$ group with its 37 generators seems to be just one among many possible SM gauge group extensions. However, it is actually not such an arbitrary choice, as it can be uniquely specified by postulating 4 reasonable requirements on the gauge group $G \supseteq SMG$:

1. G should transform the presently known (left-handed, say) Weyl particles into each other, so that $G \subseteq U(45)$. Here $U(45)$ is the group of all unitary transformations of the 45 species of Weyl fields (3 generations with 15 in each) in the SM.

2. No anomalies, neither gauge nor mixed. We assume that only straightforward anomaly cancellation takes place and, as in the SM itself, do not allow for a Green-Schwarz type anomaly cancellation [2].

3. The various irreducible representations of Weyl fields for the SM group remain irreducible under G.

4. G is the maximal group satisfying the other 3 postulates.

With these four postulates a somewhat complicated calculation shows that, modulo permutations of the various irreducible representations in the Standard Model fermion system, we are led to our gauge group $SMG^3 \times U(1)_f$. Furthermore it shows that the SM group is embedded as the diagonal subgroup of SMG^3, as required in our AGUT model. The AGUT group breaks down an order of magnitude or so below the Planck scale to the SM group. The anomaly cancellation constraints are so tight that, apart from various permutations of the particle names, the $U(1)_f$ charge assignments are uniquely determined up to an overall normalisation and sign convention. In fact the $U(1)_f$ group does not couple to the left-handed particles or any first generation particles, and the $U(1)_f$ quantum numbers can be chosen as follows:

$$Q_f(\tau_R) = Q_f(b_R) = Q_f(c_R) = 1 \qquad (1)$$

$$Q_f(\mu_R) = Q_f(s_R) = Q_f(t_R) = -1 \qquad (2)$$

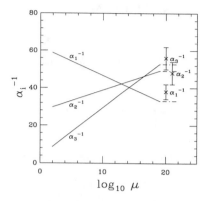

Figure 1: Evolution of the Standard Model fine structure constants α_i (α_1 in the SU(5) inspired normalisation) ¿from the electroweak scale to the Planck scale. The anti-GUT model predictions for the values at the Planck scale, $\alpha_i^{-1}(M_{Planck})$, are shown with error bars.

The SM gauge coupling constants do not, of course, unify in the AGUT model, but their values have been successfully calculated using the Multiple Point Principle [3]. According to the MPP, the coupling constants should be fixed such as to ensure the existence of many vacuum states with the same energy density; in the Euclideanised version of the theory, there is a corresponding phase transition. So if several vacua are degenerate, there is a multiple point. The couplings at the multiple points have been calculated in lattice gauge theory for the groups $SU(3)$, $SU(2)$ and $U(1)$ separately. We imagine that the lattice has a truly physical significance in providing a cut-off for our model at the Planck scale. The SM fine structure constants correspond to those of the diagonal subgroup of the SMG^3 group and, for the non-abelian groups, this gives:

$$\alpha_i(M_{Planck}) = \frac{\alpha_i^{Multiple\ Point}}{3} \qquad i = 2,\ 3 \qquad (3)$$

The situation is more complicated for the abelian groups, because it is possible to have gauge invariant cross-terms between the different $U(1)$ groups in the Lagrangian density such as:

$$\frac{1}{4g^2} F_{\mu\nu}^{gen\ 1}(x) F_{gen\ 2}^{\mu\nu}(x) \qquad (4)$$

So, in first approximation, for the SM $U(1)$ fine structure constant we get:

$$\alpha_1(M_{Planck}) = \frac{\alpha_1^{Multiple\ Point}}{6} \qquad (5)$$

The agreement of these AGUT predictions with the data is shown in figure 1.

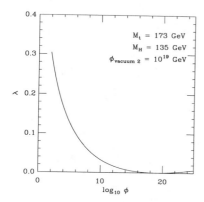

Figure 2: Plot of λ as a function of the scale of the Higgs field ϕ for degenerate vacua with the second Higgs VEV at the Planck scale $\phi_{vac\,2} = 10^{19}$ GeV.

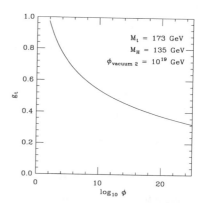

Figure 3: Plot of g_t as a function of the scale of the Higgs field ϕ for degenerate vacua with the second Higgs VEV at the Planck scale $\phi_{vac\,2} = 10^{19}$ GeV.

3 The MPP Prediction for the Top Quark and Higgs masses in the Standard Model

The application of the MPP to the pure Standard Model (SM), with a cut-off close to M_{Planck}, implies that the SM parameters should be adjusted, such that there exists another vacuum state degenerate in energy density with the vacuum in which we live. This means that the effective SM Higgs potential $V_{eff}(|\phi|)$ should, have a second minimum degenerate with the well-known first minimum at the electroweak scale $\langle |\phi_{vac\,1}| \rangle = 246$ GeV. Thus we predict that our vacuum is barely stable and we just lie on the vacuum stability curve in the top quark, Higgs particle (pole) mass (M_t, M_H) plane. Furthermore we expect the second minimum to be within an order of magnitude or so of the fundamental scale, i.e. $\langle |\phi_{vac\,2}| \rangle \simeq M_{Planck}$. In this way, we essentially select a particular point on the SM vacuum stability curve and hence the MPP condition predicts precise values for M_t and M_H.

For the purposes of our discussion it is sufficient to consider the renormalisation group improved tree level effective potential $V_{eff}(\phi)$. We are interested in values of the Higgs field of the order $|\phi_{vac\,2}| \simeq M_{Planck}$, which is very large compared to the electroweak scale, and for which the quartic term strongly dominates the ϕ^2 term; so to a very good approximation we have:

$$V_{eff}(\phi) \simeq \frac{1}{8}\lambda(\mu = |\phi|)|\phi|^4 \qquad (6)$$

The running Higgs self-coupling constant $\lambda(\mu)$ and the top quark running Yukawa coupling constant $g_t(\mu)$ are readily computed by means of the renormalisation group equations, which are in practice solved numerically, using the second order expressions for the beta functions.

The vacuum degeneracy condition is imposed by requiring:

$$V_{eff}(\phi_{vac\,1}) = V_{eff}(\phi_{vac\,2}) \qquad (7)$$

Now the energy density in vacuum 1 is exceedingly small compared to $\phi_{vac\,2}^4 \simeq M_{Planck}^4$. So we basically get the degeneracy condition, eq. (7), to mean that the coefficient $\lambda(\phi_{vac\,2})$ of $\phi_{vac\,2}^4$ must be zero with high accuracy. At the same ϕ-value the derivative of the effective potential $V_{eff}(\phi)$ should be zero, because it has a minimum there. Thus at the second minimum of the effective potential the beta function β_λ also vanishes:

$$\beta_\lambda(\mu = \phi_{vac\,2}) = \lambda(\phi_{vac\,2}) = 0 \qquad (8)$$

which gives to leading order the relationship:

$$\frac{9}{4}g_2^4 + \frac{3}{2}g_2^2 g_1^2 + \frac{3}{4}g_1^4 - 12g_t^4 = 0 \qquad (9)$$

between the top quark Yukawa coupling and the electroweak gauge coupling constants $g_1(\mu)$ and $g_2(\mu)$ at the scale $\mu = \phi_{vac\,2} \simeq M_{Planck}$. We use the renormalisation group equations to relate the couplings at the Planck scale to their values at the electroweak scale. Figures 2 and 3 show the running coupling constants $\lambda(\phi)$ and $g_t(\phi)$ as functions of $\log(\phi)$. Their values at the electroweak scale give our predicted combination of pole masses [4]:

$$M_t = 173 \pm 5 \text{ GeV} \quad M_H = 135 \pm 9 \text{ GeV} \qquad (10)$$

4 Fermion Mass Hierarchy in AGUT

The $SMG^3 \times U(1)_f$ gauge group is broken by a set of Higgs fields S, W, T and ξ down to the SM gauge group.

Together with the Weinberg Salam Higgs field, ϕ_{WS}, they are responsible for breaking the quark-lepton mass protection by the chiral AGUT quantum numbers. We have the freedom of choosing the abelian quantum numbers of the Higgs fields, which we can express as charge vectors of the form:

$$\vec{Q} \equiv \left(\frac{y_1}{2}, \frac{y_2}{2}, \frac{y_3}{2}, Q_f\right), \qquad (11)$$

where $y_i/2$ $(i = 1, 2, 3)$ are the $U(1)_i$ weak hypercharges. However we fix their non-abelian representations by imposing a natural generalisation of the SM charge quantisation rule

$$y_i/2 + d_i/2 + t_i/3 = 0 \quad (\text{mod} \quad 1) \qquad (12)$$

and requiring that they be singlet or fundamental representations. The duality, d_i, and triality, t_i, here are given by $d_i = +1, 0$ for the doublet and singlet representations respectively of $SU(2)_i$, and $t_i = +1, -1, 0$ for the $3, \bar{3}, 1$ representations of $SU(3)_i$.

By requiring a realistic charged fermion spectrum (with ϕ_{WS} giving an unsuppressed top quark mass), we are led to the following choice:

$$\vec{Q}_{\phi_{WS}} = (0, \frac{2}{3}, -\frac{1}{6}, 1) \ , \quad \vec{Q}_W = (0, -\frac{1}{2}, \frac{1}{2}, -\frac{4}{3}),$$

$$\vec{Q}_T = (0, -\frac{1}{6}, \frac{1}{6}, -\frac{2}{3}) \ , \quad \vec{Q}_\xi = (\frac{1}{6}, -\frac{1}{6}, 0, 0),$$

$$\vec{Q}_S = (\frac{1}{6}, -\frac{1}{6}, 0, -1) \qquad (13)$$

The orders of magnitude for the effective SM Yukawa coupling matrix elements are then given by:

$$Y_U \simeq \begin{pmatrix} S^\dagger W^\dagger T^2 (\xi^\dagger)^2 & W^\dagger T^2 \xi & (W^\dagger)^2 T \xi \\ S^\dagger W^\dagger T^2 (\xi^\dagger)^3 & W^\dagger T^2 & (W^\dagger)^2 T \\ S^\dagger (\xi^\dagger)^3 & 1 & W^\dagger T^\dagger \end{pmatrix}, \qquad (14)$$

$$Y_D \simeq \begin{pmatrix} SW(T^\dagger)^2 \xi^2 & W(T^\dagger)^2 \xi & T^3 \xi \\ SW(T^\dagger)^2 \xi & W(T^\dagger)^2 & T^3 \\ SW^2(T^\dagger)^4 \xi & W^2(T^\dagger)^4 & WT \end{pmatrix}, \qquad (15)$$

$$Y_E \simeq \begin{pmatrix} SW(T^\dagger)^2 \xi^2 & W(T^\dagger)^2 (\xi^\dagger)^3 & (S^\dagger)^2 WT^4 \xi^\dagger \\ SW(T^\dagger)^2 \xi^5 & W(T^\dagger)^2 & (S^\dagger)^2 WT^4 \xi^2 \\ S^3 W(T^\dagger)^5 \xi^3 & (W^\dagger)^2 T^4 & WT \end{pmatrix} (16)$$

Here W, T, ξ, S should be interpreted as the vacuum expectation values (VEVs) of the Higgs fields in units of M_{Planck} We have used the Higgs fields W, T, ξ, S and the fields $W^\dagger, T^\dagger, \xi^\dagger, S^\dagger$ (with opposite charges) equivalently here, which we can do in non-supersymmetric models. In our fit below we do not need any suppression from the Higgs field S and so we set its VEV to be $S = 1$. This means that the quantum numbers of the other Higgs fields $\vec{Q}_{\phi_{WS}}, \vec{Q}_W, \vec{Q}_T$ and $\vec{Q}_\xi$ are only determined modulo $\vec{Q}_S$.

Table 1: Best fit to experimental data. All masses are running masses at 1 GeV except the top quark mass which is the pole mass.

	Fitted	Experimental
m_u	3.6 MeV	4 MeV
m_d	7.0 MeV	9 MeV
m_e	0.87 MeV	0.5 MeV
m_c	1.02 GeV	1.4 GeV
m_s	400 MeV	200 MeV
m_μ	88 MeV	105 MeV
M_t	192 GeV	180 GeV
m_b	8.3 GeV	6.3 GeV
m_τ	1.27 GeV	1.78 GeV
V_{us}	0.18	0.22
V_{cb}	0.018	0.041
V_{ub}	0.0039	0.0035

Since the diagonals of Y_U, Y_D and Y_E are equal we expect to have the approximate relations

$$m_b \approx m_\tau, \quad m_s \approx m_\mu \qquad (17)$$

at M_{Planck}, since these masses come from the diagonal elements. There are no such relations involving the top or charm quark masses, since they come from off-diagonal elements which dominate Y_U. We also note that we expect

$$m_d \gtrsim m_u \approx m_e \qquad (18)$$

at M_{Planck}, since there are two approximately equal contributions to the down quark mass.

The VEVs of the three Higgs fields W, T and ξ are taken to be free parameters in a fit[5] to the 12 experimentally known charged fermion masses and mixing angles. The results of the best fit, which reproduces all the experimental data to within a factor of 2, are given in table 1 and correspond to the parameters

$$\langle W \rangle = 0.179 \quad \langle T \rangle = 0.071 \quad \langle \xi \rangle = 0.099, \qquad (19)$$

in Planck units. This fit is as good as we can expect in a model making order of magnitude predictions.

5 Neutrino Mass and Mixing Problem

There is now strong evidence that the neutrinos are not massless as they would be in the SM. Physics beyond the SM can generate an effective light neutrino mass term

$$\mathcal{L}_{\nu-mass} = \sum_{i,j} \psi_{i\alpha} \psi_{j\beta} \epsilon^{\alpha\beta} (M_\nu)_i j \qquad (20)$$

in the Lagrangian, where $\psi_{i,j}$ are the Weyl spinors of flavour i and j, and $\alpha, \beta = 1, 2$. Fermi-Dirac statistics mean that the mass matrix M_ν must be symmetric.

In models with chiral flavour symmetry we typically expect the elements of the mass matrices to have different orders of magnitude. The charged lepton matrix is then expected to give only a small contribution to the lepton mixing. As a result of the symmetry of the neutrino mass matrix and the hierarchy of the mass matrix elements it is natural to have an almost degenerate pair of neutrinos, with nearly maximal mixing[6]. This occurs when an off-diagonal element dominates the mass matrix.

The recent Super-Kamiokande data on the atmospheric neutrino anomaly strongly suggests large $\nu_\mu - \nu_\tau$ mixing with $\Delta m^2_{\nu_\mu \nu_\tau} \sim 10^{-3}$ eV^2. Large $\nu_\mu - \nu_\tau$ mixing is given by the mass matrix

$$M_\nu = \begin{pmatrix} \times & \times & \times \\ \times & \times & A \\ \times & A & \times \end{pmatrix} \qquad (21)$$

and we have

$$\Delta m^2_{23} \ll \Delta m^2_{12} \sim \Delta m^2_{13} \qquad (22)$$
$$\sin^2 \theta_{23} \sim 1 \qquad (23)$$

However, this hierarchy in Δm^2's is inconsistent with the small angle (MSW) solution to the solar neutrino problem, which requires $\Delta m^2_{12} \sim 10^{-5}$ eV^2.

Hence we need extra structure for the mass matrix such as having several elements of the same order of magnitude. e.g.

$$M_\nu = \begin{pmatrix} a & A & B \\ A & \times & \times \\ B & \times & \times \end{pmatrix} \qquad (24)$$

with $A \sim B \gg a$. This gives

$$\frac{\Delta m^2_{12}}{\Delta m^2_{23}} \sim \frac{a}{\sqrt{A^2 + B^2}}. \qquad (25)$$

The mixing is between all three flavours. and is given by the mixing matrix

$$U_\nu \sim \begin{pmatrix} \frac{1}{\sqrt{2}} & -\frac{1}{\sqrt{2}} & 0 \\ \frac{1}{\sqrt{2}} \cos\theta & \frac{1}{\sqrt{2}} \cos\theta & -\sin\theta \\ \frac{1}{\sqrt{2}} \sin\theta & \frac{1}{\sqrt{2}} \sin\theta & \cos\theta \end{pmatrix} \qquad (26)$$

where $\theta = \tan^{-1} \frac{B}{A}$. So we have large $\nu_\mu - \nu_\tau$ mixing with $\Delta m^2 = \Delta m^2_{23}$, and nearly maximal electron neutrino mixing with $\Delta m^2 = \Delta m^2_{12}$. However the AGUT model naturally gives a structure like eq. (21) rather than eq. (24).

There is also some difficulty in obtaining the required mass scale for the neutrinos. In models such as the AGUT the neutrino masses are generated via superheavy intermediate fermions in a see-saw type mechanism. This leads to too small neutrino masses:

$$m_\nu \lesssim \frac{\langle \phi_{WS} \rangle^2}{M_F} \sim 10^{-5} \text{ eV}, \qquad (27)$$

for $M_F = M_{Planck}$ (in general m_ν is also supressed by the chiral charges). So we need to introduce a new mass scale into the theory. Either some intermediate particles with mass $M_F \lesssim 10^{15}$ GeV, or an $SU(2)$ triplet Higgs field Δ with $\langle \Delta^0 \rangle \sim 1$eV is required. Without further motivation the introduction of such particles is *ad hoc*.

6 Conclusions

We presented two applications of the Multiple Point Principle, according to which nature should choose coupling constants such that the vacuum can exist in degenerate phases. Applied to the AGUT model, it successfully predicts the values of the three fine structure constants, as illustrated in figure 1. In the case of the pure SM, it leads to our predictions for the top quark and Higgs pole masses: $M_t = 173 \pm 5$ GeV and $M_H = 135 \pm 9$ GeV.

The maximal AGUT group $SMG^3 \times U(1)_f$ assigns a unique set of anomaly free chiral gauge charges to the quarks and leptons. With an appropriate choice of Higgs field quantum numbers, the AGUT chiral charges naturally give a realistic charged fermion mass hierarchy. An order of magnitude fit in terms of 3 Higgs VEVs is given in table 1, which reproduces all the masses and mixing angles within a factor of two. On the other hand, the puzzle of the neutrino masses and mixing angles presents a challenge to the model.

References

1. C.D. Froggatt and H.B. Nielsen, *Nucl. Phys.* B **147**, 277 (1979).
2. M.B. Green and J. Schwarz, *Phys. Lett.* B **149**, 117 (1984).
3. D.L. Bennett, C.D. Froggatt and H.B. Nielsen in *Proceedings of the 27th International Conference on High Energy Physics* (Glasgow, 1994), eds. P. Bussey and I. Knowles, (IOP Publishing Ltd, 1995) p. 557; *Perspectives in Particle Physics '94*, eds. D. Klabučar, I. Picek and D. Tadić, (World Scientific, 1995) p. 255, hep-ph/9504294.
4. C.D. Froggatt and H.B. Nielsen, *Phys. Lett.* B **368**, 96 (1996).
5. C.D. Froggatt, M. Gibson, H.B. Nielsen and D.J. Smith, *Int. J. Mod. Phys.* A (to be published), hep-ph/9706212.
6. C.D. Froggatt and H.B. Nielsen *Nucl. Phys.* B **164**, 114 (1979)

GRAND UNIFICATION FROM GAUGE THEORY IN $M_4 \times Z_N$

M. KUBO, Z. MAKI, M. NAKAHARA

Department of Physics, Kinki University, Higashi-Osaka 577-8502, Japan
E-mail: ziromaki@phys.kindai.ac.jp

T. SAITO

Department of Physics, Kwansei Gakuin University, Nishinimiya 662-0891, Japan
E-mail: tsaito@jpnyitp.yukawa.kyoto-u.ac.jp

The $SU(5)$ grand unified theory (GUT) is constructed from the geometrical viewpoint of gauge theory on a three-sheeted space-time, i.e., the $M_4 \times Z_3$ manifold. The Higgs potential is derived with some new features which are not seen in the usual phenomenological potential.

1 Introduction

It has become clear in recent years that spontaneously broken gauge theories based on the Higgs mechanism are all gauge theories in $M_4 \times Z_N$, where M_4 is the 4-dimensional Minkowski space and Z_N is the discrete space with N points. In these theories the Higgs fields are regarded as gauge fields along the discrete space Z_N.

The Weinberg-Salam (WS) model for electroweak interactions is also one such gauge theory with the manifold $M_4 \times Z_2$. In this case the Higgs potential takes a form which necessarily leads to spontaneous gauge symmetry breaking. Thus we have a possible explanation of the guage-theoretical origin of the Higgs field and the symmetry breaking mechanism.

Gauge theories in $M_4 \times Z_N$ were first formulated in terms of noncommutative geometry (NCG) by Connes[1] with its various alternatives. Recently one of the authors (T.S.), in collaboration with Konisi,[2] constructed another gauge theory in $M_4 \times Z_N$ from purely geometrical point of view, without employing the entire context of NCG. The theory has been applied to the WS model, $N = 2$ and $N = 4$ super Yang-Mills theories and the Brans-Dicke theory by one of the present authors (T.S.) and his colleagues.[3]

The motivation of the present work is to construct $SU(5)$ GUT based on this 'geometric' formalism. Although $SU(5)$ GUT in its simplest form is ruled out as a realistic model, the reconstruction of $SU(5)$ GUT in our framework is still important since this is a typical pattern of GUTs.

In the present case we need two kinds of Higgs fields, H_5 and ϕ_{24}, belonging to 5- and 24-dimensional representations of $SU(5)$, respectively. Then we have to prepare at least a 3-sheeted space-time $M_4 \times Z_3$, where Z_3 has three discrete points g_0, g_1 and g_2. Hence we introduce three kinds of $SU(5)$ gauge fields $A_\mu(x, g_p)$ $(p = 0, 1, 2)$

independently.

2 $SU(5)$ GUT from gauge theory in $M_4 \times Z_3$

Let us consider a manifold $M_4 \times Z_3$, where M_4 is the 4-dimensional space-time and Z_3 the discrete space with three points g_p $(p = 0, 1, 2)$, which are subject to the algebraic relations

$$g_p + g_q = g_{p+q} \pmod 3, \tag{1}$$

where additive notation is used for the group product. Let us attach a complex 5-dimensional internal space $V[5, x, g_p]$ to each point $(x, g_p) \in M_4 \times Z_3$. The fermionic fields $\psi(x, g_p)$ are chosen as

$$\psi(x, g_0) = \psi_5(x) = {}^t\!\left(d^1, d^2, d^3, e^+, \bar{\nu}_e\right)_R, \tag{2}$$

and

$$\psi(x, g_1) = \psi_{10}(x) = \begin{pmatrix} 0 & u_3^c & -u_2^c & -u^1 & -d^1 \\ -u_3^c & 0 & u_1^c & -u^2 & -d^2 \\ u_2^c & -u_1^c & 0 & -u^3 & -d^3 \\ u^1 & u^2 & u^3 & 0 & -e^+ \\ d^1 & d^2 & d^3 & e^+ & 0 \end{pmatrix}_L, \tag{3}$$

where ψ_5 and ψ_{10} represent the 5-dimensional fundamental representation and the antisymmetric 10-dimensional representation of $SU(5)$, respectively. We also choose

$$\psi(x, g_2) = \psi'_{10}(x), \tag{4}$$

which is another field independent of $\psi_{10}(x)$ defined in Eq.(3).

Three $SU(5)$ gauge fields $A_\mu(x, g_p)$ $(p = 0, 1, 2)$ are introduced on the three sheets of M_4. Two Higgs fields H_5 and ϕ_{24} belong to the 5- and 24-dimensional representations of $SU(5)$ respectively. H_5 plays the rôle of the mapping function from g_0 to g_1, while ϕ_{24} from g_1 to g_2.

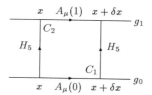

Figure 1: Two paths by which the curvature $F_{\mu g_1}$ is defined.

We also need another Higgs field H_5', which is a mapping function from g_0 to g_2.

The fermionic field $\psi_5(0)$ is mapped from g_0 to g_1 as

$$\psi_{10,H}^{ij}(1) \equiv H^{ij}{}_k(1,0)\psi_5^k(0), \qquad (5)$$

where

$$H^{ij}{}_k(1,0) = \frac{1}{2}\left(H_5^i\delta_k^j - \delta_k^i H_5^j\right) \qquad (6)$$

We would like to stress that the mapping function (6) can be identifie as a gauge field. This is because it transforms as, like in lattice gauge theories,

$$H(1,0) \to \tilde{H}(1,0) = U(1) \otimes U(1)H(1,0)U^{-1}(0) \qquad (7)$$

under the $SU(5)$ gauge transformations

$$\begin{aligned}\psi_5(0) &\to \tilde{\psi}_5(0) = U(0)\psi_5(0),\\ \psi_{10}(1) &\to \tilde{\psi}_{10}(1) = U(1) \otimes U(1)\psi_{10}(1).\end{aligned} \qquad (8)$$

Another reason why $H(0,1)$ be a gauge field comes from "covariant differences" defined by

$$D_q\psi(p) \equiv \psi(p) - H(p,p+q)\psi(p+q) \pmod 3, \quad (q=1,2) \qquad (9)$$

Let us consider $D_2\psi(1) = \psi(1) - H(1,0)\psi(0)$ for example. One easily finds from Eqs. (7) and (8) that

$$D_2\psi(1) \to \tilde{D}_2\tilde{\psi}(1) = U(1) \otimes U(1)D_2\psi(1), \qquad (10)$$

implying $D_2\psi(1)$ transforms covariantly under gauge transformations.

Now we calculate the curvature associated with Fig. 1. The fermion field $\psi_5(x,g_0)$ is mapped from (x,g_0) to $(x+\delta x, g_1)$ through paths C_1 and C_2. The resulting images are denoted by ψ_{C_1} and ψ_{C_2} respectively, whose difference defines a curvature $F_{\mu g_1}(x,g_0)$;

$$\psi_{C_1} - \psi_{C_2} = F_{\mu g_1}(x,g_0)\delta x^\mu \psi_5(x,g_0), \qquad (11)$$

where

$$\begin{aligned}F_{\mu g_1}(x,g_0) &= D_\mu H_5\\ &\equiv \partial_\mu H_5 + ig\left(H_5 A_\mu(0) - A_\mu(1)H_5\right)\end{aligned} \quad (12)$$

In the same way, one obtains

$$\begin{aligned}F_{\mu g_2}(x,g_1) &= D_\mu \phi_{24}\\ &\equiv \partial_\mu \phi_{10} + ig\left(\phi_{24}A_\mu(1) - A_\mu(2)\phi_{24}\right)\end{aligned} \quad (13)$$

It should be noted here that guage fields $A_\mu(0), A_\mu(1)$ and $A_\mu(2)$ at this stage are independent of each other. After constructing the Lagrangian, however, we may identify these fields with the help of action principle;

$$A_\mu(0) = A_\mu(1) = A_\mu(2). \qquad (14)$$

To see this, we first notice that the action I should be of the Yang-Mills type since Higgs fields are gauge fields. Accordingly the Euclidean action contains

$$I \sim \left|\partial_\mu H_5 + ig\left(H_5 A_\mu(0) - A_\mu(1)H_5\right)\right|^2. \qquad (15)$$

Let $|0\rangle$ be the vacuum on g_0, which satisfies

$$\langle 0|H_5|0\rangle = {}^t(0,0,0,0,\beta), \quad (\beta = \text{real const.}) \qquad (16)$$

Then one obtains

$$\begin{aligned}I &\sim \sum_{j=1}^{5}|A_\mu^{5j}(0) - A_\mu^{j5}(1)|^2\\ &= \sum_{j=1}^{4}|A_\mu^{5j}(0) - A_\mu^{j5}(1)|^2 + |A_\mu^{55}(0) - A_\mu^{55}(1)|^2,\end{aligned} \quad (17)$$

which takes a minimum at

$$A_\mu^{55}(1) = A_\mu^{55}(0). \qquad (18)$$

This equality is achieved if we choose the gauge such that

$$\begin{aligned}A_\mu(1) \to \tilde{A}_\mu(1) &= U(1)A_\mu(1)U^{-1}(1) - \frac{i}{g}\partial_\mu U(1)U^{-1}(1)\\ &= A_\mu(0).\end{aligned} \quad (19)$$

Note that

$$\begin{aligned}\langle 0|\tilde{H}(1,0)|0\rangle &= \langle 0|U(1) \otimes U(1)H(1,0)|0\rangle\\ &= \langle 0|H(1,0)|0\rangle,\end{aligned} \quad (20)$$

since the vacuum $|0\rangle$ on g_0 is invariant under $U(1)$.

In the same way we can show that

$$\tilde{A}_\mu(2) = \tilde{A}_\mu(1) = A_\mu(0). \qquad (21)$$

This observation leads to the conclusion that, once the Lagrangian is introduced and the action principle is invoked, we may identify

$$\begin{aligned}\psi_{10}' &= \psi_{10}, \quad H_5' = H_5,\\ A_\mu(x,g_2) &= A_\mu(x,g_1) = A_\mu(x,g_0).\end{aligned} \quad (22)$$

3 Higgs Potential and Fermionic Interactions

Taking the above arguments on identification problems of fermionic and/or gauge fields for granted, we can easily write down our resulting Lagrangian. For example, the Higgs potential $V(\phi, H)$ is expressed as

$$
\begin{aligned}
V(\phi, H) = &-\frac{1}{2}\mu^2 \mathrm{tr}\phi^2 + \frac{1}{4}a \left[\left(\mathrm{tr}\phi^2\right)^2 - \mathrm{tr}\phi^4 \right] - \frac{1}{2}v^2 H^\dagger H \\
&+\frac{1}{4}\lambda \left(H^\dagger H\right)^2 + \alpha \left[\mathrm{tr}\phi^2 \left(H^\dagger H\right) + H^\dagger \phi^2 H\right] \\
&+\beta \left(H^\dagger \phi H\right) + \text{h.c.}, \quad\quad (23)
\end{aligned}
$$

where we put $\phi \equiv \phi_{24}$ and $H \equiv H_5$. We see that the second and the last term in Eq.(23) obey new features not seen in the "phenomenological potential" treated in many literatures.[4] Spontaneous breaking of $SU(5)$ symmetry is not automatic unlike the case of $M_4 \times Z_2$.

The fermionic part of our Lagrangian $\mathcal{L}_F$ takes the usual form. Namely, we have

$$
\begin{aligned}
\mathcal{L}_F = &i\bar{\psi}_5 \gamma^\mu D_{mu}\psi_5 + i\bar{\psi}_{10}\gamma^\mu D_\mu \psi_{10} \\
&-\kappa \left\{ \bar{\psi}_{10} H_5 \psi_5 + \bar{\psi}_5 H_5^* \psi_{10} \right\}, \quad\quad (24)
\end{aligned}
$$

where D_μ is the conventional covariant derivative in M_4.

Concluding Remarks

We have described briefly our construction of $SU(5)$-GUT as the guage theory in the $M_4 \times Z_3$ manifold from geometrical viewpoint. The readers should refer to hep-th/9804161 and *Prog. Theor. Phys.* **100** (1998), 165.

We thank G. Konisi for useful discussions.

References

1. A. Connes, *Noncommutative Geometry* (Academic Press, London, 1994).
2. G. Konishi and T. Saito, *Prog. Theor. Phys.* **95**, 657 (1996).
3. T. Saito and K. Uehara! $Phys. Rev.* D **56**, 2390 (1997). B. Chen, T. Saito, H-B. Teng, K. Uehara and K. Wu, *Prog. Theor. Phys.* **95**, 1173 (1996). A. Kokado, G. Konisi, T. Saito and Y. Tada, *Prog. Theor. Phys.* **99**, 293 (1998).
4. R. N. Mohapatra and M. K. Parida, *Phys. Rev.* D471993264.

SU(N) SUSY GUTS WITH STRING REMNANTS:
MINIMAL SU(5) AND BEYOND

J. L. CHKAREULI

Institute of Physics, Georgian Academy of Sciences
380077 Tbilisi, Georgia

A new superstring-motivated framework for a treatment of $SU(N)$ SUSY GUTs is argued. We show that all the present difficulties of the minimal supersymmetric $SU(5)$ model can successfully be overcome with a new renormalizable reflection-invariant superpotential with two-adjoint scalars, one of which is interpreted as a massive string mode essentially decoupled from the low-energy particle spectra. This superpotential is proved to properly fix a mass ratio of the basic adjoint scalar moduli, while its gauge-type reflection symmetry essentially protects the model from gravitational smearing. The significant heavy threshold effect related with the generic mass splitting of adjoint moduli is shown to alter appropriately the running of gauge couplings towards the realistic string-scale grand unification. Furthermore, the extension of the superpotential to some $SU(N)$ GUTs gives rise to, among many degenerate vacua, the missing VEV vacuum configuration for the basic adjoint scalar, thus providing a further clue to a doublet-triplet splitting problem, on the one hand, and a family symmetry $SU(N-5)_F$ for quarks and leptons, on the other. We predict the existence on a TeV scale two or three families of pseudo-Goldstone bosons of type $(5+\bar{5}) + SU(5)$-singlets depending on $SU(7)$ or $SU(8)$ GUT selected.

1 Introduction

Presently, leaving aside the non-supersymmetric Grand Unified Theories which certainly contradict the experiment unless some special extension of particle spectrum at intermediate scale(s) is made [1], one can see that even SUSY GUTs seem to be far from perfection. The problems, as they appear for the prototype supersymmetric $SU(5)$ model (with a minimal content for matter, Higgs and gauge bosons) [2], can conventionally be qualified as phenomenological and conceptual.

The phenomenological ones include:

(1) The large value of the strong coupling $\alpha_s(M_Z)$ predicted, $\alpha_s(M_Z) > 0.126$ for the effective SUSY scale $M_{SUSY} < 1$ TeV [3] in contrast to the world average value [4] $\alpha_s(M_Z) = 0.119(2)$;

(2) The proton decay due to the color-triplet $H_c(\bar{H}_c)$ exchange [2] at a rate largely excluded by a combination of the detailed renormalization group (RG) analysis for the gauge couplings [5] and the improved lower limits on proton decay mode $p \to \bar{\nu}K^+$ from Super-Kamiokande and on the superparticle masses from LEP2 [6];

(3) An absence of the sizeable neutrino masses $m_\nu \geq 10^{-2}$ eV, as it can explicitly be derived from the atmospheric neutrino deficit data reported [7], at least for one of neutrino species.

Furthermore, for conceptual reasons, the present status of the minimal $SU(5)$ appears to be inadequate as well:

(4) The first one is, of course, the doublet-triplet splitting problem entirely underlying the gauge hierarchy phenomenon in SUSY GUTs [2];

(5) The next point is an absence in the minimal theory of any flavor or family symmetry mechanism which can guarantee together with rather peculiar masses of quarks and leptons a nearly uniform mass spectrum for their superpartners with a high degree of flavor conservation in SUSY theories [8];

(6) Then a low unification scale M_U whose value lies one order of magnitude below than the typical string scale $M_{STR} \simeq 5 \cdot 10^{17}$ GeV [9];

(7) And lastly, the gravitational smearing of its principal predictions (particularly for $\alpha_s(M_Z)$) due to the uncontrollable high-dimension operators induced by gravity in the kinetic terms of the basic gauge SM bosons [10] that makes the ordinary $SU(5)$ model to be largely untestable.

The question arises: Where the possible solution to those problems could come from? According to a superstring theory [9] we seem to say "no" for turning into play of any new states other than the massive string modes, or any mass scales besides the string one. The only exception could be made for the adjoint scalar Σ moduli states $\Sigma_8(8_c, 1)$ and $\Sigma_3(1, 3_w)$ (in a self-evident $SU(3)_C \otimes SU(2)_W$ notation) appearing in many string models at a well-motivated intermediate scale $m \sim M_P^{2/3} M_{SUSY}^{1/3}$ [11]. That means, as an example, the starting mass of the basic adjoint scalar supermultiplet $\Sigma(24)$ of $SU(5)$ theory (remaining just the non-Goldstone remnants Σ_8 and Σ_3 after symmetry breaking) may be taken at a scale m. Thus, the problems listed above remain to be addressed to some new superpotential provided there is the "better" one that develops a proper vacuum configurations when including extra massive ($\sim M_P$) states and/or increasing its symmetry. Strange as it may seem, those simple string-motivated rules of an "allowed" generalization of the minimal $SU(5)$ are turned out to work.

Towards this end, let us consider a general $SU(N)$ invariant renormalizable superpotential of two adjoint scalars Σ and Ω satisfying also the gauge-type reflection symmetry ($\Sigma \to -\Sigma$, $\Omega \to \Omega$) inherited from superstrings

$$W = \frac{1}{2}m\Sigma^2 + \frac{1}{2}M_P\Omega^2 + \frac{1}{2}h\Sigma^2\Omega + \frac{1}{3}\lambda\Omega^3 + W_H \quad (1)$$

where the second adjoint Ω can be considered as a state originated from the massive string mode with the (conventionally reduced) Planck mass $M_P = (8\pi G_N)^{-1/2} \simeq 2.4 \cdot 10^{18}$ GeV, while the basic adjoint Σ remains (relatively) light when one goes from the string scale to lower energies, $m \sim M_P^{2/3} M_{SUSY}^{1/3}$ [11]. The superpotential also includes the ordinary Higgs-doublet containing fundamental chiral supermultiplets H and $\bar{H}$ presented in W_H which is unessential for the moment.

We show below that the superpotential (1) entirely constructed according to our simple rules lead to the natural string scale unification even in a case of the minimal $SU(5)$, whereas its new missing VEV vacuum configurations in the higher $SU(N)$ symmetry cases can give a further clue to other problems mentioned above: doublet-triplet splitting, family symmetry, neutrino masses etc. At the same time, due to the gauge reflection symmetry of the superpotential W, the operators gravitationally induced at a Planck scale for the basic adjoint Σ (developing the principal VEV in the model, see below) in the kinetic terms of the SM gauge bosons should have dimension 6 and higher. Thus, their influence on our predictions seems to be negligible in contrast to the standard $SU(5)$ where they can largely be smeared out [10].

2   The SU(5) model

We start by recalling that, to one-loop order, gauge coupling unification is given by the three RG equations relating the values of the gauge couplings at the Z-peak $\alpha_i(M_Z)$ ($i = 1, 2, 3$), and the common gauge coupling α_U [1]:

$$\alpha_i^{-1} = \alpha_U^{-1} + \sum_p \frac{b_i^p}{2\pi} \ln\frac{M_U}{M_p} \quad (2)$$

where b_i^p are the three b-factors corresponding to the $SU(5)$ subgroups $U(1)$, $SU(2)$ and $SU(3)$, respectively, for the particle labeled by p. The sum extends over all the contributing particles in the model, and M_p is the mass threshold at which each decouples. All of the SM particles and also the second Higgs doublet of MSSM are already presented at the starting scale M_Z. The next is assumed to be supersymmetric threshold associated with the decoupling of the supersymmetric particles at some single effective scale M_{SUSY} [3]; we propose thereafter the relatively low values of M_{SUSY}, $M_{SUSY} \sim M_Z$ to keep sparticle masses typically in a few hundred GeV region. The superheavy states, such as the adjoint fragments Σ_8 and Σ_3 at the masses M_8 and M_3, respectively, and the color-triplets H_c and $\bar{H}_c$ at a mass M_c are also included

in the evolution equations (2). As to the superheavy gauge bosons and their superpartners (X-states), they do not contribute to the Eq.(2), for they are assumed to lie on the GUT scale M_U ($M_X = M_U$), above which all particles fill complete $SU(5)$ multiplets.

Now, by taking the special combination of Eqs.(2) we are led to the simple relation between gauge couplings and the logarithms of the neighboring threshold mass ratios

$$12\alpha_2^{-1} - 7\alpha_3^{-1} - 5\alpha_1^{-1} =$$
$$\frac{3}{2\pi}\left(-2\ln\frac{M_X}{M_c} + \ln\frac{M_c}{M_3} - 7\ln\frac{M_3}{M_8} - \frac{19}{6}\ln\frac{M_{SUSY}}{M_Z}\right) \quad (3)$$

which can be viewed as the basis for giving the qualitative constraints to the $\alpha_s(M_Z)$ depending on the present (very precise) measurement of $\sin^2\theta_W$ [4] and superheavy mass splitting, when one goes beyond the MSSM limit ($M_X = M_c = M_3 = M_8$). One can see from Eq.(3) that α_s increases with $\frac{M_c}{M_3}$ and decreases with $\frac{M_X}{M_c}$, $\frac{M_{SUSY}}{M_Z}$ and, especially, with $\frac{M_3}{M_8}$ (the largest coefficient before logarithm). Unfortunately, in the standard $SU(5)$ case [2] with the degenerate adjoint moduli Σ_3 and Σ_8 ($M_3 = M_8$ at the GUT scale M_X) one is inevitably lead to unacceptably high values of $\alpha_s(M_Z)$ for the allowed M_c region [5,6] and subTeV M_{SUSY} area [3].

However, a drastically different unification picture appears when a generically large mass splitting between Σ_3 and Σ_8 that follows from a new superpotential (1) is taken into account. Actually, one can see that in a basic vacuum which breaks $SU(5)$ to $SU(3)_C \otimes SU(2)_W \otimes U(1)_Y$ the VEVs of the adjoints Σ and Ω are given by

$$\Sigma(\Omega) = \frac{\sqrt{8mM_P}}{h}\left(\frac{2m}{h}\right)diag(1, 1, 1, -3/2, -3/2) \quad (4)$$

respectively, with the hierarchically large VEV ratio being inverse to their masses, $\Sigma/\Omega = (2M_P/m)^{1/2}$. As this takes place, the (physical) mass ratio of the survived adjoint moduli Σ_3 and Σ_8 is turned out to be fixed (at a GUT scale)

$$M_3 = 10m, \quad M_8 = \frac{5}{2}m, \quad \frac{M_3}{M_8} = 4 \quad (5)$$

(just as in a non-supersymmetric $SU(5)$ case) in contrast to $M_3/M_8 = 1$ in the standard one-adjoint superpotential [2].

So, with the observations made we are ready now to carry out the standard two-loop analysis (with conversion from $\overline{MS}$ scheme to $\overline{DR}$ one included) [1,12] for gauge (α_1, α_2, α_3) and Yukawa (α_t, α_b and α_τ in a self-evident notation for top- and bottom-quarks and tau-lepton) coupling evolution depending on, apart from the single-scale (M_{SUSY}) supersymmetric threshold corrections mentioned above, the heavy Σ moduli threshold

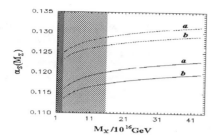

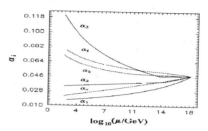

Figure 1: The predictions in the present model (the solid lines) and in the standard supersymmetric $SU(5)$ model (the dotted lines) of $\alpha_s(M_Z)$ as a function of the grand unification scale M_X for the two cases of small $tan\beta$ values (a) with top-Yukawa coupling $\alpha_t(M_X) = 0.3$ taken and large $tan\beta$ (b) values corresponding to top-bottom unification under $\alpha_t(M_X) = \alpha_b(M_X) = 0.05$. The unification mass M_X varies from the adjoint-moduli threshold-degeneration point ($M_X = M_\Sigma$) to the string scale ($M_X = M_{STR}$ with a level $k = 1$), while the color-triplet mass is assumed to be at unification scale in all cases ($M_c = M_X$). The all-shaded areas on the left are generally disallowed by the present bound[4] on nucleon stability for both cases ((a), dark) and ((b), light), respectively.

Figure 2: The superunification of gauge (α_1, α_2, α_3) and Yukawa (α_t, α_b, α_τ) couplings at the string scale (the solid and dotted lines, respectively).

only. This varies, in turn, from the GUT scale M_X ($M_3 = M_X$, $M_8 = \frac{1}{4} M_X$) down to some well-motivated intermediate value $O(10^{14})$ GeV[11] pushing thereafter the M_X up to the string scale M_{STR}. The mass splitting between weak triplet Σ_3 and color octet Σ_8 in themselves noticeably decreases, while M_3 and M_8 run from M_X down to the lower energies, as it results from their own two-loop RG evolution, which is also included in the analysis. On the other hand, the color triplets $H_c(\bar{H}_c)$ are always taken at M_X, for the strings seem to say nothing why any extra states, other than the adjoint moduli Σ_3 and Σ_8, could left relatively light.

Our results, as appeared after numerical integration of all the RG equations listed above, are largely summarized in Figure 1. One can see that the $\alpha_s(M_Z)$ values predicted (with a percent accuracy due to the very precise value of $\sin^2\theta_W = 0.2313(3)$[4] and top-Yukawa coupling appropriately fixed at M_X), are in a good agreement with the world average value (see above) in contrast to the standard SUSY $SU(5)$ taken under the same conditions.

Remarkably enough, the presently testable (SUSY threshold neglecting) top-bottom unification[13] turned out to work well in the model, thus giving the good prediction of the top-quark mass. Furthermore, the low starting values of α_t and α_b at M_X in this case, as well as the closeness of the unification mass M_X to the string scale, allow one to make a next step towards the most symmetrical case which can be realized in the present string-motivated $SU(5)$ - Yukawa and gauge coupling su-

perunification at a string scale:

$$\alpha_t(M_X) = \alpha_b(M_X) = \alpha_\tau(M_X) = \alpha_U \ ,$$
$$M_X = M_{STR} \tag{6}$$

This conjecture certainly could concern the third-family Yukawa couplings solely, since those ones could naturally arise from the basic string-inspired interactions, whereas masses and mixing of the other families seemed to be caused by some more complex and model-dependent dynamics showing itself at lower energies. Due to a crucial reduction of a number of the fundamental parameters the gauge-Yukawa coupling unification leads immediately to a series of the very distinctive predictions (of α_s, in general, and masses in absence of any large supersymmetric threshold corrections)

$$\alpha_s(M_Z) = 0.119 \pm 0.001 \ , \quad m_t = 180 \pm 1 \ ,$$
$$\frac{m_b}{m_\tau} = 1.79 \pm 0.01 \ , \quad tan\beta = 52 \pm 0.2 \tag{7}$$

in a surprising agreement with experiment[4]. In Figure 2 the superunification of gauge and Yukawa couplings is demonstrated.

This is how a new superpotential (1) seems for the first time to open the way to the natural string-scale grand unification in the supersymmetric $SU(5)$, as prescribed at low energies by the gauge coupling values and the minimal particle content[14].

3   Beyond the SU(5)

As a general analysis of the superpotential W (1) shows [15] that possible VEV patterns of the adjoints Σ and Ω include the following four cases only: (i) the trivial symmetry unbroken case, $\Sigma = \Omega = 0$; (ii) the single-adjoint condensation, $\Sigma = 0$, $\Omega \neq 0$; (iii) the "parallel" vacuum configurations, $\Sigma \propto \Omega$ and (iv) the "orthogonal" vacuum configurations, $Tr(\Sigma\Omega) = 0$. While the Planck-mass mode Ω having a cubic term in W develops in all non-trivial cases only a "standard" VEV pattern which breaks the starting $SU(N)$ symmetry to $SU(k) \otimes S(N-k) \otimes U(I)$ the basic adjoint Σ develops a radically new missing VEV vacuum configuration in a case (iv), thus giving a "double" breaking of $SU(N)$ to $SU(k/2) \otimes SU(k/2) \otimes SU(N-k) \otimes U(I)_1 \otimes U(I)_2$ in this case:

$$\Omega = \omega\, diag\Big(\overbrace{1 \ldots 1}^{k} \;,\; \overbrace{-\frac{k}{N-k} \cdots -\frac{k}{N-k}}^{N-k} \Big)$$

$$\Sigma = \sigma\, diag\Big(\overbrace{1 \ldots 1}^{k/2} \;,\; \overbrace{-1 \ldots -1}^{k/2} \;,\; \overbrace{0 \ldots 0}^{N-k} \Big) \qquad (8)$$

with $\omega = (-m/h)(\frac{N-k}{N-2k})$ and $\sigma = \sqrt{mM_P}/h(\frac{2N}{N-k})^{1/2}$, respectively.

Now, if it is granted that the "missing VEV subgroup" $SU(N-k)$ is just the weak symmetry group $SU(2)_W$, we come to a conclusion that the numbers of fundamental colors and flavors (or families) must be equal ($n_C = n_F = k/2$). They all are entirely unified in the framework of the starting $SU(8)$ symmetry [15]. The quark-lepton families (and their superpartners) having been properly assigned to $SU(8)$ multiplets [15] appear as the fundamental triplet of the chiral family symmetry $SU(3)_F$ [16] that meets a natural conservation of flavor both in the particle and sparticle sectors, respectively [8]. Another possibility is when $SU(N-k)$ identifying with the color symmetry group $SU(3)_C$, thus giving interrelation between the weak and flavor groups ($n_W = n_F = k/2$) in the framework of the starting $SU(7)$ symmetry. In the latter case the quark-lepton (squark-slepton) families are doublet plus singlet under the family symmetry $SU(2)_F$ providing the valuable mass matrices for quarks and leptons [15] as well as a flavor conservation in the sparticle sector [8]. The higher $SU(N)$ groups, if considered, are based solely on those principal possibilities mentioned.

Let us see now how this missing VEV mechanism works to solve doublet-triplet splitting problem in $SU(8)$ or $SU(7)$ GUT due to the superpotential W (1). There the W_H part is, in fact, the only reflection-invariant coupling of the basic adjoint Σ with a pair of the ordinary Higgs-boson containing supermultiplets H and $\bar{H}$

$$W_H = f\bar{H}\Sigma H \quad (\Sigma \to -\Sigma,\; \bar{H}H \to -\bar{H}H) \qquad (9)$$

having the zero VEVs, $\bar{H} = H = 0$, during the first stage of the symmetry breaking. Thereupon W_H turns to the mass term of H and $\bar{H}$ depending on the missing VEV pattern (8). This vacuum, while giving generally heavy masses (of the order of M_{GUT}) to them, leaves

their weak components strictly massless. To be certain we must specify the multiplet structure of H and $\bar{H}$ in the both cases of the weak-component and color-component missing VEV vacuum configurations, that is, in $SU(8)$ and $SU(7)$ GUTs, respectively. In the $SU(8)$ case H and $\bar{H}$ are fundamental octets whose weak components (ordinary Higgs doublets) do not get masses from the basic coupling (9). In the SU(7) case H and $\bar{H}$ are the 2-index antisymmetric 21-plets which (after the proper projecting out of extra states) contain just a pair of the massless Higgs doublets. Thus, there certainly is a natural doublet-triplet splitting in the both cases although we drive at the vanishing μ term on this stage. However, one can argue that the right order μ term always appears from the radiative corrections on the next stage when SUSY breaks [15].

Inasmuch as the missing VEV vacua appear only in the higher than $SU(5)$ symmetry cases and extra flavor symmetry should break

$$SU(N-5)_F \otimes U(I)_1 \otimes U(I)_2 \to U(1)_Y \qquad (10)$$

at the GUT scale as well (not to spoil the gauge coupling unification), a question arises: How can those vacua survive so as to be subjected at most to the weak scale order shift? This requires, in general, that a superpotential (1) to be strictly protected from any large influence of $N-5$ scalars $\varphi^{(n)}$ ($n = 1, ..., N-5$) providing the flavor symmetry breaking (10). Technically, such a custodial symmetry could be the superstring-inherited anomalous $U(1)_A$ [17] which can naturally get untie those two sectors and induce the right-scale flavor symmetry breaking (10) through the Fayet-Iliopoulos D-term [2]. Anyway, as it takes place in the supersymmetric $SU(N)$ theory considered, the accidental, while radiatively broken, global symmetry $SU(N)_I \otimes SU(N)_{II}$ appears and $N-5$ families of the pseudo-Goldstone states of type

$$5 \;+\; \bar{5} \;+\; SU(5) - \text{singlets} \qquad (11)$$

are produced at a TeV scale where SUSY softly breaks. Together with ordinary quarks and leptons and their superpartners the two or three families of PG states (11), depending on $SU(7)$ or $SU(8)$ GUT selected, determine the whole particle environment at low energies.

4 Conclusion

The recent Super-Kamiokande data [7] arise a question about a modification of the SM to get a mass of order 0.1 eV at least for one of neutrino species. That means, in general, a particle content of the SM or the minimal $SU(5)$ should be extended to include new states, that is, the properly heavy right-handed neutrinos or even light sterile left-handed ones. We have found that the proper missing VEV vacuum configurations require the extended GUTs $SU(7)$ or $SU(8)$ where such states, together with ordinary quarks and leptons, naturally appear, thus providing on this stage at least a qualitative explanation [15] for data [7].

Meanwhile, despite the common origin there is a principal difference between the $SU(7)$ and $SU(8)$ cases that manifests itself not only in number of PG families (11). The point is the basic adjoint Σ moduli mass ratio M_3/M_8 (which, as we could see in the $SU(5)$, essentially determines a high-energy behavior of gauge couplings) appears according to the missing VEV vacua (8) to be 2 and 1/2 for $SU(7)$ and $SU(8)$, respectively. That means the unification scale in $SU(7)$ can be pushed again to the string scale, while scale in $SU(8)$ always ranges closely to the standard unification value [15]. The detailed RG analysis [15] carried along with that of $SU(5)$ (Section 2) lives up to our expectations. So, the $SU(7)$ seems to be the only GUT that can give a solution to all seven problems which we have started from.

Acknowledgments

I would like to acknowledge the stimulating conversations with many colleagues, especially with Alexei Anselm, Riccardo Barbieri, Zurab Berezhiani, Gia Dvali, Colin Froggatt, Mike Green, S. Randjbar-Daemi, Alexei Smirnov, David Sutherland, and not least, with my collaborators Ilia Gogoladze and Archil Kobakhidze. Financial support by INTAS Grants No. RFBR 95-567 and 96-155 are also gratefully acknowledged.

References

1. W. de Boer, *Prog. Part. Nucl. Phys.* 33, 201 (1994).

2. S. Dimopoulos and H. Georgi, *Nucl. Phys.* B 193, 150 (1981); N. Sakai, *Z. Phys.* C 11, 153 (1981); S. Weinberg, *Phys. Rev.* D 26, 287 (1982); N. Sakai and T. Yanagida, *Nucl. Phys.* B 197, 533 (1982).

3. J. Bagger, K. Matchev and D. Pierce, *Phys. Lett.* B 348, 443 (1995); P. Langacker and N. Polonsky, *Phys. Rev.* D 52, 3081 (1995).

4. Particle Data Group, *Eur. Phys. J. C.* 3, 1 (1998).

5. H. Hisano, T. Moroi, K. Tobe and T. Yanagida, *Mod. Phys. Lett.* A 10, 2267 (1995).

6. H. Murayama, *Nucleon Decay in GUT and Non-GUT SUSY models* , LBNL-39484; hep-ph/9610419.

7. The Super-Kamiokande Collaboration (Y.Fukuda *et al.*), *Phys. Rev. Lett.* 81, 1562 (1998).

8. M. Dine, R. Leigh and A. Kagan, *Phys. Rev.* D 48, 4269 (1993); L.J. Hall and H. Murayma, *Phys.*

Rev. Lett. 75, 3985 (1995); Z. Berezhiani, *Phys. Lett.* B 417, 287 (1998).

9. For a recent discussion and extensive references, see K. Dienes, *Phys. Rep.* 287, 447 (1997).

10. C.T. Hill, *Phys.Lett.* 135 B, 47 (1984); L.J. Hall and U. Sarid, *Phys. Rev. Lett.* 70, 2673 (1993); S. Urano, D. Ring and R. Arnowitt, *Phys. Rev. Lett.* 76, 3663 (1996).

11. A.E. Faraggi, B. Grinstein and S. Meshkov, *Phys. Rev.* D 47, 5018 (1993); C. Bachas, C. Fabre and T. Yanagida, *Phys. Lett.* B 370, 49 (1996); M. Bastero-Gil and B. Brahmachari, *Phys. Lett.* B403, 51 (1997).

12. V. Barger, M.S. Berger and P. Ohmann, *Phys. Rev.* D 47, 1093 (1993); P. Martin and M.T. Vaughn, *Phys. Rev.* D 50, 2282 (1994).

13. L.J. Hall, R. Rattazzi and U. Sarid, *Phys. Rev.* D 50, 7048 (1994); M. Carena, M. Olechowski, S. Pokorski and C.E.M. Wagner, *Nucl. Phys.* B 426, 269 (1994).

14. J.L. Chkareuli and I.G. Gogoladze, *Phys. Rev.* D 58, 551 (1998).

15. J.L. Chkareuli and A.B. Kobakhidze, *Phys. Lett.* B 407, 234 (1997); J.L. Chkareuli, I.G. Gogoladze and A.B. Kobakhidze, *Phys. Rev. Lett.* 80, 912 (1998); *Phys. Lett.* B (in press).

16. J.L. Chkareuli, *JETP Letters* 32, 671 (1980); Z.G. Berezhiani, *Phys. Lett.* B 129, 99 (1983); B 150, 117 (1985); J.L. Chkareuli, *Phys. Lett.* B 246, 498 (1990); B 300, 361 (1993).

17. M. Green and J. Schwarz, *Phys. Lett.* B 149, 117 (1998).

THE DUALIZED STANDARD MODEL AND ITS APPLICATIONS

CHAN Hong-Mo

Rutherford Appleton Laboratory, Chilton, Didcot, Oxon, OX11 0QX, United Kingdom
E-mail: chanhm @v2.rl.ac.uk

José BORDES

Dept. Fisica Teorica, Univ. de Valencia, c. Dr. Moliner 50, E-46100 Burjassot (Valencia), Spain
E-mail: jose.M.bordes @uv.es

TSOU Sheung Tsun

Mathematical Institute, University of Oxford, 24-29 St. Giles', Oxford, OX1 3LB, United Kingdom
E-mail: tsou @maths.ox.ac.uk

The Dualized Standard Model offers a natural explanation for Higgs fields and 3 generations of fermions plus a perturbative method for calculating SM parameters. By adjusting only 3 parameters, 14 quark and lepton masses and mixing parameters (including ν oscillations) are calculated with general good success. Further predictions are obtained in post-GZK air showers and FCNC decays.

In this article, we summarize some work which has occupied us for some years. The material has been summarized in 5 papers submitted to this conference (Papers 607, 610, 611, 613, 636), and this talk is but a summary of these summaries.

The Dualized Standard Model [1] is a scheme which aims to answer some of the questions left open by the Standard Model (such as why Higgs fields or fermion generations should exist) and to explain the values of some the Standard Model's many parameters (such as fermion masses and mixing angles). In contrast to most schemes with similar aims, the DSM remains entirely within the Standard Model framework, introducing neither supersymmetry nor higher space-time dimensions. That it is able to derive results beyond the Standard Model while remaining within its framework is by exploiting a generalization of electric-magnetic duality to nonabelian Yang-Mills theory found a couple of years ago [2].

The concept of duality is best explained by recalling the well-known example in electromagnetism. There a dual transform (the Hodge star) is defined: $^*F_{\mu\nu} = -(1/2)\epsilon_{\mu\nu\rho\sigma}F^{\rho\sigma}$, where for both the Maxwell field $F_{\mu\nu}$ and its dual $^*F_{\mu\nu}$ potentials A_μ and $\tilde{A}_\mu$ exist, so that the theory is invariant under both A_μ and $\tilde{A}_\mu$ gauge transformations. The theory has thus in all a $U(1) \times \tilde{U}(1)$ gauge symmetry with $U(1)$ corresponding to electricity and $\tilde{U}(1)$ to magnetism. Magnetic charges are monopoles in $U(1)$, while electric charges are monopoles of $\tilde{U}(1)$.

The same statements do not hold for nonabelian Yang-Mills theory under the dual transform [3] (star). However, it was shown [2] that there exists a generalized dual transform for which similar results apply. Its exact

form, in the language of loop space [4,5], need not here bother us. What matters, however, is that given this generalized transform, a potential can again be defined for both the Yang-Mills field and its dual, and that the theory is invariant under both the gauge transformations:

$$A_\mu \to A_\mu + \partial_\mu\Lambda + ig[\Lambda, A_\mu], \quad \tilde{A}_\mu \to \tilde{A}_\mu + \partial_\mu\tilde{\Lambda} + i\tilde{g}[\tilde{\Lambda}, \tilde{A}_\mu], \tag{1}$$

with $g, \tilde{g}$ satisfying the (generalized) Dirac quantization condition [6]: $g\tilde{g} = 4\pi$. As a result, the theory has in all the gauge symmetry $SU(N) \times \widetilde{SU}(N)$ with $SU(N)$ corresponding to (electric) 'colour' and $\widetilde{SU}(N)$ to (magnetic) 'dual colour'. And again, dual colour charges are monopoles in $SU(N)$, while colour charges appear as monopoles in $\widetilde{SU}(N)$ [7].

Applied to colour in the Standard Model, this nonabelian duality [2] gives two new interesting features. First, dual to the $SU(3)$ symmetry of colour, the theory possesses also an $\widetilde{SU}(3)$ symmetry of dual colour. Then, by a well-known result of 't Hooft [8], since colour is confined, it follows that this $\widetilde{SU}(3)$ of dual colour has to be broken via a Higgs mechanism[a]. Hence, the theory already contains within itself a broken 3-fold gauge symmetry which could play the role of the 'horizontal' symmetry wanted to explain the existence in nature of the 3 fermion generations. Second, in the generalized dual transform [2], the frame vectors ('dreibeins') in the gauge group take on a dynamical role [7] which suggests that they be promoted to physical fields. If so, then they possess the properties

---

[a]It has been shown [6] the duality introduced [2] indeed satisfies the commutation relations of the order-disorder parameters used by 't Hooft to define his duality.

that one wants for Higgs fields for symmetry breaking (as in electroweak theory): space-time scalars belonging to the fundamental representation having classical values (vev's) with finite lengths.

The basis of the Dualized Standard Model is just in making this bold assumption of identifying the dual colour $SU(3)$ as the 'horizontal' generation symmetry and of the frame vectors in it as the Higgs fields for its breaking. We note that according to nonabelian duality [2], the niches already exist in the original theory in the form of the dual symmetry and the 'dreibeins'. One could thus claim that the DSM offers an explanation for the existence both of exactly 3 fermion generations and of Higgs fields necessary for breaking this generation symmetry.

This identification further suggests the manner in which the symmetry ought to be broken. As a result, the fermion mass matrix at tree-level takes the form [1]:

$$m = m_T \begin{pmatrix} x \\ y \\ z \end{pmatrix} (x, y, z), \qquad (2)$$

where m_T is a normalization factor depending on the fermion-type T, namely whether U- or D-type quarks, charged leptons (L) or neutrinos (N), and x, y, z are vacuum expectation values of Higgs fields (normalized for convenience: $x^2 + y^2 + z^2 = 1$), which are independent of the fermion-type T. Because m is factorizable it has only one nonzero eigenvalue so that at tree-level there is only one massive generation (fermion mass hierarchy). Further, because (x, y, z) is independent of the fermion-type, the state vectors of, say, the U- and D-type quarks in generation space have the same orientation, so that the CKM matrix is the unit matrix. These are already not a bad first approximation to the experimental situation.

One can go further, however. With loop corrections, it is seen that the mass matrix m' remains factorizable [1], with the same form as (2), but the vector (x', y', z'), in which the relevant information of m' is encoded, now rotates with the energy scale, tracing out a trajectory on the unit sphere. Hence, the lower generation fermions acquire small finite masses via 'leakage' from the highest generation. Furthermore, the vector (x', y', z') depends now on the fermion-type, giving rise to a nontrivial CKM matrix. The result is a perturbative method for calculating fermion mass and mixing parameters.

In a 1-loop calculation [9,10] it is found that out of the many diagrams only the Higgs loop diagram dominates, involving thus only a few adjustable parameters. The present score is as follows. By adjusting 3 parameters, namely a Yukawa coupling strength ρ and the 2 ratios between the Higgs vev's x, y, z, one has calculated the following 14 of the 'fundamental' SM parameters:
- the 3 parameters of the quark CKM matrix $|V_{rs}|$,
- the 3 parameters of the lepton CKM matrix $|U_{rs}|$,

Table 1: Fermion Masses

	$Calculation$	$Experiment$
m_c	1.327GeV	1.0 – 1.6GeV
m_s	0.173GeV	100 – 300MeV
m_μ	0.106GeV	105.7MeV
m_u	235MeV	2 – 8MeV
m_d	17MeV	5 – 15MeV
m_e	7MeV	0.511MeV
m_{ν_1}	10^{-15}eV	< 10eV
B	400TeV	?

- $m_c, m_s, m_\mu, m_u, m_d, m_e$,
- the masses m_{ν_1} of the lightest and B of the right-handed neutrinos,

there being no CP-violation at 1-loop order.

First, for the quark CKM matrix $|V_{rs}|$, where $r = u, c, t$ and $s = d, s, b$, one obtains for a sample fit [11]:

$$|V_{rs}| = \begin{pmatrix} 0.9752 & 0.2215 & 0.0048 \\ 0.2210 & 0.9744 & 0.0401 \\ 0.0136 & 0.0381 & 0.9992 \end{pmatrix}, \qquad (3)$$

as compared with the experimental values [12]:

$$\begin{pmatrix} 0.9745 - 0.9760 & 0.217 - 0.224 & 0.0018 - 0.0045 \\ 0.217 - 0.224 & 0.9737 - 0.9753 & 0.036 - 0.042 \\ 0.004 - 0.013 & 0.035 - 0.042 & 0.9991 - 0.9994 \end{pmatrix}. \qquad (4)$$

All the calculated values are seen to lie roughly within the experimental bounds.

Second, for the lepton CKM matrix $|U_{rs}|$, one obtains with the same 3 input parameters:

$$|U_{rs}| = \begin{pmatrix} 0.97 & 0.24 & 0.07 \\ 0.22 & 0.71 & 0.66 \\ 0.11 & 0.66 & 0.74 \end{pmatrix}, \qquad (5)$$

where $r = e, \mu, \tau$ and $s = 1, 2, 3$ label the physical states of the neutrinos. The empirical values of $|U_{rs}|$ for leptons are much less well-known. Collecting all the information so far available from neutrino oscillation experiments, one arrives at the following tentative arrangement:

$$|U_{rs}| = \begin{pmatrix} \star & 0.4 - 0.7 & 0.0 - 0.15 \\ \star & \star & 0.45 - 0.85 \\ \star & \star & \star \end{pmatrix}. \qquad (6)$$

which is seen to be in very good agreement with the prediction (5) for U_{e3} and $U_{\mu3}$, but not for U_{e2}.

Lastly, from the same calculation with the same 3 parameters, one obtains the fermion masses listed in Table 1. The second generation masses agree very well with experiment. Those of the lowest generation were obtained by extrapolation on a logarithmic scale and should be regarded as reasonable if of roughly the right magnitude. As for the 2 neutrino masses, the experimental bounds are so weak that there is essentially no check.

In summary, out of the 14 quantities calculated, 8 are good ($|V_{rs}|, |U_{\mu 3}|, |U_{e3}|, m_c, m_s, m_\mu$), 2 are reasonable ($m_d, m_e$), 2 are unsatisfactory ($|U_{e2}|, m_u$), and 2 are untested, which is not a bad score for a first-order calculation with only 3 parameters.

One interesting feature for the calculation outlined above is that the trajectories traced out by the vector (x', y', z') for the 4 different fermion-types U, D, L, N all coincide to a very good approximation, only with the 12 physical fermion states at different locations (Figure 1). The points $(1, 0, 0)$ and $\frac{1}{\sqrt{3}}(1, 1, 1)$ are fixed points so that

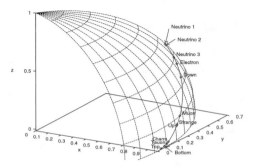

Figure 1: The trajectory of (x', y', z') as the energy scale varies.

the rate of flow is slower near the ends of the trajectory than in the middle. For this reason, the states t and b are close together in spite of their big mass difference. This observation will be of relevance later.

Since neutrino oscillations [13] are of particular interest at this conference, let us take a closer look [10]. The element $U_{\mu 3}$ of the lepton CKM matrix giving the mixing between the muon neutrino ν_μ and the heaviest neutrino ν_3 is constrained mainly by the data on atmospheric neutrinos. From the old Kamiokande data [14] an analysis by Giunti et al.[15] gives the bounds on $U_{\mu 3}$ shown in Figure 2. In the DSM scheme, with parameters already fixed by the fit to the quark sector [9], the elements of $|U_{rs}|$ are calculable given the masses of ν_3 and ν_2. Then, with m_{ν_2} taken in the range $10^{-11}\text{eV}^2 < m_2^2 < 10^{-10}\text{eV}^2$ as suggested by the Long Wave-Length Osicillation (LWO) (or the 'vacuum' or 'just-so') solution to the solar neutrino problem [16,17], the predicted band of values of $|U_{\mu 3}|$ for a range of input values of m_{ν_3} is shown in Figure 2, passing right through the middle of the allowed region. No similar detailed analysis of the new SuperKamiokande data [13] has yet been performed, but the predicted band can be seen to remain well within the allowed region:

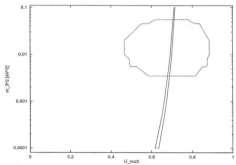

Figure 2: 90 % CL limits on $U_{\mu 3}$ compared with DSM calculation.

$.53 < U_{\mu 3} < .85,\ 5 \times 10^{-4} < m_3^2 < 6 \times 10^{-3}\text{eV}^2.$

The same calculation gives the prediction shown in Figure 3 for the element U_{e3} representing the mixing between the electron neutrino ν_e and ν_3, which is constrained mainly by the reactor data from CHOOZ[18] and Bugey [19]. If m_3^2 is higher than $2 \times 10^{-3}\text{eV}^2$, as favoured by the old Kamiokande data [14,15] and the new data from Soudan reported in this conference [20], then the negative result from CHOOZ restricts U_{e3} to quite small values, as indicated in Figure 3 and quoted in (6). The new Su-

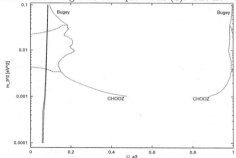

Figure 3: 90 % CL limits on U_{e3} compared with DSM calculation.

perKamiokande data [13] gives a best-fit value for m_3^2 of 2.2×10^{-3}, still implying by CHOOZ a small value for U_{e3}, but do not exclude lower values of m_3 and hence much larger values of U_{e3}. In any case, as seen in Figure 3, the band of values predicted by the DSM calculation falls always comfortably within the allowed region.

The DSM results summarized above for neutrino oscillations were obtained assuming m_2^2 of order 10^{-11} to 10^{-10}eV^2, as suggested by the LWO solution [16,17]. The alternative MSW [21] solutions for solar neutrinos require[22] $m_2^2 \sim 10^{-5}\text{eV}^2$, for which no sensible DSM solution was found [10]. It is thus intriguing to hear in this conference that the new SuperKamiokande data on the day-night variation and energy spectrum of solar neutrinos [13] also favours the LWO solution.

Further, generation being identified with dual colour in DSM, one expects only 3 generations of neutrinos. Thus, the result from Karmen [23] reported in this conference against the existence of another neutrino with mass of order eV, as previously suggested by the LSND experiment [24], is also in the DSM's favour.

It is particularly instructive to compare the CKM matrices for leptons and quarks. Both the empirical (4), (6) and the calculated (3), (5) matrices show the following salient features:

- that the 23 element for leptons is much larger than that for quarks,
- that the 13 elements for both quarks and leptons are much smaller than the rest,
- that the 12 element is largish and comparable in magnitude for quarks and leptons.

These features, all so crucial for interpreting existing data, not only are all correctly reproduced by DSM calculation, but also can be understood within the scheme using some classical differential geometry as follows [25].

First, it turns out [1,9] that in the limit when the separation between the top 2 generations is small on the trajectory traced out by (x', y', z'), which is the case for all 4 fermion-types as seen in Figure 1, then the vectors for the 3 generations form a Darboux triad [26] composed of (i) the radial vector (x', y', z') for the heaviest generation, (ii) the tangent vector to the trajectory for the second generation, and (iii) the vector normal to both the above for the lightest generation. The CKM matrix is thus just the matrix which gives the relative orientation between the Darboux triads for the two fermion-types concerned. Secondly, by the Serret–Frenet–Darboux formulae, it follows that the CKM matrix can be written, to first order in the separation Δs on the trajectory between t and b for quarks and between τ and ν_3 for leptons, as

$$CKM \sim \begin{pmatrix} 1 & -\kappa_g \Delta s & -\tau_g \Delta s \\ \kappa_g \Delta s & 1 & \kappa_n \Delta s \\ \tau_g \Delta s & -\kappa_n \Delta s & 1 \end{pmatrix}, \quad (7)$$

where κ_n and κ_g are respectively the normal and geodesic curvature and τ_g is the geodesic torsion of the trajectory. Lastly, for our case of a curve on the unit sphere, $\kappa_n = 1$ and $\tau_g = 0$, from which it follows that :

- the 23 element equals roughly Δs,
- the 13 element is of second order in Δs,
- the 12 element depends on the details of the curve.

In Figure 1, Δs between τ and ν_3 is much larger than that between t and b, hence also the 23 element of the CKM matrix. Indeed, measuring the actual separations in Figure 1, one obtains already values very close to the actual CKM matrix elements in (3) and (5) or in (4) and (6). The 13 elements should be small in both cases, as already noted. As for the 12 elements, they depend on both the locations and details of the curve, which

explains why they need not differ much between quarks and leptons in spite of the difference in separation, and also why the DSM prediction in (5) is not as successful for this element as for the others.

To test DSM further, one seeks predictions outside the Standard Model framework. These are not hard to come by. Identifying generation to dual colour, which is a local gauge symmetry, makes it imperative that any particle carrying a generation index can interact via the exchange of the dual colour gauge bosons, leading to flavour-changing neutral current (FCNC) effects. Given the calculations on the CKM matrices outlined above, all low energy FCNC effects can now be calculated in terms of a single mass parameter ζ related to the vev's of the dual colour Higgs fields [27]. Inputting the mass difference $K_L - K_S$ which happens to give the tightest bound on $\zeta \sim 400$ TeV, one obtains bounds on the branching ratios of various FCNC decays. In the following paragraph, an argument will be given which converts these bounds into actual order-of-magnitude estimates. In particular, the mode $K_L \to \mu^\pm e^\mp$ has a predicted branching ratio of around 10^{-13}, less than 2 orders away from the new BNL bound of 5.1×10^{-12} reported in this conference [28].

Since neutrinos carry a generation index, it follows that they will also acquire a new interaction through the exchange of dual colour bosons. At low energy, this interaction will be very weak, being suppressed by the large mass parameter ζ. However, at C.M. energy above ζ, this new interaction will become strong. With an estimate of at least 400 TeV, the predicted new interaction is not observable in laboratory experiments at present or in the foreseeable future, but it may be accessible in cosmic rays. For a neutrino colliding with a nucleon at rest in our atmosphere, 400 TeV in the centre of mass corresponds to an incoming energy of about 10^{20} eV. Above this energy, neutrinos could thus in principle acquire a strong interaction and produce air showers in the atmosphere. Now air showers at and above this energy have been observed. They have long been a puzzle to cosmic ray physicists since they cannot be due to proton or nuclear primaries which would be quickly degraded from these energies by interaction with the 2.7 K microwave background [29]. Indeed, the GZK cutoff [30] for protons is at around 5×10^{19} eV. Neutrinos, on the other hand, are not so affected by the microwave background. Hence, if they can indeed produce air showers via the new interaction predicted by the DSM, they can give a very neat explanation for the old puzzle of air showers beyond the GZK cut-off [31]. Further tests for the proposal have been suggested [32]. The proposal also gives a rough upper bound on the mass parameter ζ governing FCNC effects which is close to the lower bound obtained in the preceding paragraph. It was on the basis of this coincidence that the above FCNC bounds were converted into actual order-of-magnitude estimates.

The conclusions are summarized in Figure 4.

It is a pleasure for us to acknowledge our profitable and most enjoyable collaboration with Jacqueline Faridani and Jakov Pfaudler. TST also thanks the Royal Society for a travel grant to Vancouver.

References

1. Chan Hong-Mo and Tsou Sheung Tsun, Phys. Rev. D57, 2507, (1998).
2. Chan Hong-Mo, Jacqueline Faridani, and Tsou Sheung Tsun, Phys. Rev. D53, 7293, (1996).
3. CH Gu and CN Yang, Sci. Sin. 28, 483, (1975); TT Wu and CN Yang, Phys. Rev. D12, 3843 (1975).
4. A.M. Polyakov, Nucl. Phys. 164, 171, (1980).
5. Chan Hong-Mo and Tsou Sheung Tsun, *Some Elementary Gauge Theory Concepts* (World Scientific, Singapore, 1993).
6. Chan Hong-Mo and Tsou Sheung Tsun, Phys. Rev. D56, 3646, (1997).
7. Chan Hong-Mo, Jacqueline Faridani, and Tsou Sheung Tsun, Phys. Rev. D51, 7040 (1995).
8. G. 't Hooft, Nucl. Phys. B138, 1, (1978); Acta Phys. Austriaca. Suppl. 22, 531, (1980).
9. J Bordes, Chan H-M, J Faridani, J Pfaudler, Tsou ST, Phys. Rev. D58, 013004, (1998)
10. José Bordes, Chan Hong-Mo, Jakov Pfaudler, Tsou Sheung Tsun, Phys. Rev. D58, 053003 (1998).
11. Chan Hong-Mo and Tsou Sheung Tsun, Acta Phys. Polonica, B28, 3041, (1997)
12. Particle Physics Booklet, (1996), RM Barnett et al., Phys. Rev. D54, 1, (1996); web updates.
13. SuperKamiokande data, presented by C McGrew, M Vagins, M Takita at ICHEP'98.
14. K.S. Hirata et al., Phys. Letters, B205, 416, (1988); B280, 146, (1992); Y. Fukuda et al. Phys. Letter B335, 237, (1994).
15. C Giunti, CW Kim, M Monteno, hep-ph/9709439.
16. V. Barger, R.J.N. Phillips, and K. Whisnant, Phys. Rev. Letters, 69, 3135, (1992).
17. PI Krastev, ST Petcov, PRL 72, 1960, (1994).
18. CHOOZ collaboration, M. Apollonio et al., Phys. Lett. B420, 397, (1997).
19. B. Ackar et al., Nucl. Phys. 434, 503, (1995).
20. Soudan II, presented by H Gallagher at ICHEP'98.
21. L Wolfenstein, Phys. Rev. D17, 2369, (1978); SP Mikheyev, AYu Smirnov, Nuo. Cim. 9C, 17 (1986).
22. For example see G.L. Fogli, E. Lisi, and D. Montanino, Phys. Rev. D54, 2048 (1995).
23. Karmen, presented by J Kleinfeller at ICHEP'98.
24. C Athanassopoulos et al, PRL 75, 2650 (1995).
25. José Bordes, Chan Hong-Mo, Jakov Pfaudler, Tsou Sheung Tsun, Phys. Rev. D58, 053006 (1998).
26. See e.g. L.P. Eisenhart, *A Treatise on the Differential Geometry of Curves and Surfaces*, Ginn and Company 1909, Boston.
27. J Bordes, Chan H-M, J Faridani, J Pfaudler, Tsou ST, hep-ph/9807277, (1998).
28. BNL data in review by D Bryman at ICHEP'98.
29. Murat Boratav, astro-ph/9605087, Proc. 7th Int. Workshop on Neutrino Telescopes, Venice 1996.
30. K Greisen, PRL 16 (1966) 748; GT Zatsepin and VA Kuz'min, JETP Letters, 4 (1966) 78.
31. J Bordes, Chan HM, J Faridani, J Pfaudler, Tsou ST, hep-ph/9705463; hep-ph/9711438, to appear in Proc. of the Int. Workshop on Physics Beyond the Standard Model (1997) Valencia.
32. J Bordes, Chan HM, J Faridani, J Pfaudler, Tsou ST, Astroparticle Phys. J., 8, 135 (1998).

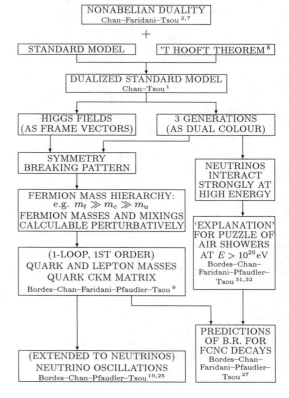

Figure 4: Summary flow-chart

PREDICTIONS FOR SUSY PARTICLE MASSES FROM ELECTROWEAK BARYOGENESIS

J. M. CLINE

McGill University, Dept. of Physics, 3600 University St., Montréal (Qc) H3A 2T8, Canada
E-mail: jcline@physics.mcgill.ca

In collaboration with G.D. Moore,[1] the electroweak phase transition in the minimal supersymmetric standard model is studied using the two-loop effective potential. We make a comprehensive search of the MSSM parameter space consistent with electroweak baryogenesis, taking into account various factors: the latest experimental constraints on the Higgs boson mass and the ρ parameter, the possibility of significant squark and Higgs boson mixing, and the exact rate of bubble nucleation and sphaleron transitions. Most of the baryogenesis-allowed regions of parameter space will be probed by LEP 200, hence the Higgs boson is likely to be discovered soon if the baryon asymmetry was indeed created during the electroweak phase transition.

1 Introduction

An attractive possibility for explaining the baryon asymmetry of the universe,

$$\frac{n_B - n_{\bar{B}}}{n_\gamma} \sim 10^{-10} \qquad (1)$$

is electroweak baryogenesis, combined with supersymmetry. This is one of the only pictures which makes fairly definite low-energy predictions, which are in effect being tested now at LEP and the Tevatron by their searches for the Higgs boson and the top squark.[1,2]

Let us review the basic picture of electroweak baryogenesis. It assumes that the electroweak phase transition (EWPT) is strongly first order, hence bubbles of the true vacuum, with nonzero Higgs VEV $\langle H \rangle$, nucleate inside the false vacuum, at a critical temperature T_c near 100 GeV (fig. 1). Outside the bubble, anomalous baryon-violating interactions (sphalerons), present in the Standard Model, are occuring much faster than the Hubble expansion rate. Inside the bubble, if $\langle H \rangle / T_c$ is larger than ~ 1, these interactions are out of equilibrium. In addition there must be CP-violating interactions at the bubble wall, which cause an asymmetry in the reflection probability for particles versus antiparticles (and left-handed versus right-handed particles) from the wall. This causes a build-up of CP asymmetry in front of the wall, which the sphalerons attempt to erase. But in so doing, they create a baryon asymmetry in front of the wall, which gradually falls behind the wall due to the steady expansion of the latter into the plasma, and collisions of the reflected particles with other particles in the plasma. This baryon excess survives to become the baryonic matter of the present-day universe.

There have been attempts to implement electroweak baryogenesis in other ways, for example, using cosmic strings instead of bubbles. The idea is that $\langle H \rangle$ can be suppressed inside the strings; then strings could work similarly to bubbles except that the baryon (B) violation is going on inside of the strings as they sweep through

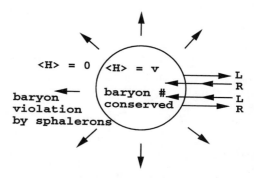

Figure 1: Expanding bubble during the electroweak phase transition. Left-handed fermions reflect into right-handed ones and vice versa, but with different probabilities.

space, while B is conserved outside because the temperature is assumed to be much less than T_c. We have made a quantitative study of all aspects of this proposal [3] (except for the possibility of superconducting strings which carry an enormous current), and concluded that it falls short of being able to produce the required B asymmetry by 10 orders of magnitude. One reason for the failure is that the CP violation scales with the string velocity (v) squared, whereas the density of a string network scales like v^{-2}. It is therefore impossible to tune the density to enhance the asymmetry. Another reason for the failure is that sphalerons are typically large compared to strings, and their energy, E_{sph}, is increased if they are squashed so as to fit inside a string. This suppresses their likelihood, hence the rate of B violation, by a Boltzmann factor, $e^{-E_{sph}/T}$.

Thus the expanding bubble picture is the most likely realization of electroweak baryogenesis. And it is also highly constrained, since it is not easy to get a large enough asymmetry. One of the major challenges is in getting the phase transition to be strong enough so that $\langle H \rangle / T_c \geq 1$.

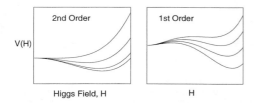

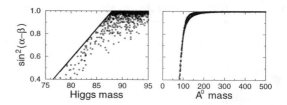

Figure 2: (a) Higgs effective potential for several temperatures, with a second order transition. (b) Same, but with a first order transition.

Figure 4: (a) Scatter plot from Monte Carlo in plane of Higgs mixing angles versus lightest Higgs mass (in GeV); (b) Same, but with the mass of the A^0 Higgs boson.

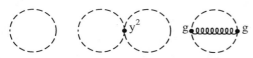

Figure 3: The most important one- and two-loop stop diagrams contributing to the effective potential.

is strong enough for successful baryogenesis, while still compatible with other experimental constraints.

3 Important Parameters and Constraints

Now I will discuss the parameters which have the strongest effect on the EWPT.

Lightest Higgs mass, $\mathbf{m}_h$. The transition is strongest if m_h is small. Moreover, the light Higgs field h is related to the flavor eigenstates H_i by a mixing angle α,

$$h = \sin\alpha\, H_1 + \cos\alpha\, H_2, \qquad (3)$$

and the phase transition is also strongest when $\sin^2(\alpha - \beta) \approx 1$, where $\tan\beta = \langle H_2\rangle/\langle H_1\rangle$. This is illustrated in fig. 4a, which shows the distribution of allowed points from our Monte Carlo of the parameter space, in the plane of $\sin^2(\alpha - \beta)$ and m_h. The region to the left of the line is experimentally excluded by LEP.

CP-odd Higgs mass, $\mathbf{m}_{A^0}$. A strong phase transition also favors the pseudoscalar Higgs boson, A^0, being heavy. In fact, this explains the preference for $\sin^2(\alpha - \beta) \approx 1$, because there is a strong correlation between m_{A^0} and $\sin^2(\alpha - \beta)$, as shown in fig. 4b. In the limit where both become large, the Higgs sector of the MSSM becomes SM-like, with all the heavy Higgs bosons decoupling.

$\tan\beta$ and $\mathbf{m}_U^2$. We find a lower limit on the value of $\tan\beta$ which gives a strong enough transition:

$$\tan\beta > 2.1 \qquad (4)$$

This is coming largely from the fact that m_h depends on $\tan\beta$,

$$m_h^2 = \tfrac{1}{2}\left[m_A^2 + m_Z^2 - \sqrt{(m_A^2 + m_Z^2)^2 - 4m_Z^2 m_A^2 \cos^2 2\beta}\right]$$
$$+ O\left[(m_t^4/v^2)\ln\left(m_{\tilde{t}_1} m_{\tilde{t}_2}/m_t^2\right)\right], \qquad (5)$$

in such a way that its tree-level value vanishes when $\tan\beta = 1$. Thus to satisfy the experimental constraints

2 Strength of the Phase Transition

In the Standard Model, the EWPT is second order, as illustrated in figure 2a. The effective potential, $V(H)$, never develops a barrier between the high-temperature, symmetric phase minimum ($H = 0$) and the low-T, broken phase one ($H \neq 0$). Bubbles do not form in this case, so baryogenesis as outlined above cannot occur. However, the EWPT can be first order in the Minimal Supersymmetric Standard Model (MSSM) if one of the top squarks (the mostly right-handed one) is sufficiently light,[2,4,5,6] giving an effective potential of the form shown in figure 2b.

In the presence of light stops (light compared to the temperature), $V(H)$ gets finite-temperature contributions from vacuum loop diagrams containing stops and possibly gluons, like those of figure 3. The propagators are evaluated at an arbitrary value of the background Higgs fields, of which there are two in the MSSM. The squark masses appearing in the propagators are the Higgs-field-dependent eigenvalues of the mass matrix

$$\mathcal{M}_{\tilde{t}}^2 \cong \begin{pmatrix} m_Q^2 + y^2 H_2^2 + O(m_Z^2) & y(A_t\, H_2 - \mu\, H_1) \\ y(A_t\, H_2 - \mu\, H_1) & m_U^2 + y^2 H_2^2 + O(m_Z^2) \end{pmatrix}. \qquad (2)$$

Because the top Yukawa coupling y is large, the stops couple strongly to the Higgs field H_2, and this is why they have a potentially strong effect on $V(H)$.

We have computed $V(H)$ to two loops, including mixing effects between the stops and between the heavy and light Higgs fields. We then searched the MSSM parameter space to find regions where the phase transition

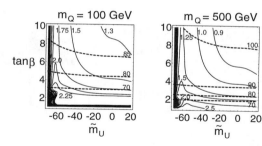

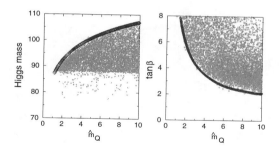

Figure 5: Contours of v/T (solid) and Higgs mass (dashed) in the plane of $\tan\beta$ and $\tilde{m}_U \equiv m_U^2/|m_U|$ (in GeV), for $m_Q = 100$ and 500 GeV, respectively, at zero squark mixing ($\mu = A_t = 0$). The potential has color-breaking minima in the black regions near $\tilde{m}_U = -70$ GeV.

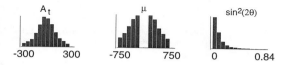

Figure 6: Histograms for squark mixing parameters and mixing angle from Monte Carlo. Mass parameters are in GeV.

on the Higgs mass, one must take $\tan\beta$ to be greater than some minimum value. As for larger values of $\tan\beta$, these tend to weaken the phase transition. However this effect can be counteracted by enhancing the stop contribution to $V(H)$, *i.e.*, by making the stop lighter. To make the right-handed stop sufficiently light, we must in fact take its mass parameter m_U^2 to be negative! This combines with the term m_t^2 in eq. (2) to give an overall positive mass squared, except before the EWPT when $m_t = 0$. Figure 5 shows how the strength of the phase transition (measured as $\langle H \rangle / T$, where $\langle H \rangle = \sqrt{\langle H_1 \rangle^2 + \langle H_2 \rangle^2}$) depends on $\tan\beta$ and on $\tilde{m}_U \equiv m_U^2/|m_U|$.

Squark mixing parameters, μ and A_t. Although the EWPT is generally stronger for small values of the squark mixing parameters (appearing in the off-diagonal elements of the mass matrix (2)), the preference is rather weak. Histograms for μ, A_t and the squark left-right mixing angle θ are shown in figure 6. Small values of $|\mu|$ are experimentally excluded by chargino and neutralino searches, explaining the absence of points near $\mu = 0$.

Left-handed stop mass. The left-handed stop mass parameter is bounded from below,

$$m_Q > 130 \text{ GeV}. \qquad (6)$$

This comes from the requirement that the contributions to the ρ parameter from the squarks not exceed precision electroweak bounds (we take $\Delta\rho < 1.5 \times 10^{-3}$). Because the light Higgs mass depends on m_Q through radiative

Figure 7: Scatter plots from Monte Carlo for m_h and $\tan\beta$ versus $\hat{m}_Q \equiv m_Q/(100 \text{ GeV})$.

corrections, it has a noticeable effect on the baryogenesis limits on these parameters. The maximum allowed light Higgs mass, and the minimum allowed values of $\tan\beta$, depend on $\hat{m}_Q \equiv (m_Q/100 \text{ GeV})$ as

$$m_h \leq 85.9 + 9.2\ln(\hat{m}_Q) \text{ GeV}$$
$$\tan\beta \geq (0.03 + 0.076\,\hat{m}_Q - 0.0031\,\hat{m}_Q^2)^{-1}. \qquad (7)$$

The corresponding scatter plots from the Monte Carlo, for these quantities versus m_Q are shown in figure 7.

4 Strong Phase Transition at Small m_{A^0}?

We have seen that the majority of accepted points in our Monte Carlo are those with large values of m_{A^0}. There are some rare exceptions, as can be seen in fig. 7a: the sparse points below $m_h = 87$ GeV are those with $m_{A^0} \sim 100$ GeV. Although they infrequently give a strong enough transition, they could be important for the following reason. Most of the contributions to the CP asymmetry that gives baryogenesis are proportional to the amount by which $\tan\beta$ changes inside the bubble wall, and this is strongly correlated with the value of m_{A^0}.

The concept of $\Delta\beta$, the deviation in β, is illustrated in fig. 8. The path in field space taken by the Higgs field as it goes from inside the bubble ($\langle H \rangle > 0$) to outside ($\langle H \rangle = 0$) is shown for the case of $m_{A^0} = \infty$, where it is a straight line, and for $m_{A^0} \sim 100$ GeV, which gives some curvature to the path. The maximum angular deviation from straightness can be called $\Delta\beta$, and computed from the effective potential. (For technical reasons we define $\Delta\beta \equiv \max_v [v(\beta(v) - \beta(v_c))]/v_c$, where v is the value of $\langle H \rangle$ at any point inside the bubble wall.) In figure 9 we show how $\Delta\beta$ is correlated with m_{A^0} and the frequency of $\Delta\beta$ values. Typically $\Delta\beta$ is quite small, 10^{-3}, and only very rarely reaches 10^{-2}, as also found by others.[7]

Since most electroweak baryogenesis mechanisms have $n_B \propto \Delta\beta$, this gives an additional suppression in the

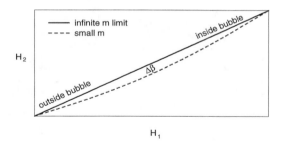

Figure 8: Trajectories in Higgs field space for going through the bubble wall, in the limits of large and small m_{A^0}.

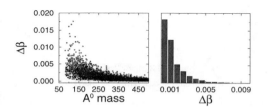

Figure 9: (a) Maximum deviation in weighted Higgs VEV orientation, $\Delta\beta \equiv \max_v[v(\beta(v) - \beta(v_c))]/v_c$, inside bubble wall, as a function of m_{A^0}; (b) distribution of $\Delta\beta$ values.

baryon asymmetry that can be produced in electroweak mechanisms, which has been overlooked by some authors. Interestingly, there is one contribution to the CP asymmetry which has been shown to be unsuppressed [8] in the limit of vanishing $\Delta\beta$. Charginos, which have the Dirac mass matrix

$$\mathcal{M}_{\tilde{W}\tilde{h}} = \begin{pmatrix} m_2 & gH_2/\sqrt{2} \\ gH_1/\sqrt{2} & \mu \end{pmatrix}, \qquad (8)$$

can experience a CP-violating force while traversing the bubble wall if the μ parameter is complex, even if H_1/H_2 remains constant in the wall.

5 Squark Masses and Mixing

Previous studies of the EWPT have emphasized the weakening effect that squark mixing has on the phase transition strength. We have already pointed out that the Monte Carlo, although favoring small mixing between the left- and right-handed stops, does not strongly exclude large mixing: see figure 6 and eq. (2). Since nonzero values of μ are needed to get CP violation, this is fortunate!

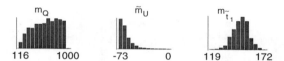

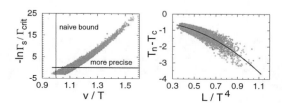

Figure 10: Monte Carlo distributions for the mass parameters of (a) the left stop and (b) the right stop; and (c) the actual mass of the right stop, in GeV.

Figure 11: (a) Correlation of the sphaleron rate with v/T; below the line would be ruled out baryon preservation; (b) $\Delta T \equiv T_n - T_c$ in GeV versus $\Lambda \equiv$ (latent heat)$/T^4$, fit by the function $\Delta T = -0.5 + 0.5\Lambda - 2.9\Lambda^2$.

As for the stop masses, the distributions of the relevant parameters can be seen in figure 10. The left-handed stop (which in the limit of large m_Q has mass approximately equal to m_Q) is usually much heavier than the right-handed one, but can be as light as 116 GeV. The right-handed one is always relatively light, in the range 119 GeV $< m_{\tilde{t}_1} < 172$ GeV. Such a light squark is potentially discoverable at the Tevatron, but this depends strongly on the value of gluino mass, $m_{\tilde{g}}$. The discovery potential is greatly suppressed if $m_{\tilde{g}} > 300$ GeV.

6 Sphaleron Rate versus Higgs VEV; Latent Heat

In this work we have also made a detailed study of the exact criterion for a strong enough phase transition. Recall that the requirement is that sphaleron interactions inside the bubbles must be slow enough so that the baryon number produced does not get erased afterwards. The naive condition for fulfilling this is that $\langle H \rangle / T_c \geq 1$, but the more exact criterion is that the rate of sphaleron interactions, Γ_s, must exceed some critical lower value, Γ_{crit}, derived in ref. [1]. By plotting $-\ln \Gamma_s/\Gamma_{crit}$ versus v/T (fig. 11a), we can see how far off the naive condition is. The figure shows that a significant number of trial parameter sets are excluded by the exact bound, although they may pass the $\langle H \rangle / T > 1$ condition. However, we have tried to correct for the fact that the effective potential approach underestimates the strength of the phase transition by about 10%, compared to nonperturbative lattice results.[6] Because of this, our threshold for acceptance is somewhat looser than $\Gamma_s = \Gamma_{crit}$.

We have also studied the issue of reheating after the onset of the phase transition, which is related to the latent heat L. L is defined to be the difference in $dV/d\ln T$ between the symmetric ($\langle H \rangle = 0$) and broken ($\langle H \rangle \neq 0$) phases. Although the phase transition becomes energetically possible starting at the critical temperature T_c, where the two phases are degenerate in energy, stable bubbles do not start to appear until the somewhat lower nucleation temperature, T_n. If the transition is strong enough, entropy production at the bubble walls could conceivably reheat the universe all the way back to T_c. In general the reheat temperature (T_r) is given by

$$\frac{T_c - T_r}{T_c - T_n} = 1 - \frac{30L}{g_* \pi^2 (T_c^4 - T_n^4)}$$
$$\cong 1 - \frac{15L}{2g_* \pi^2 T_c^3 (T_c - T_n)}, \qquad (9)$$

which approaches zero if $T_r \to T_c$ (here g_* is the number of relativistic degrees of freedom in the plasma). We find a correlation between L and $T_c - T_n$ (fig. 11b) so that the right hand side can be thought of as being a function of L alone, roughly. We also find that $(T_c - T_r)/(T_c - T_n)$ is always in the range [0.6, 0.8], so that reheating to T_c is never achieved. One reason to be interested in this is that complete reheating to T_c tends to slow the growth of the bubbles significantly, and most baryogenesis mechanisms predict that the baryon asymmetry is enhanced by $1/v$ if the bubble wall velocity v is small.

7 Will Electroweak Baryogenesis be Ruled Out (or In) by LEP?

The most pressing question confronting electroweak baryogenesis is whether it is really testable at LEP. Fig. 12 shows the regions in the $\tan\beta$-m_h plane which will be excluded in runs near 200 GeV center of mass energy. At first sight the experimentally inaccessible region $m_h > 95$ GeV, $\tan\beta > 10$ might look worrisome, since these values appear to be allowed by our Monte Carlo. Closer examination shows that these points only escape detection by LEP if m_{A^0} and $\sin^2(\alpha - \beta)$ are too small to be compatible with a strong phase transition. The real worry is whether $m_h > 107$ GeV, which is above the discovery potential of LEP 200, but still compatible with baryogenesis if the left-handed stop is heavy enough. In this case we will have to hope for the slim possibility of the Higgs being discovered at Tevatron.

8 Conclusions

The most promising experimental signal for electroweak baryogenesis is a light Higgs, with $m_h < 86$ GeV if the

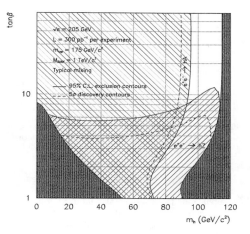

Figure 12: Discovery potential of LEP 2.

left-stop mass parameter m_Q is 100 GeV, and $m_h < 116$ GeV if $m_Q = 2$ TeV. It is also possible that the mostly right-handed top squark will be observed, since its mass is constrained to lie in the range 119 GeV $< m_{\tilde{t}} < 172$ GeV. This extreme lightness of the stop can only be achieved by taking its mass parameter m_U^2 to be dangerously negative. The danger is that the universe will get stuck in a color-breaking minimum with a nonzero stop condensate, and not be able to tunnel into the normal vacuum state where we live. Although it has been previously investigated,[5] this issue probably deserves closer scrutiny.

References[a]

1. J.M. Cline and G.D. Moore, hep-ph/9806354, to appear in Phys. Rev. Lett. (1998).
2. M. Carena, M. Quiros and C.E.M. Wagner, hep-ph/9710401, Nucl. Phys. B524, 3 (1998).
3. J.M. Cline, J.R. Espinosa, G.D. Moore and A. Riotto, preprint MCGILL-98/24 (1998).
4. J.R. Espinosa, , hep-ph/9604320, Nucl. Phys. B475 (1996) 273
5. D. Bödeker, P. John, M. Laine and M.G. Schmidt, hep-ph/9612364, Nucl. Phys. B497 (1997) 387.
6. M. Laine and K. Rummukainen, hep-ph/9804255, Phys. Rev. Lett. 80 (1998) 5259 ; hep-lat/9804019, preprint CERN-TH-98-122 (1998)
7. J.M. Moreno, M. Quiros and M. Seco, hep-ph/9801272, Nucl. Phys. B526 (1998) 489.
8. J.M. Cline, M. Joyce and K. Kainulainen, hep-ph/9708393, Phys. Lett. B417 (1998) 79.

---

[a]see ref. 1 for a more complete list of references.

SELECTED LOW-ENERGY SUPERSYMMETRY PHENOMENOLOGY TOPICS

J. F. Gunion

Department of Physics, University of California at Davis, Davis, CA 95616, USA
E-mail: jfgucd@ucdhep.ucdavis.edu

I review selected topics in supersymmetry, including: effects of non-universality, high $\tan\beta$ and phases on SUSY signals; a heavy gluino as the LSP; gauge-mediated SUSY signals involving delayed decays; R-parity violation and the very worst case for SUSY discovery; some topics regarding Higgs bosons in supersymmetry; and doubly-charged Higgs and higgsinos in supersymmetric left-right symmetric models. I emphasize scenarios in which detection of supersymmetric particles and/or the SUSY Higgs bosons might require special experimental and/or analysis techniques.

1 Non-universality, $\tan\beta \gg 1$ and phases

Many deviations from universal boundary conditions at the unification or string scale are now being actively considered. Neither the gaugino masses nor the scalar masses are required to be universal. One well-motivated model with non-universal gaugino masses is the O-II orbifold model,[1] in which supersymmetry breaking (SÜSY) is dominated by the overall size modulus (as opposed to the dilaton). It is the only string model where the limit of pure modulus SÜSY is possible without charge and/or color breaking. One finds.

$$M_3 : M_2 : M_1 \overset{O-II}{\sim} -(3+\delta_{GS}) : (1-\delta_{GS}) : (\frac{33}{5} - \delta_{GS}),$$ (1)

The phenomenology of this model changes dramatically as a function of the Green-Schwarz parameter, δ_{GS}; indeed, a heavy gluino is the LSP when $\delta_{GS} \sim -3$ (a preferred range for the model). Another class of models with non-universal gaugino masses are those where SÜSY arises due to F-term breaking with $F \neq$ SU(5) singlet.[2] Possible representations for F include:

$$F \in (\mathbf{24 \times 24})_{\text{symmetric}} = \mathbf{1} \oplus \mathbf{24} \oplus \mathbf{75} \oplus \mathbf{200},$$ (2)

leading to $\langle F \rangle_{ab} = c_a \delta_{ab}$, with c_a depending on the representation. Results for the gauginos masses at the grand-unification scale M_U and at m_Z are given in Table 1.

F	M_U			m_Z		
	M_3	M_2	M_1	M_3	M_2	M_1
1	1	1	1	~ 6	~ 2	~ 1
24	2	-3	-1	~ 12	~ -6	~ -1
75	1	3	-5	~ 6	~ 6	~ -5
200	1	2	10	~ 6	~ 4	~ 10
$O-II$ $\delta_{GS}=-4$	1	5	$\frac{53}{5}$	~ 6	~ 10	$\sim \frac{53}{5}$

Table 1: M_a at M_U and m_Z for the four F irreducible representations and in the O-II model with $\delta_{GS} \sim -4$.

Both $F \in \mathbf{200}$ and the O-II model allow for the possibility that $m_{\tilde\chi_1^\pm} \simeq m_{\tilde\chi_1^0}$, since $m_{\tilde\chi_1^0} \sim \min(M_1, M_2)$,

$m_{\tilde\chi_1^\pm} \sim M_2$ and $M_2 < M_1$. In this situation, there are two possibilities. (1) The degeneracy is so extreme ($\Delta m_{\tilde\chi} \equiv m_{\tilde\chi_1^\pm} - m_{\tilde\chi_1^0} \lesssim 0.1$ GeV) that the $\tilde\chi_1^\pm$ is long-lived. In this case, one searches for heavily-ionizing charged tracks. (2) The degeneracy is still small, but large enough that the $\tilde\chi_1^\pm$ is not pseudo-stable: $0.3 \leq \Delta m_{\tilde\chi} \leq 3$ GeV. One must search for $\tilde\chi_1^+ \tilde\chi_1^-$ production at an e^+e^- collider using a photon tag: $e^+e^- \to \gamma\tilde\chi_1^+\tilde\chi_1^-$. In case (1) [(2)], a DELPHI analysis[3] yields $m_{\tilde\chi_1^\pm} \geq 84$ GeV [≥ 54 GeV, provided $m_{\tilde\nu}$ is large]. In general (but not preferred in the GUT context), there is also a third possibility: $M_2, M_1 \gg |\mu|$. In this case, the $\tilde\chi_1^\pm$ and $\tilde\chi_1^0$ are again nearly degenerate, but the $\gamma\tilde\chi_1^+\tilde\chi_1^-$ cross section is smaller and no limits (beyond the LEP $m_{\tilde\chi_1^\pm} \geq 45$ GeV limit) are possible unless the $\tilde\chi_1^\pm$ is pseudo-stable.[3]

Scalar mass non-universality can emerge from many sources; a particularly popular source is D-term contributions to scalar mass, especially from an anomalous U(1). A typical model is one[4] which employs U(1)$_Y$. The result is a Fayet-Illiopoulos D-term contribution to the scalar masses at M_U: $\tilde m_i^2 = m_0^2 + Y_i D_Y$, where m_0 is the usual mSUGRA universal mass. (The other mSUGRA parameters are denoted $m_{1/2}$, A_0, $\tan\beta$, sign(μ).) As D_Y is turned on, the scalar masses are altered and the value of $|\mu|$ required for RGE electroweak symmetry breaking (in which $m_{H_2^0}^2$, the scalar mass-squared associated with the Higgs boson that couples to the top quark, becomes negative at low energy scales) to give the correct value of m_Z changes. The 'normal' mSUGRA relation between gaugino masses, scalar masses and $|\mu|$ is altered so that the LSP need not be the $\tilde\chi_1^0$. As D_Y is changed, it becomes possible for the LSP to be: the $\tilde\tau_R$ ($|\mu| > |\mu|_{mSUGRA}$); a higgsino ($|\mu| < |\mu|_{mSUGRA}$); or a sneutrino (in a small band with $|\mu| < |\mu|_{mSUGRA}$). Cosmology suggests these latter are disfavored, but reheating can obviate such constraints and even a stable LSP=$\tilde\tau_R$ would then be allowable.

Clearly, such scalar non-universality leads to drastic changes in collider phenomenology. In particular, if the

$\tilde{\tau}_R$ is the LSP one should look for a stable $\tilde{\tau}_R$, whereas if a higgsino is the LSP then $m_{\tilde{\chi}_1^\pm} \simeq m_{\tilde{\chi}_1^0} \simeq m_{\tilde{\chi}_2^0}$ and LEP2 constraints will be weakened (see above). Further, in collider events there will be much less missing transverse momentum ($\not{p}_T$) than for mSUGRA boundary conditions.

Let us next mention the phenomenological implications of high $\tan\beta$ for superparticle discovery. RGE equations cause $\tilde{\tau}$ to decline in mass relative to $\tilde{e}$, $\tilde{\mu}$ (but the $\tilde{\chi}_1^0$ is still the LSP). This leads to dominance of τ's in cascade decays and in the 'tri-lepton' signal. Tevatron signals for SUSY become more difficult; it definitely takes TeV33 to probe SUSY if gluino and squarks are $\gtrsim 1$ TeV with corresponding mass scales for other sparticles. [5]

Normally, the possible phases for the soft-SUSY-breaking parameters have been neglected in studying SUSY collider phenomenology. For example, in mSUGRA, A_0 and μ can have phases. More generally, there are 79 masses and real mixing angles and 45 CP-violating phases in the MSSM. These phases appear in mass matrices as well as couplings. EDM and CP-violation constraints do not require that these phases be small; cancellations among different contributions to CP-violating observables are possible. [6] Extraction of all SUSY parameters from experiment becomes considerably more complex in general, [7] even at an e^+e^- collider.

2 A heavy gluino as the LSP

There are several attractive models in which the gluino is heavy and yet is the LSP. These models include: the O-II model discussed earlier [1] when $\delta_{GS} \sim -3$ (the preferred range); and the GMSB model of Raby. [8]

A detailed study of the phenomenology of a $\tilde{g}$-LSP has appeared. [9] First, one must consider constraints coming from the relic density of $R^0 = \tilde{g}g$ (almost certainly the lightest) bound states. Taking into account annihilations that continue after freezeout, and allowing for non-perturbative contributions to the annihilation cross section, it is found [9] that the relic density can be small enough, even at very large $m_{\tilde{g}}$ and even without including late stage inflation (as might be needed for the Polonyi problem), to avoid all constraints from stable isotope searches, underground detectors, etc. Certainly, the R^0's are very unlikely to be the primary halo constituent.

Next, one must consider how the $\tilde{g}$-LSP manifests itself in a detector and in relevant experimental analyses. This is sensitively dependent upon several ingredients. First, there is the question of how the $\tilde{g}$ hadronizes. In general, it can pick up quarks and/or a gluon to form either charged $R^\pm$ (e.g. $\tilde{g}u\bar{d}$) or neutral R^0 (e.g. $\tilde{g}g$, $\tilde{g}u\bar{u}$, ...) bound states with probabilities P and $1 - P$, respectively. ($R^\pm$ states that are not pseudo-stable between hadronic collisions are not counted in P.) These probabilities are assumed to apply to a heavy $\tilde{g}$ both as it exits from the initial hard interaction and also after each hadronic collision. (The picture is that the light quarks and gluons are stripped away in each hadronic collision and that the heavy $\tilde{g}$ is then free to form the $R^\pm$ and R^0 bound states in the same manner as after initial production.) In any reasonable quark-counting model $P < 1/2$, in which case the $\tilde{g}$ spends most of its time as an R^0 as it passes through the detector. The second critical ingredient is the $\langle \Delta E \rangle$ deposited in a hadronic collision; several models that bracket the known result for a pion are employed. Since, a heavy $\tilde{g}$ is typically not produced with an ultra-relativistic velocity, it does not deposit very much energy even in its first few hadronic collisions; indeed, it can often penetrate the detector unless it is in an $R^\pm$ state a large fraction of the time ($P > 1/2$) and is slowed down by ionization energy deposits. Third, the net hadronic energy deposit depends on λ_T, the path length in iron given by the $\tilde{g}$ total cross section. One popular model [10] suggests $\lambda_T \sim 2\lambda_T(\pi)$. Fourth, the effective Fe thickness of instrumented and uninstrumented portions of the relevant detectors (OPAL and CDF) must be known. Fifth, one must account for how a calorimeter treats ionization energy deposits as compared to hadronic collision energy deposits; the latter are measured correctly when the calorimeter is calibrated for a light hadron, but the former are over-estimated by a factor of roughly 1.6 in an iron calorimeter (as employed by OPAL and CDF). Thus, when a calorimeter is calibrated to give correct π energy, calorimeter response after one λ_T is $E_{\text{calorimeter}} = rE_{\text{ionization}} + E_{\text{hadronic}}$, where $r \sim 1.6$ for an iron calorimeter. Sixth, it is necessary to determine if the $\tilde{g}$-jet is charged at appropriate points in the detector, and other analysis-dependent criteria are satisfied, such that the $\tilde{g}$-jet is identified as containing a muon. 'Muonic' jets are discarded in the CDF jets + missing energy analysis, but retained in the corresponding OPAL analysis. In the latter, the jet energy of a jet that is 'muonic' is computed as:

$$E_{\text{jet}} = p_{\text{jet}} = E_{\text{calorimeter}}^{\text{tot}} + \theta(\mu\text{id})(p_{\text{tracker}} - 2 \text{ GeV}), \quad (3)$$

where $\theta(\mu\text{id}) = 1$ or 0, $p_{\text{tracker}} = m_{\tilde{g}}\beta\gamma$ is the momentum as measured by the tracking system, and the 2 GeV subtraction is the energy that would have been deposited by a muon in the calorimeter.

In the end, the $E_{\text{jet}} = p_{\text{jet}}$ as defined by the experiments is normally quite different from the true gluino jet momentum, and most events will be associated with large missing momentum. Further, for moderately large P (but not too close to 1), there are large fluctuations on an event-by-even basis in how the $\tilde{g}$-jets are treated. Thus, one [9] employs an event-by-event model of $\tilde{g}$ passage

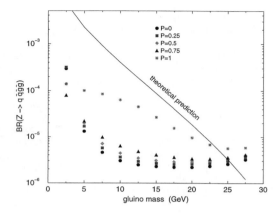

Figure 1: We compare the prediction of $BF(Z \to q\bar{q}\tilde{g}\tilde{g})$ compared to the extracted [9] OPAL 95% CL upper limit as a function of $m_{\tilde{g}}$ for $P = 0, 1/4, 1/2, 3/4, 1$. Both smearing and fragmentation effects are included. Results are for the standard λ_T and $\langle \Delta E \rangle$ (SC1) case. [9]

through the detector accounting for P at each hadronic collision and the associated calorimeter responses.

Sample results for OPAL and CDF are illustrated in Figs. 1 and 2. These figures illustrate that, for 'standard' choices [9] of λ_T and $\langle \Delta E \rangle$, one can use the jets + missing energy OPAL and CDF analyses to exclude any $m_{\tilde{g}}$ from ~ 3 GeV up to $\sim 130 - 150$ GeV, regardless of the charged fragmentation probability P. For $P > 1/2$, there is some sensitivity to the λ_T and $\langle \Delta E \rangle$ scenario choices: limits could be weaker (or stronger). For choices that yield weak limits when $P > 1/2$, one can use the OPAL and CDF searches for tracks corresponding to a heavily-ionizing charged particle to eliminate all $m_{\tilde{g}}$ values up to $\sim 130 - 150$ GeV except in the interval $23 \lesssim m_{\tilde{g}} \lesssim 50$ GeV, which is the gap between the OPAL analysis and the current version of the CDF analysis. A refined CDF heavily-ionizing-track analysis should be able to eliminate this gap.

3 Delayed decay signals for gauge-mediated supersymmetry breaking (GMSB)

The two canonical GMSB possibilities are: $\tilde{\tau}_R$=NLSP, with $\tilde{\tau}_R \to \tau \tilde{G}$; and $\tilde{\chi}_1^0$=NLSP followed by $\tilde{\chi}_1^0 \to \gamma \tilde{G}$, where the $\tilde{G}$ is the Goldstino. In either case, the NLSP decay can be either prompt or delayed. In the $\tilde{\tau}_R$-NLSP case, detection of SUSY will be easy, either using heavily ionizing tracks [11] for long path length of the $\tilde{\tau}_R$ or τ signals [12] if the decay is prompt. However, if the $\tilde{\chi}_1^0$ is the

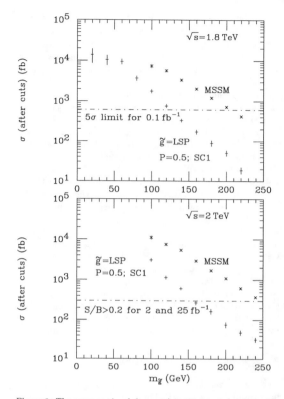

Figure 2: The cross section (after cuts) in the jets + $\not{p}_T$ channel is compared to (a) the 95% CL upper limit for $L = 19$ pb^{-1} (which is the same as the 5σ signal level for $L = 100$ pb^{-1}) at $\sqrt{s} = 1.8$ TeV and (b) the $S/B = 0.2$ level at Run-II ($L \geq 2$ fb^{-1}, $\sqrt{s} = 2$ TeV) as a function of $m_{\tilde{g}}$ for $P = 1/2$. Standard choices for λ_T and $\langle \Delta E \rangle$ are employed.

NLSP, detection of a SUSY signal can be much more challenging, and measurement of the SUSY scale, $\sqrt{F}$, requires special attention. [13] In fact, it is quite possible, and required in some models, that $\sqrt{F} \sim 1000 - 5000$ TeV, in which case

$$(c\tau)_{\tilde{\chi}_1^0 = \tilde{B} \to \gamma \tilde{G}} \sim 130 \left(\frac{100 \text{ GeV}}{m_{\tilde{\chi}_1^0}} \right)^5 \left(\frac{\sqrt{F}}{100 \text{ TeV}} \right)^4 \mu\text{m} \tag{4}$$

is typically quite large. In particular, in GMSB models with a hidden sector communicating at two-loops with a messenger sector, we have [14,15] (to within a factor of 5 or less) $\sqrt{F} > \Lambda\sqrt{f}$, where $f \sim 2.5 \times 10^4/g_m^4$ and Λ is the parameter that sets the scale of soft-susy-breaking masses. Roughly, 40 TeV $\lesssim \Lambda \lesssim 150$ TeV is required (see below), implying $\sqrt{F} \gtrsim 1000 - 5000$ TeV. Meanwhile, the gravitino has mass $m_{\widetilde{G}} = \frac{F}{\sqrt{3}M_{\text{Planck}}} \sim 2.5 \left(\frac{\sqrt{F}}{100 \text{ TeV}}\right)^2$ eV, and $m_{\widetilde{G}} \lesssim 1$ keV is preferred by cosmology, implying $\sqrt{F} \lesssim 3000$ TeV. Thus, we should take seriously $1000 \lesssim \sqrt{F} \lesssim 5000$ TeV and the possibility of delayed $\widetilde{\chi}_1^0$ decays. At the very least, one should explore the phenomenology of the model for the full range of possible $\sqrt{F}$ values.

A recent study [13] has explored Tevatron phenomenology for the full range of $\sqrt{F}$ in a sample model in which the superparticle masses have the relative magnitudes typical of the simpler GMSB models with minimal messenger sector content. In the model employed, the $\widetilde{\chi}_1^0$ is the LSP and the sparticle masses are: $m_{\widetilde{\chi}_1^0} \sim 1.35$ GeV $\cdot$ Λ, $m_{\widetilde{\chi}_1^+} \sim 2.7$ GeV $\cdot \Lambda$ ($\sim 2m_{\widetilde{\chi}_1^0}$), $m_{\widetilde{g}} \sim 8.1$ GeV $\cdot \Lambda$ ($\sim 6m_{\widetilde{\chi}_1^0}$), $m_{\widetilde{\ell}_R} \sim 1.7$ GeV $\cdot \Lambda$, $m_{\widetilde{\ell}_L} \sim 3.5$ GeV $\cdot \Lambda$ ($\sim 2m_{\widetilde{\ell}_R}$), $m_{\widetilde{q}} \sim 11$ GeV $\cdot \Lambda$ ($\sim 6m_{\widetilde{\ell}_R}$), with Λ in TeV. From these mass formula, we see that if $\Lambda \lesssim 40$ TeV then the $\widetilde{\ell}_R$ would have been seen at LEP or LEP2, while if $\Lambda \gtrsim 150$ TeV the $\widetilde{g}$ and the $\widetilde{q}$'s becomes so heavy that naturalness problems for the Higgs sector would certainly be substantial. For the above hierarchy of masses, the primary normal SUSY signal at the Tevatron is the tri-lepton signal. It is found [13] that this signal is viable for $\Lambda \lesssim 65$ TeV for any $\sqrt{F}$; but it does not distinguish a SUGRA-like model from a GMSB model. In order to distinguish between the two model possibilities, one must detect the photon(s) that result from the $\widetilde{\chi}_1^0$ decays.

One possibility is to detect a prompt photon in association with the tri-lepton signal. One finds that this will be possible only if $\sqrt{F}$ is not very large. Additional associated-photon signals that can be considered include: observation of a photon with non-zero impact-parameter (b); decay of the $\widetilde{\chi}_1^0$ leading to an isolated energy deposit in an outer-hadronic-calorimeter cell (OHC); a photon signal in a specially designed roof-array detector placed on the roof of the detector building (RA); and the appearance of two prompt (emergence before the electromagnetic calorimeter) photons (2γ). The first three are present only if the $\widetilde{\chi}_1^0 \to \gamma\widetilde{G}$ decay is delayed, while the latter signal will be very weak if the decay is substantially delayed. After imposing strong cuts that hopefully reduce backgrounds to a negligible level (detailed detector studies being needed to confirm), the regions in $(\sqrt{F}, \Lambda)$ parameter space for which these signals are viable at the D0 detector for Run-I, Run-II and Tev33 luminosities at the Tevatron are illustrated in Fig. 3.

We can summarize as follows. If both $\sqrt{F}$ and Λ are large, then we will not see either the tri-lepton signal or the prompt 2γ signal. However, Fig. 3 shows that the large impact parameter photon signal from delayed decays of the $\widetilde{\chi}_1^0$ can cover most of the preferred parameter space region not accessible by the former two modes. The roof-array detector also provides an excellent signal at large $\sqrt{F}$. Putting all the signals together, the portion of $(\sqrt{F}, \Lambda)$ parameter space for which a GMSB $\widetilde{\chi}_1^0$ =NLSP SUSY signal can be seen at the Tevatron is greatly expanded by delayed $\widetilde{\chi}_1^0$ decay signals. We re-emphasize the fact that one needs the b, RA and/or OHC delayed-decay signals to distinguish GMSB from a SUGRA model with GMSB-like boundary conditions when $\sqrt{F}$ is too large for a viable prompt photon signal. Finally, if delayed-decay signatures are seen, an absolutely essential goal will be to determine $\sqrt{F}$ (the most fundamental SUSY parameter of all): the b and RA signals provide the information needed to do so.

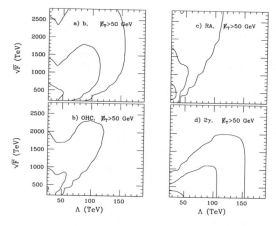

Figure 3: σ contours in the $(\sqrt{F}, \Lambda)$ parameter space for the (a) impact-parameter (b), (b) outer-hadronic (OHC), (c) roof-array (RA), and (d) prompt-two-photon (2γ) signals. Contours are given at $\sigma = 0.16, 2.5,$ and 50 fb — these correspond to 5 events at $L = 30, 2, 0.1$ fb^{-1} as for Tev33, Run-II and Run-I, respectively.

4   The nightmare R-parity violating (RPV) scenario

This scenario [16] is designed as a warning against complacency regarding SUSY discovery. The first ingredient in the nightmare scenario is a non-zero B-violating RPV coupling (often denoted λ''), which leads to LSP decay to three jets: $\widetilde{\chi}_1^0 \to 3j$. This means that the $\widetilde{\chi}_1^0$'s, resulting from chain decay of a pair of produced supersymmetric particles, do not yield missing energy. The standard jets + $\not{p}_T$ signal is absent. Of course, for universal gaugino masses at the GUT scale this is not a problem since $m_{\widetilde{\chi}_1^\pm} \sim 2m_{\widetilde{\chi}_1^0}$ (at energy scales < 1 TeV). For such large $m_{\widetilde{\chi}_1^\pm} - m_{\widetilde{\chi}_1^0}$, a robust signal for supersymmetry is provided [17] by the like-sign dilepton signal that arises when two $\widetilde{\chi}_1^+$'s (or two $\widetilde{\chi}_1^-$'s) are produced in the decay chains and both decay leptonically: *e.g.* $\widetilde{\chi}_1^+ \widetilde{\chi}_1^+ \rightarrow \ell^+ \ell^+ \nu_\ell \nu_\ell \widetilde{\chi}_1^0 \widetilde{\chi}_1^0$. Since the leptons have significant momentum and the neutrinos yield some missing momentum, the like-sign lepton events are typically quite easily isolated at the LHC, and for lower SUSY mass scales, also at the Tevatron. [17,18]

However, as discussed in an earlier section, if gaugino masses are not universal it is very possible to have $M_2 < M_1 \ll |\mu|$ (at energy scales < 1 TeV), leading to both $\widetilde{\chi}_1^\pm$ and $\widetilde{\chi}_1^0$ being SU(2) winos with $m_{\widetilde{\chi}_1^\pm} \simeq m_{\widetilde{\chi}_1^0}$. Also, in the models with $|\mu| \ll M_{1,2}$ the lightest *two* neutralinos and the lightest chargino are *all* closely degenerate: $m_{\widetilde{\chi}_1^0} \simeq m_{\widetilde{\chi}_1^\pm} \simeq m_{\widetilde{\chi}_2^0}$. In either case, the leptons in $\widetilde{\chi}_1^\pm \rightarrow \ell^\pm \nu \widetilde{\chi}_1^0$ decays are very soft. The like-sign dilepton signal would be very weak (after necessary cuts requiring reasonable momenta for the leptons). The implications of these scenarios are the following. At LEP2 one would need to use the $e^+ e^- \rightarrow \gamma \widetilde{\chi}_1^+ \widetilde{\chi}_1^-$ photon-tag signal, but, unlike the case where the $\widetilde{\chi}_1^0$'s from $\widetilde{\chi}_1^\pm$ decays yield missing energy, the $\widetilde{\chi}_1^+ \widetilde{\chi}_1^-$ would decay to a final state containing six (relatively soft) jets. Perhaps, such a signal can be shown to be viable over backgrounds up to some reasonable value of $m_{\widetilde{\chi}_1^\pm}$. To go beyond this value would require a viable signal at the Tevatron and/or LHC. However, at the hadron colliders, leptonic signals will be very weak. Aside from W decays, energetic leptons can emerge only from decays of the heavy gauginos (*e.g.* the higgsino states in the $M_2 < M_1 \ll |\mu|$ case) that are present by virtue of either being directly produced or arising in decays of still heavier produced supersymmetric particles. If the leptonic signals turn out to be too weak, the only signal with a substantial rate will be spherical events containing an extra large number of jets. This signal might prove very difficult to isolate from backgrounds.

5 Two topics regarding Higgs bosons in SUSY.

The first topic concerns the use of experimental limits on the lightest SUSY Higgs boson, h^0, to exclude parameter regions in various SUSY models. The second topic is the construction of a truly difficult scenario for SUSY (or any) Higgs detection.

5.1 Model constraints from limits on the h^0

To illustrate the possibilities, I present a brief discussion of two representative papers. The first is that of de Boer *et al.*. [19] They assume universal mSUGRA CMSSM (constrained MSSM) boundary conditions and impose radiative electroweak symmetry breaking and

gauge coupling unification with $\alpha_s(m_Z) = 0.122$ and $m_t = 173.9 \pm 5.2$ GeV. They also require $b - \tau$ Yukawa unification, with $m_b(m_b) = 4.2 \pm 0.15$ GeV (do we really know it so well?). Additional input data is the current combined ALEPH/CLEO result for $b \rightarrow s\gamma$ (including combined errors) and Higgs mass limits from LEP2. Regarding the latter, the CMSSM approach with RGE electroweak symmetry breaking implies that m_{A^0} is large and that the h^0 is very SM-like. Thus, they require $m_{h^0} \gtrsim 89$ GeV at 95% CL. Finally, they require $\Omega h^2 < 1$ for the relic neutralinos of the model. With this input, including a systematic treatment of experimental errors, they compute the χ^2 for different parameter choices in the CMSSM context.

They find significant constraints on the allowed parameter space. It is convenient to think of the allowed parameter regions as follows. First, by imposing $b - \tau$ unification they end up with only 4 good $\tan\beta$ and sign(μ) solution scenarios: two at low $\tan\beta$ and two at high $\tan\beta$. These 4 possibilities are then restricted by other constraints as shown in the following Table.

Constraint	$\tan\beta = 1.65$ $\mu < 0$	$\tan\beta = 1.65$ $\mu > 0$
$b \rightarrow s\gamma$	OK	OK
$m_{h^0} > 90$?	No	$m_{1/2} > 400$
$\Omega h^2 < 1$?	$m_0 < 300$	$m_0 < 300$
Constraint	$\tan\beta = 35$ $\mu < 0$	$\tan\beta = 64$ $\mu > 0$
$b \rightarrow s\gamma$	$m_{1/2} > 700$	$m_{1/2} > 500$
$m_{h^0} > 90$?	Yes	Yes
$\Omega h^2 < 1$?	$m_0 > 300$	$m_0 > 300$

The Higgs limits are most restrictive for the low-$\tan\beta$ solutions. First, for low $\tan\beta$, $\mu < 0$ is pretty much excluded unless one allows $m_0, m_{1/2} > 1$ TeV. Second, low $\tan\beta$ and $\mu > 0$ will soon be excluded if no Higgs is seen at LEP200: for $\mu > 0$ and $m_0, m_{1/2} < 1$ TeV, the other constraints imply $m_{h^0}^{max} = 97 \pm 6$ GeV (error dominated by uncertainty in m_t). If $\tan\beta$ is large, then $m_{h^0}^{max} = 120 \pm 2$ GeV, which will hopefully be testable at TeV33. The best χ^2 solutions have large squark masses > 1 TeV and fine-tuning problems.

Many other studies, especially in the fixed-point context, reach very similar conclusions. For example, Carena *et al.* [20] show that $m_{h^0} > 89$ GeV implies a lower bound on $\tan\beta$ well above the perturbativity bound unless the stop mass matrix is carefully chosen. In particular, the fixed-point value of $\tan\beta \sim 1.5$ is allowed only if the heavier stop is not too heavy (*i.e.* there is an implicit upper bound on $m_{\widetilde{t}_2}$). If the bound on m_{h^0} increases to $m_{h^0} \gtrsim 103$ GeV (as expected at LEP200), the low-$\tan\beta$ fixed point scenario will be ruled out. Of course, one should keep in mind that the low-tan β fixed point solution is ruled out in mSUGRA and CMSSM if we require $\Omega h^2 < 1$ [21] and/or no charge/color breaking. [22] This means, that we can only have a low-tan β fixed-point solution with $m_{h^0} > 89$ GeV that is consistent with $\Omega h^2 < 1$ if we disconnect the slepton, Higgs and squark soft-supersymmetry-breaking mass parameters by not requiring a universal value at the GUT scale. Finally, we recall that adequate electroweak baryogenesis in the MSSM requires m_{h^0} in the LEP2 range, a light $\tilde{t}_1$, and small stop mixing. Imposing these constraints in conjunction with the fixed-point low-tan β solution requires $m_{\tilde{t}_2} > 2$ TeV.

Of course, all these constraints depend greatly upon the fact that the MSSM contains exactly two doublets and no other Higgs representations. For example, the constraints found in these studies are obviated if one adds extra singlet(s) S with $\lambda S H_1 H_2$ style coupling.

5.2 A very difficult Higgs scenario: is there a no-lose theorem?

An interesting question that has emerged in several different models is the question of whether there is a no-lose theorem for Higgs discovery at a $\sqrt{s} = 500$ GeV e^+e^- collider. Typically, [23,24] if one adds just one or two Higgs bosons to the spectrum the answer is yes: one or more of the scalar Higgs bosons will be discovered in the Zh production mode. However, the situation could be much more complicated. A very difficult case [24] is one in which there are many Higgs bosons, as could arise in a string model with many U(1)'s, [25] and they share the ZZ-Higgs coupling-strength-squared (g^2_{ZZh}) fairly uniformly. Further, assume these Higgs bosons are spread out such that the experimental resolution is insufficient to resolve the separate peaks, in which case the only signal is an unresolved continuum excess over background. Finally, assume the Higgs bosons all decay into a variety of channels, including invisible decays, various $q\bar{q}$ channels, etc., in which case identification of the h decay final state would not be useful because of the large background in any one channel. In particular, in $e^+e^- \rightarrow Zh$, there would be no guarantee we can use $Z \rightarrow q\bar{q}$ or $\nu\bar{\nu}$ decays because of the large number of possible channels in the recoil state and, thus, small signal relative to background in any one channel. The only clearly reliable signal would be an excess in the recoil M_X distribution in $e^+e^- \rightarrow ZX$ (with $Z \rightarrow e^+e^-$ and $\mu^+\mu^-$).

To describe this scenario quantitatively, one [24] employs a continuum description in which there are Higgs bosons from m_h^{min} to m_h^{max}. Defining $K(m_h)$ as the g^2_{ZZh} strength (relative to SM strength) as a function of m_h, one then can write two sum rules:

$$\int dm_h K(m_h) \geq 1 \qquad (5)$$

$$\int dm_h K(m_h) m_h^2 \leq \langle M^2 \rangle, \qquad (6)$$

where the former becomes an equality if only Higgs singlet and doublet representations are involved. The key to a no-lose theorem is to limit $\langle M^2 \rangle$. In the context of supersymmetry one can write $\langle M^2 \rangle \equiv m_B^2 = \lambda v^2$, where $v = 246$ GeV and λ, a typical quartic Higgs coupling at low energy scales, is limited by requiring perturbativity for λ up to some high scale Λ. In the most general SUSY model, one finds $m_B \leq 200$ GeV for $\Lambda \sim 10^{17}$ GeV. Alternatively, independently of a SUSY context, the success of fits to precision electroweak data using $m_{h_{SM}} \lesssim 200$ GeV implies, in a multi-Higgs model, that the Higgs bosons with large $K(m_h)$ must have average mass $\lesssim 200$ GeV, which would imply $m_B \lesssim 200$ GeV.

Taking $K(m_h) = K$, a constant, Eq. (5) leads to $K = 1/(m_h^{max} - m_h^{min})$ (assuming only singlet and doublet representations), and Eq. (6) implies

$$m_B^2 \equiv \langle M^2 \rangle \geq \frac{1}{3} \left([m_h^{max}]^2 + m_h^{max} m_h^{min} + [m_h^{min}]^2 \right) (7)$$

The maximal spread is achieved for $m_h^{min} = 0$, in which case Eq. (7) requires $m_h^{max} \leq \sqrt{3} m_B \leq 340$ GeV.

To analyze this situation, [24] assume $\sqrt{s} = 500$ GeV, for which $\sigma(Zh)$ for a SM-like h is substantial out to $m_h \sim 200$ GeV. [$\sigma(Zh_{SM})$ falls from 70 fb at low m_h to 42 fb at $m_h \sim 200$ GeV.] Confining the signal region to $70 \leq m_h \leq 200$ GeV, a fraction $f \sim 0.4$ of the uniform $K(m_h)$ spectrum would lie in this region. If LEP2 data can eventually be used to show that $K(m_h)$ is small for $m_h \lesssim 70$ GeV (i.e. $m_h^{min} = 70$ GeV) then $m_h^{max} = 300$ GeV [from Eq. (7) with $m_B = 200$ GeV] and a fraction $f \sim 0.55$ would lie in the $70 - 200$ GeV region. Alternatively, one can consider only the $100 - 200$ GeV interval (to avoid the large background in the vicinity of $m_h \sim m_Z$), in which case $f \sim 0.3$ for $m_h^{min} = 0$ and $f \sim 0.43$ for $m_h^{min} = 70$ GeV, respectively.

The results for the overall excess in Zh, with $Z \rightarrow e^+e^- + \mu^+\mu^-$, integrated over the $70 - 200$ GeV and $100 - 200$ GeV intervals, assuming $\int K(m_h) = 1$ over the interval, are given in Table 2, assuming an integrated luminosity of $L = 500$ fb^{-1} (which is very optimistic). Including the factor f, one finds $S \sim 1350f$ with a background of either $B = 6340$ or $B = 2700$, for the $70 - 200$ GeV or $100 - 200$ GeV windows, respectively. Correspondingly, one must detect the presence of a broad $\sim 21\%f$ or $\sim 50\%f$ excess over background, respectively. For $f \sim 0.4 - 0.55$ in the 1st case and $f \sim 0.3 - 0.43$ in the 2nd case, this would probably be possible. Nominally, $S/\sqrt{B} \sim 17f$ and $\sim 26f$ for the $70 - 200$ GeV and $100 - 200$ GeV windows in M_X, respectively. However, if $L \lesssim 200$ fb, the detection of the excess will become quite marginal. As an aside, we note [24] that $e^+e^- \rightarrow e^+e^-h$ via ZZ-fusion is not useful because of very small S/B.

Table 2: Approximate S, B and $S/\sqrt{B}$ values for Zh (with $Z \to e^+e^- + \mu^+\mu^-$) after integrating the M_X recoil mass spectrum from (a) 70 GeV to 200 GeV and (b) 100 GeV to 200 GeV, assuming that many Higgs bosons are distributed evenly throughout the interval with uniform $K(m_h)$. Results are for $\sqrt{s} = 500$ GeV, $L = 500$ fb^{-1}.

| M_X | Zh, $Z \to e^+e^- + \mu^+\mu^-$ | | |
Interval	S	B	$S/\sqrt{B}$
$70 - 200$	1350	6340	17
$100 - 200$	1356	2700	26

Table 3: Approximate S, B and $S/\sqrt{B}$ values for Zh (with $Z \to e^+e^- + \mu^+\mu^-$) in each of the thirteen 10 GeV bins in M_X from 70 to 200 GeV, assuming that $S \sim 1350$ events are distributed equally among these bins. We assume $\sqrt{s} = 500$ GeV, $L = 500$ fb^{-1}.

Bin No.	1	2	3	4
S	104	104	104	104
B	1020	1560	1440	734
$S/\sqrt{B}$	3.3	2.6	2.7	3.8
Bin No.	5	6	7	8–13
S	104	104	104	104
B	296	162	125	~ 130
$S/\sqrt{B}$	6.0	8.2	9.3	9.1

Of course, if an excess is observed, the next interesting question is whether we can analyze the amount of this excess on a bin-by-bin basis. The situation is illustrated in Table 3 assuming that the roughly 1350 (i.e. $f = 1$ for the moment) signal events are distributed equally in the thirteen 10 GeV bins from 70 to 200 GeV. Table 3 gives S for $f = 1$, B and the corresponding $S/\sqrt{B}$ for each bin. Both S and $S/\sqrt{B}$ must be reduced by f. One sees that $L = 500$ fb^{-1} would yield $S/\sqrt{B} > 3$ only for the $M_X \gtrsim 120$ GeV bins when $f \sim 0.5$. Further, with only $L = 100$ fb^{-1} (as might be achieved after a few years of running at a 'standard' luminosity design), this bin-by-bin type of analysis would not be possible for 10 GeV bins if $f \sim 0.5$; one really needs $L = 500$ fb^{-1}.

A final question is how many Higgs force us into the continuum scenario? In the inclusive $e^+e^- \to ZX$ mode, with $Z \to e^+e^-, \mu^+\mu^-$, the electromagnetic calorimeter and tracking resolutions planned for electrons and muons imply $\Delta m \sim 20$ GeV at $\sqrt{s} = 500$ GeV. As a result, something like five Higgs bosons distributed from 70 to 200 GeV would put us into the continuum scenario unless a specific Higgs decay final state (for which resolutions are expected to be below 10 GeV and backgrounds would be smaller) could be shown to be dominant.

6 Doubly-charged Higgs and higgsinos in supersymmetric L-R models

In supersymmetric L-R symmetric models, the Lagrangian cannot contain terms that explicitly violate R-parity. The presence or absence of RPV is determined by whether or not there is spontaneous RPV. There are two generic possibilities. [26]

If certain higher dimensional operators are small or absent, then the scalar field potential must be such that L-R symmetry breaking induces RPV through some combination of non-zero $\langle \tilde{\nu}_i^c \rangle$'s. In this case, the W_R mass scale is low and, of course, there are lots of new phenomena associated with RPV. In this scenario, the triplet Higgs and higgsinos, including $\Delta_{L,R}^{--}$ and their fermionic partners, are not necessarily light. Considerable phenomenological discussion of the resulting RPV signatures for this case has appeared. [27]

If the above-mentioned higher-dimensional operators are present and are of full strength (but, of course, $\propto 1/M_U$ or $1/M_{\text{Planck}}$), then L-R symmetry breaking does not require RPV. In this case, the W_R mass scale must be very large. Further, the Δ_L triplet members and their superpartners must be very heavy unless one removes the (naturally present) parity-odd singlet from the theory (which is normally included in order to avoid $v_L \neq 0$ vacua). However, when the R-sector Higgs mechanism comes in at high scale (assumed to be above the SUSY breaking scale) to give $v_R \neq 0$ and generate W_R mass, one is breaking a U(3) symmetry and there are 4 surviving massless (goldstone) fields, which are the Δ^{--} superfield and its charge conjugate, whose component fields only become massive via the higher-dimensional operators. In this case, it is natural for the mass scales of the Δ_R^{--} and $\tilde{\Delta}_R^{--}$, $\sim v_R^2/M_{\text{Planck}}$, to be at the ~ 100 GeV level.

The phenomenology of doubly charged Higgs bosons has a long history. [28] The above Δ_R^{--} (hereafter we drop the R subscript) would generally be narrow. Noting that $\Delta^{--} \to W^-\Delta^-$ is expected to be kinematically forbidden, its primary decay modes would most probably be via the Majorana couplings associated with the see-saw mechanism for neutrino mass generation:

$$\mathcal{L}_Y = ih_{ij}\psi_i^T C\tau_2 \Delta \psi_j + \text{h.c.}, \tag{8}$$

where $i, j = e, \mu, \tau$ are generation indices, and Δ is the 2×2 matrix of Higgs fields:

$$\Delta = \begin{pmatrix} \Delta^+/\sqrt{2} & \Delta^{++} \\ \Delta^0 & -\Delta^+/\sqrt{2} \end{pmatrix}. \tag{9}$$

Limits on the h_{ij} by virtue of the $\Delta^{--} \to \ell^-\ell^-$ couplings include: Bhabha scattering, $(g-2)_\mu$, muonium-antimuonium conversion, and $\mu^- \to e^-e^-e^+$. Adopting the convention

$$|h_{\ell\ell}^{\Delta^{--}}|^2 \equiv c_{\ell\ell}m_{\Delta^{--}}^2 \, (\text{ GeV}), \qquad (10)$$

one finds $c_{ee} < 10^{-5}$ (Bhabbha) and $\sqrt{c_{ee}c_{\mu\mu}} < 10^{-7}$ (muonium-antimuonium) are the strongest of the limits. There are no limits on $c_{\tau\tau}$ which is, naively, expected to be the largest. If all the c's are very tiny, virtual versions of $\Delta^{--} \to \Delta^- W^-$ could be important.

Regarding production, because of the very large W_R mass, the doubly-charged Higgs bosons would be primarily produced at hadron colliders via $\gamma^*, Z^* \to \Delta^{--}\Delta^{++}$. At an e^-e^- or $\mu^-\mu^-$ collider they could be produced directly as an s-channel resonance via the lepton-number-violating couplings h_{ee} and $h_{\mu\mu}$, respectively. The strategy for discovering and studying the Δ^{--} would be the following. First, one would discover the Δ^{--} in $p\bar{p} \to \Delta^{--}\Delta^{++}$ with $\Delta^{--} \to \ell^-\ell^-, \Delta^{++} \to \ell^+\ell^+$ ($\ell = e, \mu, \tau$) at TeV33 or LHC.[29] One finds that Δ^{--} detection at the Tevatron ($\sqrt{s} = 2$ TeV, $L = 30$ fb^{-1}) is possible for $m_{\Delta^{--}}$ up to 300 GeV for $\ell = e$ or μ and up to 180 GeV for $\ell = \tau$. At the LHC, discovery is possible up to roughly 925 GeV (1.1 TeV) for $\ell = e, \mu$ and 475 GeV (600 GeV) for $\ell = \tau$, for $L = 100$ fb^{-1} ($L = 300$ fb^{-1}). Thus, TeV33 + LHC will tell us if such a Δ^{--} exists in the mass range accessible to the next linear collider or a first muon collider, and, quite possibly, its decays will indicate if it has significant coupling to e^-e^- and/or $\mu^-\mu^-$ (unless $\tau^-\tau^-$ is completely dominant, as is possible). Whether or not these decays are seen, we will wish to determine the strength of these couplings by studying e^-e^- and $\mu^-\mu^-$ s-channel production of the Δ^{--}. We note that if the Δ^{--} is observed at the LHC, we will know ahead of time what final state to look in and have a fairly good determination of $m_{\Delta^{--}}$.

At the NLC, taking $L = 50$ fb^{-1} and defining R to be the beam energy spread in percent,

$$N(\Delta^{--}) \sim 3 \times 10^{10} \left(\frac{c_{ee}}{10^{-5}}\right)\left(\frac{0.2\%}{R}\right), \qquad (11)$$

implying an enormous event rate if c_{ee} is near its upper bound. The ultimate sensitivity to c_{ee} when $\Gamma_{\Delta^{--}}$ is much smaller than the beam energy spread can be estimated by supposing that 100 events are required. From Eq. (11), we predict 100 Δ^{--} events for

$$c_{ee}|_{100 \text{ events}} \sim 3.3 \times 10^{-14}(R/0.2\%), \qquad (12)$$

independent of $m_{\Delta^{--}}$, which is dramatic sensitivity. Because of the much smaller R values possible at a $\mu^-\mu^-$ collider ($R \sim 0.003\%$ is possible), comparable or greater sensitivity to $c_{\mu\mu}$ could be achieved there despite the lower expected integrated luminosity.

In the L-R symmetric models the phenomenology of the doubly-charged Higgsinos would be equally interesting.[30] The basic experimental signatures always involve

τ's. In non-GMSB SUSY, if $h_{\tau\tau}$ is full strength (~ 0.5) then it influences the RGE's so that the $\tilde{\tau}$'s (especially $\tilde{\tau}_R$) are lighter than $\tilde{e}$ and $\tilde{\mu}$, even if $\tan\beta$ is not large. Further, starting with a common mass at the v_R scale, evolution leads to $m_{\tilde{\Delta}^{--}} < m_{\tilde{\Delta}^{--}}$ and the $\tilde{\Delta}^{--}$ would be easily visible as described above. Less attention has been paid to $\tilde{\Delta}^{--}, \tilde{\Delta}^{++}$, which could be produced at the Tevatron in pairs. Indeed, for $m_{\Delta^{--}} = m_{\tilde{\Delta}^{--}}$, the $\tilde{\Delta}^{--}\tilde{\Delta}^{++}$ pair cross section is bigger than that for $\Delta^{--}\Delta^{++}$ due to the fact that the former is not p-wave suppressed. Normally, $\tilde{\Delta}^{--} \to \tilde{\tau}_R\tau$ is kinematically allowed and will dominate over all other lepton channels because of larger coupling. The dominant $\tilde{\tau}_R$ decay would be $\tilde{\tau}_R \to \tilde{\chi}_1^0\tau$. Thus, a typical signature would be $p\bar{p} \to \tilde{\Delta}^{--}\tilde{\Delta}^{++} \to \tau^-\tau^-\tau^+\tau^+ + \not{p}_T$. Note that the presence of $\not{p}_T$ would make reconstruction of the Δ^{--} and Δ^{++} masses difficult.

In the GMSB context there are some alterations to the above scenario. First, one finds that the $\tilde{\Delta}^{--}$ is now lighter than the Δ^{--}. In fact, the $\tilde{\Delta}^{--}$ could even be the NLSP. If not, the $\tilde{\tau}_R$ very probably is (even for minimal messenger sector content), with $\tilde{\tau}_R \to \tau\tilde{G}$ being its dominant decay. The typical signature would be the same as above except the $\not{p}_T$ would now be due to the $\tilde{G}$'s rather than $\tilde{\chi}_1^0$'s. In the small portion of parameter space where the $\tilde{\chi}_1^0$ is the NLSP, the signature for $\tilde{\Delta}^{--}\tilde{\Delta}^{++}$ production changes to $4\tau2\gamma + \not{p}_T$, where the γ's come from the $\tilde{\chi}_1^0 \to \gamma\tilde{G}$ decays.

Overall, a supersymmetric L-R symmetric model would give rise to a very unique phenomenology with many exciting ways to explore the content and parameters of the model.

7 Conclusion

I have tried to give an overview of recent results in supersymmetry phenomenology with emphasis on unusual scenarios that one might encounter, especially ones for which detection of supersymmetric particles and/or the SUSY Higgs bosons might require special experimental/analysis techniques. Experimentalists should pay attention to these special cases to make sure that their detector designs, triggering algorithms and analysis techniques do not discard these possibly important signals.

Acknowledgements

This work was supported in part by the U.S. Department of Energy.

References

1. The phenomenology is discussed in C.H. Chen, M. Drees and J.F. Gunion, *Phys. Rev. Lett.* **76**, 200 (1996); *Phys. Rev.* D **55**, 330 (1997).
2. G. Anderson, C.H. Chen, J.F. Gunion, J. Lykken, T. Moroi, Y. Yamada, in *New Directions for High-Energy Physics*, Proceedings of the 1996 DPF/DPB Summer Study on High Energy Physics, Snowmass '96, edited by D.G. Cassel, L.T. Gennari and R.H. Siemann (Stanford Linear Accelerator Center, Stanford, CA, 1997) pp. 669–673, hep-ph/9609457.
3. DELPHI Collaboration, http://delphiwww.-cern.ch/delfig/figures/search/chadeg172/-char_dege.html.
4. A. de Gouvea, A. Friedland and H. Murayama, hep-ph/9803481.
5. H. Baer, C.H. Chen and X. Tata, *Phys. Rev.* D **35**, 075008 (1998).
6. This is exemplified by the work of T. Ibrahim and P. Nath, hep-ph/9807501, in the 'mSUGRA' context.
7. M. Brhlik and G. Kane, hep-ph/9803391.
8. S. Raby, *Phys. Rev.* D **56**, 2852 (1997); *Phys. Lett.* B **422**, 158 (1998); S. Raby and K. Tobe, hep-ph/9807281.
9. H. Baer, K. Cheung and J.F. Gunion, hep-ph/9806361.
10. J.F. Gunion and D. Soper, *Phys. Rev.* D **15**, 2617 (1977).
11. J. Feng and T. Moroi, *Phys. Rev.* D **58**, 035001 (1998).
12. B. Dutta, D.J. Muller and S. Nandi, hep-ph/9807390; K. Cheung and D. Dicus, *Phys. Rev.* D **58**, 057705 (1998).
13. C.-H. Chen and J.F. Gunion, *Phys. Lett.* B **420**, 77 (1998) and *Phys. Rev.* D **58**, 075005 (1998).
14. J.F. Gunion, in *Future High Energy Colliders*, Proceedings of the ITP Symposium, U.C. Santa Barbara, October 21–25, 1996, AIP Press, ed. Z. Parsa, pp. 41–64.
15. A. de Gouvea, T. Moroi and H. Murayama, *Phys. Rev.* D **56**, 1281 (1997).
16. J.F. Gunion, under study.
17. P. Binetruy and J.F. Gunion, in *Heavy Flavors and High Energy Collisions in the 1—100 TeV Range*, Proceedings of the INFN Eloisatron Project Workshop, Erice, Italy, June 10–27, 1988, edited by A. Ali and L. Cifarelli (Plenum Press, New York, 1989) p. 489.
18. H. Dreiner and G.G. Ross, *Nucl. Phys.* B **365**, 597 (1991); H. Dreiner, M. Guchait and D.P. Roy, *Phys. Rev.* D **49**, 3270 (1994); V. Barger, M.S. Berger, P. Ohmann, R.J.N. Phillips, *Phys. Rev.* D **50**, 4299 (1994); H. Baer, C. Kao and X. Tata, *Phys. Rev.* D **51**, 2180 (1995); H. Baer, C.-H. Chen and X. Tata, *Phys. Rev.* D **55**, 1466 (1997); A. Bartl *et al.*, *Nucl. Phys.* B **502**, 19 (1997).
19. Example: W. de Boer *et al.*, hep-ph/9805378.
20. M. Carena, P. Chankowski, S. Pokorski and C. Wagner, hep-ph/9805349.
21. J. Ellis, T. Falk, K. Olive and M. Schmitt, *Phys. Lett.* B **413**, 355 (1977).
22. S.A. Abel and B.C. Allanach, *Phys. Lett.* B **431**, 339 (1998).
23. J. Kamoshita, Y. Okada and M. Tanaka, *Phys. Lett.* B328 67 1994 ; B.R. Kim, S.K. Oh and A. Stephan, *Proceedings of the 2nd International Workshop on "Physics and Experiments with Linear e^+e^- Colliders"*, edited by F. Harris, S. Olsen, S. Pakvasa and X. Tata, Waikoloa, HI, (World Scientific, Singapore, 1993) p. 860; B.R. Kim, G. Kreyerhoff and S.K. Oh, hep-ph/9711372.
24. J.R. Espinosa and J.F. Gunion, hep-ph/9807275.
25. K.R. Dienes, *Phys. Rep.* **287**, 447 (1997).
26. See, for example, K. Kuchimanchi and R.N. Mohapatra, *Phys. Rev. Lett.* **75**, 3989 (1995); and Z. Chacko and R.N. Mohapatra, *Phys. Rev.* D **58**, 015001 (1998).
27. For a brief review and references, see K. Huitu, J. Maalampi and K. Puolamaki, hep-ph/9708491, in Proceedings of the 5th International Conference on *Physics Beyond the Standard Model*, Balholm, Norway, 1997, p. 500.
28. J.F. Gunion, *Int. J. Mod. Phys.* A **11**, 1551 (1996) and *Int. J. Mod. Phys.* A **13**, 2277 (1998).
29. J.F. Gunion, C. Loomis and K. Pitts, in *New Directions for High-Energy Physics*, Proceedings of the 1996 DPF/DPB Summer Study on High Energy Physics, Snowmass '96, edited by D.G. Cassel, L.T. Gennari and R.H. Siemann (Stanford Linear Accelerator Center, Stanford, CA, 1997) p. 603, hep-ph/9610237.
30. B. Dutta and R. Mohapatra, hep-ph/9804277.

Tau Leptons as the New Signal for Supersymmetry

S. Nandi

Department of Physics, Oklahoma State University, Stillwater, OK 74078

For a large class of supersymmetry breaking theories, tau leptons may be the only observable signal for SUSY at the present and forthcoming colliders. At LEP-2, the dominant signals are events having four tau leptons and large missing energy. At the upgraded Tevatron, the signal is inclusive high p_T 2τ and 3τ events accompanied by large missing transverse energy. The observation of these signals will require good tau detection efficiencies. In this talk, I review the recent works on the observability of the τ signals at the LEP-2 and the upgraded Tevatron.

1 Introduction

If the supersymmetry at the Standard Model (SM) sector is broken effectively at the weak scale, then some of the lighter SUSY particles are expected to be produced at LEP-2 or the upgraded Tevatron. The subsequent decays of this particles will give rise to the observable SUSY signals. The important question is what these signals are, and what is the prospect of detecting such signals. In the past decade, usual SUSY searches have mostly concentrated on looking for events having high p_T jets and/or e,μ accompanied by large missing energy ($\not{E}_T$) . What if the dominant signal involves τ leptons only? Is this a theoretically sound possibility? In this talk, I argue that the answer to this later question is "Yes", and discuss the prospects of observing these τ signals at LEP-2 and Tevatron. The outline of this talk is as follows: In section 2, I discuss the theoretical motivations regarding why the τ leptons are the important signals and may be the only signal at LEP-2 and Tevatron Run 2. In sections 3,4,and 5 , I discuss what the important SUSY productions processes are, and the possibility of detecting these signals at LEP-2 and the upgraded Tevatron . The conclusions are summerized in section 5.

2 Theoretical Motivation

Supersymmetry signals at the colliders depend very much on the mechanism of the SUSY breaking and what the lightest SUSY particle (LSP) is.There are three major scenarios for effective weak scale supersymmetry breaking: Gauge Mediated Supersymmetry Breaking (GMSB) [1,2], Gravity Mediated Supersymmetry Breaking (Usual Supergravity, USG) [3],and the More Minimal Supersymmetric Standard Model (MMS) [4]. In the traditional GMSB, gravitino,$\tilde{G}$ is the LSP with a mass usually in the range of a few eV to a few keV. In the USG and MMS, the LSP is model dependent, but is usually the lightest neutralino, χ_1^0. The Next to Lightest SUSY Particle (NLSP) is the one whose production and decay will give rise to SUSY signals at the colliders. In GMSB, NLSP can either be the χ_1^0, or the lighter of the scalar taus, $\tilde{\tau}_1$

[5]. χ_1^0 will decay to a photon and a gravitino, while $\tilde{\tau}_1$ will decay to a tau and a gravitino. In USG with a χ_1^0 as LSP, $\tilde{\tau}_1$ can also be the NLSP for the parameter space having large values of $\tan\beta$ and A_τ, and will decay to a τ and a χ_1^0 [6]. In MMS, $\tilde{\tau}_1$ can also be the NLSP decaying to τ and χ_1^0. Thus, in all of the three SUSY breaking scenario, the possibility exists that we will get only the τ signals for SUSY at the colliders. In this talk, we will discuss only the τ signals arising from the GMSB case.

In GMSB with the messenger sector in $5+\bar{5}$ of SU(5), , there are five parameters, M, Λ, $\tan\beta$, n, and Sign(μ) which determine all the SUSY particle masses and mixing angles. M is the messenger scale. Λ is equal to $< F_s > / < S >$, and is related to the SUSY breaking scale. The parameter n is the number of $5+\bar{5}$ in the messenger sector. n take take the values 1,2,3, or 4 in order to satisfy the constraint of the perturbative unification of the gauge couplings in MSSM. The parameter $\tan\beta$ is the usual ratio of the up and down type Higgs VEVs. Sign(μ) is negative in our convention [7] from the $b \to s\gamma$ constraints. In our calculation, we demand electroweak radiative symmetry breaking, which then determines the other parameter, μ^2 and B . The soft SUSY breaking gaugino and the scalar masses at the messenger scale are given by[1,8]

$$\tilde{M}_i(M) = n \, g \left(\frac{\Lambda}{M} \right) \frac{\alpha_i(M)}{4\pi} \Lambda \qquad (1)$$

and

$$\tilde{m}^2(M) = 2 \, n \, f \left(\frac{\Lambda}{M} \right) \sum_{i=1}^{3} k_i C_i \left(\frac{\alpha_i(M)}{4\pi} \right)^2 \Lambda^2. \qquad (2)$$

where α_i are the three SM gauge coupling constants, and $k_i = 1,1,$and 3/5 for SU(3), SU(2), and $U(1)_Y$, respectively. The constants C_i are zero for gauge singlets, and 4/3, 3/4 , and $(Y/2)^2$ for the fundamental representation of SU(3), SU(2), and $U(1)_Y$ respectively. f(x) and g(x) are messenger scale threshold functions. The SUSY mass spectrum is calculated using the appropriate RGE equations [7]with the boundary conditions given by equations (1) and (2). Detail investigation shows that Λ lies between about 20 and 200 TeV for weak scale supersymmetry. The mass of the gravitino is given by

$$m_{3/2} = \frac{F}{3^{1/2} M_*} \qquad (3)$$

where F is the VEV of the auxiliary component of the hidden sector superfield which breaks supersymmetry, and $M_* = M_P/\sqrt{8\pi}$, is the reduced Planck's mass. In the scalar tau NLSP scenario, $\tilde{\tau}_1$ decays to a tau and a gravitino. The decay width of $\tilde{\tau}_1$ is given by

$$\Gamma_{\tilde{\tau}_1} = \frac{1}{16\pi} \frac{m^5_{\tilde{\tau}_1}}{F^2} \qquad (4)$$

and its decay length is given by

$$L \simeq 10km \times <\beta\gamma> \left[\frac{F^{1/2}}{10^7 \, GeV}\right]^4 \left[\frac{100 GeV}{m_{\tilde{\tau}_1}}\right]^5 \qquad (5)$$

where β is the speed of $\tilde{\tau}_1$, and $\gamma \equiv \left(1 - \beta^2\right)^{-1/2}$.

From equation (5), we see that there are three possibilities for the $\tilde{\tau}_1$ signal depending on the value of F. $\tilde{\tau}_1$ can decay promptly within the detector giving rise to the τ signal. Its decay can be delayed resulting in kink, or a displaced vertex [9]. It may also be long-lived and thus not decaying within the detector. In the last case, we will have highly ionizing tracks from the long-lived slow moving $\tilde{\tau}_1$, or an excess over the SM dimuon events from the less massive fast moving $\tilde{\tau}_1$. The discovery reach at the Tevatron Run 2 for this long-lived case has been studied in detail [10]. In this talk, I will discuss mainly the prompt decay case. However, in section 5, I also include a brief review for the long-lived case as was suggested to me by the session organizer.

3 $\tilde{\tau}$ NLSP, Prompt decay signals at LEP-2

The dominant production process at LEP-2 is the pair production of the lightest neutralino, χ^0_1 [11]. Each neutralino decays to a τ and a $\tilde{\tau}_1$ followed by the decay $\tilde{\tau}_1 \rightarrow \tau \tilde{G}$ [11]. The resulting final state involves four tau leptons plus missing energy due to the escaping gravitinos [11]. Such a final state has negligible SM background. Because of the Majorana nature of the neutralinos, two of the most energetic taus can have the same sign of electric charge. Also, two of the four taus (arising from the decays of the $\tilde{\tau}_1$) will have higher p_T and higher energy. A recent analysis by the DELPHI collaboration has found two such candidate four tau events, [12] where their expected background is one. From this , DELPHI collaboration set a lower bound of 72 GeV for the stau mass [12]. Analysis by the other LEP-2 collaborations are in progress, and we will soon know if these events are real signal for SUSY.

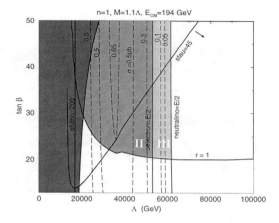

Figure 1: Cross section contours for the neutralino pair production in the Λ-$\tan\beta$ plane at the LEP-2 energy of $E_{CM} = 194$ GeV.

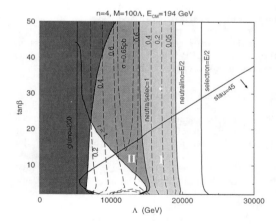

Figure 2: Cross section contours for the neutralino pair production in the Λ-$\tan\beta$ plane at the LEP-2 energy of $E_{CM} = 194$ GeV.

The direct production of opposite sign stau pair and their subsequent dacays can also give rise to observable signal at LEP-2, although this cross section is much smaller than the χ_1^0 pair productions [13]. In this case, the final state is 2τ plus $\not{E}_T$ due to the escaping gravitinos. This signal has background from the SM W pair productions, although the angular distributions of the τ pair in the two cases are different. Similar signals are also expected from the charged Higgs pair productions if it is kinematically allowed. The third possibility is the scalar electron or muon pair productions and their decays to χ_1^0 e (or $\chi_1^0 \mu$) and the subsequent decay of χ_1^0. In this case, the final states will be four taus or six charged leptons [14]. At the LEP-2 energy of 194 GeV, the cross section contours for the process, $e^+ e^- \to \chi_1^0 \chi_1^0$ in the Λ-$\tan\beta$ plane is shown in Fig. 1 and Fig. 2. The region **I** corresponds to the hierarchy $E > M_{\chi_1^0} > m_{\tilde{e}_R} > m_{\tilde{\tau}_1}$, region **II** corresponds to $E > m_{\tilde{e}_R} > M_{\chi_1^0} > m_{\tilde{\tau}_1}$, while the region **III** corresponds to $m_{\tilde{e}_R} > E > M_{\chi_1^0}$. Note that the region **I** gives rise to four charged leptons in the final state, region **II** to four taus or six charged leptons, while region **III** to four tau leptons. The detailed discussions for the different final states , and also for the other LEP-2 energies can be found in ref. [14].

4 $\tilde{\tau}$ NLSP, Prompt decay signals at the Upgraded Tevetron

In GMSB, the gluinos and the squarks are very heavy, and can not be produced at the upgraded Tevatron energy. Here the dominant production processes (giving rise to SUSY signals) are the chagino pair ($\chi^+ \chi^-$) and the chargino, 2nd neutralino ($\chi_1^\pm \chi_2^0$) [15,10]. $\chi_1^\pm$ decay dominantly to a $\tilde{\tau}_1$ and a ν_τ, while the χ_2^0 decays dominantly to $\tau \tilde{\tau}_1$. $\tilde{\tau}_1$ decays subsequently to $\tau\tilde{G}$. Thus the signal is inclusive 2τ or 3τ with high p_T accompanied by large missing transverse energy ($\not{E}_T$) [15]. The major question is whether these events are observable in the hadronic environment [16]. In table 1, we give the relevent masses, $\chi^+ \chi^-$ and $\chi_1^\pm \chi_2^0$ production cross sections, and the inclusive 2τ and 3τ -jet production rates for two illustrative set of parameters at the upgraded Tevatron ($\sqrt{s} = 2$ TeV). The τ leptons has been decayed to produce the τ-jets. In performing this analysis, the cuts employed are that final state charged leptons must have $p_T > 10$ GeV and a pseudorapidity, $\eta \equiv -\ln(\tan\frac{\theta}{2})$ (where θ is the polar angle with respect to the proton beam direction), of magnitude less than 1. Jets must have $E_T > 10$ GeV and $|\eta| < 2$. In addition, hadronic final states within a cone size of $\Delta R \equiv \sqrt{(\Delta\phi)^2 + (\Delta\eta)^2} = 0.4$ are merged to a single jet. Leptons within this cone radius of a jet are discounted. For a τ-jet to be counted as such, it must have $|\eta| < 1$. The most energetic τ jet is required to have

Table 1: Relevent masses and branching ratios for the two examples of the GMSB Prompt $\tilde{\tau}_1$ decay case where the ordering of the masses is $M_{\chi_2^0} \approx M_{\chi_1^\pm} > m_{\tilde{\nu}} > M_{\chi_1^0} > m_{\tilde{e}_1, \tilde{\mu}_1} > m_{\tilde{\tau}_1}$.

	Example 1	Example 2
	$\Lambda = 32$ TeV	$\Lambda = 22$ TeV
	$M = 4.69\Lambda$	$M = 40\Lambda$
	$n = 3$	$n = 4$
	$\tan\beta = 12$	$\tan\beta = 18$
m_h (GeV)	118	117
$m_{\chi_1^0}$	127	116
$m_{\chi_2^0}$	224	206
$m_{\chi_{1,2}^\pm}$	222, 355	205, 341
$m_{\tilde{\tau}_{1,2}}$	101, 220	73, 194
$m_{\tilde{e}_{1,2}}$	108, 218	94, 186
$m_{\tilde{\nu}_\tau}$	202	168
$m_{\tilde{\nu}_e}$	202	168
$\sigma_{p\bar{p} \to \chi_1^+ \chi_1^-}$ (fb)	22.9	37.2
$\sigma_{p\bar{p} \to \chi_1^\pm \chi_2^0}$ (fb)	26.3	45.1
$\sigma \cdot BR_{2\tau-jets}$ (fb)	9.07	15.7
$\sigma \cdot BR_{3\tau-jets}$ (fb)	2.19	5.24

$E_T > 20$ GeV. In addition, a missing transverse energy cut of $\not{E}_T > 30$ GeV is imposed. With an integrated luminosity of 2,10 and 30 fb^{-1}, the number of 2τ (3τ) events for the parameter set in example 1 are 18(4),90(20) and 270(60) respectively, while the corresponding numbers for the the the parameter set of example 2 are 30(10), 150((50) and 450(150) respectively. Fig. 3 gives the E_T distribution of the highest E_T tau-jet, whereas fig. 4 gives the missing transverse energy ($\not{E}_T$) distributions of the events involving $\chi^+\chi^-$ /$\chi_1^\pm\chi_2^0$ productions. Both the E_T of the τ jets, and the $\not{E}_T$ of the events are large enough to be detected at the upgraded Tevatron. However, a careful calculation of the associated background, as well as a good tau jet efficiency for the detectors will be required to establish these signals. More details of this study, and the results for the other final states can be found in ref. [16,17].

5 Long-lived $\tilde{\tau}_1$ NLSP Signals

If the hidden sector SUSY breaking scale, $\sqrt{F}$, is larger than about 10^7 GeV, the $\tilde{\tau}_1$ NLSP will be long-lived, and will not decay inside the detector. In this scenario, the usual 2τ and 3τ plus $\not{E}_T$ signal discussed above will be absent. In this case, if the produced stau NLSP's are very massive, they will be slow, and will appear as a highly ionozing tracks. Alternatively, if they are highly relativistic, they can not be distinguished from muons, and their productions may give rise to excesses in the dimuon or multilepton productions over the usual SM productions.

Current LEP-2 searches, upto $\sqrt{s} = 172$ GeV , yield a lower bound of of 75 GeV for such a long-lived $\tilde{\tau}_1$. Tevatron Run 2 will be able to significantly improve over the lower bound at the highest LEP-2 energy.

Long-lived very massive staus, if produced at the upgraded Tevatron, will be slow, and will lose lot of energy via ionization, resulting in the highly ionizing tracks in the detector. Their production processes are the pair productions of the opposite sign $\tilde{\tau}_1$. Each $\tilde{\tau}_1$ will produce a highly ionizing track with no hadronic activity within some isolated cone. The resulting 5 events discovery reaches at the upgraded Tevatron ($\sqrt{s} = 2$ TeV) are $m_{\tilde{\tau}_1} = 110$, 180, and 230 GeV for an integrated luminosity of 2, 10, and 30 fb^{-1} respectively. The other production process are the opposite sign chargino pair productions, as well as the chargino-second neutralino productions, and their subsequent decays to $\tilde{\tau}_1$. In this case, the 5 event discovery reaches, including a 75% experimental efficiency, at the upgraded Tevatron are $M_{\chi_1^\pm}$ about 220, 270, and 310 for an integrated luminosity of 2, 10, and 30 fb^{-1} respectively. If the staus are not very massive, their productions will contribute to an excess over the SM dimuon rates. In this case, the discovery reach for a 3σ excess at the upgraded Tevatron are $m_{\tilde{\tau}_1} = 70$, 90, and 110 GeV for an intregrated luminosity of 2, 10, and 30 fb^{-1} respectively.

6 Conclusions

LEP-2 and upgraded Tevatron have good potential to discover supersymmetry by looking for the tau signals. This may be the only signal for a large class of SUSY breaking theories for a good part of the parameter space. At LEP-2, the dominant signal is 4τ plus $\not{E}_T$. Two such candidates have been observed by thr DELPHI collaboration. At the upgraded tevatron, the signal is inclusive high p_T 2τ and 3τ events with large $\not{E}_T$. Detailed analysis with the appropriate cuts shows that LEP-2 can explore the GMSB scale Λ upto about 70 TeV, while the upgraded Tevatron can go upto Λ of about 90 TeV. Detectors with good τ detection efficiency will play a very important part in such a discovery.

Acknowledgement

The work reported in the sections 3 and 4 of this talk was done in various collaborations with K. Cheung, D. A. Dicus, B. Dutta and D. J. Muller. I thank them for many important discussions. Thanks are also due to J. Heussman and D. J. Muller for some technical help during the preparation of this manuscript This work was supported in part by the U.S. Department of Energy Grant Number DE-FG03-98ER41076.

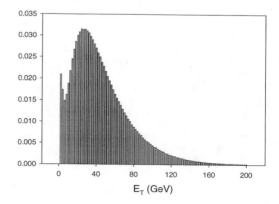

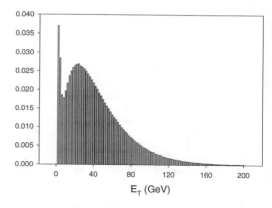

Figure 3: E_T distribution of the highest E_T τ-jet from $\chi_1^+\chi_1^-/\chi_1^\pm\chi_2^0$ production for the case where $\tan\beta = 12$, $M = 150$ TeV, $\Lambda = 32$ TeV, and $n = 3$. In the upper figure no cuts are imposed, while in the lower figure a pseudorapidity cut of $\eta < 1$ is imposed on the τ-jets.

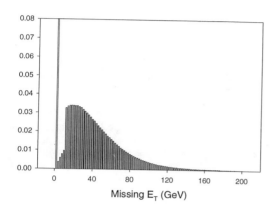

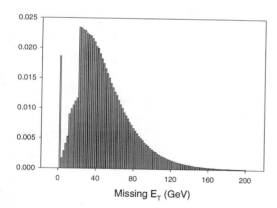

Figure 4: $\not{E}_T$ distribution for $\chi_1^+\chi_1^-/\chi_1^\pm\chi_2^0$ production where $\tan\beta = 12$, $M = 150\,\text{TeV}$, $\Lambda = 32\,\text{TeV}$, and $n = 3$. The lower figure includes all the cuts, while the upper figure omits the $E_T > 20\,\text{GeV}$ cut on the highest E_T τ-jet.

References

1. M. Dine and A. Nelson, *Phys. Rev.* D **47**, 1277 (1993); M. Dine, A. Nelson and Y. Shirman, *Phys. Rev.* D **51**, 3188 (1997); M. Dine, A. Nelson, Y. Nir, Y. Shirman,*Phys. Rev.* D **53**, 2658 (1996); M. Dine, Y. Nir and Y. Shirman, *Phys. Rev.* D **55**, 1501 (1997).

2. I. Affleck, M. Dine and N. Seiberg, *Nucl. Phys.* B **256**, 557 (1997); R. N. Mohapatra and S. Nandi,*Phys. Rev. Lett.* **79**, 181 (1997); B. Dobrescu, *Phys. Lett.* B **403**, 285 (1997); Z. Chako, B. Dutta, R. N. Mohapatra and S. Nandi, *Phys. Rev.* D **56**, 5466 (1997).

3. Examples of recent reviews areM. Dine, hep-ph/9612389; X. Tata, hep-ph/9712464; S. P. Martin, hep-ph/9612389; S. Dawson, hep-ph/9712464.

4. A. G. Cohen, D. B. Kaplan, and A. E. Nelson, *Phys. Lett.* B **388**, 588 (1996); S. Ambrosanio, A. E. Nelson, *Phys. Lett.* B **411**, 283 (1997); See also the plenary talk by A. Nelson at this conference.

5. K. S. Babu, C. Kolda and F Wilczek,*Phys. Rev. Lett.* **77**, 3070 (1996); J. Bagger, D. Pierce, K. Matchev and R. J. Zhang,*Phys. Rev.* D **55**, 3188 (1997); S, Dimopoulos, S. Thomas and J. D. Wells,*Nucl. Phys.* B **488**, 39 (1997).

6. V. Barger, C. Kao, and T-J Li, *Phys. Lett.* B **433**, 328 (1998); H. Baer, C-H Chen, M. Drees, F. Paige, and X. Tata, *Phys. Rev.* D **58**, 075008 (1998).

7. V. Barger, M. S. Berger, and P. Ohmann, *Phys. Rev.* D **49**, 4908 (1994)

8. S. Dimopoulos, G. F. Giudice, and A. Pomerol, *Phys. Lett.* B **389**, 37 (1996); S. P. Martin,*Phys. Rev.* D **55**, 3177 (1997).

9. S. Dimopoulos and G. F. Giudice,*Phys. Lett.* B **379**, 105 (1996).

10. J. L. Feng and T Moroi,*Phys. Rev.* D **58**, 035001 (1998)

11. D. A. Dicus, B. Dutta and S. Nandi,*Phys. Rev. Lett.* **78**, 3055 (1997).

12. G. Wolf, DELPHI Collaboration, Talk at this Conference.

13. D. A. Dicus. B. Dutta and S. Nandi,*Phys. Rev.* D **56**, 5748 (1997).

14. K. Cheung, D. A. Dicus, B. dutta and S. Nandi,*Phys. Rev.* D **56**, 5748 (1998)

15. B. Dutta and S. Nandi, hep-ph/9709511.

16. B. Dutta, D. J. Muller and S. Nandi, hep-ph/9807390, (submitted for publication in Nuclear Physics B)

17. D. J. Muller and S. Nandi (in preparation).

RESONANT SLEPTON PRODUCTION AT HADRON COLLIDERS IN R-PARITY VIOLATING MODELS

J. L. HEWETT and T. G. RIZZO

Stanford Linear Accelerator Center, Stanford University, Stanford, CA 94309, USA
E-mail: hewett,rizzo@slac.stanford.edu

Single s-channel production of sleptons, such as $\tilde{\tau}$'s and/or $\tilde{\nu}_\tau$, with their subsequent decay into purely leptonic or dijet final states is possible in hadronic collisions via R-parity violating couplings. We examine the impact of slepton production on bump searches in both the Drell-Yan and dijet channels and examine whether the lepton charge asymmetry in the $\ell\nu$ channel provides for additional search sensitivity. As a consequence, search reaches in the slepton mass-R-parity violating coupling plane are obtained for both the Tevatron and LHC. The possibility of using the leptonic angular distributions and the lepton charge asymmetry to distinguish slepton resonances from new gauge bosons is also analyzed.

1 Introduction

As is well known, the conventional gauge symmetries of the supersymmetric extension of the Standard Model(SM) allow for the existence of additional terms in the superpotential that violate Baryon(B) and/or Lepton(L) number. One quickly realizes that simultaneous existence of such terms leads to rapid proton decay. These phenomenologically dangerous terms can be written as

$$W_R = \lambda_{ijk}L_iL_jE_k^c + \lambda'_{ijk}L_iQ_jD_k^c + \lambda''_{ijk}U_i^cD_j^cD_k^c + \epsilon_iL_iH \,, \tag{1}$$

where i,j,k are family indices and symmetry demands that $i < j$ in the terms proportional to either the λ or λ'' Yukawa couplings. In the MSSM, the imposition of the discrete symmetry of R-parity removes by brute force all of these 'undesirable' couplings from the superpotential. However, it easy to construct alternative discrete symmetries that allow for the existence of either the L- or B-violating terms [1] in W_R (but not both kinds). As far as we know there exists no strong theoretical reason to favor the MSSM over such R-parity violating scenarios. Since only B- or L-violating terms survive when this new symmetry is present the proton now remains stable in these models. Consequently, various low-energy phenomena then provide the only significant constraints [2] on the Yukawa couplings λ, λ' and λ''.

If R-parity is violated much of the conventional wisdom associated with the MSSM goes by the wayside, *e.g.*, the LSP (now not necessarily a neutralino!) is unstable and sparticles may now be produced singly. In particular, it is now possible that some sparticles can be produced as s-channel resonances, thus appearing as bumps in cross sections if kinematically accessible. In the case of the two sets of trilinear L-violating terms in W_R, which we consider below, an example of such a possibility is production of a $\tilde{\nu}_\tau$ via $d\bar{d}$ annihilation at the Tevatron or LHC (through λ' couplings). If this sneu-

trino decays to, *e.g.*, opposite sign leptons (through the λ couplings) then an event excess, clustered in mass, will be observed in the Drell-Yan channel similar to that expected for a Z'. Similarly, the corresponding process $u\bar{d} \to \tilde{\tau} \to \ell\nu$ may also occur through these couplings and mimics a W' signature. In addition to these leptonic final states, both $\tilde{\tau}$ and $\tilde{\nu}$ resonances may decay hadronically via the same vertices that produced them, hence leading to potentially observable peaks in the dijet invariant mass distribution. Thus resonant slepton production, first discussed in Ref. [3], clearly offers a unique way to explore the R-parity violating model parameter space. It is important to note that R-parity violation also allows for other SUSY particles, such as $\tilde{t}$ and/or $\tilde{b}$, to be exchanged in the non-resonant $t,u-$channels and contribute to Drell-Yan events. However, it can be easily shown that their influence on cross sections and various distributions will be quite small if the low energy constraints on the Yukawa couplings are satisfied [4].

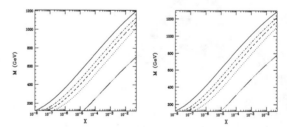

Figure 1: Discovery regions (lying below the curves) in the mass-coupling plane for R-parity violating resonances in the neutral (left) and charged (right) Drell-Yan channels at the Run II Tevatron. From top to bottom the curves correspond to integrated luminosities of 30, 10, 5 and 2 fb^{-1}. The estimated reach for Run I is given by the lowest curve. The parameter X is defined in the text.

The questions we address in this analysis are: (i) what are the mass and coupling reaches for slepton resonance searches at the Tevatron and LHC in the Drell-Yan and dijet channels and (*ii*) how can slepton resonances, once discovered, be distinguished from Z', W' production. Below we will mainly concern ourselves with the third generation sleptons but our analysis is easily extended to those of the first and second generation as well.

2 The Drell-Yan Channel

In the case of Drell-Yan production the search reach analysis is straightforward being nearly identical to that used for new gauge boson production, apart from acceptance issues, *i.e.*, we now have spin-0 and not spin-1 resonances. Since sleptons are expected to be narrow, the narrow width approximation is adequate and we can directly follow the analysis presented in Ref. [5]. In addition to the slepton mass itself, the only other parameter in this calculation is the product of the appropriate Yukawa couplings, λ', from the initial state $d\bar{u}$ or $d\bar{d}$ coupling vertex, and the slepton's leptonic branching fraction, B_ℓ. Calling this product $X = (\lambda')^2 B_\ell$, we can obtain the search reach as a function of X in the charged and neutral channels for both the Tevatron and LHC; these results are shown in Figures 1 and 2. Not only is it important to notice the very large mass reach of these colliders for sizeable values of $X \sim 10^{-3}$, but we should also observe the small X reach, $X \sim 10^{-(5-7)}$ and below, for relatively small slepton masses. Clearly these results show the rather wide opportunity available to discover slepton resonances over extended ranges of masses and couplings at these hadron colliders. Note that for fixed values of X the search reach is greater in the charged current channel due to the higher parton luminosities.

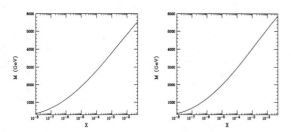

Figure 2: Same as the previous figure but now for the LHC with an integrated luminosity of 100 fb^{-1}.

Our next issue concerns identifying the resonance (or Jacobian peak) as a slepton instead of a new gauge boson. One immediate difference which would signal $\tilde{\nu}$ production would be the observation of the very unusual $e\mu$ final states which are allowed by the generational structure of the superpotential; such final states are not expected to

occur for a Z' and would be a truly remarkable signature for R-parity violation. Clearly if the $e\mu$ or SUSY decay modes of the slepton dominate there will be no identification problem. If the R-parity violating modes dominate it is best to look for universality violations, *e.g.*, if the resonance decays to only one of e^+e^- or $\mu^+\mu^-$ or if these two rates are substantially different. Most new gauge bosons which are kinematically accessibly are not anticipated to have substantially different couplings to the first two fermion generations. In the case of a $\tilde{\nu}$ versus a Z', it is well known that most Z' bosons have parity violating fermionic couplings which would lead to a forward-backward asymmetry, A_{FB}, in their leptonic decay distributions. The $\tilde{\nu}$, being spin-0, would always produce a null asymmetry. A_{FB} is more easily measured and requires less statistical power than does the reconstruction of the complete angular distribution. This is important since, whereas only 10 or so background free events would constitute a discovery many more, $\sim 100 - 200$ are required to determine the asymmetry. This would imply that the reach for performing this test is somewhat if not substantially less than the discovery reach. For example, the Tevatron may discovery a $\tilde{\nu}$ with a mass of 700 GeV for a certain value of X but only for masses below 500 GeV would there be enough statistics to extract A_{FB} for this same X value.

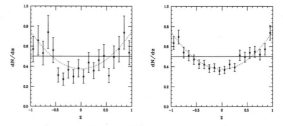

Figure 3: Comparison of the Monte Carlo generated normalized angular distribution for the leptons in Z' decay with that for a $\tilde{\nu}$ assuming a (left)400 event sample or a (right)2000 event sample; the errors are purely statistical. The Z' is assumed to have no forward-backward asymmetry due to its fermionic couplings. The $1 + z^2$ angular distribution is also shown. The effect of potential acceptance losses in the two outer bins has not been included.

A more complex and interesting situation arises when the Z' naturally has $A_{FB} = 0$ as in, *e.g.*, some E_6 models; [6] in this case the on-resonance asymmetry data alone is insufficient. If A_{FB} could be measured throughout the resonance region, it would be possible to deduce through detailed line-shape studies whether or not the new contribution interferes with the SM amplitude (something that does not occur in the case of $\tilde{\nu}$ production). Besides requiring substantial statistics, finite dilepton mass resolution, especially for the $\mu^+\mu^-$ final state, may disrupt this program.

Of course, with a plethora of statistics the complete angular distribution can be obtained as shown in Fig.3. Here we compare Monte Carlo generated data for a Z' with a zero forward-backward asymmetry with both the flat distribution and the $\sim 1 + z^2$ distribution hypotheses($z = \cos\theta$) and ignore complications due to possible acceptance losses arising from rapidity cuts in the forward and backward directions. Such a distribution has been measured by CDF both on the Z and above. [7] Both analyses would seem to indicate that of order ~ 1000 events are required to make a clean measurement, a sample approximately 100 times larger than that required for discovery. Although such measurements would be conclusive as to the identity of the resonance, the required statistics results in a significant loss in the mass range over which it can be performed. In our Tevatron example above where the search reach was 700 GeV we would find that the angular distribution could only be determined for masses below ~ 400 GeV assuming the same X value.

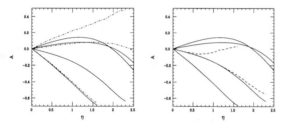

Figure 4: The lepton charge asymmetry in the charged current Drell-Yan production channel at the 2 TeV Tevatron for the SM (solid curves)and with 250(700) GeV $\tilde\tau$ exchange (the dash-dotted or dashed curves) assuming $\lambda, \lambda' = 0.15$ for purposes of demonstration. From top to bottom in the center of the figure, the SM curves correspond to M_T bins of 50-100, 100-200, 200-400 and > 400 GeV, respectively. Note that for M_T in the 50-100 GeV range there is no distinction between the SM result and that with a $\tilde\tau$.

The angular distribution approach cannot be used to separate between the $\tilde\tau$ and W' cases due to the missing energy in the event. However, there are two useful observables in this situation. First, one can examine the transverse mass (M_T) distribution associated with the new Jacobian peak region to see if interference with SM amplitudes is occurring. This is far more difficult than in the $\tilde\nu$ case again due to the missing energy and mass smearing. A second possibility is to examine the leptonic charge asymmetry, $A(\eta)$, for the electrons or muons in the final state as a function of their rapidity. We remind

the reader that $A(\eta)$ is defined as

$$A(\eta) = \frac{dN_+/d\eta - dN_-/d\eta}{dN_+/d\eta + dN_-/d\eta}, \qquad (2)$$

where $N_\pm$ are the number of positively/negatively charged electrons of a given rapidity, η. In the SM, the charge asymmetry is sensitive to the ratio of u-quark to d-quark parton densities and the $V - A$ production and decay of the W. [8] Since the coupling structure of the W has been well-measured, any deviations in this asymmetry within the M_T bin surrounding the W have been attributed to modifications in the parton distributions (PDF's). Here, we are more interested in events with larger M_T. Note that $A(-\eta) = -A(\eta)$ if CP is conserved(which we assume) so that we will only need to deal with $\eta \geq 0$ in the following discussion.

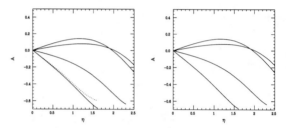

Figure 5: Same as the previous figure but now for 1000(1500) GeV $\tilde\tau$ exchange corresponding to the the dotted or dashed curves in the left(right) panel.

Consider the case for $\tilde\tau$ production at the Tevatron. Fig.4 shows the lepton charge asymmetry, within four M_T bins corresponding to $50 < M_T < 100$ GeV, $100 < M_T < 200$ GeV, $200 < M_T < 400$ GeV, and $400 < M_T < 1800$ GeV for the SM and how it is modified by the presence of a 250(700) GeV $\tilde\tau$ with, for purposes of demonstration, $\lambda, \lambda' = 0.15$. In particular we observe that the lepton charge asymmetry can be significantly altered for larger values of M_T in the bins associated with the new Jacobian peak. Note, however, that there is essentially no deviation in the asymmetry in the transverse mass bin associated with the W peak, $50 < M_T < 100$ GeV, so that this M_T region can still be used for determination of the PDFs. Fig.4 also shows that the presence of the $\tilde\tau$ tends to drive the asymmetry to smaller absolute values as perhaps might be expected due to the presence of a spin-0 resonance which provides a null 'raw' asymmetry. In Fig.5 we see that the asymmetry is still visible in the last M_T bin for the case of a 1 TeV $\tilde\tau$ but becomes essentially non-existent for these values of the Yukawa couplings when the mass is raised to 1.5 TeV.

Fig.6 shows the corresponding modifications in the leptonic charge asymmetry due to an 800 GeV W'

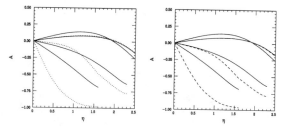

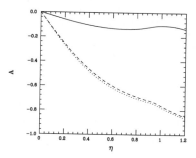

Figure 6: Same as the previous two figures but now for the case of a 800 GeV W' with purely left-handed(left panel) or purely right-handed(right panel) couplings.

with either purely left-handed(LH) or purely right-handed(RH) fermionic couplings. Note that the W' with purely RH couplings, unlike the LH W', does not interfere with the SM amplitude, similar to the case of $\tilde{\tau}$ production. The deviation in the asymmetry due to either type of W' is very different than that for a $\tilde{\tau}$. Here we see that the W' substantially increases the magnitude of the asymmetry for both coupling types and that RH and LH W' bosons are *themselves* potentially distinguishable by using the data in the M_T bin below but not containing the Jacobian peak.

The M_T bins we have taken in this analysis are rather broad. We might expect that if we compress the width of the M_T bin around the W' or $\tilde{\tau}$ Jacobian peak we will increase the purity of the resonant contribution and have an even better separation of the two possibilities, at the price of reduced statistics. (Of course as we narrow this bin we will no longer be able to distinguish LH from RH W' bosons since this information comes from SM–W' interference.) These expectations come to fruition in Fig.7 which shows a more direct comparison of the lepton charge asymmetries for a $\tilde{\tau}$ and W' of the same mass (800 GeV) and narrowing the width of the M_T bin surrounding the Jacobian peak to only 300 GeV. Note that the LH and RH W' cases are no longer separable. Clearly such measurements will allow the production of W' and $\tilde{\tau}$ to be distinguished.

It is interesting to note that lepton asymmetry deviations can be used to probe indirectly for the exchange of $\tilde{\tau}$ through R-parity violating couplings. To demonstrate this let us fix the $\tilde{\tau}$ width to mass ratio to be $\Gamma/m = 0.004$ and subdivide each of the four M_T bins discussed above into rapidity intervals of $\Delta\eta = 0.1$. For a given $\tilde{\tau}$ mass we can then ask down to what value of the product of the Yukawa couplings, $\lambda\lambda'$, will the asymmetry differ significantly from SM expectations. For a fixed mass and integrated luminosity we generate Monte Carlo data for various values of the Yukawas and then perform

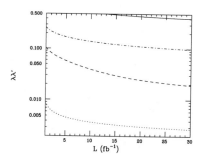

Figure 7: Direct comparison of the charge asymmetry induced by a 800 GeV $\tilde{\tau}$(solid) and a LH or RH W'(dot and dash) of the same mass at the 2 TeV Tevatron. For this comparison, a narrow bin in M_T was chosen: $600 < M_T < 900$ GeV. The Yukawa couplings are as in the earlier figures.

a χ^2 analysis to obtain the sensitivity. The results of this analysis are shown in Fig.8 and one can see that the search reaches obtained in this manner are rather modest.

Figure 8: Search reach for $\tilde{\tau}$ exchange as a function of the 2 TeV Tevatron integrated luminosity assuming $\Gamma/m = 0.004$ for masses of 1500, 1000, 750 and 250 GeV(from top to bottom).

3 The Dijet Channel

Since $d\bar{d}$ and/or $u\bar{d}$ annihilation are responsible for producing the slepton resonances, it is obvious that the resonance must also decay into these same fermion pairs. This means that $\tilde{\tau}$ or $\tilde{\nu}$ will decay to dijets and may appear as observable peaks above the conventional QCD backgrounds. This hope will very hard to fulfill at the LHC where QCD backgrounds are expected to be severe for searches for narrow resonances which are not strongly produced. At the Tevatron, one can be much more optimistic. In fact, searches for such narrow dijet resonances have already been performed [9] at the Tevatron by both CDF and D0 during Run I. Using their results and scaling by appropriate factors of beam energy and integrated luminosities we may estimate the probable search reaches for CDF and D0 from Run II. (These estimates conform to the expectations given in Ref. [10].) The cross sections themselves are immediately calculable in the narrow width approximation in terms of the product $Y = (\lambda')^2 B_{2j}$, where λ' is the familiar Yukawa coupling and B_{2j} is the dijet branching fraction. The results of these calculations are shown in Fig.9. Here we clearly see that for values of $Y \sim 0.001 - 0.01$ or greater, the Tevatron will have a substantial mass reach for slepton induced dijet mass bumps during Run II. Note that as in the case of Drell-Yan, larger cross sections for fixed Y occur in the CC channel than in the NC channel due to the larger parton luminosities. Unfortunately, if such a bump is observed it will not be straightforward to identify it as a slepton resonance.

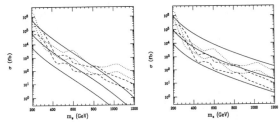

Figure 9: Cross sections for narrow dijet resonances(solid) at the 2 TeV Tevatron arising from $\tilde{\nu}$(left) or $\tilde{\tau}$(right) production in comparison to the anticipated search reaches of CDF(dots) and D0(dashes). The upper(lower) curve for each experiment assumes an integrated luminosity of 2(30) fb^{-1}. The three solid curves from top to bottom correspond to slepton resonance predictions for Y=0.1, 0.01 and 0.001, respectively, where Y is defined in the text.

4 Conclusion

As we have seen from the analysis above, resonant s-channel production of $\tilde{\tau}$ and/or $\tilde{\nu}$ with their subsequent decay into purely leptonic or dijet final states is observable over a wide range of parameters in hadronic collisions via R-parity violating couplings. We have obtained the corresponding search reaches in the slepton mass-R-parity violating coupling plane for both the Tevatron and LHC. If this signature is observed, we have demonstrated that the leptonic angular distributions and the lepton charge asymmetry can be successfully used to distinguish slepton resonances from those associated with new gauge bosons.

This process provides a clean and powerful probe of R-parity violating supersymmetric parameter space.

References

1. H. Dreiner, hep-ph/9707435, to be published in *Perspectives on Supersymmetry*, ed. by G.L. Kane. See also P. Roy, hep-ph/9712520.
2. V. Barger, G.F. Giudice and T. Han, *Phys. Rev.* D **40**, 2987 (1989); D. Choudhury and P. Roy, *Phys. Lett.* B **387**, 153 (1996); G. Bhattacharyya, hep-ph/9709395 .
3. R. Kalinowski *et al.*, *Phys. Lett.* B **406**, 314 (1997) and *Phys. Lett.* B **414**, 297 (1997); J.Erler, J.L. Feng and N. Polonsky, *Phys. Rev. Lett.* **78**, 3063 (1997); B.C. Allanach *et al.*, hep-ph/9708495; S.Bar-Shalom, G.Eilam and A. Soni, hep-ph/9804339 and hep-ph/9802251; J.L. Feng, J.F. Gunion and T. Han, hep-ph/9711414.
4. J.L Hewett and T.G. Rizzo, *Phys. Rev.* D **56**, 5709 (1997) and *Phys. Rev.* D **58**, 055005 (1998).
5. T.G. Rizzo, hep-ph/9609248, published in the *New Directions for High High Energy Physics, Proceedings of the 1996 DPF/DPB Summer Study on High Energy Physics*, eds. D.G Cassel, L.T. Gennari and R.H. Siemann, Snowmass, CO 1996.
6. J.L. Hewett and T.G. Rizzo, *Phys. Rep.* **183**, 193 (1989).
7. F. Abe *et al.*, CDF Collaboration, *Phys. Rev. Lett.* **77**, 2616 (1996).
8. F. Abe *et al.*, CDF Collaboration, *Phys. Rev. Lett.* **74**, 850 (1995) and hep-ex/9809001.
9. F. Abe *et al.*, CDF Collaboration, *Phys. Rev. Lett.* **77**, 5336 (1996), Erratum, *ibid.*, **78**, 4307 (1997); R. Harris, in *Proceedings of the 10th Topical Workshop on Proton-Antiproton Collider Physics*, Batavia, IL 1995, ed. R. Raja and J. Yoh; S. Abachi *et al.*, D0 Collaboration, in the *Proceedings of the 28th International Conference on High Energy Physics*, Warsaw, Poland, July 1996, FERMILAB-CONF-96-168-E and hep-ph/9807014.
10. *Future Electroweak Physics at the Fermilab Tevatron*, eds D. Amidei and R. Brock, Fermilab-Pub-96/082.

EXTRACTING CHARGINO/NEUTRALINO MASS PARAMETERS FROM PHYSICAL OBSERVABLES

G. MOULTAKA

Physique Mathématique et Théorique
UMR-CNRS,
Université Montpellier II, F34095 Montpellier Cedex 5, France
E-mail: moultaka@lpm.univ-montp2.fr

I report on two papers [1,2] where complementary strategies are proposed for the determination of the chargino/neutralino sector parameters, M_1, M_2, μ and $\tan\beta$, from the knowledge of some physical observables. This determination and the occurrence of possible ambiguities are studied as far as possible analytically within the context of the unconstrained MSSM, assuming however no CP-violation.

1 Introduction

The gauge bosons and Higgs bosons superpartners have every chance to play, in the minimal version of the supersymmetric standard model (MSSM), an important part in the first direct experimental evidence for supersymmetry, if the latter happens to be linearly realized in nature around the electroweak scale. This would go through the study of the direct production of the light states and their subsequent decays, eventually cascading down to leptons (or jets) and missing energy [3][4][5].

The chargino/neutralino sector is an overconstrained system in the sense that only a few basic parameters in the Lagrangian are needed to determine all the six physical masses and the mixing angles of the various states. The latter determine the couplings to gauge bosons, Higgs bosons and matter fermions, so that various phenomenological tests could be in principle envisaged in the process of experimental identification. Alternatively, one might hope that a partial experimental knowledge of this sector would be sufficient to allow a reasonably unequivocal reconstruction of the full set of parameters; at stakes, on one hand the determination of the magnitude of the fermion soft susy breaking parameters, on the other, the existence of a heavy neutral stable particle, of prime importance to the cold dark matter issue [6]. Furthermore, the sensitivity to $\tan\beta$, the ratio of the two vacuum expectation values of the Higgs fields, and to the supersymmetric parameter μ, brings in a further correlation with the other sectors of the MSSM.

Hereafter we describe two strategies: the first deals with the extraction of M_2, μ and $\tan\beta$ form the study of the lightest chargino pair production and decay in e^+e^- collisions [1], the second with the extraction of M_1, M_2 and μ form the knowledge of any three ino masses and $\tan\beta$ [2]. We start by stating the common features to these complementary approaches as well as their specific assumptions. We then highlight the main ingredients of each of them and illustrate some of their results. Finally we show in what sense they eventually complement one another. [Obviously, the reader is referred to [1] and [2] for more details and references. Still, we add some comments at various places of the ongoing presentation, which differ slightly from, and hopefully complete, the latter references.]

The reconstruction of the basic parameters of the theory involves generically two steps which can be sketched as follows:

$$
\begin{array}{c}
\text{Experimental Observables} \\
\updownarrow \ (I) \\
\text{Physical Parameters} \\
\updownarrow \ (II) \\
\text{Lagrangian parameters}
\end{array}
\tag{1}
$$

Each of these steps can suffer from equivocal reconstructions due to partial experimental knowledge or to theoretical ambiguities. In the present report we concentrate on the theoretical aspects of both steps.

2 CDDKZ and KM common features

The ino sector is considered in both [1] (referred to as CDDKZ) and [2] (KM) with the following assumptions:

- No reference to model-dependent assumptions about physics at energies much higher than the electroweak scale, like the GUT scale, and their possible implication on the parameters of this sector. [Thus the study is mainly carried out in the unconstrained MSSM, but any model-assumptions can be easily overlaid.]

- R-parity conservation;

- CP-conservation in the ino sector; This assumption is here only for practical reasons and should be eventually removed in future studies in order to cope with the possibility to deal with (complex) phases [7];

- CDDKZ and KM choose $M_2 > 0$. This is of course a mere convention due to the partial phase freedom through redefinition of fields, the only physical signs being the relative ones among M_1, M_2 and μ as one can easily see from the relevant terms in the Lagrangian. (also $\tan\beta$ is taken positive and the μ term convention is that of ref.[8].)

Let us now recall briefly the basic ingredients of the ino mass matrices. The physical charginos (resp. neutralinos) are mixtures of charged (resp. neutral) higgsino and gaugino components. The chargino mass matrix reads:

$$\mathcal{M}_C = \begin{pmatrix} M_2 & \sqrt{2}m_W\sin\beta \\ \sqrt{2}m_W\cos\beta & \mu \end{pmatrix} \qquad (2)$$

It has a supersymmetric contribution coming from the μ term in the superpotential, the higgsino component, a contribution from the soft susy breaking wino mass term, and off-diagonal terms due to the electroweak symmetry breaking. Since $\mathcal{M}_C$ is not symmetric one needs two independent unitary matrices for the diagonalization. This is but the reflection of the fact that there are two independent mixings involving separately the two higgsino $SU(2)_L$ doublets. The eigenvalues are most easily obtained from the diagonalization of $\mathcal{M}_C{}^\dagger\mathcal{M}_C$ giving the squares of the chargino masses:

$$M_{\chi_{1,2}^\pm}^2 = \tfrac{1}{2}[M_2^2 + \mu^2 + 2m_W^2 \\ \mp\sqrt{(M_2^2 + \mu^2 + 2m_W^2)^2 - 4(M_2\mu - m_W^2\sin 2\beta)^2}] \qquad (3)$$

On the other hand, the angles ϕ_L, ϕ_R defining the two independent left- and right- chiral mixings among the winos and higgsinos in the four component Dirac representation are given by

$$\cos 2\phi_L = \frac{M_2^2 - \mu^2 - 2m_W^2\cos 2\beta}{2(M_{\chi_1}^2 - m_W^2) - M_2^2 - \mu^2}$$

$$\sin 2\phi_L = \frac{2\sqrt{2}m_W(M_2\cos\beta + \mu\sin\beta)}{2(M_{\chi_1}^2 - m_W^2) - M_2^2 - \mu^2}$$

$$\qquad (4)$$

$$\cos 2\phi_R = \frac{M_2^2 - \mu^2 + 2m_W^2\cos 2\beta}{2(M_{\chi_1}^2 - m_W^2) - M_2^2 - \mu^2}$$

$$\sin 2\phi_R = \frac{2\sqrt{2}m_W(M_2\sin\beta + \mu\cos\beta)}{2(M_{\chi_1}^2 - m_W^2) - M_2^2 - \mu^2}$$

where M_{χ_1} is the lightest chargino mass given by eq.(3). This form of the mixing angles is such that the

eigenvalues of $\mathcal{M}_C$ are always positive definite.

The neutralino mass matrix corresponds to bilinear terms in the photino, zino and neutral higgsino two-component fields. It receives contributions from the μ term, the soft mass terms of the gaugino $SU(2)_L$ triplet (M_2) and singlet (M_1), while the mixing among states is triggered by the electroweak symmetry breaking:

$$\mathcal{M}_N =$$

$$\begin{pmatrix} M_1 & 0 & -m_Z s_w\cos\beta & m_Z s_w\sin\beta \\ 0 & M_2 & m_Z c_w\cos\beta & -m_Z c_w\sin\beta \\ -m_Z s_w\cos\beta & m_Z c_w\cos\beta & 0 & -\mu \\ m_Z s_w\sin\beta & -m_Z c_w\sin\beta & -\mu & 0 \end{pmatrix}$$

$$\qquad (5)$$

In contrast with $\mathcal{M}_C$, $\mathcal{M}_N$ is symmetric so that it can be diagonalized with one unitary matrix. On the other hand the eigenmasses are not positive definite[a]. Finally we note that in general the diagonalization of $\mathcal{M}_N$ cannot be achieved through a similarity transformation, unless all three parameters M_1, M_2 and μ are real. This will be a key point in the algorithm we present for the reconstruction of the parameters in the neutralino sector.

3 Specific features

3.1 CDDKZ

The lightest chargino χ_1^+ can be produced in pairs in e^+e^- collisions, at LEPII[4] or NLC[5] energies, through γ and Z s-channel exchange as well as sneutrino t-channel exchange. The production cross section will thus depend on the chargino mass m_{χ_1}, the sneutrino mass $m_{\tilde{\nu}}$ and the mixing angles, eq.(4), which determine the couplings of the chargino states to the Z and the sneutrino. The unpolarized total cross section is illustrated in fig.1 with three representative cases of higgsino, gaugino or mixed content of the lightest chargino mass. The sharp rise near threshold should allow a precise determination of the chargino mass. Also the sensitivity to the sneutrino mass with the typical destructive interference in the gaugino and mixed cases necessitates the knowledge of this parameter.

Subsequently the chargino will decay directly to a pair of matter fermions (leptons or quarks) and the (stable) lightest neutralino, through the exchange of a W boson (charged Higgs exchange is suppressed for light fermions) or scalar partners of leptons or quarks. Of course the decay matrix elements will depend on further

[a]For more details about the ino sector see for instance [8],[9] and references therein.

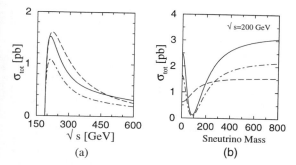

(a)   (b)

Figure 1: Total cross section for the charginos pair production for a representative set of M_2, μ, solid line gaugino case, dashed line higgsino case, dot-dashed line mixed case. In (a) $m_{\tilde{\nu}} = 200 GeV$. (taken from ref.[1])

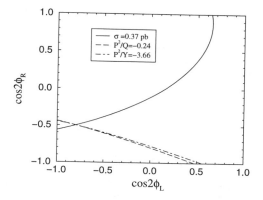

Figure 2: Contours for the "measured values" of the total cross section (solid line), $\frac{\mathcal{P}^2}{\mathcal{Q}}$, and $\frac{\mathcal{P}^2}{\mathcal{Y}}$ (dot-dashed line) for $m_{\chi_1^\pm} = 95 GeV$ $[m_{\tilde{\nu}} = 250 GeV]$. (taken from ref.[1])

parameters like the susy scalar masses and couplings to the neutralino. However, CDDKZ propose that, looking at the total production cross section and some polarization components and spin-spin correlations of the final state charginos, one can define measurable combinations for which the details of the chargino decay products cancel out. This allows to isolate to a large extent the chargino system from the neutralino system and thus extract the mixing angles and chargino mass from these observables (step (I) in eq.(1)). In fact, for step (I) to work completely for the chargino system, one needs to know, besides the sneutrino mass, *also the lightest neutralino mass*, as will become clear later on. Once the chargino mass $m_{\chi_1^+}$ and $\cos 2\phi_L, \cos 2\phi_R$ are known one can determine M_2, μ and $\tan\beta$ up to possible ambiguities, [step (II) of eq.(1).]

Before going further let us first describe in some detail the basic ingredients of step (I). The presence of invisible neutralinos, in the final state of the process $e^+e^- \rightarrow \chi_1^+\chi_1^- \rightarrow \chi_1^0\chi_1^0(f_1\bar{f}_2)(\bar{f}_3 f_4)$, makes it impossible to measure directly the chargino production angle in the laboratory frame. From now on we will thus concentrate on observables where this angle is integrated out. Integrating also over the invariant masses of the fermionic systems $(f_1\bar{f}_2)$ and $(\bar{f}_3 f_4)$ one can write the differential cross section in the following form:

$$\frac{d^4\sigma(e^+e^- \rightarrow \chi_1^+\chi_1^- \rightarrow \chi_1^0\chi_1^0(f_1\bar{f}_2)(\bar{f}_3 f_4))}{d\cos\theta^* d\phi^* d\cos\bar{\theta}^* d\bar{\phi}^*} = \frac{\alpha^2\beta}{124\pi s} Br_{\chi^- \rightarrow \chi^0 f_1\bar{f}_2} Br_{\chi^+ \rightarrow \chi^0 \bar{f}_3 f_4} \Sigma(\theta^*, \phi^*, \bar{\theta}^*, \bar{\phi}^*) \quad (6)$$

where α is the fine structure constant, β the velocity of the chargino in the c.m. frame, θ^* ($\bar{\theta}^*$) denotes the polar angle of the $f_1\bar{f}_2$ ($\bar{f}_3 f_4$) system in the χ_1^- (χ_1^+) rest frame with respect to the charginos flight direction

in the laboratory frame, and ϕ^* ($\bar{\phi}^*$) the corresponding azimuthal angle with respect to a canonical production plane. $\Sigma(\theta^*, \phi^*, \bar{\theta}^*, \bar{\phi}^*)$ is made out of combinations of helicity amplitudes which lead to an unpolarized term plus fifteen other contributions from polarization components and spin-spin correlations. We reproduce here only those components which are relevant to our discussion.

$$\Sigma = \Sigma_{unpol} + (\kappa - \bar{\kappa})\cos\theta^*\mathcal{P} + \cos\theta^*\cos\bar{\theta}^*\kappa\bar{\kappa}\mathcal{Q}$$
$$+ \sin\theta^*\sin\bar{\theta}^*\cos(\phi^* + \bar{\phi}^*)\kappa\bar{\kappa}\mathcal{Y} + \dots \quad (7)$$

Actually, among the sixteen terms which contribute to Σ only ten survive because of CP-invariance (when violation of CP from the Z-boson width or radiative corrections is neglected). Of these ten, three are redundant being CP eigenstates. Of the remaining seven independent components, only those which can be extracted from experimentally measurable angular distributions are explicitly written in eq.(7). This means that the others will be integrated out through appropriate projections.

In eq.(7) Σ_{unpol} corresponds to the unpolarized cross section for the chargino pair production and is given in terms of helicity amplitudes by

$$\Sigma_{unpol} = \frac{1}{4}\int d\cos\Theta \sum_{\sigma=\pm}[|\langle\sigma; ++\rangle|^2 + |\langle\sigma; +-\rangle|^2 + |\langle\sigma; -+\rangle|^2 + |\langle\sigma; --\rangle|^2] \quad (8)$$

$\mathcal{P}$ is a polarization component coming separately from the χ^- (or χ^+) system

$$\mathcal{P} = \tfrac{1}{4} \int d\cos\Theta \sum_{\sigma=\pm} [|\langle\sigma;++\rangle|^2 + |\langle\sigma;+-\rangle|^2 \\ -|\langle\sigma;-+\rangle|^2 - |\langle\sigma;--\rangle|^2] \quad (9)$$

while $\mathcal{Q}$ and $\mathcal{Y}$ describe the spin correlations between the two chargino systems and have the following structure

$$\mathcal{Q} = \tfrac{1}{4} \int d\cos\Theta \sum_{\sigma=\pm} [|\langle\sigma;++\rangle|^2 - |\langle\sigma;+-\rangle|^2 \\ -|\langle\sigma;-+\rangle|^2 + |\langle\sigma;--\rangle|^2] \quad (10)$$

$$\mathcal{Y} = \tfrac{1}{2} \int d\cos\Theta \sum_{\sigma=\pm} \mathrm{Re}\{\langle\sigma;--\rangle\langle\sigma;++\rangle^*\}$$

where σ is the initial state electron helicity. The strategy of CDDKZ is based on the following two observations:

i) The three angular contributions, $\cos\theta^*$, $\cos\bar{\theta}^*$ and $\sin\theta^* \sin\bar{\theta}^* \cos(\phi^*+\bar{\phi}^*)$ are fully determined by the measurable parameters $E, |\vec{P}|$ (the energy and momentum of each of the decay systems $f_i\bar{f}_j$ in the laboratory frame), and the chargino mass $m_{\chi_1^+}$;

ii) The three quantities $\Sigma_{unpol}, \mathcal{P}^2/\mathcal{Q}$ and $\mathcal{P}^2/\mathcal{Y}$ lead to κ free observables, where κ (and $\bar{\kappa} = -\kappa$)) measures the asymmetry between left- and right-chirality form factors in the decay products of the chargino;

Here the kinematic configuration is similar to that of a τ lepton pair production with successive decays in light leptons or quarks plus missing energy. However in the present context the invisible particle has a non negligible mass whose knowledge is necessary to relate the energy of the $f_i\bar{f}_j$ system in the chargino rest frame to that in the laboratory frame. Thus the neutralino mass is actually necessary in the reconstruction of the angular contributions in i). The crucial point in ii) is that the dependence on the final state decay fermions through the asymmetry in the left- and right- chiral structure cancels out. Thus $\Sigma_{unpol}, \mathcal{P}^2/\mathcal{Q}$ and $\mathcal{P}^2/\mathcal{Y}$ allow to study the chargino sector independently of the details of the decay products. In the same time, their extraction from the experimental data, via convolution with appropriate moments, requires the measurement of the energies and momenta of the two $f_i\bar{f}_j$ systems, the chargino mass (ex. via threshold effects, see fig.1), as well as the neutralino mass (ex. from the energy distribution of the final particles). Once extracted, one can combine $\Sigma_{unpol}, \mathcal{P}^2/\mathcal{Q}$ and $\mathcal{P}^2/\mathcal{Y}$ which depend on the c.m. energy $\sqrt{s}$, the sneutrino mass $m_{\bar{\nu}}$, and $\cos 2\phi_L, \cos 2\phi_R$ to determine the latter cosines. An illustration is given in fig. 2 of a unique consistent solution corresponding to the intersection point of the

contour plots at $(\cos 2\phi_L = -0.8, \cos 2\phi_R = -0.5)$. The requirement that the three curves should meet in one point offers clearly a very stringent consistency check of the model. On the other hand, while Σ_{unpol} is a quadratic polynomial in $\cos 2\phi_L$, $\cos 2\phi_R$, the two other observables are quartic in these variables. A potential ambiguity in the determination of $(\cos 2\phi_L, \cos 2\phi_R)$ will be, however, very unlikely, especially if $m_{\bar{\nu}}$ is fixed independently and the c.m. energy varied. We do not dwell here on further aspects of step (I) which can be found in [1].

We now go to step (II) of eq.(1) and describe briefly how to determine M_2, μ and $\tan\beta$. Starting from eq.(4) and $m_{\chi_1^+}$ in eq.(3), CDDKZ give closed expressions for $M_2, \mu, \tan\beta$ in terms of the quantities $p = \cot(\phi_R - \phi_L), q = \cot(\phi_R + \phi_L)$. They considered all possible cases and concluded to the existence of at most a two-fold ambiguity in the determination of the Lagrangian parameters, traceable to a sign ambiguity in $\sin 2\phi_{L,R}$, (see [1] for details). Here we only sketch an equivalent discussion which shows that, when it occurs, this two-fold ambiguity is *always associated with opposite μ sign solutions*. This can be most easily seen as follows: from $\cos 2\phi_L, \cos 2\phi_R$ in eq.(4) one easily determines M_2 uniquely (remember that M_2 is positive in our convention) and μ with a global sign ambiguity, as functions of $m_{\chi_1^+}, \tan\beta, \cos 2\phi_L$ and $\cos 2\phi_R$. Plugging those functions in the $m_{\chi_1^+}$ part of eq.(3) one gets, thanks to some cancellations, a simple quadratic equation in $\tan\beta$. The two solutions encompass automatically the sign of $\sin 2\phi_L \sin 2\phi_R$. Furthermore, each of them is consistent only with (at most) one μ sign reproducing the correct $m_{\chi_1^+}$, since eq.(3) is not invariant under $\mu \to -\mu$ for a given $M_2, \tan\beta$. As a numerical illustration, taking the input of fig.2, $\sigma_{tot} = 0.37$pb, $\mathcal{P}^2/\mathcal{Q} = -0.24, \mathcal{P}^2/\mathcal{Y} = -3.66$, CDDKZ find the following two-fold solution

$$[\tan\beta; M_2, \mu] = \begin{cases} (A) & [1.06; \ 83\text{GeV}, \ -59\text{GeV}] \\ (B) & [3.33; \ 248\text{GeV}, \ 123\text{GeV}] \end{cases} \quad (11)$$

We see that the two-fold ambiguity comes with a sign change for μ in accord with the general pattern just described. To eliminate this discrete ambiguity one would clearly need an independent information about any of the three parameters. Finally, the reconstruction obviously depends on the quality of the experimental accuracy with which the needed observables can be determined. This requires among other things:

- running at different c.m. energies: at threshold for a good determination of $m_{\chi_1^+}$, away from threshold to increase the sensitivity to chargino polarization;

- a good reconstruction of the final state fermion systems for a good determination of the neutralino mass;

- identification of the chargino electric charge, necessary for the extraction of $\mathcal{P}$;

- an independent knowledge of the sneutrino mass, to avoid a three parameter fit to the observables;

3.2 KM

In this section we describe another strategy for extracting the Lagrangian parameters[2]. It consists in assuming that only ino masses are known. Among other things, this strategy will be complementary to the one described in the previous section in the sense that it provides (within the CDDKZ strategy) an algorithm for the determination of M_1, the only parameter which was not reconstructed in ref.[1]. KM concerns mainly step (II) of eq.(1). The emphasis is put on the extent to which the reconstruction can be made through a controllable analytical procedure including all possible ambiguities, if three ino masses and $\tan\beta$ were known[b]. This is particularly relevant for the neutralino sector where the analytical reconstruction is far from trivial.

The next aim in [2] is to provide a numerical code which uses as much of the analytical solutions as possible and allows a direct reconstruction of M_1, M_2 and μ from the physical ino masses. We do not address here the more realistic issues when only mass differences are measured [3], however it is clear that the study provides a useful building block even in this case, and practically allows to avoid parameter scanning numerical procedures as well as model-dependent assumptions. KM distinguish two cases:

S_1: The two charginos and one neutralino masses are input;

S_2: One chargino and two neutralino masses are input;

Although S_1 is phenomenologically less compelling than S_2 as far as the generic pattern of low lying states is concerned, it turns out that it leads to a full analytical reconstruction. In contrast, S_2 needs partly a purely numerical algorithm which is, however, minimized through the use of the S_1 solutions. The bottom line is that the resulting algorithms are very fast, the first being fully analytical and the second needing seldom more than a few iterations to reach numerical convergence (see [2] for more details).
Let us now describe briefly the solutions for S_1.

---

[b]The fact that $\tan\beta$ needs to be an input is actually a marginal point here, as one can assume that this parameter has been determined from elsewhere, like for instance in [1] or from the study of yet another sector of the MSSM.

Chargino sector:
Starting from eq.(3) one can determine analytically μ^2 and M_2 in terms of $M_{\chi_1^+}, M_{\chi_2^+}, \tan\beta$ and m_W. Without further information in the chargino sector, μ and M_2 will be determined, but up to a $|\mu| \leftrightarrow M_2$ ambiguity (that is, one cannot determine uniquely at this level the Higgsino and Gaugino content of the charginos). On the other hand the global sign ambiguity in μ, due to the fact that only μ^2 is known, is actually lifted by the relation

$$M_2\,\mu = m_W^2\sin 2\beta \pm M_{\chi_1^+} M_{\chi_2^+} \qquad (12)$$

since M_2 is positive by definition. Nonetheless there remains a two-fold ambiguity coming from the relative $\pm$ sign in eq.(12). On the other hand, some constraints will come from the requirement of real-valuedness of M_2 and μ. All these aspects are analytically delineated in [2] in terms of domains of $\tan\beta$ and the sum and difference of the input chargino masses.

Neutralino sector:
Let us now assume that $M_2, \mu, \tan\beta$ and one neutralino mass have been determined, and address the question of reconstructing M_1 and thus the three remaining neutralino masses. It should be clear that the answer to this question is not straightforward independently of whether it can be phrased analytically or not. Indeed, with all parameters but M_1 fixed in eq.(5), and the knowledge of the mass of just one neutralino state (say the lightest), it could well be that multiple branch solutions exist which would be lifted only through extra information about the couplings in this sector. It turns out, however, not to be the case (at least when phases are ignored): there is basically a unique solution, apart from the fact that one should allow for negative and positive values for the input neutralino mass since $\mathcal{M_N}$ can have negative eigenvalues. (This sign liberty is actually the only ambiguity which can be eventually fixed through the study of the couplings and will not be discussed further here.)
The trick is to write down the four independent combinations of the entries of $\mathcal{M_N}$ which are invariant under similarity transformations, and thus relate them simply to the four eigenvalues of $\mathcal{M_N}$. One can then express the correlations between the eigenvalues and the basic parameters in the following form:

$$M_1 = -\frac{P_{ij}^2 + P_{ij}(\mu^2 + m_Z^2 + M_2 S_{ij} - S_{ij}^2) + \mu m_Z^2 M_2 s_w^2 \sin 2\beta}{P_{ij}(S_{ij} - M_2) + \mu(c_w^2 m_Z^2 \sin 2\beta - \mu M_2)} \qquad (13)$$

$$M_2 =$$

$$\frac{S_{ij} P_{ij}^2 + P_{ij} m_Z^2 \mu \sin 2\beta - (P_{ij}^2 + (\mu^2 + m_w^2)P_{ij} + S_{ij} m_Z^2 \mu \sin 2\beta)M_1}{P_{ij}^2 + P_{ij}(\mu^2 + s_w^2 m_Z^2) + \mu S_{ij}(s_w^2 m_Z^2 \sin 2\beta - \mu M_1)}$$

$$\qquad (14)$$

1707

where

$$S_{ij} \equiv \tilde{M}_{N_i} + \tilde{M}_{N_j}$$
$$P_{ij} \equiv \tilde{M}_{N_i} \tilde{M}_{N_j}$$

and $i \neq j$ index any neutralino mass parameter. (The tilde denotes the fact that the mass parameters can be negative valued) The nice thing about the above equations is that if any of the neutralino masses is taken as input (say $\tilde{M}_{N_2}$), then the other three are determined analytically through a simple cubic equation. A unique value for M_1 is then determined from eq.(13) after plugging any of these solutions. Eqs.(13, 14) express in a specially convenient way the various correlations among the four eigenvalues and the basic parameters. It is also noteworthy that the input set $(M_2, \mu, \tan\beta)$ plus one neutralino mass is optimal for a fully analytical determination. In particular this is precisely the input set required in ref.[1]. We illustrate here the complementarity of the two approaches by taking the two sets of numbers (A) and (B) in eq.(11) and a lightest neutralino $M_{N_1} = 30GeV$, to reconstruct M_1 and the remaining neutralino masses from eqs.(13, 14):

$$[\; M_1, \qquad \tilde{M}_{N_2}, \qquad \tilde{M}_{N_3}, \qquad \tilde{M}_{N_4}]$$

$$(A) \begin{cases} (+);\; [30GeV,\; 59GeV,\; -107GeV,\; 122GeV] \\ \\ (-);\; [-52GeV,\; 58GeV,\; -119GeV,\; 120GeV] \end{cases}$$

$$(B) \begin{cases} (+);\; [46GeV,\; 110GeV,\; -130GeV,\; 284GeV] \\ \\ (-);\; [-25GeV,\; 101GeV,\; -132GeV,\; 284GeV] \end{cases}$$

(15)

Here ($\pm$) refer to the two possible signs of the $30GeV$ lightest eigenmass input. The effect of this sign tends to be more important for M_1 than for the neutralino masses. Of course a minus sign should be accompanied with the appropriate sign change in the Feynman rules involving neutralinos (see [8]). A further study of the left and right form factors in the chargino decay system could then lift partially the four-fold ambiguity in eq.(15). Lifting the remaining two-fold ambiguity will still necessitate further measurements from the ino sector.

Back to the S_1 strategy, we give in fig.3 an illustration of the sensitivity to a chargino mass, fixing the other two masses of chargino and neutralino. The behavior of the reconstructed (M_1, M_2, μ) turns out to be fairly simple, up to the two-fold ambiguity induced by μ in the chargino sector. A simple behavior shows as well for the

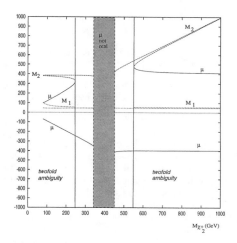

Figure 3: μ, M_1 and M_2 (with the "higgsino-like" convention $|\mu| \leq M_2$) as functions of $M_{\chi_2^+}$ for fixed $M_{\chi_1^+}$ ($= 400$ GeV); M_{N_2} ($= 50$ GeV), and $\tan\beta$ ($= 2$). (taken from ref.[2])

remaining three neutralino masses (see [2]), when the input neutralino mass is varied. This behavior which is fully controlled analytically can be used to discuss the generic gross features of the spectrum even when one deviates from the present input strategy. For instance the sensitivity to $\tan\beta$ turns out to be rather mild, and the effect of the sign change in the input neutralino mass tends to be negligible apart from well localized regions (see ref.[2] for further illustrations, including a reconstruction of the parameters at the GUT scale).

In fig.4 we illustrate the S_2 strategy. The input set in this case requires a partial numerical algorithm since the output becomes controlled by high degree polynomials. However using eqs.(13, 14) in conjunction with the chargino sector relations allows an optimized iterative algorithm. Fig.4 shows a rather intricate behavior of (M_1, M_2, μ) when one chargino and two neutralino masses are taken as input, a reflection of the above mentioned high degree polynomials, which nevertheless boils down (at least in our numerical trials) to at most a four-fold ambiguity. The regions of many-fold ambiguities or no ambiguity at all are separated by domains where the output parameters become complex valued (the shaded areas). Furthermore the singular behavior in some small regions is generically traced back to zeroing some parameters (see ref.[2] for more details).

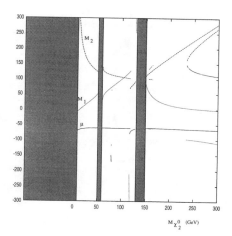

Figure 4: μ, M_2 and M_1 as function of $M_{\chi_2^0}^0$ for fixed M_{N_3} ($= -100\text{GeV}$), $M_{\chi_1^+}$ (= 80 GeV) and $\tan\beta(= 2)$. (taken from ref.[2])

4 Final comments

In this talk we presented two possible theoretical strategies for the extraction of the ino sector parameters from physical observables. The first relied on the study of the production and decay of the lightest chargino in e^+e^- collisions, the second on the knowledge of some ino masses. We also illustrated a full reconstruction of the ino sector when the two complementary approaches are brought together. Generally speaking, these approaches provide with efficient tools for the study of the ino sector. In the same time, they suggest the need in some cases for further experimental information due to the occurrence of possible discrete ambiguities in the reconstruction. Furthermore, although we only considered real-valued parameters, some of the material presented here goes through unaltered if phases are allowed. This is the case in CDDKZ for the chargino sector, even though extra information will still be needed to determine those phases. The inclusion of phases is less obvious in KM, especially in the neutralino sector, and deserves a separate study by itself. One should, however, keep in mind that the above strategies can give indirect information about the need for phases, whenever the experimental data place the parameters in the forbidden regions delineated in KM. In any case, a by-product of the analytical study would have been the construction of fast and flexible algorithms which can be used in various ways when reconstructing the ino parameters.

Finally, it should be stressed that the strategies we

presented here are just at the theoretical level. Obviously a more realistic examination of the experimental extraction of observables and related errors is still needed to assess their degree of efficiency. In addition, these strategies should eventually be placed in a wider context involving the other sectors of the MSSM, taking into account plausible discovery scenarios of the susy partners. The inter-correlations between these sectors, endemic to supersymmetry, will then hopefully allow a unique determination of all the parameters of the model.

Acknowledgements

I would like to thank Jean-Loïc Kneur and Peter Zerwas for valuable discussions on their contributed papers. Thanks go also to Seongyoul Choi and Abdelhak Djouadi for providing me with useful material for the talk.

References

1. S.Y. Choi, A. Djouadi, H. Dreiner, J. Kalinowski and P.M. Zerwas, hep-ph/9806279, (to appear in Eur.Phys.J. C)
2. J.L. Kneur and G. Moultaka, hep-ph/9807336, (to appear in Phys.Rev. D)
3. I. Hinchliffe et al., Phys. Rev. D55 (1997) 5520; D. Denegri, W. Majerotto and L. Rurua, CMS NOTE 1997/094, HEPHY PUB 678/97, hep-ph/9711357; CMS Collaboration (S. Abdullin et al.), CMS-NOTE-1998-006, hep-ph/9806366;
4. *Physics at LEP II*, Report No. CERN-96-01, eds. G. Altarelli, T. Sjöstrand, and F. Zwirner;
5. E. Accomando *et al.* Physics Reports **299** (1998) 1;
6. see for instance, G. Jungman, M. Kamionkowski and K. Griest, Phys.Rep. 267 (1996) 195;
7. M. Brhlik and G.L. Kane, hep-ph/9803391; T. Ibrahim and P. Nath, hep-ph/9807501; S.Y. Choi, J.S. Shim, H.S. Song, W.Y. Song hep-ph/9808227;
8. J.F. Gunion and H.E. Haber Nucl. Phys. B272 (1986) and Erratum hep-ph/9301205;
9. H.E. Haber and G.L. Kane Phys.Rep. 117 (1985) 75;

RADIATIVE CORRECTIONS IN THE MSSM BEYOND ONE-LOOP: PRECISION OBSERVABLES AND NEUTRAL HIGGS MASSES

W. HOLLIK

Theoretical Physics Division, CERN, CH-1211 Geneva 23, Switzerland
and
Institut für Theoretische Physik, Universität Karlsruhe, D-76128 Karlsruhe, Germany
E-mail: Wolfgang.Hollik@physik.uni-karlsruhe.de

Recent results on two-loop QCD corrections to electroweak precision observables and to the neutral Higgs boson masses in the Minimal Supersymmetric Standard Model are presented. Main emphasis is on the QCD corrections to electroweak precision observables via squark contributions to the ρ-parameter, and on the Feynman-diagrammatic calculation of the masses of the MSSM Higgs bosons h^0, H^0 yielding the currently most precise prediction of the lightest Higgs mass m_h within the Feynman-graph approach.

1 Precision observables

Supersymmetry (SUSY) predicts the existence of scalar partners to each standard model (SM) fermion, and spin–1/2 partners to the gauge and Higgs bosons. So far, the direct search of SUSY particles at colliders has not been successful, and under some assumptions one can only set lower bounds on their masses [1].

An alternative way to probe SUSY is to search for the virtual effects of the additional particles. one can use the high-precison measurements of electroweak observables to search for the quantum effects of the SUSY particles. In the Minimal Supersymmetric Standard Model (MSSM) one finds two types of potentially large effects: The first possibility is that charginos and scalar top quarks are light enough to affect the decay width of the Z boson into b–quarks [2]; however, for masses beyond the LEP2 or Tevatron reach, these effects become too small to be observable [3].

The second possibility is the contribution of the top and bottom squark loops to the electroweak gauge–boson self–energies [4]: if there is a large splitting between the masses of these particles, the contribution will grow with the mass of the heaviest scalar quark and can be sizable. This is similar to the SM case, where the top/bottom weak isodoublet generates a quantum correction that grows as m_t^2. This contribution enters the electroweak observables via the ρ parameter [5], which influences the normalization of the NC Z couplings, the effective electroweak mixing angle at the Z–boson resonance and the $W - Z$ mass correlation.

The ρ parameter, in terms of the transverse parts of the $W-$ and $Z-$boson self–energies at zero momentum–transfer, is given by

$$\rho = \frac{1}{1 - \Delta\rho} \; ; \; \Delta\rho = \frac{\Pi_{ZZ}(0)}{M_Z^2} - \frac{\Pi_{WW}(0)}{M_W^2} \; . \quad (1)$$

In the SM, the contribution of a fermion isodoublet (u, d) to $\Delta\rho$ reads at one–loop order

$$\Delta\rho_0^{\mathrm{SM}} = \frac{N_c G_F}{8\sqrt{2}\pi^2} F_0\left(m_u^2, m_d^2\right) \; , \quad (2)$$

with the color factor N_c and the function F_0 given by

$$F_0(x, y) = x + y - \frac{2xy}{x - y} \log\frac{x}{y} \; . \quad (3)$$

The only relevant SM contribution is due to the top/bottom weak isodoublet. Because $m_t \gg m_b$, one obtains $\Delta\rho_0^{\mathrm{SM}} = 3G_F m_t^2/(8\sqrt{2}\pi^2)$. The two-loop QCD corrections corrections have been calculated long ago [6], yielding: $\Delta\rho_1^{\mathrm{SM}} = -\Delta\rho_0^{\mathrm{SM}} \cdot \frac{2}{3}\frac{\alpha_s}{\pi}(1 + \pi^2/3)$. Meanwhile the three-loop term is also avialable [7]. The QCD corrections decrease the one–loop result by approximately 10%, corresponding to a shift of $\sim +10$ GeV in the indirect determination of the top mass.

In order to treat the SUSY loop contributions to the electroweak observables at the same level of accuracy as the standard contribution, higher order corrections should be incorporated. In particular the QCD corrections, which because of the large value of the strong coupling constant can be rather important, have to be included. They have been derived in ref [8], and will be briefly discussed here.

In SUSY theories, the scalar partners of each SM quark will induce additional contributions. The current eigenstates, $\tilde{q}_L$ and $\tilde{q}_R$, mix to give the mass eigenstates. The mixing angle is proportional to the quark mass and therefore is important only in the case of the third generation scalar quarks [9]. In particular, due to the large value of m_t, the mixing angle $\theta_{\tilde{t}}$ between $\tilde{t}_L$ and $\tilde{t}_R$ can be very large and lead to a scalar top quark $\tilde{t}_1$ much lighter than the t–quark and all the scalar partners of the light quarks [9]. The mixing in the $\tilde{b}$–quark sector can be sizable only in a small area of the SUSY parameter space.

Neglecting the mixing in the $\tilde{b}$ sector, $\Delta\rho$ is given at one–loop order by the simple expression

$$\Delta\rho_0^{\rm SUSY} = \frac{3G_F}{8\sqrt{2}\pi^2}\left[-\sin^2\theta_{\tilde{t}}\cos^2\theta_{\tilde{t}}F_0\left(m_{\tilde{t}_1}^2, m_{\tilde{t}_2}^2\right)\right.$$
$$\left. + \cos^2\theta_{\tilde{t}}F_0\left(m_{\tilde{t}_1}^2, m_{\tilde{b}_L}^2\right) + \sin^2\theta_{\tilde{t}}F_0\left(m_{\tilde{t}_2}^2, m_{\tilde{b}_L}^2\right)\right]. \quad (4)$$

In a large area of the parameter space, the scalar top mixing angle is either very small $\theta_t \sim 0$ or maximal, $\theta_t \sim -\pi/4$. The contribution can be at the level of several per mille and is therefore within the range of the experimental observability. Relaxing the assumption of a common scalar quark mass, the corrections can become even larger [4,8].

At $\mathcal{O}(\alpha\alpha_s)$, the two–loop Feynman diagrams contributing to the ρ parameter in SUSY consist of two sets which are separately ultraviolet finite and gauge–invariant: diagrams involving only gluon exchange, and diagrams involving scalar quarks, gluinos, and quarks. The two–loop Feynman diagrams have to be supplemented by the corresponding one-loop counterterm insertions. The mass renormalization is performed in the on–shell scheme, where the mass is defined as the pole of the propagator. The mixing angle renormalization is performed in such a way that all transitions from $\tilde{q}_i \leftrightarrow \tilde{q}_j$ which do not depend on the loop–momenta in the two–loop diagrams are canceled.

The contribution of the $(\tilde{t}, \tilde{b})$ doublet to the ρ parameter, including the two–loop gluon exchange and pure scalar quark diagrams in the case where the $\tilde{b}$ mixing is neglected, is given by an expression similar to Eq. (4):

$$\Delta\rho_1^{\rm SUSY} = \frac{G_F\alpha_s}{4\sqrt{2}\pi^3}\left[-\sin^2\theta_{\tilde{t}}\cos^2\theta_{\tilde{t}}F_1\left(m_{\tilde{t}_1}^2, m_{\tilde{t}_2}^2\right)\right.$$
$$\left. + \cos^2\theta_{\tilde{t}}F_1\left(m_{\tilde{t}_1}^2, m_{\tilde{b}_L}^2\right) + \sin^2\theta_{\tilde{t}}F_1\left(m_{\tilde{t}_2}^2, m_{\tilde{b}_L}^2\right)\right]. \quad (5)$$

The two–loop function $F_1(x,y)$ is given in terms of dilogarithms by

$$F_1(x,y) = x + y - 2\frac{xy}{x-y}\log\frac{x}{y}\left[2 + \frac{x}{y}\log\frac{x}{y}\right]$$
$$+ \frac{(x+y)x^2}{(x-y)^2}\log^2\frac{x}{y} - 2(x-y)\text{Li}_2\left(1 - \frac{x}{y}\right). \quad (6)$$

In the case of large mass splitting it increases with the heavy scalar quark mass squared: $F_1(x,0) = x(1+\pi^2/3)$.

The two–loop gluonic SUSY contribution to $\Delta\rho$ is shown in Fig. 1 as a function of the common scalar mass $m_{\tilde{q}}$, for the two scenarios discussed previously: $\theta_{\tilde{t}} = 0$ and $\theta_{\tilde{t}} \simeq -\pi/4$. The two–loop contribution is of the order of 10 to 15% of the one–loop result. Contrary to the SM case where the two–loop correction screens the one–loop contribution, $\Delta\rho_1^{\rm SUSY}$ has the same sign as $\Delta\rho_0^{\rm SUSY}$. In

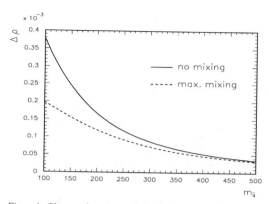

Figure 1: Gluon exchange contribution to the ρ parameter at two-loop as a function of the common mass $m_{\tilde{q}}$, for $\theta_{\tilde{t}} = 0$ and $\theta_{\tilde{t}} \sim -\pi/4$ [with $\tan\beta = 1.6$ and $M_t^{\rm LR} = 0$ and 200 GeV, respectively, where $M_t^{\rm LR}$ is the off–diagonal term in the $\tilde{t}$ mass matrix].

the case of degenerate $\tilde{t}$ quarks with masses $m_{\tilde{t}} \gg m_{\tilde{b}}$, the result is the same as the QCD correction to the (t, b) contribution in the SM, but with opposite sign. The gluonic correction to the contribution of scalar quarks to the ρ parameter will therefore enhance the sensitivity in the search of the virtual effects of scalar quarks in high-precision electroweak measurements.

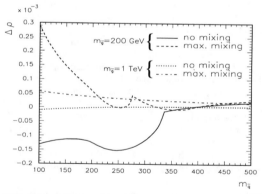

Figure 2: Contribution of the gluino exchange diagrams to $\Delta\rho_1^{\rm SUSY}$ for two values of $m_{\tilde{g}}$ in the scenarios of Fig. 1.

The gluino exchange diagrams cannot be cast into a simple anlytical expression. In general they give smaller contributions compared to gluon exchange. Only for gluino and scalar quark masses close to the experimental lower bounds they compete with the gluon exchange contributions. In this case, the gluon and gluino contributions add up to $\sim$ 30% of the one–loop value for maximal mixing [Fig. 2]. For larger values of $m_{\tilde{g}}$, the contribution decreases rapidly since the gluinos decouple for high masses.

2 Masses of the neutral Higgs bosons

The search for the lightest Higgs boson provides a direct and very stringent test of SUSY. A precise prediction for the mass of the lightest Higgs boson in terms of the relevant SUSY parameters hence is crucial in order to determine the discovery and exclusion potential of LEP2 and the upgraded Tevatron and also for physics at the LHC, where a high-precision measurement of the mass of this particle might be possible.

In the MSSM the mass of the lightest Higgs boson, m_h, is restricted at the tree level to be smaller than the Z-boson mass. This bound, however, is strongly affected by the inclusion of radiative corrections. The dominant one-loop corrections arise form the top and scalar-top sector via terms of the form $G_F m_t^4 \ln(m_{\tilde{t}_1} m_{\tilde{t}_2}/m_t^2)$ [10]. They increase the predicted values of m_h and yield an upper bound of about 150 GeV. These results have been improved by performing a complete one-loop calculation in the on-shell scheme, which takes into account the contributions of all sectors of the MSSM [11,12,13]. Beyond one-loop order renormalization group (RG) methods have been applied in order to obtain leading logarithmic higher-order contributions [14,15,16,17], and a diagrammatic calculation of the dominant two-loop contributions in the limiting case of vanishing $\tilde{t}$-mixing and infinitely large M_A and $\tan\beta$ has been carried out [18].

Recently a Feynman-diagrammatic calculation of the leading two-loop corrections of $\mathcal{O}(\alpha\alpha_s)$ to the masses of the neutral $\mathcal{CP}$-even Higgs bosons has been performed [19]. They have been combined with the complete one-loop diagrammatic calculation [20] to obtain in this way the currently most precise prediction for m_h within the Feynman-diagrammatic approach for arbitrary values of the parameters of the Higgs and scalar top sector of the MSSM. Further refinements concerning the leading two-loop Yukawa corrections of $\mathcal{O}(G_F^2 m_t^6)$ [15,21] and of leading QCD corrections beyond two-loop order are also included. The results will be discussed in this section.

In the MSSM two Higgs doublets are needed, decomposed as follows:

$$H_1 = \begin{pmatrix} H_1^1 \\ H_1^2 \end{pmatrix} = \begin{pmatrix} v_1 + (\phi_1^0 + i\chi_1^0)/\sqrt{2} \\ \phi_1^- \end{pmatrix},$$

$$H_2 = \begin{pmatrix} H_2^1 \\ H_2^2 \end{pmatrix} = \begin{pmatrix} \phi_2^+ \\ v_2 + (\phi_2^0 + i\chi_2^0)/\sqrt{2} \end{pmatrix}. \quad (7)$$

The Higgs sector can be described with the help of two independent parameters (besides the gauge couplings g, g'), usually chosen as $\tan\beta = v_2/v_1$ and the mass M_A of

the $\mathcal{CP}$-odd A boson. The $\mathcal{CP}$-even neutral mass eigenstates, h^0, H^0 are obtained from $\phi_{1,2}^0$ by a rotation yielding the tree-level masses $m_{h,H,\text{tree}}$.

In the Feynman-diagrammatic approach the one-loop corrected Higgs masses are derived by finding the poles of the $h - H$-propagator matrix whose inverse is given by

$$\begin{pmatrix} q^2 - m_{H,\text{tree}}^2 + \hat{\Sigma}_H(q^2) & \hat{\Sigma}_{hH}(q^2) \\ \hat{\Sigma}_{hH}(q^2) & q^2 - m_{h,\text{tree}}^2 + \hat{\Sigma}_h(q^2) \end{pmatrix}, \quad (8)$$

where the $\hat{\Sigma}$ denote the full one-loop contributions to the renormalized Higgs-boson self-energies. For these self-energies we take the result of the complete one-loop on-shell calculation of [12]. The agreement with the result obtained in [11] is better than 1 GeV for almost the whole MSSM parameter space.

As mentioned above the dominant contribution arises from the $t - \tilde{t}$-sector. The current eigenstates of the scalar quarks, $\tilde{q}_L$ and $\tilde{q}_R$, mix to give the mass eigenstates $\tilde{q}_1$ and $\tilde{q}_2$. The non-diagonal entry in the scalar quark mass matrix is proportional to the mass of the quark and reads for the $\tilde{t}$-mass matrix

$$m_t M_t^{LR} = m_t(A_t - \mu \cot\beta), \quad (9)$$

where we have adopted the conventions used in [8]. Due to the large value of m_t mixing effects have to be taken into account. Diagonalizing the $\tilde{t}$-mass matrix one obtains the eigenvalues $m_{\tilde{t}_1}$ and $m_{\tilde{t}_2}$ and the $\tilde{t}$ mixing angle $\theta_{\tilde{t}}$.

The leading two-loop corrections have been obtained in [19] by calculating the $\mathcal{O}(\alpha\alpha_s)$ contribution of the $t - \tilde{t}$-sector to the renormalized Higgs-boson self-energies at zero external momentum from the Yukawa part of the theory. At the two-loop level the matrix (8) consists of the renormalized Higgs-boson self-energies

$$\hat{\Sigma}_s(q^2) = \hat{\Sigma}_s^{(1)}(q^2) + \hat{\Sigma}_s^{(2)}(0), \quad s = h, H, h\dot{H}, \quad (10)$$

where the momentum dependence is neglected only in the two-loop contribution. The Higgs-boson masses at the two-loop level are obtained by determining the poles of the matrix Δ_{Higgs} in Eq. (8).

We have implemented two further steps of refinement into the prediction for m_h, which are shown separately in the plots below. The leading two-loop Yukawa correction of $\mathcal{O}(G_F^2 m_t^6)$ is taken over from the result obtained by renormalization group methods [15,21]. The second step of refinement concerns leading QCD corrections beyond two-loop order, taken into account by using the $\overline{MS}$ top mass, $\overline{m}_t = \overline{m}_t(m_t) \approx 166.5$ GeV, for the two-loop contributions instead of the pole mass, $m_t = 175$ GeV. In the $\tilde{t}$ mass matrix, however, we continue to use the pole mass as an input parameter. Only when performing the comparison with the RG results we use $\overline{m}_t$ in the $\tilde{t}$ mass matrix for the two-loop result, since in the RG results the running masses appear everywhere. This three-loop effect gives rise to a shift up to 1.5 GeV in the prediction for m_h.

For the numerical evaluation we have chosen two values for $\tan\beta$ which are favored by SUSY-GUT scenarios [22]: $\tan\beta = 1.6$ for the $SU(5)$ scenario and $\tan\beta = 40$ for the $SO(10)$ scenario. Other parameters are $M_Z = 91.187$ GeV, $M_W = 80.375$ GeV, $G_F = 1.16639 \, 10^{-5}$ GeV^{-2}, $\alpha_s(m_t) = 0.1095$, and $m_t = 175$ GeV. For the figures below we have furthermore chosen $M = 400$ GeV (M is the soft SUSY breaking parameter in the chargino and neutralino sector), $M_A = 500$ GeV, and $m_{\tilde{g}} = 500$ GeV as typical values (if not indicated differently). The scalar top masses and the mixing angle are derived from the parameters $M_{\tilde{t}_L}$, $M_{\tilde{t}_R}$ and M_t^{LR} of the $\tilde{t}$ mass matrix (our conventions are the same as in [8]). In the figures below we have chosen $m_{\tilde{q}} \equiv M_{\tilde{t}_L} = M_{\tilde{t}_R}$.

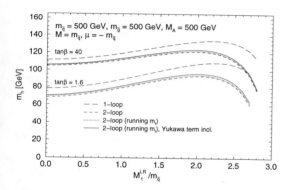

Figure 3: One- and two-loop results for m_h as a function of $M_t^{LR}/m_{\tilde{q}}$ for two values of $\tan\beta$.

The plot in Fig. 3 shows the result for m_h obtained from the diagrammatic calculation of the full one-loop and leading two-loop contributions. The two steps of refinement discussed above are shown in separate curves. For comparison the pure one-loop result is also given. The results are plotted as a function of $M_t^{LR}/m_{\tilde{q}}$, where $m_{\tilde{q}}$ is fixed to 500 GeV. The qualitative behavior is the same as in [19], where the result containing only the leading one-loop contribution (and without further refinements) was shown. The two-loop contributions give rise to a large reduction of the one-loop result of 10–20 GeV. The two steps of refinement both increase m_h by up to 2 GeV. A minimum occurs for $M_t^{LR} = 0$ GeV which we refer to as 'no mixing' (differently from section 1). A maximum in the two-loop result for m_h is reached for about $M_t^{LR}/m_{\tilde{q}} \approx 2$ in the $\tan\beta = 1.6$ scenario as well as in

the $\tan\beta = 40$ scenario. This case we refer to as 'maximal mixing'. The maximum is shifted compared to its one-loop value of about $M_t^{LR}/m_{\tilde{q}} \approx 2.4$. The two steps of refinement have only a negligible effect on the location of the maximum.

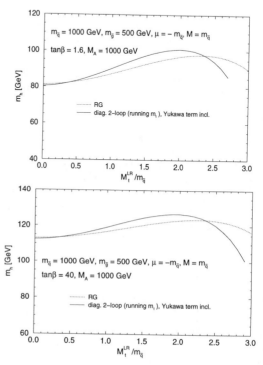

Figure 4: Comparison between the Feynman-diagrammatic calculations and the results obtained by renormalization group methods [16]. The mass of the lightest Higgs boson is shown for the two scenarios with $\tan\beta = 1.6$ and $\tan\beta = 40$ for increasing mixing in the $\tilde{t}$-sector and $m_{\tilde{q}} = M_A$.

We now turn to the comparison of our diagrammatic results with the predictions obtained via renormalization group methods. We begin with the case of vanishing mixing in the $\tilde{t}$ sector and large values of M_A, for which the RG approach is most easily applicable and is expected to work most accurately. In order to study different contributions separately, we have first compared the diagrammatic one-loop on-shell result [12] with the one-loop leading log result (without renormalization group improvement) given in [17] and found very good agreement, typically within 1 GeV. We then performed a leading log expansion of our diagrammatic result (which corresponds to the two-loop contribution in the RG approach) and also found agreement with the full two-loop result within about 1 GeV. Finally, we have compared our diagrammatic result for the no-mixing case including the refinement terms with the RG results [a] obtained in [16]. After the inclusion of the refinement terms the diagrammatic result for the no-mixing case agrees very well with the RG result. The deviation between the results exceeds 2 GeV only for $\tan\beta = 1.6$ and $m_{\tilde{q}} < 150$ GeV. For smaller values of M_A the comparison for the no-mixing case looks qualitatively the same. For $\tan\beta = 1.6$ and values of M_A below 100 GeV slightly larger deviations are possible. Since the RG results do not contain the gluino mass as a parameter, varying $m_{\tilde{g}}$ gives rise to an extra deviation, which in the no-mixing case does not exceed 1 GeV. Varying the other parameters μ and M in general does not lead to a sizable effect in the comparison with the corresponding RG results.

We now consider the situation when mixing in the $\tilde{t}$ sector is taken into account. We have again compared the full one-loop result with the one-loop leading log result used within the RG approach [17] and found good agreement. Only for values of M_A below 100 GeV and large mixing deviations of about 5 GeV occur. In Fig. 4 our diagrammatic result including the refinement terms is compared with the RG results [16] as a function of $M_t^{LR}/m_{\tilde{q}}$ for $\tan\beta = 1.6$ and $\tan\beta = 40$. For larger $\tilde{t}$-mixing sizable deviations between the diagrammatic and the RG results occur, which can exceed 5 GeV for moderate mixing and become very large for large values of $M_t^{LR}/m_{\tilde{q}}$. As already stressed above, the maximal value for m_h in the diagrammatic approach is reached for $M_t^{LR}/m_{\tilde{q}} \approx 2$, whereas the RG results have a maximum at $M_t^{LR}/m_{\tilde{q}} \approx 2.4$, i.e. at the one-loop value. Varying the value of $m_{\tilde{g}}$ in our result leads to a larger effect than in the no-mixing case and shifts the diagrammatic result relative to the RG result within ± 2 GeV.

So far, the results of our diagrammatic on-shell calculation and the RG methods have been compared in terms of the parameters $M_{\tilde{t}_L}$, $M_{\tilde{t}_R}$ and M_t^{LR} of the $\tilde{t}$ mixing matrix, since the available numerical codes for the RG results [16,17] are given in terms of these parameters. However, since the two approaches rely on different renormalization schemes, the meaning of these (non-observable) parameters is not precisely the same in the two approaches starting from two-loop order. Indeed we have checked that assuming fixed values for the physical parameters $m_{\tilde{t}_1}$, $m_{\tilde{t}_2}$, and $\theta_{\tilde{t}}$ and deriving the corresponding values of the parameters $M_{\tilde{t}_L}$, $M_{\tilde{t}_R}$ and M_t^{LR} in the on-shell scheme as well as in the $\overline{MS}$ scheme, sizable differences occur between the values of the mixing parameter M_t^{LR} in the two schemes, while the parameters $M_{\tilde{t}_L}$,

[a]The RG results of [16] and [17] agree within about 2 GeV with each other.

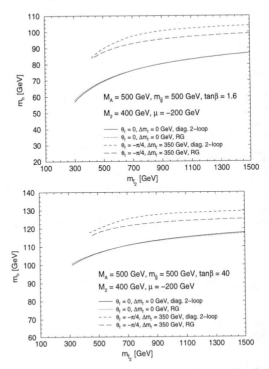

Figure 5: Comparison between the Feynman-diagrammatic calculations and the results obtained by renormalization group methods [16]. The mass of the lightest Higgs boson is shown for the two scenarios with $\tan\beta = 1.6$ and $\tan\beta = 40$ as a function of the heavier physical $\tilde{t}$ mass $m_{\tilde{t}_2}$. For the curves with $\theta_{\tilde{t}} = 0$ a mass difference $\Delta m_{\tilde{t}} = 0$ GeV is assumed whereas for $\theta_{\tilde{t}} = -\pi/4$ we chose $\Delta m_{\tilde{t}} = 350$ GeV, for which the maximal Higgs masses are achieved.

$M_{\tilde{t}_R}$ are approximately equal in the two schemes. Thus, part of the different shape of the curves in Fig. 4 may be attributed to a different meaning of the parameter M_t^{LR} in the on-shell scheme and in the RG calculation.

For the purpose of comparing results obtained in different renormalization schemes it is very desirable to express the prediction for the Higgs-boson masses in terms of physical observables, i.e. the physical masses and mixing angles of the model instead of unphysical parameters. As a step into this direction we compare in Fig. 5 the diagrammatic results and the RG results as a function of the physical mass $m_{\tilde{t}_2}$ and with the mass difference $\Delta m_{\tilde{t}} = m_{\tilde{t}_2} - m_{\tilde{t}_1}$ and the mixing angle $\theta_{\tilde{t}}$ as parameters. In the context of the RG approach the running $\tilde{t}$ masses, derived from the $\tilde{t}$ mass matrix, are considered as an approximation for the physical masses. The range of the $\tilde{t}$ masses appearing in Fig. 5 has been constrained by requiring that the contribution of the third generation of scalar quarks to the ρ-parameter [8] does not exceed the value of $1.3 \cdot 10^{-3}$. As in the comparison performed above, in Fig. 5 very good agreement is found between the results of the two approaches in the case of vanishing $\tilde{t}$ mixing. For the maximal mixing angle $\theta_{\tilde{t}} = -\pi/4$, however, the diagrammatic result yields values for m_h which are higher by about 5 GeV.

References

1. D. Treille, plenary talk at ICHEP98, these proceedings
2. A. Djouadi, G. Girardi, W. Hollik, F. Renard, C. Verzegnassi, *Nucl. Phys.* B **349**, 48 (1991); M. Boulware, D. Finnell, *Phys. Rev.* D **44**, 2054 (1991)
3. W. de Boer et al., *Z. Phys.* C **75**, 627 (1997)
4. R. Barbieri et al., *Nucl. Phys.* B **224**, 32 (1983); M. Drees, K. Hagiwara, *Phys. Rev.* D **42**, 1709 (1990); P. Chankowski et al., *Nucl. Phys.* B **417**, 101 (1994)
5. M. Veltman, *Nucl. Phys.* B **123**, 89 (1977)
6. A. Djouadi, C. Verzegnassi, *Phys. Lett.* B **195**, 265 (1987); A. Djouadi, *Nuovo Cim.* A **100**, 357 (1988)
7. K. Chetyrkin, J. Kühn, M. Steinhauser, *Phys. Lett.* B **351**, 331 (1995); L. Avdeev et al., *Phys. Lett.* B **336**, 560 (1994), E: B **349**, 597 (1995)
8. A. Djouadi, P. Gambino. S. Heinemeyer, W. Hollik, C. Jünger, G. Weiglein, *Phys. Rev. Lett.* **78**, 3626 (1997); *Phys. Rev.* D **57**, 4179 (1998)
9. J. Ellis, S. Rudaz, Phys. Lett. **B128**, 248 (1983)
10. H. Haber and R. Hempfling, *Phys. Rev. Lett.* **66**, 1815 (1991); Y. Okada, M. Yamaguchi and T. Yanagida, *Prog. Theor. Phys.* **85**, 1 (1991); J. Ellis, G. Ridolfi and F. Zwirner, *Phys. Lett.* B **257**, 83 (1991); *Phys. Lett.* B **262**, 477 (1991); R. Barbieri and M. Frigeni, *Phys. Lett.* B **258**, 395 (1991)
11. P. Chankowski, S. Pokorski and J. Rosiek, *Nucl. Phys.* **423**, 437 (1994)
12. A. Dabelstein, *Nucl. Phys.* B **456**, 25 (1995); *Z. Phys.* C **67**, 495 (1995)
13. J. Bagger, K. Matchev, D. Pierce and R. Zhang, *Nucl. Phys.* B **491**, 3 (1997)
14. J. Casas, J. Espinosa, M. Quirós and A. Riotto, *Nucl. Phys.* B **436**, 3 (1995), E: *ibid.* B **439**, 466 (1995)
15. M. Carena, J. Espinosa, M. Quirós and C. Wagner, *Phys. Lett.* B **355**, 209 (1995)
16. M. Carena, M. Quirós and C. Wagner, *Nucl. Phys.* B **461**, 407 (1996)
17. H. Haber, R. Hempfling and A. Hoang, *Z. Phys.* C **75**, 539 (1997)
18. R. Hempfling and A. Hoang, *Phys. Lett.* B **331**, 99 (1994)
19. S. Heinemeyer, W. Hollik and G. Weiglein, *Phys. Rev.* D **58**, 091701 (1998)
20. S. Heinemeyer, W. Hollik, G. Weiglein, hep-ph/9807423 (*Phys. Lett.* B , to appear)
21. M. Carena, P. Chankowski, S. Pokorski and C. Wagner, FERMILAB-PUB-98/146-T, hep-ph/9805349.
22. M. Carena, S. Pokorski and C. Wagner, *Nucl. Phys.* B **406**, 59 (1993); W. de Boer et al., *Z. Phys.* C **71**, 415 (1996)

BOSONIC DECAYS OF $\tilde{t}_2$ AND $\tilde{b}_2$ INCLUDING SUSY–QCD CORRECTIONS

W. MAJEROTTO[a], A. BARTL[b], H. EBERL[a], K. HIDAKA[c], S. KRAML[a], T. KON[d], W. POROD[b], and
Y. YAMADA[e]

[a] *Institut für Hochenergiephysik der Österreichischen Akademie der Wissenschaften,*
A–1050 Vienna, Austria
[b] *Institut für Theoretische Physik, Universität Wien, A–1090 Vienna, Austria*
[c] *Department of Physics, Tokyo Gakugei University, Koganei, Tokyo 184–8501, Japan*
[d] *Faculty of Engineering, Seikei University, Musashino, Tokyo 180, Japan*
[e] *Department of Physics, Tohoku University, Sendai 980–8578, Japan*

We discuss in detail the decays of the heavier top and bottom squarks ($\tilde{t}_2$ and $\tilde{b}_2$). We show that the bosonic decay modes, as for instance, $\tilde{t}_2 \to \tilde{t}_1 + (h^0, H^0, A^0 \text{ or } Z^0)$, $\tilde{t}_2 \to \tilde{b}_{1,2} + (H^+ \text{ or } W^+)$, can dominate over the decays into charginos and neutralinos in a wide region of the parameter space. We have also calculated the SUSY–QCD corrections to these decays. The individual decay widths and branching ratios can get corrections up to 50%.

1 Introduction

Most studies on squark ($\tilde{q}$) decays concentrated on the decays into fermions

$$\tilde{q}_i \to q^{(\prime)} + (\tilde{\chi}_k^0, \tilde{\chi}_j^\pm, \text{ or } \tilde{g}), \qquad (1)$$

with $i, j = 1, 2$, and $k = 1, \ldots 4$. $\tilde{\chi}_j^\pm$ denotes the charginos, $\tilde{\chi}_k^0$ the neutralinos, and $\tilde{g}$ the gluino. However, in the Minimal Supersymmetric Standard Model (MSSM) the squarks of the 3rd generation ($\tilde{t}, \tilde{b}$) can have a large mass splitting $\Delta m = m_{\tilde{q}_2} - m_{\tilde{q}_1}$ due to $\tilde{q}_L - \tilde{q}_R$ mixing. Therefore, the heavier squark can also decay into bosons:

$$\tilde{q}_2 \to \tilde{q}_i' + (W^\pm, \text{ or } H^\pm), \qquad (2)$$
$$\tilde{q}_2 \to \tilde{q}_1 + (Z^0, h^0, H^0, \text{ or } A^0). \qquad (3)$$

We have performed a detailed study of these decays[1]. We have shown that the decays of Eqs.(2) and (3) can dominate in a large region of the MSSM parameter space due to large $\tilde{t}$ (or $\tilde{b}$) mixing parameters and large top (or bottom) Yukawa couplings.

The mass matrix in the basis ($\tilde{q}_L, \tilde{q}_R$) is:

$$\mathcal{M}_{\tilde{q}}^2 = \begin{pmatrix} m_{\tilde{q}_L}^2 & a_q\,m_q \\ a_q\,m_q & m_{\tilde{q}_R}^2 \end{pmatrix}, \qquad (4)$$

with

$$m_{\tilde{q}_L}^2 = M_{\tilde{Q}}^2 + m_Z^2 \cos 2\beta \left(I_{qL}^3 - e_q \sin^2 \theta_w \right) + m_q^2, \quad (5)$$
$$m_{\tilde{q}_R}^2 = M_{\{\tilde{U},\tilde{D}\}}^2 + e_q\, m_Z^2 \cos 2\beta \, \sin^2 \theta_w + m_q^2, \qquad (6)$$
$$a_q m_q = \begin{cases} (A_t - \mu \cot \beta)\, m_t & (\tilde{q} = \tilde{t}) \\ (A_b - \mu \tan \beta)\, m_b & (\tilde{q} = \tilde{b}) \end{cases}, \qquad (7)$$

with I_{qL}^3 the third component of the weak isospin and e_q the electric charge of the quark q. $M_{\tilde{Q}}, M_{\tilde{U}}, M_{\tilde{D}}$, and A_q

are SUSY soft–breaking parameters, μ is the Higgsino mass parameter, and $\tan \beta = v_2/v_1$ with v_1 and v_2 the vacuum expectation values of the two Higgs doublets. Diagonalizing the matrix (3) one gets the mass eigenstates $\tilde{q}_1 = \tilde{q}_L \cos \theta_{\tilde{q}} + \tilde{q}_R \sin \theta_{\tilde{q}}$, $\tilde{q}_2 = -\tilde{q}_L \sin \theta_{\tilde{q}} + \tilde{q}_R \cos \theta_{\tilde{q}}$ with the masses $m_{\tilde{q}_1}, m_{\tilde{q}_2}$ ($m_{\tilde{q}_1} < m_{\tilde{q}_2}$) and the mixing angle $\theta_{\tilde{q}}$.

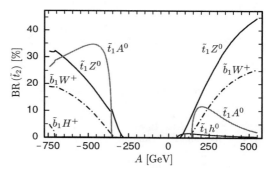

Figure 1: Branching ratios of $\tilde{t}_2$ decays into bosons as a function of $A \equiv A_t = A_b$, for $M_{\tilde{Q}} = 320\,\text{GeV}$, $M_{\tilde{U}} = 288\,\text{GeV}$ $M_{\tilde{D}} = 352\,\text{GeV}$, $\mu = -320\,\text{GeV}$, $\tan \beta = 3$, $M = 140\,\text{GeV}$, $m_{A^0} = 130\,\text{GeV}$.

2 Tree–level branching ratios

The width of the squark decays into Higgs and gauge bosons, Eqs.(2) and (3), are given by[2]:

$$\Gamma(\tilde{q}_i \to \tilde{q}_j^{(\prime)} V) = \frac{\kappa_{ijV}^3}{16\pi\, m_V^2\, m_{\tilde{q}_i}^3} \, (c_{ijV})^2, \qquad (8)$$

$$\Gamma(\tilde{q}_i \to \tilde{q}_j^{(\prime)} H_k) = \frac{\kappa_{ijk}}{16\pi\, m_{\tilde{q}_i}^3} \, (G_{ijk})^2. \qquad (9)$$

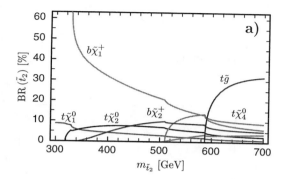

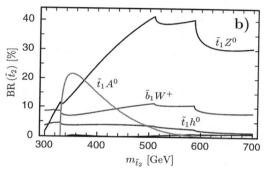

Figure 2: Branching ratios of $\tilde{t}_2$ decays as a function of $m_{\tilde{t}_2}$, for $m_{\tilde{t}_1} = 200$ GeV, $\cos\theta_{\tilde{t}} = -0.6$, $M = 140$ GeV, $\mu = -320$ GeV, $\tan\beta = 3$, $m_{A^0} = 130$ GeV, $M_{\tilde{D}} = 1.1\,M_{\tilde{Q}}$, $A_t = A_b$; (a) decays into fermions and (b) decays into bosons.

with $V = \{Z^0, W^\pm\}$ and $H_k = \{h^0, H^0, A^0, H^\pm\}$, $\kappa_{ijX} \equiv \kappa(m_{\tilde{q}_i}^2, m_{\tilde{q}_j^{(\prime)}}^2, m_X^2)$, $\kappa(x,y,z) = (x^2 + y^2 + z^2 - 2xy - 2xz - 2yz)^{1/2}$. The complete expressions for c_{ijV} and G_{ijk} are given in [2]. The leading terms of some of the stop couplings to vector and Higgs bosons are:

$$\tilde{t}_2 \tilde{t}_1 Z^0 \sim g \sin 2\theta_{\tilde{t}}, \tag{10}$$

$$\tilde{t}_1 \tilde{t}_2 h^0 \sim h_t \left(\mu \sin\alpha + A_t \cos\alpha\right) \cos 2\theta_{\tilde{t}}, \tag{11}$$

$$\tilde{t}_1 \tilde{t}_2 H^0 \sim h_t \left(\mu \cos\alpha - A_t \sin\alpha\right) \cos 2\theta_{\tilde{t}}, \tag{12}$$

$$\tilde{t}_1 \tilde{t}_2 A^0 \sim h_t \left(\mu \sin\beta + A_t \cos\beta\right), \tag{13}$$

$$\tilde{t}_1 \tilde{b}_2 H^\pm \sim -h_t \left(\mu \sin\beta + A_t \cos\beta\right) \sin\theta_{\tilde{t}} \sin\theta_{\tilde{b}} \tag{14}$$
$$+ h_b \left(\mu \cos\beta + A_b \sin\beta\right) \cos\theta_{\tilde{t}} \cos\theta_{\tilde{b}},$$

$$\tilde{t}_2 \tilde{b}_1 W^+ \sim -g \sin\theta_t \cos\theta_{\tilde{b}}, \tag{15}$$

where $h_{t,b}$ are the Yukawa couplings

$$h_t = \frac{g m_t}{\sqrt{2} m_W \sin\beta} \quad , \quad h_b = \frac{g m_b}{\sqrt{2} m_W \cos\beta}.$$

One can see, for example, that the $\tilde{t}_2 \tilde{t}_1 Z^0$ coupling is largest for maximal mixing $\theta_{\tilde{t}} = \pi/4$ because the Z^0 couples only to $\tilde{t}_L \tilde{t}_L$ and $\tilde{t}_R \tilde{t}_R$. On the other hand, the Higgs bosons couple mainly to $\tilde{t}_L \tilde{t}_R$ combinations with the coupling proportional to the Yukawa coupling h_t. Notice that the coupling $\tilde{t}_2 \tilde{t}_1 A^0$ does not depend explicitly on the mixing angle.

In the numerical analysis we have taken into account the constraints from LEP2 and Tevatron: $m_{\tilde{\chi}_1^+} > 90$ GeV, $m_{\tilde{\chi}_1^0} > 30$ GeV, $m_{h^0} > 72$ GeV, $m_{\tilde{q}_1} > 80$ GeV, $\Delta\rho(\tilde{t} - \tilde{b}) < 0.0012$, and the necessary conditions to avoid a colour or electric charge breaking minimum: $A_{t,b}^2 < 3(M_{\tilde{Q}}^2 + M_{\tilde{U},\tilde{D}}^2 + m_{H_{2,1}}^2)$, with $m_{H_2}^2 = (m_A^2 + m_Z^2)\cos^2\beta - \frac{1}{2}m_Z^2$, $m_{H_1}^2 = (m_A^2 + m_Z^2)\sin^2\beta - \frac{1}{2}m_Z^2$.

In Fig. 1 we show the branching ratios of $\tilde{t}_2$ decays into bosons as a functions of $A_t = A_b \equiv A$ for $M_{\tilde{Q}} = 320$ GeV, $M_{\tilde{U}} : M_{\tilde{Q}} : M_{\tilde{D}} = 0.9 : 1 : 1.1$, $\tan\beta = 3$, $\mu = -320$ GeV, $M = 140$ GeV, $m_{A^0} = 130$ GeV. We see that the bosonic decays play an important rôle. In Fig. 2 we show the branching ratios of $\tilde{t}_2$ decays (a) into fermions as well as (b) into bosons as a function of $m_{\tilde{t}_2}$ for $m_{\tilde{t}_1} = 200$ GeV, $\cos\theta_{\tilde{t}} = -0.6$, $M_{\tilde{D}} = 1.1 M_{\tilde{Q}}$, $A_t = A_b$, $\mu = -320$ GeV, $M = 140$ GeV, $\tan\beta = 3$, $m_{A^0} = 130$ GeV. It is interesting to see the step–wise appearance of the various decay channels as a function of $m_{\tilde{t}_2}$.

Figs. 3a and 3b exhibit the $\tan\beta$ dependence of the bosonic $\tilde{t}_2$ and $\tilde{b}_2$ decay branching ratios for $M_{\tilde{Q}} = 500$ GeV, $M_{\tilde{U}} = 444$ GeV, $M_{\tilde{D}} = 556$ GeV, $M = 300$ GeV, $\mu = -700$ GeV, $m_{A^0} = 150$ GeV, $A_t = A_b = 600$ GeV. Notice that the $\tilde{t}_2(\tilde{b}_2)$ decays into $\tilde{b}_1$ and a gauge or Higgs boson become increasingly important for larger $\tan\beta$.

3 SUSY–QCD corrections

The SUSY–QCD corrections to $\mathcal{O}(\alpha_s)$ were calculated in [3,4,5] for squark decays into charginos/neutralinos, Eq.(1), in [6] for decays into vector bosons, Eqs.(2), and (3), in [7,8] for the decays into Higgs bosons, Eqs.(2), and (3), and in [9] for the decays into gluinos, Eq.(1). The corrected decay widths get the following contributions:

$$\Gamma = \Gamma^0 + \delta\Gamma^{(v)} + \delta\Gamma^{(w)} + \delta\Gamma^{(c)} + \delta\Gamma^{(\text{real gluon})}, \tag{16}$$

where Γ^0 is the tree–level width, $\delta\Gamma^{(v)}$ is the vertex correction (g, $\tilde{g}$, q, and $\tilde{q}$ exchange), $\delta\Gamma^{(w)}$ is the wave–function correction due to the self–energy graphs of $\tilde{q}$ and q. $\delta\Gamma^{(c)}$ is due to the shift from the bare to the on–shell coupling. $\delta\Gamma^{(\text{real gluon})}$ is due to real gluon radiation. We use the on–shell renormalization scheme and SUSY conserving dimensional reduction regularization ($\overline{\text{DR}}$).

There are some subtleties in the calculation:

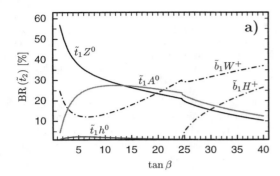

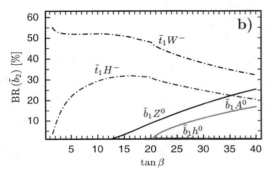

Figure 3: Branching ratios of (a) $\tilde{t}_2$ and (b) $\tilde{b}_2$ decays into bosons as a function of $\tan\beta$, for $M_{\tilde{Q}} = 500\,\text{GeV}$, $M_{\tilde{U}} = 444\,\text{GeV}$ $M_{\tilde{D}} = 556\,\text{GeV}$, $A_t = A_b = 600\,\text{GeV}$, $\mu = -700\,\text{GeV}$, $M = 300\,\text{GeV}$, $m_{A^0} = 150\,\text{GeV}$.

(i) $\delta\Gamma^{(c)}$ is due to the counterterms to the couplings, for instance, in the case of $\tilde{q}_i \to \tilde{q}_j^{(\prime)} H_k$ due to δG_{ijk}. The couplings G_{ijk}, Eq.(9), depend in general on m_q, A_q, and the mixing angle $\theta_{\tilde{q}}$, hence $\delta G_{ijk} = \delta G_{ijk}(\delta m_q, \delta(m_q A_q), \delta\theta_{\tilde{q}})$. This implies that also the mixing angle must be renormalized. We have fixed $\delta\theta_{\tilde{q}}$ by $e^+e^- \to \tilde{q}_i\bar{\tilde{q}}_j$ such that the off-diagonal squark self-energies cancel [10]. Similar conditions have been used in [4,5,11].

(ii) A further subtlety concerns the parameter $M_{\tilde{Q}}$ in the on-shell scheme. At tree-level and in the $\overline{\text{DR}}$ renormalization scheme SU(2) symmetry requires that the parameter $M_{\tilde{Q}}$ in the $\tilde{t}$ and $\tilde{b}$ matrices have the same value. This is, however, not the case at loop-level in the on-shell scheme due to different shifts $\delta M_{\tilde{Q}}^2$ in the $\tilde{t}$ and $\tilde{b}$ sectors [12]. We have in the on-shell scheme $M_{\tilde{Q}}^2(\tilde{b}) = M_{\tilde{Q}}^2(\tilde{t}) + \delta M_{\tilde{Q}}^2(t) - \delta M_{\tilde{Q}}^2(\tilde{b})$, where $M_{\tilde{Q}}^2$ and $\delta M_{\tilde{Q}}^2$ is calculated from $M_{\tilde{Q}}^2(\tilde{q}) = m_{\tilde{q}}^2 \cos^2\theta_{\tilde{q}} + m_{\tilde{q}_2}^2 \sin^2\theta_{\tilde{q}} - m_Z^2 \cos 2\beta(I_{qL}^3 -$

$e_q \sin^2\theta_W) - m_q^2.$

(iii) Moreover, we have found points in the parameter space, especially for large $\tan\beta$, where the corrected width becomes negative. A consideration of higher order effects seems to be necessary.

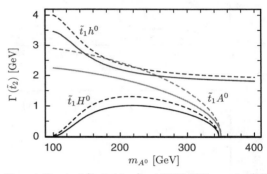

Figure 4: Tree–level (dashed lines) and SUSY–QCD corrected (full lines) widths of $\tilde{t}_2 \to \tilde{t}_1 + (h^0, H^0, A^0)$ decays as a function of m_{A^0}, for $m_{\tilde{t}_1} = 250\,\text{GeV}$, $m_{\tilde{t}_2} = 600\,\text{GeV}$, $\cos\theta_{\tilde{t}} = 0.26$, $\mu = 550\,\text{GeV}$, $\tan\beta = 3$, $m_{A^0} = 130\,\text{GeV}$, $m_{\tilde{g}} = 600\,\text{GeV}$.

In Fig. 4 we show the tree–level and the SUSY–QCD corrected decay widths of $\tilde{t}_2 \to \tilde{t}_1 + (h^0, H^0, A^0)$ as a function of m_{A^0} for $m_{\tilde{t}_1} = 250\,\text{GeV}$, $m_{\tilde{t}_2} = 600\,\text{GeV}$, $\cos\theta_{\tilde{t}} = 0.26, \mu = 550\,\text{GeV}$, $\tan\beta = 3$. The corrections can go up to 40%.

Fig. 5 shows the tree–level and the SUSY–QCD corrected decay branching ratios of $\tilde{t}_2$ decays as a function of $\cos\theta_{\tilde{t}}$ (a) into neutralinos/charginos and (b) into bosons as a function of $\cos\theta_{\tilde{t}}$ for $m_{\tilde{t}_1} = 200$ GeV, $m_{\tilde{t}_2} = 400$ GeV, $M = 140$ GeV, $\mu = -320$ GeV, $\tan\beta = 3, m_{A^0} = 130$ GeV, $M_{\tilde{D}} = 1.1\,M_{\tilde{Q}}, A_t = A_b$.

Finally, Fig. 6 exhibits the $m_{\tilde{t}_2}$ dependence of the branching ratios of $\tilde{t}_2$ decays into bosons for $m_{\tilde{t}_1} = 200$ GeV, $\cos\theta_{\tilde{t}} = 0.6$ and the other parameters as in Fig. 5. Especially for smaller branching ratios the corrections can go up to 50%.

In conclusion, we have shown that the decays of $\tilde{t}_2$ and $\tilde{b}_2$ into Higgs and gauge bosons play an important rôle in a large region of the parameter space. The SUSY–QCD corrections can be significant in some cases.

Acknowledgements

The work of A.B., H.E., S.K., W.M., and W.P. was supported by the "Fonds zur Förderung der wissenschaftlichen Forschung" of Austria, project no. P10843–PHY. The work of Y.Y. was supported in part by the Grant–in–aid for Scientific Research from the

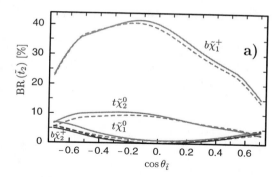

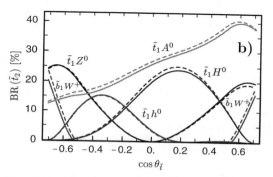

Figure 5: Tree–level (dashed lines) and SUSY–QCD corrected (full lines) branching ratios of $\tilde{t}_2$ decays (in %) as a function of $\cos\theta_{\tilde{t}}$, for $m_{\tilde{t}_1} = 200\,\text{GeV}$, $m_{\tilde{t}_2} = 400\,\text{GeV}$, $M = 140\,\text{GeV}$, $\mu = -320\,\text{GeV}$, $\tan\beta = 3$, $m_{A^0} = 130\,\text{GeV}$, $M_{\tilde{D}} = 1.1\,M_{\tilde{Q}}$, $A_t = A_b$; (a) decays into fermions and (b) decays into bosons.

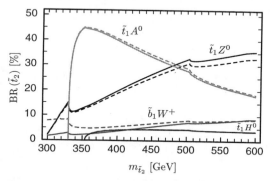

Figure 6: Tree–level (dashed lines) and SUSY–QCD corrected (full lines) branching ratios of $\tilde{t}_2$ decays (in %) as a function of $m_{\tilde{t}_2}$, for $m_{\tilde{t}_1} = 200\,\text{GeV}$, $\cos\theta_{\tilde{t}} = 0.6$, $M = 140\,\text{GeV}$, $\mu = -320\,\text{GeV}$, $\tan\beta = 3$, $m_{A^0} = 130\,\text{GeV}$, $M_{\tilde{D}} = 1.1\,M_{\tilde{Q}}$, $A_t = A_b$.

Ministry of Education, Science, and Culture of Japan, No. 10740106, and by Fuuju–kai Foundation.

References

1. A. Bartl, H. Eberl, K. Hidaka, S. Kraml, T. Kon, W. Majerotto, W. Porod, and Y. Yamada, hep–ph/9804265, to be published in *Phys. Lett.* B.
2. A. Bartl, W. Majerotto, and W. Porod, *Z. Phys.* C **64**, 499 (1994); *Z. Phys.* C **68**, 518(E) (1995).
3. S. Kraml, H. Eberl, A. Bartl, W. Majerotto, and W. Porod, *Phys. Lett.* B **386**, 175 (1996).
4. A. Djouadi, W. Hollik, and C. Jünger, *Phys. Rev.* D **55**, 6975 (1997).
5. W. Beenakker, R. Höpker, T. Plehn, and P.M. Zerwas, *Z. Phys.* C **75**, 349 (1997).
6. A. Bartl, H. Eberl, K. Hidaka, S. Kraml, W. Majerotto, W. Porod, and Y. Yamada, *Phys. Lett.* B **419**, 243 (1998).
7. A. Arhib, A. Djouadi, W. Hollik, and C. Jünger, *Phys. Rev.* D **57**, 5860 (1998).
8. A. Bartl, H. Eberl, K. Hidaka, S. Kraml, W. Majerotto, W. Porod, and Y. Yamada, hep–ph/9806299.
9. W. Beenakker, R. Höpker, and P.M. Zerwas, *Phys. Lett.* B **378**, 159 (1996).
10. H. Eberl, A. Bartl, and W. Majerotto, *Nucl. Phys.* B **472**, 481 (1996).
11. A. Djouadi, W. Hollik, and C. Jünger, *Phys. Rev.* D **54**, 5629 (1996).
12. A. Bartl, H. Eberl, K. Hidaka, T. Kon, W. Majerotto, and Y. Yamada, *Phys. Lett.* B **402**, 303 (1997).

DISCOVERY LIMITS FOR TECHNI-OMEGA PRODUCTION IN $e\gamma$ COLLISIONS

STEPHEN GODFREY, PAT KALYNIAK

Ottawa-Carleton Institute for Physics, Department of Physics, Carleton University, Ottawa CANADA, K1S 5B6
E-mail: kalyniak@physics.carleton.ca, godfrey@physics.carleton.ca

TAO HAN

Department of Physics, University of Wisconsin, Madison, WI 53706
E-mail: than@pheno.physics.wisc.edu

In a strongly-interacting electroweak sector with an isosinglet vector state, such as the techni-omega, ω_T, the direct $\omega_T Z\gamma$ coupling implies that an ω_T can be produced by $Z\gamma$ fusion in $e\gamma$ collisions. This is a unique feature for high energy e^+e^- or e^-e^- colliders operating in an $e\gamma$ mode. We consider the processes $e^-\gamma \to e^- Z\gamma$ and $e^-\gamma \to e^- W^+W^- Z$, both of which proceed via an intermediate ω_T. We find that at a 1.5 TeV e^+e^- linear collider operating in an $e\gamma$ mode with an integrated luminosity of 200 fb^{-1}, we can discover an ω_T for a broad range of masses and widths.

1 Introduction

The mechanism for electroweak symmetry breaking remains the most prominent mystery in elementary particle physics. In the Standard Model (SM), a neutral scalar Higgs boson is expected with a mass (m_H) to be less than about 800 GeV. In the weakly coupled supersymmetric extension of the SM, the lightest Higgs boson should be lighter than about 140 GeV. Searching for the Higgs bosons has been the primary goal for current and future collider experiments [1]. However, if no light Higgs boson is found for $m_H < 800$ GeV, one would anticipate that the interactions among the longitudinal vector bosons become strong [2]. This is the case when strongly interacting dynamics is responsible for electroweak symmetry breaking, such as in Technicolor models [3].

Without knowing the underlying dynamics of the strongly-interacting electroweak sector (SEWS), it is instructive to parametrize the physics with an effective theory for the possible low-lying resonant states. This typically includes an isosinglet scalar meson (H) and an isotriplet vector meson (ρ_T) [4]. However, in many dynamical electroweak symmetry breaking models there exist other resonant states such as an isosinglet vector (ω_T), and isotriplet vector (a_{1_T}) [5,6]. In fact it has been argued that to preserve good high energy behavior for strong scattering amplitudes in a SEWS, it is necessary for all the above resonant states to coexist [7]. It is therefore wise to keep an open mind and to consider other characteristic resonant states when studying the physics of a SEWS at high energy colliders.

Among those heavy resonant states, the isosinglet vector state ω_T has rather unique features. Due to the isosinglet nature, it does not have strong coupling to two gauge bosons. It couples strongly to three longitudinal gauge bosons $W_L^+ W_L^- Z_L$ (or equivalently electroweak

Goldstone bosons $w^+ w^- z$) and electroweakly to $Z\gamma$, ZZ, and W^+W^-. It may mix with the $U(1)$ gauge boson B, depending on its hypercharge assignment in the model. The signal for ω_T production was studied for pp collisions at 40 TeV and 17 TeV [5,6]. It appears to be difficult to observe the ω_T signal at the LHC. On the other hand, the direct $\omega_T Z_L \gamma$ coupling implies that an ω_T can be effectively produced by $Z_L\gamma$ fusion in $e\gamma$ collisions. This is a unique feature for high energy e^+e^- or e^-e^- colliders operating in an $e\gamma$ mode. In this paper we concentrate on the ω_T production at $e\gamma$ linear colliders. We first describe a SEWS model in Sec. II in terms of an effective Lagrangian involving ω_T interactions. We then present our results in Sec. III for the production and decay of ω_T in $e\gamma$ colliders. We show that a high energy $e\gamma$ linear collider will have great potential to discover ω_T with a mass of order 1 TeV. We conclude in Sec. IV.

2 ω_T Interactions

For an isosinglet vector, a Techniomega-like state ω_T, the leading strong interaction can be parameterized by

$$\mathcal{L}_{\text{strong}} = \frac{g_\omega}{v\Lambda^2}\, \varepsilon_{\mu\nu\rho\sigma}\, \omega_T^\mu\, \partial^\nu w^+ \partial^\rho w^- \partial^\sigma z \qquad (1)$$

where $v = 246$ GeV is the scale of electroweak symmetry breaking and Λ is the new physics scale at which the strong dynamics sets in. The effective coupling g_ω is of strong coupling strength, and is model-dependent. It governs the partial decay width $\Gamma(\omega_T \to w^+ w^- z) \equiv \Gamma_{WWZ}$. To study the ω_T signal in a model-independent way, we will take the physical partial width as an input parameter to give the value for the factor $(g_\omega/v\Lambda^2)$.

The effective Lagrangian describing the electroweak interactions of the ω_T with the gauge bosons can be writ-

Table 1: Feynman rules for the effective interactions of ω_T

Vertex	Feynman rule
$\omega_T w^+ w^- z$	$i\frac{g_{\omega_T}}{v\Lambda^2}\varepsilon_{\mu\nu\rho\sigma}\epsilon^\mu(\omega)q_1^\nu q_2^\rho q_3^\sigma$
$\omega_T Z\gamma$	$i\frac{4\chi e^2}{\sin\theta_w\cos\theta_w}\varepsilon_{\mu\nu\rho\sigma}\epsilon^\mu(\omega)\epsilon^\nu(Z)p_\gamma^\rho\epsilon^\sigma(\gamma)$
$\omega_T ZZ$	$i4e^2\chi(\cot^2\theta_w-\tan^2\theta_w)\varepsilon_{\mu\nu\rho\sigma}\epsilon^\mu(\omega)\epsilon_1^\nu p_2^\rho\epsilon_2^\sigma$
$\omega_T W^+W^-$	$i\frac{2e^2\chi}{\sin^2\theta_w}\varepsilon_{\mu\nu\rho\sigma}\epsilon^\mu(\omega)\epsilon_-^\nu(p_+^\rho-p_-^\rho)\epsilon_+^\sigma$

ten as [5]:

$$\mathcal{L}_{e.w.} = \chi\varepsilon_{\mu\nu\rho\sigma}[g'\mathrm{Tr}(\frac{\sigma^3}{2}B^{\rho\sigma}\{\Sigma^\dagger D^\nu\Sigma,\omega_T^\mu\})$$
$$-g\mathrm{Tr}(\frac{\vec{\sigma}}{2}\cdot\vec{W}^{\rho\sigma}\{\Sigma D^\nu\Sigma^\dagger,\omega_T^\mu\})] \quad (2)$$

where the covariant derivative is defined by

$$D^\mu\Sigma = \partial^\mu\Sigma - ig\ \vec{W}^\mu\cdot\frac{\vec{\sigma}}{2}\Sigma + ig'\ \Sigma B^\mu\frac{\sigma_3}{2} \quad (3)$$

g and g' are the $SU(2)_L$ and $U(1)_Y$ coupling constants, Σ is the non-linearly realized representation of the Goldstone boson fields and transforms like $\Sigma \to L\Sigma R^\dagger$. In the Unitary gauge $\Sigma \to 1$. With these substitutions $\mathcal{L}_{e.w.}$ leads to

$$\mathcal{L}_{e.w.} \sim 2ie^2\ \chi\varepsilon_{\mu\nu\rho\sigma}\ \omega_T^\mu\left[\frac{2}{\sin\theta_w\cos\theta_w}\partial^\rho\gamma^\sigma Z^\nu\right.$$
$$+\left(\frac{\cos^2\theta_w}{\sin^2\theta_w}-\frac{\sin^2\theta_w}{\cos^2\theta_w}\right)\partial^\rho Z^\sigma Z^\nu \quad (4)$$
$$\left.+\frac{1}{\sin^2\theta_w}(\partial^\rho W^{+\sigma}W^{-\nu}+\partial^\rho W^{-\sigma}W^{+\nu})\right] + \cdots$$

where θ_w is the electroweak mixing angle. The first term gives the $\omega_T Z\gamma$ vertex, of importance for production of the ω_T in $e\gamma$ colliders. It involves the unknown coupling parameter χ, where $e^2\chi$ may be of electroweak strength. Similarly, the $\omega_T ZZ$ and $\omega_T W^+W^-$ vertices, corresponding to the second and third terms, respectively, are proportional to χ. Hence, we take the partial width of the ω_T into these two body states, $\Gamma_{2-body} = \Gamma(\omega_T \to Z\gamma) + \Gamma(\omega_T \to W^+W^-) + \Gamma(\omega_T \to ZZ)$ as input, to determine this coupling.

The Feynman rules for the effective interactions of the ω_T, represented in Fig. 1, are given in Table 2.

3 Calculation and Results

We consider the two signal processes which proceed via an intermediate ω_T:

$$e^-\gamma \to e^-\omega_T \to e^-Z\gamma \quad (5)$$
$$e^-\gamma \to e^-\omega_T \to e^-W^+W^-Z. \quad (6)$$

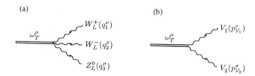

Figure 1: Effective interactions of ω_T with (a) w^+w^-z and (b) two vector bosons.

Depending upon the $\omega_T Z\gamma$ coupling, the signal cross section can be fairly large. The cross section expressions are lengthy and we will not present them here. We choose to look at these channels based on the distinctive signature of the first and the potential enhancement of the second arising from its dependence on the strong coupling g_ω.

The SM background to the process $e\gamma \to eZ\gamma$ is the bremsstrahlung of photons and Z's from the electron. The background contribution to the $e^-W^+W^-Z$ final state has a complicated structure, mainly from the subprocesses $e^+e^- \to W^+W^-, \gamma\gamma \to W^+W^-$ with a radiated Z. We use the MADGRAPH package[8] to evaluate the full SM amplitudes for the background processes.

In calculating the total cross sections for the $eZ\gamma$ signal and the backgrounds, we impose the following "basic cuts" to roughly simulate the detector coverage:

$$170° > \theta_e > 10°, \quad 165° > \theta_\gamma > 15°, \quad \theta_{e\gamma} > 30°,$$
$$E_\gamma > 50\ \text{GeV}, \quad (7)$$

where θ_e is the polar angle with respect to the e^- beam direction in the lab (e^+e^- c. m.) frame and $\theta_{e\gamma}$ is the angle between the outgoing e^- and γ. Only the cut on θ_e is relevant to the $e^-W^+W^-$ process. The cuts on photons also regularize the infrared and collinear divergences in the tree-level background calculations.

We have also implemented the back-scattered laser spectrum for the photon beam[9]. For simplicity, we have ignored the possible polarization for the electron and photon beams, although an appropriate choice of photon beam polarization may enhance the signal and suppress the backgrounds.

We present the total cross section for the signal and background versus the e^+e^- c. m. energy $\sqrt{s_{e^+e^-}}$ in Fig. 2 with various choices of ω_T mass and partial widths. We have taken representative values for M_{ω_T} of 0.8 (1.0) TeV. In Fig. 2(a), we use partial widths $\Gamma_{2-body} = 5$ (20) GeV and $\Gamma_{WWZ} = 20$ (80) GeV, setting $\Gamma_{WWZ} = 4\Gamma_{2-body}$. In Fig. 2(b), we take $\Gamma_{2-body} = 15$ (40) GeV and $\Gamma_{WWZ} = 30$ (80) GeV, such that $\Gamma_{WWZ} = 2\Gamma_{2-body}$. As noted above, the values for the couplings χ and g_ω are obtained using these partial widths as input. In Fig. 2(b), the 2-body decay modes represent a larger fraction of the total width and, hence, the cross sections for the signal processes, which go via

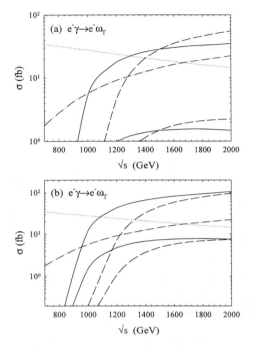

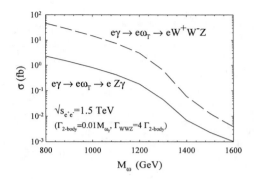

Figure 3: Cross sections versus the M_{ω_T} for the signal $e^-\gamma \to e^-\omega_T$ for $\sqrt{s_{e^+e^-}} = 1.5$ TeV with $\omega_T \to W^+W^-Z$ (dashed line) and $Z\gamma$ (solid line).

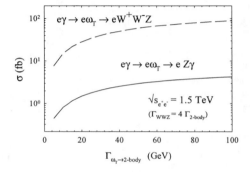

Figure 4: Cross sections versus the partial width Γ_{2-body} for the signal $e^-\gamma \to e^-\omega_T$ for $\sqrt{s_{e^+e^-}} = 1.5$ TeV and $M_{\omega_T} = 1$ TeV with $\omega_T \to W^+W^-Z$ (dashed line) and $Z\gamma$ (solid line).

Figure 2: Cross sections versus the e^+e^- c. m. energy for the signal $e^-\gamma \to e^-\omega_T$ with $\omega_T \to W^+W^-Z$ and $Z\gamma$ and the SM backgrounds. The solid lines are for $M_{\omega_T} = 0.8$ TeV and the long-dashed lines for $M_{\omega_T} = 1.0$ TeV In each case the curve with the larger cross section is for the W^+W^-Z final state and the lower for the $Z\gamma$ final state. The dash-dot line is for the $e^-W^+W^-Z$ SM background and the dotted line is for the $e^-Z\gamma$ SM background. In (a), $\Gamma_{WWZ} = 4\Gamma_{2-body}$ and in (b), $\Gamma_{WWZ} = 2\Gamma_{2-body}$. The choices of partial widths are given in the text.

$Z\gamma$ fusion, are enhanced due to the larger value of χ. We see that, for the parameters considered, the signal cross sections for the $e^-W^+W^-Z$ channel are about $10 - 100$ fb once above the mass threshold and overtake the background rates by as much as an order of magnitude. Such high production rates imply that the linear collider would have great potential to discover and study the ω_T. The cross sections for the $Z\gamma$ final state are lower as expected and lie below the background with only the cuts of Eq. (7) imposed.

The reason that the cross section for $M_{\omega_T} = 1.0$ TeV becomes larger than that for 0.8 TeV in Fig. 2(a) is because we have chosen relatively larger couplings for the 1.0 TeV ω_T, based on the input partial widths.

In Fig. 3, we show the total signal cross sections versus M_{ω_T} for $\sqrt{s_{e^+e^-}} = 1.5$ TeV. For simplicity, the couplings are the values obtained by taking $\Gamma_{2-body} = $

$1\% M_{\omega_T}$ and $\Gamma_{WWZ} = 4\Gamma_{2-body}$. As expected, below the mass threshold $\sqrt{s_{e\gamma}} < M_{\omega_T}$, the signal cross section drops sharply. However, depending on the broadness of the resonance, there is still non-zero signal cross section.

We have chosen partial decay widths of ω_T as input parameters to characterize its coupling strength. It is informative to explore how the cross section changes with the widths. Figure 4 demonstrates this point, for $\sqrt{s_{e^+e^-}} = 1.5$ TeV and $M_{\omega_T} = 1$ TeV, where we vary Γ_{2-body} and take $\Gamma_{WWZ} = 4\Gamma_{2-body}$. The signal cross section rate and relative branching fractions for the two channels would reveal important information for the underlying SEWS dynamics.

Although the SM backgrounds seem to be larger or, at best comparable, to the signal rate for the $Z\gamma$ channel, the final state kinematics is very different between

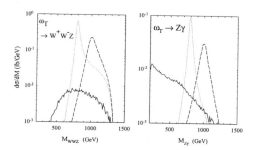

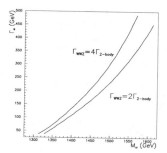

Figure 5: Differential cross sections for $\sqrt{s_{e^+e^-}} = 1.5$ TeV as a function of the invariant mass of the ω_T decay products $M(WWZ)$ and $M(Z\gamma)$ for the signal $e^-\gamma \to e^-\omega_T$ with $\omega_T \to W^+W^-Z$ and $Z\gamma$. The dotted lines are for $M_{\omega_T} = 0.8$ TeV ($\Gamma_{\omega_T} = \Gamma_{2-body} + \Gamma_{WWZ} = 15 + 30$ GeV) and the dashed lines for $M_{\omega_T} = 1.0$ TeV ($\Gamma_{\omega_T} = 40 + 80$ GeV). The SM backgrounds relevant to each case are given by the solid lines.

Figure 6: Contours representing 10 signal events in the parameter space for M_{ω_T} and Γ_{ω_T} with $\sqrt{s_{e^+e^-}} = 1.5$ TeV and an integrated luminosity of 200 fb^{-1}. Two choices of the ratio of Γ_{WWZ} to Γ_{2-body} are shown.

them. Because the final state vector bosons in the signal are from the decay of a very massive particle, they are generally very energetic and fairly central. We thus impose further cuts to reduce the backgrounds at little cost to the signal:

$$15° < \theta_{\gamma,Z,W} < 165°, \quad E_{\gamma,Z,W} > 150 \text{ GeV}. \quad (8)$$

The most distinctive feature for the signal is the resonance in the invariant mass spectrum for W^+W^-Z and $Z\gamma$ final states. We demonstrate this in Fig. 5 for both W^+W^-Z and $Z\gamma$ modes. The cuts in both Eqs. (7) and (8) are imposed. We see that a resonant structure at M_{ω_T} is evident and the SM backgrounds after cuts (8) are essentially negligible for the particular choice of parameters shown.

To further assess the discovery potential, we explore the parameter space for M_{ω_T} and Γ_{ω_T} at a 1.5 TeV linear collider. The signal for the WWZ mode consists of both W's decaying hadronically and the Z decaying hadronically or into electrons or muons. The same Z decay modes provide the signal for the $Z\gamma$ channel, along with the detected photon. We assume an 80% detection efficiency for each of the W, Z, and γ and an integrated luminosity of 200 fb^{-1}. The contours given in Fig. 6 represent 10 signal events. For these results, both cuts (7) and (8) are imposed. For the cases where the cuts reduce the background to an insignificant level, this gives a reasonable estimate of the discovery potential. We are undertaking a more detailed analysis for that part of the parameter space where the background is significant.

4 Conclusions

A high energy $e\gamma$ collider is unique in producing an isosinglet vector state such as ω_T. We calculated the signal cross sections for processes (5) and (6) in an effective Lagrangian framework. We found that signal rates can be fairly large once above the M_{ω_T} threshold, although the determining factor is the effective electroweak coupling of the ω_T, χ. The signal characteristics are very different from the SM backgrounds, making the discovery and further study of ω_T physics very promising at the linear collider. With an integrated luminosity of 200 fb^{-1} at $\sqrt{s_{e^+e^-}} = 1.5$ TeV, one may discover an ω_T for a broad range of masses and widths.

Acknowledgements

This research was supported in part by the Natural Sciences and Engineering Research Council of Canada, and in part by the U. S. Department of Energy under Grant No. DE-FG02-95ER40896. Further support for T.H. was provided by the University of Wisconsin Research Committee, with funds granted by the Wisconsin Alumni Research Foundation.

References

1. For a recent review on weakly coupled electroweak sector, see *e.g.*, H. E. Haber, T. Han, F. Merritt and J. Womersley, hep-ph/9703391, 1996 DPF/DPB Summer Study on *New Directions for High-Energy Physics* Snowmass, CO, 25 Jun - 12 Jul 1996.

2. For recent reviews on strongly coupled electroweak sector, see *e.g.*, R.S. Chivukula, M.J. Dugan, M. Golden, E.H. Simmons, Ann. Rev. Nucl. Part. Sci. **45** 255 (1995); T. Barklow *et al.*, hep-

ph/9704217, 1996 DPF/DPB Summer Study on *New Directions for High-Energy Physics* Snowmass, CO, 25 Jun - 12 Jul 1996; T. Han, hep-ph/9704215, in *Tegernsee 1996, The Higgs puzzle*, p.197, Ringberg, Germany, 8-13 Dec 1996.

3. For a modern review on technicolor theories, see *i.e.* K. Lane, hep-ph/9610463, published in ICHEP 96: p.367.

4. J. Bagger, V. Barger, K. Cheung, J. Gunion, T. Han, G. Ladinsky, R. Rosenfeld and C.P. Yuan, Phys. Rev. **D49**, 1246 (1994); Phys. Rev. **D52**, 3878 (1995).

5. R.S. Chivukula and M. Golden, Phys. Rev. **D41**, 2795 (1990).

6. J. Rosner and R. Rosenfeld, Phys. Rev. **D38**, 1530 (1988); R. Rosenfeld, Phys. Rev. **D41**, (1989).

7. T. Han, Z. Huang and P.Q. Hung, Mod. Phys. Lett. **A11**, 1131 (1996).

8. T. Stelzer and W. Long, Comput. Phys. Commun. **81**, 357 (1994).

9. I.F. Ginzburg *et al.*, Nucl. Instrum. Methods, **205**, 47 (1983); **219**, 5 (1984). C. Akerlof, Ann Arbor report UM HE 81-59 (1981; unpublished).

MAJORANA NEUTRINO TRANSITION MAGNETIC MOMENTS IN LEFT-RIGHT SYMMETRIC MODELS

M. CZAKON, J. GLUZA and M. ZRALEK

Department of Field Theory and Particle Physics
Institute of Physics, University of Silesia, Uniwersytecka 4
PL-40-007 Katowice, Poland
E-mails:czakon,gluza,zralek@us.edu.pl

Transition magnetic moments of Majorana neutrinos are discussed in the frame of the most natural version of the LR model (with left- and right-handed triplets and a bidoublet in the Higgs sector). We show that their largest values could be at most $6 \cdot 10^{-13} \mu_B$ from diagrams with W_L in the loop. This could happen for specific models where (i) neutrino-charged lepton mixing is maximal and (ii) $\kappa_1 \simeq \kappa_2$ (VEVs for neutral Higgs fields in the bidoublet ϕ are equal). Contributions from diagrams with charged Higgses in the loop are smaller than those in the SM with right-handed neutrinos.

1 Introduction

1.1 Bounds on ν-magnetic moments

The existence of a nonzero neutrino magnetic moment is a theoretically interesting issue in neutrino physics which is even strengthened by the first indication that neutrinos are massive particles [1]. Whether it is also an experimentally relevant quantity depends obviously on its magnitude.

Measurements of the $\nu_i e^- \rightarrow \nu_i e^-$ ($\nu_i = \bar{\nu}_e, \nu_e, \bar{\nu}_\mu$) and the $\nu_\tau e^- \rightarrow \nu_\tau e^-$ cross sections give the following limits [a2]

$$\mu_{\nu_e} \leq 1.8 \cdot 10^{-10} \mu_B, \tag{1}$$

$$\mu_{\nu_\mu} \leq 7.4 \cdot 10^{-10} \mu_B, \tag{2}$$

$$\mu_{\nu_\tau} \leq 5.4 \cdot 10^{-7} \mu_B. \tag{3}$$

There are also astrophysical bounds in addition to direct laboratory limits given above. In particular red giant luminosity and helium stars cooling by neutrinos emission impose ν-magnetic moments smaller than $10^{-12} \mu_B$ [3]. However, this limit is not as reliable as the terrestrial [4]. We expect that values in the range $\mu \sim 10^{-10} \div 10^{-12} \mu_B$ could have practical implications for the Sun [5], Supernova and/or neutron star physics [6].

Unfortunately, the Standard Model (SM) with its massless neutrinos and sole left-handed currents leaves no space for a nonvanishing neutrino magnetic moment. It is easily understandable. Since it arises from the operator $\sigma_{\mu\nu} q^\nu$ and as

$$\bar{\Psi}_f \sigma_{\mu\nu} \Psi_i = \bar{\Psi}_{fL} \sigma_{\mu\nu} \Psi_{iR} + \bar{\Psi}_{fR} \sigma_{\mu\nu} \Psi_{iL}, \tag{4}$$

---

[a]These limits are quoted by PDG. The authors know of no terrestrial experimental limits on transition magnetic moments (Majorana neutrinos), although there is a general consensus that they are at most of the same order.

we can see that there is a chirality change which makes it necessary to have both left and right handed particle states.

The easiest way to avoid this shortcoming of the SM is to add right-handed singlets with additional mass terms. The latter also change chirality.

As it was found such a theory yields [7]

$$\mu_{\nu_\alpha \nu_\beta} = \frac{3 e G_F}{16\sqrt{2}\pi^2}(m_\alpha + m_\beta) \sum_{l=e,\mu,\tau} Im\left(U^\dagger_{\beta l} U_{l\alpha}\right) \left(\frac{m_l}{M_W}\right)^2$$

$$\simeq 1.6 \times 10^{-19} \left(\frac{m_{\nu_\alpha} + m_{\nu_\beta}}{1 eV}\right)$$

$$\times \sum_{l=e,\mu,\tau} Im\left(U^\dagger_{\beta l} U_{l\alpha}\right) \left(\frac{m_l}{M_W}\right)^2. \tag{5}$$

Recent Superkamiokande as well as solar neutrino results and cosmological arguments suggest that neutrino masses are much smaller than present terrestrial bounds. Since obviously

$$\sum_{l=e,\mu,\tau} Im\left(U^\dagger_{\beta l} U_{l\alpha}\right) \left(\frac{m_l}{M_W}\right)^2 < 10^{-4}$$

we can safely assume that in the very best of the cases $\mu_{\nu_{e(\mu)}\nu_\tau} \leq 10^{-16} \mu_B$ for m_{ν_τ} of order of a few MeV.

This discouraging conclusion shows that we need more sophisticated models than the SM alone (with right-handed neutrinos) to get experimentally viable magnetic moments.

1.2 Models with large neutrino magnetic moments

Plenty of models with large ν-magnetic moments have emerged in literature during past decades [4,7]. Those that remain interesting from the phenomenological point of view can be subdivided into two categories with reference to our problem (i) renormalizable ones (charged scalars in the SM, left-right symmetric models,...) and (ii) finite ones (MSSM, supersymmetric left-right model,...). In the latter class there is a direct connection between the mass of the neutrino and its magnetic moment which requires special treatment (see the case of MSSM in [8]). The situation turns up to be much simpler in the former class where corrections to the mass are divergent. Then renormalization makes it a free parameter. As the magnetic moment contribution is always finite we can safely consider it alone.

So, how to make the magnetic moment contribution larger than in the SM?

First we need to generate a term containing no external neutrino mass. To one loop order this can only be done by left-right transition in vertices which is possible by adding

1. charged Higgs scalars, whose Yukawa couplings contain left-right transition from neutrinos to charged leptons,

and/or

2. right currents

Both of these are present in the left-right symmetric models, on which we shall concentrate in the present article.. Although many estimations have been done by other authors, too rough treatments led to quite contrary conclusions [10,11]. Here we attempt to clarify the situation by extending the calculation to the Higgs sector and using more phenomenological arguments.

2 ν-transition magnetic moments in Left-Right symmetric models

We concentrate on the popular version of the L-R symmetric model with Higgs bidoublet ϕ and two Higgs triplets $\Delta_{L,R}$. In such a model neutrinos have a Majorana character meaning only transition magnetic moments are allowed. Other Higgs sectors may lead to Dirac neutrinos, however the magnetic moment is very small [12].

Until now only diagrams with gauge bosons in the loop (see Fig.1) have been considered in the literature, with the dominant W_1 gauge boson contribution.

However, as the charged Higgs - leptons ($H^\pm l\nu$) coupling is proportional to heavy neutrino masses, even if $H^\pm$ masses are very large the contribution of diagrams with exchanged Higgs particles to μ_ν seems to be interesting.

The contribution to the $\mu_{\alpha\beta}$ can be described by the following diagrams[b] (1) with gauge bosons $W_{1,2}^\pm$ exchange (Fig.1), (2) with charged Higgs bosons $H_{1,2}$ exchange (Fig.2) , (3) with both $W_{1,2}^\pm$ and $H_{1,2}$ exchange (Fig.3).

---

[b]The calculation has been performed in the unitary gauge.

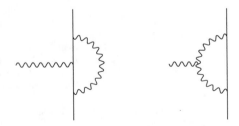

Figure 1: Diagrams for the neutrino magnetic moment with $W_{1,2}$ gauge bosons and charged leptons (wavy and solid lines, respectively) in the loop. External wavy(solid) line(s) stands for the photon(neutrinos).

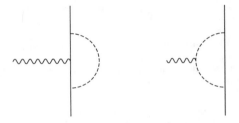

Figure 2: Diagrams for the neutrino magnetic moment with $H_{1,2}$ scalars (dashed lines) in the loop. External wavy(solid) line(s) stands for the photon(neutrinos).

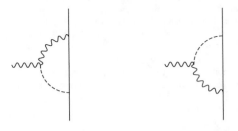

Figure 3: Diagrams for the neutrino magnetic moment with $H_{1,2}$ (dashed lines) and $W_{1,2}$ (wavy lines) charged particles in the loop. External wavy(solid) line(s) stands for the photon(neutrinos).

2.1 Diagrams with gauge bosons in the loop

As $M_{W_2} >> M_{W_1}$ the diagrams with W_1 exchange dominate. Their contribution to the $\mu_{\alpha\beta}$ can be classified into two categories

(1) diagrams with neutrino mass insertion on the external neutrino legs,

and

(2) a diagram with charged lepton mass insertion on the internal lepton leg.

Diagrams from class (1) are proportional to the sum of neutrino masses $m_\alpha + m_\beta$ and are the same as in Eq.(5). The diagram (2) is proportional to the $W_L - W_R$ mixing angle ξ, namely [7]

$$\mu_{\nu_a \nu_b} \simeq \frac{\sqrt{2} G_F}{\pi^2} \sin\xi \cos\xi \cdot m_e$$
$$\sum_{\alpha=e,\mu,\tau} m_\alpha Im\left[(K_R)_{a\alpha}(K_L^\dagger)_{\alpha b} + (K_L)_{a\alpha}(K_R^\dagger)_{\alpha b}\right]\mu_B.$$
$$(6)$$

The numerical value of $\mu_{\alpha\beta}$ depends on the mixing angle ξ and the $K_{R(L)}$ matrix elements. If we assume that the K_L matrix is diagonal (lepton number conservation), the K_R matrix elements are given by the see-saw mechanism

$$K_R \sim O\left(\frac{< m_D >}{M_N}\right) \quad (7)$$

and $K_R \sim 0.01$ for $< m_D >\sim 1$ GeV and $m_N = 100$ GeV. Taking also that (limits from fits to low energy data [13])

$$M_{W_2} \geq 477 \text{ GeV}$$
$$\xi \leq 0.031 \text{ rad} \quad (8)$$

we obtain (for e(μ) transition to τ neutrino)

$$\mu_{ab} \leq 6 \cdot 10^{-13}\mu_B \quad (9)$$

This value is at the edge of physical interest. However, (9) is very optimistic. For $M_{W_2} >> M_{W_1}$ we have

$$\xi \simeq \epsilon \left(\frac{M_{W_1}}{M_{W_2}}\right)^2 \quad (10)$$

¿From the above and (8) it follows that $\epsilon \simeq 1$ but this value (which is equivalent to $\kappa_1 \simeq \kappa_2$, $\kappa_{1,2}$ are VEV for the bidoublet ϕ) is very unprobable [14,15].

2.2 Diagrams with scalars in the loop

To calculate the diagrams of Fig.(2) and (3) we need couplings of charged Higgs particles to gauge bosons and leptons. To this end we must define the Yukawa and the Higgs sector. The latter is the most general potential [14] with a vanishing VEV for the neutral field of the left-handed triplet Δ_L, $v_L = 0$. All necessary couplings can be found in [16].

We found that all couplings vanish in the $\epsilon = \frac{2\kappa_1\kappa_2}{\kappa_1^2+\kappa_2^2} \to 0$ limit. For $\epsilon \neq 0$ the magnitude of the Higgs diagram contribution to $\mu_{\alpha\beta}$ is also very small. For example diagrams of Fig.(3) for H_2 and W_2 exchange yield

$$\mu_{ab} \simeq \frac{1}{\sqrt{2} \cdot 4\pi^2} f(\epsilon) \cdot \left(m_a^N + m_b^N\right) Im(K_L^\dagger K_L)$$
$$\times \frac{m_e}{M_{H_2}^2 - M_{W_2}^2}\left(\frac{2M_{H_2}^2}{M_{H_2}^2 - M_{W_2}^2} ln\left(\frac{M_{H_2}}{M_{W_2}}\right) - 1\right)\mu_B.$$
$$(11)$$

where $f(\epsilon) < 1$ and $f(\epsilon) \to 0$ for $\epsilon \to 0$. Taking $M_{W_2(H_2)} = 1(1.6)$ TeV we have

$$\mu_{ab} \leq f(\epsilon) \cdot 10^{-22}\left(\frac{m_a^N + m_b^N}{eV}\right)\mu_B. \quad (12)$$

Let's note that the expectation that the Higgs diagrams contain terms proportional to the heavy neutrino masses was incorrect. Only light masses remain after cancellation due to the Majorana nature of neutrinos [12]. Taking into account all diagrams (Figs.1-3) we can see that the dominant contribution to Majorana neutrino transition magnetic moment $\mu_{\alpha\beta}$ is given by Eq.(6).

We also see that $\mu_{\alpha\beta} \neq 0$ if $K_{L(R)}$ matrices are complex which is the case of broken CP symmetry. When CP is conserved the magnetic transition is possible only for neutrinos of opposite CP eigenvalues. For three light neutrinos this leaves two nonzero values at most.

3 Conclusion

We have recalculated the transition magnetic moments of Majorana neutrinos in the Left-Right symmetric model with a bidoublet and two triplets. We found that the contribution of the diagrams with Higgs scalar exchange is very small, smaller than the SM contribution (Eq.5). Only one diagram (with W_1 exchange) gives an interesting result which for $\mu_{\nu_e\nu_\mu}$ transition magnetic moment is

$$\mu_{\nu_e\nu_\mu}/\mu_B \simeq 7 \cdot 10^{-12}|(K_R)_{e\nu_e}|\left(\frac{0.5 TeV}{M_{W_2}(TeV)}\right)^2 \cdot \epsilon \quad (13)$$

For acceptable values of the K_R matrix elements, the mass of the heavy W_2 particle and the ϵ parameter, the value of $\mu_{\nu_e\nu_\mu}$ is however much too small to be interesting from experimental and astrophysical point of view.

Acknowledgements

This work was supported by Polish Committee for Scientific Researches under Grant No. 2P03B08414 and 2P03B04215.

References

1. Y. Fukuda et al. (Super-Kamiokande Collab.), ICRR-REPORT-418-98 and hep-ex/9805006; Y. Kajita, talk at 'Neutrinos 98', June 1998.
2. A.V. Derbin, Phys. Atom. Nucl., **57** 222 (1994); D.A. Krakauer et al., Phys. Lett. **B252**, 177 (1990); A.M. Cooper-Sarkar et al., Phys. Lett. **B280**, 153 (1992).
3. G. Raffelt, D. Dearborn, J. Silk, Astroph. J. **336** (1989); G. Raffelt, Phys. Rev. Lett. **64** 2856 (1990).
4. P. Pal, Int. J. Mod. Phys. **A7** (1992) 5387
5. E. Kh. Akhmedov, hep-ph/9705451.
6. K. Fujikawa, R. Shrock, Phys. Rev. Lett. **45**, 963 (1980); P.B. Pal, astro-ph/9803312 and references therein.
7. M. Fukugita, A. Suzuki (eds.) 'Physics and Astrophysics of Neutrinos', Springer-Verlag, Tokyo, 1994, 1 and references therein.
8. K.S. Babu, R.N.Mohapatra, Phys. Rev. Lett. **64**, 1705 (1990).
9. J.C. Pati and A. Salam, Phys. Rev. **D10**, 275 (1974); R. N. Mohapatra and J.C. Pati, ibid. **D11**, 566 (1975); 2558 (1975); G. Senjanovic and R.N. Mohapatra, ibid. **D12**, 1502 (1975); G. Senjanovic, Nucl. Phys. **B153**, 334 (1979).
10. M.A.B. Beg, W.J. Marciano, M. Ruderman, Phys. Rev. **D17**, 1395 (1978); M.B. Voloshin, M.I. Vysotskii and L.B. Okun, Sov. J. Nucl. Phys. **44**, 440(1986); M.B. Voloshin and M.I. Vysotskii, Sov. J. Nucl. Phys. **44**, 544 (1986).
11. P. Pal, R. N. Mohapatra 'Massive neutrinos in Physics and Astrophysics', World Scientific, Singapore, 1991; M. Fukugita, T. Yanagida, Phys. Rev. Lett. **58**, 1807 (1987).
12. M. Czakon, J. Gluza, M. Zrałek, to appear in Phys. Rev. **D**.
13. J. Polak, M. Zrałek, Nucl. Phys. **B363**, 385 (1991).
14. N.G. Deshpande, J.F. Gunion, B. Kayser and F. Olness, Phys. Rev. **D44**, 837 (1991); J.F. Gunion et al., Phys. Rev. **D40**, 1546 (1989).
15. G. Ecker, W. Grimus, W. Konetschny, Phys. Lett. **B94**, 381 (1980); Nucl. Phys. **B250**, 517 (1985).
16. J. Gluza and M. Zrałek, Phys. Rev. **D51**, 4695 (1995).

LIGHT RAYS OF NEW PHYSICS: CP VIOLATION IN $B \to X_s\gamma$ DECAYS

M. NEUBERT

Theory Division, CERN, CH-1211 Geneva 23, Switzerland
E-mail: Matthias.Neubert@cern.ch

The observation of a sizable direct CP asymmetry in the inclusive decays $B \to X_s\gamma$ would be a clean signal of New Physics. In the Standard Model, this asymmetry is below 1% in magnitude. In extensions of the Standard Model with new CP-violating couplings, large asymmetries are possible without conflicting with the experimental value of the $B \to X_s\gamma$ branching ratio. In particular, large asymmetries arise naturally in models with enhanced chromo-magnetic dipole transitions. Some generic examples of such models are explored and their implications for the semileptonic branching ratio and charm yield in B decays discussed.

1 Introduction

Studies of rare decays of B mesons have the potential to uncover the origin of CP violation and provide hints to physics beyond the Standard Model. The measurements of several CP asymmetries will make it possible to test whether the CKM paradigm is correct, or whether additional sources of CP violation are required. In order to achieve this goal, it is necessary that the theoretical calculations of CP-violating observables are, to a large extent, free of hadronic uncertainties. This can be achieved, e.g., by measuring time-dependent asymmetries in the decays of neutral B mesons into particular CP eigenstates. In many other cases, however, the theoretical predictions for direct CP violation in exclusive decays are obscured by strong-interaction effects [1-5].

Inclusive decay rates of B mesons, on the other hand, can be reliably calculated in QCD using the operator product expansion. Up to small bound-state corrections these rates agree with the parton model predictions for the underlying decays of the b quark. The disadvantage that the sum over many final states partially dilutes the CP asymmetries in inclusive decays is compensated by the fact that, because of the short-distance nature of these processes, the strong phases are calculable using quark–hadron duality. In this talk, I report on a study [6] of direct CP violation in the rare radiative decays $B \to X_s\gamma$, both in the Standard Model and beyond. These decays have already been observed experimentally, and copious data samples will be collected at the B factories. The theoretical analysis relies only on the weak assumption of global quark–hadron duality, and the leading nonperturbative corrections are well understood.

We perform a model-independent analysis of CP-violating effects in terms of the effective Wilson coefficients $C_7 \equiv C_7^{\rm eff}(m_b)$ and $C_8 \equiv C_8^{\rm eff}(m_b)$ multiplying the (chromo-) magnetic dipole operators $O_7 = e\, m_b\, \bar{s}_L \sigma_{\mu\nu} F^{\mu\nu} b_R$ and $O_8 = g_s m_b\, \bar{s}_L \sigma_{\mu\nu} G^{\mu\nu} b_R$ in the effective weak Hamiltonian, allowing for generic New Physics contributions to these coefficients. Several extensions of the Standard Model in which such contributions

arise have been explored, e.g., in Refs. 7–11. We find that in the Standard Model the direct CP asymmetry in $B \to X_s\gamma$ decays is very small (below 1% in magnitude) because of a combination of CKM and GIM suppression, both of which can be lifted in New Physics scenarios with additional contributions to the dipole operators containing new weak phases. We thus propose a measurement of the inclusive CP asymmetry in radiative B decays as a clean and sensitive probe of New Physics. Studies of direct CP violation in the inclusive decays $B \to X_s\gamma$ have been performed previously by several authors, both in the Standard Model [12] and in certain extensions of it [13,14]. In all cases rather small asymmetries were obtained. We generalize and extend these analyses in various ways. Besides including some contributions neglected in previous works, we investigate a class of New Physics models with enhanced chromo-magnetic dipole contributions, in which large CP asymmetries of order 10–50% are possible. We also employ a next-to-leading order analysis of the CP-averaged $B \to X_s\gamma$ branching ratio in order to derive constraints on the parameter space of the New Physics models considered.

2 Direct CP violation in $B \to X_s\gamma$ decays

The starting point in the calculation of the inclusive $B \to X_s\gamma$ decay rate is provided by the effective weak Hamiltonian renormalized at the scale $\mu = m_b$. Direct CP violation in these decays may arise from the interference of non-trivial weak phases, contained in CKM parameters or in possible New Physics contributions to the Wilson coefficient functions, with strong phases provided by the imaginary parts of the matrix elements of the operators in the effective Hamiltonian [15]. These imaginary parts first arise at $O(\alpha_s)$ from loop diagrams containing charm quarks, light quarks or gluons. Using the formulae of Greub et al. for these contributions [16], we calculate at next-to-leading order the difference $\Delta\Gamma = \Gamma(\bar{B} \to X_s\gamma) - \Gamma(B \to X_{\bar{s}}\gamma)$ of the CP-conjugate, inclusive decay rates. The contributions to $\Delta\Gamma$ from virtual corrections arise from interference of the one-loop diagrams with insertions of the operators O_2 and O_8 with the tree-level diagram containing O_7. Here $O_2 = \bar{s}_L \gamma_\mu q_L \, \bar{q}_L \gamma^\mu b_L$ with $q = c, u$ are the usual current–current operators in the effective Hamiltonian. There are also contributions to $\Delta\Gamma$ from gluon bremsstrahlung diagrams with a charm-quark loop. They can interfere with the tree-level diagrams for $b \to s\gamma g$ containing an insertion of O_7 or O_8. Contrary to the virtual corrections, for which in the parton model the photon energy is fixed to its maximum value, the gluon bremsstrahlung diagrams lead to a non-trivial photon spectrum, and so the results depend on the experimental lower cutoff on the photon energy. We define a quantity δ by the requirement that $E_\gamma > (1 - \delta) E_\gamma^{\mathrm{max}}$. Combining the two contributions and dividing the result by the leading-order expression for twice the CP-averaged inclusive decay rate, we find for the CP asymmetry

$$
\begin{aligned}
A_{\mathrm{CP}}^{b \to s\gamma}(\delta) &= \left. \frac{\Gamma(\bar{B} \to X_s\gamma) - \Gamma(B \to X_{\bar{s}}\gamma)}{\Gamma(\bar{B} \to X_s\gamma) + \Gamma(B \to X_{\bar{s}}\gamma)} \right|_{E_\gamma > (1-\delta)E_\gamma^{\mathrm{max}}} \\
&= \frac{\alpha_s(m_b)}{|C_7|^2} \left\{ \frac{40}{81} \operatorname{Im}[C_2 C_7^*] - \frac{4}{9} \operatorname{Im}[C_8 C_7^*] \right. \\
&\quad - \frac{8z}{9} \left[v(z) + b(z, \delta) \right] \operatorname{Im}[(1 + \epsilon_s) C_2 C_7^*] \\
&\quad \left. + \frac{8z}{27} b(z, \delta) \operatorname{Im}[(1 + \epsilon_s) C_2 C_8^*] \right\},
\end{aligned}
\tag{1}
$$

where $z = (m_c/m_b)^2$, and the explicit expressions for the functions $g(z)$ and $b(z, \delta)$ can be found in Ref. 6. The quantity ϵ_s is a ratio of CKM matrix elements given by

$$
\epsilon_s = \frac{V_{us}^* V_{ub}}{V_{ts}^* V_{tb}} \approx \lambda^2 (i\eta - \rho) = O(10^{-2}),
\tag{2}
$$

where $\lambda = \sin\theta_{\mathrm{C}} \approx 0.22$ and $\rho, \eta = O(1)$ are the Wolfenstein parameters. An estimate of the C_2–C_7 interference term in (1) was obtained previously by Soares [12], who neglected the contribution of $b(z, \delta)$ and used an approximation for the function $v(z)$. The relevance of the C_8–C_7 interference term for two-Higgs-doublet models, and for left–right symmetric extensions of the Standard Model, was explored in Refs. 13,14.

In the Standard Model, the Wilson coefficients take the real values $C_2 \approx 1.11$, $C_7 \approx -0.31$ and $C_8 \approx -0.15$. The imaginary part of the small quantity ϵ_s is thus the only source of CP violation. All terms involving this quantity are GIM suppressed by a power of the small ratio $z = (m_c/m_b)^2$, reflecting the fact that there is no non-trivial weak phase difference in the limit where $m_c = m_u = 0$. Hence, the Standard Model prediction for the CP asymmetry is suppressed by three small factors: $\alpha_s(m_b)$ arising from the strong phases, $\sin^2\theta_{\mathrm{C}}$ reflecting the CKM suppression, and $(m_c/m_b)^2$ resulting

Table 1: Values of the coefficients a_{ij} in %

δ	E_γ^{min} [GeV]	a_{27}	a_{87}	a_{28}
1.00	0.00	1.06	-9.52	0.16
0.30	1.85	1.23	-9.52	0.10
0.15	2.24	1.40	-9.52	0.04

from the GIM suppression. The numerical result for the asymmetry depends on the values of the strong coupling constant and the ratio of the heavy-quark pole masses, for which we take $\alpha_s(m_b) \approx 0.214$ (corresponding to $\alpha_s(m_Z) = 0.118$ and two-loop evolution down to the scale $m_b = 4.8\,\mathrm{GeV}$) and $\sqrt{z} = m_c/m_b = 0.29$. This yields $A_{\mathrm{CP,SM}}^{b \to s\gamma} \approx (1.5\text{–}1.6)\% \, \eta$ depending on the value of δ. With $\eta \approx 0.2\text{–}0.4$ as suggested by phenomenological analyses, we find a tiny asymmetry of about 0.5%, in agreement with the estimate obtained in Ref. 12. Expression (1) applies also to the decays $B \to X_d\gamma$, the only difference being that in this case the quantity ϵ_s must be replaced with the corresponding quantity $\epsilon_d = (V_{ud}^* V_{ub})/(V_{td}^* V_{tb}) \approx (\rho - i\eta)/(1 - \rho + i\eta) = O(1)$. Therefore, in the Standard Model the CP asymmetry in $B \to X_d\gamma$ decays is larger by a factor of about -20 than that in $B \to X_s\gamma$ decays. However, experimentally it is difficult to distinguish between $B \to X_s\gamma$ and $B \to X_d\gamma$ decays. If only their sum is measured, the CP asymmetry vanishes by CKM unitarity [12].

From (1) it is apparent that two of the suppression factors operative in the Standard Model, z and λ^2, can be avoided in models where the effective Wilson coefficients C_7 and C_8 receive additional contributions involving non-trivial weak phases. Much larger CP asymmetries then become possible. In order to investigate such models, we may to good approximation neglect the small quantity ϵ_s and write

$$
A_{\mathrm{CP}}^{b \to s\gamma}(\delta) = a_{27}(\delta) \operatorname{Im}\left[\frac{C_2}{C_7} \right] + a_{87} \operatorname{Im}\left[\frac{C_8}{C_7} \right] + a_{28}(\delta) \frac{\operatorname{Im}[C_2 C_8^*]}{|C_7|^2}.
\tag{3}
$$

The values of the coefficients a_{ij} are shown in Table 1 for three choices of the cutoff on the photon energy: $\delta = 1$ corresponding to the (unrealistic) case of a fully inclusive measurement, $\delta = 0.3$ corresponding to a restriction to the part of the spectrum above 1.85 GeV, and $\delta = 0.15$ corresponding to a cutoff that removes almost all of the background from B decays into charmed hadrons. In practice, a restriction to the high-energy part of the photon spectrum is required for experimental reasons. Note, however, that the result for the CP asymmetry is not very sensitive to the choice of the cutoff. Whereas the third term in (3) is generally very small, the first two terms can give rise to sizable effects. Assume, e.g., that there is a New Physics contribution to C_7 of similar magnitude as the Standard Model contribution (so as not to spoil the prediction for the $B \to X_s\gamma$ branching ratio) but with a non-trivial weak phase. Then the first term in (3) may give a contribution of up to about 5% in magnitude. Similarly, if there are New Physics contributions to C_7 and C_8 such that the ratio C_8/C_7 has a non-trivial weak phase, the second term may give a contribution of up to about 10% × $|C_8/C_7|$. In models with a strong enhancement of $|C_8|$ with respect to its Standard Model value, there is thus the possibility of generating a large direct CP asymmetry in $B \to X_s\gamma$ decays.

The impact of nonperturbative power corrections on the rate ratio defining the CP asymmetry is very small, since most of the corrections cancel between the numerator and the denominator. Potentially the most important bound-state effect is the Fermi motion of the b quark inside the B meson, which determines the shape of the photon energy spectrum in the endpoint region. This effect is included in the heavy-quark expansion by resumming an infinite set of leading-twist contributions into a "shape function", which governs the momentum distribution of the heavy quark inside the meson [17,18]. The physical decay distributions are obtained from a convolution of parton model spectra with this function. In the process, phase-space boundaries defined by parton kinematics are transformed into the proper physical boundaries defined by hadron kinematics. Details of the implementation of this effect can be found in Refs. 6,19. We note that the largest coefficient, a_{87}, is not affected by Fermi motion, and the impact on the other two coefficients is rather mild. As a consequence, the predictions for the CP asymmetry are quite insensitive to bound-state effects, even if a restriction on the high-energy part of the photon spectrum is imposed.

Below we explore the structure of New Physics models with a potentially large inclusive CP asymmetry. A non-trivial constraint on such models is that they must yield an acceptable result for the total, CP-averaged $B \to X_s\gamma$ branching ratio, which has been measured experimentally. Taking a weighed average of the results reported by the CLEO and ALEPH Collaborations [20,21] at this Conference gives $B(B \to X_s\gamma) = (3.14 \pm 0.48) \times 10^{-4}$. The complete theoretical prediction for the $B \to X_s\gamma$ branching ratio at next-to-leading order has been presented for the first time by Chetyrkin et al. [22], and subsequently has been discussed by several authors [23-25]. It depends on the Wilson coefficients C_2, C_7 and C_8 through the combinations $\mathrm{Re}[C_i C_j^*]$. Recently, we have extended these analyses in several aspects, including a discussion of Fermi motion effects and a conservative analysis of perturbative uncertainties [19]. In contrast to the case of the CP asymmetry, Fermi motion ef-

fects are very important when comparing experimental data for the $B \to X_s\gamma$ branching ratio with theoretical predictions. With our choice of parameters, we obtain for the total branching ratio in the Standard Model $B(B \to X_s\gamma) = (3.29 \pm 0.33) \times 10^{-4}$, which is in very good agreement with the experimental findings.

3 CP asymmetry beyond the Standard Model

In order to explore the implications of various New Physics scenarios for the CP asymmetry and branching ratio in $B \to X_s\gamma$ decays, we use the renormalization group to express the Wilson coefficients $C_7 = C_7^{\mathrm{eff}}(m_b)$ and $C_8 = C_8^{\mathrm{eff}}(m_b)$ in terms of their values at the high scale m_W, for which we write $C_{7,8}(m_W) = C_{7,8}^{\mathrm{SM}}(m_W) + C_{7,8}^{\mathrm{new}}(m_W)$. The first term corresponds to the Standard Model contributions, which are functions of the mass ratio $x_t = (m_t/m_W)^2$. Numerically, one obtains

$$C_7 \approx -0.31 + 0.67\, C_7^{\mathrm{new}}(m_W) + 0.09\, C_8^{\mathrm{new}}(m_W) ,$$
$$C_8 \approx -0.15 + 0.70\, C_8^{\mathrm{new}}(m_W) . \qquad (4)$$

We choose to parametrize our results in terms of the magnitude and phase of the New Physics contribution $C_8^{\mathrm{new}}(m_W) \equiv K_8\, e^{i\gamma_8}$ as well as the ratio

$$\xi = \frac{C_7^{\mathrm{new}}(m_W)}{Q_d\, C_8^{\mathrm{new}}(m_W)} , \qquad (5)$$

where $Q_d = -\frac{1}{3}$. A given New Physics scenario predicts these quantities at some large scale M. Using the renormalization group, it is then possible to evolve these predictions down to the scale m_W. Typically, $\xi \equiv \xi(m_W)$ tends to be smaller than $\xi(M)$ by an amount of order -0.1 to -0.3, depending on how close the New Physics is to the electroweak scale [6]. We restrict ourselves to cases where the parameter ξ in (5) is real; otherwise there would be even more potential for CP violation. This happens if there is a single dominant New Physics contribution, such as the virtual exchange of a new heavy particle, contributing to both the magnetic and the chromomagnetic dipole operators.

Ranges of $\xi(M)$ for several illustrative New Physics scenarios are collected in Table 2. For a detailed discussion of the model parameters which lead to the ξ values quoted in the table the reader is referred to Ref. 6. Our aim is not to carry out a detailed study of each model, but to give an idea of the sizable variation that is possible in ξ. It is instructive to distinguish two classes of models: those with moderate (class-1) and those with large (class-2) values of $|\xi|$. It follows from (4) that for small positive values of ξ it is possible to have large complex contributions to C_8 without affecting too much the

Table 2: Ranges of $\xi(M)$ for various New Physics contributions to C_7 and C_8, characterized by the particles in penguin diagrams

Class-1 models	$\xi(M)$
neutral scalar–vectorlike quark	1
gluino–squark ($m_{\tilde{g}} < 1.37 m_{\tilde{q}}$)	$-(0.13\text{--}1)$
techniscalar	≈ -0.5

Class-2 models	$\xi(M)$
scalar diquark–top	4.8–8.3
gluino–squark ($m_{\tilde{g}} > 1.37 m_{\tilde{q}}$)	$-(1\text{--}2.9)$
charged Higgs–top	$-(2.4\text{--}3.8)$
left–right W–top	≈ -6.7
Higgsino–stop	$-(2.6\text{--}24)$

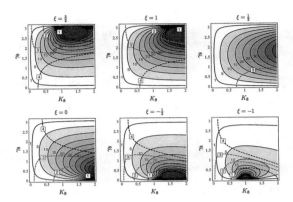

Figure 1: Contours for the CP asymmetry $A_{\mathrm{CP}}^{b \to s\gamma}$ for various class-1 models. We show contours only until values $A_{\mathrm{CP}} = 50\%$; for such large values, the theoretical expression (3) for the asymmetry would have to be extended to higher order to get a reliable result.

magnitude and phase of C_7, since

$$\frac{C_8}{C_7} \approx \frac{0.70 K_8\, e^{i\gamma_8} - 0.15}{(0.09 - 0.22\xi) K_8\, e^{i\gamma_8} - 0.31}. \tag{6}$$

This is also true for small negative values of ξ, albeit over a smaller region of parameter space. New Physics scenarios that have this property belong to class-1 and have been explored in Ref. 8. They allow for large CP asymmetries resulting from the C_7–C_8 interference term in (3). Figure 1 shows contour plots for the CP asymmetry in the (K_8, γ_8) plane for six different choices of ξ between $\frac{3}{2}$ and -1, assuming a cutoff $E_\gamma > 1.85\,\mathrm{GeV}$ on the photon energy. For each value of ξ, the plots cover the region $0 \leq K_8 \leq 2$ and $0 \leq \gamma_8 \leq \pi$ (changing the sign of γ_8 would only change the sign of the CP asymmetry). The contour lines refer to values of the asymmetry of 1%, 5%, 10%, 15% etc. The dashed lines indicate contours where the $B \to X_s\gamma$ branching ratio takes values between 1×10^{-4} and 4×10^{-4}, as indicated by the numbers inside the squares. The Standard Model prediction with this choice of the photon-energy cutoff is about 3×10^{-4}. The main conclusion to be drawn from the figure is that in class-1 scenarios there is a great potential for having a sizable CP asymmetry in a large region of parameter space. Any point to the right of the 1% contour for $A_{\mathrm{CP}}^{b \to s\gamma}$ cannot be accommodated by the Standard Model. Note that quite generally the regions of parameter space that yield large values for the CP asymmetry are not excluded by the experimental constraint on the CP-averaged branching ratio. To have a large CP asymmetry the products $C_i C_j^*$ are required to have large imaginary parts, whereas the total branching ratio is sensitive to the real parts of these quantities.

There are also scenarios in which the parameter ξ takes on larger negative or positive values. In such cases, it is not possible to increase the magnitude of C_8 much over its Standard Model value, and the only way to get a large CP asymmetry from the C_7–C_8 or C_7–C_2 interfer-

ence terms in (3) is to have C_7 tuned to be very small; however, this possibility is constrained by the fact that the total $B \to X_s\gamma$ branching ratio must be of an acceptable magnitude. That this condition starts to become a limiting factor is already seen in the plots corresponding to $\xi = -\frac{1}{2}$ and -1 in Figure 1. For even larger values of $|\xi|$, the C_7–C_8 interference term becomes ineffective, because the weak phase tends to cancel in the ratio C_8/C_7. Then the C_2–C_7 interference term becomes the main source of CP violation; however, as discussed in Section 2, it cannot lead to an asymmetry exceeding the level of about 5% without violating the constraint that the $B \to X_s\gamma$ branching ratio not be too small. Models of this type belong to the class-2 category. The branching-ratio constraint allows larger values of C_8 for positive ξ. For example, for $\xi \approx 5$, which can be obtained from scalar diquark–top penguins, asymmetries of 5–20% are still consistent with the $B \to X_s\gamma$ bound. On the other hand, for $\xi \approx -(2.5\text{--}5)$, which includes the multi-Higgs-doublet models, asymmetries of only a few percent are attainable, in agreement with the findings of previous authors[13,14,25].

The class-1 scenarios explored in Figure 1 have the attractive feature of a possible large enhancement of the magnitude of the Wilson coefficient C_8. This would have important implications for the phenomenology of the semileptonic branching ratio and charm yield in B decays, through enhanced production of charmless hadronic final states induced by the $b \to sg$ transition[7–9]. At $O(\alpha_s)$, the $B \to X_{sg}$ branching ratio is proportional to $|C_8|^2$. The left-hand plot in Figure 2 shows contours for this branching ratio in the (K_8, γ_8) plane. In the Standard Model, $\mathrm{B}(B \to X_{sg}) \approx 0.2\%$ is very small; however,

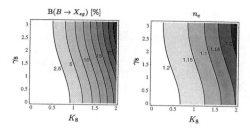

Figure 2: Contours for the $B \to X_{sg}$ branching ratio (left) and for the charm yield n_c in B decays (right)

in scenarios with $|C_8| = O(1)$ sizable values of order 10% for this branching ratio are possible, which simultaneously lowers the theoretical predictions for the semileptonic branching ratio and the charm production rate n_c by a factor of $[1 + B(B \to X_{sg})]^{-1}$. The current value of n_c reported by the CLEO Collaboration is [26] 1.12 ± 0.05. Although the systematic errors in this measurement are large, the result favours values of $B(B \to X_{sg})$ of order 10%. This is apparent from the right-hand plot in Figure 2, which shows the central theoretical prediction for n_c as a function of K_8 and γ_8. Note that there is an overall theoretical uncertainty in the value of n_c of about 6% resulting from the dependence on quark masses and the renormalization scale [27]. The theoretical prediction for the semileptonic branching ratio would have the same dependence on K_8 and γ_8, with the normalization $B_{SL} = (12 \pm 1)\%$ fixed at $K_8 = 0$. A large value of $B(B \to X_{sg})$ could also help in understanding the η' yields in charmless B decays [28,29]. For completeness, we note that the CLEO Collaboration has presented a preliminary upper limit on $B(B \to X_{sg})$ of 6.8% (90% CL) [30]. It is therefore worth noting that large CP asymmetries of order 10–20% can also be attained at smaller $B \to X_{sg}$ branching ratios of a few percent, which would nevertheless represent a marked departure from the Standard Model.

4   Conclusions

I have reported on a study of direct CP violation in the inclusive, radiative decays $B \to X_s\gamma$. From a theoretical point of view, inclusive decay rates entail the advantage of being calculable in QCD, so that a reliable prediction for the CP asymmetry can be confronted with data. From a practical point of view, it is encouraging that $B \to X_s\gamma$ decays have already been observed experimentally, and high-statistics measurements will be possible in the near future. We find that in the Standard Model the CP asymmetry in $B \to X_s\gamma$ decays is strongly suppressed by three small parameters: $\alpha_s(m_b)$ arising from the ne-

cessity of having strong phases, $\sin^2\theta_C \approx 5\%$ reflecting a CKM suppression, and $(m_c/m_b)^2 \approx 8\%$ resulting from a GIM suppression. As a result, the asymmetry is only of order 1% in magnitude – a conclusion that cannot be significantly modified by long-distance contributions. The latter two suppression factors are inoperative in extensions of the Standard Model for which the effective Wilson coefficients C_7 and C_8 receive additional contributions involving non-trivial weak phases. Much larger CP asymmetries are therefore possible in such cases.

A model-independent analysis of New Physics scenarios in terms of the magnitudes and phases of the Wilson coefficients C_7 and C_8 shows that sizable CP asymmetries are predicted in large regions of parameter space. Asymmetries of 10–50% are possible in models which allow for a strong enhancement of the coefficient of the chromo-magnetic dipole operator. They are, in fact, natural unless there is a symmetry that forbids new weak phases from entering the Wilson coefficients. Quite generally, having a large CP asymmetry is not in conflict with the observed value for the CP-averaged $B \to X_s\gamma$ branching ratio. Indeed, it may help to lower the theoretical predictions for the semileptonic branching ratio and charm multiplicity in B decays, thereby bringing these observables closer to their experimental values.

The fact that a large inclusive CP asymmetry in $B \to X_s\gamma$ decays is possible in many generic extensions of the Standard Model, and in a large region of parameter space, offers the possibility of looking for a signature of New Physics in these decays using data sets that will become available during the first period of operation of the B factories. A negative result of such a study would impose constraints on many New Physics scenarios. A positive signal, on the other hand, would provide interesting clues about the nature of physics beyond the Standard Model. In particular, a CP asymmetry exceeding the level of 10% would be a strong hint towards enhanced chromo-magnetic dipole transitions.

We have restricted our analysis to the case of inclusive radiative decays since they entail the advantage of being theoretically very clean. However, if there is New Physics that induces a large inclusive CP asymmetry in $B \to X_s\gamma$ decays it will inevitably also lead to sizable asymmetries in some related processes. In particular, since we found that the inclusive CP asymmetry remains almost unaffected if a cut on the high-energy part of the photon spectrum is imposed, we expect that a large asymmetry will persist in the exclusive decay mode $B \to K^*\gamma$, even though a reliable theoretical analysis would be much more difficult because of the necessity of calculating final-state rescattering phases [31]. Still, it would be worthwhile searching for CP violation in this channel.

Acknowledgments

The work reported here has been done in a most pleasant collaboration with Alex Kagan, which is gratefully acknowledged.

1. B. Blok and I. Halperin, Phys. Lett. B **385**, 324 (1996); B. Blok, M. Gronau and J.L. Rosner, Phys. Rev. Lett. **78**, 3999 (1997).

2. J.M. Gérard and J. Weyers, Preprint UCL-IPT-97-18 [hep-ph/9711469].

3. M. Neubert, Phys. Lett. B **424**, 152 (1998).

4. A.F. Falk, A.L. Kagan, Y. Nir and A.A. Petrov, Phys. Rev. D **57**, 4290 (1998).

5. D. Atwood and A. Soni, Phys. Rev. D **58**, 036005 (1998).

6. A.L. Kagan and M. Neubert, Preprint CERN-TH/98-1 [hep-ph/9803368], to appear in Phys. Rev. D.

7. B.G. Grzadkowski and W.-S. Hou, Phys. Lett. B **272**, 383 (1991).

8. A.L. Kagan, Phys. Rev. D **51**, 6196 (1995).

9. M. Ciuchini, E. Gabrielli and G.F. Giudice, Phys. Lett. B **388**, 353 (1996) [E: **393**, 489 (1997)].

10. A. Abd El-Hady and G. Valencia, Phys. Lett. B **414**, 173 (1997).

11. G. Barenboim, J. Bernabeu and M. Raidal, Phys. Rev. Lett. **80**, 4625 (1998).

12. J.M. Soares, Nucl. Phys. B **367**, 575 (1991).

13. L. Wolfenstein and Y.L. Wu, Phys. Rev. Lett. **73**, 2809 (1994).

14. H.M. Asatrian and A.N. Ioannissian, Phys. Rev. D **54**, 5642 (1996); H.M. Asatrian, G.K. Yeghiyan and A.N. Ioannissian, Phys. Lett. B **399**, 303 (1997).

15. M. Bander, D. Silverman and A. Soni, Phys. Rev. Lett. **43**, 242 (1979).

16. C. Greub, T. Hurth and D. Wyler, Phys. Rev. D **54**, 3350 (1996).

17. M. Neubert, Phys. Rev. D **49**, 3392 and 4623; T. Mannel and M. Neubert, Phys. Rev. D **50**, 2037 (1994).

18. I.I. Bigi, M.A. Shifman, N.G. Uraltsev and A.I. Vainshtein, Int. J. Mod. Phys. A **9**, 2467 (1994); R.D. Dikeman, M. Shifman and N.G. Uraltsev, Int. J. Mod. Phys. A **11**, 571 (1996).

19. A.L. Kagan and M. Neubert, Preprint CERN-TH/98-99 [hep-ph/9805303], to appear in Eur. Phys. J. C.

20. T. Skwarnicki (CLEO Collaboration), Talk no. 702 presented at this Conference.

21. R. Barate et al. (ALEPH Collaboration), Phys. Lett. B **429**, 169 (1998).

22. K. Chetyrkin, M. Misiak and M. Münz, Phys. Lett. B **400**, 206 (1997) [E: **425**, 414 (1998)].

23. A.J. Buras, A. Kwiatkowski and N. Pott, Phys. Lett. B **414**, 157 (1997); Nucl. Phys. B **517**, 353 (1998).

24. M. Ciuchini, G. Degrassi, P. Gambino and G.F. Giudice, Nucl. Phys. B **527**, 21 (1998).

25. F.M. Borzumati and C. Greub, Preprint ZU-TH 31/97 [hep-ph/9802391].

26. P. Drell, CLNS-97-1521 [hep-ex/9711020], to appear in the Proceedings of the 18th International Symposium on Lepton–Photon Interactions, Hamburg, Germany, July 1997.

27. M. Neubert and C.T. Sachrajda, Nucl. Phys. B **438**, 235 (1995).

28. W.-S. Hou and B. Tseng, Phys. Rev. Lett. **80**, 434 (1998).

29. A.L. Kagan and A.A. Petrov, Preprint UCHEP-97/27 [hep-ph/9707354].

30. T.E. Coan et al. (CLEO Collaboration), Phys. Rev. Lett. **80**, 1150 (1998).

31. C. Greub, H. Simma and D. Wyler, Nucl. Phys. B **434**, 39 (1995).

LESSONS FROM $\overline{B} \to X_s\gamma$ IN TWO HIGGS DOUBLET MODELS

F.M. BORZUMATI

Laboratoire de Physique Mathématique et Théorique, Université Montpellier II, F-3405 Montpellier Cedex 5, France
E-mail: francesc@lpm.univ-montp2.fr

C. GREUB

Institut für theoretische Physik, Universität Bern, Sidlerstrasse 5, 3012 Bern, Switzerland
E-mail: greub@itp.unibe.ch

The next–to–leading order predictions for the branching ratio BR($\overline{B} \to X_s\gamma$) are given in a generalized class of two Higgs doublets models. Included are the recently calculated leading QED corrections. It is shown that the high accuracy of the Standard Model calculation is in general not shared by these models. Updated lower limits on the mass of the charged Higgs boson in Two Higgs Doublet Models of Type II are presented.

1 Introduction

Two Higgs Doublet Models (2HDMs) are conceptually among the simplest extensions of the Standard Model (SM). They contain additional sources of flavour change due to their extended Higgs sectors. Studies of the $\overline{B} \to X_s\gamma$ decay in this class of models, therefore, can already test how unique is the accuracy of the SM result for the branching ratio BR($\overline{B} \to X_s\gamma$) at the next–to–leading (NLO) order in QCD[1], even requiring calculations at the same level of complexity as the SM one. They can obviously provide also important indirect bounds on the new parameters contained in these models. In spite of their apparent simplicity, indeed, 2HDMs have not been correctly constrained in ongoing experimental searches[2,3].

The well–known Type I and Type II models are particular examples of 2HDMs, in which the same or the two different Higgs fields couple to up– and down–type quarks. The second one is especially important since it has the same couplings of the charged Higgs H^+ to fermions that are present in the Minimal Supersymmetric Standard Model (MSSM). The couplings of the neutral Higgs to fermions, however, have important differences from those of the MSSM[2,3]. Since, beside the W, only charged Higgs bosons mediate the decay $\overline{B} \to X_s\gamma$ when additional Higgs doublets are present, the predictions of BR($\overline{B} \to X_s\gamma$) in a 2HDM of Type II give, at times, a good approximation of the value of this branching ratio in some supersymmetric models[4].

It is implicit in our previous statements that we do not consider scenarios with tree–level flavor changing couplings to neutral Higgs bosons. We do, however, generalize our class of models to accommodate Multi–Higgs Doublet models, provided only one charged Higgs boson remains light enough to be relevant for the process $\overline{B} \to X_s\gamma$. This generalization allows a simultaneous study of different models, including Type I and Type II, by a continuous variation of the (generally complex) charged Higgs couplings to fermions. It allows also a more complete investigation of the question whether the measurement of BR($\overline{B} \to X_s\gamma$) closes the possibility of a relatively light $H^\pm$ not embedded in a supersymmetric model.

This summer (1998), a new (preliminary) measurement of this decay rate was reported by the CLEO Collaboration[5] BR($\overline{B} \to X_s\gamma$) = $(3.15 \pm 0.35 \pm 0.32 \pm 0.26) \times 10^{-4}$, which, compared to older results, is based on 53% more data (3.3×10^6 $B\overline{B}$ events). The upper limit allowed by this measurement, reported in the same paper, is 4.5×10^{-4} at 95% C.L.. The ALEPH Collaboration has measured[6] BR($\overline{B} \to X_s\gamma$) = $(3.11 \pm 0.80 \pm 0.72) \times 10^{-4}$ from a sample of b hadrons produced at the Z–resonance.

The theoretical prediction for BR($\overline{B} \to X_s\gamma$) has a rather satisfactory level of accuracy in the SM. The main uncertainty, slightly below $\pm 10\%$, comes from the experimental error on the input parameters. The more genuinely theoretical uncertainty, due to the unknown value of the renormalization scale μ_b and the matching scale μ_W, which was unacceptably large at the leading–order (LO) level, was reduced to roughly $\pm 4\%$ when the NLO QCD corrections to the partonic decay width $\Gamma(b \to X_s\gamma)$ were completed[7]. (See ref.[1] for the milestone papers which brought to the complete LO calculation.) In addition, non–perturbative contributions to BR($\overline{B} \to X_s\gamma$), scaling like $1/m_b^2$[8] and $1/m_c^2$[9], were computed. Very recently, the leading QED and some classes of electroweak corrections were also calculated[10,11,12].

Following the procedure described in refs.[1,13], and including QED corrections as in ref.[12], we obtain a theoretical prediction[13] in agreement with the existing data:

$$\mathrm{BR}(\overline{B} \to X_s\gamma) = (3.32 \pm 0.14 \pm 0.26) \times 10^{-4}. \quad (1)$$

The first error in (1) is due to the μ_b and μ_W scale uncertainties; the second, comes from the experimental uncertainty in the input parameters.

A detailed study of $\overline{B} \to X_s \gamma$ at the NLO in QCD[1] in 2HDMs, on the contrary, shows that the NLO corrections and scale dependences in the Higgs contributions to BR$(\overline{B} \to X_s \gamma)$ are very large, irrespectively of the value of the charged Higgs couplings to fermions. This feature remains undetected in Type II models, where the SM contribution to BR$(\overline{B} \to X_s \gamma)$ is always larger than, and in phase with, the Higgs contributions. In this case, a comparison between theoretical and experimental results for BR$(\overline{B} \to X_s \gamma)$ allows to conclude that values of $m_{H^\pm} = O(m_W)$ can be excluded. Such values are, however, still allowed in other 2HD models.

These issues are illustrated in Sec. 4, after defining in Sec. 2 the class of 2HDMs considered, and presenting the NLO corrections at the amplitude level in Sec. 3. A brief discussion on the existing lower bounds on $m_{H^\pm}$ coming from direct searches at LEP is included in Sec. 3. This rests on observations brought forward in refs.[2,14] contributed to this conference.

2   Couplings of Higgs bosons to fermions

Models with n Higgs doublets have generically a Yukawa Lagrangian (for the quarks) of the form:

$$-\mathcal{L} = h_{ij}^d \, \overline{q}'_{Li} \, \phi_1 \, d'_{Rj} + h_{ij}^u \, \overline{q}'_{Li} \, \widetilde{\phi}_2 \, u'_{Rj} + \text{h.c.}, \qquad (2)$$

where q'_L, ϕ_i, $(i = 1, 2)$ are SU(2) doublets $(\widetilde{\phi}_i = i\sigma^2 \phi_i^\star)$; u'_R, d'_R are SU(2) singlets and h^d, h^u denote 3×3 Yukawa matrices. To avoid flavour changing neutral couplings at the tree–level, it is sufficient to impose that no more than one Higgs doublet couples to the same right–handed field, as in eq. (2).

After a rotation of the quark fields from the current eigenstate to the mass eigenstate basis, and an analogous rotation of the charged Higgs fields through a unitary $n \times n$ matrix U, we assume that only one of the $n - 1$ charged physical Higgs bosons is light enough to lead to observable effects in low energy processes. The n–Higgs doublet model then reduces to a generalized 2HDM, with the following Yukawa interaction for this charged physical Higgs boson denoted by H^+:

$$\mathcal{L} = \frac{g}{\sqrt{2}} \left\{ \frac{m_{di}}{m_W} X \, \overline{u}_{Lj} V_{ji} \, d_{Ri} + \frac{m_{ui}}{m_W} Y \, \overline{u}_{Ri} V_{ij} \, d_{Lj} \right\} H^+. \qquad (3)$$

In (3), V is the Cabibbo–Kobayashi–Maskawa matrix and the symbols X and Y are defined in terms of elements of the matrix U (see citations in ref.[1]). Notice that X and Y are in general complex numbers and therefore potential sources of CP violating effects. The ordinary Type I

and Type II 2HDMs (with $n = 2$), are special cases of this generalized class, with $(X, Y) = (-\cot\beta, \cot\beta)$ and $(X, Y) = (\tan\beta, \cot\beta)$, respectively.

We do not attempt to list here the generic couplings of fermions to neutral Higgs fields. In a 2HDM of Type II, the neutral physical states are two CP–even states h and H ($m_h < m_H$) and a CP–odd state A. In this case, only one additional rotation matrix is needed, parametrized by the angle α, which is independent of the rotation angle β of the charged sector. This independence stops to be true when this model is supersymmetrized since supersymmetry induces a relation between $\tan 2\alpha$ and $\tan 2\beta$.

3   NLO corrections at the amplitude level

The NLO corrections are calculated using the framework of an effective low–energy theory with five quarks, obtained by integrating out the the t–quark, the W–boson and the charged Higgs boson. The relevant effective Hamiltonian for radiative B–decays

$$\mathcal{H}_{eff} = -\frac{4G_F}{\sqrt{2}} V_{ts}^\star V_{tb} \sum_{i=1}^{8} C_i(\mu) \mathcal{O}_i(\mu) \qquad (4)$$

consists precisely of the same operators $\mathcal{O}_i(\mu)$ used in the SM case, weighted by the Wilson coefficients $C_i(\mu)$. The explicit form of the operators can be seen elsewhere[1].

Working to NLO precision means that one is resumming all the terms of the form $\alpha_s^n(m_b) \ln^n(m_b/M)$, as well as $\alpha_s(m_b) (\alpha_s^n(m_b) \ln^n(m_b/M))$. The symbol M stands for one of the heavy masses m_W, m_t or m_H which sets the order of magnitude of the matching scale μ_W. This resummation is achieved through the following 3 steps: 1) matching the full standard model theory with the effective theory at the scale μ_W. The Wilson coefficients are needed at the $O(\alpha_s)$ level; 2) evolving the Wilson coefficients from $\mu = \mu_W$ down to $\mu = \mu_b$, where μ_b is of the order of m_b, by solving the appropriate renormalization group equations. The anomalous dimension matrix has to be calculated up to order α_s^2; 3) including corrections to the matrix elements of the operators $\langle s\gamma | \mathcal{O}_i(\mu) | b \rangle$ at the scale $\mu = \mu_b$ up to order α_s.

Only step 1) gets modified when including the charged Higgs boson contribution to the SM one. The new contributions to the matching conditions have been worked out independently by several groups[15,16,1], by simultaneously integrating out all heavy particles, W, t, and H^+ at the scale μ_W. This is a reasonable approximation provided $m_{H^\pm}$ is of the same order of magnitude as m_W or m_t.

Indeed, the lower limit on $m_{H^\pm}$ coming from LEP I, of 45 GeV, guarantees already $m_{H^\pm} = O(m_W)$. There exists a higher lower bound from LEP II of 55 GeV for any value of $\tan\beta$[17] for Type I and Type II models, which has been recently criticized in ref. [3]. This criticism is based on the fact that there is no lower bound on m_A and/or m_h coming from LEP [2]. As it was already mentioned, in 2HDMs (unlike in the MSSM), the two rotation angles of the neutral and charged Higgs sector, α and β, are independent parameters. Therefore, the pair–production process $e^+e^- \to Z^* \to hA$ and the Bjorken process $e^+e^- \to Z^* \to Zh$, which are sufficient in the MSSM to put lower limits on m_h and m_A separately, imply in this case only that $m_h + m_A \gtrsim 100\,\text{GeV}$[18]. The other two production mechanisms possible at LEP I (they require larger numbers of events than LEP II can provide) are the decay $Z \to h/A\gamma$ and the radiation of A off $b\bar{b}$ and $\tau^+\tau^-$ pairs. The latter, allows for sizable rates only for large values of $\tan\beta$ and it yields the constraint $\tan\beta \leq 40$ for $m_A = 15\,\text{GeV}$, in a 2HDM of Type II [19]. The former one, limits weekly $\tan\beta$ to be in the range $\{0.2, 100\}$ for $m_{h/A} \approx 10\,\text{GeV}$ [14]. Indirect searches from the anomalous magnetic moment of the muon also lead to constraints on $\tan\beta$ for a light h and A. For a pseudoscalar this limit is stronger than that obtained from the radiation mechanism at LEP I only for $m_A < 2\,\text{GeV}$ [2].

Therefore, one of the neutral Higgs bosons can still be light. Charged Higgs bosons pair–produced at LEP II can then decay as $H^+ \to AW^+$ and/or hW^+ with an off–shell W boson. The rate is not negligible and invalidates the limit of 55 GeV in Type I models and models of Type II with modest $\tan\beta$, obtained considering only $c\bar{s}$ and $\tau^+\nu_\tau$ as possible decay modes of $H^\pm$. The unescapable limit of LEP I is however already large enough for a simultaneous integration out of H^+, W and t.

After performing steps *1)*, *2)*, and *3)*, it is easy to obtain the quark level amplitude $A(b \to s\gamma)$. As the matrix elements $\langle s\gamma|\mathcal{O}_i|b\rangle$ are proportional to the tree–level matrix element of the operator $\mathcal{O}_7$, the amplitude A can be written in the compact form

$$A(b \to s\gamma) = \frac{4G_F}{\sqrt{2}} V_{tb}V_{ts}^* \overline{D} \langle s\gamma|\mathcal{O}_7|b\rangle_{tree} \quad . \quad (5)$$

(It should be noticed that a subset of Bremsstrahlung contributions was transferred to (5), as described in ref. [1].) For the following discussion it is useful to decompose the reduced amplitude $\overline{D}$ in such a way that the dependence on the couplings X and Y (see eq. (3)) becomes manifest:

$$\overline{D} = \overline{D}_{SM} + XY^*\overline{D}_{XY} + |Y|^2\overline{D}_{YY} \quad . \quad (6)$$

In Fig. 1 the individual $\overline{D}$ quantities are shown in LO (dashed) and NLO (solid) order, for $m_{H^\pm} = 100\,\text{GeV}$, as a function of μ_b; all the other input parameters are taken at their central values, as specified in ref. [1]. To explain the situation, one can concentrate on the curves for $\overline{D}_{XY}$. Starting from the LO curve (dashed), the final NLO prediction is due to the change of the Wilson coefficient C_7,

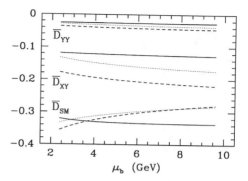

Figure 1: LO (dashed) and NLO (solid) predictions of the various pieces of the reduced amplitude $\overline{D}$ for $m_{H^\pm} = 100\,\text{GeV}$ (see text).

shown by the dotted curve, and by the inclusion of the virtual QCD corrections to the matrix elements. This results into a further shift from the dotted curve to the solid curve. Both effects contribute with the same sign and with similar magnitude, as it can be seen in Fig. 1. The size of the NLO corrections in the term $\overline{D}_{XY}$ in (6) is

$$\frac{\Delta\overline{D}_{XY}}{\overline{D}_{XY}^{LO}} \equiv \frac{\overline{D}_{XY}^{NLO} - \overline{D}_{XY}^{LO}}{\overline{D}_{XY}^{LO}} \sim -40\% \, ! \quad (7)$$

A similarly large correction is also obtained for $\overline{D}_{YY}$. For the SM contribution $\overline{D}_{SM}$, the situation is different: the corrections to the Wilson coefficient C_7 and the corrections due to the virtual corrections in the matrix elements are smaller individually, and furthermore tend to cancel when combined, as shown in Fig. 1.

The size of the corrections in $\overline{D}$ strongly depends on the couplings X and Y (see eq. (6)): $\Delta\overline{D}/\overline{D}$ is small, if the SM dominates, but it can reach values such as -50% or even worse, if the SM and the charged Higgs contributions are similar in size but opposite in sign.

4 Results and Conclusions

The branching ratio $\text{BR}(\overline{B} \to X_s\gamma)$ can be schematically written as

$$\text{BR}(\overline{B} \to X_s\gamma) \propto |\overline{D}|^2 + \cdots \quad , \quad (8)$$

where the ellipses stand for Bremsstrahlung contributions, electroweak corrections and non–perturbative effects. As required by perturbation theory, $|\overline{D}|^2$ in eq. (8) should be understood as

$$|\overline{D}|^2 = |\overline{D}^{LO}|^2 \left[1 + 2\text{Re}\left(\frac{\Delta\overline{D}}{\overline{D}^{LO}}\right) \right] \quad , \quad (9)$$

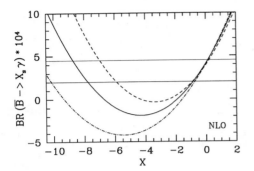

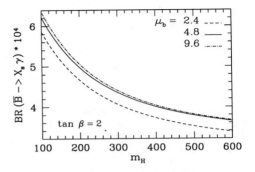

Figure 2: BR($\overline{B} \to X_s\gamma$) for Y = 1, $m_{H\pm} = 100$ GeV as a function of X, for $\mu_b = 4.8$ GeV (solid), $\mu_b = 2.4$ GeV (dahed) and $\mu_b = 9.6$ GeV (dash-dotted). Superimposed is the range of values allowed by the CLEO measurement.

Figure 3: BR($\overline{B} \to X_s\gamma$) in a Type II model with $\tan\beta = 2$, for various values of μ_b. The leading QED corrections are included (see text).

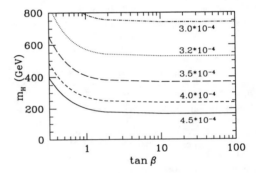

Figure 4: Contour plot in ($\tan\beta, m_{H\pm}$) in Type II models, obtained by using the NLO expression for BR($\overline{B} \to X_s\gamma$) and possible experimental upper bounds. The leading QED corrections are also included (see text). The allowed region is above the corresponding curves.

i.e., the term $|\Delta\overline{D}/\overline{D}^{LO}|^2$ is omitted. If Re($\Delta\overline{D}/\overline{D}^{LO}$) is larger than 50% in magnitude and negative, the NLO branching ratio becomes negative, i.e. the truncation of the perturbative series at the NLO level is not adequate for the corresponding couplings X and Y. This happens also for modest values of X and Y, as it is illustrated in Fig. 2, where only real couplings are considered. The values X = 1 and X = −1 in this figure, correspond respectively to the predictions of a the Type II and a Type I 2HDM with $\tan\beta = 1$.

Theoretical predictions for the branching ratio in Type II models stand, in general, on a rather solid ground. Fig. 3 shows the low–scale dependence of BR($\overline{B} \to X_s\gamma$) for matching scale $\mu_W = m_{H\pm}$, for $m_{H\pm} > 100$ GeV. It is less than ±10% for any value of $m_{H\pm}$ above the LEP I lower bound of 45 GeV. Such a small scale uncertainy is a generic feature of Type II models and remains true for values of $\tan\beta$ as small as 0.5. In this, as in the following figures where reliable NLO predictions are presented, the recently calculated leading QED corrections are included in the way discussed in the addendum [13] of ref. [1]. They are not contained in the result shown in Figs. 1 and 2, which have only an illustrative aim.

In Type II models, the theoretical estimate of BR($\overline{B} \to X_s\gamma$) can be well above the experimental upper bound of 4.5×10^{-4} (95% C.L.), reported by the CLEO Collaboration at this conference [5], leading to constraints in the ($\tan\beta, m_{H\pm}$) plane. The region excluded by the CLEO bound, as well as by other hypothetical experimental bounds, is given in Fig. 4. These contours are obtained minimizing the ratio BR($\overline{B} \to X_s\gamma$)/BR($b \to cl\nu_l$), when varying simultaneously the input parameters within their errors as well as the two scales μ_b and μ_W. For $\tan\beta = 0.5, 1, 5$, we exclude re-

spectively $m_{H\pm} \leq 280, 200, 170$ GeV, using the present upper bound from CLEO. Notice that the flatness of the curves shown in Fig. 3 towards the higher end of $m_{H\pm}$, causes a high sensitivity of these bounds on all details of the calculation (see ref. [1]). These details can only alter the branching ratio at the 1% level, i.e. well within the estimated theoretical uncertainty, but they may produce shifts of several tens of GeV, in either direction, in the lower bounds quoted above.

Also in the case of complex couplings, the results for BR($\overline{B} \to X_s\gamma$) range from ill–defined, to uncertain, up to reliable. One particularly interesting case in which the perturbative expansion can be safely truncated at the NLO level, is identified by: Y = 1/2, X = 2 exp($i\phi$), and $m_{H\pm} = 100$ GeV. The corresponding branching ratios, shown in Fig. 5, are consistent with the CLEO measurement, even for a relatively small value of $m_{H\pm}$ in a large

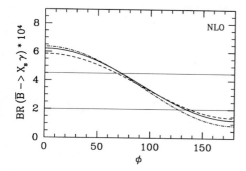

Figure 5: BR($\overline{B} \to X_s\gamma$) as a function of ϕ, where ϕ parametrizes $X = 2\exp(i\phi)$, for $Y = 1/2$, $m_{H\pm} = 100$ GeV. Solid, dashed and dash–dotted lines correspond to $\mu_b = 4.8, 2.4, 9.6$ GeV. The leading QED corrections are included (see text). Superimposed is the range of values allowed by the CLEO measurement.

range of ϕ. Such a light charged Higgs can contribute to the decays of the t–quark, through the mode $t \to H^+b$.

The imaginary parts in the X and Y couplings induce –together with the absorptive parts of the NLO loop-functions– CP rate asymmetries in $\overline{B} \to X_s\gamma$. A priori, these can be expected to be large. We find, however, that choices of the couplings X and Y which render the branching ratio stable, induce in general small asymmetries, not much larger than the modest value of 1% obtained in the SM.

We conclude with the most important lessons which can be extracted out of the calculation presented here. The high accuracy reached in the theoretical prediction of BR($\overline{B} \to X_s\gamma$) at the NLO for the SM, is not a general feature. In spite of its potential sensitivity to new sources of chiral flavour violation, the $\overline{B} \to X_s\gamma$ decay may turn out to be unconstraining for many models, because of the instability of NLO calculations. This is not only a temporary situation, since it is higly unlikely that a higher order QCD improvement is carried out. Nevertheless, there are scenarios in which BR($\overline{B} \to X_s\gamma$) can be reliably predicted at the NLO level as in 2HDMs of Type II. In these models, $m_{H\pm}$ can then be safely excluded up to values which depend on the experimentally maximal allowed value of BR($\overline{B} \to X_s\gamma$). Today, we find $m_{H\pm} \gtrsim 165$ GeV. It has to be stressed, however, that in the more general 2HDMs discussed in this article, H^+ can be much lighter and may still be detected as a decay product of the t–quark.

Acknowledgements

We thank M. Krawczyk for helpful discussions. This work was partially supported by CNRS and the Swiss National Foundation.

References

1. F. Borzumati and C. Greub, *Phys. Rev.* D 58 (1998) 07004.
2. M. Krawczyk and J. Zochowski, *Phys. Rev.* D 55 (1997) 6968.
3. F. Borzumati and A. Djouadi, hep-ph/9806301.
4. see for example: F. Borzumati, hep-ph/9702307.
5. M.S. Alam, CLEO CONF 98-17, contribution to this conference; J. Alexander, contribution to this conference.
6. R. Barate et al. (ALEPH Collab.), CERN-EP/98-044 (1998).
7. A. Ali and C. Greub, *Zeit. für Physik* C 49 (1991) 431; *Phys. Lett.* B 361 (1995) 146; N. Pott, *Phys. Rev.* D 54 (1996) 938; C. Greub, T. Hurth, and D. Wyler, *Phys. Lett.* B 380 (1996) 385; *Phys. Rev.* D 54 (1996) 3350; K. Adel and Y.P. Yao, *Phys. Rev.* D 49 (1994) 4945; C. Greub and T. Hurth, *Phys. Rev.* D 56 (1997) 2934; K. Chetyrkin, M. Misiak, and M. Münz, *Phys. Lett.* B 400 (1997) 206; A.J. Buras, A. Kwiatkowski, and N. Pott, *Nucl. Phys.* B 517 (1998) 353.
8. I.I. Bigi, M. Shifman, N.G. Uraltsev, and A.I. Vainshtein, *Phys. Rev. Lett.* 71 (1993) 496; A. Falk, M. Luke, and M. Savage, *Phys. Rev.* D 49 (1994) 3367, *Phys. Rev.* D 53 (1996) 2491; A.V. Manohar and M.B. Wise, *Phys. Rev.* D 49 (1994) 1310.
9. M.B. Voloshin, *Phys. Lett.* B 397 (1997) 295; Z. Ligeti, L. Randall, and M.B. Wise, *Phys. Lett.* B 402 (1997) 178; A.K. Grant, A.G. Morgan, S. Nussinov, and R.D. Peccei, *Phys. Rev.* D 56 (1997) 3151; G. Buchalla, G. Isidori, and S.J. Rey, *Nucl. Phys.* B 511 (1998) 594.
10. A. Czarnecki and W.J. Marciano, hep-ph/9804252.
11. A. Strumia, hep-ph/9804274.
12. A.Kagan and M.Neubert, hep-ph/9805303.
13. F. Borzumati and C. Greub, hep-ph/9809438.
14. M. Krawczyk, P. Mättig, and J. Zochowski, IFT 98/8, in preparation.
15. P. Ciafaloni, A. Romanino, and A. Strumia, *Nucl. Phys.* B 524 (1998) 361.
16. M. Ciuchini, G. Degrassi, P. Gambino, and G.F. Giudice, *Nucl. Phys.* B 527 (1998) 21.
17. P. Janot, in "International Europhysics Conference on HEP 1997", Jerusalem (Israel), August 1997.
18. K. Desch, contribution to this conference.
19. ALEPH Collab., contribution PA13-027 at ICHEP96, Warsaw 1996.

SCALAR-MEDIATED FLAVOR-CHANGING NEUTRAL CURRENTS

MARC SHER

Physics Dept., College of William and Mary, Williamsburg VA 23187 USA
E-mail: sher@physics.wm.edu

The simplest extension of the standard model involves adding a scalar doublet–the so-called two-Higgs model. In general, the additional scalar will mediate tree-level flavor-changing neutral currents. Although one can arbitrarily impose a discrete symmetry to avoid these, it isn't necessary to do so; reasonable assumptions about the size of the flavor-changing couplings can make them sufficiently small as to avoid problems in the kaon sector. However, these same assumptions give much larger effects in the third-family case. We discuss the model, the "reasonable assumptions" on the size of the couplings, and examine phenomenological bounds on the couplings, showing that the most promising signatures are from $B_s \to \mu\tau$, $\mu \to e\gamma$, etc. We then include the newest result which shows potentially significant effects on the anomalous magnetic moment of the muon.

One of the simplest, and most-often studied, extensions of the standard model is the "two-Higgs doublet" model, in which an additional scalar doublet is included. A potentially serious problem with such models is the existence of tree-level flavor-changing neutral currents. Although one can eliminate such currents with a discrete symmetry, such a symmetry is ad hoc, and may have cosmological problems. In this talk, I'll point out that the problem of flavor-changing neutral currents (FCNC) may not be as serious as often believed, and that reasonable assumptions about the structure of Yukawa coupling matrices allow sufficient suppression of FCNC couplings between the first two generations, while simultaneously predicting much bigger effects in coupling the second and third generations. The phenomenological bounds will be reviewed and the most promising decays examined. Finally, a new bound, from the anomalous magnetic moment of the muon, will be discussed.

It is easy to see why FCNC do not appear in the standard model. The Yukawa interactions in the standard model are of the form (focusing on the neutral fields only)

$$\mathcal{L}_Y = -\frac{h_{ij}}{\sqrt{2}}\overline{\psi}_i\psi_j\phi \tag{1}$$

After spontaneous symmetry breaking, this gives a mass matrix

$$M_{ij} = \frac{h_{ij}}{\sqrt{2}}\langle\phi\rangle \tag{2}$$

Clearly, when the mass matrix is diagonalized, the Yukawa coupling matrix will also be diagonalized, and thus there will be no neutral flavor-changing couplings involving the Higgs boson.

In the two doublet model, this is not the case. The Yukawa interactions are of the form

$$\mathcal{L}_Y = -\frac{f_{ij}}{\sqrt{2}}\overline{\psi}_i\psi_j\phi_1 - \frac{g_{ij}}{\sqrt{2}}\overline{\psi}_i\psi_j\phi_2 \tag{3}$$

which gives a mass matrix

$$M_{ij} = \frac{f_{ij}}{\sqrt{2}}\langle\phi_1\rangle + \frac{g_{ij}}{\sqrt{2}}\langle\phi_2\rangle \tag{4}$$

It is clear that diagonalizing M_{ij} will **not** generally diagonalize f_{ij} and g_{ij}, leading automatically to flavor-changing neutral couplings. These models will thus have a tree-level contribution to, for example, $K^o - \overline{K}^o$ mixing.

There are three solutions to this problem that have been discussed. The first two make use of the Glashow-Weinberg theorem[7], which states that neutral flavor-changing couplings will be absent if all of the quarks of a given charge couple only to a single Higgs doublet. In Model I, one imposes a discrete symmetry to couple only one doublet to the fermions. In Model II, one imposes a discrete symmetry to couple the $Q = 2/3$ quarks to one doublet and the $Q = -1/3$ quarks to the other (in supersymmtry, this discrete symmetry arises automatically). These both eliminate tree-level FCNC. However, they are *ad hoc*, with no justification other than eliminating FCNC, and discrete symmetries may have problems with cosmological domain walls. The third option, Model III, will be the focus of this talk. In this model, no discrete symmtries are introduced, and one looks at the phenomenological constraints on the flavor changing couplings.

To define the flavor-changing couplings, one first chooses a basis so that $\langle\phi_2\rangle = 0$. In that case, $M_{ij} = f_{ij}\langle\phi_1\rangle/\sqrt{2}$, and then diagonalization of M_{ij} will automatically diagonalize the couplings of ϕ_1. The couplings of ϕ_2 remain arbitrary:

$$\begin{aligned}\xi_{ij}^U\overline{Q}_{i,L}\tilde{\Phi}_2 U_{j,R} + \text{h.c.}\\ \xi_{ij}^D\overline{Q}_{i,L}\Phi_2 D_{j,R} + \text{h.c.}\end{aligned} \tag{5}$$

with a similar term for leptons. Note that the choice of basis will not, in general, diagonalize the Higgs mass matrices, and thus one should replace Φ_2 with each mass eigenstate times an appropriate mixing angle. Since the lightest mass eigenstate will dominate in the phenomenological effects, and since the couplings are arbitrary, this angle will be ignored.

The biggest danger from the above couplings is the contribution to $K^o - \overline{K}^o$ mixing, due to tree-level ϕ_2 exchange. Prior to 1987, it was claimed that the observed value of the mixing would be inconsistent with the presence of this effect unless the mass of the ϕ_2 was greater than 100 TeV. Since that is far beyond the electroweak scale, Model III was excluded.

However, as pointed out by Cheng and Sher[7], this result was based on the assumption that ξ_{ds}^D was of the order of the gauge coupling, which is certainly not reasonable for a Yukawa coupling involving the lightest generations. They noted that in a wide variety of models, if one assumes that there is no fine-tuning (in which several large numbers add or subtract to form a small number), then the flavor-changing couplings will be of the order of the geometric mean of the Yukawa couplings of the two generations. Thus, writing

$$\xi_{ij} = \lambda_{ij} \frac{\sqrt{m_i m_j}}{\langle \phi_2 \rangle} \tag{6}$$

they argued that the most natural value of the λ_{ij} is unity. This ansatz then strongly suppresses the value of ξ_{ds}, weakening the lower bound on ϕ_2 into the hundreds of GeV range.

Since we do not yet have a theory of flavor, it would be premature to make any firm assumptions about the magnitude of the λ_{ij}, rather, one should look for experimental signatures and bounds on these parameters, keeping in mind that in a variety of models, their value should be approximately of order one.

The most extensive discussion of the bounds on the λ_{ij} was given in a recent paper by Atwood, Reina and Soni[7]; most of these results are discussed in more detail there. The first process to examine is $F - \overline{F}$ mixing, where $F = (K, D, B)$. For a pseudoscalar mass of 500 GeV (the pseudoscalar gives the strongest bounds, since the matrix element is largest), the bounds are $\lambda_{ds} < 0.2$, $\lambda_{db} < 0.25$ and $\lambda_{uc} < 0.6$. One can see that the bounds involving the first generation are *very* stringent.

What about processes involving the third generation? Atwood, Reina and Soni show that the most important are $B_s \to \mu^+ \mu^-$ and $B \to X_s \mu^+ \mu^-$. The former gives a bound $\lambda_{sb}\lambda_{\mu\mu} < 9$ and the latter gives a bound $\lambda_{sb}\lambda_{\mu\mu} < 4$. The former bound depends on the pseudoscalar mass, and the latter on the scalar mass (a value of 200 GeV was chosen). Thus, these processes are getting into the interesting region. It was shown by Aliev, et al.[7] that the bound from $B \to X_s\gamma$ is not useful unless λ_{bb} is very large.

It should be pointed out that the above ansatz for the couplings indicate that the $tc\phi_2$ coupling could be very large. In papers by Bar-Shalom, et al.[7], the processes $e^+e^- \to \overline{t}c\nu_e\overline{\nu}_e$ and $\mu^+\mu^- \to \overline{t}c$ are examined; Han, et al.[7] examined the processes $t \to (g, Z, \gamma)$, and Hou[7] looked at the unusual process $e^+e^- \to tt\overline{cc}$. All of these processes, of course, require substantial top quark production.

In all of the above analyses, it was assumed that the flavor-changing couplings occurred in the quark sector. There is no reason, of course, why they should not also occur in the lepton sector. As pointed out by Sher and Yuan[7], standard unification arguments should give relationships between, say, λ_{sb} and $\lambda_{\mu\tau}$. They examined these relationships, and found bounds on some rather unusual (and therefore experimentally distinct) processes.

For example, from $B_s \to \mu\tau$ and $B \to K\mu\tau$, they found that $\sqrt{\lambda_{sb}\lambda_{\mu\tau}} < 10$; and from $B_s \to \tau\tau$ and $B \to K\tau\tau$, $\sqrt{\lambda_{sb}\lambda_{\tau\tau}} < 30$. Although these bounds are somewhat weaker than those involving just muons, they (especially $B_s \to \mu\tau$) are very experimentally distinct, and can thus be significantly improved at a B-factory. In fact, the current bound on $B_s \to \mu\tau$ was guessed from a published bound on $B \to \mu\mu X$; as yet there are **no published bounds** on the **most** distinct and interesting process!

In the purely leptonic sector, the strongest bound[7] comes from $\mu \to e\gamma$, with a τ intermediate state. This gives $\sqrt{\lambda_{\mu\tau}\lambda_{e\tau}} < 5$. They did not calculate the contribute to muon-electron conversion off a nucleus, which could give stronger bounds. Processes involving rare τ decays give extremely weak bounds.

Note that all of the bounds above depend on the product of two different λ's, with the exception of $F - \overline{F}$ mixing (which bounds λ's which have a first generation index). Since one might expect the biggest contribution to come from coupling between the second and third generations ($\lambda_{sb}, \lambda_{\mu\tau}, \lambda_{ct}$), it is desirable to look for processes which only depend on a single coupling. Such a process was recently discussed by Nie and Sher[7]. They noted that one can look at $g - 2$ for the muon. If one looks at the vertex correction, where the muons exchange a scalar and turn into τ's, then this contribution will only depend on $\lambda_{\mu\tau}$. Using current bounds, they find that $\lambda_{\mu\tau} < 50$; but this will be improved by a factor of 5 in an upcoming experiment. One might think that $\lambda_{\mu\tau} < 10$ is not an impressive bound, but recall that the above ansatz is just a guess, and that the actual bound on $\xi_{\mu\tau} \equiv \lambda_{\mu\tau}(\sqrt{m_\mu m_\tau}/v)$ would be $\xi_{\mu\tau} < \frac{1}{40}$, which is certainly in the interesting range.

In summary, one of the simplest extensions of the standard model is the addition of a scalar doublet. In general, this yields tree-level FCNC. Defining couplings as $\xi_{ij} \equiv \lambda_{ij}(\sqrt{m_i m_j}/v)$, theorists would generally expect λ_{ij} to be of order 1, however in view of our lack of understanding of flavor physics, one should look at the phenomenological bounds on these FCNC processes. Bounds from $F - \overline{F}$ mixing show that couplings involving the first generation are quite small. Couplings involving the second and third generations can be much larger. The strongest and most intriguing bound (because it has no one-loop counterpart in the standard model) is from $B_s \rightarrow \mu\tau$, which is not experimentally bounded in the literature. This bound, and others, depends on the product of two λ's. A recent analysis of the anomalous magnetic moment of the muon showed that it is only sensitive to $\lambda_{\mu\tau}$, and not to a product of two couplings.

References

1. S.L. Glashow and S. Weinberg, *Phys. Rev.* D **15**, 1958 (1977).
2. , T.P. Cheng and M. Sher, *Phys. Rev.* D **35**, 3484 (1987).
3. D. Atwood, L. Reina and A. Soni, *Phys. Rev.* D **55**, 3156 (1977).
4. T.M. Aliev and E.O. Iltan, hep-ph/9803272.
5. S. Bar-Shalom, G. Eilam and A. Soni, *Phys. Rev.* D **57**, 2957 (1998); *Phys. Rev. Lett.* **79**, 1217 (1997).
6. T. Han, K. Whisnant, B.L. Young and X. Zhang, *Phys. Lett.* B **385**, 311 (1996).
7. W.-S. Hou, G.-L. Lin and C.-Y. Ma, *Phys. Rev.* D **56**, 7434 (1997).
8. M. Sher and Y. Yuan, *Phys. Rev.* D **44**, 1461 (1991).
9. S. Nie and M. Sher, *Phys. Rev.* D **58**, 11xxxx (1998).

ANOMALOUS HIGGS COUPLINGS AT COLLIDERS[1]

M. C. Gonzalez-Garcia

Instituto de Física Corpuscular - C.S.I.C./Univ. de València. 46100 Burjassot, València, SPAIN

I summarize our results on the attainable limits on the coefficients of dimension–6 operators from the analysis of Higgs boson phenomenology using data taken at Tevatron RUNI and LEPII. Our results show that the coefficients of Higgs–vector boson couplings can be determined with unprecedented accuracy. Assuming that the coefficients of all "blind" operators are of the same magnitude, we are also able to impose bounds on the anomalous vector–boson triple couplings comparable to those from double gauge boson production at the Tevatron and LEPII

1 Introduction

Despite the impressive agreement of the Standard Model (SM) predictions for the fermion–vector boson couplings with the experimental results, the couplings among the gauge bosons are not determined with the same accuracy. The $SU_L(2) \times U_Y(1)$ gauge structure of the model completely determines these self–couplings, and any deviation can indicate the existence of new physics beyond the SM.

Effective Lagrangians are useful to describe and explore the consequences of new physics in the bosonic sector of the SM [2,3,4,5]. After integrating out the heavy degrees of freedom, anomalous effective operators can represent the residual interactions between the light states. Searches for deviations on the couplings WWV ($V = \gamma, Z$) have been carried out at different colliders and recent results [6] include the ones by CDF [7], and DØ Collaborations [8,9]. Forthcoming perspectives on this search at LEP II CERN Collider [10,11], and at upgraded Tevatron Collider [12] were also reported.

In the framework of effective Lagrangians respecting the local $SU_L(2) \times U_Y(1)$ symmetry linearly realized, the modifications of the couplings of the Higgs field (H) to the vector gauge bosons (V) are related to the anomalous triple vector boson vertex [3,4,5,13]. Here, I summarize our results on the attainable limits on the coefficients of dimension–6 operators from the analysis of Higgs boson phenomenology using data taken at Tevatron RUNI and LEPII. Our results show that the coefficients of Higgs-vector boson couplings can be determined with unprecedented accuracy. Assuming that the coefficients of all "blind" operators are of the same magnitude, we are also able to impose bounds on the anomalous vector–boson triple couplings comparable to those from double gauge boson production at the Tevatron and LEPII.

2 Effective Lagrangians

A general set of dimension–6 operators that involve gauge bosons and the Higgs scalar field, respecting local $SU_L(2) \times U_Y(1)$ symmetry, and C and P conserving, contains eleven operators [3,4]. Some of these operators either affect only the Higgs self–interactions or contribute to the gauge boson two–point functions at tree level and can be strongly constrained from low energy physics below the present sensitivity of high energy experiments [4,5]. The remaining five "blind" operators can be written as [3,4,5],

$$\mathcal{L}_{\text{eff}} = \sum_i \frac{f_i}{\Lambda^2} \mathcal{O}_i = \frac{1}{\Lambda^2} \Big[f_{WWW} \, Tr[\hat{W}_{\mu\nu} \hat{W}^{\nu\rho} \hat{W}_\rho^\mu]$$
$$+ f_W (D_\mu \Phi)^\dagger \hat{W}^{\mu\nu} (D_\nu \Phi) + f_B (D_\mu \Phi)^\dagger \hat{B}^{\mu\nu} (D_\nu \Phi) \quad (1)$$
$$+ f_{WW} \Phi^\dagger \hat{W}_{\mu\nu} \hat{W}^{\mu\nu} \Phi + f_{BB} \Phi^\dagger \hat{B}_{\mu\nu} \hat{B}^{\mu\nu} \Phi \Big]$$

where Φ is the Higgs field doublet, and

$$\hat{B}_{\mu\nu} = i(g'/2) B_{\mu\nu} \quad \hat{W}_{\mu\nu} = i(g/2) \sigma^a W_{\mu\nu}^a$$

with $B_{\mu\nu}$ and $W_{\mu\nu}^a$ being the field strength tensors of the $U(1)$ and $SU(2)$ gauge fields respectively.

Anomalous $H\gamma\gamma$, $HZ\gamma$, and HZZ and HWW and couplings are generated by (1), which, in the unitary gauge, are given by

$$\mathcal{L}_{\text{eff}}^{\text{H}} = g_{H\gamma\gamma} H A_{\mu\nu} A^{\mu\nu} + g_{HZ\gamma}^{(1)} A_{\mu\nu} Z^\mu \partial^\nu H$$
$$+ g_{HZ\gamma}^{(2)} H A_{\mu\nu} Z^{\mu\nu} + g_{HZZ}^{(1)} Z_{\mu\nu} Z^\mu \partial^\nu H$$
$$+ g_{HZZ}^{(2)} H Z_{\mu\nu} Z^{\mu\nu} + g_{HWW}^{(2)} H W_{\mu\nu}^+ W_-^{\mu\nu}$$
$$+ g_{HWW}^{(1)} \left(W_{\mu\nu}^+ W_-^\mu \partial^\nu H + h.c. \right) \quad (2)$$

where $A(Z)_{\mu\nu} = \partial_\mu A(Z)_\nu - \partial_\nu A(Z)_\mu$. The effective couplings $g_{H\gamma\gamma}$, $g_{HZ\gamma}^{(1,2)}$, and $g_{HZZ}^{(1,2)}$ and $g_{HWW}^{(1,2)}$ are related to the coefficients of the operators appearing in (1) through,

$$g_{H\gamma\gamma} = -\left(\frac{g M_W}{\Lambda^2} \right) \frac{s^2 (f_{BB} + f_{WW})}{2} ,$$
$$g_{HZ\gamma}^{(1)} = \left(\frac{g M_W}{\Lambda^2} \right) \frac{s(f_W - f_B)}{2c} ,$$
$$g_{HZ\gamma}^{(2)} = \left(\frac{g M_W}{\Lambda^2} \right) \frac{s[2s^2 f_{BB} - 2c^2 f_{WW}]}{2c} , \quad (3)$$

$$g_{HZZ}^{(1)} = \left(\frac{gM_W}{\Lambda^2}\right)\frac{c^2 f_W + s^2 f_B}{2c^2},$$

$$g_{HZZ}^{(2)} = -\left(\frac{gM_W}{\Lambda^2}\right)\frac{s^4 f_{BB} + c^4 f_{WW}}{2c^2},$$

$$g_{HWW}^{(1)} = \left(\frac{gM_W}{\Lambda^2}\right)\frac{f_W}{2},$$

$$g_{HWW}^{(2)} = -\left(\frac{gM_W}{\Lambda^2}\right)f_{WW},$$

with g being the electroweak coupling constant, and $s(c) \equiv \sin(\cos)\theta_W$.

Equation (1) also generates new contributions to the triple gauge boson vertex. Using the standard parametrization for the C and P conserving vertex [2]

$$\mathcal{L}_{WWV} = g_{WWV}\left\{ g_1^V\left(W_{\mu\nu}^+ W^{-\mu}V^\nu - W_\mu^+ V_\nu W^{-\mu\nu}\right) \right.$$
$$\left. + \kappa_V W_\mu^+ W_\nu^- V^{\mu\nu} + \frac{\lambda_V}{M_W^2}W_{\mu\nu}^+ W^{-\nu\rho}V_\rho{}^\mu \right\}, \quad (4)$$

where $V = Z, \gamma$, the coupling constants are $g_{WW\gamma} = e$ and $g_{WWZ} = e/(s\,c)$. The field-strength tensors include only the Abelian parts, i.e. $W^{\mu\nu} = \partial^\mu W^\nu - \partial^\nu W^\mu$ and $V^{\mu\nu} = \partial^\mu V^\nu - \partial^\nu V^\mu$, and

$$g_1^Z = 1 + \frac{1}{2}\frac{M_Z^2}{\Lambda^2}f_W,$$

$$\kappa_\gamma = 1 + \frac{1}{2}\frac{M_W^2}{\Lambda^2}\left(f_W + f_B\right), \quad (5)$$

$$\kappa_Z = 1 + \frac{1}{2}\frac{M_Z^2}{\Lambda^2}\left(c^2 f_W - c^2 f_B\right),$$

$$\lambda_\gamma = \lambda_Z = \frac{3}{2}s^2\frac{M_W^2}{\Lambda^2}f_{WWW}. \quad (6)$$

As seen above, the operators $\mathcal{O}_W$ and $\mathcal{O}_B$ give rise to both anomalous Higgs–gauge boson couplings and to new triple and quartic self–couplings amongst the gauge bosons, while the operator $\mathcal{O}_{WWW}$ solely modifies the gauge boson self–interactions. The operators $\mathcal{O}_{WW}$ and $\mathcal{O}_{BB}$ only affect HVV couplings, like HWW, HZZ, $H\gamma\gamma$ and $HZ\gamma$, since their contribution to the $WW\gamma$ and WWZ tree–point couplings can be completely absorbed in the redefinition of the SM fields and gauge couplings [13]. Therefore, one cannot obtain any constraint on these couplings from the study of anomalous trilinear gauge boson couplings. These anomalous couplings were studied in electron–positron collisions [13,15,16]. In particular, limits on f_{WW} and f_{BB} can be established via the reactions $e^+e^- \to \gamma\gamma\gamma$ and $e^+e^- \to b\bar{b}\gamma$ at LEP II, and read $|f/\Lambda| \leq 10^2$ TeV^{-2}, for $f_{WW} = f_{BB} = f$, and $80 \leq M_H \leq 100$ GeV.

3 New Higgs Signatures

In this talk I will review our results on Higgs production at the Fermilab Tevatron collider and at LEPII with its subsequent decay into two photons [1]. This channel in the SM occurs at one–loop level and it is quite small, but due to the new interactions (1), it can be enhanced and even become dominant. I will summarized our results on the signatures:

$$
\begin{aligned}
p\bar{p} &\to j\,j\,\gamma\,\gamma \\
p\bar{p} &\to \gamma\,\gamma + \not{E}_T \\
p\bar{p} &\to \gamma\,\gamma\,\gamma \\
e^+e^- &\to j\,j\,\gamma\,\gamma \\
e^+e^- &\to \gamma\,\gamma\,\gamma
\end{aligned}
\quad (7)
$$

We have included in our calculations all SM (QCD plus electroweak), and anomalous contributions that lead to these final states. The SM one–loop contributions to the $H\gamma\gamma$ and $HZ\gamma$ vertices were introduced through the use of the effective operators with the corresponding form factors in the coupling [18]. Neither the narrow–width approximation for the Higgs boson contributions, nor the effective W boson approximation were employed. We consistently included the effect of all interferences between the anomalous signature and the SM background. As an example of I quote here that 1928 SM amplitudes plus 236 anomalous ones, contribute to the process $p\bar{p} \to j\,j\,\gamma\,\gamma$ [19]. The SM Feynman diagrams corresponding to the background subprocess were generated by Madgraph [20] in the framework of Helas [21]. The anomalous couplings arising from the Lagrangian (1) were implemented in Fortran routines and were included accordingly. For the $p\bar{p}$ processes, we have used the MRS (G) [22] set of proton structure functions with the scale $Q^2 = \hat{s}$.

All processes listed in (7) have been the object of direct experimental searches. In our analysis we have closely followed theses searches in order to make our study as realistic as possible. In particular when studying the $\gamma\gamma jj$ final state we have closely followed the results recently presented by DØ Collaboration for $p\bar{p} \to \gamma\gamma jj$ events with high two–photon invariant mass [14].

For events containing two photons plus large missing transverse energy ($\gamma\gamma\,\not{E}_T$) as well as three photons in the final state we have used the results from DØ and CDF collaborations [23,24,27]. These events represent an important signature for some classes of supersymmetric models and in Refs.[23,24,27] the experimental collaborations use their results to set limits in some of the SUSY parameters. However, as we pointed out [1], these final states can also be a signal of Higgs production and subsequent decay into photons and can be used to place limits on the coefficients of the anomalous operators (1).

Finally, in order to obtain constraints on the anomalous couplings described above, we have also used the OPAL data [25,26] for the reactions,

$$e^+e^- \to \gamma\gamma ,\qquad\qquad (8)$$

$$e^+e^- \to \gamma\gamma + \text{hadrons} .\qquad\qquad (9)$$

As an example, I describe below more in detail our analysis of the $\gamma\gamma jj$ final state.

4 The process $p\bar{p} \to jj\gamma\gamma$: An Example

As mentioned before when studying the $\gamma\gamma jj$ final state we have closely followed the results presented by DØ Collaboration for $p\bar{p} \to \gamma\gamma jj$ events with high two–photon invariant mass [14]. The cuts applied on the final state particles are:

For the photons

$$|\eta_{\gamma 1}| < 1.1 \text{ or } 1.5 < |\eta_{\gamma 1}| < 2 \quad p_T^{\gamma 1} > 20 \text{ GeV}$$
$$|\eta_{\gamma 2}| < 1.1 \text{ or } 1.5 < |\eta_{\gamma 2}| < 2.25 \, p_T^{\gamma 2} > 25 \text{ GeV}$$
$$\sum \vec{p}_T^{\,\gamma} > 10 \text{ GeV}$$

For the $l\nu\gamma\gamma$ final state

$$|\eta_e| < 1.1 \text{ or } 1.5 < |\eta_e| < 2 \qquad |\eta_\mu| < 1$$
$$p_T^{e,\mu} > 20 \text{ GeV} \qquad\qquad \not{p}_T > 20 \text{ GeV}$$

For the $jj\gamma\gamma$ final state

$$|\eta_{j1}| < 2 \qquad\qquad\qquad p_T^{j1} > 20 \text{ GeV}$$
$$|\eta_{j2}| < 2.25 \qquad\qquad\qquad p_T^{j2} > 15 \text{ GeV}$$
$$\sum \vec{p}_T^{\,j} > 10 \text{ GeV} \qquad\qquad R_{\gamma j} > 0.7$$
$$40 \le M_{jj} \le 150 \text{ GeV}$$

We also assumed an invariant–mass resolution for the two–photons of $\Delta M_{\gamma\gamma}/M_{\gamma\gamma} = 0.15/\sqrt{M_{\gamma\gamma}} \oplus 0.007$ [17]. Both signal and background were integrated over an invariant–mass bin of $\pm 2\Delta M_{\gamma\gamma}$ centered around M_H. Finally, we isolate the majority of events due to associated production, and the corresponding background, by integrating over a bin centered on the W or Z mass, which is equivalent to the invariant mass cut listed above.

After imposing all the cuts, we get a reduction on the signal event rate which depends on the Higgs mass. For instance, for the $jj\gamma\gamma$ final state the geometrical acceptance and background rejection cuts account for a reduction factor of 15% for $M_H = 60$ GeV rising to 25% for $M_H = 160$ GeV. We also include in our analysis the particle identification and trigger efficiencies. For leptons and photons they vary from 40% to 70% per particle [8,9]. For the $jj\gamma\gamma$ final state we estimate the total effect of these efficiencies to be 35%. We therefore obtain an overall efficiency for the $jj\gamma\gamma$ final state of 5.5% to 9% for $M_H = 60\text{--}160$ GeV in agreement with the results of Ref. [14].

Dominant backgrounds are due to missidentification when a jet fakes a photon. The probability for a jet to fake a photon has been estimated to be of a few times 10^{-4} [8]. Although this probability is small, it becomes the main source of background for the $jj\gamma\gamma$ final state because of the very large multijet cross section. In Ref. [14] this background is estimated to lead to 3.5 ± 1.3 events with invariant mass $M_{\gamma\gamma} > 60$ GeV and it has been consistently included in our derivation of the attainable limits.

5 Results and Conclusion

I now present our results on the attainable limits on the coefficients of the anomalous operators. In order to establish these bounds on the coefficients in each process, we imposed an upper limit on the number of signal events based on Poisson statistics. In the absence of background this implies $N_{\text{signal}} < 1\,(3)$ at 64% (95%) CL. In the presence of background events, we employed the modified Poisson analysis. We are currently working on the statistical combination of the information from the different final states [28].

The coupling $H\gamma\gamma$ derived from (2) involves f_{WW} and f_{BB} [13]. In consequence, the anomalous signature $f\bar{f}\gamma\gamma$ is only possible when those couplings are not vanishing. The couplings f_B and f_W, on the other hand, affect the production mechanisms for the Higgs boson. In Fig. 1.a we present our results for the excluded region in the f_{WW}, f_{BB} plane from the different channels studied [1] for $M_H = 100$ GeV assuming that these are the only non-vanishing couplings. Since the anomalous contribution to $H\gamma\gamma$ is zero for $f_{BB} = -f_{WW}$, the bounds become very weak close to this line, as is clearly shown in Fig. 1. In Fig. 1.b we show the preliminary results for the same plot after combining all the channels. As seen in the figure, one expects a clear improvements of the individual bounds, when the information from all channels is combined.

These bounds depend on the Higgs mass and became weaker as the Higgs boson becomes heavier. In Table 1 we display the allowed values for f/Λ^2, at 95% CL, from $\gamma\gamma jj$ Tevatron D0 data analysis assuming that $f_{WW} = f_{BB}$ and $f_W = f_B = 0$ for different Higgs masses. For the sake of completeness we also show the accessible bound for future Tevatron Upgrades. We should remind that this scenario will not be restricted by data on W^+W^- production since there is no trilinear vector boson couplings involved. Therefore the limits here presented are the only existing direct bounds on these operators.

One may wonder how reasonable are these bounds, or how they compare with other existing limits on the coefficients of other dimension-six operator. In order to address this question one can make the assumption that

Table 1: Allowed range of f/Λ^2 in TeV^{-2} at 95% CL, assuming that $(f_{BB} = f_{WW} \gg f_B, f_W)$ for the different final states, and for different Higgs boson masses

M_H (GeV)		100	150	200	250
$jj\gamma\gamma$	RunI	$(-20 - 49)$	$(-26 - 64)$	$(-96 -> 100)$	$(< -100 -> 100)$
	RunII	$(-8.4 - 26)$	$(-11 - 31)$	$(-36 - 81)$	$(-64 -> 100)$
	TeV33	$(-4.2 - 6.5)$	$(-4.5 - 12)$	$(-19 - 40)$	$(-28 - 51)$

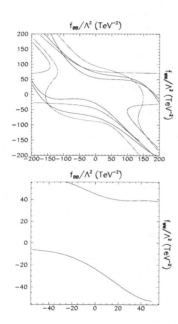

Figure 1: a)Exclusion region outside the curves in the $f_{BB} \times f_{WW}$ plane, in TeV^{-2}, based on the CDF analysis of $\gamma\gamma\gamma$ production (most external black lines), on the D0 analysis of $\gamma\gamma jj$ production (most internal black lines), on the D0 analysis of $\gamma\gamma$ E_T (blue lines), and on the OPAL analysis of $\gamma\gamma\gamma$ production (red lines), always assuming $M_H = 100$ GeV. The curves show the 95% CL deviations from the SM total cross section. b) Same as a) for all processes combined.

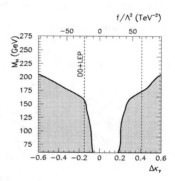

Figure 2: Excluded region in the $\Delta\kappa_\gamma \times M_H$ plane from the combined analysis from the combined results of the $\gamma\gamma\gamma$ production at LEP, $\gamma\gamma\gamma$, $\gamma\gamma + E_T$, and $\gamma\gamma jj$ production at Tevatron, assuming that all f_i are equal (see text for details).

all blind operators affecting the Higgs interactions have a common coupling f [5], i.e.

$$f_W = f_B = f_{WW} = f_{BB} = f , \qquad (10)$$

In this scenario, $g_{HZ\gamma}^{(1)} = g_{HZZ}^{(3)} = 0$, and we can relate the Higgs boson anomalous coupling f with the conventional parametrization of the vertex WWV $(V = Z, \gamma)$

$$\Delta\kappa_\gamma = \frac{M_W^2}{\Lambda^2} f = c^2 \Delta g_1^Z \quad, \Delta\kappa_Z = \frac{(1 - 2s^2)}{c^2} \Delta\kappa_\gamma . \qquad (11)$$

In Fig. 2, we show the region in the $\Delta\kappa_\gamma \times M_H$ that can be excluded through the combined analysis of the $\gamma\gamma\gamma$ production at LEP, $\gamma\gamma\gamma$, $\gamma\gamma + E_T$, and $\gamma\gamma jj$ production at Tevatron [28]. For the sake of comparison, we also show in Fig. 2 the best available experimental limit on $\Delta\kappa_\gamma$ from double gauge boson production at Tevatron and LEP II [29]. In all cases the results were obtained assuming the HISZ scenario. We can see that, for $M_H \leq 170$ GeV, the limit that can be established at 95% CL from the Higgs production analysis is tighter than the present limit coming from gauge boson production.

In conclusion, we have shown that the analysis of an anomalous Higgs boson production at the Fermilab

Tevatron and the CERN LEP II collider may be used to impose strong limits on new effective interactions. Under the assumption that the coefficients of the four "blind" effective operators contributing to Higgs–vector boson couplings are of the same magnitude, the study can give rise to a significant indirect limit on anomalous $WW\gamma$ couplings. Furthermore, this analysis is able to set constraints on those operators contributing to new Higgs interactions for Higgs masses far beyond the kinematical reach of LEP II.

References

1. Based on the work by F. de Campos, M.C. Gonzalez-Garcia y S.F.Novaes, Phys. Rev. Lett. **79**, 5210 (1997); M.C. Gonzalez-Garcia, S.M. Lietti, S.F. Novaes, Phys. Rev. **D57**, 7045 (1998); O.J.P. Eboli, M.C. Gonzalez-Garcia, S.M. Lietti, S.F. Novaes, hep-ph/9802408, To appear in Physics Letters B; F. de Campos, M.C. Gonzalez-Garcia, S.M. Lietti, S.F. Novaes, R. Rosenfeld hep-ph/9806307, To appear In Physics Letters B.

2. K. Hagiwara, H. Hikasa, R. D. Peccei and D. Zeppenfeld, Nucl. Phys. **B282**, 253 (1987).

3. C. J. C. Burguess and H. J. Schnitzer, Nucl. Phys. **B228**, 464 (1983); C. N. Leung, S. T. Love and S. Rao, Z. Phys. **31**, 433 (1986); W. Buchmüller and D. Wyler, Nucl. Phys. **B268**, 621 (1986).

4. A. De Rujula, M. B. Gavela, P. Hernandez and E. Masso, Nucl. Phys. **B384**, 3 (1992); A. De Rujula, M. B. Gavela, O. Pene and F. J. Vegas, Nucl. Phys. **B357**, 311 (1991).

5. K. Hagiwara, S. Ishihara, R. Szalapski and D. Zeppenfeld, Phys. Lett. **B283**, 353 (1992); *idem*, Phys. Rev. **D48**, 2182 (1993); K. Hagiwara, T. Hatsukano, S. Ishihara and R. Szalapski, Nucl. Phys. **B496**, 66 (1997).

6. For a review see: T. Yasuda, report FERMILAB-Conf–97/206–E, and hep-ex/9706015.

7. F. Abe *et al.*, CDF Collaboration, Phys. Rev. Lett. **74**, 1936 (1995); *idem* **74**, 1941 (1995); *idem* **75**, 1017 (1995); *idem* **78**, 4536 (1997).

8. S. Abachi *et al.*, DØ Collaboration, Phys. Rev. Lett. **75**, 1023 (1995); *idem* **75**, 1028 (1995); *idem* **75**, 1034 (1995); *idem* **77**, 3303 (1996); *idem* **78**, 3634 (1997); *idem* **78**, 3640 (1997).

9. B. Abbott *et al.*, DØ Collaboration, Phys. Rev. Lett. **79**, 1441 (1997).

10. For a review see: Z. Ajaltuoni *et al.*, "Triple Gauge Boson Couplings", in *Proceedings of the CERN Workshop on LEP II Physics*, edited by G. Altarelli, T. Sjöstrand, and F. Zwirner, CERN 96–01, Vol. 1, p. 525 (1996), and hep-ph/9601233.

11. T. Barklow *et al.*, Summary of the Snowmass Sub-group on Anomalous Gauge Boson Couplings, to appear in the *Proceedings of the 1996 DPF/DPB Summer Study on New Directions in High-Energy Physics*, June 25 — July 12 (1996), Snowmass, CO, USA, and hep-ph/9611454.

12. D. Amidei *et al.*, *Future Electroweak Physics at the Fermilab Tevatron: Report of the TeV-2000 Study Group*, preprint FERMILAB-PUB-96-082 (1996).

13. K. Hagiwara, R. Szalapski and D. Zeppenfeld, Phys. Lett. **B318**, 155 (1993).

14. B. Abbott *et al.*, DØ Collaboration, FERMILAB-CONF-97/325-E, contribution to the Lepton-Photon Conference, Hamburg, July 1997.

15. K. Hagiwara, and M. L. Stong, Z. Phys. **C62**, 99 (1994); B. Grzadkowski, and J. Wudka, Phys. Lett. **B364**, 49 (1995); G. J. Gounaris, J. Layssac and F. M. Renard, Z. Phys. **C65**, 245 (1995); G. J. Gounaris, F. M. Renard and N. D. Vlachos, Nucl. Phys. **B459**, 51 (1996).

16. S. M. Lietti, S. F. Novaes and R. Rosenfeld, Phys. Rev. **D54**, 3266 (1996); F. de Campos, S. M. Lietti, S. F. Novaes and R. Rosenfeld, Phys. Lett. **B389**, 93 (1996).

17. A. Stange, W. Marciano, and S. Willenbrock, Phys. Rev. **D49**, 1354 (1994); Phys. Rev. **D50**, 4491 (1994).

18. J. F. Gunion, H. E. Haber, G. Kane, S. Dawson, *The Higgs Hunter's Guide* (Addison–Wesley, 1990).

19. V. Barger, T. Han , D. Zeppenfeld, and J. Ohnemus, Phys. Rev. **D41**, 2782 (1990).

20. T. Stelzer and W. F. Long, Comput. Phys. Commun. **81**, 357 (1994).

21. H. Murayama, I. Watanabe and K. Hagiwara, KEK report 91-11 (unpublished).

22. A. D. Martin, W. J. Stirling, R. G. Roberts Phys. Lett. **B354**, 155 (1995).

23. S. Abachi *et al.*, DØ Collaboration, Phys. Rev. Lett. **78**, 2070 (1997).

24. B. Abbott *et al.*, DØ Collaboration, Phys. Rev. Lett. **80**, 442 (1998).

25. OPAL Collaboration, K. Ackerstaff *et al.*, Eur. Phys. J. **C1** (1998) 31.

26. OPAL Collaboration, K. Ackerstaff *et al.*, Eur. Phys. J. **C1** (1998) 21.

27. F. Abe *et al.*, CDF Collaboration, Phys. Rev. Lett. **81**, 1791 (1998).

28. M. C. Gonzalez–Garcia, S. M. Lietti, and S. F. Novaes, in preparation.

29. See for instace, talks by T. Diehl and H. Phillips in these proceedings.

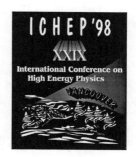

Parallel Session 17

Superstrings

Convenors: Costas Kounnas (ENS, CPT)

STABLE NON-BPS STATES IN STRING THEORY

Ashoke Sen

Mehta Research Institute of Mathematics and Mathematical Physics,
Chhatnag Road, Jhusi, Allahabad 211019, INDIA
E-mail: sen@mri.ernet.in

In this talk I discuss examples of stable non-BPS states in string theory and show how string duality helps us calculate their mass at strong coupling. Special emphasis is given on the non-BPS states transforming in the spinor representation of the gauge group in type I string theory.

1 Introduction and Summary of Results

During the last year we have learned a lot about BPS states in string theory. They are special states which are

- annihilated by a subset of the supersymmetry generators of the theory,

- satisfy a specific relation between mass and charge which does not receive quantum corrections (in theories with at least 16 unbroken supersymmetry generators):

$$m = f(g), \qquad (1)$$

$f(g)$ being a known function of the coupling constant g, and,

- are stable as a consequence of this relationship.

Many string theories contain states in their perturbative spectrum which are stable due to charge conservation, but are not BPS states. For these states, we still expect a definite functional relationship between the mass of the state and the string coupling constant:

$$m = f(g). \qquad (2)$$

For small g, $f(g)$ can be evaluated using string perturbation theory, and has the form:

$$f(g) = \sum_n c_n g^n, \qquad (3)$$

where c_n are computable coefficients. However, the general form of f, valid for finite g, is not known. In this talk we shall use duality symmetries of string theory to compute $f(g)$ for large g in some special examples. The details of the computation relevant for this talk are given in refs. [1,2].

Before I discuss specific examples, let me give some motivation for such a study.

1. Stable particles in the observed universe are not BPS states. Thus understanding properties of stable non-BPS states in string theory may be a small step towards understanding the properties of stable particles in our universe.

2. This analysis illustrates the power of duality symmetries in string theory.

3. In string theory progress has often come from unexpected directions. Thus one should explore every direction which might lead to new results.

The first example that we shall consider is that of SO(32) heterotic string theory. The spectrum of this theory at weak coupling contains states in the spinor representation of SO(32). These states are known to be non-BPS. Since the lightest state in the spinor representation of SO(32) must be stable due to charge conservation, we can ask: what will be the mass of these states at strong coupling? We shall find that the answer is:[a]

$$m = K g^{1/2} \qquad (4)$$

where K is a known constant.

The second example involves a configuration of parallel Dirichlet p-brane and orientifold p plane in type II string theory.[3] The ground state of the string stretched between the D-p-brane and its image under the orientifold operation is known to be non-BPS. However, it is charged under the U(1) gauge field living on the D-p-brane, and hence is stable. At weak coupling the mass of this state can be computed using string perturbation theory, but we are interested in finding its mass in the strong coupling limit. I shall now quote the results for different values of p:

p	Mass in string metric for $g \gg 1$	Overall coefficient C_p
6	$C_6 g$	Known
5	$C_5 g^{\frac{1}{2}}$	Known
4	$C_4 g^{\frac{1}{3}}$	Unknown
3	Unknown	Unknown

The details of the computation can be found in refs. [1], but I shall not discuss them here.

---
[a]We have set $\hbar = c = \alpha' = 1$.

Note that in this table we have restricted the value of p in the range $3 \le p \le 6$. For $p \ge 7$ the presence of the brane modifies the fields around it to such an extent that asymptotic value of the string coupling constant is not a well defined quantity. For $p \le 2$ the mass of a state carrying U(1) gauge charge becomes ill defined since the integral of the electromagnetic energy density ($\int d^p x F^2$) receives a divergent contribution from infinity.

2 A Specific Example: SO(32) Spinors in Heterotic String Theory

In this part of the talk I shall discuss the computation of the mass of the lightest SO(32) spinor state in heterotic string theory in the strong coupling limit. For this we use the fact that the heterotic string theory in the strong coupling limit is equivalent to weakly coupled type I string theory.[4,5] Thus we need to identify the SO(32) spinor states in weakly coupled type I string theory. Our starting point will be type I string theory compactified on a circle S^1 of (large) radius R, and a D-string anti-D-string pair wrapped on this circle. Due to the existence of a Z_2 gauge symmetry on the world volume of the D-string, we have a choice of putting either periodic or anti-periodic boundary condition on the open strings with one end lying on the D-string (anti-D-string) and the other end free (*i.e.* lying on the D9-brane). We choose periodic boundary condition on open string with one end on the D-string, and anti-periodic boundary condition on the open string with one end lying on the anti-D-string. Such a state can be shown to transform in the spinor representation of SO(32) without carrying any other conserved charge.[2] Thus this is a candidate for the state we are looking for.

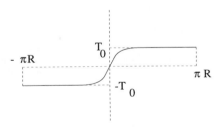

Figure 1: Lowest energy configuration for the tachyon field on the D-string anti-D-string pair. Here $\pm T_0$ denote the locations of the minimum of the tachyon potential.

Since both the D-string and the anti-D-string have tension $T_D = (2\pi \tilde{g})^{-1}$ measured in the type I metric, the mass of this state is given by

$$2 \cdot 2\pi R \cdot T_D = 2R/\tilde{g}. \tag{5}$$

Here $\tilde{g}$ is the type I coupling constant, related to the heterotic coupling through the relation:

$$\tilde{g} = g^{-1}. \tag{6}$$

(5) blows up as $R \to \infty$. Thus the state that we have found is not quite the right state. The situation improves upon noting that there is a tachyonic mode on the D-string anti-D-string pair. This signals that the true ground state of the system has energy lower than that given in eq.(5). Since the tachyon comes from an open string with one end on the D-string and the other end on the anti-D-string, it must satisfy antiperiodic boundary condition along S^1. Thus if $\pm T_0$ denote the locations of the minimum of the tachyonic potential $V(T)$, then for large R the lowest energy configuration will be given by a tachyonic kink solution interpolating between $-T_0$ and T_0, as displayed in Fig.1. The configuration where the tachyon is constant and equal to T_0 (or $-T_0$) is not allowed, as this is not anti-periodic under translation by $2\pi R$ along the compact direction. It can be shown that

$$V(\pm T_0) + 2T_D = 0, \tag{7}$$

so that at the minimum of the tachyon potential the negative contribution to the energy density from the tachyon potential exactly cancels the tension of the D-string anti-D-string pair. This implies that the configuration shown in Fig.1 has finite mass in the $R \to \infty$ limit, since far away from the core of the soliton, the energy density vanishes.

The mass of the state can be computed as follows. Since the world volume action of the D-string anti-D-string pair depends on the type I coupling constant $\tilde{g}$ via an overall multiplicative factor of $\tilde{g}^{-1}$, it is clear that the mass of the tachyonic kink solution, as measured in the type I metric, will be of the form:

$$C\tilde{g}^{-1}, \tag{8}$$

where C is a numerical constant. Using the standard relationship between the type I and the heterotic metric[4] we can reexpress this mass in the heterotic metric. The result is

$$Kg^{1/2}, \tag{9}$$

where K is another numerical constant related to C. This establishes the result quoted in the introduction. As mentioned in the introduction, the value of C (and hence K) can be computed, but we shall not give the details here. The result is

$$C = \sqrt{2}. \tag{10}$$

References

1. A. Sen, JHEP06(1998)007 [hep-th/9803194]; JHEP08(1998)010 [hep-th/9805019]; JHEP08(1998)012 [hep-th/9805170].
2. A. Sen, JHEP09(1998)023 [hep-th/9808141].
3. E. Gimon and J. Polchinski, Phys. Rev. **D54** (1996) 1667 [hep-th/9601038].
4. E. Witten, Nucl. Phys. **B443** (1995) 85 [hep-th/9503124].
5. J. Polchinski and E. Witten, Nucl. Phys. **B460** (1996) 525 [hep-th/9510169].

DUALITY IN STRING COSMOLOGY

RAM BRUSTEIN

Department of Physics, Ben-Gurion University, Beer-Sheva 84105, Israel
E-mail: ramyb@bgumail.bgu.ac.il

Scale factor duality, a truncated form of time dependent T-duality, is a symmetry of string effective action in cosmological backgrounds interchanging small and large scale factors. The symmetry suggests a cosmological scenario ("pre-big-bang") in which two duality related branches, an inflationary branch and a decelerated branch are smoothly joined into one non-singular cosmology. The use of scale factor duality in the analysis of the higher derivative corrections to the effective action, and consequences for the nature of exit transition, between the inflationary and decelerated branches, are outlined. A new duality symmetry is obeyed by the lowest order equations for inhomogeneity perturbations which always exist on top of the homogeneous and isotropic background. In some cases it corresponds to a time dependent version of S-duality, interchanging weak and strong coupling and electric and magnetic degrees of freedom, and in most cases it corresponds to a time dependent mixture of both S-, and T-duality. The energy spectra obtained by using the new symmetry reproduce known results of produced particle spectra, and can provide a useful lower bound on particle production when our knowledge of the detailed dynamical history of the background is approximate or incomplete.

1 Introduction

Our starting point is the effective action (in the so-called "string-frame"),

$$S_{eff} = \int d^4x \left\{ \sqrt{-g} e^{-\phi} \left[\frac{1}{16\pi\alpha'} (R + \partial_\mu\phi\partial^\mu\phi) \right. \right.$$
$$\left. - \frac{1}{4} F_{\mu\nu} F^{\mu\nu} + \bar{\Psi} D\!\!\!/ \Psi + \cdots \right] \tag{1}$$
$$\left. + \text{ higher orders in } \alpha'\partial^2 + \text{ higher orders in } e^\phi \right\}.$$

The dilaton ϕ, is a "Brans-Dicke-like" scalar with $\omega_{BD} = -1$.

The basic length scales of the theory are the string scale $\alpha' \equiv \ell_S^2$ and the Planck scale $e^\phi \alpha' = G_N \equiv \ell_P^2$, related by $e^\phi \equiv g_{string}^2 \simeq \frac{1}{4\pi}\alpha_{GUT}(1/\ell_S) = \left(\frac{\ell_P}{\ell_S}\right)^2$. These relations are modified for strongly coupled string theory. We will assume that the theory is weakly coupled throughout the evolution.

At early times fields may have been displaced from their present state, so we look for general FRW type solutions $g_{\mu\nu} = diag(-1, a^2(t)dx_i dx^i)$, $\phi = \phi(t)$, and if other fields are present we assume that they have only time dependence. We obtain equations for the Hubble parameter $H(t) = \dot{a}(t)/a(t)$, $\phi(t)$, and additional fields, in particular moduli, and matter (if present), and solve them, requiring that at late times the evolution has to be that of standard cosmology.

In a general effective action we may represent any contributions in addition to lowest order action by a "matter" Lagrangian [1,2,3]

$$S_{eff} = \int d^4x \left\{ \sqrt{-g} \left[\frac{e^{-\phi}}{16\pi\alpha'} (R + \partial_\mu\phi\partial^\mu\phi) \right] \right.$$

$$\left. + \frac{1}{2}\mathcal{L}_m(\phi, g_{\mu\nu}, ...) \right\}. \tag{2}$$

The equations of motion are the following

$$\dot{\phi} = 3H_S \pm \sqrt{3H_S^2 + e^\phi\rho_S}$$
$$\dot{H}_S = \pm H_S\sqrt{3H_S^2 + e^\phi\rho_S} + \frac{1}{2}e^\phi(p_S + \Delta_\phi\mathcal{L}_m) \tag{3}$$
$$\dot{\rho}_S + 3H_S(\rho_S + p_S) = -\Delta_\phi\mathcal{L}_m\dot{\phi},$$

where

$$T_{\mu\nu} = \frac{1}{\sqrt{-g}}\frac{\delta\mathcal{L}_m}{\delta g^{\mu\nu}}$$
$$\Delta_\phi\mathcal{L}_m = \frac{1}{2}\frac{1}{\sqrt{-g}}\frac{\delta\mathcal{L}_m}{\delta\phi} \tag{4}$$
$$T^\mu_{\ \nu} = diag(\rho, -p, p, -p).$$

As a result of scale factor duality (SFD) which will be discussed in more detail below, the solutions come in pairs or branches, the $(+)$ branch vacuum (without any sources) satisfies

$$\dot{H}_S = +H_S\sqrt{3H_S^2},$$

$H_S = \frac{1}{\sqrt{3}}\frac{1}{t_0-t}$, $t < t_0$ and is characterized by a future singularity. If the universe starts expanding according to the $(+)$ branch vacuum solution, H is positive, and therefore its time derivative is also positive. This branch cannot connect smoothly to radiation dominated (RD) FRW with constant dilaton, $\dot{\phi} = 3H_S + \sqrt{3H_S^2 + e^\phi\rho_S} \Rightarrow \dot{\phi} > 0$. The $(-)$ branch vacuum satisfies

$$\dot{H}_S = -H_S\sqrt{3H_S^2},$$

$H_S = \frac{1}{\sqrt{3}}$, $\frac{1}{t-t_0}t > t_0$ and is characterized by a past singularity. This branch can connect smoothly to RD

FRW with constant dilaton

$$\dot{\phi} = 3H_S - \sqrt{3H_S^2 + e^\phi \rho_S} \Rightarrow \dot{\phi} = 0 \text{ if } 6H_S^2 = e^\phi \rho_S.$$

A cosmological scenario ("pre-big-bang") in which two duality related branches, an inflationary branch and a decelerated branch are smoothly joined into one cosmology has been proposed [1,2]. In this scenario the universe quickly becomes homogeneous, isotropic, and spatially flat. The transition between the inflationary and decelerated branches, the so-called graceful exit transition, is expected to occur when the universe reaches string scale curvature. The use of scale factor duality in the analysis of the higher derivative corrections to the effective action, and consequences for the nature of exit transition are outlined.

Quantum fluctuations superimpose on top of the smooth classical background inhomogeneity perturbations, which are then amplified by the accelerated expansion of the universe and materialize as particles with specific energy spectra later on [4].

A new duality symmetry is obeyed by the lowest order equations for inhomogeneity perturbations [5]. In some cases it corresponds to a time dependent version of T-duality, interchanging small and large scale factors, in some cases it corresponds to a time dependent version of S-duality, interchanging weak and strong coupling and electric and magnetic degrees of freedom, and in most cases it corresponds to a time dependent mixture of both. As in other applications, duality turns out to be a powerful tool for obtaining results that are inaccessible otherwise. In particular, lower bounds on the energy density of the produced particles. The energy spectrum obtained by truncating the solutions of the perturbation equations to the constant modes, is characterized by a residual duality symmetry, reproduces known results of produced particle spectra, and can provide a useful lower bound on particle production when our knowledge of the detailed dynamical history of the background is approximate or incomplete.

2 Scale Factor Duality

2.1 Lowest order

To introduce SFD we look at the lowest order 4-d effective dilaton-gravity action

$$S_{LO} = \int d^4x \ \sqrt{-g} \left[\frac{e^{-\phi}}{16\pi\alpha'} \left(R + \partial_\mu \phi \partial^\mu \phi \right) \right]. \quad (5)$$

Integration by parts leads to

$$S_{LO} = -\frac{1}{16\pi\alpha'} \int d^4x a^3 e^{-\phi} \left(6H^2 - 6H\dot{\phi} + \dot{\phi}^2 \right)$$

$$= -\frac{1}{16\pi\alpha'} \int d^4x e^{-\bar{\phi}} \left(3H^2 - \dot{\bar{\phi}}^2 \right), \quad (6)$$

where $\bar{\phi} \equiv \phi - 3\ln a$, and we have set the lapse function $n(t)$ to unity.

The action (6) is invariant under the symmetry transformation

$$a(t) \rightarrow 1/a(t),$$
$$\phi(t) \rightarrow \phi - 6\ln a(t),$$

$\bar{\phi}$ and H^2 are invariant,

$$\bar{\phi}(t) \rightarrow \bar{\phi}(t),$$

$$H(t) \rightarrow -H(t),$$

and the equations of motion are covariant. The two branches describing an expanding universe are related to each other SFD× Time reversal. In general, for more complicated cosmological backgrounds the symmetry of the action is more complicated [6].

2.2 Leading corrections

The transition between the two duality related branches is called the graceful exit transition. It is known that to lowest order the two branches are separated by a singularity [7]. The emerging scenario for the exit transition requires classical α' corrections which can bound the curvature below the string scale, as well as quantum corrections [8,9]. The leading classical corrections determine whether the solution can reach a "good" region in $\dot{\bar{\phi}}, H$ phase space. A model for the exit transition has been presented [9], and therefore we know that a transition is possible. The question is whether string theory actually determines the coefficients such that a transition occurs.

We have investigated [10] effective classical corrections, and demanded that the action will really be an effective dilaton-gravity action, without additional new degrees of freedom [11]. This means that we have to use actions that produce equations without higher derivatives. Field redefinitions can change that but, it is better to use a "frame" with no higher derivatives ensuring numerical stability and control. We also require an action that reproduces whatever string theoretic information available such as scattering amplitudes, perturbative beta-function calculations etc. As we will see this is not enough to obtain the full corrected action. If scale factor duality could be imposed in a practical way it would help, however, the situation is more complicated.

The action including leading classical corrections is given by

$$S_{LCC} = \int d^4x \sqrt{g} e^{-\phi} \{ R + (\nabla\phi)^2 +$$

$$\frac{1}{2}[A(\nabla\phi)^4 + B R^2{}_{GB} +$$

$$C\left(R^{\mu\nu} - \frac{1}{2}g^{\mu\nu}R\right)\nabla_\mu\phi\,\nabla_\nu\phi + D\,\nabla^2(\phi)\,(\nabla\phi)^2]\}.$$

In covariant variables, it takes the following form

$$S_{LCC} = e^{-\bar\phi}\left\{\frac{3\,H(t)^2}{n(t)} - \frac{\dot{\bar\phi}(t)^2}{n(t)}\right.$$
$$+\frac{3\,(27\,A + 8\,B + 9\,C + 27\,D)\,H(t)^4}{2\,n(t)^3} +$$
$$\frac{(54\,A + 4\,B + 9\,C + 45\,D)\,H(t)^3\,\dot{\bar\phi}(t)}{n(t)^3}$$
$$+\frac{3\,(18\,A + C + 12\,D)\,H(t)^2\,\dot{\bar\phi}(t)^2}{2\,n(t)^3} +$$
$$\frac{3\,(2\,A + D)\,H(t)\,\dot{\bar\phi}(t)^3}{n(t)^3}$$
$$\left.+\frac{(3\,A + D)\,\dot{\bar\phi}(t)^4}{6\,n(t)^3}\right\},$$

where we have performed integration by parts to get rid of the $h'(t)$, $\phi''(t)$ and $n'(t)$. This is possible in general due to the 'no higher derivatives' condition.

Perturbative string calculation can provide two of the coefficients, one additional coefficient sets the overall scale at which the leading corrections kick in, so one coefficient remains unknown. If some symmetry principle, such as SFD could be used, the leading corrections action would be determined completely. For example, to impose naive SFD we would have to eliminate the odd parts of the action. We can set, for example,

$$D = -2A$$
$$C = 4A - \frac{4}{9}B.$$

However, this is only possible in a homogeneous background [12]. In an inhomogeneous background we get more equations and the only consistent solution is $A = B = C = D = 0$. The conclusion is that, at the very least, the SFD transformation has to be modified at this order, making it less useful for determination of the one remaining coefficient. If we insist on naive SFD in the homogeneous case, then it is possible to show [10] that if a stable algebraic fixed point exists, another non-stable fixed point will also exist, and that the generic solution will encounter the unstable fixed point first and run into a singularity.

2.3 All orders

As we have seen, additional input is required to determine the behaviour of solutions when classical stringy corrections are included. The best would be to establish the existence of an exact conformal field theory solution corresponding to the algebraic fixed point. In general this requires working with 2-d conformal field theories rather than with effective actions. I outline here some of the possibilities to achieve progress in that direction [10].

For highly symmetric backgrounds, such as the linear-dilaton deSitter background, it is possible to use the isometries of the background to impose additional symmetries on world-sheet operators, and constrain the beta-function coefficients. Another possibility is to use conformal perturbation theory to add (1,1) operators to an established conformal field theory, a linear dilaton flat-space background [10]. Yet another possibility is to start with known exact solutions in higher dimensions and compactify down to 4-d [13].

3 "S"-Duality

The quadratic action for perturbations of any tensor field expanded around a cosmological dilaton-gravity background is given by

$$S_{pert} = \frac{1}{2}\int d^3x\,d\eta\,S(\eta)\left[\psi'^2 - (\nabla\psi)^2\right].\qquad(7)$$

The prefactor $S(\eta)$ is given by $a^{2m}e^{\ell\phi}$, where m and ℓ depend on the type of field. For example, gravitons, dilatons and moduli have $m = 1$, $\ell = -1$, model independent axions have $m = 1$, $\ell = 1$, while Ramond-Ramond axions have $m = 1$, $\ell = 0$, and so on. We would like to compute the evolution of perturbation and eventually compute an important physical observable quantity: the spectrum of produced particles at late times.

3.1 "S"-Duality symmetry

To discuss duality symmetry of the action (7) it is more convenient to to use the Hamiltonian formalism. The Hamiltonian density corresponding to (7) is given by

$$H = \frac{1}{2}\int d\eta\left\{S^{-1}\Pi^2 + S(\nabla\Psi)^2\right\},\qquad(8)$$

where the momentum conjugate to Ψ is given by

$$\Pi = S\Psi'.\qquad(9)$$

The Hamilton equations of motion are first order

$$\Pi' = -\frac{\delta H}{\delta\Psi} = S\nabla^2\Psi$$
$$\Psi' = \frac{\delta H}{\delta\Pi} = S^{-1}\Pi,\qquad(10)$$

and lead to second order equations

$$\Pi'' - \frac{S'}{S}\Pi' - \nabla^2\Pi = 0 \qquad (11)$$

$$\Psi'' + \frac{S'}{S}\Psi' - \nabla^2\Psi = 0. \qquad (12)$$

The second equation (12) is commonly used in analysis of perturbation spectra.

In Fourier space the Hamiltonian density is given by

$$H = \frac{1}{2}\int d\eta \left\{ S^{-1}\Pi_{\vec{k}}\Pi_{-\vec{k}} + Sk^2\Psi_{\vec{k}}\Psi_{-\vec{k}} \right\}, \qquad (13)$$

and the equations of motion are given by

$$\Pi'_{\vec{k}} = -Sk^2\Psi_{-\vec{k}}$$
$$\Psi'_{\vec{k}} = S^{-1}\Pi_{-\vec{k}}. \qquad (14)$$

"S"-duality exchanges the variables and momenta and at the same time sends S to its inverse,

$$\Pi_{\vec{k}} \to \widetilde{\Pi}_{\vec{k}} = k\Psi_{\vec{k}}$$
$$k\Psi_{\vec{k}} \to k\widetilde{\Psi}_{\vec{k}} = -\Pi_{\vec{k}}$$
$$S \to \widetilde{S} = S^{-1}, \qquad (15)$$

leaving the Hamiltonian, equations of motion and Poisson brackets invariant.

We are interested in a situation in which the initial conditions correspond to zero-point vacuum fluctuations of the field Ψ, and therefore

$$\langle S^{-1}\Pi^2 \rangle = \langle S(\nabla\Psi)^2 \rangle, \qquad (16)$$

where $\langle \cdots \rangle$ denotes ensemble average.

The duality (15), contains strong-weak coupling duality as a special case. For perturbative heterotic 4-d gauge bosons the function S is given simply by $S(\eta) = e^{\phi(\eta)}$. Recall that $e^{\phi(\eta)} = g_{string}$, so the transformation $S \to \widetilde{S} = S^{-1}$ is, at each time η, simply the celebrated strong-weak coupling duality $g_{string} \to g_{string}^{-1}$, which appears as a part of the $SL(2, Z)$ group, usually called S-duality. The transformation (15) exchanges in this case electric and magnetic degrees of freedom.

3.2 Approximate solutions

To construct approximate solutions define $\widehat{\Psi}$, $\widehat{\Pi}$, whose Fourier modes are given by

$$\widehat{\Psi}_k = S^{1/2}\,\Psi_k$$
$$\widehat{\Pi}_k = S^{-1/2}\,\Pi_k. \qquad (17)$$

The new variables have simple transformation law under "S"-duality

$$k\widehat{\Psi}_k \to k\widetilde{\widehat{\Psi}}_k = -\widehat{\Pi}_k$$
$$\widehat{\Pi}_k \to \widetilde{\widehat{\Pi}}_k = k\widehat{\Psi}_k$$
$$S \to \widetilde{S} = S^{-1}. \qquad (18)$$

The variables $\widehat{\Psi}$, $\widehat{\Pi}$ satisfy the following Schrödinger-like equations

$$\widehat{\Psi}_k{}'' + \left(k^2 - (S^{1/2})''S^{-1/2} \right)\widehat{\Psi}_k = 0$$
$$\widehat{\Pi}_k{}'' + \left(k^2 - (S^{-1/2})''S^{1/2} \right)\widehat{\Pi}_k = 0. \qquad (19)$$

Since $S(\eta) \sim \eta^\alpha$, the potentials $V_\Psi = (S^{1/2})''S^{-1/2}$, $V_\Pi = (S^{-1/2})''S^{1/2}$, if non-vanishing, are proportional to $1/\eta^2$. For $k^2 > V_\Psi$, V_Π, or equivalently $(k\eta)^2 > 1$ (inside the horizon), we look for WKB-like approximate solutions

$$\widehat{\Psi}_k(\eta) = \left(k^2 - V_\Psi\right)^{-1/4} e^{-i\int_{\eta_0}^{\eta} d\eta' \left(k^2 - V_\Psi\right)^{1/2}}$$
$$\widehat{\Pi}_k(\eta) = k\left(k^2 - V_\Pi\right)^{-1/4} e^{-i\int_{\eta_0}^{\eta} d\eta' \left(k^2 - V_\Pi\right)^{1/2}}. \qquad (20)$$

The advantage of looking at solutions (20) is that they manifestly preserve the "S"-duality symmetry of the equations, because the potentials V_Ψ, V_Π get interchanged under $S \to S^{-1}$.

For very large k^2, $k^2 \gg V_\Psi$, V_Π solutions (20) reduce to correctly normalized vacuum fluctuations

$$\widehat{\Psi}_k(\eta) = k^{-1/2} e^{-ik\eta + i\varphi_0}$$
$$\widehat{\Pi}_k(\eta) = k^{+1/2} e^{-ik\eta + i\varphi_0'}, \qquad (21)$$

where φ_0, φ_0' are random phases, originating from the random initial conditions. Note that because of the random phases, "S"-duality holds only on the average in the sense of eq.(16).

For $k^2 < V_\Psi$, V_Π, or equivalently $(k\eta)^2 < 1$ (outside the horizon), it is possible to write "exact" solutions [5]. It is convenient to define the functions $T\cos(S^{-1}, S)$, $T\sin(S^{-1}, S)$

$$T\cos(S^{-1}, S) = 1 - k\int_{\eta_{ex}}^{\eta} d\eta_1 S^{-1}(\eta_1)\, k\int_{\eta_{ex}}^{\eta_1} d\eta_2 S(\eta_2) + \cdots$$

$$+ (-1)^{n+1}k^{2(n-1)}\prod_{n-1\ times}\int S^{-1}\int S \cdots \int S^{-1}\int S + \cdots$$

$$T\sin(S^{-1}, S) = k\int_{\eta_{ex}}^{\eta} d\eta_1 S^{-1}(\eta_1) - \cdots \qquad (22)$$

$$+ (-1)^{n+1}k^{2n-3}\int S^{-1}\prod_{n-2\ times}\int S\int S^{-1}\cdots \int S\int S^{-1} + \cdots,$$

in terms of which the "exact" solutions take the following form

$$\widehat{\Psi}_k(\eta) = \sqrt{S}\left\{ A_k\; T\cos(S^{-1}, S) + B_k\; T\sin(S^{-1}, S) \right\}$$

$$\widehat{\Pi}_k(\eta) = \frac{k}{\sqrt{S}}\left\{ B_k\; T\cos(S, S^{-1}) - A_k\; T\sin(S, S^{-1}) \right\}. \quad (23)$$

Using the relations

$$\left[T\cos(S^{-1}, S) \right]' = -\frac{k}{S}\; T\sin(S, S^{-1})$$

$$\left[T\sin(S^{-1}, S) \right]' = \frac{k}{S}\; T\cos(S, S^{-1}), \quad (24)$$

and similar relations for $\left[T\cos(S, S^{-1}) \right]'$ and $\left[T\sin(S, S^{-1}) \right]'$, it is possible to verify explicitly that $\widehat{\Psi}_k$, $\widehat{\Pi}_k$ in eq.(23) are indeed solutions of eqs.(19). Formally, these solutions are valid also inside the horizon, but the functions $T\cos$, $T\sin$ are not well defined there.

We need to match the solutions inside and outside the horizon and do it such that "S"-duality is respected. One way of doing so is to use solutions (20) inside the horizon, and (23) outside the horizon and match them at some time near horizon exit time η_{ex}, for which $k\eta_{ex} \sim 1$. Taking advantage of the phenomenon of "freezing of perturbations" outside the horizon we obtain the following result,

$$\widehat{\Psi}_k(\eta) = \frac{1}{\sqrt{k}}\left[\left(\frac{S_{ex}}{S_{re}}\right)^{-1/2}\cos(k\eta) + \left(\frac{S_{ex}}{S_{re}}\right)^{1/2}\sin(k\eta) \right]$$

$$\widehat{\Pi}_k(\eta) = \sqrt{k}\left[\left(\frac{S_{ex}}{S_{re}}\right)^{1/2}\cos(k\eta) - \left(\frac{S_{ex}}{S_{re}}\right)^{-1/2}\sin(k\eta) \right] \quad (25)$$

where $S_{re} = S(\eta_{re})$. The reentry time η_{re} is the second time at which $k\eta_{re} \sim 1$.

3.3 Energy Spectrum

We compute an important physical observable, the Hamiltonian density,

$$\langle H_k \rangle = \frac{1}{2}\left(\langle |\widehat{\Pi}_k|^2 \rangle + k^2 \langle |\widehat{\Psi}_k|^2 \rangle \right) \quad (26)$$

Using the approximate solutions (25) we obtain

$$\langle H_k \rangle = k\left(\frac{S_{ex}}{S_{re}} + \frac{S_{re}}{S_{ex}} \right). \quad (27)$$

It is invariant under $S_{ex} \to S_{ex}^{-1}$, $S_{re} \to S_{re}^{-1}$, and overall rescaling of S. Note that, for a given k, $\langle H_k \rangle$ depends only on S_{ex} and S_{re}, and not on the whole evolution.

The spectral energy distribution, $d\rho_k/d\ln k = (k^3/a^4)\langle H_k \rangle$ $(\omega = k/a)$ is given by

$$\frac{d\rho(\omega)}{d\ln\omega} = \omega^4\left[\frac{S_{ex}(\omega)}{S_{re}(\omega)} + \frac{S_{re}(\omega)}{S_{ex}(\omega)} \right]$$

$$\simeq \omega^4\; \mathrm{Max}\left\{ \frac{S_{ex}}{S_{re}}, \frac{S_{re}}{S_{ex}} \right\}. \quad (28)$$

It has the same invariance properties as the Hamiltonian density.

From eq.(28) we obtain model independent lower bound on energy density of cosmologically produced particles. The spectrum (28) is a sum of two terms, one being the inverse of the other. Therefore it is not possible to decrease the contribution of one term without increasing the contribution of the other. The physical origin of this lower bound is indeed the uncertainty principle. Recall that one term originates from the contribution of the perturbation conjugate momentum and the other from the contribution of the perturbation itself. The uncertainty principle says that it is not possible to decrease both without limits. For specific cases, the lower bound may be improved using some particular properties of the background.

The result (28) provides an easy and a very general way of computing $\frac{d\rho(\omega)}{d\ln\omega}$. The prescription is simple. Once the function $S(\eta)$ is known for all times, substitute for $\eta_{ex} \to k^{-1}$, and for η_{re} substitute the properly redshifted k^{-1}. The results obtained using this simple method reproduce known results obtained by explicit complicated calculations. [14,15,16,17,18,19]

Acknowledgements

I would like to thank my collaborators Maurizio Gasperini, Merav Hadad, Dick Madden and Gabriele Veneziano. This work is supported in part by the Israel Science Foundation administered by the Israel Academy of Sciences and Humanities.

1. G. Veneziano, *Phys. Lett.* B265 (1991) 287.
2. M. Gasperini and G. Veneziano, *Astropart. Phys.* 1 (1993) 317.
3. R. Brustein and R. Madden, *Phys. Lett.* B410 (1997) 110.
4. N.D. Birrell and P.C.W. Davies, Quantum fields in curved space, Cambridge University Press, 1984; V. F. Mukhanov, A. H. Feldman and R. H. Brandenberger, *Phys. Rep.* 215 (1992) 203.
5. R. Brustein, M. Gasperini and G. Veneziano, *Phys. Lett.* B431 (1998) 277.
6. K.A. Meissner and G. Veneziano, *Phys. Lett.* B267 (1991) 33; A.A. Tseytlin *Mod. Phys. Lett.* A6 (1991) 1721; A. Sen, *Phys. Lett.* B271 (1991) 295;

J. Maharana and J. H. Schwarz, *Nucl. Phys.* B390 (1993) 3.

7. R. Brustein and G. Veneziano, *Phys. Lett.* B329 (1994) 429; N. Kaloper, R. Madden and K.A. Olive, *Nucl. Phys.* B452 (1995) 677.

8. M. Gasperini, M. Maggiore and G. Veneziano, *Nucl. Phys.* B494 (1997) 315.

9. R. Brustein and R. Madden, *Phys. Rev.* D57 (1998) 712.

10. R. Brustein and R. Madden, to appear.

11. K. Forger, B. A. Ovrut, S. Theisen and D. Waldram, *Phys. Lett.* B388 (1996) 512.

12. N. Kaloper and K. Meissner, *Phys. Rev.* D56 (1997) 7940; M. Maggiore, *Nucl. Phys.* B525 (1998) 413.

13. C. Kounnas, private communication.

14. M. Gasperini, *Phys. Lett.* B327 (1994) 214; M. Gasperini and G. Veneziano, *Phys. Rev.* D50 (1994) 2519.

15. R. Brustein, M. Gasperini, M. Giovannini and G. Veneziano, *Phys. Lett.* B361 (1995) 45.

16. M. Gasperini, M. Giovannini and G. Veneziano, *Phys. Rev. Lett.*75 (1995) 3796; D. Lemoine and M. Lemoine, *Phys. Rev.* D52 (1995) 1955.

17. E. J. Copeland, R. Easther and D. Wands, *Phys. Rev.* D56 (1997) 874; E. J. Copeland, J. E. Lidsey and D. Wands, *Nucl. Phys.* B506 (1997) 407.

18. R. Brustein and M. Hadad *Phys. Rev.* D57 (1998) 725; A. Buonanno, K. Meissner, C. Ungarelli and G. Veneziano, J.High Energy Phys. 01 (1998) 004.

19. R. Durrer, M. Gasperini, M. Sakellariadou and G. Veneziano, astro-ph/9806015.

ZERO TEMPERATURE PHASE TRANSITIONS IN GAUGE THEORIES

L.C.R.Wijewardhana

Physics Department, University Of Cincinnati, Cincinnati, Ohio, 45221
E-mail: rohana@physics.uc.edu

SU(N) gauge theories undergo a chiral transition when the number of quark flavors N_f is varied. In this lecture we discuss the recent progress made in understanding such transitions.

1 Introduction

In an $SU(N)$ gauge theory with N_f massless quark flavors in the fundamental representation , it is expected that both confinement and spontaneous chiral symmetry breaking take place providing that N_f is not too large. If, on the other hand, N_f is large enough, the theory is expected to neither confine nor break chiral symmetry. For example, if N_f is larger than $11N/2$ for quarks in the fundamental representation, asymptotic freedom (and hence confinement and chiral symmetry breaking) is lost. Even for a range of N_f below $11N/2$, the theory should remain chirally symmetric and deconfined. The reason is that an infrared fixed point is present[1], determined by the first two terms in the renormalization group (RG) beta function. By an appropriate choice of N and N_f, the coupling at the fixed point, α_*, can be made arbitrarily small, making a perturbative analysis reliable. Such a theory is massless and conformally invariant in the infrared. It is asymptotically free, but without confinement or chiral symmetry breaking.

As N_f is reduced, α_* increases. At some critical value of N_f (N_f^c) there will be a phase transition to the chirally asymmetric and confined phase. It is an important problem in the study of gauge field theories to determine N_f^c and to characterize the nature of the phase transition.

It has ben suggested[2], that the phase transition takes place at a large enough value of N_f^c so that the infrared fixed point α_* reliably exists and governs the phase transition. The transition was then analyzed using the ladder expansion of a gap equation, or equivalently the CJT effective potential[3]. It was argued that confinement effects can be neglected to estimate N_f^c and to determine the nature of the transition. It was then shown that the chiral order parameter vanishes continuously at $N_f \to N_f^c$ from below, but that the phase transition is not conventionally second order in that there is no effective, low energy Landau-Ginzburg Lagrangian, i.e. the correlation length does not diverge as the critical point is approached.

Once chiral symmetry breaking sets in, the quarks decouple at momentum scales below the dynamical mass leaving the pure gauge theory behind. The effective cou-

pling then grows, leading to confinement at a scale on the order of the quark mass. Thus for N_f just below N_f^c, the fixed point is only an approximate feature of the theory governing momentum scales above the dynamically generated mass. This is adequate, however, since it is this momentum range that determines N_f^c and the character of the transition.

The analysis of this phase transition paralleled an analysis of the chiral transition in 2+1 dimensional gauge theories with N_f quarks[4]. Using a large N_f expansion it was found[5] that the effective infrared coupling runs to a fixed point proportional to $1/N_f$. As N_f is lowered this coupling strength exceeds the critical coupling necessary to produce spontaneous symmetry breaking. It was argued that this critical $1/N_f$ coupling lies in a range where the large N_f expansion is reliable[6]. These conclusions were also supported by lattice simulations[7]. It was then noted that as in the case of the 3+1 dimensional $SU(N)$ theory, this phase transition is not conventionally second order[4].

For QCD the study of the chiral phase transition as a function of N_f is of theoretical interest, but is unlikely to shed direct light on the physics of the real world. There remains the possibility, however, that if technicolor is the correct framework for electroweak symmetry breaking, the transition could be physically relevant. In a recent letter[8], it was pointed out that in an $SU(2)$ technicolor theory, a single family of techniquarks ($N_f = 8$) leads to an infrared fixed point near the critical coupling for the chiral phase transition. This can provide a natural origin[9] for walking technicolor[10] and has other interesting phenomenological features.

In a recent paper[11] , the features of the chiral phase transition as function of N_f has further been clarified. It has been observed that the transition takes place at a relatively large value of N_f ($N_f^c \approx 4N$) where the infrared coupling is determined by a fixed point accessible in the loop expansion of the β function, and that the transition can be studied using a ladder gap equation. Our higher order estimates suggest that the estimate of N_f^c is good to about 20%. To phrase things in physical terms, the effect of the light quarks is to screen the long range force, eventually disordering the system and taking it to the symmetric phase. That the transition takes place at a relatively large value of N_f means that the quarks are relatively ineffective at long range screening.

With an infrared fixed point governing the transition, the order parameter vanishes in a characteristic exponential fashion and all physical scales vanish in the same way[12], [13]. There is no finite set of light degrees of freedom that can be identified to form an effective, Landau-Ginzburg theory. In the symmetric phase ($N_f > N_f^c$) , no light degrees of freedom are formed as $N_f \to N_f^c$. Thus the transition is continuous but not conventionally second order. The validity of the approach is considered by estimating higher order terms in both the β function and the anomalous dimension of the mass operator.

In Ref. [14], it was noted that single instanton effects in a theory with an infrared fixed point seem capable of triggering a chiral phase transition at similarly large values of N_f/N. A detailed computation was carried only out for an $SU(2)$ gauge theory but the analysis indicated that this could be the case at larger values of N as well.

It is interesting to compare our results with the phase structure of supersymmetric $SU(N)$ theories where exact results are available [15]. In such theories there is also a large range of N_f where the theory is asymptotically free and an infrared fixed point occurs. A transition to a strongly coupled phase occurs at $N_{SUSY}^c = 3N/2$. Thus it seems plausible that infrared fixed points are fairly generic in asymptotically free gauge theories with a large number of flavors. One prominent difference between the supersymmetric and non-supersymmetric cases is that the strongly coupled phase for $N_f^c - 1$ flavors does not have chiral symmetry breaking or confinement for $N > 2$. However a class of supersymmetric chiral gauge theories (with antisymmetric tensor fields) have been found [16] where the theory does go from an infrared fixed point to confinement upon the removal of one flavor.

The results of this paper can be contrasted with preliminary lattice work [17] and the instanton liquid model [18] suggesting that the chiral transition takes place at much smaller values of N_f contrary to earlier lattice results [19]. The transition would then be an intrinsically strong coupling phenomenon inaccessible to the methods used here. The quarks would have to be much more effective at long range screening than indicated by the gap equation, disordering the system even in the presence of a strong, attractive long range force. Further work on all these approaches will be required to help to resolve this difference.

There is some evidence from lattice studies that approximate Parity doubling may occur in confining SU(N) theories when one reaches the critical number of flavors from the broken side. This could substantially reduce the leading positive contribution to the S parameter if such theories are utilized in building technicolor models. The leading contribution to S comes from the techni ρ and a_1 sectors with opposite signs. Thus approximate parity doubling could makes the value of S fall within the experimentally acceptable range. The reader is refered to the recent work by Appelquist and Sanino for a discussion of the possibility of parity doubling[20].

The work reported here is done in collaboration with T.Appelquist, A.Ratnaweera and J.Terning. I thank them for valuable discussions. My research was partially supported by the U.S DOE under contract #DE-FG-02-84ER40153. Part of this work was done at the Institute of Fundamental Studies in Kandy Sri Lanka. I thank the director professor K.Tennakone for hospitality and financial support.

References

1. T. Banks and A. Zaks, *Nucl. Phys.* **B196** (1982) 189.

2. T. Appelquist, J. Terning, and L.C.R. Wijewardhana, *Phys. Rev. Lett.* **77** (1996) 1214 hep-ph/9602385.

3. J.M. Cornwall, R. Jackiw, and E. Tomboulis, *Phys. Rev.* **D10** (1974) 2428.

4. T. Appelquist, J. Terning, and L.C.R. Wijewardhana, *Phys. Rev. Lett.* **75** (1995) 2081, hep-ph/9402320 .

5. T. Appelquist, D. Nash, and L.C.R. Wijewardhana, *Phys. Rev. Lett.* **60** (1988) 2575; D. Nash *Phys. Rev. Lett.* **62** (1989) 3024; T. Appelquist and D. Nash, *Phys. Rev. Lett.* **64** (1990) 721.

6. D. Nash *Phys. Rev. Lett.* **62** (1989) 3024; T. Appelquist and D. Nash, *Phys. Rev. Lett.* **64** (1990) 721.

7. E. Dagotto, J.B. Kogut, and A.Kocic, *Phys. Rev. Lett.* **62**, 1083 (1989); *Nucl. Phys.* **B334**, 279 (1990).

8. T. Appelquist, J. Terning, and L.C.R. Wijewardhana, *Phys. Rev. Lett.* **79** (1997) 2767, hep-ph/9706238.

9. K. Lane, M.V. Ramana *Phys. Rev.* **D44** (1991) 2678.

10. B. Holdom, *Phys. Lett.* **B150** (1985) 301; K. Yamawaki, M. Bando, and K. Matumoto, *Phys. Rev. Lett.* **56** (1986) 1335; T. Appelquist, D. Karabali, and L.C.R. Wijewardhana, *Phys. Rev. Lett.* **57** (1986) 957; T. Appelquist and L.C.R. Wijewardhana, *Phys. Rev* **D35** (1987) 774; T. Appelquist and L.C.R. Wijewardhana, *Phys. Rev* **D36** (1987) 568.

11. T.Appelquist, A,Ratnaweera, J.Terning and L.C.R.Wijewardhana, Phys.Rev.D58,10525(98).

12. V.A. Miransky and K. Yamawaki *Phys. Rev.* **D55** (1997) 5051, hep-th/9611142; erratum ibid. **D56**

(1997) 3768; V.A. Miransky, hep-ph/9703413.

13. S. Chivukula, *Phys. Rev.* **D55** (1997) 5238, hep-ph/9612267.

14. T. Appelquist and S. Selipsky, *Phys. Lett.* **B400** (1997) 364, hep-ph/9702404.

15. For a recent review see K. Intriligator and N. Seiberg, *Nucl. Phys. Proc. Suppl.* **45BC** (1996)1; *Nucl. Phys. Proc. Suppl.* **55B** (1996) 200; hep-th/9509066.

16. J. Terning, *Phys. Lett.* **B422** (1998)149; hep-th/9712167.

17. D. Chen, R.D. Mawhinney, *Nucl. Phys. Proc. Suppl.* **53** (1997)216 hep-lat/9705029; R.D. Mawhinney, hep-lat/9705030; see also Y. Iwasaki, K. Kanaya, S. Kaya, S. Sakai, T. Yoshie, *Nucl. Phys. Proc.Suppl.* **53** (1997) 449, hep-lat/9608125.

18. M. Velkovsky, E. Shuryak, hep-ph/9703345.

19. J. B. Kogut and D. K. Sinclair, *Nucl. Phys.* **B295** [FS21] (1988) 465, F. Brown et. al., *Phys. Rev.* **D46** (1992) 5655.

20. T.Appelquist and A.Sannino, hep-ph/9806409.

Newtonian Dynamics in an Infinite Momentum Frame

Charles B. Thorn

Institute for Fundamental Theory, Department of Physics, University of Florida, Gainesville, FL 32611, USA
E-mail: thorn@phys.ufl.edu

The recent conjectured equivalence between classical string theory on certain Anti de Sitter backgrounds and large N_c gauge theories motivates a reassessment of more direct approaches to 't Hooft's large N_c limit. One such approach, the direct summation of planar diagrams in an infinite momentum frame, can be mapped to a Galilei invariant linear chain problem. However, this is a true Newtonian system only in a "wee-parton" approximation. We address here the problem of dealing with valence partons. This means confronting the full light-front dynamics of field theory in which an infinite number of species of particles with varying Newtonian mass P^+ can transform into each other by the usual vertices of Feynman diagrams. Motivated by the fact that field theories can be obtained as infinite tension limits of string theories, we suggest a way to cast field theory as a standard Newtonian system, with only a finite number of particle species. We use a simple quantum mechanical system to display the essence of the proposal, and we discuss prospects for applying these ideas to large N_c QCD.

1 Introduction

String theory was originally devised as a theory of the strong interactions. In the seventies, after the consensus developed that QCD was the true theory of strong interactions but before string theory was touted as a theory of everything including quantum gravity, it was hoped that string theory might provide an avenue for understanding the hadronic spectrum of QCD, especially in 't Hooft's large N_c limit[1] (here N_c = the number of colors). Recent work in string theory initiated by Maldacena[2] has revived this hope. Maldacena has conjectured an exact equivalence between $\mathcal{N} = 4$ super Yang-Mills theory at $N_c \to \infty$ and classical type IIB String theory compactified on the manifold $AdS_5 \times S_5$. The detailed nature of the conjecture was subsequently sharpened.[3,4]

In the detailed correspondence between the two sides of the equivalence, the radii of both AdS_5 and S_5 are proportional to $(N_c g^2)^{1/4}$. Since string theory on the requisite manifold is so far problematic, firm tests of the conjecture exist only for $N_c g^2 \to \infty$ where the radii are large and string theory can be replaced by IIB supergravity. But the qualitative nature of the string theory is revealed by examining the bosonic part of the world sheet action

$$S_{w.s.} = \int d\tau d\sigma \{e^\phi (\partial x)^2 + (\partial \phi)^2\} \qquad (1)$$

which has the form of Polyakov's[5] "confining string." The Liouville field ϕ is just the extra dimension of AdS_5. Evidently, the dynamical tension $T_0(\phi) = e^\phi$ equilibrates to 0 at $\phi \to -\infty$, so we have a "tensionless" string, and therefore no confinement. This is of course to be expected since the $\mathcal{N} = 4$ Yang-Mills theory is invariant under the full superconformal group, with no room for a mass scale. To apply these ideas to real QCD both supersymmetry and conformal invariance must be broken, for example, in the manner proposed by Witten.[6]

So far the pursuit of these ideas has exploited the string or gravity side of the conjectured equivalence to yield insight into the strong 't Hooft coupling limit of Yang-Mills theory. In this talk, I would like to explore how the physics of such a QCD/string equivalence might look entirely within the framework of the field theory. This will involve a return to ideas on summing the planar diagrams of QCD in an infinite momentum frame, proposed in the late 70's.[7,8,9,10,11,12] In the next section I will review some of this work, especially recalling the technical obstacles that stalled our progress. Then in Sec. 3 I will describe a recent proposal I have developed for dealing with these obstacles. In Sec. 4 I will illustrate how this proposal plays out in a simple quantum mechanics example. Then in the concluding section I will discuss prospects for its application to real QCD.

2 Review of Early Work and the Difficulties

In the late seventies, my collaborators and I tried to exploit the most basic simplification of infinite momentum frame or light-front quantization: that it turns the dynamics of field theory into those of a Newtonian many particle system, albeit one in which the Newtonian mass of each particle is variable. In this interpretation the Newtonian time is x^+ where $x^\pm \equiv (x^0 \pm x^1)/\sqrt{2}$, and the transverse coordinates x^k, $k = 2, \ldots, D - 1 = 3$ describe the Newtonian space. Then $P^+ = (P^0 + P^1)/\sqrt{2}$ plays the role of Newtonian mass, which can assume every value from 0 to ∞. By virtue of quantum mechanics knowledge about P^+ is equivalent to knowledge about x^-.

2.1 The Gluon Field on a Light-front

Now consider the gluon field of QCD. In light-cone gauge, $A^+ = 0$, only the transverse components are dynamically independent and can be resolved into components of definite P^+:

$$\mathbf{A}_\alpha^\beta(\mathbf{x}, x^-) = \int_0^\infty \frac{dP^+}{\sqrt{4\pi P^+}} [\mathbf{a}_\alpha^\beta(\mathbf{x}, P^+)e^{-iP^+x^-} + h.c.]$$

$$\to \sum_{M=1}^\infty \frac{1}{\sqrt{4\pi M}} [\mathbf{a}_{M\alpha}^\beta(\mathbf{x})e^{-iMmx^-} + h.c.] \quad (2)$$

where in the last line we have discretized $P^+ \to Mm$. The creation and annihilation operators then obey the algebra

$$[a_{M\alpha}^{k\ \beta}(\mathbf{x}), a_{M'\gamma}^{\dagger l\ \delta}(\mathbf{y})] = \delta_{kl}\delta_{MM'}\delta_\gamma^\beta\delta_\alpha^\delta\delta(\mathbf{x} - \mathbf{y}). \quad (3)$$

2.2 Light-front Dynamics

The Newtonian Hamiltonian is just P^- which has the zero coupling (free gluon) limit:

$$H_0 \equiv P_0^- = \int d\mathbf{x} \sum_{M=1}^\infty \text{Tr } \mathbf{a}_M^\dagger(\mathbf{x}) \frac{-\nabla^2}{2Mm} \cdot \mathbf{a}_M(\mathbf{x}), \quad (4)$$

which is the kinetic energy operator of a Newtonian many body system with an infinite number of species of elementary Newtonian particles labeled by Newtonian mass Mm; spin or polarization k; and color α, β. The interactions of the field theory can be divided into terms which represent standard Newtonian mass conserving (but spin dependent) potential energy and all the non-Newtonian terms which involve at least some redistribution of Newtonian mass amongst the gluons.

Newtonian Interactions. These all involve $P^+ = 0$ exchange. They occur directly as part of the quartic gluon term $\text{Tr}[A_k, A_l]^2$ in the "bare" Hamiltonian. The corresponding potential energy will have the form

$$\frac{g^2}{MNm}\delta(\mathbf{x} - \mathbf{y})(\text{Color Spin Matrix}). \quad (5)$$

But they can also be induced in the process of integrating out the $P^+ = 0$ modes of the gluon field, which have no Newtonian interpretation. These induced potentials will be nonlocal and not restricted to two body interactions. Whether or not one explicitly includes these terms can be regarded as an ambiguity of the discretization of P^+, since $P^+ = 0$ is just a set of measure zero in the continuum limit. However, including them to some degree could dramatically reduce lattice artifacts and make coarse discretizations more accurate than they would otherwise be. In practice one can obtain their short distance behavior order by order in weak coupling perturbation theory, but since this is not a closed process, it is probably best to keep them generic, with the hope that much important physics will depend only on

qualitative properties such as whether they are attractive, repulsive, can form bound states, etc.

Non-Newtonian Interactions. Most of the interactions in the "bare" Hamiltonian are of this type. They include all of the cubic terms, which of necessity change the number of gluons by ± 1. Also among the quartic interactions are terms which change the number of gluons by ± 2. Changing the number of gluons of necessity rearranges the P^+ distribution amongst the gluons. But there are also quartic interactions which conserve the number of gluons but cause P^+ to be exchanged, and we include these in the non-Newtonian interactions.

These non-Newtonian P^+ rearranging interactions are characteristic field theoretic interactions, and in fact give a direct description of things that actually occur (e.g. gluon production and annihilation) in time-dependent phenomena such as scattering. However, they represent an annoying complication to the problem of describing bound states – glueballs in the case of the QCD without quarks. If the P^+ of each gluon were fixed, the bound state dynamics would be just that of a standard Newtonian many-body system. It is our goal to try to achieve such a description, but first we discuss how 't Hooft's large N_c limit separates the bound state problem from time dependent phenomena.

2.3 $N_c \to \infty$

The wonderful thing about 't Hooft's $N_c \to \infty$ limit is that *if* there is confinement, then *all* glueball scattering amplitudes vanish in the extreme limit. For example, the 2 glueball elastic scattering amplitude is of order $1/N_c^2$. Thus in this limit the non-trivial physics resides in the glueball masses and their wave-functions as seen by weak probes such as leptons, photons and weak bosons. Let us recall how the glueball spectrum problem can be formulated in this limit.[12]

Consider the subspace of states spanned by those of the form

$$\frac{1}{N_c^{k/2}} \text{Tr } a_{M_1}^{\dagger i_1}(\mathbf{x}_1) \cdots a_{M_k}^{\dagger i_k}(\mathbf{x}_k)|0\rangle, \quad (6)$$

with the trace over color indices. These states are normalized to $O(1)$ as $N_c \to \infty$. Consider a general superposition of such states with fixed total P^+, so $\sum_k M_k = M$. Let $\Psi_k(\mathbf{x}_1 i_1 M_1, \cdots, \mathbf{x}_k i_k M_k)$ be the coefficient of Eq. (6) in this superposition. Note that each Ψ_k is cyclically symmetric under $(12 \cdots k) \to (k12 \cdots k-1)$, but the relative cyclic position of each gluon is fixed by its location in the trace. The light front Hamiltonian is a sum of terms each of which is a single trace of a product of up to four creation and annihilation operators. In 't Hooft's large N_c limit $g^2 = O(1/N_c)$, so that only color structures in the Hamiltonian that supply excess factors of N_c will survive the limit. The color structure of the interaction terms in the light-front Hamiltonian are

$$g\text{Tr}[a^\dagger a^\dagger a], \quad g\text{Tr}[a^\dagger aa], \quad g^2\text{Tr}[a^\dagger a^\dagger a^\dagger a],$$
$$g^2\text{Tr}[a^\dagger a^\dagger aa], \quad g^2\text{Tr}[a^\dagger aaa], \quad g^2\text{Tr}[: a^\dagger aa^\dagger a :].$$

The terms with a single annihilation operator, acting on a state of the form Eq. (6), replace one of the gluons by 2 or 3 gluons at the *same* cyclic location. They survive the limit because the factors of N_c shown in Eq. (6) need to be changed. The terms with 2 or more annihilation operators will survive only if they contract against neighboring gluons in the state Eq. (6). The color structure $g^2\text{Tr}[: a^\dagger aa^\dagger a :]$ will not survive the limit at all. Thus in the limit $N_c \to \infty$ the energy eigenvalue equation implies the following equations relating the various Ψ_k, $k = 1 \cdots M$:

$$\left(\sum_{l=1}^{k} \frac{\mathbf{p}_l^2}{2mM_l}\right)\Psi_k + V_{k,k}\Psi_k + V_{k,k-2}\Psi_{k-2} + V_{k,k-1}\Psi_{k-1}$$
$$+ V_{k,k+1}\Psi_{k+1} + V_{k,k+2}\Psi_{k+2} = E\Psi_k. \quad (7)$$

In this equation the V's are, in general, matrices in the spin and P^+ labels, functions of the transverse coordinates, and up to first order in transverse derivatives.

Every interaction term in Eq. (7) involves *only* neighboring gluons in the trace. Thus the system described by Eq. (7) resembles a long ring molecule (polymer) in which the gluons play the role of atoms. The dynamics is non-relativistic quantum mechanics, but it is exotic in that rings with different numbers of gluons ("atoms") mix quantum mechanically. If we could neglect this mixing, the molecular analogy would be almost perfect. In particular the component described by Ψ_M, containing the maximal number of gluons ("wee" gluons), would in the absence of mixing be a solution of the Schrödinger equation

$$\left[\sum_{l=1}^{M} \frac{\mathbf{p}_l^2}{2m} + \sum_{l=1}^{M} \mathcal{V}(\mathbf{x}_{l+1} - \mathbf{x}_l)\right]\Psi_M = E\Psi_M. \quad (8)$$

If the potential is a simple attractive well and stong enough to form bonds, the physics of this system in the limit $M \to \infty$ reproduces that of a light-cone quantized relativistic string. The detailed shape of the potential would not matter because all of the important excitations would involve energies of $O(1/M)$ and hence very mild deformations of each gluon's wave function. One can argue that the components with large numbers of gluons dominate at strong 't Hooft coupling $N_c g^2 \to \infty$ in order to take advantage of as many strong interactions as possible. But to say that only the component Ψ_M is present at strong coupling is a very discretization sensitive conclusion. However, the universal behavior of

large systems might justify replacing the mixing of the components for $k \sim M$ by a new degree of freedom (the Liouville field?) living on a chain with a fixed number of gluons. In any case it appears that dealing with the full coupled system of equations Eq. (7) in a brute force manner is pretty hopeless.

3 A Proposal

We have shown how light-front dynamics reduced the problem of finding the glueball mass spectrum to finding the energy spectrum of a ring of "atoms" or gluons, each with a variable Newtonian mass. In particular this variability of mass allows quantum mechanical mixing between rings of different numbers of gluons. Thus we do not yet have a good physical analogy to ordinary molecules in non-relativistic quantum mechanics. If we could find a truer analogy, we might learn properties of glueballs from the known physical behavior of molecules.

Our proposal is to regard the variable Newtonian mass of each gluon as an indicator that the gluon is a tightly bound composite particle of gluon "bits," each bit possessing the minimal amount of mass m. Then a gluon with Newtonian mass mM is to be interpreted as a tightly bound composite of M bits. To take this proposal beyond words, we must replace the original field theory of gluons with a new theory of "bits," which includes the necessary dynamics for forming these tightly bound clusters. The idea is that the number of bits in a glueball will be fixed, and that the coupled equations of Eq. (7) will be replaced by a Schrödinger equation of the standard Newtonian form Eq. (8). The Yang-Mills theory must then be regained as an effective field theory describing the behavior of these composites at energies much lower than their excitation energy.

Note that one realization of this proposal occurs if the original Yang-Mills theory is just the infinite tension limit of some string theory. Each gluon would then "really" be a short piece of string consisting of string bits, all of which are "wee." Our suggestion in this context is that it might be more tractable to think of the glueball mass spectrum from the vantage point of the string theory (T_0 large but finite) than to take the infinite tension limit and work only with the effective field theory.

4 Quantum Mechanics Example

Here we take a bare bones quantum system exhibiting the kind of mixing encountered in our large N_c glueball problem, and find an equivalent system that doesn't have that mixing[13]. Consider a system of two particles each of mass m mixing with a one particle system of mass $2m$.

The coupled Schrödinger equations are

$$\left[\frac{-\nabla_1^2 - \nabla_2^2 + 2\mu^2 - 2mE}{2m}\right]\psi(\mathbf{x}_1, \mathbf{x}_2)$$
$$+ \frac{g}{8m\sqrt{2\pi}}\delta(\mathbf{x}_1 - \mathbf{x}_2)\chi(\mathbf{x}_1) = 0$$
$$\left[\frac{-\nabla^2 + \mu_2^2 - 4mE}{4m}\right]\chi(\mathbf{x}) + 2\frac{g}{8m\sqrt{2\pi}}\psi(\mathbf{x}, \mathbf{x}) = 0. \quad (9)$$

Here μ is the Lorentzian mass to be distinguished from the Newtonian mass m. We have let the Lorentzian mass of the χ particle $\mu_2 \neq \mu$ to allow for mass renormalization $\mu_2^{ren} = \mu$. Because of Galilei invariance we can work in the center of mass system, where $\psi = f(\mathbf{x}_1 - \mathbf{x}_2)$ and $\chi = \text{constant}$. Then χ can be eliminated leading to the single particle equation

$$[-\nabla^2 + \mu^2 - mE]f(\mathbf{x}) - \frac{g^2}{8\pi(\mu_2^2 - 4mE)}\delta(\mathbf{x})f(\mathbf{x}) = 0. \quad (10)$$

Note that although this is a single equation, the presence of mixing is signaled by the presence of E in the coefficient of the potential term. A convenient way to regulate the singular delta potential is to replace $\delta(\mathbf{x})$ by $\delta(|\mathbf{x}| - a)/a^{d-1}\Omega_d$ on s-waves and by zero on non-s-waves. Here $\Omega_d = 2\pi^{d/2}/\Gamma(d/2)$ is the volume of a unit $d-1$ sphere.

Solving the Schrödinger equation we first require a bound state with $E = \mu^2/4m$ which is identified with the one particle state with Newtonian mass $2m$ and renormalized Lorentzian mass μ. This determines μ_2:

$$\mu_2^2 = \mu^2 + \frac{g^2 K_\nu(\kappa a) I_\nu(\kappa a)}{8\pi\Omega_d a^{d-2}}, \quad (11)$$

where $\nu \equiv (d-2)/2$ and $\kappa \equiv \sqrt{\mu^2 - mE} = \mu\sqrt{3}/2$. The remaining physics in this model is contained in the s-wave phase shift:

$$\cot \delta =$$
$$\frac{N_\nu(ka)}{J_\nu(ka)} + \frac{2K_\nu(\kappa a)I_\nu(\kappa a)}{\pi J_\nu^2(ka)} + \frac{16\Omega_d a^{d-2}(\mu^2 - 4mE)}{g^2 J_\nu^2(ka)}. \quad (12)$$

Now we can remove the temporary uv cutoff by sending $a \to 0$ keeping κ fixed:

$$k^{2\nu}\cot \delta = k^{2\nu}\cot \pi\nu - \kappa^{2\nu}\csc \pi\nu$$
$$- \frac{\Gamma(1 + \nu)^2 2^{2\nu+6}\Omega_d}{g^2}(k^2 + \kappa^2). \quad (13)$$

Not too surprisingly this is precisely the effective range formula in the case of $d = 2\nu + 2$ spatial dimensions. The key qualitative feature to note is that the effective range, the coefficient of k^2, is *negative*.

This then is the baby version of the effective field theory. Our proposal in this case is to find a quantum

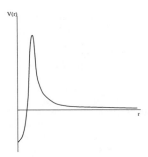

Figure 1: Potential Energy for the Quantum Mechanics Model

mechanical 2 particle system in which the one particle state is a composite, not an independent degree of freedom. That is, we must find a potential $V(\mathbf{x}_1 - \mathbf{x}_2)$ to insert in the two particle Schrödinger equation

$$\frac{1}{2m}[-\nabla_1^2 - \nabla_1^2 - 2mE]\psi + V\psi = 0 \quad (14)$$

that exactly reproduces at low energies the physics of the model with mixing. Clearly, V must (1) support a bound state and (2) at low energies display an effective range approximation with negative effective range. In our paper we exhibit an explicit potential of the form $V(r) = -\gamma\delta(r - b) + \lambda\delta(r - a)$ with $0 < b < a$ and $\gamma, \lambda > 0$. This is an example of a more generic shape shown in Fig. 1.

Qualitatively the short range attraction enables a very tight bound state and the barrier inhibits the coupling of the composite to the two particle continuum. Indeed, the effective coupling of the baby field theory is inversely related to the height of the barrier, $g^2 \sim 1/\lambda^2$. One can show[13] that the physics of that baby field theory is exactly reproduced for low energies: $ka \ll 1$.

5 Discussion and Conclusion

The example of Sec. 4 shows, in a highly simplified context, how an apparently independent species of Newtonian particle can be regarded as a composite of 2 "elementary" particles. The quantum mechanical mixing between the one particle system and the two particle system is simply the process of barrier penetration in the underlying theory. To conclude let us discuss the prospects for applying these ideas to large N_c QCD.

As described in Sec. 2, the problem of the glueball mass spectrum in the large N_c limit reduces, in discretized light-front quantization, to M coupled Schrödinger equations for polymers of gluons that describe how components of the glueball wave function

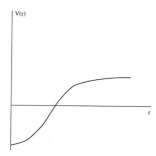

Figure 2: Potential Energy for a String-Bit Model

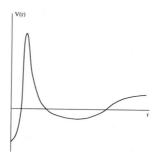

Figure 3: Potential Energy for a QCD-like Model.

with all numbers of gluons, from a small number of "valence" gluons to a large number of "wee" gluons, mix with each other. At weak 't Hooft coupling $N_c g^2 << 1$, ordinary perturbation theory gives tractable information about the interaction of small numbers of valence gluons. At large 't Hooft coupling one can arguably deal with large numbers of "wee" gluons, although any literal strong coupling expansion would be highly sensitive to the discretization of P^+. In fact the coupling parameter is scale dependent and a true solution of the problem must simultaneously include all scales and therefore a range of coupling constants from arbitrarily small values to values at least of $O(1)$.

The polymer analogy offers a way to capture the essential physics of glueballs by comparison with known molecular physics. However the mixing of states with different numbers of gluons ("atoms" in the molecular analogy) spoils the analogy with real molecules. In this talk we have suggested how these complicated mixing effects may be found as standard quantum mechanical tunneling phenomena within a completely standard Newtonian many-body system. To describe further how this observation might be helpful, recall the simplicity of the "wee" gluon approximation to the glueball dynamics Eq. (8). It is in fact a string-bit model of fundamental string, if the potential $\mathcal{V}$ is taken to have a shape as shown in Fig. 2. Such a potential is physically equivalent to $T_0^2 \mathbf{x}^2$ for $M \to \infty$. This limit gives a string, but with *no hard (valence) partons*. It is therefore infinitely far from large N_c QCD.

However if the potential is changed to one with a sharp barrier, as pictured in Fig. 3, many wee partons can cluster into a tightly bound subsystem that acts like a valence parton. We hope that these clustering effects can accurately simulate the mixing effects present in the original M coupled Schrodinger equations. We do not dare to hope that such a replacement Newtonian theory is literally soluble. Rather, the hope is that because the

replacement theory is a standard Newtonian many-body system, its relevant physics can be accurately extracted in the large system limit $M \to \infty$.

Acknowledgments

This work was supported in part by the Department of Energy under grant DE-FG02-97ER-41029.

References

1. G. 't Hooft, *Nucl. Phys.* **B72** (1974) 461.
2. J. M. Maldacena, "The large N Limit of Superconformal Field Theories and Supergravity," hep-th/9711200.
3. E. Witten, "Anti-de Sitter Space and Holography," hep-th/9802150.
4. S. S. Gubser, I. R. Klebanov, and A. M. Polyakov, "Gauge Theory Correlators from Noncritical String Theory," hep-th/9802109.
5. A. M. Polyakov, "String Theory and Quark Confinement," Talk given at Strings 97, Amsterdam, The Netherlands, 16-21 Jun 1997, hep-th/9711002.
6. E. Witten, "Anti-de Sitter Space, Thermal Phase Transition, and Confinement in Gauge Theories," hep-th/9803131.
7. R. Giles and C. B. Thorn, *Phys. Rev.* **D16** (1977) 366.
8. C. B. Thorn, *Phys. Lett.* **70B** (1977) 85; *Phys. Rev.* **D17** (1978) 1073.
9. R. Giles, L. McLerran, and C. B. Thorn, *Phys. Rev.* **D17** (1978) 2058.
10. R. Brower, R. Giles, and C. Thorn, *Phys. Rev.* **D18** (1978) 484.
11. C. B. Thorn, *Phys. Rev.* **D19** (1979) 639.
12. C. B. Thorn, *Phys. Rev.* **D20** (1979) 1435.
13. C. B. Thorn, "Quantum Newtonian Dynamics on a Light-Front," hep-th/9807151.

NON–PERTURBATIVE RESULTS IN GLOBAL SUSY AND TOPOLOGICAL FIELD THEORIES

D. BELLISAI

Università di Roma "Tor Vergata", Via della Ricerca Scientifica, I-00133 Roma, Italy
E-mail: bellisai@roma2.infn.it

F. FUCITO, A. TANZINI, G. TRAVAGLINI

I.N.F.N.- Sezione di Roma II "Tor Vergata", Via della Ricerca Scientifica, I-00133 Roma, Italy
E-mail: fucito@roma2.infn.it, tanzini@roma2.infn.it, travaglini@roma2.infn.it

In this lecture we briefly review the Seiberg–Witten model and explore some of its topological aspects.

1 Introduction

Many progresses have been made in recent years in the understanding of non–perturbative phenomena in globally supersymmetric (SUSY) gauge theories. In a seminal paper [1], Seiberg and Witten calculated all the non–perturbative contributions to the holomorphic effective action for an $N = 2$ Super Yang–Mills (SYM) theory. The low–energy action can in fact be expressed in terms of a unique holomorphic function $\mathcal{F}$, called the effective prepotential. Arguing that the different phases of the theory are described by a *polymorphic* prepotential, Seiberg and Witten were able to translate a set of physical conditions in terms of the number of singularities and the monodromies of $\mathcal{F}$. From this knowledge, it was thus possible to reconstruct the entire prepotential. The only physical information concerning non–perturbative effects fed to $\mathcal{F}$ was the surviving quantum symmetry of the theory which, for gauge group $SU(2)$, is $\mathbb{Z}_2$. To check the reliability of this framework, the non–perturbative contributions to $\mathcal{F}$ were computed directly in the Coulomb phase of the theory (for gauge instantons of winding number one and two), by using a saddle point approximation for certain correlators of the relevant fields [2,3,4]. The results were in agreement with those of Seiberg and Witten. These successful checks raise a number of questions which are the motivations of our investigation. The most compelling of them is: How come that a saddle point approximation, in which only quadratic terms are retained in the expansion of the action, is able to give the correct result? Why are higher–order corrections exactly zero? In the presence of SUSY a certain number of welcome simplifications allow the computation of some Green's functions. The most remarkable simplification is that these correlators are given by a constant times the appropriate power of the renormalization group invariant (RGI) scale. This fact was exploited by Witten [5] to argue that these computations are relevant for the determination of a class of invariants of four dimensional manifolds.

In the same paper it was shown that in the topological twisted version of $N = 2$ SYM the semiclassical expansion is exact. Here we suggest that these guidelines could help us give an answer to the problems previously raised.

2 A Review of the Seiberg–Witten Model

The Lagrangian density for the microscopic $N = 2$ SYM theory, in the $N = 2$ supersymmetric formalism is given by

$$L = \frac{1}{16\pi} \text{Im} \int d^2\theta d^2\tilde{\theta} \; \mathcal{F}(\Psi) \; . \tag{1}$$

The chiral superfield Ψ, which describes the vector multiplet of the $N = 2$ SUSY, transforms in the adjoint representation of the gauge group G (which will be $SU(2)$ from now on). Re–expressing the Lagrangian density in the $N = 1$ formalism, we have

$$L = \frac{1}{16\pi} \text{Im} \left[\int d^2\theta d^2\bar{\theta} K(\Phi, \bar{\Phi}, V) + \int d^2\theta f_{ab}(\Phi) W^a W^b \right] , \tag{2}$$

where a, b are indices of the adjoint representation of G. The Kähler potential $K(\Phi, \bar{\Phi}, V)$ and the holomorphic function $f_{ab}(\Phi)$ are given, in terms of $\mathcal{F}$, by $K(\Phi, \bar{\Phi}, V) = (\bar{\Phi} e^{-2V})^a \partial \mathcal{F}/\partial \Phi^a$ and $f_{ab}(\Phi) = \partial^2 \mathcal{F}/\partial \Phi^a \partial \Phi^b$. The classical action is obtained by choosing for the holomorphic prepotential $\mathcal{F}$ the functional form

$$\mathcal{F}_{cl}(\Psi) = \frac{\tau_{cl}}{2}(\Psi^a \Psi^a) \; , \tag{3}$$

where we define $\tau_{cl} = \frac{\theta}{2\pi} + \frac{4\pi i}{g^2}$. The classical action of the theory contains the scalar potential $S_{pot} = \int d^4x \; \text{Tr}[\phi, \phi^\dagger]^2$. The most general (supersymmetric) classical vacuum configuration is then

$$\phi_0 = a \left(\Omega \frac{\sigma_3}{2} \Omega^\dagger \right) \; , \; a \in \mathbb{C} \; , \; \Omega \in SU(2) \; . \tag{4}$$

When $a \neq 0$ the $SU(2)$ gauge symmetry is spontaneously broken to $U(1)$. The classical vacuum "degeneracy" is lifted neither by perturbative nor by non–perturbative quantum corrections [6] so that one has a fully quantum moduli space. The effective Lagrangian for the massless $U(1)$ fields Φ, W_α, will be the $U(1)$ version of Eq. (1) and reads, in $N = 1$ notation,

$$L_{eff} = \frac{1}{16\pi} \text{Im} \left[\int d^2\theta \mathcal{F}''(\Phi) W^\alpha W_\alpha + \int d^2\theta d^2\bar\theta \bar\Phi \mathcal{F}'(\Phi) \right] \quad . \tag{5}$$

The low–energy dynamics are then governed by the effective prepotential $\mathcal{F}(\Phi)$, whose crucial property is holomorphicity. The effective coupling constant is given by $\tau(\Phi) = \mathcal{F}''(\Phi)$. The description of the low–energy dynamics in terms of Φ and W_α is not appropriate for all vacuum configurations. The quantum moduli space $\mathcal{M}_{SU(2)}$ is better described in terms of the variable a and its dual $a_D = \partial_a \mathcal{F}$. When the gauge group is $SU(2)$, we can describe $\mathcal{M}_{SU(2)}$ in terms of the gauge–invariant coordinate $u = <\text{Tr}\phi^2>$. Then $\mathcal{M}_{SU(2)}$ is the Riemann sphere with punctures at $u = \infty$ and $u = \pm\Lambda^2$, where Λ is the RGI scale in the normalization of Seiberg and Witten [1]. At the classical level

$$\mathcal{F}_{cl}(a) = \frac{\tau_{cl}}{2} a^2 \quad ; \tag{6}$$

however perturbative as well as non–perturbative effects modify the expression of the prepotential, so that

$$\mathcal{F}(a) = \mathcal{F}_{pert}(a) + \mathcal{F}_{np}(a) \quad . \tag{7}$$

The perturbative term has been calculated by Seiberg [7] and is

$$\mathcal{F}_{pert}(\Psi) = \frac{i}{2\pi} \Psi^2 \ln \frac{\Psi^2}{M^2} \quad , \tag{8}$$

where M can be fixed by assigning the value of the coupling constant at some subtraction point. The R–symmetry of the theory constrains the non–perturbative prepotential to be

$$\mathcal{F}_{np}(a) = \sum_{k=1}^{\infty} \mathcal{F}_k \left(\frac{\Lambda}{a} \right)^{4k} a^2 \quad , \tag{9}$$

and similarly

$$u(a) = \frac{1}{2} a^2 + \sum_{k=1}^{\infty} \mathcal{G}_k \left(\frac{\Lambda}{a} \right)^{4k} a^2 \quad . \tag{10}$$

The 1–instanton contribution to $u(a)$ was found to be [2,4]

$$< \text{Tr}\phi^2 >_{k=1} = \frac{\Lambda_{PV}^4}{g^4 a^2} \tag{11}$$

where $\Lambda_{PV} = \Lambda/\sqrt{2}$ is the Pauli–Villars RGI scale, which naturally arises when performing instanton calculations.

Matone's relation [8] expresses then the $\mathcal{F}_k$'s as functions of the $\mathcal{G}_k$'s,

$$2i\pi k \mathcal{F}_k = \mathcal{G}_k \quad . \tag{12}$$

By making some hypotheses on the structure of the moduli space and on the monodromies of τ around its singularities, Seiberg and Witten obtained the expressions of $a(u)$ and $a_D(u)$,

$$a(u) = \frac{\sqrt{2}}{\pi} \int_{-\Lambda^2}^{\Lambda^2} dx \frac{\sqrt{x-u}}{\sqrt{x^2 - \Lambda^4}} \quad , \tag{13}$$

$$a_D(u) = \frac{\sqrt{2}}{\pi} \int_{\Lambda^2}^{u} dx \frac{\sqrt{x-u}}{\sqrt{x^2 - \Lambda^4}} \quad . \tag{14}$$

From Eqs. (13), (14) one can derive the $\mathcal{F}_k$'s and check whether they agree with those computed via instanton calculations, which we study in the next Section.

3 Instanton Calculus in SUSY Gauge Theories

We now briefly review the strategy to perform semiclassical computations in supersymmetric gauge theories, in the context of the constrained instanton method [9].

We expand the action functional around a properly chosen field configuration, which is the solution of the equations

$$D_\mu(A) F_{\mu\nu} = 0 \quad , \tag{15}$$

$$D^2(A)\phi_{cl} = 0 \quad , \quad \lim_{|x|\to\infty} \phi_{cl} \equiv \phi_\infty = a \frac{\sigma_3}{2i} \quad . \tag{16}$$

The boundary condition in Eq. (16) is dictated by Eq. (4). Equation (15) admits instanton solutions. When $a = 0$, Eq. (16) has only the trivial solution $\phi = 0$. On the other hand, when $a \neq 0$, we shall decompose the fields ϕ, $\phi^\dagger$ as

$$\phi = \phi_{cl} + \phi_Q \quad , \quad \phi^\dagger = (\phi_{cl})^\dagger + \phi_Q^\dagger \quad , \tag{17}$$

and integrate over the quantum fluctuations ϕ_Q, $\phi_Q^\dagger$. We want to remark that Eqs. (15), (16) are just approximate saddle point equations of the $N = 2$ action functional. This is a tricky point of the constrained instanton method.

The integration over the bosonic zero–modes can be traded for an integration over the collective coordinates, at the cost of introducing the corresponding Jacobian. The existence of fermion zero–modes is the way by which Ward identities related to the group of chiral symmetries of the theory come into play. When $a = 0$, the anomalous $U(1)_R$ symmetry

$$\lambda \longrightarrow e^{i\alpha}\lambda \quad , \quad \phi \longrightarrow e^{2i\alpha}\phi \tag{18}$$

and gauge invariance allow a nonzero result for the Green's functions with n insertions of the gauge–invariant quantity $(\phi^a\phi^a)(x)$ only when $n = 2k$. These correlators possess the right operator insertions needed to saturate the integration over the Grassmann parameters and, by supersymmetry, they are also position-independent. On the other hand, when $a \neq 0$, the correlator $\langle \phi^a \phi^a \rangle$ admits a complete expansion in terms of instanton contributions.

Fermion zero-modes are found by solving the equation $D_\mu \bar{\sigma}_\mu^{\dot{\alpha}\beta} \lambda_{\beta \dot{A}} = 0$, where $\dot{A} = 1, 2$ is the supersymmetry index. For instantons of winding number $k = 1$ the whole set of solutions of this equation is obtained via SUSY and superconformal transformations, which yield

$$\lambda_{\alpha \dot{A}}^a = \frac{1}{2} F_{\mu\nu}^a (\sigma_{\mu\nu})_\alpha^\beta \zeta_{\beta \dot{A}} \quad , \tag{19}$$

where $\zeta = \xi + (x - x_0)_\mu \sigma_\mu \bar{\eta} / \sqrt{2}\rho$, ξ, $\bar{\eta}$ being two arbitrary quaternions of Grassmann numbers.

The correct fermionic integration measure is given by the inverse of the determinant of the matrix whose entries are the scalar products of the fermionic zero-mode eigenfunctions and it reads $d^4\xi d^4\bar{\eta} \left(\frac{g^2}{32\pi^2}\right)^4$, where $d^4\xi d^4\bar{\eta} \equiv d^2\xi_{\dot{1}} d^2\xi_{\dot{2}} d^2\bar{\eta}_{\dot{1}} d^2\bar{\eta}_{\dot{2}}$.

As an example of instanton calculation we now consider the correlator $< \phi^a \phi^a >$ in the semiclassical approximation around an instanton background of winding number $k = 1$,

$$< \phi^a \phi^a > =$$

$$\int d^4x_0 d\rho \left(\frac{2^{10}\pi^6\rho^3}{g^8}\right) e^{-\frac{8\pi^2}{g^2} - 4\pi^2|a|^2\rho^2}$$

$$\int [\delta Q \delta \lambda] \delta \bar{\lambda} \delta \phi_Q^\dagger \delta \phi_Q \delta \bar{c} \delta c$$

$$\exp\left[-S_H[\phi_Q, \phi_Q^\dagger, A^{cl}] - S_F[\lambda, \bar{\lambda}, A^{cl}] + \right.$$

$$\left. -\frac{1}{2} \int d^4x \, Q_\mu M_{\mu\nu} Q_\nu - \int d^4x \, \bar{c} D^2(A^{cl})c\right]$$

$$\int d^4\xi d^4\bar{\eta} \left(\frac{g^2}{32\pi^2}\right)^4$$

$$\exp\left[-S_Y[\phi_{cl} + \phi_Q, (\phi_{cl})^\dagger + \phi_Q^\dagger, \lambda^{(0)}, \bar{\lambda} = 0]\right]$$

$$(\phi_{cl} + \phi_Q)^a (\phi_{cl} + \phi_Q)^a(x) \quad . \tag{20}$$

Let us now explain where the different terms in Eq. (20) come from:

1. $d^4x_0 d\rho \left(\frac{2^{10}\pi^6\rho^3}{g^8}\right)$ is the bosonic measure after the integration over $SU(2)/\mathbb{Z}_2$ global rotations in color space has been performed. x_0 and ρ are the center and the size of the instanton.

2. $S_H[\phi_{cl}, (\phi_{cl})^\dagger, A^{cl}] = 4\pi^2|a|^2\rho^2$, is the contribution of the classical Higgs action.

3. The third and fourth lines include the quadratic approximation of the different kinetic operators for the quantum fluctuations of the fields, and the symbol $[\delta\lambda\delta Q]$ denotes integration over nonzero-modes. $\bar{c}$ and c are the usual ghost fields, $\int d^4x \, \bar{c} D^2(A^{cl})c$ being the corresponding term in the action.

4. $S_Y[\phi, \phi^\dagger, \lambda^{(0)}, \bar{\lambda} = 0]$ is the Yukawa action calculated with the complete expansion of the fermionic fields replaced by their projection over the zero-mode subspace. It reduces to $\sqrt{2}g\epsilon^{abc} \int \phi^{a\dagger}(\lambda_{\dot{1}}^{(0)b} \lambda_{\dot{2}}^{(0)c})$.

After the integration over ϕ, $\phi^\dagger$ and the nonzero-modes, the ϕ_Q insertions get replaced by ϕ_{inh}, where

$$\phi_{inh}^a = \sqrt{2}g\epsilon^{bdc}[(D^2)^{-1}]^{ab}(\lambda_{\dot{1}}^{(0)d} \lambda_{\dot{2}}^{(0)c}) \quad , \tag{21}$$

and the determinants of the various kinetic operators cancel against each other. The r.h.s. of Eq. (20) now reads

$$\Lambda_{PV}^4 \int d^4x_0 d\rho \left(\frac{2^{10}\pi^6\rho^3}{g^8}\right) e^{-4\pi^2|a|^2\rho^2}$$

$$\int d^4\xi d^4\bar{\eta} \left(\frac{g^2}{32\pi^2}\right)^4 \exp\left[-\sqrt{2}g\epsilon^{abc} \int \phi_{cl}^\dagger(\lambda_{\dot{1}}^{(0)b} \lambda_{\dot{2}}^{(0)c})\right]$$

$$(\phi_{cl} + \phi_{inh})^a (\phi_{cl} + \phi_{inh})^a(x) \quad , \tag{22}$$

where $\Lambda_{PV}^4 = \mu^{8 - \frac{1}{2}(4+4)} e^{-\frac{8\pi^2}{g^2}}$. μ comes from the Pauli-Villars regularization of the determinants and the exponent is $b_1 k = (n_B - n_F/2)$ where n_B, n_F, b_1 are the number of bosonic, fermionic zero-modes and the first coefficient of the β-function of the theory.

The Yukawa action does *not* contain the Grassmann parameters of the zero-modes coming from SUSY transformations. As a consequence the only nonzero contributions are obtained by picking out the terms in the ϕ_{inh} insertions which contain the SUSY solutions of the Dirac equation. This amounts to say[a]

$$(\phi_{cl} + \phi_{inh})^a (\phi_{cl} + \phi_{inh})^a \longrightarrow -\xi_{\dot{1}}^2 \xi_{\dot{2}}^2 (F_{\mu\nu}^a F_{\mu\nu}^a). \tag{23}$$

Equation (20) now becomes

$$< \phi^a \phi^a > = \Lambda_{PV}^4 \int d\rho \left(\frac{2^{10}\pi^6\rho^3}{g^8}\right) e^{-4\pi^2|a|^2\rho^2}$$

$$\int d^4x_0 \, (F_{\mu\nu}^a F_{\mu\nu}^a)$$

$$\frac{g^2}{2}(a^*)^2 \int d^4\xi \left(\frac{g^2}{32\pi^2}\right)^2 \xi_{\dot{1}}^2 \xi_{\dot{2}}^2 \quad . \tag{24}$$

---

[a]This property generally holds for all the multi-instanton contributions to $u(a)$.

We can then immediately integrate over x_0 remembering that $\int d^4x \, F_{\mu\nu}^a F_{\mu\nu}^a = 32\pi^2/g^2$. The remaining integrations over ξ and ρ in Eq. (24) are trivial and yield [2,4]

$$< \phi^a \phi^a > = \frac{2}{g^4} \frac{\Lambda_{PV}^4}{a^2} \quad . \tag{25}$$

This result agrees with the Seiberg–Witten prediction.

4 Topological Yang–Mills Theory and Instanton Moduli Spaces

The relationship between supersymmetric and topological theories shows up when one observes that in the former there exists a class of position–independent Green's functions. When formulated on a generic manifold M by redefining the generators of the Lorentz group in a suitably twisted fashion, the $N = 2$ SYM theory gives rise to the so–called Topological Yang–Mills theory (TYM) [5]. *All* the observables, included the partition function itself, are in this case topological invariants, in the sense that they are independent of the metric on M.

With respect to the twisted Lorentz group, SUSY charges decompose as a scalar Q, an antisymmetric tensor $Q_{\mu\nu}$ and a vector Q_μ:

$$\bar{Q}_{\dot{\alpha}}^{\dot{A}} \to Q \oplus Q_{\mu\nu} \quad ,$$
$$Q_\alpha^A \to Q_\mu \quad . \tag{26}$$

Moreover, the twist transforms the $N = 2$ SYM fields as

$$A_\mu \to A_\mu \quad ,$$
$$\bar{\lambda}_{\dot{\alpha}}^{\dot{A}} \to \eta \oplus \chi_{\mu\nu} \quad ,$$
$$\lambda_\alpha^A \to \psi_\mu \quad ,$$
$$\phi \to \phi \quad , \tag{27}$$

where the anticommuting fields $\eta, \chi_{\mu\nu}, \psi_\mu$ are respectively a scalar, a self–dual two–form and a vector. The scalar supersymmetry charge of TYM plays a major rôle, in that it is preserved on any (differentiable) four–manifold M, and has the crucial property of being nilpotent modulo gauge transformations. This allows one to interpret it as a BRST–like charge. Actually, in order to have a strictly nilpotent BRST charge, one needs to include gauge transformations with the appropriate ghost c [10]. The BRST transformations are then

$$sA = \psi - Dc \quad ,$$
$$s\psi = -[c, \psi] - D\phi \quad ,$$
$$sc = -[c, c] + \phi \quad ,$$
$$s\phi = -[c, \phi] \quad . \tag{28}$$

This algebra can be read as the definition and the Bianchi identities for the curvature $\hat{F} = F + \psi + \phi$ of the connection $\hat{A} = A + c$ of the universal bundle $P \times \mathcal{A}/\mathcal{G}$ ($P, \mathcal{A}, \mathcal{G}$

are respectively the principal bundle over M, the space of connections and the group of gauge transformations). The exterior derivative on the base manifold $M \times \mathcal{A}/\mathcal{G}$ is given by $\hat{d} = d + s$.

The TYM action can be interpreted as a pure gauge–fixing action which localizes the universal connection to

$$\hat{A} = A + c = U^\dagger(d + s)U \quad . \tag{29}$$

A is then an anti–self–dual (ASD) connection which we have written in the ADHM formalism[b]. Once $\hat{A}$ is given, the components F, ψ, ϕ of $\hat{F}$ are in turn determined. F is anti–self–dual ($F = F^-$), and ψ is an element of the tangent bundle $T_A \mathcal{M}^-$, where $\mathcal{M}^-$ is the instanton moduli space. Moreover, the scalar field ϕ is the solution to the equation $D^2\phi = [\psi, \psi]$. The explicit expression for ϕ is then given by the twisted version of Eq. (21). For the following discussion it is important to point out that the ϕ field has trivial boundary conditions ($\phi = 0$) at spatial infinity.

In this geometrical framework, the BRST operator s has a very nice explicit realization as the exterior derivative on $\mathcal{M}^-$. This leads us to compute correlators of s–exact operators as integrals of forms on $\partial \mathcal{M}^-$ [11]. For example, we can write

$$\mathrm{Tr}\phi^2 = sK_c \quad , \quad K_c = \mathrm{Tr}\left(csc + \frac{2}{3}ccc\right) \quad , \tag{30}$$

an expression which parallels the well–known relation

$$\mathrm{Tr}F^2 = sK_A \quad , \quad K_A = \mathrm{Tr}\left(AdA + \frac{2}{3}AAA\right) \quad . \tag{31}$$

For winding number $k = 1$, the top form on the (eight–dimensional) instanton moduli space is $\mathrm{Tr}\phi^2(x_1)\mathrm{Tr}\phi^2(x_2)$, and one can compute [11]

$$\int_{\mathcal{M}^-} \mathrm{Tr}\phi^2 \mathrm{Tr}\phi^2 = \int_{\partial \mathcal{M}^-} \mathrm{Tr}\phi^2 K_c = \frac{1}{2} \quad . \tag{32}$$

5 Topological Aspects of the Seiberg–Witten Model

From the comparison between the Seiberg–Witten ansatz for the $N = 2$ low–energy effective action and explicit instanton calculations, we learn two important lessons. On one hand, this comparison provides us with a consistency check of the scenario proposed by Seiberg and Witten[1]. On the other hand, it strongly suggests that the semiclassical approximation around the instanton background saturates the non–perturbative sector. This calls for an explanation. A key property of the TYM theory is the exactness of the semiclassical limit [5]: we are

---

[b]We use the definitions and conventions of Sec. II of [4].

then naturally led to explore the Coulomb phase of the $N = 2$ SYM theory starting from its topological twisted counterpart.

The first problem one must face is that nontrivial boundary conditions for the scalar field are not compatible with the BRST algebra in Eq. $(28)^c$. This is because a nonzero v.e.v. for ϕ implies the existence of a (nonzero) central charge Z in the SUSY algebra which acts on the fields as a $U(1)$ transformation with gauge parameter ϕ. This new symmetry has to be included in an appropriate extension of the BRST operator, and it is implemented through the introduction of a new *global* ghost field Λ. The resulting algebra is then [11]

$$
\begin{aligned}
sA &= \psi - D(c + \Lambda) \ , \\
s\psi &= -[c + \Lambda, \psi] - D\phi \ , \\
s(c + \Lambda) &= -[c + \Lambda, c + \Lambda] + \phi \ , \\
s\phi &= -[c + \Lambda, \phi] \ .
\end{aligned} \tag{33}
$$

Once the universal connection is projected onto Eq. (29) by the gauge–fixing TYM action, the above extended algebra includes scalar fields with nonvanishing boundary conditions. Therefore, we can see that in this picture the field configurations dictated by the constrained instanton method (see Eqs. (15), (16)) naturally come into play, without resorting to any approximation procedure.

The TYM action gets now a nonzero boundary contribution from the term

$$
S_{inst} = \int_{R^4} d^4x \ \partial^\mu s \text{Tr}(\phi^\dagger \psi_\mu) \ . \tag{34}
$$

The explicit ADHM expressions for the fields in Eq. (34) can be derived from the extended algebra in Eq. (33) starting from the universal connection $U^\dagger(d + s)U^d$. It is easy to see [11] that, when inserted into Eq. (34), they yield the multi–instanton action[3] for the Seiberg–Witten model. This provides a natural and simplifying framework for further studies of non–perturbative effects in $N = 2$ theories.

Acknowledgements

It is a pleasure to thank D. Anselmi, C.M. Becchi, S. Giusto, C. Imbimbo, M. Matone, G.C. Rossi and S.P. Sorella for many stimulating discussions.

References

1. N. Seiberg and E. Witten, *Nucl. Phys.* B **431**, 19 (1994).

2. D. Finnell and P. Pouliot, *Nucl. Phys.* B **453**, 225 (1995).

3. N. Dorey, V.V. Khoze and M.P. Mattis, *Phys. Rev.* D **54**, 2921 (1997).

4. F. Fucito and G. Travaglini, *Phys. Rev.* D **55**, 1099 (1997).

5. E. Witten, *Commun. Math. Phys.* **117**, 353 (1988).

6. N. Seiberg, *Phys. Lett.* B **328**, 469 (1993).

7. N. Seiberg, *Phys. Lett.* B **206**, 75 (1988).

8. M. Matone, *Phys. Lett.* B **357**, 342 (1995).

9. I. Affleck, *Nucl. Phys.* B **191**, 429 (1981); I. Affleck, M. Dine and N. Seiberg, *Phys. Rev. Lett.* **51**, 1026 (1983); V. Novikov, M. Shifman, A. Vainshtein and V. Zakharov, *Nucl. Phys.* B **260**, 157 (1985).

10. L. Baulieu and I.M. Singer, *Nucl. Phys. Proc. Suppl.* **B5** 12 (1988).

11. D. Bellisai, F. Fucito, A. Tanzini and G. Travaglini, *in preparation.*

---

^{c}Recall that c localizes to $U^\dagger sU$.

^{d}In this context, the global ghost Λ is necessary to ensure invariance under local reparametrizations in the moduli space.

SUPERGRAVITY PREDICTIONS
FOR DARK MATTER

R. ARNOWITT

Center for Theoretical Physics, Department of Physics
Texas A&M University
College Station, TX 77843-4242, USA

PRAN NATH

Department of Physics, Northeastern University
Boston, MA 02115, USA

An analysis of the preliminary DAMA data indicating the possible existence of an annual modulation effect in dark matter detection event rates is given within the framework of supergravity (SUGRA) grand unification. For minimal SUGRA models, the theory is consistent with the 90% C.L. DAMA bands for $\tan\beta \gtrsim 30$. Non-universal soft breaking of appropriate signs can greatly enhance the theoretical signal, making the theory consistent with the data for $\tan\beta \gtrsim 7$. In both cases, the central value of the Wimp (lightest neutralino) mass is about 60 GeV, in accord with the DAMA result.

1 Introduction

Supersymmetry allows one to treat a wide variety of topics, which prior to its inception would have been thought to be disparate subjects. These include accelerator physics, the nature and detection of dark matter (DM), and proton stability. In supersymmetry, information or experimental bounds in one of these areas affects predictions for the others.

We consider here models based on supergravity (SUGRA) grand unification (GUTs) where supersymmetry is broken in a hidden sector by gravity with gravity as the messenger field to the physical sector [1]. Such models can discuss all three of the above phenomena.

The simplest SUGRA GUT model is the "minimal" model, mSUGRA, which possesses two Higgs fields $H_{1,2}$ and universal soft breaking parameters. This model depends on only four additional parameters and one sign. These may be chosen as m_0 (the universal scalar soft breaking mass at the GUT scale $M_G \cong 2 \times 10^{16} \, GeV$); $m_{1/2}$ (the universal gaugino mass at M_G or alternately the gluino ($\tilde{g}$) mass $m_{\tilde{g}} \cong (\alpha_3/\alpha_G)m_{1/2} = 2.83m_{1/2}$ where $\alpha_G \cong 1/24$ is the GUT scale coupling constant); A_0 (the universal cubic soft breaking mass at M_G, or alternately A_t the t-quark parameter at M_Z); and $\tan\beta = <H_2>/<H_1>$ (where $<H_2>$ gives rise to u-quark masses and $<H_1>$ to d-quark and lepton masses). In addition, the sign of the Higgs mixing parameter μ in the superpotential ($W \sim \mu H_1 H_2$) is a priori arbitrary.

Both theoretical and experimental constraints limit this parameter space. Thus, the renormalization group equations (RGE) determine μ^2 in terms of M_Z and the other parameters, so that $SU(2) \times U(1)$ breaking occurs at the electroweak scale, and experiments at LEP, SLC, the Tevatron and CESR limit the parameter space.

While mSUGRA is the simplest possibility (and consequently the most predictive), non-universal soft breaking may also occur. Further, some phenomena are sensitive to even moderate amounts of non-universality. In the following, we will maintain universality in the first two generations [to satisfy the suppression of flavor changing neutral currents (FCNC)] and allow non-universality in the Higgs and third generation masses. Thus at M_G we assume

$$m_{H_1}^2 = m_0^2(1 + \delta_1); \; m_{H_2}^2 = m_0^2(1 + \delta_2) \tag{1}$$

$$m_{q_L}^2 = m_0^2(1 + \delta_3); \; m_{u_R}^2 = m_0^2(1 + \delta_4);$$

$$m_{e_R}^2 = m_0^2(1 + \delta_5) \tag{2}$$

where $q_L = (\tilde{t}_L, \tilde{b}_L)$, $u_R = \tilde{t}_R$, $e_R = \tilde{\tau}_R$. In addition, there are three cubic parameters A_{0t}, A_{0b}, $A_{0\tau}$ at M_G. If supersymmetry is broken above M_G (e.g. at the Planck scale $M_{P\ell} = 2.4 \times 10^{18} \, GeV$), and the GUT group contains an SU(5) subgroup (e.g. SU(N), SO(2N), $N \geq 5$, E_6 etc.) with matter embedded in the SU(5) in the usual fashion, one has $\delta_3 = \delta_4 = \delta_5$ and $A_{0b} = A_{0\tau}$.

In the following, we will limit the SUSY parameters by $|\delta_i| \leq 1$, m_0, $m_{\tilde{g}} \leq 1 \, TeV$ and $|A_t/M_o| \leq 7$.

2 Annual Modulation Effect

SUGRA models of the above type which possess an R-parity invariance, automatically possess a cold dark matter (CDM) candidate, which over most of the parameter space is the lightest neutralino, $\tilde{\chi}_1^0$. Further, the predicted relic density of the $\tilde{\chi}_1^0$ falls within the astronomically measured range, $0.1 \leq \Omega_{\chi_1^0} h^2 \leq 0.4$, for a significant amount of the parameter space. (Here $\Omega_{\chi_1^0} = \rho_{\chi_1^0}/\rho_c$, $\rho_{\chi_1^0}$ is the relic density of $\tilde{\chi}_1^0$, $\rho_c = 3H^2/8\pi G_N$, and $H = h(100\, kms^{-1} Mpc^{-1})$ is the Hubble constant.) Thus, it may be possible to detect the $\tilde{\chi}_1^0$ which reside in the halo of the Milky Way and are incident on the earth, from their scattering by quarks in a nuclear target.

Last year, the DAMA NaI ($T\ell$) experiment in the Gran Sasso National Laboratory found preliminary evidence for the existence of Milky Way Wimps using the annual modulation signal [2]. This signal arises from the motion of the Earth around the Sun which varies the velocity of the detector on the Earth relative to the halo Wimps. Thus, the Earth's velocity is $v_E = v_S + v_0 \cos\gamma \cos\omega(t - t_0)$, where $v_S \cong 232 km/s$ is the Sun's velocity around the Galaxy, $v_0 \cong 30 km/s$ is the Earth's velocity around the Sun, $\gamma \cong 60°$ is the inclination of the plane of the Earth's orbit relative to the Galactic plane, $\gamma = 2\pi/T$, $T = 1$ year, and $t_0 =$ June 2. This motion predicts about a 7% modulation in the expected event rate.

Based on 4549 kg-day of data, the DAMA observations were consistent with a Wimp mass M_w and Wimp-proton cross section σ_{w-p} of

$$M_w = (59^{+36}_{-19})\, GeV;\ \ \xi\sigma_{w-p} = (1.0^{+0.1}_{-0.4}) \times 10^{-5} pb \quad (3)$$

where $\xi = \rho_{local}/0.3 GeV cm^{-3}$ and ρ_{local} is the local Milky Way mass density of Wimps.

3 mSUGRA Analysis

While the data is preliminary in nature, it is of interest to analyze its significance with respect to supersymmetry predictions. Previous analysis has been done using the MSSM [3]. We examine here what might be expected theoretically using supergravity grand unified models.

We consider first the mSUGRA model described above, and we assume in the following a central value of ρ_{local} of $4.5 GeV cm^{-3}$. Fig. 1 plots the maximum theoretical $\tilde{\chi}_1^0 - p$ cross section for $\tan\beta = 30$, as the other parameters are varied over their ranges. The DAMA 90% C.L. bound is a band ranging from between about $(1 \times 10^{-6}$ to $5 \times 10^{-6})$ pb at $m_{\chi_1^0} = 50$ GeV and between about $(2 \times 10^{-6}$ to $6 \times 10^{-6})$ pb at $m_{\chi_1^0} = 90$ GeV. (We have here reduced the upper experimental bound to coincide with the 90% C.L. sensitivity obtained by DAMA from previous data [2].) Since the theoretical cross section increases with $\tan\beta$, Fig. 1 implies that $\tan\beta \gtrsim 30$ is required for the mSUGRA model to lie within the

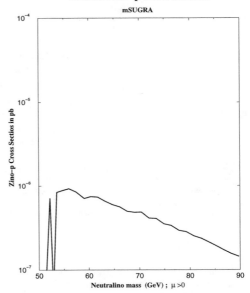

DAMA Zino–p Cross Sections

Figure 1: Maximum $\tilde{\chi}_1^0 - p$ cross section vs. $m_{\chi_1^0}$ for $\tan\beta = 30$, $\mu > 0$, for mSUGRA. The DAMA 90% C.L. is approximately a band between $(1 \times 10^{-6}$ to $5 \times 10^{-6})$ pb at $m_{\chi_1^0} = 50 GeV$ and between $(2 \times 10^{-6}$ to $6 \times 10^{-6})$ pb at $m_{\chi_1^0} = 90\, GeV$.

90% C.L. band of the DAMA data. This result restricts the particle spectrum expected at accelerators. Thus for $\tan\beta = 30$, one finds: $52\, GeV \lesssim m_{\chi_1^0} \lesssim 62\, GeV$; $95\, GeV \lesssim m_{\chi_1^\pm} \lesssim 110\, GeV$; $100\, GeV \lesssim m_h \lesssim 110\, GeV$; $380\, GeV \lesssim m_{\tilde{g}} \lesssim 480 GeV$ at the 90% C.L. Hence for $\tan\beta = 30$, most likely the $\chi_1^\pm$ and h will not be observable at LEP, but should be accessible to the TeV(33) (with $20 fb^{-1}$ of data). However, the gluino would require the LHC to be seen. Note that $m_{\chi_1^0}$ falls in the center of the allowed DAMA range given in Eq. (3). For $\tan\beta > 30$ the above spread in the masses would increase.

4 Non-Universal Soft Breaking

The situations changes significantly if one allows for non-universal soft breaking as in Eqs. (1,2). One of the major effects occur in the μ parameter determined by the renormalization group equations. One finds for $\tan\beta \lesssim 25$ [4]

DAMA Zino–p Cross Sections

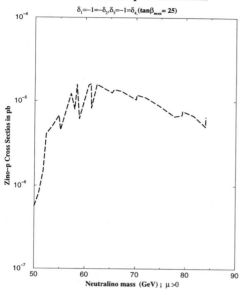

Figure 2: Maximum $\tilde{\chi}_1^0 - p$ cross section vs. $m_{\chi_1^0}$ for $\tan\beta = 25$, $\mu > 0$, for non-universal soft breaking with $\delta_1 = \delta_3 = \delta_4 = -1$, $\delta_2 = 1$. The 90% C.L. DAMA bands are stated in Fig. 1.

DAMA Zino–p Cross Sections

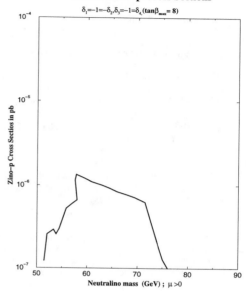

Figure 3: Same as Figure 2 with $\tan\beta = 8$.

$$\mu^2 \cong \frac{t^2}{t^2 - 1}[\frac{1}{t^2}\delta_1 - \frac{1 + D_0}{2}\delta_2 + \frac{1 - D_0}{2}(\delta_3 + \delta_4)]$$

$$m_0^2 + (\mu^2)_{univ} \qquad (4)$$

where $D_0 \cong 1 - (m_t/200\sin\beta)^2$, $t \equiv \tan\beta$ and $(\mu^2)_{univ}$ is the universal soft breaking contribution. Reducing μ^2 generally increases the $\tilde{\chi}_1^0 - p$ cross section (since it increases the Higgsino-gaugino interference in the spin-independent cross section). Thus, the choice $\delta_1, \delta_3, \delta_4 < 0$ and $\delta_2 > 0$ will reduce μ^2 from its mSUGRA value. Fig. 2 shows this effect for the case $\tan\beta = 25, \delta_1 = \delta_3 = \delta_4 = -1 = -\delta_2$. Here the theoretical cross section can exceed the DAMA 90% C.L. bound for a considerable part of the parameter space, in contrast to the universal soft breaking result of Fig. 1 (with $\tan\beta = 30$). Thus, it is possible to reduce $\tan\beta$ and still be consistent with the DAMA data. Fig. 3 shows that $\tan\beta = 8$ leads to a theoretical cross section within the 90% C.L. DAMA bands similar to what required $\tan\beta = 30$ for mSUGRA. (The opposite choice of signs of δ_i would reduce $\sigma_{\chi_1^0 - p}$

below the DAMA bounds.) Models allowing proton decay generally require a small $\tan\beta$ to be consistent with the Kamiokande data, and so such models in general will require non-universal soft breaking to be consistent with the DAMA results.

5 Conclusions

We have analyzed here the preliminary DAMA data of 4549 kg-day for the annual modulation effect [2] within the framework of supergravity grand unified models with R-parity invariance. The minimal supergravity model (mSUGRA) with universal soft breaking is consistent with this data provided $\tan\beta \gtrsim 30$. It predicts a neutralino mass of $m_{\chi_1^0} \simeq (50 - 60)\ GeV$, in accord with the data, and a $\chi_1^\pm$ and h accessible to the TeV(33), but a gluino requiring the LHC.

Models with non-universal soft breaking in the Higgs and third generation can more easily account for the data with a wider range of parameters, ie. $\tan\beta \gtrsim 7$, provided the deviations from universality have specific signs.

The DAMA data is still quite preliminary and more data is accumulating [5]. We have seen here, however, that SUGRA grand unified models imply the possibility of $\chi_1^0 - p$ cross sections within the range sensitivity of current NaI detectors.

Acknowledgment

This work was supported in part by National Science Foundation grants PHY-9722090 and PHY-9602074.

1. A.H. Chamseddine, R. Arnowitt and P. Nath, Phys. Rev. Lett **49**, 970 (1982). For reviews see P. Nath, R. Arnowitt and A. H. Chamseddine, Applied N=1 Supergravity (World Scientific, Singapore, 1984); H. P. Nilles, Phys. Rep. **110**, 1 (1984); R. Arnowitt and P. Nath, Proc. of VII J. A. Swieca Summer School, ed. E. Eboli and V. O. Rivelles (World Scientific, Singapore, 1994).
2. R. Bernabei et al., Phys. Lett. **B424**, 195 (1998).
3. A. Bottino, F. Donato, N. Fornengo and S. Scopel, Phys. Lett. **B423**, 109 (1998).
4. P. Nath and R. Arnowitt, Phys. Rev. **D56**, 2820 (1997).
5. R. Bernabei et al., ROM2F/98/34. An additional 14,962 kg-day of data has now been obtained. An analysis of this data within the framework of SUGRA models is presented elsewhere.

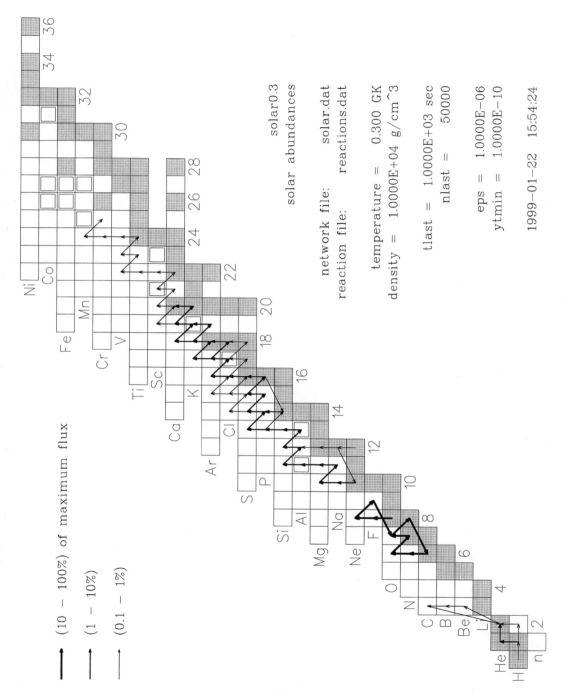

(10 – 100%) of maximum flux

(1 – 10%)

(0.1 – 1%)

network file: solar.dat
reaction file: reactions.dat

temperature =   0.300 GK
density   = 1.0000E+04 g/cm^3

tlast = 1.0000E+03 sec
nlast =   50000

eps = 1.0000E-06
ytmin = 1.0000E-10

1999-01-22 15:54:24

solar0.3
solar abundances

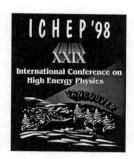

Parallel Session 18

Lattice Gauge Theory

Convenors: Kenneth Bowler (Edinburgh)

Keh-Fei Liu (Kentucky)

CHIRALITY ON THE LATTICE

H. NEUBERGER

Department of Physics and Astronomy, Piscataway, NJ 08855, USA
E-mail: neuberg@physics.rutgers.edu

During the last several years a non-perturbative formulation of exact chiral symmetry on the lattice has been developed. I shall outline the main ideas of these developments and discuss prospects for the future. The focus will be on the basic concepts enciphered in a new jargon consisting of terms like "infinite number of fermions", "domain wall fermions", "the overlap", "the Ginsparg-Wilson relation". Technical details will be omitted.

1 Introduction

As far as we know the basic constituents of matter are chiral fermions. Their interactions are described by an effective theory, the minimal standard model. This model is experimentally established at relatively long scales. The scales where field theory will cease being an adequate description of Nature are much shorter. Thus, there is a large range of scales where field theory is reliable. A complete theory would contain all scales. We are led[1] to ask: Is it plausible for a complete theory to contain such a large separation of scales, given that the long scales are chiral? If we restrict the question to renormalized, perturbative field theory, the answer is positive. But, if we now consider the same question in a non-perturbative context, the answer is less clear.

To simply ignore the non-perturbative aspects of the question would be wrong: Non-perturbative considerations have proven to be relevant before, for example providing an intrinsic upper bound on the Higgs mass (the triviality bound)[2]. Chiral gauge theories pose difficulties to non-perturbative analysis. The most famous of these is encountered in lattice field theory. For a long time it was believed that chiral gauge theories cannot be defined in any reasonable way on the lattice[3]. Developments over the last six years are poised to falsify this belief.

The main physics question is then what can be learned from the difficulties on the lattice and their resolution: Are the problems general or just a specific lattice quirk? Does the recent resolution point to a generic structure, relevant also off the lattice, in the real world? Most experts would agree that the problem is not just a lattice quirk; nevertheless, most field theorists work in the continuum, and implicitly assume otherwise. Accepting that the problem is generic we would expect a resolution that is not limited to the lattice. The recent resolution indeed is not.

In a nutshell, it works by postulating an infinite number of extra heavy fermions per unit four-volume. All the extra fermions are as heavy as we wish. Their masses can be kept naturally at energies many orders of magnitude above the typical energy-momentum scales of the chiral gauge theory governing the long scales. They can be integrated out fully, leaving a consistent low energy chiral theory. The required infinite number of fermions is suggestive of one or more extra dimensions, beyond the four we know. It is tantalizing that there exist other speculations about the fundamental laws of Nature that also use extra dimensions.

This solution was developed in collaboration with Narayanan[4,5,6]. We started to work on the problem in response to independently conceived ideas that appeared in two papers. The most widely known paper[7] is by Kaplan and has roots in earlier work[8] by Callan and Harvey. The other paper[9] was authored by Frolov and Slavnov. These two papers seemed very different but Narayanan and I identified a common mechanism. We pursued these insights to a concrete realization. Recently it has become clear that some germs of the idea can be found in an earlier paper by Ginsparg and Wilson[10].

During the last year many more lattice field theorists have started working on implementations of exact chiral symmetry on the lattice. Progress has been particularly rapid for vector-like field theories, where the exact chiral symmetries are global. This is important for QCD. Although in QCD the quarks are massive, there are chiral symmetries that are approximate and their incorporation in effective descriptions of the low energy properties of QCD has been extremely successful phenomenologically. Thus, numerical work on lattice QCD would benefit enormously from a practical lattice formulation of exact global chiral symmetry.

2 Different views of infinite numbers of fermions

Callan and Harvey were looking for a physical setup to find connections between anomalies in dimensions that differ by one or two units. They considered a gauge theory in five dimensions with a mass term that depended on the fifth dimension going monotonically from a positive value to a negative one. The five dimensional free Dirac equation has solutions that propagate with the speed of light along the wall but are exponentially suppressed in directions perpendicular to the wall. These solutions also turn out to be unpaired eigenstates of the γ_5-matrix. They represent chiral fermions confined to the four dimensional domain wall located at the point x_5 where the mass vanishes. The effective action of the fermions on the wall is chiral, so can be anomalous. The associated charge must leak into the fifth dimension since from the five dimensional point of view the charge is conserved. This charge has to be allowed to disappear at the $\pm$ infinities in the fifth direction; one cannot compactify that dimension, as charge would then have no way of disappearing. The number of fermions is infinite not only because the fifth direction is continuous, but, more importantly, because it is open at the two ends.

Kaplan showed that the Callan-Harvey phenomenon could be realized on a lattice using a Wilson mass term to create the domain-wall, the "defect". He understood that the continuity of the fifth direction (in addition to that of the first four) was not essential. But, he abandoned the openness of the ends of the fifth direction in order to make the whole setup finite. This added an anti-domain-wall housing fermions of opposite chirality. Kaplan speculated that one could devise a gauge action which could keep the chiral fermions at their respective walls and the two chiralities would not communicate even in the presence of arbitrary gauge fields. So, one still had a vector-like theory but the two chiral components of the fermions were "physically separated" in the fifth direction and, if it so happened that there were no anomalies at one of the walls, one could hope that the walls would dynamically decouple. It still could happen that fermion number would be violated at one of the walls. Such violations had to be exactly compensated at the other wall - the decoupling of the walls was not complete. Before Kaplan, Boyanowski et al. [11] discussed Callan-Harvey phenomena on the lattice, but their suggestions were less concrete and attracted no attention.

Frolov and Slavnov considered the regularization of $SO(10)$ gauge theory with one 16-plet of Weyl fermions. They introduced an infinite number of heavy fermion fields of normal and abnormal statistics. Their scheme made use of specific $SO(10)$ properties. The choice of statistics of the fermi fields ensured that the real part of the effective action induced by the fermions came in with the required weight, half of that induced by Dirac fermions. The ordering implied by the masses implied that the infinity was one dimensional. On the other hand, the gauge fields were just four dimensional. The entire treatment was in the continuum.

The papers by Kaplan and Frolov and Slavnov appeared at a time that the importance of topology induced fermion number violating processes was recognized. In a bilinear action different multiplets of fermions do not couple to each other. Therefore, one can imagine integrating

over the multiplets independently. But, in an instanton background, something very strange must be allowed to happen: a single ψ field may acquire a nonzero expectation value. This is impossible if the kernel of the bilinear form is a finite square matrix. In the continuum, this is possible because of a non-zero index. The index of a linear operator can be understood as a measure of the difference between the number of rows and the number of columns of a finite matrix approximating the operator. This difference cannot be frozen: it must change as the gauge background changes. This can be achieved only if the kernel is an infinite matrix, in other words, the number of fermions is infinite.

Narayanan and I interpreted the papers of Kaplan and Frolov and Slavnov in a new way. Our setup also accommodated the above described instanton phenomena. We start with [4] an apparently vectorial gauge theory:

$$\mathcal{L}_\psi = \bar{\psi}\gamma_\mu D_\mu \psi + \bar{\psi}\left(\frac{1+\gamma_5}{2}\mathcal{M} + \frac{1-\gamma_5}{2}\mathcal{M}^\dagger\right)\psi \quad (1)$$

The operator $\mathcal{M}$ acts in a new space and has an analytic index there (do not confuse this index with the one associated with instantons): while $\dim(\mathrm{Ker}\mathcal{M})=1$, $\dim(\mathrm{Ker}\mathcal{M}^\dagger)=0$. This property is radiatively stable, but can be realized only for $\dim\mathcal{M}=\infty$. The infinite size of $\mathcal{M}$ and its index make it impossible to use a standard bi-unitary transformation on the fermi fields to replace a non-hermitian $\mathcal{M}$ by a hermitian operator. If in Kaplan's original approach one made the gauge fields four dimensional his scheme looked just as Frolov and Slavnov's only the mass matrices were different. In his case the masses approached asymptotic values, while Frolov and Slavnov chose them to increase without bound. But, in both cases one had an index.

To control the infinity we interpreted the fifth direction as an imaginary time coordinate. So, we had to calculate the projectors on the vacua of the many-fermion problems associated with asymptotic translations in the fifth direction. Starting from the $\pm$ infinities, two different states propagate inwards, towards the defect. The fermion-induced action of interest is localized at the interface and is contained in the overlap of the two states - hence the term "overlap" [5].

The projectors are onto rays: thus, while the absolute value of the overlap is well defined its phase is not. This is just right: gauge invariance is preserved at all steps and because of anomalies it has to be impossible to achieve a complete definition for arbitrary fermion representations. We have a $U(1)$ bundle over the space of gauge fields and the bundle itself can be reduced to the space of gauge orbits. But, there is no guarantee that reasonable sections of the bundle can be found: We ended up with a mathematical structure able to reproduce the subtle features of chiral fermions [6].

Ginsparg and Wilson proposed in 1982 a renormalization group approach to representing exact global chiral symmetries on the lattice. The renormalization group is designed to bring out quantum versions of scale invariance which are anomalous. Massless QCD is classically scale invariant and consequently has exact global chiral symmetry. Therefore, a fixed point for massless QCD should exhibit chiral symmetry and also chiral anomalies. For concreteness, Ginsparg and Wilson imagined starting the renormalization group iteration from a chirally symmetric action. The iteration has to break chiral symmetry to make anomalies possible, but this is the single source of breaking. The initial symmetry constrains the resulting effective action. This action has to satisfy a remnant of chiral symmetry, the Ginsparg-Wilson relation. From it, Ginsparg and Wilson derive the anomalous $U(1)_A$ Ward identity. In footnote 11 they observe that all non-anomalous Ward identities could be derived in the same manner. Such identities hold when more than one quark flavor is present ($N_f > 1$). In footnote 7 they comment that the fixed point action cannot be bilinear in the fermion fields when $N_f > 1$ [12].

Their paper was forgotten because, in the presence of gauge fields, they had no explicit solution to their relation. That solutions should exist for $N_f = 1$ was plausible if one accepted the existence of renormalization group transformations with acceptable bilinear fixed points. To reach a fixed point an infinite number of iterations are required. Each iteration removes a slice of short ranged fermion modes and compensates by dilating the remainder. The chirally symmetric starting point is then separated from the actual action by the elimination of an infinite number of fermionic (and bosonic) degrees of freedom.

I have mentioned already that the infinite matrix $\mathcal{M}$ can realize both Kaplan's and Frolov and Slavnov's ideas and in its Kaplan-version leads to the overlap. In the vector-like context, if one does not insist on explicit decoupling of left and right Weyl spinors in the action, the kernel of the bilinear fermion can have an unrestricted square shape for any gauge background. Thus, one can hope for a sequence of kernels of finite square shape that converge to the infinite kernel representing the strictly massless case. Such a sequence was known since Kaplan: it is nothing but his domain wall construction, slightly modified [11,13]. Viewed differently, it consists of a light Dirac fermion coupled to many heavy Dirac fermions [14] by a mass matrix of seesaw type. The seesaw suppression produces strictly massless fermions only after an infinite amplification. But, the mass decreases exponentially fast.

I have emphasized the infinite number of fermions because it is a physically appealing picture. Mathematically however, the overlap construction [5] was recast [6] in a way that made no references to anything infinite. As a result, some technical simplifications were achieved. Moreover, the scheme is very flexible. This flexibility amounts to freedom to choose from large classes of lattice H-operators. Adapting the form of H to the specific problem at hand turned out very useful for efficient numerical simulations in two dimensions and also in analytical work. This flexibility will very likely be further exploited in future applications.

3 Finite number of fermions

In the vector-like case the overlap admits an explicit form involving a finite square matrix, the overlap-Dirac operator [15], which couples left and right Weyl components, and is therefore of the type that should obey the Ginsparg Wilson relation. It should do so because it was obtained by infinite iteration from an explicitly chiral starting point. It indeed does [14,16], and thus we finally have an explicit solution to this relation and the properties established by Ginsparg and Wilson hold.

More in line with the original work it has been recently claimed that a true fixed point to full massless QCD can be replaced by a classical approximation, called "a perfect action" [17], and the Ginsparg Wilson relation still holds. The perfect action is bilinear in the fermions for any number of flavors, but no explicit expression is known. An explicit definition is provided only for the "perfect" renormalization group transformation and any fixed point of this map is a "perfect action". The map contains a minimization step which introduces a nonanalyticity in the background. Some singularities are necessary, because of instantons [16]. In the overlap-Dirac operator the singularities are in direct correspondence with exact zero eigenvalues of a finite, local and analytic matrix.

At present it is unclear whether another class of explicit solutions to the Ginsparg Wilson relation will be found. There are obvious variations on the overlap itself, but it seems hard to find something explicit and really new. There also are indications [18,19] that all acceptable solutions to the Ginsparg Wilson relation have an overlap "flavor".

The infinite number of fermions imply a dependence only on the *rays* making up the overlap. On a finite lattice these rays are points in a $CP(\mathcal{N})$ or $RP(\mathcal{N})$ space and their Berry phases are the mathematical vehicle bringing in unavoidable anomalies and the need for their cancelation [20]. Anomalies can show up because the infinite number of fermions introduces a lack of determinacy. Without it, the regulated chiral determinant would be a function of gauge fields, rather than a quantity defined up to phase. The phase freedom is restricted by requiring the states in the overlap to depend smoothly on the gauge field. This requirement cannot be made compatible with gauge invariance if anomalies do not cancel. On the other hand, the real part of the induced action is gauge invariant, no matter what happens with anomalies.

If one wishes to work within a scheme that makes no reference to infinite numbers of fermions, the needed rays can be introduced by hand. However, solutions to the Ginsparg Wilson relation, unlike the overlap Hamiltonians or true fixed point actions, cannot be smoothly dependent on the gauge fields. Therefore, even if one extracts the relevant subspaces, it becomes unclear why one should care about Berry phases, given that the associated operators depend non-analytically on the gauge fields. To consider Berry phases we should allow ourselves to be aware of the smooth overlap-Hamiltonians. What distinguishes overlap solutions to the Ginsparg Wilson relation is that these come from analytic Hamiltonians and consequently have only singularities of a certain type. It seems contrived to eliminate these Hamiltonians but keep the singularity structure. In any case, an approach ostensibly based solely on the Ginsparg Wilson relation, but exploiting spectral representations of the fermion kernel ends up being equivalent to the overlap construction.

Ginsparg and Wilson knew nothing about the overlap and simply assumed the existence of an explicitly chirally symmetric starting point which had to be left somewhat ill defined. It is clear now that one can proceed this way and get a well defined scheme in the end. But, should we really ignore the starting point of the iteration and just focus on the relation observed by the fixed point? I think not: Nature has no reason to first come up with a Ginsparg Wilson relation and then find a solution - this relation is either obeyed or not, but the reason must be elsewhere, at a deeper level.

4 Main achievements of the overlap

The overlap has been extensively tested both on and off the lattice.

Explicit computations [6,21] in perturbative gauge backgrounds confirmed that the fermions were indeed chiral. In perturbation theory the generality of the overlap structure was made evident by calculations in non-lattice regularization schemes. These continuum schemes also simplified the needed algebra. Perturbation theory was used to compute anomalies, the vacuum polarization, and to check [22] the radiative stability of masslessness. It produced both consistent and covariant anomalies, pinpointed the source for their difference, and provided insights that could later be abstracted outside perturbation theory. Also, even in the vector-like context [23] it was necessary to see how various "no-go" theorems were avoided at the level of perturbation theory.

Numerical work established that instanton effects were correctly reproduced [6,24,25]. Rather vexing questions had to be answered. Was it indeed true that a single fermion could acquire an expectation value in an instanton background? How could one have explicit violation of $U(1)_A$ in the vector-like context without violating any other global axial symmetry? Eventually it became clear that a fully regulated version was available where 't Hooft's solution to the QCD $U(1)_A$ problem was valid.

For chiral models instanton effects are more dramatic. In two dimensions it was shown that fermion number violation is reproduced in the overlap [26]. The model that was investigated also has composite massless fermions and provides a simple example where 't Hooft's consistency conditions are non-trivially respected. Using the overlap, this model was simulated numerically, and the success of this experiment constitutes the most subtle test of the overlap to date. The test also shows that exact gauge invariance on the lattice is not needed so long as the model makes sense in the continuum [27]. A slight breaking of gauge invariance amounting to short ranged correlations between the gauge degrees of freedom $g(x)$ is irrelevant in the continuum limit where the $g(x)$ become independent and decouple from observables.

Non-perturbative anomalies are relatively subtle in continuum physics. In a non-perturbative setting they should become simpler to understand than perturbative anomalies. This was shown for the overlap in three and four dimensions [28,29]. In both dimensions the mechanism behind the anomaly is simple: the non-perturbative anomalies reflect Berry phase obstructions to choosing global phases of the states making up the overlap. These obstructions are directly related to overlap-Hamiltonian level crossings over an extended parameter space.

Three dimensions is a promising area of applications for the overlap [28,30]. Although there is no chirality in any odd dimensions, there exists an analogue, and the overlap machinery can be extended to three dimensions with ease. A particular set of three dimensional models that were studied by Appelquist [31] and collaborators, admit an overlap formulation. ¿From it, I derived a three dimensional generalization of the Ginsparg Wilson relation. Three dimensional models will be instructive.

5 Prospects

The initial apathetic reaction of dominant factions in the lattice community to the progress on the chiral fermion front has all but disappeared. The most convincing sign of change are the emerging priority squabbles.

If the overlap is correct and practical difficulties are overcome, the way lattice QCD is currently being done will undergo a revolution. Currently, a large fraction of the numerical QCD effort, which commands the bulk of support, manpower and visibility in lattice field theory, is invested in diminishing and controlling chirality violating effects. Following initial work on a two dimensional vector-like model [32], serious studies [33] of the domain wall truncation of the overlap have been undertaken and consequently the largest computer resources in the US will be applied to the domain-wall-fermion seesaw-approximation of the overlap [34].

Even more recently, progress has been made also on the direct implementation of the overlap-Dirac operator on the lattice [30,35], relegating all truncations to their natural place: numerical algorithms. It is too early to say whether the direct approach or the domain wall one will ultimately prove more efficient. In principle, the direct approach is cleaner.

6 Summary

A new and exciting lattice methodology is emerging and, possibly, we are witnessing a big step forward in numerical QCD. At the base of this progress lies a world consisting of an infinite number of fermions, all but one having very large masses. This infinity is fully under control and consequently can be completely eliminated. It is natural to speculate that we have discovered more than just a trick designed for computers, namely, that we have obtained a valuable hint about chirality in Nature.

Acknowledgements

This research was supported in part by the DOE under grant #DE-FG05-96ER40559.

References

1. H. B. Nielsen, S. E. Rugh, *Nucl. Phys. B (Proc. Suppl.)* **29B,C**, 200 (1992).
2. H. Neuberger, U. M. Heller, M. Klomfass and P. Vranas, *Proceedings of the XXVI International Conference on High Energy Physics*, **Vol. II**, 1360 (1993), editor James R. Stanford, (American Institute of Physics).
3. J. Smit, *Nucl. Phys. B (Proc. Suppl.)* **17**, 3 (1990).
4. R. Narayanan, H. Neuberger, *Phys. Lett. B* **302**, 62 (1993).
5. R. Narayanan, H. Neuberger, *Nucl. Phys. B* **412**, 574 (1994).
6. R. Narayanan, H. Neuberger, *Nucl. Phys. B* **443**, 305 (1995).
7. D. B. Kaplan, *Phys. Lett. B* **288**, 342 (1992).
8. C. Callan, J. Harvey, *Nucl. Phys. B* **250**, 427 (1984).
9. S. A. Frolov, A. A. Slavnov, *Phys. Lett. B* **309**, 344 (1993).
10. P. Ginsparg, K. Wilson, *Phys. Rev. D* **25**, 2649 (1982).
11. D. Boyanowski, E. Dagoto, E. Fradkin, *Nucl. Phys. B* **285**, 340 (1987).
12. E. Eichten, J. Preskill, *Nucl. Phys. B* **268**, 179 (1986).
13. Y. Shamir, *Nucl. Phys. B* **406**, 90 (1993).
14. H. Neuberger, *Phys. Rev. D* **57**, 5417 (1998).
15. H. Neuberger, *Phys. Lett. B* **417**, 141 (1998).
16. H. Neuberger, *Phys. Lett. B* **427**, 353 (1998).
17. P. Hasenfratz, V. Laliena, F. Niedermayer, *Phys. Lett. B* **427**, 125 (1998); M. Lüscher, *Phys. Lett. B* **428**, 342 (1998).
18. R. Narayanan, hep-lat/9802018.
19. T-W Chiu, C-W Wang, S. Zenkin, hep-lat/9806031.
20. H. Neuberger, hep-lat/9802033.
21. S. Randjbar-Daemi, J. Strathdee, *Phys. Lett. B* **348**, 543 (1995), *Nucl. Phys. B* **443**, 386 (1995); *Phys. Lett. B* **402**, 134 (1997); *Nucl. Phys. B* **461**, 305 (1996); *Nucl. Phys. B* **466**, 335 (1996).
22. A. Yamada, *Nucl. Phys. B* **514**, 399 (1998); Y. Kikukawa, H. Neuberger, A. Yamada, *Nucl. Phys. B* **526**, 572 (1998).
23. H. Neuberger, hep-lat/9807009.
24. R. Narayanan, H. Neuberger, *Phys. Rev. Lett.* **71**, 3251 (1993).
25. R. Narayanan, P. Vranas, *Nucl. Phys. B* **506**, 373 (1997).
26. Y. Kikukawa, R. Narayanan, H. Neuberger, *Phys. Rev. D* **57**, 1233 (1998); *Phys. Lett. B* **105**, 399 (1997). R. Narayanan, H. Neuberger, *Phys. Lett. B* **393**, 360 (1997); *Nucl. Phys. B* **477**, 521 (1996); *Phys. Lett. B* **348**, 549 (1995).
27. D. Förster, H. B. Nielsen, M. Ninomiya, *Phys. Lett. B* **94**, 135 (1980).
28. Y. Kikukawa, H. Neuberger, *Nucl. Phys. B* **513**, 735 (1998).
29. H. Neuberger, hep-lat/9805027.
30. H. Neuberger, hep-lat/9806025.
31. T. Appelquist, D. Nash, *Phys. Rev. Lett.* **64**, 721 (1990).
32. R. Narayanan, H. Neuberger, P. Vranas, *Phys. Lett. B* **353**, 507 (1995); *Nucl. Phys. B (Proc. Suppl.)*, **47** 596 (1995).
33. P. M. Vranas, *Phys. Rev. D* **57**, 1415 (1998).
34. T. Blum, A. Soni, *Phys. Rev. Lett.* **79**, 3595 (1997); P. Chen et al., hep-lat/9807029.
35. R. G. Edwards, U. M. Heller, R. Narayanan, hep-lat/9807017.

GAUGE INVARIANT PROPERTIES OF ABELIAN MONOPOLES

S. THURNER, H. MARKUM

Institut für Kernphysik, TU-Wien, Wiedner Hauptstraße 8-10, A-1040 Vienna
E-mail: thurner@kph.tuwien.ac.at

E.-M. ILGENFRITZ

Institute for Theoretical Physics, Kanazawa University, Japan

M. MÜLLER-PREUSSKER

Institut für Physik, Humboldt Universität zu Berlin, Germany

Using a renormalization group motivated smoothing technique, we investigate the large scale structure of lattice configurations at finite temperature, concentrating on Abelian monopoles identified in the maximally Abelian, the Laplacian Abelian, and the Polyakov gauge. Monopoles are mostly found in regions of large action and topological charge, rather independent of the gauge chosen to detect them. Gauge invariant properties around Abelian monopoles, the local non-Abelian action and topological density, are studied. We show that the local averages of these densities along the monopole trajectories are clearly above the background, which supports the existence of monopoles as physical objects. Characteristic changes of the vacuum structure at the deconfinement transition can be attributed to the corresponding Abelian monopoles, to an extent that depends on the gauge chosen for Abelian projection. All three Abelian projections reproduce the full $SU(2)$ string tension within 10 % which is preserved by smoothing.

1 Introduction

Over the last two decades a variety of attempts in field theory have been aiming for a qualitative understanding and modeling of two basic properties of QCD: quark confinement and chiral symmetry breaking. The most prominent schemes are the instanton liquid model [1] and the dual superconductor picture of the QCD vacuum.[2] While the first model explains chiral symmetry breaking and solves the $U_A(1)$ problem, the second one provides a simple idea for the confinement mechanism. In this scenario, where the vacuum is viewed as a dual superconductor, condensation of color magnetic monopoles leads to confinement of color charges through a dual Meissner effect. The superconductor picture was substantiated by a large number of lattice simulations over the last years. So it was shown that in the confinement phase monopoles percolate through the $4D$ volume [3] and are responsible for the dominant contribution to the string tension.[4] At present, more and more groups characterize their lattice vacuum in accordance to the instanton liquid picture.[5]

Both models rest on the existence of very different kinds of topological excitations, instantons and color magnetic monopoles. For a long time they have been treated independently, only recently some deeper connection among those different objects has been pointed out, both on the lattice and in the continuum.[6] Instantons are localized solutions of the Euclidean equations of motion in Yang-Mills theory carrying action and integer topological charge. Even though it is difficult to detect instantons and antiinstantons among quantum fluctuations,

there is no problem to study these well-defined objects in classical or semiclassical (heated) configurations on the lattice. The situation for monopoles is more difficult due to two reasons. First, there are no monopole solutions in QCD as, for example, in the Georgi-Glashow model where 't Hooft-Polyakov monopoles with finite extension and mass exist. Second, following 't Hooft, monopoles should be searched for as pointlike singularities of some gauge transformation dictated by a local, gauge covariant composite field. The standard prescription, however, is localizing monopoles in QCD as Abelian monopoles via an Abelian projection from some gauge (for example the maximally Abelian gauge). This leads to monopole trajectories which are dependent on the gauge chosen. Note however, that the condensation mechanism of monopoles itself seems to be gauge independent.[7]

In this contribution we relate gauge invariant observables to monopole trajectories, with the intention to further understand the semiclassical vacuum structure in terms of monopoles and lumps of topological charge and their role for the confinement problem. We will comment on a new way of monopole identification on the lattice which evades serious problems of previous methods, and which might have a close formal relationship to 't Hooft-Polyakov monopoles.

2 Smoothing

To resolve semiclassical structures in gauge field configurations provided by lattice simulations, the cooling method has been used, which locally minimizes the ac-

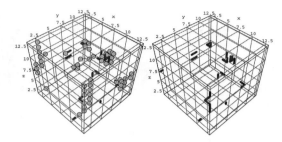

Figure 1: Regions of low modulus of the auxiliary Higgs field (dots), which *should* mark the trajectories of monopoles according to LAG, are found very close to the trajectories of DGT monopoles obtained by Abelian projection. For clarity DGT monopoles are also shown alone (right).

tion. However, even improved versions of cooling rapidly destroy monopole percolation and reduce the string tension. From the instanton point of view cooling is known to destroy small instantons and instanton-antiinstantons pairs, such that the true topological structure is accessible at best by a backward extrapolation to zero cooling steps. Up to now, most lattice studies are performed with the Wilson action, for which the lattice definitions of the topological charge Q are known to violate the bound valid for the continuum action: $S \geq 8\pi^2 |Q|/g^2$, where g is the coupling constant. The Wilson action is known to decrease with smaller size ρ of an instanton, such that isolated instantons are unstable under cooling. In contrast to this, improved cooling finds instantons stabilized within a size interval $\rho > 2\ a$ (a is the lattice spacing). Other methods like APE smearing let instantons grow.

To avoid these ambiguities we have used a method of 'constrained smoothing'[8] which is based on the concept of perfect actions.[9] These actions respect the above bound for the topological charge and lead to a theoretically consistent 'inverse blocking' operation. Inverse blocking is a method to find a smooth interpolating field on a fine lattice by constrained minimization of the perfect action, provided a configuration is given on a coarse lattice. This makes an unambiguous definition of topological charge possible. Constrained smoothing is a renormalization group motivated method which first blocks fields $\{U\}$, sampled on a fine lattice with lattice spacing a, to a coarse lattice $\{V\}$ with lattice spacing $2\ a$ by a standard blockspin transformation. Then inverse blocking is used to find a smoothed field $\{U^{\mathrm{sm}}\}$ replacing $\{U\}$.

An important feature of this method is that it does not drive configurations into classical fields as unconstrained minimization of the action would do. It saves the long range structure of the Monte Carlo configuration in $\{V\}$, such that the smooth background contains semiclassical objects deformed by classical and quantum

interaction. The upper blocking scale roughly defines the border line between 'long and short range'[a]. In this work we used a simplified fixed-point action[11] for Monte Carlo sampling and for constrained smoothing before the configurations were analyzed.

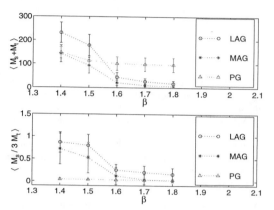

Figure 2: Total monopole length (top) and space-time asymmetry (bottom) as a function of β for monopoles obtained in different gauges. $\beta_c = 1.545(10)$ is the deconfinement point.

3   Gauge Fixing

The most popular gauge to study monopoles on the lattice is the maximally Abelian gauge (MAG).[12] This gauge is enforced by an iterative minimization procedure, which can get stuck in local minima, so-called *technical* Gribov copies. The Laplacian Abelian gauge (LAG)[13] is not afflicted by this problem. MAG and LAG can be understood along the same lines. The gauge functional of the MAG can be written:

$$F(\Omega) = \sum_{x,\mu} (1 - \frac{1}{2} tr\, (\sigma_3 U_{x,\mu}^{(\Omega)} \sigma_3 U_{x,\mu}^{(\Omega)\dagger}))$$

$$= \sum_{x,\mu,a} (X_x^a - \sum_b R_{x,\mu}^{a,b} X_{x+\hat{\mu}}^b)^2 \to \int_V (D_\mu X)^2 ,(1)$$

with the gauge transformation Ω_x acting on $\{U\}$

$$U_{x,\mu}^{(\Omega)} = \Omega_x U_{x,\mu} \Omega_{x+\hat{\mu}}^\dagger$$

encoded in an *auxiliary* adjoint Higgs field

$$\Phi_x = \Omega_x^\dagger \sigma_3 \Omega_x = \sum_a X_x^a \sigma_a$$

---

[a]The iterative application of this method, 'cycling',[10] obscures the idea of a definite blocking scale while it still preserves rather well features of long range physics as the string tension.

subject to local constraints $\sum_a (X_x^a)^2 = 1$ and with adjoint links

$$R_{x,\mu}^{a,b} = \frac{1}{2} tr\left(\sigma_a U_{x,\mu} \sigma_b U_{x,\mu}^\dagger\right).$$

In LAG the local constraints are relaxed and replaced by a global normalization: $\sum_{x,a}(X_x^a)^2 = V$, such that Eq. (1) can be further written:

$$\int_V (D_\mu X)^2 \to \sum_{x,a} \sum_{y,b} X_x^a \{-\square_{x,y}^{a,b}(R)\} X_y^b. \qquad (2)$$

Then the minimization reduces to a search for the lowest eigenmode of the covariant lattice Laplacian. LAG is unambiguously defined, except for degenerate lowest eigenmodes, which correspond to *true* Gribov copies. For both MAG and LAG, the gauge transformation is finally performed by diagonalization of the field Φ_x. Quite similarly we enforce the Polyakov gauge (PG) by diagonalization of Polyakov loops.

After the Abelian gauge of choice has been fixed one extracts the Abelian degrees of freedom (Abelian projection). The Abelian link angles can then be used for the identification of monopoles, like in compact $U(1)$ theory, searching for the ends of Dirac strings. Monopoles identified in this manner are generally referred to as DeGrand-Toussaint (DGT) monopoles. The Higgs field introduced in the LAG provides an alternative for monopole identification which is more satisfactory from a physical point of view. Lines of $\rho_x = |\Phi_x| = 0$ where ρ_x is defined as

$$X_x^a = \rho_x \hat{X}_x^a \;, \quad \rho_x = \sqrt{\sum_{a=1}^{3}(X_x^a)^2}, \qquad (3)$$

directly define lines of gauge fixing singularities (monopoles), more in the original spirit of 't Hooft.[b] Note here that this way of monopole identification does not require to perform the actual gauge fixing and Abelian projection!

In Fig. 1 we show that both methods of monopole identification turn out to be quite related. Regions of small ρ are highly correlated with trajectories of monopoles identified by the DGT method.

4 Physical Properties of Monopoles

The following results were obtained from simulations of pure $SU(2)$ theory on a $12^3 \times 4$ lattice. Observables were computed on 50 independent configurations per β. Different β values were considered to study the behavior

[b]For the 't Hooft-Polyakov monopole regions with $\rho = 0$ of the *physical* Higgs field are identified with the centers of such monopoles.

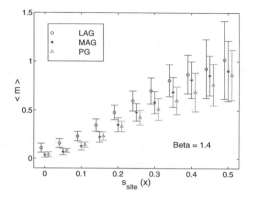

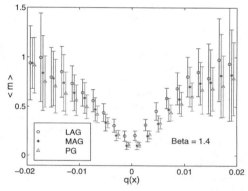

Figure 3: Average occupation number of monopoles $< m >$ nearest to sites with action density s_{site} (top) and topological charge density q (bottom) in the confinement phase.

slightly below and above the deconfinement phase transition. For the particular action used,[11] $\beta_c = 1.545(10)$. Global properties like the total loop length and the space-time asymmetry are shown in Fig. 2. DGT monopoles extracted from the MAG and the LAG behave qualitatively similar. Those from the PG show no change at the deconfinement phase transition. This reflects the fact that PG monopoles should be static.

In Fig. 3 we present the average occupation number of monopoles on dual links nearest to a given site as a function of the local action $s_{site}(x)$ and charge $q(x)$, for different gauges. One observes that the probability of finding monopoles increases with the amount of action/charge density at the same lattice position. This result is practically independent of the gauge used to define the (DGT) monopoles.

If monopoles are physical objects, one expects that

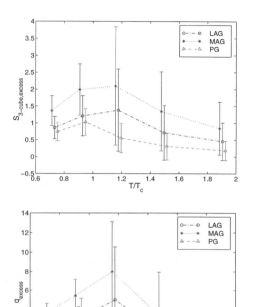

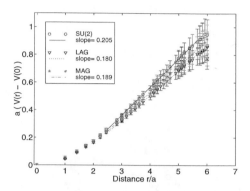

Figure 5: Static quark-antiquark potentials obtained from Polyakov-line correlators after smoothing. The slope of the Abelian potential in the MAG after Abelian projection is about 5% less than that of $SU(2)$ gauge field. Another 5% are lost in the LAG. Still the LAG carries 90% of the original string tension, indicating Abelian dominance for the LAG.

In Fig. 5 we display static quark-antiquark potentials obtained from Polyakov-Antipolyakov correlators, for the $SU(2)$ fields and, in the case of LAG and MAG, after Abelian projection. The Abelian string tension of the MAG is about 5% less than that the $SU(2)$ field. The Abelian string tension of LAG is a little smaller than for the MAG but still exhibits Abelian dominance. The Abelian string tension of PG is trivially identical with that measured on the smoothed $SU(2)$ configurations.

Finally we present an intuitive argument, that monopoles also should carry electric charge, that they are dyons. Consider the selfduality equations $F_{\mu\nu} = \pm \tilde{F}_{\mu\nu}$. From the trivial relation

$$\int d^4x \text{Tr}[(F_{\mu\nu} \pm \tilde{F}_{\mu\nu})^2] \geq 0 \qquad (5)$$

it immediately follows that

$$S \geq \frac{8\pi}{g^2}|Q| \qquad (6)$$

and (anti)selfdual fields saturate the identity. Fig. 6 depicts the probability distribution of topological charge density for a given local action and shows that for $s_{site} > 0.3$ the local action obeys a local version of Eq. (6) near to saturation, $s_{site}(x) \sim \frac{8\pi}{g^2}|q(x)|$. The plot was obtained after one constrained smoothing step, and exhibits that the gauge fields are already sufficiently smooth to expose semi-classical structure. This is suggested by the relatively clear ridges indicating approximate local selfduality for large enough action density. In Fig. 3 we provided

Figure 4: Excess action (top) and charge (bottom) as a function of temperature.

they can be characterized by a local excess of the (gauge invariant) action. We define such an excess action of monopoles by

$$S_{\text{ex}} = \frac{< S_{\text{monopole}} - S_{\text{nomonopole}} >}{< S_{\text{nomonopole}} >} , \qquad (4)$$

where S_{monopole} is the action contained in a three-dimensional cube which corresponds to the dual link occupied by a monopole. Replacing the action in the above expression by the modulus of the topological charge density according to the Lüscher method we obtain the charge excess q_{ex}. For details of the definition of the local operators see Ref. 11. Fig. 4 shows that just below T_c the excess action and charge for the MAG and LAG monopoles are clearly above one, indicating an excess of action of more than a factor of two compared to the bulk average (background). The large error bars above T_c reflect the fact that the topological activity diminishes in the deconfinement phase. These results are somewhat

enhanced in comparison to a $T = 0$ study with Wilson action without cooling or smoothing.[14]

evidence that monopoles are found predominantly in regions of large action. We thus conclude that monopoles also carry electric charge and should be interpreted as dyons. Note that this way of argumentation is a short-cut, to be more precise, one would have to test for local selfdualtiy along individual monopole trajectories.

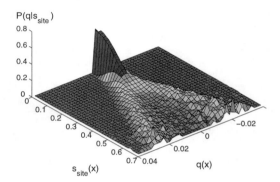

Beta = 1.40

Figure 6: Probability distribution for finding a topological charge density $q(x)$ at a lattice site x if the local action density equals $s_{site}(x)$. The ridges follow the lines $s_{site}(x) = \frac{8\pi}{g^2}|q(x)|$ where local (anti)selfduality is satisfied.

5 Conclusions

We have demonstrated that the renormalization group smoothing technique with an (approximate) classically perfect action provides a powerful tool to investigate the semiclassical vacuum structure. Analyzing trajectories of monopoles identified in various gauges we found that monopoles appear preferably in regions which are characterized by enhanced action and topological charge density. We showed that in exactly those regions local (anti)selfduality of the gauge fields is prevailing. This is further evidence that monopoles should be addressed as dyons. We demonstrated that almost the complete string tension can be recovered from the Abelian projected field corresponding to various Abelian gauges, indicating Abelian dominance also for the LAG. This is trivially true for the PG, but the corresponding monopoles do not change at the deconfinement transition. We have shown that monopole trajectories carry an excess action of about twice the background action density of smoothed gauge fields. Similarly, monopoles also carry excess topological charge. In the confinement phase this observation is rather independent of the gauge chosen for identifying Abelian monopoles, but the behavior

of PG monopoles is different in the deconfinement. We therefore conclude that MAG and LAG monopoles behave similar physically and can be interpreted as physical objects which carry action and topological charge.

This work was supported in part by FWF, No. P11456.

References

1. E. Shuryak, *Nucl. Phys.* B **203**, 93 (1982); T. Schäfer and E. Shuryak, *Rev. Mod. Phys.* **70**, 323 (1998).
2. G. 't Hooft, *Nucl. Phys.* B **190**, 455 (1981); S. Mandelstam, *Phys. Rep.* C **23**, 245 (1976).
3. V.G. Bornyakov, V.K. Mitrjushkin and M. Müller-Preussker, *Phys. Lett.* B **284**, 99 (1992).
4. H. Shiba and T. Suzuki, *Phys. Lett.* B **333**, 461 (1994).
5. J. Negele, talk presented at *Lattice 98*.
6. M.N. Chernodub and F.V. Gubarev, *JETP Lett.* **62**, 100 (1995); S. Thurner, H. Markum and W. Sakuler, in *Proceedings of Confinement 95* (World Scientific, Singapore, 1995); S. Thurner, M. Feurstein, H. Markum and W. Sakuler, *Phys. Rev.* D **54**, 3457 (1996); H. Suganuma, S. Sasaki, H. Ichie, F. Araki and O. Miyamura, *Nucl. Phys.* B (Proc. Suppl.) **53**, 528 (1997); R.C. Brower, K.N. Orginos and C.-I. Tan, *Phys. Rev.* D **55**, 6313 (1997); H. Reinhardt, *Nucl. Phys.* B **503**, 505 (1997).
7. M.N. Chernodub, M.I. Polikarpov and A.I. Veselov, *Nucl. Phys.* B (Proc. Suppl.) **49**, 307 (1996); A. Di Giacomo and G. Paffuti, *Phys. Rev.* D **56**, 6816 (1997).
8. M. Feurstein, E.-M. Ilgenfritz, M. Müller-Preussker and S. Thurner, *Nucl. Phys.* B **511**, 421 (1998) (hep-lat/9611024).
9. P. Hasenfratz and F. Niedermayer, *Nucl. Phys.* B **414**, 785 (1994).
10. T. DeGrand, A. Hasenfratz and T.G. Kovacs, *Nucl. Phys.* B **505**, 417 (1997).
11. E.-M. Ilgenfritz, H. Markum, M. Müller-Preussker and S. Thurner, *Phys. Rev.* D, in print (hep-lat/9801040).
12. A.S. Kronfeld, G. Schierholz and U.-J. Wiese, *Nucl. Phys.* B **293**, 461 (1987).
13. A. van der Sijs, *Nucl. Phys.* B (Proc. Suppl.) **53**, 535 (1997); *Prog. Theor. Phys. Suppl.*, in print (hep-lat/9803001).
14. B.L.G. Bakker, M.N. Chernodub and M.I. Polikarpov, *Phys. Rev. Lett.* **80**, 30 (1998).

THEORETICAL ASPECTS OF TOPOLOGICALLY UNQUENCHED QCD

S. Dürr

University of Washington, Physics Department, Box 351560, Seattle, WA 98195, U.S.A.
E-mail: durr@phys.washington.edu

I give an outline of my recent proposal to take the QCD functional determinant in lattice simulations partially into account: The determinant is split into two factors, the factor referring to a standard background in each topological sector is kept exactly, the factor describing the effect of the smooth deviation of the actual configuration from the reference background is replaced by one. The issue of how to choose the reference configurations is discussed and it is argued that "topologically unquenched QCD" is an interesting starting point to study full QCD in lattice simulations as it gets the main qualitative features right from the beginning.

1 What's wrong with quenched QCD ?

In QCD the fermion functional determinant is a nonlocal contribution to the gluon effective action. This nonlocality has an unpleasant effect in lattice calculations as it slows down present numerical algorithms dramatically when the quark-masses get small.

In order to elude this problem most numerical simulations in the past have been done in the "quenched approximation" where the determinant is simply replaced by one [1]. More recently, simulations in the "partially quenched approximation" have become available [2], where the determinant gets evaluated at quark masses which are higher than those in the propagators. Thus (partial) quenching amounts to suppressing the contribution of all internal fermion loops in QCD by giving the quarks unphenomenological high or infinite masses.

Attempts to introduce the corresponding modifications in the low-energy theory artificially in order to learn how to correct for them – the results being "quenched" and "partially quenched Chiral Perturbation Theory" – have shown that the (partially) quenched approximation is, in some aspects, fundamentally different from the full theory: First of all, numerical results won in quenched or partially quenched simulations should (at least in principle) be corrected for the occurrence of "enhanced chiral logarithms" [3,4]. In addition, the η' was found to be a pseudo-Goldstone boson in quenched QCD (as opposed to the situation in QCD, where it is heavier than the lowest-lying octet of pseudo-scalar pseudo-Goldstone mesons by more than a factor $\sqrt{3}$) and its propagator shows – in case the low-energy analysis is correct – a pole of order two (which spoils any field-theoretic interpretation) right at the same position in the p^2-plane where its first-order pole is [3,4]. Even without particle-interpretation, quenched QCD is strictly confined to Euclidean space-time; there is no continuation of its Green's functions to Minkowski space-time [5].

Thus it seems legitimate to search for an alternative starting point for approaching full QCD which gets some qualitative aspects of full QCD right from the beginning.

2 What is "topologically unquenched QCD" ?

The aim is to identify a part of the functional determinant which is cheap from the computational point of view but essential from the field-theoretical point of view.

We start from the generating functional of (euclidean) QCD in the form where the fermionic degrees of freedom have been integrated out

$$Z_\theta^{\mathrm{QCD}}[\bar{\eta}, \eta] = N \cdot \int DA \, \frac{\det(\slashed{D}+M)}{\det(\slashed{\partial}+M)}$$
$$e^{\bar{\eta}(\slashed{D}+M)^{-1}\eta} \, e^{-\int \frac{1}{4}GG \, + i\theta \int \frac{g^2}{32\pi^2}G\tilde{G}} \quad (1)$$

where $\slashed{D} = \gamma_\mu(\partial_\mu - igA_\mu)$ is the (euclidean) Dirac operator and $\tilde{G}_{\mu\nu} = \frac{1}{2}\epsilon_{\mu\nu\sigma\rho}G_{\sigma\rho}$ the dual of the field-strength operator and where the measure DA includes gauge-fixing and Faddeev Popov terms. In (1) a factor which does not depend on the gauge field to be integrated over has been pulled out of the normalizing factor and the convention is that the quark mass matrix M is diagonal and of rank N_f with the CP-violation stemming entirely from θ (if $\theta \neq \pi\mathbf{Z}$); the shorthand-notation to be used here and in subsequent formulas is $\det(\slashed{D}+M) = \prod_{i=1}^{N_f} \det(\slashed{D}+m_i)$ in the determinant and $\bar{\eta}(\slashed{D}+M)^{-1}\eta = \sum_{i=1}^{N_f} \bar{\eta}_{(i)}(\slashed{D}+m_i)^{-1}\eta_{(i)}$ in the propagator.

QCD is known to show a topological structure [6]: The $SU(3)$-gauge-field configurations on $\mathbf{R}^4$ with finite action boundary condition or on the torus $\mathbf{T}^4$ fall into inequivalent topological classes (labeled by an index $\nu = g^2/32\pi^2 \cdot \int G\tilde{G} \, dx \in \mathbf{Z}$). In a given sector ν, any two configurations may be continuously deformed into each other but not into any configuration with a different index ν. Due to this topological structure, the integral in (1) may be rewritten as a sum of integrals over the individual sectors $\int DA \to \sum_{\nu \in \mathbf{Z}} \int DA^{(\nu)}$ and the determinant factorizes ($\slashed{D} = \slashed{D}^{(\nu)}$)

$$\frac{\det(\slashed{D}+M)}{\det(\slashed{\partial}+M)} = \frac{\det(\slashed{D}_{\mathrm{std}}^{(\nu)}+M)}{\det(\slashed{\partial}+M)} \cdot \frac{\det(\slashed{D}^{(\nu)}+M)}{\det(\slashed{D}_{\mathrm{std}}^{(\nu)}+M)} \quad (2)$$

where the first factor depends on ν only and may be

pulled out of the integral. In this form it is obvious that quenched QCD actually does two modifications: It sets both determinant factors equal one, whereas partially quenched QCD keeps both of them at the price of using unrealistically high quark masses.

In "topologically unquenched QCD" the first factor in (2) is kept exact (with the same quark mass as in the propagator between $\bar{\eta}, \eta$) and only the second factor is replaced by one [7], i.e. the "theory" is defined through

$$
Z^{\mathrm{TU-QCD}}_{\theta, \{A^{(\nu)}_{\mathrm{std}}\}}[\bar{\eta}, \eta] = N \cdot \sum_{\nu \in \mathbf{Z}} \frac{\det(\mathbb{D}^{(\nu)}_{\mathrm{std}}+M)}{\det(\mathbb{D}^{(0)}_{\mathrm{std}}+M)} \; e^{i\nu\theta} \cdot
$$
$$
\int DA^{(\nu)} \; e^{\bar{\eta}(\mathbb{D}^{(\nu)}+M)^{-1}\eta} \; e^{-\int \frac{1}{4}GG} \; . (3)
$$

At this point, the motivation to treat the two determinant factors in (2) on unequal footing is just an economic one: The "topological" factor is universal for all configurations within one class and bears the knowledge about the nontrivial topological structure of QCD. On the other hand, the "continuous" factor in (2) causes a dramatic slowdown in numerical simulations as this part of the determinant (or its change) has to be computed for each configuration individually.

Note that, in contrast to the full-QCD generating functional (1), the "topologically unquenched" truncation (3) does depend on the choice of reference-backgrounds. Later, two alternative strategies of how to choose, in a given sector, the reference-configuration are described and such a choice of strategy, once it is done, is considered part of the definition of the theory. Obviously there is no a-priori evidence that – with any of these two choices – the approximation (3) should be particularly good. Nevertheless, known analytical knowledge about QCD in a finite box seems to indicate that including the "topological" part of the determinant is sufficient to get some basic features of full QCD qualitatively right.

3 How to implement "topologically unquenched QCD" ?

Obviously, for "topologically unquenched QCD" to be practically useful, at least two requirements have to be fulfilled: First, a recipe of how to make a good choice for the reference backgrounds in (3) is needed. Second, the costs in terms of CPU-time for separating the "topological" part of the determinant from the remainder have to be smaller than the costs would be to compute the determinant as a whole. These issues shall be discussed.

3.1 Upgrading a quenched sample

A quenched sample may get modified to be representative in the sense of "topologically unquenched QCD":

1. Use a method you consider both trustworthy and efficient to compute for each configuration its topological index ν.

2. Use the gauge action you trust to compute, in each class, the gauge-action of every configuration as well as the class-average $\bar{S}^{(\nu)}$ and choose the reference-configuration according to one of the following two prescriptions:

    (i) Choose – out of the class ν – the configuration with minimal gauge-action as the representative $A^{(\nu)}_{\mathrm{std}}$.

    (ii) Choose – out of the class ν – the configuration for which the gauge-action is closest to the class-average $\bar{S}^{(\nu)}$ as the representative $A^{(\nu)}_{\mathrm{std}}$.

3. Use the fermion action and the method you consider both trustworthy and efficient to compute the determinants $\det((\mathbb{D}^{(\nu)}_{\mathrm{std}}+M)/((\mathbb{D}^{(0)}_{\mathrm{std}}+M))$.

4. For any higher topological sector either include the corresponding determinant computed in step 3 into the measurement or eliminate the corresponding fraction of configurations from that sector.

A few points need immediate clarification:

The first problem is that on the lattice, the topological structure of the continuum-theory is washed out; simply computing $\nu = g^2/32\pi^2 \cdot \int G\tilde{G} \, dx$ gives a value for ν which is, in general, not an integer. A solution to this problem is so crucial to the overall performance of the "topologically unquenched" approximation that it shall be discussed in a separate subsection.

The second point is that for computing the determinant ratio in step 3 one cannot rely on any method which is tantamount to an expansion in δA, since the two backgrounds are far from each other. The necessary ab-initio computation is achieved e.g. by the eigenvalue method: typically the first few hundred eigenvalues of the Dirac operator on a given background may be determined.

Finally: What's the difference in terms of physics between the two strategies of how to choose the reference-backgrounds in step 2 ? We emphasize that for either choice there is a sound theoretical motivation. Strategy (i) – choose the configuration which minimizes the gauge action – is nothing but the semiclassical ansatz being pushed to account for topology: Within each sector, the determinant is exact for the configuration having least gauge-action, i.e. for the one which, in a semiclassical treatment, gives the dominant contribution to that sector to the path-integral. Strategy (ii) – choose the configuration which, in its gauge-action, is closest to the class-average of that sector – takes into account that the

Monte Carlo simulation as a whole doesn't try to minimize the total action density but rather the free-energy density: The configuration which is most typical in a certain sector is not the one with minimal action but the one which has additional instanton-antiinstanton pairs plus topologically trivial excitations such as to find an optimum between the additional amount of action to be paid and the additional amount of entropy to be gained. It is the very aim of the second strategy to choose in each sector a "most-typical" background (which realizes such an optimum pay-off) as reference-configuration. It should be stressed that even though the two strategies end up selecting reference-backgrounds which look highly different (exceedingly smooth in the first case versus pretty rough in the second case) the final results may still be close to each other – the only thing which matters is the (strategy-intrinsic) determinant ratio computed in step 3.

3.2 Generating a "topologically unquenched" sample

The fact that the "topological" determinant in (3) is a number which depends only on the total topological charge of the actual configuration but not on its other details suggests that one could try to precompute these "topological" determinants on artificially constructed backgrounds prior to running the simulation.

Within the strictly semiclassical strategy which is reflected by choice (i) the reference-backgrounds are gotten in a rather simple way: Place ν copies (for $\nu > 0$) of a single-instanton solution with typical radius (i.e. $\rho \simeq 0.3$fm, cf.[8]) on the lattice and minimize the action with respect to variation of their relative orientations and positions. Within the more realistic strategy of choice (ii) which accounts for the competing effects of increased action versus increased entropy the reference-backgrounds are constructed as follows: Place ν instantons (for $\nu > 0$) with typical sizes (i.e. $\rho \simeq 0.3$fm, cf.[8]) randomly on the lattice plus additional instanton-antiinstanton pairs such as to achieve a total instanton density of 1fm^{-4} (cf.[8]). Optionally, this background may be dressed with thermal fluctuations by applying a reasonable number of heating-steps (monitoring ν in order to guarantee that it stays unchanged).

Having defined the standard backgrounds in this way a pure Metropolis algorithm generating a "topologically unquenched" sample employs the following steps:

1. Make use of an updating procedure to propose a new configuration and determine its topological index (via the method you trust).

2. If the configuration realizes a previously unseen ν: Evaluate the functional determinant on the standard background constructed for that ν according to your choice of strategy.

3. Base the decision on whether to accept the proposed configuration on

$$\Delta S = S_{\text{new}}^{(\nu)} - S_{\text{old}}^{(\nu)} - \log\left(\frac{\det(\not{D}_{\text{std}}^{(\nu,\text{new})} + M)}{\det(\not{D}_{\text{std}}^{(\nu,\text{old})} + M)}\right) \quad (4)$$

where $S_{\text{new/old}}^{(\nu)}$ denotes the gluonic action of the newly proposed / last accepted configuration.

There is one conceptual deficit this algorithm suffers from: The procedure for constructing the standard-backgrounds makes use of knowledge about the size-distribution and partly about the density of (anti-) instantons which was won in previous lattice-studies. In other words: The "topologically unquenched" simulation as outlined above is not entirely from first principles. Moreover, the quality of the final sample depends on how appropriate the artificial backgrounds are which were used in the computation of the standard determinants. Constructing these reference backgrounds is particularly demanding within strategy (ii) as it means that one has to do an a-priory guess which configuration, within a given sector, is "most-typical" in the sense of full QCD. Even if "most typical" is translated into a technical criterion (e.g. the configuration which, in its total effective action, is closest to the corresponding class-average of a finite full-QCD sample) there is no other algorithmic solution to this problem than by doing a full-QCD simulation. Thus it is reasonable to construct the reference backgrounds as indicated above, thereby making use of existing knowledge concerning size-distribution and density of instantons in QCD – in particular as there is, for strategy (ii), a final check concerning the quality of the artificial backgrounds: In case the guess would have been just perfect, the a-priori ratios of determinants evaluated on these backgrounds (i.e. the ratios which got used in the "topologically unquenched" simulation) would perfectly agree with the analogous a-posteriori determinant ratios evaluated on the "most-typical" configurations as produced by the run. Accordingly, a procedure one might think of trying in case the agreement turns out to be less than satisfactory is just to start over with the simulation – but this time using the "most-typical" configurations found in the first run rather than the artificial guesses.

3.3 Measuring topological indies

Determining the topological index of a newly proposed background in a way which is fast and reliable is so crucial to the overall-performance of "topologically unquenched QCD" as to justify few remarks about this point.

There are several methods[9] to determine, for a given configuration, its topological index ν. Some of them were recently compared and found to give – when implemented with sufficient care – to comparable results for the topological susceptibility [10]. Nevertheless, it is clear that in the present case, where nothing is known about the spectrum of the Dirac operator on the background at hand, the field-theoretic method is likely to determine ν in the quickest possible way. However, the fact that on the lattice the relevant operator undergoes thermal renormalization provides a challenge: Simply integrating the Chern density, i.e. computing $g^2/(32\pi^2) \int G^a_{\mu\nu} \tilde{G}^a_{\mu\nu} \, dx$ gives a value which is, in general, not close to an integer; in fact, a histogram-plot over many configurations tends to reveal accumulations near regularly displaced, non-integer values, e.g. near $0, \pm 0.7, \pm 1.4$ etc. There are two options of how to deal with this situation:

The first, simplistic, approach is just to define a "confidence interval" – e.g. ± 0.2 – around each of the values $0, \pm 0.7, \pm 1.4, \ldots$ and to assign the configurations lying within these bounds the indices $\nu = 0, \pm 1, \pm 2 \ldots$ etc. The remaining configurations which didn't get an index assigned are then simply tossed away.

The second, more sophisticated, approach is to make use of the fact that cooling a configuration is able to remove the effect brought in by thermal renormalization: Cooling a set of gluon-configurations results in the peaks (in the histogram plot) being shifted closer to the corresponding integers and the valleys between the peaks getting thinned out under each sweep.

The problem is that these two methods do not necessarily agree in their results for a given configuration – a fact which can be understood on rather simple grounds: Under repeated cooling with the naive (Wilson) action, a single-instanton solution shrinks monotonically until it finally falls through the grid. In order to prevent the cooling algorithm at least from loosing the large instantons one has to modify the action w.r.t. which cooling is done in such a way that all instantons with a radius ρ above a certain ρ_{thr} tend to get blown up ("over-improved action") or stay constant ("perfect action") under a sweep, where typically $\rho_{\text{thr}} \simeq 2.3a$. From the evidence given in [11] how quickly cooling with an "over-improved" action tends to pin down $g^2/(32\pi^2) \int G^a_{\mu\nu} \tilde{G}^a_{\mu\nu} \, dx$ near an integer (say 5 sweeps to be within 2.99 and 3.01, etc.) performing $O(3)$ "over-improved" cooling sweeps seems to be sufficient to get an unambiguous assignment. The price to pay, however, is that the small instantons ($\rho < \rho_{\text{thr}}$) get compressed and finally pushed through the grid even more efficiently than under cooling with an unimproved action [11]. From these consideration we conclude that the field-theoretic method with cooling yields, once it has stabilized, a correct assignment for the latter cooled configuration which, however, isn't necessarily appropriate for the initial configuration which may have contained small ($\rho < 2.3a$) instantons. On the other hand, determining the topological index by the first (simplistic) approach (no cooling being involved) has an inferior signal-to-noise ratio (about half of the configurations can't get assigned an index and have to be tossed away) but for the remaining ones the procedure is sensitive to all instantons the lattice can support (i.e. $\rho > 0.7a$).

3.4 Rudimentary cost analysis

The overhead as compared to a quenched simulation results from the CPU-time spent on determining ν for every newly proposed configuration and from the determinants which get evaluated. Preparing the reference-backgrounds and computing the determinants is a fixed investment which is given by L, a, m only (i.e. independent of the length of the simulation) and evaluating the $O(10)$ standard determinants (for nowadays typical values of m and L) is pretty cheap [7]. On the other hand, determining for each configuration its index ν gives rise to costs which grow linearly in simulation-time and thus provide the main overhead (as compared to Q-QCD) in a long run. As a consequence, the method for determining the topological index will have the greatest impact on the overall-performance in TU-QCD. We have advocated choosing a field-theoretical definition with little or no cooling at all, which means that either $O(3)$ cooling-sweeps are performed or 50% of the configurations have to be tossed away. Accordingly, in an approximation where a cooling-sweep is considered twice as expensive as a complete update, the overhead from ν-determinations is roughly a factor 2...6 over a quenched simulation. Thus doing a "topologically unquenched" run might be considered an alternative to a high-statistics quenched run.

4 What about qualitative features of "topologically unquenched QCD" ?

In "topologically unquenched QCD" a determinant is introduced which – as is seen from (4) – only influences the relative weight of the different topological sectors; within each sector there is no difference to quenched QCD.

For QCD in a finite box Leutwyler and Smilga have shown that in the regime [a] $V\Sigma m \gg 1$ the distribution of topological indices is gaussian with width [12]

$$\langle \nu^2 \rangle = V\Sigma m / N_f \quad . \tag{5}$$

In quenched QCD the corresponding distribution is much broader as there is no determinant which suppresses the higher sectors. In "topologically unquenched QCD" the standard determinants result in the higher sectors being suppressed as compared to a quenched sample but the

---

[a] $\Sigma = \lim_{m \to 0} \lim_{V \to \infty} |\langle \overline{\psi}\psi \rangle|$, where $m_i = m \ \forall i$ for simplicity; note that $V\Sigma m \to \infty$ when $m \to 0$ as the box has to be scaled accordingly: $L \simeq 1/M_\pi, M_\pi^2 \simeq \Lambda_{\text{had}} m$.

amount of suppression strongly depends on the strategy for selecting or constructing the reference backgrounds.

In strategy (i) a sectorial determinant is introduced which is exact for the background which – from the classical point of view – dominates that sector. The point is that this semiclassical treatment is indeed justified for sufficiently small coupling-constant, i.e. in a ridiculously small box where the topological distribution in QCD is known to be extremely narrow [12]. As the box-volume increases the effective coupling gets stronger and strategy (i) is unable to account for this change. To see this more clearly we stipulate the validity of the index theorem on the lattice [9] which allows us to rewrite the two factors in (2) using the Vafa-Witten representation [13]

$$\frac{\det(D\!\!\!/_{\rm std}^{(\nu)}+M)}{\det(D\!\!\!/_{\rm std}^{(0)}+M)} = \prod_{i=1}^{N_f} m_i^{|\nu|} \cdot \frac{\prod_{\lambda>0} (\lambda_{\rm std}^{(\nu)\,2} + m_i^2)}{\prod_{\lambda>0} (\lambda_{\rm std}^{(0)\,2} + m_i^2)} \quad (6)$$

$$\frac{\det(D\!\!\!/^{(\nu)}+M)}{\det(D\!\!\!/_{\rm std}^{(\nu)}+M)} = \prod_{i=1}^{N_f} \frac{\prod_{\lambda>0} (\lambda^{(\nu)\,2} + m_i^2)}{\prod_{\lambda>0} (\lambda_{\rm std}^{(\nu)\,2} + m_i^2)} \quad . \quad (7)$$

Strategy (i) retains a determinant which is appropriate in a small volume and thus strongly suppresses the higher topological sectors. As it comes to larger volumes, the semiclassical treatment breaks down and the quantum fluctuations packed into the "continuous" determinant (7) prove able to milder the suppression – in full QCD, but not within strategy (i). The virtue of strategy (ii) is that this change is accounted for by successively redefining the standard-backgrounds used in (6). In other words: Within strategy (ii) parts which would belong to (7) in (i) are gradually reshuffled into the "topological" part (6) as the box-volume increases. As a consequence, either strategy is supposed to be trustworthy as long as $V\Sigma m \leq 1$, but only strategy (ii) may give a reasonable approximation to full QCD in the regime $V\Sigma m \gg 1$.

Comparing the two factors (6) and (7) one ends up realizing that the "topological" determinant (6) has exactly the same structure as its QCD counterpart (the latter comes without the subscript "std" in the numerator): The essential ingredient is the prefactor $m^{|\nu|}$. In QCD, this prefactor is known [12] to cause the strong suppression of nonzero indices in the limit $V\Sigma m \ll 1$. The fact that it is still around in the "topologically unquenched" approximation (with either choice for the reference-backgrounds) means that TU-QCD (unlike Q-QCD) shows the phenomenon of chiral symmetry restoration if the chiral limit is performed in a finite box.

Finally, the fact that the number of virtual quark-loops is not restricted in TU-QCD means that there is an infinite number of diagrams contributing to the η'-propagator (not just the connected and the hairpin diagram as in Q-QCD) and this propagator may even be well-defined in the field-theoretic sense.

In summary, the fermions in "topologically unquenched QCD" are fully dynamical, but they interact in a way which does not pay attention to the details of the gluon background configuration but to its topological index only and this seems to be sufficient to get a number of basic features of full QCD qualitatively right.

Acknowledgements

The author is supported by the Swiss National Science Foundation (SNF).

References

1. H. Hamber, G. Parisi, *Phys. Rev. Lett.* **47**, 1792 (1981); E. Marinari, G. Parisi, C. Rebi, *Phys. Rev. Lett.* **47**, 1795 (1981); D. Weingarten, *Phys. Lett.* B **109**, 57 (1982).
2. N. Eicker *et al.*, *Phys. Lett.* B **407**, 290 (1997).
3. C. Bernard, M. Golterman, *Phys. Rev.* D **46**, 853 (1992) and *Nucl. Phys. Proc. Suppl.* **26**, 360 (1992); S. Sharpe, *Phys. Rev.* D **46**, 3146 (1992) and *Nucl. Phys. Proc. Suppl.* **30**, 213 (1993).
4. C. Bernard, M. Golterman, *Phys. Rev.* D **49**, 486 (1994); S. Sharpe, *Phys. Rev.* D **56**, 7052 (1997); M. Golterman, K. Leung, *Phys. Rev.* D **57**, 5703 (1998).
5. A. Morel, *J. Physique* **48**, 111 (1987).
6. A. Belavin, A. Polyakov, A. Schwartz, Y. Tyupkin, *Phys. Lett.* B **59**, 85 (1975); G. t'Hooft, *Phys. Rev. Lett.* **37**, 8 (1976) and *Phys. Rev.* D **14**, 3432 (1976); C. Callan, R. Dashen, D. Gross, *Phys. Lett.* B **63**, 334 (1976); R. Bott, *Bull. Soc. Math. France* **84**, 251 (1956).
7. S. Dürr, hep-lat/9801005.
8. T. Schäfer, E.V. Shuryak, *Rev. Mod. Phys.* **70**, 323 (1998).
9. J. Smit, J.C. Vink, *Nucl. Phys.* B **286**, 485 (1987); R. Narayanan, P. Vranas, *Nucl. Phys.* B **506**, 373 (1997); C.R. Gattringer, I. Hip, C.B. Lang, *Phys. Lett.* B **409**, 371 (1997) and *Nucl. Phys.* B **508**, 329 (1997); C.R. Gattringer, I. Hip, hep-lat/9712015; P. Hasenfratz, V. Laliena, F. Niedermayer, *Phys. Lett.* B **427**, 125 (1998); P. Hernandez, hep-lat/9801035.
10. B. Allés *et al*, hep-lat/9711026.
11. P. deForcrand, M. GarciaPérez, I-O. Stamatescu, *Nucl. Phys.* B **499**, 409 (1997) and *Nucl. Phys. Proc. Suppl.* **63**, 549 (1998).
12. H. Leutwyler, A. Smilga, *Phys. Rev.* D **46**, 5607 (1992); A. Smilga, hep-th/9503049.
13. C.Vafa, E.Witten, *Nucl. Phys.* B **234**, 173 (1984) and *Comm. Math. Phys.* **95**, 257 (1984).

Quantum Hall Dynamics on von Neumann Lattice

K. ISHIKAWA, N. MAEDA, T. OCHIAI, and H. SUZUKI

Department of Physics,Graduate School of Science,Hokkaido University,060-0810 Sapporo,Japan
E-mail:name@particle.sci.hokudai.ac.jp

Quantum Hall Dynamics is formulated on von Neumann lattice representation where electrons in Landau levels are defined on lattice sites and are treated systematically like lattice fermions. We give a proof of the integer Hall effect, namely the Hall conductance is the winding number of the propagator in the momentum space and is quantized exactly as integer multiple of $\frac{e^2}{h}$ in quantum Hall regime of the system of interactions and disorders. This shows that a determination of the fine structure constant from integer quantum Hall effect is in fact possible. We present also a unified mean field theory of the fractional Hall effect for the in-compressible quantum liquid states based on flux condensation and point out that the known Hofstadter butterfly spectrum of the tight binding model has a deep connection with the fractional Hall effect of the continuum electrons. Thus two of the most intriguing and important physical phenomena of recent years, the integer Hall effect and fractional Hall effect are studied and are solved partly by von Neumann lattice representation.

1 von Neumann lattice representation

1.1 Quantum Hall system

Quantum Hall system is a system of two dimensional electrons in strong perpendicular magnetic field,and is realized in semiconductors. The system shows intriguing physical phenomena such as the integer Hall effect and the fractional Hall effect. The Hall conductance agrees with $N \times e^2/h$ or $p/q \times e^2/h$ in finite parameter regions, hence ,quantum Hall effect is used as a standard of resistance and precise determination of the fine structure constant provided the above relations are exact.It is a theoretical issue to find out if the above relations are exact or not.

The fractional Hall effect shows that the ground state of many electron systems of certain fractional filling is unique and is a quantum liquid of having large energy gap. The energy gap vanishes and the ground state has an enormous degeneracy in the absence of interactions. To find the mechanism of forming this kind of incompressible liquid by interactions is another issue for the theorists.

Von Neumann lattice is a spatial lattice which is defined from a complete set of coherent states, i.e., eigenstates of annihilation operator. Von Neumann lattice representation preserves a spatial symmetry in lattice form and is useful in studying the above problems. We have used it for studying the quantum Hall systems and for solving the quantum Hall dynamics.The proof of the integer Hall effect and others have been given. [1,2,3,4,6]

In quantum Hall system, it is convenient to decompose the electron coordinates (x,y) into two sets of variables,guiding center variables and relative coordinates variables. Guiding center variables (X,Y) stand for the center coordinates of cyclotron motion and their commutation relation becomes an imaginary number that is inversely proportional to the external magnetic field. A minimum complete set of coherent states in this space has discrete complex eigenvalues and is known as von Neumann lattice representation. Relative coordinates (ξ, η) satisfy the equivalent commutation relation and the one body free Hamiltonian is proportional to the summation of squares of relative coordinates. Hence the electron has a discrete eigenvalue of energy. Each energy level is known as Landau level and its degeneracy is specified by the center variables. From commutation relation, the degeneracy per area is proportional to the magnetic field. The coherent states are defined by

$$(X + iY)|\alpha_{mn}\rangle = z_{mn}|\alpha_{mn}\rangle, \tag{1}$$
$$z_{mn} = (m\omega_x + n\omega_y)a,$$

where m, n are integers and ω_x, ω_y are complex numbers which satisfy $\text{Im}[\omega_x^*\omega_y] = 1$ and z_{mn} is a point on the lattice site in the complex plane; an area of the unit cell is a^2. We call this lattice the magnetic von Neumann lattice. With a spacing $a = \sqrt{\frac{2\pi\hbar}{eB}}$, the completeness of the set $\{|\alpha_{mn}\rangle\}$ is ensured. [7] Fourier transformed states denoted by

$$|\alpha_{\mathbf{p}}\rangle = \sum_{m,n} e^{ip_x m + ip_y n}|\alpha_{m,n}\rangle, \tag{2}$$

are orthogonal, that is,

$$\langle\alpha_{\mathbf{p}}|\alpha_{\mathbf{p}'}\rangle = \alpha(\mathbf{p})\sum_N (2\pi)^2\delta(\mathbf{p} - \mathbf{p}' - 2\pi\mathbf{N}). \tag{3}$$

Here, $\mathbf{N} = (N_x, N_y)$ is a vector with integer values and $\mathbf{p} = (p_x, p_y)$ is a momentum in the Brillouin zone (BZ), that is, $|p_x|, |p_y| \leq \pi$. The function $\alpha(\mathbf{p})$ is calculated by using the Poisson resummation formula as follows:

$$\alpha(\mathbf{p}) = \beta(\mathbf{p})^*\beta(\mathbf{p}), \tag{4}$$
$$\beta(\mathbf{p}) = (2\text{Im}\tau)^{\frac{1}{4}} e^{i\frac{\tau}{4\pi}p_y^2}\vartheta_1(\frac{p_x + \tau p_y}{2\pi}|\tau), \tag{5}$$

where $\vartheta_1(z|\tau)$ is a theta function and the moduli of the von Neumann lattice is defined by $\tau = -\omega_x/\omega_y$. The magnetic von Neumann lattice is parameterized by τ. To indicate the dependence on τ, we sometimes use a notation such as $\beta(\mathbf{p}|\tau)$. For $\tau = i$, the von Neumann lattice becomes a square lattice. For $\tau = e^{i2\pi/3}$, it becomes a triangular lattice. Some properties of the above functions are presented in References .Whereas $\alpha(\mathbf{p})$ satisfies the periodic boundary condition, $\beta(\mathbf{p})$ obeys a nontrivial boundary condition

$$\beta(\mathbf{p} + 2\pi\mathbf{N}) = e^{i\phi(p,N)}\beta(\mathbf{p}), \qquad (6)$$

where $\phi(p, N) = \pi(N_x + N_y) - N_y p_x$. We can define the orthogonal state which is normalized with δ-function as follows:

$$|\beta_\mathbf{p}\rangle = \frac{|\alpha_\mathbf{p}\rangle}{\beta(\mathbf{p})}. \qquad (7)$$

It should be noted that the state $|\alpha_0\rangle$ is a null state, that is, $\sum_{m,n}|\alpha_{mn}\rangle = 0$, because $\beta(0) = 0$.

The Hibert space of one-particle states is also spanned by the state $|f_l \otimes \beta_\mathbf{p}\rangle$. We call the state $|f_l \otimes \beta_\mathbf{p}\rangle$ the momentum state of the von Neumann lattice. The wave function of the state $|f_l \otimes \beta_\mathbf{p}\rangle$ in the spatial coordinate space is given in references.

The probability density $|\langle\mathbf{x}|f_l \otimes \beta_\mathbf{p}\rangle|^2$ is invariant under the translation $(\tilde{x}, \tilde{y}) \rightarrow (\tilde{x} + aN_x, \tilde{y} + aN_y)$. Thus, the momentum state is an extended state.

1.2 Field Theoretical Formalism and Topological Formula of Hall Conductance

In the preceding section we obtain one-particle states based on the von Neumann lattice, that is, the coherent state $|f_l \otimes \alpha_{mn}\rangle$ and the momentum state $|f_l \otimes \beta_\mathbf{p}\rangle$ .Now we develop the field theoretical formalism based on the momentum state . From now on, we denote $|f_l \otimes \beta_\mathbf{p}\rangle$ as $|l, \mathbf{p}\rangle$. We expand the electron field operator in the form

$$\psi(\mathbf{x}) = \int_{\mathrm{BZ}} \frac{d^2p}{(2\pi)^2} \sum_{l=0}^{\infty} b_l(\mathbf{p})\langle\mathbf{x}|l, \mathbf{p}\rangle. \qquad (8)$$

$b_l(\mathbf{p})$ satisfies the anti-commutation relation

$$\{b_l(\mathbf{p}), b_{l'}^\dagger(\mathbf{p}')\} = \delta_{l,l'}\sum_N (2\pi)^2\delta(\mathbf{p} - \mathbf{p}' - 2\pi\mathbf{N})e^{i\phi(p',N)}, \qquad (9)$$

and the same boundary condition as $\beta(\mathbf{p})$. $b_l^\dagger$ and b_l are creation and annihilation operators which operate on the many-body states. The free Hamiltonian is given by

$$\mathcal{H}_0 = \int d^2x \psi^\dagger(\mathbf{x})\hat{H}_0\psi(\mathbf{x}) = \sum_l \int_{\mathrm{BZ}} \frac{d^2p}{(2\pi)^2} E_l b_l^\dagger(\mathbf{p})b_l(\mathbf{p}). \qquad (10)$$

The density and current operators in the momentum space $j_\mu = (\rho, \mathbf{j})$ become

$$j_\mu(\mathbf{k}) = \int_{\mathrm{BZ}} \frac{d^2p}{(2\pi)^2} \sum_{l,l'} b_l^\dagger(\mathbf{p})b_{l'}(\mathbf{p} - a\hat{\mathbf{k}})$$
$$\langle f_l|\frac{1}{2}\{v_\mu, e^{ik\cdot\xi}\}|f_{l'}\rangle e^{-\frac{i}{4\pi}a\hat{k}_x(2p - a\hat{k})_y}, \quad (11)$$

Here, $v^\mu = (1, -\omega_c\eta, \omega_c\xi)$, and $\hat{k}_i = W_{ij}k_j$. The explicit form of $\langle f_l|e^{ik\cdot\xi}|f_{l'}\rangle$ and W_{ij} are given in references.

The free Hamiltonian $\mathcal{H}_0$ is diagonal in the above basis. However, the density operator is not diagonal with respect to the Landau level index. This basis, which we call the energy basis, is convenient to describe the energy spectrum of the system. In another basis, $\mathcal{H}_0$ is not diagonal and the density operator is diagonal. This basis, which we call the current basis, is convenient to describe the Ward-Takahashi identity and the topological formula of Hall conductance. There is no basis in which both the Hamiltonian and the density are diagonal. This is one of peculiar features in a magnetic field.

The current basis is constructed as follows. Using a unitary operator, we can diagonalize the density operator in the Landau level indices. We define the unitary operator

$$U_{ll'}^\dagger(\mathbf{p}) = \langle f_l|e^{i\tilde{p}\cdot\xi/a - \frac{i}{4\pi}p_x p_y}|f_{l'}\rangle. \qquad (12)$$

By introducing a unitary transformed operator $\tilde{b}_l(\mathbf{p}) = \sum_{l'} U_{ll'}(\mathbf{p})b_l(\mathbf{p})$, the density operator is written in the diagonal form and the current operator becomes a simple form:

$$\rho(\mathbf{k}) = \int_{\mathrm{BZ}} \frac{d^2p}{(2\pi)^2} \sum_l \tilde{b}_l^\dagger(\mathbf{p})\tilde{b}_l(\mathbf{p} - a\hat{\mathbf{k}}). \qquad (13)$$

$\tilde{b}_l$ and $\tilde{b}_l^\dagger$ satisfy the anti-commutation relation and boundary condition.

Here we show the Ward-Takahashi identity and the topological formula of Hall conductance using the current basis. The one-particle irreducible vertex part $\tilde{\Gamma}^\mu$ is connected with the full propagator by the Ward-Takahashi identity. The identity has crucial roles in the following derivation of the topological formula of Hall conductance. The Ward-Takahashi identity in this case becomes

$$\tilde{\Gamma}_\mu(p, p) = \frac{\partial\tilde{S}^{-1}(p)}{\partial p^\mu}. \qquad (14)$$

In a theory without a magnetic field, Ward-Takahashi identity gives a relation that the state of the dispersion $\epsilon(p)$ moves with the velocity $\frac{\partial\epsilon(p)}{\partial p_i}$. However in a magnetic field, we can not diagonalize both the current and the energy simultaneously. Therefore, the Ward-Takahashi identity does not imply the relation.

In a gap region, it was proven that the Hall conductance is obtained not only from the retarded product of the current correlation function (Kubo formula), but also from the time-ordered product of it. From the time-ordered product of the current correlation function, the Hall conductance is given by the slope of $\pi^{\mu\nu}(q)$ at the origin and is written as

$$\sigma_{xy} = \frac{e^2}{3!}\epsilon^{\mu\nu\rho}\partial_\rho\pi_{\mu\nu}(q)|_{q=0}. \qquad (15)$$

If the derivative ∂_ρ acts on the vertex with the external line attached, its contribution becomes zero owing to the epsilon tensor. Therefore, the case that the derivative acts on the bare propagator is survived.

By the Ward-Takahashi identity, σ_{xy} is written as a topologically invariant expression of the full propagator:

$$\sigma_{xy} = \frac{e^2}{h}\frac{1}{24\pi^2}\int_{BZ\times S^1} d^3p\,\epsilon_{\mu\nu\rho} \qquad (16)$$
$$\text{tr}\left(\partial_\mu\tilde{S}^{-1}(p)\tilde{S}(p)\partial_\nu\tilde{S}^{-1}(p)\tilde{S}(p)\partial_\rho\tilde{S}^{-1}(p)\tilde{S}(p)\right)$$

Here, the trace is taken over the Landau level index and the p_0 integral is a contour integral on a closed path.

Thus, we denote S^1 as the integration range. The integral $\frac{1}{24\pi^2}\int tr(d\tilde{S}^{-1}\tilde{S})^3$ gives a integer value under general assumptions and in fact counts the number or Landau bands bellow the Fermi energy. Thus, the Hall conductance is proved to be a integer times e^2/h. The impurities and interactions dot not modify the value of the σ_{xy} if the Fermi energy is located in the gap region or in the localized state region.

2 Flux state mean field theory

2.1 flux state on von Neumann lattice

We propose a new mean field theory based on the flux state on von Neumann lattices in this section.

The dynamical flux which is generated by interactions plays an important role in our mean field theory. Dynamics is described by a lattice Hamiltonian, which is due to the external magnetic field, and by the induced magnetic flux due to interaction, although the original electrons are defined on the continuum space. Consequently, our mean field Hamiltonian is close to the Hofstadter Hamiltonian, which is a tight-binding model with uniform constant flux. For this reason there are similarities between their solutions.

Due to the two scales of periodicity, the Hofstadter Hamiltonian exhibits interesting structure as is seen in Figure. [5] The largest gap exists along a line $\Phi = \nu\Phi_0$ with a unit of flux Φ_0. The ground state energy becomes minimum also with this flux. These facts may suggest that the Hofstadter problem has some connection with the fractional Hall effect. We pursue a mean field theory of the condensed flux states in the quantum Hall system and point out that the Hofstadter problem is actually connected with the fractional Hall effect.

From the induced magnetic field, new Landau levels are formed . If integer number of these Landau levels are filled completely, the integer quantum Hall effect occurs. The ground state has a large energy gap and is stable against perturbations, just as in the case of the ordinary integer quantum Hall effect. We identify them as fractional quantum Hall states.

We postulate, in the quantum Hall system of the filling factor ν, the induced flux per plaquette and magnetic field of the following magnitudes:

$$\Phi_{\text{ind}} = \nu\Phi_0, \quad \Phi_0 = \Phi_{\text{external flux}},$$
$$B_{\text{ind}} = \nu B_0, \quad B_0 = B_{\text{external magnetic field}}, \qquad (17)$$

where ν is the filling factor measured with the external magnetic field. We obtain a self-consistent solution with this flux. Then the integer quantum Hall effect due to the induced magnetic field could occur just at the filling factor ν, because the density satisfies the integer Hall effect condition,

$$\frac{eB_{\text{ind}}}{2\pi}N = \frac{eB_0}{2\pi}\nu,$$
$$N = 1. \qquad (18)$$

The ground state has a large energy gap, generally.

At the half-filling $\nu = 1/2$, the half-flux $\Phi_0/2$ is induced. We first study the state of $\nu = 1/2$, and next the states of $\nu = p/(2p \pm 1)$. At $\Phi = \Phi_0/2$, the band structure is that of a massless Dirac field and has doubling symmetry. When an even number of Landau levels of the effective magnetic field, $B_{\text{ind}} - B_0/2$, is filled, ground states have large energy gaps. This occurs if the condition of the density,

$$\frac{e}{2\pi}|\nu - \frac{1}{2}|B_0 \cdot 2p = \frac{eB_0}{2\pi}\nu,$$
$$\nu = \frac{p}{2p \pm 1} \; ; \; p, \text{ integer}, \qquad (19)$$

is satisfied. The factor of 2 on the left-hand side is due to the doubling of states and will be discussed later. We study these states in detail based on the von Neumann lattice representation.

The action and density operator show that there is an effective magnetic field in the momentum space. The total flux in the momentum space is in fact a unit flux. In the thermodynamic limit, in which the density in space is finite, the density in momentum space is infinite. Consequently, it is possible to make this phase factor disappear using a singular gauge transformation in the momentum space with infinitesimally small coupling.

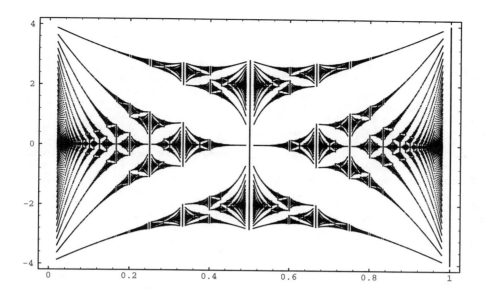

Figure 1: The Hofstadter butterfly. Horizontal axis represents magnetic flux per plaquette. Vertical axis represents energy.

We make a singular gauge transformation of the field in the momentum space and we have the commutation relation and the charge density.

By a Chern-Simons gauge theory in momentum space, the gauge transformation is realized. Here, the coupling constant $\tilde{e}$ is infinitesimally small, and hence fluctuations of the Chern-Simons gauge field have a small effect, and we ignore them.

The most important part in the action in the coordinate representation is obtained and the corresponding Hamiltonian in the lowest Landau level space is given by

$$H = -\tfrac{1}{2}\sum v(\mathbf{R}_2 - \mathbf{R}_1)c_0^\dagger(\mathbf{R}_1)c_0(\mathbf{R}_2)c_0^\dagger(\mathbf{R}_2)c_0(\mathbf{R}_1),$$
$$v(\mathbf{R}) = \tfrac{\pi}{a}e^{-\frac{\pi}{2}\mathbf{R}^2}I_0(\tfrac{\pi}{2}\mathbf{R}^2), \qquad (20)$$

where I_0 is zero-th order modified Bessel function. We study a mean field solution of this Hamiltonian.

(i) Half-filled case, $\nu = 1/2$

At half-filling $\nu = 1/2$, the system has a half flux, $\Phi = \Phi_0/2$. The system, then, is described equivalently with the two-component Dirac field by combining the field at even sites with that at odd sites. We obtained self-consistent solutions numerically. As is expected, the spectrum has two minima and two zeros corresponding to doubling of states.

Around the minima, the energy eigenvalue is approximated as,

$$E(p) = E_0 + \frac{(\mathbf{p} - \mathbf{p}_0)^2}{2m^*}, \quad \mathbf{p}_0 = (0,0), (0,\pi/a), \quad (21)$$

The m^* in Eq.(21) is the effective mass, and is computed numerically:

$$m^* = 0.225\sqrt{\frac{B}{B_0}}m_e, \ B_0 = 20\text{Tesla}, \ \kappa = 13, \ \gamma = 0.914\frac{e^2}{\kappa}. \tag{22}$$

The $\nu = 1/2$ mean field Hamiltonian is invariant under a kind of Parity, P, and anti-commutes with a chiral transformation, α_5. If the parity is not broken spontaneously, there is a degeneracy due to the parity doublet. The doubling of the states also appears at $\nu \neq 1/2$ and plays an important role when we discuss the states away from $\nu = 1/2$ in the next part. When an additional vector potential with the same gauge, $A_x = 0$, $A_y = Bx$, is added, the Hamiltonian satisfies the properties under the above transformations, and doubling due to the parity doublet also appears. Thus, the factor of 2 is necessary in Eq.(19) and leads the principal series at $\nu = p/(2p\pm1)$ to have the maximum energy gap.

(ii) $\nu = \frac{p}{2p\pm1}$

If the filling factor, ν, is slightly away from $1/2$, the total system can be regarded as a system with a small magnetic field of magnitude $(\nu-1/2)B_0$. The band structure may be slightly modified. It is worthwhile to start from the band of $\nu = 1/2$ as a first approximation and to make iteration in order to obtain self-consistent solutions at arbitrary $\nu = p/(2p \pm 1)$. We solve the mean field Hamiltonian

$$H_{\mathrm{M}} = \sum U_0^{(\frac{1}{2}+\delta)}(\mathbf{R}_1 - \mathbf{R}_2)e^{i\int(\mathbf{A}^{(\frac{1}{2})}+\delta\mathbf{A})\cdot d\mathbf{x}}v(\mathbf{R}_1 - \mathbf{R}_2)$$
$$c^\dagger(\mathbf{R}_1)c(\mathbf{R}_2),$$

$$\langle c^\dagger(\mathbf{R}_1)c(\mathbf{R}_2)\rangle_{1/2+\delta} = U_0^{(\frac{1}{2}+\delta)} e^{i\int (\mathbf{A}^{(\frac{1}{2})}+\delta\mathbf{A})\cdot d\mathbf{x}} \quad (23)$$

under the self-consistency condition at $\nu = 1/2 + \delta$. Here we solve, instead, a Hamiltonian which has the phase of Eq.(23) but has the magnitude of the $\nu = 1/2$ state.

The integer quantum Hall state has an energy gap of the Landau levels due to $\delta\mathbf{A}$. This occurs when an integer number of Landau levels is filled completely. The Landau level structure is determined by the phase factor. Magnitudes of the physical quantities may be modified, nevertheless.

$U_0^{\frac{1}{2}}(\mathbf{R}_1 - \mathbf{R}_2)$ was obtained in the previous part, and it is approximated with the effective mass formula.

From Eq.(19), p Landau levels are completely filled, and the integer quantum Hall effect occurs at $\nu = p/(2p\pm 1)$. The energy gap is given by the Landau level spacing

$$\Delta E_{\text{gap}} = \frac{e\delta B_{\text{eff}}}{m^*} = \frac{eB_0}{m^*}|\nu - \frac{1}{2}|. \quad (24)$$

These equations were solved numerically, and the energy gaps and the widths of excited bands are obtained. Some bands are narrow and some bands are wide. Near $\nu = 1/2$, the effective magnetic field approaches zero, and the Landau level wave functions have large spatial extension. The lattice structure becomes negligible, and the spectrum shows simple Landau levels of the continuum equation in these regions. Near $\nu = 1/3$, the lattice structure is not negligible, and bands have finite widths. There are non-negligible corrections from those of continuum calculations.

Due to the energy gap of the integer Hall effect caused by the induced dynamical magnetic field, the states at $\nu = p/(2p \pm 1)$ are stable, and fluctuations are weak. Invariance under P, moreover, ensures these states to have uniform density. In systems with impurities, localized states with isolated discrete energies are generated by impurities and have energies in the gap regions. These states contribute to the density but do not contribute to the conductance. If the Fermi energy is in one of these gap regions, the Hall conductance σ_{xy} is given by a topological formula, Eq.(16), and remains constant, at $\frac{e^2}{h} \cdot \frac{p}{2p\pm 1}$. The fractional Hall effect is realized.

At a value of ν smaller than $1/3$, the Hofstadter butterfly exhibits other kinds of structures. They may be connected with the Wigner crystal.

2.2 Comparison with experiment

In the previous section we presented our mean field theory based on flux condensation, where lattice structure generated by the external magnetic field and condensed flux due to interaction are important ingredients. Consequently, our mean field Hamiltonian becomes very similar

to that of Hofstadter, which is known to show a large energy gap zone along the $\Phi = \nu\Phi_0$ line. The line $\Phi = \nu\Phi_0$ is special in the Hofstadter problem and hence in our mean field Hamiltonian. This explains why the experiments of the fractional quantum Hall effects show characteristic behavior at $\nu = p/(2p \pm 1)$. The ground states at $\nu = p/(2p \pm 1)$ have the lowest energy and the largest energy gap, and hence these states are stable. In this section we compare the energy gaps of the principal series in the lowest order, with experiments and with Laughlin variational wave function.

The effective mass m^* of Eq.(22) was obtained from the curvature of the energy dispersion and should show a characteristic mass scale of the fractional Hall effect. The Landau level energy in the lowest approximation is calculated, and the gap energy is given in Eq.(24).

These values are compared with the experimental values and with the composite fermion mean field values. Our mean field values are close to the experimental values. For example, at $\nu = 1/3$, the composite fermion mean field theory gives E_{gap}=380[K], which is a factor of 40 larger than the experimental value. Whereas, our effective mass formula gives E_{gap}=24[K], and the other approximation gives E_{gap}=36[K], which are a factor of two or three larger than the experimental value and close to the value of the Laughlin wave function, E_{gap}=26[K]. The agreement is not perfect, but should be regarded as good as the lowest mean field approximation. Near $\nu = 1/3$, the bands have finite widths, and near $\nu = 1/2$, the widths are infinitesimal. The dependence of the width upon the filling factor, ν, and the entire structure of the bands are characteristic features of the present mean field and should be tested experimentally.

3 Summary

We formulated the quantum Hall effect, integer Hall effect and fractional Hall effect with the von Neumann lattice representation of two-dimensional electrons in a strong magnetic field. The von Neumann lattice is a subset of the coherent state. The overlap of states is expressed with a elliptic theta function. They allow for a systematic method of expressing the quantum Hall dynamics.

A topological invariant expression of the Hall conductance was obtained in which compactness of the momentum space is ensured by the lattice of the coordinate space. Because the lattice has an origin in the external magnetic field, the topological character of the Hall conductance is ensured by the external magnetic field. The conductance is quantized exactly as $(e^2/h) \cdot N$ in the quantum Hall regime. The integer N increases monotonically with the chemical potential.

A new mean field theory of the fractional Hall effect

that has dynamical flux condensation was proposed . In our mean field theory, lattice structure is introduced from the von Neumann lattice and, flux is introduced dynamically. The mean field Hamiltonian becomes a kind of tight-binding model, and the rich structure of the tight-binding model is seen as characteristic features of the fractional Hall effect in our mean field flux states of having a liquid property with an energy gap. These states satisfy the self-consistency condition of having the lowest energy and the largest energy gap. The physical quantities of our mean field theory are close to the experimental values in the lowest order at $\nu = p/(2p \pm 1)$.

Acknowledgements

This work was partially supported by the special Grant-in-Aid for Promotion of Education and Science in Hokkaido University provided by the Ministry of Education, Science, Sports and Culture, the Grant-in-Aid for Scientific Research on Priority Area(Physics of CP violation) (Grant No.10140201), and the Grant-in-Aid for International Scientific Research (Joint Research Grant No.10044043) from the Ministry of Education, Science, Sports and Culture, Japan.

References

1. N. Imai, K. Ishikawa, T. Matsuyama and I. Tanaka, Phys. Rev. **B42** (1990), 10610.
2. K. Ishikawa, Prog. Theor. Phys. Supple. No.107 (1992), 167.
3. K. Ishikawa, N. Maeda and K. Tadaki, Phys. Rev. **B51** (1995), 5048; **B54** (1996), 17819. K. Ishikawa et al., Phys. Lett. **210A** (1996), 321.
4. K. Ishikawa and N. Maeda, Prog. Theor. Phys. **97** (1997),507 .
5. D.R.Hofstadter,Phys.Rev.**B14**,(1976),2239
6. K. Ishikawa, N. Maeda,T.Ochiai and H.Suzuki, Phys. Rev. **B58** (1998), 1088;to appear in Physica E(1998);to appear in Phys.Rev.B(1998).
7. A. M. Perelomov, Teor. Mat. Fiz. **6**, 213 (1971); V. Bargmann, P. Butera, L. Girardello, and J. R. Klauder, Rep. Math. Phys. **2**, 221 (1971).

TOWARD THE CHIRAL LIMIT OF QCD:
Quenched and Dynamical Domain Wall Fermions

PING CHEN, NORMAN CHRIST, GEORGE FLEMING, ADRIAN KAEHLER, CATALIN MALUREANU,
ROBERT MAWHINNEY, GABRIELE SIEGERT, CHENGZHONG SUI, YURI ZHESTKOV

Department of Physics, Columbia University, New York, NY 10027

PAVLOS VRANAS

Physics Dept., University of Illinois, Urbana, IL 61801

A serious difficulty in conventional lattice field theory calculations is the coupling between the chiral and continuum limits. With both staggered and Wilson fermions, the chiral limit cannot be realized without first taking the limit of vanishing lattice spacing. In this talk, we report on extensive studies of the domain wall formulation of lattice fermions, which avoids this difficulty at the expense of requiring that fermion propagators be computed in five dimensions. A variety of results will be described for quenched and dynamical simulations at both zero and finite temperature. Conclusions about the benefits of this new method and some new physical results will be presented. These results were obtained on the *QCDSP* machine recently put into operation at Columbia and the RIKEN Brookhaven Research Center.

1 Introduction

Important theoretical and algorithmic advances have opened a new approach to the numerical study of chiral symmetry in QCD. The two widely studied lattice fermion formalisms, staggered and Wilson fermions, both present serious difficulties to the numerical simulation of chiral symmetry in lattice QCD. Not only do both of these lattice descriptions explicitly break most or all of the chiral symmetries present in massless QCD, but they also obscure the underlying relationship between topology in the gauge sector and zero modes for the fermions. As a result, lattice calculations will typically fail to show the full consequences of the Goldstone theorem, current algebra, or the 't Hooft solution to the $U_A(1)$ problem, without explicit extrapolation to the continuum limit. Thus, the physics of chiral symmetry discovered in both the 60's and the 70's is corrupted by the usual lattice discretization of QCD.

However, by considering Wilson fermions formulated in five dimensions with a large negative mass, $m_W = -m_0$, Kaplan [1] showed that it is possible to avoid the fermion doubling problem while still achieving complete chiral symmetry in the much simpler limit of large lattice extent in the new fifth dimension. A further important advance was achieved by Narayanan and Neuberger [2] who generalized Kaplan's approach and recognized that this limit of large fifth dimension not only realized the desired continuum chiral symmetry but also created a structure of exact fermion zero modes with a close relation to the continuum Atiyah-Singer theorem and the physics of the axial anomaly. An efficient and well-elaborated version of this domain wall method was developed and analyzed by Shamir and by Shamir and Furman,[3,4] providing a

practical and attractive method for large-scale numerical calculation.[5] There has now been considerable exploratory numerical work suggesting that this method lives up to its promise for both the Schwinger model [6] and quenched QCD [7] and that fermion zero-modes with the desired properties are realized.[8,9]

We will describe these issues in somewhat more detail in Section 2 below and then describe a series of calculations [10,11,12,13,14,15] underway on the recently completed computers at Columbia (8,196-nodes 0.4Tflops) and the RIKEN Brookhaven Research Center (12,288-nodes, 0.6Tflops).[16] These calculations are designed to i) establish the extent to which zero-mode effects can be seen in practical simulations with fifth-dimension extent $L_s \approx 10$ (Section 3), ii) study the chiral symmetry of quenched QCD with careful control over the effects of L_s, examining both the hadron spectrum and the chiral condensate (Section 4) and iii) explore QCD thermodynamics using domain wall fermions both as a probe of the properties of the pure gauge theory and as a method to realize complete flavor symmetry in a lattice study of the full QCD phase transition. This final calculation represents the best-controlled approach yet available to examine the role of the axial anomaly in the full QCD, chiral phase transition.

2 Domain Wall Fermions

2.1 Formulation

Our formulation of domain wall fermions follows closely that proposed by Shamir.[3,4] The starting point is the standard lattice treatment of the gauge variables as 3×3 special unitary matrices $U_\mu(n)$, associated with each link in a four-dimensional space-time lattice. Here n locates the site from which the corresponding link extends in the positive μ^{th} direction. These link variables enter the Wilson gauge action in the usual way. The new features of the domain wall treatment appear in the fermion action. Here the fermion field, $\psi(n, s)$ is 4-spinor and 3-component color vector as usual but now depends on both the normal lattice coordinate n and a new fifth co-ordinate s, lying in the range $0 \le s \le L_s - 1$. The action takes the form:

$$\mathcal{A}_{\mathrm{DWF}} = \sum_{n,s;n',s'} \bar{\psi}(n, s)\{(D_{\mathrm{W}})_{n,n'}\delta_{s,s'} + m_0$$
$$+ (D_5)_{s,s'}\delta_{n,n'}\}\psi(n', s') \tag{1}$$

Here D_{W} is the usual Wilson Dirac operator:

$$(D_{\mathrm{W}})_{n,n'} = \frac{1}{2}\sum_{\mu}\{(1 + \gamma^\mu)U(n)_\mu \delta_{n,n'-\hat{\mu}} \tag{2}$$
$$+ (1 + \gamma^\mu)U(n')_\mu^\dagger \delta_{n,n'+\hat{\mu}} - 2\delta_{n,n'}\},$$

m_0 the Wilson mass (but with an unconventional neg-ative sign), and D_5 an additional fifth-dimension piece, diagonal in color:

$$(D_5)_{s,s'} = \frac{1}{2}\{(1 + \gamma^5)\delta_{s,s'-1} + (1 - \gamma^5)\delta_{s,s'+1} - 2\delta_{s,s'}$$
$$-m_f(1 - \gamma^5)\delta_{s,0}\delta_{s',L_s-1}$$
$$-m_f(1 + \gamma^5)\delta_{s,L_s-1}\delta_{s',0}\}. \tag{3}$$

The combination of the boundaries at $s = 0$ and $s = L_s - 1$ and the negative Wilson mass, $-m_0$, allows un-usual massless boundary states to form, localized on the $s = 0$ and $s = L_s - 1$ four-dimensional boundaries of the five-dimensional problem. The result is two types of states, decaying exponentially as one moves away from 0 or $L_s - 1$: massless, right-handed particles bound to the right-hand wall and massless, left-handed particles bound to the left-hand wall. The mixing between these two boundary states, given by the small overlap of the ex-ponentially decreasing 5-dimensional wave functions, im-plies that these states will actually form a massive four-component fermion. They will become literally massless only in the limit of infinite separation between the walls, $L_s \to \infty$, a limit in which this exponentially small over-lap vanishes.

The extra mass term with coefficient m_f in Eq. 3 explicitly couples the right and left walls and gives these boundary states an additional mass proportional to m_f. This extra term provides an explicit mass which is easily adjusted in contrast to the overlap-induced mass whose dependence on L_s is not so precisely known. We will attempt to choose L_s and m_f so that the effects of the mixing are much smaller than those of m_f, so the mass of the domain-wall states is explicitly proportional to m_f.

2.2 Chiral Symmetry

The domain wall fermion formulation described above should become a theory of massless fermions in the limit $L_s \to \infty$. This is easy to see in the free field case and can be argued to be true for the interacting theory as well [4]. Heuristically, we observe that if the gauge coupling is suf-ficiently small, the relevant gauge configurations will be smooth on the lattice scale and interact primarily with the low-energy, domain wall states in exactly the fash-ion of four-dimensional QCD. Of course, the extent to which this is actually true in a particular calculation for a specific value of L_s, must be studied numerically.

In the usual Wilson formulation of lattice fermions, the axial part of the underlying $SU(N_f) \times SU(N_f)$ flavor symmetry is strongly broken by the dimension-five, "Wil-son term" added to make the doublers heavy. A chiral theory is expected to be found in the low energy portion of the theory provided the mass-related hopping param-eter κ is tuned to its critical value. However, in practice, the effects of this Wilson term have been so large as to obscure the character of the chiral phase transition and to serious impede the extraction of a variety of weak ma-trix elements whose efficient calculation requires the use of chiral symmetry.

The staggered fermion formulation preserves a sin-gle chiral symmetry, making the chiral phase transition much easier to simulate. However, this approach con-tains lattice artifacts which break the normal vector fla-vor symmetries making many quantities more difficult to interpret than in the Wilson formulation.

Both the Wilson and staggered formulations of lat-tice fermions are expected to show full, physical flavor symmetry only as the continuum limit is approached. In contrast, N_f species of domain wall fermions, as formu-lated above, support a complete $SU(N_f) \times SU(N_f)$ flavor symmetry which becomes exact in the possibly much less demanding $L_s \to \infty$ limit.

2.3 Index Theorem on the Lattice

An important part of the relativistic quantum physics of fermions is the axial anomaly which is required for proper understanding of the $\pi^0 \to \gamma\gamma$ decay, the η' mass and the order of the QCD phase transition. With both Wilson and staggered fermions, the symmetry generated by the anomalously non-conserved current is explicitly broken by order a or order a^2 terms in the lattice fermion action. Thus, without a careful study of the limit of vanishing lattice spacing, $a \to 0$, one is unable to be certain whether a given anomalous effect results from a lattice artifact or will survive in the continuum limit.

Happily, this fundamental physical phenomona is also represented in a new and potentially more reliable fashion by the domain wall fermion formulation. As developed by Narayanan and Neuberger, the domain wall formulation supports a variant of the Atiyah-Singer index theorem in the $L_s \to \infty$ limit, even for finite lattice spacing. In the continuum, this theorem relates the winding number of the gauge field background to the number of zero modes of the Dirac operator and provides the explicit mechanism for gauge field topology to effect the physics of quarks.

To understand the connection between gauge field topology and exact zero modes of the Dirac operator, for domain wall fermions one begins by viewing the five-dimensional fermion path integral in a fixed gauge background and for a fixed L_s as representing both the determinant of D_{DWF} and as a quantum mechanical expectation value of the transfer matrix T_5 for unit translation in the fifth dimension in the presence of the s-independent background gauge field:

$$Z(L_s) = \det\{D_{\text{DWF}}\} = \langle 0|T_5^{L_s}|0\rangle. \qquad (4)$$

Here, the fermion state $|0\rangle$ is a particular four-dimensional fermionic state, determined by the boundary conditions with half of the available states filled. In the limit $L_s \to \infty$, the right-most factor of Eq. 4 projects onto the eigenstate of T_5 with the largest eigenvalue, $|0'\rangle$. Thus, we can conclude that

$$\lim_{L_s \to \infty} \det\{D_{\text{DWF}}\} \propto |\langle 0'|0\rangle|^2 \qquad (5)$$

The Dirac operator of our five dimensional formulation will then have exact zero modes whenever the states $|0\rangle$ and $|0'\rangle$ have a different number of occupied states (and are then, necessarily, orthogonal). This connection of exact Dirac zero modes with integers defined from the gauge fields is in close analogy to the continuum Atiyah-Singer theorem. In the continuum limit, such zero modes will correspond to topologically non-trivial gauge configurations. Furthermore, these zero modes should continue to exist as lattice corrections and quantum fluctuation are added [8].

In the next Section, we will examine the question of how large L_s must be for these expected near-zero modes to be recognized.

3 Fermion Zero Modes

In order to investigate the degree to which expected fermion zero modes will actually be visible for finite L_s and non-zero lattice spacing, we have looked for zero mode effects in the background of a discretized, instanton-like gauge field.[17] We do this by evaluating the volume-averaged, chiral condensate as a function of the explicit quark mass, m_f. In the continuum, the chiral

condensate is easily related to the spectrum of Dirac eigenvalues through the Banks-Casher formula:

$$\langle \bar{\psi}\psi \rangle = -m \int_{-\infty}^{\infty} \frac{\rho(\lambda, m)d\lambda}{\lambda^2 + m^2} \qquad (6)$$

where $\rho(\lambda)$ is the density of Dirac eigenvalues corresponding to the gauge configuration or ensemble of configurations used to compute $\bar{\psi}\psi$ and m is the bare quark mass.

Eq. 6 implies that the presence of a zero mode, $\rho(\lambda) = Z\delta(\lambda) + \dots$ implies a Z/m divergence in $\bar{\psi}\psi$ as m approaches zero. Such zero-mode effects can be easily studied [11,15] by computing $\bar{\psi}\psi$ for a fixed gauge configuration, constructed to be close to a Belavin, Polyakov, Schwartz, Tyupkin instanton. Such a configuration will have an exact zero mode for the continuum Dirac operator. In Figure 1, we show $\bar{\psi}\psi$ as a function of quark mass computed in this fixed, instanton-like background using the staggered Dirac operator. Instead of the desired $1/m_f$ divergence as $m_f \to 0$, one sees a weak inflection around $m_f \approx 10^{-2}$ suggesting that the lattice spacing has shifted the zero mode from zero to approximately this value. Even this weak signal essentially disappears if even 10% noise is superimposed on the original instanton-like background.

The behavior for the domain wall Dirac operator in this same background, shown in Figure 2, is dramatically different. There one sees the expected decrease in $\bar{\psi}\psi$ as m_f decreases, until $m_f \approx 2\ 10^{-2}$, at which point a clear $1/m_f$ signal is seen, even for the curve corresponding to L_s as small as 6. For finite L_s this $1/m_f$ behavior is cut off and $\bar{\psi}\psi$ becomes constant as m_f falls below the residual mass introduced by direct mixing between the two overlapping domain wall states. As can be seen in the figure, this effect moves to much smaller m_f as L_s increases. Equally important, the appealing picture shown in Figure 2 changes very little as noise is superimposed on the instanton-like, background solution.

Thus, it appears that the domain wall formulation will allow us to simulate fermions at fixed lattice spacing and not-too-large L_s which show both the full $SU(2) \times SU(2)$ flavor/chiral symmetry and also the important connection between lattice topology and fermion zero-modes which underlies the physics of the axial anomaly.

4 QCD at Zero Temperature

4.1 Hadronic Spectrum

As our first test of these ideas we have carried out a series of careful, quenched calculations on small $8^3 \times 32$ lattices for a variety of values of L_s and m_f.[12] Using physical quark operators whose left- and right-handed components were obtained by evaluating the 5-dimensional fermion fields on the $s = 0$ and $s = L_s - 1$ hyperplanes,

1804

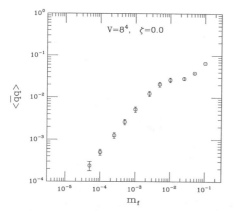

Figure 1: The chiral condensate computed on an instanton-like background evaluated using staggered fermions. The small inflection interrupting the steady decrease of $\langle \bar\psi\psi \rangle$ with decreasing m_f is caused by the expected "zero-mode", shifted far from zero by this coarse 8^4 lattice.

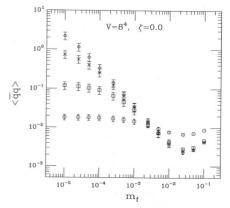

Figure 2: The chiral condensate computed on the instanton-like background of Figure 1, now evaluated using domain wall fermions. Results for four different choices of L_s are shown: $L_s = 4$ (circles), 6 (squares), 8 (crosses) and 10 (diamonds). Now the expected, zero-mode induced $1/m_f$ behavior can be easily seen for $L_s > 4$ and $m_f \leq 10^{-2}$.

we had little difficulty computing hadron masses following the usual methods. In Figure 3, we show the resulting ρ and nucleon masses, extrapolated to $m_f = 0$, as a function of L_S. Although we have examined quite large domain wall separations, up to $L_s = 48$, it is clear from the figure, that we could have extracted accurate results from the smaller $L_s = 10$ or 16 calculations.

The corresponding L_s behavior of the pion mass is shown in Figure 4. In contrast to the case of the nucleon and the ρ, we see significant dependence on L_s, even at the largest values of L_s. This contrast can be easily explained if we hypothesize that the largest effect of finite L_s is to give an additional L_s-dependent contribution to the quark mass. Such an effect will be much more visible for the very light pion than for the heavier ρ or nucleon. Using the observed m_f sensitivity (not shown here) of m_ρ and m_π, one can conclude that the 100% change in $m_\pi^2|_{m_f=0}$ seen as L_s increases from 16 to 48, would amount to only a 4% effect on $m_\rho|_{m_f=0}$, an effect hidden by our errors. Of greater potential interest is the failure of m_π to vanish as $L_s \to \infty$. While this may be a new feature of the quenched approximation made visible by the domain wall formalism, it is very likely a simple, finite-volume effect. This non-vanishing contribution to m_π is an appreciable fraction of m_π only for very small quark masses, $m_f \approx 0.02$. However, for such light quarks the product $m_f \langle \bar\psi\psi \rangle V = 0.02 \cdot 0.0019 \cdot 8^3 \times 32 = 0.7$, a clear warning that the Goldstone phenomena will begin to be influenced by our finite $8^3 \times 32$, space-time volume.

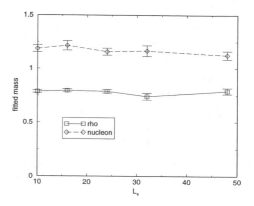

Figure 3: The ρ and nucleon masses, extrapolated to $m_f = 0$, plotted as a function of L_s, the lattice extent in the fifth dimension. These masses were computed with domain wall height $m_0 = 1.65$ from quenched, $\beta = 5.85$, $8^3 \times 32$ configurations.

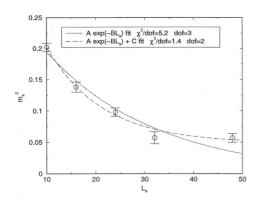

Figure 4: The π mass, extrapolated to zero quark mass, plotted as a function of L_s, again for quenched, $\beta = 5.85$, $8^3 \times 32$ configurations. The failure of m_π to approach zero as $L_s \to 0$, as required by the Goldstone theorem, is most likely a simple, finite-volume effect.

4.2 Chiral Condensate

Lacking the fermion determinant, the quenched approximation can potentially yield gauge configurations with small or vanishing Dirac eigenvalues. As was discussed in Section 3, such configurations could lead to an unphysical, $1/m$ divergence in the chiral condensate. However, such behavior could be easily suppressed by lattice artifacts for staggered fermions and may be reduced when "exceptional" configurations are discarded for Wilson fermions.

We have explicitly searched for such effects in a $\beta = 5.85$, quenched simulation and found them quite easily.[13] Our results for a $8^3 \times 32$ lattice are shown in Figure 5. One sees a clearly $1/m_f$ divergence for small m_f. The $1/m_f$ coefficient is $3.8(3)\ 10^{-6}$. If this effect comes only from exact zero modes, whose number increases as $V^{1/2}$, we should expect this behavior to be less pronounced on larger volumes. In fact, a similar, $16^3 \times 32$ calculation also shows this $1/m_f$ behavior, but with a much smaller coefficient, $0.6(1)\ 10^{-6}$. Thus, while one should worry that conventional quenched chiral extrapolations do not allow for such behavior, this new effect may not be important for quenched calculations performed on large physical volumes.

5 QCD near T_c

Because of the importance of chiral symmetry for the QCD phase transition, one of the most interesting applications of domain wall fermions may be to studies of QCD thermodynamics.

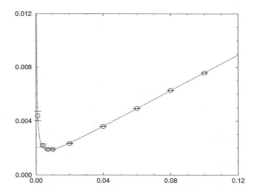

Figure 5: The chiral condensate is plotted as a function of quark mass from the same quenched, $\beta = 5.85$ $8^3 \times 32$ calculation. The clear $1/m_f$ behavior is a new, unphysical feature of the quenched approximation, resulting from the absence of the fermion determinant.

5.1 The quenched Chiral Condensate

We will first continue the discussion above of possible quenched, $1/m_f$ behavior for $T \geq T_c$.[13] Here, one expects to see such singular behavior and a $1/m_f$ term that may be non-zero, even in the limit of infinite volume.[18] While earlier staggered fermion calculations have not seen this effect,[18,19] it is very visible in the $\beta = 5.71$ results shown in Figure 6. A companion calculation on a larger $32^3 \times 4$ lattice shows $1/m_f$ behavior with the same coefficient. Thus, by using a fermion formulation more sensitive to the effects of zero modes, we have discovered that the quenched chiral condensate shows quite striking behavior. Rather than vanishing for small m_f in the deconfined region, as would be expected if quenched chiral symmetry were restored above β_c, the chiral condensate diverges as $1/m_f$. In addition, as can be seen in Figure 6, there is a non-zero, $m_f = 0$ intercept even if the $1/m_f$ term is removed. Clearly the old picture of quenched chiral symmetry restoration above β_c is seriously incomplete.

5.2 $SU(2) \times SU(2)$ Chiral Transition

Now let us examine the physical, 2-flavor QCD phase transition using domain wall fermions.[14] We have done a series of calculations on $8^3 \times 4$ lattices with a variety of values of β, L_s and domain wall heights. In Figure 7, we show the Wilson line expectation value and the chiral condensate computed with a domain wall height $m_0 = 1.90$, $L_s = 12$ and quark mass $m_f = 0.1$. One sees the expected cross-over behavior as β increases, suggesting that a full QCD, thermodynamic calculation using domain wall fermions is may well be possible. We have seen very similar behavior for m_0 varying between 1.65

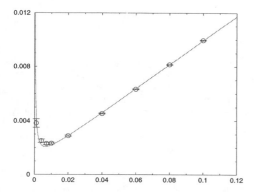

Figure 6: The chiral condensate is plotted as a function of quark mass, computed using pure gauge configurations on a $16^3 \times 4$ lattice, just above the deconfining phase transition for $\beta = 5.71$. There is an new $1/m_f$ divergence visible for small mass as well as a non-zero constant piece in $\bar{\psi}\psi$. These appear to be new, volume-independent terms, first visible when domain wall fermions are used.

and 2.15, suggesting that we are seeing stable, 2-flavor behavior for $m_0 = 1.9$, although more study of this question is warranted.

Of special interest is the degree of chiral symmetry present on either side of this transition region. This is examined in Figure 8 where the limit $m_f \to 0$ is displayed using a series of full QCD calculation with quark masses $m_f = 0.06, 0.10, 0.14$ and 0.18. The domain wall formulation appears to have allowed a calculation, for the first time, in which a transition is seen between a phase with spontaneously broken chiral symmetry ($\beta = 5.20$) and one with what should be full, restored, $SU(2) \times SU(2)$ symmetry ($\beta = 5.45$). Calculations are now underway to study this system further, examining the properties of the QCD phase transition in greater detail for larger spatial volumes.

5.3 Anomalous Symmetry Breaking

As a final topic, we attempt to exploit the sensitivity of the domain wall fermion formulation to topological effects, and examine the degree of anomalous symmetry breaking slightly above the phase transition.[14] The current results of this study are shown in Figure 9. We plot the difference of two screening masses, computed at $\beta = 5.40$, just above the transition region, evaluated on a $16^3 \times 4$ lattice, using $L_s = 16$ and $m_0 = 1.9$. Here we are subtracting the screening mass m_δ, extracted from the exponential damping of correlation functions computed from the operator $(\bar{\psi}\vec{\tau}\psi)(x)$, and m_π computing using the operator $(i\bar{\psi}\vec{\tau}\gamma^5\psi)(x)$. The 2×2 matrices, $\vec{\tau}$, are the usual Pauli, isospin generators.

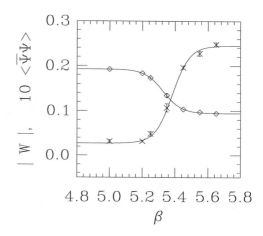

Figure 7: The chiral condensate and Wilson line are plotted as a function of β. These values were obtained in a full QCD, $N_f = 2$, $m_f = 0.1$ calculation on an $8^3 \times 4$ lattice. Behavior characteristic of a phase transition with $\beta_c \approx 5.35$ is seen.

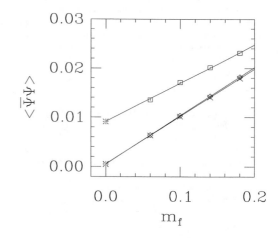

Figure 8: The chiral limit of the $\langle\bar{\psi}\psi\rangle$ in full, 2-flavor QCD for $L_s = 16$ and $m_0 = 1.9$ Shown are points below the transition, $\beta = 5.20$, $8^3 \times 4$(squares) and points above the transition, $\beta = 5.45$, $8^3 \times 4$(diamonds) and $16^3 \times 4$(crosses).

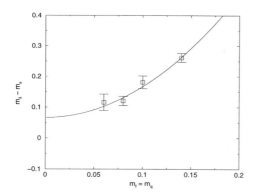

Figure 9: The anomalous difference of the screening masses for the π and δ are plotted as a function of quark mass. These calculations were performed with two flavors of dynamical quarks on a $16^3 \times 4$ lattice, at $\beta = 5.40$, just of above the transition, using $L_s = 16$ and $m_0 = 1.9$.

Because the δ and π correlation functions are related by the anomalous chiral symmetry, any difference between these two screening masses in the limit $m_f \to 0$, is an unambiguous measure of anomalous chiral symmetry breaking. Since these screening masses are both about 1 in lattice units, Figure 9 shows a small ($\approx 5\%$), but possibly significant effect. The present data, shown in the figure, give an intercept of 0.067(1.8) when fitted to the expected quadratic m_f dependence. While we cannot be certain we have seen the sought-after evidence of anomalous symmetry breaking (the sort required to predict a second-order QCD phase transition), we can assert with confidence that the effect is quite small.

6 Conclusion

We have demonstrated that the domain wall formulation of lattice fermions can be used to study both zero temperature, quenched hadron spectroscopy and the QCD phase transition for both zero and two flavors. This new method appears to realize accurate chiral symmetry at fixed, non-zero lattice spacing, and does not require unreasonably large domain wall separation in the fifth dimension. This method offers a very promising new approach to the study of chiral symmetry using lattice methods.

Acknowledgements

This work was supported in part by the U.S. Department of Energy. We have benefited from discussions with a number of people included T. Blum, R. Edwards, R. Narayanan, H. Neuberger, and Y. Shamir.

References

1. D.B. Kaplan, Phys. Lett. **B288** (1992) 342; Nucl. Phys. **B30** (Proc. Suppl.) (1993) 597.
2. R. Narayanan, H. Neuberger, Phys. Lett. **B302** (1993) 62; Phys. Rev. Lett. **71** (1993) 3251; Nucl. Phys. **B412** (1994) 574; Nucl. Phys. **B443** (1995) 305.
3. Y. Shamir, Nucl. Phys. **B406** (1993) 90;
4. V. Furman, Y. Shamir, Nucl. Phys. **B439** (1995) 54.
5. Alternative, promising methods have also been proposed. See H. Neuberger, Phys. Lett. **B417** (1998) 141; H. Neuberger, Phys. Rev. **D57** (1998) 5417; H. 'Neuberger, RU-98-03, hep-lat/9801031; H. Neuberger, RU-98-28, hep-lat/9806025; R. G. Edwards, U. M. Heller and R. Narayanan, hep-lat/9807017; A. A. Slavnov hep-lat/9807040
6. P.M. Vranas, Nucl. Phys. **B53** (Proc. Suppl.) (1997) 278; P.M. Vranas Phys. Rev. **D57** (1998) 1415.
7. T. Blum and A. Soni, Phys. Rev. **D56** (1997) 174; Phys. Rev. Lett. **79** (1997) 3595.
8. R.Narayanan, P. Vranas, Nucl. Phys. **B506** (1997) 373
9. R.G. Edwards, U.M. Heller and R. Narayanan, hep-lat/9801015. R.G. Edwards, U.M. Heller and R. Narayanan, hep-lat/9802016; R.G. Edwards, U.M. Heller and R. Narayanan, hep-lat/9806011;
10. Columbia lattice group contributions to the RIKEN-BNL workshop on Fermion Frontiers on Vector Lattice Gauge Theory held May 6-9, 1998, at the Brookhaven National Laboratory;
11. P. Chen, et al., CU-TP-906, hep-lat/9807029.
12. P. Chen, et al., Quenched QCD with domain wall fermions, CU-TP-915.
13. P. Chen, et al., The domain wall fermion chiral condensate in quenched QCD, CU-TP-916.
14. P. Chen, et al., CU-TP-914, hep-lat/9809159.
15. P. Chen, et al., The anomaly and topology in quenched QCD above T_c, CU-TP-913.
16. D. Chen, et al., CU-TP-911, hep-lat/9810004.
17. A. Kaehler, in preparation.
18. S. Chandrasekharan, D. Chen, N. Christ, W. Lee, R. Mawhinney and P. Vranas, CU-TO-902, hep-lat/9807018.
19. A. Kaehler, Nucl. Phys. B (Proc. Suppl.) 63A-C (1998) 823.

GAUGE INVARIANT FIELD STRENGTH CORRELATORS IN QCD

presented by A. DI GIACOMO

Dip. Fisica Università and INFN, Via Buonarroti 2 ed.B 56126 PISA, ITALY
E-mail: digiacomo@pi.infn.it

M. D'ELIA, H. PANAGOPOULOS

Department of Natural Sciences University of Cyprus, PO Box 537 Nicosia, CY-1678, Cyprus
E-mail: delia@dirac.ns.ucy.ac.cy
E-mail: haris@earth.ns.ucy.ac.cy

E. MEGGIOLARO

Institut für Theoretische Physik Universität Heidelberg Philosophenweg 16 D-69120 Heidelberg Germany
E-mail: e.meggiolaro@sns.it

Gauge invariant correlators in QCD are studied on the lattice. A systematic determination of the correlation lengths for gluon field strength correlators and quark correlators is made. The measurement of the gluon and quark condensates is discussed.

1 Introduction

The gauge invariant gluon field strength correlators are defined as

$$\mathcal{D}_{\mu\nu\rho\sigma}(x,C) = \langle 0| T \left(G^a_{\mu\nu}(x) S^{ab}_{adj,C}(x,0) G^b_{\rho\sigma}(0) \right) |0\rangle \quad (1)$$

with $S^{ab}_{adj,C}(x,0)$ the parallel transport from 0 to x along the path C, in the adjoint representation

$$S_{adj,C}(x,0) = P \exp \left(i \int_{0,C}^{x} A^a_\mu(y) T^a_{adj} \, dy^\mu \right) \quad (2)$$

T^a_{adj} are the group generators in the adjoint representation.

$\mathcal{D}_{\mu\nu\rho\sigma}(x,C)$ depends on the choice of C: in what follows we will take for C a straight line, and drop the dependence on C.

Higher correlators are usually defined by parallel transport to a fixed point x_0.

Fermion correlators are defined as

$$S_i(x) = \langle 0| T \left(\bar{\psi}(x) S_{fund}(x,0) M^i \psi(0) \right) |0\rangle \quad (3)$$

M^i is a generic element of the Clifford algebra of the γ matrices. S_{fund} is the analog of the transport (2) in the fundamental representation, and again we have once and for all assumed for the path C a straight line.

By use of general covariance arguments[1,2] $\mathcal{D}_{\mu\nu\rho\sigma}$ can be parametrized in terms of two independent invariant form factors $\mathcal{D}(x^2)$ and $\mathcal{D}_1(x^2)$

$$\mathcal{D}_{\mu\nu\rho\sigma}(x) = (g_{\mu\rho}g_{\nu\sigma} - g_{\mu\sigma}g_{\nu\rho}) \left[\mathcal{D}(x^2) + \mathcal{D}_1(x^2) \right] + \quad (4)$$
$$+ (x_\mu x_\rho g_{\nu\sigma} - x_\mu x_\sigma g_{\nu\rho} - x_\rho x_\nu g_{\mu\sigma} + x_\nu x_\sigma g_{\mu\rho}) \frac{\partial \mathcal{D}_1}{\partial x^2}$$

Similarly one can prove that all the correlators (3) vanish by T, P invariance, except the correlator S_0 corresponding to $M^i = I$, the identity matrix

$$S_0(x) = \langle 0| T \left(\bar{\psi}(x) S_{fund}(x,0) \psi(0) \right) |0\rangle \quad (5)$$

and the vector correlator $(M^i = \gamma^\mu)$, with γ^μ in the direction of x

$$\frac{x^\mu}{|x|} S_V(x) = \langle 0| T \left(\bar{\psi}(x) S_{fund}(x,0) \gamma^\mu \psi(0) \right) |0\rangle \quad (6)$$

The physical interest of the above correlators stems from the basic idea of the ITEP sum rules[3]: the long distance modes of QCD are described by a slowly varying background, made e.g. of instantons, on which high momentum perturbative fluctuations are superimposed. In the O.P.E. low modes generate the condensates, and high frequency modes the corresponding coefficient functions $C_n(x)$

$$T \left(j_\mu(x) j_\nu(0) \right) \simeq \sum_n C_n(x) O_n \quad (7)$$
$$= C_I(x) I + C_G(x) G_{\mu\nu}(0) G_{\mu\nu}(0) +$$
$$+ C_\psi(x) \sum_f m_f \bar{\psi}_f(0) \psi_f(0) + \ldots$$

As is well known expressing the left hand side in terms of a dispersive integral, relates masses and widths of resonances in $e^+ e^- \to hadrons$ to the condensates

$$G_2 = \frac{\beta(g)}{g} \langle 0| G^a_{\mu\nu}(0) G^a_{\mu\nu}(0) |0\rangle \quad , \quad \langle 0| m_f \bar{\psi}_f(0) \psi_f(0) |0\rangle$$

(SVZ sum rules).

By use of this idea it was proposed in ref.[4,5] that the gluon condensate G_2 could be determined from the spectrum of bound states of heavy $Q\bar{Q}$ systems. If the correlation length of the slow varying field, λ, is much bigger than the typical time of the bound system, then its effect is in all respects a static Stark effect on the levels, and the gluon condensate can be extracted from it.

A more detailed analysis[6] involves $\mathcal{D}_{\mu\nu\rho\sigma}(x)$, and gives a shift depending on the parameter

$$\rho = \lambda \frac{m_q \alpha_s^2}{4} \qquad (8)$$

where $4/m_q\alpha_s^2$ is the typical time of the low lying levels of the system, and λ is the correlation length defined as

$$\mathcal{D}(x) \underset{|x|\to\infty}{\simeq} G_2 \exp(-x/\lambda) \qquad (9)$$

Measuring λ was the motivation to investigate for the first time $\mathcal{D}_{\mu\nu\rho}(x)$ on the lattice[7]. The computation was done in quenched $SU(2)$ and gave a surprisingly small value of λ

$$\lambda \simeq 0.16\,\text{fm} \qquad (10)$$

shaking the very bases of the SVZ approach.

A stochastic model of the vacuum was subsequently developed[1,2], in which observables are expressed in terms of invariant field strength correlators, and a cluster expansion is made. The basic assumption of the model is that higher order clusters are negligible.

The quark correlator S_0 instead, known also as "non local fermion condensate" enters in the construction of the wave functions of hadrons, in particular in the computation of the pion form factor[8].

A technical breakthrough, the use of cooling or smearing procedures to polish short distance fluctuations, leaving long distance physics unchanged[9], allowed a better determination of correlators[10,11,12].

The physical motivations to study correlators on the lattice are in conclusion

(i) understanding λ, and the basis of SVZ sum rules.

(ii) measuring condensates from first principles.

(iii) more generally providing inputs from first principles to the community of stochastic vacuum practitioners.

2 Cooling-smearing correlators

Short range fluctuations in lattice configurations can be smoothed off by a local cooling procedure which consists in replacing a link by the sum of the inverse "staples" attached to it

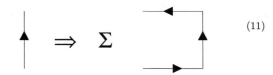

$$(11)$$

Since the action density is

$$S \sim \sum_{\mu\nu}\left(1 - \frac{1}{N_c}Tr\Pi_{\mu\nu}\right) \quad \text{with } \Pi_{\mu\nu} = \qquad (12)$$

this procedure makes locally $S = 0$. In the euclidean region S plays the role of energy and the replacement (11) locally minimizes S, whence the terminology "cooling".

Like any local procedure cooling n_t times affects distances d by a diffusion process, with

$$d^2 \sim n_t \qquad (13)$$

According to eq.(13), n_t can be made sufficiently large to eliminate short range fluctuations but not enough to modify long range correlations[9]. Fluctuations are thus reduced by orders of magnitude without changing long distance physics.

Gauge invariant correlators can be represented on lattice by the following operators[7]

$$\mathcal{D}^L_{\mu\nu\rho\sigma} = \left\langle \underset{\Pi_{\mu\nu}}{\rule{0pt}{0pt}} \underset{\Pi_{\rho\sigma}}{\rule{0pt}{0pt}} - \frac{1}{N_c}\underset{\Pi_{\mu\nu}}{\rule{0pt}{0pt}}\ \underset{\Pi_{\rho\sigma}}{\rule{0pt}{0pt}}\right\rangle$$

A series expansion in a gives

$$\mathcal{D}^L_{\mu\nu\rho\sigma}(d) \simeq Z^2 a^4 \mathcal{D}_{\mu\nu\rho\sigma}(d) + \mathcal{O}(a^6) \qquad (14)$$

By the cooling procedure $\mathcal{O}(a^6)$ terms disappear and possible renormalizations Z of the field $G_{\mu\nu}$ tend to 1. So finally

$$\mathcal{D}^L_{\mu\nu\rho\sigma}(d) \simeq a^4 \mathcal{D}_{\mu\nu\rho\sigma}(da) \qquad (15)$$

In order to use cooling profitably the distance d has to be $\sim 3 - 4$ lattice spacings at least. At a given β this corresponds to some physical length l_{min}. If we want l_{min} to be small, say 0.1 fm, since the lattice must be at least 1 fm across, the lattice size must be $(10l_{min})^4$. Going to small distances requires big lattices.

3 Correlators, OPE and renormalons

The SVZ sum rules are based on the OPE of the correlator

$$\Pi_{\mu\nu}(q) = \int d^4x\, e^{iqx}\langle 0|T\left(j_\mu(x)j_\nu(0)\right)|0\rangle \qquad (16)$$

$$= \Pi(q^2)\left(g_{\mu\nu}q^2 - q_\mu q_\nu\right)$$

The OPE gives

$$\Pi(q^2) \simeq c_1 I + c_2 \frac{G_2}{q^4} + \dots \qquad (17)$$

The first term corresponds to the perturbative expansion. That expansion, however, is not Borel summable, and is ambiguous by terms of order μ^4/q^4, with μ the renormalization scale (Renormalons). As a consequence the second term in eq.(17) is intrinsically undefined. This is a basic and unavoidable drawback of perturbation theory, reflecting the fact that perturbative vacuum is not the ground state[13].

However, keeping the first few terms in the perturbative expansion of c_1 and c_2 gives a consistent phenomenology[3].

The same happens for the correlators. The OPE of the invariant form factors has the form

$$\mathcal{D}(x^2) \underset{x^2 \to 0}{\simeq} \frac{c_1}{x^4} + c_2 G_2 + \mathcal{O}(x^2) \qquad (18)$$

The second term is again undetermined by renormalons coming from the perturbative expansion of c_1. As in the SVZ sum rules we shall assume that the first few terms of the perturbative expansion work, and use it to determine G_2. We shall parametrize the lattice determination of $\mathcal{D}(x^2)$ and $\mathcal{D}^{(1)}(x^2)$ as

$$\frac{1}{a^4} \mathcal{D}_L(x^2) = \frac{a}{|x|^4} e^{-|x|/\lambda_a} + A_0 e^{-|x|/\lambda} \qquad (19)$$

$$\frac{1}{a^4} \mathcal{D}_L^{(1)}(x^2) = \frac{a_1}{|x|^4} e^{-|x|/\lambda_a} + A_1 e^{-|x|/\lambda} \qquad (20)$$

Eq.(19), (20) obey the OPE eq.(18), and reflect the existence of a mass gap in the theory.

In ref.[7] the first term of eq.(18), (19), (20) was computed in perturbation theory and subtracted. The residual term was an exponential and λ could be extracted from it, giving $\lambda = 0.16$ fm.

In ref.[10,11] quenched $SU(3)$ was studied. The above parametrization gave a good fit to the data (fig.1) with $\lambda = 0.22$ fm and $G_2 = (0.14 \pm 0.02)$ GeV^4, a value larger by an order of magnitude than the phenomenological value[3]. In ref[12] full QCD with 4 staggered fermions was studied, at quark masses $am_q = 0.01$, $am_q = 0.02$. The results are in this case[12]

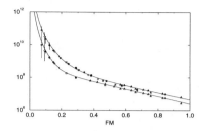

Fig.1 $D_{||}^L/a^4 = \mathcal{D} + \mathcal{D}_1 + x^2 \partial \mathcal{D}_1/\partial x^2$ and $D_\perp/a^4 = \mathcal{D} + \mathcal{D}_1$ versus x. The lines correspond to the best fit to eq.(19),(20).

$$am_q = .01 \quad \lambda = (.34 \pm .02)\,\text{fm} \quad G_2 = .015 \pm .003\,\text{GeV}^4$$
$$am_q = .02 \quad \lambda = (.29 \pm .02)\,\text{fm} \quad G_2 = .031 \pm .005\,\text{GeV}^4$$

A common feature to all the determinations is that $|A_1| \simeq A_0/10$. In full QCD the correlation length is bigger, and could agree with the basic philosophy of ref.[3]. Also the value of the condensate G_2 is smaller than the quenched value and agrees with phenomenology.

By use of the relation[14]

$$\frac{d}{dm_f} G_2 = -\frac{24}{b_0} \langle \bar{\psi}\psi \rangle \qquad b_0 = 11 - \frac{2}{3} N_f$$

one can extrapolate in m_f to the physical value of G_2 getting

$$G_2 \simeq 0.022 \pm .006\,\text{GeV}^4 \qquad (21)$$

in agreement with sum rules determination[15].

Similar arguments allow to extract G_2 from the measurement of the average value of the density of action (plaquette): again the level of rigour is the same as for SVZ sum rules[3], at least if only a few terms are kept of the perturbative expansion of the coefficients in the OPE[16,17,18].

A detailed analysis of the behaviour of the correlators at finite temperature, in the vicinity of the deconfining transition T_c was made in ref.[11]. There $O(4)$ invariance is lost reducing to $O(3)$ and 5 independent form factors exist.

The main result is that magnetic correlators are unchanged across T_c, while electric correlators have a sharp drop (fig.2,fig.3).

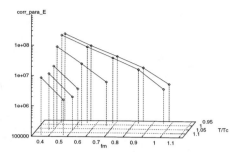

Fig.2 The electric longitudinal correlator versus distance, for different values of T/T_c.

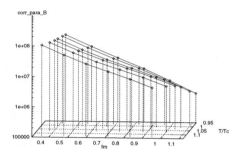

Fig.3 The magnetic longitudinal correlator versus distance, for different values of T/T_c.

The scalar quark correlator

$$S_0(x) = -\langle 0|T(\bar{\psi}(x)S\psi(0))|0\rangle \qquad (22)$$

has recently been determined in full QCD and in quenched QCD[19].

A comparison has been made between the determinations in full QCD, at given values of the quark masses $am_q = 0.01$, $am_q = 0.02$ and of the lattice spacing a, and in quenched QCD at the same values of these physical parameters. No difference has been found, within errors, indicating that quark loops do not affect appreciably the quark correlator.

A sensible parametrization for the lattice regulator S_0^L is

$$S_0^L(x) = a^3 A_0 \exp(-x/\lambda_f) + \frac{B_0 a^3}{x^2} \qquad (23)$$

Simulations have been performed[19] at $am_q = 0.01$ in full QCD at $\beta = 5.35$ and quenched QCD at $\beta = 6.0$, which correspond both to a lattice spacing $a \simeq 0.10$ fm. No appreciable difference is found and in both cases (fig.4).

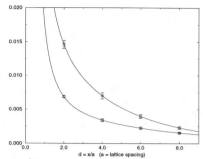

Fig.4 $S_0^L(x)$ versus x. The curve is the best fit to eq.(23).

$$\frac{a}{\lambda_f} = 0.16 \pm 0.04 \qquad am_\pi = 0.26 \pm 0.01 \qquad (24)$$

A similar determination at $am_q = 0.02$ $\beta = 5.35$ full QCD, $\beta = 5.91$ quenched, where $a \simeq 0.12$ fm, give indistinguishable results:

$$\frac{a}{\lambda_f} = 0.26 \pm 0.04 \qquad am_\pi = 0.37 \pm 0.01 \qquad (25)$$

Putting the two determinations together gives

$$\lambda_f m_\pi = 1.5 \pm 0.3 \qquad (26)$$

The typical correlation length is now $\sim 1/m_\pi$ in agreement with the approach of ref.[3].

A determination of A_0 and a study of its relation to the quark condensate $\langle \bar{\psi}\psi \rangle$ is on the way.

4 Discussion

The typical correlation length is rather small for gluon correlators: $\lambda = 0.16$ fm for quenched $SU(2)$, 0.22 fm for quenched $SU(3)$, 0.32 fm for full QCD with 4 flavours. It is bigger for fermion correlators where

$$\lambda_f \simeq (1.5 \pm 0.3)m_\pi^{-1}$$

Condensates can be determined, despite the presence of renormalons, by the same philosophy used for SVZ sum rules. $G_2 = (0.022 \pm 0.006)$ GeV^4 in full QCD is consistent with the determination by sum rules. It is an order of magnitude bigger in quenched QCD, $G_2 = (0.14 \pm 0.02)$ GeV^4.

The behaviour of correlators at the deconfining transition is consistent with expectations.

Our determinations are useful to phenomenology and to test models of QCD vacuum.

Acknowledgements

Partially supported by EC TMR Program ERBFMRX-CT97-0122, and by MURST, project: "Fisica Teorica delle Interazioni Fondamentali".

References

1. H.G. Dosch, *Phys. Lett.* B **190**, 177 (1987).
2. H.G. Dosch, Yu. A. Simonov, *Phys. Lett.* B **205**, 339 (1988).
3. M.A. Shifman, A.I. Vainshtein, V.I. Zakharov, *Nucl. Phys.* B **147**, 385,448,519 (1979).
4. H. Leutwyler, *Phys. Lett.* B **98**, 447 (1981).
5. M.B. Voloshin, *Nucl. Phys.* B **154**, 365 (1979).
6. M. Campostrini, A. Di Giacomo, Š. Olejník, *Z. Phys.* C **31**, 5 (1986).
7. M. Campostrini, A. Di Giacomo, G. Mussardo, *Z. Phys.* C **25**, 173 (1984).
8. S.V. Mikhailov, A.V. Radjushkin, *JETP Lett.* **43**, 713 (1986).
9. M. Campostrini, A. Di Giacomo, M. Maggiore, H. Panagopoulos, E. Vicari, *Phys. Lett.* B **225**, 403 (1989).
10. A. Di Giacomo, H. Panagopoulos, *Phys. Lett.* B **285**, 133 (1992).
11. A. Di Giacomo, E. Meggiolaro, H. Panagopoulos, *Nucl. Phys.* B **483**, 371 (1997).
12. M. D'Elia, A. Di Giacomo, E. Meggiolaro, *Phys. Lett.* B **408**, 315 (1997).
13. A.H. Müller, *Nucl. Phys.* B **250**, 327 (1985).
14. V.A. Novikov, M.A. Shifman, A.I. Vainshtein, V.I. Zakharov, *Nucl. Phys.* B **198**, 301 (1981).
15. S. Narison, *Phys. Lett.* B **387**, 162 (1996).
16. A. Di Giacomo, G.C. Rossi, *Phys. Lett.* B **100**, 481 (1981).
17. A. Di Giacomo, G. Paffuti, *Phys. Lett.* B **108**, 327 (1982).
18. M. Campostrini, A. Di Giacomo, Y. Gündüc, *Phys. Lett.* B **225**, 393 (1983).
19. M. D'Elia, A. Di Giacomo, E. Meggiolaro, in preparation.

GAUGE FIXING ON THE LATTICE AND THE GIBBS PHENOMENON

JEFFREY E. MANDULA

U. S. Department of Energy, Washington, DC 20585, USA
E-mail: mandula@hep2.er.doe.gov

We discuss global gauge fixing on the lattice, specifically to the lattice Landau gauge, with the goal of understanding the question of why the process becomes extremely slow for large lattices. We construct an artificial "gauge-fixing" problem which has the essential features encountered in actuality. In the limit in which the size of the system to be gauge fixed becomes infinite, the problem becomes equivalent to finding a series expansion in functions which are related to the Jacobi polynomials. The series converges slowly, as expected. It also converges non-uniformly, which is an observed characteristic of gauge fixing. In the limiting example, the non-uniformity arises through the Gibbs phenomenon.

1 Introduction

In very broad terms, there are two classes of applications of lattice gauge theory. One is the extraction of the physical predictions of QCD at low energies, where perturbation theory is not trustworthy. The other is learning about the mathematical structure of quantum field theory, especially the Green's functions of QCD, again outside the region in which we can safely trust perturbative analysis. For many applications of the first type, most notably the calculation of the hadronic spectrum, there is no need to specify a gauge. In fact, the original proposal of Wilson was to calculate the Feynman path integral by explicitly summing over all gauges. For problems of the second type, by contrast, the specification of a gauge is intrinsic. Also, the most straightforward way to evaluate the non-perturbative renormalization constants of operators is to work in a fixed gauge. These renormalization constants are physically relevant. They are needed to calculate the absolute scale of the strong interaction matrix elements that multiply the fundamental Standard Model couplings in weak decays and in $K^0 \bar{K}^0$ and $B^0 \bar{B}^0$ mixing.

It is well known that gauge fixing is afflicted with a host of problems. Gribov copies[1] — both the analogues of continuum copies and new lattice artifacts — occur on the lattice. We continue to lack a good understanding of how to treat them in principle, and we also have no general understanding of their practical significance, although in some examples the manner in which they are treated affects the results of calculations.[2] Presumably performing the path integral for a gauge theory within a "Fundamental Modular Domain" is an appropriate method, but we have no operational procedure for finding such a domain.[3]

Gauge fixing to the Landau gauge is also a notoriously slow process, especially for large lattices. While there are a number of techniques that work adequately well on small or moderate sized lattices, for the size lattices currently in common use, all algorithms are unfortunately very inefficient.[4] The goal of the work described here is to get some insight into this algorithmic inefficiency.

2 The Lattice Landau Gauge

On the lattice, a general gauge transformation has the structure

$$U_\mu(x) \rightarrow U_\mu^{(G)}(x) \equiv G(x) U_\mu(x) G(x + \hat{\mu})^\dagger$$

Any gauge condition $f(U) = 0$ is implemented following the Fade'ev-Popov procedure:

$$\int DU \, e^{-S} \, O(U) \bigg|_{f(U) = 0}$$

$$= \int DU \int DG \, e^{-S} \, O(U) \, \Delta_{FP}(U) \, \delta(f(U^{(G)}))$$

$$= \int DU \, e^{-S} \, O(U^{(G)})$$

where here $U^{(G)}$ denotes the gauge transform of U to the gauge $f(U^{(G)}) = 0$.

Note that on the lattice the path integral measure remains the same as without gauge fixing. The effect of the Fade'ev-Popov determinant is incorporated not by keeping only configurations of links satisfying $f(U) = 0$ gauge and modifying their relative weights, but by keeping all configurations weighted only by the action, and transforming each to the $f(U) = 0$ gauge.

For the lattice Landau gauge, one formulates the gauge condition as a maximization so as to avoid some lattice artifacts

$$\underset{G}{Max} \sum_{x\mu} Re \, Tr \, U_\mu^{(G)}(x)$$

At each site, a gauge transformation affects 8 links, so the maximization condition at site x is

$$\underset{G(x)}{Max} \, Re \, Tr \, G(x) \sum_\mu (U_\mu(x) + U_\mu(x - \hat{\mu})^\dagger)$$

If $G(x) = 1$ satisfies the maximization condition, then we recover the lattice form of the differential condition

$$\Delta_\mu A_\mu(x) \equiv \sum_\mu (A_\mu(x) - A_\mu(x - \hat{\mu})) = 0$$

where the gauge potential is

$$A_\mu(x) = \frac{U_\mu(x) - U_\mu^\dagger(x)}{2iag} - Tr\frac{U_\mu(x) - U_\mu^\dagger(x)}{6iag}$$

The most naive implementation of the Landau gauge condition is to cycle through the lattice, imposing the maximization requirement one site at a time, and repeating the process until the configuration has relaxed sufficiently into a global maximum.[5] Two classes of improvements on this method are widely used. One is overrelaxation[6], that is the replacement $G_{max}(x) \rightarrow G_{max}^w(x)$, $(1 \le w \le 2)$, either exactly or stochastically. Another is Fourier preconditioning[7].

The conventional wisdom about the inefficiency of gauge fixing is that the lattice configuration develops "hot spots", that is exceptional poorly fixed points which move from site to site under the effect of repeated gauge fixing sweeps, but which are strongly persistent. Another part of the conventional wisdom, not actually compatible with the idea of "hot spots", is that it is the longest wavelength modes that relax most slowly.

If one looks in detail at the distribution of the values of the deviation from Landau gauge at each site, using $Tr(\Delta_\mu A_\mu)^2$ as a measure of the distance from perfect fixing, one sees not just a few very poorly fixed sites, but a broad range of deviations with no notable gaps. If one looks in Fourier space, one finds that there is a broad range of relaxation times, with many modes decaying slowly, including, but not limited to, the longest wavelengths.

3 Gauge Fixing Convergence

If the link variables $U_\mu(x)$ are close to satisfying the Landau gauge condition, we may expand the gauge transformation needed to satisfy it exactly $G(x) = \exp(i\,g(x))$ to second order, and express the Landau gauge condition as the maximization of a quadratic form

$$\underset{g(x)}{Max}\ Re\ Tr\ i\,[\,g(x)\,U_\mu(x) - U_\mu(x - \hat{\mu})\,g(x)\,]$$

$$- \frac{1}{2}Re\ Tr\,[\,g^2(x)\,U_\mu(x) - U_\mu(x - \hat{\mu})\,g^2(x)\,]$$

$$+ Re\ Tr\,[\,g(x)\,U_\mu(x)\,g(x + \hat{\mu})\,]$$

Setting the variation of the quadratic form with respect to the generator of gauge transformations $g(x)$ to zero gives a matrix

inversion problem with a nearest neighbor structure. The eigenvalues of this matrix control the convergence of relaxation methods, and a two dimensional example illustrates the typical character of the eigenvalues.

The eigenvalues of the tridiagonal $N \times N$ matrix

$$-\begin{pmatrix} 1 & 1 & 0 & \cdot & \cdot & \cdot & \cdot \\ 1 & 2 & 1 & \cdot & \cdot & \cdot & \cdot \\ 0 & 1 & 2 & \cdot & \cdot & \cdot & \cdot \\ \cdot & \cdot & \cdot & \cdot & \cdot & \cdot & \cdot \\ \cdot & \cdot & \cdot & \cdot & 2 & 1 & \cdot \\ \cdot & \cdot & \cdot & \cdot & 1 & 2 & 1 \\ \cdot & \cdot & \cdot & \cdot & \cdot & 1 & 1 \end{pmatrix}$$

are

$$\lambda_m = 2\left(\cos\frac{m\pi}{N} - 1\right) \qquad m = 0, 1, ..., N - 1$$

The important points to note are that there is always a zero eigenvalue, and that as N becomes large, many small eigenvalues develop.

$$\lambda_m \cong -\frac{m^2\pi^2}{2N^2} \qquad m \ll N$$

It is not the single vanishing eigenvalue which is troublesome — its eigenvector could always be projected out — but the multiplicity of small ones.

4 Conjugate Gradient Gauge Fixing

Landau gauge fixing is a maximization problem, and there are many methods for addressing such problems. Rather than simply sweeping through the lattice, we may elect to use a global method, which act on all sites at once. Conjugate gradient is one such. It is a well known, efficient method for minimizing (or maximizing) functions of many variables. In lattice work it is the method of choice for inverting the Dirac operator.

Conjugate gradient is a recipe for finding a sequence of mutually conjugate vectors, that is vectors satisfying

$$(h_i, Mh_j) = 0 \qquad (i \neq j)$$

where M is the quadratic form to be extremized. Each vector in the sequence constructed from its predecessors

$$h_{i+1} = Mh_i + \sum_{j=0}^{j=i} c_j h_j$$

The initial vector is taken to be in direction of steepest descent from the starting point.

The power of conjugate gradient lies in the fact that the values of calculated coefficients do not change as the number of terms in successive approximations is increased. For minimization problems which are exactly, and not just approximately quadratic forms, conjugate gradient converges exactly in a finite number of steps.

A tractable problem which nonetheless retains the characteristic features of Landau gauge fixing is the minimization of an N-dimensional quadratic form with eigenvalues evenly distributed between 0 and 1, starting from a point whose displacement from the exact minimum has an equal projection on each eigenvector of the quadratic form. We know the exact solution, of course, but is it quite instructive to study the convergence of the conjugate gradient algorithm to the exact solution.

The clearest basis in which to analyze the problem is that in which the quadratic form to be minimized is diagonal. This is just a choice of basis, and neither accelerates nor retards the convergence.

$$Min \sum_{i,j=1}^{N} v_j M_{ij} v_j \qquad M_{ij} = \lambda_i \delta_{ij} = \frac{i}{N} \delta_{ij}$$

In this basis the sequence of approximations is

$$v_i^{[n]} = \sum_{m=0}^{n} c_m h_i^{[m]}$$

$$h_i^{[m+1]} = M_{ij} h_j^{[m]} + \sum_{m'=1}^{m} d_{m'} h_i^{[m']} \qquad h^{[0]} = \begin{pmatrix} 1 \\ 1 \\ 1 \\ \cdot \\ \cdot \\ \cdot \\ 1 \end{pmatrix}$$

$$c_0 = 1$$

5   The $N \rightarrow \infty$ Limit

We are concerned with the convergence of the conjugate gradient method as the dimension of the quadratic form becomes very large. Conveniently, in the limit where the dimension goes to infinity we can transform the problem of minimizing the quadratic form into a problem of expanding a function in a series of polynomials. This is accomplished by rescaling the index labeling directions into a continuous variable on the interval $[0,1]$ and replacing sums over indices by finite integrals. The mutually conjugate vectors are replaced by polynomials $P_n(x)$, which are mutually orthogonal with respect to the weight function $w(x) = x$ on this interval. The initial point, which by convention we take as the origin, is the constant function. Explicitly, the mapping is

$$x = i/N$$
$$h_i^{[n]} \rightarrow P_n(x)$$
$$M_{ij} = \frac{i}{N} \delta_{ij} \rightarrow x\delta(x - y)$$
$$v_i M_{ij} u_j \rightarrow \int_0^1 x f(x) g(x)\, dx$$

$$\sum_{i=1}^{N} \rightarrow \int_0^1 dx$$
$$v_i \rightarrow f(x)$$
$$M_{ij} v_j \rightarrow x f(x)$$
$$v^{[0]} \rightarrow f_0 = 1$$

6   The Sturm-Liouville Expansion

6.1 Statement and Solution

The convergence of the conjugate gradient minimization becomes the convergence of the sequence of approximations

$$f_n(x) \equiv \sum_{m=1}^{n} c_m P_m(x) \rightarrow f(x) = 1$$

The polynomials $P_n(x)$ are determined by the two properties that they mutually orthogonal on the interval $[0,1]$ with respect to the weight function $w(x) = x$, and that each successive polynomial is a linear combination of all the foregoing polynomials and $w(x) = x$ times its predecessor.

Such a sequence of functions sounds like the solution of a Sturm-Liouville eigenvalue-eigenvector problem. The equation must have regular singular points at 0 and 1, and a factor of x multiplying the eigenvalue. Such an equation need not have solutions which are polynomials at all. However it is simple to verify that with the following choice of exponents and coefficients:

$$\left[\frac{d}{dx} x^2 (1 - x) \frac{d}{dx} - 2 \right] P_n(x) = -n(n + 2) x P_n(x)$$

there are solutions which are analytic at both ends of the interval $[0,1]$ and whose power series terminate. These functions are related to Jacobi polynomials, and one can work out all their properties and expansions using standard analysis methods.

The reader may object that we are not implementing conjugate gradient at all, and so we may learn nothing about the convergence of the method. While it is true that we are avoiding the explicit use of conjugate gradient, we have nonetheless determined a sequence of polynomials which have all the properties of the conjugate gradient sequence. Since there can only be one such, we are using the same expansion vectors as would have been given by a direct use of conjugate gradient. That is, the convergence of the Sturm-Liouville expansion is exactly the convergence of the conjugate gradient expansion, because they are the same expansion.

A convenient normalization for these polynomials is

$$P_n(x) = \sum_{m=1}^{n} \frac{(-1)^{m-1}(n-1)!}{(m-1)!(n-m)!} \frac{3!(n+m+1)!}{(n+2)!(m+2)!} x^m$$

Note that the sequence starts at $n = 1$, not $n = 0$. The constant function is not in the sequence. This is what makes the problem of expanding the exact solution of the minimization problem, the constant function, other than trivial. It is straightforward to derive the integrals for single functions and for a product of two:

$$\int_0^1 x\, P_n(x)\, dx = \frac{12}{[n(n+1)(n+2)]^2}$$

$$\int_0^1 x\, P_m(x)\, P_n(x)\, dx = \frac{18}{n^2(n+1)^3(n+2)^2} \delta_{mn}$$

These integrals are used in evaluating the conjugate gradient expansion coefficients and examining the convergence in the norm of the partial sums. Explicitly, the successive approximations $f_n(x)$ to $f(x) = 1$ are given by

$$f_n(x) = \frac{2}{3} \sum_{m=1}^{n} (n+1)\, P_m(x)$$

$$= 1 - 2 \sum_{m=0}^{n} \frac{(n!)^2 (-x)^m (1-x)^{n-m}}{m!\,(m+2)!((n-m)!)^2}$$

6.2 Convergence Properties

The convergence of the partial sums in the norm is only a power of the number of terms,

$$\int_0^1 x\, (f_n(x) - 1)^2 = \frac{2}{[(n+1)(n+2)]^2}$$

The number of terms needed to successively reduce the error by a fixed finite factor grows like the $-1/4$ power of the residual error, so that the relaxation time of the sequence is infinite.

The pointwise convergence is both slow and non-uniform. At fixed x in the interior of $[0,1]$ one has

$$f_n(x) - 1 \sim \frac{\cos(n\theta + c)}{n^{-5/2}} \qquad (\cos\theta = 1 - 2x)$$

At the upper end of the interval the convergence is different:

$$f_n(1) = \frac{2(-1)^n}{(n+1)(n+2)}$$

This is a marginally slower rate of convergence. At the lower end of the interval, pointwise convergence actually fails.

6.3 The Gibbs Phenomenon

No matter how many terms n are kept in the expansion, there remains a region in which the difference between the partial sum and the exact result does not approach 0. The region shrinks as n grows, of course. Specifically, for small values of $x \propto 1/n^2$, one derives

$$\lim_{n \to \infty} f_n\!\left(\frac{a}{n^2}\right) = 1 - \frac{2}{a} J_2(2\sqrt{a})$$

where, J_2 denotes the ordinary (oscillatory) Bessel function. The maximum value of this expression is about 1.059, which is attained at $a \approx 1.8$

This is an example of the Gibbs phenomenon. The series overshoots and oscillates about true value. Even though this behavior is restricted to an ever smaller range, $x \propto 1/n^2$, it saturates the convergence in the norm. Explicitly,

$$\lim_{b \to \infty} \frac{1}{n^4} \int_0^b a\, da \lim_{n \to \infty} \left(f_n\!\left(\frac{a}{n^2}\right) - 1 \right)^2 = \frac{2}{n^4}$$

which is the leading asymptotic behavior of the full deviation in the norm from $f(x) = 1$.

6.4 Role of the Small Eigenvalues

The appearance of the Gibbs phenomenon, and the turgid convergence in general, are a result of the multiplicity of small eigenvalues. It is instructive to consider what would be the convergence of conjugate gradient if one chose the matrix M_{ij} not to have any small eigenvalues. In the $N \to \infty$ limit, the effect would be to replace the polynomials $P_n(x)$ with polynomials $K_n(x)$ whose orthogonality relation involves a weight function that does not take arbitrarily small values. As an example, let us consider polynomials orthogonal with respect to

$$w(x) = (x+1)$$

These polynomials are not related to any of the classical polynomials, and so we have no short cuts to assist the evaluation of the expansion coefficients or the convergence properties of the partial sums. Nonetheless, it is a simple computer algebra exercise to compute the first dozen or two terms exactly. The result shows numerically that the successive approximations to

$f(x) = 1$ converge exponentially with n. The following table shows the contrast between the singular expansion in the P_n polynomials and the smooth expansion in the K_n polynomials

Table 1: Convergence of $\int_0^1 w(x) (f_n(x) - 1)^2 \, dx$

Order	P - Expansion	K - Expansion
0	1.	1.
1	.2592593	0.03670782
2	.1203704	0.001182794
3	.07	3.597242e-5
4	.04592593	1.077474e-6
5	.03250189	3.206815e-8
6	.02423469	9.511775e-10
7	.01877572	2.815558e-11
8	.01497942	8.32333e-13
9	.0122314	2.458333e-14
10	.01017753	7.256163e-16

7   Conclusions

The principal lesson of this analysis is qualitative. It is that the slowness of the gauge fixing process to the lattice Landau gauge for large lattices is not likely to be overcome. Even global maximization methods, and even neglecting their limitations imposed by round-off errors, do not seem to help. To a lesser extent, the same observation could be made about the lattice Coulomb gauge; lesser simply because the number of sites to be fixed is smaller by a factor of the number of sites in the time direction.

One way to try to deal with difficulty of gauge fixing on very large lattices would be via the use of perfect, or at least improved actions. That is, one may try to use lattices with many fewer sites, for the same level of precision. This will not be automatic to implement however. Just as additional operators in the action are needed to eliminate the leading finite lattice spacing errors, similar terms will be required in the gauge condition, to improve the convergence of the lattice Landau gauge to the continuum version.

Another necessary observation about extending the improvement program to gauge-variant quantities is that all the evidence for the viability of that program comes from the study of gauge-invariant quantities. It will be necessary to check carefully the extension of improved actions and operators to the unphysical sectors of gauge theory. A most non-trivial test of those ideas will be whether the improvement of the action and operators with gauge-invariant improvement terms suffices, or whether non-gauge-invariant improvement terms will need to be introduced.

References

1. V. N. Gribov, Nucl. Phys. B **139** (1978) 1; I. M. Singer, Comm. Math. Phys. **60** (1978)7.

2. Ph. de Forcrand, J. E. Hetrick, A. Nakamura, and M. Plewnia, in *Proceedings of the International Symposium on Lattice Gauge Theory 1990*, Tallahassee (1991).

3. D. Zwanziger, Nucl. Phys. B **209** (1982) 336; Phys. Letts. B **257** (1991) 168; G. Dell'Antonio and D. Zwanziger, Nucl. Phys. B **326** (1989) 336; A. Cucchieri, Nucl. Phys. B **521** (1998) 365.

4. A. Cucchieri and T. Mendes, Nucl. Phys. B **471** (1996) 263; H. Suman and K. Schilling, hep-lat/9306018.

5. K. Wilson, in Proc. NATO Adv. Study Inst. (Cargèse, 1979), Plenum Press; J. E. Mandula and M. Ogilvie, Phys. Letts. B **139** (1987) 127.

6. J. E. Mandula and M. Ogilvie, Phys. Letts. B **248** (1990) 156; Ph. de Forcrand and R. Gupta, Nucl. Phys. B (Proc. Suppl.) **9** (1989) 516.

7. C.T.H. Davies *et al.*, Phys. Rev. D **37** (1988) 1581.

IMPROVED LATTICE ACTIONS AND CHARMED HADRONS

B. H. J. McKELLAR

School of Physics, University of Melbourne, Parkville, Vic 3052, AUSTRALIA
and
Special Research Centre for the Subatomic Structure of Matter, University of Adelaide, SA 5005, AUSTRALIA
E-mail:mckellar@physics.unimelb.edu.au

J. P. MA

School of Physics, University of Melbourne, Parkville, Vic 3052, AUSTRALIA
and
Institute for Theoretical Physics, Academica Sinica, PO Box 2735, Beijing 100080, CHINA
E-mail:majp@itp.ac.cn

We report on calculations of hadron masses using improved lattice actions for glue, light quarks and heavy quarks. We find that the results for charmed hadrons show that charmed quarks are an awkward intermediate position in that they are not well approximated by either the light quark or the heavy quark ansatz.

1 The Lattice Actions

1.1 The gluonic action

We take the one-loop improved action for glue [1], where the action consists of plaquette, rectangle and parallelogram terms and is accurate up to errors of $O(\alpha_s^2 a^2, a^4)$. Implementing tadpole improvement the action becomes [2]

$$S(U) = \beta \sum_{pl} \frac{1}{3} \mathrm{ReTr}(1 - U_{pl}) + \beta_{rt} \sum_{rt} \frac{1}{3} \mathrm{ReTr}(1 - U_{tr})$$
$$+ \beta_{pg} \sum_{pg} \frac{1}{3} \mathrm{ReTr}(1 - U_{pg}), \qquad (1)$$

with

$$\beta_{rt} = -\frac{\beta}{20u_0^2}(1 + 0.4805\alpha_s),$$

$$\beta_{pg} = -\frac{\beta}{u_0^2}0.03325\alpha_s,$$

$$u_0 = \left(\frac{1}{3}\mathrm{ReTr}\langle U_{pl}\rangle\right)^{\frac{1}{4}},$$

$$\alpha_s = -\frac{\ln(\frac{1}{3}\mathrm{ReTr}\langle U_{pl}\rangle)}{3.06839} \qquad (2)$$

We used this action to generate gluonic configurations at $\beta = 7.4$, on a lattice whose size is $8^3 \times 16$, and also at $\beta = 6.8$ with the size $6^3 \times 12$. The parameter u_0 is determined by self-consistency. It is 0.8631 and 0.8267 at $\beta = 7.4$ and $\beta = 6.8$ respectively. The pseudo heat bath method [3] was used to update the links, and the three $SU(2)$ subgroups were updated 3 times in each overall update step. For each lattice we generated 100 configurations for our mass calculations.

1.2 The improved action for light quarks

The improved action proposed in [4] can be written as:

$$S_{\text{light}} = -\sum_x \left\{ m\bar{\psi}(x)\psi(x) \right.$$
$$+ \sum_\mu \bar{\psi}(x)\gamma_\mu\Delta_\mu(1 - c_1\Delta_\mu^{(2)})\psi(x)$$
$$\left. + r\sum_\mu \bar{\psi}(x)\Delta_\mu^{(2)}\Delta_\mu^{(2)}\psi(x) \right\}, \qquad (3)$$

where Δ_μ and $\Delta_\mu^{(2)}$ are lattice derivatives with the gauge link $U_\mu(x)$:

$$\Delta_\mu\psi(x) = \frac{1}{2}(U_\mu(x)\psi(x+\hat{\mu}) - U_\mu^\dagger(x-\hat{\mu})\psi(x-\hat{\mu})),$$
$$\Delta_\mu^{(2)}\psi(x) = U_\mu(x)\psi(x+\hat{\mu}) + U_\mu^\dagger(x-\hat{\mu})\psi(x-\hat{\mu})$$
$$-2\psi(x). \qquad (4)$$

The parameter c_1 is determined to be $1/6$ at tree-level to remove lattice effects at $O(a^2)$, m is the mass parameter for a quark. The last term in the action with the parameter r is introduced in analogy to the Wilson term in the Wilson action to solve the doubling problem of lattice fermions. However, this device does not totally solve the doubling problem. In the free case one can solve the equation of motion determined by the action to see whether the doublers are removed or not. An analysis in ref [5] and the analysis in ref [6] for the D234 action, show that in the low-energy regime the actions describe one particle in the sense that the propagator has only one pole in that regime. In the high-energy regime there are additional "unphysical" poles. As we are only interested in the low-energy regime as we are in this work, we can expect that the effect from these "unphysical poles" will be negligible. We will take $r = 1/6$. With this choice the action is the same as employed in ref[7] in which it is shown that there is indeed no effect in the low-energy regime which can be related to the "unphysical poles".

In our work we used the stabilized biconjugate gradient algorithm[8] to calculate propagators for light quarks. We find in our calculations that this algorithm is at least three times faster than the conventional conjugate gradient algorithm.

1.3 The Action of Lattice NRQCD

Heavy quarks whose mass is larger than 1 in lattice units cannot be simulated directly as described in the previous section with reliable results. To simulate them one uses the heavy quark effective theory, HQET. The formulation of HQET for hadrons with zero velocity is equivalent to that of NRQCD on the lattice, except that the expansion parameters are different. As we will only create hadronic states on lattice with zero space-momenta, we may use lattice NRQCD for heavy quarks like the b- and c-quarks. On the lattice one needs to calculate the propagator of heavy quarks as the solution to the appropriate evolution equation. We take the evolution equation proposed in ref[9,10]. The propagator $G(t)$ for a heavy quark of mass M_Q (where $G(t) = 0$ for $t \leq 0$) is calculated on the lattice as:

$$G(1) = \left(1 - \frac{H_0}{2n}\right)^n U_4^\dagger \left(1 - \frac{H_0}{2n}\right)^n \delta_{\mathbf{x},\mathbf{0}},$$

$$G(t+1) = \left(1 - \frac{H_0}{2n}\right)^n U_4^\dagger \left(1 - \frac{H_0}{2n}\right)^n$$
$$\times (1 - \delta H) G(t), \qquad (5)$$

where

$$H_0 = -\frac{\Delta^{(2)}}{2M_Q},$$

$$\delta H = -\frac{g}{2M_Q}\sigma \cdot \mathbf{B} + \frac{\Delta^{(4)}}{24M_Q} - \frac{(\Delta^{(2)})^2}{16nM_Q^2}. \qquad (6)$$

In (6) $\Delta^{(4)}$ is the lattice version of the continuum operator $\sum_i D_i^4$. The first term in δH, in which $\mathbf{B}$ is the chromomagnetic field, is responsible for spin-splitting in the mass spectrum. The last two terms are the correction terms which make (6) accurate to order $O(a^2)$. We use the definition of $\mathbf{B}$ in terms of gauge links given in ref[9], which is also improved up to errors of order $O(a^4, g^2a^2)$. The parameter n is introduced to avoid numerical instability when high-momentum modes occur. With propagators calculated with Eq.(5) we reach an accuracy of order $\frac{1}{M_Q}$ in the mass spectrum. With lattice NRQCD we

have calculated the mass spectrum of quarkonium at the above β-values[11], and have determined that at $\beta = 7.4$ the mass parameters for b- and c-quarks are $aM_b = 4.6$ and $aM_c = 1.4$ respectively. We will now use these parameters for our calculations of mas spectra of b- and c-flavored hadrons. We find that $n = 3$ for c-quark calculations, and $n = 1$ for b-quark calculations, give stable results.

We have implemented tadpole improvement in our calculations of quark propagators with the action in Eq.(3) and with the action of Eq.(5).

2 The Mass Spectrum of c- and b- flavored Hadrons

Our results for the mass spectrum of hadrons containing a c or b quark are based on our earlier results for simulations of quarkonia[11]. There we calculated, from simulations on a $8^3 \times 10$ lattice, using 200 configurations, the spin-averaged mass of S- and P-wave charmonium and bottonium, and matched the results with the experimental data to find the mass parameters of the c- and b- quarks in lattice units. We found that $aM_c = 1.4$ and $aM_b = 4.6$ for $\beta = 7.4$. Using the mass splitting between S- and P- wave quarkonia we then determined the lattice spacing to be

$$a_c^{-1} = 0.749(4)\text{GeV}, \quad \text{from charmonium}, \qquad (7)$$
$$a_b^{-1} = 0.861(5)\text{GeV}, \quad \text{from bottonium}. \qquad (8)$$

and obtain the mass for S-wave charmonium and bottonium:

$$M_{1S} = 3.00(2)\text{GeV}, \quad M_{2S} = 3.67(5)\text{GeV}, \qquad (9)$$
$$M_{1S} = 9.5(3)\text{GeV}, \quad M_{2S} = 10.1(3)\text{GeV}, \qquad (10)$$

in good agreement with experiment.

With the configurations at $\beta = 7.4$ described in our light hadron calculations of[12], and the heavy quark mass parameters reported above, we have calculated masses of the hadrons B^0, B^*, B_s, B_s^* and Λ_b and D^0, D^*, D_s, D_s^* and Λ_c. For the hadrons we use operators of the form

$$O_D(x) = \bar{u}(x)\gamma_5 c(x), \qquad (11)$$
$$O_{D^*}^\mu(x) = \bar{u}(x)\gamma^\mu c(x), \qquad (12)$$
$$O_{\Lambda_c}(x) = \varepsilon_{abc} c_a(x) \left[u_b^T(x) C \gamma_5 d_c(x)\right] \qquad (13)$$

to create the hadronic states on lattice. Note that the heavy quark field $Q(x)$ has only two nonzero components.

The propagation of the heavy quarks is determined as described in the previous section, using a source smeared by a gaussian:

$$F(\mathbf{x},\mathbf{y},t) = \left[1 + \epsilon \sum_i \Delta_i^{(2)}\right]^{n_s} (\mathbf{x},\mathbf{y},t), \qquad (14)$$

but the sink is left unsmeared. For light quarks we use the propagators of[12]. The smearing parameter n_s is fixed at $n_s = 10$, while the parameter ϵ is adjusted so that the smearing radius is about the half that of the simulated system. Using a smeared source is essential for our simulations. With a local source for heavy quark propagators the signal of hadron correlations is overwhelmed by the statistical noise after 3 or 4 time slices from the source. The measured propagators for hadron H are fitted to the form

$$C_H(t) \sim a_H e^{-E_H t}, \tag{15}$$

giving the the hadron mass M_H from

$$M_H = \Delta_Q + E_H. \tag{16}$$

Δ_Q is the difference between the renormalized mass of the heavy quark and the zero point energy of the heavy quark on lattice, and it does not depend on the type of hadron.

Even with a smeared source for heavy quark propagator, the signal is buried in statistical noise after 10 or 11 time slices from the source for mesons and after 8 or 9 time slices for baryons. For the meson we already see a plateau in the effective mass. We will take the fit-window for meson propagators from 6 to 9. For baryons we take the fit-window from 5 to 7.

Our results for E_H of various hadrons with nonzero mass of light quarks are given in Table 1.

For c- and b-flavored hadrons with s-quarks we have:

$$E_{B_s} = 0.920(11), \tag{17}$$
$$E_{B_s^*} = 0.952(9), \tag{18}$$
$$E_{D_s} = 0.937(9), \tag{19}$$
$$E_{D_s^*} = 1.018(9). \tag{20}$$

To extrapolate to the zero light quark mass we assume the dependence of E_H on the mass parameter m to be:

$$E_H = E_H^{(0)} + E_H^{(1)}(m - m_0), \tag{21}$$

and use this relation to fit our data. ¿From our results with varying values of the light quark mass parameter we extrapolate to zero mass for the light quarks and obtain:

$$
\begin{aligned}
E_{B^0}^{(0)} &= 0.796(4), \\
E_{B^*}^{(0)} &= 0.836(3), \\
E_{\Lambda_b}^{(0)} &= 1.218(7), \\
E_{D^0}^{(0)} &= 0.800(3), \\
E_{D^*}^{(0)} &= 0.889(3), \\
E_{\Lambda_c}^{(0)} &= 1.428(5).
\end{aligned} \tag{22}
$$

$$ \tag{23} $$

It is interesting to note that the energies given in Eq.(22) for b-flavored hadrons are close to those given

in Eq.(23) of c-flavored hadrons. These energies can be expanded in the inverse of M_Q with a leading order M_Q^0 and are calculated with an accuracy of M_Q^{-1} in this work. The above fact indicates that the effect from the next-to-leading order and from higher orders is small in these spin averaged masses. However, as we will see, this is not true for spin-splittings.

With the above results for the E_H, and the value of a determined from the bottonium system in Eq.(8) we are able to predict the mass differences of the b-hadrons. For doing this we take the lattice spacing determined by the bottonium spectrum in Eq.(8). We obtain

$$M_{B^*} - M_{B^0} = 34(6)\text{MeV}, \tag{24}$$
$$M_{B_s} - M_{B^0} = 107(13)\text{MeV}, \tag{25}$$
$$M_{B_s^*} - M_{B_s} = 27(17)\text{MeV}, \tag{26}$$
$$M_{\Lambda_b} - M_{B^0} = 363(9)\text{MeV}. \tag{27}$$

These results should be compared with the experimental results

$$M_{B^*} - M_{B^0} = 46\text{MeV}, \tag{28}$$
$$M_{B_s} - M_{B^0} = 91\text{MeV}, \tag{29}$$
$$M_{B_s^*} - M_{B_s} = 47\text{MeV}, \tag{30}$$
$$M_{\Lambda_b} - M_{B^0} = 363\text{MeV}. \tag{31}$$

We find that our results for $M_{\Lambda_b} - M_{B^0}$ and $M_{B_s} - M_{B^0}$ agree well with experimental results, while the spin-splittings $M_{B^*} - M_{B^0}$ and $M_{B_s^*} - M_{B_s}$ are not in such good agreement, but differ by less than 2 standard deviations from the observed values. Although the value of $M_{B^*} - M_{B^0}$ is 28% lower than the experimental value, it agrees with the result from a large and fine lattice[13].

For c-flavored hadrons we take the lattice spacing determined by the charmonium system in Eq.(7) and obtain

$$M_{D^*} - M_{D^0} = 67(5)\text{MeV}, \tag{32}$$
$$M_{D_s} - M_{D^0} = 103(9)\text{MeV}, \tag{33}$$
$$M_{D_s^*} - M_{D_s} = 66(14)\text{MeV}, \tag{34}$$
$$M_{\Lambda_c} - M_{D^0} = 471(6)\text{MeV}. \tag{35}$$

The experimental results are:

$$M_{D^*} - M_{D^0} = 143\text{MeV}, \tag{36}$$
$$M_{D_s} - M_{D^0} = 104\text{MeV}, \tag{37}$$
$$M_{D_s^*} - M_{D_s} = 144\text{MeV}, \tag{38}$$
$$M_{\Lambda_c} - M_{D^0} = 420\text{MeV}. \tag{39}$$

Our value for $M_{D_s} - M_{D^0}$ agrees well with the experimental value, but the predicted $M_{\Lambda_c} - M_{D^0}$ is 12% larger the experimental value. Worse are the spin-splittings which

Table 1: Masses of hadrons (in lattice units) for various values of the light quark mass parameter.

	$m = -0.5$	$m = -0.6$	$m = -0.65$	$m = -0.7$	$m = -0.75$
E_{D^0}	1.069(10)	0.990(10)	0.950(9)	0.909(9)	0.863(9)
E_{D^*}	1.143(8)	1.067(9)	1.030(9)	0.991(10)	0.948(10)
E_{Λ_c}	1.976(16)	1.810(17)	1.727(17)	1.643(15)	1.561(17)
E_{B^0}	1.043(14)	0.967(12)	0.928(12)	0.894(12)	0.856(13)
E_{B^*}	1.066(11)	0.999(9)	0.962(9)	0.927(9)	0.891(10)
E_{Λ_b}	1.892(12)	1.713(12)	1.617(14)	1.504(16)	1.381(22)

are much lower than the experimental values. Many reasons can advanced be for such discrepancies, such as the effects of quenching. However, the two most likely reasons are:

1. the use of lattice NRQCD upto order of $1/M_Q$. For c-quarks the effect from higher orders is very significant because the charm mass is not very large.

2. the use of tree level perturbative values for the coefficients of the action of NRQCD, where one may expect loop corrections to be large at the energy scale of M_c.

A recent study of quarkonium systems [14] shows that the effect of higher order terms in $1/M_Q$ is large in that system. One should take the loop corrections for the coeffients of Eq.(6) into account when they become available.

To obtain the absolute mass of a b- or c-flavored hadron we need to know Δ_Q in Eq.(26) or in Eq.(33). This quantity can be calculated perturbatively. It can also be extracted by studying the mass spectrum of the quarkonium system. We extracted this number from our previous study at $\beta = 7.4$ [11], and find

$$\Delta_c = 1.59(11), \quad \Delta_b = 5.25(18). \quad (40)$$

It should be recalled that effects from the spin-dependent interaction are neglected in that calculation. Adding the spin-dependent interaction has little effect on these numbers, which have quite a large uncertainty. This gives, from the above results and the appropriate lattice spacing,

$$M_{D^0} = 1.79(8)\text{GeV}, \quad M_{B^0} = 5.20(15)\text{GeV} \quad (41)$$

These predictions are in good agreement with the experimental values:

$$M_{D^0} = 1.864\text{GeV}, \quad M_{B^0} = 5.278\text{GeV}. \quad (42)$$

However, the accuracy of our predictions is not good because of the large statistical error of Δ_Q.

3 Conclusion

Our results, and those of Trottier [14], show that one needs to "improve" the action for calculations of charmed hadrons, not only to render the lattice action accurate up to high order terms in a, the lattice spacing, but also accuare to higher terms in M_c^{-1}. The fundamental difficulty is that aM_c is of order unity, and an expansion in either aM_c or in $(aM_c)^{-1}$ will converge slowly if at all.

It is interesting to note that a similar difficulty with charmed quarks arises in a completely different context — when one estimates the effective weak Hamiltonian for strange particle decays, or for parity violating nucleon- nucleon interactions. One would like to regard the charmed quark as heavy and integrate it out, and although that has become standard practice it is not really justified because the parameter μM_c (μ is the QCD scale) is not large but is of order unity. It seem that nature has provided us with a charmed quark whose mass is such as to make life difficult for us.

Acknowledgments

This work is supported by Australia Research Council. We thank our computer manager at the University of Melbourne, Dr. M. Munro, for help with the implementation of our calculations.

BHJMcK wishes to thank the Special Research Centre for the Subatomic Structure of Matter of the University of Adelaide, and the Abdus Salam International Center for Theoretical Physics in Trieste, where this work was written up for publication, for their hospitality.

References

1. M. Lüscher and P. Weisz, Comm. Math. Phys. 97 (1985) 59
2. M. Alford et. al. Phys. Lett. B361 (1995) 87
3. N. Cabibbo and E. Marinari, Phys. Lett. B119 (1982) 387
4. H. Hamer and C.M. Wu, Phys. Lett. B133 (1983) 351
   W. Wetzel, Phys. Lett. B136 (1984) 407

T. Eguchi and N. Kawamoto, Nucl. Phys. B237 (1984) 609

5. B. Sheikholeslami and R. Wohlert, Nucl. Phys. B259 (1985) 572

6. M. Alford, T. Klassen and G.P. Lepage, Nucl. Phys.(Proc. Suppl.) B47 (1996) 370

7. H.R. Fiebig and R.M. Woloshyn, Phys. Lett. B385 (1996) 273

8. A. Frommer et.al. Int. J. Mod. Phys. C5 (1994) 1073

9. G.P. Lepage et al, Phys. Rev. D46 (1992) 4052

10. C.T.H. Davies et al, Phys. Rev. D50 (1994) 6963

11. J.P. Ma and B.H.J. McKellar, Melbourne Preprint UM-P-96/55

12. J.P. Ma and B.H.J. McKellar, Phys Rev D57 (1998) 6723

13. A. Ali Khan et al., Phys. Rev. D53 (1996) 6433

14. H.D. Trottier, Phys. Rev. D55 (1997) 6844

CONFERENCE PARTICIPANTS

H. ABRAMOWICZ	Tel Aviv	Marco BATTAGLIA	Helsinki
Gerald ABRAMS	LBNL	Daniel BAUER	UCSB
Pedro ABREU	LIP	Ulrich J. BAUR	SUNY
Iris ABT	MPI Munich	Aurelio BAY	Lausanne
Raghunath ACHARYA	Arizona	James J. BEATTY	Penn State
Darin ACOSTA	Florida	Ulrich BECKER	MIT
Todd ADAMS	FNL	Jules BECKERS	U Liege, Sart Tilman
M.S. ALAM	Albany (SUNY)	Eugene W. BEIER	Pennsylvania
Mike ALBROW	Fermilab	Leo BELLANTONI	Fermilab
Jim ALEXANDER	Cornell	Bruce BERGER	Cornell
Gideon ALEXANDER	Tel Aviv	Mike BERGER	Indiana
Ahmed ALI	DESY	K. BERKELMAN	Cornell
James ALLABY	CERN	Gregorio BERNARDI	LPNHE-Paris
Attilio ANDREAZZA	CERN	W. BERNREUTHER	Aachen
Valery ANDREEV	Louisiana	Daniel BERTRAND	Bruxelles
Harald ANLAUF	U Siegen	Botio BETEV	CERN
Antonella ANTONELLI	Frascati	S. BETHKE	RWTH Aachen
Thomas APPELQUIST	Yale	Sampa BHADRA	York
S. ARCELLI	Maryland	Helena BIALKOWSKA	Warsaw
Katsushi ARISAKA	UCLA	Mikhail BILENKY	Prague
N. ARKANI-HAMED	SLAC	Ewart BLACKMORE	TRIUMF
Richard ARNOWITT	Texas A&M	Reiner BLAES	IUT de Colmar
Marina ARTUSO	Syracuse	Frederic BLANC	Fribourg
Alan ASTBURY	TRIUMF	Gerald BLAZEY	Illinois
David ATWOOD	Iowa State	Brigitte BLOCH-DEVAUX	CEA-Saclay
Jean-jacques AUBERT	CPPM	C. BLOISE	INFN
Dario AUTIERO	CERN	Alain BLONDEL	CERN
David AXEN	U British Columbia	A. BLOTZ	LANL
Georges AZUELOS	TRIUMF/U Montreal	Edward C. BLUCHER	Chicago
K.S. BABU	Princeton	Hans BLUEMER	CERN
Mark BACHMAN	UC Irvine	Barry BLUMENFELD	Johns Hopkins
Barbara BADELEK	Warsaw	G.J. BOBBINK	CERN
Giuseppe BAGLIESI	CERN	F. BOBISUT	INFN Padova
Ken BAIRD	SLAC	Peter BOCK	Heidelberg
Troy BAKER	Saskatchewan	Arie BODEK	Rochester
A. BALLESTRERO	INFN Torino	Hans BOGGILD	NBI Copenhagen
Andreas BAMBERGER	Freiburg	M. BOHM	Wuerzburg
M. BANDER	U California	Tim BOLTON	Kansas SU
Marcel BANNER	U Paris VI-VII	Alain BONISSENT	CPPM
Maria del Carmen BANULS	Valencia	Edward E. BOOS	Moscow State
Robert BARATE	LAPP	Paula BORDALO	LIP
Elisabetta BARBERIO	CERN	Jose Manuel BORDES	Valencia
Emanuela BARBERIS	Fermilab	Guennadi BORISSOV	Orsay
Barry C. BARISH	Caltech	Csaba BOROS	Adelaide
Anthony BARKER	Colorado	Daniela BORTOLETTO	Purdue
K.J. BARNES	Southampton	Martine BOSMAN	Altas Energias
Bruce BARNETT	Johns Hopkins	G. BOTTAZZI	Milan
David BARNEY	CERN	Jacques BOUCROT	Paris-Sud
Giles BARR	CERN	V. BOUDRY	Ecole Polytechnique
S. BARWICK	UC Irvine	Claire BOURDARIOS	LAL
M. BASILE	Bologna	Arthur BOWLING	Agnes Scott College
Ross T. BATES	Selkirk College	Ivanka BOZOVIC	VINCA

Sylvie BRAIBANT	CERN
Uwe BRATZLER	MPI Munich
Jean-Claude BRIENT	LPNHE-Paris
Paul BRIGHT-THOMAS	Birmingham
Carl BROMBERG	Michigan State
R.M. BROWN	RAL
Antje BRUELL	DESY
R. BRUSTEIN	Ben-Gurion U
Douglas BRYMAN	TRIUMF
Michel BUENERD	IN2P3/CNRS
Antonio BUENO	CERN
William M. BUGG	Tennessee
Andrzej BURAS	TU Munchen
Toby BURNETT	Washington
Philip BURROWS	Oxford
Gerd BUSCHHORN	MPI Munich
Jerry BUSENITZ	Alabama
P.J. BUSSEY	Glasgow
Joel BUTLER	FNAL
Bruce CAMPBELL	Alberta
Tiziano CAMPORESI	CERN
Jean-Noel CAPDEVIELLE	College de France
G. CAPON	Frascati
R. CAREY	Boston
P. CARLSON	RIT
Robert CARNEGIE	Carleton
Joao CARVALHO	Coimbra
David CASSEL	Cornell
Maria-Gabriella CATANESI	CERN
V. CAVASINNI	INFN
Claudia CECCHI	Geneva
Fabrizio CEI	Michigan
Filippo CERADINI	Roma Tre
F. CERVELLI	INFN Pisa
H.M. CHAN	RAL
Darwin CHANG	National Tsing Hua U
Lay Nam CHANG	Virginia Tech
D.G. CHARLTON	Birmingham
Vladimir CHEKELIAN	MPI Munich
Mark CHEN	Princeton
Hai-Yang CHENG	Taiwan
Elliott C. CHEU	Arizona
Jon CHKAREULI	Tbilisi
Frank CHLEBANA	Fermilab
Yong-Min CHO	Seoul National
Norman CHRIST	Columbia
Jiri CHUDOBA	Charles U
Robert CLARE	MIT
James CLINE	McGill
Thomas COAN	Southern Methodist
Guy COIGNET	LAPP
Nicanor COLINO	CIEMAT
Johann COLLOT	ISN
Janet CONRAD	Columbia
John CONWAY	Rutgers
Oliver COOKE	CERN
Stephane COUTU	Pennsylvania SU
Gilles COUTURE	UQAM
Mary Anne CUMMINGS	Illinois
Donald CUNDY	CERN
David CUSSANS	CERN
G. D'AMBROSIO	INFN
Yuanben DAI	CAS
John DAINTON	DESY
Carlo DALLAPICCOLA	Maryland
M.V. DANILOV	ITEP
Mourad DAOUDI	Stanford
W. Dieter DAU	Kiel
Aharon DAVIDSON	Ben Gurion
Michel DAVIER	Paris-Sud
C. DE CLERCQ	Brussels
Nicolo DE GROOT	SLAC
Aldo DEANDREA	CNRS
Pascal DEBU	Saclay
Marc DEJARDIN	Saclay
Paul DERWENT	Fermilab
Klaus DESCH	CERN
Abhay DESHPANDE	Yale
N.G. DESHPANDE	Oregon
Asoka DESILVA	TRIUMF
Claude DETRAZ	IN2P3/CNRS
A. DI GIACOMO	INFN Pisa
E. DI SALVO	INFN Genova
Cristinel DIACONU	CPPM
Thomas DIEHL	FNAL
T.V. DIMOVA	Budker
Carlo DIONISI	CERN
Madhu DIXIT	Carleton
Antonio DOBADO	Complutense
Niels DOBLE	CERN
Yuri DOKSHITZER	INFN
Brian DOLAN	Ireland
N. DOMBEY	Sussex
Jonathan DORFAN	SLAC
T. DORIGO	Padova
Peter J. DORNAN	Imperial College
Michael DOSER	SLAC
Tony DOYLE	Glasgow
Marcos DRACOS	Louis Pasteur
J. DREES	Wuppertal
Robert DRUCKER	Oregon
Dominique DUCHESNEAU	Annecy
Stephen DUERR	Washington
Wolfgang DUNNWEBER	Munchen
Friedrich DYDAK	CERN
Oscar Jose Pinto EBOLI	U Sao Paulo (IFUSP)
Karl ECKLUND	Cornell
Reinhard ECKMANN	Austin

Klaus EHRET	Dortmund	Ekkehard GERNDT	DESY
Gerald EIGEN	Bergen	Christoph GEWENIGER	Heidelberg
Martin EINHORN	Michigan	Claude GHESQUIERE	College de France
Y. EISENBERG	Weizmann	Paolo GIACOMELLI	INFN Bologna
Tord J.C. EKELOF	Uppsala	Lawrence GIBBONS	Cornell
Victor ELIAS	U Western Ontario	Gary GLADDING	Illinois
Eckhard ELSEN	DESY	Hank GLASS	Fermilab
Markus ELSING	CERN	Stephen GODFREY	Carleton
Martin ERDMANN	Karlsruhe	Nelli GOGUITIDZE	H1 Collaboration
Pavel F. ERMOLOV	Moscow SU	Jacques GOLDBERG	TECHNION
Carlos ESCOBAR	UNICAMP	Marvin GOLDBERG	NSF
Pierre ESPIGAT	College de France	Steven GOLDFARB	Lausanne
E. ETZION	Tel Aviv	Eugene GOLOWICH	Massachusetts
Harold EVANS	Columbia	V. GOLUBEV	Budker
M. FABBRICHESI	INFN	M.C. GONZALEZ-GARCIA	Valencia
S. FAHEY	Colorado	G. GONZALEZ-SPRINGBERG	Uruguay
Svjetlana FAJFER	Inst. Jozef Stefan	Maury GOODMAN	Argonne
Alvise FAVARA	Caltech	Alfred GOSHAW	Duke
J. FAY	Lyon	H. GRAESSLER	TH Aachen
Th. FELDMANN	Wuppertal	F. GRANCAGNOLO	INFN Lecce
Thomas FERBEL	Rochester	Paul GRANNIS	SUNY
Enrique FERNANDEZ	Altas Energias	F. GRARD	U Mons-Hainaut
Peter FILIP	Slovak Acad. Sci.	Stephen GRAY	Cornell
Margret FINCKE-KEELER	Victoria	E. GRAZIANI	INFN
Raffaele FLAMINIO	LAPP	M. GREEN	Royal Holloway College
Brenna FLAUGHER	Fermilab	T.J. GREENSHAW	Liverpool
Manfred FLEISCHER	Dortmund	Christoph GREUB	Bern
Robert FLEISCHER	CERN	Guenter GRINDHAMMER	MPI Munich
R. FLOREANINI	INFN	Marc GRISARU	Brandeis
Lorenzo FOA	CERN	Eilam GROSS	Weizmann
Zoltan FODOR	Eotvos Lorand	J. GROSSE-KNETTER	Oxford U
Yiota FOKA	IKF Frankfurt	Carla GROSSO-PILCHER	U Chicago
Roger FORTY	CERN	Martin GRUNEWALD	Humboldt U
B. FOSTER	Bristol	John GUNION	UC Davis
P. FRANZINI	Frascati	Eugene GURVICH	Tel Aviv
William FRAZER	LBNL	Laszlo GUTAY	Purdue
Klaus FREUDENREICH	ETH Zurich	Claude GUYOT	Saclay
Stefano FRIXIONE	ETH Zurich	Geoff HALL	Imperial College
Colin FROGGATT	Glasgow	Kurt HALLER	U Connecticut
F. FUCITO	INFN Rome	A.W. HALLEY	CERN
Emidio GABRIELLI	Depart. de Fisica Teorica	Klaus HAMACHER	Wuppertal
Pauline GAGNON	CRPP, Ottawa	Odile HAMON	U Paris VI-VII
T.K. GAISSER	Delaware	Cherif HAMZAOUI	UQ Montreal
H. GALLAGHER	Oxford	Tao HAN	U Wisconsin
E. GALLO	INFN Firenze	J. Dines HANSEN	NBI Copenhagen
Angela GALTIERI	LBNL	John Renner HANSEN	NBI Copenhagen
Gerardo GANIS	MPI Munich	Peter HANSEN	NBI Copenhagen
Theodore GARAVAGLIA	Dublin	Gail HANSON	Indiana
Carmen GARCIA	IFIC	Michael HAPKE	Westfield College
Pablo GARCIA-ABIA	CERN	David HARDTKE	LBNL
Arthur GARFINKEL	Purdue	John HAUPTMAN	Iowa State
Ugo GASPARINI	INFN	Masaki HAYASHI	Tokyo
Chao-Qiang GENG	National Tsing Hua U	Yoshinari HAYATO	KEK
Ulrich GENSCH	DESY	Mao HE	Shandong

Zoltan KUNSZT	Inst. f Theor. Physik, Zürich	Vera LUTH	SLAC
Tibor KURCA	DESY	Louis LYONS	Oxford
Piret KUUSK	Tartu	Hong MA	BNL
Masahiro KUZE	KEK/IPNS	Daniela MACINA	Geneva
Manuel Torres LABANSAT	U Nacional Mexico	Bogdan Castle MAGLICH	HiEnergy Microdevices, Inc.
Luis LABARGA	Madrid	K.T. MAHANTHAPPA	Colorado
Didier LACOUR	U Paris VI-VII	L. MAIANI	INFN
R. (Bob) LADENDORFF	Cyprus Miami Mining	E. MAINA	INFN Torino
Harry C.S. LAM	McGill	Walter MAJEROTTO	Austria
Jodi LAMOUREUX	Brandeis	Akihiro MAKI	KEK/IPNS
Rolf LANDUA	CERN	Ziro MAKI	Kinki U
Francisco LARIOS	CINVESTAV	A.I. MALAKHOV	JINR
Mary-Lu LARSEN	Towson	Adolfo MALBOUISSON	LAFEX/CBPF
Amitabh LATH	Rutgers	Russell MALCHOW	Colorado SU
Kwong LAU	Houston	Luca MALGERI	CERN
Luis LAVOURA	IST, Lisboa	Jeffrey MANDULA	DOE, HEP Div.
I. LAZZIZZERA	INFN	Rainer MANKEL	Humboldt U
P. LEBRUN	U. Claude Bernard	I. MANNELLI	INFN Pisa
Pierre LECOMTE	Swiss Fed. Inst. Tech.	Philip D. MANNHEIM	Connecticut
G. LEDER	Austria	Pierre MARAGE	Brussels
Fabienne LEDROIT	ISN	Chiara MARIOTTI	CERN
Alfred LEE IV	Duke	Marvin MARSHAK	Minnesota
J. LEE-FRANZINI	Frascati	Eric MARTELL	U Illinois at Chicago
Michel LEFEBVRE	U Victoria	F. MARTELLI	Urbino
Jacques LEFRANCOIS	Paris-Sud	Cecile MARTIN	U Paris Vi-VII
George LEIBBRANDT	Guelph	John MARTIN	Toronto
Lars LEISTAM	EST Division	Silvia MASCIOCCHI	MPI Munich
David LEITH	SLAC	Francisco MATORRAS	Cantabria
J. LEMONNE	Brussels	Masahisa MATSUDA	Lund
Rod W. LESSARD	Purdue	Satoshi MATSUDA	Kyoto
A. LEVY	Tel Aviv	Takayuki MATSUI	KEK
Bing An LI	Kentucky	Takashi MATSUSHITA	KEK/IPNS
Guey-Lin LIN	Chiao-Tung U	C. MATTEUZZI	INFN Milano
Willis T. LIN	Central U	Victor MATVEEV	INR
Ludger LINDEMANN	DESY	Chris MAXWELL	Durham
Manfred LINDNER	MPI	Patricia MCBRIDE	Fermilab
Steve LINN	Florida SU	Robert MCCARTHY	SUNY
V.A. LIOUBIMOV	DESY	Gary MCCARTOR	Southern Methodist U
Harry LIPKIN	Weizmann	W. John MCDONALD	U Alberta
Alan LITKE	UC Santa Cruz	Clark MCGREW	SUNY
Keh-Fei LIU	Kentucky	B.H.J. MCKELLAR	Melbourne
Lianshou LIU	Huazhong Normal U	Peter MCNAMARA	CERN
C.H. LLEWELLYN SMITH	CERN	Robert MCPHERSON	U Victoria
Marcelo LOEWE	Chile	David MEASDAY	U British Columbia
Georges LONDON	Saclay	Thomas C. MEHEN	Caltech
Nat LONGLEY	Swarthmore College	Michael MELLES	Durham
Michael LOSTY	Carleton	Oliver MELZER	NIKHEF
William LOUIS	LANL	Ramon MENDEZ-GALAIN	Uruguay
Sherwin LOVE	Purdue	Ta-chung MENG	Freie U Berlin
Adolph LU	UC Santa Barbara	Bernard MERKEL	Paris-Sud
Jose Luis LUCIO-MARTINEZ	Guanajuato	W.J. METZGER	Nijmegen
Dieter LUEKE	Dortmund	Arnd MEYER	DESY
Bengt LUND-JENSSEN	KTH	Joachim MEYER	DESY
E. LUPPI	INFN, Ferrara	Douglas MICHAEL	Caltech

Bernard MICHEL	U Blaise Pascal	Sun Kun OH	Konkuk
Jouko MICKELSSON	KTH	Thorsten OHL	Darmstadt
Pasquale MIGLIOZZI	CERN	Tokio OHSKA	KEK
Roger MIGNERON	U Western Ontario	Toru OKUSAWA	Osaka City
G. MIKENBERG	Weizmann	Arne P. OLSON	Argonne
Andy MILLER	TRIUMF	Jan OLSSON	Desy
David MILLER	U College London	Christopher J. ORAM	TRIUMF
David MILSTEAD	DESY	Risto ORAVA	Helsinki
K. MILTON	Oklahoma	Shuji ORITO	Tokyo
Marie-Noelle MINARD	LAPP	Lynne ORR	Rochester
Allen MINCER	New York U	Per OSLAND	Bergen
Ramon MIQUEL	Barcelona	Eivind OSNES	Oslo
Lluisa MIR	Barcelona	Ilmar OTS	Tartu
Hironari MIYAZAWA	Kanagawa	P. PAGANINI	LPNHE
Aldo MOENZIONE	INFN	Philip PAGE	LANL
Rasmus MOLLER	NBI Copenhagen	Reynald PAIN	U Paris VI-VII
Myriam MONDRAGON	Uruguay	P.N. PAKHLOV	ITEP
Lorenzo MONETA	Geneva	William PALMER	Ohio SU
Stephane MONTEIL	IN2P3-U Blaise Pascal	James T. PANTALEONE	Alaska
Masahiro MORII	SLAC	L. PAOLUZI	INFN Rome
Kjell MORK	Norwegian U	Vaia PAPADIMITRIOU	Texas Tech U
Hans-Guenther MOSER	MPI Munich	Luc PAPE	CERN
Leszek MOTYKA	Jagellonian	Fabrizio PARODI	Genoa
Gilbert MOULTAKA	U Montpellier II	Zohreh PARSA	BNL
David MULLER	SLAC	Richard PARTRIDGE	Brown
Thomas MULLER	Karlsruhe	Marc PATERNO	Fermilab
D.J. MUNDAY	Cambridge	Ritchie PATTERSON	Cornell
Yorikiyo NAGASHIMA	Osaka	Thomas PATZAK	College de France
Tatsuya NAKADA	CERN	Ewald PAUL	Bonn
Satyanarayan NANDI	Oklahoma SU	Manfred PAULINI	LBNL
M. NAPOLITANO	INFN Napoli	Christoph PAUS	CERN
G. NARDULLI	U Bari and INFN Bari	Felicitas PAUSS	ETH Zurich
Beate NAROSKA	Hamburg	K.J. PEACH	RAL
Jordan NASH	SLAC	L.S. PEAK	Sydney
Ann NELSON	Washington	Roberto PECCEI	UCLA
Charles NELSON	SUNY at Binghamton	Thomas PEITZMANN	Münster
Donald NELSON	Washington	Vladimir PENEV	JINR
Kenneth NELSON	U. Virginia	Jen-Chieh PENG	LANL
Herbert NEUBERGER	Rutgers	John PEOPLES	FNAL
N. NEUMEISTER	Vienna	Miguel Angel PEREZ	CINVESTAV
P. NEWMAN	Birmingham	C. PERONI	INFN Torino
Barbara NICOLESCU	U Paris VI-VII	S. PETRERA	INFN
Bogdan B. NICZYPORUK	TJNAF	Yem Xuan PHAM	U Paris VI-VII
Friedrich NIEBERGALL	Hamburg	H. PHILLIPS	RAL
Carsten NIEBUHR	DESY	L. PIEMONTESE	INFN Ferrara
V.A. NIKITINE	JINR	M. PIERI	INFN Florence
Masao NINOMIYA	Kyoto	Francois PIERRE	Saclay
Jouni NISKANEN	Helsinki	Massimo PIETRONI	INFN
Seishi NOGUCHI	NARA Women's U	Herbert PIETSCHMANN	Wien
Sergio Ferraz NOVAES	UNESP	James PILCHER	Chicago
Patrick J. O'DONNELL	Toronto	Mario PIMENTA	LIP
John O'FALLON	DOE	James PINFOLD	U Alberta
Horst OBERLACK	MPI Munich	Krzysztof PIOTRZKOWSKI	DESY
Vladimir OBRAZTSOV	Protvino	David PLANE	CERN

George POCSIK	Eotvos Lorand	Roy RUBINSTEIN	FNAL
G. POELZ	Hamburg	James RUSS	Carnegie Mellon U
Paul POFFENBERGER	U Victoria	Jean SACTON	U Libre de Bruxelles
Ron POLING	Minnesota	Willis SAKUMOTO	Rochester
Matjaz POLJSAK	Inst. Jozef Stefan	Julien SALGADO	U Paris VI-VII
Bernard POPE	Michigan SU	Per SALOMONSON	Inst. Theor. Phys., CTH
Frank PORTER	Caltech	Peter SANDERS	Imperial College
Jean-Michel POUTISSOU	TRIUMF	M.G. SAPOJNIKOV	JINR
Michael PROCARIO	Carnegie Mellon	Rodolfo SASSOT	Buenos Aires
Helenka PRZYSIEZNIAK	Saclay	Patrick SAULL	DESY
Kai PUOLAMAKI	Helsinki Inst. Physics	Gilles SAUVAGE	LAPP
T.A. PURLATZ	Budker	Alexandre SAVINE	DESY
Modesto PUSTERLA	Padua	H. Lee SAWYER	Louisiana
Arnulf QUADT	Oxford	David H. SAXON	Glasgow
K. RABBERTZ	Aachen	Carla SBARRA	U British Columbia
A. RADYUSHKIN	TJNAF	Terry SCHALK	UC Santa Cruz
John P. RALSTON	Kansas	Heidi SCHELLMAN	Northwestern U
Sergio RAMOS	LIP	Phil SCHLABACH	Fermilab
Peter RATOFF	Lancaster	Stefan SCHLENSTEDT	DESY Zeuthen
S. RATTI	INFN Pavia	Peter SCHLEPER	Heidelberg
Pierpaolo REBECCHI	CERN	Helga SCHMAL	CERN
George REDLINGER	TRIUMF	William SCHMIDKE	DESY
Duncan REID	CERN	Diethard SCHMIDT	Wuppertal
Laura REINA	Wisconsin - Madison	Ivan SCHMIDT	U Santa Maria
Peter B. RENTON	Oxford	L. SCHMITT	Munich
Jean-Paul REPELLIN	IN2P3/CNRS	Olivier SCHNEIDER	CERN
Jose REPOND	Argonne	D.J. SCHOTANUS	KUN
James (Jim) REVILL	IOP, Publishing	B. SCHREMPP	U Kiel
R. RIAZUDDIN	King Fahd U	Fridger SCHREMPP	DESY
Alberto RIBON	Padova	Henning SCHROEDER	DESY
Stefania RICCIARDI	CERN	Adolf SCHWIND	DESY
Francois RICHARD	Orsay	Felix SEFKOW	DESY
Burton RICHTER	SLAC	Ashoke SEN	Mehta Research Inst.
Michael RIJSSENBEEK	SUNY - Stony Brook	Ron SETTLES	MPI Munich
Keith RILES	Michigan	Michael SHAEVITZ	Columbia
Jack RITCHIE	Texas	Stephen SHARPE	Washington
Thomas RIZZO	SLAC	Tara SHEARS	CERN
B. Lee ROBERTS	Boston	Paul D. SHELDON	Vanderbilt
Maria ROCO	Fermilab	Claire SHEPHERD-THEMISTOCLEOUS	CERN
Jorge RODRIGUEZ	Hawaii	Marc SHER	College of William & Mary
Francois ROHRBACH	CERN	Takao SHINKAWA	KEK/IPNS
Ewa RONDIO	Soltan Inst. Nucl. Studies	I. SHIPSEY	Purdue U
J.M. RONEY	U Victoria	L.E. SINCLAIR	DESY
Gang RONG	IHEP	Pekka SINERVO	Toronto
Jerome ROSEN	Northwestern U	Virendra SINGH	Tata Inst.
Peter ROSEN	US DOE,	AplenarySessions.pslexandre SINGOVSKI	CERN
Rob ROSER	FNAL	Rahul SINHA	Inst. of Math Sci., Tarami
Jonathan L. ROSNER	Chicago	Tomasz SKWARNICKI	Syracuse
Philippe ROSNET	Blaise Pascal IN2P3/CNRS	A.J. Stewart SMITH	Princeton
L. ROSSI	INFN	Alasdair SMITH	CERN
Andrei ROSTOVTSEV	ITEP	Wesley H. SMITH	Wisconsin
Myron ROSVICK	TRIUMF	Hakan SNELLMAN	RIT
Jean ROY	Colorado	G. SNOW	Nebraska
Probir ROY	Tata Inst.		

Cristina SOAVE	Torino	Georgios TSIPOLITIS	ETH Zurich
Randall SOBIE	U Victoria	Sheung Tsun TSOU	Oxford
Volker SOERGEL	MPI Munich	Wu-Ki TUNG	Michigan State U
Michael SOKOLOFF	Cincinnati	Nikolai E. TYURIN	IHEP
Stefan SOLDNER-REMBOLD	Freiburg	Ekaterini TZAMARIUDAKI	DESY
E.P. SOLODOV	Budker	George TZANAKOS	Athens
Hee Sung SONG	Seoul National	Mark VAGINS	UC Irvine
Amarjit SONI	BNL	Andrea VALASSI	LAL
Lee SORRELL	SLAC	Claude VALLEE	U Marseille II
Michel SPIRO	Saclay	Juan VALLS	FNAL
Achim STAHL	Bonn	W. VAN DONINCK	Brussels
Doug STAIRS	McGill	Rick VAN KOOTEN	Indiana
L. STANCO	INFN Padova	G. VAN MIDDELKOOP	NIKHEF
Kenneth C. STANFIELD	FNAL	K.E. VARVELL	Sydney
Steinar STAPNES	Oslo	Mayda VELASCO	CERN
Howard STEELE	Washington	G. VENTURI	INFN Bologna
Tom STEELE	Saskatchewan	Walter VENUS	CERN
Johannes STEUERER	DESY	Piero VERDINI	INFN Pisa
Frederic STICHELBAUT	SUNY	Wouter VERKERKE	NIKHEF
Jurgen STIEWE	Heidelberg	Michel VETTERLI	TRIUMF
Sheldon STONE	Syracuse	Jean-Pierre VIALLE	LAPP/IN2P3
Raimund STROEHMER	U Munchen	Daniel VIGNAUD	Saclay
Mark STROVINK	UC Berkeley/LBNL	Manuella VINCTER	U Alberta
R. STROYNOWSKI	Southern Methodist U	Oscar VIVES	Valencia
Hirotaka SUGAWARA	KEK	Erich VOGT	TRIUMF
Larry SULAK	Boston	Harald VOGT	DESY
Padmanabha N. SWAMY	Southern Illinois U	Rudiger VOSS	CERN
Eiichi TAKASUGI	Osaka	Dejan VUCINIC	MIT
Hiroshi TAKEDA	Kobe	Albrecht WAGNER	DESY
Masato TAKITA	Osaka	Roland WALDI	Dresden
Mossadek TALBY	SLAC	Kameshwar WALI	Syracuse
Reisaburo TANAKA	Ecole Polytechnique	Chris WALTHAM	U British Columbia
G.F. TARTARELLI	CERN	Shu-hong WANG	IHEP
Toshiaki TAUCHI	KEK	Yi Fang WANG	Stanford
Richard TAYLOR	SLAC	Bennie WARD	Tennessee
Frederic TEUBERT	CERN	Steven WASSERBAECH	Washington
Richard TEUSCHER	CERN	Nigel WATSON	CERN
Jon THALER	U Illinois	A. WEBER	CERN/EP
John THOMPSON	RAL	Harry WEERTS	Michigan SU
Mark THOMSON	CERN	Pippa WELLS	CERN
Charles THORN	Florida	Dennis WEYGAND	TJNAF
Stefan THURNER	TU Vienna	Richard WIGMANS	Texas Tech U
Charles TIMMERMANS	Nijmegen	Bjorn WIIK	DESY
J. TIMMERMANS	NIKHEF	L.C.R. WIJEWARDHANA	Cincinnati
Paul TIPTON	Rochester	Guy WILKINSON	CERN
Kirsten TOLLEFSON	MIT	Raymond WILLEY	Pittsburgh
Ian TOMALIN	CERN	David WILLIAMS	CERN
Eric TORRENCE	CERN	Jimmy WILLIAMS	Guelph
Fumihiko TOYODA	Kinki U	Stephane WILLOCQ	SLAC
Minh-Tam TRAN	Inst. Phys. Nucl.	Brian WINER	Ohio SU
Michael TREICHEL	CERN	Peter WINTZ	PSI
Daniel TREILLE	CERN	Mark WISE	Caltech
Howard TROTTIER	SFU	Markus WOBISCH	RWTH Aachen
Jeff TSENG	MIT	Stephen WOLBERS	FNAL

Gustavo WOLF	CERN
Knut WOLLER	DESY
Dennis WRIGHT	TRIUMF
S.L. WU	Wisconsin
Yue-Liang WU	CAS
Stephan WYNHOFF	RWTH Aachen
Sakue YAMADA	IPNS/KEK
Yoshio YAMAGUCHI	IUPAP
Taku YAMANAKA	Osaka
Narendra YAMDAGNI	Stockholm
Un Ki YANG	Rochester
Weiming YAO	LBNL
Minghan YE	CAS
G.P. YEH	Fermilab/Academia Sinica
John YELTON	Florida
Jae YU	FNAL
F. ZACH	U. Claude Bernard
Cosmas ZACHOS	Argonne
A. ZALLO	Frascati
Guenter ZECH	Siegen
Peter ZERWAS	DESY
Wolfram ZEUNER	DESY
Mehmet T. ZEYREK	METU
Amina ZGHICHE	U. Louis Pasteur
Zhiqing ZHANG	LAL
Zhengguo ZHAO	IHEP
Marek ZRALEK	U Silesia